The New Palgrave

A Dictionary of Economics

THE NEW
PALGRAVE
A DICTIONARY OF
ECONOMICS

EDITED BY

JOHN EATWELL

MURRAY MILGATE

PETER NEWMAN

Volume 2
E to J

THE MACMILLAN PRESS LIMITED, LONDON
THE STOCKTON PRESS, NEW YORK

The New Palgrave: A Dictionary of Economics
Edited by John Eatwell, Murray Milgate and Peter Newman
in four volumes, 1987

First published in hardback, 1987
ISBN 0-333-372352
Reprinted 1988 (twice), 1991 (with corrections), 1994, 1996

First published in paperback, 1998
ISBN 0-333-740408

Published in the United Kingdom by
MACMILLAN REFERENCE LTD, 1998
25 Eccleston Place, London, SW1W 9NF
and Basingstoke

Associated companies throughout the world

A catalogue record for this book is available from the British Library

Published in the United States and Canada by
STOCKTON PRESS LTD, 1998
345 Park Avenue South, 10th Floor, New York
NY 10010-1707

ISBN 1-56159-197-1

The New Palgrave is a trademark of Macmillan Reference Ltd

Typeset by Morton Word Processing Limited, Scarborough, Great Britain
Printed and bound in Hong Kong.

Contents

LIST OF ENTRIES   A – Z

One-line cross-references are shown in *italics*.

E

Easterlin hypothesis. The Easterlin, or 'cohort size', hypothesis posits that, other things constant, the economic and social fortunes of a cohort (those born in a given year) tend to vary inversely with its relative size, approximated by the crude birth rate in the period surrounding the cohort's birth (Easterlin, 1980). The linkage between higher birth rates and adverse economic and social effects arises from what might be termed 'crowding mechanisms' operating within three major social institutions – the family, school and labour market. Empirically, the most important application of the hypothesis has been to explain the varying experience of young adults in the United States since World War II, although there is some evidence of its relevance to the experience of developed countries more generally in this period. The hypothesis has led to recognition of the possibility of a self-generating fluctuation in current and future United States experience with a period of around four to five decades.

Crowding effects within the family may best be understood by recognizing that a sustained upsurge in the birth rate is likely to entail an increase in the average number of siblings, higher average birth order, and a shorter average birth interval, and that there is a substantial literature in psychology, sociology, and, more recently, in economics linking child development to one or more of these magnitudes (Behrman et al., 1980; Ernst and Angst, 1983; Heer, 1985; Wray, 1971). Typically, negative effects are associated with more siblings, closer birth spacing, and a higher birth order. The effects that have been investigated range over a wide variety of phenomena, such as mental and physical health, intelligence, educational achievement, and personality. With regard to mental health, for example, there is evidence that problem behaviours such as fighting, breaking rules, and delinquency are associated with increased family size. Adverse effects on morbidity and mortality of children have been found to be associated with increased family size and shorter birth spacing. A negative association between IQ and number of siblings has been found in a number of studies, and, controlling for IQ, between educational attainment and family size.

One of the mechanisms underlying such developments, and perhaps the most pervasive one, is the dilution of parental time and energy per child and family economic resources per child associated with increased family size, though parental preferences may also play a part (Behrman et al., 1982; Heer, 1985). Higher order children, given birth spacing, are likely to receive less parental attention than lower order, and suffer from a lower input of family economic resources. Closer birth spacing, given family size, results in more concentrated demands on parental time and income, and corresponding dilution in the amounts of these available per child. Human ecology studies of the adverse effects of household crowding, as measured, say, by persons per room, are doubtless relevant in this connection, but so far little has been done to explore linkages to the cohort size hypothesis.

The family mechanisms just discussed imply that, on the average, a larger cohort is likely to perform less well in school.

If, for example, average IQ is lower and the probability greater of disruptive behaviour in the classroom, this would make for poorer educational achievement. But even in the absence of any adverse effects within the family, a large cohort is likely in the course of its schooling to experience crowding phenomena that would react unfavourably on its average educational performance (Freeman, 1976; Waring, 1975). At any given time the human and physical capital stock comprising the school system tends either to be fixed in amount or expanding at a fairly constant rate. Because of this, a surge in entrants into the school system tends to be accompanied by a reduction in physical facilities and teachers per student. Crowding in schools is likely to entail a reduced probability for a given student of success both in academic pursuits and extra-curricular activities (e.g. making an athletic team or winning a leading role in a school play). True, if planning of educational expansion were geared fully to anticipate a large cohort's entry, then such crowding effects might be largely avoided. In the United States, however, school planning decisions are divided among numerous local governments and private institutions, and expansion has tended to occur in reaction to, rather than anticipation of, a large cohort's entry. Moreover, even when expansion occurs it is usually not accompanied by maintenance of curriculum standards, for one reason, because of the diminishing pool of qualified teachers available to supply the needs of educational expansion. The proliferation in the 1960s of two-year community colleges has been seen by some as little more than a holding device for easing the impact of a sharp increase in cohort size on four year colleges and the labour market (Waring 1975).

Other things constant, then, crowding within the school system arising from large cohort size tends to lower the amount and quality of schooling obtained. Psychologically, it is likely to give rise to frustration and a diminished self-image associated with lessened academic and extracurricular achievement. Such circumstances would reinforce similar effects arising from crowding within the family.

The prior socialization experience of a large cohort both in the family and in school is likely, in turn, to leave the cohort less well-prepared, on reaching adulthood, for success in the labour market. But even if there were no prior effects, the entry of a large proportion of young and relatively inexperienced workers into the labour market creates a new set of crowding phenomena, because the expansion of complementary factor inputs is unlikely to be commensurate with that of the youth labour force. Additions to the physical capital stock tend to be dominated by considerations other than the relative supply of younger workers, and the growth in older, experienced, workers is largely governed by prior demographic conditions. Growth in the relative supply of younger workers results, in consequence, in a deterioration of their relative wage rates, unemployment conditions, and upward job mobility (Anderson, 1982; Berger, 1985). The adverse effects of labour market crowding, in turn, tend to reinforce those of crowding within the school and family. For example, the deterioration in relative wage rates of the young translates into

1

lower returns to education and a consequent adverse impact on school dropout rates and college enrolment (Freeman, 1976). Also, the problems encountered in finding a good job may reinforce feelings of inadequacy or frustration already stirred up by prior experiences at home or in school.

In some cases, such as parental time per child, the adverse impact of large cohort size may take the form of absolute declines; in others, for example real wages or years of schooling, where the long-term trend is upward, of reductions in the rate of increase. In either case, if a large cohort is the offspring of a prior small cohort, then a cross-section comparison on a given attribute at a point in time will tend to show relative deterioration for the large cohort. For example, real wages may be rising over time for all age groups in the working population, but in a period when younger workers are plentiful relative to older, the shortfall of younger adults' wage rates relative to older will be greater than in a period when younger workers are scarce relative to older.

A comparison between younger and older adults of the type just given translates largely into a comparison of children with their parents. Thus, the relative standing of successive generations at a point in time may be altered systematically by fluctuations in relative cohort size. In turn, if parents' living levels play an important role in setting their children's material aspirations, as socialization theory leads one to believe, then an increase in the shortfall of children's wage rates relative to parents, will cause the children to feel relatively deprived and under greater pressure to keep up. The importance of relative status influences of this type in affecting attitudes or behaviour has been widely recognized in social science theory (Durkheim, 1897; Duesenberry, 1949; Merton, 1968; Modigliani, 1949; Stouffer, 1949).

As has already been suggested, the mechanisms that have been described, implying relative or absolute deterioration in the circumstances of large cohorts, result in greater psychological stress in a large cohort. Disappointment or lack of success – at home, in school, in the labour market, in pursuit of material aspirations – tends to give rise to feelings of inadequacy, anxiety, frustration and bitterness. Increased psychological stress, in turn, induces a wide range of responses in a large cohort, some of which are compensatory. Confronted with the prospect of a deterioration in its living level relative to that of its parents, a large young adult cohort may make a number of adaptations in an attempt to preserve its comparative standing. Foremost among these are changes in behaviour related to family formation and family life (Espenshade, 1985; Moffitt, 1982). To avoid the financial pressures associated with family responsibilities, marriage may be deferred. If marriage occurs, wives are more likely to work and to put off childbearing. If the wife bears children, she is more likely to couple labour force participation with childrearing, and to have a smaller number of children more widely spaced. Note that the fertility response induced by the effects of large cohort size itself tends to reverse subsequent cohort size.

Other reactions to the psychological stresses inducted by large cohort size may be viewed as socially dysfunctional (Ahlburg and Schapiro, 1984; Easterlin, 1980; Preston and McDonald, 1979). Feelings of inadequacy and frustration, for example, may lead to disproportionate consumption of alcohol and drugs, to mental depression, and, at the extreme, a higher rate of suicide. Feelings of bitterness, disappointment and rage may induce a higher incidence of crime. If men are unusually reluctant to undertake the financial pressures associated with family life, then there is likely to be less legitimization via marriage of premarital conceptions, and illegitimacy and/or

abortion rates may rise. Within marriage, the stresses of conflicting work and motherhood roles for women, and feelings of inadequacy as a breadwinner for men, are likely to result in a higher incidence of divorce. In the political sphere, the disaffection felt by a large cohort because of its lack of success may make it more responsive to the appeals of those who are politically alienated.

Adaptive mechanisms are likely to function with regard to psychological attitudes as well as behaviour. The theory of cognitive dissonance, for example, suggests that as conflicts emerge between a cohort's norms and behaviour, goals will be redefined and attitudes revised to rationalize divergent behaviour (Festinger, 1957). Thus, young adults may start with the view that a mother of pre-school children should be at home raising her children, rather than in the labour market. But as economic pressures increasingly lead to the coupling of childrearing with labour force participation, attitudes shift in favour of such behaviour, and growing support emerges for the notion that pre-school children are not adversely affected by their mothers' working.

Institutional adaptations to the stresses arising from large cohort size may also occur. Reference has already been made to responses to a surge in cohort size that may be made by the school system. Business firms and nonprofit organizations often have age related rules regulating hiring and promotion, and a surge in cohort size may force relaxation of the rules. In large families, older children may help care for younger, thus substituting the input of children's time for parents'.

Despite adaptive mechanisms, the stresses and strains associated with large cohort size are likely to continue throughout a cohort's life course. In the labour market, the disadvantage of excessive supply persists at all stages in the life cycle. The heritage of frustrations and disappointments that a large cohort carries with it will continue to make it relatively unhappy and prone to dysfunctional behaviour. When the cohort reaches retirement age, its returns from public pension support are likely to be relatively diminished by virtue of its disproportionate size.

The effects of cohort size discussed here should not be confused with 'age distribution' effects of the type commonly identified in the demographic literature. For example, a rise in the national crime rate is sometimes explained in terms of an increase in the proportion of the population in young adult age groups, because it is in those groups that the incidence of crime is typically highest. The present set of effects, however, relates to what demographers term 'age-specific' effects. Thus a rise in relative cohort size is hypothesized to raise the frequency of crime within the age groups whose relative size has increased.

Recently in social science literature, a fair amount of attention has been directed to the development of mathematical models for formally distinguishing cohort from period and age effects (Hobcraft et al., 1982). The impact of cohort size as described here is not necessarily the same as cohort effects as identified in these models. For one, the hypothesis here relates not to all influences subsumed under the term 'cohort effects', but only to those influences specifically associated with cohort size (cf. Mason et al., 1976). Also, because of what might be termed 'social contagion', the impact of cohort size is not necessarily confined to a given cohort. As was pointed out, at any given time a young adult cohort is tied to older cohorts by child-parent relationships. Because the attitudes and well-being of parents are likely to be influenced by the well-being of their children, then disaffection among young adults arising from the effects of large cohort size may precipitate similar reactions among their parents. Such interdependence among age groups would undermine an attempt to distinguish cohort from age effects by purely statistical means.

The fortunes of a cohort depend, of course, on a number of factors in addition to cohort size alone. A cohort's labour market experience will depend on conditions of labour demand as well as labour supply. Moreover, labour supply itself depends, not only on prior birth rates, but also on exogenous changes in immigration and labour force participation rates (though the latter may, of course, reflect endogenous influences as well). In addition, government policies may modify significantly the impact of free market forces on different age groups. For these reasons, one would expect cohort size effects to dominate a cohort's experience only when other factors are relatively constant.

These considerations explain why, in empirical applications, the cohort size hypothesis has come to the fore only recently, and particularly in the United States. With the implementation of governmental fiscal and monetary policies in the post-World War II period, growth of aggregate demand was largely stabilized. Also, the relative impact of immigration on the labour force was greatly reduced as a result of restrictive legislation adopted in the 1920s. Before World War II the economic experience of a native born cohort was dominated by major swings in aggregate demand and directly associated movements in immigration, so-called 'Kuznets cycles' of 15 to 25 years in length (Abramovitz, 1961; Easterlin, 1968). After World War II, however, movements in the growth of aggregate demand and immigration were comparatively much smaller. In contrast, swings in cohort size due to birth rate changes became much greater. In the century and a half before 1940 the birth rate trended downward with only intermittent intervals of slowing or levelling off. Then, in the 1940s and 1950s, the birth rate surged dramatically upward, reaching a 1957 peak almost 40 per cent above its all-time low in 1936. Subsequently, in the 1960s and 1970s, in just as startling fashion, the birth rate turned downward, plunging to a new low in 1975–76, before levelling off.

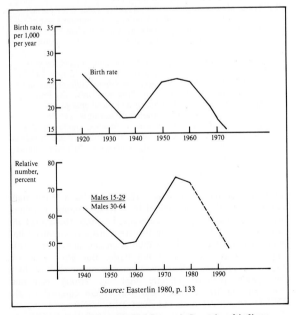

Figure 1 The Lagged Effect of the Birth Rate on the Proportion of the Young Adult Population, United States, 1920-1990

The swing in the United States birth rate from 1930s low to 1950s peak and back to a 1970s low caused with about a twenty-year lag a corresponding swing in the proportion of the adult population reaching working and family forming age, approximated empirically by the proportion of those aged 15–29 to 30–64 (Figure 1). As a result, for the young adult population, the post-World War II period witnessed decreasing relative cohort size to about 1960; then, increasing size to around 1980; and, thereafter, a return to decreasing size. Associated with these movements were corresponding changes in the economic and social behaviour of young adults of the type anticipated by the hypothesis (Easterlin, 1980). For example, the 1940s and 1950s saw the emergence of relatively favourable conditions in the labour market for young adults, a shift toward accelerated family formation (the 'baby boom'), cessation of the prior uptrend in labour force participation of young women, and comparatively favourable changes in social indicators relating to illegitimacy, divorce, suicide and crime. Then, in the 1960s and 1970s as the size of the young adult population soared upward relative to the older, younger workers' relative wage rates, unemployment rates, and upward job mobility deteriorated. Marriage was deferred and childbearing reduced, while labour force participation of younger women rose dramatically. At the same time rates of illegitimacy, crime, divorce, and suicide among young adults turned sharply upward. Subsequently, in the 1980s as relative cohort size for the young adult population levelled off and then started to decrease, a number of these conditions began to show signs once again of reversal.

As was noted, the adaptation in fertility behaviour implied by large cohort size implies that a self-correcting mechanism is at work. Because a large 'baby boom' cohort tends to have fewer children, and more widely spaced, it tends to be succeeded by a smaller 'baby bust' cohort, for whom crowding effects in the family, school, and labour market are disproportionately small. The good fortunes of this 'baby bust' cohort would be expected to give rise, in turn, to an upsurge in fertility and a subsequent large cohort. Assuming this cohort size mechanism were to dominate fertility behaviour over time, the result would be a self-generating fertility cycle of around four to five decades in length with associated changes in a wide range of social and economic phenomena induced by the fluctuations in relative cohort size. Pre-World War II Kuznets cycles of 15 to 25 years due to aggregate demand fluctuations would be replaced by post-World War II swings of double the length originating in fertility movements operating with a lagged effect on the age structure of labour supply (Ahlburg, 1982). The possibility of self-generating fertility waves has spurred the development of mathematical models (Lee, 1979; Samuelson, 1976).

In a number of industrialized countries in addition to the United States, the post-World War II era saw the adoption of macroeconomic policies aimed at stabilizing the growth of aggregate demand, along with a baby boom. Thus, there is some reason to think that cohort size effects like those in the United States might be observed, allowing, of course, for differences in the relative magnitude of cohort size vis-à-vis other factors, and for disparate institutional conditions such as government policies that, intentionally or not, might mitigate the effects of free market forces. Although no comprehensive investigation has been undertaken, there are some indications that this has been the case. For example, findings supportive of the hypothesis have been reported for labour force participation in Canada (Robertson and Roy, 1982), wage rates in Japan (Mosk and Nakata, 1985), and fertility in various developed countries (O'Connell, 1978). Some fertility

3

studies, however, have reached sceptical or mixed conclusions (Butz and Ward, 1979; Devaney, 1983).

The empirical studies cited above uniformly relate to time series data. Some critics of the cohort size hypothesis have maintained that the expected effects are not evident in micro level data or, if present, are, at best, mixed or not strong (Olneck and Wolfe, 1978; MacDonald and Rindfuss, 1978), though there are also micro studies reporting consistent findings (Elder, 1981; Coombs and Zumeta, 1970). Contradictions between cross-section and time series findings are not new to social science and there is a long history of attempts to resolve them – perhaps the best-known example in economics is the savings–income relationship. Whether in the present case a contradiction does, in fact, exist, and, if so, why, is as yet undecided.

Over the last few decades, demographic concepts have encroached modestly on economic theory, as evidenced by the appearance of life cycle, overlapping generations and vintage models. The cohort size hypothesis might be viewed as another in this sequence. Its roots, however, extend beyond economics, reaching out into sociology, demography and psychology, and it seeks to encompass a wider range of attitudinal and behavioural phenomena than is traditionally considered economic.

RICHARD A. EASTERLIN

See also DEMOGRAPHIC TRANSITION; FAMILY; FERTILITY; KUZNETS' SWINGS; POPULATION CYCLES.

BIBLIOGRAPHY

Abramovitz, M. 1961. The nature and significance of Kuznets cycles. *Economic Development and Cultural Change* 9(3), April, 225–48.

Ahlburg, D.A. 1982. The new Kuznets cycle: a test of the Easterlin–Wachter–Wachter hypothesis. *Research in Population Economics* 4, 93–115.

Ahlburg, D.A. and Schapiro, M.O. 1984. Socioeconomic ramifications of changing cohort size: an analysis of US postwar suicide rates by age and sex. *Demography* 21(1), February, 97–108.

Anderson, J. 1982. An economic–demographic model of the United States labor market. *Research in Population Economics* 4, 117–53.

Behrman, J., Hrubec, Z., Taubman, P. and Wales, T., 1980. *Socioeconomic Success: A Study of the Effects of Genetic Endowments, Family Environment and Schooling.* Amsterdam: North-Holland.

Behrman, J.R., Pollak, R. and Taubman, 1982. Parental preferences and provision for progeny. *Journal of Political Economy* 90(1), February, 52–73.

Berger, M.C. 1985. The effect of cohort size on earnings growth: a reexamination of the evidence. *Journal of Political Economy* 93(3), June, 561–73.

Butz, W. and Ward, M. 1979. The emergence of countercyclical US fertility. *American Economic Review* 69(3), June, 318–28.

Coombs, L.C. and Zumeta, Z. 1970. Correlates of marital dissolution in a prospective fertility study: a research note. *Social Problems* 18(1), Summer, 92–101.

Devaney, B. 1983. An analysis of variations in U.S. fertility and female labor force participation trends. *Demography* 20(2), May, 147–61.

Duesenberry, J.S. 1949. *Income, Saving, and the Theory of Consumer Behavior.* Cambridge, Mass.: Harvard University Press.

Durkheim, E. 1897. *Le suicide: étude de sociologie,* Paris: Felix Alcan. Trans. as *Suicide: A Study in Sociology,* New York: Free Press, 1951.

Easterlin, R.A. 1968. *Population, Labor Force, and Long Swings in Economic Growth.* New York: National Bureau of Economic Research.

Easterlin, R.A. 1980. *Birth and Fortune.* New York: Basic Books.

Elder, G.H. 1981. Scarcity and prosperity in postwar childbearing: explorations from a life course perspective. *Journal of Family History* 6(4), Winter, 410–33.

Ernst, C. and Angst, J. 1983. *Birth Order: Its Influence on Personality.* Berlin: Springer.

Espenshade, T.J. 1985. Marriage trends in America: estimates, implications, and underlying causes. *Population and Development Review* 11(2), June, 193–245.

Festinger, L. 1957. *A Theory of Cognitive Dissonance.* Evanston, Ill.: Row, Peterson.

Freeman, R.B. 1976. *The Overeducated American.* New York: Academic Press.

Heer, D.M. 1985. Effects of sibling number on child outcome. *Annual Review of Sociology* 11, 27–47.

Hobcraft, J., Menken, J. and Preston, S. 1982. Age, period, and cohort effects in demography: a review. *Population Index* 48(1), Spring, 4–43.

Lee, R. 1974. The formal dynamics of controlled populations and the echo, the boom and the bust. *Demography* 11(4), November, 563–85.

MacDonald, M.M. and Rindfuss, R.R. 1978. Relative economic status and fertility: evidence from a cross-section. *Research in Population Economics* 1, 291–308.

Mason, W.M., Mason, K.O. and Winsborough, H.H. 1976. Reply to Glenn. *American Sociological Review* 41(5), October, 904–5.

Merton, R.K. 1968. *Social Theory and Social Structure.* New York: Free Press.

Modigliani, F. 1949. Fluctuations in the saving–income ratio: a problem in economic forecasting. In Conference on Research in Income and Wealth, *Studies in Income and Wealth* No. 11, New York: National Bureau of Economic Research.

Moffitt, R.A. 1982. Postwar fertility cycles and the Easterlin hypothesis: a life-cycle approach. *Research in Population Economics* 4, 237–52.

Mosk, C. and Nakata, Y.-F. 1985. The age-wage profile and structural change in the Japanese labor market for males, 1964–1982. *Journal of Human Resources* 20(1), 100–16.

O'Connell, M. 1978. The effect of changing age distribution on fertility: an international comparison. *Research in Population Economics* 1, 233–46.

Olneck, M. and Wolfe, B. 1978. A note on some evidence on the Easterlin hypothesis. *Journal of Political Economy* 86(5), October, 953–8.

Preston, S.H. and McDonald, J. 1979. The incidence of divorce within cohorts of American marriages contracted since the Civil War. *Demography* 16(1), February, 1–25.

Robertson, M. and Roy, A.S. 1982. Fertility, labor force participation and the relative income hypothesis: an empirical test of the Easterlin–Wachter model on the basis of Canadian experience. *American Journal of Economics and Sociology* 41(4), October, 339–50.

Samuelson, P.A. 1976. An economist's non-linear model for self-generated fertility waves. *Population Studies* 30(2), July, 243–7.

Stouffer, S.A. et al. 1949. *The American Soldier: Adjustment During Wartime Life.* Vol. 1, Princeton: Princeton University Press.

Waring, J. 1975. Social replenishment and social change. *American Behavioral Scientist* 19(2), November–December, 237–56.

Wray, J.D. 1971. Population pressure on families: family size and child spacing. In National Academy of Sciences, *Rapid Population Growth: Consequences and Policy Implications,* Baltimore: Johns Hopkins Press.

East–West economic relations. The decade 1966–75 is usually considered as the golden age of East–West economic relations. Already during the previous decade, i.e. since the end of the cold war, the USSR and the Eastern European countries had increased their trade with the West at an annual rate of growth slightly higher than their total trade. But after 1966 the expansion of trade and cooperation was sustained both by a favourable political climate and by strong economic complementarities between the West (here equated to the OECD countries) and the East (the USSR and the six European countries that are members of the CMEA, or Council for Mutual Economic Assistance; hereafter we shall mention them as CPEs or centrally planned economies, for the sake of brevity).

These years were marked by *détente*, initiated in 1966 with the triumphal visit to the USSR of the French President General de Gaulle. This was not only a bilateral event, but it set the stage for diversified and institutionalized links between Eastern and Western European economies. Later on, in 1972, US President Nixon's visit to Moscow opened the shorter phase of bright US–USSR economic relations which ended in 1975. At the beginning of that year, the Soviet Union unilaterally repudiated the Soviet–American treaty of commerce, as a retaliation for the deprivation of the most favoured nation clause; according to the American legislation just introduced, the clause could not be granted to a country restricting the rights for its citizens to emigrate. Before *détente* came altogether to its end, it was symbolically magnified in the final Act of the Conference for Security and Co-operation in Europe, signed in Helsinki in August 1975. The economic 'basket' of this text was meant to appear as the Charter of East–West mutually profitable relations.

From the economic point of view, the 1966–75 decade was indeed a time of converging interests. The USSR and Eastern European countries had just engaged in economic reforms. They needed to modernize their industries. The Western firms found new markets for selling equipment and turnkey plants. High rates of economic growth, both in the West and in the East, sustained the prospects for increased exports from the East to the West, once the new capacities acquired from the West were put into operation. An era of deepening industrial cooperation, based upon technology imports and reverse flows of manufactured goods, seemed to open.

It was then almost forgotten that even in such a favourable context, East–West trade accounted for less then 3 per cent of world trade. While in 1975 it amounted to slightly under 30 per cent of total trade for the CPEs (slightly more for the USSR and less for the six smaller CPEs taken together), it never exceeded 5 per cent of total trade for the Western countries, except for some non-typical cases (such as Austria or Finland).

The following decade, ending in 1985, has witnessed a general shrinking of East–West trade. There was a conspicuous deterioration of the political climate with the invasion of Afghanistan by the Soviet troops in December 1979 and, two years later, martial law in Poland. The world economic crisis exerted some adverse effects as well. True, it benefited the USSR as an oil exporter. But the Western recession hampered the export drive of the smaller CPEs. The manufactured goods which they intended to export so as to repay their imports of equipment became less saleable in the East. Thus the imbalance between imports and exports, which had been steadily growing since 1970, could not be corrected through expanded sales. An easier way out was to borrow on Western financial markets. The CPEs were still creditworthy, and the level of international liquidity was high as a result of the inflow of petro-dollars. The total indebtedness of the CPEs culminated in 1981. The subsequent adjustments conducted in 1981–3 (though a decrease in imports and domestic investment) ended up with a marked improvement in the CPEs external financial position and with a decrease in their foreign debt (except for Poland). But the general slowdown of growth in the East, partly due to these adjustments, does not allow for a steep upward trend in East–West trade.

The outlook for East–West economic relations is to be evaluated through the combination of two opposed sets of factors. On the one hand, there are strong interests on both sides pressing for the expansion of trade and cooperation. On the other, equally strong obstacles are hindering such a development. The outcome is probably to be seen in a stabilization of those relations, below the level reached during the 'golden age' decade.

ECONOMIC INTERESTS. East–West trade is sometimes said to be a one-way street. As the magnitudes of shares in total trade show, these relations are several times more important for the East than for the West. However, dependencies are to be found on both sides, with an uneven distribution.

In the West, European countries are the main group of partners. They account for roughly 75 per cent of sales to the East and 90 per cent of imports from the East (figures of 1983). This pattern has been stable since the end of the seventies. In 1970 the share of Western Europe was very similar on the import side, but larger on the export side (about 10 points more). Since then, two major exporters have emerged outside Europe, Japan (for technology) and the United States (for grain, mainly to the USSR).

In the East, the USSR gained a growing share of East–West trade after 1970. From two-fifths of the total trade of the European CPEs with the West, it reached 50 per cent in the mid-1970s and over 60 per cent in the 1980s. This is mainly due to the increase in oil prices after 1973; it allowed the Soviet Union to secure a higher rate of growth of its trade with the West compared with the other CPEs up to 1980, and to avoid the decrease in trade which the other CPEs experienced in the beginning of the 1980s.

This growing concentration of East–West trade on the Soviet Union is an expression of stronger interdependences.

For Western Europe, especially for the large industrial corporations, the USSR emerged in the 1970s as a major purchaser of heavy equipment, whose orders helped to sustain the level of activities and jobs during the recession years. The controversial multi-billion dollars gas pipeline deal concluded in 1981 is a clear demonstration of such interests. When in 1982 the US government tried to oppose the supply of tubes and other equipment for the pipeline, as a retaliation for the Soviet role in the Polish crisis, and also as an attempt to reduce the export capacities of natural gas of the USSR, the European governments backed their firms. Even though the Soviet orders for equipment substantially declined after then, the Soviet Union remains a huge market.

On the other side, the Soviet Union has become a significant supplier of energy to Western Europe. Fuels now account for about 80 per cent of its sales to the West, from about half that share in the beginning of the 1970s. The major Western European energy importers (Germany, France, Italy) are now dependent for 6–7 per cent of their total energy imports on the Soviet Union. For natural gas alone, their dependence may be above the 30 per cent mark at the end of the 1980s, from about 15–20 per cent a decade earlier. The Soviet market is a means of achieving a diversification in energy imports; it is a cheaper supplier for oil and gas because of the distance factor, and may be considered as a more reliable one, than the Third World.

Regarding trade with the United States, the major link is grain. The Soviet Union began to buy large quantities of American grain in 1975–76 and has been the largest single customer of the United States since then. US sales never again reached the 70 per cent share of Soviet grain imports which they formed in 1979. However, the strength of economic versus political interests is clearly demonstrated by the failure of the grain embargo, which had to be lifted under the pressure of American farmers. The long-term grain sales agreement linking the two countries, first signed in 1975, has not only been renewed but also supplemented with an anti-embargo clause (in 1983).

The Western trade of Eastern Europe lacks these powerful interdependences. The smaller CPEs taken together are on average less involved in trade with the West than the USSR. In 1984, the share of Western trade in their total trade was about 25 per cent (against 30 per cent for the USSR) and had been declining since 1980. But while Bulgaria and Czechoslovakia, much more oriented toward trade within CMEA, have a very low share of their total trade with the West (12–15 per cent), Hungary (35 per cent), Poland, GDR and Romania (30 per cent) are potentially interested in expanding their trade with the West. However, opportunities for that are low. Their supply is made of sensitive goods (steel, chemicals, textiles, manufactured goods, agricultural products), the demand for which is sluggish in the West – and they complain of growing protectionism. For these goods competition is growing on Western markets from the new industrializing countries of the Third World, which in addition are more advanced in some high technology fields (electronics). They can hardly expect concessions from Western countries, for which they provide less promising markets than the USSR. The development of compensation deals is only a marginal way of securing outlets for their goods.

OBSTACLES. In the background of these differentiated economic interests, specific obstacles hinder East–West trade, in the political, institutional (systemic) and financial fields, to which must be added the 1986 developments on the world oil market.

Is East–West trade *political* in essence? In Western Europe, politics and economic relations are regarded as distinct by governments and firms. The lasting failure to find an agreement between EEC and the CMEA, since the beginning of official talks in 1976, is mainly due to the lack of institutional competence of the CMEA in matters of trade as appraised by the EC Commission (even if on the side of the Commission there is a political concern to avoid strengthening the Soviet-dominated CMEA as an organization). The major involvement of politics in East–West economic relations is related to US policy. The 'linkage' concept of tying economic advantages to Soviet concessions in the political sphere was associated in the 1970s with commercial policies (the granting of the MFN clause) or financial conditions (for access to bank credits). Since the end of that decade it has evolved into a policy of sanctions, first as a retaliation for the Soviet invasion of Afghanistan in 1979 (the grain embargo against the USSR, which was lifted in April 1981, and a tighter control of high technology sales); then as a response to the martial law imposed in December 1981 in Poland. In this last case the sanctions hit Poland (though credit and export restrictions, a suspension of the MFN clause), and the USSR (through attempts to stop the Eurosiberian pipeline deal by preventing the Western European countries from selling equipment to the USSR and from concluding the agreements for the purchases of gas). They were also extended to the other CPEs through a very severe credit squeeze. All these measures culminated in 1982. They proved largely ineffective but generated conflicts within the Western Alliance. The major and lasting field of political pressure is to be found in the embargo on high technology sales to the CPEs, conducted through the Cocom (Coordinating Committee), an informal organization set up in 1949 and including the NATO countries plus Japan. Very active during the years of the cold war, it seemed to be withering in the late 1970s but regained momentum from 1980 on. The present rationale of the Cocom restrictions is threefold: to impose sanctions; to prevent the Soviet bloc from acquiring dual-use technologies (for military as well as civilian ends); to enlarge the scope of controls by restricting high-technology exports of non-Cocom members (Sweden, Switzerland, Austria, and even some Third World countries such as India).

The *systemic* obstacles in trade are related to the specific organization of state trading in the CPEs. The monopoly of foreign trade and the related planning of trade flows remain very rigid in the Soviet Union. Increased flexibility has been introduced in the trade mechanisms of all the other CPEs, where enterprises are gaining easier access to foreign trade transactions. Direct interfirm contacts have been stimulated through industrial cooperation. In all these countries except for GDR, it is now possible to create joint enterprises with foreign equity capital (the experiences remain limited). The state trading system, however reformed, still prevents the CPEs from successfully adjusting to the market requirements in the West.

The *financial* problems of East–West relations are less dramatic than in 1980–81, when the total indebtedness of the USSR and Eastern Europe combined exceeded $80 billion, more than four times its level of the end of 1974. Two countries, Poland and Romania, entered in 1981 a process of rescheduling, which is still going on for Poland. Two others, GDR and Hungary, successfully managed to restore their external accounts in 1982–4. Since then, the Western banks have again been ready to expand their loans not only to the Soviet Union, which has always remained a good risk, but also to the other CPEs, which by all accounts seem more creditworthy than the Third World.

East–West economic relations are finally to be replaced in the broader context of the CPEs' foreign economic relations, including intra-CMEA trade. The move toward closer integration, advocated by the Soviet Union at the Summit meeting of the CMEA in June 1984 and based upon the heavy requirements of the USSR regarding its imports from its partners, might well appear as an additional constraint to the expansion of East–West relations for the smaller CPEs.

The fall in oil prices, since the end of 1985, may have strong adverse effects on East–West trade. If the average price for oil is for some time stabilized at half its 1985 level, the Soviet Union will lose at least one third of its export gains in its trade with the West. These losses may be compensated for, in the short run, by cuts in imports and increased borrowing, together with a stronger pressure on the smaller CMEA countries. The latter will thus have to divert to the Soviet market goods exportable to the West. In addition, they too will lose as sellers of refined oil products, with the same consequences as for the USSR. The 'golden age' of East–West trade is definitely not to be renewed.

MARIE LAVIGNE

See also CONVERGENCE HYPOTHESIS; CYCLES IN SOCIALIST ECONOMIES; SOCIALIST ECONOMIES.

BIBLIOGRAPHY

Bornstein, M., Gitelman, Z., Zimmerman, W. (eds) 1981. *East–West Relations and the Future of Eastern Europe: Politics and Economics.* London: Allen and Unwin.

Economic Bulletin for Europe. 1949 onwards. Geneva: Economic Commission for Europe, United Nations. Each volume contains developments on East–West trade.

Fallenbuchl, Z. and McMillan, C. (eds) 1980. *Partners in East–West Relations, the Determinants of Choice.* New York: Pergamon Press.

Holzman, F. 1976. *International Trade Under Communism, Politics and Economics.* New York: Basic Books.

Lavigne, M. 1979. *Les relations économiques Est–Ouest.* Paris: Presses Universitaires de France.

Lavigne, M. 1985. *Economie internationale des pays socialistes.* Paris: Armand Colin.

Levcik, F. (ed.) 1978. *International Economics – Comparisons and Interdependencies. Essays in honour of F. Nemschak.* Vienna: Springer Verlag.

Marer, P. and Montias, J.M. (eds) 1980. *East European Integration and East–West Trade.* Bloomington: Indiana University Press.

Eckstein, Otto (1927–1984).

Eckstein was an entrepreneur who moved a whole technology from the research community into the market place. Until he founded Data Resources, Inc., macro-econometric models were research vehicles and not vehicles for aiding business decision making. Under his direction Data Resources came to dominate the market place for this type of information, but more importantly it changed the nature of the game. To be taken seriously after his innovation, all economic forecasts had to be buttressed with econometric equations and no large firm would attempt to begin its decision-making processes without an understanding of the national and international economic forecasts emanating from such models.

Born in Ulm, Germany, in 1927, Dr Eckstein fled to England in 1938 and came to the United States in 1939. He graduated from Stuyvesant High School in New York City and served in the United States Army Signal Corps from 1946 to 1947. He received and AB degree from Princeton University in 1951 and a PhD from Harvard University in 1955.

In 1968, he and Donald B. Marron founded Data Resources, Inc., which has grown into the largest economic information company in the world. The firm became a subsidiary of McGraw-Hill, Inc. in 1979. He directed the development of the Data Resources Model of the US economy, and was responsible for its forecasting operations.

As an immigrant to the United States from Nazi Germany, Otto Eckstein wanted to contribute something to America's future success. Better economic policies that would lead to a higher American standard of living were not an abstraction to him. They were the centre of his professional life.

His professional career began with the analysis of large scale multi-year water resources projects and how one might better allocate national resources in such projects. In the late 1950s he was the principal intellectual director of a Joint Economic Committee study on how the United States might break out of what was then seen as the stagnation of the mid-1950s. His study on growth, full employment and price stability laid the basis for the successful economic policies that were followed in the first two-thirds of the 1960s. But he went on to implement those intellectual foundations as a member of the President's Council of Economic Advisers under President Johnson.

No one who knew the enthusiasm of Otto Eckstein for studying, teaching, and practising economics could thereafter think of economics as the dismal science.

LESTER C. THUROW

SELECTED WORKS

1958a. *Water Resources Development: The Economics of Project Evaluation.* Cambridge, Mass.: Harvard University Press.

1958b. (With J.V. Krutilla.) *Multiple Purpose River Development.* Baltimore: Johns Hopkins Press.

1964a. *Public Finance.* New York: Prentice-Hall. 4th edn, 1979.

1964b. (With E.S. Kirschen and others.) *Economic Policy in Our Time.* Amsterdam: North-Holland.

1967 (ed.) *Studies in the Economics of Income Maintenance.* Washington, DC: Brookings.

1970 (ed.) *The Econometrics of Price Determination.* Washington, DC: Board of Governors of the Federal Reserve System and Social Science Research Council.

1976 (ed.) *Parameters and Policies in the U.S. Economy.* Amsterdam: North-Holland.

1978. *The Great Recession.* Amsterdam: North-Holland.

1981. *Core Inflation.* New York: Prentice-Hall.

1983. *The DRI Model of the U.S. Economy.* New York: McGraw-Hill.

1984. (With Christopher Caton, Roger Brinner and Peter Duprey.) *The DRI Report on U.S. Manufacturing Industries.* New York: McGraw-Hill.

Ecole Nationale des Ponts et Chaussées.

French School of Civil Engineering, located at 28 rue des Saint-Pères, Paris. Established in 1747 by Daniel Trudaine, Finance Minister to Louis XV, the Ecole has traditionally produced economists of exceptional talent and originality. Isnard, Dupuit and Cheysson were students there and at various times its faculty included the likes of Henri Navier, Joseph Minard, Joseph Garnier, Henri Baudrillart, Charles Gide, Clément Colson and François Divisia.

The idea of an institution dedicated to the professionalization of French engineers had its roots in the 17th century. In 1690 Vauban created the Corps of Military Engineers, which was to serve as a model for future public bodies of this sort. He even went so far as to propose a public examination to test the scientific knowledge of young people aspiring to become engineers. After an inauspicious beginning, the Ecole slowly acquired more scholarly aspirations. It was directed in its formative years by J.R. Perronet, who established the high standards and pedagogical technique responsible for the ultimate success of the school, so much so that French engineers became the envy of the world. Although a formal course in economics was not established until 1847 (receiving impetus from Dupuit's pioneer researches in 1844), engineers were 'officially' exhorted to study economics as early as 1792.

The Revolution of 1789 brought sweeping changes to higher education in France. For a time it seemed as though the Ecole would be swept away as a vestige of the *ancien régime*, but Mirabeau successfully defended its existence, and by the time Napoleon came to power, a major expansion of faculty, students and curriculum was under way. With the establishment of the Ecole Polytechnique in 1794, the nature of the Ecole des Ponts et Chaussées changed from an undergraduate to a postgraduate institution, offering admission by competitive examination and specialized training for polytechnicians seeking to become civil engineers. These civil engineers became problem-solvers in the areas of flood control, municipal water distribution and sewage disposal, canal building, railway construction, road building and myriad other matters of concern to engineers.

The 19th century was the 'golden age' of the Ecole, a time when the faculty was upgraded and the curriculum was stretched to include stereotomy (1799), modern languages (1806), mineralogy and geology (1826), administrative law (1831), political economy (1847), thermodynamics (1851), and applied chemistry (1864). The role of the Ecole was pivotal and international in both engineering and economics. Henri Navier, for example, was sent in 1821 and in 1823 by the Director General of the Corps to study British achievements in suspension bridge design and construction. Upon his return Navier, who wrote a number of essays on the economic value of public works, offered a *Mémoire sur les ponts suspendus* which brought the French to the forefront of such technology for much of the 19th century. Jules Dupuit entered the Ecole in 1824, where he reacted to both Navier's engineering and economic studies, later producing a theory of marginal utility and a full scale welfare analysis of markets and market structure. In 1830 an American student, Charles Ellet, Jr., entered the Ecole, returning home as the premier suspension bridge builder and designer of his age *and* as one of the most creative American economic theorists of the century. In short, the 19th century is the period when economic inquiry at the Ecole burgeoned, easily outdistancing the policy squabbles that occupied French academic economists at the universities and in academic journals. It was the era of Dupuit, Cheysson and Colson, the unrecognized giants of 19th century French economics.

Today the Ecole des Ponts et Chaussées stands as the oldest of France's *grandes écoles*. Perched at the top of a rigid and highly centralized educational system, it persists in admitting the country's intellectual elite and in providing them with solid training in economics.

ROBERT B. EKELUND, JR. AND ROBERT F. HÉBERT

BIBLIOGRAPHY

Ekelund, R.B., Jr. and Hébert, R.F. 1973. Public economics at the Ecole des Ponts et Chaussées: 1830–1850. *Journal of Public Economics* 2(3) July, 241–56.

econometrics.

1 WHAT IS ECONOMETRICS? Econometrics is a rapidly developing branch of economics which, broadly speaking, aims to give empirical content to economic relations. The term 'econometrics' appears to have been first used by Pawel Ciompa as early as 1910; although it is Ragnar Frisch, one of the founders of the Econometric Society, who should be given the credit for coining the term, and for establishing it as a subject in the sense in which it is known today (see Frisch, 1936, p. 95). Econometrics can be defined generally as 'the application of mathematics and statistical methods to the analysis of economic data', or more precisely in the words of Samuelson, Koopmans and Stone (1954),

... as the quantitative analysis of actual economic phenomena based on the concurrent development of theory and observation, related by appropriate methods of inference (p. 142).

Other similar descriptions of what econometrics entails can be found in the preface or the introduction to most texts in econometrics. Malinvaud (1966), for example, interprets econometrics broadly to include 'every application of mathematics or of statistical methods to the study of economic phenomena'. Christ (1966) takes the objective of econometrics to be 'the production of quantitative economic statements that either *explain* the behaviour of variables we have already seen, or *forecast* (i.e. predict) behaviour that we have not yet seen, or both'. Chow (1983) in a more recent textbook succinctly defines econometrics 'as the art and science of using statistical methods for the measurement of economic relations'.

By emphasizing the quantitative aspects of economic problems, econometrics calls for a 'unification' of measurement and theory in economics. Theory without measurement, being primarily a branch of logic, can only have limited relevance for the analysis of actual economic problems. While measurement without theory, being devoid of a framework

necessary for the interpretation of the statistical observations, is unlikely to result in a satisfactory explanation of the way economic forces interact with each other. Neither 'theory' nor 'measurement' on their own is sufficient to further our understanding of economic phenomena. Frisch was fully aware of the importance of such a unification for the future development of economics as a whole, and it is the recognition of this fact that lies at the heart of econometrics. This view of econometrics is expounded most eloquently by Frisch (1933a) in his editorial statement and is worth quoting in full:

... econometrics is by no means the same as economic statistics. Nor is it identical with what we call general economic theory, although a considerable portion of this theory has a definitely quantitative character. Nor should econometrics be taken as synonymous with the application of mathematics to economics. Experience has shown that each of these three view-points, that of statistics, economic theory, and mathematics, is a necessary, but not by itself a sufficient, condition for a real understanding of the quantitative relations in modern economic life. It is the *unification* of all three that is powerful. And it is this unification that constitutes econometrics.

This unification is more necessary today than at any previous stage in economics. Statistical information is currently accumulating at an unprecedented rate. But no amount of statistical information, however complete and exact, can by itself explain economic phenomena. If we are not to get lost in the overwhelming, bewildering mass of statistical data that are now becoming available, we need the guidance and help of a powerful theoretical framework. Without this no significant interpretation and coordination of our observations will be possible.

The theoretical structure that shall help us out in this situation must, however, be more precise, more realistic, and, in many respects, more complex, than any heretofore available. Theory, in formulating its abstract quantitative notions, must be inspired to a larger extent by the technique of observation. And fresh statistical and other factual studies must be the healthy element of disturbance that constantly threatens and disquiets the theorist and prevents him from coming to rest on some inherited, obsolete set of assumptions.

This mutual penetration of quantitative economic theory and statistical observation is the essence of econometrics (p. 2).

Whether other founding members of the Econometric Society shared Frisch's viewpoint with the same degree of conviction is, however, debatable, and even today there are no doubt economists who regard such a viewpoint as either ill-conceived or impractical. Nevertheless, in this survey I shall follow Frisch and consider the evolution of econometrics from the unification viewpoint.

2 EARLY ATTEMPTS AT QUANTITATIVE RESEARCH IN ECONOMICS. Empirical analysis in economics has had a long and fertile history, the origins of which can be traced at least as far back as the work of the 16th-century Political Arithmeticians such as William Petty, Gregory King and Charles Davenant. The political arithmeticians, led by Sir William Petty, were the first group to make systematic use of facts and figures in their studies. (See, for example, Stone (1984) on the origins of national income accounting.) They were primarily interested in the practical issues of their time, ranging from problems of taxation and money to those of international trade and finance. The hallmark of their approach was undoubtedly

quantitative and it was this which distinguished them from the rest of their contemporaries. Political arithmetic, according to Davenant (1698, Part I, p. 2) was 'the art of reasoning, by figures, upon things relating to government', which has a striking resemblance to what might be offered today as a description of econometric policy analysis. Although the political arithmeticians were primarily and understandably preoccupied with statistical measurement of economic phenomena, the work of Petty, and that of King in particular, represented perhaps the first examples of a unified quantitative/theoretical approach to economics. Indeed Schumpeter in his *History of Economic Analysis* (1954) goes as far as to say that the works of the political arithmeticians 'illustrate to perfection, what Econometrics is and what Econometricians are trying to do' (p. 209).

The first attempt at quantitative economic analysis is attributed to Gregory King, who is credited with a price–quantity schedule representing the relationship between deficiencies in the corn harvest and the associated changes in corn prices. This demand schedule, commonly known as 'Gregory King's law', was published by Charles Davenant in 1699. The King data are remarkable not only because they are the first of their kind, but also because they yield a perfectly fitting cubic regression of price changes on quantity changes, as was subsequently discovered independently by Whewell (1850), Wicksteed (1889) and by Yule (1915). An interesting account of the origins and nature of 'King's law' is given in Creedy (1986).

One important consideration in the empirical work of King and others in this early period seems to have been the discovery of 'laws' in economics, very much like those in physics and other natural sciences. This quest for economic laws was, and to a large extent still is, rooted in the desire to give economics the status that Newton had achieved for physics. This was in turn reflected in the conscious adoption of the method of the physical sciences as the dominant mode of empirical enquiry in economics. The Newtonian revolution in physics, and the philosophy of 'physical determinism' that came to be generally accepted in its aftermath, had far-reaching consequences for the method as well as the objectives of research in economics. The uncertain nature of economic relations only began to be fully appreciated with the birth of modern statistics in the late 19th century and as more statistical observations on economic variables started to become available. King's law, for example, was viewed favourably for almost two centuries before it was questioned by Ernest Engel in 1861 in his study of the demand for rye in Prussia (see Stigler, 1954, p. 104).

The development of statistical theory at the hands of Galton, Edgeworth and Pearson was taken up in economics with speed and diligence. The earliest applications of simple correlation analysis in economics appear to have been carried out by Yule (1895, 1896) on the relationship between pauperism and the method of providing relief, and by Hooker (1901) on the relationship between the marriage-rate and the general level of prosperity in the United Kingdom, measured by a variety of economic indicators such as imports, exports, and the movement in corn prices. In his applications Hooker is clearly aware of the limitations of the method of correlation analysis, especially when economic time series are involved, and begins his contribution by an important warning which continues to have direct bearing on the way econometrics is practised today:

> The application of the theory of correlation to economic phenomena frequently presents many difficulties, more especially where the element of time is involved; and it by no means follows as a matter of course that a high correlation coefficient is a proof of causal connection between any two variables, or that a low coefficient is to be interpreted as demonstrating the absence of such connection (p. 485).

It is also worth noting that Hooker seems to have been the first to use time lags and de-trending methods in economics for the specific purpose of avoiding the time-series problems of spurious or hidden correlation that were later emphasized and discussed formally by Yule (1926).

Benini (1907), the Italian statistician, according to Stigler (1954) was the first to make use of the method of multiple regression in economics. He estimated a demand function for coffee in Italy as a function of coffee and sugar prices. But as argued in Stigler (1954, 1962) and more recently detailed in Christ (1985), it is Henry Moore (1914, 1917) who was the first to place the statistical estimation of economic relations at the centre of quantitative analysis in economics. Through his relentless efforts, and those of his disciples and followers Paul Douglas, Henry Schultz, Holbrook Working, Fred Waugh and others, Moore in effect laid the foundations of 'statistical economics', the precursor of econometrics. Moore's own work was, however, marred by his rather cavalier treatment of the theoretical basis of his regressions, and it was therefore left to others to provide a more satisfactory theoretical and statistical framework for the analysis of economic data. The monumental work of Schultz, *The Theory and the Measurement of Demand* (1938), in the United States and that of Allen and Bowley, *Family Expenditure* (1935), in the United Kingdom, and the pioneering works of Lenoir (1913), Wright (1915, 1928), Working (1927), Tinbergen (1930) and Frisch (1933b) on the problem of 'identification' represented major steps towards this objective. The work of Schultz was exemplary in the way it attempted a unification of theory and measurement in demand analysis; whilst the work on identification highlighted the importance of 'structural estimation' in econometrics and was a crucial factor in the subsequent developments of econometric methods under the auspices of the Cowles Commission for Research in Economics.

Early empirical research in economics was by no means confined to demand analysis. Another important area was research on business cycles, which in effect provided the basis of the later development in time-series analysis and macroeconometric model building and forecasting. Although, through the work of Sir William Petty and other early writers, economists had been aware of the existence of cycles in economic time series, it was not until the early 19th century that the phenomenon of business cycles began to attract the attention that it deserved. (An interesting account of the early developments in the analysis of economic time series is given in Nerlove and others, 1979.) Clement Juglar (1819–1905), the French physician turned economist, was the first to make systematic use of time-series data for the specific purpose of studying business cycles, and is credited with the discovery of an investment cycle of about 7–11 years duration, commonly known as the Juglar cycle. Other economists such as Kitchin, Kuznets and Kondratieff followed Juglar's lead and discovered the inventory cycle (3–5 years duration), the building cycle (15–25 years duration) and the long wave (45–60 years duration), respectively. The emphasis of this early research was on the morphology of cycles and the identification of periodicities. Little attention was paid to the quantification of the relationships that may have underlain the cycles. Indeed, economists working in the National Bureau of Economic

Research under the direction of Wesley Mitchell regarded each business cycle as a unique phenomenon and were therefore reluctant to use statistical methods except in a non-parametric manner and for purely descriptive purposes (see, for example, Mitchell, 1928 and Burns and Mitchell, 1947). This view of business cycle research stood in sharp contrast to the econometric approach of Frisch and Tinbergen and culminated in the famous methodological interchange between Tjalling Koopmans and Rutledge Vining about the roles of theory and measurement in applied economics in general and business cycle research in particular. (This interchange appeared in the August 1947 and May 1949 issues of *The Review of Economics and Statistics*.)

3 THE BIRTH OF ECONOMETRICS. Although, as I have argued above, quantitative economic analysis is a good three centuries old, econometrics as a recognized branch of economics only began to emerge in the 1930s and the 1940s with the foundation of the Econometric Society, the Cowles Commission in the United States, and the Department of Applied Economics (DAE) under the directorship of Richard Stone in the United Kingdom. (A highly readable blow-by-blow account of the founding of the first two organizations can be found in Christ (1952, 1983), while the history of the DAE is covered in Stone, 1978.) The reasons for the lapse of more than two centuries between the pioneering work of Petty and the recognition of econometrics as a branch of economics are complex, and are best understood in conjunction with, and in the light of, histories of the development of theoretical economics, national income accounting, mathematical statistics, and computing. Such a task is clearly beyond the scope of the present paper. However, one thing is clear: given the multi-disciplinary nature of econometrics, it would have been extremely unlikely that it would have emerged as a serious branch of economics had it not been for the almost synchronous development of mathematical economics and the theories of estimation and statistical inference in the late 19th century and the early part of the 20th century. (An interesting account of the history of statistical methods can be found in Kendall, 1968.)

Of the four components of econometrics, namely, *a priori* theory, data, econometric methods and computing techniques, it was, and to a large extent still is, the problem of econometric method which has attracted most attention. The first major debate over econometric method concerned the applicability of the probability calculus and the newly developed sampling theory of R.A. Fisher to the analysis of economic data. As Morgan (1986) argues in some detail, prior to the 1930s the application of mathematical theories of probability to economic data was rejected by the majority in the profession, irrespective of whether they were involved in research on demand analysis or on business cycles. Even Frisch was highly sceptical of the value of sampling theory and significance tests in econometrics. His objection to the use of significance tests was not, however, based on the epistemological reasons that lay behind Robbins's and Keynes's criticisms of econometrics. He was more concerned with the problems of multicollinearity and measurement errors which he believed, along with many others, afflicted all economic variables observed under non-controlled experimental conditions. By drawing attention to the *fictitious determinateness created by random errors* of observations, Frisch (1934) launched a severe attack on regression and correlation analysis which remains as valid now as it was then. With characteristic clarity and boldness Frisch stated:

As a matter of fact I believe that a substantial part of the regression and correlation analyses which have been made on economic data in recent years is nonsense for this very reason [the random errors of measurement] (1934, p. 6).

In order to deal with the measurement error problem Frisch developed his confluence analysis and the method of 'bunch maps'. Although his method was used by some econometricians, notably Tinbergen (1939) and Stone (1945), it did not find much favour with the profession at large. This was due, firstly, to the indeterminate nature of confluence analysis and, secondly, to the alternative probabilistic rationalizations of regression analysis which were advanced by Koopmans (1937) and Haavelmo (1944). Koopmans proposed a synthesis of the two approaches to the estimation of economic relations, namely the error-in-variables approach of Frisch and the error-in-equation approach of Fisher, using the likelihood framework; thus rejecting the view prevalent at the time that the presence of measurement errors *per se* invalidates the application of the 'sampling theory' to the analysis of economic data. In his words

It is the conviction of the author that the essentials of Frisch's criticism of the use of Fisher's specification in economic analysis may also be formulated and illustrated from the conceptual scheme and in the terminology of the sampling theory, and the present investigation is an attempt to do so (p. 30).

The formulation of the error-in-variables model in terms of a probability model did not, however, mean that Frisch's criticisms of regression analysis were unimportant, or that they could be ignored. Just the opposite was the case. The probabilistic formulation helped to focus attention on the reasons for the indeterminacy of Frisch's proposed solution to the problem. It showed also that without some *a priori* information, for example, on the relative importance of the measurement errors in different variables, a determinate solution to the estimation problem would not be possible. What was important, and with hindsight path-breaking, about Koopmans's contribution was the fact that it demonstrated the possibility of the probabilistic characterization of economic relations, even in circumstances where important deviations from the classical regression framework were necessitated by the nature of the economic data.

Koopmans did not, however, emphasize the wider issue of the use of stochastic models in econometrics. It was Haavelmo who exploited the idea to the full, and argued forcefully for an explicit probability approach to the estimation and testing of economic relations. In his classic paper published as a supplement to *Econometrica* in 1944, Haavelmo defended the probability approach on two grounds: firstly, he argued that the use of statistical measures such as means, standard errors and correlation coefficients for inferential purposes is justified only if the process generating the data can be cast in terms of a probability model: 'For *no tool developed in the theory of statistics has any meaning* – except, perhaps, for descriptive purposes – *without being referred to some stochastic scheme*' (p. iii). Secondly, he argued that the probability approach, far from being limited in its application to economic data, because of its generality is in fact particularly suited for the analysis of 'dependent' and 'non-homogeneous' observations often encountered in economic research. He believed what is needed is

to assume that the *whole set* of, say n, observations may be considered as *one* observation of n variables (or a 'sample point') following an n-dimensional *joint* probability law, the 'existence' of which may be purely hypothetical. Then, one can test hypotheses regarding this joint probability law, and draw inference as to its possible form, by means of *one* sample point (in n dimensions) (p. iii).

Here Haavelmo uses the concept of joint probability distribution as a tool of analysis and not necessarily as a characterization of 'reality'. The probability model is seen as a convenient *abstraction* for the purpose of understanding, or explaining or predicting events in the real world. But it is not claimed that the model represents reality in all its minute details. To proceed with quantitative research in any subject, economics included, some degree of formalization is inevitable, and the probability model is one such formalization. This view, of course, does not avoid many of the epistemological problems that surround the concept of 'probability' in all the various senses (subjective, frequentist, logical, etc.) in which the term has been used, nor is it intended to do so. As Haavelmo himself put it:

The question is not whether probabilities *exist* or not, but whether – if we proceed *as if* they existed – we are able to make statements about real phenomena that are 'correct for practical purposes' (1944, p. 43).

The attraction of the probability model as a method of abstraction derives from its generality and flexibility, and the fact that no viable alternative seems to be available.

Haavelmo's contribution was also important as it constituted the first systematic defence against Keynes's (1939) influential criticisms of Tinbergen's pioneering research on business cycles and macroeconometric modelling. The objective of Tinbergen's research was twofold. Firstly, to show how a macroeconometric model may be constructed and then used for simulation and policy analysis (Tinbergen, 1937). Secondly, 'to submit to statistical test some of the theories which have been put forward regarding the character and causes of cyclical fluctuations in business activity' (Tinbergen, 1939, p. 11). Tinbergen assumed a rather limited role for the econometrician in the process of testing economic theories, and argued that it was the responsibility of the 'economist' to specify the theories to be tested. He saw the role of the econometrician as a passive one of estimating the parameters of an economic relation already specified on *a priori* grounds by an economist. As far as statistical methods were concerned he employed the regression method and Frisch's method of confluence analysis in a complementary fashion. Although Tinbergen discussed the problems of the determination of time lags, trends, structural stability and the choice of functional forms, he did not propose any systematic methodology for dealing with them. In short, Tinbergen approached the problem of testing theories from a rather weak methodological position. Keynes saw these weaknesses and attacked them with characteristic insight (Keynes, 1939). A large part of Keynes's review was in fact concerned with technical difficulties associated with the application of statistical methods to economic data. Apart from the problems of the 'dependent' and 'non-homogeneous' observations mentioned above, Keynes also emphasized the problems of misspecification, multi-collinearity, functional form, dynamic specification, structural stability, and the difficulties associated with the measurement of theoretical variables. In view of these technical difficulties and Keynes's earlier warnings against 'inductive generalisation' in his *Treatise on Probability* (1921), it was not surprising that he focussed his attack on

Tinbergen's attempt at *testing* economic theories of business cycles, and almost totally ignored the practical significance of Tinbergen's work on econometric model building and policy analysis (for more details, see Pesaran and Smith, 1985a).

In his own review of Tinbergen's work, Haavelmo (1943) recognized the main burden of the criticisms of Tinbergen's work by Keynes and others, and argued the need for a general statistical framework to deal with these criticisms. As we have seen, Haavelmo's response, despite the views expressed by Keynes and others, was to rely more, rather than less, on the probability model as the basis of econometric methodology. The technical problems raised by Keynes and others could now be dealt with in a systematic manner by means of formal probabilistic models. Once the probability model was specified, a solution to the problems of estimation and inference could be obtained by means of either classical or of Bayesian methods. There was little that could now stand in the way of a rapid development of econometric methods.

4 EARLY ADVANCES IN ECONOMETRIC METHODS. Haavelmo's contribution marked the beginning of a new era in econometrics, and paved the way for the rapid development of econometrics on both sides of the Atlantic. The likelihood method soon became an important tool of estimation and inference, although initially it was used primarily at the Cowles Commission where Haavelmo himself had spent a short period as a research associate.

The first important breakthrough came with a formal solution to the identification problem which had been formulated earlier by E. Working (1927). By defining the concept of 'structure' in terms of the joint probability distribution of observations, Haavelmo (1944) presented a very general concept of identification and derived the necessary and sufficient conditions for identification of the entire system of equations, including the parameters of the probability distribution of the disturbances. His solution, although general, was rather difficult to apply in practice. Koopmans, Rubin and Leipnik, in a paper presented at a conference organized by the Cowles Commission in 1945 and published later in 1950, used the term 'identification' for the first time in econometrics, and gave the now familiar rank and order conditions for the identification of a single equation in a system of simultaneous *linear* equations. The solution of the identification problem by Koopmans (1949) and Koopmans, Rubin and Leipnik (1950), was obtained in the case where there are *a priori* linear restrictions on the structural parameters. They derived rank and order conditions for identifiability of a single equation from a complete system of equations without reference to how the variables of the model are classified as endogenous or exogenous. Other solutions to the identification problem, also allowing for restrictions on the elements of the variance–covariance matrix of the structural disturbances, were later offered by Wegge (1965) and Fisher (1966). A comprehensive survey of some of the more recent developments of the subject can be found in Hsiao (1983).

Broadly speaking, a model is said to be identified if all its structural parameters can be obtained from the knowledge of its underlying joint probability distribution. In the case of simultaneous equations models prevalent in econometrics the solution to the identification problem depends on whether there exists a sufficient number of *a priori* restrictions for the derivative of the structural parameters from the reduced-form parameters. Although the purpose of the model and the focus of the analysis on explaining the variations of some variables in terms of the unexplained variations of other variables is an important consideration, in the final analysis the specification of a minimum number of identifying restrictions was seen by researchers at the Cowles Commission to be the function and the responsibility of 'economic theory'. This attitude was very much reminiscent of the approach adopted earlier by Tinbergen in his business cycle research: the function of economic theory was to provide the specification of the econometric model, and that of econometrics to furnish statistically optimal methods of estimation and inference. More specifically, at the Cowles Commission the primary task of econometrics was seen to be the development of statistically efficient methods for the estimation of structural parameters of an *a priori* specified system of simultaneous stochastic equations.

Initially, under the influence of Haavelmo's contribution, the maximum likelihood (ML) estimation method was emphasized as it yielded consistent estimates. Koopmans and others (1950) proposed the 'information-preserving maximum-likelihood method', more commonly known as the Full Information Maximum Likelihood (FIML) method, and Anderson and Rubin (1949), on a suggestion by M.A. Girshick, developed the Limited Information Maximum Likelihood (LIML) method. Both methods are based on the joint probability distribution of the endogenous variables and yield consistent estimates, with the former utilizing all the available *a priori* restrictions and the latter only those which related to the equation being estimated. Soon other computationally less demanding estimation methods followed, both for a fully efficient estimation of an entire system of equations and for a consistent estimation of a single equation from a system of equations. The Two-Stage Least Squares (2SLS) procedure, which involves a similar order of magnitude of computations as the least squares method, was independently proposed by Theil (1954, 1958) and Basmann (1957). At about the same time the instrumental variable (IV) method, which had been developed over a decade earlier by Reiersol (1941, 1945), and Geary (1949) for the estimation of errors-in-variables models, was applied by Sargan (1958) to the estimation of simultaneous equation models. Sargan's main contribution consisted in providing an asymptotically efficient technique for using surplus instruments in the application of the IV method to econometric problems. A related class of estimators, known as k-class estimators, was also proposed by Theil (1961). Methods of estimating the entire system of equations which were computationally less demanding than the FIML method also started to emerge in the literature. These included the Three-Stage Least Squares method due to Zellner and Theil (1962), the iterated instrumental variables method based on the work of Lyttkens (1970), Brundy and Jorgenson (1971), Dhrymes (1971); and the system k-class estimators due to Srivastava (1971) and Savin (1973). An interesting synthesis of different estimators of the simultaneous equations model is given by Hendry (1976). The literature on estimation of simultaneous equation models is vast and is still growing. Important contributions have been made in the areas of estimation of simultaneous non-linear models, the seemingly unrelated regression model proposed by Zellner (1962), and the simultaneous rational expectations models which will be discussed in more detail below. Recent studies have also focused on the finite sample properties of the alternative estimators in the simultaneous equation model. Interested readers should consult the relevant entries in this Dictionary, or refer to the excellent survey articles by Hausman (1983), by Amemiya (1983) and by Phillips (1983).

While the initiative taken at the Cowles Commission led to a rapid expansion of econometric techniques, the application of these techniques to economic problems was rather slow. This was partly due to a lack of adequate computing facilities at the time. A more fundamental reason was the emphasis of all the Cowles Commission on the simultaneity problem almost to the exclusion of other problems that were known to afflict regression analysis. Since the early applications of the correlation analysis to economic data by Yule and Hooker, the serial dependence of economic time series and the problem of spurious correlation that it could give rise to had been the single most important factor explaining the profession's scepticism concerning the value of regression analysis in economics. A satisfactory solution to the spurious correlation problem was therefore needed before regression analysis of economic time series could be taken seriously. Research on this important topic began in the mid-1940s under the direction of Richard Stone at the Department of Applied Economics (DAE) in Cambridge, England, as a part of a major investigation into the measurement and analysis of consumers' expenditure in the United Kingdom (see Stone and others, 1954a). Stone had started this work during the 1939–45 war at the National Institute of Economic and Social Research. Although the first steps towards the resolution of the spurious correlation problem had been taken by Aitken (1934/35) and Champernowne (1948), the research in the DAE introduced the problem and its possible solution to the attention of applied economists. Orcutt (1948) studied the autocorrelation pattern of economic time series and showed that most economic time series can be represented by simple autoregressive processes with similar autoregressive coefficients, a result which was an important precursor to the work of Zellner and Palm (1974) discussed below. Subsequently in their classic paper, Cochrane and Orcutt (1949) made the important point that the major consideration in the analysis of stationary time series was the autocorrelation of the error term in the regression equation and not the autocorrelation of the economic time series themselves. In this way they shifted the focus of attention to the autocorrelation of disturbances as the main source of concern. Secondly, they put forward their well-known iterative method for the computation of regression coefficients under the assumption that the errors followed a first order autoregressive process.

Another important and related development at the DAE was the work of Durbin and Watson (1950, 1951) on the method of testing for residual autocorrelation in the classical regression model. The inferential breakthrough for testing serial correlation in the case of observed time-series data had already been achieved by von Neumann (1941, 1942), and by Hart and von Neumann (1942). The contribution of Durbin and Watson was, however, important from a practical viewpoint as it led to a bounds test for residual autocorrelation which could be applied irrespective of the actual values of the regressors. The independence of the critical bounds of the Durbin–Watson statistic from the matrix of the regressors allowed the application of the statistic as a general diagnostic test, the first of its type in econometrics. The contributions of Cochrane and Orcutt and of Durbin and Watson under the leadership of Stone marked the beginning of a new era in the analysis of economic time-series data and laid down the basis of what is now known as the 'time-series econometrics' approach.

The significance of the research at the DAE was not confined to the development of econometric methods. The work of Stone on linear expenditure systems represented one of the first attempts to use theory *directly* and *explicitly* in applied econometric research. This was an important breakthrough. Previously, economic theory had by and large been used in applied research only indirectly and as a general method for deciding on the list of the variables to be included in the regression model and, occasionally, for assigning signs to the parameters of the model. (For an important exception, see Marschak and Andrews, 1944.) In his seminal paper in the *Economic Journal*, Stone (1954b) made a significant break with this tradition and used theory not as a substitute for common sense, but as a formal framework for deriving 'testable' restrictions on the parameters of the empirical model. This was an important move towards the formal unification of theory and measurement that Frisch had called for and Schultz earlier had striven towards.

5 CONSOLIDATION AND FURTHER DEVELOPMENTS. The work at the Cowles Commission on identification and estimation of the simultaneous equation model and the development of appropriate techniques in dealing with the problem of spurious regression at the DAE paved the way for its widespread application to economic problems. This was helped significantly by the rapid expansion of computing facilities, the general acceptance of Keynesian theory and the increased availability of time-series data on national income accounts. As Klein (1971) put it, 'The Keynesian theory was simply "asking" to be cast in an empirical mold' (p. 416). The IS–LM version of the Keynesian theory provided a convenient and flexible framework for the construction of macroeconomic models for a variety of purposes ranging from pedagogic to short- and medium-term forecasting and policy analysis. In view of Keynes's criticisms of econometrics, it is perhaps ironic that his macroeconomic theory came to play such a central role in the advancement of econometrics in general and that of macroeconometric modelling in particular.

Inspired by the Keynesian theory and the pioneering work of Tinbergen, Klein (1947, 1950) was the first to construct a macroeconometric model in the tradition of the Cowles Commission. Soon others followed Klein's lead: prominent examples of early macroeconometric models included the Klein–Goldberger and the Brookings–SSRC models of the US economy, and the London Business School and the Cambridge Growth Project models of the UK economy. Over a short space of time macroeconometric models were built for almost every industrialized country, and even for some developing and centrally planned economies. Macroeconometric models became an important tool of *ex ante* forecasting and economic policy analysis, and started to grow both in size and sophistication. The relatively stable economic environment of the 1950s and 1960s was an important factor in the initial success enjoyed by macroeconometric models. Whether the use of macroeconometric models in policy formulation contributed towards the economic stability over this period is, of course, a different matter.

The construction and use of large-scale models presented a number of important computational problems, the solution of which was of fundamental significance not only for the development of macroeconometric modelling, but also for econometric practice in general. In this respect advances in computer technology were clearly instrumental, and without them it is difficult to imagine how the complicated computational problems involved in the estimation and simulation of large-scale models could have been solved. The increasing availability of better and faster computers was also instrumental as far as the types of problems studied and the types of solutions offered in the literature were concerned. For example, recent developments in the area of microeconometrics (see section 6.3 below) could hardly have been possible if it were not for the very important recent advances in computing facilities.

The development of economic models for policy analysis, however, was not confined to macroeconometric models. The inter-industry input–output models originating from the seminal work of Leontief (1936, 1941, 1951), and the microanalytic simulation models pioneered by Orcutt and his colleagues (1961), were amongst the other influential approaches which should be mentioned here. But it was the surge of interest in macroeconometric modelling which provided the single most important impetus to the further development of econometric methods. I have already mentioned some of the advances that took place in the field of estimation of the simultaneous equation model. Other areas where econometrics witnessed significant developments included dynamic specification, latent variables, expectations formation, limited dependent variables, discrete choice models, random coefficient models, disequilibrium models, and non-linear estimation. The Bayesian approach to econometrics was also developed more vigorously, thanks to the relentless efforts of Zellner, Drèze and their colleagues. (See Drèze and Richard (1983), and Zellner (1984, 1985) for the relevant references to theoretical and applied Bayesian econometric studies.) It was, however, the problem of dynamic specification that initially received the greatest attention. In an important paper, Brown (1952) modelled the hypothesis of habit persistence in consumer behaviour by introducing lagged values of consumption expenditures into an otherwise static Keynesian consumption function. This was a significant step towards the incorporation of dynamics in applied econometric research and allowed the important distinction to be made between the short-run and the long-run impacts of changes in income on consumption. Soon other researchers followed Brown's lead and employed his autoregressive specification in their empirical work.

The next notable development in the area of dynamic specification was the distributed lag model. Although the idea of distributed lags had been familiar to economists through the pioneering work of Irving Fisher (1930) on the relationship between the nominal interest rate and the expected inflation rate, its application in econometrics was not seriously considered until the mid 1950s. The geometric distributed lag model was used for the first time by Koyck (1954) in a study of investment. Koyck arrived at the geometric distributed lag model *via* the adaptive expectations hypothesis. This same hypothesis was employed later by Cagan (1956) in a study of demand for money in conditions of hyperinflation, by Friedman (1957) in a study of consumption behaviour and by Nerlove (1958a) in a study of the cobweb phenomenon. The geometric distributed lag model was subsequently generalized by Solow (1960), Jorgenson (1966) and others, and was extensively applied in empirical studies of investment and consumption behaviour. At about the same time Almon (1965) provided a polynomial generalization of Fisher's (1937) arithmetic lag distribution which was later extended further by Shiller (1973). Other forms of dynamic specification considered in the literature included the partial adjustment model (Nerlove, 1958b; Eisner and Strotz, 1963) and the multivariate flexible accelerator model (Treadway, 1971) and Sargan's (1964) work on econometric time series analysis which we discuss below in more detail. An excellent survey of this early literature on distributed lag and partial adjustment models is given in Griliches (1967).

Concurrent with the development of dynamic modelling in econometrics there was also a resurgence of interest in time-series methods, used primarily in short-term business forecasting. The dominant work in this field was that of Box and Jenkins (1970), who, building on the pioneering works of Yule (1921, 1926), Slutsky (1927), Wold (1938), Whittle (1963) and others, proposed computationally manageable and asymptotically efficient methods for the estimation and forecasting of univariate autoregressive-moving average (ARMA) processes. Time-series models provided an important and relatively cheap benchmark for the evaluation of the forecasting accuracy of econometric models, and further highlighted the significance of dynamic specification in the construction of time-series econometric models. Initially univariate time-series models were viewed as mechanical 'black box' models with little or no basis in economic theory. Their use was seen primarily to be in short-term forecasting. The potential value of modern time-series methods in econometric research was, however, underlined in the work of Cooper (1972) and Nelson (1972) who demonstrated the good forecasting performance of univariate Box–Jenkins models relative to that of large econometric models. These results raised an important question mark over the adequacy of large econometric models for forecasting as well as for policy analysis. It was argued that a properly specified structural econometric model should, at least in theory, yield more accurate forecasts than a univariate time-series model. Theoretical justification for this view was provided by Zellner and Palm (1974), followed by Trivedi (1975), Prothero and Wallis (1976), Wallis (1977) and others. These studies showed that Box–Jenkins models could in fact be derived as univariate final form solutions of linear structural econometric models so long as the latter were allowed to have a rich enough dynamic specification. In theory, the pure time-series model could always be embodied within the structure of an econometric model and in this sense it did not present a 'rival' alternative to econometric modelling. This literature further highlighted the importance of dynamic specification in econometric models and in particular showed that econometric models that are out-performed by simple univariate time-series models most probably suffer from serious specification errors.

The response of the econometrics profession to this time-series critique was rather mixed and has taken different forms. On the one hand a full integration of time-series methods and traditional econometric analysis has been advocated by Zellner and Palm, Wallis and others. This blending of the econometric methods which Zellner has called the SEMTSA (structural econometric modelling times-series analysis) approach is discussed in some detail in Zellner (1979). The SEMTSA approach emphasizes that dynamic linear structural models are a special case of multivariate time-series processes, and argues that time-series methods should be utilized to check the empirical adequacy of the final equation forms and the distributed lag (or transfer function) forms implicit in the assumed structural model. The modelling process is continued until the implicit estimates of the final equation forms and the distributed lag forms of the structural model are empirically compatible with the direct time-series estimates of these equations.

An alternative 'marriage' of econometric and time-series techniques has been developed by Sargan, Hendry and others largely at the London School of Economics (LSE). This marriage is based on the following two premises:

(i) Theoretical economic considerations can at best provide the specification of equilibrium or long-run relationships between variables. Little can be inferred from *a priori* reasoning about the time lags and dynamic specification of econometric relations.

(ii) The best approach to identification of lags in econometric models lies in the utilization of time-series methods, appropriately modified to allow for the existence of long-run relations among economic variables implied by economic theory.

Although the approach is general and in principle can be applied to systems of equations, in practice it has been primarily applied to modelling one variable at a time. The origins of this approach can be found in the two highly influential papers by Sargan (1964) on the modelling of money wages, and by Davidson and others (1978) on the modelling of non-durable consumption expenditures. By focusing on the modelling of one endogenous variable at a time, the LSE approach represents a partial break with the structural approach advocated by the Cowles Commission. But in an important sense the LSE approach continues to share with the Cowles Commission the emphasis it places on *a priori* economic reasoning, albeit in the form of equilibrium or long-period relationships.

6 RECENT DEVELOPMENTS. With the significant changes taking place in the world economic environment in the 1970s, arising largely from the breakdown of the Bretton Woods system and the quadrupling of oil prices, econometrics entered a new phase of its development. Mainsteam macroeconometric models built during the 1950s and 1960s, in an era of relative economic stability with stable energy prices and fixed exchange rates, were no longer capable of adequately capturing the economic realities of the 1970s. As a result, not surprisingly, macroeconometric models and the Keynesian theory that underlay them came under severe attack from theoretical as well as from practical viewpoints. While criticisms of Tinbergen's pioneering attempt at macroeconometric modelling were received with great optimism and led to the development of new and sophisticated estimation techniques and larger and more complicated models, the more recent bout of disenchantment with macroeconometric models prompted a much more fundamental reappraisal of quantitive modelling as a tool of forecasting and policy analysis. At a theoretical level it is argued that econometric relations invariably lack the necessary 'microfoundations', in the sense that they cannot be consistently derived from the optimizing behaviour of economic agents. At a practical level the Cowles Commission approach to the identification and estimation of simultaneous macroeconometric models has been questioned by Lucas and Sargent and by Sims, although from different viewpoints. There has also been a move away from macroeconometric models and towards microeconometric research where it is hoped that some of the pitfalls of the macroeconometric time-series analysis can be avoided. The response of the econometric profession as a whole to the recent criticism has been to emphasize the development of more appropriate techniques, to use new data sets and to call for a better quality control of econometric research with special emphasis on model validation and diagnostic testing.

What follows is a brief overview of some of the important developments of the past two decades. Given space limitations and my own interests there are inevitably significant gaps. These include the important contributions of Granger (1969), Sims (1972) and Engle and others (1983) on different concepts of 'causality' and 'exogeneity', and the vast literature on disequilibrium models (Quandt, 1982; Maddala, 1983, 1986), random coefficient models (Chow, 1984), continuous time models (Bergstrom, 1984), non-stationary time series and testing for unit roots (Dickey and Fuller, 1979, 1981; Evans and Savin, 1981, 1984; Phillips, 1986, 1987; Phillips and Durlauf, 1986) and small sample theory (Phillips, 1983; Rothenberg, 1984), not to mention the developments in the area of policy analysis and the application of control theory of econometric models (Chow, 1975, 1981; Aoki, 1976).

6.1 Rational expectations and the Lucas critique. Although the Rational Expectations Hypothesis (REH) was advanced by Muth in 1961, it was not until the early 1970s that it started to have a significant impact on time-series econometrics and on dynamic economic theory in general. What brought the REH into prominence was the work of Lucas (1972, 1973), Sargent (1973), Sargent and Wallace (1975) and others on the new classical explanation of the apparent breakdown of the Phillips curve. The message of the REH for econometrics was clear. By postulating that economic agents form their expectations *endogenously* on the basis of the *true* model of the economy and a *correct* understanding of the processes generating exogenous variables of the model, including government policy, the REH raised serious doubts about the invariance of the structural parameters of the mainstream macroeconometric models in the face of changes in government policy. This was highlighted in Lucas's critique of macroeconometric policy evaluation. By means of simple examples Lucas (1976) showed that in models with rational expectations the parameters of the decision rules of economic agents, such as consumption or investment functions, are usually a mixture of the parameters of the agents' objective functions and of the stochastic processes they face as historically given. Therefore, Lucas argued, there is no reason to believe that the 'structure' of the decision rules (or economic relations) would remain invariant under a policy intervention. The implication of the Lucas critique for econometric research was not, however, that policy evaluation could not be done, but rather than the traditional econometric models and methods were not suitable for this purpose. What was required was a separation of the parameters of the policy rule from those of the economic model. Only when these parameters could be identified separately given the knowledge of the joint probability distribution of the variables (both policy and non-policy variables), would it be possible to carry out an econometric analysis of alternative policy options.

There have been a number of reactions to the advent of the rational expectations hypothesis and the Lucas critique that accompanied it. The least controversial has been the adoption of the REH as *one* of several possible expectations formation hypotheses in an otherwise conventional macroeconometric model containing expectational variables. In this context the REH, by imposing the appropriate cross-equation parametric restrictions, ensures that 'expectations' and 'forecasts' generated by the model are consistent. The underlying economic model is in no way constrained to have particular Keynesian or monetarist features, nor are there any presumptions that the relations of the economic model should necessarily correspond to the decision rules of economic agents. In this approach the REH is regarded as a convenient and effective method of imposing cross-equation parametric restrictions on time series econometric models, and is best viewed as the 'model-consistent' expectations hypothesis. The econometric implications of such a model-consistent expectations mechanism have been extensively analysed in the literature. The problems of identification and estimation of linear RE models have been discussed in detail, for example, by Wallis (1980), Wickens (1982) and Pesaran (1987). These studies show how the standard econometric methods can in principle be adapted to the econometric analysis of rational (or consistent) expectations models.

Another reaction to the Lucas critique has been to treat the problem of 'structural change' emphasized by Lucas as one

more potential econometric 'problem' (on this see Lawson, 1981). It is argued that the problem of structural change resulting from intended or expected changes in policy is not new and had been known to the economists at the Cowles Commission (Marschak, 1953), and can be readily dealt with by a more careful monitoring of econometric models for possible changes in their structure. This view is, however, rejected by Lucas and Sargent and other proponents of the rational expectations school who argue for a more fundamental break with the traditional approach to macroeconometric modelling.

The optimization approach of Lucas and Sargent is based on the premise that the 'true' structural relations contained in the economic model and the policy rules of the government can be obtained *directly* as solutions to well-defined dynamic optimization problems faced by economic agents and by the government. The task of the econometrician is then seen to be the disentanglement of the parameters of the stochastic processes that agents face from the parameters of their objective functions. As Hansen and Sargent (1980) put it,

> Accomplishing this task [the separate identification of parameters of the exogenous process and those of taste and technology functions] is an absolute prerequisite of reliable econometric policy evaluation. The execution of this strategy involves estimating agents' decision rules jointly with models for the stochastic processes they face, subject to the cross-equation restrictions implied by the hypothesis of rational expectations (p. 8).

So far this approach has been applied only to relatively simple set-ups involving aggregate data at the level of a 'representative' firm or a 'representative' household. One important reason for this lies in the rather restrictive and inflexible econometric models which emerge from the strict adherence to the optimization framework and the REH. For analytical tractability it has often been necessary to confine the econometric analysis to quadratic objective functions and linear stochastic processes. This problem to some extent has been mitigated by recent developments in the area of the estimation of the Euler equations (see Hansen and Singleton, 1982). But there are still important technical difficulties that have to be resolved before the optimization approach can be employed in econometrics in a flexible manner. In addition to these technical difficulties, there are fundamental issues concerning the problem of aggregation across agents, information heterogeneity, the learning process, and the effect that these complications have for the implementation of the Lucas–Sargent research programme (cf. Pesaran, 1987).

6.2 Atheoretical macroeconometrics. The Lucas critique of mainstream macroeconometric modelling has also led some econometricians, notably Sims (1980, 1982), to doubt the validity of the Cowles Commission style of achieving identification in econometric models. The view that economic theory cannot be relied on to yield identification of structural models is not new and has been emphasized in the past, for example, by Liu (1960). The more recent disenchantment with the Cowles Commission's approach has its origins in the REH, and the unease with a priori restrictions on lag lengths that are needed if rational expectations models are to be identified (see Pesaran, 1981). Sims (1980, p. 7) writes: 'It is my view, however, that rational expectations is more deeply subversive of identification than has yet been recognized.' He then goes on to say that 'In the presence of expectations, it turns out that the crutch of a priori knowledge of lag lengths is indispensable, even when we have distinct strictly exogenous variables shifting supply and demand schedules' (p. 7). While it is true that the REH complicates the necessary conditions for the identification of structural models, the basic issue in the debate over identification still centres on the validity of the classical dichotomy between exogenous and endogenous variables. Whether it is possible to test the 'exogeneity' assumptions of macroeconometric models is a controversial matter and is very much bound up with what is in fact meant by exogeneity. In certain applications exogeneity is viewed as a property of a proposed model (à la Koopmans, 1950), and in other situations it is defined in terms of a group of variables for purposes of inference about 'parameters of interest' (Engle and others, 1983). In the Cowles Commission approach exogeneity was assumed to be the property of the structural model, obtained from a priori theory and testable only in the presence of maintained restrictions. Thus it was not possible to test the identifying restrictions themselves. They had to be assumed a priori and accepted as a matter of belief or on the basis of knowledge extraneous to the model under consideration.

The approach advocated by Sims and his co-researchers departs from the Cowles Commission methodology in two important respects. It denies that a priori theory can ever yield the restrictions necessary for identification of structural models, and argues that for forecasting and policy analysis, structural identification is not needed (Sims, 1980, p. 11). Accordingly, this approach, termed by Cooley and LeRoy (1985) 'atheoretical macroeconometrics', maintains that only unrestricted vector-autoregressive (VAR) systems which do not allow for a priori classification of the variables into endogenous and exogenous are admissible for macroeconometric analysis. The VAR approach represents an important alternative to conventional large-scale macroeconometric models and has been employed with some success in the area of forecasting (Litterman, 1985). Whether such unrestricted VAR systems can also be used in policy evaluation and policy formulation exercises remains a controversial matter. Cooley and LeRoy (1985) in their critique of this literature argue that even if it can be successfully implemented, it will still be of limited relevance except as a tool for ex ante forecasting and data description (on this also see Leamer, 1985a). They argue that it does not permit direct testing of economic theories, it is of little use for policy analysis and, above all, it does not provide a structural understanding of the economic system it purports to represent. Sims and others (Doan, Litterman and Sims, 1984; Sims, 1986), however, maintain that VAR models can be used for policy analysis, and the type of identifying assumptions needed for this purpose are no less credible than those assumed in conventional or RE macroeconometric models.

6.3 Microeconometrics. Emphasis on the use of micro-data in the analysis of economic problems is not, of course, new and dates back to the pioneering work of Ruggles and Ruggles (1956) on the development of a micro-based social accounting framework and the work of Orcutt and his colleagues already referred to above, and the influential contribution of Prais and Houthakker (1955) on the analysis of family expenditure surveys. But it is only recently, partly as a response to the dissatisfaction with macroeconometric time-series research and partly in view of the increasing availability of micro-data and computing facilities, that the analysis of micro-data has started to be considered seriously in the econometric literature. Important micro-data sets have become available especially in the United States in such areas as housing, transportation, labour markets and energy. These data sets include various longitudinal surveys (e.g. University of Michigan Panel Study of Income Dynamics and Ohio State NLS Surveys), cross-sectional surveys of family expenditures, and the population and labour force surveys. This increasing availability of micro-data, while opening up new possibilities for analysis, has also raised a number of new and interesting econometric issues primarily originating from the nature of the data. The errors of measurement are more likely to be serious in the case of micro- than macro-data. The problem of the heterogeneity of economic agents at the micro level cannot be assumed away as readily as is usually done in the case of macro-data by appealing to the idea of a 'representative' firm or a 'representative' household. As Griliches (1986) put it

> Variables such as age, land quality, or the occupational structure of an enterprise, are much less variable in the aggregate. Ignoring them at the micro level can be quite costly, however. Similarly, measurement errors which tend to cancel out when averaged over thousands or even millions of respondents, loom much larger when the individual is the unit of analysis (p. 1469).

The nature of micro-data, often being qualitative or limited to a particular range of variation, has also called for new econometric models and techniques. The models and issues considered in the micro-econometric literature are wide-ranging and include fixed and random effect models (e.g. Mundlak, 1961, 1978), discrete choice or quantal response models (Manski and McFadden, 1981), continuous time duration models (Heckman and Singer, 1984), and micro-econometric models of count data (Hausman and others, 1984 and Cameron and Trivedi, 1986). The fixed or random effect models provide the basic statistical framework. Discrete choice models are based on an explicit characterization of the choice process and arise when individual decision makers are faced with a finite number of alternatives to choose from. Examples of discrete choice models include transportation mode choice (Domenich and McFadden, 1975), labour force participation (Heckman and Willis, 1977), occupation choice (Boskin, 1974), job or firm location (Duncan 1980), etc. Limited-dependent variables models are commonly encountered in the analysis of survey data and are usually categorized into truncated regression models and censored regression models. If all observations on the dependent as well as on the exogenous variables are lost when the dependent variable falls outside a specified range, the model is called *truncated*, and, if only observations on the dependent variable are lost, it is called *censored*. The literature on censored and truncated regression models is vast and overlaps with developments in other disciplines, particularly in biometrics and engineering. The censored regression model was first introduced into economics by Tobin (1958) in his pioneering study of household expenditure on durable goods where he explicitly allowed for the fact that the dependent variable, namely the expenditure on durables, cannot be negative. The model suggested by Tobin and its various generalizations are known in economics as Tobit models and are surveyed in detail by Amemiya (1984).

Continuous time duration models, also known as survival models, have been used in analysis of unemployment duration, the period of time spent between jobs, durability of marriage, etc. Application of survival models to analyse economic data raises a number of important issues resulting primarily from the non-controlled experimental nature of economic observations, limited sample sizes (i.e. time periods), and the heterogeneous nature of the economic environment within which agents operate. These issues are clearly not confined to duration models and are also present in the case of other microeconometric investigations that are based on time series or cross section or panel data. (For early literature on the analysis of panel data, see the error components model developed by Kuh, 1959 and Balestra and Nerlove, 1966.) A satisfactory resolution of these problems is of crucial importance for the success of the microeconometric research programme. As aptly put by Hsiao (1985) in his recent review of the literature:

> Although panel data has opened up avenues of research that simply could not have been pursued otherwise, it is not a panacea for econometric researchers. The power of panel data depends on the extent and reliability of the information it contains as well as on the validity of the restrictions upon which the statistical methods have been built (p. 163).

Partly in response to the uncertainties inherent in econometric results based on non-experimental data, there has also been a significant move towards 'social experimentation', especially in the United States, as a possible method of reducing these uncertainties. This has led to a considerable literature analysing 'experimental' data, some of which has been recently reviewed in Hausman and Wise (1985). Although it is still too early to arrive at a definite judgement about the value of social experimentation as a whole, from an econometric viewpoint the results have not been all that encouraging. Evaluation of the Residential Electricity Time-of-Use Experiments (Aigner, 1985), the Housing-Allowance Program Experiments (Rosen, 1985), and the Negative-Income-Tax Experiments (Stafford, 1985) all point to the fact that the experimental results could have been equally predicted by the earlier econometric estimates. The advent of social experimentation in economics has nevertheless posed a number of interesting problems in the areas of experimental design, statistical methods (e.g. see Hausman and Wise (1979) on the problem of attrition bias), and policy analysis that are likely to have important consequences for the future development of micro-econometrics. (A highly readable account of social experimentation in economics is given by Ferber and Hirsch, 1982.)

Another important aspect of recent developments in microeconometric literature relates to the use of microanalytic simulation models for policy analysis and evaluation to reform packages in areas such as health care, taxation, social security systems, and transportation networks. Some of this literature is covered in Orcutt and others (1986).

6.4 Model evaluation. While in the 1950s and 1960s research in econometrics was primarily concerned with the identification and estimation of econometric models, the dissatisfaction with econometrics during the 1970s caused a shift of focus from problems of estimation to those of model evaluation and testing. This shift has been part of a concerted effort to restore confidence in econometrics, and has received attention from Bayesian as well as classical viewpoints. Both these views reject the 'axiom of correct specification' which lies at the basis of most traditional econometric practices, but differ markedly as how best to proceed.

Bayesians, like Leamer (1978), point to the wide disparity that exists between econometric method and the econometric practice that it is supposed to underlie, and advocate the use of 'informal' Bayesian procedures such as the 'extreme bounds analysis' (EBA), or more generally, the 'global sensitivity analysis'. The basic idea behind the EBA is spelt out in Leamer and Leonard (1983) and Leamer (1983) and has been the subject of critical analysis in McAleer, Pagan and Volker (1985). In its most general form, the research strategy put forward by Leamer involves a kind of grand Bayesian sensitivity analysis. The empirical results, or in Bayesian terminology the posterior distributions, are evaluated for 'fragility' or 'sturdiness' by checking how sensitive the results are to changes in prior distributions. As Leamer (1985b) explains:

> Because no prior distribution can be taken to be an exact representation of opinion, a global sensitivity analysis is carried out to determine which inferences are fragile and which are sturdy (p. 311).

The aim of the sensitivity analysis in Leamer's approach is, in his words, 'to combat the arbitrariness associated with the choice of priori distribution' (Leamer, 1986, p. 74).

It is generally agreed, by Bayesians as well as by non-Bayesians, that model evaluation involves considerations other than the examination of the statistical properties of the models, and personal judgements inevitably enter the evaluation process. Models must meet multiple criteria which are often in conflict. They should be relevant in the sense that they ought to be capable of answering the questions for which they are constructed. They should be consistent with the accounting and/or theoretical structure within which they operate. Finally, they should provide adequate representations of the aspects of reality with which they are concerned. These criteria and their interaction are discussed in Pesaran and Smith (1985b). More detailed breakdowns of the criteria of model evaluation can be found in Hendry and Richard (1982) and McAleer and others (1985). In econometrics it is, however, the criterion of 'adequacy' which is emphasized, often at the expense of relevance and consistency.

The issue of model adequacy in mainstream econometrics is approached either as a model selection problem or as a problem in statistical inference whereby the hypothesis of interest is tested against general or specific alternatives. The use of absolute criteria such as measures of fit/parsimony or formal Bayesian analysis based on posterior odds are notable examples of model selection procedures, while likelihood ratio, Wald and Lagrange multiplier tests of nested hypotheses and Cox's centred log-likelihood ratio tests of non-nested hypotheses are examples of the latter approach. The distinction between these two general approaches basically stems from the way alternative models are treated. In the case of model selection (or model discrimination) all the models under consideration enjoy the same status and the investigator is not committed *a priori* to any one of the alternatives. The aim is to choose the model which is likely to perform best with respect to a particular loss function. By contrast, in the hypothesis-testing framework the null hypothesis (or the maintained model) is treated differently from the remaining hypotheses (or models). One important feature of the model-selection strategy is that its application always leads to one model being chosen in preference to other models. But in the case of hypothesis testing, rejection of all the models under consideration is not ruled out when the models are non-nested. A more detailed discussion of this point is given in Pesaran and Deaton (1978).

While the model-selection approach has received some attention in the literature, it is the hypothesis-testing framework which has been primarily relied on to derive suitable statistical procedures for judging the adequacy of an estimated model. In this latter framework, broadly speaking, three different strands can be identified, depending on how specific the alternative hypotheses are. These are the *general specification tests*, the *diagnostic tests*, and the *non-nested tests*. The first of these, introduced in econometrics by Ramsey (1969) and Hausman (1978), and more recently developed by White (1981, 1982) and Hansen (1982), are designed for circumstances where the nature of the alternative hypothesis is kept (sometimes intentionally) rather vague, the purpose being to test the null against a *broad* class of alternatives. Important examples of general specification tests are Ramsey's regression specification error test (RESET) for omitted variables and/or misspecified functional forms, and the Hausman–Wu test of misspecification in the context of measurement error models, and/or simultaneous equation models. Such general specification tests are particularly useful in the preliminary stages of the modelling exercise.

In the case of diagnostic tests, the model under consideration (viewed as the null hypothesis) is tested against more specific alternatives by embedding it within a general model. Diagnostic tests can then be constructed using the likelihood ratio, Wald or Lagrange multiplier (LM) principles to test for parametric restrictions imposed on the general model. The application of the LM principle to econometric problems is reviewed in the papers by Breusch and Pagan (1980), Godfrey and Wickens (1982) and Engle (1984). Examples of the restrictions that may be of interest as diagnostic checks of model adequacy include zero restrictions, parameter stability, serial correlation, heteroskedasticity, functional forms, and normality of errors. As shown in Pagan and Hall (1983), most existing diagnostic tests can be computed by means of auxiliary regressions involving the estimated residuals. In this sense diagnostic tests can also be viewed as a kind of residual analysis where residuals computed under the null are checked to see whether they can be explained further in terms of the hypothesized sources of misspecification. The distinction made here between diagnostic tests and general specification tests is more apparent than real. In practice some diagnostic tests such as tests for serial correlation can also be viewed as a general test of specification. Nevertheless, the distinction helps to focus attention on the purpose behind the tests and the direction along which high power is sought.

The need for non-nested tests arises when the models under consideration belong to separate parametric families in the sense that no single model can be obtained from the others by means of a suitable limiting process. This situation, which is particularly prevalent in econometric research, may arise when models differ with respect to their theoretical underpinnings and/or their auxiliary assumptions. Unlike the general specification tests and diagnostic tests, the application of non-nested tests is appropriate when *specific* but rival hypotheses for the explanation of the same economic phenomenon have been advanced. Although non-nested tests can also be used as general specification tests, they are designed primarily to have high power against specific models that are seriously entertained in the literature. Building on the pioneering work of Cox (1961, 1962), a number of such tests for single equation models and systems of simultaneous equations have been proposed (see the entry on NON-NESTED HYPOTHESIS in this Dictionary for further details and references).

The use of statistical tests in econometrics, however, is not a straightforward matter and in most applications does not admit of a clear-cut interpretation. This is especially so in circumstances where test statistics are used not only for checking the adequacy of a *given* model but also as guides to model construction. Such a process of model construction involves specification searches of the type emphasized by Leamer and presents insurmountable pre-test problems which in general tend to produce econometric models whose 'adequacy' is more apparent than real. As a result, in evaluating econometric models less reliance should be placed on those indices of model adequacy that are used as guides to model construction, and more emphasis should be given to the performance of models over other data sets and against rival models. The evaluation of econometric models is a complicated process involving practical, theoretical and econometric considerations. Econometric methods clearly have an important contribution to make to this process. But they should not be confused with the whole activity of econometric modelling which, in addition to econometric and computing skills, requires data, considerable intuition, institutional knowledge and, above all, economic understanding.

7 APPRAISALS AND FUTURE PROSPECTS. Econometrics has come a long way over a relatively short period. Important advances have been made in the compilation of economic data and in the development of concepts, theories and tools for the construction and evaluation of a wide variety of econometric models. Applications of econometric methods can be found in almost every field of economics. Econometric models have been used extensively by government agencies, international organizations and commercial enterprises. Macroeconometric models of differing complexity and size have been constructed for almost every country in the world. Both in theory and practice econometrics has already gone well beyond what its founders envisaged. Time and experience, however, have brought out a number of difficulties that were not apparent at the start.

Econometrics emerged in the 1930s and 1940s in a climate of optimism, in the belief that economic theory could be relied on to identify most, if not all, of the important factors involved in modelling economic reality, and that methods of classical statistical inference could be adapted readily for the purpose of giving empirical content to the received economic theory. This early view of the interaction of theory and measurement in econometrics, however, proved rather illusory. Economic theory, be it neoclassical, Keynesian or Marxian, is invariably formulated with *ceteris paribus* clauses, and involves unobservable latent variables and general functional forms; it has little to say about adjustment processes and lag lengths. Even in the choice of variables to be included in econometric relations, the role of economic theory is far more limited than was at first recognized. In a Walrasian general equilibrium model, for example, where everything depends on everything else, there is very little scope for *a priori* exclusion of variables from equations in an econometric model. There are also institutional features and accounting conventions that have to be allowed for in econometric models but which are either ignored or are only partially dealt with at the theoretical level. All this means that the specification of econometric models inevitably involves important auxiliary assumptions about functional forms, dynamic specifications, latent variables, etc. with respect to which economic theory is silent or gives only an incomplete guide.

The recognition that economic theory on its own cannot be expected to provide a complete model specification has important consequences both for testing economic theories and for the evaluation of econometric models. The incompleteness of economic theories makes the task of testing them a formidable undertaking. In general it will not be possible to say whether the results of the statistical tests have a bearing on the economic theory or the auxiliary assumptions. This ambiguity in testing theories, known as the Duhem–Quine thesis, is not confined to econometrics and arises whenever theories are conjunctions of hypotheses (on this, see for example Cross, 1982). The problem is, however, especially serious in econometrics because theory is far less developed in economics than it is in the natural sciences. There are, of course, other difficulties that surround the use of econometric methods for the purpose of testing economic theories. As a rule economic statistics are not the results of designed experiments, but are obtained as by-products of business and government activities often with legal rather than economic considerations in mind. The statistical methods available are generally suitable for large samples while the economic data (especially economic time-series) have a rather limited coverage. There are also problems of aggregation over time, commodities and individuals that further complicate the testing of economic theories that are micro-based.

The incompleteness of economic theories also introduces an important and unavoidable element of data-instigated searches into the process of model construction, which creates important methodological difficulties for the established econometric methods of model evaluation. Clearly, this whole area of specification searches deserves far greater attention, especially from non-Bayesians, than it has so far attracted.

There is no doubt that econometrics is subject to important limitations, which stem largely from the incompleteness of the economic theory and the non-experimental nature of economic data. But these limitations should not distract us from recognizing the fundamental role that econometrics has come to play in the development of economics as a scientific discipline. It may not be possible conclusively to reject economic theories by means of econometric methods, but it does not mean that nothing useful can be learned from attempts at testing particular formulations of a given theory against (possible) rival alternatives. Similarly, the fact that econometric modelling is inevitably subject to the problem of specification searches does not mean that the whole activity is pointless. Econometric models are important tools of forecasting and policy analysis, and it is unlikely that they will be discarded in the future. The challenge is to recognize their limitations and to work towards turning them into more reliable and effective tools. There seem to be no viable alternatives.

M. Hashem Pesaran

See also ESTIMATION; HYPOTHESIS TESTING; MACROECONOMETRIC MODELS; SPECIFICATION PROBLEMS IN ECONOMETRICS; TIME SERIES ANALYSIS.

BIBLIOGRAPHY

Aigner, D.J. 1985. The residential electricity time-of-use pricing experiments: what have we learned? In *Social Experimentation*, ed. J.A. Hausman and D.A. Wise, Chicago: University of Chicago Press.

Aitken, A.C. 1934–5. On least squares and linear combinations of observations. *Proceedings of the Royal Society of Edinburgh* 55, 42–8.

Allen, R.G.D. and Bowley, A.L. 1935. *Family Expenditure*. London: P.S. King.

Almon, S. 1965. The distributed lag between capital appropriations and net expenditures. *Econometrica* 33, 178–96.

Amemiya, T. 1983. Nonlinear regression models. In *Handbook of Econometrics*, ed. Z. Griliches and M.D. Intriligator, Vol. 1, Amsterdam: North-Holland.

Amemiya, T. 1984. Tobit models: a survey. *Journal of Econometrics* 24, 3–61.

Anderson, T.W. and Rubin, H. 1949. Estimation of the parameters of a single equation in a complete system of stochastic equations. *Annals of Mathematical Statistics* 20, 46–63.

Aoki, M. 1976. *Dynamic Economic Theory and Control in Economics*. New York: American Elsevier.

Balestra, P. and Nerlove, M. 1966. Pooling cross section and time series data in the estimation of a dynamic model: the demand for natural gas. *Econometrica* 34, 585–612.

Basmann, R.L. 1957. A generalized classical method of linear estimation of coefficients in a structural equation. *Econometrica* 25, 77–83.

Benini, R. 1907. Sull'uso delle formole empiriche a nell'economia applicata. *Giornale degli economisti*, 2nd series, 35, 1053–63.

Bergstrom, A.R. 1984. Continuous time stochastic models and issues of aggregation over time. In *Handbook of Econometrics*, ed. Z. Griliches and M.D. Intriligator, Vol. 2, Amsterdam: North-Holland.

Boskin, M.J. 1974. A conditional logit model of occupational choice. *Journal of Political Economy* 82, 389–98.

Box, G.E.P. and Jenkins, G.M. 1970. *Time Series Analysis: Forecasting and Control*. San Francisco: Holden-Day.

Breusch, T.S. and Pagan, A.R. 1980. The Lagrange multiplier test and its applications to model specification in econometrics. *Review of Economic Studies* 47, 239–53.

Brown, T.M. 1952. Habit persistence and lags in consumer behaviour. *Econometrica* 20, 355–71.

Brundy, J.M. and Jorgenson, D.N. 1971. Efficient estimation of simultaneous equations by instrumental variables. *Review of Economics and Statistics* 53, 207–24.

Burns, A.F. and Mitchell, W.C. 1947. *Measuring Business Cycles*. New York: Columbia University Press for the National Bureau of Economic Research.

Cagan, P. 1956. The monetary dynamics of hyperinflation. In *Studies in the Quantity Theory of Money*, ed. M. Friedman, Chicago: University of Chicago Press.

Cameron, A.C. and Trivedi, P.K. 1986. Econometric models based on count data: comparisons and applications of some estimators and tests. *Journal of Applied Econometrics* 1, 29–53.

Champernowne, D.G. 1948. Sampling theory applied to autoregressive sequences. *Journal of the Royal Statistical Society* Series B, 10, 204–31.

Chow, G.C. 1975. *Analysis and Control of Dynamic Economic Economic Systems*. New York: John Wiley.

Chow, G.C. 1981. *Econometric Analysis by Control Methods*. New York: John Wiley.

Chow, G.C. 1983. *Econometrics*. New York: McGraw-Hill.

Chow, G.C. 1984. Random and changing coefficient models. In *Handbook of Econometrics*, ed. Z. Griliches and M.D. Intriligator, Vol. 2, Amsterdam: North-Holland.

Christ, C.F. 1952. *Economic Theory and Measurement: a twenty year research report, 1932–52*. Chicago: Cowles Commission for Research in Economics.

Christ, C.F. 1966. *Econometric Models and Methods*. New York: John Wiley.

Christ, C.F. 1983. The founding of the Econometric Society and Econometrica. *Econometrica* 51, 3–6.

Christ, C.F. 1985. Early progress in estimating quantitative economic relations in America. *American Economic Review*, December, 39–52 (supplementary information and statistical summaries).

Cochrane, P. and Orcutt, G.H. 1949. Application of least squares regression to relationships containing autocorrelated error terms. *Journal of the American Statistical Association* 44, 32–61.

Cooley, T.F. and Leroy, S.F. 1985. Atheoretical macroeconometrics: a critique. *Journal of Monetary Economics* 16, 283–368.

Cooper, R.L. 1972. The predictive performance of quarterly econometric models of the United States. In *Econometric Models of Cyclical Behaviour*, ed. B.G. Hickman, Studies in Income and Wealth 36, Vol. 2, 813–925.

Cox, D.R. 1961. Tests of separate families of hypotheses. *Proceedings of the Fourth Berkeley Symposium on Mathematical Statistics and Probability*, Vol. 1, Berkeley: University of California Press, 105–23.

Cox, D.R. 1962. Further results of tests of separate families of hypotheses. *Journal of the Royal Statistical Society*, Series B, 24, 406–24.

Creedy, J. 1986. On the King–Davenant 'law' of demand. *Scottish Journal of Political Economy* 33, August, 193–212.

Cross, R. 1982. The Duhem–Quine thesis, Lakatos and the appraisal of theories in macroeconomics. *Economic Journal* 92, 320–40.

Davenant, C. 1698. *Discourses on the Publick Revenues and on the Trade of England*, Vol. 1. London.

Davidson, J.E.H., Hendry, D.F., Srba, F. and Yeo, S. 1978. Econometric modelling of the aggregate time-series relationship between consumers' expenditure and income in the United Kingdom. *Economic Journal* 88, 661–92.

Dhrymes, P. 1971. A simplified estimator for large-scale econometric models. *Australian Journal of Statistics* 13, 168–75.

Dickey, D.A. and Fuller, W.A. 1979. Distribution of the estimators for autoregressive time series with a unit root. *Journal of the American Statistical Association* 74, 427–31.

Dickey, D.A. and Fuller, W.A. 1981. The likelihood ratio statistics for autoregressive time series with a unit root. *Econometrica* 49, 1057–72.

Doan, T., Litterman, R. and Sims, C.A. 1984. Forecasting and conditional projection using realistic prior distributions. *Econometric Reviews* 3, 1–100.

Domenich, T. and McFadden, D. 1975. *Urban Travel Demand: A Behavioural Analysis*. Amsterdam: North-Holland.

Drèze, J.H. and Richard, J-F. 1983. Bayesian analysis of simultaneous equation systems. In *Handbook of Econometrics*, ed. Z. Griliches and M.D. Intriligator, Vol. 1 Amsterdam: North-Holland.

Duncan, G. 1980. Formulation and statistical analysis of the mixed continuous/discrete variable model in classical production theory. *Econometrica* 48, 839–52.

Durbin, J. and Watson, G.S. 1950. Testing for serial correlation in least squares regression I. *Biometrika* 37, 409–28.

Durbin, J. and Watson, G.S. 1951. Testing for serial correlation in least squares regression II. *Biometrika* 38, 159–78.

Eisner, R. and Strotz, R.H. (1963). Determinants of business investment. In Commission on Money and Credit, *Impacts of Monetary Policy*, Englewood Cliffs, NJ: Prentice-Hall, 59–337.

Engle, R.F. 1984. Wald likelihood ratio and Lagrange multiplier tests in econometrics. In *Handbook of Econometrics*, Vol. 2, ed. Z. Griliches and M.D. Intriligator, Amsterdam: North-Holland.

Engle, R.F., Hendry, D.F. and Richard, J.-F. 1983. Exogeneity. *Econometrica* 51, 277–304.

Evans, G.B.A. and Savin, N.E. 1981. Testing for unit roots I. *Econometrica* 49, 753–77.

Evans, G.B.A. and Savin, N.E. 1984. Testing for unit roots II. *Econometrica* 52, 1241–70.

Ferber, R. and Hirsch, W.Z. 1982. *Social Experimentation and Economic Policy*. Cambridge: Cambridge University Press.

Fisher, F.M. 1966. *The Identification Problem in Econometrics*. New York: McGraw-Hill.

Fisher, I. 1930. *The Theory of Interest*. New York: Macmillan. Reprinted, Philadelphia: Porcupine Press, 1977.

Fisher, I. 1937. Note on a short-cut method for calculating distributed lags. *Bulletin de l'Institut International de Statistique* 29, 323–7.

Friedman, M. 1957. *A Theory of the Consumption Function*. Princeton: Princeton University Press.

Frisch, R. 1933a. Editorial. *Econometrica* 1, 1–4.

Frisch, R. 1933b. *Pitfalls in the Statistical Construction of Demand and Supply Curves*. Leipzig: Hans Buske Verlag.

Frisch, R. 1934. *Statistical Confluence Analysis by Means of Complete Regression Systems*. Oslo: University Institute of Economics.

Frisch, R. 1936. Note on the term 'Econometrics'. *Econometrica* 4, 95.

Geary, R.C. 1949. Studies in relations between economic time series. *Journal of the Royal Statistical Society*, Series B 10, 140–58.

Godfrey, L.G. and Wickens, M.R. 1982. Tests of misspecification using locally equivalent alternative models. In *Evaluation and Reliability of Macro-economic Models*, ed G.C. Chow and P. Corsi, New York: John Wiley.

Granger, C.W.J. 1969. Investigating causal relations by econometric models and cross-spectral methods. *Econometrica* 37, 424–38.

Griliches, Z. 1967. Distributed lags: a survey. *Econometrica* 35, 16–49.

Griliches, Z. 1986. Economic data issues. In *Handbook of Economet-*

rics, ed. Z. Griliches and M.D. Intriligator, Vol. 3, Amsterdam: North-Holland.

Haavelmo, T. 1943. Statistical testing of business cycle theories. *Review of Economics and Statistics* 25, 13–18.

Haavelmo, T. 1944. The probability approach in econometrics. *Econometrica* 12, Supplement, 1–118.

Hansen, L.P. 1982. Large sample properties of generalized method of moments. *Econometrica* 50, 1029–54.

Hansen, L.P. and Sargent, T.J. 1980. Formulating and estimating dynamic linear rational expectations models. *Journal of Economic Dynamics and Control* 2, 7–46.

Hansen, L.P. and Singleton, K.J. 1982. Generalized instrumental variables estimation of non-linear rational expectations models. *Econometrica* 50, 1269–86.

Hart, B.S. and von Neumann, J. 1942. Tabulation of the probabilities for the ratio of mean square successive difference to the variance. *Annals of Mathematical Statistics* 13, 207–14.

Hausman, J.A. 1978. Specification tests in econometrics. *Econometrica* 46, 1251–72.

Hausman, J.A. 1983. Specification and estimation of simultaneous equation models. In *Handbook of Econometrics*, Vol. 1, ed. Z. Griliches and M.D. Intriligator, Amsterdam: North-Holland.

Hausman, J. and Wise, D.A. 1979. Attrition bias in experimental and panel data: the Gary income maintenance experiment. *Econometrica* 47, 455–73.

Hausman, J.A. and Wise, D.A. 1985. *Social Experimentation*. Chicago: University of Chicago Press for the National Bureau of Economic Research.

Hausman, J.A., Hall, B.H. and Griliches, Z. 1984. Econometric models for count data with application to the patents – R & D relationship. *Econometrica* 52, 909–1038.

Heckman, J.J. and Singer, B. 1984. Econometric duration analysis. *Journal of Econometrics* 24, 63–132.

Heckman, J.J. and Willis, R. 1977. A beta-logistic model for the analysis of sequential labour force participation by married women. *Journal of Political Economy* 85, 27–58.

Hendry, D.F. 1976. The structure of simultaneous equations estimators. *Journal of Econometrics* 4, 51–88.

Hendry, D.F. and Richard, J-F. 1982. On the formulation of empirical models in dynamic econometrics. *Journal of Econometrics* 20, 3–33.

Hooker, R.H. 1901. Correlation of the marriage rate with trade. *Journal of the Royal Statistical Society* 44, 485–92.

Hsiao, C. 1983. Identification. In *Handbook of Econometrics*, Vol. 1, ed. Z. Griliches and M.D. Intriligator, Amsterdam: North-Holland.

Hsiao, C. 1985. Benefits and limitations of panel data. *Econometric Reviews* 4, 121–74.

Jorgenson, D.W. 1966. Rational distributed lag functions. *Econometrica* 34, 135–49.

Kendall, M.G. 1968. The history of statistical methods. In *International Encyclopedia of the Social Sciences*, ed. D.L. Sills, Vol. 15, New York: Macmillan and The Free Press, 224–32.

Keynes, J.M. 1921. *A Treatise on Probability*. London: Macmillan.

Keynes, J.M. 1939. The statistical testing of business cycle theories. *Economic Journal* 49, 558–68.

Klein, L.R. 1947. The use of econometric models as a guide to economic policy. *Econometrica* 15, 111-51.

Klein, L.R. 1950. *Economic Fluctuations in the United States 1921–1941*. Cowles Commission Monograph No. 11, New York: John Wiley.

Klein, L.R. 1971. Whither econometrics? *Journal of the American Statistical Society* 66, 415–21.

Klein, L.R. and Goldberger, A.S. 1955. *An Econometric Model of the United States, 1929–1952*. Amsterdam: North-Holland.

Koopmans, T.C. 1937. *Linear Regression Analysis of Economic Time Series*. Haarlem: De Erven F. Bohn for the Netherlands Economic Institute.

Koopmans, T.C. 1949. Identification problems in economic model construction. *Econometrica* 17, 125–44.

Koopmans, T.C. 1950. When is an equation system complete for statistical purposes? In *Statistical Inference in Dynamic Economic Models*, ed. T.C. Koopmans, Cowles Commission Monograph No. 10, New York: John Wiley.

Koopmans, T.C., Rubin, H. and Leipnik, R.B. 1950. Measuring the equation systems of dynamic economics. In *Statistical Inference in Dynamic Economic Models*, ed. T.C. Koopmans, Cowles Commission Monograph No. 10, New York: John Wiley.

Koyck, L.M. 1954. *Distributed Lags and Investment Analysis*. Amsterdam: North-Holland.

Kuh, E. 1959. The validity of cross-sectionally estimated behaviour equations in time series applications. *Econometrica* 27, 197–214.

Lawson, T. 1981. Keynesian model building and the rational expectations critique. *Cambridge Journal of Economics* 5, 311–26.

Leamer, E.E. 1978. *Specification Searches: Ad Hoc Inference with Non-experimental Data*. New York: John Wiley.

Leamer, E.E. 1983. Let's take the con out of Econometrics. *American Economic Review* 73, 31–43.

Leamer, E.E. 1985a. Vector autoregressions for causal inference. In *Carnegie-Rochester Conference Series on Public Policy* 22, ed. K. Brunner and A.H. Meltzer, Amsterdam: North-Holland, 255–304.

Leamer, E.E. 1985b. Sensitivity analyses would help. *American Economic Review* 85, 308–13.

Leamer, E.E. 1986. A Bayesian analysis of the determinants of inflation. In *Model Reliability*, ed. D.A. Belsley and E. Kuh, Cambridge, Mass: MIT Press.

Leamer, E.E. and Leonard, H. 1983. Reporting the fragility of regression estimates. *Review of Economics and Statistics* 65, 306–17.

Lenoir, M. 1913. *Etudes sur la formation et le mouvement des prix*. Paris: Giard et Brière.

Leontief, W.W. 1936. Quantitative input–output relations in the economic system of the United States. *Review of Economic Statistics* 18, 105–25.

Leontief, W.W. 1941. *The Structure of American Economy, 1919–1929*. Cambridge, Mass: Harvard University Press.

Leontief, W.W. 1951. *The Structure of American Economy, 1919–1939*. 2nd edn, Oxford: Oxford University Press.

Litterman, R.B. 1985. Forecasting with Bayesian vector autoregressions: five years of experience. *Journal of Business and Economic Statistics* 4, 25–38.

Liu, T.C. 1960. Underidentification, structural estimation and forecasting. *Econometrica* 28, 855–65.

Lucas, R.E. 1972. Expectations and the neutrality of money. *Journal of Economic Theory* 4, 103–24.

Lucas, R.E. 1973. Some international evidence on output-inflation tradeoffs. *American Economic Review* 63, 326–34.

Lucas, R.E. 1976. Econometric policy evaluation: a critique. In *The Phillips Curve and Labor Markets*, ed. K. Brunner and A.M. Meltzer, Carnegie-Rochester Conferences on Public Policy, Vol. 1, Amsterdam: North-Holland, 19–46.

Lyttkens, E. 1970. Symmetric and asymmetric estimation methods. In *Interdependent Systems*, ed. E. Mosback and H. Wold, Amsterdam: North-Holland.

McAleer, M., Pagan, A.R. and Volker, P.A. 1985. What will take the con out of econometrics? *American Economic Review* 75, 293–307.

Maddala, G.S. 1983. *Limited Dependent and Qualitative Variables in Econometrics*. Cambridge: Cambridge University Press.

Maddala, G.S. 1986. Disequilibrium, self-selection, and switching models. In *Handbook of Econometrics*, Vol. 3, ed. Z. Griliches and M.D. Intrilligator, Amsterdam: North-Holland.

Malinvaud, E. 1966. *Statistical Methods of Econometrics*. Amsterdam: North-Holland.

Manski, C.F. and McFadden, D. 1981. *Structural Analysis of Discrete Data with Econometric Applications*. Cambridge, Mass: MIT Press.

Marschak, J. 1953. Economic measurements for policy and prediction. In *Studies in Econometric Method*, ed. W.C. Hood and T.C. Koopmans, Cowles Commission for Research in Economics Monograph No. 14, New York: John Wiley.

Marschak, J. and Andrews, W.H. 1944. Random simultaneous equations and the theory of production. *Econometrica* 12, 143–205.

Mitchell, W.C. 1928. *Business Cycles: the Problem in its Setting*. New York: National Bureau of Economic Research.

Moore, H.L. 1914. *Economic Cycles: Their Law and Cause*. New York: Macmillan Press.

Moore, H.L. 1917. *Forecasting the Yield and the Price of Cotton*. New York: Macmillan Press.

Morgan, M.S. 1986. Statistics without probability and Haavelmo's revolution in econometrics. In *The Probabilistic Revolution: Ideas in the Social Sciences*, ed. L. Kruger, G. Gigerenzer and M. Morgan, Cambridge, Mass: MIT Press.

Mundlak, Y. 1961. Empirical production function free of management bias. *Journal of Farm Economics* 43, 44–56.

Mundlak, Y. 1978. On the pooling of time series and cross section data. *Econometrica* 46, 69–85.

Muth, J.F. 1961. Rational expectations and the theory of price movements. *Econometrica* 29, 315–35.

Nelson, C.R. 1972. The prediction performance of the FRB–MIT–Penn model of the US economy. *American Economic Review* 62, 902–17.

Nerlove, M. 1958a. Adaptive expectations and the cobweb phenomena. *Quarterly Journal of Economics* 72, 227–40.

Nerlove, M. 1958b. *Distributed Lags and Demand Analysis*. USDA, Agriculture Handbook No. 141, Washington, DC.

Nerlove, M., Grether, D.M. and Carvalho, J.L. 1979. *Analysis of Economic Time Series: A Synthesis*. New York: Academic Press.

Neumann, J. von. 1941. Distribution of the ratio of the mean square successive difference to the variance. *Annals of Mathematical Statistics* 12, 367–95.

Neumann, J. von. 1942. A further remark on the distribution of the ratio of the mean square successive difference to the variance. *Annals of Mathematical Statistics* 13, 86–8.

Orcutt, G.H. 1948. A study of the autoregressive nature of the time series used for Tinbergen's model of the economic system of the United States, 1919–1932. *Journal of the Royal Statistical Society*, Series B 10, 1–45 (Discussion, 46–53).

Orcutt, G.H., Greenberger, M., Korbel, J. and Rivlin, A.M. 1961. *Microanalysis of Socioeconomic Systems: A Simulation Study*. New York: Harper & Row.

Orcutt, G.H., Merz, J. and Quinke, H. (eds) 1986. *Microanalytic Simulation Models to Support Social and Financial Policy*. Amsterdam: North-Holland.

Pagan, A.R. and Hall, A.D. 1983. Diagnostic tests as residual analysis. *Econometric Reviews* 2, 159–218.

Paris, S.J. and Houthakker, H.S. 1955. *The Analysis of Family Budgets*. Cambridge: Cambridge University Press.

Pesaran, M.H. 1981. Identification of rational expectations models. *Journal of Econometrics* 16, 375–98.

Pesaran, M.H. 1987. *The Limits to Rational Expectations*. Oxford: Basil Blackwell.

Pesaran, M.H. and Deaton, A.S. 1978. Testing non-nested nonlinear regression models. *Econometrica* 46, 677–94.

Pesaran, M.H. and Smith, R.P. 1985a. Keynes on econometrics. In *Keynes' Economics: methodological issues*, ed. T. Lawson and M.H. Pesaran, London: Croom Helm.

Pesaran, M.H. and Smith, R.P. 1985b. Evaluation of macroeconometric models. *Economic Modelling*, April, 125–34.

Phillips, P.C.B. 1983. Exact small sample theory in the simultaneous equations model. In *Handbook of Econometrics*, ed. Z. Griliches and M.D. Intrilgator, Vol. 1, Amsterdam: North-Holland.

Phillips, P.C.B. 1986. Understanding spurious regressions in econometrics. *Journal of Econometrics* 33, 311–40.

Phillips, P.C.B. and Durlauf, S.N. 1986. Multiple time series regression with integrated processes. *Review of Economic Studies* 53, 473–95.

Phillips, P.C.B. 1987. Time series regression with unit roots. *Econometrica* (forthcoming).

Prothero, D.L. and Wallis, K.F. 1976. Modelling macroeconomic time series. *Journal of the Royal Statistical Society*, Series A 139, 468–86.

Quandt, R.E. 1982. Econometric disequilibrium models. *Econometric Reviews* 1, 1–63.

Ramsey, J.B. 1969. Tests for specification errors in classical linear least squares regression analysis. *Journal of the Royal Statistical Society*, Series B 31, 350–71.

Reiersol, O. 1941. Confluence analysis by means of lag moments and other methods of confluence analysis. *Econometrica* 9, 1–24.

Reiersol, O. 1945. *Confluence Analysis by Means of Instrumental Sets of Variables*. Stockholm.

Rosen, H.S. 1985. Housing behaviour and the experimental housing-allowance program: what have we learned? In *Social Experimentation*, ed. J.A. Hausman and D.A. Wise, Chicago: University of Chicago Press.

Rothenberg, T.J. 1984. Approximating the distributions of econometric estimators and test statistics. In *Handbook of Econometrics*, ed. Z Griliches and M.D. Intriligator, Vol. 2, Amsterdam: North-Holland.

Ruggles, R. and Ruggles, N. 1956. *National Income Account and Income Analysis*. 2nd edn, New York: McGraw-Hill.

Samuelson, P.A., Koopmans, T.C. and Stone, J.R.N. 1954. Report of the evaluative committee for Econometrica. *Econometrica* 22, 141–6.

Sargan, J.D. 1958. The estimation of economic relationships using instrumental variables. *Econometrica* 26, 393–415.

Sargan, J.D. 1964. Wages and prices in the United Kingdom: a study in econometric methodology. In *Econometric Analysis for National Economic Planning*, ed. P.E. Hart, G. Mills and J.K. Whitaker, London: Butterworths.

Sargent, T.J. 1973. Rational expectations, the real rate of interest and the natural rate of unemployment. *Brookings Papers on Economic Activity* No. 2, 429–72.

Sargent, T.J. and Wallace, N. 1976. Rational expectations and the theory of economic policy. *Journal of Monetary Economics* 2, 169–84.

Savin, N.E. 1973. Systems k-class estimators. *Econometrica* 41, 1125–36.

Schultz, M. 1938. *The Theory and Measurement of Demand*. Chicago: University of Chicago Press.

Schumpeter, J.A. 1954. *History of Economic Analysis*. London: George Allen & Unwin.

Shiller, R.J. 1973. A distributed lag estimator derived from smoothness priors. *Econometrica* 41, 775–88.

Sims, C.A. 1972. Money, income and causality. *American Economic Review* 62, 540–52.

Sims, C.A. 1980. Macroeconomics and reality. *Econometrica* 48, 1–48.

Sims, C.A. 1982. Policy analysis with econometric models. *Brookings Papers on Economic Activity* No. 1, 107–64.

Sims, C.A. 1986. Are forecasting models usable for policy analysis? *Federal Reserve Bank of Minneapolis Review* 10, 2–16.

Slutsky, E. 1927. The summation of random causes as the source of cyclic processes. In *Problems of Economic Conditions*, Vol. 3, Moscow. English trans. in *Econometrica* 5, (1937), 105–46.

Solow, R.M. 1960. On a family of lag distributions. *Econometrica* 28, 393–406.

Srivastava, V.K. 1971. Three-stage least-squares and generalized double k-class estimators: a mathematical relationship. *International Economic Review* 12, 312–16.

Stafford, F.P. 1985. Income-maintenance policy and work effort: learning from experiments and labour-market studies. In *Social Experimentation*, ed. J.A. Hausman and D.A. Wise, Chicago: University of Chicago Press.

Stigler, G.J. 1954. The early history of empirical studies of consumer behaviour. *Journal of Political Economy* 62, 95–113.

Stigler, G.J. 1962. Henry L. Moore and statistical economics. *Econometrica* 30, 1–21.

Stone, J.R.N. 1945. The analysis of market demand. *Journal of the Royal Statistical Society* Series A 108, 286–382.

Stone, J.R.N. et al. 1954a. *Measurement of Consumers' Expenditures and Behaviour in the United Kingdom, 1920–38*, Vols 1 and 2. London: Cambridge University Press.

Stone, J.R.N. 1954b. Linear expenditure systems and demand analysis: an application to the pattern of British demand. *Economic Journal* 64, 511–27.

Stone, J.R.N. 1978. Keynes, political arithmetic and econometrics. British Academy, Seventh Keynes Lecture in Economics.

Stone, J.R.N. 1984. The accounts of society. Nobel Memorial lecture. Reprinted in *Journal of Applied Econometrics* 1, (1986), 5–28.

Theil, H. 1954. Estimation of parameters of econometric models. *Bulletin of International Statistics Institute* 34, 122–8.

Theil, H. 1958. *Economic Forecasts and Policy*. Amsterdam: North-Holland; 2nd edn, 1961.

Tinbergen, J. 1929–30. Bestimmung und Deutung von Angebotskurven: ein Beispiel. *Zeitschrift für Nationalökonomie* 1, 669–79.

Tinbergen, J. 1937. *An Econometric Approach to Business Cycle Problems*. Paris: Herman & Cie Editeurs.

Tinbergen, J. 1939. *Statistical Testing of Business Cycle Theories*. Vol. 1: *A Method and its Application to Investment activity*; Vol. 2: *Business Cycles in the United States of America, 1919–1932*. Geneva: League of Nations.

Tobin, J. 1958. Estimation of relationships for limited dependent variables. *Econometrica* 26, 24–36.

Treadway, A.B. 1971. On the multivariate flexible accelerator. *Econometrica* 39, 845–55.

Trivedi, P.K. 1975. Time series analysis versus structural models: a case study of Canadian manufacturing behaviour. *International Economic Review* 16, 587–608

Wallis, K.F. 1977. Multiple time series analysis and the final form of econometric models. *Econometrica* 45, 1481–97.

Wallis, K. 1980. Econometric implications of the Rational Expectations Hypothesis. *Econometrica* 48, 49–73.

Wegge, L.L. 1965. Identifiability criteria for a system of equations as a whole. *Australian Journal of Statistics* 3, 67–77.

Whewell, W. 1850. *Mathematical Exposition of some Doctrines of Political Economy: Second Memoir*. Cambridge: Cambridge Philosophical Society, 1856, Transaction 9, Pt. I. *Philosophical Society*.

White, H. 1981. Consequences and detection of misspecified nonlinear regression models. *Journal of the American Statistical Association* 76, 419–33.

White, H. 1982. Maximum likelihood estimation of misspecified models. *Econometrica* 50, 1–26.

Whittle, P. 1963. *Prediction and Regulation by Linear Least-squares Methods*. London: English Universities Press.

Wickens, M. 1982. The efficient estimation of econometric models with rational expectations. *Review of Economic Studies* 49, 55–68.

Wicksteed, P.H. 1889. On certain passages in Jevons's theory of political economy. *The Quarterly Journal of Economics* 3, 293–314.

Wold, H. 1938. *A Study in the Analysis of Stationary Time Series*. Stockholm: Almqvist and Wiksell.

Working, E.J. 1927. What do statistical 'demand curves' show? *Quarterly Journal of Economics* 41, 212–35.

Wright, P.G. 1915. Review of economic cycles by Henry Moore. *Quarterly Journal of Economics* 29, 631–41.

Wright, P.G. 1928. *The Tariff on Animal and Vegetable Oils*. London: Macmillan for the Institute of Economics.

Yule, G.U. 1895, 1896. On the correlation of total pauperism with proportion of out-relief. *Economic Journal* 5, 603–11; 6, 613–23.

Yule, G.U. 1915. Crop production and price: a note on Gregory King's law. *Journal of the Royal Statistical Society* 78, March, 296–8.

Yule, G.U. 1921. On the time-correlation problem, with special reference to the variate-difference correlation method. *Journal of the Royal Statistical Society* 84, 497–526.

Yule, G.U. 1926. Why do we sometimes get nonsense correlations between time-series? A study in sampling and the nature of time-series. *Journal of the Royal Statistical Society* 89, 1–64.

Zellner, A. 1962. An efficient method of estimating seemingly unrelated regressions and tests for aggregation bias. *Journal of the American Statistical Association* 57, 348–68.

Zellner, A. 1979. Statistical analysis of econometric models. *Journal of the American Statistical Association* 74, 628–43.

Zellner, A. 1984. *Basic Issues in Econometrics*. Chicago: University of Chicago Press.

Zellner, A. 1985. Bayesian econometrics. *Econometrica* 53, 253–70.

Zellner, A. and Palm, F. 1974. Time series analysis and simultaneous equation econometric models. *Journal of Econometrics* 2, 17–54.

Zellner, A. and Theil, H. 1962. Three-stage least squares: simultaneous estimation of simultaneous equations. *Econometrica* 30, 54–78.

economic anthropology. The establishment of the Society of Economic Anthropology in 1971 symbolized the coming-of-age of an important sub-discipline in Anthropology.

The rise of economic anthropology has been co-terminous with the rise of development economics. They were both products of the post-World War II changes in the nature of the world economy. The process of de-colonization focused attention on the problems of development and underdevelopment in ex-colonial countries and generated a need to know more about the particular economic problems facing them. In spite of this common purpose, economic anthropology and economic development have developed independently of one another. This is partly because the former usually adopts an empirically oriented, descriptive, microeconomic approach while the latter tends to adopt a more conceptually oriented, prescriptive, macroeconomic approach. Perhaps the main reason, though, is that in development economics the neoclassical paradigm is overwhelmingly predominant. This makes development economists somewhat inward looking and exclusive in their approach to development problems: they simply have little interest in what others outside their field are doing, as some economic anthropologists who have tried to bridge the gap have found to their dismay (Hill, 1985). Economic anthropologists, on the other hand, are pluralistic in their theoretical approach and have been quite open to ideas developed outside the discipline of anthropology: the theories and concepts from different schools of economic thought – Marxist, institutional and neoclassical – have been particularly influential.

Economic anthropology has a short history, with the term 'economic anthropology' only becoming common currency after the publication of Herskovits's *Economic Anthropology: A Study of Comparative Economics* (1952). This was the title of the second edition of his book *The Economic Life of Primitive Peoples* (1940), which he revised in the light of a critical review by the neoclassical economist Frank Knight (1941). This dialogue, together with the publication in 1957 of *Trade and Market in the Early Empires* by Polanyi, Arensberg and Pearson sparked off a debate that still persists today between the 'formalists' (i.e. neoclassical economists) and the 'substantivists' (i.e. the institutionalists – see Firth, 1967; Le Clair and Schneider, 1968; Dalton, 1971). The controversy became a three-way debate in the 1970s when the Marxist-inspired theories of Meillassoux (1960, 1975), Godelier (1966, 1973) and other French Marxists began to be translated into English (see Seddon, 1978; Clammer, 1978; Kahn and Llobera, 1981).

There is no point in rehearsing the arguments of the principal protagonists to this debate as this has been done many times before (see Ortiz, 1983). In any case the literature has become so vast over the past decade that no short review could do justice to it. Instead, I want to take a somewhat longer view of the relationship between the two disciplines and in particular to consider the relationship between *political economy* and anthropology. The debate within economic anthropology has become arid and sterile – witness Dalton and Kocke's (1983) espousal of the Polanyi group's cause on the grounds that its members share a paradigm, whereas Marxist economic anthropology only shares a name. By taking a broader perspective it is possible to highlight the contribution of the respective disciplines to each other and at the same time bring the debate back to the central issue, namely, the analysis of inequality and poverty in the ex-colonial countries. In order to narrow the scope of this vast topic I will limit myself to a comparative analysis of fundamental issues concerning technology, land and labour in agrarian communities. One subject of fundamental importance that has been excluded is exchange, but this is covered elsewhere (*see* GIFTS).

The intensive micro study of tribal and peasant societies by the use of the 'participant observation' method is the hallmark of the anthropologist. This method involves a process of developing generalizations about social relations from the analysis of particular cases over an extended period of time. It is based on the assumption that relevant questions and hypotheses only begin to emerge during the fieldwork process when the researcher begins to see social phenomena in their totality. This method is sharply contrasted with the large-scale sample survey method of the sociologist. Here questions and hypotheses are formulated prior to fieldwork and the researcher only stays in a village for as long as it takes to collect answers to a questionnaire.

The fieldwork tradition is primarily associated with the British empiricist school of anthropology and for any academic anthropologist who works in this tradition an extended fieldwork trip of 12–18 months is an obligatory *rite de passage*. As a result, literally thousands of monographs now exist describing the social, economic and religious life of tribal and peasant societies from many different parts of the world. These ethnographies provided the data for the development of another important tradition within anthropology. This was primarily developed by the French rationalist school. It involves library–based study of ethnographies and the development of general theories by use of the comparative method. These two methods of anthropological research are not mutually exclusive. There is a feedback relationship between them and nowadays most anthropologists would use a combination of both methods.

The sum total of these two traditions has been to revolutionize our understanding of non-capitalist economies at the empirical, conceptual and theoretical level. In the early stages of the development of anthropology the pristine tribal society was the principal object of analysis. However, this is not the case today because the expansion of the world economy has incorporated even the remotest tribe into its orbit and most contemporary ethnographies attempt to come to terms with this fact empirically and theoretically. Peasant economies and societies became a major focus of attention after World War II and ethnographers of these societies have always situated them historically in the context of the wider society. Anthropology, then, has made a significant contribution to the theory of comparative economic systems and the theory of development and change.

In order to assess the nature of this contribution it is necessary to review briefly some of the literature of classical political economy on peasant economy to see what questions were being posed. It will then be possible to see how anthropologists have answered these questions, reformulated them, and posed new questions.

British economic thought has made little contribution to the theory of peasant economy because the 'peasant question' was of little interest to Adam Smith and his successors. The reason for this is to be found in the particular conditions of English agriculture. By the beginning of the 19th century capitalist farming predominated in England and policy makers had no 'tribal' or 'peasant' problem to deal with. England has been unique in this respect: every other country in the world has had, or continues to have, some part of its rural population engaged in non-capitalist forms of farming. The problems of capitalist development in this context pose questions of a kind fundamentally different from those with which Smith, Ricardo and the other British economists of their day were primarily concerned. Their European contemporaries, on the other hand, were deeply concerned with these questions and it is in their writings that we find the origins of many of the present day debates in economic anthropology.

Consider the French Physiocrats for example. They were primarily concerned to analyse the nature and causes of wealth and poverty in the French countryside of the 18th century. They distinguished between large-scale (*la grande*) and small-scale farming (*la petite culture*). For Quesnay (1756) the difference was basically technological: large-scale farmers were wealthy and ploughed with horses, small-scale farmers were poor and ploughed with oxen. Horses, argued Quesnay, were superior to oxen and enabled greater yields to be produced, but they cost more to buy. Quesnay did not see the issue as a simple problem of choice of technique, but rather as one of access to capital. He argued that the cause which forced peasants to cultivate with oxen – namely their poverty – did not allow them to use horses. Turgot (1766) developed this distinction by placing more emphasis on its sociological basis. Both kinds of farmer were alike in that both were tenants. However, the large-scale farmer was a capitalist farmer who advanced his own capital in the form of livestock, tools and the seed and paid a money rent to the landlord. He was a modern entrepreneur of agriculture – an accumulator of capital – and as such presented a sharp contrast to the small scale pre-capitalist share-cropper who advanced only his own labour and shared the physical product of his labour with the landlord. Turgot also distinguished an intermediary category, the small farmer, who paid a fixed rent to the landlord. The policy conclusion the Physiocrats arrived at was that capitalist farming was technologically superior to other forms of farming and as such should be encouraged if the wealth of the state was to be increased.

Their critics disputed the claim that capitalist farming was technologically superior and argued for quite different policies for agrarian change which had quite different political consequences. This debate persisted in western Europe late into the 19th century, moved to eastern Europe where it was debated until the 1930s, then moved to Asia and Latin America where it is still being debated today.

The eastern European version of the debate, the controversy between Lenin and Chayanov, in particular, has had a significant impact on contemporary anthropological thought. The recent English translation of Chayanov's *Theory of Peasant Economy* (1923–5) has done much to popularize his ideas. Sahlins, for example, in his influential *Stone Age Economics* (1972), has applied Chayanov's theories on the dynamics of peasant household economy in an ahistorical analysis of data on tribal economies in Melanesia and elsewhere. Chayanov's theory has two central pillars: a biological determinist theory of household size and composition, and a marginal utility theory of household output determination. Sahlins' development of Chayanov's theory builds on the first pillar but rejects the second. Some anthropologists have rejected Chayanov's approach altogether and have attempted to develop Lenin's (1899) class conflict theory of the peasantry. Perhaps the most successful example of this is Harriss's careful and scholarly analysis of technological change in a South Indian peasant economy (1982). Harriss is sensitive to the historical context of the European 'agrarian question' debate and the historical context of the peasant economy he studied, and he combines theory, history and ethnographic data in a masterly way with many interesting results. There are other anthropologists who reject both the Leninist and Chayanov's approaches. Polly Hill's writings (1963, 1970, 1972, 1977, 1982) – based on some thirteen years field work in Africa and India – combine an iconoclastic approach to received theory with meticulous attention to empirical detail. If Hill has a classical economist as a predecessor then it is Richard Jones, who argued for an empirically based comparative and historical approach to agrarian relations.

Before considering some of the contemporary findings of Harriss, Hill and others on the question of the relative efficiency of capitalist and peasant farms it is important to appreciate what the participant observation method amounted to in terms of the quality of data collected.

In the early part of the 19th century casual empirical observations, and/or travellers' tales provided the data base for many of the debates on the peasantry. For example, John Stuart Mill – who was one of the few 19th-century English economists to consider the peasant question seriously – based his pro-peasant argument on lengthy quotations from the books of travellers (1848, chs VI–VIII); European economists, such as Quesnay (1756), Turgot (1766) and Sismondi (1847), based their arguments on their own casual observations of the European farming practices of their time. Statistical data on peasant production, consumption and exchange were simply not available or, if they were, of very dubious quality. The debate, then, was largely a matter of one unsubstantiated opinion versus another. Russia was probably the first country where data on peasants were collected on a systematic basis. In 1861 provincial and district assemblies (*zemstvo*) were set up to implement land reform and they launched a vast programme of economic and statistical investigation into peasant economic problems. Over 4000 volumes of data were collected in the period up to World War I. These data – which were collected using large scale sample survey methods – radically transformed the way in which the debate was conducted, as differing opinions could now be examined statistically. Both Lenin and Chayanov made extensive use of these data in their studies of the peasant question.

This revolutionary development in peasant data collection was immediately followed by another – the participant observation method of the anthropologists. This method was developed by Malinowski (1922) in his study of a tribal economy in Papua New Guinea during World War I. However, within a decade American anthropologists, such as Redfield (1930) were applying it to the study of peasant societies in Mexico. Following World War II the number of anthropological studies of peasant societies grew exponentially and their number would now easily rival the *zemstvo* data in terms of the number of volumes. These data represent a quantum leap in our understanding because they are concerned with the description of relations between people as well as with census data on household size, landholdings and livestock numbers. An anthropologist who spends one-and-a-half to two years in a single village, observing the daily life of the peasants and conversing with them in the vernacular is obviously going to be better informed than, for example, the Russian official who spent one-and-a-half to two days in a village with a translator administering a complicated questionnaire consisting of 677 questions (Kerblay, 1966). A valid criticism of early ethnographies was that they only gave a sketch of a society at a point in time. However, repeat studies of villages by the same anthropologist or by 'second generation' anthropologists is now increasingly common. This has provided an historical depth to anthropological research, a check on the accuracy of the data, and much controversy (Lewis, 1951; Freeman, 1983).

Of course, the availability of data is only one of the many necessary conditions for the development of knowledge. It is not a sufficient condition and certainly not independent of the concepts, theories or methods of the people who collect it or use it. The quality of the data then, whilst obviously important, cannot by itself resolve a controversy. The fact that the debate initiated by the physiocrats about the alleged technological superiority of capitalist farming is still unresolved today illustrates this point.

Harriss (1982, p. 153), for example, found that the relationship between farm size and yield per acre for the village he studied in South India in 1973–4 was 'generally positive'. Farms less than 1.5 acres in size yielded 1.39 tons of paddy per irrigated acre, farms in the size range 1.5 to 5 acres yield 1.50 tons, 5–10 acre farms 1.51 tons, 10–20 acre farms 2.25 tons, and farms over 20 acres 2.12 tons. Taussig (1978) by way of contrast, found that small farmers in the Cauca Valley in Colombia use resources more efficiently than big farmers. This latter finding is in keeping with the general, though far from unanimous, conclusion reached by development economists (Lipton, 1977; Schultz, 1964). Perhaps the general conclusion we should reach is, as economist-turned-anthropologist Polly Hill (1982; p. 168) has argued, is that the question to which the debate has been directed – Do peasant farmers use their land more efficiently than large farmers? – is oversimplified. Hill, for example, objects to the world 'peasant' because, among other things, it entails the concept of an amorphous, undifferentiated class of cultivators. If the peasantry is conceptualized as consisting of poor agricultural workers who own virtually no land, tenant farmers who farm small plots of land for a rent payable in kind or produce, and relatively wealthy joint households who own small holdings of scattered plots farmed with the assistance of bonded labourers, then it is obvious that the question is indeed oversimplified.

One long–standing controversy within the political economy literature to which anthropologists have made a significant contribution concerns the general questions of land tenure, inheritance and population growth. Here, not only have they been able to answer some old questions, but, more importantly, they have managed to pose an array of completely new questions of great importance. This is because land tenure and inheritance in peasant economies raises the question of kinship, and anthropological data and theories have transformed our understanding of this subject.

Peasant land tenure presents a striking contrast to that of a capitalist farm. Whereas a capitalist farmer typically owns or rents a large consolidated block of land and employs wage-labour, the peasant farmer usually owns or rents a smallholding of scattered plots and uses family labour. Many 19th–century political economists blamed inheritance laws for this state of affairs: the large consolidated plots of the wealthy English capitalist farmers were, it was argued, a product of the English law of primogeniture whereas the poverty of the European peasantry was seen to be a result of the Napoleonic codes of partible inheritance. Peasant farming, according to this argument, had no future; peasants were destined to a life of poverty and misery and the only policy solution was to encourage large-scale farming by changing the inheritance laws. The apologists for peasantry (e.g. Sismondi, 1847) did not deny that peasant land was fragmented and scattered. What they did deny, however, was that sub-division was a growing evil and they argued that consolidations of land through purchase and marriage balanced the subdivisions. The key role of marriage in peasant economy was only noted by the 19th-century economists; their analytical attention focused on the role of the land market because it was observed that peasants pay more for their land than the capitalized value of differential rent. For the pro-peasant lobby this was simply an expression of the technical superiority of peasant farming: they were more efficient in their use of land so they could afford to pay more for it.

Marx (1894, ch. 47) had a different explanation. He argued that the peasant proprietor is simultaneously capitalist, landlord and worker. The commodity economy, of which he was inextricably part, forced him to pay himself (and his family members) extra low wages, to accept a less than average rate of return on his capital, and to capitalize rent at a less than average interest rate. The sum total of all of these effects was under-consumption, over–production and high prices of land. It also enabled the market price of agricultural produce to fall below prices of production. The latter fact explained why peasant farms were able to compete with capitalist farms. But Marx saw the persistence of peasant farming as a transitional phase: in the long run the technological superiority of capitalist farming would bring about the downfall of peasant farming.

The European peasantry did not disappear as fast as predicted by Marx and its persistence in some parts of Europe until today has enabled anthropologists to study it. Numerous studies from various parts of Europe are available. These studies highlight the central role of kinship and marriage, neglected by the 19th-century observers (see e.g. Davis, 1973; Cutiliero, 1971). Furthermore, many of these studies have taken an historical perspective and provide much fascinating data relevant to the agrarian question that has relevance to contemporary debates about agrarian change in Africa and Latin America. However, European ethnographers show little awareness of the 19th-century debates. Historians of economic thought concerned with the peasant question, on the other hand, are similarly unaware of the ethnographic literature (Dewey, 1974). A synthetic work that combines these two bodies of data and uncovers the lessons of the European experience has therefore yet to be written.

The wealth of data on kinship, marriage, and land tenure gathered by anthropologists working in non-European countries has also been largely ignored by economists and policy-makers. As a result many policies have been based on misconceptions about the nature of peasant economy. Leach's (1961) classic study of kinship and land tenure in a Sri Lankan village demonstrates this point well. Land tenure in this irrigated rice-growing area of Sri Lanka means rights to a particular quantity of *water* and individual property rights are expressed in terms of relations of descent and affinity. British land policy in Sri Lanka was based on the false assumption that in farming communities land tenure meant rights to a particular piece of *land* and the misconception that property rights were communal.

Leach's book is a study of 146 individuals in a village with a total area of 135 acres. It is a classic example of the anthropological method of developing generalizations from the particular case. The data he presents on the relationship between kinship, marriage and the transmission of land and the conclusions he draws challenge many preconceptions economists hold about property transmission, labour relations, and technology in peasant economies in South Asia. There are now literally hundreds of similar studies from different parts of South Asia (e.g. Bailey, 1957; Epstein, 1962; Macfarlane, 1976; Mencher, 1978; Gough, 1981; Hill, 1982; Harriss, 1982 and references therein), not to mention those from other parts of the world. Recent work by feminist-inspired writers has focused attention on women, work and property and has stimulated much rethinking of earlier theories (Jeffery, 1979; Sharma, 1980).

The ethnographic literature on peasant society is therefore exceptionally large and, in many cases, highly controversial. It defies simple summary. The message it carries for economists is that non-market transfers of labour and property are extremely important in peasant economies, and that cultural variables such as honour, prestige, caste, etc. play a significant role in determining the size and composition of these transfers.

Anthropologists' concern with kinship relations in peasant societies developed in earnest only after World War II. Prior to this their concern was mainly with kinship relations among the 'tribal' societies of Africa and Oceania. In these societies there are no private property rights in land of the type found in the peasant societies found in Europe or Asia. Instead the clan or lineage is usually the principal land owning group and an individual's access to land is mediated by a complex hierarchy of rights and obligations.

Lewis Henry Morgan, in his *Ancient Society* (1877), was one of the first anthropologists to analyse clan organization and to contrast it with the European system of private property and classes. Morgan's theory was based on his own fieldwork among the American Indians as well as comparative ethnographic and historical data from around the world. His theory was couched in evolutionary terms and it is now no longer seriously entertained. However, his work – because it was empirically based – contained many profound insights and many important analytical distinctions. As such it has served as the basic building block for much subsequent anthropological research.

While contemporary anthropological research has shown that clan organization is nothing like as simple or as uniform from one place to another as Morgan thought it was, few anthropologists would deny Morgan's proposition that kinship relations and property rights in land are intricately related in tribal economies in a way that differs markedly from a capitalist economy. From the market perspective this difference in land tenure assumes the form of inalienable land rights. This is not to say that tribal land tenure is rigid and unchanging. Land, where it is in short supply, is fixed but the social structure of the people who occupy it necessarily changes over the generations. Clan rights to land, then, are the subject of continuous negotiation and political manoeuvre and beyond the control of any individual who may like to assert private property rights.

Among the Tiv of Nigeria, for example, land tenure is expressed in terms of a 'segmentary lineage system'. The land map is a spatial version of the relations between segments of a lineage, whose apical ancestor is some three to six generations removed from living elders. As one moves down the lineage, sons and grandsons of the apical ancestor became in turn apical ancestors of maximal and minimal lineages. These segments describe major and minor boundaries within the lineage land. These boundaries assume importance for intra-lineage disputes over land but in confrontation with other tribes over land the boundaries disappear. The Tiv land map then is not a rigid and precise surveyor's map; it constantly changes with the dynamics of kinship organization and political disputes. Land for the Tiv is a unique category and nothing is exchangeable for it. No one can buy land in a lineage on which they are living because they have rights to sufficient land there already; on the other hand, no one can buy land in a lineage in which they are not living (see Bohannan and Bohannan, 1968).

In the Papua New Guinea highlands the land tenure system is quite different. Here the basic unit is a territorial clan whose members only have a shallow genealogical link of a few generations. One of the reasons for this is the high population density in the highlands. Clans were, and still are, engaged in violent struggles for land. Defeated clans lose their land and

become refugees but are readily adopted by other weaker clans wishing to build up their membership for defence purposes. Kinship links in this situation are created just as much by the sharing of food and residence as by genealogical links (Strathern, 1972; Meggitt, 1965). Kinship relations, then, are political relations just as much as blood relations in tribal socieities and it is these relations rather than private property relations that control the distribution of land.

What the anthropological literature has demonstrated is the tenacity and adaptability of these 'traditional' systems of economic organization in the face of quite dramatic social and political change during the colonization and decolonization process (Gregory, 1982). Among the Tolai of Papua New Guinea, for example, a century of colonization has brought many far reaching changes in life style: half their land was alienated and a copra plantation economy established; villagers have become highly educated and participate in the modern cash economy as senior public servants, entrepreneurs and cash crop producers; the population has grown rapidly and pressure on land is acute. Yet these changes have had little effect on the traditional system of land tenure: land still cannot be bought and sold (Salisbury, 1970, p. 70).

The persistence of these traditional forms of economic organization is often seen by development economists as a barrier to development. Such assertions are often made in complete ignorance of the facts as has been brilliantly demonstrated by Polly Hill in her classic study *The Migrant Cocoa-Farmers of Southern Ghana* (1963). The creation of Ghana cocoa growing – 'one of the great events of the recent economic history of Africa south of the Sahara' – was due primarily to the migration of 'traditional' African farmers. Hill shows how the migratory process derived much of its strength and impetus from the fact that it was based firmly on 'traditional orgnization'. It is also interesting to note that the farmers established their farms in uninhabited forests they purchased from other tribal groups. In other words, the whole enterprise was premised on a very active market in land. It is clear from this case that tribes are quite prepared to alienate 'inalienable' land if it suits their interests. The task of explaining contradictions such as these requires much rethinking of our Eurocentric economic theories. What is needed, Hill argues, is empirically based concepts, which can only be developed by engaging in fieldwork. She argues that we are so ignorant of the economic condition of people in rural tropical regions that we are unaware of just how ignorant we are. Much economic development theory, she argues, is based on profound ignorance; many early hypotheses of development theorists, which were based on little or no evidence, have hardened into truths (see Hill, 1982). Here then, is perhaps another reason why development economics has shunned economic anthropology: it has everything to lose by taking it seriously.

This brief review of the history of economic anthropology has not mentioned many important contributions anthropologists have made to our understanding of tribal and peasant economics in the areas of economic history (Geertz, 1963), agricultural technology and choice of technique (Conklin, 1957; Miles, 1979), consumption (Douglas and Isherwood, 1978) rural marketing (Skinner, 1964–5; Bohannan and Dalton, 1962), fishing communities (Firth, 1946; Alexander, 1982), contemporary hunters and gatherers (Woodburn, 1982; Morris, 1982), pastoralists (Proceedings, 1979), collective farming (Humphrey, 1983; Hann, 1980), and development (Hart, 1982), among others. The scope of economic anthropology is wide and the theoretical predilections of its practitioners diverse, but it is clear that anthropology and political economy have contributed much to each other in the development of out understanding of the problems of poverty and inequality in peasant and tribal economies. The particular contribution of economic anthropology is not so much in the answers it has provided but in the stimulating, empirically informed, novel questions it has posed.

C.A. GREGORY

See also ECONOMIC THEORY AND THE HYPOTHESIS OF RATIONALITY; HUNTING AND GATHERING ECONOMIES; PEASANTS; PRIMITIVE CAPITALIST ACCUMULATION.

BIBLIOGRAPHY

Alexander, P. 1982. *Sri Lankan Fishermen: Rural Capitalism and Peasant Society*. Canberra: ANU Monographs on South Asia No. 7.

Bailey, F.G. 1957. *Caste and the Economic Frontier*. Manchester: Manchester University Press.

Bohannan, P. and Bohannan, L. 1968. *Tiv Economy*. Evanston: Northwestern University Press.

Bohannan, P. and Dalton, G. (eds) 1962. *Markets in Africa*. Evanston: Northwestern University Press.

Chayanov, A.V. 1923–5. *The Theory of Peasant Economy*. Ed. D. Thorner, B. Kerblay and R.E.F. Smith, Homewood: Irwin, 1966.

Clammer, J. (ed.) 1978. *The New Economic Anthropology*. London: Macmillan.

Conklin, H. 1957. *Hanunoo Agriculture in the Philippines*. Rome: FAO.

Cutiliero, J.A. 1971. *A Portuguese Rural Society*. Oxford: Clarendon.

Dalton, G. 1971. *Economic Anthropology and Development*. New York: Basic Books.

Dalton, G. and Köcke, J. 1983. The work of the Polanyi group: past, present, and future. In Ortiz (1983).

Davis, J. 1973. *Land and Family in Pisticci*. London: Athlone.

Dewey, C.J. 1974. The rehabilitation of the peasant proprietor in nineteenth-century economic thought. *History of Political Economy* 6, 17–47.

Douglas, M. and Isherwood, B. 1978. *The World of Goods*. Harmondsworth: Penguin, 1980.

Epstein, T.S. 1962. *Economic Development and Social Change in South India*. Manchester: Manchester University Press.

Firth, R. 1946. *Malay Fishermen: Their Peasant Economy*. London: Kegan Paul.

Firth, R. (ed.) 1967. *Themes in Economic Anthropology*. London: Tavistock.

Freeman, D. 1983. Inductivism and the test of truth: a rejoinder to Lowell D. Holmes and others. *Canberra Anthropology* 6(2), 101–92.

Geertz, C. 1963. *Agricultural Involution: The Process of Ecological Change in Indonesia*. Berkeley: University of California Press.

Godelier, M. 1966. *Rationality and Irrationality in Economics*. London: New Left Books, 1972.

Godelier, M. 1973. *Perspectives in Marxist Anthropology*. Cambridge: Cambridge University Press, 1977.

Gough, K. 1981. *Rural Society in Southeast India*. Cambridge: Cambridge University Press.

Gregory, C.A. 1982. *Gifts and Commodities*. London: Academic Press.

Hann, C. 1980. *Tazlar: A Village in Hungary*. Cambridge: Cambridge University Press.

Harriss, J. 1982. *Capitalism and Peasant Farming: Agrarian Structure and Ideology in Northern Tamil Nadu*. Bombay: Oxford University Press.

Hart, K. 1982. *The Political Economy of West African Agriculture*. Cambridge: Cambridge University Press.

Herskovits, M.J. 1940. *The Economic Life of Primitive Peoples*. New York: Alfred A. Knopf.

Herskovits, M.J. 1952. *Economic Anthropology: A Study in Comparative Economics*. New York: Alfred A. Knopf.

Hill, P. 1963. *The Migrant Cocoa-Farmers of Southern Ghana: A Study in Rural Capitalism*. Cambridge: Cambridge University Press.

Hill, P. 1970. *Studies in Rural Capitalism in West Africa*. Cambridge: Cambridge University Press.

Hill, P. 1972. *Rural Hausa: A Village and a Setting*. Cambridge: Cambridge University Press.

Hill, P. 1977. *Population, Prosperity and Poverty: Rural Kano, 1900 and 1970*. Cambridge: Cambridge University Press.

Hill, P. 1982. *Dry Grain Farming Families: Hausaland (Nigeria) and Karnataka (India) Compared*. Cambridge: Cambridge University Press.

Hill, P. 1985. The gullibility of development economists. *Anthropology Today* 1(2), 10–12.

Humphrey, C. 1983. *Karl Marx Collective*. Cambridge: Cambridge University Press.

Jeffery, P. 1979. *Frogs in a Well: Indian Women in Purdah*. London: Zed Books.

Kahn, J.S. and Llobera, J.R. 1981. *The Anthropology of Pre-Capitalist Societies*. London: Macmillan.

Kerblay, B. 1966. A.V. Chayanov: Life, Career, Works. See Chayanov (1923–5).

Knight, F.H. 1941. Anthropology and economics. *Journal of Political Economy* 49(2), April, 547–68. Reprinted in *Economic Anthropology*, ed. M.J. Herskovits, New York: Alfred A. Knopf, 1952.

Leach, E.R. 1961. *Pul Eliya: A Village in Ceylon: A Study of Land Tenure and Kinship*. Cambridge: Cambridge University Press.

LeClair, E.E. and Schneider, N.K. (eds) 1968. *Economic Anthropology: Readings in Theory and Analysis*. New York: Holt Rinehart and Winston.

Lenin, V.I. 1899. *The Development of Capitalism in Russia*. Vol. III, *Collected Works*, London: Lawrence and Wishart.

Lewis, O. 1951. *Life in a Mexican Village: Tepoztlan Restudied*. Urbana: University of Illinois Press.

Lipton, M. 1977. *Why Poor People Stay Poor: A Study of Urban Bias in World Development*. London: Temple Smith.

Macfarlane, A. 1976. *Resources and Population: a Study of the Gurungs of Nepal*. Cambridge: Cambridge University Press.

Malinowski, B. 1922. *Argonauts of the Western Pacific*. New York: E.P. Dutton and Co., 1961.

Marx, K. 1894. *Capital*. Vol. III. Moscow: Progress Publishers.

Meggitt, M.J. 1965. *The Lineage System of the Mae-Enga of New Guinea*. London: Oliver & Boyd.

Meillassoux, C. 1960. Essai d'interprétation du phénomène économique dans les sociétés traditionelles d'auto-subsistance. *Cahiers d'Etudes Africains* 38–67. Translated and reprinted in Seddon (ed.) 1978.

Meillassoux, C. 1975. *Maidens, Meal and Money: Capitalism and the Domestic Community*. Cambridge: Cambridge University Press, 1981.

Mencher, J. 1978. *Agriculture and Social Structure in Tamil Nadu: Past Origins, Present Transformations and Future Prospects*. New Delhi: Allied Publishers.

Miles, D. 1979. The Finger Knife and Ockham's Razor: a problem in Asian culture history and economic anthropology. *American Ethnologist* 6(2), 223–43.

Mill, J.S. 1848. *Principles of Political Economy*. 9th edn. New York: Longmans, 1923.

Morgan, L.H. 1877. *Ancient Society*. London: Macmillan.

Morris, B. 1982. *Forest Traders: A Socio-Economic Study of the Hill Pandaram*. London: Athlone Press.

Ortiz, S. (ed.) 1983. *Economic Anthropology: Topics and Theories*. New York: University Press of America.

Proceedings, 1979. *Pastoral Production and Society*. Proceedings of the 1976 International Meeting on Nomadic Pastoralism. Cambridge: Cambridge University Press.

Polanyi, K., Arensberg, C.M. and Pearson, H.W. (eds) 1957. *Trade and Market in the Early Empires*. Glencoe: Free Press.

Quesnay, F. 1756. Farmers. *Reprints of Economic Classics*, Series 2, Number 2. Edited, translated and introduced by P.D. Groenewegen, Sydney: University of Sydney Department of Economics, 1983.

Redfield, R. 1930. *Tepoztlan: A Mexican Village*. Chicago: University of Chicago Press.

Sahlins, M. 1972. *Stone Age Economics*. Chicago: Aldine–Atherton.

Salisbury, R.F. 1970. *Vunamani*. Melbourne: Melbourne University Press.

Seddon, D. (ed.) 1978. *Relations of Production: Marxist Approaches to Economic Anthropology.* London: Frank Cass.

Schultz, T.W. 1964. *Transforming Traditional Agriculture.* New Haven: Yale University Press.

Sharma, U. 1980. *Women, Work and Property in North-West India.* London: Tavistock.

Sismondi, M. de. 1847. *Political Economy and the Philosophy of Government: Selected Essays.* London: Chapman.

Skinner, G.W. 1964–5. Marketing and Social Structure in Rural China. *Journal of Asian Studies* 24, 3–43, 195–228, 363–99.

Strathern, A.J. 1972. *One Father One Blood: Descent and Group Structure Among the Melpa People.* Canberra: Australian National University Press.

Taussig, M. 1978. Peasant economics and the development of capitalist agriculture in the Cauca Valley, Colombia. *Latin American Perspectives* 18, 62–90.

Turgot, A.R.J. 1766. On the characteristics of La Grande and La Petite Culture. *Reprints of Economic Classics*, Series 2, Number 2. Edited, translated and introduced by P.D. Groenewegen, Sydney: University of Sydney Department of Economics, 1983.

Woodburn, J. 1982. Egalitarian Societies. *Man* 17(3), 431–52.

economic calculation in socialist economies. The basic method of economic calculation used in the state socialist countries is that of incrementalism, or as it is known in the USSR, 'planning from the achieved level'. The starting point of all economic plans is the actual or expected outcome of the previous period. The planners adjust this by reference to anticipated growth rates, current economic policy, shortages and technical progress. For nearly all products, the planned output for next year will be the anticipated output for this year plus a few per cent added on. The advantages of incrementalism as a method of economic calculation are its simplicity, realism and compatibility with the functioning of a hierarchical bureaucracy. Its disadvantages are that it provides no method for making technically efficient or consistent decisions, nor does it ensure that the population will derive maximum satisfaction from the resources available.

PLANNING AND COUNTERPLANNING. A widely used method of economic calculation is that of planning and counterplanning. If the plan were simply handed down to the enterprises from above, in accordance with the planners' view of national economic requirements but in ignorance of the real possibilities of each enterprise, then it would be unfeasible (if it was too high) or wasteful (if it was too low) or both at the same time (i.e. unfeasible for some products and wasteful for others). Conversely, if plans were simply drawn up by each enterprise, they might fail to use resources in accordance with national economic requirements. The process of planning and counter-planning involves a mutual submission and discussion of planning suggestions, designed to lead to the adoption of a plan which is feasible for the enterprise and ensure that the resources of each enterprise are used in accordance with national requirements.

Unfortunately, the bureaucratic complexity of this procedure militates against both efficiency and consistency.

INPUT NORMS. The main method of economic calculation used to ensure efficiency is that of input norms. An input norm is simply a number assumed to describe an efficient process of transformation of inputs into outputs. For example, suppose that the norm for the utilization of coal in the production of one ton of steel is x tons. Then the efficient production of z tons of steel is assumed to require zx tons of coal.

The method of norms is widely used in Soviet planning, and considerable effort is devoted to updating them. Very detailed norm fixing takes place for expenditures of fuel and energy.

Much attention is devoted to the development of norms for the expenditure of metal, cement, and timber in construction. All this work is directed by the department of norms and normatives of Gosplan. Responsibility for elaborating and improving the norms lies with Gosplan's Scientific Research Institute of Planning and Norms.

Nevertheless, the method of norms is incapable of ensuring efficiency. The norms used in planning calculations are simply averages of input requirements, weighted somewhat in favour of efficient producers. Actual technologies show a wide dispersion in input–output relations. Furthermore, given norms take no account of the possibilities of substitution of inputs for one another in the production process, non-constant returns to scale, and the results of technical progress. Thus in general, the method of norms does not make it possible to calculate efficient input requirements, and plans calculated in this way are always inefficient.

The method of norms is not only used in interindustry planning, it is also used in consumption planning. In calculating the volume of particular consumer goods and services required, the planners use two main methods. One is forecasts of consumer behaviour, based on extrapolation, expenditure patterns of higher income groups, income and price elasticities of demand and consumer behaviour in the more advanced countries. The other method is that of consumption norms. The first method attempts to foresee consumer demand, the latter to shape it.

An example of the method of norms, and its policy implications, is set out in Table 1.

TABLE 1. The Soviet diet

	Norm (kgs/head/year)	Per capita consumption in 1976 as % of norm
Bread and bread products	120	128
Potatoes	97	123
Vegetables and melons	37	59
Vegetable oil and margarine	7	85
Meat and meat products	82	68
Fish and fish products	18	101
Milk and milk products	434	78
Eggs	17	72

Source: P. Weitzman, Soviet long term consumption planning: distribution according to rational need, *Soviet Studies*, July 1974, and E. M. Agababyan and Ye. N. Yakovleva (eds), *Problemy raspredeleniya i rost narodnogo blagosostoyaniya* (Moscow 1979) p. 142.

The table makes clear the logic of the Soviet policy of expanding the livestock sector, and also importing fodder and livestock products. Since the consumption of livestock products is below the norm level, the government seeks to make possible an increase in their consumption.

The method of consumption norms is an alternative to the price mechanism for the determination of output. It is, however, also used in Western countries. It is used there in those cases where distribution on the basis of purchasing power has been replaced by distribution on the basis of need. Examples are, the provision of housing, hospitals, schools and parks. Calculations of the desirable number of rooms, hospital beds and school places per person are a familiar tool of planning in welfare states.

There are two main problems with the norm method of consumption planning. The first is that of substitution between products. Although consumers may well have a medically necessary need for x grams of protein per day, they can obtain these proteins from a wide variety of foods. Secondly, consumers may choose to spend their money 'irrationally', e.g. to buy spirits instead of children's shoes.

MATERIAL BALANCES. A material balance is a balance sheet for a particular commodity showing, on the one hand, the economy's resources and potential output, and on the other, the economy's need for a particular product. Material (and labour) balances are the main methods used in calculating production and distribution plans for goods, supply plans and labour plans. Soviet planners take great pride in the balance method and consider it one of the greatest achievements of planning theory and practice. Material balances are drawn up for different periods (e.g. for annual or five year periods), by different organizations (e.g. Gosplan, Gossnab, the ministries) and at different levels (e.g. national and republican). The material balances are also drawn up with different degrees of aggregation. Highly aggregated balances are drawn up for the Five Year Plans, and highly disaggregated balances by the chief administrations of Gossnab for annual supply planning. The aim of the material balance method is to ensure the consistency of the plans.

Normally, at the start of the planning work, the anticipated availability of a commodity is not sufficient to meet anticipated requirements. To balance the two, the planners seek possibilities of economizing on scarce products and substituting for scarce materials; they investigate the possibilities of increasing production or importing raw materials or equipment, or in the last resort they determine the priority needs to be fulfilled by the scarce commodity. Even with great efforts, achieving a balance is difficult. The complexity of an economy in which a great variety of goods are produced by different processes, all of which are subject to continuous technological change, is often too great for anything more than a balance that balances only on paper. Hence it is normal, during the planned period, for the plan to be altered, often repeatedly, as imbalances come to light. Particularly important problems with the use of material balances are the highly aggregated nature of the balances and their interrelated nature.

INPUT–OUTPUT. A wide variety of input–output tables are regularly constructed in socialist countries. Ex post national tables in value terms, planning national tables in value and physical terms, regional tables and capital stock matrices are widely constructed and used. An interesting and important use concerns variant calculations of the structure of production in medium term planning.

Because an input–output table can be represented by a simple mathematical model, and because of the assumption of constant coefficients, an input–output table can be utilized for variant calculations.

$$X = (I - A)^{-1} Y$$

Assuming that A is given, X can be calculated for varying values of Y. Variant calculations of the structure of production were not undertaken with material balances because of their great labour intensity. Variant calculations have a useful role to play in medium-term planning because they enable the planners to experiment with a wide range of

possibilities. The first major use of variant calculations of the structure of production in Soviet national economic planning was in connection with the 1966–70 five year plan. Gosplan's economic research institute analysed the results of various possible shares of investment in the national income for 1966–70. It became clear that stepping up the share of investment in the national income would increase the rate of growth of the national income, but that this would have very little effect on the rate of growth of consumption (because almost all of the increased output would be producer goods). The results of the calculations are set out in Tables 2 and 3.

TABLE 2. Output of steel on various assumptions

	Variants				
	I	II	III	IV	V
Production of steel in 1970 (millions of tonnes)	109	115	121	128	136

A sharp increase in the share of investment in the national income in the five year plan 1966–70 would have led to a sharp fall in the share of consumption in the national income, and only a small increase in the rate of growth of consumption (within a five year plan period). What is very sensitive to the share of investment in the national income is the output of the producer goods industries, as Tables 2 and 3 show.

These results are along the lines of what one would expect on the basis of Feldman's model, but the input–output technique improves on Feldman's model since it enables the effect of different strategies to be seen at industry level rather than merely in terms of macroeconomic aggregates.

Another example of the use of input–output for economic calculations concerns the statistical data about the relations between industries contained in the national ex post tables in value terms. In his controversial 1968 book *Mezhotraslevye svyazi sel'skogo khozyaistva*, M. Lemeshev, then deputy head of the sector for forecasting the development of agriculture of the USSR Gosplan's Economic Research Institute, used the Soviet input–output table for 1959 as the basis for a powerful plea for more industrial inputs to be made available to agriculture.

TABLE 3. Average growth rates of selected industries, 1966–1970

	Variants				
	I	II	III	IV	V
Engineering and metal working	7.1	8.2	9.3	10.4	11.4
Light industry	6.3	6.6	6.8	7.0	7.2
Food industry	7.1	7.3	7.4	7.5	7.6

Source: M. Ellman, *Planning problems in the USSR: the contribution of mathematical economics to their solution 1960–1971* (Cambridge University Press, Cambridge 1973) p. 71.

He began by observing that from the 1959 input–output table it is clear that of the current material inputs into agriculture in that year only 23.4 per cent came from industry, while 54.7 per cent came from agriculture itself (feed, seed etc.). He argued that this was most unsatisfactory. In the section on the relationship between agriculture and engineering Lemeshev argued that the supply to agriculture of agricultural machinery was inadequate, in the section on the relationship between agriculture and the chemical industry he argued that the supply of fertilizers was inadequate, and in the section on agriculture and electricity he argued that the supply of electricity to the villages for both productive and unproductive needs was inadequate, and in the section on the relationship between agriculture and the processing industry he argued that the latter was not helping agriculture as it should do, for example, it was sometimes impossible to accept vegetables (although the consumption of these in the towns was well below the norms) because of inadequate processing and distribution facilities. In addition, he argued that the supply of concentrated feed was inadequate and the processing of milk wasteful. In view of the inadequate development of the food processing industry, he argued for the development of processing enterprises by the farms themselves.

The chapter on the productive relations between agriculture and the building industry is an extensive critique of the practice of productive, and housing and communal, building in the villages. Lemeshev argued that the state should take on responsibility for building on the collective farms. The chapter on the relationship between agriculture and transport is critical of the shortage of river freight boats. The chapter on agriculture and investment in agriculture argued that investment in agriculture was inadequate, that in the period 1959–65 there was an unwarranted increase in the proportion of investment in the collective farms which they had to finance themselves, that a greater proportion of agricultural investment should be financed by bank loans, and that as a criterion of investment efficiency the recoupment period is satisfactory. The concluding chapter is concerned with improving the productive relations between agriculture and the rest of the economy. The author argued for improving central planning by the use of input–output, for replacing procurement plans by free contracts between farms and the procurement organs (if a shortage of a particular product threatens then its price can be raised), the elimination of the supply system (i.e. the rationing of producer goods) which hinders farms from receiving the goods they want and sometimes supplies them with goods that they do not want, higher pay in agriculture and the reorganization of the labour process within state and collective farms on the basis of small groups which are paid by results.

This book was a good example of the use of input–output to provide statistical data which can be used, alongside other information, to provide a description of important economic relations and to support a case for important institutional and policy changes.

PROJECT EVALUATION. In the USSR of the 1930s, it was officially considered that there was no problem of project evaluation to which economists could contribute. The sectoral allocation of investment was a matter for the central political leadership to decide. It was they who decided in which sectors and at which locations production should be expanded. These decisions were based on the experience of the more advanced countries, the traditions of the Russian state (e.g. stress on railway building) and of the Bolshevik movement (e.g. stress on electrification and on the metal-using industries) and on the needs of defence. As far as decisions within sectors were concerned, here the main idea was to fulfil the plan using the world's most advanced technology.

The practical study of methods for choosing between variants within sectors was begun by engineers in the electricity and railway industries. The problem analysed was that of comparing the cost of alternative ways of meeting particular plan targets. A classic example of the type of problem considered was the choice between producing electricity by a hydro station or a thermal station.

During Stalin's lifetime, the elaboration by orthodox economists and the adoption by the planners of economic criteria for project evaluation were impossible because they were outside Stalin's conception of the proper role of economists (apologetics). When economists did make a contribution in this area, as was done by Novozhilov, it was ignored. After Stalin's death, however, it became possible for Soviet economists to contribute to the elaboration of methods of economic calculation for use in the decision-making process. An early and important example was in the field of project evaluation. An official method for project evaluation was adopted in 1960, and revised versions in 1964, 1966, 1969 and 1981. In a very abbreviated and summary form, the 1981 version is as follows.

In evaluating investment projects, a wide variety of factors have to be taken into account, e.g. the effect of the investment on labour productivity, capital productivity, consumption of current material inputs (e.g. metals and fuel), costs of production, environmental effects, technical progress, the location of economic activity and so on. Two indices which give useful synthetic information about economic efficiency (but are not necessarily decisive in choosing between investment projects) are the coefficient of absolute economic effectiveness and the coefficient of relative economic effectiveness.

At the national level, the coefficient of absolute effectiveness is defined as the incremental output–capital ratio.

$$E_p = \frac{\Delta Y}{I}$$

where

E_p is the coefficient of absolute effectiveness for a particular project,
ΔY is the increase in national income generated by the project, and
I is the investment cost.

The value of E_p calculated in this way for a particular investment, has to be compared with E_a, the normative coefficient of absolute effectiveness, which is fixed for each Five Year Plan and varies between sectors. In the 11th Five Year Plan (1981–85) it was 0.16 in industry, 0.07 in agriculture, 0.05 in transport and communications, 0.22 in construction and 0.25 in trade.

$$\text{If } E_p > E_a$$

then the project is considered efficient.

For calculating the criterion of absolute effectiveness at the level of individual industries, net output is used in the numerator instead of national income. At the level of individual enterprises and associations, in particular when a firm's own money or bank loans are the source of finance, profit is used instead of national income.

The coefficient of relative effectiveness is used in the comparison of alternative ways of producing particular products. In the two products case

$$E = \frac{C_1 - C_2}{K_2 - K_1}$$

where

E is the coefficient of relative effectiveness,
C_i is the current cost of the ith variant, and
K_i is the capital cost of the ith variant.

If $E > E_n$, where E_n is the officially established normative coefficient of relative economic efficiency, then the more capital intensive variant is economically justified. In the 11th Five Year Plan, E_n was in general 0.12, but exceptions were officially permitted in the range 0.08/0.10–0.20/0.25.

In the more than two variants case, they should be compared according to the formula

$$C_i + E_n K_i \rightarrow \text{minimum}$$

i.e. choose that variant which minimizes the sum of current and capital costs.

It is important not to adopt the rationalist misinterpretation of socialist planning according to which a planned economy is one in which rational decisions are made after a dispassionate analysis by omniscient and all-powerful planners of all the alternative possibilities. In such a system, the adoption of rational criteria for project evaluation would be of enormous importance. Socialist planning, however, is just one part of the social relations between individuals and groups in the course of which decisions are taken, all of which are imperfect and many of which produce results quite at variance with the intentions of the top economic and political leadership.

A good example of the factors actually influencing investment decisions under state socialism is the notorious Baoshan steel plant near Shanghai. The site was apparently chosen because of the political influence of a high-ranking Shanghai party official. The location decision ignored the fact that because of the swampy nature of the site, necessitating large expenditures on the foundations, this was in fact the most expensive of the sites considered. Very expensive, dogged with cost overruns, involving major pollution problems, the whole project was kept alive for some time by a powerful steel lobby. In due course, as a result of a national policy reversal in Beijing, the second phase was deferred and those involved publicly criticized. Judging by its costs of production, it produced gold rather than steel.

In general, the choice of projects owes more to inter-organization bargaining in an environment characterized by investment hunger than it does to the detached choice of a cost minimizing variant. The development of new and better criteria for project evaluation has turned out to be no guarantee that project evaluation will improve since the criteria are often not in fact used to evaluate projects. Their main function is to provide an acceptable common language in which various bureaucratic agencies conduct their struggles. Agencies adopt projects on normal bureaucratic grounds and then try to get them adopted by higher agencies, or defend them against attack, by presenting efficiency calculations using the official methodology but relying on carefully selected data.

LINEAR PROGRAMMING AND EXTENSIONS. Linear programming was discovered by the Soviet mathematician Kantorovich in the late 1930s. Its relevance for Soviet planning was widely discussed in the USSR in the 1960s and it was widely introduced in Soviet planning in the 1970s. Three examples of its use follow.

PRODUCTION SCHEDULING IN THE STEEL INDUSTRY. Linear programming was discovered by Kantorovich in the course of solving the problem, presented to him by the Laboratory of the all-Union Plywood Trust, of allocating productive tasks between machines in such a way as to maximize output given the assortment plan. From a mathematical point of view, the problem of optimal production scheduling for tube mills and rolling mills in the steel industry, which was tackled by Kantorovich in the 1960s, is very similar to the Plywood Trust problem, the difference being its huge dimensions.

The problem arises in the following way. As part of the planning of supply, Soyuzglavmetal (the department of Gossnab concerned with the metal industries), after the quotas have been specified, has to work out production schedules and attachment plans in such a way that all the orders are satisfied and none of the producers receives an impossible plan. In the 1960s an extensive research programme was initiated by the department of mathematical economics (which was headed by Academician Kantorovich) of the Institute of Mathematics of the Siberian branch of the Academy of Sciences, to apply optimizing methods to this problem. The chief difficulties were the huge dimensions of the problem and the lack of the necessary data. About 1,000,000 orders, involving 60,000 users, more than 500 producers and tens of thousands of products, are issued each year for rolled metal. Formulated as a linear programming problem it had more than a million unknowns and 30,000 constraints. Collecting the necessary data took about six years. Optimal production scheduling was first applied to the tube mills producing tubes for gas pipelines (these are a scarce commodity in the USSR). In 1970 this made possible an output of tubes 108,000 tons greater than it would otherwise have been, and a substantial reduction in transport costs was also achieved.

The introduction of optimal production scheduling into the work of Soyuzglavmetal was only part of the work initiated in the late 1960s on creating a management information and control system in the steel industry. This was intended to be an integrated computer system which would embrace the determination of requirements, production scheduling, stock control, the distribution of output and accounting. Such systems were widely introduced in Western steel firms in the late 1960s. Work on the introduction of management information and control systems in the Soviet economy was widespread in the 1970s but by the 1980s there was widespread scepticism in the USSR about their usefulness. This largely resulted from the failure to fulfil the earlier exaggerated hopes about the returns to be obtained from their introduction in the economy.

INDUSTRY INVESTMENT PLANS. In the state socialist countries investment plans are worked out for the country as a whole, and also for industries, ministries, departments, associations, enterprises, republics, economic regions and cities. An important level of investment planning is the industry. Industry investment planning is concerned with such problems as the choice of products, of plants to be expanded, location of new plants, technology to be used, and sources of raw materials.

The main method used at the present time in the CMEA countries for processing the data relating to possible investment plans into actual investment plans is mathematical programming. After extensive experience in this field, in 1977 a Standard Methodology for doing such calculations was adopted by the Presidium of the USSR Academy of Sciences. The use of mathematical programming for calculating optimal investment plans is an example of the possibilities for efficient control of national economies which the scientific-technical revolution in the field of management and control of large systems is bringing about.

The Soviet Standard Methodology presents models for three

standard problems. They are: a static multiproduct production problem with discrete variables, a multiproduct dynamic production problem with discrete variables, and a multiproduct static problem of the production-transport type with discrete variables. The former can be set out as follows:

Let $i = 1, \ldots, n$ be the finished goods or resources, $j = 1, \ldots, m$ be the production units, $r = 1, \ldots, R_j$ be the production technique in a unit, a_{ij}^r be the output of good $i = 1, \ldots, n'$ or input of resource $i = n' + 1, \ldots, n$, using technique r of production in unit j; C_j^r are the costs of production using technique r in unit j; D_i is the given level of output of good i, $i = 1, \ldots, n'$; P_i is the total use of resource i, $i = n' + 1, \ldots, n$ allocated to the industry; Z_j^r is the unknown intensity of use of technique r at unit j.

The problem is to find values of the variables Z_j^r that minimise the objective function

$$\sum_{j=1}^{m} \sum_{r=1}^{R_j} C_j^r Z_j^r \qquad (1)$$

i.e. minimize costs of production subject to

$$\sum_{j=1}^{m} \sum_{r=1}^{R_j} a_{ij}^r Z_j^r \geqslant D_i, \qquad i = 1, \ldots, n' \qquad (2)$$

i.e. each output must be produced in at least the required quantities

$$\sum_{j=1}^{m} \sum_{r=1}^{R_j} a_{ij}^r Z_j^r \leqslant P_i, \qquad i = n' + 1, \ldots, n \qquad (3)$$

i.e. the total use of resources cannot exceed the level allocated to the branch

$$\sum_{r=1}^{R_j} Z_j^r \leqslant 1, \qquad j = 1, \ldots, m \qquad (4)$$

$$Z_j^r = 0 \text{ or } 1, \qquad j = 1, \ldots, m, \qquad r = 1, \ldots, R_j \qquad (5)$$

i.e. either a single technique of production for unit j is included in the plan or unit j is not included in the plan.

In order to illustrate the method, an example will be given which is taken from the Hungarian experience of the 1950s in working out an investment plan for the cotton weaving industry for the 1961–65 Five Year Plan. The method of working out the plan can be presented schematically by looking at the decision problems, the constraints, the objective function and the results.

The decision problems to be resolved were:

(a) How should the output of fabrics be increased, by modernizing the existing weaving mills or by building new ones?

(b) For part of the existing machinery, there were three possibilities. It could be operated in its existing form, modernized by way of alterations or supplementary investments, or else scrapped. Which should be chosen?

(c) For the other part of the existing machinery, either it could be retained or scrapped. What should be done?

(d) If new machines are purchased, a choice has to be made between many types. Which types should be chosen, and how many of a particular type should be purchased?

The constraints consisted of the output plan for cloth, the investment fund, the hard currency quota, the building quota and the material balances for various kinds of yarn. The objective function was to meet the given plan at minimum cost.

The results provided answers to all the decision problems. An important feature of the results was the conclusion that it was cheaper to increase production by modernizing and expanding existing mills than by building new ones.

It would clearly be unsatisfactory to optimize the investment plan of each industry taken in isolation. If the calculations show that it is possible to reduce the inputs into a particular industry below those originally envisaged, then it is desirable to reduce planned outputs in other industries, or increase the planned output of the industry in question, or adopt some combination of these strategies. Accordingly, the experiments in working out optimal industry investment plans, begun in Hungary in the 1950s, led to the construction of multi-level plans linking the optimal plans of the separate industries to each other and to the macroeconomic plan variables. Multi-level planning of this type was first developed in Hungary, but has since spread to the other CMEA countries. Extensive work on the multi-level optimization of investment planning was undertaken in the USSR in connection with the 1976–90 long-term plan. (The 1976–90 plan, like all previous Soviet attempts to compile a long-term plan, was soon overtaken by events. The plan itself seems never to have been finished and was replaced by ten year guidelines for 1981–90.)

THE DETERMINATION OF COSTS IN THE RESOURCE SECTOR. In view of the wide dispersion of production costs in the resource sector, the use of average costs (and of prices based on average costs) in allocation decisions is likely to lead to serious waste. An important outcome of the work of Kantorovich and his school for practical policy has been (after a long lag) official acceptance of this proposition and of linear programming as a way of calculating the relevant marginal costs. For example, in 1979 in the USSR the State Committee for Science and Technology and the State Committee for Prices jointly approved an official method for the economic evaluation of raw material deposits. This was a prescribed method for the economic evaluation of exploration and development of raw material deposits. What was new in principle about this document was that it permitted the output derived from the deposits to be evaluated either in actual (or forecast) wholesale prices or in marginal costs. For the fuel-energy sector, a lot of work has been done to calculate actual (and forecast) marginal costs for each fuel at different locations throughout the country and for different periods. These figures are regularly calculated on optimizing models (they are the dual variables to the output maximizing primal) and have been widely used in planning practice for many years.

COMPARISON WITH THE WEST. An important method of economic calculation in socialist countries is comparison with the West. If a particular product or method of production has already been introduced (or phased out) in the West, this is generally considered a good argument to introduce it (or phase it out) in the socialist countries, subject to national priorities and economic feasibility. Obtaining advanced technology from overseas has always been an integral part of socialist planning. Comparisons with the West are particularly important in an economic system which lags behind the leading countries, lacks institutions which automatically introduce innovations into production (i.e. profit seeking business firms), and finds it difficult (because of the ignorance of the planners, stable cost plus prices and the self-interest of rival bureaucratic agencies) to notice, appraise realistically when noticed, and adopt, innovations.

ECONOMIC CALCULATION AND ECONOMIC RESULTS. It is important not to exaggerate the influence of methods of economic calculation on the performance of an economy. The performance of an economy is largely determined by external factors (e.g. the world market), economic policy (e.g. the decision to import foreign capital or to declare a moratorium), economic institutions (e.g. collective farms) and the behaviour of the actors within the system (e.g. underestimation of investment costs by initiators of investment projects). It is entirely possible for an improvement in the methods of economic calculation to coincide with a worsening of economic performance (as happened in the USSR after 1978). Realization of these facts led in the 1970s to a shift from the traditional normative approach (which concentrates on the methods of economic calculation and which regards their improvement as the main key to improved economic performance and the main role of the economist) in the study of planned economies, to the systems and behavioural approaches.

MICHAEL ELLMAN

See also CENTRAL PLANNING; DECENTRALIZATION; LANGE–LERNER MECHANISM; MARKET SOCIALISM; PRICES AND QUANTITIES; SOCIALIST ECONOMIES.

BIBLIOGRAPHY

Birman, I. 1978. From the achieved level. *Soviet Studies* 30(2), April, 153–72.
Boltho, A. 1971. *Foreign Trade Criteria in Socialist Economies.* Cambridge: Cambridge University Press.
Cave, M. 1980. *Computers and Economic Planning – the Soviet Experience.* Cambridge: Cambridge University Press.
Ellman, M. 1973. Changing views on central planning: 1958–1983. *ACES Bulletin* 25(1), Spring, 11–34.
Ellman, M. and Simatupang, B. 1982. *Odnowa* in statistics. *Soviet Studies* 34(1), January, 111–17.
Gács, J. and Lackó, M. 1973. A study of planning behaviour on the national-economic level. *Economics of Planning* 13(1–2), 91–119.
Giffen, J. 1981. The allocation of investment in the Soviet Union. *Soviet Studies* 33(4), October, 593–609.
Kantorovich, L.V., Cheshenko, N.I., Zorin, Iu.M. and Shepelev, G.I. 1979. On the use of optimization methods in automated management systems for economic ministries. *Matekon* 15(4), Summer, 42–66.
Kornai, J. 1967. *Mathematical Planning of Structural Decisions.* Amsterdam: North-Holland.
Kornai, J. 1980. *Economics of Shortage.* 2 vols, Amsterdam: North-Holland.
Kushnirsky, F.I. 1982. *Soviet Economic Planning, 1965–1980.* Boulder, Colorado: Westview, ch. 4.
Lee Travers, S. 1982. Bias in Chinese economic statistics. *China Quarterly* 91, September, 478–85.
Levine, H.S. 1959. The centralized planning of supply in Soviet industry. In *Comparisons of the United States and Soviet Economies,* Washington, DC: Joint Economic Committee, US Congress.
Stalin, J. 1952. *Economic Problems of Socialism in the USSR.* Moscow: Foreign Languages Publishing House.
Standard methodology for calculations to optimize the development and location of production in the long run. 1978. *Matekon* 15(1), Fall, 75–96.
Tretyakova, A. and Birman, I. 1976. Input–output analysis in the USSR. *Soviet Studies* 28(2), April, 157–86.

economic freedom. Economic freedom describes a particular condition in which the individual finds himself as a result of certain characteristics in his economic environment. Taking a simple formulation of decision-making in which it is assumed that the individual maximizes his satisfaction both as a consumer of private and government goods and services and as a supplier of factor services, his position may be depicted as follows:

$$\text{Max } U^i = U^i(x^i, q_k, a^i) \tag{1}$$

Subject to

$$p_k^c \cdot x^i + T_k^i = Y^i = \varphi(a^i) = p_k^a \cdot a^i \tag{2}$$

where x_i is a vector of 'private goods', q_k is a vector of goods supplied by government, a^i is a vector of factor inputs, p_k^c is a vector of product prices for private goods, p_k^a is a vector of factor prices, T_k^i is net tax liability of individual i (tax obligation *less* transfers), Y^i is personal income before tax of individual i and subscript k denotes an exogenously determined variable.

Assuming the budget constraint (2) is exactly satisfied, the individual maximizes his satisfaction solving for the vector of private-goods consumption in terms of their prices, disposable income and predetermined levels of public goods available for consumption, where goods prices and factor prices, quantities of factor inputs and tax liabilities are either known or predicted by the individual.

Economic freedom requires that the various terms in the budget constraint reflect the absence of 'preference or restraint' (Adam Smith) on the individual. Therefore p_k^c is a vector of product prices which result from the operation of competitive market forces with the individual being free to choose between alternatives. Similarly, p_k^a must be characterized by competition in the factor market with the individual being 'free to bring both his industry and capital into competition with those of any other man or order or men' (Adam Smith). There is less certainty concerning the constraints placed on T_k and q_k. Some writers would argue that economic freedom requires a pre-established limit on the values of T_k and q_k either expressed or implied in a country's constitution (Nozick's 'minimal state'; see Nozick, 1974). Others would argue that within a system of democratic government it should be possible to devise voting systems through which individuals express their preferences for values of T_k and q_k which simulate if they do not replicate the competitive market in the private sector (see Buchanan, 1975). All agree, however, that economic freedom is not compatible with large values of T and q in relation to values of x^i, mainly because a large public sector increases the monopoly power of public servants both as suppliers of public goods and factor services to produce them and encourages the growth of private monopolies as a defence against public monopsony buying.

ECONOMIC FREEDOM AND LIBERTARIAN PHILOSOPHY. There are features of this attempt at a 'technical' definition which may be called in question and which must be considered later, but it will be recognizable to those economists who have elevated economic freedom to an important goal in its own right and have claimed that it is the most important means for ensuring that the economy develops at the right 'tempo'. Discussion of the usefulness of the concept of economic freedom, therefore, centres in these two libertarian propositions.

The first proposition is contained in a striking passage in Book III of his *Essay on Liberty*: J.S. Mill wrote:

> He who lets the world, or his own portion of it, choose his plan of life for him, has no need of any other faculty than the ape-like one of imitation. He who chooses to plan for himself, employs all his faculties. He must use observation to see, reasoning and judgment to foresee, activity to gain materials for decision, discrimination to decide and, when he has decided, firmness and self-control to hold to his deliberate decision ... It is possible that he might be guided

on some good path, and kept out of harm's way, without any of these things. But what will be his comparative worth as a human being? It really is of importance, not only what men do, but also what manner of men they are that do it (Mill, 1859).

The passage captures the essence of the libertarian view of the good society, clearly implying that it requires that individuals should accept the necessity for choosing and for recognizing their responsibility for making choices. It must simultaneously require that, to develop the capacity for choosing, individuals must have the widest possible freedom of choice in the acquisition and disposal of resources. Two further conclusions follow.

The only restriction on economic freedom experienced by the individual should be when such freedom harms others.

The individual is not accountable to society for his actions and this, together with the different and changing preferences of individuals, makes libertarians distance themselves from attempts to establish a 'social welfare function' (cf. Rowley and Peacock, 1975).

The second proposition maintains that economic freedom brings the added bonus of promoting the economic welfare of both the individual and of society. Economic freedom encourages the individual to 'better his condition' (Smith, 1776) by exploiting opportunities for specialization and gains from trade which will be fully realized through the spontaneous emergence of markets. Not only is economic freedom regarded as the only material condition compatible with human dignity but it is also a necessary condition for the economic growth of the economy and for its adjustment to the changing preference structures of its members in response to market forces. The market is a 'discovery process' (Hayek, 1979) in which participants adjust to change giving rise to the notion of the 'invisible hand' which coordinates human economic actions automatically without recourse to government intervention. *Pace* Hahn (1982) and others, libertarians do not attach importance to a general equilibrium solution, attained by the operation of competitive market forces (cf. Barry, 1985). Indeed, though some exceptions will be noted below, it is claimed by supporters of the doctrine of economic freedom that disturbance of the natural process of exchange by government intervention assumes knowledge of the intricacies of the economy which is vouchsafed to no one, but there is no guarantee that officials, who maximize their private interests like everyone else, would be willing to maximize some social optimum even if they knew how to do so.

It was clearly recognized, by Hume and Smith for example, that for markets to work efficiently there must be a well-defined system of property rights and that costs of contracting between individuals in order to benefit from gains-from-trade would need to be minimized. The promotion of market efficiency was therefore bound to require some government intervention. No specialization or gains-from-trade would take place in a society in which there was no machinery for settling disputes and for preserving law and order. Acceptance of coercive intervention, however, requires that the 'rule of law' prevails. The law must be prospective and never retrospective in its operation, the law must be known and, as far as possible, certain, and the law must apply with equal force to all individuals without exception or discrimination. The state could also have a role in reducing the costs of contracting both by the removal of barriers to trade and to factor mobility and by the positive encouragement to the reduction in the costs of transport. In this latter respect Adam Smith supported reduction in the 'expense of carriage' by state

financing of road building and supervision of financial methods to promote road maintenance and improvement.

At no stage therefore in the development of the doctrine of economic freedom, as understood by economists, was it regarded as synonymous with 'laissez-faire'. At the same time, the role of the state in respect of the promotion of economic freedom was and has remained strictly limited in libertarian thinking. Indeed, some modern libertarians devote much discussion to the possibilities of 'privatizing' even such traditional functions of the state as the maintenance of law and order.

SOME PROBLEMS RAISED BY THE CONCEPT OF ECONOMIC FREEDOM. The most obvious question posed to libertarians by those who are sceptical of their position is that the system of economic freedom is silent on the question of the distribution of property rights. In terms of our simple model, what principle should determine the values of $Y^1, \ldots, Y^i, \ldots, Y^n$ which, when aggregated, would describe some initial distribution of income as measured, say, by the shape of the Lorenz Curve? What reason have we for supposing that the 'optimal' distribution of income would emerge from the process of economic exchange between individuals?

The answer to this question does not find libertarians speaking with one voice. The problem is not one of principle, for the ultimate test to them is how far any government intervention represents a restriction of freedom. The problem is one of interpretation. It would be difficult today to find libertarians who would object to government intervention designed to assure protection to those who are severely deprived. Thus Hayek has argued that so long as 'a uniform minimum income is provided outside the market to all those who, for any reason, are unable to earn in the market an adequate maintenance, this need not lead to a restriction of freedom, or conflict with the Rule of Law'. This still leaves room for much disagreement among libertarians as to the precise level of the minimum and how to decide on who is entitled to receive it. Some supporters of the libertarian position, including the present author, would go much further and argue, along with J.S. Mill, that concentrations of wealth sustained over lengthy time periods can endanger economic freedom, not to speak of political freedom, by the association of such concentrations with the concentration of power of wealthy individuals over the less fortunate.

If the concept of economic freedom cannot embrace some precise guidance about the extent to which economic exchanges should be interfered with, it certainly places limits on the form of that interference. Thus libertarians, to the extent that they accept the need for a state-guaranteed minimum standard of living, prefer the use of money transfers to individuals rather than the provision of social services below or at zero cost, that is to say the economic condition of individuals in receipt of state support should be reflected in reduction in T_k^i (whose value may have to be negative) rather than an increase in q_k. Thus it is argued that individuals then retain responsibility for the purchase of goods and services designed to promote their own welfare and that the power of the state over the individual by bureaucratic dictatorship of preferences and by the lack of incentives in the public sector to economize in resource use is circumscribed.

A more severe test for the practicality of libertarian measures, designed to permit some redistribution without increasing the power of the state, arises in the case of any attack on the concentration of wealth. Clearly, a system of inheritance taxation which results in the transfer of capital from the private to the public sector would not conform to libertarian thinking, not only because this would discourage private saving but also because it would build up the power of the state. A system of taxation would have to be devised which not only did not discourage accumulation of private capital but also simultaneously encouraged legators to disseminate capital in favour of those with little capital. It is a long time since libertarians have plucked up the courage to try to develop such a system, given that eminent public finance specialists have failed in their attempts to fulfil these requirements.

The second major question arises from the persistent objection of Marxists and other Socialist writers that the system of economic freedom, as depicted by the libertarians, fails to solve the problem of 'worker alienation'. It may be that the system of economic freedom can allow employees alone or in combination with others to influence the price of factor inputs (p_k^e) and the work/leisure combination (a^i), variables which play a crucial part in individual welfare. The fact remains that the system of property rights, which libertarians support, includes the individual ownership of capital and the use of capitalistic methods of production which imply an authority relationship between employer and worker. The hierarchical order at the place of work seems at complete variance with the independence of economic action attributed to the individual by the supporters of economic freedom.

Reactions to this argument by libertarians are sometimes reminiscent of the Scots preacher who, on recognizing a theological difficulty in his sermon, recommended his congregation to look the difficulty squarely in the face and pass it by. However, even Socialist writers, notably the prominent Marxist Ota Sik (1974), have recognized that the alternative to market capitalism – collectivist production – does not solve the problem for it is not synonymous with democratization at the shop-floor level. In other words, the basis of alienation is technological and not institutional. Some libertarians, notably Mill, have made common cause with Socialists by arguing that alienation must not be taken to be an inevitable consequence of productive activity. Mill sought one solution in the encouragement of firms owned and managed by the labour force, but still subject to competition. Utopian Socialists have claimed that the only solution is to reject altogether the technology which imposes hierarchical relations in the first place. Both 'solutions' are still the subject of living debate in both the professional and political arena.

ALAN PEACOCK

See also ECONOMIC HARMONY; SELF-INTEREST; UTILITARIANISM.

BIBLIOGRAPHY
Barry, N.P. 1985. In defense of the invisible hand. *The Cato Journal* 5(1), Spring, 133–48.
Buchanan, J.M. 1975. *The Limits of Liberty: Between Anarchy and Leviathan.* Chicago and London: University of Chicago Press.
Hahn, F. 1982. Reflections on the invisible hand. *Lloyds Bank Review* 144, April, 1–21.
Hayek, F.A. 1979. *Law, Legislation and Liberty*, Vol. 3. London: Routledge & Kegan Paul.
Mill, J.S. 1859. *Essay on Liberty.* Oxford: Oxford University Press, 1942.
Mill, J.S. 1871. *Principles of Political Economy*, Book IV, ch. VII. Toronto: University of Toronto Press, 1965.
Nozick, R. 1974. *Anarchy, State and Utopia.* Oxford: Basil Blackwell.
Peacock, A. 1979. *The Economic Analysis of Government and Related Themes.* Oxford: Martin Robertson, chs 5 and 6.
Rowley, C. and Peacock, A. 1975. *Welfare Economics: A Liberal Re-Interpretation.* London: Martin Robertson.
Sik, O. 1974. The shortcomings of the Soviet economy as seen in Communist ideologies. *Government and Opposition* 9(3), 263–76.
Smith, A. 1776. *An Inquiry into the Nature and Causes of the Wealth of Nations*, Book IV, ch. 9 and Book V. Ed. R.H. Campbell and A.J. Skinner, Oxford: Clarendon Press, 1976.

economic goods. *See* FREE GOODS; GOODS AND COMMODITIES.

economic growth. *See* ACCUMULATION OF CAPITAL; CLASSICAL GROWTH MODELS; HARROD–DOMAR GROWTH MODEL; MEASUREMENT OF ECONOMIC GROWTH; NEOCLASSICAL GROWTH THEORY.

economic harmony. This term has been introduced frequently into economic discussion, and especially into discussions concerning the history of economic thought. Yet there seems to be a good deal of ambiguity as to what it is to mean. Moreover, there has developed considerable disagreement concerning the centrality of the 'harmony' idea to the development of economic thought, and similar disagreement concerning the extent to which the classical economists, in particular, are to be seen as harmony-theorists. We will return a little later to distinguish various different senses that have been attached to the term 'harmony' in economics. For each of these different senses, however, acceptance of the harmony thesis has been held to imply a favourable stance towards a policy of laissez-faire. It is thus not surprising that 18th-century precursors of the notion of harmony have been discovered in Cantillon and in Quesnay (Schumpeter, 1954, p. 234). And we are not surprised to find some writers emphasizing the harmony ideas they see in the classical economists, especially in Adam Smith (Halévy, 1901–4, p. 89: Heimann, 1945, p. 65), while others vehemently question the unqualified identification of these writers with harmony theories (Robbins, 1952, pp. 22–9; Samuels, 1966, pp. 6–8; Sowell, 1974, pp. 16f). It was in the middle of the 19th century that the best-known writings appeared concerning economic harmony. The term appeared in the title of two books by the American economist Henry C. Carey (Carey, 1836, 1852). These works were followed by a general treatise stressing the same theme (Carey, 1858–60). The term also appeared in the title of a book by the French economic writer Frédéric Bastiat (1850). For a (muted) defence of Bastiat against widespread 19th-century charges that his work in this respect was a crude plagiarism of Carey, see Teilhac (1936, pp. 100–113), who points to the inspiration that both Carey and Bastiat received from J.B. Say. Subsequent references to harmony theories in economics generally tended to be critical, as economists began to argue (from the latter decades of the 19th century into the 20th century) for greater state intervention in market economies on perceived grounds of economic efficiency or economic justice. During most of the 20th century economists, even when they have defended the efficiency and justice of markets, have generally not couched their arguments explicitly in terms of harmony theory. Even Ludwig von Mises who, as we shall see, was an important exception to this last generalization, relegated the notion of harmony to a distinctly subsidiary role in his system. Recent re-awakened attention to 18th-century theories of spontaneous order, especially as rediscovered and expanded in the work of Hayek, has not had the effect of reintroducing the term 'economic harmony' to current usage. We turn now to take notice of the several different (although certainly interrelated) senses in which this term has been used during the history of economics.

HARMONY AS FLOWING FROM DIVINE PROVIDENCE. A harmony 'theory' is not, in this sense, one that flows out of economic science; rather it represents an attitude of (usually religious) optimism and faith, which itself suggests and guides the course of scientific investigation.

> Just as Kepler was inspired by the doctrine of harmony in the spheres to discover the laws which govern the orbits of the planets, so the early economists were inspired by the doctrine that there is a harmony of interests in a society to formulate economic laws (Streeten, 1954, p. 208).

It was from this sense of the term that Lord Robbins vigorously dissociated the classical school. It was this optimistic doctrine that came to be referred to contemptuously by the German term 'Harmonielehre'. Archbishop Whately, who in 1832 set up a chair of political economy at Trinity College, Dublin, was an influential harmony theorist in this sense. He saw the purpose of the chair as that of combatting the irreligious implications, as he saw them, of Ricardian economics. The early Dublin professors 'were under pressure to present an optimistic or harmonious picture of how the market economy operates' and the resulting critical attitude towards Ricardian theory reflected 'these extrascientific concerns' (Moss, 1976, p. 153). A variant of this approach to the harmony doctrine was the Enlightenment view, in which Deistic philosophy perceived a natural order as responsible for 'predetermined harmony' (Mises, 1949, p. 239; Heimann, 1945, p. 49).

HARMONY THEORY AS THE DOCTRINE OF MAXIMUM SATISFACTION. When major neoclassical economists such as Marshall (1920, p. 470) and Wicksell (1901, p. 73) referred to harmony theorists, they evidently had in mind those who believed that economic theory demonstrates that free competitive markets generate maximum total satisfaction for society as a whole. 'Harmony theory' thus referred to a very specific conclusion of economic science, a conclusion central to welfare economics, but a conclusion whose validity both Marshall and Wicksell were concerned to refute. Of special concern, in this context, was the issue of whether the new marginal utility doctrines had been successfully deployed by Jevons, or by Walras, to arrive at 'harmony' conclusions similar to those that had been reached, on other grounds, by Bastiat.

Parallel to this sense of harmony was that which attributed *ethical* virtues to the distributive results of competitive markets. Thus J.B. Clark's demonstration of the justice of marginal-productivity incomes is seen as 'harmony doctrine' (Myrdal, 1932, p. 148).

HARMONY DOCTRINE AS THE DENIAL OF CLASS CONFLICT. One sense in which harmony doctrines have been understood throughout the history of economics is that in which it is sought to demonstrate the mutual compatibility of the interests of the various individuals and groups in society . In particular, such doctrines tend to dismiss the notion of inherent class conflict under capitalism. A 20th-century economist who has himself emphasized this idea of harmony of interests in the market society, put the genesis of this idea as follows:

> When the classical economists [asserted 'the theorem of the harmony of the rightly understood interests of all members of the market society' they were stressing] two points: First, that everybody is interested in the preservation of the social division of labour, the system that multiplies the productivity of human efforts. Second, that in the market

society consumers' demand ultimately directs all production activities (Mises, 1949, p. 674).

Mises, indeed, saw these ideas as important results of economic science, having wide application. 'There is no conflict between the interests of the buyers and those of the sellers, between the interest of the producers and those of the consumers' (Mises, 1949, p. 357). Only in the special case of resource monopoly ownership may it happen that the 'emergence of monopoly prices ... creates a discrepancy between the interests of the monopolist and those of the consumers' (Mises, 1949, p. 680).

HARMONY AND THE SPONTANEOUS ORDER TRADITION. Since the early 1940s F.A. Hayek has succeeded in drawing the attention of economists and others to a line of social analysis since the 18th century, an approach often termed the 'spontaneous order tradition'. The emphasis, in this tradition, is on the evolution of institutions and social outcomes 'which are indeed the results of human action, but not the execution of any human design' (Ferguson, 1767, p. 187, cited in Hayek, 1967, p. 96). There is no doubt that the term 'economic harmony' has often been applied as an expression of belief in the *possibility and social benignity of undesigned social outcomes*. To some extent, of course, this sense of the term overlaps those listed above, but the emphasis here is not in the denial of conflict, not on any particular welfare theorem, certainly not on any religiously based optimism, but on the counter-intuitive possibility of orderly results emerging without deliberate design from the spontaneous interplay of independently acting individuals. 'Order' in this context has come to mean 'mutually reinforcing expectations'. The following reference to this notion of harmony expresses this usage of the term:

> The great general rule governing human action at the beginning, namely that it must conform to fair expectations, is still the scientific rule. All the forms of conduct complying with this rule are consistent with each other and become the recognized customs. The body of custom therefore tends to become a harmonious system (Carter 1907, p. 331, cited in Hayek, 1973, p. 169).

The above survey has been confined to notions of economic harmony believed to be achieved spontaneously, 'naturally', without design. For the sake of completeness it should perhaps be noted that the term 'harmony' has occasionally been used to describe the objective of *deliberate* social policy. Thus a well-known debate was initiated by E. Halévy in his claim that Bentham and the philosophical Radicals subscribed to two partly contradictory principles: the 'economic' principle of 'natural identity' (i.e. harmony) of interests, and the 'juristic' principle of the 'artificial identification of interests' (Halévy, 1901–4, pp. 15, 17, 489). Lord Robbins, in disputing Halévy concerning any contradiction in the Benthamite position, refers to the juristic principle as contending it to be 'the function of the legislator to bring about an artificial harmonization of interest' (Robbins, 1952, pp. 190f). While occasional references may be found to harmony sought to be artificially accomplished, the term has, in general, been associated almost invariably with harmony achieved undeliberately in a decentralized system.

ISRAEL M. KIRZNER

See also BASTIAT, CLAUDE FRÉDÉRIC; SCOTTISH ENLIGHTENMENT.

BIBLIOGRAPHY
Bastiat, F. 1850. *Les harmonies économiques.* Paris: Guillaumin.
Carey, H.C. 1836. *The Harmony of Nature.* Philadelphia: Carey, Lea & Blanchard.
Carey, H.C. 1852. *The Harmony of Interests, Agricultural, Manufacturing, and Commercial.* 2nd edn, New York: Myron Finch.
Carey, H.C. 1858–60. *Principles of Social Science.* Philadelphia: J.B. Lippincott.
Carter, J.C. 1907. *Law, Its Origin, Growth and Function.* New York and London: G.P. Putnam's Sons.
Ferguson, A. 1767. *An Essay on the History of Civil Society.* London.
Halévy, E. 1901–4. *The Growth of Philosophic Radicalism.* Translated from the French by M. Morris, 1928, Boston: Beacon, 1955.
Hayek, F.A. 1967. *Studies in Philosophy, Politics and Economics.* Chicago: University of Chicago Press.
Hayek, F.A. 1973. *Law, Legislation and Liberty.* Vol. I: *Rules and Order,* Chicago: University of Chicago Press.
Heimann, E. 1945. *History of Economic Doctrines, An Introduction to Economic Theory.* New York: Oxford University Press.
Marshall, A. 1920. *Principles of Economics.* 8th edn, London: Macmillan, 1936.
Mises, L. von. 1949. *Human Action: A Treatise on Economics.* 3rd edn, Chicago: Regnery, 1966.
Moss, L.S. 1976. *Mountifort Longfield: Ireland's First Professor of Political Economy.* Ottowa, Ill.: Green Hill.
Myrdal, G. 1932. *The Political Element in the Development of Economic Theory.* Translated from the German by P. Streeten, Cambridge, Mass.: Harvard University Press, 1954.
Robbins, L. 1952. *The Theory of Economic Policy in English Classical Political Economy.* London: Macmillan, 1965.
Samuels, W.J. 1966. *The Classical Theory of Economic Policy.* Cleveland and New York: World.
Schumpeter, J.A. 1954. *History of Economic Analysis.* New York: Oxford University Press.
Sowell, T. 1974. *Classical Economics Reconsidered.* Princeton: Princeton University Press.
Streeten, P. 1954. Recent controversies. Appendix to Myrdal (1932).
Teilhac, E. 1936. *Pioneers of American Economic Thought in the Nineteenth Century.* Translated from the French by E.A.J. Johnson (1936), reprinted, New York: Russell and Russell, 1967.
Wicksell, K. 1901. *Lectures on Political Economy.* Vol. I, Translated from the Swedish by E. Classen, London: Routledge and Kegan Paul, 1934.

economic history. This essay focuses on the relationship between economic history and mainstream economics. As such, although it provides a review of major developments in economic history and gives examples to illustrate the chief points in the arguments, it does not purport to offer a comprehensive history of the discipline.

THE DEVELOPMENT OF ECONOMIC HISTORY AS A SUBJECT. Economic history has changed enormously in the years since World War II. Forty years ago American economic history was 'with a few brilliant exceptions, neither good economics nor good history, a dim echo of the American institutionalists' (McCloskey, 1976, p. 435) whilst British economic history had concentrated disproportionately on the Industrial Revolution and its links to the social improvement tradition in adult education (Barker, 1985, p. 36). The subsequent development of the subject has embraced a wide variety of problems, methodologies and datasets, and a somewhat uneasy co-existence, in Britain at least, with social and business history.

The early postwar period was notable for the historical work prompted by the National Bureau of Economic Research and by Keynesian preoccupations with economic fluctuations and for the painstaking construction of historical national income accounts which by the mid–1960s provided a remarkable increase in economic historical knowledge concerning many OECD countries. By 1960 the stage was set for the New Economic History and the first Cliometrics Meeting was held at Purdue. From this point on American economic historians turned increasingly to the formal use of economic modelling and hypothesis testing and thus to a much more quantitative economic history. As has always tended to be the case, the subject found itself 'problem-driven' (Mathias, 1971, p. 371) and in this case there were, of course, three particularly interesting controversies which dominated the early years, namely, the social savings from 19th-century railways, the economics of slavery in the American south and entrepreneurial failure in late 19th-century Britain.

1960 also saw the publication of Rostow's *A Stage Theory of Economic Growth,* which capitalized on the increasing evidence on incomes in the past and which created a large industry of comment and further research. Indeed, the 1960s literature was permeated by the language of 'take-off' even though most professional economic historians soon became sceptical of Rostow's claims. Economic historians were at the time extremely keen to discuss the lessons of historical experience for developing countries.

The last ten years have seen a period of greater 'maturity' in the subject. The brash claims of the 1960s have given way to a more sober and less controversialist American literature which has a much broader scope, looks more to 20th-century topics and has to some extent started to rediscover the virtues of the older, descriptive tradition.

European economic historians have been much more reluctant to follow this research path. Even in Britain which, given the strong Anglo-American tradition in economics, might have been expected to go down this route, economic history has remained largely in a descriptive, historical mode. Separate departments of economic history flourished in the years of academic expansion, adding significantly to knowledge in an increasingly wide number of areas but on the whole producing relatively little of interest to economists. Recent developments have seen a boom in social history, the turning away by the majority of British economic historians from the new economic history (Supple, 1981) and a rapid contraction in the number of chairs and separate departments of economic history.

WHAT DOES ECONOMIC HISTORY OFFER TO ECONOMISTS? A session at the 1984 American Economic Association meetings was devoted to this question; the papers and discussion can be found in Parker (1986). The context is, of course, that in the postwar years successive vintages of professional economists seemingly embody ever greater mathematical and econometric sophistication together with more profound scorn for and ignorance of economic history. The pragmatic value of the subject, such as it, is can be assessed under the five headings suggested by McCloskey (1976) in his stimulating contribution on this issue: (a) more economic facts, (b) better economic facts, (c) better economic theory, (d) better economic policy and (e) better economists.

Most obviously economic history offers a much wider range of evidence and the possibility of working with quite large time series of data. Research in economic history has now significantly expanded the database available in readily accessible form. Encouragingly there are important publications by economists aware of some of the possibilities; perhaps even more importantly techniques of time-series econometrics have become much more powerful and thus more advantage can be obtained from this data. Thus we have recently seen not only Friedman and Schwartz's monumental study (1982), which uses amongst other sources Feinstein's (1972) study for UK incomes, output and prices and which for the US is based on their own meticulous construction of time series on the money stock, but we have also witnessed Hendry and Ericsson's devastating econometric critique (1983). Moreover the pre-1914 monetary time series themselves for the UK have now been improved upon by Capie and Webber (1985).

It is also true that economic history provides data which *cannot* be obtained for the present – for example, 19th-century censuses provide a rich source of panel data – and also offers examples of the operation of institutional arrangements instructive to present policy concerns but not extant. White's (1984) study of free banking in the 19th century is a good example and perhaps of wider interest is the recent NBER volume on the workings of the classical Gold Standard (Bordo and Schwartz, 1984).

At the same time a respect for history necessarily limits its value as a source of direct lessons for the economist or of tests of economic models. Consider for instance the problem of unemployment where interwar experience has recently been the subject of great interest. Many of the standard series used in the work of UK economists like Layard and Nickell (1986) either do not exist (e.g. union/non-union wage mark-up) or were collected on a quite different basis (e.g. unemployment rates). National accounts estimates are annual not quarterly and there are only eighteen observations including two massive recessions. However, it is clear that use of the models and proxy variables commonly relied on to estimate NAIRU in the 1970s imply that NAIRU would have been far below *any* unemployment figure observed between the wars (Crafts et al., 1984). Thus it seems that both the structure of the interwar economy and its database are very different from the territory familiar to present-day economists and also that a satisfactory macromodel of the period is virtually impossible to construct. It is easy enough to generate simple-minded accounts of unemployment based on a tight prior and a single equation, as did Benjamin and Kochin (1979) in claiming that the experience was one of voluntary unemployment based on high unemployment benefit to wage rate ratios, or founded on an unquestioning assertion that demand deficiency was the problem as in Beveridge (1944) and many subsequent textbooks, but it is very hard to understand the workings of the interwar labour market through a macromodel.

Another much-studied topic where it might be thought that historical experience would be instructive to the present is the 'Demographic Transition' from high to low fertility, which the now advanced countries experienced between 1870 and the 1930s and about which a substantial body of data exists. Thanks in particular to the Princeton project we do indeed now know a lot about the determinants of fertility during the period of decline (Coale and Watkins, 1986). Yet the timing of the onset of the transition from the 'natural fertility' of earlier centuries remains unexplained and occurred more or less simultaneously in very different economic circumstances. It is at least arguable that the neoclassical optimizing model is appropriate for the past hundred years of Western European fertility but not earlier when group rules predominated (Wrigley, 1978, p. 148).

To some extent then, the importance of the past to the economist lies in the information that it was in some respects different from today. Thus the proper choice of a model depends on institutional and social circumstances and the use of a policy instrument can have a very different outcome as times change; for example, in Britain the use of bank rate in 1913, 1925 and 1931 or devaluation in 1949 and 1967 (Cairncross and Eichengreen, 1983). Not all historical experience is directly relevant to today's policy problems but contextual awareness is another important contribution that economic history can make to better economics.

THE CONTRIBUTION OF ECONOMICS TO ECONOMIC HISTORY. The past twenty years has seen a considerable expansion in the use of theoretical models and quantitative techniques in economic history – the majority of these studies can be said to use economic analysis and/or econometrics. It was noted earlier that British economic history had been much less influenced than that of the United States, yet expansion of 'new economic history' articles on Britain saw an increase from five a year in the late 1960s to 28 a year in 1979–81 (Lee, 1983, p. 47). It is, of course, debatable how successful the influx of economists' methodology into economic history has been. The doubters include not only many traditional economic historians but also economists such as Solow (1986, p. 26):

> As I inspect some current work in economic history, I have the sinking feeling that a lot of it looks exactly the kind of economic analysis I have just finished caricaturing: the same integrals, the same regressions, the same substitutions of t-ratios for thought!

Assessment of the contribution of economics to economic history should perhaps adopt similar categories to those proposed by McCloskey for examining the value of economic history to economics. Suitable headings might be (a) more facts, (b) better facts, (c) better hypotheses, (d) better interpretation of the data, and (e) better historians.

Economic analysis used in conjunction with historical records has indeed significantly expanded the database available to economic historians as has also the use of quantitative techniques in general, perhaps most notably in the case of Wrigley and Schofield's (1981) path-breaking reconstruction of the population history of England over 1541–1871. To a considerable extent, as in the case of national income or money stock series, this is because the organizing concepts are rooted in economics but data in readily accessible form tend to be a natural result of the type of enquiries economists pursue in historical endeavours, and this trait was exhibited early on: for example, in the price data constructed by Gayer, Rostow and Schwartz (1953) and in the information on farm inputs and outputs in the Parker–Gallman sample of the United States manuscript consensus of 1860 (Parker, 1970).

Perhaps the chief contribution in terms of 'better facts' comes in the areas affected by index number problems which, not surprisingly, abound in economic history. Traditional historians have on the whole rather too readily used crude index numbers for purposes to which they are ill-suited – most infamously perhaps Clapham (1926) in his discussion of real wages during the industrial revolution, which used the wholly inadequate Silberling price index. Recent work by 'new' economic historians has produced, for example, much better weighted indices of both industrial production and the cost of living during the British industrial revolution than existed hitherto.

More controversial are the achievements under the headings (c), (d) and (e). Obviously the chief source of hypotheses has been neoclassical economics, particularly since the end of the Keynesian era. The interpretative techniques used in historical analysis are by now very numerous and include for example, Hatton et al.'s (1983) use of Granger-causality to examine 18th-century English trade, Williamson's (1985) use of Jones-type general equilibrium models to examine the sources of income inequality in 19th-century Britain, Fogel's (1964) use of cost–benefit analysis to examine social savings on American railroads, Temin's (1976) implementation of IS–LM analysis to gain insight into the onset of the Great Depression, Thomas's (1983) investigation of the employment effects of 1930s rearmament using an input–output/social accounting matrix approach together with scores of routine applications of two stage least squares, growth accounting etc. Both the questions posed and the methodology employed are quite foreign to the economic history of the 1930s and 1940s and the technical apparatus can seem an insuperable barrier to entry to traditional historians. As a result effective critiques of this kind of work most often come from within the new economic history camp – for example, the brilliant essay by Sutch (1975) which responded to Fogel and Engerman's (1974) interpretation of slavery in the American South – and perhaps come too infrequently.

It should be stressed that a number of long-standing issues have been illuminated by the application of formal economic analysis. An obvious case in point is the question of the standard of entrepreneurial performance in the pre-1914 British economy. Neoclassical analysis formalized the problem in terms of profit-maximization and showed that in many cases – for example, retention of mule-spinning in cotton (Sandberg, 1974) and slow adoption of the basic process in steel (McCloskey, 1973) – the choices were correct at British relative factor costs. It soon became clear that blanket accusations of amateurism, conservatism and plain incompetence were vastly exaggerated.

Even more decisive was the demonstration that slavery in the American South was profitable (and would not have collapsed simply through lack of viability in the absence of the Civil War) together with evidence that slaves were competitively priced in relation to other possible investments and that slave owners on the whole responded rationally to economic incentives (Aitken, 1971).

Yet this research eventually led to a most acrimonious debate among new economic historians, with the critics of Fogel and Engerman claiming that they had misinterpreted slavery from a desire to make everything fit comfortably into a neoclassical model in which 'each and every slave owner regarded slaves solely as productive instruments and used them for a single transcendent purpose: the maximization of pecuniary gain' (David et al., 1976, p. 341). Thus the critics claimed that a priori notions had led to a selective reading of the evidence and too cheerful a view of slavery, and also to unwarranted inferences about the motives of slave owners as a whole.

Similarly, controversy arose with McCloskey's (1970) attempt to build from successful defence of particular choices of technique to a general claim that markets worked well and that the late Victorian economy was growing as fast as was possible. Again McCloskey's argument can be seen as relying heavily on a priori reasoning and an ordering of facts based on a rather simple kind of neoclassical model. Thus growth was independent of the investment rate in the long run; income foregone from inefficiency in capital markets could only have been a small fraction of national income; product markets were competitive and thus productivity levels were as high as possible and the idiosyncratic structure of the British economy was a result of comparative advantage. Later writers, perhaps with weaker priors, have not accepted this picture: thus Hannah (1974) and Kennedy (1984) have revealed failures of information and the takeover mechanism which reduced the stringency of the market check on inefficient management; Allen (1979) demonstrated a worrying shortfall in total factor

productivity in iron and steel; Crafts and Thomas (1986) showed comparative advantage to be in low wage labour intensive exports and raised questions about the accumulation of human capital, etc.

These examples reflect well-known problems of methodology in economics. There has, of course, been a long-running debate over instrumentalism, i.e. judging theories on the basis of their predictions. Given that economists are often concerned with prediction rather than explanation up to a point this may be an acceptable criterion and, for example, firms may be regarded as if they maximize profits if their actions are not inconsistent with profit maximization. For an economic historian this may be a dangerous oversimplification leading to an erroneous belief that motives have been understood or that all decisions are based simply on profit maximization. In each of the above cases this type of difficulty is an important element in the minds of the critics.

Additional but not unrelated problems come from the use of tight priors. Recent work has suggested that economists of different schools generally regard certain models as requiring less severe tests than others – for example, Chicago economists tend to accept empirical results consistent with standard price theory much more readily than those that are not and to believe in market failure much less readily than others (Reder, 1982). In fact, testing in economics has rarely amounted to serious falsificationism and microeconomic research proceeds on the agreed but largely untested assumption that the marginal equivalences of the neoclassical model are tolerably achieved (McClelland, 1975, pp. 26–7). Obviously economists despite their differences are much more likely to share each other's priors than those of historians and equally obviously most work applying economics to history does not involve tests of competing hypotheses.

If the new economic history is to convince historians as well as economists, much higher standards of proof than those evident generally so far will be required – and they will provide the verstehen complement to regression analysis and deductive reasoning. This much is clearly indicated by historians' reaction to Time on the Cross (Stampp, 1976). Similarly, historians will (rightly) demand more of a research programme into entrepreneurship in the steel industry than an exoneration of entrepreneurial decisions based on examination of one (important) technical change and an ingenious attempt to show that price-cost margins were low so as to infer that managers were constrained not to fail (McCloskey, 1973). In the case of the more sophisticated modelling exercises such as Williamson's, historians will need to be persuaded that the game is worth playing even though they find the assumptions highly restrictive and have no feel as to whether competing models may do better in explaining the historical record.

Perhaps it should be said therefore that economics has made an important contribution as a source of hypotheses in economic history but that the 'testing' of these hypotheses is more likely to convince a true believer in the neoclassical paradigm than a sceptical historian. The usefulness of economic models as a method of ordering facts has been shown as far afield as uncovering the rationality of common field agriculture (Fenoaltea, 1986) or in understanding the Domesday tax assessments (McDonald and Snooks, 1985). On the other hand the degree of commitment to prior beliefs which professional training in economics inculcates and the limited range of evidence which economists typically consider, are better suited to prediction than explanation and may act as an obstacle to a fuller understanding of historical events. The danger then is that 'economic history is as much corrupted as enriched by economic theory' (Solow, 1986, p. 21).

ECONOMIC GROWTH AND ECONOMIC HISTORY. Traditionally the closest links between economics and economic history have been in the study of long run economic growth and development and it is therefore appropriate to review this area in more detail than others. Sadly, it must be said straightaway that economic history has had little influence upon and has been relatively little affected by growth theory of the postwar variety. More disappointingly and surprisingly, economic history has contributed hardly at all to major questions of applied economics of growth such as the relative decline of the UK economy in the long run.

One close and productive point of contact between growth theory and economic history has come through growth accounting in the Denison mould. The results for the United States show strikingly how different the 20th century has been compared with earlier in terms of total factor productivity growth (Abramovitz and David, 1973). This impression of relatively late acceleration to rapid overall productivity growth is confirmed by Crafts's (1985) finding that the Industrial Revolution saw productivity growth confined to a small advanced subset of the economy with the result that macroadvance was perhaps only 0.4 per cent per year. These findings are not especially surprising but do provide a framework for future research.

Indeed, it can be argued that the most important contribution economics has made to the understanding of long-run economic development recently is through better measurement, a research programme begun by Kuznets and still progressing vigorously. Thus while our understanding of the reasons for the onset of modern economic growth has advanced little our knowledge of its quantitative dimensions has grown substantially. Unfortunately the data now accumulated on long-run growth of national income offer little support for ambitious 'grand theories' of economic growth.

The general outline explanation of how the West became rich has not changed much over the years despite the attention of many economists and economic historians and would be well understood by an economist of a century ago. Thus it is widely agreed that the development of private property rights, the possibility of pursuing business relatively free of political or religious interference, the move to a more experimentalist (scientific) approach to the world, all stimulated by expansion of trade and commercial activity eventually provoked a marked acceleration in technological progress in the 18th and 19th centuries embodied in a much expanded quantity and variety of capital (Rostow, 1960; Landes, 1969; Hicks, 1969; North and Thomas, 1973; Rosenberg and Birdzell, 1986). In turn the 20th century has seen a further acceleration of productivity growth based on more education and research and development. This persistence and vigour of technological progress would be the surprise for Jevons and Ricardo in today's accounts of the rise of the West.

Within this general framework economic historians trying to apply economic analysis have sought to advance more detailed, predictive models still at the grand level. Within the past generation three attempts stand out and have for a while intrigued both economists and historians, namely the schema put forward by Rostow (1960), Gerschenkron (1962) and North and Thomas (1973).

Rostow's stages theory of growth, memorable in particular for its notion of 'take-off into sustained growth', dominated the literature for a decade or so to the extent of conditioning the thinking even of its many opponents. At its heart is a warranted growth rate equation based on the investment rate and the capital to output ratio and the idea of 'leading sectors' with powerful impacts on demand and costs through linkage

effects, i.e. through the input–output matrix of the economy. Take-off characterized by the presence of leading sectors, an appropriate institutional framework, a sharp rise in the rate of productive investment and a dramatic acceleration of the economy would occur when necessary prerequisites in terms of agricultural productivity and infrastructure had been achieved. The theory was sketched out for Britain, which was seen as a model for other countries following the same sequence of stages to the modern world and, based on then available national income data, Rostow suggested dates for take-off in various countries.

Rostow's work sparked off a stream of critical comment and extremely fruitful research responses. At the theoretical level critics like Kuznets (1963) and Fishlow (1965) objected that the stages and conditions for movement between stages were ambiguously defined and that the aggregate growth equation lacked behavioural content while at the empirical level examples were found of countries like France where a take-off could not be identified (Marczewski, 1963) and quantitative examination of railroads dealt harshly with the notion of the leading sector (Fogel, 1964). Quantitative research on Britain by Deane and Cole (1962) quickly established that Rostow's model case of take-off did not fit the data particularly well – a view which subsequent research has reinforced (Crafts, 1985) – in terms of investment, leading sectors or turning points in growth.

A further criticism of Rostow's 'uniform' model was that follower countries would differ in their development from pioneers perhaps because of different factor endowments, technological or trading opportunities. This type of argument was systematized by Gerschenkron (1962) who argued against the idea of prerequisites for take-off and suggested the 'backward' countries would be likely to develop differently in a process characterized by a great spurt in manufacturing output with heavy investment, much pressure on living standards, large-scale plant concentrated in heavy industry, coercive political institutions and a lagging agricultural sector. In this case Russia was the model economy.

Although the critical reaction has taken longer to emerge than in the case of Rostow, in the end the evidence has seemed not to support this view either. Gregory (1983) has shown that the quantitative data on Russia does not support Gerschenkron's claims on living standards, agriculture or heavy industry. Crafts (1984) for a sample of 17 European countries was able to show that the evidence does not confirm Gerschenkron's beliefs on correlations between backwardness and industrial output growth, consumption or investment.

The general tendency in the literature now is to accept that we do not have a typology of Western growth. Countries are perceived as pursuing *different* paths to the modern world (O'Brien and Keyder, 1978) and showing a high variance of structural characteristics at given income levels (Crafts, 1984). In particular, Britain appears to be very atypical with its early rundown of the agricultural labour force, its high agricultural productivity and its low rates of investment in physical and human capital. Perhaps reflection on the basis of international trade theory suggests that this is not surprising – production structures surely will vary according to *relative* factor endowments and countries did indeed play very different parts in the expanding trade of the 19th century. Differing population growth patterns, which as was noted earlier are not yet fully understood, impinge heavily on development paths especially in cases as different as France and Russia yet they are not a part of the grand theories. In retrospect, the predictive models of Gerschenkron and Rostow appear not only to fail empirically, but also to be misconceived.

The third grand schema, that put forward by North and Thomas, relates to the period prior to the onset of modern economic growth and sees as *the* prerequisite the establishment of an appropriate set of private property rights which ensure a reasonable degree of equality between the social and private rate of return to innovation. The key novelty of this approach relates to the attempt to endogenize institutional structure and in particular to relate institutional innovations to population growth, changing land/labour ratios and relative prices. Crucial, however, in the outcome was the strength of the state and its response to fiscal problems and the degree of political collusion between lords.

North and Thomas identify an important question, namely the economic origins of the institutional structure that all agree was conducive to modern economic growth. In doing so they wish to relax an assumption of conventional neoclassical general equilibrium models, namely the exogeneity of rules whilst retaining a neoclassical utility maximizing approach to modelling. Success in this area would be a major advance and would represent an important addition to the variety of models and range of societal variations normally considered by economists, i.e. a potentially important contribution to economics from the study of historical problems.

Unfortunately, in order to explain variations between countries and centuries North and Thomas are forced back on a variety of *ad hoc* (political) explanations such that their model appears to lack an unambiguous ability to predict the consequences of changing population (Field, 1981, p. 190). As North himself later reflected, his enterprise though stimulating was ultimately unsuccessful in that it lacked amongst other things a theory of demographic change, a theory of conflict resolution, a supply function of new institutional arrangements and above all 'a theory of ideology to account for ... deviations from the individualistic calculus of neoclassical theory' (1981, p. 12). This is an important, if negative, lesson to have learnt and should give food for thought to economists who fail to recognize that the past was different from today.

The other vital area in the study of long-run growth, which is less than adequately dealt with by the above approaches and where economic historians would hope to be able to use economic theory, is technological change. Here there are limited gains which have been realized. In particular, the diffusion process is somewhat better understood as are the gains from innovations. Thus Hyde's study of the adoption of coke-smelting (1977), David's of the mechanical reaper (1975) and Greasley's (1982) of the coal-cutting machine all show the importance of relative costs and the economic nature of the decision to adopt and have proved fruitful applications of basic neoclassical methodology. Also measurements of social savings from innovations has revealed the relative unimportance of even the most spectacular changes relative to overall income levels (Fogel, 1964; Von Tunzelmann, 1978) and has effectively banished the notion of indispensability together with hyperbolic accounts of technological advance.

In other aspects, however, the record is much less satisfactory – most notably in terms of induced innovation, where advance seems to be crucial in terms of understanding the accelerations in productivity growth which the quantitative evidence indicates. The most notable contribution by economic historians has obviously been that of Habakkuk (1962). He argued for the importance of 'labour-scarcity' in the United States' manufacturing in the early 19th century as an inducement both to more machinery and better machinery – in terms of technological change both to a faster overall rate of improvement and to more bias in the labour-saving direction as compared with Britain. David (1975) in an extensive review of the literature using the concept of the innovation possibility frontier shows that to sustain a belief in this, the only game in town, leads to extremely paradoxical results, especially that the UK was on the *higher* innovation possibility frontier, contrary to everyone's prior beliefs. He argues for a solution to the problem in terms of localized technological progress arising from learning by doing in the vicinity of the techniques chosen as a result of initial factor prices but this remains no more than an intriguing speculation.

The implications of this review are somewhat pessimistic. They are as follows. First, the requirements of economic history outstrip the present capabilities of economic theory but this situation has not induced any important theoretical developments. Second, in the key areas of institutional and technological change neoclassical economics has major limitations which are revealed by the difficulty of the questions history poses. Third, we are no nearer to a precise answer to the basic issues which stimulated the development of this research field as posed by Landes,

> (1) why did this first breakthrough to a modern industrial system take place in Western Europe? and (2) why, within this European experience, did change occur when and where it did?' (1969, p. 12).

Economists should learn something of the narrowness of their basic theoretical approach, new economic historians that they have been much more successful at measuring than at explaining modern economic growth.

THE FUTURE OF ECONOMIC HISTORY. It should be clear that the application of economic analysis and quantitative methods in economic history has produced very significant achievements if not living up to quite all the hopes of twenty years ago. Easy publications making valid and important points but giving less than satisfactory overall interpretations characterized the early years – unfortunately often alienating traditional historians by a combination of apparently simplistic arguments based on tight priors and a lack of real accessibility to the 'layperson'. The profession now shows every sign of recognizing and seeking to remedy these traits (Field, 1986).

In many respects there is reason therefore to be optimistic about future relationships between economics and economic history. Econometrics is more powerful and easier to carry out, interest in long-run rise and decline of nations has probably never been higher in both US and UK and the teething problems of new economic history are in the past. And yet, economists appear unreceptive on the whole to historical study and economic historians are in retreat everywhere as financial stringency impinges more severely on university budgets. As McCloskey put it a decade ago, 'for fifteen years or so Cliometricians have been explaining to their colleagues in history the wonderful usefulness of economics. It is time they began explaining to their colleagues in economics the wonderful usefulness of history' (1976, p. 455).

N.F.R. CRAFTS

See also CLIOMETRICS; CONTINUITY IN ECONOMIC HISTORY; ECONOMIC INTERPRETATION OF HISTORY; INDUSTRIAL REVOLUTION.

BIBLIOGRAPHY

Abramovitz, M. and David, P.A. 1973. Reinterpreting economic growth: parables and realities. *American Economic Review, Papers and Proceedings* 63(2), 428–39.

Aitken, H.G.J. (ed.) 1971. *Did Slavery Pay?* Boston: Houghton Mifflin.

Allen, R.C. 1979. International competition in iron and steel. *Journal of Economic History* 39(4), 911–37.

Barker, T.C. 1985. What is economic history? *History Today*, 36–8.

Bordo, M.D. and Schwartz, A. (eds) 1984. *A Retrospective on the Classical Gold Standard, 1821–1931*. Chicago: University of Chicago Press.

Cairncross, A. and Eichengreen, B. 1983. *Sterling in Decline: The Devaluations of 1931, 1949 and 1967*. Oxford: Blackwell.

Capie, F. and Webber, A. 1985. *A Monetary History of the United Kingdom 1870–1982*, Vol. I. London: Allen & Unwin.

Clapham, J.H. 1926. *An Economic History of Modern Britain*. Vol. I: *The Early Railway Age*. Cambridge: Cambridge University Press.

Coale, A.J. and Watkins, S.C. (eds) 1986. *The Decline of Fertility in Europe*. Princeton: Princeton University Press.

Crafts, N.F.R. 1984. Patterns of development in nineteenth-century Europe. *Oxford Economic Papers* 36(3), 438–58.

Crafts, N.F.R. 1985. *British Economic Growth During the Industrial Revolution*. Oxford: Clarendon Press.

Crafts, N.F.R., Thomas, M.F. and Mackinnon, M.E. 1984. International trade and structural unemployment in interwar Britain. Paper presented to the American Economic Association, Dallas.

Crafts, N.F.R. and Thomas, M.F. 1986. Comparative advantage in UK manufacturing trade, 1910–1935. *Economic Journal* 96(383), September, 629–45.

David, P.A. 1975. *Technical Choice, Innovation and Economic Growth*. Cambridge: Cambridge University Press.

David, P.A. et al. 1976. *Reckoning With Slavery*. New York: Oxford University Press.

Deane, P. and Cole, W.A. 1962. *British Economic Growth, 1688–1959*. Cambridge: Cambridge University Press.

Feinstein, C.H. 1972. *National Income, Expenditure and Output of the United Kingdom, 1855–1965*. Cambridge: Cambridge University Press.

Fenoaltea, S. 1986. The economics of the common fields: the state of the debate. Paper presented to the International Symposium on Property Rights, Organizational Forms and Economic Behaviour, Uppsala, Sweden.

Field, A.J. 1981. The problem with neoclassical institutional economics: a critique with special reference to the North/Thomas model of pre-1500 Europe. *Explorations in Economic History* 18(2), 174–98.

Field, A.J. 1986. The future of economic history. In *The Future of Economic History*, ed. A.J. Field, Boston: Kluwer-Nijhoff.

Fishlow, A. 1965. Empty economic stages? *Economic Journal* 75, 112–25.

Fogel, R.W. 1964. *Railroads and American Economic Growth: Essays in Econometric History*. Baltimore: Johns Hopkins University Press.

Fogel, R.W. and Engerman, S.L. 1974. *Time on the Cross*. Boston: Little, Brown.

Friedman, M. and Schwartz, A. 1982. *Monetary Trends in the United States and the United Kingdom: their Relation to Income, Prices and Interest Rates, 1867–1975*. Chicago: Chicago University Press.

Gayer, A.D. et al. 1953. *The Growth and Fluctuation of the British Economy, 1790–1850*. Oxford: Clarendon Press.

Gerschenkron, A. 1962. *Economic Backwardness in Historical Perspective*. Cambridge, Mass.: Harvard University Press.

Greasley, D. 1982. The diffusion of machine cutting in the British coal industry, 1902–1938. *Explorations in Economic History* 19(3), 246–68.

Gregory, P. 1983. *Russian National Income, 1885–1913*. Cambridge: Cambridge University Press.

Habakkuk, H.J. 1962. *American and British Technology in the Nineteenth Century*. Cambridge: Cambridge University Press.

Hannah, L. 1974. Takeover bids in Britain before 1950: an exercise in business pre-history. *Business History* 16(1), 65–77.

Hatton, T. et al. 1983. Eighteenth-century British trade: home spun or Empire made? *Explorations in Economic History* 20(2), 163–82.

Hendry, D.F. and Ericsson, N. 1983. Assertion with empirical basis: an econometric appraisal of 'Monetary Trends in ... the United Kingdom' by Milton Friedman and Anna Schwartz. *Bank of England Panel of Academic Consultants Paper* No. 22, 45–101.

Hicks, J.R. 1969. *A Theory of Economic History*. Oxford: Clarendon Press.

Hyde, C.K. 1977. *Technological Change and the British Iron Industry, 1700–1870*. Princeton: Princeton University Press.

Kennedy, W.P. 1984. Notes on economic efficiency in historical perspective: the case of Britain, 1870–1914. *Research in Economic History* 9, 109–41.

Kuznets, S.S. 1963. Notes on the take-off. In *The Economics of Take-Off Into Sustained Growth*, ed. W.W. Rostow, London: Macmillan.

Landes, D.S. 1969. *The Unbound Prometheus*. Cambridge: Cambridge University Press.

Lee, C.H. 1983. *Social Science and History*. London: Social Science Research Council.

McClelland, P.D. 1975. *Causal Explanation and Model Building in History, Economics and the New Economic History*. Ithaca: Cornell University Press.

McCloskey, D.N. 1970. Did Victorian Britain fail? *Economic History Review* 23(3), December, 446–59.

McCloskey, D.N. 1973. *Economic Maturity and Entrepreneurial Decline: British Iron and Steel, 1870–1913*. Cambridge, Mass.: Harvard University Press.

McCloskey, D.N. 1976. Does the past have useful economics? *Journal of Economic Literature* 14(2), June, 434–61.

McDonald, J. and Snooks, G.D. 1985. Statistical analysis of Domesday Book (1086). *Journal of the Royal Statistical Society* 148, 147–60.

Marczewski, J. 1963. The take-off and French experience. In *The Economics of Take-Off Into Sustained Growth*, ed. W.W. Rostow, London: Macmillan.

Mathias, P. 1971. Living with the neighbours: the role of economic history. In *The Study of Economic History*, ed. N.B. Harte, London: Cass.

North, D.C. 1981. *Structure and Change in Economic History*. New York: Norton.

North, D.C. and Thomas, R.P. 1973. *The Rise of the Western World: A New Economic History*. Cambridge: Cambridge University Press.

O'Brien, P.K. and Keyder, C. 1978. *Economic Growth in Britain and France, 1780–1914*. London: Allen & Unwin.

Parker, W.N. (ed.) 1970. *The Structure of the Cotton Economy of the Antebellum South*. Berkeley: University of California Press.

Parker, W.N. (ed.) 1986. *Economic History and the Modern Economist*. Oxford: Blackwell.

Reder, M.W. 1982. Chicago economists: permanence and change. *Journal of Economic Literature* 21, March, 1–38.

Rosenberg, N. and Birdzell, L.E. 1986. *How the West Grew Rich*. London: Tauris.

Rostow, W.W. 1960. *The Stages of Economic Growth*. Cambridge: Cambridge University Press.

Sandberg, L.G. 1974. *Lancashire in Decline*. Columbus: Ohio State University Press.

Solow, R. 1986. Economics: is something missing? In *Economic History and the Modern Economist*, ed. W.N. Parker, Oxford: Blackwell.

Stampp, K.M. 1976. Introduction. In *Reckoning With Slavery*, ed. P.A. David et al., New York: Oxford University Press.

Supple, B.E. 1981. Old problems and new directions. *Journal of Interdisciplinary History* 12(2), 199–205.

Sutch, R. 1975. The treatment received by American slaves; a critical review of the evidence presented in *Time on the Cross*. *Explorations in Economic History* 12(4), 335–438.

Temin, P. 1976. *Did Monetary Forces Cause the Great Depression?* New York: Norton.

Thomas, M. 1983. Rearmament and economic recovery in the late 1930s. *Economic History Review* 36, November, 552–79.

Von Tunzelmann, G.N. 1978. *Steam Power and British Industrialization to 1860*. Oxford: Clarendon Press.

White, L.H. 1984. *Free Banking in Britain: Theory, Experience and Debate, 1800–1845*. Cambridge: Cambridge University Press.

Williamson, J.G. 1985. *Did British Capitalism Breed Inequality?* London: Allen & Unwin.

Wrigley, E.A. 1978. Fertility strategy for the individual and the group. In *Historical Studies of Changing Fertility*, ed. C. Tilly, Princeton: Princeton University Press.

Wrigley, E.A. and Schofield, R. 1981. *The Population History of England, 1541–1871*. London: Arnold.

economic integration. In everyday parlance, integration is defined as bringing together of parts into a whole. In the economic literature, the term 'economic integration' does not have such a clear-cut meaning. At one extreme, the mere existence of trade relations between independent national economies is considered as a form of economic integration; at the other, it is taken to mean the complete unification of national economies.

Economic integration is defined here as process and as a state of affairs. Considered as a process, it encompasses measures designed to eliminate discrimination between economic units that belong to different national states; viewed as a state of affairs, it represents the absence of various forms of discrimination between national economies.

Economic integration may take several forms that represent various degrees of integration. In a free trade area, tariffs (and quantitative import restrictions) among participating countries are eliminated, but each country retains its own tariffs against non-members. Establishing a customs union involves, apart from the suppression of intra-area trade barriers, equalizing tariffs on imports from non-member countries.

A common market goes beyond a customs union, inasmuch as it also entails the free movement of factors of production. In turn, an economic union combines the suppression of restrictions on commodity and factor movements with some degree of harmonization of national economic policies, so as to reduce discrimination owing to disparities in these policies. Finally, total economic integration means the unification of economic policies, culminating in the establishment of a supra-national authority whose decisions are binding for the member states.

HISTORY. The first important case of economic integration was the German Zollverein in the 19th century, which subsequently led to total economic integration through the unification of the German states with the establishment of the Deutsches Reich. In the 20th century, the creation of the Benelux customs (1948) and subsequently economic (1949) union, comprising Belgium, Luxemburg, and the Netherlands, represented the first step towards European economic integration. It was followed by the establishment of the European Coal and Steel Community (1953) and the European Economic Community or EEC (1958), both comprising Belgium, France, Italy, Luxemburg, the Netherlands, and West Germany.

Austria, Denmark, Norway, Portugal, Sweden, Switzerland, and the United Kingdom founded the European Free Trade Association or EFTA in 1960, with Finland participating first as an associate and later as a full member. In turn, Denmark and the United Kingdom left EFTA and, together with Ireland, entered the European Economic Community in 1968; Greece became a member of the EEC in 1978, and Portugal and Spain joined in 1986.

In Eastern Europe, the Council for Mutual Economic Assistance or CMEA was established in 1948, with the participation of the Soviet Union, Bulgaria, Czechoslovakia, Hungary, Poland, and Romania. Albania and East Germany joined shortly thereafter; subsequently, Cuba and Mongolia became full members while Albania ceased to participate in CMEA activities.

There have been a number of attempts at economic integration in developing countries. Some were to involve the establishment of a free trade area, such as the Latin American Free Trade Association (1960) comprising Argentina, Bolivia, Brazil, Chile, Colombia, Ecuador, Mexico, Peru, Uruguay, and Venezuela; others were designed to become customs unions, such as the West African Customs Union (1959), including the Ivory Coast, Mali, Mauritania, Niger, Senegal, and Upper Volta. In 1960, the Central American Common Market was established, with Costa Rica, Guatemala, Honduras, Nicaragua, and El Salvador as members; in turn, the East African Common Market, comprising Kenya, Tanzania, and Uganda and subsequently transformed into the East African Economic Community (1967), was designed to become an economic union. None of these attempts has come to fruition, however, as barriers to intra-area trade have not been fully eliminated or have subsequently been restored.

TRADE CREATION AND TRADE DIVERSION. Viner's *The Customs Union Issue* (1950) was the first important contribution to the theory of economic integration. Viner investigated the impact of a customs union on trade flows and distinguished between the 'trade-creating' and the 'trade-diverting' effects of a union. In the first case, there is a shift from domestic to partner country sources of supply of a particular commodity; in the second case, the shift occurs from non-member country to partner country sources of supply.

Trade creation increases economic welfare, inasmuch as higher-cost domestic sources of supply are replaced by lower-cost imports from partner countries that were previously excluded by the tariff. In turn, trade diversion has a welfare cost since tariff discrimination against non-member countries, attendant on the establishment of the customs union, leads to the replacement of lower-cost sources of supply in these countries by higher-cost partner country sources.

The net welfare effects of the customs union will depend on the amount of trade created and diverted as well as on differences in unit costs. In a partial equilibrium framework, under constant costs, there will be a welfare gain (loss) if the amount of trade created, multiplied by differences in unit costs between the home and the partner countries, exceeds (falls short of) the amount of trade diverted, multiplied by differences in unit costs between the partner and the non-member countries.

Meade (1955) further considered the effects of a customs union on intercommodity substitution, involving the replacement of domestic products by partner country products (trade creation) and the replacement of products of non-member countries by partner country products (trade diversion). As in the case of substitution among the sources of supply of a particular commodity (production effects), trade creation involves a welfare improvement, and trade diversion the deterioration of welfare, in the event of substitution among commodities (consumption effects).

The separation of production and consumption effects does not imply the absence of interaction between the two. Substitution among sources of supply will affect the pattern of consumption through changes in the prices paid by the consumer. Also, intercommodity substitution will lead to modifications in the pattern of production by changing the prices received by producers.

At the same time, as Lipsey and Lancaster (1956–57) first noted, production and consumption effects – and the theory of customs unions in general – should be considered as special cases of the theory of the second best. Assuming that the usual conditions for a Pareto optimum are fulfilled, free trade will lead to efficient resource allocation while pre-union, as well as the post-union, situations are sub-optimal because tariffs exist in both cases. In the abstract, then, one cannot make a judgement as to whether establishing a customs union will increase or reduce welfare. Nevertheless, a consideration of certain factors may provide a presumption as to the possible direction of the welfare effects of a union.

FACTORS INFLUENCING THE WELFARE EFFECTS OF A CUSTOM UNION. Lipsey (1960) suggested that the welfare effects of a customs union will depend on the relative importance in home consumption of goods produced domestically and imported from non-member countries prior to the establishment of the union. Ceteris paribus, the larger the share of domestic goods and the smaller the share of goods imported from non-member countries, the greater is the likelihood of an improvement in welfare following the union's establishment. Such will be the case since substitution of partner country products for domestic products entails trade creation and their substitution for the products of non-member countries involves trade diversion.

These propositions are consistent with Tinbergen's (1957) conclusion that increases in the size of a customs union will augment the probability of favourable welfare effects; in the limiting case, the customs union includes the entire world, which is equivalent to free trade. Applying the argument that gains are obtained through the enlargement of a union because of increased possibilities for the reallocation of production, it also follows that the gains are positively correlated with increases in the market size of the participating countries (e.g. small countries will gain more from participation in a customs union than large countries).

Viner further considered the implications that differences in production structures among the member countries have for the welfare effects of a customs union. He suggested that the more competitive (the less complementary) is the production structure of the member countries, the greater is the chance that a customs union will increase welfare.

This proposition reflects the assumption that countries with similar production structures tend to replace domestic goods by competing imports from partner countries following the establishment of a customs union, while differences in the production structure within the union lead to substitution of partner country products for lower-cost products originating in non-member countries (the latter conclusion does not hold if the union includes the low-cost producer).

The welfare effects of a customs union will also depend on transportation costs. Ceteris paribus, the lower are transport costs among the member countries, the greater will be the gains from their economic integration. Thus, the participation of neighbouring countries in a union, with greater possibilities for trade creation across their borders, will offer advantages over the participation of faraway countries that tends to promote trade diversion.

The height of tariffs will further affect the potential gains and losses derived from a customs union. High pre-union tariffs against the future member countries will increase the possibility of trade creation, and hence gains in welfare, following the establishment of the union while low tariffs against non-member countries will reduce the chances for trade diversion. But, these conclusions have little relevance under the application of the most-favoured-nation clause that entails providing equal tariff treatment to all countries before the customs union is established.

CUSTOMS UNIONS VS. UNILATERAL TARIFF REDUCTIONS. In the Viner–Meade–Lipsey analysis, participation in a trade-creating customs union was considered as a means to reduce the distorting effects of the country's own tariffs. This argument was carried to its logical conclusion in contributions by Cooper and Massell (1965a) and Johnson (1965) who suggested that participation in a customs union is inferior to the unilateral elimination of tariffs, which leads to greater trade creation without giving rise to trade diversion.

The same authors claimed that the reasons for the establishment of customs unions lie in the gains participating countries may obtain in furthering non-economic objectives, and considered preference for industry as such an objective. They further assumed that this objective can be pursued at a lower cost in the framework of the larger market of a customs union than in the country's own domestic market.

As Johnson noted, the formation of a customs union in the pursuit of the stated objective presupposes that the member countries are at a comparative disadvantage in the production of industrial goods vis-à-vis the rest of the world. Cooper and Massell (1965b) identified such countries with developing countries, further suggesting that the economic planners of these countries are willing to accept some reduction in national income in order to assure increases in industrial production.

The question remains as to why there is a preference for industry. Johnson (1965) expressed the view that such preference may reflect nationalist aspirations and rivalry with other countries; the power of industrial firms and workers to increase their incomes; or the belief that industrial activity involves beneficial externalities. The last point, however, implies that there is no need to introduce non-economic considerations to obtain the Cooper–Mansell–Johnson result; the desirability of a customs union may be established in economic terms, provided that it permits obtaining externalities that cannot be achieved otherwise.

A further question is if unilateral tariff reductions will be superior to a customs union in the absence of a preference for industry or beneficial externalities. The Wonnacotts (1981) showed that this may not be the case if one admits the existence of tariffs in partner and in non-member countries prior to the formation of the customs union.

The elimination of tariffs by partner countries will provide benefits to the home country as it can now sell at a higher price in partner country markets. This gain will be larger the higher is the pre-union tariff in the partner countries and will further be affected by tariffs in the non-member countries. This is because, in selling in partner country markets free of duty, home country producers avoid paying the tariff in non-member countries.

Finally, Cooper and Massell (1965b) noted that a subsidy-union, with each participating country subsidizing its own industrial production, is superior to a customs union. This conclusion follows since the consumption cost of the tariff can be avoided if the prices of industrial products in the union are maintained at the world market level through subsidies. However, production subsidization may be done by each country individually, with the attendant welfare benefits, without participating in a union.

MULTI-COUNTRY ANALYSIS OF A CUSTOMS UNION. Traditionally, the welfare effects of a customs union were considered from the point of view of a single country. Yet, these effects may differ among member countries, depending on their production structure, location, the height of pre-union tariffs, and other characteristics. In fact, one member country may obtain a gain and another a loss, when any attempt to aggregate gains and losses encounters the well-known difficulties of international welfare comparisons.

The distribution of welfare gains and losses in a customs union will be further affected by changes in the terms of trade. The establishment of a union may give rise to price changes in trade between the member countries, even if the prices at which trade takes place with non-member countries remain unchanged (the case of the 'small' union).

In the more general case, prices in trade with non-member countries will also vary. Now, while trade diversion involves a welfare loss to the member countries of a customs union under unchanged terms of trade, this loss may be offset by a welfare gain due to improvements in the terms of trade attendant on trade diversion. Conversely, whereas under the assumption of unchanged terms of trade the welfare of non-member countries is unaffected by the establishment of a customs union, non-member countries will lose owing to the adverse impact of trade diversion on their terms of trade. This may be interpreted as the result of a shift in the union members' reciprocal demand curve for products originating in non-member countries.

Improvements in the terms of trade thus provide reasons for the establishment of a customs union even in the absence of non-economic objectives and beneficial externalities. Such improvements also favour a customs union over unilateral tariff reductions, which would lead to the deterioration of the terms of trade of the country concerned.

Other things being equal, the larger the union the greater will be its gain, and hence the loss to non-member countries, through terms of trade changes. This is because, ceteris paribus, the larger the union the higher will be the elasticity of its reciprocal demand for foreign products and the lower the elasticity of reciprocal demand on the part of non-member countries for the union's products.

The extent of terms-of-trade effects will further depend on the height of tariffs before and after the establishment of a customs union. As Vanek (1965) first showed, a customs union will not involve a loss to non-member countries, while benefiting its own members, if the union's external tariff level is sufficiently lower than the pre-union tariffs of the member countries.

Vanek's proposition was formulated in a three-country, two-commodity (3×2) model. It has subsequently been extended to a general case, under which compensatory payments to non-member countries were also introduced (Kemp and Wan, 1976). At the same time, these propositions indicate a theoretical possibility rather than a likely outcome, since customs unions have shown little inclination to compensate non-member countries for losses attendant upon the union's establishment.

3×3 models represent an intermediate case between 3×2 and $m \times n$ models. They permit introducing a greater number of possible trade patterns, differential tariffs, complementarity and substitution in consumption, with a large number of marginal conditions in production and consumption, as well as intermediate products (Lloyd, 1982). The 3×3 model is thus richer in content than the 3×2 model. Despite attempts made at introducing new terminology (Collier, 1979), however, adding a third commodity does not appear to have materially affected the basic propositions of customs union theory. This conclusion may also find application to $m \times n$ models.

FREE TRADE AREAS. In a free trade area, maintaining different tariffs among member countries on the products of non-members introduces the possibility of trade deflection. Furthermore, production and investment deflection may occur if one admits trade in intermediate products.

There will be trade deflection if imports enter the free trade area via the member country which applies the lowest tariff. Transportation costs apart, this is equivalent to adopting a tariff equal to the lowest tariff for each commodity in any of the member countries. Under the assumption of unchanged terms of trade, the deflection of trade will increase welfare in the member countries by limiting the extent of trade diversion.

Removing this assumption, trade deflection will affect the distribution of welfare between member and non-member countries by reducing the terms of trade gain (loss) for the former (latter).

Production deflection will occur if the manufacture of products containing imported inputs shifts to countries which have lower tariffs on these inputs, because differences in tariffs outweigh differences in production costs. The deflection of production will have unfavourable effects on welfare, since the pattern of productive activity will not follow lines of comparative advantage but rather differences in duties.

The deflection of production may also affect the pattern of investment. Other things being equal, investors will establish factories in countries with lower tariffs on imported inputs. Again, adverse welfare effects will ensue because investments respond to tariff differences rather than to differences in production costs.

The deflection of trade, production, and investment represent unintended effects of free trade areas. To avoid such an eventuality member countries of free trade areas have imposed country of origin rules. These rules limit the freedom of intra-area trade to commodities that incorporate a certain proportion of domestic products or undergo a particular process of transformation in one of the member countries. The application of origin rules limits, but does not entirely eliminate, trade, production and investment deflection in a free trade area. Other things being equal, then, their self-interest would tend to encourage member countries to reduce their own tariffs.

FACTOR MOVEMENTS. The deflection of investment may occur within a country or may involve international capital movements. In the first case, it affects the allocation of the country's own capital among industries; in the second case, it influences the international allocation of capital.

The last point leads to the case of common markets where, by definition, the full mobility of factors is assured. Meade (1953) first analysed the welfare effects of the movement of factors of production in an integrated area. He concluded that free factor movement will increase the gains obtained in a union by reducing the relative scarcities of the factors of production. This conclusion reflected the assumption that the conditions for factor price equalization through trade are not fulfilled.

If factors of production were not free to move between member and non-member countries, there will be no welfare loss due to factor movements among member countries to correspond to trade diversion in commodity trade. In the event of such factor movements, however, an analogous case to trade creation and trade diversion occurs if the movement of factors were subject to taxes prior to the establishment of a union and these taxes have been removed among union members. And, in any case, there will be indirect effects on welfare to the extent that factor movements substitute for trade. These effects may involve welfare losses to non-member countries as the newly-established productions substitute for imports from them.

ECONOMIES OF SCALE. Economic integration may lead to lower costs through increases in the volume of plant output. For various types of equipment, such as containers, pipelines, and compressors, cost is a function of the surface area whereas capacity is related to volume; per unit costs decline with increases in output in the case of bulk transactions as well as for nonproportional activities such as design production planning, research, and the collection and channelling of information; inventory holdings do not need to increase proportionately with output; larger output warrants the application of technological methods that call for the use of specialized equipment or assembly-line production; and large-scale production may be necessary to ensure the optimum use of various kinds of indivisible equipment.

Corden (1972) showed that the traditional concepts of trade creation and trade diversion will be relevant in the case of economies of scale on the plant level but new concepts are added: the cost-reduction effect and the trade-suppression effect. The former refers to reductions in average unit costs as domestic output expands following the establishment of the union; the latter refers to the replacement of cheaper imports from non-member countries by domestic production under economies of scale. In Corden's view, a net benefit is likely to ensue as the cost-reduction effect tends to outweigh the trade-suppression effect.

Plant size and unit costs are not necessarily correlated in the case of multiproduct firms. In such instances, costs may be lowered by reducing product variety through specialization in an integrated area, which permits lengthening production runs for individual products.

The advantages of longer production runs derive from improvements in manufacturing efficiency along the 'learning curve' as cumulated output increases; the lowering of expenses involved in moving from one operation to another that involves the resetting of machines, the shifting of labour, and the reorganization of the work process; and the use of special-purpose machinery in the place of general purpose machinery.

Apart from product or horizontal specialization, there are possibilities for vertical specialization by subdividing the production process among individual establishments in an integrated area. As the sales of the final product increase, parts, components, and accessories may be manufactured in separate plants, each of which enjoys economies of scale, thereby resulting in cost reductions.

COMPETITION AND TECHNOLOGICAL CHANGE. Economic integration will also create the conditions for more effective competition (Scitovsky, 1958). By increasing the number of firms each producer considers as his competitors, the opening of national frontiers will contribute to the loosening of monopolistic and oligopolistic market structures in the individual countries. At the same time, there is no contradiction between gains from economies of scale and increased competition, since a wider market can sustain a larger number of efficient units (Balassa, 1961).

Greater competition may have beneficial effects through improvements in manufacturing efficiency as well as through technological change. While the former has no place in traditional theory, which postulates the choice of the most efficient production methods among those available to the firm, it may assume considerable importance in countries whose markets have been sheltered from foreign competition.

The stick and the carrot of competition also provides inducement for technological progress in the member countries. In particular, increased competition may stimulate research activity aimed at developing new products and improving production methods. Finally, economic integration may contribute to the transmission of technological knowledge by increasing the familiarity of producers with new products and technological processes originating in the partner countries.

It has been suggested, however, that gains from competition, and from economies of scale, may be obtained through unilateral trade liberalization and that the gains are predicated on the response of economic agents to the stimulus provided by competition (Krauss, 1972). While the validity of the second point depends on factors which are particular to each country, the first neglects the gains obtained through the increases in output associated with sales in the markets of partner countries.

POLICY HARMONIZATION. Policy differences among the member countries may influence trade flows and factor movements, thereby modifying the welfare effects of economic integration. Industrial policies, social policies, fiscal policies, monetary policies, and exchange rate policies are relevant in this context (Balassa, 1961).

Industrial policies may involve granting credit preferences and/or tax benefits across the board or to particular activities. 'Horizontal' policies that are applied across-the-board do not create distortions, unless the conditions under which they are provided favour one activity over another. By contrast, 'vertical' measures are granted to particular activities and thereby introduce distortions, which may counteract the effects of the elimination of intra-area tariffs.

Intercountry differences in social policies will not give rise to distortions, provided that social benefits are financed from the contributions of employers and employees. Nor are these conclusions affected if factor mobility is introduced into the analysis as long as the employees regard the resulting social benefits as part of their compensation.

The situation is different if social benefits are financed from general tax revenue. This case is equivalent to a wage subsidy that favours labour-intensive activities. Correspondingly, differences in the mode of financing social security among the member countries will introduce distortions in resource allocation. This conclusion is strengthened if consideration is given to factor movements that respond to international differences in labour costs.

The elimination of vertical measures of industrial policy and the equalization of the conditions of financing social security will reduce distortions in resource allocation as well as differences in tax burdens among the member countries. Differences in the tax burden may remain, however, owing to national preferences as to the provision of collective goods. The effects of such differences on factor movements will depend on the spending of the tax proceeds. But, there may be 'supply-side' effects, with a lower tax burden providing incentives for work effort and risk taking.

A further question is if, for a given tax burden, intercountry differences in reliance on indirect taxes and income taxes will distort competition. Under the destination principle, indirect taxes are rebated on exports and imposed on imports without such adjustments occurring in regard to income taxes. Nevertheless, distortions in the conditions of competition will not ensue as flexibility in exchange rates will offset differences in rates of indirect taxes.

The application of the origin principle, with indirect taxes levied on production irrespective of the country of sale, in one country and that of the destination principle in another will similarly be offset through exchange rate flexibility. Such will not be the case, however, if cascade-type taxation applied in one country and value added taxation in another, with the former raising the tax burden on industries that go through several stages of fabrication, each of which is subject to tax. Eliminating this source of distortion would necessitate the adoption of value added taxation in all member countries of a union.

While exchange rate flexibility is necessary to offset intercountry differences in systems of taxation, it has been proposed that fixed exchange rates be established following the creation of a union. But such an action is predicated on the coordination – and eventual unification – of monetary and fiscal policies, since otherwise pressures are created for exchange rate changes. The fixity of exchange rate should thus be considered as the final outcome of policy coordination rather than an intermediate step in economic integration (Balassa, 1975).

BELA BALASSA

See also CUSTOMS UNIONS; LIST, FRIEDRICH.

BIBLIOGRAPHY

Balassa, B. 1961. *The Theory of Economic Integration.* Homewood, Ill.: Richard D. Irwin.

Balassa, B. 1975. Monetary integration in European Common Market. In *European Economic Integration*, ed. B. Balassa, Amsterdam: North-Holland, 175–220.

Collier, P. 1979. The welfare effects of a customs union: an anatomy. *Economic Journal* 83, 84–7.

Cooper, C.A. and Massell, B.F. 1965a. A new look at customs union theory. *Economic Journal* 75, 742–7.

Cooper, C.A. and Massell, B.F. 1965b. Towards a general theory of customs unions for developing countries. *Journal of Political Economy* 73, 461–76.

Corden, W.M. 1972. Economies of scale and customs union theory. *Journal of Political Economy* 80, 465–75.

Johnson, H.G. 1965. An economic theory of protectionism, tariff bargaining, and the formation of customs unions. *Journal of Political Economy* 73, 256–83.

Kemp, M.C. and Wan, H.Y., Jr. 1976. An elementary proposition concerning the formation of customs unions. *Journal of International Economics* 6, 95–7.

Krauss, M.B. 1972. Recent developments in customs union theory: an interpretative survey. *Journal of Economic Literature* 10, 413–36.

Lipsey, R.G. 1960. The theory of customs unions: a general survey. *Economic Journal* 70, 496–513.

Lipsey, R.G. and Lancaster, K.J. 1956–7. The general theory of second best. *Review of Economic Studies* 24, 11–32.

Lloyd, P.J. 1982. The theory of customs unions. *Journal of International Economics* 12, 41–63.

Meade, J.E. 1953. *Problems of Economic Union.* London: Allen & Unwin.

Meade, J.E. 1955. *The Theory of Customs Union.* Amsterdam: North-Holland.

Scitovsky, T. 1958. *Economic Theory and Western European Integration.* London: Allen & Unwin.

Tinbergen, J. 1957. Customs unions: influence of their size on their effect. *Zeitschrift der gesamten Staatswissenschaft* 113, 404–14.

Vanek, J. 1965. *General Equilibrium of International Discrimination. The Case of Customs Unions.* Cambridge, Mass.: Harvard University Press.

Viner, J. 1950. *The Customs Union Issue.* New York: Carnegie Endowment for International Peace.

Wonnacott, P. and Wonnacott, R. 1981. Is unilateral tariff reduction preferable to a customs union? The curious case of the missing foreign tariffs. *American Economic Review* 71, 704–14.

economic interpretation of history. Marxism does not possess a monopoly of the economic interpretation of history. Other theories of this kind can be formulated – for instance that which can be found in the very distinguished work of Karl Polanyi, dividing the history of mankind into three stages, each defined by a different type of economy. If Polanyi is right in suggesting that reciprocity, redistribution and the market each defined a different kind of society, this is, in a way, tantamount to saying that the economy is primary, and thus his work constitutes a species of the economic interpretation of history. Nevertheless, despite the importance of Polanyi's work and the possibility of other rival economic interpretations, Marxism remains the most influential, the most important, and perhaps the best elaborated of all theories, and we shall concentrate on it.

One often approaches a theory by seeing what it denies and what it repudiates. This approach is quite frequently adopted in the case of Marxism, where it is both fitting and misleading. We shall begin by adopting this approach, and turn to its dangers subsequently.

Marxism began as the reaction to the romantic idealism of Hegel, in the ambience of whose thought the young Karl Marx reached maturity. This no doubt is the best advertised fact about the origin of Marxism. The central point about Hegelianism was that it was acutely concerned with history and social change, placing these at the centre of philosophical attention (instead of treating them as mere distractions from the contemplation of timeless objects, which had been a more frequent philosophical attitude); and secondly, it taught that history was basically determined by intellectual, spiritual, conceptual or religious forces. As Marx and Engels put it in *The German Ideology*, 'The Young Hegelians are in agreement with the Old Hegelians in their belief in the rule of religion, of concepts, of an abstract general principle in the existing world' (Marx and Engels, [1845–6], p. 5).

Now the question is – why did Hegel and followers believe this? If it is interpreted in a concrete sense, as a doctrine claiming that the ideas of men determined their other activities, it does not have a great deal of plausibility, especially when put forward as an unrestricted generalization. If it is formulated – as it was by Hegel – as the view that some kind of abstract principle or entity dominates history, the question may well be asked: what evidence do we have for the very existence of this mysterious poltergeist allegedly manipulating historical events? Given the fact that the doctrine is either implausible or obscure, or indeed both, why were intelligent men so strongly drawn to it?

The answer to this may be complex, but the main elements in it can perhaps be formulated simply and briefly. Hegelianism enters the scene when the notion of what we now call *culture* enters public debate. The point is this: men are not machines. When they act, they do not simply respond to some kind of push. When they do something, they generally have an idea, a concept, of the action which they are performing. The idea or conception in turn is part of a whole system. A man who goes through the ceremony of marriage has an *idea* of what the institution means in the society of which he is part, and his understanding of the institution is an integral part of his action. A man who commits an act of violence as part of a family feud has an idea of what family and honour *mean*, and is committed to those ideas. And each of these ideas is not something which the individual had excogitated for himself. He took it over from a corpus of ideas which differ from community to community, and which *change* over time, and which are now known as *culture*.

Put in this way, the 'conceptual' determination of human conduct no longer seems fanciful, but on the contrary is liable to seem obvious and trite. In various terminologies ('hermeneutics', 'structuralism', and others) it is rather fashionable nowadays. The idea that conduct is concept-saturated and that concepts come not singly but as *systems*, and are carried not by individuals but by on-going historic communities, has great plausibility and force. Admittedly, those who propose it, in Hegel's day and in ours, do not always define their position with precision. They do not always make clear whether they are merely saying that culture in this sense is important (which is hardly disputable), or claiming that it is the prime determinant of other things and the ultimate source of change, which is a much stronger and much more contentious claim. Nonetheless, the idea that culture is important and pervasive is very plausible and suggestive, and Hegelianism can be credited with being one of the philosophies which, in its own peculiar language, had introduced this idea. It is important to add that Hegelianism often speaks of 'Spirit' in the singular; our suggestion is that this can be interpreted as *culture*, as the spirit of the age. This made it easy for Hegelianism to operate as a kind of surrogate Christianity: those no longer able to believe in a personal god could tell themselves that this had been a parable on a kind of guiding historical spirit. For those who wanted to use it in that way, Hegelianism was the continuation of religion by other means.

But Hegelianism is not exhausted by its sense of culture, expressed in somewhat strange language. It is also pervaded by another idea, fused with the first one, and one which it shares with many thinkers of its period: a sense of *historical plan*. The turn of the 18th and 19th centuries was a time when men became imbued with the sense of cumulative historical change, pointing in an upward direction – in other words, the idea of Progress.

The basic fact about Marxism is that it retains this second idea, the 'plan' of history, but aims at inverting the first idea, the romantic idealism, the attribution of agency to culture. As the two founders of Marxism put it themselves in *The German Ideology* (pp. 14–15),

In direct contrast to German philosophy which descends from heaven to earth, here we ascend from earth to heaven ... We set out from real active men, and on the basis of their real life-process we demonstrate the development of the ideological reflexes and echoes of this life-process ... Morality, regligion, metaphysics, all the rest of ideology and their corresponding forms of consciousness, thus no longer retain the semblance of independence. They have no history, no development; but men, developing their material production and their material intercourse, alter, along with their real existence, their thinking and the products of their thinking. Life is not determined by consciousness, but consciousness by life.

Later on in the same work, the two founders of Marxism specify the recipe which, according to them, was followed by those who produced the idealistic mystification. First of all, ideas were separated from empirical context and the interests of the rulers who put them forward. Secondly, a set of logical connections was found linking successive ruling ideas, and their logic is then meant to explain the pattern of history. (This links the concept-saturation of history to the notion of historic *design*. Historic pattern is the reflection of the internal logical connection of successive ideas.) Thirdly, to diminish the mystical appearance of all this, the free-floating, self-transforming concept was once again credited to a person or group of persons.

If this kind of theory is false, what then is true? In the same work a little later, the authors tell us:

This sum of productive forces, forms of capital and social forms of intercourse, which every individual and generation finds in existence as something given, is the real basis of ... the ... 'essence of man'.... These conditions of life, which different generations find in existence, decide also

whether or not the periodically recurring revolutionary convulsion will be strong enough to overthrow the basis of all existing forms. And if these material elements of a complete revolution are not present ... then, as far as practical developments are concerned, it is absolutely immaterial whether the 'idea' of this revolution has been expressed a hundred times already ... (p. 30).

The passage seems unambiguous: what is retained is the idea of a plan, and also the idea of primarily internal, endogenous propulsion. What has changed is the identification of the propulsion, of the driving force of the transformation. Change continues to be the law of all things, and it is governed by a plan, it is not random; but the mechanism which controls it is now identified in a new manner.

From then on, the criticisms of the position can really be divided into two major species: some challenge the identification of the ruling mechanism, and others the idea of historic *plan*. As the most dramatic presentation of Marxist development, Robert Tucker's *Philosophy and Myth in Karl Marx* (1961, p. 123) puts it:

Marx founded Marxism in an outburst of Hegelizing. He considered himself to be engaged in ... [an] ... act of translation of the already discovered truth ... from the language of idealism into that of materialism.... Hegelianism itself was latently or esoterically an economic interpretation of history. It treated history as 'a history of production' ... in which spirit externalizes itself in thought-objects. But this was simply a mystified presentation of *man* externalizing himself in *material* objects.

This highlights both the origin *and* the validity or otherwise of the economic interpretation of history. Some obvious but important points can be made at this stage. The Hegel/Marx confrontation owes much of its drama and appeal to the extreme and unqualified manner in which the opposition is presented. This unqualified, unrestricted interpretation can certainly be found in the basic texts of Marxism. Whether it is the 'correct' interpretation is an inherently undecidable question: it simply depends on which texts one treats as final – those which affirm the position without restriction and without qualification, or those which contain modifications, qualifications and restrictions.

The same dilemma no doubt arises on the Hegelian side, where it is further accompanied by the question as to whether the motive force, the spirit of history, is to be seen as some kind of abstract principle (in which case the idea seems absurd to most of us), or whether this is merely to be treated as a way of referring to what we now term culture (in which case it is interesting and contentious).

One must point out that these two positions, the Hegelian and the Marxist, are contraries, but not contradictories. They cannot both be true, but they can perfectly well both be false. A world is easily conceivable where neither of them is true: a world in which social changes sometimes occur as a consequence of changes in economic activities, and sometimes as a consequence of strains and stresses in the culture. Not only is such a world conceivable, but it does really rather look as if that is the kind of world we do actually live in. (Part of the appeal of Marxism in its early days always hinged on presenting Hegel-type idealism and Marxism as two contradictories, and 'demonstrating' the validity of Marxism as a simple corollary of the manifest absurdity of strong versions of Hegelianism.) In this connection, it is worth noticing that by far the most influential (and not unsympathetic) sociological critic of Marx is Max Weber, who upholds precisely this kind of position. Strangely enough, despite explicit and categorical denials on his own part, he is often misrepresented as offering a return to some kind of idealism (without perhaps the mystical idea of the agency of abstract concepts which was present in Hegel). For instance, Michio Morishima, in *Why has Japan 'Succeeded'?* (1982, p. 1), observes: 'Whereas Karl Marx contended that ideology and ethics were no more than reflections ... Max Weber ... made the case for the existence of quite the reverse relationship.' Weber was sensitive to both kinds of constraint; he merely insisted that on occasion, a 'cultural' or 'religious' element might make a crucial difference.

Connected with this, there is another important theoretical difference to be found in Weber and many contemporary sociologists. The idea of the inherent historical plan, which had united Hegel and Marx, is abandoned. If the crucial moving power of history comes from one source only, though this does not strictly speaking entail that there should be a plan, an unfolding of design, it nevertheless does make it at least very plausible. If that crucial moving power had been *consciousness*, and its aim the arrival at self-consciousness, then it was natural to conclude that with the passage of time, there would indeed be more and more of such consciousness. So the historical plan could be seen as the manifestation of the striving of the Absolute Spirit or humanity, towards ever greater awareness. Alternatively, if the motive force was the growth of the forces of production, then, once again, it was not unreasonable to suppose that history might be a series of organizational adjustments to expanding productive powers, culminating in a full adjustment to the final great flowering of our productive capacity. (Something like that is the essence of the Marxist vision of history.)

If on the other hand the motive forces and the triggers come from a *number* of sources, which moreover are inherently diverse, there is no clear reason why history should have a pattern in the sense of coming ever closer to satisfying some single criterion (consciousness, productivity, congruence between productivity and social ethos, or whatever). So in the Weberian and more modern vision, the dramatic and unique developments of the modern industrial world are no longer seen as the inevitable fulfilment and culmination of a potential that had always been there, but rather as a development which only occurred because a certain set of factors happened to operate at a given time simultaneously, and which would otherwise not have occurred, and which was in no way *bound* to occur. Contingency replaces fatality.

So much for the central problem connected with the economic interpretation of history. The question concerning the relative importance of conceptual (cultural) and productive factors is the best known, most conspicuous and best advertised issue in this problem area. But in fact, it is very far from obvious that it is really the most important issue, the most critical testing ground for the economic theory of history. There is another problem, less immediately obvious, less well known, but probably of greater importance, theoretically and practically. That is the relative importance of productive and *coercive* activities.

The normal associations which are likely to be evoked by the phrase 'historical materialism' do indeed imply the downgrading of purely conceptual, intellectual and cultural elements as explanatory factors in history. But it does not naturally suggest the downgrading of force, violence, coercion. On the contrary, for most people the idea of coercion by threat or violence, or death and pain, seems just as 'realistic', just as 'materialistic' as the imperatives imposed by material need for sustenance and shelter. Normally one assumes that the difference between coercion by violence or the threat of violence, and coercion by fear of destitution, is simply that the former is more immediate and works more quickly. One might even argue that *all* coercion is ultimately coercion by violence: a man or a group in society which coerces other members by controlling the food supply, for instance, can only do it if they control and defend the store of food or some other vital necessity by force, even if that force is kept in reserve. Economic constraint, it could be argued (as Marxists themselves argue in other contexts), only operates because a certain set of rules is enforced by the state, which may well remain in the background. But economic constraint is in this way parasitic on the ultimate presence of enforcement, based on the monopoly of control of the tools of violence.

The logic of this argument may seem persuasive, but it is contradicted by a very central tenet of the Marxist variant of the economic theory of history. Violence, according to the theory, is not fundamental or primary, it does not initiate fundamental social change, nor is it a fundamental basis of any social order. This is the central contention of Marxism, and at this point, real Marxism diverges from what might be called the vulgar image possessed of it by non-specialists. Marxism stresses economic factors, and downgrades not merely the importance of conceptual, 'superstructural' ones, but equally, and very significantly, the role of coercive factors.

A place where this is vigorously expressed is Engels's 'Anti-Dühring' (1878):

> ... historically, private property by no means makes its appearance as the result of robbery or violence. ... Everywhere where private property developed, this took place as the result of altered relations of production and exchange, in the interests of increased production and in furtherance of intercourse – that is to say, as a result of economic causes. Force plays no part in this at all. Indeed, it is clear that the institution of private property must be already in existence before the robber can *appropriate* another person's property ... Nor can we use either force or property founded on force to explain the 'enslavement of man for menial labour' in its most modern form – wage labour.... The whole process is explained by purely economic causes; robbery, force, and the state of political interference of any kind are unnecessary at any point whatever (Burns, 1935, pp. 267–9).

Engels goes on to argue the same specifically in connection with the institution of slavery:

> Thus force, instead of controlling the economic order, was on the contrary pressed into the service of the economic order. *Slavery* was invented. It soon became the predominant form of production among all peoples who were developing beyond the primitive community, but in the end was also one of the chief causes of the decay of that system (ibid., p. 274).

Engels a little earlier in the same work was on slightly more favourable ground when he discussed the replacement of the nobility by the bourgeoisie as the most powerful estate in the land. If physical force were crucial, how should the peaceful merchants and producers have prevailed over the professional warriors? As Engels puts it: 'During the whole of this struggle, political forces were on the side of the nobility ...' (ibid., p. 270).

One can of course think of explanations for this paradox: the nobility might have slaughtered each other, or there might be an alliance between the monarchy and the middle class (Engels himself mentioned this possibility, but does not think it constitutes a real explanation) and so forth. In any case, valid or not, this particular victory of producers over warriors would seem to constitute a prima facie example of the non-dominance of force in history. The difficulty for the theory arises when the point is generalized to cover all social orders and all major transitions, which is precisely what Marxism does.

Engels tries to argue this point in connection with a social formation which one might normally consider to be the very paradigm of the domination by force: 'oriental despotism'. (In fact, it is for this very reason that some later Marxists have maintained that this social formation is incompatible with Marxist theory, and hence may not exist.) Engels does it, interestingly enough, by means of a kind of functionalist theory of society and government: the essential function, the essential role and duty, of despotic governments in hydraulic societies is to keep production going by looking after the irrigation system. As he puts it:

> However great the number of despotic governments which rose and fell in India and Persia, each was fully aware that its first duty was the general maintenance of irrigation throughout the valleys, without which no agriculture was possible (Burns, 1935, p. 273).

It is a curious argument. He cannot seriously maintain that these oriental despots were always motivated by a sense of duty towards the people they governed. What he must mean is something like this: unless they did their 'duty', the society in question could not survive, and they themselves, as its political parasites, would not survive either. So the real foundation of 'oriental despotism' was not the force of the despot, but the functional imperatives of despotically imposed irrigation systems. Economic need, as in the case of slavery, makes use of violence for its own ends, but violence itself initiates or maintains nothing. This interpretation is related to what Engels says a little further on. Those who use force can either aid economic development or accelerate it, or go against it, which they do rarely (though he admits that it occasionally occurs), and then they themselves usually go under: 'Where ... the internal public force of the country stands in opposition to economic development ... the contest has always ended with the downfall of the political power' (Burns, 1935, p. 277).

We have seen that Engels's materialism is curiously functional, indeed teleological: the economic potential of a society or of its productive base somehow seeks out available force, and enlists it on its own behalf. Coercion is and ought to be the slave of production, he might well have said. This teleological element is found again in what is perhaps the most famous and most concise formulation of Marxist theory, namely certain passages in Marx's preface to *A Contribution to 'The Critique of Political Economy'* (1859):

> A social system never perishes before all the productive forces have developed for which it is wide enough; and new, higher productive relationships never come into being before the material conditions for their existence have been brought to maturity within the womb of the old society itself. Therefore, mankind always sets itself only such problems as it can solve; for when we look closer we will always find that the problem itself only arises when the material conditions for its solution are already present, or at least in the process of coming into being. In broad outline, the Asiatic, the ancient, the feudal, and the modern bourgeois mode of production can be indicated as progressive epochs in the economic system of society (Burns, 1935, p. 372).

The claim that a new order does not come into being before

the conditions for it are available, is virtually a tautology: nothing comes into being unless the conditions for it exist. That is what 'conditions' mean. But the idea that a social system never perishes before it has used up all its potential is both strangely teleological and disputable. Why should it not be replaced even before it plays itself out to the full? Why should not some of its potential be wasted?

It is obvious from this passage that the purposive, upward surge of successive modes of production cannot be hindered by force, nor even aided by it. Engels, in 'Anti-Dühring', sneers at rulers such as Friedrich Wilhelm IV, or the then Tsar of Russia, who despite the power and size of their armies are unable to defy the economic logic of the situation. Engels also treats ironically Herr Dühring's fear of force as the 'absolute evil', the belief that the 'first act of force is the original sin', and so forth. In his view, on the contrary, force simply does not have the capacity to initiate evil. It does however have another 'role in history, a revolutionary role'; this role, in Marxist words, is midwifery:

> ... it is the midwife of every old society which is pregnant with the new,... the instrument by the aid of which social movement forces its way through and shatters the dead, fossilized, political forms ... (Burns, 1935, p. 278).

The midwifery simile is excellent and conveys the basic idea extremely well. A midwife cannot create babies, she can only aid and slightly speed up their birth, and once the infant is born the midwife cannot do much harm either. The most one can say for her capacity is that she may be necessary for a successful birth. Engels seems to have no fear that this sinister midwife might linger after the birth and refuse to go away. He makes this plain by his comment on the possibility of a 'violent collision' in Germany which 'would at least have the advantage of wiping out the servility which has permeated the national consciousness as a result of the humiliation of the Thirty Years War'.

There is perhaps an element of truth in the theory that coercion is and ought to be the slave of production. The element of truth is this: in pre-agrarian hunting and gathering societies, surrounded by a relative abundance of sustenance but lacking means of storing it, there is no persistent, social, economic motive for coercion, no *sustained* employment for a slave. By contrast, once wealth is systematically produced and stored, coercion and violence or the threat thereof acquire an inescapable function and become endemic. The surplus needs to be guarded, its socially 'legitimate' distribution enforced. There is some evidence to support the view that hunting and gathering societies were more peaceful than the agrarian societies which succeeded them.

One may put it like this: in societies devoid of a stored surplus, no surplus needs to be guarded and the principles governing its distribution do not need to be enforced. By contrast, societies endowed with a surplus face the problem of protecting it against internal and external aggression, and enforcing the principles of its distribution. Hence they are doomed to the deployment, overt or indirect, of violence or the threat thereof. But all of this, true though it is, does not mean that surplus-less societies are necessarily free of violence: it only means that they are not positively obliged to experience it. Still less does it mean that within the class of societies endowed with a surplus, violence on its own may not occasionally or frequently engender changes, or inhibit them. The argument does not preclude coercion either from initiating social change, or from thwarting change which would otherwise have occurred. The founding fathers of Marxism directed their invective at those who raised this possibility, but they never succeeded in establishing that this possibility is not genuine. All historic evidence would seem to suggest that this possibility does indeed often correspond to reality.

Why is the totally unsubstantiated and indeed incorrect doctrine of the social unimportance of violence so central to Marxism?

The essence of Marxism lies in the retention of the notion of an historical plan, but a re-specification of its driving force. But the idea of a purposive historical plan is not upheld merely out of an intellectual desire for an elegant conceptual unification of historical events. There is also a deeper motive. Marxism is a salvation religion, guaranteeing not indeed individual salvation, but the collective salvation of all mankind. Ironically, its conception of the blessed condition is profoundly bourgeois. Indeed, it constitutes the ultimate apotheosis of the bourgeois vision of life. The bourgeois preference for peaceful production over violent predation is elevated into the universal principle of historical change. The wish is father of the faith. The work ethic is transformed into the essence, the very species-definition of man. Work is our fulfilment, but work patterns are also the crucial determinants of historical change. Spontaneous, unconstrained work, creativity, is our purpose and our destiny. Work patterns also determine the course of history and engender patterns of coercion, and *not* vice versa. Domination and the mastery of techniques of the violence is neither a valid ideal, nor ever decisive in history. All this is no doubt gratifying to those imbued with the producer ethic and hostile to the ethic of domination and violence: but is it true?

Note that, were it true, Marxism is free to commend spontaneously cooperative production, devoid of ownership and without any agency of enforcement, as against production by competition, with centrally enforced ground rules. It is free to do it, without needing to consider the argument that only competition keeps away centralized coercion, and that the attempt to bring about propertyless and total cooperation only engenders a new form of centralized tyranny. *If* tyranny *only* emerges as a protector of basically pathological forms or organization of work, then a sound work-pattern will on its own free us for ever from the need for either authority or checks on authority. Man is held to be alienated from his true essence as long as he works for extraneous ends: he finds his true being only when he indulges in work for the sake of creativity, and choses his own form of creativity. This is of course precisely the way in which the middle class likes to see its own life. It takes pride in productive activity, and chooses its own form of creativity, and it understands what it does. Work is not an unintelligible extraneous imposition for it, but the deepest fulfilment.

On the Marxist economic interpretation of history, mankind as a whole is being propelled towards this very goal, this bourgeois-style fulfilment in work without coercion. But the guarantee that this fulfilment will be reached is only possible if the driving force of history is such as to ensure this happy outcome. If a whole multitude of factors, economic, cultural, coercive, could all interact unpredictably, there could hardly be any historic plan. But if on the other hand only one factor is fundamental, and that factor is something which has a kind of vectorial quality, something which increases over time and inevitably points in one direction only (namely the augmentation of the productive force of man), then the necessary historical plan does after all have a firm, unprecarious base. This is what the theory requires, and this is what is indeed asserted.

The general problem of the requirement, ultimately, of a *single-factor theory*, with its well-directed and persistent factor,

is of course related to the problems which arise from the plan that Marxists discern in history. According to the above quotation from Marx, subsequent to primitive communism, four class-endowed stages arise, namely the Asiatic, the ancient, the feudal, and the modern bourgeois, which is said to be the last 'antagonistic' stage (peaceful fulfilment follows thereafter). Marxism has notoriously had trouble with the 'Asiatic' stage because, notwithstanding what Engels claimed, it *does* seem to exemplify and highlight the autonomy of coercion in history, and the suspension of progress by a stagnant, self-maintaining social system.

But leaving that aside, in order to be loyal to its basic underlying intuition of a guaranteed progression and a final happy outcome, Marxism is not committed to any particular number or even any particular sequence of stages. The factual difficulties which Marxist historiography has had in finding all the stages and all the historical sequences, and in the right order, are not by themselves necessarily disastrous. A rigid unilinealism is not absolutely essential to the system. What it *does* require (apart from the exclusiveness, in the last analysis, of that single driving force) is the denial of the possibility of stagnation, whether in the form of absolute stagnation and immobility, or in the form of circular, repetitive developments. If this possibility is to be excluded, a number of things need to be true: all exploitative social forms must be inherently unstable; the number of such forms must be finite; and circular social developments must not be possible. *If* all this is so, then the alienation of man from his true essence – free fulfilment in unconstrained work – *must* eventually be attained. But if the system can get stuck, or move in circles, the promise of salvation goes by the board. This would be so even if the system came to be stuck for purely economic reasons. It would be doubly disastrous for it if other factors, such as coercion, were capable of freezing it. The denial of any autonomous role for violence in history is the most important, and most contentious, element in the Marxian economic theory of history.

So what the Marxist economic interpretation of history really requires is that no non-economic factor can ever freeze the development of society, that the development of society itself be pushed forward by the continuous (even if on occasion slow) growth of productive forces, that the social forms accompanying various stages of the development of productive forces should be finite in number, and that the last one be wholly compatible with the fullest possible development of productive forces and of human potentialities.

The profound irony is that a social system marked by the prominence and pervasiveness of centralized coercion, should be justified and brought about by a system of ideas which denies autonomous historical agency both to coercion and to ideas. The independent effectiveness both of coercion and of ideas can best be shown by considering a society built on a *theory*, and one which denies the effectiveness of either.

ERNEST GELLNER

BIBLIOGRAPHY

Burns, E. 1935. *A Handbook of Marxism*. London: Victor Gollancz.

Engels, F. 1878. 'Anti-Dühring'. In Burns (1935).

Marx, K. 1859. A Contribution to 'The Critique of Political Economy'. In Burns (1935).

Marx, K. and Engels, F. 1845–6. *The German Ideology*. London: Lawrence & Wishart, 1940.

Morishima, M. 1982. *Why has Japan 'Succeeded'? Western Technology and the Japanese Ethos*. Cambridge: Cambridge University Press.

Tucker, R.C. 1961. *Philosophy and Myth in Karl Marx*. Cambridge: Cambridge University Press.

economic laws. The social sciences, and economics in particular, separated from moral and political philosophy in the second half of the 18th century when the results of the myriad of intentional actions of people were perceived to produce regularities resembling the laws of a system. Both physiocratic thought and Smith's *Wealth of Nations* reflect this extraordinary discovery: scientific laws thought to be found only in nature could also be found in society. This extension poses several problems. A serious one refers to the tension of combining individuals' freedom of action with the scientists' desire to discover the systematic aspects of the unintended and quite often unpredictable consequences of human action. i.e. the desire to arrive at laws characterized by a certain degree of generality and permanence.

In the history of economic thought this fundamental tension has been solved in different ways. In the 18th century, the mechanistic ideal of the natural sciences, combined with the natural law idea of a harmonious order of nature, determined the way social phenomena were treated. There was a desire to discover the 'natural laws' of economic life and to formulate the natural precepts which rule human conduct. The classical economists upheld the notion that natural laws are embedded in the economic process as beneficial laws, along with the belief in the existence of rules of nature capable of being discovered. Thus the belief that things could follow the beneficial 'natural course' only in a rationally organized society which it was a duty to create according to the precepts of nature. The economic system is the mechanism by which the individual is driven to fostering the prosperity of society while pursuing his private interest. Hence the automatic operation of the economic system may be combined with freedom of individual action. This is the core of the doctrine of economic harmony. Besides being causal laws of a mechanical type, the laws of nature are providentially imposed norms of conduct. In such a setting it would have been pointless to separate means and ends, since the implementation of natural laws is both an end and a means, and even more pointless to think of a tension between 'explaining' and 'understanding' economic behaviour. Causal and teleological, positive and normative, theoretical and practical started being seen as separate categories only when the economic discourse freed itself from the philosophy of natural law and all its implications.

Post-classical economics set out to be a science of the laws regulating the economic order and of the conditions allowing these laws to operate. It became the basis of a theory that, in Jevons's own terms, proposed to construct a 'social physics'. The view of a social world ordered according to transcendent ends was abandoned in favour of an ideal of objective knowledge of economic phenomena gained through a 'positive' study of the laws that regulate market activities. In so doing, neoclassical 'positive' economics solves the aforementioned tension by extrapolating the theoretical model of natural sciences to economics: economics is to produce the laws of motion similar to those of physics, chemistry, astronomy.

But what is a scientific law and which role do laws play within the logical positivist's perspective adopted by neoclassical economics? Laws provide the foundation of a deductive scientific method of inquiry. According to the deductive–nomological conception of explanation, due to C. Hempel, laws are universal statements not requiring reference to any one particular object or spatio–temporal location. To be valid, laws are constrained neither to finite populations nor to particular times and places; they are, in effect, expressions of natural stationarities. This interpretation of the notion of law provides the so-called covering-law model

of explanation with an unquestionably firm inferential foundation. Deductive logic is employed to ensure the truth status of propositions and since the deductions are (by hypothesis) predicated on true universal statements (laws), the empirical validity of these statements may be ascertained. However, what sort of constraints on economic discourse are imposed by this positivistic structure? On the one hand this structure constitutes its object; on the other hand it generates specific economic questions together with their method of solution. Following the model of natural sciences and its success in controlling a natural world made up of objects and unvarying relations among them expressed in the form of laws, the neoclassical approach arrives at a study of regularities conceived of as specifying the nature of its objects.

To capture the different interpretations of the notion of law by classical and neoclassical economists let us refer to one of the most famous of economic laws: the law of diminishing returns, also known as the law of variable proportions. Studying agricultural production, Ricardo had noted that different quantities of labour, assisted by certain quantities of other inputs (farm tools, fertilizers, etc.), could be employed on a given piece of land, i.e. it was possible to vary the proportions in which land and complex labour (labour assisted by other inputs) are employed. He accordingly arrived at the law which states that production increases resulting from equal increments in the employment of complex labour, while the quantity of land farmed remains constant, will initially be increasing and then decreasing. (To be sure, the first statement of the law is due to the physiocratic economist Turgot.)

Three points deserve attention. First, Ricardo and classical authors in general offer no formal demonstration of this law. To them, it is basically an empirical law, on which no functional association between output and variable inputs can be built. Second, the classics' use of the law refers to their theories of distribution and development: as the supply of land in the whole system is fixed, sooner or later a point will be reached at which economic growth will come to halt, notwithstanding any countervailing effects due to technical progress. Finally, the law presupposes a comparative statics framework: the pattern of the marginal products of complex labour refers to different observable equilibrium positions and not to hypothetical or virtual variations.

With the advent of the marginalist revolution, two subtle changes in the interpretation of the law took place. (a) The *de facto* elimination of the distinction between the extensive case (the case of the simultaneous cultivation of pieces of land of different fertility) and the intensive case (the application of successive doses of capital and labour to the same piece of land) with an over-evaluation of the latter. Classical economists, being interested in the explanation of rent, concentrated on the extensive case; they took also the intensive case into consideration but with many qualifications. Indeed, whereas the various levels of productivity of different qualities of land is a circumstance which may be directly observed in a given situation, the marginal productivity of a given input is related to a virtual increment in output and therefore to a virtual change in the situation. (b) The change in the method of analysis – it was preferred to reason in terms of hypothetical rather than observable changes – brought about by the shift of interest towards the intensive margin, supported the thesis of the symmetrical nature of land and other inputs. This in turn favoured the extension of the substitutability between land and complex labour from agricultural production to all kinds of production, including those in which land does not figure as a direct input. It so happened that whereas in classical economics the substitutability between land and complex labour presupposes that simple labour and equipment are strictly complementary, in neoclassical economics this substitutability is applied to all inputs indiscriminately.

However, the neoclassical interpretation of the law poses serious problems. In the first place, there is the problem of justifying, on empirical grounds, the general applicability of the substitution principle. Secondly, and more importantly, in order to allow the substitution of inputs to take place, a certain lapse of time is required during which the required modifications to the productive structure can be made. (It is certainly true that coal can replace oil to provide heating, but before this can happen it will be necessary to change the heating system.) The well-known distinction between the short-run and the long-run is a partial and indirect way to take the temporal element into consideration. In the short-run the plant is fixed by definition. It is therefore the fixed input which, in the neoclassical interpretation of the law, plays the same role as land in the classical interpretation. Now, neoclassical theory correctly states the law of diminishing returns with respect to the short-run; however it is in the long-run that the substitutability of inputs becomes actually feasible. One is therefore confronted with a dilemma: the neoclassical interpretation of the law seems to be more plausible in a long-run framework when there exists the necessary time to accommodate input adjustments; on the other hand, fixed inputs cannot, by definition, exist in the long-run so that the law of variable proportions cannot be stated in such a context.

This dilemma is the price neoclassical theory has to pay for its interpretation of the law in accordance with the positivistic statute. Indeed, the power of deductive, truth-preserving rules of scientific inference is not purchased without a cost. A school of economic thought which is not prepared to sustain such a cost is the neo-Austrian. The neo-Austrian economists solve what has been called the fundamental tension by arguing economics cannot and should not provide general laws since, by its very nature, it is an idiographic and not a nomothetical discipline. The general target of economics is 'understanding' grounded in *Verstehen* doctrine: by introspection and empathy, the study of the economic process should aim at explaining individual occurrences, not abstract classes of phenomena. It follows that if by a scientific law one should mean a universal conditional statement of type 'for all *x*, if *x* is *A*, then *x* is *B*', statements regarding unique events cannot by definition express any regularity for the simple reason that any regularity presupposes the recurrence of what is defined as regular. In the words of L. von Mises, who shares with F. von Hayek the paternity of the neo-Austrian school, what assigns economics its peculiar and unique position in the orbit of pure knowledge '... is the fact that its particular theorems are not open to any verification or falsification on the ground of experience ... the ultimate yardstick of an economic theorem's correctness or incorrectness is solely reason unaided by experience' (von Mises, 1949, p. 858).

There is indeed a place for economic 'laws' in the framework of Austrian economics. The familiar 'laws' of economics (diminishing marginal utility, supply and demand, diminishing returns to factors, Say's law and so on) are seen as 'necessary truths' which explain the essential structure of the economic world but with no predictive worth. In other words, economic laws are not generalizations from experience, as it is the case within the positivistic paradigm, but are theorems which enable us to understand the economic world. It is ironic that Mises' position of radical apriorism joined to Hayek's attack on scientism and methodological monism are completely at variance with the position taken by the father of the Austrian

school, Carl Menger (1883), who announced that in economic theories exact laws are defined which are just as rigorous as in fact are the laws of nature.

Between the extreme positions of neoclassical positive economic and neo-Austrian economics are those who, without denying that economics is in search for laws in the same sense in which natural sciences are and that laws perform an explanatory as well as a predictive function, underline that the explicative structure of economics, albeit nomothetical, substantially differs from that of natural sciences. This intermediate position can be traced back to Keynes's methodology which considers the conditions of truth and universality of the positivistic conception of scientific laws as far too rigid for a discipline such as economics. Two main reasons account for the different epistemological status of laws in natural sciences and in economics. First, the knowledge of economic phenomena is itself an economic variable, i.e. it changes, along with the process of its own acquisition, the economic situation to which it refers. The formulation of a new physical law does not change the course of physical processes; it does not influence the truth or falsity of the prognosis. This is not the case in economics where the prognosis, say, that in two years time there will be a boom can cause overproduction and a resulting recession. In turn, this specific aspect is strictly connected to the fact that the object of study of economics posseses an historical dimension. Economics is in time in a way that natural sciences are not. The ensuing mutability of observed regularities is well expressed by Keynes when he writes, 'As against Robbins, economics is essentially a moral science and not a natural science. That is to say it employs introspection and judgements of value' (1973, p. 297) to which he adds, 'It deals with motives, expectations, psychological uncertainties. One has to be constantly on guard against treating the material as constant and homogenous' (p. 300).

Second, the role played by *ceteris paribus* clauses in natural sciences and in economics is substantially different. The modern economists appeal to the 'other things being equal' clause – which according to Marshall is invariably attached to any economic law – in all those cases where the classical economists were talking of 'disturbing causes'. J.S. Mill's discussion of inexact sciences is suggestive here:

> When the principles of Political Economy are to be applied to a particular case then it is necessary to take into account all the individual circumstances of that case ... These circumstances have been called *disturbing causes*. This constitutes the only uncertainty of Political Economy (1836, p. 300).

Also in natural sciences we find *ceteris paribus* clauses. Indeed, a scientific theory that could dispense with them would in effect achieve perfect closure, which is a rarity. So where lies the difference? The example of the science of tides used by Mill is revealing. Physicists know the laws of the greater causes (the gravitational pull of moon) but do not know the laws of the minor causes (the configuration of the sea bottom). The 'other things' which scientists hold equal are the lesser causes. So could we conclude, that just about all generalizations in both natural sciences and economics express in fact *tendency laws*, in the sense that these 'laws' truly capture only the functioning of 'greater causes' within some domain? Certainly not, since there is a world of difference between the two cases. Galileo's law of falling bodies certainly presupposes a *ceteris paribus* clause, so much so that he had to employ the idealization of a 'perfect vacuum' to get rid of the resistance of air. However, he was able to give estimates of the magnitudes of the amount of distortion that friction and the other 'accidents' would determine and which the law ignored. In other words, whereas in natural sciences the 'disturbing causes' have their own laws, this is not the case in economics where we find tendency statements with unspecified *ceteris paribus* clauses or, if specified, specified only in qualitative terms. In economics it is generally impossible to list all the conceivable inferences implied in a lawlike statement and to replace the *ceteris paribus* clause with precise conditions. So, for example, the law that 'less will be bought at a higher price' is not refuted by panic buying, nor is it confirmed by organized consumer boycotts. No test is decisive unless *ceteris* are really *paribus*.

These remarks help to understand the role acknowledged by Keynes to laws in economic inquiry. Besides general laws, there are also rules and norms which are significant in the explanation of economic behaviour. To Keynes, it makes no sense to reduce all forms of explanation in economics to that of the covering-law model. Indeed, whereas to justify a law one has to show that it is logically derivable from some other more general statements, often called principles or postulates, the justification of rules occurs through the reference to goals and the justification of norms through the reference to values which are not general sentences, but rather intended singular patterns or even ideal entities. Since no scientific law, in the natural scientific sense, has been established in economics, on which economists can base predictions, what are used and have to be used to explain or to predict are tendencies or patterns expressed in empirical or historical generalizations of less than universal validity, restricted by local and temporal limits. Recently, Arrow has amazed orthodox economists when raising doubts about the mechanistically inspired understanding of economic processes: 'Is economics a subject like physics, true for all time or are its laws historically conditioned?' (Arrow, 1985, p. 322).

The list of generally accepted economic laws seems to be shrinking. The term itself has come to acquire a somewhat old-fashioned ring and economists now prefer to present their most cherished general statements as theorems or propositions rather than laws. This is no doubt a healthy reaction: for too long economists have been under the nomological prejudice, of positivistic origin, that the only route towards explanation and prediction is the one paved with laws, and laws as forceful as Newton's laws. Images in science are never innocent: wrong images can have disastrous effects.

STEFANO ZAMAGNI

BIBLIOGRAPHY

Arrow, K. 1985. Economic history: a necessary though not sufficient condition for an economist. *American Economic Review, Papers and Proceedings* 75(2), May, 320–23.

Keynes, J.M. 1973. *The General Theory and After. Part II: Defence and Development*. In *The Collected Writings of John Maynard Keynes*, Vol. XIV, London: Macmillan.

Menger, C. 1883. *Unterschungen über die Methode der Sozialwissenschaften*. Leipzig: Duncker & Humblot.

Mill, J.S. 1836. On the definition of political economy and the method of investigation proper to it. Reprinted in *Collected Works of John Stuart Mill, Essays on Economy and Society*, ed. J.M. Robson, Toronto: University of Toronto Press, 1967, Vol. 4.

Mises, L. von. 1949. *Human Action. A Treatise on Economics*. London: William Hodge.

economic man. Among the many different portrayals of economic agents, the title of *homo economicus* is usually reserved for those who are rational in an instrumental sense. Neoclassical economics provides a ready example. In its ideal-type case the agent has complete, fully ordered preferences (defined over the domain of the consequences of his feasible actions), perfect information and immaculate computing power. After deliberation he chooses the action which satisfies his preferences better (or at least no worse) than any other. Here rationality is a means-to-ends notion, with no questions raised about the source or worth of preferences. The rational economic man is a bargain-hunter, who never pays more than he needs or gets less than he could at the price.

This basic model is then made more sophisticated. The theory of risk allows for the point that an action may have several possible consequences. The agent assesses its expected utility by discounting the utility of each consequence by how likely it is to be the actual one. That requires him to have a probability distribution for the consequences, even if only a subjective one. Other refinements include allowance for costs of information, of its processing and of action. Then there are complexities, perhaps best illustrated by the theory of games, when other agents are introduced into the story. The basic vision remains, however, one of agents who are rational in the sense that they maximize an objective function subject to constraints.

This vision is not unique to neoclassical economics. For example, Marx's profit-maximizing capitalist fits the same instrumental model of rationality. Institutionalist accounts of, for instance, banks or trade unions often conceive economic bodies as unitary rational agents similarly. Nor is the vision confined to any specific motivating desire in agents, like a selfish pleasure-maximizing drive. There is scope for allowing ethical preferences alongside the symptomatic textbook desires for apples and oranges. Agents are, however, regarded as self-interested, in the looser sense that they are moved to satisfy whatever preferences they happen to have. Furthermore, granted that *de gustibus non disputandum*, this modest base is enough to ground a full-blown social theory on a model of agency which can be exported to other social sciences.

Such a social theory is individualist and contractarian, with a pedigree which includes Hobbes's *Leviathan* and Benthamite utilitarianism. The satisfaction of individual preference, aided by felicific calculation, is what makes the social world go round. Social relations become instrumental, in the sense that they embody exchanges in the service of individual preferences. For instance, marriage has been analysed in this spirit as an arrangement to secure the mutual benefit of exchange between two agents with different endowments. Crime has been claimed to occur because calculation of costs and benefits proves it to be the action which maximizes expected utility. Meanwhile, institutions, which feature in elementary microeconomics as constraints on individual choice, become deposits left by earlier transactions, often deliberately deposited as devices to prevent preferences being frustrated by situations of the prisoner's dilemma type. Government policies are explained on the hypothesis that the political arena is also peopled by individuals maximizing expected utility, who form coalitions to market policies which will secure re-election. In this sort of way *homo economicus* turns into a universal *homo sapiens*.

Such a full-blown social theory may be too ambitious, because assumptions which are plausible for simple market transactions become suspect when scaled up. For example, the ideal-type case makes agents, so to speak, transparent to themselves, and does not allow for history occurring behind men's backs. Freudians would object to transparency of preferences and Marxians would invoke theories of false consciousness. (Although Marx's capitalists are instrumentally rational, their desire to maximize profit is an alienated one, 'forced' on them by a competitive capitalist system.) Many other social theorists would object to the treatment of norms and social relations as instrumental, on the grounds that norms are prior to preferences. For instance, cultural forms like the rules of orchestral composition are a source of musical preferences rather than a solution to *a priori* problems of maximizing musical enjoyment. In general it can be argued that ambition overreaches itself at the point where it reduces the parameters, needed for analysing economic choice as instrumentally rational, to outcomes of rational choice.

Meanwhile, however, such objections need not affect the more modest enterprise of explaining economic transactions within the parameters of a market. But even here *homo economicus* has his critics. Philosophically, it is not plain that preferences can be taken as given in a sense which makes them impervious to the agent's beliefs about the moral quality of his actions. In supposing that only desires can motivate agents, the economist is taking sides in a continuing philosophical dispute between Humeans, who regard reason as the slave of the passions, and Kantians, who make place for the rational monitoring of desire. This dispute surfaces plainly in welfare economics, when it is asked whether all preferences should count equally, but bears on the elementary model of action too. There are also methodological doubts about the empirical standing of the model. What would falsify the claim that economic agents seek the most effective means to satisfy their preferences? Apparent counter-examples can always be dealt with by treating them as evidence that preferences have changed. Indeed, since preferences are unobservable, they can be identified only if the correctness of the model is presupposed. In other words, there is room for deeper dispute about the foundations of orthodox microeconomics than is always realized.

Even within economics there are critics. The most substantial attack comes from those who think that perfect information is not a useful limiting case of imperfect information. Granted that there is often no way of calculating the likely marginal costs and benefits of acquiring extra information (short of actually acquiring it), how shall the agent decide rationally when to stop? Simon (1976) uses the question to argue for 'satisficing' models, in place of maximizing ones, and for 'procedural' rationality. Rationality, he suggests, is a matter of following a procedure which halts with a good solution, and should not be defined in terms of best solutions. Instead of grounding explanations in ideal-type cases, economists should direct their attention to the procedures which businesses and consumers in fact follow and, hence, regard *homo economicus* as more of an Organization Man than an abstract maximizer.

On the other hand, *homo economicus* has proved to be a mathematically tractable assumption of a fertile kind. To portray action as optimizing a reasonably well-behaved objective function is to make possible the analytic insights got by use of calculus, set theory and other powerful mathematical tools. By comparison, satisficing models have fared less well by having to rely on simulation techniques. It is computationally cumbersome to explore the equivalent fo the comparative static properties of these models, and the generality of the results obtained is open to suspicion.

A different approach to the information issue comes from the Rational Expectations School. They argue that a rational agent who is short of information should not use an information-generating mechanism which gives rise to systematic errors. If errors are systematic, the agent should be able to learn how to eliminate them by amending the mechanism. There is an incentive to do so, because improved estimates of future variables will be profitable. On the face of it this makes rational expectations the natural ally of economic-man models. Economic Man can proceed much as before, in the assurance that inadequate information involves nothing more systematic than 'white noise' and with the benefit of fresh analytic results which flow from a rational expectations hypothesis.

But this is to sidestep the informational problem set earlier, unless one sees how rational agents will learn to remove systematic errors. In some simple learning situations a Bayesian updating procedure turns a rational expectations generating process into an approximation of adaptive expectations, which could be construed as a procedural rule of thumb. But no general rapprochement between maximizing and procedural models of rationality follows. In more general learning situations the rational agent is trying to learn the rational expectations equilibrium relationship between variables – the one which, if used by agents to form their expectations, would reproduce itself in experience (white noise apart). This sounds easy, in that repeated experience of a particular relationship should lead to convergence on accurate parameter estimates. However, ignorance of the rational expectations equilibrium values produces behaviour which departs from those values. So observed values of variables embody a distortion which agents cannot correct without knowing the dimensions of their own ignorance. To know this, however, they would have to know the rational expectations equilibrium values already. To put it as the procedural critics might, learning would be feasible only if there were nothing to learn. The information question has been begged.

Yet Economic Man remains a powerful model of action not only in neoclassical theories, where insights in comparative statics have been especially notable, but elsewhere too. How powerful it finally is depends, within economics, on what becomes of the informational difficulties and on whether a procedural model can come up with rival results of equal scope and elegance. On the export front it offers a tempting analysis of social behaviour at large both for transactions in other social arenas and for the emergence of the institutions which govern those arenas. But the greater its ambitions, the more serious become the unresolved doubts about the origin of preferences and their relation to norms and institutions.

SHAUN HARGREAVES-HEAP AND MARTIN HOLLIS

See also ALTRUISM; ECONOMIC THEORY AND THE HYPOTHESIS OF RATIONALITY; HEDONISM; RATIONAL BEHAVIOUR; SELF-INTEREST; UTILITARIANISM.

BIBLIOGRAPHY

Simon, H.A. 1976. From substantive to procedural rationality. In *Method and Appraisal in Economics*, ed. S. Latsis, Cambridge: Cambridge University Press.

economic organization and transaction costs. One important extension of the Coase Theorem states that, if all costs of transactions are zero, the use of resources will be similar no matter how production and exchange activities are arranged. This implies that in the absence of transaction costs, alternative institutional or organizational arrangements would provide no basis for choice and hence could not be interpreted by economic theory. Not only would economic organization be randomly determined; there actually would not be any organization to speak of: production and exchange activities would simply be guided by the invisible hand of the market.

But organizations or various institutional arrangements do exist, and to interpret both their presence and their variation, they must be treated as the results of choice subject to the constraints of transaction costs.

In the broadest sense transaction costs encompass all those costs that cannot be conceived to exist in a Robinson Crusoe economy where neither property rights, nor transactions, nor any kind of economic organization can be found. This breadth of definition is necessary because it is often impossible to separate the different types of cost. So defined, transaction costs may then be viewed as a spectrum of institutional costs including those of information, of negotiation, of drawing up and enforcing contracts, of delineating and policing property rights, of monitoring performance, and of changing institutional arrangements. In short, they comprise all those costs not directly incurred in the physical process of production. Apparently these costs are weighty indeed, and to term them 'transaction costs' may be misleading because they may loom large even in an economy where market transactions are suppressed, as in a communist state.

By definition, an organization requires someone to organize it. In the broadest sense, all production and exchange activities not guided by the invisible hand of the market are organized activities. Thus, any arrangement that requires the use of a manager, a director, a superviser, a clerk, an enforcer, a lawyer, a judge, an agent, or even a middleman implies the presence of an organization. These professions would not exist in the Crusoe economy, and payments for their employment are transaction costs.

When transaction costs are defined to include all costs not found in a Crusoe economy, and economic organizations are defined equally broadly to include any arrangement requiring the service of a visible hand, a corollary appears: all organization costs are transaction costs, and vice versa. That is why during the past two decades economists have striven to interpret the various forms of organizational arrangements in terms of the varying costs of transactions.

Some obvious examples will illustrate the point. A worker in a factory (an organization) may be paid by a piece rate or by a wage rate. If the costs of measuring and enforcing performance (one type of transaction cost) are zero, then either arrangement will yield the same result. But if these costs are positive, the piece-rate contract will more likely prevail if the costs of measuring outputs are relatively low, whereas the wage contract will more likely be chosen if the costs of measuring hours and enforcing performance are low relative to the costs of measuring outputs. As another example, some restaurants (again an organization) measure the quantity of food sold; others serve buffet dinners, allowing customers to eat as much as they please at a fixed price per head. The cost of metering and quantifying food consumption relative to the basic cost of the food will determine which arrangement is chosen. In the total absence of transaction costs, the factory or the restaurant would not exist in the first place, because consumers would buy directly from the input owners who produce the goods and services.

As early as 1937, R.H. Coase interpreted the emergence of the firm (an organization) in light of the costs of determining market prices (transaction costs). When these costs are substantial because of the difficulties of measuring separate contributions by workers and of negotiating prices for separate components of a product, a worker may choose to work in a factory (a firm); he surrenders the right to use his labour by contract and voluntarily submits to direction by a visible hand, instead of personally selling his services or contributions to customers through the invisible hand of the market. The firm is therefore said to supersede the market. As the supersession progresses, the saving in the costs of determining prices will be countered by the rising costs of supervision and of management in the firm. Equilibrium is reached when, at the margin, the cost saving in the former equals the rising cost in the latter.

The firm superseding the market may be regarded as a factor market superseding a product market. If all costs of transactions were zero, the two markets would be inseparable in that a payment made by a customer to the owner of a factor of production would be the same as payment made to a product seller. In such a world it would be a fallacy to speak of the factor market and the product market as coexisting entities.

The presence of transaction costs is a prelude to separate the factor market from the product market. However, in some arrangements, such as the use of certain piece rates, it may become impossible to separate the one market from the other. Therefore, instead of viewing the firm as superseding the market, or the factor market as superseding the product market, it is more correct to view the organizational choice as one type of contract superseding another type. In these terms, the choice of organizational arrangements is actually the choice of contractual arrangements.

When organizational choices are viewed as contractual choices, it becomes evident that it is often impossible to draw a clear dividing line separating one organization from another. Take the firm, for example. It is often the case that the entrepreneur who holds employment contracts (and it is not clear whether it is the entrepreneur who employs the workers or the workers who employ the entrepreneur) may contract with other firms; a contractor may subcontract; a subcontractor may sub-subcontract further; and a worker may contract with a number of 'employers' or 'firms'. If the chain of contracts were allowed to spread, the 'firm' might encompass the whole economy. With this approach the size of the firm becomes indeterminate and unimportant. What are important are the choice of contracts and the costs of transactions that determine this choice.

Traditional economic analysis has been confined to resource allocation and income distribution. Contractual arrangements as a class of observations have been slighted in that tradition. In a world complicated by transaction costs, this neglect not only leaves numerous interesting observations unexplained, but actually obscures the understanding of resource allocation and income distribution. The economics of organization or institution or, for that matter, the workings of various economic systems, were never placed in the proper perspectives under the traditional approach. For generations students were told that various kinds of 'imperfections' were the cause of seemingly mysterious observations: policies were 'misguided', or antitrust specialists were barking up the wrong trees.

The costs of introducing new and more valid ideas must have been enormous. Even today textbooks still discuss marginal productivity theory only with reference to fixed wage and rental payments. Yet economists have known all along that (for labour alone) payments may be in the peripheral forms of piece rates, bonuses, tips, commissions, or various sharing arrangements; moreover, even wage rates may assume a number of forms. Each type of contract implies different costs of supervision, of measurement, and of negotiation, and the form of economic organization, along with the function of the visible hand, changes whenever a different contractual arrangement is chosen.

The choice of contractual arrangements is not, of course, confined to the factor markets. In the product markets, pricing arrangements such as tie-in sales, full-line forcing, or membership fees associated with clubs, may similarly be interpreted in light of transaction costs. Further, business organizations in mergers, franchises, and various forms of integration are now beginning to be viewed as transaction-cost phenomena. Indeed, close inspection of department stores and shopping centres reveals pricing and contractual arrangements between a central agent and individual sellers, as well as among the sellers themselves, which could not be explained by textbook economics.

Transaction costs are often difficult to measure and, as noted earlier, difficult to separate by type. However, the measurement problem can be avoided if only we are able to specify how these costs vary under different observable circumstances, and their different types are separable if viewed in terms of changes at the margin. These two conditions are requisite in the derivation of testable implications for the interpretation of organizational behaviour.

The use of transaction costs to analyse institutional (organizational) choice is superior to three other approaches. One approach would focus on incentives. However, incentives are not in principle observable, and we will do better in deriving testable propositions if the same problem is viewed in terms of the costs of enforcing performance. A second approach adopts risk. However, it is difficult to ascertain how risk is altered under different circumstances. Many risk problems, such as the uncertainty of whether an agreement will be honoured, are also problems of transaction costs, and it is easier to deal directly with those. Finally, some recent advances in transaction-cost analysis have called attention to the costs embodied in dishonesty, cheating, shirking, and opportunistic behaviour. Yet these are loose terms and, whatever they describe, to some extent are always to be found. To the degree that we can identify the particular costs of transactions that promote dishonesty, that shadowy explanation is no longer needed. After all, in what sense can we say a person is 'increasingly dishonest' or 'increasingly opportunistic'?

The transaction-cost approach to analysis of economic organizations can be extended upward from a few participants to the 'government' or even the nation itself. At the lower level, the owners of condominium units almost as a rule form associations with specific by-laws and elect committees to act on matters of common concern, the decisions being determined by majority vote. The transaction costs of ballot voting are less than those of using prices and dollar votes in certain circumstances, and trivial matters may even be delegated to a 'dictatorial' manager to further reduce the cost of voting. Similarly, residents in a particular location may choose to incorporate into a city, selecting their own mayor, with a committee setting up the building codes, hiring firemen and policemen, and deciding other matters of common concern.

Private property rights offer the unique advantage of allowing individual property owners the option of *not* joining an organization. This choice is an effective restraint against the adoption of an organization with higher transaction costs. It is true that a home-owner in a given region may, by majority vote, lose his option of not joining in a city corporation (unlike a worker who, in a free enterprise economy, always has the option of not joining a 'firm'). But with private property rights the majority vote aims at cost saving, and a reluctant resident may exercise his own judgement by selling his house and moving elsewhere.

Private property rights further reduce transaction costs under competition. An entrepreneur or agent who wants to recruit other resource owners to join his organization must, under competition, offer attractive terms, and this can be achieved only if his organization can effectively reduce transaction costs. On the other hand, the resource owner competing to join an organization will be more inclined to deliver a good performance when at risk of losing his job.

The option of not joining an organization and the cost-reducing function of competition are, of course, restrained when an organization is extended to encompass an entire nation. When citizenship is dictated by birth, the option of not joining is restrained, and competition among nations to recruit members is decidedly less than among organizations within a nation. This relative lack of cost-reducing mechanisms is all the more evident in a communist state, where a citizen does not have the option of choosing an organization within that state.

A communist state may be regarded as a 'superfirm' in which comrades lack the option of not joining. Each worker is assigned to a particular job supervised and directed by the visible hands of comrade officials of varying ranks. In this aspect the communist state is remarkably similar to what Coase calls a 'firm', where workers are told what to do instead of being directed by market prices. But the lack of market prices in the communist state is not due to the costs of determining prices; rather, in the absence of private property rights market prices simply do not exist, and visible supervision by a hierarchy ranking becomes the remaining alternative to chaos.

The transaction costs of operating an organization are necessarily higher in a communist state than in a free enterprise economy, due to the lack of option of not joining and the lack of competition both to recruit members among organizations and to induce members to perform well.

If the transaction costs of operating organization were zero, resource allocation and income distribution would be the same in a communist state as in a free enterprise state: consumer preferences would be revealed without cost; auctioneers and monitors would provide freely all the services of gathering and collating information; workers and other factors of production would be directed free of cost to produce in perfect accord with consumer preference; each consumer would receive goods and services in conformity with his preferences; and the total income received by each worker, as determined costlessly by an arbitrator, would equal his marginal productivity plus a share of the rents of all resources other than labour, according to any of a number of criteria costlessly agreed upon. But such an ideal situation is obviously not to be found.

We therefore conclude that the poor economic performance of a communist state is attributable to the high transaction costs of operating that organization. Under the postulate of constrained maximization, the communist state survives for the same reason that any 'inefficient' organization survives: namely, the transaction costs of *changing* an organizational (institutional) arrangement are prohibitive. Such costs include those of obtaining information about the workings of alternative institutions, and of using persuasive or coercive power to alter the status of the privileged groups whose incomes might be adversely affected by the institution of a different form of economic organization.

STEVEN N.S. CHEUNG

See also COASE THEOREM: TRANSACTIONS COSTS; VERTICAL INTEGRATION.

BIBLIOGRAPHY

Alchian, A.A. and Demsetz, H. 1972. Production, information costs, and economic organization. *American Economic Review* 62, 777–95.

Barzel, Y. 1982. Measurement costs and the organization of markets. *Journal of Law and Economics* 25, 27–48.

Cheung, S.N.S. 1969. Transaction costs, risk aversion, and the choice of contractual arrangements. *Journal of Law and Economics* 12, 23–42.

Cheung, S.N.S. 1982. *Will China Go 'Capitalist'?*. Hobart Paper 94, London: International Economic Association.

Cheung, S.N.S. 1983. The contractual nature of the firm. *Journal of Law and Economics* 26, 1–21.

Coase, R.H. 1937. The nature of the firm. *Economica* 4, 386–405.

Jensen, M.C. and Meckling, W.H. 1976. Theory of the firm: managerial behavior, agency costs and ownership structure. *Journal of Financial Economics* 3, 305–60.

Klein, B., Crawford, R.G. and Alchian, A.A. 1978. Vertical integration, appropriable rents, and the competitive contracting process. *Journal of Law and Economics* 21, 297–326.

Knight, F.H. 1921. *Risk Uncertainty and Profit*. Boston: Houghton Mifflin.

McManus, J.C. 1975. The costs of alternative economic organizations. *Canadian Journal of Economics* 8, 334.

Williamson, O.E. 1975. *Markets and Hierarchies: Analysis and Anti-Trust Implications*. Glencoe, Ill.: Free Press.

economic science and economics. The terms 'economy' and 'economic' or 'economical', are now used chiefly in two meanings, which it is well to distinguish clearly; since, though divergent in their history, they are liable to fusion, and therefore in some degree to confusion.

'Economy' originally meant, in Greek, the management of the affairs of a household, especially the provision and administration of its income. But since both in the acquisition and in the employment of wealth it is fundamentally important to avoid waste either of labour or of its produce, 'economy' in modern languages has come to denote generally the principle of seeking to attain, or the method of attaining, a desired end with the least possible expenditure of means; and the words 'economy', 'economic', 'economical', are often used in this sense, even without any direct relation to the production, distribution, or consumption of wealth. Thus we speak of 'economy of force' in a mechanical arrangement without regard to its utility, and of 'economy of time' in any employment whether productive of wealth or not.

On the other hand, as there is an obvious analogy between the provision for the needs of a state and the provision for the needs of a household, 'political economy', in Greek, came to be recognized as an appropriate term for the financial branch of the art or business of government. It is found in this sense in a treatise translated as Aristotle's in the 13th century; and so, when, in the transition from medieval to modern history, the question of ways and means obtrusively claimed the attention of statesmen, 'political economy' was the name naturally given to that part of the art of government which had for its aim the replenishment of the public treasury, and – as a means to this – the enrichment of the community by a provident regulation of industry and trade. And the term retained this meaning till the latter part of the 18th century without perceptible change – except that, towards the end of this period, the enrichment of the people came to be less exclusively regarded from the point of view of public finance, and more sought as a condition of social well-being.

But in the latter part of the 18th century, under the influence primarily of the leading French 'Économistes' or 'Physiocrats' – Quesnay, De la Rivière, and others – the conception of political economy underwent a fundamental change, in consequence of a fundamental change in the kind of answer which these thinkers gave to the question 'how to make a nation wealthy'. The physiocrats proclaimed to France, and through France to the world, that a statesman's true business was not to *make* laws for industry and trade in the hope of increasing wealth; but merely to ascertain and protect from encroachment the simple and immutable laws of nature, under which the production of wealth would regulate itself in the best possible way if governments would abstain from meddling. A view broadly similar to this, but less extreme, and, partly for this reason, more directly influential, was expounded in Adam Smith's *Wealth of Nations*. Instead of showing the statesman how to 'provide a plentiful revenue or subsistence for the people' – which was one of the two main objects of political economy, according to the traditional view – Adam Smith aims at showing him how nature, duly left alone, tends in the main to attain this end better than the statesman can attain it by governmental interference. Accordingly, so far as the widespread influence of Adam Smith's teaching went, that branch of the statesman's art which aimed at 'providing a plentiful revenue for the people' tended almost – though not altogether – to shrink to the simple maxim of *laisser faire*: leaving in its place a scientific study of the processes by which wealth is produced, distributed, and exchanged, through the spontaneous and partly unconscious division of labour among the members of human society, independently of any governmental interference beyond what is required to exclude violence or fraud. A part, indeed, of the old art of political economy – that which aimed at 'supplying the state with a revenue sufficient for the public service' – remained indispensable to the statesman; but it was held that this traditional art required to be renovated by being rationally based on the doctrines of the new-born science just described. It is, then, this scientific study of the department of social activity that most writers on the subject now primarily mean by the term 'political economy': such part of the old governmental art so called, as the doctrine of the new science is held to admit, being commonly regarded as 'applied political economy'. In consequence of this change the adjective 'economic', instead of the too cumbrous 'politico-economic', has come to denote the matters investigated by the science of political economy, and the propositions and arguments relating to them.

By thinkers and duly-instructed students this distinction between 'science' and 'art' – between the study of 'what is' and the study of 'what ought to be' – is usually regarded as simple and clear; and accordingly when such persons speak of the 'laws of political economy' they mean not rules by which the process of the social production and distribution of wealth *ought* to be governed, but general relations of co-existence and sequence among phenomena of this class, ascertained by a scientific study of this process as it actually takes place. This distinction, however, has been found difficult to establish in common thought: even well-educated persons still occasionally speak of the 'laws of political economy' as being 'violated' by the practice of statesmen, trades-unions, and other individuals and bodies. It is partly in order to prevent this confusion that the terms 'economic science' and 'economics' have recently come more and more into use, as a preferable alternative for political economy, so far as it is the name of a science. As to the scope of this science – it would be generally agreed that it is a branch of a larger science, dealing with man in his social relations; that it is to an important extent, but not altogether, capable of being usefully studied in separation from other branches of this science; and that it is mainly concerned with the social aspect – as distinct from the special technical aspect – of such human activities as are directed towards the production, appropriation, and application of the material means of satisfying human desires, so far as such means are capable of being exchanged. It would also be generally agreed that the method of economic science is partly deductive, partly inductive and historico-statistical. But to attempt a more precise determination of its method and scope, and especially of its relation to the art or system of practical rules which should guide the action of governments or private individuals in economic matters, would require us to enter into questions of a highly controversial kind; which will be more conveniently discussed when we come to deal with the older and wider term Political Economy.

[Henry Sidgwick]
Reprinted from *Palgrave's Dictionary of Political Economy*.

economics and philosophy. *See* PHILOSOPHY AND ECONOMICS.

economics and politics. *See* POLITICS AND ECONOMICS.

economics and psychology. *See* PSYCHOLOGY AND ECONOMICS.

economics and race. *See* RACE AND ECONOMICS.

economics libraries and documentation. Libraries for a discipline are formed and characterized by that discipline. It is quite easy for a visitor to recognize that a library is for scientists or for humanists or for social scientists just by a glance at the types of books, periodicals and other material on the shelves. Libraries for economists reflect the distinctive and changing sources and documentation of economics, which are in turn products of changes in the discipline itself. As economists have successively widened the scope of their enquiries and added weapons to their methodological armoury, so the types of material they have needed to consult have multiplied. As the immediate communication of their results has become more and more pressing, so the types of publication they have favoured have evolved in response. The library providing effective service to an econometrician today would have been as irrelevant to a 17th-century mercantilist as the literature of econometrics would have been incomprehensible. This article will deal with economics libraries in their natural context of economics documentation.

The first economics collections were in the private libraries of 17th- and 18th-century scholars. Adam Smith, renowned for his forgetfulness and carelessness in dress, was able to defend himself with the claim that at least 'I am a beau in my books'. In fact the majority of his 3000 titles were in the miscellaneous topics a gentleman scholar might have been expected to cultivate, with only about 100 directly on economic topics. Nevertheless, his economic method was a book-based one, using material from his predecessors and building it into his own system. About 100 authors are quoted in the *Wealth of Nations* (1776), though not always by name. By Smith's day there was already a large monograph literature the economist could draw on in several languages: pamphlets advocating trading schemes, tracts on farm management, alarming arguments on the growth or decline of population, accounts of particular industries, countries, cities, etc. The first economists wrote in monograph form too, either the pamphlet directed at some particular case or instance, or, from the 18th century onwards, the full length treatise summing up a whole theory of economics. Such was to remain the pattern of economics publishing until the later 19th century.

Not all economists relied on published sources; Malthus and others travelled widely in search of facts which they blended with information derived from extensive reading. There were other economists like Ricardo, whose theorizing worked outwards from his own inner store of business experience and personal insights, for whom access to a library was of less significance. An economist whose method did rely on books was usually forced to be a collector himself since libraries well-stocked with suitable publications were few. The 19th-century British economist, Richard Jones, in the laborious progress of his treatise on *Rent* (1831) made frequent calls on the kindness of his friend William Whewell to provide him with books from the 'Public Library' (in fact the predecessor of Cambridge University Library). As a country clergyman with little money for book buying and in comparative isolation from access to libraries, it was a difficult assignment for him to develop a theory of rent based on worldwide evidence. Indeed, as the complexity and specialization of the economics discipline grew during the 19th century it became less and less possible for an economist to function away from well-stocked libraries.

One reason for the increased dependence was the question of priority. As the literature of economics grew, so did the possibility of the prior publication of a supposed original idea. W.S. Jevons, for instance, developed and published his marginal utility theory in virtual isolation from other ideas on the topic. His subsequent realization that Menger and Walras had arrived at the same theoretical point in the same year of 1871 was disturbing, but as a voracious collector of economics literature he began to appreciate that there were also predecessors, most notably Hermann Heinrich Gossen, whose work published in 1854, needed to be acknowledged. It was becoming clear in various inescapable ways that the economist needed a library. Great general libraries such as that of the British Museum, could and still can provide much for the economist. Karl Marx, for instance, laboured there to great effect for years, and was able to display an encyclopedic knowledge of the relevant literature as a result. Nevertheless, by the end of the 19th century, the discipline was ready for specialized economics libraries.

Estimates of the numbers of practising economists at a given time are even today fraught with problems of definition. However, it is fairly safe to say that until the end of the 19th century and the early 20th century brought the creation of schools and faculties of economics in the universities of Europe and North America, and governments and business began to hire people with the degrees awarded by these institutions, practising economists were very few in number and more significantly were thinly scattered geographically. A few, most notably Smith himself but others such as Jean-Baptiste Say, were professors in faculties of law or other marginally related disciplines. Most were first and foremost otherwise employed: as a businessman like Ricardo, an official like J.S. Mill, a clergyman like the above-mentioned Richard Jones, an engineer like Dupuit, or a landowner like von Thünen. Libraries to serve such a scattered and heterogenous group were not practical. With the rise of Cambridge as a centre of economics excellence in the second half of the 19th century, the creation of the London School of Economics in 1895, and the emergence of great faculties of economics at

Harvard, Columbia and Chicago by the beginning of the 20th century in the USA, there were for the first time concentrations of economics teachers, researchers and students, whose needs gave rise to specialized economics libraries.

The source materials that such libraries might stock were multiplying fast. The business world was the chief generator of publications of use to economists. Journals from many countries in specific fields such as mining, insurance and banking, prospectuses and annual reports of railway companies, histories of firms or industries, biographies of businessmen, documents from international fairs and exhibitions, all burgeoned during the 19th century. For the first time, governments became significant publishers of economics-related literature: British Parliamentary Papers and United States Congressional Documents began to appear with increased volume and regularity almost immediately the 19th century began. In other countries the quantity, if not the regularity, of government publications also increased swiftly as the century progressed. Much of their content was statistical; and the US Census of 1790 and the British Census of 1801 were milestones in the practice of number gathering. Statistical societies, such as that of London (founded 1825, now the Royal Statistical Society) arose to take advantage of this material and in the process created a new layer of publication which economists could exploit.

The growth of economics as a discipline not only brought about the multiplication of materials but also created a need for more immediate channels of communication. The monograph which had been the chief avenue of publication for so long, began to lose some of its importance to the periodical. During the 19th century, economists had written for the great literary reviews, for general interest magazines, for newspapers, and for the one or two specialized publications such as the *Economist*. It was only after 1886, however, when the *Quarterly Journal of Economics* was founded by Harvard University, followed quite swiftly by the Royal Economic Society's *Economic Journal* (1891) and Chicago's *Journal of Political Economy* (1892), that there was a serious rival medium for economic writings. When the *American Economic Review* joined them in 1911, a main core of prestigious journals was taking shape, and the nature of the communication of economic ideas was completely altered. The need to convince a publisher of the validity and interest of one's ideas, or alternatively to publish at one's own expense, was replaced by interaction with an editor or editorial board drawn from one's peers. The possibility of quite swift and direct communication of one's ideas, even on highly technical matters, was opened up. Economics publication became more focused and more isolated from public debate.

This rich growth of sources permitted economists to build more and more elaborate structures of argument and evidence. W.S. Jevon's perilous but gallant attempts to link cyclical commercial fluctuations back through the business statistics to agricultural yields, to weather data, and finally to sunspot activity in a chain of causality, was just a particularly bold exploitation of the wealth that was becoming available. Even a collector of books as enthusiastic as Jevons could not reasonably hope to acquire materials of the number and type required for work of this kind. It is no coincidence that Jevons was an enthusiastic member of the Library Association (UK) and published papers on the topic of libraries. Whilst the 18th-century or early 19th-century economist urgently needed to be a collector of books, an economist of the late 19th and early 20th century like F.Y. Edgeworth needed to own hardly any books and could rely on libraries entirely.

The economics libraries which were created to cope with the growing need of economic researchers for specialized materials were, in the first place, historical collections designed to allow the reconstruction of past theory and the recovery of past knowledge. Two of the greatest have their roots in the personal collecting of one man. H.S. Foxwell (1849–1936) bridged the age of the great amateur collectors and that of specialised professional libraries. He amassed two great collections during a lifetime of acquisition carried on at a level of bibliomania. His first collection, sold in 1901 to the Goldsmiths Company, is now the basis of London University Library's Goldsmiths' Collection. His second collection, begun immediately the Goldsmiths sale had put his finances to rights, was sold in 1929 to Harvard Business School, which took possession on Foxwell's death to form the Kress Library. Kress is in turn now only part of the Baker Library of Harvard Business School.

Other great academic libraries also follow this pattern of a core historical collection alongside a fast-growing and fast-changing current collection. Columbia University, with its Seligman Collection, Johns Hopkins with the Hutzler Collection (assembled by Jacob Hollander) and the University of Illinois with its Hollander Collection, show a similar pattern. In Britain, the historical riches of Goldsmiths' are complemented within the London University system by the London School of Economics' British Library of Political and Economic Science, founded in 1896. Other parts of the world also have libraries of similar scope – Japan, for instance, where the Menger, Schumpeter and Burt Franklin collections of Hitotsubashi University provide a basis of old economics material to underpin its modern collections. With business history now an accepted discipline in graduate business schools in North America, the older titles in these libraries' core collections are attracting renewed attention from researchers.

Other types of economic research library have grown up in the 20th century to supplement the provisions of the academic libraries. Departments of Government and their associated agencies are the most common alternative source of economics materials. In the USA in particular, excellent libraries of this type are numerous; the US Department of Commerce Library, founded in 1913, for instance, has extensive collections including much statistical material, and agricultural economics material is one of the chief strengths of the National Library of Agriculture, originally established as the Department of Agriculture Library in 1862. The Federal Trade Commission, the Treasury, the Federal Reserve System Board of Governors, and other US Government agencies have fine economics libraries. The Library of Congress too, in its Social Science Reading Room, gives access to an enormous wealth of economics material. Other countries also have government economics libraries, the UK Board of Trade Library being one particularly fine example. International organizations also support economics collections; the International Labour Organisation (ILO) in Geneva, maintains an extremely large library, accessible via a computerized retrieval system. The Chambre de Commerce et d'Industrie Library in Paris is an example of a large, modern economics library supported by a trade association. The greatest economics collection in the public library sector is that of the New York Public Library, whose Economic Division has since 1919 produced the weekly *Bulletin of the Public Affairs Information Service*, the foremost index to the world's English language literature in the related fields of economics, finance, business, labour and public affairs.

Only a comparatively few libraries have the resources to devote the necessary attention to the full potential range of material required by economists and the techniques that will permit its efficient acquisition and retrieval. The Library of the Instituts für Weltwirtschaft in Kiel, Federal Republic of Germany, regarded by many as the premier economics library in the world, has been able to do this consistently since its foundation in 1914. Its catalogue of persons gives access to material not only via authors of books, articles and chapters, editors of books, symposia and journals, writers of prefaces and introductions, but also to material whose subject is a person or persons connected with economics. Its title catalogue includes not merely books, but also annual reports, serials, newspapers, collections, and the corporate bodies, congresses and conferences which publish material. The subject catalogue has geographical entries in addition to conventional subject headings. Cross reference cards are hardly ever needed, for as many copies of the full entry card as are necessary are entered in the relevant places in the catalogue. The work of the library is largely in the hands of professional economists whose subject expertise ensures the accuracy of subject cataloguing. Because the provision of good catalogues is never enough to ensure that the user obtains documents that relate to his interests, Kiel like many other special libraries alerts users to pertinent new acquisitions. Mark Perlman (1973) has justly called its methods 'the acme of the traditional approach to economics literature retrieval'.

In addition to dealing with the complexities of the multiplying sources for economic research, libraries have had to come to terms with yet another shift in the forms of economic communication. The urgency to establish the priority of ideas or to publish research before it becomes obsolete has placed strains on book and journal publishing with which they have been unable to cope. The average lag between submission of an article and its publication in a journal has increased over the years, and the maximum wait may be in excess of two years. This is despite the introduction of submission fees intended to reduce the number of submissions to some journals, the growth in the number of specialized journals and the introduction of a journal, *Economics Letters*, specifically designed to ensure swift publication of material. Academic monograph publishing is currently under the severest financial strains, with spiralling costs leading to higher price per copy to the consumer, with consequently reduced numbers of sales driving the unit cost up still further. In their editorial decisions, therefore, publishers are putting increasing weight on the market value of proposed titles.

The solution to this problem has been the increasing use of semi-published forms usually referred to as working papers. This form of distribution for an individual paper, reproduced by some inexpensive method and circulated via the writer's own institution's mailing list, causes confusion and distress to some librarians. Some libraries only acquire working papers if they are free, many do not catalogue them, some bind them in series, others do not. Some libraries avoid them altogether because of the difficulties they cause, and in the conviction that anything worthwhile which appears as a working paper will eventually be published in more permanent form. This is a serious disservice to economic scholarship. Roy Harrod (1969) said 'Mimeographed essays issued in advance of publication, if any, by the research unit of one university to the professors of other universities all over the world have come to constitute the main matter for reading, at least among theoretical economists.' Indeed, some of the difficulty of knowing all the writings of an economist as distinguished as the Norwegian Ragnar Frisch stems from the fact that he published so frequently in working paper form. The collection of working

papers at Warwick University Library (UK), their published *Economics Working Papers Bibliography* and their microform service are a major contribution to this problem area.

In the last quarter of the 20th century, the forms of source material for the economist have begun to challenge the service capacities of traditional librarianship. Most particularly, published statistics are no longer adequate for the purposes of many economists: they are out-of-date when published; they are in summary rather than comprehensive form; and they require recording in computerized form so that they can be arranged, shuffled and manoeuvred into revealing forms by the researcher. The electronic publication of statistics, making them available originally in computerized form, is growing rapidly. Increasingly, economic research depends on access to suitable computer hardware, availability of programmes which will perform the required tasks, and tapes of the data, rather than on books or other paper formats. To some extent this trend tends to exclude the library from the research process, but that is not necessarily always the case.

The issue of whether an economics library should merely confine itself to searching bibliographic databases on behalf of its economist clients or whether it should go further in identifying and making available numeric databases, has already begun to be explored, for instance in the Economics Library of the Ministry of Agriculture Fisheries and Food in the UK (O'Sullivan, 1982). The Baker Library at Harvard has long been involved with the acquisition of computerized data bases, but now has professional staff members who are also actively involved in reviewing, publicising, and manipulating numerical databases. What is more, to add the exploitation of numerical databases, whether commercial, such as the many provided by Evans Economics, Inc. (EEI), or government created, such as the US Bureau of Labor Statistics' LABSTAT, or the various services of the Bureau of Economic Analysis, to the functions of a library which is already managed by computerized systems often does not seem like a major problem. The availability of statistics in published or database form, the relative costs, types of series available in the alternative forms, the compatibility of computerized data with the systems available to the economist, are undeniably difficult issues. However, librarians in their new roles as information specialists are proving themselves capable of lending invaluable assistance to their clients with this type of material.

The potential of computerized systems for the swift transmission of information is likely to be further exploited. Electronic mail, for instance, already permits the flexible exchange of messages, long or short, amongst individuals or groups, by users of computer systems linked by telephone lines. The potential of electronic mail for the almost instant communication of research findings amongst a group of interested experts, an 'invisible college', is obvious. The electronic journal, which is at present being developed at a number of centres, seems likely to be an answer to the problem of the chronic pressure on economics journals. The electronic journal exists originally as a database controlled by an editor. Authors send their 'manuscripts' to the editor on-line from wherever their computer is located, the text can be refereed, edited and amended on-line, and then the subscribers read the journal on-line, printing out text on demand. The journal can be altered daily if new material is received and old material can be relegated to storage files. Within a group of specialists, united by the necessary machinery and associated financial arrangements, accurate, easily modified information can be available in a form more swift and convenient than any previously used. This new communication medium need not necessarily circumvent specialized libraries, which in future might actually provide facilities for such operations, for instance in the archiving of material from the electronic journal.

With their computer expertise, their staffs of specialists, and their vested interest in the storage and dissemination of information, there are very obviously roles which libraries are developing in the management of electronic information systems. Though there seems little indication of the large-scale return to conventional library-based scholarship in economics prophesied by Harry Johnson (1977), a total disregard of the value of historic and rare books collections already existing should not be contemplated. No discipline can safely neglect its past and it is to be hoped that economics libraries will cherish their treasures for the use of the minority of scholars whose approach is more reflective and retrospective. Nonetheless, the cutting edge of economics librarianship in the 21st century seems likely to be as different from that of the 20th or the 19th as will be its forms of documentation, and indeed the economics discipline which generates them.

P. STURGES

BIBLIOGRAPHY

Cole, A.H. 1957. *The Historical Development of Economic and Business Literature.* Kress Library Publication No. 12, Boston: Baker Library, Harvard Business School.
Fletcher, J. 1972. A view of the literature of economics. *Journal of Documentation* 28(4), December, 283–95.
Harrod, R. 1969. How can economists communicate? *Times Literary Supplement,* 24 July, 805–6.
Johnson, H.G. 1977. Methodologies of economics. In *The Organisation and Retrieval of Economic Knowledge,* ed. M. Perlman, London: Macmillan, 496–509.
Kindleberger, C. 1977. The use of libraries by economists: a personal view. In M. Perlman, *The Organisation and Retrieval of Economic Knowledge,* London: Macmillan.
Koch, J.E. and Pask, J.M. 1980. Working papers in academic business libraries. *College and Research Libraries* 41(6), November, 517–23.
O'Sullivan, S.D.A. 1982. Numeric and bibliographic databases for agricultural statistics: conflict or co-operation? *Sixth International Online Information Meeting,* Oxford: Learned Information.
Perlman, M. 1972. Economic libraries and collections. In *Encyclopedia of Library and Information Science,* Vol. 7, New York: Marcel Dekker.
Perlman, M. 1973. Editor's comment [on Kiel Instituts für Weltwirtschaft Library]. *Journal of Economic Literature* 11(1), March, 56–8.
Ruokonen, K. 1981. BILD–integrated online system for economic and business literature. *Tidskrift für Dokumentation* 37(3), 62–7.
Yohe, G.W. 1980. Current publication lags in economics journals. *Journal of Economic Literature* 18(3), September, 1050–55.

economics of sports. See SPORTS.

economic surplus and the equimarginal principle. Marginal analysis is actually only a particular case of a more general theory, the theory of surpluses and the economy of markets, which, if considered first, facilitates the discussion of the equimarginal principle.

THE GENERAL THEORY OF SURPLUSES AND THE ECONOMY OF MARKETS – FUNDAMENTAL CONCEPTS AND THEOREMS

To simplify the exposition, it is assumed that one good (U), enters all preference and production functions, and that its quantity can vary continuously. Except for the hypothesis of continuity with respect to this good (U), the discussion in this first part is free of any restrictive hypothesis of continuity, differentiability or convexity for the goods (V), ... , (W) considered, and the preference indexes and production functions. (For an exposition of the following theory in the case where no one good plays a particular role, see Allais, 1985, Section II, pp. 139–41.)

Structural conditions. The needs of every unit of consumption, individual or collective, can be entirely defined by considering a preference index

$$I_i = f_i(U_i, V_i, \ldots, W_i) \tag{1}$$

increasing as it passes from a given situation to one it finds preferable. Every quantity V_i is counted positively if it refers to a consumption, negatively if it refers to a service supplied.

The set of feasible techniques for a unit of production j can be represented by a condition of the form

$$f_j(U_j, V_j, \ldots, W_j) \geqslant 0$$

where every quantity V_j is considered as representing a consumption or an output depending on whether it is positive or negative. The extreme points corresponding to the boundary between possible and impossible situations represent states of maximum efficiency for the production unit considered. They may be represented by the condition

$$f_j(U_j, V_j, \ldots, W_j) = 0 \tag{2}$$

The function f_j may be called the production function. It is defined up to any transformation which leaves its sign unchanged.

From a technical point of view, maximum efficiency implies quite specific conditions. If for instance, one considers a production technique $A = A(X, Y, \ldots, Z)$ and if n production units are technically preferable to a single one, we should have (Allais, 1943, pp. 187–8; 1981, pp. 319–22)

$$\sum_j A(X_j, Y_j, \ldots, Z_j) > A\left[\sum_j X_j, \sum_j Y_j, \ldots, \sum_j Z_j\right]. \tag{3}$$

In the opposite case we have

$$A\left[\sum_j X_j, \sum_j Y_j, \ldots, \sum_j Z_j\right] > \sum_j A(X_j, Y_j, \ldots, Z_j). \tag{3*}$$

An industry is referred to as differentiated if the use of distinct production units is technically more advantageous than the concentration of all production operations into a single production unit. It is called non-differentiated in the opposite case. Conditions (3) and (3*) are two particular illustrations of differentiation (Allais, 1943, p. 637).

From inequality (3) it is possible to show that the whole production function of a differentiated industry is asymptotically homogeneous. In this case $(n \gg 1)$ there is quasi-homogeneity (Allais, 1943, pp. 201–6; 1974b).

Distributable surplus corresponding to a given modification of the economy. The distributable surplus σ_u relative to a good (U) and to a realizable modification of the economy which leaves all preference indexes unchanged is defined as the quantity of that good which can be released following this shift (Allais, 1943, pp. 610–16). The surplus considered here differs essentially from the concepts of consumer surplus as normally considered in the literature (e.g. Samuelson, 1947, pp. 195–202; Blaug, 1985, pp. 355–70; Allais, 1981, pp. 297–8, and 1985, nn. 12–13).

Let us consider an initial state $(\mathscr{E}_1)$ characterized by consumption values $U_i, V_i, \ldots, W_i$ and $U_j, V_j, \ldots, W_j$ (positive or negative) of the different units of consumption and production. We have

$$\sum_i U_i + \sum_j U_j = U_0;$$

$$\sum_i V_i + \sum_j V_j = V_0; \ldots; \quad \sum_i W_i + \sum_j W_j = W_0 \quad (4)$$

where $U_0, V_0, \ldots, W_0$ designate available resources. Let $(\delta\mathscr{E}_1)$ be a feasible modification of $(\mathscr{E}_1)$ characterized by finite variations $\delta U_i, \delta V_i, \ldots, \delta W_i, \delta U_j, \delta V_j, \ldots, \delta W_j$, and let

$$(\mathscr{E}_2) = (\mathscr{E}_1) + \delta(\mathscr{E}_1)$$

represent the new state.

According to (4) we naturally have

$$\sum_i \delta V_i + \sum_j \delta V_j = 0$$

for every good $(U), (V), \ldots, (W)$. From (2) we also have for every unit of production j

$$f_j(U_j + \delta U_j, V_j + \delta V_j, \ldots, W_j + \delta W_j) = 0$$

According to (1) the preference indexes become

$$I_i + \delta I_i = f_i(U_i + \delta U_i, V_i + \delta V_i, \ldots, W_i + \delta W_i)$$

The δI_i can be positive, zero, or negative.

Let us now define a third state $(\mathscr{E}_3)$ by the condition that by the modification $-\delta\sigma_{ui}$ of just the quantities $U_i + \delta U_i$ all the preference indexes return to their initial values.

We then have the conditions

$$f_i(U_i + \delta U_i - \delta\sigma_{ui}, V_i + \delta V_i, \ldots, W_i + \delta W_i)$$
$$= f_i(U_i, V_i, \ldots, W_i) \quad (5)$$

The state $(\mathscr{E}_3)$ can be termed 'isohedonous' with the state $(\mathscr{E}_1)$. In passing from $(\mathscr{E}_1)$ to $(\mathscr{E}_3)$ the quantity

$$\delta\sigma_u = \sum_i \delta\sigma_{ui} \quad (6)$$

of the good (U) is released, as all the units of consumption find themselves again in situations which they consider equivalent, since their preference indexes return to the same values (Allais, 1943, pp. 637–8).

The surplus $\delta\sigma_u$ has been released during the passage from $(\mathscr{E}_1)$ to $(\mathscr{E}_3)$. It may then be considered that in the situation $(\mathscr{E}_1)$ this surplus was both realizable and distributable. It may further be considered that in passing from $(\mathscr{E}_1)$ to $(\mathscr{E}_2)$, it has in effect been distributed.

The distributable surplus thus defined covers the whole economy, but this definition can be used for any group of agents. It is necessary only to consider the functions f_i and f_j and the resources relating to this group in the preceding relations.

Any exchange system, with the corresponding production operations it implies, is deemed 'advantageous' when a distributable surplus is achieved and distributed, so that the preference index of any consumption unit concerned increases. If an exchange and production system is advantageous, there must be at least one system of prices which allows it, the prices used by each pair of agents being specific to them. The distribution of the realized surplus between agents is determined by the system of prices used in the exchanges between them.

Conditions of equilibrium and maximum efficiency. In essence all economic operations of whatever type may be considered as

reducing to the search for, the achievement of, and the distribution of surpluses. Thus stable general economic equilibrium exists if, and only if, in the situation under consideration, there is no realizable surplus, which means

$$\delta\sigma_u \leqslant 0 \quad (7)$$

for all feasible modifications of the economy (Allais, 1943, pp. 606–12).

In such a situation the distributable surplus is zero or negative for all possible modifications of the economy compatible with its structural relations, and it is impossible to find any set of prices that would permit effective bilateral or multilateral exchanges (accompanied by the implied production operations) which are advantageous to all the agents concerned.

A situation of maximum efficiency can be defined as a situation in which it is impossible to improve the situation of some people without undermining that of others, i.e. to increase certain preference indices without decreasing others. The set of states of maximum efficiency represents the boundary between the possible and the impossible (Figure 1).

From those definitions of the situations of maximum efficiency and stable general economic equilibrium, it follows, with the greatest generality and without any restrictive hypothesis of continuity, differentiability or convexity, except for the common good (U), that:

Any state of stable general economic equilibrium is one of maximum efficiency (*First theorem of equivalence*).

Any state of maximum efficiency is one of stable general economic equilibrium (*Second theorem of equivalence*).

Since there can be no stable general economic equilibrium if there is any distributable surplus, every state of stable general economic equilibrium is a state of maximum efficiency. Conversely, if there is maximum efficiency, there is no realizable surplus which could be used to increase at least one preference index without decreasing the others, and consequently, every state of maximum efficiency is a state of stable general economic equilibrium.

Because of the theorems of equivalence, the terms 'conditions of stable general economic equilibrium' and 'conditions of maximum efficiency' are used interchangeably below.

The dynamic process of the economy: decentralized search for surpluses. In their essence all economic operations, whatever they may be, can be thought of as boiling down to the pursuit, realization and allocation of distributable surpluses. The corresponding model is the Allais model of the economy of markets (1967), defined by the fundamental rule that every agent tries to find one or several other agents ready to accept at specific prices a bilateral or multilateral exchange (accompanied by corresponding production decisions) which will release a positive surplus that can be shared out, and which is realized and distributed once discovered. Thus the evolution of the market's economy is characterized by the condition

$$\delta I_i \geqslant 0$$

for every consumption unit.

Since in the evolution of an economy of markets, surpluses are constantly being realized and allocated, the preference indexes of the consumption units are never decreasing, at the same time as some are increasing. This means that for a given structure, that is to say, for given preferences, resources, and technical know-how, the working of an economy of markets tends to bring it nearer and nearer to a state of stable general economic equilibrium, hence a state of maximum efficiency (Figure 1), which is the third fundamental theorem.

Naturally such evolution takes place only if sufficient information exists about the actual possibilities of realizing surpluses.

To any given initial situation whatsoever, assumed not to be a situation of equilibrium, there corresponds an infinite number of possible equilibrium situations, each corresponding to a particular path and each satisfying the general condition that no index of preference should take on a lower value than in the initial situation (Figure 1).

Economic loss. The loss σ_u^* which is associated with a given situation is defined as the greatest quantity of the good (U) which can be released in a transformation of the economy for which all the preference indexes remain unchanged (Figure 1): (Allais, 1943, pp. 638–49).

It is a well determined function

$$\sigma_u^* = F[I_1, I_2, \ldots, I_n, U_0, V_0, \ldots, W_0] \qquad (8)$$

of the preference indexes I_i and of the resources V_0 which characterize this situation. The loss σ_u^* is an indicator of inefficiency, and $-\sigma_u^*$ an indicator of the efficiency of the economy as a whole.

The loss is minimum and nil in every state of maximum efficiency, and positive in every feasible situation which is not a state of maximum efficiency. It decreases in any modification of the economy, whereby some preference indexes increase, others remaining unchanged, or whereby some surpluses are released with no decline in some preference indexes.

Paths to states of economic equilibrium and maximum efficiency. Since the preference indices I_i are continuous functions of the quantities U_i of the common good (U), the boundary between the possible and the impossible situations in the hyperspace of preference indexes is constituted by a continuous surface. On this surface the loss σ_u^* is nil. This representation allows an immediate demonstration by simple topological considerations of propositions whose proof would otherwise be very difficult. (The paternity of this representation has been unduly attributed to P. Samuelson, 1959, but it was in fact published for the first time in Allais, 1943, and systematically used by Allais in later years especially 1945 and 1947; see Allais, 1971, n.11, p. 385; and 1974a, n.18, pp. 176–7.)

For every feasible situation which is not a state of maximum efficiency, represented by a point such as M_0, there are an infinity of realizable displacements M_0M enabling a situation of maximum efficiency M^* to be approached, such that all the preference indexes have greater values than in the initial situation M_0.

Figure 1 presents an illustration of the process of dynamic evolution by releasing and sharing out of surpluses during which the loss σ_u^* is constantly decreasing (Allais 1943, 1974b, and 1981, p. 121).

The changing structure of the economy. As psychological patterns vary, as techniques are improved, or as new resources are discovered (or existing resources depleted), the set of situations of maximum efficiency relative to the indexes of preference constantly undergoes change over time. Consequently, situations of equilibrium and maximum efficiency are never reached, and what is really important is to determine the rules of the game which must be applied to come constantly closer to them as rapidly as possible. At a given time t, if information is sufficient and if the adjustments are sufficiently

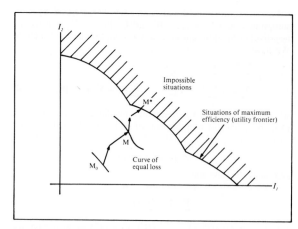

Figure 1 Process of dynamic evolution. Illustrative diagram

rapid, the point representing the economy will never be very far from the maximum efficiency surface of that time t.

General comment. An economy of markets can be defined as one in which the agents – consumption, production, and arbitrage units – coexist and are free to undertake any exchange transaction or production operation which can result in rendering some distributable surplus available. The principle of the markets economy is that any surplus realized is shared among the operators involved. How the surpluses achieved are shared out depends on the specific systems of prices used in the exchanges between the agents concerned. The prices used are always specific to the exchange and production operations considered and there is never a unique system of prices used in common by all the agents.

Diagrammatic representation like that of Figure 1 reveals clearly three basic facts:

(1) There is an infinity of situations of maximum efficiency corresponding to a given initial situation characterized by some distribution of property.

(2) To each situation of maximum efficiency there corresponds a final distribution of property.

(3) This final distribution depends on the initial situation and the distribution of surpluses in the course of the transition.

Thus there is a very strong inter-dependence between the point of view of efficiency corresponding to the discovery and realization of surpluses and the ethical point of view corresponding to their sharing.

In any event, since only what is produced can be shared, the incentive stemming from the partial or total appropriation of the surpluses by the various agents appears as a fundamental factor for the functioning of the economy of markets.

On the general theory of surpluses and the economy of markets in the general case; and on the fundamental theorems see: Allais (1943, pp. 112–77; 181–211; 604–56), (1967, § 8–65), (1968a, vol. II), (1968b), (1971), (1974a), (1981, pp. 27–48), (1985), and (1986).

THE EQUIMARGINAL PRINCIPLE

Continuity and differentiability. The preceding definitions and theorems are very general and do not make any hypothesis of continuity, derivability or convexity, except the hypothesis of continuity for the common good (U).

We now assume in addition only that all the quantities and functions considered are continuous and that all functions have first and second order derivatives, the following developments being totally independent of any hypothesis of general convexity.

From the sign conventions adopted earlier it follows that for any i, j and V

$$f'_{iv} = \partial f_i / \partial V_i \geqslant 0, \qquad f'_{jv} = \partial f_j / \partial V_j \geqslant 0.$$

The second partial derivatives are written

$$f''_{ivw} = \partial^2 f_i / \partial V_i \delta W_i, \qquad f''_{jvw} = \partial^2 f_j / \partial V_j \delta W_j.$$

In the following, the symbol $\overline{d^2 g}$ represents the second differential

$$\overline{d^2 g} = \sum_U^W g''_{v^2} \, dV^2 + 2 \sum_{U,V} g''_{vw} \, dV \, dW$$

of a function $g(U, V, \ldots, W)$ when all parameters in that function are taken as independent, while the symbol $\overline{d^2 g_u}$ represents what this second differential becomes after du has been replaced by its expression derived from

$$dg = \sum_U^W g'_v \, dV = 0$$

(Allais, 1968a, vol. II, pp. 77–8; 1973b, pp. 151–5; 1981, pp. 688–9).

Convexity and concavity. The local properties of diminishing or increasing marginal returns are related to local conditions of convexity or concavity. Convexity is defined as follows:

Ordinal fields of preference: A field of choice is said to be convex in the whole space (postulate of general convexity) if, at all points of the field, the condition

$$I(M_0) \leqslant I(M_1)$$

entails

$$I(M_0) \leqslant I(M)$$

with

$$M = \lambda M_0 + (1 - \lambda) M_1 \qquad 0 < \lambda < 1$$

There is local convexity at M_0 if this condition is satisfied only for

$$|\mathbf{M}_0 \mathbf{M}_1| < \epsilon$$

where ϵ is a given positive number.

When differentiability is assumed local convexity implies

$$\overline{d^2 f_{iu}} \leqslant 0 \qquad \text{for } df_i = 0.$$

Fields of production: A field of production is said to be convex over the whole space (postulate of general convexity) if, for any two possible points M_0 and M_1, the centre of gravity defined by the relation

$$\mathbf{M} = \lambda \mathbf{M}_0 + (1 - \lambda) \mathbf{M}_1$$

is likewise a possible point for

$$0 < \lambda < 1.$$

Local convexity obtains at M_0 if the preceding condition is satisfied only for

$$|\mathbf{M}_0 \mathbf{M}_1| < \epsilon$$

where ϵ is a given positive number.

When differentiability is assumed, local convexity implies

$$\overline{d^2 f_{ju}} \leqslant 0 \qquad \text{for } df_j = 0.$$

In fact there is no production operation that does not begin by providing increasing marginal returns, and it is only beyond a certain threshold that diminishing marginal returns are observed. That is a general physical law of nature (Allais, 1943, pp. 193–5; 1968a, II, pp. 68–96; 1971, pp. 362–4; 1974a, pp. 153–7). Similarly it can be considered as an introspective datum that psychological returns begin by increasing but in the end always decrease beyond certain threshold values. That is a general psychological law (Allais, 1968a, II, pp. 109–38; 1971, pp. 360–62; 1974a, pp. 153–5). These are two fundamental properties of fields of choice and production. They rule out the postulate of general convexity which is generally accepted in the contemporary literature.

Generation of distributable surplus. Consider any economic state ($\mathscr{E}$) and a realizable modification ($\delta \mathscr{E}$) such that all the preference indexes I_i remain constant (isohedonous modification). Let the conditions of constancy of these indexes and the conditions corresponding to the production functions be written in the same general form

$$g_k(U_k, V_k, \ldots, W_k) = 0 \qquad (9)$$

where U_k, V_k, $\ldots$, W_k represent the consumption of both consumption and production units. By convention, any quantity V_k, if positive, represents consumptions, either by a consumption or a production unit. For any production or consumption unit, any parameter V_k, if negative, represents production of a good or a service.

Let dU_k, dV_k, $\ldots$, dW_k, be the first order differentials of the variations δU_k, δV_k, $\ldots$, δW_k of consumptions U_k, V_k, $\ldots$, W_k in the displacement ($\delta \mathscr{E}$). From (9), we have

$$g'_{ku} \, dU_k + g'_{kv} \, dV_k + \cdots + g'_{kw} \, dW_k = 0 \qquad (10)$$

Let δV_{kl} be the quantity of (V) received by the consumption or the production unit k from the consumption or production unit l. By definition, we have

$$\delta V_k = \sum_{k \neq l} \delta V_{kl} \qquad (11)$$

$$\delta V_{lk} = -\delta V_{kl}. \qquad (12)$$

Assuming that the displacement ($\delta \mathscr{E}$) is such that

$$\sum_k \delta V_k = 0, \ldots, \sum_k \delta W_k = 0. \qquad (13)$$

Let

$$\epsilon_{v,u}^k = g'_{kv} / g'_{ku}. \qquad (14)$$

The ratio $\epsilon_{v,u}^k$ is the coefficient of marginal equivalence (or marginal rate of substitution) of goods (V) and (U) for agent k (Allais, 1943, pp. 609–10, and 617–21).

From (10) and (14) we have the relation

$$dU_k = -[\epsilon_{vu}^k \, dV_k + \cdots + \epsilon_{wu}^k \, dW_k] \qquad (15)$$

between the first order differential dU_k, dV_k, $\ldots$, dW_k.

If dU_k is positive, agent k receives a quantity dU_k to within the second order. If dU_k is negative, agent k supplies a quantity $-dU_k$ to within the second order.

From the condition (13), it follows that the displacement considered releases a global distributable surplus

$$\delta \sigma_u = -\sum_k \delta U_k$$

representing the excess of the quantities supplied over the quantities received of good (U) whose first order differential is

$$d\sigma_u = -\sum_k dU_k.$$

65

From (11) and (15)

$$dU_k = -\sum_v^w \left[\epsilon_{vu}^k \sum_{\substack{k,l \\ k<l}} dV_{kl} \right]$$

and from (12), we have (Allais, 1952c, p. 31; 1968a, II, p. 174; 1981, p. 88)

$$d\sigma_u = \sum_v^w \sum_{\substack{k,l \\ k<l}} (\epsilon_{vu}^k - \epsilon_{vu}^l) \, dV_{kl}. \tag{16}$$

According to definitions (5) and (6) $d\sigma_u$ is the first differential of the global distributable surplus $\delta\sigma_u$ released in the displacement considered. For all economic agents the unit of value is defined by condition $u_k = u = 1$. The marginal values $v_k, \ldots, w_k$ of goods $(V), \ldots, (W)$ for unit k are defined with respect to the u_k by the relations

$$\frac{g'_{ku}}{u_k} = \frac{g'_{kv}}{v_k} = \cdots = \frac{g'_{kw}}{w_k} \tag{17}$$

$$u_k = u = 1. \tag{18}$$

Under the adopted sign convention, all the v_k are positive. We have from (14) and (18)

$$\epsilon_{vu}^k = v_k \tag{19}$$

and relation (16) is written

$$d\sigma_u = \sum_v^w \sum_{\substack{k,l \\ k<l}} (v_k - v_l) \, dV_{kl} \tag{20}$$

where v_k and v_l are the marginal values of good (V) for units k and l. This summation covers all agents, both consumption and production units. It can thus be seen that all the differences between the marginal values in the situation $(\mathscr{E})$ can give rise to the release of potential surpluses which can be released and distributed.

The meaning of relation (20) is immediate. Thus if $v_k > v_l$ the relative value of good (V) is higher for agent k than for agent l. The transfer of a positive quantity dV_{kl} of good (V) from agent l to agent k therefore creates an additional positive value

$$d\sigma_{ukl} = (v_k - v_l) \, dV_{kl}$$

If in this 'isohedone' transformation surpluses are released, all positive, they can be distributed in such a way as to increase all preference indexes. In such a modification of the economy, the maximum distributable surplus diminishes, and the point representing the economic situation considered moves closer to the surface of maximum efficiency in the hyperspace of preference indexes. Naturally, for this condition to obtain, the corresponding exchanges and the changes of the consumptions and productions they imply in the production system, must effectively occur.

Psychological values and marginal psychological values. Naturally, the v_k are only marginal values for the agents. The psychological values v_i^* of the consumption V_i of a subject i is defined by the relation

$$f_i(U_i + v_i^* V_i, 0, \ldots, W_i) = f(U_i, V_i, \ldots, W_i)$$

where $v_i^* V_i$ is the sum he would accept to receive to offset the drop in his consumption V_i to zero. The unit value v_i^* is

generally much higher than the marginal value v_i corresponding to relations (17), (18) and (19).

In any event, a consumption is only advantageous when its psychological value is higher than its marginal value, because, if this were not so, it would be in the subject's interest to reduce his consumption V_i.

Conditions of stable general economic equilibrium and maximum efficiency of the economy. From condition (7) it follows that the necessary and sufficient condition for a situation $(\mathscr{E})$ to be of stable equilibrium and maximum efficiency is that the distributable surplus $\delta\sigma_u$ defined by (5) and (6) be negative or zero for every feasible modification $(\delta\mathscr{E})$, i.e. every modification that is compatible with the constraint conditions, i.e., the structural relations of the economy (2) and (4) above.

Condition (7) implies the two conditions (Allais, 1943, p. 612)

$$d\sigma_u = 0 \qquad \text{(first order condition)} \quad (21)$$

$$d^2\sigma_u \leqslant 0 \qquad \text{(second order condition)} \quad (22)$$

for any realizable and reversible modification $(\delta\mathscr{E})$ in which the expressions of $d\sigma_u$ and $d^2\sigma_u$ represent the first and second differential of $\delta\sigma_u$.

Thus we have according to (21) and (22) using the above notations

$$d\sigma_u = \sum_i d\sigma_{ui} = \sum_i dI_i/I'_{iu} = 0 \tag{23}$$

$$d^2\sigma_u = \sum_i \overline{\frac{d^2 f_{iu}}{f'_{iu}}} + \sum_j \overline{\frac{d^2 f_{ju}}{f'_{ju}}} \leqslant 0 \qquad \text{for } d\sigma_u = 0. \tag{24}$$

Actually, and according to relation (20), *the first order condition (23) implies that when the quantities V_k are not nil, all the marginal values v_k are equal to a same value v and a same system of prices $u, v, \ldots, w$ then exists for all the agents k concerned, such that*

$$\frac{g'_{ku}}{u} = \frac{g'_{kv}}{v} = \cdots = \frac{g'_{kw}}{w}. \tag{25}$$

These equalities condense the general equimarginal principle into a single formulation. They express the fact that in a situation of equilibrium and maximum efficiency, the psychological (or objective) value v_k of the last dollar is the same, for any agent (consumption or production unit), whatever use it is put to.

For the quantities V_k which are nil (terminal equilibria), we necessarily have

$$v_k \leqslant v$$

since, if this were not true, the operator's interest would be to increase V_k from the value $V_k = 0$; he could indeed do this because of the existence of other operators who are in a situation of tangential equilibrium for good (V).

The second order condition (24) holds whether or not the df_i are equal to zero. It is only subject to the constraint (21). If we consider only the modifications of the economy involving units k and l, condition (24) is written

$$d^2\sigma_u = \overline{\frac{d^2 f_{ku}}{f'_{ku}}} + \overline{\frac{d^2 f_{lu}}{f'_{lu}}} \leqslant 0 \qquad \text{for } d\sigma_{uk} + d\sigma_{ul} = 0.$$

shows that when in a situation of maximum efficiency consumption or production units consume (or produce) the same goods, one unit at most is in a situation of local concavity, i.e., in a situation of marginal increasing returns (Allais, 1968a, pp. 196–9; 1974a, n.125, p. 184; 1981, p. 65).

Consequently, when maximum efficiency obtains, most operators are in a situation of local convexity and marginal decreasing returns. However, this condition cannot be interpreted as meaning that all fields of choice and production are convex everywhere, this hypothesis being totally contradicted by observed data.

When local convexity obtains for a consumption unit, its index of preference is effectively at a maximum, subject to the budgetary constraint, equilibrium prices being taken as given. Similarly if local convexity obtains for a production unit, the unit's income is effectively at a maximum, equilibrium prices again being taken as given. However, these two principles, which in any case could be valid only for a situation of maximum efficiency, cannot be considered as corresponding in all cases to optimum behaviour, and they cannot be taken to be of general value. As a matter of fact and for instance, if, in a situation of maximum efficiency, a production unit is in a situation of local concavity, its income is minimum, the equilibrium prices being considered as given.

Conditions (25) and (24) show the total symmetry of the implications of the psychological and technical structures of the economy.

Approximate value of the economic loss corresponding to the non-equality of marginal values in the neighbourhood of a situation of maximum efficiency. The integration of equation (20) along a path leading to a state of maximum efficiency leads to the following approximate estimate to within third order accuracy of the global loss involved in the initial situation (relation 8)

$$\sigma_u^* \sim \frac{1}{2u} \sum_v \sum_{\substack{k,l \\ k<l}} (v_k - v_l)\delta V_{kl}^*. \tag{26}$$

In this relation, the quantities $v_k - v_l$ represent the differences of marginal values in the initial state considered, and the δV_{kl}^* are the quantities of the good (V) received by operator k from operator l in the transition from the initial to the final state. Relation (26) is of the broadest generality, and holds whatever the initial state (Allais, 1952, pp. 31–2, n8; 1968a, II, p. 207; 1981, p. 110).

Its simplicity is really extraordinary in view of the complexity of the concept it represents, namely the maximum of the distributable surplus for all the modifications which the economy can undergo while leaving the preference indexes unchanged.

In the neighbourhood of a situation of maximum efficiency, the $(v_k - v_l)$ and δV_{kl}^* are first order quantities, whereas the loss σ_u^* is only of the second order. However, since the δV_{kl}^* are of the first order, the variations δI_i of the preference indexes are also of the first order. As a result, and for instance, in the neighbourhood of a situation of maximum efficiency, taxes have major first order effects on the distribution of income but only second order effects on the efficiency of the economy.

On the theoretical foundations of the equimarginal principle, see: Allais (1943, pp. 604–56), (1945), (1952a, pp. 28–32), (1967), (1968a, vol. II), (1971), (1973a), (1973b), (1974a), (1974b), (1981) and (1986). Illustrative models: Allais (1943, Annexe I, pp. 4–24), (1945, pp. 57–69). On its extension see: cases of perfect and imperfect foresight: Allais (1943, pp. 343–84), (1947, pp. 23–228), (1964), (1967), (1968a, II).

Illustrative models: (1947, pp. 631–771). Capitalistic optimum theory: Allais (1947, pp. 179–228), (1962), (1963). Demographic optimum theory: Allais (1943, pp. 749–85). Case of risk: Allais (1952b). Application of marginal analysis to transport: Allais (1964) and (1986). For a general overview on the meaning, limits, generalizations, and history of the equimarginal analysis see Allais (1986).

Theory of surpluses and marginal analysis. As a matter of fact a single relation, the relation (20) (or the equivalent relation (16)) condenses the whole marginal approach as it has developed for over a century. Subject only to the hypotheses of continuity and derivability implied by any marginal theory, it applies in all cases, and its simplicity is really extraordinary.

It also shows that equilibrium and maximum efficiency can obtain only when all marginal values are equal, which is the equimarginal principle.

The equimarginal principle was discovered first by Gossen (1854), and rediscovered, broadened and introduced independently into economics by Jevons (1871), Menger (1871) and Walras (1874–7). In the following years numerous new developments of the principle have been presented by their immediate successors, especially by Edgeworth (1881), Irving Fisher (1892) and Vilfredo Pareto (1894–1911). Particularly striking illustrations of the role of differences in marginal equivalences are Ricardo's theory of comparative costs (1817) and Dupuit's theory of economic losses (1844–53).

This principle corresponds to the outcome of the dynamic process of the economy induced by differences in marginal equivalences. According to Irving Fisher (1892), with whose judgement I agree fully, 'No idea has been more fruitful in the history of economic science.' Its applications and generalizations dominate all economic analysis in real terms.

From the foregoing a double conclusion emerges: the classical theory of marginal equivalences is irreplaceable to make understandable the underlying nature of all economic phenomena; the general theory of surplus, of which classical marginal theory is only a special case, allows one to extend the propositions of marginal analysis to the most general case of discrete variations and indivisibilities.

As important as the analysis of the conditions of general equilibrium and maximum efficiency may be, the analysis of the dynamic processes which enable surpluses to be generated from a given situation, is much more important. From this point of view the analyses by Dupuit, Jevons, Edgeworth, Pareto, and the marginal school and its predecessors in general, appears much more realistic that the contributions which rest only upon the consideration of Walras's general model of equilibrium.

In fact, what is really important is not so much the knowledge of the properties of a state of maximum efficiency as the rules of the game which have been applied to the economy effectively to move nearer to a state of maximum efficiency.

The decentralized search for surpluses is truly the dynamic principle from which a thorough and yet very simple conception of the operation of the whole economy can be derived. Whereas in the market economy model the search for efficiency is essentially focused on the determination of a certain set of prices, the analysis of the model of the economy of markets is based on the search for potential surpluses and their realization. Not only is the economy of markets model much more realistic than the market economy model while lending itself to much simpler proofs, but also these proofs are not subordinated to any restrictive assumptions relating to continuity, differentiability of functions, or convexity. All of economic dynamics is reduced to a single principle: the search for and realization of potential surpluses, which leads to the minimization of loss for the economy as a whole.

On all these points see especially: Allais (1971) and (1974a).

The tendencies of the contemporary literature. From Walras on, the literature became progressively – and unduly – concentrated on equilibrium analysis which, however interesting it could be, is less so than the analysis of the processes by which the economy tends at any time towards situations of equilibrium which in fact are never reached.

Today there is a tendency to neglect the dynamic marginal approach based on the consideration of differences in marginal equivalences; and in the name of a so-called rigour it has been replaced by new theories. A fortiori, the general theory of surpluses which generalizes marginal analysis is simply ignored. This development, which in reality, and despite the too-widely held belief to the contrary, represents an immense step backward, basically stems from the unquestioning acceptance of 'established truths' taught by the dominant 'establishments', whose only real basis is their incessant repetition.

As a matter of fact the guiding principles of the contemporary theories descending from Walras: the adoption of the market economy model; the hypothesis that a common price system applicable to all operators prevails at each instant; the assumption of general convexity; and the exaltation of mathematical formalism of the theory of sets to the detriment of conformity with actual facts, constitute an impediment to any genuine progress in analysis of the economy in real terms.

The essential difference between the market economy model and the model of the economy of markets is that, in the latter, the exchanges leading to equilibrium take place successively at different prices, and that, at any given moment, the price sets used by different operators are not necessarily the same. Whereas in the first model the final situation is determined totally by the initial situation, which correspondingly plays a privileged role without any real justification, in the second the final situation depends both on the initial situation and the path taken from it to the final situation (Figure 1).

Whereas the market economy model postulates perfect competition and a large number, if not an infinity, of operators, the model of the economy of markets applies just as well to the cases of monopoly as to the cases of competition.

Not only is the market economy model unrealistic, but it also gives rise to considerable mathematical difficulties when an attempt is made to demonstrate the above three fundamental theorems. Whether differential calculus or set theory is used, the theorems can only be demonstrated under extremely restrictive conditions, and the difficulties they imply are, from an economic standpoint, completely artificial, for they arise solely from the unrealistic nature of the model used. Paradoxically, whereas these restrictive assumptions are totally unrealistic, most of the theoretical difficulties encountered disappear, as shown above, once they are discarded.

The market economy approach leads to imposing on any economic model, for it to be considered satisfactory, conditions which actually apply to a particular model, which are generally not fulfilled in reality, and for which, at all events, no rigorous justification can be found.

By departing from the great tradition of marginal theory and by adopting an unrealistic model and unrealistic assumptions,

the contemporary theories, purely mathematical, have doomed themselves to sterility as regards the understanding of reality.

On the contemporary theories see especially: Samuelson (1947); Arrow (1968); Debreu (1959) and (1985); Blaug (1970 and 1985); Arrow and Hahn (1971); Hutchison (1977, pp. 62–97 and 161–70); Woo (1985); and Allais (1952b, 1968b, 1968e, 1971, 1974a, and 1981).

MAURICE ALLAIS

See also ALLAIS, MAURICE; EFFICIENT ALLOCATION; GENERAL EQUILIBRIUM; OPTIMALITY AND EFFICIENCY; SURPLUS APPROACH TO VALUE AND DISTRIBUTION.

BIBLIOGRAPHY
Allais, M. 1943. *A la recherche d'une discipline économique. Première partie, l'économie pure.* 2 vols, Paris: Ateliers Industria. 2nd edn, *Traité d'économie pure,* 5 vols, Paris: Imprimerie Nationale, Paris, 1952. (The second edition is identical to the first, except for a new Introduction, 1–63.)

Allais, M. 1945. *Economie pure et rendement social.* Paris: Sirey.

Allais, M. 1947. *Economie et intérêt.* 2 vols, Paris: Imprimerie Nationale and Librairie des Publications Officielles.

Allais, M. 1952a. Introduction to the 2nd edn of Allais (1943). Paris: Imprimerie Nationale.

Allais, M. 1952b. L'éxtension des théories de l'équilibre économique général et du rendement social au cas du risque. *Colloques Internationaux du Centre National de la Recherche Scientifique* 40, *Econométrie,* 81–120. A summarized version was published under the same title in *Econometrica,* 21 April 1953, 269–90.

Allais, M. 1962. The influence of the capital output ratio on real national income. *Econometrica* 30, 700–28. Republished in American Economic Association, *Readings in Welfare Economics,* Vol. XII, 1969, 682–714, with an additional Note, 711–14.

Allais, M. 1963. The role of capital in economic development. In *Study Work on the Econometric Approach to Development Planning* (Oct. 7–13, 1963), Pontificiae Academiae Scientiarum Scripta Varia 28, Pontifica Academia Scientiarum, Amsterdam: North-Holland and Chicago: Rand McNally, 1965. 697–1002.

Allais, M. 1964. La théorie économique et la tarification optimum de l'usage des infrastructures de transport. *La Jaune et la Rouge* (publication of the Société Amicale des Anciens Eléves de l'Ecole Polytechnique), special issue *Les Transports,* Paris, 1964, 108–14.

Allais, M. 1967. Les conditions de l'éfficacité dans l'économie. Fourth International Seminar, Centro Studi e Ricerche su Problemi Economico-Sociali, Milan. Italian translation: 'Le condizioni dell' efficienza nell' economia' in *Programmazione E Progresso Economico,* Milan: Franco Angeli, 1969, 13–303. Original French text in M. Allais, *Les Fondements du Calcul Economique,* Paris: Ecole Nationale Supérieure des Mines de Paris, Vol. I, 1967.

Allais, M. 1968a. *Les fondements du calcul économique* 3 vols, Paris: Ecole Nationale Supérieure des Mines, Paris, Vol. I, 1967, and Vols II and III, 1968.

Allais, M. 1968b. The conditions of efficiency in the economy. *Economia Internazionale,* 399–420.

Allais, M. 1968c. Pareto, Vilfredo: contributions to Economics. In *International Encyclopedia of the Social Sciences,* New York: Macmillan and Free Press, Vol. 11, 399–411.

Allais, M. 1968d. Fisher, Irving. In *International Encyclopedia of the Social Sciences,* New York: Macmillan and Free Press, 1968, Vol. 5, 475–85.

Allais, M. 1968e. L'économique en tant que science. *Revue d'Economie Politique,* January–February, 5–30. Trans. as 'Economics as a Science', *Cahiers Vilfredo Pareto,* 1968, 5–24.

Allais, M. 1971. Les théories de l'équilibre économique général et de l'éfficacité maximale – impasse récentes et nouvelles perspectives. Congrès des Economistes de Langue Française, 2–6 June 1971. *Revue d'Economie Politique* 3, May–June, 331–409. Spanish translation: 'Las theorias del equilibrio economico general y de la eficacia maxima – recientes callejones sin salida y nuevas perspectivas', *El Trimestre Economico* (Mexico), 39(3), 1972. 557–633; English translation: see Allais (1974a).

Allais, M. 1973a. La théorie générale des surplus et l'apport fondamental de Vilfredo Pareto. *Revue d'Economie Politique* 6, November–December, 1044–97.

Allais, M. 1973b. The general theory of surplus and Pareto's fundamental contribution. *Convegno Internazionale Vilfredo Pareto* (Roma, 25–27 October 1973), Rome: Accademia Nazionale dei Lincei, 1975, 109–63. (English trans. of Allais, 1973a.)

Allais, M. 1974a. Theories of general economic equilibrium and maximum efficiency. Institute for Advanced Studies, Vienna, 1974. In *Equilibrium and Disequilibrium in Economic Theory*, ed. G. Schwödiauer. Dordrecht: Reidel, 1977, 129–201. (English version of Allais, 1971, with some additions.)

Allais, M. 1974b. Les implications de rendements croissants et décroissants sur les conditions de l'équilibre economique général et d'une éfficacité maximale. In *Hommage à François Perroux*, Grenoble: Presses Universitaires de Grenoble, 1978, 605–74.

Allais, M. 1981. *La théorie générale des surplus. Economies et Sociétés*, 2 vols, January–May 1981, Institut de Sciences mathématique et économie appliquées.

Allais, M. 1985. The concepts of surplus and loss and the reformulation of the theories of stable general economic equilibrium and maximum efficiency. In *Foundations and Dynamics of Economic Knowledge*, ed. M. Baranzini and R. Scazzieri, Oxford: Basil Blackwell, 135–74.

Allais, M. 1987. *The Equimarginal Principle, Meaning, Limits, and Generalisations*. Centre d'Analyse Economique. *Revista internazionale di scienze economiste e commerciale*.

Arrow, K.J. 1968. Economic equilibrium. In *International Encyclopedia of the Social Sciences*, New York: Macmillan and Free Press, Vol. 4, 376–89.

Arrow, K.J. and Hahn, F.H. 1971. *General Competitive Analysis*. San Francisco: Holden-Day; Edinburgh: Oliver.

Blaug, M. 1979. *Economic Theory in Retrospect*. 4th edn, 1985. London: Heinemann Educational Books.

Debreu, G. 1959. *Theory of Value*. New York: Wiley.

Debreu, G. 1985. *Theoretic Models: Mathematical Form and Economic Content*. Frisch Memorial Lecture, Fifth World Congress of the Econometric Society, MIT, 17–24 August 1985.

Dupuit, J. 1844. De la mesure de l'utilité des travaux publics. *Annales des Ponts et Chaussées*, 2nd series, Mémoires et Documents no. 116, vol. VIII, 332–75.

Dupuit, J. 1849. De l'influence des péages sur l'utilité des voies de communication. *Annales des Ponts et Chaussées*, 2nd series, 170–248.

Dupuit, J. 1853. De l'utilité et de sa mesure. *Journal des Economistes* 36(147), 1–28.

Edgeworth, F.Y. 1881. *Mathematical Psychics: An Essay on the Application of Mathematics to the Moral Sciences*. London: Kegan Paul. Reprinted, New York: Kelley, 1953.

Fisher, I. 1892. *Mathematical Investigations in the Theory of Value and Prices*. New Haven: Yale University Press, 1925.

Gossen, H.H. von 1854. *Entwickelung der Gesetze des menschlichen Verkehrs und der daraus fliessender Regeln für menschliches Handeln*. Third edn, 1927, introduction by Friedrich Hayek, Berlin: Präger.

Hutchison, T.W. 1977. *Knowledge and Ignorance in Economics*. Oxford: Blackwell.

Jevons, W.S. 1871. *The Theory of Political Economy*. London: Macmillan. 5th edn trans. as *La théorie de l'économie politique*, Paris: Giard, 1909.

Menger, C. 1871. *Grundsätze der Volkswirtschaftslehre*. Vienna: Braumneller.

Pareto, V. 1896–7. *Cours d'économie politique*. 2 vols, Lausanne: Rougé. Reprinted, Geneva: Droz, 1964.

Pareto, V. 1901. Anwendungen der Mathematik auf Nationalökonomie. *Encyklopädie der Mathematichen Wissenschaften*, Leipzig, Vol. I, 1094–120.

Pareto, V. 1906. *Manuale d'economia politica*. Milan. Trans. as *Manuel d'économie politique*, Paris: Giard et Brière, 1909, and Geneva: Droz, 1966; and as *Manual of Political Economy*, reprinted, New York: Kelley, 1971.

Pareto, V. 1911. Economie mathématique. *Encyclopedie des Sciences Mathématiques*, Paris: Gauthier-Villars, 1911, 591–641; also in *Statistique et économie mathématique*, Geneva: Droz, 1966,

319–76; published in English as 'Mathematical economics', *International Economic Papers* No. 5, New York, 1955, 58–102.

Ricardo, D. 1817. *On the Principles of Political Economy and Taxation*. Vol. I of *The Works and Correspondence of David Ricardo*, ed. Piero Sraffa, Cambridge: Cambridge University Press, 1951–1955.

Samuelson, P.A. 1947. *Foundations of Economic Analysis*. 2nd edn, Cambridge, Mass.: Harvard University Press, 1948.

Samuelson, P.A. 1950. Evaluation of real national income. *Oxford Economic Papers* NS, January, 1–29.

Walras, L. 1874–7. *Eléments d'économie politique pure – théorie de la richesse sociale*. 6th edn, Paris: Guillaumin; reprinted, Paris: Pichon et Durand-Auzias, 1952. English translation of the 6th edition as *Elements of Pure Economics*, ed. W. Jaffé, London: Allen & Unwin, 1954.

Woo, H.K.H. 1985. *What's Wrong with Formalization in Economics? – An Epistemological Critique*. Hong Kong: Hong Kong Institute of Economic Science.

economic theory and the hypothesis of rationality. In this paper, I want to disentangle some of the senses in which the hypothesis of rationality is used in economic theory. In particular, I want to stress that rationality is not a property of the individual alone, although it is usually presented that way. Rather, it gathers not only its force but also its very meaning from the social context in which it is embedded. It is most plausible under very ideal conditions. When these conditions cease to hold, the rationality assumptions become strained and possibly even self-contradictory. They certainly imply an ability at information processing and calculation that is far beyond the feasible and that cannot well be justified as the result of learning and adaptation.

Let me dismiss a point of view that is perhaps not always articulated but seems implicit in many writings. It seems to be asserted that a theory of the economy must be based on rationality, as a matter of principle. Otherwise, there can be no theory. This position has even been maintained by some who accept that economic behaviour is not completely rational. John Stuart Mill (1848, bk. 2, ch. 4) argued that custom, not competition, governs much of the economic world. But he adds that the only possible theory is that based on competition (which, in his theories, includes certain elements of rationality, particularly shifting capital and labour to activities that yield higher returns); 'Only through the principle of competition has political economy any pretension to the character of science', ([1848] 1909, p. 242).

Certainly, there is no general principle that prevents the creation of an economic theory based on hypotheses other than that of rationality. There are indeed some conditions that must be laid down for an acceptable theoretical analysis of the economy. Most centrally, it must include a theory of market interactions, corresponding to market clearing in the neoclassical general equilibrium theory. But as far as individual behaviour is concerned, any coherent theory of reactions to the stimuli appropriate in an economic context (prices in the simplest case) could in principle lead to a theory of the economy. In the case of consumer demand, the budget constraint must be satisfied but many theories can easily be devised that are quite different from utility maximization. For example, habit formation can be made into a theory; for a given price–income change, choose the bundle that satisfies the budget constraint and that requires the least change (in some suitably defined sense) from the previous consumption bundle. Though there is an optimization in this theory, it is different from utility maximization; for example, if prices and income return to their initial levels after several alterations, the final bundle purchased will not be the same as the initial. This theory would strike many lay observers as plausible, yet it is not rational as economists have used that term. Without belabouring the point, I simply observe that this theory is not only a logically complete explanation of behaviour but one that is more powerful than standard theory and at least as capable of being tested.

Not only is it possible to devise complete models of the economy on hypotheses other than rationality, but in fact virtually every practical theory of macroeconomics is partly so based. The price- and wage-rigidity elements of Keynesian theory are hard to fit into a rational framework, though some valiant efforts have been made. In the original form, the multiplier was derived from a consumption function depending only on current income. Theories more nearly based on rationality make consumption depend on lifetime or 'permanent' income and reduce the magnitude of the multiplier and, with it, the explanatory power of the Keynesian model. But if the Keynesian model is a natural target of criticism by the upholders of universal rationality, it must be added that monetarism is no better. I know of no serious derivation of the demand for money from a rational optimization. The loose arguments that substitute for a true derivation, Friedman's economizing on shoe leather or Tobin's transaction demand based on costs of buying and selling bonds, introduce assumptions incompatible with the costless markets otherwise assumed. The use of rationality in these arguments is ritualistic, not essential. Further, the arguments used would not suggest a very stable relation but rather one that would change quickly with any of the considerable changes in the structure and technology of finance. Yet the stability of the demand function for money must be essential to any form of monetarism, not excluding those rational expectations models in which the quantity theory plays a major role.

I believe that similar observations can be made about a great many other areas of applied economics. Rationality hypotheses are partial and frequently, if not always, supplemented by assumptions of a different character.

So far, I have argued simply that rationality is not in principle essential to a theory of the economy, and, in fact, theories with direct application usually use assumptions of a different nature. This was simply to clear the ground so that we can discuss the role of rationality in economic theory. As remarked earlier, rationality in application is not merely a property of the individual. Its useful and powerful implications derive from the conjunction of individual rationality and the other basic concepts of neoclassical theory – equilibrium, competition, and completeness of markets. The importance of all these assumptions was first made explicit by Frank Knight (1921, pp. 76–79). In the terms of Knight's one-time student, Edward Chamberlin (1950, pp. 6–7), we need not merely pure but perfect competition before the rationality hypotheses have their full power.

It is largely this theme on which I will expand. When these assumptions fail, the very concept of rationality becomes threatened, because perceptions of others and, in particular, of their rationality become part of one's own rationality. Even if there is a consistent meaning, it will involve computational and informational demands totally at variance with the traditional economic theorist's view of the decentralized economy.

Let me add one parenthetic remark to this section. Even if we make all the structural assumptions needed for perfect competition (whatever is needed by way of knowledge, concavity in production, absence of sufficient size to create market power, etc.), a question remains. How can equilibrium

be established? The attainment of equilibrium requires a disequilibrium process. What does rational behaviour mean in the presence of disequilibrium? Do individuals speculate on the equilibrating process? If they do, can the disequilibrium be regarded as, in some sense, a higher-order equilibrium process? Since no one has market power, no one sets prices; yet they are set and changed. There are no good answers to these questions, and I do not pursue them. But they do illustrate the conceptual difficulties of rationality in a multiperson world.

RATIONALITY AS MAXIMIZATION IN THE HISTORY OF ECONOMIC THOUGHT. Economic theory, since it has been systematic, has been based on some notion of rationality. Among the classical economists, such as Smith and Ricardo, rationality had the limited meaning of preferring more to less; capitalists choose to invest in the industry yielding the highest rate of return, landlords rent their property to the highest bidder, while no one pays for land more than it is worth in product. Scattered remarks about technological substitution, particularly in Ricardo, can be interpreted as taking for granted that, in a competitive environment, firms choose factor proportions, when they are variable, so as to minimize unit costs. To be generous about it, their rationality hypothesis was the maximization of profits by the firm, although this formulation was not explicitly achieved in full generality until the 1880s.

There is no hypothesis of rationality on the side of consumers among the classicists. Not until John Stuart Mill did any of the English classical economists even recognize the idea that demand might depend on price. Cournot had the concept a bit earlier, but neither Mill nor Cournot noticed – although it is obvious from the budget constraint alone – that the demand for any commodity must depend on the price of all commodities. That insight remained for the great pioneers of the marginalist revolution, Jevons, Walras, and Menger (anticipated, to be sure, by the Gregor Mendel of economics, H.H. Gossen, whose major work, completely unnoticed at the time of publication [1854], has now been translated into English [1983]). Their rationality hypothesis for the consumer was the maximization of the utility under a budget constraint. With this formulation, the definition of demand as a function of all prices was an immediate implication, and it became possible to formulate the general equilibrium of the economy.

The main points in the further development of the utility theory of the consumer are well known. (1) Rational behaviour is an ordinal property. (2) The assumption that an individual is behaving rationally has indeed some observable implications, the Slutsky relations, but without further assumptions, they are not very strong. (3) In the aggregate, the hypothesis of rational behaviour has in general no implications; that is, for any set of aggregate excess demand functions, there is a choice of preference maps and of initial endowments, one for each individual in the economy, whose maximization implies the given aggregate excess demand functions (Sonnenschein, 1973; Mantel, 1974; Debreu, 1974; for a survey, see Shafer and Sonnenschein, 1982, sec. 4).

The implications of the last two remarks are in contradiction to the very large bodies of empirical and theoretical research, which draw powerful implications from utility maximization for, respectively, the behaviour of individuals, most especially in the field of labour supply, and the performance of the macroeconomy based on 'new classical' or 'rational expectations' models. In both domains, this power is obtained by adding strong supplementary assumptions to the general model of rationality. Most prevalent of all is the assumption that all individuals have the same utility function (or at least that they differ only in broad categories based on observable magnitudes, such as family size). But this postulate leads to curious and, to my mind, serious difficulties in the interpretation of evidence. Consider the simplest models of human capital formation. Cross-sectional evidence shows an increase of wages with education or experience, and this is interpreted as a return on investment in the form of foregone income and other costs. But if all individuals are alike, why do they not make the same choice? Why do we observe a dispersion? In the human capital model (a particular application of the rationality hypothesis), the only explanation must be that individuals are not alike, either in ability or in tastes. But in that case the cross-sectional evidence is telling us about an inextricable mixture of individual differences and productivity effects. Analogously, in macroeconomic models involving durable assets, especially securities, the assumption of homogeneous agents implies that there will never be any trading, though there will be changes in prices.

This dilemma is intrinsic. If agents are all alike, there is really no room for trade. The very basis of economic analysis, from Smith on, is the existence of differences in agents. But if agents are different in unspecifiable ways, then remark (3) above shows that very few, if any, inferences can be made. This problem, incidentally, already exists in Smith's discussion of wage differences. Smith did not believe in intrinsic differences in ability; a porter resembled a philosopher more than a greyhound did a mastiff. Wage differences then depended on the disutilities of different kinds of labour, including the differential riskiness of income. This is fair enough and insightful. But, if taken seriously, it implies that individuals are indifferent among occupations, with wages compensating for other differences. While there is no logical problem, the contradiction to the most obvious evidence is too blatant even for a rough approximation.

I have not carried out a scientific survey of the uses of the rationality hypothesis in particular applications. But I have read enough to be convinced that its apparent force comes only from the addition of supplementary hypotheses. Homogeneity across individual agents is not the only auxiliary assumption, though it is the deepest. Many assumptions of separability are frequently added. Indeed, it has become a working methodology to start with very strong assumptions of additivity and separability, together with a very short list of relevant variables, to add others only as the original hypotheses are shown to be inadequate, and to stop when some kind of satisfactory fit is obtained. A failure of the model is attributed to a hitherto overlooked benefit or cost. From a statistical viewpoint, this stopping rule has obvious biases. I was taught as a graduate student that data mining was a major crime; morality has changed here as elsewhere in society, but I am not persuaded that all these changes are for the better.

The lesson is that the rationality hypothesis is by itself weak. To make it useful, the researcher is tempted into some strong assumptions. In particular, the homogeneity assumption seems to me to be especially dangerous. It denies the fundamental assumption of the economy, that it is built on gains from trading arising from individual differences. Further, it takes attention away from a very important aspect of the economy, namely, the effects of the distribution of income and of other individual characteristics on the workings of the economy. To take a major example, virtually all of the literature on savings behaviour based on aggregate data assumes homogeneity. Yet there have been repeated studies that suggest that saving is not proportional to income, from which it would follow that distributional considerations matter. (In general, as data have improved, it has become increasingly difficult to find any

simple rationally based model that will explain savings, wealth, and bequest data.)

The history of economic thought shows some other examples and difficulties with the application of the rationality hypothesis. Smith and the later classicists make repeated but unelaborated references to risk as a component in wage differences and in the rate of return on capital (e.g., Mill, [1848] 1909, pp. 385, 406, 407, 409). The English marginalists were aware of Bernoulli's expected-utility theory of behaviour under uncertainty (probably from Todhunter's *History of the Theory of Probability*) but used it only in a qualitative and gingerly way (Jevons, [1871] 1965, pp. 159–60; Marshall, 1920, pp. 842–3). It was really not until the last 30 years that it has been used systematically as an economic explanation, and indeed its use coincided with the first experimental evidence against it (see Allais, 1979). The expected-utility hypothesis is an interesting transition to the theme of the next section. It is in fact a stronger hypothesis than mere maximization. As such it is more easily tested, and it leads to stronger and more interesting conclusions. So much, however, has already been written about this area that I will not pursue it further here.

RATIONALITY, KNOWLEDGE, AND MARKET POWER. It is noteworthy that the everyday usage of the term 'rationality' does not correspond to the economist's definition as transitivity and completeness, that is, maximization of something. The common understanding is instead the complete exploitation of information, sound reasoning, and so forth. This theme has been systematically explored in economic analysis, theoretical and empirical, only in the last 35 years or so. An important but neglected predecessor was Holbrook Working's random-walk theory of fluctuations in commodity futures and securities prices (1953). It was based on the hypothesis that individuals would make rational inferences from data and act on them; specifically, predictability of future asset prices would be uncovered and used as a basis for current demands, which would alter current prices until the opportunity for gain was wiped out.

Actually, the classical view had much to say about the role of knowledge, but in a very specific way. It emphasized how a complete price system would require individuals to know very little about the economy other than their own private domain of production and consumption. The profoundest observation of Smith was that the system works behind the backs of the participants; the directing 'hand' is 'invisible'. Implicitly, the acquisition of knowledge was taken to be costly.

Even in a competitive world, the individual agent has to know all (or at least a great many) prices and then perform an optimization based on that knowledge. All knowledge is costly, even the knowledge of prices. Search theory, following Stigler (1961), recognized this problem. But search theory cannot easily be reconciled with equilibrium or even with individual rationality by price setters, for identically situated sellers should set identical prices, in which case there is nothing to search for.

The knowledge requirements of the decision may change radically under monopoly or other forms of imperfect competition. Consider the simplest case, pure monopoly in a one-commodity partial equilibrium model, as originally studied by Cournot in 1838. The firm has to know not only prices but a demand curve. Whatever definition is given to complexity of knowledge, a demand curve is more complex than a price. It involves knowing about the behaviour of others. Measuring a demand curve is usually thought of as a job for an econometrician. We have the curious situation that scientific analysis imputes scientific behaviour to its subjects.

This need not be a contradiction, but it does seem to lead to an infinite regress.

From a general equilibrium point of view, the difficulties are compounded. The demand curve relevant to the monopolist must be understood *mutatis mutandis*, not *ceteris paribus*. A change in the monopolist's price will in general cause a shift in the purchaser's demands for other goods and therefore in the prices of those commodities. These price changes will in turn affect by more than one channel the demand for the monopolist's produce and possibly also the factor prices that the monopolist pays. The monopolist, even in the simple case where there is just one in the entire economy, has to understand all these repercussions. In short, the monopolist has to have a full general equilibrium model of the economy.

The informational and computational demands become much stronger in the case of oligopoly or any other system of economic relations where at least some agents have power against each other. There is a qualitatively new aspect to the nature of knowledge, since each agent is assuming the *rationality* of other agents. Indeed, to construct a rationality-based theory of economic behaviour, even more must be assumed, namely, that the rationality of all agents must be *common knowledge*, to use the term introduced by the philosopher David Lewis (1969). Each agent must not only know that the other agents (at least those with significant power) are rational but know that each other agent knows every other agent is rational, know that every other agent knows that every other agent is rational, and so forth (see also Aumann, 1976). It is in this sense that rationality and the knowledge of rationality is a social and not only an individual phenomenon.

Oligopoly is merely the most conspicuous example. Logically, the same problem arises if there are two monopolies in different markets. From a practical viewpoint, the second case might not offer such difficulties if the links between the markets were sufficiently loose and the monopolies sufficiently small on the scale of the economy that interaction was negligible; but the interaction can never be zero and may be important. As usually presented, bargaining to reach the contract curve would, in the simplest case, require common knowledge of the bargainer's preferences and production functions. It should be obvious how vastly these knowledge requirements exceed those required for the price system. The classic economists were quite right in emphasizing the importance of limited knowledge. If every agent has a complete model of the economy, the hand running the economy is very visible indeed.

Indeed, under these knowledge conditions, the superiority of the market over centralized planning disappears. Each individual agent is in effect using as much information as would be required for a central planner. This argument shows the severe limitations in the argument that property rights suffice for social rationality even in the absence of a competitive system (Coase, 1960).

One can, as many writers have, discuss bargaining when individuals have limited knowledge of each other's utilities (similarly, we can have oligopoly theory with limited knowledge of the cost functions of others: see, e.g., Arrow, 1979). Oddly enough, it is not clear that limited knowledge means a smaller quantity of information than complete knowledge, and optimization under limited knowledge is certainly computationally more difficult. If individuals have private information, the others form some kind of conjecture about it. These conjectures must be common knowledge for there to be a rationality-based hypothesis. This seems to have as much informational content and to be as unlikely as

knowing the private information. Further, the optimization problem for each individual based on conjectures (in a rational world, these are probability distributions) on the private information of others is clearly a more difficult and therefore computationally more demanding problem than optimization when there is no private information.

RATIONAL KNOWLEDGE AND INCOMPLETE MARKETS. It may be supposed from the foregoing that informational demands are much less in a competitive world. But now I want to exemplify the theme that perfect, not merely pure, competition is needed for that conclusion and that perfect competition is a stronger criterion than Chamberlin perhaps intended. A complete general equilibrium system, as in Debreu (1959), requires markets for all contingencies in all future periods. Such a system could not exist. First, the number of prices would be so great that search would become an insuperable obstacle; that is, the value of knowing prices of less consequence, those of events remote in time or of low probability, would be less than the cost so that these markets could not come into being. Second, markets conditional on privately observed events cannot exist by definition.

In any case, we certainly know that many – in fact, most – markets do not exist. When a market does not exist, there is a gap in the information relevant to an individual's decision, and it must be filled by some kind of conjecture, just as in the case of market power. Indeed, there turn out to be strong analogies between market power and incomplete markets, though they seem to be very different phenomena.

Let me illustrate with the rational expectations equilibrium. Because of intertemporal relations in consumption and production, decisions made today have consequences that are anticipated. Marshall (1920, bk 5, chs 3–5) was perhaps the first economist to take this issue seriously. He introduced for this purpose the vague and muddled concepts of the short and long runs, but at least he recognized the difficulties involved, namely, that some of the relevant terms of trade are not observable on the market. (Almost all other accounts implicitly or explicitly assumed a stationary state, in which case the relative prices in the future and between present and future are in effect current information. Walras (1874, lessons 23–25) claimed to treat a progressive state with net capital accumulation, but he wound up unwittingly in a contradiction, as John Eatwell has observed in an unpublished dissertation. Walras's arguments can only be rescued by assuming a stationary state.) Marshall in effect made current decisions, including investment and savings, depend on expectations of the future. But the expectations were not completely arbitrary; in the absence of disturbances, they would converge to correct values. Hicks (1946, chs 9–10) made the dependence of current decisions on expectations more explicit, but he had less to say about their ultimate agreement with reality.

As has already been remarked, the full competitive model of general equilibrium includes markets for all future goods and, to take care of uncertainty, for all future contingencies. Not all of these markets exist. The new theoretical paradigm of rational expectations holds that each individual forms expectations of the future on the basis of a correct model of the economy, in fact, the same model that the econometrician is using. In a competitive market-clearing world, the individual agent needs expectations of prices only, not of quantities. For a convenient compendium of the basic literature on rational expectations, see Lucas and Sargent (1981). Since the world is uncertain, the expectations take the form of probability distributions, and each agent's expectations are conditional on the information available to him or her.

As can be seen, the knowledge situation is much the same as with market power. Each agent has to have a model of the entire economy to preserve rationality. The cost of knowledge, so emphasized by the defenders of the price system as against centralized planning, has disappeared; each agent is engaged in very extensive information gathering and data processing.

Rational expectations theory is a stochastic form of perfect foresight. Not only the feasibility but even the logical consistency of this hypothesis was attacked long ago by Morgenstern (1935). Similarly, the sociologist Robert K. Merton (1957) argued that forecasts could be self-denying or self-fulfilling; that is, the existence of the forecast would alter behaviour so as to cause the forecast to be false (or possibly to make an otherwise false forecast true). The logical problems were addressed by Grunberg and Modigliani (1954) and by Simon (1957, ch. 5). They argued that, in Merton's terms, there always existed a self-fulfilling prophecy. If behaviour varied continuously with forecasts and the future realization were a continuous function of behaviour, there would exist a forecast that would cause itself to become true. From this argument, it would appear that the possibility of rational expectations cannot be denied. But they require not only extensive first-order knowledge but also common knowledge, since predictions of the future depend on other individuals' predictions of the future. In addition to the information requirements, it must be observed that the computation of fixed points is intrinsically more complex than optimizing.

Consider now the signalling equilibrium originally studied by Spence (1974). We have large numbers of employers and workers with free entry. There is no market power as usually understood. The ability of each worker is private information, known to the worker but not to the employer. Each worker can acquire education, which is publicly observable. However, the cost of acquiring the education is an increasing function of ability. It appears natural to study a competitive equilibrium. This takes the form of a wage for each educational level, taken as given by both employers and workers. The worker, seeing how wages vary with education, chooses the optimal level of education. The employer's optimization leads to an 'informational equilibrium' condition, namely, that employers learn the average productivity of workers with a given educational level. What dynamic process would lead the market to learn these productivities is not clear, when employers are assumed unable to observe the productivity of individual workers. There is more than one qualitative possibility for the nature of the equilibrium. One possibility, indeed, is that there is no education, and each worker receives the average productivity of all workers (I am assuming for simplicity that competition among employers produces a zero-profit equilibrium). Another possibility, however, is a dispersion of workers across educational levels; it will be seen that in fact workers of a given ability all choose the same educational level, so the ability of the workers could be deduced from the educational level ex post.

Attractive as this model is for certain circumstances, there are difficulties with its implementation, and at several different levels. (1) It has already been noted that the condition that, for each educational level, wages equal average productivity of workers is informationally severe. (2) Not only is the equilibrium not unique, but there is a continuum of possible equilibria. Roughly speaking, all that matters for the motivation of workers to buy education are the relative wages at different educational levels; hence, different relations between wages and education are equally self-fulfilling. As will be seen below, this phenomenon is not peculiar to this model.

On the contrary, the existence of a continuum of equilibria seems to be characteristic of many models with incomplete markets. Extensive non-uniqueness in this sense means that the theory has relatively little power. (3) The competitive equilibrium is fragile with respect to individual actions. That is, even though the data of the problem do not indicate any market power, at equilibrium it will frequently be possible for any firm to profit by departing from the equilibrium.

Specifically, given an equilibrium relation between wages and education, it can pay a firm to offer a different schedule and thereby make a positive profit (Riley, 1979). This is not true in a competitive equilibrium with complete markets, where it would never pay a firm to offer any price or system of prices other than the market's. So far, this instability of competitive equilibrium is a property peculiar to signalling models, but it may be more general.

As remarked above, the existence of a continuum of equilibria is now understood to be a fairly common property of models of rational market behaviour with incomplete information. Thus, if there were only two commodities involved and therefore only one price ratio, a continuum of equilibria would take the form of a whole interval of price ratios. This multiplicity would be nontrivial, in that each different possible equilibrium price ratio would correspond to a different real allocation.

One very interesting case has been discussed recently. Suppose that we have some uncertainty about the future. There are no contingent markets for commodities; they can be purchased on spot markets after the uncertainty is resolved. However, there is a set of financial contingent securities, that is, insurance policies that pay off in money for each contingency. Purchasing power can therefore be reallocated across states of the world. If there are as many independent contingent securities as possible states of the world, the equilibrium is the same as the competitive equilibrium with complete markets, as already noted in Arrow (1953). Suppose there are fewer securities than states of the world. Then some recent and partly still unpublished literature (Duffie, 1985; Werner, 1985; Geanakoplos and Mas-Colell, 1986) shows that the prices of the securities are arbitrary (the spot prices for commodities adjust accordingly). This is not just a numéraire problem; the corresponding set of equilibrium real allocation has a dimensionality equal to the number of states of nature.

A related model with a similar conclusion of a continuum of equilibria is the concept of 'sunspot' equilibria (Cass and Shell, 1983). Suppose there is some uncertainty about an event that has in fact no impact on any of the data of the economy. Suppose there is a market for a complete set of commodity contracts contingent on the possible outcomes of the event, and later there are spot markets. However, some of those who will participate in the spot markets cannot participate in the contingent commodity markets, perhaps because they have not yet been born. Then there is a continuum of equilibria. One is indeed the equilibrium based on 'fundamentals,' in which the contingencies are ignored. But there are other equilibria that do depend on the contingency that becomes relevant merely because everyone believes it is relevant. The sunspot equilibria illustrate that Merton's insight was at least partially valid; we can have situations where social truth is essentially a matter of convention, not of underlying realities.

THE ECONOMIC ROLE OF INFORMATIONAL DIFFERENCES. Let me mention briefly still another and counterintuitive implication of thoroughgoing rationality. As I noted earlier, identical individuals do not trade. Models of the securities markets based on homogeneity of individuals would imply zero trade;

all changes in information are reflected in price changes that just induce each trader to continue holding the same portfolio. It is a natural hypothesis that one cause of trading is difference of information. If I learn something that affects the price of a stock and others do not, it seems reasonable to postulate that I will have an opportunity to buy or sell it for profit.

A little thought reveals that, if the rationality of all parties is common knowledge, this cannot occur. A sale of existing securities is simply a complicated bet, that is, a zero-sum transaction (between individuals who are identical apart from information). If both are risk averters, they would certainly never bet or, more generally, buy or sell securities to each other if they had the same information. If they have different information, each one will consider that the other has some information that he or she does not possess. An offer to buy or sell itself conveys information. The offer itself says that the offerer is expecting an advantage to himself or herself and therefore a loss to the other party, at least as calculated on the offerer's information. If this analysis is somewhat refined, it is easy to see that no transaction will in fact take place, though there will be some transfer of information as a result of the offer and rejection. The price will adjust to reflect the information of all parties, though not necessarily all the information.

Candidly, this outcome seems most unlikely. It leaves as explanation for trade in securities and commodity futures only the heterogeneity of the participants in matters other than information. However, the respects in which individuals differ change relatively slowly, and the large volume of rapid turnover can hardly be explained on this basis. More generally, the role of speculators and the volume of resources expended on informational services seem to require a subjective belief, at least, that buying and selling are based on changes in information.

SOME CONCLUDING REMARKS. The main implication of this extensive examination of the use of the rationality concept in economic analysis is the extremely severe strain on information-gathering and computing abilities. Behaviour of this kind is incompatible with the limits of the human being, even augmented with artificial aids (which, so far, seem to have had a trivial effect on productivity and the efficiency of decision making). Obviously, I am accepting the insight of Herbert Simon (1957, chs 14, 15), on the importance of recognizing that rationality is bounded. I am simply trying to illustrate that many of the customary defences that economists use to argue, in effect, that decision problems are relatively simple break down as soon as market power and the incompleteness of markets are recognized.

But a few more lessons turned up. For one thing, the combination of rationality, incomplete markets, and equilibrium in many cases leads to very weak conclusions, in the sense that there are whole continua of equilibria. This, incidentally, is a conclusion that is being found increasingly in the analysis of games with structures extended over time; games are just another example of social interaction, so the common element is not surprising. The implications of this result are not clear. On the one hand, it may be that recognizing the limits on rationality will reduce the number of equilibria. On the other hand, the problem may lie in the concept of equilibrium.

Rationality also seems capable of leading to conclusions flatly contrary to observation. I have cited the implication that there can be no securities transactions due to differences of information. Other similar propositions can be advanced, including the well-known proposition that there cannot be any money lying in the street, because someone else would have picked it up already.

The next step in analysis, I would conjecture, is a more consistent assumption of computability in the formulation of economic hypotheses. This is likely to have its own difficulties because, of course, not everything is computable, and there will be in this sense an inherently unpredictable element in rational behaviour. Some will be glad of such a conclusion.

KENNETH J. ARROW

Reprinted from *Journal of Business*, 1986, vol. 59, no. 4, pt. 2.

See also BOUNDED RATIONALITY; MODELS AND THEORY; PREFERENCES; RATIONAL BEHAVIOUR.

BIBLIOGRAPHY

Allais, M. 1979. The so-called Allais paradox and rational decisions under uncertainty. In *Expected Utility Hypothesis and the Allais Paradox*, ed. M. Allais and O. Hagen, Boston: Reidel.

Arrow, K.J. 1953. Le rôle des valeurs boursières dans la répartition la meilleure des risques. In *Économétrie*, Paris: Centre National de la Recherche Scientifique.

Arrow, K.J. 1979. The property rights doctrine and demand revelation under incomplete information. In *Economics and Human Welfare*, ed. M.J. Boskin, New York: Academic Press.

Aumann, R.J. 1976. Agreeing to disagree. *Annals of Statistics* 4, 1236–9.

Cass, D. and Shell, K. 1983. Do sunspots matter? *Journal of Political Economy* 91, 193–227.

Chamberlin, E. 1950. *The Theory of Monopolistic Competition*. 6th edn, Cambridge, Mass.: Harvard University Press.

Coase, R. 1960. The problem of social cost. *Journal of Law and Economics* 3, 1–44.

Cournot, A.A. 1838. *Researches into the Mathematical Principles of the Theory of Wealth*. Translated by N.T. Bacon, New York: Macmillan, 1927.

Debreu, G. 1959. *Theory of Value*. New York: Wiley.

Debreu, G. 1974. Excess demand functions. *Journal of Mathematical Economics* 1, 15–23.

Duffie, J.D. 1985. Stochastic equilibria with incomplete financial markets. Research Paper No. 811, Stanford: Stanford University, Graduate School of Business.

Geanakoplos, J. and Mas-Colell, A. 1986. Real indeterminacy with financial assets. Paper No. MSRI 717–86, Berkeley: Mathematical Science Research Institute.

Gossen, H.H. 1983. *The Laws of Human Relations*. Cambridge, Mass.: MIT Press.

Grunberg, E. and Modigliani, F. 1954. The predictability of social events. *Journal of Political Economy* 62, 465–78.

Hicks. H.R. 1946. *Value and Capital*. 2d edn, Oxford: Clarendon.

Jevons, W.S. 1871. *The Theory of Political Economy*. 5th edn; reprinted, New York: Kelley, 1965.

Knight, F. 1921. *Risk, Uncertainty, and Profit*. Boston: Houghton Mifflin.

Lewis, D. 1969. *Convention*. Cambridge, Mass.: Harvard University Press.

Lucas, R. and Sargent, T. 1981. *Rational Expectations and Econometric Practice*. 2 vols, Minneapolis: University of Minnesota Press.

Mantel, R. 1974. On the characterization of excess demand. *Journal of Economic Theory* 6, 345–54.

Marshall, A. 1920. *Principles of Economics*. 8th edn; reprinted, New York: Macmillan, 1948.

Merton, R.K. 1957. The self-fulfilling prophecy. In R.K. Merton, *Social Theory and Social Structure*, revised and enlarged edn, Glencoe, Ill.: Free Press.

Mill, J.S. 1848. *Principles of Political Economy*. London: Longmans, Green, 1909.

Morgenstern, O. 1935. Vollkommene Voraussicht und wirtschaftliches Gleichgewicht. *Zeitschrift für Nationalökonomie* 6, 337–57.

Riley, J.G. 1979. Informational equilibrium. *Econometrica* 47, 331–60.

Shafer, W. and Sonnenschein, H. 1982. Market demand and excess demand functions. In *Handbook of Mathematical Economics*. Vol. 2, ed. K.J. Arrow and M. Intriligator, Amsterdam: North-Holland.

Simon, H. 1957. *Models of Man*. New York: Wiley.

Spence, A.M. 1974. *Market Signaling*. Cambridge, Mass.: Harvard University Press.

Sonnenschein, H. 1973. Do Walras's identity and continuity characterize the class of community excess demand functions? *Journal of Economic Theory* 6, 345–54.

Stigler, G.J. 1961. The economics of information. *Journal of Political Economy* 69, 213–25.

Walras, L. 1874. *Elements of Pure Economics*. Translated by W. Jaffé, London: Allen & Unwin, 1954.

Werner, J. 1985. Equilibrium in economies with incomplete financial markets. *Journal of Economic Theory* 36, 110–19.

Working, H. 1953. Futures trading and hedging. *American Economic Review* 43, 314–43.

economic theory of the state. The basic forms, social functions, institutional boundaries and legitimating principles of states vary across historical epochs and also differ among specific regimes in the same epoch. This makes it difficult (some would even say impossible) to develop a theory which applies to all states – whether in general or simply in their economic aspects. This entry limits itself to some economic aspects of the capitalist state.

THE CAPITALIST TYPE OF STATE. Capital accumulation has occurred under the most divergent state forms, but not all state forms are equally supportive of capital accumulation. Various attempts have been made to construct theoretically an ideal type of state which is both possible and particularly appropriate under capitalism without claiming, however, that this 'capitalist type of state' exists always and everywhere in capitalist societies. Among other characteristics of this state form, three institutional features are worth noting here: it has an effective monopoly of coercive power, its resources are purchased with money derived from taxation, and its activities are subject to the rule of law. Each of these features is not only compatible with but also potentially supportive of the capitalist economic order.

Firstly, the state is able to monopolize coercion because capital appropriates the surplus labour of workers through the wage-relation rather than through extra-economic compulsion. This monopoly is also functional since it prevents particular economic agents from using direct force to subvert the free play of market forces. Secondly, state resources can be purchased because capitalism involves generalized commodity production and money mediates the exchange of all commodities (including labour-power). The state should raise monetary taxation because it cannot meet its reproduction needs by selling its own output, and cannot expropriate them forcibly only from those who happen to produce them without undermining the formal equality and property rights which underpin capitalism. Thirdly, the rule of law can exist because capitalism presupposes the formal freedom and equality of all economic agents. Only if it exists can such agents rely on a stable and impartial legal and political environment for their long-term economic activities.

These three institutional traits of the state facilitate capital accumulation. But they are neither logically nor historically necessary; nor, where they occur, do they guarantee accumulation. This is not simply because economic factors themselves engender recurrent crises within capitalism. There are also distinct political reasons. These are rooted in the institutional form of the state and in the struggles which occur around the nature and purposes of state power. To take only three examples. The institutional separation between the state and economy is crystallized above all in the state's legitimate coercive monopoly and its incarnation of national-popular unity vis-à-vis the antagonistic private interests of civil society. This means that the state has the political and ideological capacities to disturb as well as to promote capital accumulation. Nor does the tax form have any self-evident limits. It can produce fiscal crises and/or disproportions between state expenditure and the requirements of capital accumulation. Thirdly, because the rule of law implies formal neutrality towards particular economic agents, it is correspondingly inadequate as a steering mechanism. But more purposive, ad hoc, discretionary interventions can produce bureaucratic overload and also disrupt the labour process and capitalist market forces. Whether such problems occur depends not only on the form of the state and its integration into the circuit of capital but also on the changing balance of political forces.

ECONOMIC ASPECTS OF THE CAPITALIST STATE. Nowhere are economic systems self-reproducing, self-regulating and self-sufficient. They always depend on other institutional systems and the contingent support of non-economic forces. The capitalist state clearly has a key role in securing such institutional preconditions; and it is also the nodal site for political support. This does not mean, however, that one can enumerate a set of essential economic functions which must be performed by the capitalist state. Indeed, paraphrasing Max Weber's more general comment on the modern state, one could say: there are no economic activities which capitalist states have not at some time undertaken and none which they undertake invariably and exclusively. In particular the capitalist state is neither confined to producing 'public goods' nor is it the sole producer of such goods. Instead, even if certain broad developmental tendencies can be identified, its precise economic activities are always conjunctural. They are always influenced, furthermore, by political and ideological as well as economic factors.

ECONOMIC PERIODIZATION OF THE CAPITALIST STATE. The structural relations between state and economy and the forms of state intervention typically vary across time as well as nations. This has encouraged attempts at periodization. Although labels vary, four phases are often identified: mercantilism, liberal capitalism, (simple) monopoly capitalism and late (or state monopoly) capitalism. Without necessarily endorsing these attempts at periodization, the basic features of each stage can be presented as follows.

Under mercantilism state power is used to establish the dominance of the capital relation and market forces. This is the period of primitive accumulation and capitalist manufacture and is associated historically with the absolutist state. Once this dominance is secured, a liberal phase is said to follow. This involves the nightwatchman state which is restricted to securing the general external conditions of production and has no significant directly economic role. The third phase is linked to the dominance of monopoly capital and the rise of imperialism. In this stage the state serves to regulate the economic dominance of monopoly capital, assumes an active role of managing the economic and political relations between organized capital and the labour movement, and also employs extra-economic coercion abroad in inter-imperialist competition. Next comes the state monopoly capitalist stage. State management of the domestic economy through taxation, state credit, public enterprise and/or the so-called military-industrial complex now have an increasingly important role; the welfare state system and collective consumption become central to the reproduction of labour-power and to political management; and international and transnational state organizations have a key role in managing the world economy. Not all national economies have experienced all four stages and much depends on the timing of their capitalist development and on their place within the international division of labour.

Such changes in the state's economic role also involve reorganizing its overall institutional form. Growing state intervention is typically associated with the strengthening of the executive at the expense of the legislative branch, the rise of functional (as opposed to territorial) representation closely tied to the administration, the increased importance of the state economic apparatus and the growing dominance of economic criteria within non-economic departments, and the decline of the substantive rule of law (as opposed to the simple maintenance of legal forms) in favour of more discretionary forms of intervention. Thus the growth of state economic intervention leaves neither the economy nor the state unchanged. The circuit of capitals is socialized through the state and the state is reorganized to reflect economic needs (cf. Poulantzas, 1978).

EXPLANATIONS FOR THE ECONOMIC ROLE OF THE STATE. Various explanations have been offered for the state's assumption of economic functions and for their general developmental tendencies. Broadly speaking these comprise two main groups: explanations which focus on the essential structure and laws of motion of capitalism and explanations which focus on the social relations which obtain between class forces. Included among the former are explanations which emphasize the inability or failure of individual capitals, market forces or the law of value to secure all the institutional and economic conditions needed for capital accumulation. These conditions are frequently said to include: (a) 'general external conditions' such as bourgeois law or a formally rational monetary system; (b) public goods such as fire services, sea walls or statistical services which facilitate production in all branches; and (c) material factors productively consumed in all or most branches, such as labour-power or energy supplies. The more 'class-theoretical' explanations focus either on the state's instrumentalization by particular (capitalist) class interests and/or on its relatively autonomous role in managing the balance of class forces both within the economic sphere and in society more generally. In turn the relative strength of class forces is sometimes attributed to changes in the mode of production and sometimes to broader social and political factors ranging from unionization to wars.

The reasons advanced for increasing state intervention can be used to illustrate such arguments. Some theorists highlight changes in the forces of production (e.g their increasing socialization, growing capital intensity or lengthening turnover time of capital). Others emphasize changes in the relations of production (e.g. the shift from absolute to relative surplus value, growth of monopoly capital, increased importance of banking or financial capital, the internationalization of production, or changing forms of economic crisis). Yet others have stressed an increased importance of the tendency of the rate of profit to fall. Whatever reasons are advanced, however, the same conclusion is drawn. In the course of capital accumulation there is a growing need for state intervention to socialize the forces of production (e.g. infrastructural provision, manpower training, technological innovation)

and/or the relations of production (e.g. state credit, economic management or collective consumption) to compensate for the failures of market forces and competition adequately to coordinate and integrate the circuit of capital. Whilst such explanations often identify important structural changes in capitalism, they do not provide a satisfactory explanation for the political response to such changes. Nor does an emphasis on the mediating role of class struggle help much here unless attention is paid to the full range and forms of political forces.

THE GENERAL LIMITS TO STATE INTERVENTION. There has been considerable interest in the limits as well as the reasons for state intervention. Again we find both general explanations and arguments relating to various stages of intervention. The following factors are frequently cited here: (a) the exclusion of the state from the heart of the production process – which means it must react *a posteriori* to events it cannot directly control or engage in ineffective *a priori* planning; (b) its tendency to respond to economic problems and crises in terms of surface appearances (e.g. inflation, unemployment, trade deficits) which have no obvious or consistent relationship to the real course of capital accumulation – which means that state policies often have limited or perverse effects; (c) the inherent limitations of law and money as steering mechanisms for a constitutional tax-state – since both mechanisms operate at a distance from real economic agents and processes; (d) the contradictions involved in the expansion of non-commodity forms of provision – they may promote capital accumulation but they also withdraw money from the circuit of capital, they can promote fiscal crises, and they suggest that the commodity form is neither natural nor necessary; and (e) the *sui generis* interests of state managers which can conflict with the supposed needs of capital. Most of these difficulties are aggravated by the co-existence of an effective world economy and a multiplicity of nation-states.

POLITICAL AND IDEOLOGICAL COMPLICATIONS. The state's economic role is always affected by its other tasks. These include its own organizational reproduction, maintaining domestic political order and territorial integrity, and defining and interpreting national unity. Thus economic policies are typically inserted into more general political strategies and influenced by political and ideological struggles. This affects the inputs, 'withinputs' and outputs of the state system.

On the input side economic needs must be translated into political demands through whatever organizational and institutional channels are available; and they must be coupled with political values and legal norms which are often only indirectly relevant to economic considerations. Within the state system it is the balance of political forces which determines how these economic demands are expressed in economic policies. This will vary with the individual forms of policy production (e.g. bureaucratic, purposive programming, participation, delegation to professionals) and with the manner in which some basic unity is imposed on the state's manifold activities. Each mode of policy-production contains its own limitations; moreover, problems of internal unity often preclude the flexible responses needed for economic management. All this is aggravated because political forces are generally most immediately concerned with other political forces and only indirectly with the economic sphere. Accordingly, it is the political repercussions of economic events and crises which matter more than their inherent economic form or substance. Finally, the outputs of the state are generally mediated in and through its own forms of intervention which operate at one or more removes from the real economy.

Even the increasingly dominant state economic apparatus must operate in this environment and it is also prey to muddling through, administrative inertia, political pressures and ideological thinking. State-owned industries and central banks typically operate in a political environment which shapes their economic activities and distinguish them from private industrial or financial enterprises. In general, state intervention reflects the balance among all political forces and these extend well beyond the classes, fractions and strata defined by the circuit of capital. This helps to explain the incoherence of economic policies and the difficulties of rational economic planning.

Indeed the state's current expanded role involves two double-binds: the one economic, the other political. Firstly, when the state intervenes to alleviate structural economic crises, it must substitute its own policies for the purgative effects of market-mediated reorganization. Thus it typically changes the forms in which economic crises operate rather than eliminating them and even internalizes such crises within the state. Here they can take such forms as fiscal crises, legitimacy crises, representational crises, crises of internal unity and crises of governmental effectiveness or overload. But, since the state's role has now become vital for accumulation, it cannot solve economic crises simply by withdrawing or refusing to intervene. At best it can reorganize how it intervenes. Moreover, in so far as economic crises are seen to follow from such withdrawal, refusal or reorganization, they can also precipitate new forms of political crisis. Secondly, in attempting to resolve crises on behalf of capital, it faces a political dilemma. If its crisis-management deliberately favours one fraction of capital at the expense of others, it is liable to aggravate economic problems for capital as a whole and to weaken its own legitimacy. But even if it succeeds in winning support for policies in the collective interests of capital, it cannot thereby avoid favouring some capitals more than others. This will modify the balance of forces and could disturb the initial alliance which sustained such policies.

FURTHER RESEARCH. A general economic theory of the capitalist state is impossible because national economies and nation-states are too varied and because economic issues are always influenced by non-economic factors. But a theoretically informed account of the economic aspects of particular capitalist states is certainly possible. In this context it would be worth exploring the following issues. What forms are taken by the institutional separation of the state from the economic realm and what do these forms imply for the nature and limits of state intervention? How can one identify the collective interests of capital when these are always overdetermined by contingent political and ideological factors and when alternative paths and strategies are followed in different national economies? What difference do the various forms of political representation and intervention make to the economic role of the state in capitalist societies? What scope is there for international state organizations to regulate or manage economic crises? In answering such questions one must recognize that, despite the above-mentioned limitations to the state's capacities to manage capitalism, some states and regimes are more successful than others. This suggests the need for much more detailed historical analyses and for taking seriously the 'political' moment of political economy.

R. JESSOP

See also KEYNESIANISM; MARX, KARL HEINRICH; NATIONALIZATION; WELFARE STATE.

BIBLIOGRAPHY
Alford, R.R. and Friedland, R. 1986. *Powers of Theory: The State, Capitalism, and Democracy.* Cambridge: Cambridge University Press.
Badie, B. and Birnbaum, P. 1983. *The Sociology of the State.* Chicago: Chicago University Press.
de Brunhoff, S. 1978. *The State, Capital, and Economic Policy.* London: Pluto.
Galbraith, J.K. 1967. *The New Industrial State.* London: André Deutsch.
Jessop, B. 1982. *The Capitalist State.* Oxford: Martin Robertson; New York: New York University Press.
Kraetke, M. 1985. *Kritik der Finanzwissenschaft.* Frankfurt: VSA.
Luhmann, N. 1982. *Politische Theorie im Wohlfahrtstaat.* Munich: Olzog.
O'Connor, J. 1973. *The Fiscal Crisis of the State.* London: Macmillan; New York: St Martin's.
Offe, C. 1984. *Contradictions of the Welfare State.* London: Hutchinson.
Offe, C. 1985. *Disorganized Capitalism.* Oxford: Polity Press.
Poggi, G. 1978. *The Development of the Modern State.* London: Hutchinson; Stanford: Stanford University Press.
Poulantzas, N. 1978. *State, Power, Socialism.* London: New Left Books.

economic war. Economic war constitutes all economic measures taken, before, during or instead of a military war, to harm an enemy. Compare protectionism, which is all the measures taken to 'defend' the national economy. These latter are often precisely the same measures. The subjective perception of how they do defend our own long-run economic interests is very often incorrect, and always controversial: for free trade lies at the root of Western economics. By contrast there is little theory about economic war, and (or so?) most of the measures taken seem by common admission well fitted to their time and place.

In view of the paucity of 'embargological' writing this entry must be of a frankly introductory character. First, it is well to establish some key definitions:

Embargo – a state's (or alliance's) prohibition to all its (their) citizens to sell to, or buy from, a named party, even when the price is right. An embargo is not an act of military war, and one on imports is little different from protectionism, except that its motive is to harm the foreign seller not benefit his domestic competitor.

Blockade – the prohibition by a state upon third states to trade with the second state, its enemy; a blockade must be enforced by military means and so is an act of war, possibly even against third states.

Both embargoes and blockades normally list specific goods and services. Note that the embargo of a sufficiently wide alliance is as good as a blockade, but is still no act of war.

Boycott – an embargo, usually popular or informal, on purchases alone. Typically the state machine is not involved, but some social group.

Contraband – goods on such a list that a third state tries to smuggle through a blockade.

Sanctions – the League of Nations' word for its members' punishment of an aggressor by a combined official blockade (the wording of Article 16 is vague in all original and amended versions, so the word 'blockade' is a little strong).

Transport strangleholds – when one country's transport system monopolizes, or nearly so, access to another. The great case recently is Mozambique over Southern Rhodesia (the Beira railway, see below). A near case used to be the Arab League's use of the Suez Canal against Israel.

Black List – when the state imposing an embargo (or blockade) seeks to enforce it by a secondary embargo directly on specific firms within a third (capitalist) country, that are 'violating' the original embargo as they are of course entitled by international law and the law of their own state. The Arab League runs such a blacklist. The USA enforces its stricter view of CoCom in the same way. Communist enterprises are of course 'unblacklistable' – apart from their states.

Hostile Planner – the external authority who intervenes in the market (his, ours or the world's) in order to do us harm. This concept is necessary to remind us that interventions are not always benevolent. However, just as those of the friendly (and so mainly internal) planner may be mistaken and so maleficent, those of the hostile planner may be mistaken and so beneficent.

Bottleneck effect – when an unsubstitutable import is successfully embargoed, and some activity must, at least in the short run, be shut down.

CoCom – the Co-ordinating Committee of the NATO powers plus Japan, Australia and New Zealand. Administers an embargo of militarily significant industrial products. Is consultative only, each member remaining sovereign.

Dual Use – services like rail freight and goods like special steel and aero engines have dual military and civilian use. Thus an embargo on military goods *must* hit some civilian ones. Economic war has a very long history indeed. Its variety is best appreciated by considering its first in the Mercantilist era. In the 17th and 18th centuries the state was broadly proto-Keynesian. It sought to expand the quantity of money, in order to increase employment, encourage development and – above all – collect a gold stock in case of war. Economic war was the normal condition of Mercantilist international relations, interrupted only by military war. Although it certainly had military implications, it was not, as in the 20th century, a sign of extreme hostility or easily distinguished from mere protectionism.

But how does a bankless state acquire money? If, as was normal, it had no gold or silver mines and could not steal any in its colonial conquests it could only run a balance of payments surplus. This would not only bring in money, it would also set off the foreign-trade multiplier – a concept dimly perceived but not analysed; i.e. the new money would not be hoarded. So trade was a zero-sum game – the international division of labour dates only from Smith – and indeed a war of all against all in search of gold. Therefore one embargoed imports and encouraged exports, for one's own good. In peace time one did this *contra mundum*; in wartime one concentrated on one's enemy, doing oneself good and him harm all at once. Exporting to the enemy (except technology) was very patriotic, since it harmed him. Even military supplies were allowed (British cloth for the Grande Armée), though not actual arms. This was all an essentially monetary, not an input/output, view of economic war.

Banks and paper money added to but did not modify these policies, notably in the Napoleonic Wars. Paper money was regarded with extreme suspicion – a sign of national weakness even if convertible, since clearly the authorities had already failed to gather enough gold for all purposes. So we add to our goals the destruction of the enemy's convertibility. When Pitt went off gold it was a Napoleonic victory, due to France's superior exports (of wheat). The nature of this victory was that it was a blow to the morale of an enemy with a weak balance of payments. Drained externally of the means of internal

payment, Pitt was faced with severe unemployment in Yorkshire, and a budget deficit if he wished to do something about it (in the absence of an existing and functioning welfare state). He therefore went off gold and printed the money – a defeat in all itself.

Let us jump to the 19th century, during which 'embargology' declined as free trade doctrine spread, and the notion spread that war is an epiphenomenon on the real, freely trading, peaceful, liberal, capitalist, democratic world of the planetary economy. In such an environment, where also in practice few wars were fought between major powers, there was no incentive even to consider economic measures short of war.

The 20th century has not forgotten the 19th, and it is only shamefacedly that it has reverted to the practice of the 17th and 18th centuries. Not accidentally protectionism has grown back too, but the two are not mixed up as under Mercantilism. All modern states have banks and paper money, and the monetary peculiarities of late Mercantilism have dropped away. With the welfare state and fiscal/monetary policy the modern state can sufficiently mitigate external crises to retain domestic political stability. Inconvertibility and inflation will not alter its warlike stance. Economic war has therefore – again very rationally – become an input–output matter, though the state of our enemy's gold reserve continues to be a preoccupation since gold is fungible into any input.

But the main change, surely due to 19th century example, is that economic war is no longer waged for economic ends (make him economically weaker so that I can be economically stronger, trade being a zero-sum game), but only for 'political' ends (make him economically weaker so that I can be militarily stronger). We may even infer that civilization has advanced: dirty tricks are no longer played by states merely for civilian gain. Let us examine a few examples of the new, more purely military economic war.

In the simplest case a specific export is embargoed to the enemy. If it is not a finished good, like a weapon, but an input (e.g. special steel) or both (e.g. refined petrol), our enemy must shut down some activity because of the bottleneck effect. This is the main weapon of modern economic war. If, however, his gold reserve is low and his balance of payments strained we may also embargo his exports, quite in general. This will force him to cut an import of his choice, and so suffer a mild bottleneck.

The practical complications are illuminating. Should the USA embargo the sale of wheat to the USSR, or should France embargo the purchase of Soviet gas? Provided that France does not become dependent on this gas (e.g. above five per cent of all fuel consumption) she clearly has a better economic case for doing what she prefers. For the USA, wheat relieves a serious and immediate bottleneck: that of fodder, leading to the immediate slaughter of Soviet livestock.

By a simple and well tried 'iteration', the livestock slaughter first raises, then lowers the supply of meat, the great crucial consumer good shortage that has already lead to very serious rioting and many deaths (Novocherkassk, 1962), not to mention a huge consumer subsidy. For comparison, in 1801 Britain imported French wheat to avoid a serious food shortage and despite Mercantilist doctrine. The mad Tsar Paul suggested a wheat embargo, this being his period of alliance with France. But that would have been to embargo an export, so everyone pointed out that he was only the mad Tsar Paul. Napoleon, of course, supported by current doctrine, had no qualms about his export. Anyway had not low farm prices contributed to the Vendée? Similarly Reagan fears, or feared, low farm prices, and brought Carter's wheat embargo to an end.

Yet again, wheat is a perfectly competitive commodity, and so much less suitable to be embargoed (though sometimes easy enough to blockade). In fact under President Carter the USSR bought wheat from Argentina instead. But the price was higher, the docking facilities worse and the delay considerable. All this imposed external costs the USSR, while USA, selling elsewhere in the world, had very minor external losses. Her losses were internal, indeed mainly only transfers, embarrassing the government but not much impoverishing the people: price support outlays, storage costs and electoral shifts.

Nevertheless it is part of the conventional wisdom of modern 'embargology' to count as far as possible in physical terms. The embargo deprived USSR of scarcely any bushels of wheat, so it is accounted a failure. The notion of a discriminatory export tax, of depressing the enemy's terms of trade, has achieved no recognition: the intellectual world of modern economic war is one of input-output and, seemingly, fixed co-efficients Mercantilism knew better. Even the export of money itself (long-term loans) is not taxed, but simply subjected to administrative control. But it has eventually been agreed, among the NATO powers, no longer to subsidize loans to Warsaw Pact countries; i.e. not to operate export credit guarantees in favour of even non-embargoed exports. At least, like machinery, large long-term loans are not perfectly competitive and so much easier to control.

If Mercantilism knew little about foreign lending, it knew as well as we do about technology transfer. Technology, like gold itself, was an exception: it must never be exported. For with better technology 'we' beat 'them', both in war and in the exportation of ordinary goods and services. In modern times technological levels differ much more, and the subject has become more important. Although no one country has a monopoly, the advanced have become very advanced, and it has become much more difficult to absorb their output; their active help is needed. There has also grown up an unduly sharp distinction between civilian and military technology – as if dual use were inconceivable. Moreover, military R&D bulks much larger in the total.

It was the beginning of the end of Mercantilism when David Hume declared that,

> In opposition to his narrow and malignant opinion, I will venture to assert, that the increase of riches and commerce in any one nation, instead of hurting, commonly promote (sic) the riches and commerce of all its neighbours; and that a state can scarcely carry its trade and industry very far, where all the surrounding states are buried in ignorance, sloth and barbarism (*Three Essays ... II: On the Jealousy of Trade*, Josiah Tucker's edn, London 1787, first page).

Economic war contributes much less than nothing if we only want to prosper. In the circumstances of the Cold War, the sole long-term economic war that the world now knows, this is clearly still true, but irrelevant. The great question is purely, will this new political system – opposed to 'us' on principle, and both expecting and working for 'our' total defeat – become more friendly just because it is richer? Or will it spend the extra resources on yet more arms?

The political aims of an economic war are seldom clear. Do we want (i) to incapacitate our enemy, (ii) to dissuade him, or (iii) much more ambitiously, to change his policy and aims? And with which economic instruments should we proceed in each case? In the absence of good theory modern political leaders enter upon economic war in permanent ignorance and temporary passion; their Mercantilist predecessors were far better served.

Case (iii) is bimodal. It includes, as a valid 'offensive' tactic, bringing the enemy into our group, transferring technology to him, lending him money at a discount and so enriching him: 'stab with a sausage'. In a basically economic analysis we need only say, this is absolutely correct, and the best policy by far, but only if it is sure to work, and within reasonable time. If not, case (iii) means, bimodally, that very severe measures indeed are appropriate: conversion through fear.

Case (ii) implies short slaps on the wrist, with valid threats of worse to come. It implies that we have *some* ability to change policy, at least in small matters, and are therefore prepared to 'fine-tune' our measures and to agree with each other on tactics.

Case (i) implies despair over ultimate friendship, and accepts a 'peace that is no peace' as a long-term goal: the establishment of military superiority by permanently slowing up the enemy's economic growth, without fine-tuning. One cannot after all fine-tune so diverse and fractious a coalition as the CoCom.

Modern economic war concerns mainly military and dual-use goods. This is an unnecessary restraint: if our enemy can make wheat with difficulty and rifles with ease we should deprive him of wheat. The logic is irrefragable in Case (iii) strategies, indeed hard to beat in Case (ii). Lipstick, therefore, is a highly strategic commodity if our enemy taxes it heavily and his comparative cost situation makes its production for any reason expensive for him. The concentration of embargoes in military goods serves however a good electoral purpose; ordinary people do not understand the lipstick argument but do agree that we should not deliver weapons (a not wholly correct proposition!).

Do the initiators of economic warfare always fail in their aims? This is often stated these days, by those who wish to end the CoCom and widen embargoes and with it (unilaterally) the Cold War against USSR. There is, however, no truth in 'always'; at most one can say, politicians initiate military war with far more thought, and it is not the fault of economic war, but of those who wage it, that its record is so spotty.

We list the main disputed or forgotten incidents since 1919:

1935-6. League of Nations sanctions against Italy, on the occasion of her Abyssinian aggression. Excessive moderation shown: neither oil imports nor use of the Suez Canal embargoed, but these were the only two serious bottlenecks. Reasons: fear of war in Mediterranean, and of Fascist-Nazi alliance.

1940. Anglo-American partial embargo on oil for Japan. Japanese general staff estimate military action will shortly become impossible. Pearl Habor results. This catastrophe for the initiators shows, at any rate, the effectiveness of their threat.

1976. Ian Smith, leader of the illegal white government of Southern Rhodesia, was forced to go to the negotiating table with his black enemies by Samora Machel's closure of the Beira railway. In power since 1974, Machel had hesitated because of the huge loss of invisible earnings. The effect of this was to divert all traffic to the South African network, which is about five times as far to the sea, and so very expensive; overloading it was also very unpopular with the South African government (but to South African pressure was added greater guerrilla activity). So the success of the Mozambican embargo redeemed the failure of the British. The latter was of course grossly mis-conceived. Even if better administered it could not have worked before the Portuguese Revolution.

The beginning of East-West Détente in 1970 merits longer treatment. First Brezhnev offered the German Treaty, then the

Helsinki Declaration and then, more informally, the emigration of Jews. These were, in their original form, substantial concessions, and the quid pro quo was to be technology transfer, and access to Western capital markets. In 1972 the deal was in place: the frontiers of West Berlin were recognized, the European Security Conference had begun (to end in 1975 with the Helsinki Declaration on human rights, communications, etc.), and the Jews were coming out.

But in the same year, 1972, Senator Jackson boasted during elections too much of how he had literally bargained the loans against the emigration. Sheer pride forced Brezhnev to hold back his emigrant Jews and the deal turned sour. This however does not alter the US fact that the original détente was made possible by the US embargoes on technology and capital: the very Soviet political concessions basic to the earlier Détente, which the Western enemies of CoCom and the renewed Cold War wish to bring back, were themselves the product of the relaxation of the still earlier embargoes.

Economic war against South Africa since about 1946 has been, until September 1985, mainly a matter of private boycott; except that the Communist powers have embargoed her (save Mozambique, which is much more dependent than upon Southern Rhodesia; and the USSR which has co-operated in the international diamond duopoly). Ideologically motivated private groups in the advanced capitalist democracies have refused to buy this or that export; but since they have never fully controlled any enterprises this has affected only consumer goods. States have embargoed the sale of weapons and police equipment (except Israel and Brazil). All this is standard stuff – and was very ineffective.

Much more novel was the 'extra-territorial' use of shareholder power. Much as the US government forces its firms to boycott Swedish firms that have been blacklisted for ignoring CoCom, so have ideological groups of shareholders forced enterprises with branches in South Africa to raise black wages above the market level, recognize black unions and even to evade local laws. This has been achieved more by bad publicity than by serious voting blocks at shareholders' meetings. The role of the churches, both as shareholders and as propagandists, has been considerable. Such interference is known as extra-territoriality: the state (Sweden or South Africa) on whose territory the enterprise produces or sells, or the trade union organizes, loses the degree of control over events that is normal in a capitalist state owing to foreign bodies with their own political will. This is not the case if it has merely to deal with a profit seeking headquarters abroad. Non-profit seekers are much more formidable, once in full control.

Disinvestment runs clean contrary to this. Anti-Apartheid campaigners have divided into pragmatists wishing to use such little powers as extra-territoriality confers, and extremists wishing to keep, above all, their hands clean. Disinvestment is no weapon at all against a company that does not want to borrow more, and the refusal to recognize this simple fact shows us again at what a low intellectual level economic war is ordinarily discussed. But disinvestment has a corollary of very great potency indeed: the refusal to buy new issues. This refusal rubs off on the bonds and bills of the South African government. It was of course the disinvestment controversy, and the spreading of the consciousness of what Apartheid really means, that made conservative Western banking circles refuse to 'put together a package' during the debt crisis of September 1985, turning them into a sort of moralized IMF. It will be observed that the more monetary, Mercantilist view of economic war has lost little validity.

Let us conclude with a mixed bag of applications of economic theory, for war and trade have many parallels:

(a) small countries are seldom in a position to make economic war, but are ideal victims of it:

(b) even large ones are not often well placed. Countries should form alliances, or coalitions as one says in oligopoly theory.

(c) even before size comes factor endowment. To be the monopolist of a raw material is great, but to possess an irreplaceable transport artery is still greater. And factor endowment is always largely historical chance.

(d) trade unions make economic war and throw up many parallels.

(e) to a most curious extent there is little notion of compensation for the losses caused to one's side by economic war. Once's image is of rich corporations losing small sums by not selling, or delaying the sale of, high technology. So the issue only arises domestically when small enterprises (e.g. farmers) are hit. As to international burden sharing, say with CoCom, the diplomacy of it would be horrendously complicated and divisive. But could not Britain have subsidized Mozambique, already in 1975, to close the Beira railway?

P.J.D. WILES

See also BEGGAR-THY-NEIGHBOUR; CONFLICT AND SETTLEMENT; DUMPING; FREE TRADE AND PROTECTION; INTERNATIONAL TRADE; OPTIMAL TARIFFS; TARIFFS.

economies and diseconomies of scale. 1. CONCEPTUAL ISSUES; 2. ECONOMIES OF SCALE AND MARKET STRUCTURE; 3. NORMATIVE ANALYSIS.

1. CONCEPTUAL ISSUES

1.1. Definitions. We consider the unit costs of producing a (single or composite) output under a given technology (no technical change). We say that there are *economies* (or *diseconomies) of scale* in some interval of output if the average cost is decreasing (or increasing) there. This definition focuses on economies and diseconomies of a technical character. It is sometimes extended to cover business activities other than production (such as marketing, financing, training: see Scherer, 1980).

Note that, in the case of a composite output, the proportions among the goods produced are kept constant. (A different notion, that of 'economies of scope' contemplates variations in cost as the output mix varies.) The definition of cost may, on the other hand, imply that the input proportions are adjusted in order to minimize expenditures. A related idea is that of returns to scale: here both the output and input proportions are kept fixed, and one compares the amount of (the simple or composite) output $f(x)$ produced by a given input vector x with the amount produced by vector λx, for $\lambda > 1$. *Increasing* (or *decreasing*, or *constant*) *returns to scale* are said to prevail if $f(\lambda x)$ is greater than (or smaller than, or equal to) $\lambda f(x)$. Under some conditions (see, e.g., Fuss–McFadden, 1978, p. 48) increasing (or decreasing) returns to scale are equivalent to economies (or diseconomies) of scale.

If f is a strictly concave function and $f(0) \geqslant 0$, or if f is homogeneous of degree less than one, then decreasing returns to scale prevail. Conversely, homogeneity of degree greater than one is a sufficient condition for increasing returns to scale.

1.2. Internal and external economies and diseconomies. It is sometimes useful (see, e.g., section 2.1 below) to consider economies of scale that appear only at the aggregate level and not at the level of the individual firm. For example (see Chipman, 1970) let there be two firms with cost functions $C_j(y_j) = k_j y_j$, $j = 1, 2$. Firm j treats k_j as a parameter, and in this sense its technology displays constant returns to scale. But suppose that k_j actually depends on the amount of output of the other firm, say $k_j = [y_i]^\beta$. Then the aggregate cost is $[y_2]^\beta y_1 + [y_1]^\beta y_2$. We have *external economies* if $\beta < 0$ and *external diseconomies* if $\beta > 0$.

1.3. Explaining diseconomies and economies of scale. We consider diseconomies first. Decreasing returns imply that duplicating *all* inputs yields less than twice the amount of output. But an exact clone of a production process that exhaustively lists all factors of production should give exactly the same output. The failure to double the output suggests the presence of an extra input, not listed among the arguments in the production function, that cannot be duplicated. This idea goes back to Ricardo's rent as based on the impossibility of duplicating agricultural land of a given quality. Alternatively, the extra input can be interpreted as managerial skill.

Consider (see McKenzie, 1959) a strictly concave production function $f(x)$, where x is an L-dimensional input vector. One can associate to it a constant returns to scale technology with $L + 1$ inputs $F(x; z)$ and a fixed level of the extra input, say $z = 1$, such that $f(x) = F(x; 1)$, i.e., f describes the amounts of output obtainable by varying the first L inputs when the 'managerial skill' is kept at the constant level $z = 1$. To this end, define $F(x; z) = zf(x/z)$. It is easy to check that F is quasi-concave and homogeneous of degree one, i.e., constant returns to scale. Moreover, competitive profits can be viewed as the competitive reward to the 'managerial skill' (at $z = 1$, $z(\partial F/\partial z) = f(x) - \nabla f(x) \cdot x)$.

A similar notion can be applied to the case of external diseconomies of scale: the extra input can then be identified with a common pool resource (say, clean water), available in a limited amount. Conversely, the extra public input may be created by the activity of the industry (say, information or specific training of the labour pool): this will generate external economies of scale.

We turn now to internal economies of scale. Koopmans (1957) reviews some controversies on this issue and remarks (p. 152 fn.3), 'I have not found one example of increasing returns to scale where there is not some indivisible commodity in the surrounding circumstances.' The following ideas have appeared in the literature.

(a) *Indivisible input.* Assume for instance that the only input is some specific capital good (a machine, plant, ship or pipeline) which is indivisible in the sense that it becomes useless if physically divided. It has a given maximal capacity $\bar{y}$, but it can be underutilized to produce amounts of output less than $\bar{y}$. Then $C(y)$ looks like Figure 1, and there are economies of scale in each of the intervals $[0, \bar{y}]$, $[\bar{y}, 2\bar{y}], \ldots,$ $[(n-1)\bar{y}, n\bar{y}], \ldots.$

(b) *Set-up cost.* Take the only input to be labour time and assume that a certain amount of time has to be spent in preparation for the task (the set-up cost can be given several interpretations, as time spent in: (1) concentrating and getting psychologically ready for the task; (2) learning how to do it; (3) preparing the tools needed). Once the set-up cost is paid, the amount of output is proportional to the extra labour spent. This looks like Figure 2, where increasing returns to scale prevail. Set-up costs can here be viewed as a form of indivisibility: 'readiness' (or 'information' or 'preparation') is indivisible: a 'half-ready' worker is useless.

(c) The above examples can be extended to more than one capital good (or type of set-up cost). Consider, for instance, pipelines ten miles long. Only metal sheet is used in their production: the amount needed is proportional to the radius of the pipeline. Output y (flow of oil between two points ten miles apart) is proportional to the section area, a quadratic function of the radius. A pipeline of a given radius is indivisible, but one can build pipelines of any radius. The cost function looks like Figure 3. The vertical coordinate can be interpreted as the minimal dollar outlay of a firm that buys pipelines and sells y, or as the amount of the input 'square yards of metal sheet' used by a vertically integrated firm that produces its own pipelines and sells output y.

(d) *Adam Smith's division of labour.* The *Wealth of Nations* attributes to the 'division of labour' the increase in output per worker. The main argument seems to be based on the set-up costs of (b) above. Smith's notion is related to another fundamental idea: the Ricardian gains from specialization and trade. But, in Arrow's (1979) words, 'the Ricardian idea of specialization lacks some characteristics of the Smithian; in Ricardo's system the abilities to produce are given. In Smith's view, specialization is more a matter of deliberate choice.'

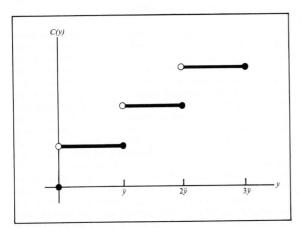

Figure 1

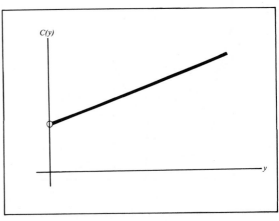

Figure 2

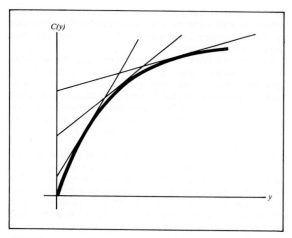

Figure 3

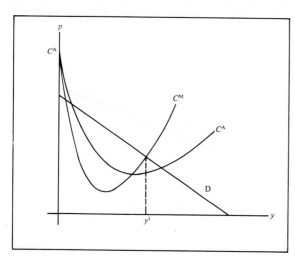

Figure 4

Figure 5

2. ECONOMIES OF SCALE AND MARKET STRUCTURE

2.1. Perfect competition as price taking behaviour. A perfectly competitive firm is often defined as one that faces a horizontal demand curve. It is clear that, as long as $C(0) = 0$, no such firm can be at equilibrium at a level of output at which the average cost faced by the firm is decreasing.

This in particular implies that competition cannot prevail under the presence of internal economies of scale at all levels of output. The argument allows for economies of scale which are external to the competitive firm. (But the *laissez-faire* equilibrium will then typically be suboptimal.) The idea of constant (or increasing) average cost at the firm level but of decreasing average cost at the aggregate level (industry economies of scale) did play an important role in the controversies on the compatibility between economies of scale and competition (see Chipman, 1965). This idea is illustrated in the example of Section 1.2 above: the reader is referred to Chipman (1970) for a rigorous study.

We now focus on internal economies. Let the market demand curve be as in Figure 4 where the average and marginal curves of a typical firm are also drawn. This situation does not *per se* violate the price taking rule: one could have, for instance, a single price taking firm operating at y^1. But such a combination of demand and cost does not fit well with the idea of perfect competition.

First, it is graphically clear that at most two firms may operate in this market. But it is then unrealistic to assume that each firm will take the market price (or the price charged by the other firm) as given.

Second, with two firms the aggregate supply curve would look like the discontinuous curve in Figure 5, where supply at no price equals demand. Difficulties with the existence of competitive equilibrium will in general appear as soon as the average cost is somewhere decreasing.

These difficulties become more severe when considering unrestricted entry (with identical cost curves for all incumbents and entrants), a natural attribute of perfect competition: it is then required that at a 'long run' equilibrium no potential entrant have incentives to enter. But if potential entrants are themselves price takers, a long-run equilibrium implies zero profits, and typically none will exist if the cost curves are U-shaped. The entry model of Baumol–Panzar–Willig (1982) faces the same existence difficulties.

2.2. Explaining the number of firms. The previous discussion suggests that imperfect competition will prevail under economies of scale. On the other hand, one would expect the number of firms in an industry to be inversely related to the degree of scale economies relative to the extent of the market. Novshek's (1980) approach yields a rigorous version of this idea. Novshek considers a model of Cournot oligopoly with free entry where potential entrants adopt themselves the Cournot conjecture that active firms will keep their output constant (as proposed by Bain, 1956, and Sylos-Labini, 1962).

An example will illustrate Novshek's method. Consider a market where the aggregate (inverse) demand is given by $p = a - bY$. The number of firms in the industry is not given, but all firms, incumbent or potential, have access to the same cost function: $C(y) = g + cy$. The positive parameter g is a set-up cost: the larger g, the stronger the economies of scale. Similarly, the larger a or the smaller b, the larger the extent of the market.

Write $\hat{y}(n)=(a-c)/(n+1)b$. This is the output of a firm at the (unique and symmetric) Cournot equilibrium for n active firms. For this to be a Novshek (or Long-run Cournot) Equilibrium we require that: (a) the price be not less than average cost; this is a *no exit* condition. (b) No potential entrant has incentives to enter, i.e., $a-b(n\hat{y}(n)+y)\leqslant c+g/y$ for all $y>0$: this is a *no entry* condition.

It can be checked that (a) and (b) impose lower and upper bounds on the number of firms n^* that can prevail at a Novshek Equilibrium. The no exit condition implies that $n^*+1\leqslant(a-c)/\sqrt{(bg)}$, and the no entry condition implies that $n^*+1\geqslant(1/2)(a-c)/\sqrt{(bg)}$. The expression $(a-c)/\sqrt{(bg)}$ can be viewed as an index of the 'extent of the market relative to the scale economies', since it is increasing in a and decreasing in b and g. Both bounds increase with this index, and in this sense the number of firms in an industry increases with it.

2.3. Perfect competition as a limit. Section 2.1 above discussed the existence difficulties that appear when production functions are not concave. These difficulties, serious for competitive equilibrium, are attenuated when considering noncompetitive equilibria: existence results for general equilibrium models can be found in Arrow–Hahn (1971) and Silvestre (1977, 1978). We focus now on partial equilibrium markets of the Novshek type (as in section 2.2 above, but perhaps with U-shaped costs). Consider a sequence of such markets each with the same technology but with increasing size of the consumer sector (say, the parameter b of section 2.2 above tends to zero). Equilibria turn out to exist at least for all but a finite number of markets in such a sequence.

Moreover, the equilibria of such a sequence converge to an optimal state (zero welfare loss) where the price equals the marginal cost. Such a limiting state can motivate an alternative definition of a (long-run) competitive equilibrium: this approach has the virtue of providing a justification (from the noncooperative, Cournot viewpoint) of the price taking postulate (see Mas-Colell, 1980, 1981).

The convergence to long-run competitive equilibrium obtains both in the case of U-shaped cost curves and in the case of everywhere decreasing average costs (see Guesnerie and Hart, 1985). This suggests a certain degree of compatibility between increasing returns and competition, in the sense that if the economy is sufficiently large, the price will be approximately equal to marginal cost in either case (even when price taking behaviour is ruled out). But Guesnerie and Hart also show that the per capita welfare loss tends to zero much more rapidly in the U-cost case. Thus, in their words, 'there remains a sense in which everywhere increasing returns do cause greater problems for the competitive model than do increasing returns which are eventually exhausted' (p. 541).

3. NORMATIVE ANALYSIS

Consider a commodity that is produced with economies of scale and that is sold in a market at a uniform price. Efficiency requires that price be equal to marginal cost. But the marginal cost is lower than the average cost. Hence, efficient pricing requires that the producing firm suffer losses. This is a basic obstacle to efficiency under increasing returns to scale. (When the commodity is not easily transferable among buyers efficiency can sometimes be achieved by means of price discrimination or nonlinear pricing schedules.)

One institutional arrangement that can in principle resolve the conflict is the public ownership of the firm. The firm can then be instructed to set prices equal to marginal costs and be subsidized for the resulting losses (see Hotelling, 1938). This motivates the concept of Marginal Cost Pricing Equilibrium (see Guesnerie, 1975; Beato, 1982), an extension of the notion of general competitive equilibrium where firms with increasing returns to scale must follow the marginal cost pricing rule instead of profit maximization. Any efficient allocation can be attained as a marginal cost pricing equilibrium for some redistribution of income. But setting prices equal to marginal cost only guarantees the first order conditions for efficiency, not sufficient here. Thus, one should not expect that all marginal cost pricing equilibria will be efficient. A weaker desideratum is the existence of at least one efficient marginal cost pricing equilibrium compatible with a given income distribution. This turns out to obtain in some special cases (e.g., when there is a representative consumer) but not in general (see Guesnerie, 1975; Brown and Heal, 1979, 1980 and Beato and Mas-Colell, 1985).

There are, on the other hand, practical obstacles to the achievement of efficiency by a publicly owned firm. First, the absence of a profit maximization target may reduce the incentives for cost minimization. Second, the implied redistribution from taxpayers to buyers may be ethically objectionable if, say, only the wealthy are buyers. Or the public firm may find itself legally or politically constrained to break even, because it may in practice be hard to distinguish the losses mandated by marginal cost pricing from those caused by mismanagement. First best efficiency is then unattainable.

An interesting second best problem for a publicly owned or regulated firm with economies of scale is the following one (see Ramsey, 1927, and Boiteux, 1956). Let the firm be constrained to break even, and let it sell the same commodity in two separate markets. Efficiency would require setting the prices in both markets equal to the (common) marginal cost, but this violates the break-even constraint. The second best solution requires charging different prices: the market with less elastic demand is charged a higher price.

JOAQUIM SILVESTRE

See also EXTERNAL ECONOMIES AND DISECONOMIES; FIXED FACTORS; INCREASING RETURNS; INDIVISIBILITIES; INTERNAL ECONOMIES; RETURNS TO SCALE.

BIBLIOGRAPHY

Arrow, K.J. 1979. The division of labor in the economy, the polity and society. In *Adam Smith and Modern Political Economy*, ed. Gerald P. O'Driscoll, Jr., Ames: Iowa State University Press.

Arrow, K.J. and Hahn, F.H. 1971. *General Competitive Analysis*. San Francisco: Holden–Day.

Bain, J.S. 1956. *Barriers to New Competition*. Cambridge, Mass.: Harvard University Press.

Baumol, W.J., Panzar, J.C. and Willig, R.D. 1982. *Contestable Markets and the Theory of Industry Structure*. New York: Harcourt Brace Jovanovich.

Beato, P. 1982. The existence of marginal cost pricing with increasing returns. *Quarterly Journal of Economics* 97, 669–89.

Beato, P. and Mas-Colell, A. 1985. On marginal cost pricing with given tax-subsidy rules. *Journal of Economic Theory* 37, 356–65.

Boiteux, M. 1956. Sur la gestion des monopoles publics astreints à l'équilibre budgétaire. *Econometrica* 24, 22–40.

Brown, D.J. and Heal, G. 1979. Equity, efficiency and increasing returns. *Review of Economic Studies* 46, 571–85.

Brown, D.J. and Heal, G. 1980. Two-part tariffs, marginal cost pricing and increasing returns in a general equilibrium model. *Journal of Public Economics* 13, 25–49.

Chipman, J.S. 1965. A survey of the theory of international trade: part 2, the neo-classical theory. *Econometrica* 33, 685–760.

Chipman, J.S. 1970. External economies of scale and competitive equilibrium. *Quarterly Journal of Economics* 84(3), 347–85.

Fuss, M. and McFadden, D. (eds) 1978. *Production Economics: A Dual Approach to Theory and Applications*. Amsterdam: North-Holland.

Guesnerie, R. 1975. Pareto optimality in non-convex economies. *Econometrica* 43, 1–29.

Guesnerie, R. and Hart, O. 1985. Welfare losses due to imperfect competition: asymptotic results for Cournot–Nash equilibria with and without free entry. *International Economic Review* 26(3), 525–45.

Hotelling, H. 1938. The general welfare in relation to problems of taxation and of railway and utility rates. *Econometrica* 6, 242–69.

Koopmans, T.C. 1957. *Three Essays on the State of Economic Science*. New York: McGraw-Hill.

McKenzie, L. 1959. On the existence of general equilibrium for a competitive market. *Econometrica* 27, 54–71.

Mas-Colell, A. (ed.) 1980. *Noncooperative Approaches to the Theory of Perfect Competition*. Symposium issue, *Journal of Economic Theory. Reprinted, New York: Academic Press, 1982.*

Mas-Colell, A. 1981. Cournotian foundations of Walrasian equilibrium theory: an exposition of recent theory. In *Advances in Economic Theory*, ed. W. Hildenbrand, Cambridge: Cambridge University Press.

Novshek, W. 1980. Cournot equilibrium with free entry. *Review of Economic Studies* 47, 473–86.

Ramsey, F. 1927. A contribution to the theory of taxation. *Economic Journal* 37, 47–61.

Scherer, F.M. 1980. *Industrial Market Structure and Economic Performance*. 2nd edn, Boston: Houghton Mifflin.

Silvestre, J. 1977. General monopolistic equilibrium under non-convexities. *International Economic Review* 18, 425–34.

Silvestre, J. 1978. Increasing returns in general non-competitive analysis. *Econometrica* 46(2), 397–402.

Sylos-Labini, P. 1962. *Oligopoly and Technical Progress*. Trans. Elizabeth Henderson, Cambridge, Mass.: Harvard University Press.

Economistes. *See* PHYSIOCRATS.

Eden, Frederick Morton (1766–1809). The son of Sir Robert Eden, F.M. Eden was educated at Oxford, gaining a Master's degree in 1789. A co-founder of the Globe Insurance Company, he published in 1797 the three volumes of his investigation into the conditions of the labouring poor, *The State of the Poor*. This work was perhaps the most detailed appraisal of social legislation and its actual workings that had appeared, and the findings provided ample material for ensuing debate on the best form of dealing with poverty and pauperism. In the years that followed Eden wrote a number of pamphlets on related issues.

The greater part of *The State of the Poor* records Eden's findings relating to the actual conditions prevailing in the parishes of England. Stimulated by the high prices prevailing in 1794–5, Eden initially set out to study the condition of the poor, but later extended this to the labouring classes. He encountered at times great resistance from local parish authorities, but despite this he was able to gather a considerable amount of information on wage levels, diet and prices. This was linked to an appraisal of the nutritional value of available foodstuffs, such that it was possible to arrive at some kind of comparative assessment of levels of poverty and want. It emerged from his empirical findings that the actual conditions and treatment of the poor varied greatly from parish to parish, this in part reflecting the patchwork of legislation that had grown up over the years in relation to the pauper and the workless. He argued however that existing legislation implied a policy of support for the indigent, and

that in general a civilized society had an obligation to make such provision.

K. Tribe

Edgeworth, Francis Ysidro (1845–1926). Giving the Sidney Ball Lecture in May 1929, A.C. Pigou remarked that 'During some thirty years until their recent deaths in honoured age, the two outstanding names in English economics were Marshall at Cambridge and Edgeworth here in Oxford' (Pigou and Robertson, 1931, p. 3). That the names were presented in that non-alphabetical order was not just Cambridge insularity but a universal perception. In a letter to Edgeworth from Bari in November 1890, Pantaleoni wrote that '... you are the closest approximation of a match for Marshall in England. You know that to my mind, Marshall is simply a new Ricardo who has appeared in the field – and to be second to him is as great an honour as a scientific man can wish for, in our time.' The modest Edgeworth would have concurred: 'Marshall was at the Council to-day; it was as if Achilles had come back,' (reported by Bonar, 1926, p. 650).

In our time the issue of the relative merit of these two great scientists is not so clear cut. 'Edgeworth, the tool-maker, gloried in his tools' (Pigou, ibid.). The general form of the utility function, indifference curves, the convexity of those curves, Pareto optimality, the contract curve; where would we be without them? But Marshall too forged some powerful and durable tools: the three-fold division of time for adjustment (market, short and long periods), consumer's surplus, elasticity, they are all now second nature to us. On tools, it is a tie.

The universal use of mathematics in economics and the necessity of statistical inference in econometrics clearly tell against Marshall and for Edgeworth. The relative shift of interest from the partial to the general equilibrium of competitive markets has also been unkind to Marshallian analysis, though marking only a marginal gain for Edgeworth, who clearly never understood such important aspects of Walrasian thought as the capital theory.

But it is above all Edgeworth's profound analysis of the relation between non-market and market forms of economic organization, with its emphasis on contract and the core, that resonates in the modern mind in a way that Marshallian ideas do not. This very modern-seeming aspect of Edgeworth's work may of course be a temporary phenomenon, simply a reflection of currently fashionable preoccupations with contract theory and game theory. But that seems unlikely. It took almost 80 years for the profession to grasp the meaning and significance of the core, Edgeworth's greatest invention, and we are not going to let go now.

Edgeworth was a prolific writer, but regrettably there has never been a complete bibliography of his work. An incomplete and slightly inaccurate compilation was made under Harry Johnson's direction in 1953–4 but was never published, although a mimeographed version was circulated from Chicago sometime after 1960. A partial enlargement and correction of this yields four books; 172 articles, pamphlets and notes, several of them in multiple parts; at least 173 book reviews; and 132 entries in Palgrave's original *Dictionary of Political Economy*, with 77 in Volume I, 38 in II, 17 in III, plus 4 new items in Higgs's second edition (1925).

One of the books and 89 of the articles were on probability and statistics and have not been reprinted. Of the rest, 32 were reprinted in whole or in part in Edgeworth (1925), as were 69 distinct review essays. Some 41 of his non-statistical articles appeared in the *Economic Journal*, of which he was editor for so long, as did 132 of his book reviews. The reader should be warned that the aged Edgeworth severely edited many of the items appearing in his collected *Papers*, so that to avoid error one must refer to the article as originally published.

Each quotation from Edgeworth given here is intended to be exact; in particular, all italicizing of words and sentences is his, not mine.

With such profusion in his publications, this relatively short essay cannot begin to do justice to the variety and subtlety of Edgeworth's work. However, references to and discussions of his seminal contributions to the theory of monopoly and duopoly, of international trade, of taxation and public utility pricing, of index numbers and of distribution will be found throughout this Dictionary, under the appropriate headings. Rather than attempt any wide coverage (which would at best be merely an annotated bibliography) this essay concentrates instead on his biography and on two early works, *New and Old Methods of Ethics* (1877) and *Mathematical Psychics* (1881). Since the first of these has never been reprinted and seems almost never to have been read, disproportionate attention is quite deliberately paid to it here; but it is of course the better-known Economic Calculus in *Mathematical Psychics* which decisively changed economic theory, albeit some 80 years later.

I. LIFE

Edgeworth's remark that 'The life of Jevons was not eventful' (1886, p. 355) applies just as well to his own. He was born at Edgeworthstown, Co. Longford in Ireland on 8 February 1845 and actually named Ysidro Francis (Butler and Butler, 1928, Dedication and p. 244; Kendall, 1968, p. 269). Educated at home by tutors (and so avoiding the rigours of Victorian public school life) he entered Trinity College, Dublin in 1862 and specialized in classics. There he earned the highest honours and great praise from his teachers, among whom was J.K. Ingram. In 1867 he moved from Dublin to Oxford, without a degree but still reading classics and ancient philosophy, and as a student of Balliol achieved in 1869 a First Class in *Literae Humaniores*, the degree itself not being awarded until 1873 (Bowley, 1934, 113).

In his successful application for the Drummond Professorship in Political Economy at Oxford in 1890, Edgeworth recalled that 'After leaving Oxford, I studied mathematics for some years', apparently on his own in Hampstead and living obscurely on a modest private income. Either then or soon after he read law as well and was called to the Bar by the Inner Temple in 1877, a year which also saw his first substantial publication (at his own expense) *New and Old Methods of Ethics*, and apparently his first regular (if temporary) academic appointment as teacher of Greek at Bedford College for Women in London. Like his grandfather R.L. Edgeworth, he never practised the law. Instead, struggling through many unsuccessful applications for teaching posts in classics and 'moral science', he gradually made his way up through the margins of Victorian academic life, helped along by an increasing stream of publications.

The benchmarks of success were a lectureship in logic in 1880, a professorship in political economy in 1888, and the Tooke Professorship of Economic Science and Statistics in 1890, all at King's College, London and all badly paid. In the Tooke Chair he succeeded the economic historian Thorold Rogers and did so again in 1891 with the Drummond Chair at Oxford, since Rogers had held both posts simultaneously. With the latter appointment Edgeworth was elected a Fellow of All Souls which 'became his home for the rest of his life' (Keynes, 1926, p. 143), although he continued to rent the two small rooms near Hampstead Heath that he had taken when first in London.

Edgeworth received many academic honours: twice President of Section F of the British Association (1889 and 1922), President of the Royal Statistical Society (1912), Vice-President of the Royal Economic Society, and one of the original Fellows of the British Academy (1903). One valued prize escaped him: although his case was often urged by his strong admirer Francis Galton, Edgeworth was never elected a Fellow of the Royal Society (Stephen Stigler, 1978, p. 309), mainly because, as John Venn wrote to Galton in 1896, 'it is difficult to point to any single essay of his which,..., is decisive at once in the way of power and originality' – this of the author of *Mathematical Psychics*, which Venn certainly read! He became emeritus at Oxford in 1922.

The most important honour was his Editorship of the *Economic Journal*, offered to him after J.N. Keynes declined appointment as its first editor. Edgeworth served from 1890 until 1911, when he resigned to make way for J.M. Keynes and became Chairman of the Editorial Board (Harrod, 1951, pp. 158–9). At Versailles in 1919 Maynard Keynes found himself more than usually busy and so Edgeworth became active again as Joint Editor with Keynes, continuing until the day he died, 13 February 1926, at the age of 81. Keynes 'received a final letter from him about its business after the news of his death' (1926, p. 140).

Family background. Edgeworth cannot be understood apart from his extraordinary family. The tradition in that family was that its Irish branch came originally from Edgeworth in Middlesex (now Edgeware, a London suburb). R.L. Edgeworth alleged that it descended from a monk named Roger Edgeworth, who sermonizing against Henry the Eighth fell in love 'with the bright eyes of beauty' and like Henry changed his faith in order to marry (1820, I, p. 5; this story should however be taken 'with a grain of salt', Butler and Butler, p. 7fn). Roger's two sons Edward and Francis crossed to Ireland in 1583, probably under the auspices of the Earl of Essex, and successfully sought their fortune. Edward became Bishop of Down and Connor, and Francis married into the Irish squirearchy.

For the next 250 years the family prospered as part of the Protestant Ascendancy, their proclivity for gambling and conspicuous consumption sufficiently balanced by talents for marrying wealthy wives and widows and for obtaining the favours offered by a place-seeking society. Edgeworth's grandfather Richard Lovell Edgeworth (1744–1817) was an eccentric in the grand tradition, inventor of many odd mechanical devices and a devoted member of the Lichfield circle of savants around Erasmus Darwin (the grandfather of Charles), which included such luminaries of the Industrial Revolution as Boulton, Watt and Wedgwood. Enthusiastic in all things, Richard raised his eldest son by the precepts of *Emile* and in Paris had the boy inspected by Rousseau himself (1820, I, pp. 177–9, 258–9). Energetic in all things, he married four times and had 22 children over a span of 48 years, of whom seven sons and eight daughters survived him; the last of these 'died at the age of 92 in 1897, a hundred and fiftythree years after her father's birth' (Butler and Butler, 1928, p. 250).

His chief favourite was his second child and eldest daughter Maria (1767–1849), with whom he collaborated in works on education and who became a celebrated and pioneering novelist in her own right. Father and daughter travelled widely and made many friends, among them David Ricardo and

Dugald Stewart, Jeremy Bentham and his early disciple and translator Etienne Dumont (1820, II, pp. 275–6). With such friends, and visitors like Sir Walter Scott, Edgeworthstown became 'the most illustrious of all the country houses' (H. Butler, *Social Life in Ireland, 1800–1845*, quoted in Hankins, 1980, p. 404 n52). In 1824, when her father was dead and the glory years past, still Maria and her Edgeworthstown could so impress the 19-year-old William Rowan Hamilton (already acclaimed as a second Newton and forever commemorated in the word 'Hamiltonian') that the rest of his life became interwoven with theirs. It was 'a house full of "children" of all ages. Maria lived in the midst of constant commotion, writing her books in the drawing room with the rest of the family about her' (Hankins, 1980, p. 35).

Edgeworth's father Francis Beaufort Edgeworth (1809–1846) was Richard's sixth son, by his last wife Anne Beaufort. She was of Huguenot extraction and eldest sister of the man who invented the Beaufort Wind Scale. Francis was an engaging child, a favourite both of his elderly father and his half-sister Maria (R.L. Edgeworth, 1820, II, p. 359; Graves, 1882, pp. 330–31). Sardonic and sensible and his senior by 42 years, Maria was more Wodehousian Aunt than sister and took a genuine though clear-eyed interest in his education (Graves, 1882, pp. 287–94) and marriage (Hankins, p. 414 n56). The contrary assertion by Mozley (quoted by Keynes) seems mainly to reflect that author's tory prejudice against Maria and her 'sensational novels,..., written to damage the character of the statesmen, the aristocracy, and the Church of this country' (1882, pp. 41–2).

Francis grew to be an affectionate though ineffectual man with many friends, living the kind of nomadic literary life common among those with some talent and less money. Unusual for an Edgeworth in being educated at Cambridge, he wilted in the dry cool atmosphere of Trinity College's mathematics and science, subjects for which he had no aptitude at all. While Keynes (1926, p. 146n) was perhaps mistaken in citing a partial passage from Mozley to imply that the boy Francis believed in perpetual motion – the full passage making more sense if the credulity was that of his friend David Reid – the adult Francis was just as credulous, and on the same subject (Hankins, pp. 411–12 n3). Hamilton lamented this lack of scientific judgement but shared his friend's passion for Plato and Kant and lengthy philosophical and poetical discussion.

Failing with a private tutoring establishment at Eltham near London, Francis returned home some time after 1836 to help manage the family estate, a task at which – if the sneering Carlyle is to be believed – 'it was said he shone ... and had become a taciturn grim landmanager' (1851, 173). After a long illness he died at the age of 37 on 12 October 1846, in the worst season of the Irish Famine (Graves, 1882, p. 331n).

The most decisive act in his life was his marriage. In 1831, aged 22 and in matrimonial mood made possible by a private income, he seriously considered two of Hamilton's sisters as possible partners. He wrote poetry for the elder and in Dublin came within 'one quarter of an inch' of proposing to her. However, much to Maria's relief – 'I ... did not feel that I could have *loved* either of these sisters' – he fled instead to London and his former lodgings near the British Museum, where quite by accident he soon met a beautiful Catalan refugee, Rosa Florentina Eroles. She was quite poor and only 16, from a family that apparently was well-educated and whose precise identity is the subject of some speculation by Hicks (1984, pp. 160–62). Like so many of his ancestors Francis fell in love at once, writing to her 'a poem of remarkable beauty' (Graves, 1882, p. 510) and to his mother a

letter describing Rosa as 'fat, voluptuous and made for love' (Hankins, p. 414,n56). Within three weeks she was married and carried off to Florence, where the young couple were reported to be 'very happy' (Graves 1882, p. 558). Meeting her in Dublin three years, later, Hamilton wrote of his liking for Francis's 'foreign wife' who 'has conquered ... his dislike to learning modern languages' (Graves, 1885, p. 93).

Ysidro Francis was the couple's fifth and last son, born just 18 months before his father died. Rosa continued to live at Edgesworthstown and survived only until 1864, outlasted by her mother-in-law who died the next year aged 96. Death and emigration (Keynes, 1951, p. 218 n3) took a quite remarkable toll of Richard Edgeworth's numerous progeny, so much so that Francis Ysidro's death in 1926 marked the end of the male line in Europe. He inherited Edgeworthstown in 1911 but never lived there.

Character and style. Edgeworth's character was elusive and contradictory. 'On anyone who knew Edgeworth he must have made a strong individual impression as a person. But it is scarcely possible to portray him to those who did not' (Keynes, 1926, p. 156). Nevertheless Keynes made a wonderful try, his vivid impressionistic skill bringing the old gentleman marvellously alive, in a way reminiscent of his sketches of Lloyd George and other protagonists at Versailles (Keynes, 1933). Sparkling as it was, however, his memoir remains a portrait of the economist as an old man.

Indeed, none of Edgeworth's biographers actually met him before 1888, by which time he was 43 and well on his way to becoming (at least on the surface) almost a caricature of the popular idea of an unworldly professor. He had in abundance all the usual traits of that stereotype. He was extraordinarily absent-minded, in New York in 1902 quite literally missing the boat back to Britain (Fisher, 1956, p. 92) – although, because he 'was a bad sailor even on Swiss lakes' (Bonar, 1926, p. 647), that may simply have been a subconscious protest. He was an incoherent lecturer, hilariously described as such by several witnesses; 'when, after many hours ... he at last made the supply curve intersect the demand curve ... One knew it was a great moment. He wagged his beard and muttered inaudible things into it. He seemed to be in a kind of ecstasy' (Harrod, 1951, p. 373). He was adept at 'avoiding conversational English' (Graves, 1958, p. 267) using instead the long Latinate words that he loved. 'Was it very caliginous in the Metropolis?' he once asked T.E. Lawrence at All Souls' Gate, receiving the grave reply: 'Somewhat caliginous but not altogether inspissated.'

He was both diffident and a pickfault, a maddening combination. L.L. Price catches well if haltingly a reaction shared by anyone who knows his work:

> hesitating and tentative, [he] was always seeking shelter behind deference to multiplied authority, and yet was not displeased to find, and was punctilious in exhibiting, minute discrepancies in the numerous texts consulted, ending, as a result, to all appearance, in more, rather than less, unstable ambiguity than that with which he started (1926, pp. 371–2).

The diffidence was chronic. 'Edgeworth lacked the force that produces impressive treatises and assembles adherents ... unleaderly is, I think, the word' (Schumpeter, 1954, p. 831). A sadder and more personal consequence was his lifelong bachelorhood. Himself just married, Keynes wrote that Edgeworth 'did not have as much happiness as he might have had' although 'it was not for want of susceptibility' that he did not marry. Indeed, in June 1889 Beatrice Potter (passionately in love a year earlier with Joseph Chamberlain and to be married three years later to Sidney Webb) was being 'halfheartedly courted by a middle-aged economist named Edgeworth' and 'shamefacedly observed [to her diary] that "relations with men stimulate and excite one's lower nature" and that "that part of a woman's nature dies hard" ' (N. and J. MacKenzie, 1977, p. 134). Later, Edgeworth 'collected his friends' opinions on the subject of matrimony and told me he was disappointed: "They were all so happily married" ' (Bonar, 1926, p. 649). Remembering his father's urgent courtship of Rosa, such wretched diffidence!

To set against these infirmities were many positive qualities. 'The kindest and most courteous of men' (Butler and Butler, 1928, p. 244 n1), 'with a natural inclination to encourage the youthful and the unknown' (Keynes, 1926, p. 152). Sanguine by nature, 'few visitors were so beloved as he' (Bonar, 1926, p. 652). His enormous scientific productivity bears witness that he worked very hard, being physically strong and mentally alert well into his seventies, 'his iron frame' fond of walking, cycling, golfing, boating, swimming and mountaineering. Rather surprisingly, in the administration of the *Economic Journal* this archetypal abstracted academic displayed that 'great practical sense of the Edgeworth kind' for which his Aunt Maria had been famous (Mill, 1873, 56).

Few economists have read so widely, not only economics and philosophy but also science and mathematics and the literature of many languages alive and dead. He reviewed for the *Economic Journal* books in French, Italian, German, Dutch and modern Greek, and 'appears to have looked critically at every book that reached the *Journal*'s office' (Bowley, 1934, p. 123). Like his fellow Irish outsiders and younger contemporaries Shaw and Wilde he was cosmopolitan in outlook, warmly hospitable to visiting foreign scholars and 'the most accessible of the English economists' (ibid., p. 122), a marked contrast to the insularity that prevailed elsewhere. His sense of humour was no less acute for being expressed often by apt quotation from foreign languages ancient and modern, or by allusion to dry donnish English prose and poetry.

Which brings us to his literary style, the most searching appraisal of which may be found in Stephen Stigler (1978, pp. 292–3); see, for example, his enlightening analysis of Edgeworth's constant recourse to 'authority'. He justly remarks that 'investigators of Edgeworth's work must at some point come to grips with his unique style of writing' and means precisely that – one of a kind. Especially idiosyncratic in his early writings, with their abundance of free and apposite quotation from the classics so that one 'can scarcely tell whether it is a line of Homer or a mathematical abstraction which is in course of integration' (Keynes, 1926, p. 145), there is nothing like Edgeworth's style in the entire literature of economics and statistics, nor scarcely in general literature. To some people (Keynes, Schumpeter, Stigler *père et fils*) that style is charming and addictive, to many others the reverse.

Only Stephen Stigler seems to have observed how appropriate was Edgeworth's peculiar style to his peculiar purposes. Particularly in those early works, the extravagant and sustained metaphors were clearly meant to help the reader grasp ideas whose depth and originality required that hardest response of all, a wrenching of the mind from its old familiar routines. While his success in this was at best partial (the rediscovery of the core did after all have to wait for almost 80 years) that was probably due as much to the sheer newness and profundity of the ideas as it was to the 'strange but charming amalgam of poetry and pedantry, science and art,

wit and learning, of which he had the secret;' (Keynes, 1926, p. 146).

That style was already 'all ... there full-grown' (Keynes again) at the age of 32 in *New and Old Methods of Ethics* and we do not know why. This is just one of the many questions about the young Edgeworth to which there are no clear answers. On the personal side, why did he switch from TCD to Oxford and take so long to graduate; why did he study law and then not practise it; and, to be mundane, what was his income during the many years of study in London? On the intellectual side, what gave rise to his interest in ethics and why did he become so strong a utilitarian; why did he study mathematics to such depth; why was he drawn to economics; and why then to statistics, so soon after the fundamental contributions to economic theory of *Mathematical Psychics*?

Intellectual influences. The interest in ethics is perhaps sufficiently accounted for by his immersion in Plato and Aristotle, caused by *Lit. Hum.* if nothing else; but his passion for utilitarianism is not so readily explained. The early Edgeworth was the most exact of utilitarians, utility itself being for him a stuff 'as real as his morning jam' (Samuelson, 1947, p. 206, who however later referred to Edgeworth's All Souls' breakfasts of pheasant and champagne: 1951, p. 53n). While it was reported that his father 'showed an early and strong revolt against the hollowness, callousness and deadness of utilitarianism' (Mozley, 1882, p. 41), that may again speak more to the reporter's prejudices than to the facts; and the youthful revolt may not have lasted. Keynes suggests that in 'his early adherence to Utilitarianism Edgeworth reacted back from his father's reaction against Maria Edgeworth's philosophy in these matters' (1926, p. 148n), a Bloomsbury-like speculation which seems to ignore how little direct intellectual influence either of them could have had on the infant Edgeworth.

It is almost as hard to explain his longer-lasting attachment to mathematics, though perhaps here the speculation can be more solidly based. We have already seen that William Rowan Hamilton was for decades a close and affectionate friend to both Edgeworth's aunt and father. Writing to the widowed Rosa Edgeworth early in 1847 Hamilton referred to 'our lost Francis' (Graves, 1885, p. 554) and in a letter to his Scottish disciple Peter Tait, to 'Francis Edgeworth ... who was during his life a great friend of mine, and for whose memory I retain respect and love' (Graves, 1889, p. 188). In 1858 he was invited to Edgeworthstown to take part in a tenants' festival honouring the return of one of Francis's sons, William Edgeworth, from army service in India. This retying of old bonds so overcame Hamilton that he was moved to write in Maria's memory what is apparently his last poem extant, a sonnet of superb Victorian sentimentality, and to present it on the spot to Edgeworth's sister Mary (Graves, 1889, pp. 102–3).

'One would guess' says Hicks (1984, p. 162) 'that when the boy went to Trinity, he would have been given an introduction to Hamilton.' There is no need to guess. Writing from Dunsink Observatory in Dublin on 6 December 1864, nine months before he died at the age of 60, Hamilton wrote to a correspondent that 'We all enjoyed much a recent visit from Francis Edgeworth – who permits me in conversation to call him "Frank" – for to me there can be no second "Francis" ' (Graves, 1889, p. 168). That Edgeworth knew Hamilton is confirmed by his review of the third volume of Graves's biography (Edgeworth, 1889), which also makes clear his unbounded admiration and respect for Hamilton, 'not merely the Irish Lagrange ... [but] ... the Pascal or Descartes of his country. What Leibnitz said of himself, that his mind could not be satisfied by one species of study, may be said with equal truth of the Irish polymath.'

Given this evidence it seems reasonable to speculate that the young fatherless Edgeworth, discovering in himself both a comparative lessening of interest in the classics (Stephen Stigler, 1978, p. 289) and an increasing enthusiasm for mathematics, should have taken as model and inspiration this close family friend, only four years older than his father, who as an undergraduate had gained a rare *optime* in both mathematics and Greek, and who happened to be one of the greatest men of science of his or any other age. If true, this hypothesis would explain not only the fact but also the direction of his studies in mathematics, which is otherwise rather puzzling. 'His first substantial effort, *New and Old Methods of Ethics*,... already showed a confident and creative mastery of the calculus of variations, not to mention some knowledge of mathematical physics' (ibid. p. 290). Playing as it did so essential a role in Hamilton's path-breaking work on dynamics and optics, it would then not be surprising that 'the wily charms of the calculus of variations', 'the most sublime branch of analysis' (1881, pp. 93, 109) – became the focus of Edgeworth's mathematical studies.

It would also help to explain Edgeworth's delight in the aesthetic aspects of mathematics, a response so depressingly missing in Marshall. Hamilton's masters were Lagrange and Laplace, 'the mathematics of the Ecole Polytechnique', and for him Lagrange's *Mecanique analytique* was quite literally a 'scientific poem', a point of view 'that Hamilton had worked out in his discussions with [FYE's father] Francis Edgeworth' (Hankins, 1980, pp. 23 and 104–5). Contrast this with the system that produced Marshall, the Cambridge Mathematical Tripos of the 19th century, when 'True to their sporting instincts the English had contrived to turn even the university examinations into an athletic contest' (Annan, 1952, p. 24).

> It was an examination in which the questions were usually of considerable mechanical difficulty – but unfortunately did not give any opportunity for the candidate to show mathematical imagination or insight or any quality that a creative mathematician needs (Snow, 1967, p. 22).

The result was a system that according to G.H. Hardy effectively ruined serious mathematics in England for a hundred years, so that when in 1908 he came to write an elementary text on analysis it was like 'a missionary talking to cannibals' (Hardy, 1938, Preface).

Edgeworth's passionate vision of 'the double-sided height of one maximum principle, the supreme pinnacle of moral as of physical science' (1881, p. 12) is also traceable to Hamilton, since it is to him we owe 'the most general energetical principle governing all dynamical motions' (Lindsay and Margenau, 1957, p. 195), and 'the cornerstone of modern physics' (Erwin Schrodinger, quoted in Hankins, 1980, p. 64). The striking combination of scientific power and aesthetic appeal in Hamilton's mathematics must have made it impossible for Edgeworth to resist.

'Clearly, the superior genius who reduced the general dynamical problem to the discovery of a single action-function was as much affected by the ideal beauty of "one central idea," as by the practical consequences of his discovery' (1881, p. 94; see also p. 11); a footnote here cites Hamilton's two great papers on dynamics in the *Philosophical Transactions* of the Royal Society of 1834 and 1835.

Edgeworth's attraction to economics was almost certainly due to Jevons alone. In the mid-1870s Edgeworth became friends with the psychologist James Sully (1842–1923), a fellow member of the recently founded Savile Club (Sully, 1918, 164). Their common interest in the 'psychophysics' of Weber and Fechner and in moral science was reflected in *New and Old Methods of Ethics*, whose only acknowledgement was to Sully. At about that time the latter moved to Hampstead, in part to be near his 'bachelor chum' (ibid., pp. 177–9; Howey, 1960, pp. 98–100). Although Sully coyly does not identify this chum, his detailed description fits Edgeworth like a glove, even down to Edgeworth's love (reported also by Keynes) of swimming in ice-cold water.

Sully was also a neighbour and great friend of Jevons (Black, 1977, Vol. IV, p. 239 n2), and 'A common interest drew him and my chum together, and so we made a trio in many a pleasant walk and skating excursion' (Sully, 1918, p. 181). We also know from a perceptive testimonial written for Edgeworth by H. S. Foxwell on 6 October 1881, in connection with an unsuccessful application for a professorship at Liverpool, that

> Between the publication of these brilliant and suggestive papers [Edgeworth (1879) and (1881)], I had the pleasure of making the acquaintance of Mr. Edgeworth, through our common friend Professor Jevons, at whose house and elsewhere we have had many discussions, chiefly upon economics, but frequently also upon ethical and philosophical subjects ... [in] these conversations ... he showed great speculative ability, singularly wide and various acquirements, and marked originality and vigour of expression.

Edgeworth's article 'Hedonical Calculus' in *Mind*, July 1879 was reprinted essentially unchanged as pages 56 to 82 of *Mathematical Psychics* and is itself almost free of economics. It seems likely, then, that at least until 1879 he had learned very little of the subject. If so, the appearance of *Mathematical Psychics* in the middle of 1881 was a truly stunning performance (to use a favourite word of Schumpeter). The economics of the book are strongly Jevonian rather than Classical, with very few references to Smith, Ricardo or Malthus, some to Mill, and many to Cournot, Jevons, Marshall and Walras.

Perhaps because his father died so young Edgeworth seems always to have had a need for heroes. Hamilton was probably one such, Jevons certainly another. In an affecting letter to Jevons' widow written three days after her husband had drowned in August 1882, Edgeworth lamented

> the loss of my venerated friend ... and a peculiar intellectual sympathy which he extended ... It is difficult to realize that I shall never more meet Mr. Jevons on the ice or heath be fascinated by his philosophic smile and drink in his words. I shall always regard it as one of the privileges of my life to have come under the influence of his serene and lofty intellect (Black, V, pp. 201–2).

This severe intellectual loss could help to account for his lack of research in economics in the years after *Mathematical Physics*, although a possible further explanation is a perceived need for more reading in the subject, especially in classical economics.

With the loss of Jevons a new hero was needed. Sidgwick was an obvious candidate, since *New and Old Methods of Ethics* had been variations on a theme in that author's *Methods of Ethics*. Moreover, 'Partly ... from his strong sense of humour, Sidgwick came near to rivalling Marshall for first place in Edgeworth's admiration' (Bonar, 1925, pp. 649–50). But 'on the whole Marshall was the great Apollo, oracle, or highest authority' (ibid.). The change from Jevons to Marshall was in fact soon made, perhaps accelerated by Marshall's review of *Mathematical Psychics* (Marshall, 1881), which was

marginally less uncomprehending than that of Jevons (1881). .
As early as September 1882 Marshall was writing to 'My dear
Edgeworth' and ending 'Yours very sincerely', while Mary
Marshall sent her 'very kind regards'.

In spite of some strains such as that over barter (recorded in
Marshall, 1961, II, pp. 791–8) the closeness thus established
between Edgeworth and Marshall was not broken until the
latter's death in 1924. There was genuine respect and affection
on each side, even though 'there can seldom have been a
couple whose conversational methods were less suited to one
another' (Keynes, 1926, p. 144). The relationship was
asymmetric, however. It was understood that Edgeworth was
to make strong (and sometimes unsuccessful) efforts to
comprehend the other's modes of thought, his 'arms of
mathematics ... under the garb of literature' (*Mathematical
Psychics*, p. 138; the order is inverted here), but that Marshall
need not reciprocate those efforts; for he seems never to have
entered seriously into Edgeworth's economics or statistics.

Edgeworth and Hobson. Edgeworth was 'entirely incapable of
intentional discourtesy. It was against his nature to inflict an
insult' (Bonar, 1926, p. 650). So it is a shock to encounter the
allegation that his disapproval of J.A. Hobson's heterodox
views led to the latter's exclusion from offering courses in
political economy as an Extension Lecturer at London (see e.g.
the entry on J.A. Hobson in this Dictionary). The allegation
appears to originate with Hobson himself (1938, p. 30).'This
was due, I learned, to the intervention of an economic
Professor who had read my book [*The Physiology of Industry*]
and considered it as equivalent in rationality to an attempt to
prove the flatness of the earth.' Hobson did not name the
professor, and the connection to Edgeworth appears to have
been made first by Hutchison (1953, pp. 118–19), who
observed that Edgeworth, then a professor at London, indeed
reviewed the book in the journal *Education* in 1890. That
seems to be the extent of the proof.

Even if it could be firmly established that it was Edgeworth
who was responsible for the refusal to permit Hobson to teach
economics, it does not follow that his reason was the latter's
heterodoxy – it may have been his ignorance of some
important parts of economic analysis. Thus in his 1904 paper
on Distribution (1925, pp. 19–20, fn3), Edgeworth took
Hobson severely to task for the latter's lack of understanding
of marginal productivity theory in *The Economics of
Distribution*, where in effect he confused (according to
Edgeworth) Δx or dx with x. Schumpeter said of Hobson (1954,
pp. 832–3 fn2) that he

> was by no means averse to the comfortable explanation
> that his Marshallian opponents were actuated by an
> inquisitorial propensity to crush dissent, if not by class
> interest: the possibility that, owing to his inadequate
> training, many of his propositions, especially his criticisms,
> might be provably wrong and due to nothing but failure to
> understand never entered his head, however often it was
> pointed out to him.

II. WORKS

A. EXACT UTILITARIANISM. Edgeworth's first known publication
was a short note on Matthew Arnold's interpretation of
Bishop Butler, in the first volume of *Mind* (1876). His next was
New and Old Methods of Ethics, a quite extraordinary
monograph of 92 pages that remains almost unread to the
present day (see however Howey, 1960, ch. XI; and Creedy,
1986, ch. 2). Half of the book (the first 34 pages plus 12 pages
of Notes) is concerned with an elaborate philosophical critique

of the recently published work of Barratt and Sidgwick on
ethics and need not detain us here, except to note Howey's
interesting suggestion that Edgeworth's style owed something
to 'that of the brilliant and precocious Alfred Barratt, whose
writings read quite like Edgeworth's' (1960, p. 107).

For the economist the book takes off abruptly on p. 35,
which is headed 'Meaning of Utilitarianism'. *Exact utilitarianism* is defined by Edgeworth to be 'The doctrine of Fechner
and Sidgwick', which is 'the greatest quantity of happiness of
sentients, exclusive of number and distribution – an end to
which number and distribution are but means'. He then
analyses various mathematical meanings that can be given to
the formula 'the greatest happiness of the greatest number',
and in so doing uses a Lagrange multiplier for apparently the
first time ever in social science, in a variational (rather than
finite-dimensional) problem and in very throwaway of-course-
everyone-knows-this fashion (p. 38, (4)). He concludes: 'it
appears impossible to assign any intelligible or tenable
meaning to the formula ... but that of exact utilitarianism'
(p. 39).

He next calls upon the work of the psychophysicists, such as
Weber, Fechner, Wundt, Delboeuf and Helmholtz (for useful
brief accounts of which see Howey, 1960, pp. 95–8 and
Stigler, 1965, pp. 113–17). '[A]greeably to the Fechnerian law,
the pleasure of each sentient element is represented by $k(\log \gamma
- \log \beta)$, where γ is the stimulus, and k and β are constants.'
Then 'if a given amount of "stimulus" ... is to be distributed
among a given set of sentients, the distribution favourable to
the production of the greatest quantity of happiness is
equality' (p. 40).

However, this Fechnerian law had been criticized 'by many
high authorities' so Edgeworth modifies it to keep its desirable
qualitative properties but dispense with its objectionable
quantitative aspects. These qualitative properties are 'that the
first differential is positive, the second differential negative for
all values of the variable, or at least all with which we are
concerned' (p. 40). In more modern language, he wishes to
make the pleasure an increasing *concave* function of the
stimulus. (Later, he actually used that language in referring to
'concave functions, as they might be called' in (1881, p. 130),
which is the more interesting since the perception that separate
theory is possible for such functions is quite recent. 'The
foundations of the theory of convex functions are due to
Jensen [1906]' (Hardy, Littlewood and Polya, 1934, p. 71)).
Edgeworth admits that the curve might have an initially
convex part, as in the standard Knightian diagram of
diminishing returns, and frequently adverts to this possibility;
but by reference to second order conditions (a constant
preoccupation in all this early work) he argues that 'the *upper
part* of the curve is alone that capable of being employed in
cases of maximum pleasure, and therefore alone concerning us
here' (p. 42).

His generalization is the 'quasi-Fechnerian law
$\pi = k|f(y) - f(\beta)|$', where '$\pi\, dt$ represents, or is proportional
to, the pleasure of a sentient element during an element of
time'; f is any function having the two properties listed above;
and β and k are two co-efficients, the former denoting 'the
"threshold", the lowest value of stimulus for which the sentient
has any pleasure at all', it being understood, with Wundt, that
'The threshold[s] of sensation and pleasure are, . . . , identical].
It follows that β^{-1} is a measure of the sentient element's
'sensibility'.

The interpretation of the co-efficient k is subtle, and
important for later developments. On p. iii ·of the Contents at
the beginning of the book we find an interesting double
criterion (reminiscent of topologies on spaces of differentiable

functions) for one sentient to be more productive of pleasure than another:

> A sensory element is said to have greater capacity for pleasure, when it not only affords a greater quantity of pleasure, pleasure arising from simple sensations, for the same quantity of stimulus, *ceteris paribus*; but also a greater increment of pleasure for the same increment of stimulus, *ceteris paribus*.

It is immediate from the quasi-Fechnerian formula, with its stimulus function f common to all, that for two elements i and j such that $\beta_i = \beta_j$, both $\pi_i > \pi_j$ and $[\mathrm{d}\pi_i(y_i)/\mathrm{d}y] > [\mathrm{d}\pi_j(y_j)/\mathrm{d}y]$ if and only if $k_i > k_j$. Thus 'The co-efficient k in the Fechnerian formula perhaps corresponds to "capacity for pleasure"' (p. iii). In principle, both co-efficients β and k 'vary with different elements' (42).

The Calculus of Hedonics. Edgeworth formulates 'two main problems of the Calculus of Hedonics', the first posed as 'a certain quantity of stimulus to be distributed among a given set of sentients ... to find the law of distribution productive of the greatest quantity of pleasure', subject to the proviso that 'every element is to have *some* stimulus' (p. 43). The second main problem drops this proviso.

He attacks the first problem under the assumption that the set of sentients is finite, which leads at once to a standard Lagrangean problem of finite multidimensional calculus. Assuming that k and β are the same for all sentients, he finds that 'the law of distribution is equality' though he does not interpret the Lagrange multiplier. Allowing β but not k to vary across sentients does not change this conclusion, unlike the converse assumption, which leads to the proposition that 'Unto him that hath greater capacity for pleasure shall be added more of the means of pleasure.' When both k and β vary across sentients, this last conclusion remains unaffected.

In the next step he argues that it is 'more appropriate' to treat each sentient as

> infinitesimal with regard to the whole ... since, strictly speaking, each pleasure element consists of the indefinitely small pleasure afforded during a given[i] time, by each indefinitely small element of the whole sensory tract to which the stimulus is applied. (The whole sensory tract is generally considered as made up of tracts belonging to different sentient individuals, e.g. different animals). The latter conception (that of infinitesimals) will therefore generally be adopted in the sequel,... the conclusions are equally deducible on either supposition [i.e. finite or infinitely many sentients] (p. 44).

This passage seems to be the earliest use of the idea of a continuum of economic actors, which was introduced into the modern literature in the classic paper of Aumann (1964). Moreover, it makes explicit that the units of feeling which he calls sentients (the term seems borrowed from Sidgwick, e.g. 1877, p. 382) comprise more than humankind. In this Edgeworth follows Bentham and Mill, and their practice finds an echo in current arguments concerning animal rights (see e.g. Singer, 1985, pp. 4–6).

To keep the second main problem as simple as possible, Edgeworth first supposes that k and β each depend on the same variable x. Hence if y is the required function that expresses the law of distribution over sentients, the interval $[x_0, x_1]$ the support of this distribution, and D the available stimulus, Problem II may be expressed: Find y, x_0 and x_1 to

$$\max \int_{x_0}^{x_1} k\{f(y) - f(\beta)\}\,\mathrm{d}x \text{ subject to } \int_{x_0}^{x_1} y\,\mathrm{d}x = D \quad (1)$$

Although Edgeworth's treatment of (1) and similar problems is quite competent and careful by the standards of rigour of English mathematics of the 19th century, and even by the standards of many latter-day economists (for whom simple Euler necessary conditions suffice), still much of his variational analysis is suspect. He was himself far from satisfied with his analysis; for example,

> It is not pretended that even the mathematical reasoning of the preceding theory is free from objection. Thus it might be objected that account is taken only of *some* solutions. ... Again, it might be objected that a *maximum* found by the Calculus of Variations ... may not be the *greatest possible* value of a proposed integral (pp. 48–9).

Although that of a self-professed 'amateur' mathematician (1889, p. 262) his discussion was far from naive.

As with the first problem, he considers four cases: where k and β are constant across sentients; where β alone varies; where k but not β varies; and where they both vary. In the first case the conclusion is that y gives equality of distribution and that $[x_1 - x_0]$ is determinate in length though not position. In the second case, the first two results remain and the indeterminacy as a position is removed, the location comprising 'a region of maximum sensibility'. The solution of the third case results in a law of distribution like that of the corresponding case of the first problem, with a determinate interval $[x_1 - x_0]$ whose location 'corresponds to the greatest capacity for pleasure' (p. 47). The solution of the fourth case is similar.

He next tackles the case where the support of y may not be an interval but instead a disconnected set. The analysis is heuristic rather than formal, 'not mathematically very elegant' (p. 47), and illustrated in typical Edgeworth fashion by a 'snake–argument' (pp. 53, 54), a 'somewhat amphibious investigation'. This is a long and fanciful (but accurate) story of an unfortunate snake which crawls up a range of mountains, in the process being dismembered and rearranged in order to ensure that it occupies the largest possible total height. It is argued that the basic structure and conclusions of the previous arguments remain in this new, possibly disconnected situation.

Extensions. The argument is then extended in several directions, allegedly all without loss of the main thread of reasoning or conclusion. The first is to the case where k and β are functions of not one but several variables. It is pointed out (p. 52) that the 'sentient elements' of this calculus are not individuals but parts of individuals. One individual may be composed of a small number of elements with high-ranking capacity for pleasure (k) and/or sensibility (β^{-1}) and a large number of low-ranking elements, while another individual may consist entirely of medium-ranking elements and so contribute more pleasure to the social total than the high-variance individual.

The next extension is to where the distribuend D is not a constant but can be written $D = \phi(x_1 - x_0)$, where $\mathrm{d}\phi/\mathrm{d}x > 0$ and $\mathrm{d}^2\phi/\mathrm{d}x^2 < 0$.

> To excite interest in this case, it may be well to make the premature remark, that these are precisely the relations which Malthus correctly supposed to exist between the quantity of the food and the number of population, and which he illustrated by the properties of the logarithm (p. 53).

This is the first and essentially the only intrusion of standard political economy into the book.

Another extension, 'an important modification of the preceding theory', is given towards the end of the book but is conveniently treated here, even though that involves the anticipation of some of his later arguments and language. His introduction of the problem is succinct:

> Up to this, sentients being regarded as so many lamps of different lighting power, the questions have been what lamps shall be lit, and how much material shall be supplied to each lamp, in order to produce the greatest quantity of light. And the answers, neither unexpected, nor yet distinctly foreshadowed by common sense, are, that a limited number of the best burners are to be lit, and that most material is to be given to the best lamp. But the conception more appropriate to the real phenomena is, that a large portion of the material to be distributed is applied not to be burned by the lamp, but to construct and repair it (p. 74).

The consequences of incorporating these needs for reproduction are straightforward but serious. Because '*minus* pain is sweeter than *plus* pleasure ... where the curve of pain descends to ever lower depths ... if there is not enough for all, no one should get very much ... [so that] ... the very rich will go away sorrowful from utilitarian teaching' (p. 75). In giving an apt example, Edgeworth cannot resist some characteristically donnish humour:

> Prior to [this] consideration, it might normally have been argued (p. 48), that out of the same quantity of material a hundred philosophers would elicit more happiness than a hundred capuchin monkeys. But perhaps the material which would keep a hundred little monkeys in health and happiness would not feed twenty philosophers! (p. 76).

Fechner to Bentham to Spencer. Edgeworth's overriding concern is with exact utilitarianism, the allocation of scarce means to secure the maximum of pleasure over all sentients and all time. But, strictly speaking, the conclusions reached so far (i.e. through p. 53) apply only to the psychophysical entities of stimulus and sensation. So two separate amendments need to be made in order to transit from the world of Fechner to that of Bentham, one at each 'end' of the Calculus of Hedonics, as it were. 'Stimulus' must be replaced by 'means to stimulus', and 'sensation' by 'emotion' or 'pleasure', and the wording of the two main problems of the Calculus amended accordingly (no matter that such terms as 'capacity for pleasure' have already been used; 'The reader is requested to make allowances and corrections, if the language has already appeared hovering, preparing for this hypothetical flight above the region of sensation' (p. 54, fn1).

In making this change Edgeworth takes the opportunity to generalize the stimulus-sensation function yet further. Pleasure π is a function of β, k and stimulus, as before, but now stimulus σ is a function ϕ of y, the 'share of distribuend'. Hence,

$$\pi = f(\beta, k, \ldots, \phi(y)) \qquad (2)$$

He argues that if the following four conditions hold, then 'The preceding conclusions [i.e. those prior to p. 54] are reducible by reasoning analogous to that ... [for] ... the simple cases of isolated sensations; subject to the same mathematical difficulties' (p. 56).

(I) β and k vary with the individual, but ϕ is the same across all individuals.

(II) $\partial^2\pi(\ )/\partial y^2$ is always negative; 'or, at any rate, there must exist satisfactory criteria of a maximum'.

(III) $\partial\phi(\ )/\partial y$ always increases with k.

(IV) If both k and β depend on another variable x, then $\partial\pi(\ )/\partial x$ is always positive.

The next stage is to argue, at some length, that '*None* of these conditions are fulfilled' (p. 57) but that in effect they are a good place from which to start. Thus, he admits that ϕ depends not on y alone, and not even only on y and the individual's characteristics k and β, but also on the y,k and β of 'his fellow-creatures'. He attempts some mathematical analysis of this complicated situation, which leaves the tentative impression that under further plausible assumptions the original conclusions are not much affected.

Rather separable from this discussion of Condition I is an interesting but brief analysis (pp. 63–6) of a society 'homogeneous as to tastes and pleasures' which, to carry out various consumption activities, forms into individuals, couples, triples, ..., such that in each of these groups each individual has the same (partial) utility function for that activity. Not surprisingly, it turns out by the usual Lagrange multiplier analysis that the distribution of this society's means should be equal for each group of a given size:

> on each individual, for his self-regarding pleasures; on each couple, for their *egoisme à deux*; on each triplet, and more numerous association, for their social pleasures. The theory does not pronounce upon the manner in which the distribuend should thus be equally applied; whether, for example, the expenditure of a household should be defrayed directly out of the common store, or should pass through the hands of the members of the household (p 64).

He also allows for some, rather systematic, heterogeneity in this society of groups.

The case for Condition (II) is simply that if it did not hold then 'the greatest quantity of pleasure is obtained ... by assigning the whole of ... [a] ... given quantity ... [of material] ... to *one* individual' (p. 69). So $\partial^2\pi(\ )/\partial y^2$ 'must be negative for the higher values of y at least'.

The reasonableness of Conditions III and IV really turn on what interpretation is to be given to k, heretofore the 'capacity for pleasure'. Sadly, it is here that Edgeworth's analysis begins to slide into Victorian eugenics (a word not coined until 1833), which is so alien in spirit and conclusion to modern readers. Up to now, Edgeworth's exact utilitarianism and its blithe reckoning of the net pleasure of every living being into the accounts of common happiness may have provoked smiles, but after all differ only in shamelessness and universality from such everyday exercises of modern welfare economics as normative growth theory. Now, with k given 'the second intention of order of evolution' (p. 57), his happy utilitarian world switches abruptly to the deadly earnestness of Herbert Spencer and Social Darwinism, of Galton and eugenics. What was before merely a matter of individual differences in capacity for pleasure is transmuted into *systematic* variation of such capacity with the creature's alleged position in the alleged evolutionary order, so that indeed $\partial\pi(\ )/\partial y$ increases with k. In *New and Old Methods of Ethics* the potentially harmful consequences of this new meaning for k are kept in check (as shown by the next two extensive quotations) but in 'The Hedonical Calculus' of 1879 they come perilously close to disaster.

Edgeworth next applies the third of the extensions listed above, where a substantial proportion of available means is devoted to the reproduction of society, to analyse Mill's advocacy of a stationary state. 'To the [reproductive] necessities of the individual must be added his contribution to the necessities of the social state to which he belongs' [public goods?], in order to arrive at the 'threshold' T which 'must be deducted from the individual's share (y) before the quasi-Fechnerian theory can be applied' (p. 76). Next, total happiness being $\Sigma\pi$, the number of the population n, 'the order of evolution' (monkeys, apes, men, etc.) x, and the distribuend function ϕ (now dependent on n and x), an implicit definition of F is given by

$$\sum\pi = nF(x, [\phi(n, x)/n - T]) \tag{3}$$

where $\partial F(\ )/\partial z > 0$. (cf. the deeply interesting discussion by Yaari (1981, pp. 14–21) of the Hedonical Calculus). Edgeworth observes that from this equation,

> As long as T does not increase with x, an indefinite progress in evolution is desirable ... . But if the threshold increased with evolution, then we should tend to a 'stationary state', not only wealth and number, but also, what Mill hardly contemplated, cultivation, evolution, stationary (p. 78).

The next quotation shows the Edgeworth of 1877 very far from crasser versions of eugenics, exemplifying it does, all at once, his style of exposition, subtlety of thought, liberality of spirit, allusiveness of reference, and diffidence (not to say slyness) of manner. He first remarks (pp. 75–6), 'Behind the first problem, to distribute over the present generation, looms the second problem, to select from posterity.' This prompts the following footnote:

> Not only from the second generation, but from 'the innumerable multitude of living beings present and to come.' This extension of view is not always favourable to privilege. For example, *prima facie*, unequal legislation directed against the influx of Chinese labour might be justified, on the supposition that, if on a large scale Chinese competed successfully with Aryans, an inferior race would inherit the earth. But this *prima facie* correspondence between exact utilitarianism and commercial selfishness would disappear, *if*, it were probable that the inferior race, not retarded by unequal laws, would catch up the superior in the race of evolution, and become ultimately as highly civilized,
>
>   'as completely so,
>
> As who began a thousand years ago.'
>
> The difference of civilization during a short interval of such pursuit might be neglected, or rather, would counterbalanced by the invidiousness and deteriorating tendencies of unequal legislation.
>
> Of course, it will be understood that examples are put forward in these pages τυπω [in outline] and παχυλὼς [coarsely], and without practical qualifications. It has been sought, not to clothe generality in circumstances, but to exhibit boldly the conception of exact utilitarianism – with a not unfelt sacrifice of delicacy to clearness.

He concludes the whole investigation of the consequences of exact utilitarianism as follows:

> With regard to the theory of distribution, there is no indication that, at any rate between classes so nearly in the same order of evolution as the modern Aryan races, a law of distribution other than equality is to be wished ... . With regard to the theory of population, there should be a limit to the number. As to the quality ... if number and quality should ultimately come into competition, ..., then the indefinite improvement of quality is no longer to be wished ... . Not the most cultivated coterie, not the most numerous proletariate [sic], but a happy middle class shall

inherit the earth. – It is submitted that these conclusions are acceptable to common utilitarianism, and, if not to common, at least to good, sense (p. 78).

The Hedonical Calculus. 'The Hedonical Calculus' of 1879, reprinted in *Mathematical Psychics* (whose pagination is used here), has been the primary source for scholarly discussion of Edgeworth's utilitarianism. This is unfortunate, since it is both less subtle and more tarnished with eugenics than the analysis of *New and Old Methods of Ethics*. Since its basic approach is the same as that of the earlier work, however, the discussion here can be brief, concentrating only on points of difference.

(i) Capacity for happiness is defined as before, and a similar double criterion is given for *capacity for work*, 'when for the same amount whatsoever of work done he incurs a less amount of fatigue, *and also* for the same increment (to the same amount) whatsoever of work done a less increment of fatigue' (p. 59; fn.1 here is *not* in the article). As before, pleasure increases at a diminishing rate with means, and now fatigue increases at an increasing rate with work. It is postulated 'that capacity for pleasure and capacity for work generally speaking go together; that they both rise with evolution' (p. 68). Indeed, these capacities are *identified* with evolution (p. 70). Unsurprisingly, eugenics creeps in fast: 'it is probable that the highest in the order of evolution are most *capable of education* and improvement. In the general advance the most advanced should advance most' (p. 68).

(ii) The variational problem (1) above is generalized to include 'the pains of production' as well as 'the pleasures of consumption' (p. 66), and conclusions similar to those of the earlier work are found. Possibly because of the addition of the capacity for work, it is not made clear that the discussion includes the whole of the animal kingdom and not just humankind; perhaps, in this new version, it does not.

(iii) In exploring the solution for optimal growth of 'sections' of the population, it is assumed 'that the issue of each (supposed endogamous) section ranges on either side of the parental capacity' according to the normal distribution (pp. 69–70). This is apparently Edgeworth's first use of his beloved 'Law of Error'. Consideration of the problem of optimum population structure leads to the following 'outlandish' solution: 'the average issue shall be as large as possible for all sections above a determinate degree of capacity, but zero for all sections below that degree' (p. 70). Yaari (1981, pp. 18–19) argues that the bizarre aspects of this bang-bang solution to the control problem spring not from Edgeworth's method but from its application to population planning.)

(iv) Edgeworth poses the Rawls-like question: '*What is the fortune of the least favoured class in the Utilitarian community?*' (p. 72), and concludes with relief that '*the condition of the least favoured class is positive happiness*' (p. 73). If nothing else, this shows that not only did he consider utility cardinally measurable, both individually and collectively, but also that for him such measurement does not, like height, temperature and von Neumann–Morgenstern utility, have an arbitrary origin and scale. 'The zero-point of pleasure corresponding to a minimum of means' (p. 72) implies that at most only scale can be arbitrary; and even there any permitted scale change must be the same for everyone.

(v) *New and Old Methods of Ethics* urged utilitarians to be aware that sentients may differ in capacity for pleasure, before making their felicific calculations. The later conjunction of capacity for pleasure with position in the evolutionary order, however, threatened an unappetising brew of utilitarianism and eugenics. In 'The Hedonical Calculus' the potion turns poisonous.

> But Equality is not the whole of distributive justice ... in the minds of many good men among the moderns and the wisest of the ancients, there appears a deeper sentiment in favour of aristocratical privilege – the privilege of man above brute, of civilised above savage, or birth, of talent, and of the male sex. This sentiment of right has a ground of utilitarianism in supposed differences of *capacity*, (p. 77).

Luckily, Edgeworth at last draws back from such sophistry:

> Pending a scientific hedonimetry, the principle 'Every man, and every woman, to count for one,' should be very cautiously applied (p. 81).

(vi) After all is said and done and integrated, 'the Hedonical Calculus supplies less a definite direction than a general bias' (p. 81), though towards what is not clear. Its deductions are 'of a very abstract, perhaps, only negative, character; negativing the assumption that *Equality* is necessarily implied in Utilitarianism' (p. vii). Thus the bright shining stream of *New and Old Methods of Ethics* finally stagnates into the muddy delta of '*Nothing indeed appears to be certain from a quite abstract point of view*' (pp. 74–5).

Such a sad conclusion did not keep the irrepressible Edgeworth down for long. Newly fired by the stimulus of Jevons, his own powerful capacity for work and intellectual pleasure soon produced one of the greatest achievements of all economic theory, the Economical Calculus of *Mathematical Psychics*.

B. MATHEMATICAL PSYCHICS. Short as it is, the book may be divided into three parts:

(A) An account and defence of the role of mathematics in economics [pp. 1–15, plus Appendices I (pp. 83–93), II (pp. 94–8) and VI (pp. 116–25)];

(B) The Economical Calculus [pp. 15–56, plus Appendices V (pp. 104–116) and VII (pp. 126–48)];

(C) The Utilitarian Calculus [pp. 56–82, plus Appendices III (pp. 98–102) and IV (pp. 102–4)].

Part C, which is essentially 'The Hedonical Calculus' of 1879, has already been discussed.

(A) *Mathematical Economics*. Edgeworth's defence of mathematical economics consists of two related lines of argument, buttressed in the three Appendices by some vigorous counter-attacks against the ἀγεωμετρητοι (a term which Edgeworth uses frequently, and which means literally, 'those who have not studied geometry'; presumably the reference is to the inscription on the door at the entrance to Plato's Academy: 'no one who has not studied geometry may enter').

'Loose Indefinite Relations'. In the first line of argument, 'it is attempted to illustrate the possibility of Mathematical reasoning without *numerical* data' (p. v). Keynes observed that it was 'a thesis which at the time it was written was of much originality and importance' (1926, p. 146) but did not elucidate the nature of its originality. In fact most of the more obvious points about the non-necessity of actual numerical data and actual functional forms for economic analysis had already been made in 1838 by Cournot, 'the father of Mathematical Economics' (p. 83), and Edgeworth was never one to repeat what was already clear, at least to him. 'We must take for granted that our intelligent inquirer understands what is intelligible to the intelligent' (p. 126).

His new arguments are more subtle. First, it is suggested that indeed

our social problems ... [have] ... *some* precise data: for example, the property of *uniformity of price* in a market; ... the *fulness* [sic] of the market: that there *continues* to be up to the conclusion of the dealing an indefinite number of dealers; ... the *fluidity* of the market, or infinite dividedness of the dealers' interests (p. 5).

Secondly, although very often we only know 'a certain *loose quantitative relation*, the *decrease-of-the-rate-of-increase* of a quantity' this is often enough, since 'The criterion of a *maximum* turns not upon the *amount* but upon the *sign* of a certain quantity.' It is precisely this quantitative relation that is 'given in such data as the *law of diminishing returns to capital and labour*, the *law of diminishing utility*, the *law of increasing fatigue*: the very same irregular, unsquared material which constitutes the basis of the Economical and the Utilitarian Calculus' (p. 6). For problems of optimum it is 'the loose, indefinite relations of positive and negative, convex or concave' (p. 91) that are important, and it is 'this very relation of concavity, not a whit more indefinite in psychics than in physics,.. [which is] ..quarried from such data as the law of decreasing utility, of increasing fatigue, of diminished returns to capital and labour;' (p. 92).

Nor is the use of such relations confined to economics. 'The great Bertrand–Thompson maximum–minimum principles and their statical analogues present abundant instances of mathematical reasoning about loose, indefinite relations' (pp. 89–90). The constant repetition of the phrase 'loose indefinite [or quantitative] relations' – at least 16 times in these pages and in Appendix I – achieves at last an almost incantatory effect.

'*Mécanique sociale*'. The earlier discussion of Hamilton's possible influences on the young Edgeworth referred to the latter's passionate comparison of Social Mechanics with Celestial Mechanics in *Mathematical Psychics*, in a section which is perhaps that most frequently quoted, possibly because it comes early on in a difficult book, more likely because it inspired his most rhapsodic writing.

> Imagine a material Cosmos, a mechanism as composite as possible, and perplexed with all manner of wheels, pistons, parts, connections, and whose mazy complexity might far transcend in its entanglement the webs of thought and wiles of passion; nevertheless, if any given impulses be imparted ... each part of the great whole will move off with a velocity such that the energy of the whole may be the greatest possible. (p. 9)

> Mécanique sociale' may one day take her place along with 'Mécanique Celeste', throned each upon the double-sided height of one maximum principle, the supreme pinnacle of moral as of physical science (p. 12).

Such analogies seem to have had an almost mystical significance for Edgeworth, undoubtedly reinforcing his faith in 'the employment of mechanical terms and Mathematical reasoning in social science' (p. 15).

The ἀγεωμετρητοι

Appendix II is chiefly notable for setting a very Oxbridge-like 'examination paper' containing seven tough questions for the mathematically unblessed, such as: '6. It has been said that the *distribution of net produce* between cooperators (labourers and capitalists associated) is arbitrary and *indeterminate*. Discuss this question' (p. 96). It is admitted that some scholars may of course turn 'away contemptuously from such questions', but 'Are they not all quantitative conceptions, best treated by means of the science of quantity?' (p. 98).

Appendix VI carries the attack further, listing mathematical errors by those eminent but ageometrical social scientists Bentham, J.S. Mill, Cairnes, Spencer and Sidgwick. It is remarkable chiefly for Edgeworth's quite deliberate singling out of Cairnes for a savage attack, apparently provoked by the latter's 'amazing blindness' (p. 119) to Jevons's theory of exchange.

(B) *Economical Calculus.* Edgeworth now enters quite precipitately upon the work on which his main claim to immortality as an economist must rest. It is best told predominantly in his own subtle words.

First (pp. 16–19) come some important definitions and assumptions, which set the table very precisely for what is to follow:

(a) '[E]very agent is actuated only by self-interest', and may act '*without*, or *with*, the consent of others affected by his actions ... the first species of action may be called *war*; the second, *contract.*

(b) '[E]conomic *competition* ... is both, *pax* or *pact* between contractors during contract, *war*, when some of the contractors *without the consent of others recontract.*'

(c) 'The *field of competition* ... consists of all the individuals who are willing and able to recontract about the articles under consideration.'

(d) Any individual may contract or 'recontract with any out of an indefinite number' of agents, 'without the consent being required of, any third party'; 'e.g., any [individual] *X* (and similarly *Y*) may deal with any number of *Y*s.'

(e) '[I]f any *X* deal with an indefinite number of *Y*s he must given each an indefinitely small portion of *x*', from which there follows 'the indefinite divisibility of each *article* of contract.'

(f) 'A *settlement* is a contract which cannot be varied with the consent of all the parties to it.'

(g) 'A *final settlement* is a settlement which cannot be varied by recontract within the field of competition.'

(h) 'Contract is *indeterminate* when there are an indefinite number of *final settlements*.' (A copy of the book annotated by Edgeworth himself replaces 'are' by 'may be'; I am indebted to Stephen Stigler for this information).

Notice the lawyerly language employed, most probably a consequence of Edgeworth's legal training. In particular, note the stress on the word 'contract', which is used in simple or compound form over 200 times in this Economical Calculus and its two Appendices. As Howey, writing in 1960 (and not today!) pointed out: 'No one had used it in this sense earlier in economics and no one followed Edgeworth in his use' (p. 240, fn48).

More importantly, notice the subtlety of (f) and (g). If a contract cannot be varied by the consent of *all* parties to it, then that must be because at least one will be hurt by a given change and would therefore block any such proposed move. Hence, a settlement is *precisely* a Parto-optimal point for the parties to the contract; and this many years before Pareto considered the matter. Moreover, a *final* settlement is one which cannot be varied, even by negotiation with any or all of the parties outside the contract. Taken in conjunction with (d), this implies that a final settlement is *precisely* one that, as we would now say, lies in the core.

The indeterminacy of contract. Most probably inspired by Cournot's passage from monopoly to pure competition (1838, ch. VII), and 'Going beyond ... [him] ..., not without trembling' (p. 47), Edgeworth at once raises the question: '*How far contract is indeterminate?*' (p. 20). He begins its answer by considering first two individuals *X* and *Y*,

exchanging two goods in the amounts *x* and *y* from initial holdings $(a, 0)$ and $(0, b)$, respectively. Hence the post-trade amounts held are $(a - x, y)$ and $(x, b - y)$. At first he assumes additive Jevonian utility functions but immediately switches, 'more generally' to the form $P = F(x, y)$ and $\Pi = \Phi(x, y)$ for *X* and *Y*, respectively—the first use of this general form in economics, unless one includes his formulations in *New and Old Methods of Ethics.*

We now have to find 'To what *settlement* they [i.e. *X* and *Y*] will consent; the answer is in general that contract by itself does not supply sufficient conditions to determinate the solution;' (p. 20), but 'will supply only *one* condition (for the two variables), namely' (p. 21).

$$F_x(x,y)\phi_y(x,y) - F_y(x,y)\phi_x(x,y) = 0 \qquad (4)$$

where $F_x(x,y) = \partial F(x,y)/\partial x$, etc. The locus of points (x,y) which satisfy (4) 'it is proposed here to call the *contract-curve*' (p. 21), 'along which the pleasure-forces of the contractors are mutually antagonistic' (p. 29).

Edgeworth arrives at (4) by several different routes. First, '*X* will step only on one side of a certain line, the *line of indifference*, as it might be called;' (p. 21). (The name probably comes not from Jevons's Law of Indifference but from *New and Old Methods of Ethics* (p. 9, fn1), where 'indifferent actions' are 'those which are supposed equally to tend to the general good', and/or from Sidgwick (e.g. 1877, p. 125): 'a scale of desirability, measured positively and negatively from a zero of perfect indifference').

> And ... *X* will *prefer* to move ... perpendicular to the line of indifference ... *X* and *Y* will consent to move together ... in any direction between their respective lines of indifference ... they will refuse to move at all ... When their *lines of indifference* are coincident ... whereof the *necessary* (but not *sufficient*) condition is [(4)] (p. 22).

The second route is via consideration of the total differential of *F*, expressed in polar form, but a more interesting route is the third:

> motion is possible so long as, one party not losing, the other gains. The point of equilibrium, therefore, may be described as a *relative maximum*, the point at which e.g. Π being constant, *P* is a maximum. Put $P = P - c(\Pi - \Pi')$, where *c* is a constant [i.e. a Lagrange multiplier] and Π' is the supposed given value of Π (p. 23).

Eliminating *c* leads to (4) again.

The interest of this third route is of course that Edgeworth, giving here the standard alternative definition of a Pareto-optimal point, is making it quite clear exactly what a settlement means, namely, an efficient allocation for the parties to the contract.

The argument is then 'extended to several persons and several variables' (pp. 26–8). Pages 28–9 illustrate the two-person two-good case by a diagram (*not* a box, the usual historical ascription being inaccurate) which measures along the abscissa the wages paid by Crusoe and along the ordinate the labour given up by Friday. Because of this the indifference curves run from southwest to northeast, and the contract curve from northwest to southeast, the reverse of the usual box diagram.

> [S]ettlements are represented by an *indefinite number of points*, a locus, the *contract-curve* CC', or rather, a certain portion of it which ... lies between two points ... which are respectively the intersections with the contract-curve of the *curves of indifference* [the first such naming' for each party drawn through the [no-trade] origin (p. 29).

Perfect and imperfect competition.

With this clogged and underground procedure [of bilateral exchange] is contrasted . . . the smooth machinery of the open market . . . You might suppose each dealer to write down[3] [the reference is to Walras, 1874, Article 50; see Walras, 1954, p. 93] his *demand*, how much of an article he would take at each price, without attempting to conceal his requirements; and these data having been furnished to a sort of market-machine, the *price* to be passionlessly evaluated. That contract in a state of perfect competition is determined by demand and supply is generally accepted, but is hardly to be fully understood without mathematics. . . . The familiar pair of equations is [i.e. will be] deduced by the present writer from the first principle: Equilibrium is attained when the existing contracts can neither be varied without recontract with the consent of the existing parties, nor by contract within the field of competition. The advantage of this general method is that it is applicable to the particular cases of imperfect competition; where the conceptions of *demand and supply at a price* are no longer appropriate (pp. 30–31).

In other words, equilibrium of perfect competition is *defined* to be attained when the contract is in the core.

Edgeworth starts the analysis of 'imperfect competition', the 'limitation of numbers' (p. 42), by supposing that there is introduced a second X and a second Y, each of them with 'the same requirements, the same nature as the old' X and Y, respectively (p. 35). For this four-member economy,

there cannot be an equilibrium unless (1) all the field is collected at one point; (2) that point is on the *contract-curve*. For (1) if possible let one couple be at one point, and another couple at another point. It will generally be the interest of the X of one couple and the Y of the other to rush together, leaving their partners in the lurch. And (2) if the common point is not on the contract-curve, it will be the interest of *all parties* to descend to the contract-curve.

However, the contract-curve appropriate to the new, quadrilateral situation is *not* that belonging to the old, but a strict subset of it. In particular, the end-points of the old bilateral contract-curve, determined by the respective indifference curves through the initial no-trade point, will no longer be sustainable. Take either such end-point, say that closer to the X-origin. Then 'it will in general be possible for *one* of the Ys (without the consent of the other) to *recontract* with the two Xs, so that for all those three parties the recontract is more advantageous than the previously existing contract' (p. 35), i.e. than that which was at the X-end of the original contract-curve. The detailed geometric argument that establishes this is presented in pp. 35–7 (for a somewhat simpler discussion see Newman, 1965, pp. 111–15), and depends crucially on the property that the indifference curves of each agent are strictly convex.

It is sometimes said (e.g. by Stigler, 1965, p. 103) that Edgeworth assumed this property of convex indifference curves, but in fact he *proves* it on the basis of earlier assumptions, viz: that $F_x < 0$, $F_y > 0$, $F_{xx} < 0$, $F_{yy} < 0$, and $F_{xy} < 0$. From these it readily follows that the standard sufficient conditions for indifference curves to be convex (as given for example in Allen (1938, p. 375)) are satisfied, and that in fact they are equivalent to the conditions presented by Edgeworth on page 36. Note that $F_x < 0$ because x is the amount of good given up by X, while F_{xx} and F_{yy} are both negative by the law of diminishing marginal utility. The unusual assumption is

$F_{xy} < 0$, which in the later definition of Auspitz and Lieben (1889, p. 482; 1914, Texte, p. 318) would imply that x and y are substitutes. Edgeworth chooses to be coy about this: 'Attention is solicited to the interpretation of the third condition [i.e. $F_{xy} < 0$]' (p. 34), and so almost deliberately loses the chance for yet another explicit first, a formal definition of substitutes and complements; but he is quite clear about its role in establishing convexity.

'If now a *third X* and third Y (still equal-natured) be introduced into the field' (p. 37) then the original bilateral contract-curve shrinks yet further, as now $2Y$s can recontract with $3X$s, and so on.

[I]n general for any number short of the *practically infinite* (if such a term be allowed) there is a finite length of contract-curve, . . . , at any point of which if the system is placed, it cannot by contract or recontract be displaced; . . . there are *an indefinite number of final settlements*, a quantity continually diminishing as we approach a perfect market (p. 39),

since competitive allocations always remain in that finite length (p. 37).

There seems to be no warrant for the claim (e.g. in Creedy, 1986, pp. 65, 69) that Edgeworth considered competitive equilibrium to be unique. Indeed, speaking of one application (to personal service) he says that 'there is no *determinate*, and very generally[1], *unique*, arrangement towards which the system tends' (p. 46), the footnote referring explicitly to Marshall and Walras on multiple equilibria. Elsewhere in the book he makes frequent and delighted reference to the (very different) demonstrations by Walras and Marshall of the possibility of such multiple equilibria.

Finally, even a passing remark sounds startlingly modern, especially in the example given

[D]ifferent final settlements would be reached if the system should run down from different *initial positions* or contracts. The sort of difference which exists between[1] Dutch and English auction, theoretically unimportant in *perfect competition*, does correspond to different results, *different final settlements* in imperfect competition. And in general, and in the absence of imposed conditions, the said final settlements are not on the demand-curve, but on the contract–curve (pp. 47–8),

i.e. they are core allocations but not competitive allocations. The footnote mentions (but does not give a reference to) W.T. Thornton, who first pointed to the possibility of such a discrepancy between the results of English and Dutch auctions (see 1870, pp. 56ff): 'Now we believe', Edgeworth remarks acidly, 'but not because that unmathematical writer has told us.' Down with the ἀγεωμέτρητοι!

Determinateness and arbitration. Towards the end of the Economical Calculus there is an indictment of competition as sharp as any penned by an anti-neoclassical, and more powerful than most since it is based on exact analysis.

To impair, it may be conjectured, the reverence paid to *competition*; in whose results – as if worked out by a play of physical forces, impersonal, impartial – economists have complacently acquiesced. Of justice and humanity there was no pretence; but there seemed to command respect the majestic neutrality of Nature. But if it should appear that the field of competition is deficient in that *continuity of fluid*, that *multiety of atoms* which constitute the foundations of the uniformities of Physics; if competition is found wanting, not only the regularity of law, but even the impartiality of chance – the throw of a die loaded with villainy – economics would be indeed a 'dismal science', and the reverence for competition would be more no more.

There would arise a general demand for a *principle of arbitration* (pp. 50–51).

Edgeworth is ready to hand with just the right principle for the occasion.

Equity and 'fairness of division' are charming ... but how would they be applicable to the distribution of a joint product between cooperators? ... *Justice* requires to be informed by some more definite principle (p. 51).

This principle is – exact utilitarianism.

Now, it is a circumstance of momentous interest ... that *one* of the in general indefinitely numerous *settlements*[1] between contractors is ... the contract tending to the greatest possible utility of the contractors. In this direction ... is to be sought the required principle (p. 53).

The footnote first offers a proof of the proposition in the text, a proof which is incomplete since it shows neither the uniqueness of the utilitarian point nor that it is on the contract-curve, a settlement, and not merely on the (larger) efficiency-locus. It then goes on to consider the case of altruistic agents, Xs whose utility functions are not F but $F + \lambda\phi$, 'where λ is a *coefficient of effective sympathy*' and similarly for any Y, whose utility is now given by $\phi + \mu F$. For these 'modified contractors', it turns out that their contract-curve is '*The old contract-curve between narrower limits ... As the coefficients of sympathy increase, utilitarianism becomes more pure, ... the contract-curve narrows down to the utilitarian point.*'

The argument has come, designedly, full circle. After the Utopias of *New and Old Methods of Ethics* and 'The Hedonical Calculus', the positive economics of the Economical Calculus leads us after all back to Utilitarianism, considered now as a practical way of overcoming the evils of imperfect competition and indeterminacy. But, to paraphrase a harsh (though just) remark that Edgeworth once made (1889a, p. 435) about Walras' *tâtonnement*, this utilitarian principle of arbitration 'indicate[s] *a* way, not the *way*,' to justice. Moreover, it is 'not a very good [way]' (ibid.) since not only does it require cardinality of utility measurement (as does, for example, the bargaining solution of Nash, 1950), but its cardinality is quite strict, not allowing (as does Nash) independent changes of origin and scale in each of the agents' utility functions. Edgeworth's attempt to move from his 'weak' bargaining solution, the contract-curve, to a 'strong' solution, the utilitarian point, cannot be judged a success. But what a road he travelled before arriving at this Oz-like destination!

AFTER MATHEMATICAL PSYCHICS

It must have been deeply disappointing to Edgeworth that in their reviews of his book neither of his two great heroes, Jevons and Marshall, displayed the slightest understanding of his profound analysis of contract and its indeterminacy. Both were quite laudatory in general terms, but both were patronizing, Jevons in his criticism of the difficulty of Edgeworth's style – 'an uncouth and even clumsy piece of literary work' (1881, p. 583), Marshall in his worry 'to see how far he succeeds in preventing his mathematics from running away with him,... out of sight of the actual facts of economics' (1881, p. 457).

However, it was some consolation to Edgeworth to receive a very encouraging letter from Galton (see Stephen Stigler, 1978, pp. 290–91), who strongly disagreed with Jevons. From the narrow point of view of economics, encouragement from such a quarter was perhaps not to be wished, for it may have confirmed Edgeworth's attraction to eugenics. More probably and more usefully, it may well have helped to set him on the road to his distinguished contributions to probability and statistics, which began in 1883 (see the companion entry on his contributions to this subject).

Apart from a brief note (1884), in which he drew a rather forced analogy between the limit processes that lead to competitive allocations and the limiting processes that lead to the Central Limit Theorem, Edgeworth never again returned to the theory of contract. Why this should be so is a puzzle. The unperceptive reviews must have been a factor, and so perhaps was his consciousness that his 'proofs' of the main results were '*Conclusions*, rather, the mathematical demonstration of which is not fully exhibited' (1881, p. 20, fn1). Perhaps the failure to return was fortuitous. It is not uncommon for a scholar, on completion of a major work, to turn to something quite different. So Edgeworth, tempted by statistics, may have found its attractions overwhelming, to the exclusion not only of the theory of contract but of all other work in economics. Between 1881 and 1888 he published essentially nothing in economics, apart from the note already alluded to, during a period for which over 30 publications in statistics and probability are recorded.

It is clear that he must have continued to read widely and deeply in economics during this time – those 77 entries in the first volume of Palgrave's *Dictionary* did not come out of the air. Moreover, he took an active part in economic discussions, not only in such formal settings as the British Association but also in informal clubs like those in intellectual Hampstead that were frequented by Wicksteed, Shaw and Webb. With his election to the Drummond Chair at Oxford in 1890, Edgeworth settled down to that 'long stream of splinters ... split off from his bright mind to illumine (and obscure) the pages of the *Statistical* and *Economic Journals*' (Keynes, 1926, p. 149).

If its creator did not languish, the theory of contract certainly did. The silence was extraordinary. Keynes's biography was superb about the man but again rather patronizing about his work, especially about 'Mathematical Psychics [which] has not ... fulfilled its early promise' (1926, p. 149). It cannot be said that Keynes's criticism was very acute, being mainly of that kind of Platonist pseudo-profundity which can be gained by capitalizing such words as 'Organic Unity, ... Discrete-ness, ... Discontinuity' (p. 150).

A faint rustle stirred the silence in an article by the young Hicks (1930; see also 1932; p. 26), while Stigler (1942, p. 81) observed that 'It can be demonstrated that the length of the contract curve decreases as the number of bargainers increases', though without direct reference to *Mathematical Psychics*. That was apparently the entire literature on the subject. I remember vividly a conversation in 1952–3 with a distinguished economic theorist, subsequently a Nobel Prize winner, in which he (and I, come to that) expressed puzzlement about what this passage of Stigler's could possibly mean.

The silence was broken by Martin Shubik (1959), who made the essential connection between Edgeworth's contract theory, coalitions and game theory. After nearly 80 years in the wilderness Edgeworth's theory finally moved, all at once, on to centre stage. Thereafter progress was rapid, with the theorems of Scarf (1962), Scarf and Debreu (1963), Aumann (1964) and many others. Edgeworth's faith in the importance of the theory of contract has been triumphantly vindicated.

ENVOI. Following the strategy of this essay, which has been to let Edgeworth speak for himself, let the final word concerning the significance of his work be left to him. If this eloquent passage is reminiscent of another and much better known peroration (Keynes, 1936, pp. 383–4), remember that Edgeworth was always ahead of his time.

> Considerations so abstract it would of course be ridiculous to fling upon the flood-tide of practical politics. But they are not perhaps out of place when we remount to the little rills of sentiment and secret springs of motive where every course of action must be originated. It is at a height of abstraction in the rarefied atmosphere of speculation that the secret springs of action take their rise, and a direction is imparted to the pure fountains of youthful enthusiasm whose influence will ultimately affect the broad current of events (1881, Appendix VII, 'On the Present Crisis in Ireland', pp. 128–9).

PETER NEWMAN

SELECTED WORKS

Books
1877. *New and Old Methods of Ethics*. Oxford: James Parker & Co.
1881. *Mathematical Psychics*. London: C. Kegan Paul & Co.
1887. *Metretike*. London: The Temple Company.
1925. *Papers Relating to Political Economy*. 3 vols, London: Macmillan for the Royal Economic Society.

Articles
1876. Mr Matthew Arnold on Bishop Butler's Doctrine of Self-Love. *Mind* I, 570–71 (signed T.Y. Edgeworth).
1879. The Hedonical Calculus. *Mind* IV, 349–409.
1884. The rationale of exchange. *Journal of the Statistical Society* 47, 164–6.

Reviews
1886. Journal and Letters of W. Stanley Jevons, edited by his Wife. *The Academy*, 22 May, No. 733, 355–6.
1889a. (Review of) L. Walras: *Eléments d'économie politique pure*, 2nd edition. *Nature* 40, 434–36.

BIBLIOGRAPHY
Works referring to the life of Edgeworth and his family
Black, R.D.C. 1977. *Papers and Correspondence of William Stanley Jevons*, Vols. IV and V. London: Macmillan, in association with the Royal Economic Society.
Bonar, J. 1926. Memories of F.Y. Edgeworth. *Economic Journal* 36, 647–53.
Bowley, A.L. 1934. Francis Ysidro Edgeworth. *Econometrica* 2, 113–24.
Butler, H.J. and H.E. 1928. *The Black Book of Edgeworthstown and other Edgeworth Memories 1585–1817*. London: Faber & Gwyer.
Carlyle, T. 1851. *The Life of John Sterling*. Boston: Phillips, Sampson.
Creedy, J. 1986. *Edgeworth and the Development of Neoclassical Economics*. Oxford: Basil Blackwell.
Edgeworth, R.L. 1820. *Memoirs of Richard Lovell Edgeworth, Esq.; begun by himself and concluded by his daughter Maria Edgeworth*. 2 vols, London: R. Hunter and Baldwin, Cradock and Joy.
Fisher, I.N. 1956. *My Father Irving Fisher*. New York: Comet Press.
Graves, R. 1958. *Goodbye to All That*. New edn, revised, London: Cassell & Co. Originally published in England in 1929; the first American edition (1930) omitted the Edgeworth–Lawrence anecdote.
Graves, R.P. 1882, 1885, 1889. *Life of Sir William Rowan Hamilton*. Vol. I (1882), Vol. II (1885), and Vol. III (1889), Dublin: Hodges, Figgis.
Hankins, T.L. 1980. *Sir William Rowan Hamilton*. Baltimore: Johns Hopkins Press. (Contains portraits of Edgeworthstown House and Francis Beaufort Edgeworth.)
Harrod, R.F. 1951. *The Life of John Maynard Keynes*. London: Macmillan.

Hicks, J.R. 1984. Francis Ysidro Edgeworth. In *Economists and the Irish Economy: from the Eighteenth Century to the Present Day*, ed. Dublin: Irish Academic Press in association with *Hermathena*, Trinity College Dublin, 157–74.
Hildreth, C. 1968. Edgeworth, Francis Ysidro. In *International Encyclopaedia of the Social Sciences*, ed. D.E. Sills, New York: Macmillan and Free Press, Vol. 4, 506–9.
Hobson, J.A. 1938. *Confessions of an Economic Heretic*. London: Allen & Unwin. Reprinted Brighton: Harvester Press, 1976.
Howey, R.S. 1960. *The Rise of the Marginal Utility School 1870–1889*. Lawrence: University of Kansas Press.
Hutchison, T.W. 1953. *A Review of Economic Doctrines 1870–1929*. Oxford: Clarendon Press.
Kendall, M.G. 1968. Francis Ysidro Edgeworth, 1845–1926. *Economic Journal* 36, March, 140–53.
Keynes, J.M. 1933. *Essays in Biography*. London; Macmillan. New edn, ed. Geoffrey Keynes, London; Rupert Hart-Davis, 1951. Reprinted as Vol. X of *The Collected Writings of John Maynard Keynes*, London: Macmillan for the Royal Economic Society, 1972. (The footnote referred to was in neither the 1926 nor (the first printing of) the 1933 version of Keynes' essay.)
MacKenzie, N. and MacKenzie, J. 1977. *The Fabians*: New York; Simon & Schuster.
Marshall, A. 1961. *Principles of Economics*. 9th (Variorum) edn, ed. C.W. Guillebaud, 2 vols, London: Macmillan.
Mill, J.S. 1873. *Autobiography*. London: Longmans, Green, Reader & Dyer.
Mozley, T. 1882. *Reminiscences, chiefly of Oriel College and the Oxford Movement*. 2 vols, Boston: Houghton Mifflin.
Pigou, A.C. and Robertson, D.H. 1931. *Economic Essays and Addresses*. London: P.S. King & Son.
Price, L.L. 1926. Francis Ysidro Edgeworth. *Journal of the Royal Statistical Society* 89, March, 371–7.
Samuelson, P.A. 1947. *Foundations of Economic Analysis*. Cambridge, Mass.: Harvard University Press.
Samuelson, P.A. 1951. Schumpeter as a teacher and economic theorist. In *Schumpeter: Social Scientist*, ed. S.E. Harris, Cambridge, Mass.: Harvard University Press.
Schumpeter, J.A. 1954. *History of Economic Analysis*. New York: Oxford University Press.
Stigler, S. 1978. Francis Ysidro Edgeworth, statistician. *Journal of the Royal Statistical Society*, Series A 141(3), 287–322.
Sully, J. 1918. *My Life and Friends*. New York: E.P. Dutton & Co.

Other references
Allen, R.G.D. 1938. *Mathematical Analysis for Economists*. London: Macmillan.
Annan, N. 1952. *Leslie Stephen*. Cambridge, Mass.: Harvard University Press.
Aumann, R. 1964. Markets with a continuum of traders. *Econometrica* 32, 39–50.
Auspitz, R. and Lieben, R. 1889. *Untersuchungen über die Theorie des Preises*. Leipzig; Duncker & Humblot. French trans. in 2 vols (Texte and Album), Paris: Giard, 1914.
Cournot, A.A. 1838. *Recherches sur les principes mathématiques de la théorie des richesses*. Paris: Hachette. New edn., ed. G. Lutfalla, Paris: Rivière, 1938.
Hardy, G.H. 1938. *Pure Mathematics*. 7th edn, Cambridge: Cambridge University Press.
Hardy, G.H., Littlewood, J.E. and Polya, G. 1934. *Inequalities*. Cambridge: Cambridge University Press.
Hicks, J.R. 1930. Edgeworth, Marshall, and the indeterminateness of wages. *Economic Journal* 40, 215–31.
Hicks, J.R. 1932. *The Theory of Wages*. London: Macmillan.
Jensen, J.L.W.V. 1906. Sur les fonctions convexes et les inégalités entre les valeurs moyennes. *Acta Mathematica* 30, 175–93.
Jevons, W.S. 1881. Review of *Mathematical Psychics*. *Mind* 6, 581–3.
Keynes, J.M. 1936. *The General Theory of Employment, Interest and Money*. London: Macmillan.
Lindsay, R.B. and Margenau, H. 1957. *Foundations of Physics*. New York: Dover.
Marshall, A. 1881. Review of *Mathematical Psychics*. *Academy* 476, June, 457.
Nash, J.F. 1950. The bargaining problem. *Econometrica* 18, 155–62.

Newman, P. 1965. *The Theory of Exchange*. Englewood Cliffs, NJ: Prentice-Hall.

Scarf, H. 1962. An analysis of a market with a large number of participants. In H. Scarf, *Recent Advances in Game Theory*, Princeton: Princeton University Press.

Scarf, H. and Debreu, G. 1963. A limit theorem on the core of an economy. *International Economic Review* 4, 235–47.

Shubik, M. 1959. Edgeworth market games. In *Contributions to the Theory of Games*, Vol. IV, ed. A.W. Tucker and R.D. Luce, Annals of Mathematics Studies No. 40, Princeton: Princeton University Press.

Sidgwick, H. 1874. *The Methods of Ethics*. London: Macmillan. 2nd edn, 1877.

Singer, P. (ed.) 1985. *In Defence of Animals*. Oxford: Basil Blackwell.

Snow, C.P. 1967. Foreword to new edition of G.H. Hardy: *A Mathematician's Apology*. Cambridge: Cambridge University Press.

Stigler, G.J. 1942. *The Theory of Competitive Price*. New York: Macmillan.

Stigler, G.J. 1965. *Essays in the History of Economics*. Chicago: University of Chicago Press.

Thornton, W.T. 1870. *On Labour*. 2nd edn, London: Macmillan. Rome: Edizioni Bizzarri, 1969.

Walras, L. 1874. *Eléments d'économie politique pure*. Lausanne: L. Corbaz. English translation by W. Jaffé of the Edition Definitive (1926), London: Allen & Unwin, 1954.

Yaari, M.E. 1981. Rawls, Edgeworth, Shapley and Nash: theories of distributive justice re-examined. *Journal of Economic Theory* 24, 1–39.

Edgeworth as a statistician. Francis Edgeworth was the leading theorist of mathematical statistics of the latter half of the 19th century, though his influence was diminished by the difficulty of his exposition. He is most frequently remembered today for his work on the Edgeworth Series, but in fact he touched on nearly every sphere of modern statistics, from the analysis of variance to stochastic models, to multivariate analysis, to the asymptotic theory of maximum likelihood estimates, to inventory theory. In some areas such as correlation, his work was decisive in the development of all that followed.

Edgeworth's first purely statistical work was published in 1883, when he began a series of papers examining the methods, rationale and philosophical foundations of probability and its application to the analysis of observational data. Most of this work appeared in the *Philosophical Magazine, Mind*, or the *Journal* of the London (later the Royal) Statistical Society. Between 1883 and 1890 he published over 30 separate papers on a wide selection of statistical topics; these works are best viewed as the tracks left by a first-rate mind as it took an excursion through territory that had already been explored. He found much that was new, but his principal occupation re-examining past works, particularly those of Laplace, to see how they might be used in social science. A major (and under appreciated) accomplishment of this period was Edgeworth's explanation of how simple significance tests could be used to compare averages. The mathematical technique was not new, but the conceptual framework was subtly different from that of the early astronomers, and while Edgeworth's (1885b) explanation may today seem elementary, it had a lasting widespread impact. In subsequent work (Edgeworth, 1885c; Stigler, 1978) he developed what might now be viewed as an analysis for an additive effects model for a two-way classification, and he was sensitive to the effect non-normality or serial dependence could have upon the procedures.

Edgeworth's main orientation in his inferential work was Bayesian, and he presented both philosophical and mathematical investigations of this approach. To Edgeworth, a prior distribution was based in a rough way upon experience. A uniform prior was often justified because, Edgeworth observed, we do not find a pattern in nature that tends to favour one set of values for its constants over another set. Edgeworth tempered this with a realization that inferences would frequently not be very sensitive to the prior specification (Edgeworth, 1885a). When evaluating the significance of differences, however, Edgeworth reverted to a sampling theory viewpoint. One of his 1883 works includes a derivation of Student's t-distribution as the posterior distribution for a normal mean. From 1890 to 1893 Edgeworth, reacting to work by Galton, gave the first fully developed mathematical examination of correlation and its relation to the multivariate normal distribution (Edgeworth, 1892a, b). Edgeworth showed how the constants of a multivariate normal distribution could be expressed in terms of pairwise correlation coefficients (and hence how the conditional expectation of one variable given others could be expressed in terms of correlation coefficients), and he investigated how a correlation coefficient could be estimated from data. His work gave what may be the earliest version of what has come to be called Pearson's product moment estimate (or Pearson's r). Incidentally, it was Edgeworth who coined the term 'coefficient of correlation', as Galton had used 'index'.

Edgeworth's work on correlation had an immense influence upon Karl Pearson, and through him upon all 20th-century work on this topic. In the 1890s, Edgeworth's statistical work became increasingly occupied by a competition with Karl Pearson as to who could best model skew data. Pearson, with his family of skew curves that included gamma distributions and a scheme (the method of moments) for selecting a curve within this family, is generally conceded to have won the contest. Edgeworth at one time or another tried three different approaches. One of these (the 'method of translation', or fitting a normal curve to transformed data) has become popular in more recent times. Another (fitting separate half-normal curves to the left and right sides of the distribution) has been largely forgotten. The third was based upon what we now call the Edgeworth Series. The essence of Edgeworth's approach was to generalize the central limit theorem by the inclusion of correction terms, terms that appeared in the derivation of the distribution of sums but which became negligible if the number of terms in the sum was large. The idea was that skew distributions found in nature were skew because they were aggregates of relatively small numbers of non-normal components. Edgeworth was thus taking a theoretical approach, one that he felt was more appealing than Pearson's more ad hoc approach. The Edgeworth Series was foreshadowed in his work as early as 1883 (when he found it as a series solution to the heat equation), but the full development came later (Edgeworth, 1905), and the labour he put into it after 1895 was immense, and largely unrewarded. His attempts to provide a methodology for fitting the series to data attracted few followers, Arthur Bowley being the only important one. Bowley's brave attempt to explain the method in his assessment of Edgeworth's work (Bowley, 1928) was only marginally more readable than Edgeworth's own many efforts on this. Ironically, later statisticians (notably Harold Cramér, see Cramér, 1972) have found that Edgeworth's mode of arranging correction terms was far superior to alternatives proposed by Bruns, Gram and Charlier, and the Edgeworth Series has become an important technique for approximating sampling distributions (rather than data distributions, as Edgeworth had intended).

In addition to these major themes, Edgeworth's work abounds in minor nuggets. The largest of these may be a series of papers in 1908–9 that we can now recognize as containing the germ of a proof of the asymptotic efficiency of maximum likelihood estimates. In a contentious 1935 meeting of the Royal Statistical Society this work was pointed out to R.A. Fisher by Bowley as an unacknowledged predecessor, although it seems doubtful that it had any influence on Fisher (see Pratt, 1976). Of more importance was Edgeworth's work on index numbers and on the theory of banking. While his work on index numbers is more properly treated with his economic work, it is worth noting here that he was a pioneer in the application of probability to the analysis and choice of index numbers. In regard to banking, based upon statistical considerations, he promulgated in 1888 the rule that the reserves of a bank need only be proportional to the square root of its liabilities (Edgeworth, 1888).

In all Edgeworth's work one is constantly coming upon minor, often paradoxical observations (see for example, Stigler, 1980) that reveal the depth of his understanding, the subtlety of his thoughts, and a grasp of mathematics that seems quite at odds with his lack of formal training in the subject. Edgeworth was an independent thinker upon statistical matters, though he was perhaps the earliest to appreciate and follow up on Galton's innovative concepts of regression and correlation. Edgeworth's most important influence was upon Karl Pearson, though Pearson was chary in his recognition of this influence. Taken together, Galton, Edgeworth and Pearson shaped modern statistics to a greater degree than any other individual or group before R.A. Fisher. Edgeworth's works on statistics number at least 75, and it is rare to find one that is self-contained. Bowley (1928) made an attempt to summarize all of Edgeworth's statistical work, and he gave a bibliography of most of it. Stigler (1978, 1986) gives a more recent assessment, and comments upon different aspects of Edgeworth's work can be found in papers by Kendall (1968, 1969) and Pratt (1976).

STEPHEN M. STIGLER

BIBLIOGRAPHY

Bowley, A.L. 1928. *F.Y. Edgeworth's Contributions to Mathematical Statistics*. London: Royal Statistical Society. Reprinted, New York: Augustus M. Kelley, 1972.

Cramér, H. 1972. On the history of certain expansions used in mathematical statistics. *Biometrika* 59, 205–7.

Edgeworth, F.Y. 1885a. Observations and statistics. An essay on the theory of errors of observation and the first principles of statistics. *Transactions of the Cambridge Philosophical Society* 14, 138–69.

Edgeworth, F.Y. 1885b. Methods of statistics. *Jubilee Volume of the Statistical Society*, 181–217.

Edgeworth, F.Y. 1885c. On methods of ascertaining variations in the rate of births, deaths, and marriages. *Journal of the Royal Statistical Society* 48, 628–49.

Edgeworth, F.Y. 1888. The mathematical theory of banking. *Journal of the Royal Statistical Society* 51, 113–27.

Edgeworth, F.Y. 1892a. Correlated averages. *Philosophical Magazine* (Fifth Series) 34, 190–204.

Edgeworth, F.Y. 1892b. The law of error and correlated averages. *Philosophical Magazine* (Fifth Series) 34, 429–38, 518–26.

Edgeworth, F.Y. 1905. The law of error. *Transactions of the Cambridge Philosophical Society* 20, 36–65, 113–41.

Edgeworth, F.Y. 1908–9. On the probable errors of frequency-constants. *Journal of the Royal Statistical Society* 71, 381–97, 499–512, 651–78; 72, 81–90.

Kendall, M.G. 1968. Francis Ysidro Edgeworth, 1845–1926. *Biometrika* 55, 269–75.

Kendall, M.G. 1969. The early history of index numbers. *Review of the International Statistical Institute* 37, 1–12.

Pratt, J. 1976. F.Y. Edgeworth and R.A. Fisher on the efficiency of maximum likelihood estimation. *Annals of Statistics* 4, 501–14.

Stigler, S.M. 1978. Francis Ysidro Edgeworth, statistician (with discussion). *Journal of the Royal Statistical Society*, Series A 141, 287–322.

Stigler, S.M. 1980. An Edgeworth curiosum. *Annals of Statistics* 8, 931–4.

Stigler, S.M. 1986. *The History of Statistics: The Measurement of Uncertainty before 1900*. Cambridge, Mass.: Belknap Press of the Harvard University Press.

Edgeworth, Maria (1767–1849). Born in England of an Irish land-owning family, Maria Edgeworth began her career as amanuensis and co-author to her father Richard Lovell Edgeworth, the educator and amateur inventor. Her first publications were a series of moral tales for children (*The Parents' Assistant*, 1796, and *Early Lessons*, 1802) which aimed to instil the virtues she saw as essential to a 'good' individual and so a 'good' society: honesty, frugality and hard work. These characteristics match rather precisely those of Adam Smith's 'prudent man' in the *Wealth of Nations*. Her tales teach the value of a work ethic, sharply contrasting the evils of sloth and idleness with the pleasures of diligence and achievement. Indeed, her attitude towards this aspect of labour did not exclude her own privileged class of landowners, who, as she witnessed in her own country, frequently abused the landlord–tenant contract.

In 1800 she published the work which is, perhaps, of most interest to economists, *Castle Rackrent*. Through the character of Thady Quirk, an ancient retainer of the Rackrent family, she recounts the history of three generations of absentee landlords, of their tenants and of the depths to which the Rackrent fortunes had fallen through successive generations of dissolute lifestyle. The book not only influenced prominent literary figures of the time (for example, Turgenev and Walter Scott) but also established a literary precedent for the development of fictional characters within the context of a realistic historical, social and economic setting – an approach which, in England, could be said to reach its peak with George Eliot's *Middlemarch*. In the 19th century the name Rackrent came to stand for the embodiment of the vices of the landed aristocracy and was freely used as such by writers like Carlyle and, later, her nephew F.Y. Edgeworth.

Maria Edgeworth continued her critical examination of the landlord–tenant relationship in novels like *The Absentee* (1812) and *Ennui* (1825) where she addressed issues such as leases, population and economic progress and the impact of manufacture on a traditional agricultural economy. Her letters to David Ricardo confirm her interest in the poverty and distress among the Irish agricultural peasantry. She initiated and engaged in a vigorous correspondence with Ricardo over the potato question and the effects of famines in the 1820s. On this subject she differed with both Ricardo and Malthus arguing that the essential cause of the difficulty lay in mismanagement. She rather amusingly suggested that instead of theorizing from afar, Ricardo should travel to Ireland and see for himself.

J.P. CROSHAW

education, economics of. *See* HUMAN CAPITAL.

effective demand. This is the term used by Keynes in his *General Theory* (1936) to represent the forces determining changes in the scale of output and employment as a whole.

Keynes attributed the first discussions of the determinants of the supply and demand for output as a whole to the classical economists, in particular the debate between Ricardo and Malthus concerning the possibility of 'general gluts' of commodities, or what has come to be known as Say's Law of Markets. Indeed, Keynes's theory was intended to replace Say's Law, although the emergence of effective demand from his *Treatise on Money* (1930) critique of the quantity theory of money, and his insistence on its application in what he originally called a 'monetary production economy', suggests that it should also be seen in antithesis to classical monetary theory. For Adam Smith (1776, p. 285), 'A man must be perfectly crazy who ... does not employ all the stock which he commands, whether it be his own or other peoples' on consumption or investment. As long as there was what Smith called 'tolerable security', economic rationality implied that it was impossible for demand for output as a whole to diverge from aggregate supply. Although Smith (p. 73) did call the demand 'sufficient to effectuate the bringing of the commodity to the market', the 'effectual demand' 'of those who are willing to pay the natural price' of the commodity, the idea referred to divergence of market from natural price of particular commodities and the process of gravitation of prices to their natural values. J.B. Say's discussion of the problem of the 'disposal of commodities' adopted Smith's position. Against those who held that 'products would always be abundant, if there were but a ready demand, or market for them,' Say's 'law of markets' argued 'that it is production which opens a demand for products' (1855, pp. 132–3); if production determined ability to buy, then demand could not be deficient. While excesses in particular markets were admitted, they would always be offset by deficiencies in others. Ricardo used similar arguments against Malthus, who responded by suggesting that:

> from the want of a proper distribution of the actual produce, adequate motives are not furnished to continued production,... the grand question is whether it [actual produce] is distributed in such a manner between the different parties concerned as to occasion the most effective demand for future produce ... (Malthus, 1821).

Malthus argues that the composition of output affects its quantity by producing doubts in the minds of Smith's rational entrepreneurs concerning the 'security' of their future profit.

The final word in the classical debate was J.S. Mill's 'On the Influence of Consumption on Production', which sought exceptions to the proposition that 'All of which is produced is already consumed, either for the purpose of reproduction or enjoyment' so that 'There will never, therefore, be a greater quantity produced, of commodities in general, than there are customers for' (1874, pp. 48–9). Mill accused those who argued that demand limits output of a fallacy of composition, for the individual shopkeeper's failure to sell is due to a disproportion of demand which cancels out for the nation as a whole. Mill also notes that the argument that every purchaser must be a seller presumes barter, for money enables exchange 'to be divided into two separate acts' so one 'need not buy at the same moment when he sells' (p. 70). To avoid this problem 'money must itself be considered as a commodity', for 'there cannot be an excess of all other commodities, and an excess of money at the same time' (p. 71). Mill admits that if money were 'collected in masses', there might be an excess of all commodities, but this would mean only a temporary fall in the value of all commodities relative to money. Similarly to Smith's 'tolerable security', Mill explains an excess of

commodities in general by 'a want of commercial confidence', which he denies may be caused by an overproduction of commodities (p. 74).

Mill's defence of Say's Law highlights the importance of the classical quantity theory, which was originally formulated to oppose the undue emphasis given to precious metals as components of national wealth by the mercantilists. Hume noted that labour, not gold, produced the commodities which composed national wealth; that gold was only as good as the labour it commanded to produce output. Thus the classical position that the velocity of circulation of money was independent of its quantity was built on the view that money would only be held to be spent. Money could at best cause temporary general gluts; in the long term, 'rational' men would not choose to hold money rather than spend it.

On the eve of the marginal revolution, classical theory thus admitted the temporary occurrence of general gluts explained by cyclical disproportions in demand for money and commodities due to crises of confidence. It is paradoxical that while the marginal revolution was motivated by the failure of classical theory to give sufficient attention to the role of demand in value theory, it failed to extend its analysis of demand to output as a whole in either the long or the short period. Indeed, the emphasis on individual equilibrium produced by the subjective theory of value which replaced the classical theory, made separate discussion of aggregate supply and demand redundant. Thus Keynes's reference to 'the disappearance of the theory of demand and supply for output as a whole, that is the theory of employment *after* it has been for a quarter of a century the most discussed thing in economics' (Keynes, 1936c).

But it was discussion, not Say's Law, which disappeared from neoclassical economics. Thus Keynes classed economists from Smith and Ricardo to Marshall and Pigou as 'Classical', for despite antagonistic theories of value and distribution, they all held a similar theory of supply and demand for output as a whole.

Keynes suggests that this was due more to the failure of neoclassical economists to heed Mill's warning concerning the extension of the conditions faced by the individual to the economy as a whole, than to positive analysis. If consumers (producers) maximize utility (profit) subject to an income (cost) constraint, reaching the maximum by substituting in consumption (production) goods (inputs) which were cheaper per unit of utility (output), then excess supply of any good (resource) is due to its price exceeding its marginal utility (productivity). Market competition would lead to relative price adjustments which eliminate excess supply. Since it was impossible for any single good (resource) to be unsold (unemployed), it was natural to extend this analysis to the aggregate level to deny the possibility of general gluts without further analysis.

Any divergence from this position was explained, not by reference to hoarding money due to crises of confidence, but by temporary impediments to the automatic adjustment of relative prices in competitive markets. Thus, despite their new marginal theory of value, Keynes's contemporaries reached a similar result that divergence of employment from its full employment level would be determined by temporary non-persistent causes eliminated in the long run.

From 1921 to 1939 the unemployment rate in the United Kingdom never fell below 10 per cent, peaking in 1932 at 22.5 per cent (over 2.7 million). This exceeded the limits that most economists attributed to short-period frictions. The self-adjusting nature of the neoclassical version of Say's Law that Keynes chose to criticize was thus contradicted by reference to economic events as well as by Keynes's conception of effective demand.

Keynes was not concerned with impediments to the equality of the supply and demand, but with the

> problem of the equilibrium of supply and demand for output as a whole, in short, of effective demand ... When one is trying to discover the volume of output and employment, it must be this point of equilibrium for which one is searching.

While the Classics solved the problem by assuming the identity of savings and expenditure on investment goods, neoclassical theory presumed Say's Law 'without giving the matter the slightest discussion' (1936b, p. 215).

Keynes's theory of effective demand thus had to replace Say's Law. To do this Keynes departed from the Classical position on two points. The first was to assume that wages exceed subsistence so that expenditure on consumption goods does not exhaust factor incomes. As expressed in Keynes's psychological law of consumption, this implied that as output increased, the gap between aggregate expenditure and factor costs increased, so that unless investment expenditure expanded to fill the gap, entrepreneurs would experience losses.

The second departure was from the assumption that rationality dictated that entrepreneurs' savings represented productive investment expenditure. If investment could produce losses, or changes in interest rates change capital values, then greater future enjoyment might be assured by not investing; holding money might be 'rational' in such conditions. Further, in a monetary economy, nothing guarantees that maximization of returns in money will maximize either productive capacity or the demand for labour.

In Keynes's theory the propensity to consume and the multiplier produce the proposition that it is the level of output which adjusts saving to investment, rather than the rate of interest, while the explanation of the decisions over the level of investment in a monetary economy requires an explanation of rates of interest in money terms. The two factors are closely related.

In a 1934 letter to Kahn, Keynes gives a 'precise definition of what is meant by effective demand' (1934a, p. 422). If O is the level of output, W the marginal prime cost of production for that output, and P the expected selling price, 'Then OP is effective demand'. The classical theory that 'supply creates its own demand' assumes that OP equals OW, irrespective of the value of O, 'so that effective demand is incapable of setting a limit to employment which consequently depends on the relation between marginal product in wage-goods industries and marginal disutility of employment'. Thus, what Keynes later called (1936a, ch. 2) the two 'classical' postulates limit O at full employment. In contrast,

> On my theory OW ≠ OP for *all* values of O, and entrepreneurs have to choose a value of O for which it is equal – otherwise the equality of price and marginal prime cost is infringed. This is the real starting point of everything.

The key point was thus the impact of different levels of O on the difference between costs and prices, that is on entrepreneurs' profits. Keynes took up this question, in an undated exchange with Sraffa of about the same time (1934b, pp. 157ff). Keynes notes that a non-unitary marginal propensity to consume implies OP ≠ OW for any O, and generates

the general principle that *any* expansion of output gluts the market unless there is a *pari passu* increase of investment appropriate to the community's marginal propensity to consume; and any contraction leads to windfall profits to producers unless there is an appropriate *pari passu* contraction of investment.

The level of O at which OP = OW will be determined by the level of investment and the propensity to consume. Changes in the rate of investment, based on entrepreneurs' expectations of their future profits, will determine O.

In an early draft of the *General Theory* Keynes put it this way:

> Effective demand is made up of the sum of two factors based respectively on the expectation of what is going to be consumed and on the expectation of what is going to be invested (1973a, p. 439).

Thus the theory of effective demand required, in addition to explanation of consumption based on the propensity to consume, an explanation of variations in the level of investment. Since neoclassical theory resolved this problem by presuming that investment was brought into balance with full employment saving by means of the rate of interest, Keynes located the 'flaw being largely due to the failure of the Classical doctrine to develop a satisfactory theory of the rate of interest' (1934c, p. 489).

Keynes concentrated his efforts to produce a theory of interest compatible within this theory of effective demand within what he called a monetary production economy. The *Treatise on Money* (1930) had explained changes in prices in terms of households' consumption decisions relative to entrepreneurs' production decisions. If these decisions were incompatible, investment diverged from saving and prices of consumption goods adjusted producing windfall profits or losses. The prices of investment goods were determined separately from this process, by means of the interaction of the bearishness of the public reflecting their decisions to hold bank deposits or securities on the one hand, and the monetary policy of the banking system on the other.

Investment goods are held because their present costs or supply prices are lower than the present value of their anticipated future earnings or demand prices; the larger this difference, the higher the expected rate of return. Since any change in the price of a durable capital asset will influence its rate of return, a theory that explains the price of capital assets also explains rates of return (which Keynes called marginal efficiency). With the demand price of an asset based on the value of expected future earnings discounted by the rate of interest, it is clear why a satisfactory theory of interest is crucial to the explanation of effective demand.

But money was a durable asset like any other, and as such it has a spot or demand price and a supply price or forward price, which determine the money rate of interest. Keynes thus transformed his concept of bearishness into liquidity preference which, together with banking policy, would determine the rate of interest. For Keynes, 'the money rate of interest ... is nothing more than the percentage excess of a sum of money contracted for forward delivery ... over what we may call the "spot" or cash price of the sum thus contracted for forward delivery' (1936a, p. 222), it is:

> the premium obtainable on current cash over deferred cash ... No one would pay this premium unless the possession of cash served some purpose, that is had some efficiency. Thus we may conveniently say that interest on money measures the marginal efficiency of money measured in terms of itself as a unit (1937a, p. 101).

Since both money and capital assets had marginal efficiencies representing their rates of return, profit-maximizing individuals in a monetary economy would demand money and capital assets in proportions which equated their respective returns. The equilibrium level of output chosen by entrepreneurs would then be represented by equality of the marginal efficiency of capital and the rate of interest (the marginal efficiency of money). The question of the effect of an increase in output on profit raised by a propensity to consume less than unity can now be seen as the effect of an increase in investment on the marginal efficiency of money relative to the marginal efficiencies of capital assets. Since these marginal efficiencies reflect pairs of spot and forward asset prices, the question can also be put as the effect of an increase in investment on relative money prices. Thus Keynes's independent variables, the propensity to consume, the efficiency of capital and liquidity preference, given expectations and monetary policy, interact to determine effective demand.

Since this equilibrium could be described by S = I, or equality between the rate of interest and the marginal efficiency of capital, the level of output which equates aggregate demand and supply also equates marginal efficiency with the rate of interest. To complete his theory of effective demand, Keynes faced the question first raised by Wicksell of the causal relation between the natural and the money rate of interest. Just as Keynes rejected the determination of the level of O at which OP = OW by the equality of the marginal productivity and disutility of labour, he rejected marginal productivity as the determinant of marginal efficiency and the real rate of interest determining the money rate because it was based on 'circular reasoning' (1937b, p. 212).

Keynes argues instead that it is the marginal efficiency of capital assets which adapts to the money rate of interest rather than vice versa. These two points of departure are discussed in chapters 16 and 17 of the *General Theory*, where Keynes points out that the money rate of return to be expected from a capital asset depends on the relation of anticipated money receipts relative to expected money costs, and that there is no reason to believe that these will be related in any predictable way to the asset's physical productivity. Wicksell's natural rate, derived from physical relations of production and exchange, has no application in a monetary economy; Keynes thus substitutes the concept of marginal efficiency.

Keynes also notes that increased investment in particular capital assets increases supply prices and reduces demand prices, causing a decline in marginal efficiencies; an increase in output thus leads to investment in assets with lower rates of return. At some point the marginal efficiency of money will make investment in money as profitable as the purchase of capital assets. At this point the rate of interest equals the marginal efficiency of capital, and any further increase in output would confirm Keynes's 'general principle' that any further expansion in output gluts the market, for increased income is not spent but held in the form of money which becomes a 'generalised sink for purchasing power'.

The question that distinguishes Keynes's theory is thus why money's liquidity premium does not fall as output expands, for this is what prevents investment from rising by just the amount to fill the gap created by the propensity to consume being less than one. To describe these 'essential properties of interest and money', Keynes departs from Mill's position that money is just another commodity. When money is the debt of the banking system its price and quantity behaviour will differ from physical commodities, for it has no real costs of production nor real substitutes. Thus an asset which has a negligible elasticity of production and substitution with respect

to a change in effective demand, will have a rate of return which responds less rapidly to an expansion in demand. As long as the rate of interest falls less rapidly than the marginal efficiencies of capital assets, its rate will be the one which sets the point at which further expansion creates losses.

Thus the propensity to consume shows that investment will have to increase by the amount of the gap between incomes and expenditures as incomes rise if entrepreneurs are not to make losses, while the marginal efficiency of capital and liquidity preference in a monetary production economy explain why the behaviour of the rate of interest relative to the marginal efficiency of capital makes it unlikely that the rate of investment should adjust by just that amount. Since entrepreneurs maximize monetary returns, not employment or physical output, there is no reason why their investment decisions should lead to an equilibrium at full employment. Keynes's explanation of the limit to the level of employment permits any level as a stable equilibrium, including full employment; it is thus more general than the classical Say's Law position, in which the only stable equilibrium was the limit set by full employment as given in the labour market.

J.A. KREGEL

See also SAY'S LAW.

BIBLIOGRAPHY
Keynes, J.M. 1930. *A Treatise on Money*. Reprinted in Keynes (1971).
Keynes, J.M. 1934a. Letter to R.F. Kahn, 13 April. Reprinted in Keynes (1973b).
Keynes, J.M. 1934b. Letter to P. Sraffa, undated. Reprinted in Keynes (1979).
Keynes, J.M. 1934c. Poverty in plenty: is the economic system self-adjusting? Reprinted in Keynes (1973b).
Keynes, J.M. 1936a. *The General Theory of Employment, Interest and Money*. Reprinted in Keynes (1973a).
Keynes, J.M. 1936b. Letter to A. Lerner, 16 June. Reprinted in Keynes (1979).
Keynes, J.M. 1936c. Letter to R.F. Harrod, 30 August. Reprinted in Keynes (1973c).
Keynes, J.M. 1937a. The theory of the rate of interest. Reprinted in Keynes (1973c).
Keynes, J.M. 1937b. Alternative theories of the rate of interest. Reprinted in Keynes (1973c).
Keynes, J.M. 1971–83. *The Collected Writings of John Maynard Keynes*. Ed. D. Moggridge. London: Macmillan for the Royal Economic Society: 1971. Vols. V and VI. *A Treatise on Money* (1930). 1973a. Vol. VII. *The General Theory of Employment, Interest and Money* (1936). 1973b. Vol. XIII. *The General Theory and After: Part I – Preparation*. 1973c. Vol. XIV. *The General Theory and After: Part II – Defence and Development*. 1979. Vol. XXIX. *The General Theory and After – A Supplement*.
Malthus, T.M. 1821. Letter from Malthus to Ricardo, 7 July. Reprinted in Ricardo (1952) 9–10.
Mill, J.S. 1874. On the influence of consumption on production. In J.S. Mill, *Essays on Some Unsettled Questions of Political Economy*, 2nd edn, reprinted Clifton, NJ: A.M. Kelley, 1974.
Ricardo, D. 1952. *Works and Correspondence of David Ricardo*, Vol. IX. Ed. P. Sraffa with the collaboration of M. Dobb, Cambridge: Cambridge University Press.
Say, J.B. 1855. *A Treatise on Political Economy*. 6th American edn, Philadelphia: J.B. Lippincott.
Smith, A. 1776. *An Inquiry into the Nature and Causes of the Wealth of Nations*. Oxford: Oxford University Press, 1976.

effective protection. The effective rate of protection is the rate of protection provided to the value added in the production of a product. Let the *effective price* be defined as the domestic price of a unit of value added. Then the effective rate of protection (henceforth, ERP) is the proportional increase in the effective price made possible by tariffs and other measures. It is to be contrasted with the *nominal* tariff and (more generally) nominal rate of protection, which refers to the proportional increase in the *nominal price*. If the only policy instruments are tariffs, the ERP depends not only on the nominal tariff on the commodity concerned but also on the tariffs on the inputs and on the input coefficients.

Consider the simple case of an importable product, j, which has only a single input, also an importable, i. There are no taxes and subsidies affecting j and i other than the import tariffs. The formula for the ERP for the activity producing j is then

$$g_j = \frac{t_j - a_{ij} t_i}{1 - a_{ij}}$$

where g_j is the ERP, t_j is the tariff on j, t_i is the tariff on i, and a_{ij} is the share of i in the cost of j in the absence of tariffs.

This shows that if $t_j = t_i$, then $g_j = t_j$. It is common for input tariffs to be low relative to final goods tariffs, that is, $t_j > t_i$, and in that case $g_j > t_j$, an important result, since it shows that effective rates tend to be higher than nominal rates. A rise in the input tariff clearly reduces effective protection for the using industry, even though it raises protection for the input-producing activity.

Actual measurements involve using a'_{ij}, which is the input share that results after the tariffs have raised both the domestic final good price and the domestic input price. The connection between the input share before tariffs are imposed (a_{ij}) and after (a'_{ij}) is as follows:

$$a'_{ij} = a_{ij} \frac{1 + t_i}{1 + t_j}$$

and from this, and the formula for the ERP given above, one can obtain the formula which is commonly used in empirical studies, namely:

$$g_j = \frac{1 - a'_{ij}}{\dfrac{1}{1 + t_j} - \dfrac{a'_{ij}}{1 + t_i}} - 1.$$

The effective protection concept can be extended to allow for all taxes and subsidies affecting tradeable goods, i.e. all importables and exportables. An export subsidy raises the domestic price of an exportable; if the input is an exportable, then t_i represents the rate of subsidy, and if the activity for which the ERP is being calculated is an exportable, then t_j can represent the subsidy. Similarly, export taxes, production taxes, consumption taxes, and production and consumption subsidies can be allowed for. A production tax or subsidy for the final product will affect t_j, while a consumption tax or subsidy for the input will affect t_i. Thus the ERP measure allows a single figure to sum up the net result of various trade and non-trade taxes and subsidies affecting any particular activity.

The ERP measurements revealed at an early stage the high protection that developed countries provided for final processing of primary products even in cases where nominal tariffs were low, the reason being the duty-free entry of the basic materials. The measurements also bring out the negative effective protection provided for exports in many countries: the exports receive no subsidies or other assistance, i.e. $t_j = 0$

for most exports, but import tariffs on their inputs make their t_is positive. It was also noted that tariff reductions are not always what they seem: an offer of tariff cuts at an international negotiation may actually raise the ERP for some domestic industries.

Much attention has been given in the literature to the discovery of *negative value added*, a discovery which was a by-product of effective protection calculations. There are cases where the free trade price of the final product is less than the free trade price of its inputs, so that under free trade the effective price would be negative. One possible reason for this phenomenon is that transport costs on inputs may be much higher than those on the final product. There would then be no production of the final good under free trade. But a sufficiently high tariff on the final good relative to the tariff on the input could make the effective price domestically positive, so that domestic production begins. The rate of protection is then infinite, and algebraically the calculation of the ERP will yield a negative figure. Clearly domestic production of a product where the cost of imported inputs exceeds the free trade price of the final product is an extreme form of waste.

GENERAL EQUILIBRIUM. The next step is to put ERPs into a general equilibrium framework. One can imagine a *scale of effective rates*, which will include ERPs for all traded activities, including both exportables and importables. The scale will give some indication of the direction in which resources have been pulled by the protective structure. Of course, actual resource movements will also depend on production substitution elasticities, that is, on the whole general equilibrium system, so that the scale is only indicative of resource movement effects. The crucial point is that, in general equilibrium, *relative* ERPs matter, not absolute rates. This is simplest to see in a model with only two activities where both may be obtaining positive ERPs (if one is an export, this implies it is getting an export subsidy), and resources will then tend to move into the activity with the relatively higher ERP.

There are complications, and it has been shown in the literature that one can produce paradoxes. For example, in a three-activity model, with A and B complementary in a general equilibrium sense, protection of A may expand B even though B may get a lower ERP than C. It must also be remembered that relative nominal rates will determine the direction in which consumption is pulled or distorted.

The idea of the scale of ERPs was first presented in Corden (1966), where it was said that

> Assuming normal non-zero substitution elasticities in production, [the scale of effective rates] tells us the *direction* in which this structure causes resources to be pulled as between activities producing traded goods. Domestic production will shift from low to high effective-protective-rate activities.

This was too strong. It is true in a rather special model, later set out formally in Jones (1975), but more generally, there are various 'paradoxical' circumstances where it need not be true. Several articles have explored such possibilities, and examples are expounded in Corden (1971).

At the general equilibrium level the important and somewhat complex issue also arises of whether particular traded goods activities are protected relative to non-tradeables, and what the role of the exchange rate is. The basic point is that the imposition of a protective structure which is generally positive will tend to draw resources out of non-tradeables, and if the nominal rates are also mainly positive, divert consumption towards non-tradeables. Assuming balance initially, protection

103

then results in excess demand for non-tradeables. Balance would be restored by a rise in the price of non-tradeables, relative to the free trade prices of tradeables, this being a *real* appreciation. It could be brought about with a fixed nominal exchange rate combined with an absolute rise in the price of non-tradeables, or by a nominal appreciation when the price of non-tradeables is constant, possibly because the nominal wage is given.

The usual expositions assumed the average price-level of non-tradeables constant, and stressed the exchange rate adjustment that then needs to be associated with a change in protection levels. When this adjustment is taken into account one can obtain a *net* protective rate which shows whether a particular activity is protected relative to non-tradeables. For example, a particular activity may obtain an ERP of 10 per cent but, if the protection for it and all other activities were removed, there might have to be a devaluation which is equivalent to a uniform tariff and export subsidy of, say, 15 per cent. In that case this activity would have obtained a higher effective price under free trade, so that the system of protection has provided it with negative *net* effective protection. Relative to non-tradeables, it has been anti-protected. If its resources were primarily mobile into and out of non-tradeables, it would expand as a result of a movement towards free trade.

KEY ASSUMPTIONS. The theory of effective protection, at least in its more formal version, makes a number of assumptions. The first is the *small country assumption*, namely the assumption that the country concerned faces given prices of its exports and imports (the terms of trade being exogenous). The second assumption is that for all tradeable goods some trade remains, so that domestic prices are determined by the given world prices as modified by tariffs, export subsidies and other interventions. Thirdly, imports are assumed to be perfect substitutes for the import-competing goods for which the ERPs are calculated. Finally, it is assumed that there are fixed coefficients between final outputs and traded inputs, even though substitution between the domestic factors that contribute to value added can be allowed.

Much theoretical work has gone into exploring the implications of removing the last assumption, though all the others are also important. It does not follow that calculations of ERPs are meaningless when these assumptions do not hold, but figures must be interpreted with care, as discussed in Corden (1971).

NORMATIVE IMPLICATIONS. Do ERPs have normative significance? The formal theory of effective protection and tariff structure was developed with the focus on a question of positive economics: namely, how does a protective structure affect the allocation of resources? But the great interest in the theory and the widespread activity in making calculations has been motivated by a concern with normative issues. It must be stressed again that only *relative* effective rates matter. Knowing a single effective rate on its own sheds no light on either positive or normative implications. The frequent assumption, often only implicit, has been that free trade with appropriate exchange rate adjustment would be the optimal situation, and that non-uniform effective rates therefore impose a *production cost* of protection – that is, a welfare loss through a distortion in resource use. Large divergences are then an indication of a high cost of protection. Furthermore, the structure of effective protection gives then a guide to the welfare (or efficiency) effects of tariff changes: a change that reduces a divergence between effective rates is likely to reduce the cost of protection.

A practical implication is that if there is to be gradual tariff reduction without extra costs being imposed during the process, any increase of such divergences should be avoided. This will be so, for example, if high effective rates are always reduced first. In a three-product model, with industry A getting 0 per cent, industry B 20 per cent and industry C 50 per cent, a reduction in B's effective rate first would increase the divergences between the ERPs on B and C, so that the ERP of C should be reduced first. This is the *concertina method* of tariff reduction, but may have quite complicated implications in terms of nominal tariffs. *Radial* (uniform across-the-board) reductions would also avoid divergences being increased. Finally, it must be remembered that nominal tariffs affect the pattern of consumption whether by final users, or in use of inputs, so that divergences in nominal tariffs determine the *consumption cost* of protection.

If there are other (non-trade) distortions in the economy, a tariff distortion may actually be offsetting. Thus, if an industry is established on the basis of a very high tariff (relative to other industries), so that a positive cost of protection might be expected, there may be a gain if (for example) the industry uses labour for which it has to pay a wage that exceeds its opportunity cost owing to distortions in the labour market. When such non-trade distortions are prevalent one cannot use effective rates on their own as indicators of which activities should expand and which decline if resource allocation is to improve. The broader concept of *domestic resource cost* has been developed to take all distortions into account.

PRACTICAL PROBLEMS. The calculation of ERPs and their use as a guide to policy has become very widespread, especially in developing countries. But all sorts of practical problems arise in the calculations, essentially because the assumptions of the formal theory do not hold, and many ways have been devised to deal with these problems. The problems can only be listed here, but they are important for practitioners. For more details, see Corden (1975) and Balassa and associates (1982).

When quotas are the principal method of protection, comparisons between domestic and world market prices must be made in order to obtain the implicit nominal rates of protection which must be the starting point for any calculations. When tariffs alone are relevant there may be *tariff redundancy*, so that, again, price comparisons must be made; a difficulty here is that the quality of the local product and the import may differ. Available input–output coefficients in most countries are rarely sufficiently disaggregated for the ERP calculations. There is a need for tariff averaging, and this has built-in biases.

A decision has to be made as to how to treat non-traded inputs into the tradeable products for which ERPs are calculated. This last issue has given rise to much theoretical discussion (on which see Corden, 1971, 1975). The correct method appears to be very complicated: lump the non-traded and primary-factor content of non-traded inputs with value added, but group the traded-input content of non-traded inputs with traded inputs. Tariffs on traded inputs into non-traded inputs then reduce the ERP for the final product.

THE SUBSTITUTION PROBLEM. By far the most sophisticated theoretical work has gone into the 'substitution problem'. This involves removing the assumption of fixed coefficients between the final output and the produced traded inputs. Thus substitution between traded inputs and the primary factor content of value added is allowed for. Two distinct issues then arise.

First, suppose that the production functions are separable, so that substitution between traded inputs and the various primary factors, for example, labour and capital, is 'unbiased'. In that case, the concept of value added retains a clear meaning. One can think of a 'value added product' which is combined with traded inputs in varying proportions (depending on the input tariff and the final good tariff, among other things), to make a final product. Since the ERP is the proportional increase in the effective price, which is the price of this 'value added product'. ERP also then has a clear meaning. But the problem remains that measurements based on the coefficients after tariffs have been imposed (which is what the data yield) will have a bias, reflecting the substitution effects. It can be shown that the tendency will always be to overstate the 'true' ERPs. The problem is then one of inevitable measurement error. Since one is interested in the relative position in the scale of effective rates and in the divergences between ERPs, it is relevant that the measurement error will differ between ERPs, depending on production functions and relationships between final goods and input tariffs. Fortunately, there is some possibility that this complication may not be important in practice.

The second issue is more fundamental. If production functions are not separable, so that substitution is 'biased', the whole concept of the 'value added product' and hence of ERP is thrown into doubt. The question is whether 'value added' has a meaning. One really needs to assume that, on a probability basis, the bias is generally zero.

ARE GENERAL EQUILIBRIUM MODELS PREFERABLE? Another basic criticism of the ERP concept and of all the resources that have gone into calculations of ERPs can be made. It has been pointed out that a scale of ERPs is an imperfect and possibly misleading indicator of resource allocation movements. Actual resource pulls also depend on supply elasticities, on production functions, and on a whole lot of complex interactions which have been analysed in the literature, but which deprive the scale of effective rates of any simple significance. Various paradoxes have been shown to be possible – for example, that resources will be drawn into a low ERP activity out of a high ERP one under particular factor-intensity and relative tariff conditions. The conclusion of some critics has been either that no measurements are any use or that one might as well use only nominal rates. Another view is that the best approach is to use computable general equilibrium models, and these make ERPs redundant.

The answer must be that if the data and estimates for complete general equilibrium models are available – and sufficiently disaggregated with respect to activities or industries to be policy-relevant – there is indeed no need to calculate ERPs. The latter contain some information, taking into account input tariffs, and so on, but pause half-way to the complete answer. The case for ERP calculations and their use for policy must be that the data and estimated functional relationships required for complete and detailed general equilibrium models do not usually exist, and certainly not in sufficiently disaggregated form, so that ERPs, which *are* feasible to calculate, give some indication of possible resource pulls and costs of protection. The extensive theoretical work is designed to indicate the direction of various probable biases and to bring out the stringent assumptions required for firm conclusions to be reached from the data.

THE LITERATURE. Until the mid-1960s the vertical relationships between tariff rates derived from the input–output relationships between products were completely neglected in the literature of trade theory. In fact tariff theory was either narrowly partial equilibrium, focusing on just one vertically integrated product, or consisted of two-sector general equilibrium models. A major feature of the theory of tariff structure was not just to bring out the relevance of input tariffs but also to focus on the horizontal general equilibrium relationships when there are more than two products.

With regard to the ERP concept itself, while there were early precursors, the first extended exposition was in Barber (1955), the first systematic theoretical papers were a 1965 paper of Johnson's, reprinted in Johnson (1971), and Corden (1966). The latter paper opened up various general equilibrium issues, the significance of the scale of effective rates, the problem of non-traded inputs, the substitution problem, and so on, and later a systematic and more complete exposition was presented in Corden (1971), which also contains a history of the ERP concept and references to various precursors. Pioneering empirical work was done in Balassa (1965) and Basevi (1966). Later Balassa became a sponsor of major multi-country empirical studies (Balassa and associates, 1971, 1982), and these volumes also contain extensive reviews by Balassa of theoretical and measurement issues.

The central theoretical issues of the meaning of ERPs have been discussed in numerous papers subsequent to the early work. Particularly to be noted are Jones (1975) and Ethier (1977). In addition, there have been several articles on the 'substitution problem', beginning with Jones (1971), a paper reprinted in Corden (1971), and Ethier (1972), followed by papers by Bruno, by Khang and by Bhagwati and Srinivasan, all in the *Journal of International Economics* of 1973.

W.M. CORDEN

See also FREE TRADE AND PROTECTION; INTERNATIONAL TRADE; QUOTAS AND TARIFFS.

BIBLIOGRAPHY

Balassa, B. 1965. Tariff protection in industrial countries: an evaluation. *Journal of Political Economy* 73(6), December, 573–94.

Balassa, B. and associates. 1971. *The Structure of Protection in Developing Countries*. Baltimore: Johns Hopkins Press.

Balassa, B. and associates. 1982. *Development Strategies in Semi-industrial Countries*. Baltimore: Johns Hopkins Press.

Barber, C.L. 1955. Canadian tariff policy. *Canadian Journal of Economics and Political Science* 21, November, 513–30.

Basevi, G. 1966. The US tariff structure: estimation of effective rates of protection of US industries and industrial labor. *Review of Economics and Statistics* 48, 147–60.

Bruno, M. 1973. Protection and tariff change under general equilibrium. *Journal of International Economics* 3(3), August, 205–25.

Bruno, M., Khang, C., Ray, A., Bhagwati, J.N. and Srinivasan, T. 1973. The theory of effective protection in general equilibrium: a symposium. *Journal of International Economics* 3(3), August, 205–81.

Corden, W.M. 1966. The structure of a tariff system and the effective protective rate. *Journal of Political Economy* 74, June, 221–37.

Corden, W.M. 1971. *The Theory of Protection*. Oxford: Oxford University Press.

Corden, W.M. 1975. The costs and consequences of protection: a survey of empirical work. In *International Trade and Finance: Frontiers for Research*, ed. P.B. Kenen, Cambridge: Cambridge University Press.

Ethier, W.J. 1972. Input substitution and the concept of the effective rate of protection. *Journal of Political Economy* 80(1), January/February, 34–47.

Ethier, W.J. 1977. The theory of effective protection in general equilibrium: effective-rate analogues of nominal rates. *Canadian Journal of Economics* 10(2), May, 233–45.

Grubel, H.G. and Johnson, H.G. (eds) 1971. *Effective Tariff Protection*. Geneva: Graduate Institute of International Studies.

Johnson, H.G. 1971. *Aspects of the Theory of Tariffs*. London: George Allen & Unwin.

Jones, R.W. 1971. Substitution and effective protection. *Journal of International Economics* 1(1), February, 59–81.

Jones, R.W. 1975. Income distribution and effective protection in a multicommodity trade model. *Journal of Economic Theory* 11(1), August, 1–15.

'effectual demand' in Adam Smith. Smith's notion of 'effectual demand' is still the subject of several discussions dealing with the role of demand in classical and neoclassical theories of price and distribution and with the influence of demand on 'division of labour' and economic progress. Smith defined 'effectual demand' as the 'demand of those who are willing to pay the natural price of the commodity, or the whole value of rent, labour and profit, which must be paid in order to bring it thither' (Smith, 1776, vol. 1, p. 58). According to him, when the quantity of any commodity brought to market falls short of the effectual demand, those who demand it

> cannot be supplied with the quantity they want. Rather than want it altogether, some of them will be willing to give more. A competition will immediately begin among them, and the market price will rise more or less above the natural price (ibid.).

On the other hand, 'when the quantity brought to market exceeds the effectual demand,... the market price will sink more or less below the natural price' (p. 59), whereas 'when the quantity brought to market is just sufficient to supply the effectual demand and no more, the market price naturally comes to be ... the same with the natural price' (ibid.)

'Effectual demand' is thus defined as the demand for any *individual* commodity, corresponding to the natural price for it. It was a *long-period* concept, since it was associated with those prices which allow the payment of wages, rents and profits at their natural levels, and which hold when in all industries productive capacity is fully adjusted and a uniform rate of profits is earned (see Smith, 1776, vol. 1, pp. 59–65).

The definition of 'effectual demand' was introduced in dealing with the adjustment process between demand and supply. This process was conceived to occur on a *single* market assuming as known the natural prices of that and all other commodities. The process of adjustment implies, therefore, a *prior* determination of distributive variables and of all natural prices, *associated with given levels of effectual demand in each industry*. Smith's notion of 'effectual demand' thus refers as much to a specific industry as to the whole economy: it can be seen as a 'micro' and a 'macroeconomic' concept.

The study of effectual demand involves a description of how the working of competition enforces natural prices but does not constitute a theory of what determines them. Smith never derived demand-functions for any commodity. 'Effectual demand' represented a point, and no attempt was made to determine the magnitude of the rise (fall) in demand when the price falls below (rises above) its natural level. He thus used a different notion from that implied by demand-curves in neoclassical theory, which

> requires a specific ordering between *each* price–quantity point. ... The theory does not regard these points as results of accidental and temporary deviations of the quantity supplied from the 'normal' level, but rather as determinate

points likely to emerge from a repetition of events (Garegnani, 1983, p. 310).

Smith's notion of effectual demand has been recalled by those who, following Sraffa's rehabilitation of the surplus approach of the classical political economists (Sraffa, 1960), have proposed to separate the analysis of price and distribution from that of the levels of output and demand. Within this approach, *given the level and the composition of output* and one distributive variable, it is possible to determine the 'socially necessary' technique, the other distributive variables and natural prices. The levels of output and demand in each industry, taken as given, represent long-period values, since they are associated with fully adjusted productive capacity and uniform rates of profit in all industries.

The analysis of the classical tradition is characterized by integration between historical, institutional and economic factors. This approach is applied to the analysis of the level and composition of demand. The analysis of the aggregate level is related to Say's law, whose acceptance is an open option in classical political economy. Among the elements affecting the composition of demand, two groups of factors appear to emerge in Smith's writings. First of all, *objective* factors, like the degree of development of the economy and the distribution of income among different classes of society. Secondly, *subjective* factors, which are influenced by customs, social rules and fashion. The limited attention paid to substitution within the bundle of commodities demanded by different income groups suggests a minor role attributed to this factor, without denying the possibility of its further analysis, carried out case by case.

Marx analysed the factors influencing demand in a similar way. His stress was on objective factors, that is on the ratio of total surplus-value to wages and the proportions in which the surplus-value is split up among profits, interests, ground rents, taxes, etc. (see Marx, 1894, pp. 181–2). Given the historically achieved degree of development of the economy (whose analysis is not based on the acceptance of Say's law) and the distribution of income, it is possible to determine the average level of demand for different commodities from each class or social group. The total consumption expenditure of each class is an increasing function of the income earned (Marx, 1894, pp. 188–9), while the composition of its consumption is influenced by habits and rules which, over a certain historical period, are dominant within that class. Limited possibilities of substitution within the bundle of commodities demanded by each class are recognized, and again appear left to be studied case by case.

> The working class must find at least the same quantity of necessities on hand if it is to continue living in its accustomed average way, although they may be more or less differently distributed among the different kinds of commodities ... The same, with more or less modification, applies to other classes (Marx, 1972a, pp. 188).

Besides, Marx pointed out that the analysis of demand has to recognize the distinction between the part coming from consumers and that coming from entrepreneurs requiring means of production in order to meet what he called the need for commodities in the market, depending on the 'actual social needs of the different classes and on the income available to them' (Marx, 1894, pp. 188–9).

Some remarkable similarities can be found between this approach and that followed by Keynes in the *General Theory*. In chapters 8 and 9 of this work, the factors affecting aggregate consumption are examined in an analysis which is separate from that determining prices and distribution, and which pays hardly any attention to substitution within the bundle of commodities demanded for consumption, a factor to which a secondary role appears to be attributed. According to Keynes (see 1936, pp. 90–95), total consumption depends partly on total income, partly on other objective circumstances, like the interest rate, and partly on subjective factors, which 'include those psychological characteristics of human nature and those social practices and institutions' (p. 91), which are unlikely to change over limited periods of time except in abnormal or revolutionary circumstances, and which it is necessary to consider 'in an historical inquiry or in comparing one social system with another of a different type' (ibid.). Talking of the interest rate, Keynes concluded that its influence on consumption

> is open to a great deal of doubt. ...[Its influence] is complex and uncertain, being dependent on conflicting tendencies ... Substantial changes in the rate of interest tend to modify social habits considerably, thus affecting the subjective propensity to spend – though in which direction it would be hard to say, except in the light of actual experience (p. 93).

Thus, as in classical tradition, the actual influence of the factors considered is evaluated by Keynes according to the historical circumstances considered, taking into account that their influence may be uncertain in its intensity and direction. The integration between economic and institutional and social factors also emerges in the analysis of the influence of subjective factors (pp. 107–112), whose relative strength

> will vary enormously according to the institutions and organisation of the economic society which we presume, according to habits formed by race, education, convention, religion and current moral, according to present hopes and past experience, according to the scale and technique of capital equipment, and according to the prevailing distribution of wealth and the established standards of life (p. 109).

However, the principle of substitution and that of diminishing marginal returns play a primary role in Keynes's analysis of investment in the *General Theory*. In this respect, Keynes said, 'I am simply accepting the usual theory of the subject' (Keynes, 1973, p. 615), 'meaning exactly the same as Marshall ... means' (p. 630). Yet, alongside this neoclassical element, Keynes referred to other factors influencing investment, like the present and expected level of effective demand (see Keynes, 1936, p. 147), which may come from the private or the public sector and may affect what he called 'the state of long-term expectation'. The analysis of investment of the *General Theory* may thus suggest some elements to develop a theory of demand within classical tradition.

One element is that 'the state of long-term expectation is often steady' (Keynes, 1936, p. 162), since factors like the institutional environment and government policies do not only influence it, but also 'exert their compensating effects' on its fluctuations, together with factors related to the maintenance of the efficiency of capital goods. Within this line, government policies, and industrial policy in particular, relations between industry and finance, industrial relations and the history of competitiveness and technological changes are to be seen as relevant factors affecting the prevailing state of long-term expectation (see Eatwell, 1983, p. 283).

Another element is that there may be 'short-period changes in the state of long-term expectation' (Keynes, 1936, p. 164) due, among other things, to reactions of investors during the transition process from one state of long-term expectation, 'which has its definite corresponding level of long-period employment' (ibid., p. 48) with fully adjusted capacity, to another to which a new long-period position corresponds. This process was described by Keynes in chapter 5 of the *General Theory* (1936, pp. 46–50), where he concluded that 'a mere change in expectation is capable of producing an oscillation of the same kind of a shape as a cyclical movement, in the course of working itself out' (p. 49). This chapter points out the possibility of presenting a long-period analysis of demand and output, which is integrated with an analysis of the cyclical movements of the economy.

Smith's notion of 'effectual demand' thus appears a fruitful concept linking the classical theory of prices and distribution and that of output and demand. The historical elements present in the latter theory underline an outstanding feature of Smith's and of classical political economists' work, i.e. that the analysis of output and demand is part of the analysis of concrete 'historical processes' of accumulation which, as said above, can show cyclical fluctuations around the main trend.

Carlo Panico

BIBLIOGRAPHY

Eatwell, J.L. 1983. The long-period theory of employment. *Cambridge Journal of Economics* 7, 269–85.

Garegnani, P. 1983. The classical theory of wages and the role of demand schedules in the determination of relative prices. *American Economic Review, Papers and Proceedings* 73, 309–13.

Keynes, J.M. 1936. *The General Theory of Employment, Interest and Money*. London: Macmillan.

Keynes, J.M. 1973. *The General Theory and After*. Part I: *Preparation*. Vol. XIII of *The Collected Writings of J.M. Keynes*, ed. D. Moggridge, London: Macmillan.

Marx, K. 1894. *Capital*, Vol. 3. London: Lawrence & Wishart, 1972.

Marx, K. 1910. *Theories of Surplus Value*, Vol. 3. London: Lawrence & Wishart, 1972.

Smith, A. 1776. *An Inquiry into the Nature and the Causes of the Wealth of Nations*. 2 vols, ed. E. Cannan, London: Methuen, 1930.

Sraffa, P. 1960. *Production of Commodities by Means of Commodities*. Cambridge: Cambridge University Press.

efficient allocation. Analysis of efficiency in the context of resource allocation has been a central concern of economic theory from ancient times, and is an essential element of modern microeconomic theory. The ends of economic action are seen to be the satisfaction of human wants through the provision of goods and services. These are supplied by production and exchange and limited by scarcity of resources and technology. In this context efficiency means going as far as possible in the satisfaction of wants within resource and technological constraints. This is expressed by the concept of Pareto optimality, which can be stated informally as follows: a state of affairs is Pareto optimal if it is within the given constraints and it is not the case that everyone can be made better off in his own view by changing to another state of affairs that satisfies the applicable constraints.

Because knowledge about wants, resources and technology is dispersed, efficient outcomes can be achieved only by coordination of economic activity. Hayek (1945) pointed out the role of knowledge or information, particularly in the context of prices and markets, in coordinating economic activity. Acquiring, processing and transmitting information are costly activities themselves subject to constraints imposed by technological and resource limitations. Hayek pointed out that the institutions of markets and prices function to communicate information dispersed among economic agents so as to bring about coordinated economic action. He also drew attention to motivational properties of those institutions, or incentives. In this context, the concept of efficiency takes account of the organizational constraints on information processing and transmission in addition to those on production of ordinary goods and services. The magnitude of resources devoted to business or governmental bureaucracies, and to some of the functions performed by industrial salesmen, attests to the importance of these constraints. Economic analysis of efficient allocation has formally imposed only the constraints on production and exchange, and until recently recognized organizational constraints only in an informal way. But it is these constraints that motivate the pervasive and enduring interest in decentralized modes of economic organization, particularly the competitive mechanism.

It is necessary to limit the scope of this essay so that it is not coextensive with microeconomic theory. The main limitation imposed here is to confine attention to models in which either the role of information is ignored, or in which agents do not behave strategically on the basis of private information. In so doing, a large and important class of models involving problems of efficient allocation in the presence of incentive constraints is excluded.

The main ideas of efficient resource allocation are present in their simplest form in the linear activity analysis model of production. We begin with that model.

EFFICIENCY OF PRODUCTION: LINEAR ACTIVITY ANALYSIS

The analysis of production can to some extent be separated from that of other economic activity. The concept of efficiency appropriate to this analysis descends from that of Pareto optimality, which refers to both productive and allocative efficiency in the full economy in which production is embedded. It is useful to begin with a model in which technological possibilities afford constant returns to scale, that is, with the (linear) activity analysis model of production pioneered by Koopmans (1951a, 1951b, 1957), and closely related to the development of linear programming associated with Dantzig (1951a, 1951b) and independently with the Russian mathematician Kantorovitch (1939, 1942) and Kantorovitch and Gavurin (1949).

The two primitive concepts of the model are *commodity* and *activity*. A list of n commodities is postulated; a commodity *bundle* is given by specifying a sequence of n numbers $a_1, a_2, \ldots, a_n$. Technological possibilities are thought of as knowledge of how to transform commodities. Such knowledge may be described in terms of collections of activities called *processes*, much as knowledge of how to prepare food is described by recipes. A recipe commonly has two parts, a list of ingredients or inputs and of the output(s) of the recipe, and a description of how the ingredients are to be combined to produce the output(s). In the activity analysis model the description of productive activity is suppressed. Only the specification of inputs and outputs is retained; this defines the production process.

Commodities are classified into 'desired', 'primary' and 'intermediate' commodities. Desired commodities are those whose consumption or availability is the recognized goal of production; they satisfy wants. Primary commodities are those available from nature. (A primary commodity that is also desired is listed separately among the desired commodities and

must be transformed by an act of production into its desired form.) Intermediate commodities are those that merely pass from one stage of production to another. Each commodity can exist in any non-negative amount (*divisibility*). Addition and subtraction of the numbers measuring the amount of a commodity represent joining and separating corresponding amounts of the commodity.

An activity is characterized by a *net output number* for each commodity, which is positive if the commodity is a net output, negative if it is a net input and zero if it is neither. The term *input–output vector* is also used for this ordered array of numbers. Activity analysis postulates a finite number of basic activities from which all technologically possible activities can be generated by suitable combination. Allowable combinations are as follows. If two activities are known to be possible, then the activity given by their algebraic sum is also possible, i.e. if $a = (a_1, a_2, \ldots, a_n)$ and $b = (b_1, b_2, \ldots, b_n)$, then $a + b = (a_1 + b_1, a_2 + b_2, \ldots, a_n + b_n)$ is also possible. Thus, additivity embodies an assumption of non-interaction between productive activities, at least at the level of knowledge. Furthermore, if an activity is possible, then so is every non-negative multiple of it (*proportionality*), i.e. if $a = (a_1, a_2, \ldots, a_n)$ is possible, then so is $\mu a = (\mu a_1, \mu a_2, \ldots, \mu a_n)$ for any non-negative real number μ. This expresses the assumption of constant returns to scale. The family of activities consisting of all non-negative multiples of a given one forms a process. Since there is a finite number of basic activities, there is also a finite number of basic processes, each intended to describe a basic method of production capable of being carried out at different levels, or intensities.

The assumptions of additivity and proportionality determine a linear model of technology that can be given the following form. Let A be an n by k matrix whose jth column is the input–output vector representing the basic activity that defines the jth basic process, and let $x = (x_1, x_2, \ldots, x_n)$ be the vector whose jth component x_j is the scale (level or intensity) of the jth basic process. Let $y = (y_1, y_2, \ldots, y_n)$ be the vector of commodities. Technology is represented by a linear transformation mapping the space of activity levels into the commodity space, i.e.

$$y = Ax \qquad x \geq 0.$$

With the properties assumed, a process can be represented geometrically in the commodity space by a halfline from the origin including all non-negative multiples of some activity in that process. The finite number of halflines representing basic processes generate a convex polyhedral cone consisting of all activities that can be expressed as sums of activities in the basic processes, or equivalently, as non-negative linear combinations of the basic activities, sometimes called a *bundle of basic activities*. This cone is called the *production set*, or set of *possible productions*.

Two other assumptions are made about the production set itself, rather than just the individual activities. First, there is no activity, whether basic or derived, in the production set with a positive net output of some commodity and non-negative net outputs of all commodities. This excludes the possibility of producing something from nothing, whether directly or indirectly. Second, it is assumed that the production set contains at least one activity with a positive net output of some commodity.

If the availability of primary commodities is subject to a bound, the technologically possible productions described by the production set are subject to another restriction; only those possible productions that do not require primary inputs in amounts exceeding the given bounds can be produced. Furthermore, because intermediate commodities are not desired in themselves, their net output is required to be zero. (Strictly speaking, the technological constraint on intermediate commodities is that their net output be non-negative. The requirement that they be zero can be viewed as one of elementary efficiency, excluding accumulation or necessity to dispose of unwanted goods.) With these restrictions the model can be written

$$y = Ax, \ x \geqslant 0, \ y_i = 0$$

if i is an intermediate commodity, and

$$y_i \geqslant r_i \quad \text{if } i \text{ is a primary commodity,}$$

where r_i is the (non-positive) limit on the availability of primary commodity i. This leads to the concept of an *attainable* activity.

A bundle of basic activities is *attainable* if the resulting net outputs are non-negative for all desired commodities, zero for intermediate commodities and non-positive for primary commodities, and if the total inputs of primary commodities do not exceed (in absolute amount) the prescribed bounds of availability of those commodities. The set of activities satisfying these conditions is a truncated convex polyhedral cone in the commodity space called the *set of attainable productions*.

The concept of productive efficiency in this model is as follows. An activity (a bundle of basic activities) is *efficient* if it is attainable and if every activity that provides more of some desired commodity and no less of any other is not attainable.

This concept can be seen to be a specialization of Pareto optimality. If for each desired commodity there is at least one consumer who is not satiated in that commodity, at least in the range of production attainable within the given resource limitations, then increasing the amount of any desired commodity without decreasing any other can improve the state of some non-satiated consumer without worsening that of any other.

CHARACTERIZING EFFICIENT PRODUCTION IN TERMS OF PRICES

Efficient production can be characterized in terms of *implicit prices*, also called *shadow prices*, or in the context of linear programming, *dual variables*. Efficient activities are precisely those that maximize profit for suitably chosen prices. The profit returned by a process carried out at the level x is

$$x \sum_i p_i a_i,$$

where the prices are $p = (p_1, \ldots, p_n)$, and $a = (a_1, \ldots, a_n)$ is the basic activity defining the process; the profit on the bundle of activities Ax at prices p is given by the inner product $py = pAx$.

This characterization is the economic expression of an important mathematical fact about convex sets in n-dimensional Euclidean space, namely that through every point of the space not interior to the convex set in question there passes a hyperplane that contains the set in one of its two halfspaces (Fenchel, 1950; Nikaido, 1969, 1970). (A hyperplane in n dimensional space is a level set of a linear function of n variables, and thus is a translate of an $n - 1$ dimensional linear subspace. A hyperplane is given by an equation of the form $c_1 x_1 + c_2 x_2 + \cdots + c_n x_n = k$, where the x's are variables, the c's are coefficients defining the linear function and k is a constant identifying the level set. A hyperplane divides the space into two halfspaces corresponding to the two inequalities $c_1 x_1 + c_2 x_2 + \cdots + c_n x_n \gtrless k$ respectively.) It can also be seen that a point of a convex set is a boundary point if and only if it maximizes a linear function on the (closure of the) set. These

facts can be used to characterize efficient production because the attainable production set is convex and efficient activities are boundary points of it. Because the efficient points are those, roughly speaking, on the 'north-east' frontier of the set, the linear functions associated with them have non-negative coefficients, interpreted as prices. On the other hand, if a point of the attainable set maximizes a linear function with strictly positive coefficients (prices), then it is on the 'north-east' frontier of the set.

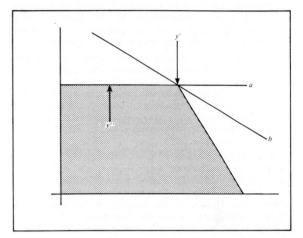

Figure 1

In Figure 1 the set enclosed by the broken line and the axes is the projection of the attainable set on the output coordinates; inputs are not shown. The point y' in the figure is efficient; the point y'' is not; both y' and y'' maximize a linear function with non-negative coefficients (the level set containing y' is labelled a and also contains y''). However, y' maximizes a linear function with positive coefficients (one such, whose level set through y' is labelled b, is shown), while y'' does not.

These implicit, or efficiency prices arise from the logic of efficiency or maximization when the relevant sets are convex, not from any institutions such as markets or exchange. An important reason for interest in them is the possibility of achieving efficient performance by decentralized methods. As described above, under the assumptions of additivity and constant returns to scale the production set can be seen to be generated by a finite number of basic processes, each of which consists of the activities that are non-negative multiples of a basic activity, the multiple being the scale (level, or intensity) at which the process is operated. Following the presentation of Koopmans (1957), each basic process is controlled by a manager, who decides on its level. The manager of a process is assumed to know only the input–output coefficients of his process. Each primary resource is in the charge of a resource holder, who knows the limit of its availability. Efficiency prices are used to guide the choices of managers and resource holders. (Under constant returns to scale, if an activity yields positive profit at a given system of prices, then increasing the scale of the process containing that activity increases the profit. Since the scale can be increased without bound, if the profitability of a process is not zero or negative, then, in the eyes of its manager, who does not know the aggregate resource constraints, it can be made infinite. Therefore, the systems of prices that can be considered for the role of efficiency prices must be restricted to those *compatible with the given technology,* namely prices such that no process is profitable and at least one process breaks even.) Two propositions characterize efficient production by prices and provide the basis for an interpretation in terms of decentralized control of production.

In a given linear activity analysis model, if there is a given system of prices compatible with the technology, in which the prices of all desired commodities are positive, then any attainable bundle of basic activities selected only from processes that break even and which utilizes all positively priced primary commodities to the limit of their availability and does not use negatively priced primary commodities at all, is an efficient bundle of activities.

In a given linear activity analysis model, each efficient bundle of activities has associated with it at least one system of prices compatible with the technology such that every activity in that bundle breaks even and such that prices of desired commodities are positive, and the price of a primary commodity is non-negative, zero or non-positive, according as its available supply is full, partly, or not used at all (Koopmans, 1957).

These propositions are stated in a static form. There is no reference to managers raising or lowering the levels of the processes they control, or to resource holders adjusting prices. A dynamic counterpart of these propositions would be of interest, but because of the linearity of the model such dynamic adjustments are unstable (Samuelson, 1949).

It should also be noted that the concept of decentralization is not explicitly defined in this literature; the interpretation is by analogy with the competitive mechanism. Nevertheless, the interest in characterizing efficiency by prices and their interpretation in terms of decentralization is an important theme in the study of efficient resource allocation.

The linear activity analysis model has been generalized in several directions. These include dropping the assumption of proportionality, dropping the restriction to a finite number of basic activities, dropping the restriction to a finite number of commodities and dropping the restriction to a finite number of agents. Perhaps the most directly related generalization is to the nonlinear activity analysis, or nonlinear programming, model.

EFFICIENCY OF PRODUCTION: NONLINEAR PROGRAMMING

In the nonlinear programming model there is, as in the linear model, a finite number of basic processes. Their levels are represented by a vector $x = (x_1, x_2, \ldots, x_k)$, where k is the number of basic processes. Technology is represented by a nonlinear transformation from the space of process levels to the commodity space (still assumed to be finite dimensional), written

$$y = F(x), x \geq 0.$$

The production set in this model is the image in the commodity space of the non-negative orthant of the space of process levels. Under the assumptions usually made about F, the production set is convex, though, of course, not a polyhedral cone.

In this model as in the linear activity analysis model a central result is the characterization of efficient production in terms of prices. The simplest case to begin with is that of one desired commodity, say, one output, with perhaps several inputs. In this case the (vector-valued) function F can be written

$$F(x) = [f(x), g_1(x), g_2(x), \ldots, g_m(x)],$$

where the value of f is the output, and $g_1, \ldots, g_m$ correspond

to the various inputs. Resource constraints are expressed by the conditions

$$g_j(x) \geq 0, \quad \text{for} \quad j = 1, 2, \ldots, m,$$

and non-negativity of process levels by the condition, $x \geq 0$. (Here the resource constraints $r_j \leq h_j(x) \leq 0$ are written more compactly as $h_j(x) - r_j = g_j(x) \geq 0$.)

In this model the definition of efficient production given in the linear model amounts to maximizing the value of f subject to the resource and non-negativity constraints just mentioned.

Problems of constrained maximization are intimately related to saddle-point problems. Let L be a real valued function defined on the set $X \times Y$ in R^n. A point (x^*, y^*) in $X \times Y$ is a *saddle point* of L if

$$L(x, y^*) \leq L(x^*, y^*) \leq L(x^*, y),$$

for all x in X and all y in Y.

The concept of a concave function is also needed. A real valued function f defined on a convex set X in R^n is a *concave function* if for all x and y in X and all real numbers $0 \leq a \leq 1$

$$f(ax + (1 - a)y) \geq af(x) + (1 - a)f(y).$$

The following mathematical theorem is fundamental.

Theorem (Kuhn and Tucker, 1951; Uzawa, 1958): Let f and g_1, $g_2, \ldots, g_m$ be real valued concave functions defined on a convex set X in R^n. If f achieves a maximum on X subject to $g_j(x) \geq 0$, $j = 1, 2, \ldots, m$ at the point x^* in X, then there exist non-negative numbers $p_0^*, p_0^*, \ldots, p_m^*$, not all zero, such that $p_0^* f(x) + p^* g(x) \leq p_0^* f(x^*)$ for all x in X, and furthermore, $p^* g(x^*) = 0$. (Here the vectors $p^* = (p_1^*, p_2^*, \ldots, p_m^*)$, and $g(x) = [g_1(x), g_2(x), \ldots, g_m(x)]$.) The vector p^* may be chosen so that

$$\sum_0^m p_j^* = 1.$$

An additional condition (Slater, 1950) is important. (It ensures that the coefficient p_0 of f is not zero.)

Slater's Condition: There is a point x' in X at which $g_j(x') > 0$ for all $j = 1, 2, \ldots, m$.

If attention is restricted to concave functions, as in the Kuhn–Tucker–Uzawa Theorem, the relation between constrained maxima and saddle points can be summarized in the following theorem.

Theorem: If f and g_j, $j = 1, 2, \ldots, m$ are concave functions defined on a convex subset X in R^n, and if Slater's Condition is satisfied, then x^* in X maximizes f subject to $g_j(x) \geq 0$, $j = 1, 2, \ldots, m$, if and only if there exists $\lambda^* = (\lambda_1^*, \lambda_2^*, \ldots, \lambda_m^*)$, $\lambda_j^* \geq 0$ for $j = 1, 2, \ldots, m$, such that (x^*, λ^*) is a saddle point of $L(x, \lambda) = f(x) + \lambda g(x)$ on $X \times R_+^n$.

This theorem is easily seen to cover the case where some constraints are equalities, as in the case of intermediate commodities. The sufficiency half of this theorem holds for functions that are not concave.

The auxiliary variables $\lambda_1, \lambda_2, \ldots, \lambda_m$, called *Lagrange multipliers*, play the role of efficiency prices, or shadow prices; they evaluate the resources constrained by the condition $g(x) \geq 0$. The maximum characterized by the theorem is a global one, as in the case of linear activity analysis.

If the functions involved are differentiable, a saddle point of the Lagrangean can be studied in terms of first-order conditions. The first-order conditions are necessary conditions for a saddle point of L. If the functions f and the g's are concave on a convex set X, then the first-order conditions at a point

(x^*, λ^*) are also sufficient; that is, they imply that (x^*, λ^*) is a saddle point of L. Thus,

Theorem: If $f, g_1, g_2, \ldots, g_m$ are concave and differentiable on an open convex set X in R^n, and if Slater's Condition is satisfied, then x^* maximizes f subject to $g_j(x) \geqslant 0$ for $j = 1, 2, \ldots, m$ if and only if there exists numbers $\lambda_1^*, \lambda_2^*, \ldots, \lambda_m^*$ such that the first-order conditions for a saddle point of $L(x, \lambda) = f(x) + \lambda g(x)$ are satisfied at (x^*, λ^*).

If there are non-negativity conditions on the x's,

$$g_j(x) \geqslant 0, \quad x \geqslant 0, \quad x \text{ in } R^n$$

and the first-order conditions can be written

$$f_x^* + \lambda^* g_x^* \leqslant 0, (f_x^* + \lambda^* g_x^*)x^* = 0,$$
$$\lambda^* g(x^*) = 0, g(x^*) \geqslant 0, g(x^*) \geqslant 0,$$
$$\lambda^* \geqslant 0 \quad \text{and} \quad \lambda^* g(x^*) = 0,$$

where f_x^* denotes the derivative of f evaluated at x^*. In more explicit notation, the conditions $f_x^* + \lambda^* g_x^* = 0$ can be written as

$$\partial f / \partial x_i + \sum_{j=1}^m \lambda_j^* \partial g_j / \partial x_i = 0, \qquad i = 1, 2, \ldots, n$$

When the assumption of concavity is dropped, it is no longer possible to ensure that the local maximum is also a global one. However, it is still possible to analyse local constrained maxima in terms of local saddle-point conditions. In this case a condition is needed to ensure that the first-order conditions for a saddle point are indeed necessary conditions. The Kuhn–Tucker Constraint Qualification is such a condition. Arrow, Hurwicz and Uzawa (1961) have found a number of conditions, more useful in application to economic models, that imply the Constraint Qualification.

The case of more than one desired commodity leads to what is called the *vector maximum problem*, Kuhn and Tucker (1951). This may be defined as follows. Let $f_1, f_2, \ldots, f_k$ and $g_1, g_2, \ldots, g_m$ be real valued functions defined on a set X in R^n. We say x^* in X achieves a (global) *vector maximum* of $f = (f_1, f_2, \ldots, f_k)$ subject to $g_j(x) \geqslant 0, j = 1, 2, \ldots, m$ if,

(I) $g_j(x^*) \geqslant 0, j = 1, 2, \ldots, m$,

(II) there does not exist x' in X satisfying $f_i(x') \geqslant f_i(x^*)$ for $i = 1, 2, \ldots, k$ with $f_i(x') > f_i(x^*)$ for some value of i, and $g_j(x') \geqslant 0$ for $j = 1, 2, \ldots, m$.

This is just the concept of an efficient point expressed in the present notation.

A vector maximum has a saddle-point characterization similar to that for a scalar valued function.

Theorem: Let $f_1, f_2, \ldots, f_k$ and $g_1, g_2, \ldots, g_m$ be real valued concave functions defined on a convex X set in R^n. Suppose there is x^0 in X such that $g_j(x^0) > 0, j = 1, 2, \ldots, m$ (Slater's Condition). If x^* achieves a vector maximum of f subject to $g(x) \geqslant 0$ then there exist $a = (a_1, a_2, \ldots, a_k)$ and $\lambda^* = (\lambda_1^*, \lambda_2^*, \ldots, \lambda_m^*)$ with $a_j \geqslant 0$ for all j, $a \neq 0$ and $\lambda \geqslant 0$ such that (x^*, λ^*) is a saddle point of the Lagrangean $L(x, \lambda) = af(x) + \lambda g(x)$.

Several different 'converses', to this theorem are known. One states that if x^* maximizes $L(x, \lambda^*)$ for some strictly positive vector a and non-negative λ^*, and if $\lambda^* g(x^*) = 0$ and $g(x^*) \geqslant 0$, then x^* gives a vector maximum of f subject to $g(x) \geqslant 0$, and x in X. Another, parallel to the result for the case of one desired commodity, is the following.

Theorem: Let f and g be functions as in the theorem above. If there are positive real numbers $a_1, a_2, \ldots, a_k$ and if (z^*, λ^*) is a saddle point of the Lagrangean L (defined as above) then (I) x^* achieves a maximum of f subject to $g(x) \geqslant 0$ on X, and (II) $\lambda^* g(x^*) = 0$.

The positive numbers $a_1, \ldots, a_k$ are interpreted as prices of desired commodities, and the non-negative numbers λ_j^* are prices of the remaining commodities. The condition $\lambda^* g(x^*) = 0$ which arises in these theorems states that the value of unused resources at the efficiency prices λ^* is zero; that is, resources not fully utilized at a vector maximum have a zero price.

The connection between vector maxima and Pareto optima is as follows. Because a vector maximum is an efficient point (for the vectorial ordering of the commodity space), it is a Pareto optimum for appropriately specified (non-satiated) utility functions, as was already pointed out in the case of the linear activity analysis model. Furthermore, if the functions $f_1, \ldots, f_k$ are themselves utility functions, and the variable x denotes allocations, with the constraints g defining feasibility, then a vector maximum of f subject to the constraints $g(x) \geqslant 0$ and x in X is a Pareto optimum, and vice versa. Hence the saddle-point theorems give a characterization of Pareto optima by prices. The interpretation of prices in terms of decentralized resource allocation described in the linear activity analysis model also applies in this nonlinear model. The proofs of these theorems reveal an important logical role played by the principle of marginal cost pricing.

The basic theorems of nonlinear programming, especially the Kuhn–Tucker–Uzawa Theorem in the setting of the vector maximum problem, have been extended to the case of infinitely many commodities. (Hurwicz, 1958, first obtained the basic results in this field.) Technicalities aside, the theorems carry over to certain infinite dimensional spaces, namely linear topological spaces, or in the case of first-order conditions, Banach spaces.

Dropping the restriction to a finite number of basic processes leads to classical production or transformation function models of production, whose properties depend on the detailed specifications made.

Samuelson (1947) used Lagrangean methods to analyse interior maxima subject to equality constraints in the context of production function models, as well as that of optimization by consumers. He also gave the interpretation of Lagrange multipliers as shadow prices.

EFFICIENT ALLOCATION IN AN ECONOMY WITH CONSUMERS AND PRODUCERS

In an economy with both consumption and production decisions, efficiency is concerned with distribution as well as production. Data about restrictions on consumption and the wants of consumers must be specified in addition to the data about production. The elements of the models are as follows.

The commodity space is denoted X; it might be l-dimensional Euclidean space, or a more abstract space such as an additive group in which, for example, some coordinates are restricted to have integer values. There is a (finite) list of consumers, $1, 2, \ldots, n$, and a similar list of producers, $1, 2, \ldots, m$. A *state* of the economy is an array consisting of a commodity bundle for each agent in the economy, consumer or producer. This may be written $(\langle x^i \rangle, \langle y^j \rangle)$, where $\langle x^i \rangle = (x^1, x^2, \ldots, x^n)$ and $\langle y^j \rangle = (y^1, y^2, \ldots, y^m)$ and x^i and y^j are commodity bundles. Absolute constraints on consumption are expressed by requiring that the allocation $\langle x^i \rangle$ belong to a specified subset X of the space X^n of allocations.

Examples of such constraints are:

1. The requirement that the quantity of a certain commodity be non-negative.

2. The requirement that a consumer requires certain minimum quantities of commodities in order to survive.

Each consumer i has a preference relation, denoted $\succsim_i$, defined on X. This formulation admits externalities in consumption, including physical externalities and externalities in preferences; for example, preferences that depend on the consumption of other agents, termed non-selfish preferences. The consumption set of the ith consumer is the projection X^i of X onto the space of commodity bundles whose coordinates refer to the holdings of the ith consumer.

Technology is specified by a production set Y, a subset of X^m, consisting of those arrays $\langle y^j \rangle$ of input–output vectors that are jointly feasible for all producers. The production set of the jth producer, denoted Y^j, is the projection of Y onto the subspace of X^m whose coordinates refer to the jth producer.

The (aggregate) initial endowment of the economy is denoted by w, a commodity bundle in X.

These specifications define an *environment*, a term introduced by Hurwicz (1960) in this usage and according to him suggested by Jacob Marschak. This term refers to the primitive or given data from which analysis begins. Each environment determines a set of *feasible* states. These are the states $(\langle x^i \rangle, \langle y^j \rangle)$ such that $\langle x^i \rangle$ is in X, $\langle y^j \rangle$ is in Y and $\Sigma x^i - \Sigma y^j \leqslant w$.

An environment determines the set of states that are Pareto optimal for that environment. Explicitly, they are the states $(\langle x^{*i} \rangle, \langle y^{*j} \rangle)$ that are feasible in the given environment, and such that if any other state $(\langle x^i \rangle, \langle y^j \rangle)$ has the property that $\langle x^i \rangle \succsim_i \langle x^{*i} \rangle$ for all i with $\langle x^{i'} \rangle \succ_{i'} \langle x^{*i'} \rangle$ for some i', then $(\langle x^i \rangle, \langle y^j \rangle)$ is not feasible in the given environment.

It is important to note that the set of feasible states and the set of Pareto optimal states are completely determined by the environment; specification of economic organization is not involved.

At this level of generality, where externalities in consumption and production are admitted as possibilities, and where commodities may be indivisible, no general characterization of Pareto optima in terms of prices is possible. (Indeed, Pareto optima may not exist. Conditions that make the set of feasible allocations non-empty and compact and preferences continuous suffice to ensure the existence of Pareto optima.) In environments with externalities, or other non-neoclassical features, Pareto optima are generally not attainable by decentralized processes (Hurwicz, 1966).

If the class of environments under consideration is restricted to the neoclassical environments, the fundamental theorems of welfare economics provide a characterization of Pareto optimal states via efficiency prices. That characterization has a natural interpretation in terms of a decentralized mechanism for allocation of resources.

The framework for these results is obtained by restricting the class of environments specified above as follows. The commodity space is to be Euclidean space of l dimensions, i.e. $X = R^l$. The consumption set for the economy is to be the product of its projections, i.e. $X = X^1 \times X^2 \times \cdots \times X^n$. This expresses the fact that if each agent's consumption is feasible for him, the total array is jointly feasible. Furthermore, each agent is restricted to have selfish preferences; that is, agent i's preference relation depends only on the coordinates of the allocation that refer to his holdings. In that case the preference relation $\succsim_i$ may be defined only on X^i, for each i. Similarly, externalities are ruled out in production, i.e. $Y = Y^1 \times Y^2 \times \cdots \times Y^m$.

The concept of an *equilibrium relative to a price system* (Debreu, 1959) serves to characterize Pareto optima by prices. A price system, denoted p, is an element of R^l; the environment $e = [(X^i), (\succsim_i), (Y^j), w]$ is of the restricted type specified above (free of externalities and indivisibilities).

A state $[(x^{*i}), (y^{*j})]$ of e is an *equilibrium relative to price system p* if:

1. For every consumer i, x^{*i} maximizes preference $\succsim_i$ on the set of consumption bundles whose value at the prices p does not exceed the value of x^{*i} at those prices, i.e., if x^i is in $\{x^i$ in $X^i : px^i \leqslant px^{*i}\}$ then $x^i \succsim_i x^{*i}$.

2. For every producer j, y^{*j} maximizes profit py^j on Y^j.

3. Aggregate supply and demand balance, i.e.

$$\sum_i x^{*i} - \sum_j y^{*j} = w.$$

An equilibrium relative to a price system differs from a competitive equilibrium (see below) in that the former does not involve the budget constraints applying to consumers in the latter concept. In an equilibrium relative to a price system the distribution of initial endowment and of the profits of firms among consumers need not be specified.

The first theorem of neoclassical welfare economics states, subject only to the exclusion of externalities and a mild condition that excludes preferences with thick indifference sets, that a state of an environment e that is an equilibrium relative to a price system p is a Pareto optimum of e (Koopmans, 1957).

The second welfare theorem is deeper and holds only on a smaller class of environments, sometimes referred to in the literature as the *classical environments* (called neoclassical above). One version of this theorem is as follows. Let $e = [(X^i), (\succsim_i), (Y^j), w]$ be an environment such that for each i

1. X^i is convex.
2. The preference relation $\succsim_i$ is continuous.
3. The preference relation $\succsim_i$ is convex.
4. The set $\Sigma_j Y^j$ is convex.

Let $[(x^{*i}), (y^{*j})]$ be a Pareto optimum of e such that there is at least one consumer who is not satiated at x^{*i}. Then there is a price system p, with not all components equal to 0, such that – except for Arrow's (1951) 'exceptional case', where p is such that for some i the expenditure px^{*i} is a minimum on the consumption set X^i – the state $[(x^{*i}), (y^{*j})]$ is an equilibrium relative to p.

(The condition that preferences are convex and not satiated is sufficient to exclude 'thick' indifference sets. A preference relation on X^i is convex if whenever x' and x'' are points of X^i with x' strictly preferred to x'' then the line segment connecting them (not including the point x'') is strictly preferred to x'. The consumption set X^i must be convex for this property to make sense. A preference relation is not satiated if there is no consumption preferred to all others.)

Hurwicz (1960) has given an alternative formalization of the competitive mechanism in which Arrow's exceptional case presents no difficulties.

If the exceptional case is not excluded, then it can still be said that:

1. x^{*i} minimizes expenditure at prices p on the upper contour set of x^{*i}, for every i, and

2. y^{*j} maximizes 'profit' py^j on the production set Y^j, for every j.

The state (x^*, y^*) together with the prices p, constitute a *valuation equilibrium* (Debreu, 1954).

As in the case of efficiency prices in pure production models, these prices have in themselves no institutional significance. They are, however, in the same way as other efficiency prices, suggestive of an interpretation in terms of decentralization.

If, in addition to the restriction to classical environments, the economic organization is specified to be that of a system of markets in a private ownership economy, and if agents are assumed to take prices as given, then the welfare theorems can translate into the assertion that the set of Pareto optima of an environment e and the set of competitive equilibria for e (subject to the possible redistribution of initial endowment and ownership shares) are identical. More precisely, the specification of the environment given above is augmented by giving each consumer a bundle of commodities, his initial endowment, denoted w^i. The total endowment is $w = \Sigma_i w^i$. Furthermore, each consumer has a claim to a share of the profits of each firm; the claims for the profit of each firm are assumed to add up to the entire profit. When prices and the production decisions of the firms are given, the profits of the firms are determined and so is the value of each consumer's initial endowment. Therefore, the income of each consumer is determined. Hence, the set of commodity bundles a consumer can afford to buy at the given prices, called his *budget set*, is determined; this consists of all bundles in his consumption set whose value at the given prices does not exceed his income at the given prices. Competitive behaviour of consumers means that each consumer treats the prices as given constants and chooses a bundle in his budget set that maximizes his preference: that is, a bundle x^i that is in X^i and such that if any other bundle x'^i is preferred to it, then x'^i is not in his budget set.

Competitive behaviour of firms is to maximize profits computed at the given prices p, regarded by the firms as constants; that is, a firm chooses a production vector y^j in its production set with the property that any other vector affording higher profits than py^j is not in the production set of firm j.

A *competitive equilibrium* is a specification of a commodity bundle for each consumer, a production vector for each firm, and a price system, together denoted $[(x^{*i}), (y^{*j}), p^*]$, where p^* has no negative components, satisfying the following conditions:

1. For each consumer i the bundle x^{*i} maximizes preference on the budget set of i.
2. For each firm j the production vector y^{*j} maximizes profit $p^* y^j$ on the production set Y^j.
3. For each commodity, the total consumption does not exceed the net total output of all firms plus the total initial endowment, i.e. $\Sigma_i x^{*i} - \Sigma_j y^{*j} \leqslant w = \Sigma_i w^i$;
4. For those commodities k for which the inequality in 3 is strict; that is, the total consumption is less than initial endowment plus net output, the price p_k^* is zero.

The welfare theorems stated in terms of equilibrium relative to a price system translate directly into theorems stated in terms of competitive equilibrium. Briefly, every competitive equilibrium allocation in a given classical environment is Pareto optimal in that environment, and every Pareto optimal allocation in a given classical environment can be made a competitive equilibrium allocation of an environment that differs from the given one only in the distribution of the initial endowment. (Arrow (1951), Koopmans (1957), Debreu (1959) and Arrow and Hahn (1971) give modern and definitive treatment of the classical welfare theorems.)

It should be noted that the equilibria involved must exist for these theorems to have content. Sufficient conditions for existence of competitive equilibrium, which, since a competitive equilibrium is automatically an equilibrium relative to a price system, are also sufficient for existence of an equilibrium relative to a price system, include convexity and continuity of consumption sets and preferences and of production sets, as well as some assumptions which apply to the environment as a whole, restricting the ways in which individual agents may fit together to form an environment (Arrow and Debreu, 1954; Debreu, 1959; McKenzie, 1959).

The second welfare theorem involves redistribution of initial endowment. This is essential because the set of competitive equilibria from a given initial endowment is small (essentially finite) (Debreu, 1970), while the set of Pareto optima is generally a continuum. The set of Pareto optima cannot in general be generated as competitive allocations without varying the initial point. If redistribution is done by an economic mechanism, then it should be a decentralized one to support the interpretation given of the second welfare theorem. No such mechanism has been put forward as yet. Redistribution of initial endowment by lump-sum taxes and transfers has been discussed. A customary interpretation views these as brought about by a process outside economics, perhaps by a political process; no claim is made that such processes are decentralized. Some economists consider dependence on redistribution unsatisfactory because information about initial endowment is private; only the individual agent knows his own endowment. Consequently the expression of that information through political or other action can be expected to be strategic. The theory of second-best allocations has been proposed in this context. Redistribution of endowment is excluded, and the mechanism is restricted to be a price mechanism, but the price system faced by consumers is allowed to be different from that faced by producers; all agents behave according to the rules of the (static) competitive mechanism. The allocations that satisfy these conditions, when the price systems are variable, are maximal allocations in the sense that they are Pareto optimal within the restricted class just defined. These are so-called *second-best* allocations. This analysis was pioneered by Lipsey and Lancaster (1956) and Diamond and Mirrlees (1971).

EFFICIENT ALLOCATION IN NON(NEO)CLASSICAL ENVIRONMENTS

The term *nonclassical* refers to those environments that fail to have the properties of classical ones; there may be indivisible commodities, nonconvexities in consumption sets, preferences or production sets, or externalities in production or consumption. An example of nonconvex preference would arise if a consumer preferred living in either Los Angeles or New York to living half the time in each city, or living half-way between them, depending on the way the commodity involved is specified. A production set representing a process that affords increasing returns to scale is an example of nonconvexity in production. A large investment project such as a road system is an example of a significant indivisibility. Phenomena of air or water pollution provide many examples of externalities in consumption and production.

The characterization of optimal allocation in terms of prices provided by the classical welfare theorems does not extend to nonclassical environments. If there are indivisibilities, equilibrium prices may fail to exist. Lerner (1934, 1947) has proposed a way of optimally allocating resources in the presence of indivisibilities. It would typically require adding up consumers' and producers' surplus.

Increasing returns to scale in production generally results in non-existence of competitive equilibrium, because of unbounded profit when prices are treated as given. Nash equilibrium, a concept from the theory of games, can exist even in cases of increasing returns. The difficulty is that such equilibria need not be optimal. Similar difficulties occur in cases of externalities.

Failure of the competitive price mechanism to extend the properties summarized in the classical welfare theorems to nonclassical environments has led economists to look for alternative ways of achieving optimal allocation in such cases. Such attempts have for the most part sought institutional arrangements that can be shown to result in optimal allocation. Ledyard (1968, 1971) analysed a mechanism for achieving Pareto optimal performance in environments with externalities. The use of taxes and subsidies advocated by Pigou (1932) to achieve Pareto optimal outcomes in cases of externalities is such an example. In a similar spirit Davis and Whinston (1962) distinguish externalities in production that leave marginal costs unaffected from those that do change marginal costs. In the former case they propose a pricing scheme, but one that involves lump-sum transfers. Marginal cost pricing, including lump-sum transfers to compensate for losses, which was extensively discussed as a device to achieve optimal allocation in the presence of increasing returns (Lerner, 1947; Hotelling, 1938; and many others) is another example of a scheme to realize optimal outcomes in nonclassical environments in a way that seeks to capture the benefits associated with decentralized resource allocation. In the case of production under conditions of increasing returns, the use of nonlinear prices has been suggested in an effort to achieve optimality with at least some of the benefits of decentralization. (See Arrow and Hurwicz, 1960; Heal, 1971; Brown and Heal, 1982; Brown, Heal, Khan and Vohra, 1985; Jennergren, 1971; Guesnerie, 1975.)

In the case of indivisibilities, and in the context of productive efficiency, integer programming algorithms exist for finding optima in specific problems, but a general characterization in terms of prices such as exists for the classical environments is not available. A decentralized process, involving the use of randomization, whose equilibria coincide with the set of Pareto optima has been put forward by Hurwicz, Radner and Reiter (1975). This process has the property that the counterparts of the classical welfare theorems hold for environments in which all commodities are indivisible, and the set of feasible allocations is finite, or in which there are no indivisible commodities, or externalities, but there may be nonconvexities in production or consumption sets, or in preferences. This, of course, includes the possibility of increasing returns to scale in production.

The schemes and processes that have been proposed, including many not described here, are quite different from one another. If attention is confined to pricing schemes without additional elements, such as lump-sum transfers, it may be satisfactory to proceed on the basis of an informal intuitive notion of decentralization. This amounts in effect to identifying decentralization with the competitive mechanism, or more generally with price or market mechanisms. If a broader class of processes is to be considered, including some already mentioned in this discussion, then a formal concept of decentralized resource allocation process is needed.

EFFICIENT ALLOCATION THROUGH INFORMATIONALLY DECENTRALIZED PROCESSES

A formal definition of a concept of *allocation process* was first given by Hurwicz (1960). He also gave a definition of *informational decentralization* applying to a broad class of allocation mechanisms, based in part on a discussion by Hayek (1945) of the advantages of the competitive market

mechanism for communicating knowledge initially dispersed among economic agents so that it can be brought to bear on the decisions that determine the allocation of resources. Hurwicz's formulation is as follows.

There is an initial dispersion of information about the environment; each agent is assumed to observe directly his own characteristic, e^i, but to know nothing directly about the characteristics of any other agent. In the absence of externalities, specifying the array of individual characteristics specifies the environment, i.e. $e = (e^1, \ldots, e^n)$. When there are externalities, an array of individual characteristics, each component of which corresponds to a possible environment, may not together constitute a possible environment. In more technical language, when there are externalities the set of environments is not the Cartesian product of its projections onto the sets of individual characteristics.

The goal of economic activity, whether efficiency, Pareto optimality or some other desideratum such as fairness, can be represented by a relation between the set of environments and the set of allocations, or outcomes. This relation assigns to each environment the set of allocations that meet the criterion of desirability. In the case of the Pareto criterion, the set of allocations that are Pareto optimal in a given environment is assigned to that environment. Formally, this relation is a correspondence (a set-valued function) from the set of environments to the set of allocations.

An allocation process, or mechanism, is modelled as an explicitly dynamic process of communication, leading to the determination of an outcome. In formal organizations standardized forms are frequently used for communication; in organized markets like the Stock Exchange, these include such things as order forms; in a business, forms on which weekly sales are reported; in the case of the Internal Revenue Service, income tax forms. A form consists of entries or blanks to be filled in a specified way. Thus, a form can be regarded as an ordered array of variables whose values come from specified sets. In the Hurwicz model, each agent is assumed to have a *language*, denoted M^i for the ith agent, from which his (possibly multi-dimensional) *message*, m^i, is chosen. The *joint message* of all the agents, $m = (m^1, \ldots, m^n)$ is in the *message space* $M = M^1 \times \cdots \times M^n$. Communication takes place in time, which is discrete; the message $m_t = (m_t^1, \ldots, m_t^n)$ denotes the message at time t. The message an agent emits at time t can depend on anything he knows at that time. This consists of what the agent knows about the environment by direct observation, by assumption, (*privacy*) his own characteristics, e^i for agent i, and what he has learned from others via the messages received from them. The agents' behaviour is represented by *response functions*, which show how the current message depends on the information at hand. Agent i's message at time t is

$$m_t^i = f^i(m_{t-1}, m_{t-2}, \ldots; e^i), \quad i = 1, \ldots, n, \quad t = 0, 1, 2, \ldots$$

If it is assumed that memory is finite, and bounded, it is possible without loss of generality to take the number of past periods remembered to be one. (If memory is unbounded, taking the number of periods remembered to be one excludes the possibility of a finite dimensional message space.) In that case the response equations become a system of first order temporally homogeneous difference equations in the messages. Thus:

$$m_t^i = f_i(m_{t-1}; e^i) \quad i = 1, \ldots, n, \quad t = 0, \ldots,$$

which can be written more compactly as

$$(*) \quad m_t = f(m_{t-1}; e).$$

(This formulation can accommodate the case of directed communication, in which some agents do not receive some messages; if agent i is not to receive the message of j, then f^i is independent of m^j, although m^j appears formally as an argument.) Analysis of informational properties of mechanisms is to begin with separated from that of incentives. When the focus is on communication and complexity questions, the response functions are not regarded as chosen by the agent, but rather by the designer of the mechanism.

The iterative interchange of messages modelled by the difference equation system (*) eventually comes to an end, by converging to a stationary message. (It is also possible to have some stopping rule, such as to stop after a specified number of iterations.) The stationary message, which will be referred to as an *equilibrium message*, is then translated into an outcome, by means of the *outcome function*:

$$h: M \to Z,$$

where Z is the space of outcomes, usually allocations or trades. An allocation mechanism so modelled is called an *adjustment process*; it consists of the triple (M, f, h). Since no production or consumption takes place until all communication is completed, these processes are *tâtonnement* processes.

A more compact and general formulation was given by Mount and Reiter (1974) by looking only at message equilibria when attention is restricted to static properties. A correspondence is defined, called the *equilibrium message correspondence*. It associates to each environment the set of equilibrium messages for that environment. In order to satisfy the requirement of privacy, namely that each agent's message depend on the environment only through the agent's characteristic, the equilibrium message correspondence must be the intersection of individual message correspondences, each associating a set of messages acceptable to the individual agent as equilibria in the light of his own characteristic. Thus the equilibrium message correspondence

$$\mu: E \to M,$$

is given by

$$\mu(e) = \bigcap_i \mu^i(e^i),$$

where $\mu^i: E^i \to M$ is the individual message correspondence of agent i. Note that here the message space M need not be the Cartesian product of individual languages. In the case of an adjustment process, the equilibrium message correspondence is defined by the conditions

$$\mu^i(e^i) = \{m \text{ in } M \,|\, f^i(m; e^i) = m^i\}, \qquad i = 1, \ldots, n$$

together with the condition that μ is the intersection of the μ^i. Specification of the outcome function $h: M \to Z$ completes the model, (M, μ, h).

The performance of a mechanism of this kind can be characterized by the mapping defined by the composition of the equilibrium message correspondence μ and the outcome function h. The mapping $h\mu$; $E \to Z$, possibly a correspondence, specifies the outcomes that the mechanism (M, μ, h) generates in each environment in E. A mechanism, whether in the form of an adjustment process, or in the equilibrium form, is called *Pareto-satisfactory* (Hurwicz, 1960) if for each environment in the class under consideration, the set of outcomes generated by the mechanism coincides with the set of Pareto optimal outcomes for that environment. Allowance must be made for redistribution of initial endowment, as in the case of the second welfare theorem. (A formulation in the framework of mechanisms is given in Mount and Reiter, 1977).

The competitive mechanism formalized as a static mechanism is as follows. (Hurwicz, 1960, has given a different formulation, and Sonnenschein, 1974, has given an axiomatic

characterization of the competitive mechanism from a somewhat different point of view.) The message space M is the space of prices and quantities of commodities going to each agent (it has dimension $n(l-1)$ when there are n agents and l commodities, taking account of budget constraints and Walras' Law), the individual message correspondence μ^i maps agent i's characteristic e^i to the graph of his excess demand function. The equilibrium message is the intersection of the individual ones, and is therefore the price–quantity combinations that solve the system of excess demand equations. The outcome function h is the projection of the equilibrium message onto the quantity components of M. Thus $h\mu(e)$ is a competitive equilibrium allocation (or trade) when the environment is e. The classical welfare theorems state that for each e in E_c, $h[\mu(e)] = P(e)$, where E_c denotes the set of classical environments and P is the Pareto correspondence. (Allowance must be made for redistribution of initial endowment in connection with the second welfare theorem. Explicit treatment of this is omitted to avoid notational complexity. The decentralized redistribution of initial endowment is, as in the case of the second welfare theorem, not addressed.) The welfare theorems can be summarized in the Mount–Reiter diagram (Figure 2) (Reiter, 1977).

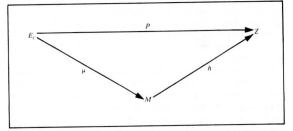

Figure 2

The welfare theorems state that this diagram *commutes* in the sense that starting from any environment e in E_c one reaches the same allocations via the mechanism, that is, via $h\mu$, as via the Pareto correspondence P.

With welfare theorems as a guide, the class of environments E_c can be replaced by some other class E, and the Pareto correspondence can be replaced by a correspondence, P, embodying another criterion of optimality, and one can ask whether there is a mechanism, (M, μ, h) that makes the diagram commute, or, in other words, *realizes P?* Without further restrictions on the mechanism, this is a triviality, because one agent can act as a central agent to whom all others communicate their environmental characteristics; the central agent then has the information required to evaluate P.

The concept of an *informationally decentralized mechanism* defined by Hurwicz (1960) makes explicit intuitive notions underlying the view that the price mechanism is decentralized.

Informationally decentralized processes are a subclass of so-called *concrete processes*, introduced by Hurwicz (1960). These are processes that use a language and response rules that allow production and distribution plans to be specified explicitly. The informationally decentralized processes are those whose response rules permit agents to transmit information only about their own actions, and which in effect require each agent to treat the rest of the economy either as one aggregate, or in a symmetrical way that, like the aggregate, gives anonymity to the other agents.

In the case of static mechanisms, the requirements for informational decentralization boil down to the condition that the message space have no more than a certain finite dimension, and in some cases only that it be of finite dimension. In the case of classical environments this can be seen to include the competitive mechanism, and to exclude the obviously centralized one mentioned above.

Without going deeply into the matter, an objective of this line of research is to analyse explicitly the consequences of constraints on economic organization that come from limitations on the capacity of economic agents to observe, communicate and process information. One important result in this field is that there is no mechanism (M, μ, h) where μ preserves privacy, that uses messages smaller (in dimension) than those of the competitive mechanism (Hurwicz, 1972b; Mount and Reiter, 1974; Walker, 1977; Osana, 1978). Similar results have been obtained for environments with public goods, showing that the Lindahl mechanism uses the minimal message space (Sato, 1981). Another objective is to analyse effects on incentives arising from private motivations in the presence of private information; that is, information held by one agent that is not observable by others, except perhaps at a cost. (There is a large literature on this subject under the rubric 'incentive compatibility', or 'strategic implementation' (Dasgupta, Hammond and Maskin, 1979; Hurwicz, 1971, 1972a). The informational requirements of achieving a specified performance taking some aspects of incentive compatibility into account have been studied by Hurwicz (1976), Reichelstein (1984a, 1984b) and by Reichelstein and Reiter (1985).

Some important results for non-neoclassical environments can be mentioned. Hurwicz (1960, 1966, 1972a) has shown that there can be no informationally decentralized mechanism that realizes Pareto optimal performance on a class of environments that includes those with externalities. Calsamiglia (1977, 1982) has shown in a model of production that if the set of environments includes a sufficiently rich class of those with increasing returns to scale in production, then the dimension of the message space of any mechanism that realizes efficient production cannot be bounded.

EFFICIENT ALLOCATION WITH INFINITELY MANY COMMODITIES

An infinite dimensional commodity space is needed when it is necessary to make infinitely many distinctions among goods and services. This is the case when commodities are distinguished according to time of availability and the time horizon in the model is not bounded or when time is continuous, or according to location when there is more than a finite number of possible locations; differentiated commodities provide other examples, and so does the case of uncertainty with infinitely many states. The bulk of the literature deals with the infinite horizon model of allocation over time, though recently more attention is given to models of product differentiation. Ramsey (1928) studied the problem of saving in a continuous time infinite horizon model with one consumption good and an infinitely lived consumer. He used as the criterion of optimality the infinite sum (integral) of undiscounted utility. Ramsey's contribution was largely ignored, and rediscovered when attention returned to problems of economic growth. A model of maximal sustainable growth based on a linear technology with no unproduced inputs was formulated by von Neumann (1937 in German; English translation, 1945–6). This contribution was unknown among English-speaking economists until after World War II. Study of intertemporal allocation by Anglo-American economists effectively began with the contributions of Harrod (1939) and Domar (1946). These models were concerned with stationary growth at a constant sustainable rate (stationary growth paths) rather than full intertemporal efficiency. Malinvaud (1953) first addressed this problem in a pioneering model of intertemporal allocation with an infinite horizon.

Efficient allocation over (discrete) time would be covered by the finite dimensional models described above if the time horizon were finite. It might be thought that a model with a sufficiently large but still finite horizon would for all practical purposes be equivalent to one with an infinite horizon, while avoiding the difficulties of infinity, but this is not the case, because of the dependence of efficient or optimal allocations on the value given to final stocks, a value that must depend on their uses beyond the horizon.

Malinvaud (1953) formulated an important infinite horizon model, which is the infinite dimensional counterpart of the linear activity analysis model of Koopmans. In Malinvaud's model time is discrete. The time horizon consists of an infinite sequence of time periods. At each date there are finitely many commodities. All commodities are desired in each time period, and no distinction is made between desired, intermediate and primary commodities. As in the activity analysis model, there is no explicit reference to preferences of consumers. Productive efficiency over time is analysed in terms of the output available for consumption, rather than the resulting utility levels.

Technology is represented by a production set X^t for each time period $t = 1, 2, \ldots$, an element of X^t being an ordered pair (a^t, b^{t+1}) of commodity bundles where a^t represents inputs to a production process in period t, and b^{t+1} represents the outputs of that process available at the beginning of period $t + 1$. Here both a^t and b^{t+1} are non-negative. The set X^t is the aggregate production set for the economy during period t. The net outputs available for consumption are given by

$$y^t = b^t - a^t, \qquad \text{for } t \geqslant 1,$$

where b^1 is the initial endowment of resources available at the beginning of period 1. A *programme* is an infinite sequence $\langle (a^t, b^{t+1}) \rangle$; it is a *feasible programme* if (a^t, b^{t+1}) is in X^t, and $b^t - a^t \geqslant 0$ for each $t \geqslant 1$, given b^1. The sequence $y = \langle y^t \rangle$ is called the *net output programme* associated with the given programme; it is a *feasible net output programme* if it is the net output programme of a feasible programme. A programme is *efficient* if it is (1) feasible and (2) there is no other programme that is feasible, from the same initial resources b^1, and provides at least as much net output in every period and a larger net output in some period. This is the concept of efficient production, already seen in the linear activity analysis model, now extended to an infinite horizon model. The main aim of this research is to extend to the infinite horizon model the characterization of efficient production by prices seen in the finite model. This goal is not quite reached, as is seen in what follows.

The main difficulties presented by the infinite horizon are already present in a special case of the Malinvaud model with one good and no consumers. Let Y be the set of all non-negative sequences $y = (y_t)$ that satisfy $0 \leqslant y_t = f(a_{t-1}) - a_t$ for $t \geqslant 1$, and $0 \leqslant y^0 = b^1 - a^0$, $b^1 > 0$, where f is a real-valued continuous concave function on the non-negative real numbers (the production function), $f(0) = 0$, and b^1 is the given initial stock. The set Y is the set of all feasible programmes. A programme $y' - y > 0$. A price system is an infinite sequence $p = (p^t)$ of non-negative numbers. Denote by P the set of all price systems.

Malinvaud recognized the possibility that an efficient net output programme (y^t) need not have an associated system of non-zero prices (p^t) relative to which the production

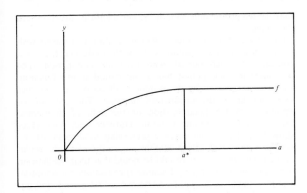

Figure 3

programme generating y satisfies the condition of intertemporal profit maximization, namely that

$$p^{t+1}f(a^t) - p^t a^t \geqslant p^{t+1}f(a) - p^t a$$

for all t and every $a \geqslant 0$. (Here (a^t) is the sequence of inputs producing y.) A condition introduced by Malinvaud, called *nontightness*, is sufficient for the existence of such nonzero prices. Alternative proofs of Malinvaud's existence theorem were given by Radner (1967) and Peleg and Yaari (1970). (An example showing the possibility of non-existence given by Peleg and Yaari (1970) is as follows. Suppose f is as shown in figure 3. At an interior efficient, and therefore value maximizing, programme the first-order necessary conditions for a maximum imply $p^{t+1}f'(a^t) = p^t$. If there is a time at which $a^t = a^*$, in an efficient programme, then, since $f'(a^*) = 0$, it follows that prices at all prior and future times are 0. Nontightness rules out such examples.)

On the side of sufficiency, Malinvaud showed that intertemporal profit maximization relative to a strictly positive price system p is not enough to ensure that a feasible programme is efficient. An additional (transversality) condition is needed. In the present model the following is such a condition;

$$\lim_{t \to \infty} p^t y^t = 0.$$

Cass (1972) has given a criterion that completely characterizes the set of efficient programmes in a one good model with strictly concave and smooth production technology that satisfies endpoint conditions $0 \leqslant f'(\infty) < 1 < f'(x) < \infty$ for some $x > 0$. Cass's criterion, states that a programme is *inefficient* if and only if the associated competitive prices – that is, satisfying $p^{t+1}f'(a^t) = p^t$ – also satisfy $\sum_{t=1}^{\infty}(1/p^t) < \infty$. This criterion may be interpreted as requiring the terms of trade between present and future to deteriorate sufficiently fast. Other similar conditions have been presented (Benveniste and Gale, 1975; Benveniste, 1976; Majumdar, 1974; Mitra, 1979). It is hard to see how any transversality condition can be interpreted in terms of decentralized resource allocation.

An alternate approach to characterizing efficient programmes was taken by Radner (1967), based on value functions as introduced in connection with valuation equilibrium by Debreu (1954). (Valuation equilibrium was discussed in connection with Arrow's exceptional case, above.) The value function approach was followed up by Majumdar (1970, 1972) and by Peleg and Yaari (1970). A price system defines a continuous linear functional, (a real-valued linear function) on the commodity space. This function assigns to a programme its present value. The present value may not be

well-defined, because the infinite sequence that gives it diverges. This creates certain technical problems passed over here. A more important difficulty is that linear functionals exist that are not defined by price systems. Radner's approach was to characterize efficient programmes in terms of maximization of present value relative to a linear functional on the commodity space. Radner showed, technical matters aside, that:

1. If a feasible programme maximizes the value of net output (consumption) relative to a strictly positive continuous linear functional, then it is efficient.

2. If a given programme is efficient, then there is a nonzero non-negative continuous linear functional such that the given programme maximizes the value of net output relative to that functional on the set of feasible programmes.

These propositions seem to be the precise counterparts of the ones characterizing efficiency in the finite horizon model. Unfortunately, a linear functional may not have a representation in the form of the inner product of a price sequence with a net output sequence. (The production function $f(a) = a^\beta$, with $0 < \beta < 1$ provides an example. It is known that the programme with constant input sequence $x_t = (1/\beta)^{\beta/\beta - 1}$ and output sequence $y_t = (1/\beta)^{\beta/\beta - 1} - (1/\beta)^{1/\beta - 1}t = 1, 2, \ldots$, is efficient, and therefore there is a continuous linear functional relative to which it is value maximizing. But there is no price sequence (p^t) that represents that linear functional.) This presents a serious problem, because in the absence of such a representation it is unclear whether this characterization has an interpretation in terms of decentralized allocation processes; profit in any one period can depend on 'prices at infinity'.

This approach has the advantage that it is applicable not only to infinite horizon models, but to a broader class in which the commodity space is infinite dimensional. Bewley (1972), Mas-Colell (1977) and Jones (1984) among others discuss Pareto optimality and competitive equilibrium in economies with infinitely many commodities. Hurwicz (1958) and others analysed optimal allocation in terms of nonlinear programming in infinite dimensional spaces. Theorems of programming in infinite dimensional spaces are also used in some of the models mentioned in this discussion.

The basic difficulties encountered in the one-good model, apart from the numerous technical problems that tend to make the literature large and diverse as different technical structures are investigated, are on the one hand the fact that transversality conditions are indispensable, and on the other the possibility that linear functionals, even when they exist, may not be representable in terms of price sequences. These problems raise strong doubt about the possibility of achieving efficient intertemporal resource allocation by decentralized means, though they leave open the possibility that some other decentralized mechanism, not using prices, might work. Analysis of this possibility has just begun, and is discussed below.

The difficulties seen in the one-good production model persist in more elaborate ones, including multisectoral models with efficiency as the criterion, and models with consumers in which Pareto optimality is the criterion. McFadden, Mitra and Majumdar (1980) studied a model in which there are firms, and overlapping generations of consumers, as in the model first investigated by Samuelson (1958). Each consumer lives for a finite time and has a consumption set and preferences like the consumers in a finite horizon model. A model with overlapping generations of consumers presents the fundamental difficulty that consumers cannot trade with future consumers as yet unborn. This difficulty can appear even in a finite horizon model if there are too few markets. The economy is closed in the sense that there are no nonproduced resources; the von Neumann growth model is an example of such a model. Building on the results of an earlier investigation (Majumdar, Mitra and McFadden, 1976), these authors introduced several notions of price systems, of competitive equilibrium, efficiency and optimality, and sought to establish counterparts of the classical welfare theorems. To summarize, in the 1976 paper they strengthen an earlier result of Bose (1974) to the effect that the problem of proper distribution of goods in essentially a short-run problem, and that the only long-run problem, one created by the infinite horizon, is that of inefficiency through overaccumulation of capital. In the 1980 paper the focus is on the relationships among various notions of equilibrium and Pareto optimality. The force of their results is, as might be expected, that the difficulties already seen in one-good model without consumers persist in this model. A transversality condition is made part of the definition of competitive equilibrium in order to obtain the result that an equilibrium is optimal. A partial converse requires some additional assumptions on the technology (reachability) and on the way the economy fits together (nondecomposability). These results certainly illuminate the infinite horizon model with overlapping generations of consumers and producers, but the possibility of efficient or optimal resource allocation by decentralized means is not different from that in the one-good Malinvaud model.

Recently, Hurwicz and Majumdar in an unpublished manuscript dated 1983, and later Hurwicz and Weinberger (1984), have addressed this issue directly, building on the approach of mechanism theory.

Hurwicz and Majumdar have studied the problem of efficiency in a model with an infinite number of periods. In each period there are finitely many commodities, one producer who is alive for just one period, and no consumers' choices. The criterion is the maximization of the discounted value of the programme (well-defined in this model). The producer alive in any period knows only the technology in that period. The question is whether there is a (static) privacy preserving mechanism using a finite dimensional message space whose equilibria coincide with the set of efficient programmes. The question can be put as follows. In each period a message is posted. The producer alive in that period responds 'Yes' or 'No'. If every producer over the entire infinite horizon answers 'Yes', the programme is an outcome corresponding to the equilibrium consisting of the infinite succession of posted messages. Since each producer knows only the technology prevailing in the period when he is alive, the process preserves privacy. If in addition the message posted in each period is finite dimensional, the process is informationally decentralized. Period-by-period profit maximization using period-by-period prices is a mechanism of this type; the message posted in each period consists of the vector of prices for that period, and the production plan for that period, both finite dimensional. The object is to characterize all efficient programmes as equilibria of such a mechanism. This would be an analogue of the classical welfare theorems, but without the restriction to mechanisms that use prices in their messages.

The main result is in the nature of an impossibility theorem. If the technology is constant over time, and that fact is common knowledge at the beginning, the problem is trivial since knowledge of the technology in the first period automatically means knowledge of it in every period. On the other hand, if there is some period whose technology is not known in the first period, then there is no finite dimensional message that can characterize efficient programmes, and in

that sense, production cannot be satisfactorily decentralized over time.

Hurwicz and Weinberger (1984) have studied a model with both producers and consumers. As with producers, there is a consumer in each period, who lives for one period. The consumer in each period has a one-period utility function, which is not known by the producer; similarly the consumer does not know the production function. The criterion of optimality is the maximization of the sum of discounted utilities over the infinite horizon. Hurwicz and Weinberger show that there is no privacy preserving mechanism of the type just described whose equilibria correspond to the set of optimal programmes. It should be noted that their mechanism requires that the first-period actions (production, consumption and investment decisions) be made in the first period, and not be subject to revision after the infinite process of verification is completed. (On the other hand, under tâtonnement assumptions it may be possible to decentralize. In this model tâtonnement entails reconsideration 'at infinity'.)

If attention is widened to efficient programmes, and if technology is constant over time, there is an efficient programme with a fixed ratio of consumption to investment. This programme can be obtained as the equilibrium outcome of a mechanism of the specified type. However, this corresponds to only one side of the classical welfare theorems. It says that the outcome of such a mechanism is efficient; but it does not ensure that every efficient programme can be realized as the outcome of such a mechanism. The latter property fails in this model.

STANLEY REITER

See also INCENTIVE COMPATIBILITY; LINEAR PROGRAMMING; ORGANIZATION THEORY; WELFARE ECONOMICS.

BIBLIOGRAPHY

Arrow, K. 1951. An extension of the basic theorems of classical welfare economics. In *Proceedings of the Second Berkeley Symposium on Mathematical Statistics and Probability*, ed. J. Neyman, Berkeley: University of California Press.

Arrow, K. and Debreu, G. 1954. Existence of an equilibrium for a competitive economy. *Econometrica* 22, July, 265–90.

Arrow, K. and Hahn, F. 1971. *General Competitive Analysis*. San Francisco: Holden-Day.

Arrow, K. and Hurwicz, L. 1960. Decentralization and computation in resource allocation. In *Essays in Economics and Econometrics*, ed. R.W. Pfouts, Chapel Hill: University of North Carolina Press, 34–104.

Arrow, K., Hurwicz, L. and Uzawa, H. 1961. Constraint qualifications in maximization problems. *Naval Research Logistics Quarterly* 8(2), June, 175–91.

Benveniste, L. 1976. Two notes on the Malinvaud condition for efficiency of infinite horizon programs. *Journal of Economic Theory* 12, 338–46.

Benveniste, L. and Gale, D. 1975. An extension of Cass' characterization of infinite efficient production programs. *Journal of Economic Theory* 10, 229–38.

Bewley, T. 1972. Existence of equilibria in economies with infinitely many commodities. *Journal of Economic Theory* 4, 514–40.

Bose, A. 1974. Pareto optimality and efficient capital accumulation. Discussion Paper No. 74–4, Department of Economics, University of Rochester.

Brown, D. and Heal, G. 1982. Existence, local-uniqueness and optimality of a marginal cost pricing equilibrium in an economy with increasing returns. Cal. Tech. Social Science Working Paper No. 415.

Brown, D., Heal, G., Ali Khan, M. and Vohra, R. 1986. On a general existence theorem for marginal cost pricing equilibria. Cowles Foundation Working Paper No. 724. Reprinted in *Journal of Economic Theory* 38, 1986, 111–19.

Calsamiglia, X. 1977. Decentralized resource allocation and increasing returns. *Journal of Economic Theory* 14, 263–83.

Calsamiglia, X. 1982. On the size of the message space under non-convexities. *Journal of Mathematical Economics* 10, 197–203.

Cass, D. 1972. On capital over-accumulation in the aggregative neoclassical model of economic growth: a complete characterization. *Journal of Economic Theory* 4(2), April, 200–23.

Dantzig, G.B. 1951a. The programming of interdependent activities. In *Activity Analysis of Production and Allocation*, ed. T. Koopmans, Cowles Commission Monograph No. 13, New York: Wiley, ch. 2, 19–32.

Dantzig, G.B. 1951b. Maximization of a linear function of variables subject to linear inequalities. In *Activity Analysis of Production and Allocation*, ed. T. Koopmans, Cowles Commission Monograph No. 13, New York: Wiley, ch. 21, 339–47.

Dasgupta, P., Hammond, P. and Maskin, E. 1979. The implementation of social choice rules: some general results on incentive compatibility. *Review of Economic Studies* 46, 185–216.

Davis, O.A. and Whinston, A.B. 1962. Externalities welfare and the theory of games. *Journal of Political Economy* 70, 214–62.

Debreu, G. 1954. Valuation equilibrium and Pareto optimum. In *Proceedings of the National Academy of Sciences of the USA* 40(7), 588–92.

Debreu, G. 1959. *Theory of Value*. New York: Wiley.

Debreu, G. 1970. Economies with a finite set of equilibria. *Econometrica* 38(3), May, 387–92.

Diamond, P. and Mirrlees, J. 1971. Optimal taxation and public production. I: Production efficiency; II: Tax rules. *American Economic Review* 61, 8–27; 261–78.

Domar, E. 1946. Capital expansion, rate of growth, and employment. *Econometrica* 14, April, 137–47.

Fenchel, W. 1950. Convex cones, sets, and functions. Princeton University (hectographed).

Guesnerie, R. 1975. Pareto optimality in non-convex economies. *Econometrica* 43, 1–29.

Harrod, R.F. 1939. An essay in dynamic theory. *Economic Journal* 49, 14–33.

Hayek, F. von. 1945. The use of knowledge in society. *American Economic Review* 35, 519–53. Reprinted in F. von Hayek, *Individualism and Economic Order*, Chicago: University of Chicago Press, 1949, 77–92.

Heal, G. 1971. Planning, prices and increasing returns. *Review of Economic Studies* 38, 281–94.

Hotelling, H. 1938. The general welfare in relation to problems of taxation and of railway and utility rates. *Econometrica* 6, 242–69.

Hurwicz, L. 1958. Programming in linear spaces. In *Studies in Linear and Non-Linear Programming*, ed. K. Arrow, L. Hurwicz and H. Uzawa, Stanford: Stanford University Press.

Hurwicz, L. 1960. Optimality and informational efficiency in resource allocation processes. In *Mathematical Methods in the Social Sciences, 1959*, ed. K.J. Arrow, S. Karlin and P. Suppes, Stanford: Stanford University Press.

Hurwicz, L. 1971. Centralization and decentralization in economic processes. In *Comparison of Economic Systems: Theoretical and Methodological Approaches*, ed. A. Eckstein, Berkeley: University of California Press, ch. 3.

Hurwicz, L. 1972a. On informationally decentralized systems. In *Decision and Organization*, ed. C. McGuire and R. Radner, Amsterdam, London: North-Holland, ch. 14, 297–336.

Hurwicz, L. 1972b. On the dimensional requirements of informationally decentralized Pareto-satisfactory processes. Presented at the Conference Seminar in Decentralization, Northwestern University. In *Studies in Resource Allocation Processes*, ed. K.J. Arrow and L. Hurwicz, Cambridge: Cambridge University Press, 1977.

Hurwicz, L. 1976. On informational requirements for nonwasteful resource allocation systems. In *Mathematical Models in Economics: Papers and Proceedings of a US–USSR Seminar, Moscow*, ed. S. Shulman, New York: National Bureau of Economic Research.

Hurwicz, L., Radner, R. and Reiter, S. 1975. A stochastic decentralized resource allocation process. *Econometrica* 43: Part I, 187–221; Part II, 363–93.

Hurwicz, L. and Weinberger, H. 1984. Paper presented at IMA seminar in Minneapolis.

Jennergren, L. 1971. Studies in the mathematical theory of decentralized resource-allocation. PhD dissertation, Stanford University.

Jones, L. 1984. A competitive model of commodity differentiation. *Econometrica* 52, 507–30.

Kantorovitch, L. 1939. *Matematicheskie metody organizatii i planirovania proizvodstva* (Mathematical methods in the organization and planning of production). Izdanie Leningradskogo Gosudarstvennogo Universiteta, Leningrad. Trans. in *Management Science* 6(4), July 1960, 363–422.

Kantorovitch, L. 1942. On the translocation of masses. (In English.) *Comptes Rendus (Doklady) de l'Academie des Sciences d l'URSS* 37(7–8).

Kantorovitch, L. and Gavurin, M. 1949. Primenenie matematicheskikh metodov v voprosakh analiza grusopotokov (The application of mathematical methods to problems of freight flow analysis). In *Problemy Povysheniia Effektivnosty Raboty Transporta* (Problems of raising the efficiency of transportation), ed. V. Zvonkov, Moscow and Leningrad: Izdatel'stvo Akademii Nauk SSSR.

Koopmans, T.C. 1951a. Analysis of production as an efficient combination of activities. In *Activity Analysis of Production and Allocation*, ed. T. Koopmans, Cowles Commission Monograph No. 13, New York: Wiley, ch. 3, 33–97.

Koopmans, T.C. 1951b. Efficient allocation of resources. *Econometrica* 19, 455–65.

Koopmans, T.C. 1957. *Three Essays on the State of Economic Science*. New York: McGraw-Hill, 66–104.

Kuhn, H. and Tucker, A. 1951. Nonlinear programming. In *Proceedings of the Second Berkeley Symposium on Mathematical Statistics and Probability*, ed. J. Neyman, Berkeley: University of California Press, 481–92.

Ledyard, J. 1968. Resource allocation in unselfish environments. *American Economic Review* 58, 227–37.

Ledyard, J. 1971. A convergent Pareto-satisfactory non-tâtonnement adjustment process for a class of unselfish exchange environments. *Econometrica* 39, 467–99.

Lerner, A. 1934. The concept of monopoly and measurement of monopoly power. *Review of Economic Studies* 1(3), June, 157–75.

Lerner, A. 1944. *The Economics of Control*. New York: Macmillan.

Lipsey, R. and Lancaster, K. 1956. The general theory of second best. *Review of Economic Studies* 24, 11–32.

McFadden, D., Mitra, T. and Majumdar, M. 1980. Pareto optimality and competitive equilibrium in infinite horizon economies. *Journal of Mathematical Economics* 7, 1–26.

McKenzie, L. 1959. On the existence of general equilibrium for a competitive market. *Econometrica* 27(1), January, 54–71.

Majumdar, M. 1970. Some approximation theorems on efficiency prices for infinite programs. *Journal of Economic Theory* 2, 399–410.

Majumdar, M. 1972. Some general theorems of efficiency prices with an infinite dimensional commodity space. *Journal of Economic Theory* 5, 1–13.

Majumdar, M. 1974. Efficient programs in infinite dimensional spaces: a complete characterization. *Journal of Economic Theory* 7, 355–69.

Majumdar, M., Mitra, T. and McFadden, D. 1976. On efficiency and Pareto optimality of competitive programs in closed multisector models. *Journal of Economic Theory* 13, 26–46.

Malinvaud, E. 1953. Capital accumulation and efficient allocation of resources. *Econometrica* 21, 233–68.

Mas-Colell, A. 1977. Regular nonconvex economies. *Econometrica* 45, 1387–404.

Mitra, T. 1979. On optimal economic growth with variable discount rates: existence and stability results. *International Economic Review* 20, 133–45.

Mount, K. and Reiter, S. 1974. The informational size of message spaces. *Journal of Economic Theory* 8, 161–92.

Mount, K. and Reiter, S. 1977. Economic environments for which there are Pareto satisfactory mechanisms. *Econometrica* 45, 821–42.

Neumann, J. von. 1937. A model of general economic equilibrium. *Ergebnisse eines mathematischen Kolloquiums*, No. 8. Trans. from German, *Review of Economic Studies* 13(1), (1945–6), 1–9.

Nikaido, H. 1969. *Convex Structures and Economic Theory*. New York: Academic Press.

Nikaido, H. 1970. *Introduction to Sets and Mappings in Modern Economics*. Trans. K. Sato, Amsterdam: North-Holland (Japanese original, Tokyo, 1960).

Osana, H. 1978. On the informational size of message spaces for resource allocation processes. *Journal of Economic Theory* 17, 66–78.

Peleg, B. and Yaari, M. 1970. Efficiency prices in an infinite dimensional commodity space. *Journal of Economic Theory* 2, 41–85.

Pigou, A. 1932. *The Economics of Welfare*. 4th edn, London: Macmillan.

Radner, R. 1967. Efficiency prices for infinite horizon production programs. *Review of Economic Studies* 34, 51–66.

Ramsey, F. 1928. A mathematical theory of saving. *Economic Journal* 38, 543–59.

Reichelstein, S. 1984a. Dominant strategy implementation, incentive compatibility and informational requirements. *Journal of Economic Theory* 34(1), October, 32–51.

Reichelstein, S. 1984b. Information and incentives in economic organizations. PhD dissertation, Northwestern University.

Reichelstein, S. and Reiter, S. 1985. Game forms with minimal strategy spaces. Discussion Paper No. 663, The Center for Mathematical Studies in Economics and Management Science, Northwestern University, Evanston, Ill.

Reiter, S. 1977. Information and performance in the (new)[2] welfare economics. *American Economic Review* 67, 226–34.

Samuelson, P. 1947. *Foundations of Economic Analysis*. Cambridge, Mass.: Harvard University Press.

Samuelson, P. 1949. Market mechanisms and maximization, I, II, III. Hectographed memoranda, The RAND Corporation, Santa Monica.

Samuelson, P. 1958. An exact consumption-loan model of interest with or without the social contrivance of money. *Journal of Political Economy* 66, December, 467–82.

Sato, F. 1981. On the informational size of message spaces for resource allocation processes in economies with public goods. *Journal of Economic Theory* 24, 48–69.

Slater, M. 1950. Lagrange multipliers revisited: a contribution to non-linear programming. *Cowles Commission Discussion Paper*, Math. 403, also RM-676, 1951.

Sonnenschein, H. 1974. An axiomatic characterization of the price mechanism. *Econometrica* 42, 425–34.

Uzawa, H. 1958. The Kuhn–Tucker Theorem in concave programming. In *Studies in Linear and Non-Linear Programming*, ed. K. Arrow, L. Hurwicz and H. Uzawa, Stanford: Stanford University Press.

Walker, M. 1977. On the informational size of message spaces. *Journal of Economic Theory* 15, 366–75.

efficient market hypothesis. A capital market is said to be efficient if it fully and correctly reflects all relevant information in determining security prices. Formally, the market is said to be efficient with respect to some information set, ϕ, if security prices would be unaffected by revealing that information to all participants. Moreover, efficiency with respect to an information set, ϕ, implies that it is impossible to make economic profits by trading on the basis of ϕ.

It has been customary since Roberts (1967) to distinguish three levels of market efficiency by considering three different types of information sets:

(1) The weak form of the Efficient Market Hypothesis (EMH) asserts that prices fully reflect the information contained in the historical sequence of prices. Thus, investors cannot devise an investment strategy to yield abnormal profits on the basis of an analysis of past price patterns (a technique known as technical analysis). It is this form of efficiency that is associated with the term 'Random Walk Hypothesis'.

(2) The semi-strong form of EMH asserts that current stock prices reflect not only historical price information but also all publicly available information relevant to a company's securities. If markets are efficient in this sense, then an analysis of balance sheets, income statements, announcements of dividend changes or stock splits or any other public information about a company (the technique of fundamental analysis) will not yield abnormal economic profits.

(3) The strong form of EMH asserts that all information that is *known* to any market participant about a company is fully reflected in market prices. Hence, not even those with privileged information can make use of it to secure superior investment results. There is perfect revelation of all private information in market prices.

WEAK FORM MARKET EFFICIENCY AND THE RANDOM WALK HYPOTHESIS. If markets are efficient, the (technical) analysis of past price patterns to predict the future will be useless because any information from such an analysis will already have been impounded in current market prices. Suppose market participants were confident that a commodity price would double next week. The price will not gradually approach its new equilibrium value. Indeed, unless the price adjusted immediately, a profitable arbitrage opportunity would exist and could be expected to be exploited immediately in an efficient market. Similarly, if a reliable and profitable seasonal pattern for equity prices exists (e.g. a substantial Christmas rally) speculators will bid up prices sufficiently prior to Christmas so as to eliminate any unexploited arbitrage possibility. Samuelson (1965) and Mandelbrot (1966) have proved rigorously that if the flow of information is unimpeded and if there are no transactions costs, then tomorrow's price change in speculative markets will reflect only tomorrow's 'news' and will be independent of the price change today. But 'news' by definition is unpredictable and thus the resulting price changes must also be unpredictable and random.

The term 'random walk' is usually used loosely in the finance literature to characterize a price series where all subsequent price changes represent random departures from previous prices. Thus, changes in price will be unrelated to past price changes. (More formally, the random walk model states that investment returns are serially independent, and that their probability distributions are constant through time.) It is believed that the term was first used in an exchange of correspondence appearing in *Nature* in 1905. The problem considered in the correspondence was the optimal search procedure for finding a drunk who had been left in the middle of a field. The answer was to start exactly where the drunk had been placed. That point is an unbiased estimate of the drunk's future position since he will presumably stagger along in an unpredictable and random fashion.

The earliest empirical work on the random walk hypothesis was performed by Bachelier (1900). He concluded that commodities prices followed a random walk, although he did not use that term. Corroborating evidence from other time series was provided by Working (1934 – various time series), Cowles and Jones (1937 – US stock prices) and Kendall (1953 – UK stock and commodities prices). These studies generally found that the serial correlation between successive price changes was essentially zero. Roberts (1959) found that a time series generated from a sequence of random numbers had the same appearance as a time series of US stock prices. Osborne (1959) found that stock price movements were very similar to the random Brownian motion of physical particles. He found that the logarithms of price changes were independent of each other.

More recent empirical work has used alternative techniques and data sets and has searched for more complicated patterns in the sequence of prices in speculative markets. Granger and Morgenstern (1963) used the powerful technique of spectral analysis but were unable to find any dependably repeatable patterns in stock price movements. Fama (1965) not only looked at serial correlation coefficients (which were close to zero) but also corroborated his investigation by examining a series of lagged price changes as well as by performing a number of nonparametric 'runs' tests. Fama and Blume (1966) examined a variety of filter techniques – trading techniques where buy (sell) signals are generated by some upward (downward) price movements from recent troughs (peaks) – and found they could not produce abnormal profits. Other investigators have done computer simulations of more complicated techniques of technical analysis of stock price patterns and found that profitable trading strategies could not be employed on the basis of these techniques. Solnik (1973) measured serial correlation coefficients for daily, weekly, and monthly price changes in nine countries and also concluded that profitable investment strategies could not be formulated on the basis of the extremely small dependencies found.

While the empirical data are remarkably consistent in their general finding of randomness, equity markets do not perfectly conform to the statistician's ideal of a random walk. As noted above, while serial correlation coefficients are always found to be small, there are some small dependencies that have been isolated. While 'runs' tests found only slight departures from randomness, there is a slight tendency for runs in daily price changes to persist. Merton (1980) has shown that changes in the variance of a stock's return (price) can be predicted from its variance in the recent past. Such departures from a pure random walk do not violate the weak form of EMH, which states only that unexploited trading opportunities should not exist in an efficient market. Still, the formal random walk model does not strictly hold. The probability distributions of stock returns are not constant through time and thus the appropriate model for stock prices may be a submartingale rather than a random walk.

In addition, some disturbing seasonal patterns have recently been found in stock price series. Keim (1983) and others have documented a January effect, where stock returns are abnormally higher during the first few days of January (especially for small firms) and French (1980) and others have also documented a so-called 'weekend effect' where average returns to stocks are negative from the close of trading on Friday to the close of trading on Monday. Seasonals appear to exist in several international markets as documented by the Gultekins (1983) and by Jaffe and Westerfield (1984). But departures from randomness are generally remarkably small and an investor who pays transaction costs cannot choose a profitable investment strategy on the basis of these anomalies. Thus, while the random-walk hypothesis is not strictly upheld, the departures from randomness that do exist are not large enough to leave unexploited investment opportunities. Consequently, the empirical evidence presents strong evidence in favour of the weak form of the efficient market hypothesis. The history of stock price movements does not offer investors any information that permits them to outperform a simple buy-and-hold investment strategy.

SEMI-STRONG FORM EFFICIENCY. The weak form of EMH has found general acceptance in the financial community, where technical analysts have never been held in high repute. The stronger assertion that all publicly available information has already been impounded into current market prices has proved far more controversial among investment professionals, who practise 'fundamental' analysis of publicly available information as a widely accepted mode of security analysis. In general, however, the empirical evidence suggests that public information is so rapidly impounded into current market prices that fundamental analysis is not likely to be fruitful.

A variety of tests have been performed to ascertain the speed of adjustment of market prices to new information. Fama, Fisher, Jensen and Roll (1969) looked at the effect of stock splits on equity prices. While splits themselves provide no economic benefit, splits are usually accompanied or followed by dividend increases that do convey to the market information about management's confidence about the future progress of the enterprise. Thus, while splits usually do result in higher share prices, the market appears to adjust to the announcement fully and immediately. Substantial returns can be earned prior to the split announcement, but there is no evidence of abnormal returns after the public announcement. Indeed, in cases where dividends were not raised following the split, firms suffered a loss in price, presumably because of the unexpected failure of the firm to increase its dividend. Similarly, while merger announcements, especially where premiums are being paid to the shareholders of the acquired firm, can raise market prices substantially, it appears that the market adjusts fully to the public announcements. Dodd (1981) finds no evidence of abnormal price changes following the public release of the merger information.

Scholes (1972) studied the price effects of large secondary offerings. The general belief among market professionals is that such offerings will depress prices temporarily so as to facilitate a large distribution relative to normal trading volume. Such a *temporary* decline would be inconsistent with market efficiency. Scholes hypothesized that the decline would be permanent, however, reflecting release of privileged information (since block traders are usually insiders) of an expected decline in the company's performance. Scholes found that the declines were permanent, especially when sales were by insiders, and thus inconsistent with the temporary price-pressure hypothesis. However, Kraus and Stoll (1972) used intraday prices and did find some evidence of a price reversal and an arbitrage opportunity. But these reversals took place within a 15-minute period – a speed of adjustment that suggests the market is remarkably efficient.

While the vast majority of studies support the semi-strong version of EMH, there have been some that do not. Ball (1978) found that stock-price reactions to earnings announcements are not complete. Abnormal risk-adjusted returns are systematically non-zero in the period following the announcement. Ball attributed this to inadequacies in the capital asset pricing model (CAPM) used to adjust for risk differentials and suggested several steps to reduce the estimation bias. Watts (1978), however, performed the steps suggested by Ball and still found systematic abnormal returns. Rendleman, Jones and Latané (1982) also find a relationship between unexpected quarterly earnings and excess returns for common shares subsequent to the announcement date. Roll (1984) found that orange-juice futures prices were made informationally inefficient over short periods by the existence of exchange-imposed maximum daily price moves. Apart from this constraint, however, prices did fully reflect all known information. Moreover, the other abnormalities have not been shown to exist *consistently* over time, and when they did occur, they have usually been small enough that only a professional broker-dealer could have earned economic profits. Thus, it remains to be seen how robust these anomalies are as compared with the vast body of evidence supporting the semi-strong EMH. The evidence in favour of the market's rapid adjustment to new information is sufficiently pervasive that it is now a generally, if not universally, accepted tenet of financial econometric research.

THE STRONG FORM OF THE EFFICIENT MARKET HYPOTHESIS. As the previous studies indicated, stock splits, dividend increases and merger announcements can have substantial impacts on share prices. Consequently insiders trading on such information can clearly profit prior to making the announcement, as has been documented by Jaffe (1974). While such trading is generally illegal, the fact that the market often at least partially anticipates the announcements suggests that it is certainly possible to profit on the basis of privileged information. Thus, the strongest form of the EMH is clearly refuted. Nevertheless, there is considerable evidence that the market comes reasonably close to strong-form efficiency.

Several studies have been performed on the records of professional investment managers. In general they show that randomly selected portfolios or unmanaged indices do as well or better than professionally managed portfolios after expenses. Cowles (1933) examined the records of selected financial services and professional investors. He failed to find any evidence of performance superior to that which could be achieved by investing in the market as a whole. Friend et al. (1962) concluded that the performance of the average mutual fund was insignificantly different from the performance of an unmanaged portfolio with similar asset composition. Jensen (1969) measured the risk-adjusted performance of mutual funds utilizing the capital asset pricing model to measure the appropriate risk return trade-off. Jensen found that while the funds tended to earn *gross* positive abnormal returns, any relative advantage of the professional managers was lost in management fees. Note that the EMH would not rule out small gross abnormal returns as an incentive to acquire information. Grossman and Stiglitz (1980) and Cornell and Roll (1981) have shown that a sensible market equilibrium should leave some incentive for analysis. Those who acquire costly information would have superior gross returns but only average net returns. And the overwhelming evidence on the performance of professional investors is that net returns are only average or below average. For example, during the 20 years to 1984, two-thirds of US professional pension fund managers were out-performed by the unmanaged Standard and Poor's 500 stock index. Moreover, there seems to be little consistency to whatever exceptional performance one finds. It appears that a professional manager who has achieved exceptional performance in one period is just as likely to underperform the market in the next period. It is clear that while superior investment managers may well exist, they are extremely rare.

SOME FURTHER ANOMALIES. In general, the empirical evidence in favour of EMH is extremely strong. Probably no other hypothesis in either economics or finance has been more extensively tested. Thus, it is not surprising that along with general support for EMH there have been scattered pieces of anomalous evidence inconsistent with the hypothesis in its strongest forms. Basu (1977, 1983) found that stocks with low price-earnings (P/E) multiples have higher average risk-adjusted returns than stocks with high P/E's. Banz (1981) found that substantial abnormal (risk-adjusted) long-run rates of return could be earned by investing in portfolios of smaller firms. As was noted above, a large part of this higher return

occurs early in January. We know that transactions costs are higher for smaller firms but this factor does not seem to explain the size effect. This size effect appears to persist in varying degrees over time and is related to the evidence regarding higher returns for stocks with low P/E multiples. Of course, we must always keep in mind that these findings of abnormal returns are always joint tests of market efficiency and the particular form of the asset pricing model involved. Thus, it is impossible to distinguish if the abnormal returns are truly due to inefficiencies or result instead because of inadequacies of the capital asset pricing model as a method of measuring risk.

In another empirical study rejecting the concept of market efficiency, Shiller (1981) argued that variations in aggregate stock market prices are much too large to be justified by the variation in subsequent dividend payments. This apparent rejection of the Efficient Market Hypothesis for the entire stock market goes far beyond the narrow issue of whether or not some investors or some trading schemes can beat the market. Shiller's tests, however, are joint tests of market efficiency and the correctness of his model of the dividend process. Marsh and Merton (1983) derive an alternative model for dividend and stock price behaviour. They conclude that Shiller's findings that stock prices are 'too volatile' is a result of his misspecification of the dividend process rather than a result of market inefficiency. A similar conclusion was reached by Kleidon (1986). Nevertheless, the history of fads and excesses in speculative markets I have reviewed (Malkiel, 1985), from tulip bulbs to blue-chip growth stocks, gives me some doubts that we should always consider that the current tableau of market prices represents the best estimates available of appropriate discounted present value.

There have been other scattered instances of inefficiencies as summarized by Jensen (1979) and Ball (1978). I have argued (Malkiel, 1980), that closed-end funds (even those holding essentially 'market' portfolios) were inefficiently priced over many years so that they would provide investors with abnormal returns over and above those involved in buying and holding directly the well-diversified portfolios owned by the funds.

But this last illustration, rather than convincing me of substantial areas of market inefficiency, actually drives me to the opposite conclusion. If there is *truly* some area of pricing inefficiency that can be discovered by the market and dependably exploited, then profit-maximizing traders and investors will eventually through their purchases and sales bring market prices in line so as to eliminate the possibility of extraordinary return. In time investors recognized that closed-end funds at discounts represented extraordinary value and the discounts on these funds were eventually largely eliminated.

So we are again driven back to the position of the EMH. Pricing irregularities may well exist and even persist for periods of time, and markets can at times be dominated by fads and fashions. Eventually, however, any excesses in market valuations will be corrected. Undoubtedly with the passage of time and with the increasing sophistication of our data bases and empirical techniques, we will document further departures from efficiency and understand their causes more fully. But I suspect that the end result will not be an abandonment of the profession's belief that the stock market is remarkably efficient in its utilization of information.

BURTON G. MALKIEL

See also ARBITRAGE; FINANCE; FINANCIAL MARKETS.

BIBLIOGRAPHY

Bachelier, L. 1900. *Théorie de la speculation. Annales de l'Ecole normale Supérieure*, 3rd series, 17, 21–86. Trans. by A.J. Boness in *The Random Character of Stock Market Prices*, ed. P.H. Cootner, Cambridge, Mass.: MIT Press, 1967.

Ball, R. 1978. Anomalies in relationships between securities' yields and yield-surrogates. *Journal of Financial Economics* 6(2–3), 103–26, June/September, 1981.

Banz, R. 1981. The relationship between return and market value of common stocks. *Journal of Financial Economics* 9(1), March 3–18.

Basu, S. 1977. Investment performance of common stocks in relation to their price earnings ratios: a test of the efficient markets hypothesis. *Journal of Finance* 32(3), June, 663–82.

Basu, S. 1983. The relationship between earnings' yield, market value and the return of NYSE common stocks: further evidence. *Journal of Financial Economics* 12(1), June, 129–56.

Cornell, B. and Roll, R. 1981. Strategies for pairwise competitions in markets and organizations. *Bell Journal of Economics* 12(1), Spring, 201–13.

Cowles, A. 1933. Can stock market forecasters forecast? *Econometrica* 1(3), July, 309–24.

Cowles, A. and Jones, H. 1937. Some posteriori probabilities in stock market action. *Econometrica* 5(3), July, 280–94.

Dodd, P. 1981. *The Effect on Market Value of Transactions in the Market for Corporate Control, Proceedings of Seminar on the Analysis of Security Prices, CRSP*. Chicago: University of Chicago, May.

Fama, E. 1965. The behavior of stock market prices. *Journal of Business* 38(1), January, 34–105.

Fama, E. and Blume, M. 1966. Filter rules and stock market trading. *Security Prices: A Supplement, Journal of Business* 39(1), January, 226–41.

Fama, E., Fisher, L., Jensen, M. and Roll, R. 1969. The adjustment of stock prices to new information. *International Economic Review* 10(1), February, 1–21.

French, K. 1980. Stock returns and the weekend effect. *Journal of Financial Economics* 8(1), March, 55–69.

Friend, I., Brown, F., Herman, E. and Vickers, D. 1962. *A Study of Mutual Funds*. Washington, DC: US Government Printing Office.

Granger, D. and Morgenstern, O. 1963. Spectral analysis of New York Stock Market prices. *Kyklos* 16, January, 1–27.

Grossman, S. and Stiglitz, J. 1980. On the impossibility of informationally efficient markets. *American Economic Review* 70(3), June, 393–408.

Gultekin, M. and Gultekin, N. 1983. Stock market seasonality, international evidence. *Journal of Financial Economics* 12(4), December, 469–81.

Jaffe, J. 1974. The effect of regulation changes on insider trading. *Bell Journal of Economics and Management Science* 5(1), Spring, 93–121.

Jaffe, J. and Westerfield, R. 1984. The week-end effect in common stock returns: the international evidence. Unpublished manuscript, University of Pennsylvania, December.

Jensen, M. 1969. Risk, the pricing of capital assets, and the evaluation of investment portfolios. *Journal of Business* 42(2), April, 167–247.

Jensen, M. 1978. Some anomalous evidence regarding market efficiency. *Journal of Financial Economics* 6(2–3), June–September, 95–101.

Keim, D. 1983. Size related anomalies and stock return seasonality: further empirical evidence. *Journal of Financial Economics* 12(1), June, 13–32.

Kendall, M. 1953. The analysis of economic time series. Part I: Prices. *Journal of the Royal Statistical Society* 96(1), 11–25.

Kleidon, A. 1986. Variance bounds tests and stock price valuation models. *Journal of Political Economy* 94(5), October, 953–10001.

Kraus, A. and Stoll, H. 1972. Price impacts of block trading on the New York Stock Exchange. *Journal of Finance* 27(3), June, 569–58.

Malkiel, B. 1980. *The Inflation-Beater's Investment Guide*. New York: Norton.

Malkiel, B. 1985. *A Random Walk Down Wall Street*. 4th edn, New York: Norton.

Mandelbrot, B. 1966. Forecasts of future prices, unbiased markets, and martingale models. *Security Prices: A Supplement, Journal of Business* 39(1), January, 242–55.

Marsh, T. and Merton, R. 1983. Aggregate dividend behavior and its implications for tests of stock market rationality. Working Paper, Sloan School of Management, September.

Merton, R. 1980. On estimating the expected return on the market: an exploratory investigation. *Journal of Financial Economics* 8(4), December, 323–61.

Osborne, M. 1959. Brownian motions in the stock market. *Operations Research* 7(2), March/April, 145–73.

Pearson, K. and Rayleigh, Lord. 1905. The problem of the random walk. *Nature* 72, 294, 318, 342.

Rendleman, R., Jones, C. and Latané, H. 1982. Empirical anomalies based on unexpected earnings and the importance of risk adjustments. *Journal of Financial Economics* 10(3), November, 269–87.

Roberts, H. 1959. Stock market 'patterns' and financial analysis: methodological suggestions. *Journal of Finance* 14(1), March, 1–10.

Roberts, H. 1967. Statistical versus clinical prediction of the stock market. Unpublished manuscript, CRSP, Chicago: University of Chicago, May.

Roll, R. 1984. Orange juice and weather. *American Economic Review* 74(5), December, 861–80.

Samuelson, P. 1965. Proof that properly anticipated prices fluctuate randomly. *Industrial Management Review* 6(2), Spring, 41–9.

Scholes, M. 1972. The market for securities; substitution versus price pressure and the effects of information on share prices. *Journal of Business* 45(2), April, 179–211.

Shiller, R.J. 1981. Do stock prices move too much to be justified by subsequent changes in dividends? *American Economic Review* 71(3), June, 421–36.

Solnik, B. 1973. Note on the validity of the random walk for European stock prices. *Journal of Finance* 28(5), December, 1151–9.

Thompson, R. 1978. The information content of discounts and premiums on closed-end fund shares. *Journal of Financial Economics* 6(2–3), June/September, 151–86.

Watts, R. 1978. Systematic 'abnormal' returns after quarterly earnings announcements. *Journal of Financial Economics* 6(2–3), June/September, 127–50.

Working, H. 1934. A random difference series for use in the analysis of time series. *Journal of the American Statistical Association* 29, March, 11–24.

egoism. *See* ALTRUISM; ECONOMIC MAN; INDIVIDUALISM; RATIONAL BEHAVIOUR; SELF-INTEREST.

Einaudi, Luigi (1874–1961).   An outstanding Italian economist and influential figure on the broader political and cultural scene, Einaudi was born in Carru (Piedmont) on 24 March 1874 and died in Rome on 30 October 1961. He graduated in law from Turin in 1895 and then, whilst continuing with this studies, embarked on a career in journalism. The success he achieved in both fields underlined his rare talent and his endless capacity for work. In fact, his academic progress was so rapid that in 1907 he was appointed as professor of public finance at the University of Turin. Meanwhile, he wrote articles for the most influential Italian daily newspaper of the period, the *Corriere delle Serra*, which not only brought him national recognition but also earned him the reputation of 'educator' of the entire country. He became a member of the Senate in 1919, but retired from all political and public activity with the advent of fascism. Towards the end of the World War II he went into exile in Switzerland. On his return, he was appointed Governor of the Bank of Italy (1945), Vice-President of the Cabinet and Minister in charge of the Budget (1947), and was finally elected President of the Republic of

Italy (1948–1955). At the end of his seven-year presidential term of office, he was made a life member of the Senate.

The most important aspect of Einaudi's achievements is the use he made of his academic and journalistic ability, as foundations for his activity as a statesman and politician. In addition, close study of his strictly scientific works reveals the extent to which he drew on the wealth of knowledge and experience which he had gained also in other fields. The 3800 recorded items of Einaudi's works cover such a wide range of interests that it is necessary here to concentrate on his contributions to the study of public finance and his ideas on economic policy. Einaudi's main contributions to the study of public finance were investigations, based on the classical ideas of John Stuart Mill, which gave a solid logical basis to the principle of the exclusion of savings from taxable income; his research into the theory of capitalization of taxation; his critical and constructive contributions on the effects of certainty and stability of fiscal principles; his important analysis of the concept of taxable income which he identified with normal income, or, in other words, with the average income potentiality of the person subject to taxation.

Einaudi's position vis-à-vis public intervention in the economy was not hostile in principle, though he undoubtedly took a limited view of state interference in economic life. Since, for Einaudi, 'All liberties were jointly liable', autonomous sources of income were a necessity to avoid men to be subjected to a single centralizing order of the state. He asserted this during the 20 years of fascism, when he continued to teach with the same independence of mind and without compromising his fidelity to economic liberalism. Even though Einaudi had been stressing the usefulness of productive public expenditure since 1919, he showed a singular lack of comprehension of the Keynesian contribution, in the belief that it would be an inevitable cause of inflation.

F. Caffè

SELECTED WORKS
On Luigi Einaudi himself there is a *Bibliografia degli scritti* edited by Luigi Firpo under the auspices of the Bank of Italy, Turin, 1971. Even though Einaudi outlined a plan for reprinting his works, this reprint has still not been completed. It is, nevertheless, still useful to divide his work into the three main areas which he outlined: theory, politics and history. Representative works of the three sections are as follows:

1912. *Intorno al concetto di reddito imponibile e di un sistema di imposte sul reddito consumato.* Turin: V. Bona.
1919. *Osservazioni critiche intorno alla teoria dell'ammortamento dell'imposta e teoria delle variazioni nei redditi e nei valori capitali susseguenti all'imposta.* Turin: Fratelli Bocca.
1929. *Contributo all ricerca della 'ottima imposta'.* Milan: Bocconi.
1938. *Miti e paradossi delli giustizia tributaria.* Turin: Luigi Einaudi.
The following handbooks are available:
1914. *Corso di scienza delle finanze.* Turin: Tip. e Bono.
1932–66. *Principi di scienza delle finanze.* Turin: La Riforma Sociale.
1932. *Il sistema tributario italiano.* Turin: La Riforma Sociale. With reference to the history of finance and the history of ideas see:
1908. *La finanza sabauda all'aprirsi del secolo XVIII e durante la guerra di successione spagnola.* Turin: Società Tip. Editrice Nazionale.
1927. *La guerra e il sistema tributario italiano.* Bari: Laterza.
1953. *Saggi bibliografici e storici intorno alle dottrine economiche.* Rome: Ediz. Storia e Litteratura. Einaudi's journalistic work has to a large extent been collected in 8 volumes comprising the *Cronache economiche e politiche di un trentennio* (1893–1925), Turin: Ed. Einaudi, 1959–65, and in *Lo scrittoio del Presidente 1948–1955*, Turin: Ed. Einaudi, 1956. For many years Einaudi was Italian correspondent for the *Economist*.

Einzig, Paul (1897–1973). Einzig was born in Brasov, Transylvania (Austria-Hungary). He was both a prolific, widely read author on international monetary topics and a renowned journalist. Educated in Hungary and France, he received his PhD from the University of Paris. In 1919, Einzig settled in the UK. Soon he became the Paris correspondent of the *Financial News* and was appointed its political editor in 1929. When the *Financial News* was bought by the *Financial Times*, Einzig became the political editor of the latter newspaper. Also he wrote the daily 'Lombard Street Column' during the mid- and late 1930s and many feature articles on currency questions. One of his top 'scoops' as a journalist was the revelation in 1943 of how the Swiss National Bank was buying looted gold from the Reichsbank on a huge scale. Already in 1939, Einzig's book *The Bloodless Invasion* had provided an original account of how Nazi Germany in its exchange rate policies exploited South-East Europe.

Einzig wrote more than 50 books on financial topics. Perhaps *A Dynamic Theory of Forward Exchange* (1961) is the best example of his powerful combination of economic, practical and historical knowledge. The book has a section describing the methods of intervention by central banks in forward exchange markets in the interwar period – and also by the Austrian and Russian central banks in the late 19th century. Einzig takes issue with the 'static theory of forward exchange' in which forward rates are shown as determined by given international interest rate differentials. He stresses that these themselves are influenced by speculation in the forward market. Einzig showed that except in the case of perfect arbitrage, forward markets have to be considered explicitly in an analysis of international short-term capital movements.

In *Primitive Money, in its Ethnological, Historical and Economic Aspects*, Einzig looks at how different commodities came to be used as money in primitive and ancient society. He refutes the hypothesis that money developed primarily through the progress of division of labour and the resulting complexity of trade, which made barter increasingly cumbersome. Much more important was the designation of a commodity for use in non-commercial payments (religious sacrifices, blood money, bride prices etc.).

Brendan Brown

SELECTED WORKS
1931. *Behind the Scenes of International Finance.* London: Macmillan.
1939. *The Bloodless Invasion.* London: Macmillan.
1940. *Europe in Chains.* London: Penguin.
1949. *Money in its Ethnological Aspects.* London: Macmillan.
1954. *Monetary Policy.* London: Penguin.
1960. *In the Centre of Things.* London: Hutchinson.
1961. *Theory of Foreign Exchange.* London: Macmillan.
1961. *A Dynamic Theory of Forward Exchange.* London: Macmillan.
1962. *A History of Foreign Exchange.* London: Macmillan.
1967. *Foreign Exchange Crisis.* London: Macmillan.

elasticities approach to the balance of payments. The substance of a theory is independent of the manner in which it is dressed. In particular, it is a matter of style only whether or not formulae are expressed in terms of elasticities of demand and supply, or in terms of ordinary derivatives. To speak of an 'elasticities approach' to the balance of payments is therefore to speak no sense at all.

However, behind the nonsensical label there hides a coherent and distinctive theory of what determines the response of a country's balance of payments to parametric changes in its rate of exchange, that is, to changes in the terms on which its currency exchanges for other currencies. The theory goes back to a paper published by Charles Bickerdike (1920).

Consider a simplified world containing just two countries (the 'home' country and the 'foreign') and producing and trading just two commodities. Let R be the price of foreign currency in terms of home currency, let p_i be the home price of the ith commodity in terms of home currency (so that, in arbitrage equilibrium, $p_i^* \equiv p_i/R$ is the foreign price of the commodity in terms of foreign currency), and let B be the home balance of trade in terms of foreign currency. Then, writing $z_i(p_i)$ and $z_i^*(p_i^*)$ as the home and foreign excess demands for the ith commodity, Bickerdike's model of the balance of payments reduces to the system of three equations

$$z_i(p_i) + z_i^*(p_i/R) = 0 \qquad (i = 1,2)$$
$$B = -(1/R)[p_1 z_1(p_1) + p_2 z_2(p_2)] \qquad (1)$$

In this system the rate of exchange R is treated as a parameter and p_1, p_2 and B as variables to be determined. Differentiating (1) with respect to R, solving for dB and the dp_i, and converting to elasticities, we obtain

$$dB = \left\{ -p_2^* z_2^* \left[\frac{\eta_1^*(1+\eta_1)}{\eta_1^* - \eta_1} - \frac{\eta_2^*(1+\eta_2)}{\eta_2^* - \eta_2} \right] - B \right\} \frac{dR}{R} \qquad (2)$$

and

$$\frac{dp_i}{p_i} = \frac{\eta_i^*}{\eta_i^* - \eta_i} \frac{dR}{R}, \qquad i = 1, 2 \qquad (3)$$

where $\eta_i \equiv (dz_i/dp_i)(p_i/z_i)$ and $\eta_i^* \equiv (dz_i^*/dp_i^*)(p_i^*/z_i^*)$. In the special case in which B is initially zero, (2) takes the simpler form

$$dB = -p_2^* z_2^* \left[\frac{\eta_1^*(1+\eta_1)}{\eta_1^* - \eta_1} - \frac{\eta_2^*(1+\eta_2)}{\eta_2^* - \eta_2} \right] \frac{dR}{R}. \qquad (2')$$

Equation (2) is often referred to as the Bickerdike–Robinson–Metzler formula: however, the role of Robinson (1947) and of Metzler (1949) was that of expositor only.

Suppose for concreteness that the home country exports the first commodity and imports the second, so that η_1 and η_2^* are export-supply elasticities and η_2 and η_1^* import-demand elasticities. Suppose further that all marginal propensities to buy are positive, so that η_1 and η_2^* are positive, η_2 and η_1^* negative. Then for the balance of payments to improve in response to devaluation it suffices that the sum of the two import demand elasticities exceed one in magnitude, that is, that the Marshall–Lerner condition be satisfied. Thus equation (2') can be rewritten as

$$dB = -p_2^* z_2^* \left[\frac{\eta_1 \eta_2^*(1+\eta_1^*+\eta_2) - \eta_1^* \eta_2(1+\eta_1+\eta_2^*)}{(\eta_1 - \eta_1^*)(\eta_2 - \eta_2^*)} \right] \frac{dR}{R}$$

with all terms of known sign except $(1 + \eta_1^* + \eta_2)$. For a positive response of the balance of payments to devaluation it suffices also that the terms of trade improve, or at least that they not worsen. For changes in the terms of trade are indicated by changes in p_1/p_2 and, from equation (3),

$$\frac{d(p_1/p_2)}{p_1/p_2} = \frac{dp_1}{p_1} - \frac{dp_2}{p_2} = \left(\frac{\eta_1^*}{\eta_1^* - \eta_1} - \frac{\eta_2^*}{\eta_2^* - \eta_2} \right) \frac{dR}{R}$$

If this expression is non-negative then, from (2'), dB must be positive.

Bickerdike's theory is very special in that the excess demand for each commodity depends on the money price of that commodity only. Implicitly, all 'cross' price elasticities are set

equal to zero. For more general theories and, in particular, more general versions of (2'), the reader is referred to Negishi (1968), Kemp (1970), Dornbusch (1975) and Kyle (1978).

Murray C. Kemp

See also ABSORPTION APPROACH TO THE BALANCE OF PAYMENTS; MONETARY APPROACH TO THE BALANCE OF PAYMENTS.

BIBLIOGRAPHY

Bickerdike, C.F. 1920. The instability of foreign exchange. *Economic Journal* 30, March, 118–22.

Dornbusch, R. 1975. Exchange rates and fiscal policy in a popular model of international trade. *American Economic Review* 65(5), December, 859–71.

Kemp, M.C. 1970. The balance of payments and the terms of trade in relation to financial controls. *Review of Economic Studies* 37(1), January, 25–31.

Kyle, J.F. 1978. Financial assets, non-traded goods and devaluation. *Review of Economic Studies* 45(1), February, 155–63.

Metzler, L.A. 1949. The theory of international trade. In *A Survey of Contemporary Economics*, ed. H.S. Ellis, Philadelphia: Blakiston.

Negishi, T. 1968. Approaches to the analysis of devaluation. *International Economic Review* 9, June, 218–27.

Robinson, J. 1947. *Essays in the Theory of Employment*. 2nd edn, Oxford: Basil Blackwell.

elasticity. One day in the winter of 1881–2 Alfred Marshall came down from the sunny rooftop of his hotel in Palermo 'highly delighted', for he had just invented elasticity of demand (Keynes, 1925, pp. 39,n3, 45,n2). So delighted was he that within a mere four years he had introduced the word *elasticity* into the technical literature of economics (Marshall, 1885), which by his own standards was rushing pell-mell into print. But if the speed of its introduction was uncharacteristic the manner of it was not, tucked away as it was at the end of a lecture dull even for its time, and giving no hint that elasticity was new and exciting (ibid, p. 187).

The notion that demand varies less or more than price can of course be found rather often in classical economics, especially in John Stuart Mill (Edgeworth, 1894, p. 691). But to turn that trite idea into something useful requires a firm grip on the prior idea of quantity demanded *at a price*. So it is not surprising that the only ancient who came close to Marshall's idea was Cournot himself, the inventor of (among much else) the demand function.

In fact Cournot came so close that it is hard to understand, first, why he did not go all the way, and second, why Marshall gave him no credit for showing that way. Such lack of generosity is the more puzzling since we know that between the time when (according to Mrs Marshall) he invented elasticity, and the late spring of 1882 when he first drafted the chapter on Elasticity for the *Principles*, Marshall reread Cournot (Whitaker, 1975, Volume I, p. 85).

Starting with the demand function $D = F(p)$, Cournot pointed out that $pF(p)$ is total revenue, so that for maximum revenue the price p must be such that $F(p) + pF'(p) = 0$ (1838, p. 56). Thus total revenue will increase or decrease with increase in price according as $\Delta D/\Delta p$ is larger or smaller than D/p, where ΔD is the absolute value of the change in quantity demanded.

Commercial statistics should therefore be required to separate articles of high economic importance into two categories, according as their current prices are above or below the value which makes a maximum of $pF(p)$. We shall see that many economic problems have different solutions, according as the article in question belongs to one or other of these two categories. (Bacon's translation, 1897, p. 54).

Let f be a real-valued nonzero differentiable function whose domain is some open interval I of the real line. In conformity with Marshall's Mathematical Appendix (1890, Note IV, pp. 738–40), the *elasticity of f at the point x*, denoted by $\eta_f(x)$, is defined here to be the *number* $xf'(x)/f(x)$. The *function* η_f defined by this formula is called the *elasticity of f*. To define the elasticity of *demand*, some authors prefer to follow the convention $f(x) = -xf'(x)/f(x)$, which is not used here. Unfortunately there is no standard notation for elasticity, since the obvious candidates are already taken, e for e and E for the expectations operator.

Cournot's critical value of p, his criterion for sorting out commodities, is simply that p^* for which $\eta_f(p^*) = -1$; he was close indeed. However, unlike Marshall (who is crystal clear on the point) there is no trace in Cournot of the crucial property that the elasticity measure is *invariant* to changes in units of measurement of quantities and prices, and it is this property alone that makes it so important in pure and applied economics.

A little calculus will prove such invariance, but is more enlightening to apply the dimensional analysis of Jevons and Wicksteed. Let the dimension of x be X and that of $f(x) = y$ be Y, so that $f'(x)$ has dimension YX^{-1}. The dimension of $\eta_f(x)$ is then $X \cdot YX^{-1} \cdot Y^{-1}$ and everything cancels. The elasticity of f at x is a pure number, unaffected by change in the units of either x or y. (This application is so obvious that the most plausible explanation of why it was not included in Wicksteed (1894) is that his entry was actually written before Marshall's *Principles* appeared.) Although invariance to transformation of units is the key property of elasticities, partly as a consequence the measure has a number of other agreeable properties. For example, it is easily seen that $\eta_f(x) = d \log f(x)/d \log x$, which paves the way for a whole calculus of elasticities in terms of logarithmic derivatives (Champernowne, 1935; Allen, 1938, pp. 251–4). One simple application of this calculus is the formula $\eta_{fg}(x) = \eta_f(x) + \eta_g(x)$, where fg is the product of f and g (with a corresponding formula for the quotient function f/g), while another is the characterization of constant elasticity functions as those which are linear in logarithms, i.e. of Wicksell–Cobb–Douglas type. Incidentally, Douglas's paper of 1927 was apparently intended to introduce elasticity of supply, which is odd since it had already appeared twenty years before (and rather late at that) in the Fifth Edition of the *Principles* (see Marshall, 1961, Vol. II, p. 521).

The extension of elasticity to functions of more than one variable is easy—one simply uses the partial derivatives f_i rather than the derivative f'—and is staple fare in textbooks (see e.g. Allen, 1938, pp. 310–12). However, many of those textbooks underplay another useful property of elasticities of strictly monotonic functions (such as the usual demand and supply curves) which follows from the inverse function theorem. Considering just functions of one variable, if we write $\Phi = f^{-1}$ then from that theorem $\Phi' = f^{-1}$, so from this and the definition of elasticity,

$$\eta_{\Phi}(y) = y\Phi'(y)/\Phi(y) = f(x)/xf'(x) = (\eta_f(x)^{-1}),$$

i.e. the elasticity of the inverse function is the inverse of the elasticity. Two obvious applications of this to the elementary theory of the firm are:

(i) Since the revenue function is $R(q) = pq = q\Phi(q)$,

marginal revenue $(mr) = \Phi(q) + q\Phi'(q)$

$$= \Phi(q)[1 + (q\Phi'(q)/\Phi(q)] = \Phi(q)(1 + \eta_{\Phi}(q)),$$

from which one can derive the more usual but less intuitive formula $mr = p[1 + (1/\eta_f(p))]$; and (ii) since at the firm's profit maximizing output marginal cost $mc = mr$, the Lerner (1934) measure of monopoly power $(p - mc)/p$ may be written $[\Phi(q) - mr]/\Phi(q) = 1 - [\Phi(q)(1 + \eta_{\Phi}(q)]/\Phi(q) = -\eta_{\Phi}(q)$.

Arc elasticity, which is really ordinary elasticity with the index number problem thrown in, was introduced quite early by Dalton (1920, pp. 192–7). But the heyday of elasticities of all kinds came later, in the 1930s, so much so that it is small wonder that in the immediate post-war period Samuelson (1947, pp. 4–5) used elasticity statements to exemplify what he meant both by 'meaningful theorems' and by non-meaningful theorems in economics. A peculiar aspect of some of the elasticity measures introduced then was their definition not in terms of the properties of a given *function* f (as here), but rather as the ratio of proportionate change in one variable to proportionate change in another, allegedly causative, variable, without any explicit functional relationship intervening. Thus with Hicks's 'elasticity of expectations' (1939, p. 205) there *is* no 'expectation function' of which it is an elasticity, as that term is defined above. Similarly, although the elasticity of substitution (σ) invented by Hicks (1932) and Robinson (1933) immediately provoked many articles in response (e.g. Lerner, 1933), at no time was a 'substitution function' introduced whose elasticity it was. The lack of a generating function for σ might help to explain why its use often occasions technical difficulty.

It is of some interest to apply duality theory to the problem of deriving simple formulas for entities like σ (cf. Woodland, 1982, p. 31). Consider the elasticity of substitution σ between two consumer's goods x and y, with no restriction being placed on preferences apart from the smoothness conditions implicit at this level of analysis. First, take advantage of homogeneity in both the ordinary and compensated demand functions to write the former function as $f(p, m)$ and the latter as $h(p, t)$, where p is the price of x in terms of y, m is the consumer's income in terms of y, and t is the *maximized* level of utility for the price-income situation (p, m). Put $x^* = f(p, m)$. Finally, observe that σ is wholly determined by the price slope corresponding to p together with the indifference curve corresponding to t, so that we may write $\sigma = \sigma(p, t)$.

From a modern version of the Fundamental Equation of Value Theory (Hicks, 1939, p. 309),

$$f_p(p, m) = h_p(p, t) - x^* f_m(p, m) \qquad (1)$$

where f_p, h_p and f_m are, in sequence, the partial derivatives of f and h with respect to p, and of f with respect to m. Multiplying (1) by $p/f(p, m)$ and writing η_{fp}, η_{fm} for the two partial elasticities of f, we obtain

$$\eta_{fp}(p, m) = ph_p(p, t)/x^* - px^* mf_m(p, m)/(mf(p, m))$$
$$= ph_p(p, t)/x^* - k\eta_{fm}(p, m) \qquad (2)$$

where $k = px^*/m$, i.e. the fraction of m spent on x. Now since t is the maximized level of utility, given local non-satiation $x^* = h(p, t)$ (see COST MINIMIZATION AND UTILITY MAXIMIZATION). Hence, the first term on the right-hand side of (2) is $\eta_{hp}(p, t)$, the partial elasticity of h with respect to p, and (2) becomes

$$\eta_{fp}(p, m) = \eta_{hp}(p, t) - k\eta_{fm}(p, m) \qquad (3)$$

A standard result of Hicks and Allen (1934; see Hicks, 1981,

p. 20) for the two-good case can be written in the present notation as

$$-\eta_{fp}(p, m) = k\eta_{fh}(p, m) + (1 - k)\sigma(p, t) \qquad (4)$$

so from (3) and (4),

$$(k - 1)\sigma(p, t) = \eta_{hp}(p, t) \qquad (5)$$

Let the cost (expenditure) function for this problem be $c(p, t)$, and denote its partial derivative with respect to p by c_p. Then, writing η_{cpp} for the partial elasticity of c_p with respect to p, since Shephard's Lemma implies $c_p = h$ we have

$$\eta_{hp}(p, t) = \eta_{cpp}(p, t) \qquad (6)$$

Now $k = px^*/m = ph(p, m)/m = pc_p(p, t)/m$. Because t is the maximized level of utility $m = c(p, t)$, so $k = pc_p(p, t)/c(p, t) = \eta_{cp}(p, t)$, where η_{cp} is the partial elasticity of c with respect to p. Substituting from this and (6) into (5),

$$\sigma(p, t) = \eta_{cpp}(p, t)/(\eta_{cp}(p, t) - 1) \qquad (7)$$

Thus the elasticity of substitution in this two-good case can be expressed entirely in terms of the cost function.

PETER NEWMAN

See also MARSHALL, ALFRED.

BIBLIOGRAPHY

Allen, R.G.D. 1938. *Mathematical Analysis for Economists*. London: Macmillan.
Champernowne, D.G. 1935. A mathematical note on substitution. *Economic Journal* 15, 246–58.
Cournot, A.A. 1838. *Recherches sur les principes mathématiques de la théorie des richesses*. Paris: Hachette. New edn, ed. G. Lutfalla, Paris: Rivière, 1938. English trans. by N.T. Bacon, 1897. Reprinted, New York: A.M. Kelley, 1960.
Dalton, H. 1920. *Some Aspects of the Inequality of Incomes in Modern Communities*. London: Routledge.
Douglas, P.H. 1927. Elasticity of supply as a determinant of distribution. In *Economic Essays Contributed in Honor of John Bates Clark*, ed. J.H. Hollander, New York: Macmillan.
Edgeworth, F.Y. 1894. Elasticity. In *Dictionary of Political Economy*, Vol. I, ed. R.H.I. Palgrave, London: Macmillan, 691.
Hicks, J.R. 1932. *The Theory of Wages*. London: Macmillan.
Hicks, J.R. 1939. *Value and Capital*. Oxford: Clarendon Press.
Hicks, J.R. 1981. *Collected Essays on Economic Theory*. Vol. I: Wealth and Welfare. Cambridge, Mass.: Harvard University Press.
Hicks, J.R. and Allen, R.G.D. 1933. A reconsideration of the theory of value. *Economica* 1, 52–76, 196–219. Reprinted in Hicks (1981, Volume I).
Keynes, J.M. 1925. Alfred Marshall, 1842–1924. In Pigou (1925).
Lerner, A.P. 1933. The diagrammatical representation of elasticity of substitution. *Review of Economic Studies*. Reprinted in Lerner (1953).
Lerner, A.P. 1934. The concept of monopoly and the measurement of monopoly power. *Review of Economic Studies* 1, June, 157–75. Reprinted in Lerner (1953).
Lerner, A.P. 1953. *Essays in Economic Analysis*. London: Macmillan.
Marshall, A. 1885. The graphic method of statistics. *Journal of the Royal Statistical Society*. Reprinted in Pigou (1925).
Marshall, A. 1890. *Principles of Economics*, Vol. I. London: Macmillan.
Marshall, A. 1961. *Principles of Economics*. 9th (Variorum) edn, with annotations by C.W. Guillebaud. 2 vols, London: Macmillan.
Pigou, A.C. (ed.) 1925. *Memorials of Alfred Marshall*. London: Macmillan. Reprinted, New York: A.M. Kelley, 1966.
Robinson, J.V. 1933. *The Economics of Imperfect Competition*. London: Macmillan.
Samuelson, P.A. 1947. *Foundations of Economic Analysis*. Cambridge, Mass.: Harvard University Press.
Whitaker, J.K. (ed.) 1975. *The Early Economic Writings of Alfred Marshall, 1867–1890*. 2 vols, New York: Free Press.
Wicksteed, P.H. 1894. Dimensions of economic quantities. In *Dictionary of Political Economy*, Vol. I, ed. R.H.I. Palgrave, London: Macmillan, 583–5.
Woodland, A.D. 1982. *International Trade and Resource Allocation*. Amsterdam: North-Holland.

elasticity of substitution. The concept of the elasticity of substitution, developed by Joan Robinson and John Hicks separately in the 1930s, represented an important addition to the marginal theory of the 1870s, in the tradition of Marshall, Edgeworth and Pareto. It brought together two concepts which were already well established in the literature – the ideas of elasticities (which derive from Mill) and those of substitution (which go back to Smith). The relationship defined by the concept is a mathematical one relating to utility and production functions, with considerable economic implications. It has two applications: to the theory of production, and in particular the isoquant relationship between factor inputs, and to consumer behaviour and the indifference curve. Let us look at each in turn.

The two inventors of the concept – Joan Robinson, in her *Economics of Imperfect Competition* (1933), and John Hicks in his *Theory of Wages* (1932) – each developed Marshall's formula for the elasticity of derived demand. Each defined the concept somewhat differently. For Hicks, the definition was the percentage change in the relative amount of the factors employed resulting from a given percentage change in the relative marginal products or relative prices, i.e. (following Samuelson, 1968):

$$\sigma = \sigma_{12} = (F_1 F_2 / FF_{12}) = \sigma_{21},$$

where $F(V_1, V_2)$ is a standard neoclassical production function, and the subscripts are the partial derivatives. This is sometimes called the direct elasticity of substitution. For Joan Robinson, on the other hand, concerned with relative shares and hence distributional issues, the elasticity of substitution was defined as 'the proportionate change in the ratio of the amounts of the factors employed divided by the proportionate change in the ratio of their prices' (p. 256):

$$\sigma = -\frac{\partial(V_1/V_2)/(V_1/V_2)}{\partial(W_1/W_2)/(W_1/W_2)}$$

where W_1 is the price of the V_1 factor.

These two definitions of the concept gave rise to a considerable debate in the early issues of the *Review of Economic Studies*, with in particular a notable contribution from Kahn (1933) concerned to identify how these concepts related to each other. It turns out that these two original definitions are identical when the production function is confined to two factors of production, where the partial derivatives of the production function are the marginal productivities of the factor inputs and yield the relevant factor prices. In addition, the contributors to the debate attempted to identify the implications of these somewhat abstract concepts. Amongst these were the joint determination by the elasticity of substitution and the factor supplies of the relative shares of the factor reward (wages and profits), and implications for the definition of imperfect competition with increasing returns to scale.

It is not surprising that it is with the cases where the restrictive neoclassical assumptions for the production function are not met that most interest arises. Two important developments are where production function involves three or more factors and in extending from Cobb–Douglas to Constant Elasticity of Substitution (CES) production functions. But although considerable emphasis has been placed on the elasticity of substitution in production, it remains a technical concept concerning factor substitutability. It has no direct allocation consequence. Diminishing elasticity of substitution does not imply diminishing returns to scale, since for returns we must have prices. Thus it is restricted to describing the technical conditions of production. But, being a technical concept, it can be generalized to all forms of transformation. Thus as we noted above, along with a number of other concepts, these tools developed for production were taken over to consumer theory. Because of the implications the concept had for the development of consumer behaviour, and because of the insight which the resulting difficulties threw up concerning the concept more generally, this application is of special interest.

It was Hicks (and Allen) who made that step. While Joan Robinson's development of the concept was closely related to her extension of Marshall's theory of the industry, Hicks was familiar with a very different approach to value theory, that of Edgeworth, Pareto and Walras. While Joan Robinson had focused on production substitutions, and hence isoquants, Hicks took the idea developed in that domain, and translated it across to consumer theory, and to the indifference curves which he had got from Edgeworth. In the two goods case, price elasticity could be represented in terms of his fundamental formula, according to which:

$$\text{Price elasticity} = k \text{ (income elasticity)} + (1 - k) \text{ (e.s.)}$$

where k is the total expenditure that is spent on the commodity. Thus, with income elasticity, consumer theory led into a representation of the effect of a price change in terms of the income and substitution effects, with elasticity being thus of prime importance in classifying goods by their demand characteristics.

But whereas the elasticity concept in production theory naturally led on to the possibility of measurement, that step in consumer theory was more contentious. For although this technical concept represented one important step in the development of the marginalist approach to the theory of value, the theory of demand behaviour requires a behavioural theory of choice. The elasticity of substitution with respect to the indifference curve is one technical component. But, as with production theory, prices, and in this case the budget line, are also required.

Technical concepts thus aided the formulation of modern consumer theory as outlined in Hicks and Allen's 'A reconsideration of the theory of value' (1934) and the opening chapters of *Value and Capital* (1939), a path from which it has scarcely deviated. But despite the mathematical elegance of this construction, it may be argued that it disguised many of the important underlying questions. The increased power of the indifference curve analysis begged the question of whether consumer preferences could in reality be represented in this abstract way. Ultimately, whether consumer behaviour is well described by concepts like the elasticity of substitution, depends upon whether preferences can be represented by complete, transitive, utility functions. Much recent evidence from psychologists and decision theorists suggests otherwise. Likewise for production theory, the concepts of capital and labour may be themselves ambiguous.

D.R. HELM

See also CES PRODUCTION FUNCTION; COBB–DOUGLAS FUNCTIONS; PRODUCTION FUNCTIONS.

BIBLIOGRAPHY
Hicks, J.R. 1932. *The Theory of Wages*. London: Macmillan.
Hicks, J.R. 1939. *Value and Capital*. Oxford: Clarendon Press.
Hicks, J.R. 1970. Elasticity of substitution again: substitutes and complements. *Oxford Economic Papers* 22(3), November, 289–96.
Hicks, J.R. and Allen, R.G.D. 1934. A reconsideration of the theory of value I–II. *Economica* 1, Pt 1, February, 52–76; Pt II, May, 196–219.
Kahn, R.F. 1933. The elasticity of substitution and the relative share of a factor. *Review of Economic Studies* 1, October, 72–8.
Robinson, J. 1933. *The Economics of Imperfect Competition*. London: Macmillan.
Samuelson, P.A. 1968. Two generalizations of the elasticity of substitution. In *Value, Capital and Growth*, ed. J.N. Wolfe, Edinburgh: Edinburgh University Press.

elections. *See* VOTING.

Ellet, Charles, Jr. (1810–1862). American engineer and economic theorist, Ellet was born on 1 January 1810 at Penn's Manor, Pennsylvania, and died on 21 June 1862, a victim of the Civil War. Ellet grew up on a family farm but showed little inclination for agriculture: at age 17 he joined a surveying crew. With no formal education or training, he soon became an assistant engineer to Benjamin Wright, chief engineer of the Chesapeake and Ohio Canal. With ability and hard work Ellet taught himself mathematics and French, earning the respect of influential engineers. Letters of introduction to Lafayette and the American ambassador helped secure Ellet a place at the Ecole des Ponts et Chaussées, Dupuit's alma mater, in 1830. On his return to America in 1832 Ellet became the premier suspension bridge designer in America, building in 1849 the (then) longest suspension bridge in the world across the Ohio River at Wheeling. Colonel Ellet designed, constructed and commanded the ram fleet of the Union forces at the Naval Battle at Memphis, Tennessee. He died as a result of a wound received in the heat of that battle.

Ellet spent most of his professional life as an engineer, but in one major work and in a number of contributions to the *Journal of the Franklin Institute* between 1840 and 1844, he significantly advanced the economic theory of monopoly, input selection, spatial economics, benefit-cost theory and econometric estimation. All Ellet's contributions were facilitated by the use of the differential calculus, which permitted him to express the simple theory of the firm, and some of its extensions, in mathematical terms. In his *Essay on the Laws of Trade* (1839) Ellet established the demand curve for a monopoly railroad with distance as a variable. Utilizing first order conditions and solving for the gross toll on passenger traffic, Ellet demonstrated that the profit-maximizing toll would be equal to one–half the costs of transportation added to a constant quantity, a well–known result.

Ellet considered not one monopoly model but a multiplicity of them, including those dealing with freight transport, duopoly conditions and the principles of monopoly price discrimination. Further, Ellet's particular insights into simple and discriminatory pricing systems led him to provide, with distance as a variable, an amazingly complete mathematical and graphical analysis of the impact of changes in the pricing system upon the market area served by a profit–maximizing railroad (1840a). In this important contribution to market area analysis Ellet argued that a set of (constrained) discriminatory tolls inverse to distance, in contrast to tolls proportional to distance, could be devised whereby all interested parties (management, shippers, the State) could all be made better off.

In a series of papers (1842–4) Ellet extended his theoretical analysis of inputs and input selection (1839) to one of the earliest attempts to develop, empirically specify and test a theoretical cost function. Utilizing a 'law' of costs which included his selected determinants of annual total railway costs, Ellet estimated the empirical dimensions from data collected from the mid–1830s. He then reaffirmed the power of his initial equation with new and supplementary data.

In all, the calibre and completeness of Ellet's theoretical and empirical inventions would not compare unfavourably with those of von Thünen, Cournot, Dupuit or Lardner. Ellet, who was primarily an engineer, was America's best representative among the pioneer contributors to scientifically oriented economics in the 19th century.

ROBERT B. EKELUND, JR.

SELECTED WORKS

1839. *An Essay on the Laws of Trade in Reference to the Works of Internal Improvement in the United States.* Richmond. Reprint from the 1st edn, New York: Augustus Kelley, 1966.
1840a. The laws of trade applied to the determination of the most advantageous fare for passengers on railroads. *Journal of the Franklin Institute.*
1840b. A popular exposition of the incorrectness of the tariffs of tolls in use on the public improvements of the United States. *Journal of the Franklin Institute.*
1842–4. Cost of transportation on railways. *Journal of the Franklin Institute*, various issues.

BIBLIOGRAPHY

Baumol, W. and Goldfeld, S.M. (eds) 1968. *Precursors in Mathematical Economics: An Anthology.* London: London School of Economics and Political Science.
Calsoyas, C.D. 1950 The mathematical theory of monopoly in 1839: Charles Ellet, Jr. *Journal of Political Economy* 58, April, 162–70.
Ekelund, R.B., Jr. and Hooks, D. 1972. Joint demand, discriminating two-part tariffs and location theory: an early American contribution. *Western Economic Journal* 10(1), March, 84–94.
Viner, J. 1958. *The Long View and the Short: Studies in Economic Theory and Policy.* New York: Glencoe.

Ely, Richard Theodore (1854–1943). Ely was born in Ripley, New York, on 13 April 1854 and died at Old Lyme, Connecticut, on 4 October 1943.

Ely's long and vigorous career epitomizes the general proposition that an economist can exert a major constructive influence on his subject and profession even though his original contribution to economic theory is negligible. A highly effective teacher and maker of careers for his former students; prolific author of popular articles, scholarly volumes, and publications series; organizer and fund-raiser for major research projects; founder of various academic institutes and associations; leader or participant in numerous reform societies; and centre of innumerable controversies, Ely was the most widely known, even notorious, economist in the USA around the turn of the century.

After a brief spell as a country schoolteacher and a preliminary year at Dartmouth College, Ely graduated from Columbia College in 1876 and was awarded a three-year fellowship to study philosophy in Germany. He soon switched to political economy, came under the influence of Karl Knies at Heidelberg, where he obtained a PhD summa cum laude, in 1878, and later attended Adolph Wagner's lectures in Berlin. Returning to the USA he was unemployed for over a year before his appointment, initially on a half-time basis, at Johns Hopkins, where he taught from 1881 to 1892. He then moved to Wisconsin, founding an outstanding school of Economics, Political Science and History including such luminaries as F.J. Turner, E.A. Ross, and J.R. Commons. A unique collaboration developed between the social scientists and the state legislators, especially under the La Follette governorship, which pioneered major social and economic reform legislation. In 1925 Ely took his Institute for Research in Land Economics and Public Utilities, founded in 1920, from Madison to Northwestern University, and remained there until 1932, when he launched a new, but impoverished Institute for Economic Research in New York City. Eventually hit by the depression, Ely was forced to depend on the support of friends and former students as he completed his autobiography and failed to complete a massive history of American economic thought initiated fifty years earlier.

An ardent Christian Socialist and outspoken critic of laissez-faire individualism and 'old school' English classical economics, Ely delighted social reformers and outraged conservatives by his writings on such controversial current topics as socialism and the American labour movement. Prone to emotional overstatement and careless in exposition, his public pronouncements and reputation frequently embarrassed the aspiring young professional economists with whom he founded the American Economic Association, in 1885, and for a time discouraged some moderate and conservative economists from joining. Although Ely's original draft prospectus had been rejected, and the Association's original constitution was toned down, and then dropped, the organization hovered uneasily between missionary evangelism and scholarly objectivity until he was obliged to relinquish his Secretaryship in 1892.

Two years later, at Wisconsin, Ely's fellow professionals rallied around him when he was denounced for preaching socialism and encouraging strikes, and although he was completely exonerated in a 'trial' that attracted national attention, Ely gradually became more conservative. Ironically, in the 1920s his Institute was attacked, no doubt unfairly, as a tool of the public utilities, and was referred to disparagingly in a report on professional ethics by a committee of the American Association of University Professors, in 1930.

During his long lifetime Ely wrote extensively on an extraordinarily wide variety of topics, often in a popular and journalistic fashion. Nevertheless he repeatedly opened up new research topics that were developed by his colleagues and former students – for example in labour history, state taxation, land economics, natural resources etc. – and his various textbooks, especially the multi-edition *Outlines of Economics* which sold 350,000 copies, were both widely used, and highly regarded.

At Wisconsin he helped to launch the American Association for Labor Legislation, of which he became President, and raised private resources to finance John R. Commons's massive *Documentary History of American Society* (11 vols, 1910–11). He served as President of the American Economic Association in 1900–1901.

Ely was a stimulating teacher whose ideas formed a direct link between the doctrines of the German historical school and American institutionalism, a link most clearly evident in his neglected two volume study of *Property and Contract in their Relations to the Distribution of Wealth* (1914). Many of his students went on to distinguished careers in academic and/or public life. He was undoubtedly an outstanding academic entrepreneur, and his contribution to the American Economic Association is recognized in its annual invited Richard T. Ely lecture, which was inaugurated in 1963.

A.W. COATS

SELECTED WORKS

1883. *French and German Socialism in Modern Times.* New York: Harper & Brothers. Reprinted, 1911.

1884a. *The Past and the Present of Political Economy.* Baltimore: N. Murray for Johns Hopkins.

1884b. *Recent American Socialism.* Baltimore: N. Murray for Johns Hopkins. Reprinted, 1885.

1886. *The Labour Movement in America.* New York: T.Y. Crowell Co. New edn, revised and enlarged, New York: Macmillan Co, 1905.

1888a. *Taxation in American States and Cities.* New York: T.Y. Crowell Co.

1888b. *Social Aspects of Christianity.* Boston: W.L. Greene and Co.

1889a. *An Introduction to Political Economy.* New York: Chautauqua Press. New and revised edn, New York: Eaton and Mains; Cincinnati: Jennings and Pye, 1901.

1889b. *Social Aspects of Christianity and other Essays.* New York: T.Y. Crowell Co. Reprinted, 1895.

1893. *Outlines of Economics.* Meadville, Pennsylvania and New York: Flood and Vincent. 6th edn with Ralph Hess, New York: Macmillan Co., 1938.

1894. *Socialism: An Examination of its Nature, its Strength, its Weakness. With Suggestions for Social Reform.* London: S. Sonnenschein Co.; New York and Boston: T.Y. Crowell Co.

1900. *Monopolies and Trusts.* New York: Macmillan Co. Reprinted, 1912.

1903. *Studies in the Evolution of Industrial Society.* New York and London: Macmillan Co. Reprinted, 1918.

1914. *Property and Contract in their Relations to the Distribution of Wealth.* 2 vols. New York: Macmillan Co. Reprinted, 1922.

1924. (With Edward W. Morehouse.) *Elements of Land Economics.* New York: Macmillan Co. Reprinted, 1932.

1928. (With George S. Wehrwein.) *Land Economics.* Ann Arbor, Michigan: Edwards Bros. Revised edn, Madison, Wisconsin: University of Wisconsin Press, 1964.

1938. *Ground Under Our Feet: an Autobiography.* New York: Macmillan Co.

BIBLIOGRAPHY

Rader, B.G. 1966. *The Academic Mind and Reform: the influence of Richard T. Ely in American life.* Lexington: University of Kentucky Press.

embargo. *See* ECONOMIC WAR.

emigration. *See* INTERNATIONAL MIGRATION.

empiricism. *See* ENGLISH HISTORICAL SCHOOL; GERMAN HISTORICAL SCHOOL; METHODENSTREIT; METHODOLOGY; PHILOSOPHY AND ECONOMICS.

employer. *See* PRINCIPAL AND AGENT.

employment, full. *See* FULL EMPLOYMENT.

employment, theories of. Theories of employment are actually concerned with involuntary *unemployment*. They deal with the definition, nature, and causes of such unemployment, and also with economic policies to reduce or alleviate it. They consider such questions as:

How serious is the problem of involuntary unemployment, both in the short and in the longer run?

Is such unemployment a feature of economic equilibrium, or is it an exclusively disequilibrium phenomenon?

Is there a 'natural' or 'normal' or 'non-inflation-accelerating' unemployment rate?

Is there a trade-off between unemployment and inflation rates? If so, what are the terms of trade-off, and are they stable?

Under what circumstances, if any, can assurance of long-term high employment be combined with assurance of price-level stability (or non-accelerating inflation)?

OPPOSED POSITIONS. Two basic and opposed positions of economists on employment theory may be summarized as follows:

(1) Applying standard supply-and-demand analysis to labour markets makes unemployment a disequilibrium phenomenon, resulting from the prevalence and persistence of real and money wage rates higher than the demand for labour will support. Its solution is the lowering of real wage rates to market-clearing levels, rather than any arbitrary removal of certain classes of workers from the labour supply. Public support for the unemployed may perhaps subsidize search for desirable jobs, but it should not subsidize withdrawal from the labour force. Intentional stimulation of labour demand, as by monetary expansion or fiscal deficits, is apt to kindle or accelerate inflation, and/or to raise interest rates and discourage investment. Limitation of labour-saving technical progress will slow economic growth at the expense of future generations.

(2) Unemployment results from equilibrium between aggregate supply and demand for the national output at a level too low to require the productive services of the full labour supply. The appropriate remedy is expansion of demand by fiscal and monetary measures – increased public spending, lower taxes, accelerated monetary growth, lower interest rates. Removal of particular groups from the labour market – youth, the elderly, secondary workers in families with employed breadwinners – and moratoria on labour-saving innovations may be legitimate devices to reduce unemployment in the short run, as may the export of unemployment by export subsidies and protection against imports.

Variants of the first of these positions – sometimes called *neoclassical,* but actually much older than the 'neoclassical revolution' of the 1870s – dominated economic orthodoxy in Western countries prior to the 1930s. Variants of the second position, articulated by John Maynard Keynes's *General Theory of Employment Interest and Money* (1936, esp. ch. 19) are called *Keynesian,* although none of the basic ideas was precisely new in 1936. (Marx, for example, had gone far beyond Keynes in regarding a 'reserve army of the unemployed' as highly functional in capitalism, its function being to hold wage rates at an established subsistence level.) But much as persistent depression unemployment threatened neoclassicism, persistent and accelerating inflation has later threatened Keynesianism. Revolts against Keynesian neo-orthodoxy have taken two opposite tacks; towards reformulated neoclassicism on one side, and towards 'incomes policies' of employment guarantees (with regulated prices and usually also wages) at the opposite end of the spectrum. The current (mid-1980s) situation is variously described as fluid, as chaotic, and as 'in shambles'. It cannot be called 'cut and dried'!

Conflict between neoclassicals and Keynesians is exacerbated by denunciation in each group of the other's policy proposals as dangerously harmful. To the confirmed neoclassicist, artificial demand stimulus, repeated and anticipated, soon raises society's (unofficial) 'discomfort index' by raising the inflation rate more than it lowers the measured unemployment rate. To the confirmed Keynesian, the immediate effect of any real or money wage cut is to deepen recession by shifting purchasing power from 'spenders' (the working class) to 'savers' (capitalists and corporate treasuries).

NEOCLASSICAL EMPLOYMENT THEORY. Mature neoclassical employment theory, as represented by A.C. Pigou's *Theory of Unemployment* (1933), draws its analysis from Alfred Marshall's *Principles of Economics* (1890). There had been little formal employment theory in Marshall himself, the most nearly relevant materials being the treatment of derived demand (Book V, ch. vi) with reference to the building trades, rather than the 'wages' chapters of Book VI. Book V includes Marshall's famous 'four laws' governing the extent to which a rise in the demand and price of an output (houses) causes a rise in the demand and wage of an input (building workers). In today's economic terminology and Marshall's order, these laws state that the rise in demand for the input will increase more, the lower the elasticity of substitution between that input and other inputs, the lower the elasticity of demand for the output, the less the importance of that input in the production process for the output, and the less the elasticities of supply of substitute inputs. This seems far removed from employment theory, but Pigou, in successive editions of his *Economics of Welfare* (1920, 1st edition entitled *Wealth and Welfare*, 1912) restated Marshall's laws as conditions under which workers might obtain higher wages, presumably after union organization, with minimum losses of employment. A full mathematical statement, combining all four laws in a single formula for the elasticity of the derived demand for a labour input, dates from J.R. Hicks's *Theory of Wages* (1932). Denoting by E, η, σ, e the respective elasticities of labour demand, output demand, substitution between labour and 'capital', and supply of 'capital', the Hicks equation is:

$$E = \frac{\sigma(\eta + e) + ke(\eta - \sigma)}{(\eta + e) + k(\eta - \sigma)}$$

with k representing the relative importance of labour in production as measured by the proportionate share of wage payments in total cost. (The equation ignores shifts of consumer demand between more and less labour-intensive commodities.) From Hicks's equation, Marshall's laws follow immediately, with the possible exception of the third one on 'the importance of being unimportant'.

We come now to Pigou's *Theory of Unemployment*. Based on the Marshallian structure and appearing in mid-Depression, this volume is remembered chiefly as the fuse that lit Keynes's *General Theory*, but deserves a better fate. Much of it can be interpreted as standing Pigou's earlier argument on its head, so as to provide us an exposition of conditions under which employment can be *restored* most rapidly in a depression, and with *minimum* cuts in real wages. These conditions are embodied in devices to *raise* the elasticity of demand for labour. These several devices involve shifts in aggregate demand, private and (especially) public from what Pigou calls 'centres' where (1) σ is low to others where it is high, (2) from centres where η is low to others where it is high, (3) from centres where k is low to others where it is high – surely the most important quantitatively – and finally (4) from centres where e is low to others where it is high.

On the more aggregative plane, Pigou seems wavering and inconsistent in the light of fifty years' additional development of macroeconomic theory. Over the long term, he argues in the second chapter of Part V, increases in aggregate demand serve only to raise prices and wages without increasing employment. In the short run, however, they can be helpful if not carried too far. This apparently implies that Say's Law is valid as a long-run proposition, but inoperative in the short run.

Pigou's approach may seem naive in transferring microeconomic analysis to the macroeconomic plane, and in its failure to examine Say's Law more intensively than it does. But it is far from the labour- and union-bashing that neoclassical economic argumentation is often supposed to represent.

KEYNESIAN EMPLOYMENT THEORY. Despite its title, Lord Keynes's *General Theory* is a treatise on the macroeconomic theory of income determination. Its employment theory is confined largely to attacks on the Marshall–Pigou tradition, under the assumption (too obvious to require either stress or detailed development) that aggregate real income and the unemployment rate are inversely related under any given state of technology.

Strengthening and specifying of the Keynesian relation between income on the one hand, and employment and unemployment on the other, came only after advances in econometrics and in computer technology. A standard specification has been contributed by Arthur Okun. 'Okun's Law' or the 'Okun curve', as it is variously known, may be written in disguised differential-equation form. For example, let us denote by $\hat{U}$ the percentage change in the measured unemployment rate over a given period and by $\hat{Y}$ the percentage change in a real-income measure like deflated GNP or GDP over the same period.

$$-\hat{U} = a(\hat{Y} - \bar{Y}_0) \quad \text{or} \quad \hat{U} + a(\bar{Y} - \hat{Y}_0) = 0$$

Here a is a statistical parameter, while $\bar{Y}_0$ is the income growth rate estimated to be required if the unemployment rate is not to rise. Neither real nor nominal wage rate movements are taken into explicit account. (The estimates of $\bar{Y}_0$ are based on regressions of high-employment points only, and the slope of this regression is called the country's potential growth rate. The area between such a regression and the country's actual growth path is sometimes called an 'Okun gap'.)

The $(a, \bar{Y}_0)$ values may of course vary widely, both across countries and over time. For the US 1949–60, Okun's estimate of a was about 0.3, and his estimate of $\bar{Y}_0$ about 3.75 per cent. The usefulness of the widely used Okun analysis is questionable, however, in the presence of supply shocks, wage 'explosions', and similar disturbances. (A productivity jump, for example, might be expected to lower a and raise $\bar{Y}_0$. The combination of these shifts increases the income growth rate required to reduce an existing unemployment rate or to keep that rate from rising.)

Shifting from the empirical to the analytical side, a fruitful development of Keynesian unemployment theory was the distinction between high and low full employment, introduced by A.P. Lerner's *Economics of Employment* (1951, ch. 13). By 'low full employment' Lerner meant essentially what was later called NAIRU in the United Kingdom, namely the non-inflation-accelerating rate of unemployment. By 'high full employment', on the other hand, Lerner meant 100 per cent of the labour force, minus only the frictional lacunae consequent upon job changing. Lerner set low rather than high full employment as the preferred target of employment policy, and outlined a detailed (but probably impractical) scheme of wage controls for attaining and maintaining it. (Most professed Keynesians, more ambitious than Lerner, would strive for high full employment.)

The 'expository Keynesianism' of the textbooks stresses primarily the *shapes* and *parameters* of such relations as the consumption, marginal efficiency, and liquidity functions which determine the level of income in the Keynesian scheme. At a more advanced level, the emphasis appears to be changing, to stress rather the *volatility* of these functions as expectations fluctuate. Leijonhufvud (1968) calls the newer view 'the economics of Keynes' as distinguished from the 'Keynesian economics' of the textbooks. The change in emphasis may also result from the defence of Keynesian macroeconomics, with its employment-theory appendage, against its critics, whose argument can be paraphrased: 'The aggregate demand for the national output depends inversely on the general price level, as the demand for a single commodity depends upon its price. If human wants are insatiable at a zero price level, it follows that there exists a positive price (and likewise wage) level at which full-employment output can be absorbed.' (The critics could not, however, prove that the market-clearing set of full-employment wage rates was at or above 'subsistence', however defined.)

The Keynesian rebuttal, due largely to Clower (1965), introduced the distinction between actual and 'notional' demands for output and especially for labour. In a recession, the demand function for output is weaker than the 'notional' full-employment demand function would be. At the same time, the actual demand for labour is less than the notional one which would prevail were potential employers reasonably sure of selling full-employment output at profitable prices. The impasse or vicious circle could be broken when the demand for output (see Figure 1) could rise from *AD* to *AD'* by public policies which improved the state of confidence – without reference to the multiplier mechanism of the Keynesian textbooks. The equilibrium position could move from point *E*, possibly all the way to point *E'* at the full-employment income level *F*, with money wages remaining the same and with no need to press aggregate supply *AS* vertically downward to *AS'* as by a wage cut.

Now suppose, in the same recession, there was to be a complete 'hands-off' policy. As unemployment continued with nothing done about it beyond calls for wage reductions, notional aggregate demand would fall, perhaps as far as *AD''* if the policy were to lead to budget-balancing and monetary

contraction. Even if hard times and wage concessions from labour eventually forced aggregate supply to the *AS'* position, the result would be hyper-deflation rather than recovery. (The Hoover debacle of 1931–2 in the United States and the contemporaneous Brüning one in Germany are cases in point.) By the time aggregate supply had reached *AS'*, the equilibrium point *E''* would prevail, with output (and therefore employment) below those at *E*, even though deflationary cost-cutting would have restored high employment had aggregate demand remained at *AD*.

Equilibrium is then not unique. Depending upon the state of notional as well as actual demand, there are an infinite number of possible equilibria at as many levels of income and employment. We cannot be sure *a priori* that wage deflation will produce increases in employment.

CONCLUSION. The British Broadcasting Company featured in December 1944 a series of postwar-planning programmes entitled *Jobs for All*. These were inspired not only by Lord Keynes's *General Theory* but by the extensions and applications proposed by Michal Kalecki in the Oxford Institute of Statistics' *Economics of Full Employment* (Burchardt et al., 1944). (Kalecki, who would later assume temporary leadership of British Keynesianism after Keynes's own death in 1946, had proposed 'incomes policies' and income redistribution as preferable to either deficit finance or stimulation of private investment as a route to full employment.) A junior member of that Oxford team, and a speaker in the BBC *Jobs for All* series, was G. D. N. Worswick. In July 1984, the same G.D.N. Worswick, now Professor at Oxford and President of the Royal Economic Society, delivered his presidential address on 'Jobs for All?', later published in the *Economic Journal* (Worswick, 1985). It was the same subject, but note the question mark.

The substance of Worswick's address was that, unfortunately, that question mark belonged in his new title, and could not be expunged even after forty additional years of planning, theorizing, and experimentation. We quote from his final paragraph (p. 14):

> When it comes to action, [*The Economics of Full Employment*] was already too optimistic. We assumed that trade unions would readily accept some limitations on free collective bargaining as a small price to pay for ending unemployment. There was too little recognition that it is my restraint which is necessary to secure your employment. Is it possible to devise schemes which are not only of advantage for the national economy, or for workers as a whole, but can also be seen to be to the advantage ... of members of trade unions who are already in employment? This is a task for the new generation of economists to undertake. Until a lasting solution is found, the question mark after my title must remain.

M. BRONFENBRENNER

See also AGGREGATE DEMAND AND SUPPLY ANALYSIS; EFFECTIVE DEMAND; FULL EMPLOYMENT; INVOLUNTARY EMPLOYMENT; KEYNES, JOHN MAYNARD; KEYNES'S GENERAL THEORY; OUTPUT AND EMPLOYMENT; PHILLIPS CURVE.

BIBLIOGRAPHY

Burchardt, F.A., et al. 1944. *The Economics of Full Employment.* Oxford: Basil Blackwell.

Clower, R. 1965. The Keynesian counter-revolution, a theoretical appraisal. In *The Theory of Interest Rates*, ed. F. Hahn and F. Brechling. London: Macmillan.

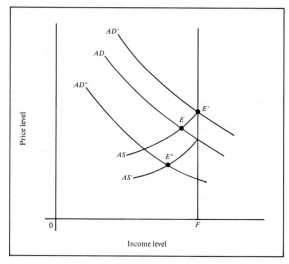

Figure 1   Investment and saving in a 2-period model

Hicks, J.R. 1932. *The Theory of Wages*. London: Macmillan.
Keynes, J.M. 1936. *The General Theory of Employment, Interest and Money*. London: Macmillan.
Leijonhufvud, A. 1968. *On Keynesian Economics and the Economics of Keynes*. New York, and Oxford: Oxford University Press.
Lerner, A.P. 1951. *Economics of Employment*. New York: McGraw-Hill.
Marshall, A. 1890. *Principles of Economics*. London: Macmillan.
Okun, A. 1970. *The Political Economy of Prosperity*. Washington, DC: Brookings Institution; New York: Norton.
Patinkin, D. 1956. *Money, Interest, and Prices*. Evanston, Ill.: Row, Peterson.
Pigou, A.C. 1920. *The Economics of Welfare*. London: Macmillan.
Pigou, A.C. 1933. *The Theory of Unemployment*. London: Macmillan.
Worswick, G.D.N. 1985. Jobs for all? *Economic Journal* 93, March, 1–14.

empty boxes. 'Empty Economic Boxes' is the title of a famous article in the *Economic Journal* of 1922 and a phrase which has subsequently entered the language of economics as a shorthand for 'abstract theory without practical relevance'. The paper was written by J.H. Clapham, the leading British economic historian of the interwar years and first Professor of Economic History at Cambridge University.

As an economic historian, Clapham showed a 'special predilection for hard and tangible facts' (Postan, 1946, p. 57). His classic text, *An Economic History of Modern Britain* (3 vols.: 1926, 1932, 1938) is notable for its pioneering insistence on presenting facts as far as possible in detailed quantitative form. Clapham is especially remembered for his statement on the methodology of economic history subsequently eagerly embraced by 'new economic historians':

> Every economic historian should, however, have acquired what might be called the statistical sense, the habit of asking in relation to any institution, policy, group or movement the questions: how large? how long? how often? how representative? (1931, p. 328).

Clapham was not (nor was any of his contemporaries), however, a new economic historian, in that his work was not characterized by the use of formal economic analysis or econometrics. His approach to economic history echoes his 1922 article in finding little of practical relevance in the economic theory of his day. In effect, Clapham revealed a preference for a shift in the balance of economists' research programmes away from pure, abstract theory towards collection of better economic data.

The 1922 article was directed particularly towards Pigou's (1920) proposals for taxation of decreasing returns and subsidy of increasing returns industries. Clapham argued that 'the Laws of Returns have never been attached to specific industries; that the boxes are, in fact, empty' (1922, p. 312). In effect, Clapham criticized the lack of evidence on long-run cost curves for both firm and industry and on learning effects. He also expressed scepticism about the prospects for empirical work in this area. Pigou's reply (1922) expressed the belief that evidence would be forthcoming and that, in any case, economic thinking was useful as a method of analysis in policy questions.

Ironically, Pigou's work gave rise to a considerable body of theoretical rather than empirical literature, which left relatively little of his initial proposals intact. Notable articles by Knight (1924), Robertson (1924), Viner (1931) and Ellis and Fellner (1943) made clear, for example, the distinctions between external diseconomies and transfer payments to fixed factors and between externalities from irreversible learning effects and minimum efficient scale on a given long-run average cost curve.

Already by the end of the interwar period Pigou's hopes for empirical research were starting to be fulfilled by pioneering investigations into production functions, cost curves and learning effects, which are reviewed in Walters (1963), Johnston (1960) and Alchian (1963). It should be said that econometric work has by now achieved far more than Clapham believed possible, and the results are seriously considered in the context of antitrust policy (e.g. Cmnd. 7198, 1978), if not in taxation policy. At the same time, industrial economics textbooks still offer very substantial reservations about the precision of available estimates (Hay and Morris, 1979, ch. 2).

In economic history also, there has been progress in applying Pigou's ideas as revised by the subsequent theoretical literature. By far the most interesting study is that of David (1975, ch. 2), which found econometric evidence of external economies from irreversible learning effects in the American cotton textile industry before 1825 and *also* that thereafter there was no justification for infant-industry protection.

David's paper is an excellent example of the 'new economic history', with its stress on use of theory and hypothesis testing. In fact, new economic history has generally used models based on mainstream neoclassical economics and, in general, this has undoubtedly proved fruitful – much more so than a sceptic like Clapham would have imagined. For example, discussions of the slave economy in the United States and entrepreneurial behaviour in late Victorian Britain have been substantially enriched.

Nevertheless, there is cause for concern about the one-way relationship which has developed in the past quarter-century between economics and economic history. There is a danger not so much of inevitably empty boxes as of forcing historical examples into particular boxes; that is, of operating with priors that are too tight. In particular, as McClelland (1975, p. 125) has emphasized, new economic historians should be wary of automatically believing that the marginal equivalences of the neoclassical model are tolerably achieved in all situations.

Perhaps, also, economists have more to learn from economic history than they seem presently to believe. Obviously, the past offers a much wider array of facts and institutions than the present, evidence which at present is underutilized. In addition, the study of history necessarily involves seeking to understand particular events, and exposure to the difficulties of this can give an interesting perspective on modern economic analysis. Economic historians like David would argue that the past is characterized by pervasive learning effects and technical interrelatedness so as to produce path-dependent sequences of economic changes. If in fact the increasing-returns box is as full as initial investigations in technological history suggest it could be, then critics of orthodox theory like Kaldor (1972) will find their positions strengthened.

Moreover, in such a world the past matters in ways that neoclassical theory ignores, and the balance of research in economic history should be different; for example, less emphasis in research on the rationality of individual entrepreneurs in Victorian Britain and more on the impact of an 'early start' on subsequent economic performance.

A modern-day Clapham would still say that we do not know enough about 'increasing returns', but rather than turning his back on the concept, he would surely insist on the importance of economic history in establishing how full this box is and would now recognize that to some extent the importance of economic history depends on the answer. He would also find common cause with applied economists and econometricians like Leontief (1971) and Hendry (1980) in wishing that more resources were devoted to gathering information on economies past and present.

N.F.R. CRAFTS

See also CLAPHAM, JOHN HAROLD; FIRM, THEORY OF THE; INCREASING RETURNS.

BIBLIOGRAPHY

Alchian, A. 1963. Reliability of progress curves in airframe production. *Econometrica* 31(4), 679–93.

Clapham, J.H. 1922. Of empty economic boxes. *Economic Journal* 32, 305–14.

Clapham, J.H. 1926. *An Economic History of Modern Britain*. Vol. 1: *The Early Railway Age*. Cambridge: Cambridge University Press.

Clapham, J.H. 1931. Economic history as a discipline. In *Encyclopaedia of the Social Sciences*, ed. E.R.A. Seligman and A. Johnson, London: Macmillan, Vol. 5.

Clapham, J.H. 1932. *An Economic History of Modern Britain*. Vol. 2: *Free Trade and Steel*. Cambridge: Cambridge University Press.

Clapham, J.H. 1938. *An Economic History of Modern Britain*. Vol. 3: *Machines and National Rivalries*. Cambridge: Cambridge University Press.

Command Paper no. 7198. 1978. *A Review of Monopolies and Mergers Policy*. London: HMSO.

David, P.A. 1975. *Technical Choice, Innovation and Economic Growth*. Cambridge: Cambridge University Press.

Ellis, H.S. and Fellner, W. 1943. External economies and diseconomies. *American Economic Review* 33(3), 493–511.

Hay, D.A. and Morris, D.J. 1979. *Industrial Economics: Theory and Evidence*. Oxford: Oxford University Press.

Hendry, D.F. 1980. Econometrics: alchemy or science? *Economica* 47(188), 387–406.

Johnston, J. 1960. *Statistical Cost Analysis*. New York: McGraw-Hill.

Kaldor, N. 1972. The irrelevance of equilibrium economics. *Economic Journal* 82, 1237–55.

Knight, F.H. 1924. Some fallacies in the interpretation of social cost. *Quarterly Journal of Economics* 38, 582–606.

Leontief, W. 1971. Theoretical assumptions and nonobserved facts. *American Economic Review* 61(1), 1–7.

McClelland, P.D. 1975. *Causal Explanation and Model Building in History, Economics and the New Economic History*. Ithaca: Cornell University Press.

Pigou, A.C. 1920. *The Economics of Welfare*. London: Macmillan.

Pigou, A.C. 1922. Empty economic boxes: a reply. *Economic Journal* 32, 458–65.

Robertson, D.H. 1924. Those empty boxes. *Economic Journal* 34, 16–30.

Viner, J. 1931. Cost curves and supply curves. *Zeitschrift für Nationalökonomie* 3, 23–46.

Walters, A.A. 1963. Production functions and cost functions: an econometric survey. *Econometrica* 31(1), 1–66.

enclosures. *See* COMMON PROPERTY RIGHTS; COMMON LAND; OPEN FIELD SYSTEM.

endogeneity and exogeneity. Endogeneity and exogeneity are properties of variables in economic or econometric models. The specification of these properties for respective variables is an essential component of the entire process of model specification. The words have an ambiguous meaning, for they have been applied in closely related but conceptually distinct ways, particularly in the specification of stochastic models. We consider in turn the case of deterministic and stochastic models, concentrating mainly on the latter.

A deterministic economic model typically specifies restrictions to be satisfied by a vector of variables **y**. These restrictions often incorporate a second vector of variables **x**, and the restrictions themselves may hold only if **x** itself satisfies certain restrictions. The model asserts

$$\forall \mathbf{x} \in R, \qquad G(\mathbf{x}, \mathbf{y}) = 0.$$

The variables **x** are exogenous and the variables **y** are endogenous. The defining distinction between **x** and **y** is that **y** may be (and generally is) restricted by **x**, but not conversely. This distinction is an essential part of the specification of the functioning of the model, as may be seen from the trivial model,

$$\forall \mathbf{x} \in \mathbf{R}^1, \qquad x + y = 0.$$

The condition $x + y = 0$ is symmetric in x and y; the further stipulation that x is exogenous and y is endogenous specifies that in the model x restricts y and not conversely, a property that cannot be derived from $x + y = 0$. In many instances the restrictions on **y** may *determine* **y**, at least for $\mathbf{x} \in R^* \subset R$, but the existence of a *unique* solution has no bearing on the endogeneity and exogeneity of the variables.

The formal distinction between endogeneity and exogeneity in econometric models was emphasized by the Cowles Commission in their pathbreaking work on the estimation of simultaneous economic relationships. The class of models they considered is contained in the specification

$$\mathbf{B}(L)\mathbf{y}(t) + \boldsymbol{\Gamma}(L)\mathbf{x}(t) = \mathbf{u}(t);$$

$$\mathbf{A}(L)\mathbf{u}(t) = \boldsymbol{\epsilon}(t);$$

$$\text{cov}[\boldsymbol{\epsilon}(t), \mathbf{y}(t - s)] = \mathbf{O}, \, s > O;$$

$$\text{cov}[\boldsymbol{\epsilon}(t), \mathbf{x}(t - s)] = \mathbf{O}, \, \text{all } s;$$

$$\boldsymbol{\epsilon}(t) \sim \text{IIDN}(\mathbf{O}, \boldsymbol{\Sigma}).$$

The vectors $\mathbf{x}(t)$ and $\mathbf{y}(t)$ are observed, whereas $\mathbf{u}(t)$ and $\boldsymbol{\epsilon}(t)$ are underlying disturbances not observed but affecting $\mathbf{y}(t)$. The lag operator L is defined by $L\mathbf{x}(t) = \mathbf{x}(t - 1)$; the roots of $|\mathbf{B}(L)|$ and $|\mathbf{A}(L)|$ are assumed to have modulus greater than 1, a stability condition guaranteeing the nonexplosive behaviour of **y** given any stable path for **x**. The Cowles Commission definition of exogeneity in this model (Koopmans and Hood, 1953, pp. 117–120) as set forth in Christ (1966, p. 156) is

> An exogenous variable in a stochastic model is a variable whose value in each period is statistically independent of the values of all the random disturbances in the model in all periods.

All other variables are endogenous. In the prototypical model set forth above **x** is exogenous and **y** is endogenous.

The Cowles Commission distinction between endogeneity and exogeneity applied to a specific class of models, with linear relationships and normally distributed disturbances. The exogenous variables **x** in the prototypical model have two important but quite distinct properties. First, the model may be solved to yield an expression for $\mathbf{y}(t)$ in terms of current and past values of **x** and ϵ,

$$\mathbf{y}(t) = \mathbf{B}(L)^{-1}\boldsymbol{\Gamma}(L)\mathbf{x}(t) + \mathbf{B}(L)^{-1}\mathbf{A}(L)\boldsymbol{\epsilon}(t).$$

Given suitably restricted $\mathbf{x}(t)$ (e.g. all **x** uniformly bounded, or being realizations of a stationary stochastic process with finite variance) it is natural to complete the model by specifying that it is valid for all **x** meeting the restrictions, and this is often done. The variables **x** are therefore exogenous here as **x** is exogenous in a deterministic economic model. A second, distinct property of these variables is that in estimation $\mathbf{x}(t)$ $(-\infty < t < \infty)$ may be regarded as fixed, thus extending to the environment of simultaneous equation models methods of statistical inference initially designed for experimental settings. It was generally recognized that exogeneity in the prototypical model was a sufficient but not a necessary condition to justify treating variables as fixed for purposes of inference. If $\mathbf{u}(t)$ in the model is serially independent (i.e. $\mathbf{A}(L) = \mathbf{I}$) then lagged values of $\mathbf{y}$ may also be treated as fixed for purposes of the model; this leads to the definition of 'predetermined variables' (Christ, 1966, p. 227) following Koopmans and Hood (1953, pp. 117–121):

> A variable is predetermined at time t if all its current and past values are independent of the vector of current disturbances in the model, and these disturbances are serially independent.

These two properties were not explicitly distinguished in the prototypical model (Koopmans, 1950; Koopmans and Hood, 1953) and tended to remain merged in the literature over the next quarter-century (e.g. Christ, 1966; Theil, 1971; Geweke, 1978). By the late 1970s there had developed a tension between the two, due to the increasing sophistication of estimation procedures in nonlinear models, treatment of rational expectations, and the explicit consideration of the respective dynamic properties of endogenous and exogenous variables (Sims, 1972, 1977; Geweke, 1982). Engle, Hendry and Richard (1983), drawing on this literature and discussions at the 1979 Warwick Summer Workshop, formalized the distinction of the two properties we have discussed. Drawing on their definitions 2.3 and 2.5 and the discussions in Sims (1977) and Geweke (1982), $\mathbf{x}$ is *model exogenous* if given $\{\mathbf{x}(t), t \leq T\} \in R(T)$ the model may restrict $\{\mathbf{y}(t), t \leq T\}$, but given

$$\{\mathbf{x}(t), t \leq T + J\} \in R(T + J)$$

there are no further restrictions on $\{\mathbf{y}(t), t \leq T\}$, for any $J > 0$. If the model in fact does restrict $\{\mathbf{y}(t), t \leq T\}$, then $\mathbf{y}$ is model endogenous. As examples consider

Model 1:

$$y(t) = ay(t - 1) + bx(t) + u(t),$$
$$x(t) = cx(t - 1) + v(t);$$

Model 2:

$$y(t) = ay(t - 1) + bx(t) + u(t),$$
$$x(t) = cx(t - 1) + dy(t) + v(t);$$

Model 3:

$$y(t) = ay(t - 1) + b\{x(t) + E[x(t)|x(t - s), s > 0]\} + u(t),$$
$$x(t) = cx(t - 1) + v(t).$$

In each case $u(t)$ and $v(t)$ are mutually and serially independent, and normally distributed. The parameters are assumed to satisfy the usual stability restrictions guaranteeing that x and y have normal distributions with finite variances. In all three models y is model endogenous, and x is model exogenous in Models 1 and 3 but not 2. For estimation the situation is different. In Model 1, treating $x(t)$, $x(t-1)$ and $y(t-1)$ as fixed simplifies inference at no cost; $y(t-1)$ is a classic predetermined variable in the sense of Koopmans and Hood (1953) and Christ (1966). Similarly in Model 2, $x(t-1)$ and $y(t-1)$ may be regarded as fixed for purposes of inference despite the fact that x and y are both model endogenous. When Model 3 is re-expressed

$$y(t) = ay(t - 1) + bx(t) + bcx(t - 1) + u(t),$$
$$x(t) = cx(t - 1) + v(t),$$

it is clear that $x(t)$ cannot be treated as fixed if the parameters are to be estimated efficiently since there are cross-equation restrictions involving the parameter c. Model exogeneity of a variable is thus neither a necessary nor a sufficient condition for treating that variable as fixed for purposes of inference.

The condition that a set of variables can be regarded as fixed for inference can be formalized, following Engle, Hendry and Richard (1983) along the lines given in Geweke (1984). Let

$$\mathbf{X} \equiv [\mathbf{x}(1), \ldots, \mathbf{x}(n)] \quad \text{and} \quad \mathbf{Y} \equiv [\mathbf{y}(1), \ldots, \mathbf{y}(n)]$$

be matrices of n observations on the variables $\mathbf{x}$ and $\mathbf{y}$ respectively. Suppose the likelihood function $L(\mathbf{X}, \mathbf{Y} | \boldsymbol{\Theta})$ can be reparameterized by $\lambda = F(\boldsymbol{\Theta})$ where F is a one-to-one transformation; $\lambda' = (\lambda_1, \lambda_2)'$, $(\lambda_1, \lambda_2) \in \Lambda_1 X \Lambda_2$; and the investigator's loss function depends on parameters of interest λ_1 but not nuisance parameters λ_2. Then $\mathbf{x}$ is *weakly exogenous* if

$$L(\mathbf{X}, \mathbf{Y} | \lambda_1, \lambda_2) = L_1(\mathbf{Y} | \mathbf{X}, \lambda, \lambda_1) \cdot L_2(\mathbf{X} | \lambda_2),$$

and in this case $\mathbf{y}$ is *weakly endogenous*. When this condition is met the expected loss function may be expressed using only $L_1(\mathbf{Y} | \mathbf{X}, \lambda_1)$, i.e. $\mathbf{x}$ may be regarded as fixed for purposes of inference.

The concepts of model exogeneity and weak exogeneity play important but distinct roles in the construction, estimation, and evaluation of econometric models. The dichotomy between variables that are model exogenous and model endogenous is a global property of a model, drawing in effect a logical distinction between the inputs of the model $\{\mathbf{x}(t), t \leq T\} \in R(T)$ and the set of variables restricted by the model $\{\mathbf{y}(t), t \leq T\}$. Since model exogeneity stipulates that $\{\mathbf{x}(t), t \leq T + J\}$ places no more restrictions on $\{\mathbf{y}(t), t \leq T\}$ than does $\{\mathbf{x}(t), t \leq T\}$, the global property of model exogeneity is in principle testable, either in the presence or absence of other restrictions imposed by the model. When conducted in the absence of most other restrictions this test is often termed a 'causality test' and its use as a test of specification was introduced by Sims (1972). The distinction between weakly exogenous and weakly endogenous variables permits a simplification of the likelihood function that depends on the subset of the model's parameters that are of interest to the investigator. It is a logical property of the model: the same results would be obtained using $L(\mathbf{X}, \mathbf{Y} | \lambda_1, \lambda_2)$ as using $L(\mathbf{Y} | \mathbf{X}, \lambda_1)$. The stipulation of weak exogeneity is therefore not, by itself, testable.

JOHN GEWEKE

See also CAUSAL INFERENCE; CAUSALITY IN ECONOMIC MODELS; IDENTIFICATION; SIMULTANEOUS EQUATIONS MODELS.

BIBLIOGRAPHY

Christ, C.F. 1966. *Econometric Models and Methods*. New York: Wiley.

Engle, R.F., Hendry, D.F. and Richard, J.-F. 1983. Exogeneity. *Econometrica* 51(2), March, 277–304.

Geweke, J. 1978. Testing the exogeneity specification in the complete dynamic simultaneous equation model. *Journal of Econometrics* 7(2), April, 163–85.

Geweke, J. 1982. Causality, exogeneity, and inference. Ch. 7 in *Advances in Econometrics*, ed. W. Hildenbrand, Cambridge: Cambridge University Press.

Geweke, J. 1984. Inference and causality. Ch. 19 in *Handbook of Econometrics*, Vol. 2, ed. Z. Griliches and M.D. Intriligator, Amsterdam: North-Holland.

Koopmans, T.C. 1950. When is an equation system complete for statistical purposes? In *Statistical Inference in Dynamic Economic Models*, ed. T.C. Koopmans, New York: Wiley.

Koopmans, T.C. and Hood, W.C. 1953. The estimation of simultaneous economic relationships. In *Studies in Econometric Method*, ed. W.C. Hood and T.C. Koopmans, New York: Wiley.

Sims, C.A. 1972. Money, income, and causality. *American Economic Review* 62(4), September, 540–52.

Sims, C.A. 1977. Exogeneity and causal ordering in macroeconomic models. In *New Methods in Business Cycle Research*, ed. C.A. Sims, Minneapolis: Federal Reserve Bank of Minneapolis.

Theil, H. 1971. *Principles of Econometrics*. New York: Wiley.

endogenous and exogenous money. The issue of endogeneity or exogeneity of money is one that runs through the history of monetary theory, with prominent authors appearing to hold views on either side. Narrowly put, those who plug for the exogeneity view take one or all among the cluster of variables – price level, interest rate or real output – as being determined by movements in the stock of money. Those who hold the endogeneity view consider that the stock of money in circulation is determined by one or all of the variables mentioned above. This narrow definition begs several questions. The variables price level (P), interest rate (R), real output (Y) and money stock (M) are all at the macroeconomic level, i.e. in the context of a one-good economy. Some part of the continuing debate can be traced to the view held by various participants in the controversy about whether such a high level of aggregation is appropriate, e.g. is there *a* rate of interest? Another part of the debate refers to the choice of money stock variable. Is it commodity money (gold), fiat (paper) money, bank deposits or a larger measure of liquidity that is to stand for *the* money stock? The problem can be dealt with even at a one-good level either in the context of a closed economy or an open economy and either in an equilibrium or a disequilibrium context, static or dynamic, short run or long run. The basic issue is about the direction of causality-money to other variables or other variables to money. But as our understanding of the underlying statistical theory concerning causality and exogeneity has advanced in recent years, it must also be added that participants in the controversy conflate the exogeneity of a variable (especially of money) with its *controllability* by policy. Strictly speaking one can have exogeneity without any presumption that the variable can be manipulated by policy, for example rainfall. Also once posed in a dynamic context, we should distinguish between weak exogeneity, which allows for feedback from the endogenous to the exogenous variables over time, and strong exogeneity, which does not allow such a feedback (Hendry, Engle and Richard, 1983). Endogeneity or exogeneity are notions that only make sense in the context of a model. Frequently in the past, there has been a failure to specify such a model, which has then allowed the controversy to continue.

SOME DEFINITIONS. To simplify matters, at the risk of putting off readers, let us begin by specifying a small model within whose context endogeneity and exogeneity can be defined. This macroeconomic model will consist of four variables P, Y, R and M whose exogenous/endogenous status is at debate. We subdivide them into the three non-monetary variables P, Y, R labelled X and money M. There are of course other truly exogenous variables – tastes, technology, international variables – which we label Z. Now we observe that the variables X and M are correlated, i.e. jointly distributed conditional upon the set of variables Z. The question of endogeneity or exogeneity of money is as to whether the correlation between X and M can be written in terms of X being a function of M and Z, or M being a function of X and Z. In econometric terms, can we partition the joint distribution of X and M into a *conditional* distribution of X on M,Z and a marginal distribution of M on Z (the exogenous money case) or a conditional distribution of M on X and Z and a marginal distribution of X on Z. Thus when we say money is exogenous it is exogenous with respect to X variables but it could still be determined by Z variables; symmetrically for the X variables being exogenous. If M is influenced by the past values of X as well as by Z though not by the current values of X, then M is said to be weakly exogenous. Thus M may be controlled by monetary authorities but they may be reacting to past behaviour of X variables. Then M is determined by a reaction function and is only weakly exogenous. The same definition of weak exogeneity extends to the Z variables. Thus even international variables, such as capital inflow, may be determined by past values of X variables in which case they are weakly exogenous (for further detail, see Desai, 1981). The best way to consider the issue of exogeneity of money is to specify the type of money economy envisaged – commodity money, paper money, credit money and look at the variables likely to influence the supply of money and its relation with other variables.

COMMODITY MONEY. Historically the argument about exogeneity is constructed around the Quantity Theory of Money, which stated that the amount of money in circulation at any time determined the volume of trade and if the amount went on increasing it would lead sooner or later to an increase in price. In the context of commodity money, the proposition concerned attempts by coining authorities to debase coinage by clipping or alloying it with inferior metal. These were ways in which the amount of money could be altered by policy manipulation and then exogenously act upon prices. But in a commodity money regime, the stock of money could also be altered by influx of precious metal through gold discoveries and greater influx. These were exogenous variations not susceptible to policy manipulation but presumed an open economy. The first statement of the quantity theory of money by David Hume starts with an illustration of an influx of gold from outside and traces its effects first on real economic activity and eventually on prices. In Hume's quantity theory, money is exogenous but not subject to policy manipulation. The opposite view (argued by James Steuart for instance) was that it was the volume of activity that elicited the matching supply of money. This could be done partly by dis-hoarding on the parts of those who now expected a better yield on their stock. It could also be altered if banks were willing to 'accommodate' a larger volume of bills (see Desai, 1981). Dis-hoarding implies that a portion of the money supply *in circulation* is endogenously determined in a commodity money economy. It could be argued that even the influx of gold could have been caused by the discrepancy between the domestic and the world gold price, which in the 18th century before a world gold market existed could be substantial. In the latter case money would be weakly exogenous as long as there were lags between the appearance of discrepancy and the inflow of gold.

INSIDE MONEY. Once however one introduces banks into the scheme of things, the issue of exogeneity becomes complex. Till very recently we have lacked a theory of banking behaviour of any degree of sophistication, although in terms of institutional description we have much knowledge. If banks are willing to 'accommodate' a greater volume of trade, this can only be because they find it profitable to do so. This increased profitability may be actual or perceived but it must be a result of an increase in differential between the interest (discount) rate borrowers are willing to pay and the rate at which banks can acquire liquidity. Banks can then choose to expand the ratio of credit to the cash base and sustain a higher volume. Banks create inside money and inside money can only be regarded as endogenous. But the extent to which a single bank can create money will depend on the behaviour of the *banking system*. The banking system can by the cloakroom mechanism choose any ratio of credit to cash base. It is conceivable though not likely that in such a system of inside money, banks could arbitrarily, i.e. exogenously, increase money supply. They must however base such an action on considerations of expected profitability. We can envisage a situation in which banks guided by 'false' expectations can sustain a credit boom by a bootstraps mechanism. This is the way in which a Wicksellian cumulative process could sustain itself. An arbitrary, exogenous increase in inside money by the banking system though possible is not very likely. It runs into the problems caused by the leakage of cash either internally (finite limits to the velocity of circulation of cash) or abroad. It was the international leakage that was normally regarded as the most likely constraint since it caused outflow of gold – the International Gold Standard which provided the context for 19th-century theories in this imposed exogenous constraints on money supply by imposing a uniform gold price in all countries. In such a case, money is exogenous and not subject to policy manipulation. In as much as gold movements are triggered by internal variables, it is weakly exogenous.

OUTSIDE FIAT MONEY. It is the case of fiat money printed as the state's liability, i.e. as outside money, that provides the best illustration of exogenous money not subject to any constraint. In a world where only paper currency was used and it was printed by the monetary authorities, the stock of money could be exogenously determined. This would be additionally so even if there was inside money as long as the monetary authorities could insist that banks obeyed a strict cash to deposit ratio and there were no substitutes for cash available beyond the control of the monetary authorities. It is this view of money that most closely corresponds to Keynes's assumptions in the *General Theory* and it is also in the monetarist theory of Milton Friedman. The banking system is a passive agent in this view and given the cash base is always fully loaned up. Thus given the amount of high powered money in the system providable only by the monetary authorities, the supply of money is determined. Even if the stock of money were exogenous, its impact on the non-monetary variables X can be variable. This is because the velocity of circulation which translates the stock of money into money in circulation need not be constant but variable. If the velocity of circulation were not only a variable but also a function of the X variables, then although the monetary authorities can determine the stock of money the influence of money on real variables is not as predicted by the Quantity Theory. Thus it is not the exogeneity of money issue that divides monetarists and Keynesians but the determinants of the velocity of circulation. For the monetarist, the velocity of circulation ($M/P \cdot Y$) has to be independent of P, Y, R and M. For Keynesians, the demand for money depends on the rate of interest crucially and the interest elasticity of demand for money is a variable tending to infinity in a liquidity trap.

MODERN CREDIT ECONOMY. In a world with inside and outside money with a sophisticated banking system as well as a non-banking financial sector, the question of exogeneity is the most complex. In the previous case of outside fiat money we assumed, that the cash ratio was fixed and adhered to by banks. It is when the banks' reserve base contains government debt instruments – treasury bills, bonds, etc. – that the profit-maximizing behaviour of the banks renders a greater part of the money stock endogenous. Thus while the narrow money base – currency in circulation and in central bank reserves – can be regulated by the monetary authority, the connection between money base and total liquidity in the economy becomes highly variable. Banks will expand their loan portfolio as long as the cost of replenishing their liquidity does not exceed the interest rate they can earn on loans. The relation between broad money (M_3) and narrow money (M_0) becomes a function of the funding policy concerning the budget deficit and the structure of interest rates. Thus the stock of narrow money can be exogenous and policy determined. But the stock of broad money is endogenous. A crucial recent element has been the financial revolution of the last decade (De Cecco, 1987). A variety of financial instruments – credit cards, charge cards, money market funds, interest-bearing demand deposits, electronic cash transfer – has made the ratio of cash to volume of financial transactions variable though with a steep downward trend. It has also increased the number of money substitutes and made the cost of liquidity lower. The non-banking financial system thus can create liquidity by 'accommodating' a larger volume of business, advancing trade credit, allowing consumer debt to increase etc. The velocity of circulation of cash increases very sharply in such a world and liquidity, a broader concept than even broad money, becomes endogenous. Here again profitability of liquidity creation becomes the determining variable. But the financial revolution has also integrated world financial markets and economies are increasingly open. Thus capital flows are rapid and respond to minute discrepancies in the covered interest parity. In such a world money is at best weakly exogenous but more usually endogenous. The issue of exogeneity or endogeneity of money thus crucially depends on the type of money economy that one is considering – commodity money, paper money, credit (mobile) money. It also depends on the sophistication of the banking and financial system within which such money is issued. Debates over the last two hundred years have used the word money to cover a variety of situations. It has also not been clarified whether the issue is exogeneity of money or its controllability and whether it is merely the stock of money or its velocity as well which is being considered. Once these issues have been clarified, the notion of exogeneity needs to be defined in the modern econometric fashion, relative to a model in order to decide whether money can be exogenous. It seems likely that the narrower the definition of money stock, the more likely is it to fulfil the requirement of (weak) exogeneity. Such exogeneity is necessary but not sufficient to demonstrate that money determines the price level or the real economy.

MEGHNAD DESAI

See also CAPITAL, CREDIT AND MONEY MARKETS; HIGH-POWERED MONEY AND THE MONETARY BASE; MONETARY BASE; MONEY SUPPLY; THEORY OF MONEY.

BIBLIOGRAPHY
De Cecco, M. 1987. *Changing Money: Financial Innovations in Developed Countries.* Oxford: Blackwell.
Desai, M. 1981. *Testing Monetarism.* London: Frances Pinter.
Hendry, D., Engle, R. and Richard, J.L. 1983. Exogeneity. *Econometrica* 51(2), March, 227–304.

endowments. *See* GENERAL EQUILIBRIUM.

energy economics. Political instability in the Middle East caused dramatic oil price increases in late 1973 and during a 17-month period over 1979–80. These events created a natural laboratory for economists to study the response of markets and economies to shifts in the relative price of an important input. Prior to the decade of energy shocks, economists specializing in this area usually focused on market structure of the critical oil market (Adelman, 1973) or the structure and regulation of specific fuel markets. While these issues continued to receive attention, new empirical issues were intensely studied, of which three key areas are highlighted here: energy demand and its response to price, the connection between energy and other factors of production, and the response of the aggregate economy to input price shocks. Analysis of cartel behaviour in the world oil market also attracted much interest and is briefly reviewed. The economic research on energy supply remained largely theoretical, focusing on the formal aspects of modelling depletion. In general, these contributions provided fewer empirical insights than those in the above areas. Additionally, contributions to the study of the market structure and industrial organization of other fuel markets were not as noteworthy, although the shift away from oil and the recent deregulation trends in natural gas and electricity markets have regenerated new interests in these topics.

ENERGY DEMAND

Economists were more optimistic than others on the response of aggregate energy demand to a change in its relative price. And in hindsight, this optimism appears justified. Most studies found that while energy demand was responsive to price, it tended to be price inelastic, even in the long run. During the late 1970s, many estimates placed the price elasticity of aggregate energy in the -0.3 to -0.7 range when measured at the wholesale (or secondary) level (Energy Modeling Forum, 1982). The oil price collapse in 1986, as well as a more modest price decline in 1983, are providing another real-world experiment for testing the symmetry of energy demand responses to rising and falling prices. Price elasticities at the end-use level could be some 50 per cent higher due to differences between end-use and wholesale price levels. Elasticities for individual fuels and electricity would also be higher, reflecting the potential for interfuel substitution within aggregate energy.

Interfuel substitution. The shift away from petroleum products accelerated after the 1979–80 oil price shocks, particularly for fuel oil where ready substitutes are available. These economic incentives, combined with environmental concerns about coal and nuclear power, have intensified the interest in the relationships between energy sources. In contrast to the finding about aggregate energy, the econometric evidence appears to suggest considerably *less* substitution potential between fuels than is deduced from engineering studies of specific technologies. The econometric estimates may change, however, as competitive fuel markets become less regulated and as equipment that can burn multiple fuels penetrates the market more widely.

Energy-using capital. Energy is a derived demand that depends upon the mix of final goods and services desired by households and firms. Even for residential energy use, households are viewed as combining energy-using capital stock and energy inputs to produce service flows that provide utility to the decisionmaker. This approach provides a clear distinction

between short and long-run energy demand responses. Working with a fixed capital stock, agents can alter their rate of utilization to changes in price, income and weather in the short run. In the long run, the demand for energy is tantamount to the demand for energy-using capital stock. The development of reliable capital stock estimates for empirical application is a decidedly more complex issue. Usually researchers have serious reservations about the quality of such estimates and seek ways to avoid them. Their recourse is to represent implicitly the capital stock adjustment by eliminating the capital stock variable in the theoretical model. The flow-adjustment model (Houthakker and Taylor [1966], 1970) is also a tractable way for separating the short and long-run energy demand responses in the absence of explicit capital stock estimates. The approach of using explicit capital stock estimates has been applied most successfully to the study of the demand for gasoline in the transportation sector (Sweeney, 1979), where information on the planned fuel efficiencies of different makes and vintages of passenger cars is available. This approach has been applied with somewhat less success to the residential sector. Some US studies have used state-level estimates of energy-using appliances over time (Taylor et al., 1982), while others have employed household surveys (Cowing and McFadden, 1984) to provide a cross-sectional view of appliance ownership. Estimates for the other sectors are unavailable. Commercial floor space does not commit the agent to any particular fuel; hence, the short versus long-run distinction is lost. And industrial capital stock is too heterogeneous to be meaningful for empirical studies of manufacturing. Industrial energy demand appears to be particularly influenced by the significant shift in economic structure from more to less energy-intensive sectors. At least one-third of the reduction in fossil fuel use per dollar of output may be due to the compositional shift in output in this sector. The sources of shift are uncertain and require additional research; energy prices, the cost of capital, capital obsolescence, the business cycle, technological change, and the appreciating dollar have all been mentioned as potential contributing factors.

Price variables. Some ambiguity has arisen about the price variable in energy demand studies as well. The frequent use of ex post average prices introduces simultaneity biases because this price variable is determined in part by total demand. Moreover, electricity and natural gas are sold on a declining block basis, in which the marginal price falls as consumption increases. The substitution of marginal for average prices may be insufficient to capture adequately the full effects of rate schedules on fuel demand. Marginal prices will reflect changes in the slope of the budget line for households, but the budget line itself can be shifted inward by an increase in the fixed charge. This consideration would argue, at least conceptually, for the inclusion of both marginal and fixed charges in demand studies for such fuels (Taylor, 1975). Fuel availability can be an important issue in some demand studies. Not all households in the US have had access to natural gas pipelines, even when price controls were not binding. This access problem affects the demand for substitutes (e.g., fuel oil and electricity) as well as for the fuel itself (Blattenberger et al., 1983). Moreover, price regulations have had similar distorting effects. Binding price regulations have prevented the estimation of the demand for natural gas during the 1970s in the US. Natural gas shortages have also induced greater demand for substitute fuels than would be the case if all markets were clearing. For the most part, these issues have been insufficiently analysed in traditional demand studies.

ENERGY AND OTHER INPUTS

Energy-economy linkages. The price elasticity of aggregate energy demand measures the proportional change in the use of all energy for a one per cent change in the aggregate price of energy. It is closely akin to the concept of the elasticity of substitution between energy and non-energy inputs. If the supplies of non-energy inputs are held fixed, the aggregate elasticity will shape the long-run, energy-economy linkage. Energy's relatively low historical value share of GNP may not be an appropriate indicator of energy's importance to the economy, if limited flexibility in substituting capital and labour for energy greatly influence the future value share of energy (Hogan and Manne, 1977). A higher elasticity implies less economic loss resulting from a reduction in energy availability or from a change in the cost of imported energy. However, if energy costs are raised by a domestic tax on energy that keeps the higher energy expenditures within the country, the economic loss becomes greater as the aggregate elasticity increases.

Energy and capital. More in-depth study of the energy substitution issue was made possible by significant conceptual advancements during the 1960s that provided general flexible forms for estimating production and cost functions. In particular, the translogarithmic and generalized Leontief functions were advanced during this period. These functions are appropriate for analysing substitution and complementarity between three or more inputs. This theoretical development was propitious for economists who were interested in probing the relationship between energy inputs and other factors. At the same time, the energy price shocks provided a real-world problem for testing these functions. The application of these general flexible forms became widespread during the 1970s. While there was consensus on the methodology for estimating functions, there was sharp disagreement on the principal findings. The empirical dispute centred around whether capital and energy were substitute or complementary inputs in manufacturing. Some researchers found that energy and capital were net complements (Hudson and Jorgenson, 1974; Berndt and Wood, 1976). Thus, higher capital costs would reduce energy demand. More importantly, higher energy prices would reduce capital investment and hence the long-run GNP through a lower capital stock. The initial studies supporting this view used cost functions that were estimated on national time-series data and that included the prices of four inputs: capital, labour, energy and materials. Other researchers concluded that energy and capital were substitutes for each other. Energy demand would be discouraged by lower capital costs, while higher energy prices would encourage the substitution towards more capital, contributing positively to long-run potential GNP. These studies often used cost functions that were estimated by pooling time-series data across several countries and that excluded the price of materials (Griffin and Gregory, 1976; Pindyck, 1979). Intuitive explanations for each finding were advanced. Within individual technologies, energy conservation could be realized by substituting more capital as the real price of capital declined. However, with more than two inputs, this effect could be offset or even reversed by the substitution towards more automated processes using more of both capital and energy and less of labour and materials. While some think that this empirical issue revolves around differences in methodologies, for example, pooled versus time series, there is some evidence that the results are sensitive to data construction, such as the calculation of a capital price series that accounts for tax variables in countries other than the US, or reliable estimates of the capital stock that incorporate the effect of energy shocks on capital obsolescence.

ECONOMIC IMPACTS OF HIGHER ENERGY PRICES

Energy economists shifted their attention from the long run to the short run after the 1979–80 disruption. Oil and macroeconomics became inextricably tied, as issues of long-run dependence received less attention in favour of short-run vulnerability concerns. Perhaps no other issue has revealed so clearly the schism within the profession between microeconomists and macroeconomists. Microeconomists estimated the effects of disruptions and energy security programmes by assuming full employment of resources, while macroeconomists often assumed exogenous oil prices. Some economists emphasized the adjustment in real prices while others focused on nominal price stickiness. With some exceptions, there was little bridging of this gap.

Types of losses. A permanent increase in the real resource cost of energy has a direct and lasting effect on a nation's real income. This can happen either with greater depletion of domestic energy resources or with an increase in the foreign price of energy. When foreign prices are the cause, this reduction in real income is also called the terms of trade effect. It is approximately equal to the change in real oil prices times the level of oil imports. The terms of trade losses are excluded from GNP as conventionally measured with national income accounts. Higher domestic or foreign energy prices will also reduce physical output as measured by GNP. Much of the observed reduction in output during the 1970s appears to reflect macroeconomic adjustment costs (or indirect costs) as a result of the sudden oil price shocks. Quantitatively, these output effects tend to dominate the terms of trade effects. The adjustment costs appear to depend upon the change in energy prices, the stickiness in wages and prices, and the relative importance of energy use (rather than just imports) in the economy. Macroeconomic adjustment costs from an oil shock are experienced by both oil importers and oil exporters.

Policies. In contrast to a foreign price shock for final goods, an energy shock reduces real output in the short run while increasing the aggregate price level. Since the problem originates on the supply side, demand-oriented policies alone are usually considered insufficient. Policy prescriptions usually include supply-oriented policies that reduce prices while augmenting output. Some policies like oil stockpile releases are aimed at directly lowering the oil price, while others like reductions in payroll or excise taxes operate on lowering production costs generally.

Aggregate demand and supply. There are alternative and not mutually exclusive explanations for this adjustment to a lower level of activity. Greater unemployment can be attributed to either aggregate demand or supply conditions, and it is difficult to distinguish empirically between the effects. Aggregate demand will be reduced by an oil price shock if prices of other goods exhibit nominal stickiness, resulting in a higher aggregate price level. Unless money demand is allowed to grow faster, real money supply will contract, interest rates will rise, and expenditures will decline. Real domestic income will also be drained by the OPEC tax, which shifts income overseas where it is unlikely to be fully respent in the oil-importing country. While this effect is often emphasized, its importance can be limited by several conditions. First, export prices for some countries can also increase in real terms, as might be the case for energy-intensive exports. Moreover, the tax multiplier effect may be relatively modest, as has been exhibited by empirical models of the US economy (Energy Modelling Forum, 1984). Aggregate supply conditions can also create unemployment during an oil shock as well. If wages are rigid in real terms, reductions in the demand for labour after an oil shock will result in classical unemployment in which wages are too high for full employment. The unemployment emerges from an imbalance in the labour market rather than from insufficient demand in the product market. These conditions appear to be important for explaining the response of European countries to the oil price instability of the 1970s (Sachs, 1979).

WORLD OIL MARKETS

Analysing the market. Empirical studies of the world oil market generally decompose the market into at least three sectors: demanders, cartel producers (OPEC), and residual producers. OPEC's behaviour is critical to determining the price and output in such a system. The standard approach in economics is to represent OPEC as a cartel that maximizes its wealth subject to the constraint of an exhaustible resource. Following the Hotelling principle, the cartel adopts a pricing path that allows marginal rents (marginal revenue minus the marginal extraction cost) to increase over time at the rate of interest. Since the monopolist's price exceeds the competitive price, the cartel encourages more conservation than otherwise, for well-behaved demand functions. Additional complications can be introduced through various assumptions about the reactions of a group of producers to the pricing strategy of another group; both Nash–Cournot and Stackelberg models have been applied to this market. Generated price paths will be sensitive to these assumptions about the optimizing behaviour of all sectors of the market, including the strategy of oil consumers. These models appear to be most suitable for gaining insights about long-run price trends over a period of several decades. Applied models have been less satisfactory for shorter periods, for example during the 1970s. They could not explain the price doubling in 1979–80 and do not address many important short-run phenomena due to the absence of lags in the supply, demand and pricing relationships. As a result, applied wealth-maximization models have generally been replaced by simulation models using *ad hoc* pricing relationships based upon the experiences of the 1970s (Gately, 1984). In these models OPEC sets its price on the basis of a desired capacity-utilization target. The cartel increases price when the market becomes tighter and lowers it when excess capacity increases.

Policy. The presence of market power in international oil has some interesting policy implications as well. Although economists have a strong predilection towards keeping government out of domestic energy markets, they often retain a role for public intervention in the world crude oil market. Economic inefficiencies result from cartel prices that are artificially high. Policies that make an oil-importing economy less dependent upon oil in undisrupted markets reduce the demand for OPEC oil as well as increase the price elasticity for OPEC oil. Both factors will place downward pressure on the cartel's price, providing economic benefits by lowering the price on the oil that the economy does import. This argument is based upon the optimal tariff literature in international trade economics. Oil-demand reduction policies in stable markets may also provide additional benefits during a disruption by reducing the losses in real income caused by the oil price shock. These benefits must be compared with the resource costs of the policy action itself. Moreover, a number of qualifications apply, including the assumption of no retaliation by cartel producers.

A CONCLUDING COMMENT. The oil decade has given the economist a very rich experience with which to test his analytical tools. Even so, research on energy demand must often bend to the dictates of available data rather than to theory. General agreement exists on the role of markets with limited government intervention, the importance of the price elasticity of demand, and broad policy responses to sudden oil price shocks. Less consensus emerges, however, on the more technical issues, like energy-capital complementarity and the relative importance of aggregate demand and supply factors in the economy during oil disruptions.

HILLARD G. HUNTINGTON

See also DEPLETION; EXHAUSTIBLE RESOURCES; NATURAL RESOURCES; SUPPLY SHOCKS.

BIBLIOGRAPHY

Adelman, M. 1973. *The World Petroleum Market.* Baltimore: Johns Hopkins Press for Resources for the Future.

Berndt, E. and Wood, D. 1975. Technology, prices and the derived demand for energy. *Review of Economics and Statistics* 57(3), August, 259–68.

Blattenberger, G., Taylor, L. and Rennhack, R. 1983. Natural gas availability and residential demand for energy. *Energy Journal* 4(1), January, 23–45.

Cowing, T. and McFadden, D.L. 1984. *Microeconomic Modeling and Policy Analysis: Studies in Residential Energy Demand.* Orlando, Florida: Academic Press.

Energy Modeling Forum. 1982. *Aggregate Elasticity of Energy Demand.* Stanford University, California.

Energy Modeling Forum. 1984. *Macroeconomic Impacts of Energy Shocks.* Stanford University, California.

Gately, D. 1984. A ten-year retrospective: OPEC and the world oil market. *Journal of Economic Literature* 22(3), September, 1100–14.

Griffin, J. and Gregory, P. 1976. An inter-country translog model of energy substitution responses. *American Economic Review* 66(5), December, 845–57.

Hogan, W. and Manne, A. 1977. Energy-economic interactions: the fable of the elephant and the rabbit? In *Energy and the Economy*, Vol. 2, Energy Modeling Forum Report 1, September 1977, Stanford University, California; also in *Modeling Energy-Economy Interactions: Five Approaches*, ed. C. Hitch, Washington, DC: Resources for the Future.

Houthakker, H. and Taylor, L. 1966. *Consumer Demand in the United States.* Cambridge, Mass.: Harvard University Press. 2nd edn, 1970.

Hudson, E.A. and Jorgenson, D.W. 1974. US energy policy and economic growth. *Bell Journal of Economics* 5(2), Autumn, 461–514.

Pindyck, R. 1979. Interfuel substitution and industrial demand for energy: an international comparison. *Review of Economics and Statistics* 61(2), May, 169–79.

Sachs, J. 1979. Wages, profits, and macroeconomic adjustment: a comparative study. *Brookings Papers on Economic Activity* No. 2, 269–319.

Sweeney, J. 1979. Effects of Federal policy on gasoline consumption. *Resources and Energy* 2(1), September, 3–26.

Taylor, L.D. 1975. The demand for electricity: a survey. *Bell Journal of Economics* 6(1), Spring, 74–110.

Taylor, L., Blattenberger, G. and Rennhack, R. 1982. *Residential Demand for Energy*, Vol. 1. Palo Alto, California: Electric Power Research Institute, April.

enforcement. Enforcement, with its usual connotations of agents being compelled to behave in ways that are at variance with how they would like to act, seems a long way removed from the conventional neoclassical approach of laissez-faire inherent in a decentralized economic system. If an enforcement mechanism is viewed as a method by which rule-breaking may be discouraged then laissez-faire attempts to be a set of rules that are self-enforceable so that no formal enforcement mechanism is required (see Stigler, 1970 for a useful discussion of enforcement). But it is clear that decentralizability is not, in itself, enough. For instance, the fact that tax evasion is kept in check only by legal pressure shows how the addition of a tax system to a decentralized competitive structure may destroy self-enforceability. The possibility of gain by the exercise of monopoly power shows how the rule of price-taking behaviour is not immune to the problems of enforcement.

Except when self-enforceability holds, so that agents wish to follow the given set of rules – the rules are incentive compatible – an enforcement mechanism will be necessary to prevent rule-breaking. A formal enforcement mechanism has two components. First, there must be a method of monitoring agents so that it is possible to observe, perhaps imperfectly, rule-breaking. Second, there must exist a sanction or punishment which can be imposed on rule-breakers. If perfect observability is costless and there is no limit to the disutility that a punishment can impose then there is no difficulty in enforcing a rule and, importantly, the enforcement procedure never needs to be exercised. Thus, the existence of an enforcement procedure alters the incentives to obey rules and can be used as a component of the incentive structure within an economic system. It is clear that this is in part the role of a legal system.

To examine enforcement procedures in greater detail, first consider the monitoring mechanism. If monitoring is impossible then agents will abide with a rule only if it is self-enforceable. If monitoring is costly then there may be a decision to be taken with regard to the best monitoring mechanism. Two dimensions of choice seem natural. First, there is the quality of the monitoring mechanism. After monitoring, an inference will be made with regard to whether a rule has been broken. The better the quality of the monitoring mechanism, the more accurate will be this inference. Accuracy will involve both a low probability of inferring rule-breaking when it has not occurred and a low probability of inferring rule-compliance when rule-breaking has occurred.

The second component of the monitoring mechanism is the intensity with which it is applied. The same agent may be monitored several times to improve the accuracy of the test; agents may be monitored on a random basis so that any one agent will be monitored with a probability of less than unity. The overall mechanism is likely to be more effective if agents must decide upon their behaviour with regard to rule-compliance before they know the intensity with which they will be monitored.

A high punishment level will deter rule-breaking more than a low punishment level. To see what an optimal enforcement mechanism may look like, consider a planner who wishes to deter rule-breaking (which, for instance, imposes significant negative externalities on other agents). Assume that the planner wishes to maximize the expected utility of agents whilst deterring rule-breaking – rule-breaking is assumed to be privately optimal. If a is the accuracy of the test used then let $p(a)$ be the probability of a rule-breaking inference if rule-breaking has not occurred. Similarly, let $q(a)$ be the probability of such an inference if rule-breaking has occurred.

Thus, $p(a) < q(a)$, $p(a)$ is falling in a and $q(a)$ is rising in a. Let U be the utility that an agent gets from rule-compliance before the introduction of an enforcement mechanism and V be the utility that would result from rule-breaking. Then if m is the probability of being monitored and f is the punishment cost then, taking the simplest specification of linear utility, the expected utility of an agent who complies with the rule is given by

$$U - mp(a)f - am \tag{1}$$

where am is taken to be the cost of operating the enforcement mechanism, the money being raised through taxation, say. The punishment f is assumed to be a deadweight loss rather than a monetary fine which could be used to finance the monitoring mechanism. An agent compares (1) with the expected utility from rule-breaking:

$$V - mq(a)f - am \tag{2}$$

so that rule-breaking can be deterred if

$$mf[q(a) - p(a)] \geqslant V - U. \tag{3}$$

If the planner wishes to maximize (1) subject to (3) then the first point to notice is that (1) can always be increased while still satisfying (3) if mf is held constant but m is reduced. Thus, it is desirable to increase the punishment whilst reducing the rate of monitoring and it is optimal to punish as severely as possible (this argument is credited to Becker, 1968). Once the worst punishment is chosen, the optimal monitoring mechanism satisfies

$$\frac{fp' + 1}{fp + a} = \frac{q' - p'}{q - p} \tag{4}$$

which gives rise to the comparative statics results that would be expected. If f can be chosen without limit (punishment can reduce utility to minus infinity) then m will be set arbitrarily small and, keeping mf fixed, (1) can be increased by increasing a if $p'(a)$ is strictly negative. If perfect accuracy is attainable ($p(a) = 0$ for some a) then the expected utility of the agent will be U – the first-best is achievable – and the enforcement mechanism will take the form of an infinitesimally small amount of monitoring using a very accurate procedure and infinitely large punishments being imposed on rule-breakers. This strong result depends upon the strong assumptions that have been imposed on the model. However, Nalebuff and Scharfstein (1985) obtain a similar result in a richer setting than that of the above model. For a model where finite levels of punishment are optimal because of the risk aversion of agents, see Polinsky and Shavell (1979).

The structure outlined above has the property that the enforcement mechanism is operated by a planner. One reason for assuming this is that it is unnecessary to consider the planner's incentives to operate the mechanism. In some situations, it is necessary for a group of agents to operate an enforcement mechanism that will be imposed upon themselves. The classic example of this is the operation of a cartel (Stigler, 1964). A major problem introduced by not having an external agency is that costs may be imposed on firms if they choose to punish a firm that breaks the cartel's rules. An example of this is when a firm is punished by all firms entering into a price-cutting war. To provide an incentive to punish, it may be necessary to have a mechanism to enforce the operation of the original enforcement mechanism, and so on. For an insightful analysis of this problem, see Abreu (1986). Considerations of this sort show the potential richness of the structure of enforcement mechanisms in general.

KEVIN ROBERTS

See also CARTELS; COOPERATIVE EQUILIBRIA; COOPERATIVE GAMES; GAME THEORY; INDUSTRIAL ORGANIZATION.

BIBLIOGRAPHY

Abreu, D. 1986. Extremal equilibria of oligopolistic supergames. *Journal of Economic Theory* 39(1), June, 191–225.

Becker, G. 1968. Crime and punishment: an economic approach. *Journal of Political Economy*, March-April, 169–217.

Nalebuff, B. and Scharfstein, D. 1985. Self-selection and testing. Forthcoming in *Review of Economic Studies*.

Polinsky, A. and Shavell, S. 1979. The optimal tradeoff between the probability and magnitude of fines. *American Economic Review* 69(5), December, 880–91.

Stigler, G. 1964. A theory of oligopoly. *Journal of Political Economy* 72(1), February, 44–61.

Stigler, G. 1970. The optimum enforcement of laws. *Journal of Political Economy* 78(3), March–April, 526–35.

Engel, Ernst (1821–1896). Born in Dresden, Engel was a German statistician best known for the discovery of the Engel curve and of Engel's Law. In his early years he was associated with the French sociologist Frédéric Le Play, whose interest in the family led him to conduct household surveys. The expenditure data collected in these surveys convinced Engel that there was a relation between a household's income and the allocation of its expenditures between food and other items. This was one of the first functional relations ever established quantitatively in economics. Furthermore, he observed that households with higher incomes tended to spend more on food than poorer households, but that the share of food expenditures in the total budget tended to vary inversely with income. From this empirical regularity he went on to infer that in the course of economic development agriculture would decline relative to other sectors of the economy (Engel, 1857). From 1860 to 1882 Engel was director of the Prussian statistical bureau in Berlin, in which capacity he did much to expand and strengthen official statistics. His resignation resulted from his opposition to Bismarck's protectionist policies. In his own research he dealt particularly with the value of human life (Engel, 1877), which he approached from the cost side. He also investigated the influence of price on demand. His influence on official statistics extended well beyond Germany, and in 1885 he was among the founders of the International Statistical Institute. He died in Radebeul in 1896.

H.S. HOUTHAKKER

SELECTED WORKS

1857. Die Productions- und Consumptionsverhaeltnisse des Koenigsreichs Sachsen. Reprinted with Engel (1895), *Anlage* I, 1–54.

1877. *Der Kostenwerth des Menschen*. Berlin.

1895. Die Lebenskosten Belgischer Arbeiter-Familien frueher und jetzt. Reprinted in *International Statistical Institute Bulletin* 9, 1–124.

Engel curve. By tabulating data from a survey of Belgian working-class families, Engel (1857) was the first to show that a household's expenditure on food and other items depended on its income or total expenditure. The graphical representation of this dependence, particularly as it appears in cross-section data, soon became known as the Engel curve. Efforts to demonstrate its relevance to other countries met with considerable success; for a centennial review see

Houthakker (1957). Although discovered at a time when economists thought in terms of price rather than income, the Engel curve gradually came to be recognized as a cornerstone of demand analysis. The Keynesian consumption function may be considered an extension of the Engel curve; it is outside the scope of this article. In what follows the basic relation will be written

$$x_i = f_i(y, z), \tag{1}$$

where x_i is a household's expenditure (in money terms) on the ith commodity, y is some indicator of the household's overall resources, and z stands for a vector of other variables influencing x_i. Engel curves may be conveniently characterized by the *income elasticity*, which is the partial derivative of log x_i with respect to log y. Expenditure items are called luxuries, necessities and inferior goods depending on whether the income elasticity is greater than one, between zero and one, or less than zero. In general these elasticities are not independent of income, however, and an item may be a luxury in a certain income range and a necessity or an inferior good in another. The extensive research on the Engel curve, which became active after the emergence of econometrics in the 1930s, will be reviewed under four headings: the dependent variable x_i, the independent variable y, the mathematical form of f_i, and the nature of the catchall variable z.

THE DEPENDENT VARIABLE. In empirical work the classification of consumption into separate items is usually determined by the data, which may distinguish anywhere from three or four to many hundreds of items. Some of these items, especially durables, are bought infrequently and consequently may not be represented adequately in a survey of expenditures in a short time interval. To analyse durables effectively it is desirable to use time series or to combine two or more surveys taken at different times. Panel surveys, in which the same household is observed in several periods, have recently been introduced in a number of countries; they also shed light on such phenomena as habit formation. Most of the interest in one-period surveys has centred on relatively broad categories such as food, clothing and housing. Strictly speaking, however, the economic theory of consumer demand deals with more narrowly specified items, for instance a 16oz. can of Delmonte peaches bought in a certain store. Data in such detail are rarely available in household surveys. Moreover this theory postulates the existence of a utility function

$$u(q_1, q_2, \ldots, q_n)$$

whose arguments are quantities q_i, not expenditures. Thus if in (1) the index i refers to (say) fruit we should bear in mind that

$$x_i = \sum_j p_{ij} q_{ij}.$$

A total quantity q_i for fruit may be obtained by summing the q_{ij}. Although the meaning of a 'price' p_i for all fruit is not equally clear, it is often calculated by simply dividing x_i by q_i. In empirical studies where quantity data are available it has been found that the average price so calculated is usually an increasing function of the 'income' variable y. Prais and Houthakker (1955) interpreted this phenomenon as showing that richer households buy items of a better 'quality' than poorer ones. A theoretical discussion of this general topic may be found in Deaton and Muellbauer (1980, ch. 10.) In a related vein Wold (1953) has suggested that a higher income allows households to buy a larger variety of items. The same idea was expressed and demonstrated by Frazier (1984) as the 'Engel curve for variety'.

THE INDEPENDENT VARIABLE. In the older literature the variable y in (1) was usually identified with the income or total expenditure reported during the survey period. This identification was put into question by the *life-cycle hypothesis* of Modigliani and Brumberg (1955), according to which consumers maximize utility over their lifetime, not over a year or some other short period. Although addressed primarily to the Keynesian consumption function, the life-cycle hypothesis clearly also has a bearing on the Engel curve. So does the closely related *permanent income hypothesis* (Friedman, 1957), which divides both current income and current consumption into a permanent and a transitory component. According to Friedman's theory permanent consumption, at least of nondurables, depends on permanent income rather than on current income. Since permanent income, which is essentially the same as expected lifetime income, cannot be observed, these ideas might appear to be very damaging to cross-section analysis. Fortunately this is not the case. Friedman pointed out that by appropriate grouping of households the transitory components of income and consumption can be made to cancel out. Such grouping had long been customary in the study of household expenditures; in fact Engel himself derived the first Engel curves from grouped data. Friedman's analysis also gave support to another long-standing practice, namely the use of total expenditure rather than current income as an indicator of household resources. The empirical relevance of the life-cycle and permanent income hypotheses to cross-section data (or for that matter to time series) has not been fully established. The weakness of both theories is that they take no account of liquidity constraints, which may force households to pay more attention to current income than these hypotheses suggest.

THE SHAPE OF ENGEL CURVES. The systematic study of the mathematical form of the Engel curve started with Allen and Bowley (1935), who assumed it to be linear. When this assumption was shown to be unrealistic a variety of other forms, usually involving logarithms, was investigated. At present many researchers favour the following form first proposed by Working (1943):

$$x_i/y = a_i + b_i \log y. \qquad (2)$$

This expression is consistent with utility theory, and more particularly with the Almost Ideal Demand System of Deaton and Muellbauer (1980). Utility theory also suggests, however, that the shape of the Engel curve depends on the level of aggregation at which expenditures are analysed. While (2) may be a good approximation for broad categories, it cannot be generally valid because the ratio on the left is negative for certain income ranges. Taking non-negativity into account, Houthakker (1953) suggested that for a narrowly specified item (such as the can of peaches mentioned earlier) the Engel curve coincides with the horizontal axis for low values of income; at some initial income the item enters the budget as a luxury; for higher incomes the income elasticity declines until the expenditure again becomes zero after the item has become an inferior good. Wales and Woodland (1983) incorporated the non-negativity constraint in empirical work.

OTHER VARIABLES. Income is prominent in the cross-section analysis of consumption, yet it explains only a small fraction of the enormous variation in expenditure patterns among individual households first analyzed by Allen and Bowley (1935). This diversity incidentally contradicts the widespread belief that consumers are merely slaves to advertising and social pressures. Chief among the variables that have been used to reduce this variation is family size, or more generally family composition. To quantify the effect of family size Engel (1895) introduced the 'equivalent adult scale', in which persons were given weights according to their age and sex regardless of their expenditures. Sydenstricker and King (1921) suggested that there should be a different scale for every expenditure category, and in addition a 'general scale' applied to income. This idea was implemented empirically by Prais and Houthakker (1955) and put on a firmer theoretical basis by Barten (1964), Pollak and Wales (1981) and others. In addition to family composition, variables representing location, ethnic origin, social class and homeownership have been studied.

H.S. HOUTHAKKER

See also CONSUMERS' EXPENDITURE; DEMAND THEORY.

BIBLIOGRAPHY

Allen, R.G.D. and Bowley, A.L. 1935. *Family Expenditure. A study of its variation.* London: P.S. King & Son.

Barten, A.P. 1964. Family composition, prices and expenditure patterns. In *Econometric Analysis for National Economic Planning*, ed. P.E. Hart et al., London: Butterworths.

Deaton, A. and Muellbauer, J. 1980. *Economics and Consumer Behaviour.* Cambridge: Cambridge University Press. (Contains extensive bibliography.)

Engel, E. 1857. Die Productions- und Consumptionsverhaeltnisse des Koenigreichs Sachsen. Reprinted with Engel (1895), Anlage I, 1–54.

Engel, E. 1895. Die Lebenskosten Belgischer Arbeiter-Familien frueher und jetzt. *International Statistical Institute Bulletin* 9, 1–124.

Friedman, M. 1957. *A Theory of the Consumption Function.* Princeton: Princeton University Press.

Houthakker, H.S. 1953. La forme des courbes d'Engel. *Cahiers du Seminaire d'Econometrie* 2, 59–66.

Houthakker, H.S. 1957. An international comparison of household expenditure patterns commemorating the centennial of Engel's Law. *Econometrica* 25(4), October, 532–51.

Jackson, L.F. 1984. Hierarchic demand and the Engel curve for variety. *Review of Economics and Statistics* 66(1), February, 8–15.

Modigliani, F. and Brumberg, R. 1954. Utility analysis and the consumption function: an interpretation of cross-section data. In *Post-Keynesian Economics*, ed. K.K. Kurihara, London: George Allen & Unwin.

Pollak, R. and Wales, T.J. 1981. Demographic variables in demand analysis. *Econometrica* 49(6), November, 1533–51.

Prais, S.J. and Houthakker, H.S. 1955. *The Analysis of Family Budgets.* Cambridge: Cambridge University Press.

Sydenstricker, E. and King, W.I. 1921. The measurement of the relative economic status of families. *Quarterly Publication of the American Statistical Association* 17(135), September, 842–57.

Wales, T.J. and Woodland, A.D. 1983. Estimation of consumer demand systems with binding non-negativity constraints. *Journal of Econometrics* 21(3), April, 263–85.

Wold, H.L. with Jureen. 1953. *Demand Analysis.* New York: John Wiley & Sons; Stockholm: Almqvist & Wiksell.

Working, H. 1943. Statistical laws of family expenditures. *Journal of the American Statistical Association* 38, March, 43–56.

Engel's Law. Engel's Law states that the share of food in total expenditures is inversely related to the household's income (or some other measure of its total resources). This implies that the income elasticity of food expenditure is less than one. Of all empirical regularities observed in economic data, Engel's Law is probably the best established; indeed it holds not only in the cross-section data where it was first observed, but has often been confirmed in time-series analysis as well. Like most economic laws, however, it holds *ceteris paribus*; prices, among other things, are assumed constant. There is also evidence that the income elasticity of food, like the budget share, is inversely related to income; the elasticity may be as high as 0.8 or 0.9 at very low income levels, and close to zero for high incomes. As Engel himself emphasized, his law has profound consequences for economic development. Food is the main product of agriculture; therefore a declining share of food in aggregate consumption implies a declining share of agriculture in aggregate production. 'Balanced growth', in which all sectors grow at the same rate, is impossible. A full analysis of the effects of Engel's Law would require explicit consideration of relative prices and of productivity in the farm and nonfarm sectors; foreign trade may also have an effect. Under plausible assumptions the number of farmers and farm workers will decline not only relatively but also absolutely, and population will flow from rural to urban areas. If these adjustments do not occur quickly enough, per capita income in the farm sector will fall behind its nonfarm equivalent. In developed countries this phenomenon is known as the 'farm problem'. It can be shown (Houthakker, 1967) that the severity of this problem depends on the growth rate of farm productivity: by becoming more efficient farmers tend to work themselves (or more precisely their less efficient competitors) out of a job. The root cause, however, is Engel's Law, which is peculiar to food; attempts to demonstrate similar laws for other categories of expenditure have had less success. It should be added that when applying to agricultural development the decomposition of the expenditure elasticity into a quantity and a quality elasticity is relevant. The reason is that, by and large, agriculture produces quantity, while food processing and trade produce quality. In the case of total food Bunkers and Cochrane (1957) have shown that the quantity elasticity is much less than the expenditure elasticity.

H.S. HOUTHAKKER

See also HOUSEHOLD BUDGETS.

BIBLIOGRAPHY

Bunkers, E.W. and Cochrane, W.W. 1957. On the income elasticity of food services. *Review of Economics and Statistics* 39(2), May, 211–17.

Houthakker, H.S. 1967. *Economic Policy for the Farm Sector.* Washington, DC: American Enterprise Institute for Public Policy Research.

Engels, Friedrich (1820–1895). Born in Barmen, the eldest son of a textile manufacturer in Westphalia, Engels was trained for a merchant's profession. From school onwards however, he developed radical literary ambitions which eventually brought him into contact with the Young Hegelian circle in Berlin in 1841. In 1842, Engels left for England to work in his father's Manchester firm. Already converted by Moses Hess to a belief in 'communism' and the imminence of an English social revolution, he used his two-year stay to study the conditions which would bring it about. From this visit, came two works which were to make an important contribution to the formation of Marxian socialism: 'Outlines of a Critique of Political Economy' (generally called the 'Umrisse') published in 1844 and *The Condition of the Working Class in England*, published in Leipzig in 1845.

Returning home via Paris in 1844, Engels had his first serious meeting with Marx. Their life-long collaboration dated from this point with an agreement to produce a joint work (*The Holy Family*), setting out their positions against other tendencies within Young Hegelianism. This was followed by a second unfinished joint enterprise, (*The German Ideology*, 1845–7), where their materialist conception of history was expounded systematically for the first time.

Between 1845 and 1848, Engels was engaged in political work among German communist groups in Paris and Brussels. In the 1848 revolution itself, he took a full part, first as a collaborator of Marx on the *Neue Rheinische Zeitung* and subsequently in the last phase of armed resistance to counter-revolution in the summer of 1849.

In 1850, Engels returned once more to Manchester to work for his father's firm and remained there until he retired in 1870. During this period, in addition to numerous journalistic contributions, including attempts to publicize Marx's *Critique of Political Economy* (1859) and *Capital*, Volume One (1867), he first developed his interest in the relationship between historical materialism and the natural sciences. These writings were posthumously published as *The Dialectics of Nature* (1925). In 1870 Engels moved to London.

As Marx's health declined, Engels took over most of his political work in the last years of the First International (1864–72) and took increasing responsibility for corresponding with the newly founded German Social Democratic Party and other infant socialist parties. Engels's most important work during this period was his polemic against the positivist German socialist, Eugen Dühring. The *Anti-Dühring* (1877) was the first comprehensive exposition of a marxian socialism in the realms of philosophy, history and political economy. The success of this work, and in particular of extracts from it like *Socialism, Utopian and Scientific*, represented the decisive turning point in the international diffusion of Marxism and shaped its understanding as a theory in the period before 1914.

In his last years after Marx's death in 1883, Engels devoted most of his time to the editing and publishing of the remaining volumes of *Capital* from Marx's manuscripts. Volume Two appeared in 1885, Volume Three in 1894, a year before his death. Engels had also hoped to prepare the final volume dealing with the history of political economy. But the difficulty of deciphering Marx's handwriting, his own failing eyesight and the formidable editorial problems encountered in constructing Volumes Two and Three, induced him to hand over this task to Karl Kautsky, who subsequently published it under the title *Theories of Surplus Value*.

Engels's work was of importance, both in the construction and interpretation of Marxian economic theory and in the laying down of important guidelines in the subsequent development of marxist economic policy.

In the realm of theory, his contribution is of particular significance in three respects.

First, and of real importance in the formation of a distinctively marxian stance towards political economy was Engels's 'Outlines of a Critique of Political Economy' (the *Umrisse*), published in 1844. In 1859 in his own *Critique of Political Economy*, Marx acknowledged this sketch as 'brilliant' (Marx, 1859) and its impact is discernible in Marx's 1844 writings. The *Umrisse* represented the first systematic confrontation between the 'communist' strand of Young Hegelianism and political economy. The communist aspiration was expressed in Feuerbachian language, while the mode of analysis was Hegelian. But, as has recently been demonstrated (Claeys, 1984), the content of Engels's critique was first and foremost a product of his early stay in Manchester. For, apart from some indebtedness to Proudhon's *What is Property?* (1841), the main source of Engels's essay was John Watts, *The Facts and Fictions of Political Economy* (1842), a resumé of the Owenite case against the propositions of political economy. At this stage, Engels's own acquaintance with the work of political economists seems to have been mainly at second hand.

The *Umrisse* was an attempt to demonstrate that all the categories of political economy presupposed competition which in turn presupposed private property. He began with an analysis of value, which juxtaposed a 'subjective' conception of value as utility ascribed to Say with an 'objective' conception as cost of production attributed to Ricardo and McCulloch. Reconciling these two definitions in Hegelian fashion, Engels defined value as the relation of production costs to utility. This was the equitable basis of exchange, but one impossible to implement on the basis of competition which was responsive to market demand rather than social need. (Engels still adhered to this definition of value thirty years later in the *Anti-Dühring*. Discussing the disappearance of the 'law of value' with the end of commodity production, he wrote:

> As long ago as 1844, I stated that the above mentioned balancing of useful effects and expenditure of labour would be all that would be left, in a communist society, of the concept of value as it appears in political economy... . The scientific justification for this statement, however,... was only made possible by Marx's *Capital* (Engels, 1877, pp. 367–8).

This shows how much greater continuity of thought there was between the young and the old Engels than is normally imagined.)

He next analysed rent, counterposing a Ricardian notion of differential productivity to one attributed to Smith and T.P. Thompson based upon competition. Interestingly, in this analysis Engels differed both from Watts and Proudhon, in denying the radical form of the labour theory – the right to the whole product of labour – both by citing the case of the need to support children and in querying the possibility of calculating the share of labour in the product.

Finally, after an attack on the Malthusian population theory, which closely followed Alison and Watts, Engels attacked competition itself, both because it provided no mechanism of reconciling general and individual interest, and because it was argued to be self-contradictory. Competition based on self-interest bred monopoly. Competition as an immanent law of private property led to polarization and the centralization of property. Thus private property under competition is self-consuming.

What particularly impressed Marx was the argument that all the categories of political economy were tied to the assumption of competition based on private property. This, for him, represented an important advance over Proudhon whose notion of equal wage would lead to a society conceived as 'abstract capitalist' and whose conception of labour right presupposed private property. Proudhon had not seen that labour was the essence of private property. His critique was of 'political economy from the standpoint of political economy'. He had not 'considered the further creations of private property, e.g. wages, trade, value, price, money etc. as forms of private property in themselves' (Marx, 1844, p. 312). The Umrisse suggested a new means of underpinning the marxian ambition to transcend the categorical world of political economy and private property altogether. Moreover, by representing competition as a law which would produce its opposite, monopoly, the elimination of private property and

revolution, Engels preceded Marx in positing the 'free trade system' as a process moving towards self-destruction through the operation of laws immanent within it.

These conclusions were amplified in Engels's other major work of this period, *The Condition of the Working Class in England*. Here, the law of competition by engendering 'the industrial revolution' had created a revolutionary new force, the working class. The single thread underlying the development of the working class movement had been the attempt to overcome competition. Such an analysis prefigured the famous statement in the *Communist Manifesto* that the capitalists were begetting their own gravediggers (Stedman Jones, 1977).

Between the mid-1840s and the mid-1870s, Engels played no discernible part in the elaboration of *Capital* beyond supplying Marx with practical business information. His vital contributions to the pre-history of the theory were forgotten and it was only in his better-known role as interpreter and publicist of Marx's work that his writings received widespread attention. During the Second International period, these writings attained almost canonical status, but in the 20th century they have generally provided a polemical target for all those attempting to retheorize Marx in the light of the publication of his early writings.

In the realm of political economy more narrowly conceived, Engels helped to set up the 'transformation' debate by his dramatization of Marx's switch from value to production price in his introductions to Volumes Two and Three of *Capital*. Engels's own contribution to this debate in his last published article in *Neue Zeit* in 1895 (now published as 'Supplement and Addendum' to Volume Three of *Capital*) was to argue that the shift from value to production price was not merely a logical development entailed by the enlargement of the scope of investigation to include circulation and the 'process of capitalist production as a whole', but also reflected a real historical transition from the stage of simple commodity production to that of capitalism proper. 'The Marxian law of value has a universal economic validity for an era lasting from the beginning of the exchange that transforms products into commodities down to the fifteenth century of our epoch' (Marx, 1894, p. 1037).

Leaving aside the empirical question whether during the pre-capitalist era commodities were exchanged in accordance with the amount of labour embodied in them, commentators as diverse as Bernstein and Rubin, have objected that this makes no sense in terms of Marx's theory, since during this epoch, there existed 'no mechanism of the general equalisation of different individual labour expenditures in separate economic units on the market' and that consequently it was not appropriate to speak of 'abstract and socially necessary labour which is the basis of the theory of value' (Rubin, 1928, p. 254). They have further objected, appealing to Marx's 1857 'Introduction to the Critique of Political Economy', that there is no necessary connection between the logical and historical sequence of concepts, and that the order of appearance of concepts in *Capital* is determined simply by the logical place they occupy in an exposition of the theory of the capitalist mode of production.

Engels could certainly claim explicit textual support from Volume Three for his historical interpretation of value ('It is also quite apposite to view the value of commodities not only as theoretically prior to the prices of production, but also as historically prior to them. This applies to those conditions in which the means of production belong to the worker...': Marx 1894, p. 277.) It should also be stressed that there was nothing new in Engels's representation of the character of Marx's theory. Back in 1859, in a review of Marx's *Critique of Political Economy*, Engels stated, 'Marx was, and is, the only one who could undertake the work of extracting from the Hegelian Logic the kernel which comprised Hegel's real discoveries ... and to construct the dialectical method divested of its idealistic trappings' (Engels 1859, pp. 474–5); and in characterizing that method as a form of identity between logical and historical progression, he continued, 'the chain of thought must begin with the same thing that this history begins with, and its further course will be nothing but the mirror image of the historical course in abstract and theoretically consistent form...' (ibid., p. 475). It is implausible to suppose that Marx at this time should have sanctioned a fundamental distortion of his method and it is suggestive that he himself, describing his relationship to Hegel should have endorsed the metaphor of discovering 'the rational kernel in the mystical shell' in his 1873 Postface to the Second Edition of *Capital* (Marx, 1873, p. 103). Perhaps the real difficulty lies not in Engels, but in Marx himself. It may be, as Louis Althusser has claimed, that Marx did not find a suitable language in which to characterize the distinctiveness of his approach, or it may be more simply that Marx remained ambivalent about how to characterize the theory. In any event, it is not difficult to establish disjunctions between the way he proceeds and the descriptions he gives of his procedures. Engels stuck fairly closely to Marx's descriptions of his procedures and can hardly be reproached for taking Marx at his word.

The problem of Engels's role as an interpreter of Marx's theory debouches onto a third and potentially yet more contentious aspect of Engels's legacy, his role as editor of *Capital*, Volumes Two and Three. Engels's work was not confined to the transcription of Marx's illegible handwriting. He had to make active editorial choices. The published versions of these volumes contain over 1300 pages, but the original manuscripts amount to almost twice as many. For Volume Two for instance, Marx had composed eight versions of his treatment of the process of circulation, from which Engels made a collation. In the absence of an independent transcription and publication of the manuscripts, from which Engels worked, it is impossible to assess whether the emphasis and meaning of the published Volumes differ in any significant way from the original. What seems clear, is that in his cautious desire to reproduce as much of the original material as possible, Engels produced a much bulkier and more repetitive version than Marx originally intended. Marx, it seems, always hoped that *Capital* should consist of two volumes and a further volume on the history of political economy (Rubel 1968, Levine, 1984). From a detailed comparison of Volume Two, Part 1, with the original manuscripts, it appears that Engels also occasionally committed inaccuracies in the citation of the manuscripts he had used (Levine, 1984). Much more doubtful, given all we know of Engels's caution as an editor, is the further suggestion that Engels's editing procedures may have shifted the meaning of the text in ways that lent support to a 'collapse theory' of capitalism (*Zusammenbruchstheorie*) (Levine, 1984). Apart from the smallness of the sample and Engels's own reservations about such a theory, the fact is that proponents of such a position already had sufficient ammunition from *Capital*, Volume One. Moreover, it simply begs the question whether Marx's attitude to the collapse of capitalism was any more or less apocalyptic than that of Engels.

This discussion by no means exhausts Engels's importance in the history of economic theory or policy. A fuller treatment would have to discuss his analysis of the 'peasant question'

which included the important prescription that collectivisation must be by example rather than force, his definition of political economy in the *Anti-Dühring*, his interpolations in Capital, Volume Three, on banks, the stock exchange and cartels which set the agenda for the early 20th-century discussion of finance capital, his various writings on the relationship between the state and economic forces and his later surveys of English developments since 1844 which prepared the way for later marxist theories of labour aristocracy. These are only some of the more salient examples.

Finally, at a time when it seems that the technical debate on value seems to have reached a moment of exhaustion, it is perhaps worth going back to Engels if only to remind us of the anti-economic purpose underlying Marx's attempt to construct a theory of value in the first place.

GARETH STEDMAN JONES

SELECTED WORKS

1843. *Outlines of a Critique of Political Economy*. In Karl Marx and Frederick Engels, *Collected Works* [MECW], Vol. III, London: Lawrence & Wishart, 1975.

1845. *The Condition of the Working Class in England*. MECW, Vol. IV, London: Lawrence & Wishart, 1975.

1859. Karl Marx, *A Contribution to the Critique of Political Economy*. MECW, Vol. XVI, London: Lawrence & Wishart, 1976. iSN*Anti-Dühring*. Moscow: Foreign Languages Publishing House, 1954.

1894. *The Peasant Question in France and Germany*. In Karl Marx and Frederick Engels, *Selected Works*, Vol. 3, Moscow: Progress Publishers, 1970.

n.d. *Engels on Capital*. London: Lawrence & Wishart.

BIBLIOGRAPHY

Claeys, G. 1984. Engels' *Outlines of a critique of political economy* (1843) and the origins of the Marxist critique of capitalism. *History of Political Economy* 16(2), Summer, 207–32.

Levine, N. 1984. *Dialogue within Dialectics*. London: Allen & Unwin.

Marx, K. and Engels, F. 1844. *The Holy Family*. In *Collected Works*, Vol. IV.

Marx, K. 1959. *Contribution to a Critique of Political Economy: preface*. In Marx–Engels, *Collected Works* (MECW), vol. XV.

Marx, K. 1873. *Capital*, Vol. I, 2nd edn. Harmondsworth: Penguin, 1976.

Marx, K. 1894. *Capital*, Vol. III. Harmondsworth: Penguin, 1981.

Rubel, M. (ed.) 1968. *Karl Marx, Oeuvres*, Vol. II. Paris: Gallimard.

Rubin, I. 1928. *Essays on Marx's Theory of Value*. Detroit: Black & Red, 1972.

Stedman Jones, G. 1977. Engels and the history of Marxism. In *The History of Marxism*, ed. E.J. Hobsbawm, Hassocks: Harvester, 1983.

English historical school. A group of economists whose heyday was from 1875 to 1890 and whose major figures were John Kells Ingram (1823–1907), James E. Thorold Rogers (1822–1890), T.E. Cliffe Leslie (1827–1882), William Cunningham (1849–1919), Arnold Toynbee (1852–1883), William Ashley (1860–1927) and W.A.S. Hewins (1865–1931). H.S. Foxwell (1849–1936) was sympathetic to their approach but outside the group's mainstream. All were united by an inductive approach to economics, a determination to stress that no economic theory or policy could be appropriate to all times and places, and a conviction that classical and neoclassical economics alike were already too abstract to give state or citizen much practical help, and were getting worse.

The movement's most important forerunner was Richard Jones (1790–1855), whose criticisms of Ricardian economics – both for its hyper-deductive character and its pretensions to universality – enjoyed intelligent public attention without much persuasive power. Jones offered neither a historically relative political economy to put in Ricardianism's place nor even any substantial contribution to economic history. But, in any case, the time was not right for Jones's ideas to take hold. By the 1870s a number of factors had combined to prepare the ground for a far more influential historical critique of orthodox economics. There was the influence of John Stuart Mill, who in his later years both practised and lent his philosophical authority to a more inductive approach to political economy. Yet when Mill's influence was removed by his death in 1873, silencing the most authoritative voice in economics, the collapse of classical orthodoxy was further accelerated. And of its two main potential heirs, marginalism and historicism, it was the historicists who were more in tune with the general intellectual climate of the time.

As Darwinian ideas were absorbed into social science, the call went up for an evolutionary (and hence relativistic) science of political economy. (No one was to call for it more loudly than Marshall.) The Comtean critique of overspecialization within social science was still near its zenith, and applied with especial force to the increasingly narrow world of neoclassical economics. 'Straight' history was increasingly emphasizing its economic aspects in the work of F.W. Maitland, F. Seebohm and P. Vinogradoff. And, for those who were prepared to listen, Karl Marx was reiterating the potential scope and grandeur of economic dynamics.

The representatives of the English historical school drew on such influences with varying degrees of emphasis. Ingram used his presidency of Section F of the British Association (the social science section) to mount an explicitly Comtean attack on political economy's 'narrowness' in 1878. Ashley painstakingly catalogued the aspects of Marxism with which he was and was not in agreement. The one conditioning factor which, oddly enough, was of limited influence was the work of the German historical school of economists. English historicists might invoke the authority of their German contemporaries; Ashley and Hewins had important contacts with the later German historical school; but it is hard to point to any German historicist as a major formative influence on any English counterpart.

What, then, was the detailed message of the historical school? (In answering this question we shall be able to throw light on how far it should be regarded as a distinct 'school' at all.) First, as has already been mentioned, they were reacting against the narrow scope of orthodox economics. Thus Ingram's address of 1878, while accepting the arguments in favour of doing 'one thing at a time' warned that the social sciences were still branches of one subject 'and the relations of the branches may be precisely the most important thing to be kept in view respecting them'. Ingram saw the narrow intellectual vision of orthodox economists as both cause and consequence of their neglect of moral issues, and further argued that once it was accepted that 'the idea of forming a true theory of the economic frame and working of society apart from its other sides is illusory' it necessarily followed that 'the economic structure of society and its mode of development cannot be deductively foreseen but must be ascertained by direct historical investigation' (Ingram, 1879).

But should one's methodological stance in fact depend on one's assessment of the appropriate intellectual boundaries of economics? J.A. Hobson was later to argue that the two issues had nothing whatever to do with one another. However, historicists to a man – albeit with different degrees of emphasis – followed Ingram's lead in using their calls for a broader-based discipline to buttress their onslaught on

unbalanced deductivism. The link was 'economic man', seen by historicists as an unreal psychological stereotype wholly unable to support the pyramids of deductive logic burdened upon him by Ricardians and Jevonians alike. Whether it was wealth or utility that he was supposed to maximize, he turned out very much the same, 'an abstraction compounding a great variety of different and heterogeneous motives which have been mistaken for a single homogeneous force' (Leslie, 1888). Other Ricardian propositions which, in Leslie's view, contradicted actual experience included the quantity theory of money and the contention that competition operated so as to equalize rates of profit across the economy.

Leslie's suggestion that the whole edifice of Ricardian economics be levelled to the ground, prior to economists making a fresh and cautious start, marked the high point of historicist iconoclasm. There were a number of different stopping-places (most of them inhabited by Ashley at one time or another) along the road from orthodoxy to this extreme point. Yet the historicists hang together as a school because of their common emphasis on factual and statistical thoroughness, on the relativity of economic doctrines, and on entering unfamiliar territory with an open mind and doing painstaking research before allowing the first tentative inductive generalizations to filter through. The most orthodox of the school, Thorold Rogers, made the most impressive statistical contribution with his *History of Agriculture and Prices in England* (1866) which, among other objectives, sought to marshall the figures needed to refute Ricardian rent and wage theory. Ashley's verdict, however, that Rogers' practice of merely illustrating his preconceived opinions with historical material was alien to a genuine historical method has been endorsed by modern commentators.

It would be wrong to conclude from the above that the historical school was hostile to deduction as such. 'Deduction', said Ingram, 'is a legitimate process when it sets out not from *a priori* assumptions, but from proved generalisations.' The historicist position, in effect, was that one had to ascertain by factual investigation exactly how amenable to deductive analysis different economic phenomena actually were. That the calculating maximizing spirit (where it existed) was amenable to Ricardian treatment was conceded on all sides. This point had been heavily stressed by Walter Bagehot (in his centenary essay on *The Wealth of Nations*) in the hope of rendering orthodox economics more plausible by demarcating its boundaries as those of the modern commercial world. Ashley's inaugural lecture at Harvard in 1893 endorsed this point; Cunningham's *Modern Civilisation in its Economic Aspects* (1896) asserted that deductive analysis was coming into its own because 'business of a modern type is being extended over a larger and larger area'. That this last tendency was – on balance – welcomed by Ashley and regretted by Cunningham may help explain the difference in their attitudes to Marshallian economics. Ashley (who was to become professor of commerce at Birmingham in 1901) shared Marshall's enthusiasm for most of what the modern businessman represented. In Cunningham, by contrast, distaste for the modern world and nostalgia for the Middle Ages predominated. But personal temperament counted for just as much in explaining the contrast between Ashley's relatively placatory attitude to Marshall and Cunningham's violently hostile one.

Marshall's inaugural lecture at Cambridge in 1885 had met, head-on, the historicist assertion that the forces of custom and habit in economic life were strong enough to make orthodox economics, with its basic postulate of maximization, widely redundant. Marshall predicted that 'economic science' would soon be even more successful than it was already in 'break[ing] up and explain[ing] economic customs'; asserted that statements that this or that economic arrangement was due to custom were little more than confessions of ignorance of true causes; and entrusted economic analysis with the illumination of such ignorance – the demonstration, for example, that 'rents seldom diverge much for a long time from their Ricardian level in the East' (Marshall, 1885). Cunningham, while regarding the whole lecture as a personal and public affront, fastened especially onto this last point, telling the British Association (1889) that 'Professor Marshall, instead of accepting the description of mediaeval or Indian economic forms as they actually occur, sets himself to show that the accounts of them can be so arranged and stated as to afford illustrations of Ricardo's law of rent.' Marshall's *Principles of Economics*, published the following year, opened with a long historical introduction which Ashley saw as a conciliatory gesture and Cunningham as a further provocation. (Today it reads as neither.) In 'The Perversion of Economic History' (*Economic Journal*, September 1892), Cunningham joyously rebuked what he saw as Marshall's hasty and amateurish style of historiography. It would all have read more convincingly if Cunningham had refrained from grotesquely out-of-context quotation, even at one point inserting a rogue word into Marshall's text to make it sound marginally more implausible.

Marshall's reply to Cunningham's criticisms (it took Cunningham three years and seven polemics to induce it) was seen in most quarters as the final statement in the dispute (if only because the *Economic Journal* refused Cunningham the space for a counter-riposte.) Ashley, in his Harvard inaugural the following year, praised the historical chapters in the *Principles* and claimed that 'to most of us the recent exchange of hostilities between two distinguished English economists has seemed almost an anachronism'.

The methodological debate, then, subsided after the early 1890s. But the Protectionist controversy which began when Joseph Chamberlain disavowed free trade in 1903 saw survivors of the old historicists grouping reconstituted for a new battle. The episode is best approached via a general look at historicist attitudes to policy questions.

It is no coincidence that the entire historical school, regardless of whether as individuals they were of the 'left' or the 'right', favoured an acceleration of the existing trend towards increased state intervention in the economy. Irish social reform, the recognition and legal protection of the trades unions, and the conditions of industrial and agricultural workers were all seen as urgent areas of responsibility for the state. The general view was well summarized by Foxwell (1885):

> we have been suffering for a century from an acute outbreak of individualism unchecked by the old restraints and invested with almost a religious sanction by a certain soul-less school of writers. The narrowest selfishness has been recommended as public virtue.

Ingram praised the German historical school for upholding the power of the state as 'the organ of the nation for all ends which cannot be adequately effected by voluntary individual effort'. Cunningham's *Politics and Economics* (1885) introduced his readers to 'National Husbandry', Cunningham's scheme for an economic policy holistic in its inspiration and nationalistic in its objectives: 'the duty we owe to posterity [is] to make the future of our nation as great and noble as lies within our power.'

The link between holism (refusal to isolate the individual as a unit of analysis) and historical relativism was an irreproachably logical one: only if an individual can be isolated from his social context can a theory involving him be isolated from time and place. And Cunningham for one kept his readers' eyes firmly on the fact that policy recommendations were as historically relative as economic principles, even suggesting at one point that the fact that a measure had worked well in very different circumstances was a consideration *against* proposing it here and now. Such pragmatism characterized much of the protectionist campaign. If free-trading economists were to be charged with inflexible dogmatism, intellectual arrogance and subservience to abstractions it was essential that no such taint could be thought to cling to the Protectionist cause. Ashley, indeed, never went beyond recommending temporary and selective tariffs for purposes of retaliation, and stressed that 'with England as she has been for some centuries the notion that imports are paid for by money which might otherwise be spent at home is the crudest of popular fallacies.' Cunningham – eventually – did arrive at a more thoroughly protectionist stance than this, but it took him until 1910 to do so. And by 1910 the steam was running out of the protectionist campaign anyway, at least as far as the historical school was concerned. Ashley's administrative responsibilities at Birmingham and Hewins's parliamentary ones virtually terminated their contributions to serious economic debate; Cunningham turned his attention to the relations between Christianity, political practice and social science. The historical school's achievements were complete by 1914.

How significant were they? Today their part in the foundation of economic history as a subject in its own right is more obvious than their contribution to economics. Their lack of facility with marginal analysis – no historicist tried to master the neoclassical 'paradigm' and it must be doubted whether most of them would have been able to handle it even if they had tried – relegated them to outsiders' roles once the dominance of neoclassicism was secured. Could they have prevented this dominance? The answer depends on whether one thinks that the inductive, historically based economics which they demanded but ostentatiously failed to supply could ever have been a feasible project. As it was, their lack of solid achievement inevitably weakened their position even as critics. Yet they forced both Marshall and his disciples to change both their thoughts and their presentation of these thoughts in a number of ways. Economic concepts were more carefully defined, and the bounds of their applicability more precisely demarcated. Policy recommendation became more cautious and less likely to be accompanied by exaggerated statements of the contributions of pure theory. The modern economist, said L.L. Price (1906),

> evinces a readiness to recognise without reserve those qualifications of subtle delicate theory which a comparison with rough, unyielding facts must necessarily require. This reasonable attitude is largely due to the abiding influence of the vigorous controversy in which Cliffe Leslie bore a leading part.

J. MALONEY

See also GERMAN HISTORICAL SCHOOL.

BIBLIOGRAPHY
Ashley, W.J. 1903. *The Tariff Problem*. London: P.S. King & Son.
Coats, A.W. 1954. The historicist reaction in English political economy, 1870–90. *Economica* 21, May, 143–53.
Cunningham, W. 1885. *Politics and Economics*. London: Kegan Paul, Trench & Co.
Foxwell, H.S. 1885. What is political economy? *The Eagle*, No. 79.

Ingram, J.K. 1878. *The Present Position and Prospects of Political Economy*. Dublin and London: Longmans. Reprinted in *Essays in Economic Method*, ed. R.L. Smith, London: Duckworth, 1962.

Koot, G. 1980. English historical economics and the emergence of economic history in England. *History of Political Economy* 12(2), Summer, 174–205.

Marshall, A. 1885. *The Present Position of Economics*. London: Macmillan.

Rogers, J.E.T. 1866–1902. *A History of Agriculture and Prices in England*. Oxford: Clarendon Press.

Semmel, B. 1960. *Imperialism and Social Reform*. London: George Allen & Unwin.

Enlightenment. *See* SCOTTISH ENLIGHTENMENT.

entitlements. In the strong sense, an entitlement is something owed by one set of persons to another. The thing owed is either a performance of a certain kind, such as a dental extraction, or a forbearance from interfering from some aspect of the title-holder's activity or enjoyment, such as not trespassing on someone's land. Strong entitlements imply the presence of a right in the person entitled and a corresponding or *correlative* obligation in the person owing the performance or forbearance. Typically, the person entitled is further vested with ancillary powers to waive the obligation or, alternatively, to initiate proceedings for its enforcement. A secondary (and contested) instance of a strong entitlement arises with respect to the position of a third-party beneficiary of a right–obligation relation between two other parties, such as the beneficiary of an insurance policy. Third parties usually lack powers of waiver and enforcement, for it is not strictly to them that fulfilment of the obligation is owed.

A weaker form of entitlement may be said to pertain to those of a person's activities which, while not specifically protected by obligations in others not to interfere, are nevertheless indirectly and extensively protected by their other forbearance obligations. Thus, while persons may be under no obligation specifically to allow someone to use a pay telephone, they probably do have forbearance obligations with respect to assault, theft, property damage, etc., the joint effect of which is to afford some high (but incomplete) degree of protection to someone using a pay telephone. However, such an entitlement amounts to less than the full protection afforded by a right inasmuch as it does not, for example, avail against anyone who may already be using that telephone.

Beyond strong and weak entitlements, one may also possess many largely unprotected liberties. These consist in those activities from which one has no obligation to refrain but with which, equally, no direct or extensive indirect claims to non-interference. So, broadly speaking, persons' strong entitlements may be construed as conjunctively constituting their spheres of ownership, while their weak entitlements and their unprotected liberties constitute the fields of activity within which they exercise the powers and privileges of ownership. Normally, it is persons' strong entitlements that are of primary normative concern, with weak entitlements and unprotected liberties being determined residually.

Entitlements may be either legal or moral. Sets of legal entitlements tend to reflect the multifarious demands of various customs, moral principles, judicial decisions and state policy. A set of moral entitlements, on the other hand, is commonly derived from some basic principle embedded in a moral code. The nature of this derivation varies with the type of code involved. In many single-value codes (such as utilitarianism), entitlements are instrumental in character:

whether and what sort of an obligation is owed, by one person to another, depends upon the relative magnitude of the contribution that fulfilment of that obligation would make to realizing that value. Changing causal conditions of maximization warrant alterations in the content and distribution of entitlements. Codes containing a plurality of independent values characteristically generate entitlements from a principle of justice. The set of entitlements thus derived possesses intrinsic and not merely instrumental value, though its normative status depends upon the ranking of justice in relation to the code's other values. In such codes, the chief distinction between moral obligations that (like kindness) are not correlative to any entitlement and those of justice that are, lies in the fact that only the latter are waivable and permissibly enforceable.

Much of the philosophical treatment of entitlements is located in discussions of rival theories of justice. These theories differ according to the various norms they propose for determining who owes what to whom. Endorsing the classical formal conditions of justice – 'rendering to each what is due to him' and 'treating like cases alike' – they diverge widely in their interpretations of what is due to a person and what count as like cases. Procedural and substantive criteria that have been offered for determining individuals' entitlements include: relative need, productivity, equal freedom, equal utility, personal moral worth, interpersonal neutrality, personal inviolability, initial contract and so forth. As is immediately obvious, the nature and distribution of the entitlements mandated by each of these criteria are by no means self-evident, and their identification thus requires supplementary postulates that are variously drawn from psychological theories, from theories of moral and rational choice, and from conceptual analyses of the criteria themselves. It is also true that not all of these criteria are mutually exclusive: given a plausible set of premises, some can be derived from others.

There are other dimensions, apart from their distributive norms, in which theories of just entitlements differ. Some of these differences are logically implied by the nature of the norms themselves, while others are independent of them. One such dimension is the kinds of object to be distributed in conformity with a proposed criterion. Proffered items include all utility-producing goods, means of production, natural resources, the rents of superior skills or talents, and even human body parts. What one may do with the things to which one has strong entitlements – what weak entitlements and unprotected liberties one possesses – is largely a function of the sorts of thing to which others are strongly entitled. The intricate structure of permissibility, jointly formed by the rights one has against others and the rights others have against oneself, constitutes the fields of activity within which each person exercises those rights. It thereby also determines the respective spheres of market, state and charitable activities.

A third differentiating dimension is the range of subjects to be counted as having entitlements. Generally accepting the membership of all adult human beings in the class of title-holders, theories differ over whether their distributive norms extend to minors, members of other societies, deceased persons (in respect of bequest), persons conceived but not yet born (in respect of abortion), persons not yet conceived (in respect of capital accumulation and environmental conservation) and non-human animals. Again, the nature and interpretation of a theory's distributive criterion often work to delimit its class of title-holders.

In the light of this multiplicity of differentiating dimensions, the classification – let alone assessment – of theories is no simple task. One, but by no means the only, important respect in which many of them can be compared is in terms of the scope they allow for unconstrained individual choice. Thus theories might be ranged along a spectrum from those that prescribe only an initial set of entitlements (permitting persons thereafter to dispose of these as they choose), to those that require constant enforceable adjustment of the content and distribution of entitlements to conform to certain norms. However, even this way of arraying competing theories is somewhat underspecified, inasmuch as it fails to capture the varied ramifications of the restrictions implied by different initial entitlements.

Hence it is an open question as to where on this spectrum one would locate theories that (via a unanimity requirement) construe each person's initial entitlement as a veto on a social or constitutional contract. Such an entitlement may in turn be derived from some interpretation of equal freedom, personal inviolability or interpersonal neutrality. Or it may itself be taken as an intuitively acceptable foundational postulate for deriving a more complex set of entitlements. Whether an initial contract theory is permissive or restrictive of wide individual choice depends upon its account of the terms of that contract. The derivation of these terms usually proceeds from some conception of human nature – of human knowledge and motivation – along with some meta-ethical theory about the nature of moral reasoning. Contractual terms generated by these premises may extend only to the design of political institutions, thereby leaving the determination of individuals' substantive entitlements to the legislative process. Alternatively, such contracts may stipulate a set of basic individual rights that are immune to legislative encroachment. In either case, the resultant scope for individual choice remains underdetermined. In the first case it depends upon the extent of legislation, while in the second it depends upon the size and nature of the stipulated set of rights. Laws and constitutional rights imply both restrictions on each person's conduct but also, *ipso facto*, restrictions on the extent of permissible interference with others' conduct.

Dispensing with the initial contract device and hypothetical unanimous agreement, some theories derive a set of entitlements directly (non-procedurally) from a substantive foundational value. Among such theories, one type assigns entitlements according to the differential incidence of some stipulated variable in the population of title-holders. Need and productivity are particularly prominent variables in this field, often acquiring their normative import from the values of welfare equalization and maximization. Clearly, applications of these distributive criteria respectively presuppose accounts of essential human requirements and of economic value. Although, for such theories, any shift in the incidence of the stipulated variable occasions a corresponding adjustment of entitlements, the issue of whether this adjustment must be imposed or occurs spontaneously partly turns on the model of interactive behaviour employed. In general, models indicating spontaneous adjustment generate that conclusion by ascribing dominance to altruistic (need) or income-maximizing (productivity) behaviour. To the extent that these ascriptions are empirically unrealistic, such theories mandate enforceable restrictions on the scope for individual choice.

Another type of directly derived (non-contract-based) entitlement set is drawn from foundational values like equal freedom, personal inviolability or interpersonal neutrality, which, by definition, are of uniform non-differential incidence in the population of title-holders. Varying interpretations of these concepts tend nonetheless to converge on the Kantian injunction that persons must be treated as ends in themselves and, more specifically, that no person's ends may be systematically subordinated to those of another. Here the theoretical task is to design a set of entitlements that is independent of any particular conception of 'the good' – independent of particular preferences and (other) moral values – and that is such as to ensure that the consequences of persons' actions, whether harmful or beneficial, are not imposed on others. A typical, though by no means invariable, structural feature of such an entitlement set is its extensive use of a threefold classification of things in the world as selves, raw natural resources and objects which are combinations of these. While title-holders are each vested with ownership of themselves (their bodies and labour), such theories often contain some sort of egalitarian constraint on individual entitlements to raw natural resources. The precise form of this constraint determines the nature of the encumbrances that may be imposed on the ownership of objects in the third category. But since these encumbrances exhaust the restrictions on what persons may do with what they own, such theories are presumed to allow considerable scope for individual choice.

It is hardly worth remarking that many theories of entitlement combine aspects of the three types outlined above. The assessment of competing theories – a complex task, as stated previously – commonly consists in testing for internal coherence and in appraising the interpretations placed on core concepts in the theory. Thus, if it is supposed that the moral principle underpinning a set of entitlements is that of justice, and that justice is analytically linked to the concept of rights, there is room for dispute as to whether the first (initial contract) and second (needs, productivity) types of theory are properly viewed as theories of entitlement. A distinctive normative feature of rights is that they are held non-contingently to confer an element of individuated discretion on their owners. It is unclear whether possession of a veto in a collective-choice procedure amounts to a sufficiently individuated sphere of discretion. On the other hand, the entitlements generated by considerations of need or productivity, while sufficiently individuated, appear to lack any necessarily discretionary character. A difficulty besetting the first and third types of theory arises with regard to the notion of initial entitlements. Specifically, it seems clear that the identification of each person's initial entitlement – either in a collective-choice procedure or under an egalitarian constraint on natural resource ownership – cannot be interpreted as an historically 'one-off' determination, in the face of an undecidable number and size of partially concurrent future generations. These are among the more salient problems commanding attention in current work on theories of entitlement.

HILLEL STEINER

See also ECONOMIC FREEDOM; EQUALITY; INEQUALITY; JUSTICE; POVERTY; PROPERTY RIGHTS; REDISTRIBUTION OF INCOME AND WEALTH.

BIBLIOGRAPHY

Buchanan, J.M. 1974. *The Limits of Liberty*. Chicago: University of Chicago Press.

Demsetz, H. 1964. Toward a theory of property rights. *American Economic Review, Papers and Proceedings* 57, 347–59.

Dworkin, R. 1981. What is equality? *Philosophy and Public Affairs* 10, 185–246 and 283–345.

Hohfeld, W.N. 1919. *Fundamental Legal Conceptions*. New Haven: Yale University Press.

Lyons, D. (ed.) 1979. *Rights*. Belmont: Wadsworth Publishing Company.

Nozick, R. 1974. *Anarchy, State and Utopia*. Oxford: Blackwell.

Rawls, J. 1971. *A Theory of Justice*. Oxford: Oxford University Press, 1972.

Sen, A.K. 1981. Rights and agency. *Philosophy and Public Affairs* 11, 3–39.

Steiner, H. 1987. *An Essay on Rights*. Oxford: Blackwell.

entrepreneur. There are several theories of the entrepreneur, but very few mathematical models which formally analyse entrepreneurial behaviour within a closed economic system. Indeed, it is often argued that by its very nature entrepreneurial behaviour cannot be predicted using deterministic models. Entrepreneurship, it is claimed, is essentially a spontaneous and evolutionary phenomenon.

The term 'entrepreneur' seems to have been introduced into economic theory by Cantillon (1755), but the entrepreneur was first accorded prominence by Say (1803). It was variously translated into English as 'merchant', 'adventurer' or 'employer', though the precise meaning is the undertaker of a project. John Stuart Mill (1848) popularized the term in England, though by the turn of the century it had almost disappeared from the theoretical literature (though see Marshall, 1890).

The 'disappearance' of the entrepreneur is associated with the rise of the neoclassical school of economics. The entrepreneur fills the gap labelled 'fixed factor' in the neoclassical theory of the firm. Entrepreneurial ability is analogous to a fixed factor endowment because it sets a limit to the efficient size of the firm. The static and passive role of the entrepreneur in the neoclassical theory reflects the theory's emphasis on perfect information – which trivializes management and decision-making – and on perfect markets – which do all the coordination that is necessary and leave nothing for the entrepreneur (cf. Baumol, 1968).

According to Schumpeter (1934), the entrepreneur is the prime mover in economic development, and his function is to innovate, or 'carry out new combinations'. Five types of innovation are distinguished: the introduction of a new good (or an improvement in the quality of an existing good), the introduction of a new method of production, the opening of a new market – in particular an export market in new territory – the 'conquest of a new source of supply of raw materials or half-manufactured goods' and the creation of a new type of industrial organization – in particular the formation of a trust or some other type of monopoly. Schumpeter is also very clear about what the entrepreneur is *not*: he is not an inventor, but someone who decides to allocate resources to the exploitation of an invention; nor is he a risk-bearer: risk-bearing is the function of the capitalist who lends funds to the entrepreneur. Essentially, therefore, Schumpeter's entrepreneur has a managerial, or decision-making role.

This view receives qualified support from Hayek (1937) and Kirzner (1973), who emphasize the role of the entrepreneur in acquiring and using information. The entrepreneur's alertness to profit-opportunities, and his readiness to exploit them through arbitrage-type operations, makes him the key element in the 'market process'. Hayek and Kirzner regard the entrepreneur as responding to change – as reflected in the information he receives – whilst Schumpeter emphasized the role of the entrepreneur as a source of change. These two views are not incompatible though: a change effected by one entrepreneur may cause spill-over effects which alter the environment of other entrepreneurs. Hayek and Kirzner do not insist on the novelty of entrepreneurial activity, however, and it is certainly true that a correct decision is not always a decision to innovate; premature innovation may be commer-cially disastrous. Schumpeter begs the question of whether someone who is the first to evaluate an innovation, but decides (correctly) not to innovate, qualifies as an entrepreneur.

Leibenstein (1968) regards the entrepreneur as someone who achieves success by avoiding the inefficiencies to which other people – or the organization to which they belong – are prone. The main virtue of Leibenstein's approach is that it emphasizes that, in the real world, success is exceptional and failure is the norm.

Knight (1921) insists that decision-making involves uncertainty. Each business situation is unique, and the relative frequencies of past events cannot be used to evaluate the probabilities of future outcomes. According to Knight, measurable risks can be diversified – or 'laid off' – through insurance markets, but uncertainties cannot. Those who take decisions in highly uncertain environments must bear the full consequences of those decisions themselves. These people are entrepreneurs: they are the owners of businesses and not the salaried managers that make day-to-day decisions.

It is not clear, at first sight, whether there is any common thread which runs through these various theories of the entrepreneur. Casson (1982) attempts to identify a shared element by introducing the concept entrepreneurial judgement. The entrepreneur is defined as someone who specializes in taking judgemental decisions about the allocation of scarce resources. The essence of a judgemental decision is that there is no decision rule that can be applied that is both obviously correct and involves using only freely available information. Suppose, for example, that a decision rule, is used; then there must be some initial judgement that the chosen rule, and not some other rule, is the appropriate one. No rule can ever be fully self-justifying: there is no definitive model which demonstrates that one rule is always superior to another. Ultimately, the justification for a rule must be some property of the environment, which in many cases cannot be observed.

It is evident that this concept of judgemental decision-making rejects the 'naive neoclassical' view that all decision-making merely involves marginalist calculations based upon public information supplied by the price system. It recognizes that not only is information costly, but that the costs of acquiring information are different for different people. Furthermore, because their access to information differs, different people will make different decisions in the same situation. The essence of judgemental decision-making is that the outcome depends upon *who* makes the decision.

When judgements differ, confident individuals can back their own judgement by taking up speculative positions against other people who hold a conventional view. The confident individuals 'bet' against others by acquiring assets that they believe other people have under-valued, disposing of assets that they believe other people have over-valued, undertaking projects that other people do not consider profitable, and so on. Using this approach, the arbitraging activity described by Hayek and Kirzner, and the innovative activity described by Schumpeter, are seen to be special cases of the general concept of entrepreneurial speculation based upon self-confident judgement.

In a market economy, individuals who lack confidence in their judgement can delegate decisions to entrepreneurs. The individual entrusts his wealth to an entrepreneur, who allocates this wealth in accordance with his own judgement. In practice, the individual will often diversify his risks by using a 'portfolio' of different entrepreneurs. The delegation of decision-making can be effected in various ways. An individual may supply capital at fixed interest to an entrepreneur who is self-employed or is the owner-manager of a firm. He may own an equity stake in a firm where the entrepreneur acts partly as a salaried employee; or he may deposit his funds in a bank whose managers advance loans to firms and self-employed entrepreneurs.

To overcome the principal–agent problems involved in the delegation of decisions, it is normally necessary for the supplier of finance not only to have confidence in the entrepreneur's judgement, but also to trust the entrepreneur to exercise this judgement in pursuit of maximum profit. Unless the entrepreneur has an established reputation, he has a strategic problem in obtaining the confidence of others. Because of the differences in judgement mentioned earlier, the entrepreneur will normally be more optimistic about a project than are his potential financial backers. His backers will therefore perceive higher risks, and set a higher cost of capital than the entrepreneur believes is warranted. If, however, he persuades his backers to share his optimism, then they may preempt his project, since they already have the finance to proceed with it and he does not.

This leads directly to the question of trust. Just as the financiers must trust the entrepreneur to use his funds in their interests, so the entrepreneur must trust his financial backers not to preempt his project for themselves. Part of the problem can be solved by using an 'honest broker' such as a bank, which vets entrepreneurial projects on behalf of investors but ties its own hands by not entering into entrepreneurial projects on its own account. In countries where the banking system is underdeveloped, the extended family often fulfils a similar function of 'honest broking' between the older generation who are potential investors and the younger generation who are potential entrepreneurs. Another method of building trust is to supply finance in a sequence of small instalments so that both parties have an incentive to behave honourably in order not to put future relations between them at risk.

Much of the information required for decision-making is not merely costly to obtain, but is not available by direct observation at all. Another way of saying this is that decisions are governed not only by objective information but also by subjective beliefs. An individual's beliefs originate with his culture and his religion as well as with his direct experience of life. Some cultures appear to give greater encouragement to entrepreneurship than others. A culture which stresses individuality rather than conformity encourages an individual to form an independent judgement of a situation. A culture which emphasizes human autonomy rather than fatalistic submission to nature encourages the kind of self-confidence required of the entrepreneur. A culture which emphasizes the heroic aspects of leadership rather than the corrupting effects of power-seeking encourages individuals to undertake ambitious projects which call for a high degree of organization, and so on. Cultural values have always been emphasized in the literature on entrepreneurship: Schumpeter, for example, refers to the dream and will to found a private dynasty, the will to conquer and the joy of creating, while Weber (1930) emphasizes the Protestant ethic and the concept of calling and Redlich (1956) the militaristic values of the 'captains of industry'. Writers on business history almost invariably stress the influence of culture and personality on the behaviour of the entrepreneur.

A common criticism of theories which place considerable weight on cultural characteristics and personality traits is that they are difficult to test. Indeed, it is often suggested that because the behaviour of individual entrepreneurs tends to be unpredictable, theories of entrepreneurship are untestable. It is, however, quite possible that while the behaviour of individual entrepreneurs cannot be predicted, the behaviour of

entrepreneurs as a group is predictable. Furthermore, a theory of entrepreneurship may generate propositions relating to other social phenomena besides the behaviour of entrepreneurs themselves. With certain qualifications, it is possible to develop a model in which both the level of entrepreneurial activity and the functional distribution of income between entrepreneurship and other factors are simultaneously determined.

Given that the entrepreneur specializes in judgemental decision-making, it is possible to formulate a derived demand for entrepreneurial services which varies according to the demands which the business environment places upon judgement. The more complex the environment, the faster the pace of change, and the more radical the structural adjustments that these changes call for, the greater will be the demand for entrepreneurs. A large demand for entrepreneurs will be reflected in substantial profit opportunities for people who can anticipate changes and correctly foresee their consequences. Individual profit opportunities will not be competed away so long as each opportunity can be preempted by a single entrepreneur before others come to form the same judgement as he has done. In the short run, therefore, the successful entrepreneur can earn a monopolistic rent to superior judgement.

In the long run, however, entry into entrepreneurship will tend to compete away any expected return to entrepreneurial activity which exceeds the expected return to non-entrepreneurial activity (after due allowance for different levels of risk and for the non-pecuniary net benefits of the two kinds of activity). The competing away of entrepreneurial rents may not be complete, however, because access to capital may prove a barrier to entry for the reasons explained above.

Long-run entry corresponds to a movement along a long-run supply curve for entrepreneurs. The total supply of entrepreneurs is measured by the number of individuals whose principal activity is to exercise their judgement to allocate resources. The people concerned may be senior salaried managers or the self-employed – given the definition above, it is impossible to identify the entrepreneur simply by his contractual status in employment. An increase in the supply of entrepreneurs is effected by individuals transferring out of manual work and non-entrepreneurial decision-making (i.e. routine management), and from unemployment and leisure, and by net inward migration of entrepreneurs from abroad. The position and elasticity of the entrepreneurial supply curve depends upon the expected return to non-entrepreneurial activity abroad, the distribution of judgemental ability within the indigenous population, cultural attitudes, and barriers to entry and exit which reduce mobility between the entrepreneurial and non-entrepreneurial groups.

Given both the long run supply and long run demand for entrepreneurs, it is possible to visualize a long equilibrium in which the marginal entrepreneur earns an approximately normal return, intra-marginal entrepreneurs earn a quasi-rent to superior judgement, and intra-marginal non-entrepreneurs earn quasi-rents for their non-entrepreneurial abilities. The equilibrium return to entrepreneurship, and the equilibrium number of entrepreneurs, depend upon the parameters of the demand and supply curves, as described above. This equilibrium is a partial equilibrium, conditional upon the returns to non-entrepreneurial activity within the economy. It is also possible to derive a general equilibrium by endogenizing the return to non-entrepreneurial activity.

It should be emphasized, however, that any kind of 'equilibrium' in a 'market for entrepreneurs' is essentially an analytical fiction because the adjustment of this market to an equilibrium is itself an entrepreneurial task. The decision whether to hire an entrepreneur, and the decision whether to become one, are both entrepreneurial decisions. It is difficult for entrepreneurs to intermediate in the market for entrepreneurs because it is difficult to buy and sell 'human capital' of this kind. To introduce a Walrasian auctioneer to coordinate supply and demand decisions in the market for entrepreneurs would be self-contradictory, for it is only because of the absence of the Walrasian auctioneer that entrepreneurs are required in the first place. Thus while the concept of a market equilibrium for entrepreneurs is a useful analytical device, it is erroneous to suppose that the market for entrepreneurs is ever in a full equilibrium.

Twenty years ago, the study of the entrepreneur was regarded as a 'gap' in economic theory. It is now recognized that this gap cannot be filled without radically changing the nature of the theory itself. The entrepreneur can only be understood properly in the context of an economic model which does full justice to the structural complexity and the evolutionary nature of the economy (Nelson and Winter, 1982). Within such a model the 'equilibrium' concept remains a useful analytical device, but one of limited practical relevance. The study of the entrepreneur leads to a vision of economics much wider than that of a subject which parsimoniously derives a consistent set of price and quantity equations. Aspects of human personality – such as self-confidence – acquire a crucial role – and so too does the malleability of the personality under the influence of cultural attitudes. The theory of the entrepreneur, therefore, is not the last step which renders the conventional theory of value complete, but the first step towards an economic theory which forms part of a wider integrated body of social science.

MARK CASSON

See also CODETERMINATION AND PROFIT-SHARING; CORPORATE ECONOMY; INTEREST AND PROFIT; PROFIT AND PROFIT THEORY.

BIBLIOGRAPHY

Baumol, W.J. 1968. Entrepreneurship in economic theory. *American Economic Review, Papers and Proceedings* 58, 64–71.

Cantillon, R. 1755. *Essai sur la nature du commerce en général*. Ed. H. Higgs, London: Macmillan, 1931.

Casson, M.C. 1982. *The Entrepreneur: An Economic Theory*. Oxford: Martin Robertson.

Hayek, F.A. von. 1937. Economics and knowledge. *Economica*, NS 4, 33–54.

Kirzner, I.M. 1973. *Competition and Entrepreneurship*. Chicago: University of Chicago Press.

Knight, F.H. 1921. *Risk, Uncertainty and Profit*. Ed. G.J. Stigler, Chicago: University of Chicago Press, 1971.

Leibenstein, H. 1968. Entrepreneurship and development. *American Economic Review* 58, 72–83.

Marshall, A. 1890. *Principles of Economics*. 9th edn, 2 vols, ed. G.W. Guillebaud, London: Macmillan, 1961.

Mill, J.S. 1848. *Principles of Political Economy*. New edn, ed. W.J. Ashley, London: Longmans, 1909.

Nelson, R.R. and Winter, S.G. 1982. *An Evolutionary Theory of Economic Change*. Cambridge, Mass: Harvard University Press.

Redlich, F. 1956. The military enterpriser: a neglected area of research. *Explorations in Entrepreneurial History*, Series 1, 8, 252–6.

Say, J.B. 1803. *A Treatise on Political Economy: or the Production, Distribution and Consumption of Wealth*. New York: Augustus M. Kelley, 1964.

Schumpeter, J.A. 1934. *The Theory of Economic Development*. Trans. R. Opie, Cambridge, Mass.: Harvard University Press.

Weber, M. 1930. *The Protestant Ethic and the Spirit of Capitalism*. Trans. T. Parsons, London: George Allen & Unwin.

entropy. A concept of momentous importance for our understanding of physical reality though it is, entropy is one of the most poorly understood even by many physicists as a keen thermodynamicist, D. ter Haar, opined. A 'far-fetched' notion, 'obscure and difficult of comprehension' judged J. Willard Gibbs, the architect of statistical thermodynamics. It was apposite for Lord Snow to argue that some familiarity with the law of entropy, the second law of thermodynamics, separates the educated into two cultures. But this condition is quite curious given that the fountainhead of thermodynamics is anthropomorphic in a far more pronounced degree than that of any other branch of physics. No other physical concept belongs to our ordinary experience as inherently as heat and work or temperature and pressure. Indeed, thermodynamics is at bottom a physics of economic value as Sadi Carnot initiated it in his famous 1824 memoir about our efficient use of energy (Georgescu-Roegen, 1971).

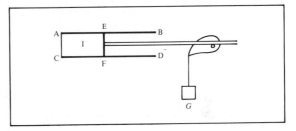

Figure 1

An explication of entropy at the ground level being beset with unusual difficulties, over the years the concept has received several, not always logically related, definitions. However, there is an original nature of that concept which can be grasped only by a punctilious description of the so-called Carnot cycle (Georgescu-Roegen, 1986). This cycle – the pillar of thermodynamic theory – is a model of unmatched idealization of an engine that converts thermal into potential energy. It consists of a piston-and-cylinder (Figure 1). The cylinder, ABCD, as well as the piston diaphragm, EF, are made of a perfect thermal insulator; the head of the cylinder, AC, instead is a perfect thermal conductor. The piston-rod turns a noncircular cam which raises or lowers a suspended weight, G. The space AEFC is filled with an ideal gas whose simple properties enable us to represent analytically the working of the engine (Figure 2a). The model involves several other idealizations. First, the piston moves reversibly, i.e., with an infinitesimally slow speed, an assumption which does away with any friction. And since any such motion would require an infinite time, the point exposes one of the basic anthropomorphic limitations. Second, the piston performs no other work than turning the cam. And the cam on turning does not change its potential energy. Third, when the cycle begins, the volume of the gas is at, say, V_0, and its absolute temperature at T_1. A hot reservoir (a virtually limitless source of thermal energy at the same temperature T_1) is already attached at AC. On expanding, the gas absorbs thermal energy from the reservoir at a rate that keeps its temperature constant throughout. The work thus produced raises the weight from its initial position G_0. Because the internal energy of an ideal gas remains constant if its temperature does not change and because the work only raises the weight, the thermal energy, Q, absorbed by the gas from the reservoir must at any time be equal to the work, W.

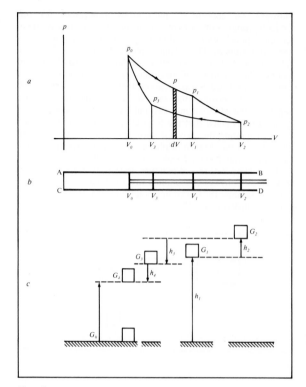

Figure 2

Two observations call now for undivided attention. First, the phase considered so far proves that *it is possible to convert thermal energy (heat) from a single source into potential energy (work)*. Hence, in principle, we could sail by tapping the immense energy of the ocean water. Lord Kelvin (1882, I) was not exact therefore when he said in 1852 that '*It is impossible by means of an inanimate material agency, to derive mechanical effect from any portion of matter by cooling it below the temperature of the coldest of the surrounding objects.*' The true reason is not technical (as we have just seen), but still another anthropomorphic limitation: no human can work with a piston that expands beyond any limit; we must bring it back to begin another conversion. But to bring it back we need some energy. Of course, we could use the potential energy of the weight to push the piston back reversibly. That would not do for everything would be just as at the beginning.

There is, however, a technical means to bring the piston back by using less energy than Q. Sadi Carnot (p. 36) revealed the secret in a rather unnoticed law: *The fall of the caloric [thermal energy] produces more motive power [work] at inferior than at superior temperatures.* Let us then assume that the gas expands only to V_1 which raises the weight to G_1 (Figure 2). The performed work, W_1 ($=Q_1$), is represented by the area below the segment $p_0 p_1$ of the curve pV = constant which represents the isothermal expansion of the ideal gas. The reason for using a cam is that the work $p\,dV$ is not uniform as V increases.

At V_1, an idealized operation takes place: instantaneously the hot reservoir is replaced by a perfect insulator covering AC and a different cam replaces the old. During this second phase the gas expands adiabatically (i.e. without exchanging heat with the outside) to V_2, converting Q_{12} of its internal energy into work, $W_{12} = Q_{12}$. This raises the weight further to G_2. By

V_2 the temperature of the ideal gas has decreased to, say, T_2. Another idealized operation now replaces the thermal insulator with a cold reservoir of temperature T_2 and a different cam replaces the previous one. From V_2 on, Carnot's law applies: an energy Q_2 obtained by lowering G_2 to G_3 is transmitted to the reservoir through the intermediary of the gas which contracts so that its temperature and hence its energy remain constant. At V_3 again the cam is exchanged and the reservoir replaced by a perfect insulator over AC. The gas continues then to contract this time adiabatically, from V_3 to V_0. The work W_{34} supplied by the lowering of the weight from G_3 to G_4 is converted into internal energy $Q_{34} = Q_{12}$ because equal temperature changes cause equal changes of the internal energy of an ideal gas.

To sum up: the cycle of the piston is complete, from V_0 back to V_0. Yet the weight is not back at G_0, but at G_4. For the puzzling contrast we should recall the role of the cams in the model. It could not be otherwise since the final work of the cycle, W, is positive, being represented by the area $p_0 p_1 p_2 p_3$; accordingly,

$$W = Q_1 + Q_{12} - Q_2 - Q_{34} = Q_1 - Q_2 \qquad (1)$$

The foregoing analysis proves the correctness of the so-called Carnot's principle, that any heat engine needs two sources of energy of different temperatures. The point is explicit in Planck's amended form of the second law of thermodynamics as given by Lord Kelvin:

'*It is impossible to construct an engine which will work in a complete cycle, and produce no effect except the raising of a weight and the cooling of a heat-reservoir*' (Planck, 1897).

Planck thus negated what Wilhelm Ostwald called perpetual motion of the second kind.

From the equations of the curves of Figure 2(a) there follows an expression of Carnot's law

$$Q_2/T_2 - Q_1/T_1 = 0. \qquad (2)$$

This surprising relation led Rudolf Clausius (in an 1854 essay) to interpret Q/T as *the transformation equivalent-value of Q* at the temperature T. And, as Lord Kelvin had done a few months earlier, he proved that for a reversible cycle

$$N = \sum Q_i/T_i = 0 \quad \text{or} \quad N = \oint dQ/T = 0. \qquad (3)$$

Both based the *theorem* on the first law of thermodynamics, $W = \Sigma Q/T = 0$ for an isolated system, and on a formulation of the second law. Clausius', which is equivalent to that of Kelvin–Planck, is still the most transparent one: *Heat cannot by itself pass from a colder to a warmer body.*

Clausius also proved the epochal result that, with the convention for signs as in (2), for an *irreversible* cycle

$$\oint dQ/T > 0 \qquad (4)$$

In 1865, on the basis of (3) he defined a new thermodynamic function, S, by

$$S_a - S_b = \int_a^b dQ/T \qquad (5)$$

for any reversible path from a state a to b. Then, by (3) and (4), always

$$\Delta S \geqslant \oint dQ/T, \qquad (6)$$

the equality prevailing only for reversible transformations.

At that juncture, Clausius replaced the term, 'transformation' by its Greek equivalent, *entropí*, and ended the memoir by the famous stanza:

1. *The energy of the universe is constant.*
2. *The entropy of the universe tends to a maximum.*

Although S was then defined only for thermal equilibrium, it was one of the greatest novelties ever thought up by a scholarly mind.

Once the idea that heat is not an indestructible fluid but a kind of molecular motion gained credence, to explain thermodynamic phenomena by the laws of mechanics became a vital programme. The most ambitious attempt was that of an 1872 epochal, albeit hard-going, essay by Ludwig Boltzmann (Brush, 1966). He argued that if the collisions between molecules follow some (apparently innocuous) rules, the distribution, $f(x, t)$, of their velocities at any time, t, is such that

$$H(t) = \int f \log f \, dx \qquad (7)$$

is never increasing,

$$dH(t)/dt \leqslant 0, \qquad (8)$$

the equality prevailing only for thermal equilibrium. Naturally, Boltzmann went on to claim not only that $S = -H$, but that (7) defined entropy for any thermodynamic system as well.

Since to derive an irreversible property from a completely reversible axiomatic basis was as incredible a feat as claiming that the angles of a triangle in an Euclidean plane add to more than two right angles, protests had to come. In 1876, a Boltzmann colleague, Joseph Loschmidt, pointed out that if at any time the velocities of a system satisfying (8) are reversed, we obtain a system for which H increases. And twenty years later, Ernst Zermelo, a pupil of Max Planck, recalled that in 1890 Henri Poincarè had proved that any finite mechanical system must eventually return as close as we wish to any of its *previous* positions (cf. the elementary illustrations in Georgescu-Roegen, 1971, pp. 154f). Hence, if H decreases for some time, it must necessarily increase before the return. Boltzmann was at a loss to defend his strong position of 1872 (see Brush, 1966). He asserted that if velocities are reversed the entropy most probably would still increase, since his theorem was not based on 'the nature of the forces' but on the immensely greater probability of the initial conditions that yield (8). But there is no reason for these conditions – known as *Stossenanzahl* (statistical postulate) – to be perpetuated from one collision to the next (Georgescu-Roegen, 1971, App. C). As to Zermelo's entropy recurrence, Boltzmann dismissed it as irrelevant in practice because of the extremely long time for the return of present conditions (Brush, 1966). Loschmidt's objection, however, had a weak point, namely, that systems with reversed velocities do not exist always in actuality. And in favour of Zermelo's, one could imagine that we may be in the middle of a recurrence period begun eons ago.

Pressed by persistent criticism Boltzmann abandoned his purely mechanistic basis of entropy, to concede that his *H*-theorem 'can never be proved from the equations of motions alone, [it] can only be deduced from the laws of probability' (Boltzmann, 1895). Already in 1877, he anchored the law of increasing entropy on the postulate that every state always passes to one of greater probability. Reasonable though the postulate may seem, it is not supported by probability laws: the occurrence of highly improbable bridge hands is subject to no sequential condition (Georgescu-Roegen, 1971, VI.1 and App. F). Then, observing that any

state ultimately reaches one of the molecular structures corresponding to thermal equilibrium, and that the number, Ω, of those structures is far greater than that of any other state, Boltzmann proposed that for thermal equilibrium $S = \log \Omega$, which after the dimensional correction by Max Planck became

$$S = k \log \Omega, \qquad (9)$$

where k is now registered as Boltzmann's constant. As has been observed by Planck (1897), the logarithm function must be used in (9) because for a thermal equilibrium composed of two such states we know that $S = S_1 + S_2$ whereas $\Omega = \Omega_1 \times \Omega_2$. Later, Boltzmann (1896/8) connected (9) with (7) and (8) by observing that n_i, being the number of molecules in some state i, with $n = \Sigma n_i$, then the number of possible structural combinations for that system is

$$W = n!/n_1!n_2! \ldots n_m! \qquad (10)$$

Hence, granting that every n_i is sufficiently great, by Stirling's asymptotic formula for $n!$ we obtain $\log W = \Sigma n_i \log (n_i/n)$, where by putting $f_i = n_i/n$, by analogy to (7) and (9) Boltzmann put

$$S = -nk \sum f_i \log f_i = -nkH, \qquad (11)$$

still another expression for entropy. At least, this formula fared well with the advent of Planck's quantum theory, which explains why Planck, who at first opposed Boltzmann's position, ultimately changed his mind (1897, seventh edn).

To buttress the probabilistic interpretation of entropy, Boltzmann (1896/8) ultimately brought in the ideas of ordered and disordered states. Herman von Helmholtz had already defined (in 1892) entropy as the measure of disorder, an unfortunate connection as disorder certainly cannot be defined analytically. Curiously, this definition has had amazing success: the entry for 'entropy' in *The Encyclopedia of Philosophy* opens with it.

Because the endeavours to systematize the various ways of looking at entropy could not arrive at a simple, natural explanation of that notion, writers have moved deeper and deeper into purely formal lucubrations, so that Jacques Hadamard in his 1910 review of J. Willard Gibbs's treatise could say that statistical thermodynamics is only mathematics (cited in Georgescu-Roegen, 1971). Yet as early as 1882 Helmholtz placed entropy in an accessible cadre as he showed that the internal energy, U, of an isolated system consists of free (or available) energy, F, and of bound (or unavailable) energy measured by TS. Hence, if we represent U by a rectangle of base T, the entropy would indicate the height for separating the *bound* energy. It was with the introduction of the relation $F = U - TS$ that the importance of the singular concept, entropy, was brought to the surface. It was only thereafter that one could translate Clausius's entropy law into

The free energy of any isolated system continuously degrades into bound energy.

It is this form that pinpoints the reason for the supreme role of entropy in all nature, as has been recognized by great luminaries such as Sir Arthur Eddington and Albert Einstein. Interestingly, Lord Kelvin who first spoke (in 1852) of 'The Universal Tendency in Nature to the Dissipation of Energy', hardly ever used the term 'entropy'. And Walter Nernst, another illuminator of thermodynamics, even decided not to have recourse to it.

We should next recall that just as the foundation of classical thermodynamics was being laid, Herbert Spencer came forth with some tenets that presaged Darwin's own theory. One was that 'the homogeneous is the hotbed of the heterogeneous', which looks like a characterization of living systems. Lord Kelvin (quoted earlier) as well as Helmholtz thus were not prepared to admit that the entropic degradation applies to animated matter, too. Later Henri Bergson even claimed that life opposes that degradation. Of course, life does not violate the entropy law, for as Erwin Schrödinger was to put it not very long ago, a living creature is not an isolated system: it exchanges entropy with its environment. Yet from the fact that such a phenomenon is not impossible thermodynamically, it does not follow at all that it should also exist. That is why several scholars have argued that in nature there must also be at work an anti-entropy principle – *ektropy*, coined by G. Hirth and adopted by Felix Auerbach, or *anti-chance*, Sir Arthur Eddington's term (Georgescu-Roegen, 1971). The dissipative structures, recently set forth by Ilya Prigogine (1980) to portray the entropic process specific to living organisms imply as Prigogine recognizes, a new crowning of the Spencerian tenet.

A real imbroglio involving the entropy concept grew from the seminal work of Claude H. Shannon (1948) on the purely technical problem of communication, which is to find out how many distinct sequences (messages) of a given length can be formed by a code, a set of different single signals. For the communication engineer it is totally irrelevant what each message may mean. However, that meaning must be understood by both the originator and the intended receiver. Knowledge thus passes from the former to the latter; while in transit, it is information (a distinction analogous to the heat being thermal energy in transit).

For messages transmitted in a vernacular language, Shannon found that the ratio of messages per letter is given by the same formula as $-H$ in (11) if the f_i's represent the statistical frequencies of the corresponding alphabet. Seeking a name for his new formula (a measure of the efficiency of the code), Shannon accepted the suggestion of John von Neumann: 'entropy'. But the mere coincidence of formulae was not a basis for justifying the terminological transfer. We would not call 'kinetic energy' the second moment of a distribution just because both formulae are a sum of squares: $\Sigma \eta_i X_i^2$. With that transfer, the concept of entropy started travelling from one domain to another with hardly any discrimination. One now speaks of the income distribution in country A being greater than in B, although the student interested in B will certainly think otherwise. And the reader is dizzied by the frequent phrases in which 'information', 'knowledge', 'negentropy', intervene pell-mell. Further, O, Onicescu suggested that Σf_i^2 can also serve as 'informational energy', which, of course, could not be related to the real entropy. Most important, in a vignetted article, 'The Bandwagon', Shannon himself protested without delay the use of 'entropy' beyond the domain of technical communication (Georgescu-Roegen, 1975).

The idea that the entropic degradation of anything is in fact a 'loss of information' was set forth earlier by a consummate thermodynamicist, G.N. Lewis (*Science*, 1930). But it was E.T. Jaynes who after the spread of Shannon's theorem set out to erect thermodynamics on that basis alone. In spite of its bizarreness, or perhaps because of it, the idea is still running in some circles. So, in his recent primer (*The Second Law*, 1984), P.W. Atkins was in good order to deliberately omit any reference to entropy and information because of the 'muddleheadedness' of the idea that entropy is not a property, of an engine but of the engineer's mind.

While the concept of entropy was thus converted almost at will, a vital issue found no place in thermodynamics, namely, the macroscopic role of matter. Matter is mentioned but only indirectly, as friction. Prigogine (1955) did extend the domain of thermodynamics from closed (impermeable to matter) systems to open systems, but he considered matter only as a vehicle of energy – the heat carried by a red-hot iron, for instance. No one seems to have derived the important object lesson from Gibbs's proof that the interdiffusion of two gases of the *same* temperature increases entropy. The increase is due to the entropic degradation of matter. To fill this lacuna, a new law of thermodynamics states that

Perpetual motion of the third kind is impossible,

which means that no closed (not to be confused with 'isolated') system can perform work indefinitely at constant rate. The reason is that macroscopic matter also degrades entropically (Georgescu-Roegen, 1980).

With its exotic name and its complicated fate even within the evolution of thermodynamics, entropy has become a word of great alluring power. Occasionally, littérateurs have used it manifestly as a selling point, 'Entropy' by Thomas Pynchon in 1960, *Against Entropy* by Michael Frayn (1967). Clausius certainly did not foresee this development from his coinage of the bizarre word.

NICHOLAS GEORGESCU-ROEGEN

See also INFORMATION THEORY.

BIBLIOGRAPHY

Boltzmann, L. 1895. On certain questions of the theory of gases. *Nature* 51, February.

Boltzmann, L. 1896–98. *Lectures on Gas Theory*. Trans. from German by S.G. Brush, Berkeley: University of California Press, 1964.

Brush, S. 1966. *Kinetic Theory*, Vol. 2. Oxford: Pergamon Press.

Carnot, S. 1824. *Reflections on the Motive Power of Fire, and on the Machines Fitted to Develop that Power*. Trans. and ed. R.H. Thurston, New York: Dover, 1960.

Clausius, R. 1867. *The Mechanical Theory of Heat with Its Application to the Steam Engine and to the Physical Properties of Bodies*. Ed. T.A. Archer, London: John van Voorst.

Georgescu-Roegen, N. 1971. *The Entropy Law and the Economic Process*. Cambridge, Mass.: Harvard University Press.

Georgescu-Roegen, N. 1975. The measure of information: a critique. *Proceedings of the Third International Congress of Cybernetics and Systems*, Bucharest, August 25–29 1975, ed. J. Rose and C. Bilciu, New York: Springer-Verlag.

Georgescu-Roegen, N. 1980. Matter: a resource ignored by thermodynamics. In *Future Sources of Organic Raw Materials*, ed. L.E. St-Pierre and G.R. Brown, Oxford: Pergamon Press.

Georgescu-Roegen, N. 1987. *The Promethean Destiny of Mankind's Technology*. Brighton: Wheatsheaf.

Helmholtz, H. von. 1882. On the thermodynamics of chemical processes. In H. von Helmholtz, *Physical Memoirs*, Vol. I, London: Taylor & Francis, 1888–90.

Planck, M. 1897. *Treatise on Thermodynamics*. Translated from the 7th German edn by A. Ogg, New York: Dover, 1925.

Prigogine, I. 1955. *Thermodynamics of Irreversible Processes*. 3rd edn, New York: Interscience Publishers, 1967.

Prigogine, I. 1980. *From Being to Becoming*. San Francisco: W.H. Freeman.

Rankine, W.J.M. 1881. *Miscellaneous Scientific Papers*. Ed. W.J. Millar, London: Charles Griffin.

Shannon, C. and Weaver, W. 1948. *The Mathematical Theory of Communication*. Urbana: University of Illinois Press.

Thomson, W., Baron Kelvin. 1882. *Mathematical and Physical Papers*. Cambridge: Cambridge University Press.

entry and market structure. Entry – and its opposite, exit – have long been seen to be the driving forces in the neoclassical theory of competitive markets. Long-run equilibrium in such a market requires that no potential entrant finds entry profitable, and that no established firm finds exit profitable. In conjunction with the price-taking assumption, the first condition requires that price be no greater than minimum average cost, A$\hat{C}$, and the second that price be no less than A$\hat{C}$. Hence, in equilibrium, price is equal to A$\hat{C}$. There is very little more to the theory of equilibrium in a competitive market than this simple yet powerful story of no-entry and no-exit.

Surprisingly, considerations of entry and potential competition, so central to the economist's view of competitive markets, played almost no role in oligopolistic and monopolistic markets until the work of Bain (1956) and Sylos-Labini (1957) in the mid-1950s. One important strand of the modern literature on entry combines, in essence, their insights into the role of potential competition in oligopolies with Schelling's (1956) ideas on commitment. This essay focuses on this strand of literature.

In reviewing Bain (1956) and Sylos-Labini (1957), Modigliani (1958) proposed what has come to be known as the *limit-output* (more commonly, *limit-price*) model, which is a formalization of one of the key ideas in these books. One version of this model has been the focal point of much of the recent literature on entry. Consider a market for some undifferentiated good, currently served by *one established firm*, in which demand and cost conditions are unchanging over an infinite time horizon. Now suppose that all potential entrants take the established firm's output today, denoted by X, as the output which it will produce tomorrow and forever – the so-called Sylos postulate. Let $\underline{X}$ denote the smallest value of X such that the maximized profit of a representative potential entrant is non-positive. This value, $\underline{X}$, is called the *limit output* (and the corresponding price the *limit price*) because, given the Sylos postulate, there will be no entry if and only if $X \geqslant \underline{X}$. (We assume for convenience that zero profit does not induce entry.)

Two important insights are already clear. First, potential competition constrains the ability of the established firm to exploit its position of market power since the no-entry condition ($X \geqslant \underline{X}$) places a lower bound on industry output in long-run equilibrium. Second, by producing at least $\underline{X}$ units of output, the established firm can deter entry. In one case central to the evolution of the literature, entry deterrence is also always profitable.

Suppose that the established firm's average cost function is nowhere upward-sloping, and that the interest rate is not too high. In this situation, the established firm will always choose to deter entry by producing at least $\underline{X}$. There are two sorts of solutions. If the ordinary monopoly output, M, is greater than or equal to $\underline{X}$, then the monopolist will produce M, and we have what we could call *natural monopoly*, in the positive sense of the term. If $M < \underline{X}$, the monopolist produces $\underline{X}$ to deter entry strategically, a case of *artificial monopoly*. When $M < \underline{X}$ the monopolist must choose between deterring and accommodating entry. Relative to the deterrence strategy – produce $\underline{X}$ today and forever – accommodation produces larger profit for the one established firm today (since today's output will be M) but smaller profit tomorrow and forever (since price will be no higher than the limit price and the established firm's output will be less than $\underline{X}$). If the rate of interest is not too high, it is obvious that the established firm will choose to deter entry.

The essence of both solutions is a message which the established firm wants to communicate to potential entrants. 'If you enter, then my output will be (no smaller than) $\underline{X}$.' This 'deterrence message' has the property that it deters entry – if it is believed. It raises the obvious credibility question, 'Would the established firm really produce the promised output post-entry'? Much of the recent literature on entry has implicitly or explicitly focused on this question. Four interrelated insights have emerged.

First, to answer the credibility question we can use Schelling's distinction between threats and commitments (1956). The deterrence message is a commitment if, entry having occurred, it is in the established firm's self-interest to produce $\underline{X}$, or if the production of $\underline{X}$ follows automatically. Otherwise the message is a threat and is not credible. To implement this approach we need a model of oligopoly. Given such a model, all interested parties can compute the post-entry equilibrium, providing a direct answer to the credibility question.

Second, by virtue of being there first, the established firm has the opportunity to make *irreversible decisions* which alter its real economic circumstances in any post-entry oligopoly game. These irreversible decisions can sometimes make the established firm a more aggressive competitor post-entry, and therefore serve to make the deterrence message more believable. That is, the established firm has the opportunity to do some things prior to entry which cannot be undone subsequent to entry, and which affect the profitability of entry. It is convenient to refer to these irreversible decisions as *commitments*. As Spence (1977) observed, since the rate at which output is produced is reversible, producing the limit output prior to entry is not a commitment and therefore has no bearing on the credibility of the deterrence message. However, holding the capacity or capital to do so is – provided that it is specific. By acquiring specific capital, the established firm reduces its marginal cost, making it more aggressive in any post-entry oligopoly game. Inventory (analysed by Ware, 1985) is a particularly illuminating commitment since it puts the firm in the position of having zero marginal cost post-entry (until its inventory is exhausted).

The third insight, which arises from the attempt to implement the first two, concerns the *form* of the game which established firms and potential entrants play, and the appropriate *equilibrium concept* which this form seems to imply. Even in the simplest of circumstances, any game involving established firms and potential entrants is a *multi-stage game* played out in real time with two important features: (1) commitments, which inevitably involve *sunk costs*, are made in earlier stages of the game; (2) the net revenues which justify these sunk costs are generated only in later stages. As Brander and Spencer (1983) observe, this form is an unavoidable feature of economic reality. Product development costs, for example, must be sunk prior to production; the costs associated with specific capital goods must be sunk prior to production, and so on.

A rational firm must therefore think of commitments in the way it thinks of other investment decisions. In particular, it must form expectations about how decisions with respect to today's commitments will alter its net revenues tomorrow, and thereafter. In the presence of sunk costs, *rational* or *consistent* expectations are a desirable feature of any equilibrium concept. If firms' expectations are not constrained to be rational, then any outcome is possible – that is, given an outcome, there is a set of expectations which will produce it. Rational expectations are then necessary to constrain results. See, for example, the discussions in Eaton and Lipsey (1979, 1980) and Dixit (1980) on this point. Given this view, Selten's (1975) notion of sub-game perfection is the appropriate equilibrium concept in these entry games.

To convey the flavour of modern theories of entry and to see how rational expectations enter the analysis, it is instructive to write down a simple entry game and to consider the way in which one finds the perfect equilibrium. Most of the recent envelope theorem

literature on entry has focused on exercises which involve one established firm and one potential entrant, and which are not a great deal more complex than the following illustration.

Consider an entry game played in three stages. In stage 1, firm 1 (the established firm) chooses the value of some commitment, c_1. In stage 2, knowing the value of c_1 which firm 1 chose in stage 1, firm 2 chooses a value for its commitment, c_2. By appropriate choice of units, we can interpret c_1 and c_2 as costs, which once they are incurred are sunk. These sunk costs might, for example, be expenditures on advertising or on cost-reducing research and development. In stage 3, the two firms play a market game in which goods are produced and sold and the net revenues which justify the upstream sunk costs are realized.

To find the perfect equilibrium of this game we work backwards.

Stage 3: Given an oligopoly model which determines the equilibrium of the market game, the net revenue to each firm in stage 3 is determined by c_1 and c_2. Denote these net-revenue functions by $\Pi_1(c_1, c_2)$ and $\Pi_2(c_1, c_2)$. If, for example, the oligopoly model is the Cournot model, then in stage 3 each firm chooses its own quantity to maximize its revenues minus its *avoidable* costs. The net-revenue functions are simply revenues minus avoidable costs in the Cournot equilibrium.

Stage 2: If firm 2 is to have rational expectations, it must know $\Pi_2(c_1, c_2)$. This, of course, means that it knows the oligopoly model which determines the equilibrium of the market game. In stage 2, knowing its net-revenue function and the value of c_1, firm 2 chooses c_2 to maximize $[\Pi_2(c_1, c_2) - c_2]$. The solution to this maximization problem determines c_2 as a function of c_1: $c_2 = g(c_1)$.

Stage 1: Rational expectations for firm 1 means that it knows both $g(c_1)$ and $\Pi_1(c_1, c_2)$. In stage 1 it chooses c_1 to maximize $[\Pi_1(c_1, c_2) - c_1]$ subject to $c_2 = g(c_1)$. The only endogenous variable in this maximization problem is c_1 and the solution to it therefore determines a value for c_1 say c_1^*. Firm 2 then chooses $c_2^* = g(c_1^*)$, and in the third stage of the game the firms realize $\Pi_1(c_1^*, c_2^*)$ and $\Pi_2(c_1^*, c_2^*)$.

In this sort of game the established firm may or may not be able to deter entry, and if it is able, it may or may not choose to. Duopoly solutions will, however, be asymmetric. The established firm will rig the duopoly market structure to its own advantage.

Using this approach, or one that is in the spirit of this one, the recent literature on entry has focused on many of the commitments which established firms can and, indeed, must make. Advertising (Cubbin, 1981), brand proliferation (Schmalensee, 1978), the location of retail outlets (Eaton and Lipsey, 1979), patenting (Gilbert and Newbery, 1982), learning-by-doing (Spence, 1981), the durability of specific capital (Eaton and Lipsey, 1980), the exercise of monopsony power (Salop and Scheffman, 1983) and, of course, specific capital (Spence, 1977; Dixit, 1980 and Ware, 1984) are just some of the vehicles for commitment which have been considered.

This rich set of games and possible solutions brings us to the fourth insight in this literature. Implicit in this way of thinking about oligopolistic markets, and the role which entry plays in those markets, is much more than a theory of how one established firm strategically positions itself with respect to one potential entrant. There is, in this paradigm, a theory of market structure, a theory which remains largely unexplored.

B. Curtis Eaton

See also CONTESTABLE MARKETS; LIMIT PRICING; NATURAL MONOPOLY; PREDATORY PRICING.

BIBLIOGRAPHY

Bain, J.S. 1956. *Barriers to New Competition*. Cambridge, Mass.: Harvard University Press.
Brander, J.A. and Spencer, B.J. 1983. Strategic commitment with R&D: the symmetric case. *Bell Journal of Economics* 14(1), Spring, 225–35.
Cubbin, J. 1981. Advertising and the theory of entry barriers. *Economica* 48, August, 289–98.
Dixit, A. 1980. The role of investment in entry deterrence. *Economic Journal* 90, March, 95–106.
Eaton, B.C. and Lipsey, R.G. 1979. The theory of market pre-emption: the persistence of excess capacity and monopoly in growing spatial markets. *Economica* 46(182), May, 149–58.
Eaton, B.C. and Lipsey, R.G. 1980. Exit barriers are entry barriers: the durability of capital as a barrier to entry. *Bell Journal of Economics* 11(2), Autumn, 721–9.
Gilbert, R.J. and Newbery, D.M.G. 1982. Preemptive patenting and the persistence of monopoly. *American Economic Review* 72(3), June, 514–26.
Modigliani, F. 1958. New developments on the oligopoly front. *Journal of Political Economy* 66, June, 215–32.
Salop, S.C. and Scheffman, D.T. 1983. Raising rivals' costs. *American Economic Review* 73(2), May, 267–71.
Schelling, T.C. 1956. An essay on bargaining. *American Economic Review* 46, June, 281–306.
Schmalensee, R. 1978. Entry deterrence in the ready-to-eat breakfast cereals industry. *Bell Journal of Economics* 9(2), Autumn, 305–27.
Selten, R. 1975. Re-examination of the perfectness concept for equilibrium points in extensive games. *International Journal of Game Theory* 4(1), January, 25–55.
Spence, A.M. 1977. Entry, capacity, investment and oligopolistic pricing. *Bell Journal of Economics* 8(2), Autumn, 534–44.
Spence, A.M. 1981. The learning curve and competition. *Bell Journal of Economics* 12(1), Spring, 49–70.
Sylos-Labini, P. 1956. *Oligopolio e progresso technico*. Milan: Giuffre.
Ware, R. 1984. Sunk costs and strategic commitment: a proposed three-stage equilibrium. *Economic Journal* 94, June, 370–78.
Ware, R. 1985. Inventory holding as a strategic weapon to deter entry. *Economica* 52, February, 93–101.

envelope theorem. The origin of this famous theorem is the discussion between Jacob Viner (1931) and his draftsman Y.K. Wong concerning the relationship between short and long run average cost curves. Viner had apparently reasoned that since in the long run average costs should be at a minimum, the long run average cost curve should not only always be below the short run curves, but should also pass through the minimum points of each short run curve. Wong pointed out the impossibility of this joint occurrence, but Viner opted to draw the long run curve through the minimum points, thereby necessarily passing above sections of the short run curves. The puzzle was solved by Samuelson (1947) who showed in a general way why the long run curve would be the 'envelope' curve to the set of short run curves. The envelope theorem has since become one of the fundamental tools of modern economic analysis.

Consider any two-variable model, maximize $y = f(x_1, x_2, \alpha)$, where x_1 and x_2 are the choice variables and, for the moment, α is a single parameter representing some constraint on the maximizing agent's behaviour. The first-order equations are $f_1 = f_2 = 0$. By solving the first-order equations simultaneously, assuming unique solutions, explicit choice functions $x_1 = x_1^*(\alpha)$, $x_2 = x_2^*(\alpha)$ are implied. That is, if the parameter α changes, both choice variables will in general change. The refutable propositions in economics consist of predictions of the directions of change in some or all x_i^*'s as α changes. Define the 'indirect objective function' $\phi(\alpha)$ as the maximum value of f for given α. Then $\phi(\alpha) = f(x_1^*(\alpha), x_2^*(\alpha), \alpha)$. The envelope theorem con-

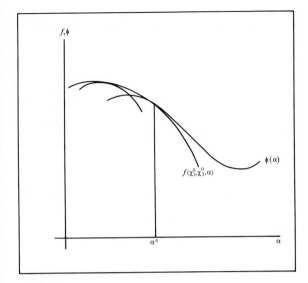

Figure 1

From the above, therefore,

$$\phi_{\alpha\alpha} - f_{\alpha\alpha} = f_{1\alpha}\partial x_1^*/\partial\alpha + f_{2\alpha}\partial x_2^*/\partial\alpha > 0 \qquad (4)$$

This analysis is readily generalized to the n-variable case, producing the condition $\Sigma\partial x_i^*/\partial\alpha > 0$.

If the parameter α enters only one first-order condition $f_i = 0$, that is $f_{j\alpha} = 0$ for $j \neq i$, then the condition reduces to $f_{i\alpha}\partial x_i^*/\partial\alpha > 0$, the famous 'conjugate pairs' result. When a parameter disturbs only one first-order equation, that choice variable responds in the same direction as the disturbance. In many economic models the objective functions contain expressions linear in the parameters, for example the expression for (negative) total cost when constant per-unit costs are asserted, $-\Sigma p_i x_i$. In this case $f_{p_i x_j} = -1$ if $i = j$ and zero otherwise; hence $\partial x_i^*/\partial p_i < 0$. In the case where α is a vector of parameters $\alpha = (\alpha_1, \ldots, \alpha_m)$, condition (4) says that the matrix $F_{\alpha\alpha} = f_{\alpha\alpha} - \phi_{\alpha\alpha}$ is negative (semi)definite; the usual comparative statics results follow from the negativity of the diagonal elements.

The envelope theorem also yields the non-intuitive 'reciprocity' conditions. Suppose there are two parameters α and β. Then from invariance of second partial derivatives to the order of differentiation (Young's theorem), $\phi_{\alpha\beta} = \phi_{\beta\alpha}$. Using equation (3) above, $\Sigma f_{i\alpha}\partial x_i^*/\partial\beta = \Sigma f_{i\beta}\partial x_i^*/\partial\alpha$. When the objective function contains a linear expression $p_1 x_1 + p_2 x_2 + \cdots + p_n x_n$, $f_{p_1 x_1} = 1, f_{p_1 x_2} = 0$, etc., and therefore $\partial x_1^*/\partial p_2 = \partial x_2^*/\partial p_1$, etc. This result occurs in consumer theory for the compensated demands and in the profit maximization and cost minimization models of production theory.

The extension of these results to models involving one or more side-conditions (constraints) depends on whether the parameters enter only the objective function or whether they enter the constraints also (or exclusively). If the parameters enter only the objective function, the only effect is to reduce the dimensionality of the x_is; no changes in any of the above results occur. However, if a parameter enters a constraint, as that parameter changes, the constraint space also changes, destroying the relation $f_{\alpha\alpha} < \phi_{\alpha\alpha}$. In this case $\phi_\alpha = L_\alpha$, where L is the associated Lagrangian of the constrained maximization model, and the second-order conditions are not sufficient to restrict the signs of comparative statics equations.

Eugene Silberberg

See also COMPENSATED DEMAND; COST FUNCTIONS; DUALITY; LE CHATELIER PRINCIPLE.

BIBLIOGRAPHY

Samuelson, P.A. 1947. *Foundations of Economic Analysis*. Cambridge, Mass.: Harvard University Press.

Silberberg, E. 1978. *The Structure of Economics*. New York: McGraw-Hill.

Viner, J. 1931. Cost curves and supply curves. *Zeitschrift für Nationalökonomie* 3, 23–46. Reprinted in American Economic Association, *Readings in Price Theory*, Homewood, Ill.: Irwin, 1952.

environmental economics. Environmental economics is a new field, essentially the creation of the present generation of economists. But its roots are in the externality theories of Marshall and Pigou, the public goods theories of Wicksell and Bowen, the general equilibrium theory of Walras, and the applied field of benefit-cost analysis, foreshadowed by Dupuit but cultivated to maturity by economists in the water resource agencies of the US Government.

cerns the rates of change of $\phi(\alpha)$ versus $f(x, \alpha)$ with respect to changes in α.

In Figure 1, a typical $\phi(\alpha)$ is plotted. For an arbitrary α^0 some $x_1^0 = x_1^*(\alpha^0)$ and $x_2^0 = x_2^*(\alpha^0)$ are implied. Consider the behaviour of $f(x_1, x_2, \alpha)$ when x_1 and x_2 are held fixed at x_1^0 and x_2^0 as opposed to when they are variable. When $\alpha = \alpha^0$, the 'correct' x_is are chosen, and therefore $\phi(\alpha) = f(x_1^0, x_2^0, \alpha)$ at that one point. However, both to the left and to the right of α^0, the 'wrong' (i.e. non-maximizing) x_is are chosen, and since $\phi(\alpha)$ is the *maximum* value of f for given α, $f(x_1^0, x_2^0, \alpha) < \phi(\alpha)$ in any neighbourhood around α^0. This implies that ϕ and f must be tangent at α^0 (assuming differentiability), and, moreover, f must be either more concave or less convex than ϕ there. Since this must happen for arbitrary α, similar tangencies occur at other values of α. It is apparent from figure 1 that $\phi(\alpha)$ is the *envelope* of the $f(x_1, x_2, \alpha)$'s for each α. Most important, but not so obvious, is that *all comparative statics theorems in maximization models are consequences of the relative curvatures of ϕ and f.*

Consider a new function, the difference between the actual and the maximum value of f for given α,

$$F(x_1, x_2, \alpha) = f(x_1, x_2, \alpha) - \phi(\alpha).$$

Since $f < \phi$ for $x \neq x^*$ and $f = \phi$ for $x_i = x_i^*$, F has a maximum (of zero) when $x_i = x_i^*(\alpha)$. Therefore, $F(x_1, x_2, \alpha)$ has zero partial derivatives with respect to the original variables x_1 and x_2, and also α:

$$F_i = f_i = 0, \qquad i = 1, 2 \qquad (1)$$

and

$$F_\alpha = f_\alpha - \phi_\alpha = 0 \qquad (2)$$

Equations (1) are simply the original maximum conditions. Equation (2) is the traditional 'envelope' result, $\phi_\alpha = f_\alpha$. Moreover, the sufficient second-order conditions imply $F_{\alpha\alpha} = f_{\alpha\alpha} - \phi_{\alpha\alpha} < 0$. It is from this last result that all known comparative statics results flow. Since

$$\phi_\alpha(\alpha) \equiv f_\alpha[x_1^*(\alpha), x_2^*(\alpha), \alpha], \qquad (3)$$

differentiating both sides with respect to α yields

$$\phi_{\alpha\alpha} \equiv f_{\alpha 1}\partial x_1^*/\partial\alpha + f_{\alpha 2}\partial x_2^*/\partial\alpha + f_{\alpha\alpha}$$

External effects of economic transactions have been recognized by economists for a long time. The 1920s saw a series of articles in the *Economic Journal* dealing with the possible external economies that were associated with investment in decreasing cost industries. In restrospect this discussion appears rather arid and concern with the issue has long since mostly faded away.

But, in addition, economic theory has long recognized, at least in a limited way, the existence of 'common property' problems and resource misallocations associated with them. It was early appreciated that when property rights to a valuable resource were not or could not be parcelled out in such a way that one participant's activities in the use of that resource would leave the others unaffected, except through market exchange, unregulated private exchange would lead to external effects and therefore to inefficiencies. For instance, the individual crude petroleum producer, pumping from a common pool, has no market incentive to take account of the increased cost imposed on others because of reduced gas pressure resulting from his own pumping. While this problem was recognized with respect to such resources as petroleum, fisheries and groundwater, and there was rather sophisticated theorizing with respect to it, private property and exchange have still been regarded by the profession as the keystones of an efficient allocation of resources.

To quote the famous early welfare economist Pigou:

> When it was urged above, that in certain industries a wrong amount of resources is being invested because the value of the marginal social net product there differs from the value of the marginal private net product, it was tacitly assumed that in the main body of industries these two values are equal (Pigou, 1932, p. 199).

And Scitovsky, another important early student of externalities, after having described his cases two and four, which deal with externalities affecting consumers and producers respectively, says:

> The second case seems exceptional, because most instances of it can be and usually are eliminated by zoning ordinances and industrial regulations concerned with public health and safety ... . The fourth case seems unimportant, simply because examples of it seem to be few and exceptional (Scitovsky, 1954, pp. 143–51).

Even as late as 1957 Bator could write. 'The very pastoral quality of the example suggests that in a statical context such direct interaction among producers-interaction that is not reflected by prices-is probably rare' (p. 42). In the early 1960s the economics profession began to take the problem of externalities more seriously, as indicated by the appearance of a spate of articles in major economics journals. Articles by Coase (1960), Buchanan and Tullock (1962) and Turvey (1963) went some distance toward systematizing definitions and illuminating policy issues. However, even these contributions dealt with externality as a comparatively minor aberration from Pareto optimality in competitive markets, and focused upon externalities between two parties. Mishan, after a careful review of the literature, commented on this as follows:

> The form in which external effects have been presented in the literature is that of partial equilibrium analysis; a situation in which a single industry produces an equilibrium output, usually under conditions of perfect competition, some form of intervention being required in order to induce the industry to produce an 'ideal' or 'optimal' output. If the point is not made explicitly, it is

tacitly understood that unless the rest of the economy remains organized in the conformity with optimum conditions, one runs smack into Second Best problems (Mishan, 1965, pp. 1–34).

By the end of the decade it had become clear that at least one class of externalities – those associated with the disposal of residuals resulting from modern consumption and production activities – would have to be viewed quite differently. In reality they are a normal, indeed inevitable, part of these processes. Their economic significance tends to increase as economic development proceeds, and the ability of the natural environment to receive and assimilate them is an important natural resource of increasing value. It appears that the common failure up to that time to recognize these facts in economic theory may have resulted from viewing the production and consumption processes in a manner which is at variance with the fundamental physical law of conservation of mass/energy.

Modern welfare economics concludes that if (1) preference orderings of consumers and production functions of producers are independent and their shapes appropriately constrained, (2) consumers maximize utility subject to given income and price parameters, and (3) producers maximize profits subject to these price parameters, then a set of prices exists such that no individual can become better off without making some other individual worse off. For a given distribution of income this is an efficient state. Given certain further assumptions concerning the structure of markets, this 'Pareto optimum' can be achieved via a pricing mechanism and voluntary decentralized exchange.

But if the capacity of the environment to assimilate residuals is scarce, the decentralized voluntary exchange process cannot be free of uncompensated technological external diseconomies unless (1) all inputs are fully converted into outputs, with no unwanted material and energy residuals along the way, and all final outputs are utterly destroyed in the process of consumption, or (2) property rights are so arranged that all relevant environmental attributes are in private ownership and these rights are exchanged in competitive markets. Neither of these conditions can be expected to hold in an actual economy, and they do not.

Nature does not permit the destruction of matter except by annihilation with antimatter, and the means of disposal of unwanted residuals which maximizes the internal rate of return of decentralized decision units is by discharge to the environment, watercourses, land and the atmosphere. Water and air were traditionally examples of free goods in economics. But in reality in developed economies they are common property resources of great and increasing value, which present society with important and difficult allocation problems that exchange in private markets cannot solve. These problems loom larger as increased population and industrial production put more pressure on the environment's ability to dilute, chemically degrade, and simply accumulate residuals from production and consumption processes. Moreover, technological means for processing or purifying one or another type of residuals do not destroy the residuals but only alter their form. Thus, given the level, patterns and technology of production and consumption, the recycling of materials into productive uses or discharge into an alternative medium are the only general operations for protecting a particular environmental medium such as water. Residual problems must be seen in a broad regional or economy-wide context rather than as separate and isolated problems of disposal of gaseous, liquid, solid, and energy waste products.

Frank Knight perhaps provided a key to why these elementary facts played so small a role in earlier economic theorizing and empirical research:

> The next heading to be mentioned ties up with the question of dimensions from another angle, and relates to the second main error mentioned earlier as connected with taking food and eating as the type of economic activity. The basic economic magnitude (value or utility) is service, not good. It is inherently a stream or flow in time ... (Knight, 1921, p. 92).

Standard economic allocation theory is in truth concerned with services. Material objects are merely the vehicles which carry some of these services, and they are exchanged because of consumer preferences for the services associated with their use or because they can help to add value in the manufacturing process. Yet economists persisted in referring to the 'final consumption' of goods as though material objects such as fuels, materials, and finished goods somehow disappear into a void – a practice which was comparatively harmless only so long as air and water were both literally 'free goods'. Of course, residuals from both the production and consumption processes remain, and they usually render disservices (like killing fish, increasing the difficulty of water treatment, reducing public health soiling and deteriorating buildings, etc.) rather than services. These disservices flow to consumers and producers whether they are wanted or not, and except in unusual cases they cannot be controlled by engaging in individual exchanges.

A 1970 publication formalized these ideas by incorporating mass balance into a Walrasian-type general equilibrium model (Kneese, Ayres and d'Arge, 1970). This work provided the seed for a large amount of further theoretical and, by then, also applied work in environmental economics in the 1970s. Much of this work was conducted by researchers at Resources for the Future or by persons affiliated with RFF but located at other institutions. A summary of the methods and results of these lines of inquiry can be found in Kneese and Bower (1979). A major theoretical publication of this period, and still today the standard reference is Mäler (1974).

Subsequent applied and theoretical work has taken many directions. It is useful in exposition, however, to classify the research under a few broad heads. The set that seems most useful to us is: How to make decisions about environmental quality? How to measure benefits and costs of changing environmental quality? How to implement decisions about changing or maintaining environmental quality? How do decisions about environmental quality affect the larger economy?

Before proceeding to brief summaries of work under each heading, we note two things. First, the categories are very closely interrelated – more so than may be obvious. For example, worrying about cost-benefit analysis assumes a particular answer to the first question. Choice of implementation method helps to determine the costs of any particular decision. And those costs, in turn, are central to macroeconomic assessments of policy effects.

Second, hallmark of this applied work in environmental economics is its inherently interdisciplinary character. To achieve the needed depth in addressing real problems it is necessary to draw on the skills of engineers, biologists, physicists, lawyers, and political scientists, as well as economists. To many, this is an important appeal of the subject.

HOW TO DECIDE ENVIRONMENTAL ISSUES? As has already been pointed out, the essence of environmental issues is that they involve externalities and public goods. The combination means that only in rare cases can we appeal to familiar theorems about the splendid welfare results produced by the free market. And because the market has no particular claim to our attention, we face the urgent question how can (and how should) society decide what amount of environmental quality it should purchase. To fail to provide an answer is to leave matters by default to the market, a non-decision that has been shown to lead to the provision of a lower quality than would be optimal if individuals' willingness to pay could be summed and balanced against the cost of provision. An explicit decision in favour of a non-market mechanism opens a wide range of alternatives for consideration, none of which has a completely persuasive claim to our allegiance.

One possibility, which might be called the straightforward welfare economics answer, is to use cost-benefit analysis. Abstracting for the moment from the difficulties of measurement to be discussed below, notice that this answer says the collectivity should decide on environmental quality by adding up the benefits to whomsoever they may accrue and comparing the total to the costs; with a similar qualifier. Because willingness to pay is constrained by income, the result is heavily conditioned by the status quo. This criterion, or something very similar, is implicit in the judgement about the undesirability of market solutions referred to above.

One way to move away from income and cardinal welfare measures generally is to look to voting for solutions. Here, each person counts exactly the same, and the extent of an individual's prospective gain or loss is not important to the ultimate decision. All that matters is whether the individual is for or against the proposition.

Economists, political scientists and philosophers have concerned themselves with examining the range of possible decision mechanisms and criteria. Two principal sorts of examination have been performed. In one, the criteria have been matched up against ethical systems. For example, a version of the utilitarian system of ethics has been shown to be equivalent to the use of cost benefit analysis. This line of work is summarized and examined in Kneese and Schulze (1985).

The other major form of research into environmental decision making has been built on the subdiscipline known as public or social choice theory (see Plott, 1976, for a review of the basic theory). Here interest has been in the behaviour of the decision-making processes themselves rather than the normative nature of the decisions they might lead to. The ornery result about the potential cyclicity of majority decisions over three or more alternatives when individual preferences are not single-peaked is the most familiar example. But more recent research has gone far beyond this, examining the effect of such 'technical' or constitutional issues as (a) the regulatory process and its relation to larger social choice issues (e.g. Russell and Shelton, 1974; Fiorina and Noll, 1978; Fiorina, 1982); (b) the design of geographical jurisdictions and the relationships among different jurisdictional levels (e.g. McMillan, 1976; Rose-Ackerman, 1981); (c) the design of specific institutions (Klevorick and Kramer, 1973); (d) the politics of particular projects (e.g. Plotkin, 1972). Further, the methods of public choice have been used to examine such questions as why citizens join environmental groups that seek public good, when naive economic theory might predict free riding (see for example Hardin, 1982).

For some, however, the methods of social choice and the questions dealt with therein remain fundamentally unsatisfying because firm prescriptive results can never be obtained. All practical institutions for collective decision making share some version of the original sin captured in Arrow's famous impossibility theorem. Being unable either to prescribe the 'correct' constitution or to judge among the outcomes thrown up by alternative decision making arrangements is sufficiently frustrating that welfare economics, for all its warts, seems more enticing. This leads back to consideration of benefit-cost analysis and its practical implementation.

HOW TO MEASURE BENEFITS AND COSTS? Again, because of the nature of environmental quality, the benefits to be expected from proposed changes in prescribed quality are difficult to measure. This is simply the other side of the observation that markets do not work well to produce decisions because willingness to pay is not registered in any market or markets. In these circumstances, the ingenuity of economists has been strained to the limit to devise ways of teasing benefit estimates out of the record of other transactions. The resulting methods may conveniently be referred to as *indirect*, and will be contrasted below to *direct* methods in which individuals are asked to say what they are willing to pay.

Estimating benefits using one of the indirect methods usually involves, at least in principle, understanding and being able to characterize quantitatively not only the human valuation behaviour that gives rise to the actual benefit, but the prior linkages – from proposed policy to change in pollution (often discharges) and from that change to the resulting change in ambient environmental quality. It is the latter change to which humans and other organisms or materials react, either as a matter of conscious choice or of biological or chemical necessity.

Of particular interest in the context of the indirect methods is to sort out what can be measured from existing (or plausible future) data and whether dollar values may be inferred from that same data or must be supplied from some other source. Thus, for example, the best–known indirect method is probably that of using travel cost incurred to infer willingness to pay for the experience or site travelled to (as historical background, see Clawson and Knetsch, 1966). A neat theoretical proof that the travel cost method does what it is supposed to do is given by Bowes and Loomis (1980). Here the method rests on a decision to travel, producing a single 'purchase' from which value may be inferred via valuation of distance and time – albeit the methods for the valuation are subject to dispute. Applying this to environmental quality benefit estimation is far from trivial though the idea is easy to see: improving environmental quality *at a place* should shift outwards the demand for that place or the experience it provides.

But if environmental quality is changing everywhere at once it may be practically impossible to infer the benefits from a travel cost model. A second indirect method based on an observable decision focuses on participation in an activity dependent on the natural environment (such as recreation fishing) rather than travel to a place. Here the argument is that benefits arise because quality improvements increase the supply (lower the price) of the requisite resource. This effect can be sought by matching data on participation and availability (e.g. see Vaughan and Russell, 1984).

The problem, however, is that no expense is uniquely attached to the participation decision. Travel may occur but we will not, by assumption, know to where. Equipment will be purchased but may be of any degree of elaborateness and, indeed, necessity. In these circumstances, a secondary valuation question arises: how to value a unit of participation. No altogether satisfactory answer exists, but such tricks as use

of a subsidiary travel-cost model have been tried (Vaughan and Russell, 1982).

A third major type of indirect benefit estimation technique might be called the biological response approach. Here human (or plant or animal) physiology responds to stress that is associated with environmental quality. For example, air pollution by sulphates and particulates appears to be related to a variety of acute and chronic respiratory tract conditions (Lave and Seskin, 1977; for a more recent and correct modern treatment, Portney and Mullahy, 1985).

Again, no valuation is subsumed in the response. It must be supplied by inference; hence the indirect label. For plants or animals involved in commercial fishing, forestry or agriculture, such inferences are relatively straightforward. For 'aesthetic' species, for situations involving general ecological health, for arguments about the future 'gene pool', and most importantly, for human health, the inferences are anything but straightforward. Human mortality effects have, for example, been valued on the basis of property values (Portney, 1981b) and occupational risks and wages (e.g. Thaler and Rosen, 1976; for a survey see Bailey, 1980). For effects involving sickness but not death, even greater ingenuity has been required. A recent published example is Cropper (1981).

Rather than dwell further on the details of the indirect methods, however, we shall make a few generalizations. First, the understanding and data necessary to capturing the several linkages referred to above are almost never available. Thus, the estimates of physical effect, whether it be in the ambient environment or in the human body, are always imprecise and possibly biased as well. Second, the chains of inference and the data, on which values of particular decision units or physical effects are to be based, admit of substantial additional imprecision. Thus, while the methods, for the most part, are respectable economically, they do not in general lead to numbers one would want to claim were correct even within a factor of 5 or 10.

In this circumstance a method that is less respectable but also a good deal less inflexible and expensive, has commended itself to the profession's attention. This is the method known, by reason of bureaucratic accident, as contingent valuation or, less grandly, willingness to pay surveys. In this method individuals are asked, in more or less sophisticated ways, what they would be willing to pay for some carefully specified environmental good. The technique has not been considered respectable because as everyone has known since Samuelson's original papers on public goods, respondents in such situations have incentives to answer strategically (more bluntly, to lie). Recently, however, the profession has been examining evidence rather than relying on a priori argument and has begun to find that things are not so bad. (A recent state of the art assessment appears in Cummings, Brookshire and Schulze, 1985.) This method seems likely to be the basis of future benefit estimates directed to practical policy decisions.

On the other side of the benefit-cost problem, the estimation of costs of pollution control policies has been dominated by engineers, since most discharge reductions are achieved by process change or insertion of some new process into a waste stream. The contributions of economists have been of two sorts. First, there have been cautions about the difficulties that lurk beneath the apparently unthreatening surface of the cost-estimation problem (see, for example, Portney, 1981a). Second, models that integrate air, water, and solid waste forms of the problem and preserve the materials balance – at least in its important aspects – have been constructed to help policy makers avoid the pitfall of assuming that treatment equals destruction (that removal of a substance from stack gas for example, means the end of that problem). Several of these models are described in Kneese and Bower (1979).

HOW TO IMPLEMENT DECISIONS? Making a collective decision to attain this or that ambient environmental quality does not mark the end of society's problem. To transform that decision into physical reality the behaviour of individual procedures and consumers must be changed. (In the commonest case, pollution discharges must be reduced.) One of the great continuing debates in environmental economics concerns the relative merits of alternative ways of achieving those changes. In overly simple, but commonly used terms, the debate has been between one or another version of command and control regulations and one or another version of economic incentives. The former are distinguished by their form as orders to do or not do something (to install and operate a piece of equipment, or reduce discharges by some per cent from a base case, or discharge no more than some amount). The latter require no action but payment, but encourage the economically rational discharger to change behaviour to the extent that reduced payments more than balance increased costs incurred in reducing discharges.

Early work in this field (e.g. Kneese, 1964) stressed the desirability of the economic incentive approach. This result was based on considerations of static efficiency and information economy, but rested on some very special assumptions (linear total benefit functions and no effect of a source's location on its damage causing potential). Subsequent analysis has relaxed the assumptions and extended the criteria of judgement to include: dynamic incentives, ease of monitoring and enforcement, flexibility in the face of exogenous changes, and political considerations (especially ethical connotations and distributional impact). These analyses have shown that the case for incentives is considerably less clear cut than original drawn, though in the encouragement of environment-saving technological development the incentives do have unambiguous superiority.

Similar early enthusiasm and analogous subsequent qualification were both lavished on the other major form of economic incentive, the marketable permit, either to pollute a place in the ambient environment or to discharge. Both types of instrument as well as others such as deposit-refund and liability systems are discussed at some length in Bohm and Russell (1985).

The practical manifestations of the economic incentives idea are few and far between despite the enthusiasm of the economic profession. The only serious system of incentive emission charges seems to be the German national system described by Brown and Johnson (1984). The major trial with marketable permits, though hedged about with constraints, is the offset-bubble-emission reduction credit system being put in place for air pollution control in the United States (see Tietenberg, 1984).

HOW TO PREDICT MACROECONOMIC EFFECTS? There is considerable practical interest in predicting the effects of environmental policies on the wider economy. For the most part, cost benefit analysis stops with the initial incidence of the implied cost and does not trace how that cost gets translated through the economy into changed input use and output mix. Two routes to the extra information may usefully be distinguished.

The first and most common depends on running projected first round costs through a macroeconomic summary model. This sort of analysis is done routinely by the econometric consulting firms such as Data Resources, Inc. and Wharton Associates. It is analogous to any exercise involving an assumed cost increase for producers (for an early report on such an effort, see Evans, 1972; for a more recent discussion of issues and results see Peskin, Portney and Kneese, 1981).

A different approach, and one still far from routine, is to use computable general equilibrium models of Walrasian type. This application of such models represents a significant extension from their more familiar use in the analysis of tax and trade problems (see the review by Shoven and Whalley, 1984). The problem for environmental application is how to enter the models, since in contrast to the first approach, simple dollar cost numbers are not sufficient. Rather, the changes in technologies of the producers subject to environmental regulation must be captured and available during the iterative solution, because the technology based standards force the producers off their efficiency frontiers. Development of practical tricks to accomplish this is an activity on the cutting edge of this subfield.

CONCLUDING COMMENT. From its fledging years in the early 1960s the field of environmental economics has grown to maturity. It is now complete with textbooks, monographs, journals, and courses in most colleges and universities. As the nature of modern economies and societies changes and evolves, new challenges appear. To meet them, environmental economics must now reach out increasingly to its sister social disciplines. Whereas, in its earlier years, effective environmental economics research had to call mostly on related natural sciences, such as engineering, ecology and meteorology, now the need is to build better bridges to collective choice theory, sociology, psychology, and formal ethics. The eclectic nature of environmental economics is to many a major part of its fascination.

ALLEN V. KNEESE AND CLIFFORD S. RUSSELL

See also EXTERNALITIES; WATER RESOURCES; USER FEES.

BIBLIOGRAPHY

Bailey, M.J. 1980. *Reducing Risks of Life: Measurement of the Benefits.* Washington, DC: American Enterprise Institute.

Bator, F. 1957. The simple analytics of welfare maximization. *American Economic Review* 47(1), March, 22–59.

Bohm, P. and Russell, C.S. 1985. Comparative analysis of alternative policy instruments. In *Handbook of Natural Resource and Energy Economics,* ed. A.V. Kneese and J.L. Sweeney, Amsterdam: North-Holland.

Bowes, M.D. and Loomis, J.B. 1980. A note on the use of travel costs models with unequal zonal populations. *Land Economics* 56, November, 465–70.

Brown, G. and Johnson, R.W. 1984. Pollution control by effluent charges: it works in the Federal Republic of Germany why not in the United States. *Natural Resources Journal* 24(4), October, 929–66.

Buchanan, J.M. and Tullock, G. 1962. *The Calculus of Consent. Logical Foundations of Constitutional Democracy.* Ann Arbor: University of Michigan Press.

Clawson, M. and Knetsch, J.L. 1966. *Economics of Outdoor Recreation.* Baltimore: Johns Hopkins University Press for Resources for the Future.

Coase, R.H. 1960. The problem of social cost. *Journal of Law and Economics* 3(1), October, 1–44.

Cropper, M.L. 1981. Measuring the benefits from reduced morbidity. *American Economic Review* 71(2), May, 235–40.

Cummings, R.G., Brookshire, D.S. and Schulze, W.D. 1985. *Valuing Environmental Goods: A State of the Art Assessment of the Contingent Valuation Method.* New York: Rowman and Allanheld.

Evans, M.K. 1972. A forecasting model applied to pollution control costs. *American Economic Review* 43(2), 244–56.

Fiorina, M.P. 1982. Legislative choice of regulatory forms: legal process or administrative process? *Public Choice* 39(1), 33–66.

Fiorina, M.P. and Noll, R. 1978. Voters, bureaucrats and legislators. *Journal of Public Economics* 9, 239–54.

Hardin, R.G. 1982. *Collective Action*. Baltimore: Johns Hopkins University Press for Resources for the Future.

Klevorick, A.K. and Kramer, G.H. 1973. Social choice on pollution management: the Genossenschaften. *Journal of Public Economics* 2, 101–46.

Kneese, A.V. 1964. *The Economics of Regional Water Quality Management*. Baltimore: Johns Hopkins University Press for Resources for the Future.

Kneese, A.V. and Schulze, W. 1985. Ethics and environmental economics. In *Handbook of Natural Resource and Energy Economics*, ed. A.V. Kneese and J.L. Sweeney, Amsterdam: North-Holland.

Kneese, A.V. and Bower, B.T. 1979. *Environmental Quality and Residuals Management*. Baltimore: Johns Hopkins University Press for Resources for the Future.

Kneese, A.V., Ayres, R.U. and d'Arge, R.C. 1970. *Economics and the Environment: A Materials Balance Approach*. Washington, DC: Resources for the Future.

Knight, F.H. 1921. *Risk, Uncertainty and Profit*. Boston: Houghton Mifflin Company.

Lave, L. and Seskin, E. 1977. *Air Pollution and Human Health* Baltimore: Johns Hopkins University Press for Resources for the Future.

Mäler, K.-G. 1974. *Environmental Economics*. Baltimore: Johns Hopkins University Press for Resources for the Future.

McMillan, M. 1976. Criteria for jurisdictional design. *Journal of Environmental Economics and Management* 3, 46–68.

Mishan, E.J. 1965. Reflections on recent developments in the concept of external effects. *Canadian Journal of Economics and Political Science* 31, February, 1–34.

Peskin, H.M., Portney, P.R. and Kneese, A.V. 1981. *Environmental Regulation and the US Economy*. Washington, DC: Resources for the Future.

Pigou, A.C. 1932. *The Economics of Welfare*. 4th edn, London: Macmillan.

Plotkin, N. 1972. A Social choice model of the California Feather River Project. *Public Choice* 12, Spring, 69–87.

Plott, C.R. 1976. Axiomatic social choice theory: an overview and interpretation. *American Journal of Political Science* 20(3), August, 511–96.

Portney, P.R. 1981a. The macroeconomic impacts of federal environmental regulation. In Henry Peskin et al., *Environmental Regulation and the US Economy*, Washington, DC: Resources for the Future.

Portney, P.R. 1981b. Housing prices, health effects, and valuing reductions in risk of death. *Journal of Environmental Economics and Management* 8(1), 72–8.

Portney, P.R. and Mullahy, J. 1985. Urban air quality and acute respiratory illness. *Journal of Urban Economics* 20(1), 21–38.

Rose-Ackerman, S. 1981. Does federalism matter? Political choice in a federal republic. *Journal of Political Economy* 89(1), 152–65.

Russell, M. and Shelton, R.B. 1974. A model of regulatory agency behavior. *Public Choice* 20, Winter, 47–62.

Scitovsky, T. 1954. Two concepts of external economies. *Journal of Political Economy* 62, April, 70–82.

Shoven, J.B. and Whalley, J. 1984. Applied general-equilibrium models of taxation and international trade. *Journal of Economic Literature* 22(3), September, 1007–51.

Thaler, R.H. and Rosen, S. 1976. The value of saving a life: evidence from the labor market. In *Household Production and Consumption*, ed. N.E. Terleckyj, New York: Columbia University Press.

Tietenberg, T.H. 1984. *Emission Trading*. Washington, DC: Resources for the Future.

Turvey, R. 1963. On divergences between social cost and private cost. *Economica* 30(119), August, 309–13.

Vaughan, W.J. and Russell, C.S. 1984. The role of recreation resource availability variables in recreation participation analysis. *Journal of Environmental Management* 19, December, 185–96.

envy. Envy is a deadly sin, but then so is avarice or greed, and greed seems not to trouble economists. Envy does, however, perhaps because it is an externality. Different economists have also used the term in different senses. Veblen (1899) avoids the word 'envy', but one feels that some of the pleasure of conspicuous consumption may come from the malicious belief that it induces envy in others. Brennan (1973) uses the term 'malice' to indicate negative altruism – a distaste for the income of others – and 'envy' to indicate that the marginal disutility of another's income increases as their income increases. For other concepts of envy, see Nozick (1974) and Chaudhuri (1985).

Most economists now use the word 'envy' in a narrow technical sense due to Foley (1967), who was much more interested, however, in finding an adequate concept of 'equity'. First, however, one should turn to Rawls (1971), whose *Theory of Justice* has 12 pages of very pertinent discussion.

RAWLS ON ENVY AND JUSTICE. Rawls (1971) defines envy in a way which he attributes to Kant, and he is careful to distinguish 'envy' from 'jealousy', which can be thought of as a protective response to envy:

> we may think of envy as the propensity to view with hostility the greater good of others even though their being more fortunate than we are does not detract from our advantages. We envy persons whose situation is superior to ours ... and we are willing to deprive them of their greater benefits even if it is necessary to give up something ourselves. When others are aware of our envy, they may become jealous of their better circumstances and anxious to take precautions against the hostile acts to which our envy makes us prone. So understood envy is collectively disadvantageous: the individual who envies another is prepared to do things that make them both worse off, if only the discrepancy between them is sufficiently reduced (p. 532).

This is in section 80, on 'the problem of envy' in which Rawls asks whether his theory of justice is likely to prove impractical because 'just institutions ... are likely to arouse and encourage these propensities [such as envy] to such an extent that the social system becomes unworkable and incompatible with human good' (p. 531). This is a positive question; a normative one arises when one recognizes the possibility of 'excusable envy' because 'sometimes the circumstances evoking envy are so compelling that, given human beings as they are, no one can reasonably be asked to overcome his rancorous feelings' (p. 534). In the following section 81, on 'envy and equality', Rawls argues carefully that his 'principles of justice are not likely to arouse excusable ... envy ... to a troublesome extent' (p. 537). Thereafter, he discusses the conservative contention 'that the tendency to equality in modern social movements is the expression of envy' (p. 538), and Freud's lamentable suggestion that an egalitarian sense of justice is but an adult manifestation of childish feelings of envy and jealousy. Recently, indeed, a particular progressive tax in West Germany has been labelled an 'envy tax' (*Neidsteuer*), as noted by Bös and Tillmann (1985). Anyway, Rawls is careful to distinguish 'rancorous' envy from the justifiable feelings of resentment at being treated unjustly. While envy may often form the basis of an appeal to justice, the claims that all appeals to justice rely on envy often fail to distinguish envy from resentment. I shall return to this later.

EQUITY AS ABSENCE OF 'ENVY'. More recently, however, 'envy' has acquired a precise technical sense in economic theory, following the (apparently independent) lead taken by Feldman and Kirman (1974) and by Varian (1974) in analysing a concept of 'equity' due to Foley (1967, p. 75). Apparently Foley was the first to use the term 'envy' in this sense, though only informally: Feldman, Kirman and Varian include it in their titles.

Consider any allocation (x_g^i) $(g = 1$ to n; $i = 1$ to $m)$ of n goods to each of m individuals. Suppose these individuals have preferences represented by ordinal utility functions $U^i(x^i)$ ($i = 1$ to m) of each individual i's own (net) consumption vector x^i. Then individual i is said to envy j if $U^i(x^j) > U^i(x^i)$, so that i prefers j's allocation to his own. Notice that this is a purely technical definition; it tells us nothing about i's emotional or psychological state, whether i is unhappy because he prefers what j has, or whether i's 'envy' makes him want to harm j. There is no sin in this unemotional economists' concept of envy, but no particular ethical appeal either. Indeed, it might be better to say that 'i finds j's position to be enviable', to minimize the suggestions of emotion.

Nevertheless, Foley was concerned to introduce a concept of equity of welfare which overcomes the deficiencies of equality of after-tax income – deficiencies which are obvious when there are different public goods in different areas, different preferences for leisure as against consumption, and different needs as well. Thus Foley proposes the absence of 'envy' as a test of whether an allocation is equitable. Formally, (x_g^i) is equitable if $U^i(x^i) \geqslant U^i(x^j)$ for all pairs of individuals i and j.

Foley (1967) was careful to qualify this test. First, lifetime consumption plans must be considered so that the prodigal do not envy the higher later consumption enjoyed by the thrifty. Second, as he says:

> if a gas station attendant has the desire to be a painter but not the ability, it may be necessary to make the painter's life very unattractive in other ways before the gas station attendant will prefer his own; so unattractive, perhaps, that the painter will envy the attendant while the attendant is still envying him. These cases must be interpreted flexibly; either equivalents to the talents must be postulated which the gas station attendant does possess, or reasonable alternatives framed that abstract from the glamour and prestige of certain activities (p. 75).

Foley (1967, p. 76) concludes his discussion of 'equity' as follows:

> If tastes differ greatly, there is very little gained by the analysis since a very wide range of allocations will meet the equity criterion. The definition is offered only as a tentative contribution to a difficult and murky subject and concludes the sketchy discussion this paper will make of welfare economics.

FAIRNESS AND OTHER EXTENSIONS OF EQUITY. In a one-good problem of dividing a cake, procedures for achieving 'fair' allocations, without envy and with no cake wasted, were discussed in works such as Steinhaus (1948, 1949), and Dubins and Spanier (1961) before Foley. Fairness with many goods was considered by Schmeidler and Vind (1972), by adding Pareto efficiency to the requirement that nobody should envy anybody else's net trade vector (as opposed to the consumption vector, which includes the endowment). Pazner and Schmeidler (1974) and Varian (1974) then came up with examples of economies with production in which there are no fair allocations, because Pareto efficiency requires skilled workers to supply more hours of labour than unskilled workers, and tastes are such that no allocation of consumption then avoids envy. Feldman and Kirman (1974) considered reducing the

degree of envy, whereas Pazner and Schmeidler (1978) weakened the notion of equity to 'egalitarian equivalence, – finding an allocation (x_g^i) in which each individual i is indifferent between x^i and a consumption in an 'egalitarian' allocation $(\bar{x}_g^i)$ with $\bar{x}^i$ independent of i. Of course, in this egalitarian allocation there is no envy. These later developments all seem like attempts to rescue a dubious notion of equality without giving up first-best Pareto efficiency, even though that is surely unattainable anyway in economies with private information.

ENVY AND RESENTMENT RECONSIDERED. In the definition of envy, each individual i compares the consumption vector x^j of another with his own, x^i, using i's own utility function U^i. But $U^i(x^j) > U^i(x^i)$ is insufficient for i's envy to be excusable, in Rawls's sense. Indeed, if x^j is preferable for i because j has some special needs met, i's envy is quite unjustifiable. As Sen (1970, ch. 9) points out in his discussion of Suppes's (1966) grading principles of justice, comparisons of x^j and x^i must allow for differences in tastes, needs, and so on. Thus the appropriate comparison in determining what is inequitable is rather whether $U^j(x^j) > U^i(x^i)$. If this is true, we might say that i resents j. Absence of resentment then requires all individuals to have equal utility levels; of course, this requires interpersonal comparisons of utility levels, of the kind used to make decisions 'in an original position', behind the 'veil of ignorance', before each individual knows his tastes. The technical sense of envy defined earlier differs from this technical notion of resentment precisely because it ignores the original position; not surprisingly, then, envy has no moral force, whereas resentment may well have.

Complete absence of resentment in this sense is probably too strong; but one should look for there to be no resentment in the weaker sense that no individual can legitimately feel treated unjustly by the institutions that determine his welfare. That, of course, reverts to Rawls (1971), though not necessarily to his particular concept of justice.

PETER J. HAMMOND

See also ALTRUISM; EQUITY; FAIRNESS.

BIBLIOGRAPHY

Bös, D. and Tillmann, G. 1985. An 'Envy Tax': theoretical principles and applications to the German surcharge on the rich. Public Finance/Finances Publiques 40, 35–63.

Brennan, G. 1973. Pareto desirable redistribution: the case of malice and envy. Journal of Public Economics 2, 173–83.

Chaudhuri, A. 1985. Formal properties of interpersonal envy. Theory and Decision 18, 301–12.

Dubins, L.E. and Spanier, E.H. 1961. How to cut a cake fairly. American Mathematical Monthly 68, 1–17.

Feldman, A. and Kirman, A. 1974. Fairness and envy. American Economic Review 64, 995–1005.

Foley, D.K. 1967. Resource allocation and the public sector. Yale Economic Essays 7, 45–198.

Nozick, R. 1974. Anarchy, State, and Utopia. New York: Basic Books.

Pazner, E. and Schmeidler, D. 1974. A difficulty in the concept of fairness. Review of Economic Studies 41, 441–3.

Pazner, E. and Schmeidler, D. 1978. Egalitarian equivalent allocations: a new concept of economic equity. Quarterly Journal of Economics 92, 671–87.

Rawls, J. 1971. A Theory of Justice. Cambridge, Mass: Harvard University Press; Oxford: Clarendon Press.

Schmeidler, D. and Vind, K. 1972. Fair net trades. Econometrica 40, 637–42.

Sen, A.K. 1970. Collective Choice and Social Welfare. San Francisco: Holden-Day.

Steinhaus, H. 1948. The problem of fair division. *Econometrica* 16, 101–4.

Steinhaus, H. 1949. Sur la division pragmatique. *Econometrica* 17, 315–19.

Suppes, P. 1966. Some formal models of grading principles. *Synthèse* 16, 284–306.

Varian, H.R. 1974. Equity, envy and efficiency. *Journal of Economic Theory* 9, 63–91.

Veblen, T. 1899. *The Theory of the Leisure Class.* New York: Macmillan and Viking, 1967.

Ephémérides du citoyen ou chronique de l'ésprit national. French economic periodical issued in three series under different names from 1766 to 1772, 1774 to 1776 and in 1788. Published first as a bi-monthly by its founder and first editor, l'Abbé Baudeau, it became a monthly as from January 1767 after Baudeau's conversion to Physiocracy by Mirabeau and Le Trosne. Its contents included contributed articles on economic and political subjects, book reviews, comments and letters to the editor, together with a chronicle of public events of interest to its readership. This provided its format from January 1769, when Du Pont de Nemours took over the editorship. Although censorship problems troubled the journal persistently (as disclosed in the Turgot–Du Pont correspondence, for this reason many issues appeared well after the ostensible month of publication) the first series was terminated by l'Abbé Terray in November 1772, presumably because it contained much vigorous criticism of his abolition of domestic free trade in grain. The first series produced therefore six issues in 1766 as a bi-monthly and 63 monthly issues from January 1767 to March 1772 inclusive. Under the title *Nouvelles Ephémérides ou Bibliothèque raisonnée de l'histoire, de la morale et de la politique,* it was revived by Baudeau after Turgot became Contrôleur-général in 1774, publishing 18 issues in all from January 1775 to June 1776, that is, the month after Turgot's dismissal from the ministry. A third series, *Nouvelles Ephémérides économiques* published three issues from January to March 1788, again under Baudeau's editorship, but his failing mental powers were presumably the reason why this final series ended so quickly.

Although initially set up by Baudeau in imitation of the English *Spectator,* within a year of its inception economics began to dominate its contents and many of the leading Physiocrats, in particular Mirabeau, Baudeau and Du Pont de Nemours, contributed most of the articles. A detailed discussion of its contents is given in Bauer (1894) and in Coquelin and Guillaumin (1854, I, pp. 710–12). Perhaps the most important piece it contained is Turgot's *Réflexions sur la formation et distribution des richesses* in serial form (*Ephémérides,* 1769, No. 11, pp. 12–56, No. 12, pp. 31–98 and 1770, No. 1, pp. 113–73), although with considerable unauthorized alterations and notes by Du Pont (see Groenewegen, 1977, pp. xix-xxi). It also published foreign contributions in French translation, including Beccaria's inaugural lecture with copious notes and comments by Du Pont (*Ephémérides,* 1769, No. 6, pp. 57–152) and a contribution by Franklin on the increasing troubles between England and her American colonies (*Ephémérides,* 1768, No. 8, pp. 159–92). As an early if not the first, economic journal, the *Ephémérides* remains an important part of economic literature and an indispensable source for those interested in the study of Physiocracy.

PETER GROENEWEGEN

See also PHYSIOCRATS.

BIBLIOGRAPHY

Bauer, S. 1894. Ephémérides. In *Dictionary of Political Economy,* ed. R.H.I. Palgrave, London: Macmillan, Vol. I.

Coquelin C. and Guillaumin (eds). 1854. Ephémérides. In *Dictionnaire de l'économie politique,* Paris: Guillaumin, Hachette, Vol. I.

Groenewegen, P.D. 1977. *The Economics of A.R.J. Turgot.* The Hague: Martinus Nijhoff.

epistemological issues in economics. Economics has raised such high hopes with its sophisticated techniques that the lack of agreed findings sets a puzzle. It will be suggested here that this lack of agreement reflects an epistemological puzzle, partly of how to recognize a causal law and partly of whether the search should be for causal laws in any case.

Start with the hypothetico-deductive method and the standard idea that economic theory advances when hypotheses confront the empirical evidence, as laid out in textbooks of positive economics, like Lipsey (1980). The backdrop is an empiricist picture of the natural world as an ordered realm, independent of our concepts, beliefs, hypotheses and conjectures about it. Science captures the order by testing our conjectures so as to arrive at causal laws or hypotheses describing what happens next in various initial conditions. Whether or not the world is governed by underlying forces and mechanisms, our knowledge of it, formulated in terms of causal laws, has only the empirical warrant conferred when observation and experiment uphold our generalizations. Scientific method is a matter of generalizing either from a known pattern to the next case (prediction) or from a particular case to a pattern which subsumes it (explanation). Prediction and explanation are thus two sides of the only epistemic coin, experience generalized. The economic world and economic knowledge are then construed on this natural science model.

This very basic empiricism needs supplementing in two ways, if it is to carry conviction. One is to give theory a more explicit role, which it plainly has in the practice of economics, without undermining the claims of prediction to be the only test of truth. A neat source is Friedman (1953), who associates economic theory with 'a language' and 'a body of substantive hypotheses'. The former is 'a set of tautologies' and 'its function is to act as a filing system'. The latter is 'designed to abstract essential features of a complex reality'. Whether the right features have been abstracted and included in the filing system depends solely on the success of the resulting predictions. This echoes the Logical Positivists' distinction between analytic statements, whose truth relies on the meaning of terms and tells us nothing about the world, and synthetic statements, whose truth depends on the facts. Friedman's 'substantive hypotheses' serve to link pure theory (the 'filing system') to the world, while making sure that empirical facts wear the trousers.

The other supplement is Popper's (1963) account of science as conjectures and refutations. Reflecting that circular theories, like those of Freud and Marx, are always confirmed by experience in the eyes of their holders, he weakens the claim of empirical confirmation as a test of truth. Instead, it is falsifiability which separates science from pseudo-science. Formally, if hypothesis H implies observation O, then O does not prove H, but *not-O* refutes H. So a genuinely scientific theory must state possible empirical conditions in which it would be refuted. A causal law is an empirical hypothesis of sufficient scope and generality, which has risked refutation. In Popper's own eyes, he has made a radical break with traditional empiricism by undermining the value of induction.

However, epistemologically, claims to knowledge still face the same old test and the facts of observation are still trumps. (This paragraph refers to Popper's best-known, classic account and not to his more recent writings.)

Later philosophy of science has raised serious difficulties for this empiricist approach to judging a theory. The Quine–Duhem hypothesis has it that a scientific theory is a web, which includes observation sentences and which can only be understood as a whole. Quine (1961), evoking Duhem (1914), argues that traditional empiricism relies on two untenable dogmas. One is that our five senses supply us with 'unvarnished news' as an objective and independent test of hypotheses. The other is that meaningful statements divide cleanly into analytic and synthetic (as defined above). Both dogmas are to be rejected. Our beliefs form a 'seamless web', a 'field of force which touches experience only at the edges'. At the moment of testing they 'face the tribunal of experience together' and, in assessing the verdict, we always have a choice of what to accept, revise or reject (including our own observations). In place of a single hypothesis H implying an unvarnished observation O, we have a set H_1H_2 ... etc. linked to another theory-laden set O_1O_2 ... etc. What we do about it, when we find that we cannot keep both sets, may be governed by criteria like parsimony, elegance or fertility, whose epistemic warrant is open to question and is far removed from any which relies on there being brute facts.

Economics is full of illustrations of the Quine–Duhem hypothesis. The 'money supply' or 'the general price level' are not brute facts. There are a number of different definitions of the money supply and ways of aggregating prices, and the choice of one rather than another will reflect theoretical considerations. The economist works with descriptions of data which already have theoretical order built into them, and this seems inescapable. Likewise, the joint testing of hypotheses is notoriously recognized in economics as complicated by the role of *ceteris paribus* conditions. Predictions are always issued subject to certain *ceteris paribus* clauses, and if the prediction is falsified, the economist is often able to claim that this is because the *ceteris paribus* conditions were violated, particularly when, as in consumer theory, there are unobservable variables like preference orderings or utility functions to reckon with. So we find that empirical tests in economics are often indecisive. Few neoclassical economists, for example, seem willing to forsake the homogeneity assumption (the absence of money illusion) in consumer theory despite its frequent 'falsification'. Indeed, it looks rather as if, despite the common gestures of respect for Popper, economics is full of those circular theories which are always confirmed by experience in the eyes of their holders. The Quine–Duhem hypothesis is descriptively plausible.

This conclusion would not surprise Kuhn (1970), who argues that even the natural sciences have not progressed in the way expected by an empiricist methodology. Instead, their history is best understood as a discontinuous series of paradigms. By 'paradigms' he usually seems to mean the definitive current practices of the dominant scientific community and sometimes a more loosely specified set of currently shared presuppositions or world views. Paradigms rise and fall for many reasons, but empirical testing is never decisive. Rather, paradigms simply acquire more and more anomalies as contrary empirical evidence accumulates, and it is only when a new paradigm surfaces, which can incorporate the anomalies, that the anomalies become regarded as counter-examples. A paradigm shift then occurs. Feyerabend (1975) has continued this process of dismantling empiricism, until it is unclear whether theory choice can be a rational process at all. This can be read as an invitation to a sociology of knowledge approach. If there is no good intellectual reason behind theory selection, then we must study the social pressures on and within the scientific communities, which influence the evolution of theory. In economics one might cite the suspicion in some quarters that the dominance of neoclassical economics owes much to its apparent support for a free enterprise system and that its mathematical sophistication is best understood as an exclusionary device typical of a closed profession. But a step into the sociology of knowledge is not the only option.

Lakatos (1978) tries to reaffirm the core of Popper by accepting that the units at stake are whole research programmes rather than single hypotheses, and that the core of the research programme is often defended against falsification by suitable adjustments to the auxiliary hypotheses or 'protective belt'. A research programme is to be judged, however, by whether these revisions to the protective belt are progressive or degenerating. Degenerating ones are *ad hoc* and cover only the anomaly which has precipitated the adjustment, whereas progressive ones provide additional and novel areas of application for the theory. The progressive/degenerating distinction sounds a promising way of reintroducing empirical criteria into the evaluation of a theory, but in practice this designation, like falsification earlier, is prone to vary with the eyes of the beholder. What looks progressive from one theoretical perspective can become *ad hoc* when viewed from another. For a setting to these developments the reader might usefully consult Harré (1972); for a chart of the options, Chalmers (1976); and for a robust post-empiricist philosophy of science, Hesse (1980).

Recent philosophy of science thus makes it unsurprising that economic methodologies inspired by empiricism are alive with controversy. But there are other sources of methodological inspiration. Weber (1922) draws an appealing but cloudy distinction between explaining (*erklären*) and understanding (*verstehen*). Natural sciences seek to explain by means of causal laws; the humanities seek to understand by reconstructing the actors' world from within. The crux is the idea that the agent's own point of view matters in the social world in a way which does not hold for the natural sciences. Unlike the natural world, the social is *not* an ordered realm independent of our concepts, beliefs, hypotheses and conjectures about it. Our beliefs influence our actions and hence the outcomes we observe in any empirical investigation, whereas subatomic particles have no beliefs to affect outcomes in the natural world. This sets a puzzle for an empiricist methodology, which seeks the causal laws governing an independent realm. How is the social scientist to investigate the world from within and to relate these findings to the demands of objectivity?

Weber's answer is that some of his work is to be done by a process of '*verstehen*' – a key term from the German idealist, 'hermeneutic' (interpretative) tradition. The guiding thought is that what turns behaviour into action is its inward meaning and that institutions similarly are meaningful practices. But 'meaning' is an elusive concept, which threatens to let in more subjective variety than a social scientist can welcome. Weber tries to render *verstehen* more precise by stressing the rationality of action seen through the actor's eyes. He borrows the neoclassical economic concept of rational action. To the objection that real actions are not always rational, even when seen from within, he replies that, by establishing what would be fully rational, one can identify departures from the ideal type as *explananda*. *Verstehen* is not the only method, however. The social sciences seek both adequacy at the level of meaning and adequacy at the causal level, with the 'causal level' said to be one of statistically significant correlations.

Verstehen is thus finally less of an alternative to *erklären* and more of a heuristic device; but an interesting line has been opened.

Among examples of the use of *verstehen* Weber cites pure mathematics. This suggests a more ambitious thought, one which finds echoes in economics itself. Von Mises (1949, 1960) presents economics as the science of human action and economic theory as the construction *a priori* of ideal types of rational allocation. This yields an element of *a priori* knowledge contrary to empiricism. Similarly, Hayek (1960) and the new Austrians appeal to Kant and *a priori* knowledge when making their commitment to a methodological individualism grounded in the preconditions of the possibility of free choice.

Although a shift to a Kantian epistemology of pure reason would create more problems than it answers, it could have interesting implications. Consider, for example, Arrow–Debreu general equilibrium theory. It is not plausibly regarded either as a set of empirical hypotheses or as a mere filing system. Its proponents would not accept a sociology-of-knowledge suggestion that it belongs to the initiation rites of the profession. It functions very much as an 'ideal type', used for judging economic performance and making policy recommendations. What claim has to be made for it, if it is to be a reliable benchmark for performance? A Kantian reply is that it has to lay claim to truth, in the sense of stating correctly what would be the outcome of a fully rational allocation of resources throughout an economy. As in mathematics, the Kantian adds, *a priori* truths are hard to come by but are nonetheless what theory aims for. Arrow–Debreu general equilibrium theory makes sense on no other terms.

This is, of course, a contentious approach even to pure mathematics. It is doubly so for economics, owing to the role of rationality in its 'ideal types'. Not everyone would accept that Arrow–Debreu general equilibrium theory embodies an ideally rational allocation. Whether it does depends in part on how 'rationality' is defined. It is standardly given an instrumental definition, with a rational allocation being one which adopts the most efficient means to a given end. One reason for this definition is that it is believed to keep on the safe side of the positive/normative distinction and so to preserve the basic parts of economic theory from value judgements. This does not satisfy advocates of other approaches, especially political economists, who allege that Arrow–Debreu general equilibrium theorem has ideological commitments. A Kantian comment would be that 'ideal type' rationality is bound to involve the rationality of ends as well as of means. An ideally rational allocation of resources is the one which a just or good society would display, and an ideally rational choice is, in the last analysis, a moral choice. A switch to a Kantian epistemology may breach the positive/normative distinction and restore economics to the position of a moral science.

A challenge to the positive/normative distinction opens up deep epistemological questions. But some economists are willing to take the plunge, inspired by Rawls (1971). By defining a just society as one whose allocation of resources rational agents would agree upon in advance of knowing what they would each get out of it personally, Rawls connects economic rationality to moral choice. That makes it possible to ask precise questions about, for instance, the rationality of preferences and which kind of preferences should be satisfied in cases of conflict. The hope might be said to be an *a priori*, normative science with implications both for efficiency and for moral advance. This is to take seriously the thought that to

know what would be ideally rational is to know how to do better.

Rawls's epistemological novelty lies in the use of a thought-experiment, inviting readers to think themselves into the shoes of fully rational, self-interested agents, who do not yet know whether they themselves will gain or lose from any arrangement proposed. Their knowledge of what would be rational comes from a proof that each does best to settle for equal basic rights and a maximum distribution of goods. The proof is controversial, but Rawls has certainly given economists new thoughts, especially in welfare economics, and a new line of defence against the charge that pure theorizing about reflective equilibria is only a parlour game.

The article began by suggesting that the lack of many results in economics comparable to the major discoveries in natural science reflected puzzles about causal laws and the value of seeking them. Some of the puzzles are general for all sciences; witness the unfinished arguments started by Quine, Kuhn and others. An initial empiricism seems peculiarly difficult to uphold in economics, however, because one is trying to predict what will be done by agents, who themselves have beliefs and whose world depends on their expectations.

That makes rationality an epistemologically important yet special concept in economics. But it is unlikely to be a gateway to a simple rationalist epistemology. Whereas an 'ideal type' of frictionless motion is simply a limiting case with a zero coefficient of friction, an 'ideal type' of rational choice is not an abstraction from normal behaviour but a solution to a theoretical problem with a likely normative dimension. Even if economics is still regarded as the search for a different kind of causal law, rather than as an alternative to causal thinking, this difference in kind is great enough to set epistemological problems. Changes in belief change the course of economic events and introduce discontinuities, which make ideal types of rational action unlike timeless models of causal regularities. Rational belief is part of the concept of rationality. So the normative element of the concept of rationality will remain important for economics, even if only for prediction.

SHAUN HARGREAVES-HEAP AND MARTIN HOLLIS

See also ARROW–DEBREU MODEL; ECONOMIC MAN; MODELS AND THEORY; PHILOSOPHY AND ECONOMICS; RHETORIC.

BIBLIOGRAPHY

Chalmers, A.F. 1976. *What is this Thing called Science?* St Lucia, Queensland: University of Queensland Press.

Duhem, P. 1914. *The Aim and Structure of Physical Theory*. Princeton: Princeton University Press, 1954.

Feyerabend, P.K. 1975. *Against Method: An Outline of an Anarchistic Theory of Knowledge*. London: New Left Books.

Friedman, M. 1953. On the methodology of positive economics. In M. Friedman, *Essays in Positive Economics*, Chicago: University of Chicago Press.

Harré, R. 1972. *The Philosophies of Science*. Oxford: Oxford University Press.

Hayek, F.A. 1960. *The Constitution of Liberty*. Chicago: University of Chicago Press.

Hesse, M. 1980. *Revolutions and Reconstructions in the Philosophy of Science*. Brighton: Harvester Press.

Kuhn, T.S. 1970. *The Structure of Scientific Revolutions*. 2nd edn, Chicago: University of Chicago Press.

Lakatos, I. 1978. The methodology of scientific research programmes. In *Philosophical Papers*, Vol. I, ed. J. Worral and G. Currie, Cambridge: Cambridge University Press.

Lipsey, R.G. 1980. *Introduction to Positive Economics*. London: Weidenfeld & Nicolson.

Popper, K.R. 1963. *Conjectures and Refutations: The Growth of Scientific Knowledge*. London: Routledge & Kegan Paul.

Quine, W. 1961. Two dogmas of empiricism. In *From a Logical Point of View*, ed. W. Quine, Cambridge, Mass.: Harvard University Press.

Rawls, J. 1971. *A Theory of Justice*. Cambridge, Mass.: Belknap Press of Harvard University Press.

von Mises, L. 1949. *Human Action*. London: William Hodge.

von Mises, L. 1960. *Epistemological Problems of Economics*. Princeton: Princeton University Press.

Weber, M. 1921. *Economy and Society*. New York: Bedminster Press, 1968.

equality. The very use of the term 'equality' is often clouded by imprecise and inconsistent meanings. For example, 'equality' is used to mean equality before the law (equality of treatment by authorities), equality of opportunity (equality of chances in the economic system), and equality of result (equal distribution of goods), among other things. These different meanings often conflict, and are almost never wholly consistent. See Hayek (1960, p. 85; 1976, pp. 62–4) for a discussion of equality before the law and equality of result, and Rawls (1971) for a discussion of equality of opportunity within a theory of distributive justice. Elsewhere I have discussed the difference between equality of opportunity and equality of result in education (Coleman, 1975). See also Pole (1978) for a detailed examination of the changing conceptions of equality in American history.

Some order can be brought into the confusion among the different uses of the term 'equality' by first conceiving of a system that constitutes an abstraction from reality. The system consists of:

(a) a set of positions which have two properties:

    (i) when occupied by persons, they generate activities which produce valued goods and services;

    (ii) the persons in them are rewarded for these activities, both materially and symbolically;

(b) a set of adult persons who are occupants of positions;

(c) children of these adults;

(d) a set of normative or legal constraints on certain actions.

What is ordinarily meant by equality under the law has to do with (b), (c), and (d): that the normative or legal constraints on actions depend only on the nature of the action, and not on the identity of the actor. That is, the law treats persons in similar positions similarly, and does not discriminate among them according to characteristics irrelevant to the action.

What is ordinarily meant by equality of opportunity has to do with (a), (b), and (c): that the processes through which persons come to occupy positions give an equal chance to all. More particularly, this ordinarily means that a child's opportunities to occupy one of the positions (a) do not depend on which particular adults from set (b) are that child's parents. What is ordinarily meant by equality of result has to do with (a.ii): that the rewards given to the position occupied by each person are the same, independent of the activity.

These three conceptions, equality under the law, equality of opportunity, and equality of result can also be seen as involving different relations of the State to the inequalities that exist or spontaneously arise in ongoing social activities. Equality before the law implies that the laws of the State do not recognize distinctions among persons that are irrelevant to the activities of the positions they occupy, but otherwise make no attempt to eliminate inequalities that arise. Equality of opportunity implies that the State intervenes to insure that inequalities in one generation do not cross generations, that children have opportunities unaffected by inequalities among their parents. Equality of result implies a continuous or periodic intervention and redistribution by the State to insure that the inequalities which arise through day-to-day activities are not accumulated, but are continuously or periodically eliminated.

The relations between the first two kinds of equality differ according to how close a society is to a legally minimalist society or a legally maximalist society. In a society that is legally minimalist, equality before the law is compatible with a high degree of inequality of opportunity – depending on the distribution of opportunity provided by other institutions in society, such as the family. In a legally maximalist society, in which many functions of traditional institutions have been taken over by institutions that are creatures of the State (e.g., functions of the family taken over by the public school), equality before the law implies a high degree of equality of opportunity. Only in a society in which the law was far more intrusive than found anywhere, and children were taken from their families to be raised 'with equal opportunity' by the State, could it be said that equality before the law would coincide with equality of opportunity.

The relation between equality of opportunity and equality of result is somewhat different, for it implies two different kinds of interventions of the State. Equality of opportunity implies intervention to provide each person with resources that give equal *chances* to obtain the material and symbolic rewards that arise from productive activity, while equality of result implies intervention in the distribution of these rewards, to provide each person with equal *amounts*. The two concepts become indistinguishable only when the State intervenes to insure that each position (in (a) above) provides the same set of material and symbolic rewards; and in such a circumstance, 'opportunity' loses meaning altogether.

IS EQUALITY 'NATURAL'? There are certain philosophical positions that take equality of result as a 'natural' point, from which all others are deviations. Isaiah Berlin probably states this as well as any other

> No reason need be given for ... an equal distribution of benefits for that is 'natural', self evidently right and just, and needs no justification, since it is in some sense conceived as being self justified ... The assumption is that equality needs no reasons, only inequality does so; that uniformity, regularity, similarity, symmetry,... need not be specially accounted for, whereas differences, unsystematic behavior, changes in conduct, need explanation and, as a rule, justification. If I have a cake and there are ten persons among whom I wish to divide it, then if I give exactly one tenth to each, this will not, at any rate automatically, call for justification; whereas if I depart from this principle of equal division I am expected to produce a special reason. It is some sense of this, however latent, that makes equality an idea which has never seemed intrinsically eccentric ... (1961, p. 131).

This quotation describes a view with which Berlin does not necessarily identify himself. In the same paper, he states that 'equality is one value among many ... it is neither more nor less rational than any other ultimate principle ... rational or non-rational' It is, however, the position implicitly taken by John Rawls in his *Theory of Justice*, for the book is addressed to the question, 'When can inequalities (of result) be regarded as just?' Rawls's answer can be paraphrased as 'Only those inequalities are just which make the least well off person better off than that person would be (other things being equal) in the absence of the inequalities.'

Whether equality of result is 'natural' or not, and whether the position of Berlin and Rawls is correct or incorrect, would appear to depend on how the distribution of goods occurs: If goods are initially the property of a single central source (e.g., 'the State'), then Berlin's position and that of Rawls appear correct. If all rights and resources originate with the State (or with the king, as in early political theory), than an equal distribution has some claim to be seen as natural. (If, for example, the revenue from oil discovered on public lands is a major component of GNP, as in some Middle Eastern states, equal distribution constitutes a natural point.) But if goods are seen to arise from the activities of a set of independent actors each with certain initial property rights, and each with a certain amount of zeal and skill, 'equality' (meaning equality of result) is hardly natural, and is inconsistent with the distribution of property rights including rights to the fruits of one's own activity.

EQUALITY, ENVY AND RESENTMENT. The idea of equality as 'natural' appears also to derive in part from the ubiquity of envy and resentment in society, with the demand for 'equality' as an expression of these feelings which carries legitimacy. A number of sociologists have pointed to this connection. For example, Simmel writes (1922, translated in Schoeck, 1969, p. 236–7):

> Characteristically, no one is satisfied with his position in relation to his fellow beings, but everyone wishes to achieve a position that is in some way an improvement. When the needy majority experiences the desire for a higher standard of living, the most immediate expression of this will be a demand for equality in wealth and status with the upper ten thousand.

Simmel follows with an anecdote: at the time of the 1848 revolution, a woman coal-carrier remarked to a richly dressed lady, 'Yes, madam, everything's going to be equal now; I shall go in silks and you'll carry coal.'

Helmut Schoeck, in an extensive examination of the role of envy in society, argues that

> social philosophers have largely failed to see how little the individual is concerned with being *equal* to someone else. For very often his sense of justice is outraged by the very fact that he is denied the measure of inequality which he considers to be right and proper (1969, p. 234).

Feelings of envy and resentment constitute a challenge to the existing distribution of *rights* in society, between those held collectively and those held individually. In particular, it is a challenge to the existence of individual property rights. The centrality of property rights for conceptions of equality is seen most clearly in neoclassical economic theory, which assumes a distribution of property rights among a set of independent actors, accompanied by a free market. (See Meade, 1964, for a discussion of property rights and the market in relation to equality.) It is to economic theory that I now turn.

THE ROLE OF 'EQUALITY' IN ECONOMIC THEORY. The concept of 'equality' has no place in positive economic theory. In this it is unlike the concept of 'liberty', for economic theory is predicated on the assumption of liberty, that is, free choice (subject only to resource constraints) among alternative actions. There is, in the concept of free choice, however, something closer to the idea of equality before the law than to

equality of opportunity, and closer to the latter than to equality of result. Equality of result implies a distribution process that is the antithesis of the market.

But normative economics, that is, welfare economics, makes up for the absence of 'equality' from positive economic theory, for the idea of equality of result is a part of the very atmosphere surrounding welfare economics. The question of what policies will maximize social welfare is not often answered directly in terms of equality in the distribution of valued goods, but the idea seems always to hover nearby. The most direct expression of the central importance of equality in welfare economics was probably that of Pigou (1938; see also Bergson, 1966, ch. 9) who reasoned that because money, like everything else, had declining marginal utility, and thus a dollar was worth much less to a person when he had a million others than when it was the only one he had, then the maximum of social welfare could only be achieved when incomes were made equal. (Neither Pigou nor any other welfare economist followed this implication with actual policy recommendations for equality of income, thus raising the question: if the criterion is correct, then why not recommend implementing it?)

The rock on which Pigou's argument is often regarded as foundering is that of interpersonal comparison of utility. To move from the relative importance for one person of a dollar when he is rich and when he is poor to its relative importance to different persons is a move which, as has been often reiterated, cannot be justified on positive grounds. Perhaps the most widely quoted statement to this effect is that of Lionel Robbins (1938):

> But, as time went on, things occurred which began to shake my belief in the existence between so complete a continuity between politics and economic analysis ... I am not clear how these doubts first suggested themselves; but I well remember how they were brought to a head by my reading somewhere – I think in the work of Sir Henry Maine – the story of how an Indian official had attempted to explain to a high-caste Brahmin the sanctions of the Benthamite system. 'But that,' said the Brahmin, 'cannot possibly be right – I am ten times as capable of happiness as that untouchable over there.' I had no sympathy with the Brahmin. But I could not escape the conviction that, if I chose to regard men as equally capable of satisfaction and he to regard them as differing according to a hierarchial schedule, the difference between us was not one which could be resolved by the same methods of demonstration as were available in other fields of social judgement ... 'I see no means,' Jevons had said, 'whereby such comparison can be accomplished.'

Edgeworth expressed the same point, 'The Benthamite argument that equality of means tends to maximum happiness, presupposes a certain equality of natures; but if the capacity for happiness of different classes is different, the argument leads not to equal, but to unequal distribution' (1897, p. 114).

Such arguments are ordinarily taken as conclusive within the domain of economics, and with their acceptance, the very programme of welfare economics – not to speak of the foundations for a policy designed to bring equality – is emasculated.

A philosopher might argue, of course, that there is no logical difference between the comparison of utilities of two persons and the comparison of utilities of one person at two different times. Neither, by this argument, is warranted. See, for example, Parfit (1984).

However, Pigou's conclusion has, quite apart from problems of interpersonal comparison of utility, another deficiency. It assumes that each person is an island, and contributes nothing to the welfare of others, nor has his welfare contributed to by others. Yet is is the essence of social and economic systems that there is interdependence, that one person's activities do affect the welfare of others, whether intended or not. One person spends money on loud radios that cause disturbance, while another plants flowers that others enjoy. Or one uses income for training which is productive, benefiting general welfare, while another uses income on drink and becomes alcoholic, requiring public-expense hospitalization.

But if this is so, then maximization of welfare one time period into the future would require that these interdependencies be taken into account. Maximization would occur only if resources were distributed among persons in accordance with the positive impact of their activities on those events which bring welfare to others. But in general persons do not capture the full benefits of their welfare-generating activities, nor do persons pay the full costs of their welfare–diminishing activities.

The matter can also be seen as a problem in input–output economics: What current allocation of resources among productive activities (i.e., among positions in the system as described earlier) will achieve some desired distribution of final consumption? If the aim is to maximize the sum of final consumption ('maximizing welfare'?), it is quite unlikely that either the current allocation necessary to achieve that, or the distribution of final consumption itself, will approach equality. Even if the desired final distribution is equality, and even if that is achievable within the system of activities, it is highly unlikely that the allocation at time O necessary to achieve that at time t will be equal. And it may well be that the only distribution at time O that would achieve equality at time t would do so at a low level of welfare, with each having less than if there were inequality at time t resulting from a different distribution at time O. If Pareto optimality is taken as a self-evident necessary condition for optimal policies, then because of the processes described above, a criterion of equal distribution (either initially or subsequently) would violate the condition. This suggests that Rawls's question was misdirected, and should have been 'when (assuming non-violation of constitutional rights) is *equality* of distribution justified?' and should have been answered, 'Only when there is no unequal distribution that would subsequently make each better off.'

Thus even if Pigou's point that maximizing welfare requires equalizing marginal utilities is accepted, and noncomparability of utilities is ignored, the policy implication of equalizing incomes appears shortsighted in the extreme. Another way of seeing so is by use of Robert Nozick's Wilt Chamberlain example, an example designed to argue against theories of distributive justice which, like that of Rawls, use the resulting distribution of goods ('end state theories', to use Nozick's term) as a criterion.

Now suppose that Wilt Chamberlain is greatly in demand by basketball teams, being a great gate attraction. (Also suppose contracts run only for a year, with players being free agents.) He signs the following sort of contract with a team: In each home game, twenty five cents from the price of each ticket of admission goes to him. (We ignore the question of whether he is 'gouging' the owners, letting them look out for themselves.) The season starts, and people cheerfully attend his team's games; they buy their tickets, each dropping a separate twenty five cents for their

admission price into a special box with Chamberlain's name on it. They are excited about seeing him play; it is worth the total admission price to them. Let us suppose that in one season one million persons attend his home games, and Wilt Chamberlain winds up with $250,000, larger even than anyone else has. Is he entitled to this income? (Nozick, 1974, p. 161)

Thus as Nozick points out, an equal distribution at one point will lead to an unequal distribution at a later point, due to the very system of activities through which persons satisfy their interests.

There are only three ways to prevent this, all of which, carried to their limit, can be shown to reduce welfare. One is to prevent the economic exchange through which persons spend their quarters as they see fit, for such exchanges may lead to a large accumulation in the hands of the Wilt Chamberlains.

A second is to attack the system of activities itself, the system which generates that matrix of coefficients that transform equality into inequality – that is, shutting down professional basketball, which redistributes income from those with low incomes to those with high incomes. The third way is to allow the exchange, but then to tax the high incomes back down to equality. This effectively eliminates the activity, because if income is an incentive to carry out the activity that is paid for, the Wilt Chamberlains lose all incentive to carry the activity.

Indeed, unless there is a perfect positive correspondence of those activities which are intrinsically pleasurable with those which produce benefits for others, and a perfect negative correspondence with those that produce harm for others, the absence of any extrinsic incentives will lower the welfare for all. The more interrelated the activities of individuals, the greater the reduction in social welfare when extrinsic incentives are absent.

It is true that taxation which is not carried to the limit, but is merely 'progressive', does not eliminate the incentive for activities that bring high income, for these activities continue in societies that have progressive taxation. But this taxation may lead to underprovision of welfare-generating activities. That is, efficiency may be sacrificed to achieve some distributional goals. The potential conflicts between efficiency and equality are discussed in the literature on optimal taxation (e.g., Atkinson and Stiglitz, 1980, part II). (A device which is informally used in social systems to reduce the disincentive effect of regimes of taxation and redistribution that shift incomes in the direction of equality is the attachment of social stigma to the receiving of income thus redistributed, for example, stigma associated with being 'on welfare'. The existence of this stigma constitutes a means of informally reconstituting the differential incentives that are reduced by redistribution.)

All three approaches to preventing inequalities from arising out of equality give, at their extreme, the same result: elimination of the very system of activities that generates welfare in the first place; for it is these activities which not only generate welfare, but also transform equality at one time into inequality at a later time.

Thus it becomes clear that the source of inequalities is embedded in the very matrix of social and economic activities through which individuals increase the welfare of themselves and one another. If, through technology for example, this matrix changes in such a way that individuals' satisfaction of wants is more concentrated in a few hands (e.g., by the invention and development of television), then inequalities will necessarily increase.

More generally, the degree of inequality seems related to the degree of interdependence in this matrix of social and economic activities. In a social system that has very low interdependence (e.g., a social system composed largely of subsistence farmers, a condition that was once the case for nearly all societies), the welfare of each in future periods depends largely on his own initial distribution of resources (including zeal and skill). If that distribution is near equality, then near equality is perpetuated into the future, modified only by random events. More important, even if the initial distribution is unequal, the low interdependence of the system of activities means that these inequalities (also modified by random events) are merely carried forward into the future. In a system with a high degree of interdependence, however, there are a great many configurations which constitute 'inequality-generating' activity structures. In such activity structures, initial distribution of equality will lead to highly unequal distributions. This inequality in turn will lead in the next generation to inequality of opportunity, constrained only by random processes or explicit policies towards non-inheritance of position, i.e., toward equality of opportunity. (In a system in which attention to basketball was directed not to televised professional teams, but to games of the local high school, both the material and nonmaterial rewards among basketball player would be more equally distributed. There would be greater equality of results, which would arise not through a change in the set of persons (b), the distribution of children (c), or the normative and legal constraints (d), but only through a change in the distribution of positions.)

Does this mean that there tends to be a negative relation between the interdependence of activities in a social system (and thus the total social product) and the equality with which the activities of the system distribute the product? If so, this is a discouraging result for those who would prefer a social system in which incomes are not increasingly unequal, for it specifies an opposition between two goals both regarded as desirable.

This question has two parts, a within-generation part and a between-generation part. Within generations, it appears likely that there is a negative relation, that increased interdependence does, except in unlikely activity structure, increase inequality. It is possible that this negative relation is responsible for the rise in redistributive actions of governments as interdependence of economic activities increases.

Between generations, the answer would appear to hinge largely upon the relative rates of increase of interdependence of activities and of equality of opportunity (i.e., non-inheritance of position). The latter can occur through regression to the mean as well as through explicit policy intervention (see Becker and Tomes, 1986, for a discussion). If equality of opportunity increases more slowly than interdependence of activities, then (except for unlikely configurations of the activity matrix) there will be a decrease in equality of result among lineages of persons. If equality of opportunity increases more rapidly than the increase in interdependence of activities, there will be an increase in equality of result among lineages, even with a decrease in equality of result within generations.

Altogether, there has been little investigation of the matters discussed above, that is, just how the structure of social and economic activities itself affects inequalities. Such investigations would lead toward taking work on equality partly out of the realm of normative theory, bringing it partly into the realm of positive theory.

JAMES S. COLEMAN

See also ENVY; FAIRNESS; INEQUALITY; POVERTY; DISTRIBUTIVE JUSTICE.

BIBLIOGRAPHY

Atkinson, A.B. and Stiglitz, J.E. 1980. *Lectures on Public Economics.* New York: McGraw-Hill.

Becker, G. and Tomes, N. 1986. Inequality, human capital, and the rise and fall of families. In *Approaches to Social Theory*, ed. S. Lindenberg, J. Coleman, and S. Nowak, New York: Russell Sage.

Bergson, A. 1966. *Essays in Normative Economics.* Cambridge, Mass: Harvard University Press.

Berlin, I. 1961. Equality. In *Justice and Social Policy*, ed. F.A. Olafson, Englewood Cliffs: Prentice Hall, 131.

Coleman, J. 1975. What is meant by 'an equal educational opportunity'? *Oxford Review of Education* 1(1), 27–9.

Edgeworth, F.Y. 1897. *Papers Relating to Political Economy.* London: Macmillan for the Royal Economic Society, 1925.

Hayek, F.A. 1960. *The Constitution of Liberty.* Chicago: University of Chicago Press.

Hayek, F.A. 1976. *Law, Legislation and Liberty*, Vol. 2. Cambridge: Cambridge University Press.

Meade, J.E. 1964. *Efficiency, Equality and the Ownership of Property.* London: Allen & Unwin.

Nozick, R. 1974. *Anarchy, State and Utopia.* New York: Basic Books.

Parfit, D. 1984. *Reasons and Persons.* Oxford: Oxford University Press.

Pigou, A.C. 1938. *Economics of Welfare.* 4th edn, New York: Macmillan.

Pole, J.R. 1978. *The Pursuit of Equality in American History.* Cambridge: Cambridge University Press.

Rawls, J. 1971. *A Theory of Justice.* Cambridge, Mass.: Harvard University Press.

Robbins, L. 1938. Interpersonal comparisons of utility. *Economic Journal* 48, 635–41.

Schoeck, H. 1969. *Envy: A Theory of Social Behavior.* New York: Harcourt, Brace and World.

Simmel, G. 1962. *Soziologie.* 2nd edn, Munich and Leipzig: Duncker & Humblot.

equal rates of profit. The concept of equality, or its opposite inequality, implies a comparison, and a comparison must be based on the consideration of a population of cases. Therefore equality or inequality has different implications according to the definition of that population. This general observation applies in particular to rates of profit.

Three different types of comparison of rates of profit will be examined:

(i) We may compare the rates of profit in terms of a fixed *numéraire*, particularly money, which can be obtained over a certain period of time from investment of funds in different lines of activity. We shall refer to equality in this sense as *sectoral equality* of rates of profit. Or,

(ii) We may compare the rate of profit obtainable over a certain period of time in terms of one *numéraire* with that obtainable over the same period of time in terms of another *numéraire*. In a famous chapter of his *General Theory* (Keynes, 1936, ch. 17), Keynes employs the term 'own rates of interest' to describe these rates of return in terms of different *numéraires*, and we shall borrow the same term and call equality of the rates of return in different numeraires *own rates* equality. Finally,

(iii) We may compare the rate of profit obtainable in terms of the fixed *numéraire*, which may again be money, during one period of time with that obtainable during another later period of time. This comparison include the historically important question of the long-term trend of the rate of interest, whether it will tend to constancy, to increase, or to decline and, if not constant, what will be its eventual limit. We shall refer to equality in this sense as *temporal equality* of rates of profit. (In common with many writers, particularly in the past, we ignore in the present discussion distinctions between the rate of interest and the rate of profit. The main cause of a persistent difference between the two must be sought in the uncertainty from which our analysis abstracts.)

While it is convenient to have discussions of respectively sectoral, own-rate and temporal equality of rates of profit collected together in one article, it will be clear that these are distinct notions and that the investigation of the conditions required for another.

THE THEORY OF PROFIT. An argument concerning equality of rates of profit might depend importantly on which theory of the rate of profit is invoked. Such is inescapably the case where temporal equality of rates of profit is concerned. However a good deal of our argument concerning sectoral and own rate equality of rates of profit is independent of the exact theory of the determination of rates of profit in general. This unexpected possibility might be realized because equality of rates of profit depends above all upon arbitrage, the tendency for capital to seek the highest return. Indeed in some cases an arbitrage condition alone suffices to demonstrate that rates of profit must be equal.

We shall refer to a state of the economy in which all possibilities of profitable arbitrage have been put into effects, which is a kind of short-period equilibrium, as an *arbitrage equilibrium*. It has sometimes been claimed that profit (where what is intended is a part of profit distinct from a normal rate of return) is essentially a phenomenon of disequilibrium. On this account an arbitrage equilibrium would not only exhibit equal rates of profit, all rates of profit would equal zero. Only the normal rate of return would be realised in an arbitrage equilibrium.

To argue about terminology where weighty issues are involved shows poor judgement. Even if profit is defined to be an excess of return to capital above the return generally available, and even if we exclude temporary rents, it remains to show that no sector can enjoy a permanent profit advantage against which arbitrage is for some reason powerless. If, on the other hand, profit is taken to include temporary rents it is evident that there is really no case for equality. Hence the only interesting question to decide is whether rates of profit defined as net returns to capital divided by the values of capital employed (on average or at the margin) are equal in an arbitrage equilibrium.

SECTORAL EQUALITY OF RATES OF PROFIT. Nowhere is the power of arbitrage, together with its limitations, better illustrated than in the case of comparisons of profit rates across sectors. The desire of every investor to obtain the highest possible rate of return may reasonably be assumed to equalize the equivalent rates of return on different bonds. Will not a similar principle ensure the equalization of rates of profit in different activities, be they regions or industries?

The answer depends on two important points. First, we must decide how to compare two rates of return, what are the principles of equivalence? Secondly, arbitrage may encounter obstacles. This is true even where bonds are concerned, and is more important still where we are concerned with different sectors.

Clearly rates of return should be true economic rates including allowances for capital gains, etc. Moreover, two apparently different rates of return may not excite arbitrage if they represent different risks, or different liabilities to taxation, or if the difference is too small to overcome transactions costs. Although they are important in empirical investigations, these detailed considerations may be neglected for our purposes. So we are left with structural obstacles to arbitrage.

When economic theorists assume equal rates of profit in different sectors they are implicitly ignoring questions of industrial structure. (For an excellent treatment of the concept of industrial structure and it implications for profitability, see Hay and Morris, 1979, ch. 7.) It is typically supposed, for example, that capital may be shifted from one sector to another in arbitrarily small quantities. If increasing returns to scale imply that operation at a very small scale will be costly, the putative entrant must choose between staying out of the sector of fighting his way into what must be an oligopolistic market. There is naturally no reason to suppose that the rate of profit enjoyed by those already inside may not exceed that obtainable in a competitive sector of small-scale units.

It would not be necessary to reiterate the foregoing point if it had not apparently been challenged by the late Piero Sraffa in the oft-quoted foreword to his *Production of Commodities by Means of Commodities* (1960). Sraffa's model for the determination of prices is striking for its simplicity and for the fewness of its assumptions. In his forward the author warned his readers against the temptation to assume that his argument depended upon assuming constant returns to scale. In a sense it does not, as that assumption is never directly employed. However equality of rates of profit, sectoral equality according to our present terminology, is assumed. We cannot of course claim that sectoral equality requires constant returns to scale. However it requires some assumptions about the environment, specifically the market environment, in which firms operate, and constant returns to scale and free entry are obvious sufficient conditions for sectoral equality of profit rates.

EQUALITY OF OWN RATES OF INTEREST. Consider a price system extending through time so that for each period t there is a present price for each of N goods. Such a price system may be represented thus:

$$
\begin{array}{cccccc}
p_{11} & p_{12} & \cdots & p_{1t} & \cdots & p_{1T} \\
p_{21} & p_{22} & \cdots & p_{2t} & \cdots & p_{2T} \\
\cdots & \cdots & \cdots & \cdots & \cdots & \cdots \\
p_{N1} & p_{N2} & \cdots & p_{Nt} & \cdots & p_{NT}
\end{array}
\qquad (4.1)
$$

As problems raised by infinite price systems need not concern us here, we suppose that the prices only extend forward to period T. If we imagine that good 1 is money, it will be seen that the money rate of interest in period 1 for a t-period loan may be calculated as follows. One unit of present money costs p_{11} and one unit of money bought now for delivery in period t costs p_{1t}. Hence one unit of money surrendered now buys p_{11}/p_{1t} units of money at t. This corresponds to a rate of interest equal to $p_{11}/p_{1t}-1$, or $(p_{11} - p_{1t})/p_{1t}$. What was denoted above by the term the money rate of interest can equally be designated the *own rate of interest* on money, in this case for a t-period loan.

The money rate of interest measures the extra money obtainable by postponing payment as a proportion of the payment deferred. This notion generalises to any good. We may for example measure the extra wheat obtainable by postponing delivery as a proportion of the quantity of wheat delivery deferred. Suppose that wheat prices occupy the second row of (4.1) above. Then the t period own rate of interest for wheat will be equal to $p_{21}/p_{2t} - 1$, or $(p_{21}-p_{2t})/p_{2t}$, which is exactly analogous to the expression of the money rate of interest already derived.

Turning from the rows of (4.1), which correspond to different goods, consider the columns, which correspond to different periods of time. It is easily shown that if the columns are proportional to each other, which is the same as saying that relative prices are the same in all periods, then the own rate of interest for a given duration of loan is the same for all goods. Suppose that the own rate of interest for good 1 for a deferment from period 1 to period t is $r_1 t$. Then, as we have already seen:

$$r_1 t = (p_{11} - p_{1t})/p_{1t} \tag{4.2}$$

However, by assumption:

$$p_{1t}/p_{1t} = p_{11}/p_{1t} \tag{4.3}$$

so that:

$$(p_{11} - p_{1t})/p_{1t} = (p_{11} - p_{1t})/p_{1t} \tag{4.4}$$

Or,

$$r_1 t = r_1 t \tag{4.5}$$

Here, constancy of relative prices implies equality of own rates of return, as required. Conversely, variations in relative prices will be reflected in differences in own rates of return.

Under what circumstances is it reasonable to assume constancy of relative prices over time? We shall certainly require the assumption that the economy is stationary in some sense. Suppose for example that as time passes timber becomes more and more scarce relative to demand as forests are depleted or demand grows. Then we would expect the price of timber to rise through time relative to other goods. Similarly, technical progress, unless it be of the simplest labour-augmenting kind, will typically imply changes in relative prices. The transistor, the microchip and other innovations, to cite another example, have certainly caused electric goods to become relatively cheaper.

Consider therefore a stationary state, which may be growing economy, but which is stationary in the sense that in each period it is technically exactly the same as in every other period, except perhaps for scale. As the economy is essentially the same at every moment of time, it makes intuitive sense to suppose that relative prices might be the same at each moment of time, and this intuition is valid in so far as it can be shown that any development of the economy which is stationary, in the sense just described, may be supported by a price system which is itself stationary, in the sense that relative prices are invariant over time[3].

Stationarity of the real economy is sufficient for stationarity of a price system that will support production activities, but does not imply that any such price system will be stationary. Indeed it is an implication of the multiplicity of price systems and interest rates which goes under the name of 'double-switching'. That prices which support stationary production will frequently be neither unique nor themselves stationary. Their non-uniqueness is an immediate implication of double-switching. The existence of non-stationary price systems for these equilibria follows when we note that the average of two systems of equilibrium prices must themselves be equilibrium prices. However the average of two price systems based on different rates of interest produces a rate of interest variable over time, and varying relative prices.

The importance of these findings may be questioned because the price system is required to support not only production (supply) but also consumption (demand). This will make the observation of non-unique prices, and in particular of a history including double-switching, much less probable than a considering of the production side alone might suggest.

It remains to briefly mention Keynes's use of own rates of interest in his *General Theory*, if only to point out that it is not in fact particularly germane to the present discussion of equality of own rates of interest. Keynes's extraordinary argument is concerned with the comparison of money rates of return at the margin to accumulating various assets, which is something like the question of sectoral equality.

We may imagine that as the various assets are accumulated the money rates of return to further accumulation for each of them is forced down, and that the quantities accumulated are such that these marginal returns on all assets are equalized. If we could conceive of the elasticity of the money rate of return for each asset to the stock accumulated (which we may call the return-stock elasticity) as a value independent of other accumulations, which Keynes in effect does, then assets with low return-stock elasticities will accumulate rapidly relatively to assets with higher return-stock elasticities. Keynes's argument claims that money is eventually the asset with the lowest return-stock elasticity, and that this has the implication that, in an economy with a limited supply of money, the money rate of return (which of course is the own rate of interest of money) will eventually rise to a level which discourages the further accumulation of real assets.

5. TEMPORAL EQUALITY OF RATES OF PROIT. We now turn to the equality, or inequality as the case may be, of the rates of profit which prevail at different moments of time. There is a longer tradition among economic theorists, which goes back to the classical writers, of explaining the long-run tendency for the rate of profit to fall. This was largely a response to a supposed fall in the rate of interest which the classical economists 'took to be an indisputable fact'. For these older theories the reader is referred to entries on Adam Smith, Marx, Mill, Ricardo and Say. Here we consider only a modern view of the problem. A justification for this division of labour may be sought in the fact that modern theories of the rate of profit are radically different from classical views.

The main source of the difference between modern and classical theories (which in this context should be taken to exclude Marx) is that the former treat technical progress as having regular and continuous effects on the economy, where the latter typically do not. Thus the characteristic classical argument for a falling rate of profit is stagnationist in nature. The decline in the rate of profit is part of the grinding to a halt of a previously progressive economy. In contrast, the modern neoclassical approach locates the explanation of a falling rate of profit in the character of a technical progress conceived as an indefinitely continuing process.

To demonstrate the theoretical issues involved we first show when a declining rate of profit would arise in a neoclassical model with aggregate capital and a constant saving propensity, and then discuss some of the shortcomings of that model as an account of capital accumulation.

Let output, Y, depend upon the input of labour, L, and a capital stock which is homogeneous with the output flow, K, according to a constant returns production function as:

$$Y = F(K, L, t). \tag{5.1}$$

Let partial derivatives be denoted by subscripts so that, for example, the marginal product of capital is denoted $F_K(K, L, t)$. We denote the rate of profit by r, so that:

$$r = F_K(K, L, t). \tag{5.2}$$

Time derivatives are shown by a dot over the variable concerned. Differentiating $F_K(K, L, t)$ totally with respect to time we obtain an expression for the time rate of change of the rate of profit as:

$$\dot{r} = F_{KK} \cdot \dot{K} + F_{KL} \cdot \dot{L} + F_{Kt}, \tag{5.3}$$

Hence for constancy of the rate of profit we must have:

$$F_{KK} \cdot \dot{K} + F_{KL} \cdot \dot{L} + F_{Kt} = 0. \tag{5.4}$$

which on rearrangement yields:

$$\frac{F_{KK} \cdot K}{F_K} \cdot k + \frac{F_{KL} \cdot L}{F_K} \cdot l + \frac{F_{Kt}}{F_K} = 0, \tag{5.5}$$

where k and l are respectively the logarithmic rates of growth of capital and labour. Now (5.5) can be expressed more simply as:

$$\sigma_K \cdot k + \sigma_L \cdot l + \gamma = 0; \tag{5.6}$$

where σ_K and σ_L are respectively the elasticity of the marginal product of capital with respect to K and L, and γ is the proportional change in the marginal product of capital due to the passage of time alone.

We know that $F_K(K, L, t)$ is homogeneous of degree zero in K and L. Hence:

$$\sigma_K + \sigma_L = 0, \tag{5.7}$$

and (5.6) reduces to:

$$\sigma_K \cdot (k - 1) + \gamma = 0. \tag{5.8}$$

This last expression has an intuitive interpretation. As σ_K is the elasticity of the marginal product of capital with respect to capital, it will be negative. It is weighted by $k - 1$, the rate of growth of capital per unit of labour, which will be positive under normal economic growth. Thus $\sigma_K \cdot (k - 1)$ measures the rate at which capital accumulation is pushing down the rate of profit due to the substitution of capital for labour at constant technical knowledge. The second term represents the rate at which technical progress is tending to raise the rate of profit at constant factor proportions, which must be positive term if technical progress is beneficial. Now, unsurprisingly, (5.8) says that, for the rate of profit to remain constant, these two effects must exactly offset.

As it is known that a production function with aggregate capital cannot be derived rigorously except for simple or special production technologies, it may reasonably be asked how fare the above account, of a downward pressure on the rate of profit due to accumulation being offset by an upward pressure due to technical progress, generalizes. In particular, is it generally true that accumulation with constant technical knowledge exerts a downward pressure on the rate of profit?

Given the enormous literature on the theory of capital which has been produced in recent years, it is perhaps surprising that this question remains relatively under investigated. Many discussions of capital accumulation simply beg the question by assuming that the rate of interest would fall continuously through time. Indeed double-switching is most at variance with the traditional neoclassical view of capital accumulation when that assumption is made. However there is no guarantee of a continuous fall of the rate of profit through time, and the demand side of the economy is likely to prohibit a return to a previous and lower income state.

On the other hand, linear models of the type that have been used to illustrate simple stories of capital accumulation can lead to quite eccentric time profiles of consumption being associated with the accumulation of capital (where this is defined simply as an increase in long-term consumption). Hence there is no possibility in general of ruling out erratic developments in the rate of interest over time.

CHRISTOPHER BLISS

See *also* CAPITAL PERVERSITY; INTEREST AND PROFIT; SURPLUS APPROACH TO VALUE AND DISTRIBUTION; SRAFFIAN ECONOMICS.

BIBLIOGRAPHY

Bliss, C.J. 1975. *Capital Theory and the Distribution of Income.* Amsterdam: North-Holland

Harcourt, G.C. 1972. *Some Cambridge Controversies in the Theory of Capital.* Cambridge: Cambridge University Press.

Hay, D.A. and Morris, D.J. 1979. *Industrial Economics: Theory and Evidence.* Oxford: Oxford University Press.

Keynes, J.M. 1936. *The General Theory of Employment, Interest and Money.* London: Macmillan.

Schumpeter, J. 1954. *History of Economic Analysis.* New York: Oxford University Press.

Sraffa, P. 1960. *Production of Commodities by Means of Commodities.* Cambridge: Cambridge University Press.

equation of exchange. The equation of exchange (often referred to as the quantity equation) is one of the oldest formal relationships in economics, early versions of both verbal and algebraic forms appearing at least in the 17th century. Perhaps the best known variant of the equation of exchange is that expressed by Irving Fisher (1922):

$$MV = PT \tag{1}$$

Equation (1) represents a simple accounting identity for a money economy. It relates the circular flow of money in a given economy over a specified period of time to the circular flow of goods. The left-hand side of equation (1) stands for money exchanged, the right-hand side represents the goods, services and securities exchanged for money during a specified period of time. M is defined as the total quantity of money in the economy, T as the total physical volume of transactions, where a transaction is defined as any exchange of goods, including physical capital, services and securities for money, P is an appropriate price index representing a weighted average of the prices of all transactions in the economy. Finally, to make the stock of money comparable with the flow of the value of transactions (PT), and to make the two sides of the equation balance, it is multiplied by V, the transactions velocity of circulation, defined as the average number of times a unit of currency turns over (or changes hands) in the course of effecting a given year's transactions.

An alternative variant of the Equation of Exchange is the income version by Pigou (1927). Empirical difficulties in measuring an index of transactions, and the special price index related to it, led, with the development of national income accounting, to the formulation of equation (2):

$$MV = PY \tag{2}$$

where y represents national income expressed in constant dollars, P the implicit price deflator and V the income velocity of circulation defined as the average number of times a unit of currency turns over in the course of financing the year's final activity.

Equations (1) and (2) differ from each other because the volume of transactions in the economy includes intermediate goods and the exchange of existing assets, in addition to final goods and services. Thus vertical integration and other factors which affect the ratio of transactions to income would also alter the ratio of transactions velocity to income velocity.

A third version of the Equation of Exchange, the Cambridge Cash Balance Approach (Pigou, 1917: Marshall, 1923; Keynes, 1923), converts the flow of spending into units comparable to the stock of money

$$M = kPY \tag{3}$$

where $k = 1/V$ is defined as the time duration of the flows of goods and services money could purchase, for example, the average number of weeks income held in the form of money balances.

Equations (2) and (3) are arithmetically equivalent to each other but they rest on fundamentally different notions of the role of money in the economy. Both equations (2) and (1) view money primarily as a medium of exchange and the quantity of money is represented as continually 'in motion' – constantly changing hands from buyer to seller in the course of a time period. Equation (3) views money as a temporary abode of purchasing power (an asset) forming part of a cash balance 'at rest'. Consequently, the items included in the definition of money in the transactions and income versions of the Equation of Exchange are assets used primarily to effect exchange – currency and checkable deposits, whereas the Cash Balance approach includes, in addition to these items, non-checkable deposits and possibly other liquid assets.

The Equation of Exchange is useful both as a classification scheme for analysing the underlying forces at work in a money economy and as a building block or engine of analysis for monetary theory and in particular for the Quantity Theory of Money.

As a classification scheme, the equation as a basic accounting identity of a money economy demonstrates the two-sided nature of the circular flow of income – that the sum of expenditures must equal the sum of receipts. The left-hand side of the equation shows the market value of goods and services purchased (dollar value of goods exchanged) and the money received. The equation also relates the stock of money to the circular flow of income by multiplying M by its velocity. Finally, the equation is useful in creating definitional categories – M, V, P, T – amenable both to empirical measurement and to theoretical analysis.

The Equation of Exchange is best known as a building block for the Quantity Theory of Money. The traditional approach has been to make behavioural assumptions about each of the variables in the equation, converting it from an identity to a theory. The simplest application, dubbed the 'Naive Quantity Theory' (Locke, 1691) treated V and T in equation (1) as constants, with P varying in direct proportion to M.

A more sophisticated version (Fisher, 1911) treats each of M, V and T as being normally determined by independent sets of forces, with V as determined by slowly changing factors such as those affecting the payments process and the community's money holding habits.

The Cambridge Cash Balance approach, based on equation (3), views the Quantity Theory as encompassing both a theory of money demand and money supply. In this approach the nominal money supply is determined by the monetary standard and the banking system while the nominal quantity of money demanded is proportional to nominal income, with k the factor of proportionality, representing the community's desired holding of real cash balances. k in turn is determined by economic variables such as the rate of interest in addition to the factors stressed by the Fisher approach. The price level (value of money) is then determined by the equality of money supply and demand.

The Equation of Exchange can also be regarded as a building block for a macro theory of aggregate demand and supply (Schumpeter, 1966). If we view MV as aggregate demand and T or y as aggregate supply, then P would be determined in the familiar Marshallian way.

Finally, the equation can be used to construct a theory of nominal income. According to this approach (Friedman and Schwartz, 1982), nominal income is determined by the interaction of the money supply and a stable demand for real cash balances. The decomposition of a given change in nominal income into a change in the price level and in real output is determined in the short run by inflation (deflation) forecast errors and in the long run by the natural rate of output.

The Equation of Exchange both as a classification scheme and as a building block for the Quantity Theory of Money can be traced back to the earliest development of economic science.

The pre-Classical writers of the 17th and 18th centuries viewed the Equation in both senses. Locke (1691), Hume (1752) and Cantillon (1735) each organized his approach to monetary issues using the Equation. Locke had a clear statement of the naive quantity theory assuming both V and T to be immutable constants. Hume followed Locke but made a clear distinction between long run statics and short run dynamics. In the long run the price level would be proportional to M but in the short run or transition period, changes in M would produce changes in T. Cantillon had a clear understanding of the relationship between the stock of money and the circular flow of income. Indeed, he was the first to define explicitly the concept of velocity of circulation, viewing V not as a constant but as a variable influenced in a stable way by both technological and economic variables. Furthermore, like Hume, Cantillon distinguished between the long run equilibrium nature of the quantity theory and short run disequilibrium. Both Locke and Hume viewed the Equation from the perspective of money 'at rest' forming a cash balance whereas Cantillon viewed money as continuously in 'motion'.

John Law (1705) understood the Equation of Exchange but used it to derive a link between changes in the quantity of M and changes in T.

The Classical economists, Thornton, Ricardo, Mill, Senior and Cairnes followed the Locke/Hume/Cantillon tradition of the quantity theory of money using a verbal version of the Equation of Exchange in their monetary analysis.

Algebraic versions of the Equation first appeared in the 17th and 18th centuries (see Marget, 1942; Humphrey, 1984). The British writers Briscoe (1694) and Lloyd (1771) both expressed a rudimentary version of equation (1), unfortunately omitting a term for velocity. Turner (1819) formulated the equation without breaking PT into separate components. The most complete early statement of the equation was by Sir John Lubbock (1840) who not only included all the items of the Equation but (preceding Fisher) distinguished between the quantities and velocities of hard currency, bank notes and bills of exchange. Similar complete algebraic statements of the Equation were made by the German writers Lang (1811) and Rau (1841); the Italian Pantaleoni (1889); the Frenchmen Levasseur (1858), Walras (1874) and de Foville (1907); and the Americans Newcomb (1885), Hadley (1896), Norton (1902) and Kemmerer (1907). Of this group Newcomb presented the clearest statement. Newcomb started with the concept of exchange as involving the transfer of money for wealth. Summing up all exchanges in the economy he arrived at his Equation of Societary Circulation:

$$VR = KP \qquad (4)$$

where V represents the total value of currency, R the rapidity (velocity) of circulation, K the volume of real transactions, P a price index.

The clearest and best known algebraic expressions of the Equation were by the neoclassical economists Irving Fisher

(1922) and A.C. Pigou (1917). Fisher (1911), directly following Newcomb, defined the Equation of Exchange as

> a statement, in mathematical form, of the total transaction effected in a certain period in a given community. ...[I]n the grand total of all exchanges for a year, the total money paid is equal to the total value of goods bought. The equation thus has a money side and a goods side. The money side is the total money paid, and may be considered as the product of the quantity of money multiplied by its rapidity of circulation. The goods side is made up of the products of quantities of goods exchanged multiplied by their respective prices (pp. 15–17).

This statement expressed as in equation (1) or in an expanded version distinguishing between currency and deposits payable by check,

$$MV + M'V' = PT \qquad (5)$$

where M' is defined as checkable deposits and V' their velocity, Fisher then used to analyse the forces determining the price level.

Fisher's approach followed the 'motion' theory tradition of Cantillon with velocity determined primarily by technological and institutional factors. In contrast Pigou (1917) and other writers in the Cambridge tradition, Marshall (1923) and Keynes (1923), followed the 'rest' approach of Locke and Hume expressing the Equation as

$$1/P = kR/M \qquad (6)$$

where R represents total resources enjoyed by the community, k the proportion of resources the community chooses to keep in the form of titles to legal tender, M the number of units of legal tender and P a price index. For Pigou the fundamental difference between his approach and that of Fisher was that by focusing

> attention on the proportion of their resources that people *choose* to keep in the form of titles to legal tender instead of focusing on the 'velocity of circulation' ... it brings us ... into relation with *volition* – an ultimate *cause of demand* – instead of with something that seems at first sight *accidental and arbitrary* (p. 174, emphasis added).

The Cambridge Cash Balance Version of the Equation of Exchange, by focusing on the demand for money and volition rather than emphasizing mechanical aspects of the circular flow of money, can be viewed as the starting point for the Keynesian approach to the demand for money (Keynes, 1936), for modern choice theoretic approaches to money demand (Hicks, 1935) and for the Modern Quantity Theory of Money (Friedman, 1956).

MICHAEL D. BORDO

See also NEWCOMB, SIMON; QUANTITY THEORY OF MONEY.

BIBLIOGRAPHY
Bordo, M.D. 1983. Some aspects of the monetary economics of Richard Cantillon. *Journal of Monetary Economics* 12, 234–58.
Briscoe, J. 1694. *Discourse on the Late Funds...* . London.
Cantillon, R. 1755. *Essai sur la nature du commerce en général*. Ed. H. Higgs, 1931, London: Macmillan; reprinted New York: Augustus M. Kelley, 1964.
Fisher, I. 1911. *The Purchasing Power of Money*. 2nd edn, 1922. Reprinted, New York: Augustus M. Kelley, 1963.
Foville, A. de. 1907. *La monnaie*. Paris.
Friedman, M. 1956. The quantity theory of money – a restatement. In *Studies in the Quantity Theory of Money*, ed. M. Friedman, Chicago: University of Chicago Press.
Friedman, M. and Schwartz, A.J. 1982. *Monetary Trends in the United States and the United Kingdom: their relation to income, prices and interest rates, 1867–1975*. Chicago: University of Chicago Press for the National Bureau of Economic Research.
Hadley, A.T. 1896. *Economics*. New York.
Hicks, J.R. 1935. A suggestion for simplifying the theory of money. *Economica* 2, February, 1–19.
Holtrop, M.W. 1929. Theories of the velocity of circulation of money in earlier economic literature. *Economic Journal* 39, January, 503–24.
Hume, D. 1752. Of money. In *Essays, Moral, Political and Literary*, Vol. I of *Essays and Treatises*, a new edition, Edinburgh: Bell and Bradfute; Cadell and Davies, 1804.
Humphrey, T.M. 1984. Algebraic quantity equations before Fisher and Pigou. *Federal Reserve Bank of Richmond Economic Review* 70(5), September/October, 13–22.
Kemmerer, E.W. 1907. *Money and Credit Instruments in Their Relation to General Prices*. New York: H. Holt & Co.
Keynes, J.M. 1923. *A Tract on Monetary Reform*. Reprinted, London: Macmillan for the Royal Economic Society, 1971.
Keynes, J.M. 1936. *The General Theory of Employment, Interest and Money*. Reprinted, London: Macmillan for the Royal Economic Society, 1973.
Lang, J. 1811. *Grundlinien der politischen Arithmetik*. Kharkov.
Levasseur, E. 1858. *La question de l'or: les mines de Californie et d'Australie*. Paris.
Lloyd, H. 1771. *An Essay on the Theory of Money*. London.
Locke, J. 1691. *The Works of John Locke*, Vol. 5. London, 1823.
Lubbock, J. 1840. *On Currency*. London.
Marget, A.W. 1942. *The Theory of Prices*. New York: Prentice-Hall.
Marshall, A. 1923. *Money, Credit and Commerce*. London: Macmillan. Reprinted, New York. Augustus M. Kelley, 1965.
Newcomb, S. 1885. *Principles of Political Economy*. New York: Harper & Brothers.
Norton, J.P. 1902. *Statistical Studies in the New York Money Market*. New York.
Pantaleoni, M. 1889. *Pure Economics*. Trans. T.B. Bruce, London: Macmillan, 1898.
Pigou, A.C. 1917. The value of money. *Quarterly Journal of Economics* 32, November. Reprinted in *Readings in Monetary Theory*, ed. F.A. Lutz and L.W. Mints for the American Economic Association, Homewood, Ill.: Irwin, 1951.
Pigou, A.C. 1927. *Industrial Fluctuations*. 2nd edn, London: Macmillan, 1929.
Rau, K.H. 1842. *Grundsatze der Volkswirtsaftslehre*. 4th edn, Leipzig and Heidelberg.
Schumpeter, J.A. 1954. *History of Economic Analysis*. New York: Oxford University Press.
Turner, S. 1819. *A Letter Addressed to the Right Hon. Robert Peel with Reference to the Expediency of the Resumption of Cash Payments at the Period Fixed by Law*. London.
Walras, L. 1874–7. *Eléments d'économie politique pure*. Lausanne: Corbaz.

equilibrium: an expectational concept. Economic equilibrium, at least as the term has traditionally been used, has always implied an outcome, typically from the application of some inputs, that conforms to the expectations of the participants in the economy. Many theorists, especially those employing the 'economic man' postulate, have also required the further condition for equilibrium that every participant be optimizing in relation to those correct expectations. However it is the former condition, correct expectations, that appears to be the essential property of equilibrium at least in the orthodox use of the term. Economic equilibrium is therefore not defined in the same terms as physical equilibrium. The rest positions or damped oscillations of pendulums cannot be economic equilibria nor disequilibria since pendulums have no expectations.

Yet it is natural and obvious that the first applications of the equilibrium idea identified some position of rest, or stationary state, as being the .equilibrium in the problem at hand. Undoubtedly the term equilibrium, referring to an 'equal weight' of forces pushing capital or what-not *in* as pulling it *out*, owes its origins to the balance of forces prevailing in a stationary situation. But there can also be a *sequence* of positions in which there is a new balance with each new position. There was no reason why equilibria might exist only among stationary states or balanced-growth paths.

Once efforts began to extend economic theory to the case of moving equilibrium paths the expectational meaning of equilibrium began to be explicit. Two of the pioneers here are Myrdal and Hayek. In his 1927 book on price determination and anticipations (in Swedish) Myrdal addresses the two-way interdependency arising in a dynamic analysis of an on-going economy: present disturbances influence future prices and anticipations of future disturbances affect present prices (the latter relation being Myrdal's main subject). In a 1928 article (in German) on what he called intertemporal equilibrium, Hayek drew the analogy between intertemporal trade and international (or interspatial) trade: prices of the same thing at two different places or times are not generally equal, though they may be pulled up or down together. In a 1929 article (in Swedish) Lindahl studied what is considered to be the first mathematical model of intertemporal equilibrium. This literature is surveyed in Milgate (1979).

The English-speaking world was slow to take up the new line of research. In his *General Theory* of 1936, Keynes speaks grandly of having shown the existence of an (implicitly moving) equilibrium with underemployment, and he does argue that the expectation of falling wages and thus prices makes the slump worse, which suggests he had an expectational notion of equilibrium in mind; but he gives no clues as to what he means by equilibrium, so both the nature and the basis of his claim are left unclear. The new topic of intertemporal equilibrium and the explicit expectational treatment of equilibrium make their English debut in Hicks's *Value and Capital* in 1939. (In the same year Harrod's expectational notion of 'warranted growth', alias equilibrium, and the translation of Lindahl's writings appear.) Hicks makes clear the analytical problem that the analyst and the economic agents alike must solve to find equilibrium: in view of the dependence of future endogenous variables, such as next period's price, on present actions of firms and households, and the dependence of such actions on expectations of those future variables, what expectation would cause the actual outcome to coincide with the expectation? For example, if the actual price P is a function f of the expected price P^e find the value of P^e such that $P^e = f(P^e)$. Thus the fixed-point character of equilibrium from a mathematical standpoint has a human, or real, interpretation. One might say, semi-jocularly, that pendulums have no economic equilibria since their motions, unlike those of trapeze artists, are not a function of expectations, if they have any.

In the postwar period the notion of equilibrium turns up in contexts quite different from that of the inter-war economic theorists. In game theory, begun by von Neumann and Morgenstern, the term equilibrium is used to refer to the theoretical solution to the policies, or play, of two or more players in strategic interaction. If the model postulates optimizing, or expected-utility-maximizing, behaviour by all players, as game theorists' models invariably do, the equilibrium necessarily has the feature that no player can do better acting alone; but lying behind this feature is the essential property that each player has correctly expected the strategy of the others and hence optimized relative to those correct expectations.

In the late 1960s the notion of equilibrium begins to take root in the new territory of non-classical markets – markets without costless and thus complete information. An economy may have markets – the resort hotel market is perhaps a suitable example – in which there are costs in the acquisition or processing of information about prices (and perhaps product specifications) so that arbitrage tendencies are delayed and the classical law of one price operates only with a lag. One well-known portrait of such a market imagines that the national market is composed of Phelpsian islands lacking current-period information about one another's prices. Another image visualizes each firm as an island unto itself with its own stock of customers, who are not knowledgeable about the policies (and perhaps even the whereabouts or existence) of other firms. In such non-Walrasian markets the prevailing prices can be (and usually are) supposed to be market-clearing: no buyer or seller is subjected to rationing (sometimes called non-price rationing by overfastidious writers). However the market will be in *equilibrium* if and only if the prices (and other variables) reflect correct expectations on the part of suppliers and buyers about the prices prevailing elsewhere – at other islands or other firms; otherwise there is *dis*equilibrium.

An economy may also have markets – one may think of labour markets or markets for rental housing – in which, although information is immediate, the wage or rental setters have to make decisions of some durability, however short-lived, and without advance information about the similar decisions of the other firms. In such quasi-Walrasian markets there may be reasons – having to do with incentives, or efficiency – why wages tend to exceed and rentals lie below the market-clearing level. Yet the market will be in *equilibrium* in the case (if such exists) in which no wage setter or rental setter experiences surprise at the corresponding decisions being made simultaneously (or perhaps somewhat later within the period of the commitment) by the other wage or rental setters; otherwise the market must be in disequilibrium, however long or brief (see Phelps et al., 1970).

Thus the analogy between intertemporal equilibrium and interspatial equilibrium, which was drawn by Hayek and others in their analysis of the former, now seems deeper than it could have at first. The expectational meaning of equilibrium, which is so unavoidably clear in the context of intertemporal equilibrium, where future prices are generally expected future prices, turns out to be just as natural and inevitable in the interspatial context as soon as one gives up the fictive device of the Walrasian auctioneer and thus admit that there are 'other' prices elsewhere, about which there must be expectations, not merely a single market-wide price.

The 1970s witnessed the formal analysis of equilibrium in terms of expectations, or forecasts, of the probability distributions of prices. Lucas, adopting the device of separate market-clearing islands, analysed a model in which there is non-public, or local, information (later called asymmetric information), namely local prices, and these price observations are used to update people's conditional forecasts of the currently unobserved prices elsewhere. There may exist a *rational-expectations* equilibrium in which everyone knows and uses the correct *conditional* expectations of the unobserved prices – that is, the statistically optimal forecasts conditional upon his particular information set. This is equilibrium with a qualification.

In surveying the meaning of equilibrium Grossman has remarked that, in Hicks, 'perfect foresight is an equilibrium concept rather than a condition of individual rationality'. A similar comment applies, with even greater weight, to statistical equilibrium and to its rational-expectations variant. The agents of equilibrium models are not simply rational creatures; they have somehow come to possess fantastic knowledge. The equilibrium premise raises obvious problems of knowledge: why should it be supposed that all the agents have hit upon the true model, and how did they manage to estimate it and conform to it more and more closely? There has always been a strand of thought, running from Morgenstern in the 1930s to Frydman in the present, that holds that we cannot hope to understand the major events in the life of an economy, and perhaps also its everyday behaviour, without entertaining hypotheses of disequilibrium.

EDMUND S. PHELPS

See also ARROW–DEBREU MODEL; CONJECTURAL EQUILIBRIA; DISEQUILIBRIUM ANALYSIS; GENERAL EQUILIBRIUM; SUNSPOT EQUILIBRIA; UNCERTAINTY AND GENERAL EQUILIBRIUM.

BIBLIOGRAPHY

Frydman, R. and Phelps, E.S. (eds) 1983. *Individual Forecasts and Aggregate Outcomes.* Cambridge: Cambridge University Press.

Grossman, S.J. 1981. An introduction to the theory of rational expectations under asymmetric information. *Review of Economic Studies* 54, June, 541–60.

Hayek, F.A. 1928. Das intertemporale Gleichgewichtssystem der Preise und die Bewegungen des Geldwertes. *Weltwirtschaftliches Archiv* 28(1), July, 33–76.

Hicks, J.R. 1939. *Value and Capital.* Oxford: Clarendon Press.

Keynes, J.M. 1936. *General Theory of Employment, Interest and Money.* London: Macmillan.

Lindahl, E. 1929. Prisbildningproblemets uppläggning från kapitalteoretisk synpunkt. *Ekonomisk Tidskrift* 2.

Lucas, R.E., Jr. 1972. Expectations and the neutrality of money. *Journal of Economic Theory* 4(2), April, 103–24.

Milgate, M. 1979. On the origin of the notion of 'intertemporal equilibrium'. *Economica* 46(1), February, 1–10.

Morgenstern, O. 1935. Vollkommene Voraussicht und wirtschaftliches Gleichgewicht. *Zeitschrift für Nationalökonomie* 6(3), 337–57.

Myrdal, G. 1927. *Prisbildningsproblemet och Föränderligheten.* Uppsala: Almqvist and Wiksell.

Phelps, E.S. et al. 1970. *Microeconomic Foundations of Employment and Inflation Theory.* New York: W.W. Norton.

von Neumann, J. and Morgenstern, O. 1944. *The Theory of Games.* Princeton: Princeton University Press.

equilibrium: development of the concept. From what appears to have been the first use of the term in economics by James Steuart in 1769, down to the present day, equilibrium analysis (together with its derivative, disequilibrium analysis) has been the foundation upon which economic theory has been able to build up its not inconsiderable claims to 'scientific' status. Yet despite the persistent use of the concept by economists for over two hundred years, its meaning and role have undergone some quite profound modifications over that period.

At the most elementary level, 'equilibrium' is spoken about in a number of ways. It may be regarded as a 'balance of forces', as when, for example, it is used to describe the familiar idea of a balance between the forces of demand and supply. Or it can be taken to signify a point from which there is no endogenous 'tendency to change': stationary or steady states exhibit this kind of property. However, it may also be thought of as that outcome which any given economic process might be said to be 'tending towards', as in the idea that competitive processes tend to produce determinate outcomes. It is in this

last guise that the concept seems first to have been applied in economic theory. Equilibrium is, as Adam Smith might have put it (though he did not use the term), the centre of gravitation of the economic system – it is that configuration of values towards which all economic magnitudes are continually tending to conform.

There are two properties embodied in this original concept which when taken into account begin to impart to it a rather more precise meaning and a well-defined methological status. Into this category enters the formal definition of 'equilibrium conditions' and the argument for taking these to be a useful object of analysis.

There are few better or more appropriate places to isolate the first two properties of 'equilibrium' in this original sense than in the seventh chapter of the first book of Adam Smith's *Wealth of Nations.* The argument there consists of two steps. The first is to define 'natural conditions':

> There is in every society … an ordinary or average rate of both wages and profits… . When the price of any commodity is neither more nor less than what is sufficient to pay … the wages of the labour and the profits of the stock employed … according to their natural rates, the commodity is then sold for what may be called its natural price (Smith, 1776, I.vii, p. 62).

The key point here is that 'natural conditions' are associated with a general rate of profit – that is, uniformity in the returns to capital invested in different lines of production under existing best-practice technique. In the language of the day, this property was thought to be the characteristic of the outcome of the operation of the process of 'free competition'.

The second step in the argument captures the analytical status to be assigned to 'natural conditions':

> The natural price … is, as it were, the central price, to which the prices of all commodities are continually gravitating. Different accidents may sometimes keep them suspended a good deal above it, and sometimes force them down even somewhat below it. But whatever may be the obstacles which hinder them from settling in this center of repose and continuance, they are constantly tending towards it (I.vii, p. 65).

This particular 'tendency towards equilibrium' was held to be operative in the *actual* economic system at any given time. It is not to be confused with the familiar question concerning the stability of competitive equilibrium in modern analysis. There the question about convergence to equilibrium is posed in some *hypothetical* state of the world where none but the most purely competitive environment is held to prevail. It is also essential to observe that in defining 'natural conditions' in this fashion, nothing has yet been said (nor need it be said) about the forces which act to determine the natural rates of wages and profits, or the natural prices of commodities. It will therefore be possible to refrain from discussing the *theories* offered by various economists for the determination of these variables in most of what follows. Treatment of these matters may be found elsewhere in the Dictionary. Similarly, there will be no discussion here of existence or uniqueness of equilibrium – matters which are the subject of separate entries.

'Natural conditions' so defined and conceived are the formal expression of the idea that certain systematic or persistent forces, regular in their operation, are at work in the economic system. Smith's earlier idea, that 'the co-existent parts of the universe … contribute to compose one immense and connected system' (1759, VII. ii, 1.37), is translated in this later formulation into an analytical device capable of generating conclusions with a claim to general (as opposed to a particular, or special) validity. These general conclusions were customarily referred to as 'statements of tendency', or 'laws', or 'principles' in the economic literature of the 18th and 19th centuries. It is worth emphasizing that there was no implication that these general tendencies were either swift in their operation or that they were not subject at any time to interference from other obstacles. Like sea level, 'natural conditions' had an unambiguous meaning, even if subject to innumerable cross-currents.

To put it another way, the distinction between 'general' and 'special' cases (like its counterpart, the distinction between 'equilibrium' and 'disequilibrium'), refers neither to the immediate practical relevance of these kinds of cases to actual existing market conditions, nor to the prevalence, frequency, or probability of their occurrence. In fact, as far as simple observation is concerned, it might well be that 'special' cases would be the order of the day. John Stuart Mill expressed this idea especially clearly when he held that the conclusions of economic theory are only applicable 'in the *abstract*', that is, 'they are only true under certain suppositions, in which none but general causes – causes common to the *whole class* of cases under consideration – are taken into account' (Mill, 1844, pp. 144–5). Marshall, of course, understood their application as being subject not only to this qualification (which he spoke about in terms of 'time'), but also to the condition that 'other things are equal' (1890, I.iii, p. 36). There will be cause to return to this matter below.

To unearth these regularities, one had to inquire behind the scene, so to speak, to reveal what otherwise might remain hidden. Adam Smith had set out the basis of this procedure in an early essay on 'The Principles which Lead and Direct Philosophical Enquiries':

> Nature, after the largest experience that common observation can acquire, seems to abound with events which appear solitary and incoherent ... by representing the invisible chains which bind together all these disjointed objects, [philosophy] endeavours to introduce order into this chaos of jarring and discordent appearances (Smith, 1795, p. 45).

In short, 'equilibrium', if we may revert to the modern terminology for a moment, became the central organizing category around which economic theory was to be constructed. It is no accident that the formal introduction of the concept into economics is associated with those very writers whose names are closely connected with the foundation of 'economic science'. It could even be argued that its introduction marks the foundation of the discipline itself, since its appearance divides quite neatly the subsequent literature from the many analyses of individual problems which dominated prior to Smith and the Physiocrats.

Cementing this tradition, Ricardo spoke of fixing his 'whole attention on the permanent state of things' which follows from given changes, excluding for the purposes of general analysis 'accidental and temporary deviations' (1817, p. 88). Marshall, though substituting the terminology 'long-run normal conditions' for the older 'natural conditions', excluded from this category results upon which 'accidents of the moment exert a preponderating influence' (1890, p. vii). J.B. Clark followed suit and held that 'natural or normal' values are those to which 'in the long run, market values tend to conform' (1899, p. 16). Jevons (1871, p. 86), Walras (1874–7, p. 380), Böhm-Bawerk (1899, II, p. 380) and Wicksell (1901, I, p. 97) all followed the same procedure.

Not only was the status of 'equilibrium' as the centre of

gravitation of the system (the benchmark case, so to speak) preserved, but it was defined in the manner of Smith. The primary theoretical object of all these writers was to explain that situation characterized by a uniform rate of profit on the supply price of capital invested in different lines of production. Walras, whose argument is quite typical, stated the nature of the connection forcefully:

> uniformity of ... the price of net income [rate of profit] on the capital goods market ... [is one] condition by which the universe of economic interests is governed (1874–7, p. 305).

From an historical point of view, the novelty of these arguments which were worked out in the 18th century by Smith and the Physiocrats, is not that they recognized that there might be situations which could be described as 'natural', but that they associated these conditions with the outcome of a specific process common to market economies (free competition) and utilized them in the construction of a general economic analysis of market society. Earlier applications of 'natural order' arguments were little more than normative pronouncements about some existing or possible state of society. They certainly made no 'scientific' use of the idea of systematic tendencies, even if these might have been involved. This is particularly apparent in the case of the 'natural law' philosophers, but is also true of the early liberals like Locke and Hobbes. Even Hume, who to all intents and purposes had in his possession all of the building blocks of Smith's position, drew back from the one crucial step that would have led him to Smith's 'method' – he was just not prepared to admit that thinking in terms of regularities, however useful it might prove to be in dispelling theological and other obfuscations (and thus in advancing 'human understanding'), was anything more than a convenient and satisfying way of thinking. The question as to whether the social and economic world was actually governed by such regularities, so central to Smith and the Physiocrats, just did not concern Hume.

Yet the earlier normative connotations of ideas like 'natural conditions', 'natural order', and the like, quite rapidly disappeared when the terminology was appropriated by economic theory. Nothing was 'good' simply by virtue of its being 'natural'. This, of course, is not to say that once the theoretical analysis of the natural tendencies operating in market economies had been completed, and the outcomes of the competitive process had been isolated in abstract, an individual theorist might not at that stage wish to draw some conclusions about the 'desirability' of its results (a normative statement, so to speak). But such statements are not implied by the concept of equilibrium – they are value judgements about the characteristics of its outcomes.

Indeed, contrary to the view sometimes expressed, even Smith's use of Deistic analogies and metaphors in the *Theory of Moral Sentiments*, where we read about God as the creator of the 'great machine of the universe', and where we encounter for the first time the famous 'invisible hand', is no more than the extraneous window-dressing which surrounds a well-defined *theoretical* argument based upon the operation of the so-called 'sympathy' mechanism. Thus, as W.E. Johnson noted when writing for the original edition of Palgrave's *Dictionary*, 'the confusion between scientific law and ethical law no longer prevails', and he observes that 'the term normal has replaced the older word natural – to be understood by this terminology as 'something which presents a certain empirical uniformity or regularity' (1899, p. 139).

While 'natural conditions' or 'long-run normal conditions' represent the original concept of 'equilibrium' utilized in economic theory, John Stuart Mill's *Political Economy* seems to have been the source from which the actual term equilibrium gained widespread currency (though, like so much else, it is also to be found in Cournot's *Recherches*). More significant, however, is the fact that in Mill's hands the meaning and status of the concept undergoes a modification. While maintaining the idea of equilibrium as a long-period position, Mill introduces the idea that the equilibrium theory is essentially 'static'. The relevant remarks appear at the beginning of the fourth book:

> We have to consider the economical condition of mankind as liable to change ... thereby adding a theory of motion to our theory of equilibrium – the Dynamics of political economy to the Statics (Mill, 1848, IV.i, p. 421).

Since he retained the basic category of 'natural and normal conditions', Mill's claim had the effect of adding a property to the list of those associated with the concept of equilibrium. However, over the question of whether this additional property was necessary to the concept of equilibrium, there was to be less uniformity of opinion. Indeed, this matter gave rise to a debate in which at one time or another (until at least the 1930s) almost all theorists of any repute became contributors. The problem was a simple one – are natural or long-period normal conditions the same thing as the 'famous fiction' of the stationary or steady state. Much hinged upon the answer; a 'yes' would have limited the application of equilibrium to an imaginary stationary society in which no one conducts the daily business of life.

On this question, as might be expected, Marshall vacillated. The thrust of his argument (as well as those of his major contemporaries, with the important exception of Pareto) seems to imply that such a property was not essential to his purpose, but as was his habit on so many occasions, in a footnote he qualified that position (1890, p. 379, n.1). In the final analysis, the answer seems to have depended rather more on the explanation given for the determination of equilibrium values, than upon the concept of equilibrium proper. It was not until the 1930s that the issue seems to have been resolved to the general satisfaction of the profession. But then its 'resolution' required the introduction of a new definition of equilibrium (the concept of intertemporal equilibrium) due in the main to Hicks.

However, some further embellishments and modifications were worked upon the concept of equilibrium before the 1930s. Here, two developments stand out. The first concerns the distinction between partial equilibrium analysis and general equilibrium analysis. The second concerns a trend that seems to have developed consequent upon Marshall's treatment of the element of time, which led him to his threefold typology of periods ('market', 'short', and 'long' – we shall leave to one side the further category of 'secular movement'). The upshot of this trend which is decisive, is that it became common to speak of the possibility of 'equilibrium' in each of these Marshallian periods.

The analytical basis for partial equilibrium analysis was laid down in 1838 by Cournot in his *Recherches*. Mathematical convenience, more than methodological principle, seems to have been responsible for his adopting it (see, for example, 1838, p. 127). Though this small volume failed to exercise any widespread influence on the discipline much before the present century, it was known and read by Marshall (who spoke of Cournot as his 'gymnastics master'), from whose *Principles* the popularity of partial equilibrium analysis is largely derived (though it would be remiss to overlook Auspitz, Lieben and von Mangoldt). Unlike the case of Cournot, however, it would

be difficult to argue that Marshall came across the method in anything other than a roundabout way (though some have argued that its principal attraction for him lay in its facility in allowing him to express his theory in a manner which required little recourse to mathematics).

When Marshall first introduced the idea of assuming 'other things equal' in the *Principles*, the *ceteris paribus* condition which is taken as the hallmark of the partial equilibrium approach, he seems to have done so not in order to justify the procedure of analysing 'one bit at a time', but in order to make a quite different point – that a long-run normal equilibrium would only *actually* emerge if none but the most general causes were allowed to operate without interference (see, for example, 1890, p. 36, p. 366, and pp. 369–70). In other words, the 'other things' that were being held 'equal' were the given data of the theory and the external environment – if the data remained the same and the external environment was freely competitive, then a long-run normal equilibrium would result. Indeed, Walrasian general equilibrium holds 'other things equal' in this sense. To put it another way, in Marshall's initial argument nothing was said about the possibility of assuming the interdependencies between long-run variables themselves to be of secondary importance, as is customary in partial equilibrium analysis.

This latter requirement of Marshallian analysis, the idea of the negligibility of indirect effects when one looks at individual markets (1919, p. 677ff.), seems to have sprung from his habit of presenting equilibrium *theory* in terms of *particular* market demand and supply curves (with their attendant notions of representative consumers and firms). It is here, in fact, that Marshall's presentation of demand and supply theory differs so markedly from its presentation by Walras. To the extent that this is so, it would seem to be better to recognize that the idea of 'partial' versus 'general' equilibrium has more to do with the presentation of a particular theory, and Marshall's propensity to consider markets one at a time, than it has to do with the abstract category of equilibrium with which this discussion is concerned. This view would accord, incidentally, with the fact that the great disputes over the relative merits of these two modes of analysis (for example, that between Walras on the one hand, and Auspitz and Lieben on the other) were fought over the specification of demand and cost functions.

Another modification to the concept of equilibrium that has become more significant in recent literature also makes an appearance in Marshall; though it is not carried as far as it has been in recent literature. The second, third and fifth chapters of the fifth book of Marshall's *Principles* set out the conditions for the determination of what he calls the 'temporary equilibrium', the 'short-run equilibrium' and the 'long-run equilibrium' of demand supply. The last of these categories, as Marshall makes perfectly clear in the text, corresponds to Adam Smith's 'natural conditions' (1890, p. 347). The first two are to a greater or lesser degree 'more influenced by passing events, and by causes whose action is fitful and short lived' (p. 349). What is striking about Marshall's terminology is the fact that situations which from an analytical point of view would traditionally have been regarded as 'deviations' from long-period normal equilibrium (that is, disequilibria) are explicitly referred to as different cases of 'equilibrium'. This trend has taken on an entirely new significance in recent literature, and has had dramatic consequences for the meaning and status of the concept of equilibrium in economic theory. But just as important in comprehending this development is the introduction of the notion of intertemporal equilibrium into theoretical discourse.

The notion of intertemporal equilibrium (introduced by

Hayek, Lindahl and Hicks in the inter-war years and developed in the 1950s by Malinvaud, Arrow and Debreu) warrants special consideration since 'equilibrium conditions' under this notion are defined quite differently from 'natural' or 'long-run normal' conditions. Intertemporal equilibrium defines as its object the determination of nt market-clearing prices (for n commodities over t elementary time periods commencing from an arbitrary short-period starting point). The chief implication of this definition of equilibrium conditions, and that which sets it apart from long-run normal conditions, is that not only will the price of the same commodity be different at different times but also that the stock of capital need not yield a uniform return on its supply price.

This fundamental change in the concept of equilibrium did not mean that intertemporal equilibrium positions were immediately divested of the status that had been given to 'equilibrium' ever since Adam Smith. In certain circles they continued to be regarded as positions towards which the economic system could actually be said to be 'tending' (or as benchmark cases).

However, once the *sequential* character of this equilibrium concept came to be better understood, it became apparent that there could be no 'tendency' towards it – at least not in the former meaning of that idea. One was either in it, in which case the sequence was 'inessential', or one was not, in which case the sequence was 'essential' (see Hahn, 1973, p. 16). And the probabilities overwhelming suggested the latter. Attention was thus turned to the individual points in the sequence; the temporary equilibria, as Hicks had dubbed them (applying the terminology of Marshall in a new context). A whole new class of cases, disequilibrium cases from the point of view of full intertemporal equilibrium, began to be examined. The discipline has now accumulated so many varieties that it is impossible to document them all here. Instead, two broad features of this development may be noted here, the first concerning the role that expectations were thereby enabled to play, the second the common designation now uniformly applied to all such cases: 'equilibrium'.

When equilibrium is interpreted as a solution concept in the sense that *all* solutions to *all* models (for which solutions exist) enjoy equal analytical status and differ only in that they become 'significant', as von Neumann and Morgenstern put it, when they are 'similar to reality in those respects which are essential in the investigation at hand' (1944, p. 32), it is sometimes said that economics has availed itself of a very powerful notion of equilibrium. On this line of argument, Walrasian equilibrium and, say, conjectural equilibrium compete with one another not for the title 'general' (since, in the traditional sense at least, there is no such category), but for the title 'significant'. Furthermore, at any given time they are competing for this title with as many other models as are available to the profession.

It seems to be the case that the status of equilibrium in economic analysis has come full circle since its introduction in the late 18th century. From being derived from the idea that market societies were governed by certain systematic forces, more or less regular in their operation in different places and at different times, it now seems to be based on an opinion that nothing essential is 'hidden' behind the many and varied situations in which market economies might actually find themselves. In fact, it seems that these many cases are to be thought of as being more or less singular from the point of view of modern theory. From being the central organizing category around which the whole of economic theory was constructed, and therefore the ultimate basis upon which its

practical application was premissed, equilibrium has become a category with no meaning independent of the exact specification of the initial conditions for *any* model. Instead of being thought of as furnishing a theory applicable, as Mill would have said, to the whole class of cases under consideration, it is increasingly being regarded by theorists as the solution concept relevant to a particular model, applicable to a limited number of cases. The present fashion for replacing economic theory proper by game theory, an approach which could be regarded by no less a theorist than Professor Arrow as contributing only 'mathematical tools' to economic analysis not many years ago (1968, p. 113), seems to exemplify the trend of modern economics.

MURRAY MILGATE

See also ARROW–DEBREU MODEL OF GENERAL EQUILIBRIUM; CENTRE OF GRAVITATION; COMPETITION; CLASSICAL CONCEPTIONS OF; CONJECTURAL EQUILIBRIUM; EQUILIBRIUM; AN EXPECTATIONAL CONCEPT; GENERAL EQUILIBRIUM; MONETARY DISEQUILIBRIUM AND MARKET CLEARING; NATURAL AND NORMAL CONDITIONS; STABILITY; STATIONARY STATE; TEMPORARY EQUILIBRIUM.

BIBLIOGRAPHY

Arrow, K.J. 1968. Economic equilibrium. In *International Encyclopedia of the Social Sciences*, as reprinted in *The Collected Papers of Kenneth J. Arrow*, Volume 2, Cambridge, Mass: Harvard University Press.

Böhm-Bawerk, E. von. 1899. *Capital and Interest*. 3 vols; reprinted, Illinois: Libertarian Press, 1959.

Clark, J.B. 1899. *The Distribution of Wealth*. London: Macmillan.

Cournot, A.A. 1838. *Researches into the Mathematical Principles of the Theory of Wealth*. Translated by N.T. Bacon with an introduction by Irving Fisher, 1897; 2nd edn, London and New York: Macmillan, 1927.

Garegnani, P. 1976. On a change in the notion of equilibrium in recent work on value. In *Modern Capital Theory*, ed. M. Brown et al., Amsterdam: North–Holland.

Hahn, F.H. 1973. *On the Notion of Equilibrium in Economics*. Cambridge: Cambridge University Press.

Hicks, J.R. 1939. *Value and Capital*. 2nd edn, Oxford: Clarendon Press, 1946.

Jevons, W.S. 1871. *Theory of Political Economy*. Edited from the 2nd edition (1879) by R.D.C. Black, Harmondsworth: Penguin, 1970.

Marshall, A. 1890. *Principles of Economics*. 9th (variorum) edition, taken from the text of the 8th edition, 1920. London: Macmillan.

Marshall, A. 1919. *Industry and Trade*. 2nd edn, London: Macmillan.

Mill, J.S. 1844. *Essays on Some Unsettled Questions of Political Economy*. 2nd edn, 1874; reprinted, New York: Augustus M. Kelley.

Mill, J.S. 1848. *Principles of Political Economy*. 6th edn, 1871 (reprinted 1909), London: Longmans, Green & Company.

Palgrave, R.H.I. (ed.) 1899. *Dictionary of Political Economy*, Vol. III. London: Macmillan.

Pareto, V. 1909. *Manual of Political Economy*. Translated from the French edition of 1927 and edited by A.S. Schwier and A.N. Page, New York: Augustus M. Kelley, 1971.

Ricardo, D. 1817. *The Principles of Political Economy and Taxation*. Edited from the 3rd edition of 1821 by P. Sraffa with the collaboration of M. Dobb, Vol. I of *The Works and Correspondence of David Ricardo*, 11 vols, Cambridge: Cambridge University Press, 1951–73.

Smith, A. 1759. *The Theory of Moral Sentiments*. Edited by D.D. Raphael and A.L. Macfie from the 6th edn of 1790, Oxford: Oxford University Press, 1976.

Smith, A. 1776. *An Inquiry into the Nature and Causes of the Wealth of Nations*. 2 vols, ed. E. Cannan, London: Methuen, 1961.

Smith, A. 1795. *Essays on Philosophical Subjects*. Edited by W.P.D. Wrightman and J.C. Bryce, Oxford: Oxford University Press, 1980.

Von Neumann, J. and Morgenstern, O. 1944. *Theory of Games and Economic Behavior*. 3rd edn, Princeton: Princeton University Press, 1953.

Walras, L. 1874–7. *Elements of Pure Economics*. Translated and edited by W. Jaffé from the definitive edition of 1926, London: Allen & Unwin, 1954.

Wicksell, K. 1901. *Lectures on Political Economy*. 2 vols, ed. L. Robbins, London: Routledge and Kegan Paul, 1934.

equity. Depending on the user's inclinations, 'equity' can mean almost anything; this user will adopt a meaning which has been followed by economists and other social scientists since the late 1960s (see particularly Foley, 1967), a meaning close to equality or fairness.

Although 'equality' is less ambiguous than 'equity', it too has many definitions: Jefferson's adage that 'all men are created equal' clearly does not mean that they all have the same talents, skills, inherited and acquired wealth; it only means that they share, or ought to share, certain narrowly defined legal rights and political powers. However, in a simple economic model, equality can be made simple. If we assume that society is comprised of a certain set of n individuals who produce among themselves certain quantities of various goods, we can speak of an equal division of the goods: an allocation that would give each person exactly $1/n$ of the total of each good. Economists would agree that this is equality (at least on the consumption side). Most would also agree that it is an undesirable state of affairs, if for no other reason than that no two people would ever want to consume exactly the same bundle of goods. They would be equal, but not especially happy. Moreover, getting society to that equal allocation would require transferring wealth from the more productive individuals to the less productive, and the transfer mechanism itself would destroy incentives to produce.

So equality in its extreme form – an equal consumption bundle for every consumer – is an obviously unworkable idea, and needs to be weakened. We shall say in this assay that individual i *envies* individual j if i would rather have j's consumption bundle than his own. Formally, let $u_i(\cdot)$ represent individual i's utility function, and x_i represent his consumption bundle. (For now, production is ignored.) Then i envies j if $u_i(x_j) > u_i(x_i)$. This is now a more-or-less standard usage by economists, who have ignored wiser and older counsel, for example, J. S. Mill, who calls envy 'that most odious and anti-social of all passions' (*On Liberty*, ch. 4). Mill would presumably not endorse an economic analysis founded on envy.

Following Varian (1974) we define an allocation as *equitable* if under it no individual envies another; that is, if

$$u_i(x_i) \geqslant u_i(x_j) \quad \text{for all } i \text{ and } j.$$

Obviously, the equal allocation is equitable. But equity does not share equality's obvious disadvantage of forcing all to consume the same no matter what their tastes. If Adam loves apples and Eve loves oranges, and if God has endowed them with a total of one apple and one orange, then the equal allocation (half an apple and half an orange for each) is clearly foolish, but the equitable allocation (one apple for Adam and one orange for Eve) makes good sense.

But the notion of equity has an obvious disadvantage, aside from its being founded on that odious passion. For instance, the economist's model, which reduces person i to a utility function $u_i(\cdot)$ and a bundle of goods x_i, ignores the fact that life is full of things not captured in $u_i(\cdot)$ or x_i, for instance, non-transferable attributes like beauty, health and family. Even if the division of economic goods is equitable, i will probably envy j his looks, or his good health. This problem was alluded to by Kolm (1972). A well-meaning economist who follows his equity theory to its bitter end will conclude that the beautiful should be disfigured, and the well made sick.

Less obvious disadvantages of the idea of equity require references to Pareto efficiency, the foundation of modern welfare economics. An allocation y is *Pareto superior* to an allocation x if all individuals prefer y to x. (This assumes, of course, a constant set of individuals who are making the judgement.) If y is Pareto superior to x, the move from x to y is a *Pareto move*. An allocation x is *Pareto optimal* if there is no y that is Pareto superior to x.

Several authors (e.g. Kolm, 1972) have established that in an economy where there is no production, there exist allocations that are both equitable and Pareto optimal. To find one, start at the equal allocation and move the economy to a competitive equilibrium. By the first fundamental theorem of welfare economics, a competitive equilibrium is Pareto optimal. Since the equilibrium is based on the equal allocation, every individual has the same budget. But if i has the same budget as j, he cannot envy the bundle j buys since he could have bought it himself. So this theorem creates a link between equity and the more traditional, more fundamental notion of Pareto optimality.

But it is a weak link. Pazner and Schmeidler (1974) and Varian (1974) consider an economy with production, where i's utility depends not only on his consumption bundle x_i, but also on the number of hours he works q_i. However, production attributes are non-transferable. If person i is ten times as productive as j, there may be no Pareto optimal distribution of consumption goods and of work hours that is also equitable. Think of an economy of which you are a part and Luciano Pavarotti is a part. You would have to train for 10 lifetimes before you could sing an aria like he does, and therefore there may be no possibility of arriving at an allocation of consumption and work effort among all that is both equitable and Pareto optimal.

Various possible solutions to this quandary have been suggested (e.g. in Pazner, 1976, and Pazner and Schmeidler, 1978). For instance, consider an economy where 'everybody shares an equal property right in everybody's time'. This may lead to the existence of allocations that are both equitable and optimal, but it makes Pavarotti a slave to everyone who is less gifted. Or, as another possible solution, consider an *egalitarian equivalent* allocation. This is one such that the utility distribution it produces could be generated by a theoretical economy in which all consumers are assigned identical consumption bundles. Pazner and Schmeidler (1978) show that egalitarian equivalent allocations that are also Pareto optimal exist, even in economies with production. But this idea is also unworkable; it is simply too airy.

Turn back to an economy without production. It is true that there will exist, under general assumptions, allocations that are both equitable and Pareto optimal in the pure exchange economy. But Feldman and Kirman (1974) show two disturbing facts: First, even if traders start at the equal allocation, and they make a Pareto move to the core (the solution set for frictionless barter), they may end up at an inequitable allocation. Second, if traders start at an equitable allocation, and make a Pareto move to a competitive equilibrium they may end up at an allocation where someone envies someone else. The 'green sickness' springs up where once there was equity.

The Edgeworth box diagram below illustrates the second possibility. In the figure, x_{I1} and x_{I2} represent quantities of goods 1 and 2 belonging to trader I; x_{J1} and x_{J2} represent quantities belonging to J. Also, i_1 and i_2 are two of trader I's indifference curves: j_1 and j_2 and two of trader J's indifference curves; $w = (w_i, w_j)$ is the initial allocation; $w^{-1} = (w_j, w_i)$ is the allocation which switches the bundles between I and J. Note that w^{-1} is found by reflecting w through the centre of the box. Now w is equitable since the indifference curves through it pass above w^{-1}, and the move from w to x is a competitive equilibrium trade that makes both better off. But $x = (x_i, x_j)$ is not equitable, since i_2 passes below $x^{-1} = (x_j, x_i)$, which means that trader I envies J when they are at x.

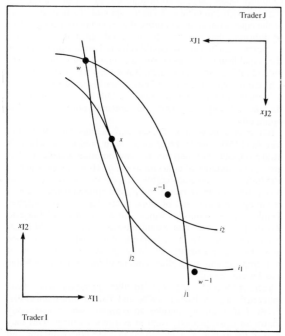

Figure 1

In an interesting extension of the Feldman and Kirman result, Goldman and Sussangkarn (1978) show with generality that in 2 person, 2 good exchange economies there exist allocations x such that (a) x is equitable in the non-envy sense but (b) x is not Pareto optimal and (c) every y which is Pareto superior to x is inequitable! This is formal proof of Johnson's assertion (*The Rambler*, No.183) that 'envy is almost the only vice which is practicable at all times, and in every place; the only passion which can never lie quiet from want of irritation'.

The concept of equity as non-envy is still alive among prominent economists; for instance, Baumol (1982) applies non-envy to an analysis of rationing. This in spite of the fact that recent history suggests the average man fares better under regimes that are less committed to elimination of envy through redistribution of goods, and in spite of the serious theoretical objections raised to the concept as outlined above. Should we care about equity? The temptation to pronounce judgement on

what is equitable and what is not may be irresistible. But economic theory suggests that the pursuit of equity in the sense of non-envy will lead to some peculiar and unpalatable results.

ALLAN M. FELDMAN

See also FAIRNESS.

BIBLIOGRAPHY

Baumol, W. 1982. Applied fairness theory and rationing policy. *American Economic Review* 72(4), 639–51.

Feldman, A. and Kirman, A. 1974. Fairness and envy. *American Economic Review* 64(6), December, 995–1005.

Foley, D. 1967. Resource allocation and the public sector. *Yale Economic Essays* 7(1), Spring, 45–98.

Goldman, S. and Sussangkarn, C. 1978. The concept of fairness. *Journal of Economic Theory* 19(1), 210–16.

Kolm, S.-C. 1972. *Justice et équité* Paris: Editions du Centre de la Recherche Scientifique.

Pazner, E. 1976. Recent thinking on economic justice. *Journal of Peace Science.*

Pazner, E. and Schmeidler, D. 1974. A difficulty in the concept of fairness. *Review of Economic Studies* 41(3), 441–3.

Pazner, E. and Schmeidler, D. 1978. Egalitarian equivalent allocations: a new concept of economic equity. *Quarterly Journal of Economics* 92(4), November, 671–87.

Schmeidler, D. and Vind, K. 1972. Fair net trades. *Econometrica* 40(4), July, 637–42.

Varian, H. 1974. Equity, envy and efficiency. *Journal of Economic Theory* 9(1), 63–91.

equivalent income scales. *See* CHARACTERISTICS; HOUSEHOLD BUDGETS; SEPARABILITY.

ergodic theory. To begin in the middle; for that is where ergodic theory started, in the middle of the development of statistical mechanics, with the solution, by von Neumann and Birkhoff, of the problem of identifying space averages with time averages. This problem can be formulated as follows: If $x_t(-\infty < t < \infty)$ represents the trajectory (orbit) passing through the point $x = x_0$ at time $t = 0$ of a *conservative dynamical system*, when can one make the identification

$$(*) \quad \lim_{T \to \infty} (1/T) \int_0^T f(x_t)\, dt = \int_\Omega f\, dm/m(\Omega)$$

for suitable functions defined on the *phase space* Ω of the system?

There are many things to be explained here. For example one might imagine a 'large' number of particles contained in a box, which collide with one another and with the sides of the box according to the usual laws of elastic collision. Each of these particles has three coordinates of position and three coordinates of velocity so that the state of the system is describable by $6n$ coordinates if n is the number of particles. Newtonian laws, of course, provide a history and future for each of these points in $6n$ dimensional space. The same laws imply the law of conservation of energy, so that in principle dynamical systems may be studied with the assumption that energy is constant for each trajectory of a conservative system. Thus in (*) we take the phase space Ω to be that hypersurface of $6n$ dimensional space where the total energy has a given (constant) value, and m is the hypersurface volume (measure) associated with the Liouville invariant volume whose existence is guaranteed by the conservativity of the system. In general $m(\Omega)$ is a finite quantity.

The left-hand side of (*) is the time average along a trajectory for a function (observable) f and the right-hand side is the phase or space average.

Von Neumann proved a mean convergence version of (*) and shortly after G. D. Birkhoff proved (*) as stated, for *almost all* states, in both cases under the assumption that the system (restricted to Ω) is *ergodic*, a notion, we shall explain presently. (Cf. von Neumann, 1932a, and Birkhoff, 1931.) It was soon realized that both versions of (*) (the ergodic theorems) could be formulated and proved in a more abstract setting and indeed one can say that this abstraction and the subsequent mathematics thereby generated is ergodic theory proper.

Let (Ω, m) represent an abstract space with a finite measure. (There is no loss in generality in assuming $m(\Omega) = 1$, as we shall do.) Let T_t represent a family of transformations indexed by time (in various contexts, the real numbers, the integers) such that $T_{t+s} = T_t \circ T_s$. Assume that this family is measure-preserving ($mT_t B = mB$, for all 'measurable' sets). The study of T_t as t varies through its index set, provides a model for an evolutionary system, such as the dynamics in phase space described earlier, in which measure (volume) is preserved. The system is said to be *ergodic* if Ω cannot be decomposed into two disjoint invariant measurable sets

$$A, B (A \cup B = \Omega, A \cap B = \varnothing, T_t A = A, T_t B = B \text{ all } t)$$

of positive measure.

In a strict sense, the time-average space-average problem was not *solved* by von Neumann and Birkhoff, as far as the classical dynamical system given at the outset is concerned, for the question of whether this system *is* ergodic was left open and it is only recently (Sinai, 1963) that progress has been made in this direction.

Most workers in ergodic theory concern themselves with measure-preserving transformations T_t indexed by the integers, so that with $T_1 = T$, T_t is the iteration of T repeated t times. Results in this context invariably lead to results for real continuous time.

Having freed itself from a particular (albeit important) dynamical system, ergodic theory or more particularly the theory of measure-preserving transformations began to encounter a rich diversity of problems:

(i) When does a measurable transformation, non-singular with respect to a given measure, preserve an equivalent finite (or even σ-finite) measure?

(ii) Are there analogues of the ergodic theorems for Markov processes?

(iii) Where do we find examples of measure-preserving (or non-singular) transformations in other branches? If they are non-singular answer question (i). If they are measure-preserving are they ergodic? If so, interpret the ergodic theorems for them.

(iv) Is it possible to (at least partially) classify the myriad examples coming from other branches of mathematics?

One should notice that in posing these problems ergodic theory became a *global* analysis in two senses: The phase space dynamical system described at the beginning of this entry is global in that *all* solutions of a differential equation are involved. Ergodic theory then moves on to treat all other problems having a dynamical character in which an invariant measure appears.

Concerning (ii) one should note that a measure-preserving transformation T gives rise to an isometric operator $Lf = f \circ T$ on various Banach spaces, the most important being $L^1(m)$. In a similar way a Markov process gives rise to a semi-group of

positive contractions. For such operators there is a variety of ergodic theorems generalizing the classical results of Birkhoff and von Neumann. As an example there is the powerful general ergodic theorem (Chacon and Ornstein, 1960): If L is a positive contraction on $L^1(m)$ and $f, g \in L^1(m)$ then

$$\sum_{k=0}^{n} L^n f \bigg/ \sum_{k=0}^{n} L^n g$$

converges almost everywhere on the set where the denominator is persistently positive.

Here we have an instance of ergodic theory providing a powerful tool for statistics. This should hardly be surprising, however, as even the classical Birkhoff ergodic theorem has an immediate impact on stochastic processes, for one can always associate a measure-preserving transformation with, say, a sequence of independent and identically distributed random variables in such a way that the strong law of large numbers is an easy corollary of Birkhoff's theorem.

Markov and other stochastic processes have played and continue to play a central role in the development of ergodic theory. In recent years a modelling procedure for understanding hyperbolic dynamical systems based on Markov chains has led to profound results in the area of differentiable statistical mechanics. Thus statistical ideas are exchanged, measure for measure, with those of ergodic theory.

Concerning (iii) here are some examples:

(a) *An 'irrational flow'*. Here $\Omega = \{(z, w): z, w \text{ complex } |z| = |w| = 1\}$, $T_t(z, w) = (e^{2\pi i \alpha t} z, e^{2\pi i \beta t} w)$, α, β are real with α/β irrational. m is an ordinary Lebesgue measure.

(b) *A skew product*. Here (Ω, m) is the same as in (a).

$$T(z, w) = (e^{2\pi i \alpha} z, zw), \quad \alpha \text{ irrational}.$$

(c) *An automorphism of a torus*. Again (Ω, m) is the same as in (a).

$$T(z, w) = (z^2 w, zw).$$

(d) *A translation of a homogeneous space*. G is a locally compact Lie group and H is a closed subgroup such that the homogeneous space $G/H = \{gH : g \in G\}$ is compact. $\Omega = G/H$ and m is a Haar measure. The transformation T is defined as a translation.

$$T(g, H) = agH$$

for a given element $a \in G$.

(e) *A geodesic flow*. Here we consider an n-dimensional Riemannian manifold M with unit length tangent vectors v located at points of M. Such a vector v defines a unique geodesic curve on M. Ω is the totality of such v and $T_t v$ is the unit tangent vector obtained by allowing v to flow along its geodesic at unit speed after time t. The measure m may be taken to be the natural one associated with Liouville's measure.

(f) *A Hamiltonian dynamical system*. Instead of defining this we mention that (e) above and the n particle phase space system at the beginning of this article are both examples of such a system.

(g) *The evolutionary shift associated with a Markov chain or more particularly of a Bernoulli (independent) sequence of trials*.

For the Bernoulli case Ω consists of points $w = \{w_n\}_{-\infty}^{\infty}$ where w_n represents the outcome of an experiment (heads or tails, for example, in the tossing of a coin) at time n. m is the probability which guarantees the independence of these trials, and T is the shift in time $Tw = w'$ where $w'_n = w_{n+1}$.

(h) *A stationary Gaussian (normal process)*.

(i) *The continued fraction transformation*. Here Ω consists of the irrational numbers between 0 and 1. m is 'Gauss's' measure whose density is $1/\log 2(1 + x)$ and $Tx = 1/x \bmod 1$.

An alternative account of ergodic theory, which admittedly ignores the history of the subject, could be given which is based on the above examples (and many others). It would motivate the subject by the questions: What do these examples have in common? What concepts underlie them? However, only a posteriori would these questions lose their artificiality.

The first four examples (a), (b), (c) and (d), all arise from algebraic or homogeneous space structures and even (e) falls into this category under certain conditions on the curvature of the manifold. In general, (e) arises from differential geometry. The Bernoulli example (g) (or more generally a Markov chain) arises from probability theory as does (h). The example (i) occurs in the study of continued fractions.

These examples (under suitable conditions) are flows and transformations which display varying degrees of ergodicity or *mixing* and ergodic theoretical techniques reveal important information about them. For example, in Furstenberg, (1961) (b) was used to give a proof of the famous theorem of Weyl that $\alpha n^2 + \beta n + \gamma$ mod 1 is uniformly distributed in the unit interval $[0,1)$ as n varies (as long as α or β is irrational). The example (c) was closely analysed as a prototype of *hyperbolicity* prior to the development of Anosov and Axiom A dynamical systems. The examples covered by (e) (and the related *horocycle flows*) are central to the study of hyperbolic geometry and to the theory of unitary representations of semi-simple Lie groups. The examples (g), (h) provide the most important classes of stationary stochastic processes and are intimately related to Brownian motion. Example (i) is of vital importance in number theory.

Question (iv) was first approached (von Neumann, 1932b; von Neumann and Halmos, 1942) using spectral techniques. Two measure-preserving transformations S, T are said to be (spatially) *isomorphic* if there is an invertible measure-preserving transformation ϕ between their respective spaces such that $\phi S = T \phi$ a.e. (almost everywhere). Isomorphism implies that the unitary spectral characteristics are indistinguishable (i.e. *spectral* equivalence), but not vice versa. The main result obtained characterized all ergodic measure-preserving transformations with a pure point spectrum. For such transformations S, T identity of point spectrum implies spatial isomorphism and such transformations are (isomorphically) precisely the ergodic translations of compact metric abelian groups.

A similar theory was developed in Abramov (1962) for so-called transformations with quasi-discrete spectrum. Example (b) provides an example of this type of transformation. They had been studied earlier by Anzai. The works of Auslander, Green and Hahn (1963) and Parry (1971) provide further developments in this direction. A completely analogous theory is modelled on the 'rigid' examples of *nilflows* and *unipotent affines* on nil manifolds. The rigidity here refers to the phenomenon of measure isomorphisms *necessarily* being algebraic in character. The most recent work concerning rigidity in ergodic theory (Ratner, 1982) finds this, and related phenomena, in horocycle flows.

So far we have given a condensed account of only one strand in isomorphism theory. The most active work has occurred in connection with examples of an entirely different and *random* character.

This work began with the problem of deciding whether two Bernoulli shifts (which are necessarily spectrally isomorphic) are spatially isomorphic. The first breakthrough occurred with Kolmogorov's introduction of entropy theory into the subject (Kolmogorov, 1958). As modified in (Sinai, 1959) entropy is a numerical invariant of isomorphism (i.e. if S, T are isomorphic then their entropies $h(S)$, $h(T)$ coincide). This fact provides a multitude of Bernoulli transformations which are not isomorphic. The basic ideas originate with Shannon and McMillan, but they required significant adaptation before they could be used in ergodic theory. The new entropy theory developed apace in the hands of, principally, Russian mathematicians in the 1960s and received its biggest impetus from the American mathematician Ornstein, who in 1968 proved that two Bernoulli transformations with the same entropy are isomorphic (cf. Ornstein, 1970). From that time the subject has grown exponentially, with ever more transformations shown to be (isomorphic to) Bernoulli transformations. Such transformations have to have (to say the least) positive entropy and their *intrinsic* random character is in marked contrast to the rigid examples referred to earlier which are *deterministic* (with zero entropy).

Entropy plays very little role in the looser classification theory which allows velocities (along trajectories) to vary. There are continuous (real) time and discrete versions of this theory and as early as 1943 Kakutani had conjectured that all ergodic systems are *Kakutani equivalent*, using the current nomenclature for this loose equivalence (Kakutani, 1943). Although this conjecture turned out to be false (in fact entropy ensures the existence of at least three Kakutani inequivalent systems), Feldman (1976) and Katok (1977) showed that remarkably dissimilar systems are equivalent according to this notion. In Ornstein and Weiss (1984) it is shown that a modification of Kakutani's conjecture is true. In this connection a grand theory of equivalence relations in ergodic theory has been developed in Rudolph (1984). This is ergodic theory with its head in the clouds.

From a more earthly point of view ergodic theory in the 1960s, through the developments of entropy theory and stimulated by Anosov (1967) and Smale (1967), began to connect with the newly flourishing field of differentiable dynamical systems.

Examples (c) and (f) are, respectively, prototypes of Anosov diffeomorphisms and flows. Their principal feature here is their global hyperbolic structure. Dynamicists are particularly interested in such systems as they are *structurally stable*, a concept which became something of a dogma in the 1960s and 1970s, as some mathematicians went so far as to assert that any real and persistent system must be structurally stable. (A structurally stable system is, roughly speaking, one which retains its principal features after a small perturbation.) This important concept was modified by Smale when he introduced Axiom A systems and proved that the latter are Ω-stable (structural stability relative to non-wandering sets). Smale thereby axiomatized a vast category of new dynamical systems and presented us with an approach which unified Anosov systems, gradient like dynamical systems and his so-called 'horse shoes'. For these systems Smale proved his spectral decomposition theorem, which describes the non-wandering set of an Axiom A system in much the same way as one describes the irreducible block behaviour of a non-negative matrix, in the theory of Markov chains (Smale, 1970). The *basic* sets of an Axiom A system received further scrutiny in terms of Markov partitions by Sinai (Anosov case) and Bowen (Axiom A case).

Bowen was a key figure in the fruitful convergence of ergodic theory and differentiable dynamical systems because of his profound expertise in both subjects. In a series of papers he provided deep analyses of Axiom A diffeomorphisms and flows (roughly speaking, hyperbolic dynamics) from the point of view of symbolic dynamics and periodic orbits (Bowen, 1977). His work connected happily with the direction Ruelle and Sinai were taking in statistical mechanics (Sinai, 1972; Ruelle, 1978). Together they laid the foundations for statistical mechanics on manifolds.

The subject then has gone full circle to its origins, but on the way it encountered a dazzling variety of iteration problems from other areas, viz. maps of the unit interval (Collet and Eckmann, 1980), boundary measures associated with Fuchsian groups (Patterson, 1976; Sullivan (1979), analytic maps of the Riemann sphere or complex plane (Rees, 1982), to name but three.

As to recent developments in statistical mechanics the one-dimensional lattice gas has received the most attention. Here one considers a shift transformation (as in the case of a Markov chain) initially in the absence of any probability but supplemented with a natural topology which reflects the connectivity of the transformation. Such a shift is called a *topological Markov chain* (or *shift of finite type*).

Then one considers an action potential describable in terms of a continuous function. Under a stronger (Lipschitz) condition it turns out that there is always a unique shift invariant probability (called an *equilibrium state*) given by a variational principle involving the *pressure* of the potential.

A key tool in this theory is the *transfer* matrix or operator associated with the potential and under suitable (aperiodic and irreducible) conditions, iterations of this operator will force arbitrary probabilities to converge to the equilibrium state.

The one-dimensional lattice gases provide models for simple gases and also for statistical mechanics on manifolds. The results above have analogues (when appropriate conditions are imposed) for differential or even topological dynamical systems (Pesin, 1977). Moreover, at least for hyperbolic systems, one can view topological Markov chains with their potentials, equilibrium states, closed orbits, transfer operators and pressures as (in a technical sense) building schemes for these systems.

A recent new area of ergodic theory which stands outside the developments just sketched is concerned with the application of ergodic theory and topological dynamics to combinatorial number theory. The motivation for this recent work was Szemeredi's proof of a conjecture of Erdos and Turan. The conjecture, which emanated from a result of van der Waerden's, states that if $a_1 < a_2 < \cdots$ is an increasing sequence of positive integers and if θ_N denotes the number of these integers less than N, then for every $k > 0$ there is an arithmetic progression in the sequence of length k, as long as $\theta_N/N > \epsilon$ infinitely often (for some $\epsilon > 0$).

Furstenberg (1977) provided an ergodic theoretical proof of this result and of many other results with the same 'flavour'. His technique involved building a sequence $\{x_n\}$ of zeros and ones ($x_n = 1$ when and only when n is in the sequence), and embedding this sequence in a shift space. The details, which are quite intricate, involve the proof of a multiple recurrence theorem:

If T is a measure-preserving transformation on (X, m) and if $m(A) > 0$ then for every positive integer k

$$m(A \cap T^n A \cap T^{2n} A \cap \cdots \cap T^{kn} A) > 0$$

for infinitely many integers n.

Although this area is somewhat askew to the other developments outlined above, it needs to be mentioned because of the great research potential it possesses.

Where are the likely growing points for the subject? Here is a list of guesses. Some of them are wild; others are safe; and they are not all of equal weight:

(1) Applications to combinatorial number theory (Furstenberg, 1977);

(2) Problems involving a mixture of prime number theory and ergodic theory inspired perhaps by Vinogradov's theorem that $p\alpha$ mod 1 is uniformly distributed when p runs through the primes and α is irrational;

(3) Greater understanding of the connections between the prime number theorem and the prime orbit theorem (Hejhal, 1976; Parry and Pollicott, 1983);

(4) Further developments of cohomology theory in ergodic theory (Schmidt, 1977);

(5) Developments in *restricted* classification theories of processes; in particular a solution of Williams's problem (Williams, 1973); in particular a solution of the stochastic version of the theory of Adler and Marcus (1979);

(6) Developments from Ornstein's and Weiss's modified Kakutani problem in the theory of von Neumann algebras;

(7) Which ergodic translations, affines and flows are rigid?

(8) A greater understanding of turbulence (Ruelle and Takens, 1971).

WILLIAM PARRY

See also CONTINUOUS AND DISCRETE TIME MODELS; CONTINUOUS-TIME STOCHASTIC MODELS; CONTINUOUS-TIME STOCHASTIC PROCESSES.

BIBLIOGRAPHY

Abramov, L.M. 1962. Metric automorphisms with quasi-discrete spectrum. *Izvestiya Akademii Nauk Ser. Mat.* 26, 513–30; *American Mathematical Society Translations* 2(39), 37–56.

Adler, R.L. and Marcus, B. 1979. Topological entropy and equivalence of dynamical systems. *Memoirs of the American Mathematical Society* 219, 1–84.

Anosov, D.V. 1967. Geodesic flows on closed Riemannian manifolds with negative curvature. *Trudy. Mat. Inst. Steklova* 90, 1–209; *Proceedings of the Steklov Institute of Mathematics (American Mathematical Society Translations)*, 1969, 1–235.

Auslander, L., Green, L. and Hahn, F. 1963. Flows on homogeneous spaces. *Annals of Mathematics Studies* 53, Princeton.

Birkhoff, G.D. 1931. Proof of the ergodic theorem. *Proceedings of the National Academy of Sciences of the USA* 17, 656–60.

Bowen, R. 1977. On Axiom A diffeomorphisms. *American Mathematical Society Regional Conference Series* 35, 1–45.

Chacon, R.V. and Ornstein, D.S. 1960. A general ergodic theorem. *Illinois Journal of Mathematics* 4, 153–60.

Collet, P. and Eckmann, J.P. 1980. *Iterated Maps on the Interval as Dynamical Systems. Progress in Physics*, Vol. 1, Boston: Birkhauser.

Feldman, J. 1976. Non-Bernoulli K-automorphisms and a problem of Kakutani. *Israel Journal of Mathematics* 24, 16–37.

Furstenberg, H. 1961. Strict ergodicity and transformations of the torus. *American Journal of Mathematics* 83, 573–601.

Furstenberg, H. 1977. Ergodic behaviour of diagonal measures and a theorem of Szemeredi on arithmetic progressions. *Journal d'analyse mathématique* 31, 2204–56.

Halmos, P.R. and von Neumann, J. 1942. Operator methods in classical mechanics II. *Annals of Mathematics* 43, 332–50.

Hejhal, D.A. 1976 The Selberg trace formula and the Riemann zeta function. *Duke Mathematical Journal* 43, 441–82.

Kakutani, S. 1943. Induced measure-preserving transformations. *Proceedings of the Imperial Academy of Tokyo* 19, 635–41.

Katok, A. 1977. Monotone equivalence in ergodic theory. *Izvestiya Akademii Nauk Ser. Mat.* 41, 104–157.

Kolmogorov, A.N. 1958. A new metric invariant of transient dynamical systems and automorphisms of Lebesgue spaces. *Doklady Akademii Nauk SSSR* 119, 8561–864 (Russian).

Ornstein, D.S. 1970. Bernoulli shifts with the same entropy are isomorphic. *Advances in Mathematics* 4, 337–52.

Ornstein, D.S. and Weiss 1984. Any flow is the orbit factor of any other. *Ergodic Theory and Dynamical Systems* 4, 105–16.

Parry, W. 1971. Metric classifications of ergodic nil flows and unipotent affines. *American Journal of Mathematics* 93, 819–28.

Parry, W. and Pollicott, M. 1983. An analogue of the prime number theorem for closed orbits of Axiom A flows. *Annals of Mathematics* 118, 573–91.

Patterson, S.J. 1976. The limit set of a Fuchsian group. *Acta Mathematica* 136, 241–73.

Pesin, J. 1977. Characteristic Lyapunov exponents and smooth ergodic theory. *Russian Mathematical Surveys* 32(4), 55–114.

Ratner, M. 1982. Rigidity of horocycle flows. *Annals of Mathematics* 115, 597–614.

Rees, M. 1982. Positive measure sets of ergodic rational maps. University of Minnesota Mathematics Report.

Rudolph, D. 1984. *Restricted Orbit Equivalence*. Reprinted, Baltimore University of Maryland.

Ruelle, D. 1978. *Thermodynamic Formalism*. Reading, Mass.: Addison-Wesley.

Ruelle, D. and Takens, F. 1971. On the nature of turbulence. *Communications in Mathematical Physics* 20, 167–92.

Schmidt, K. 1977. *Cocycles on Ergodic Transformation Groups*. London: Macmillan.

Sinai, J.G. 1959. On the concept of entropy of a dynamical system. *Doklady Akademii Nauk SSSR* 124, 768–71.

Sinai, J.G. 1963. On the foundations of the ergodic hypothesis for a dynamical system of statistical mechanics. *Doklady Akademii SSSR* 153, 1261–4; *Sov. Math. Dokl.* 4 (1963), 1818–22.

Sinai, J.G. 1972. Gibbsian measures in ergodic theory. *Usphehi Matematiceskich Nauk* 27, No. 4, 21–64; *Russian Mathematical Surveys* 27(4), 21–69.

Smale, S. 1967. Differentiable dynamical systems. *Bulletin of the American Mathematical Society* 73, 747–817.

Sullivan, D. 1979. The density at infinity of a discrete group of hyperbolic motions. *Publications mathématiques* 50, 419–450.

Von Neumann, J. 1932a. Proof of the quasi-ergodic hypothesis. *Proceedings of the National Academy of Sciences of the USA* 18, 70–82.

Von Neumann, J. 1932b. Zur operatoren Methode in der klassischen Mechanik. *Annals of Mathematics* 33, 587–642.

Walters, P. 1973. A variational principle for the pressure of continuous transformations. *American Journal of Mathematics* 97, 937–71.

Williams, R.F. 1973. Classification of subshifts of finite type. *Annals of Mathematics* 88, 120–93.

Erhard, Ludwig (1897–1977). Erhard was a man who had his moment in history and grasped it. As head of the Economic Department of the administration which preceded the creation of the Federal Republic of Germany, he was the author of the decision to combine the currency reform of 1948 with the abolition of rationing, and of restrictive regulations concerning production, distribution and capital movements. Many have argued that Germany's 'economic miracle' (and not less the political miracle) owes much to these decisions which at the time were regarded as either unrealistic or indefensible by many, including the Occupation Powers.

In a sense, Erhard's life before 1948 was a preparation for this moment, and his career afterwards a continuation of its theme. Born in Fürth in Franconia into a small business family, Erhard studied economics after World War I and joined an economic research institute. His teachers were, on the one hand, Wilhelm Rieger, first director of the Nuremberg Commercial College, and, on the other, Franz Oppenheimer, economist and sociologist in Frankfurt, whose influence on Erhard went much deeper. In the Sixties Erhard described Oppenheimer's importance for him in this way: his own economic policy was in a sense the redirection of Oppenheimer's 'liberal socialism' to 'social liberalism'. During World War II he wrote a memorandum sketching his project for a market economy in ways which left no doubt that he foresaw and wished for the defeat of the Nazis. This was one reason why he was appointed Bavarian Minister of Economic Affairs in 1945, and in 1947, head of the small special unit which prepared the currency reform of 1948. When Konrad Adenauer formed the first Federal Government, Erhard became Minister of Economic Affairs, a post which he held until he succeeded Adenauer as Federal Chancellor in 1963. It was as Economics Minister that Erhard preached and implemented the concept of 'social market economy', a market economy tempered by basic social policies, for which the Federal Republic has become famous. Erhard's Chancellorship was undistinguished; in 1966, his own party, the Christian Democratic Union (CDU) forced him to resign. However, his effect on Germany's economic institutions and the prevailing mould of economic thought is profound and lasting.

RALF DAHRENDORF

BIBLIOGRAPHY
Caro, M.K. 1965. *Der Volkskanzler – Ludwig Erhard*. Cologne and Berlin: Kiepenheuer & Witsch.
Lukomski, J.M. 1965. *Ludwig Erhard – Der Mensch und der Politiker*. Dusseldorf: Econ.

Erlich, Alexander (1913–1985). Alexander Erlich was born in St. Petersburg on 6 December 1913 and died on 7 January 1985. He moved to Poland with his family in 1918. In 1914, his father Henryk Erlich, a leader in the Socialist movement in Poland, was executed. In the same year, after university studies in Berlin and Warsaw, Erlich emigrated to the US, where he earned a PhD at the New School for Social Research and joined the faculty of Columbia University in 1955. From 1966 until his retirement in 1981, Erlich was professor of economics at Columbia, teaching in the Economics department, the Russian Institute and the Institute for East Central Europe. Professor Erlich was revered by his students for his unstinting help and encouragement and respected by his colleagues for his breadth of knowledge and understanding of socialist economics.

Alexander Erlich's main contribution to the economics of socialism is his work on the critical issue of industrialization policy in the USSR in the 1920s. To this issue, Erlich brought an unusual blend of sophisticated economic reasoning and penetrating political analysis. His major thesis concerning Soviet policy in this period is that the structural disproportions in the Soviet economy were so deep that virtually any policy would have had negative side-effects on reconstruction. Specifically, Erlich argued throughout his career that the economic policies of both the left *and* the right opposition were equally problematic. While the left analysis was correct in pointing out that future growth was limited after 1925 by the existing high capacity utilization and scarce investment funds, Preobrazhenskii and others were wrong in underestimating the reaction of the peasantry to an industrialization policy that would squeeze peasant incomes. On the other hand, the right opposition did not appreciate the implications of high capacity utilization for continued growth through small profit margins and high turnover of consumer goods and light manufacturers. The right, and Bukharin in particular, were seen by Erlich to be naive on the intensity of the conflict between consumption and investment once existing capacity was fully utilized. This, his major work, exhibits a detailed knowledge of the Soviet experience and a dispassionate and rigorous analysis of policy choices that set the standard for such work in the field.

DIANE FLAHERTY

SELECTED WORKS

1950. Preobrazhenskii and the economics of Soviet industrialization. *Quarterly Journal of Economics* 64, February, 57–88.

1959. The Polish economy after October 1956: background and outlook. *American Economic Review, Papers and Proceedings* 49, May, 94–112.

1960. *Soviet Industrialization Debate, 1924–1928*. Cambridge, Mass.: Harvard University Press.

1967a. Development strategy and planning: the Soviet experience. In *National Economic Planning*, ed. M.F. Millikan, New York: Columbia University Press.

1967b. Notes on a Marxian model of capital accumulation. *American Economic Review, Papers and Proceedings* 57, May, 599–616.

1973. A Hamlet without the Prince of Denmark. *Politics and Society* 4(1), Fall, 35–53.

1977. Stalinism and Marxian growth models. In *Stalinism: Essays in Historical Interpretation*, ed. Robert C. Tucker, New York: Norton.

1978. Dobb and the Marx–Fel'dman model: a problem in Soviet economic strategy. *Cambridge Journal of Economics* 2(2), June, 203–14.

BIBLIOGRAPHY

Desai, P. (ed.) 1983. *Marxism, Central Planning and the Soviet Economy: economic essays in honour of Alexander Erlich*. Cambridge, Mass. and London: MIT Press.

error, law of. *See* RANDOM VARIABLES.

errors in specification. *See* SPECIFICATION PROBLEM IN ECONO-METRICS.

errors in variables. 1. THE HISTORICAL AMBIVALENCE; 2. EARLY ECONOMETRIC DEVELOPMENTS; 3. RECENT ECONOMETRIC DEVELOPMENTS; 4. PROSPECTS.

1. THE HISTORICAL AMBIVALENCE

This entry surveys the history and recent developments on economic models with errors in variables. These errors may arise from the use of substantive unobservables, such as permanent income, or from ordinary measurement problems in data collection and processing. The point of departure is the classical regression equation with random errors in variables:

$$y = X^*\beta + u$$

where y is a $n \times 1$ vector of observations on the dependent variable, X^* is a $n \times k$ matrix of unobserved (latent) values on the k independent variables, β is a $k \times 1$ vector of unknown coefficients, and u is a $n \times 1$ vector of random disturbances. The matrix of observed values on X^* is

$$X = X^* + V$$

where V is the $n \times k$ matrix of measurement errors. If some variables are measured without error, the appropriate columns of V are zero vectors. In the conventional case the errors are uncorrelated in the limit with the latent values X^* and the disturbances u; and the errors have zero means, constant variances, and zero autocorrelation. In observed variables the model becomes

$$y = X\beta + (u - V\beta).$$

Since the disturbance $(u - V\beta)$ is correlated with X, ordinary least squares estimates of β are biased and inconsistent. The errors thus pose a potentially serious estimation problem. In

regard to systematic errors in variables, they will not be discussed, since they raise complex issues of model misspecification which lie outside the scope of this entry.

Errors in variables have a curious history in economics, in that economists have shown an ambivalent attitude toward them despite the universal awareness that economic variables are often measured with error and despite the commitment to economics as a science. Griliches (1974) suggested that much of the ambivalence stems from the separation in economics between data producers and data analysers. If so, why have not economists made a greater effort to cross the breach? Griliches (p. 975) further suggested that 'another good reason for ignoring errors in variables was the absence of any good cure for this disease'. If so, why have not economists made greater use of the econometric techniques developed since Griliches wrote his survey?

We propose an alternative explanation: the way of economic thinking, epitomized by utility theory and consumer maximization, has promoted a neglect of measurement errors. Bentham (1789, ch. IV) was a pioneer of measurement theory in the social sciences in his attempt to provide a theory for the measurement of utility. He went so far as to recommend that the social welfare of a given policy be computed by summing up *numbers* expressive of the 'degrees of good tendency' across individuals. Bentham's notion of cardinal utility met rightfully with great resistance. Pareto (1927, ch. III) pressed the dominant view: the economic equilibrium approach, by producing empirical propositions about consumer demand in terms of observables (quantities, prices, incomes), is to be favoured over theories connecting prices to utility, a metaphysical entity. Thus, theory – here optimization by rational consumers in a competitive market – overcame a fundamental measurement problem.

We do not wish to quarrel with the neoclassical equilibrium approach to the study of demand, although some economists wonder whether the assumptions of the theory have sufficient validity to warrant their acceptance (as part of the maintained hypothesis) in so many empirical demand studies. Rather, our point is that the great successes of this theory, and analogous successes of similar theories about other economic behaviour, have implanted a subconscious bias toward the substitution of economic theory (assumptions) for difficult measurement. In consequence, many economic models do not have an adequate empirical basis, cf. Leontief (1971) and Koopmans (1979).

Whatever the reasons for their neglect, errors in variables have been hard to keep down. Substantive unobservables such as permanent income, expected price and human capital continue to work their way into economic models and raise measurement issues. Friedman's (1957) permanent income model has served as a prototype for the errors-in-variables setup:

$$c_p = ky_p$$
$$y = y_p + y_t$$
$$c = c_p + c_t$$

where c = consumption, y = income, subscript 'p' = permanent, subscript 't' = transitory, and k is a behavioural parameter. Friedman (p. 36) clearly recognized the connection of the model expressed in observed c and y to the errors-in-variables setup; in his words, 'The estimation problem is the classical one of "mutual regression" or regression "when both variables are subject to error" '.

In the next two sections, early and recent developments on economic models with random errors in variables will be surveyed. Then, we will speculate on the future use of errors-in-variables methods.

2. EARLY ECONOMETRIC DEVELOPMENTS

Frisch (1934) was the first econometrician to face squarely the problem of errors in variables. In a brave book addressing model search, multicollinearity, simultaneity, and errors in variables, he decomposed the observed variables into a systematic (latent) part and a random disturbance part. His complicated correlation approach, sometimes resembling common factor analysis, did not satisfactorily resolve the errors-in-variables problem, but he raised fruitful questions. While Koopmans (1937), Geary (1942), Hurwicz and Anderson (1946), Reiersöl (1950), and a few others followed up on Frisch's endeavour, interest in the problem waned by the start of the 1950s. The famous Cowles Commission may have unintentionally buried the errors-in-variables problem when the chief investigators put it aside in order to make progress on the simultaneity problem. Applied economists, in their zeal to employ the new simultaneous equations model, ignored the limitations in their data despite the warning cry of Morgenstern (1950). Sargan (1958), Liviatan (1961), Madansky (1959), and a few others made contributions in the 1950s and 1960s, but for the most part errors in variables lay dormant. Widely used econometrics textbooks aggravated matters by highlighting the lack of identification of the classical regression equation in the absence of strong prior information such as known ratios of error variances. Neglect by the theorists led to the widespread use of *ad hoc* proxies in practice.

3. RECENT ECONOMETRIC DEVELOPMENTS

Zellner (1970) sparked a revival of interest in errors in variables. He attained identification of the permanent income prototype by appending a measurement relation that predicted unobservable permanent income in terms of *multiple causes* (e.g. education, age, housing value), to accompany the natural *indicator* relation in which observed current income is a formal proxy for permanent income. Goldberger (1971, 1972b) stimulated the revival by showing how models with substantive unobservables could be identified and estimated by combining all of the measurement information in a set of multiple equations that arise from multiple indicators or multiple causes. He also drew out the connections among the errors-in-variables model of econometrics, the confirmatory factor analysis model of psychometrics, and the path analysis model of sociometrics. On the applications side, Griliches and Mason (1972) and Chamberlain and Griliches (1975) studied the important socioeconomic problem of estimating the economic returns to schooling, with allowance for unobservable 'ability'. With Goldberger and Griliches leading the way, the econometric literature on errors in variables flourished in the 1970s.

Multiple equations. For the permanent income prototype, Zellner's multiple cause relation and indicator relation formed a two-equation measurement system that could be appended to the structural consumption equation. For this three-equation model, Zellner (1970) provided an efficient generalized least squares estimator, and Goldberger (1972a) added a maximum likelihood estimator. This errors-in-variables framework, which can be applied to many situations in which an unobservable appears as an independent variable in an otherwise classical regression equation, has been very useful.

Example applications have included Aigner's (1974) study of labour supply in which the wage is an unobservable, Lahiri's (1977) study of the Phillips curve in which price expectations is an unobservable, and Geraci and Prewo's (1977) study of international trade in which transport costs is an unobservable.

Jöreskog and Goldberger (1975) generalized the framework to situations in which there are more than two observed dependent variables. Their model combined prior constraints on the reduced-form coefficients (of the type that arise in econometric simultaneous equations models) with prior constraints on the reduced-form disturbance covariance matrix (of the type that arise in psychometric factor analysis models). For this model which contains multiple indicators and multiple causes (MIMIC) for a single unobservable, they developed a maximum likelihood estimator. Applications of the MIMIC framework have included Kadane et al's (1977) study of the effects of environmental factors on changes in unobservable intelligence over time, and Robins and West's (1977) study of unobservable home value. The framework could be extended to endogenous causes, as Robinson and Ferrara (1977) demonstrated.

Simultaneity. The MIMIC model assumes unidirectional causation. Suppose instead that unobservables appear in a simultaneous equations model. This case calls for less 'outside information' than the single-equation case, since coefficient overidentification, as it would exist in the hypothetical absence of measurement errors, may compensate for the underidentification associated with errors. This idea had appeared in early unpublished works by Hurwicz and Anderson (1946); Goldberger resurrected it in 1971. Geraci (1974), Hausman (1977), and Hsiao (1976) subsequently developed identification analyses and estimators for the simultaneous equations model with errors, taking account of the 'disturbance' covariance restrictions induced by the error structure as well as the usual coefficient restrictions. Many of their results have an instrumental variables interpretation. For illustration, an indicator for an unobservable in a simultaneous equations system is a valid instrumental variable for a given structural equation if the unobservable either (a) does not appear in that equation or (b) appears but has an associated error variance that is identified using information from some other part of the system. Once the latter linkage is recognized, the model well may be identifiable and hence estimable. A notable application has been Griliches and Chamberlain's series of studies on the economic returns to schooling. They employed a triangular model with structural disturbances specified as a function of unobservables (common factors). In various ways their models incorporated simultaneity, multiple indicators, and multiple causes.

Dynamics. Maravall and Aigner (1977), Hsiao (1977), and Hsiao and Robinson (1978) extended the errors-in-variables analysis to dynamic economic models. Among their findings, dynamics are a 'blessing' for identification in that autocorrelation of exogenous variables may provide additional information; and, upon taking a discrete Fourier transform of the data, many of the results for the contemporaneous model can be carried over to the dynamic model. In the same vein, Geweke (1977) developed a maximum likelihood estimator for the dynamic factor analysis model by reprogramming, in complex arithmetic, Jöreskog's (1970) maximum likelihood algorithm which had developed into the widely used LISREL software package. Geweke applied this estimator to investigate manufacturing sector adjustments to unobservable product demand. Singleton (1977) extended this factor analysis approach to study the cyclical behaviour of the term structure of interest rates. His framework allowed estimation of the model without specifying the causes of the unobservable real rate of interest and price expectations, thus isolating the classic Fisher hypothesis for testing.

As the preceding survey indicates, the recent econometric literature contains many theoretical results on the identification and estimation of structural models that contain substantive unobservables and measurement errors. (For a further survey, see Aigner et al., 1984.) The literature also contains some interesting applications, but not many. Are more forthcoming?

4. PROSPECTS

We have an uneasy feeling about the state of empirical economics. The development of formal economic theory and associated econometric technique has proceeded at an extraordinary pace. At the same time, what do economists know empirically? Many reported inferences hinge upon model assumptions whose validity remains to be assessed, and the gap between econometric technique and available data seems to be growing. None the less, there are grounds for some optimism.

Although few in number, the applications of errors-in-variables methods in the 1980s have been striking in their relevance to central economic issues. For example, Attfield (1980) has made the permanent income model a more complete explanation of consumption by incorporating unobservable liquid assets, rateable value, and windfall income. His model is a special simultaneous equations model with errors, in which identification of the individual structural equations can be established on a recursive basis. Geweke and Singleton (1981) also have taken up the permanent income model, adapting the classical latent variables model to this time series context and thereby generating some new tests of the permanent income hypothesis. As another example, Garber and Klepper (1980) have defended the competitive model of short-run pricing in concentrated industries through an explicit accounting for errors in measuring cost and output changes. They concluded that short-run price behaviour may appear to be related to market structure primarily because of estimation biases due to the measurement errors. As a final example, Stapleton (1984) has shown that the symmetry restrictions on structural parameters imposed by demand theory can be used to identify a linear model's parameters when measurement errors in price perceptions exist. This study is noteworthy in two respects. First, it shows how price, that bedrock of economic theory, may be measured erroneously. Second, economic theory is used to permit the explicit treatment of errors in variables.

These recent empirical works indicate the potential of errors-in-variables methods to lend fresh insights into important economic issues, and should stimulate more use of these methods. There are other encouraging signs as well. Recent studies using micro data have shown increasing attention to measurement error problems. In the macro area the rational expectations hypothesis has raised economists' consciousness of the difference between key conceptual variables of economic theory (i.e. permanent income, expected price, ex ante real rate of interest) and the available measurements. With respect to applications of errors-in-variables methods in economics, the stock is not great but the flow is encouraging.

VINCENT J. GERACI

See also ECONOMETRICS; LATENT VARIABLES; REGRESSION AND CORRELATION ANALYSIS.

BIBLIOGRAPHY

Aigner, D.J. 1974. An appropriate econometric framework for estimating a labor-supply function from the SEO file. *International Economic Review* 15(1) February, 59–68.

Aigner, D.J., Hsiao, C. Kapteyn, A. and Wansbeek, T. 1984. Latent variable models in econometrics. In *Handbook of Econometrics*, ed. Z. Griliches and M.D. Intriligator, Amsterdam: Elsevier Science, ch. 23.

Attfield, C.L.G. 1980. Testing the assumptions of the permanent-income model. *Journal of the American Statistical Association* 75, March, 32–8.

Bentham, J. 1789. *An Introduction to the Principles of Morals and Legislation.* London: Clarendon Press, 1907.

Chamberlain, G. and Griliches, Z. 1975. Unobservables with a variance-components structure: ability, schooling, and the economic success of brothers. *International Economic Review* 16(2), June, 422–49.

Friedman, M. 1957. *A Theory of the Consumption Function.* Princeton: Princeton University Press.

Frisch, R. 1934. *Statistical Confluence Analysis by Means of Complete Regression Systems.* Oslo: University Institute of Economics.

Garber, S. and Klepper, S. 1980. Administered pricing' or competition coupled with errors of measurement. *International Economic Review* 21(2), June, 413–35.

Geary, R.C. 1942. Inherent relationships between random variables. *Proceedings of the Royal Irish Academy* 47, 63–76.

Geraci, V.J. 1974. *Simultaneous Equation Models with Measurement Error.* PhD dissertation. New York: Garland, 1982.

Geraci, V.J. and Prewo, W. 1977. Bilateral trade and transport costs. *Review of Economics and Statistics* 59(1), February, 67–74.

Geweke, J.F. 1977. The dynamic factor analysis of economic time-series models. In *Latent Variables in Socioeconomic Models*, ed. D.J. Aigner and A.S. Goldberger, Amsterdam: North-Holland, ch. 19.

Geweke, J.F. and Singleton, K.J. 1981. Latent variable models for time series: a frequency domain approach with an application to the permanent income hypothesis. *Journal of Econometrics* 17(3), December, 287–304.

Goldberger, A.S. 1971. Econometrics and psychometrics. *Psychometrika* 36, June, 83–107.

Goldberger, A.S. 1972a. Maximum-likelihood estimation of regressions containing unobservable independent variables. *International Economic Review* 13(1), February, 1–15.

Goldberger, A.S. 1972b. Structural equation methods in the social sciences. *Econometrica* 40(6), November, 979–1001.

Griliches, Z. 1974. Errors in variables and other unobservables. *Econometrica* 42(6), November, 971–98.

Griliches, Z. and Mason, W.M. 1972. Education, income and ability. *Journal of Political Economy* 80(3), Pt II, May–June, 74–103.

Hausman, J. 1977. Errors in variables in simultaneous equation models. *Journal of Econometrics* 5(3), May, 389–401.

Hsiao, C. 1976. Identification and estimation of simultaneous equation models with measurement error. *International Economic Review* 17(2), June, 319–39.

Hsiao, C. 1977. Identification of a linear dynamic simultaneous error-shock model. *International Economic Review* 18(1), February, 181–94.

Hsiao, C. and Robinson, P.M. 1978. Efficient estimation of a dynamic error-shock model. *International Economic Review* 19(2), June, 467–79.

Hurwicz, L. and Anderson, T.W. 1946. Statistical models with disturbances in equations and/or disturbances in variables. Unpublished memoranda. Chicago: Cowles Commission.

Jöreskog, K.G. 1970. A general method for the analysis of covariance structures. *Biometrika* 57(2), August, 239–51.

Jöreskog, K.G. and Goldberger, A.S. 1975. Estimation of a model with multiple indicators and multiple causes of a single latent variable. *Journal of the American Statistical Association* 70, Pt I, September, 631–9.

Kadane, J.B., McGuire, T.W., Sanday, P.R. and Stuelin, P. 1977. Estimation of environmental effects on the pattern of I.Q. scores over time. In *Latent Variables in Socioeconomic Models*, ed. D.J. Aigner and A.S. Goldberger, Amsterdam: North-Holland, ch. 17.

Koopmans, T.C. 1937. *Linear Regression Analysis of Economic Time Series*. Haarlem: De Erven F. Bohn N.V.

Koopmans, T.C. 1979. Economics among the sciences. *American Economic Review* 69(1), March, 1–13.

Lahiri, K. 1977. A joint study of expectations formation and the shifting Phillips curve. *Journal of Monetary Economics* 3(3), July, 347–57.

Leontief, W. 1971. Theoretical assumptions and nonobserved facts. *American Economic Review* 61(1), March, 1–7.

Liviatan, N. 1961. Errors in variables and Engel curve analysis. *Econometrica* 29, July, 336–62.

Madansky, A. 1959. The fitting of straight lines when both variables are subject to error. *Journal of the American Statistical Association* 54, March, 173–205.

Maravall, A. and Aigner, D.J. 1977. Identification of the dynamic shock-error model. In *Latent Variable Models in Socioeconomic Models*, ed. D.J. Aigner and A.S. Goldberger, Amsterdam: North-Holland, ch. 18.

Morgenstern, O. 1950. *On the Accuracy of Economic Observations*. 2nd edn, Princeton: Princeton University Press, 1963.

Pareto, V. 1927. *Manual of Political Economy*. New York: Augustus M. Kelley, 1971.

Reiersl, O. 1950. Identifiability of a linear relation between variables which are subject to error. *Econometrica* 18, October, 375–89.

Robins, P.K. and West, R.W 1977. Measurement errors in the estimation of home value. *Journal of the American Statistical Association* 73, June, 290–94.

Robinson, P.M. and Ferrara, M.C. 1977. The estimation of a model for an unobservable variable with endogenous causes. In *Latent Variable Models in Socioeconomic Models*, ed. D.J. Aigner and A.S. Goldberger, Amsterdam: North-Holland, ch. 9.

Sargan, J.D. 1958. The estimation of economic relationships using instrumental variables. *Econometrica* 26, July, 393–415.

Singleton, K.J. 1977. The cyclical behavior of the term structure of interest rates. Unpublished PhD dissertation, Madison: University of Wisconsin.

Stapleton, D. 1984. Errors-in-variables in demand systems. *Journal of Econometrics* 26(3), December, 255–70.

Zellner, A. 1970. Estimation of regression relationships containing unobservable independent variables. *International Economic Review* 11, October, 441–54.

estimation.   Point estimation concerns making inferences about a quantity that is unknown but about which some information is available, e.g., a fixed quantity θ for which we have n imperfect measurements $x_1, \ldots, x_n$. The theory of estimation deals with how best to use the information (combine the values $x_1, \ldots, x_n$) to obtain a single number, estimate, for θ, say $\hat\theta$. Interval estimation does not reduce the available information to a single number and is a special case of hypothesis testing. This entry deals only with point estimation.

Justification for any particular way of combining the available information can be given only in terms of a *model* connecting the x's to θ. For example, in the case of imperfect measurements $x_1, \ldots, x_n$, we could regard the *errors*, $x_i - \theta$, $i = 1, \ldots, n$ as independent outcomes of a random process so that the joint distribution of the x's depends on θ:

$$p(x_1, \ldots, x_n | \theta) = \prod_1^n f(x_i - \theta).$$

In general, a *statistical model* represents the data, observations $x_1, \ldots, x_n$, where the x's may be vectors of quantities, as having arisen as a drawing from a joint distribution depending on some unknown parameters $\theta = (\theta_1, \ldots, \theta_k)'$. For example, consider $x_1, \ldots, x_t$, where x_t is identically and independently distributed according to a univariate normal distribution with mean μ and variance σ^2 (Cramer, 1946). The "location parameter," μ, and

the "scale parameter," σ^2, are unknown but, because they determine the distribution from which the data are supposed to arise, the latter may be used to form a point estimate of the vector $\theta = (\mu, \sigma^2)'$, e.g., $\hat\theta = (\bar{x} \Sigma_1^T x_i / T, s^2 = \Sigma_1^T (x_i - \bar{x})^2 / T)'$, the properties of which may be discussed in terms of various criteria and the properties of the family of probability distributions $p(x | \theta)$ from which the data are assumed to come. An *estimator* is a *function* of the observations; an *estimate* is the value of such a function for a particular set of observations. The theory of point estimation concerns the justification for estimators in terms of the properties of the estimates which they yield relative to specified criteria.

General treatments of the theory of point estimation may be found in Lehmann (1983), Cox and Hinkley (1974), Rao (1973) and Zellner (1971), *inter alia*.

Econometric estimation problems usually concern inferences about the parameters of conditional rather than unconditional distributions. For example, if the observations $(y_1, x_1), \ldots, (y_n, x_n)$ are assumed to represent a drawing from a multivariate normal distribution with mean vector μ and variance-covariance matrix Σ, then the *conditional* distribution of y given x, $p(y | x, \theta)$, is univariate normal with mean $\theta_1 = \mu_1 + \sigma_{12} \sigma_{22}^{-1} (x - \mu_2)$ and variance $\theta_2 = \sigma_{11} - \sigma_{12} \sigma_{22}^{-1} \sigma_{21}$, where

$$\mu = (\mu_1, \mu_2) \quad \text{and} \quad \Sigma = \begin{bmatrix} \sigma_{11} & \sigma_{12} \\ \sigma_{21} & \sigma_{22} \end{bmatrix}.$$

Note that θ_1 is a linear function of x which depends upon the parameters of the originally assumed joint distribution; this function is called the *regression* of y on x. *regression analysis* deals with the general problem of estimating such functions which characterize conditional distributions, usually those derived from normal distributions.

A standard method, and the one most common in econometrics, for obtaining estimators is the method of *maximum likelihood*. Consideration of this method provides a good introduction to alternative principles of estimation. Let the data $x = (x_1, \ldots, x_n)'$ be fixed and regard $p(x | \theta)$ as a function of θ; it is then called the *likelihood*. The value of $\hat\theta = \hat\theta(x_1, \ldots, x_n)$ which maximizes $p(x | \theta)$, if it exists and is unique, is called the maximum-likelihood estimator, or estimate (MLE). (For a general survey, see Norden, 1972–73, or Lehmann, 1983.) The MLE of a continuous function $g(\theta)$ is $g(\hat\theta)$ where $\hat\theta$ is the MLE of θ. Other desirable properties of the MLE are asymptotic as $n \to \infty$. Under regularity conditions: (1) The MLE is weakly *consistent*, i.e., $\lim n \to \infty \Pr(|\hat\theta_n - \theta| < \epsilon) = 1$ for all $\epsilon > 0$. (2) The MLE is *asymptotically normal*, i.e. the distribution of $\hat\theta$ appropriately normalized, $\sqrt{n}(\hat\theta_n - \theta)$, tends to the normal distribution, with mean 0 and variance-covariance matrix $[I(\theta)]^{-1}$ where

$$I(\theta) = -E[\partial^2 \log p(x | \theta) / \partial\theta \, \partial\theta'].$$

$I(\theta)$ is called the information matrix and shows the information a single observation contains about the parameter θ. (3) The MLE is *asymptotically efficient* in the sense that if θ^* is any other estimator such that $\sqrt{n}(\theta_n^* - \theta)$ tends in distribution to the normal with mean zero and variance-covariance matrix $\Sigma(\theta)$, the matrix $[\Sigma(\theta) - I^{-1}(\theta)]$ is positive semi-definite. For example, in the case of one parameter this means that no other asymptotically normal estimator has, as $n \to \infty$, a smaller variance than the MLE. The conditions for asymptotic normality do ensure, with probability tending to one, a solution to the *likelihood equation* $\partial \log p(x | \theta) / \partial\theta = 0$, which is consistent and asymptotically normal and efficient. The problem is that there may be more than one solution, but only one can be the MLE. When the number of parameters to be estimated (elements of the vector θ) tends to infinity with n, the MLE's for some may exist but may not be consistent (Neyman and Scott, 1948).

Solutions to the likelihood equation are not the only estimators which may be consistent, asymptotically normal and efficient, but comparison with the MLE, assuming correct specification of $p(x|\theta)$, is facilitated by the fact that all have a normal distribution as $n \to \infty$. For fixed n, the distributions of different estimators are difficult to determine and may, indeed, be quite different. Moreover, when the distributions underlying the data are misspecified, the MLE's generally no longer have these optimal properties (White, 1982; Gourieroux, Monfort and Trognon, 1984), although other, weaker, optimality properties remain. Apart from specification problems, however, the likelihood function provides an important and useful summary of the data, and point estimates and hypothesis testing procedures based on it are often justified in this way (Fisher, 1925; Barnard, Jenkins, and Winsten, 1962; Edwards, 1972).

The 'accuracy' of an estimator $\hat{\theta}$ of a scalar parameter θ may be measured (defined) in a variety of ways: by its expected squared or absolute error, relative error, or by $\Pr\{|\hat{\theta} - \theta| \le \alpha\}$ for some α. Any choice is arbitrary; for convenience expected squared error is the usual choice. Some justification for a particular choice may be provided in terms of a *loss function* $L(\theta, \hat{\theta})$ or the expected loss $EL(\theta, \hat{\theta})$ or *risk function* of *statistical decision theory*. Choice of estimators may be justified in terms of the extent to which the choice minimizes risk or some aspect thereof. Both the *sampling theory* and *Bayesian* approaches to estimation can be interpreted in these terms.

A very weak property that any estimator should have is that no other estimator exists which dominates it in the sense that the latter leads to estimates having uniformly lower expected loss irrespective of θ. Estimators satisfying this criterion are called *admissible*.

In the sampling theoretic approach, emphasis is placed on finding estimators which have desirable properties in terms of relative frequencies in hypothetically repeated samples. For example, we might require that the distribution of an estimator be centred on the true parameter value, i.e., $E(\hat{\theta} - \theta) = 0$. Such estimators are called *unbiased*. Among all unbiased estimators we presumably would prefer one yielding estimates with a distribution concentrated about the mean. Such minimum variance unbiased estimators (MVU) play a key role in the theory of estimation. Specifically, the famous Rao–Blackwell Theorem states that if an unbiased estimator $\hat{\theta}$ is a function of a complete sufficient statistic for θ then it is MVU. A statistic, say T, is said to be sufficient for θ if the conditional distribution of the observations given T is independent of θ. Completeness is also a property of the distribution functions for the observations; (a family P of distributions (of T) indexed by a parameter θ is said to be complete if there is no 'unbiased estimator of zero' other than $\phi(x) \equiv 0$.) Note that choosing an estimator so as to minimize the expected squared error of the estimate it yields is equivalent to minimizing the unweighted sum of the variance and the squared bias. From a decision theoretic point of view, it may be better to accept an estimator with a small bias if such an estimator has a smaller risk.

In the sampling theoretic approach, emphasis is given to the distribution of estimates yielded by a specified estimator. The likelihood approach, on the other hand, emphasizes the distribution of the observations, given a parametrically specified distribution, under alternative values of these parameters. Concern is primarily with the maximum value of the likelihood function with respect to the parameters and its curvature near the point at which the global maximum occurs,

but some approaches stress the relevance of the likelihood function in other neighbourhoods (Barnard et al., 1962; Edwards, 1972). The Bayesian approach carries concern with the entire likelihood function further: estimation and inference are based on the posterior density of the unknown parameters of the distribution generating the observations. This posterior density is proportional to the likelihood function multiplied by a prior density of the parameters, i.e., a weighted average of likelihoods for different parameter values where the weights are determined by prior (subjective) beliefs. (See BAYESIAN INFERENCE.)

In the Bayesian approach, both observations and parameters are taken to be stochastic. Let $p(x_1, \theta)$ be the joint probability density function for an observation vector, x, and a parameter vector θ; then $p(x_1, \theta) = p(x|\theta)p(\theta) = p(\theta|x)p(x)$, where $p(\xi|\eta)$ denotes the conditional density of ξ given η and $p(\xi)$ denotes the marginal density of ξ. Thus $p(\theta|x)$ is proportional to $p(\theta)p(x|\theta)$ by the factor

$$p(x) = \int p(\theta)p(x|\theta)\,d\theta.$$

$p(\theta|x)$ is the *posterior* distribution of θ, after having observed the data; $p(\theta)$ is the *prior* distribution of θ; and $p(x|\theta)$ is the *likelihood*. Alternatively, consider the weighted average risk (as defined above):

$$\int EL(\theta, \hat{\theta})\,w(\theta)\,d\theta,$$

with weights $w(\theta)$ such that

$$\int w(\theta)\,d\theta = 1.$$

When $L(\theta, \hat{\theta}) = (\hat{\theta} - \theta)^2$, the estimator which minimizes such a weighted average risk is

$$\hat{\theta}(x) = \int \theta w(\theta) p(x|\theta)\,d\theta \bigg/ \int w(\theta) p(x|\theta)\,d\theta.$$

If the weights $w(\theta)$ are taken to be the values of the marginal density $p(\theta)$, the mean of the posterior Bayes distribution minimizes the expected squared error of the estimates when both the variation of data and the uncertainty with respect to θ are taken into account: $\hat{\theta}$ is the expected value of θ based on the *posterior* distribution of θ. As $n \to \theta$, it may be shown that the influence of the prior distribution diminishes until in the limit it disappears; then, under general circumstances, the minimization of mean square error in the Bayesian framework yields the MLE. The principal difficulty in the Bayesian approach is the choice of a reasonable prior for θ, $p(\theta)$. (For a comprehensive discussion, see Zellner, 1971.)

Instead of minimizing the expected loss, one may minimize the maximum loss. Estimators which do are called *minimax*; the theory is developed in Wald (1950).

There are three general approaches to choice of a prior in Bayesian analysis. First, the prior may be obtained empirically (Maritz, 1970). For example, suppose that the problem is to estimate the percentage of defective items in a particular batch. Assuming such batches were produced in the past suggests a prior based on the proportion of defective items observed in previous batches. This kind of 'updating' forms the basis for the celebrated Kalman filter. Second, the prior may be viewed as representing a 'rational degree of belief' (Jeffreys, 1961). What represents a 'rational degree' is not specified, but the idea leads directly to the use of priors that represent knowing little or nothing, so-called *non-informative priors*. However, total ignorance has proved difficult to capture in many cases. A third approach is that the prior represents a subjective degree of belief (Savage, 1954; Raiffa and Schlaifer, 1961). But, of whom? and how arrived at? Minimax-estimation theory offers one possible approach for it leads to the minimum mean-square-error Bayes estimator, i.e., the mean of the posterior distribution of the parameters, when the prior is least favourable in the sense of making expected loss the largest for whatever class of priors is chosen.

Related to this problem is the more general question of *robust estimation*. In order to make sense of any data, it is necessary to assume something. For example, the justification for using the sample mean to estimate the mean of the distribution generating the data is often the assumption that that distribution is normal or nearly so. In that case, the sample mean is not only asymptotically efficient but uniformly MVU, minimax, admissible, etc. But suppose that the distribution is Cauchy (having roughly the same shape as the normal but with very thick tails); then, the sample mean has the same distribution as any individual observation, its accuracy does not improve with n and it is not even a consistent estimator. At least, within the class of distributions which include the Cauchy, the properties of the sample mean, and similarly ordinary least squares, are quite sensitive to the true nature of the underlying distribution of the data. We say that such estimators are not *robust*. Complete discussions are contained in Huber (1981) and Hampel, Ronchetti, Rousseeuw and Stahel (1985).

To conclude, three estimation problems of special concern in economics are discussed: (1) classical linear regression; (2) non-linear regression, and (3) estimation of simultaneous structural equations.

The classical theory of linear regression deals with the following problem: Let X be an $n \times k$ matrix of nonstochastic observations (n for each variable ($x_1, \ldots, x_k$), β be a $k \times 1$ vector of parameters (one of which becomes an intercept if $x_1 \equiv 1$, say), and y be an $n \times 1$ vector of stochastic variables such that $y - x\beta = \epsilon \sim N(0, \Sigma)$. The *ordinary least-squares estimates* (OLS), $\hat{\beta} = (X'X)^{-1}X'y$ are MLE and MVU when $\Sigma = \sigma^2 I$. When this is not true, although the OLS estimates are unbiased and consistent, they are not asymptotically efficient or minimum variance. The *generalized least squares estimates* (GLS), $\hat{\beta} = (X'\Sigma^{-1}X)^{-1}X'\Sigma^{-1}$ are efficient, but of course Σ, and therefore Σ^{-1}, is generally unknown. Often, however, a consistent estimate of Σ is available, leading to *feasible*, or *estimated*, GLS estimates.

Many problems in economics lead to non-linear relationships. Linear regression may be a good (local) approximation to such relationships if the data do not vary too widely. Moreover, many non-linear relationships may be transformed into linear ones (e.g., the Cobb–Douglas production function). Often, however, the data are sufficiently variable to make a linear relationship a poor approximation and no linearizing transformation exists. The general non-linear regression model is $y = f(X, \beta, \epsilon)$ or more frequently $y = f(X, \beta) + \epsilon$. Least-squares or maximum-likelihood estimates may still be obtained, but the first-order conditions for a minimum or a maximum will generally be non-linear, frequently ruling out analytic expressions for the estimates. Consider the problem of minimizing the sum of squared residuals, $(y - f(X, \beta))'(y - f(X, \beta))$, with respect to β (non-linear least squares); numerical methods for solving this problem are of the general form: $\hat{\beta}_{i+1} = \hat{\beta}_i - s_i P_i V_i$, where $\hat{\beta}_i$ = the value of the estimator parameter vector at iteration i, s_i = the step size at iteration i, P_i = the direction matrix at iteration i, and V_i = the gradient of the objective function at iteration i. The matrix P_i determines the direction in which the parameter vector is changed at each iteration; it is generally taken to be the Hessian matrix evaluated at the current value of the parameter vector or some approximation to it. Let $g(\beta)$ be the objective function; then

$$P_i = [\partial^2 g(\beta)/\partial\beta\partial\beta' | \beta = \hat{\beta}_i]^{-1}$$

is the Hessian. A justification for this choice is obtained from the second-order (quadratic) approximation to the objective function in the neighborhood of the current estimate. For a detailed treatment of this problem as well as constrained non-linear estimation see Quandt (1983). The statistical properties of non-linear estimators are discussed by Amemiya (1983).

Economic theory teaches us that the values of many economic variables are often determined simultaneously by the joint operation of several economic relationships, for example, supply and demand determine price and quantity. This leads to a representation in terms of a system of simultaneous structural equations (simultaneous equations model, or SEM). The problem of how to estimate the parameters of an SEM has occupied a central place in econometrics since Haavelmo (1943). A linear SEM is given by, $By_t \times \Gamma x_t = u_t$, $t = 1, \ldots, T$ where B is $G \times G$, Γ is $G \times K$, y_t is $G \times 1$, x_t is $K \times 1$, and u_t is $G \times 1$. u_t is assumed to be zero mean with variance-covariance matrix Σ, often normally distributed, independently and identically for each t. Thus the u_t are serially independent. It is also assumed that $\text{plim} \Sigma_1^T x_{it} u_{jt}/T = 0$ all $i = 1, \ldots, K$ and $j = 1, \ldots, G$ and $\text{plim } X'X/T$ is a positive definite matrix, where $X = (x_1, \ldots, x_T)'$. If B is non-singular this system of *structural equations*, as they are called, may be solved for the so called 'endogenous' variables, y_t, in terms of the 'exogenous' variables x_t: $y_t = \Pi x_t + v_t$ where $\Pi = -B^{-1}\Gamma$, $v_t = B^{-1}u_t$, so that $Ev_t = 0$ and $Ev_t v_t' = B^{-1}\Sigma(B^{-1})' = \Omega$. It is, in general, not possible to determine B, Γ and Σ from knowledge of the *reduced form* (RF) parameters Π and Ω; there are, in principle, many structural systems compatible with the same RF. Given sufficient restrictions on the structural system, however, knowledge of the RF parameters can be used, together with the assumed restrictions, to determine the structural parameters. The SEM is then said to be *identified*.

For *linear* structural equations with *normally distributed* disturbances, the conditions for identification may be derived from the condition that for any system $B^* y_t + \Gamma^* x_t = u_t^*$ for which u_t^* and u_t are identically distributed, where $B^* = FB$, $\Gamma^* = F\Gamma$ and $u_t^* = Fu_t$, then $F \equiv fI$ is implied by the restrictions, where f is any positive scalar (Hsiao, 1983).

Methods of estimating the parameters of SEMs may be put into two categories: (1) *limited-information* methods which estimate parameters of a subset of the equations, usually a subset consisting of a single equation, taking into account only the identifying restrictions on the parameters of equations in that subset, and (2) *full-information* methods which estimate all of the identifiable parameters in the system simultaneously and therefore take into account all identifying restrictions. Full- or limited-information methods may be based on either least-squares or maximum-likelihood principles. ML-based methods yield estimates which are invariant according to the normalization rule (choice of f).

For systems or single equations in SEMs for which there are restrictions just sufficient to identify the parameters of interest, estimates may be based on *indirect least squares*, that is, derived directly from the reduced form parameters estimated by applying OLS to each equation of the RF; such estimates are ML. If the restrictions are just sufficient to identify the parameters of each equation, the resulting estimates are *full-information maximum-likelihood* (FIML) estimates. When an equation is over-identified, in the sense that there are more than enough restrictions to identify it, *two-stage least squares* (2SLS) or *limited-information maximum likelihood* (LIML) may be applied equation by equation to each equation which is identified. Provided the model is correctly specified, such estimates are consistent and asymptotically unbiased but not asymptotically efficient, because some restrictions are neglected in the estimation of some parameters. An analog of 2SLS, *three-stage least squares* (3SLS), yields estimates which are asymptotically equivalent to FIML and therefore efficient.

Amemiya (1983) extends all of these methods to non-linear systems. Sargan (1980) discusses identification in non-linear systems.

MARC NERLOVE AND FRANCIS X. DIEBOLD

See also BAYESIAN INFERENCE; LEAST SQUARES; LIKELIHOOD; REGRESSION AND CORRELATION ANALYSIS; RESIDUALS; STATISTICAL DECISION THEORY; STATISTICAL INFERENCE.

BIBLIOGRAPHY

Amemiya, T. 1983. Nonlinear regression models. In *Handbook of Econometrics*, Vol. 1, ed. Z. Griliches and M.D. Intriligator, Amsterdam: North-Holland.

Barnard, G.A., Jenkins, G.M. and Winsten, C.B. 1962. Likelihood inference and time series (with discussion). *Journal of the Royal Statistical Society*, Series A, 125, 321–72.

Cox, D.R. and Hinkley, D.V. 1974. *Theoretical Statistics*. London: Chapman & Hall.

Cramer, H. 1946. *Mathematical Methods of Statistics*. Princeton: Princeton University Press.

Edwards, A.W.F. 1972. *Likelihood*. Cambridge: Cambridge University Press.

Fisher, R.J. 1925. Theory of statistical estimation. *Proceedings of the Cambridge Philosophical Society* 22, 700–25.

Gourieroux, C., Monfort, A. and Trognon, A. 1984. Pseudo maximum likelihood methods: theory. *Econometrica* 52, 681–700.

Haavelmo, T. 1944. The probability approach in econometrics. *Econometrica* 12, Supplement, July, 1–115.

Hampel, F.R., Ronchetti, E.M., Rousseeuw, P.J. and Stahel, W.A. 1985. *Robust Statistics*. New York: John Wiley.

Hsiao, C. 1983. Identification. In *Handbook of Econometrics*, Vol. 1, ed Z. Griliches and M. Intriligator, Amsterdam: North-Holland, 223–83.

Huber, P.J. 1981. *Robust Statistics*. New York: John Wiley.

Jeffreys, H. 1961. *Theory of Probability*. 3rd edn, Oxford: Clarendon Press.

Lehmann, E.L. 1983. *The Theory of Point Estimation*. New York: John Wiley.

Maritz, J.S. 1970. *Empirical Bayes Analysis*. London: Methuen.

Neyman, J. and Scott, E.L. 1948. Consistent estimates based on partially consistent observations. *Econometrica* 16, 1–32.

Norden, R.H. 1972–3. A survey of maximum-likelihood estimation. *Review of the International Institute of Statistics* 40, 329–54; 41, 39–58.

Quandt, R.E. 1983. Computational problems and methods. In *Handbook of Econometrics*, Vol. 1, ed. Z. Griliches and M. Intriligator, Amsterdam: North-Holland, 699–764.

Raiffa, H. and Schlaifer, R. 1961. *Applied Statistical Decision Theory*. Boston: Harvard Business School.

Rao, C.R. 1973. *Linear Statistical Inference and its Applications*. 2nd edn, New York: John Wiley.

Sargan, J.D. 1980. Identification and lack of identification. Paper presented to 4th World Congress of the Econometric Society, 28 August–2 September 1980, Aix-en-Provence, France.

Savage, L.J. 1954. *The Foundations of Statistics*. New York: John Wiley.

Wald, A. 1950. *Statistical Decision Functions*. New York: John Wiley.

White, H. 1982. Maximum-likelihood estimation of misspecified models. *Econometrica* 50, 1–25.

Zellner, A. 1971. *An Introduction to Bayesian Inference in Econometrics*. New York: John Wiley.

Eucken, Walter (1891–1950). Head of the Freiburg School of German Neo-liberalism and founder of the yearbook *Ordo*, Eucken was born at Jena on 17 January 1891. Eucken earned his doctoral degree at Bonn (1913). After the *Habilitation* in Berlin (1921), he was professor of economics at Tübingen (1925) and Freiburg (1927–50). He died on 20 March 1950 in London, during a lecture series at the London School of Economics.

Eucken's works mark the return to (neo)classical theory in German economics after the dominance of the Historical School. He stressed, however, the theorist's task to explain reality and rejected model-building if it was purely an intellectual game. Eucken's outstanding analytical contributions include a masterly explanation of the German inflation and currency depreciation on quantity-theoretical grounds (1923), a capital theory (1934) building on Böhm-Bawerk and Wicksell and, in particular, his theory of economic systems (1940) and of economic policy (1952).

Eucken's theory of economic policy starts from the distinction between the *economic order*, the legal and institutional framework of economic activity, and the *economic process*, the daily transactions of economic agents. Under laissez-faire the state neither shapes the economic order nor intervenes in the economic process; in a centrally planned economy the state dominates both. Eucken conceived a *Wettbewerbsordnung* (competitive system) different from both systems: Government should abstain from directly intervening into market processes, but it has to shape the economic order by guaranteeing, through *Ordnungspolitik*, the 'constituent principles' of the market economy (monetary stabilization, free entry, private property, freedom of contract, liability, consistency in economic policy and, primarily, maintaining competition). Subsidiary are the 'regulatory principles': monopoly regulation, social policy, process stabilization policy. Eucken's theory laid the ground for West Germany's 'social market economy'.

JOSEF MOLSBERGER

SELECTED WORKS

1923. *Kritische Betrachtungen zum deutschen Geldproblem*. Jena: Gustav Fischer.

1932. Staatliche Strukturwandlungen und die Krise des Kapitalismus. *Weltwirtschaftliches Archiv* 36.

1934. *Kapitaltheoretische Untersuchungen*. Jena: Gustav Fischer. 2nd enlarged edn, Tübingen: Mohr; Zürich, Polygraphischer Verlag, 1954.

1940. *Die Grundlagen der Nationalökonomie*. Jena: Gustav Fischer. 8th edn, Berlin, Heidelberg and New York: Springer, 1965. Trans. as *The Foundations of Economics*, London, 1950.

1948a. On the theory of the centrally administered economy: an analysis of the German experiment. Pts I–II, *Economica* 15; Pt I, May, 79–100; Pt II, August, 173–93.

1948b. Das ordnungspolitische Problem. *Ordo* 1.

1949. Die Wettbewerbsordnung und ihre Verwirklichung. *Ordo* 2.

1951. *Unser Zeitalter der Misserfolge: 5 Vorträge zur Wirtschaftspolitik*. Tübingen: Mohr.

1952. *Grundsätze der Wirtschaftspolitik*. Ed. Edith Eucken and K. Paul Hensel, Bern: Francke; Tübingen: Mohr. 5th edn, Tübingen: Mohr, 1975.

BIBLIOGRAPHY

Böhm, F. 1950. Die Idee des Ordo im Denken Walter Euckens. *Ordo* 3, xv–lxiv.

Jöhr, W.A. 1950. Walter Euckens Lebenswerk. *Kyklos* 4(4), 257–78.

Lenel, H.O. 1975. Walter Euckens ordnungspolitische Konzeption, die wirtschaftspolitische Lehre in der Bundesrepublik und die Wettbewerbstheorie von Heute. *Ordo* 26, 22–78.

Lutz, F.A. 1961. Eucken, Walter. In *Handwörterbuch der Sozialwissenschaften*, Vol. 3, Stuttgart: Fischer; Tübingen: Mohr; Göttingen: Vandenhoeck & Ruprecht.

Schmidtchen, D. 1984. German 'Ordnungspolitik' as institutional choice. *Zeitschrift für die gesamte Staatswissenschaft* 140(1), March, 54–70.

Stackelberg, H. von. 1940. Die Grundlagen der Nationalökonomie. *Weltwirtschaftliches Archiv* 51(2), March, 245–86.

Welter, E. 1965. Walter Eucken. In *Lebensbilder grosser Nationalökonomen: Einführung in die Geschichte der Politischen Ökonomie*, ed. H.C. Recktenwald, Cologne and Berlin: Kiepenheuer & Witsch.

Euler's Theorem. Euler's Theorem on homogeneous functions is one of those useful pieces of multivariable calculus that has tended not to receive the attention in mathematical textbooks that its importance in economic theory warrants. An analogous case is Lagrange multipliers, though there the analysis in most textbooks falls far short of the rigour and depth that are needed for fruitful economic applications, as it often does of Euler's other discovery of direct importance in economics, the so-called Euler equations in the calculus of variations (for a critical discussion, see Young, 1969). With Euler's Theorem there are no such worries, however, and the discussion in a work like that of Courant (1936, Vol. II, pp. 108–10) is quite adequate.

Once the necessary notation and terminology is established the statement of the theorem follows easily. Given any real number k, let F be a real-valued function defined on some non-empty subset S of vectors $x \in R^n$. Then F is said to be *homogeneous of degree* k (h.d.k.) if the equation

$$F(tx) = t^k F(x) \qquad (1)$$

holds for every $x \in S$ and every real number t.

Let f be a differentiable function defined on a non-empty open subset $G \subset R^n$, and denote the *gradient* of f at x, i.e. the n-dimensional vector of its partial derivatives f_i evaluated at x, by $\nabla f(x)$. The inner product of any two vectors a and b is written $\langle a, b \rangle$.

Euler's Theorem. The differentiable function f is homogeneous of degree k if and only if the following Euler relation holds for every $x \in G$,

$$\langle x, \nabla f(x) \rangle = k f(x) \qquad (2)$$

For a proof, see Courant (1936). Notice that this theorem *characterizes* homogeneous functions, i.e. any function satisfying (2) for all x must satisfy (1), hence be h.d.k. A simple but often useful corollary of the theorem is that if f is r-times differentiable and $m \leqslant r$, then each of its partial derivatives of order m is homogeneous of degree $k - m$, so that each f_i is h.d.$(k-1)$, each f_{ij} is h.d.$(k-2)$, and so on. Since homogeneous functions crop up almost everywhere in economics, Euler's Theorem is a standard tool with innumerable applications. So it is slightly odd that what was apparently its first use occurred so late, and that it was not by an established mathematical economist. In his review of Wicksteed (1894), A. W. Flux (1894) pointed out that Wicksteed could have saved himself a great deal of trouble if he had simply cited Euler's Theorem instead of, in essence, proving it all over again. It was indeed in the controversy over the so-called Adding-Up Problem in the theory of distribution that Euler's Theorem first gained notoriety, and the reader is referred to the entry on that Problem for details; here only a few of the main points will be lightly sketched in.

Assume that the firm wishes to minimize the cost of producing a scalar output η by the use of factors $x_1, x_2, \ldots, x_n = x$ bought at competitive prices $p_1, p_2, \ldots, p_n = p$. Under standard assumptions the first order conditions for this minimization yield

$$p_i = \lambda f_i(x) \qquad (3)$$

where $i = 1, 2, \ldots, n$ and λ is the associated Lagrange multiplier. This multiplier is of course the marginal cost of output, a fact which can be guessed at from (3) on purely dimensional grounds alone. Assume now that the production function f, where $\eta = f(x)$, is homogeneous of some unknown degree k.

Then, substituting from (3) into (2) and remembering the meaning of $\nabla f(x)$,

$$\lambda^{-1} \langle x, p \rangle = k f(x) \qquad (4)$$

If there is competition in the product market as well, i.e. free entry, then in long-run equilibrium marginal cost will equal the price q of the product, so that (4) becomes

$$\langle x, p \rangle = k q f(x) \qquad (5)$$

The left hand side of (5) is total factor payments. If constant returns prevail f is h.d.1 and so $k = 1$. All is well, since the right hand side is then total revenue, equal to the sum of factor payments. If on the other hand $k < 1$, so that returns to scale are decreasing, (5) shows that there will be something left over after all the purchased inputs have been paid.

What this residual really means is not clear. Some writers have interpreted it to be the returns (rent) to some non-marketed factor internal to the firm. But in that case why isn't the factor sold by its owner to the firm (after all we are in long-run equilibrium, so that quasi-rents do not apply)? Or is there no external market for the factor?

This is not the place to go into such questions, but it may be suggested that the incompleteness resides more in the theory than in either the markets or the factor payments. If $k > 1$ there are increasing returns to scale, and (5) then suggests that there will not be enough revenue to meet total factor payments. But with increasing returns the hypothesis of perfect competition in the product market has to be abandoned, so the passage from (4) to (5) is illegitimate and (5) does not hold.

PETER NEWMAN

BIBLIOGRAPHY

Courant, R. 1936. *Differential and Integral Calculus*. 2 vols, London: Blackie & Son.

Flux, A.W. 1894. Review of Wicksteed (1894). *Economic Journal* 4, 305–8.

Wicksteed, P.H. 1894. *An Essay on the Co-ordination of the Laws of Distribution*. London: Macmillan.

Young, L.C. 1969. *Lectures on the Calculus of Variations and Optimal Control Theory*. Philadelphia: W.B. Saunders Co.

Eurodollar Market. The Eurodollar Market is the term still commonly used to describe international financial activities by intermediaries whereby deposits are accepted and loans made in currencies other than the currency of the country in which the participating intermediary is located. The geographical focus was originally in Europe, especially London, but activities are now located in centres around the world.

The origins of the Eurodollar Market lay in exchange control restrictions on the use of national currencies for capital transactions, including the financing of trade between third parties. Banks and other financial intermediaries having longstanding involvement in the financing of such international trade, sought other means of maintaining their position in this sphere. Funds were canvassed in the one capital market where exchange controls did not apply, namely the United States. Thus dollars were borrowed and used to finance exports and imports between countries other than the country in which the financing intermediary was located. Given this circumstance it was inevitable that the US dollar became the currency of denomination for these transactions. Thus the term Eurodollar Market bears witness to the origins of this international financing arrangement. Nevertheless it remains a fair description as more than 70 per cent of all transactions continue to be denominated in that currency.

London remains the largest centre for Eurodollar activity, followed by New York, Frankfurt and Zurich. Transactions may be determined in these centres. The formal completion of transactions may, however, be registered in 'off-shore' places such as the Cayman Islands and the Bahamas in the Caribbean. By completing transactions in various centres external to the United States, such as those in the Caribbean, US banks avoided restrictions on their portfolios such as the locking up of assets in non-income earning reserve requirements. The US authorities bowed to the realities of this situation when in 1981, they established International Banking Facilities in the United States. Extra-territorial recognition to these activities was given while keeping them on-shore.

With the origins of eurodollar activity in a period of rigorous exchange controls, it might have been expected that the market would stagnate, and possibly wither, when exchange controls were whittled down or abandoned altogether as was the case by the late 1970s in many industrial economies. This was not the case. However, distinctions between the Eurodollar Market and transactions in the national currency of the country in which intermediaries were located, were blurred.

SIZE AND STRUCTURE. Estimates of the size of the Eurodollar Market or more correctly the Eurocurrency Market are compiled by the Bank for International Settlement, Morgan Guaranty Trust Company of New York and the International Monetary Fund. The IMF series was not published until 1984 so that most attention has been given to the other two. The series of estimates provided by the Bank for International Settlements (BIS) has been the basis for most analysis. The differences between the BIS series and those provided by Morgan Guaranty are explained largely by timing and coverage.

The BIS series is compiled on gross and net bases, the difference being depositing and lending between participating intermediaries. As is evident from the estimates in Table 1, the growth of lending was spectacular for many years during the 1970s. Only in recent times, with the onset of the debt crisis in 1981/82, has lending slowed.

The difference between the gross and net series for loans outstanding reflects mainly transactions between intermediaries in the Eurodollar Market, often referred to as 'interbank transactions'. The significance of these transactions lies in the provision of liquidity within the market. Intermediaries, by borrowing and lending amongst themselves, could meet liquidity needs arising from the mismatching of maturities between liabilities and assets. Loss of confidence in the asset quality of intermediaries brought illiquidity from 1982 onwards. Many intermediaries were not prepared to place funds with other intermediaries, a feature evident in the different rates of growth of gross and net lending.

TABLE 1: Eurofinance Lending, 1972 to 1983 (U.S. $ billions)

| Year | Amounts outstanding at the end of year | | | |
| | Gross | | Net | |
	Value	% increase	Value	% increase
1972	203.7	—	120.6	—
1973	291.4	43.1	172.1	42.7
1974	363.6	24.8	214.6	24.7
1975	442·4	21.7	254.6	18.6
1976	548.0	23.9	324.6	27.5
1977	671.3	22.5	435.0	34.0
1978	856.4*	28.9	530.0*	21.8
1979	1120.3	30.8	665.0	25.5
1980	1335.4	19.2	810.0	21.8
1981	1550.2*	14.1	945.0*	16.0
1982	1694.5	9.3	1020.0	7.9
1983:I	1757.0	0.4	1085.0	6.4
1983:II	2097.9	—	1240.0	—
1984	2153.9	2.7	1280.0	3.2

Source: Bank for International Settlements, various annual reports and reviews.

Note: *Change in series due to alteration in coverage of countries and transactions. Where the series breaks the estimated rate of growth is based on the old series for that year while the new series is the base for estimating growth in the subsequent year. This is illustrated in 1983 where both series are shown.

Estimates of the maturity structure of liabilities and assets are provided by the Bank of England for activities in London. About 40 per cent of liabilities have had a maturity of less than a month with over 20 per cent less than eight days. Assets have a longer maturity, about 32 per cent with maturities of less than a month and more than 22 per cent over one year. Maturity mismatching increased during the 1970s and 1980s.

Vital to any understanding of the impact of this market is the short maturities of these deposit liabilities. The financial intermediaries are constantly seeking new deposits or the rolling-over of existing deposits. Given the size of the Eurodollar Market and the frequency with which new funding is required, foreign exchange transactions have come to be dominated by these capital transactions. A modest estimate would be a turnover of US $150 billions per day and, in all likelihood, closer to double that estimate. The scale of these transactions far outweighs transactions related to trade in goods and services.

MECHANISMS. Analyses of the workings of the Eurodollar market remain controversial. Initially the market was treated as an extension of a national monetary system; given the predominance of transactions denominated in US dollars, this was viewed as an adjunct of American banking. This general approach for explaining the growth of the Eurodollar Market was matched by the belief that the function of this market was to 'recycle petrodollars'. What that expression meant was simply the functioning of the Eurodollar Market to take up the balance of payments surpluses of the oil-exporting countries after 1973, and then again in 1980, to fund the deficits of mainly oil-importing countries.

These interpretations do not stand inspection. The Eurodollar Market is not bound by the actions of any one national supervisory authority. Equally, it is not supported by the activities of any central bank. The participating intermediaries, even though they may be called banks, do not have recourse to lender of last resort facilities. They cannot, as yet, write cheques on themselves; rather they must write cheques on accounts in banks within a national banking system. They accept deposits with a specified term to maturity, often very short, and lend for specified periods with provisions for adjusting interest rates. Lending rates are most frequently quoted as some margin over LIBOR – the London Inter-Bank Offered Rate. The participating intermediaries bid for deposits from business, governments, banks, monetary authorities and wealthy individuals all being resident within national monetary systems, and from other participating intermediaries. They lend amongst themselves and to final users of funds, predominantly governments, banks and business in various countries, most often those with trade and payments deficits.

The Eurodollar Market is not an extension of national banking systems. It is a market in debt, not money. The functions performed by the participating intermediaries are central to an understanding of the impact of that market for not only the substantial debts incurred by many countries but also the balance of payments adjustment and exchange rate relativities.

ISSUES. The stability of the Euromarket mechanism rested upon the capacity of borrowers to meet their obligations. But in providing a means whereby deficit countries could maintain those deficits and not adjust to worsening balance of payments, the participating intermediaries accumulated an increasing proportion of their assets in the obligations of a relatively few countries. Portfolio risk could not be spread.

Impetus to expansion in the Euromarket was maintained during the late 1970s and early 1980s by the narrowing of margins over the cost of funds to participating intermediaries and slender capital/assets ratios. Although the dangers of such developments were recognised quite early in the 1970s, the Bank for International Settlements was unable to make effective its efforts to get coordination amongst national bank supervisory authorities about activities being pursued in the Euromarket. Efforts by the Bank of England and the Swiss authorities, while valuable within their national boundaries, did not gain that wider recognition which, with hindsight, was all too obviously needed.

An explanation for this failure to gain international coordination is the lack of recognition of problems likely to arise with the system. Attention was focused on individual failure either of a participating intermediary or a borrowing country. Only in the 1980s did the systemic problem become clear. By then modest arrangements for coordination proved inadequate, most obviously with the collapse of Banco Ambrosiano in Italy with repercussions for affiliates in Luxembourg and Switzerland.

The rapid escalation of debt problems for chronic borrowing countries in Eastern Europe, Latin America and Africa brought Euromarket activities to a virtual halt by late 1982. Debt renegotiation strained the capital structure of many participating intermediaries and their parent banks. Most were forced to improve capital ratios and hence liquidity, a not surprising development in view of debt rescheduling stretching the maturity structure of assets.

Direct lending by the participating intermediaries meant that the quality of the borrower was not subject to market tests. That lending activity was opaque, not transparent. By maintaining the flow of funds to chronic deficit countries, adjustment problems for those countries were deferred, but the strains were transferred in part to the participating intermediaries. The inherent weakness of this financing system was revealed when rising interest rates, shifts in exchange rate

relativities and weak commodity markets in the early 1980s found the chronic debtor countries unable to service their borrowings and, in many instances, meet repayments.

One repercussion of the harsh strains on participating intermediaries has been to restore activity in international bond financing. That financing, being directly subject to market tests, is confined to countries not facing chronic debt problems. Moreover, governments and companies from those countries are superior credit risks quite often to many participating intermediaries now bearing the penalties of lending to chronic debtors. Those same intermediaries have been willing to foster new techniques of financing, such as note issuance facilities and revolving credits, to maintain participation in international capital transactions. These new techniques offer possibilities for future strains no less than what emerged through direct lending.

W.P. HOGAN

See also INTERNATIONAL MONETARY POLICY.

Evans, Griffith Conrad (1887–1973). A distinguished American mathematician and pioneer mathematical economist, Evans was born on 11 May 1887 in Boston, Massachusetts. Educated in mathematics at Harvard University (AB, 1907; MA, 1908; PhD, 1910), he spent two years as a post-doctoral fellow studying with Vito Volterra at the University of Rome, then joined the faculty of the Rice Institute in Houston, Texas, where he taught from 1912 to 1934. In 1934, he became chairman of the mathematics department at the University of California, Berkeley, retaining that position until his retirement in 1954. He died on 8 December 1973, at the age of 86.

Evans's important contributions to mathematics, especially in functional analysis and potential theory, earned him membership in the National Academy of Sciences in 1933, as well as numerous other professional honours. His interest in mathematical economics became evident about 1920, when he gave his first series of lectures on that subject at the Rice Institute, and it continued up to the time of his retirement, his last publication on the subject appearing in 1954. It is likely that his initial contact with mathematical economics took place in Italy and France, for he shows great familiarity with the work of such writers as Pareto, Amoroso and Divisia, who were flourishing during and after his early Continental sojourn. Among earlier writers in mathematical economics, he mainly cites Cournot and Jevons; among his contemporaries, Irving Fisher, Henry Schultz and Henry Moore.

Evans's most important work in economics is his *Mathematical Introduction to Economics* (1930), which also contains materials from his earlier papers. In the book and his other publications, he applied the calculus and the calculus of variations to problems of monopoly, duopoly and competition, and to a whole range of problems of comparative statics, including the incidence of taxes and the effects of tariffs. His approach was quite different from that of Walras (to whom he does not refer in his book), in that most of his models dealt with one or a few actors in a single market, or a small number of markets. In his 'Maximum production studied in a simplified economic system' (1934, p. 37), he gave clear expression to his attitude toward general equilibrium models: 'Large numbers of simultaneous equations in a large number of variables convey little information ... about an economic system.'

In Evans's models, supply was generally described in terms of a cost function rather than a production function. To deal with macroeconomic problems, he constructed aggregate models, and, in order to provide a rationale for such models, he and his students made a deep study of the problem of index number construction. Their starting point was the work of Irving Fisher and François Divisia.

Evans's books and his articles were an important resource for early American students with an appetite for mathematical economics, who prior to their publication found an extremely sparse literature on which to graze. Samuelson, for example, mentions Evans as one whose works he 'pored over' when working on the *Foundations of Economic Analysis*. Moreover, Evans's methods of modeling economic situations gave new impetus to the approach of comparative statics, especially in application to macroeconomic problems. He constructed an early (perhaps the first) two-sector aggregative model containing a consumption good and a capital good; and he saw the power of second-order conditions of stability in reasoning about comparative statics, thereby anticipating by more than a decade Samuelson's important contributions to that topic.

Evans did only a little work in nonequilibrium dynamics, although he saw clearly the need for further development of that subject. In his 'Simple theory of economic crises' (1931, p. 61), he complained that 'the fact of lack of equilibrium in economic systems continually, and practically, stares us in the face; yet the principal discussion from a theoretical point of view has been of equilibrium, and thus at one stroke has eliminated a major issue'.

Evans was a fellow of the Econometric Society, and one of its founders. His principal influence upon the progress of economics came through the methodologies employed in his book, and through the work of his students, among whom were Francis W. Dresch, Kenneth May, C.F. Roos and Ronald W. Shephard, and one step removed, Lawrence W. Klein and Herbert A. Simon, who were colleagues or pupils of these students.

HERBERT A. SIMON

SELECTED WORKS

1922. A simple theory of competition. *American Mathematical Monthly* 29, 371–80.

1924. The dynamics of monopoly. *American Mathematical Monthly* 31, 77–83.

1925a. Economics and the calculus of variations. *Proceedings of the National Academy of Sciences of the USA* 11, 90–95.

1925b. The mathematical theory of economics. *American Mathematical Monthly* 32, 104–10.

1929. Cournot on mathematical economics. *Bulletin of the American Mathematical Society* 35, 269–71.

1930. *Mathematical Introduction to Economics*. New York: McGraw-Hill.

1931. A simple theory of economic crises. *American Statistical Association Journal* 26 (supplement), March, 61–68.

1932a. Stabilité et dynamique de la production dans l'économie politique. *Mémorial des Sciences Mathématiques* 61. Paris: Gauthier-Villars.

1932b. The role of hypothesis in economic theory. *Science* 75, 321–4.

1934. Maximum production studied in a simplified economic system. *Econometrica* 2, 37–50

1937. Indices and the simplified system. In, Cowles Commission for Research in Economics, *Report of Third Annual Research Conference on Economics and Statistics*, Chicago: University of Chicago Press, 56–9.

1939. (With K. May.) Stability of limited competition and cooperation. *Reports of a Mathematical Colloquium*. Notre Dame University, 2nd series, 3–15.

1950. Mathematics for theoretical economists. *Econometrica* 18, 203–4.

1952. Note on the velocity of circulation of money. *Econometrica* 20, 1.

1954. Subjective values and value symbols in economics. In *From Symbols and Values, an Initial Study, Thirteenth Symposium of the Conference on Science, Philosophy, and Religion*, New York: Harper & Row.

evolution. *See* NATURAL SELECTION AND EVOLUTION.

examples. Examples in economics, as elsewhere, are simply cases, real or fictitious, or partly both, supposed to embody a general principle. They may be classified as follows: (1) *Real but general*, as Ricardo's hunters (*Principles*), and Adam Smith's bricklayers, carpenters, and men of letters (*Wealth of Nations*). The examples are taken from a known genus but not from known individuals. Where the genus is perfectly well known, no cavil is possible. Adam Smith's illustration of division of labour could hardly have been improved by a reference to a particular pin-making establishment in a specified place. But, in exposition, the more concrete the genus the more telling the example; e.g. 'blacksmith' seems nearer life than 'workman'. (2) *Real and particular*, as in Cairnes's illustration of the theory of international trade from the Australian gold discoveries. Adam Smith, where he does not use the real and general, uses the real and particular, and falls back on fiction only for his similes (as 'the highway', 'the waggonway through the air', the 'wings', and 'the pond and the buckets', *Wealth of Nations*, II, ii), or his metaphors ('wheel of circulation', 'channel of circulation'). Ricardo and his immediate followers have preferred, as a rule, (3) *Fictitious* examples. These may be illustrations of which the component elements are generically well known, even the favourite 'man on the desert island', but the combining of the elements is the work of the writer, and is more or less arbitrary, as de Quincey's 'man with the musical box on Lake Superior', and Bastiat's 'plank and plane'. There is also a risk that the construction of the example may involve a begging of the question to be proved. 'Suppose that there are but two nations in the world living side by side, with a population of one million souls in each' (Barbour, *Bimetallism*). 'My object was to elucidate principles, and to do this I imagined strong cases that I might show the operation of those principles' (Ricardo, *Letters*). There is no necessary fallacy in this method of exposition any more than in illustrating the law of gravitation by the action of bodies *in vacuo*. Concrete cases must necessarily exemplify much more than one principle, and, even if they suggested a particular generalization, they may perhaps not clearly illustrate it without a fictitious simplification. The lawfulness of such a method of exposition or, it may be, of proof is discussed elsewhere.

[JAMES BONAR]

Reprinted from *Palgrave's Dictionary of Political Economy*.

excess capacity. *See* CAPITAL UTILIZATION.

ex ante and ex post. The concepts of *ex ante* and *ex post* are the most popular terminological innovations developed by the famous so-called Stockholm School in the 1930s. The terminology was introduced into macroeconomic theory, especially with regard to the savings–investment relation by Gunnar Myrdal (1933, 1939) and clarified and incorporated into sequence or period analysis by Erik Lindahl (1934, 1939b), whose conceptual system of 'prospective' and 'retrospective' values achieved 'world-citizenship' as a method for drawing up national budgets (Hansen, 1951, p. 27). The popularization of the method of *ex ante* and *ex post* is due to Ohlin's seminal articles on the Stockholm School (1937) which made it 'generally accepted over the whole world with a rapidity unusual to economics' (Palander, 1941, p. 34).

The significance of the distinction between *ex ante* and *ex post* 'as one of the most transforming insights that theoretical economics has had' (Shackle, 1972, p. 440) does not follow so much, as often stressed in the literature, from the simple fact that there exist always two alternative definitions of flow-related economic magnitudes like income, production etc., depending on whether they are looked at 'from before' or 'from after'. The central idea of the necessity to distinguish between *ex ante* and *ex post* stems rather from the recognition of the fundamental difference, originally expressed by Frank Knight (1921, pp. 35f.) and definitely formulated by Myrdal (1939, pp. 59f.), between 'foreseen' and 'unforeseen' changes where only the latter result in 'gains and losses' which, as shown by Lindahl (1939b, pp. 103f.), have to be 'windfalls'. Therefore, in the analysis of expectations under uncertainty time has to be included in an essential way by two alternative methods of calculation of economic variables: (i) an *ex ante* computation or business calculation which refers to a point of time at the beginning of a period and (ii) an *ex post* computation or bookkeeping referring to the development in time at the end of the period (Myrdal, 1939, pp. 45–7). As a consequence, economic analysis can be divided into (i) an *ex ante* analysis explaining how expectations determine an economic magnitude and (ii) an *ex ante/ex post* analysis explaining the possible divergence between the expected and realized value of this variable.

The emergence of the concepts of *ex ante* and *ex post* can be dated to Lindahl's and Myrdal's early writings in the 1920s. In his first treatise in macroeconomic theory (1924, ch. 3) Lindahl stressed the time factor for economic analysis and used the notion of 'subjective calculations of the future' as well as the term *ex post* when he discussed 'a negative investment recognized only ex post' (p. 33). A more coherent analysis of these concepts was given in Myrdal's dissertation on expectations and price changes (1927, pp. 67f.) where he showed, emphasizing Knight's idea of the difference between certain and uncertain changes, how divergences between incomes and costs of an investment calculated '*before*' will be balanced by gains and losses calculated '*after*'.

The first application of these ideas to macroeconomic problems was made by Lindahl in *Penningpolitikens medel* (1930; cf. 1939a). However, the dynamic method of temporary equilibrium used in this treatise, i.e. 'an analysis dividing time into a number of short equilibrium periods during which no changes occur' led to a 'theoretically inadmissible mixture of the *ex ante* and *ex post* analysis' (Myrdal, 1939, p. 122). As shown by Myrdal in the original Swedish version of *Monetary Equilibrium* (1932, pp. 228–30), this mixture was especially obvious in Lindahl's discussion of the relation between investment and saving, where he could not demonstrate in a satisfactory way how an initial discrepancy due to a shift in the rate of interest will always be balanced by changes in the distribution of income between borrowers and lenders. If, however, Lindahl would give up his method of temporary equilibrium, i.e. allow for disequilibrium during a period, his analysis could be interpreted in an *ex ante/ex post* framework.

It was exactly this disequilibrium analysis which enabled Myrdal to clarify the relation between investment and saving in his three different versions of *Monetary Equilibrium*, where he introduced the notions of *ex ante* and *ex post* first in the German edition (1933, § 29). In his discussion of these concepts Myrdal (1939, pp. 59–62, 116–25; cf. 1933, §§ 32, 55–6) allowed for a discrepancy between investment and saving *ex ante* based on 'anticipatory calculations' at a point of time demarcating the beginning of a period, while at its end their values were constructed by 'a subsequent "bookkeeping" ' in such a way that there is *always* an *ex post* balance 'regardless of how short the period'. Therefore, it is not 'this meaningless balance' which is of interest to economic analysis but '*the very changes during the period which are required to bring about this ex post balance*'. Myrdal assumed that these balancing factors arise out of 'unanticipated changes' in 'revenues and costs', i.e. in incomes, during the period for which they can be calculated only *ex post*: gains and losses. As later shown by Lindahl (1939b, pp. 103 f.; cf. 1958), these values have to be windfalls and must not be confused, as sometimes in Myrdal's analysis (1939, p. 65), with entrepreneurial gains or losses, which are already included in the *ex ante* values and which, therefore, can not serve as balancing factors.

Although Myrdal always spoke of 'income changes' as the balancing factor *ex post* of discrepancies *ex ante* between investment and saving, his examples implied almost exclusively '*price changes*' (1939, p. 60; cf. Palander, 1941, pp. 42f.; Hansson, 1982, p. 149). It was left to Lindahl (1934, 1939b) to demonstrate how the *ex post* equality was achieved in a disequilibrium process via a change in quantities.

Lindahl presented his solution in an aggregate demand and supply framework, where demand was identified with the purchase plans of consumers and producers and supply with the expectations of the sellers of production and consumption goods. In his analysis he made the fundamental assumption that the purchase plans as well as the supply prices of the sellers at the beginning of a period 'have been actually realized during the period' (1934, p. 207, cf. 1939b, p. 92). Under this assumption a possible deviation between expected and realized sales, which Lindahl took as given if the future is not foreseen with certainty, must be considered as a result of a difference between investment and saving *ex ante* which in turn will cause a divergence between expected and realized total real income. These changes in income represent gains or losses to the producers which in a form of 'unintentional', not 'forced', saving or dissaving equalize investment and saving *ex post*.

The purpose of Lindahl's rigid assumption that purchase plans are always fulfilled, which made it impossible to apply his analysis to the conditions of full employment (Hansen, 1951, pp. 29–32), was to demonstrate that, once prices are given, the *actions* of the economic subjects during the period 'can be directly deduced from the plans at the beginning of the period' (1939b, p. 92). With this demonstration Lindahl had taken the first step to a sequence analysis, i.e. a single-period analysis where *ex ante* plans determine *ex post* results. The second step consisted of a continuation analysis where the *ex post* events of the current period lead to revisions of the *ex ante* plans for the consecutive period at the transition point between these periods, 'especially as regards the supply prices and the producers' and consumers' demand' (1934, p. 211). However, as Lindahl 'never succeeded in formulating 'laws of motion' for revisions of plans ... this promising branch of dynamic theory became abortive' (Hansen, 1966, p. 3).

Of greater influence for economic analysis was Lindahl's second contribution to the development of the *ex ante/ex post* method, his discussion of the relations between 'prospective' and 'retrospective' values of micro- and macro-economic variables (1939b) which contained 'the germ of many lines in later works' on the methodology of national accounting

(Ohlsson, 1953, p. 266). Although Lindahl's accounting structure was criticized as 'deficient' (Ohlsson) in the treatment of government accounting, it has been emphasized recently that Lindahl's system 'does have some continuing merits ... for the accounting of the public sector': 'In this field at least, it may still be contended, *ex ante ex post* remains respectable' (Hicks, 1985, p. 80).

For a long time it was argued that the exposition of the Keynesian system 'requires the language of *ex ante* and *ex post*' (Shackle, 1972, p. 172), with the emphasis placed on the possible divergences, due to uncertainty, between disappointed *ex ante* expectations and *ex post* results as one of the relevant factors in determining the level of employment. However, the posthumous publication of Keynes's 1937 lecture notes have shown that Keynes (1937, p. 183) emphasized that even under the assumption of an 'identity of *ex post* and *ex ante*', i.e. with expectations always fulfilled but without having to assume for this case, as did Myrdal and Lindahl, the absence of uncertainty, 'the theory of effective demand is substantially the same' (p. 181). Moreover, as Keynes regarded the '*time relationship*' between the concepts of *ex ante* and *ex post* as 'incapable of being made precise' (p. 179), he rejected this method as an inadequate tool in handling the problems of uncertainty and time.

OTTO STEIGER

See also MYRDAL, GUNNAR.

BIBLIOGRAPHY

Hansen, B. 1951. *A Study in the Theory of Inflation.* London: Allen & Unwin.
Hansen, B. 1966. *Lectures in Economic Theory. Part I: General Equilibrium Theory.* Lund: Studentlitteratur.
Hansson, B. 1982. *The Stockholm School and the Development of Dynamic Method.* London: Croom Helm.
Hicks, J. 1985. *Methods of Dynamic Economics.* Oxford: Clarendon Press.
Keynes, J.M. 1937, Ex post and ex ante. Notes from the 1937 lectures. In *The General Theory and After*, Part II: *Defence and Development*, Vol. XIV of *The Collected Writings of John Maynard Keynes*, ed. D. Moggridge. London: Macmillan, 1973, 179–83.
Knight, F.H. 1921. *Risk, Uncertainty, and Profit.* Chicago: University of Chicago Press, 1971.
Lindahl, E. 1924. *Penningpolitikens mål och medel. Del I.* Lund: Gleerup; Malmö: Försäkringsaktiebolaget.
Lindahl, E. 1930. *Penningpolitikens medel.* Lund: Gleerup; Malmö: Försäkringsaktiebolaget; enlarged version of 1st edn, 1929; revised version trans. as Lindahl (1939a).
Lindahl, E. 1934. A note on the dynamic pricing problem. Mimeo, Gothenburg, 13; quoted from the corrected version published in Steiger (1971), 204–11.
Lindahl, E. 1939a. The rate of interest and the price level. In Lindahl (1939c), 139–260; revised version of Lindahl (1930).
Lindahl, E. 1939b. Algebraic discussion of the relations between some fundamental concepts. In Lindahl (1939c), 74–136.
Lindahl, E. 1939c. *Studies in the Theory of Money and Capital.* London: Allen & Unwin.
Lindahl, E. 1958. The concept of gains and losses. In *Festskrift til Frederik Zeuthen*, Copenhagen: Nationalkonomisk Forening, 208–19.
Myrdal, G. 1927. *Prisbildningsproblemet och föränderligheten.* Uppsala and Stockholm: Almqvist & Wiksell.
Myrdal, G. 1932. Om penningteoretisk jämvikt. En studie över den 'normala räntan' i Wicksells penninglära. *Ekonomisk Tidskrift* 33(5–6), (1931; printed 1932), 191–302; revised version trans. as Myrdal (1933).
Myrdal, G. 1933. Der Gliechgewichtsbegriff als Instrument der geldtheoretischen Analyse. In *Beiträge zur Geldtheorie*, ed. F.A. Hayek. Vienna: J. Springer, 361–487; 1st revised version of Myrdal (1932); 2nd revised version trans. as Myrdal (1939).
Myrdal, G. 1939. *Monetary Equilibrium.* London: Hodge; revised version of Myrdal (1933).
Ohlin, B. 1937. Some notes on the Stockholm theory of savings and investment I–II. *Economic Journal* 47, 53–69, 221–40.
Ohlsson, I. 1953. *On National Accounting.* Stockholm: Konjunkturinstitutet.
Palander, T. 1941. Om 'Stockholmsskolans' begrepp och metoder. Metodologiska reflexioner kring Myrdals 'Monetary Equilibrium'. *Ekonomisk Tidskrift* 43(1), March, 88–143; quoted from and trans. as: On the concepts and methods of the 'Stockholm School'. Some methodological reflections on Myrdal's 'Monetary Equilibrium'. *International Economic Papers* No. 3, 1953, 5–57.
Shackle, G.L.S. 1972. *Epistemics & Economics. A Critique of Economic Doctrines.* Cambridge: Cambridge University Press.
Steiger, O. 1971. *Studien zur Entstehung der Neuen Wirtschaftslehre in Schweden. Eine Anti-Kritik.* Berlin: Duncker & Humblot.

excess demand and supply. In general equilibrium theory the economy may be represented by the function which specifies the aggregate excess demands (positive) or excess supplies (negative) which are expressed at all possible price systems.

While the concept of excess demand is implicit in Walras's (1874) framework it is first introduced explicitly in Hicks's (1946) treatment of the general equilibrium system. In the present discussion we first examine how the excess demand function is obtained from the underlying parameters of the economy, that is the preferences and endowments of the various agents in the economy. We then discuss why the concept is useful, and what restrictions economic theory imposes on the excess demand function, and, equally importantly, what it does not.

An economy with n commodities consists of a number of agents, each with given preferences for and endowments of these commodities. An agent's preferences are represented by a complete preordering P on the commodity space R_+^n.

The preordering has the following properties:

Continuity: the set of all x in R_+^n such that xPy and the set such that yPx are both closed for all y.

Monotonicity: if $x \geqslant y$ but xy then xPy but not yPx.

Convexity: if xPy and z is a proper linear combination of x and y then zPy but not yPz.

The agent's endowment is represented simply by a point e in $S = R_+^n - 0$.

Given an agent's preferences P and endowment e, and given any price system p in S, there is one and only one $x(P)$ in R_+^n such that $x(p)Py$ for all y such that $p \cdot y \geqslant p \cdot e$; existence of this element can be shown using Weierstrass's theorem, while uniqueness follows immediately from convexity (Debreu, 1959). The difference between this $x(p)$ and his endowment that is $x(p) - e$, is his excess demand at the price system p, denoted $g(p)$.

One property which follows immediately from the definition of excess demand is that $g(p)$ is homogeneous (of degree zero), that is $g(tp) = g(p)$ for all positive t. A second property which follows immediately from the definition and monotonicity is that $p \cdot g(p) = 0$ for all p.

A third important property which is less immediate is that g is continuous provided that the value of the agent's endowment, $p \cdot e$, is positive (Debreu, 1959). The reason why we need this proviso can be seen by an example. Let $n = 2$ and $e = (1, 0)$, and let p tend to $(0, 1)$; for all positive $p, g_1(p)$ is non-positive but in the limit when $p = (0, 1)g_1(p)$ is infinite because of monotonicity.

A further property of excess demand is the revealed preference property. Consider two distinct prices p and q, with the corresponding excess demands $g(p)$ and $g(q)$, also assumed to be distinct. If $g(p)$ is available at the price system q, that is if $q \cdot g(p) \leqslant 0$, then $g(q)$ must not be available at the price system p, that is $p \cdot g(q) > 0$. We thus have $q \cdot g \cdot (p) \leqslant 0$ implies $p \cdot g(q) > 0$.

Excess demands of individual agents are mainly of interest in that they determine aggregate excess demands. The aggregate excess demand function $f \colon S \to R^n_+$ is obtained simply by defining $f(p)$ as the sum of each agent's $g(p)$.

It is immediate that $f(p)$ is homogeneous, that is $f(tp) = f(p)$ for all positive t, and also that $p \cdot f(p) = 0$ for all p, a property known as Walras's Law. It is also clear that f is continuous at any strictly positive p, for then the value of each agent's endowment must be positive, so that each g is continuous. In fact f is continuous everywhere on S, essentially because at any p there must be some agent whose endowment has positive value.

The revealed preference property, which applies to individual agents' excess demands, does not, however, carry over to aggregate excess demands. If all agents are identical, that is, have identical preferences and endowments, then the property carries over, but it may be that individual agents' excess demands are related in a perverse way, which means that the property is lost in aggregation. If the property does apply in aggregate then we may consider the economy to behave as an individual agent.

The excess demand function f provides us with a simple way to define equilibrium, and various of its properties. A price system p in S is an equilibrium price if $f(p) = 0$. Further, for example, an equilibrium price p is unique if, for any q in S, $f(q) = 0$ implies that $q = tp$ for some positive t. Equilibrium and its properties may be defined without the concept of excess demand, but the concept provides a useful framework for their investigation.

For example, with two commodities (and thus effectively one price and one excess demand function) it is intuitively clear that equilibrium will be unique, and indeed stable, if excess demand is everywhere downward-sloping. More generally, uniqueness and stability are typically investigated using generalizations of this idea of downward-sloping excess demand.

We have noted that aggregate excess demands have the properties of homogeneity, Walras's Law and continuity. It is also important to note what is in effect the converse of this: that any function with these properties may be an excess demand function. This is to say that economic theory places no restrictions on excess demand functions other than these three properties.

This property was first investigated by Sonnenschein (1973), but the most powerful result is due to Debreu (1974), who showed that there is an economy with precisely n agents which generates any continuous homogeneous excess demand function obeying Walras's Law, at least on the set of strictly positive prices. This is proved by first decomposing f into n individual excess demand functions, each having the revealed preference property, and then showing that these individual functions are the result of preference maximization subject to a wealth constraint.

Indeed, if f is restricted to being twice differentiable then there is an economy with n agents, each with homothetic preferences, which generates f as its excess demand function, again on the set of positive prices (Mantel, 1979). This is shown by creating indirect utility functions for each agent and then using Roy's Identity to obtain the excess demand functions. However, the extension of this result, and also

Debreu's result, to the whole of S, rather than only the set of strictly positive prices, remains open.

Not surprisingly, these results imply that the set of equilibrium prices has little structure. Indeed, it may be any non-empty compact subset of S (Mas-Colell, 1977). Thus without further restrictions we can say very little about the set of equilibria.

MICHAEL ALLINGHAM

BIBLIOGRAPHY

Debreu, G. 1959. *Theory of Value*. New York: Wiley.
Debreu, G. 1974. Excess demand functions. *Journal of Mathematical Economics* 1(1), 15–21.
Hicks, J.R. 1946. *Value and Capital*. 2nd edn, Oxford: Clarendon Press.
Mantel, R. 1976. Homothetic preferences and community excess demand functions. *Journal of Economic Theory* 12(2), April, 197–201.
Mas-Colell, A. 1977. On the equilibrium price set of an exchange economy. *Journal of Mathematical Economics* 4(2), 117–26.
Sonnenschein, H. 1973. Do Walras' identity and continuity characterize the class of community excess demand functions? *Journal of Economic Theory* 6(4), August, 345–54.
Walras, L. 1874–7. *Eléments d'économie politique pure*. Definitive edn, Lausanne: Corbaz, 1926; trans. by W. Jaffé as *Elements of Pure Economics*, London: George Allen & Unwin, 1954.

exchange. The accepted purview of economics is the allocation of scarce resources. Allocation comprises production and exchange, according to a division between processes that transform commodities and those that transfer control. For production or consumption, exchange is essential to efficient use of resources. It allows decentralization and specialization in production; and for consumption, agents with diverse endowments or preferences require exchange to obtain maximal benefits. If two agents have differing marginal rates of substitution then there exists a trade benefiting both. The advantages of barter extend widely, e.g. to trade among nations and among legislators ('vote trading'), but it suffices here to emphasize markets with enforceable contracts for trading private property unaffected by externalities. In such markets, voluntary exchange involves trading bundles of commodities or obligations to the mutual advantage of all parties to the transaction.

In a market economy using money or credit, terms of trade are usually specified by prices. Besides purchases at prices posted by producers and distributors, exchange occurs in bargaining, auctions, and other contexts with repeated or competitive offers. In institutionalized 'exchanges' for trading commodities, brokers offer bid and ask prices; and for trading financial instruments, specialists cross orders and maintain markets continually by trading for their own accounts.

Records of transaction prices and quantities are the raw data of many empirical studies of economic activity, and explanation of these data is a main purpose of economic theory. Theories of exchange attempt to predict the terms of trade and the resulting transactions, depending on the market structure and the agents' attributes, including such features as each agent's endowment, productive opportunities, preferences, and information. Also relevant are the markets accessible, the trading rules used, and the contracts available; these may depend on property rights, search or transaction costs, and on events observable or verifiable to enforce contracts. If a particular trading rule is used, it specifies the actions allowed each agent in each contingency, and the trades resulting from each combination of the agents' actions. These features are the ingredients of experimental designs to test theories, and they motivate models used for empirical estimates of market behaviour. Normative considerations are also relevant, and welfare analyses emphasize the distributional consequences of alternative trading procedures and contracts.

Most theories hypothesize that each agent acts purposefully to maximize (the expected utility of) gains from trade. Some behaviour may be erratic, customary, or reflect dependency on a *status quo*, but experimental and empirical evidence substantially affirms the hypothesis of 'rational' behaviour, at least in the aggregate. Although more general theories are available, the main features are explained by preferences that are quite regular, as assumed here: monotone, convex, as smooth as necessary, and possibly allowing risk aversion.

Typically there are many efficient allocations of a fixed endowment: any allocation that equates all agents' marginal rates of substitution is efficient. In the case of risk sharing, for example, an allocation is efficient if all agents achieve the same marginal rates of substitution between income in every two states. The distribution of endowments among agents evidently matters, however, and a major accomplishment has been the identification of a small set of salient efficient allocations. Named for Léon Walras, this set is a focus of nearly all theories, in the sense that other allocations are explained by departures from the Walrasian model. A major theme is to elaborate the special role of the Walrasian allocations.

An allocation is Walrasian if it is obtainable by trading at prices such that it would cost each agent more to obtain a preferable allocation. That is, items are bought at uniform prices available to all, and each agent chooses a preferred trade within a budget constraint imposed by the value of goods sold. A Walrasian allocation is necessarily efficient to the extent that markets are complete; for another allocation preferred by every agent would cost each agent more at the current prices and therefore more in total, which cannot be true if the preferred allocation is a redistribution of the present one. Conversely, each efficient allocation is Walrasian without further trade, since the agents' common marginal rates of substitution serve as the price ratios. The basic formulation considers trade for delivery in all future contingencies, but refined formulations elaborate the realistic case that markets reopen continually and trade is confined to a limited variety of contracts for spot and contingent future delivery.

Sufficient conditions for Walrasian allocations to exist have been established. Mainly these require that agents' preferences be convex and insatiable, and that each agent has an endowment sufficient to obtain a positive income. For 'most' economies the number of Walrasian allocations is finite; strong assumptions on substitution and income effects are required to ensure uniqueness.

Walrasian allocations and prices for specified models can be computed by solving a fixed-point problem and general methods have been devised. The task is complex (e.g. linear models with integer data can yield irrational prices) but an important simplifying feature is that the Walrasian prices depend only on the distribution of agents' attributes, and in particular only on the aggregate excess demand function. Essentially, any continuous function satisfying Walras's Law and homogeneity in prices is the excess demand for some economy.

The key requirement for a Walrasian allocation is that each agent's benefit be maximized within the budget imposed by the assigned prices; and that markets clear at those prices. Complete exploitation of all gains from trade may be precluded however by incomplete markets, pecuniary externalities (such as absence of necessary complementary goods), or insufficient contracts; or by strategic behaviour. If producers with monopoly power restrain output to elevate prices, or practise any of the myriad forms of price discrimination, then the resulting allocation is not Walrasian. Much discrimination segments markets via quality differentiation or bundling, but equally common is discriminatory pricing of the various conditions of delivery (e.g. spatial, temporal, service priority) or nonlinear pricing of quantities (e.g. two-part and block declining tariffs) if purchases can be monitored and resale markets are absent.

The Walrasian model of exchange is substantially defined by the absence of such practices affecting prices. It also relies on a fixed specification of agents, products, markets, and contracts. The theory of economies with large firms having power to influence prices and to choose product designs is significantly incomplete. The deficiencies derive partly from inadequate formulations, and partly from technical considerations: characterizations and even the existence of equilibria (in non-randomized strategies) depend on special structural features. For example, the simplest models positing simultaneous choices of qualities and prices by several firms lack equilibria; models with sequential choices encounter similar obstacles but to lesser degrees. In addition, if firms have fixed costs and must avoid losses then efficiency may require nonlinear pricing and other discriminatory practices if lump-sum assessments imposed on customers are precluded.

Market clearing is also essential to the Walrasian model and prices are determined entirely by the required equality of demand and supply. In contrast, successive markets with overlapping generations of traders need not clear 'at infinity'. Such markets can exhibit complicated dynamics even if the underlying data of the economy are stationary. Similarly, continually repeated markets where buyers and sellers arrive at the same rate that others depart from after completing transactions admit non-Walrasian prices or may have persistent excess on one side of the market if search or dispersed bargaining prevents immediate clearing.

When it is that among the feasible allocations the best prediction might be one of the Walrasian allocations, has been answered in several ways.

Competition is the first answer. On the supply side, for instance, with many sellers each one's incentive to defect from collusive pricing arrangements is increased. Absent collusion, if prices reflect supplies offered on the market and each seller chooses an optimal supply in response to anticipations of other sellers' supplies, then each seller's optimal percentage profit margin declines inversely with the number of sellers offering substitutes. Price discrimination, such as nonlinear pricing, is inhibited if there are many sellers, resale markets are available, or customers' purchases are difficult to monitor. Without capacity limitations, direct price competition among close or perfect substitutes erodes profits since undercutting is attractive. Although these conclusions are weakened to the extent that buyers incur search or switching costs, easy entry incurring no sunk costs remains important to ensure that markets are contestable and monopoly rents are eliminated. Monopoly rents are often substantially dissipated in entry deterrence, price wars, and other competitive battles to retain or capture monopoly positions. This is true both when entrants bring perfect substitutes and more generally, since entrants tend to fill in the spectra of quality attributes and conditions of delivery.

Arbitrage is important in commodity markets with standardized qualities, and especially in financial markets; to the extent that the contingent returns from one asset replicate those from a bundle of other assets, or from some trading strategy, its price is linked to the latter. Also, repeated opportunities to trade contingent on events enable a few securities to substitute for a much wider variety of absent contingent contracts.

One form of the competitive hypothesis emphasizes the option that each subset of the traders has to redistribute their endowments among themselves. For example, a seller and those who purchase from him are a coalition redistributing their resources among themselves. A core allocation is one such that no coalition can redistribute its endowments to the greater advantage of each member. The core allocations include the Walrasian allocations. A basic result first explored by F.Y. Edgeworth establishes that as the economy is enlarged by adding replicates of the original traders, the set of core allocations shrinks to the set of Walrasian allocations.

Another form emphasizes the view that in an economy so large that each agent's behaviour has an insignificant effect on the terms of trade, every trader's best option is to maximize the gains from trade at the prevailing prices. For example, any one trader's potential gain from behaviour that influences the terms of trade becomes insignificant (generically) as the set of traders expands, providing the limit distribution is 'atomless'. Similar results obtain for various models of markets with explicit price formation via auctions. Generally, an efficient allocation is necessarily Walrasian if each agent is unnecessary to attainment of others' gains from trade. An idealized formulation considers an atomless measure space of agents in which only measurable sets of agents matter and the behaviour of each single agent is inconsequential. In this case the Walrasian allocations are the only core allocations. Similarly, the allocation obtained from the Shapley value, in which each agent shares in proportion to his expected marginal contribution to a randomly formed coalition, is a Walrasian allocation.

Structural features of trading processes suggest an alternative hypothesis. Matching problems (e.g. workers seeking jobs) admit procedural rules that with optimal play yield core allocations, and for a general exchange economy an appropriately designed auction yields a core allocation. Other games have been devised for which optimal play by the agents produces a Walrasian allocation. Continual bilateral bargaining among dispersed agents (with sufficiently diverse preferences), in which agents are repeatedly matched randomly and one designated to offer some trade to the other, also results in a Walrasian allocation with optimal play. In a related vein, several methods of selecting allocations create incentives for agents to falsify reports of their preferences, and if they do this optimally then a Walrasian allocation results. Quite generally, any process that is fair in the sense that all agents enjoy the same opportunities for net trades yields essentially a Walrasian allocation. In one axiomization, some 'signal' is announced publicly and then based on his preferences each agent responds with a message that affects the resulting trades: if a core allocation is required, and each signal could be the right signal for some larger economy, then the signal must be essentially equivalent to announcement of a Walrasian price to which each agent responds with his preferred trade within his budget specified by the price.

Traders' impatience can also affect the terms of trade. In the simplest form of impatience, agents discount delayed gains from trade. Dynamic play is assumed to be sequentially rational in the sense that a strategy must specify an optimal continuation from each contingency; this is a strong requirement and severely restricts the admissible equilibria. For example, if a seller and a buyer alternate proposing prices

for trading an item, then in the *unique* equilibrium trade occurs immediately at a price dependent on their discount rates. As the interval between offers shrinks the seller's share of the gains from trade becomes proportional to the relative magnitude of the buyer's discount rate; for example, equal rates yield equal division. Extensions to multilateral contexts produce analogous results. A monopolist with an unlimited supply selling to a continuum of buyers might plausibly extract favourable terms, but actually in any equilibrium in which the buyers' strategies are stationary, as the interval between offers shrinks the seller's profit disappears and all trade occurs immediately at a Walrasian price. Similarly, a durable-good manufacturer lacking control of resale or rental markets has an incentive to increase the output rate as the production period shrinks; or, to pre-commit to limited capacity. This emphasizes that monopoly power depends substantially on powers of commitment stemming from increasing marginal costs, capacity limitations, or other sources.

Impatience and sequential rationality can, however, produce inefficiencies in product design, as in the case of a manufacturer's choice of durability, or in market structure, as in a preference to rent rather than sell durable goods.

Complete information is a major factor justifying predictions of Walrasian prices. With complete information and symmetric trading opportunities among agents, many models predict a Walrasian outcome, but incomplete information often produces departures from the Walrasian norm.

Although information may be productively useful, in an exchange economy the arrival of information may be disadvantageous to the extent that risk-averse agents forgo insurance against its consequences. A basic result considers an exchange economy that has reached an efficient allocation before some agents receive further private information, and this fact is common knowledge: the predicted response is no further trade, though prices may change.

Each efficient allocation has 'efficiency prices' that reflect the marginal rates of substitution prevailing; in the Walrasian case all trades are made at these prices. They summarize a wealth of information about technology, endowments and preferences. Prices (and other endogenous observables) are therefore not only sufficient instruments for decentralization but also carriers of information. If information is dispersed among agents then Walrasian prices are signals, possibly noisy, that can inform an agent's trading. Models of 'temporary equilibrium' envision a succession of markets, in each of which prices convey information about future trading opportunities. 'Rational expectations' models assume that each agent maximizes an expected utility conditioned on both his private information and the informational content of prices. In simple cases prices are sufficient statistics that swamp an agent's private information, whereas in complex real economies the informational content of prices may be elusive; nevertheless, markets are affected by inferences from prices (e.g. indices of stock and wholesale prices) and various models attempt to include these features realistically. Conversely, responses of prices to events and disclosures by firms are studied empirically.

The privacy of each agent's information about his preferences and endowment affects the realized gains and terms of trade. Many procedures require that the relative prices of 'qualities' provide incentives for self-selection. An example is a product line comprised of imperfect substitutes, in which price increments for successive quality increments induce customers to select according to their preferences. Several forms of discrimination in which prices depend on the quality (e.g. the time, location, priority or other circumstances of delivery) or, if resale is prevented, the quantity purchased, operate similarly.

Absence of the relevant contingent contracts is implicitly a prime source of inefficiencies and distributional effects. Trading may fail if adverse selection precludes effective signalling about product quality: without quality assurances or warranties, each price at which some quality can be supplied attracts sellers offering lesser qualities. Investments in signals, possibly unproductive ones, that are more costly for sellers supplying inferior qualities induce signal-dependent schedules in which the price paid depends on the signal offered. For example, to signal his ability a worker may over-invest in education or work in a job for which he would be underqualified on efficiency grounds. If buyers make repeat purchases based on the quality experienced from trying a product, then the initial price itself, or even dissipative expenditures such as uninformative advertising, can be signals used by the seller to induce initial purchases.

Principal–agent relationships in which a risk-averse agent has superior information and his actions cannot be monitored completely by the principal require complex contracts. For example, in a repeated context with perfect capital markets and imperfect insurance, the optimal contract provides the agent with a different reward for each measurable output, and the total remuneration is the accumulated sum of these rewards. Contracting is generally affected severely by limited observation of contingencies (either events or actions relevant to incentives) and in asymmetric relationships nonlinear pricing is often optimal. Insurance premia may vary with coverage, for example, to counter the effects of adverse selection or moral hazard.

Labour markets are replete with complex incentives and forms of contracting, partly because workers cannot contract to sell labour forward and partly because labour contracts substitute for imperfect loan markets and missing insurance markets (e.g. against the risk of declining productivity). Workers may have superior information about their abilities, technical data, or effort and actions taken; and firms may have superior information about conditions affecting the marginal product of labour. Incentives for immediate productivity may be affected by conditioning estimates of ability on current output, or by procedures selecting workers for promotion to jobs where the impact of ability is multiplied by greater responsibilities. The complexity of the resulting incentives and contracts reflects the multiple effects of incomplete markets and imperfect monitoring.

In the context of trading rules that specify price determination explicitly, analyses of agents' strategic behaviour emphasize the role of private information. The trading rule and typically the probability distribution of agents' privately known attributes are assumed to be common knowledge; consequently, formulations pose games of incomplete information. An example is a sealed-bid auction in which the seller awards an item to the bidder submitting the highest price: suppose that each bidder observes a sample, independently and identically distributed (i.i.d.) conditional on the unknown value of the item. With equilibrium bidding strategies, as the number of bidders increases the maximal bid converges in probability to the expectation of the value conditional on knowing the maximum of all the samples; for the common distributions this implies convergence to the underlying value. Alternative auction rules are preferred by the seller according to the extent that the procedures dilute the informational advantages of bidders (e.g. progressive oral bidding has this effect), and exploit any risk aversion. Rules can be constructed that maximize the seller's expected revenue:

if bidders' valuations are i.i.d., then for the common distributions, awarding the item to the highest bidder at the first or second-highest price is optimal, subject to an optimal reservation price set by the seller. In such a second-price or oral progressive auction with no reservation price, bidders offer their valuations, so the price is Walrasian.

Another example is a double auction, used in the London gold and Japanese stock markets, in which multiple buyers and sellers submit bid and ask prices and then a clearing price is selected from the interval obtained by intersecting the resulting demand and supply schedules. For a restricted class of models, requiring sufficiently many buyers and sellers with i.i.d. valuations, a double auction is incentive efficient, in the sense that there is no other trading rule that is sure to be preferred by every agent; also, as the numbers increase the clearing price converges to a Walrasian price.

The effects of privileged information held by some traders have been studied in the context of markets mediated by brokers and specialists, as in most stock markets. The results show that specialists' strategies impose all expected losses from adverse selection on uninformed traders. On the other hand, specialists may profit from knowledge of the order book and immediate access to trading opportunities.

Private information severely affects bargaining. With alternating offers even the simplest examples have many equilibria, plausible criteria can select different equilibria, and a variety of allocations are possible. In most equilibria, delay in making a serious offer (one that has some chance of acceptance) is a signal that a seller's valuation is not low or a buyer's is not high; or the offers made limit the inferences the other party can make about one's valuation. When both valuations are privately known, signalling must occur in some form to establish that gains from trade exist. Typically all gains from trade are realized eventually, but significant delay costs are incurred.

In a special case, a seller with a commonly known valuation repeatedly offers prices to a buyer with a privately known valuation: assume that the buyer's strategy is a stationary one that accepts the first offer less than a reservation price depending on his valuation. As mentioned previously for the monopoly context, as the period between offers shrinks the seller's offers decline to a price no more than the least possible buyer's valuation and trade occurs quickly: the buyer captures most of the gains. Even with alternating offers, the buyer avoids serious offers if his valuation is high and the periods short. Thus, impatience, frequent offers, and asymmetric information combine to skew the terms of trade in favour of the informed party.

The premier instance of exchange is the commodity trading 'pit' in which traders around a ring call out bid and ask prices or accept others' offers. These markets operate essentially as multilateral versions of bargaining but with endogenous matching of buyers and sellers: delay in making or accepting a serious offer can again be a signal about a trader's valuation, but with the added feature that 'competitive pressure' is a source of impatience. That is, a trader who delays incurs a risk that a favourable opportunity is usurped by a competing trader. These markets have been studied experimentally with striking results: typically most gains from trade are realized, at prices eventually approximating a Walrasian clearing price, especially if the subjects bring experience from prior replications. However, if 'rational expectations' features are added, subjects may fail to make the required inferences from information revealed by offers and transactions.

Trading rules can be designed to maximize the expected

realized gains from trade, using the 'revelation principle'. Each trading rule and associated equilibrium strategies induce a 'direct revelation game' whose trading rule is a composition of the original trading rule and its strategies; in equilibrium each agent has an incentive to report accurately his privately known valuation. In the case that a buyer and a seller have valuations drawn independently according to a uniform distribution, the optimal revelation rule is equivalent to a double auction in which trade occurs if the buyer's bid exceeds the seller's offer, and the price used is halfway between these. More generally, with many buyers and sellers and an optimal rule, the expected unrealized gains from trade declines quickly as the numbers of buyers and sellers increase. Such static models depend, however, on the presumption that subsequent trading opportunities are excluded.

Enforceable contracts facilitate exchange, and most theories depend on them, but they are not entirely essential. Important in practice are 'implicit contracts' that are not enforceable except via threats of discontinuing the relationship after the first betrayal. Similarly, in an infinitely repeated situation, if a seller chooses a product's quality (say, high or low) and price before sale, and a buyer observes the quality only after purchasing, then the buyer's strategy of being willing to pay currently only the price associated with the previously supplied quality suffices to induce continual high quality.

Studies of exchange without enforceable contracts focus on the Prisoners' Dilemma game: both parties can gain from exchange but each has an incentive to defect from his half of the agreement. In any finite repetition of this game with complete information the equilibrium strategies predict no agreements, since each expects the other to defect. Infinite repetitions can sustain agreements enforced by threats of refusal to cooperate later. With incomplete information, reputational effects can sustain agreements until near the end. For example, if one party thinks the other might automatically reciprocate cooperation, then he has an incentive to cooperate until first betrayed, and the other has an incentive to reciprocate until defection becomes attractive near the end. Reputations are important also in competitive battles among firms with private cost information: wars of attrition select the efficient survivors.

Continuing studies of exchange are likely to rely on game-theoretic methods. This approach is useful to study strategic behaviour in dynamic contexts; to elaborate the roles of private information, impatience, risk aversion, and other features of agents' preferences and endowments; to describe the consequences of incomplete markets and contracting limited by monitoring and enforcement costs; and to establish the efficiency properties of the common trading rules. It also integrates theories of exchange with theories of product differentiation, discriminatory pricing, and other strategic behaviour by producers. Technically, the game-theoretic approach enables a transition from theories of a large economy with a specified distribution of agents' attributes, to theories of an economy with few agents having private information but commonly known probability assessments; further realism may depend on reducing the assumed common knowledge and developing better formulations of competition among large firms. Grand theories of general economic equilibrium incorporating all these realistic aspects are unlikely until the foundations are established.

In sum, the Walrasian model remains a paradigm for efficient exchange under 'perfect' competition in which equality of demand and supply is the primary determinant of the terms of trade. Further analysis of agents' strategic behaviour with private information and market power elaborates the causes of incomplete or imperfectly competitive markets that impede efficiency, and it delineates the fine details of endogenous product differentiation, contracting, and price formation essential to the application of the Walrasian model.

ROBERT B. WILSON

See also CORES; GAME THEORY; GENERAL EQUILIBRIUM; INCOMPLETE CONTRACTS; INCOMPLETE MARKETS; PERFECTLY AND IMPERFECTLY COMPETITIVE MARKETS.

BIBLIOGRAPHY

Arrow, K.J. and Debreu, G. 1954. Existence of an equilibrium for a competitive economy. *Econometrica* 22, 265–90.

Arrow, K.J. and Hahn, F.H. 1971. *General Competitive Analysis*. San Francisco: Holden-Day.

Aumann, R.J. 1964. Markets with a continuum of traders. *Econometrica* 32, 39–50.

Debreu, G. 1959. *Theory of Value*. New York: John Wiley & Sons.

Debreu, G. 1970. Economies with a finite set of equilibria. *Econometrica* 38, 387–92.

Debreu, G. and Scarf, H. 1963. A limit theorem on the core of an economy. *International Economic Review* 4, 235–46.

Gresik, T. and Satterthwaite, M.A. 1984. The rate at which a simple market becomes efficient as the number of traders increases: an asymptotic result for optimal trading mechanisms. Discussion Paper 641, Northwestern University; *Journal of Economic Theory* (1987).

Grossman, S.J. and Perry, M. 1986. Sequential bargaining under asymmetric information. *Journal of Economic Theory* 39, 120–54.

Gul, F., Sonnenschein, H. and Wilson, R.B. 1986. Foundations of dynamic monopoly and the Coase conjecture. *Journal of Economic Theory* 39, 155–90.

Hildenbrand, W. 1974. *Core and Equilibria of a Large Economy*. Princeton: Princeton University Press.

Hölmstrom, B.R. and Milgrom, P.R. 1986. Aggregation and linearity in the provision of intertemporal incentives. Report Series D, No. 5, School of Organization and Management, Yale University.

Hölmstrom, B.R. and Myerson, R.B. 1983. Efficient and durable decision rules with incomplete information. *Econometrica* 51, 1799–820.

Kreps, D.M., Milgrom, P.R. Roberts, D.J. and Wilson, R.B. 1982. Rational cooperation in the finitely repeated prisoners' dilemma. *Journal of Economic Theory* 27, 245–52.

McKenzie, L. 1959. On the existence of general equilibrium for a competitive market. *Econometrica* 27, 54–71.

Milgrom, P.R. 1979. A convergence theorem for competitive bidding with differential information. *Econometrica* 47, 679–88.

Milgrom, P.R. 1985. The economics of competitive bidding: a selective survey. In *Social Goals and Social Organization*, ed. L. Hurwicz, D. Schmeidler and H. Sonnenschein, Cambridge: Cambridge University Press.

Milgrom, P.R. and Stokey, N. 1982. Information, trade, and common knowledge. *Journal of Economic Theory* 26, 17–27.

Myerson, R.B. and Satterthwaite, M.A. 1983. Efficient mechanisms for bilateral trading. *Journal of Economic Theory* 29, 265–81.

Radner, R. 1972. Existence of equilibrium of plans, prices and price expectations in a sequence of markets. *Econometrica* 40, 289–303.

Roberts, D.J. and Postlewaite, A. 1976. The incentives for price-taking behavior in large exchange economies. *Econometrica* 44, 115–28.

Roberts, D.J. and Sonnenschein, H. 1977. On the foundations of the theory of monopolistic competition. *Econometrica* 45, 101–13.

Rubinstein, A. 1982. Perfect equilibrium in a bargaining model. *Econometrica* 50, 97–109.

Scarf, H.(With T. Hansen.) 1973. *The Computation of Economic Equilibria*. New Haven: Yale University Press.

Schmeidler, D. 1980. Walrasian analysis via strategic outcome functions. *Econometrica* 48, 1585–93.

Schmeidler, D. and Vind, K. 1972. Fair net trades. *Econometrica* 40, 637–42.

Smith, V. 1982. Microeconomic systems as experimental science. *American Economic Review* 72, 923–55.

Sonnenschein, H. 1972. Market excess demand functions. *Econometrica* 40, 549–63.

Sonnenschein, H. 1974. An axiomatic characterization of the price mechanism. *Econometrica* 42, 425–34.

Spence, A.M. 1973. *Market Signalling: Information Transfer in Hiring and Related Process.* Cambridge, Mass.: Harvard University Press.

Wilson, R.B. 1985. Incentive efficiency of double auctions. *Econometrica* 53, 1101–16.

exchangeable value. *See* ABSOLUTE AND EXCHANGEABLE VALUE.

exchange control. In a rather narrow sense, we refer to exchange control when monetary institutions (governments, central banks or specialized institutions) impose strictly defined limitations on international transactions or on the exchange of national currency into foreign currency. So exchange control occupies the middle ground between unrestricted convertibility into foreign exchange and the total ban on convertibility which is practised in a number of developing countries and in the socialist countries.

In dealing with balance of payments, these restrictions serve different purposes. The most frequent objectives consist in balancing trade (the export and import of goods) or the current account.

In all instances, exchange control measures aim at preserving the autonomy of domestic policies threatened by trade deficits, foreign debts, or a switch in control of the national productive capital. At all times, the primary purpose of exchange controls, as well as that of the non-convertibility of certain currencies, has consisted in preserving the national autonomy of a country from outside interference. The Sparta of Lycurgus, which in turn inspired Plato's project on non-convertible fiduciary currency, provides us with the first famous example (see Einzig, 1962). Here the object was to reduce the possibility of corruption by foreign agents. In times of crisis or war, such measures represent a means to control trade or to guarantee the supply of primary strategic materials. But such practices are relatively recent. Einzig (1934) rightly points out that only after 1917 did the World War I belligerents attempt to control exchange. A return to unlimited convertibility occurred only during the second half of the 1920s, after the return to the gold standard in 1925. But the experience gained during World War I facilitated a quick return to control measures during the crises of the 1930s and especially during World War II.

During the decades following the end of World War II, the purpose of controls consisted in limiting any imbalance which in some countries went hand in hand with the development of trade. Basically, controls are introduced when monetary exchange systems fail to fulfil their role as regulators in the international market.

Due to the diversity in exchange control measures, it is difficult to measure their impact. The role of most controls is to keep the current balance of payments in check. In this sense, controls affect the financing of imports (prior demand, necessary deposits, specific rates of exchange), terms of payment (fixed payment delays for export or imports) or limits to travel spending. The exercise of some form of control is the norm rather than the exception.

In the first half of the 1980s – a period characterized by the liberalization of capital movements – only the United States,

Switzerland, Britain (since 1979) and the Federal Republic of Germany (since 1984) allowed the free circulation of capital. Japan, which applied vigorous exchange controls until the end of the 1970s, turned to a gradual liberalization spurred by the US–Japanese negotiations in 1984. While improving convertibility, France, Italy and Belgium have kept certain restrictions relating to the circulation of capital.

The rationale underlying such restrictive practices can first of all be found in the limitations imposed by the two traditional regimes of exchange at fixed and flexible rates. We cannot expect such systems to lead to unique equilibrium rates of exchange (a fantasy already denounced by Joan Robinson in 1937). The purpose of such systems is to prevent the development of strong creditor or debtor positions. From time to time these limitations force the countries in question to inhibit the application of free convertibility regimes.

But this is not the sole reason for restricting exchange. Certain control measures aim to shield the development of domestic industries from foreign competition. Here it is useful to distinguish measures bearing on the exchange of commodities and those having effect on the movement of capital, in order to perceive the different stages in a policy geared towards protecting national economies.

In order to understand the advantages or disadvantages of such measures, an attempt will be made to define the limits to exchange regimes and the diversity in control measures.

1. EXCHANGE CONTROLS AS A REACTION TO LIMITATIONS OF THE TWO CLASSICAL EXCHANGE REGIMES. For some countries, standard exchange regimes lead to long-lasting imbalances (excessively high debts, inflation triggered by depreciation of the national currency) which can inhibit the autonomy of domestic policies, and so may lead to control measures. The character of these disturbances varies according to the current exchange regime, as demonstrated by the experience of the principal market economies since 1945; at the beginning of the 1970s, these market economies passed from a system of fixed exchange to a system of floating exchange.

In a system of fixed exchange, the defence of parities leads to vast movements of currency reserves by the central banks, inciting speculative movements by private capital, in turn staking on the sudden realignment of parities. The destabilizing effect produced by credit balance or imbalance parity adjustments or drastic economic measures

In a system of flexible exchange, where financial markets are largely integrated, domestic economic policies are bounded in terms of prices and rates of interest. This leads to instability in the adjustment of the balance of payments. On the one hand, the increase in the cost of imports in the case of depreciation of the national currency stimulates inflationary pressures and inhibits the re-establishment of foreign debt. On the other hand, the integration of financial markets forces national real interest rates to align with international levels. Speculation then is a measure of the ability of an economic policy to accommodate any pressures on prices and on rates of interest.

The fixed rate regime followed by the flexible rate experience demonstrated the instability generated by free convertibility regimes for medium sized countries, faced with either the risk associated with the issue of a standard currency, i.e. the dollar, or with the erratic speculative movements resulting from its decline.

The gold convertibility of the American currency (at a fixed price of 35 dollars an ounce) constituted one of the pillars of the system of fixed exchange inaugurated at Bretton Woods in 1944. The United States then held eighty per cent of the world gold stock. From the 1960s onwards, military expenses abroad and investment abroad, as well as the balance of trade deficit, led to a sharp decrease in gold reserves and to accumulation by European and Japanese central banks. In early 1971, gold reserves corresponded to only a third of foreign holdings in dollars and the flight of capital increased. In March 1973, an intensification in speculative movements – encouraged by the announcement by the US Secretary of State of the future total liberalization of capital movements – forced the central banks to endorse the general free floating of currencies under the auspices of the International Monetary Fund.

For current critics of the system of fixed parities (Johnson, 1969; Mundell, 1968), only a system of flexible exchange could allow for a degree of autonomy and stability in domestic policies. Yet after more than a decade, the system of floating exchange seems conducive to remarkable instability and interdependence between national policies.

With the change in the hegemonic role of the dollar (see Parboni, 1981), the monetary uncertainty which characterized the beginning of the 1970s soon led to inflationary pressures. In terms of domestic policy, it was manifest in the large price increase of raw materials together with the explosion of wage conflicts towards the end of the 1960s. In terms of economic policy, inflation represented the main preoccupation of the decade. In general terms, the new symmetry between internationally used currencies made 'financial markets very sensitive to waves of anticipations, polarised alternatively by political and financial events' (Aglietta, 1984).

These arbitrary movements reinforced constraints upon national monetary policies. The development of a whole literature using game theory (under the influence of Hamada, 1974) highlights the development of this interdependence.

When confronted with such uncertain situations, a country whose currency has no international role tends to preserve the autonomy of its domestic policy either by participating in the erection of monetary blocs, or by limiting the convertibility of its currency.

The creation in 1977 of the European Monetary System in order to reduce pressures on the monetary policies of the European countries induced by variations in American interest rates, represents a reaction of the first type. In fact, during the 1960s Mundell and McKinnon advocated the creation of a zone of fixed parities in a system of flexible exchange.

The permanency of exchange control measures (when the growth in trade by nature tends towards the dismantling of such measures) together with recommendations by economists such as Tobin (1978) in favour of some control on short-term movements of capital, represents a reaction of the second type.

In exchange regimes which permit the development of important imbalances or which cannot cope with sudden crises of confidence, restriction on free convertibility appears as one of the only means at the disposal of an isolated country which aims at preserving a degree of autonomy in the elaboration of its economic policy. No exchange regime can dispose of the plurality of currencies and the predominance of one of these currencies, such as the dollar. This is a continuous source of conflict (Brunhoff, 1986). The attempt by monetarists such as Friedman to present a flexible exchange system as a means to merge the diversity of national currencies into one neutral international currency turned out to be irrelevant.

It is against this background, and in the face of imbalances in terms of employment or external payments, that the relative maintenance of control measures has to be seen, especially since in the postwar years organizations such as the International Monetary Fund or OECD have concurrently been aiming at the liberalization of trade and the convertibility of currencies.

This retention of restrictive practices is all the more remarkable in view of the fact that since the postwar period trade has been more multilateral and diversified than at any other time. In the more segmented world trade system illustrated by the three trade blocs – dollar, pound, gold – in the late 1930s, or in a trade system involving economies at very different stages of development, the limitations to free convertibility appear even more obvious. In fact, there is a strong relationship between circulation of commodities and means of payment. This link plays a decisive role in the evolution in exchange practices.

2. THE DOUBLE NATURE OF EXCHANGE CONTROLS. Commercial and financial aspects of international economic relations are largely dissociated in a tradition which ex-post can be qualified as monetarist-inspired. Bergsten and Williamson (1983) attribute this bias mainly to the implicit (and erroneous) hypothesis of an automatic and regulating adjustment of parities. This dichotomous approach to international economic relations is also apparent in the distribution of roles between international institutions in the immediate postwar period. While the role of GATT consists in reducing barriers to trade, the role of the IMF consists of reducing limitations on the free convertibility of currencies (according to Article 1 of its 1944 Statute).

Yet it is clear that these two aspects – commercial and monetary – of international relations are linked. The orientation and the level of commercial exchange is influenced by financial conditions and conditions of payment. Reciprocally, the development of commercial exchange stimulates the extension of banking networks and financial innovation. In the case of so-called invisible trade, the two areas tend to merge.

This is evident in the case of exchanges of investment revenues realized externally. But it is also applicable to specialized services calling for the establishment of subsidiaries abroad (in the case of financial and insurance activities, consultancies, chambers of commerce etc.). The freedom of capital movements and the right to settle are thus pre-conditions for more advanced specialization by developed countries in the exchange of services. But these specialized services also play a strategic role in commercial exchanges: they are mainly produced by multinational firms and mainly used by other multinational firms which occupy a dominant position in world trade (see Clarimonte and Cavanagh, 1984).

The liberalization of invisible trade is therefore not automatic since it implies ample liberalization of capital movements. Since the Tokyo Round the United States has led a campaign within GATT for the liberalization of invisible trade. OECD, which already in the 1950s defined a liberalization code for invisible trade, has noted its ineffectiveness, particularly in the case of the right to do business in a foreign country. Opposition to such policies is manifest in developing countries, where control over the exchange of services is considered strategic (in the early 1980s, liberalization in the area of transport is still a very touchy question).

To this basic complementarity between exchanges of goods and capital movements must be added an interdependence between the different forms of control on monetary exchange and on commercial transactions. Monetary authorities and private agents use the former in order to complement or to thwart the latter. To this end, juggling terms of payment goes hand in hand with fictional or falsified commercial transactions in order to avoid measures that control exchange. Quotas imposed on commercial transactions are sometimes aimed principally at reducing monetary transfers (as illustrated by French experience in the 1930s according to Einzig, 1934).

So it is possible like Krueger (1978), Bhagwati (1978) and McKinnon (1979) to consider together all the different forms of controls on commercial and monetary relations between economies in order to define the different forms of foreign trade 'regimes'. On the basis of national studies of practices which inhibit the freedom of trade (and exchange) in eleven developing countries, Krueger (1978) and Bhagwati (1978) have identified five (not necessarily consecutive) stages. The first stage is characterized by the establishment of generalized and undifferentiated control over imports; this kind of control often follows an unbearable balance of payments deficit. During the second stage there is a large differentiation in the measures of control according to the way in which imports are utilized. A third stage implements a reduction of direct controls (or quotas) together with a sharp devaluation of the national currency. Following this stage there is either a return to the situation described in the second stage or an attempt to liberalize trade by replacing some quotas by tariffs. Finally, the fifth stage is characterized by free convertibility of the national currency for current transactions, which now are only subject to customs duties. So it is only during this last stage that the commercial and monetary aspects of foreign relations are dissociated.

To this must be added the possibility of controlling capital movements (and therefore of controlling exchange) that influence the trade dynamics mentioned above in respect to developed countries.

Krueger and Bhagwati measure the degree of liberalization of trade according to the disappearance of all types of quotas – despite an increase in customs duties – but nevertheless stress the damaging effects of restrictive practices. Such a position raises two paradoxes.

First, according to a classical theorem in international economics, quotas and customs dues have equivalent effects. Second, why are practices that are hardly 'optimal' so widespread? Krueger and Bhagwati clearly stress that conditions required by the theorem of equivalence so to speak never coincide. The distribution of import licences in the case of quotas, with endogenous effects on supply and demand, modifies resource allocation resulting from custom taxation. The authors also stress the great variety of distributive criteria which lead to distortions. Finally we must consider the question of the rationale for control measures which, according to McKinnon (1979), is hardly considered in the theoretical literature or in case studies. McKinnon brings up the question of political power created by controlling foreign trade in developing countries. Such discretionary power in the distribution of licenses favours 'clientisme', especially if the informal character of domestic activities limits the possibility for internal control. But the importance of such controls is principally derived from the possibility of selecting productive activities. An import licence ensures the viability of an enterprise, in turn giving it access to other limited resources such as capital. The absence or weakness of a financial market, capable of directing sufficient funds toward activities having priority, is according to McKinnon (1979) an essential factor in the origin of trade control practices. The volatility of domestic capital is a sign of this weakness in the financial market.

This explanation for protectionist policies appears to be applicable to the analysis of exchange controls practised in developed countries. The weakness or narrowness of financial markets seems to be one of the major causes for restrictions on the free circulation of currencies. This is indeed suggested by the history of exchange controls.

3. PAST AND PRESENT REASONS FOR EXCHANGE CONTROL POLICIES. It is seldom asked what imperatives lead to the application of such obviously inconvenient exchange control practices, which simultaneously contain vast opportunities for fraud, potentially ad hoc measures and weak global coherence. Indeed, it is too often a question of faith on the part of opponents of exchange control, who blame a pernicious propensity to bureaucratize to explain the choice of easy direct controls in the place of rigorous but unpopular policies. Their position is not without foundation. But it remains secondary in the face of the risks involved in massive movements of capital. The financial pressures mentioned by McKinnon (1979) do not constitute a marginal phenomenon. No financial market is safe from the flight of capital, feeding itself rapidly to the extent of changing the policies pursued or putting an end to the free circulation of capital. It has been stressed that standard exchange systems do not have the stabilizing effects necessary to prevent speculative moves. On the contrary, there is clear evidence that orthodox exchange systems allow for the maintenance of 'over-evaluation' or 'under-evaluation' of parities. It is within these limits to exchange regimes that major necessary conditions for restrictive practices must be found.

The importance of financial markets and the voluntarist and innovative character of current economic policies constitute sufficient conditions for the elaboration of these control policies. Here margins are fixed by the amplitude and stability of the domestic financial market; the need for autonomy is defined by the type of policies pursued.

The history of exchange controls emphasizes this double aspect: the impact of current economic policies, and the importance of the financial market with its international links.

One of the first experiences of exchange control in modern times took place in 17th-century England. The Royal Exchange was then introduced, together with the Navigation Acts, in order to secure a basis for growing British power in the face of the decline of Spain and the Netherlands, at a time when the City did not yet carry the weight required for such a rise to power. In Germany in the 1930s, rigorous exchange controls were due to the autarkic tendencies pursued by the national-socialist government. Yet in France, the introduction of such controls after 1936 when the tripartite alliance (with the US and Britain) had failed, calls into question the ability of the financial market to withstand speculative moves such as those following the 1930s crisis.

In the immediate postwar period, generalized controls in Europe revealed the fragility of financial markets in a period when reconstruction absorbed most resources. The progressive and partial liberalization of capital movements (see Einzig, 1962) had to rely on massive and conditional Marshall Plan aid and on the regulatory action of institutions such as the IMF, OECD and GATT.

This attempt to insulate a fragile financial market from competition by foreign capital (without the evolution of the rates of exchange correcting this distortion) can explain the relative continuity of restrictions on the movement of capital in France (see Claassen and Wyplosz, 1982).

In the growing integration of financial markets there might be seen a stabilizing factor which enables the opening up of relevant economies to the free circulation of capital. The development of new information techniques in the area of communication has largely contributed to the acceleration of this integration of financial markets (initiated in the 1930s by the opening of the first transatlantic communication line): a world market in currency transactions was established in the 1970s; a securities market was in turn established during the

1980s. But the extension of these information networks has also considerably increased the amplitude and scope for short-term speculative movements, thereby increasing global instability in the financial international system. The resulting prospect of international crisis renders unlikely a definite liberalization of monetary movements. If crises can break out more rapidly than in the past, the possibility of introducing rapid and efficient exchange control can play an important role in deterring speculation and the development of a major exchange crisis.

PASCAL PETIT

See also CAPITAL FLIGHT; EXTERNAL DEBT; FIXED EXCHANGE RATES; FUNDAMENTAL DISEQUILIBRIUM; INTERNATIONAL CAPITAL FLOWS; INTERNATIONAL FINANCE.

BIBLIOGRAPHY
Aglietta M. 1984. Les régimes monétaires de crise. *Critiques de l'Economie Politique* 26–7, January-June.
Bergsten, F. and Williamson, J. 1983. Exchange rates and trade policy. In Cline (1983).
Bhagwati, J. 1978. *Anatomy and Consequences of Exchange Control Regimes*. Vol XI of *Foreign Trade Regimes and Economic Development*, Cambridge, Mass.: Ballinger for the National Bureau of Economic Research
Claassen, E.M. and Wyplosz, C. 1982. Capital controls: some principles and the French experience. *Annales de l'Insée* 47–8.
Clarimonte, P. and Cavanagh, J.H. 1984. Transnational corporations and services: the final frontier. In UNCTAD (1984).
Cline, W.R. (ed.) 1983. *Trade Policy in the 1980s*. Cambridge, Mass.: MIT Press.
De Brunhoff, S. 1986. *L'heure du marché*. Paris: Presses Universitaires de France.
Einzig, P. 1934. *Exchange Control*. London: Macmillan.
Einzig, P. 1962. *The History of Foreign Exchange*. London: Macmillan. Reprinted 1979.
Einzig, P. 1968. *Leads and Lags*. London: Macmillan.
Johnson, H.G. 1969. The case for flexible exchange rates. *Federal Reserve Bank of St Louis Review* 51.
Kindleberger, C.P. 1984. *A Financial History of Western Europe*. London: George Allen & Unwin.
Krueger, A.O. 1978. *Liberalization Attempts and Consequences*. Vol X of *Foreign Trade Regimes and Economic Development*, Cambridge, Mass.: Ballinger for the Nationae Bureau of Economic Research.
McKinnon, R. 1963. Optimum currency areas. *American Economic Review* 53, September, 717–25.
McKinnon, R. 1979. Foreign trade regimes and economic development: a review article. *Journal of International Economics* 9(3), August, 429–52.
Mundell, R.A. 1961. A theory of optimum currency areas. *American Economic Review* 51, September, 657–65.
Mundell, R.A. 1968. *International Economics*. New York: Macmillan.
Parboni, R. 1981. *The Dollar and its Rivals*. London: Verso.
Robinson, J. 1937. The foreign exchanges. In J. Robinson, *Essays in the Theory of Employment*, London: Macmillan.
Tobin, J. 1978. A proposal for international monetary reform. Cowles Foundation Discussion Paper No. 506, Yale University.
UNCTAD. 1984. *Trade and Development*. An Unctad Review, Geneva.

exchange rate policy. *See* INTERNATIONAL FINANCE; INTERNATIONAL MONETARY POLICY.

exchange rates. The large movements in the price of the US dollar in terms of the German mark, the Japanese yen, the British pound, and various other currencies since the breakdown of the Bretton Woods system in the early 1970s again raises the question of how exchange rates are determined. This question tends to be dormant when the major countries peg their currencies – and then reappears when their currencies float. Approaches to explaining the movement of the exchange rate must recognize that the range of variation in the price of the US dollar in terms of the currencies of various other countries has been substantially larger than the contemporary change in the differential between the increase in the US price level and the increase in the price levels in these other countries. Moreover, deviations between market exchange rates and real (or price-level adjusted) exchange rates have been substantially larger with the floating exchange rate system than with the pegged exchange rate system of the 1950s and the 1960s. A second observation is that at times countries with strong and appreciating currencies have had large trade deficits (the United States in the early 1980s) and the countries with weak and depreciating currencies have had large trade surpluses (the United States in the late 1970s) – even though large trade deficits frequently are associated with weak or depreciating currencies, and large trade surpluses with strong or appreciating currencies. A third observation is that the range of variation in the price of the US dollar in terms of various foreign currencies does not appear to have declines during the first decade of experience with the floating exchange rate system. A fourth observation is that the differences in interest rates on comparable assets denominated in the US dollar and various foreign currencies have proven to be poor predictors of the rate of change of the price of the US dollar in terms of each of these currencies. Similarly, forward exchange rates have not proven to be effective predictors of future spot exchange rates at the maturity of the forward contracts.

The experience with floating exchange rates has proved very different from the experience with the Bretton Wood system of pegged exchange rates in the 1950s and the 1960s. Then countries devalued their currencies when their trade deficits were excessively large; changes in currency parities generally were consistent with the changes in international competitiveness. The countries which devalued their currencies, like France in 1959 and again in 1969 and Great Britain in 1967, had experienced higher rates of inflation than their major trading partners. And the countries which revalued their currencies, like Germany in 1961 and again in 1969, generally experienced lower rates of inflation than their major trading partners.

The next section discusses five approaches toward explaining the level of the exchange rate and changes in exchange rates. Then a disequilibrium toward analysing exchange rate movements is contrasted with an equilibrium approach. Then the relation between the determinants of exchange rates under a floating exchange rate system is compared with the determinants of the exchange rate under a pegged exchange rate system.

APPROACHES TOWARD MODELLING THE DETERMINATION OF EXCHANGE RATES. Most of the approaches toward explaining changes in the exchange rate were developed to explain phenomena at particular times. Five approaches are distinguished: purchasing power parity, elasticities, absorption, portfolio balance, and the asset market approach. The purchasing power parity approach is identified with Cassel, who sought to develop a way for European governments to determine the equilibrium values for their currencies if they again were to peg them to gold after World War I. During the war, inflation rates varied extensively among countries. Either the countries with the more rapid inflation would have to accept large declines in their price levels, or they would be obliged to peg their currencies to gold at new parities that would reflect that their inflation rates had been higher than those of their major trading partners. Cassel's insight was that changes in exchange rates should conform to differences in national inflation rates – in effect an extension of the arbitrage proposition known as the Law of One Price from individual goods to national market baskets of goods (traded) and services (non-traded). The terms *undervaluation* and *overvaluation* are the layman's expression that the value for the exchange rate seems inconsistent with the relationship between the domestic price level and the price levels in the major trading partners. Subsequent analysis has been directed at whether the equilibrium exchange rate should be inferred from absolute price levels or whether instead the exchange rate should be based on changes in price levels from data when the exchange market was in equilibrium should provide the basis for determining the new equilibrium exchange rate.

The elasticities approach to the determination of the exchange rate developed the 1930s in response to the observation – or at least to the stylized fact – that countries which devalued their currencies were not successful in increasing their exports relative to their imports. The devaluation improved competitiveness in that the price of domestic goods fell relative to the price of foreign goods, but the improvement in competitiveness did not lead to the desired reduction in the trade deficit. The explanation was that the spending on imports might increase if domestic demand were price inelastic and export receipts might decline if foreign demand for domestic goods were price inelastic. This 'elasticity pessimism' view led to the conclusion that changes in exchange rates would prove ineffective in improving the trade balance, which was formalized in the Marshall–Lerner condition (that a devaluation would reduce the trade deficit if the sum of the elasticities is greater than one). An alternative interpretation for the observation that the devaluation of one country's currency would not reduce its trade deficit in the 1930s was that the subsequent devaluations of the currencies of its trading partners effectively neutralized its own devaluation.

The absorption approach to the determination of the exchange rate was developed in the period after World War II to highlight that changes in exchange rates would not lead to a permanent improvement in the trade balance unless the devaluating country adopted a sufficiently contractive monetary and fiscal policy. This approach followed the Keynesian tradition that the trade balance was the residual between domestic consumption and domestic production; a trade deficit occurred when domestic consumption exceeded domestic production, which meant that imports exceeded exports. Unless consumption declined relative to production as domestic currency was devalued, the trade deficit would persist. Thus a devaluation might be necessary to reduce a trade deficit; if excess demand remained after the devaluation, then domestic price level would rise, and the improvement in competitiveness effected by the devaluation would be negated by the subsequent increase in consumption and in imports.

Both the absorption approach and the elasticities approach provide explanations of why a devaluation would be ineffective in improving the trade balance in terms of the levels of demand. Thus the elasticities approach highlighted a deficiency of demand associated with the Great Depression, while the absorption approach reflects excess demand of the years immediately after World War II. In both cases, however, the equilibrium exchange rate was determined by the need to have national price levels aligned so that there would be equilibrium in the goods market; the change to price-level competitiveness was necessary for a permanent improvement in the trade balance, but not a sufficient condition.

The Bretton Woods decades were marked by infrequent changes in exchange parities of the industrial countries. Inflation rates in most countries were low. For most of this period, the United States incurred payments deficits; the problem was to explain the persistence of the US payments deficit despite a variety of measures adopted to reduce the imbalance. The payments surpluses of other industrial countries were explained by their demand for international reserves, or by their demand for money. The portfolio balance approach emphasized that payments balances and hence the exchange rate, reflected trade in securities as well as trade in goods. Trade in securities would involve a stock adjustment in the volume of foreign securities owned by domestic residents. The Monetary Theory of the Balance of Payments emphasized that payments surpluses and deficits reflected imbalances between the demand for money in each country and the supply of reserves that results from the monetization of domestic assets; if the demand for money increased more rapidly than the supply based on the expansion of the domestic assets owned by the central bank, then the country would realize a trade and payments surplus, since the supply of goods will exceed the demand. The inflow of gold and foreign exchange would lead to increases in the assets of the central bank, and thus lead to an increase in the money supply. Both the portfolio balance approach and the monetary theory placed payments surpluses and deficits in a general equilibrium framework. The monetary approach was generally mute on whether the payments surplus reflected the trade account or the capital account. In contrast, the portfolio balance approach could explain the US payments in terms of the desire of residents of other countries to borrow long and lend short in their transactions with the United States.

The observations about the wide variation in the price of national currencies since the early 1970s are similar to the observations about the movements in exchange rates in the early 1920s. In both cases, the movements of exchange rates were much larger than the movement that would be inferred from the changes in the relationship among national price levels. In both periods, countries with relatively high inflation rates experienced a significantly more rapid reduction in the foreign exchange value of their currencies than would be inferred from the relative price level movements alone.

The dominant explanation for the large variations in market exchange rates derives from Irving Fisher's observation in *The Theory of Interest* that the interest rate differential between bonds payable in gold and bonds payable in rupees or silver reflected the anticipated rate of change in the price of gold in terms of silver. Thus the current spot exchange rate at any moment is the anticipated spot exchange rate for various future dates discounted to the present by the differential between interest rates on domestic securities and interest rates on similar securities denominated in the foreign currency. If the current spot exchange rate differed substantially from the anticipated spot exchange rate adjusted for the interest rate differential, investors would have a virtually riskless profit opportunity. The implication is that the spot exchange rate changes whenever the anticipated spot exchange rate changes, or whenever the differential between interest rates on comparable securities denominated in the domestic currency and foreign currency changes.

The Asset Market Approach to the Exchange Rate places the Fisherian observation in a general equilibrium framework. The exchange rate is the price of two national monies. Changes in the exchange rate reflect changes in the demand for securities denominated in each currency relative to the supply. The changes in the exchange rate that occur to obtain equilibrium in the asset market may induce disequilibrium in the goods market; the goods produced by the countries subject to capital outflows will become undervalued. In effect, the capital outflow can occur only if the country can generate a current account surplus. The asset market approach slights the role of trade in goods in determining the value of the exchange rate today, on the presumption that the daily volume of transactions in assets across national borders is so much larger than the volume of trade in goods. However, the anticipated exchange rate may reflect the value that would lead to goods market equilibrium at the anticipated price levels during future years.

The exchange rate necessary to achieve asset market equilibrium leads to disequilibrium in the goods market. The anticipated spot exchange rate reflects the value that will clear the goods markets at some future date; this anticipated value may be extrapolated from current or recent movements in the domestic and foreign inflation rate. In this way, changes in the inflation rate can have a major impact on the current spot exchange rate as the revised anticipated values are discounted to the present.

A DISEQUILIBRIUM APPROACH TO EXCHANGE RATE DETERMINA-
TION. Sudden large movements of exchange rate during particular episodes have been explained in terms of extrapolation by investors of the future exchange rate from the direction of movement in the exchange rate in the recent past. For a while, at least, some participants in the exchange market rely on a 'follow-the-leader' approach; changes in exchange rates thus reflect a bandwagon effect. Hence there may be 'speculative bubbles' in the exchange rate. Momentum models of exchange rate forecasting are based on this view. In such cases, the exchange rates may move away from an equilibrium value, and eventually these sharp movements will be reversed.

The approach toward explaining changes in exchange rates raises the question whether the foreign exchange market is efficient, or whether instead period-to-period movements in the exchange rate are serially correlated. The serially correlated movements in exchange rates could explain why the exchange rate tends to overshoot the ultimate equilibrium. In a few brief episodes, exchange rate movements appear serially correlated.

The momentum approach toward the determination of the exchange rate can be reconciled with the Asset Market Approach in that new information about inflation rates may lead to sharp revisions in anticipated exchange rates and, to a lesser extent, in interest rates. Frequently this new information will develop over a period of weeks and months; each revision will lead to a new value for the current spot exchange rate. If the trend-like movement is sufficiently strong, then momentum-based approaches to forecasting exchange rates may be triggered, and the exchange rate movement may become abrupt, and overshoot its ultimate equilibrium.

EXCHANGE RATE DETERMINATION – PEGGED RATE PERIODS AND
FLOATING EXCHANGE RATE PERIODS. The amplitude and suddenness of movements in exchange rates under the floating rate period highlights the conditions necessary for the successful operation of a pegged exchange rate regime – and for a floating exchange rate system in which movements in spot exchange rates are gradual, and conform with differences in inflation rates. A pegged exchange rate system can be maintained if the anticipated exchange rates are more or less identical with the current spot exchange rate – because the authorities are committed to maintaining their parities, and manage their policies accordingly. And for most periods, this commitment was credible. In contrast, during the floating rate

periods, the authorities generally have had no such commitment about a particular value for the future spot exchange rate.

The sharp movements in the price of the US dollar in terms of various foreign currencies during the floating exchange rates reflects that various types of shocks affect the anticipated spot exchange rates, domestic interest rate and foreign interest rates. Most of these shocks reflected changes in monetary policy in the United States and in other countries.

The number and magnitude of monetary disturbances has been substantially larger during periods associated with floating exchange rates. In part, this reflects that many of the moves to floating exchange rates occur in an inflationary environment: the early 1920s and the 1970s were both periods of divergent movement in national price levels. During the 1950s and the early 1960s, inflation rates were low and similar among the industrial countries; the anticipated price of foreign exchange more or less approximates the parity. Interest rates differed modestly among countries. Capital flows occur primarily in response to national differences in savings and investment rates. The exchange rate at the level necessary to maintain the approximate balance in trade accounts flow of capital were modest in size relative to flows of goods.

ROBERT Z. ALIBER

See also CRAWLING PEG; FIXED EXCHANGE RATES; FLEXIBLE EXCHANGE RATES; INTERNATIONAL FINANCE; INTERNATIONAL MONETARY INSTITUTIONS; MONETARY APPROACH TO THE BALANCE OF PAYMENTS.

BIBLIOGRAPHY

Alexander, S.S. 1952. Effects of a devaluation on a trade balance. *IMF Staff Papers* 2, April, 263–78.

Aliber, R.Z. 1973. The interest rate parity theorem: a reinterpretation. *Journal of Political Economy* 81(6), November–December, 1451–9.

Aliber, R.Z. 1980. Floating exchange rates: the twenties and the seventies. In *Flexible Exchange Rates and the Balance of Payments*, ed. J.S. Chipman and C.P. Kindleberger, Amsterdam, Oxford: North-Holland.

Balassa, B. 1964. The purchasing-power-parity doctrine: a reappraisal. *Journal of Political Economy* 72, December, 584–96.

Bernstein, E.M. 1956. Strategic factors in balance of payments adjustment. *IMF Staff Papers* 5, August, 151–69.

Bilson, J.F. 1978. Rational expectations and the exchange rate. In *The Economics of Exchange Rates*, ed. J.A. Frenkel and H.G. Johnson, Reading, Mass.: Addison-Wesley.

Branson, W.H. and Henderson, D.W. 1985. The specification and influence of asset markets. In *Handbook of International Economics*, ed. R.W. Jones and P.B. Kenen, Amsterdam, Oxford: North-Holland.

Cassel, G. 1922. *Money and Foreign Exchange After 1914*. New York: Macmillan.

Dornbusch, R. 1976. Expectations and exchange rate dynamics. *Journal of Political Economy* 84(6), December, 1161–76.

Fisher, I. 1930. *The Theory of Interest*. New York: The Macmillan Company.

Frenkel, J.A. 1976. A monetary approach to the exchange rate: doctrinal aspects and empirical evidence. *Scandinavian Journal of Economics* 78(2), May, 200–24.

Frenkel, J.A. 1981. Flexible exchange rates, prices and the role of 'news': lessons from the 1970s. *Journal of Political Economy* 89(4), August, 665–705.

Frenkel, J.A. and Mussa, M. 1985. Asset markets, exchange rates, and the balance of payments. In *Handbook of International Economics*, ed. R.W. Jones and P.B. Kenen, Amsterdam and Oxford: North-Holland.

Friedman, M. 1953. The case for flexible exchange rates. In M. Friedman, *Essays in Positive Economics*, Chicago: University of Chicago Press.

Haberler, G. 1949. The market for foreign exchange and stability of the balance of payments: a theoretical analysis. *Kyklos* 3, 193–218.

Isard, P. 1977. How far can we push the 'law of one price'? *American Economic Review* 67(5), December, 942–8.

Mundell, R.A. 1960. The monetary dynamics of international adjustment under fixed and flexible exchange rates. *Quarterly Journal of Economics* 74, May, 227–57.

Mussa, M. 1979. Empirical regularities in the behavior of exchange rates and theories of the foreign exchange market. In *Policies for Employment, Prices, and Exchange Rates*, ed. K. Brunner and A.H. Meltzer, Amsterdam and Oxford: North-Holland.

Mussa, M. 1982. A model of exchange rate dynamics. *Journal of Political Economy* 90(1), February, 74–104.

Officer, L.H. 1976. The purchasing-power-parity theory of exchange rates: a review article. *IMF Staff Papers* 23(1), March, 1–60.

excise duties. *See* CONSUMPTION TAXES; INDIRECT TAXES.

exhaustible resources. An exhaustible resource is a term that has come to be associated with an ore such as coal, oil or platinum which does not renew itself rapidly in its natural setting. Any resource can be exhausted and most resources are being renewed but for exhaustible resources, the speed of replenishment is very slow. Forests, fishing grounds, and arable land have and are being exhausted but are generally treated as natural resources other than exhaustible. Exhaustible and non-renewable have come to be used interchangeably. Since forest and fish stocks generally do renew themselves by natural as opposed to human actions, they have come to be called renewable resources. Land as a means of producing useful product has come to be treated separately from renewable and exhaustible resources.

It was the Malthusian notion of fully exploiting the finite acreage of arable land that led to the development of a fairly clear notion of the economic return on land to the landowner, the concept of land rent. Reflection by L.C. Gray (1914) on land rent in the context of resources such as coal led to the first precise presentation of the idea of rent originating in the inherent scarcity associated with a depletable stock such as a body of ore being mined. The central question is quite clear: if land of uniform quality commands a positive price because it is in finite supply, so should a stock of useful ore. Given its scarcity value, how is this return to an ore body related to the rate of which the ore body is mined or systematically depleted? Gray worked these matters out correctly with numerical examples. He dealt with homogeneous ore in fixed supply and assumed that mining costs per period increased more than proportionately with the amount mined. The maximization of the present value of profit from extracting the ore period by period resulted in an 'optimal' schedule of extractions and an optimized present value of the original stock, a rent ascribable to the body of ore. Gray solved by careful reasoning a constrained dynamic optimization problem without the use of higher mathematics, a problem belonging to a class at the heart of the analysis of exhaustible resources. Hotelling (1931) worked out a related problem, using the appropriate mathematical methods for the first time, in his classic 'The Economics of Exhaustible Resources' and the basic efficiency condition characterizing the rent maximizing rate of ore depletion for a homogeneous stock is often called 'Hotelling's Rule'. The rule's essential import is that the marginal ton extracted in any period should yield the same rent, in present value terms. Otherwise, adjustments could be made in the sequence of extractions initially which would yield a higher total rent from the stock in present value terms. In L.C. Gray's problem, the marginal ton extracted in any period should yield the same profit, in present value terms, if the extraction programme is optimal (profit-maximizing).

A parallel stream of thought dealt with the implications of exhaustibility of an essential resource in an economy. This is a type of Malthusianism. A growing social system must face some form of awkward adjustment when exhaustible resources approach exhaustion. Bounds on food supply in the face of an exponentially growing population preoccupied Malthus. Subsequent writers have emphasized that industrial growth must cease since coal and oil are in finite supply.

Jevons (1865), in a book-length application of Malthusianism to energy resources, rather than agricultural resources, considered Britain's industrial prosperity to be based on coal and foresaw a day at which the stocks beneath Britain's soil would become exhausted. He entertained the idea that technical progress might delay the date of exhaustion but was pessimistic about disaster being delayed in the long term. He entertained the sophisticated view that stocks were not homogeneous and the coal prices would rise as exhaustibility was approached but he failed to relate his views to the matter of the market mechanisms which determined the rate of depletion of stocks. He did refer to the coal stocks as a depleting 'capital good' and suggested using revenues obtained from taxing coal sales to pay off the national debt but he did not emphasize using such revenues for the purpose of investment rather than consumption. For a theorist of great stature it is curious that he devoted no space to asking the L.C. Gray question; namely, where is rent of the Ricardian kind in the analysis of a depleting stock? He remained above the microeconomic details of a depleting stock. 'While other countries mostly subsist upon the annual and ceaseless income of the harvest, we are drawing more and more upon a capital which yields no annual interest, but once turned to light and heat and motive power, is gone forever into space' (p. 412).

In Gray (1913) there is explicit recognition that a concept of socially optimal exhaustible resource utilization plan might be defined and that market determined utilization programmes exist. Gray raises the issue that the two utilization programmes might not coincide because social and private discount rates may differ. This is a sound but partial explanation of such divergences.

About one hundred years after Jevons, Forrester (1971) and Meadows et.al. (1972) restated the general Malthusian position in a simultaneous equation dynamic system, 'solved' on computers. Simulations of possible histories on the computers provide the 'solutions' for the models. Economic collapse could occur because an exhaustible resource constraint became binding, because population grew to the maximum food supply of the earth, or because pollution became very severe and costly to control. Relative prices played no role in the models. More recent computer-based models have been developed in an attempt to predict how various economies in the world will react to constraints on growth.

That the original force of population growth on living standards could be blunted if not eliminated by technical progress appeared early in the debate over the importance of Malthus's ideas. J.S. Mill adopted a somewhat optimistic view in his *Principles of Political Economy* – Malthusian decline ultimately but not immediately. But it was only after sustained industrial growth set in towards the end of the 19th century that Malthusianism became no longer a principal topic of debate in economics. The other way to mitigate the impact of ultimate exhaustion of essential stocks of minerals is to use some of the returns to owners of the stocks to invest in capital goods which are themselves durable and provide capacity for future production. The capital need not be a perfect substitute for the services of the depleting stock (perfect substitutes being called 'backstop technologies') but could be useful new investment generally. Jevons (1865) hinted at this approach in Chapter XVII, 'Of Taxes and the National Debt', when he proposed using revenues from taxes on coal to pay off Britain's national debt.

It clearly makes no economic difference whether we save resources for tomorrow by not using them today, or instead use them today in a process that ... yields capital goods which yield their services tomorrow. The question of the ... conservation of resources in general is inseparable from the larger question of the optimum rate of capital investment in the economy (Gordon, 1958, pp. 114–15).

Solow (1974) developed formally the idea of new investment being a substitute for a currently partly depleted stock such as coal and Hartwick indicated that paths of constant consumption with stationary populations could be sustained if current investment equalled the current value of current depletion of the finite homogeneous stocks. This savings–investment rule which yields a plausible concept of equity or fairness across generations has been shown to be a quite general property of market models of capital accumulation (e.g. Dixit, Hammond and Hoel, 1980).

Many economists have come to view exhaustibility of a given stock, say petroleum reserves, as a phase in a longer programme in which unlimited energy supplies will be available from say fusion power at the end of a sequence of consecutive depletions. Fusion power is the substitute source of energy (the backstop technology of supply) for wood, coal, oil, and uranium. Nordhaus (1973) made this view of exhaustibility popular in an empirical study of world energy production and consumption. Recycling, an imperfect means of renewing a stock, of durable minerals is a possibility for forestalling the disappearance of such useful metals as copper.

GRAY'S PROBLEM OF THE EXHAUSTIBLE RESOURCE EXTRACTING FIRM. The owner of the firm faces a constant price p of mined ore, a homogeneous known stock of size S, and an extraction cost $c(q)$ with $c'(q) > 0$ and $c''(q) > 0$, r is a constant interest rate and the present value of profits $\int_0^T [pq(t) - c(q(t))]e^{-rt} dt$ is to be maximized by choice of extraction path $\{q(t)\}$ subject to $\int_0^T q(t) dt \leq S$. The solution to this problem has $(p - c'(q))e^{-rt} = \text{constant}$ and

$$q(T)[p - c'(q(T))] = pq(T) - c(q(T)),$$

for S scarce (not too large an initial stock). $[p - c'(q)]$ is defined as *rent* per ton and it obviously increases at the rate of interest r (the so-called $r\%$ or Hotelling Rule). For an increase in S, $[p - c'(q(0))]$ or the constant above declines yielding the intuition that the level of rent reflects the overall scarcity or 'degree of finiteness' of the original stock.

If the average cost of mining is U-shaped then n symmetric firms forming an industry will, in following the above path, display a jump in market price and an industry market equilibrium will not exist (Eswaran, Lewis, Heaps, 1983). If the market price above is made a random variable, the level of expected profits will exceed that level observed if the mean price is used by the firm (Hartwick and Yeung, 1985). Firms prefer price uncertainty. Similarly for interest rate uncertainty.

HOTELLING'S PROBLEM OF AN INDUSTRY EXTRACTING AN EXHAUSTIBLE STOCK. Hotelling (1931) glossed over the details of each extraction firm's decision-making situation and worked at an industry-wide level. Implicit are many small price-taking firms with extraction cost c per ton constant, possibly zero. At each instant of time the firms face a negatively sloped demand schedule $p(q)$ where Q is the sum of current extractions of all firms from a homogeneous finite industry stock S. A social planning solution for arranging an optimal industry extraction programme $(Q(t))^*$ is the maximization of the present value of social surplus

$$\int_0^T [B(Q(t)) - cQ(t)]e^{-\rho t} dt \text{ subject to } \int_0^T Q(t) dt \leq S,$$

where $B(Q(t)) = \int_0^{Q(t)} p(Q) dQ$ is the area under the demand schedule, a proxy for the social value of $Q(t)$ is an instant of time. Note $dB/dQ = p(t)$. For the case of $p(0) < \infty$ (a finite price at which demand is choked off), the solution is $(p(t) - c)e^{-\rho t} = \text{constant}$ and $(p(T) - c)Q(T) = B(Q(T)) - cQ(T)$, where the constant is a decreasing function of the stock size parameter S, $p(t) - c$ is rent per ton extracted at time t and it clearly rises along an optimal extraction programme at a rate ρ where ρ is the social discount rate. For $\rho = r$, the market rate of interest, we observe that along the optimal extraction programme the percentage increase in rent at each instant equals r.

The Hotelling Rule is a member of the class of efficiency conditions for capital accumulation. It indicates that the capital gain dp/dt on the marginal ton at t should be the same as the interest obtainable, $(p(t) - c)r$, if that marginal ton were extracted at t and the net proceeds invested in the best alternative, earning $r\%$ per annum. This is a condition for zero profit from arbitraging the marginal ton extracted at time t over consecutive periods. The planner is indifferent between extracting the marginal ton at time t and leaving it in the ground. Clearly the zero profit arbitrage condition is a market condition. At each instant no firm can make profit by re-arranging its extraction plan. Hence the planning solution mimics a market realization for profit seeking firms. For a firm to be on its best extraction path, it must be able to see how prices will behave into the indefinite future. Though this sounds demanding, one can respond that an owner of stock who is a poor at calculating the time path of future prices will, at a profit, be brought out by a more clear-eyed firm. Thus though perfect foresight is a very demanding criterion it is one which is enforced by the discipline of the market. The possibility of destabilizing speculation has been reflected on by observers such as Solow (1974a) but a definitive development of the idea has not been worked out. Yeung and Hartwick [1986] characterize equilibrium with extracting firms facing interest rate uncertainty in an industry framework. Hotelling explicitly developed the idea of a bias in extraction rates arising from a competitive industry becoming a private monopoly. The formal statement of extraction plans being pro-conservation or anti-conservation was established with the benchmark case being the social planner's solution. He also considered how taxation might affect extraction programmes.

The notion of a backstop technology, a continuous source of supply ultimately substituting for supply from a depleting stock, shifted attention from extraction paths *per se* to the relationship between extraction paths and the economic characteristics of the backstop. For example, how does uncertainty about the cost of fusion power (the backstop technology) affect the speed of exploitation of known oil reserves? (see Dasgupta and Heal, 1974). How does uncertainty about the timing of the arrival of fusion power affect the speed of exploitation of known oil reserves? (see Dasgupta and Stiglitz, 1981). Strategic interactions between owners or developers of the backstop and separate different owners of the stock and how extraction paths are affected have been much analysed (see for example Gallini, Lewis and Ware, 1983).

The analysis of stocks of declining quality has been approached by indexing layers or seams by aggregate ore lifted by a point in time. Lower quality is simply defined as a higher cost of extraction per ton lifted. Two sources of rent emerge; from quality variation at a point in time there is static Ricardian rent, while between dates an identical ton will exhibit a rent arising from the interest rate wedge, a dynamic or Hotelling rent. The arbitrage principle governing the pace of extraction remains essentially the same: the rent on the marginal ton extracted at date t must equal $1/(1+r)$ times the rent earned by this ton (now the 'most intramarginal' ton) at date $t+1$ in a discrete time formulation. Quality variation can of course be continuous or discrete as in Herfindahl's (1967) original statement. With quality *improving* as extraction proceeds, the situation becomes one of declining cost of extraction per ton, a variant of increasing returns to scale. In such situations market solutions fail to mimic optimal planning solutions, as Hartwick, Kemp and Long (1986) have made clear for the polar case of set-up costs having to be met before any extraction begins from a deposit. The redevelopment of the Hotelling model to incorporate oligopoly, the case of a dominant extractor and a competitive fringe group of extractors was done by Salant (1976) and Gilbert (1978). Newbery (1980) explained that paths which were optimal for each party at the initial date of extraction could be departed from later in the programme, an instance of a so-called dynamic inconsistency. Richer assumptions on information and strategy (Stackelberg consistency) lead to these difficulties disappearing. Many oligopoly models were constructed in order to attempt to analyse the effects of the partial cartelization of the world oil industry which became effective in 1973.

Since the stock of a mineral is below ground, its size to the extractor is a random variable. Exploration of various types can reduce uncertainty about stock size but actual depletion is the only way to reveal its true magnitude. The incorporation of stock uncertainty and possibly explicit exploration costs makes the Hotelling Rule for arranging the pace of extraction approximate at best (for example, Arrow and Chang, 1978) Unanticipated discoveries of new ore bodies will result in declines in current prices. Technical improvements in extraction techniques and exploration methods also put downward pressure on market prices for minerals. Slade (1982) has analysed long run price changes for many exhaustible resources.

THE SUBSTITUTION OF NEW CAPITAL GOODS FOR DECLINING STOCKS OF EXHAUSTIBLE RESOURCES IN ECONOMY-WIDE MODELS. In one commodity representation of economy-wide models, the single output X is produced from the flows of services from buildings and machines of stock size K, from labour services L, and flows of resources R from a finite stock S. A so-called neoclassical production function relates inputs to output: $X = f(K, L, R)$. At each instant X is divided between the demands of current consumption, C and investment, dK/dt. For a world with a constant population and labour force, the intertemporal allocation decision is how to divide X between C and dk/dt and how large to make R at each instant. One approach (e.g. Dasgupta and Heal [1974]) is to assume that a social planner optimizes

$$W = \int_0^x U(C) e^{-\rho t} \, dt$$

where $U(\cdot)$ is a concave utility indicator for consumption at a point in time and ρ is a social discount rate. Optimalization yields an 'optimal savings rule' for output and an optimal depletion rule, $[\partial f/\partial R][\partial f/\partial K] = d[\partial f/\partial R]/dt$ for stock S. This depletion rule is of course Hotelling's Rule since $\partial f/\partial R$ can be identified as the current price of the exhaustible resource and $\partial f/\partial K$ as the current rate of return on capital in the economy.

If instead of optimizing W, one imposes the savings-investment rule (i.e. invest in new capital goods the current value of exhaustible resources used up, thus setting $(df/dR) \cdot R = dK/dt$) and requires that the stock of resources be consumed according to the above Hotelling efficiency condition, one finds, given L constant, that C remains constant over time (Dixit, Hammond and Hoel, 1980). No generation experiences a higher per capita income than another and C can remain positive over infinite time if $f(\cdot)$ is assumed to be Cobb–Douglas with the coefficient on R small relative to that on K (Solow, 1974). We have then a formalization of the argument that the current generation should compensate the next generation for currently depleting finite resource stocks, by 'replacing' the amount depleted with new producible capital goods such as roads, buildings, and machines.

JOHN M. HARTWICK

See also DEPLETION; ENERGY ECONOMICS; NATURAL RESOURCES; RENEWABLE RESOURCES.

BIBLIOGRAPHY

Arrow, K.J. and Chang, S. 1978. Optimal pricing, use and exploration of uncertain resource stocks. Technical Report No. 31, Dept. of Economics, Harvard University.

Dasgupta, P. and Heal, G.M. 1974. Optimal depletion of exhaustible resources. *Review of Economic Studies*, Symposium, 3–28.

Dasgupta, P. and Stiglitz, J.E. 1981. Resource depletion under technological uncertainty. *Econometrica* 49, January, 85–104.

Dixit, A., Hammond, P. and Hoel, M. 1980. On Hartwick's rule for constant utility and regular maximum paths of capital accumulation and resource depletion. *Review of Economic Studies* 47(3), April, 347–54.

Eswaran, M., Lewis, T.R. and Heaps, T. 1983. On the non-existence of market equilibria in exhaustible resources markets with decreasing costs. *Journal of Political Economy* 91, 145–67.

Forrester, J.W. 1971. *World Dynamics*. Cambridge: Wright Allen Press.

Gallini, N., Lewis, T.R. and Ware, R. 1983. Strategic timing and pricing of a substitute in a cartelized resource market. *Canadian Journal of Economics* 16, August, 429–46.

Gilbert, R.J. 1978. Dominant firm pricing policy in a market for an exhaustible resource. *Bell Journal of Economics*, Autumn, 385–95.

Gordon, H. 1958. Economics and the conservation question. *Journal of Law and Economics* 1, October, 110–21.

Gray, L.C. 1913. The economic possibilities of conservation. *Quarterly Journal of Economics* 27, 497–519.

Gray, L.C. 1914. Rent under the assumption of exhaustibility. *Quarterly Journal of Economics* 28, 466–89.

Hartwick, J.M., Kemp, M.C. and van Long, N. 1986. Set-up costs and the theory of exhaustible resources. *Journal of Environmental Economics and Management* 13(3), 212–24.

Hartwick, J.M. and Yeung, D. 1985. Preference for output price uncertainty by the nonrenewable resource extracting firm. *Economics Letters* 19(1), 85–90.

Herfindahl, O.C. 1967. Depletion and economic theory. In *Extractive Resources and Taxation*, ed. M. Gaffney, Madison: University of Wisconsin Press.

Hotelling, H. 1931. The economics of exhaustible resources. *Journal of Political Economy* 39, 137–75.

Jevons, W.S. 1865. *The Coal Question: An Inquiry Concerning the Progress of the Nation, and the Probable Exhaustion of our Coal Mines.* New York: Augustus M. Kelley, 1965.

Meadows, D.H., Meadows, D.L., Randers, J. and Behrens, W.W. III. 1972. *The Limits to Growth.* New York: Universe Books.

Newbery, D. 1980. Oil prices, cartels, and the problem of dynamic inconsistency. *Economic Journal* 91, September, 617–46.

Nordhaus, W. 1973. The allocation of energy resources. *Brookings Papers on Economic Activity* No. 3, 529–70.

Salant, S. 1976. Exhaustible resources and industrial structure: a Nash–Cournot approach to the world oil market. *Journal of Political Economy* 84(5), 1079–93.

Slade, M.E. 1982. Trends in natural–resource commodity prices: an analysis of the time domain. *Journal of Environmental Economics and Management* 9, 122–37.

Solow, R.M. 1974a. Intergenerational equity and exhaustible resources. *Review of Economic Studies*, Symposium, 29–45.

Solow, R.M. 1974b. The economics of resources of the resources of economists. *American Economic Review, Papers and Proceedings* 64(2), May, 1–14.

Yeung, D. and Hartwick, J.M. 1986. Interest rate and output price uncertainty and industry equilibrium for nonrenewable resource extracting firms. Queen's University Economics Discussion Paper (mimeo).

existence of general equilibrium. Léon Walras provided in his *Eléments d'économie politique pure* (1874–7) an answer to an outstanding scientific question raised by several of his predecessors. Notably, Adam Smith had asked in *An Inquiry into the Nature and Causes of the Wealth of Nations* (1776) why a large number of agents motivated by self-interest and making independent decisions do not create social chaos in a private ownership economy. Smith himself had gained a deep insight into the impersonal coordination of those decisions by markets for commodities. Only a mathematical model, however, could take into full account the interdependence of the variables involved. In constructing such a model Walras founded the theory of general economic equilibrium.

Walras and his successors were aware that his theory would be vacuous in the absence of an argument supporting the existence of its central concept. But for more than half a century that argument went no further than counting equations and unknowns and finding them to be equal in number. Yet for a non-linear system this equality does not prove that there is a solution. Nor would it provide a proof even for a linear system, especially when some of the unknowns are not allowed to take arbitrary real values.

A successful attack on the problem of existence of a general equilibrium was made possible by an exceptional conjunction of circumstances in Vienna in the early 1930s. It started from the formulation of the Walrasian model in terms of demand functions which had been given by Gustav Cassel in 1918. As Hans Neisser (1932) noted, certain values of commodity quantities and prices appearing in the solutions of Cassel's system of equations might be negative in such a way as to render those solutions meaningless. Heinrich von Stackelberg (1933) also made a cogent remark. Let x_i be the quantity of the ith final good demanded by consumers, a_{ij} the fixed technical coefficient specifying the input of the jth primary resource required for a unit output of the ith final good, and r_j the available quantity of the jth primary resource. The equality of demand and supply for every resource is expressed by

$$\sum_i a_{ij}x_i = r_j \quad \text{for all } j.$$

Von Stackelberg observed that if there are fewer final goods than primary resources, the preceding linear system of equations in $(x_1, \ldots, x_m)$ has, in general, no solution. Karl Schlesinger (1933–4) then remarked that equalities should be replaced by inequalities

$$\sum_i a_{ij}x_i \leqq r_j \quad \text{for all } j,$$

with the condition that a resource for which the strict inequality holds has a zero price. This suggestion, which had already been hinted at by Frederik Zeuthen (1932) in a different context, was essential to the proper formulation of the existence problem.

The problem thus posed received its first solution from Abraham Wald (1933–4), whose work on the existence of a general equilibrium gave rise to three published articles. The first two appeared in *Ergebnisse eines mathematischen Kolloquiums* in 1933–4 and in 1934–5. The third appeared in *Zeitschrift für Nationalökonomie* (1936) and was translated into English in *Econometrica* (1951). In that body of work Wald separately studied a model of production and a model of exchange and proved the existence of an equilibrium for each one.

By the standards prevailing in economic theory at that time, his mathematical arguments were of great complexity, and the major contribution that he had made did not attract the attention of the economics profession. A two-decade pause followed, and when research on the existence problem started again after 1950 it was under the dominant influence of work done, also in the early 1930s, by John von Neumann. His article on the theory of growth, published in *Ergebnisse eines mathematischen Kolloquiums* (1935–6) and translated into English in the *Review of Economic Studies* (1945), contained in particular a lemma of critical importance. That lemma was reformulated in the following far more convenient form, and was also given a significantly simpler proof, by Shizuo Kakutani (1941). Let K be a non-empty, compact, convex set of finite dimension. Associate with every point x in K a non-empty, convex subset $\phi(x)$ of K, and assume that the graph $G = \{(x, y) \in K \times K \mid y \in \phi(x)\}$ of the transformation ϕ is closed. Then ϕ has a fixed point x^*, i.e., a point x^* that belongs to its image $\phi(x^*)$.

Kakutani's theorem was applied by John Nash, in a one-page note of 1950, to establish the existence of an equilibrium for a finite game. It can be used as well (Debreu, 1952) to prove the existence of an equilibrium for a more general system composed of n agents. The ith agent chooses an action a_i in a set A_i of *a priori* possible actions. A state of the social system is therefore described by the list $a = (a_1, \ldots, a_n)$ of the actions chosen by the n agents. The preferences of the ith agent are represented by a real-valued utility function u_i defined for every a in the set of states $A = \times_{i=1}^{n} A_i$. Moreover the ith agent is restricted in the choice of his action in A_i by the actions chosen by the other agents. Formally let N denote the set $\{1, \ldots, n\}$ of all the agents and $N \backslash i$ denote the set of the agents other than the ith. Let also $a_{N \backslash i}$ denote the list of the actions $(a_1, \ldots, a_{i-1}, a_{i+1}, \ldots, a_n)$ chosen by the agents in $N \backslash i$. The ith agent is constrained to choose his own action in a subset $\phi_i(a_{N \backslash i})$ of A_i depending on $a_{N \backslash i}$. In these conditions the ith agent, considering $a_{N \backslash i}$ as given, chooses his action in $\mu_i(a_{N \backslash i})$, the set of the elements of $\phi_i(a_{N \backslash i})$ at which the maximum of the utility function $u_i(\cdot, a_{N \backslash i})$ in $\phi_i(a_{N \backslash i})$ is attained. Consider now the transformation $a \mapsto \mu(a) = \times_{i=1}^{n} \mu_i(a_{N \backslash i})$ associating with any element a of A, the subset $\mu(a)$ of A. A state a^* is an equilibrium if and only if for every $i \in N$, the action a_i^* of the ith agent is best according to his preferences given the actions $a_{N \backslash i}^*$ of the others, that is, if and only if for every $i \in N$, $a_i^* \in \mu_i(a_{N \backslash i}^*)$, that is, if and only if

$a^* \in \mu(a^*)$. Thus the concept of an equilibrium for the social system is equivalent to the concept of a fixed point for the transformation $a \mapsto \mu(a)$ of elements of A into subsets of A. Ensuring that the assumptions of Kakutani's theorem are satisfied for the transformation μ yields a proof of existence of an equilibrium for the social system.

In the revival of interest in the problem of existence of a general economic equilibrium after 1950, the first solutions were published in 1954 by Kenneth Arrow and Gerard Debreu, and by Lionel McKenzie. The article by McKenzie emphasized international trade aspects, and the article by Arrow and Debreu dealt with an integrated model of production and consumption. Both rested their proofs on Kakutani's theorem. They were followed over the next three decades by a large number of publications (a bibliography is given in Debreu, 1982) which confirmed the concept of a Kakutani fixed point as the most powerful mathematical tool for proofs of existence of a general equilibrium.

A simple prototype of the various economies that were the subject of those numerous existence results is (following Arrow–Debreu) composed of m consumers and n producers, producing, exchanging and consuming l commodities. The consumption of the ith consumer ($i = 1, \ldots, m$) is a vector x_i in R^l whose positive (or negative) components are his inputs (or outputs) of the l commodities. Similarly the production of the jth producer ($j = 1, \ldots, n$) is a vector y_j in R^l whose negative (or positive) components are his inputs (or outputs) of the l commodities. The ith consumer has three characteristics. (1) His consumption set X_i, a non-empty subset of R^l, is the set of his possible consumptions. (2) A binary relation $\precsim_i$ on X_i defines his preferences, and '$x_i \precsim_i x_i'$' is read as 'x_i' is at least as desired as x_i by the ith consumer'. Formally the preference relation of the ith consumer is the set $\{(x, x') \in X_i \times X_i | x \precsim_i x'\}$. (3) A vector e_i in R^l describes his initial endowment of commodities. The jth producer has one characteristic, his production set Y_j, a non-empty subset of R^l defining his possible productions. Finally the number $\theta_{ij} \geqq 0$ specifies the fraction of the profit of the jth producer distributed to the ith consumer. These numbers satisfy the equality $\Sigma_{i=1}^m \theta_{ij} = 1$ for every j. In summary, the economy $\mathscr{E}$ is characterized by the list of mathematical objects

$$\left[(X_i, \precsim_i, e_i)_{i=1,\ldots,m}, (Y_j)_{j=1,\ldots,n}, (\theta_{ij})_{\substack{i=1,\ldots,m \\ j=1,\ldots,n}} \right].$$

Given a price-vector p in R^l different from 0, the jth producer ($j = 1, \ldots, n$) chooses a production y_j in Y_j that maximizes his profit, that is, such that the value $p \cdot y_j$ of y_j relative to p satisfies the inequality $p \cdot y_j \geqq p \cdot y$ for every y in Y_j. Thus the ith consumer receives in addition to the value $p \cdot e_i$ of his endowment, $\Sigma_{j=1}^n \theta_{ij} p \cdot y_j$ as the sum of his shares of the profits of the n producers. The value $p \cdot x$ of his consumption x is therefore constrained by the budget inequality $p \cdot x \leqq p \cdot e_i + \Sigma_{j=1}^n \theta_{ij} p \cdot y_j$. Under that constraint he chooses a consumption x_i in X_i that is best according to his preferences. The list $[p, (x_i)_{i=1,\ldots,m}, (y_j)_{j=1,\ldots,n}]$ of a non-zero price-vector, m consumptions and n productions forms a general equilibrium of the economy $\mathscr{E}$ if for every commodity, the excess of demand over supply vanishes,

$$\sum_{i=1}^m x_i - \sum_{j=1}^n y_j - \sum_{i=1}^m e_i = 0.$$

The existence of a general equilibrium can be proved (following Arrow–Debreu) by casting the economy $\mathscr{E}$ in the form of a social system of the type defined above. For this it suffices to introduce, in addition to the m consumers and to the n producers, a fictitious price-setting agent whose set of actions and whose utility function are now specified. Note first that the definition of a general equilibrium is invariant under multiplication of the price-vector p by a strictly positive real number. In the simple case where all prices are non-negative, one can therefore restrict p to be an element of the simplex $P = \{p \in R_+^l | \Sigma_{h=1}^l p^h = 1\}$, the set of the vectors in R^l whose components are non-negative and add up to one. The set of actions of the price-setter is specified to be P. Given the consumptions $(x_i)_{i=1,\ldots,m}$ chosen by the m consumers, and the productions $(y_j)_{j=1,\ldots,n}$ chosen by the n producers, there results an excess demand

$$z = \sum_{i=1}^m x_i - \sum_{j=1}^n y_j - \sum_{i=1}^m e_i.$$

The utility function of the price-setter is specified to be $p \cdot z$. Maximizing the function $p \mapsto p \cdot z$ over P carries to one extreme the idea that the price-setter should choose high prices for the commodities that are in excess demand, and low prices for the commodities that are in excess supply.

Some of the assumptions on which the theorems of Arrow–Debreu (1954) are based are weak technical conditions: closedness of the consumption-sets, of the production-sets and of the preference relations, existence of a lower bound in every coordinate for each consumption-set, possibility of a null production for each producer. Other assumptions were later shown to be superfluous for economies with a finite set of agents: irreversibility of production (if both y and $-y$ are possible aggregate productions, then $y = 0$), free disposal (any aggregate production $y \leqq 0$ is possible), and completeness and transitivity of preferences. Convexity of preferences can be dispensed with, and convexity of consumption-sets can be weakened, in economies with a large number of small consumers. Insatiability of consumers is an acceptable behavioural postulate. There remain, however, two overly strong assumptions. They are the hypothesis that for every i, the endowment e_i yields a possible consumption for the ith consumer (after disposal of a suitable commodity-vector if need be), and the assumption of convexity on the total production-set $Y = \Sigma_{j=1}^n Y_j$ which implies non-increasing returns to scale in the aggregate.

An alternative approach to the problem of existence of a general equilibrium, closer to traditional economic theory, is centred on the concept of excess demand function, or of excess demand correspondence. Given an economy $\mathscr{E}$ defined as before, consider a price-vector p in R_+^l different from 0. The productions $(y_1, \ldots, y_n)$ chosen by the producers, and the consumptions $(x_1, \ldots, x_m)$ chosen by the consumers in reaction to the price-vector p result in an excess demand z in the commodity-space R^l. If z is uniquely determined, the excess demand function f from $R_+^l \backslash 0$ to R^l is thereby defined. If z is not uniquely determined, the set of excess demands in R^l associated with p is denoted by $\phi(p)$, and the excess demand correspondence ϕ is thereby defined on $R_+^l \backslash 0$. Both f and ϕ are homogeneous of degree zero since $f(p)$ and $\phi(p)$ are invariant under multiplication of p by a strictly positive real number. This permits various normalizations of p. For instance, p may be restricted to the simplex P. Moreover, for every $i = 1, \ldots, m$, one has $p \cdot x_i \leqq p \cdot e_i + \Sigma_{j=1}^n \theta_{ij} p \cdot y_j$. By summation over i, one obtains

$$p \cdot \sum_{i=1}^m x_i \leqq p \cdot \sum_{i=1}^m e_i + p \cdot \sum_{j=1}^n p \cdot y_j,$$

or equivalently $p \cdot z \leqq 0$. Therefore for every p in $R_+^l \backslash 0$, one has either $p \cdot f(p) \leqq 0$ or $p \cdot \phi(p) \leqq 0$. This observation leads to the following proof of existence of a general equilibrium (Gale, 1955; Nikaidô, 1956; Debreu, 1956). Let ϕ be a correspondence transforming points of the simplex P into non-empty convex subsets of R^l. If ϕ is bounded, has a closed graph and satisfies $p \cdot \phi(p) \leqq 0$ for every p in P, then, by Kakutani's theorem, there are a point p^* in P and a point z^* in R^l such that $z^* \in \phi(p^*)$ and $z^* \leqq 0$. In economic terms, there is a price-vector p^* in P yielding an associated excess demand z^* in $\phi(p^*)$, all of whose components are negative or zero.

If all the consumers in the economy $\mathscr{E}$ are insatiable, every individual budget constraint is binding, and one has for every i, $p \cdot x_i = p \cdot e_i + \Sigma_{j=1}^n \theta_{ij} p \cdot y_j$. By summation over i, $p \cdot z = 0$. Thus in the case where the vector z associated with p is uniquely determined, the excess demand function satisfies

Walras's Law: for every p in $R_+^l \backslash 0$, $\quad p \cdot f(p) = 0$.

In geometric terms, in the commodity-price space R^l the vectors p and $f(p)$ are orthogonal. This prompts one to normalize the price-vector p so that it belongs to the positive part of the unit sphere $\bar{S} = \{p \in R_+^l \,|\, \|p\| = 1\}$, for then $f(p)$ can be represented as a vector tangent to $\bar{S}$ at p. The excess demand function is now seen as a vector field on $\bar{S}$. This in turn suggests another proof of existence of a general equilibrium (Dierker, 1974) for the particular case of an exchange economy $\mathscr{E}$ whose consumers have continuous demand functions, monotone preferences and strictly positive endowments of all commodities. In that case for every $i = 1, \ldots, m$, the consumption-set X_i of the ith consumer is R_+^l, and $x < x'$ implies $x \prec_i x'$ (if x' is at least equal to x in every component, and $x' \neq x$, then x' is preferred to x). Since the demand of a consumer with monotone preferences is not defined when some prices vanish, one must restrict the price-vector p to be strictly positive in every component, that is, to belong to $S = \{p \in \text{Interior } R_+^l \,|\, \|p\| = 1\}$. Moreover let p_q be a sequence of price-vectors in S converging to p_0 in the boundary $\bar{S} \backslash S$ of S. Thus for every q, the vector p_q is strictly positive in each component, while, in the limit, p_0 has some zero components. Then the associated sequence of excess demands $f(p_q)$ is unbounded. As a consequence, the vector field f points inward towards S near the boundary of S. In these conditions Brouwer's fixed point theorem yields the existence of an equilibrium price-vector p^* in S for which excess demand vanishes, $f(p^*) = 0$.

The preceding solutions of the problem of existence of a general equilibrium all rest directly on fixed point theorems. Three different lines of approach are provided by (1) combinatorial algorithms for the computation of approximate general equilibria, (2) differential processes converging to general equilibria, (3) the theory of the fixed point index of a mapping.

(1) The past two decades have witnessed the development of algorithms of a combinatorial nature for the computation of an approximate general equilibrium (see COMPUTATION OF GENERAL EQUILIBRIA). Given any number $\epsilon > 0$, a constructive procedure thereby yields a price-vector p such that the norm $|f(p)|$ of the associated excess demand is smaller than ϵ. A compactness argument then gives a sequence of price-vectors p_q in S converging to p_0 for which $|f(p_q)|$ tends to 0. In the limit, $f(p_0) = 0$.

(2) Global Analysis was introduced into economic theory at the beginning of the 1970s to study the set of general equilibria of an economy and the manner in which it depends on the economy. In that framework Stephen Smale proposed in (1976) a differential process which starts from a point in the boundary of the set of normalized price-vectors, and which converges to the set of equilibria provided that the initial point does not lie

in a negligible exceptional set (see GLOBAL ANALYSIS). Another constructive procedure thus gives, from a differentiable viewpoint, conditions under which the set of general equilibria is not empty.

(3) In the same differentiable framework Egbert Dierker (1972) used the theory of the fixed point index of a mapping to prove that a regular economy (as defined by him in REGULAR ECONOMIES below) whose excess demand points inward near the boundary of S has an odd (hence non-zero) number of general equilibria. The significance of this theorem rests on the fact that under its assumptions almost every economy is regular.

The previous existence results have been extended in many directions. The study of the core of an economy led to the consideration of a set of agents, all of whom are negligible relative to their totality. This concept was formalized first as an atomless measure space of agents, and later by means of non-standard analysis. In both cases the existence of a general equilibrium had to be proved for economies with infinitely many agents.

In order to specify a commodity one lists its physical characteristics, the date, the location, and the event at which it is available. As soon as one of those four variables can take infinitely many values, the analysis of general equilibrium must be set in the framework of infinite-dimensional commodity spaces. Several existence results were obtained in that context.

In yet another direction, external effects called for extensions. When the characteristics of each agent (e.g. his preferences, his production set, . . .) depend on the actions chosen by the other agents, formulating the economy as a social system of the type described earlier immediately yields an existence theorem. Still other extensions have covered economies with public goods, with indivisible commodities, and with non-convex production sets.

GERARD DEBREU

See also ARROW–DEBREU MODEL; FIXED POINT THEOREMS; GENERAL EQUILIBRIUM.

BIBLIOGRAPHY

Arrow, K. J. and Debreu, G. 1954. Existence of an equilibrium for a competitive economy. *Econometrica* 22, 265–90.
Arrow, K. J. and Intriligator, M. D. (eds) 1981–6. *Handbook of Mathematical Economics*, Vols I–III, Amsterdam: North-Holland.
Cassel, K. G. 1918. *Theoretische Sozialökonomie*. Leipzig: C. F. Winter.
Debreu, G. 1952. A social equilibrium existence theorem. *Proceedings of the National Academy of Sciences* 38, 886–93.
Debreu, G. 1956. Market equilibrium. *Proceedings of the National Academy of Sciences* 42, 876–8.
Debreu, G. 1982. Existence of competitive equilibrium. Ch. 15 in Arrow and Intriligator (1981–6).
Dierker, E. 1972. Two remarks on the number of equilibria of an economy. *Econometrica* 40, 951–3.
Dierker, E. 1974. *Topological Methods in Walrasian Economics*. Berlin, New York: Springer-Verlag.
Gale D. 1955. The law of supply and demand, *Mathematica Scandinavica* 3, 155–69.
Kakutani, S. 1941. A generalization of Brouwer's fixed point theorem. *Duke Mathematical Journal* 8, 457–9.
McKenzie, L. W. 1954. On equilibrium in Graham's model of world trade and other competitive systems. *Econometrica* 22, 147–61.
Nash, J. F. 1950. Equilibrium points in N-person games. *Proceedings of the National Academy of Sciences of the USA* 36, 48–9.
Neisser, H. 1932. Lohnhöhe und Beschäftigungsgrad im Marktgleichgewicht. *Weltwirtschaftliches Archiv* 36, 415–55.
Neumann, J. von 1935–36. Über ein ökonomisches Gleichungssystem und eine Verallgemeinerung des Brouwerschen Fixpunktsatzes.

Ergebnisse eines mathematischen Kolloquiums 8, 73–83. Trans. by G. Morgenstern as 'A model of general economic equilibrium', *Review of Economic Studies* 13(1), 1945, 1–9.

Nikaidô, H. 1956. On the classical multilateral exchange problem. *Metroeconomica* 8, 135–45.

Schlesinger, K. 1933–4. Über die Produktionsgleichungen der ökonomischen Wertlehre. *Ergebnisse eines mathematischen Kolloquiums* 6, 10–20.

Smale, S. 1976. A convergent process of price adjustment and global Newton methods. *Journal of Mathematical Economics* 3, 107–20.

Smith A. 1776. *An Inquiry into the Nature and Causes of the Wealth of Nations*. 2 vols, ed. R. H. Campbell, A. S. Skinner and W. B. Todd, Oxford: Clarendon Press, 1976.

von Stackelberg, H. 1933. Zwei kritische Bemerkungen zur Preis-theorie Gustav Cassels. *Zeitschrift für Nationalökonomie* 4, 456–72.

Wald A. 1933–4. Über die eindeutige positive Lösbarkeit der neuen Produktionsgleichungen. *Ergebnisse eines mathematischen Kolloquiums* 6, 12–20.

Wald, A. 1934–5. Über die Produktionsgleichungen der ökonomischen Wertlehre. *Ergebnisse eines mathematischen Kolloquiums* 7, 1–6.

Wald, A. 1936. Über einige Gleichungssysteme der mathematischen Ökonomie. *Zeitschrift für Nationalökonomie* 7, 637–70. Trans. by Otto Eckstein as 'On some systems of equations of mathematical economics', *Econometrica* 19(4), October 1951, 368–403.

Walras, L. 1874–7. *Éléments d'économie politique pure*. Lausanne: L. Corbaz. Trans. by William Jaffé as *Elements of Pure Economics*, Homewood, Ill.: Richard D. Irwin, 1954.

Zeuthen, F. 1932. Das Prinzip der Knappheit, technische Kombination und ökonomische Qualität. *Zeitschrift für Nationalökonomie* 4, 1–24.

exit and voice. A central place is held in economics and social science in general by principles and forces making for order or equilibrium in economic and social systems. Disorder and disequilibrium are then understood as resulting from some malfunction of these principles or forces. Explanations of order–disorder or equilibrium–disequilibrium have typically been discipline-bound, dealing with either the political or the economic world. Since the two are interrelated it would be useful to have a construct that bridges them. Such is the claim of the exit–voice perspective. It addresses the changing balance of order and disorder in the social world by pointing out that social actors who experience developing disorder have available to them two activist reactions and perhaps remedies: *exit*, or withdrawal from a relationship that one has built up as a buyer of merchandise or as a member of an organization such as a firm, a family, a political party or a state; and *voice*, or the attempt at repairing and perhaps improving the relationship through an effort at communicating one's complaints, grievances and proposals for improvement. The voice reaction belongs in good part to the political domain since it has to do with the articulation and channelling of opinion, criticism and protest. Much of the exit reaction, on the contrary, involves the economic realm as it is precisely the function of the markets for goods, services, and jobs to offer alternatives to consumers, buyers and employees who are for various reasons dissatisfied with their current transaction partners.

The exit–voice alternative was proposed and explored in *Exit, Voice, and Loyalty: Responses to Decline in Firms, Organizations, and States* (Hirschman, 1970, henceforth *EVL*). Attempts to apply the book's perspective were made over many areas of social life. In the following, the basic concepts will be recapitulated and, where necessary, reformulated. Subsequently some major applications of the exit–voice polarity will be reviewed.

BASIC CONCEPTS

Exit. Exit means withdrawal from a relationship with a person or organization. If this relationship fulfils some vital function, then the withdrawal is possible only if the same relationship can be re-established with another person or organization. Exit is therefore often predicated on the availability of choice, *competition*, and well-functioning *markets*.

Exit of customers (or employees) serves as a signal to the management of firms and organizations that something is amiss. A search for causes and remedies will then be undertaken and some plan of action designed to restore performance will be adopted. This is one way in which markets and competition work to prevent decay and to maintain and perhaps improve quality.

Exit is a powerful but indirect and somewhat blunt way of alerting management to its failings. Most of the time, those customers and members of organizations who exit have no interest in improving them by their withdrawal, so that exit does not provide management with much information on what is wrong.

Voice. The direct and more informative way of alerting management is to alert it: this is *voice*. Its role is, or should be, paramount in situations where exit is either not available at all or is difficult, costly, and traumatic. This is so for certain primordial groupings one is born into – the family, the ethnic or religious community, the nation – or for those organizations one joins with the intention of staying for a prolonged period – school, marriage, political party, firm. With regard to buying and selling, voice should take over from exit when competition is weak or nonexistent as in the case of goods and services being produced under oligopolistic or monopolistic conditions, or when exit is expensive for both parties as in certain interfirm relations.

Unlike exit in the case of well-functioning markets, voice is never easy. It can even be dangerous. Many organizations and their agents are not at all keen on being told about their shortcomings by members and the latter often expose themselves to reprisals if they utter any criticism (Birch, 1975). Even in the absence of reprisals, the cost of voice to an individual member will often exceed, in terms of time and effort, any conceivable benefit from voicing. Frequently, moreover, any effective channelling of individual voices requires a number of members to join together so that voice formation depends on the potential for collective action.

In spite of these problems, voice exists or, rather, it has come into being. Its history is to a considerable extent the history of the right to dissent, of due process, of safeguards against reprisal, and of the advance of trade unions and of consumer and many other organizations articulating the demands of individuals and groups who once were silent. Similarly, the history of exit is the history of the broadening of the market, of the right to move freely, to emigrate, to be a conscientious objector, to divorce, etc. Being two basic, complementary ingredients of democratic freedom, the right to exit and the right to voice have on the whole been enlarged or restricted jointly. Yet, there are important instances of unilateral advances or retreats of either the one or the other response mechanism (Rokkan, 1975; Finer, 1974).

Interaction of exit and voice. As noted, exit is paramount as a reaction to discontent in some circumstances and voice holds a similarly privileged position in others, but frequently both mechanisms are available jointly. In such situations they may either reinforce or undercut each other. The availability and threat of exit on the part of an important customer or group of members may powerfully reinforce their voice. On the other hand, the actual recourse to exit will often diminish the volume of voice that would otherwise be forthcoming and, should the organization be more sensitive to voice than to exit, the stage could be set for cumulative deterioration. For example, after an incipient deterioration of public schools or inner cities, the availability of private schools or suburban housing would lead, via exit, to further deterioration – a turn of events that might have been prevented if the parents sending their children to private school or the inner city residents who move to the suburbs had instead used their voice to press for reform. In their aggregate effects, the individual exit decisions are harmful – an instance of the 'tyranny of small decisions' – also because they are likely to be taken on the basis of a short-run private-interest calculus only and do not take into account the 'public bad' that will be inflicted, even on those who exit, by decaying inner cities and segregated education (Levin, 1983; Breneman, 1983).

These kinds of situations are sufficiently numerous and important to be of interest not only as curious paradoxes showing that under some circumstances the availability of exit (that is, of competition) could have undesirable effects. In this connection, *EVL* stressed the value of *loyalty* as a factor that might delay over-rapid exit. Loyalty would make a member reluctant to leave an organization upon the slightest manifestation of decline even though rival organizations were available. Provided it is not 'blind', loyalty would also activate voice as loyal members are strongly motivated to save 'their' organization once deterioration has passed some threshold.

The difficulties of combining exit and voice in an optimal manner are in a sense 'problems of the rich': they relate to situations and societies where exit and voice are both forthcoming more or less abundantly, but where, for best results, one would wish for a different mix. Historically more frequent are cases where exit and voice are both in short supply, in spite of many reasons for discontent and unhappiness. There is no doubt, as many commentators have pointed out, that passivity, acquiescence, inaction, withdrawal, and resignation have held sway much of the time over wide areas of the social world. This is largely the result of repression of both exit and voice – a repression that has flourished in spite of the fact that all human organizations could put to good use the feedback provided by the two reaction modes.

Problems in voice formation. The development of voice among customers of firms or members of organizations poses a number of problems that were not fully explored in *EVL*. Critics have asserted that, in its endeavour to present voice as a ready alternative to exit, the book understated the difficulties of voice formation. In examining this issue it is useful to start with the extreme no-voice case: the authoritarian state which is dedicated to repressing and suppressing voice. This situation has given rise to a useful distinction between *horizontal* and *vertical* voice (O'Donnell, 1986). The latter is the actual communication, complaint, petition, or protest addressed to the authorities by a citizen and, more frequently, by an organization representing a group of citizens. Horizontal voice is the utterance and exchange of opinion, concern and criticism *among* citizens: in the more open societies it is today regularly ascertained through opinion polls revealing the approval rating of presidents, prime ministers, mayors, etc. Horizontal voice is a necessary precondition for the mobilization of vertical voice. It is the earmark of the more frightful authoritarian regimes that they suppress not only vertical voice – any ordinary tyranny does that – but

horizontal voice as well. The suppression of horizontal voice is generally the side-effect of the terrorist methods used by such regimes in dealing with their enemies.

The distinction between vertical and horizontal voice is relevant to the 'free ride' argument in relation to voice formation (Barry, 1974). For *vertical* voice to come about, that is, for members of the organization to engage management in meaningful dialogue, it is frequently necessary for members to forge a tie among themselves, to create an organization which will agitate for their demands, etc. But the hoped-for result of collective voice is a freely available public good; hence, so goes the critical argument, self-interested, 'rational' individuals may well withhold their contribution to the voice enterprise in the expectation that others will take on the entire burden. Important as it is, this argument has its limitations. First of all, it is addressed only to vertical voice which it mistakenly equates (as *EVL* did) with voice in general. Horizontal voice is not subject to the strictures of the free-rider argument: it is free, spontaneous activity of men and women in society, akin to breathing. As just noted, extraordinary violence has to be deployed if it is to be suppressed. Under ordinary circumstances, horizontal voice is continuously generated and has an impact even without becoming vertical: in many environments managers of organizations cannot help noticing and reacting to critical opinions and hostile moods of the members, whether or not organized protest movements break out. That the planned economies of Eastern Europe function to the extent they do has been explained on precisely this ground (Bender, 1981, p. 30).

Another limitation of the free-rider argument lies in its assumption that individuals will always act instrumentally. Just because the desired result of collective voice is typically a *public* good – or, better, some aspect of the *public* happiness – participation in voice provides an alternative to self-centred, instrumental action. It therefore has the powerful attractions of those activities that are characterized by the fusion of striving and attaining and can be understood as investments in individual or group identity (Hirschman, 1985).

SOME AREAS OF APPLICATION

Trade unions. In economics, the major application of the exit–voice theme has been the analysis of trade unions as collective voice by Freeman and Medoff in their book *What Do Unions Do?* (1984). Instead of looking at unions as a monopolistic device raising wages for unionized workers beyond the 'market–clearing' equilibrium level or – much the same zero-sum interpretation in different language – as a tool in the class struggle serving to reduce the degree of exploitation, the book finds that a major function of unions is that of channelling information to management about workers' aspirations and complaints. Collective voice, in the form of union bargaining, is more efficient in conveying information about workers' discontent – and in doing something about it – than individual decisions to quit, as voice carries more information than exit. The presence of union voice is shown to reduce costly labour turnover. Moreover, the fringe benefits, workplace practices, and seniority rules which unions negotiate often result in offsetting labour productivity increases.

Markets and hierarchies vs. exit and voice. Renewed attention has been given in recent years to the question why some kinds of economic activities are carried on through many independent firms while others, to the contrary, are tied together through bureaucratic and hierarchical relations. In accounting for hierarchy, one approach has directed attention to such matters as uncertainty about the evolution of the market and the technology and in particular to asymmetric availability of information to buyer and seller, creating opportunities for deceitful behaviour (Williamson, 1975). Hierarchy is then seen as superior to markets whenever there is need for a sustained and frank dialogue between the contracting parties. Critics of this position have argued: (1) relations between independent firms, such as contractors and subcontractors, are often quite effective in discouraging malfeasance; (2) correlatively, hierarchy frequently leads to characteristic patterns of concealment and control evasion (Eccles, 1981; Granovetter, 1985); and (3) industry structure varies substantially from one country to another as well as within the same country over time: in Japan, for example, subcontracting is much more widely practiced than in the West and in Italy subcontracting has become more widespread in the last 10–20 years.

A formulation in terms of exit–voice is helpful here. The characteristics which are said to justify hierarchy – incomplete information, considerable apprenticing of one firm by the other, openings for 'opportunistic' (i.e., dishonest) behaviour, etc. – all make for situations in which there is need for voice: the firms contracting together must intensively consult with, and watch over, each other. *But the need for voice does not necessarily imply that hierarchy is in order.* Whether voicing is done best within the same organization or from one independent firm to another is by no means a foregone conclusion. Moreover, when the two parties are independent and resort a great deal to voice, the possibility of exit from the relationship often looms in the background. The implicit threat of exit could carry as much clout as that of sanctions in hierarchical relationships.

The argument for hierarchy in cases where voice has an important role to play may arise from thinking of market relationships only in terms of the ideal, anonymous market where voice is wholly absent. But most markets involve voice: commerce *is* communication, and is premised on frequent and close contact of the contracting parties who deliver promises, trust them, and engage in mutual adjustment of claims and complaints – all of this was implicit in the eighteenth-century notion of *doux commerce* (Hirschman, 1977, 1982). Adam Smith even conjectured that it was man's ability to communicate through speech that lies at the source of his 'propensity to truck and barter'. How odd, then, that the need for frequent and intensive communication should be adduced as a conclusive argument for hierarchy.

Public services: education, health, others. The organization of public services represents a privileged area for the application of exit–voice reasoning – significantly the exit–voice idea had its origin in the analysis of a public service in trouble, the Nigerian railroads (*EVL*, Preface). Public services are typically sold or delivered by a single public or publicly regulated supplier, for various well-known reasons.

With the production of most public services being thus deprived of the 'discipline of the market', problems of productive efficiency and quality maintenance arise necessarily. An obvious way of mitigating these problems is to attempt to reintroduce market pressures in some fashion. For example, when certain categories of goods and services are to be made available either to all citizens regardless of their income or to some deprived social groups, the state and its agencies can sometimes refrain from producing or distributing these goods directly, and instead issue special purpose money or *vouchers* enabling the beneficiaries to acquire the goods or services through ordinary market channels. In this manner the voucher system reintroduces the market and the possibility of exit. A particularly successful example of the voucher system is the distribution of Food Stamps to low-income persons in the United States. Instead of creating and administering its own food distribution network the state hands out vouchers (food stamps) which the beneficiaries can then use at existing, competitive commercial outlets.

In part because of the success of this programme and in part because of the belief in 'market solutions' as the remedy for all that ails government programmes, voucher schemes have been proposed for a large number of other public services, from education to low-cost housing to the supply of certain health services. Voucher systems are appropriate primarily under the following conditions (Bridge, 1977): (1) there are widespread differences in tastes and these differences are recognized as legitimate; (2) individuals are well informed about quality and different qualities are easily compared and evaluated; (3) purchases are recurrent and relatively small in relation to income so that buyers can learn from experience and easily switch from one brand and supplier to another.

These conditions are ideally present in the case of foodstuffs, but much less so in the case of, say, health and educational services. Hence the development of voice constitutes here an important alternative strategy for assuring and maintaining product quality. In other words, the beneficiaries of certain public services should be induced to become active on their own behalf, individually or collectively As always, development of voice is arduous because of apathy and passivity of the members, but also because it will often be resisted by the organizations that have been set up to deliver the services. A number of proposals and attempts have been made to introduce more voice into the administration of both health and educational services (Stevens, 1974; Klein, 1980).

EVL had insisted on the see-saw character of exit and voice interventions in these fields. Education and health systems seemed particularly exposed to the danger that premature exit – of the potentially most influential members – would undermine voice. The opposite relation may also occur, however, for the opening up of the exit perspective could serve to strengthen voice: parents who have been wholly passive because of feelings of powerlessness and fear of reprisals may feel empowered for the first time once they are given vouchers that could be used 'against' the schools currently attended by their children, and will be more ready than before to speak out with regard to desirable changes in those very schools.

Spatial mobility (migration) and political action. Another substantial area of exit–voice applications opens up when exit is taken in the literal, spatial sense. Here exit–voice boils down to the familiar flight or fight alternative. While often institutionalized among nomadic groups (Hirschman, 1981, ch. 11), this alternative is not necessarily available in sedentary societies. Here the traditionally available choice is fight or submit in silence. The option of removing oneself from an oppressive environment has become available on a massive scale only in modern times, with the advances in transportation and the *uneven* opening up of economic opportunity, religious tolerance, and political freedom. Where the option has existed, the interaction of exit and voice has been on display in three principal types of migration: (1) that from the countryside to the city, the oldest and no doubt largest of the modern migrations; (2) the migration from the city to the suburbs, which was most intense in the United States during the fifties and sixties, owing to the spread of the automobile and also to the large-scale migration of blacks and Hispanics *into* the cities; (3) finally, of course, international migration with its numerous economic and political determinants and constraints. Under this rubric, the international movement of capital also deserves attention.

Looking at the varieties of exit–voice interplay in these diverse settings, it is possible, on the basis of the numerous studies now available, to distinguish the following patterns:

(1) In accordance with the basic hypothesis of *EVL*, exit-migration deprives the geographical unit which is left behind (countryside, city, nation) of many of the more activist residents, including potential leaders, reformers, or revolutionaries. Exit weakens voice and reduces the prospects for advance, reform, or revolution in the area that is being left.

Something of this pattern can be observed in all three types of migration. Massive rural–urban migration could obviously reduce the potential as well as the need for land reforms which the voice of the countryside might otherwise have precipitated (Huntington and Nelson, 1976, pp. 103ff.). The large outward migration from Europe to the United States in the 19th century up to World War I probably functioned as a political safety-valve for the rapidly industrializing European societies of that period, as has been shown for Italy (MacDonald, 1963–4). In a similar vein, the possibility of westward migration within the United States has been invoked as an explanation for the lack of a militant working-class movement in that country. Finally, the city-to-suburbs migration in the United States has led, at least initially, to cumulative deterioration in the urban areas affected by out-migration in spite of, and in some cases because of, reduced density. At times, the voice-weakening effect of exit is consciously utilized by the authorities: permitting, favouring, or even ordering the exit of enemies or dissidents has long been one – comparatively civilized – means for autocratic rulers to rid themselves of their critics, a practice revived on a large scale by Castro's Cuba and, on a more selective basis, by the Soviet Union.

(2) But the basic see-saw pattern – the more exit the less voice – does not exhaust the rich historical material. The mechanism through which voice is strengthened rather than weakened as a result of exit is distinctive in the case of migration. In some societies the accumulated social pressures could be so high that authoritarian political controls will only be relaxed if a certain amount of out-migration takes place concurrently. This is what happened in the fifty years prior to World War I when the franchise was extended in many European states from which large contingents of people were departing. In other words, the state accommodated some of the pressures toward democratization because it could be reasonably surmised, in part as a result of out-migration, that opening the door slightly to voice would not blow away the whole structure. A similar positive relation between exit and voice may exist today with regard to such southern European countries as Spain, Portugal, and Greece: here the large-scale emigration to northern Europe may also have eased the transition to a more democratic (more vociferous) order.

(3)Exit–voice theory posits remedial or preventive responses to any large-scale out-migration on the part of the entity that is being left. A firm losing customers or a party losing members will normally undertake a search for the reasons of such declines in fortune and then determine upon a strategy for recovery. For out-migration such reactions are not easy to identify. In the case of massive rural–urban migration, for example, there is usually no organized entity such as the 'countryside' that registers the flight from it and can undertake corrective action. With regard to migration from the city to the suburbs, the situation is not too different. Here entities exist – city administrations – but they have generally been ineffective in modifying the individual decisions of millions of people to move into their own homes in the suburbs.

The analogy to the firm is – or should be – most applicable when the geographic entity losing residents is the State, which is after all a highly organized, self-reflective body with considerable means of action. There is, of course, the already noted possibility that out-migration relieves economic or political stress in a country, is therefore *welcome*, and may even be encouraged by the state. But massive emigration is at some point bound to be viewed as dangerous. Just like a business firm, the state may then take measures to make itself more attractive to its citizens. One example of this reaction is the national plan for economic recovery and industrialization adopted by Ireland in 1958, in the midst of very high levels of emigration, mostly to England (Burnett, 1976). It has also been shown that the pioneering welfare state measures of the late 19th and early 20th century, starting in Bismarck's Germany in the eighties and then spreading to the Scandinavian countries and Great Britain, were all taken in countries with high rates of overseas migration. These measures can be seen as attempts of states to make themselves more attractive to their citizens (Kuhnle, 1981).

The international movement of capital was first commented upon from the exit perspective in the 18th century. Montesquieu and Adam Smith both thought that the threat of exit on the part of movable capital could play a useful role in preventing arbitrary and confiscatory measures against the legitimate interests of commerce and industry. The threat of exit or exit itself was expected to function, like the customer's exit, as a curb on misconduct, this time on the part of the state. While this relationship is still pertinent, exit of capital often plays a less constructive role today. In the more peripheral capitalist countries the owners of capital have become fully alive to the possibility of removing part of their holdings to the United States or other reliable places in case they become unhappy about the 'investment climate'. In this manner, capital exit (or flight) will often be practised on a large scale as soon as the state undertakes some, perhaps long overdue, reforms with respect to such matters as land tenure or fiscal equity. Instead of preventing arbitrary and ill-considered policies, exit can thus complicate and render more hazardous certain *needed* reforms. Moreover, exit undercuts voice: as long as the capitalists are able to remove their patrimony to a safe place, they will have that much less incentive to raise their voice for the purpose of making a responsible contribution to national problem-solving. Capital mobility and propensity to exit may thus be a major reason for the instability of states in the capitalist periphery (Hirschman, 1981, ch. 11).

Political parties. Two principal propositions were put forward by *EVL* with regard to the dynamics of political parties in a democracy:

(1)In a two-party system, the tendency of the parties to move toward the non-ideological centre in order to capture the (allegedly) voluminous middle-of-the-road vote is countered by those party members and militants who are on the parties' ideological fringes, have 'nowhere else to go', but just because of that are maximally motivated to exert influence inside the party, by forceful uses of voice.

(2)In a multi-party system, with the ideological distance from one party to the next being presumably shorter than in two-party systems, dissatisfaction with party performance is more likely to lead to exit than in two-party systems; in the latter, voice will play the more important role as switching to the other party requires too big an ideological jump. One inference is that parties in two-party systems may be expected to exhibit more internal divisions, but also more internal democracy and less bureaucratic centralism than parties in multiple-party systems.

The first of these propositions has been strongly supported by events subsequent to the publication of the book. At that time, only the nomination of Barry Goldwater to be the standard bearer of the Republican Party in 1964 could be cited in support. Since them, additional evidence has accumulated: from the nomination by the Democrats of George McGovern to contend the Presidential elections in 1972 to the increasing power of the more radical wing of the Labour party and the ascendancy of Margaret Thatcher within the Conservative party and of Ronald Reagan among the Republicans. The theory that in a two-party system the two parties would increasingly converge toward some middle ground has been amply disconfirmed.

The second proposition on political parties which was deduced from the *EVL* framework has undergone several qualifications. For example, in democracies with old cleavages along ethnic, linguistic, and religious lines, the distances between the several parties rooted in ethnic, etc., identities could actually be wider than that between the parties of two-party systems. Under these conditions, the exit–voice logic would in fact predict that member participation (voice) in parties of multi-party systems would also be vigorous and exit infrequent (Lorwin, 1971; Hirschman, 1981, ch. 9).

A more serious complication is being stressed in a work by S. Kernell still in progress. In two-party systems, exit is a particularly powerful move for dissatisfied members as by casting their vote for the other party they are doubling its impact, something they cannot be sure of in multi-party systems. Hence, in case of disappointment with the performance of one's own party, there could arise a special temptation in two-party systems to switch to the other party so as to *punish* one's own. Such a preference for exit is likely to come to the fore primarily when a party in power is perceived as having seriously mishandled its mandate. Under the circumstances, the prospect of being able to punish that party retrospectively could overcome party loyalty and past ideological commitment. This constellation was an important factor in the sharp defeat of the Democratic ticket in the 1980 Presidential elections in the United States.

The family: marriage and divorce. Modern marriage is one of the simplest illustrations of the exit–voice alternative. When a marriage is in difficulty, the partners can either make an attempt, usually through a great deal of voicing, to reconstruct their relationship or they can divorce. The complexities of the interplay between exit and voice are well in evidence here. Just as the threat of strike in labour–management relations, so is the threat of divorce important in inducing the parties to 'bargain seriously'; but as exit becomes ever easier and less costly (and perhaps even profitable to one of the parties – see Weitzman, 1985), its availability will undermine voice: rather than being an action of last resort, divorce could become the automatic response to marital difficulty with less and less effort made at communication and reconciliation.

This is exactly what appears to have happened in the United States during the last fifteen years, i.e. since *EVL* stated that 'the expenditure of time, money and nerves' necessitated by complicated divorce procedures serves the useful, if unintended purpose of 'stimulating voice in deteriorating, yet recuperable organizations which would be prematurely destroyed through free exit' (p. 79). In 1970 California adopted a new 'no-fault'

law on divorce which spread, though often in attenuated form, to most other states (Weitzman, 1985). The California law drastically altered divorce procedures: instead of requiring proof that one of the parties was guilty of some specific type of behaviour constituting grounds for divorce, the new law permitted divorce when both *or just one* of the two parties asserted that the marriage had irretrievably broken down. The possibility of a unilateral decision, of just 'walking out', is symbolic of the way in which the California law undercuts the recourse to voice.

With the new regime, the pendulum has swung quite far in the direction of facilitating exit and of thereby weakening voice. It was of course a reaction to the many abuses of the older fault-based system which required costly and degrading adversarial proceedings, and in effect discriminated against the poor. But the framers of the new legislation probably did not realise the extent to which the earlier obstacles to divorce indirectly encouraged attempts at mending the so easily frayed conjugal relationship and how much the new freedom to exit would torpedo such attempts, with the results that one of every two new marriages now ends in divorce.

The family: adolescent development. This is another family situation for whose analysis a formulation in terms of exit and voice has been found useful (Gilligan, 1986). Adolescent development has often been portrayed as a process through which the 'dependent' child becomes an 'independent' adult through progressive 'detachment' from the parents. Freud saw this as 'one of the most significant, but also one of the most painful psychic accomplishments of the pubertal period ... a process that alone makes possible the opposition, which is so important for the progress of civilization, between the new generation and the old' (1905, p. 227). Here is a celebration of exit; Freud's statement neglects a complementary aspect and task of adolescent development which is to maintain and enrich the bond with the older generation through continued, if conflict-ridden, communication. In other words, voice has an important role to play in transforming the adolescent's relationship to the parents. The peculiar poignancy of the adolescent–parents conflict resides in fact in the impossibility of relying *wholly* on voice in resolving it: given the closeness of the relationship, a full accord that would be the outcome of successful voicing risks ending up in incest, as the 'meeting of minds would suggest a meeting of bodies' (Gilligan, 1986). It is because of the incest taboo that exit must be part of the solution, but different generations of adolescents are likely to achieve emancipation by practising very different characteristic mixes of exit and voice. Moreover, as Gilligan stresses, the balance of exit and voice differs according to gender. Girls place a greater value than boys on continued attachment to the family, and are therefore less attracted to the masculine ideal of independence–isolation. Hence they experience a greater tension between exit and voice.

With this imaginative use of the exit–voice concept, the outer limits of its sphere of influence may have been reached.

ALBERT O. HIRSCHMAN

See also CONTESTABLE MARKETS; FAMILY; TIEBOUT HYPOTHESIS.

BIBLIOGRAPHY

Barry, B. 1974. Review article: 'Exit, Voice, and Loyalty'. *British Journal of Political Science* 4, February, 79–107.

Bender, P. 1981. *Das Ende des ideologischen Zeitalters*. Berlin: Severin und Siedler.

Birch, A.H. 1975. Economic models in political science: the issue of 'Exit, Voice, and Loyalty'. *British Journal of Political Science* 5, January, 69–82.

Breneman, D.W. 1983. Where would tuition tax credit take us? Should we agree to go? In *Public Dollars for Private Schools*, ed. T. James and H.M. Levin, Philadelphia: Temple University Press.

Bridge, G. 1977. Citizen choice in public services: voucher systems. In *Alternatives for Delivering Public Services*, ed. E.S. Savas, Boulder, Colorado: Westview.

Burnett, N.R. 1976. Emigration and modern Ireland. Unpublished PhD dissertation, School of Advanced International Studies, Johns Hopkins University.

Eccles, R.G. 1981. The quasifirm in the construction industry. *Journal of Economic Behavior and Organization* 2(4), December, 335–57.

Fainstein, N.I and Fainstein, S.S. 1980. Mobility, community, and participation: the American way out. In E.G. Moore and W.A.V. Clark, *Residential Mobility and Public Policy*, Beverly Hills: Sage.

Finer, S.E. 1974. State-building, state boundaries and border control in the light of the Rokkan–Hirschman model. *Social Science Information* 13(4–5), 79–126.

Freeman, R.B. and Medoff, J.L. 1984. *What Do Unions Do?* New York: Basic Books.

Freud, S. 1905. Three essays on the theory of sexuality. In *Complete Psychological Works*, Vol. 7, London: Hogarth, 1953.

Gilligan, C. 1986. Exit–voice dilemmas in adolescent development. In *Development, Democracy and the Art of Trespassing: Essays in Honor of A.O. Hirschman*, ed. A. Foxley et al., Notre Dame, Indiana: University of Notre Dame Press.

Granovetter, M. 1985. Economic action and social structure: a theory of embeddedness. *American Journal of Sociology* 91, November, 481–510.

Hirschman, A.O. 1970. *Exit, Voice, and Loyalty: Responses to Decline in Firms, Organizations, and States*. Cambridge, Mass.: Harvard University Press.

Hirschman, A.O. 1977. *The Passions and the Interests: Political Arguments for Capitalism before its Triumph*. Princeton: Princeton University Press.

Hirschman, A.O. 1981. *Essays in Trespassing: Economics to Politics and Beyond*. Cambridge: Cambridge University Press.

Hirschman, A.O. 1982. *Shifting Involvements: Private interest and public action*. Princeton: Princeton University Press.

Hirschman, A.O. 1985. Against parsimony: three easy ways of complicating some categories of economic discourse. *Economics and Philosophy* 1, 7–21.

Huntington, S.P. and Nelson, J.M. 1976. *No Easy Choice: political participation in developing countries*. Cambridge, Mass.: Harvard University Press.

Kernell, S. 1987. *Retrospective Voting and Contemporary Macrodemocracy*. Washington, DC: Brookings Institution.

Klein, R. 1980. Models of man and models of policy: reflections on *Exit, Voice and Loyalty* ten years later. *Milbank Memorial Fund Quarterly* 58(3), Summer, 413–29.

Kuhnle, S. 1981. Emigration, democratization, and the rise of the European welfare states. In *Mobilization, Center-Periphery Structures, and Nation-Building* (a volume in commemoration of Stein Rokkan), ed. P. Torsvik, Bergen: Universitetsforlaget.

Levin, H.M. 1983. Educational choice and the pains of democracy. In *Public Dollars for Private Schools: the case for tuition tax credits*, ed. T. James and H.M. Levin, Philadelphia: Temple University Press.

Lorwin, V. 1971. Segmented pluralism: ideological cleavages and political cohesion in the smaller European democracies. *Comparative Politics* 3(2), January, 141–75.

MacDonald, J.S. 1963–64. Agricultural organization, migration and labour militancy in rural Italy. *Economic History Review* 16, August, 61–75.

O'Donnell, G. 1986. On the convergences of Hirschman's *Exit, Voice and Loyalty* and *Shifting Involvements*. In *Development, Democracy and the Art of Trespassing: Essays in Honor of A.O. Hirschman*, ed. A. Foxley et al., Notre Dame, Indiana: University of Notre Dame Press.

Rokkan, S. 1975. Dimensions of state formation and nation-building: a possible paradigm for research on variations in Europe. In *The Formation of National States in Western Europe*, ed. C. Tilly, Princeton: Princeton University Press.

Stevens, C.M. 1974. Voice in medical care markets: 'consumer participation'. *Social Science Information* 13(3), 33–48.

Weitzman, L.J. 1985. *The Divorce Revolution: The Unexpected Social and Economic Consequences for Women and Children in America*. New York: Free Press.

Williamson, O.E. 1975. *Markets and Hierarchies: analysis and antitrust implications*. New York: Free Press.

exogeneity. *See* ENDOGENEITY AND EXOGENEITY.

exogenous money. *See* ENDOGENOUS AND EXOGENOUS MONEY.

expectational equilibrium. *See* EQUILIBRIUM: AN EXPECTATIONAL CONCEPT.

expectation of life. *See* LIFE TABLES.

expectations. Most decisions that economic agents must make involve uncertainty about the future. Thus, any economic model that is intended to be descriptive of human behaviour is likely to involve human expectations about uncertain future economic variables. Areas in economics that involve expectations in fundamental ways include the theories of intertemporal consumption or labour supply decisions, theories of firms' pricing, sales, investment, or inventory decisions, theories of financial markets and money, theories of insurance, and of search behaviour, signalling, agency, and bidding. If our purpose is to describe human behaviour, then the study of human expectations is inseparable from the study of the behavioural models in which these expectations are imbedded. Only a few general observations can be discussed here.

ECONOMIC EXPECTATIONS, SURVEYS AND PROXIES. Applied econometric research often relies on simple models involving *expectations variables* that represent the expectations of economic agents for some specified economic variables. For example, the total savings of individuals may be related to a variable purporting to measure their expectations for their pension benefits on the date of their retirement, years in the future. Or, an individual's decision whether to purchase a long-term or short-term bond may be related to a variable representing expectations as to the course of future short-term interest rates over the life of the long-term bond.

Expectations variables included in such models are often referred to as measuring, perhaps imperfectly, some idealized *economic expectations*. What economic expectations actually represent is usually not spelled out. Different people have different perceptions about the outlook for future variables, and so there is an index number problem in reducing their divergent opinions into a single measure. Moreover, people when asked for their expectation for some economic variable may answer that they have no expectation. If pressed, they may hazard a guess. Certainly, most individuals make some economic decisions without making an effort to learn about relevant economic variables. From time to time, circumstances require making difficult or important decisions, and then people may trouble themselves more to find out about economic variables. Economic models that speak of 'the' expectation of an economic variable are presumably talking about some average of the expectations of some people and guesses other people would make if pressed, or about averages of the better or worse forecasts of the same people at different times.

The expectations variables used in econometric work to measure economic expectations may be *survey expectations*, representing the average expectation respondents reported on a public opinion survey, or they may be *expectations proxies* consisting of transformations of other variables that appear to the econometrician to be plausible guesses as to the economic expectations. For example, a moving average of lagged inflation rates may serve as an expectations proxy for future inflation.

Expectations surveys commonly take two forms: those that survey individuals representative of the general population and those that sample experts. The former provide measures of expectations that are relevant to decisions, like individual decisions how much to save in a given month or whether to put money in a savings account or in corporate bonds, to which decision-makers do not attach great importance. The latter provide measures of expectations that are relevant to decisions, like firms' decisions on whether to market a new product or invest in a new plant, on which decision-makers are likely to spend the resources to obtain informed forecasts.

DAY-TO-DAY EXPECTATIONS OF THE GENERAL POPULATION. Survey research finds, according to Katona (1975), that most people can be induced to make a guess as to the direction of change in the near future of major macroeconomic variables, but are reluctant to give quantitative estimates of the extent of the change. The information on which most people base their expectations is fragmentary. Based on decades of survey research on the general public in the United States, Katona concluded that the majority knew whether unemployment had increased or decreased in the preceding months, whether profits or retail sales had gone up or down, and also whether interest rates had risen or fallen, but did not know how much larger or smaller any of these magnitudes were. The extent of knowledge about macroeconomic variables is generally greater the more important or dramatic the recent changes in these variables, and, of course, the more the variable has been emphasized in the mass media.

Since we generally want to incorporate expectations variables in an economic model that describes human behaviour, we are likely to want any variable measuring economic expectations to represent the actual thoughts of individuals *before* they were forced to sit down at a questionnaire and carefully think about how to forecast an economic variable. In modelling, say, income expectations for the purpose of studying the saving decision, we want to get into the individual's frame of mind at the times when saving decisions are made.

When we try to characterize a person's frame of mind at these times, we should recognize that the expectations are likely to differ through time qualitatively as well as quantitatively. For example, an expectation of a future rise in income may become more vividly impressed on individuals' consciousness by some public event that reminds that person of the reasons to expect income to rise. At the same time, the expectation as measured on a survey may be unchanged. Psychologists who study the saving decision have emphasized the importance of changing *aspirations* as distinct from changing expectations.

Individuals who are not thinking much about economic theories and who are merely confronted with economic variables whose stochastic properties are difficult to comprehend, may be modelled fairly well in terms of simple expectations proxies like that proposed by Fisher (1930). In Fisher's proxy, expected inflation is a *distributed lag* with coefficients that decline linearly with the lag of actual inflation (i.e. a weighted average of current and past values with weights that decline linearly with time into the past until the weight reaches zero). A variation on Fisher's expectation mechanism is *adaptive expectations* (Cagan, 1956), in which expectations are formed as a distributed lag, with coefficients that decline exponentially, rather than linearly, with the lag of actual past inflation, and that sum to one. The rate at which coefficients decline might be determined by the rate at which human memory decays. It may be natural to form expectations of a future variable (e.g. inflation) by thinking back over the recent past of experience of the variable, and hence such memory decay may result in a distributed lag pattern like that hypothesized by Cagan.

With adaptive expectations, the change in the expectations variable is proportional to the difference between its previous value and the latest value of the variable to be forecasted. This construction resembles that of the *error-learning hypothesis*, Meiselman (1962). However, in the error learning hypothesis the change in the expectation for a variable at a specific future date is proportional to the error just discovered in the forecast for the variable for today's date.

Alternatives to adaptive expectations are *regressive expectations*, in which variables are expected to return gradually to a fixed level independent of their recent past behaviour (this term has also been used as a synonym for adaptive expectations), and *extrapolative expectations* in which the recent direction of change in the variables is expected to continue (see, for example, Modigliani and Sutch, 1966). Any of these expectations mechanisms may be consistent with optimal forecasts of the future variables under certain special circumstances (see Sargent and Wallace, 1973; Shiller, 1978).

We may not want to use such simple models of expectations in periods when individuals may think a great deal about economic theories. During a period of hyperinflation, for example, it is perhaps unlikely that people will form expectations adaptively, since the inflation affects them so noticeably. They may seek out the opinions of experts at such times.

EXPECTATIONS OF EXPERTS. It is often the case that it is much easier for surveyors to find the expectations of randomly sampled individuals than the expert opinions. Expert opinions may be generated only at the time a crucial decision is made, and not when an expert is asked to fill out a questionnaire. Moreover, experts may feel that their time is too valuable to merit attending carefully to a questionnaire.

It is now the case that economic forecasting has become a profession in which practitioners regularly publish their forecasts of macroeconomic variables, and thereby open themselves up to systematic evaluation by outsiders. Usually these forecasts have some basis in econometric models subject to judgementally introduced 'add factors'. Professional forecasters now make available regularly tabulated forecasts of macroeconomic variables for the succeeding few years. The success of these forecasts are now regularly computed by independent evaluators, and this provides a genuine incentive to forecast well. The market-place will tend to reduce the numbers of those who do not forecast well.

Professional forecasts made in organizations, when not made in anticipation of the kind of 'forecasting race' judged by outside evaluators, may not be serious individual attempts to predict. Instead, they may be 'conventional' forecasts using methods and information that are perceived as having sanction in the organization. Organizations may stipulate what information a forecaster is to use and how the information is to be translated into a forecast. The aim of such sanctions may be to produce uniformity in the organization as to factual premises on which decisions are made, but they may also lead to forecasts that are not as accurate as they could be. The costs to individuals of violating the assumptions of the organization may be very large relative to the possible benefits of forecasting well.

The distinction between day-to-day expectations of individuals and the expectations of experts may in practice not be important. The advantages that experts have, of access to data, understanding of economic theory and use of statistical methods, may confer little advantage in circumstances when the structure of the economy is changing. Then the data may be viewed as of little help, as it is generated by a different model, and statistical analysis also may be of little help. Experts may then fall back on adaptive expectations or other methods of producing guesses like those of ordinary individuals.

RATIONAL EXPECTATIONS VERSUS MECHANICAL EXPECTATIONS PROXIES. Why does the idea sound plausible that economic expectations of future inflation may be proxied fairly well by adaptive expectations or some other distributed lag on actual inflation? Is it just because of the theory of psychologists that human memory decays gradually through time and the notion that casual guesses of future inflation would correspond to recent memories of inflation? Perhaps instead it is that a distributed lag on inflation is not a bad way to forecast inflation.

Suppose people were asked each month to forecast the rate of increase of the price of some seasonal commodity, let us say, fresh tomatoes. Certainly, many of them would be aware that fresh tomatoes are more expensive in the winter, when they must be grown in hothouses or brought in from greater distances. Not all people would know this, and many who did know about the seasonality in price would not know its magnitude. But certainly a distributed lag with smoothly declining coefficients on actual tomato price changes is not what we would think of first to model their expectations. Such a distributed lag would imply some seasonality in expectations but would also generally imply that people misforecast the month of highest price.

If there is any doubt as to the value of simple expectations proxies for modelling the expectations of tomato consumers, there is certainly no doubt that it would be inappropriate to use such proxies to model the expectations of tomato producers. Some producers specialize in producing hothouse tomatoes, and time their production for the winter months. Surely *they* know in which month prices are higher, and by how much they tend to be higher.

How then should we build a model that describes the supply of tomatoes over time? Since tomatoes must be planted months in advance of the anticipated demand for them, the supply function for tomatoes must depend on expectations formed at this time by producers, as well as on seasonal factors affecting the cost of production. We might then model the supply on tomatoes by finding a good way to predict the price of tomatoes (using, say, seasonal dummies and other information) and substituting the prediction in place of the expectation in the supply function. The result would be a *rational expectations model*.

One could use such a model to predict the supply response to some variable that has been found to predict price. If, let us say, we found that bad weather in Mexico, which might later reduce the supply of winter tomatoes to the United States, tended to cause the seasonal peak in tomato prices in the United States to be higher than usual, then we might in these circumstances forecast the supply of domestic tomatoes in the United States to be higher than usual. A rational expectations model would produce such a forecast if the model was based on an empirical forecasting relation for price that used the weather variable as an explanatory variable.

Of course, for the purpose of forecasting supply we might also have used the 'naive' approach of estimating a forecasting equation directly for supply (without the intermediate step of developing a forecasting equation for price) depending on such variables as earlier weather and on seasonal dummies. Such a method may also predict supply satisfactorily, but it might not do as well since it would not make use of the information contained in economic theory, that weather affects supply only through its effect on rationally expected price. For example, suppose we had a long time series of data on various weather variables and prices but only a few observations on quantities supplied. We could not include all the weather variables directly in a 'naive' forecasting equation for supply, since we would thereby exhaust degrees of freedom. But we could first find how these weather variables predict price and then use a single price expectations variable to predict supply.

RATIONAL EXPECTATIONS SIMULTANEOUS EQUATIONS MODELS. The above example of the use of a rational expectations model was very special in that the model consisted only of a single equation relating supply to an earlier expectation of price. Moreover, the equation was used only to forecast supply in a situation where we expect the correlations observed in the past with explanatory variables to continue. Very often we wish instead to predict the effect on supply of some change in government policy or other structural change that is expected to change the correlations with other variables.

Suppose, for example, we wish to know the effect on the seasonal pattern of tomato supply in the United States of a government policy of blocking further international trade in tomatoes. Here, the naive forecasting model that related tomato supply to weather and seasonal dummy variables would be of no value. An estimated rational expectations model relating supply to expected price might still be of value. We need only to model the determination of expected price.

Suppose we then also estimate (using a sample period in which some tomatoes were imported) a domestic demand function for tomatoes, relating, say, total quantities demanded in the United States to contemporaneous price. Consider a two-equation model consisting of this demand equation and the rational expectations domestic supply equation for tomatoes described above. In the sample period domestic demand did not equal domestic supply because of imports. After the policy change the domestic supply and demand will be equal. Can we now predict how the seasonal pattern of quantities supplied may be changed by the government policy?

To answer this question, we cannot just solve the two-equation model with the two endogenous variables, quantity and price, because both price and expectations appear separately in the model. However, the expectation of price, if it is a rational expectation, ought to be determined by the very model in which it appears. How can we find the rational expectation of price?

One approach is first to guess a function relating expected price to the exogenous variables in the model, in our example, the seasonal dummies and weather variable. If one substitutes this guess into the model in place of the expected price, one then has an ordinary simultaneous equation model in price and quantity in terms of exogenous variables. However, unless one made a lucky guess, one would then find that the model that resulted from the guess was inconsistent with the guess, in that the model implies that a different way of forecasting price is optimal, given the expectations function.

What we need to find is an equation defining the expectation of price which, on substituting into the model, produces a model in which that equation gives the optimal forecast of price. Muth (1961) showed how this can be done if the simultaneous equations model is linear and if rational expectations are defined as mathematical expectations conditioned on variables in the model that are in the public's information set.

Using such a solution method, we might find how the seasonality of both quantities and price will be changed under the new government policy. In this simple example, doing this would seem to be preferable to using a model with an expectations proxy for price that did not take into account how the changing seasonal pattern of price would change the way expectations are formed.

MATHEMATICAL EXPECTATIONS. *Mathematical expectations*, conditioned on the information set available to agents, are convenient to use to represent economic expectations in simple linear rational expectations models of the kind described above. But of course, we should recognize that the term 'expectation' used in economics does not necessarily conform to the term 'expectation' used by probability theorists. There are other candidates to represent economic expectations, for example, other measures of central tendency such as the median or mode, or measures of central tendency applied to transformations of the random variable.

Ultimately, many economic models that involve mathematical expectations as economic expectations derive from the assumption of maximization of the mathematical expectation of a utility function. The mathematical expectations operator is initially brought into the assumptions of the model because such expected utility maximization is viewed as a good way to represent human behaviour. Expected utility maximization has been shown to follow from some plausible axioms representing an idealization of 'rational' human behaviour. But it is only in certain special cases that maximization of expected utility produces simple behavioural relations involving mathematical expectations as 'economic expectations' of the kind that many applied econometricians have been using.

Linear utility functions representing risk neutral agents may give rise to models in which agents care only about the mathematical expectations of variables, as in the models in finance in which the mathematical expectations of returns on various assets are equalized. A quadratic expected utility function may also produce models that depend on mathematical expectations. It is a result of Simon (1956) that if there are no terms of degree higher than two in control variables and exogenous stochastic processes then optimal behaviour depends linearly on a 'certainty equivalent' equal to the conditional expected value of future values of the stochastic process, and not on any other characteristics of their conditional distribution. Simon set up a problem in which there was nothing that could be done by the maximizing agent about the variance of the outcome. In contrast, in the capital asset pricing model in finance, a utility function quadratic in wealth (but where there are terms of degree higher than two in control variables and wealth) yields a behavioural relation that involves both a mathematical expectation and a variance matrix of the underlying stochastic variables. More generally, expected utility function models that are not linear or quadratic will produce Euler–equation type first-order conditions involving the mathematical expectation operator and economic variables, but usually some non-linear transformations of economic variables.

Many models, like our simple tomato rational expectations model described above, start from behavioural relations involving mathematical expectations, and do not derive these from the hypothesis of expected utility maximization. In these cases, the popularity of mathematical expectations as representations of economic expectations may derive from some intuitively desirable and convenient properties of mathematical expectations, properties that are not shared by other measures of central tendency.

The mathematical expectation of the sum of two random variables is equal to the sum of their mathematical expectations whether or not the two variables are independent, a property not shared by the median or mode, even if the variables are independent. If we have a joint distribution of two random variables, x and y, and we define the conditional distribution of x given y, then the mathematical expectation $E(x|y)$ of x in the conditional distribution is a function of y. The *law of iterated projections* states that the mathematical expectation of the mathematical expectation of x, $E(E(x|y))$ equals the mathematical expectation of x, $E(x)$. In simple terms, this law might be described as saying that people do not expect to change their expectations. Again, this law does not hold in general for the mode or median of x. On the other hand, the median has the desirable property that the median of any monotonic transformation of a random variable is the transformation of the median, a property not shared by the mathematical expectation.

CRITICISMS OF RATIONAL EXPECTATIONS MODELS. The simple supply and demand model for tomatoes described above was chosen as an ideal example of the application of rational expectations models. In this example there is substantial seasonal variation in price, which ought to be forecastable. Moreover, as the model was set up, only producers' expectations entered the model, and producers are far more likely than others to have rational expectations about price. But few of the applications of the theory of rational expectations have dealt with such ideal examples.

The best-known application of rational expectations models has been to an interpretation of the observed relation between unemployment in inflation. A.W. Phillips (1958) noted a negative relation between the unemployment rate and the wage inflation rate in the United Kingdom between 1861 and 1957. A similar relation was also found in the United States for much of the same sample period. Since then, the negative relation has broken down. Lucas (1973) and Sargent and Wallace (1973) offered interpretations of the Phillips relation and its subsequent breakdown. In its simplest terms, this interpretation asserts that there may be a stable relation between unemployment and *unexpected* inflation. Unexpected inflation may cause job seekers to misperceive the real value of wage offers they have received, and thus to accept offers that they wouldn't have accepted if they had known the true real wage they were getting. By accepting these jobs, they lower the unemployment rate. In the period Phillips studied, the price level might have been well-enough approximated by a random walk that actual inflation may have approximately equalled unexpected inflation. Since then, when inflation has become much more serially correlated, actual and expected inflation may have diverged widely.

In its general idea, the Lucas–Sargent–Wallace theory of the Phillips curve sounds like an appealing possibility. The question for econometric testing of the theory is whether we want to assume that expectations of unemployed workers are fully rational.

The tests Sargent (1976) made of the model are illustrative of the manner in which rational expectations models are often tested. Sargent tested whether the model holds under the assumption that unemployed workers are making optimal use in their forecasts of inflation of current and lagged values of the real government surplus, real and money government expenditures, the price level, the money supply, and a wage index. It is commonplace today in the rational expectations literature to see similar extravagant assumptions made about the information sets of ordinary individuals.

The most basic criticism of many rational expectations models is that they make implausible claims for individual economic agents' ability and willingness to compute. But the criticism of these models goes beyond that: see for example Friedman (1979), Shiller (1978) or Tobin (1980).

The rational expectations models assume that economic agents behave as if they know the structure of the economy so that they can compute the optimal forecasts that represent their expectations. But the structure of the economy is always changing, as technology, tastes, and government interventions change. These changes themselves vary qualitatively from time to time, and so it may not be possible for economic agents to group instances in such a way as to allow dealing with the changes in statistical terms. If these changes occur frequently relative to the speed at which people can assess the economy, it may never be appropriate to assume that their forecasts are optimal forecasts.

In most rational expectations models, the behaviour of the economic variables that individual economic agents must forecast is itself affected by the way in which the economic agents form expectations. This fact was noted above in connection with our efforts to solve the supply and demand rational expectations model for tomatoes. Thus, if economic agents learn something about how to forecast an economic variable, the random properties of the economic variable may change in consequence. A rational expectations equilibrium is achieved only when people have adopted a way of forecasting that is consistent with the implications for the economy of their own way of forecasting. How do they find such a way of forecasting? Achievement of a rational expectations equilibrium might take place as a consequence of a long iterative process, each step representing the learning by economic agents of how to forecast in the preceding step, and thereby necessitating the next step of learning anew how to forecast. In models that are more complicated than the simple supply and demand models (for example, models of the entire macroeconomy), the time required for each step may need to be enormous. The problem of convergence of forecasting methods to a rational expectations equilibrium recalls the problem in mathematical economics of the convergence of a price vector to Walrasian equilibrium. However, the former problem has received much less attention. Moreover, convergence may well be orders of magnitude slower in the former. It would appear likely, given the complexity of the macroeconomy, that economic agents learn very slowly about how to forecast given the present structure of the economy. Each step in the iteration requires sifting through large amounts of data and learning how these are related statistically.

Despite these criticisms, rational expectations models may well be useful for some applications when compared with alternative models based on expectations proxies. As regards the assumptions in the models for the ability and willingness of economic agents to process information, there is no alternative for model builders to that of judging for plausibility on a case by case basis.

RATIONAL EXPECTATIONS MODELS, STOCHASTIC PROCESSES AND OPTIMAL CONTROL. The advent of rational expectations in econometric models has marked a revolution in economic thinking that is comparable in the magnitude of its impact on the economics profession to the Keynesian revolution of a half century ago.

Muth (1961) and those who carried on the rational expectations literature have borrowed heavily from another literature once outside economics: the theory of stochastic processes and optimal control. What is substantially new about the rational expectations models derives ultimately from these theories, which were developed for the most part over the last half century. The implications of these theories were so profound that it was inevitable that they should make themselves felt in economics, as well as in many fields in science and engineering.

The rational expectations revolution is not primarily the result of any failure of conventional econometric models to forecast well, as some (e.g. Lucas and Sargent, 1981) have argued. It is true that initial optimism for the forecasting ability of such models has been tempered by experience, but it has not been established that shortcomings of the expectations modelling methods have been the major fault. It has certainly not been established empirically that rational expectations models can predict better.

Interest by economists in optimal control and the theory of stochastic processes was initially expressed in their efforts to apply control methods to existing econometric models, to achieve their stabilization. However, the optimal control of conventional 'Keynesian' econometric models involving expectations proxies like adaptive expectations has never become as influential in the profession as its developers had hoped. Perhaps the general profession thought that the methods of control were too refined for the crude models to which they were applied. More concern was felt for improving the models themselves.

The idea that optimal control might be applied to conventional Keynesian econometric models did have the effect of generating hopes that the macroeconomy might be controlled very well, 'fine tuned' so to speak, and thus great importance was placed on the structural stability of these models. Much of the polemic against 'Keynesian' economics waged by those who promoted rational expectations models as alternatives were really directed against these efforts to apply optimal control systematically to the models (see, for example, Sargent and Wallace, 1981). The fault of these models – in that they relied heavily on crude expectations proxies like adaptive expectations – became central to the criticism of them.

The rational expectations models applied stochastic optimal control theory by assuming, in effect, that human behaviour could be modelled as if everyone all along had been applying the principles of optimal control to their own economic decisions. Given the natural interest of economists in rational behaviour, the optimal filtering and extrapolation that was developed as part of the theory of stochastic processes would naturally be used in modelling how individuals forecast.

Of course, there are strict limits to the extent to which people's actual behaviour can be described in such terms.

Rational expectations models thus often sacrifice descriptive accuracy in the hope that the models would exhibit stability in the presence of interventions of the kind envisioned by makers of government macroeconomic policy. The models may not be generally well-suited to forecasting when the policy regime is unchanged. They are most appropriately considered policy analysis tools.

ROBERT J. SCHILLER

See also ADAPTIVE EXPECTATIONS; INTERTEMPORAL EQUILIBRIUM AND EFFICIENCY; RATIONAL EXPECTATIONS; UNCERTAINTY; UNCERTAINTY AND GENERAL EQUILIBRIUM.

BIBLIOGRAPHY

Cagan, P. 1956. The monetary dynamics of hyperinflation. In *Studies in the Quantity Theory of Money*, ed. M. Friedman, Chicago: University of Chicago Press.

Fisher, I. 1930. *The Theory of Interest*. New York: Macmillan.

Friedman, B.M. 1979. Optimal expectations and the extreme information assumptions of 'rational' expectations models. *Journal of Monetary Economics* 5, 23–41.

Katona, G. 1975. *Psychological Economics*. New York: Elsevier.

Lucas, R.E., Jr. 1973. Some international evidence on output–inflation tradeoffs. *American Economic Review* 63, 326–34.

Lucas, R.E., Jr. and Sargent, T.J. 1981. After Keynesian macroeconomics. In *Rational Expectations and Econometric Practice*, ed. Lucas and Sargent, Minneapolis: University of Minnesota Press.

Meiselman, D. 1962. *The Term Structure of Interest Rates*. Englewood Cliffs, New Jersey: Prentice-Hall.

Modigliani, F. and Sutch, R. 1966. Innovations in interest rate policy. *American Economic Review, Papers and Proceedings* 56, 178–97.

Muth, J.F. 1961. Rational expectations and the theory of price movements. *Econometrica* 29, 315–35.

Phillips, A.W. 1958. The relation between unemployment and the rate of change of money wage rates in the United Kingdom, 1861–1957. *Economica* 25, 283–99.

Sargent, T.J. 1976. A classical macroeconomic model for the United States. *Journal of Political Economy* 84, 207–37.

Sargent, T.J. and Wallace, N. 1973. Rational expectations and the dynamics of hyperinflation. *International Economic Review* 14, 328–50.

Sargent, T.J. and Wallace, N. 1976. Rational expectations and the theory of economic policy. *Journal of Monetary Economics* 2, 169–84.

Sargent, T.J. and Wallace, N. 1981. 'Rational' expectations, the optimal monetary instrument, and the optimal money supply rule. In *Rational Expectations and Econometric Practice*, ed. Robert E. Lucas and Thomas J. Sargent, Minneapolis: University of Minnesota Press.

Shiller, R.J. 1978. Rational expectations and the dynamic structure of rational expectations models: a critical review. *Journal of Monetary Economics* 4, 1–44.

Simon, H.A. 1956. Dynamic programming under uncertainty with a quadratic objective function. *Econometrica* 24, 74–81.

Tobin, J. 1980. *Asset Accumulation and Economic Activity*. Yrjö Jahnsson Lectures, Oxford: Basil Blackwell.

expected utility and mathematical expectation. 1. Expected utility theory deals with choosing among acts where the decision-maker does not know for sure which consequence will result from a chosen act. When faced with several acts, the decision-maker will choose the one with the highest 'expected utility', where the expected utility of an act is the sum of the products of probability and utility over all possible consequences.

The introduction of the concept of expected utility is usually attributed to Daniel Bernoulli (1738). He arrived at this concept as a resolution of the so-called St Petersburg paradox. It involves the following gamble: A 'fair' coin is flipped until the first time heads up. If this is at the kth flip, then the gambler receives $\$2^k$. The question arose how much to pay for participation in this gamble. Since the probability that heads will occur for the first time in the kth flip is 2^{-k} (assuming independence of the flips), and the gain then is $\$2^k$, the 'expected value' (i.e. the *mathematical expectation* of the gain) of the gamble is infinite. It has been observed though that gamblers were not willing to pay more then $\$2$ to $\$4$ to participate in such a gamble. Hence the 'paradox' between the mathematical expectation of the gain, and the observed willingness to pay.

Bernoulli suggested that the gambler's goal is not to maximize his expected gain, but to maximize the expectation of the logarithm of the gain which is $\Sigma_{j=1}^{\infty} 2^{-j} \log 2^j$, i.e. 2 log $2(= \log 4)$. Then the gambler is willing to pay $\$4$ for the gamble. The idea that *homo economicus* considers the expected utility of the gamble, and not the expected value, is a cornerstone of expected theory.

In the next section the approach of Savage to decisions under uncertainty is presented. In section 3 the von Neumann–Morgenstern characterization of expected utility maximization for the context of decisions under risk is given. Section 4 briefly mentions some related approaches. Section 5, the Appendix, defines (mathematical) expectation.

2. EXPECTED UTILITY WHEN APPLIED TO DECISIONS UNDER UNCERTAINTY; SAVAGE'S APPROACH

2.1 The main ingredients of a decision problem under uncertainty are acts consequences and states of nature. Suppose that a decision-maker has to choose one of three feasible acts f, g, h. Act f leads to one (only) of the two consequences a and b. Act g leads to a or c, act h to b or d. Thus the set of consequences, C, is in this example {a, b, c, d}.

The matching of feasible acts to consequences is expressed by the concept of 'state of nature', or 'state' for short. More precisely, a given state of nature indicates for each feasible act what the resulting consequence will be. In the above example, there are three feasible acts f, g, h, each leading to one of two possible consequences. See Table 2.1.

TABLE 2.1 The eight logically possible matchings of feasible acts to consequences

Acts	States							
	s_1	s_2	s_3	s_4	s_5	s_6	s_7	s_8
f	a	a	a	a	b	b	b	b
g	a	a	c	c	a	a	c	c
h	b	d	b	d	b	d	b	d

A state of nature completely resolves the uncertainty relating acts to consequences. If the decision-maker would know for sure which state of nature is the true one, then he would choose an act which results in a most desirable consequence. The desirability of a consequence neither depends on the act nor on the state of nature leading to it.

In constructing a table like Table 2.1 some of the states of nature may be deleted if the decision-maker is certain that they cannot occur.

The next step in the process of selecting the best act is to construct 'conceivable' acts, which are not feasible. Thus the set of acts, F, in Savage's set-up consists of all functions from the set of states of nature, S, to the set C of consequences. In our example there are 4^8 acts. Of these, three acts f, g and h are actually feasible: The additional 65533 acts are only conceivable. The construction of the conceivable acts and the possibility of ranking all acts of F is a basic assumption of the present approach. For the sake of presentation we will in the next subsection assume the validity of the expected utility theory and then we will return to the rationale of our construction.

2.2 Suppose for the present that the decision-maker, in choosing between acts, indeed computes the expected utility of each act, and selects a feasible act with the highest expected utility. Thus we are assuming that he has assigned probability $P(s)$ to every state of nature s in S, and the utility $U(c)$ to every consequence c in C. So, given an act f in F, the expected utility $EU(f)$ of f equals $\Sigma_{s \in S} P(s) U[f(s)]$. More generally, if the set S is infinite, then P is a finitely additive probability measure defined on all events (i.e. subsets of S), and $EU(f)$ equals $\int U[f(s)] \, dP(s)$ (assuming the integral to exist; say U is bounded; see the Appendix, on Mathematical Expectation, section 5). So in fact in this case the decision-maker has a well-defined 'preference relation' (i.e. binary relation) $\geqslant$ on the set of acts F, with, for all f, g in F:

$$(2.3) \qquad f \succsim g \qquad \text{iff } EU(f) \geqslant EU(g)$$

It is easily seen that the preference relation, defined in (2.3), is not affected when the utility function $U : C \mapsto R$ is replaced by any positive linear transformation of it (say $\bar{U} : c \mapsto \alpha U(c) + \beta$, for some real β and positive α).

2.4 If a preference relation, $\succsim$, over acts is derived from comparisons of expected utility as in (2.3), then it must satisfy several properties. We follow the terminology and order of Savage (1954). He listed seven postulates, five of which (P1 up to P4, and P7) are implied by (2.3). Postulate P1 says that the preference relation is complete ($f \succsim g$ or $g \succsim f$ for all acts f, g) and transitive. Postulate P2 is referred to as the sure-thing principle. It says that, when comparing two acts, only those states of nature matter, on which these acts differ. In other words, for the comparison between two acts, if they coincide on an event A, it really does not matter what actually the consequence is for each state in A. Thus P2 makes it possible to derive a preference relation over acts, conditioned on the event A^c; this for any event A.

Postulate P3 entails that the desirability of a consequence does not depend on the combination of state and act that lead to it; hence the possibility to express the desirability of consequences by a utility function on C.

P4 guarantees that the preference relation over acts induces a qualitative probability relation ('at least as probable as') over events, which is transitive and complete. P7 is a technical monotonicity condition.

P5 and P6 are Savage's only postulates which are not a necessary implication of (2.3). P5 simply serves to exclude the trivial case where the decision-maker is indifferent between any two acts. P6 implies some sort of continuity of the preference relation, and non-atomicity of the probability measure; the last term means that any non-impossible event can be partitioned into two non-impossible events. Hence there must be an infinite number of states.

Savage's great achievement was not to *assume* (2.3), but to show that his list of postulates P1–P7 *implies* that the preference relation over acts has an expected utility representation as in (2.3). Savage argued compellingly for the appropriateness of his postulates. Furthermore, Savage showed that the probability measure in (2.3) is uniquely determined by the preference relation $\geqslant$, and that the utility function is unique up to a positive linear transformation.

2.5. The significance of Savage's achievement is that it gives the first, and until today most complete, conceptual foundation to expected utility. Savage's conclusion, to use expected utility for the selection of optimal acts, can be used even if we do not have the structure and the seven postulates of Savage. Indeed, the assumption needed on consequences, states, acts, and preferences, is that they can be extended so as to satisfy all requirements of Savage's model. Also other models, as mentioned in section 4, can be used to obtain expected utility representations.

Given a decision problem under uncertainty, if we assume that it can be embedded in Savage's framework, then it is not necessary to actually carry out this embedding. In other words, if the decision-maker is convinced that in principle it is possible to construct the conceivable acts as in subsection 2.1 and the ranking of all acts in accordance with the postulates, then this construction does not have to be made. Instead one can directly try to assess probabilities and utilities, and apply the expected utility criterion. As an example, suppose a market-vendor has to decide whether to order 50 portions of ice-cream (f), or not (g). One portion costs \$1, and is sold for \$2. If the weather will be nice the next day, the school nearby will allow the children to go to the market, and all 50 portions, if ordered, will be sold, yielding a profit of \$50. If the weather is not nice, no portion will be sold. We assume that the ice-cream cannot be kept in stock and hence bad weather will yield a 'gain' of \$$-50$ if the portions have been ordered.

Instead of embedding the above example into Savage's framework, the market salesman may immediately assess P_1 (or $1 - P_1$), the probability for good (bad) weather; next assess the utilities of gaining \$50, \$0 and $-$\$50; finally order the 50 portions if $P_1 U(\$50) + (1 - P_1) U(-\$50) > U(\$0)$.

Theoretical conclusions can be derived from the mere assumption of expected utility maximization, without an actual assessment of the probabilities and utilities. Examples are the theories of attitudes towards risk, with applications to insurance, portfolio choice, etc. The validity of these applications depends on expected utility theory, which in turn depends on the plausibility of Savage's model (or other derivations of expected utility).

Another important theoretical application of Savage's model is to neo-Bayesian statistics. For applied statistics, in this vein, the availability of a 'prior distribution', as proved by Savage's approach, is essential.

3. EXPECTED UTILITY WHEN APPLIED TO DECISIONS UNDER RISKS; THE VON NEUMANN–MORGENSTERN APPROACH

Special and extreme cases of decisions under uncertainty are decisions in 'risky' situations. In decisions under uncertainty, as exposited in the previous section, the decision-maker who follows the dictum of expected utility has to assign utilities to the consequences and probabilities to the states. He can do it by mimicking the proof of Savage's theorem, or more directly by organizing his information, as the case may be.

Decision-making under risk considers the special case where the formulation of the problem for the decision-maker includes probabilities for the events, so that he only has to derive the utilities of consequences. As an example, consider a gambler in a casino who assumes that the roulette is really unbiased, so that each number has probability $1/37$ (or $1/38$). Another example is the St Petersburg paradox, described in subsection 1.2.

Within the framework of expected utility theory, for the evaluation of an act, only its probability distribution over the consequences has to be taken into account. Thus, for decision-making under risk, with probabilities known in advance, one may just as well describe acts as probability distributions over consequences instead of as functions from the states to the consequences.

3.1. Let us denote by L the set of probability distributions over C with finite support. We refer to them as lotteries. Von Neumann and Morgenstern (1947, Appendix) suggested conditions on a preference relation $\geqslant$ between lotteries, necessary and sufficient for the existence of a real-valued utility function U on C, such that for any two lotteries P and Q in L:

$$(3.2) \qquad P \geqslant Q \qquad \text{iff} \sum_{c \in C} P(c)U(c) \geqslant \sum_{c \in C} Q(c)U(c)$$

It is easy to see that the utility function, U, is unique up to positive linear transformations. Before we present a version of von Neumann–Morgenstern's theorem, recall that for any $0 \leqslant \alpha \leqslant 1$, and for any two lotteries P and W, $R := \alpha P + (1-\alpha)Q$ is again a lottery, assigning probability $R(c) = \alpha P(c) + (1-\alpha)Q(c)$ to any c in C. Also note that the assumption that all lotteries are given, is sometimes as heroic as Savage's assumption that all functions from S to C are conceivable acts.

The first axiom of von Neumann–Morgenstern, NM1, says that the preference relation over the lotteries is complete and transitive. NM2, the continuity axiom, says that, if $P \succ Q \succ R$, then there are α, β in $]0, 1[$, such that $\alpha R + (1-\alpha)P \succ Q \succ \beta P + (1-\beta)R$. Here the strict preference relation $\succ$ is derived from $\gtrsim$ in the usual way: $P \succ Q$ if $P \gtrsim Q$ and not $Q \gtrsim P$.

The third axiom NM3 is the independence axiom. It says that for α in $]0,1]$, P is preferred to Q iff $\alpha P + (1-\alpha)R$ is preferred to $\alpha Q + (1-\alpha)R$. This condition is the antecedent of Savage's sure-thing principle, and is the most important innovation of the above axioms.

3.3 Von Neumann and Morgenstern originally stated their theorem for more general sets than L. They did it for so-called mixture spaces, i.e. spaces endowed with some sort of convex combination operation. This has been done more precisely by Herstein and Milnor (1953).

Von Neumann and Morgenstern introduced their theory of decision-making under risk as a normative tool for playing zero-sum games in strategic form. There the 'player' (i.e. decision-maker) can actually construct any lottery he wishes over his pure-strategies (but not over his consequences).

The theorem of von Neumann and Morgenstern, stated above, is a major step in the proof of Savage's theorem.

Recently there has been much research on decision making under risk for its own end. Some of this research is experimental, subjects are asked to express their preferences between lotteries. These experiments, or polls, reveal violations of most of the axioms. They lead to representations different from expected utility.

4. OTHER APPROACHES AND BIBLIOGRAPHICAL REMARKS

The first suggestion for expected utility theory in decision-making under uncertainty in the vein of Savage was Ramsey's (1931). His model was not completely formalized. The work of Savage was influenced by de Finetti's approach to probabili-

ties, as in de Finetti (1931, 1937). The decision theoretic framework to which Savage's expected utility model owes much is that of Wald (1951), who regards a statistician as a decision-maker.

A model which can be considered intermediate between those of Savage and von Neumann and Morgenstern is that considered by Anscombe and Aumann (1963). Formally it is a special case of a mixture set, but like Savage it introduces states of nature, and gives a simultaneous derivation of probabilities for the states, and of utilities for the consequences. A consequence in this model consists of a lottery over deterministic outcomes; this involves probabilities known in advance, as in the approach of von Neumann and Morgenstern. The Anscombe and Aumann theory, as well as most of the technical results up to 1970, are presented in detail in Fishburn (1970).

In the expected utility theory, described above, the desirability (utility) of consequences does not depend on acts or states of nature. This is a restriction in many applications. For example the desirability of family income may depend on whether the state of nature is 'head of family alive' or 'head of family deceased'. Karni (1985) summarized and developed the expected utility theory without the restrictive assumption of state-independent preferences over consequences.

Ellsberg (1961) argued against the expected utility approach of Savage by proposing an example, inconsistent with it. A way of resolving the inconsistency is to relax the additivity property of the involved probability measures. Schmeidler (1984) formulated expected utility theory with non-additive probabilities for the framework of Anscombe and Aumann (1963). Gilboa (1985) did the same for the original framework of Savage. Wakker (1986) obtained expected utility representation, including the non-additive case, for a finite number of states of nature and non-linear utility.

5. APPENDIX: MATHEMATICAL EXPECTATION

5.1 *Expectation with respect to finitely additive probability.* A non-empty collection Σ of subsets (called events) of a non-empty set S is said to be an algebra if it contains the complement of each set belonging to it, and it contains the union of any two sets belonging to it. A (finitely additive) probability P on Σ assigns to every event in Σ a number between 0 and 1 such that $P(S) = 1$ and for any two *disjoint* events A and B, $P(A \cup B) = P(A) + P(B)$.

A random variable X is a real-valued function on S such that, for any open or closed (bounded or unbounded) interval I, $\{s \in S | X(s) \in I\}$ (or $[X \in I]$ for short) is an event i.e., in Σ. Given such a random variable X, its (mathematical) expectation is:

$$(5.2) \quad E(X) = \int_0^\infty P[X \geqslant \alpha] \, d\alpha - \int_{-\infty}^0 (1 - P[X \geqslant \alpha]) \, d\alpha,$$

where the integration above is Riemann-integration and it is assumed that the integral exist. The integrands in (5.2) are monotonic, so $E(X)$ exist if X is bounded. If the random variable X has finitely many values, say $x_1, \ldots, x_n$ then (5.2) reduces to

$$(5.3) \qquad\qquad E(X) = \sum_{i=1}^n P(X = x_i)x_i,$$

However, an equation like that above may not hold if the random variable obtains countably many different values. An example will be provided in subsection 5.7.

5.4 σ-*additive probability.* Kolmogorov (1933) imposed an additional continuity assumption on probability P on Σ: To

231

simplify presentation he first assumed that Σ is a σ-algebra, i.e., an algebra such that for every sequence of events $(A_i)_{i=1}^{\infty}$ it contains its union $\bigcup_{i=1}^{\infty} A_i$. He then required that $P(\bigcup_{i=1}^{\infty} A_i) = \Sigma_{i=1}^{\infty} P(A_i)$ if the A_i's are pairwise disjoint.

This last property is referred to as σ-additivity of the probability P. In this way Kolmogorov transformed large parts of probability theory into (a special case of) measure theory. Thus an expectation of a random variable X is

$$(5.5) \qquad E(X) = \int_S X(s)\,dP(s)$$

where the right side is a Lebesgue integral (if it exists...), defined as a limit of integrals of random variables with countably many values. Let Y be such a random variable with values $(y_i)_{i=1}^{\infty}$, then

$$(5.6) \qquad E(Y) = \sum_{i=1}^{\infty} P(Y = y_i) y_i$$

if the right side is absolutely convergent.

5.7 *An example* will now be introduced of a finitely additive probability, i.e. a probability for which (5.3) holds but (5.6) does not hold. Let S be the set of rational numbers in the interval $[0, 1]$ and let Σ be the algebra of all subsets of S. (It is in fact a σ-algebra.) For $0 \leqslant \alpha \leqslant \beta \leqslant 1$ define $P(S \cap [\alpha, \beta]) = \beta - \alpha$ and extend P to all subsets of Σ. For each s in S, $P(s) = 0$. Since S is countable we can write $S = \{s_1, s_2, \ldots\}$ and $1 = P(S) > \Sigma_{i=1}^{\infty} P(s_i) = 0$. Defining $Y(s_i) = 1/i$ for all i, we get a contradiction to (5.6). The finitely additive probability P has also the property implied by Savage's P6 (see 2.4): If $P(A) > 0$ then there is an event $B \subset A$ such that $0 < P(B) < P(A)$.

5.8 *Distributions.* A non-decreasing right continuous function on the extended real line is called a distribution function if $F(-\infty) = 0$ and $F(\infty) = 1$. Given a random variable X, its distribution function F_X is defined by $F_X(\alpha) = P(X \leqslant \alpha)$ for all real α. Then

$$(5.9) \qquad E(X) = \int_0^{\infty} [1 - F_X(\alpha)]\,d\alpha - \int_{-\infty}^0 F(\alpha)\,d\alpha$$

which is the dual of formula (5.2). If the distribution F_X is smooth we say that the random variable X has a density $f_X : R \to R$, which is the derivative of F_X. In this case

$$(5.10) \qquad E(X) = \int_{-\infty}^{\infty} \alpha f(\alpha)\,d\alpha$$

5.11 *Non-additive probability.* A function $P : \Sigma \to [0, 1]$ is said to be *non-additive* probability (or capacity) if $P(S) = 1$, $P(\phi) = 0$ and for $A \subset B, P(A) \leqslant P(B)$. Choquet (1954) suggested to integrate a random variable with respect to non-additive probability by formula (5.2).

DAVID SCHMEIDLER AND PETER WAKKER

See also ALLAIS PARADOX; MEAN VALUES; RISK; SUBJECTIVE PROBABILITY; UNCERTAINTY; UTILITY THEORY AND DECISION-MAKING

BIBLIOGRAPHY

Anscombe, F.J. and Aumann, R.J. 1963. A definition of subjective probability. *Annals of Mathematical Statistics* 34, 199–205.
Bernoulli, D. 1738. Specimen theoriae novae de mensura sortis. *Commentarii Academiae Scientiarum Imperialis Petropolitanae* 5, 175–92. Translated into English by L. Sommer (1954) as: Exposition of a new theory on the measurement of risk, *Econometrica* 12, 23–36; or in *Utility Theory: A Book of Readings*, ed. A.N. Page, New York: Wiley, 1986.

Choquet, G. 1953–54. Theory of capacities. *Annales de l'Institut Fourier* (Grenoble), 131–295.
de Finetti, B. 1931. Sul significato soggettivo della probabilita. *Fundamenta Mathematicae* 17, 298–329.
de Finetti, B. 1937. La prévision: ses lois logiques, ses sources subjectives. *Annales de l'Institut Henri Poincaré* 7, 1–68. Translated into English in *Studies in Subjective Probability*, ed. H.E. Kyburg and H.E. Smokler, 1964, New York: Wiley.
Ellsberg, D. 1961. Risk, ambiguity, and the Savage axioms. *Quarterly Journal of Economics* 75, 643–69.
Fishburn, P.C. 1970. *Utility Theory for Decision Making.* New York: Wiley.
Gilboa, I. 1986. Non-additive probability measures and their applications in expected utility theory. PhD Thesis submitted to Tel Aviv University.
Herstein, I.N. and Milnor, J. 1953. An axiomatic approach to measurable utility. *Econometrica* 21, 291–7.
Karni, E. 1985. *Decision-Making under Uncertainty: The Case of State-Dependent Preferences.* Cambridge, Mass.: Harvard University Press.
Kolmogorov, A.N. 1933. *Grundbegriffe der Wahrscheinlichkeitsrechnung.* Berlin. Translated into English by Nathan Morrison (1950, 2nd edn, 1956), New York: Chelsea Publishing Company.
Loeve, M. 1963. *Probability Theory.* 3rd edn, Princeton: Van Nostrand.
von Neumann, J. and Morgenstern, O. 1947. *Theory of Games and Economic Behavior.* 2nd edn, Princeton: Princeton University Press.
Ramsey, F.P. 1931. Truth and probability. In *The Foundations of Mathematics and Other Logical Essays*, ed. R.B. Braithwaite, New York: Harcourt, Brace.
Savage, L.J. 1954. *The Foundations of Statistics.* New York: Wiley, 2nd edn, 1972.
Schmeidler, D. 1984. Subjective probability and expected utility without additivity. CARESS, University of Pennsylvania and IMA University of Minnesota, mimeo.
Wald, A. 1951. *Statistical Decision Functions.* New York: Wiley.
Wakker, P.P. 1986. Representations of choice situations. PhD thesis, University of Tilburg, Department of Economics.

expected utility hypothesis. The expected utility hypothesis of behaviour towards risk is essentially the hypothesis that the individual decision-maker possesses (or acts as if possessing) a 'von Neumann–Morgenstern utility function' $U(\cdot)$ or 'von Neumann–Morgenstern utility index' $\{U_i\}$ defined over some set of outcomes, and when faced with alternative risky prospects or 'lotteries' over these outcomes, will choose that prospect which maximizes the expected value of $U(\cdot)$ or $\{U_i\}$. Since the outcomes could represent alternative wealth levels, multidimensional commodity bundles, time streams of consumption, or even non-numerical consequences (e.g. a trip to Paris), this approach can be applied to a tremendous variety of situations, and most theoretical research in the economics of uncertainty, as well as virtually all applied work in the field (e.g. optimal trade, investment or search under uncertainty) is undertaken in the expected utility framework.

As a branch of modern consumer theory (e.g. Debreu, 1959, ch. 4), the expected utility model proceeds by specifying a set of objects of choice and assuming that the individual possesses a preference ordering over these objects which may be represented by a real-valued maximand or 'preference function' $V(\cdot)$, in the sense that one object is preferred to another if and only if it is assigned a higher value by this preference function. However, the expected utility model differs from the theory of choice over non-stochastic commodity bundles in two important respects. The first is that since it is a theory of choice under uncertainty, the objects of choice are not deterministic outcomes but rather probability distributions over these outcomes. The second difference is that, unlike in the non-stochastic case, the expected utility model imposes a very specific restriction on the functional form of the preference function $V(\cdot)$.

The formal representation of the objects of choice, and hence of the expected utility preference function, depends upon the structure of the set of possible outcomes. When there are a finite number of outcomes $\{x_1, \ldots, x_n\}$, we can represent any probability distribution over this set by its vector of probabilities $P = (p_1, \ldots, p_n)$ (where $p_i = \mathrm{prob}(\tilde{x} = x_i)$), and the preference function takes the form

$$V(P) = V(p_1, \ldots, p_n) \equiv \Sigma U_i p_i.$$

When the outcome set consists of the real line or some subset of it, probability distributions are represented by their cumulative distribution functions $F'(\cdot)$ (where $F(x) = \mathrm{prob}(\tilde{x} \leqslant x)$), and the expected utility preference function takes the form $V(F) \equiv \int U(x)\, dF(x)$. (When $F(\cdot)$ possesses a density function $f(\cdot) \equiv F'(\cdot)$ this integral can be equivalently written as $\int U(x)f(x)\, dx$.) When the outcomes are multivariate commodity bundles of the form $(z_1, \ldots, z_n)$, $V(\cdot)$ takes the form $\int \ldots \int U(z_1, \ldots, z_n)\, dF(z_1, \ldots, z_n)$ over multivariate cumulative distribution functions $F(\cdot, \ldots, \cdot)$. The expected utility model derives its name from the fact that in each case, the preference function $V(\cdot)$ consists of the mathematical expectation of the von Neumann–Morgenstern utility function $U(\cdot)$, $U(\cdot, \ldots, \cdot)$, or utility index $\{U_i\}$ with respect to the probability distribution $F(\cdot)$, $F(\cdot, \ldots, \cdot)$, or P.

Mathematically, the hypothesis that the preference function $V(\cdot)$ takes the form of a statistical expectation is equivalent to the condition that it be 'linear in the probabilities'; that is, either a weighted sum of the components of P (i.e. $\Sigma U_i p_i$) or else a weighted integral of the functions $F(\cdot)$ or $f(\cdot)[\int U(x)\, dF(x)$ or $\int U(x)f(x)\, dx]$. Although this still allows for a wide variety of attitudes towards risk, depending upon the shape of the von Neumann–Morgenstern utility function $U(\cdot)$ or index $\{U_i\}$, the restriction that $V(\cdot)$ be linear in the probabilities is the primary empirical feature of the expected utility model and provides the basis for many of its observable implications and predictions.

It is important to distinguish between the preference function $V(\cdot)$ and the von Neumann–Morgenstern utility function $U(\cdot)$ (or index $\{U_i\}$) of an expected utility maximizer, in particular with regard to the prevalent though mistaken belief that expected utility preferences are somehow 'cardinal' in a sense which is not exhibited by preferences over non-stochastic commodity bundles. As with any real-valued representation of a preference ordering, an expected utility preference function $V(\cdot)$ is 'ordinal' in that it may be subject to any increasing transformation without affecting the validity of the representation; thus, for example, if $V(F) \equiv \int U(x)\, dF(x)$ represents the preferences of some expected utility maximizer, so will the (nonlinear) preference function $Y(F) \equiv [\int U(x)\, dF(x)]^3$. On the other hand, the von Neumann–Morgenstern utility functions which generate these preference functions are 'cardinal' in the sense that a function $U^*(\cdot)$ will generate an ordinally equivalent linear preference function $V^*(F) \equiv \int U^*(x)\, dF(x)$ if and only if it satisfies the cardinal relationship $U^*(x) \equiv a \cdot U(x) + b$ for some $a > 0$ (in which case $V^*(\cdot) = a \cdot V(\cdot) + b$). However, such situations also occur in the theory of preferences over non-stochastic commodity bundles: the Cobb–Douglas preference function $\alpha \cdot \ln(x) + \beta \cdot \ln(y) + \gamma \cdot \ln(z)$ (written here in its additive form) can be subject to any increasing transformation and is clearly ordinal, even though a vector of parameters $(\alpha^*, \beta^*, \gamma^*)$ will generate an ordinally equivalent additive form $\alpha^* \cdot \ln(x) + \beta^* \cdot \ln(y) + \gamma^* \cdot \ln(z)$ if and only if it satisfies the cardinal relationship $(\alpha^*, \beta^*, \gamma^*) = \lambda \cdot (\alpha, \beta, \gamma)$ for some $\lambda > 0$.

In the case of a simple outcome set of the form $\{x_1, x_2, x_3\}$, it is possible to illustrate the 'linearity in the probabilities'

property of an expected utility maximizer's preferences over lotteries. Since every probability distribution (p_1, p_2, p_3) over this set must satisfy the condition $\Sigma p_i = 1$, we may represent each such distribution by a point in the unit triangle in the (p_1, p_3) plane, with p_2 given by $p_2 = 1 - p_1 - p_3$ (Figures 1 and 2). Since they represent the loci of solutions to the equations

$$U_1 p_1 + U_2 p_2 + U_3 p_3 = U_2 - [U_2 - U_1] \cdot p_1 + [U_3 - U_2] \cdot p_3$$
$$= \text{constant}$$

for the fixed utility indices $\{U_1, U_2, U_3\}$, the indifference curves of an expected utility maximizer consist of parallel straight lines in the triangle of slope $[U_2 - U_1]/[U_3 - U_2]$, as illustrated by the solid lines in Figure 1. An example of indifference curves which do *not* satisfy the expected utility hypothesis (i.e. are not linear in the probabilities) is given by the solid curves in Figure 2.

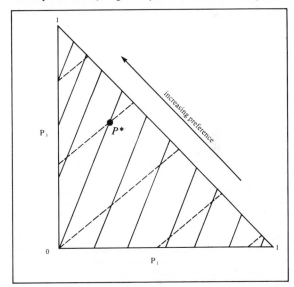

Figure 1   Expected Utility Indifference Curves

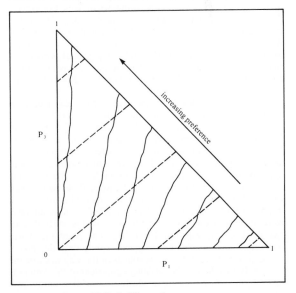

Figure 2   Non-Expected Utility Indifference Curves

When the outcomes $\{x_1, x_2, x_3\}$ represent different levels of wealth with $x_1 < x_2 < x_3$, this diagram can be used to illustrate other possible aspects of an expected utility maximizer's attitudes toward risk. On the general principle that more wealth is better, it is typically postulated that any change in a distribution (p_1, p_2, p_3) which increases p_3 at the expense of p_2, increases p_2 at the expense of p_1, or both, will be preferred by the individual: this property is known as 'first-order stochastic dominance preference'. Since such shifts of probability mass are represented by north, west or north-west movements in the diagram, first-order stochastic dominance preference is equivalent to the condition that indifference curves are upward sloping, with more preferred indifference curves lying to the north-west. Algebraically, this is equivalent to the condition $U_1 < U_2 < U_3$.

Another widely (though not universally hypothesized aspect of attitudes towards risk is that of 'risk aversion' (e.g. Arrow, 1974, ch. 3; Pratt, 1964). To illustrate this property of preferences, consider the dashed lines in Figure 1, which represent loci of solutions to the equations

$$x_1 p_1 + x_2 p_2 + x_3 p_3 = x_2 - [x_2 - x_1] \cdot p_1 + [x_3 - x_2] \cdot p_3$$

$$= \text{constant}$$

and hence may be termed 'iso-expected *value* loci'. Since north-east movements along any of these loci consist of increasing the tail probabilities p_1 and p_3 at the expense of middle probability p_2 in a manner which preserves the mean of the distribution, they correspond to what are termed 'mean preserving increases in risk' (e.g. Rothschild and Stiglitz, 1970, 1971). An individual is said to be 'risk averse' if such increases in risk always lead to less preferred indifference curves, which is equivalent to the graphical condition that the indifference curves be steeper than the iso-expected value loci. Since the slope of the latter is given by $[x_2 - x_1]/[x_3 - x_2]$, this is equivalent to the algebraic condition that $[U_2 - U_1]/[x_2 - x_1] > [U_3 - U_2]/[x_3 - x_2]$. Conversely, individuals who *prefer* mean preserving increases in risk are termed 'risk loving': such individuals' indifference curves will be flatter than the iso-expected value loci, and their utility indices will satisfy $[U_2 - U_1]/[x_2 - x_1] < [U_3 - U_2]/[x_3 - x_2]$.

Note finally that the indifference map in Figure 1 indicates that the lottery P^* is indifferent to the origin, which represents the degenerate lottery yielding x_2 with certainty. In such a case the amount x_2 is said to be the 'certainty equivalent' of the lottery P^*. The fact that the origin lies on a lower iso-expected value locus than P^* reflects a general property of risk averse preferences, namely that the certainty equivalent of any lottery will always be less than its mean. (For risk lovers, the opposite is always the case.)

When the outcomes are elements of the real line, it is possible to represent the above (as well as other) aspects of preferences in terms of the shape of the von Neumann–Morgenstern utility function $U(\cdot)$, as seen in Figures 3 and 4. In each figure, consider the lottery which assigns the probabilities 2/3: 1/3 to the outcome levels x' and x'', respectively. The expected value of this lottery (i.e. the value $\bar{x} = 2/3 \cdot x' + 1/3 \cdot x''$) is seen to lie between these two values, two-thirds of the way towards x'. The expected *utility* of this lottery – i.e. the value $\bar{u} = 2/3 \cdot U(x') + 1/3 \cdot U(x'')$ – is similarly seen to lie between $U(x')$ and $U(x'')$ on the vertical axis, two-thirds of the way towards $U(x')$. The point $(\bar{x}, \bar{u})$ will accordingly lie on the line segment connecting the points $(x', U(x'))$ and $(x'', U(x''))$, two-thirds of the way towards the former. In each figure, the certainty equivalent of this lottery is given by that sure outcome c which also yields a utility level of $\bar{u}$.

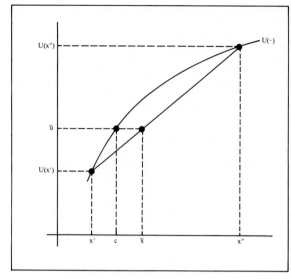

Figure 3 Von Neumann-Morgenstern Utility Function of a Risk Averse Individual

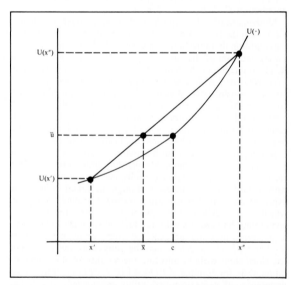

Figure 4 Von Neumann-Morgenstern Utility Function of a Risk Loving Individual

It is clear from our definition of first-order stochastic dominance preference above that this property of preferences can be extended to the case of density functions $f(\cdot)$ or cumulative distribution functions $F(\cdot)$ over the real line (e.g. Quirk and Saposnik, 1962), and that it is equivalent to the condition that $U(x)$ be an increasing function of x, as in Figures 3 and 4. It is also possible to generalize the notion of a mean preserving increase in risk to density functions or cumulative distribution functions (e.g. Rothschild and Stiglitz, 1970, 1971), and our earlier algebraic condition for risk aversion generalizes to the condition that $U''(x) < 0$ for all x, i.e. that the von Neumann–Morgenstern utility function $U(\cdot)$ be concave, as in Figure 3. As before, the property of risk aversion implies that the certainty equivalent c of any lottery will always lie below its mean, as seen in Figure 3, and once again, the opposite is true for the convex utility function of a risk lover, as seen in Figure 4. Two of the earliest and most important graphical analyses of risk attitudes in terms of the shape of the von Neumann–Morgenstern utility function are those of Friedman and Savage (1948) and Markowitz (1952).

The tremendous analytic capabilities of the expected utility model for the study of behaviour towards risk derive largely from the work of Arrow (1974) and Pratt (1964). Roughly speaking, these researchers showed that the 'degree' of concavity of the von Neumann–Morgenstern utility function can be used to provide a measure of an expected utility maximizer's 'degree' of risk aversion. Formally, the Arrow–Pratt characterization of comparative risk aversion is the result that the following conditions on a pair of (increasing, twice differentiable) von Neumann–Morgenstern utility functions $U_a(\cdot)$ and $U_b(\cdot)$ are equivalent:

$U_a(\cdot)$ is a concave transformation of $U_b(\cdot)$ (i.e.

$U_a(x) \equiv \rho[U_b(x)]$ for some increasing concave function $\rho(\cdot)$), $-U_a''(x)/U_a'(x) \geq -U_b''(x)/U_b'(x)$, for each x, and

if c_a and c_b solve

$$U_a(c_a) = \int U_a(x)\,\mathrm{d}F(x) \quad \text{and} \quad U_b(c_b) = \int U_b(x)\,\mathrm{d}F(x)$$

for some distribution $F(\cdot)$, then $c_a \leq c_b$,

and if $U_a(\cdot)$ and $U_b(\cdot)$ are both concave, these conditions are in turn equivalent to:

if $r > 0$, $E[\tilde{z}] > r$, and α_a and α_b maximize

$$\int U_a[(I-\alpha)r + \alpha z]\,\mathrm{d}F(z) \quad \text{and} \quad \int U_b[(I-\alpha)r + \alpha z]\,\mathrm{d}F(z)$$

respectively, then $\alpha_a \leq \alpha_b$.

The first two of these conditions provide equivalent formulations of the notion that $U_a(\cdot)$ is a more concave function than $U_b(\cdot)$. In particular, the curvature measure $R(x) \equiv -U''(x)/U'(x)$ is known as the 'Arrow–Pratt index of (absolute) risk aversion', and plays a key role in the analytics of the expected utility model. The third condition states that the more risk averse utility function $U_a(\cdot)$ will never assign a higher certainty equivalent to any lottery $F(\cdot)$ than will $U_b(\cdot)$. The final condition pertains to the individuals' respective demands for risky assets. Specifically, assume that each of them must allocate $\$I$ between two assets, one yielding a riskless (gross) return of r per dollar, and the other yielding a risky return $\tilde{z}$ with a higher expected value. This condition thus says that the less risk averse utility function $U_b(\cdot)$ will generate at least as great a demand for the risky asset than the more risk averse utility function $U_a(\cdot)$. It is important to note that it is the *equivalence* of the above certainty equivalent and asset demand conditions which makes the Arrow–Pratt characterization such an important result in expected utility theory. (See Ross, 1981, however, for an alternative and stronger characterization of comparative risk aversion.)

Although the applications of the expected utility model extend to virtually all branches of economic theory (e.g. Hey, 1979), much of the flavour of these analyses can be gleaned from Arrow's (1974, ch. 3) analysis of the portfolio problem of the previous paragraph: rewriting $(Ir - \alpha)r + \alpha z$ as $Ir + \alpha \cdot (z - r)$, the first-order condition for this problem can be

expressed as:

$$\int z \cdot U'[Ir + \alpha \cdot (z - r)]\,\mathrm{d}F(z)$$
$$- r \cdot \int U'[Ir + \alpha \cdot (z - r)]\,\mathrm{d}F(z) = 0,$$

that is, the marginal *expected* utility of the last dollar allocated to each asset is the same. The second-order condition can be written as:

$$\int (z - r)^2 \cdot U''[Ir + \alpha \cdot (z - r)]\,\mathrm{d}F(z) < 0$$

and is ensured by the property of risk aversion [i.e. $U''(\cdot) < 0$].

As usual, we may differentiate the first-order condition to obtain the effect of a change in some parameter, say initial wealth I, on the optimal level of investment in the risky asset (i.e. on the optimal value of α). Differentiating the first-order condition (including α) with respect to I, solving for $\mathrm{d}\alpha/\mathrm{d}I$, and invoking the second-order condition and the positivity of r yields that this effect possesses the same sign as:

$$\int (z - r) \cdot U''[Ir + \alpha \cdot (z - r)]\,\mathrm{d}F(z).$$

Making the substitution $U''(\cdot) \equiv -R(\cdot) \cdot U'(\cdot)$ and subtracting $R(Ir)$ times the first-order condition yields that this term is equal to:

$$- \int (z - r) \cdot \{R[Ir + \alpha \cdot (z - r)] - R(I)\}$$
$$\times U'[Ir + \alpha \cdot (z - r)]\,\mathrm{d}F(z).$$

On the assumption that α is positive and $R(\cdot)$ is monotonic, the expression $(z - r) \cdot [R(Ir + \alpha \cdot (z - r)) - R(Ir)]$ will possess the same sign as $R'(\cdot)$. This implies that the derivative $\mathrm{d}\alpha/\mathrm{d}I$ will always be positive (negative) whenever the Arrow–Pratt index $R(x)$ is a decreasing (increasing) function of the individual's wealth level x. In other words, an increase in initial wealth will always increase (decrease) the demand for the risky asset if and only if $U(\cdot)$ exhibits decreasing (increasing) absolute risk aversion in wealth. Further examples of the analytics of risk and risk aversion in the expected utility model may be found in the above references as well as the surveys of Hirshleifer and Riley (1979), Lippman and McCall (1981) and Machina (1983b).

Finally, in addition to the case of preferences over probability distributions, it is also possible to refer to expected utility preferences over alternative 'state-payoff bundles' (e.g. Hirshleifer, 1965, 1966). This approach postulates a (typically finite) set of 'states of nature' (i.e. a mutually exclusive and exhaustive partition of the set of observable occurrences) and the objects of choice consist of state-payoff bundles of the form $(x_1, \ldots, x_n)$, where x_i denotes the outcome the individual will receive should state i occur. An expected utility maximizer whose subjective probabilities of the n states are given by the values $(\bar{p}_1, \ldots, \bar{p}_n)$ will rank such bundles according to the preference function $V(x_1, \ldots, x_n) \equiv \Sigma U(x_i)\bar{p}_i$, or in the event that the utility of wealth function $U_i(\cdot)$ itself depends upon the state of nature, according to the 'state-dependent' preference function $V(x_1, \ldots, x_n) \equiv \Sigma U_i(x_i)\bar{p}_i$ (e.g. Karni, 1985). One of the advantages of the general 'state-preference' approach is that it does not require that we be able to observe the individual's probabilistic beliefs, or that different individuals share the same probabilistic beliefs.

AXIOMATIC DEVELOPMENT. Although there exist dozens of formal axiomatizations of the expected utility model in its

235

different contexts, most proceed by specifying an outcome space and postulating that the individual's preferences over probability distributions on this outcome space satisfy the following four axioms: completeness, transitivity, continuity and the independence axiom. Although it is beyond the scope of this entry to provide a rigorous derivation of the expected utility model in its most general setting, it is possible to illustrate the meaning of the axioms and sketch a proof of the expected utility representation theorem in the simple case of a finite outcome set of the form $\{x_1, \ldots, x_n\}$.

Recall that in such a case the objects of choice consist of all probability distributions $P = (p_1, \ldots, p_n)$ over $\{x_1, \ldots, x_n\}$, so that the following axioms refer to the individuals' weak preference relation $\succsim$ over this set, where $P^* \succsim P$ is read 'P^* is weakly preferred (i.e. preferred or indifferent) to P' (the associated strict preference relation $\succ$ and indifference relation $\sim$ are defined in the usual manner):

Completeness: For any two distributions P and P^* either $P^* \succsim P$, $P \succsim P^*$, or both.

Transitivity: If $P^{**} \succsim P^*$ and $P^* \succsim P$, then $P^{**} \succsim P$,

Mixture Continuity: If $P^{**} \succsim P^* \succsim P$, then there exists some $\lambda \in [0, 1]$ such that $P^* \sim \lambda P^{**} + (1 - \lambda)P$, and

Independence: For any two distributions P and P^*, $P^* \succsim P$ if and only if $\lambda P^* + (1 - \lambda)P^{**} \succsim \lambda P + (1 - \lambda)P^{**}$ for all $\lambda \in (0, 1]$ and all P^{**},

where $\lambda P + (1 - \lambda)P^*$ denotes the 'probability mixture' of P and P^*, i.e., the lottery with probabilities

$$(\lambda p_1 + (1 - \lambda)p_1^*, \ldots, \lambda p_n + (1 - \lambda)p_n^*).$$

The notion of a probability mixture is closely related (though not identical) to that of a 'compound lottery', in the sense that the probability mixture $\lambda P + (1 - \lambda)P^*$ yields the same probabilities of ultimately obtaining the outcomes $\{x_1, \ldots, x_n\}$ as would a compound lottery yielding a $\lambda : (1 - \lambda)$ chance of obtaining the respective lotteries P or P^*.

The completeness and transitivity axioms are completely analogous to their counterparts in the standard theory of the consumer (in particular, transitivity of $\succsim$ can be shown to imply transitivity of both $\succ$ and $\sim$). Mixture continuity states that if the lottery P^{**} is weakly preferred to P^*, and P^* is weakly preferred to P, then there will exist some probability mixture of the most and least preferred lotteries which is indifferent to the intermediate one.

As in standard consumer theory, completeness, transitivity and continuity serve essentially to establish the existence of a real-valued preference function $V(p_1, \ldots, p_n)$, which represents the relation $\succsim$, in the sense that $P^* \succsim P$ if and only if $V(p_1^*, \ldots, p_n^*) \geq V(p_1, \ldots, p_n)$. It is the independence axiom which, besides forming the basis of its widespread normative appeal, gives the theory its primary empirical content by implying that the preference function must take the linear form $V(p_1, \ldots, p_n) \equiv \Sigma U_i p_i$. To see the meaning of this axiom, assume that one is always indifferent between a compound lottery and its probabilistically equivalent single-stage lottery, and that P^* happens to be weakly preferred to P. In that case, the choice between the mixtures $\lambda P^* + (1 - \lambda)P^{**}$ and $\lambda P^* + (1 - \lambda)P^{**}$ is equivalent to being presented with a coin that has a $(1 - \lambda)$ chance of landing tails (in which case the prize will be P^{**}) and being asked *before the flip* whether one would rather win P or P^* in the event of a head. The normative argument for the independence axiom is that either the coin will land tails, in which case the choice would not have mattered, or it will land heads, in which case one is 'in effect' back to a choice between P and P^* and one 'ought' to have the same preferences as before. Note finally that the above statement of

the axiom in terms of the weak preference relation $\succsim$ also implies its counterparts in terms of strict preference and indifference.

In the following sketch of the expected utility representation theorem, expressions such as '$x_i \succsim x_j$' should be read as saying that the individual weakly prefers the degenerate lottery yielding x_i with certainty to that yielding x_j with certainty, and '$\lambda x_i + (1 - \lambda)x_j$' will be used to denote the $\lambda : (1 - \lambda)$ probability mixture between these two degenerate lotteries, and so on.

The first step in the argument is to define the von Neumann–Morgenstern utility index $\{U_i\}$ and the expected utility preference function $V(\cdot)$. Without loss of generality, we may reorder the outcomes so that $x_n \succsim x_{n-1} \succsim \cdots \succ x_2 \succsim x_1$. Since $x_n \succsim x_i \succsim x_1$ for each outcome x_i, we have by mixture continuity that there will exist scalars $\{U_i\} \subset [0, 1]$ such that $x_i \sim U_i x_n + (1 - U_i)x_1$ for each i (note that we can define $U_1 = 0$ and $U_n = 1$). Given this, define $V(P)$ to equal $\Sigma U_i p_i$ for all P.

The second step is to show that each lottery $P = (p_1, \ldots, p_n)$ is indifferent to the mixture $\lambda x_n + (1 - \lambda)x_1$ where $\lambda = \Sigma U_i p_i$. Since $(p_1, \ldots, p_n)$ can be written as the n-fold probability mixture $p_1 \cdot x_1 + p_2 \cdot x_2 + \cdots + p_n \cdot x_n$ and each outcome x_i is indifferent to the mixture $U_i x_n + (1 - U_i)x_1$, an n-fold application of the independence axiom yields that $(p_1, \ldots, p_n)$ is indifferent to the mixture

$$p_1 \cdot [U_1 x_n + (1 - U_1)x_1] + p_2 \cdot [U_2 x_n + (1 - U_2)x_1] + \cdots$$
$$\cdots + p_n \cdot [U_n x_n + (1 - U_n)x_1],$$

which is equal to $(\Sigma U_i p_i) \cdot x_n + (1 - \Sigma U_i p_i) \cdot x_1$.

The third step is to demonstrate that the mixture $\lambda x_n + (1 - \lambda)x_1$ is weakly preferred to the mixture $\gamma x_n + (1 - \gamma)x_1$ if and only if $\lambda \geq \gamma$. This follows immediately from the independence axiom and the fact that $\lambda \geq \gamma$ implies that these two lotteries may be expressed as the respective mixtures

$$(\lambda - \gamma) \cdot x_n + (1 - \lambda + \gamma) \cdot Q$$

and

$$(\lambda - \gamma) \cdot x_1 + (1 - \lambda + \gamma) \cdot Q,$$

where Q is defined as the mixture

$$[\gamma/(1 - \lambda + \gamma)] \cdot x_n + [(1 - \lambda)/(1 - \lambda + \gamma)] \cdot x_1.$$

The completion of the proof is now simple. For any two distributions P^* and P, we have by transitivity and the second step that $P^* \succsim P$ if and only if

$$(\Sigma U_i p_i^*) \cdot x_n + (1 - \Sigma U_i p_i^*) \cdot x_1 \succsim (\Sigma U_i p_i) \cdot x_n + (1 - \Sigma U_i p_i) \cdot x_1,$$

which by the third step is equivalent to the condition that $\Sigma U_i p_i^* \geq \Sigma U_i p_i$, or in other words, that $V(P^*) \geq V(P)$.

As mentioned, the expected utility model has been axiomatized many times and in many contexts. The most comprehensive account of the axiomatics of the model is undoubtedly Fishburn (1982).

HISTORY. The hypothesis that individuals might maximize the expectation of 'utility' rather than of monetary value was first proposed independently by the mathematicians Gabriel Cramer and Daniel Bernoulli, in each case as the solution to a problem posed by Daniel's cousin Nicholas Bernoulli (see Bernoulli, 1738). This problem, which has since come to be known as the 'St Petersburg Paradox', considers the gamble which offers a 1/2 chance of $1.00, a 1/4 chance of $2.00, a 1/8 chance of $4.00, and so on. Although the expected value of this prospect is

$$(1/2) \cdot \$1.00 + (1/4) \cdot (\$2.00) + (1/8) \cdot (\$4.00) + \cdots$$
$$\cdots = \$0.50 + \$0.50 + \$0.50 + \cdots = \$\infty,$$

common sense suggests that no one would be willing to forgo a very substantial certain payment in order to play it. Cramer and Bernoulli proposed that instead of looking at expected value, individuals might evaluate this and other lotteries by their 'expected utility', with utility given by a function such as the natural logarithm or the square root of wealth, in which case the certainty equivalent of the St Petersburg gamble becomes a moderate (and plausible) amount.

Two hundred years later, the St Petersburg Paradox was generalized by Karl Menger (1934), who noted that whenever the utility of wealth function was unbounded (as with the natural logarithm or square root functions), it would be possible to construct similar examples with infinite expected utility and hence infinite certainty equivalents (replace the payoffs \$1.00, \$2.00, \$4.00, ... in the above example by x_1, x_2, x_3, ... where $U(x_i) = 2^i$ for each i). In light of this, von Neumann–Morgenstern utility functions are typically (though not universally) postulated to be bounded functions of wealth.

The earliest formal axiomatic treatment of the expected utility hypothesis was developed by Frank Ramsey (1926) as part of his theory of subjective probability or individuals' 'degrees of belief' in the truth of various alternative propositions. Starting from the premise that there exists an 'ethically neutral' proposition whose degree of belief is 1/2 and whose validity or invalidity is of no independent value, Ramsey proposed a set of axioms on how the individual would be willing to stake prizes on its truth or falsity in a manner which allowed for the derivation of the 'utilities' of these prizes. He then used these utility values and betting preferences to determine the individual's degrees of belief in other propositions. Perhaps because it was intended as a contribution to the philosophy of belief rather than the theory of risk bearing, Ramsey's analysis did not have the impact upon the economics literature that it deserved.

The first axiomatization of the expected utility model to receive widespread attention was that of John von Neumann and Oskar Morgenstern, which was presented in connection with their formulation of the theory of games (von Neumann and Morgenstern, 1944, 1947, 1953). Although both these developments were recognized as breakthroughs, the mistaken belief that von Neumann and Morgenstern had somehow mathematically overthrown the Hicks–Allen 'ordinal revolution' led to some confusion until the difference between 'utility' in the von Neumann–Morgenstern and ordinal (i.e. non-stochastic) senses was illuminated by writers such as Ellsberg (1954) and Baumol (1958).

Another factor which delayed the acceptance of the theory was the lack of recognition of the role played by the independence axiom, which did not explicitly appear in the von Neumann–Morgenstern formulation. In fact, the initial reaction of researchers such as Baumol (1951) and Samuelson (1950) was that there was no reason why preferences over probability distributions must *necessarily* be linear in the probabilities. However the independent discovery of the independence axiom by Marschak (1950), Samuelson (1952) and others, and Malinvaud's (1952) observation that it had been implicitly invoked by von Neumann and Morgenstern, led to an almost universal acceptance of the expected utility hypothesis as both a normative and positive theory of behaviour toward risk. Practically the only dissenting voice was that of Maurice Allais, whose famous paradox (see below) and other empirical and theoretical work (e.g. Allais, 1952) has provided the basis for the resurgence of interest in alternatives to expected utility in the late 1970s and 1980s. This period also saw the development of the elegant axiomatization of Herstein and Milnor (1953) as well as Savage's (1954) joint

axiomatization of utility and subjective probability, which formed the basis of the state-preference approach described above.

While the 1950s essentially saw the completion of foundational work on the expected utility model, the 1960s and 1970s saw the flowering of its analytic capabilities and its application to fields such as portfolio selection (Merton, 1969), optimal savings (Levhari and Srinivasan, 1969), international trade (Batra, 1975), and even the measurement of inequality (Atkinson, 1970). This movement was spearheaded by the development of the Arrow–Pratt characterization of risk aversion (see above) and the characterization, by Rothschild–Stiglitz (1970, 1971) and others, of the notion of 'increasing risk'. This latter work in turn led to the development of a general theory of 'stochastic dominance' (e.g. Whitmore and Findlay, 1978), which further expanded the analytical powers of the model.

Although the expected utility model received a small amount of experimental testing by economists in the early 1950s (e.g. Mosteller and Nogee, 1951; Allais, 1952) and continued to be examined by psychologists, interest in the empirical validity of the model waned from the mid-1950s through the mid-1970s, no doubt due to both the normative appeal of the independence axiom and model's analytical successes. However, the late 1970s and 1980s have witnessed a revival of interest in the testing of the expected utility model; a growing body of evidence that individuals' preferences *systematically* depart from linearity in the probabilities; and the development, analysis and application of alternative models of choice under risk (see below). It is fair to say that today the debate over the descriptive (and even normative) validity of the expected utility hypothesis is more extensive than it has been in 30 years, and the outcome of this debate will have important implications for the direction of research in the economic theory of individual behaviour towards risk.

EVIDENCE AND ALTERNATIVE HYPOTHESES. As mentioned above, the current body of experimental evidence suggests that individual preferences over lotteries are typically *not* linear in the probabilities, but rather depart systematically from this property. The earliest, and undoubtedly best-known, example of this is the so-called 'Allais paradox' (Allais, 1952), in which the individual is asked to rank each of the following pairs of prospects (where \$1M = \$1,000,000):

$$a_1: \{1.00 \text{ chance of } \$1M \quad \text{versus} \quad a_2: \begin{cases} 0.10 \text{ chance of } \$5M \\ 0.89 \text{ chance of } \$1M \\ 0.01 \text{ chance of } \$0, \end{cases}$$

and:

$$a_3: \begin{cases} 0.10 \text{ chance of } \$5M \\ 0.90 \text{ chance of } \$0 \end{cases} \quad \text{versus} \quad a_4: \begin{cases} 0.11 \text{ chance of } \$1M \\ 0.89 \text{ chance of } \$0. \end{cases}$$

Since each of these lotteries involves outcomes in the set $\{x_1, x_2, x_3\} = \{\$0, \$1M, \$5M\}$, they may be plotted in the (p_1, p_3) triangle diagram, as illustrated in Figures 5 and 6. The fact that the four prospects form a parallelogram in this triangle makes this problem a useful test of linearity (i.e. the expected utility hypothesis), since it implies that an expected utility maximizer will prefer a_1 to a_2 if and only if he or she prefers a_4 to a_3 (algebraically, this is in turn equivalent to the inequality $(0.10 \cdot U(\$5M) - 0.11 \cdot U(\$1M) + 0.01 \cdot U\$(0) < 0)$.

However, experimenters such as Allais (1952), Morrison (1967), Moskowitz (1974), Raiffa (1968), Slovic and Tversky (1974) and others, have found that the modal if not majority choice was for a_1 in the first pair and a_3 in the second pair, as would be chosen by an individual whose indifference curves 'fanned out' as in Figure 6. Subsequent studies by Hagen (1979), Karmarkar (1974), MacCrimmon and Larsson (1979), McCord and de Neufville (1983) and others, using both similar and qualitatively different types of examples, have also revealed systematic departures from linearity in the direction of 'fanning out' (see Machina, 1983a, 1983b).

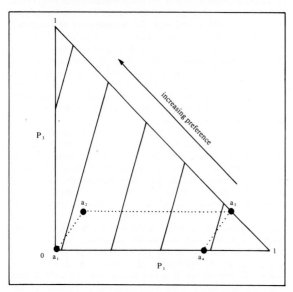

Figure 5   Allais Paradox with Expected Utility Indifference Curves

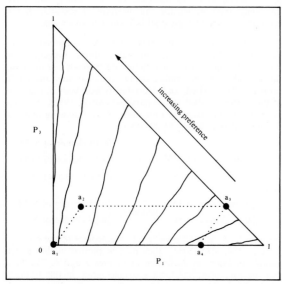

Figure 6   Allais Paradox with Non-Expected Utility Indifference Curves that "Fan Out"

In light of this evidence, researchers have begun to develop alternatives to the expected utility model (typically generalizations of it) which are capable of exhibiting this form of nonlinearity as well as other standard properties of risk preferences such as first-order stochastic dominance preference and risk aversion. (A set of non-expected utility indifference curves which exhibits these three properties, for example, is

given in Figure 2.) Specific nonlinear functional forms for preference functions which have been proposed include those of Edwards (1955) and Kahneman and Tversky (1979) $(\Sigma U(x_i)\pi(p_i))$; Chew and MacCrimmon (1979) and Chew (1983) $\{[\int U(x)\,dF(x)]/[\int W(x)\,dF(x)]\}$; and Quiggin (1982) $\{\int U(x)\,dG[F(x)]\}$. A general framework for the analysis of differentiable non-expected utility preference functions in terms of their local linear approximations, which can be interpreted as local 'expected utility' approximations, is developed in Machina (1982, 1983a). Finally, the findings by Lichtenstein and Slovic (1971), Grether and Plott (1979) and others of systematic intransitivities in preferences over lotteries (but see Karni and Safra, 1984), have led to the development of non-transitive models by researchers such as Bell (1982), Fishburn (1983), and Loomes and Sugden (1982). (For a more complete survey of the experimental evidence on the expected utility hypothesis as well as alternative models of behaviour towards risk, see Machina, 1983b.)

MARK J. MACHINA

See also BERNOULLI, DANIEL; DECISION THEORY; RAMSEY, FRANK PLUMPTON; REPRESENTATION OF PREFERENCES; RISK UNCERTAINTY; UTILITY THEORY AND DECISION-MAKING.

BIBLIOGRAPHY

Allais, M. 1952. Fondements d'une théorie positive des choix comportant un risque et critique des postulats et axiomes de l'école Américaine. *Colloques Internationaux du Centre National de la Recherche Scientifique* 40, (1953), 257–332. Trans. as: The foundations of a positive theory of choice involving risk and a criticism of the postulates and axioms of the American School, in Allais and Hagen (1979).

Allais, M. and Hagen, O. (eds) 1979. *Expected Utility Hypotheses and the Allais Paradox.* Dordrecht: D. Reidel.

Arrow, K. 1974. *Essays in the Theory of Risk-Bearing.* Amsterdam: North-Holland.

Atkinson, A. 1970. On the measurement of inequality. *Journal of Economic Theory* 2(3), September, 244–63.

Batra, R. 1975. *The Pure Theory of International Trade under Uncertainty.* London: Macmillan.

Baumol, W. 1951. The Neumann–Morgenstern utility index: an ordinalist view. *Journal of Political Economy* 59(1), February, 61–6.

Baumol, W. 1958. The cardinal utility which is ordinal. *Economic Journal* 68, December, 665–72.

Bell, D. 1982. Regret in decision making under uncertainty. *Operations Research* 30, September–October, 961–81.

Bernoulli, D. 1738. Specimen theoriae novae de mensura sortis. *Commentarii Academiae Scientiarum Imperialis Petropolitanae.* Trans. as: Exposition of a new theory on the measurement of risk, *Econometrica* 22, January 1954, 23–36.

Chew, S.H. 1983. A generalization of the quasilinear mean with applications to the measurement of income inequality and decision theory resolving the Allais paradox. *Econometrica* 51(4), July, 1065–92.

Chew, S. and MacCrimmon, K. 1979. Alpha–Nu choice theory: a generalization of expected utility theory. University of British Columbia Faculty of Commerce and Business Administration Working Paper No. 669, July.

Debreu, G. 1959. *Theory of Value: An Axiomatic Analysis of Economic Equilibrium.* New Haven: Yale University Press.

Edwards, W. 1955. The prediction of decisions among bets. *Journal of Experimental Psychology* 50(3), September, 201–14.

Ellsberg, D. 1954. Classical and current notions of 'measurable utility'. *Economic Journal* 64, September, 528–56.

Fishburn, P. 1982. *The Foundations of Expected Utility.* Dordrecht: D. Reidel.

Fishburn, P. 1983. Nontransitive measurable utility. *Journal of Mathematical Psychology* 26(1), August, 31–67.

Friedman, M. and Savage, L. 1948. The utility analysis of choices involving risk. *Journal of Political Economy* 56, August, 279–304. Reprinted in *Readings in Price Theory*, ed. G. Stigler and K. Boulding, London: George Allen & Unwin, 1953.

Grether, D. and Plott, C. 1979. Economic theory of choice and the preference reversal phenomenon. *American Economic Review* 69(4), September, 623–38.

Hagen, O. 1979. Towards a positive theory of preferences under risk. In Allais and Hagen (1979).

Herstein, I. and Milnor, J. 1953. An axiomatic approach to measurable utility. *Econometrica* 21, April, 291–7.

Hey, J. 1979. *Uncertainty in Microeconomics*. Oxford: Martin Robinson; New York: New York University Press.

Hirshleifer, J. 1965. Investment decision under uncertainty: choice theoretic approaches. *Quarterly Journal of Economics* 79, November, 509–36.

Hirshleifer, J. 1966. Investment decision under uncertainty: applications of the state-preference approach. *Quarterly Journal of Economics* 80, May, 252–77.

Hirshleifer, J. and Riley, J. 1979. The analytics of uncertainty and information – an expository survey. *Journal of Economic Literature* 17(4), December, 1375–421.

Kahneman, D. and Tversky, A. 1979. Prospect theory: an analysis of decision under risk. *Econometrica* 47(2), March, 263–91.

Karmarkar, U. 1974. The effect of probabilities on the subjective evaluation of lotteries. Massachusetts Institute of Technology Sloan School of Management Working Paper No. 698–74, February.

Karni, E. 1985. *Decision Making under Uncertainty: the Case of State-Dependent Preferences*. Cambridge, Mass.: Harvard University Press.

Karni, E. 1985. Increasing risk with state dependent preferences. *Journal of Economic Theory* 35(1), 172–7.

Karni, E. and Safra, Z. 1984. 'Preference reversal' and the theory of choice under risk. Johns Hopkins University Working Papers in Economics No. 141.

Levhari, D. and Srinivasan, T.N. 1969. Optimal savings under uncertainty. *Review of Economic Studies* 36–2, April, 153–64.

Lichtenstein, S. and Slovic, P. 1971. Reversals of preferences between bids and choices in gambling decisions. *Journal of Experimental Psychology* 89(1), July, 46–55.

Lippman, S. and McCall, J. 1981. The economics of uncertainty: selected topics and probabilistic methods. In *Handbook of Mathematical Economics*, ed. K. Arrow and M. Intriligator, Vol. 1, Amsterdam: North-Holland.

Loomes, G. and Sugden, R. 1982. Regret theory: an alternative theory of rational choice under uncertainty. *Economic Journal* 92. (368), December, 805–24.

McCord, M. and de Neufville, R. 1983. Empirical demonstration that expected utility analysis is not operational. In Stigum and Wenstøp (1983).

MacCrimmon, K. and Larsson, S. 1979. Utility theory: axioms versus 'paradoxes'. In Allais and Hagen (1979).

Machina, M. 1982. 'Expected utility' analysis without the independence axiom. *Econometrica* 50(2), March, 277–323.

Machina, M. 1983a. Generalized expected utility analysis and the nature of observed violations of the independence axiom. In Stigum and Wenstøp (1983).

Machina, M. 1983b. The economic theory of individual behavior toward risk: theory, evidence and new directions. Institute for Mathematical Studies in the Social Sciences Technical Report No. 433, Stanford University, October.

Malinvaud, E. 1952. Note on von Neumann–Morgenstern's strong independence axiom. *Econometrica* 20(4), October, 679.

Markowitz, H. 1952. The utility of wealth. *Journal of Political Economy* 60, April, 151–8.

Marschak, J. 1950. Rational behavior, uncertain prospects, and measurable utility. *Econometrica* 18, April, 111–41 (Errata, July 1950).

Menger, K. 1934. Das Unsicherheitsmoment in der Wertlehre. *Zeitschrift für Nationalökonomie*. Trans. as: The role of uncertainty in economics, in *Essays in Mathematical Economics in Honor of Oskar Morgenstern*, ed. M. Shubik, Princeton: Princeton University Press, 1967.

Merton, R. 1969. Lifetime portfolio selection under uncertainty: the continuous time case. *Review of Economics and Statistics* 51(3), August, 247–57.

Morrison, D. 1967. On the consistency of preferences in Allais' paradox. *Behavioral Science* 12(5), September, 373–83.

Moskowitz, H. 1974. Effects of problem representation and feedback on rational behavior in Allais and Morlat-type problems. *Decision Sciences* 2.

Mosteller, F. and Nogee, P. 1951. An experimental measurement of utility. *Journal of Political Economy* 59, October, 371–404.

Pratt, J. 1964. Risk aversion in the small and in the large. *Econometrica* 32, January–April, 122–36.

Quiggin, J. 1982. A theory of anticipated utility. *Journal of Economic Behavior and Organization* 3(4), December, 323–43.

Quirk, J. and Saposnick, R. 1962. Admissibility and measurable utility functions. *Review of Economic Studies* 29, February, 140–46.

Raiffa, H. 1968. *Decision Analysis: Introductory Lectures on Choice under Uncertainty*. Reading, Mass.: Addison Wesley.

Ramsey, F. 1926. Truth and probability. In *The Foundations of Mathematics and Other Logical Essays*, ed. R. Braithwaite, New York: Harcourt, Brace and Co., 1931. Reprinted in *Foundations: Essays in Philosophy, Logic, Mathematics and Economics*, ed. D. Mellor, New Jersey: Humanities Press, 1978.

Ross, S. 1981. Some stronger measures of risk aversion in the small and in the large, with applications. *Econometrica* 49(3), May, 621–38.

Rothschild, M. and Stiglitz, J. 1970. Increasing risk I: a definition. *Journal of Economic Theory* 2(3), September, 225–43.

Rothschild, M. and Stiglitz, J. 1971. Increasing risk II: its economic consequences. *Journal of Economic Theory* 3(1), March, 66–84.

Safra, Z. 1985. Existence of equilibrium for Walrasian endowment games. *Journal of Economic Theory* 37(2), 366–78.

Samuelson, P. 1950. Probability and attempts to measure utility. *Economic Review* 1, July, 167–73. Reprinted in Stiglitz (1965).

Samuelson, P. 1952. Probability, utility, and the independence axiom. *Econometrica* 20, October, 670–78. Reprinted in Stiglitz (1965).

Savage, L. 1954. *The Foundations of Statistics*. New York: John Wiley & Sons. Enlarged and revised edn, New York: Dover, 1972.

Slovic, P. and Tversky, A. 1974. Who accepts Savage's Axiom? *Behavioral Science* 19(6), November, 368–73.

Stiglitz, J. (ed.) 1965. *Collected Scientific Papers of Paul A. Samuelson*, Vol. 1. Cambridge, Mass.: MIT Press.

Stigum, B. and Wenstøp, F. (eds) 1983. *Foundations of Utility and Risk Theory with Applications*. Dordrecht: D. Reidel.

von Neumann, J. and Morgenstern, O. 1944. *Theory of Games and Economic Behavior*. Princeton: Princeton University Press. 2nd edn, 1947; 3rd edn, 1953.

Whitmore, G. and Findlay, M. (eds) 1978. *Stochastic Dominance: An Approach to Decision Making Under Risk*. Lexington, Mass.: D.C. Heath.

expenditure functions. *See* COST FUNCTIONS; DUALITY.

expenditure tax. The idea of an expenditure tax has a long ancestry, dating back at least to Hobbes, who argued that people should be taxed according to the resources of the community they absorb not according to what they contribute. The case was later taken up by J.S. Mill, Marshall, Pigou and Irving Fisher. In modern times, the advocacy of an expenditure tax is most associated with the Cambridge economist Nicholas Kaldor (1955). Recently it has been espoused by the Meade Committee (Meade, 1978), and separately by two members of that Committee, Kay and King (1978).

There are efficiency and equity arguments for considering an expenditure tax as an alternative to income taxation. As far as efficiency is concerned, there is a commonly held view that because income tax involves the double taxation of saving, and therefore lowers the rate of return on saving below the rate of return on investment, this distorts the choice between consumption and saving, and represents a wasteful distortion. However, recent work, using optimal tax theory, has questioned this conclusion. The theory of the second best teaches that alternative tax systems cannot be evaluated according to the number of distortions: the magnitude of the distortions and their interaction also needs to be taken into account. Using optimal tax theory in an intertemporal context, Atkinson and Sandmo (1980) have shown that no firm conclusions can be reached on the relative efficiency of expenditure and income taxes from a welfare point of view: it all depends on the form of the social welfare function, what other instruments governments can use to achieve a desired intertemporal allocation of consumption, and on crucial parameters of the model such as the interest elasticity of labour supply.

As far as equity is concerned, the case for an expenditure tax may be stronger. The principle of progressive income taxation rests on the concept of taxable capacity or ability to pay. The question is, does 'income' approximate to this concept? There are three main difficulties. First, income is only one measure of taxable capacity. Secondly, income itself is not an unambiguous concept. Thirdly, the actual definition of income for tax purposes can introduce inequities into the system by some receipts being treated as income and others not. Income is taken as a proxy for 'spending power', but there are other sources of spending power (e.g. wealth) and it is not easy to express them all in a single measure of taxable capacity. There are particular problems associated with irregular receipts and capital gains. It can be argued that many of the problems created by the non-comparability of different forms of income, wealth and capital gains would be resolved by taxing expenditure rather than income. The individual himself would declare his spending power when he spends. Since there is no objective definition of income that can provide a true measure of spending power, there can be no presumption that any income tax system would be superior from an equity point of view to an expenditure tax.

Kaldor was the first to argue in a comprehensive way that the measurement of income as a measure of taxable capacity is inevitably ambiguous and is likely to be a bad proxy for the measurement of spending power, so that the taxation of spending as such may be regarded at least as equitable as income tax, if not more so, with other positive advantages – particularly, an expenditure tax would be a more efficient instrument for controlling the economy, so that there is no necessary conflict between an egalitarian system of taxation, efficiency and growth. The Meade Committee, as well as mentioning the traditional arguments concerning the difficulty of defining income and measuring accruals, placed most emphasis on the elimination of capital market distortions, particularly those associated with various concessions in the existing income tax system which have differential and distorting effects on rates of return to saving outlets, and with having to correct nominal capital gains and losses for inflation. Such problems automatically disappear with an expenditure tax.

Kaldor also discussed at length the effect that a switch to an expenditure tax is likely to have on risk bearing, the supply of effort, saving and economic progress. It turns out that it is impossible to predict the net effects with any degree of certainty. As far as risk bearing is concerned it is impossible to say whether an expenditure tax is better or worse than an income tax yielding the same revenue, since on the one hand it is less discriminating against risk in so far as part of taxable income is saved, but on the other hand is more discriminating

in so far as part of the capital gain is spent. Likewise, as far as the supply of effort is concerned, it is possible to reach different conclusions according to the assumptions made concerning the relative stability of income and consumption, and whether taxation is progressive or proportional. Kaldor did not pay much attention to the argument that an expenditure tax would avoid the double taxation of saving under income tax, and therefore avoid distortions and encourage saving, but it has other advantages relating to enterprise and economic progress. For example, without a capital gains tax, income tax puts a premium on speculation compared with an expenditure tax where both yield and capital gains are equally taxed if spent, or equally exempt if saved. An expenditure tax which discouraged speculation would enhance the supply of risk capital.

Several theoretical objections have been raised against the expenditure tax, but none is very convincing. The main difficulty concerns the practical implementation of the tax. In evidence to the 1929 Colwyn Committee on National Debt and Taxation, Keynes had earlier described the expenditure tax idea as theoretically sound but 'practically impossible'. This was the prevailing view (see also Pigou, 1928) largely because of the difficulty of getting taxpayers to keep accurate records of personal expenditure and checking returns. It was Irving Fisher (1937) who first showed that this would not be necessary since a person's expenditure is the difference between what he has available for spending and what he has left at the end of the accounting period. Thus in theory the only information required is the size of a person's bank balance at the beginning of the year plus income and other receipts, and from this is then deducted net investments, exempted expenditure and the size of the bank balance at the end of the year, and the difference is chargeable expenditure. The major problems concern the definition of chargeable expenditure, and evasion through the avoidance of the use of bank accounts.

In his original exposition of the expenditure tax, Kaldor perhaps overestimated the drawbacks of income tax and understated the difficulties of the expenditure tax. Conceptually the dividing line between what is consumption and what is saving may be said to be as arbitrary and fraught with difficulties, as in answering the question, when does income accrue? In defence, he admitted, however, that comparing the expenditure tax with a more *comprehensive* income tax, the balance between the two is much more finely poised: the conclusion which Prest (1979) also comes to in his review of the Meade Committee Report.

A.P. Thirlwall

See also CONSUMPTION TAXES; DIRECT TAXES; PUBLIC FINANCE; TAXATION OF INCOME; TAXATION OF WEALTH.

BIBLIOGRAPHY
Atkinson, A.B. and Sandmo, A. 1980. Welfare implications of the taxation of savings. *Economic Journal* 90, September, 529–49.
Fisher, I. 1937. Income in theory and income taxation in practice. *Econometrica* 5, January, 1–55.
Kaldor, N. 1955. *An Expenditure Tax*. London: Allen and Unwin.
Kay, J.A. and King, M.A. 1978. *The British Tax System*. London: Oxford University Press.
Meade, J. 1978. *The Structure and Reform of Direct Taxation*. London: George Allen and Unwin.
Pigou, A.C. 1928. *A Study in Public Finance*. London: Macmillan.
Prest, A.R. 1979. The structure and reform of direct taxation. *Economic Journal* 89, June, 243–60.

expense curve. *See* CUNYNGHAME, HENRY; MARSHALL, ALFRED.

experience. *See* LEARNING BY DOING.

experimental methods in economics (i). Experiment in the scientific sense has been well described as 'putting in action causes and agents over which we have control, and purposely varying their combinations and noticing what effects take place' (Herschel, *Study of Natural Philosophy*, p. 76). In sciences such as physics and chemistry, in which the phenomena are amenable to arrangement, it is by far the most potent instrument of discovery. Where, however, there is not the same facility for easy manipulation, the inquirer is compelled to fall back on the less effective method of simple observation. Instead of creating instances for himself, he has to find them in nature, or wait till they are presented spontaneously to his view.

Economics, in common with the other social sciences, clearly belongs to the latter class. The phenomena of wealth are closely inter-connected, and are besides affected by the other forms of social activity. Hardly any economic event can be said to be the result of a single cause, it is rather the product of several contributory causes. Nor are the total effects of any one agency easily separable; they are combined with those of others in a whole which cannot be analysed. In technical language 'plurality of causes' and 'intermixture of effects', the two great hindrances to the use of experiment (Mill, *Logic*, Book iii, ch. x), are generally present in economic facts. To secure the requisite isolation of any phenomenon selected for study is rarely possible. The most rigorous form of inquiry, known as the 'method of difference', the essence of which 'is the comparison of two instances, which resemble one another in all material respects, except that in one a certain cause is present, while in the other it is absent' (Keynes, *Scope and Method of Political Economy*, p. 170), is plainly excluded, since we cannot introduce a single cause that will have only a measurable effect, nor can we be sure that the surrounding conditions remain unaltered. The 'method of agreement' in which the instances compared resemble each other in only one particular is not merely inferior as an experimental resource, but is inapplicable to social phenomena. Two countries or periods that had one common feature would have more than one. In two classes of cases, however, experiment may be sometimes used, viz. (1) in reference to the premises or data of economic science, thus the 'law of diminishing returns' admits of experimental proof; (2) More important than the preceding exception, which is rather apparent than real, are those cases in which, by deductive reasoning, it can be shown that the action of an economic force is limited, and then its working within those limits can be experimentally ascertained.

These exceptions notwithstanding, it may be said that scientific experiments (*experimenta lucifera*) are a very slight resource in economics.

The case is somewhat different with regard to practical questions. Legislative measures and individual actions are, if so intended, so many experiments on the social system. Thus if several countries, widely differing in other respects, have established a system of peasant proprietary with good results, while several other countries, also widely differing *inter se*, are without that system and show inferiority, we may argue that peasant proprietary is experimentally justified. The same reasoning would be applicable to commercial policy, and has actually been used in reference to the case of Victoria and New

South Wales, but illogically, as a number of cases are required to exclude other influences.

Again, by applying special legislation, e.g. a particular kind of land tenure, to one part of a country, we can ascribe to its influence the special effects noticed in that district. Practical experiments (*experimenta fructifera*) may also be employed by means of (1) permissive legislation, or (2) temporary legislation.

Private persons also carry out practical economic experiments, as in the case of profit-sharing (Leclaire), and the recent eight hours day experiment at Sunderland (*Economic Journal*, ii, pp. 755, 756). A large accumulation of instances may even give a very near approach to rigorous scientific proof.

A vaguer use of the term 'experimental method' is common in continental and especially in French writers. J.B. Say, for example, declares that the true method of political economy is *La méthode expérimentale qui consiste essentiellement à n'admettre comme vrais que les faits dont l'observation et expérience ont démontre la réalité* (*Traité, Discours préliminaire*, p. x, 5th edn, 1826). Here 'experiment' is used as synonymous with 'experience'; it therefore includes observation and experiment in the strict sense.

[C.F. BASTABLE]

Reprinted from *Palgrave's Dictionary of Political Economy*.

BIBLIOGRAPHY

Donnat, L. 1891. *La Politique Expérimentale*. Paris.
Jevons, W.S. 1883. *Methods of Social Reform and other papers*. London.
Keynes, J.N. 1891. *The Scope and Method of Political Economy*. London.
Lewis, G.C. 1852. *Methods of Observation and Reasoning in Politics*. London.
Mill, J.S. 1848. *A System of Logic*. London.
Say, J.B. 1826. *Traité d'économie politique*. 5th edn, Paris: Rapilly.

experimental methods in economics (ii). Historically, the method and subject matter of economics have presupposed that it was a non-experimental (or 'field observational') science more like astronomy or meteorology than physics or chemistry. Based on general, introspectively 'plausible', assumptions about human preferences, and about the cost and technology based supply response of producers, economists have sought to understand the functioning of economies, using observations generated by economic outcomes realized over time. The data of the astronomer is of this same type, but it would be wrong to conclude that astronomy and economics are methodologically equivalent. There are two important differences between astronomy and economics which help to illuminate some of the methodological problems of economics. First, based upon parallelism (the maintained hypothesis that the same physical laws hold everywhere), astronomy draws on all the relevant theory from classical mechanics and particle physics – theory which has evolved under rigorous laboratory tests. Traditionally, economists have not had an analogous body of tested behavioural principles that have survived controlled experimental tests, and which can be assumed to apply with insignificant error to the microeconomic behaviour that underpins the observable operations of the economy. Analogously, one might have supposed that there would have arisen an important area of common interest between economics and, say, experimental psychology, similar to that between astronomy and physics, but this has only started to develop in recent years.

Second, the data of astronomy are painstakingly gathered by professional observational astronomers for scientific purposes, and these data are taken seriously (if not always non-controversially) by astrophysicists and cosmologists. Most of the data of economics has been collected by government or private agencies for non-scientific purposes. Hence astronomers are directly responsible for the scientific credibility of their data in a way that economists have not been. In economics, when things appear not to turn out as expected the quality of the data is more likely to be questioned than the relevance and quality of the abstract reasoning. Old theories fade away, not from the weight of falsifying evidence that catalyses theoretical creativity into developing better theory, but from lack of interest, as intellectual energy is attracted to the development of new techniques and to the solution of new puzzles that remain untested.

At approximately the mid–20th century, professional economics began to change with the introduction of the laboratory experiment into economic method. In this embryonic research programme economists (and a psychologist, Sidney Siegel) became directly involved in the design and conduct of experiments to examine propositions implied by economic theories of markets. For the first time this made it possible to introduce *demonstrable* knowledge into the economist's attempt to understand markets.

This laboratory approach to economics also brought to the economist direct responsibility for an important source of scientific data generated by controlled processes that can be replicated by other experimentalists. This development invited economic theorists to submit to a new discipline, but also brought an important new discipline and new standards of rigour to the data gathering process itself.

An untested theory is simply a hypothesis. As such it is part of our *self*-knowledge. Science seeks to expand our knowledge of *things* by a process of testing this type of self-knowledge. Much of economic theory can be called, appropriately, 'ecclesiastical theory'; it is accepted (or rejected) on the basis of authority, tradition, or opinion about assumptions, rather than on the basis of having survived a rigorous falsification process that can be replicated.

Interest in the replicability of scientific research stems from a desire to answer the question 'Do you see what I see?'. Replication and control are the two primary means by which we attempt to reduce the error in our common knowledge of economic processes. However, the question 'Do you see what I see?' contains three component questions, recognition of which helps to identify three different senses in which a research study may fail to be replicable:

(1) *Do you observe what I observe?* Since economics has traditionally been confined to the analysis of non-experimental data, the answer to this question has been trivially, 'yes'. We observe the same thing because we use the same data. This non-replicability of our traditional data sources has helped to motivate some to turn increasingly to experimental methods. We can say that you have replicated my experiments if you are unable to reject the hypothesis that your experimental data came from the same population as mine. This means that the experimenter, his/her subjects, and/or procedures are not significant treatment variables.

(2) *Do you interpret what we observe as I interpret it?* Given that we both observe the same, or replicable data, do we put the same interpretation on these data? The interpretation of observations requires theory (either formal or informal), or at least an empirical interpretation of the theory in the context that generated the data. Theory usually requires empirical interpretation either because (i) the theory is not developed

directly in terms of what can be observed (e.g. the theory may assume risk aversion which is not directly observable), or (ii) the data were not collected for the purpose of testing, or estimating the parameters of a theory. Consequently, failure to replicate may be due to differences in interpretation which result from different meanings being ascribed to the theory. Thus two researchers may apply different transformations to raw field data (e.g. different adjustments for the effect of taxes), so that the results are not replicable because their theory interpretations differ.

(3) *Do you conclude what I conclude from our interpretation?* The conclusions reached in two different research studies may be different even though the data and their interpretation are the same. In economics this is most often due to different model specifications. This problem is inherent in non-experimental methodologies in which, at best, one usually can estimate only the parameters of a prespecified model and cannot credibly test one model or theory against another. An example is the question of whether the Phillips' curve constitutes a behavioural trade-off between the rates of inflation and unemployment, or represents an equilibrium association without causal significance.

I. MARKETS AND MARKET EXPERIMENTS. Markets and how they function constitute the core of any economic system, whether it is highly decentralized – popularly, a 'capitalistic' system, or highly centralized – popularly, a 'planned' system. This is true for the decentralized economy because markets are the spontaneous institutions of exchange that use prices to guide resource allocation and human economic action. It is true for the centralized economy because in such economies markets always exist or arise in legal form (private agriculture in Russia) and clandestine or illegal form (barter, bribery, the trading of favours, and underground exchange in Russia, Poland and elsewhere). Markets arise spontaneously in all cultures in response to the human desire for betterment (to 'profit') through exchange. Where the commodity or service is illegal (prostitution, gambling, the sale of liquor under Prohibition or of marijuana, cocaine, etc.) the result is not to prevent exchange, but to raise the risk and therefore the costs of exchange. This is because enforcement is itself costly, and it is never economical for the authorities (whether Soviet or American) even to approximate perfect enforcement. The spontaneity with which markets arise is perhaps no better illustrated than when (1979–80) US airlines for promotional purposes issued travel vouchers to their passengers. One of these vouchers could be redeemed by the bearer as a cash substitute in the purchase of new airline tickets. Consequently vouchers were of value to future passengers. Furthermore, since (as Hayek would say) the 'circumstances of time and place' for the potential redemption of vouchers were different for different individuals, there existed the preconditions for the active voucher market that was soon observed in all busy airports. Current passengers with vouchers who were unlikely to be travelling again soon held an asset worth less to themselves than to others who were more certain of their future or impending travel plans. The resulting market established prices that were discounts from the redemption or 'face' value of vouchers. Sellers who were unlikely to be able to redeem their vouchers preferred to sell them at a discount for cash. Buyers who were reasonably sure of their travel plans could save money by purchasing vouchers at a discount. Thus the welfare of every active buyer and seller increased via this market. Without a market, many – perhaps most – vouchers would not have been exercised and would thus have been 'wasted'.

The previous paragraph illustrates a fundamental hypothesis (theorem) of economics: the ('competitive') market process yields welfare improving (and, under certain limiting ideal conditions, welfare maximizing) outcomes. But is the hypothesis 'true', or at least very probably true? (Lakatos (1978) would correctly ask 'Has it led to an empirically progressive research programme?') I think it is 'true', but how do I know this? Do you see what I see? A Marxist does not see what I see in the above interpretation of a market. The young student studying economics does not see what I see, although if they continue to study economics eventually they (predictably) come to see what I see (or, at least, they say they do). Is this because we have inadvertently brainwashed them? The gasoline consumer does not see what I see. They see themselves in a *zero* sum game with an oil company: any increase in price merely redistributes wealth from the consumer to the company, which is not 'fair' since the company is richer. What I see in a market is a *positive* sum

game yielding gains from exchange, which constitutes the fundamental mechanism for creating, not merely redistributing wealth. The traditional method by which the economist gets others to see this 'true' function of markets is by logical arguments (suppose it were not true, then ...), examples, and 'observations', such as are contained in my description of the voucher market, in which what is 'observed' is hortatively described and interpreted in terms of the hypothesis itself. But if this knowledge of the function of markets is 'true', can it be demonstrated? Experimentalists claim that laboratory experiments can provide a uniquely important technique of demonstration for supplementing the theoretical interpretation of field observations.

I conducted my first experiment in the spring of 1956. Since then hundreds of similar, as well as environmentally richer experiments have been conducted by myself and by others. In 1956, my introductory economics class consisted of 22 science and engineering students, and although this might not have

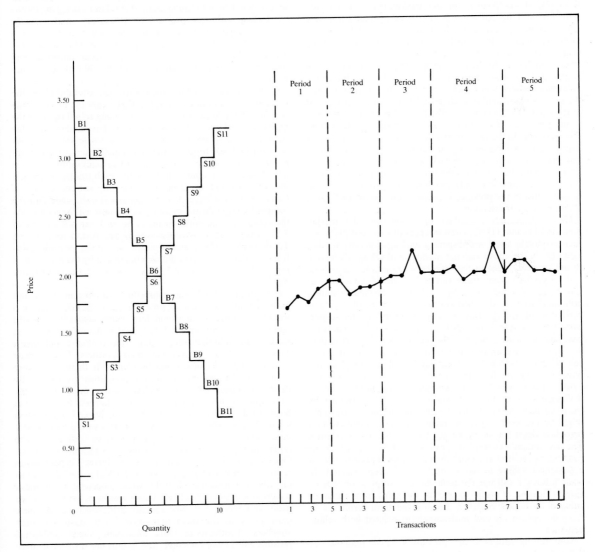

Figure 1

243

been the 'large number' traditionally thought to have been necessary to yield a competitive market, I though it was large enough for a practice run to initiate a research programme capable of falsifying the standard theory. I conducted the experiment before lecturing on the theory and 'behaviour' of markets in class so as not to 'contaminate' the sample. The 22 subjects were each assigned one card from a well-shuffled deck of 11 white and 11 yellow cards. The white cards identified the sellers, and the yellow cards identified the buyers. Each white card carried a price, known only to that seller, which represented that seller's minimum selling price for one unit, and each yellow card identified a price, known only to that buyer, representing that buyer's maximum buying price for one unit. On the left of Figure 1 is listed these so-called 'limit' prices, identified by buyer, B1, B2 etc. (in descending order, D) and by seller, S1, S2 etc. (in ascending order, S). To keep things simple and well controlled each buyer (seller) was informed that he/she was a buyer (seller) of at most one unit of the item in each of several trading periods. Thus demand, D(supply, S) was 'renewed' in each trading period as a steady state flow, with no carry-over in unsatisfied demand (or unsold stock), from one period to the next. In the airline voucher example, imagine the vouchers being issued, followed by trading; the vouchers then expire, new vouchers are issued, traded and so on. In the experiment, suppose real motivation is provided by promising to pay (in cash) to each buyer the difference between that buyer's assigned limit buying price and the price actually paid in each period that a unit is purchased in the market. Thus suppose seller 5 sells their unit to buyer 2 at the price 2.25. Then buyer 2 earns a 'profit' of $0.75 from this exchange. In this way we induce on each buyer a value (or hypothesized willingness-to-pay) equal to the assigned limit buy price. Similarly, suppose each seller is paid the difference between that seller's actual sales price and assigned limit price ('cost', or willingness-to-sell)in each trading period that a unit is sold. Thus in the previous exchange example, seller 5 earns $0.50 from the transaction.

This experimental procedure operationalizes the market preconditions that (1) 'the circumstances of time and place' for each economic agent are dispersed and known only to that agent (as in the above voucher market) and (2) agents have a secure property right in the objects of trade and the private gains ('profits') from trade (an airline travel voucher was transferable and redeemable by any bearer). The reader should note that 'profit' is identified as much with the act of buying as with that of selling. This is because 'profit' is the surplus earned by a buyer who buys for less than his willingness-to-pay, just as a seller's 'profit' is the surplus earned when an item is sold for more than the amount for which they are willing to sell. Willingness-to-sell need not have, and usually does not have anything to do with accounting 'cost', or production 'cost', from which one computes accounting profit. Willingness-to-sell, like willingness-to-buy, is determined by the immediate circumstances of each agent. Hence, a passenger might be prepared to pay the regular full fare premium on a first-class ticket for an emergency trip to visit a sick relative. The accountant's concept of profit cannot be applied to the passenger's decision any more than it can be applied to that of a passenger willing to sell a voucher at a deep discount. In what follows I will use the term 'buyer's surplus' or 'seller's surplus' instead of 'profit' to refer to the gains from exchange enjoyed by buyers or sellers because the term 'profit' is so strongly, exclusively and misleadingly associated with selling activities.

Now let us interpret the previously cited fundamental theorem of economics in the context of the experimental design contained in Figure 1. We note first that the ordered set of seller (buyer) limit prices defines a supply (demand) function (Figure 1). A supply (demand) function provides a list of the total quantities that sellers (buyers) would be willing to sell (buy) at corresponding hypothetical fixed prices. Neither of these functions is capable of being observed, scientifically, in the field. This is because the postulated limit prices are inherently private and not publicly observable. We could poll every potential seller (buyer) of vouchers in Chicago's O'Hare airport on 20 December 1979 to get each person's reported limit price, but we would have no way of validating the 'observations' thus obtained. Referring to Figure 1, we see that in my 1956 experiment, sellers (hypothetically) were just willing to sell three units at price 1.25, nine units at 2.75 and so on. Similarly buyers (hypothetically) were just willing to buy four units at 2.50, seven units at 1.75 and so on. If seller 3 is indifferent between selling and not selling at 1.25, and if every seller (buyer) is likewise indifferent at his/her limit price, then any particular unit may not be sold (purchased) at this limit price. One means of dealing with this problem in laboratory markets is to promise to pay a small 'commission', say 5 cents, to each buyer and seller for each unit bought or sold. Thus seller 3 has a small inducement to sell at 1.25 if he can do no better, and buyer 6 has a small inducement to buy at 2.00 if she can do no better.

Economic theory defines the competitive equilibrium as the price and corresponding quantity that clears the market; that is, it sets the quantity that sellers are willing to sell equal to the quantity that buyers are willing to buy. This assumes that the subjective cost of transacting is zero; otherwise any units with limit prices equal to the competitive equilibrium price will not exchange. In Figure 1 this competitive equilibrium price is 2.00. If the 5 cent 'commission' paid to each trading buyer and seller is sufficient to compensate for any subjective cost of transacting, then buyer 6 and seller 6 will each trade and the competitive equilibrium quantity exchanged will be 6 units. At the competitive equilibrium price, buyer 1 earns a surplus of $3.25 - 2.00 = 1.25$ (plus commission) per period and so on. Total surplus, which measures the maximum possible gains from exchange, or maximum wealth *created* by the existence of the market institution, is 7.50 per period, at the competitive equilibrium.

If by some miracle the competitive equilibrium price and exchange quantity were to prevail in this market, sellers 1–6 would sell, buyers 1–6 would buy, while sellers 7–11 would make no sales and buyers 7–11 would make no purchases. It might be thought that this is unfair – the market should permit some or all of the 'submarginal' buyers (sellers) 7–11 to trade – or that more wealth would be created if there were more than six exchanges. But these interpretations are wrong. By definition, buyer 10 is not willing to pay more than 1.00. Consequently, it is a peculiar notion of fairness to argue that buyer 10 should have as much priority as buyer 1 in obtaining a unit. In the airline voucher example, this would mean that a buyer who is unlikely to redeem a voucher should have the same priority as a buyer who is likely to redeem a voucher. One can imagine a market in which, say, buyer 1 is paired with seller 9 at price 3.00, buyer 2 with seller 8 at price 2.75, and so on with nine units traded. If this were to occur it would mean buyers 7–9, who are less likely to use vouchers, have purchased them, and sellers 7–9, who initially held vouchers, and were more likely to use them than buyers 7–9, have sold their vouchers. Furthermore, this allocation yields additional possible gains from exchange, and is thus *not sustainable*, even if it were thought to be desirable. That is, buyer 9, who bought from seller 1 at price 1.00, could resell the unit to seller 9 (who sold her unit to buyer 2), at price (say) 2.00. Why? Because, by definition a voucher is worth 2.75 to seller 9 and only 1.25 to buyer 9. Similar additional trades can be made by buyers (sellers) 7 and 8. The end result would be that buyers 1–6 and sellers 7–11 would be the terminal holders of vouchers, just as if the competitive equilibrium had been reached initially.

Hence, either the competitive equilibrium prevails, or if inefficient trades occur at dispersed prices, then further 'speculative' gains can be made by some buyers and sellers. If these gains are fully captured the end result is the same allocation as would occur at the competitive equilibrium price and quantity.

Having specified the environment (individual private values) of our experimental market, what remains is to specify an exchange institution. In my 1956 experiment I elected to use trading rules similar to those that characterize trading on the organized stock and commodity exchanges. These markets use the 'double oral auction' procedure. In this institution as soon as the market 'opens' any buyer is free to announce a bid to buy and any seller is free to announce an offer to sell. In the experimental version each bid (offer) is for a single unit. Thus a buyer might say 'buy, 1.00', while a seller might say 'sell, 5.00', and it is understood that the buyer bids 1.00 for a unit and the seller offers to sell one unit for 5.00. Bids and offers are freely announced and can be modified. A contract occurs if any seller accepts the bid of any buyer, or any buyer accepts the offer of any seller. In the simple experimental market, since each participant is a buyer or seller of at most one unit per trading period, the contracting buyer and seller drop out of the market for the remainder of the trading period, but return to the market when a new trading 'day' begins. The experimenter announces the close of each trading period and the opening of the subsequent period, with each trading period timed to extend, say, five minutes. Each contract price is plotted on the right of Figure 1 for the five trading periods of the experiment. This result was not as expected. The conventional view among economists was that a competitive equilibrium was like a frictionless ideal state which could not be conceived as actually occurring, even approximately. It could be conceived of occurring only in the presence of an abstract 'institution' such as a Walrasian *tâtonnement* or an Edgeworth recontracting procedure. It was for teaching, not believing.

From Figure 1 it is evident that in the strict sense the competitive equilibrium was not attained in any period, but the accuracy of the competitive equilibrium theory is easily comparable to that of countless physical processes. Certainly, the data clearly do not support the monopoly, or seller collusion model. The total return to sellers is maximized when four units are sold at price 2.50. Similarly, the monopsony, or buyer collusion model requires four units to exchange at price 1.50.

Since 1956, several hundred experiments using different supply and demand conditions, experienced as well as inexperienced subjects, buyers and sellers with multiple unit trading capacity, a great variation in the numbers of buyers and sellers, and different trading institutions, have established the replicability and robustness of these results. For many years at the University of Arizona and Indiana University we have been using various computerized (the PLATO system) versions of the double 'oral' auction, developed by Arlington Williams, in which participating subjects trade with each other through computer terminals. These experiments establish that the 1956 results are robust with respect to substantial reductions in the number of buyers and sellers. Most such experiments use only four buyers and four sellers, each capable of trading several units. Some have used only two sellers, yet the competitive equilibrium model performs very well under double auction rules. Figure 2 shows the supply and demand design and the market results for a typical experiment in which subjects trade through PLATO computer terminals under computer-monitored double auction rules.

In addition to its antiquarian value, Figure 1 illustrates the problem of monitoring the rules of a 'manual' experiment. Observe that in period 4 there were seven contracts which are recorded as occurring in the price range between $1.90 and $2.25. This is not possible since there are only six buyers with limit buy prices above $1.90. Either a buyer violated his budget constraint, or the experimenter erred in recording a price in his first experiment. In Figure 2 there is plotted each contract (an accepted bid if the contract line passes through a 'dot'; an accepted offer if the line passes through a 'circle') and the bids ('dots') and offers ('circles') that preceded each accepted bid or offer. One of the several advantages of computerized experimental markets is that the complete data of the market (all bids, offers, and contracts at their time of execution) are recorded accurately and non-invasively, and all experimental rules are enforced perfectly. In particular the violation of a budget constraint revealed in Figure 1, which is a perpetual problem with manually executed experiments, is not a problem when trading is perfectly computer monitored.

The rapid convergence shown in Figures 1 and 2 has not always extended to trading institutions other than the double auction. For example, the 'posted offer' pricing mechanism (associated with most retail markets), in which sellers post take it or leave it non-negotiable prices at the beginning of each period, yields higher prices and less efficient allocations than the double auction. This difference in performance becomes smaller with experienced subjects and with longer trading sequences in a given experiment (Ketcham et al., 1984). Similarly, a comparison of double auction with a sealed bid-offer auction finds the latter to be less efficient and to deviate more from the competitive equilibrium predictions (Smith et al., 1982). Thus, institutions have been demonstrated to make a difference in what we observe. The data and analysis strongly suggest that institutions make a difference because the rules (legal environment) make a difference, and the rules make a difference because they affect individual incentives.

II. BRIEF INTERPRETIVE HISTORY OF THE DEVELOPMENT OF EXPERIMENTAL ECONOMICS. The two most influential early experimental studies represent the two primary poles of experimental economics: the study of individual preference (choice) under uncertainty (Mosteller and Nogee, 1951) and of market behaviour (Chamberlin, 1948). The investigation of uncertainty and preference has focused on the testing of von Neumann–Morgenstern–Savage subjective expected utility theory. Battalio, Kagel and others have pioneered in the testing of the Slutsky–Hicks commodity demand and labour supply preferences using humans (1973) and animals (1975). A series of large-scale field experiments in the 1970s extended the experimental study of individual preference to the measurement of the effect of the negative income tax and other factors on labour supply and to the measurement of the demand for electricity, housing and medical services.

Since the human species has been observed to participate in market exchange for thousands of years, the experimental study of market behaviour is central to economics. Preferences are not directly observable, but preference theory, as an

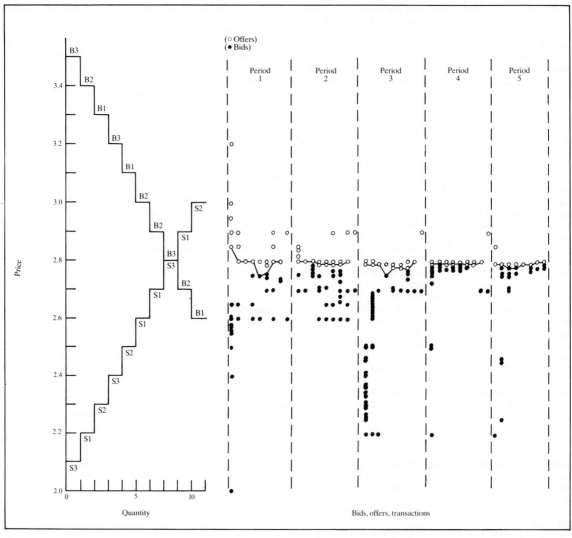

(○ Offers)
(● Bids)

Period 1   Period 2   Period 3   Period 4   Period 5

Price

Quantity

Bids, offers, transactions

Figure 2

abstract construct, has been *postulated* by economists to be fundamental to the explanation and understanding of market behaviour. In this sense the experimental study of group market behaviour depends upon the study of individual preference behaviour. But this intellectual history should not obscure the fact that the study of markets and the study of preferences need not be construed as inseparable. Adam Smith clearly viewed the human 'propensity to truck, barter and exchange' (and *not* the existence of human preferences) as axiomatic to the scientific study of economic behaviour. Obversely, the work of Battalio and Kagel showing that animals behave as if they had Slutsky–Hicks preferences makes it plain that substitution behaviour is an important cross species characteristic, but that such phenomena need not be associated with market exchange.

A significant feature of Chamberlin's (1948) original work is that it concerned the study of behaviourally complete markets; that is all trades, including purchases as well as sales, were

executed by active subject agents. This feature has continued in the subsequent bilateral bargaining experiments of Siegel and Fouraker (1960) and in market experiments (Smith, 1962, 1982; Williams and Smith, 1984) such as those discussed in section I. This feature was not present in the early and subsequent experimental oligopoly literature (Hoggatt, 1959; Sauermann and Selten, 1959; Shubik, 1962; Friedman, 1963), in which the demand behaviour of buyers was simulated, that is, programmed from a specified demand function conditional on the prices selected in each 'trading' period by the sellers. This simulation of demand behaviour is justified as an intermediate step in testing models of seller price behaviour that assume passive, simple maximizing, demand-revelation behaviour by buyers. But the conclusions of such experimental studies should not be assumed to be applicable, even provisionally, to any observed complete market without first showing that the experimental results are robust with respect to the substitution of subject buyers for simulated buyers.

III. THE FUNCTIONS OF MARKET EXPERIMENTS IN MICROECONOMIC ANALYSIS. A conceptual framework for clarifying some uses and functions of experiments in microeconomics can be articulated by suitable modification and adaptation (Smith, 1982) of the concepts underlying the adjustment process, as in the welfare economics literature (see references to Hurwicz and Reiter in Smith, 1982). In this literature a microeconomic environment consists of a list of agents $\{1, \ldots, N\}$, a list of commodities and resources $\{1, \ldots K\}$, and certain characteristics of each agent i, such as the agent's preferences (utility) u_i, technological (knowledge) endowment T^i, and commodity endowment w_i. Thus agent i is defined by the triplet of characteristics $E^i = (u^i, T^i, w^i)$ defined on the K-dimensional commodity space. A microeconomic *environment* is defined by the collection $E = (E^1, \ldots, E^N)$ of these characteristics. This collection represents a set of primitive circumstances that condition agents' interaction through institutions. The superscript i, besides identifying a particular agent, also means that these primitive circumstances are in their *nature* private: it is the individual who likes, works, knows and makes.

There can be no such thing as a credible institution-free economics. Institutions define the property right rules by which agents communicate and exchange or transform commodities within the limits and opportunities inherent in the environment, E. Since markets require communication to effect exchange, property rights in messages are as important as property rights in goods and ideas. An institution specifies a language, $M = (M^1, \ldots, M^N)$, consisting of message elements $m = (m^1, \ldots, m^N)$, where M^i is the set of messages that can be sent by agent i (for example, the range of bids that can be sent by a buyer). An institution also defines a set of allocation rules $h = (h^1(m), \ldots, h^N(m))$ and a set of cost imputation rules $c = (c^1(m), \ldots, c^N(m))$, where $h^i(m)$ is the commodity allocation to agent i and $c^i(m)$ is the payment to be made by i, each as a function of the messages sent by all agents. Finally, the institution defines a set of adjustment process rules (assumed to be common to all agents), $g(t_0, t, T)$, consisting of a starting rule, $g(t_0, \cdot, \cdot)$, a transition rule, $g(\cdot, t, \cdot)$, governing the sequencing of messages, and a stopping rule, $g(\cdot, \cdot, T)$, which terminates the exchange of messages and triggers the allocation and cost imputation rules. Each agent's property rights in communication and exchange is thus defined by $I^i = (M^i, h^i(m), c^i(m), g(t_0, t, T))$. A microeconomic *institution* is defined by the collection of these individual property right characteristics, $I = (I^1, \ldots, I^N)$.

A microeconomic *system* is defined by the conjunction of an environment and an institution, $S = (E, I)$. To illustrate a microeconomic system, consider an auction for a single indivisible object such as a painting or an antique vase. Let each of N agents place an independent, certain, monetary value on the item $v_1, \ldots, v_N$, with agent i knowing his own value, v_i, but having only uncertain (probability distribution) information on the values of others. Thus $E^i = (v_i; P(v), N)$. If the exchange institution is the 'first price' sealed-bid auction, the rules are that all N bidders each submit a single bid any time between the announcement of the auction offering at t_0, and the closing of bids, at T. The item is then awarded to the maker of the highest bid at a price equal to the amount bid. Thus, if the agents are numbered in descending order of the bids, the first price auction institution $I_1 = (I_1^1 = [h^1(m) = 1, \ c^1(m) = b_1]$ and $I_1^i = [h^i(m) = 0, c^i(m) = 0], i > 1$, where $m = (b_1, \ldots, b_N)$ consists of all bids tendered. That is, the item is awarded to the high bidder, $i = 1$, who pays b_1, and all others receive and pay nothing. This contrasts with the 'second price' sealed-bid auction $I_2 = (I_2^1, \ldots, I_2^N)$ in which $I_2^1 = [h^1(m) = 1, \ c^1(m) = b_2]$ and $I_2^i = [h^i(m) = 0, c^i(m) = 0], i > 1$; that is, the highest bidder

receives the allocation but pays a price equal to the second highest bid submitted.

Another example is the English or progressive oral auction, whose rules are discussed under the entry AUCTIONS. It should be noted that the 'double oral' auction, used extensively in stock and commodity trading and in the two experimental markets discussed in section I, is a two-sided generalization of the English auction.

A microeconomic system is activated by the behavioural choices of agents in the set M. In the static, or final outcome, description of an economy, agent *behaviour* can be defined as a function (or correspondence) $m^i = \beta^i(E^i | I)$ carrying the characteristics E^i of agent i into a message m^i, conditional upon the property right specifications of the operant institution I. If all exchange-relevant agent characteristics are included in E^i, then $\beta \equiv \beta^i$ for all i. Given the message-sending behaviour of each agent, $\beta(E|I)$, the institution determines the outcomes

$$h^i(m) = h^i[\beta(E^1|I), \ldots, \beta(E^N|I)]$$

and

$$c^i(m) = c^i[\beta(E^1|I), \ldots, \beta(E^N|I)].$$

Within this framework we see that agents do not choose allocations directly; agents choose messages with institutions determining allocations under the rules that carry messages into allocations. (You cannot choose to 'buy' an auctioned item; you can only choose to raise the standing bid at an English auction or submit a particular bid in a sealed bid auction.) However, the allocation and cost imputation rules may have important incentive effects on behaviour, and therefore messages will in general depend on these rules. Hence, market outcomes will result from the conjunction of institutions' and agents' behaviour.

A proper theory of agents' behaviour allows one to deduce a particular β function based on assumptions about the agent's environment and the institution, and his motivation to act. Auction theory is perhaps the only part of economic theory that is fully institution specific. For example, in the second price sealed bid auction it is a dominant strategy for each agent simply to bid his or her value; that is

$$b_i = \beta(E^i | I_2) = \beta(v_i | I_2) = v_i, \qquad i = 1, \ldots, N.$$

The resulting outcome is that $b_1 = v_1$ is the winning bid and agent 1 pays the price v_2. Similarly, in the English auction, agent 1 will eventually exclude agent 2 by raising the standing bid to v_2 (or somewhat above), and obtain the item at this price. In the first price auction Vickrey proved that if each agent maximizes expected surplus $(v_i - b_i)$ in an environment with $P(v) = v$ (the v_i are drawn from a constant density on [0, 1]), then we can deduce the noncooperative equilibrium bid function, $b_i = \beta(E^i | I_1) = \beta[v_i; P(v), N | I_1] = (N - 1)v_i | N$ (see the entry on AUCTIONS for a more complete discussion).

With the above framework it is possible to explicate the roles of theory and experiment, and their relationship, in a progressive research programme (Lakatos, 1978) of economic analysis. But to do this we must first ask two questions:

(1) 'Which of the elements of a microeconomic system are not observable?' The nonobservable elements are (i) preferences, (ii) knowledge endowments, and (iii) agent message behaviour, $\beta(E|I)$. Even if messages are available and recorded, we still cannot observe message behaviour functions because we cannot observe, or vary, preferences. The best we can do with field observations of outcomes is to interpret them in terms of models based on assumptions about preferences (Cobb–Douglas, constant elasticity of substitution,

homothetic), knowledge (complete, incomplete, common), and behaviour (cooperative, noncooperative). Any 'tests' of such models must necessarily be joint tests of all of these unobservable elements. More often the econometric exercise is parameter estimation, which is conditional upon these same elements.

(2) 'What would we like to know?' We would like to know enough about how agents' behaviour is affected by alternative environments and institutions so that we can classify them according to the mapping they provide into outcomes. Do some institutions yield Pareto optimal outcomes, and/or stable prices, and, if so, are the results robust with respect to alternative environments?

These two questions together tell us that what we want to know is inaccessible in natural experiments (field data) because key elements of the equation are unobservable and/or cannot be controlled. If laboratory experiments are to help us learn what we want to know, certain precepts that constitute proposed sufficient conditions for a valid controlled microeconomic experiment must be satisfied:

(1) *Non-satiation* (or monotonicity of reward). Subject agents strictly prefer any increase in the reward medium, π; that is $U_i(\pi_i)$ is monotone increasing for all i.

(2) *Saliency.* Agents have the unqualified right to claim rewards that increase (decrease) in the good (bad) outcomes, x_i, in an experiment; the institution of an experiment renders these rewards salient by defining outcomes in terms of the message choices of agents.

In both the field and the laboratory it is the institution that induces value on messages, given each agent's (subjective) value of commodity outcomes. In the laboratory we use a monetary reward function to induce utility value on the abstract accounting outcomes ('commodities') of an experiment. Thus, agent i is given a concave schedule, $V_i(x_i)$, defining the 'redemption value' in dollars for x_i units purchased in an experimental market, and is assured of receiving a net payment equal to $V_i(x_i)$ less the purchase prices of the x_i units in the market. If the x_i units are all purchased at price p (which is the assumption used to derive a hypothetical demand schedule) the agent is paid $\pi_i = V_i(x_i) - px_i$, with utility $u^i(x_i) = U_i(\pi_i(x_i))$. In defining demand it is assumed that the agent directly chooses x_i (that is $x_i = m_i$). Therefore, if i maximizes $u^i(x_i) = U_i[V_i(x_i) - px_i]$, then at a maximum we have $U_i' \cdot [V_i'(x_i) - p] = 0$, giving the demand function $x_i = V_i'^{(-1)}(p)$ if $U_i' > 0$, where $V_i'^{(-1)}$ is the inverse of i's marginal redemption value of x_i units. (The same procedure for a seller using a cost function $C_j(x_j)$ and paying $px_j - C_j(x_j)$ allows one to induce a marginal cost supply of j.) This illustration generalizes easily: if the joint redemption value is $V_i(x_i, y_i)$ for two abstract commodities (x_i, y_i), $u^i = U_i[V^i(x_i, y_i)]$ induces an indifference map given by the level curves of $V^i(x_i, y_i)$, on (x_i, y_i), with marginal rate of substitution $U_i' V_x^i / U_i' V_y^i = V_x^i / V_y^i$, if $U_i' > 0$. If $V^i(x_i, X)$ is the reward function, with x_i a private and X a common (public) outcome good, we are able to control preferences in the study of public good allocation mechanisms, or if

$$X = \sum_{i=1}^{N} x_i$$

we are poised to study allocation with an 'atmospheric' externality (Coursey and Smith, 1985).

The first two precepts are sufficient to allow us to assert that we have created a microeconomic sysem $S = (E, I)$ in the laboratory. But to assure that we have created a controlled microeconomy, we need two additional precepts:

(3) *Dominance.* Own rewards dominate any subjective costs of transacting (or other motivation) in the experimental market.

As with any person, subject agents may have variables other than money in their utility functions. In particular, if there is cognitive and kinesthetic (observe the traders on a Stock Exchange floor) disutility associated with the message-transaction process of the institution, then utility might be better written $U_i(\pi_i, m^i)$. To the extent that this is so we induce a smaller demand on i with the payoff $V_i(x_i)$ than was computed above, and we lose control over preferences. As a practical matter experimentalists think the problem can usually be finessed by using rewards that are large relative to the complexity of the task, and by adopting experimental procedures that reduce complexity (e.g. using the computer to record decisions, perform needed calculations, provide perfect recall, etc.). Another approach, as noted in section I, is to pay a small commission for each trade to compensate for the subjective transaction costs.

(4) *Privacy.* The subjects in an experiment each receive information only on his/her own reward schedule.

This precept is used to provide control over interpersonal utilities (payoff externalities). Real people may experience negative or positive utilities from the rewards of others, and to the extent that this occurs we lose control over induced demand, supply and preference functions. Remember that the reward functions have the same role in an experiment that preference functions have in the economy, and the latter preferences are private and non-observable.

If our interest is confined to testing hypotheses from theory, we are done. Precepts (1)–(4) are sufficient to provide rigorous tests of the theorist's ability to model individual and market behaviour. But one naturally asks if replicable results from the laboratory are transferable to field environments. This requires,

(5) *Parallelism.* Propositions about behaviour and/or the performance of institutions that have been tested in one microeconomy (laboratory or field) apply also to other microeconomies (laboratory or field) where similar *ceteris paribus* conditions hold.

Astronomy, meteorology, biology and other sciences use the maintained hypothesis that the same physical laws hold everywhere. Economics postulates that when the environment and institution are the same, behaviour will be the same; that is, behaviour is determined by a relatively austere subset of life's parameters. Whether this is 'true' is an empirical question. Hence, when one experimentalist studies variations on the treatment variables of another it is customary to replicate the earlier work to check parallelism. Similarly, one must design field experiments, or devise econometric models using non-experimental field data, that provide tests of the transferability of experimental results to any particular market in the field. Only in this way can questions of parallelism be answered. They are not answered with speculations about alleged differences between the experimental subject's behaviour and (undefined) 'real world' behaviour. The experimental laboratory *is* a real world, with real people, real institutions, real payoffs and commodities just as real as stock certificates and airline travel vouchers, both of which have utility because of the claim rights they legally bestow on the bearer.

IV. CLASSIFYING THE APPLICATION OF EXPERIMENTAL METHODS. There are many types of experiments and many fields of economic study to which experimental methods have been applied.

The experimental study of auctions makes the most extensive use of models of individual behaviour based explicitly on the message requirements of the different institutions. This literature provides test comparisons of predicted behaviour, $m^i = \beta(E^i|I)$, with observations on individual choice, $\hat{m}_i = \beta(\hat{E}_i|I)$ for given realizations, $\hat{E}_i$ (such as values, $\hat{v}_i$, where they are assigned at random). The large literature on experimental double auctions makes no such individual comparisons, because the theoretical literature had not yielded tractable models of individual bid-offer behaviour (but recent contributions by Friedman (1984), and Wilson (1984) are providing such models). Here as in most other areas of experimental research the comparisons are between the predicted price–quantity outcomes of static theory (such as competitive, monopoly, and Cournot models), and observed outcomes. But double auctions have been studied (see references in Smith, 1982) in a variety of environments; for example, the effect of price floors and ceilings have been examined (see references in Plott, 1982). In all cases these studies are making comparisons. In *nomotheoretical* experiments one compares theory and observation, whereas in *nomoempirical* experiments one compares the effect of different institutions and/or environments as a means of documenting replicable empirical 'laws' that may stimulate modelling energy in new directions. The idea that formal theory must precede meaningful observation does not account for most of the historical development of science. *Heuristic* or exploratory experiments that provide empirical probes of new topics and new experimental methods should not be discouraged.

In industrial organization, and antitrust economics, experimental methods have been applied to examine the effects of monopoly, conspiracy, and alleged anticompetitive practices, and to study the concept of natural monopoly and its relation to scale economics, entry cost and the contestable markets hypothesis (see references in Plott, 1982; Smith, 1982; Coursey et al., 1984).

An important development in the experimental study of allocation processes has been the extension of experimental market methods to majority rule (and other) committee processes, and to market-like group processes for the provision of goods which have public or common outcome characteristics (loosely, public goods). These studies have examined public good allocation under majority (and Roberts') rules for committee including the effect of the agenda (see the references to Fiorina and Plott, and Levine and Plott in Smith, 1982), and under compensated unanimity processes suggested by theorists (see the references in Coursey and Smith, 1985). Generally, this literature reports substantial experimental support for the theory of majority rule outcomes, the theory of agenda processes (the sequencing of issues for voting decisions), and for incentive compatible models of the provision of public goods.

VERNON L. SMITH

See also ALLAIS PARADOX; EFFICIENT ALLOCATION OF RESOURCES; PREFERENCE REVERSALS; PSYCHOLOGY AND ECONOMICS.

BIBLIOGRAPHY

Battalio, R., Kagel, J., Winkler, R., Fisher, E., Basmann, R. and Krasner, L. 1973. A test of consumer demand theory using observations of individual consumer purchases. *Western Economic Journal* 11(4), December, 411–28.

Chamberlin, E. 1948. An experimental imperfect market. *Journal of Political Economy* 56, April, 95–108.

Coursey, D., Isaac, M., Luke, M. and Smith, V. 1984. Market contestability in the presence of sunk (entry) costs. *Rand Journal of Economics* 15(1), Spring, 69–84.

Coursey, D. and Smith, V. 1985. Experimental tests of an allocation mechanism for private, public or externality goods. *Scandinavian Journal of Economics* 86(4), 468–84.

Friedman, D. 1984. On the efficiency of experimental double auction markets. *American Economic Review* 74(1), March, 60–72.

Friedman, J. 1963. Individual behavior in oligopolistic markets: an experimental study. *Yale Economic Essays* 3(2), 359–417.

Hoggatt, A. 1959. An experimental business game. *Behavioral Science* 4(3), July, 192–203.

Kagel, J., Battalio, R., Rachlin, H., Green, L., Basmann, R. and Klemm, W. 1975. Experimental studies of consumer behavior using laboratory animals. *Economic Inquiry* 13(1), March, 22–38.

Ketcham, J., Smith, V. and Williams, A. 1984. A comparison of posted-offer and double-auction pricing institutions. *Review of Economic Studies* 51(4), October, 595–614.

Lakatos, I. 1978. *The Methodology of Scientific Research Programmes, Philosophical Papers*, Vol. 1, Ed. J. Worrall and G. Currie, Cambridge: Cambridge University Press.

Mosteller, F. and Nogee, P. 1951. An experimental measurement of utility. *Journal of Political Economy* 59, October, 371–404.

Plott, C. 1982. Industrial organization theory and experimental economics. *Journal of Economic Literature* 20(4), December, 1485–527.

Sauermann, H. and Selten, R. 1959. Ein Oligopolexperiment. *Zeitschrift für die Gesamte Staatswissenschaft* 115(3), 427–71.

Shubik, M. 1962. Some experimental non zero sum games with lack of information about the rules. *Management Science* 81(2), January, 215–34.

Siegel, S. and Fouraker, L. 1960. *Bargaining and Group Decision Making*. New York: McGraw-Hill.

Smith, V. 1962. An experimental study of competitive market behavior. *Journal of Political Economy* 70, April, 111–37.

Smith, V. 1982. Microeconomic systems as experimental science. *American Economic Review* 72(5), December, 923–55.

Smith, V., Williams A., Bratton, K. and Vannoni, M. 1982. Competitive market institutions: double auctions versus sealed bid-offer auctions. *American Economic Review* 72(1), March, 58–77.

Williams, A. and Smith, V. 1984. Cyclical double-auction markets with and without speculators. *Journal of Business* 57(1) Pt 1, January, 1–33.

Wilson, R. 1984. Multilateral exchange. Working Paper No. 7, Stanford University, August.

exploitation. In the most general sense, to exploit something means to make use of it for some particular end, as in the exploitation of natural resources for social benefit or for private profit. Insofar as this use takes advantage of other people, exploitation also implies something unscrupulous. If the other people are endemically powerless, as in the case of the poor in relation to their landlords, creditors and the like, then the term exploitation takes on the connotation of oppression.

Marx uses the word exploitation in all the above senses. But he also defines a new concept, the *exploitation of labour*, which refers specifically to the extraction of the surplus labour upon which class society is founded. In this latter sense, exploitation becomes one of the basic concepts of the Marxist theory of social formations.

EXPLOITATION AND CLASS. Society consists of people living within-and-through complex networks of social relations which shape their very existence. Marx argues that the relations which structure the social division of labour lie at the base of social reproduction, because the division of labour simultaneously accomplishes two distinct social goals: first, the production of the many different objects which people use in their myriad activities of daily life; and second, the reproduction of the basic social framework under which this production takes place, and hence of the social structures which rest on this foundation. Social reproduction is always the reproduction of individuals as *social individuals*.

Class societies are those in which the rule of one set of people over another is founded upon a particular kind of social division of labour. This particularity arises from the fact that the dominant class maintains itself by controlling a process through which the subordinate classes are required to devote a portion of their working time to the production of things needed by the ruling class. The social division of labour within a class society must therefore be structured around the extraction of *surplus labour*, i.e. of labour time over and above that required to produce for the needs of the labouring classes themselves. In effect, it is the subordinate classes which do the work for the reproduction of the ruling class, and which therefore end up *working to reproduce the very conditions of their own subordination*. This is why Marx refers to the extraction of surplus labour in class societies as the exploitation of labour (Marx, 1867, Part 3 and Appendix). It should be clear from this, incidentally, that the mere performance of labour beyond that needed to satisfy immediate needs does not in itself constitute exploitation. Robinson Crusoe, labouring away in his solitude in order to plant crops for future consumption or to create fortification against possible attacks, is merely performing some of the labour necessary for his own needs. He is neither exploiter nor exploited. But all this changes once he manages to subordinate the man Friday, to 'educate' him through the promise of religion and the threat of force to his new place in life, and to set him to work building a proper microcosm of English society. Now it is Robinson who is the exploiter, and Friday the exploited whose surplus labour only serves to bind him ever more tightly to his new conditions of exploitation (Hymer, 1971).

Although the exploitation of labour is inherent in all class societies, the form it takes varies considerably from one mode of production to another. Under slavery, for instance, the slave belongs to the owner, so that the whole of his or her labour and corresponding net product (i.e. product after replacement of the means of production used up) is ostensibly appropriated by the slave owner. But in fact the slave too must be maintained out of this very same net product. Thus it is the surplus product (the portion of the net product over that needed to maintain the slaves), and hence the surplus labour of the slaves, which in the end sustains the slave-owning class. In a similar vein, under feudalism the surplus labour of the serf and tenant supports the ruling apparatus. But here, the forms of its exaction are many and varied: sometimes direct, as in the case of the quantities of annual labour and/or product which the serf or tenant is required to hand over to Lord, Church and State; and sometimes indirect, as in the payment of money rents, tithes and taxes which in effect require the serf or tenant to produce a surplus product and sell it for cash in order to meet these imposed obligations.

The material wealth of the dominant class is directly linked to the size of the surplus product. And this surplus product is in turn greater the smaller the standard of living of the subordinate classes, and the longer, more intense or more productive their working day. Both of these propositions translate directly into a higher ratio of surplus labour time to the labour time necessary to reproduce the labourers themselves, that is, into a higher *rate of exploitation* of labour: given the productivity of labour and length and intensity of the working day, the smaller the portion of the product

consumed by the producing class, the greater the portion of their working day which is in effect devoted to surplus labour; similarly, given the consumption level of the average peasant or worker, the longer, more intense and/or more productive their labour, the smaller the portion of their working day which has to be devoted to their own consumption needs, and hence the greater the portion which corresponds to surplus labour.

Because the magnitude of the surplus product can be raised in the above ways, it is always in the direct interest of the ruling class to try and push the rate of exploitation towards its social and historical limits. By the same token, it is in the interest of the subordinate classes not only to resist such efforts but also to fight against the social conditions which make this struggle necessary in the first place. The exploitative base of class society makes it a fundamentally antagonistic mode of human existence, marked by a simmering hostility between rulers and ruled, and punctuated by periods of riots, rebellions and revolutions. This is why class societies must always rely heavily on ideology to motivate and rationalize the fundamental social cleavage upon which they rest, and on force to provide the necessary discipline when all else fails.

CAPITALISM AND EXPLOITATION. Capitalism shares the above general attributes. It is a class society, in which the domination of the capitalist class is founded upon its ownership and control of the vast bulk of the society's means of production. The working class, on the other hand, is made up of those who have been 'freed' of this self-same burden of property in means of production, and who must therefore earn their livelihood by working for the capitalist class. As Marx so elegantly demonstrates, the *general social condition* for the reproduction of these relations is that the working class as a whole be induced to perform surplus labour, because it is this surplus labour which forms the basis of capitalist profit, and it is this profit which in turn keeps the capitalist class willing and able to reemploy workers. And as the history of capitalism makes perfectly clear, the whole process is permeated by the struggle between the classes about the conditions, terms and occasionally even about the future, of these relations.

The historical specificity of capitalism arises from the fact that its relations of exploitation are almost completely hidden behind the surface of its relations of exchange. At first glance, the transaction between the worker and capitalist is a perfectly fair one. The former offers labour power for sale, the latter offers a wage rate, and the bargain is struck when both sides come to terms. But once this phase is completed, we leave the sphere of freedom and apparent equality and enter into 'the hidden abode of production' within which lurks the familiar domain of surplus labour (Marx, 1867, ch. 6). We find here a world of hierarchy and inequality, of orders and obedience, of bosses and subordinates, in which the working class is set to work to produce a certain amount of product for its employers. Of this total product, a portion which corresponds to the materials and depreciation costs of the total product is purchased by the capitalists themselves, in order to replace the means of production previously used up. A second portion is purchased by the workers with the wages previously paid to them by their employers. But if these two portions happen to exhaust the total product, then the capitalists will have succeeded in producing only enough to cover their own (materials, depreciation and wage) costs of production. *There would be no aggregate profit*. It follows, therefore, that for capitalist production to be successful, i.e. for it to create its own profit, workers must be induced to work longer than the time required to produce their own means of consumption.

They must, in other words, perform surplus labour time in order to produce the surplus product upon which profit is founded.

The above propositions can be derived analytically (Morishima, 1973, ch. 7). More importantly, they are demonstrated *in practice* whenever working time is lost through labour strikes or slowdowns. Then, as surplus labour time is eroded, the normally hidden connection between surplus labour and profit manifests itself as a corresponding fall in profitability. Every practising capitalist must learn this lesson sooner or later.

Orthodox economics, encapsulated within its magic kingdom of production functions, perfect competition, and general equilibrium, usually manages to avoid such issues. Indeed, it concerns itself principally with the construction and refinement of an idealized image of capitalism, whose properties it then investigates with a concentration so ferocious that it is often able to entirely ignore the reality which surrounds it. Within this construct, production is a disembodied process undertaken by an intangible entity called the firm. This firm hires 'factors of production' called capital and labour in order to produce an output, paying for each factor according to its estimated incremental contribution to the total output (i.e. according to the value of its marginal product). If all goes well, the sum of these payments turn out to exhaust exactly the net revenues actually received by the firm, and the ground is set for yet another round.

Notice that this conception puts a thing (capital) and a human capacity (labour power) on equal footing, both as so-called factors of production. This enables the theory to deny any class difference between capitalists and workers by treating all individuals as essentially equal because they are all owners of at least one factor of production. The fact that 'factor endowments' may vary considerably across individuals is then merely a second-order detail whose explanation is said to lie outside of economic theory. Next, by treating production as some disembodied process, the human labour process is reduced to a mere technical relation, to a production function which 'maps' things called inputs (which include labour power) into a thing called output. All struggle over the labour process thus disappears from view. Finally, since capital and labour are mere things, they cannot be said to be exploited. However, to the extent that the payment for some factors falls short of equality with its particular marginal product, the *owner* of this factor may be said to be exploited. In this sense, exploitation is defined as a discrepancy between an actual and an ideal 'factor payment' (it can be established that a very similar construction underlies notions of unequal exchange such as those in Emmanuel, 1969). More importantly, exploitation as defined above can in principle apply just as well to profits as to wages. Capitalism thus emerges as a system in which capitalists are just as liable to be exploited by workers as vice versa (Hodgson, 1980, section 2). With this last step, the very notion of exploitation is reduced to utter triviality.

EXPLOITATION, GENDER AND RACE. We have focused on the notion of exploitation as the extraction of surplus labour because this relation is the foundation upon which class society is built, in the sense that the other legal, political and personal relations within the society are structured and limited by this central one. This does not mean that these other relations lack a history and logic of their own. It only means that within any given mode of production, they are bound to the system by the force field of this central relation, and characteristically shaped by its ever present gravitational pull.

In the same vein, the notion that class society is marked by oppression along class lines obviously does not exclude other equally egregious forms of subjugation. It is evident, for instance, that the oppression of women by men is common to all known societies, and to all classes within them. Thus any proper understanding of the oppression of workers by capitalists must also encompass the oppression of working-class women by men of all classes, as well as the oppression of ruling-class women by men of their own class.

But even this is not enough. It is not sufficient to say that class and patriarchy are coexistent forms of oppression. We need to know also how they relate to one another. And it is here that Marxists generally give preeminence to class, not because class oppression is more grievous, but because of the sense that it is the nature of the class relation which modulates and shapes the corresponding form of patriarchy. That is to say, Marxists argue that capitalist patriarchy is distinct from feudal patriarchy precisely because capitalist relations of production are characteristically different from feudal ones.

Needless to say, there is still considerable controversy about the exact relationship between patriarchy and class (Barret, 1980), as there is about the relation of race to either of them (Davis, 1981). These are issues of great theoretical significance. Most importantly, a united struggle against these various forms of oppression has truly revolutionary potential.

ANWAR SHAIKH

See also CAPITAL AS A SOCIAL RELATION; LABOUR POWER; MARXIAN VALUE ANALYSIS; RATE OF EXPLOITATION; SURPLUS VALUE; VALUE AND PRICE.

BIBLIOGRAPHY
Barret, M. 1980. *Women's Oppression Today: Problems in Marxist Feminist Analysis.* London: Verso.
Davis, A.Y. 1981. *Women, Race and Class.* New York: Vintage, 1983.
Emmanuel, A. 1969. *Unequal Exchange: A Study of the Imperialism of Trade.* New York: Monthly Review Press.
Hodgson, G. 1980. A theory of exploitation without the labor theory of value. *Science and Society* 44(3), Fall, 257–73.
Hymer, S. 1971. Robinson Crusoe and the secret of primitive accumulation. *Monthly Review* 23 (4), September, 11–36.
Marx, K. 1867. *Capital*, Vol. I. London: Penguin Books, 1976.
Morishima, M. 1973. *Marx's Economics.* Cambridge: Cambridge University Press.

export-led growth. *See* IMPORT SUBSTITUTION AND EXPORT-LED GROWTH.

exports. *See* INTERNATIONAL TRADE.

extended family. It has long been assumed that extended families are typical of pre-capitalist or non-capitalist societies, while the nuclear family form is the product of industrialization and urbanization. Modernization theories, deriving ultimately from 19th-century thinkers such as the French social reformer Frédéric Le Play (e.g., 1871), and finding different forms of expression in the Chicago School of urban sociology (e.g., Wirth, 1938) and Parsonian functionalism (e.g., Parsons and Bales, 1955), was articulated in a moderate form by W. Goode:

Whenever the economic system expands through industrialisation, family patterns change. Extended kinship ties weaken, lineage patterns dissolve, and a trend toward some form of the conjugal system generally begins to appear – that is, the nuclear family becomes a more independent kinship unit (1963, p. 6).

While Goode himself recognizes that the conjugal family was prevalent in Western Europe long before the Industrial Revolution and limits himself to stating how functionally suited it is to the industrial system, it has long been assumed that the nuclear family emerged as a result of the development of capitalism (e.g., Tawney, 1912). How this supposed transformation is interpreted depends on ideological positions: for those critical of the effects of capitalism the extended family evokes a world of solidarity and human values, while for the opposite tradition which finds its decisive expression in liberalism as a political doctrine, the extended family serves to maintain dependency between kin and to prevent the development of the entrepreneurial spirit.

What is meant by the extended family? The terms is ambiguous in the same way as the concept of the family itself: that is, it can refer either to a *co-resident group*, consisting of a wider group of kin than the single nuclear family, or to a network of genealogically and affinally related kin who cooperate and interact closely. However, in either case it is used especially to contrast with the dissolving of kin ties and their replacement by various types of voluntary association and contractual ties, which are said to be typical of capitalist society.

The extended family is frequently defined by the criterion of co-residence, not only because large residential groupings contrast so strikingly with the small units of Western capitalist society, but also because there is a normative assumption contained within the word 'family' that close kin *should* share their resources and if possible live together.

However, even the criterion of co-residence is not as clear-cut as might at first appear (Goody, 1972). Available historical and anthropological evidence reveals that there are many different ways in which kin can share domestic space, from maintaining virtually independent budgets, to operating as a close-knit single economic unit with a strong head (usually known as a 'patriarch'). Moreover, while most authors try to maintain a distinction between the terms family and household, the terms frequently become elided or confused. The word family of course derives from the Latin *familia*, which referred to a whole complex household enterprise, including slaves. This broader definition of 'family' was only restricted to genealogical kin from the beginning of the 19th century (Flandrin, 1976, pp. 4–10). The ambiguity arises from the contemporary ideology that co-resident domestic groups (i.e. households) *should* be based on close kin relationships, and that the intimacy, cooperation and pooling of resources found within a household are only appropriate between close kin. More specifically, Western familial ideology assumes that the co-resident kin group is the social unit within which sharing and pooling take place, while exchange takes place *between* households (Harris, 1982).

Such assumptions have two problems: first, that they render us blind to the presence within households of people who are not related to the household core, that is, servants, lodgers and others. This has had damaging consequences for European family history, which only in recent years has begun to appreciate the significance of what is clearly a major pattern of European family organization and domestic life (Macfarlane, 1970; Harris, 1982; Smith, 1984). Secondly, such assumptions place undue emphasis on the individual household unit, at the expense of adequate consideration of movement and cooperation *between* households or their members.

Apart from methodological problems, the view that extended families are 'pre-capitalist' while the nuclear family is typical of capitalism has turned out to be inaccurate even for English history and the early development of capitalism. Macfarlane has recently summarized a large body of historical research to argue that the English peasant economy was not based on extended families from at least the 13th century. English rural society was mobile, with a developed market in land and labour; children were as likely to work as servants for a wage and buy land when they reached maturity, as to inherit a family farm (Macfarlane, 1978). Moreover, a study of 19th-century Lancashire proposed that industrialization actually *increased* the number of extended (i.e. three-generation) households (Anderson, 1971; see also Tilly and Scott, 1978; Hareven, 1982).

On a more general level, the work of Laslett and his associates has used extensive historical demographic research to argue that 'the nuclear family predominates numerically almost everywhere, even in underdeveloped parts of the world' (Laslett, 1972, p. 9). Laslett's arguments were particularly directed against the orthodoxy established by Le Play that the three-generational 'stem family' (*famille souche*) was the dominant family form of the European peasantry, derived from factors such as the inalienability of the land belonging to a particular house and patriline, the buying out of siblings by a chosen successor, and extensive provisions for retirement.

Certain difficulties can be found with Laslett's influential and important arguments; in particular, Berkner (1972) has demonstrated the existence of stem families in 18th-century Austria by emphasizing an essential feature of households ignored by Laslett, namely that they change over time in accordance with individual life cycles and mortality rates. Thus, even in areas where the 'stem family' is the basic principle of organization, only a minority of actual households will conform to this type. This approach has been extended by Wolf (1984) to include the notion of 'family cycle' in a discussion of rural Taiwan in the 20th century. Others have argued that Laslett's use of the 'community' as his basic unit of analysis is inappropriate, since there are significant variations between households according to class and socio-economic position. Overall there is a problem in assessing how far majority household forms in terms of *statistical frequency* reflect what each society considers to be the ideal. The discrepancy can be illustrated by contemporary industrial Britain, where research has revealed that a surprisingly large percentage of households do not conform to the ideal nuclear family type, for all that it is enshrined in legislation, welfare policies and religious belief (McIntosh, 1979). The problems are obviously magnified when we turn to scanty historical data.

In recent years, detailed historical research on European families has revealed a complexity and variation that modifies Laslett's early argument but also refuses any simple correlation of family types with particular modes of production, economic stages, or even countries. Even regarding England, opinions differ as to how far and in what circumstances the nuclear family was the dominant type; taking a broader European perspective, Anderson summarizes the debate as follows:

> the European pre-industrial household was a regionally diverse one with England, northern France, North America and possibly the Low Countries ... being unique in both their low proportion of complex households and their

overall homogeneity of household patterns. By contrast, areas of much greater complexity predominated in the east and south ... while in Northern Europe a more locally diverse pattern was found (1980, p. 29).

Various explanations have been proposed for variation in household and family forms; the influence of different systems of distributing productive property has rightly been considered of major importance. Goody (1976) offers a global theory of the formation of domestic groups in terms of land distribution which in turn he derives from agricultural technology (see also Goody, Thirsk and Thompson, 1976). However, it is too general to be applicable to the understanding of local variations; a recent exhaustive discussion of the evidence from English history from 1250 to 1800 concludes that it would be hard to maintain that the relationship between landed property and the family's development cycle was the sole or even the most important determinant of rural family forms (Smith, 1984, p. 86).

In modernization theories, one of the structural explanations for the replacement of the extended family by the nuclear or conjugal family was precisely the shift away from agrarian production with land as the basic means of subsistence, to an industrial system in which the majority owned no means of production except their own labour power. However, some have argued that the organization of labour can be a major determinant of household size in peasant societies. Conversely, studies of proto-industrialization and factory production show how family ties can be strengthened and household size increased in order to maximize cooperation and the pooling of labour (Anderson, 1971; Medick, 1976). The same pattern can be documented outside of Europe in different contexts: some of the classic examples of large extended family households, for example, the Indian joint family, or the Japanese *dozuku*, have shown remarkable resilience in adapting to the various processes of urbanization and industrialization (Yanagisako, 1979; see also Smith, Wallerstein and Evers, 1984).

Large, extended family households, although not as generally found in pre-capitalist societies as was once supposed, do occur in many different contexts. However, there are major problems of definition. Is it really appropriate to include within a single category an Amazonian longhouse (*maloca*) inhabited by a variety of agnatically related families who cooperate in consumption, and the famous Balkan *zadruga*, where an older man might run a unit consisting of up to fifty people all his direct descendants, who operate as a single production unit?

Thus the whole notion of the extended family household needs substantial modification, whether one considers its historical distribution, its determinants, or its status as an individual unit.

In the broader sense of extended family as a network of kin, too, debates centre on how far such kin ties are typical of 'pre-capitalist' societies, and how far they disintegrate with the development of capitalism. Shorter (1975) presents a provocative version of the modernization thesis, arguing that the 'modern family' is private and more independent of wider kin and social ties than the 'traditional' family. But there is substantial disagreement: students of European history cite evidence to argue that neighbourhood ties have long been more significant in everyday life than kin ties (e.g., Macfarlane, 1970). Goody (1983) argues that European family structures are unusual in this respect, because of policies of the early Church to proscribe marriage between close kin. Conversely, case studies from the nations of the economic periphery, and of non-European migrant groups in metropol-

itan regions, indicate how suited extended family networks are to business success (e.g., the classic studies for West Africa of Cohen, 1969, and Okali, 1983). Overall, the emphasis of modernization theories on linear change determined by the economy cannot be sustained, both because the pattern supposed to be typical of industrial society is found much earlier in European history, and because it is tied too closely to a supposed functional fit with industrial production. With the current restructuring of the world economy away from industry, we can expect extended family forms to thrive in many economic situations.

OLIVIA HARRIS

See also DEMOGRAPHIC TRANSITION; ECONOMIC ANTHROPOLOGY; FAMILY; INHERITANCE.

BIBLIOGRAPHY

Anderson, M. 1971. *Family Structure in Nineteenth Century Lancashire*. Cambridge: Cambridge University Press.

Anderson, M. 1980. *Approaches to the History of the Western Family 1500–1914*. London: Macmillan.

Berkner, L. 1972. The stem family and developmental cycle of the peasant household: an eighteenth-century Austrian example. *American Historical Review* 77, 398–418.

Flandrin, J-L. 1976. *Families in Former Times*. Trans., Cambridge: Cambridge University Press, 1979.

Goode, W. 1963. *World Revolution and Family Patterns*. New York and Glencoe, Ill.: Free Press.

Goody, J. 1972. The evolution of the family. In *Household and Family in Past Time*, ed. P. Laslett and R. Wall, Cambridge: Cambridge University Press.

Goody, J. 1976. *Production and Reproduction: A Comparative Study of the Domestic Domain*. Cambridge: Cambridge, University Press.

Goody, J. 1983. *The Development of the Family and Marriage in Europe*. Cambridge: Cambridge University Press.

Goody, J., Thirsk, J. and Thompson, E.P. (eds) 1976. *Family and Inheritance*. Cambridge: Cambridge University Press.

Hareven, T. 1982. *Family Time and Industrial Time*. Cambridge: Cambridge University Press.

Harris, O. 1982. Households and their boundaries. *History Workshop Journal* 13, 143–52.

Laslett, P. 1972. Introduction. In *Household and Family in Past Time*, ed. P. Laslett and R. Wall, Cambridge: Cambridge University Press.

Le Play, F. 1871. *L'organisation de la famille selon le vrai modèle signalé par l'histoire de toutes les races et de tous les temps*. Paris.

Macfarlane, A. 1970. *The Family Life of Ralph Josselin*. Cambridge: Cambridge University Press.

Macfarlane, A. 1978. *The Origins of English Individualism*. Oxford: Basil Blackwell.

McIntosh, M. 1979. The welfare state and the needs of the dependent family. In *Fit Work for Women*, ed. S. Burman, London: Croom Helm.

Medick, H. 1976. The proto-industrial family economy. *Social History* 1(3), 291–315.

Netting, R., Wilk, R. and Arnould, E. (eds) 1984. *Households. Comparative and Historical Studies of the Domestic Group*. Berkeley: University of California Press.

Parsons, T. and Bales, R. 1955. *Family. Socialization and Interaction Process*. Glencoe, Ill.: Free Press.

Shorter, E. 1975. *The Making of the Modern Family*. New York: Basic Books.

Smith, J., Wallerstein, I. and Evers, H.D. (eds) 1984. *Households and the World Economy*. California: Sage Publications.

Smith, R. (ed.) 1984. *Land, Kinship and Life Cycle*. Cambridge: Cambridge University Press.

Tawney, R.H. 1912. *The Agrarian Problem of the Sixteenth Century*. London: Longmans.

Tilly, L. and Scott, J. 1978. *Women, Work and Family*. New York: Holt, Rinehart & Winston.

Wirth, L. 1938. Urbanism as a way of life. *American Journal of Sociology* 44, July, 1–24.

Wolf, A. 1984. Family life and the life cycle in rural China. In Netting, Wilk and Arnold (1984).

Yanagisako, S. 1979. Family and household: the analysis of domestic groups. *Annual Review of Anthropology* 8, 161–205.

extended reproduction. *See* SIMPLE AND EXTENDED REPRODUCTION.

extensive and intensive rent. The distinction between extensive and intensive rent appears clearly in the history of economic thought with Ricardo, even though a number of economists discussed these concepts previously on various occasions (e.g. Anderson, 1777). After Ricardo, until the end of the century, every economist understood the concept of rent to mean the possibility of obtaining an income from the ownership of scarce natural resources, such as land and mines. But that notion of rent changed progressively and substantially after the so-called 'marginalist revolution'. It may be, therefore, useful to examine the transformation of the notion of rent from classical to marginalist economics.

To understand the concept of rent in classical political economy proper, it is essential to relate it to the notion of surplus, that is the quantity of commodities which at the end of the production cycle (usually the year) are left for consumption, net investment (or waste) after the means of production are replaced so that the new production cycle can begin again on at least the same scale. Quesnay was the first to show clearly that rent is a component of surplus. In the *Tableau économique* (1758) rent appears as a share of the agricultural product – agriculture is the only productive sector – which is paid every year by farmers to landlords. In the same way, Adam Smith spoke of rent as a revenue belonging to landlords as a surplus. As soon as land becomes private property [says Smith] the landlord demands a share of almost all the produce which the labourer can either raise, or collect from it. His rent makes the first deduction from the produce of the labour which is employed upon land (Smith, 1776, p. 58). But neither Quesnay, nor Smith gave a satisfactory explanation of the causes affecting the level of the rates of rent. For Quesnay, landlords had a feudal right to ask farmers for a rent. And Smith considered rent as a 'monopoly price'.

Only with Ricardo, who published in 1815, at the same time as Malthus's and West's works on the same subject, *An Essay on the Influence of a Low Price of Corn on the Profits of Stock*, does the classical theory of rent take on a clear and precise shape. In a certain sense, the *Principles*, published two years later, can be considered, as far as rent is concerned, only as an application of the labour theory of value of the specific case already worked on in the *Essay*. 'Rent', says Ricardo, 'is that portion of the produce of the earth, which is paid to the landlord for the use of the original and indestructible powers of the soil' (Ricardo, 1817, p. 67). The landlord can raise a rent only when land becomes scarce, that is when the demand for agricultural produce cannot be satisfied without putting into cultivation lands of inferior quality.

> When in the progress of society [says Ricardo] land of the second degree of fertility is taken into cultivation, rent immediately commences on that of the first quality, and the amount of that rent will depend on the difference in the quality of these two portions of land (Ricardo, 1817, p. 70).

This is extensive rent.

But Ricardo recognized a second kind of rent. Before inferior land is cultivated, 'capital can be employed more productively on those lands which are already in cultivation'. This is intensive rent, 'for rent is always the difference between the produce obtained by the employment of two equal quantities of capital and labour' (Ricardo 1817, p. 71).

This viewpoint changed considerably in the marginalist theory of value and distribution. The distinction between extensive and intensive rent was maintained, but progressively the notion of rent as a surplus disappeared and a new explanation of the law of decreasing returns was introduced into economic theory. In Jevons, for instance, rent is again a surplus, even if its explanation is founded on the new mathematical marginalist techniques. 'The accepted theory of rent', says Jevons, referring to Ricardo's and J.S. Mill's doctrine, 'needs little or no alteration to adapt it to expression in mathematical symbols'. And supposing that a worker is employed on a given area of land, rent will be 'the excess of produce which can be exacted from him ... if he be not himself the owner of the land' (Jevons, 1871, pp. 220, 223). A few years later, Walras protested over such a treatment of rent. Lesson 39 of his *Eléments d'économie politique pure* (1874) is devoted to an 'Exposition and refutation of the English theory of rent'. His starting point is Ricardo's and Mill's distinction between extensive and intensive rent on which he is in agreement, but after a mathematical restatement of these theories, he asks himself 'why this school does not try to formulate a unified general theory to determine the prices of all productive services in the same way'. Walras' aim is to include rent theory in a system of general economic equilibrium in which all prices of commodities and services are interdependent.

> All that remains of Ricardo's theory after a rigorous analysis is that rent is not a component part, but a result, of the price of products. But the same thing can be said of wages and interest. Hence, rent, wages, interest, the prices of products, and the coefficients of production are all unknowns within the same problem; they must always be determined together and not independently of one another (Walras, 1874, p. 416, 418).

Marshall considerably extended the concept of rent. 'The rent of land', affirms Marshall, 'is no unique fact, but simply the chief species of a large genus of economic phenomena' (Marshall, 1890, p. 523). All income can include an element of rent. 'There is an element of true rent in the composite product that is commonly called wages, an element of true earnings in what is commonly called rent and so on' (Marshall, 1890, p. 350). If the supply of a certain factor is scarce, and cannot increase in a certain period of time, then it is possible to gain an income which may be properly called rent. We can imagine cases of 'pure rent' and cases of 'quasi-rent'. An example of quasi-rent is that of incomes gained on old investments of capital. There is no sharp line of division between pure rent and quasi-rent. Commodities which are in short supply in the short run can be produced in a greater quantity in the long run, so that any possibility of obtaining rent disappears.

Moreover, Marshall softened the distinction between extensive and intensive rent, which he calls differential and scarcity rent. 'In a sense', says Marshall, 'all rents are scarcity rents, and all rents are differential rents'. We can say that differential rent arises because the land of a single quality comes to be, at a certain point in time, in short supply.

> In this connection [concludes Marshall], it may be noted that the opinion that the existence of inferior land, or other

agents of production, tends to raise the rents of the better agents is not merely untrue. It is the reverse of the truth. For, if the bad land were to be flooded and rendered incapable of producing anything at all, the cultivation of other land would need to be more intensive; and therefore the price of the product would be higher, and rents generally would be higher, than if that land had been a poor contributor to the total stock of produce (Marshall, 1890, pp. 351–2).

If Marshall was the economist who gave the greatest contribution to the extension of the concept of rent, Wicksteed was the one who gave it a precise and probably definitive arrangement inside the marginalist theory of distribution. According the Wicksteed the marginalist, or as he calls it, the differential theory of distribution, when fully grasped 'must destroy the very conception of separate laws of distribution such as the law of rent, the law of interest, or the law of wages' (Wicksteed, 1914, p. 789). The possibility of coordinating the distribution shares, in order that their sum amounts to the total net product, rests on the fact that the differential service to production of every factor is always the same, even if the way may differ from factor to factor. For example, for land the quality which is relevant is extension, for labour skill and dexterity, etc. 'The law of distribution', affirms Wicksteed, 'is one, and is governed not by the differences of nature in the factors, but by the identity of their differential effect' (Wicksteed, 1914, p. 789). In other words, what is important for the marginalist theory is not the heterogeneity of factors, but the differential effect of a different quantity of a factor of the same quality.

The consequences of this observation are quite radical.

> Ricardo's celebrated law of rent really asserts nothing except that the superior article fetches the superior price, in proportion to its superiority; and it is obvious that all 'superiorities' in land, whether arising from 'inalienable' properties or from expenditure of capital, tell in exactly the same way upon the rent (Wicksteed, 1914, p. 790).

When we consider the usual diagram in which different qualities of land are represented on the abscissas and the different fertilities obtained form a given 'dose' of labour and capital on the ordinate we must be aware of the fact that that curve is not a *functional curve*. We simply arranged the different kinds of land in a descending order according to their fertility. On the contrary, if we increase the quantity of a certain factor in relation to a fixed quantity of another factor we may construct a curve which shows a *functional* relation between the 'doses' of the factor and the marginal product, which has a behaviour depending on the quantity of the variable factor employed. In the first case considered, there is no law of rent at all 'but the tacit assumption that the differential theory of distribution is true of every factor of production except land, and that rent is what is left after everything that is not rent is taken away'. Only with a functional curve do we have a true theory of rent and distribution, but in such a case 'we must understand that when the differential distribution is affected there is no surplus or residuum at all' (Wicksteed, 1914, pp. 791–2). All the product is therefore distributed to the production factors according to their marginal product.

To conclude our short description of the place of rent in the marginalist theory of distribution, it is clear that with Wicksteed rent is no longer a share of the annual surplus, but an income, the nature of which is perfectly symmetrical with

that of capital and labour: they are all paid according to their marginal product. But to reach that result it was essential to reduce the classical law of decreasing returns to the 'law of variable proportions'. As Wicksteed states firmly, only the expansion or contraction of a homogeneous factor in relation to a given quantity of another one gives rise to a marginal product.

The marginalist theory of rent and, more generally, of distribution were widely accepted until recent times. Only with the publication in 1960 of *Production of Commodities by Means of Commodities* by Piero Sraffa were increasing doubts raised by economists on the reasonableness of its assumptions. Sraffa resumed the classical point of view of the problem of value and distribution, placing the notion of surplus at the centre of his inquiry. Inside that theoretical framework it is again quite natural to consider rent as a surplus, that is, a share of the net income distributed to landlords (or other owners of scarce natural resources). Indeed, Sraffa states that 'it is hardly necessary to dwell on the doctrine that 'taxes on rent fall wholly on landlords and thus cannot affect the prices of commodities or the rate of profits' (Sraffa, 1960 p. 74), which was precisely the point of view defended by Ricardo.

Sraffa draws a clear-cut distinction between extensive and intensive rent. If we consider a system of production equations, we can take into consideration two sectors: industry and agriculture. The equations of the industrial sector enable us to determine the rate of profits, given the wage rate, and all industrial prices. In the agricultural sector only corn is produced (which is not utilized as a means of production in the industrial sector). Let us suppose now that *n* different qualities of land are disposable. If the quantity of corn required can be raised on the more fertile land, land will be redundant and there will be no rent. The price of corn is determined by its production equation, since the costs of its means of production (reckoned on the basis of the industrial prices) are known and the rate of profits and wages are equally known. Only when the need arises to put less fertile lands into cultivation will a rent be possible for the owners of the more fertile farms. But the marginal land will pay no rent.

Intensive rent is possible when land of a single quality is in short supply. In such a case, two different processes or methods of cultivation can be used side by side determining a uniform rent per acre. The existence of two methods adopted simultaneously on the same land should be considered as the result of a process of intensive diminishing returns.

> The existence side by side, of two methods can be regarded as a phase in the course of a progressive increase of production on the land. The increase takes place through the gradual extension of the method that produces more corn at a higher unit cost, at the expense of the method that produces less (Sraffa, 1960, p. 76).

It is worthwhile here noticing that Sraffa's analysis is based on the assumption that the quantities of commodities which should be produced and brought to the market are given. Any inquiry regarding the factors affecting these magnitudes should be undertaken at a further stage: here relative prices, the levels of personal income, consumption habits, and so on should be considered as independent variables affecting the quantities of commodities required. But if it is agreed, as a first step, to consider the quantities to be produced as given, some interesting results will emerge:

(a) In the case of extensive rent, the extension of production from a more fertile to a less fertile land, owing to the need to produce more agricultural products, will cause a lowering of the general rate of profits (if corn is a means of production), an increase in the price of the agricultural commodity in relation the industrial ones and a consequent rise in rents on the more fertile lands. In a similar way, in the case of intensive rent, an increase in the quantity produced of a certain agricultural good will cause a change in the methods of production – the old pair of techniques will give place to a new more efficient one – with a consequent increase in the agricultural price and rent, and the reduction of the general rate of profits. Both results (Montani, 1972) are perfectly in agreement with the Ricardian doctrine of rent.

(b) The order of fertility of the various kinds of land is not given, once and for all, by nature. The more fertile lands (i.e. lands which are put into cultivation first because of the rate of profits they give to the agricultural entrepreneur) do not coincide with the lands paying higher rents. Ricardo's opinion is not correct on this point:

> When land of the third quality is taken into cultivation, rent immediately commences on the second. At the same time, the rent of the first quality will rise, *for that must always be above the rent of the second,* by the difference between the produce which they yield with a given quantity of capital and labour (Ricardo, p. 70; our italics).

Generally speaking, if lands 1, 2 and 3 are already cultivated and rent on land 1 is higher than rent on land 2, it may happen that when land 4 is put into cultivation the rate of rent on land 2 may become greater than the rate of rent on land 1. The reversal of the order of rents is possible both for the rate of rent per unit of land and for rate of rent per unit of product (Montani, 1972).

(c) The scarcity of land does not depend only on the quantity of the agricultural product to be produced, but on the distribution of income between profits and wages too. Given certain methods of production, some natural resources, as land, mineral deposits, etc, become 'scarce' or 'redundant' according to the quantity produced of a given commodity and to the relative level of the rate of profits in relation to wages. This may happen both for extensive and intensive rent. Therefore, since no change in the proportions of the 'production factors' occurs when the natural resource becomes scarce or redundant owing to the change in distribution, it is obvious that the meaning of 'decreasing returns' inside the classical theory of rent (and distribution) is different from that connected with the 'law of variable proportions' (Montani, 1975).

The vicissitudes of the theory of rent in the history of economic thought are, therefore, strictly connected to the meaning of the law of decreasing returns; this can be appreciated from what has been said above about how the content of this law changed considerably during the transition from the classical to the marginalist paradigm. Marshall was well aware of the diversity of the two points of view, and stated quite clearly that

> the diminishing return which arises from an ill-proportioned application of the various agents of production into a particular task has little in common with that broad tendency to the pressure of a crowded and growing population on the means of subsistence. The great classical Law of Diminishing Return has its chief application, not to any one particular crop, but to all the chief food crops (Marshall, p. 338).

This classical meaning of the law was progressively forgotten by the economists owing to the over-narrow view imposed by the marginalist theory of distribution.

GUIDO MONTANI

See also ABSOLUTE RENT; LAND RENT; RENT; RICARDO, DAVID.

BIBLIOGRAPHY

Anderson, J. 1777. *An Inquiry into the Nature of the Corn-Laws; with a View to the New Corn-Bill proposed for Scotland.* Edinburgh.

Jevons, W.S. 1871. *The Theory of Political Economy.* Harmondsworth: Penguin Books, 1970.

Marshall, A. 1890. *Principles of Economics.* 8th edn. London: Macmillan, 1972.

Montani, G. 1972. La teoria ricardiana della rendita. *L'Industria* 3/4, 221–43.

Montani, G. 1975. Scarce natural resources and income distribution. *Metroeconomica* 27, 68–101.

Ricardo, D. 1817. *On the Principles of Political Economy and Taxation.* In *The Works and Correspondence of David Ricardo,* Vol. I, ed. P. Sraffa, Cambridge: Cambridge University Press, 1960. Reprinted, 1966.

Smith, A. 1776. *An Inquiry into the Nature and Causes of the Wealth of Nations.* London: Everyman's Library, 1964.

Sraffa, P. 1960. *Production of Commodities by Means of Commodities.* Cambridge: Cambridge University Press.

Walras, L. 1874. *Eléments d'économie politique pure.* Trans. as *Elements of Pure Economics,* ed. W. Jaffé, 1954. Reprinted, Fairfield: A.M. Kelley, 1977.

Wicksteed P. H. 1914. The scope and method of political economy in the light of the 'marginal' theory of value and distribution. *Economic Journal* 24, March. Reprinted in the *The Common Sense of Political Economy and Selected Papers and Reviews on Economic Theory,* Vol. II, ed. L. Robbins, 1933; reprinted, New York: A.M. Kelley, 1967.

extensive form games. The most general model used to describe conflict situations is the extensive form model, which specifies in detail the dynamic evolution of each situation and thus provides an exact description of 'who knows what when' and 'what is the consequence of which'. The model should contain all relevant aspects of the situation; in particular, any possibility of (pre)commitment should be explicitly included. This implies that the game should be analysed by solution concepts from noncooperative game theory, that is, refinements of Nash equilibria. The term extensive form game was coined in von Neumann and Morgenstern (1944) in which a set theoretic approach was used. We will describe the graph theoretical representation proposed in Kuhn (1953) that has become the standard model. For convenience, attention will be restricted to finite games.

The basic element in the Kuhn representation of an n-person extensive form game is a rooted tree, that is, a directed acyclic graph with a distinguished vertex. The game starts at the root of the tree. The tree's terminal nodes correspond to the endpoints of the game and associated with each of these is an n-vector of real numbers specifying the payoff to each player (in von Neumann–Morgenstern utilities) that results from that play. The nonterminal nodes represent the decision points in the game. Each such point is labelled with an index i ($i \in \{0, 1, \ldots, n\}$) indicating which player has to move at that point. Player O is the chance player who performs the moves of nature. A maximal set of decision points that a player cannot distinguish between is called an information set. A choice at an information set associates a unique successor to every decision point in this set, hence, a choice consists of a set of edges, exactly one edge emanating from each point in the set.

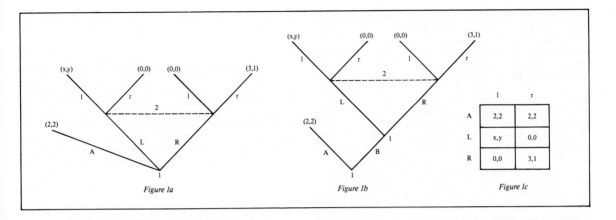

Figure 1a *Figure 1b* *Figure 1c*

Information sets of the chance player are singletons and the probability of each choice of chance is specified. Formally then, an extensive form game is a sixtuple $\Gamma = (K, P, U, C, p, h)$ which respectively specify the underlying tree, the player labelling, the information sets, the choices, the probabilities of chance choices and the payoffs.

As an example, consider the 2-person game of Figure 1a. First player 1 has to move. If he chooses A, the game terminates with both players receiving 2. If he chooses L or R, player 2 has to move and, when he is called to move, this player does not know whether L or R has been chosen. Hence, the 2 decision points of player 2 constitute an information set and this is indicated by a dashed line connecting the points. If the choices L and ℓ are taken, then player 1 receives x, while player 2 gets y. The payoff vectors at the other endpoints are listed similarly, that is, with player 1's payoff first. The game of Figure 1b differs from that in Figure 1a only in the fact that now player 1 has to choose between L and R only after he has decided not to choose A. In this case, the game admits a proper subgame starting at the second decision point of player 1. This subgame can also be interpreted as the players making their choices simultaneously.

A strategy is a complete specification of how a player intends to play in every contingency that might arise. It can be planned in advance and can be given to an agent (or a computing machine) who can then play on behalf of the player. A pure strategy specifies a single choice at each information set, a behaviour strategy prescribes local randomization among choices at information sets and a mixed strategy requires a player to randomize several pure strategies at the beginning of the game. The normal form of an extensive game is a table listing all pure strategy combinations and the payoff vectors resulting from them. Figure 1c displays the normal form of Figure 1a, and, up to inessential details, this also represents the game of Figure 1b. The normal form suppresses the dynamic structure of the extensive game and condenses all decision-making into one stage. This normalization offers a major conceptual simplification, at the expense of computational complexity: the set of strategies may be so large that normalization is not practical. Below we return to the question of whether essential information is destroyed when a game is normalized.

A game is said to be of perfect recall if each player always remembers what he has previously known or done, that is, if information is increasing over time. A game may fail to have perfect recall when a player is a team such as in bridge and in this case behaviour strategies may be inferior to mixed strategies since the latter allow for complete correlation between different agents of the team. However, by modelling different agents as different players with the same payoff function one can restore perfect recall, hence, in the literature attention is usually restricted to this class of games. In Kuhn (1953) and Aumann (1964) is has been shown that, if there is perfect recall, the restriction to behaviour strategies is justified.

A game is said to be of perfect information if all information sets are singletons, that is, if there are no simultaneous moves and if each player always is perfectly informed about anything that happened in the past. In this case, there is no need to randomize and the game can be solved by working backwards from the end (as already observed in Zermelo, 1913). For generic games, this procedure yields a unique solution which is also the solution obtained by iterative elimination of dominated strategies in the normal form. The assumption of the model that there are no external commitment possibilities implies that only this dynamic programming solution is viable; however, this generally is not the unique Nash equilibrium. In the game of Figure 2, the roll-back procedure yields (R, r), but a second equilibrium is (L, ℓ). The latter is a Nash equilibrium since player 2 does not have to execute the threat when it is believed. However, the threat is not credible: player 2 has to move only when 1 has chosen R and facing the *fait accompli* that R has been chosen, player 2 is better off choosing r. Note that it is essential that 2 cannot commit himself: If he could we would have a different game of which the outcome could perfectly well be $(2, 2)$.

A major part of noncooperative game theory is concerned with how to extend the backwards induction principle to games with imperfect information, that is how to exclude intuitively

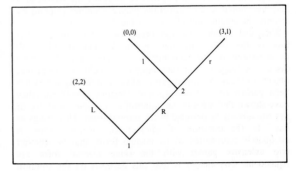

Figure 2

257

unreasonable Nash equilibria in general. This research originates with Selten (1965) in which the concept of subgame perfect equilibria was introduced, that is, of strategies that constitute an equilibrium in every subgame. If $y < 0$, then the unique equilibrium of the subgame in Figure 1b is (R, r) and, consequently, (BR, r) is the unique subgame perfect equilibrium in that case. If $x < 2$, however, then (AL, ℓ) is an equilibrium that is not subgame perfect. The game of Figure 1a does not admit any proper subgames; hence, any equilibrium is subgame perfect, in particular (A, ℓ) is subgame perfect if $x > 2$. This shows that the set of subgame perfect equilibria depends on the details of the tree and that the criterion of subgame perfection does not eliminate all intuitively unreasonable equilibria.

To remedy the latter drawback, the concept of (trembling hand) perfectness was introduced in Selten (1975). The idea behind this concept is that with a small probability players make mistakes, so that each choice is taken with an infinitesimal probability and, hence, each information set can be reached. If $y \leqslant 0$, then the unique perfect equilibrium outcome in Figures 1a and 1b is $(1, 3)$: player 2 is forced to choose r since L and R occur with positive probability.

The perfectness concept is closely related to the sequential equilibrium concept proposed in Kreps and Wilson (1982). The latter is based on the idea of 'Bayesian' players who construct subjective beliefs about where they are in the tree when an information set is reached unexpectedly and who maximize expected payoffs associated with such beliefs. The requirements that beliefs be shared by players and that they be consistent with the strategies being played (Bayesian updating) imply that the difference from perfection is only marginal. In Figure 1a, only (R, r) is sequential when $y < 0$. When $y = 0$, then (A, ℓ) is sequential, but not perfect: choosing ℓ is justified if one assigns probability 1 to the mistake L, but according to perfectness R also occurs with a positive probability.

Unfortunately, the great freedom that one has in constructing beliefs implies that many intuitively unreasonable equilibria are sequential. In Figure 1a, if $y > 0$, then player 2 can justify playing ℓ by assigning probability 1 to the 'mistake' L, hence, (A, ℓ) is a sequential equilibrium if $x \leqslant 2$. However, if $x < 0$, then L is dominated by both A and R and thinking that 1 has chosen L is certainly nonsensical. (Note that, if $x < 0$, then (AL, ℓ) is not a sequential equilibrium of the game of Figure 1b, hence, the set of sequential (perfect) equilibrium outcomes depends on the details of the tree.) By assuming that a player will make more costly mistakes with a much smaller probability than less costly ones (as in Myerson's (1978) concept of proper equilibria) one can eliminate the equilibrium (A, ℓ) when $x \leqslant 0$ (since then L is dominated by R), but this does not work if $x > 0$. Still, the equilibrium (A, ℓ) is nonsensical if $x < 2$: If player 2 is reached, he should conclude that player 1 has passed off a payoff of 2 and, hence, that he aims for the payoff of 3 and has chosen R. Consequently, player 2 should respond by r: only the equilibrium (R, r) is viable.

What distinguishes the equilibrium (R,r) in Figure 1 is that this is the only one that is stable against all small perturbations of the equilibrium strategies, and the above discussion suggests that such equilibria might be the proper objects to study. An investigation of these stable equilibria has been performed in Kohlberg and Mertens (1984) and they have shown that whether an equilibrium outcome is stable or not can already be detected in the normal form. This brings us back to the question of whether an extensive form is adequately represented by its normal form, that is, whether two extensive games with the same normal form are equivalent. One answer is that this depends on the solution concept employed: it is affirmative for Nash equilibria, for proper equilibria (van Damme, 1983, 1984) and for stable equilibria, i.e. for the strongest and the weakest concepts, but it is negative for the intermediate concepts of (subgame) perfect and sequential equilibria. A more satisfactory answer is provided by a theorem of Thompson (1952) (see Kohlberg and Mertens, 1984) that completely characterizes the class of transformations that can be applied to an extensive form game without changing its (reduced) normal form: The normal form is an adequate representation if and only if these transformations are inessential. Nevertheless, the normal form should be used with care, especially in games with incomplete information (cf. Harsanyi, 1967–8; Aumann and Maschler, 1972), or when communication is possible (cf. Myerson, 1986).

ERIC VAN DAMME

See also COOPERATIVE GAMES; GAMES WITH INCOMPLETE INFORMATION; GAME THEORY; NASH EQUILIBRIUM; NON-COOPERATIVE GAMES.

BIBLIOGRAPHY

Aumann, R.J. 1964. Mixed and behavior strategies in infinite extensive games. In *Advances in Game Theory*, ed. M. Dresher, L.S. Shapley and A.W. Tucker, Princeton: Princeton University Press.

Aumann, R.J. and Maschler, M. 1972. Some thoughts on the minimax principle. *Management Science* 18(5), January, 54–63.

Harsanyi, J.C. 1967–8. Games with incomplete information played by 'Bayesian' players. *Management Science* 14; Pt I, (3), November 1967, 159–82; Pt II, (5), January 1968, 320–34; Pt III, (7), March 1968, 486–502.

Kohlberg, E. and Mertens, J-F. 1984. On the strategic stability of equilibria. Mimeo, Harvard Graduate School of Business Administration. Reprinted in *Econometrica* 53, 1985, 1375–85.

Kreps, D.M. and Wilson, R. 1982. Sequential equilibria. *Econometrica* 50(4), July, 863–94.

Kuhn, H.W. 1953. Extensive games and the problem of information. *Annals of Mathematics Studies* 28, 193–216.

Myerson, R.B. 1978. Refinements of the Nash equilibrium concept. *International Journal of Game Theory* 7(2), 73–80.

Myerson, R.B. 1986. Multistage games with communication. *Econometrica* 54(2), March, 323–58.

Selten, R. 1965. Spieltheoretische Behandlung eines Oligopolmodells mit Nachfragetragheit. *Zeitschrift für die gesamte Staatswissenschaft* 121, 301–24; 667–89.

Selten, R. 1975. Reexamination of the perfectness concept for equilibrium points in extensive games. *International Journal of Game Theory* 4(1), 25–55.

Thompson, F.B. 1952. Equivalence of games in extensive form. RAND Publication RM-769.

Van Damme, E.E.C. 1983. *Refinements of the Nash equilibrium concept*. Lecture Notes in Economics and Mathematical Systems, No. 219, Berlin: Springer-Verlag.

Van Damme, E.E.C. 1984. A relation between perfect equilibria in extensive form games and proper equilibria in normal form games. *International Journal of Game Theory* 13(1), 1–13.

Von Neumann, J. and Morgenstern, O. 1944. *Theory of Games and Economic Behavior*. Princeton: Princeton University Press.

Zermelo, E. 1913. Über eine Anwendung der Mengenlehre auf die Theorie des Schachspiels. *Proceedings, Fifth International Congress of Mathematicians* 2, 501–4.

external debt. The term 'external debt' refers to financial obligations incurred by individuals or, more commonly, institutions resident in one country vis-à-vis those resident in another. In other words, the obligations cross the borders of sovereign states. Usually different nationalities or citizenships are involved, as well as different residencies, but this is not strictly necessary. For instance, a US corporation in the

United States may borrow from a US bank branch on London, which in turn funds the loan by taking deposits from American residents of the United States. The involvement of a financial intermediary in London means that this chain of transactions among American nationals is governed in part by English law and regulations rather than, or as well as, US law (and of course the laws may conflict). In fact, legal or regulatory differences are likely to be the reason why such all-American transactions move 'off-shore' in the first place. For example, reserve or liquidity requirements applying to US banks in London may be less severe than those applying in New York, thus enabling the London branch to offer more attractive terms than its New York counterpart to depositors and borrowers alike.

In the above example US residents have both a debt and a claim vis-à-vis non-residents. So the net external asset position of the United States is the same as it would be if neither of these particular transactions had taken place. The same would apply if the bank in London were British or Japanese rather than American, and if US residents had placed their overseas assets not with this bank but with some other institution in a third country. Nonetheless, with countries as with individuals or firms, it is not only the net (external) financial position which matters. A country (or a firm) may have substantial net (external) assets and still face financial difficulties – a liquidity crisis or a run on the currency – if its liabilities are more liquid than its assets, and external creditors demand repayment; or if domestic residents wish to send more capital abroad and foreign creditors refuse to grant additional loans. Conversely, insofar as a debtor country is readily able to meet the interest due on its foreign debt from export proceeds or other normal receipts, its net debtor position is not a source of embarrassment.

These considerations point the way to theoretical and practical criteria for external borrowing and lending by a country; but a number of other points have first to be clarified. Aside from debts, external liabilities (and assets) may take the form of claims on real property (ordinary shares, land, buildings, etc). Unlike such claims, debts carry legal obligations to pay interest and/or amortization over some stated period; but, unless specifically indexed, these obligations are in nominal money terms, not in real terms. The rate of interest may be either fixed or fluctuating according to some formula, but will normally depend in large measure upon the currency denomination of the loan. The faster the real-terms purchasing power of a currency is expected to depreciate in the future, the higher the interest rate that it will carry. Creditors therefore lose, and debtors gain, to the extent that the currency in question loses real value during the period of the loan at a faster pace than financial markets had anticipated at the time the loan was made; and conversely when it loses real value more slowly. Monetary authorities in countries with longer-term borrowings denominated in their own currency are accordingly exposed to a degree of temptation (moral hazard) to encourage faster depreciation of their currencies in order to reduce the burden of such debts; and conversely for monetary authorities of countries with longer-term external lending in their own currency (though this has not been viewed as a significant risk in practice).

External debt may take the form either of securities that are in principle marketable (bonds or bills) or of non-marketable claims such as bank loans, trade credit between firms or government-to-government loans. Before 1914, international lending was conducted overwhelmingly by way of securities; subsequently other claims have played a larger role, most especially so in the 1970s, which saw a huge wave of lending to less developed countries (notably in Latin America) in the form of medium-term credits, denominated mainly in US dollars, put together by syndicates of major commercial banks.

EXTERNAL BORROWING AND THE MACRO-ECONOMY. Starting with a clean sheet, a country acquires *net* external liabilities (assets) by spending abroad out of current income more (less) than it is currently earning abroad – in other words, by running a deficit (surplus) on the current account of its balance of payments. Once external assets and liabilities exist, the net position is also affected by valuation changes, i.e. appreciation and depreciation, of the various items. Another way of looking at a current payments deficit is to say that it represents external dis-saving by the country in question, and comes about as a result of the combined saving and spending decisions of all the country's residents. Macroeconomic policies exercise a crucial influence, both directly (the government being a major spending unit) and indirectly (through the impact of fiscal, monetary and exchange-rate policies upon private-sector outlays). It is important to distinguish between external saving and total saving. Most of a country's total saving is matched by expenditure on domestic capital formation, i.e. investment in buildings, machinery, etc., whether by the private or public sectors. External dis-saving and hence the current payments deficit equals the *excess* of domestic investment over national saving (and conversely for external saving and a current payments surplus). In symbols $X - M = S - I$, where X is export receipts, including earnings on overseas assets as well as sales of goods and services, M is import payments, S is national saving and I is domestic investment. Clearly the net decline in the external asset/liability position which matches a current account deficit may take many different forms – such as acquisition of domestic factories by foreign firms, sale of overseas assets by domestic residents, drawing down of official foreign-exchange reserves, etc. – and need not involve external borrowing in the strict sense of the term.

CRITERIA. Some external borrowing is undertaken in emergency or disequilibrium situations, as an alternative to depletion of foreign exchange reserves or, beyond that, to rapid elimination of a current account deficit by means of restraint on domestic expenditure which would cause significant disruption or hardship. Thus, a domestic investment boom may be sustained and allowed to run its course rather than be cut short by balance-of-payments pressures. Or the impact of an unwelcome event, such as a sudden and unexpected drop in export receipts in a primary-product-exporting country, may be cushioned. In this instance the argument for emergency borrowing is most evident when the drop in export income is expected for good reason to be short-lived. When it is expected to be of long duration, some external borrowing may still be appropriate, so that the country may adapt to its new circumstances gradually. In sum, borrowing is rational whenever the discounted present value of its cost is judged to be less than that of the expenditures which would be foregone if the borrowing did not take place.

Analogous criteria apply to external borrowing along an equilibrium growth path. Assume for simplicity that there is a single world-wide interest rate at which funds can be both lent and borrowed. A country should undertake all domestic investment projects expected to be profitable in foreign-currency terms at this interest rate, arranging the structure and timing of its investment programme so as to maximize the present-value difference between total return and total cost. At the same time the country's total saving will depend on

259

population size, income level, income distribution and thriftiness or time preference. The country will be a net borrower or net lender abroad in any particular phase of its development, depending on whether domestic investment is running ahead of or behind national saving. More explicit and precise formulations of this principle can be derived in the context of the mathematical theory of optimum growth (see Shell, 1967).

UNCERTAINTY, DEFAULT RISK AND THE BANKING SYSTEM. The principles outlined so far largely abstract from problems of imperfect foresight and information. Any credit market has to cope with such problems, but special issues arise in relation to international lending.

Although a borrower promises to service his debt, circumstances or dishonesty may lead him to default on or repudiate the obligation. In the face of this possibility, various kinds of safeguard are available to actual or potential creditors. First, private creditors should not merely lend to one or a few borrowers, but should spread risks by holding a market portfolio of claims. Secondly, the market itself will generate differential pricing of loans, with debtors judged riskier having to pay higher interest rates and/or being rationed in the amounts that they can borrow at any price. The role of rationing is emphasised by the theory of adverse selection (Stiglitz and Weiss, 1981), which suggests that higher interest rates tend to drive away the prudent and good-quality borrowers, leaving mainly the more doubtful prospects still in the market. Thirdly, borrowers who default on their obligations are liable to forfeit property either automatically or after legal proceedings. Such proceedings may, however, be costly and troublesome to pursue, especially in the international context, where action may be required in more than one country. Fourthly, default or bankruptcy carries a stigma which will prevent or hamper the possessor's access to credit and other markets for a greater or lesser period of time. By the same token, creditors may be able to enforce prudent policies of their choosing upon debtors as a condition of prolonging credit rather than declaring the debtor in default. In practice such power can be exercised only by banks or other financial institutions; and in the case of government ('sovereign') borrowers mostly only by other governments or supranational institutions such as the International Monetary Fund (IMF).

The point about continued access to international markets has to carry a lot of weight in the case of sovereign borrowers, where the mechanism of foreclosure (property forfeiture) is inapplicable. With a commercial (non-sovereign) loan, lenders will wish to ensure that the market value of the assets available as security in the event of default or insolvency is at least equal to the discounted present value of the lost interest and repayments. The nearest comparable condition in the case of a sovereign loan is to ensure that the borrower believes his self-interest to lie in maintaining debt service, because default would bring him more losses than gains. If his losses are defined simply in terms of denial of future loan inflows (a plausible albeit somewhat narrow definition), then the incentive to default arises if and only if the discounted present value of the potential future inflows is less than that of the interest and amortization payments avoided by defaulting. To prevent this condition being met, lenders must put appropriate limits on the growth of their loans to sovereign borrowers, the limits depending at any one time on the existing level of debt, the real rate of interest and the debtor economy's growth rate (see, for instance, Niehans, 1986; Eaton and Gersovitz, 1981).

Suppose, however, that lenders miscalculate and allow external debt to build up to a point where sovereign borrowers are seriously tempted to default. To lure them away from this option, lenders must then continue to re-lend debt service payments as they fall due and, in addition, ensure that the real interest rate being charged is at least no higher, and probably somewhat lower, than the borrower's long-term economic growth rate. Depending on circumstances, lenders may prefer to cut their losses by accepting default; and some may have no effective choice in the matter – for instance, widely scattered small bondholders with no practical means of influencing the scale or terms of new lending.

The foregoing optimal lending strategies to sovereign borrowers may not be achievable by competitive private lenders acting on their own. Financial markets, being characterized by incompleteness of information about the creditworthiness of borrowers and about future economic trends, are prone to be heavily influenced by herd instinct and fashion, even if decisions are rationalized in analytical terms. When circumstances look favourable, it is difficult to prevent excessive growth of lending to favoured sectors or customers, including sovereign borrowers – especially with financial institutions also competing for positions of market leadership and size. Then, when the climate turns sour, private lenders left to themselves may precipitate a financial crash by recalling loans when debtors are in no position to repay. Observing these tendencies at work in 19th-century British credit markets, Walter Bagehot (1873; see also Hirsch, 1977) argued that stabilization of the financial system requires a degree of restraint on competition through some mixture of oligopolistic market structure and central-bank guidance or approval, the latter as a *quid pro quo* for protecting banks against runs by means of last-resort lending facilities.

A similar lesson emerged from global lending developments in the 1970s and 1980s. When less developed countries in Latin America and elsewhere became major international borrowers after 1970, and especially after the quadrupling of oil prices in 1973–4, most of the sovereign lending was handled by the commercial bank network of the major industrial countries, which was far more flexible than official credit channels centred on the IMF in intermediating to new patterns of international capital flows. In the process, however, lending banks, especially in the United States, made themselves vulnerable by lending amounts greatly in excess of their capital and reserves to a handful of sovereign borrowers. From 1979 onwards monetary restraint imposed by the US Federal Reserve System triggered a shift in the world financial climate whose extent and duration could scarcely have been foreseen. In particular, international real interest rates rose from a negative figure to around 7 per cent, and the terms of trade moved heavily against primary producers. After two-to-three years the burden of debt service had become unsustainable in relation to borrowers' export receipts. During 1982–3 some 35 countries were obliged to seek rescheduling of their external debts. The fact that this happened in an orderly manner, and without involving a serious chain of bank failures, was due to the action of the principal OECD-country central banks, led by the Federal Reserve, in conjunction with the IMF and the Bank for International Settlements. They buttressed commercial bank lending with official credits and, more important, exercised moral suasion to induce the banks to renew sovereign loans as they fell due, rather than seek a large-scale withdrawal of funds which would not have been feasible. This cooperative 'crisis management' was the international equivalent of last-resort lending in a domestic banking system.

PETER M. OPPENHEIMER

See also DEPENDENCY; INTERNATIONAL FINANCE; INTERNATIONAL INDEBTEDNESS.

BIBLIOGRAPHY

Bagehot, W. 1873. *Lombard Street*. London: H.S. King & Co.

Eaton, J. and Gersovitz, M. 1981. *Poor Country Borrowing in Financial Markets and the Repudiation Issue*. Princeton Studies in International Finance No. 47, Princeton: Princeton University Press.

Hirsch, F. 1977. The Bagehot problem. *Manchester School of Economics and Political Science* 45(3), September, 241–57.

Niehans, J. 1986. In *Strategic Planning in International Banking*, ed. P. Savona and G. Sutija, London.

Shell, K. (ed.) 1967. *Essays on the Theory of Optimum Economic Growth*. Cambridge, Mass.: MIT Press.

Stiglitz, J. and Weiss, A. 1981. Credit rationing in markets with imperfect information. *American Economic Review* 71(3), June, 393–410.

external economies. 'The concept of external economies is one of the most elusive in economic literature'. This is how Tibor Scitovsky began his article 'Two Concepts of External Economies' (1954). His statement is still true, and it may be added that there are at least two such concepts.

The meaning of external economies and its counterpart, external diseconomies, has changed over time. Nowadays, it is essentially synonymous with externality or external effects in the sphere of production. That is, external economies (diseconomies) or positive (negative) external effects in production are unpaid side-effects of one producer's output or inputs on other producers. (As an illustrative example, we can take the case where a dam constructed by a hydroelectric power plant eliminates flooding of farmers' crop fields (external economy) or reduces the catches of fishermen downstream (external diseconomies); a producer's pollution which increases the costs of, *inter alia*, other producers is perhaps the most important case of externalities.) Sometimes, external economies also refer to unpaid side-effects of or on consumption activities, but this meaning is disregarded here.

External economies in this modern sense imply as a rule that market prices in a competitive market economy will not reflect marginal social costs of production. Hence, a 'market failure' arises, meaning that the market economy cannot attain a state of efficiency on its own. Specifically, in an otherwise 'perfect' market economy, a producer who has external economies (positive external effects) on other producers would not extend his externality-generating activity, say, his output, to the point where marginal cost of production equals marginal social benefits of production, which amounts to the market value of his marginal output *plus* the market value of the side-effect on the output of other producers.

At an earlier stage, external economies in the meaning now given were called *technological external economies*, reflecting the fact that the effects were transmitted outside the market mechanism and altered the technological relationship between the recipient firm's output and the inputs under its control. Formally, we have that the output q_i of the ith producer is affected not only by changes in his control variables, a vector x_i, but also by e_j, a variable controlled by some other producer j. This gives us the following production function:

$$q_i = f_i(x_i; e_j).$$

The reason for specifying such effects as technological is that the concept of 'external economies' has been given a broad meaning ever since its introduction. During the early part of the 20th century, external economies (diseconomies) were defined so as to include beneficial (detrimental) *price effects* of producer activities. Thus, in principle, the concept included cases where increases in factor inputs by one firm lowered or raised input prices for other firms. However, much of the discussion centred on the case where increases in *industry* output lowered or raised input prices for the individual member firm. The case of reduced input prices presupposes that the supply side of the market for inputs is characterized either by imperfect competition (say, a profit-maximizing monopoly producing at decreasing marginal costs, decreasing at a rate sufficient for an increase in demand to lower price) or by a competitive industry having a downward-sloping 'supply' curve, which in turn reflects external economies in this industry. Supply conditions in the original industry, as well as in the industry producing inputs for the first industry, are shown in Figure 1. Here, $\Sigma s_i(Q_0)$ is the aggregate supply of the firms in the industry when actual industry output is Q_0. When industry output increases to Q_1, input prices drop (or technological external economies arise), causing a downward shift in individual cost and supply curves and hence in the aggregate supply ($\Sigma s_i(Q_1)$). The curve M, the downward-sloping 'supply' curve, is actually a market equilibrium curve showing the equilibrium price/output combinations at different levels of demand (see Bohm, 1967).

The price effects between firms or between industry and its firms were termed *pecuniary external economies (diseconomies)* by Viner (1931). Before him, A.C. Pigou (1920) had argued (although he phrased it differently) that external economies and diseconomies, both technological and pecuniary, would call for government intervention in order for the industry to attain a socially efficient level of output. Specifically, Pigou argued that if expansion of a competitive industry would increase prices of inputs sold to the industry, thus creating pecuniary external diseconomies for the individual firms in the industry, the aggregate 'supply' (or market equilibrium) curve would not reflect social marginal costs.

Pigou's argument can be illustrated in Figure 2, where M is the upward-sloping long-run 'supply' curve for the industry due to rising input prices as a consequence of increasing industry output. SMC is the curve showing the total marginal outlay on inputs, where the difference to M is the increased outlay for *intra*-marginal inputs. Pigou contended (but later rescinded this position) that the SMC curve indicated the true social marginal costs and hence, that the price/output combination attained by the market, as shown by the intersection of market 'supply' M and market demand D, was suboptimal. He argued that the optimal level could not be attained unless a tax were levied on this industry so that, in equilibrium, price and output would be those shown by the intersection of SMC and D. (Similarly, a bounty would be

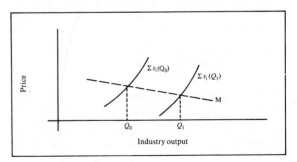

Figure 1

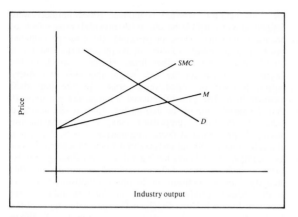

Figure 2

required in the case of pecuniary external economies (see Figure 1), where SMC would be downward-sloping and steeper than the M curve.)

It was demonstrated by F.H. Knight (1924) and D.H. Robertson (1924) and further elaborated by Ellis and Fellner (1943) that the total marginal outlay is irrelevant as an indicator of marginal social costs. The effect on outlay for intramarginal units of inputs represents an increment to rent on these units and hence, it is not part of the real costs of increased output; the real marginal costs are only those which have to be paid for the required marginal inputs. In other words, the rising costs for marginal inputs are those shown by the industry 'supply' curve M. Hence, pecuniary external economies or diseconomies would not call for government intervention. The case is different, however, for technological economies and diseconomies. Assume, for example, that output of a geographically concentrated industry pollutes the area where it is located and hence reduces the productivity of labour inputs of the individual firms in the industry. Here, all effects on input productivity, for marginal as well as intramarginal units, constitute real costs. So if the rising M and SMC reflect such effects only, Pigou's argument holds.

When the debate had arrived at this point, pecuniary external economies could be dropped as a cause of market failure and hence, the concept lost its specific economic interest. But, now, what do the *technological* external economies and diseconomies of industry output on the individual firms in the industry actually represent? Alfred Marshall (1890) introduced the term external economies when analysing industry production costs as a function of output:

> We may divide the economies arising from an increase in the scale of production of any kind of goods, into two classes – firstly, those dependent on the general development of the industry; and secondly, those dependent on the resources of individual houses of business engaged in it, on their organization and the efficiency of their management. We may call the former *external economies*, and the latter *internal economies* (Marshall, 8th edn, 1920, p. 266).

The latter concept is now recognized as economies of scale in the individual firm. Marshall elaborated the meaning of the former concept using scattered examples, the most explicit of which is perhaps the increased knowledge accompanying the expansion of industry output materialized in the publication of trade journals and other forms of improved information about markets and technology in the industry. But, in addition, he argued that industry growth, especially when concentrated to a particular region, might create a market for skilled labour, advance subsidiary industries or give rise to specialized service industries as well as improve railway communication and other infrastructure.

Thus, external economies emerged here essentially as cost reductions for individual firms as a consequence of industry growth, that is, as economies external to the firm but internal to the industry. To remain firmly within the framework of static analysis, these economies should be thought of as being reversible. But some of Marshall's examples alluded to irreversible phenomena and to dynamic effects of industry growth. This was particularly obvious when he at times referred to external economies as being dependent on the 'general progress of industrial environment'.

Marshall's claim that external economies were important in the long run – in fact, more important than internal economies – seemed to have little immediate impact on the thinking of his fellow economists. For example, the discussion of 'empty economic boxes' in the early 1920s (Clapham, 1922; Pigou, 1922; Robertson, 1924) centred on questioning the relevance of external economies and diseconomies, the latter concept deriving from issues raised by Pigou (1920). Actually, the significance of technological external economies and diseconomies in the *static* analysis of an industry or among producers in general escaped most economists all the way up to the postwar years when traffic congestion (although already mentioned by Pigou) and the common pool problem of interdependent producers in an oil field or a fishing area became the leading examples of the industry case and environmental pollution the leading example of the general case. Eventually, diseconomies emerged as the important case and economies as the exceptional case, whereas earlier hardly any importance was attached to technological *dis*economies (see e.g. Robertson, 1924).

If static external effects turned out to be the most important legacy of Marshall's original contribution, external economies as a *dynamic* concept, which seems closer to Marshall's own ideas, also came to play a role in economic analysis. Dynamic external economies refer to increased division of labour resulting from industry growth, the emergence of firms specializing in new activities, some of which aim at developing capital equipment for, or servicing, other firms. An early elaboration of these ideas was made by Young (1928). Later, the concept of external economies came to play a prominent role in development planning, primarily for underdeveloped countries or regions. The general idea here was hardly at all related to the non-market interdependence of technological external economies, but rather to the market interdependence of which pecuniary external economies were part, now on an economy-wide basis. It was argued, in particular by Rosenstein-Rodan (1943) and – with a somewhat different focus – by Scitovsky (1954), that development must be planned so that supply and demand relationships among different sectors of the economy are taken into account. Specifically, it was pointed out that major investments in industrial capacity in a poor economy risk being a failure unless a required increase in the supply of inputs to meet the need of the expanding industry, as well as a sufficient stimulus of demand for the output of the expanding industry, occur at the same time. This doctrine of 'balanced growth' called for a simultaneous expansion of investment in several sectors of the economy, a 'Big Push' (Rosenstein-Rodan, 1943) determined by input–output relations or general market interdependence. Scitovsky argued specifically that indivisibilities in capital formation may call for investment criteria, which – in contrast to those of individual firms – take into account the indirect supply and demand effects on profitability elsewhere in the economy.

Here, external economies came to be synonymous with what was later called linkage effects (Hirschman, 1958), where backward linkages refer to the supply to the investing sector and forward linkages to the demand for the output of the investing sector. Hirschman's own analysis led him to recommend a strategy for economic development in poor countries that was opposite to that of a 'balanced growth'. Focusing on the shortage of entrepreneurial capacity and the small impact of subtle market signals in backward areas, he advocated a strategy of 'unbalanced growth', according to which investments should be undertaken so as to reinforce market signals and create pressure for investments in sectors related to the original investments via strong linkage effects.

That linkage effects offered a more precise terminology than dynamic external economies is one reason why the latter concept now has lost ground. Another is, of course, that this meaning of external economies referred only or primarily to market interdependence, which is a general economic phenomenon. Thus, with technological external economies and diseconomies now most often replaced by the well-defined concept of external effects, and with pecuniary external economies and diseconomies being synonymous with general market interdependence, external economies no longer have much of a role to play in economic analysis. Aside from occasional use as a synonym for external effects, the concept now stands for interdependence that does not clearly fall into any of the categories mentioned here. That is, when firms affect one another in a way not covered by static equilibrium analysis or by interdependence among existing markets in the context of the dynamic analysis of economic development, external economies are still used by economists as a convenient catchall.

Again, one may ask – as did economists in the interwar period – what do these external economies actually stand for? At least some examples can be given. Growth of an industry may create a supply of new skills which turn out to provide a starting point for an altogether new line of business. Or, growth of a technologically advanced industry in a particular region leading to the location of a school of higher learning to this region may in turn stimulate – or reduce – growth of other activities. These cases are awkward to handle in traditional, well-structured economic analysis. So the main characteristic of these external economies, very much like most of those suggested by Marshall, is that we cannot yet say in any systematic way exactly what they represent.

PETER BOHM

See also ECONOMIES AND DISECONOMIES OF SCALE; EXTERNALITY; LINKAGES; RISING SUPPLY PRICE; YOUNG, ALLYN ABBOTT.

BIBLIOGRAPHY

American Economic Association (AEA). 1952. *Readings in Price Theory*, Vol. VI. Homewood, Ill.: Irwin.

Arndt, H.W. 1955. External economies in economic growth. *Economic Record* 31, November, 192–214.

Arrow, K. and Scitovsky, T. (eds) 1969. *Readings in Welfare Economics*. Homewood, Ill.: Irwin.

Bohm, P. 1967. *External Economies in Production*. Stockholm: Almqvist & Wiksell.

Clapham, J.H. 1922. On empty economic boxes. *Economic Journal* 32, 305–14; repr. in AEA (1952).

Ellis, H. and Fellner, W. 1943. External economies and diseconomies.

American Economic Review 33, 493–511. Reprinted in AEA (1952).

Heller, W.P. and Starrett, D.A. 1976. On the nature of externalities. In *Theory and Measurement of Economic Externalities*, ed. S. Lin, New York: Academic Press.

Hirschman, A.O. 1958. *The Strategy of Economic Development*. New Haven: Yale University Press.

Knight, F.H. 1924. Some fallacies in the interpretation of social cost. *Quarterly Journal of Economics* 38, 582–. Reprinted in Arrow and Scitovsky (1969).

Marshall, A. 1890. *Principles of Economics*. 8th edn, London: Macmillan, 1920.

Meade, J. 1952. External economies and diseconomies in a competitive situation. *Economic Journal* 62, 54–67. Reprinted in Arrow and Scitovsky (1969).

Pigou, A.C. 1920. *The Economics of Welfare*. London: Macmillan.

Pigou, A.C. 1922. Empty economic boxes. *Economic Journal* 32, 458–65. Reprinted in AEA (1952).

Robertson, D.H. 1924. Those empty boxes. *Economic Journal* 34, 16–30. Reprinted in AEA (1952).

Rosenstein-Rodan, P.N. 1943. Problems of industrialization of Eastern and South-Eastern Europe. *Economic Journal* 53, 202–11.

Scitovsky, T. 1954. Two concepts of external economies. *Journal of Political Economy* 62, 70–82. Reprinted in Arrow and Scitovsky (1969).

Viner, J. 1931. Cost curves and supply curves. *ZeitschriftfürNationalÖkonomie* 3, 23–46. Reprinted in AEA (1952).

Young, A. 1928. Increasing returns and economic progress. *Economic Journal* 38, 527–42. Reprinted in Arrow and Scitovsky (1969).

externalities. Competitive equilibria are Pareto optimal when they exist if preferences are locally non-satiated and if externalities are not present in the economy. Why externalities upset the first fundamental theorem of welfare economics and which economic policies can remedy this failure are the major questions addressed below.

TECHNOLOGICAL EXTERNALITIES. Let us call technological externality, the indirect effect of a consumption activity or a production activity on the consumption set of a consumer, the utility function of a consumer or the production function of a producer. By indirect, we mean that the effect concerns an agent other than the one exerting this economic activity and that this effect does not work through the price system.

Externalities may be positive or negative and are quite diverse. Major examples include pollution activities (air pollution, water pollution, noise pollution ...), malevolence and benevolence, positive interaction of production activities. From a practical point of view the most significant are negative pollution activities, so that we can say that the theory of technological externalities is essentially the foundation of environmental economics.

The formalization of technological externalities is achieved in microeconomics by making production sets, utility functions and production sets (or functions) affected by externalities functionally dependent on the activities of the other agents creating these indirect effects.

For example, the utility function of a consumer is made dependent on the level of production of a firm polluting the air breathed by the consumer. This modelling option that we will implicitly adopt here is right as long as the link between production and air pollution is not alterable.

If de-polluting activities are possible the link between the level of pollution and the economic activities generating them must be made explicit. An important difficulty in analysing these activities is due to the non-convexities which they usually introduce.

PECUNIARY EXTERNALITIES. During the 1930s, a confused debate occurred between economists on the relevance of pecuniary externalities, i.e. on externalities which work through the price system. A quite general consensus was that pecuniary externalities are irrelevant for welfare economics: the fact that by increasing my consumption of whisky I affect your welfare through the consequent increase in price does not jeopardize the Pareto optimality of competitive equilibria.

This is true when all the assumptions required for the competitive equilibria to be Pareto optimal are satisfied. In such a framework prices only equate supply and demand and pecuniary externalities do not matter. As soon as we move away from this set of assumptions prices generally play additional roles. For example, in economies with incomplete contingent markets, prices span the subspace in which consumption plans can be chosen. In economies with asymmetric information, prices transmit information. When agents affect prices, they affect the welfare of the other agents by altering their feasible consumption sets or their information structures. Pecuniary externalities matter for welfare economics.

In what follows we focus only on technological externalities.

COMPETITIVE EQUILIBRIUM WITH EXTERNALITIES. How is the characterization of Pareto optima in convex economies affected by externalities? Very simply, as Pigou early understood. The classical equality of marginal rates of substitution and marginal rates of transformation must now be expressed using *social* marginal rates and not only *private* marginal rates as in an economy without externalities. Social marginal rates must be computed taking into account direct *and* indirect effects of economic activities. For example, the marginal cost of a polluting activity must include not only the direct marginal cost of production, but also the marginal cost imposed on the environment.

Note that Pareto optima do not exclude polluting activities, but set them at levels such that their social marginal benefit equates their social marginal cost well computed.

It is now easy to understand that in a private competitive economy, equilibria will not be in general Pareto optimal since the private decentralized optimizations of economic agents lead them to the equalization of *private and not social* marginal rates through the price system.

MARKETS FOR POLLUTION RIGHTS. Consider for concreteness a firm polluting a consumer. One potential solution is to create a market for this externality. Before producing, the firm must buy from the consumer the right to pollute. If both actors were behaving competitively with respect to the price of this right, the competitive equilibrium in the economy with an extended price system would be Pareto optimal, since there is no externality left.

A number of difficulties exist with this approach. In general we cannot expect agents to behave competitively unless we are in the special case of impersonal externalities. Then, there is a fundamental non-convexity in the case of negative externalities since as a negative externality increases the production set shrinks, but there is a limit to this effect which is the zero production level. Competitive equilibria cannot then exist unless bounds are set on supplies of pollution rights. (For a positive price, a firm would like to offer an infinite amount of pollution rights and close down.)

In the above set-up, the implicit status quo was the absence of externalities. The initial rights are a clean environment: 'Polluters must pay.' We can instead give to the polluting firm the right to pollute and then ask the consumer to buy from the

firm a decrease of his pollution. This different allocation of initial rights does not upset the Pareto optimality of the competitive equilibrium, but of course has distributional effects.

TAXATION OF EXTERNALITIES. The likely strategic behaviour by agents on markets of pollution rights makes taxation of externalities the most common policy tool. The polluter must then pay for each unit of a polluting activity a tax which equals the marginal cost imposed by this activity on the other agents. The polluter then internalizes the externality and Pareto optimality is restored. If the externality is positive he must be similarly subsidized.

Note that nothing is said about the amount of taxes so obtained by the Government. There is no presumption that it is given to the polluters. In fact, the implicit assumption is that it is redistributed through lump sum transfers which do not affect agents' behaviours (in the sense of their first order conditions). From the point of view of Pareto optimality, the important goal is to modify polluters' behaviours.

If lump sum transfers are not available, the budget of the Government must be balanced and then goods different from the polluting activities must be taxed or subsidized to solve the ensuing second best problem.

The major difficulty with this solution is informational.

IMPERFECT INFORMATION. The traditional theory of externalities has proceeded as if the regulators had complete knowledge of the economy and were therefore able to compute optimal taxes, or as if agents were not behaving strategically with respect to their private information. Very often this is not the case and the problem is to elicit this private information and use it to compute taxes, a more difficult problem.

Intuitively, the solution of what is now a second best problem is to have taxes which depend nonlinearly on polluting activities. This nonlinearity may sometimes take the extreme form of a zero tax up to a given amount and a very large tax above, a mechanism which is equivalent to a quota.

PLANNING AND EXTERNALITIES. Externalities are not only a problem of market economies with an insufficient number of markets. One way to suppress an externality between two agents is to have them integrate into a single agent. All externalities would be internalized if the whole economy was integrated.

Leaving aside imperfect information and the associated strategic behaviours, the planning problem of these integrated agents is more complicated than if externalities were not present. Planning procedures appropriated to externalities have been provided.

EXTERNALITIES AND COOPERATIVE GAME THEORY. Suppose we attempt to represent the outcome of cooperation in an economy with externalities by the core. The core is the set of allocations which are not blocked by any coalition. A coalition blocks an allocation if it can do better for all its members than this allocation.

Externalities introduce a difficulty in the definition of a blocking coalition. When a group of agents envision forming a coalition they must conjecture what will be the behaviour of the complementary coalition since it is affected by the externalities of this complementary coalition.

Two extreme notions have been proposed. In the α-core a coalition is said to block an allocation if it can do better, whatever the actions of the complementary coalition. This is extremely prudent. In the β-core a coalition is said to block an allocation if, for any action of the complementary coalition, it can do better. The β-core is of course included in the α-core.

Results depends a lot on these conjectures about the actions of the complementary coalition, an unsatisfactory feature. One lesson, however, is that the core may be empty, i.e. that externalities introduce an element of instability in economic games.

HISTORICAL NOTE. Following the pioneering work by Sidgwick (1887) and Marshall (1890), Pigou (1920) has provided the basic theory of static technological externalities. Coase (1960) has explained how initial rights could be assigned in various ways. Arrow (1969) has explained how externalities could be internalized by the creation of additional markets. Starrett (1972) has pointed out the associated problem of non-convexity. The first theorem of existence of an equilibrium with externalities has been provided by McKenzie (1955). Shapley and Shubik (1969) studied the core with externalities. A large number of authors have studied various second best problems associated with externalities (Buchanan, 1969; Plott, 1966; Diamond, 1973; Sandmo, 1975).

J.J. LAFFONT

See also CLUBS; COASE THEOREM; EXTERNAL ECONOMIES AND DISECONOMIES; INCENTIVE COMPATIBILITY; INTERDEPENDENT PREFERENCES; NON-CONVEXITY; RISING SUPPLY PRICE; TRANSACTIONS COSTS AND ECONOMIC ORGANIZATION.

BIBLIOGRAPHY
Arrow, K. 1969. The organization of economic activity: issues pertinent to the choice of market versus non-market allocation. In Joint Economic Committee, *The Analysis and Evaluation of Public Expenditures: the PPB System*, Washington, DC: Government Printing Office, 47–64.
Buchanan, J.M. 1969. External diseconomies, corrective taxes and market structure. *American Economic Review* 59, 174–6.
Coase, R.H. 1960. The problem of social cost. *Journal of Law and Economics* 3, 1–44.
Diamond, P. 1973. Consumption of externalities and imperfect corrective pricing. *Bell Journal of Economics and Management*, Autumn, 526–38.
McKenzie, L. 1955. Competitive equilibrium with dependent consumer preferences. In *Proceedings of the Second Symposium in Linear Programming*, ed. H.A. Antosiewicz, Washington, DC.
Marshall, A. 1890. *Principles of Economics*. London: Macmillan.
Pigou, A.C. 1920. *The Economics of Welfare*. London: Macmillan.
Plott, C.R. 1966. Externalities and corrective taxes. *Economica* 33, 84–7.
Sandmo, A. 1975. Optimal taxation in the presence of externalities. *Swedish Journal of Economics* 77, 96–8.
Shapley, L. and Shubik, M. 1969. On the core of an economic system with externalities. *American Economic Review* 59, 687–9.
Sidgwick, H. 1887. *Principles of Political Economy*. 2nd edn, London: Macmillan.
Starrett, D. 1972. Fundamental non-convexities in the theory of externalities. *Journal of Economic Theory* 4, 180–99.

extortion. *See* BRIBERY.

F

Fabian economics. Despite the Webbs' disdain for abstract economics ('sheer waste of time'), economic arguments have always held a central place in the Fabian case for socialism. As in most matters the Fabian Society has approached the dismal science eclectically. Some members have accepted market economics, others have rejected it; some embraced the Keynesian revolution, others remained sceptical; some have believed in market pricing, others have been convinced that controls are essential for centralized planning. There is no consistent body of thought which could properly be described as Fabian economics. There is nonetheless a distinctive Fabian approach to economics, which this essay identifies while tracing the significant shifts in its key elements.

THE FABIANS AND THE MARGINAL REVOLUTION. When the small group including Sydney Olivier, Bernard Shaw and Sidney Webb first started to meet at Mrs Charlotte Wilson's house in Hampstead, they set themselves the task of reading Marx's *Das Kapital* chapter by chapter. Graham Wallas, who joined the group in February 1885, later recalled that we were astonished to find 'that we did not believe in Karl Marx at all' (Wallas, 1923). Webb, Wallas and Shaw were also members of the Economic Circle, an offshoot of the Bedford Chapel Debating Society, where Professor Edgeworth helped to expound the principles of the new marginal economics with another economist, Philip Wicksteed. Thus, according to Wallas, under Webb's leadership the group thrashed out 'the Jevonian anti-Marx value theory as the basis of our socialism'. Shaw apparently needed more convincing than others. In the *Fabian Essays* he later described how he had been converted from his earlier Marxist faith that the working class revolution would take place in Britain by 1889 'at latest' (Shaw, [1908] 1967, pp. 218–19). Instead of manning the barricades that year, Shaw was busily explicating the new Fabian economic basis for socialism.

In his preface to the essays, Shaw explained that the writers were all social democrats, 'with a common conviction of the necessity of vesting the organisation of industry and the material of production in a State identified with the whole people by complete Democracy'. In his contributions he propounded the theory of marginal productivity, demonstrating that Ricardian economic rent, or 'surplus value', can accrue to all the factors of production, to land and to labour, and not just to capital as in the Marxist version. Similarly, he rejected the labour theory of value, and advanced the neoclassical version, which he called 'exchange value'; in other words value was determined by the interaction of supply and demand in the marketplace. Shaw concluded:

> What the achievement of Socialism involves economically, is the transfer of rent from the class which now appropriates it to the whole people. Rent being that part of the produce which is individually unearned, this is the only equitable method of disposing of it (p. 220).

The method proposed to accomplish the transition was the common ownership of property, or as Webb put it: 'the gradual substitution of organized operation for the anarchy of competitive struggle' (ibid., p. 62).

The original essayists all shared Marx's moral outrage at the evils of capitalism, particularly as a cause of hopeless poverty, inhuman working conditions and excessive inequality, and they also identified the institution of private property as its prime motivating force. However, they did not share the Marxist belief that capitalism must inevitably collapse. Although they recognized that periodic slumps were endemic to the system, they were more struck by its spectacular long-run growth and saw no reason to suppose that it would not continue to reap the benefits of technological change. Thus, as Schumpeter later explained, they were the kind of socialists who believed in the productive success of capitalism while they deplored its distributive results (Schumpeter, 1942, pp. 61–2). They thought that through the gradual extension of public property socialism would evolve from democratic efforts to mitigate the effects of industrialization. Indeed, Webb provided an extraordinary two-page catalogue of socialism's accomplishments to date, which ranged from the army and navy to public baths and cow meadows (Shaw, [1908] 1967, pp. 66–7). William Clarke described the growth of joint stock companies, and more recently of 'rings' and 'trusts', through which ownership became ever more divorced from entrepreneurial function and capitalism ever more inconsistent with democracy and the public interest. These changes provided the other main Fabian justifications for the public ownership of industry.

Their views on the actual operations of a socialist system were hazy. Shaw and Webb both imply that socialism will have arrived when the entire market operation is administered through nationalization, municipalization and government regulation. Shaw described the aim of social democracy:

> to gather the whole people into the state, so that the state may be trusted with the rent of the country, and finally with the land, the capital, and the organisation of the national industry – with all sources of production, in short, which are now abandoned to the cupidity of irresponsible private individuals (ibid., p. 224).

Yet, in other Fabian Tracts, Shaw extolled the virtues of competition and of individual freedom, asserting that the latter was 'as highly valued by the Fabian Society as Freedom of Speech, Freedom of Press, or any other article in the charter of popular liberties' (Shaw, 1896, p. 327).

Later, of course, the Webbs provided a far more detailed view of their ideas for the organization of a Social Parliament to decide economic policy and to administer public enterprises. Beatrice herself remained ambivalent as to whether unemployment was caused by personal failings or 'the disease of industry'; their apparently countercyclical unemployment scheme only shifted existing projects without requiring fundamental changes in government policy (Harris, 1972, pp. 42–3). The Webbs' ideas about state planning were based on administrative principles, not economic science.

In the next Fabian generation, Hugh Dalton, a student of

Pigou, used Pigou's revised version of neoclassical theory to demonstrate the critical differences between factor incomes and personal incomes. He introduced and defined the nature of inheritance and its role in maintaining wealth differentials; he broadened its concept to include educational opportunity, access to public services and institutional customs (Dalton, 1920). According to Gaitskell's later assessment of the British tradition, Dalton's work was a decisive influence in shifting socialist thought from the 'sterile, out-of-date, somewhat academic arguments of earlier writers' to the practical issues of progressive taxation and educational reform (Gaitskell, 1955, pp. 936–7). Although still grounded in neoclassical criteria of allocative efficiency, Dalton's analysis dealt directly with income equality, opening up ways to achieve socialism other than through Webbian public ownership. Thus, Gaitskell believed that the case for socialist equality could be stated on 'straightforward ethical principles', rather than on 'complicated arguments about economic abstractions'.

THE FABIANS, THE KEYNESIAN REVOLUTION AND ECONOMIC PLANNING IN THE 1930s. The great depression threatened both the political and economic stability of capitalist systems. Inspired by the Russian revolution and its apparent success in replacing capitalism and avoiding mass unemployment, many leftist sympathisers turned to Marxism. They struggled through *Das Kapital*, they visited the Soviet Union, and they recommended the Soviet political philosophy and economic system. The Webbs fell in love with Russia; in their last major work, *Soviet Communism: a new civilization?*, they advocated a totally controlled economy, visualising Soviet planning as the ultimate Fabian collective. In *New Fabian Essays* Crossman argued that they had simply superimposed Marxism on their basic utilitarianism; he believed that only John Strachey successfully re-thought the entire system 'in Anglo-Saxon terms' (Crossman, 1970, p. 5).

It fell to the younger generation to restate the traditional Fabian case against Marxist economic thought and revolutionary methods and to redefine the democratic socialist alternative. Hugh Gaitskell and Evan Durbin organized the Economic Section of the New Fabian Research Bureau, which had been founded by G.D.H. Cole in March 1931 and merged with the parent Fabians in 1938; their purpose was to explore the implications of the theoretical economic controversies for socialism and to make policy recommendations to the Labour Party (Durbin, 1985). At the same time the obvious failures of the market system were challenging economists to rethink the role of government intervention and to redesign their tool kit. Keynesian macroeconomics, the economics of imperfect competition and the principles of economic planning embodied in the new 'market socialism' were first developed during the 1930s. After the war they were incorporated into the orthodox case for the mixed economy.

In pointed contrast to official policy, Keynes had begun pressing British governments to expand, not to contract, public expenditure to cope with unemployment. In the early 1930s his position was largely intuitive; *The General Theory* published in 1936 was the first systematic exposition of his theoretical case. Until then the most fundamental cleavage on the unemployment issue was between those who advocated government intervention in the market, and those who did not. Socialists were naturally allied with the interventionists on social and political grounds, as well as economic, and thus were sympathetic to Keynes's policy efforts: but they were suspicious of his political ties to the Liberal Party, and some of the professional; economists were sceptical about his expansionist policies. James Meade and Colin Clark, who were

working alongside Keynes, were convinced expansionists by August 1931. Together they were responsible for converting the New Fabians well before 1936. Amongst the sceptics were Gaitskell and Durbin, who were strongly influenced by Hayek's trade cycle theories and who were deeply concerned to demolish 'treasured dogma' within the Labour party, namely the myths that capitalism was collapsing and that socialism could easily replace it and automatically solve the unemployment problem. As early as 1932 Gaitskell explained why, although 'prosperity' was an important socialist goal, it was not 'the distinguishing characteristic of the Socialist ideal' (Gaitskell, 1932).

Meade also played an important role in converting Douglas Jay, whose influential book, *The Socialist Case*, published in 1937, was the first to propose that Keynesian fiscal and monetary measures to control output and employment be explicitly incorporated as part of socialist planning methods. Cole, who thought that the *General Theory* was the most important economics book published since Marx's *Das Kapital* and Ricardo's *Principles*, was quick to point out that because Jay gave such a low priority to nationalization his book contained very little of 'what most people habitually think of as socialism' (Cole, 1937). Thus, the introduction of Keynesian methods also served to weaken the case for public ownership as the basis of the socialist economic alternative.

By the late 1930s most democratic socialists in Britain had recognized the importance of the Keynesian message for socialism, and by the end of the war the Labour Party had officially adopted a Keynesian full employment policy. The new macroeconomic analysis provided an obvious answer to the problem of dealing with capitalist collapse. It also reinforced distributive goals, since lower income families had a higher propensity to consume, and it underscored the importance of central planning to control the economy, since only the government had the power to offset insufficient private spending. So compelling were these arguments that they also converted at least one influential Marxist, John Strachey, to the Fabian cause.

Yet Fabian acceptance of Keynes's economics and of Keynes's basic individualism is often overstated, particularly in the prewar context. Anthony Wright (in Pimlott, 1984) has suggested that the Tawney approach to equality is fundamentally different from the liberal philosophy behind Beveridge's welfare state. A similar contrast can be made between Fabian conceptions about economic planning in the 1930s and Keynesian macroeconomic management. Fabians were explicit about their opposition to the capitalist system, which Keynes wanted to repair, but which they wanted to replace. They were emphatic about the need for major reform of Britain's financial institutions and for substantial growth of the public sector; indeed, they believed that both were essential to implement a successful full employment policy. At least one Fabian economist, Evan Durbin, never accepted *The General Theory* model as the solution to all macroeconomic problems; he believed that it failed to explain the trade cycle, and was therefore unsuitable for the long-term growth problems which the socialist state must solve in order to improve upon capitalism's record.

The principles of market socialism grew out of work initiated by Durbin and Gaitskell, who undertook a systematic reconsideration of the Marshallian microeconomic grounds for intervention and the implications for socialist planning. Together with H.D. Dickinson they demonstrated that the market system by definition could neither price collective goods nor reflect the true social value of externalities, and, therefore, that it could not determine the appropriate allocation of resources for their production. They also incorporated the new economics of imperfect competition associated with Joan Robinson to restate the objections to the existing system, which they termed 'monopoly capitalism'. A planning authority would be able to correct these deficiencies and use the principles of optimal allocation to guide its decisions; in other words, neoclassical criteria should serve as the handmaiden to collective decision-making. In the 1930s and 1940s, many Fabians contributed to the further elaboration of these ideas into a socialist economic system based on free choices in the labour market, consumers' sovereignty through market pricing and marginal cost pricing in nationalized industries. The importance of this analysis was that it added strong theoretical arguments for a mixed economy as an explicit complement to the macroeconomic Keynesian ones.

There were, however, other Fabians who found such arguments hard to take and/or to follow. Barbara Wootton, whose planning schemes were an updated version of the Webbian administrative structure, was clear that prices would have to be controlled in the public interest. Even Dalton, who recognized that planning was not necessarily socialist, still maintained the early Fabian belief that 'Socialism is primarily a question of ownership' (Dalton, 1935, p. 247). With more appreciation for the problems of allocative efficiency under socialism, Cole attempted to fashion a different socialist economics, one which was neither Marxist nor neoclassical (Cole, 1935). Although his own system remained a rather sketchy attempt to incorporate socialist distributional goals into decisions about production, he had some telling arguments against his neoclassical comrades, pointing out that market prices reflected the existing income distribution, and thus could not provide the proper signals for socialist allocation. His efforts are particularly interesting for the light they throw on the need to mesh social policy with economic planning, and on the problem of applying neoclassical analysis to meet essentially political goals.

By the end of the 1930s, most Fabians had come to accept the necessity for a mixed economy, if only on practical grounds, because the legislation necessary to secure socialism by parliamentary methods could not be accomplished by one Labour government. Government planning was necessary to ensure aggregative and allocative efficiency and to redistribute income and wealth. Control of what were later known as 'the commanding heights' of the economy was essential to implement the planning alternative, and a central authority was required to make sure that sectional interests, such as bankers, business and trade unionists, did not subvert the public good. However, in an important change of emphasis, Durbin and Gaitskell were explicit that their objections to capitalism and to the Marxist alternative were social and political, not economic (Gaitskell, 1935; Durbin, 1940). The essence of their socialism was social justice as Tawney defined it. In short, the mixed economy was not simply politically expedient, it was central to the economic operation of the democratic socialist state.

THE FABIANS AND THE MIXED ECONOMY IN PRACTICE. As authors in the New Fabian Essays later pointed out, the war substantially altered the balance of power between the government and the private sector. And in comprehensive plans for recovery, the wartime coalition laid the foundations for bipartisan support of full employment, a unified system of social services and educational reform. Thus, when the Labour government took over in 1945, there was not much resistance to its programme or to its Fabian philosophy.

In 1948 the Fabian Society commissioned W. Arthur Lewis to write a pamphlet on 'the economic perplexities of the moment'. These turned out to be so numerous that Lewis ended up writing a short book, The Principles of Economic Planning, an influential statement of the revised conception of market socialism. Like Meade in Planning and the Price Mechanism, published in the same year, he argued the case for planning on general interventionist grounds, implicitly rejecting the Durbin/Gaitskell notion that only a socialist government could run the economy efficiently, although one might still believe only a socialist government would. To paraphrase Lewis, socialism was not about the state, any more than it was about property; 'socialism is about equality'. There could be many ways to handle property and to plan the economy, which were not inconsistent with socialism (Lewis, 1949, pp. 10–11). Lewis argued that the crucial issue was whether the state should operate 'through the price mechanism or in supersession of it'; the real choice was 'between planning by inducement, and planning by direction'. Lewis himself was neutral on the issue, believing that Britain needed some of both. Although insistent that there must be free consumer goods and labour markets, he argued that demand was not sacred and that it should be manipulated in specific markets and in the aggregate to achieve policy goals. Similarly, he did not believe that nationalization should be taken on its merits. Lewis wanted 'more than we have already got' (steel, banking and chemicals were his candidates), but in no circumstances the whole economy; 'a country whose people love freedom will not wish the state to become the sole employer' (ibid., p. 104).

Shortly after this book was published in 1949, Cole as chairman of the Society organized a conference to begin to rethink the way forward now that the main components of the first Fabian stage to socialism were in place. New Fabian Essays published in 1952 was the end result of this effort to take account of important societal changes and the Keynesian revolution. The essayists were all agreed that the British version of the mixed economy was a permanent Fabian accomplishment, and that the Tories would not dismantle the welfare state nor renege on full employment. Yet, despite the enormous gains, substantial inequities remained and new problems emerged: in particular, the great concentrations of bureaucratic power in the public and private sectors which threatened individual freedom. In general terms the way forward was to continue to pursue equality, to improve labour/management relations and to disperse power as much as possible.

However, the Fabians were still united in their dissatisfaction with that system. Although they were clear that the postwar version of welfare capitalism did not meet their conception of socialism, many of the essayists were vague about what they did want. Writing about equality in New Fabian Essays, Roy Jenkins explained that a classless society was one 'in which men will be separated from each other less sharply by variations in wealth and origin than by differences in character', but it was impossible to describe 'the exact shape of the goal'. Of contributors to New Fabian Essays, only Crosland was willing to be explicit in the negative sense that he specified four policies which would not achieve equality; the continued extension of free social services, more nationalization, the proliferation of controls and further redistribution of income by direct taxation. In an important shift, many Fabians had come not only to believe in the mixed economy, but also to accept its current structural form.

Crosland outlined the main features of what he called 'post-capitalist society': he concluded that it was more equal and more planned than before, but that it was still based on unacceptable class divisions. While individual property rights were no longer the essential basis of economic and social power, they still affected the distribution of wealth. He felt that the power of the state had been expanded sufficiently to exert control over the economy: if anything, physical controls should be reduced as they were unpopular and inefficient. Similarly, nationalization had secured government power in the central sectors of the economy, social legislation had ensured a national minimum welfare level, and full employment policies had removed insecurity and demonstrated that central planning could be directed to meet social ends. Keynesian policies were crucial to maintaining this system, but as these were now well understood, Crosland argued that 'the new society may prove to be a very enduring one'. In *The Future of Socialism* (1956), Crosland spelled out his ideas on planning in more detail; he believed its 'essential role' was Keynesian economic management, that the techniques were no longer controversial nor the preserve of any one party, and that political will, not planning theory, were required to plan effectively; 'if socialists want bolder planning, they must choose bolder ministers'.

One lone dissenter from the general Fabian romance with Keynes was G.D.H. Cole. Although enthusiastic about the *General Theory* when it was published, he had become increasingly concerned about these new directions after the war. Indeed, this was precisely why he had initiated the process of rethinking, and why, as the discussions progressed, he resigned his position as chairman of the Fabian Society. In 1950 he published a short book, *Socialist Economics*, which spelled out his disagreements with the new Fabian approach. First, he thought that Keynesian economics was too involved with aggregates and not sufficiently concerned with the structural problems necessary for a socialist economy to replace the capitalist system. As far as he was concerned the new direction provided a diluted form of socialism, which was 'little more than Keynesian Liberalism with frills'. Second, although Cole had advocated using a wide range of industry controls as early as 1929 and was opposed to total public ownership, he was also explicit in rejecting the current version of the mixed economy 'as a permanent resting place'.

Elizabeth Durbin

See also DURBIN, EVAN FRANK MOTTRAM; SOCIAL DEMOCRACY; WEBB, BEATRICE AND SIDNEY.

BIBLIOGRAPHY
Clarke, P. 1978. *Liberals and Social Democrats*. Cambridge: Cambridge University Press.
Cole, G.D.H. 1935. *Principles of Economic Planning*. London: Macmillan.
Crossman, R.H. (ed.) *New Fabian Essays*. London: J.M. Dent; 3rd impression, 1970.
Dalton, H. 1920. *Some Aspects of the Inequality of Incomes in Modern Communities*. London: Routledge & Kegan Paul.
Dalton, H. 1935. *Practical Socialism for Britain*. London: George Routledge.
Durbin, E.F.M. 1940. *The Politics of Democratic Socialism*. London: Routledge & Kegan Paul.
Durbin, E. 1985. *New Jerusalems: The Labour Party and the Economics of Democratic Socialism*. London: Routledge & Kegan Paul.
Gaitskell, H.T.N. 1932. Socialism and wage policy. Fabian Society Papers, Box J24/2 in Nuffield College, Oxford.
Gaitskell, H.T.N. 1935. Financial policy in the transition period. In *New Trends in Socialism*, ed. G.E.G. Catlin, London: Lovat Dickson & Thompson.
Gaitskell, H.T.N. 1955. The ideological development of democratic socialism in Britain 1955. *Socialist International Information* 5, 52–3.
Harris, J. 1972. *Unemployment and Politics*. Oxford: Oxford University Press.
Lewis, W.A. 1949. *The Principles of Economic Planning*. London: Dobson.
Pimlott, B. (ed.) 1984. *Fabian Essays in Economic Thought*. London: Gower Publishing.
Schumpeter, J.A. 1942. *Capitalism, Socialism, and Democracy*. New York: Harper & Row; Torchbook edition, 1962.
Shaw, G. Bernard. 1884, 1896. *Fabian Tract No. 2*. (1884) *Tract No. 70* (1896). Quoted in C.A.R. Crosland, *The Future of Socialism*, London: Jonathan Cape, abridged edition, 1964.
Shaw, G. Bernard. (ed.) *Fabian Essays in Socialism*. New York: Doubleday edn, 1967.
Wallas, G. 1923. Article in *Morning Post*, 1 January. See Clarke (1978) for further details.
Wright, A.W. 1979 *G.D.H. Cole and Socialist Democracy*. Oxford: Clarendon Press.

Fabricant, Solomon (born 1906). Fabricant was born in Brooklyn, New York, on 15 August 1906. He began his association with the National Bureau of Economic Research in 1930, serving as director of research from 1953 to 1965 and continuing as a member of the Board. From 1944 to 1973 he was on the economics faculty at New York University. His economic studies range across a wide field, including productivity and economic growth, national income and capital formation, trends in government activity, and economic accounting under conditions of inflation.

Fabricant's initial work on productivity demonstrated that in industries with large productivity gains, the resulting cost and price reductions have usually been sufficient to cause output and employment to rise faster than in other industries – a conclusion at variance with the common contention that technology, which is often a source of rapid productivity growth, deprives workers of jobs. Fabricant's research also clarified the understanding of productivity gains and losses during business cycles, with systematic effects on the movements in costs and profits, which in turn play an important role in generating recessions and recoveries.

In his investigation of trends in government activity (1952), he showed how economic development in the United States during the first half of the 20th century had fostered a rise in the relative importance of government. Thus, for example, urbanization promoted the demand for municipal services, advances in transportation technology led to government building of roads and airfields, and increases in family income supported government activities in education, public health, welfare, and old-age assistance. By carefully assembling the facts on government functions, types of organization, and use of labour and capital, and developing a reasoned account of the factors that led to their growth or decline over the past 50 years, Fabricant cast a bright light over what was to happen over the following 30 years.

G.H. Moore

SELECTED WORKS
(Except as noted, all were published in New York by the National Bureau of Economic Research.)
1938. *Capital Consumption and Adjustment*.
1940. *The Output of Manufacturing Industries, 1899–1937*.
1942. *Employment in Manufacturing, 1899–1939: An Analysis of Its Relation to the Volume of Production*.
1952. *The Trend of Government Activity in the United States since 1900*.
1958. *Investing in Economic Knowledge*.
1959. *The Study of Economic Growth*.
1959. *Basic Facts on Productivity Change*. Occasional paper 63.
1969. *Primer on Productivity*.

1976. (With others.) *Economic Calculation under Inflation.* Indianapolis: Liberty Press.

1984. *Toward a Firmer Basis of Economic Policy: The Founding of the National Bureau of Economic Research.* Cambridge, Mass.: National Bureau of Economic Research.

factor analysis. Factor analysis is a branch of analysis of variance used to investigate the structure of a data set. Consider a data set x_{ij} resulting from the observation of several variables j on several objects i. If the data set arises from a complex multidimensional process about which little is known a priori statistical analysis of the data itself might profitably be used to gain insights into various characteristics of the processes which generated the data set. In particular, statistical techniques can be used to: (1) search for a simpler representation of the underlying processes which generated the data by reducing the dimension of the variable space in which the objects are represented; (2) look for the interactions among the variables by forming linear clusters of variables; and (3) seek characterizations of the clusters of variables which relate them to the underlying processes which generated the data set being analysed. Factor analysis performs all three functions.

A variety of factor analytic methods has been introduced. They differ in estimation procedures (least squares or maximum likelihood); fitting equation (original data matrix, covariance or correlation matrix); scaling assumption (original or normalized data, type of normalization and in whether the scaling is performed prior to the estimation or as part of the estimation procedure); and in the normalization principles applied to the factor matrix. For a discussion of the relationship between them see Kruskal (1978). Following Kruskal, we start from the original data, derive the covariance matrix and then discuss the procedures applied to it. The basic technical references are Hotelling (1933), Bartlett (1938), Lawley (1940), Lawley and Maxwell (1971), Joreskog (1967), and Joreskog, and Goldberger (1972).

Let the variables j characterizing the objects i be measured as deviations from their means. Assume further that the data set x_{ij} was generated by an r-dimensional linear process, with r significantly smaller than the original number of variables J. We are then seeking a representation of x of the form

$$x_{ij} = \sum_r a_{ir} b_{rj} + v_{ij} \qquad (1)$$

which, in some sense, comes closest to representing the original data set. In (1) the a_{ir} represent the coefficients, known as 'factor scores', which indicate the 'regression coefficients' of the objects upon each of the r clusters of variables; the b_{rj} represent the coefficients of the variables in each of the r clusters, known as 'factor loadings' or 'factor patterns'. The r clusters of variables are known as factors or components, and represent the coordinates of the lower-dimensional space onto which the data matrix is mapped. In matrix notation, we can write (1) as (2)

$$X = AB + \Sigma \qquad (2)$$

where A is the matrix of a_{ir}, B is the matrix of b_{rj} and Σ is a diagonal disturbance matrix with typical element σ_j^2.

One can fit (2) directly, by least squares or by maximum likelihood, or one can form the sample covariance matrix $C = X'X/N$, where N is the number of objects, and fit it instead. If one assumes that: (1) the a_{ij} are random, identically distributed, with mean 0, and independent both of each other

and of the disturbances and (2) applies the normalization T that sets

$$\frac{N-1}{N}(AT^{-1})'(AT^{-1}) = I \qquad (3)$$

then the expected value of the sample covariance matrix C is

$$E(C) = B'B + \Sigma^2 \qquad (4)$$

This equation can be fitted either by least squares (Hotelling, 1933; Anderson, 1958; Harman, 1960; Joreskog and Goldberger, 1972) or by maximum likelihood methods (Lawley, 1940; Joreskog, 1967), to obtain estimates for b_{rj} and σ_j^2. Once these estimates have been obtained, a_{ir} can be estimated by regression methods from eqn (2) keeping B fixed.

In the least squares approach the matrix B is estimated by extracting the successive eigenvectors of

$$(C - \lambda_r I)b_r = 0 \qquad (5)$$

where λ_r is the tth characteristic root and b_r is the rth eigenvector. The rth column of B, b_r, represents the makeup of the rth component in terms of the original, observable variables. Goodness of prediction measures analogous to significance intervals can be derived for the estimates of B by using Stone–Geisser or Tukey-jack-knife methods (Wold, 1982).

In the maximum likelihood approach, we form the likelihood function,

$$L = \tfrac{1}{2}(N-1)\ln|C| - \tfrac{1}{2}(N-1)\sum_{i,j} x_{ji}x_{ij}C^{ij}/N - 1 \qquad (6)$$

where $|C|$ is the determinant of C, and C^{ij} is the ijth element of C^{-1}. To find the maximum likelihood estimators of B and Σ, we differentiate (6) with respect to the elements of B and Σ and set the resulting equations equal to zero. The maximum equations are then solved simultaneously for B and Σ by applying techniques such as Fletcher–Powell (1963) for the simultaneous optimization of nonlinear equation systems. The maximum likelihood approach was first developed by Lawley (1940); practical estimation techniques for it were developed by Joreskog (1967). The use of maximum likelihood has both advantages and disadvantages: it requires stringent assumptions about the distributions of the parameter set B and the disturbances Σ but it also enables one to estimate confidence intervals on the parameters of B and on the goodness of fit (Lawley and Maxwell, 1971; and Jennrich and Thayer, 1973).

Both the least squares approach and the maximum likelihood approach yield estimates of B which are not unique since a rigid rotation of B yields the same estimating equations. Several approaches have been proposed for deriving unique estimates. These include normalization assumptions on $A'A$ or $B'B$ and rotation assumptions aimed at increasing ease of interpretability such as the varimax rotation (Kaiser, 1958).

The first applications of factor analysis in the social sciences were in psychology, for which the technique was first developed by Spearman (1904), and used to analyse mental abilities (see Bolton et al., 1973 for a survey). In economics, the first application was to demand analysis (Stone, 1945). Stone hypothesized that demand for commodities is explained by three types of influences: national income and own and other prices; social influences affecting tastes and market conditions; and forces peculiar to a particular community. He used a three-factor confluence analysis model, similar to factor analysis, to identify the factors affecting consumer demand. A recent study of market demand employing modern factor analysis is Huang et al. (1980). Stone (1947) and Geary (1948) used factor analysis to study interaction patterns among time series. Using time series representing the components of national income and product in the US, Stone showed that 97.5 per cent of their total variance could be represented by three factors. Banks (1954) used factor analysis in agriculture to predict overall agricultural productivity from crop productivity data on a small number of crops.

The most numerous applications of factor analysis to economics have been in economic development (Adelman and Morris, 1967; Rayner, 1970; Schilderinck, 1969). In a series of studies, Adelman and Morris investigated the interdependence of economic, social and political phenomena in the development process. Their observations were 74 countries; their variables were typologies representing various aspects of economic, social and political structure. Four factors explained most of the covariance: a modernization factor, which includes indicators of economic and social development; a political development factor; a political leadership factor; and a social and political stability factor. They found that the relative importance of these factors in explaining intercountry differences in growth rates changes systematically with country development levels, with social forces declining in importance and political leadership increasing. Other applications have been to the economics of education (Aigner and Goldberger, 1977) and to stock market prices (King, 1966).

Recent uses of factor analysis have been in the estimation of the parameters of unobservable variables, defined as variables whose measurable quantities differ from their theoretical counterparts and to error-in-variables models. Other recent advances have been in nonlinear factor analysis (McDonald, 1967) and in the dynamic analysis of factor structures (Geweke, 1977).

IRMA ADELMAN

See also ARBITRAGE PRICING THEORY; PRINCIPAL COMPONENTS; STONE, JOHN RICHARD NICHOLAS.

BIBLIOGRAPHY

Adelman, I. and Morris, C.T. 1967. *Society, Politics, and Economic Development: A Quantitative Approach.* Baltimore: Johns Hopkins Press.
Aigner, D.J. and Goldberger, A.S. (eds) 1977. *Latent Variables in Socioeconomic Models.* Amsterdam: North-Holland.
Anderson, T.W. 1958. *An Introduction to Multivariate Statistical Analysis.* New York: Wiley.
Banks, C. 1954. The factorial analysis of crop productivity: a reexamination of professor Kendall's data. *Journal of the Royal Statistical Society,* Series B 16, 100–111.
Bartlett, M.S. 1938. Methods of estimating mental factors. *Nature* 141, 609–10.
Bolton, B., Hinman, S. and Tuft, S. 1973. *Annotated Bibliography: Factor Analytic Studies 1941–1970.* 4 vols, Fayetteville: University of Arkansas, Arkansas Rehabilitation Research and Training Center. (Tuft did not collaborate on vols 3 and 4.)
Fletcher, R. and Powell, M.J.D. 1963. A rapidly convergent descent method for minimization. *Computer Journal* 6, 163–8.
Geary, R.C. 1948. Studies in relation between economic time series. *Journal of the Royal Statistical Society,* Series B 10, 140–58.
Geweke, J. 1977. The dynamic factor analysis of economic time-series models. In *Latent Variables in Socioeconomic Models,* ed. D.J. Aigner and A.S. Goldberger, Amsterdam: North-Holland.
Harman, H.H. 1960. *Modern Factor Analysis.* 3rd edn, revised, Chicago: University of Chicago Press, 1976.
Hotelling, H. 1933. Analysis of a complex of statistical variables into principal components. *Journal of Educational Psychology* 24, 417–41, 498–520.
Huang, C.-L., Raunika, R. and Fletcher, S.M. 1980. Estimation of demand parameters based on factor analysis. Paper presented at

the American Agricultural Economics Association Meetings in Urbana, Illinois.
Jennrich, R.I. and Thayer, D.T. 1973. A note on Lawley's formulas for standard errors in maximum likelihood factor analysis. *Psychometrika* 38, 571–80.
Jöreskog, K.G. 1963. *Statistical Estimation in Factor Analysis: A New Technique and its Foundation.* Stockholm: Almqvist & Wiksell.
Jöreskog, K.G. 1967. Some contributions to maximum likelihood factor analysis. *Psychometrika* 32, 443–82.
Jöreskog, K.G. 1984. *Advances in Factor Analysis and Structural Equation Models.* Lanham: University Press of America.
Jöreskog, K.G. and Goldberger, A.S. 1972. Factor analysis by generalized least squares. *Psychometrika* 37, 243–60.
Kaiser, H.F. 1958. The varimax criterion for analytic rotation in factor analysis. *Psychometrika* 23, 187–200.
King, B. 1966. Market and industry factors in stock price behavior. *Journal of Business* 39, Supplement, 139–90.
Kruskal, J.B. 1978. Factor analysis: bilinear methods. In *International Encyclopedia of Statistics,* New York: Macmillan, 307–30.
Lawley, D.N. 1940. The estimation of factor loadings by the method of maximum likelihood. Royal Society of Edinburgh, Section A, *Proceedings* 60, 64–82.
Lawley, D.N. and Maxwell, A.E. 1963. *Factor Analysis as a Statistical Method.* 2nd edn, London: Butterworth, 1971.
McDonald, R.P. 1967. Factor interaction in nonlinear factor analysis. *British Journal of Mathematical and Statistical Psychology* 20, 205–15.
Rayner, A.C. 1970. The use of multivariate analysis in development theory: a critique of the approach used by Adelman and Morris. *Quarterly Journal of Economics* 84, 639–47.
Schilderinck, J.H.F. 1969. *Factor Analysis Applied to Developed and Developing Countries.* Rotterdam: Rotterdam University Press.
Spearman, C.E. 1904. 'General intelligence' objectively determined and measured. *American Journal of Psychology* 15, 201–293.
Stone, R. 1945. The analysis of market demand. *Journal of the Royal Statistical Society,* Series A 108, 286–382.
Stone, R. 1947. On the interdependence of blocks of transactions. *Journal of Royal Statistical Society,* Series B 9, 1–45.
Thurstone, L.L. 1935. *The Vectors of Mind: Multiple-factor Analysis for the Isolation of Primary Traits.* Chicago: University of Chicago Press.
Wold, H. 1982. Soft modeling and some extensions. In *Systems under Indirect Observation,* ed. K.G. Jreskog and H. Wold, Amsterdam: North-Holland, II, 1–54.

factor-price equilization theorem. See HECKSCHER–OHLIN TRADE THEORY.

factor price frontier. The constraint binding changes in the distributive variables, in particular the real wage rate (w) and the rate of profit (r), was discovered (though not consistently demonstrated) by Ricardo: 'The greater the portion of the result of labour that is given to the labourer, the smaller must be the rate of profits, and vice versa' (Ricardo, 1971, p. 194). He was thus able to dispel the idea, generated by Adam Smith's notion of price as a sum of wages and profits, that the wage and the rate of profit are determined *independently* of each other. Ever since the inverse relationship between the distributive variables played an important role in long-period analysis of both classical and neoclassical descent. In more recent times it was referred to by Samuelson (1957), who later dubbed it 'factor price frontier' (cf. Samuelson, 1962). Hicks (1965, p. 140, n.1) objected that this term is unfortunate, since it is the earnings (quasi-rents) of the (proprietors of) capital goods rather than the rate of profit which is to be considered the 'factor price' of capital (services). A comprehensive treatment of the problem under consideration within a classical framework of the analysis, including joint production proper, fixed capital and scarce natural resources, such as land, was provided by Sraffa (1960). The relationship is also known as the 'wage frontier' (Hicks, 1965), the 'optimal transformation frontier' (Bruno, 1969) and the 'efficiency curve' (Hicks, 1973). The duality of the $w–r$ relationship and the $c–g$ relationship, that is, the relationship between the level of consumption output per worker (c) and the rate of growth (g) in steady-state capital theory has been demonstrated by the latter two authors and in more general terms by Burmeister and Kuga (1970); for a detailed account, see Craven (1979).

To begin with, suppose for simplicity that there are only single-product industries with labour as the only primary input and that only one (indecomposable) system of production is known (cf. Sraffa, 1960, Part I). Then, with gross outputs of the different products all measured in physical terms and made equal to unity by choice of units and with wages paid at the end of the uniform production period, we have the price system.

$$p = (1 + r)ap + wa_0, \qquad (1)$$

where p is the column vector of normal prices, a is the square matrix of material inputs, which is assumed to be productive, and a_0 is the column vector of direct labour inputs. Using the consumption basket d as standard of value or *numéraire*,

$$dp = 1, \qquad (2)$$

we can derive from (1) and (2) the $w–r$ relationship for system (a, a_0)

$$w = \{d[I - (1 + r)a]^{-1}a_0\}^{-1} \qquad (3)$$

The relationship is illustrated in Figure 1. At $r = 0$ the real wage in terms of d is at its maximum value W; it falls monotonically with increases in r, approaching zero s r approaches its maximum value R. (The $w–r$ relationship can be shown to be a straight line if Sraffa's Standard commodity s is used as *numéraire*, where s is a row vector such that $s = (1 + R)\, sa$; cf. Sraffa, 1960, chap. IV.)

Let us now assume that several systems are available for the production of the different commodities and that all the production processes exhibit constant returns to scale. We call the set of all the alternative methods (or processes) of production known the *technology* of the economic system. From this set a series of alternative *techniques* can be formed by grouping together these methods of production, one for each commodity. Hence there is the question of the *choice of technique*. Under competitive conditions this choice will be exclusively grounded

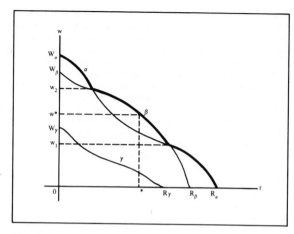

Figure 2

on cheapness, that is, the criterion of choice is that of *cost-minimization*. In the case depicted, it can be shown that the competitive tendency of entrepreneurs to adopt whichever technique is cheapest in the existing price situation, will for a given w (or, alternatively, r) lead to the technique yielding the highest $r(w)$, whereas techniques yielding the same $r(w)$ for the same $w(r)$ are equiprofitable and can co-exist (cf. Garegnani, 1970, p. 411).

What has just been said is illustrated in Figure 2. It is assumed that only three alternative techniques, α, β and γ, are available, each of which is represented by the associated $w–r$ relationship; since w is always measured in terms of the consumption basket d, all three relationships can be drawn in the same diagram. Obviously, technique γ is inferior and will not be adopted. Technique α will be chosen for $0 < w < w_1$ and $w_2 < w \leqslant w_\alpha$, while technique β dominates at $w_1 < w < w_2$; there are two *switch points* (at $w = w_1$ and $w = w_2$, respectively) at which both techniques are equiprofitable. The heavy line represents the economy's $w–r$ *frontier* (or 'factor price frontier') and is the outer envelope of the $w–r$ relationships. At a level of the wage rate w^*, for example, technique β will be adopted giving a rate of profit r^*. (For a discussion of more general cases of single production, see Pasinetti, 1977, ch. VI; for a reformulation of some results in capital theory in terms of the so-called 'dual' cost and profit functions, see Salvadori and Steedman, 1985; on the maximum number of switch points between two production systems, see Bharadwaj, 1970.)

Figure 2 shows that the same technique (α) may be cost-minimizing at more than one level of the wage rate (rate of profit) even though other techniques (here β) dominate at wage rates in between. The implication of this possibility of the *re-switching* of techniques (and of the related possibility of *reverse capital deepening*) is that the direction of change of input proportions cannot be related unambiguously to changes in the distributive variables. This can be demonstrated by making use of the duality between the $w–r$ and the $c–g$ frontier. Denoting the value of net output per labour unit by y and the value of capital per labour unit by k, we have in steady-state equilibrium

$$y = w + rk = c + gk. \qquad (4)$$

Solving for k we get

$$k = (c - w)/(r - g) \qquad (5)$$

except in golden rule equilibrium ($g = r$), where k can be shown

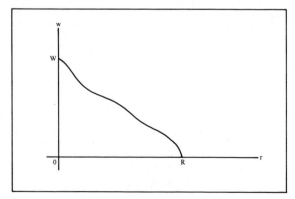

Figure 1

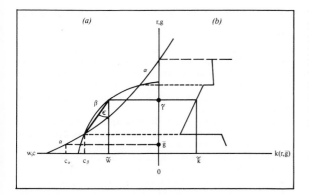

Figure 3

to be (minus) the slope of the golden rule $w-r$ relationship at the going level of r. In Figure 3(a) the frontier built of two techniques, α and β, is depicted. The rate of growth is fixed at the level $\bar{g}$, to which correspond c_α and c_β. For values of $r \geqslant \bar{g}$, that is, on the right side of the golden rule, Figure 3(b) gives the corresponding value of $k(r, \bar{g})$. For example, at $\tilde{w}$ technique β will be chosen, yielding a rate of profit $\tilde{r}$; the associated capital intensity is given by

$$\tan \epsilon = (c_\beta - \tilde{w})/(\tilde{r} - \bar{g}) = \tilde{k}.$$

Figure 3(b) shows that the capital–labour ratio need not be inversely related to the rate of profit as neoclassical long-period theory maintained. In more general terms, it cannot be presumed that input uses, per unit of output, are related to the corresponding 'factor prices' in the conventional way (see Metcalfe and Steedman, 1972, and Steedman, 1985). This result calls in question the validity of the traditional demand and supply approach to the determination of quantities, prices and income distribution.

The results stated above essentially carry over to the more general case with fixed capital, pure joint production and several primary inputs, such as land and labour of different qualities, provided the formalization of the problem is appropriately adapted to the specific case under consideration. Here it suffices to point out a few additional aspects of the choice of technique problem.

With fixed capital there is always such a problem to be solved. This concerns both the choice of the system of operation of plant and equipment, that is, for example, whether a single or a double-shift system is to be adopted; and the choice of the economic lifetimes of fixed capital goods. During the capital theory debates of the 1960s and early 1970s attention focussed on the latter aspect of the use of capital. It was shown that with decreasing or changing efficiency of the durable capital good cost minimization implies that for a given level of the rate of profit premature truncation is advantageous as soon as the price (book value) of the partly worn out item becomes negative. While the $w-r$ relationship for a given truncation may slope upwards over some range of r, the $w-r$ frontier consists only of those parts of the $w-r$ relationships that are downward-sloping. Moreover, it was demonstrated that the frontier can display the *return of the same truncation* (cf., for example, Hagemann and Kurz, 1976). As to the other aspect of capital utilization, a similar possibility can be shown to exist: the *return of the same system of operation of plant and equipment* (cf. Kurz, 1986). Both phenomena are of course variants of the reswitching of techniques.

In systems with pure joint production a choice of technique is inherent, even where the number of processes available does not exceed the number of products. Sraffa's approach to joint production is in terms of 'square' systems of production, that is, systems where the number of processes operated is equal to the number of commodities (i.e. positively-priced products). However, as Salvadori (1982) has shown, in such a framework a cost-minimizing system does not need to exist. A way out of this impasse may be seen in a formalization of joint production that is similar to von Neumann's. In such a formalization the free disposal assumption plays a crucial role. It can be shown that the $w-r$ frontier is downward-sloping, even though individual $w-r$ relationships may have positive ranges.

HEINZ D. KURZ

See also RESWITCHING OF TECHNIQUE; SRAFFIAN ECONOMICS; TWO-SECTOR MODEL.

BIBLIOGRAPHY

Bharadwaj, K. 1970. On the maximum number of switches between two production systems. *Schweizerische Zeitschrift für Volkswirtschaft und Statistik* 106, December, 409–29.

Bruno, M. 1969. Fundamental duality relations in the pure theory of capital and growth. *Review of Economic Studies* 36, January, 39–53.

Burmeister, E. and Kuga, K. 1970. The factor price frontier, duality and joint production. *Review of Economic Studies* 37, January, 11–19.

Craven, J. 1979. Efficiency curves in the theory of capital: a synthesis. In *The Measurement of Capital, Theory and Practice*, ed. K.D. Patterson and K. Schott, London: Macmillan.

Garegnani, P. 1970. Heterogeneous capital, the production function and the theory of distribution. *Review of Economic Studies* 37, July, 407–36.

Hagemann, H. and Kurz, H.D. 1976. The return of the same truncation period and reswitching of techniques in neo-Austrian and more general models. *Kyklos* 29, December, 678–708.

Hicks, J.R. 1965. *Capital and Growth*. Oxford: Oxford University Press.

Hicks, J.R. 1973. *Capital and Time, A Neo-Austrian Theory*. Oxford: Oxford University Press.

Kurz, H.D. 1986. 'Normal' positions and capital utilisation. *Political Economy* 2, May, 37–54.

Metcalfe, J.S. and Steedman, I. 1972. Reswitching and primary input use. *Economic Journal* 82, March, 140–57.

Pasinetti, L.L. 1977. *Lectures on the Theory of Production*. London: Macmillan.

Ricardo, D. 1971. *The Works and Correspondence of David Ricardo*, Vol. VIII. Edited by P. Sraffa in collaboration with M.H. Dobb, Cambridge: Cambridge University Press.

Salvadori, N. 1982. Existence of cost-minimizing systems within the Sraffa framework. *Zeitschrift für Nationalökonomie* 42, September, 281–98.

Salvadori, N. and Steedman, I. 1985. Cost functions and produced means of production: duality and capital theory. *Contributions to Political Economy* 4, March, 79–90.

Samuelson, P.A. 1957. Wages and interest: a modern dissection of Marxian economic models. *American Economic Review* 47, December, 884–912.

Samuelson, P.A. 1962. Parable and realism in capital theory: the surrogate production function. *Review of Economic Studies* 29, June, 193–206.

Sraffa, P. 1960. *Production of Commodities by Means of Commodities*. Cambridge: Cambridge University Press.

Steedman, I. 1985. On input 'demand curves'. *Cambridge Journal of Economics* 9, June, 165–72.

factor reversals. *See* HECKSCHER–OHLIN TRADE THEORY.

factory system. *See* INDUSTRIAL REVOLUTION; LABOUR PROCESS.

facts. *See* MODELS AND THEORY; STYLIZED FACTS.

fair division. The theory of fair division is concerned with the design of procedures for allocating a bundle of goods among n persons who are perceived to have equal rights to the goods. Both equity (according to criteria discussed below) and efficiency are sought. The theory is of interest primarily because its approach to allocation problems enjoys some important advantages over the alternative approach suggested by neoclassical welfare economics, and because studying the sense in which procedures actually in use are equitable is a good way to learn about popular notions of equity.

The modern theory of fair division has it origins in papers by Steinhaus (1948) and Dubins and Spanier (1961), who described methods (attributed by Steinhaus in part to S. Banach and K. Knaster) for sharing a perfectly divisible 'cake' among n people. In the method described by Steinhaus, the people are ordered (randomly, if desired) and the first person cuts a slice from the cake. Then each other person, in turn, may diminish the slice if he wishes. The last person to diminish the slice must take it as his share, with the slice reverting to the first person if no one chooses to diminish it. The process then continues, sharing the remainder of the cake in the same way among those people who have not yet received a share.

In the closely related method described by Dubins and Spanier, one person passes a knife continuously over the cake, at each instant determining a well-defined slice, which grows over time. The first other person to indicate his willingness to accept the slice then determined by the knife's location receives it as his share. The process then continues as before.

These n-person fair-division schemes are in the spirit of the classical two-person method of divide and choose, in which one person divides the cake into two portions and the other then chooses between them. Neither n-person scheme, however, is a true generalization of the two-person method. Steinhaus (1950) proposed a three-person scheme (formalized and generalized to n persons by Kuhn, 1967) that is a true generalization. In this scheme, one person divides the cake into n portions and the others announce which of the portions are acceptable to them. Then, if it is possible to give each of the others a share acceptable to him, this is done. Otherwise, it is possible to assign a share to the divider in such a way that it is still feasible to give each other person $1/n$th of the cake in his own estimation. This share is assigned, and the process then continues as before.

Each of these schemes is fair in the sense that, under reasonably general conditions (see Kuhn, 1967), it allows each person to ensure, independent of the others' behaviour, that he will obtain at least $1/n$th of the total value of the cake in his estimation. In the Steinhaus (1948) method, if a person is called upon to cut, he takes a slice with $1/n$th the value of the original cake; and a person given an opportunity to diminish a slice reduces it to $1/n$th value, if possible, or does nothing if it already has value $1/n$th or less. In the method described by Dubins and Spanier, each person indicates his willingness to accept any slice whose value reaches $1/n$th of the total value of the cake. Finally, in the method of Steinhaus (1950), the divider divides the cake into n portions, each acceptable to him, and the others declare acceptable all portions they deem to have at least $1/n$th of the value of the entire cake.

These results are of considerable interest, but are incomplete in several ways. First, they ignore the question of efficiency, which is central to the problem of designing allocation mechanisms.

Second, although it does not involve interpersonal comparisons, the notion of fairness employed is inherently cardinal, and therefore difficult to make operational. This obscures a major advantage of the fair-division approach over that of neoclassical welfare economics.

Finally, when operationally meaningful notions of fairness are employed in an environment with nontrivial efficiency issues, the fact that each person has a strategy that ensures him at least his share of the cake does not guarantee that allocations resulting from strategic behaviour are fair: a person might give up the social desideratum of fairness to get more of the goods he desires.

The modern theory of fair division answers these criticisms by studying the implications of rational behaviour and employing a different concept of equity. A fair procedure is defined as one that always yields a fair allocation, in the sense formalized by Foley (1967): an allocation is *fair* if and only if no person prefers any other person's share to his own.

Kolm (1972) and Crawford (1977) (see also Luce and Raiffa, 1957, and Crawford and Heller, 1979) use this notion to formalize the sense in which the two-person method of divide and choose is fair. They characterize the perfect-equilibrium strategies when the divider (D) knows the preferences of the chooser (C) and show that in equilibrium, D divides so that he is indifferent about C's choice and C then chooses as D would prefer. The resulting allocation is fair, in Foley's sense, but not generally efficient unless D and C have identical preferences. The allocation is, however, efficient in the set of fair allocations.

These results establish an operationally meaningful sense in which the two-person divide-and-choose method is fair, and show that it has some tendency toward efficiency. However, when preferences are common knowledge, the role of divider is an advantage, in the sense that the divider always weakly prefers his allocation to what he would receive if he were chooser. This follows from the facts that the game always yields a fair allocation and the divider can divide so that any desired fair allocation is the result. Further, n-person versions of the divide-and-choose method need not even yield fair allocations.

Crawford (1979) and Crawford (1980) study schemes that improve upon the classical divide-and-choose method while preserving its good points. In the two-person scheme studied in Crawford (1980), D offers C a choice between a proposal of D's choosing and equal division, instead of making him choose between a proposal and its complement. The resulting perfect-equilibrium outcomes are both fair and efficient, under reasonable assumptions; the role of divider is still an advantage, but less so than in the classical divide-and-choose method. These results extend, in part, to the n-person case.

In the n-person scheme studied in Crawford (1979), the role of divider in the scheme of Crawford (1980) is auctioned off. This completely eliminates the asymmetry of roles, and yields perfect-equilibrium allocations that are both efficient and egalitarian-equivalent, in the sense of Pazner and Schmeidler (1978): an allocation is *egalitarian-equivalent* if and only if it is indifferent, for all people, to equal division of some (not necessarily feasible) bundle of goods. However, although egalitarian-equivalence shares many of fairness's advantages as an equity notion, egalitarian-equivalent allocations need not be fair.

Despite their flaws, the schemes just described share several advantages over the traditional approach of choosing an allocation that maximizes a neoclassical social welfare function.

First, they deal with notions of equity that (like efficiency) do not involve interpersonal comparisons and have an objective meaning.

Second, their prescriptions are implementable in a stronger

sense than those of neoclassical welfare economics. The classical welfare theorems establish that a competitive equilibrium is efficient and that, under reasonable assumptions, any efficient allocation can be obtained as a competitive equilibrium for suitably chosen initial endowments. But finding the endowments that yield the allocation that maximizes social welfare is informationally virtually equivalent to computing the entire optimal allocation. By contrast, the fair-division approach often allows the specification of procedures that are independent of the details of the environment but still yield equitable and efficient allocations.

Finally, most of the procedures studied in the literature on fair division are self-administered, in the sense that they can be implemented without a referee. This is difficult to formalize, but clearly important in practice.

VINCENT P. CRAWFORD

See also ENVY; EQUALITY; FAIRNESS.

BIBLIOGRAPHY

Crawford, V. 1977. A game of fair division. *Review of Economic Studies* 44(2), June, 235–47.

Crawford, V. 1979. A procedure for generating Pareto-efficient egalitarian-equivalent allocations. *Econometrica* 47(1), January, 49–60.

Crawford, V. 1980. A self-administered solution of the bargaining problem. *Review of Economic Studies* 47(2), January, 385–92.

Crawford, V. and Heller, W. 1979. Fair division with indivisible commodities. *Journal of Economic Theory* 21(1), August, 10–27.

Dubins, L. and Spanier, E. 1961. How to cut a cake fairly. *American Mathematical Monthly* 68(1), January, 1–17.

Foley, D. 1967. Resource allocation and the public sector. *Yale Economic Essays* 7(1), Spring, 45–98.

Kolm, S. 1972. *Justice et équité.* Paris: Editions du Centre National de la Recherche Scientifique.

Kuhn, H. 1967. On games of fair division. Ch. 2 in *Essays in Honor of Oskar Morgenstern*, ed. M. Shubik, Princeton: Princeton University Press.

Luce, R. and Raiffa, H. 1957. *Games and Decisions: Introduction and Critical Survey.* New York: John Wiley.

Pazner, E. and Schmeidler, D. 1978. Egalitarian equivalent allocations: a new concept of economic equity. *Quarterly Journal of Economics* 92(4), November, 671–87.

Steinhaus, H. 1948. The problem of fair division. *Econometrica* 16(1), January, 101–4.

Steinhaus, H. 1950. *Mathematical Snapshots.* New York: Oxford University Press.

fairness. The issues of equity and efficiency are central aspects of most economic problems. In the political domain it often seems that concerns with equity – or at least distribution – often outweigh concerns with economic efficiency in discussion of policy alternatives. Despite this, most economic analysis has paid much more attention to issues of efficiency than to equity.

The notion of efficiency has been repeatedly refined in economics, and today the concept of a Pareto efficient allocation has a firm place in the economist's tool-kit. There is no similar agreement about the proper concept of 'equitable' or 'fair' allocations. This is not to say that proposals are lacking, and in this essay I will examine a few of the ideas concerning economic definitions of fairness and equity. Since I have provided a more detailed survey of contributions in this area elsewhere (Thomson and Varian, 1985), I will focus more on the conceptual underpinnings, rather than the technical results.

Suppose that you had a bundle of goods to divide in a 'fair' way among *n* economic agents. How would you do it? In the absence of any further information, the natural choice is equal division. But even if equal division is a fair way to divide the bundle *initially*, it may not remain fair. If agents have different tastes, they will generally desire to trade the goods among themselves. Even though the initial allocation is symmetric, the final allocation will not necessarily inherit this desirable property of the original division.

What would be an economic definition of 'symmetry'? One proposal, due to Duncan Foley (1967), goes as follows: an agent *i* is said to *envy* another agent *j* if *i* prefers *j*'s bundle to his own. An allocation in which no agent envies any other agent is known as an *envy-free* allocation. Equal division is, of course, envy-free, but there will typically be many other allocations that satisfy this symmetry property. Allocations that are both Pareto efficient and envy-free are particularly interesting since they are allocations that will not be disturbed by voluntary trade. An envy-free allocation is sometimes referred to as an 'equitable' allocation. An envy-free Pareto efficient allocation is often called a 'fair' allocation. The term 'envy-free' seems to me to be both more descriptive and less misleading.

But do Pareto efficient envy-free allocations necessarily exist? It is too much to ask for allocations that are both equitable and efficient? As it turns out, it is possible to show that a competitive equilibrium from equal division is necessarily an envy-free and efficient allocation. It is efficient by the First Theorem of Welfare Economics, and the envy-free property follows from the fact that equal division guarantees that all agents will have the same wealth.

Other sorts of allocative mechanisms may not necessarily preserve the symmetry of equal division. For example, it is easy to exhibit allocations in the core of an equal division market game in which some agent envies another. The particular feature of trade on a competitive market that is important is the fact that all agents have the same trading opportunities, and hence cannot in equilibrium prefer some other agent's choices to their own. This insight has been examined in detail by Schmeidler and Vind (1972) using the notion of 'fair net trades'.

The concept of envy-free allocations has been generalized in many different ways. For example, there is the idea of a 'coalitionally envy-free allocation', which requires that there is no *group* of agents that unanimously prefers some other group's bundle to their own. A closely related idea is that of an *egalitarian equivalent* allocation, which is one in which every agent is indifferent between the bundle he holds in that allocation and a bundle in some (hypothetical) equal division allocation.

There will typically exist envy-free Pareto efficient allocations that are not competitive equilibria with equal wealths, but an equal-wealth allocation turns out to be especially interesting in a number of ways. For example, only in an equal-wealth allocation does each agent have the same budget set, and thus have *equal trading opportunities*. Furthermore, it can be shown that when preferences vary continuously across the population, the *only* Pareto efficient envy-free allocations are those with equal wealth.

The concept of envy-free allocations seems to work quite well as a formalization of the concept of symmetry when agents are themselves more or less symmetrically situated. However, when the agents are not themselves symmetric, the envy-free concept becomes somewhat forced. Consider, for example, the case of agents with severe handicaps. Do they not deserve some kind of special compensation for these handicaps in a 'fair' allocation? Shouldn't a diabetic's demand for insulin take precedence over a gourmet's demand for truffles?

These questions arise naturally when we consider models of production. For in this case agents with different abilities are like agents with different degrees of being handicapped. As Ronald Dworkin (1981) puts it: 'someone who cannot play basketball like Wilt Chamberlain ... suffers from an (especially common) handicap.' How does the concept of an envy-free allocation generalize to production economies? First we should consider what we mean by stating that one agent envies another agent in a production context. Since one agent cannot directly consume another agent's leisure, the extension of the concept to production is not immediate. More formally, if one agent's consumption set is not identical with another's, the concept of envy-free allocation is not necessarily well defined.

The natural thing to do here is to consider what would happen if agents swapped not only consumption bundles but also labour commitments – in order to envy another agent, you not only have to desire his consumption, but you also have to be willing to work as much as he does.

But this definition has a serious problem which was first discovered by Pazner and Schmeidler (1974): it may be that there are no Pareto efficient envy-free allocations by this definition. The problem is that just because one agent is willing to work as much as another doesn't mean that he will be able to produce as much output as the other. When abilities are different, the concept of 'envy' needs some refinement.

One suggestion, made by Varian (1974), is to have agents compare their consumption-output bundles, not their consumption-input bundles. Thus in order to 'envy' another agent, I must be willing to produce as much as he produces, or, more generally, I have to produce output with the same value that he produces. This sort of envy comparison, happily, is consistent with Pareto efficiency. Another suggestion, due to Pazner and Schmeidler (1978), is that we consider allocations in which each agent has a consumption-leisure bundle that has equal value at the efficiency prices. Again, it can be shown that such allocations will always exist. In some sense these two proposals are at opposite extremes: Varian's suggestion favours the able, while Pazner and Schmeidler's favours the unable. Is there a natural intermediate concept that is in some sense more balanced? The answer is not known.

An area that is closely related to that of envy-free allocations is that of *games of fair division*. Everyone is familiar with the classic scheme of 'I divide and you choose' as a solution to two person division games. But what do you do if you want to divide a good (or a bundle of goods) among more than two agents? There have been several schemes proposed; Kuhn (1967) provides a nice survey of the early literature. Since this survey, there have been some further study of games of fair division and an increasing interest in the implementability of some of the equity concepts described above.

HAL R. VARIAN

See also EQUITY; FAIR DIVISION; JUSTICE; SOCIAL JUSTICE; WELFARE ECONOMICS.

BIBLIOGRAPHY

Dworkin, R. 1981. What is equality? Part 2: Equality of resources. *Philosophy and Public Affairs* 10, 283–345.
Foley, D. 1967. Resource allocation and the public sector. *Yale Economic Essays* 7, 45–98.
Kuhn, H. 1967. On games of fair division. In *Essays in Honor of Oskar Morgenstern*, ed. M. Shubik, Princeton: Princeton University Press.
Pazner, E. and Schmeidler, D. 1974. A difficulty in the concept of fairness. *Review of Economic Studies* 41, 441–3.
Pazner, E. and Schmeidler, D. 1978. Decentralization and income distribution in socialist economies. *Economic Inquiry* 16(2), April, 257–64.
Schmeidler, D. and Vind, K. 1972. Fair net trades. *Econometrica* 40, 637–47.
Thomson, W. and Varian, H. 1985. Theories of justice based on symmetry. In *Social Goals and Social Organization: Essays in Honor of Elisha Pazner*, ed. Leo Hurwicz et al., New York: Oxford University Press.
Varian, H. 1974. Equity, envy, and efficiency. *Journal of Economic Theory* 9, 63–91.

fairs and markets. *See* MARKETPLACES; PRODUCERS' MARKETS.

falling rate of profit. Adam Smith, David Ricardo, Karl Marx, John Stuart Mill and John Maynard Keynes all expected the rate of profit to decline in the longest of long runs. It goes without saying that their reasons differ, with the result that we have several theories which point to this possibility.

ADAM SMITH. Smith is generally regarded as an optimist who saw more potential for progress in real wages and labour productivity than several of his successors. He expected productivity to be constant in agriculture, while in his *Lectures* he claimed that the division of labour in industry would permit twenty million workers to produce one hundred times the output of two million (p. 392). Capital accumulation would inevitably lead to population growth which should enable these potential benefits from the division of labour to be realized, and the extra population would allow the division of labour to be further extended and permit yet higher productivity. If productivity is constant in agriculture and rising in industry, its average in industry and agriculture together, Q_y, will have a persistent tendency to rise (as Hollander (1973) shows).

At first sight most of the benefits from this rising productivity trend should go to profits and rents. The *level* of wages will be higher in a fast than in a slow-growint economy, but Smith does not expect the *high* wages of a fast growing economy to rise each year. There will be one particular *level* of wages, W, in a stationary state, a higher *level*, $W+$, in a slow growing economy, and a still higher *level*, $W++$, where capital and population are growing rapidly. As capital and employment can grow rapidly without any need for wages to rise above $W++$, all the gains in Q_y can be added to profits and rent. This ought to produce a *rising* rate of profit, for if Q_y is rising while the wage is stuck at $W++$, then the surplus for profits and rent per worker, $(Q_y - W++)$, and the share of profits and rent in output, $(Q_y - W++)/Q_y$ will all the time increase. Unless rents take an ever growing share of 'profits and rents', this continual rise in the proportion of output which can go to profits and rent should allow the share of profits and therefore the rate of profit to keep on rising. So it is at first sight puzzling that Smith should insist in *The Wealth of Nations* (1776) that:

> In a country which had acquired that full complement of riches which the nature of its soil and climate, and its situation with respect to other countries allowed it to acquire; which could, therefore, advance no further, and which was not going backwards, both the wages of labour and the profits of stock would probably be very low (p. 111).

The cause of the declining rate of profit which takes Smith's economy gradually to a stationary state where wages and profits are both low is most easily understood if (to follow

Eltis, 1984) attention is focused on agriculture, and the corn harvest in particular.

Smith suggests that each corn harvest is produced with unchanging labour productivity, 'In every different stage of improvement ... the raising of equal quantities of corn in the same soil and climate, will, at an average, require nearly equal quantities of labour' (p. 206), while the wage is also just sufficient to buy a given quantity of corn for,

> the money price of labour ... must always be such as to enable the labourer to purchase a quantity of corn sufficient to maintain his family either in the liberal, moderate, or scanty manner in which the advancing, stationary or declining circumstances of the society oblige the employer to maintain him (p. 509).

In a rapidly progressing economy where the wage is $W + +$, this will be a sufficient sum of money to purchase a fixed quantity of corn of $W_a + +$. If the constant output of corn per worker is Q_a, while the wage represents $W_a + +$ of corn, the surplus that is available for profits and rent will be the constant $(Q_a - W_a + +)$ of corn per worker. Therefore, if we measure output per worker and the wage in corn, there is no tendency for profits plus rent per workers to rise. If this constant share of surplus is divided equally between profits and rents, then Smith would predict an approximately constant *share* of profits in agriculture.

Smith envisages that an economy will become increasingly capital intensive.

> As the division of labour advances ... in order to give constant employment to an equal number of workmen, an equal stock of provision, and a greater stock of materials and tools than what would have been necessary in a ruder state of things must be accumulated beforehand (p. 277).

In the case of agriculture this increase in capital intensity takes the form of a growing use of oxen ('labouring cattle') and increasing sums will be spent on fertilization and improvements to the soil. So there will be a continual tendency for the agricultural capital–output ratio to rise. With a constant share of profits (P/Y), and a rising capital–output ratio (K/Y), the rate of profit $((P/K)$ which is $(P/Y) \div (K/Y))$ will tend to fall. Entrepreneurs can choose whether to deploy their capital in agriculture, industry, or commerce, so the rate of profit cannot fall in agriculture without similar falls elsewhere. Hence as the agricultural rate of profit falls, the capital withdrawn from agriculture will be transferred, and the increase in competition that this causes will also force industrial and commercial profits down for,

> When the stocks of many rich merchants are turned into the same trade, their mutual competition naturally tends to lower its profit; and when there is a like increase of stock in all the different trades carried on in the same society, the same competition must produce the same effect in them all (p. 105).

The general fall in the rate of profit will gradually reduce capital accumulation, and as this diminishes, wages will fall from $W_a + +$ to $W_a +$ and subsequently to W_a. At this lower wage, profits will recover a little, but the same cause, a rising K/Y in agriculture while P/Y is constant, will cause a resumption of the falling trend which will continue until the stationary state where wages and profits are both 'very low' is reached.

DAVID RICARDO. During the Napoleonic Wars, high food prices caused British farmers to cultivate inferior land, and this led Ricardo and his great contemporary, Malthus, to attribute a major role to agricultural diminishing returns. The simplest representation of Ricardo's theory of income distribution which follows from this is also a 'corn-model' (as Sraffa (1951) and Eatwell (1975) suggest; Hollander (1979) dissents). Ricardo himself published a table in his initial statement of his new theory, *An Essay on the Influence of a Low Price of Corn on the Profits of Stock* (1815), where the wage, output and capital per agricultural worker are all expressed as quantities of corn. Because landlords receive no rent from marginal land, its entire corn output, Q_a, goes either to wages or profits. If the equilibrium or natural wage is fixed as a specific quantity of corn, W_a, then the equilibrium profits earned from the employment of a marginal agricultural worker will be $(Q_a - W_a)$. If the capital required to employ him can be expressed as a quantity of corn, K_a, then the rate of profit at the margin will be $(Q_a - W_a)/K_a$. In the *Essay* table, corn output per worker, Q_a, falls as a growing demand for food forces the margin of cultivation onto inferior land; capital per worker, K_a, rises because extra transport costs are involved in farming inferior land (which is further from the market), while W_a, the natural wage expressed as a quantity of corn, is constant and independent of the extent to which inferior agricultural land has to be used. The continual tendency for marginal agricultural productivity to diminish, while the capital cost of employing an agricultural worker increases, persistently reduces the agricultural rate of profit, $(Q_a - W_a)/K_a$. As with Smith, if the rate of profit falls in agriculture, then competition must reduce it equally in industry and commerce.

Ricardo moved on from the 'corn-model' of the *Essay* table to a more general theory in *Principles of Political Economy and Taxation* (1817). There (as Hicks (1972) suggests) the natural wage is expressed as specific quantities of food-and-manufactures. The food items in this 'basket' of consumer goods become more expensive as agriculture is driven onto inferior land where more workers are needed to produce the food workers require, while the manufactured items included in the natural wage become cheaper as technical progress, the division of labour and a growing use of machinery reduce labour requirements. Ricardo believed that the tendency for food to require more labour will have a stronger influence on the real cost of the basket of goods that constitute the natural wage than the tendency for manufactures to require less. In consequence the aggregate labour required to produce wage goods rises all the time, so a marginal worker will spend a higher fraction of his week producing the wage goods that his constant wage requires. Then the fraction of his output that is surplus to wages and available for profits (*marginal* output never goes to rent) will have a continual tendency to fall, so there will be a declining trend in P/Y and in the rate of profit:

> The natural tendency of profits then is to fall; for, in the progress of society and wealth, the additional quantity of food required is obtained by the sacrifice of more and more labour. This tendency, this gravitation as it were of profits, is happily checked at repeated intervals by the improvements in machinery, connected with the production of necessaries, as well as by discoveries in the science of agriculture which enable us to relinquish a portion of labour before required, and therefore to lower the price of the prime necessary of the labourer (p. 120).

In this statement of Ricardo's argument in the *Principles*, the rate of profit is influenced by developments in both agriculture and industry (as Hollander (1979) emphasizes), for anything which causes workers to spend a higher fraction of time producing wage goods must increase the proportion of marginal production that goes to wages, while any increase in productivity in the manufacture of industrial *necessities* will reduce the proportion of workers' time required to produce wage goods, and so increase the fraction which can go to profits. If the tendency for real agricultural productivity to fall has more influence than the tendency for industrial productivity to rise, then P/Y, the fraction of marginal production which is surplus to wages will have a continual tendency to fall. In the *Principles* Ricardo does not repeat the proposition (from the *Essay*) that capital per worker rises as the margin of cultivation moves onto inferior land, so the tendency for the rate of profit to fall is dominated by the influence of declining agricultural productivity upon P/Y, while K/Y plays a neutral rôle.

In the *Principles* as in Smith, wages fall (from $W + +$ to $W +$ and then to W) as capital accumulation and population growth diminish, and (as Hicks and Hollander (1977) show) this reduces the rate at which profits decline, without affecting the proposition that they must fall eventually to the minimum stationary state level.

In 1820 five years after the conclusion of the Napoleonic Wars, Ricardo wrote an essay for the *Encyclopedia Brittanica* on 'Funding Systems' (which Dobb (1973) considers significant) in which he modified the proposition that declining agricultural productivity *in an individual country* will inevitably cause a continual decline in its rate of profit. A country such as Britain could avoid the influence of agricultural diminishing returns by importing its marginal food and paying with exports of manufactures:

> a country could go on for an indefinite time increasing in wealth and population, for the only obstacle to this increase would be the scarcity, and consequently high value, of food and other raw produce. Let these be supplied from abroad in exchange for manufactured goods, and it is difficult to say where the limit is at which you would cease to accumulate wealth and to derive profit from its employment (p. 179).

Ricardo did not go on to say, though it is implicit in this statement, that global diminishing returns would force profits down in the end. If marginal productivity fell at a world level, food and necessary minerals would only be obtainable at a rising real marginal cost, and wages would absorb a growing fraction of marginal production and leave a diminishing fraction over for profits, so that P/Y would persistently fall. A country importing food in such circumstances would face deteriorating terms of trade, and wages would have to rise in its manufacturing industries to pay the ever rising cost of imported food with the result that wages would absorb an increasing fraction of the revenues that manufacturers obtained, and so force P/Y downwards in precisely the manner set out in Ricardo's *Principles*.

JOHN STUART MILL. Mill went on to develop the economic analysis of Smith and Ricardo (as Hollander (1985) shows). He agreed that the rate of profit will be strongly influenced by population growth, capital accumulation and techniques of production which he refers to as 'the arts of production', and that these will generally advance together. But 'Agricultural skills are of slow growth', and inventions occur only occasionally, so that, as with Ricardo, agricultural improvements are no more than an intermittent counteracting tendency which temporarily relieves the adverse pressure of growing population on agricultural productivity. 'The economical progress of a society constituted of landlords, capitalists, and labourers, tends to the progressive enrichment of the landlord class; while the cost of the labourer's

subsistence tends on the whole to increase, and profits to fall' (1848, pp. 731–2).

The fall in the rate of profit will continue until an eventual stationary state is reached. The minimum to which the rate of profit will then fall will be made up of two elements. There must first be a sufficient reward for the postponement of consumption to ensure the maintenance of the capital stock. This will determine the riskless rate of interest that lenders will receive from financially sound governments. The rate of profit will exceed this minimal interest rate for there will inevitably be risks of default in commercial undertakings and entrepreneurs must earn more than the rates at which they borrow if they are to be persuaded to organize production in circumstances where each faces risk. Mill believed that the minimum rate of profit set by these considerations will have a tendency to fall because a growing security of property rights would continually improve incentives to accumulate and at the same time reduce the risks involved:

> a change which has always hitherto characterized, and will assuredly continue to characterize the progress of every civilized society, is a continual increase of the security of person and property. The people of every country in Europe, the most backward as well as the most advanced, are, in each generation, better protected against the violence and rapacity of one another, both by a more efficient judicature and police for the suppression of private crime, and by the decay and destruction of those mischievous privileges which enabled certain classes of the community to prey with impunity upon the rest. They are also, in every generation, better protected, either by institutions or by manners and opinion, against arbitrary exercise of the power of government (p. 707).

For these and similar reasons, 'The risks attending the investment of savings in productive employment require, therefore, a smaller rate of profit to compensate for them than was required a century ago' (p. 737). As civilization advances, mankind becomes less the slave of the moment, and more habituated to carry their desires forward into a distant future which is 'a natural result of the increased assurance with which futurity can be looked forward to' (p. 738). All this will 'diminish the amount of profit which people absolutely require as an inducement to save and accumulate'.

Hence the minimum rate of profit required to sustain a stationary state should fall all the time. Because there has been

> a diminution of risk and increase of providence, a profit or interest of three or four per cent is as sufficient a motive to the increase of capital in England at the present day, as thirty or forty per cent in the Burmese Empire, or in England at the time of King John.

Mill actually envisaged a time when this minimal interest rate might fall as low as one per cent.

He believed that opulent societies like 19th-century England were continually close to this minimum. If all British saving was suddenly invested at home, 'Few persons would hesitate to say, that there would be great difficulty in finding remunerative employment every year for so much new capital', and 'if the present annual amount of savings were to continue, without any of the counteracting circumstances which now keep in check the natural influence of those savings in reducing profit, the rate of profit would speedily attain the minimum, and all further accumulation of capital would for the present cease' (p. 741).

Counteracting tendencies which prevent the rate of profit from actually attaining the minimum are the diversion of a good deal of saving overseas where a higher rate of profit can be earned, and technical progress in the manufacture of wage goods which adds new opportunities for profitable investment, but Mill believed that the adverse influence on the rate of profit of the pressure to accumulate would exercise the dominant influence. Diminishing returns would even set in in North America, for as its population rose 'unless great improvements take place in agriculture' there would need to be increases in capital per worker which would gradually produce the same effects on profitability as in Europe (p. 745).

Like Ricardo, Mill did not envisage that technical progress in the production of workers' necessities would be sufficient to overcome the influence of population growth and agricultural diminishing returns, so profits would continually fall towards the level set by the returns which savers and entrepreneurs must receive, which would itself diminish.

KARL MARX. Marx did not follow Ricardo and Mill in attributing particular significance to agricultural diminishing returns. In *Capital* the production of food is not singled out in relation to the other goods that workers buy, and there is no tendency for the real cost of workers' consumer goods to rise. Marx actually argued the contrary, for he attributed great significance to the favourable effects of industrial mechanization and the division of labour. Because of these, there is a falling trend in the real cost of the goods workers buy in order to achieve the equilibrium wage, 'the value of labour in exchange'. Analogously with Smith and Ricardo, this has to provide a standard of living sufficient to sustain the population – or as Marx puts it, 'to ensure the reproduction of the working class'.

Measured in hours of labour time, his preferred unit of value, each worker labours for $(V + S)$ hours a day, of which V suffice for the production of the wage goods required for the equilibrium wage, while the product of the remaining S hours is surplus to workers' subsistence requirements and belong to the capitalist employers. Marx describes the ratio of S, the total hours workers labour for others, to V, the hours they labour for their own subsistence needs, as 'the rate of exploitation'. Because of continuing productivity growth as a result of increasing mechanization and extensions of the division of labour, workers' subsistence needs can be met in fewer hours, so V has a persistent tendency to fall. As Marx sees no tendency for total hours of work to fall, S can rise as V falls with the result that there is a persistent tendency for S/V, the rate of exploitation, to rise. As S/V rises, so will the ratio of profits to wages and therefore the share of profits in output. Given this prediction of a rising P/Y, Marx can only arrive at the conclusion that P/K, the rate of profit, has a persistent tendency to fall, if K/Y, the capital–output ratio, rises still more persistently than P/Y.

Marx believed that there are strong historical tendencies for capital per worker and the capital–output ratio to rise. The total capital tied up in the employment of a worker consists of means of production, namely physical capital equipment and raw materials, of C, and advance payments of wages of V per worker, in return for which the employer obtains the worker's 'labour power'. Total capital per worker is $(C + V)$, and Marx refers to C, raw materials and machinery, as constant capital, and V, the advance purchase of 'labour power', as variable capital. He believed that there is a persistent tendency for C/V which he refers to as 'the organic composition of capital' to rise. As the division of labour advances, a 'greater mass of raw material and auxiliary substances enter into the labour process', while increases in the mass of machinery, furnaces, means of transport and the means of production concentrated

in buildings, are 'a condition of the increasing productivity of labour'. A 'growing extent of the means of production, as compared with the labour-power incorporated with them, is an expression of the growing productiveness of labour', and the 'law of the progressive increase in constant capital, in proportion to the variable, is confirmed at every step ... whether we compare different economic epochs or different nations in the same epoch' (Vol. 1, pp. 583–4). Now the rate of profit is the ratio of total profit, that is, surplus-value, S, to total capital, $(C + V)$, and $S/(C + V)$ can be written as $(S/V)/(C/V + 1)$. The continual tendency for C/V, the organic composition of capital, to rise will all the time reduce the rate of profit, but the tendency for S/V, the rate of exploitation, to rise, will continually raise the rate of profit. Marx believed he had demonstrated a continual tendency for the rate of profit to decline, but (as Meek shows) there is no presumption that $(S/V)/(CV + 1)$ will decline if there are upward tendencies in both the organic composition of capital (C/V), and the rate of exploitation (S/V).

But Marx's conclusion of a declining rate of profit can be established if these trends are pushed to their ultimate limits. The upper limit to total surplus value per worker, S, cannot exceed one working day, while the upper limit to C, constant capital per worker (or the 'dead labour' with which workers are equipped) can become indefinitely high if the tendency for C/V to rise is continual. Thus, if the historical tendency is for S to rise to the maximum hours in a working day, S_{max}, and for C to rise without limit, then the rate of profit, $S_{max}/(C + V)$, will become indefinitely small.

Several modern commentators (e.g., Fine and Harris, 1976, and Shaikh, 1978) have underlined this interpretation, by adding that the upward boundary to $S/(C + V)$, which is set by the profit rate where the wage (V) is zero is S_{max}/C, which will fall continually as C rises. If the *upper limit* to the rate of profit has a continual tendency to fall, then it is a reasonable presumption that there will be a declining trend in the actual rate of profit, despite fluctuations associated with vicissitudes in wage bargaining.

Marx himself emphasized that his 'law of the tendency of the rate of profit to fall' is no more than a *tendency* which can and will be counteracted by a variety of developments over considerable periods. Wage costs may fall for a time and permit the rate of profit to rise if imported workers' consumer goods can be produced more cheaply overseas. New industries may begin to produce with low capital intensity (and therefore a low C/V): in Marx's words they begin by employing mainly living labour. But as these industries develop, capital intensity will rise and the ratio of dead to living labour increase, so that C/V rises in the same way as in older industries. Another possibility is that capital equipment may fall in price relative to consumer goods, and in this case industry will become more capital intensive in technical terms without any necessary tendency for the organic composition of capital to rise. The 'technical composition of capital' $(C/V$ measured in technical units) would still be rising, but not its 'organic composition' which is C/V in Marx's labour units.

But the tendency for growing capital intensity to reduce the rate of profit would dominate any secular trend, for these helpful developments could only operate for a time.

JOHN MAYNARD KEYNES. There is an echo of Mill's theory in *The General Theory of Employment, Interest and Money* (1936). Keynes believed that if a country could reduce its rate of interest to a level compatible with full employment, and then invest its full employment saving, it would be 'comparatively easy to make capital-goods so abundant that the marginal efficiency of capital is zero'. He believed that 'a properly run community equipped with modern technical resources, of which the population is not increasing rapidly, ought to be able to bring down the marginal efficiency of capital in equilibrium approximately to zero within a single generation' (pp. 220–21).

Thus Keynes, like Mill, believed that a modern economy's potential to accumulate greatly transcended the rate at which new investment opportunities would arise, with the result that the rate of profit would rapidly fall towards the stationary state level if its full potential for accumulation could ever be realized.

CONCLUSION. The theories of these great economists have rested on three general predictions about the future development of capitalist economies which have not been borne out empirically.

Smith and Marx both believed that capital would have a persistent tendency to grow faster than output, which could be expected to produce a declining tendency in the rate of profit, and the trend in the capital–output ratio was indeed upward prior to 1776 and 1867. But the British capital–output ratio has been approximately stable since 1867 (Matthews, Feinstein and Odling-Smee, 1982), while the United States capital–output ratio has been falling (Klein and Kosobud, 1961). Few now speak of a long term tendency for the capital–output ratio to rise, so this line of argument finds little echo in 20th-century economics.

Ricardo and Mill were much influenced by a belief that the adverse influence of agricultural diminishing returns would inevitably outweigh any favourable effects from technical progress, with the result that the real cost of workers' necessities would rise continuously and squeeze the rate of profit. But since they wrote, there has been no tendency for the real cost of food and raw materials to rise faster than manufactures. The terms of trade have fluctuated a good deal, but technical progress has raised productivity enormously in both industry and agriculture, and there has been no tendency for a rising relative cost of food to squeeze profits in the manner that Ricardo and Mill expected. Some futurologists predict a gradual depletion of the world's natural resources with inevitable Ricardian (and Malthusian) consequences, but the 20th century itself has provided no empirical support for their pessimism.

Mill and Keynes were impressed by the proposition that continuing capital accumulation would exhaust opportunities for profit faster than new investment opportunities can be created. But since World War II technical progress has accelerated, and there have been decades when new investment opportunities providing enormous scope for profitable investment have emerged. It is rarely argued now that there is any *necessary* tendency for the new investment opportunities created by technical advance to fall short of actual investment so that the marginal efficiency of investment must tend to fall.

So there is little late 20th-century support for the theories which have been outlined. There is however a further hypothesis which is germane to the general direction of Marx's political and social thought. If there is a continual increase in the power of workers in wage bargaining in comparison with the power of capitalists to resist their influence, then the share and rate of profit will have a tendency to fall. This will be reinforced if workers' political representatives exercise a growing legislative influence over wage bargaining and price formation. If workers become immune from dismissal or redundancy without compensation, while the prices companies set are increasingly subject to public scrutiny, then there will

be an accompanying tendency for the rate and share of profits to decline. In the 1970s in several countries the political power of the working class appeared to rise with accompanying shifts in income distribution, but political developments have been in the other direction in the early 1980s, so as with previous hypotheses, there is no particular reason to anticipate any clear future trend.

WALTER ELTIS

See also CLASSICAL ECONOMICS; MARXIST ECONOMICS: STATIONARY STATE.

BIBLIOGRAPHY
Primary
Keynes, J.M. 1936. *The General Theory of Employment, Interest and Money.* Republished as Vol. VII of *The Collected Writings of John Maynard Keynes,* London: Macmillan, 1973.
Marx, K. 1867–94. *Capital.* 3 vols. Reprinted, Moscow: Progress Publishers for Lawrence and Wishart, 1974.
Mill, J.S. 1848. *Principles of Political Economy with Some of Their Applications to Social Philosophy.* 2 vols. Reprinted in *Collected Works of John Stuart Mill,* Vols II and III, ed. J.M. Robson, Toronto: University of Toronto Press, 1965.
Ricardo, D. 1815. *An Essay on the Influence of a Low Price of Corn on the Profits of Stock.* Reprinted in Vol. IV of *The Works and Correspondence of David Ricardo,* 11 vols, ed. P. Sraffa, Cambridge: Cambridge University Press, 1951–73.
Ricardo, D. 1817. *On the Principles of Political Economy and Taxation.* Reprinted as Vol. I of *The Works and Correspondence of David Ricardo,* ed. P. Sraffa, Cambridge: Cambridge University Press, 1951–73.
Ricardo, D. 1820. Funding systems. In *Encyclopedia Brittanica,* Supplement to 4th edn. Reprinted in Vol. IV of *The Works and Correspondence of David Ricardo,* ed. P. Sraffa, Cambridge: Cambridge University Press, 1951.
Smith, A. 1763. *Lectures on Jurisprudence.* Ed. R.L. Meek, D.D. Raphael and P.G. Stein, Oxford: Oxford University Press, 1978.
Smith, A. 1776. *An Inquiry into the Nature and Causes of the Wealth of Nations.* 2 vols. Reprinted, ed. R.H. Campbell, A.S. Skinner and W.B. Todd, Oxford: nOxford University Press, 1976.
Secondary
Dobb, M. 1973. *Theories of Value and Distribution since Adam Smith.* Cambridge: Cambridge University Press.
Eatwell, J. 1975. The interpretation of Ricardo's *Essay on Profits. Economica* 42(166), May, 182–7.
Eltis, W. 1984. *The Classical Theory of Economic Growth.* London: Macmillan.
Fine, B. and Harris, L. 1976. Controversial issues in Marxist economic theory. *Socialist Register.*
Hicks, J. 1972. Ricardo's theory of distribution. Reprinted in *Classics and Moderns,* Oxford: Blackwell, 1983.
Hicks, J. and Hollander, S. 1977. Mr Ricardo and the Moderns. *Quarterly Journal of Economics* 91, August, 351–69.
Hollander, S. 1973. *The Economics of Adam Smith.* Toronto: University of Toronto Press.
Hollander, S. 1979. *The Economics of David Ricardo.* Toronto: University of Toronto Press.
Hollander, S. 1985. *The Economics of John Stuart Mill.* 2 vols, Oxford: Blackwell.
Klein, L.R. and Kosobud, R.F. 1961. Some econometrics of growth: great ratios of economics. *Quarterly Journal of Economics* 75(2), May, 173–98.
Matthews, R.C.O., Feinstein, C.H. and Odling-Smee, J.C. 1982. *British Economic Growth, 1856–73.* Oxford: Oxford University Press.
Meek, R.L. 1960. The falling rate of profit. *Science and Society* 24, Winter, 36–52.
Shaikh, A. 1978. Political economy and capitalism: notes on Dobb's theory of crisis. *Cambridge Journal of Economics* 2, June, 233–51.
Sraffa, P. 1951. Introduction to D. Ricardo, *On the Principles of Political Economy and Taxation.* Cambridge: Cambridge University Press.

false trading. *See* TÂTONNEMENT AND RECONTRACTING.

family. In virtually every known society – including ancient, primitive, developing, and developed societies – families have been a major force in the production and distribution of goods and services. They have been especially important in the production, care, and development of children, in the production of food, in protecting against illness and other hazards, and in guaranteeing the reputation of members. Moreover, parents have frequently displayed a degree of self-sacrifice for children and each other that is testimony to the heroic nature of men and women.

Of course, families have radically changed over time. The detailed kinship relations in primitive societies traced by anthropologists contrast with the predominance of nuclear families in modern societies, where cousins often hardly know each other, let alone interact in production and distribution. The obligations in many societies to care for and maintain elderly parents is largely absent in modern societies, where the elderly either live alone or in nursing homes.

Nevertheless, families are still much less prominent in economic analysis than in reality. Although the major economists have claimed that families are a foundation of economic life, neither Marshall's *Principles of Economics*, Mill's *Principles of Political Economy*, Smith's *Wealth of Nations* nor any of the other great works in economics have made more than casual remarks about the operation of families.

One significant exception is Malthus's model of population growth. Malthus was concerned with the relation between fertility, family earnings, and age at marriage, and he argued that couples usually do (or should) marry later when economic circumstances are less favourable. However, this important insight (see Wrigley and Schofield, 1981, for evidence that prior to the 19th century, marriage rates in England did increase when earnings rose) had no cumulative effect on the treatment of the family by economists.

During the last 40 years, economists have finally begun to analyse family behaviour in a systematic way. No aspect of family life now escapes interpretation with the calculus of rational choice. This includes such esoteric subjects as why some contraceptive techniques are preferred to others, and why polygamy declined, as well as more 'traditional' subjects such as what determines age at marriage, number of children, the amount invested in the human capital of children, and the amount spent by children on the care of elderly parents. This essay sets out the 'economic approach' to various aspects of family behaviour. Detailed discussions of particular aspects can be found in the bibliography.

I FERTILITY. Let us start with the Malthusian problem: how is the number of children, or fertility, of a typical family determined? Crucial to any discussion is the recognition, taken for granted by Malthus, that men and women strongly prefer their own children to children produced by others. This preference to produce one's children eventually helped stimulate economists to recognize that families, and households more generally, are important producers as well as consumers.

The desire for own children means that the number of children in a family is affected by supply conditions. Supply is determined by knowledge of birth control techniques, and by the capacity to produce children, as related to age, nutrition, health, and other variables.

The demand side emerges through maximization of the utility of a family that depends on the quantity of children (n) and other commodities (z), as in

$$U = U(n, z). \tag{1}$$

Utility is maximized subject not only to household production functions for children and other commodities, but also to constraints on family resources. Money income is limited by wage rates and the time spent working, and the time available for household production is limited by the total time available. These constraints are shown by the following equations where λ is the marginal utility of family income. The total net cost of rearing a child (Π_n) equals the value of the goods and services that he consumes, plus the value of the time spent on him by family members ($\Sigma w_i(t_{n_i})$), minus his earnings that contribute to family resources.

$$\left. \begin{array}{l} p_n n + p_z z = \sum w_i t_{w_i} + v \\ t_{n_i} + t_{z_i} + t_{w_i} = t \end{array} \right\} \text{ all } i \in f, \tag{2}$$

where t_{w_i} is the hours worked by the ith family member, w_i is his or her hourly wage, v is non-wage family income, t_{n_i} and t_{z_i} are the time allocated to children and other commodities by the ith member, and t is the total time available per year or other time unit.

By substituting the time constraints into the income constraint, one derives the family's full income (S):

$$\left(p_n + \sum w_i t_{n_i} \right) n + \left(p_z + \sum w_i t_{z_i} \right) Z = \sum w_i t + v = S,$$

$$\Pi_n n + \Pi_z Z = S. \tag{3}$$

If utility is maximized subject to full income, the usual first order conditions follow:

$$\frac{\partial U}{\partial n} = \lambda \Pi_n, \tag{4}$$

and

$$\frac{\partial U}{\partial Z} = \lambda \Pi_z. \tag{5}$$

The basic theorem of demand states that an increase in the relative price of a good reduces the demand for that good when real income is held constant. If the qualification about income is ignored, then, in particular, an increase in the relative price of children would reduce the children desired by a family. The net cost of children is reduced when opportunities for child labour are readily available, as in traditional agriculture. This implies that children are more valuable in traditional agriculture than in either cities or modern agriculture, and explains why fertility has been higher in traditional agriculture (see the evidence in Jaffe, 1940; Gardner, 1973).

Production and rearing of children have usually involved a sizeable commitment of the time of mothers, and sometimes also that of close female relatives, because children tend to be more time intensive than other commodities, especially in mother's time (i.e. in equation (3), $p_n/\Pi_n < p_z/\Pi_z$). Consequently, a rise in the value of mother's time would reduce the demand for children by raising the relative cost of children. In many empirical studies for primitive, developing, and developed societies, the number of children has been found to be negatively related to various measures of the value of mother's time (see e.g. Mincer, 1962; Locay, 1987).

Women with children have an incentive to engage in activities that are complementary to child care, including work in a family business based at home, and sewing or weaving at home for pay. Similarly, women who are involved in complementary activities are encouraged to have children because children do not make such large demands on their time. This explains why women on dairy farms have more children than women on grain farms: dairy farming inhibits off-farm work because that is not complementary with children.

During the past one hundred years, fertility declined by a remarkable amount in all Western countries; as one example, married women in the US now average a little over two live births compared with about five-and-a-half live births in 1880 (see US Bureau of the Census, 1977). Economic development raised the relative cost of children because the value of parents' time increased, agriculture declined, and child labour became less useful in modern farming. Moreover, parents substituted away from number of children toward expenditures on each child as human capital became more important not only in agriculture, but everywhere in the technologically advanced economies of the 20th century (for a further discussion, see Becker, 1981, ch. 5).

II 'QUALITY' OF CHILDREN. The economic approach contributes in an important way to understanding fertility by its emphasis on the 'quality' of children. Quality refers to characteristics of children that enter the utility functions of parents, and has been measured empirically by the education, health, earnings, or wealth of children. Although luck, genetic inheritance, government expenditures, and other events outside the control of a family help determine child quality, it also depends on decisions by parents and other relatives.

The quality and quantity of children interact not because they are especially close substitutes in the utility function of parents, but because the true (or shadow) price of quantity is partly determined by quality, and vice versa. To show this, write the utility function in equation (1) as

$$U = U(n, q, Z), \tag{6}$$

where q is the quality of children. Also write the family budget equation in equation (3) as

$$\Pi_n n + \Pi_q q + \Pi_c nq + \Pi_z Z = S, \tag{7}$$

where Π_n is the fixed cost of each child, Π_q is the fixed cost of a unit of quality, and Π_c is the variable cost of children.

By maximizing utility subject to the family income constraint, one derives the following first order conditions:

$$\frac{\partial U}{\partial n} = \lambda(\Pi_n + \Pi_c q) = \lambda \Pi_n^*, \tag{8}$$

$$\frac{\partial U}{\partial q} = \lambda(\Pi_q + \Pi_c n) = \lambda \Pi_q^*, \tag{9}$$

$$\frac{\partial U}{\partial Z} = \lambda \Pi_z. \tag{10}$$

Quantity and quality interact because the shadow price of quantity (Π_n^*) is positively related to the quality of children, and the shadow price of quality (Π_q^*) is positively related to the quantity of children.

To illustrate the nature of this interaction, consider a rise in the fixed cost of quantity (Π_n) that raises the shadow price of quantity (Π_n^*), and thereby reduces the demand for quantity. A reduction in quantity however, lowers the shadow price of quality (Π_q^*), which induces an increase in quality. But the increase in quality, in turn, raises further the shadow price of quantity, which reduces further the quantity of children, which induces a further increase in quality, and so on until a new equilibrium is reached. Therefore, a modest increase in the fixed cost of quantity could greatly reduce the quantity of children, and greatly increase their quality, *even when quantity and quality are not good substitutes in the utility function.*

The interaction between quantity and quality can explain why large declines in fertility are usually associated with large increases in the education, health, and other measures of the quality of children (see the evidence in Becker, 1981, ch. 5). It also explains why quantity and quality are often negatively related among families: evidence for many countries indicates that years of schooling and the health of children tend to be negatively related to the number of their siblings (see e.g. De Tray, 1973; Blake, 1981).

The influence of parents on the quality of their children links family background to the achievements of children, and hence links family background to inequality of opportunity and intergenerational mobility. Sociologists have dominated discussions of intergenerational mobility, but in recent years economists have emphasized that the relation between the occupations, earnings, and wealths of parents and children depends on decisions by parents to spend time, money, and energy on children. Economists have used the concepts of investment in human capital and bequests of nonhuman wealth to model the transmission of earnings and wealth from parents and children (see e.g. Conlisk, 1974; Loury, 1981; Becker and Tomes, 1986). These models show that the relation between say the earnings of parents and children depends not only on biological and cultural endowments 'inherited' from parents, but also on the interaction between these endowments, government expenditures on children, and investments by parents in the education and other human capital of their children.

III ALTRUISM IN THE FAMILY. I have followed the agnostic attitude of economists to the formation of preferences, and have not specified how quality of children is measured. One analytically tractable and plausible assumption is that parents are altruistic toward their children. By 'altruistic' is meant that the utility of parents depends on the utility of children, as in

$$U_p = U(z_p, U_1, \ldots, U_n), \tag{11}$$

where z is the consumption of parents, and $U_i, i = 1, \ldots, n$ is the utility of the ith child.

Economists have generally explained market transactions with the assumption that individuals are selfish. In Smith's famous words,

> It is not from the benevolence of the butcher, the brewer, or the baker, that we expect our dinner, but from their regard to their own interest. We address ourselves, not to their humanity but to their self-love, and never talk of our own necessities but of their advantages.

The assumption of selfishness in market transactions has been very powerful, but will not do when trying to understand families. Indeed, the main characteristic that distinguishes family households from firms and other organizations is that allocations within families are largely determined by altruism and related obligations, whereas allocations within firms are largely determined by implicit or explicit contracts. Since families compete with governments for control over resources, totalitarian governments have often reached for the loyalties of their subjects by attacking family traditions and the strong loyalties within families.

The preference for *own* children mentioned earlier suggests special feelings toward one's children. Sacrifices by parents to help children, and vice versa, and the love that frequently binds husbands and wives to each other, are indicative of the highly personal relations within families that are not common in other organizations (see also Ben-Porath, 1980; Pollak, 1985).

Although altruism is a major integrating force within families, the systematic analysis of altruism is recent, and many of its effects have not yet been determined. One significant result has been called (perhaps infelicitously) the Rotten Kid theorem, and explains the coordination of decisions among members when altruism is limited. In particular, if one member of a family were sufficiently altruistic toward other members to spend time or money on each of them, they would have an incentive to consider the welfare of the family as a whole, *even when they are completely selfish.*

The proof of this theorem is simplest when the utility of an altruist (called the 'head') depends on the combined resources of all family members. Consider a single good (x) consumed by all members: the head and n beneficiaries (not only children but possibly also a spouse and other relatives). The head's utility function can be written as

$$U_h = U(x_h, x_1, \ldots, x_n). \tag{12}$$

The budget equation would be

$$x_h + \sum_{i=h}^{n} g_i = I_h, \tag{13}$$

where I_h is the head's income, g_i is the gift to the ith beneficiary, and the price of x is set at unity. With no transactions costs, each dollar contributed would be received by a beneficiary, so that

$$x_i = I_i + g_i, \tag{14}$$

where I_i is the income of the ith beneficiary. By substitution into equation (13),

$$x_h + \sum x_i = I_h + \sum I_i = S_h. \tag{15}$$

The head can then be said to maximize the utility in (12), subject to family income (S_h).

To illustrate the theorem, consider a parent who is altruistic toward her two children, Tom and Jane, and spends say $200 on each. Suppose Tom can take an action that benefits him by $50, but would harm Jane by $100. A selfish Tom would appear to take that action if his responsibility for the changed circumstances of Jane were to go undetected (and hence not punished). However, the head's utility would be reduced by Tom's action because family income would be reduced by $50. If altruism is a 'superior good', the head will reduce the utility of each beneficiary when her own utility is reduced. Therefore, should Tom take this action, she would reduce her gift to him from $200 to less than $150, and raise her gift to Jane to less than $300. As a result, Tom would be made worse off by his actions.

Consequently, a selfish Tom who anticipates correctly the response from his parent will not take this action, even though the parent may not be trying to 'punish' Tom because she may not know that Tom is the source of the loss to Jane and the gain to herself. This theorem requires only that the head know the outcomes for both Tom and Jane and has the 'last word' (this term is due to Hirshleifer, 1977).

The head has the 'last word' when gifts depend (perhaps only indirectly) on the actions of beneficiaries. In particular, if gifts to the ith beneficiary depend both on his income and on family income, as in

$$g_i = \psi_i(S_h) - I_i, \quad \text{with} \quad \frac{d\psi_i}{dS_h} > 0 \tag{16}$$

then by substitution into equation (14),

$$x_i = I_i + g_i = \psi_i(S_h). \tag{17}$$

The head would then have the 'last word' because x_i would be maximized by maximizing S_h; for further discussion of the Rotten Kid theorem, see Becker (1981, ch. 5), Hirshleifer (1977), and Pollak (1985).

Although this theorem is applicable even when beneficiaries are envious of each other or of the head, it does not rule out conflict in families with altruistic heads. Sibling rivalry, for example, is to be expected when children are selfish because they each want larger gifts from the head, and each would try to convince the head of his or her merits. Conflict also arises when several members are altruistic to the same beneficiaries, but not to each other. For example, if parents are altruistic to their children but not to each other, each benefits when the other spends more on the children. Married parents might readily work out an agreement to share the burden, but divorced parents have more serious conflict. Noncustodial parents (usually fathers) fall behind in their child support payments partly to shift the burden of support to custodial parents (see the discussion in Weiss and Willis, 1985).

Altruism provides many other insights into the behaviour of families. For example, an efficient division of labour is possible in altruistic families without the usual principal–agent conflict because selfish as well as altruistic members consider the interests of other members. Or contrary to some opinion, bequests and gifts to children are not perfect substitutes even in altruistic families. Bequests not only transfer resources to children but also give parents the last word, which induces children to take account of the interests of elderly parents (see Becker, 1981, ch. 5; and also Bernheim, Schleiffer and Summers, 1986). Moreover, if public debt or social security were financed by taxes on succeeding generations that are anticipated by altruistic parents who make bequests, they would raise their bequests to offset the higher taxes paid by their children. Such compensatory reactions negate the effect of debt or social security on consumption and savings (see the detailed analysis in Barro, 1974).

IV THE SEXUAL DIVISION OF LABOUR. A sharp division of labour in the tasks performed by men and women is found in essentially all societies. Women have had primary responsibility for child care, and men have had primary responsibility for hunting and military activity; even when both men and women engaged in agriculture, trade, or other market activities, they generally performed different tasks (see the discussion in Boserup, 1970).

Substantial division of labour is to be expected in families, not only because altruism reduces incentive to shirk and cheat (see section III), but also because of increasing returns from investments in specific human capital, such as skills that are especially useful in child rearing or in market activities. Specific human capital induces specialization because investment costs are partially (or entirely) independent of the time spent using the capital. For example, a person would receive a higher return on his medical training when he puts more time into the practice of medicine. Similarly, a family is more efficient when members devote their 'working' time to different activities, and each invests mainly in the capital specific to his or her activities (see Becker, 1981, 1985; for developments of this argument outside families, see Rosen, 1981).

The advantages of a division of labour within families do not alone imply that women do the child rearing and other household tasks. However, the gain from specialized investments implies the traditional sexual division of labour if women have a comparative advantage in childbearing and child rearing, or if women suffer discrimination in market activities. Indeed, since a sexual division of labour segregates the activities of men and women, and since segregation is an effective way to avoid discrimination (see Becker, 1981), even small differences in comparative advantage, or a small amount of discrimination against women, can induce a sharp division of labour.

Until recently, the sexual division of labour in Western countries was extreme; for example, in 1890, less than five per cent of married women in the United States were in the labour force. In 1981, by contrast, over 50 per cent even of married women with children under six were in the labour force (see Smith and Ward, 1985). However, the occupations of employed men and women are still quite different, and women still do most of the child rearing and other household chores (see *Journal of Labor Economics*, January 1985).

The large growth in the labour force participation of married women during the 20th century is mainly explained by the economic development that transformed Western economies. Substitution toward market work was induced by the rise in the potential earnings of women (see Mincer, 1962). Moreover, the growth in clerical jobs and in the services sector generally, gave women more flexibility in combining market work and child rearing (see Goldin, 1983). In addition, the large decline in fertility during this period (see section I) greatly facilitated increased labour force participation by married women. The converse is also true, however, because the rise in participation of women discouraged child-bearing.

V DIVORCE. Since women specialize in child care, they have been economically vulnerable to divorce and the death of their mates. All societies recognized this vulnerability by requiring long term contracts, called 'marriage', between men and women legally engaged in reproduction. In Christian societies, these contracts often could not be broken except by adultery, abandonment or death. In Islam and Asia they could be broken for other reasons as well, but husbands were required to pay compensation to their wives when they divorced without cause.

The growth of divorce during this century in Western countries has been remarkable. Essentially no divorces were granted in England prior to the 1850s (see Hollingsworth, 1965), whereas now almost 30 per cent of marriages there will terminate by divorce, and the fraction is even larger in the United States, Sweden and some other Western countries (see US Bureau of the Census, 1977). What accounts for this huge growth in divorce over a relatively short period of time?

The utility-maximizing rational choice perspective implies that a person wants to divorce if the utility expected from remaining married is below the utility expected from divorce, where the latter is affected by the prospects for remarriage; indeed, most persons divorcing in Western countries now do remarry eventually (see e.g. Becker, Landes and Michael, 1977). This simple criterion is not entirely tautological because several determinants of the gain from remaining married can be evaluated.

Some persons become disappointed because their mates turn out to be less desirable than originally anticipated. That new information is an important source of divorce is suggested by the large fraction occurring during the first few years of marriage. Although disappointment is likely to be involved in most divorces, the large growth in divorce rates, especially the acceleration during the last 20 years, is not to be explained by any sudden deterioration in the quality of information. Instead, we look to forces that reduced the advantages from remaining in an imperfect marriage.

The strong decline in fertility over time discouraged divorce because the advantages from staying married are greater when young children are present. Conversely, fertility declined partly because divorce became more likely since married couples are less likely to have children when they anticipate a divorce (see Becker, Landes and Michael, 1977, for supporting evidence). The rise in the labour force participation of married women also lowered the gain from remaining married because the sexual division of labour was reduced, and women became more independent financially. At the same time, the labour force participation of married women increased when divorce became more likely since married women want to acquire skills that would raise their incomes if they must support themselves after a divorce.

Legislation certainly eased the legal obstacles to divorce, but empirical investigations have not found significant permanent effects on the divorce rate (see e.g. Peters, 1983). Moreover, economic analysis suggests that even no-fault divorce and other radical changes in divorce legislation would not significantly affect the rate of divorce because bargaining between husbands and wives about the terms of staying married or divorcing offsets even sharp changes in divorce laws.

To show this, let income be I_h^d and I_w^d respectively, if h and w decide to divorce, and I_h^m and I_w^m, respectively, if they remain married. The budget equation is

$$x_h^d + x_w^d = I_h^d + I_w^d = I^d \qquad (18)$$

when divorced, and

$$x_h^d + x_w^m = I_h^m + I_w^m = I^m \qquad (19)$$

when married. I suggest that the decision to divorce is largely independent of divorce laws, and depends basically on whether $I^d \gtreqless I^m$, because both h and w can be made better off by divorce when $I^d > I^m$ and by remaining married when $I^m > I^d$.

Consider, for example, a comparison between unilateral or no-fault divorce, and divorce only by mutual consent. Assume that the husband appears to gain from divorce ($I_h^d > I_h^m$), but the apparent loss to the wife is greater, so that $I^d < I^m$. If divorce were unilateral, he might be tempted to seek a divorce even when she would be greatly harmed. However, she could change his mind by offering a bribe (b_h) that would make both of them better off by staying married:

$$x_h^m + b_h > I_h^d, \quad \text{and} \quad I_w^m - b_h > I_w^d. \qquad (20)$$

This bribe is feasible because $x_h^m + x_w^m = I^m > I^d$. He would then prefer to remain married, even if he could divorce without her consent. Note that they would also decide to remain married if divorce required mutual consent because at least one of them must be made worse off by divorce.

Divorce rates have been affected less by legislation that has regulated the conditions for divorce than by legislation that has affected the gains from divorce. For example, aid to mothers with dependent children and negative income taxes encourage divorce by providing poorer women with child support and 'alimony' (see Hannan, Tuma and Groeneveld, 1977).

VI MARRIAGE. Marriages can be said to take place in a 'market' that 'assigns' men and women to each other or to remain single until better opportunities come along. An optimal assignment in an efficient market with utility-maximizing participants has the property that persons not assigned to each other could not be made better off by marrying each other.

In all societies, couples tend to be of similar family background and religion, and are positively sorted by education, height, age, and many other variables. The theory of assignments in efficient markets explains positive assortative mating by complementarity, or 'superadditivity', in household production between the traits of husbands and wives. Efficient assignments also partly explain altruism between husbands and wives: persons 'in love' are likely to marry because, at the detached level of formal analysis, love can be considered one source of 'complementarity'.

Associated with optimal assignments are imputations that determine the division of incomes or utilities in each marriage. Equilibrium incomes have the property that

$$I_{ii}^m + I_{ii}^f = I_{ii}, \qquad (21)$$

and

$$I_{ii}^m + I_{ij}^f \geqslant I_{ij}, \qquad i \neq j, \qquad (22)$$

where I_{ij} is the output from a marriage of the ith man (m_i) to the jth women (f_j), and I_{ii}^m and I_{ii}^f are the incomes of m_i and f_j, respectively. The inequality in equation (22) indicates the $\{ii\}$ is an optimal assignment because m_i and f_j, $j \neq i$, could not be made better off by marrying each other instead of their assigned mates (f_i and m_j, respectively). Equilibrium incomes include dowries, bride prices, leisure and 'power' (further discussion can be found in Becker, 1974, 1981; the analysis of optimal assignments in Gale and Shapley, 1962; and Roth, 1984, is less relevant to marriage because equilibrium prices – i.e. incomes – are not considered).

Many of the forces in recent decades that reduce the gain from remaining married (see section V) have also raised the gain from delaying first marriage and remarriage. These include the decline in fertility and the rise in labour force participation of married women. The reduced incentive to marry in Western societies is evident from the rapid increase in the number of couples living together without marriage, and in the number of births to unmarried women. Nevertheless, even in Scandinavia, where the trend toward cohabitation without marriage has probably gone furthest, married persons are still far more likely to remain together and to produce children than are persons who cohabit without marriage (for Swedish evidence, see Trost, 1975).

VII SUMMARY AND CONCLUDING REMARKS. Families are important producers as well as spenders. Their primary role has been to supply future generations by producing and caring for children, although they also help protect members against ill health, old age, unemployment, and other hazards of life.

Families have relied on altruism, loyalty, and norms to carry out these tasks rather than the contracts found in firms. Altruism and loyalty are concepts that have not been utilized extensively to analyse market transactions, and our understanding of their implications is only beginning. Yet a much more complete understanding is essential before the behaviour and evolution of families can be fully analysed.

Firms and families compete to organize the production and distribution of goods and services, and activities have passed from one to the other as scale economies, principal–agent problems, and other forces dictated. Agriculture and many retailing activities have been dominated by family firms that combine production for the market with production for members. Presumably, such hybrid organizations are important when altruism and loyalty are more effective than contracts in organizing market production (see Becker, 1981, ch. 8; Pollak, 1985), and when the production and care of children complements production for the market.

Families in Western countries have changed drastically during the past thirty years; fertility declined below replacement levels, the labour force participation of married women and divorce soared, cohabitation and births to

unmarried women became common, many households are now headed by unmarried women with dependent children, a large fraction of the elderly either live alone or in nursing homes, and children from first and second, sometimes even third, marriages frequently share the same household.

Nevertheless, obituaries for the family are decidedly premature. Families are still crucial to the production and rearing of children, and remain important protectors of members against ill-health, unemployment, and many other hazards. Although the role of families will evolve further in the future, I am confident that families will continue to have primary responsibility for children, and that altruism and loyalty will continue to bind parents and children.

GARY S. BECKER

See also ALTRUISM; FAMILY PLANNING; FERTILITY; GENDER; HOUSEHOLD PRODUCTION; HUMAN CAPITAL; INEQUALITY BETWEEN SEXES; VALUE OF TIME; WOMEN'S WAGES.

BIBLIOGRAPHY

Barro, R.J. 1974. Are government bonds net wealth? *Journal of Political Economy* 82(6), November–December, 1095–117.

Becker, G.S. 1974. A theory of marriage: Part II. *Journal of Political Economy* 82(2), part II, S11–26.

Becker, G.S. 1981. *A Treatise on the Family.* Cambridge, Mass.: Harvard University Press.

Becker, G.S. 1985. Human capital, effort, and the sexual division of labor. *Journal of Labor Economics* 3(1), Part II, 533–58.

Becker, G.S., Landes, E.M., and Michael, R.T. 1977. An economic analysis of marital instability. *Journal of Political Economy* 85(6), December, 1141–87.

Becker, G.S. and Tomes, N. 1986. Human capital and the rise and fall of families. *Journal of Labor Economics* 4(2, pt. 2), S1–39.

Ben-Porath, Y. 1980. The F-connection: families, friends, and firms and the organization of exchange. *Population and Development Review* 6(1), 1–30.

Bernheim, B.I., Schleiffer, A. and Summers, L.H. 1986. Bequests as a means of payment. *Journal of Labor Economics* 4(3), pt. 2, S151–82.

Blake, J. 1981. Family size and the quality of children. *Demography* 18(4), 421–42.

Boserup, E. 1970. *Woman's Role in Economic Development.* London: Allen & Unwin.

Conlisk, J. 1974. Can equalization of opportunity reduce social mobility? *American Economic Review* 64(1), March, 80–90.

De Tray, D.N. 1973. Child quality and the demand for children. *Journal of Political Economy* 81(2), Pt II, Mar–Apr, S70–95.

Gale, D. and Shapley, L.S. 1962. College admissions and the stability of marriage. *American mathematical Monthly* 69(1), January, 9–15.

Gardner, B. 1973. Economics of the size of North Carolina rural families. *Journal of Political Economy* 81(2), Part II, March–April, S99–122.

Goldin, C. 1983. The changing economic role of women: a quantitative approach. *Journal of Interdisciplinary History* 13(4), 707–33.

Hannan, M.T., Tuma, N.B. and Groeneveld, L.P. 1977. Income and marital events: evidence from an income maintenance experiment. *American Journal of Sociology* 82(6), 611–33.

Hirshleifer, J. 1977. Shakespeare vs. Becker on altruism: the importance of having the last word. *Journal of Economic Literature* 15(2), 500–502.

Hollingsworth, T.H. 1965. *The Demography of the British Peerage.* Supplement to *Population Studies* 18(2).

Jaffe, A.J. 1940. Differential fertility in the white population in early America. *Journal of Heredity* 31(9).

Locay, L. 1987. *Population Density of the North American Indians.* Cambridge, Mass.: Harvard University Press.

Loury, G.C. 1981. Intergenerational transfers and the distribution of earnings. *Econometrica* 49(4), 843–67.

Malthus, T.R. 1798. *An Essay on the Principle of Population.* Reprinted, London: J.M. Dent, 1958.

Marshall, A. 1890. *Principles of Economics.* London: Macmillan.

Mill, J.S. 1848. *Principles of Political Economy, with some of their applications to Social Philosophy.* Reprinted, New York: Colonial Press, 1899.

Mincer, J. 1962. Labor force participation of married women. In *Aspects of Labor Economics,* Princeton: Princeton University Press.

Peters, E. 1983. The impact of state divorce laws on the marital contract: marriage, divorce, and marital property settlements. Discussion Paper No. 83-19. Economics Research Center/NORC.

Pollak, R.A. 1985. A transactions cost approach to families and households. *Journal of Economic Literature* 23(2), 581–608.

Rosen, S. 1981. Specialization and human capital. *Journal of Labor Economics* 1(1), 43–9.

Roth, A. 1984. The evolution of the labor market for medical interns and residents: a case study in game theory. *Journal of Political Economy* 92(6), 991–1016.

Smith, A. 1776. *An Inquiry into the Nature and Causes of the Wealth of Nations.* Reprinted, New York: Modern Library, 1937.

Smith, J.P. and Ward, M.P. 1985. Time series growth in the female labor force. *Journal of Labor Economics* 3(1) Part II, 559–90.

Trost, J. 1975. Married and unmarried cohabitation: the case of Sweden and some comparisons. *Journal of Marriage and the Family* 37(3), 677–682.

US Bureau of the Census, 1977. *Current Population Reports.* Series P-20, No. 308, Fertility of American Women: June, 1976.

Weiss, Y. and Willis, R. 1985. Children as collective goods and divorce settlements. *Journal of Labor Economics* 3(3), 268–92.

Wrigley, E.A. and Schofield, R.S. 1981. *The Population History of England 1541–1871.* Cambridge, Mass.: Harvard University Press.

family allowance. *See* POVERTY; SOCIAL SECURITY; WELFARE STATE.

family planning. The phrase 'family planning' has come to mean the set of institutions, policies and programmes whose principal objective is to alter the family size decisions of households. Family planning institutions, private or public, attempt to influence fertility choices by (a) direct persuasion of couples to adopt socially 'appropriate' family size goals; (b) the dissemination of information on techniques of birth or conception prevention, and (c) the provision of birth or conception control services or inputs at subsidized cost. In addition, governments may adopt policies that directly alter the incentives for bearing and rearing children. Such policies may include income tax exemptions or direct transfers which vary by the number of children and/or economic and social sanctions related to family size, such as restrictions on parental work opportunities or restrictions on schooling or consumption privileges when those are principally supplied by the public sector.

Of course, to the extent that fertility decisions are responsive to changes in relative prices and to income changes, all governmental policies (tax, transfer, expenditures) indirectly influence the family size goals of households. What principally distinguishes family planning interventions from other government programmes is their attempt to affect fertility outcomes by influencing the means by which households achieve their family size goals.

FAMILY PLANNING AND THE ECONOMIC THEORY OF FERTILITY. Economic models of fertility that incorporate the technology of reproduction provide a general framework with which to analyse the influence of family planning programmes on the family size plans of families (Easterlin et al., 1980; Rosenzweig and Schultz, 1985). In these models, births or conceptions are viewed as byproducts of sexual activity. These byproducts can be averted by the employment of methods of birth control or contraceptive techniques. The set of relationships between sexual and other behaviours, contraceptive practices, and conceptions or births is the reproductive technology, analogous to the technology of production in firms, which describes the effects of inputs on outputs. Couples thus determine their fertility through the use of reproduction inputs. And just as firms adjust output when either input prices or the technology of production change, given demand for the firm's product, couples alter their fertility in response to changes in the costs of reproductive inputs or to changes in the technology of reproduction, given their family size goals.

Family planning initiatives that lower the costs of averting births through subvention of reproduction inputs or information provision have price and income effects. The lowering of the costs of averting births induces couples to avert more births (the own price effect); but couples' real incomes are also higher as a consequence and they may decide to spend some of that income by having larger families. If income effects are small relative to price effects (more likely the smaller the share of contraceptive costs in the family budget), such family planning activities should lower fertility, whatever the motivations of couples for having children.

The degree to which a couple benefits from or is influenced by programmatic family planning activities depends on its family size goals and on the type of family planning activity. If family planning interventions make birth reduction less costly, those couples who desire smaller families (avert more births) benefit most. If the poorest households have the largest families it is thus not clear that non-selective contraceptive *subsidy* programmes benefit the poor relative to the rich. To the extent, however, that family planning initiatives are characterized chiefly by information dissemination, the distribution of the benefits will depend on the pre-programme distribution of such information in the population. If more educated or wealthier couples are better able to acquire information in the absence of such programmes than are other couples, the programmes will benefit such couples least. Fertility reductions associated with contraceptive information dissemination will be larger in poorer, less-educated families.

Economic theory also suggests that the effects of family planning, by altering the costs of fertility, will not be confined to changes in family size. As noted, the increase in income associated with the subsidy may be spread among other family activities. But there are also substitution or cross price effects. In models (Willis, 1973; Becker and Lewis, 1973) in which couples care about the average 'quality' of children, reductions in the cost of fertility control and thus reduced family size make the provision of resources to children less costly, as such resources need be allocated among less children. If family size and child quality are substitutes in the usual consumer demand sense, then it is likely that the reduction in fertility induced by family planning activities will also result in increased investments by families in each child born even if there are no direct biological links between birth order, birth intervals and the characteristics of children.

RATIONALES FOR FAMILY PLANNING INTERVENTIONS. Rigorous theoretical justifications for the public subvention of family planning activities are surprisingly scarce. As for all public interventions, a rationale on efficiency grounds should be based on a demonstration that the costs incurred by private agents making fertility decisions diverge from the social costs of those decisions. The exact nature of the market failure or market incompleteness or the direct negative externalities associated with the production of children that might render family planning programmes appropriate instruments for achieving more efficient outcomes in an economy have not been clearly identified. In growth models incorporating optimal fertility decision-making, the results appear to depend critically on the assumed degree of altruism parents have for children (and vice versa), the allocation of property rights over parental investments in children, and the completeness of intertemporal markets. In the absence of clear resolutions of such issues, a number of other justifications for publicly supported family planning activities have been put forth. One rationale is based on the existence of positive externalities associated with human capital investment (Rosenzweig and Wolpin, 1986). If investments in health or in schooling by households directly benefit other households such that public subventions of such activities are optimal, then it may be efficient to subsidize fertility control (a) if reductions in family size induce greater investments in human capital and/or (b) since reductions in the number of children make less costly public subsidization of investments in children. This argument suggests that health, schooling and family planning programmes are complementary and would tend to be positively correlated over time within countries and across areas.

Two other rationales for family planning interventions are based on information problems. The rise in incomes accompanying economic development and the use of newer medical technologies have contributed to the dramatic fall in infant and child death rates in low income countries over the past decades without a concomitant decline in fertility in many countries. If parents do not correctly foresee the future drop in the risk of death for their children associated with the health externalities of economic growth and development (infection reduction), then subsidization of fertility control may be warranted to reduce fertility to appropriate levels.

Technological innovation has also characterized the control of fertility. If the market provision of information about new methods of contraception is problematic, then publicly funded information dissemination about innovations in this technology may be warranted. Family planning services are then analogous to extension services in agriculture.

EVALUATING FAMILY PLANNING PROGRAMMES. The conceptual experiment needed to ascertain how and to what extent family planning subsidies or information provision actually influence fertility and other behaviours is straightforward – randomly select an area or set of areas for intervention and compare the fertility and other relevant outcomes there with those in non-intervention areas. Since dynamic models of fertility (e.g. Heckman and Willis, 1975) have as yet little to say about how reductions in the costs of fertility control influence the timing and spacing of births, it may not be appropriate to measure the effects of such programmes over short intervals of time. Couples with less costly and/or improved means of controlling fertility may choose to have their children earlier or later; the short-run response of fertility to a family planning intervention may be quite different from the response in terms of completed family size.

Information from appropriate randomized experiments involving family planning activities is scarce. Most estimates of the impact of family planning interventions have come from non-experimental data, chiefly cross-sectional data. The best of the cross-sectional studies of the effects of public expenditures on family planning or measures of access to family planning institutions examine as well the natalist effects of other programmes (health programmes, for example). Since theory suggests that health and family planning interventions are complementary and are likely to be distributed similarly, failure to take into account the existence and distribution of other programmes when evaluating the impact of family planning interventions may yield misleading estimates of family planning efforts. Multivariate studies combining spatial information on programmes and household data from rural and urban Colombia and rural India (Rosenzweig and Schultz, 1982; Rosenzweig and Wolpin, 1982) indicate that family planning and health institutions (clinics) are associated with both lower fertility and lower rates of child mortality, although no effects of these programmes were found in rural areas of Colombia. Results from the urban Colombia data, moreover, indicated that the effects of the programmes were significantly greater among households with less-educated mothers. This result is consistent with the notion that family planning (and health) programmes principally serve to disseminate information, this function being of less value for the more educated (and better informed) households.

A study using longitudinal information on the nutritional status of children and information on the dates of initiation of health and family planning programmes (Rosenzweig and Wolpin, 1986) tested whether the timing of public programme interventions across areas was correlated with unmeasured area factors associated with child health. The results suggested the spatial distribution of both health and family planning programmes was not random, with both types of programmes tending to be similarly placed (in low health areas), and that once non-random programme placement was taken into account (but not before), both the family planning and health programmes appeared to improve significantly the nutritional status of children.

These empirical studies thus suggest that family planning activities have succeeded in lowering fertility and in augmenting human capital investment, in at least some countries, but that more attention to the rules by which public programmes are distributed and initiated may be needed to obtain more accurate estimates of the effects of such programmes. Improved estimates of the consequences of family planning initiatives are thus a byproduct of a better understanding of the rationale for such programmes and of public sector behaviour.

MARK R. ROSENZWEIG

See also FECUNDITY; FERTILITY.

BIBLIOGRAPHY

Becker, G.S. and Lewis, H.G. 1973. On the interaction between quantity and quality of children. *Journal of Political Economy* 82, April/May, S279-S288.

Easterlin, R.A., Pollak, R.A. and Wachter, M.L. 1980. Toward a more general economic model of fertility determination. In *Population and Economic Change in Developing Countries*, ed. R.A. Easterlin, Chicago: University of Chicago Press.

Heckman, J.J. and Willis, R.J. 1978. Estimation of a stochastic model of reproduction: an econometric approach. In *Household Production and Consumption*, ed. N.E. Terlecky, New York: Columbia University Press.

Rosenzweig, M.R. and Schultz, T.P. 1982. Child mortality and fertility in Colombia: individual and community effects. *Health Policy and Education*, February, 125–51.

Rosenzweig, M.R. and Schultz, T.P. 1985. The demand for and supply of births: fertility and its life-cycle consequences. *American Economic Review* 75, December, 992–1015.

Rosenzweig, M.R. and Wolpin, K.I. 1982. Governmental interventions and household behavior in a developing country: anticipating the unanticipated consequences of social programs. *Journal of Development Economics* 10(2), April, 209–25.

Rosenzweig, M.R. and Wolpin, K.I. 1986. Evaluating the effects of optimally distributed public programs: child health and family planning interventions. *American Economic Review* 76, June, 470–82.

Willis, R.J. 1973. A new approach to the economic theory of fertility behavior. *Journal of Political Economy*, March/April, S14–64.

famine. Attempts to formulate a precise definition of famine are fraught with difficulties. In commonsense terms, a famine refers to a sudden event involving large-scale deaths from starvation within a short period. In reality, the existence of starvation has been and remains a persistent characteristic of many societies. This means that a famine has to be defined in terms of a significant deviation from a 'norm' and it is hard to avoid ambiguities in such an enterprise. Moreover, in many situations, it is virtually impossible to distinguish between deaths from starvation and those from disease. Since starvation reduces the human body's resistance to disease, deaths from starvation can be easily confused with deaths from disease and famines are typically accompanied by epidemics. In spite of these difficulties of definition, however, a famine, when it occurs, rarely fails to be recognized. There is usually a sudden increase in mortality and there are obvious signs (e.g. begging, eating of inedibles, food riots, sharp increases in petty crimes, unusual scales of migration, etc.) of a desperate search for food by a sizeable section of the population.

In recorded history, nearly all societies have periodically suffered the devastating consequences of famines. The earliest recorded famine, which occurred in ancient Egypt, dates back to the fourth millennium BC. The most recent famines occurred in Ethiopia and Sudan as late as 1985–86. While a complete list of famines is unlikely ever to be compiled, even an incomplete reckoning strongly suggests that very few societies managed to escape them (see the list provided in the 1985 edition of the *Encyclopaedia Britannica*, vol. 4, p. 675).

Unfortunately, in spite of its ubiquity in human history, famine remains a misunderstood or at best an inadequately understood phenomenon. It is true, of course, that some societies (e.g. the European countries) which repeatedly suffered famines in the past are now free of this threat. But this should be seen as a fortuitous result of varied historical processes rather than as an outcome of purposive endeavours. For misconceptions about the causes of famine persist and hinder scientific investigations into past famines even today. The fact that many societies do not as yet possess the ability to anticipate, prevent or even adequately respond to famine is in part attributable to the wide acceptance of views based on such misconceptions.

In section 1, the most influential of these views are briefly discussed and are confronted with factual evidence drawn from some of the major famines of the past. Apart from highlighting some common fallacies with respect to causation, this discussion also brings into focus some important facts which a theory of famine must seek to explain. An approach to famine analysis is then outlined in section 2. Finally, some policy measures needed to counter the threat of famine in contemporary developing countries are briefly discussed in section 3.

1. THE QUESTION OF CAUSATION. If wars are not taken into account, it would be true to say that natural disasters, principally droughts and floods, preceded most of the major famines in known history (the exceptions are the Bengal famine of 1943 which was not preceded by any natural disaster

and the Irish famine of the 1840s where the precipitating factor was a plant disease in the form of potato blight). This fact gave rise to a remarkable fallacy of the *post hoc ergo propter hoc* type. Until very recently, the major famines were widely believed to have been caused by food shortages generated by natural disasters (see, for example, Masefield, 1963 and Aykroyd, 1974). Thus famines were viewed as natural rather than social phenomena. Descriptions of famine as the 'extreme and persistent shortage of food' (1985 edition of the *Encyclopaedia Britannica*, vol. 4, p. 674) abound in the literature on the subject.

The proponents of the 'food shortage' hypothesis sometimes sought support in simple-minded interpretations of Adam Smith's view that market forces achieved the best results in social production and distribution and of Malthus's view that famines constituted a 'positive check' on population growth. If market forces indeed ensure the optimal production and distribution of food in all situations, then famines must imply food shortages due to natural calamities. Alternatively, if it is societies' failure to keep population growth in line with the growth of food supply which invites nature's wrath in the form of famine, food shortages are again implied.

The view that famines are caused by food shortages was held with such certainty that some observers, when confronted with evidence that food was not in short supply during certain famines, suggested that famines were partly explained by people's reluctance to change their food habits in times of scarcity. Thus one commentator on the Irish famine insinuated that the Irish people's refusal to eat anything other than potatoes had worsened the situation (C.E. Trevelyan quoted in Woodham-Smith, 1962). Another commentator on the Bengal famine of 1943 suggested that many Bengalis could in fact have saved thier lives had they been prepared to eat wheat (Moraes, 1975). Yet, it has been reported time and again that during famines people eat all sorts of inedibles including human flesh (*Encyclopaedia Britannica*, vol. 4, p. 675; Mallory, 1926).

On reflection, the emphasis on food shortage in explaining famine appears rather puzzling. For, even on purely logical grounds, it is implausible to suppose that famines are universally caused by food shortages. While most famines were indeed preceded by reductions in food output, it cannot be deduced that such reductions necessarily led to food shortages. And while food shortages can certainly cause famines, it does not follow that all famines must necessarily be caused by food shortages. Famine implies that some people do not have adequate access to food, it does not imply that food itself is in short supply.

The explanation of the puzzle may well lie in the fact that habits of thought often fail to change *pari passu* with changing reality. It is possible to imagine situations where famines must necessarily be caused by food shortages. In a country where transport systems and trade relationships are undeveloped, a localized crop failure can cause a localized famine even though there may not be a serious food shortage in the country as a whole. Some of the famines of the remote past were undoubtedly of this type. Such circumstances, however, have long ceased to exist. The development of transport systems and trade relationships has virtually eliminated the possibility of localized crop failures generating localized food shortages.

It is arguable, however, that famines are to be explained not in terms of food shortages but in terms of 'food availability decline'. In a society with a given pattern of distribution of food among the population, a certain level of food supply is required to ensure that even the most disadvantaged sections do not starve. If the food supply falls below this level, the poorest must starve unless the decline in the food supply itself leads to a reduction in the inequality of food distribution. Since it can be plausibly argued that a decline in food supply typically leads to a worsening of food distribution, reductions in food supply can clearly cause famine. But, firstly, it still cannot be argued that famines are necessarily caused by reductions in food supply. Secondly, since changes in food distribution in consequence of reductions in food supply are of crucial importance in determining the overall effects, reductions in food supply in themselves do not have much predictive power. A ten per cent decline in food supply may lead to a major famine in one situation but to only a minor squeeze on some people's food consumption in another.

It is useful, at this point, to look at some facts. A careful scrutiny of the available evidence from some of the major famines of the past yields the following conclusions. First, some major famines were not in fact preceded by any significant decline in food production or availability. The outstanding example is the Bengal famine of 1943, but there are others (Sen, 1981). Second, many of those famines which were preceded by food availability decline did not in fact involve absolute shortages of food. The examples include the Indian famines which occurred between 1860 and 1910 and the Irish famine of the 1840s (Ghose, 1982; Woodham-Smith, 1962). Third, even during severe famines, food was often exported not only from the countries concerned but also from the famine-affected regions themselves. For example, throughout the period 1860-1910, when twenty major famines and scarcities were experienced, India was a significant net exporter of food-grains; moreover, there is evidence to show that some of the regions worst affected by famines exported food in the famine years (Ghose, 1982). Ireland also was a net exporter of food during the 1840s (Woodham-Smith, 1962). Fourth, only the people belonging to particular social classes died of starvation during famines. In India, for example, landlords and merchants often prospered during famines while agricultural labourers, artisans, barbers, washermen, etc. died in their thousands (Ghose, 1982; see also Alamgir, 1980; Sen, 1981; Woodham-Smith, 1962 for further evidence).

These facts argue strongly against the 'food shortage' hypothesis. The 'food availability decline' hypothesis does much better, but even this fails to explain a number of major famines. Clearly, famines do not have a single cause. A recognition of this helps focus attention on the determinants of people's access to food. Food supply may indeed be one of these determinants, but it certainly is not the only one.

2. FAMINE AND ECONOMIC ANALYSIS. Perhaps because famines were believed to be caused by natural disasters, economists had rarely concerned themselves with famine analysis; this was left, until very recently, largely to journalists, administrators and historians. The classical economists noted the distressing poverty of the labouring classes; Malthus and Marx in particular attempted to identify the causes of persistent misery and provided some valuable insights (Malthus, 1798 and 1800; Marx, 1867, ch. 25). But Malthus negated his own insights by insisting that famines constituted a 'positive check' on population growth and Marx was primarily concerned with the long-term dynamic of capitalist economies. On the whole, analysis of poverty and famine remained outside the framework of classical political economy. The neoclassical framework of general equilibrium analysis, currently in vogue, assumes away possibilities of starvation; it supposes that individuals possess adequate resources to survive above starvation levels even without entering exchange relations (Koopmans, 1957). Even the early development theorists did not concern themselves with problems of starvation and famines (consider the works of Mandelbaum, Nurkse and Lewis).

In the early 1970s, the problems of poverty and malnutrition in developing countries began to attract economists' attention. Subsequently, following the pioneering work of Sen (1977), famine analysis was brought into the domain of economists' interests. Much remains to be done and, in particular, famine analysis is yet to be integrated into the body of development theory. But a new approach to famine analysis has emerged and this is outlined below.

A household's (or an individual's) access to food is determined, firstly, by its asset-holding and, secondly, by the possibilities of transforming assets into food. The universally held asset is labour. But households may hold, in addition, land, other instruments of production or special skills. The transformation possibilities depend both on the employability of the assets for production (including production of services) and on the exchange ratios between assets and food or between yields of assets and food. At the level of individual households, employability of assets is not always guaranteed; labour may remain unemployed, artisanal products may remain unsold and services may not be demanded. Exchange ratios can be complex and unstable. For a peasant household which owns land and produces its own food requirements, productivities of land and labour are the relevant exchange ratios and these obviously depend on the quality of land as well as on weather conditions. For a peasant household which produces a non-food crop, an additional relevant exchange ratio is that between the non-food crop and food. For a landless labourer, the ratio of exchange between labour and food may be direct (when wages are paid in food) or indirect (when wages are paid in money). A variety of other institutional arrangements such as tenancy and credit relations may have important consequences for the ratio of exchange between labour and food for both peasants and landless labourers. The important point to note in all this is that assetholding, employability of assets and exchange ratios, which determine individual households' access to food, are subject to rules set by societies.

It takes only a small step to see that in any society, the distribution of food among the population depends on the distribution of productive assets (in the case of agrarian societies, on the structure of property rights in land), on the rules governing their employability and on a host of exchange relations and ratios. Changes in one or more of these parameters and variables imply changes in the distribution of food and may mean a drastic reduction in some people's access to it. Depending upon the particular characteristics of a society, such changes can be caused by a variety of factors including reductions in food output due to natural disasters. A serious analysis of famine requires an adequate understanding of the relevant institutional and economic characteristics of the society in question as well as identification of those variables which are liable to drastic changes.

To see these arguments in more concrete terms, consider an agrarian society which is composed of three classes: food producers (i.e. landowners), agricultural labourers and other workers (artisans and service workers such as barbers, washermen, etc.). Food flows principally from the landowners to the other classes in exchange for labour, non-food products and services, and these exchanges are mediated by money. Thus the ability of the non-landowners to acquire food depends on three types of variables: the landowners' demand for labour, non-food products and services; the prices of these items and the price of food. A crop failure may affect all these variables: the landowners' demand for agricultural labour, artisanal products and services may decline and the relative prices of these items vis-à-vis food may decline simultaneously. A crop failure can thus lead not only to a decline in food supply but also to a more unequal distribution of the available food. But the relative prices of labour, artisanal products and services vis-à-vis food can decline for other reasons too, as a consequence of a sharp rise in the government's demand for food, for instance. Thus even when food supply remains unaltered, inequality of food distribution may sharply rise leading to a famine. A focus on employment and prices is adequate for analysing both of these cases; a focus on food supply is both inadequate and misleading.

There is in fact something more to the story. In the situation described above, it has been supposed that the landowners produce only food, but they may be producing both food and non-food crops in some combination. In general, the cropping pattern depends in an important way on the structure of landownership. For example, other things being equal, food production is likely to be more emphasized in a situation where a majority of the landowners are peasant proprietors than in a situation where a majority are large landlords. The level of food production itself is in part socially determined. This also means that the extent of decline in food production in the wake of a natural disaster is not independent of the structural characteristics of an economy. While in one situation, all efforts may be made to limit the damage to food crops, such efforts may well be directed largely to limiting the damage to non-food crops in another situation.

In short, famine fundamentally is a social phenomenon which can be understood only by focusing on the institutions and arrangements which determine the access to food of different classes and groups in a society. In this perspective, the links between persistent mass poverty and periodic famines are rather obvious. The same set of forces generate mass poverty over time and famines in certain periods. Mass poverty results from long-term changes in social production and distribution mechanisms; famines result from violent short-term changes in the same mechanisms. Growth of mass poverty increases vulnerability to famine through raising the percentage of population surviving on the margin of subsistence in normal periods. Famine increases mass poverty by permanently altering the distribution of productive assets in favour of the richer section of the population (see Ghose, 1982 for some evidence). Emphasis on such simple indicators as food supply frustrates efforts to develop effective policies and instruments for combating both mass poverty and the ever-present threat of famine in the less developed parts of the world.

3. FAMINE AND STATE POLICY. In societies where universal social security systems are unlikely to be in place for a long time to come, anticipation, relief and prevention of famine must be among the central objectives of state policy. And it is absolutely essential to ensure that state policy is based on sound judgements. The potential cost of judgemental error is high; millions of lives may be at stake. These, of course, are normative statements. It certainly cannot be assumed that governments everywhere will adopt policies whose desirability is manifest from a humanitarian standpoint. It is naive to suppose that many of the past famines would not have occurred had a correct theory existed; saving lives was not always a high priority for governments. This, however, is no place to dwell on these issues.

As already noted, the index traditionally used to anticipate famine has been per capita food output (or availability). It should be clear from the arguments presented above that this index is quite inadequate for the purpose and can be totally misleading. In fact, no single index is adequate. For correct anticipation, changes in the pattern of asset distribution, in employment possibilities, in wages and prices and in per capita food output must all be carefully monitored.

As for famine relief, strictly non-interventionist policies were sometimes advocated in the past. During the Indian famines of the 19th century and the Irish famine of the 1840s, the colonial administrators invoked Adam Smith's doctrines to justify non-interventionist policies (Ambirajan, 1978; Aykroyd, 1974; Rashid, 1980; Woodham-Smith, 1962). Repeated failures of the market forces to cure famines eventually proved to be persuasive arguments in favour of intervention. For some time now, the standard method of intervention has been free distribution of food to the famine-affected population. This undoubtedly is the best option once a famine is already in course. If, however, an approaching famine can be anticipated well in advance, implementation of appropriately designed and timed food-for-work programmes is a far better method. This is not only because assets can be created at the same time as starvation deaths are prevented, but also because this method allows people to retain their dignity.

For either of these methods of intervention to be feasible, food must be available. It has been argued above that famines generally do not involve food shortages. It would be wrong to suppose, however, that governments can acquire enough food for purposes of relief through purchases from the market or that relief can be provided in the form of cash payments. Governments' attempts to purchase food from the market in the pre-famine or famine periods will increase the relative price of food and thus risk worsening the situation. Relief in the form of cash payments will also increase the relative price of food and is most likely to redistribute food among the poor. For effective relief, governments must either draw upon accumulated food stocks or stop food exports or increase food imports. A viable policy of stock management in food is an essential weapon in the fight against famine.

Elimination of the threat of famine calls for policies which are essentially the same as those needed to eradicate mass poverty. These policies fall into two categories: those designed to improve the security of access to food of the vulnerable groups and those designed to reduce the dependence of food output on the forces of nature. The first category includes such policies as agrarian reforms, creation of stable employment outside agriculture, promotion of effective organizations of the poor, development of state-sponsored social security systems for the poorest, etc. The second category includes measures to improve technological conditions of production in the food sector (e.g. development of effective irrigation and drainage systems). However, though conceptually it is convenient to separate them, the two types of policies are interactive in practice. Implementation of agrarian reforms, for example, often facilitates the development of irrigation and drainage systems. Improvement in the security of access to food of the vulnerable groups and reduction in the degree of dependence on nature generally reinforce each other.

The entire discussion in this essay has been concerned with economic processes and policies within countries. These certainly should be accorded primacy, but it needs to be recognized that they are not altogether independent of the nature and dynamics of international economic relations. To take an example, the policy of encouraging production of export crops to the neglect of food crops (a policy which has relevance for food security issues), pursued in many developing countries, is directly linked to the nature of their involvement in international economic relations. From another standpoint, it can be said that the international economy displays certain features which are remarkably similar to those observed in national economies. The level of food production in the world today is such that there need be no starvation and famine in any part of the globe. But just as adequate food availability in a country does not guarantee access to adequate food for all, the existence of food surplus at the global level does not do much to resolve food security problems in individual countries. The causes are similar and so are the remedies.

A.K. GHOSE

See also AGRICULTURAL GROWTH AND POPULATION CHANGE; AGRICULTURE AND ECONOMIC DEVELOPMENT; ENTITLEMENTS; MORTALITY; NUTRITION.

BIBLIOGRAPHY

Alamgir, M. 1980. *Famine in South Asia: Political Economy of Mass Starvation*. Cambridge, Mass.: Oelgeschlager, Gunn and Hain.
Ambirajan, S. 1978. *Classical Political Economy and British Policy in India*. Cambridge: Cambridge University Press.
Aykroyd, W.R. 1974. *The Conquest of Famine*. London: Chatto & Windus.
Ghose, A.K. 1982. Food supply and starvation: a study of famines with reference to the Indian sub-continent. *Oxford Economic Papers* 34(2), July, 368–84.
Koopmans, T.C. 1957. *Three Essays on the State of Economic Science*. New York: McGraw-Hill.
Mallory, W.H. 1926. *China: Land of Famine*. New York: American Geographical Society.
Malthus, T.R. 1798. *An Essay on the Principle of Population*. London. Harmondsworth: Penguin, 1970.
Malthus, T.R. 1800. *An Investigation of the Cause of the Present High Price of Provision*. London.
Marx, K. 1867. *Capital*, Vol. 1. Reprinted, Harmondsworth: Penguin, 1976.
Masefield, G.B. 1963. *Famine: Its Prevention and Relief*. Oxford: Oxford University Press.
Moraes, D. 1975. The dimensions of the problem: comment. In *Hunger, Politics and Markets: The Real Issues of the Food Crisis*, ed. S. Aziz, New York: New York University Press.
Rashid, S. 1980. The policy of laissez-faire during scarcities. *Economic Journal* 90, September, 493–503.
Sen, A.K. 1977. Starvation and exchange entitlements: a general approach and its application to the Great Bengal Famine. *Cambridge Journal of Economics* 1(1), 33–59.
Sen, A.K. 1981. *Poverty and Famines: An Essay on Entitlement and Deprivation*. Oxford: Clarendon Press.
Woodham-Smith, C. 1962. *The Great Hunger: Ireland 1845–49*. London: Hamish Hamilton.

'Famous Fiction'. *See* STATIONARY STATE.

Fanno, Marco (1878–1965). Fanno was a most distinguished Italian economist who became Professor of Political Economy in 1909 and taught at the universities of Sassari, Cagliari, Messina and Padua.

His work places him between the Italian tradition of General Equilibrium and the macrodynamic theories developed during the 1930s. From this perspective, Fanno was unique among the scholars who shaped Italy's economic thought until the end of World War II. Indeed, most economists were reared in the General Equilibrium school of Pareto and Pantaleoni and did not absorb the new formulations of the 1930s.

Fanno's contributions range from the theory of joint costs (1914) to the analysis of the elasticity of demand (1929, 1933)

and monetary issues (1913, 1937). Yet it is a study on economic fluctuations that constitutes Fanno's most important work (1947). This study is characterized by a systematic sifting of the major theoretical literature on the subject, as well as of a large amount of historical and empirical material. Analytically, his approach to the trade cycle reflects Ragnar Frisch's model of the propagation of impulses in economic activity. In his book, Fanno discusses in detail the role of credit in determining the duration of the cycle. In this respect he departed from the theories of the real trade cycle and moved closer to Keynes's *Treatise on Money*.

JOSEPH HALEVI

SELECTED WORKS

1913. *Le banche e il mercato monetario*. Rome: Loescher.
1914. Contributo alla teoria dell'offerta a costi congiunti. *Giornale degli Economisti* 49, Supplement, October, 1–143.
1929. Die Elastizität der Nachfrage nach Ersatzgütern. *Zeitschrift für Nationalökonomie* 1(1), May, 51–74.
1933. Interrelation des prix et courbes statistiques de demande et d'offre. *Econometrica* 1(2), April, 162–71.
1937. *Lezioni di economia e legislazione bancaria*. Padua.
1947. *La teoria delle fluttuazioni economiche*. Turin: Unione Tipografico–Editrice Torinese.

farm economics. *See* AGRICULTURAL ECONOMICS.

farming. *See* AGRICULTURAL GROWTH AND POPULATION CHANGE; COMMON LAND; PEASANTS.

Farr, William (1807–1883). William Farr, born in Kenley, Shropshire on 30 November 1807, died in London on 14 April 1883, was a statistician in the General Register Office who had been appointed in 1840 as 'compiler of abstracts' and was two years later made Statistical Superintendent, a post he held until his retirement in 1880. He pioneered the quantitative study of morbidity and mortality and in the process became one of Victorian England's most prominent figures in the public health and reform movements (Cullen, 1975). He made major contributions in the fields of data collection, being largely responsible for the introduction of a cause of death classification which was linked with his derivation of the 'zymotic' theory of epidemic disease (Eyler, 1979; Pelling, 1978). As an Assistant Census Commissioner for each of the censuses of 1851, 1861 and 1871, he was largely responsible for the development of reliable procedures for the recording of occupations (McDowall, 1983). He is, however, best known as a statistical analyst, for in 1843 he constructed the first English Life Table based on deaths in 1841 linked to the census of that year. At the same time he established the formula for deriving from a rate of mortality by age m the probability of survival p at the initial age. In 1850 and 1864 Farr produced his second and third English Life Tables, the last mentioned being used as the actuarial basis for the life insurance scheme set up by the Post Office for its employees. Farr in his work on occupational mortality was the first to make extensive use of the standard mortality rate, allowing comparisons of the mortality of different groups by means of a summary statistic which took account of differences in the age structure of the groups being compared. A recurring theme in his work was the identification of variation in mortality in different urban areas of the country. Such differential mortality was viewed as an index of human welfare. For example, in 1850 one-tenth of the registration districts, those he named 'healthy districts', had average mortality rates not exceeding 17 per 1,000, a rate he thought indicative of the 'natural' mortality which, when exceeded, would indicate those deaths attributable to unnatural and preventable diseases. An underlying aim in much of his work was to discover statistical laws or numerical expressions of regularities such as he proposed in the laws of recovery and death in smallpox, the elevation law for cholera mortality in London (Lewes, 1983) and the law of the relation between population density and mortality. He was also an early contributor to human-capital theory (Kiker, 1968) arguing, in particular, that the economic value of men varied with age as well as social class, and this he used as powerful publicity for urban reform by drawing attention to the financial losses that followed from diseases that were the causes of death and illness in society at large.

R.M. Smith

SELECTED WORKS

1843a. Causes of the high mortality in town districts. *5th Annual Report of the Registrar General*, 406–35 (*Parliamentary Papers* XXI, 200–15).

1843b. English Life Table No. 1. *5th Annual Report of the Registrar General*, 354–8, 366–7 (*Parliamentary Papers* XXI, 168–71, 178) and *6th Annual Report of Registrar General*, 517–666 (*Parliamentary Papers*, 1844, XIX, 290–358).

1850. English Life Table No. 2: Males. *12th Annual Report of the Registrar General*, Appendix, 73–152.

1852a. Influence of elevation on the fatality of cholera. *Journal of the Statistical Society* 15, 155–83.

1852b. *Report on the Mortality of Cholera in England, 1848–49*. London: HMSO.

1854. Vital statistics. In *A Descriptive and Statistical Account of the British Empire: Exhibiting Its Extent, Physical Capacities, Population, Industry, and Civil and Religious Institutions*, ed. J.R. McCulloch, 4th edn, London: Longman, Brown, Green and Longmans.

1859. English Life Table No. 2: Females. *20th Annual Report of the Registrar General*, Appendix, 177–203 (*Parliamentary Papers*, 1859, sess. 2 XII).

1864. *English Life Table: Tables of Lifetimes, Annuities and Premiums, with an Introduction by William Farr, M.D., F.R.S., D.C.L.* London.

1866. Mortality of children in the principal states of Europe. *Journal of the Statistical Society* 29, 1–35.

1867–8. *Report of the Cholera Epidemic of 1866 in England: Supplement 29th Annual Report on the Registrar General (Parliamentary Papers* XXXVI).

1885. *Vital Statistics: a Memorial Volume of Selections from the Reports and Writings of William Farr, M.D. D.C.L. C.B. F.R.S.* Ed. Noel A. Humphreys, London.

BIBLIOGRAPHY

Cullen, M.J. 1975. *The Statistical Movement in Early Victorian Britain: The Foundations of Empirical Social Research*. Brighton: Harvester Press.

Eyler, J.M. 1979. *Victorian Social Medicine: The Ideas and Methods of William Farr*. Baltimore and London: Johns Hopkins University Press.

Kiker, B.F. 1968. *Human Capital: In Retrospect*. Essays in Economics No. 16, Columbia, South Carolina: University of South Carolina, Bureau of Business and Economic Research.

Lewes, F. 1983. William Farr and cholera. *Population Trends* 31, Spring, 8–12.

McDowall, W. 1983. William Farr and the study of occupational mortality. *Population Trends* 31, Spring, 21–4.

Pelling, M. 1978. *Cholera, Fever and English Medicine*. Oxford: Oxford University Press.

Farrell, Michael James (1926–1975). M.J. Farrell was born in 1926 and read Politics, Philosophy and Economics at New College, Oxford, graduating with First Class Honours. He moved to Cambridge in 1949 to work with Richard Stone at the Department of Applied Economics. He became a Fellow of Gonville and Caius College and the University made him Lecturer in Economics and eventually Reader. He was Editor of the *Review of Economic Studies* and a Fellow of the Econometric Society. In 1957 Farrell contracted poliomyelitis which left him dependent on crutches to get about. He died in 1975.

The bibliography of Farrell's work provided by Fisher (1976) lists 25 journal papers, about one a year in a cruelly shortened academic life. The quality of these papers is remarkable. They reveal the clarity of their author's mind and an outstanding creativity. Farrell often answered questions that others had hardly considered.

As a young man Farrell was influenced by Phillip Andrews, the author of *Manufacturing Business*, and they shared a dissatisfaction concerning the prevailing theory of the firm: 'They [economists in the 1920s and 1930s] reduced the theory of the firm to a maximization problem soluble by the most elementary application of the differential calculus ...' and 'Unfortunately these conclusions did not fit the regrettably complex facts well ...' (Farrell, 1971, p. 10). Farrell's work on the theory of the firm displayed an acute understanding of the subtlety of profit maximization as a strategy. In (1954) he provided one of the first applications of linear programming to this field. Farrell believed that the case for profit maximization eventually depended in part on the operation of a selection process. His (1970) paper remains to this day one of the best papers ever written on that topic.

Farrell wrote on the measurement of productive efficiency, on the consumption function, and on welfare economics. On some topics he produced a single paper – his last was on social choice theory.

In (1959) Farrell made two observations which were important innovations at the time. First, he exposed what he called 'the fallacy'. This is a confusion between sufficient and necessary conditions for competitive equilibrium to be efficient. Convexity, as Farrell neatly demonstrated, is *sufficient* for existence of equilibrium but is *necessary* neither for existence nor for efficiency. Secondly, '... concavities in individual indifference maps disappear when one aggregates over a large enough number of individuals' (p. 381).

This deep aggregation result, which gave rise to an extensive literature (see, e.g., Arrow and Hahn, 1971, chs 7 and 8), is based on a simple point. To illustrate it consider consumers and let them all have the same tastes, which may be represented by $U(x)$, where x is a vector of consumptions. Suppose that $U(x_1) = U(x_2)$ and let there be N consumers. We now wish to see whether a convex combination of $N \cdot x_1$ and $N \cdot x_2$, that is $\lambda \cdot N \cdot x_1 + (1 - \lambda) \cdot N \cdot x_2$, can be distributed so as to make each consumer at least as well off as with x_1 or x_2. If it can, community indifference curves will be convex even if those derived from $U(\ )$ are not.

If consumers were indefinitely divisible we could achieve this result by giving x_1 to $\lambda \cdot N$ consumers and x_2 to $(1 - \lambda) \cdot N$ consumers. However, as $\lambda \cdot N$ may not be an integer this exact procedure is inadmissible. Nevertheless, as N becomes large an integer $M < N$ will eventually emerge such that M/N approximates λ to any desired degree of accuracy. Hence Farrell's result follows.

Farrell treated the often sloppily discussed question whether speculation could be destabilizing and still profitable, in (1966). His demonstration within a very general framework that linearity of demand functions is required to exclude this possibility greatly advanced the general understanding of this problem.

In (1962) Farrell considered the well-known problem of the yield gap, the observation that equities at certain times show a different rate of return from that obtained from bonds. He provided some calculations which showed that there had been yield gaps in the past even when returns were corrected for capital gains. In considering what light these *ex post* observations throw on investors' *ex ante* decisions Farrell asked '... what do we mean by perfect knowledge in a market where uncertainty is present?' (1962, p. 835) This lead him to analyse what he called 'accurate' expectations: '... an individual's expectation is 'accurate' if his subjective probability distribution is the same as the hypothetical frequency distribution by which we represent the real world' (p. 836). Long before the idea of rational expectations became fashionable, Farrell saw its relevance to the analysis of securities markets. However the careful student of profit maximization and selection processes found no reason to assume that expectations would necessarily be 'accurate'.

CHRISTOPHER BLISS

SELECTED WORKS

1954. An application of activity analysis to the theory of the firm. *Econometrica* 22, July, 291–302.
1959. The convexity assumption in the theory of competitive markets. *Journal of Political Economy* 67, August, 377–91.
1962. On the structure of the capital market. *Economic Journal* 72, December, 830–44.
1966. Profitable speculation. *Economica* 33, May, 183–93.
1970. Some elementary selection processes in economics. *Review of Economic Studies* 37(3), July, 305–19.
1971. Philip Andrews and manufacturing business. *Journal of Industrial Economics* 20(1), November, 10–13.

BIBLIOGRAPHY

Arrow, K.J. and Hahn, F.H. 1971. *General Competitive Analysis.* Amsterdam: North-Holland.
Fisher, M.R. 1976. The economic contribution of Michael James Farrell. *Review of Economic Studies* 43(3), October, 371–82. (Includes a complete discussion of Farrell's works.)

fascism. The term fascism can be applied to historical reality only as an approximation, because the differences between what are called fascist movements and regimes seem to be greater than the similarities, and leave room for many contrary interpretations (cf. de Felice, 1969; Gregor, 1974). Given this restriction the term is applied to both radical populistic mass movements, primarily of the middle classes, and, where they attained power, to the political regimes they created between the two world wars.

The fascist movements emerged as a result of the political, economic and social crisis of the bourgeois societies in European countries after World War I. They propagated an extreme anti-liberal, anti-socialist, nationalist and imperialist (and, in Germany, racial) ideology, and above all, they struggled with militancy and terror against the labour organizations. Where these movements came to power (Italy and Germany) it was by coalition with the bourgeois upper class and thanks to the simultaneous failure of labour organizations to present any effective resistance. The political structure of the fascist regimes was, on the surface, marked by the dictatorial leader, the single party system, the total control of the press and all information sources, massive propaganda campaigns, tendencies toward the coordination of all political, economical, social and cultural institutions from above, and the power of the party militia, the police and the secret police. But behind this surface of strictly hierarchical dictatorship the fascist leaders' disregard for administration, their glorification of struggle and competition as an ideological expression of Social Darwinism led to a lack of constitutionality, to a deficient division of spheres of control and influence between the agencies, and, especially in the later years, to a multiplication of hurriedly erected ad hoc Commissariats without any proper plan of coordination. That, in turn, left much room for constant quarrels and boundary disputes between the party leaders, representatives of special party organizations (e.g. the SS, the Arbeitsfront in Germany), the army, the state machinery (traditionally the realm of the conservative bourgeoisie) and big industry as rival power blocs. This disintegration of the regime's power structure often made political decision procedures very ineffective. (With regard to Germany, see Fraenkel, 1941; Neumann, 1944; Broszat, 1969; Hirschfeld and Mommsen (eds), 1980.)

FASCISM AND THE ECONOMY

Fascism did not lead to any original contributions to economic theory except for some elements in the theory of corporatism added by Italian fascists. Positing the primacy of national over individual welfare, the fascist state was to direct economic activities for these purposes. In principle national interests meant economic strength on the basis of private ownership of the means of production, military power as a precondition for imperialistic expansion, independence in the world and autarky. These objectives implied in turn the necessity of rearmament. Thus in fascism the economy became ultimately an instrument of rearmament and autarky objectives; in Germany soon after fascism came to power (1934–5), in Italy during the World Depression that followed a period of relatively liberal economic policy (until 1926–7), in which a free-trade and a deflationary fiscal policy (to balance the budget) was implemented.

To revive the economy after the Depression the fascist regimes utilized deficit-financed government expenditures partly for infrastructural investments (like the Autobahnbau in Germany) but mainly for rearmament. Thus in Germany the total government expenditures as a proportion of gross national product doubled from 1932 to 1938. The armament expenditures as a proportion of GNP rose in the same time from nearly 1 per cent to more than 15 per cent, which in 1938 was 50 per cent of total government expenditure (Erbe, 1958). In addition the regimes tried to stimulate civil economic activities – such as house renovation – by tax reductions and/or pecuniary aid.

Credit policy basically functioned as a means to finance the budget deficit. Because the public debt could not be totally financed from the private capital market, the credit institutions were obliged to absorb the public debt by accepting public treasury certificates. Thus the credit institutions lost their usual function as intermediaries in the private circulation of capital. They served instead as a collecting box of money to cover public debts. Tax credit notes and, in Germany, the so-called Mefo-bills were further financing instruments. The German Reich's debt increased from RM14 milliard in 1933 to RM42 milliard in 1938, of which RM12 milliard were raised by the Mefo-bills, showing the high proportion of short-term debts. As long as full employment had not been achieved this credit expansion had little inflationary effect.

The control over the volume of investment by prohibiting the distribution of dividends above a fixed level (in Germany, six per cent, by subjecting new issues of shares to the permission of the state and by obliging firms to lend the government all their non-invested excess capital were supportive measures to the management of deficit spending.

Falling imports and exports as a result of the Depression and the protectionism of the time led, especially in the fascist countries, to serious tendencies towards an insulation from cyclical trade movements and the creation of a closed economy. A neomercantilistic foreign trade policy became a means of achieving these objectives. Bilaterization of foreign trade, based on clearing and barter agreements accompanied by the use of economic, political and, later, military pressure to attain favourable trade arrangements; import licences; export subsidies; fixing of quotas; control over foreign exchange and high tariff barriers: all these instruments were used to regulate foreign trade totally with regard to the programmes of autarky and rearmament.

Thus, in accordance with the old imperialist aims of big business and as a preliminary to creating the closed 'Grossraumwirtschaft', German foreign trade shifted from the western to the weak southeast European countries with their large resources of raw materials (Sohn-Rethel, 1973). The volume of German foreign trade with these countries as a proportion of total German foreign trade more than doubled between 1932 and 1938. To get special raw materials German foreign trade with Latin America and northeast European countries developed in the same direction.

Based on growing internal demand Germany experienced rapid economic revival. Full employment had been achieved by 1937–8 from a situation of over 6 million jobless in 1932–3. Although this success served to establish mass loyalty toward the fascist regime, economic development was undoubtedly more for the benefit of the propertied classes and, above all, of big industry, whose profits in 1938 were twice as high as in 1932 (Bettelheim, 1971, p. 232). As a result of the brutal destruction of all traditional independent labour organizations, the prohibition of strikes and the elimination of free wage negotiations, the degree of working class exploitation was increased, scarcely masked by some welfare services. While in Germany wages were fixed at the low level of the Depression year 1932, in Italy they were even cut. In Germany, the index of average weekly real wages reached the level of 1928 only in 1938, yet the average weekly labour time increased from 41.5 hours in 1932 to nearly 47 hours in 1938. Thus the growth of wages is to be seen as the result of rising working hours (Mason, 1977, p. 149). Wages and salaries as a proportion of national income fell from 64 per cent in 1932 to 57 per cent in 1938.

The growing profits were mostly ploughed back into investments. In Germany the gross investment as a proportion of GNP rose from 9 per cent in 1932 to more than 15 per cent in 1938. Although personal consumption increased, total consumption as a proportion of GNP fell from 81 per cent in 1932 to less than 64 per cent in 1938 (Mason, 1977, p. 149). The transformation of the production structure from consumer good industries to those of capital equipment was completely in line with the rearmament programme.

In pursuit of autarky, surrogates for imports and foreign raw materials were increasingly produced, shifting the orientation of many firms' production processes from the world to the domestic market. This often led to a loss of strong world market positions. This process was supported by a cartellization policy which was in contrast to the earlier anti-capitalist slogans of the fascist movement. Moreover, state-run factories were built up to increase the use of low-quality domestic raw materials with correspondingly high production costs. However, self-sufficiency could never be achieved. At the outbreak of the war Germany was still dependent on foreign supplies of oil, iron ore, manganese and many other raw materials (Kaldor, 1945, p. 42).

With the intensification of measures for rearmament and autarky, after full employment had been achieved, beginning in Germany with the declaration of Hitler's 'Vierjahresplan' in 1936, public finances drifted towards a ruinous situation. Inflation was only suppressed by extensive controls of prices and wages. In an attempt to manage critical shortages of raw materials, quota systems were introduced. For the same reason, the employment of the labour force was increasingly controlled and directed. However, these interventions into the running of the economy took place without any proper planning.

Although the outbreak of the war necessitated the further intensification of armaments production German war potential was never fully exploited (Kaldor, 1945). This would have meant the further extension of the average labour time, the employment of more women, the further reduction of consumer good production to the advantage of war production, and total planned economy. The reason the fascist leaders did not force the people to greater sacrifices is to be seen in their interpretation of Germany's defeat in World War I as a result of internal political instability (Mason, 1977).

WOLFGANG-DIETER CLASSEN

See also CORPORATISM; WAR ECONOMY.

BIBLIOGRAPHY

Bettelheim, C. 1971. *L'économie allemande sous le nazisme. Un aspect de la décadence du capitalisme.* Paris: Maspero.
Broszat, M. 1969. *Der Staat Hitlers. Grundlegung und Entwicklung seiner inneren Verfassung.* Munich: DTV.
De Felice, R. 1966. *Mussolini il fascista.* I: *La conquista del potere, 1921–1925,* Turin: Einaudi, 1966; II: *L'organizzazione dello Stato fascista, 1925–1929,* Turin: Einaudi, 1968.
De Felice, R. 1969. *Le interpretazioni del Fascismo.* Bari: Laterza.
Erbe, R. 1958. *Die nationalsozialistische Wirtschaftspolitik 1933–1939 im Lichte der modernen Theorie.* Zurich: Polygraph Verlag.
Fraenkel, E. 1941. *The Dual State. A contribution to the theory of dictatorship.* New York, London and Toronto: Octagon Books.
Gregor, A.J. 1974. *Interpretations of Fascism.* Morristown, NJ: General Learning Press.
Hirschfeld, G. and Mommsen, W.J. (eds) 1980. *Der Führerstaat: Mythos und Realität. Studien zur Struktur und Politik des Dritten Reiches.* Stuttgart: Klett-Verlag.
Kaldor, N. 1945. The German war economy. *Review of Economic Studies* 13(1), 33–52.
Lyttleton, A. 1973. *The Seizure of Power: Fascism in Italy 1919–1929.* London: Weidenfeld & Nicolson.
Mason, T.W. 1968. The primacy of politics – politics and economics in National Socialist Germany. In *The Nature of Fascism,* ed. S.J. Woolf, London: Weidenfeld & Nicolson.
Mason, T.W. 1977. *Sozialpolitik im Dritten Reich. Arbeiterklasse und Volksgemeinschaft.* Opladen: Westdeutscher Verlag.
Milward, A.S. 1965. *The German Economy at War.* London: Athlone Press.
Neumann, F. 1944. *Behemoth. The structure and practice of National Socialism.* 2nd edn, New York: Octagon Books.
Petzina, D. 1967. *Autarkiepolitik im Dritten Reich. Der national sozialistische Vierjahresplan.* Stuttgart: Deutsche Verlagsanstalt.
Sarti, R. 1971. *Fascism and Industrial Leadership in Italy 1919–1940.* Berkeley: University of California Press.
Schweitzer, A. 1964. *Big Business in the Third Reich.* Bloomington: Indiana University Press.
Sohn-Rethel, A. 1973. *Ökonomie und Klassenstruktur des deutschen Faschismus. Aufzeichnungen und Analysen.* Frankfurt am Main: Suhrkamp Verlag. Trans. by Martin Sohn-Rethel as *Economy and Class Structure of German Fascism,* London: CSE Books.
Turner, H.A., Jr. 1985. *German Big Business and the Rise of Hitler.* Oxford and New York: Oxford University Press.

fashion. *See* CHANGES IN TASTE; CONSPICUOUS CONSUMPTION.

Fasiani, Mauro (1900–1950). Fasiani was born in Turin and died in Genoa. Clearly the most important Italian scholar of fiscal theory to emerge in the interwar period (Buchanan, 1960, p. 36), he taught public finance in Turin, Sassari, Trieste and, from 1934, in Genoa. His career was rapid and exclusively academic. Despite his untimely death, he left important works on fiscal theory, and also on economic theory, economic policy and the history of economic thought.

Following Pareto's theory of the ruling class and Puviani's idea of fiscal illusion, which he rediscovered, Fasiani asserts that fiscal activity is to be explained on the basis of the nature of the political entity and not in terms of economic calculus or by sacrifice theories or by the ability-to-pay principle (1932a, 1941). As taxation and public expenditure are political phenomena, it is impossible to know the laws of fiscal activity. Fiscal theory can only be built through static models reflecting the different types of political societies. To De Viti de Marco's models of the 'monopolistic' state, where the ruling class governs only in its own interest, and of the 'cooperative' state, where the ruling class governs in the interest of every member of the community, Fasiani adds the model of the 'modern, nationalistic or corporative' state, in which the ruling class governs in the interest of the collectivity, considered as a whole (1941).

He dealt with the duration of the process of tax shifting (1934) and with the characteristics of intermediate positions in the transition from one state of equilibrium to another (1932b); with tax shifting in conditions of constant, increasing and decreasing costs in competition and in monopoly (1941, App. I and II) and with the effects of an excise tax under conditions of industrial concentration (1942a). He analysed the different elements determining the 'quantity of labour' and proved the impossibility of understanding the effects of taxation on labour supply assuming as variables only working hours and income (1942c). He devoted much research to the problem of the double taxation of saving (1926), confirming the validity of J.S. Mill's thesis in opposition to the theories of Einaudi and of Fisher. Fasiani also wrote important notes on the application of the Paretian indifference curve apparatus to the classical problem of the relative burden of income tax and consumption tax (1930) and on the analysis of the relationship between taxation and risk-taking (1935b).

In order to study the effects of taxation in a state of equilibrium, Fasiani re-examined and criticized some problems of economic theory. Among other things, he reasserted the hypothesis of production at constant costs and redefined the variables of the labour supply. Specifically he dealt with business cycles and stabilization policy, giving a decisive role to monetary policy (1935a, 1937a, 1942b).

His most important work in the history of economic thought is a very long essay on fiscal theory in Italy (1932c). In this work Fasiani critically examined the general theories of public finance formulated in Italy between 1880 and 1930, that is to say the economic theory, the political theory, the sociological theory, and also the theses on the effects of taxation and public debt, on tax shifting and tax incidence.

Finally, the essays on fiscal theory in the 18th century (1936) and on Francesco Fuoco (1774–1841), a forerunner of mathematical economics (1937b), are worthy of note.

MASSIMO FINOIA

SELECTED WORKS

A full bibliography of Fasiani' works is contained in: *Rivista di Diritto finanziario e Scienza delle Finanze* 9, September 1950, 216–18.

1926. Sulla teoria dell'esenzione del risparmio dall'imposta. *Memorie della Reale Accademia delle Scienze di Torino* 61, off-print.

1930. Di un particolare aspetto delle imposte sul consumo. *La Riforma Sociale* 41, January–February, 1–20.

1932a. Temi teorici ed 'exponibilia' finanziari. *La Riforma Sociale* 43, July–August, 383–425.

1932b. Velocità delle variazioni della domanda e dell'offerta e punti di equilibrio stabile e instabile. *Atti della Reale Accademia delle Scienze di Torino* 67, 383–425.

1932c. Der gegenwärtige Stand der reine Theorie der Finanzwissenschaft in Italien. *Zeitschrift für Nationalökonomie* 3(3), 651–91; 4(1), (1933), 79–107, 4(3), 357–88.

1934. Materials for a theory of the duration of the process of tax shifting. *Review of Economic Studies* 1, February, 81–101; 2, February 1935, 122–37.

1935a. Fluttuazioni economiche ed economia corporativa. *Annali di Statistica e di Economia* 3, 1–70.

1935b. Imposta e rischio. In AA.VV., *Studi in onore del prof. Salvatore Ortu Carboni*, Roma: Tipografia del Senato, 139–202.

1936. Precedenti di alcune recenti teorie finanziarie. *Annali di Statistica e di Economia* 4, 195–240.

1937a. Principi generali e politiche della crisi. *Annali di Statistica e di Economia* 12, 25–108.

1937b. Note sui 'Saggi economici' di Francesco Fuoco. *Annali di Statistica e di Economia* 5, 1–131.

1941. *Principi di Scienza delle Finanze*. 2 vols, Turin: Giappichelli; 2nd edn, 1951.

1942a. La translazione dell'imposta in regime di concentrazione industriale. *Studi Economici Finanziari Corporativi* 2, April–September, 200–225.

1942b. Potenziale di lavoro e moneta. *Annali di Statistica e di Economia* 9–10, 65–137.

1942c. Appunti critici sulla teoria degli effetti dell'imposta sull'offerta individuale di lavoro. *Annali di Statistica e di Economia* 9–10, 139–233.

BIBLIOGRAPHY

Buchanan, J.M. 1960. La scienza dell finanze: the Italian tradition in fiscal theory. In J.M. Buchanan, *Fiscal Theory and Political Economy, Selected Essays*, Chapel Hill: University of North Carolina Press.

Cosciani, C. 1950. Mauro Fasiani. *Economia Internazionale* 3, November, 913–19.

Einaudi, L. 1950. Mauro Fasiani. *Rivista di Diritto finanziario e Scienza delle Finanze* 9, September, 199–201.

Scotto, A. 1950. Gli scritti di Mauro Fasiani. *Rivista di Diritto finanziario e Scienza delle Finanze* 9, September, 202–15.

Faustmann, Martin (1822–1876). Faustmann was a German forester who spent much of his life working on the grand-ducal forests of Hesse. Between 1849 and 1865 he entered into controversies with other foresters concerning methods of forest valuations, his ideas eventually prevailing among that minority of forest economists who accepted the discipline of a positive rate of interest in making forest calculations. Although it has been said that his work was approved by such 'national economists' as Wagner and Roscher (*Allgemeine Deutsche Biographie*, 1877), it was evidently quite unknown to the more theoretically oriented German and Austrian specialists in capital and interest. Incorrect solutions to the optimum forest rotation problem were subsequently offered by such economists as Jevons, J.B. Clark and Irving Fisher, in the course of simplified expositions of the idea of the production period of a single investment. Not until the 1950s did economists working outside forestry realize that Faustmann's approach as explained to generations of resistant forestry school students contained a correct approach to the forestry question.

The economists' discovery was sparked by F. and V. Lutz, M. Gaffney, P.H. Pearse and, a few years later, Paul Samuelson. (The literature suggests that some Scandinavian and German economists, notably Ohlin, either knew of Faustmann's formula or worked it out for themselves.)

Faustmann's formula is derived from his investigations into forest values, needed at that time to guide the allocation of landowners' acres between trees and agriculture. His predecessors had consequently attempted to value the soil and the forest separately. In this they failed, partly because they confused stocks and flows. Faustmann cleared this up in 1849 by providing a single forward-looking approach for the present value of the next and future forest crops. As his professional readership required, his formulation also made it possible to take account of expected planting, husbanding, thinning and harvesting net costs during the life of each subsequent stand. He was able to solve his predecessors' problem by showing that the soil value (with which agricultural values are to be compared) is the value of the forest enterprise when it is still bare land, before a crop rotation has been commenced.

Faustmann is known today by resource economists for two by-products of his original perception. First, he showed correctly how to calculate the rotation age that is optimal for the owner in the presence of all expected costs and expected subsequent harvests. Second, by including the expected net discounted returns from subsequent rotations in his value and rotation-age formulae, he took the step that later eluded 20th-century economists, such as Fisher. He included the implicit foregone rent or shadow price of the land. He showed that the effect of doing this is that a given growth-and-harvest cycle will be shorter than economists' analyses would have predicted. Shorter rotations advance the date on which the next and all subsequent rotations will be harvested, thus reducing the effect of waiting on calculated soil values.

Faustmann made subsequent contributions to professional forestry, but they are of little interest today.

ANTHONY SCOTT

SELECTED WORKS

1849. Berechnung des Werthes, welchen Waldboden sowie noch nicht haubare Holzbestände für die Waldwirtschaft besitzen. *Allgemeine Forst und Jagd-Zeitung* 25, 441–55. Trans. by W. Linnard and included in Gane (1968). Samuelson (1976) contains an extended bibliography.
1877. Faustmann, Martin. In *Allgemeine Deutsche Biographie*. , Vol. VI, Leipzig: Duncker & Humblot.

BIBLIOGRAPHY

Bently, W.R. and Teeguarden, D. 1965. Financial maturity: a theoretical review. *Forest Science* 2, March, 76–87.
Dickson, H. 1953. Forest rotation. *Weltwirtschaftliches Archiv* 70.
Dowdle, B. (ed.) 1974. *The Economics of Sustained Yield Forestry.* Seattle: College of Forestry, University of Washington.
Fernow, B.E. 1902. *Economics of Forestry.* New York: T.Y. Crowell.
Fernow, B.E. 1911. *History of Forestry.* Revised edn, Toronto: University of Toronto Press.
Gaffney, M. 1960. *Concepts of Financial Maturity of Timber and Other Assets.* Agricultural Economics Information Series No. 62, Department of Agricultural Economics, North Carolina State College, Raleigh, December.
Gane, M. (ed.) 1968. *Martin Faustmann and the Evolution of Discounted Cash Flow; Two Articles from the original German of 1849.* Trans. by W. Linnard (includes translated articles by von Gehren and Faustmann), Oxford: Commonwealth Forestry Institute; Institute Paper No. 42.
Hiley, W.E. 1930. *The Economics of Forestry.* Oxford: Clarendon Press.
Lutz, F. and Lutz, V. 1951. *Theory of Investment of the Firm.* Princeton: Princeton University Press.
Pearse, P.H. 1967. The optimal forest rotation. *Forestry Chronicle* 43, June, 178–95.
Samuelson, P.A. 1976. Economics of forestry in an evolving society. *Economic Inquiry* 14, 466–92. This article evolved from the symposium edited by B. Dowdle (1974).
Scott, A. 1983. *Natural Resources and the Economics of Conservation.* 3rd edn, Ottawa: Carleton University Press. (See Appendix to ch. 3.)

Fawcett, Henry (1833–1884). Born on 26 August 1833, the son of a Salisbury draper, Henry Fawcett died on 6 November 1884, by which time he had been Professor of Political Economy in the University of Cambridge since 1863, a Liberal MP since 1864, and Postmaster General under Gladstone since 1880. His political career fulfilled a youthful ambition; his commitment to economics was a consequence of a shooting accident which blinded him at the age of 25. For although he was elected a Fellow of Trinity Hall soon after completing the Cambridge Mathematical Tripos in 1856, the loss of his sight forced him to abandon his studies for the Bar in favour of a professional career which could more easily dovetail with his political preoccupations. He had already begun to read himself into his parliamentary role with the aid of J.S. Mill's *Principles of Political Economy* (1848), and henceforth he depended exclusively on that text to supply the analytical and theoretical framework for his economics.

Fawcett's own textbook, *A Manual of Political Economy* (1863), expounded orthodox classical political economy in the tradition of Adam Smith as updated by Mill. Designed to provide the student (whether undergraduate, politician or general reader) with a clear, relevant, uncomplicated introduction to the state of economic knowledge, and to illustrate its applicability to a changing and complex real world, it went through six diligently revised editions in his lifetime; and his wife, Millicent Garrett Fawcett, a famous suffragette, saw two further editions through the press, the last in 1907. There was much repetition between this work and his other articles and books and the 18 lectures which were his only professorial duty. Fawcett wrote as he spoke, in the spirit of a determinedly non-doctrinaire liberal economist, pragmatically applying the principles of an established discipline to the practical policy problems currently facing government. Prevented by disability from engaging in systematic research in applied economics, he lacked the interest in abstract reasoning that might have drawn him to theoretical research, where his blindness would have been less of a handicap. Nevertheless, although he chose for himself the role of a teacher, a popularizer of classical orthodoxy, he was intelligently alive to the need to take other considerations into account when prescribing practical policies. For example, his best-seller on *Free Trade and Protection* (1878), after listing all the classical arguments in favour of free trade, went on to defend an Indian five per cent tariff on cotton imports from the United Kingdom, partly on revenue grounds and partly on grounds of natural justice.

The intellectual ferment associated with marginal revolution passed Fawcett by. Yet he did contribute to the debates of the 1860s on the labour question. Mill, for example, took into his *Principles* (with handsome acknowledgement to Fawcett) the idea that unionization was altering behaviour in the labour market by making employers and workers negotiate more rationally. But Fawcett refused to follow Mill in the latter's 1869 recantation of the wages–fund doctrine and took no interest in the 'new political economy' which was exciting the younger generation of Cambridge economists in the late 1870s and early 1880s and on which his successor Alfred Marshall was to set a distinctive personal stamp. On the other hand, his direct, realistic, unpolished attempts to explain the substance and policy implications of elementary economic analysis to non-professionals reached a much wider contemporary audience than the writings of any other late 19th-century English professor of political economy.

PHYLLIS DEANE

SELECTED WORKS

1863. *Manual of Political Economy*. Cambridge.
1865. *The Economic Position of the British Labourer*. London.
1871. *Pauperism: its causes and remedies*. London.
1878. *Free Trade and Protection. An inquiry into the causes which have retarded the general adoption of free trade*. London.

Fawcett, Millicent Garrett (1847–1929). A leading suffragist, Millicent Garrett Fawcett was also the author of a widely used elementary textbook, *Political Economy for Beginners* (1870). She married Henry Fawcett in 1867, when he was already Professor of Political Economy at Cambridge, the Member of Parliament for Brighton, and sightless (the result of a stray shot from his father's hunting gun in 1858). This led her to settle down as her husband's full-time secretary. It also brought her at the early age of twenty into close contact with a progressive intellectual circle which included among its elder statesmen Grote and Mill, and also Maurice, Sidgwick and Cairnes. Her first published article, in *Macmillan's Magazine* on Sidgwick's lectures at Cambridge to the unrecognized women students of the day (who included Mary Paley), led to a commission from Alexander Macmillan to write a primer on political economy based on her husband's *Manual of Political Economy*. While her *Political Economy for Beginners* is unremarkable in most respects, it does not follow Mill into the quick-sand of the wages-fund doctrine (see, for example, 1870, p. 25), and it was influential in accelerating that process of establishing economics as a suitable discipline for textbook writers which had been set in motion by Jane Marcet.

Nearly a quarter of a century later, she followed it with *Tales in Political Economy* (1894) which she confessed was little more than a 'plagiarism of Harriet Martineau's idea of hiding the powder of political economy in the raspberry jam of a story' (p. v). The book comprises four stories set on a desert island (thereby inculcating the view that the discipline deals in universals, which some see as having had unfortunate consequences in subsequent years), to illustrate the doctrines of free trade and division of labour, the theory of competition, and the theory of money. In the latter, coconuts serve as money, and the usual rules of the quantity theory are thereby elucidated in what is, for that theory, a rich institutional setting.

In 1872 she contributed eight of the fourteen chapters to *Essays and Lectures*, a book co-authored with Henry Fawcett. Amongst other topics, she attacked the expansion of the national debt, and opposed the extension of free elementary education on the grounds that it might remove checks to population. In two other essays she promoted the cause of higher education for women, a programme to which she helped to give more concrete form when she was later instrumental in the setting-up of Newnham Hall, Cambridge, which was incorporated as the first women's college in that city's university in 1874.

It was, however, in the area of the struggle for women's citizenship that she played her most significant role. She had joined a suffragist group as early as 1867, but it was only after Henry Fawcett's death in 1884 that she was able to allocate more time to her own political activities. From 1897 until 1918 (the year in which the suffrage was first extended to women in Britain), she was President of the National Union of Women's Societies and after her retirement she continued to campaign for full suffrage (achieved in 1928) and for professional and legal rights. She gave the movement her practical and intellectual support for better than fifty years, a measure of her dedication to the cause.

MURRAY MILGATE AND ALASTAIR LEVY

SELECTED WORKS

1868. The education of women of the middle and upper classes. *Macmillan's Magazine* 17(102), 511–17.
1870. *Political Economy for Beginners*. London: Macmillan.
1872. (With H. Fawcett.) *Essays and Lectures on Social and Political Subjects*. London: Macmillan.
1874. *Tales in Political Economy*. London: Macmillan.
1924. *What I Remember*. London: T. Fisher Unwin.

Fay, Charles Ryle (1884–1961). Lancashire-born economic historian, whose grandfather worked as a boy on the construction of the first railway coaches for the Liverpool and Manchester Railway and later invented the chain brake used for the emergency stopping of trains, Fay subscribed to a vision of the progress of industrial society towards 'happiness and beauty'. Increased specialization and improvements in the division of labour were, for him, essential to progress. Fay was not, however, unaware that the historical record of industrialization had been marred by hardship, poverty and waste. But these effects had not, in his view, been unavoidable. The exploitation of child and female labour, the appalling conditions in Britain's factories and industrial towns in the 19th century, and the recurrence of distress in agricultural communities, all received Fay's strong condemnation. His liberalism had a social conscience about it. He certainly did not number among those apostles of social laissez-faire who, on his own speculation, might well be found on the lowest ledge of Dante's Inferno (1928, p. 358).

Fay's academic career is easily summarized. He was a favourite pupil of Marshall at Cambridge, and in 1908 he was elected to a fellowship at Christ's College. The same year saw the publication of his study of co-operation in agriculture which established his credentials as an economic historian. Fay remained in Cambridge until 1921, when he removed to Canada to take up a chair in Economic History at Toronto. Nine years later, he returned to Cambridge as Reader in Economic History, where he remained until his retirement.

Some idea of Fay's humane and liberal instincts can be gained from his *Co-operation at Home and Abroad* (1908). Its central thesis was that, contrary to popular opinion at the time, there remained both a social and economic role to be filled by small cultivating ownership. Its prospects, however, rested on the ability of its participants to establish what would today be called marketing boards. Fay saw in the Canadian wheat pools and the cases of co-operation among Californian fruit growers the promise of things to come (1928, p. 250). Never losing his faith in the market, he stressed that this kind of co-operation was the antithesis of collective ownership and that, what is more, it was the only form of agricultural co-operation that the historical record suggested might work (1908, pp. 350–52). There is more than a faint echo of John Stuart Mill in this advocacy of producer co-operatives over collectivization.

Fay's *Life and Labour in the Nineteenth Century* (1920) expanded on his concern with social history and was based on his Cambridge lectures; it surveyed the main features and figures of the economic, political and social history of the period, and examined the relationship between them and theoretical discourse in economics. This project was repeated on a rather more grand scale in *Great Britain From Adam Smith to the Present Day*, a book first published in 1928 which went through five editions before Fay's death in 1961. This book embodies all the hallmarks of Fay's approach to the study of history. In particular, it reveals very clearly his attempt to trace to their basis in economic theory the practical and political ideas around which history unfolded. In a similar fashion, the subject of protection came under Fay's scrutiny in *The Corn Laws and Social England* (1932) and *Imperial Economy* (1934).

MURRAY MILGATE AND ALASTAIR LEVY

SELECTED WORKS

1908. *Co-operation at Home and Abroad: A Description and Analysis.* London: P.S. King.
1920. *Life and Labour in the Nineteenth Century.* Cambridge: Cambridge University Press.
1928. *Great Britain from Adam Smith to the Present Day.* London: Longmans, Green; 5th edn, 1950.
1932. *The Corn Laws and Social England.* Cambridge: Cambridge University Press.
1934. *Imperial Economy and Its Place in the Formation of Economic Doctrine 1600–1932.* Oxford: Oxford University Press.
1940. *English Economic History, mainly since 1700.* Cambridge: W. Heffer & Sons.

fecundity. Fecundity is defined as the ability to reproduce, whereas fertility is actual reproduction. Because differences in both unobserved fecundity and contraceptive behaviour can cause observed variation in fertility, it can be difficult to separate biological from behavioural influences on fertility. This identification problem is more troublesome in studies of individual than in aggregate fertility behaviour. Fertility trends and differentials at the aggregate level must be due primarily to socioeconomic factors since even wide variations in levels of health and nutrition have little effect on fecundity. Only in populations experiencing widespread malnutrition or a high prevalence of diseases leading to sterility (as has occurred in parts of Africa) does fecundity appear to be significantly impaired.

The treatments of fecundity in economic and demographic models of individual fertility behaviour will be compared using a framework that focuses on the stochastic process generating births. The single parameter (p) characterizing a waiting time process generating births is specified as the difference between an underlying component (n) that is exogenous to the individual decisions and the choice of contraception (c). Based on perceived costs and benefits, parents choose c (between 0 and n) to affect their probability of a birth.

Demographers who follow in the tradition of Henry (1957) model the reproductive process and the stages through which a women passes throughout her fertile period, but do not model the choice of c. Such a demographic model of a non-contracepting population can be considered a special case of a more general decision-theoretic model if that model permits the optimal choice of c to be zero. The demographic and economic approaches are therefore not to be distinguished by whether they model the decision for c, but by how they implicitly or explicitly model n, the underlying component that is exogenous to the couple's decisions. The main distinguishing features are (1) whether n is assumed to be a function of fecundity only or also of various socioeconomic variables and

(2) whether n is represented by a single (or possibly age-dependent) value or takes on distinct values corresponding to different stages throughout the interval between births. If n is solely a function of fecundity, then observed correlations of fertility with socioeconomic variables will reflect only those variables' influence on c. If not, the observed correlations will reflect a combined influence on n and c.

Economic models that focus on the price and income variables affecting fertility typically regard n as reflecting the level of fecundity, with variations in n being uncorrelated both with socioeconomic variables that explain c and with c's error term. The level of fecundity can influence the choice of c, whether or not couples can perceive their fecundity. Couples who perceive their higher fecundity may try to offset their higher n by choosing a higher c. If total contraceptive costs depend on the level chosen, the offset may not be complete and couples with higher n may have a higher probability of a birth than otherwise identical couples. The contraceptive decisions of those unable to ascertain n will also be affected, to the extent that they choose c conditional on their current number of children alive and to the extent that, at any given time, higher fecundity couples have more children.

The possible dependence of c on n does not present difficulties in estimating the determinants of n–c. However, the determinants may be estimated only after eliminating the effects of both unobserved fecundity, n, and the errors in predicting the choice of c from the likelihood function used to describe fertility histories. Provided n is uncorrelated with the socioeconomic variables, the combined error terms can be treated as a random effect. The random effect can be integrated out of the likelihood function by assuming a parametric distribution, if results do not appear sensitive to the choice of distribution. If the results are sensitive to the distributional assumption, then a nonparametric procedure may be followed.

While the expected number of births can be derived from the estimated probabilities, the usual procedure has been to regress the number of births on the socioeconomic variables that determine c. If n is uncorrelated with the latter variables, then the error term of a regression on completed family size will also be uncorrelated with them.

Two potential problems arise when the number of births to those at younger ages and with incomplete families is regressed on socioeconomic variables. The distribution of the error term in the regression may then be misspecified since the number of births reflects the outcomes of waiting-time processes. This is not likely to be a serious problem when couples have had sufficient time for their behaviour to compensate for differences in levels of fecundity.

A more serious problem arises if the observations on fertility histories are censored, as would be expected for younger women. Those couples who have chosen a lower probability of a birth are more likely to have the lengths of their births intervals truncated by the observation date. This imparts a bias to the estimated effects of socioeconomic variables on the number of births. It can be corrected by using additional information on the censored lengths of birth intervals to infer the distribution of uncensored intervals. The likelihood function describing fertility histories is amended to incorporate the probability of not observing a birth, which is equal to one minus the cumulative distribution function of the uncensored distribution. How useful the censored observations are in providing information on the uncensored observations is an issue that must be decided on empirical grounds.

In summary, economic models that assume n to be uncorrelated with socioeconomic variables will attribute an observed correlation of fertility with such variables to their influence on c. An explicit consideration of fecundity will be required in these models if one is interested in the determinants of the probability of a birth and the spacing of births or if one is interested in the determinants of births and must use observations on women who cannot be assumed to have completed their fertility.

Demographic models of birth probabilities implicitly model n as being correlated with socioeconomic variables. Under this interpretation, an observed correlation between fertility and such variables can exist even when c is always equal to zero (i.e. when couples are not trying to control their births).

The implicit dependence of n on socioeconomic variables is apparent in the analyses of natural fertility populations, defined by Henry as those populations that do not practice contraception or induced abortion. A key technique of these analyses (e.g. Leridon, 1977) is to decompose natural fertility into its underlying components which are: (1) the age at marriage and duration of marital separation, (2) the waiting time to conception for a susceptible woman (3) the time added to the birth interval by intra-uterine mortality, (4) the duration of postpartum infecundability and (5) the age at onset of permanent sterility. Differences in gestation lengths are inconsequential. This methodology of breaking down fertility outcomes into intervening components has been extended to the case of contracepting populations by considering (6) the use and effectiveness of contraception and (7) induced abortion. By definition, any determinant of fertility must act through one or more of these proximate determinants.

The first five components interact to yield substantial variations across natural fertility populations in expected mean completed family sizes for women who are married at age 20. The mean family sizes range from 5.4 under the marital fertility rates prevailing in villages near Bombay in 1954–55 to 10.9 for the Hutterite population in the USA with marriages between 1921 and 1930 (Leridon, 1977). Based on a sensitivity analysis where the natural fertility components are varied separately through their approximate ranges, Bongaarts and Potter (1983) conclude that the largest variations in simulated total fertility rates are due to changes in the age at marriage and in the duration of postpartum infecundability, both of which can be substantially affected by individual decisions.

Thus, one implication of the demographic approaches is that n is determined by a combination of factors (2) through (5). If n is represented by a single value throughout the birth interval when, in fact, distinct biological factors operate over different stages of the birth interval, the model will be misspecified. The possible specification error must be balanced against the bias that would arise if the identification of the different stages is accomplished by conditioning on a choice variable of the parents, such as the length of breastfeeding.

A second implication is that n is a function of socioeconomic variables. If both n and c are functions of the same variables, then a noncontracepting population can be identified solely on the basis of fertility data, only under a maintained hypothesis that parents initiate or alter their control after a birth. This hypothesis is maintained in the literature on natural fertility. Identifying the effects of observed socioeconomic variables requires an explict formulation of how the variables affect n and c, noting that c may also depend on n. Identification may be facilitated if either economic theory, or a biological theory of the determinants of n, specifies how n and c respond to births and deaths.

Treating the unobserved components of n–c as a random effect, as described above, will provide a reduced form estimate of the effect on fertility of a variable that affects both

n and c. Identifying the separate effect on c is possible if n can be treated as a fixed effect and eliminated from the estimating equation. A comparison of the estimated coefficients from the fixed effect model and the random effect model would provide information on the variable's effect on n.

JOHN L. NEWMAN

See also DEMOGRAPHY; FAMILY PLANNING; FERTILITY.

BIBLIOGRAPHY

Bongaarts, J. and Potter, R.G. 1983. *Fertility, Biology, and Behavior: An Analysis of the Proximate Determinants.* New York: Academic Press.

Bulatao, R.A. and Lee, R.D. (eds) 1983. *Determinants of Fertility in Developing Countries.* 2 vols, New York: Academic Press.

Henry, L. 1957. Fécondité et famille: modèles mathématiques (I). *Population* 12(3). Trans. as 'Fertility and family: mathematical models I' in *On the Measurement of Human Fertility*, ed. M.C. Sheps and E. Lapierre–Adamyck, New York: Elsevier, 1972.

Leridon, H. 1977. *Human Fertility: The Basic Components.* Chicago: University of Chicago Press.

federalism. *See* FISCAL FEDERALISM.

Fel'dman, Grigorii Alexandrovich (1884–1958). Fel'dman was one of the founders of the theory of economic growth under socialism, the economics of planning and development economics. An electrical engineer by profession, he worked in Gosplan from February 1923 to January 1931. It was in this period that his contribution to economics was made. At first he was in the department analysing and forecasting developments in the world economy (he concentrated on Germany and the USA). His first work on the theory of growth was a comparative study of the structure and dynamics of the US economy in 1850–1925 with projections of the Soviet economy between 1926/27 and 1940/41. His most important work ('On the theory of the rates of growth of the national income') was a report to Gosplan's committee for compiling a long-term plan for the development of the national economy of the USSR. It was published in two parts in Gosplan's journal in 1928. A year later Fel'dman published a paper which provides a more popular presentation of how to utilize his ideas to calculate long term plans. The ideas of Fel'dman formed the methodological basis for the preliminary draft of a long term plan worked out by the committee, then headed by N.A. Kovalevskii. This draft was discussed at meetings of Gosplan's economic research institute in February and March 1930. Apart from this serious discussion, during 1930 Fel'dman came under public attack for his ideas. His reliance on mathematics and his lack of fanaticism did not fit in well with the political fervour of 1930. The concrete numerical work of Fel'dman and Kovalevskii in 1928/30 was much too optimistic. It treated as feasible entirely unrealizable goals. The attempt to realize them had disastrous effects on the economy. Unfortunately, the political situation in the USSR prevented Fel'dman from publishing anything on economics after 1930. Even when, in 1933, he reverted from the sensitive subject of socialist industrialization to the problems of capitalist growth, his book was not published.

As far as growth theory is concerned, Fel'dman's work was much in advance of contemporary Western work. He developed a two-sector growth model and showed how different growth rates implied different economic structures.

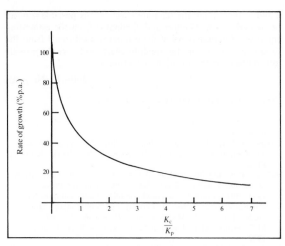

Figure 1 Fel'dman's first theorem. K_c is the capital stock in the consumer goods industry, K_p is the capital stock in the producer goods industry.

He derived two important results, one about the ratios of the capital stocks in in the two sectors, the other about the allocation of investment between the two sectors. The first result is that a high rate of growth requires that a high proportion of the capital stock be in the producer goods sector. This is illustrated in Figure 1. Fel'dman's second theorem is that, along a steady growth path, investment should be allocated between the sectors in the same proportion as the capital stock. For example, suppose that a 20 per cent rate of growth requires a K_c/K_p of 3.7. Then to maintain growth at 20 per cent p.a. requires that 3.7/4.7 of annual investment goes to the consumer goods industries and 1.0/4.7 of annual investment goes to the producer goods industries.

The interrelationship between the two theorems is shown in Table 1, in which Fel'dman explained how any desired growth rate, given the capital–output ratio, determined both the necessary sectoral composition of the capital stock and the sectoral allocation of investment.

Given the capital–output ratio, the higher the K_p/K_c ratio, i.e. the greater the proportion of the capital stock in the producer goods sector, and correspondingly the higher the $\Delta K_p/\Delta K_c + \Delta K_p$) ratio, i.e. the greater the proportion of new investment in the producer goods sector, the higher the rate of growth. With a capital–output ratio of 2.1, to raise the growth rate from 16.2 to 24.3 per cent requires raising the proportion of the capital stock in the producer goods sector from $\frac{1}{3}$ to $\frac{1}{2}$, and the share of investment in the producer goods sector from $\frac{1}{3}$ to $\frac{1}{2}$.

The conclusion Fel'dman drew from his model was that the main tasks of the planners were to regulate the capital–output ratios in the two sectors and the ratio of the capital stock in

the producer goods sector to that in the consumer goods sector. For the former task, Fel'dman recommended rationalization and multi-shift working, for the latter, investment in the producer goods sector.

As far as the economics of planning is concerned, the main lesson to be learned from the Fel'dman model is that the capacity of the capital goods industry is one of the constraints limiting the rate of growth of an economy. There may well be other constraints, such as foreign exchange, urban real wages or the marketed output of agriculture. (Indeed, it is possible that one or more of these is/are the binding constraint/s and that the limited capacity of the producer goods sector is a non-binding constraint.) Economic planning is largely concerned with the removal of constraints to rapid economic growth. Accordingly, a planned process of rapid growth may require that the planners stimulate the rapid development of the producer goods sector.

As far as development economics is concerned, Fel'dman is important because of the argument in his 1928 paper that 'an increase in the rate of growth of income demands industrialization, heavy industry, machine building, electrification ...'. When first formulated, this conclusion struck many economists as counter-intuitive and paradoxical.

Fel'dman's work, as is natural for a pioneer, suffers from serious limitations. As far as the theory of economic growth under socialism is concerned, he was an important early contributor, but his work has to be complemented by Kalecki's emphasis on the limits of growth and Kornai's emphasis on the behavioural regularities actually generating the growth process. As for the economics of planning, his arguments have to be complemented by a proper understanding of the role of agriculture, foreign trade and personal consumption and of the danger of an over-accumulation crisis. In development economics, experience in the USSR in the 1930s, India in the 1950s and China in the Maoist period has shown the limitations of a narrowly Fel'dmanite approach.

A brilliant pioneer, Fel'dman's work was ended after only a few years by the Stalinists.

MICHAEL ELLMAN

SELECTED WORKS
1927. Soobrazheniya o strukture i dinamike narodnogo khozyaistvo SSha s 1850 po 1925g i SSSR s 1926/27 po 1940/41gg (Reflections on the structure and dynamics of the national economy of the USA from 1850 to 1925 and of the USSR from 1926/27 to 1940/41). *Planovoe khozyaistvo* no. 7. Also published as a booklet.
1928. K teorii tempov narodnogo khozyaistva (On the theory of the rates of growth of the national income). *Planovoe khozyaistvo* nos. 11 and 12. English translation in *Foundations of Soviet Strategy for Economic Growth*, ed. N. Spulber, Bloomington: Indiana, 1964, pp. 174–99 and 304–31.
1929a. SSSR i mirovoe khozyaistvo na rubezhe vtorogo goda pyatletki (The USSR and the world economy on the eve of the second year of the five year plan). *Na planovom fronte* no. 2.
1929b. O limitakh industrializatsii (On the limits of industrialization). *Planovoe khozyaistvo* no. 2.
1929c. Analiticheskii metod postroeniya perspektivnykh planov (An analytical method for constructing perspective plans). *Planovoe khozyaistvo* no. 12.
1930. Problemy electrifikatsii na novom etape (Problems of electrification at a new stage). In *Na novom etape sotsialisticheskogo stroitel'stva*, vol. 1, Moscow.

BIBLIOGRAPHY Soviet evaluations Report of the discussion in Gosplan's economic research institute of February and March 1930. *Planovoe khozyaistvo*, 1930, no. 3, pp. 117–211.

TABLE 1. Fel'dman's two theorems

$\dfrac{K_p}{K_c}$	$\dfrac{dY}{dt}$ (in % p.a.) (when $K/Y = 2.1$)	$\dfrac{\triangle K_p}{\triangle K_c + \triangle K_p}$
0.106	4.6	0.096
0.2	8.1	0.167
0.5	16.2	0.333
1.0	24.3	0.500

Vainshtein, A.L. and Khanin, G.I. 1968. Pamyati vydayushchegocya sovetskogo ekonomista-matematika G.A. Fel'dmana (In memory of the outstanding Soviet mathematical economist G.A. Fel'dman). *Ekonomika i matematicheskie metody* 4(2).

English-language works

Chng, M.K. 1980. Dobb and the Marx–Fel'dman model. *Cambridge Journal of Economics* 4(4), 393–400.

Dobb, M. 1967. The question of 'Investment priority for heavy industry'. Chapter 4 of M. Dobb, *Papers on Capitalism, Development and Planning,* London: Routledge.

Domar, E.D. 1957. A Soviet model of growth. In E.D. Domar, *Essays in the Theory of Economic Growth,* New York: Oxford University Press.

Erlich, A. 1978. Dobb and the Marx-Fel'dman model: a problem in Soviet economic strategy. *Cambridge Journal of Economics* 2(2), 203–14.

Mahalanobis, P.C. 1953. Some observations on the process of growth of national income. *Sankhya* 12(4), 307–12.

Mahalanobis, P.C. 1955. The approach of operational research to planning in India. *Sankhya* 16(1 and 2), 3–130.

Sen, A.K. and Raj, K.N. 1961. Alternative patterns of growth under conditions of stagnant export earnings. *Oxford Economic Papers* 13(1), 43–52.

Tinbergen, J. and Bos, H.C. 1962. *Mathematical Models of Economic Growth.* New York: McGraw.

be preferable. He argued that price–wage expectations are critical since uncertainty could defeat Keynesian policies.

Thirty years later (Fellner, 1976), he amplified this theme. He suggested that dynamic macroeconomic equilibrium requires not only that savings–investment decisions be validated but also that price expectations be close to actual price levels. He qualified the now usual rational expectations model with the view that public predictions of government reactions are probabilistic; credibility is, therefore, critical. But government should not passively validate just any expectations. Rather, a major policy aim should be to create an environment of restraint which leads to stable rather than explosive expectations. Thus, he cast government in the same role as himself – that of a wise teacher.

IRMA ADELMAN

SELECTED WORKS

1946. *Monetary Policies and Full Employment.* Berkeley: University of California Press.

1949. *Competition Among the Few.* New York: Alfred A. Knopf.

1955. *Trends and Cycles in Economic Activity.* New York: Holt, Rinehart & Winston.

1960. *Emergence and Content of Modern Economic Analysis.* New York: McGraw-Hill.

1965. *Probability and Profit.* Homewood, Ill.: Richard D. Irwin.

1976. *Towards a Reconstruction of Macroeconomics.* Washington, DC: American Enterprise Institute.

Fellner, William John (1905–1983). Fellner was born in Budapest and received his PhD at the University of Berlin. In 1938 he moved to the United States and taught at Berkeley (1939–52) and Yale (1952–73). He was President of the American Economic Association (1969) and a member of the Council of Economic Advisers (1973–5).

His major contributions were to macroeconomic theory and policy. Those who, like myself, were fortunate enough to know him came to worship him because of his combination of nobility of spirit, profundity, subtlety, humility, deep culture and inherent humanity. His writings were shaped by all these qualities as well as by his formative experiences in interwar Europe. He was a liberal of the old school – a humanist and an anti-authoritarian. He had been traumatized by the German hyperinflation, the mass unemployment of the Great Depression, and by the Nazi totalitarianism which ensued. His teachings were committed to avoiding a repetition.

Upon re-reading his *Monetary Policies and Full Employment* (1946), one is struck by how far ahead of his time he was. A limited Keynesian, he advocated policies aimed at avoiding severe recessions and allowing small ones to run their course because he foresaw that an unconditional guarantee of full employment would lead monopolistic groups of industrialists and workers to constantly raise wages and prices and reduce quality. This is precisely what happened in the industrial countries between 1965 and 1973. The result would be that an unconditional full employment guarantee would require government controls on wages and prices and would ultimately result in a severe abrogation of both liberty and market efficiency. He argued that growth and cycles are interdependent as are price stability and employment. He emphasized the role of uncertainty, expectations and credibility of government-policy commitments. He foreshadowed both a subtler version of supply-side economics and of rational expectations. With respect to supply-side economics, he argued that fiscal expansionism should be limited to counteracting severe recessions only and that otherwise a combination of credit policies, price-cost policies, and tax policies aimed at increasing the level of private activity would

female labour force participation. *See* LABOUR SUPPLY OF WOMEN.

Ferguson, Adam (1723–1815). Ferguson was born in Perthshire in 1723 and died in Edinburgh in 1815. He was educated at St Andrew's University for the Church of Scotland and became a leading member of the 'moderate' clergy which controlled its affairs from 1752 to 1805. He was a charismatic teacher who held the Moral Philosophy chair at Edinburgh from 1764 to 1785, transforming its curriculum and laying the foundations of its international reputation. As a moralist, Ferguson was worried by the materialism inherent in modern philosophy and modern life, and was anxious to show that the classical republicanism of the Machiavellians was still of value in analysing and resolving its problems. He presented human beings as active rather than passive agents who were motivated by a natural love of perfection that seemed to be in danger of extinction in a commercial world. In the process he showed that the mechanics of social bonding in primitive societies in particular were more complex than contemporaries realized, a demonstration that continues to be admired by anthropologists.

Marx admired Ferguson's discussion of the division of labour and the apparent alienation that accompanied its progress and he thought that Smith's treatment of the subject owed much to him. In fact the resemblances are only superficial. Ferguson's treatment of the subject and of political economy generally was derivative and shaped by the classical republican's traditional concern with virtue, corruption and the place of the heroic virtues in an age of commerce. *The Wealth of Nations* was to leave no significant marks on his thought. He was a moralist who sought to tighten, not loosen the ties which bound political economy to moral philosophy.

Ferguson's contemporary reputation rested on three frequently republished and translated works, *An Essay on the History of Civil Society* (1769) and *The History of the Progress and Termination of the Roman Republic* (1783). His lectures were published as *The Principles of Moral and Political Science* (1792).

NICHOLAS PHILLIPSON

See also SCOTTISH ENLIGHTENMENT.

SELECTED WORKS
1767. *An Essay on the History of Civil Society.* Ed. D. Forbes, Edinburgh: University Press, 1966.

BIBLIOGRAPHY
Kettler, D. 1965. *The Social and Political Thought of Adam Ferguson.* Columbus: Ohio State University Press.

Ferrara, Francesco (1810–1900). Ferrara was not only an economist but also an influential figure in Italian politics, culture and journalism. He was born in Palermo on 7 December 1810 and died in Venice on 22 January 1900. His long life spanned the political unification of Italy and the country's first attempts to assert itself as a latecomer to the international scene. As a patriot, he was one of the leaders of the Sicilian revolution against the Bourbons in 1848. Although the failure of this uprising led to the return of the Bourbons and subjected Ferrara to exile in Turin, one of the most significant documents of this period is the *Letter from Malta*, which constituted a formal indictment of the Bourbon government and is attributed to him. In Turin, Ferrara became a friend of Cavour, and he was appointed Professor of Political Economy at the university there. As soon as Sicily was liberated he returned to Palermo, where he was placed in charge of indirect taxation.

In 1862 he went back to Turin to assist Quintino Sella, the founder of Italian public finance, in the formulation of fundamental laws to resolve, through harsh and unpopular measures, the financial difficulties of the time. For a brief period in 1867 he was Minister of Finance, and he subsequently became a member of the Chamber of Deputies until 1880. During this time of political and parliamentary activity he also produced a prodigious amount of important journalistic work, inspired by his guiding principle that 'economics was the new way of the necessity of freedom'. His intransigent and uncompromising liberalism placed him in direct contrast with Cavour and contributed to his position as a respected but isolated figure.

His economic achievements are such that he is highly regarded by all the greatest Italian economists (Pareto, Pantaleoni, Einaudi, Del Vecchio) as having inspired and created an Italian economic school of thought. Ferrara has a double claim to this. As the founder of an *Economists' Library*, of which he edited the first two series, he brought the greatest foreign economists within the sphere of Italian cultural life, writing perceptive prefaces to the translations of their works. Guided by his extraordinary knowledge of economic thought, he included translations of the works of authors such as Henry Carey at a time when that author's work was forgotten even by his fellow–Americans. As a theoretician, Ferrara pursued vigorous polemics against the German historical school, which rejected the theoretical method in economics. He developed the concept of cost of reproduction based on technical and psychological factors which pre-dated the marginalist theory in all but name. He was also an early forerunner of the Italian tradition in the economic foundations of public finance.

However, Ferrara has yet to achieve his rightful recognition at an international level. Praised by G.H. Bousquet, who edited the French translation of his selected works, criticized by Schumpeter for his 'ultraliberalism', he is unknown to modern proponents of neo-liberalism, even though he was the forerunner of many of their proposals for de-regulation; for example, the free creation of money without state interference, which has been recently advocated by Hayek. Another aspect of Ferrara's thinking that finds an immediate place in today's economic debates is his conception of 'generalized crowding out', which he formulated not only with regard to the predominant absorption of financial flows by the public sector (and therefore to the detriment of private enterprise) but with respect to every form of public intervention.

F. CAFFÈ

SELECTED WORKS
Part of *Le opere complete di Francesco Ferrara* was published in homage to Luigi Einaudi, on the occasion of his 80th birthday. This edition, which was edited with great philological rigour by Bruno Rossi Ragazzi, who died prematurely, and was continued by various other acedemics, has yet to be completed. In its present state it comprises ten volumes, which include his *Prefazioni alla biblioteca dell'Economista*, his *Articoli su giornali*, his *Saggi rassegne memorie economiche e finanziarie* and *Discorsi parlamentari*. Not included are the *Lezione de Economia Politica*, which were given by him and which are available in a two-volume edition edited by G. De Mauro Tesoro (1934–5), Zanichelli: Bologna. A collection of *Oeuvres économiques choisies* by Ferrara has been edited by G.H. Bousquet and J. Crisafulli (1938), Paris: Rivière. See also G.H. Bousquet (1960), *Esquisse d'une historie de la science économique en Italie – des origines à Francesco Ferrara*, Paris: Rivière.

fertility. At the aggregate level, human reproduction is the ultimate source of an economic system's labour input and of the consumers who constitute the principal destination of the economy's output. At the individual level, children are an important source of satisfaction that compete with alternatives for the limited parental resources of time, energy and money available. Despite this, reproductive behaviour has traditionally been omitted from economic theorizing, and even in the past three decades has gained only a marginal foothold.

Possibly the hesitancy of economic theory to address the determinants of childbearing reflects a sensitivity to reality. Several empirical regularities involving the relation of fertility to income have posed a formidable challenge to theoretical interpretation. First, there is the long-term trend. In an historical epoch when real income per capita and, in consequence, real consumption of almost all goods has risen at unprecedented rates for a century or more in developed countries, births per couple over the reproductive career have fallen from levels often as high as six or more to two or less. Second, the cross-sectional relation between fertility and income within countries has been found to be variable and often lacks any significant association. Third, over the business cycle a positive association between fertility and income has typically been observed. Moreover, in a number of developed countries in the post-World War II period there was an unprecedented and unanticipated 'baby boom' of a decade or more in duration followed by an almost equally startling 'baby bust'.

It is often the case that new policy concerns stimulate economic theory, and this is clearly so with regard to fertility behaviour. Although unusually low fertility in the developed countries in the Great Depression had led to some experimentation with economic incentives to childbearing, the major policy stimulus by far was the emergence in the post-World War II era of the so-called 'population problem' as a presumed obstacle to economic growth in the less-developed world. Could measures be designed to lower reproduction rates in high-fertility societies and thus reduce rates of population growth? Concern with this issue spurred a number of economists to take a fresh look at fertility behaviour.

The contemporary economic theory of fertility dates from work by Harvey Leibenstein (1957) and Gary S. Becker (1960), which sought in somewhat different ways to assimilate the explanation of fertility behaviour to the economic theory of household demand (Leibenstein (1974) and Keeley (1975) give a good review of this early history). In 1965 Becker extended his analysis to incorporate the emerging concepts of household production theory and the allocation of time (Becker, 1965; Lancaster, 1971). For a decade or so this line of work, which came to be known as the Chicago–Columbia approach, dominated the economic theory of fertility. Among the more influential contributions were those made by Mincer (1963), Nerlove (1974) and Willis (1973). A volume edited by T.W. Schultz (1974) and a survey article and subsequent book by T.P. Schultz (1976, 1981) brought together a number of attempts to apply this approach empirically to the experience of developed and less-developed countries. A valuable commentary on the evolution of this work appears in Ben-Porath (1982).

Throughout much of this period, a second line of work was in progress that came to be dubbed the 'Pennsylvania' model (Sanderson, 1976, 1980; Behrman and Wolfe, 1984). Although this work largely accepted the Chicago–Columbia view as far as it went, it sought to broaden the model to include theoretical and empirical considerations that figured prominently in the sociological and demographic literature on fertility. One set of considerations related to taste influences on the demand for children, particularly of what economists would term a 'relative income' nature (Ben-Porath, 1975; Easterlin, 1969; Leibenstein, 1975). Good theoretical expositions of the demand for children, reflecting taste considerations as well as those in the Chicago–Columbia model, are given by Lindert (1978) and Turchi (1975). The other set of considerations relates to 'natural fertility' or so-called 'supply' factors (Easterlin, 1978; Tabarrah, 1971). A formal statement of what came to be termed the 'supply–demand' approach was published in 1980 (Easterlin, Pollak and Wachter, 1980). In 1983 an interdisciplinary National Academy of Sciences panel adopted the supply–demand framework in surveying the literature on determinants of fertility in developing countries (Bulatao and Lee, 1983). (As shall become clear, in this theory supply and demand are not used in the usual economic sense.)

Recent work points to some convergence of the two lines of work. Scholars working in the Chicago–Columbia tradition have introduced into their work intergenerational influences which can be likened to the taste influences of the Pennsylvania model (Becker and Tomes, 1976), and have also started investigating supply factors (Michael and Willis, 1976; Rosenzweig and Schultz, 1985). But the two schools remain sufficiently distinct, especially in their interpretation of the empirical regularities described above, to warrant separate discussion. Using the empirical regularities as the framework for the discussion the following aims to indicate for the non-specialist the principal ideas on each side, their relations to each other, and their bearing on the interpretation of each of the empirical regularities mentioned. Needless to say, there is also variability within each school as well as work not easily classified under either head.

THE SECULAR DECLINE IN FERTILITY. Over the long term, growth in real per capita income has everywhere been accompanied by a decline in child-bearing from levels sometimes averaging as high as six or more births per woman to around two or less. In seeking to explain this development as in understanding fertility behaviour more generally, the Chicago–Columbia model focuses on changes in the demand for children.

A simple economic model analogous to that for the demand for any economic good, the original starting point for economic theorizing on fertility, would see the number of children demanded as varying directly with household income (assuming children are a 'normal' good), directly with the price of goods relative to children, and inversely with the strength of tastes for goods relative to children. In the Chicago–Columbia approach price and income are the explanatory variables featured, and especially price. The explanation of the secular fertility decline provides a typical illustration. The decline is seen as due to a decrease in the demand for children brought about by socio-economic development. This decrease in demand, in turn, is ascribed to a strong negative effect associated with an increase in the relative price of children that outweighs a weak positive effect from higher income. The most common explanation of the increased relative price of children focuses on the opportunity cost of the wife's time. In keeping with the household production function concept, children are seen as requiring inputs of goods and time, and the price of children, as depending, accordingly, on the prices of these inputs. Typically, the input of the wife's time in childbearing and raising is the central focus, and the assumption is made that children are more time-intensive with regard to the wife's time than other forms of consumption. The opportunity cost of the time input into children is then seen as increasing secularly as the wife's opportunity cost, proxied by her schooling, rises (Willis, 1973). Other factors increasing the relative price of children are sometimes cited such as the price of labour relative to capital, the prices of other child inputs, or a systematic change in the 'quality' of children demanded, where quality is identified with the quantities of inputs of time and goods into a child (Lindert, 1980, 1983; T.P. Schultz, 1979; T.W. Schultz, 1974). All of these are seen as working via a negative impact on the demand for children, as in the case of the opportunity cost of a wife's time.

Several objections based on empirical studies have been raised to a purely demand interpretation of the secular fertility decline. For one thing, a phase of increasing fertility has often preceded the secular fertility decline (Dyson and Murphy, 1985); is this to be taken as implying an initial period of increasing demand for children? Then, too, there are indications of low fertility in some sub-Saharan African societies, apparently associated with venereal disease. How is this to be treated in a demand-oriented theory? Perhaps most important, demographic surveys of contemporary premodern populations have repeatedly turned up evidence of what demographers call a 'natural fertility' regime, the absence of any attempt deliberately to limit fertility among almost all segments of the population, aside from some elite groups (Coale, 1967; Henry, 1961). If parents are choosing the number of children they have in accordance with a demand model, how is one to explain the fact that so few are doing anything to control their fertility? It seems unlikely that unregulated fertility would assure that most couples would have just as many children as they want and no more, and thus result almost uniformly throughout a population in no practice of family size limitation.

To deal with questions of this type the Pennsylvania model stresses two factors as fertility determinants in addition to the demand for children: (1) the potential supply of children, the number of surviving children parents would have if they did not deliberately limit fertility; and (2) the costs of fertility regulation, including both subjective (psychic) drawbacks and objective costs, the time and money required to learn about and use specific techniques.

The introduction of supply considerations is the most distinctive feature of the Pennsylvania model. (Regulation costs are sometimes treated in Chicago–Columbia models although usually subordinated to demand.) The most obvious example of the importance of a supply constraint in determining observed fertility is the case of a couple that has fecundity problems and is consequently unable to produce as many children as it wants. True, child adoption is a logical option in such a case, but empirically this practice is of quite limited significance. Clearly, a supply constraint due to sterility would explain the African case mentioned above. Aside from sterility problems, however, the production of children is kept down significantly in almost all pre-modern societies by various types of behaviour that have unperceived consequences for fertility (a phenomenon conceptually designated 'unperceived jointness' by Easterlin, Pollak and Wachter, 1980). The most important types of behaviour in this regard are deferment of sexual unions beyond menarche, which reduces exposure to intercourse and prolonged breast feeding, which has the effect of delaying the return of ovulation after a birth. Also, because the subject of parents' demands is not births *per se*, but surviving children, high infant and child mortality in pre-modern societies further restricts the supply of children.

The Pennsylvania model suggests two possible reasons for 'natural fertility' behaviour. First, there is the possibility of excess demand. If in most households in a pre-modern society the supply of children were less than demand, then parents would have as many children as they could. In such a situation there would be a general absence of any practice of deliberate family size limitation, and differences in observed fertility would be determined by the circumstances responsible for differences in supply.

Even if supply exceeded demand, however, deliberate family size limitation would not necessarily occur, because of the costs attaching to the various techniques of fertility control. If, for example, the disutility attaching, say, to abstinence or withdrawal, exceeds the disutility of an excess number of children, and no other contraceptive practices are known, then a couple's observed fertility would again be governed by its supply. Thus the Pennsylvania model identifies two cases in which rational behaviour would be consistent with an absence of deliberate fertility regulation – (a) an excess demand condition, and (b) high perceived costs of fertility regulation.

In interpreting the secular fertility decline, the Pennsylvania model envisages a typical pre-modern society as starting from a condition of unregulated fertility due to either or both of the circumstances just mentioned, and moving to a situation of algebraically increasing excess supply, as supply increases and demand decreases (though not necessarily concurrently) with socio-economic development. The increase in supply might reflect a decrease in breastfeeding that raises natural fertility, improved child survival, or both. The decreased demand might be due to a rise in the relative cost of children, as in the Chicago–Columbia model, or an anti-natal shift in tastes due to education or the introduction of new consumer goods (Behrman and Wolfe, 1984; Easterlin, 1978). If the disutility associated with an excess supply of children remains less than the disutility associated with use of contraception, then fertility will remain uncontrolled and an increase in actual fertility,

reflecting the growth of natural fertility, will be observed. Such circumstances could account for a phase of increasing fertility prior to the secular fertility decline. Eventually, however, as excess supply continues to rise, the pressures for adoption of deliberate control will prevail, and observed fertility will decline. Lower costs of fertility regulation due, say, to better contraceptive knowledge or improved contraceptive availability, might also contribute to the shift to lower fertility, although it is unlikely that this factor in itself could explain an initial phase of increasing fertility.

Thus, the supply–demand approach sees the supply of children and regulation costs, as well as demand, as factors that may significantly influence observed fertility in pre-modern societies and in the early stages of the secular fertility decline. Eventually, however, as modernization progresses, most households shift to a position of substantial potential excess supply and increasingly perceive costs of fertility regulation as low. Because of this, demand influences become increasingly dominant in determining fertility, although these influences may reflect taste changes in addition to anti-natal price effects (cf. Mueller and Short, 1983). Thus, in contemporary developed societies, both schools emphasize demand as the principal influence determining fertility, although they differ with regard to the underlying determinants of demand that are considered most important.

In recent work by some members of the Chicago–Columbia school, a supply–demand model has also been adopted, but sharp conceptual differences from the Pennsylvania model remain (Schultz, 1981; Rosenzweig and Schultz, 1985). In the Chicago–Columbia model supply is determined solely by biological factors and all behavioural factors operate through demand. In contrast, in the Pennsylvania model, supply is constrained by behaviour that has the unintentional effect of limiting fertility. Particularly at issue is the extent to which considerations of family size enter into individual decisions on marriage-timing, length of breastfeeding, and consumption. Based on motivational data from sample surveys as well as behavioural data of various types, the Pennsylvania model assumes such decisions to be taken largely independently of family size concerns. The positivist methodology of the Chicago–Columbia school leads to rejection of this evidence, and to a theoretical conception that stresses on *a priori* grounds the endogeneity to the fertility decision of marriage-timing and breastfeeding behaviour. In the Pennsylvania model it is the behavioural influences on supply that are principally responsible for constraining fertility to the extent that it may fall short of demand in premodern societies. Aside from cases of physiological sterility, the supply concept of the Chicago–Columbia school would be unlikely to constrain fertility effectively; hence their supply–demand model largely preserves their emphasis on demand conditions as the determinant of observed fertility behaviour in all times and places.

THE CROSS-SECTIONAL RELATION BETWEEN FERTILITY AND INCOME. As has been mentioned, cross-sectional empirical studies of the relation between fertility and income yield mixed results, and this is so even after controlling for numerous other variables (Mueller and Short, 1983; Simon, 1974). Sometimes the direction of relationship is positive, sometimes negative; sometimes the relationship is significant, sometimes not significant. Because economists of both schools working in the fertility area intuitively accept that children are a normal good, these empirical findings have provoked an extensive research for explanation, and, in particular, for price or taste factors that might vary systematically with income level. This search has again involved rather different types of emphasis by the Chicago–Columbia and Pennsylvania schools.

Consistent with the interpretation of the secular fertility decline, the variable stressed most frequently by the Chicago–Columbia school has been the opportunity cost of a wife's time, a variable first brought to the fore by Mincer, 1963). The idea is that husbands with higher permanent income are likely to have spouses with higher education and thus higher market wage rates. A general increase in the potential earnings of both sexes would then lead to a substitution effect against children due to the wife's higher wage rate that offsets a positive income effect from the husband's wage rate. Because of the absence of wage rates for nonworking wives, empirical tests of this hypothesis have usually used wife's education as a proxy for the opportunity cost of wife's time (T.W. Schultz, 1974).

A second line of explanation in the Chicago–Columbia approach stressed by Becker himself and differing from the price-of-time argument emphasizes the association between what is called 'child quality' and income (Becker and Lewis, 1974). In this case the variation with income of the quantities of child inputs, rather than their prices, is the focus. The basic idea is that as parents' income increases, they are assumed to want to increase child inputs and thus to spend more, on the average, on their children, just as they are expected to want to spend more on themselves. This positive association between desired expenditures per child and parental income causes children to be more expensive for wealthier parents than poorer ones, and is presumed to offset the positive effect of income *per se* on the demand for children.

Although the Pennsylvania school does not reject the plausibility of these hypotheses, it offers yet another one, again influenced by the demographic work of sociologists, as in the supply–demand approach. In this case, the analysis builds on the sociological notion that one's economic socialization experience early in the life cycle plays an important part in forming one's material tastes. It is assumed that the material environment surrounding young persons in the course of their upbringing leads to the formation of a socially defined subsistence level that they wish to achieve on reaching adulthood (Ahlburg, 1984; Easterlin, 1973). Only to the extent that actual income exceeds this subsistence constraint would a couple feel free to embark on family formation. Assuming that young adults with higher income come from more affluent backgrounds, then the expected positive effect of higher income on the demand for children would be offset by the higher goods aspirations of wealthier compared with poorer couples (Easterlin, 1969).

Recently there has been some convergence in the two views. On the one hand, the Chicago–Columbia school has introduced consideration of what is called 'child endowments', stressing the effect of one's family of origin on fertility behaviour, as in the Pennsylvania model (Becker, 1981; Becker and Tomes, 1976; Nerlove, 1974). On the other hand, the Pennsylvania school has added to its conception concern with parents' desires for expenditures on their children as well as for themselves (Easterlin, 1976). Both schools typically argue that because of such systematic correlates of changing income, the cross-sectional observed relation of income and fertility is uncertain. Leibenstein (1974, 1975) takes a stronger stance, asserting on *a priori* grounds that negative taste changes associated with higher income offset positive income effects and cause a negative association between income and fertility. The argument is that a higher status household must more than proportionately increase its expenditures on 'status goods' in order to maintain its relative life style and economic status.

Much of the theorizing regarding the cross-sectional income/fertility relationship assumes that the same arguments would apply in both pre-modern and developed societies. Development of the supply–demand model by the Pennsylvania school has led to reconsideration of this proposition. If, in pre-modern societies, fertility is largely determined by supply conditions, whereas in developed countries it is largely determined by demand factors, there is obviously no reason to expect the same relation between income and fertility (Behrman and Wolfe, 1984). Indeed, a systematic shift in the relation of income to fertility might be observed, as the dominant determinants shifted from supply to demand (Crimmins et al., 1984). To illustrate, from a supply viewpoint, in a pre-modern society higher income might lead to better nutrition and thereby higher fecundity of a wife, or to shorter breastfeeding as baby food substitutes become available and affordable. For both reasons the ability to produce children would be positively related to income. If a natural fertility regime prevailed, then such supply effects might yield a positive relation between observed fertility and income. With the progress of modernization and a growing predominance of demand factors in fertility determination, this initial positive cross-sectional association might change to a non-significant or (on Leibenstein reasoning) even a negative relationship. The possibility that supply factors might dominate the cross-sectional income–fertility relationship in developing countries has not been considered by the Chicago–Columbia school, presumably because of their more restricted concept of supply.

FLUCTUATIONS. Economic theorizing about fertility fluctuations has focused primarily on the experience of the developed countries, and particularly the United States. The protracted postwar baby boom and bust, from a low in the 1930s to a peak in the 1950s and then a new 1970s trough, has attracted most attention; shorter-term business cycle fluctuations, much less.

The interpretation of the United States baby boom and bust advanced by the Pennsylvania school is a relative income one that builds on the arguments about taste formation described in the preceding section (Easterlin, 1973, 1980). The basic idea is that the cohorts that were in the family forming ages in the late 1940s and 1950s were raised under the economically deprived circumstances of the Great Depression and World War II. As a result, the material aspirations formed during their economic socialization experience were low. Their labour market experience, however, was quite favourable, because of the combined circumstances of a prolonged post-World War II economic expansion and the relative scarcity of young workers, the latter echoing the unprecedentedly low fertility of the 1920s and 1930s. In consequence, these cohorts enjoyed high relative income, that is, high income relative to their material aspirations, and this led to earlier marriage and child-bearing, higher completed family size, and the baby boom that lasted through the late 1950s.

The circumstances of the subsequent cohorts tended to be the reverse – declining relative income, postponed marriage and childbearing, lower completed family size – adding up to a baby bust. On the one hand, these cohorts had formed high material aspirations as a result of their upbringing in the boom circumstances following World War II. On the other, their own labour market experience was much less favourable, partly because of some slackening in the growth of aggregate demand, and partly because of a sharply increased relative supply of workers in family forming ages, itself a consequence of the prior baby boom.

As the foregoing suggests, the relative income hypothesis can with some restrictive assumptions, be translated into a relative cohort size hypothesis. If one assumes fairly stable growth in aggregate demand and a largely closed economy, then variations in the earnings of younger compared with older workers would be dominated by variations in the relative supply of younger workers. A relatively small cohort of young adults would cause a narrowing of the shortfall of younger workers' incomes compared with older; a relatively large cohort would cause a widening of the gap. Taking older workers' incomes as a proxy for the material aspirations formed by young adults when in their parents' homes, one obtains the same type of relative income mechanism engendering fertility movements that was just described. This relative cohort size variant has been used to demonstrate the possibility of a self-generating fertility cycle (Lee, 1974; Samuelson, 1976).

In contrast to the taste formation influences emphasized in the Pennsylvania model, the Chicago–Columbia interpretation of the baby boom and bust builds on a price-of-time argument similar to that used in explaining the secular fertility decline (Butz and Ward, 1979). An increase in husband's income is thought to have a positive effect on fertility; an increase in the wife's wage rate, a negative effect due to the price-of-time effect. In the baby boom period, it is argued, the labour market for women relative to men, as indexed by wage rate movements for the two sexes, was comparatively weak; thereafter the labour market for women expanded commensurately with that for men. Thus, in the baby boom period, men's wage rates rose while women's remained relatively flat; hence a net positive impact on fertility prevailed, reflecting the dominant effect of men's compared with women's wage rate changes. Thereafter, women's wage rates rose commensurately with men's, and a negative effect dominated, due to the higher absolute magnitude of the elasticity of fertility with respect to women's wage rates than men's. The result of the disparate changes in men's and women's wages before and after 1960 was thus an upswing in fertility followed by a downswing. Young women's labour force participation moved inversely with fertility in the two periods, reflecting the differing pull of women's wage rates.

Several critiques of the econometric techniques used in this analysis have appeared (Kramer and Neusser, 1984; McDonald 1983). Also, the movement in labour force participation of older women does not fit easily into the argument. This is because older women, who are highly substitutable for younger women in most jobs, showed a marked rise in labour force participation before 1960, the period when the female labour market was presumably weak, and then much slower growth after 1960, when the female labour market was presumably stronger. In addition, the Chicago–Columbia view implies that the more favourable movement in women's wages after 1960 would have shortened birth intervals (Mincer and Polachek, 1974), whereas the opposite, in fact, occurred. Nevertheless, the Butz–Ward analysis remains the prevailing interpretation advanced by adherents of the Chicago–Columbia school. Some analysts have found that a combination of the Pennsylvania and Chicago–Columbia models is superior to either alone (Devaney, 1984; Lindert, 1978).

The Pennsylvania and Chicago–Columbia models have quite different implications for the future of fertility fluctuations. The Pennsylvania model suggests that a growing scarcity of younger workers in the 1980s and 1990s, echoing the baby bust of the 1960s and 1970s, is a factor making for a turnaround in the relative income of young adults and thus for an upturn in fertility. In contrast, the Chicago–Columbia view envisages further fertility declines on the assumption that women's wages are likely to continue to rise commensurately with men's.

The two models also differ in their predictions regarding variations in fertility over the business cycle. The Pennsylvania model anticipates a positive association of the type traditionally observed (Ben-Porath, 1973). Because of the importance of historical experience in forming tastes, one would expect tastes to remain largely invariant in periods as short as the usual business cycle. Variations in actual earnings associated with the business cycle might therefore be expected to lead to corresponding variations in fertility as income expectations were revised.

In contrast, the Chicago–Columbia analysis suggests that the short-term association between fertility and income varies with the proportion of females in the labour force. The reasoning is that the relative importance of the negative effects of women's wage changes versus the positive effect of men's wage changes will be greater, the larger the proportion of women in the labour force. Hence, as the proportion of women at work rises, the more sensitive does fertility become to fluctuations in women's wage rates. Based on the uptrend in women's labour force participation, the Chicago–Columbia model thus foresees the emergence of counter-cyclical fertility fluctuations. When women's wage rates are high, as in a boom period, the price-of-time effect will pull them into the labour market and consequently reduce fertility; when wage rates are low, the reverse will occur. This mechanism is claimed to have operated during the business cycles of the 1970s.

CONCLUSION. This survey has aimed at highlighting some of the principal differences between the two leading schools of economic theorizing about reproductive behaviour, as manifested in the interpretations offered of trends, fluctuations and cross-sectional variations in the income–fertility relationship. As the survey demonstrates, the evolution of theorizing on reproductive behaviour has been away from a simple economic model of demand emphasizing income and market price variables toward the recognition of additional constraints on behaviour. Perhaps more than in any other area of economic analysis, the constraint of time inputs has come to the forefront, both the amounts of time required in childbearing and childrearing and the prices at which these inputs should be valued. This interest has stimulated fruitful empirical inquiries by economists into the use of time within the household. Also, explicit attention has been paid to the constraint on one's behaviour arising from the way that prior experience shapes one's tastes. Thus, attention has been directed to the way that one's experience in one's family of origin and one's socialization experience more generally may shape adult preferences with regard to material aspirations and family size. In this area, economists have sometimes pushed beyond the conceptual speculations of sociologists to formulate specific empirical models of taste formation. Finally, recognition has emerged of the constraint on 'consumption' of children arising from production possibilities within the household. Whether because of biological or behavioural attributes, a couple may be unable to produce as many children as demanded, and its observed consumption would thus reflect this rationing constraint.

As the foregoing discussion shows, the introduction of these new behavioural constraints has arisen from growing awareness by economists of the intractability of empirical evidence on reproductive behaviour, and a resultant attempt to accommodate within economic analysis conceptual contributions from related disciplines. Although some progress in understanding reproductive behaviour has been made, there is still no single generally accepted theory of reproductive behaviour, and no consensus on the interpretation of the empirical regularities described above. However, if it is true that scientific breakthroughs frequently occur at the juncture of different disciplines, fertility theory is undoubtedly one of the frontiers of economic theory beckoning for more intensive exploration.

RICHARD A. EASTERLIN

See also DEMOGRAPHIC TRANSITION; DEMOGRAPHY; EASTERLIN HYPOTHESIS; FAMILY PLANNING; FECUNDITY; HISTORICAL DEMOGRAPHY; INFANT MORTALITY; STABLE POPULATION THEORY.

BIBLIOGRAPHY

Ahlburg, D.A. 1984. Commodity aspirations in Easterlin's relative income theory of fertility. *Social Biology* 31(3/4), Fall/Winter, 201–7.

Becker, G.S. 1960. An economic analysis of fertility. In *Demographic and Economic Change in Developed Countries*, Universities–National Bureau Conference Series No. 11, Princeton: Princeton University Press.

Becker, G.S. 1965. A theory of the allocation of time. *Economic Journal* 75, September, 493–517.

Becker, G.S. 1981. *A Treatise on the Family*. Cambridge, Mass.: Harvard University Press.

Becker, G.S. and Lewis, H.G. 1974. Interaction between quantity and quality of children. In *The Economics of the Family*, ed. T. W. Schultz, Chicago: University of Chicago Press.

Becker, G.S. and Tomes, N. 1976. Child endowments and the quantity and quality of children. *Journal of Political Economy* 84(4), Part 2, August, S143–S162.

Behrman, J.R. and Wolfe, B.L. 1984. A more general approach to fertility determination in a developing country: the importance of biological supply considerations, endogeneous tastes and unperceived jointness. *Economica* 51, August, 319–39.

Ben-Porath, Y. 1973. Short-term fluctuations in fertility and economic activity in Israel. *Demography* 10(2), May, 185–204.

Ben-Porath, Y. 1975. First generation effects on second generation fertility. *Demography* 12(3), August, 397–405.

Ben-Porath, Y. 1982. Economics and the family – match or mismatch? A review of Becker's 'A Treatise on the Family'. *Journal of Economic Literature* 20(1), March, 52–64.

Bulatao, R.A. and Lee, R.D. (eds) 1983. *Determinants of Fertility in Developing Countries: A Summary of Knowledge*. New York: Academic Press.

Butz, W.P. and Ward, M. P. 1979. The emergence of counter-cyclical US fertility. *American Economic Review* 69(3), June, 318–28.

Coale, A.J. 1967. The voluntary control of human fertility. *Proceedings of the American Philosophical Society* 111(3), June, 164–9.

Crimmins, E.M., Easterlin, R.A. Jejeebhoy, S.J. and Srinivasan, K. 1984. New perspectives on the demographic transition: a theoretical and empirical analysis of an Indian state, 1951–1975. *Economic Development and Cultural Change* 32(2), January, 227–53.

Devaney, B. 1984. An analysis of variations in U.S. fertility and female labor force participation trends. *Demography* 20(2), May, 147–61.

Dyson, T. and Murphy, M. 1985. The onset of fertility transition. *Population and Development Review* 11(3), September, 399–440.

Easterlin, R.A. 1969. Towards a socioeconomic theory of fertility: a survey of recent research on economic factors in American fertility. In *Fertility and Family Planning: A World View*, ed. S.J. Behrman, Leslie Corsa, Jr. and R. Freedman, Ann Arbor: University of Michigan Press.

Easterlin, R.A. 1973. Relative economic status and the American fertility swing. In *Family Economic Behavior: Problems and Prospects*, ed. E.B. Sheldon, Philadelphia: Lippincott.

Easterlin, R.A. 1976. Population change and farm settlement in the northern United States. *Journal of Economic History* 36(1), March, 45–75.

Easterlin, R.A. 1978. The economics and sociology of fertility: a synthesis. In *Historical Studies of Changing Fertility*, ed. C. Tilly, Princeton: Princeton University Press.

Easterlin, R.A. 1980. *Birth and Fortune*. New York: Basic Books.

Easterlin, R.A., Pollak, R.A. and Wachter, M.L. 1980. Toward a more general economic model of fertility determination: endogenous preferences and natural fertility. In *Population and Economic Change in Developing Countries*, ed. R.A. Easterlin, Chicago: University of Chicago Press.

Henry, L. 1961. La fécondité naturelle: observations – théorie – résultats. *Population* 16(4), October/December, 625–36.

Keeley, M. 1975. A comment on 'An Interpretation of the Economic Theory of Fertility'. *Journal of Economic Literature* 13(2), June, 461–7.

Kramer, W. and Neusser, K. 1984. The emergence of countercyclical U.S. fertility: note. *American Economic Review* 74(1), March, 201–2.

Lancaster, K.J. 1971. *Consumer Demand: A New Approach*. New York: Columbia University Press.

Lee, R. 1974. The formal dynamics of controlled populations and the Echo, the Boom and the Bust. *Demography* 11(4), November, 563–85.

Leibenstein, H. 1957. *Economic Backwardness and Economic Growth*. New York: John Wiley.

Leibenstein, H. 1974. An interpretation of the economic theory of fertility: promising path or blind alley? *Journal of Economic Literature* 12(2), June, 457–79.

Leibenstein, H. 1975. The economic theory of fertility decline. *Quarterly Journal of Economics* 89(1), February, 1–31.

Lindert, P.H. 1978. *Fertility and Scarcity in America*. Princeton: Princeton University Press.

Lindert, P.H. 1980. Child costs and economic development. In *Population and Economic Change in Developing Countries*, ed. R.A. Easterlin, Chicago: University of Chicago Press.

Lindert, P.H. 1983. The changing economic costs and benefits of having children. In *Determinants of Fertility in Developing Countries: A Summary of Knowledge*, Vol. 1, ed. R. Bulatao and R.D. Lee, New York: Academic Press.

McDonald, J. 1983. The emergence of countercyclical US fertility: a reassessment of the evidence. *Journal of Macroeconomics* 5(4), Fall, 421–36.

Michael, R. T. and Willis, R. J. 1976. Contraception and fertility: household production under uncertainty. In Conference on Research in Income and Wealth, *Household Production and Consumption*, New York: National Bureau of Economic Research.

Mincer, J. 1963. Market prices, opportunity costs, and income effects. In *Measurement in Economics*, ed. C. Christ et al., Stanford: Stanford University Press.

Mincer, J. and Polachek, S. 1974. Family investments in human capital: earnings of women. In *The Economics of the Family*, ed. T.W. Schultz, Chicago: University of Chicago Press.

Mueller, E. and Short, K. 1983. Effects of income and wealth on the demand for children. In *Determinants of Fertility in Developing Countries: A Summary of Knowledge*, Vol. 1, ed. R. Bulatao and R.D. Lee, New York: Academic Press.

Nerlove, M. 1974. Household and economy: toward a new theory of population and economic growth. *Journal of Political Economy* 82(2), Part II, March/April, S200–S218.

Rosenzweig, M.R. and Schultz, T.P. 1985. The demand for and supply of births: fertility and its life cycle consequences. *American Economic Review* 75(5), December, 992–1015.

Samuelson, P.A. 1976. An economist's non-linear model of self-generated fertility waves. *Population Studies* 30(2), July, 243–7.

Sanderson, W.C. 1976. On two schools of the economics of fertility. *Population and Development Review* 2(3–4), September/December, 469–77.

Sanderson, W.C. 1980. Comment. In *Population and Economic*

Change in Developing Countries, ed. R.A. Easterlin, Chicago: University of Chicago Press.

Schultz, T. 1976. Determinants of fertility: a micro-economic model of choice. In *Economic Factors in Population Growth*, ed. A.J. Coale, New York: Halsted Press.

Schultz, T.P. 1979. Current developments in the economics of fertility. In *International Union for the Scientific Study of Population: Economic and Demographic Change: Issues for the 1980's*, Proceedings of the Conference, Helsinki 1978, Vol. 3, Liège: IUSSP, 27–38.

Schultz, T.P. 1981. *Economics of Population*. Reading, Mass.: Addison-Wesley Co.

Schultz, T.W. (ed.) 1974. *The Economics of the Family*. Chicago: University of Chicago Press.

Simon, J.L. 1974. *The Effects of Income on Fertility*. Chapel Hill: University of North Carolina Press.

Tabarrah, R.B. 1971. Toward a theory of demographic development. *Economic Development and Cultural Change* 19(2), January, 257–77.

Turchi, B.A. 1975. *The Demand for Children: The Economics of Fertility in the United States*. Cambridge, Mass.: Ballinger.

Willis, R.J. 1973. A new approach to the economic theory of fertility behavior. *Journal of Political Economy* 81(2), Part II, March/April, S14-S64.

fetishism. *See* COMMODITY FETISHISM.

Fetter, Frank Albert (1863–1949). Fetter was born on 3 March 1863 in the town of Peru, Indiana, and died on 21 March 1949 in Princeton, New Jersey. He was educated at Indiana and Cornell Universities and received his doctorate in economics at the University of Halle in Germany in 1894; he spent most of his life teaching at Cornell (1901–11) and Princeton universities (1911–34).

In journal articles on capital, interest and rent written largely between 1900 and 1914 (Fetter, 1977), and particularly in two treatises on economic principles (Fetter, 1904 and 1915), Fetter built upon Böhm-Bawerk and the Austrian School to develop a lucid and remarkably integrated structure of economic theory. He was able to accomplish this feat by purging economics of all traces of Ricardian or other British objectivist theories of value and distribution, in particular any differential theories of rent or productivity theories of interest.

Much of Fetter's achievement rested on his insight into the ordinary language meaning of 'rent' as simply the price of any durable good per unit time. He was then able to show that the prices of consumer goods are determined by their marginal utilities, and that these values are imputed back to determining the rental prices of factors of production by their marginal value productivity in serving consumers. The capital value, or price of the whole good (whether land, capital goods, or, Fetter might have added, the labourer under slavery) is then determined by the sum of its expected future returns, or rents, discounted by the social rate of time preference, or rate of interest. Thus, Fetter went beyond Böhm-Bawerk by arriving at a pure time preference theory of interest. Productivity and time preference are both highly important, but they have very different functions: the former in determining rents, and the latter determining the rate of interest. Thus, future rents are discounted by the rate of time preference and summed up, or 'capitalized', into their present capital value. Indeed, Fetter often called his contribution the 'capitalization theory of interest'.

Fetter presented the fullest portrayal yet attained of the time market, the market for present as against future goods, as it permeates the economic system. The time market is not only the loan market, but also exists when entrepreneurs purchase or hire discounted factors of production (future goods) in return for money (a present good) and then reap a time or interest return when the product is later sold as a present good. Entrepreneurs earn profits, or suffer losses, as they lead the economy in the direction of a general equilibrium determined by marginal utility, marginal value productivity, and time preference.

While Fetter was led by his capitalization theory to arrive independently at the Mises–Hayek theory of the business cycle in 1927 (Fetter, 1977, pp. 260–316), he virtually abandoned value and distribution theory in the last two decades of his life to concentrate on the alleged monopolistic evils of basing-point pricing. He assumed that competition requires uniform pricing of products at the mill, while uniform pricing at centres of consumption is somehow monopolistic and deserves to be outlawed (Fetter, 1931). Fetter's shift of concern, coupled with a general loss of interest in economic theory in the US between the two world wars and the continuing dominance of neo-Ricardian Marshallian theory in Britain, gravely hindered the incorporation of Fetter's notable contributions into modern economics.

MURRAY N. ROTHBARD

SELECTED WORKS

1904. *The Principles of Economics*. New York: Century.
1915. *Economic Principles*. New York: Century.
1931. *The Masquerade of Monopoly*. New York: Harcourt, Brace.
1977. *Capital, Interest and Rent: essays in the theory of distribution*. Edited by M. Rothbard, Kansas City: Sheed, Andrews and McMeel.

BIBLIOGRAPHY

Coughlan, J.A. 1965. The contributions of Frank Albert Fetter (1863–1949) to the development of economic theory. Doctoral Dissertation, Washington, DC, Catholic University of America.

Hoxie, R.F. 1905. Fetter's theory of value. *Quarterly Journal of Economics*, February.

Fetter, Frank Whitson (born 1899). Fetter was born in San Francisco, California, in 1899. His published research is wide-ranging, including studies of inflation and international economic issues, but his most celebrated contributions are in the history of economic thought. These contributions were accorded special recognition in 1982, when he became a Distinguished Fellow of the History of Economics Society.

After gaining a first degree at Swarthmore (BA, 1920), Fetter went to Harvard (MA, 1924) and Princeton (MA, 1922; PhD, 1926). Thereafter, he taught economics at Princeton (to 1934) and at Haverford College (1934–48). In 1948 he was appointed Professor of Economics at Northwestern University and remained in that post until his retirement in 1967.

Fetter chose classical economics as the major focus of his research, in particular British economic thought from Adam Smith to John Stuart Mill. That thought he has characterized as, 'a time bomb under the citadels of the established order' (Fetter, 1981, p.31). In his view the core of classical economics was not the doctrine of laissez-faire. Rather, it was rationality, and this led economists to be concerned with questions such as religious discrimination and aristocratic privilege, as well with the freer operation of the forces of the market. They were advocates of social change on a broad front.

Given this understanding of classical economics, Fetter's work has not been confined to textual analysis of the treatises of the great theorists. In addition he has closely observed economists at work in the public forums of 19th–century Britain, as shown in his masterly overview of their interventions in Parliament (Fetter, 1980). This book examines the economist's role in debates concerning not only trade, working conditions, business practice, taxation and other economic matters, but also on such issues as education, church–state relations, civil rights and parliamentary reform.

Another forum for economists in the first half of the 19th century was that provided by influential periodicals such as the *Edinburgh Review*, the *Westminster Review* and *Blackwood's*. This facet of contemporary economic debate he explored in a series of pioneering papers (Fetter, 1953; 1958; 1960; 1962; 1965). Special mention is also due his work on the development of thought relating to monetary and banking policy in Great Britain (Fetter, 1955, 1965; 1973), which brings the modern reader into intimate contact with the institutions, personalities and conceptual divisions that were crucial in the evolution of a powerful monetary orthodoxy.

The contributions of Fetter are informed by the conviction that a grasp of history is a vital element in the intellectual equipment of those who would make economic judgements.

BARRY GORDON

SELECTED WORKS

1931. *Monetary Inflation in Chile*. Princeton: Princeton University Press. Spanish trans. as *La inflación monetariau en Chile*, Santiago: University of Chile, 1937.
1942. The life and writings of John Wheatley. *Journal of Political Economy* 50, June, 357–76.
1953. The authorship of economic articles in the *Edinburgh Review*, 1802–47. *Journal of Political Economy* 61, June, 232–59.
1955a. *The Irish Pound, 1797–1826*. London: Allen and Unwin.
1955b. Does America breed depressions? *Three Banks Review* 27, September, 28–41.
1957 (ed.) *The Economic Writings of Francis Horner*. London: London School of Economics and Political Science.
1958. The economic articles in the *Quarterly Review* and their authors, 1809–1852. *Journal of Political Economy* 66, Pt I, February, 47–64; Pt II, April, 154–70.
1959. The politics of the Bullion Report. *Economica* 26, May, 99–120.
1960. The economic articles in *Blackwood's Edinburgh Magazine*, and their authors, 1817–1853. *Scottish Journal of Political Economy* 7; Pt I, June, 85–107; Pt II, November, 213–31.
1962a. Robert Torrens: Colonel of Marines and political economist. *Economica* 29, May, 152–65.
1962b. Economic articles in the *Westminster Review* and their authors, 1824–51. *Journal of Political Economy* 70, December, 570–96.
1964. (ed.) *Selected Economic Writings of Thomas Attwood*. London: London School of Economics and Political Science.
1965a. *Development of British Monetary Orthodoxy, 1797–1875*. Cambridge, Mass.: Harvard University Press.
1965b. Economic controversy in the *British Review*, 1802–1850. *Economica* 32(128), November, 424–37.
1968. The transfer problem: formal elegance or historical realism? In *Essays in Money and Banking in Honour of R.S. Sayers*, ed. C.R. Whittlesey and J.S.G. Wilson, Oxford: Oxford University Press, 63–84.
1969. The rise and decline of Ricardian economics. *History of Political Economy* 1(1), 67–84.
1973. (With D. Gregory.) *Monetary and Financial Policy*. Dublin: Irish University Press.
1975. The influence of economists in Parliament on British legislation from Ricardo to John Stuart Mill. *Journal of Political Economy* 83(5), October, 1051–64.
1980. *The Economist in Parliament, 1780–1868*. Durham, North Carolina: Duke University Press.
1981. Are economists of any use? *History of Economics Society Bulletin* 3.

feudalism. Modern discussions of feudalism have been bedevilled by disagreement over the definition of that term. There are three main competing conceptualizations. (1) Feudalism refers strictly to those social institutions which create and regulate a quite specific form of legal relationship between men. It constitutes a relationship in which a freeman (vassal) assumes an obligation to obey and to provide, primarily military, services to an overlord who, in turn, assumes a reciprocal obligation to provide protection and maintenance, typically in the form of a fief, a landed estate to be held by the vassal on condition of fulfilment of obligations (Bloch, 1939–40). (2) Feudalism refers, more broadly, to a form of government or political domination. It is a form of rule in which political power is profoundly fragmented geographically; in which, even within the smallest political units, no single ruler has a monopoly of political authority; and in which political power is privately held, and can thus be inherited, divided among heirs, given as a marriage portion, mortgaged, and bought and sold. Finally, the armed forces involve, as a key element, a heavy armed cavalry which is secured through private contracts, whereby military service is exchanged for benefits of some kind (Strayer, 1965; Ganshof, 1947). (3) Feudalism refers to a type of socio-economic organization of society as a whole, a mode of production and of the reproduction of social classes. It is defined in terms of the social relationships by which its two fundamental social classes constitute and maintain themselves. Specifically, the peasants, who constitute the overwhelming majority of the producing population, maintain themselves by virtue of their possession of their full means of subsistence, land and tools, so require no productive contribution by the lords to survive. This possession is secured by means of the peasants' collective political organization into self-governing communities, which stand as the ultimate guardian of the individual peasants' land. As a result of the peasants' possession and their consequent economic independence, mere ownership of property cannot be assumed to yield an economic rent; in consequence, the lords are obliged to maintain themselves by appropriating a feudal levy by the exercise of *extra-economic* coercion. The lords are able to extract a rent by extra-economic coercion only in consequence of their political self-organization into lordly groups or communities, by means of which they exert a degree of domination over the peasants, varying in degree from enserfment to mere tribute taking (Marx, 1894; Dobb, 1946).

Though often thought to be in conflict, these conceptions are not only complementary but in fact integrally related to one another. While the lords' very existence as lords was based, as Marxists correctly insist, upon their appropriating a rent from the peasantry by extra-economic coercion, their capacity actually to exert such force in the rent relationship depended upon from their ability to construct and maintain the classically political ties of interdependence which joined overlord to knightly follower and thereby constituted the feudal groups which were the ultimate source of the lords' power. Conversely, while feudal bonds of interdependence were constructed, as the Weberians emphasize, to build highly localized governments capable at once of waging warfare, dispensing justice and keeping the peace, the *raison d'être* of the mini-states thus created was to constitute the dominant class of feudal society by establishing the instruments for extracting, redistributing and consuming the wealth upon which this class depended for their maintenance and reproduction. State and ruling class were thus two sides of the same coin. The distinctive ties which bound man to man in feudal society (not only the relations of vassalage strictly speaking, but also the more loosely defined associations structured by patronage, clientage, and family) constituted the building blocks, at one and the same time, for the peculiarly fragmented, locally based and politically competitive character of the feudal ruling class and for the peculiarly particularized nature of the feudal state. It was the lords' feudal levies which provided the material base for the feudal polity. It was the parcellized character of the feudal state, itself the obverse side of the decentralized structure of lordship through which rent was appropriated from the peasantry, which thus created the basic opportunities, set the ultimate limits and posed the fundamental problems for the lords' reproduction as a ruling class.

THE ORIGINS OF FEUDALISM. The rise of feudalism was conditioned by an extended process of political fragmentation within the old Carolingian Empire. This is understandable, in part, in terms of a tendency to decentralization inherent in patrimonial rule. The patrimonial lord, to maintain his following, had, paradoxically, to provide his followers with the means to establish their independence from him. He could counteract their tendency to assert their autonomy through successful warfare and conquest, in which the followers found it worth their while to continue to submit to his authority. But in the absence of such profitable aggression, the followers had every incentive to assert their independence. It was in this way that the devolution and dissolution of more centralized forms of authority took place within the Carolingian Empire during the 9th and 10th centuries, as the Franks and their followers ceased to be conquerors, following a long period in which the empire had expanded. Fragmentation was hastened by the contemporaneous invasions of the Northmen, Saracens and Magyars. Effective authority fell, successively, from the king to his princes, to the counts and, ultimately, to local castleholders and even manorial lords, as the newly-emerging, highly localized rulers turned their pillaging from foreign enemies to the local population (Weber, 1956; Duby, 1978, pp. 147ff).

Feudalism originally took shape in the early part of the 11th century in many parts of Western Europe, including much of France, northern Italy and western Germany. Feudal rule was first constituted through the formation of lordly political groups, initially organized around a castle and led by the castellan. The castellan's power was derived from his knightly followers. The knights possessed military training, fought on horseback wearing (increasingly elaborate) coats of armour, often lived in the castle, and, from around the second third of the 11th century, tended to be bound to the castellan through ties of vassalage. The castellan's hegemony was manifested in his capacity to exert the right of the ban over his district – whose outer limits were usually no more than half a day's ride from the central fortress. The right of the ban, traditionally in the hands of the early medieval kings and the direct expression of their authority, allowed the castellan, above all, to extract dues from the peasant households within his jurisdiction, as well as to dispense justice and keep the peace. Although the surrounding lesser lords were usually tied to a castellan, in some cases they retained their full independence, not only collecting feudal rents derived from their authority over their tenants, but imposing taxes and exerting justice within their manorial mini-jurisdictions. In any case, all these lords confirmed their membership in the dominant class by claiming exemption from fiscal exactions: freedom under feudalism thus took the form of privilege. The peasants' unfreedom in some cases originated from their ancestors' having formally commended themselves to their lord; that is, their having subjected themselves to his domination in exchange for his

assuring their safety. But, with the crystallization of feudal domination, it simply expressed the lords' having appropriated the right to extort protection money from them. The peasants' unfreedom was thus defined and constituted precisely by their subjection to arbitrary levies (Duby, 1973, 1978).

The feudal economy was thus structured, on the one hand, by a form of precapitalist property relations in which the individual peasant families, as members of a village community, *individually possessed* their means of reproduction. This contrasted with other precapitalist property forms in which the village community itself was the possessor (or more of one). On the other hand, under feudalism, the individual lords reproduced themselves by *individually appropriating* part of the peasants' product, backed up by localized communities of lords connected by various sorts of political bond, classically vassalage. This contrasted with other precapitalist property systems, in which the community, or communities, of lords appropriated the peasants' product collectively (as a tax) and shared out the proceeds among the community's, or communities', members.

FEUDAL PROPERTY RELATIONS AND THE FORMS OF INDIVIDUAL ECONOMIC RATIONALITY. The fundamental feudal property relationships of peasant possession and of lordly surplus extraction by extra-economic compulsion shaped the long-term evolution of the feudal economy. This was because these relationships were systematically maintained by the conscious actions of communities of peasants and of lords and thus constituted relatively inalterable constraints under which individual peasants and lords were obliged to choose the patterns of economic activity most sensible for them to adopt in order to maintain and improve their condition. The potential for economic development under feudalism was thus sharply restricted because both lords and peasants found it in their rational self-interest to pursue individual economic strategies which were largely incompatible with, if not positively antithetical to, specialization, productive investment and innovation in agriculture.

First, and perhaps most fundamental, because both lords and peasants were in full possession of what they needed to maintain themselves as lords and peasants, they were free from the *necessity* to buy on the market what they needed to reproduce, thus freed from dependence on the market and the necessity to produce for exchange, and thus exempt from the requirement to sell their output competitively on the market. In consequence, both lords and peasants were free from the necessity to produce at the socially necessary rate so as to maximize their rate of return and, in consequence, relieved of the requirement to cut costs so as to maintain themselves, and so of the necessity constantly to improve production through specialization and/or accumulation and/or innovation. Feudal property relations, in themselves, thus failed to *impose* on the direct producers that relentless drive to improve efficiency so as to survive, which is the *differentia specifica* of modern economic growth and required of the economic actors under capitalist property relations in consequence of their subjection to production for exchange and economic competition.

Absent the necessity to produce so as to maximize exchange values and in view of the underdeveloped state of the economy as a whole, the peasants tended to find it most sensible actually to deploy their resources so as to ensure their maintenance by producing directly the full range of their necessities; that is, *to produce for subsistence*. Given the low level of agricultural productivity which perforce prevailed, harvests and therefore food supplies were highly uncertain. Since food constituted so large a part of total consumption, the uncertainty of the food market brought with it highly uncertain markets for other commercial crops. It was therefore rational for peasants to avoid the risks attached to dependence upon the market, and to do so, they had to diversify rather than specialize, marketing only physical surpluses. In fact, beyond their concern to minimize the risk of losing their livelihood, the peasants appear to have found it desirable to carry out diversified production simply because they wished to maintain their established mode of life – and, specifically, to avoid the subjection to the market which production for exchange entails, and the total transformation of their existence which that would have meant.

To make possible ongoing production for subsistence, the peasants naturally aimed to maintain their plots as the basis for their existence. To ensure the continuance of their families into the future, they also sought to ensure their children's inheritance of their holdings. Meanwhile, they tended to find it rational to have as many children as possible, so as to ensure themselves adequate support in their old age. The upshot was relatively large families and the subdivision of plots on inheritance.

Like the peasants, the lords occupied a 'patriarchal' position, possessing all that they needed to survive and thus freed of any necessity to increase their productive capacities. Moreover, even to the extent they wished, for whatever reason, to increase the output of their estates, the lords faced nearly insuperable difficulties in accomplishing this by means of increasing the productive powers of their labour and their land. Thus, if the lords wished to organize production themselves, they had no choice but to depend for labour on their peasants, who possessed their means of subsistence. But precisely because the peasants were possessors, the lords could get them to work only by directly coercing them (by taking their feudal rent in the form of labour) and could *not* credibly threaten to 'fire' them. The lords were thereby deprived of perhaps the most effective means yet discovered to impose labour discipline in class-divided societies. Because the peasant labourers had no *economic* incentive to work diligently or efficiently for the lords, the lords found it extremely difficult to get them to use advanced means of production in an effective manner. They could force them to do so only by making costly unproductive investments in supervision.

In view of both the lords' and the peasants' restricted ability effectively to allocate investment funds to improved means of production to increase agricultural efficiency, both lords and peasants found that the only really effective way to raise their income via productive investment was by opening up new lands. Colonization, which resulted in the multiplication of units of production on already existing lines, was thus the preferred form of productive investment for both lords and peasants under feudalism.

Beyond colonization and the purchase of land, feudal economic actors, above all feudal lords, found that the best way to improve their income was by forcefully *redistributing* wealth away from the peasants or from other lords. This meant that they had to deploy their resources (surpluses) towards building up their *means of coercion* by means of investment in military men and equipment, in particular to improve their ability to fight wars. A drive to *political accumulation*, or state building, was the feudal analogue to the capitalist drive to accumulate capital.

THE LONG-TERM PATTERNS OF FEUDAL ECONOMIC DEVELOPMENT. Feudal property relations, once established, thus obliged lords and peasants to adopt quite specific patterns of individual economic behaviour. Peasants sought to produce for subsistence, to hold on to their plots, to produce large families and to provide for their families' future generations by bequeathing their plots. Both lords and peasants sought to use available surpluses funds to open new lands. Lords directed their resources to the amassing of greater and better means of coercion. Generalized on a society-wide basis, these patterns of individual economic action determined the following developmental patterns, or laws of motion, for the feudal economy as a whole:

(i) *Declining productivity in agriculture* (Bois, 1976; Hilton, 1966; Postan, 1966). The generalized tendency to adopt production for subsistence on the part of the peasantry naturally constituted a powerful obstacle to commercial specialization in agriculture and to the emergence of those competitive pressures which drive a modern economy forward. In so doing, it also posed a major barrier to agricultural improvement by the peasantry, since a significant degree of specialization was required to adopt almost all those technical improvements which would come to constitute 'the new husbandry' or the agricultural revolution (fodder crops, up-and-down farming, etc.). In addition, production aimed at subsistence and the maintenance of the plot as the basis for the family's existence posed a major barrier to those rural accumulators, richer peasants and lords, who wished to amass land or to hire wage labour, since the peasants would not readily part with their plots, which were the immediate bases for their existence, unless compelled to do so; nor could they be expected to work for a wage unless they actually needed to.

Further counteracting any drive to the accumulation of land and labour was the tendency on the part of the possessing peasants to produce large families and subdivide their holdings among their children. The peasants' parcellization of plots under population growth tended to overwhelm any tendency towards the build-up of large holdings in the agricultural economy as a whole, further reducing the potential for agriculture improvement.

Finally, individual peasant plots were, most often, integrated within a village agriculture which was, in critical ways, controlled by the community of cultivators. The peasant village regulated the use of the pasture and waste on which animals were raised, and the rotation of crops in the common fields. Individual peasants thus tended to face significant limitations on their ability to decide how to farm their plots and thus, very often, on their capacity to specialize, build up larger consolidated holdings, and so forth.

To the extent that the lords succeeded in increasing their wealth by means of improving their ability coercively to redistribute income away from the peasantry, they further limited the agricultural economy's capacity to improve. Increased rents in whatever form reduced the peasants' ability to make investments in the means of production. Meanwhile, the lords' allocation of their income to military followers and equipment and to luxury consumption ensured that the social surplus was used unproductively, indeed wasted. To the extent – more or less – that the lords increased their income, the agricultural economy was undermined.

(ii) *Population growth* (Postan, 1966). The long-term tendency to the decline of agricultural productivity thus conditioned by the feudal structure of property was realized in practice as a consequence of rising population. The peasants' possession of land allowed children to accede to plots and, on that basis, to form families at a relatively early age. Married couples, as noted, had an incentive to have many children, both to provide insurance for their old age and to assure that the line would be continued. The result was that all across the European feudal economy, we witness a powerful tendency to population growth from around the beginning of the 12th century, which led, almost everywhere, to a doubling of population over the following of two centuries.

(iii) *Colonization* (Postan, 1966; Duby, 1968). The only significant method by which the feudal economy achieved real growth and counteracted the tendency to declining agricultural productivity, was by way of opening up new land for cultivation. Indeed, economic development in feudal Europe may be understood, at one level, in terms of the familiar race between the growth of the area of settlement and the growth of population. During the 12th and 13th centuries, feudal Europe was the scene of great movements of colonization, as settlers pushed eastward across the Elbe and southward into Spain, while reclaiming portions of the North Sea in what became the Netherlands. The opening of new land did, for a time, counteract and delay the decline of agricultural productivity. Nevertheless, in the long run – as expansion continued, as less fertile land was brought into cultivation, and as the man/land ratio rose – rents rose, food prices increased, and the terms of trade increasingly favoured agricultural as opposed to industrial goods. At various points during the 13th and early 14th centuries, all across Europe, population and production appear to have reached their upper limits, and there began to ensue a process of demographic adjustment along Malthusian lines.

(iv) *Political accumulation or state building* (Dobb, 1946; Anderson, 1974; Brenner, 1982). Given the limited potential for developing the agricultural productive forces and the limited supply of cultivable land, the lordly class, as noted, tended to find the build-up of the means of force for the purpose of redistributing income to be the best route for amassing wealth. Indeed, the lords found themselves more or less *obliged* to try to increase their income in order to finance the build-up of their capacity to exert politico-military power. This was, first of all, because they could not easily escape the politico-military conflict or competition that was the inevitable consequence of the individual lords' direct possession of the means of force (the indispensable requirement for their maintenance as members of the ruling class over and against the peasants) and thus of the wide dispersal of the means of coercion throughout the society. It was, secondly, because they had to confront increasingly well-organized peasant communities and, as feudal society expanded geographically, to counteract the effects of increasing peasant mobility.

In the first instance, of course, military-politico efficacy required the collecting and organizing of followers. But to gain and retain the loyalty of their followers the overlords had to feed and equip them and, in the long run, competitively reward them. Minimally, the overlord's household had to become a focus of lavish display, conspicuous consumption and gift-giving, on par with that of other overlords. But beyond this, it was generally necessary to provide followers with the means to maintain their status as members of the dominant class – that is, a permanent source of income, requiring a grant of land with associated lordly prerogatives (classically the fief). But naturally such grants tended to increase the followers' independence from the overlords, leading to renewed potential for disorganization, fragmentation and anarchy. This was the perennial problem of all forms of patrimonial rule and at the centre of feudal concerns from the beginning. The tendency to fragmentation was, moreover, exacerbated as a result of the pressure to divide lordships and

lands among children. To an important degree, then, feudal evolution may be understood as a product of lordly efforts to counteract political fragmentation and to construct firmer intra-lordly bonds with the purpose of withstanding intra-lordly politico-military competition and indeed of carrying on the successful warfare that provided the best means to amass the wealth ultimately required to maintain feudal solidarity. This meant not only the development of better weapons and improved military organization, but also the creation of larger and more sophisticated political institutions, and naturally entailed increased military and luxury consumption.

Actually to achieve more effective political organization of lordly groups required political innovation. Speaking broadly, the constitution of military bands around a leading warlord for external warfare, especially conquest, most often provided the initial basis of intra-lordly cohesion. This served as the foundation for developing more effective collaboration within the group of lords for the protection of one another's property and for controlling the peasantry. As a further step in this direction, the overlord would establish his pre-eminence in settling disputes among his vassals (as in Norman England). Next, the leading lord might extend feudal centralization by establishing immediate relations with the undertenants of his vassals. One way this took place was through constructing direct ties of dependence with these rear vassals (as in 11th-century England). More generally, it was accomplished by the extension of central justice to ever broader layers of the lordly class, indeed the free population as a whole. Sometimes the growth of central justice was achieved through the more or less conscious collaboration of the aristocracy as a whole (as in 12th-century England). On other occasions it had to be accomplished through more conflicted processes whereby the leading lord (monarch, prince) would accept appeals over the heads of his vassals from their courts (as in medieval France). Ultimately, the feudal state could be further strengthened only by the levying of taxes, and this almost always required the constitution of representative assemblies of the lordly class.

This is not to say that a high level of lordly organization was always required. Nor is it to argue that state building took place as an automatic or universal process. At the frontiers of European feudal society, to the south and east, colonization long remained an easy option, and there was relatively little (internally generated) pressure upon the lordly class to improve its self-organization. At the same time, just because stronger feudal states might become necessary did not always determine that they could be successfully constructed. Witness the failure of the German kings to strengthen their feudal state in the 12th century, and the long-term strengthening of the German principalities which ensued. The point is that to the degree that disorganization and competition prevailed within and between groups of feudal lords, they would tend to be that much more vulnerable not only to depredations from the outside, but to the erosion of their very dominance over the peasants. The French feudal aristocracy thus paid a heavy price for their early, highly decentralized feudal organization, suffering not only significant losses of territory to the Anglo-Normans, but a serious reduction in their control over peasant communities and a consequent decline in dues. The French aristocracy's later recovery and successes may be attributed, at least in large part, to their evolution of a new, more centralized, more tightly-knit form of political organiza- tion – the tax/office state, where property in office (rather than lordship/land) gave the aristocracy rights to a share in centralized taxation (rather than feudal rent) from the peasants. In sum, the economic success of individual lords, or groups of them, does seem to have depended upon successful feudal state building, and the long-term trend throughout Europe, from the 11th through to the 17th century, appears to have been towards ever more powerful and sophisticated feudal states.

TRADE, TOWNS AND FEUDAL CRISIS. The growing requirements of the lordly class for the weaponry and luxury goods (especially fine textiles) needed to carry on intra-feudal politico-military competition were at the source of the expansion of commerce in feudal Europe. The growth of trade made possible the rise of a circuit of interdependent productions in which the artisan-produced manufactures of the towns were exchanged for peasant-produced necessities (food) and raw materials, appropriated by the lords and sold to merchant middlemen. Great towns thus emerged in Flanders and north Italy in the 11th and 12th centuries on the basis of their industries' ability to capture a preponderance of the demand for textiles and armaments of the European lordly class as a whole.

In the first instance, the growth of this social division of labour within feudal society benefited the lords, for it reduced costs through increasing specialization, thus making luxury goods relatively cheaper. Nevertheless, in the long run it meant a growing disproportion between productive and unproductive labour in the economy as a whole, for little of the output of the growing urban centres went back into production to augment the means of production or the means of subsistence of the direct peasant producers; it went instead to military destruction and conspicuous waste. Over time, increasingly sophisticated political structures and technically more advanced weaponry meant growing costs and thus increased unproductive expenditures. At the very time, then, that the agricultural economy was reaching its limits, the weight of urban society upon it grew significantly, inviting serious disruption.

Because the growth of lordly consumption proceeded in response to the requirements of intra-feudal competition in an era of increasingly well-constructed feudal states, the lords could not take into account its effect on the underlying agricultural productive structure. All else being equal, the growth of population beyond the resources to feed it could have been expected to call forth a Malthusian adjustment, and most of Europe did witness the onset of famine and the beginning of demographic downturn in the early 14th century. Nevertheless, while the decline of population meant fewer mouths to feed with the available resources, it also meant fewer rent-paying tenants and so, in general, lower returns to the lords. The decline in seigneurial incomes induced the lords to seek to increase their demands on the peasantry, as well as to initiate military attacks upon one another. The peasants were thus subjected to increasing rents and the ravages of warfare at the very moment that their capacity to respond was at its weakest, and their ability to produce and to feed themselves was further undermined. Further population decline brought further reductions in revenue leading to further lordly demands – resulting in a downward spiral which was not reversed in many places for more than a century. The lordly revenue crisis and the ensuing seigneurial reaction thus prevented the normal Malthusian return to equilibrium. A general socio-economic crisis, the product of the overall feudal class/political system, rather than a mere Malthusian downturn, gripped the European agrarian economy until the middle of the 15th century (Dobb, 1946; Hilton, 1969; Bois, 1976; Brenner, 1982).

In the long run, feudal crisis brought its own solution. With the decline of population, peasant cultivation drew back onto the better land, making for the potential of increased output per capita and growing peasant surpluses. Meanwhile, civil and external warfare seem to have abated, a reflection perhaps of the exhaustion of the lordly class, and the weight of ruling class exactions on the peasantry declined correspondingly, especially as the peasants were now in a far better position to pay. The upshot was a new period of population increase and expansion of the area under cultivation, of the growth of European commerce, industry and towns, and, ultimately, of the familiar outrunning of production by population. Meanwhile, lordly political organization continued to improve, feudal states continued to grow, intrafeudal competition continued to intensify, and, over the long run, lordly demands on the peasants continued to increase even as the capacity of the peasantry began, once again, to decline. By the end of the 16th century one witnesses, through most of Europe, a descent into the 'general crisis of the 17th century' which took a form very similar to that of the 'general crisis of the 14th and 15th centuries'. Clearly, through most of Europe, the old feudal property relations persisted, undergirding the repetition of established patterns of feudal economic non-development.

APPROACHES TO TRANSITION. It is an implication of the foregoing analysis that so long as feudal property relations persisted, the repetition of the same long term economic patterns could be expected. So long as feudal property relations obtained, lords and peasants could be expected to find it rational to adopt the same patterns of individual economic behaviour; in consequence, one could expect the same long term cyclical tendencies to declining agricultural productivity, population growth, and the opening of new land, issuing in a tendency to Malthusian adjustment but overlaid by a continuation of the secular tendency to lordly state building and growing unproductive expenditures. Generally speaking, so long as feudal property relations obtained, no inauguration of a long term pattern of modern economic growth could be expected. From these premises, it is logical to conclude that the onset of economic development depended on the transformation of feudal property relations into capitalist property relations, and that indeed is the point of departure of a long line of theorists and historians (Marx, 1894; Dobb, 1946; Hilton, 1969; Bois, 1976).

Nevertheless, beginning with Adam Smith himself, a whole school of historically-sensitive theorists have found it quite possible to ignore, or sharply to downplay, the problem of the transformation of property relations and of social relationships more generally in seeking to explain economic development. These theorists naturally refuse to go along with the Adam Smith of *Wealth of Nations* Book I in contending that the mere application of individual economic rationality will, directly and automatically, bring economic development. They nevertheless follow the Adam Smith of *Wealth of Nations* Book III in arguing that, given the appearance of certain specific, *quite-reasonable-to-expect* exogenous economic stimuli, rational self-interested individuals can indeed be expected to take economic actions which will detonate a pattern of modern economic growth. Specifically, it is their hypothesis that the growth of commerce, an enormously widespread if not universal phenomenon of human societies, systematically has led precapitalist economic actors to assume capitalist motivations or goals, to adopt capitalist norms of economic behaviour, and, eventually, to bring about the transformation of precapitalist to capitalist property relations. It is undoubtedly because Adam Smith and his followers have

313

believed that the growth of exchange will *in itself* sooner or later create the necessary conditions for modern economic growth that they have not greatly concerned themselves with these conditions or viewed their emergence as a problem which needs addressing.

Thus, Smith and a long line of followers, prominently including the economic historian of medieval Europe Henri Pirenne and the Marxist economist Paul Sweezy, have all produced analyses which follow essentially the same progression. First, merchants, emanating from outside feudal society, offer previously unobtainable products to lords and peasants who hitherto had produced only for subsistence. This is understood as a more or less epoch-making historical event, an original rise of trade. Next, the very opportunity to purchase these new commodities induces the individual economic actors to adopt businesslike attitudes and capitalist motivations, specifically to relinquish their norm of production for subsistence and to adopt the economic strategy of capitalists-in-embryo – viz., production for exchange so as to maximize returns by way of cost cutting. Third, since precapitalist property relations, marked by the producers' possession of the means of subsistence and by the lord's extraction of a surplus by means of extra-economic coercion, prevent the individual economic actors from most effectively deploying their resources to maximize exchange values, both lords and peasants move, on a unit-by-unit basis, to transform these property relations in the direction of capitalist property relations. In particular, the lords dispense with their (unproductive) military followers and military luxury expenditures; they free their hitherto-dominated peasant producers; they expropriate these peasants from the land; then, finally, they enter into contractual relations with these free, expropriated peasants. This gives rise, within each unit to the installation of free, necessarily commercialized (market dependent) tenants on economic leases, who, ultimately, hire wage labourers. The end result is the establishment of capitalist property relations and capitalist economic norms in the society as a whole and the onset of economic development (Smith, 1776; Pirenne, 1937; Sweezy, 1950).

The foregoing argument of what might be called the Smithian school is designed, implicitly or explicitly, to show how the rise of exchange in a feudal setting, in itself will create the conditions under which rational economic actors will pursue self-interested action which leads, on an economy wide basis, to modern economic growth. Nevertheless, the validity of each step in the Smithian argument can be, and has been, challenged by those who take as their point of departure the historically-established property relations. It is the essence of their position that the Smithians can sustain their argument only by failing sufficiently to understand what patterns of economic activity individual lords and peasants will find it rational to adopt in response to the rise of trade, *given* the prevalence of feudal property relations (Marx, 1894; Dobb, 1946; Bois, 1976).

In the first place, although long-distance merchants may bring to feudal lords and peasants commodities they could not previously obtain, the merchants' mere offer of these commodities cannot ensure that the lords and peasants will, in turn, put their own products on the market in order to buy them. Given the existence of feudal property relations, both lords and peasants may be assumed to have everything they need to maintain themselves. The opportunity to buy new goods may very well make it possible for the precapitalist economic actors to increase or enrich their consumption, but this does not mean that they will take advantage of this opportunity. The increased potential for exchange simply

cannot determine that exchange will increase (Luxemburg, 1913).

Secondly, even where the appearance of new goods brought by merchants does induce the lords to try to increase their consumption by raising their output and increasing the degree to which they orient their production towards exchange, this will hardly lead them to find it in their rational self-interest to dismantle, in piecemeal fashion, the existing feudal property relations by freeing and expropriating their peasants. Given the reproduction of feudal property relations by communities of feudal lords and peasants, the individual lords can hardly find it in their rational self-interests to free their peasants, for they would lose thereby their very ability to exploit them, and thus their ability to make an income. The point is that, once freed from the lord's extra-economic domination, his *possessing* peasants would have no need to pay *any* levy to him, let alone increase the quality and quantity of their work for him. Moreover, even if the lord could, at one and the same time, free *and* expropriate his peasants, he would still lose by the resulting transformation of his unfree peasant possessors into free landless tenants and wage labourers, for the newly-landless tenants or wage labourers would have no reason to stay and work for their former lord or to take up a lease from him.

To the degree, then, that lords sought to increase their output in response to trade, they appear to have found it in their rational self-interest not to transform but to intensify the precapitalist property relations. Because they found it, on the one hand, difficult to get their possessing peasants effectively to use more productive techniques on their estates, and, on the other hand, irrational to install capitalist property relations within their units, they seem to have had little choice but to try to do so within the constraints imposed by feudal property relations – by increasing their levies on the direct producers in money, kind or labour. To make this possible, they had no choice but to try to strengthen their institutionalized relationship of domination over their peasants, by investing in improved means of coercion and by improving the politico-military organization of their lordly groups. It needs to be emphasized that the lords could not be sure they could succeed in this, for the peasants would likely resist, and perhaps successfully. But in so far as the lords could dictate terms, this was the route they found most promising. Witness the growth of demesne farming in response the growth of the London market in 13th-century England or, more spectacularly, the rise of a neo-serfdom throughout later medieval and early modern Eastern Europe in response to the growth of trade with the West (Dobb, 1946).

Finally, it needs to be noted that the sort of products on the market which were most likely to stimulate the exploiters to try to increase their income for the purpose of trade were goods which 'fit' their specific reproductive needs. These were not producer goods but, on the contrary, means of consumption – specifically, materials useful for building up the exploiters' political and military strength. They were certainly not luxury goods in the ordinary sense of superfluities, for they were, in fact, necessities for the exploiters. But they were luxuries in that their production involved a subtraction from the means available to the economy to expand its fundamental productive base.

Paradoxically, then, to the extent that the rise of trading opportunities, *in itself*, can be expected to affect precapitalist economies, it is likely to bring about not the loosening but the tightening of precapitalist property forms, the growth of unproductive expenditure, and the quickening not of economic growth but of stagnation and decline.

FROM FEUDALISM TO CAPITALISM. The onset of modern economic growth thus appears to have required the break-up of precapitalist property relations characterized by the peasants' possession of their means of subsistence and the lords' surplus extraction by extra-economic compulsion. Nevertheless, neither the regular recurrence of system-wide socio-economic crisis nor the widespread growth of exchange could, in themselves, accomplish this. The problem which thus emerges is how feudal property relations could ever have been transformed?

To begin to confront this question, one can advance two basic hypotheses which follow more or less directly from the central themes of this article:

1. In so far as lords and peasants, acting either individually or as organized into communities, were able to realize their conscious goals, they succeeded, in one way or another, in maintaining precapitalist property forms. This is to say, once again, that the patterns of economic activity that individual lords and peasants found it reasonable to pursue could not aim at transforming the feudal property structure. It is also to emphasize that, because peasants and lords organized themselves into communities for the very purpose of maintaining and strengthening, respectively, peasant possession and the institutionalized relationships required for taking a feudal rent by extra-economic coercion, lords and peasants acting as communities were unlikely to aim at undermining feudal property forms. Peasants might, through collective action, conceivably have reduced to zero the lords levies and eliminated the lords' domination; but, even in this extreme case, they would have ended up constituting a community of peasants fully in possession of their means of subsistence, with all of the barriers to economic development entailed by that set of property relations. Were the lords, on the other hand, to have succeeded to the greatest extent conceivable in overcoming peasant resistance, they would only to that degree have strengthened their controls over the peasants and increased their rate of rent, thus tightening feudal property relations.

2. Where breakthroughs took place to modern economic growth in later medieval and early modern Europe, these must be understood as *unintended consequences* of the actions by individual lords and peasants and by lordly communities and peasant communities in seeking to maintain themselves as lords and peasants in feudal ways. In other words, the initial transitions from feudal to capitalist property relations resulted from the attempts by feudal economic actors, as individuals and collectivities, to follow feudal economic norms or to reproduce feudal property relations under conditions where, doing so, actually had the effect – for various reasons – of undermining those relations.

To give substance to these hypotheses would require a lengthy historical discussion. It is here possible only to note a broad contrast in the historical evolutions of the different European regions during the late medieval and early modern periods. Through most of pre-industrial Europe, East and West, varying processes of class formation brought, in one form or another, the reproduction of feudal property relations and, in turn, the repetition of long-term developmental patterns familiar from the medieval period. However, in a few European regions, feudal property relations dissolved themselves, giving rise, for the first time, to essentially modern processes of economic development.

Thus, through much of later medieval and early modern Western Europe (France and parts of Western Germany), although peasants succeeded in very much strengthening peasant possession, winning their freedom and destroying all forms of surplus extraction by extra-economic coercion by individual lords, the lords succeeded, in response, in maintaining themselves by means of constituting a new, more potent form of now-collective surplus extraction by extra-economic compulsion, the tax/office state. At the same time, throughout late medieval and early modern Eastern Europe, despite the peasants' initially very powerful rights in the land and the lords' initially very weak feudal controls, the lords ended up erecting an extremely tight form of individual lordly domination and surplus extraction by extra-economic compulsion – serf-operated demesne production. The consequence of these reconsolidations of essentially feudal property relations throughout most of Europe, East and West, was the reappearance throughout most of Europe during the early modern period of the same trends toward demographically powered expansion, toward the continued build-up of larger and more sophisticated states and, ultimately, toward socio-economic crisis as had characterized the medieval period.

The evolution of property relations in late medieval and early modern England was in some contrast to that of both Eastern and (most of) Western Europe, with epochal consequences for the long-term pattern of economic development. During this period, English lords, unlike those in Eastern Europe, failed, as did those throughout almost all of Western Europe, in their attempts to maintain, let alone intensify, their extra-economic controls over their peasantry. On the other hand, the English lords, unlike those throughout much of Western Europe, did ultimately succeed in maintaining their positions by means of preventing their customary tenants from achieving full property in their plots. They were able, in consequence, to consign these tenants to leasehold status, and thus to assert their own full property in the land.

The unintended consequence of the actions of English peasants and lords aiming to maintain themselves as peasants and lords in feudal ways was thus to introduce a new system of now-capitalist property relations in which the direct producers were free from the lords' extra-economic domination but also separated from their full means of reproduction (subsistence). In the upshot, tenants without direct access to their means of reproduction, had no choice but to produce competitively for exchange and thus, so far as possible, to specialize, accumulate and innovate. At the same time, the landlords found themselves obliged to create larger, consolidated and well-equipped farms if they wished to attract the most productive tenants. The long-run results were epoch making. Under the pressures of competition, processes of differentiation led to the emergence of an entrepreneurial class of capitalist tenant farmers who were ultimately able to employ wage labourers. Meanwhile, the drive to cut costs in agricultural production ultimately brought about an agricultural revolution, as market-dependent farmers were obliged to adopt techniques which long had been available, but long eschewed by possessing peasants who would not intentionally take the risks of specialization, let alone make the necessary capital investments. The secular decline in food costs and the secular rise in living standards which resulted underpinned the movement of population off the land and into industry and made possible the rise of the home market. Industry and agriculture, for the first time, proved mutually supporting, rather than mutually competitive, and population increase served to stimulate economic growth rather than to undermine it. England experienced unbroken industrial and demographic growth right through the 17th and 18th centuries, which ultimately issued in the Industrial Revolution.

ROBERT BRENNER

See also COMMON LAND; DOBB, MAURICE HERBERT; MODES OF PRODUCTION; OPEN FIELD SYSTEM; PEASANTS; SWEEZY, PAUL MALOR.

BIBLIOGRAPHY

Anderson, P. 1974. *Passages from Antiquity to Feudalism*. London: New Left Books.

Bois, G. 1976. *La crise du féodalisme*. Paris: Editions EHESS.

Bloch, M. 1939–40. *Feudal Society*. Trans. L.A. Manyon, Chicago: University of Chicago Press, 1961, 2 vols.

Brenner, R. 1982. The agrarian roots of European feudalism. In *The Brenner Debate: Agrarian Class Structure and Economic Development in Preindustrial Europe*, ed. T.H. Aston, Cambridge: Cambridge University Press, 1985.

Dobb, M. 1946. *Studies in the Development of Capitalism*. London: Routledge & Kegan Paul; New York: International Publishers, 1947.

Duby, G. 1968. *Rural Economy and Country Life in the Medieval West*. Trans. C. Postan, Columbia: University of South Carolina Press.

Duby, G. 1973. *The Early Growth of the European Economy*. Trans. H.B. Clarke, Ithaca: Cornell University Press, 1974.

Duby, G. 1978. *The Three Orders of Society*. Trans. T.N. Bisson, Chicago: University of Chicago Press, 1980.

Ganshof, F.L. 1947. *Feudalism*. Trans. P. Grierson, New York: Harper & Row, 1961.

Hilton, R.H. 1966. *A Medieval Society*. New York: Wiley.

Hilton, R.H. 1969. *The Decline of Serfdom*. London: Macmillan.

Luxemburg, R. 1913. *The Accumulation of Capital*. Trans. A. Schwarzschild, New York: Monthly Review Press, 1968.

Marx, K. 1894. *Capital*, Vol. III. New York: International Publishers, 1967.

Pirenne, H. 1937. *Economic and Social History of Medieval Europe*. New York: Harcourt Brace & Co.

Postan, M.M. 1966. Medieval agrarian society in its prime: England. In *The Cambridge Economic History of Europe*, Vol. 1: *The Agrarian Life of the Middle Ages*, 2nd edn, ed. M.M. Postan and H.J. Habakkuk, Cambridge: Cambridge University Press.

Strayer, J.R. 1965. *Feudalism*. New York: Van Nostrand Reinhold.

Weber, M. 1956. Patriarchalism and patrimonialism. Feudalism, Standestaat, and patrimonialism. In *Economy and Society*, 2 vols, ed. G. Roth and C. Wittich, Berkeley: University of California Press, 1978.

fiat money. The three major conventionally accepted functions of a money are as a means of exchange, unit of account and store of wealth. The most important function is as a means of exchange; the next is as a *numéraire*, and the last, the 'store of wealth', is fulfilled by many commodities. Money is a creation of law. A commodity money is a money which would have intrinsic utilitarian worth as a commodity even if it were demonetized. A fiat money is a money which if demonetized would scarcely be worth the paper it is printed on.

If there are m items of exchange only $m - 1$ ratios among m prices are needed to specify exchange rates. Without loss of generality we may select a price of one per unit for the item which is to serve as the numeraire. Conventionally, but not by logical necessity, the money used as a means of exchange is usually also selected as the numeraire.

The means of exchange feature of a money is closely related to the concept of liquidity. A *simple market* is where commodity i can be exchanged directly for j. We may define a money as an item i which has complete simple markets, that is, it is completely liquid, it can be exchanged directly for all other commodities. A perfect exchange economy with m goods and $m(m - 1)/2$ simple markets has every good completely liquid; all serve as means of payment. A near money can be loosely described as a good or instrument with almost m simple markets.

Liquidity and marketability are related but different. An asset that is highly liquid is also marketable; one that is marketable is not necessarily highly liquid. In pure exchange theory the relationship between goods i and j is symmetric. There is no strategic distinction between buyers and sellers; all are traders. In the dynamics of exchange the distinction appears.

The distinction between a money and a non-money is in essence strategic, relating to the number and nature of the markets it has. The distinction between a commodity money and a fiat money is in the intrinsic worth of the former and in the greater emphasis on legal code, custom and authority needed to back the acceptability in exchange of the latter. The store of value aspect of a fiat money is protected only by the legal and other societal backing of its means of exchange property. If prices change substantially it may be a highly imperfect store of value.

A bank cheque may be counted as a fiat money as a first approximation. A bank demand deposit is a promise to pay on demand the government's fiat money, but by the way the rules of the game have evolved the bank's IOU note itself is negotiable. Its acceptability in exchange is almost but not quite as universal as cash and in both instances the acceptability of the paper is by fiat, law, force and custom.

Another important property of a money is anonymity, a fiat money being somewhat more anonymous than a cheque. In making a purchase with the former the only name that need appear is that of the currency, such as a US dollar bill; with the latter a bank's name and the name of the signer appear.

A fiat money is best considered in terms of its system behaviour, which is delimited by the rules of the game. They in turn are determined from society to society by a highly institutional blend of law, custom, business practices and technology. Thus, although in attempting to answer many questions concerning the functions of fiat money, say in the United States and in France, for most purposes we can assume that the functions are the same but in some instances there may be a difference caused by differences in the rules.

From a somewhat more legalistic point of view it can be argued that all coinage is fiat money. Long before the invention of coinage, Egyptian tomb paintings showed precious metal being weighed in exchange (Skinner, 1967, pp. 8, 9). Thus we might wish to distinguish a precious metal used and measured in exchange from an official coin that is issued by the state. Even though the electrum coins first issued by Lydia around 635 BC by Gyges and Andys may have been traded as full-bodied monies (i.e. their precious metal content approximately determined their value), they nevertheless differed fundamentally from the equivalent nominal amount of precious metal. They had the imprimatur of the state. The state reserved the power to coin. Although in international trade foreign and clipped coins often required both weighing and reassaying, within the confines of the issuing kingdom or empire the power of the government to control the content of the coinage and to influence its acceptability set the stage for differentiating coinage from precious metal.

The first paper money known is that of China. Notes were issued as early as AD 650 in the Tang dynasty (Yang, 1952; Bereziner and Narbeth, 1973). Even earlier, around 120 BC the White Hart money of Emperor Wu (Fitzgerald, 1935, p. 166) provides an example of a fiat issue, although the notes on the hide of a white hart did not circulate but could best be regarded as a tax on the nobility.

The many desirable properties of a money include ease of identification, durability, portability, ease of protection against theft, efficiency in transfer and other features involving the technology of trade. The most abstract, durable and easily transportable funds would be a purely electronic funds transfer accounting system working essentially at the speed of light. This appears to place a technological bound on the velocity of transactions and cuts out activities such as cheque-kiting and 'the cheque is in the mail', but the upper bound on human decision-making will have occurred long before that velocity is attained. This advance in technology does not change the fundamental fact that money is defined by the rules of the game. There can be control rules (of varying efficiency) under any technology.

A discussion of the concept of a fiat money cannot be disassociated from the concept of credit. But the granting of credit is a two-party non-anonymous transaction. A gives B an IOU note in return for something of value. Among two private citizens the IOU is often non-negotiable by third parties. The rules are somewhat different for the use of credit where A is a private individual depositor and B is a bank. The rules are still different for the array of government debt including government paper money. A dollar bill is a zero interest freely negotiable perpetuity without recourse. The bankruptcy rules and who can sue whom for recovery are part of the distinctions among the various forms of highly short term credit instruments created by binary arrangements among the five major economic agents of a society, the central government, other governmental bodies, financial institutions, other corporate entities and private individuals. Fiat money is a form of credit where the issuing party is the state and the recourse of an individual creditor is negligible against the state, but by the law of the state the fiat money must be accepted in payment to extinguish other debts.

Money, other financial instruments and financial institutions are the neural network or command, control and intelligence system of an economy. Speed and efficiency in exchange and optimal coding of parsimony of information are achieved by this system which at the same time provides the means for a somewhat loosely coupled macroeconomic governmental control.

MARTIN SHUBIK

See also MONEY IN ECONOMIC ACTIVITY; QUANTITY THEORY OF MONEY.

BIBLIOGRAPHY
Bereziner, Y. and Narbeth, G. 1973. *The Story of Paper Money*. Newton Abbot: David and Charles.
Fitzgerald, C.P. 1935. *China: a short cultural history*. London: Cresset Press.
Skinner, F.G. 1967. *Weights and Measures*. London: HMSO.
Tobin, J. 1980. Explaining the existence and value of money: comment. In *Models of Monetary Economics*, ed. J.H. Kareken and N. Wallace, Minneapolis: Federal Reserve Bank of Minneapolis.
Yang, L. 1952. *Money and Credit in China*. Cambridge, Mass.: Harvard University Press.

Fichte, Johann Gottlieb (1762–1814). Fichte, though of the first importance as a philosopher, cannot be called an economist. Yet through his philosophy he has indirectly exercised great influence on economists, his system giving in outline the theory of development worked out by Hegel, and applied by certain of Hegel's followers to economic history and theory. Yet the direct influence of Fichte, through his writings on social and political questions, has been much less strong than might have been expected from the power of the writer and the brilliancy of his theories.

Fichte himself had two social ideals. (*a*) He looked forward to a condition of human society when the state and the coercion of laws would not be needed; as regards the remote future, he is what is now called an anarchist, of the type of William Godwin. (*b*) But he sees that men have, strictly speaking, no rights without the state, and conceives that they must necessarily pass through a stage of development in which the state and the laws shall educate them. He has, therefore, a proximate ideal, an ideal state. The best state is to him a 'closed state'; it is not merely to have its separate nationality and laws, but it is to be separate in its industry and wealth. It is not to be merely 'protected' against its neighbours' competition; it is to have a cordon drawn round it, and, with a few jealously-watched exceptions, it is to have no trade and hardly any intercourse with the foreigner.

The cordon once drawn, the guardians of the state can, he thinks, regulate production and trading, prices and wages. They can introduce a *Landesgeld* or peculiar national currency, valueless abroad; and they can control its value by controlling its quantity. Thus in all departments of economical life there would be hope of introducing constancy, security, and the maintenance of the chief right of man, the right to labour. Fichte means by right to labour the same sort of exclusive privilege as was secured by the old gilds to their members; and he regards this as the most important form of property. Private property in the ordinary sense of the world, family life, and even accumulation of fortunes, are not excluded; and the advantages of family life are clearly recognized. Fichte is a socialist but not a communist; and he does not try to regulate consumption.

The fire of enthusiasm always present in Fichte's writings is not wanting in the *Closed State*; but the *Characteristics*, and *Vocation of Man*, are better examples of his best manner.

His collected works were edited by J.H. Fichte, Berlin, 1845–6 (8 vols). There are passages of economic interest scattered up and down in nearly all these volumes. *Der Geschlossene Handels-Staat* (1800) was an appendix to the *Naturrecht* (1796). Both are contained in vol. III. of works.

The *Characteristics of the Present Age*, *The Vocation of Man*, and other of the more popular works of Fichte were translated into English (with much spirit) by the late Sir William Smith (Chapman, 1848, etc.). The translator published also a *Memoir* of Fichte that went through two editions. Fichte's chief philosophical treatise is *Wissenschaftslehre* (1794), vol. i. of works.

[JAMES BONAR]
Reprinted from *Palgrave's Dictionary of Political Economy*.

SELECTED WORKS
1845–6. *Collected Works*. 8 vols, ed. J.H. Fichte, Berlin.
1847. *The Characteristics of the Present Age*. Trans. W. Smith, London: Chapman.
1848. *The Vocation of Man etc.* Trans. W. Smith, London: Chapman.

BIBLIOGRAPHY
Bonar, J. 1893. *Philosophy and Political Economy*, Vol. 4. London.
Lassalle, F. 1862. *Die Philosophie Fichtes und die Bedeutung des deutschen Volksgeistes Festrede*.
Meyer, J.B. 1878. *Fichte, Lassalle, und der Sozialismus*.
Schmoller, G. 1888. *Litteraturgeschichte der Staats- und Socialwissenschaften*. Leipzig.
Smith, W. 1848. *Memoir of Fichte*. 2nd edn, London.

fictitious capital. The concept of 'fictitious capital' is rarely used by economists today. According to the rather small, though diverse, group of authors who have used the notion, it refers to the finance of productive activity by means of credit. Whatever their differences, all authors contrast 'fictitious capital' with 'real capital', where the latter usually refers to produced means of production, but may also include what Marxists call 'money-capital'. One group of authors contrasts finance by means of fictitious capital with voluntary (i.e. not forced) saving of the means of production. Hayek (1939) is a member of this group and refers to Viner's (1937) brief discussion of the use of the concept by English economists (e.g. by Lauderdale and Ricardo). On the other hand, Marx (1894), and Hilferding (1910), analyse the concept of 'fictitious capital' with respect to different forms of 'borrowed capital' and to the significance of the market value of financial titles and their relation to the value produced by labour.

Hayek (1939) argues that fictitious capital is the product of an increase in bank credit which distorts the capital market. When the plans of consumers and entrepreneurs coincide, the credit offered by the former to the latter corresponds to the placement of savings, and the stability of the capital market is assured. However, an increase in bank credit which encourages entrepreneurs to invest without a corresponding increase in saving results in what Hayek calls a crisis of 'over consumption', with, at the same time, a scarcity of capital and an excess supply of unused capital goods. Here the notion of 'fictitious capital' has a pejorative character as if it referred to counterfeit money or a *traite de cavalerie*. It is no longer solely the source of an illusory stimulus but a source of distortion and crisis.

Fictitious capital violates the necessary neutrality of money by establishing a direct relationship between banks and enterprises, in place of the banks' intermediary role. The interpretation of this relationship as illusory or harmful is related to a quantitative conception of the supply of money.

Marx (1894) discusses his quite different notion of 'fictitious capital' in the context of his theory of money and credit. According to him, productive capital, the value of which is created by labour, appears in diverse forms – first, that of money-capital, which is necessary for the payment of wages and the purchase of capital-goods. This money-capital, which is owned by a capitalist, may be loaned by a financier to an entrepreneur. Interest is payable, but this is solely a financial revenue derived from gross profit and has no 'natural' character. According to his A–A′ formula (expressing the cycle of loaned capital), 'capital seems to produce money like a pear-tree produces pears', divorced from the process of production and the exploitation of labour. This is why, according to Marx, interest-bearing capital is the most fetishized form of capital.

The notion of 'fictitious capital' derives from that of loaned money-capital. It suggests a principle of evaluation which is opposed to that which is based on labour-value: 'The formation of fictitious capital is called capitalization. Capitalization takes place by calculating the sum of capital which, at the average rate of interest, would regularly yield given receipts of all kinds.' According to Marx, financial revenues regulate the evaluation of all other receipts. It is 'totally absurd' to capitalize wages as if they were a return to 'human capital', and an 'illusion' to do the same with interest on the public debt to which there corresponds no productive investment.

Nevertheless, the issue of bonds provides the right to a part of the surplus which will be created by future work. Hilferding remains faithful to Marx when he states that 'on the stock exchange, capitalist property appears in its pure form ... outside the process of production'. Although doubly fetished, in the circuit A–A′ and on the financial markets, this fictitious

capital has some real roots – the necessity of there being money-capital, credit and the means of financial circulation as an expression of the functioning of the capitalist mode of production.

Used in these different ways the notion of 'fictitious capital' has often, for various reasons, a pejorative character. Although little used, it is at the centre of major economic problems: the relation between circulation and production, banks and enterprises and, fundamentally, the distribution of income.

S. DE BRUNHOFF

BIBLIOGRAPHY
Hayek, F.A. 1939. Price expectations, monetary disturbances and malinvestments. In Hayek, *Profits, Interest and Investment*, London: Routledge.
Hilferding, R. 1910. *Finance Capital*. London: Routledge & Kegan Paul, 1981, Pt 2.
Marx, K. 1894. *Capital*, Vol. 3, Part V. Moscow: International Publishers, 1967.
Viner, J. 1937. *Studies in the Theory of International Trade*. London: Harper.

fiducial inference. *See* FISHER, RONALD AYLMER.

fiduciary issue. The fiduciary issue of a bank (*fiduciarius* = held in trust) is that part of its note issue that is not covered by gold or by some other generally accepted means of payment, such as silver. The expression is associated especially with the Bank of England, where it dates from the Bank Charter Act 1844 (7 & 8 Vict., c.32), although this does not use the actual expression.

STATUTORY PROVISIONS. The 1844 Act, as a condition of renewing the Bank of England's charter, required it to divide its activities between an Issue Department and a Banking Department. The Issue Department, as its name implies, was responsible for the control of the Bank's note issue, and in particular for ensuring that the size of the issue complied with the Act. This prescribed that, except for a fixed amount of Government securities, the Bank's notes must be covered completely by gold coin, or by gold or silver bullion of which at least four-fifths must be gold. It is the amount of securities so fixed that is known as the fiduciary issue.

In 1844 the amount of the fiduciary issue was set at £14 million. No official reason was given for choosing this amount, but contemporaries offered a number of possible explanations. Firstly, it was probably no coincidence that the Bank's capital was, and is, £14,553,000. Secondly, an internal Bank committee, which reported while the Act was in preparation, suggested that it would be appropriate to issue £12 million of notes plus £2 million against 'unemployable deposits'. Thirdly, some commentators related the figure to the minimum actual circulation of notes, which between 1799 and 1844 had never fallen below about £15.5 million. Alternatively, it was argued that between 1826 and 1843 the average circulation of notes in excess of the Bank's holdings of bullion had been slightly above £1 million. Adding to this £3 million to replace the notes of certain country banks which had ceased to issue notes, produced a figure of £14 million. This was taken to be the amount that, characteristically, the Bank could float and the public could use. It seems probable that the decision to fix the fiduciary issue at £14 million reflected more than one of these converging considerations.

TABLE 1

Issue Department

	£		£
Notes issued	28,351,295	Government debt	11,015,100
		Other securities	2,984,900
		Gold coin and	
		bullion	12,657,208
		Silver bullion	1,694,087
	28,351,295		28,351,295

The fiduciary issue is represented by the first two items among the assets.

The Bank was required by the Act to publish weekly a Return showing how it was complying with the obligations placed upon it. In the first such Return published, for the week ended 7 September 1844, the part concerned with the Issue Department was as shown in Table 1.

The Act also restricted the issue of notes by banks other than the Bank of England. Only those banks already issuing notes on 6 May 1844 might do so in England in future, and the amount which each might issue was limited to those in circulation on that date. Two other provisions in the Act, continuing restrictions already in force, ensured that in the course of time all English note issues other than those of the Bank of England would disappear. These provisions were that no issuing bank might have more than six partners, and that no bank in London or within 65 miles of London (except of course the Bank of England itself) might issue notes.

INCREASES IN LIMIT. When, as a result of such restrictions, a country bank ceased to issue notes, the Bank of England was permitted to seek authority to increase its fiduciary issue by an amount equal to two-thirds of that which had lapsed. (The limitation to two-thirds appears to have been based on an assumption that the discontinuing bank would normally have held a reserve in gold or Bank of England notes equal to one-third of its note issue.) As a result of this provision, the Bank of England's issue not covered by coin or bullion increased by stages, eventually reaching £19,750,000 on 21 February 1923.

During World War I the issue of bank notes was supplemented by Government-issued Treasury Notes, but in 1928 the two series were amalgamated under the aegis of the Bank of England. The operative statute was the Currency and Bank Notes Act 1928 (18 & 19 Geo.V, c.13). This set the limit of the fiduciary issue at £260 million, but included provision for this to be increased or decreased on the initiative of the Bank of England. Increases, which might continue for six months at a time, were to be authorized by Treasury Minute, and were subject to a maximum of two years, after which parliamentary approval had to be obtained. Reductions, on the other hand, could be authorized by a Treasury letter. The backing for the fiduciary issue was still to be Government debt, except that silver coin to an amount not exceeding £5,500,000 might be included. Apart from the fiduciary issue, all notes had to be covered by gold coin or bullion.

The crisis of 1931, leading to Great Britain's abandonment of the Gold Standard, was accompanied by an increase in the fiduciary issue to £275 million, which was in force from August 1931 to March 1933. Thereafter the limit varied between £200 million and £260 million until 1939. In January 1939 it was temporarily increased to £400 million by a Treasury Minute. In March of that year it was altered to

£300 million by the Currency and Bank Notes Act 1939 (2 & 3 Geo.VI, c.7). In September 1939, however, practically the whole of the Bank's gold holding was transferred to the Exchange Equalization Account, and the fiduciary issue was increased to £580 million. Since then the Bank's note issue has been effectively backed only by paper. At the end of the Bank's year 1983–4 (February 1984) the notes issued totalled £11,470,000,000, while the assets of the Issue Department consisted wholly of securities. The increase above the £300 million set by the Currency and Bank Notes Act 1939 is authorized regularly by the Treasury, and is confirmed by Statutory Order placed before Parliament every second year.

PURPOSE OF LIMITATION. The philosophy underlying the limitation of the fiduciary issue was that of the Currency School. It was held that to restrict the issue of bank notes in this way would ensure that there would be no repetition of the crisis of 1836, which was believed to have been caused by an undue proliferation of notes. For this reason, proposals put forward while the Act was being deliberated, by which a relaxing clause would have been included to allow for emergencies, were held to be unnecessary, and indeed unwise. Subsequent experience, however, ensured that such a clause was included when the 1928 Act was being drafted. For, far from preventing new crises, the 1844 Act in some respects promoted them by leading the Bank to believe that it was fulfilling its responsibilities if the note issue was within the prescribed limit, without regard to the ability of the Banking Department to expand credit.

The outcome was a series of crises in 1847, 1857 and 1866. On each occasion commercial panics produced scrambles for liquidity, which led inevitably to demands for more Bank of England notes. Each time, the Bank was initially prevented from responding by the limit on its note issue, thereby exaggerating the panic. Each time, however, the Government encouraged the Bank to meet commercial requirements, even though the volume of notes issued might exceed the statutory limit and undertook to indemnify the Bank if this occurred. In practice, the limit was not exceeded in 1847 or 1866, but in 1857 the note issue was increased by £2 million above the £14,475,000 which was then the fiduciary issue; of these £2 million, some £928,000 left the Bank.

These developments, revealing the inadequacy of the Currency Theory, cast doubts on the significance of the note issue, and therefore of the limit to the fiduciary issue. Concern also shifted to the size of the Bank's gold stock in relation to the country's international commitments.

The Macmillan Committee, reporting in 1931 (paragraph 328), recommended that the fiduciary issue as such should be abolished, being replaced by a limit on the Bank's total note issue, together with an obligation to maintain a minimum stock of gold. This proposal was not adopted.

In 1959 the Radcliffe Committee, whose report stressed that it was the money supply as a whole rather than the note issue which was important, dismissed the fiduciary issue as irrelevant. The Committee further remarked (paragraph 367) that the only current use of the Bank Return, as prescribed in 1844, was to 'provide a formula for determination of the income of which the Bank has untrammelled disposal'.

Today the fiduciary issue would appear to have no other function than, through the two-year limitation upon its increase imposed in 1928, to afford Parliament a periodic reminder of the growth of the monetary base.

SCOTLAND AND IRELAND. In Scotland and Ireland the individual banks have continued to issue notes, there being no equivalent there to the provision in the Bank Charter Act 1844 extinguishing English note issues other than those of the Bank of England. However, in 1845 limits similar in effect to the Bank of England's fiduciary issue were placed upon the volume of notes which each of the 19 Scottish banks might issue, other than against a backing of legal tender (8 & 9 Vict., c.38). These limits, which totalled some £3 million for Scotland as a whole, were based on the average of each bank's actual circulation during the twelve months ended 1 May 1845. Similar legislation was passed for Ireland (8 & 9 Vict., c.37)

In 1928 parallel legislation to the Currency and Bank Notes Act restricted the fiduciary issues of the Scottish banks (then numbering eight) to a total of £2,676,350 and those of the banks in Northern Ireland to £1,634,000.

J.K. HORSEFIELD

See also BANKING SCHOOL, CURRENCY SCHOOL AND FREE BANKING SCHOOL; MONETARY BASE.

BIBLIOGRAPHY

Clapham, Sir John. 1944. *The Bank of England – A History.* Cambridge: Cambridge University Press.
Committee on Finance and Industry (Macmillan Committee). 1931. *Report* Cmd 3897. London: HMSO.
Committee on the Working of the Monetary System (Radcliffe Committee). 1959. *Report* Cmnd 827. London: HMSO.
Sayers, R.S. 1976. *The Bank of England, 1891–1944.* Cambridge: Cambridge University Press.

final degree of utility. The expression used by Jevons for the degree of utility of the last increment of any commodity secured, or the next increment expected or desired. The increments being regarded as infinitesimal, the degree of utility is not supposed to vary from the last possessed to the next expected. It will be obvious, after a study of the article on Degree of Utility that it is the *final* degree of utility of various commodities that interests us commercially, not, for instance, their initial or average degrees of utility. That is to say (Fig. 1), if a is a small unit of the commodity A, and b a small unit of the commodity B, and q_a the quantity of A I possess, and q_b the quantity of B I possess, then, in considering the equivalence of a and b I do not ask whether A or B has the greater initial degree of utility, i.e. I do not compare the lines Oa and Ob, nor do I inquire which has the greater average degree of utility, i.e. I do not compare the height of the rectangle on base Ox which shall equal the area aOxa', with the height of the rectangle on base Oy which shall equal the area bOyb', but I compare the length xa' with the length yb', and ask what are the relative rates at which increments of A and B will *now add* to my satisfaction. If xa' is twice the length of yb', then (since a and b are supposed to be small units, throughout the consumption of which the decline in the curves aa' bb' may be neglected) it is obvious that $2b$ will be equivalent to a, since either increment will yield an equal area of satisfaction.

Now suppose (Fig.2) that some other possessor of the commodities A and B, either because he possesses them in different proportions, or because his tastes and wants are different, finds that the relative final utilities of the small units a and b are not the same for him (2) as they are for me (1). Say that for him $3b$ is the equivalent of a, clearly the conditions for a mutually advantageous exchange exist. Let δ be greater than 2 and less than 3, so that $\delta - 2$ and $3 - \delta$ are both positive. Now suppose (1) exchanges with (2), giving him a and

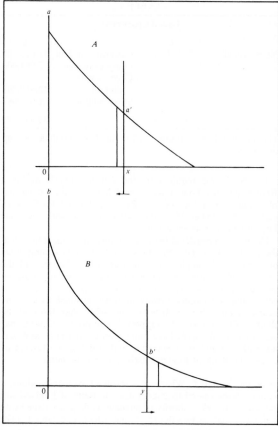

Figure 1

receiving from him δb. Then, (1) receives δb in exchange for a (worth $2b$ to him) and benefits to the extent of $(\delta - 2)b$, and by the same transaction (2) has received a (worth $3b$ to him) in exchange for δb, and has benefited to the extent of $(3 - \delta)b$. The result of this exchange will be a movement of all the verticals that indicate the amount of each commodity possessed by each exchanger, in the directions indicated by the arrow-heads; and this again will (as is obvious from inspection of the figures) tend to reduce the difference between the ratio of equivalence between a and b in the case of the two exchangers. The process of exchange will go on (δ not necessarily remaining constant) until the ratio of equivalence between a and b coincides for the two exchangers, the last exchange bringing about an equilibrium in accordance with that ratio. Such a ratio of equilibrium is a limiting ratio of exchange; that is to say, exchange constantly tends to approach such a ratio, perhaps by a series of tentative exchanges at various rates, and would cease were such a ratio actually arrived at.

Hence Jevons's fundamental theorem: 'The ratio of exchange of any two commodities will be the reciprocal of the ratio of the final degrees of utility of the quantities of commodities available for consumption after the exchange is completed', applies to an ideal ratio which would secure equilibrium at a stroke, rather than to the tentative bargains by which it is approached in the 'actual market'.

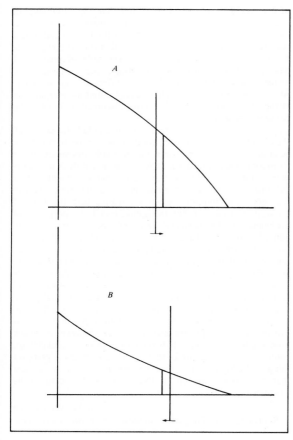

Figure 2

BIBLIOGRAPHY

Gossen, H.H. 1854. *Entwicklung der Gesetze des menschlichen Verkehrs, und der daraus fliessenden Regeln für menschliches Handeln.* Braunschweig: Vieweg.

Jevons, W.S. 1871. *Theory of Political Economy.* London. 3rd edn, London: Macmillan, 1888.

Walras, L. 1886. *Theórie de la monnaie:* Lausanne: Corbaz.

final utility. The principles and methods embodied in Jevons's doctrine of 'final utility' have received far-reaching developments in recent years. Hence a movement has arisen, variously described as 'psychological' or 'marginalist', which aims at unifying and simplifying economic theory, and at the same time affiliating its laws more closely to the principles that regulate human conduct in general.

Jevons has shown that the demand in a market in which there are no reserved prices can be represented by a collective curve. The amount of the commodity in the market is measured on the abscissa, and the equilibrating price on the ordinate. The next step is to point out that in so far as the sellers have reserved prices they ought to be regarded as themselves entering the market, with potential demands, on the same footing as the purchasers. Their intention to retain such and such quantities of their stock at such and such prices (whether for their own use or because they speculate on the demands of future purchasers) constitute *de facto* demands, and should be entered on the collective demand curve; which, together with the register of the amount of the commodity, will determine the price, as before. It follows that the cross curves of demand and supply, so often employed by economists, are really no more than two sections of the true collective curve of demand, separated out from each other, and read, for convenience, in reverse directions. This separation is irrelevant to the determination of the equilibrating price (as may easily be shown by experiment), though it enables us to read off the volume of the exchanges that will be necessary in order to bring about the equilibrium, on any given supposition as to initial holdings. These cross curves, then, as usually presented, confuse the methods by which the equilibrating price is arrived at with the conditions that determine what it is.

Passing on to the problems of production and distribution, we note that in an industrially advanced community production rests upon the cooperation of a number of heterogeneous factors, the supply of which may be controlled by a number of independent individuals or combinations; and since it is obvious that the value of a means of production must be derivative from the value of the product, we have, theoretically, to determine the principle on which the value of the product when realized will be distributed amongst the various factors which cooperated in its production. Practically the factors will generally be brought together by a series of speculative transactions based on estimates made in advance. But in any case the value of the several factors must be determined by consideration of their productive effectiveness at the margin, and their equivalence to each other in fractional substitutions. For although the nature of the productive service rendered by such factors as land, labour, and tools, for instance, is different in each case, and no main factor could be replaced in its entirety by any other, yet every manager is constantly engaged in considering alternatives and equivalences between fractional additions or subtractions of them at the margin. It is so that he determines the proportions in which to distribute his resources over the improving or extending of a site, the modification of existing buildings, the

The conceptions of 'degree of utility' and 'final degree of utility' lie at the heart of the mathematical method of political economy, and their complete history would almost coincide with the history of mathematical economics. Incidentally the idea has been struck from time to time by sundry mathematicians, and it has been worked out independently by economists no fewer than four or five times. Cournot (1838), Dupuit (1844), Gossen (1853), and Jevons (1862 and 1871) successively discovered and taught the theory, each one in ignorance of the work of his predecessors. In 1871 the Austrian Menger, and in 1874 the Swiss Walras (working on the basis laid down by Cournot), adopted essentially the same central conception, and since then the theory has not again sunk into oblivion. Many writers in Germany, Holland, Denmark, France, Italy, and England are now engaged in developing it. See the bibliographies and lists of writers in the appendix to Jevons's *Theory of Political Economy*, 3rd edn, and the Preface to Walras's *Theórie de la Monnaie*, 1886; and for far-reaching recent developments in America, England, and France see Appendix.

[Jevons's 'final degree of utility' is the *Grenznutzen* of the Austrian school, Gossen's *Werth der letzten Atome*, and Walras's *rareté*.]

[P.H. WICKSTEED]
Reprinted from *Palgrave's Dictionary of Political Economy*

replacing of machinery, the strengthening or reduction of this or that grade of labour, superintendence to reduce the waste of raw material, or the seeking of new openings, or maintenance of old ones, by advertisement. And all the time he has to convince his employers that his own skill in judging of these matters is as effectively productive as any increments in the more immediate factors of production that they could command for the salary that they pay him. The purchasers, then, in the great markets of the productive factors consider them under the uniform aspect of their relative productive efficiency at the margin, just as the purchaser in the retail market considers his heterogeneous purchases under the uniform aspect of their relative efficiency at the margin, in gratifying his desires or expressing his impulses. In a word, there are not many laws of distribution but one, and that law is the law of the market.

(Thus 'interest' is the price, reckoned in deferred payments, of present command of resources. The industrial, who expects this command actually to produce the future resources out of which he will make the payment, enters a market in which he will have to compete with the non-industrial who is willing to risk or compromise his future at the dictate of his present desires, and the ordinary consumer who, having a small revenue and no accumulations, is willing to pay a higher price for a possession, if he may spread the payment over a longer period, rather than cut deep into the quick of his other requirements at the moment.

('Rent' is a form of hire, the continuous purchase of a continuous revenue of services or enjoyments. The well-known figure of the rent curve, which represents the decreasing productive efficiency of successive applications of labour and capital to a fixed unit of land, is seen to owe its form not to any special characteristic of land but to the selection of a single factor of production which is not to increase while all the others do. The identical facts which such a curve represents, if read in the reverse order, would represent the same series of hypotheses as to the relative proportions of the several factors; but the rent would now be presented as a rectangular area, with its altitude determined by the alternative uses of land, and the return to labour and capital, as a curvilinear 'residue', determined by the decreasing yield of a fixed constant of labour, etc., when spread over more and more land.)

Thus it will be seen that the end dominates the means throughout. The direction and administration of all resources is ultimately determined by estimates of the value of some experience, or by the imperativeness of some expression of the human consciousness. If at any point the expectations based on these estimates should fail or wither, the breadth of the stream that has already flowed at their bidding is powerless to sustain their living significance. Anticipated value determines the cost and sacrifice that will be incurred in production, but the cost and sacrifice, when once incurred, cannot control the value of the product.

If we now return to our starting-point in Jevons's 'final utility' and its control of the distribution of a man's pecuniary resources, we note that the term 'final' has been generally abandoned. It seems to imply a succession of experiences, following each other in time, as when a man's hunger is gradually appeased and each morsel meets a decreasingly urgent need. It is therefore inapplicable, for instance, to the problems we have discussed under the head of 'distribution', where the units of the same factor may be indistinguishable in quality and may all be running abreast of each other in the output of a continuous stream of efficiency, but where nevertheless the withdrawal from cooperation of one unit out

of five would be a less serious matter than the withdrawal of one out of four, because it would create a less serious disturbance of the proportions between the factors and would require less serious readjustments or additions to compensate it. The term 'marginal' has been very generally adopted, but it has the disadvantage of still suggesting (especially in connection with land) some intrinsic differentiating characteristic which earmarks and individualizes a unit as 'marginal' in virtue of its own nature. The term 'fractional' may often be conveniently used.

Again, the world 'utility' so conspicuously fails to include all the objects of wise or foolish, good or bad desire, to which the economic machinery ministers, that if it still sometimes retains its place (subject to careful explanation that it does not really *mean* utility) it is only for want of general agreement as to a substitute. The anomaly becomes more glaring and extends to the term 'consumption', when we realize that the laws of political economy are but the application to a special set of problems of the universal laws of the distribution and administration of resources in general (whether of money, time, influence, powers of thought, or aught else) amongst all the objects that we deliberately pursue or to which we are spontaneously impelled, whether material or spiritual, private or social, wise or foolish. It is intolerable that 'consumption' (with its subtle suggestion of a regrettable necessity that puts a drag upon the progress of 'production') should continue to stand for the whole stream of 'actualizings', in conscious experience, of the potentialities to the development of which human effort is devoted. It is the nature of these actualizings, contemplated or realized, that is the supremely significant thing in the life of a man or a community; for it is from them that all which leads up to them derives its worth or its worthlessness.

[P.H. WICKSTEED]
Reprinted from *Palgrave's Dictionary of Political Economy.*

finance. Finance is a subfield of economics distinguished by both its focus and its methodology. The primary focus of finance is the workings of the capital markets and the supply and the pricing of capital assets. The methodology of finance is the use of close substitutes to price financial contracts and instruments. This methodology is applied to value instruments whose characteristics extend across time and whose payoffs depend upon the resolution of uncertainty.

Finance is not terribly concerned with the problems that arise in a barter economy or, for that matter, in a static and certain world. But, once the element of time is introduced, transactions develop a dual side to them. When a loan is made, the amount and the terms are recorded to insure that repayment can be enforced. The piece of paper or the computer entry that describes and legally binds the borrower to repay the loan can now trade on its own as a 'bearer' instrument. It is at the point when debts were first traded that capital markets and the subject of finance began.

The study of finance is enriched by having a large body of evolving data and market lore and some powerful and, at times, competing intuitions. These intuitions are used to structure our understanding of the data and the markets which generate it. The modern tradition in finance began with the development of well-articulated models and theories to explore these intuitions and render them susceptible to empirical testing.

While the subject of finance is anything but complete, it is now possible to recognize the broad outlines of what might be called the neoclassical theory. In the discussion which follows we will group the subjects under four main headings corresponding with four basic intuitions. The first topic is efficient markets, which was also the first area of finance that matured into a science. Next come the twinned subjects of return and risk. This leads naturally into option pricing theory and the central intuition of pricing close substitutes by the absence of arbitrage. The principle of no arbitrage is used to tie together the major subfields of finance. The fourth section looks at corporate finance from its well-developed form as a consequence of no arbitrage to its current probings. A short conclusion ends the entry.

EFFICIENT MARKETS

The word efficient is too useful to be monopolized by a single meaning in economics. As a consequence, it has a variety of related but distinct meanings. In neoclassical equilibrium theory efficiency refers to Pareto efficiency. A system is Pareto efficient if there is no way to improve the well being of any one individual without making someone worse off. Productive efficiency is an implication of Pareto efficiency. An economy is productively efficient if it is not possible to produce more of any one good or service without lowering the output of some other.

In finance the word efficiency has taken on quite a different meaning. A capital market is said to be (informationally) efficient if it utilizes all of the available information in setting the prices of assets. This definition is purposely vague and it is designed more to capture an intuition than to state a formal mathematical result. The basic intuition of efficient markets is that individual traders process the information that is available to them and take positions in assets in response to their information as well as to their personal situations. The market price aggregates this diverse information and in that sense it 'reflects' the available information.

The relation between the definitions of efficiency is not obvious, but it is not unreasonable to think of the efficient markets definition of finance as being a requirement for a competitive economy to be Pareto efficient. Presumably, if prices did not depend on the information available to the economy, then it would only be by accident that they could be set in such a way as to guarantee a Pareto efficient allocation (at least with respect to the commonly held information).

If the capital market is competitive and efficient, then neoclassical reasoning implies that the return that an investor expects to get on an investment in an asset will be equal to the opportunity cost of using the funds. The exact specification of the opportunity cost is the subject of the section on risk and return, but for the moment we can observe that investing in risky assets should carry with it some additional measure of return beyond that on riskless assets to induce risk averse investors to part with their funds. For now we will defer the measurement of this risk premium, and simply represent the opportunity cost by the letter 'r'.

In much of the early empirical work on efficient markets no attempt was made to measure risk premia, and the opportunity cost of investing was set equal to the riskless rate of interest. This can be justified either by assuming that there are risk neutral investors who are indifferent to risk (or, as we shall see, by assuming that the asset's risk is diversified away in large portfolios). Whatever the rationale, to focus on the topic of efficient markets rather than on the pricing of risk, we will let r be the riskless interest rate.

If R_t denotes the total return on the asset – capital gains as well as payouts – over a holding period from t to $t+1$, then the efficient markets hypothesis (EMH) asserts that

$$E(R_t|I_t) = (1 + r_t),\tag{1}$$

where E is the expectation taken with respect to a given information set I_t, that is available at time t (and that includes r_t). An alternative formulation of the basic EMH equation is in terms of prices. For an asset with no payouts, since

$$R_t \equiv p_{t+1}/p_t,$$

we can rewrite (1) as

$$E(p_{t+1}|I_t) = (1 + r_t)p_t,\tag{2}$$

or, equivalently, discounted prices must follow the martingale,

$$\frac{1}{(1 + r_t)} E(p_{t+1}|I_t) = p_t.$$

The EMH is given empirical content by specifying the information set that is used to determine prices. Harry Roberts (1967) first coined the terms which have come to describe the categories of information sets and, concomitantly, of efficient market theories that are employed in empirical work. Fama (1970) subsequently articulated them in the form which we now use. These categories describe a hierarchy of nested information sets. As we go up the hierarchy from the smallest to the biggest set (i.e. from coarser to finer partitions) we are requiring efficiency with respect to increasing amounts of information. At the far end of the spectrum is strong-form efficiency. Strong-form efficiency asserts that the information set, I_t, used by the market to set prices at each date t contains all of the available information that could possibly be relevant to pricing the asset. Not only is all publicly available information embodied in the price, but all privately held information as well.

A substantial notch down from strong-form efficiency is semistrong-form efficiency. A market is efficient in the semistrong sense if it uses all of the publicly available information. The important distinction is that the information set, I_t, is not assumed to include privately held information, i.e. information that has not been made public. Making this distinction precise is possible in formal models but categorizing information as publicly available or not can be subjective. Presumably, accounting information such as the income statements and the balance sheets of the firm is publicly available, as is any other information that the government mandates should be released such as the stock holdings of the top executives in the firm. Presumably, too, the true but unrevealed intention of a major stockholder would fall into the category of private information. In between these extremes is a large grey area.

The tendency in the empirical literature has been to take a purist's view of semistrong efficiency, and to adopt the position that if the information was in the public domain then it was available to the public and should be reflected in prices. This ignores the cost of acquiring the information, but the intuitive justification for this position is that the costs of acquiring such public information are small compared to the potential rewards. Thus, while the government mandated and publicly reported trades of the top executives require a bit more effort to obtain in a timely fashion than some average of their past holdings, such trades, when reported, would fall squarely within the realm of publicly available information under the semistrong version of the EMH.

If the asset is traded on an organized exchange, then of all the information that is clearly available to the public, none is as accessible and cheap as its past price history. At the bottom of the ladder in the efficiency hierarchy, weak-form efficiency requires only that the current and past price history be incorporated in the information set. If there is empirical validity to the EMH then, at the very least, the market for an asset should be weak-form efficient, that is, efficient with respect to its own past price history.

Empirical testing. The empirical implications of efficiency with respect to a particular information set are that the current price of the asset embodies all of the information in that set. Since the categories of information sets are nested, rejection of any one type, say, weak-form efficiency, implies the rejection of all stronger forms.

For example, according to weak-form efficiency, the current price of an asset embodies all of the information contained in the past price history. This implies that,

$$E(R_t | R_{t-1}, R_{t-2}, \dots) = (1 + r_t), \tag{3}$$

or, in price terms,

$$E(p_{t+1} | p_t, p_{t-1}, \dots) = (1 + r_t)p_t.$$

The most dramatic consequence of the EMH and certainly the one that receives the most attention from the public, is that it denies the possibility of successful trading schemes. If, for example, the market is weak-form efficient, then an investor who makes use of the 'technical' information of past prices can only expect to receive a return of the opportunity cost $(1 + r_t)$. No amount of clever manipulation of the past information can improve this result.

As a test of weak-form efficiency, then, we could test (although not as a simple regression) the null hypothesis that

$$H_0 : E(p_{t+1} | p_t, p_{t-1}) = \beta_0 + \beta_1 p_t + \beta_2 p_{t-1}, \tag{4}$$

where

$$\beta_0 = 0$$
$$\beta_1 = (1 + r_t).$$

and

$$\beta_2 = 0.$$

The important feature of this hypothesis is that it tells what information does *not* play a role (given r_t), namely the lagged price, p_{t-1}. If the coefficient β_2 should prove to be statistically significant, then this would constitute a rejection of the weak-form EMH.

The other empirical implication of the EMH that is often cited as a defining characteristic is that an efficient price series should 'move randomly'. The precise meaning of this in our context is that price changes should be serially uncorrelated.

Consider the serial covariance between two adjacent rates of return,

$$\text{cov}(R_{t+1}, R_t) \equiv E([R_{t+1} - E(R_{t+1})][R_t - E(R_t)]). \tag{5}$$
$$= E(R_{t+1}[R_t - E(R_t)])$$
$$= E(E(R_{t+1} | R_t)[R_t - E(R_t)])$$

In equation (5), since we have not specified the information set with respect to which the expectations are to be taken, they are unconditional expectations. Under weak-form efficiency, the information set will contain the past rates of return. Suppose that the (expected) opportunity cost, e.g. the interest rate r, is independent of past returns on the asset or that

changes are of a second order of magnitude. This would occur, for example, if we held r_t constant at r. In such a case, since weak-form efficiency implies that I_{t+1} contains R_t, we have

$$E(R_{t+1} | R_t) = E[E(R_{t+1} | I_{t+1}) | R_t]$$
$$= E[(1 + r_{t+1}) | R_t] \tag{6}$$
$$= E(1 + r_{t+1}),$$

the unconditional expectation of next period's opportunity cost. Putting (5) and (6) together yields,

$$\text{cov}(R_{t+1}, R_t) = E(1 + r_{t+1})E[R_t - E(R_t)] = 0. \tag{7}$$

which is to say that rates of return are serially uncorrelated.

Tests of the EMH are legion and by and large they have been supportive. The early tests were essentially tests of the inability of trading schemes or of the random walk nature of prices, which implies that actual rates of return are serially uncorrelated. While the EMH does not imply that prices follow a random walk, such a price process is consistent with market efficiency. Alternatively, unable to specify closely the opportunity cost, some of the early tests took refuge in the view that it must be positive, which leads to a submartingale model for prices,

$$E(p_{t+1} | I_t) \geqslant p_t. \tag{8}$$

The lack of a specification of the opportunity cost characterizes the early tests (see Cowles (1933), Granger and Morgenstern (1962) and Cootner (1964) and see Roll's (1984) study of the orange juice futures market for a modern example of such a test). Following Fama (1970), the literature shifted to a concern for specifying the opportunity cost and, in this sense, empirical tests became joint tests of the EMH and of the correct specification of the opportunity cost and its attendant theory.

In terms of the information hierarchy, the general message that emerged from the testing is that the market does appear to be consistent with weak-form efficiency. Tests of stronger forms of efficiency, though, have produced mixed results. Fama, Fisher, Jensen, and Roll (1969) introduced a new methodology to test semistrong efficiency and applied it to stock splits. They observed that the residuals from a simple regression of a stock's returns on a market index would measure the portion of the return that was not attributable to market movements. By adding the residuals over a period of time, the resulting cumulative residual measures the total return over that period that is attributable to nonmarket movements. If a stock splits, say, 2 for 1, then under semistrong efficiency its price should split in proportion, i.e., halve for a 2 for 1 split. Using this 'event study' approach, Fama, Fisher, Jensen and Roll verified that stock split data was consistent with semistrong efficiency. The event study methodology they introduced and the use of cumulative residuals (averaged over firms) has become the standard method for examining the impact of information on stock returns.

By contrast with their supportive findings, Jaffé (1974), for example, found that a rule based on the publicly released information about insider trades produced abnormal returns. These results and others like them (see the section on *Risk and Return* below) have been much debated and no final verdict on the matter is likely.

Recently a more interesting empirical challenge to the EMH has come from a different tack. Shiller (1981), has argued that the traditional statistical tests that have been employed are too weak to examine the EMH properly and, moreover, that they are misfocused. Shiller adopts the intuitive perspective that if stock prices are discounted expected dividends, then they ought not to vary over time as much as actual dividends. He argues that since the price is an expectation of the dividends and future price, what actually occurs will be this expectation *plus* the error in the forecast and should be more variable than the price. This leads him to formulate statistical tests of the EMH based on the volatility of stock prices which are claimed to be more powerful than the traditional (regression based) tests.

An alternative view has been taken by critics of this perspective, notably Kleidon (1986), Flavin (1983), and Marsh and Merton (1986). These critics have taken issue with Shiller's specification of the statistical tests of volatility and, more importantly, with his basic intuition. In particular, they contend that the single realization of dividends and prices that is observed is only one drawing from all of the random possibilities and that the price is based on the expectation taken over all of these possibilities. A little bit of information, then, can have an important influence on the current price. Furthermore, they argue that when the smoothing of dividends and the finite time horizon of the data samples are taken into account, volatility tests do not reject the EMH. The testing of the EMH is taking a new direction because of this work, but, at present, the results are still mixed.

Less cosmic in scope, but perhaps more worrisome is the discovery by French and Roll (1985) that the variance per unit time of market returns over periods when the market is closed (for example, from Tuesday's close to Thursday's close when the market was closed on Wednesday because of a backlog of paperwork) is many times smaller than when it is open. It is difficult to reconcile this result with the requirement that prices reflect information about the cash flows of the assets, unless the generation of fundamental information slows dramatically when the market closes – no matter why it is closed.

Theoretical formulations. The attempts to formalize the EMH as a consistent, analytical economic theory have met with less success than the empirical tests of the hypothesis. The theory can be broken into two parts. The first part is neoclassical and is largely formulated in terms of models in which investors share a common information set. Such models focus on the intertemporal aspects of the theory and the changing shape of the information set.

It has long been recognized that a competitive economy with a single risk neutral investor would lead to the traditional efficient market theories with respect to the information set employed by that investor. More interestingly, Cox, Ingersoll and Ross (1985a), and Lucas (1978) have developed intertemporal rational expectations models each of which is consistent with certain versions of the efficient market theories.

There is, however, an important sense in which these models fail to capture the essential intuition of efficient markets. In informationally efficient markets, prices communicate information to participants. Information possessed by one investor is communicated to another through the influence – however microscopic – that the first investor has on equilibrium prices. In models where investors have homogeneous information sets such information transfer is irrelevant.

A variety of attempts have been made to develop models of financial markets which can deal with such informational issues, but the task is formidable and a satisfactory resolution is not now in hand. This work parallels that of the neoclassical rational expectations view of macroeconomics. This is no accident since the rational expectations school of macroeconomics was very clearly influenced by the intuition of efficiency in finance. The original insight that prices reflect the available information lies at the heart of rational expectations macroeconomics. In this latter work aggregate prices, for example, not only provide the terms of trade for producers, they also inform producers about the aggregate state of production in the economy.

Perhaps the principal difficulty is that models with fully rational investors tend to break down. As investors apply the full scope of their analytical and reasoning talents, the result is an equilibrium in which they lack the incentive to engage in trade. (See Grossman, 1976; Grossman and Stiglitz, 1980; Diamond and Verrecchia, 1981; Milgrom and Stokey, 1982; and Admati, 1985.) The only way out of this bind seems to be to add a discomforting element of irrationality – or an alternative motive for trade from an equilibrium, such as insurance – to the model.

To understand this point, consider a risk-averse individual trading in a market where he or she receives information signals about the ultimate value of the asset being traded and where it is common knowledge that all investors are in the same position. That is not to say that all investors have the same information, rather, it only means that they all begin with the same information, have the same view of the world (Bayesian priors), and then receive signals from the same sort of information generating mechanism. In such a market, the offer to trade on the part of any one investor communicates information to other investors. In particular, it tells them that the individual, based upon his or her information, will be improved by the trade. If all investors are rational they will all feel similarly bettered by trade. But, if the market had been in an equilibrium prior to the receipt of new information, and if it is common knowledge that trade balances, then in the new equilibrium not all of them can be improved. This contradiction can only be resolved by having no further trade upon the receipt of information.

To put the matter in an equivalent form, consider an investor who possesses some special information. Presumably, it is by trading that this information is incorporated into the market price. The above argument implies that the mere announcement of a wish to trade results in a change in prices with no profits for the investor since none will trade at the original prices. If information is costly to acquire and impossible to profit from, then why bother? In other words, if the price reflects the available information possessed by the individual participants, then why gather information if one only needs to look at the price?

The resolution of this dilemma can take many forms, and research will proceed by altering the assumptions that lead to this result. For example, we can drop the assumption about a common prior and let investors come to the markets with different a priori beliefs. We could also drop the assumption that all investors are perfectly rational and introduce 'noisy' traders. Lastly, we could drop efficiency and complete markets or integrate insurance motives in other ways.

All of these approaches are being explored but we must leave this discussion with the theory that underlies the incorporation of asymmetric information into securities prices in an unsettled state. The traditional theory that prices reflect the available information is well understood with a representative individual. The theory with asymmetric information is not well understood at all. In short, the exact mechanism by which prices incorporate information is still a mystery and an attendant theory of volume is simply missing.

To conclude, the efficient market paradigm is the backbone of much of financial research and it continues to guide a large body of theoretical and empirical work. Its usefulness is beyond question, but its fine structure is not. In a sense, like much of economics, it remains a central intuition whose analytical representations seem less compelling than the insight itself. This presents more of a problem for theory than for empirical work, but the empirical side is also not without challenge. Although the evidence in support of the efficiency of capital markets is widespread, troublesome pockets of anomalies are growing and the power of the traditional methodology to test the theory is being seriously questioned. Nevertheless, there is currently no competitor for the basic intuition of efficient markets and few insights have proven as fruitful.

RISK AND RETURN

The theory of efficient markets leads inexorably to the second central intuition in finance, the trade-off between risk and return. It has long been recognized that risk-averse investors require additional return to bear additional risk. Indeed, this insight goes back to the earliest writings on gambling and it is as much a definition of risk aversion as it is a description of risk-averse behaviour. The contribution made by finance has been to translate this observation into a body of intuition, theory, and empirics on the workings of the capital markets.

The intuition that in a competitive market higher return is accompanied by higher risk owes at least as much to Calvin as it does to Adam Smith, but, in large part the development of capital market theory has been an attempt to explain risk premia, the difference between expected returns and the riskless interest rate. The foundations for the models that would first explain risk premia and that would become the workhorses of financial asset pricing theories were laid by Hicks (1946), Markowitz (1959), and Tobin (1958). These authors developed a rigorous micro-model of individual behaviour in a 'mean variance' world where investment portfolios were evaluated in terms of their mean returns and the total variance of their returns. They justified focusing on these two distributional characteristics by assuming either that investors had quadratic von Neumann–Morgenstern utility functions or that asset returns were normally distributed. In such a world, investors would choose mean variance efficient portfolios, i.e., portfolios with the highest mean return for a given level of variance. This observation reduced the study of portfolio choice to the analysis of the properties of the mean variance efficient set. Building on their work, Sharpe (1964), Lintner (1965), and Mossin (1969), all came to the fundamental insight that this micromodel could be aggregated into a simple model of equilibrium in the capital markets, the capital asset pricing model or CAPM.

The Mean Variance Capital Asset Pricing Model (CAPM). In neoclassical equilibrium models, an investor evaluates an asset in terms of its marginal contribution to his or her portfolio. The decision to alter the proportion of the portfolio invested in an asset will depend on whether the cost of doing so in terms of risk is greater or less than the benefit in expected return. An individual in a personal equilibrium will find the cost at the margin equal to the benefit.

We will assume that a unit addition of an asset to the portfolio can be financed at an interest rate of r. In a mean variance model the net benefit of adding an asset to a portfolio is the additional expected return it brings, E, less the cost of financing it. Such a change, Δx, will augment the expected return on the portfolio, E_p, by the risk premium of the asset, i.e. by the difference between the expected return on the asset, E_i, and the cost of the financing, r,

$$\Delta E_p = (E_i - r)\Delta x. \tag{9}$$

The marginal cost, in terms of risk, of an increase in the holding of an asset is the addition to the total variance of the portfolio occasioned by an increase in the holding of the asset. To compute this increase, let v denote the variance of returns on the current portfolio, let var(i) stand for the variance of asset i's returns, let cov(i, p) denote the covariance between the return of asset i and that of the portfolio, p, and let Δx be the addition in the holding of asset i.

The variance of the portfolio after adding Δx of asset i will be,

$$v + \Delta v = v + 2\Delta x \, \text{cov}(i, p) + (\Delta x)^2 \, \text{var}(i),$$

which means the change in the variance is given by

$$\Delta v = (\Delta x)\text{cov}(i, p) + (\Delta x)^2 \, \text{var}(i),$$

and for a small marginal change, Δx, this approximates,

$$\Delta v \approx 2(\Delta x)\text{cov}(i, p).$$

The marginal rate of transformation between return and risk, then, is given by

$$\text{MRT} = \frac{\Delta E_p}{\Delta v} = \frac{(E_i - r)\Delta x}{2(\Delta x)\text{cov}(i, p)} = \frac{(E_i - r)}{2\,\text{cov}(i, p)}. \tag{10}$$

An investor will be in a personal equilibrium when this trade-off is equal to his or her personal marginal rate of substitution between return and risk. But, if the portfolio p is an optimal one for the investor then it must also have a trade-off between return and risk that is equal to the investor's marginal rate of substitution, and this permits us to use it as a benchmark. Consider, then, the alternative possibility of changing the portfolio position not by changing the amount of asset i being held, but rather by changing the amount of the entire portfolio p being held, again financing the change by an alteration in the holding of the riskless asset. This is equivalent to leveraging the portfolio of risky assets and altering the amount of the riskless asset so as to continue to satisfy the budget constraint. Such a change will produce a trade-off between return and risk exactly analogous to the one examined above.

$$\text{MRS} = \frac{E_p - r}{2\,\text{var}(p)}, \tag{11}$$

where we have written this as the marginal rate of substitution, MRS. Since in equilibrium all of the marginal rates of transformation must equal the common marginal rate of substitution, putting these two equations together we have,

$$E_i - r = (E_p - r)\beta_{ip}, \tag{12}$$

where

$$\beta_{ip} \equiv \frac{\text{cov}(i, p)}{\text{var}(p)}, \tag{13}$$

the regression coefficient of the returns of asset i on the returns of portfolio, p. Equation (12) is the famous security market line equation, the SML. It describes the necessary and sufficient condition for a portfolio p to be mean variance efficient. It also provides a clear statement of the risk premium, asserting that it is proportional to the asset's beta, β_{ip}.

The insight of Sharpe, Lintner and Mossin was the observation that the SML and the mean variance analysis could be aggregated almost without change to a full equilibrium in the capital market. If we assume that all individuals have the same information and, therefore, see the same mean variance picture, then each individual's efficient portfolio will satisfy equation (12). Since the SML equation is linear in the portfolio holding, p, we can simply weight each individual's equation by the proportion of wealth that individual holds in equilibrium, and add up the individual SML's. The result will be an SML equation for the aggregate portfolio, m, that is the weighted average of the individual portfolios. In equilibrium, the weighted average of all of the individual portfolios, m, is the market portfolio, i.e., the portfolio of all assets held in proportion to their market valuation. In other words, each asset i, must lie on the SML with respect to the market,

$$E_i - r = (E_m - r)\beta_{im}, \qquad (14)$$

which means that the market portfolio, m, is a mean variance efficient portfolio.

The geometry of the mean variance analysis is illustrated in Figure 1. The set of mean variance efficient portfolios maps out a mean variance efficient frontier in the mean standard deviation space of Figure 1. Each investor will pick some point on this frontier and that point will be associated with a mean variance efficient portfolio that is suitable for the investor's particular degree of risk aversion. All such portfolios will themselves be portfolios of just two assets: the riskless asset, r, and a common portfolio, p, of risky assets. This fortunate simplification of the individual portfolio optimization problem is referred to as two fund separation. It implies that the only role for individual preferences lies in choosing the appropriate combination of the risky portfolio, p, and the riskless asset, r. As a consequence, when we aggregate, the market risk premium, $(E_m - r)/\text{var}(m)$, will be an average of individual measures of risk aversion.

Black (1972) showed that two fund separation would still hold in the mean variance model even if there were no riskless

asset. In such a case he found that an efficient portfolio orthogonal–the 'zero beta portfolio'–to the market portfolio could be found, and that all investors would be able to find their optimal portfolios as combinations of m and this zero beta portfolio. In the above development of the CAPM we can simply let r be the expected return on a zero beta portfolio.

The necessary and sufficient conditions on return distributions for them to have this two fund separation property – for any concave utility function – were established by Ross (1978a). Ross characterized the class of distributions whose efficient frontier, i.e. the set of portfolios that *some* investor would choose, was spanned by k funds, and showed that it extended beyond the normal distribution in the case of $k = 2$ fund separation. This work was extended by Chamberlain (1983), who found the class of distributions for which expected utility was a function of just mean and variance for any portfolio as well as for the efficient ones. Cass and Stiglitz (1970) found the conditions on investor utility functions for a similar property to hold regardless of assumptions on return distributions.

It follows immediately from two fund separation that the tangency portfolio, p, in figure 1 must be the market portfolio of risky assets since all investors hold all risky assets in the same proportions. If there is no net supply of the riskless asset then p must be the market portfolio, m, itself.

The central feature of the CAPM is the mean variance efficiency of the market portfolio and the emergence of the beta coefficient on the market portfolio as the determinant of the risk premium of an asset. Those features of an asset that contribute to its variance but do not affect its covariance with the market will not influence its pricing. Only beta matters for pricing; the idiosyncratic or unsystematic risk, i.e. that portion which is the residual in the regression of the asset's returns on the market's returns and is therefore orthogonal to the market, playing no role in pricing.

This produces some results that were at first viewed as counter-intuitive. The older view that the risk premium depended on the asset's variance was no longer appropriate, since if one asset had a higher covariance with the market than another, it would have a higher risk premium even if the total variance of its returns were lower. Even more surprising was the implication that a risky asset that was uncorrelated with the market would have no risk premium and would be expected to have the same rate of return as the riskless asset, and that assets that were inversely correlated with the market would actually have expected returns of less than the riskless rate in equilibrium.

These results for the CAPM were supposedly explicated by the twin intuitions of diversification and systematic risk. There could be no premium for bearing unsystematic risk since a large and well diversified portfolio (i.e. one whose asset proportions are not concentrated in a small subset) would eliminate it – presumably by the law of large numbers. This would leave only systematic risk in any optimal portfolio and since this risk cannot be eliminated by diversification, it has to have a risk premium to entice risk averse investors to hold risky assets. From this perspective it becomes clear why an asset that is uncorrelated with the market bears no risk premium. One that is inversely correlated with the market actually offers some insurance against the all pervasive systematic risk and, therefore, there must be a payment for the insurance in the form of a negative risk premium.

There is nothing wrong with this intuition, but it does not fit the CAPM very well. The residuals from the regression of asset returns on the market portfolios are orthogonal to the market, but they could be highly correlated with each other. In

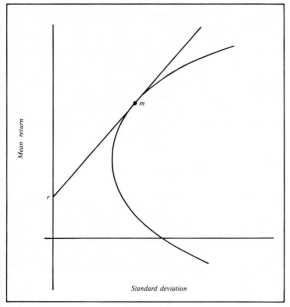

Figure 1

fact, they are linearly dependent since when they are weighted by the market proportions they sum to zero. This means the law of large numbers cannot be used to insure that large portfolios of residuals other than the market portfolio will be negligible. But, if that is the case, then the residuals could capture systematic risks not reflected in the market portfolio.

The CAPM was the genesis for countless empirical tests (see, e.g., Black, Jensen, and Scholes, 1972; and Fama and MacBeth, 1973). The latter paper developed the most widely used technique. The general structure of these tests was the combination of the efficient market hypothesis with time series and cross section econometrics. Typically some index of the market, such as the value weighted combination of all stocks would be chosen and a sample of firms would be tested to see if their excess returns, $E-r$, were 'explained' in cross-section by their betas on the index, i.e., whether the SML was rejected.

Roll (1977b, 1978) put a stop to this indiscriminate testing by calling into question precisely what was being tested. Roll's critique had two parts. First, he argued that the tests were of very low power and probably could not detect departures from mean variance efficiency. His central point, though, began by noting that tests of the CAPM were tests of the implications of the statement that the *entire* market portfolio was mean variance efficient, and were not simply tests of the efficiency of some limited index such as could be formed from the stock market. The essential role played by the market portfolio in the CAPM had been stressed by others; Ross (1977b) had shown the equivalence between the CAPM and the mean variance efficiency of the market portfolio. (Ross (1976a) had also shown that in the absence of arbitrage there was always some efficient portfolio.) Roll went beyond this simple observation, though, by stressing the essential point that the market portfolio is unmeasurable. This called into question the entire cottage industry of testing the CAPM and all of the uses to which the theory had been put, such as performance measurement.

Intertemporal models. In the aftermath of Roll's critique, attention was turned to alternative models of asset pricing and the intertemporal nature of the theory became more important. Two separate strands of development can be traced. One essentially followed the lines of the CAPM and developed the intertemporal versions of it, the ICAPM. Merton (1973a) pioneered in this. Using continuous time diffusion analysis, Merton showed that the CAPM could be generalized to an intertemporal setting. Most interestingly, though, he demonstrated that if the economic environment was described by a finite dimensional vector of state variables, x, and if asset prices were exogenously specified random variables, then a version of the SML would hold at all moments of time with the addition to the risk premium of a linear combination of the betas between the assets' returns and each of the state variables, x_j.

Ross (1975) developed a similar intertemporal extension of the CAPM, but Ross's model simplified preferences in order to close the model with an intertemporal rationality constraint and to study equilibrium price dynamics. Along the lines being developed in the modern literature on macroeconomics, intertemporal rationality and the efficient market theory required that the distribution of prices be determined endogenously. A discrete time Markov model with this feature was presented in Lucas (1978) and a full rational expectations general equilibrium in continuous time was developed in Cox, Ingersoll and Ross (1985a).

Cox, Ingersoll and Ross (1985b) applied their model to analyse and resolve some longstanding questions in the theory

of the term structure of interest rates. The theory of the term structure is one of the most important subfields of finance, and the bond markets were one of the first areas where the EMH was applied. In an efficient market, ignoring risk aversion, forward rates should be (unbiased) predictors of future spot rates and many early theories and tests of the EMH were formulated to examine this proposition (see e.g. Malkiel, 1966). Roll (1970) integrated the EMH with the CAPM and used the resulting framework to examine empirically liquidity premia in the bond markets; the work of Cox, Ingersoll and Ross (1985b) can be considered as the logical extension of his analysis to a rational intertemporal setting.

Merton's model was simplified markedly by Breeden (1979), who showed that if investors had intertemporally additive utility functions, then Merton's ICAPM and its version of the SML could be collapsed back into a single beta model, the Consumption Beta model, with all assets being priced, i.e., having their risk premiums determined, by their covariance with aggregate consumption (see also Rubinstein (1976)). If we think of returns as relative prices between wealth today and in future states of nature, then optimizing individuals will set their marginal rates of substitution between consumption today and in future states equal to the rates of return. With continuous asset prices and additive utility functions, indirect utility functions are locally quadratic in consumption and this implies that consumption plays the role of wealth in the static CAPM. This work led to a variety of attempts to measure the ability of betas on aggregate consumption to explain risk premia (see e.g. Hansen and Singleton, 1983).

Arbitrage Pricing Theory (APT). A separable but related strand of theory is the Arbitrage Pricing Theory (APT) (see e.g. Ross, 1976a, 1976b). The CAPM and the Consumption Beta model share the common feature that they explain pricing in terms of endogenous market aggregates, the market portfolio, and aggregate consumption, respectively. The APT takes a different tack.

The intuition of the CAPM (or of the Consumption Beta model) is that idiosyncratic risk can be diversified away leaving only the systematic risk to be priced. Idiosyncratic risk, though, is defined with reference to the market portfolio as the residual from a regression of returns on the market portfolio's returns. Since no further assumptions are made about the residuals, contrary to intuition a large diversified portfolio that differs from the market portfolio will not in general have insignificant residual risk. The exception is the market portfolio, but then the intuition that diversification leads to pricing by the market portfolio is circular at best.

The APT addresses this issue by assuming directly a return structure in which the systematic and idiosyncratic components of returns are defined a priori. Asset returns are assumed to satisfy a linear factor model,

$$R_i = E_i + \sum_j \beta_{ij} f_j + \epsilon_i, \quad i = 1, \ldots, n, \quad (15)$$

where E_i is the expected return, f_j is a demeaned exogenous factor influencing each asset i through its beta on the factor, β_{ij}, and ϵ_i is an idiosyncratic mean zero term assumed to be sufficiently uncorrelated across assets that it is negligible in large portfolios. An implication of the factor structure is that the ϵ terms become negligible in large well diversified portfolios and, therefore, such portfolios approximately follow an exact factor structure,

$$R_i \approx E_i + \sum_j \beta_{ij} f_j, \quad (16)$$

where i now denotes the ith well diversified portfolio. In an Arrow–Debreu state space framework, equation (16) can be interpreted as a restriction on the rank of the state-space tableaux.

An exact factor structure implies that there will be arbitrage unless the expected return on each portfolio is equal to a linear combination of the beta coefficients,

$$E_i - r = \sum_j \lambda_j \beta_{ij}, \qquad (17)$$

where λ_j is the risk premium associated with the jth factor, f_j. This equation is the APT version of the SML in the CAPM.

The APT is consistent with a wide variety of equilibrium models (including the CAPM if there is a factor structure) and it has been the object of much theoretical and empirical attention. In a sense, the APT can be thought of as a snapshot of any intertemporal model in which the factors represent innovations in the underlying state variables. This means that a rejection of the APT would imply a fairly wide ranging rejection of attempts to model asset markets with a finite set of state variables.

The original theoretical development of the APT (Ross, 1976a, 1976b) showed formally that if preferences are continuous in the quadratic mean, then the returns on a sequence of portfolios which require no wealth cannot converge to a positive return with a zero variance. This, in turn, implies that the sum of squared deviations from exact APT pricing is bounded above. These results were simplified by Huberman (1982) and extended by Ingersoll (1984) and Chamberlain and Rothschild (1983), all of whom side-stepped the issue of preferences by simply assuming that there could be no sequences converging to an arbitrage situation of a positive return with no variance. By contrast, Dybvig (1983) makes assumptions on preferences and aggregate supply to obtain a tight bound on pricing. His simple order of magnitude calculation is evidence that the pricing error is too small to be of practical significance.

By modelling the capital market explicitly as responding to innovations in exogenous variables, the APT is immediately intertemporally rational. By contrast with the CAPM and the Consumption Beta models which price assets in terms of their relation with a potentially observable and endogenous market aggregate (wealth for the CAPM and consumption for the Consumption Beta models), the APT factors are exogenous, but unspecified. Much empirical work is now underway to determine a suitable set of factors for representing systematic risk in a factor structure and to examine if they price assets successfully. (For example, see Roll and Ross, 1980; Brown and Weinstein, 1983; and Chen, Roll and Ross, 1986.)

The lack of an a priori specification for the factors has been the focus of criticism of the testability of the APT by Shanken (1982). Shanken argues that since the factors are not pre-specified, the intuitive derivation of the APT given above can be used to verify APT pricing falsely even when it does not hold, and that to prevent this some equilibrium model, such as that proposed by Connor (1984), must be used. Shanken emphasizes that his critique applies not to the theory of the APT, but rather to the way in which it has been tested. Dybvig and Ross (1985) dispute his arguments, stressing that Shanken wants to test the theory including its assumptions and approximations rather than take the positive approach of testing the model's conclusions.

Empirical testing of asset pricing models. Since Roll's critique, the methodology for testing asset pricing models has changed. There has been a retreat from testing a model per se to an explicit view that what is being tested is not the CAPM, for example, but rather whether the particular index being used for pricing is mean variance efficient. This change of focus has led to a more formal approach to the statistics of testing. Ross (1980) developed the maximum likelihood test statistic for the efficiency of a given portfolio and pointed out the analogy between this and the mean variance geometry, and Gibbons (1982) showed that the test of efficiency could be conducted by the use of seemingly unrelated regressions. These results have been extended by others. (For example, Kandel (1984) and Jobson and Korkie (1982)) and Gibbons, Ross and Shanken (1986) have developed and exploited an exact small sample test of the efficiency of a given index in the presence of a riskless asset. Similar tests of the APT have not yet been developed, and to date much of the testing of the APT has focused on comparisons between the APT and pricing using the value weighted index (see e.g. Chen, Roll, and Ross, 1986).

The most important empirical finding in asset pricing, though, has been the discovery of a wide array of phenomena that appear to be inconsistent with nearly any neoclassical model. Consider, first, the secular effects. Asset returns fall, on average, over the weekend and rise during the week (see French, 1980). Similarly, it has been found that asset returns behave differently in the first half of the month than they do in the second. The most attention, though, has been lavished on the 'small firm effect'. It appears that the average returns on small firms exceed those on large firms no matter what theory of asset pricing is used to correct for differences in the risk premium between these two categories of assets. Furthermore, the bulk of the return difference is concentrated in the first few days of January. Indeed, on average, returns in January appear to be abnormally large for all stocks (see e.g. Keim, 1983 or Roll, 1981, 1983).

Potentially these sorts of anomalies can be explained by secular changes in risk premia – perhaps due to secular patterns in the release of information – but their persistence and magnitude make them serious challenges to all the asset pricing models. When evidence of this sort appears difficult to explain by any pricing model it calls into question the efficient market hypothesis itself. Tests of an asset pricing model are usually joint tests of both market efficiency and the pricing model; rejecting a wide enough range of such models is tantamount to rejecting efficiency itself.

SUBSTITUTION AND ARBITRAGE: OPTION PRICING

The APT is the child of one of the central intuitions of finance: namely, that close substitutes have the same price. This intuition reached fruition in the path breaking paper by Black and Scholes (1973) on option pricing. Since then the theory has found myriad applications and has been significantly extended, (see, for example, Merton (1973b), Cox and Ross (1976a, 1976b), Rubinstein (1976), Ross (1976c), Ingersoll (1977), Cox, Ross and Rubinstein (1979), and Cox, Ingersoll and Ross (1985a)). The Black–Scholes model employed stochastic calculus, but a simpler framework for option pricing was presented by Cox, Ross and Rubinstein (1979) that retained its essential features and was more flexible for computational purposes. We will briefly outline this binomial approach and show its connections to the major theoretical features of option pricing.

The Binomial Model. The binomial model begins with the assumption that the price of a stock, S, follows a proportional geometric process:

$$S(t+1) = \begin{cases} aS(t) \text{ with probability } \pi \\ bS(t) \text{ with probability } 1 - \pi \end{cases}. \qquad (18)$$

In addition to the stock there is also a riskless bond with a return of $1 + r$. The basic problem of option pricing theory is to determine the value of a derivative security, i.e., a security whose payoff depends only upon the value of an underlying primitive security, the stock in this case.

Let $C(s, t)$ denote the value of the derivative security as a function of the price of the stock and the time, t. Since its value depends only upon the movement of the stock – a result that is sometimes derived as a function of other attributes such as its value at the end of some period – it will also follow a binomial process:

$$C(S, t + 1) = \begin{cases} C(aS, t) \text{ with probability } \pi \\ C(bS, t) \text{ with probability } 1 - \pi \end{cases}. \quad (19)$$

The time $t + 1$ values are illustrated in Figure 2. At any moment of time the information structure branches into relevant states, state a and state b, defined by whether the stock goes up by a or b. As the figure is drawn, $a > 1 + r > b$, and clearly $1 + r$ must lie between a and b to prevent the stock or the bond dominating. At this point there are two separate approaches to the analysis. The first is in the spirit of the original Black–Scholes model.

Suppose that at time t we form a portfolio of the riskless bond and the stock with α dollars invested in the stock and $1 - \alpha$ dollars invested in the bond. We will choose the investment proportion so that the return on the portfolio coincides with the return on the derivative security in state b. This means choosing α so that

$$\frac{C(bS, t + 1)}{C(S, t)} = \alpha b + (1 - \alpha)(1 + r), \quad (20)$$

which implies that

$$\alpha = \frac{(1 + r) - C(bS, t + 1)/C(S, t)}{(1 + r) - b}. \quad (21)$$

But, since the portfolio's return matches that of the derivative security in state b, it must also match it in state a. If it did

not, then either the portfolio or the derivative security would dominate the other, which would be an arbitrage opportunity. In other words, we must have,

$$\frac{C(aS, t + 1)}{C(S, t)} = \alpha a + (1 - \alpha)(1 + r). \quad (22)$$

Putting these two equations together produces a difference equation which is satisfied by the value of the derivative security,

$$\pi^* C(aS, t + 1) + (1 - \pi^*) C(bS, t + 1)$$
$$- (1 + r) c(S, t) = 0, \quad (23)$$

where

$$\pi^* \equiv \frac{(1 + r) - b}{a - b}. \quad (24)$$

Perhaps the most remarkable feature of this equation is that it does not involve the original probabilities for the process, π, but rather is a function of what are called the martingale probabilities, π^*.

To solve this difference equation for the value, C, of a particular derivative security we would need only to append the contractual boundary conditions that define it. For example, a European call option is specified to have the value max $(S - E, 0)$, at a specified future date, T, where E is its exercise price. Such an option gives the holder the right – but not the obligation – to buy the stock for E at time T. The dual security is a European put option which gives the holder the right, but again not the obligation, to sell the stock for E at time T. The problem is more difficult if the derivative security is of the American variety which means that the holder may exercise it any time up to and including the maturity date T and need not wait until T.

Soon after the Black–Scholes paper, Merton (1973b) examined a variety of option contracts and showed how extensive was the range of the technique. Notably, Merton was able to derive a number of qualitative results on option pricing that were relatively independent of the particular process being modelled. For example, he showed that an American call option on a stock that pays no dividends will never be exercised before its maturity date and, therefore, will have the same value as a similar European call. He also demonstrated that put/call parity, i.e. the equivalence between the positions of holding the stock and a put option and holding a bond and a call option, was not generally valid for American options. Ross (1976c) showed that the literature's emphasis on puts and calls was not misplaced since any derivative security could be composed of puts and calls.

A second approach to the valuation problem in our simple example illuminates why the original probabilities played no role in the analysis. Figure 2 displays what is essentially a two-state Arrow–Debreu model. In such a model if there are two pure contingent claims contracts paying one dollar in each state, then all securities can be valued as a function of their values, q_a and q_b. It follows, then, that any two securities which are not linearly dependent will span the space just as two pure contingent claims would and they can be used to value all securities in the space.

In our example, the value of the bond is 1 and it must satisfy,

$$1 = q_a(1 + r) + q_b(1 + r), \quad (25)$$

and the value of the stock must satisfy,

$$S = q_a(aS) + q_b(bS),$$

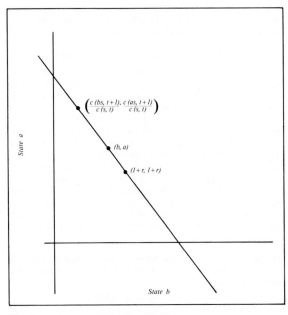

$\left(\frac{c(bs, t+1)}{c(s, t)}, \frac{c(as, t+1)}{c(s, t)} \right)$

(b, a)

$(1 + r, 1 + r)$

State a

State b

Figure 2

or

$$1 = q_a a + q_b b. \tag{26}$$

Solving these two equations we can find the implicit values of the state contingent claims,

$$q_a = \frac{(1+r) - b}{(1+r)(a-b)},$$

and

$$q_b = \frac{a - (1+r)}{(1+r)(a-b)}. \tag{27}$$

Notice that these prices do not depend on the original probability, π, since they are derived from the values of the stock and the bond. Whatever influence the probability, π, has on values is already reflected in the returns on the stock and the bond, and the derivative security value will just be a function of the implicit state prices. Using these prices, it is readily verified that the difference equation for the value of the derivative security, equation (23), is the same as,

$$q_a C(aS, t+1) + q_b C(bS, t+1) = C(S, t). \tag{28}$$

Geometrically, this means that the point,

$$\{[C(bS, t+1)/C(S, t)], [C(aS, t+1)/C(S, t)]\},$$

we plot on the same line as the return points for the bond and the stock, $(1+r, 1+r)$ and (b, a). For a call option the point will be as drawn in Figure 2 indicating that the call is more volatile than the stock.

Notice from (24) and (27) that

$$\pi^* = (1+r)q_a,$$

which means that the state space price can be interpreted as the discounted martingale probability. It is this interpretation that ties together the Cox and Ross (1976a) risk neutral approach to solving option pricing problems and the general theory of the absence of arbitrage.

Cox and Ross (1976a) argued that since the difference equation that emerged for solving option pricing problems made no explicit use of any preference information, the resulting solution must also be independent of preferences. For example, then, the resulting solution must be the same as that which would obtain in a risk neutral world. In such a world, the state probabilities must be such that the expected returns on all assets are the same,

$$\pi^* a + (1 - \pi^*)b = 1 + r,$$

where the solution for the probability, π^*, is the same martingale probability defined above. For a European call option, then, the solution will be

$$C(S, t) = \frac{1}{(1+r)^{T-t}} E^*[\max(S_T - E, 0)]$$

$$= \frac{1}{(1+r)^{T-t}} \sum_{j \geq \ln[E/Sb^{-(T-t)}]/\ln(a/b)} (\pi^*)^j$$

$$\times (1 - \pi^*)^{T-t-j}(Sa^j b^{T-t-j} - E), \tag{29}$$

where E^* is the expectation with respect to the martingale probabilities, π^* and $(1 - \pi^*)$. It is easily verified that (29) is the solution to the difference equation (23) subject to the boundary condition,

$$C(S, T) = \max(S - E, 0).$$

Contrast this formula with the original Black–Scholes formula

for the value of a call option in a continuous time diffusion model,

$$C(S, t) = SN(d_1) - e^{-r(T-t)}N(d_2), \tag{30}$$

where $N(\cdot)$ is the standard cumulative normal distribution function and,

$$d_1 \equiv \frac{\ln(S/E) + r(T-t) + \frac{1}{2}\sigma^2(T-t)}{\sigma\sqrt{(T-t)}},$$

and

$$d_2 \equiv d_1 - \sigma\sqrt{(T-t)}.$$

Equation (30) is the solution to the Black–Scholes option pricing differential equation,

$$\tfrac{1}{2}\sigma^2 S^2 C_{SS} + rSC_S - rC = -C_t, \tag{31}$$

subject to the boundary condition,

$$C(S, T) = \max(S - E, 0).$$

The Black–Scholes differential equation (31) is derived from an analogous hedging argument to that for the binomial model, applied to the continuous lognormal stock process,

$$dS/S = \mu \, dt + \sigma \, dz,$$

where z is a standard Brownian motion. In fact, as the time interval between jumps converges to zero and the jump sizes shrink appropriately, the binomial converges to the lognormal diffusion and its option pricing solution will converge to that for the lognormal diffusion. Notice, too, that in analogy with the binomial whose solution does not depend upon the state probabilities, the Black–Scholes option price (30) is independent of the expected return on the stock, μ.

The most interesting comparative statics result from these models is the observation that call or put option values increase with increasing variance, σ^2. This is a consequence of these options being convex functions of the terminal stock value, S_T (Cox and Ross, 1976b).

The general theory of arbitrage. All of the above analysis can be tied together by the general theory of arbitrage. Under quite general conditions, it can be shown that the absence of arbitrage implies the existence of a linear pricing rule that values all of the assets (see e.g., Ross, 1976a, 1978b; Harrison and Kreps, 1979). In a static model with m states of nature, this means the existence of implicit state prices, q_j, such that $q_j > 0$, and such that any asset with payoffs of x_j in the states of nature will have the value,

$$p = \sum_j q_j x_j. \tag{32}$$

The intertemporal extension of this result is most neatly displayed in terms of the martingale expectation used above. The absence of arbitrage now implies the existence of a martingale measure such that, with obvious notation,

$$p = E^* \left\{ \exp\left[-\int_0^T r(s)\,ds \right] x_T \right\}.$$

This theory permits us to tie together not only the basic results of option pricing, but also our previous analysis of asset pricing models. For example, applying it to the exact factor model,

$$R_i = E_i + \sum_j \beta_{ij} f_j, \tag{16}$$

yields the APT,

$$1 = E^*(R_i) = E^*\left(E_i + \sum_j \beta_{ij} f_j\right)$$

$$= \frac{1}{(1+r)}\left[E_i + \sum_j \beta_{ij} E^*(f_j)\right],$$

or

$$E_i = (1+r) + \sum_j \lambda_j \beta_{ij},$$

where

$$\lambda_j \equiv -E^*(f_j).$$

Similarly, in a mean variance framework the martingale analysis can be used to prove that there is always a portfolio whose covariances are proportional to the excess returns on each asset. In other words, the absence of arbitrage implies the existence of a mean variance efficient portfolio (see Ross, 1976a; Chamberlain and Rothschild, 1983).

Empirical testing. Perhaps because the option pricing theory works so well, it has generated a surprisingly small empirical literature. Some early tests, for example, Black and Scholes (1973) and Galai (1977), focused on whether the models could be used to generate successful trading rules and found that any success was easily lost to transactions costs. Most interestingly, MacBeth and Merville (1979) found that the option formulas tended to underprice 'in the money' options and overprice 'out of the money' options, but Geske and Roll (1984) have argued that this effect disappears with a reformulation of the statistics.

Given a theory that works so well, the best empirical work will be to use it as a tool rather than to test it. Chiras and Manaster (1978), for example, show that implicit volatilities, i.e. variances computed by inverting the option formulas to obtain variance as a function of the quoted option price, have strong predictive power for explaining future realized stock variances. Patell and Wolfson (1979) use the implicit variances to examine whether stock prices are more volatile around earnings announcements.

These efforts should increase; options and option pricing theory give us an opportunity to measure directly the degree of anticipated uncertainty in the markets. Financial press terms such as 'investor confidence' take on new meaning when they can actually be measured.

This does not mean, however, that there are no important gaps in the theory. Perhaps of most importance, beyond numerical results (see, for example, Parkinson, 1977; or Brennan and Schwartz, 1977), very little is known about most American options which expire in finite time. The American call option on a stock paying a dividend or the American put option are both easily solved in the infinite maturity case since the optimal exercise boundary is a fixed stock value independent of time (Merton, 1973b; Cox and Ross, 1976a, 1976b). If dividends occur at discrete points, then if the call is exercised prematurely it will only be optimal to do so just prior to a dividend payment. This permits a recursive approach to the solution of this finite maturity option (see Roll, 1977a; Geske, 1979). But, with continuous payouts, surprisingly little is known about the exercise properties of either of these options in the American case.

Despite such gaps, when judged by its ability to explain the empirical data, option pricing theory is the most successful theory not only in finance, but in all of economics. It is now widely employed by the financial industry and its impact on economics has been far ranging. At a theoretical level, we now understand that option pricing theory is a manifestation of the force of arbitrage and that this is the same force that underlies much of neoclassical finance.

THE WHOLE IS THE SUM OF THE PARTS – CORPORATE FINANCE

The use of arbitrage as a serious tool of analysis coincided with the beginning of the modern theory of corporate finance. In two seminal papers on the cost of capital, Modigliani and Miller (1958, 1963) argued that the overall cost of capital and, therefore, the value of the firm would be unaffected by its financing decision. Specifically, using arbitrage arguments, Modigliani and Miller showed that the debt/equity split would not alter a firm's value and they then argued that with the investment decision held constant, the dividend payout rate of the firm would also not affect that value. These two irrelevance propositions defined the study of corporate finance in much the same way that Arrow's Impossibility Theorem defined social choice theory. At one and the same time they propounded an irreverent theory whose central feature was the irrelevance of the topic under study. This challenge, to weaken in a useful way the assumptions of their analysis, has guided research in this area ever since.

The Modigliani–Miller analysis. Since the Modigliani-Miller (henceforth MM) irrelevance propositions are developed from the absence of arbitrage, they are quite robust to alternative specifications of the economic model. To derive the Modigliani-Miller propositions we will employ the no arbitrage theory above. Consider a firm which will liquidate all of its assets at the end of the current period, and let x denote the random liquidation value of the assets. Assume that the firm has debt outstanding with a face value of F and that the remainder of the value of the firm is owned by the stockholders who have the residual claim after the bondholders.

At the end of the period, if x is large enough the stockholders will receive $x - F$ and if x falls short of F they will receive nothing. Formally, then, the terminal payment to the stockholders is

$$\max(x - F, 0),$$

which will be recognized as the terminal payment on a call option. In other words – in a tribute to the ubiquitous nature of option pricing theory – the stockholders have a call option on the terminal value of the firm, x, with an exercise price equal to the face value of the debt, F. The bondholders can claim the entire assets if x is not sufficient to cover the promised payment of F, which means that they will receive,

$$\min(x, F).$$

The current value of the firm, V, is defined to be the value of all of the outstanding claims against its assets which in this case is the value of the stocks, S, and the bonds, B. Using the no arbitrage analysis, we find that (ignoring discounting),

$$V \equiv S + B$$

$$= E^*[\max(x - F, 0)] + E^*[\min(x, F)]$$

$$= E^*[\max(x - F, 0) + \min(x, F)]$$

$$= E^*(x),$$

which is independent of the face value, F, of the debt and, therefore, independent of the relative amounts of debt and equity. This verifies the first of the MM irrelevance propositions.

To verify the irrelevance of value to the dividend payout, consider a firm about to pay a dividend, D. The current, pre-dividend, value of the stock is $p^-(D)$ and by the no arbitrage martingale analysis this is given by,

$$p^-(D) = E^*[D + p^+(D)] = D + E^*[p^+(D)],$$

where $p^+(D)$ is the ex-dividend price. If the investment policy of the firm has been fixed, then the only impact that the current dividend payout can have on the stockholders is through its alteration of the cash in the firm. This means that changing the dividend to, say $D + \Delta D$, would necessitate a change in current assets of $-\Delta D$. From the first MM proposition the mode of financing this change in the dividend will be irrelevant to the determination of the firm's value and to simplify the analysis we will assume that it is financed by riskless debt. At an interest rate of r this would entail, say, a perpetual outflow from the firm of $r\Delta D$. Again applying the analysis and letting x_{t+s} be the cash flow at time $t + s$ given that a dividend of D is paid now, we have,

$$
\begin{aligned}
p^+(D + \Delta D) &= E^*\left[\int_0^\infty e^{-rs}(x_{t+s} - r\Delta D)\, ds\right] \\
&= E^*\left(\int_0^\infty e^{-rs}x_{t+s}\, ds\right) - E^*\left(\int_0^\infty e^{-rs}r\Delta D\, ds\right) \\
&= p^+(D) - \Delta D.
\end{aligned}
$$

Thus, we have the irrelevance proposition,

$$
\begin{aligned}
p^-(D + \Delta D) &= E^*[(D + \Delta D) + p^+(D + \Delta D)] \\
&= D + \Delta D + E^*[p^+(D + \Delta D)] \\
&= D + \Delta D + E^*[p^+(D)] - \Delta D \\
&= D + E^*[p^+(D)] \\
&= p^-(D).
\end{aligned}
$$

The MM results were startling to those who had worked in corporate finance and had taken it for granted that the way in which a firm was financed affected its value. To understand the importance of the MM results for the most practical of problems; recall that the original impetus for the study of corporate finance was the determination of the firm's opportunity cost for investments, ρ. For a marginal investment, financed by the issuance of debt and equity, the cost of capital, ρ, also known as the weighted average cost of capital, WACC, would be the weighted average cost of the debt, r, and the cost of equity, k,

$$\rho = (S/V)\, k + (B/V)\, r, \tag{33}$$

(where we have ignored tax effects).

If debt is riskless, then r is the interest rate on such debt and k, the cost of equity, will be the return required by investors for the risk inherent in the stock. Presumably k could be found by appeal to one of the asset pricing models discussed above.

Now it is tempting to think, for example, that if $k > r$, then an increase in debt relative to equity will lower ρ. If this goes too far, debt will become risky and as r rises there will be a unique optimal debt/equity ratio, $(B/S)^0$, that minimizes the cost of capital, ρ. This would be the discount rate to use for present value calculations and it would maximize the value of the firm. This was the traditional analysis of the leverage decision before MM.

By the MM theorem, though, value, V, is unaffected by leverage. This means that ρ is unaltered, since the total (expected) return to the stockholders and the bondholders, $Sk + Br$, is unaltered [see Equation (33)]. In terms of the WAAC, then, as the leverage (B/S) is increased by the substitution of debt for equity, the cost of equity changes.

$$k = \rho + (B/S)(\rho - r),$$

but not the WAAC.

Spanning arguments. The efforts to elude these results and to develop a meaningful theory of corporate finance have taken many forms. First, it has been argued that the analysis itself contains a hidden and critical assumption, namely that the pricing operator is independent of the corporate financial structure. The alternative is that the change in the debt/equity decision, for example, will also change the span of the marketed assets in the economy and, consequently, the operator used for pricing will change. The simplest such example would be a single firm in a two-state world. If the firm is an all equity firm and if there are no other traded assets, then individuals cannot adjust their consumption across the states of nature and must split it according to the equity payoff. If this firm now issues debt the two securities will span the two states of nature and complete the market. This, in turn, will generally alter pricing in the economy.

While this argument has generated a large literature, the problem of the determination of the corporate financial structure and the value of the firm is primarily a microeconomic question and it is difficult to believe that it will be resolved or even illuminated by assuming that firms have some monopoly power that enables them to alter pricing in the capital markets. At the microlevel the MM propositions are unlikely to be seriously affected by such general equilibrium arguments.

At the microlevel, too, the intuition behind the MM propositions and its conclusions is so robust as to be daunting. Consider the following argument. According to MM there can be no optimal, i.e. value maximizing, financial structure since value is independent of structure. Suppose that there was an optimal, say, debt/equity ratio, $(B/S)^0$. Any departure from this target $(B/S)^0$, however, could not lower the firm's value since it would immediately afford an arbitrage opportunity to buy the total firm at its lowered value and refinance it in the optimal target propositions, $(B/S)^0$. (This somewhat facetious argument gets the point across, but it really means that we have not fully specified the rules of the game, e.g., who moves first, what happens when no one moves, etc.)

Signalling models. A more promising route which formally exploits incomplete spanning, but does not argue that the pricing operator itself is altered by any one firm changing its financial structure, makes use of the theory of asymmetric information and signalling (see Ross, 1977a; Leland and Pyle, 1977; and Bhattacharya, 1979). If the managers of the firm possess information that is not held by the market then the market will make inferences from the actions of the firm and, in particular, from financial decisions. Changes in its financial structure or its dividend policy will alter investors' perceptions of its risk class and, therefore, its value. While the operator, $E^*(\cdot)$ does not change, the perception of the distribution of the firm's cash flows does. In an effort to maximize their value, firms will take actions, such as taking on high debt to equity ratios, which can be imitated by lesser firms only at a prohibitive cost. This will distinguish them from lesser firms that the uninformed market erroneously puts into the same class with them. In this fashion, a hierarchy of firm risk classes will emerge, and, in equilibrium, firms will signal their true situations and investors will draw correct inferences from their signals.

All of this has a nice ring to it, but the nagging question that remains is why firms use their financial decisions to accomplish all of this information transfer. Financial changes are cheap, but even cheaper might be guarantees or, for that matter, a system of legislation. These issues remain unresolved, but it is difficult to think that much will be explained by theories that argue that firms take on more debt just to show the world that 'they can do it'. There is a limit to macho-finance.

Taxes. Another line of attack has been to introduce more 'imperfections', especially taxes, into the models. Modigliani and Miller originally had noted that the presence of a corporate tax meant that firms would have an incentive to issue additional debt. Since interest payments on debt are excluded from corporate taxes, substituting debt for equity permits firms to pass returns to investors with a lowered tax cut to the government. At the limit, firms would be all debt if the tax authorities still recognized such debt payments as excludable from taxable corporate income. Presumably, the only brake to this expansion would be the real costs of dealing with the inevitable bankruptcies of high debt firms. This is logically possible, but at the expense of reducing corporate finance to the study of the tradeoff between the tax advantages of debt and the costs of bankruptcy.

Miller (1977) found a more profound brake to this tendency to increase debt. He argued that while the firm could lower its taxes by increasing its debt, the ability of investors to defer or offset capital gains implies that they pay higher taxes on interest income than on the returns from equities. With a rising tax schedule, an equilibrium is possible in which the marginal investor has a tax differential between ordinary income and equity returns that exactly offsets the firm's corporate tax advantage to debt. In such an equilibrium, investors in a higher tax bracket than the marginal investor would purchase only equity (or non-taxed bonds such as municipals for US investors) and those in lower tax brackets would purchase corporate bonds. There would be an equilibrium amount of debt for the corporate sector as a whole, but not for any individual firm (assuming the absence of inframarginal firm tax schedules).

Miller's analysis led to a large literature on the impact of taxes on pricing. Black and Scholes (1974) had made a related argument for the absence of a tax effect on dividends, arguing that stocks with relatively higher yields should not have higher gross returns to compensate investors for the additional tax burden since companies would then just cut their dividends to increase the stock price. Black and Scholes verified their results empirically, but, using a different methodology, Litzenberger and Ramaswamy (1982) found that gross returns were higher for stocks with higher dividends. Whether the supply side or the demand side dominates remains undecided.

Whatever the resolution of this and similar debates, the equilibrium tax argument initiated by Miller has changed much of the analysis of these issues. Miller and Scholes (1978), for example, argue that by employing a number of 'laundering' devices individuals can dramatically cut their taxes. Their conclusion that, in theory, taxes should be much lower than they appear to be in practice, focuses attention on the role played by informational asymmetries and the related costliness of using techniques such as investing through tax exempt intermediaries.

Agency models. The emphasis on informational asymmetries has been the cornerstone of an alternative approach to corporate finance, agency theory. Wilson (1968) and Ross (1973) developed agency models in which one party, the agent (e.g. a corporate manager) acts on behalf of another the principal (e.g. stockholders). Jensen and Meckling (1976), building on the agency theory and on Williamson's (1975) transaction cost approach, argue that corporate finance can be understood in terms of the monitoring and bonding costs imposed on stockholders and managers by such relations. The manager qua employee has an incentive to divert firm resources to his own benefit. Jensen and Meckling refer to the loss in value in restraining this incentive as the (equilibrium) agency cost of the relation.

To some extent this conflict can be resolved ex ante by the indenture agreements and covenants in financial contracts, but the cost of doing so rises with the monitoring requirements. Myers (1977), for example, has studied the implications for investment policy of the conflict between the stockholders and the bondholders. Stockholders own a call option on the assets of the firm and the value of a call increases with the variance of the asset value. Conversely, such increases will come at the expense of the bondholders. Ex ante indenture agreements can limit the ability of management and stockholders to take on additional risk, but the more precise the limits the costlier it is to write, observe and enforce them.

These trade–offs are the intuition and subject matter of the agency approach to corporate finance, but to date it is more a collection of intuitions than a well-articulated theory. The agency approach has pointed in some intriguing directions, but it fares poorly if judged by asking what it is that would be a counter observation or count as evidence against it. To the contrary, no phenomenon seems beyond the reach of 'agency costs' and at times the phrase takes on more of the trappings of an incantation than an analytical tool. The role of asymmetric information in corporate finance and in explaining the managerial and financial forces at work in the firm is self evident, but it remains fertile ground for theory.

Empirical evidence. The early empirical work examined the relation between the corporate financial structure and other characteristics of the firm. Hamada (1972), for example, studied whether the beta of a firm's equity was related to the beta of the firm's assets as would be predicted by the cost of capital, equation (33). There continues to be empirical work on these issues, but the attention of empiricists has shifted to the arena of corporate control.

A boom in merger and acquisition activity in the late 1970s and through to the present time has brought some striking and unexplained empirical regularities. On average, shareholders in firms that are the targets of tender offers gain significantly from such offers while the rewards to bidders are still ambiguous (Jensen and Ruback, 1983). For unsuccessful tenders the target firms appear to average an eventual loss and the bidders may, too. These results and the discrepancy between targets and bidders have been the object of close scrutiny.

If firms realize such abnormal gains as targets, and if it reflects the release of information about the value of their underlying assets, then that raises the question of why they were not priced correctly to begin with. On the other hand, if the returns for successful targets reflect synergies rather than simply a revaluation of their assets, why does the bidder get so little? Several game theoretic and bidding models have been built in an attempt to explain these results (see, e.g., Grossman and Hart, 1980), but a consensus has yet to emerge. Furthermore, some of the important empirical issues, such as whether bidders actually gain or lose on average remain unresolved.

Conclusion. For corporate finance, like the other major areas of finance, the neoclassical theory is now well established, but, like the other areas, the inadequacy of the neoclassical analysis is pushing researchers to begin the challenging but promising exploration of theories of asymmetric information. This work holds out the hope of explaining some of the deeper mysteries of finance that have eluded the neoclassical theory, from the embarrassing plethora of anomalies in capital markets to the basic questions of financial structure.

Perhaps the feature that truly distinguishes finance from much of the rest of economics is this constant interplay between theory and empirical analysis. The test of these new approaches will be decided less by reference to their aesthetics and more by their usefulness in explaining financial data. At the height of the subject, these two criteria become one.

STEPHEN A. ROSS

See also ARBITRAGE; CAPITAL ASSET-PRICING MODEL; DIVIDEND POLICY; EFFICIENT MARKET HYPOTHESIS; OPTIONS.

BIBLIOGRAPHY

Admati, A. 1985. A noisy rational expectations equilibrium for multi-asset securities markets. *Econometrica* 53(3), May, 629–57.

Bhattacharya, S. 1979. Imperfect information, dividend policy and the 'Bird in the Hand' fallacy. *Bell Journal of Economics and Management Science* 10(1), Spring, 259–70.

Black, F. 1972. Capital market equilibrium with restricted borrowing. *Journal of Business* 45(3), July, 444–55.

Black, F. and Scholes, M. 1972. The valuation of option contracts and a test of market efficiency. *Journal of Finance* 27(2), May, 399–417.

Black, F. and Scholes, M. 1973. The pricing of options and corporate liabilities. *Journal of Political Economy* 81(3), May–June, 637–54.

Black, F., Jensen, M. and Scholes, M. 1972. The capital asset pricing model: some empirical tests. In *Studies in the Theory of Capital Markets*, ed. M.C. Jensen, New York: Praeger.

Black, F. and Scholes, M. 1974. The effects of dividend yield and dividend policy on common stock prices and returns. *Journal of Financial Economics* 1(1), May, 1–22.

Breeden, D.T. 1979. An intertemporal asset pricing model with stochastic consumption and investment opportunities. *Journal of Financial Economics* 7(3), September, 265–96.

Brennan, M.J. and Schwartz, E.S. 1977. The valuation of American put options. *Journal of Finance* 32(2), May, 449–62.

Brown, S. and Weinstein, M. 1983. A new approach to testing asset pricing models: the bilinear paradigm. *Journal of Finance* 38(3), June, 711–43.

Cass, D. and Stiglitz, J. 1970. The structure of investor preferences and asset returns, and separatability in portfolio selection: a contribution to the pure theory of mutual funds. *Journal of Economic Theory* 2(2), June, 122–60.

Chamberlain, G. 1983. Funds, factors and diversification in arbitrage pricing models. *Econometrica* 51(5), September, 1305–23.

Chamberlain, G. and Rothschild, M. 1983. Arbitrage, factor structure and mean-variance analysis on large asset markets. *Econometrica* 51(5), September, 1281–304.

Chen, N., Roll, R. and Ross, S.A. 1986. Economic forces and the Stock market. *Journal of Business* 59, July, 383–403.

Chiras, D.P. and Manaster, S. 1978. The information content of option prices and a test of market efficiency. *Journal of Financial Economics* 6(2–3), June–September, 213–34.

Connor, G. 1984. A unified beta pricing theory. *Journal of Economic Theory* 34(1), October, 13–31.

Cootner, P. (ed.) 1964. *The Random Character of Stock Market Prices*. Cambridge, Mass.: MIT Press.

Cowles, A. 1933. Can stock market forecasters forecast? *Econometrica* 1, July, 309–24.

Cox, J.C., Ingersoll, J. and Ross, S.A. 1985a. An intertemporal general equilibrium model of asset prices. *Econometrica* 53(2), March, 363–84.

Cox, J.C., Ingersoll, J. and Ross, S.A. 1985b. A theory of the term structure of interest rates. *Econometrica* 53(2), March, 385–407.

Cox, J.C. and Ross, S.A. 1976a. The valuation of options for alternative stochastic processes. *Journal of Financial Economics* 3(1–2), January–March, 145–66.

Cox, J.C. and Ross, S.A. 1976b. A survey of some new results in financial option pricing theory. *Journal of Finance* 31(2), May, 383–402.

Cox, J.C., Ross, S.A. and Rubinstein, M. 1979. Option pricing: a simplified approach. *Journal of Financial Economics* 7(3), September, 229–63.

Diamond, D. and Verrecchia, R. 1981. Information aggregation in a noisy rational expectations economy. *Journal of Financial Economics* 9(3), September, 221–35.

Dybvig, P. 1983. An explicit bound on deviations from APT pricing in a finite economy. *Journal of Financial Economics* 12(4), December, 483–96.

Dybvig, P. and Ross, S.A. 1985. Yes, the APT is testable. *Journal of Finance* 40(4), September, 1173–88.

Fama, E.F. 1970. Efficient capital markets: a review of theory and empirical work. *Journal of Finance* 25(2), May, 383–417.

Fama, E.F., Fisher, L., Jensen, M. and Roll, R. 1969. The adjustment of stock prices to new information. *International Economic Review* 10(1), February, 1–21.

Fama, E.F. and MacBeth, J. 1973. Risk, return and equilibrium: empirical tests. *Journal of Political Economy* 81(3), May–June, 607–36.

Flavin, M. 1983. Excess volatility in the financial markets: a reassessment of the empirical evidence. *Journal of Political Economy* 91(6), December, 929–56.

French, K. 1980. Stock returns and the weekend effect. *Journal of Financial Economics* 8(1), March, 55–69.

French, K. and Roll, R. 1985. Stock return variances, the arrival of information and the reaction of traders. UCLA Working Paper.

Galai, D. 1977. Tests of market efficiency of the Chicago Board Options Exchange. *Journal of Business* 50(2), April, 167–97.

Geske, R. 1979. A note on an analytical valuation formula for unprotected American call options on stocks with known dividends. *Journal of Financial Economics* 7(4), December, 375–80.

Geske, R. and Roll, R. 1984. Isolating the observed biases in call option pricing: an alternative variance estimator. UCLA Working Paper, April.

Gibbons, M. 1982. Multivariate tests of financial models: a new approach. *Journal of Financial Economics* 10(1), March, 3–27.

Gibbons, M., Ross, S.A. and Shanken, J. 1986. A test of the efficiency of a given portfolio. Stanford University Working Paper No. 853.

Granger, C.W.J. and Morgenstern, O. 1962. Spectral analysis of New York stock market prices. Econometric Research Program, Princeton University, Research Memorandum, September.

Grossman, S.J. 1976. On the efficiency of competitive stock markets where traders have diverse information. *Journal of Finance* 31(2), May, 573–85.

Grossman, S.J. and Hart, O. 1980. Takeover bids, the free-rider problem, and the theory of the corporation. *Bell Journal of Economics and Management Science* 11(1), Spring, 42–64.

Grossman, S.J. and Stiglitz, J. 1980. The impossibility of informationally efficient markets. *American Economic Review* 70, June, 393–408.

Hamada, R.S. 1972. The effect of the firm's capital structure on the systematic risk of common stocks. *Journal of Finance* 27(2), May, 435–52.

Hansen, L. and Singleton, R. 1983. Stochastic consumption, risk aversion and the temporal behavior of asset returns. *Journal of Political Economy* 91(2), April, 249–65.

Harrison, J.M. and Kreps, D.M. 1979. Martingales and arbitrage in multiperiod securities markets. *Journal of Economic Theory* 20(3), June, 381–408.

Hicks, J.R. 1946. *Value and Capital.* 2nd edn, London: Oxford University Press.

Huberman, G. 1982. A simple approach to arbitrage pricing theory. *Journal of Economic Theory* 28(1), October, 183–91.

Ingersoll, J. 1977. A contingent-claims valuation of convertible securities. *Journal of Financial Economics* 4(3), May, 289–322.

Ingersoll, J. 1984. Some results in the theory of arbitrage pricing. *Journal of Finance* 39(4), September, 1021–39.

Jaffé, J. 1974. The effect of regulation changes on insider trading. *Bell Journal of Economics and Management Science* 5(1), Spring, 93–121.

Jensen, M.C. and Meckling, W.H. 1976. Theory of the firm: managerial behavior, agency costs and ownership structure. *Journal of Financial Economics* 3(4), October, 305–60.

Jensen, M.C. and Ruback, R.S. 1983. The market for corporate control: the scientific evidence. *Journal of Financial Economics* 11(1–4), April, 5–50.

Jobson, J.D. and Korkie, B. 1982. Potential performance and tests of portfolio efficiency. *Journal of Financial Economics* 10(4), December, 433–66.

Kandel, S. 1984. On the exclusion of assets from tests of the mean variance efficiency of the market portfolio. *Journal of Finance* 39(1), March, 63–73.

Keim, D. 1983. Size-related anomalies and stock return seasonality: further empirical evidence. *Journal of Financial Economics* 12(1), June, 13–32.

Kleidon, A. 1986. Variance bounds tests and stock price valuation models. *Journal of Political Economy* 94(5), October, 953–1001.

Leland, H. and Pyle, D. 1977. Informational asymmetries, financial structure and financial intermediation. *Journal of Finance* 32(2), May, 371–87.

Lintner, J. 1965. The valuation of risk assets and the selection of risky investments in stock portfolios and capital budgets. *Review of Economics and Statistics* 47, February, 13–37.

Litzenberger, R. and Ramaswamy, K. 1982. The effect of dividends on common stock prices: tax effects or information effects? *Journal of Finance* 37(2), May, 429–43.

Lucas, R.E., Jr. 1978. Asset prices in an exchange economy. *Econometrica* 46(6), November, 1429–45.

MacBeth, J. and Merville, L. 1979. An empirical examination of the Black–Scholes call option pricing model. *Journal of Finance* 34(5), December, 1173–86.

Malkiel, B. 1966. *The Term Structure of Interest Rates: Expectations and Behavior Patterns.* Princeton: Princeton University Press.

Markowitz, H.M. 1959. *Portfolio Selection: Efficient Diversification of Investments.* New York: Wiley.

Marsh, T. and Merton, R.C. 1986. Dividend variability and variance bounds tests for the rationality of stock market prices. *American Economic Review* 76(3), June, 483–98.

Merton, R.C. 1973a. An intertemporal capital asset pricing model. *Econometrica* 41(5), September, 867–87.

Merton, R.C. 1973b. Theory of rational option pricing. *Bell Journal of Economics and Management Science* 4(1), Spring, 141–83.

Milgrom, P, and Stokey, N. 1982. Information, trade and common knowledge. *Journal of Economic Theory* 26(1), February, 17–27.

Miller, M.H. 1977. Debt and taxes. *Journal of Finance* 32(2), May, 261–75.

Miller, M.H. and Scholes, M. 1978. Dividends and taxes. *Journal of Financial Economics* 6(4), December, 333–64.

Modigliani, F. and Miller, M.H. 1958. The cost of capital, corporation finance, and the theory of investment. *American Economic Review* 48, June, 261–97.

Modigliani, F. and Miller, M.H. 1963. Corporate income taxes and the cost of capital. *American Economic Review* 53, June, 433–43.

Mossin, J. 1966. Equilibrium in a capital asset market. *Econometrica* 34(4), October, 768–83.

Myers, S. 1977. Determinants of corporate borrowing. *Journal of Financial Economics* 5(2), November, 147–75.

Parkinson, M. 1977. Option pricing: the American put. *Journal of Business* 50(1), January, 21–36.

Patell, J.M. and Wolfson, M.A. 1979. Anticipated information releases reflected in call option prices. *Journal of Accounting and Economics* 1(2), August, 117–40.

Roll, R. 1970. *The Behavior of Interest Rates: An Application of the Efficient Market Model to U.S. Treasury Bills.* New York: Basic Books.

Roll, R. 1977a. An analytic valuation formula for unprotected American call options on stocks with known dividends. *Journal of Financial Economics* 5(2), November, 251–8.

Roll, R. 1977b. A critique of the asset pricing theory's tests. *Journal of Financial Economics* 4(2), March, 129–76.

Roll, R. 1978. Ambiguity when performance is measured by the securities market line. *Journal of Finance* 33(4), September, 1051–69.

Roll, R. 1981. A possible explanation of the small firm effect. *Journal of Finance* 36(4), September, 879–88.

Roll, R. 1983. The turn-of-the-year effect and small firm premium. *Journal of Portfolio Management* 9(2), Winter, 18–28.

Roll, R. 1984. Orange juice and weather. *American Economic Review* 74(5), December, 861–80.

Roll, R. and Ross, S.A. 1980. An empirical investigation of the arbitrage pricing theory. *Journal of Finance* 35(5), December, 1073–103.

Ross, S.A. 1973. The economic theory of agency: the principal's problem. *American Economic Review* 63(2), May, 134–9.

Ross, S.A. 1975. Uncertainty and the heterogeneous capital good model. *Review of Economic Studies* 42(1), January, 133–46.

Ross, S.A. 1976a. Return, risk and arbitrage. In *Risk and Return in Finance,* ed. I. Friend and J. Bicksler, Cambridge, Mass.: Ballinger.

Ross, S.A. 1976b. The arbitrage theory of capital asset pricing. *Journal of Economic Theory* 13(3) December, 341–60.

Ross, S.A. 1976c. Options and efficiency. *Quarterly Journal of Economics* 90(1), February, 75–89.

Ross, S.A. 1977a. The determination of financial structure: the incentive signalling approach. *Bell Journal of Economics and Management Science* 8(1), Spring, 23–40.

Ross, S.A. 1977b. The capital asset pricing model (CAPM), short-sale restrictions and related issues. *Journal of Finance* 32(1), March, 177–83.

Ross, S.A. 1978a. Mutual fund separation in financial theory – the separating distributions. *Journal of Economic Theory* 17(2), April, 254–86.

Ross, S.A. 1978b. A simple approach to the valuation of risky streams. *Journal of Business* 51(3), July, 453–75.

Rubinstein, M. 1976. The valuation of uncertain income streams and the pricing of options. *Bell Journal of Economics and Management Science* 7(2), Autumn, 407–25.

Shanken, J. 1982. The arbitrage pricing theory: is it testable? *Journal of Finance* 37(5), December, 1129–40.

Sharpe, W. 1964. Capital asset prices: a theory of market equilibrium under conditions of risk. *Journal of Finance* 19, September, 425–42.

Shiller, R. 1981. Do stock prices move too much to be justified by subsequent changes in dividends? *American Economic Review* 71(3), June, 421–36.

Tobin, J. 1958. Liquidity preference as behavior towards risk. *Review of Economic Studies* 25, February, 65–86.

Williamson, O.E. 1975. *Markets and Hierarchies: Analysis and Antitrust Implications.* New York: Free Press.

Wilson, R.B. 1968. The theory of syndicates. *Econometrica* 36(1), January, 119–32.

finance and saving.   Saving and finance are now clearly distinguished (though perhaps surprisingly this is a fairly recent development). Finance refers to monetary transactions securing the means of payment for purchases in excess of current cash flow or funding the holding of assets. Problems of finance exist for individuals or firms, not for economies as a whole except in relation to other countries; accordingly, the analysis of finance is microeconomic in character.

Saving is income not consumed. In contrast to finance it is both an action undertaken by individuals and an outcome for the economy as a whole. In the context of today's financial institutions, individual saving largely consists of money-flows to those institutions, though saving can take 'real' form as well (e.g. the purchase of houses or works of art). For an economy as a whole, apart from its interactions with the rest of the world, saving can only be net capital accumulation. Financial transactions 'consolidate out' in the aggregate balance sheet.

The problem for theory is to bridge the gap between

microeconomic money-flows, arising out of individual acts of saving and constituting potential provision of finance, and the amount of actual saving at the macroeconomic level. (Even in money terms the sum of individual attempts to save will not necessarily 'add up' to aggregate saving, for revaluations which occur as savers compete for assets are hidden in the aggregate.)

There was a time when this gap between the intentions of individuals and the macroeconomic outcome was not perceived. In Ricardo's model (1817) of an economy producing one staple good (corn), the corn not consumed was seed-corn, saving and investment in one, with no need for finance. The amount of corn so saved would depend on the expected rate of return from forgoing consumption.

Ricardo's formulation, not inappropriate to a largely non-monetized ('real') agricultural economy, survives to this day in the form of time-preference theory, in which consumption and saving are seen as two sides of an intertemporal consumption plan. The trade-off between the disutility of deferred consumption and the expected rate of return on investment determines the volume of saving and investment. Monetary factors, borrowing and lending, can be added to this theory if the rate of interest is distinguished from the rate of profit. The rate of interest in this analysis is taken as exogenous since the theory pertains to individual choice.

Amongst the classical economists with the exception of Marx, and indeed in much current theory, the distinction between interest and profit is imperfectly made. Marx (1867) was much concerned with the conditions under which finance capital could be obtained and sufficient profit on industrial capital realized to pay back finance capital, a problem also central to the work of Keynes.

The immediate background into which Keynes's work was inserted consisted of Wicksell's theory and the loanable funds theory associated with Wicksell's successors in Stockholm and with Keynes's colleague, Sir Dennis Robertson. These theories added monetary factors to the classical theory, in which the rate of interest/profit was determined by the equality of 'real' saving and investment.

Wicksell (1901) proposed the concept of the 'natural rate of interest', the rate compatible with saving–investment equality, and contrasted this with the money rate of interest. A divergence of the natural rate from the money rate of interest would result in a cumulative process of inflation or deflation caused by expansion or contraction of bank credit. The process would converge as the two rates once again became equal. Hence equality of the natural and money rates yielded price stability.

Unfortunately the concept of the natural rate is not observable nor is it determinate independently of the level of employment. It is now regarded as unhelpful.

In loanable funds theory the money rate of interest is determined by equating the demand for loanable funds, defined to include 'hoarding' (additions to idle money-holdings) as well as investment, with the supply of funds, comprising saving and additions to the money supply. In this theory as in Wicksell's, saving is implicitly equivalent to providing finance. Hoarding, clearly not a source of finance, is also not a form of saving. Saving, however, is not the only source of finance, there is also bank credit. Increases in the money supply occur when banks expand credit by more than 'prior saving' in the form of deposits. These increases were, as in Wicksell, generally held to be inflationary.

From a microeconomic perspective it would seem plausible that saving, money-income not consumed and thus available for lending to deficit spenders, automatically constitutes

finance. To Keynes (1936), however, hoarding was a form of saving, and hoarding does not provide finance. Also, it is implicit in his theory of speculative demand, in which the determination of the rate of interest is dominated by trade in existing securities, that saving does not provide finance if the pace of new issues is not adequate to absorb the savings flow. Thus for the first time the theory of savings was divorced from the theory of finance.

Until Keynes, investment was assumed to be dependent on saving as the source of finance. Keynes reversed this causal ordering, arguing that investment, financed independently of saving, created additional income adequate eventually to generate an equal volume of saving. Robertson (1940) demonstrated that the source of finance must be bank lending. This results in an increase of deposits the holding of which, on Keynes's definition but not in Robertson's, constituted saving. Much of the long debate between these two clever economists (see Keynes, 1973, pp. 201–34) rested on a misunderstanding.

Today the theory of saving has developed little beyond the debates on the relative determining roles of rates of interest and levels of income which dominated the subject in the 1930s. By contrast, the theory of finance, dealing with the appropriate portfolio choices of active managers of financial portfolios and the options open to firms in the finance of their capital, has been much developed and refined and is full of vitality.

VICTORIA CHICK

See also KEYNES, JOHN MAYNARD; LOANABLE FUNDS; SAVING EQUALS INVESTMENT.

BIBLIOGRAPHY
Keynes, J.M. 1936. *The General Theory of Employment Interest and Money.* London: Macmillan.
Keynes, J.M. 1973. *The Collected Writings of John Maynard Keynes,* Vol. XIV. London: Macmillan.
Marx, K. 1867. *Capital.* Hamburg: Otto Meisner.
Ricardo, D. 1817. *Principles of Political Economy and Taxation.* Harmondsworth: Penguin Books, 1971.
Robertson, D.H. 1940. Effective demand and the multiplier. In D.H. Robertson, *Essays in Monetary Theory,* London: P.S. King.
Wicksell, K. 1901. *Lectures in Political Economy.* London: Routledge & Kegan Paul, 1934.

finance capital. The concept of finance capital encapsulates the most theoretically significant attempt by the orthodox Marxism of the pre-1914 period to come to terms with the developments of capitalism in the late 19th century. After the Bolshevik Revolution the concept was much less frequently employed. In part this demise reflected the breakdown of orthodox Marxism as a relatively unified but developing body of doctrine, but it also reflected the inherent problems of the concept.

The term itself is not to be found in Marx's work. But subsequent formulations relied heavily on the schematic outline by Marx in Part V of Volume III of *Capital,* especially chapter 27 on 'The Role of Credit in Capitalist Production'. Marx's arguments, penned in the 1860s, but not published until 1894, focus on the two processes of the multiplication of forms of credit available to industrial capital, and the formation of joint stock companies. The two processes together he saw as heralding 'the abolition of capital as private property within the framework of capitalist production itself' (1894, p. 436).

On the basis of Marx's brief outline, Hilferding in his *Finanz Kapital* (1908), built a systematic argument, conceiving finance capital as the highest stage of capitalism. Hilferding's book presents a theoretical history of the evolution of relations between money and productive (industrial) capital. This relationship is seen as having gone through a series of historical transformations, particularly on the basis of changes in the form of credit and credit-giving institutions. Trade credit (or 'circulation credit') is seen as the initial form of credit, emerging from interruptions to the cycle of capital, and tying credit creation directly to the production and sale of commodities. This form of credit facilitated an extension of the scale of production by using funds otherwise idle.

Subsequently there developed banks which not only recycled capitalists' own idle funds but put money from other sources at the disposal of industrial capitalists. When this process of credit expansion encompassed the financing of fixed capital the relationship of the banks to industrial capital began to change, as banks came to have an enduring rather than a momentary interest in the fortunes of the industrial enterprise they lent to. So emerged the characteristically 'German' interlinking of banks and industry, with banks controlling large blocks of industrial equity and sharing large numbers of directors with industry.

The changing relationship encouraged the growth of larger banks, which could afford to tie up funds in this way, but also were enabled by expansion in size to finance lots of firms in order to spread their risks. This growing concentration of banks was seen as interacting with the growth of concentration amongst industrial firms, and is thereby closely linked with the development of the joint stock company. The growth of shares, which Hilferding stresses should be seen as another form of (irredeemable) credit, is a pre-condition of the growth of the joint stock company, which in turn is a pre-condition of a full utilization of the possibilities of technological advance (pp. 122–3).

These joint stock companies become more and more concentrated and tend to the elimination of free competition. This is paralleled by the growth of 'an ever more intimate relationship' between banks and industrial capital: 'Through this relationship ... capital assumes the form of finance capital, its supreme and most abstract expression' (p. 21).

But for Hilferding finance capital is not just a concept but a real social and political force (as indeed it was in Germany). It has its own economic policies, which are both protective of the home market and promote expansion abroad. This latter impetus leads to an intimate relationship between finance capital and the state, which is used to pursue policies of territorial aggrandizement, built partly on the desire to export commodities, but above all to facilitate the export of capital. Hence the characteristic ideology of finance capital (unlike competitive industrial capital) is aggressively expansionist and aspires to political as well as economic domination. 'Thus the ideology of imperialism arises on the ruins of the old liberal ideals, whose naivety it derides' (p. 334).

Hilferding's analysis of the structure of finance capital can be read as largely a Marxist version of the well known story of the 'divorce of ownership and control' via the development of the joint stock company. Such a parallel would not be entirely misplaced, but it would obscure some of the most important elements of Hilferding's theories.

Least surprisingly, Hilferding's analysis deploys Marx's theory of value, and this, for example, leads him to picture finance capital seizing profits originally produced by industrial capital. Such analysis simply reflects the Marxist concept of industrial capital as productive of surplus value, with other capitals obtaining their profits by redistribution from this original source. But the conceptual background of Marx's theory of value has more specific implications for Hilferding's work.

The argument that values and profits arise originally only in the industrial sector leads to the characterization of share capital as 'fictitious' capital (a term also deployed by Marx), compared with 'genuinely functioning industrial capital' (p. 111). This essentially moralistic approach cuts across the useful discussion by Hilferding of the role of share capital in making possible the joint stock company form of organization, with the progressiveness of this form for the development of production. Similarly, this allegiance to the primacy of industrial capital leads him to assert that 'the techniques of banking itself generate tendencies which affect the concentration of the banks and industry alike, but the concentration of industry is the ultimate cause of concentration in the banking system' (p. 98). Yet his analysis elsewhere makes clear that the development of the banking system, and credit system more generally, were more commonly pre-conditions of the development of forms of industrial capital than vice versa.

A problem of a rather different order is Hilferding's treatment of the relationship between banks and industry as the defining characteristic of finance capital. This leads to the view that countries such as England, where these close relations never existed, are deviants from the norm of development: '... the English system is an outmoded one and is everywhere on the decline because it makes control of the loaned-out bank capital more difficult, and hence obstructs the expansion of bank capital itself' (p. 293). But Hilferding's own arguments on the stock exchange, as the basis of a particular form of credit creation, undercuts this identification of finance capital with one particular financial institution – banks. For what is clearly at stake in Hilferding's general arguments is the development of different types of *credit*, which then impinges on forms of industrial organization, but where these types are not tied to any particular institutional form. (This is quite clear in most of his discussion of the stock exchange.)

Hilferding thus imparts a strong evolutionary element into his argument, where the normal path of development is towards the 'German' model of the relationship between banks and industry. This evolutionism is also more broadly present in Hilferding when he follows Marx in seeing the growth of finance capital and the joint stock company as a socialization of production, that is, a step towards socialist organization of the economy. This socialization is theorized as consisting of a development of a complex division of labour organized by a very few sites of decision-making. Hence the struggle for socialism in this framework is reduced to a struggle to dispossess the oligarchy who currently control production, but who have unwittingly created the 'final organizational prerequisites for socialism' (p. 368). This extraordinary line of argument implies that there is nothing specifically capitalist about the organization of large-scale capitalist industry, except who controls it – surely a *reductio ad absurdum* of the notion of the productive forces developing independently of the relations of production.

The concept of finance capital was most famously deployed by Lenin in his work on Imperialism. Lenin's aim was quite clearly to engage in a political polemic not a theoretical analysis, and he adds nothing new to the discussion of the concept. His main difference with Hilferding was to take further the stress on the aggressive tendencies of finance capital, and to argue the inescapability of imperialist war in such conditions, a conclusion not drawn by Hilferding. Whatever the merits or otherwise of Lenin's political polemic, the association of Hilferding's work with it tended to obscure the theoretical significance of *Finanz Kapital*.

After Lenin, the concept of Finance Capital has played a much lesser role in Marxist discussions. Instead, Bolshevized Marxism has tended to place more emphasis on the monopoly characteristics of modern capitalism, rather than the finance aspect; hence the common deployment of concepts of Monopoly Capital, and State Monopoly Capitalism. But even within the conceptual approaches of this post-1917 orthodox Marxism this emphasis appears misplaced. As Hussain (1976) has convincingly argued, in terms of standard Marxist categories the concept of finance capital provides a basis for the periodization of capitalism which monopoly capital cannot. It is the relationship of finance of industrial capital which largely determines the structure and size of firms, and hence finance determines the level of 'monopoly'. Starting with the total social capital, as Marx does, it is the relation of finance to industrial capital which determines the distribution of capital into firms. Within an orthodox Marxist framework, finance could in this way provide a basis for periodizing capitalism, that is, on the basis of changes in the relationship of finance to industrial capital and their implications.

Hilferding's work shares some of the defects of Marx's *Capital* in which it was so clearly grounded. Its evolutionism and its adherence to Marx's theory of value, in particular, tend to obscure what is most valuable in the analysis. Nevertheless, with the growing prominence of financial institutions and financial calculation in advanced capitalist countries, any work which provides a detailed theoretical study of the workings of finance under capitalism needs to be taken seriously. This is especially so when the study, at its best, provides analyses which avoid both the speculative character of discussion of the 'total social capital', and the empiricism of institutional description. Rather, the concept of finance capital provides an entry into analysing the nexus of relationships between financial and industrial institutions, but where these institutions are seen neither as simply representations of broader social forces, nor as complex entities knowable only through description.

More specifically, the concept of finance capital leads us to treat the industrial structure as an effect of the changes in the relationship between industrial and financial capital. Thus, for example, the well-known growth of industrial concentration in the UK and other countries in the 1950s and 1960s would be analysed primarily as an effect of the operations of the stock market, and of the credit-creating criteria deployed in that market. Equally, prediction of future trends in the industrial structure would depend upon views about the future evolution of the financial system. The development of the industrial structure, seen in this light, would neither be technologically determined, as commonly suggested, nor, as in some Marxist treatments, would it be seen as tied to the idea of the appropriation by a new class of capitalist of power over the means of production. Rather, the focus would be on the conditions of existence of the credit-giving criteria employed by financial institutions, and how these structured the forms of calculation used by firms in their deployment of means of production. In this way, forms of calculation would be seen as central to the analysis of capitalist firms, but where these forms were themselves seen as dependent upon the mechanisms of allocation of credit in the economy.

It would be an impossible project to 'revive' the orthodox Marxism of the pre-1914 period. Its theoretical presuppositions are in crucial respects no longer tenable, and its specific analyses often tied to circumstances which have changed out of all recognition. Nevertheless, this was a period when Marxism was a relatively open programme of research, and the results of that are not to be simply discarded. The concept of finance capital, shorn of some of its theoretical baggage, could be seen as a potentially fruitful legacy from that period.

J. TOMLINSON

See also HILFERDING, RUDOLPH; MONOPOLY CAPITALISM.

BIBLIOGRAPHY

Hilferding, R. 1908. *Finanz Kapital*. Translated into English, with an introduction by T.B. Bottomore, London: Routledge & Kegan Paul, 1981.

Hussain, A. 1976. Hilferding's *Finance Capital*. Bulletin of the Conference of Socialist Economists.

Marx, K. 1894. *Capital*, Vol. III. Ed. F. Engels, London: Lawrence & Wishart, 1968.

financial crisis. A financial crisis is defined as a sharp, brief, ultra-cyclical deterioration of all or most of a group of financial indicators – short-term interest rates, asset (stock, real estate, land) prices, commercial insolvencies and failures of financial institutions (Goldsmith, 1982, p. 42). Whereas a boom or bubble is characterized by a rush out of money into real or longer-term financial assets, based on expectations of a continued rise in the price of the asset, financial crisis is characterized by a rush out of the real or long-term financial asset into money, based on the expectation that the price of the asset will decline. Between the boom and a financial crisis may be a period of 'distress' in which the expectation of continued price increases has been eroded, but has not given way to the opposite expectation. Distress may be short or protracted, and may or may not end in crisis.

Bubbles or booms in a single asset, or in widely scattered assets such as Florida or California real estate, supertankers, gold in the 1980s and the like may subside slowly without crisis. The dangers lie in booms or bubbles that have spread from one asset to another, and/or one country to another, and have led to a tautness in the financial structure. In 1825, for example, the boom affected South American government bonds and mining stocks, plus English company securities, led by insurance shares. In 1847, the railway mania was paralleled by a bubble in wheat. The boom following the founding of the German Reich in 1871 affected German and Austrian railways and building, plus lending on United States railway securities. In 1890, the Baring crisis affected Argentina, Brazil, Chile, Australia, South Africa, the United States, France and Italy, which had had booms financed from London and Paris suddenly halted. In the early 1980s, financial distress came from a reversal of the expectation of continued profits in oil and in syndicated bank loans to developing countries. In addition to these loan outlets, financial distress was caused by a boom and crash in farm acreage and in California real estate, plus extensive disintermediation in thrift institutions and financial institutions that had wrongly anticipated a decline in interest rates.

Whether financial distress ends in a financial crisis depends on a variety of factors, including the fragility of the earlier extensions of credit, the speed of the reversal of expectations, the disturbance to confidence produced by some financial accident (such as a spectacular failure or the revelation of one or more swindles) and the financial community's assurance that in extreme conditions it will be rescued by a lender of last resort.

The function of a domestic lender of last resort was developed in practice in the 18th and early 19th centuries especially by the Bank of England, and rationalized by Walter Bagehot in *Lombard Street* (1873). The task is to halt the rush out of real and long-term financial assets into money by demonstrating to the financial community that there is ample money available. In the crises of 1847, 1857 and 1866 this required suspension of the Bank Act of 1844 limiting the Bank of England's note issue. The lender of last resort ostensibly lent only to solvent institutions on the basis of sound collateral; however, in practice the Bank of England, and especially the Bank of Italy in 1923, 1926 and 1930 made advances on all sorts of assets including many of dubious quality. The assets acquired by the Bank of Italy in its successive interventions to support the weak Italian capital market were consolidated in 1933 into a permanent *Istituto Riconstruzione Italiana* (IRI), patterned somewhat after the 1931 Reconstruction Finance Corporation in the United States.

Other devices historically adopted in financial crises to halt panic liquidation of assets have included: the issue of government securities to merchants against the collateral of their inventories, such securities being sold by the merchant in difficulty; the guaranteeing of the liabilities of distressed commercial or financial institutions as in Hamburg in 1857 and in London in the Baring crisis of 1890; the creation of special intermediaries to add a third signature to bills of exchange to enable them to qualify for ordinary discounting (the French *comptoirs d'escompte* and the *Golddiskontbank* in Germany in 1931); a burst of open-market operations such as undertaken by the Federal Reserve Bank of New York at the end of October and in November 1929 (Kindleberger, 1978, ch. 9).

International financial crises in which foreign asset-holders try to dump assets – usually securities or money – to escape from those denominated in a given currency have been less frequently calmed by an international lender of last resort, though this has been done. In the 19th century, the European crises of 1825, 1836, 1839, 1847, 1860, 1890 and 1907 were met by central-bank swaps of gold against silver, or loans of specie or bills of exchange. After World War II a swap network was devised among leading financial centres in which two or more central banks wrote up domestic deposits in favour of the other central bank or banks against a claim in foreign exchange (Coombs, 1976). In 1873, 1890 and 1929 international last-resort lending was either absent or inadequate, with the consequence that debt deflation proceeded further and ended in prolonged depression. In 1873 and 1890 there appears to have been no realization of the help that a lender of last resort might have provided. In the 1929 depression, and especially in 1931, Britain was financially too weak to come to the aid of Austria and Germany, and France and the United States failed to recognize the responsibility that had fallen to them (Kindleberger, 1973, ch. 14). This view, however, is not universally accepted (Moggridge, 1982).

The swap device was not adopted for the debt crises of developing countries in the early 1980s because it was instinctively understood that the resulting claims on them were not certain of ultimate satisfaction. Instead government debts were rescheduled through the so-called Paris Club, and banking claims refunded under the auspices and with the aid of credits from the International Monetary Fund. Since financial crises can occur in a matter of hours, and IMF negotiations are protracted, it has been necessary on occasion, especially for Mexico in 1982, for a 'bridging loan' from a major government or central bank until a more complete settlement could be agreed. These settlements typically required the country being aided to agree to undertake a stringent course of macroeconomic restraint, leading in some instances to internal political unrest (Williamson, 1983).

Charles P. Kindleberger

See also BANK RATE; BUBBLES; LIQUIDITY.

BIBLIOGRAPHY
Bagehot, W. 1873. *Lombard Street*. Reprinted in *The Collected Works of Walter Bagehot*, ed. N. St John Stevas, London: The Economist, 1978.
Coombs, C.A. 1976. *The Arena of International Finance*. New York: Wiley-Interscience.
Goldsmith, R.W. 1982. Comment on Hyman P. Minsky, *The Financial Instability Hypothesis*. In *Financial Crises, Theory, History and Policy*, ed. C.P. Kindleberger and J.-P. Laffargue, Cambridge: Cambridge University Press.
Kindleberger, C.P. 1973. *The World in Depression, 1929–1939*. London: Allen Lane.
Kindleberger, C.P. 1978. *Manias, Panics and Crashes: A History of Financial Crises*. New York: Basic Books.
Moggridge, D.E. 1982. Policy in the crises of 1920 and 1929. In *Financial Crises*, ed. C.P. Kindleberger and J.-P. Laffargue, Cambridge: Cambridge University Press.
Williamson, J. (ed.) 1983. *IMF Conditionality*. Washington, DC: Institute for International Economics.

financial intermediaries. The tangible wealth of a nation consists of its natural resources, its stocks of goods, and its net claims against the rest of the world. The goods include structures, durable equipment of service to consumers or producers, and inventories of finished goods, raw materials and goods in process. A nation's wealth will help to meet its people's future needs and desires; tangible assets do so in a variety of ways, sometimes by yielding directly consumable goods and services, more often by enhancing the power of human effort and intelligence in producing consumable goods and services. There are many intangible forms of the wealth of a nation, notably the skill, knowledge and character of its population and the framework of law, convention and social interaction that sustains cooperation and community.

Some components of a nation's wealth are appropriable; they can be owned by governments, or privately by individuals or other legal entities. Some intangible assets are appropriable, notably by patents and copyrights. In a capitalist society most appropriable wealth is privately owned, more than 80 per cent by value in the United States. Private properties are generally transferable from owner to owner. Markets in these properties, *capital markets*, are a prominent feature of capitalist societies. In the absence of slavery, markets in 'human capital' are quite limited.

A person may be wealthy without owning any of the assets counted in appropriable *national wealth*. Instead, a personal wealth inventory would list paper currency and coin, bank deposits, bonds, stocks, mutual funds, cash values of insurance policies and pension rights. These are paper assets evidencing claims of various kinds against other individuals, companies, institutions or governments. In reckoning personal *net worth*, each person would deduct from the value of his total assets the claims of others against him. In 1984 American households' gross holdings of financial assets amounted to about 75 per cent of their net worth, and their net holdings to about 55 per cent (Federal Reserve, 1984). If the net worths of all economic units of the nation are added up, paper claims and obligations cancel each other. All that remains, if valuations are consistent and the census is complete, is the value of the national wealth.

If the central government is excluded from this aggregation, *private net worth* – the aggregate net worth of individuals and institutions and subordinate governments (included in the 'private' sector because, lacking monetary powers, they have limited capacities to borrow) – will count not only the national-wealth assets they own but also their net claims against the central government. These include coin and currency, their equivalent in central bank deposit liabilities, and interest-bearing Treasury obligations. If these central government debts exceed the value of its real assets, *private net worth* will exceed national wealth. (However, in reckoning their net worth, private agents may subtract something for the future taxes they expect to pay to service the government's debts. Some economists argue that the subtraction is complete, so that public debt does not count in aggregate private wealth (Barro, 1974) while others give reasons the offset is incomplete (Tobin, 1980). The issue is not crucial for this essay.)

OUTSIDE ASSETS, INSIDE ASSETS AND FINANCIAL MARKETS

Private net worth, then, consists of two parts: privately owned items of national wealth, mostly tangible assets, and government obligations. These *outside* assets are owned by private agents not directly but through the intermediation of a complex network of debts and claims, *inside* assets.

Empirical magnitudes. For the United States at the end of 1984, the value of tangible assets, land and reproducible goods, is estimated at $13.5 trillion, nearly four times the Gross National Product for the year. Of this, $11.2 trillion were privately owned. Adding net claims against the rest of the world and privately owned claims against the federal government gives private net worth of $12.5 trillion, of which only $1.3 trillion represent outside financial assets. The degree of intermediation is indicated by the gross value of financial assets, nearly $14.8 trillion; even if equities in business are regarded as direct titles to real property and excluded from financial assets, the outstanding stock of inside assets is $9.6 trillion. Of these more than half, $5.6 trillion, are claims on financial institutions. The $9.6 million is an underestimate, because many inside financial transactions elude the statisticians. The relative magnitudes of these numbers have changed very little since 1953, when private net worth was $1.27 trillion, gross financial assets $1.35 trillion, $1.05 excluding equities, and GNP was $0.37 trillion (Federal Reserve, 1984).

Raymond Goldsmith, who has studied intermediation throughout a long and distinguished career and knows far more about it than anyone else, has estimated measures of intermediation for many countries over long periods of time (1969, 1985). Here is his own summary:

> The creation of a modern financial superstructure, not in its details but in its essentials, was generally accomplished at a fairly early stage of a country's economic development, usually within five to seven decades from the start of modern economic growth. Thus it was essentially completed in most now-developed countries by the end of the 19th century or the eve of World War I, though somewhat earlier in Great Britain. During this period the financial interrelations ratio, the quotient of financial and tangible assets, increased fairly continuously and sharply. Since World War I or the Great Depression, however, the ratio in most of these countries has shown no upward trend, though considerable movements have occurred over shorter periods, such as sharp reductions during inflations; and though significant changes have taken place in the

relative importance of the various types of financial institutions and of financial instruments. Among less developed countries, on the other hand, the financial interrelations ratio has increased substantially, particularly in the postwar period, though it generally is still well below the level reached by the now-developed countries early in the 20th century.

Goldsmith finds that a ratio of the order of unity is characteristic of financial maturity, as is illustrated by the figures for the United States given above (1985, pp. 2–3).

Goldsmith finds also that the relative importance of financial institutions, especially non–banks, has trended upwards in most market economies but appears to taper off in mature systems. Institutions typically hold from a quarter to a half of all financial instruments. Ratios around 0.40 were typical in 1978, but there is considerably more variation among countries than in the financial interrelations ratio. The United States, at 0.27, is on the low side, probably because of its many well-organized financial markets (1985, Table 47, p. 136).

The volume of gross financial transactions is mind-boggling. The GNP velocity of the money stock in the United States is 6 or 7 per year; if intermediate as well as final transactions for goods and services are considered, the turnover may be 20 or 30 per year. But demand deposits turn over 500 times a year, 2500 times in New York City banks, indicating that most transactions are financial in nature. The value of stock market transactions alone in the United States is one third of the Gross National Product; an average share of stock changes hands every nineteen months. Gross foreign exchange transactions in United States dollars are estimated to be hundreds of billions of dollars every day. 'Value added' in the financial services industries amounts to 9 per cent of United States GNP (Tobin, 1984).

Outside and inside money. The outside/inside distinction is most frequently applied to money. *Outside money* is the monetary debt of the government and its central bank, currency and central bank deposits, sometimes referred to as 'base' or 'high-powered' money. *Inside money*, 'low-powered', consists of private deposit obligations of other banks and depository institutions in excess of their holdings of outside money assets. Just which kinds of deposit obligations count as 'money' depends on definitions, of which there are several, all somewhat arbitrary. Outside money in the United States amounted to $186 billion at the end of 1983, of which $36 billion was held as reserves by banks and other depository institutions; the remaining $150 billion was held by other private agents as currency. The total money stock M1, currency in public circulation plus checkable deposits, was $480 billion. Thus inside M1 was $294 billion, more than 60 per cent of the total.

Financial markets, organized and informal. Inside assets and debts wash out in aggregative accounting; one person's asset is another's debt. But for the functioning of the economy, the inside network is of great importance. *Financial markets* allow inside assets and debts to be originated and to be exchanged at will for each other and for outside financial assets. These markets deal in paper contracts and claims. They complement the markets for real properties. Private agents often borrow to buy real property and pledge the property as security; households mortgage new homes, businesses incur debt to acquire stocks of materials or goods-in-process or to purchase structures and equipment. The term *capital markets* covers both financial and property markets. *Money markets* are financial markets in which short-term debts are exchanged for outside money.

Many of the assets traded in financial markets are promises to pay currency in specified amounts at specified future dates, sometimes conditional on future events and circumstances. The currency is not always the local currency; obligations denominated in various national currencies are traded all over the world. Many traded assets are not denominated in any future monetary unit of account: equity shares in corporations, contracts for deliveries of commodities – gold, oil, soy beans, hog bellies. There are various hybrid assets: preferred stock gives holders priority in distributions of company profits up to specified pecuniary limits; convertible debentures combine promises to pay currency with rights to exchange the securities for shares.

Capital markets, including financial markets, take a variety of forms. Some are highly organized auction markets, the leading real-world approximations to the abstract perfect markets of economic theory, where all transactions occurring at any moment in a commodity or security are made at a single price and every agent who wants to buy or sell at that price is accommodated. Such markets exist in shares, bonds, overnight loans of outside money, standard commodities, and foreign currency deposits, and in futures contracts and options for most of the same items.

However, many financial and property transactions occur otherwise, in direct negotiations between the parties. Organized open markets require large tradable supplies of precisely defined homogeneous commodities or instruments. Many financial obligations are one of a kind, the promissory note of a local business proprietor, the mortgage on a specific farm or residence. The terms, conditions, and collateral are specific to the case. The habit of referring to classes of heterogeneous negotiated transactions as 'markets' is metaphorical, like the use of the term 'labour market' to refer to the decentralized processes by which wages are set and jobs are filled, or 'computer market' to describe the pricing and selling of a host of differentiated products. In these cases the economists' faith is that the outcomes are 'as if' the transaction occurred in perfect organized auction markets.

FINANCIAL ENTERPRISES AND THEIR MARKETS

Financial intermediaries are enterprises in the business of buying and selling financial assets. The accounting balance sheet of a financial intermediary is virtually 100 per cent paper on both sides. The typical financial intermediary owns relatively little real property, just the structures, equipment, and materials necessary to its business. The equity of the owners, or the equivalent capital reserve account for mutual, cooperative, nonprofit, or public institutions, is small compared to the enterprises' financial obligations.

Financial intermediaries are major participants in organized financial markets. They take large asset positions in market instruments; their equities and some of their liabilities, certificates of deposit or debt securities, are traded in those markets. They are not just middlemen like dealers and brokers whose main business is to execute transactions for clients.

Fnancial intermediaries are the principal makers of the informal financial markets discussed above. Banks and savings institutions hold mortgages, commercial loans, and consumer credit; their liabilities are mainly checking accounts, savings deposits, and certificates of deposit. Insurance companies and pension funds negotiate private placements of corporate bonds and commercial mortgages; their liabilities are contracts with policy-holders and obligations to future retirees. Thus financial intermediaries do much more than participate in organized markets. If financial intermediaries confined themselves to repackaging open market securities for the convenience of their creditors, they would be much less significant actors on the economic scene.

Financial businesses seek customers, both lenders and borrowers, not only by interest rate competition but by differentiating and advertising their 'products'. Financial products are easy to differentiate, by variations in maturities, fees, auxiliary services, office locations and hours of business, and many other features. As might be expected, non-price competition is especially active when prices, in this case interest rates, are fixed by regulation or by tacit or explicit collusion. But the industry is by the heterogeneous nature of its products monopolistically competitive; non-price competition flourishes even when interest rates are free to move. The industry shows symptoms of 'wastes of monopolistic competition'. Retail offices of banks and savings institutions cluster like competing gasoline stations. Much claimed product differentiation is trivial and atmospheric, emphasized and exaggerated in advertising.

Financial intermediaries cultivate long-term relationships with customers. Even in the highly decentralized financial system of the United States, local financial intermediaries have some monopoly power, some clienteles who will stay with them even if their interest rates are somewhat less favourable than those elsewhere. Since much business is bilaterally negotiated, there are ample opportunities for price discrimination. The typical business customer of a bank is both a borrower and a depositor, often simultaneously. The customer 'earns' the right for credit accommodation when he needs it by lending surplus funds to the same bank when he has them. The same reciprocity occurs between credit unions and mutual savings institutions and some of their members. Close ties frequently develop between a financial intermediary and non-financial businesses whose sales depend on availability of credit to their customers, for example between automobile dealers and banks. Likewise, builders and realtors have funded and controlled many savings and loan associations in order to facilitate mortgage lending to home buyers.

Financial intermediaries balance the credit demands they face with their available funds by adjusting not only interest rates but also the other terms of loans. They also engage in quantitative rationing, the degree of stringency varying with the availability and costs of funds to the intermediary. Rationing occurs naturally as a by-product of lending decisions made and negotiated case by case. Most such loans require collateral, and the amount and quality of the collateral can be adjusted both to individual circumstances and to overall market conditions. Borrowers are classified as to riskiness and charged rates that vary with their classification.

United States commercial banks follow the 'prime rate convention'. One or another of the large banks acts as price leader and sets a rate on six-month commercial loans for its prime quality borrowers. If other large banks agree, as is usually the case, they follow, and the rate becomes standard for the whole industry until one of the leading banks decides another change is needed to stay in line with open-market interest rates. Loan customers are rated by the number of half-points above prime at which they will be accommodated. Of course, some applications for credit are just turned away. One mechanism of short-term adjustment to credit market conditions is to stiffen or relax the risk classifications of customers, likewise to deny credit to more or fewer applicants. Similar mechanisms for rationing help to equate demands to supplies of home mortgage finance and consumer credit.

THE FUNCTIONS OF FINANCIAL MARKETS AND INTERMEDIARY
INSTITUTIONS

Intermediation, as defined and described above, converts the outside privately owned wealth of the economy into the quite different forms in which its ultimate owners hold their accumulated savings. Financial markets alone accomplish considerable intermediation, just by facilitating the origination and exchange of inside assets. Financial intermediaries greatly extend the process, adding 'markets' that would not exist without them, and participating along with other agents in other markets, organized or informal.

What economic functions does intermediation in general perform? What do inside markets add to markets in the basic outside assets? What functions does institutional intermediation by financial intermediaries perform beyond those of open markets in financial instruments? Economists characteristically impose on themselves questions like these, which do not seem problematic to lay practitioners. Economists start from the presumption that financial activities are epiphenomena, that they create a veil obscuring to superficial observers an underlying reality which they do not affect. The celebrated Modigliani–Miller theorem (1958), generalized beyond the original intent of the authors, says so. With its help the sophisticated economist can pierce the veil and see that the values of financial assets are just those of the outside assets to which they are ultimately claims, no matter how circuitous the path from the one to the other.

However, economists also understand how the availability of certain markets alters, usually for the better, the outcomes prevailing in their absence. For a primitive illustration, consider the functions of inside loan markets as brilliantly described by Irving Fisher (1930). Each household has an inter-temporal utility function in consumptions today and at future times, a sequence of what we now would call dated 'endowments' of consumption, and an individual 'backyard' production function by which consumption less than endowment at any one date can be transformed into consumption above endowment at another date. Absent the possibility of intertemporal trades with others, each household has to do its best on its own; its best will be to equate its marginal rate of substitution in utility between any two dates with its marginal rate of transformation in production between the same dates, with the usual amendments for corner solutions. The gains from trade, i.e., in this case from auction markets in inter-household lending and borrowing, arise from differences among households in those autarkic rates of substitution and transformation. They are qualitatively the same as those from free contemporaneous trade in commodities between agents or nations.

The introduction of consumer loans in this Fisherian model will alter the individual and aggregate paths of consumption and saving. It is not possible to say whether it will raise or lower the aggregate amount of capital, here in the sense of labour endowments in the process of producing future rather than current consumable output. In either case it is likely to be a Pareto-optimal improvement, although even this is not guaranteed *a priori*.

Similar argument suggests several reasons why ultimately savers, lenders and creditors prefer the liabilities of financial intermediaries not only to direct ownership of real property but also to the direct debt and equity issues of investors, borrowers and debtors:

Convenience of denomination. Issuers of securities find it costly to cut their issues into the variety of small and large denominations savers find convenient and commensurate to their means. The financial intermediary can break up large-denomination bonds and loans into amounts convenient to small savers, or combine debtors' obligations into large amounts convenient to the wealthy. Economies of scale and specialization in financial transactions enable financial intermediaries to tailor assets and liabilities to the needs and preferences of both lenders and borrowers. This service is especially valuable for agents on both sides whose needs vary in amount continuously; they like deposit accounts and credit lines whose use they can vary at will on their own initiative.

Risk pooling, reduction and allocation. The risks incident to economic activities take many forms. Some are nation-wide or world-wide – wars and revolutions, shifts in international comparative advantage, government fiscal and monetary policies, prices and supplies of oil and other basic materials. Some are specific to particular enterprises and technologies – the capacity and integrity of managers, the qualities of new products, the local weather. A financial intermediary can specialize in the appraisal of risks, especially specific risks, with expertise in the gathering and interpretation of information costly or unavailable to individual savers. By pooling the funds of its creditors, the financial intermediary can diversify away risks to an extent that the individual creditors cannot, because of the costs of transactions as well as the inconvenience of fixed lumpy denominations.

According to Joseph Schumpeter ([1911] 1934, pp. 72–4), bankers are the gatekeepers – Schumpeter's word is 'ephor' – of capitalist economic development; their strategic function is to screen potential innovators and advance the necessary purchasing power to the most promising. They are the source of purchasing power for investment and innovation, beyond the savings accumulated from past economic development. In practice, the cachet of a banker often enables his customer also to obtain credit from other sources or to float paper in open markets.

Maturity shifting. A financial intermediary typically reconciles differences among borrowers and lenders in the timing of payments. Bank depositors want to commit funds for shorter times than borrowers want to have them. Business borrowers need credit to bridge the time gap between the inputs to profitable production and their output and sales. This source of bank business is formally modelled by Diamond and Dybvig (1983). The bank's scale of operations enables it to stagger the due dates of, say, half-year loans so as to accommodate depositors who want their money back in three months or one month or on demand. The reverse maturity shift may occur in other financial intermediaries. An insurance company or pension fund might invest short-term the savings its policy-owners or future pensioners will not claim for many years.

Transforming illiquid assets into liquid liabilities. Liquidity is a matter of degree. A perfectly liquid asset may be defined as one whose full present value can be realized, i.e., turned into purchasing power over goods and services, immediately. Dollar bills are perfectly liquid, and so for practical purposes are demand deposits and other deposits transferable to third parties by check or wire. Liquidity in this sense does not necessarily mean predictability of value. Securities traded on well organized markets are liquid. Any person selling at a given time will get the same price whether he decided and prepared to sell a month before or on the spur of the moment. But the price itself can vary unpredictably from minute to minute. Contrast a house, neither fully liquid nor predictable in value. Its selling proceeds at this moment are likely to be greater the longer it has been on the market. Consider the six-month promissory note of a small business proprietor known only to his local banker. However sure the payment on the scheduled date, the note may not be marketable at all. If the lender wants to realize its value before maturity, he will have to find a buyer and negotiate. A financial intermediary holds illiquid assets while its liabilities are liquid, and holds assets unpredictable in value while it guarantees the value of its liabilities. This is the traditional business of commercial banks, and the reason for the strong and durable relations of banks and their customers.

SUBSTITUTION OF INSIDE FOR OUTSIDE ASSETS

What determines the aggregate liabilities and assets of financial intermediaries? What determines the gross aggregate of inside assets generated by financial markets in general, including open markets as well as financial intermediaries? How can the empirical regularities found by Goldsmith, cited above, be explained?

Economic theory offers no answers to these questions. The differences among agents that invite mutually beneficial transactions, like those discussed above, offer opportunities for inside markets. Theory can tell us little a priori about the size of such differences. Moreover, markets are costly to operate, whether they are organized auction markets in homogeneous instruments or the imperfect 'markets' in heterogeneous contracts in which financial intermediaries are major participants. Society cannot afford all the markets that might exist in the absence of transactions costs and other frictions, and theory has little to say on which will arise and survive.

The macroeconomic consequence of inside markets and financial intermediaries is generally to provide substitutes for outside assets and thus to economize their supplies. That is, the same microeconomic outcomes are achievable with smaller supplies of one or more of the outside assets than in the absence of intermediation. The way in which intermediation mobilizes the surpluses of some agents to finance the deficits of others is the theme of the classic influential work of Gurley and Shaw (1960).

Consider, for example, how commercial banking diminishes the need of business firms for net worth invested in inventories, by channelling the seasonal cash surpluses of some firms to the contemporaneous seasonal deficits of others. Imagine two firms A and B with opposite and complementary seasonal zigzag patterns. A needs $2 in cash at time zero to buy inputs for production in period 1 sold for $2; the pattern repeats in 3, 4,... B needs $2 in cash at time 1 to buy inputs for production in period 2 sold for $2 in period 3, and so on in 4, 5,.... In the absence of their commercial bank, A and B each need $2 of net worth to carry on business; from period to period each alternates holding it in cash and in goods-in-process. between them the two firms always are holding $2 of currency and $2 of inventories. B enters the bank and lends A half the $2 he needs to carry his inventory in period 1; A repays the loan from sales proceeds the next period, 2; the bank now lends $1 to B, ... A and B now need only $1 of currency; each has on average net worth of $1.50 – $2 and $1 alternating; as before they are together always holding $2 of inventories. Moreover, with a steady deposit of $2 from a third party, the bank could finance both businesses completely; they would need no net worth of their own. The example is trivial, but commercial banking proper can be understood as circulation of deposits and loans among businesses and as a revolving fund assembled from other sources and lent to businesses.

As a second primitive example, consider the effects of introducing markets that enable risks to be borne by those households more prepared to take them. Suppose that of two primary outside assets, currency and tangible capital, the return on the latter has the greater variance. Individuals who are risk neutral will hold all their wealth (possibly excepting minimal transactions balances of currency) in capital as long as its expected return exceeds the expected real return on currency. If these more adventurous households are not numerous and wealthy enough to absorb all the capital, the expected return on capital will have to exceed that on currency enough to induce risk-averse wealth-owners to hold the remainder. In this equilibrium the money price of capital and its mean real return are determined so as to allocate the two assets between the two kinds of households. Now suppose that the risk-neutral households can borrow from the risk-averse types, most realistically via financial intermediaries, and that the latter households regard those debts as close substitutes for currency, indeed as inside money if intermediation by financial intermediaries is involved. The inside assets do double duty, providing the services and security of money to those who value them while enabling the more adventurous to hold capital in excess of their own net worth. As a result, the private sector as a whole will want to hold a larger proportion of its wealth in capital at any given expected real return on capital. In equilibrium, the aggregate capital stock will be larger and its expected return, equal to its marginal productivity in a steady state, will be lower than in the absence of intermediation.

Intermediation can diminish the private sector's need not just for outside money but for net worth and tangible capital. These economies generally require financial markets in which financial intermediaries are major participants, because they involve heterogeneous credit instruments and risk pooling. In the absence of home mortgages, consumer credit, and personal loans for education, young households would not be able to spend their future wages and salaries until they receive them. Constraints on borrowing against future earnings make the age-weighted average net non-human wealth of the population greater, but the relaxation of such liquidity constraints increases household welfare. Financial intermediaries invest the savings of older and more affluent households in loans to their younger and less wealthy contemporaries; otherwise those savings would go into outside assets. Likewise insurance makes it unnecessary to accumulate savings as precaution against certain risks, for example the living and medical expenses of unusual longevity. It is an all too common fallacy to assume that arrangements that increase aggregate savings and tangible wealth always augment social welfare.

DEPOSIT CREATION AND RESERVE REQUIREMENTS

The substitution of inside money for outside money is the familiar story of deposit creation, in which the banking system turns a dollar of base or 'high-powered' money into several dollars of deposits. The extra dollars are inside or 'low-powered' money. The banks need to hold only a fraction k, set by law or convention or prudence, of their deposit liabilities as reserves in base money. In an equilibrium in which they hold no excess reserves their deposits will be a multiple $1/k$ of their reserves; they will have created $(1-k)/k$ dollars of substitute money.

A key step in this process is that any bank with excess reserves makes a roughly equal amount of additional loans, crediting the borrowers with deposits. As the borrowers draw checks, these new deposits are transferred to other accounts, most likely in other banks. As deposits move to other banks, so do reserves, dollar for dollar. But now those banks have excess reserves and act in like manner. The process continues until all banks are 'loaned up', i.e. deposits have increased enough so that the initial excess reserves have become reserves that the banks require or desire.

The textbook fable of deposit creation does not do justice to the full macroeconomics of the process. The story is incomplete without explaining how the public is induced to borrow more and to hold more deposits. The borrowers and the depositors are not the same public. No one borrows at interest in order to hold idle deposits. To attract additional borrowers, banks must lower interest rates or relax their collateral requirements or their risk standards. The new borrowers are likely to be businesses that need bank credit to build up inventories of materials or goods in process. The loans lead quickly to additional production and economic activity. Or banks buy securities in the open market, raising their prices and lowering market interest rates. The lower market rates may encourage businesses to float issues of commercial paper, bonds or stocks, but the effects of investment in inventories or plant and equipment are less immediate and less potent than the extension of bank credit to a business otherwise held back by illiquidity. In either case, lower interest rates induce other members of the public, those who indirectly receive the loan disbursements or those who sell securities to banks, to hold additional deposits. They will be acquiring other assets as well, some in banks, some in other financial intermediaries, some in open financial markets. Lower interest rates may also induce banks themselves to hold extra excess reserves.

Interest rates are not the only variables of adjustment. Nominal incomes are rising at the same time, in some mixture of real quantities and prices depending on macroeconomic circumstances. The rise in incomes and economic activities creates new needs for transactions balances of money. Thus the process by which excess reserves are absorbed entails changes in interest rates, real economic activity, and prices in some combination. It is possible to describe scenarios in which the entire ultimate adjustment is in one of these variables. Wicksell's cumulative credit expansion, which in the end just raises prices, is a classic example.

Do banks have a unique magic by which asset purchases generate their own financing? Is the magic due to the 'moneyness' of the banks' liabilities? The preceding account indicates it is not magic but reserve requirements. Moreover, a qualitatively similar story could be told if reserve requirements were related to bank assets or non-monetary liabilities and even if banks happened to have no monetary liabilities at all. In the absence of reserve requirements aggregate bank assets and liabilities, relative to the size of the economy, would be naturally limited by public supplies and demands at interest rates that cover banks' costs and normal profits. If, instead of banks, savings institutions specializing in mortgage lending were subject to reserve requirements, their incentives to minimize excess reserves would inspire a story telling how additional mortgage lending brings home savings deposits to match (Tobin, 1963).

RISKS, RUNS AND REGULATIONS

Some financial intermediaries confine themselves to activities that entail virtually no risk either to the institution itself or to its clients. An open-end mutual fund or unit trust holds only fully liquid assets traded continuously in organized markets. It promises the owners of its shares payment on demand at their pro rata net value calculated at the market prices of the underlying assets – no more, no less. The fund can always meet such demands by selling assets it holds. The shareowners pay in one way or another an agreed fee from the services of the fund – the convenience and flexibility of denomination, the bookkeeping, the transactions costs, the diversification, the expertise in choosing assets. The shareowners bear the market risks on the fund's portfolio – no less and, assuming the fund is honest, no more. Government regulations are largely confined to those governing all public security issues, designed to protect buyers from deceptions and insider manipulations. In the United States regulation of this kind is the province of the federal Securities and Exchange Commission.

Most financial intermediaries do take risks. The risks are intrinsic to the functions they serve and to the profit opportunities attracting financial entrepreneurs and investors in their enterprises. For banks and similar financial intermediaries, the principal risk is that depositors may at any time demand payments the institution can meet, if at all, only at extraordinary cost. Many of the assets are illiquid, unmarketable. Others can be liquidated at short notice only at substantial loss. In some cases, bad luck or imprudent management brings insolvency; the institution could never meet its obligations no matter how long its depositors and other creditors wait. In other cases, the problem is just illiquidity; the assets would suffice if they could be held until maturity, until buyers or lenders could be found, or until normal market conditions returned.

Banks and other financial intermediaries hold reserves, in currency or its equivalent, deposits in central banks, or in other liquid forms as precaution against withdrawals by their depositors. For a single bank, the withdrawal is usually a shift of deposits to other banks or financial intermediaries, arising from a negative balance in interbank clearings of checks or other transfers to third parties at the initiative of depositors. For the banking system, as a whole, withdrawal is a shift by the public from deposits to currency.

'Withdrawals' may in practice include the exercise of previously agreed borrowing rights. Automatic overdraft privileges are more common in other countries, notably the United Kingdom and British Commonwealth nations, than in the United States. They are becoming more frequent in the United States as an adjunct of bank credit cards. Banks' business loan customers often have explicit or implicit credit lines on which they can draw on demand.

Unless financial intermediaries hold safe liquid assets of predictable value matched in maturities to their liabilities – in particular, currency or equivalent against all their demand obligations – they and their creditors can never be completely protected from withdrawals. The same is true of the banking system as a whole, and of all intermediaries other than simple mutual funds. 'Runs', sudden, massive, and contagious withdrawals, are always possible. They destroy prudent and imprudent institutions alike, along with their depositors and creditors. Of course, careful depositors inform themselves about the intermediaries to which they entrust their funds, about their asset portfolios, policies and skills. Their choices among competing depositories provide some discipline, but it can never be enough to rule out disasters. What the most careful depositor cannot foresee is the behaviour of other depositors, and it is rational for the well-informed depositor of a sound bank to withdraw funds if he believes that others are doing so or are about to do so.

Governments generally regulate the activities of banks and other financial intermediaries in greater detail than they do nonfinancial enterprises. The basic motivations for regulation appear to be the following:

It is costly, perhaps impossible, for individual depositors to appraise the soundness and liquidity of financial institutions and to estimate the probabilities of failures even if they could assume that other depositors would do likewise. It is impossible for them to estimate the probabilities of 'runs'. Without regulation, the liabilities of suspect institutions would be valued below par in check collections. Prior to 1866 banks in the United States were allowed to issue notes payable to bearers on demand, surrogates for government currency. The notes circulated at discounts varying with the current reputations of the issuers. A system in which transactions media other than government currency continuously vary in value depending on the issuer is clumsy and costly.

The government has obligation to provide at low social cost an efficient system of transactions media, and also a menu of secure and convenient assets for citizens who wish to save in the national monetary unit of account. Those transactions media and saving assets can be offered by banks and other financial intermediaries, in a way that retains most of the efficiencies of decentralization and competition, if and only if government imposes some regulations and assumes some residual responsibilities. The government's role takes several forms.

Reserve requirements. An early and obvious intervention was to require banks to hold reserves in designated safe and liquid forms against their obligations, especially their demand liabilities. Left to themselves, without such requirements, some banks might sacrifice prudence for short-term profit. Paradoxically, however, required reserves are not available for meeting withdrawals unless the required ratio is 100 per cent. If the reserve requirement is 10 per cent of deposits, then withdrawal of one dollar from a bank reduces its reserve holdings by one dollar but its reserve requirement by only ten cents. Only excess reserves or other liquid assets are precautions against withdrawals. The legal reserve requirement just shifts the bank's prudential calculation to the size of these secondary reserves. Reserve requirements serve functions quite different from their original motivation. In the systems that use them, notably the United States, they are the fulcrum for central bank control of economy-wide monetary conditions. (They are also an interest-free source of finance of government debt, but in the United States today this amounts to only $45 billion of a total debt to the public of $1700 billion.)

Last-resort lending. Banks and other financial intermediaries facing temporary shortages of reserves and secondary reserves of liquid assets can borrow them from other institutions. In the United States, for example, the well-organized market for 'federal funds' allows banks short of reserves to borrow them overnight from other banks. Or banks can gain reserves by attracting more deposits, offering higher interest rates on them than depositors are getting elsewhere. These ways of correcting reserve positions are not available to troubled banks, suspected of deep-rooted problems of liquidity or solvency or both, for example bad loans. Nor will they meet a system-wide run from liabilities of banks and other financial intermediaries into currency.

Banks in need of reserves can also borrow from the central bank, and much of this borrowing is routine, temporary, and seasonal. Massive central bank credit is the last resort of troubled banks which cannot otherwise satisfy the demands of their depositors without forced liquidations of their assets. The government is the ultimate supplier of currency and reserves in aggregate. The primary *raison d'être* of the central bank is to protect the economy from runs into currency. System-wide shortages of currency and reserves can be relieved not only by central bank lending to individual banks but by central bank purchases of securities in the open market. The Federal Reserve's inability or unwillingness – which it was is still debated – to supply the currency bank depositors wanted in the early 1930s led to disastrous panic and epidemic bank failures. No legal or doctrinal obstacles would now stand in the way of such a rescue.

Deposit insurance. Federal insurance of bank deposits in the United States has effectively prevented contagious runs and epidemic failures since its enactment in 1935. Similar insurance applies to deposits in savings institutions. In effect, the federal government assumes a contingent residual liability to pay the insured deposits in full, even if the assets of the financial intermediary are permanently inadequate to do so. The insured institutions are charged premiums for the service, but the fund in which they are accumulated is not and cannot be large enough to eliminate possible calls on the Treasury. Although the guarantees are legally limited to a certain amount, now $100,000, per account, in practice depositors have eventually recovered their full deposits in most cases. Indeed the guarantee seems now to have been extended *de facto* to all deposits, at least in major banks.

Deposit insurance impairs such discipline as surveillance by large depositors might impose on financial intermediaries; instead the task of surveillance falls on the governmental insurance agencies themselves (in the United States the Federal Deposit Insurance Corporation and the Federal Savings and Loan Insurance Corporation) and on other regulatory authorities (the United States Comptroller of the Currency, the Federal Reserve, and various state agencies). Insurance transfers some risks from financial intermediary depositors and owners to taxpayers at large, while virtually eliminating risks of runs. Those are risks we generate ourselves; they magnify the unavoidable natural risks of economic life. Insurance is a mutual compact to enable us to refrain from *sauve qui peut* behaviour that can inflict grave damage on us all. Formally, an uninsured system has two equilibria, a good one with mutual confidence and a bad one with runs. Deposit insurance eliminates the bad one (Diamond and Dybvig, 1983).

One hundred per cent reserve deposits would, of course, be perfectly safe – that is, as safe as the national currency – and would not have to be insured. Those deposits would in effect *be* currency, but in a secure and conveniently checkable form. One can imagine a system in which banks and other financial intermediaries offered such accounts, with the reserves behind them segregated from those related to the other business of the institution. That other business would include receiving deposits which required fractional or zero reserves and were insured only partially, if at all. The costs of the 100 per cent reserve deposit accounts would be met by service charges, or by government interest payments on the reserves, justified by the social benefits of a safe and efficient transactions medium. The burden of risk and supervision now placed on the insuring and regulating agencies would be greatly relieved. It is, after all, historical accident that supplies of transactions media in modern economies came to be byproducts of banking business and vulnerable to its risks.

Government may insure financial intermediaries loans as well as deposits. Insurance of home mortgages in the United States not only has protected the institutions that hold them and their depositors but has converted the insured mortgages into marketable instruments.

Balance sheet supervision. Government surveillance of financial intermediaries limits their freedom of choice of assets and liabilities, in order to limit the risks to depositors and insurers. Standards of adequacy of capital – owners' equity at risk in the case of private corporations, net worth in the case of mutual and other nonprofit forms of organization – are enforced for the same reasons. Periodic examinations check the condition of the institution, the quality of its loans, and the accuracy of its accounting statements. The regulators may close an institution if further operation is judged to be damaging to the interests of the depositors and the insurer.

Legislation which regulates financial intermediaries has differentiated them by purpose and function. Commercial banks, savings institutions, home building societies, credit unions, and insurance companies are legally organized for different purposes. They are subject to different rules governing the nature of their assets. For example, home building societies – savings and loan associations in the United States – have been required to keep most of their asset portfolios in residential mortgages. Restrictions of this kind mean that when wealth-owners shift funds from one type of financial intermediary to another, they alter relative demands for assets of different kinds. Shifts of deposits from commercial banks to building societies would increase mortgage lending relative to commercial lending. Regulations have also restricted the kinds of liabilities allowed various types of financial intermediary. Until recently in the United States, only banks were permitted to have liabilities payable on demand to third parties by check or wire. Currently deregulation is relaxing specialized restrictions on financial intermediary assets and liabilities and blurring historical distinctions of purpose and function.

Interest ceilings. Government regulations in many countries set ceilings on the interest rates that can be charged on loans and on the rates that can be paid on deposits, both at banks and at other financial intermediaries. In the United States the Banking Act of 1935 prohibited payment of interest on demand deposits. After the second world war effective ceilings on savings and time deposits in banks and savings institutions were administratively set, and on occasion changed, by federal agencies. Under legislation of 1980, these regulations are being phased out.

The operating characteristics of a system of financial intermediaries in which interest rates on deposits of various types, as well as on loans, are set by free competition are quite different from those of a system in which financial intermediary rates are subject to legal ceilings or central bank guidance, or set by agreement among a small number of institutions. For example, when rates on deposits are administratively set, funds flow out of financial intermediaries when open market rates rise and return to financial intermediaries when they fall. These processes of 'disintermediation' and 're-intermediation' are diminished when financial intermediary rates are free to move parallel to open market rates. Likewise flows between different financial intermediaries due to administratively set rate differences among them are reduced when they are all free to compete for funds.

A regime with market-determined interest rates on moneys and near-moneys has significantly different macroeconomic characteristics from a regime constrained by ceilings on deposit interest rates. Since the opportunity cost of holding deposits is largely independent of the general level of interest rates, the 'LM' curve is steeper in the unregulated regime. Both central bank operations and exogenous monetary shocks could be expected to have larger effects on nominal income, while fiscal measures and other shocks to aggregate demand for goods and services would have smaller effects (Tobin, 1983).

Entry, branching, merging. Entry into regulated financial businesses is generally controlled, as are establishing branches or subsidiaries and merging of existing institutions. In the United States, charters are issued either by the federal government or by state governments, and regulatory powers are also divided. Until recently banks and savings institutions, no matter by whom chartered, were not allowed to operate in more than one state. This rule, combined with various restrictions on branches within states, gave the United States a much larger number of distinct financial enterprises, many of them very small and very local, than is typical in other countries. The prohibition of interstate operations is now being eroded and may be effectively eliminated in the next few years.

Deregulation has been forced by innovations in financial technology that made old regulations either easy hurdles to circumvent or obsolete barriers to efficiency. New opportunities not only are breaking down the walls separating financial intermediaries of different types and specializations. They are also bringing other businesses, both financial and nonfinancial, into activities previously reserved to regulated financial institutions. Mutual funds and brokers offer accounts from which funds can be withdrawn on demand or transferred to third parties by check or wire. National retail chains are becoming financial supermarkets – offering credit cards, various mutual funds, instalment lending, and insurance along with their vast menus of consumer goods and services; in effect, they would like to become full-service financial intermediaries. At the same time, the traditional intermediaries are moving, as fast as they can obtain government permission, into lines of business from which they have been excluded. Only time will tell how these commercial and political conflicts are resolved and how the financial system will be reshaped (*Economic Report of the President*, 1985, ch. 5).

PORTFOLIO BEHAVIOUR OF FINANCIAL INTERMEDIARIES

A large literature has attempted to estimate econometrically the choices of assets and liabilities by financial intermediaries, their relationships to open market interest rates and to other variables exogenous to them. Models of the portfolio behaviour of the various species of financial intermediary also involve estimation of the supplies of funds to them, and the demands for credit, from other sectors of the economy, particularly households and nonfinancial businesses. Recent research is presented in Dewald and Friedman (1980).

Difficult econometric problems arise in using time series for these purposes because of regime changes. For example, when deposit interest rate ceilings are effective, financial intermediaries are quantity-takers in the deposit markets; when the ceilings are non-constraining or non-existent, both the interest rates and the quantities are determined jointly by the schedules of supplies of deposits by the public and of demands for them by the financial intermediary. Similar problems arise in credit markets where interest rates, even though unregulated, are administered by financial intermediaries themselves and move sluggishly. The prime commercial loan rate is one case; mortgage rates in various periods are another. In these cases and others, the markets are not cleared at the established rates. Either the financial intermediary or the borrowers are quantity-takers, or perhaps both in some proportions.

Changes in the rates follow, dependent on the amount of excess demand or supply. These problems of modeling and econometric estimation are discussed in papers in the reference above. The seminal paper is Modigliani and Jaffee (1969).

JAMES TOBIN

See also CAPITAL, CREDIT AND MONEY MARKETS; CENTRAL BANKING; DISINTERMEDIATION; FINANCE; LIQUIDITY; MONEY SUPPLY.

BIBLIOGRAPHY

Barro, R. 1974. Are government bonds net wealth? *Journal of Political Economy* 82(6), November–December, 1095–117.
Dewald, W.G. and Friedman, B.M. 1980. Financial market behavior, capital formation, and economic performance. (A conference supported by the National Science Foundation.) *Journal of Money, Credit and Banking*, Special Issue 12(2), May.
Diamond, D.W. and Dybvig, P.H. 1983. Bank runs, deposit insurance, and liquidity. *Journal of Political Economy* 91(3), June, 401–19.
Economic Report of the President. 1985. Washington, DC: Government Printing Office, February.
Federal Reserve System, Board of Governors. 1984. *Balance Sheets for the US Economy 1945–83*. November, Washington, DC.
Fisher, I. 1930. *The Theory of Interest*. New York: Macmillan.
Goldsmith, R.W. 1969. *Financial Structure and Development*. New Haven: Yale University Press.
Goldsmith, R.W. 1985. *Comparative National Balance Sheets: A Study of Twenty Countries, 1688–1978*. Chicago: University of Chicago Press.
Gurley, J.G. and Shaw, E.S. 1960. *Money in a Theory of Finance*. Washington, DC: Brookings Institution.
Modigliani, F. and Miller, M.H. 1958. The cost of capital, corporation finance and the theory of investment. *American Economic Review* 48(3), June, 261–97.
Modigliani, F. and Jaffee, D.M. 1969. A theory and test of credit rationing. *American Economic Review* 59(5), December, 850–72.
Schumpeter, J. A. 1911. *The Theory of Economic Development*. Trans. from the German by R. Opie, Cambridge, Mass.: Harvard University Press, 1934.
Tobin, J. 1963. Commercial banks as creators of 'money'. In *Banking and Monetary Studies*, ed. D. Carson, Homewood, Ill.: Richard D. Irwin.
Tobin, J. 1980. *Asset Accumulation and Economic Activity*. Oxford: Blackwell.
Tobin, J. 1983. Financial structure and monetary rules. *Kredit und Kapital* 16(2), 155–71.
Tobin, J. 1984. On the efficiency of the financial system. *Lloyds Bank Review* 153, July, 1–15.

financial journalism. Financial journalists write in the press, or talk on radio or television, about the financial markets and all the factors that move them. The term 'financial' to describe this work is traditionally used in England, but the more senior writers cover a much wider ground. They report and discuss any matter that may be of professional interest to the alert businessman. Their field includes economic analysis and comment. Their personal views on events are often expressed with a freedom rarely allowed to other journalists, and they are sometimes influential both in moving markets and in shifting public and political opinion.

There are two reasons for this state of affairs in England, one technical and one historical. The technical reason is that British Ministers and senior officials, as well as leaders of business and banking, prefer to brief journalists 'off the record' and anonymously. This is also true in France but not generally in the USA or in West Germany. The British journalist, therefore, expresses as his own view something he may have learnt from an 'inside' source, and it is often difficult for the reader of the British financial press to spot the line between official thinking and the writer's own opinion. The system has many critics, but it accounts in part for the unique role of financial journalists in England.

Historically this status goes back to the 19th century, when a number of outstanding personalities were drawn into financial and economic journalism. After the Napoleonic wars, London gradually became the chief financial centre of Europe, taking over from Amsterdam. During the industrial revolution a large, wealthy middle class had emerged which had surplus savings and looked for investment opportunities. The growing demand for financial information, analysis and advice was met by newspapers and weekly journals on the basis of advertising by company promoters, banks and (later) joint-stock companies. The London *Times* appointed its first financial editor, Thomas Massa Alsager, in 1817. He was not an economist but a businessman with wide cultural interests. He opened an office close to the stock exchange and the Bank of England, wrote the daily financial article and organized the collection of 'mercantile and foreign news'. He became a friend of the Rothschilds, made a fortune, and did not hesitate to criticize the Bank of England. For some years he stood alone in warning investors that the great boom in railway promotions was bound to collapse. The *Times* lost a great deal of advertising revenue but the proprietors were high-minded and Alsager was proved right.

Another milestone was the founding of The *Economist* in 1843 as 'a political, literary and general newspaper' by James Wilson, a banker and Member of Parliament who had been Financial Secretary to the Treasury and later became Finance Member of the Council of India. Wilson started the weekly paper, which he and his family owned, to spread the ideas of free trade, free enterprise and political reform. From the start it had a substantial statistical section, and soon monetary and banking subjects became prominent. James Wilson wrote several articles in every issue, including the important one on the 'Money Market'. In 1857 he engaged Walter Bagehot, a rising economist and banker, to write for him on banking. Bagehot soon widened his subject; he also married Wilson's daughter, and in 1859 when Wilson went to India he became sole editor and remained so until his death in 1877. Bagehot gained a high reputation as a financial expert, adviser to governments and author of books. (Keynes said Bagehot wrote *Lombard Street* in 1873 in order to 'knock two or three fundamental truths into the heads of City magnates'.) Throughout, week after week, Bagehot remained a journalist and a crusader for reform. For example, he warned of the danger of allowing the pound sterling to be used as an international currency. When that became inevitable because for a time the pound was the only currency freely convertible into gold, he demanded that the Bank of England should build up a separate gold reserve so that withdrawals of foreign deposits should not deflate the British economy. That theme has remained alive for more than 100 years.

The blend of political, economic and financial subjects has been a successful formula for The *Economist* ever since. It has also been a model for some other papers that came a little later. The first daily newspaper devoted to financial and business matters was the *Financial News*, founded in 1884, followed by the *Financial Times* in 1888. The two papers had periods of success and weakness. They merged in 1945 with the title *Financial Times* and the combined paper has greatly widened its scope and increased its circulation.

A great revival of financial journalism took place in England after World War I, when a number of gifted young university graduates were recruited, first by the *Manchester Guardian* and *The Economist* and a few years later by the *Financial News*. This group spread quickly to other publications, and though it lost many of its young members to high positions in government service, banks and universities, it raised the profession to a status not equalled in any other country. The opening was provided in the early 1920s by a surge of public demand for information and comment on a bewildering series of events. War debts, reparations, the destruction of currencies by inflation, the 1925 restoration of the gold standard, mass unemployment, the Wall Street crash of 1929 and the world-wide depression that followed, the 1931 sterling crisis – all these required some expert knowledge for proper understanding and discussion. The new type of financial journalist was picked and trained to meet this demand.

The new wave was started by Oscar Hobson (1886–1961). He went to King's College, Cambridge, partly at the same time as Maynard Keynes, whose views he sternly opposed all his life. After taking first class degrees in classics and mathematics and a brief spell in a London bank, Hobson became financial editor of the *Manchester Guardian*, where he took an active part in the public arguments of the 1920s. In 1929 he was made editor of the *Financial News*, which had just passed into new and ambitious hands. There he found the first few of the new type of financial journalists already in place, and he added quickly to them, assembling a brilliant group that became the chief nursery for British financial journalism. In 1934, after a sharp dispute over policy with his publisher (Brendan Bracken), Hobson left and became 'City Editor' of a general daily newspaper, the liberal *News-Chronicle*, where he was given less than half a page to cover financial news and comment. He managed to write each day a decisive little essay in a few square inches. One day he was summoned to the Governor of the Bank of England, Montagu Norman, who asked why he did not collect his daily essays to make a book. Hobson laughed. The Governor opened his desk, took out a pile of clippings with Hobson's daily writings and said: 'I have the book ready here, and I have arranged a publisher for it.' The book was published and went into many editions.

Hobson always kept close touch with academic economists. He was a member of the Council of the Royal Economic Society, a governor of the London School of Economics, and was made a knight. The economist Lionel Robbins wrote in his autobiography that Hobson was 'one of the creators of modern standards of professional excellence in financial journalism'.

Among these 'creators' one must certainly mention Sir Walter Layton (later Lord Layton) who arrived at *The Economist* in 1922. The editor of that paper, once appointed, is given very wide independence. Layton used his to introduce new men and new ideas. He formed a strong group of editorial writers, many of whom later left to become editors of other papers. He added greatly to the statistical and business coverage of *The Economist*. When he left in 1938 he had restored *The Economist* as a potent influence in British public life. This work was interrupted by World War II but resumed after it by Geoffrey Crowther, whose editorship gained the paper its important international readership. Crowther, too, ended his career as Lord Crowther, and he left a successful and respected enterprise behind him.

Since World War II British financial journalism has maintained its standing. Some of the leading writers have joined the profession as very young men – and more recently women – and have made their reputations. Like their predecessors they were helped by the fact that Britain is highly centralized. In contrast to the United States and West Germany (and Italy), where the political capital is separate from the financial one, London is both the seat of government and the centre of business and finance. The bank head offices, the stock exchange, the money market, the insurance and shipping markets and many others are located, with many of their professional adjuncts, in the famous 'square mile' of the 'City'. London press, radio and television dominate the country. It is true that France, too, is centralized in Paris; but the French press has not, in the past, enjoyed sufficient independence from either its proprietors or the government to build up a reputation for financial and economic authority, though a beginning has been made.

Moreover, in Britain financial journalists tend to come from the same background as people in government. In the 1970s and 1980s several of the leading economic journalists worked for a time in a government department or in the Bank of England. At one time the Treasury organized a confidential seminar for financial journalists to explain its latest forecasting methods. Some financial journalists have become Treasury officials, while Nigel Lawson was a financial journalist for some years before he went into politics and rose to become Chancellor of the Exchequer in Margaret Thatcher's second government.

It is not always necessary to have close contacts with the administration. One of the most influential financial journalists in Britain after World War II was Harold Wincott. He did not attend university but worked as a statistician for a stock exchange firm. In 1930 he joined the *Financial News* as a sub-editor, working mostly at nights. He soon began to write in various sections of the paper and attracted attention. In 1938 he was made editor of the weekly *Investor's Chronicle*, then owned by the Financial News. That was the platform he used for a number of years to comment on the financial markets. Wincott did not trouble about government secrets. He looked at what the government was doing and found it almost lunatic. In his simple, humorous style he pulled to pieces the system of regulations and controls that had been erected to keep business in its place. His weekly commentaries moved to the centre page of the *Financial Times* and became very popular. He was one of a small group of British writers who played a powerful part in restoring a belief in the market economy among the voting public. When Wincott died suddenly in his fifties some friends, led by the economist Lionel Robbins, launched an appeal for funds to form a new foundation for the spreading of these ideas. They received many times as much money as they had expected and duly set up the Wincott Foundation, which now finances a number of research projects, lectures and publications, besides awarding a prize for the 'financial journalist of the year'.

Another writer of influence was George Schwartz, an economist who had come from Vienna to the London School of Economics where he served for many years as a lecturer. He often prepared statistical and other material for J.M. Keynes. He discovered in middle age that he had a powerful gift for writing, and after a few successful attempts he left academic work to write a weekly article on economic life as he saw it for the London *Sunday Times*. Schwartz believed profoundly in the truth of liberal economics and the rightness of allowing people to strive for their own advancement. He wrote in simple terms that millions could understand, and his views gained much influence. Among his regular readers was the Queen.

Not many professional economists have made a success in financial journalism in Britain (unlike the United States). Economists are, of course, asked from time to time to contribute a comment or forecast to a daily newspaper, and some of the many economists employed by stockbrokers are often quoted for their views on specific situations. But there are no outstanding reputations. John Maynard Keynes had a close relationship with the *Manchester Guardian*. After World War I he agreed to edit for the paper a series of supplements on 'The Reconstruction of Europe'. To each of these twelve supplements (April 1922–August 1923) he made a contribution of his own; much of this material was later incorporated in his *Treatise on Money* (1930) and the *General Theory of Employment, Interest and Money* (1936). In addition Keynes persuaded many of the leading statesmen, bankers and economists of Europe to write for the supplements. They were widely read and translated into a number of languages. Later, Keynes repeatedly launched or tried out new policy ideas in newspaper articles. The brilliance of his writing alone assured these pieces a wide readership, but one could hardly call Keynes a financial journalist.

A few newspapers and journals have been mentioned above to illustrate the curious role of financial journalism in Britain. For a complete picture one would, of course, have to mention many more. 'Financial Editors' (the common description of the chief financial writer, who is usually also in charge of the reporting staff) are as a rule persons with knowledge, experience and ideas, though not all of them have had degrees in economics. The best of them are able to talk on equal terms with bank presidents, high officials and even economists. Many talented young people have been attracted to the financial services industry (banks, investment firms, stockbrokers etc) for at least a generation, and financial journalism has had its share of the recruits. Radio and television have greatly widened the scope. Financial reports form part of the main news programmes, and there are several serious analytical or discussion programmes, mainly run by former newspaper journalists.

While the standing of British financial journalists may be unique, the work itself is, of course, being done in many other countries. In the United States the *Wall Street Journal* has probably as much influence with the US administration as the *Financial Times* has in London. The *New York Journal of Commerce* has a much smaller circulation but contains much material of the highest quality. Among general newspapers one finds thorough, intelligent business sections in the *New York Times, Washington Post, Boston Globe, Christian Science Monitor, Chicago Tribune* and *Los Angeles Times*. Others like the *Miami Herald, Dallas Times Herald, Dallas Morning News* and *St Louis Post Dispatch* might claim inclusion in the list. American magazines specializing in this field include *Fortune, Forbes* (both bi-monthly), *Business Week* (weekly) and *Barrons*. Dun's *Business Month and Financial World* might be added as well as *Financier*, a monthly journal which deals seriously with policy issues involving the business and financial community. Several television programmes have gained importance. Having written this very selective list, one is left with the fact that in the United States the journalists whose names are well known and carry a certain glamour are the political commentators, not the financial journalists.

West Germany has a long tradition of good financial journalism, which goes under the broader description 'wirtschaftlich'. Many publications and their titles have changed since World War II. The chief daily business newspaper is the *Handelsblatt*, published in Düsseldorf, and for stock market subjects the daily *Börsenzeitung*. Most influential is the business (Wirtschafts-) section of the *Frankfurter Allgemeine Zeitung* which runs to 5–6 pages. Its reputation goes back to the pre-war *Frankfurter Zeitung* and it was this famous newspaper which first issued a small book

entitled 'How to read the commercial section of a daily newspaper' known to students of journalism for generations. This has now grown to a book of 550 pages and is called *So nutzt man den Wirtschaftsteil einer Tageszeitung*, edited by Jurgen Eick.

Other business and financial information is to be found in the *Wirtschaftswoche* of Düsseldorf, the *Zeitschrift für das gesamte Kreditwesen*, Frankfurt (mainly banking and financial policy) and in a number of general newspapers including *Die Welt* and *Die Zeit*. German financial journalists in the leading positions enjoy a little of the prestige that clings to their British counterparts.

The Swiss *Neue Zürcher Zeitung* occupies a special place in Europe for the quality and responsibility of its financial and economic pages. It is not surprising that Dr Franz Aschinger who for many years edited that section went on to be economic adviser of one of the three big Swiss banks and finally professor at the St Gallen university.

In France, independent financial and economic journalism is relatively new and developing slowly. Daily financial information is supplied mainly by the French-international news agency AGEFI (Agence économique et financière) which issues four editions each weekday of 12–16 pages each. It has an able staff of reporters and strong correspondents in the main financial centres. Another daily publication is *Les Echos* which concentrates on stock market information. Among general daily newspapers *Le Figaro* has an influential finance section and a weekly supplement called 'La vie économique'. The financial section of *Le Monde* and its weekly economic report are also of good quality.

The two leading weekly journals are *La Vie Française*, which now prints 120–140 pages and covers the French business scene, particularly with descriptions of companies, personalities and regions; and *Le Nouvel Economiste*, about 100 pages, covering general business subjects. Appearing twice a month is *L'Expansion*, an impressive magazine modelled in format on the American *Fortune*. It contains serious economic analysis, articles on business personalities and corporations. After a hesitant start its circulation reached 170,000 in 1984. A monthly journal specializing in banking and monetary subjects is the revue *Banque*, published by the association of French banks. It often contains serious papers on economic problems. A financial radio programme made with professional skill forms part of the nightly 'Europe No. 1 news'.

In Japan the leading daily business newspaper is *Nippon Kezai Shimbum* which combines financial and corporation news with market reports. Two general daily newspapers have substantial and respected business sections: *Asahi Shimbum* and *Yomiuri Shimbum*. There are two weekly economic journals: *Toyo Kezai Shimposha* (sometimes described as *The Economist* of Japan) and *Weekly Diamond*, published in English and dealing mainly with the investment markets. Financial journalism is a recognized profession in Japan, though its independence and social standing is not quite the same as in the West.

There are, of course, financial publications of high quality in some countries not mentioned above: Italy, India, Singapore, Hong Kong certainly have examples. Financial journalism has become a worldwide occupation.

RICHARD FRY

See also BAGEHOT, WALTER; CROWTHER, GEOFFREY; EINZIG, PAUL; LAYTON, WALTER; SHONFELD, ANDREW AKIBA; WILSON, JAMES; WITHERS, HARTLEY.

financial markets. One of the more noteworthy developments in economics over the last twenty years or so is the emergence of equilibrium models of the financial market. Included in this term is the market for financial securities such as stocks, bonds, options and insurance contracts. The chief building block and spur in this evolution has been the economics of uncertainty, which itself is of rather recent origin. The results of this new focus and the activities and synergies it has generated is often broadly referred to as financial economics. It is within this new subfield that various models of the financial market occupy the centre stage.

After a brief summary of models based on analysis in return space in section I, this essay will focus on the two-period, pure-exchange model of the financial market beginning in section II. Conditions under which full efficiency is attained in incomplete markets will be identified in section III. Finally, section IV will trace the welfare and price effects resulting from changes in the financial market.

I. RETURN SPACE ANALYSIS

Much of the earlier work in financial equilibrium focused on pay-off returns rather than total pay-offs or consumption levels. While return space is both a natural and intuitive object of concern, and in fact continues to draw much attention, it faces certain shortcomings in addressing many questions of interest where prices, endowments and consumption pay-offs play a central role. I shall therefore provide only a brief review of the main results in return space before moving on to the consumption- and wealth-oriented models.

The so-called capital asset pricing model (CAPM) was more or less independently developed by Sharpe (1964), Lintner (1965), and Mossin (1966). It studies a single-period, frictionless, competitive market of financial securities. Assuming that (a) investors' preferences are a function of only the mean and the variance of the portfolio's anticipated return (with the mean favoured and the variance disfavoured), (b) investors have homogeneous probability assessments of returns and (c) there is a risk-free asset and that unlimited borrowing is available at the lending rate, three principal results are obtained in equilibrium:

1. The expected return on an optimal portfolio is a positive linear function of its standard deviation of return.

2. The expected return on every security (and portfolio) is a positive, linear function of its (return) covariance with the market portfolio of risky assets (the portfolio which includes x per cent of the outstanding shares of all securities in the market).

3. All optimal portfolios are comprised of the market portfolio of risky assets in conjunction with either risk-free borrowing or lending.

Since the CAPM model is consistent with the von Neumann–Morgenstern (1944) theory of rational choice only under quadratic preferences and/or normally distributed returns, it has left many economists uncomfortable. Nevertheless, it has been the basis of a very large number of empirical studies, which on balance show that the CAPM model provides a rather good first approximation of observed return structures in the financial markets of various countries.

A more recent development is the so-called arbitrage pricing theory (APT) developed by Ross (1976). It posits that security returns are generated by a linear K-factor mode (with K small) in which securities' residual risks are sufficiently independent across securities for the law of large numbers to apply. APT can therefore be viewed as an extension of the single-index

model introduced by Markowitz (1952) and developed and extended by Sharpe (1963, 1967), which in turn, of course, is closely related to the CAPM. Not surprisingly, the APT appears to offer a somewhat better fit than the CAPM or single-index model.

In studying the economics of financial markets, however, the CAPM and the APT frameworks do not offer fertile ground. In the CAPM framework, for example, the capital structures of firms are a matter of indifference. To study these and other questions, we must therefore turn to more comprehensive formulations.

II. THE BASIC MODEL

The earliest models systematically incorporating uncertainty in analysing markets were those of Allais (1953), Arrow (1953), Debreu (1959, ch. 7) and Borch (1962). They may therefore be viewed as the forerunners of more comprehensive models of the financial market, including the two-period model developed below.

Assumptions. We consider a pure-exchange economy with a single commodity which lasts for two periods under the standard assumptions. That is, at the end of period 1 the economy will be in some state s where $s = 1, \ldots, n$. There are I consumer-investors indexed by i, whose probability beliefs over the states are given by the vectors $\pi_i = (\pi_{i1}, \ldots, \pi_{in})$, where, for simplicity, $\pi_{is} > 0$, all i, s. The preferences of consumer-investor i are represented by the (conditional) functions $U_{is}(c_i, w_{is})$, where c_i is the consumption level in period 1 and w_{is} is the consumption level in period 2 if the economy is in state s at the beginning of that period. These functions are defined for

$$(c_i, w_{is}) \geqq 0 \quad \text{all } i, s \tag{1}$$

and are assumed to be increasing and strictly concave.

At the beginning of period 1 (time 0), consumer-investors allocate their resources among current consumption c_i and a portfolio chosen from a set J of securities indexed by j. Security j pays $a_{js} \geqq 0$ per share at the end of period 1 and the total number of outstanding shares is Z_j. Let z_{ij} denote the number of shares of security j purchased by investor i at time 0; his portfolio $z_i = (z_{i1}, \ldots, z_{iJ})$ then yields the pay-off

$$w_{is} = \sum_{j \in J} z_{ij} a_{js}$$

available for consumption in period 2 if state s occurs at the end of period 1. Investor endowments are denoted $(\bar{c}_i, \bar{z}_i)$ and aggregate wealth or consumption in state s is given by

$$W_s \equiv \sum_{j \in J} Z_j a_{js}, \quad \text{all } s.$$

The financial markets, as is usual, are assumed to be competitive and perfect; that is, consumer-investors perceive prices as beyond their influence, there are no transaction costs or taxes, securities and commodities are perfectly divisible, and the full proceeds from short sales (negative holdings) can be invested. The number of securities, however, need not be large (although this is not ruled out). Since our focus is on the structure of the financial market, and changes therein, production decisions (and hence the vector of aggregate consumption (C, W)) are viewed as fixed.

If the rank of matrix $A = [a_{js}]$ is full (equals n), the financial market will be called *complete*; if not, it will be called *incomplete*. The significance of a complete market is that *any* pay-off pattern $w \geqq 0$ can be obtained via some portfolio z since the system $zA = w$ will always have a solution. (In incomplete markets, in contrast, some pay-offs patterns $w \geqq 0$ are infeasible.) The simplest form of a complete market is that in which $A = I$ (the identity matrix); the financial market is now said to be composed of *Arrow–Debreu* or *primitive* securities (as opposed to *complex* securities.) The main 'advantage' of an Arrow–Debreu market is that it never requires the consumer-investor to take short positions, which is generally necessary in a complete market composed of complex securities. Finally, a financial market which contains a risk-free asset, or makes it possible to construct a risk-free portfolio, is called *zero-risk compatible*.

Under our assumptions, each consumer-investor i maximizes

$$u_i \equiv \sum_s \pi_{is} U_{is}\left(c_i, \sum_{j \in J} z_{ij} a_{js}\right) \tag{2}$$

with respect to the decision vector (c_i, z_i), subject to (1) and to the budget constraint

$$c_i P_0 + \sum_{j \in J} z_{ij} P_j = \bar{c}_i P_0 + \sum_{j \in J} \bar{z}_{ij} P_j$$

as a price-taker, where P_0 is the price of a unit of period 1 consumption and P_j is the price of security j.

Equilibria and their properties. In view of our assumptions, an equilibrium will exist but need not be unique (see e.g. Hart, 1974; note also that uniqueness is with reference to the consumption allocation (c, w), not allocation (c, z)). The equilibrium conditions for any market structure A, assuming for simplicity that the non-negativity constrains on consumption are not binding may be written

$$\sum_s \pi_{is} \frac{\partial U_{is}\left(c_i, \sum_{j \in J} z_{ij} a_{js}\right)}{\partial c_i} = \lambda_i \qquad \text{all } i \tag{3}$$

$$\sum_s \pi_{is} \frac{\partial U_{is}\left(c_i, \sum_{j \in J} z_{ij} a_{js}\right) a_{js}}{\partial w_{is}} = \lambda_i P_j \qquad \text{all } i, j \tag{4}$$

$$(c_i, z_i A) \geqslant 0 \qquad \text{all } i \tag{5}$$

$$c_i + z_i P = \bar{c}_i + \bar{z}_i P \qquad \text{all } i \tag{6}$$

$$\sum_i (c_i, z_i) = (C, Z) \tag{7}$$

where the λ_i are Lagrange multipliers, (7) represents the market clearing equations, and P_0 has been chosen as numeraire, i.e. $P_0 \equiv 1$.

Any allocation (c, z) which constitutes a solution to system (3)–(7) (along with a price vector P and a vector λ) is *allocationally efficient with respect to the market structure A*—since the marginal rates of substitution for any two *securities* are the same across individuals. When (c, z) is allocationally efficient with respect to *all* conceivable allocations, whether achieved outside the existing market or not, (c, z) will be said to be *fully allocationally efficient* (FAE).

To be more precise, define the *shadow prices* R'_{is} by

$$R'_{is} \equiv \frac{1}{\lambda_i}\left(\pi_{is} \frac{\partial U_{is}\left(c_i, \sum_{j \in J} z_{ij} a_{js}\right)}{\partial w_{is}}\right).$$

It is well known that (3)–(7) plus

$$R'_{is} = R'_{1s} \qquad \text{all } i \geqslant 2, \quad \text{all } s \tag{8}$$

is a necessary and sufficient condition for the market allocation (c, z) to be FAE because (8) insures that the marginal rates of substitution of wealth between any two *states* are the same for all investors i. (4) may now be written

$$AR'_1 = P, \qquad \text{all } i. \tag{4'}$$

Implicit Prices. The equilibrium value of a feasible second-period pay-off vector w will be denoted $V(w)$; thus if w is obtainable via portfolio z, we get $w = zA$ and hence

$$V(w) = V(zA) = zP = wR = zAR.$$

In the above expression, $R = (R_1, \ldots, R_n)$ represents the not necessarily unique set of *implicit prices* of (second-period) consumption in the various states implied by P since

$$AR = P. \tag{9}$$

By Farkas' Lemma, a positive implicit price vector is always present in the absence of arbitrage and hence in equilibrium. (Arbitrage is the opportunity to obtain either a pay-off $w \geqslant 0$, $w \neq 0$, at a cost $zP \leqslant 0$, or a pay-off $w = 0$ at a cost $zP < 0$.) In view of (4') and (9), shadow prices are always implicit prices, but a set of implicit prices need not be anyone's shadow prices.

III. FULL ALLOCATIONAL EFFICIENCY IN INCOMPLETE MARKETS

When the financial market A is complete, systems (4') and (9) have only one solution, which insures that

$$R'_i = R, \qquad \text{all } i.$$

This condition, as noted, is necessary and sufficient to attain FAE. Complete financial markets, while a useful abstraction, are not an everyday occurrence, however. Securities number at most a few thousand, while the relevant set of states is no doubt much larger. This leads us to the question: under what circumstances is FAE attained in incomplete markets? One such case is trivial and will be dismissed quickly: the case when individuals are identical in their preferences, beliefs and (the value of their) endowments. We now turn to three other sets of conditions when this occurs.

Diverse Endowments. Are there any conditions under which individuals with *diverse* endowments are as well served by a single security in the market as by many? The answer is yes; beliefs must be homogeneous and preferences e.g. of the form

$$U_{is}(c_i, w_{is}) = \begin{cases} U_i^1(c_i) + \rho_s U_i^2(w_{is}) \\ \text{or} \\ U_i^1(c_i) \rho_s U_i^2(w_{is}) \end{cases} \qquad \text{all } i, s \tag{10}$$

(with $\rho_s > 0$), where

$$U_i^2(w_{is}) = (1/\gamma) w_{is}^\gamma, \qquad \gamma < 1, \qquad \text{all } i.$$

That is, preferences for second-period consumption must be separable, isoelastic and homogeneous. Everyone's optimal portfolio is now of the form

$$z_i = k_i Z, \qquad \text{all } i,$$

where the k_i are fractions. In addition, the equilibrium implicit prices R are now unique and completely independent of the market structure A.

Linear Risk Tolerance. To attain FAE with heterogeneous second-period preferences, we need at least two securities in the market. Two-fund separation occurs in every zero-risk compatible market A under homogeneous beliefs (but arbitrary return structures) when preferences are of the form (10) if and only if

$$U_i^2(w_{is}) = \begin{cases} (1/\gamma)(\phi_i + w_{is})^\gamma & \gamma < 1, \quad \text{all } i \\ \text{or} \\ -(\phi_i - w_{is})^\gamma & \gamma > 1, \quad \phi_i \text{ large, all } i \\ \text{or} \\ -\exp\{\phi_i w_{is}\} & \phi_i < 0, \quad \text{all } i \end{cases}$$

provided none of the non-negativity constraints on consumption is binding. The optimal policies are now of the form

$$z_i = k_{i1} z' + k_{i2} z'', \quad \text{all } i,$$

where the portfolio (fund) z' is risk-free and portfolio z'' is risky (see e.g. Rubinstein, 1974). It is evident that with diverse endowments, preferences must belong to a very narrow family, even when beliefs are homogeneous, in order for FAE to be attained.

Supershares. Two states s and s' such that $W_s = W_{s'}$, i.e., with equal aggregate pay-offs, are said to belong to the same superstate t (Hakansson, 1977). If the financial market is complete with respect to the superstate partition T, FAE is attained for arbitrary endowments if and only if

$$\pi_{is}/\pi_{it} = \pi_{1s}/\pi_{1t}, \quad \text{all } s \in t, \quad \text{all } i \text{ and } t \quad (11)$$

and

$$U_{is} = U_{is'}, \quad \text{all } s \text{ and } s' \in t, \quad \text{all } i \text{ and } t. \quad (12)$$

Note that (11) and (12) require only conditionally homogeneous beliefs and that preferences are insensitive to states *within* a superstate – beliefs and preferences with respect to superstates are unrestricted.

To complete the market with respect to superstates, three simple alternatives are available (Hakansson, 1978). The first is a full set of 'supershares', each share paying \$1 if and only if a given superstate occurs (superstates are readily denominated in either nominal or real terms). The second and third alternatives are a full set of (European) call options or a full set of (European) put options on the market portfolio αZ or αW, where $0 < \alpha \leqslant 1$.

It may be noted that a market in puts and calls on a crude approximation to the United States market portfolio, namely the Standard & Poors 100 Index, was opened in 1983. These options are now the most actively traded of all option instruments.

IV. CHANGES IN THE FINANCIAL MARKET

Changes in the set of securities available in the financial market are everyday occurrences. Early studies on this subject include those of Borch (1968, Ch. 8), Ross (1976) and Litzenberger and Sosin (1977). To trace fully the effects of such changes involves comparing equilibria, which is a matter of some complexity. However, using the two-period framework of this essay, it is possible to reach some general conclusions on how changes in the market structure from A' to A'', say, affect welfare, prices and other dimensions of interest in a pure exchange setting.

The Feasible Allocations. One of the critical determinants, not surprisingly, is the change in feasible allocations. Recall that a market structure A is any 'full' set of instruments; that is, any set of instruments capable of allocating, in some fashion, aggregate wealth $W = (W_1, \ldots, W_n)$. The set of feasible second-period consumption allocations $w = (w_1, \ldots, w_I)$ obtainable via market structure A will be denoted $F(A)$, i.e.

$$F(A) \equiv \left\{ w \mid w_i \geqslant 0,\ w_i = z_i A,\ \sum_i z_{ij} = Z_j, \quad \text{all } j \right\}.$$

In comparing two market structures A' and A'' with respect to feasible allocations, there are (since holding the market portfolio αZ is always feasible) three possibilities; either

$$F(A') = F(A'') \quad (\text{Type I})$$

or

$$F(A') \subset F(A'') \ (\text{or the converse}) \quad (\text{Type II})$$

or

$$\{F(A') \cap F(A'')\} \subset F(A')$$
$$\{F(A') \cap F(A'')\} \subset F(A''). \quad (\text{Type III})$$

These three types of changes will be referred to as feasibility preserving, feasibility expanding (or reducing) and feasibility altering.

A sure way to obtain a feasibility expanding change is to make a finer and finer breakdown of *existing* instruments into an ever larger set of linearly independent (or unique) securities.

Endowment Effects. Since changes in the financial market structure are generally implemented by firms or exchanges and take place when the market is closed, such changes frequently alter investors' endowments. An example would be a merger, which results in the substitution of new securities for old. It is useful to distinguish between three cases:

1. *Strong endowment neutrality.* This occurs if the endowed consumption patterns in the two markets are unaltered, i.e. if

$$(\bar{c}_i', \bar{w}_i') = (\bar{c}_i'', \bar{w}_i''), \quad \text{all } i.$$

2. *Weak endowment neutrality.* This occurs if the values of the endowments (provided there is a common implicit equilibrium price structure R) are identical in the two markets, i.e. if

$$\bar{c}_i' + \bar{z}_i' P' = \bar{c}_i' + \bar{w}_i' R = \bar{c}_i'' + \bar{w}_i'' R = \bar{c}_i'' + \bar{z}_i'' P'', \quad \text{all } i$$

where $R > 0$ satisfies $A'R = P'$ and $A''R = P''$.

3. *Non-neutral endowment changes.* While the first two cases are rather rare, strong endowment neutrality typically accompanies non-synergistic (*pro rata*) corporate spin-offs when applicable bonds remain risk-free, as well as the opening of option markets, for example.

The Welfare Dimension. As noted, in comparing different market structures, the comparison which is ultimately relevant is that which compares allocations actually attained; that is, equilibriumn allocations. Using (2), we denote investor i's equilibrium expected utility in market structure A'' by u_i'' and his equilbrium expected utility in market structure A' by u_i'. A comparison of any given equilibrium in market A'' with some equilibrium in some other market A' must then yield one of four cases:

$$u_i'' \geqslant u_i', \text{ all } i, \quad u_i'' > u_i', \text{ some } i \ (\text{Pareto dominance}) \quad (\text{i})$$

or

$$u_i'' = u_i', \text{ all } i \quad (\text{Pareto equivalence}) \quad (\text{ii})$$

or

$$u_i'' > u_i', \text{ some } i,\ u_i'' < u_i', \text{ some } i \ (\text{Pareto redistribution}) \quad (\text{iii})$$

or

$$u_i'' \leqslant u_i', \text{ all } i, \quad u_i'' < u_i', \text{ some } i \ (\text{Pareto inferiority}). \quad (\text{iv})$$

The task at hand, then, is to identify the conditions under which each of these cases, as well as combinations of these cases, will occur. All comparisons are contemporaneous in the sense that they compare welfare under market structure A'' to what it would be if A' were in use instead.

Principal results. The principal results (Hakansson, 1982) may be summarized as follows:

(1) Feasibility preserving market structure changes yield either Pareto equivalence or redistributions. To preclude Pareto redistributions we must either have efficient endow-

ments in the first market and strong endowment neutrality, or weak endowment neutrality coupled with unique equilibria. Pareto equivalence is always accompanied by value conservation.

(2) Feasibility expanding market structure changes imply either Pareto dominance, Pareto equivalence or Pareto redistributions. To preclude redistributions we must have efficient endowments in the first market and strong endowment neutrality, or weak endowment neutrality coupled with unique equilibria. Value conservation is highly unlikely.

(3) Feasibility altering changes in the market structure have unpredictable value and welfare effects.

(4) Value and welfare effects are relatively independent.

As noted by Hart (1975), the introduction of multiple commodities or more than two periods is a non-trivial step which may bring about additional complications, such as Pareto-dominated equilibria when feasibility is expanded.

Within the limits of the single-good, two-period model under pure exchange, certain tentative general conclusions concerning common market structure changes can be stated. Even under mild heterogeneity of preferences and/or beliefs, 100 per cent non-synergistic mergers tend to be welfare reducing while (non-synergistic) spin-offs and the opening of option markets tend to be beneficial. The use of risky bonds and preferred stock tends to be virtuous as well, at least apart from bankruptcy costs. Finally, value conservation is a much rarer phenomenon than suggested by Modigliani and Miller (1958) and Nielsen (1978) among others.

NILS H. HAKANSSON

See also CAPITAL, CREDIT AND MONEY MARKETS; CAPITAL ASSET PRICING MODEL; FINANCE.

BIBLIOGRAPHY

Allais, M. 1953. L'extension des théories de l'équilibre économique général et du rendement social au cas du risque. *Econometrica* 21, April, 269–90.

Arrow, K. 1953. The role of securities in the optimal allocation of risk-bearing. *Review of Economic Studies* 31, April 1964, 91–6.

Borch, K. 1962. Equilibrium in a reinsurance market. *Econometrica* 30, July, 424–44.

Borch, K. 1968. *The Economics of Uncertainty*. Princeton, NJ: Princeton University Press.

Debreu, G. 1959. *Theory of Value*. New York: Wiley.

Hakansson, N. 1977. The superfund: efficient paths toward efficient capital markets in large and small countries. *Financial Decision Making Under Uncertainty*, ed. H. Levy and M. Sarnat, New York: Academic Press.

Hakansson, N. 1978. Welfare aspects of options and supershares. *Journal of Finance* 33(3), June, 759–76.

Hakansson, N. 1982. Changes in the financial market: welfare and price effects and the basic theorems of value conservation. *Journal of Finance* 37(4), September, 977–1004.

Hart, O. 1974. On the existence of equilibrium in a securities model. *Journal of Economic Theory* 9(3), November, 293–311.

Hart, O. 1975. On the optimality of equilibrium when the market structure is incomplete. *Journal of Economic Theory* 11(3), December, 418–43.

Lintner, J. 1965. The valuation of risk assets and the selection of risky investments in stock portfolios and capital budgets. *Review of Economics and Statistics* 47, February, 13–37.

Litzenberger, R. and Sosin, H. 1977. The theory or recapitalizations and the evidence of dual purpose funds. *Journal of Finance* 32(5), December, 1433–55.

Markowitz, H. 1952. Portfolio selection. *Journal of Finance* 7, March, 77–91.

Modigliani, F. and Miller, M. 1958. The cost of capital, corporation finance, and the theory of investment. *American Economic Review* 48, June, 261–97.

Mossin, J. 1966. Equilibrium in a capital asset market. *Econometrica* 34(4), October, 768–83.

Neumann, J. von. and Morgenstern, O. 1944. *Theory of Games and Economic Behavior*. Princeton, NJ: Princeton University Press.

Nielsen, N. 1978. On the financing and investment decisions of the firm. *Journal of Banking and Finance* 2(1), March, 79–101.

Ross, S. 1976a. Options and efficiency. *Quarterly Journal of Economics* 90(1), February, 75–89.

Ross, S. 1976b. The arbitrage theory of capital asset pricing. *Journal of Economic Theory* 13(3), December, 341–60.

Rubinstein, M. 1974. An aggregation theorem for securities markets. *Journal of Financial Economics* 1(3), September, 225–44.

Sharpe, W. 1963. A simplified model for portfolio analysis, *Management Science* 9, January, 277–93.

Sharpe, W. 1964. Capital asset prices: a theory of market equilibrium under conditions of risk. *Journal of Finance* 19, September, 425–42.

Sharpe, W. 1967. Linear programming algorithm for mutual fund portfolio selection. *Management Science*, Series A 13, March, 499–510.

financial markets, spot and forward. *See* SPOT AND FORWARD MARKETS.

Finetti, Bruno de. *See* DE FINETTI, BRUNO.

fine tuning. 'Fine Tuning' was Walter Heller's phrase for fiscal and monetary actions by government aimed at countering deviations in aggregate demand – forecast or actual – from some *target* path of output and associated inflation. The idea marked an important change in doctrine. The goal was not merely to smooth out fluctuations, but to track an output-employment/inflation path chosen from the set of attainable paths according to the preferences of the policymaker (see, however, the entry on FUNCTIONAL FINANCE).

Hyperbole aside, advocates of 'tuning' believe that (1) the economy does not adequately tune itself; and (2) we know enough about its dynamic structure – the lags and multipliers – to achieve better results than a policy unresponsive to unwanted movements in aggregate demand, e.g., a regime of fixed money growth and a 'passive' fiscal policy. (To clinch the case, one has to suppose that politicians will not mess things up – that they will not produce worse results than would a policy of 'non-tuning'.)

Both technical premises have drawn sharp attack.

IF THE ECONOMY IS 'CLASSICAL'. New Classical Macroeconomics (NCM) – much in favour during the past fifteen years among young macro theorists – teaches that, if only the macroeconomic managers would stop meddling, the economy would perform about the way the stochastic version of the perfectly competitive, instantly convergent NCM model predicts it will perform: prices and wage rates would keep all markets more or less continuously cleared, and allocation would remain in the neighbourhood of its quasi-efficient Walrasian (moving) equilibrium. If that is so – an empirical question, and not a matter of methodological aesthetics or political preference – attempts by government to manage aggregate demand are at best an irrelevance, or more likely, the principal cause of macroeconomic inefficiency. Business cycles, insofar as they do not reflect feasibly efficient adjustment to changes in endowments, technology and tastes, are caused by capricious fiscal and monetary policies. Private agents make socially erroneous decisions because they are unable to decipher the behaviour of the government.

The money managers in such an NCM economy, at least in the canonical monetarist version of the story, cannot affect *real* economic magnitudes except by acting capriciously. They control the price level and only that, and should concentrate on making it behave. The fiscal managers, in turn, should stick to the neoclassical business of making the budget conform to the preferences of the electorate with respect to income redistribution and the division of output between private use and public services, present and future. As long as the government and the central bank both behave predictably, aggregate demand, total output, and employment will take care of themselves. (The meaning of efficiency in a macro context is problematic. I use the phrase quasi-efficient to allow for some microeconomic distortions, and for the virtual nonexistence of state-contingent futures markets. Quasi-efficiency is, of course, relative to given information sets.)

IF THE ECONOMY IS KEYNESIAN. Suppose, however, that prices and nominal wage rates (or their rates of change) react to excess supply and demand only sluggishly. Real disturbances give rise to cumulative, self-multiplying quantity responses that are both inefficient and slow to dissipate. Even an anticipated nominal event, for example an increase in money supply brought about by a costless airdrop of currency, causes *real* effects. Then, *in principle*, a disturbance-responsive policy could improve matters.

Not so in practice, opponents say. The coefficients (indeed, the equations) of Keynesian models are too unreliable, and the lags are too variable and too long. As a result, an activist policy – even if free of political constraint – is more likely to do harm than good. As evidence, they cite the poor performance of the US economy during the late 1960s and 1970s. (On one extreme, NCM view, Keynesian models are no good at all. What appears to be quantitative 'structure' in such models is a mirage; it reflects not durable, exploitable regularities but behaviour that is specific to private agents' expectations of government policy. Any anticipated change in policy will cause rational agents to alter their behaviour; the coefficients will shift the way the Phillips wage-inflation/unemployment relationship shifted in response to the government's attempt during 1962–8 to exploit it. On still another view, Keynesian econometric methodology is inefficient in identifying the economy's true structure. Autoregressive methods that infer structural relations among the variables entirely from the evolving pattern of leads and lags, and make no use of prior theory, are, it is alleged, more likely to reveal robust regularities.)

Pro-activists are quick to acknowledge that Keynesian econometric regularities are approximate and impermanent, and that large shifts in policy regimes may cause them to change. But they read the evidence to say that such 'structural' change is apt to be episodic or gradual or both – that the coefficients are durable enough to be *cautiously* usable. They favour large policy actions only when the gap between aggregate demand and its target is already large, or when the odds are good that it is about to become large. Against small gaps or small disturbances, they would take only small actions or none. Even then, they say, mistakes will occur. But they emphasize how singular the structure of the economy would have to be, and how special the pattern of disturbances, to justify reliance on a 'passive' policy (e.g. trying to keep the various measures of money supply growing at constant rates, and the fiscal instruments fixed in their neoclassically warranted baseline settings).

THE 1965–81 US EVIDENCE. Opponents of an activist policy make much of the American experience between 1965 and 1981. But the lesson to be learned from that experience depends critically on whether the US economy is classical or Keynesian. If in fact the economy is Keynesian, then the 1965–81 history provides little or no support for the opponents' case.

In the United States, the acceleration of inflation during 1965–8 was caused not by an overresponsive policy, but by exactly the opposite – the government's failure to heed Keynesian pleas that it counter the excessive thrust of aggregate demand by increasing taxes and making money tight. Plausibly, also, it was that failure, and the resulting rise in the pace of inflation experienced by employers and employees, that caused the Phillips unemployment/wage-inflation regularity of 1946–65 to come unstuck (thus validating the Phelps/Friedman accelerationist prediction, though not necessarily its narrowly expectations-based rationale). That the excess demand of 1965–8 was caused by a large increase in government spending, and not by an unforeseen shift in private spending propensities, made the error of non-tuning the more egregious.

To blame activist policy for the spurts of rapid inflation during the 1970s, or for the simultaneous increase in inflation and unemployment during 1973–5 and 1979–81, is to miss a crucial implication of modern Keynesian models with their lagged-inflation augmented Phillips wage equation, and raw-material price sensitive price equation. If the recently experienced rate of inflation is unacceptably high, or if the economy is subjected to a large upward supply-price shock (such as the dramatic increase in the price of oil in 1973–4 and again during 1979) then, modern Keynesian models assert, there will not exist *any* conventional fiscal and monetary actions that would produce cheerful results with respect to both (1) output and employment and (2) inflation. The entire slate of output–employment/inflation choices faced by the Federal Reserve, and Presidents Ford, Carter and Reagan was uninviting. Lacking an effective policy of direct price and wage restraint, Ford and Carter (and the Fed) could have achieved lower rates of inflation only at the cost of still more lost output and more (transient) unemployment. Reagan and Volcker could have achieved the President's ambitious 1981 output and employment objectives only at the cost of persistently rapid inflation. (The NCM model's only explanation for the acceleration of inflation during the mid- and late 1970s is that the Federal Reserve became unhinged. A determined, well publicized policy of monetary restraint could have prevented any speed-up in inflation at virtually no cost in output and employment. That same model says that the Fed can near-costlessly stop inflation. Keynesian models assert that the cure is costly, as in fact it turned out to be during 1981–4.)

REMARKS. Trade-offs involving inflation and unemployment will plague policymakers even in an accelerationist, natural rate, lagged-inflation augmented Phillips/Keynes world, especially one beset by upward supply-price shocks. The slate of inflation-unemployment choices in such a Phelps/Friedman/Phillips/Keynes economy is more complicated than in an old-fashioned Phillips/Keynes economy of the sort that Walter Heller had in mind in the early 1960s (perhaps correctly, for the range of $\dot{P}$ actually experienced during 1958–64 – there is no way to know). But only if prices instantaneously clear all markets, and, secondarily, if expectations are entirely free of inertia and strategic interdependence – that is if the economy is NCM in its structure – will the aggregate supply curve in $\dot{P}$–Q space be vertical in what may otherwise be a long-protracted short-run. (In NCM models, only capricious, unpredictable government actions give rise to an inflation–unemployment trade-off.)

One can espouse an actively responsive policy of demand management without condoning inflation. Preferences with respect to $\dot{P}$, $\ddot{P}$, ..., Q and U bear on the choice of an aggregate demand target, not on how actively responsive the government should be in pursuing that target. There is no presumption that managers instructed to minimize inflation in a cost-effective manner would enjoy a quieter life than if they were told to favour output at the expense of faster inflation.

In a non-classical, Keynesian world, policy should aim at *both* nominal and real magnitudes, in a way that recognizes their interactions. An exclusively *nominal* strategy designed to yield a given year-to-year increase in nominal GNP (ΔPQ), no matter how it divides between increased prices (ΔP) and increased output (ΔQ) makes no sense whatever. The point is especially important if supply–price disturbances are important. *Real* targeting, if interpreted to mean that one should ignore inflation, is not acceptable either, unless one simply does not care about inflation *per se*, and about whatever microeconomic inefficiency it causes.

Theoretical considerations bearing on sensible portfolio behaviour, and evidence concerning the interest-responsiveness of the demand for money, make, I think, untenable the old monetarist claim that, even in the short run, *only* money matters – that fiscal action has no independent effect on total spending. With respect to the very long run, one has to be open-minded. The answer depends on the effect of the interest rate on the demand for wealth, i.e. on saving, and the effect of wealth on the demand for money. But that long run, equilibrium-to-equilibrium outcome seems to be of no practical significance.

The selection of a policy mix – from among the many combinations of budget settings and base-money growth compatible with one's preferred output and inflation target – should reflect the community's preferences with respect to the distribution of income and the division of output between consumption and investment, private and public. In other ways, too, policy should pay attention to supply as well as demand – how to get more output out of given capital and labour, and whether and how to upgrade and augment the former, and enhance the performance and pleasure of the latter.

Sensible managers will make tactical use of *any* intermediate indicator (e.g. free reserves, help wanted ads, Michigan surveys, whatever), as long as it exhibits sufficient short-run predictive power to improve their performance. But they will never waste degrees of freedom by treating such auxiliary aiming points as though they were objectives. They will avoid shibboleth goals like budget balance. Instruments are scarce enough, even relative to true objectives.

Because the American economy has become much more 'open', demand management in the US is more complicated than it was two decades ago. The causal interconnections are more uncertain, and instruments are scarcer relative to targets. But that is not an argument for setting the controls on 'automatic'. Rather, it strengthens the case for an eclectic, regret-minimizing activism.

FRANCIS M. BATOR

See also RATIONAL EXPECTATIONS; TARGETS AND INSTRUMENTS.

BIBLIOGRAPHY

Bator, F.M. 1982. Fiscal and monetary policy: in search of a doctrine. In *Economic Choices: Studies in Tax/Fiscal Policy*, Washington, DC: Center for National Policy.

Blinder, A.S. and Solow, R.M. 1984. Analytical foundations of fiscal policy. In *Economics of Public Finance*, Washington: Brookings Institution.

Council of Economic Advisers. 1962. Annual Report of the Council of Economic Advisers. *Economic Report of the President*. Washington, DC: US Government Printing Office.

Friedman, M. 1948. A monetary and fiscal framework for economic stability. *American Economic Review* 38, June, 245–64.

Friedman, M. 1968. The role of monetary policy. *American Economic Review* 58(1), March, 1–17.

Heller, W.W. 1967. *New Dimensions of Political Economy*. New York: Norton.

Lerner, A.P. 1941. The economic steering wheel. *University Review*, Kansas City, June, 2–8.

Lucas, R. 1976. Econometric policy evaluation: a critique. *Journal of Monetary Economics BTX Supplement, Carnegie-Rochester Conference Series 1*, 19–46.

Lucas, R. 1977. Understanding business cycles. *Journal of Monetary Economics*, Supplement, Carnegie-Rochester Conference Series 5, 7–29.

Lucas, R. 1980. Methods and problems in business cycle theory. *Journal of Money, Credit and Banking* 12(4), Pt II, November, 696–715.

Modigliani, F. 1977. The monetarist controversy, or, should we foresake stabilization policies? *American Economic Review* 67(2), March, 1–19.

Okun, A.M. 1971. Rules and roles for fiscal and monetary policy. In *Issues in Fiscal and Monetary Policy: The Eclectic Economist Views the Controversy*, ed. James J. Diamond, Chicago: DePaul University Press. Reprinted in *Economics for Policymaking, Selected Essays of Arthur M. Okun*, ed. Joseph Pechman, Cambridge, Mass.: MIT Press, 1983.

Okun, A.M. 1980. Rational-expectations-with misperceptions as a theory of the business cycle. *Journal of Money, Credit and Banking* 12(4), Pt II, November, 817–25.

Phelps, E.S. 1968. Money-wage dynamics and labor-market equilibrium. *Journal of Political Economy* 76(4), Pt II, July–August, 678–711.

Samuelson, P.A. 1951. Principles and rules of modern fiscal policy: a neo-classical reformulation. In *Money, Trade and Economic Growth: Essays in Honor of John Henry Williams*, ed. Hilda L. Waitzman, New York: Macmillan.

Samuelson, P.A. and Solow, R.M. 1960. Analytical aspects of anti-inflation policy. *American Economic Review* 50, May, 177–94.

Sargent, T.J. and Wallace, N. 1975. 'Rational' expectations, the optimal monetary instrument, and the optimal money supply rule. *Journal of Political Economy* 83(2), April, 241–54.

Sims, C. 1980. Macroeconomics and reality. *Econometrica* 48(1), January, 1–48.

Solow, R.M. 1976. Down the Phillips curve with gun and camera. In *Inflation, Trade and Taxes*, ed. David A. Belsey et al., Columbus: Ohio State University Press.

Solow, R.M. 1979. Alternative approaches to macroeconomic theory: a partial view. *Canadian Journal of Economics* 12(3), August, 339–54.

Solow, R.M. 1980. What to do (macroeconomically) when OPEC comes? In *Rational Expectations and Economic Policy*, ed. Stanley Fisher, Chicago: University of Chicago Press.

Tobin, J. 1977. How dead is Keynes? *Economic Inquiry* 15(4), October, 459–68.

Tobin, J. 1980. Are new classical models plausible enough to guide policy? *Journal of Money, Credit and Banking* 12(4), Pt II, November, 788–99.

Tobin, J. 1980. Stabilization policy ten years after. *Brookings Papers on Economic Activity* No. 1, (10th Anniversary Issue), 19–71.

Tobin, J. 1982. Steering the economy then and now. In *Economics in the Public Service*, ed. Joseph A. Pechman, New York: W.W. Norton & Co.

Tobin, J. 1985. Theoretical issues in macroeconomics. In *Issues in Contemporary Macroeconomics and Distribution*, ed. George Feiwel, New York: State University of New York.

The Annual Report of the Council of Economic Advisers. 1962. *Economic Report of the President*. Washington, DC: US Government Printing Office.

Finley, Moses (1912–1986). Sir Moses Finley had an immense influence on classical studies and particularly ancient history because he brought to them the new disciplines and techniques of the modern social sciences. He was unique among ancient historians in that his early training had been in law, economics and sociology.

Born on 20 May 1912, Finley graduated (BA) from Syracuse University at the age of 15 and from Columbia (MA) at 17, his major subjects being psychology and US Constitutional Law. Westermann encouraged him to try ancient history, and he taught himself Latin and Greek, financing himself with his earnings and those of his wife Mary, a school teacher whom he married in 1932. Theirs was a childless but devoted marriage, Lady Finley dying two days before her husband.

Finley worked from 1930 to 1933 on the *Encyclopedia of Social Sciences* and was much influenced by the Frankfurt Institute for Social Research; his reading of social theory made him left-wing and at least partly Marxist. He was active on behalf of the Republicans during the Spanish Civil War and raised funds for Russian war relief in World War II. After founding the American Committee for the Defence of International Freedom against McCarthyism he was dismissed from his post as Assistant Professor of History at Rutgers University. Known by now for his lectures in England, he was given the post of Lecturer in Classics at Cambridge in 1955, and was a Fellow of Jesus College from 1957 to 1976. He became a British subject in 1962. He succeeded to the chair of Ancient History in 1960, and in 1976 became the first Master of Darwin College. Finley's doctoral thesis, *Studies in Land and Credit in Ancient Athens* (1950), gained him an international reputation. He asked questions that had not been considered before in this field, and saw the ancient world with modern eyes. Classical scholars had used the word 'economics' in its ancient and particular sense, as the management of a household and hence of a state; Finley opened up the discipline to the interests of modern social sciences, dealing with matters such as property, contracts, succession, the value of goods and coin and the laws of war. He stepped aside from the traditional track to look at the exact relationship between masters and slaves, the nature of debt bondage, the consumer society and urban and rural production. He was the first ancient historian to tackle the methodological problems implied by the new style of social history.

Finley could appear cantankerous and was famous for his feuds; he enjoyed creating shock waves in the academic world. But at his best he was a new wind blowing through an old and rather old-fashioned subject, and he changed and refreshed the classics more than any other scholar this century.

ISABEL RAPHAEL

SELECTED WORKS

1956. *The World of Odysseus.* London: Chatto & Windus.
1963. *The Ancient Greeks.* London: Chatto & Windus.
1970. *Early Greece: The Bronze and Archaic Ages.* London: Chatto & Windus.
1973a. *Democracy Ancient and Modern.* London: Chatto & Windus. Revised edn, 1985.
1973b. *The Ancient Economy.* London: Chatto & Windus.
1980. *Ancient Slavery and Modern Ideology.* London: Chatto & Windus.

firm, theory of the. It is doubtful if there is yet general agreement among economist on the subject matter designated by the title 'theory of the firm', on, that is, the scope and purpose of the part of economics so titled. There is, probably, general agreement on the subject matter of economics itself: the allocation and distribution of scarce resources. (Some economists would have us add explicitly 'and growth' to 'allocation and distribution', but traditionally growth is subsumed under 'allocation'.) Then we may take it that the purpose of the theory of the firm is to investigate the behaviour of firms as it affects allocation and distribution. We now come immediately to a fork. An economist who believes that a 'firm' is a profit-maximizing agent (whether by conscious, rational decision or otherwise), endowed with a known and given technology, and operating subject to a well-defined market constraint, will see no need for any special theory of the firm: the theory of the firm is nothing but the file of optimizing methods (and perhaps market structures). *If* firms maximize, *how* they do it is not of great interest or at least relevance to economics. The economist's job is simply to cultivate and apply optimizing techniques. Given this view, it is unnecessary to inquire further: to seek to 'inquire within' is otiose, perhaps methodologically misguided. (As we shall see, the theory of the firm has been, and perhaps still is, the battle ground for some fierce methodological warfare.) Economists who doubt any of the three critical assumptions see an urgent need to inquire within, but diverge substantially thereafter (e.g. managerial utility functions, behaviourism). Later on, I shall try to exhibit a systematic tree, although this is not easy since some of the branches are sadly tangled. Before doing that, I want to show that the first fork, referred to above, was recognized a long time ago, and to sketch some of the history of our subject. First, though, I must impose more narrow limits on it.

In most of the work on the theory of the firm it is at least implicitly assumed that the agent whose behaviour is to be examined is a capitalist firm (which may or may not be a joint-stock corporation) engaged in manufacturing, processing or perhaps extraction. Thus the study of financial intermediaries, although they are firms, is conventionally relegated to some other branch of our discipline. Partnerships and cooperatives (labour-managed firms) may be usefully examined with the techniques of the theory of the firm, as may not-for-profit organizations, but their study is conventionally filed under 'comparative systems'. For convenience and brevity, although not out of conviction, I shall respect these conventions here. It is also necessary to place some demarcation line between the theory of the firm and 'market structure' or 'industrial organization'. For the moment, at least, I think it better to let this one be implicit.

We must also ask why firms exist at all. The classic – and neoclassical – answer was provided by Coase (1937): transactions costs. I call this a 'neoclassical' answer because part of the tradition, still embodied in much contemporary general equilibrium theory, is the assumption of constant returns to scale. Some increasingness of returns may be a very good reason for the existence of firms, or at least help to explain their size, but it is obviously vastly convenient to have a sufficient reason which is not inconsistent with constant returns. Coase suggested that the firm was an area (subset of the economy) in which allocation proceeded by direction rather than via markets, because some procedures, such as the allocation of workers to tasks, could be more cheaply done that way – coordination by command rather than by price. The word 'command' suggests that some monitoring, enforcement or internal incentive structure will be required, and indeed these matters have been receiving increasing attention. Alchian and Demsetz (1972), in particular, discussed the problem of monitoring, suggesting, in effect, that the need for it explained and justified the existence of the capitalist firm.

They posed the question of who monitors the monitor, and suggested that the incentive problem is solved if the ultimate monitor is the residual claimant. O. Williamson (1980) reviewed alternative organizational structures. He suggested that the existence of firms economizes on explicit contracts which, given uncertainty and bounded rationality, are expensive instruments. He also found that ownership and hierarchy are only weakly related.

The most recent work to emphasize the reasons for the existence of firms is Aoki's (1984). He argues that if firms exist because institutional allocation is cheaper than market allocation, reasons for which he explores thoroughly, then firms must enjoy 'institutional rent'. Furthermore, not all the resources used within the firm will have prices uniquely determined by external markets. Thus the distribution of rewards is not uniquely determined, and there is room for bargaining. Aoki argues that this is best modelled as a cooperative *game*, the players of which are the stockholders and the workers. Managers are reduced to the role of technocratic mediators (which, in view of recent developments in agency theory, discussed below, is perhaps surprising). This approach proves to be very flexible: Aoki can handle as special cases the neoclassical model (shareholders get all the residual) and the labour-managed firm in which the workers get it all (and even, with some interpretation, managerial models).

In what follows, I shall take the existence of firms for granted and return later to the matter of incentives.

The first fork, referred to above, will be familiar to any careful reader of Adam Smith (1776). He relied upon the self-interest of the butcher, the baker and the brewer to provide his dinner. The 'firms' in which he had confidence were small, owner-operated (whether single owner or partnership), without limited liability. He had serious misgivings about joint-stock companies. He pointed out what has become known in this century, thanks to Berle and Means (1933), as the 'divorce between ownership and control'. And he doubted if the managers had appropriate incentives to try to maximize the owners' returns; that is, he raised the question of what is now called 'incentive compatibility'. Thus, in considering the joint-stock company, Smith went unhesitatingly down what I will call the 'troublemaker's branch': we do have to inquire within. The joint-stock company is, of course, the predominant contemporary organization.

After Smith, there is not much that can be called 'theory of the firm' in classical economics. (Ricardo's firms are Smith's butchers and bakers.) The exception, as so often, is Marx, but there is not space to discuss Marx here. (J.S. Mill, 1848, in the famous chapter 'On the Probable Futurity of the Labouring Classes', expressed concern about both the incentive structure and morality of the capitalist form of organization, and recommended a cooperative form instead.) We must notice, however, the startlingly modern work of Cournot (1838). He wrote down a demand function and, in his famous discussion of the mineral spring, employed explicit optimizing methods (and, so far as I know, was the first to do so). Not only this, he carried out a deliberate and formal exercise in comparative statics – in 1838! In applying marginal analysis to the theory of the firm he thus thoroughly anticipated the 'marginalists'. The 'marginal revolution' in due course produced a wholly desirable unification of the theories of production, allocation and distribution, creating the neoclassical branch from the fork, but with little that could be called 'theory of the firm'. The firm was, however, central in Marshall's (1890) work, and he, characteristically, put a foot on each branch. Formal, mathematical, Marshall is strictly neoclassical, as I employ the term. The informal Marshall, concerned with growth, offered suggestive literary dynamics.

Let us consider first the more formal Marshall. His distinction between the short and long runs is essential to much of his work. This distinction is, of course, the one currently in use: in the long run all factors are variable, in the short run one at least (commonly capital) is not. This allowed him to distinguish between fixed and variable costs, and between the effects of adding more labour to a fixed-capital stock and the effects of altering the scale of operations. We now have short-run diminishing returns in industry generally, while there may be increasingness in the long run. Thus Marshall was not limited to the constant coefficients case of his classical predecessors: he was able to offer a thorough analysis of the 'Laws' of Returns. This allowed him to give a fairly complete analysis of the short-run equilibrium conditions for a firm selling in a perfect market. (There is in his analysis an even shorter 'short-run', the market period in which the price of, say, a catch of herrings is determined. This does not appear to concern us here.) Marshall did not, of course, solve all the problems of the theory of production, costs, supply and distribution in competition. He left room for the important work of Viner (1931) and Stigler (1939).

A further and vital step was Marshall's generalization of Ricardo's theory of rent. He distinguished between a quasi-rent, which would in the long run be competed away, and a true rent, which definitionally could not be. (Both, of course, are any excess of rewards over opportunity cost.) If the quasi-rent is due to an increase in the demand for the product of specific capital equipment, then the long run in which it is competed away and the long run in which all factors are variable are, of course, identical. (That the period in which quasi-rent is competed away and that in which all factors are variable may differ, is noted below.) This in turn allowed Marshall to develop the long-run equilibrium conditions for a competitive industry: quasi-rent must be competed away (or negative profit eliminated by exit) so that the normal profit condition is satisfied. Here he seems to have followed Walras (1874).

Marshall made many other contributions to the theory of the firm. He noted that, if increasingness in returns (to scale, as we should say) is internal to the firm, competition is not viable, whence a downward-sloping competitive supply curve can only be attributed to economies external to the firm (internal to the industry; but he also considered economies external to the industry and internal, perhaps, only to the whole economy). He also offered a formal monopoly model some features of which require remark. The firm's demand curve coincides with the market demand curve for the 'product' (a given primitive of analysis): there is no oligopolistic interaction here. This model is still with us, although the analysis has become more elegant. In his geometry, Marshall had us finding the profit-maximizing output by looking for the biggest profit rectangle: $(AR-AC)q$. Cournot (1838) had written down the marginal revenue function in his discussion of the mineral springs case, but Marshall chose not to follow him. (The discovery of the marginal revenue curve in Cambridge in the 1930s seems to have caused great excitement.)

The less formal Marshall was concerned with growth and the intertemporal behaviour of firms. His firms were joint-stock, but otherwise rather Smithian. He had, loosely speaking, a 'clogs to clogs in three generations' model. The first entrepreneur would be vigorous and innovative, finding some source of quasi-rent. His son would be more passive and probably mistake the quasi-rent for rent itself. The spoiled and idle grandson would certainly make this mistake, the quasi-rent would be competed away, and the cycle would be over.

This is, of course, not a good description of the history of a typical (immortal) joint-stock company. What is important is the link between innovation, quasi-rent and economic growth. Now, of course, the period in which quasi-rent is competed away is not necessarily identical to that in which capital can be varied. It may be possible to copy an innovation very quickly, or necessary to wait for the expiry of a patent. And if the quasi-rent is due to exceptional managerial talent and vigour (really, a rent to ability), it does not get competed away at all, but eventually withers. It was, however, this link between innovation and quasi-rent that Schumpeter (1934) made explicit in his great vision of the source of growth in a capitalist economy: the incessant seeking for quasi-rent via innovation, each source of quasi-rent being in turn competed away by further innovation in the process of 'creative destruction'. One notes, of course, that this model does not depend on the generational cycle of Marshall's family firm: widely owned joint-stock companies can continue to play Schumpeter's game so long as they are appropriately managed.

Marshall had the task of reconciling his view of the intertemporal behaviour of firms with his short-run profit-maximizing conditions and long-run industry equilibrium conditions. His device of the 'representative firm' appears to have been designed to do this. The representative firm would not only be in short-run profit-maximizing equilibrium but would be earning precisely normal profit when the industry as a whole was in equilibrium. This means that the definition of long-run equilibrium needs to be more carefully stated. It is not 'all firms earn normal profit'. It is rather 'there is no tendency for the total number of firms in the industry to alter; the representative firm earns normal profits but others may still be expanding or already withering; in any case the net change is zero.' Here the representative firm is implicitly defined. As Newman (1960, p. 590) put it, in his discussion of Marshall's 'statistical' concept of long-run equilibrium, 'Long-run equilibrium for Marshall meant the equality of long-run demand and supply; just that and no more.' In the 1920s and 1930s there was a considerable literature on Marshall's value theory, not discussed here (see Newman, 1960, for references). Since the work of Chamberlin (1933) and Joan Robinson (1933), the notion of the representative firm has tended to disappear from the literature. It has become usual to assume that each firm is always, by choice, in short-run equilibrium, and then to consider how Marshall's long-run competitive forces will impose industry equilibrium (normal profit for all firms simultaneously). Newman and Wolfe (1961), on the other hand, followed up the 'statistical' interpretation of Marshall's long-run equilibrium. They were not the first to apply Markov-chain analysis to the behaviour of an industry; but they were the first to integrate it with value theory. (Other more or less contemporary applications of Markov-chain analysis at most appeal to 'Gibrat's Law'. Newman and Wolfe may be thought to have prepared the ground for Nelson and Winter, 1982, discussed below.)

I shall now attempt to describe some other forks and branches of the tree. To do this it is easiest to jump to the present, since so much has happened since World War II that needs to be allocated to its appropriate branch. (Chamberlin, 1933, and Joan Robinson, 1933, had, of course, made significant extensions of Marshall's formal models before the war. These contributions are discussed elsewhere.)

We encountered above a fork between what I call the smooth neoclassical branch and the rough and troublesome 'other' branch. There is another possible basis for classifica-

tion, between optimizing and other models. The advantage of the first is that it gives the neoclassical model the prominence it deserves; the advantage of the second that it brings into prominence the importance of the assumptions we make about information and computational capacity. Perhaps somewhat arbitrarily, I shall classify the models to be considered here as optimizing and 'other'. The optimizing set of models divides again, between profit maximization and the optimization of other (usually managerial) objective functions.

Let us consider some arguments concerning the classes of models we have already identified.

The advantages of an optimizing model are clear: it is analytically tractable. We have well-developed techniques to handle it, even if the economic agents considered may not. It may also be thought to have important predictive power, but this is more dubious. The programme of qualitative comparative statics (Samuelson, 1947) has been shown to be more limited than we might have hoped. The objections to optimizing models are well known, but also debatable. They are essentially two. The first is that firms, or the human beings that manage them, cannot optimize: they have neither the information nor the computational capacity, whence the most we can have is Simon's 'bounded rationality' (Simon, 1955, 1959; see also 1979). Nelson and Winter (1982) have recently made a major contribution to this approach, discussed below. The position here is not that we give up the fundamental Smithian assumption of purposeful, self-interested behaviour (with what would we replace it?) but rather that we abandon the optimizing model and consider instead how, in a world of uncertainty, firms (managers) may explore their environment and try to 'make the best of it'. It is not suggested, at least by Nelson and Winter, that we 'inquire within' for the sake of it but rather to improve our understanding of how actual firms, seeking for profit but essentially too ignorant to optimize, may try to allocate resources. The second objection to optimizing models comes from those who have enquired within and report that firms 'just don't' (see e.g. Hall and Hitch, 1939; Andrews, 1949; Cyert and March, 1963). Many critics of this behaviourist school feel that it says little more than 'firms do what they do', and fails to analyse the relationship between the observed behaviour reported and resource allocation.

An example may show the force of the criticism. It is no longer open to doubt that firms commonly adopt mark-up pricing routines. In their study of a department store, Cyert and March (1963) report their discovery of the mark-up formula in use. They then congratulate themselves on being able to predict, given the wholesale price of an article, its posted price. They also notice that if profits are not satisfactory, the firm may adjust by altering its product-mix; that is, buying better (more expensive) or cheaper stock. But it is here that the important allocational decisions are taken, and this decision process is not analysed at all. (It should be noted that Cyert and March (1963, p. 268) place on their agenda matters which do not appear to be relevant to allocation and distribution at all, and which I accordingly exclude from consideration.)

Two related arguments in favour of profit-maximizing models may usefully be noticed now. The first is the 'biological analogy': survival of the fittest (see Alchian, 1950; Penrose, 1952; Friedman, 1953; Machlup, 1946, 1967). It is suggested that in a competitive world a firm must maximize to survive. Thus, however decisions are taken, whatever routines are adopted, firms which in fact maximize will prosper and be able, in particular, to retain and attract capital, while those that do not will wither. There are three points to raise here. The first is, how competitive is the environment? (see below).

The second is that to survive, one does not have to be perfect but only good enough to handle the competition. Indeed, Charles Darwin seems to have anticipated this misuse of his argument when he wrote,

> Natural selection tends only to make each organic being as perfect as, or slightly more perfect than, the other inhabitants of the same country with which it has to struggle for existence ... Natural selection will not produce absolute perfection ... (Darwin, 1859, pp. 201–2).

The third is that, to make effective use of the *biological* analogy, one has to offer something that can serve as a *gene*. Nelson and Winter (1982) have recently suggested a candidate (see below).

The second, and related argument, is that one can maximize without consciously trying. Thus Day and Tinney (1968) show that a firm can climb to the top of a (suitably concave) profit 'hill' by use of a simple feed-back algorithm: if an action (change in output) succeeds (increases profit), repeat it; if not, back up. The notion that one may climb the hill 'driving only by the rear-vision mirror' must certainly be attractive to those who worry about the firm's information state and computational capacity. Yet obviously this simple feed-back process works only if it converges 'fast enough' relative to the stability of the environment. Otherwise, it will be necessary to improve the algorithm to speed up convergence; for example, by adding feed-forward loops. The survival argument suggests that it will then be the firms that can do this that will survive. Then the loops (routines) are identified by Nelson and Winter as the genes in the evolutionary process. Notice, however, that this identification was made in 1982, not by those who originally proposed the biological analogy (see also Winter, 1975).

We have now distinguished between optimizing models and 'other'. We have glimpsed the next two subdivisions, that between profit-maximizing and other optimizing models, and between behaviourism and other non-optimizing models. (We shall soon find another fork on the profit-maximizing branch, too; see below.) We have also noticed some relevant argument. We may now explore some developments along each of these branches.

Developments in and since World War II, some emerging from Operations Research, have extended the scope of optimizing models at a staggering rate. In a few short years, we had Linear Programming (for economic applications, see Dorfman, Samuelson and Solow, 1958), and activity analysis (see Koopmans, 1951). Optimizing techniques were extended to inventory control (Whitin, 1953; Simon, 1952). We then had what I will call the 'dynamic explosion' as the techniques of optimal control and dynamic programming were increasingly applied to the firm's problems; see, for example, Lucas (1967) and Treadway (1969) on the flexible accelerator, Mortensen (1970) and Brechling (1975) on the demand for labour.

Another major development has been the extension of optimizing models of the firm to include considerations of risk. Risk had been explicitly considered by Knight (1921), who offered an unsurpassed account of the ways in which the institutions of the capital market facilitate risk-sharing. Knight tried to distinguish between 'risk' and 'uncertainty' in a way that many have found unsatisfactory: 'risk' was insurable; 'uncertainty', any uninsurable residual. Profit was the reward for bearing uncertainty (since risk could be covered by insurance). He was, I believe, the first to make the point that entrepreneurs would have to be less risk-averse than others (their employees) with whom they entered into explicit contracts. Recent work does not, however, follow Knight. It took a new departure from the work of von Neumann and

Morgenstern (1944); see particularly Arrow (1971), and for specific applications to the theory of the firm, see for example Sandmo (1971). The main result (Sandmo) is that the risk-averse competitive firm will produce less than a risk-neutral competitive firm or one which knew with certainty that the price was going to be equal to its expected value. Drèze (1985) has used risk as a means of introducing a more realistic model of the firm into general equilibrium theory. General equilibrium theory is beyond the scope of this essay; but we should note that he does 'inquire within' and that his approach has much in common with that of Aoki (1984).

This brings us to a fork on the profit-maximizing branch. The divorce between ownership and control is explicitly recognized and the theory of agency developed to deal with it. The divorce occurs whenever an owner (or principal) submits a risky operation in which he has an interest to an operator (or agent) whose conduct he cannot monitor costlessly. Thus the theory of agency, originally developed in the discussion of share-cropping (risk-sharing) and other forms of tenancy (see Stiglitz, 1974) has the widest application, evidently to insurance, and, of particular interest in the present context, to the interior operations of firms, not only the relationship between owners and controllers but even between managers and teams (of employees) (see particularly Ross, 1973; Jensen and Meckling, 1976; Holmstrom, 1982; Grossman and Hart, 1983). It is commonly cheaper to give the operator (be he tenant, car-driver or executive) an incentive to good behaviour than to try to monitor him. This, of course, leads to less than optimal risk-sharing (collision deductible in automobile insurance). Another incentive to good behaviour in the face of costly monitoring is suggested by Eaton and White (1983): this is to give an employee a bonus, a wage above his opportunity cost, so that, in the case that misconduct is detected, dismissal is a genuine penalty (see also Shapiro and Stiglitz, 1984). Thus both carrots and sticks have been considered. When behaviour is unobservable, incentive-compatibility may require some surprising forms of contract. Thus Holmstrom has shown that the only way to avoid the free-rider problem in a team in which effort is not observable is a contract which threatens to break the budget: deliver the target, or no member gets anything (someone else takes the full value of whatever is delivered). This raises two immediate problems. First, it may pay the 'someone else' to bribe a member of the team to shirk ('just a little'). Second, if achievement of the target depends on effort and some random variable(s), how would risk-averse members of the team dare to enter into such a contract?

Above I distinguished between two approaches to the theory of the firm, that of the maximizers and of those who wished to 'inquire within'. In agency theory we see the two converging. We are 'within', but not for its own sake; the agenda is still the allocation and distribution of scarce resources. We are forced within to deal, *inter alia*, with problems raised by Adam Smith two centuries ago, in conjunction with our own better understanding of risk.

Let us now consider other optimizing models. These depend not merely upon the divorce between ownership and control but on the idea that there is 'slack' within which the controllers may play their own game without being noticed and called to account. This in turn depends on the existence of market imperfections. The usual story has been that large firms are typically in a position to make monopoly rents, and that these rents can be foregone, used up, or ploughed back at the discretion of the controllers. This story has now become considerably more sophisticated. It is acknowledged that rents usually turn out to be quasi-rents, but suggested that the large firms (conglomerates) can, by heavy R&D expenditure, enjoy a perpetual stream of quasi-rents: while one source is being competed away, another is being developed (perhaps patented). Thus there is always some room for discretionary expenditure by the controllers. This room may in turn be limited by the perspicacity of the capital market, but it is suggested (Marris, 1964) that the power of the capital market to discipline controllers is limited by the costs of information and the fact that the supply of capital to potential take-over raiders is not infinitely elastic. Suppose, however, that capital markets were perfect. So long as the divorce between ownership and control remained, so would the problem of arranging incentive-compatible contracts for managers, whoever owned the equity.

How much scope for discretionary behaviour there actually is, then, is an empirical question to which we do not have a final answer. There is, however, no shortage of models of how managers will behave if they have the room – room to maximize their own utility functions, that is. We have Baumol (1959): maximize growth subject to a minimum profit constraint. Marris (1964) and J. Williamson (1966) offer more sophisticated versions. O.E. Williamson (1964) introduces the idea of 'expense preference'. The controllers can dissipate the rents by padding costs in ways which increase their utility. These ideas (and there are others) have obvious application to regulated industries, at least in the case in which the regulatory standard is a profit ceiling. Marris and J. Williamson both take into account the financial structure of the firm. There is now a large literature on this subject which I shall not discuss here.

(The first formal application of utility maximization to the theory of the firm was probably Scitovsky's (1943). I have not listed him above because I take him to be writing of a Smithian entrepreneur taking time off to play golf rather than following the 'divorce branch'.)

The set of 'other' models may be seen to subdivide again, between behaviourism, and something more purposeful associated with the work of Herbert Simon ('don't maximize, Simonize!'). To be sure, the firms in Cyert and March wanted to make a profit: they just do not seem to have been very good at it. Along the 'Simon branch' we have purposeful, self-interested behaviour. We may call it rational too, as long as it is understood that optimization is thought to be too difficult, and it is accordingly rational not to try. It does not follow that optimization does not occur: firms may adopt a convergent process, as in Day and Tinney (1968). In a 'sufficiently stable' environment, convergence might, of course, be quite common. But convergence must be proved rather than optimization assumed. It is thought rational for the firm to adopt routines or standard operating procedures that work at least 'well enough'. The meaning of 'innovation' is now extended. The introduction of a new routine that successfully handles a complicated decision that has to be taken with limited information is as much an innovation as a new product or an improvement in the technology. (From this point of view, a new legal or financial instrument that reduces transactions costs is an innovation too.)

It would not, I think, be a good use of space to catalogue all Simon's own innovations and suggestions. (For more recent discussion of bounded rationality, and related matters, see March, 1978.) Instead, I shall consider only the most recent contribution on this branch, the work of Nelson and Winter (1982) already referred to. These writers are much concerned with economic growth, perhaps less in static allocational problems. They inherit from Schumpeter, and Marshall, as well as Simon, and they name Cyert and March among their intellectual ancestors, as well as Alchian (1950).

Nelson and Winter argue that firms do not know the well-defined technological choice sets of standard theory. They only know how to do what they do do, and how to make at least local searches to do other things. Thus there is no sharp distinction between the choice set and the choice, and maximization is not an appropriate concept or mode of analysis. Neither is equilibrium for either firm or industry. The configuration of an industry at any time is seen as the outcome of an evolutionary process, whence the appropriate tool is a Markov process (as in Newman and Wolfe, 1961). The 'genes' required for biological analogy are the firms' routines: the standard procedures (in production, marketing, finance, etc.) that it knows how to operate. Its environment is stochastic, and the firm continually has to search for new routines (mutations). Chance enters twice. The search for a new routine may be deliberate, but its success is subject to chance. Once discovered, its application is subject to chance. Thus we have purposeful, self-interested behaviour, but success is a matter of luck. Routines are inherited, but new routines, once discovered, may also be copied by others, which allows the evolutionary process to be much faster than the biological process. There is another important point here. Nelson and Winter show that it may be more profitable to wait and to copy an innovation made by others than to incur the expenses necessary to develop it oneself. This seems to be contrary to the Schumpeterian intuition. There is also a shift in focus from the 'firm'. For Nelson and Winter the evolution of the industry is the subject of study, and the routines are the genes in the evolutionary process. The 'firm', although it is assumed to adopt purposeful, self-interested conduct (to seek profit), is not itself a matter of particular interest: it is something of a transient which happens, at any moment of time, to have inherited some routines, and may or may not succeed in developing some new, successful, ones. As in the earlier biological analogies, success will be rewarded and failure punished, but this is not advanced as an argument for 'as if' optimizing behaviour; it is part of the evolutionary process. Indeed, Nelson and Winter offer the first formal proof that, in this process, it is the profitable firms that survive. For other problems (R&D and technological change; Schumpeterian competition), they have to rely on simulation techniques which, however well handled, always leave one a little uncertain about what has been established, or, at least, at what level of generality.

It is now time to return to the question posed at the beginning of this essay: what is the scope and purpose of the theory of the firm? Indeed, is there a theory of the firm at all? Perhaps not. There is a file of optimizing models. We may include in this file the theory of agency and much recent work on information and incentives. (There are also inquiries into such organizational matters as integration and the divisional structure of large corporations, which I do not discuss here.) In the 'other' branch, profit-seeking but not optimizing, there is the recent work by Nelson and Winter, in which the focus is on the development of the industry (population), and the firm is little more than an agent (unit organism) for the transmission of genes. And there is recent work, very exciting work, exploiting the ideas of capital commitment and credible threats, much of it in the spatial literature, on the strategic behaviour of firms in small group situations. Much of this work has been associated with developments in game theory. I shall not describe it here on the possibly dubious grounds that it is better filed as 'Industrial organization' or 'theory of market structure'. Demarcation lines are not, of course, well established; it could be argued that, whenever we invoke the ubiquitous Cournot–Nash equilibrium concept, we are taking a game-theoretic approach, and some might wish to interpret theory of the firm more widely than I have done. Be that as it may, there is clearly no such thing as *a* theory of the firm. But there is a great deal in the file, subdivide it as we will, and since World War II we have seen great advances, on many different fronts, albeit differently motivated and with different methodological orientations.

G.C. ARCHIBALD

See also ADVERTISING; AVERAGE COST PRICING; BEHAVIOURAL ECONOMICS; COMPETITION AND SELECTION; CORPORATE ECONOMY; ENTREPRENEUR; ENTRY AND MARKET STRUCTURE; IDEAL OUTPUT; INCREASING RETURNS; INDUSTRIAL ORGANIZATION; MONOPOLY; OLIGOPOLY; PREDATORY PRICING; PRICE DISCRIMINATION; STRATEGIC BEHAVIOUR AND MARKET STRUCTURE.

BIBLIOGRAPHY

Alchian, A.A. 1950. Uncertainty, evolution and economic theory. *Journal of Political Economy* 58, 211–2.

Alchian, A.A. and Demsetz, H. 1972. Production, information costs and economic organization. *American Economic Review* 62, 777–95.

Andrews, P.W.S. 1949. *Manufacturing Business*. London: Macmillan.

Aoki, M. 1984. *The Co-operative Game Theory of the Firm*. Oxford: Clarendon Press.

Arrow, K.J. 1971. *Essays in the Theory of Risk-Bearing*. Chicago: Markham.

Baumol, W.J. 1959. *Business Behavior, Value and Growth*. New York: Macmillan.

Berle, A.A. and Means, G.C. 1933. *The Modern Corporation and Private Property*. New York: Macmillan.

Brechling, F.P.R. 1975. *Investment and Employment Decisions*. Manchester: Manchester University Press.

Chamberlin, E.H. 1933. *The Theory of Monopolistic Competition*. Cambridge, Mass.: Harvard University Press; London: Oxford University Press.

Coase, R.H. 1937. The nature of the firm. *Economica*, NS 4, 386–405.

Cournot, A.A. 1838. *Recherches sur les principes mathématiques de la théorie des richesses*. Paris: L. Hachette. Trans. by Nathaniel T. Bacon as *Researches into the Mathematical Principles of the Theory of Wealth*, London and New York: Macmillan, 1897.

Cyert, R.M. and March, J.G. 1963. *A Behavioral Theory of the Firm*. Englewood Cliffs, NJ: Prentice Hall.

Darwin, C. 1859. *On the Origin of Species*. London: Murray.

Day, R.H. and Tinney, E.H. 1968. How to cooperate in business without really trying: a learning model of decentralized decision making. *Journal of Political Economy* 76, 583–600.

Dorfman, R., Samuelson, P.A. and Solow, R.M. 1958. *Linear Programming and Economic Analysis*. New York: McGraw-Hill.

Drèze, J.H. 1985. Uncertainty and the firm in general equilibrium theory. *Economic Journal* 95 (Supplement), 1–20.

Eaton, B.C. and White, W.D. 1983. The economy of high wages: an agency problem. *Economica*, NS 50, 175–82.

Friedman, M. 1953. The methodology of positive economics. In M. Friedman, *Essays in Positive Economics*, Chicago: University of Chicago Press.

Grossman, S.J. and Hart, O.D. 1983. An analysis of the principal-agent problem. *Econometrica* 51, 7–46.

Grossman, S.J. and Stiglitz, J. 1980. On the impossibility of informationally efficient markets. *American Economic Review* 70(3), 393–408.

Hall, R.L. and Hitch, C.J. 1939. Price theory and business behaviour. *Oxford Economic Papers* 2, 12–45.

Hölmstrom, B. 1982. Moral hazard in teams. *Bell Journal of Economics* 13, 324–40.

Jensen, M.C. and Mecklin, W.H. 1976. Theory of the firm: managerial behaviour, agency costs and ownership structure. *Journal of Financial Economics* 3, 305–60.

Knight, F. 1921. *Risk, Uncertainty and Profit*. Boston: Houghton Mifflin.

Koopmans, T.C. (ed.) 1951. *Activity Analysis of Production and Allocation*. Cowles Commission Monograph No. 13, New York: John Wiley.

Lucas, R.E. 1967. Optimal investment policy and the flexible accelerator. *International Economic Review* 8, 78–85.

Machlup, F. 1946. Marginal analysis and empirical research. *American Economic Review* 36, 519–54.

Machlup, F. 1967. Theories of the firm; marginalist, behavioral, managerial. *American Economic Review* 57, 1–33.

March, J.G. 1978. Bounded rationality, ambiguity, and the engineering of choice. *Bell Journal of Economics* 9, 587–610.

Marris, R. 1964. *The Economic Theory of 'Managerial' Capitalism*. London: Macmillan.

Marshall, A. 1890. *Principles of Economics*. London: Macmillan.

Mill, J.S. 1848. *Principles of Political Economy, with Some of Their Applications to Social Philosophy*. London: J.W. Parker.

Mortensen, D.T. 1970. A theory of wage and employment dynamics. In *Microeconomic Foundations of Employment and Inflation Theory*, ed. Edmund S.Phelps, New York: W.W. Norton.

Nelson, R.R. and Winter, S.G. 1982. *An Evolutionary Theory of Economic Change*. Cambridge, Mass. and London: Harvard University Press.

Newman, P. 1960. The erosion of Marshall's theory of value. *Quarterly Journal of Economics* 74, 587–601.

Newman, P. and Wolfe, J.N. 1961. A model for the long-run theory of value. *Review of Economic Studies* 29, 51–61.

Penrose, E.T. 1952. Biological analogies in the theory of the firm. *American Economic Review* 42, 804–19.

Robinson, J. 1933. *The Economics of Imperfect Competition*. London: Macmillan.

Ross, S. 1973. The economic theory of agency: the principal's problem. *American Economic Review* 63, 134–9.

Samuelson, P.A. 1947. *Foundations of Economic Analysis*. Cambridge, Mass.: Harvard University Press.

Sandmo, A. 1971. On the theory of the competitive firm under price uncertainty. *American Economic Review* 61, 65–73.

Schumpeter, J.A. 1934. *The Theory of Economic Development*. Cambridge, Mass.: Harvard University Press.

Scitovsky, T. 1943. A note on profit maximization and its implications. *Review of Economic Studies* 11, 57–60.

Shapiro, C. and Stiglitz, J.E. 1984. Equilibrium unemployment as a worker discipline device. *American Economic Review* 74, 433–44.

Simon, H.A. 1952. On the application of servomechanism theory in the study of production control. *Econometrica* 20, 247–68.

Simon, H.A. 1955. A behavioral model of rational choice. *Quarterly Journal of Economics* 69, 99–118.

Simon, H.A. 1959. Theories of decision making in economics. *American Economic Review* 49, 253–83.

Simon, H.A. 1979. Rational decision making in business organizations. *American Economic Review* 69, 493–513.

Smith, A. 1776. *An Enquiry into the Nature and Causes of the Wealth of Nations*. London: W. Strahan and T. Cadell.

Stigler, G.J. 1939. Production and distribution in the short run. *Journal of Political Economy* 47, 305–27.

Stiglitz, J.E. 1974. Incentives and risk sharing in sharecropping. *Review of Economic Studies* 41, 219–55.

Treadway, A.B. 1969. On rational entrepreneurial behaviour and the demand for investment. *Review of Economic Studies* 36, 227–39.

Viner, J. 1931. Cost curves and supply curves. *Zeitschrift für Nationalökonomie* 3, 23–46.

von Neumann, J. and Morgenstern, O. 1944. *Theory of Games and Economic Behavior*. Princeton: Princeton University Press.

Walras, L. 1874. *Eléments d'économie politique pure*. Lausanne: L. Corbaz; Paris: Guillaumin; Basle: H. Georg.

Whitin, T.M. 1953. *The Theory of Inventory Management*. Princeton: Princeton University Press.

Williamson, J. 1966. Profit, growth and sales maximization. *Economica*, NS 33, 1–16.

Williamson, O.E. 1964. *The Economics of Discretionary Behavior: Managerial Objectives in a Theory of the Firm*. Englewood Cliffs, NJ: Prentice-Hall.

Williamson, O.E. 1970. *Corporate Control and Business Behavior*. Englewood Cliffs, NJ: Prentice-Hall.

Williamson, O.E. 1980. The organization of work: a comparative institutional assessment. *Journal of Economic Behaviour and Organization* 1, 5–38.

Winter, S.G. 1975. Optimization and evolution in the theory of the firm. In *Adaptive Economic Models*, ed. Richard H. Day and Theodore Groves, London and New York: Academic Press.

fiscal and monetary policies in developing countries. Economic development is a highly complex process involving not only economic but also social, political, cultural and technological changes. But here it is more narrowly defined as the process of increasing the utilization and improving the productivity of available resources, a process which stimulates the growth of national income and results in an increase in the economic welfare of the community.

A person's economic welfare can roughly be measured by his income and consumption. For individuals in different income groups it is assumed, in line with the 19th-century English Utilitarians, that the amount of economic welfare, or utility, derived from a given increment in income will be larger for those in the lower than in the higher income groups. As Marshall put it, 'A shilling is the measure of less pleasure, or satisfaction of any kind, to a rich man than to a poor one' (Marshall, 1920, p. 19). This means that in assessing the pace of development one must take into account the distribution of national income; the larger the share of the lower income groups, which constitute the bulk of the population in LDCs, in any given increment of national income, the greater the increase in the economic welfare of the community and hence the higher its rate of development.

The two most striking characteristics of LDCs, which largely account for their low per capita income, are the underutilization and the inferior productivity of their land and labour resources. These characteristics are primarily due to the inadequacy of their capital equipment in relation to the size of their labour force and to the available area of cultivable land. An additional reason is that their technology is generally backward and their labour force often lacks technical, administrative and organizational skills and suffers from poor health. Both these deficiencies can be remedied only through investment – investment in capital equipment, including infrastructure, and in human resources through education, health care and new skills.

From another point of view, LDCs can be seen as 'supply-determined', in that it is the productive capacity rather than demand that generally limits the level of their activity. This means that the long-run rate of growth in them is determined by the rate at which their productive capacity grows as a result of net investment in the economy. Leaving aside the question of the capital intensity of production, the larger the share of investment in GNP, the greater the rate of growth of productive capacity. An acceleration in the rate of growth of productive capacity and GNP will require an increase in the share of investment in GNP (Kalecki, 1976, pp. 100–103).

The role of fiscal and monetary policies and instruments in development in the non-socialist economies dealt with there is discussed under two broad headings: (I) Financing of Investment and (II) Pattern of Investment; the use that can be made of these instruments in demand management and in dealing with short-term internal and external imbalances (Eshag, 1983, pp. 41–50 and ch. 6) is not considered. The discussion will cover both the *potential* of fiscal and monetary instruments for promoting development and the *actual* use made of them by developing countries. The potential is determined by assuming that the governments of LDCs are genuinely committed and give the highest priority to the promotion of development, as defined above. This assumption is not entirely realistic: in each country there are a number of classes and income groups, often with diverging interests, exercising different degrees of influence on governments, which cannot therefore be expected to act neutrally in the interest of the economic welfare of the community as a whole. Moreover, one must take into account the relative weakness and unreliability of the administrative machinery of most LDCs, which in practice constrains their freedom of choice of policy instruments (Eshag, 1983, pp. 23–6).

I. FINANCING OF INVESTMENT

The total resources available to a country for financing domestic investment is equal to the sum of national savings and net capital receipts from abroad. The inflow of foreign capital into LDCs is determined by a number of political and economic factors in which fiscal and monetary policies play a relatively minor role; the discussion can, therefore, be confined to the use of these policies to promote savings.

Of the two sets of instruments, fiscal and monetary, it is the former that can play a significant role in determining the shares of savings and consumption in GNP. The *direct* impact of monetary measures on consumption is largely confined to sales made under hire-purchase or consumer credit schemes, which represent a relatively small proportion of total consumption in LDCs. The neoclassical assumption that a rise in the real rate of interest would have a significant positive effect on private propensity to save and vice versa is of doubtful validity (Eshag, 1983, pp. 44–6). The propensity to save is determined by a large number of factors in which the interest rate plays a relatively minor role (Keynes, 1936, ch. 8–9). Moreover, such correlation as may exist between the rate of interest and private propensity to save is as likely to be negative as positive.

Since savings equal national income *less* consumption, it follows that measures which succeed in restraining the growth of government and private consumption, without at the same time retarding the growth of production, will also raise the share of savings in national income. To ensure that the growth of production is not retarded, the authorities must be willing to offset the contractionary impact of curbing consumption by an adequate increase in public investment.

Of the various categories of *government consumption*, the one that can be reduced in most LDCs – namely, those that do not face a serious threat of external aggression – without hindering their development, is expenditure on defence. A significant part of the increment in savings resulting from a cut in defence outlay will be in the form of scarce foreign exchange resources used to import military equipment.

In a study of defence expenditure of a random sample of 24 developing countries outside the region of the Arab–Israeli conflict, it was found that the unweighted arithmetic average of the ratio of their defence expenditure to GNP between 1974 and 1978 was of the same order of magnitude as that of nine members of NATO, and some 50 per cent greater than that of four neutral European countries. This ratio was particularly high for the eight Asian countries included in the sample, which together accounted for almost two-thirds of the population of LDCs; it amounted to about 4.5 per cent, compared with a ratio of 3 per cent for NATO members and 2 per cent for the neutral countries. Moreover, in almost all LDCs, expenditure on defence was higher than on health and, in some, even larger than combined outlay on education and health. The reasons for this economically wasteful pattern of expenditure, which remains unchanged to date, include the use of military force to suppress political opposition; armament races between countries involved in territorial disputes, often of little or no economic significance; and, at times, display of sophisticated modern weapons for prestige (Eshag, 1983, pp. 81–8).

Taxation. Apart from rationing, the chief instrument for increasing savings by restraining the growth of *private consumption* is taxation. Both direct and indirect taxes have the effect of reducing the purchasing power of the private sector's real disposable income and consumption. Because private consumption accounts for a considerably larger share of GNP than investment, any reduction in its rate of growth is calculated to raise the pace of expansion of savings and investment by a much higher percentage; this explains the important role generally assigned to taxation in development.

To ensure its effectiveness in promoting development, a taxation system should possess certain basic characteristics. First, the system should ensure that the burden of taxation is primarily borne by the higher income groups; the higher the per capita income of a group, the larger its contribution to tax revenue as a proportion of its income should be. This requires the implementation of a progressive system of direct taxation on income and wealth. It also implies that indirect taxes levied for revenue purposes, should be imposed mainly on 'luxuries' – namely, goods and services largely consumed by the higher income groups – rather than on those consumed by the bulk of population, 'necessities'.

Second, taxation measures should, whenever possible, be so devised as to stimulate production and, in any case, should not significantly reduce material incentives. For this reason, in two important sectors in LDCs, agriculture and small businesses, a system of progressive lump-sum taxation is preferable to taxation of production, income or profits. In neither of these two sectors is it possible in practice to enforce a dependable system of bookkeeping which could be used for the assessment of the taxpayers' income even with an efficient and reliable fiscal machinery, which is rarely to be found among LDCs. A system of 'taxation by area', under which tax rates per acre are fixed for different regions of a country on the basis of *potential* land yields and are graduated according to the taxpayers' aggregate landholding, has the advantage of being simple and is also calculated to stimulate production. Similar advantages can be derived in the taxation of small businesses from a system of 'licence fees', under which the fees are varied according to some concrete and well-defined criteria, such as the location of a business and its size (Eshag, 1983, pp. 108–112).

Third, the taxation system should be simple and readily understood by both collectors and payers of taxes, even if simplicity is achieved at the cost of some inequity in the distribution of the tax burden.

Fourth, to permit a faster growth in investment than in consumption, tax revenue should be income-elastic. In addition to implementing a progressive system of direct taxation, lump-sum taxes, such as those proposed for agriculture and small businesses, should be raised periodically in line with inflation and growth of potential land yields.

Fifth, to ensure that total tax revenue is not subject to violent fluctuations, tax sources should be diversified. The need for stability in tax revenue is a further reason for taxation of the agricultural sector along the lines suggested, rather than through export taxes or commodity boards (Eshag, 1983, pp. 104–6).

Very few developing countries have to date made adequate

use of taxation policy to restrain the growth of private consumption. 'Tax ratio', which measures the ratio of tax revenue to GNP, provides a rough indication of governments' efforts to restrain private consumption and of their success in doing so. In a study covering a random sample of 27 LDCs, is was found that in the first half of the 1970s the unweighted arithmetic mean of their tax ratio was about 14 per cent, of which less than one-third was received from direct taxes. This compared with a tax ratio of almost 40 per cent, in which about two-thirds represented direct tax revenue, for a sample of thirteen industrial countries (Eshag, 1983, pp. 92–7).

The pronounced difference between the tax ratios of LDCs and developed economies, which persists to date, can be explained only partly by the lower per capita income of the former. The large inequality of income distribution in LDCs suggests that their taxation potential is significantly higher than is indicated by their per capita income. That per capita income by itself does not provide a sufficient explanation of the low tax ratios in LDCs is clearly demonstrated by the fact that the average tax ratio among the nine African countries was appreciably higher than that for the eight Latin American countries included in the above sample, although the average per capita income of the former was about 60 per cent lower than the latter. The available evidence shows that the unwillingness and inability of governments in LDCs to tax the richer strata of the community, who exercise a substantial influence on the formulation and implementation of tax measures, goes a long way to explain their low tax ratios. The influence of big landlords, for example, is largely responsible for the use of out-of-date land and rental valuations as a basis of tax assessments and thus for allowing inflation to erode the real value of land taxes in many countries.

II. PATTERN OF INVESTMENT

Market forces would inevitably tend to pull a large proportion of investment resources to 'inessential industries', namely those that cater to the relatively strong demand of higher income groups. This would stimulate the production of luxuries at the expense of necessities, and produce a 'lop-sided' development. Selective fiscal and monetary measures can be used to influence the allocation of resources in a way which discourages this type of development. Such measures should be directed at (a) promoting 'essential industries', which produce necessities, in order to prevent a rise in the price of wage goods, which generally results in the redistribution of income in favour of higher income group, and (b) ensuring a balanced sectoral and regional growth, so as to reduce production bottlenecks and regional inequalities of income.

It is clear that the formulation of a meaningful and coherent investment policy can only take place within the framework of a development plan. As a minimum, the plan should provide a rough outline of projected movements in the volume and pattern of production and demand as well as of requirements for productive capacities, including labour and raw materials inputs.

In most LDCs, the public sector accounts for a significant proportion (varying between about 30 and 50 per cent) of total investment. The pattern of this investment is determined by governments themselves and does not require the employment of fiscal and monetary instruments whose efficacy in influencing the allocation of resources can rarely be predicted accurately. Since the value of public investment is equal to the government's savings plus its net borrowing, it follows that, *ceteris paribus*, a rise in tax ratio would increase the share of public investment in the total. Thus, taxation, apart from raising the share of investment in GNP by restraining consumption, contributes to the efficacy of the policies concerned with regulating the pattern of investment.

Fiscal instruments. Historically, *import tariffs* have been the most popular fiscal instrument used to promote industrial development. Because the protection afforded to domestic industries by tariffs cannot be accurately estimated, they have at times been reinforced by *import quotas*. Although these protectionist policies have played an important role in encouraging import substitution in LDCs since the Great Depression of the 1930s, the pattern of industrialization induced by them has in most countries failed adequately to conform to their development needs. This is largely explained by the absence of an investment licensing system that could effectively discourage the flow of resources into inessential industries; these have thrived, often with the aid of foreign investment, behind the protective barriers of tariffs and quotas. The reason for this is to be found partly in the weakness of the administrative machinery responsible for implementing investment regulations and partly in the political influence exercised by the richer classes in the formulation of such regulations; what is a 'luxury' for the poor is often regarded as a 'necessity' by the rich.

Tax concessions and *multiple exchange rates* are two other instruments employed to influence the allocation of resources, although these, like tariffs and quotas, have in practice largely served to foster industrialization in general rather than to direct resources into essential industries. Tax concessions are provided in the form of temporary exemption from profit taxes (tax holidays) for certain new industries, and of tariff concessions on industrial inputs. Under the multiple exchange rates system differential exchange rates are applied to foreign-exchange transactions to encourage the production and export of industrial products.

Apart from credit subsidies, discussed below, a number of other *subsidies* are often used to encourage the export of manufactures and to stimulate agricultural production. The latter subsidies usually take the form of the provision of agricultural inputs to farmers at subsidized prices and/or of guaranteed price schemes under which the prices paid to farmers exceed world prices.

Monetary instruments. Although monetary instruments play no significant role in directly increasing savings and investment, they can be used to influence the pattern of investment. In theory, this can be done through the operation of special *development banks* which are charged with the provision of cheap, or subsidized, credit to selected industrial establishments and to the agricultural sector. According to the OECD, there were about 340 such banks operating in some 80 developing countries in the mid-1960s; indications are that their number has significantly increased since that date. Over half the banks were state-owned and funded by the exchequer; the remainder had a mixed ownership or were private. Mixed and private banks are given governmental subsidies to enable them to earn a normal rate of profit (Eshag, 1983, pp. 186–92).

In addition to advancing loans to selected private enterprises, *industrial development banks* also help to promote new industrial ventures, directly or in partnership with private firms, and provide technical assistance to their clients. Formally, these banks are supposed to attach a greater weight to the developmental implications of their investments than to

profitability, although in practice this is generally true only of some state-owned banks (Eshag, 1983, pp. 193–6). On the whole, the primary function of the banks, like that of the fiscal instruments discussed above, has been to promote industrialization, without much discrimination between essential and inessential industries.

Officially, the principal function of the *agricultural development banks*, most of which are state-funded, is to advance credit, at subsidized interest rates, to the agricultural sector, in particular to 'small farmers', namely those who cultivate relatively small plots of land with little fixed capital and with backward technology and who are, in consequence, very poor. Owing to their low credit rating, small farmers have virtually no access to commercial banks and have to meet the bulk of their credit requirements by borrowing from non-institutional sources, notably professional moneylenders, landlords and merchant middlemen, at rates of interest which far exceed those charged by commercial banks. The subsidized credit provided by development banks can thus, in theory, contribute to development in two important ways: (a) to improve the lot of small farmers by reducing the cost of their borrowing; and (b) to stimulate agricultural production by financing the modernization of cultivation techniques, especially among small farmers.

The operation of agricultural development banks during the early 1970s has been studied by the World Bank, FAO and many agricultural economists (Eshag, 1983, pp. 196–203, footnotes). Almost all these studies indicate that very few countries have in practice made full use of the developmental potential of these banks. It is estimated that in most countries less than 30 per cent of subsidized credit was allocated to small farmers, the remainder being appropriated by medium and large landlords. According to the World Bank, between 70 and 80 per cent of small farms had commonly no access to institutional credit. Indications are that to date no significant change has taken place in the above picture of credit distribution, which has contributed to a growing inequality of income and wealth between the rich and poor farmers and has hindered agricultural production. In the opinion of most writers, including the FAO and World Bank, the primary explanation for this is to be found in the political and social influence of big landlords, which enables them to appropriate the bulk of subsidized credit (Eshag, 1983, pp. 203–7).

CONCLUSION. Three broad conclusions emerge: (a) fiscal and monetary policies have the potential to make a significant contribution to development; (b) LDCs have so far failed adequately to exploit this potential; and (c) the principal cause of this failure has been the institutional obstacles to development, largely of socio-political nature, of which the system of land tenure deserves a special mention.

ÉPRIME ESHAG

See also CURRENCY BOARDS; EXTERNAL DEBT; INTERNATIONAL INDEBTEDNESS; TERMS OF TRADE AND ECONOMIC DEVELOPMENT.

BIBLIOGRAPHY
Eshag, E. 1983. *Fiscal and Monetary Policies and Problems in Developing Countries.* Cambridge: Cambridge University Press.
Kalecki, M. 1976. *Essays on Developing Economies.* Brighton: Harvester Press.
Keynes, J.M. 1936. *The General Theory of Employment Interest and Money.* London: Macmillan.
Marshall, A. 1920. *Principles of Economics.* 8th edn. London: Macmillan, 1964.

fiscal federalism. As an offshoot of public finance theory, fiscal federalism theory analyses the special fiscal problems which arise in federal countries, drawing on the theory of public goods, taxation and public debt incidence, public choice theories of the political process and various aspects of locational theory. More specifically, fiscal federalism theory attempts to supply answers to the following questions: reasons for adopting a federal structure, rules for the assignment of activities and revenue sources to various levels of government, the efficiency properties of free and unrestricted migration from one jurisdiction to another, and the role for inter-governmental revenue transfers and their most desirable forms in a federal structure (see Grewal et al., 1980, pp. xi–xii). With its policy orientation towards practical fiscal problems in federations, it is not surprising that this branch of public finance literature has flourished particularly in the North American federations of Canada and the United States as well as in federal countries like Australia and West Germany. Fiscal federalism theory is also clearly of relevance to less well defined federal structures like Austria, India, China, Nigeria, Spain, Switzerland and Yugoslavia. However, as King (1984) is at pains to point out, the theory can also be applied to questions about the design and operation of systems of multi-level government in non-federations like the United Kingdom because it sheds light on the public finance of local government and devolution of powers to components of the union like Scotland and Wales.

In an important sense, fiscal federalism theory can be seen as an argument about the costs and benefits of decentralization in government. Decentralization allows better matching of public goods supply to local tastes, and is said to increase social welfare for general consumer surplus reasons as well as from lower organizational and signalling costs (see Breton and Scott, 1978). Economies of scale considerations act as the major constraint on decentralized public goods supply. Following Musgrave's (1959) classification of budget operations, the theory suggests that decentralization generally increases efficiency in the allocational branch of the budget process but that scope for decentralization is either virtually non-existent, or at best, rather limited, in the work of both the distribution and stabilization branches. The possibility of externalities and inter-jurisdictional spill-overs when taken into account makes application of the theory more difficult. Similar principles have been used to explain tax assignment rules, but as McLure (1983, pp. xii–xiv) points out, assessments of the tax assignment problem based thereon, 'do not all conveniently point in identical directions' while the normative rules in many cases do not conform at all with actual practice. An exception is the almost universal assignment of property or real estate taxes to local government appropriate from the perspective of the theory since such bases have low mobility. Keeping in mind the distributional and stabilization objectives of many tax forms, the rules also tend to assign a substantial proportion of the major taxes to central government.

Migration and decentralization aspects of fiscal federalism theory have played an important role in providing an escape route from the pessimistic conclusion that due to joint-supply, non-excludability and non-revelation of preferences characteristics of public goods, the market would not generate an allocative optimum in such cases. Tiebout (1956) tried to solve this theoretical dilemma by contending that decentralized or local goods supply provided the solution since analogous to the 'private market's shopping trip' this allowed individuals to reveal their preferences for particular public goods packages by voting with their feet. Although it abstracted imperfectly

from institutional rigidities, Tiebout (1956, p. 68) claimed this solution to be the 'best that can be obtained given preferences and resource endowments'. Largely because of its restrictive assumptions, much of Tiebout's theory was subsequently criticized but its emphasis on the efficiency consequences of inter-jurisdictional migration remains an important feature of fiscal federalism theory and has been the source of substantial further literature in this area.

Fiscal federalism theory has also been concerned with the theory of grants. Undersupply of public goods in a sub-national jurisdiction due to spill-overs of its benefits to non-residents provides one reason for national government grants for specific purposes; imposition of grantor preferences provides another. Where function assignments produce disparities between revenue raising abilities and expenditure requirements, general revenue grants may be required to reduce such vertical fiscal imbalance, a problem not confined to federations but equally prominent, for example, in the substantial transfers to local government in the United Kingdom. Finally, equalization transfers may be necessary to equalize either fiscal potential or fiscal capacity when for demographic, geographic and resource endowment reasons, fiscal capacities between sub-national jurisdictions differ substantially. Prevalence of inter-governmental transfers in these various dimensions has produced a substantial fiscal federalism literature on the consequences of particular grants and hence rules for the best grant policy to suit particular circumstances (see King, 1984).

The major departures from the rules developed by the theory in actual federations give fiscal federalism theory an artificiality as an explanatory device, perhaps largely because its dominant North American orientation gives much of its findings a flavour of ex-post rationalization of American experience. Even then, the strong, but not exclusive, choice-theoretic emphasis of this literature combines uneasily with the political history which the Musgraves (1973, p. 502) see as explaining most of the actual 'structure of federal arrangements in any one country'. Despite this major shortcoming, the theory has nevertheless shed much light on issues associated with the eminently practical and important topic of decentralization of government.

PETER GROENEWEGEN

See also LOCAL PUBLIC FINANCE; REGIONAL DEVELOPMENT; REGIONAL ECONOMICS; TIEBOUT HYPOTHESIS.

BIBLIOGRAPHY
Bretton, A. and Scott, A. 1978. *The Economic Constitution of Federal States.* Canberra: Australian National University Press.
Grewal, B.S., Brennan, G.H. and Mathews, R.L. (eds) 1980. *The Economics of Federalism.* Canberra: Australian National University Press.
King, D. 1984. *Fiscal Tiers. The Economics of Multi-Level Government.* London: Allen & Unwin.
McLure, C.E. (ed.) 1983. *Tax Assignment in Federal Countries.* Canberra: Centre for Research on Federal Financial Relations and International Seminar in Public Economics.
Musgrave, R.A. 1959. *The Theory of Public Finance.* New York: McGraw-Hill.
Musgrave, R.A. and Musgrave, P.B. 1973. *Public Finance in Theory and Practice.* New York: McGraw- Hill. 4th edn, 1984.
Tiebout, C.M. 1956. A pure theory of local expenditures. *Journal of Political Economy* 64, 416–24. Reprinted in Grewal et al. (1980).

fiscal policy. *See* BUDGETARY POLICY; DEFICIT FINANCING; FUNCTIONAL FINANCE; PUBLIC FINANCE; STABILIZATION POLICY.

fiscal stance. Fiscal stance is commonly understood to denote the expansionary or contractionary implications for the economy of a government's budgetary policy. More precisely, it represents an attempt to summarize, in a single measure, the combined effect on aggregate demand, and therefore potentially on real output and income, of all the various decisions taken by government in respect of public expenditure, taxation and other sources of revenue which go to make up a national budget. As such it presupposes not only that governments can affect demand in this way, but also that it is possible to devise an indicator of this kind which is sufficiently widely accepted as to be useful. This later proposition has been questioned even by a number of self-professed Keynesian economists, who have emphasized the difficulty both of aggregating the effects of the many different items included in the budget and of disentangling these from other potential influences on demand.

The most straightforward way of producing an indicator of budgetary policy is to sum the inflows of revenue and outflows of expenditure to which they give rise and to take the difference between the two, the budget balance, as a measure of fiscal stance. Indeed any assessment of a government's macroeconomic policy commonly tends to focus on this magnitude. The difficulty with this figure, as was recognized almost as soon as proposals were first made to use the budget as a tool of economic management, is that 'it fails to distinguish the budget's influence on the economy from the economy's influence on the budget' (Okun and Teeters, pp. 77–8). In other words, it incorporates both the consequences for budgetary flows of the tax rates set by government and the public expenditure outlays authorized by it from the effects on such flows of changes in income and expenditure in the economy. If there is a downturn in economic activity and income and expenditure grow less rapidly than usual, or even contract, then the revenue produced by a given set of tax rates will be correspondingly depressed and public expenditure will tend to be pushed up insofar as unemployment increases and the financial position of state (or publically supported) enterprises deteriorates (and *vice versa* if there is an upturn in activity). The problem is to separate these autonomous consequences from the discretionary effects of policy and thereby to distinguish the injection or withdrawal of purchasing power emanating from policy decisions from other sources of demand expansion or contraction, such as private sector borrowing or net export growth.

The solution, first devised in the 1940s, is to calculate budgetary flows at full employment levels of income and expenditure and to take the budget balance which would have resulted had GDP (or GNP) continuously followed such a growth path as the measures of fiscal stance. Indeed the term 'fiscal stance' has become synonomous with figures for the budget balance adjusted or normalized in this way. (The first estimate of such an adjusted balance seems to have been made by Kaldor, 1944, for the UK for the year 1938. The concept was first proposed in the USA by the Committee of Economic Development in 1947, though the most influential was probably Brown, 1956. On the US origins of the concept, see Blinder and Solow, 1974.)

Such calculations served a dual purpose. They were used not only to indicate the expansionary or contractionary nature of budgetary policy, but also to reveal whether and to what extent the current stance of policy could be sustained in the longer term as full employment was approached. Accordingly, in a number of countries, the United States, West Germany and the Netherlands in particular, the 'full employment surplus' or 'structural budget balance' became widely accepted as a useful benchmark for assessing policy. (In the USA, it was included in the Annual Report of the Council of Economic Advisors and in the Reports of the Joint Economic Committee of Congress as well as in the Budget documents. In the Netherlands, it was included in the Budget Memorandum – see Netherlands Ministry of Finance, 1970 and Budget, 1978. In Germany, it was also used to assess policy as described in Dernberg, 1975; and Chand, 1976. For a discussion of alternative measures, see Lotz, 1971.)

As time went on, as full employment in most countries became more remote, the full employment budget concept became less meaningful as a benchmark for policy, increasingly directed at objectives other than managing demand to secure particular rates of economic growth. By then, moreover, criticism of the concept as an indicator of fiscal stance was already widespread. Among the most frequently voiced concerns were: that the measure was not independent of the level of economic activity taken as the basis for normalization; that it was affected by the composition of the budget as well as by the difference between expenditure and revenue flows; that it made no allowance for the effect of inflation; that it ignored how the budget deficit was financed and more generally what kind of monetary policy was being followed; and, more recently, that it took not account of expectations and their influence on the effect of policy on the economy.

All of these criticisms are valid in some degree. The key issue, however, concerns the degree of validity and how far it is possible to modify the measure of fiscal stance to take account of them, without making it so complicated and so model-dependent that it ceases to be widely accepted as a satisfactory indicator of policy.

Thus while the level of economic activity chosen as the benchmark for standardizing the budget clearly affects the absolute value of the figures calculated, it tends to have much less effect on changes in the balance over time (see Ward and Neild, 1978, pp. 33–7). Since fiscal stance can only be interpreted meaningfully in a comparative sense, in relation to policy in different periods, it is movements in the balance which are the relevant consideration. Nevertheless the composition of domestic income and expenditure, and therefore potentially the tax and public spending flows generated, does tend to vary as activity changes. This source of difficulty, however, can readily be minimized by choosing a benchmark level of activity which is not too different from the actual level – or even, to go one step further, by changing the benchmark each year to coincide with the actual level, a picture of the changing fiscal stance being built up by cumulating successive year-to-year movements. A further problem is that there may be disagreements about the rate of growth required to ensure that activity remains constant. These disagreements, however, are not usually so great as to give rise to marked divergences in estimates of fiscal stance, except when calculated over a number of years at a time. In this case, all that is possible is to produce a range of estimates, with the range of growth rates on which they are based made explicit.

A potentially more serious problem arises from the likelihood that different components of the budget have different effects on demand, so making simple aggregation of the revenue and expenditure flows involved inappropriate and possibly misleading. This has led many (Blinder and Solow, 1974, among others) to propose that the budget components should be weighted according to the extent to which they feed into consumption or investment expenditure rather than into savings or, at one stage removed, imports (so explicitly allowing, *inter alia*, for the possibility of a balanced budget multiplier). The difficulty with such proposals is not only the increased complexity of the calculation and the greater scope for disagreement over any which is produced, but also their focus on the initial demand effects rather than on the longer-term consequences. Thus it is clearly unrealistic to suppose that 'first round' leakages from the circular process of income and expenditure determination are in some way lost for ever. Variations in income stemming from budgetary changes may not immediately feed into spending, but to major extent they ultimately will do so unless there is a permanent change in the desire of individuals and companies to increase or reduce their net holdings of financial assets. Over the long term, therefore, there tends to be a relatively stable relationship between the private sector financial balance, or savings less investment, which is the relevant concept in this context, and private sector income. For different forms of expenditure and taxation, the speed at which spending responds may well vary but there may be little significant difference in the long-term effect.

This means that any measure of fiscal stance has to specify the period over which the effect of policy is being estimated. The shorter the period, the more important are differential leakages as between budget items likely to be, the more do underlying economic circumstances come into play and the more complicated and uncertain is the process of estimation. Indeed without a fully fledged macroeconomic model with built-in behavioural functions and sufficient disaggregation of taxation and public expenditure, it is hard to see how any satisfactory estimate of short-term budgetary effects could be constructed. Such an estimate, however, is really a measure of fiscal impact rather than of fiscal stance. (Estimates of a demand-weighted measure of fiscal impact are, for example, regularly published by the UK National Institute of Economic and Social Research in its quarterly review.) The longer the period, the less important does weighting become. If the concern is to assess the cumulative effects of policy over a time horizon of a year or two, then differential leakages into net holdings of financial assets, i.e. savings less investment, ought not to be a significant problem for most items and an unweighted measure is unlikely to give misleading results.

Nevertheless, there are certain budgetary items, though usually relatively minor in scale, for which even the long-term effect on demand is likely to be small. These are lending, asset sales or purchases and other purely financial transactions which are included in total public sector borrowing (in the US partly in the Credit Budget) but not the public sector financial balance (though this is typically not true of purchases less sales of land and existing buildings) and which tend to affect income and wealth only marginally. The most sensible and straightforward course of action is to exclude these from a measure of fiscal stance.

On the other hand, differential leakages into imports may be more of a problem. There is usually a general tendency for public expenditure to involve a lower import content than private spending, at least at the first round, and a measure of fiscal stance not adjusted for this might therefore misrepresent the scale of long-term demand effects if policy is heavily concentrated on, say, expanding public expenditure or reducing taxes.

The argument for adjusting measures of fiscal stance for inflation has two aspects. The first, and least serious, is that variations in inflation can affect tax revenue differently from public expenditure outlays, insofar as the two sides of the account are indexed to differing degrees. In most advanced economies, government revenue tends to increase in propor-

tion to nominal income, or more than in proportion where the tax structure is progressive, while public expenditure sometimes lags behind inflation because of spending authorizations, or budget allocations, being specified in cash terms. In such circumstances, the budget deficit would be reduced if inflation were to increase without any overt action on the part of government. Though perhaps not intentional, a change of this kind ought to be treated as a tightening of fiscal stance in the same way as a deliberate increase in nominal tax rates which produced the same effect. To do otherwise would be to confuse action with intent and to regard inaction as signifying no change in policy even though it might be associated with significant changes in *effective* rates of taxation. This is accomplished by taking nominal changes in revenue and public expenditure in relation to nominal national income both measured in terms of actual prices, as the appropriate basis for calculating fiscal stance. Any change in this measure can then be interpreted as indicating a discretionary change in budgetary policy irrespective of its origins.

The more substantive aspect is that inflation can affect the real value of government debt in the economy, and therefore the real wealth of holders of government securities and presumably in turn their expenditure behaviour (see Tobin and Buiter, 1976; Taylor and Threadgold, 1979; and Tanzi, 1984). If, for example, such holders are not compensated for the erosion in their wealth caused by an increase in inflation relative to interest rates on the debt, then expenditure will tend to be depressed insofar as it is a function of wealth as well as current income. Conversely if interest rates lag behind prices when inflation falls, then debt holders will enjoy an increase in their real wealth which may tend to boost demand, in the longer term if not immediately. To ignore these effects is liable to give a misleading indication of fiscal stance. The cyclically adjusted budget balance ought, therefore, to be further adjusted to allow for the impact of inflation on the real value of outstanding government debt (see Price and Muller, 1984, and OECD, 1984). The difficulty is that the inflation rate which is relevant in this context is the expected future rate rather than the present rate. Since the former is unknown, there seems little practical alternative but to use the latter even though it is less than satisfactory (but see, e.g., Buiter, 1983).

The expansionary effect of higher real interest rates on spending by holders of government debt is liable to be offset by a depressing effect on investment and consumption of durable goods from the higher cost of borrowing. The question arises as to how far this and other financial effects resulting from the monetary policy being followed by the government at the time should be taken into account in the measurement of fiscal stance. In principle, it can be argued that fiscal and monetary policy should be kept separate and the effects of the two on the economy estimated individually. In practice it is not quite so simple. Even though governments have some discretion over how to finance a budget deficit – whether by expanding the money supply or selling public sector debt to the non-bank private sector and abroad – the reaction of financial markets cannot simply be ignored and in reality the two strands of policy will be considered together.

Moreover, the possible financial and wider consequences of fiscal action, both internally and externally, might themselves affect the way that demand responds to such action. For example, in a world of floating exchange rates, the exchange value of a country's currency might itself be partly determined by the fiscal policy being followed, so that a larger budget deficit might lead to a fall in the exchange rate (or possibly a rise if interest rates are expected to go up) and a stimulus to demand from net exports as well as from fiscal policy directly.

Alternatively, anticipations about the way a government might respond to prevent such a fall through modifying its monetary policy (by raising interest rates for example, or tightening credit) might itself influence the speed and scale of the internal demand response to the fiscal measures introduced.

More generally, expectations about future developments and the close relationship between the budget and other aspects of policy and other sources of demand generation represent potentially serious problems for measuring fiscal stance in any simple, straightforward manner. Thus in addition to any effects on interest rates and exchange rates, a decision to increase the budget deficit – in the present, for example – might be taken to imply a need for higher taxes in the future to service the additional debt created and hence might generate little increase in demand, to the extent that expenditure is determined by expected income over the long-run rather than current income. The argument, in its extreme version, is that reactions to the expected consequences for future public expenditure, taxation and the budget balance of present budgetary decisions are liable to frustrate the expansionary or contractionary intentions of government more or less completely. In a highly uncertain world, however, it is hardly plausible that such anticipations would fully offset attempts by government to manage demand, though it is not implausible that they might modify the effects of policy in some degree. Nor is it implausible that the degree of influence might vary according to what else is happening in the economy at the time.

In view of these considerations, it is futile to hope that any simple, easily constructed measure of fiscal stance is likely to capture fully the effects of budgetary policy at all moments in time. The question which remains is whether the only resort is to macroeconomic models which are sufficiently detailed and reliable to enable the effects of any particular package of fiscal measures to be isolated from other influences on demand (as advocated, for example, by Buiter, 1985). But in this case, the purpose of such an exercise would be unclear since the main concern is presumably with the combined effect of government policy taken as a whole rather than with any individual part of it.

In reality it is hard to believe that a measure of fiscal stance adjusted for inflation and cyclical variations in economic activity has no useful role to play in assessing government policy, despite its drawbacks and despite the heavy qualifications which ought to surround its use. Certainly the regular publication of such measures by the IMF and OECD seems to make a valuable contribution to the policy debate. At the very least, it provides an important counterbalance to the focus on the actual budget deficit which has been a feature of policy discussion in most countries in the 1970s and 1980s, which in many cases has led to the stance of policy being seriously misrepresented and which has therefore contributed to perverse policy action being taken.

TERRY WARD

See also BUDGETARY POLICY; FULL EMPLOYMENT BUDGET SURPLUS; FUNCTIONAL FINANCE; STABILIZATION POLICY.

BIBLIOGRAPHY

Blinder, A.S. and Solow, R.M. 1974. Analytical foundations of fiscal policy. In A.S. Blinder et al., *The Economics of Public Finance*, Washington, DC: Brookings Institution.

Brown, E.C. 1956. Fiscal policy in the Thirties: a reappraisal. *American Economic Review* 46, December, 857–79.

Buiter, W.H. 1983. The theory of optimum deficits and debt. In *The Economics of Large Government Deficits*, Boston: Federal Reserve Bank of Boston.

Buiter, W.H. 1985. A guide to public sector debt and deficits. *Economic Policy* 1, November, 13–79.

Chand, S.K. 1978. Summary measures of fiscal influence. *IMF Staff Papers* 24, Washington, DC.

Committee for Economic Development. 1947. *Taxes and the Budget: A Program for Prosperity in a Free Economy*. Washington, DC.

Dernberg, T.F. 1975. Fiscal analysis in the Federal Republic of Germany: the cyclically neutral budget. *IMF Staff Papers* 22, November, 825–7.

Kaldor, N. 1944. Appendix C of W.H. Beveridge, *Full Employment in a Free Society*. London: Allen & Unwin.

Lotz, J. 1971. Techniques of measuring the effects of fiscal policy. *OECD Economic Outlook, Occasional Studies*, Paris: OECD, July.

Netherlands Ministry of Finance. 1970. *The Netherlands Budget Memorandum 1970*, Annex 2. The Hague.

OECD. 1984. *Economic Outlook*. Paris: OECD, July.

Okun, A.M. and Teeters, N.H. 1970. The full employment surplus revisited. *Brookings Papers on Economic Activity* No. 1, 770–110.

Price, R.W.R. and Muller, P. 1984. Structural budget indicators and the interpretation of fiscal policy stance in OECD 'economies'. *OECD Economic Studies*, No. 3, Autumn, 27–72.

Tanzi, V. (ed.) 1984. *Taxation, Inflation and Interest Rates*. Washington, DC: IMF.

Taylor, C.T. and Threadgold, A.R. 1979. Real national savings and its sectoral composition. *Bank of England Discussion Papers* No. 6, London: Economic Intelligence Department, Bank of England.

Tobin, J. and Buiter, W.H. 1976. Long run effects of fiscal and monetary policy on aggregate demand. In *Monetarism*, ed. J.L. Stein, Amsterdam: North-Holland.

Ward, T.S. and Neild, R. R. 1978. *The Measurement and Reform of Budgetary Policy*. London: Heinemann.

Fisher, Irving (1867–1947). Irving Fisher was born in Saugerties, New York, on 27 February 1867; he was residing in New Haven, Connecticut at the time of his death in a New York City hospital on 29 April 1947.

Fisher is widely regarded as the greatest economist America has produced. A prolific, versatile and creative scholar, he made seminal and durable contributions across a broad spectrum of economic science. Although several earlier Americans, notably Simon Newcomb, had used some mathematics in their writings, Fisher's dedication to the method and his skill in using it justify calling him America's first mathematical economist. He put his early training in mathematics and physics to work in his doctoral dissertation on the theory of general equilibrium. Throughout his career his example and his teachings advanced the application of quantitative method not only in economic theory but also in statistical inquiry. He, together with Ragnar Frisch and Charles F. Roos, founded the Econometric Society in 1930; and Fisher was its first President. He had been President of the American Economic Association in 1918.

Much of standard neoclassical theory today is Fisherian in origin, style, spirit and substance. In particular, most modern models of capital and interest are essentially variations on Fisher's theme, the conjunction of intertemporal choices and opportunities. Likewise, his theory of money and prices is the foundation for much of contemporary monetary economics.

Fisher also developed methodologies of quantitative empirical research. He was the greatest expert of all time on index numbers, on their theoretical and statistical properties and on their use in many countries throughout history. From 1923 to 1936, his own Index Number Institute manufactured and published price indexes of many kinds from data painstakingly collected from all over the world. Indefatigable and innovative in empirical research, Fisher was an early and regular user of correlations, regressions and other statistical and econometric tools that later became routine.

To this day Fisher's successors are often rediscovering, consciously or unconsciously, Fisher's ideas and building upon them. He can be credited with distributed lag regression, life cycle saving theory, the 'Phillips curve', the case for taxing consumption rather than 'income', the modern quantity theory of money, the distinction between real and nominal interest rates, and many more standard tools in economists' kits. Although Fisher was not fully appreciated by his contemporaries, today he leads other old-timers by wide and increasing margins in journal citations. In column inches in the *Social Sciences Citation Index* (1979, 1983), Fisher led his most famous contemporaries, Wesley Mitchell, J.B. Clark, and F.W. Taussig in that order, by rough ratios 5:3:1:1 in 1971–5 and 9:3:1:1 in 1976–80. Much more than the others, moreover, Fisher is cited for substance rather than for history of thought.

For all his scientific prowess and achievement, Fisher was by no means an 'ivory tower' scholar detached from the problems and policy issues of his times. He was a congenital reformer, an inveterate crusader. He was so aggressive and persistent, and so sure he was right, that many of his contemporaries regarded him as a 'crank' and discounted his scientific work accordingly. Science and reform were indeed often combined in Fisher's work. His economic findings, theoretical and empirical, would suggest to him how to better the world; or dissatisfaction with the state of the world would lead him into scientifically fruitful analysis and research. Fisher's search for conceptual clarity about 'the nature of capital and income' led him not only to lay the foundations of modern social accounting but also to argue that income taxation wrongly puts saving in double jeopardy. Fisher turned his talents to monetary theory because he suspected that economic instability was largely the fault of existing monetary institutions. His 'debt-deflation theory of depression' was motivated by the disasters the Great Depression visited upon the world.

Economics was not the only aspect of human and social life that engaged Fisher's reformist zeal. He was active and prolific in other causes: temperance and Prohibition; vegetarianism, fresh air, exercise and other aspects of personal hygiene; eugenics; and peace through international association of nations.

Fisher was an amazingly prolific and gifted writer. The bibliography compiled by his son lists some 2000 titles authored by Fisher, plus another 400 signed by his associates or written by others about him. Fisher's writings span all his interests and causes. They include scholarly books and papers, articles in popular media, textbooks, handbooks for students, tracts, pamphlets, speeches and letters to editors and statesmen. They include the weekly releases of index numbers, often supplemented by commentary on the economic outlook and policy, issued for thirteen years by Fisher and assistants from the Index Number Institute housed in his New Haven home.

Fisher was the consummate pedagogical expositor, always clear as crystal. He hardly ever wrote just for fellow experts. His mission was to educate and persuade the world. He took the trouble to lead the uninitiated through difficult material in easy stages. Whenever he was teaching or tutoring students, he wrote handbooks or texts for their benefit – in mathematics and science when he was still a student himself, in the principles of economics when he was the professor responsible for the introductory course. Fisher's economics text was published in 1910 and 1911. Its graceful exposition of sophisticated theoretical material will impress a modern

connoisseur, but it was too difficult for widespread adoption. Some of it survived in a leading introductory text of the 1920s and 1930s, by the younger Yale economists Fairchild, Furniss and Buck (1926).

A BRIEF BIOGRAPHY

Irving Fisher grew up and attended school successively in Peace Dale, Rhode Island; New Haven, Connecticut; and St Louis, Missouri. His father, a Congregational minister, died of tuberculosis just when Irving had finished high school and was planning to attend Yale College, his father's *alma mater*. Irving was now the principal breadwinner for himself, his mother and his younger brother. He did have a $500 legacy from his father for his college education. The family moved to New Haven, and together managed to make ends meet. Irving tutored fellow students during term and in summers.

Fisher was a great success in Yale College, ranking first in his class and winning prizes and distinctions not only in mathematics but across the board. He was also determined to make good in the extra-curricular college culture so important in those days. His efforts won him election to the most prestigious secret senior society, Skull and Bones, the ultimate reward senior campus leaders bestowed on members of the class behind them.

Awarded a scholarship for graduate study, he stayed on at Yale. Graduate Studies were not departmentalized in those days, and Fisher ranged over mathematics, science, social science and philosophy. His most important teachers were Josiah Willard Gibbs, the mathematical physicist celebrated for his theory of thermodynamics, William Graham Sumner, famous still in sociology but at the time also important in political economy, and Arthur Twining Hadley, a leading economist specializing in what is now known as Industrial Organization.

As the time to write a dissertation approached, Fisher had still not chosen his life work. Young Fisher's interests and talents were universal. In the seven years at Yale before he finished his doctorate, he had written and published poetry, political commentary, book reviews, a geometry text together with tables of logarithms, and voluminous notes on mathematics, mechanics and astronomy for the benefit of students he was teaching or tutoring. If he had specialized in anything in six years at Yale, it was mathematics, but even in his graduate years he had spent half his time elsewhere.

Sumner put him on to mathematical economics, and in his third year of graduate study, he finished the dissertation that won him worldwide recognition in economic theory. Fisher's 1891 PhD was the first one in pure economics awarded by Yale, albeit by the faculty of mathematics. Although the university, thanks to Sumner, Hadley and Henry W. Farnum, was strong in 'political economy', there was no distinct department for the subject, let alone for 'economics'. This was generally the case in American universities. Venturing into mathematical economic theory, Fisher was very much on his own; and his route into economics was quite different from that of most American economists of his era.

The dominant tradition in American political economy was imported from the English classical economists, mainly Smith, Ricardo and John Stuart Mill; it was just beginning to be updated by Marshall. This tradition Fisher's mentors at Yale had taught him well. But the neoclassical developments on the European continent from 1870 on, the works of Walras and Menger and Böhm-Bawerk, or even those of their English counterparts Jevons and Edgeworth, had been little noticed at Yale or elsewhere in America.

At the time, the main challenge in America to classical political economy was coming from quite a different direction. The American Economic Association was founded in 1886 by young rebels against Ricardian dogma and its *laissez-faire* political and social message. They included Richard T. Ely, J.B. Clark, Edwin R.A. Seligman and other future luminaries of American economics. Many of them had pursued graduate studies in Germany. In the German emphasis on historical, institutional and empirical studies they found welcome relief from implacable classical theory, and in the German faith in the state as an instrument of socially beneficial reform they found a hopeful antidote to the fatalism of economic competition and social Darwinism. Sumner was prominent among several elders who refused to join an Association born of such heresy; he did not relent even though the AEA very soon became sufficiently neutral and catholic to attract his Yale colleagues and other initial holdouts. Fisher, a bit younger than the founding rebels and educated solely at one American university, was not involved. It was his reconstruction, rather than their revolution, that was destined eventually to replace the classical tradition in the mainstream of American economics.

Fisher stayed at Yale throughout his career. He started teaching mathematics, evidently even before he received his doctorate and was appointed Tutor in Mathematics. His first economics teaching was under the auspices of the mathematics faculty, an undergraduate course on 'The Mathematical Theory of Prices'. In 1894–5 during his Wanderjahr in Europe, this young American star was welcomed by the leading mathematically inclined theorists in every country. On his return he became Assistant Professor of Political and Social Science and began teaching economics proper. He was appointed full Professor in 1898 and retired in 1935.

Fisher was struck by tuberculosis in 1898. He spent the first three years of his professorship on leave from Yale and from science, recuperating in more salubrious climates. His lifelong crusade for hygienic living dates from this personal struggle to regain health and vigour. The experience powerfully reinforced his determination to gain 'a place among those who have helped along my science' and his ambition 'to be a *great* man', as he wrote to his wife (I.N. Fisher, 1956, pp. 87–8). After his recovery the books and articles began flowing from his pen, never to stop until his death at the age of 80.

Fisher participated actively in teaching and in university affairs until 1920. Thereafter his writings and his myriad outside activities and crusades preoccupied him. He taught only half time and had little impact on students, undergraduate or graduate. Thus Fisher had few personal disciples; there was no Fisherian School. The student to whom Fisher was closest, personally and intellectually, was James Harvey Rogers, a 1916 PhD who returned to Yale as a professor in 1930. His career was prematurely ended by his tragic death in a plane crash in 1939 at the age of 55.

Fisher was, on top of everything else, an inventor. His most successful and profitable invention was the visible card index system he patented in 1913. In 1925 Fisher's own firm, the Index Visible Company, merged with its principal competitor to form Kardex Rand Co., later Remington Rand, still later Sperry Rand. The merger made him wealthy. However, he subsequently lost a fortune his son estimated to amount to 8 or 10 million dollars, along with savings of his wife and her sister, when he borrowed money to exercise rights to buy additional Rand shares in the bull market of the late 1920s.

More than money was at risk in the market. Fisher had staked his public reputation as an economic pundit by his persistent optimism about the economy and stock prices, even

after the 1929 crash. His reputation crashed too, especially among non-economists in New Haven, where the university had to buy his house and rent it to him to save him from eviction. Until the 1950s the name Irving Fisher was without honour in his own university. Except for economic theorists and econometricians, few members of the community appreciated the genius of a man who lived among them for 63 years.

Irving Fisher's marriage to Margaret Hazard in 1893 was a very happy one for 47 years. She died in 1940. They had two daughters and one son, his father's biographer. The death of their daughter Margaret in 1919 after a nervous breakdown was the greatest tragedy of her parents' lives. Their daughter Carol brought them two grandchildren.

GENERAL EQUILIBRIUM THEORY

Fisher's doctoral dissertation (1892) is a masterly exposition of Walrasian general equilibrium theory. Fisher, who was meticulous about acknowledgements throughout his career, writes in the preface that he was unaware of Walras while writing the dissertation. His personal mentors in the literature of economics were Jevons (1871) and Auspitz and Lieben (1889).

Fisher's inventive ingenuity combined with his training under Gibbs to produce a remarkable hydraulic-mechanical analogue model of a general equilibrium system, replete with cisterns, valves, levers, balances and cams. Thus could he display physically how a shock to demand or supply in one of ten interrelated markets altered prices and quantities in all markets and changed the incomes and consumption bundles of the various consumers. The model is described in detail in the book; unfortunately both the original model and a second one constructed in 1925 have been lost to posterity. Anyway Fisher was a precursor of a current Yale professor, Herbert Scarf (1973) and other practitioners of computing general equilibrium solutions. In his formal mathematical model-building too, Fisher was greatly impressed by the analogies between the thermodynamics of his mentor Gibbs and economic systems, and he was able to apply Gibbs's innovations in vector calculus.

Fisher expounds thoroughly the mathematics of utility functions and their maximization, and he is careful to allow for corner solutions. He uses independent and additive utilities of commodities in his first mathematical approximation and in his physical model; later he was to show how this assumption could be exploited to measure marginal utilities empirically (1927). But the general formulation in his dissertation makes the utility of every commodity depend on the quantities consumed of all commodities. At the same time, he states clearly that neither interpersonally comparable utility nor cardinal utility for each individual is necessary to the determination of equilibrium. Fisher's list of the limitations of his analysis is candid and complete. The supply side of Fisher's model is, as he acknowledges, primitive. Each commodity is produced at increasing marginal cost, but neither factor supplies and prices nor technologies are explicitly modelled.

Finally, Fisher shows his enthusiasm for his discovery of mathematical economics by appending to his dissertation as published an exhaustive survey and bibliography of applications of mathematical method to economics.

General equilibrium with intertemporal choices and opportunities. The distribution of income and wealth, and in particular the sources, determinants and social rationales, of interest and other returns to private property, were obsessive topics in economics, both in Europe and North America, at the turn of the century. One important reason, especially in Europe, was the Marxist challenge to the legitimacy of property income. Answering Marx was a strong motivation for the Austrian school, in particular for the capital theory of Böhm-Bawerk and his followers. Neoclassical economics was in a much better position than its classical precursor to respond to the Marxist challenge. The labour theory of value, which Marx borrowed from the great classical economists themselves, neither explains nor justifies functionally or ethically incomes other than wages.

These topics engaged the two leading American economists of the era, John Bates Clark and Fisher. Clark (1899) set forth his marginal productivity theory of distribution, arguing that a generalized factor of production, capital, the accumulation of past savings, has like labour a marginal product that explains and justifies the incomes of its owners.

Fisher attacked these problems in a more elegant, abstract, mathematical, general and ethically neutral manner than Clark, and than Böhm-Bawerk. At the same time, his approach was clearer, simpler and more insightful than that of Walras.

The general equilibrium system of Fisher's dissertation was a single-period model. No intertemporal choices entered; hence the theory was silent on the questions of capital and interest. But Fisher took up these subjects soon after.

His first contribution, one that should not be underestimated, was to set straight the concepts and the accounting. This he did in (1896) and (1906) with clarity and completeness that have scarcely been surpassed. It's all there: continuous and discrete compounding; nominal versus real rates; the distinction between high prices and rising prices, and its implications for observations of interest rates; the inevitable differences among rates computed in different *numéraires*; rates to different maturities and consistency among them; appreciation, expected and unexpected; present values of streams of in- and out-payments; and so on. Schumpeter calls this work 'the first economic theory of accounting' and says 'it is (or should be) the basis of modern income analysis' (1954, p. 872).

Perhaps the most remarkable feature is Fisher's insistence that 'income' is consumption, including of course consumption of the services of durable goods. In principle, he says, income is psychic, the subjective utility yielded by goods and services consumed. More practically, income could be measured as the money value, or value in some other *numéraire*, of the goods and services directly yielding utility, but only of those. Receipts saved and invested, for example in the purchase of new durable goods, are not 'income' for Fisher; they will yield consumption and utility later, and those yields will be income. To include both the initial investment and the later yields as income is, according to Fisher, as absurd as to count both flour and bread in reckoning net output. This view naturally led Fisher to oppose conventional income taxation as double taxing of saving, and to favour consumption taxation instead. His views on these matters are loudly echoed today.

Fisher published his theory of the determination of interest rates in *The Rate of Interest* (1907). A revised and enlarged version was published in 1930 as *The Theory of Interest*. One motivation for the revision was that Fisher's many critics apparently did not understand the 1907 version. They typically concentrated on the 'impatience' side of Fisher's theory of intertemporal allocation and missed the 'opportunities' side. It was there in 1907 already; the theory is much the same in both versions.

In 1930 Fisher is at pains to label his theory the 'impatience and opportunity' theory. 'Every essential part of it', he

acknowledges, 'was at least foreshadowed by John Rae in 1834.' He does claim originality for his concept of 'investment opportunity'. This turns on 'the rate of return over cost, [where] both cost and return are differences between two optional income streams' (1930, p. ix). As Keynes acknowledged, this is the same as his own 'marginal efficiency of capital' (Keynes, 1936, p. 140).

In these books Fisher extended general equilibrium theory to intertemporal choices and relationships. This strategy was different from Walras. Walras tried to extend his multi-commodity multi-agent model of exchange to allow for production, saving and investment. This maintained his stance of full generality but was also difficult to expound and to understand. Fisher saw that intertemporal dependences were tricky enough to justify isolating them from the inter-commodity complexities that had concerned him in his doctoral thesis. Therefore he proceeded as if there were just one aggregate commodity to be produced and consumed at different dates. This simplification enabled him to illuminate the subject more brightly than Walras himself.

The methodology of Fisher's capital theory is very modern. His clarifications of the concepts of capital and income lead him to formulate the problem as determination of the time paths of consumption – that is, income – both for individual agents and for the whole economy. Then he divides the problem into the two sides, tastes and technologies, that are second nature to theorists today. One need only read Böhm-Bawerk's murky mixture of the two in his list of reasons for the agio of future over present consumption to realize that Fisher's procedure was not instinctive in those times.

Fisher's theory of individual saving is basically the standard model to this day. Undergraduates learn the two-period 'Fisher diagram', where a family of indifference curves in the two commodities consumption now c_1 and consumption later c_2 confront a budget constraint $c_1 + c_2/(1+r) = y_1 + y_2(1+r)$, where the y's are exogenous wage incomes in the two periods and r is the (real) market interest rate. From the usual tangency can be read the consumption choices and present saving or dissaving. This is indeed a Fisher diagram, but of course he went much beyond it.

He stated clearly what we now call the 'life cycle' model, explaining why individuals will generally prefer to smooth their consumption over time, whatever the time path of their expected receipts. But he was not dogmatic, and he allowed room for bequests and for precautionary saving. Where Fisher differed from later theorists, and especially from contemporary model-builders, was in his unwillingness to impose any assumed uniformity on the preferences (or expectations or 'endowments' – the latter term was not familiar to him though the concept was) of the agents in his economies, and in his scruples against buying definite results by assuming tractable functional forms. In general, many of the advances claimed in present-day theory appear to depend on greater boldness in these respects.

On the side of technology, Fisher's approach was the natural symmetrical partner of his formulation of preferences, equally simple, abstract and general. He assumed that the 'investment opportunities' available to an individual (not necessarily the same for everybody) and to the society as a whole can be summarized in the terms on which consumption at any date can be traded, with 'nature', for consumptions at other dates. In modern language, we would say that Fisher postulated intertemporal production possibility frontiers, properly convex in their arguments, consumptions at various dates.

All that remained for Fisher, then, was to assume complete intertemporal loan markets cleared by real interest rates, count equations, and show that in principle the equalities of saving and investment at every date determine all interest rates and the paths of consumption and production for all individuals and for the society. Like hundreds of mathematical theorists since, he set the problem up so that it conformed to a paradigm he knew, in this case the Walrasian paradigm of his own doctoral dissertation. A more rigorous proof of the existence of the equilibria Fisher was looking for came much later, from Arrow and Debreu (1954). As we know, the problems of infinity, whether agents are assumed to have infinite or finite horizons, are much more troublesome than Fisher imagined.

In any event, Fisher had an excellent vantage point from which to comment on the controversies over capital and interest raging in his day. His formulation of 'investment opportunities' seems to allow for no factor of production one could call 'capital' and enter as argument in a production function. For that matter, he doesn't explicitly model the role of labour in production either, or of land. Strangely, in Fisher's insistence that interest is *not* a cost of production, he seems to say that labour is the only cost, evidently because labour and labour alone is a source of disutility, the loss of utility from leisure, the opportunity cost of the consumption afforded by work. Proceeding in the same spirit, he postulates that, from a position of equality of present and planned future consumption a typical individual will require more extra future consumption than present consumption as compensation for extra work. The difference, the agio, is interest, whether or not it is a 'cost'. Fisher attributes the agio to 'impatience', at the same time scorning the notion that interest is the cost of securing the services of a factor of production called 'abstinence' or 'waiting'.

In the 1890s and 1900s Knut Wicksell, discovering marginal productivity independently of Clark, was modelling production as a function of labour and land inputs with the output also depending on the lags between those inputs and the harvests (Wicksell [1911], 1934, vol. I, pp. 144–66). This is an 'Austrian' formulation, akin to Böhm-Bawerk's examples of trees and wine, in which time itself appears to be productive. Fisher rightly objects to any generalization that waiting longer increases output. His own intertemporal frontiers are, to be sure, sufficiently general to encompass such technologies. They can also accommodate Leontief input–output tables and Koopmans–Dantzig activity matrices with lags, Hayekian triangular structures with inventories of intermediate goods in process, Solow technologies with durable goods and labour jointly yielding output contemporaneously or later. The only common denominator of these and other representations of technology is that they relate consumption opportunities at different dates to one another, though not necessarily always in the convex trade-off terms Fisher assumed. There does not appear to be any summary scalar measure to which the productivity of a process is generally monotonically related, whether roundaboutness, average period of production, or replacement value of existing stocks of goods.

Fisher describes himself as an advocate of 'impatience' as an explanation of interest, although he realizes there are two sides of the saving-investment market, and although he acknowledges that real interest rates can at times be zero or negative. He does appear to believe that in a stationary equilibrium with constant consumption streams, consumers will require positive interest, and that only those technologies and investment opportunities affording a 'rate of return over cost' equal to this pure time preference rate would be used. He does not face up to Schumpeter's argument in 1911 that in such a repetitive and riskless 'circular flow', rational consumers would not care

whether a marginal unit of consumption occurs today or tomorrow (Schumpeter [1911], 1934, pp. 34–6). Like Böhm-Bawerk, Fisher appeals to the shortness and uncertainty of life as a reason for time preference. For life-cycle consumers, however, time preferences are entangled with age preferences, and it is hard to defend any generalization as to their net direction. Fair annuities take care of the uncertainty.

MONETARY THEORY: THE EQUATION OF EXCHANGE AND THE QUANTITY THEORY

Irving Fisher was the major American monetary economist of the early decades of this century; the subject occupied him until the end of his career. Here especially Fisher combined theorizing with empirical research, both historical and statistical. The problems he encountered led him to invent statistical and econometric methods – index numbers and distributed lags in particular – to apply for the purposes at hand to the data he and his assistants compiled. (He even studied the turnover of cash and checking accounts of a sample of Yale students, professors and employees.)

Money was a big subject in American economic literature in the 19th century, before Fisher came on the scene. The monetary events of the times – the inconvertible greenbacks issued during the Civil War, their redemption in gold in 1879, the demonetization of silver, the rapidly increasing importance of banks – stimulated research and controversy. Nevertheless, monetary theory was relatively undeveloped and unsystematized, both in Europe and in America. Fisher's treatise (1911a) was an ambitious attempt to organize with the help of theory a large body of historical and institutional information.

Yet for all its theory, statistics and index numbers, *The Purchasing Power of Money* is a tract supporting Fisher's proposal for stabilizing the value of money. This came to be known as the 'compensated dollar', the gold-exchange standard combined with a rule mandating periodic changes in the official buying and selling prices of gold inverse to changes in a designated commodity price index. In 1911 Fisher proposed that the gold price changes be uniform and synchronous in the currencies of all countries linked by fixed exchange parities, in proportional amounts related to an international price index. Later he was willing to accept as second best that the United States adopt the scheme on its own. Keynes proposed a similar but less formal rule for the United Kingdom (1923).

The proposal is an early example of a policy *rule*, another Fisherian idea ahead of its time, more likely to be popular among economists today than it was with Fisher's contemporaries. Indeed, some rules recently proposed are quite Fisherian, for example Hall (1985).

The 'compensated dollar' is but one of several proposals Fisher advanced over the years for stabilizing price levels or mitigating the effects of their unforeseen variation. In the 1911 book he also writes favourably of the 'tabular standard', which meant no more operationally than facilitating price-indexed contracts. In the 1920s he launched a crusade for 100 per cent reserves against checkable deposits, culminating in *100% Money* (1935). This idea is also beginning to resurface in the 1980s as a preventive defence against the monetary hazards of bank failures. In Schumpeter's view, Fisher's zeal for monetary reforms lost him some of the attention and respect his scientific contributions to monetary economics deserved, and made him come across as more monetarist than his own analysis and evidence justified (Schumpeter, 1954, pp. 872–3).

The Purchasing Power of Money is a monetarist book. Fisher asserts the quantity theory as earnestly and persuasively as Milton Friedman. There are two species of quantity theories. One is a simple implication of the 'classical dichotomy': since only relative prices and real endowments enter commodity and factor demand and supply functions, the solution values for real variables in a general equilibrium are independent of scalar variations of exogenous nominal quantities. While Fisher mentions this implication of general equilibrium theory, he does not dwell upon it as one might expect. Anyway, it does not quite apply to a commodity money system like the gold standard, which Fisher was analysing. Fisher's theory is mainly of the second kind, based on the demand for and supply of the particular nominal assets serving as media of exchange.

Fisher is usually given credit for the Equation of Exchange, although Simon Newcomb, a celebrated figure in American astronomy as well as an economist, had anticipated him (1886, pp. 315–47). The Equation is the identity $MV = PT$, where M is the stock of money; V its velocity, the average number of times per year a dollar of the stock changes hands; P is the average price of the considerations traded for money in such transactions; and T is the physical volume per year of those considerations. It is an identity because it is in principle true by definition. Actually Fisher, of course, recognized the heterogeneity of transactions by writing also $MV = \Sigma p_i Q_i$, where the p_i and Q_i are individual prices and quantities. His interest in index numbers was substantially a quest for aggregate indexes P and T derived from the individual p_i and Q_i in such a way that the two forms of the equation would be consistent. Much of the book (1911a), both text and technical appendices, is devoted to this quest.

Here and in later writings, particularly (1921) and (1922), Fisher was looking for the 'best' index number formula. He postulated certain criteria and evaluated a host of formulas, investigating their properties both *a priori* and from applications to data. Since the criteria inevitably conflict, there can be no formula that excels on all counts. Although Fisher was mainly interested in measuring movements of the aggregate price level, naturally he wanted a price index P and a quantity index T to have the property that $P_1 T_1 / P_0 T_0 = (\Sigma p_1 Q_1)/(\Sigma p_0 Q_0)$, where the subscripts represent two time periods at which observations of p's and Q's are available.

This and various other desirable consistency properties are not hard to meet. The difficult question is the choice of weights in the two indexes, especially when a whole series of consistent period-to-period comparisons is desired, not just one isolated comparison. For a price index, should the quantity weights be those of a fixed base year, yielding what we now call a 'Laspeyres' index $(\Sigma p_1 Q_0)/(\Sigma p_0 Q_0)$? Or should the weights be those of the ever-changing current period, yielding a 'Paasche' index $(\Sigma p_1 Q_1)/(\Sigma p_0 Q_1)$? The indicated correlate quantity indexes would be the opposites, respectively 'Paasche' and 'Laspeyres'. In 1911 Fisher opted for the Paasche price index. He also seemed to approve the idea of chain indexes, in which the period 0 of the above formulas is not fixed in calendar time but is always the prior period, even though these violate one possible desideratum, that the relative change between two periods should be independent of the base used. He also wrote favourably of the practical advantages of an entirely different procedure, namely taking the median of an expenditure-weighed distribution of percentage price changes from one period to the next.

In 1920, however, Fisher proposed as the 'Ideal Index' a candidate he had not ranked high in 1911, namely the geometric mean of the Laspeyres and Paasche formulas. This formula has the pleasant property that the correlate of an Ideal price index is an Ideal quantity index. Correa Walsh, another index number expert, on whose comprehensive treatise (1901) Fisher relied heavily from the beginning of his own investigations, reached the same conclusion independently at about the same time (Walsh, 1921).

These index number issues do not seem as important to present-day economists as they did to Fisher. Knowing that they are intrinsically insoluble, we finesse them and use uncritically the indexes that government statisticians provide. But Fisher's explorations have been important to those practitioners.

In Fisher's Equation of Exchange (1911a) the T and the Q_i are measures of all transactions involving the tender of money, intermediate goods and services as well as final goods and services, old goods as well as newly produced commodities, financial assets as well as goods. The corresponding velocity is likewise comprehensive, much more so than the 'income' or 'circuit' velocity preferred by some monetary theorists, notably Alfred Marshall and his followers in Cambridge (England), who count only transactions for final goods, for example for Gross National Product.

Fisher elaborated the equation to distinguish the quantities M and M' of the two media currency and checking deposits and their separate velocities V and V': $MV + M'V' = PT$. This was a bow to the rising importance of bank deposits relative to currency as transactions media. Previous practice counted only government-issued currency as money, in modern parlance high-powered or base money, and regarded bank operations as increasing its velocity rather than adding to a money stock.

How does the quantity theory come out of the Equation of Exchange? Fisher argues that the real volume of money-using transactions T is exogenous; that the velocities are determined by institutions and habits and are independent of the other variables in the equation; that the division of the currency supply, the monetary base in current terminology, between currency and bank reserves is stable and independent of the variables in the equation; that banks are fully 'loaned up' so that deposits M' are a stable multiple of reserves, determined by the prudence of banks and by regulation; that exogenous changes in currency supply itself are the principal source of shocks, which, given the preceding propositions, move price level P proportionately. The many qualifications for transitional adjustments are conscientiously presented, but the monetarist message is loud and clear.

The argument is familiar to modern readers, but certain features deserve notice:

(1) Fisher gives the most illuminating account available of the institutions and habits that generate the society's demand for transactions media relative to the volume of transactions. He rightly emphasizes the fact that, and the degree to which, receipts and payments are imperfectly synchronized. He seeks the determinants of velocity in such features of social and economic structure as the frequency of wage and bill payments and the degree of vertical integration of firms. His belief that these institutions change only slowly supports his contention that velocities are exogenous constants.

(2) Much ink has been spilled on the difference between Fisher's velocity approach to money demand and the Cambridge (England) 'k' formulation. The latter, like Walras's *encaisse désiré*, directs attention to agents' portfolio decisions. To Fisher's critics that seems behavioural, while velocity is mechanical. The issue is overblown; the same phenomena can be described in either language. If the other variables in the equation are defined and measured the same way, then V and k are just reciprocals each of the other. Fisher himself

discusses hoarding. Fisher's explicit attention, in discussing economy-wide demand for circulating media in distinction to other stores of value, to the fact that money 'at rest' soon takes 'wing' to fly from one agent to another seems to be a merit of his approach.

(3) As already noted, Fisher resolved a question current in his day, whether banks' creation of deposit substitutes for currency should be regarded as increasing the velocity of basic money or as enlarging the supply of money. His choice of the latter course compels attention to the structure, behaviour and regulation of banks. He could not be expected to foresee that the proliferation of future candidates for designation as 'money' would create the monetarist ambiguities we see today.

(4) For the most part later writers have not followed Fisher in his preference for a comprehensive concept and measure of transactions volume. It is hard to attach meaning to the *real* volume of financial transactions, and therefore to see why a *T* that includes them should be a constant or exogenous term in the equation. On the other hand, modern students of money demand tend simply to forget transactions other than those on final payments.

(5) Fisher ignores the possibility that other liquid assets can serve as imperfect substitutes for money holdings because they can be converted into means of payment as needed, though at some cost. Partly for this reason, he ignores interest rate effects on demand for transactions media. In his day there may have been more excuse for these omissions than there was later. But they are still surprising for an author who elsewhere pays so much attention to the effects of interest rates and opportunity costs on behaviour.

(6) When Fisher was writing, the United States was on the gold standard; the exchange parities of the dollar with sterling and other gold-standard currencies were fixed. Fisher discusses in detail the implications of foreign transactions for the elements of the Equation of Exchange and for the quantity theory. He recognizes that tendencies towards purchasing-power parity, even though imperfect, make money supplies in any one country endogenous, tie prices to those of other countries and enhance quantity adjustments to monetary shocks in the short run. Much of the 1911 book applies, therefore, to the gold standard economies in aggregate. Indeed, Fisher finds the increase in gold production after 1896 to be the main cause of price increases throughout the world.

MACROECONOMICS: BUSINESS FLUCTUATIONS AND THE GREAT DEPRESSION

The quantity theory by no means exhausts Fisher's ideas on macroeconomics. His views were much more subtle then straightforward monetarism, but they are scattered through his writings and not systematically integrated. Consider the following non-neutralities emphasized by Fisher:

(1) Probably Fisher's principal source of fame, especially among non-economists, is his equation connecting nominal interest i, real interest r and inflation π: $i = r + \pi$. It is frequently misused. Like the Equation of Exchange, it is first of all an identity, from which, for example, an unobservable value of r can be calculated from observations of the other two variables. More interesting, certainly to Fisher, is its use as a condition of equilibrium in financial markets; for this purpose π must be replaced by expected inflation π^e, another unobservable. In a longer run, as Fisher recognized, steady-state equilibrium would also be characterized by equality of actual and expected inflation: $\pi = \pi^e$.

The Fisher equation is frequently cited nowadays in support

of complete and prompt pass-through of inflation into nominal interest rates. Fisher's view throughout his career was quite different. For one thing, neither Fisher's theory of interest nor his reading of historical experience suggested to him that equilibrium real rates of interest should be constant. Moreover, from (1896) on he believed that adjustment of nominal interest rates to inflation takes a very long time. This he confirmed by sophisticated empirical investigations, regressions in which the formation of inflation expectations was modelled by distributed lags on actual inflation. During the transition, inflation would lower real rates; nominal rates would adjust incompletely. The effect was symmetrical; he attributed the severity of the Great Depression to the high real rates resulting from price deflation.

Moreover, Fisher was quite explicit about the effects of these movements of real interest rates on real economic variables, including aggregate production and employment. In *The Purchasing Power of Money* these transitional effects are mentioned, but minimized in the author's zeal to convince readers of the importance of stabilizing money stocks. But in Fisher's writings on interest rates, the transitions turn out to be long. In his accounts of cyclical fluctuations in business activity, and especially of the Great Depression, they play the key role.

(2) An assiduous student of price data, Fisher knew that some prices were more flexible than others, that money wages were on the sticky side of the spectrum, and that the imperfect flexibility of the price level meant that the *T* on the right-hand side of his Equation of Exchange would absorb some of the variations of the left-hand side.

In the early 1930s he came to a very modern position. Real variables like production and employment are independent of the level of prices, once the economy has adjusted to the level. But they are not independent of the rate of change of prices; they depend positively on the rate of inflation. He even calculated a 'Phillips' correlation between employment and inflation (1926). He was just one derivative short of the accelerationist position (Friedman, 1968); in a little more time he would have made that step, aware as he was of the difference between actual and expected inflation. Anyway, his policy conclusion was that stabilizing the price level would also stabilize the real economy.

(3) During the Great Depression, observing the catastrophes of the world around him, which he shared personally, Fisher came to quite a different theory of the business cycle from the simple monetarist version he had espoused earlier. This was his 'debt-deflation theory of depression' (1932), summarized in the first volume of *Econometrica*, the organ of the international society he helped to found (1933). The essential features are that debt-financed Schumpeterian innovations fuel a boom, followed by a recession which can turn into depression via an unstable interaction between excessive real debt burdens and deflation. Note the contrast to the Pigou real balance effect, according to which price declines are the benign mechanism that restores full-employment equilibrium. The realism is all on Fisher's side. This theory of Fisher's has room for the monetary and credit cycles of which he earlier complained, and for the perversely pro-cyclical real interest rate movements mentioned above.

Fisher did not provide a formal model of his latter-day cycle theory, as he probably would have done at a younger age. The point here is that he came to recognize important non-monetary sources of disturbance. These insights contain the makings of a theory of a determination of economic activity, prices, and interest rates in short and medium runs. Moreover, in his neoclassical writings on capital and interest

Fisher had laid the basis for the investment and saving equations central to modern macroeconomic models. Had Fisher pulled these strands together into a coherent theory, he could have been an American Keynes. Indeed the 'neoclassical synthesis' would not have had to wait until after World War II. Fisher would have done it all himself.

His practical message in the early 1930s was 'Reflation!' When his Yale colleagues and orthodox economists throughout the country protested against public-works spending proposals and denounced Roosevelt's gold policies, Fisher was a conspicuous dissenter. He was right. Characteristically, he crusaded vigorously for his cause – in speeches, pamphlets, letters and personal talks with President Roosevelt and other powerful policy-makers. Characteristically too, as his letters home (I.N Fisher, 1956, p. 275) disclose, he saw clearly and unapologetically that in lobbying for what was good for the country he was also hoping to rescue the Fisher family finances.

Addressing the President of Yale shortly after Fisher's death, Joseph Schumpeter and eighteen colleagues in the Harvard economics department wrote, 'No American has contributed more to the advancement of his chosen subject ... The name of that great economist and American has a secure place in the history of his subject and of his country.' According to his son, this is the eulogy that would have pleased Irving Fisher the most (I.N. Fisher, 1956, pp. 337–8). Today, four decades later, economists can confirm the judgement and prediction of that eulogy.

JAMES TOBIN

Author's Note: Fortunately Fisher's son, Irving Norton Fisher, preserved the memory of his father in two indispensable publications, a biography and a comprehensive bibliography (1956, 1961). I have also relied extensively on Professor John Perry Miller's biographical essay (1967) and Professor William Barber's account (1986) of political economy at Yale before 1900. My review of Fisher's contributions to general equilibrium theory, the theory of capital and interest, monetary theory and macroeconomics draws heavily and often literally on a recent essay of my own (Tobin, 1985).

SELECTED WORKS

1892. *Mathematical Investigations in the Theory of Value and Prices.* New Haven: Connecticut Academy of Arts and Sciences, *Transactions 9*, 1892. Reprinted, New York: Augustus M. Kelley, 1961.
1896. Appreciation and interest. *AEA Publications* 3(11), August, 331–442. Reprinted, New York: Augustus M. Kelley, 1961.
1906. *The Nature of Capital and Income.* New York: Macmillan.
1907. *The Rate of Interest.* New York: Macmillan.
1910. *Introduction to Economic Science.* New York: Macmillan.
1911a. *The Purchasing Power of Money.* New York: Macmillan.
1911b. *Elementary Principles of Economics.* New York: Macmillan.
1921. The best form of index number. *American Statistical Association Quarterly* 17, March, 533–7.
1922. *The Making of Index Numbers.* Boston: Houghton Mifflin.
1926. A statistical relation between unemployment and price changes. *International Labour Review* 13, June, 785–92.
1927. A statistical method for measuring 'marginal utility' and testing the justice of a progressive income tax. In *Economic Essays Contributed in Honor of John Bates Clark*, ed. J.H. Hollander, New York: Macmillan.
1930. *The Theory of Interest.* New York: Macmillan.
1932. *Booms and Depressions.* New York: Adelphi.
1933. The debt-deflation theory of great depressions. *Econometrica* 1(4), October, 337–57.
1935. *100% Money.* New York: Adelphi.

BIBLIOGRAPHY

Arrow, K.J. and Debreu, G. 1954. Existence of an equilibrium for a competitive economy. *Econometrica* 22(3), July, 265–90.
Auspitz, R. and Lieben, R. 1889. *Untersuchungen über die Theorie des Preises.* Leipzig: Duncker & Humblot.
Barber, W.J. 1986. Yale: the fortunes of political economy in an environment of academic conservatism. In W.J. Barber, *Economists and American Higher Learning in the Nineteenth Century*, Middletown, Conn.: Wesleyan University Press.
Clark, J.B. 1899. *The Distribution of Wealth.* New York: Macmillan.
Fairchild, F.R., Furniss, E.S. and Buck, N.S. 1926. *Elementary Economics.* 2 vols, New York: Macmillan. 5th edn, 1948.
Fisher, I.N. 1956. *My Father Irving Fisher.* New York: Comet Press.
Fisher, I.N. 1961. *A Bibliography of the Writings of Irving Fisher.* New Haven: Yale University Library.
Friedman, M. 1968. The role of monetary policy. *American Economic Review* 58(1), 1–17.
Hall, R.E. 1985. Monetary policy with an elastic price standard. In *Price Stability and Public Policy*, Federal Reserve Bank of Kansas City, 137–60.
Jevons, W.S. 1871. *The Theory of Political Economy.* London: Macmillan.
Keynes, J.M. 1923. *A Tract on Monetary Reform.* London: Macmillan.
Keynes, J.M. 1936. *The General Theory of Employment, Interest and Money.* New York: Harcourt, Brace.
Miller, J.P. 1967. Irving Fisher of Yale. In *Ten Economic Studies in the Tradition of Irving Fisher*, ed. William Fellner et al., New York: Wiley.
Newcomb, S. 1885. *Principles of Political Economy.* New York: Harper.
Rae, J. 1834. *The Sociological Theory of Capital.* Reprinted, New York: Macmillan, 1905.
Samuelson, P.A. 1967. Irving Fisher and the theory of capital. In *Ten Economic Studies in the Tradition of Irving Fisher*, ed. William Fellner et al., New York: Wiley.
Scarf, H. (With T. Hansen.) 1973. *The Computation of Economic Equilibria.* New Haven: Yale University Press.
Schumpeter, J.A. 1912. *Theory of Economic Development.* Trans. from the 2nd German edn of 1926 by R. Opie, Cambridge, Mass.: Harvard University Press, 1934.
Schumpeter, J.H. 1954. *History of Economic Analysis.* Ed. E.B. Schumpeter, New York: Oxford University Press.
Social Sciences Citation Index. 1979, 1983. *Five Year Cumulation*, 1971–5 and 1976–80. Philadelphia: Institute for Scientific Information.
Tobin, J. 1985. Neoclassical theory in America. *American Economic Review* 75(6), December, 28–38.
Walsh, C.M. 1901. *The Measurement of General Exchange Value.* New York and London: Macmillan.
Walsh, C.M. 1921. *The Problem of Estimation.* London: King & Sons.
Wicksell, K. 1911. *Lectures on Political Economy.* Trans. E. Classen (from the 2nd Swedish edn), London: George Routledge & Sons, 1934.

Fisher, Ronald Aylmer (1890–1962). R.A. Fisher was born in London on 17 February 1890, the son of a fine-art auctioneer. His twin brother was still-born. At Harrow School he distinguished himself in mathematics, despite being handicapped by poor eyesight which prevented him working by artificial light. His teachers used to instruct by ear, and Fisher developed a remarkable capacity for pursuing complex mathematical arguments in his head. This manifested itself in later life in his ability to reach a conclusion whilst forgetting the argument; to handle complex geometrical trains of thought; and to develop and report essentially mathematical arguments in English (only for students to have to reconstruct the mathematics later).

He entered Gonville and Caius College, Cambridge, as a scholar in 1909, graduating BA in mathematics in 1912. Prevented from entering war service in 1914 by his poor eyesight, Fisher held several jobs before being appointed Statistician to Rothamsted Experimental Station in 1919. In

1933 he became Galton Professor of Eugenics at University College London, and in 1943 Arthur Balfour Professor of genetics in Cambridge and a Fellow of Caius College. He retired in 1957 and spent his last few years in Adelaide, Australia, where he died from a post-operative embolism on 29 July 1962.

He married Ruth Eileen Guiness in 1917 and they had two sons and six daughters. He was elected a Fellow of the Royal Society in 1929 and was knighted in 1952 for services to science.

Fisher made a most profound contribution to applied and theoretical statistics and to genetics. He had been attracted to natural history, and especially the works of Darwin, at school, and he had bought Bateson's *Principles of Genetics*, with its translation of Mendel's paper, in his first term as an undergraduate. Before graduating he had already remarked on the surprisingly good fit of Mendel's data, published a paper introducing the method of maximum likelihood, and given a proof of the distribution of the '*t*' statistic which Student had only conjectured.

In 1915 Fisher published the distribution of the correlation coefficient; in 1918 the seminal work in biometrical genetics, 'The correlation between relatives on the supposition of Mendelian inheritance', in which he introduced the word 'variance' and foreshadowed his later development of the analysis of variance; and in 1922 'On the mathematical foundations of theoretical statistics', a paper which revolutionized statistical thought.

As Statistician at Rothamsted he founded the subject of experimental design based on randomization, pursued vigorously the development of statistical estimation theory and invented – or, at least, captured – the quixotic notion of fiducial probability. Moving to London the pace did not slacken, for in addition to pioneering genetical work, especially in connection with the human blood groups, Fisher's statistical explorations revealed the likelihood principle, conditional inference and the concept of ancillarity.

World War II found him embattled on many fronts. Unhappy at home, he found his scientific activity disrupted by wartime conditions including the evacuation of his Department from London. The profundity of his work on statistical inference was ill-appreciated in America, where preoccupation with wartime problems encouraged an excessively mathematical and operational view with which Fisher had little sympathy. In mathematical genetics there were similar difficulties as the American school, starting from his 'fundamental theorem of natural selection', developed ideas of 'adaptive topographies' with false analogies to physical systems. It was not until well after his death that in both statistical inference and mathematical genetics the criticisms which he had advanced came to be appreciated.

After the war, from the relative peace of Cambridge, Fisher saw his theoretical work in both subjects suffer further temporary eclipse. He made great, but ultimately unsuccessful, efforts to establish biochemical genetics in his Department and to secure for Cambridge the national laboratories for human blood-group work. When close to retirement, he was amongst the first to realize the significance of Watson and Crick's discovery of the structure of DNA (1953), and to apply the new computers to a biological problem (1950).

Perhaps embittered by his postwar experiences (though he never relaxed his scientific work), he found some consolation in the Presidency of Caius College from 1956 to 1959, a post second to the Master, and further happiness in retirement in Adelaide.

Fisher wrote five books and published a famous set of statistical tables jointly with F. Yates. An extremely informative and admirably objective biography was published by one of his daughters in 1978 (Box, 1978).

In the field of economics Fisher's name would be remembered for his contributions to statistics alone, so fully chronicled in Box's biography, but we may here draw attention to three other areas not emphasized in the biography but which are especially relevant.

First, the 'fundamental theorem of natural selection' (1930). Although this is specifically directed at a genetical problem, it relies on a simpler implicit theorem of widespread relevance wherever discussion centres on differential growth rates, namely 'the rate of change in the growth-rate is proportional to the variance in growth-rates'. This precise theorem, which is easily proved mathematically, captures the notion that the growth rate of the fastest-growing sub-population (or economic sector, etc.) will come to dominate the overall growth-rate.

Secondly, the modern preoccupation with 'socio-biology' has as one of its origins *The Genetical Theory of Natural Selection* (1930), a fact that only surprises those who have not studied the book and Fisher's other writings on human affairs in the two decades before World War II.

Thirdly, Fisher not only introduced the Theory of Games into evolutionary biology (at the suggestion of Dr Cavalli, later Professor Cavalli-Sforza), but he discovered and published the idea of a randomized or 'mixed' strategy as early as 1934, independently of von Neumann. The problem was the card game '*Le Her*', though if Fisher had gone to the primary source (the correspondence between Montmort and Nicholas Bernoulli, published in 1713) rather than relying only on Todhunter's *History of the Mathematical Theory of Probability* (1865), he would have found that his solution had already been given by Waldegrave.

A.W.F. EDWARDS

SELECTED WORKS

1915. Frequency distribution of the values of the correlation coefficient in samples from an indefinitely large population. *Biometrika* 10, 507–21.

1918. The correlation between relatives on the supposition of Mendelian inheritance. *Transactions of the Royal Society of Edinburgh* 52, 399–433.

1922. On the mathematical foundations of theoretical statistics. *Philosophical Transactions. Royal Society of London, Series A* 222, 309–68.

1925. *Statistical Methods for Research Workers.* Edinburgh: Oliver and Boyd.

1930. *The Genetical Theory of Natural Selection.* Oxford: Clarendon Press.

1935. *The Design of Experiments.* Edinburgh: Oliver and Boyd.

1938. With F. Yates. *Statistical Tables for Biological, Agricultural and Medical Research.* Edinburgh: Oliver and Boyd.

1949. *The Theory of Inbreeding.* Edinburgh: Oliver and Boyd.

1950. *Contributions to Mathematical Statistics.* New York: Wiley.

1956. *Statistical Methods and Scientific Inference.* Edinburgh: Oliver and Boyd.

1971–4. *Collected Papers of R.A. Fisher.* Ed. J.H. Bennett. Adelaide: University of Adelaide, 5 vols.

BIBLIOGRAPHY

Box, Joan Fisher. 1978. *R.A. Fisher: The Life of a Scientist.* New York: Wiley.

fisheries. In 1980 roughly 15 per cent of animal protein consumption by humans was supplied directly by marine and freshwater fisheries. An additional fractional percentage was obtained indirectly from fish products utilized as animal feedstuffs. Industrial products derived from fish included oils, glues, drugs and fertilizers.

Prices (1985 levels) of unprocessed fish vary over several orders of magnitude, from less than $50 (US) per ton for species used for fish meal reduction (e.g. anchoveta, pilchards, herrings, menhaden) to over $10,000 (US) per ton for luxury consumer species such as prawns, lobsters, and salmon. Elasticity of demand also tends to be high for luxury products and low for low-price, high-volume fish species. Demand schedules are determined largely by consumer tastes for fish, which are strongly culture dependent. Production of several unconventional species, such as krill and lantern fish, could be greatly increased if suitable markets could be developed.

Approximately 90 per cent of world fish production is provided by marine fisheries (1981 data), the remainder coming from fresh water. Most marine fish stocks have traditionally been exploited as common property, which has resulted in the severe over-exploitation of many of these stocks. Notable examples include several species of whales and other marine mammals (fur seals, sea otters, sirenians etc.), numerous species of fish such as herring, anchoveta, pilchard and sardine (all schooling species readily located and captured by modern vessels), sea turtles, giant clams and others.

Concern over the potential loss of valuable food resources led to the establishment of both national and international fishery management agencies, particularly following World War II. These agencies were often successful in developing and analysing data bases of catch, effort, and other fishery statistics, and to a limited extent were able to apply this information to the management of fishery resources. By the mid-1970s, however, increasingly severe depletion of many stocks made it apparent that the effectiveness of these institutions was severely hampered by jurisdictional limitations.

These questions were addressed by the Third United Nations Conference on the Law of the Sea (1973–8), which, however, failed to reach agreement on marine resource issues. This development, together with the deployment of efficient distant-water fishing fleets which rapidly depleted fish stocks in many areas, culminated in the declaration in 1977–1980 of 200-mile Coastal Zones of Exclusive Fishery Jurisdiction by virtually every coastal state. Taken together, these 200-mile zones now encompass nearly 99 per cent of marine fishery landings.

The 200-mile EFJ zones provided fishing nations with increased control over coastal fisheries, although transboundary problems still persisted in many areas. By the mid-1980s several depleted stocks had recovered to productive levels, the groundfish stocks of the Grand Banks of Newfoundland being a notable example. While some important stocks remained depressed (e.g. Peruvian anchoveta, Atlantic herring, Atlantic and Pacific salmon, Alaskan King crab), the general trend was towards recovery and a gradual increase in the level of catches.

In spite of the improved state of many fish stocks, the economic position of the fishing industry remained precarious in most countries. In Canada, the federal minister of fisheries announced in 1983 that all the country's commercial fisheries had reached a state of bankruptcy, in spite of rigorous management and generous government subsidies. To some extent the difficulties could be attributed to a cost-price squeeze, as fishing costs continued to escalate, while fish prices failed to increase proportionately.

The real problem, however, appeared to lie in the common-property nature of fishery resources, the implications of which appeared to be only partially understood by management authorities. Experience had proved, as theory had indicated (Christy and Scott, 1965), that unregulated fisheries tended to reach a 'bionomic equilibrium', with depleted stocks and near-zero net incomes to fishermen. Management was directed primarily towards stock protection, and the maximization of sustained catch levels. Fishermen were assisted by various subsidies (particularly for vessel construction), which although usually introduced as temporary measures, often became transformed into quasi-permanent institutions.

The reaction of the fishing industry to these measures at first appeared hopeful. As stocks recovered and catches increased, fishermen's income levels improved, and additional employment was generated. Ultimately, however, a 'regulated' (and possibly subsidized) bionomic equilibrium was reached, with net incomes again near zero – but with greatly increased fishing capacity and correspondingly increased difficulty and expense of management. The main symptom of this situation was the appearance of large fleets of technologically efficient fishing vessels, usually tied up in port for most of the year except for brief openings of the fishing season. In some cases, overcapacity of fishing fleets (and processing plants) has been estimated to exceed 1000 per cent.

More recently, various attempts have been made to limit the extent of overcapacity in regulated fisheries. The restriction of fishing privileges to licensed vessels and operators is now common practice. In some cases, vessel buy-back programmes have also been introduced in an attempt to reduce overcapacity. But many of these 'limited entry' schemes seem to have had the perverse effect of actually *increasing* the capacity of the fishing fleets, as licensed vessels have been progressively upgraded in both power and capacity. As long as the resource remains common property and exploitation remains competitive, the incentive for overexpansion remains dominant.

In an attempt to break through this cycle of regulation and countervailing reaction, some governments have begun to introduce quasi-property rights, in the form of allocated catch quotas. On theoretical grounds, a transferable allocated quota system would have the same beneficial influence on economic efficiency of the fishery as a system of catch royalties or taxes. At the practical level, both taxes and allocated quotas face nontrivial difficulties in implementation. Both would require rigorous monitoring and enforcement, for example. In general, the fishing industry itself seems to exhibit a limited degree of comprehension of the nature of common-property externalities (although the political nature is very well understood indeed). But without understanding and cooperation from the industry, no management programme can be very effective.

The potential economic benefits of efficient fishery operation are so great, however, that it may only be a matter of time until some form of private rights allocations become established for most commercial fisheries. Ultimately, a system combining royalty payments and quota allocations could result in an equitable distribution of the economic rents from these resources – rents which until now have been largely dissipated.

Fishery resources also provide economic services through recreational use – which frequently (in the case of marine fisheries) conflicts with commercial use. Quantitative evaluation of the economic value of recreational fishing remains a controversial topic. The two proposed valuation methods, willingness to pay and fishermen's actual expenditures, are mutually inconsistent, and normally yield quite different numerical results. The current philosophy of treating fish in the water as free goods places management authorities in the position of having to select favoured beneficiaries from among an increasing number of both commercial and recreational fishermen. This dilemma doubtlessly contributes to the persistence of laissez-faire management.

C.W. CLARK

See also COMMON PROPERTY RIGHTS; RENEWABLE RESOURCES.

Bibliography
Anderson, L.G. 1977. *The Economics of Fisheries Management.* Baltimore: Johns Hopkins University Press.
Beverton, R.J.H. and Holt, S.J. 1957. *On the Dynamics of Exploited Fish Populations.* London: Ministry of Agriculture, Fisheries, and Food, Fisheries Investment Series 2(19).
Christy, F.J., Jr., and Scott, A.D. 1965. *The Common Wealth in Ocean Fisheries.* Baltimore: Johns Hopkins University Press.
Clark, C.W. 1985. *Bioeconomic Modelling and Fisheries Management.* New York: Wiley–Interscience.

fixed capital. Fixed capital is the term traditionally used to indicate durable means of production, that is all those inputs of the productive process (such as tools, machines and equipment) that are not exhausted is one single period of production. Non-durable means of production, by contrast defined circulating capital, include raw materials, energy, direct labour, semi-finished goods, etc.

Of course, while circulating capital contributes entirely to the annual production of each commodity, the contribution of fixed capital to production in each period should be determined in relation to the wear and tear actually incurred during its utilization; a datum that in general is not possible to observe directly.

Fixed capital is therefore a complication in the theory of production and it is easy to understand the reason why economists, in their search for abstract simplification of very complex real phenomena, are often induced to assume that production requires only circulating capital.

But technical progress has continuously increased the relevance of machines and plant in industrial production and, as a consequence, a theory of production able to face the problem of fixed capital has become more and more necessary. The most interesting recent contribution in this direction does not belong to mainstream traditional neoclassical theory. It has been made by Sraffa (1960) going back to the classical tradition of determining the value of commodities according to their conditions of production.

HISTORICAL DEVELOPMENTS. Fixed capital is already present in the propositions of the early economists. The determination of its contribution to the annual product of a nation by the Physiocrats and Adam Smith (1766) is however only a description of the behavioural rules of the business world rather than an attempt to explain them. The first analytical discussion of the problem of fixed capital is associated with Ricardo (1821). He is concerned with two particular aspects of the problem.

First of all he noticed that, when the rate of profits is changed, the presence of fixed capital is one of the factors that may alter the proportionality between the ratio of prices and the ratio of the quantity of labour embodied in the corresponding commodities. This is the famous exception of time to the general rule of the labour theory of value that Ricardo put forward in reply to the criticisms raised by McCulloch.

The second aspect of the problem of fixed capital considered by Ricardo, is concerned with the effects of the dynamic substitution in production of machines for labour. He concludes that workers' fears of technological unemployment may be justified, even if the conclusion does not seem to follow logically from his model, that is based on Say' Law.

Marx (1867–94) analyses in detail the consequences of the introduction of fixed capital (machines) on the productivity of labour and strongly underlines the enormous reduction in the price of commodities that it implies ; but apparently he does not care to determine the contribution of fixed capital to the cost of production in each period. A second deeper implication that Marx draws from the substitution in time of machines for labour is the increase in the organic composition of capital, from which he derives his controversial tendency of the rate of profits to fall.

RECENT CONTRIBUTIONS. There are two distinct contributions that, in very different ways, are relevant for the modern analysis of fixed capital: von Neumann (1937) and Leontief (1941, 1953). Von Neumann spends only few words in describing the economic meaning of his mathematical model, but he explicitly remarks that capital goods should appear in both the input and in the output matrix of his model, and should be considered as different goods for each different stage of their utilization, i.e. exactly the same method of analysis later adopted by Sraffa that, nevertheless, at the moment, did not receive any particular attention.

The second contribution, Leontief's input–output model, is relevant because it has many analogies with Sraffa's scheme of production and because Leontief explicitly tries to introduce fixed capital in his model. This is therefore a good starting point to appreciate Sraffa's solution of the problem.

Leontief's (1941) input–output model is a scheme of the flows of commodities among the various industries of the economic system initially conceived to take into account only circulating capital. It determines the quantities of the commodities produced and their prices as solutions of the following two systems of equations:

$$Aq + y = q \qquad (1)$$

$$pA + v = p \qquad (2)$$

where A is the input-output matrix of technical coefficients, q and y are the vectors of total production and of final demand, p is the vector of prices and v is the vector of value added.

But, as Leontief (1953) himself later recognized, a more complete description of the economic system must also involve stocks of commodities (fixed capital) in their various forms: inventories, machines, buildings, etc. He introduces therefore a second square matrix $B = b_{ij}$ that indicates the amount of commodity i required as stock to produce one unit of commodity j. Bq is then the vector of stocks of commodities required to produce the vector of commodities q. Fixed capital stocks affect the balance equation of each period only in terms of the variations of the levels of production $\dot{q} = dq/dt$. This leads Leontief to analyse the dynamic implications of the introduction of fixed capital by means of the following system of linear differential equations:

$$y = q - Aq - B\dot{q} \qquad (3)$$

showing the interaction of stocks and flows as a generalization of the acceleration principle.

Whatever the interest of these dynamic extensions may be, the treatment of fixed capital is rather crude because the determination of depreciations (the fundamental problem with fixed capital) remains exogenous to the model. The amount of fixed capital consumed in each year is in fact predetermined by simplifying assumptions either as a share of the initial stock or as a fixed percentage rate of decay of the residual stock and it is included in the flow matrix A.

FIXED CAPITAL IN A GENERAL SCHEME OF FLOWS. Sraffa's (1960) approach allows a substantial analytical improvement on the problem of fixed capital. He does not consider machines as stocks *à la* Leontief and proposes instead to consider what remains of a machine at the end of each year of operation as a joint product together with the commodity produced. An approach that Sraffa first attributes to Torrens and that afterwards was adopted by Ricardo, Malthus and Marx and then fell into oblivion with the already mentioned exception of von Neumann.

The main interest of Sraffa is in the theory of value and distribution of income. Following the approach of the classical economists, that tried to determine prices from the conditions of production of each commodity, Sraffa formulates a scheme of the production system articulated in two stages. At the first stage of the analysis, when each industry is supposed to produce one single commodity, and the number of industries is equal to the number of commodities produced, Sraffa defines a system of equations that is usually written as follows:

$$a_n w + pA(1+r) = p. \tag{4}$$

It shows that the structure of the production system, as described by the matrix of technical coefficients $A = a_{ij}$ and by the vector of labour coefficients a_n, together with one of the two distributive variables (e.g., the uniform rate of profits r), is sufficient to determine the structure of the vector of prices p and the second residual distributive variable (for analytical details see Newman, 1962 and Pasinetti, 1977).

The meaning of these prices has nothing to do with marginal or neoclassical theory. They represent a more fundamental concept: the exchange rates which ensure the reproduction of the economic system.

The introduction of fixed capital requires the second stage of the analysis, where each industry may produce jointly more than one single commodity. The outcome of this method of dealing with fixed capital is a general scheme of flows that avoids the hybrid interplay between stocks and flows of Leontief's solution.

Obviously a scheme of general joint production is much more complicated than single production. But it is not necessary to go into all the intricacies of joint production to analyse fixed capital. Sraffa considers fixed capital as the leading species of the genus of joint products, and this has suggested an analysis of the intermediate stage where fixed capital is the only element of joint production in a system of single product industries.

At this particular intermediate stage a new system of equations substitutes for the previous one:

$$a_n w + pA(1+r) = pB \tag{5}$$

where $B = b_{ij}$ is a square matrix of outputs that indicates the quantity of each commodity produced and the quantity of old machines, as their joint products, and p is the price vector of the commodities produced, including the price of all old machines at their various ages.

By contrast with the case of single production it might well happen here that, for feasible levels of the rate of profits, some price comes out to be negative, but it is possible to show that, if fixed capital is the only element of joint production of the scheme, then, only the price of old machines might be negative. This has a precise economic meaning: it is a signal of productive inefficiency. It may be shown that, by correspondingly reducing the years of utilization of the machine, the (productive) efficiency of the system would increase (i.e. it would allow higher wages at the same rate of profits). This means that it is always possible, after a suitable truncation of the period of utilization of the machine, to eliminate all negative prices and to obtain a strictly positive solution. (Further analytical details may be found in the essays by Baldone, 1974; Schefold, 1974 and Varri, 1974.)

The method of joint production therefore leads to prices that are economically meaningful and at the same time makes it possible to determine the most efficient life time of durable means of production that turns out to depend, not necessarily in a monotonic way, on the rate of profits.

The remarkable consequence of this result is that, by considering the difference of the prices of the same machine at two subsequent years, it is always possible to obtain the *correct* depreciation quota for that machine in the year considered; correct in the sense of allowing the replacement of the means of production and the payment of profits, whatever the technical conditions of use of the machine may be over its period of utilization. A solution therefore to the problem of determining the wear and tear actually occurred during the utilization of the machine that, as was noticed at the beginning, is impossible to observe directly.

FINAL REMARKS. A remarkable property of the analysis of fixed capital outlined so far is that, though avoiding the difficulties of general joint production schemes, it is rather general and comprehensive. It concerns regular systems where machines are used in their natural sequence and it is necessary to assume that at the end of their life their residual value is zero. Moreover trade of old machines among industries producing different commodities is excluded.

But the analysis does take into account two important complementary aspects of the problem of fixed capital. The first concerns the possibility of considering sets of machines jointly utilized in production, as a unique durable means of production, let us call it a plant, avoiding the indeterminacy of the price of each single component.

The second regards the valuation of obsolete machines no longer produced, but still worth using in production, that may be obtained from the computation of quasi-rents according to the same principle that applies to the rent of lands of different qualities.

More complicated schemes of fixed capital utilization are of course possible but should be analysed within the framework of general joint production.

The most important feature of Sraffa's approach to the problem of fixed capital is that, not requiring any change in the fundamental vision of production as a circular process initially adopted to analyse circulating capital, it greatly contributes to establishing it as a general approach for the analysis of modern systems of production that is alternative to marginalism and neoclassical theory.

PAOLO VARRI

See also CAPITAL AS A FACTOR OF PRODUCTION; CAPITAL GOODS; CIRCULATING CAPITAL.

BIBLIOGRAPHY

Baldone, S. 1974. Il capitale fisso nello schema teorico di Piero Sraffa. *Studi Economici* 29, 45–106. Trans. as: 'Fixed capital in Sraffa's theoretical scheme', in Pasinetti (1980), 88–137.

Leontief, W. 1941. *The Structure of American Economy, 1919–1929.* New York: Oxford University Press.

Leontief, W. et al. 1953. *Studies in the Structure of American Economy.* New York: Oxford University Press.

Marx, K. 1867–94. *Capital.* Moscow: Progress Publishers, 1965–7.

Neumann, J. von. 1937. A model of general economic equilibrium. *Review of Economic Studies* 13, 1945–6, 1–9.

Newman, P. 1962. Production of commodities by means of commodities. *Schweizerische Zeitschrift für Volkswirtschaft und Statistik* 98, 58–75.

Pasinetti, L. 1977. *Lectures on the Theory of Production.* London: Macmillan.

Pasinetti, L. 1980. *Essays on the Theory of Joint Production.* London: Macmillan.

Ricardo, D. 1821. *On the Principles of Political Economy and Taxation.* Vol. I of *The Works and Correspondence of David Ricardo,* ed. P. Sraffa, Cambridge: Cambridge University Press, 1951.

Schefold, B. 1974. Fixed capital as a joint product and the analysis of accumulation with different forms of technical progress. Mimeo, published in Pasinetti (1980), 138–217.

Smith, A. 1776. *An Inquiry into the Nature and Causes of the Wealth of Nations.* Oxford: Clarendon Press, 1976.

Sraffa, P. (ed.) 1951–73. *The Works and Correspondence of David Ricardo.* Cambridge: Cambridge University Press.

Sraffa, P. 1960. *Production of Commodities by Means of Commodities.* Cambridge: Cambridge University Press.

Varri, P. 1974. Prezzi, saggio del profitto e durata del capitale fisso nello schema teorico di Piero Sraffa. *Studi Economici* 29, 5–44. Trans. as 'Prices, rate of profit and life of machines in Sraffa's fixed capital model', in Pasinetti (1980), 55–87.

fixed exchange rates. An exchange rate is a price of one currency in terms of others. The existence of exchange rates derives from the fact that the world is divided into a large number of currency areas, mostly coterminous with nation-states, which trade with one another and therefore exchange currencies at some point (or else confine their trade to barter or 'counter-trade'). The monetary authorities of a country, which regulate money supply and credit conditions, have by the same token a responsibility for the country's exchange rate. The precise significance of the exchange rate in relation to economic policy depends on how that responsibility is exercised; in particular on how far the authorities decide to 'fix' the rate, i.e. keep its movement within a narrow band of fluctuation (in the limit, zero) over a period of time.

There are two polar cases. At one extreme monetary authorities may commit themselves to holding the exchange rate fixed on a quasi-permanent basis. This was the case with adherents to the gold standard before 1914, who defined their currency units in terms of a physical quantity of gold which was not intended to be altered in ordinary circumstances (i.e. short of war or general political breakdown). The gold parity was underwritten by official readiness to buy and sell bullion at the declared price in terms of national currency. The currency exchange rate could then fluctuate in the market only within a narrow band around the parity, limited from above by the so-called gold-import point (at which it would be just profitable for gold traders to ship gold in from abroad for sale to the monetary authorities) and from below by the corresponding gold-export point.

At the other extreme is the case of a freely floating exchange rate. Here the authorities refrain not only from declaring any kind of exchange parity for the currency, but also from intervening in the currency market in order to stabilize or influence the rate. Their impact on the rate is then purely indirect (via the influence of monetary, fiscal and other policies on the behaviour of exchange-market participants), aside from any external transactions undertaken as part of the ordinary business of government, e.g. loans to foreign governments or expenditure on the diplomatic service.

In between the two extremes is a variety of possible exchange rate arrangements. Fixity of rates becomes a matter of degree. The International Monetary Fund (IMF) Articles of Agreement, adopted after the Bretton Woods Conference of 1944, required currencies to be given a par value in terms of gold (either directly or via the US dollar which itself was defined in terms of gold); but the par values could be altered in the event of 'fundamental disequilibrium' and thus came to be known as 'adjustable pegs'. Going down the spectrum, criteria for altering parities can be set so as to encourage more frequent and presumably smaller changes (as in the various types of 'sliding' or 'crawling' peg regimes), and the permitted margins of fluctuation around any given peg can be widened. If parities are abandoned, the authorities may still engage in extensive management of the floating rate through intervention in the currency market ('dirty floating'), as well as measures of monetary policy or exchange control.

The choice of exchange-rate arrangements for a single country is constrained by circumstances in the world at large and/or by the nature of the country's own economy. If, for example, major countries form a fixed-rate system, then an individual small country will have the choice of either participating in the system or remaining outside it and selecting its own exchange-rate regime. If, on the other hand, the major currencies are floating in relation to one another (like the US dollar, the yen and the Deutschemark after 1973), then there is no straightforward fixed-rate option for other countries. At best, they can peg their currencies to *one* of the majors, or they can stabilize the value of their own currency in terms of some 'basket', i.e. weighted average of foreign currencies.

THE PRICE LEVEL AND MONETARY STABILITY. Whether freely chosen or not, a country's exchange-rate regime affects, first, the dynamic relationship between its national price level and those of other countries, and secondly, the *modus operandi* and relative impact of monetary and fiscal policy instruments.

The more rigidly fixed a country's exchange rate, the greater is the weight of external influences in determining movements of its domestic price level. The channels through which these external influences make themselves felt are varied and complex. They are relatively direct in the case of goods, services and factors of production traded internationally. To be sure, transport and transaction costs, product differentiation and other market imperfections prevent full compliance with the 'law of one price' even for the traded goods sector; but the sum total of such obstacles to full price equalization tends to be relatively constant over time, so that any significant change in the world price of a country's imports or exports is quickly passed through.

For the change in question to be, and to remain, purely monetary in nature, i.e. to have no impact on the level or composition of output and real incomes, three further conditions must be fulfilled. First, the global price shock must itself be purely monetary, i.e. must affect all traded-goods prices equiproportionally and leave the terms of trade unaltered. Secondly, the domestic economy must be characterized by widespread price flexibility, so that the price impulse is promptly transmitted to non-traded items, thus leaving domestic relative prices (of traded and non-traded goods) also unaltered. Thirdly, there must be appropriate adjustments in macro-economic, especially monetary, policy, in order to prevent either over-financing or under-financing of a given real product as the price level changes.

These conditions will seldom be met simultaneously. Monetary and real (output) disturbances are in practice intermingled. However, the conspicuous feature of fixed exchange rates in this domain is that they enforce, or presuppose, an approximately uniform system-wide inflation rate (as under the pre-1914 gold standard, or in the adjustable-peg period of 1950–1970). By contrast, floating rates permit wide divergences in national inflation rates, which are accommodated, and in part brought about, by exchange-rate movements (as was widely seen in the 1970s). Systemic inflation in the presence of fixed exchange rates will in practice always be low; otherwise the system would not command wide acceptance.

The combination of low inflation and a fixed or pegged exchange rate constitutes a virtual definition of monetary stability in an international system, and provides major real benefits by facilitating the efficient operation of the price system and the near-optimal use of money in exchange. Nonetheless, depending on the precise constitution of a fixed-rate system (i.e. whether rates are meant to be totally rigid; or if not, in what conditions and by how much they may be altered), countries may opt out or may be forced out for either of two reasons. They may find the international inflation rate unpalatable (e.g. a rate of three per cent per annum is probably acceptable to many countries but distastefully high to a few), and see insufficient compensating attractions in exchange-rate fixity as such. Alternatively, they may find the international inflation rate unattainably low, at any rate without incurring, or appearing to incur, unacceptable (even if temporary) costs. The costs comprise lost output and employment, or social disruptions over price/wage issues such as subsidies or trade union reform.

MONETARY AND FISCAL POLICY. A pegged exchange rate calls for a certain pattern of macro-economic management by national authorities. Monetary policy, especially changes in interest rates, can play a leading role in influencing aggregate private spending only if there are narrow limits to the international mobility of funds. Otherwise, the main impact of monetary measures, at least up to the medium term, is upon the disposition of internationally mobile stocks of capital, and hence upon the financial underlay to a given volume and value of national expenditures, rather than the expenditure volume itself. Monetary tightening pulls in funds from abroad; monetary easing pushes funds out (unless the respective tightening and easing is simultaneously matched by other countries). In the limit, national interest rates are determined wholly by the international capital market and its assessment of the individual country's credit rating, rather than by national preferences or policy. By the same token, fiscal policy (government expenditure, taxation and borrowing) then has a relatively great impact on national expenditure, output and the external current-account (export/import) balance.

The division of function between policy variables is quite different with freely floating exchange rates. Here the international mobility of capital (without which floating is not feasible) means that monetary measures, instead of affecting the level of external reserves, alter the exchange rate and thus the domestic price of traded goods. This in turn, depending on circumstances as before, will affect the price of non-traded goods and/or the level and composition of output. Monetary expansion, for instance, depreciates the exchange rate and raises the domestic price of traded goods, stimulating the economy and generating some combination of higher output

and higher prices. Pure fiscal policy, on the other hand, is generally less effective than before, because higher (lower) public-sector borrowing demands lead promptly to a higher (lower) exchange rate, which tends to offset the aggregate expenditure impact of the fiscal change. Only in the special case of balanced-budget fiscal policy may an equiproportionate change in government outlays and receipts affect aggregate demand even in a floating-rate regime with perfect world capital markets (McKinnon and Oates, 1966).

The contrast between the fixed and floating rate cases is most complete for a 'small' country whose behaviour has no significant impact on global economic variables. In a 'large' country monetary tightening will influence credit conditions worldwide under both fixed and floating rates, while fiscal policy will have an impact on aggregate world expenditure under either exchange-rate regime.

THE WORLD MONETARY SYSTEM. Exchange-rate arrangements are the most important constituent of the (market-economy) world monetary system. The other principal constituents are international reserve assets and arrangements for co-operation among sovereign monetary authorities and (where appropriate) international bodies such as the IMF. Global exchange-rate arrangements are determined by the small number of major countries which at any one time constitute the core of the international economy. After 1973 the world was perceived to have abandoned the pegged-rate system in favour of floating rates, even though the vast majority of the world's 100-plus currencies remained pegged to some major currency or basket of currencies. The crucial change lay in the fact that the US dollar, the Deutschemark and the yen were now in a floating relationship to one another. In addition, a few currencies of secondary importance, such as the pound sterling and the Swiss franc, were likewise floating.

The principal focus of the story is the dollar, as the system's principal reserve currency and the currency in terms of which virtually all countries had maintained pegged exchange rates over the preceding quarter-century. The key question is why the German and Japanese authorities did not re-establish a pegged-rate relationship with the dollar, despite great concern at times over the way in which floating rates were moving. Indeed, in 1978 the German government specifically took the initiative to create within Europe a stronger bloc of mutually pegged exchange rates (the European Monetary System) as a counterweight to an unstable and at that time undervalued dollar. Evidently, pegging to the dollar was seen as courting greater risks to financial stability than other courses of action. Such risks must be rooted in the presumed determinants of US financial policy, and specifically in the belief that, if other major countries commit themselves to maintaining exchange-rate pegs vis-à-vis the dollar, the US authorities for their part will not give adequate weight to the external repercussions of their policy unless they too are committed to defending an exchange-rate peg and reserve position of their own. A pure 'dollar standard' has not been an acceptable basis for a world-wide system of pegged rates, because it would leave the United States insufficiently subject to balance-of-payments discipline.

The problem of imposing payments discipline on the centre country (or countries) of a fixed-rate system has historically been solved (or avoided) in only one way, namely by pegging that country's currency and hence the system as a whole to an 'outside' commodity asset, most successfully to gold. The market for this commodity then serves as the vehicle for reconciling the competing responsibilities and preferences of the sovereign governments which make up the international system.

The theory of the pre-1914 gold standard was that movement of gold reserves determined changes in national money stocks and hence, with a given structure of domestic payments, in money national incomes. Price and wage flexibility was relied upon to assure full employment of available productive resources and, in the process, to reconcile the resulting real national incomes with their current money values as determined by the monetary mechanism.

Further implications followed. The distribution of global increments in the stock of monetary gold (equal in any period to the excess of current mine production over net private offtake for industry, hoarding, etc.) was governed by relative growth rates of real GNP. Fast growth of an economy tended to produce a relative lowering of its price level, which tendency would be checked and the price level kept in line by relatively fast growth of its gold reserves and money supply. Finally, if global economic growth was faster (slower) than current growth of money stocks, there would be downward (upward) pressure on the world price level; with the price of gold alone fixed in money terms, this meant a rise (fall) in the real price of gold, which would sooner or later augment (diminish) the net inflow of gold to the monetary system, thereby tending to halt or reverse the original movement in world price levels.

The operation of the gold standard in practice corresponded only very partially to the theoretical model. For instance, growth of monetary gold stocks was reconciled with faster growth of national outputs less by downward pressure on price levels than by increased concentration of monetary gold at central bank reserves and a shrinkage in gold's share of money aggregates (Triffin, 1964). However, national monetary policies were governed to a large degree by balance-of-payments considerations, and a broad measure of global price stability was maintained.

The adjustable-peg system of Bretton Woods (devised chiefly by J.M. Keynes and H.D. White) was a type of gold-exchange standard, but one which ultimately subordinated changes in monetary gold stocks to the growth of money incomes rather than the other way round. This intended reversal of gold-standard relationships stemmed from the Keynesian assumptions that maintenance of full employment was a government responsibility which could not in general be delegated to market forces, and that money wages and prices were inclined to be inflexible, especially downwards; hence national authorities must be free to arrange whatever level of national purchasing power they judged appropriate for maintaining high employment and avoiding inflation. Situations might arise in which one or more countries could not achieve this overriding objective at the previously declared exchange-rate pegs ('par values') without recourse to (additional) administrative restrictions on trade and current payments. In such a situation (the 'fundamental disequilibrium' of IMF terminology) a par value could be adjusted – downwards to reduce the home country's wage level in international terms, thus boosting its competitiveness; or upwards to increase the wage level in international terms, thus fending off excessive reserve gains and inflation.

Modest reserve gains, however, were viewed as desirable, and certainly as acceptable, by many countries, particularly in a period of rapid economic expansion like the 1950s and 1960s. Equilibrium of the system as a whole therefore required a certain growth of global exchange reserves to avoid a competitive scramble among countries for balance-of-payments surpluses. The annual inflow of new monetary gold after 1945 was at no time sufficient for this

purpose. The gap was filled, at first deliberately and then involuntarily, by the United States, which ran an overall deficit on its balance of payments, thereby acting as a net supplier of reserves to other countries. The immediate supply took the form of dollars, which then constituted a potential claim on the US gold stock and were in part exchanged for gold by foreign monetary authorities.

Triffin (1960) first emphasized that this process was weakening the external liquidity position of the United States and could not continue indefinitely without calling into question the gold convertibility of the dollar at its declared par value of 0.888671 grammes of gold fine or $35 per ounce of gold. Contrary, however, to what Triffin implied, the United States could not put an end to its deficit without first altering (or abandoning) its par value. By standing ready to sell gold to foreign monetary authorities at $35 an ounce, the US Treasury was acting in effect as buffer-stock manager for an under-priced commodity – a commitment which could have only one outcome. Perception of the point was paradoxically hampered by the fact that the dollar was until near the end of the 1960s scarcely overvalued against other major currencies. The pressure on the US balance of payments to act as a net source of reserves to the outside world stemmed from the dollar's overvaluation in common with all other currencies vis-à-vis gold (Gilbert, 1968, 1980).

The IMF Articles had envisaged such a possibility. Not only did they give the United States exactly the same scope to alter its par value as any other country; in addition, they provided for 'a uniform change in all par values', i.e. a general rise in the price of gold, in order to relieve a system-wide shortage of reserves or reserve increments. The US authorities declined to avail themselves of this measure, viewing or professing to view it as unlikely to promote payments equilibrium and therefore as an unwarranted blow to the prestige of the dollar. By 1970 US gold reserves had declined from their post-World War II peak of $22 billion to little more than $10 billion, while US liquid external liabilities had risen from negligible amounts to over $20 billion. The dollar's gold convertibility was formally abrogated on 15 August 1971 and the attempt to maintain a pegged-rate system on the basis of an inconvertible dollar foundered in March 1973.

Many observers have been reluctant to accept that the demise of the pegged-rate system was due to the US refusal to increase the dollar price of gold. Instead they have claimed, on the one hand, that the Bretton Woods system would in any event have been swept away by the inflation and balance-of-payments problems of the 1970s (an unconvincing line of argument, not least because the world inflation of the 1970s was itself in large measure caused by the financial turmoil in which the pegged-rate system collapsed); and on the other hand, that a fiduciary asset such as IMF Special Drawing Rights could have replaced gold (and could still do so) at the base of a pegged-rate system, but for the fact that the vulnerability of adjustable pegs to speculative attack renders them unviable anyhow in the face of free international capital movements.

Neither leg of the latter argument is persuasive. Gold was able to function as the basis of an adjustable peg system because its availability for this purpose is regulated with the help of market forces and without the need for detailed and continuous agreement on reserve creation and exchange-rate policy among sovereign governments. Specifically, the Bretton Woods System incorporated a strong and direct link between the exchange-rate policy and the international liquidity position of the United States: a reduction in the dollar's par value could always be made large enough to produce a decisive impact on US reserves. A fixed-rate system based on a fiduciary asset such as SDRs would lack this feature, and would therefore be only a special form of currency standard, like the abortive dollar standard of 1971–73.

Currency speculation, as distinct from politically motivated capital flight, does not initiate balance-of-payments problems. Rather, it emerges as an aggravating factor when there is an evident underlying disequilibrium which the authorities are slow to tackle and which therefore presents speculators with the prospect of easy gains. Variation in the method of altering an individual par value (e.g. temporary floating, or small changes of greater frequency) may in some circumstances be a useful means of containing or discouraging speculation. Such devices, however, were quite irrelevant to the disequilibrium and breakdown of Bretton Woods, since the United States was unwilling to alter the dollar's par value by any method, and without such alteration the system could not be brought to equilibrium.

PETER M. OPPENHEIMER

See also CRAWLING PEG; EXCHANGE RATE POLICY; FLEXIBLE EXCHANGE RATES; INTERNATIONAL FINANCE; INTERNATIONAL MONETARY POLICY.

BIBLIOGRAPHY

Dornbusch, R. 1980. *Open Economy Macro-Economics.* New York: Basic Books.
Eichengreen, B. (ed.) 1985. *The Gold Standard in Theory and History.* New York and London: Methuen.
Fleming, J.M. 1962. Domestic financial policies under fixed and under floating exchange rates. *IMF Staff Papers* 9, November, 369–79.
Gilbert, M. 1968. *The Gold/Dollar System: Conditions of Equilibrium and the Price of Gold.* Princeton Essays in International Finance No. 70, Princeton: Princeton University Press.
Gilbert, M. 1980. *Quest for World Monetary Order.* New York: Wiley.
McKinnon, R. and Oates, W.E. 1966. *The Implications of International Economic Integration for Monetary, Fiscal and Exchange-Rate Policy.* Princeton Studies in International Finance No. 16, Princeton: Princeton University Press.
Meade, J.E. 1951. *The Theory of International Economic Policy: Vol. I, The Balance of Payments.* London: Oxford University Press.
Mundell, R.A. 1968. *International Economics.* New York: Macmillan.
Triffin, R. 1960. *Gold and the Dollar Crisis.* New Haven: Yale University Press.
Triffin, R. 1964. *The Evolution of the International Monetary System: Historical Reappraisal and Future Perspectives.* Princeton Studies in International Finance No. 12, Princeton: Princeton University Press.

fixed factors. In moving from one market equilibrium to another, a firm may choose to hold fixed the rate of employment of one or more factors of production. The presence of fixed factors and their associated overhead costs will affect the firm's responses to changing market conditions. The residually determined quasi-rents which constitute the returns to the fixed factors must, in the long run, cover their overhead costs; otherwise, the inputs of fixed factors have to be contracted. The importance of fixed factors and overhead costs, which varies across firms and industries, was analysed by J.M. Clark (1923) who emphasized the first of the following three questions: (1) How do fixed factors affect the behaviour of prices, outputs and inputs of variable factors? (2) What determines whether a factor of production will be fixed or variable? (3) How do the fixed employment costs of

quasi-fixed labour inputs affect contractual arrangements in labour markets?

In the short run, certain paths of adjustment are barred to the firm. The usual assumption is that the input of one or more factors is fixed. Total unit costs, which include the outlays for fixed factors, lie above average variable costs so that price, in the short run, can remain well below the minimum long-run average cost. If fixed costs in an industry are high, they can pose a barrier to entry of new firms and could result in wide short-run fluctuations in price. Further, the upper-bound constraint on inputs of fixed factors affects the firm's demand for the remaining variable inputs in a manner analogous to the theory of rationing of consumer goods analysed by E. Rothbarth (1941). An increase in the demand for the final product raises the shadow price of the fixed factor, which increases the demand for variable factors that are substitutes for the fixed factor and decreases the demand for complementary variable factors. This result could explain the greater cyclical volatility in the demand for unskilled labour relative to skilled labour if unskilled labour is a closer substitute for the fixed factor, capital. Moreover, the smaller the elasticity of substitution of labour for capital, the steeper is the slope of the marginal cost curve, implying larger cyclical swings in product prices.

A firm will fix the input rate of a factor if (a) the factor is specific to the firm in the sense that employment in this firm constitutes its highest valued use, or (b) reallocation to some higher-valued use is precluded by some contractual agreement or by a prohibitively high transaction cost. In the former case, equipment, buildings and even labour can be specialized to fit into a firm's idiosyncratic production methods. The internal values of such specialized resources are likely to exceed their external values to outside users. These resources are more likely to be owned (rather than hired or leased), because of their specificity. Long-term contracts that account for some fixed factors occur where there are gains from risk-sharing or high costs of transferring resources to other firms.

A richer theory of factor markets can be developed if the dichotomy of fixed versus variable factors is replaced by a continuum of degrees of fixity. The discipline of labour economics has now accepted the principle that labour is a quasi-fixed factor. The cost of hiring and training workers constitutes the fixed component of the full cost of labour, while the variable component is the wage paid to the employee. In long-run equilibrium, the expected marginal value product which depends on the expected product price P^* and labour's marginal physical product f_N, is equated to the full labour cost:

$$P^*f_N = W + q, \qquad \left[q = \left(\frac{F}{r} \right)(1 - e^{-rT}) \right]$$

where W is the wage, and q is the periodic rent that amortizes the fixed employment cost F at a discount rate r over the worker's expected period of employment T. The gap between the wage and labour's marginal value product will be relatively larger, the higher is the degree of fixity which can be measured by $f = q/W + q$.

The cyclical behaviour of the labour market is characterized by an uneven incidence of unemployment, a compression of occupational wage differences in the upswing, persistent differences in labour turnover rates, hiring/firing practices that smack of discrimination. The quasi-fixity of labour goes a long way in explaining these phenomena. In the downswing, the product price falls below its long-run level P^*. If labour is a completely variable input, meaning that $q = F = 0$, its marginal value product Pf_N will be equated to the wage in each period. Hence, when P falls, the demand for this grade of labour is contracted until f_N climbs to restore equilibrium in both factor and product markets. However, if labour is a quasi-fixed factor, the periodic amortization of the fixed cost drives a wedge between the wage and marginal value product. For a small decline in product price, the firm will not contract the demand for a quasi-fixed grade of labour as long as its short run MVP exceeds the wage, which is the variable cost of labour; that is, if $Pf_N > W$ even though $Pf_N < (W + q)$, the input of this grade of labour will not be reduced in the downswing. There is, for each quasi-fixed factor, a trigger price P_i at which the firm will choose to reduce employment. The trigger price which induces a decline in factor demand will be lower for factors with higher degrees of fixity. In the early stages of a downturn, labour with low degrees of fixity will become unemployed, while other workers will be retained until the drop in product price P is driven below P_T. At the trough of a cycle, most grades of labour satisfy a short-run equilibrium condition where labour's MVP is equated to its variable cost, $Pf_N = W$. As P rises in the recovery, a firm will increase its demand for a quasi-fixed factor if the price rise is such that labour's MVP exceeds its *full cost*; that is, employment is expanded if and only if $Pf_N > (W + q)$. In the upturn, the rightward shift in factor demand will be greater for factors with lower degrees of fixity. Employment will be more stable, and the incidence of unemployment will be lower for those workers in occupations with higher degrees of fixity.

Some firms find that it is profitable to incur the fixed employment costs of assembling a firm-specific workforce. Recruiting is the means by which an employer identifies more productive individuals and ascertains whether an applicant will meet prescribed hiring standards. Recruitment for high-wage positions usually entails higher costs because of the variability of individual productivities. Employers who have well-defined internal labour markets and who organize production around teams also incur higher recruiting costs. In an internal labour market, workers are hired at a limited number of ports of entry and are typically given on-the-job training to adapt them to the firm's idiosyncratic production methods. Larger investments in firm-specific human capital are indicative of the greater specialization of the labour input. Firm-specific training is less profitable when labour turnover rates are high due either to the high separation propensities of workers or the low survival odds of firms. Smaller firms spend less on recruiting and appear to invest less in formal training. The estimates reported by Oi (1962) and Parsons (1972) reveal that employers incurred substantially higher fixed employment costs for workers in higher skill levels. The degree of fixity, $f = q/(W + q)$, is positively related to the wage rate W, and this relation allows us to test the implications of a theory of labour as a quasi-fixed factor. Employees in high-wage occupations experience greater employment stability over the cycle. Occupational wage differentials widen in the downswing and narrow in the upswing. Labour turnover rates are lower, and recruiting costs are higher in large firms whose workforces exhibit a higher degree of fixity.

The persistence of unemployment and the failure of wages to clear labour markets call for an explanation. Some unemployed workers are in a state of pseudo-idleness while they look for work: 'When actively searching for work, the situation is that he is really investing in himself by working on his own account without immediate remuneration. He is prospecting' (Hutt, 1977, p. 83). The time and money spent by new entrants and disemployed workers in their search for suitable job matches constitute a fixed cost which has to be recovered over the course of the employment relation. Each job is, in a very real sense, specialized to the worker–firm attachment. In a search model, unemployment can be efficient in two senses. First, it may be the least-cost means of finding a durable job. Second, a worker on a temporary layoff may stay in a state of availability awaiting recall rather than seeking work. Labour turnover is costly, both to the employer for whom labour is a quasi-fixed factor due to the fixed investments in hiring and training, as well as to the employee for whom this job is specific due to the fixed costs of search. Both parties have incentives to form an implicit contract that can raise the returns to these fixed employment costs by lengthening the expected period of employment.

Long-term employment contracts could be the result of risk-averse workers seeking job security. An employer can reduce his full labour costs by providing a tacit agreement in which the risks of income variability are shared. Such long-term agreements end up increasing the fixity of labour. Implicit, long-term contracts may also result from an employer's desire to discourage shirking and dishonesty. Firms will incur monitoring and enforcement costs to deter dysfunctional behaviour and malfeasance. These enforcement costs can be reduced by designing compensation packages which reward workers with separation pay and pensions if they perform in accordance with prescribed work standards. Stable and durable employment relations only make sense when there are fixed costs of forging and maintaining specific jobs defined by worker–firm attachments.

When physical and human capital are specialized to a firm, it must capture any quasi-rents that it can because the fixed investments in these specialized resources cannot be reallocated to some alternative use. Fixed, firm-specific factors only make sense in a world of heterogeneous firms. In Oi (1983) I advanced the thesis that firm-specific capital was systematically related to firm size. Small firms with low survival odds do not invest in custom-made machines and specifically trained employees. They are more likely to purchase used assets and to hire inexperienced workers with general human capital. The overhead costs of fixed-capital assets are relatively larger for big firms that engage in the volume production of standardized products. Large firms also incur higher fixed employment costs to recruit and train a specialized workforce. Workers in large firms are paid higher wages and are provided with employee compensation packages that are designed to reduce labour turnover rates. These phenomena could not be explained without a formal analysis of fixed and quasi-fixed factors.

WALTER Y. OI

See also OVERHEAD COSTS; RENT.

BIBLIOGRAPHY

Clark, J.M., 1923. *Studies in the Economics of Overhead Costs.* Chicago: University of Chicago Press.

Hutt, W.H. 1977. *The Theory of Idle Resources.* Indianapolis: Liberty Press.

Oi, W.Y. 1962. Labor as a quasi-fixed factor. *Journal of Political Economy* 70, November, 538–55.

Oi, W.Y. 1983. The fixed employment costs of specialized labor. In *The Measurement of Labor Costs*, ed. J.E. Triplett, Chicago: University of Chicago Press.

Parsons, D.O. 1972. Specific human capital: an application to quit rates and layoff rates. *Journal of Political Economy* 80, November, 1120–43.

Rothbarth, E. 1941. The measurement of changes in real income under conditions of rationing. *Review of Economic Studies* 8, February, 100–107.

fixed point theorems. An economic system, which consists of a number of relationships among the relevant factors, is modelled as a system of equations or inequalities of certain unknowns, whose solution represents a specific state in which the system settles. This is typically exemplified by the Walrasian competitive economy (Walras, 1874), consisting of the interaction of manifold behaviours of many individual agents with different motivations, whose specific state is a competitive equilibrium in which certain prices of goods solve the system of equations representing the simultaneous clearing of all the markets of goods so as to make those individual behaviours mutually consistent.

An economic theory is the formulation of an economic system and the elucidation of its structural and performance characteristics. It is then a primary premise of the theory, on which all its developments are built, that its modelled system be consistent in the sense that the corresponding system of equations or inequalities has a solution. Without this consistency the theory is void. For the Walrasian competitive economy the existence of a competitive equilibrium is the premise.

Conventionally, economists satisfied themselves about the existence of a solution of the relevant system of equations on the ground that there are as many equations as unknowns, a rule of thumb, which, though it is a criterion for the consistency of a system of equations in general, is neither necessary nor sufficient for the existence of a solution of a given specific system of equations.

It was generally realized in the 1950s that to deal properly with the existence of solutions of systems of equations it is necessary to rely on more sophisticated mathematical methods, among others concepts and theorems in topology, which was dawning around the beginning of the 20th century and has by now grown to a highly advanced field in contemporary mathematics. Fixed point theorems are the most notable ones among them.

Stated in the most indefinite form, a fixed point theorem is a proposition which asserts that a mapping f that transforms each point x of a set X to a point $f(x)$ within X has a fixed point x^* that is transformed to itself, so that $f(x^*) = x^*$. An arbitrary mapping need not have a fixed point, and the truth of a fixed point theorem hinges on the behavioural properties of the mapping and the structural properties of the set on which the mapping is defined.

A fixed point theorem, the first of such theorems in topology, was formulated and proved by Brouwer (1910), a great master in mathematics. It asserts the existence of a fixed point in the case where X is the set of all those points in the n-dimensional Euclidean space whose coordinates are non-negative and add up to unity, called a *simplex*, and where the mapping f transforms points within X in a continuous way.

This theorem has been used in existence problems arising in many fields, mainly in physical sciences. In economics it was used for the first time by von Neumann (1937) to prove the existence of a balanced growth path and an associated set of equilibrium prices in his multisectoral model of an expanding economy. Later in the 1950s it was applied in its original or extended versions to prove the existence of a competitive equilibrium in the Walrasian economy almost simultaneously by several authors (Arrow and Debreu, 1954: McKenzie, 1954; Gale, 1955 and Nikaido, 1956).

Brouwer's theorem still holds good in cases where the set X is not the simplex but a set topologically equivalent to the simplex, which is by definition the image of the simplex under a one-to-one continuous mapping. Bounded, closed, convex sets in the Euclidean space are sets of the simplest structure among the sets topologically equivalent to the simplex. The boundedness of a set means that the distance between any two points within it can not be indefinitely large, but has a uniform bound. The closedness of a set means that there is no point outside it which can be approached as closely as one likes from its inside. The convexity of a set means that any two points x, y can be joined by a segment within it that is represented by $tx + (1-t)y$, $(1 \geqslant t \geqslant 0)$. These three properties are often possessed by sets of relevant variables that arise in economic models.

The procedure by which von Neumann reduced the solution of the existence problem for his model of an expanding economy to Brouwer's fixed point theorem was very involved, but suggested a generalization of the concept of a fixed point on which to formulate an extension of the theorem, that fits economic models better in dealing with the existence problems arising in them. This extension was carried out by Kakutani (1941).

Kakutani's fixed point theorem pertains to a multi-valued mapping, a generalized version of the ordinary single-valued mapping. In the generalization a mapping f associates with each point x of a set X a set $f(x)$. Such a mapping is called a *point-to-set* mapping, a *set-valued* mapping or a *correspondence*. For a set-valued mapping f a fixed point x^* is a point which is included in its image, $x^* \in f(x^*)$. Regarding the way a set-valued mapping transforms points to sets there is a concept of continuity which is a counterpart of that for ordinary single-valued mappings. f is said to be *upper semi-continuous*, or *upper hemi-continuous*, if at each point x the image sets $f(y)$ of its nearby points y are very close to $f(x)$; formally, if for any neighbourhood U of $f(x)$ at each point x there is a neighbourhood V of x such that the image $f(y)$ of any point y belonging to V is included in U.

Kakutani's fixed point theorem states that if a set-valued mapping f transforms each point x of a bounded closed convex set X in the Euclidean space to a non-empty closed convex subset of X in an upper semi-continuous way, then there is a fixed point $x^* \in f(x^*)$.

Fixed point theorems in the Brouwer-Kakutani line are very powerful in solving existence problems arising in a number of important economic models. The following are typical examples of their use in the solution of such problems.

(a) Existence of an equilibrium point in a many-person noncooperative game (Nash, 1950). Let $K_i(x_1, x_2, \ldots, x_n)$ be the payoff function of the ith player, continuous with respect to the n-tuple of the players' strategies x_i, each of which is chosen from among the strategy set X_i, closed, bounded and convex for $i = 1, 2, \ldots, n$, where n is the number of players. An equilibrium point, which is such a special n-tuple of strategies $(x_1^*, x_2^*, \ldots, x_n^*)$ that x_i maximizes the ith player's payoff given that the other players' strategies are x_j^* $(j \neq i)$, is obtained as a fixed point of a mapping. The relevant set-valued mapping f is that which transforms each n-tuple of strategies $(x_1, x_2, \ldots, x_n)$ to the cartesian product of $M_i (i = 1, 2, \ldots, n)$, where M_i is the set of the ith player's strategies that maximize his payoff given that the other players' strategies are $x_j(j \neq i)$, and is assumed to be convex. f transforms each point $(x_1, x_2, \ldots, x_n)$ to the closed convex set $M_1 \times M_2 \times \cdots \times M_n$ within the cartesian product $X_1 \times X_2 \times \cdots \times X_n$, a bounded closed convex set in the Euclidean space in an upper semi-continuous way, and has a fixed point by virtue of Kakutani's fixed point theorem.

(b) Existence of a competitive equilibrium in the Walrasian competitive economy. The Walrasian competitive economy is modelled as a national economy-wide system consisting of profit-maximizing producers and utility-maximizing households in the most articulate fashion by Arrow and Debreu (1954) and Debreu (1959). A competitive equilibrium is a specific state of the system where at some prices of goods these manifold individual price-taking maximization behaviours are mutually consistent. It can be characterized as simultaneous market clearing for all goods in terms of excess demand functions of the prices that are derived from the individual maximization behaviours, and its possibility is ensured by Kakutani's fixed point theorem (Gale, 1955; Nikaido, 1956; Debreu, 1959).

Simultaneous market clearing, which is referred to as Walras's theorem can be stated as follows: Let an excess demand function ϕ be associated with each n-tuple of prices $p_1, p_2, \ldots, p_n$ non-negative and adding up to unity a non-empty bounded, closed and convex subset $\phi(p_1, p_2, \ldots, p_n)$ of an n-dimensional Euclidean space, consisting of points with co-ordinates that represent excess demands for goods, in an upper semi-continuous way; and let Walras's Law,

$$\sum_{i=1}^{n} p_i e_i = 0$$

be satisfied for any excess demand vector $(e_1, e_2, \ldots, e_n)$ included in $\phi(p_1, p_2, \ldots, p_n)$. Then there are specific prices p^*, $p_2^*, \ldots, p_n^*$ at which the corresponding set $\phi(p_1^*, p_2^*, \ldots, p_n^*)$ includes a special excess demand vector $(e_1^*, e_2^*, \ldots, e_n^*)$ with nonpositive excess demands $e_i^* \leqslant 0$ for all goods $i = 1, 2, \ldots, n$, and $e_i^* = 0$ for goods whose prices p_i^* are positive.

The excess demand function ϕ turns out to be bounded in the sense that there is a bounded closed convex set E large enough to include $\phi(p_1, p_2, \ldots, p_n)$ for all points $p = (p_1, p_2, \ldots, p_n)$ in the simplex. Thus the competitive equilibrium is obtained by Kakutani's fixed point theorem as a fixed point of the mapping f which transforms each pair (p, d) of a point p in the simplex and an excess demand vector $d = (d_1, d_2, \ldots, d_n)$ in E to the set of all pairs (q, e), e chosen from among $\phi(p_1, p_2, \ldots, p_n)$ and $q = (q_1, q_2, \ldots, q_n)$ with the coordinates

$$q_i = [p_i + \max(d_i, 0)] / \left[1 + \sum_{j=1}^{n} \max(d_j, 0) \right], \quad (i = 1, 2, \ldots, n).$$

The mapping f transforms points to closed convex sets in an upper semi-continuous way within the cartesian product X of the simplex and E, a bounded closed convex set, so that Kakutani's fixed point theorem applies to f on X. At a fixed point (p^*, e^*) hold

$$e^* \in \phi(p_1^*, p_2^*, \ldots, p_n^*)$$

$$p_i^* = [p_i^* + \max(e_i^*, 0)] / \left[1 + \sum_{j=1}^{n} \max(e_j^*, 0) \right]$$

so that

$$\sum_{i=1}^{n} p_i^* e_i^* = 0$$

$$p_i^* \sum_{j=1}^{n} \max(e_j^*, 0) = \max(e_i^*, 0), \quad (i = 1, 2, \ldots, n),$$

which rules out the possibility that some e_i^* are positive, and implies $e_i^* = 0$ for i such as $p_i^* > 0$.

As a mathematical result Brouwer's fixed point theorem is deeply rooted in the profound topological nature of Euclidean space, which necessitates the complexity of its methods of proof. The most straightforward proof, which is due to Knaster, Kuratowski and Mazurkiewicz (1929), is based on a combinatorial theorem known as Sperner's lemma (Sperner, 1928) on subdivisions of the simplex to small simplices. Kakutani's fixed point theorem is proved by applying

Brouwer's theorem to single-valued mappings approximating the given set-valued mapping.

In the late 1960s computational algorithms were developed by Scarf (1973) to approximate as closely as one likes the fixed points in the theorems in the Brouwer–Kakutani line, based on his new combinatorial theorem, similar to but distinct from Sperner's lemma, on fine grids of small simplices lying in the simplex. They serve as efficient procedures to locate fixed points explicitly, and also effect an alternative proof of Brouwer's theorem.

The Kakutani fixed point theorem has by now been further extended to cases where the set X, on which the relevant mapping f works, is of a more complex structure. The result due to Eilenberg and Montgomery (1946), which is representative of these extensions, asserts the existence of a fixed point in the case where a mapping f transforms points to *acyclic* subsets in an upper semi-continuous way within a set X which is a *compact, acyclic, absolute neighbourhood retract*. Compactness is a property of a set that is a counterpart in general spaces to boundedness and closedness combined together in Euclidean space, meaning that any nonvoid subset has at least an accumulation point within the set. The acyclicity of a set means that the set has the same homology groups as does a set consisting of a single point, and weakens *contractibility* in the sense that the set is continuously deformable to a single point within it, a very much generalized concept of convexity. Absolute neighbourhood retracts form so broad a category of sets that they include compact acyclic sets in infinite-dimensional linear spaces. The infinite-dimensional direct counterpart results of the Kakutani fixed point theorem by Fan (1952) and Glicksberg (1952) prove special cases of the Eilenberg–Montgomery theorem in direct elementary ways. These remarkable relaxations of assumptions of the Kakutani theorem broaden its applicability to economic models in which the sets of relevant variables are of more complex structures.

HUKUKANE NIKAIDO

See also COMPUTATION OF GENERAL EQUILIBRIUM; CORRESPONDENCES; EXISTENCE OF GENERAL EQUILIBRIUM.

BIBLIOGRAPHY

Arrow, K.J. and Debreu, G. 1954. Existence of an equilibrium for a competitive economy. *Econometrica* 22, July, 265–90.

Brouwer, L.E.J. 1910. Über eineindeutige, stetige Transformationen von Flächen in sich. *Mathematische Annalen* 69, 176–80.

Debreu, G. 1959. *Theory of Value: an Axiomatic Analysis of Economic Equilibrium*, Cowles Foundation Monograph No. 17, New York: John Wiley & Sons.

Eilenberg, S. and Montgomery, D. 1946. Fixed point theorems for multi-valued transformations. *American Journal of Mathematics* 68, 214–22.

Fan, K. 1952. Fixed point and minimax theorems in locally convex topological linear spaces. *Proceedings of the National Academy of Sciences of the USA* 38, 121–6.

Gale, D. 1955. The law of supply and demand. *Mathematica Scandinavica* 3, 155–69.

Glicksberg, I.L. 1952. A further generalization of the Kakutani fixed point theorem with application to Nash equilibrium points. *Proceedings of the American Mathematical Society* 3, 170–74.

Kakutani, S. 1941. A generalization of Brouwer's fixed point theorem. *Duke Mathematical Journal* 8(3), 457–59.

Knaster, B., Kuratowski, C. and Mazurkiewicz, S. 1929. Ein Beweis des Fixpunktsatzes für n-dimensionale Simplexe. *Fundamenta Mathematica* 14, 132–37.

McKenzie, L.W. 1954. On equilibrium in Graham's model of world trade and other competitive systems. *Econometrica* 22, April, 147–61.

Nash, J.F. 1950. Equilibrium points in n-person games. *Proceedings of the National Academy of Sciences of the USA* 36, 48–49.

Neumann, J. von, 1937. Über ein ökonomisches Gleichungssystem und eine Verallgemeinerung des Brouwerschen Fixpunktsatzes, *Ergebnisse eines Mathematischen Kolloquiums* 8. Translated as 'A model of general economic equilibrium', *Review of Economic Studies* 13(33), 1945–6, 1–9.

Nikaido, H. 1956. On the classical multilateral exchange problem. *Metroeconomica* 8, August, 135–45; A supplementary note, 9, December 1957, 209–10.

Scarf, H. (with T. Hansen.) 1973. *The Computation of Economic Equilibria.* Cowles Foundation Monograph 24, New Haven and London: Yale University Press.

Sperner, E. 1928. Neuer Beweis für die Invarianz der Dimesionszahl und des Gebietes. Abhandlungen an den mathematischen, Seminar der Universität, Hamburg 6, 265–72.

Walras, L. 1874–7. *Eléments d'économie politique pure.* Lausanne: Corbaz. Trans. by W. Jaffé as *Elements of Pure Economics*, London: Allen & Unwin, 1954.

fixprice models. Modern fixprice theory (Benassy, 1975, 1976, 1982; Drèze, 1975; Younès, 1975) studies trade and production at non-Walrasian prices in general environments with possibly many agents and goods. The name and the basic logic originate in Hicks (1965) where a multiperiod economy is contemplated. Hicks defines two analytical methods: the Flexprice Method, which assumes that prices adjust within each period so that current transactions equal both demand and supply (such very short-run equilibration being, in his words, 'hard to swallow') and the Fixprice Method, where prices are given at the beginning of each period and transactions may differ from supply or demand. Both the Flexprice and the Fixprice methods are 'pure' or extreme ones. Hicks's own preference is 'for something which lies between', knowing that 'anything that does so must partake to some extent the difficulties of the two'. The general models discussed here follow Hicks's Fixprice Method. They are rather abstract, and they may alternatively be applied to the short period of *Capital and Growth* or to an atemporal economy.

In the Flexprice Method transactions take place at prices for which excess demand is zero. It may be interpreted that prices adjust very rapidly in response to excess demand and that no transactions occur before equilibrium is reached. A rigorous formulation of this idea is the Walras–Samuelson tâtonnement process. Consider an exchange economy with two commodities and two agents, agent i being initially endowed with ω_{ij} units of commodity j $(i, j = 1, 2)$. Let the aggregate Walrasian excess demand functions be $z^1(p_1, p_2|\omega)$. (Here the vector $\omega = (\omega_{ij})$ is fixed.) As long as $z^i(p_1, p_2|\omega) \neq 0$ or $z^2(p_1, p_2|\omega) \neq 0$ no transactions occur and prices adjust according to the differential equation:

$$(dp_i/dt) = z^i(p_1(t), p_2(t)|\omega), \qquad i = 1, 2.$$

The Walrasian excess demands provide the 'market signals' for the adjustment of prices.

The Walrasian excess demand functions express the plans of price taking agents, i.e., plans made under the conjecture that any quantities can be bought and sold at the going prices. If transactions at non-Walrasian prices occur then such a conjecture will be falsified, since some agents will be unable to realize their plans (see Arrow, 1959). This led Patinkin (1956) to postulate that disequilibrium transactions in a market create spillover effects on others, so that, e.g., 'the pressure of excess demand in the one market affects the price movements in all other markets' (p. 157). One could, for instance, write:

$$(dp_i/dt) = f_i(z^1(p_1(t), p_2(t)|\omega), z^2(p_1(t), p_2(t)|\omega)), \quad i = 1, 2,$$

where f_i is some function. Patinkin's formulation was imprecise (Negishi, 1965; Clower, 1965), but his search for the 'relevant market signals' motivates Clower's (1965) 'dual decision hypothesis' as a microeconomic foundation of Keynesian macroeconomics. This idea also developed by Leijonhufvud (1968) and generalized by Barro–Grossman (1971, 1976), is central to Benassy's fixprice model.

It was discovered in the late 1950s that the Walras–Samuelson tâtonnement process fails to converge unless some restrictive assumptions are imposed. This difficulty, now well understood but somehow surprising at the time, led to the formulation of the non-tâtonnement adjustment process. Here two simultaneous movements occur: the distribution of the endowments changes according to some rule for trading at non-Walrasian prices, and prices adjust in response to Walrasian excess demands at the current endowments, e.g. for some rule g_{ij},

$$(d\omega_{ij}/dt) = g_{ij}(p_1(t), p_2(t), \omega(t)),$$
$$(dp_i/dt) = z^i(p_1(t), p_2(t)|\omega(t)), \qquad i, j = 1, 2.$$

This process is hard to interpret except possibly as depicting the sequential exchange of durable goods (the mineral bourse mentioned in Smale, 1976). Moreover, it ignores the irrelevance of the Walrasian signals when transactions occur at non-Walrasian prices. But some conditions on disequilibrium trading (formally, on the functions g_{ij}, see Hahn–Negishi, 1962 and Uzawa, 1962) originally meant to guarantee the convergence of the non-tâtonnement process inspired basic features of the modern fixprice models (see Younès, 1975).

A third theoretical development of the early Sixties, namely the monopolistic general equilibrium analysis pioneered by Negishi (1960–61), led to extensions of the Fixprice model where agents on one side of the market may face nonhorizontal demand (oligopoly) or supply (oligopsony) curves and have price setting power. Such market power may be based on structural conditions of the economy as in Benassy (1976, 1977, 1982), Hart (1982), Silvestre (1987) (see also DISEQUILIBRIUM ANALYSIS and RATIONED EQUILIBRIA): prices and quantities are determined simultaneously and, usually, oligopoly (or oligopsony) is formally parallel to excess supply (or excess demand). Alternatively, an inequality between supply and demand gives temporary market power to agents on the short side who may then face nonhorizontal demand or supply curves for large enough quantities (see Arrow, 1959; Negishi, 1974, 1979; Hahn, 1978; John, 1985).

CONCEPTS

Modern fixprice analysis postulates explicit trading institutions called markets: in each market a commodity is exchanged against a common medium of exchange (money). Thus, there are $n+1$ goods (from 0 to n) in the case of n markets, the zeroth good being money. This sharply contrasts with the institutional imprecision of the models alluded to in the previous section. The analysis addresses two questions: (1) Given a price vector p (normalized with respect to money), what allocations are compatible with it? (2) Given a p and an allocation compatible with it, which is the type of disequilibrium in each market? (Question 2 is a prerequisite to the study of the 'market signals' for price adjustment.) The answers are derived from three basic principles which reflect the operation of the market institution: (i) voluntary trading; (ii) absence of market frictions, and (iii) effective demand. The latter requires the explicit recognition of the interaction among markets. The first two impose conditions on trade in a market, namely that, at the going price: (i) no trader may gain by trading less; (ii) no pair formed by a buyer and a seller may gain by trading more.

The fixprice model provides a general framework (that includes perfect competition as a special case) for price-guided allocation mechanisms (see Silvestre, 1986). It has several applications (i) *short-run analysis* that assumes, as in *Capital and Growth*, that it takes time for prices and quantities to adjust (perfect competition corresponding to very fast adjustment); (ii) *monopolistic competition* (including perfect competition as the special case where market power is nil: see previous section); (iii) *price (wage or rent) controls*; this in particular motivates Drèze's formulation; (iv) *price (or wage) negotiation* (representatives of buyers and sellers negotiate prices that are taken as given by individual traders: see Silvestre, 1987). Fixprice analysis can be viewed as abstracting from specific behaviour features (say, particular oligopoly models, adjustment paths, Government policies or negotiation procedures) and focusing instead on basic market principles common to alternative specifications.

The definitions of fixprice equilibrium due to Benassy, Drèze and Younès vary in form and motivation, but turn out to be equivalent under some assumptions (see Silvestre, 1982, 1983). Rather than reproducing them in all generality, we exemplify the common concepts in two simple but important cases.

Case 1: Differentiable exchange economics. Let there be $n + 1$ goods, indexed $0, 1, \ldots, n$ (i.e. n markets). There are m traders: trader i is endowed with an $(n + 1)$ dimensional vector of initial endowments ω_i and a differentiable utility function $u_i: R^{n+1} \to R$. A net trade allocation is an m-tuple of n-dimensional net trade vectors $(z_i) = (z_{i1}, \ldots, z_{in})$, one for each trader, satisfying: $\Sigma_i z_i = 0$. It is understood that, for $j = 1, \ldots, n$, if $z_{ij} > 0$ (or < 0) then trader i is buying (or selling) in market j. The (normalized) price vector $p = (p_1, \ldots, p_n)$ is given. The vector $\hat{x}_i(p; z_i) = (\omega_{i0} - p \cdot z_i, \omega_{i1} + z_{i1}, \ldots, \omega_{in} + z_{in})$ is then the consumption vector associated with $(p; z_i)$. We only consider situations where the vectors p and $\hat{x}_i(p; z_i)$ are strictly positive. Define i's marginal utility of trading in market j at the going price as: $u_{ij}(p; z_i) = \partial u_i / \partial x_{ij} - p_j \cdot \partial u_i / \partial x_{i0}$, with derivatives evaluated at $\hat{x}(p; z_i)$.

Definition: A net trade allocation (z_i^*) is a *Fixprice Equilibrium for p* if, writing $\mu_{ij}^* \mu_{ij}(p; z_i)$, (a) *Voluntariness:* For $i = 1, \ldots, m, z_i^* \cdot \mu_{ij}^* \geq 0$; (b) *Absence of market frictions:* For $j = 1, \ldots, n$ and for any pair of consumers $i, h, \mu_{ij}^* \cdot \mu_{hj}^* \geq 0$.

Figure 1 illustrates the case of $n = 1$ and $m = 2$ in an Edgeworth box: point A represents the (unique) fixprice equilibrium at the price vector p: there trader 1 is a buyer ($z_{11} > 0$). The straight line through points ω and A depicts the budget constraints. Allocations in the segment $[\omega, A]$ would violate condition (b). Those in the segment $[A, B]$ (in particular the Pareto efficient point D) would violate condition (a) for trader 1.

The graphically more complex case of $n = m = 2$ is illustrated in Figures 2a–d. Figure 2(a) depicts trader 1's budget set in R_3^+. Rather than drawing a three-dimensional Edgeworth box, we graph first separately (Fig. 2(b–c)) and then together (Fig. 2(d)) the two-dimensional budget triangles of the traders. Figure 2(a–b) also depicts the intersections of some indifference surfaces of trader 1 with his budget set, Q_1 being his most preferred point in the budget set. At point A he is selling in both markets (i.e. $z_{11} < 0$, $z_{12} < 0$: he gets money in exchange), $\mu_{12} < 0$ (he would like to sell more in market 2) and $\mu_{11} = 0$. Figure 2(c) corresponds to trader 2: at point A he is buying in both markets (i.e. $z_{20} > 0$, $z_{21} > 0$, $z_{22} > 0$), and $\mu_{21} > 0$, $\mu_{22} = 0$. Figure 2(d) superimposes the two graphs (with the axes corresponding to trader 2 reversed and with the initial endowment points coin-

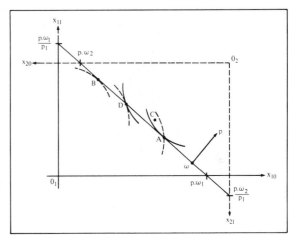

Figure 1

ciding at ω). Points A in Figures 2(b–c) have been chosen so that they also coincide in Figure 2(d), i.e., $z_{2j} + z_{2j} = 0$, $j = 1, 2$. These trades constitute a Fixprice Equilibrium.

Case 2: The Three Good Model. This by now popular model originated in Barro–Grossman (1971, 1976) and was further elaborated by Benassy (1977, 1982, 1986) and Malinvaud (1977) among others. Our presentation follows Silvestre (1986).

Let there be three goods, money, denoted by M, initially available in M_0 units, labour, denoted by L, initially available in L_0 units, and output, denoted by Y, which is initially non-available but is produced by labour according to the production function $Y = f(L)$. There are two markets, the labour market, with (nominal) wage w, and the output market, with price p. There is one firm and one consumer (the homogeneity of U makes this assumption inessential) who owns M_0 and L_0, receives all profits and is endowed with a utility function $U(Y, M)$. Note that the labour supply is fixed at L_0.

Define the marginal rate of substitution as $V'(Y) = (\partial U / \partial Y)/(\partial U / \partial M)$, with derivatives evaluated at (Y, M_0). Define the marginal cost curve as $C'_w = w(f^{-1})'(Y)$, and the full employment output as $Y_0 = f(L_0)$. Then:

Definition: The level of output Y is a *Fixprice Equilibrium output for the price–wage pair (p, w)* if:

$$Y = \min\{(V')^{-1}(p), (C'_w)^{-1}(p), Y_0\}.$$

This equality embodies in a compact way several conditions that can be interpreted as follows. First, $Y \leq Y_0$, i.e., output cannot exceed the full employment level. Second, $Y \leq (V')^{-1}(p)$, or alternatively $p \leq V'(Y)$: this means that the consumer cannot gain by buying less output at the going price: it is a condition of 'voluntary trading' for the consumer. Third, $Y \leq (C'_w)^{-1}(p)$, or, alternatively, $p \geq C'_w(Y)$, i.e. the price cannot be lower than the marginal cost: it is a condition of 'voluntary trading' for the firm (profits cannot increase by selling less at the going price). Finally, at least one of these weak inequalities must be an equality: this is the condition of frictionless markets.

Figure 3 partitions the (p, w) plane according to which one of the three possible equalities determines output (solid lines). In region E (full employment), $Y = Y_0$. In region K (Keynesian unemployment) $p = V'(Y)$, and in region C (for Classical

389

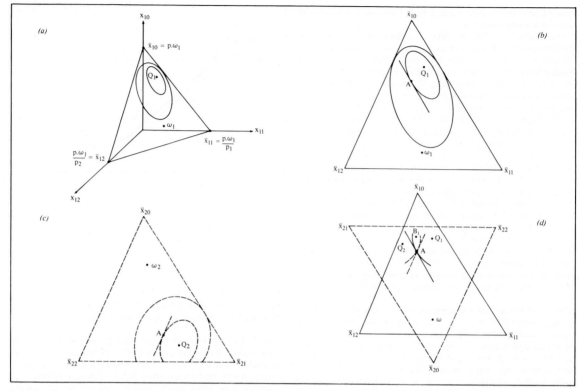

Figure 2

Unemployment of Full Capacity) $p = C'_w(Y)$. In the boundaries between regions the two relevant equalities hold. At the Walrasian point W all three equalities hold. There is full employment in Region E and unemployment outside it. The dashed lines are isoemployment loci, with the arrows indicating the directions of increasing employment.

Within the short-run interpretation (see Silvestre, 1986, for alternative ones) the labour market is in excess supply (or, excess demand) in the interior of Regions K and C (or Region E), and the output market is in excess supply (or demand) in the interior of region K (or regions C and E). At the Walrasian point W both markets are balanced. In the Keynesian region the condition for determination of output $p = V'(Y)$ can be rewritten in terms of the consumption function as in the textbook Keynesian multiplier model. The homogeneity of U implies that demand for output, as a function of p and wealth I, can be written as $h(p)I$, where the function $h(p)$ satisfies: (i) $ph(p) < 1$ and (ii) the marginal equality $(\partial U/\partial Y)/(\partial U/\partial M) = p$ whenever the consumption vector is a multiple of $(h(p), 1 - ph(p))$. By setting $I = M_0 + pY$ we obtain the effective demand for output (see the entry Disequilibrium Analysis) $C(Y) = h(p)(M_0 + pY)$: this is the traditional consumption function, with marginal propensity to consume equal to $ph(p) < 1$. The satisfaction of effective demand requires $Y = C(Y)$, i.e., $Y/M_0 = h(p)/[1 - ph(p)]$, which by the above marginal equality implies that $p = V'(Y)$.

The distinction of the two types of excess supply of labour has important implications for economic policy and for comparative statics. In Region C output is determined by the condition 'price = marginal cost'. Hence, lowering wages (nominal or real) will increase employment, but an increase in demand will have no effect on employment. In Region K, a decrease in the nominal wage has no effect on employment: only lowering the price or otherwise stimulating demand will work. This analysis also offers insights on the effects of different kinds of shocks (see Malinvaud, 1977; Silvestre, 1986): a business cycle driven by demand shocks will fluctuate between Keynesian unemployment and full employment, whereas productivity shocks will yield fluctuations between the Keynesian and the Classical types of unemployment.

INEFFICIENCIES

The budget equality and the market institution impose constraints on trades. Thus, the resulting allocation may very well be Pareto dominated by other allocations that do not satisfy these constraints. The study of such inefficiencies is important for the normative analysis of the situations covered by fixprice theory (short-run market disequilibria, price controls, monopolistic competition).

Inefficiency relative to the set of physically attainable allocations. Consider Figure 1. Note that the allocation given by A is not Pareto efficient: both traders would be better off at C, but C cannot be reached without violating some budget constraint.

A similar phenomenon may occur if there are two traders in one side of the market. Modify the example of Figure 1 by duplicating Trader 2: i.e., Traders 1 and 2 are unchanged, but now there is a Trader 3 with the same preferences and endowments as Trader 2. Let $z_{21} = (-1/4)z_{11}$, and $z_{31} = (-3/4)z_{11}$. Then there are mutually beneficial reallocations between traders 2 and 3, but they violate the budget constraint.

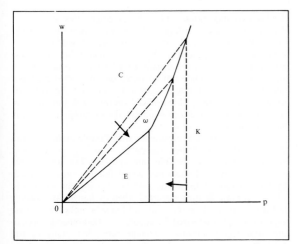

Figure 3

One can say that this type of inefficiency is caused by 'wrong prices'. Note, however, that trade at non-Walrasian prices does not per se imply inefficiency. Point *D* in Figure 1, for instance, is Pareto efficient, and all budget constraints are satisfied there. (A general treatment of allocations of this type is given by Balasko, 1979, and Keiding, 1981.) But there is forced trading at point *D*. It is the combination of non-Walrasian prices and the market institution (reflected by the voluntariness condition) that implies inefficiency. This point is made rigorous in Silvestre (1985).

Inefficiencies relative to allocation satisfying the budget constraints. When there is only one market (see Figure 1) because of the absence of frictions no allocation that satisfies the budget constraint is Pareto superior to a Fixprice Equilibrium, i.e., a Fixprice Equilibrium is efficient relative to allocations that satisfy the budget constraints. This ceases to be true with several markets: for instance, point *B* in Figure 2(d) Pareto dominates point *A* and satisfies all budget constraints. (Note that point *B* violates voluntariness.) Such inefficiencies have been studied in Benassy (1975, 1977, 1982) and Younès (1975). A particularly striking case occurs in Keynesian allocations of the Three Good Model: the markets for labour and output are in excess supply, and a direct barter of labour against output would benefit both the firm and the worker (and improve welfare). This phenomenon was viewed by Clower (1965) and Leijonhufvud (1968) as a failure of coordination among markets. It is often alleviated in modern economies by Keynesian policies rather than by the direct exchange of labour against output.

Again, the inefficiencies relative to allocations satisfying the budget constraints are based on the institutional arrangements for trade: the exchange of money against a good in each market. They could conceivably disappear under alternative trading institutions. For instance, Drèze–Müller (1980) presents a scheme where traders must satisfy, in addition to the usual budget constraint, a balance condition on 'coupons' associated with trading on each (nonmoney) commodity. The equilibria of this scheme are always efficient relative to trades that do not violate the budget constraints. (Thus, point *A* in Figure 2(d) is not an equilibrium of this kind.) Their model captures institutions that may appear under persistent excess demand, as in wartime or in planned economies.

JOAQUIM SILVESTRE

See also DISEQUILIBRIUM ANALYSIS; RATIONED EQUILIBRIA; TEMPORARY EQUILIBRIUM.

BIBLIOGRAPHY

Arrow, K.J. 1959. Towards a theory of price adjustment. In *The Allocation of Economic Resources*, ed. M. Abramovitz, Stanford: Stanford University Press.

Balasko, Y. 1979. Budget constrained Pareto-efficient allocations. *Journal of Economic Theory* 21, 359–79.

Barro, R.J. and Grossman, H.I. 1971. A general equilibrium model of income and employment. *American Economic Review* 61, 82–93.

Barro, R.J. and Grossman, H.I. 1976. *Money, Employment and Inflation*. Cambridge: Cambridge University Press.

Benassy, J.P. 1975. Neo-Keynesian disequilibrium theory in a monetary economy. *Review of Economic Studies* 42, 503–23.

Benassy, J.P. 1976. The disequilibrium approach to monopolistic price setting and general monopolistic equilibrium. *Review of Economic Studies* 43(1), 69–81.

Benassy, J.P. 1977. A Neo-Keynesian model of price and quantity determination in disequilibrium. In *Equilibrium and Disequilibrium in Economic Theory*, ed. G. Schwödiauer, Boston: Reidel.

Benassy, J.P. 1982. *The Economics of Market Disequilibrium*. New York: Academic Press.

Benassy, J.P. 1986. *Macroeconomics: An Introduction to the Non-Walrasian Approach*. New York: Academic Press.

Clower, R. W. 1965. The Keynesian counter-revolution: a theoretical appraisal. In *The Theory of Interest Rates*, ed. F. H. Hahn and F.P.R. Brechling, London: Macmillan.

Drèze, J. 1965. Existence of an exchange equilibrium under price rigidities. *International Economic Review* 16, 301–20.

Drèze, J. and Müller, H. 1980. Optimality properties of rationing schemes. *Journal of Economic Theory* 23, 131–49.

Hahan, F.H. 1978. On non-Walrasian equilibria. *Review of Economic Studies* 45(1), 1–17.

Hahn, F.H. and Negishi, T. 1962. A theorem on non-tâtonnement stability. *Econometrica* 30, 463–9.

Hahn, F.H. and Brechling, F.P.R. (eds) 1965. *The Theory of Interest Rates*. London: Macmillan.

Hart, O. 1982. A model of general equilibrium with Keynesian features. *Quarterly Journal of Economics* 97, 109–38.

Hicks, J.R. 1965. *Capital and Growth*. Oxford: Oxford University Press.

John, R. 1985. A remark on conjectural equilibria. *Scandinavian Journal of Economics* 87(1), 137–41.

Keiding, H. 1981. Existence of budget constrained Pareto-efficient allocations. *Journal of Economic Theory* 24, 393–7.

Leijonhufvud, A. 1968. *On Keynesian Economics and the Economics of Keynes*. Oxford: Oxford University Press.

Malinvaud, E. 1977. *The Theory of Unemployment Reconsidered*. Oxford: Basil Blackwell.

Negishi, T. 1960. Monopolistic competition and general equilibrium. *Review of Economic Studies* 28, 196–201.

Negishi, T. 1965. Market clearing processes in a monetary economy. In *The Theory of Interest Rates*, ed. F.H. Hahn and F.P.R. Brechling, London: Macmillan.

Negishi, T. 1974. Involuntary unemployment and market imperfection. *Economic Studies Quarterly* 25(1), 32–46.

Negishi, T. 1979. *Microeconomic Foundations of Keynesian Macroeconomics*. Amsterdam: North-Holland.

Patinkin, D. 1956. *Money, Interest and Prices*. Evanston, Ill: Row Peterson.

Silvestre, J. 1982. Fixprice analysis in exchange economies. *Journal of Economic Theory* 26, 28–58.

Silvestre, J. 1983. Fixprice analysis in productive economies. *Journal of Economic Theory* 30, 401–9.

Silvestre, J. 1985. Voluntary and efficient allocations are Walrasian. *Econometrica* 53, 807–16.

Silvestre, J. 1986. The elements of fixprice microeconomics. In *Microeconomic Theory*, ed. L. Samuelson, Boston: Kluwer Nijhoff.

Smale, S. 1976. A convergent process of price adjustment and global Newton methods. *Journal of mathematical economics* 3, 107–20.

Uzawa, H. 1962. On the stability of Edgeworth's barter process. *International Economic Review* 3(2), May, 218–32.

Younès, Y. 1975. On the role of money in the process of exchange and the existence of a non-Walrasian equilibrium. *Review of Economic Studies* 42, 489–501.

Fleming, John Marcus (1911–1976). Fleming was born on 13 March 1911 at Bathgate, Scotland, and died on 3 February 1976. He was educated at Edinburgh University, where he received the degrees of MA (honours) in history in 1932 and MA (first class honours) in political economy in 1934. He was a graduate research fellow at the Institut Universitaire des Hautes Etudes Internationales in 1934–5, and a graduate student at the London School of Economics in 1935.

At the end of 1935, he joined the Secretariat of the League of Nations, Economic Intelligence Section, as a research economist and assisted Gottfried Haberler in the latter's *Prosperity and Depression* (first published by the League in 1937). During World War II, he served with the UK Ministry of Economic Warfare from 1939 until 1942, and then joined the Economic Section of the Cabinet Office under Lord Robbins, rising eventually to the position of Deputy Director of the section. He was also a member of the UK Delegation to the San Francisco Conference in 1945; a member of the Preparatory Commission of the United Nations in 1946; a member of the International Trade Conference Preparatory Commission, 1947; and UK Representative to the Economics and Employment Commission, United Nations in 1950. From 1951 to 1954 Fleming was Visiting Professor at Columbia University, New York. He joined the International Monetary Fund in 1954 as a division chief, and in 1964 became the Deputy Director of the Research Department.

His academic contributions are mostly in the fields of welfare theory and trade and exchange policies. The most notable of his contributions was the seminal article 'On Making the Best of Balance of Payments Restrictions on Imports' (Fleming, 1951), which in James Meade's words, was 'the begetter of the analysis of the second best' (Meade, 1978), which rapidly became a fashionable new topic in welfare theory during the 1950s.

S.C. Tsiang

SELECTED WORKS

1944. (With J.E. Meade.) Price and output policy of state enterprise: a symposium. *Economic Journal* 54, December, 321–8, 337–9.

1951. On making the best of balance of payments restrictions on imports. *Economic Journal* 61, March, 48–71.

1952. A cardinal concept of welfare. *Quarterly Journal of Economics* 6, August, 366–84.

1961. International liquidity: ends and means. *IMF Staff Papers* 8, December, 439–63.

1962. Domestic financial policies under fixed and under floating exchange rates. *IMF Staff Papers* 9, November, 369–79.

1964. (With R.A. Mundell.) Official intervention in the forward exchange market. *IMF Staff Papers* 11, March, 1–17.

1971a. *Essays in International Economics*. Cambridge, Mass.: Harvard University Press.

1971b. The SDR: some problems and possibilities. *IMF Staff Papers* 18(1), March, 25–47.

1977. International aspects of inflation. In *Inflation Theory and Anti-inflation Policy*, ed. E. Lundberg, London: Macmillan.

BIBLIOGRAPHY

Meade, J.E. 1978. Commemorations. In Flemming, J. Marcus (posthumous), *Essays on Economic Policy*, New York: Columbia University Press.

flexible exchange rates. Flexible exchange rates are market determined prices of foreign exchange which move in response to supply and demand and are not pegged within narrow bands by official purchases. Flexible systems where there are no official purchases are usually called pure floating regimes, and systems with some official purchases are called managed floating regimes. The counterpart to flexible exchange rates are fixed exchange rates, where official central bank purchases or sales of foreign currencies maintain the exchange rate within narrow bands.

Until 1973, historical experience with flexible exchange rates was limited, and seemed incapable of telling much about general principles of flexible rate systems. The best known experiences involved short periods of floating brought about by the collapse of fixed rate systems. During the 1920s and 1930s, sporadic floating occurred when exchange-rate pegging failed in response to severe real and financial upheavals. These experiences, characterized by apparently destabilizing speculation and highly variable exchange rates, left many central bankers, businessmen, and a few economists, wary about the efficacy of floating rates during more tranquil periods.

In contrast to central bankers, academic economists were more sanguine about the operating characteristics of floating exchange rates. As is often the case in monetary economics Milton Friedman early and persuasively made the case for flexible exchange rates with his article, 'The Case for Flexible Exchange Rates'. He pointed out the macroeconomic benefits from flexible exchange rates, especially monetary independence, and argued that speculators, the purported villains of the interwar floating experiences, would actually help ensure the smooth working of floating rates. By the time of the breakdown of the Bretton Woods system of fixed rates, most academic economists advocated flexible rates and felt that such rates would not be very volatile.

In fact, experience since 1973 shows us that our theoretical musings about how flexible rates would work were quite wrong: exchange rates have been extremely volatile. The post-1973 events, though, have stimulated international economists to develop new theories about exchange-rate determination. These theories emphasize expectations, the role of internationally traded financial assets, and the distinctions between stock and flow phenomena. It is to these theories that we now turn.

EXCHANGE RATE DETERMINATION. A statement with which few economists would argue is that under flexible rates, the value of a currency is determined by supply and demand. Older theories, though, emphasized the supplies and demands for foreign exchange arising from flow demands for merchandise imports and exports: the trade balance was seen as the major determinant of the exchange rate. These theories were consistent with the stylized facts of the time, namely relatively free trade in goods but restricted trade in financial assets. The overriding issue on which turned the workability of flexible rates was whether import and export elasticities were 'large enough' to ensure Walrasian stability of the foreign exchange market. Whether speculators were stabilizing or destabilizing in these theories depended mostly on what sort of expectations of future exchange rate movements speculators were assumed to have.

The international monetary theoreticians of the 1970s, in an effort to explain the unforeseen volatility of floating rates, began emphasizing the demand and supply for stocks of internationally traded financial assets. Monetary factors operating through interest differentials moved to the fore as the fundamental force behind exchange-rate determination, and the trade balance and relative prices of imports and exports were pushed to the background. These theories also incorporated Muth's ideas about rational expectations, setting the stage for analyses of speculation not critically dependent on ad hoc, arbitrary specifications of how expectations of future exchange rates are formed. The emphasis of these new theories on purchases on the capital account also reflected awareness that the 1970s were fundamentally different from earlier periods in that the world had fewer controls on capital movements.

These new asset-market theories of exchange-rate determination were refined and extended to incorporate trade-balance considerations, bringing relative prices back into the picture. What has now emerged is a theory with the following implications:

(1) In the long run, a period measured in years rather than months or quarters, exchange rates are proportional to relative money supplies, so long as real factors remain roughly constant. That is, Cassel's Purchasing Power Parity Principle holds as a long-run phenomenon.

(2) The short-run behaviour of exchange rates can be highly volatile, with exchange rates deviating markedly from their long-run trends. Speculators with rational expectations play an important role in this area by exacerbating monetary shocks to the system.

We now develop a skeletal model of exchange-rate determination that captures the key features of the asset-market approach. The key building blocks are specification of a stock demand for foreign financial assets, and specification of the supply of foreign assets. Expectations are modelled as rational, letting us focus on fundamental behavioural relations rather than on the effects of various ad hoc expectational schemes. The approach here differs from more standard treatments of asset market approaches, e.g. Frankel (1983), in its emphasis on how future trade-balance effects influence current exchange rates.

The hallmark of the asset-market approach is the specification of asset demands as stock, rather than flow, demands. Elementary mean-variance portfolio theory suggests, as a first approximation, that net demand for foreign financial assets should depend on relative rates of return. Denote the stock demand for net foreign assets as F_t, where t indexes time. Let e stand for the log of the exchange rate, defined as the domestic currency value of foreign exchange, and E_t denote the mathematical expectation of any variable conditional on information available at time t. The relative rate of return on foreign assets *vis-à-vis* domestic assets is approximately:

$$E_t e_{t+1} - e_t - r_t \qquad (1)$$

where r_t is the difference between the domestic and foreign nominal interest rate. For simplicity, we specify the net demand for foreign assets as a linear function of the relative rate of return:

$$F_t = n\{E_t e_{t+1} - e_t - r_t\} \qquad (2)$$

where n is a positive constant. Portfolio theory tells us that n should depend inversely on exchange rate predictability and directly on taste for risk. If exchange rates become more unpredictable, then for any given expected return on foreign bonds the risk becomes larger. Hence, risk-averse investors will lower their holdings. Likewise, for any given expected return and degree of exchange-rate predictability, a decrease in risk-aversion by investors would lead them to increase their holdings of the risky asset.

Equilibrium in the foreign bond market means that demand equal supply. Denoting net foreign bond supply as F_t^s, this

means that the exchange rate that equilibrates demand and supply of foreign assets satisfies the following equation:

$$e_t = E_{t+1} e_t - r_t - F_t^s / n. \qquad (3)$$

That is, the current(log) exchange rate equals the current expectation of next period's rate minus the interest differential minus the stock of foreign bonds divided by the sensitivity of foreign bond demanders to expected returns. Note that if foreign bond demanders are risk neutral, i.e. if n is infinite, then the expected exchange rate change just equals the interest differential; all that investors care about is expected return, regardless of the relative riskiness of foreign bonds due to unforeseen exchange rate movements. This is the so-called efficient markets hypothesis for the foreign exchanges.

If we want to, we can think of e_t, the exchange rate, as moving to equilibrate demand and supply for net foreign bonds at each moment in time. Asset-market theorists have sometimes claimed that thinking of the exchange rate this way as opposed to thinking of it as moving to equilibrate flow supplies and demands for foreign exchange is what distinguishes the new approach from the old. Note, though, that a component of demand is $E_t e_{t+1}$, the current expectation of next period's exchange rate. Now, if the exchange rate moves to equilibrate foreign net bond demand and supply at $t + 1$, then e_{t+1} depends on F_{t+1} and $E_{t+1} e_{t+2}$, which in turn depends on F_{t+2} and $E_{t+2} e_{t+3}$, which in turn depends on $F_{t+3} e_{t+4}$, and so on. Thus, the current exchange rate depends on current expectations of all future values of net foreign bond holdings.

We can make this argument formally by iterating equation (3) forward through time, taking expectations, and substituting back in the initial value. We can then write the exchange rate as three terms: the current expectation of the long-run exchange rate, the current expectation of the sum of the current plus all future interest rate differentials, and the current expectation of the sum of current plus all future foreign bond supplies, divided by investor sensitivity to expected returns:

$$e_t = E_t e_{t+1} - E_t \sum_{i=0}^{\infty} r_{t+1} - (1/n) E_t \sum_{i=0}^{\infty} F_{t+i}^s \qquad (4)$$

What this equation highlights is how the entire future path of both interest rates and foreign bond supplies affects current exchange rates. If interest rate differentials are highly variable, perhaps reflecting variable monetary policy, then the exchange rate will be highly variable. The future values of foreign bonds highlight, indirectly, the role of the current account and relative prices in exchange rate determination.

Net additions to the stock of foreign bonds available to domestic residents can only be generated by trade balance surpluses. Symbolically, we have:

$$F_t^s - F_{t-1}^s = T_t \qquad (5)$$

where T_t denotes the trade balance at time t. The simplest specification of the behaviour of the trade balance would make it depend upon relative prices. If we denote the log of relative price levels between domestic and foreign countries as p, then we can specify the trade balance as:

$$T_t = a\{e - p\} + u_t \qquad (6)$$

where a is a parameter reflecting the responsiveness of the trade balance to relative prices, and u_t is a zero-mean serially uncorrelated random variable, capturing shocks to the underlying fundamental determinants of the trade balance, e.g. taste and technology.

393

At this point it is useful to develop the behaviour of the stock of foreign bonds through time. For expositional ease, we assume that the interest differential, r_t, is a serially uncorrelated zero-mean random variable. It turns out that F_t will follow a first-order autoregressive process through time:

$$F_t = x_1 F_{t-1} + x_2 u_t + x_3 r_t \qquad (7)$$

where x_1, x_2 and x_3 are coefficients which are functions of n and a. From the above process, it follows that

$$E_t \sum_{t=0}^{\infty} F_{t+1}^s = F_t^s[1 + x_1^2 + \cdots + x_1^i + \cdots + x_1^j + \cdots]$$
$$= F_t^s[1/(1 - x_1)] \qquad (8)$$

Hence, $e_t = E_t e_{t+1} - r_t - (1/n)F_t^s[1/(1 - x_t)]$. Assume for simplicity of exposition that relative price levels, p, are fixed. We could make any one of a variety of other standard macroeconomic assumptions about price-level determination without altering the basic lessons of this analysis. Then, $F_r = F_{t-1} + ae_t + u_t$. Using the immediately preceding expression for e_t in this equation, some simple algebra leads us to:

$$F_t = \{[n(1 - x_1)]/[n(1 - x_1) + a]\}F_{t-1}$$
$$+ \{[n(1 - x_1)]/[n(1 - x_1) + a]\}u_t$$
$$- \{[an(1 - x_1)]/[n(1 - x_1) + a]\} \qquad (9)$$

Comparing (7) and (9), we see that x_1 is implicitly defined by:

$$x_1 = \{[n(1 - x_1)]/[n(1 - x_1) + a]\} \qquad (10)$$

and there exists one unique value of x_1 between zero and one. That is, F_t^i follows a stable autoregressive process whose parameters are functions of the two structural parameters n and a.

Armed with this knowledge, we can derive the path through time of the exchange rate. Differencing the fundamental equation (3) and substituting the trade balance for $(F_t - F_{t-1})$, we get the following ARMA(1,1) process:

$$e_t = x_1 e_{t-1} - x_1 r_t + x_1 r_{t-1} + \{[1 - x_1]/a\}u_t. \qquad (11)$$

Analysis of this equation shows us how the operating characteristics of flexible exchange rates are related to whether shocks to the system are 'real', i.e., u_t, or 'monetary', i.e., r_t, and to the 'aggressiveness' of speculators, that is, to the magnitude of n. Of course, to denote r_t as the 'monetary' shock ignores the influence of real factors on nominal interest rates. In a full general equilibrium model, r_t would represent a commingling of more fundamental real and monetary shocks. First note that x_1 is monotonically increasing between zero and one as n goes between zero and infinity. Hence, when speculators are risk-neutral (n infinite), the exchange rate is completely insulated from real shocks. On the other hand, the exchange rate is least insulated from monetary shocks in this case. Whether speculators stabilize or destabilize exchange rates is thus seen to be dependent on whether shocks are real or monetary. The surprise of economists over the variability of exchange rates in the 1970s can be thought of as their lack of appreciation of how international capital mobility transmits interest-rate disturbances internationally. Their realization now that high capital mobility and variable monetary policy can lead to volatile exchange rates has led some to call for 'throwing sand in the system'.

Finally note that the long-run properties of the above asset-market model correspond to purchasing power parity. In the long run, variables are at their steady states and stocks of foreign assets are no longer changing. The trade balance, then, must be zero. Hence, on average, the exchange rate must be equal to relative price levels. Relative price levels are themselves proportional to relative money supplies, real factors remaining constant. The long-run exchange rate then will be proportional to relative money supplies. The fact that for some periods and some exchange rates this prediction has been violated implies that permanent real shocks have been important sources of variability.

The following perspectives on flexible exchange rates emerge from the preceding analysis. First, an emphasis on the role of trade in international financial assets in exchange rate determination leads to an explanation of the volatility of exchange rates under a flexible rate regime. This does not negate, though, the ability of flexible rates to provide long-run monetary independence to individual countries; countries can choose different price levels in the long run, with the exchange rate moving to equilibrate.

Second, exchange rates are endogenous variables, and exchange rate volatility is in important ways a symptom of underlying volatility, not a cause. This volatility does have real effects, though, on relative international competitiveness and associated macroeconomic dislocations.

Finally, both real and monetary factors can play an important role in exchange rate determination. Coincident with this observation, we should note that our major interest is with real exchange rates, i.e. relative prices. In the model used in this entry, our assumption of fixed relative price levels let us identify the nominal exchange rate with relative prices. Of course, in reality price levels are not fixed. Even so, we observe in the world that real and nominal exchange rate movements are highly positively correlated, probably reflecting some sort of price-level stickiness, in which case our analysis is still relevant.

R. DRISKILL

See also EXCHANGE RATES; FIXED EXCHANGE RATES; INTERNATIONAL FINANCE; PURCHASING POWER PARITY.

BIBLIOGRAPHY

Frankel, J.A. 1983. Monetary and portfolio balance models of exchange rate determination. In *Economic Interdependence and Flexible Exchange Rates*, ed. J. Bhandari and B. Putnam, Cambridge: MIT Press, 84–115.
Friedman, M. 1953. The case for flexible exchange rates. In M. Friedman. *Essays In Positive Economics*, Chicago: University of Chicago Press, 157–203.

flexprice models. *See* FIXPRICE MODELS.

floating exchange rates. *See* FLEXIBLE EXCHANGE RATES.

Florence, Philip Sargant (1890–1982). An institutional economist whose work fell broadly into the fields of regional development and industrial organization, Philip Sargant Florence was born in New Jersey (USA) on 25 June 1890 but lived and worked for the greater part of his life in England. He was educated at Rugby, Caius College, Cambridge (taking a First in the Economics Tripos in 1914), and Columbia University (PhD). From 1921 until 1929 he was a lecturer at Cambridge, and was then elected (succeeding J.F. Rees) into the Chair of Commerce in the University of Birmingham, which he occupied until his retirement in 1955. As an American citizen he travelled back to the US for appointments as visiting professor after his retirement, and this he did on at least two occasions – once visiting Johns Hopkins and on another occasion the University of Rhode Island. For the last ten years of his life he was a vice-president of the Royal Economic Society. He died on 29 January 1982 at the age of 91.

Sargant Florence's work on practical problems of industrial organization seems to have developed out of a narrower concern in his earliest writings with the internal organization of productive activities and, in particular, with the effects of fatigue (together with other sociological factors) on the productive efficiency of labour (see, for example, 1918). For Sargant Florence, the problems of industrial organization required the consideration of many issues: market structure and regulation, the direct organization of production and its management, the length of the working day, the role of competition versus combination, the education of business-men, and the efficient design of incentive systems and job ladders to name but a few (see, e.g., 1933). It is little wonder that after a lifetime of work on this kaleidoscope of questions he began to lean more and more towards the belief that the efficient organization of industry was a question that economics could only begin to answer with the help of sociology (see, for example, 1964). It should be said, however, that some of Sargant Florence's writings on these difficult subjects seem to be little more than the reflections of an educated individual with an active but not very profound interest in society, rather than the considered conclusions of a careful social scientist.

Sargant Florence's interest in regional economics seems to have taken root soon after World War II when, on returning to England from the US where he had spent the better part of the war years, he resumed his academic duties at Birmingham and began active applied research into the problems and priorities of regional development and planning in the Midlands as part of post-war reconstruction initiatives in Britain. These investigations generated a number of publications in which his role as author, editor and contributor varied: *County Town* (1946) was a study of Worcester, *English County* (1947) was an examination of the industrial infrastructure of Herefordshire, and *Conurbation* (n.d.) delineated the problems confronting attempts at regional and urban planning in Birmingham and the Black Country.

It is perhaps worth recording that Sargant Florence also wrote a 500-page treatise on the statistical method in economics and political science in 1929. He dedicated this book to all those 'who find theories unsatisfactory without the test of fact' (p. v), and in the preface to the same thanks a number of now well-known Cambridge economists for advice (including Maurice Dobb, Joan Robinson, Gerald Shove, Austin Robinson and Lavington). However, Florence's idiosyncratic understanding of the relationship between theoretical argument and empirical evidence in this work is well illustrated by his criticism and attempted refutation of Freud's theory of parapraxis (1929, pp. 196–9). Towards the end of his life, in 1975, Sargant Florence rather uncharacter-istically took up the macroeconomic question of the causes of inflation.

MURRAY MILGATE

SELECTED WORKS

1924. *The Economics of Fatigue and Unrest*. London: G. Allen & Unwin; New York: H. Holt & Co.
1926. *Over-population: theory and statistics*. London: Kegan Paul & Co.
1927. *Economics and Human Behaviour*. London: Kegan Paul & Co.; New York: W.W. Norton.
1929. *The Statistical Method in Economics and Political Science*. London: Kegan Paul, Trench and Trubner; New York: Harcourt, Brace & Co.
1930. *Uplift in Economics*. London: Kegan Paul & Co.
1933. *The Logic of Industrial Organization*. London: Kegan Paul, Trench & Trubner.
1938. (With A. Carr-Saunders et al.) *Consumers' Co-operation in Great Britain*. New York and London: Harper & Brothers.
1948. *Investment, Location, and Size of Plant*. Cambridge: Cambridge University Press.
1953. *The Logic of British and American Industry*. London: Routledge & Kegan Paul; Chapel Hill: University of North Carolina Press. Revised edn, 1971.
1957. *Industry and the State*. London: Hutchinson's University Library.
1964. *The Economics and Sociology of Industry*. London: C.A. Watts & Co. Revised edn, 1969.
1975. Stagflation in Great Britain: the role of labor. In Gardiner C. Means et al., *The Roots of Inflation*, New York: Burt Franklin & Co.

flow of capital. See INTERNATIONAL CAPITAL FLOWS.

flow of funds. See FINANCIAL INTERMEDIARIES; SOCIAL ACCOUNTING.

flows and stocks. See STOCKS AND FLOWS.

Flux, Alfred William (1867–1942). A distinguished applied economist and statistician, Flux was born in Portsmouth on 8 April 1867, the son of a journeyman cement maker. He died in Denmark, his wife's native land, on 16 July 1942. After entering St John's College, Cambridge, he was bracketed as Senior Wrangler in the Mathematics Tripos of 1887. Soon turning to economics, he came under Alfred Marshall's influence, joining Marshall as a Fellow of St John's in 1889. Leaving Cambridge in 1893 to teach economics at Owens College, Manchester, he next moved in 1901 to McGill University, Montreal. In 1908 he returned to London as statistical adviser to the Board of Trade, where he remained until retirement in 1932, being knighted in 1934.

To the pre-1908 phase belong: a steady stream of pioneering statistical studies of international trade exemplified by Flux (1894b, 1897, 1899); the first post-Marshallian British textbook (Flux, 1904), an accomplished but unoriginal exposition with an interesting geometrical appendix; a new edition of Jevons's *Coal Question* (Flux, 1906); and, though it appeared only after 1908, a study of Swedish banking for the US National Monetary Commission (Flux, 1910). But the work for which he is now best known, his only significant theoretical contribution, was his earliest publication, a review of Wicksteed's *Coordination* (Flux, 1894a) which first invoked the Euler theorem to prove that marginal productivity imputation just exhausts output given constant returns to scale in production.

After 1908, Flux found his *métier* in the development of official statistics. A series of papers given to the Royal Statistical Society, exemplified by Flux (1913, 1921, 1927, 1929), stands as monument to his important contributions. See also his Newmarch Lectures (Flux, 1924).

Flux remained a frequent reviewer for the *Economic Journal* but never matched his first performance. He also contributed to the original *Palgrave*. No comprehensive bibliography has been compiled of his many articles and pamphlets. But only three are of a theoretical or doctrinal character, none of these being especially noteworthy. For biographical details see Chapman (1942).

J.K. WHITAKER

SELECTED WORKS

1894a. Review of Wicksell, K., *Kapital und Rente nach der Neuern Nationalökonomischen Theorien,* and Wicksteed, P.H., *Essay on the Coordination of the Laws of Production and Distribution. Economic Journal* 4, June, 305–13.
1894b. The commercial supremacy of Great Britain. *Economic Journal* 4, Pt. I, September, 457–67; Pt. II, December, 595–605.
1897. British trade and German competition. *Economic Journal* 7, March, 34–45.
1899. The flag and trade: a summary review of the trade of the chief colonial empires. *Journal of the Royal Statistical Society* 62, September, 489–522.
1904. *Economic Principles.* London: Methuen; 2nd edn, 1923.
1906. (ed.) Jevons, W.S. *The Coal Question.* 2nd edn, London: Macmillan.
1910. *The Swedish Banking System.* Publications of the National Monetary Commission, Vol. XVII, No. 1, Washington, DC: Government Printing Office.
1913. Gleanings from the Census of Production Report. *Journal of The Royal Statistical Society* 76, May, 557–85.
1921. The measurement of price changes. *Journal of the Royal Statistical Society* 84, March, 167–99.
1924. *The Foreign Exchanges.* Newmarch Lectures for 1922, London: P.S. King.
1927. Indices of industrial productive activity. *Journal of the Royal Statistical Society* 90(2), 226–58.
1929. The national income. *Journal of the Royal Statistical Society* 92(1), 1–25.

BIBLIOGRAPHY

Chapman, S.J. 1942. Sir Alfred William Flux. *Economic Journal* 52, December, 400–403.

Forbonnais, François Véron Duverger de (1722–1800). French economist, industrialist and inspector of commerce, Forbonnais was born at Le Mans in 1722 and died in Paris in 1800. After initial employment in industry and trade in Nantes, his desire to obtain an official position in the government services (successful in 1756 when he was appointed general inspector of currency) inspired his career as a writer on economic and financial subjects. These all have a strong mercantilist flavour, and also display considerable antagonism to the Physiocrats. Forbonnais contributed a number of economic articles to the *Encyclopédie* and provided translations of some important writings on commerce. These include King's *The British Merchant* (1721) and Uztariz's *Theory and Practice of Commerce* (1724), the former translation according to Morellet (1821) inspired by Gournay.

Forbonnais' major works are his *Elémens du commerce* (1754) and his *Principes et observations oeconomiques* (1767). The *Elémens* has the distinction of being the first French work on economics using mathematical argument. This is his analysis of equilibrium conditions with respect to the rates of exchange between more than two countries and in situations of bimetallism where there are differences in the price ratios of gold and silver, (Theocharis, 1961). The *Principes* is a polemical work in which the major part is devoted to criticism of Quesnay's *Tableau économique* and his *Encyclopédie* articles on Farmers and Corn after an elucidation of general principles. Forbonnais' criticism of Physiocratic analysis is noteworthy because it was directed at its empirical foundations. In the discussion of general principles he develops arguments on the interdependence of production and trade, the balance of trade, the balance of trade doctrine in relation to money supply and employment, the beneficial consequences of gradual price rises, and the advantages of paper credit.

PETER GROENEWEGEN

SELECTED WORKS

1754. *Eléments du commerce.* Leyden and Paris: Briasson.
1767. *Principes et observations oeconomiques.* Amsterdam and Paris: M.M. Rey.

BIBLIOGRAPHY

King, C. 1721. *The British Merchant.* Translated freely from the English as *Le négociant anglais* by F.V.D. de Forbonnais, Dresden and Paris, 1753.
Morellet, l'Abbé de. 1821. *Mémoires de l'Abbé Morellet.* Paris: Librairie Française.
Theocharis, R.D. 1961. *Early Developments in Mathematical Economics.* London: Macmillan.
Uztariz, G. de. 1724. *Theory and Practice of Commerce (Téorica y práctica del commercio).* Translated freely from the Spanish as *Théorie et pratique du commerce et de la marine* by F.V.D. de Forbonnais, Paris, 1753.

forced labour. *See* SLAVERY.

forecasting. Decisions in the fields of economics and management have to be made in the context of forecasts about the future state of the economy or market. As decisions are so important as a basis for these fields, a great deal of attention has been paid to the question of how best to forecast variables and occurrences of interest. There are several distinct types of forecasting situations including event timing, event outcome, and time-series forecasts. Event timing is concerned with the question of when, if ever, some specific event will occur, such as the introduction of a new tax law, or of a new product by a competitor, or of a turning point in the business cycle. Forecasting of such events is usually attempted by the use of leading indicators, that is, other events that generally precede the one of interest. Event-outcome forecasts try to forecast the outcome of some uncertain event that is fairly sure to occur, such as finding the winner of some election or the level of success of a planned marketing campaign. Forecasts are usually based on data specifically gathered for this purpose, such as a poll of likely voters or of potential consumers. There clearly should be a positive relationship between the amount spent on gathering the extra data and the quality of the forecast achieved.

A time series x_t is a sequence of values gathered at regular intervals of time, such as daily stock market closing prices, interest rates observed weekly, or monthly unemployment levels. Irregularly recorded data, or continuous time sequences may also be considered but are of less practical importance. When at time n (now), a future value of the series, x_{n+h}, is a random variable where h is the forecast horizon. It is usual to ask questions about the conditional distribution of x_{n+h} given some information set I_n, available now from which forecasts will be constructed. Of particular importance are the conditional mean

$$f_{n,h} = E[x_{n+h} | I_n]$$

and variance, $V_{n,h}$. The value of $f_{n,h}$ is a point forecast and represents essentially the best forecast of the most likely value to be taken by the variable x at time $n + h$. With a normality assumption, the conditional mean and variance can be used together to determine an interval forecast, such as an interval within which $x_{n,h}$ is expected to fall with 95 per cent confidence. An important decision in any forecasting exercise is the choice of the information set I_n. It is generally recommended that I_n include at least the past and present of the individual series being forecast, $x_{n-j, j \geqslant 0}$. Such information sets are called *proper*, and any forecasting models based upon them can be evaluated over the past. An I_n that consists just of x_{n-j}, provides a univariate set so that future x_i are forecast just from its own past. Many simple time-series forecasting methods are based on this information set and have proved to be successful. If I_n includes several explanatory variables, one has a multivariate set. The choice of how much past data to use and which explanatory variables to include is partially a personal one, depending on one's knowledge of the series being forecast, one's levels of belief about the correctness of any economic theory that is available, and on data and computer availability. In general terms, the more useful explanatory variables that are included in I_n, the better the forecast that will result. However, having many series allows for a confusing number of alternative model specifications that are possible so that using too much data could quickly lead to diminishing marginal returns in terms of forecast quality. In practice, the data to be used in I_n will often be partly determined by the length of the forecast horizon. If h is small, a short-run forecast is being made and this may concentrate on frequently varying explanatory variables. Short-term forecasts of savings may be based on interest rates, for example. If h is large so that long-run forecasts are required, then slowly changing, trending explanatory variables may be of particular relevance. A long-run forecast of electricity demand might be largely based on population trends, for example. What is considered short- or long-run will usually depend on the properties of the series being forecast. For very long forecasts, allowances would have to be made for technological change as well as changes in demographics and the economy. A survey of the special and separate field of technological forecasting can be found in Martino (1983).

If decisions are based on forecasts, it follows that an imperfect forecast will result in a cost to the decision maker. For example, if $f_{n,h}$ is a point forecast made at time n, of x_{n+h}, the eventual forecast error will be

$$e_{n,h} = x_{n,h} - f_{n,h}$$

which is observed at time $n+h$. The cost of making an error e might be denoted as $C(e)$, where $C(e)$ is positive with $C(0)=0$. As there appears to be little prospect of making error-free forecasts in economics, positive costs must be expected, and the quality of a forecast procedure can be measured as the expected or average cost resulting from its use. Several alternative forecasting procedures can be compared by their expected costs and the best one chosen. It is also possible to compare classes of forecasting models, such as all linear models based on a specific, finite information set, and to select the optimum model by minimizing the expected cost. In practice the true form of the cost function is not known for decision sequences, and in the univariate forecasting case a pragmatically useful substitute for the real $C(e)$ is to assume that it is well approximated by ae^2 for some positive a. This enables least-squares statistical techniques to be used when a model is estimated and is the basis of a number of theoretical results including that the optimal forecast of x_{n+h} based on I_n is just the conditional mean of $x_{n,h}$. In the economics literature, optimum forecasts have also been called *rational expectations*, based on a specific cost function and an information set.

When using linear models and a least-square criterion, it is easy to form forecasts under an assumption that the model being used is a plausible generating mechanism for the series of interest. Suppose that a simple model of the form

$$x_t = \alpha x_{t-1} + \beta y_{t-2} + \epsilon_t$$

is believed to be adequate where ϵ_t is a zero-mean, white noise (unforecastable) series. When at time n, according to this model, the next value of x will be generated by

$$x_{n+1} = \alpha x_n + \beta y_{n-1} + \epsilon_{n+1}$$

The first two terms are known at time n, and the last term is unforecastable. Thus

$$f_{n,1} = \alpha x_n + \beta y_{n-1}$$

and

$$e_{n,1} = \epsilon_{n+1}.$$

x_{n+2}, the following x, will be generated by

$$x_{n+2} = \alpha x_{n+1} + \beta y_n + \epsilon_{n+2}.$$

The first of these terms is not known at time n, but a forecast is available for it, $\alpha f_n, \cdot$, the second term is known at time n, and the third term is not forecastable, so that

$$f_{n,2} = \alpha f_{n,1} + \beta y_n$$

and

$$e_{n,2} = \epsilon_{n+2} + \alpha(x_{n+1} - f_{n,1})$$
$$= \epsilon_{n+2} + \alpha \epsilon_{n+1}.$$

To continue this process for longer forecast horizons, it is clear that forecasts will be required for y_{n+h-2}. The forecast formation rule is that one uses the model available as though it is true, ask how a future x_{n+h} will be generated, use all known terms as they occur, and replace all other terms by optimal forecasts. For non-linear models this rule can still be used, but with the additional complication that the optimum forecast of a function of x is not the same function of the optimum forecast of x.

The central problem in practical forecasting is chosing the model from which the forecasts are derived. If a univariate information set is used, it is natural to consider the models developed in the field of time-series analysis. A class of models that have proved successful in short-run forecasting are the autoregressive-moving average models (ARMA). If a series is regressed on lagged values of itself up to p lags, and the residual is a weighted sum of a white noise series, up to q lags, the result is an ARMA (p, q) model. There is a tendency to prefer a parsimonious model so that the number of parameters (p, q) is minimized amongst models that perform equally satisfactorily. If the first difference of a series is well modelled as an ARMA process, but the level of the series is not, the basic series is said to be integrated of order one and called an integrated ARMA processes, denoted ARIMA. Box and Jenkins (1974) discussed the statistical properties of these series and relevant constraints on the parameters. They suggested three stages of analysis, (i) identification – the choice of a small subset of models for further analysis, (ii) estimation of parameters, and (iii) diagnostic checks on whether the estimated model eventually selected is fitting the data adequately by comparing with other similar models. These techniques can be generalized to modelling vector x_t series, although with some difficulties. For example, vector autoregressive models, with imprecise constraints on parameters, have been found to provide successful short-run forecasts of macroeconomic variables (Doan, Litterman, and Sims, 1984). These models are largely agnostic towards the correctness or otherwise of an economic theory. Their main competitors are the econometrics models which historically were large in that many economic variables were modelled simultaneously, occasionally several hundred to a few thousand variables, but had less emphasis on dynamics and used some economic theory, often an equilibrium theory, as a starting point for specification and analysis. They have become more dynamic in recent years whilst still emphasizing the many important interrelationships that can be expected to hold between economic variables. Both basic approaches seem to have special strengths and relative weaknesses, a good theory effectively expands the information available, but an incorrect or loosely stated theory can mislead the modelling process. Econometric models now often consider the inclusion of rational expectations as explanatory variables, and all models have considered the introduction of non-linear terms and also time-varying parameters, usually by use of the Kalman filter algorithm.

The forecasting process is generally improved by continual evaluation of previously made forecasts. The theory of optimal forecasts provides some helpful evaluation criteria, such as that h-step forecasts errors should be $MA(h-1)$, so that one-step errors from optimal forecasts should be white noise, and that generally forecasts perform less well as the horizon increases. A practical evaluation procedure is to regress the actual x_t on their forecasts, for example

$$x_{t+1} = \alpha + \beta f_{t,1} + \epsilon_{t+1}.$$

If the one-step forecast is optimal for some information set, it follows that the true values of the parameters should be $\alpha = 0$ and $\beta = 1$ and also that ϵ_t should be white noise. If any of these properties are not found for the regression, the forecasts can be immediately improved. Unfortunately this procedure does not answer the question of whether all the information in I_n is being fully utilized or if a different information set would produce better forecasts. One way that these questions can be approached is by combining forecasts from different sources or based on different information sets so that a regression is run of the form

$$x_{t+1} = \alpha + \beta_1 f_{t,1} + \beta_2 g_{t,1} + \epsilon_{t+1}$$

where f and g are the two forecasts to provide appropriate weights. This and the other topics mentioned here are discussed in greater detail in Granger and Newbold (1987).

C.W.J. GRANGER

See also MACROECONOMETRIC MODELS; PREDICTION; TIME SERIES ANALYSIS.

BIBLIOGRAPHY

Box, G. and Jenkins, G. 1970. *Time Series Analysis, Forecasting and Control*. San Francisco: Holden Day.
Doan, T., Litterman, R. and Sims, C. 1984. Forecasting and conditional projection using realistic prior distributions. *Econometric Review* 3, 131–44.
Granger, C.W.J. and Newbold, P. 1987. *Forecasting Economic Time Series*. 2nd edn, New York: Academic Press.
Martino, J. 1983. *Technological Forecasting for Decision Making*. 2nd edn, Amsterdam: North-Holland.

forced saving. The doctrine of forced saving proposes that an increase in the amount of money may be favourable to capital accumulation at the cost of a reduction in consumption of certain individuals, but the latter have not saved voluntarily and they do not receive any immediate benefit. The doctrine was developed in the early 19th century by Thornton (1802) and Bentham (1804). They used the terms 'defalcation of revenue' and 'forced frugality' respectively. It was Mises who coined the term 'forced saving' (*erzwungenes Sparen*).

Thornton published his *Paper Credit* (1802) during the debate on the suspension of gold payments by the Bank of England in 1797; the debate concerned the possible existence of a natural tendency to keep the circulation of the Bank of England within the limits which would prevent a dangerous depreciation. An excessive issue of paper money could, according to Thornton, at least temporarily increase the price level of commodities while the money wage and other fixed incomes stayed the same. This would not only lead to a general rise in prices but also to some increase in real capital, since the real consumption of the labourers and recipients of fixed incomes would be reduced, which was the meaning of 'defalcation of revenue'.

Jeremy Bentham, in the manuscript 'Institute of Political Economy' of 1804, some of which had already been written in the years 1800 and 1801, analysed the effects of an increase of paper money in a situation where all hands were employed in the most advantageous manner. If the money in the first instance were used for productive expenditure, i.e. buying inputs for producing capital goods, then it would add to real capital. In the second round the money would be exclusively used for consumption and only prices would be affected. The extra real capital was due to the 'forced frugality' of the possessors of fixed income which was engineered by the decrease in the value of money; it operated exactly like an indirect tax upon pecuniary income. But the effect of 'forced frugality' was probably quite small. It was also an unjust mechanism for increasing national wealth, and under normal circumstances voluntary sacrifices would be sufficient to augment the mass of real wealth. It is obvious in these early enquiries that the forced saving by receivers of fixed incomes came from a decrease in the amount of their real consumption, while the total amount of their money expenditures was kept the same and there was no change in the amount of hoarded funds.

During the course of the Bullionist Controversy, Malthus raised the issue in his 1811 review of Ricardo's *High Price of Bullion* (1810). Malthus proposed that if a new issue of notes came into the hands of the productive classes, (described as a change in the distribution of the circulating medium) then capital accumulation would increase. The mechanism of forced saving worked via the increase in the price level, which reduced the share of the annual produce of those classes who were only buyers and not sellers. Ricardo replied, in an appendix to the fourth edition of *The High Price of Bullion* published in 1811, that Malthus's results were based upon the assumption that those who lived on fixed incomes must consume their whole income. In the case of money saving it was possible that the issue of bank-notes and the ensuing inflation merely transferred saving from the receivers of fixed incomes to those who had borrowed from the banks. Thus Ricardo saw no reason why it should add anything to the productive classes.

Later, comments on forced saving are found in the works of J.S. Mill and Walras, but the doctrine became important once again when it was incorporated into the pre-Keynesian analysis of credit and business cycles. The analysis took off from Wicksell's brief mention that during a cumulative process rising prices might force people living on fixed money income to reduce their consumption, an 'involuntary saving' which could lead to the production of new real capital. Mises (1912) and later Hayek (1929) developed Wicksell's analysis, and forced saving was used to explain the upswing in the so-called 'over-investment' theories of cyclical movements. An overextension of credit, since the money rate of interest was too low, and the ensuing cumulative process led to a distortion of the vertical structure of production. Production of producers' goods outstripped the production of consumers' goods since means of production were transferred from the latter to the former. The increase in real capital took place because of forced saving, which worked through prices rising faster than disposable income of wage-earners and the rigidity of certain incomes. The intermediate result was the same as for voluntary saving. Consumers were forced to forego what they used to consume so as to give the entrepreneurs, who had received the additional money, command over resources for the production of extra capital goods. However, no permanent increase of real capital was possible with the help of inflationary credit expansion and forced saving, and the new capital built during the upswing would necessarily be destroyed during the downturn.

Dennis Robertson made a most detailed analysis of different forms of saving or 'lacking' in *Banking Policy and the Price Level* (1926). He introduced the term 'automatic lacking': an involuntary reduction in planned consumption, which came about when the price level increased because newly created money was added to the daily stream of money which competed for the daily stream of marketable goods.

Parts of the doctrine of forced saving were questioned with the publication of Keynes's *Treatise on Money* (1930) and his subsequent debate with Hayek and Robertson. Robertson had, according to Keynes, no distinct definition of voluntary saving, which was related to a confusion concerning the definition of income, and it implied a deficient view of the meaning of forced saving. Keynes defined saving as the difference between income or normal costs and expenditure on consumption, which could differ from investment since saving and investment were decisions taken by different agents, windfall profits and losses being the balancing figure between investment and saving. Forced saving or automatic lacking existed when investment exceeded saving and purchasing power was redistributed by the accompanying inflation; it was represented on the one hand by the increased amount of money which spenders had to pay for that part of consumption which they continued to enjoy, and on the other hand by the extra investment provided out of the windfall gains of the entrepreneurs. Hence Keynes did not challenge the fact that an increase in net investment took place via the redistribution of purchasing power, but it was not an involuntary act.

At the same time, Erik Lindahl presented a similar analysis in *The Rate of Interest and the Price Level* (1930). The rising prices during an upward cumulative process had to change the distribution in favour of those who had a strong incentive to save, until the total saving in the community corresponded to the value of real investment, which was primarily determined by the rate of interest. This saving was mainly voluntary, since an individual was free to consume as much as he liked and the only limit was his credit standing. Keynes had the same view in the *General Theory*: this type of saving was in complete agreement with the free will of the individual to save what he chose irrespective of what he or others might be investing, since no individual could be compelled to own the additional money (corresponding to the new bank-credit) unless he deliberately preferred to hold more money rather than some other form of wealth. Lindahl reserved forced saving for the possibility that the individual has to limit planned consumption out of income (defined as the rate of interest on the capital value of all capital goods including human capital) because he is not able to obtain credit, which might be explained by banking rules concerning the collateral for loans, i.e. it is not a perfect capital market.

Once the notions of *ex ante* and *ex post* were introduced all these problems could be solved. A fall in the money rate leads to an excess of planned and realised investment over planned saving (related to planned income), and the subsequent increase in prices would imply higher incomes *ex post* for the entrepreneurs, which is the same as Keynes's concept of windfall gains in the *Treatise*. This unexpected windfall, which could not be spent during the period, would contribute the extra necessary saving, since investment *ex post* had to be equal to saving *ex post*. Lindahl denoted this as 'unintentional saving' and he found 'forced saving' to be an inappropriate term. However, Keynes seemed to have changed his position slightly in *How to Pay for the War* (1940). The process could only be successful if wages lagged behind prices, for otherwise an unlimited inflation would take place. As such it was a method of compulsorily converting a part of workers' earnings, which they do not plan to save voluntarily, into the voluntary saving of the entrepreneurs. From an analytical point it was voluntary saving, but it was 'a matter of taste' whether this was a suitable name.

To sum up: there was a consensus that new credit might lead to an additional, at least temporary, investment even in a full employment situation via an increase in the price level. But the most recent contributions, e.g. Lindahl and Keynes, did not consider the extra saving to be forced. At the same time almost all of them found it unwise and unjust to rely on credit inflation as a means of increasing capital accumulation. However, after Keynes's analysis in the *General Theory* the problem seems to have disappeared from the agenda.

BJÖRN HANSSON

See also INFLATION.

BIBLIOGRAPHY

Bentham, J. 1804. Institute of Political Economy. In Vol. III of *Jeremy Bentham's Economic Writings*, ed. W. Stark, London: George Allen & Unwin, 1954.

Haberler, G. 1937. *Prosperity and Depression*. 5th edn, London: George Allen & Unwin, 1964.

Hayek, F. von. 1929. *Geldtheorie und Konjunkturtheorie*. Vienna and Leipzig: Hölder-Pichler-Tempsky. Trans. as *Monetary Theory and the Trade Cycle*, London: Jonathan Cape, 1933.

Hayek, F. von. 1931. *Prices and Production*. 2nd edn, London: George Routledge & Sons, 1935.

Hayek, F. von. 1932. A note on the development of the doctrine of forced saving. *Quarterly Journal of Economics* 47, November, 123–33.

Keynes, J.M. 1930. *A Treatise on Money*. Vol. I. Published as Vol. V of *The Collected Writings of John Maynard Keynes*, London: Macmillan, 1971.

Keynes, J.M. 1936. *The General Theory of Employment, Interest and Money*. Published as Vol. VIII of *The Collected Writings of John Maynard Keynes*, London: Macmillan, 1973.

Keynes, J.M. 1940. *How to Pay for the War*. Published in Vol. XXII of *The Collected Writings of John Maynard Keynes*, London: Macmillan, 1978.

Lindahl, E. 1930. *Penningpolitikens medel*. Lund: C.W.K. Gleerup. Trans. as 'The rate of interest and the price level' in Lindahl, *Studies in the Theory of Money and Capital*, London: George Allen & Unwin, 1939.

Machlup, F. 1943. Forced or induced saving: an exploration into its synonyms and homonyms. *Review of Economics and Statistics* 25, February, 26–39.

Malthus, T.R. 1811. Review of Ricardo's *High Price of Bullion*. *Edinburgh Review*, February.

Mill, J.S. 1844. *Essays on Some Unsettled Questions of Political Economy*. Published in Vol. IV of *Collected Works of John Stuart Mill*, London: Routledge & Kegan Paul, 1967.

Mises, L. von. 1912. *Theorie des Geldes und der Umlaufsmittel*. Munich: Dunker & Humblot. 2nd edn, 1924. Trans. as *The Theory of Money and Credit*, London: Jonathan Cape, 1934.

Ricardo, D. 1810. *The High Price of Bullion. A Proof of the Depreciation of Bank Notes*. In Vol. III of *The Works and Correspondence of David Ricardo*, ed. Piero Sraffa, Cambridge: Cambridge University Press, 1951.

Robertson, D. 1926. *Banking Policy and the Price Level*. London: P.S. King & Sons.

Thornton, H. 1802. *An Enquiry into the Nature and Effects of the Paper Credit of Great Britain*. Reprinted, London: George Allen & Unwin, 1939.

Walras, L. 1879. *Théorie mathématique du billet de banque*. Reprinted in *Etudes d'économie politique appliquée*, Lausanne and Paris, 1898.

Wicksell, K. 1935. *Lectures on Political Economy*. Vol. II. Trans. from the 3rd Swedish edn of *Förelāsningar i nationalekonomi*, Vol. II, 1929. London: George Routledge & Sons.

foreign advisers. *See* ADVISERS.

foreign aid. Foreign aid originated from the disruption of the world economy that followed World War II. Before the system of international trade and capital movements could be restored, the economies of the industrial countries had to be rebuilt and their ties with former colonies replaced by multilateral arrangements. Until these structural changes could be brought about, much of the world depended on the United States for essential imports.

Postwar reconstruction was greatly facilitated by the willingness of the United States to finance its large surplus of exports over imports on highly concessional terms under the European Recovery Program, generally known as the Marshall Plan. This arrangement was the forerunner of aid programmes for developing countries. It was built around the concept that American loans and grants could be more efficiently utilized in the context of an overall analysis of European trade and development. The Organization for European Economic Cooperation (OEEC), which was set up to administer the Marshall Plan, formulated principles and procedures that have been used in administering foreign aid programmes ever since.

The European Recovery Program had two main objectives: to restore the economies of its member countries as rapidly as possible and to develop a viable pattern of trade that would not require further concessional loans. Most participating countries achieved these objectives over the five-year period 1948–53. During this period the transfers from the United States in the peak years amounted to 2–3 per cent of total US GNP – ten times current US aid levels (Price, 1955). Since its objectives were achieved on schedule, the Marshall Plan has generally been considered a great success. This performance cannot be taken as a precedent for aid programmes for less developed countries, however, without examining the differences in their objectives and initial conditions.

As Western Europe recovered and its former colonies became independent, the OEEC was transformed into an organization for coordinating the bilateral aid programmes and trade policies of its member countries, becoming in 1961 the Organization for European Cooperation and Development (OECD). At the same time, the International Bank for Reconstruction and Development (World Bank) was shifting from reconstruction loans to industrial countries to long-term loans to developing countries. This shift was accelerated by the establishment in 1959 of a fund for soft loans to the poorest countries (the International Development Association or IDA) as an affiliate of the World Bank. This was followed by the creation of regional development banks for Latin America, Asia and Africa. By the early 1960s, therefore, the present institutional framework for government transfers of resources to less developed countries was largely in place.

PRINCIPLES OF INTERNATIONAL AID. What is foreign aid? In common usage it includes governmental resource transfers to poor countries that are mainly for development purposes; it excludes quasi-commercial transactions ('hard loans') such as export credits whose benefits to the lender approximate their cost. For most purposes, it also excludes public transfers for non-developmental objectives, such as military assistance, and private charity (Bhagwati, 1969). To be more precise, foreign aid is usually discussed using definitions adopted by the OECD, which are based on the grant element of loans for development purposes.

International aid is customarily measured in two ways: by indices of its cost to the donor country, such as the share of its GNP; and by indices of the amount of resources transferred to the recipients, such as the percentage of their imports or investment that it finances. Several of these measures are shown in Table 1 for different groups of donor countries and periods.

Although there is no general theory of international aid, there is a set of questions that regularly arises in discussing this topic. What are the benefits and costs of concessional loans and grants? How are they affected by different forms of resource transfer or by the economic structures of the participants? Is there an overall gain or loss to the world economy? While these questions must be looked at separately from the standpoint of the donor and the recipient, it is useful to start with the potential gain to the system as a whole.

If the world economy were characterized by competitive equilibrium in all countries and markets, the benefits to aid recipients would be approximately offset by costs to donors; the only net gain from aid would lie in improving income distribution (Little and Clifford, 1965, ch. 3). In the opposite case of prevailing disequilibrium in commodity and factor markets – illustrated by postwar Europe – the prevalence of bottlenecks created the possibility of achieving large increases in production in relation to the resources transferred with little reduction in consumption in the donor country. In fact, the more rapid growth of the European market for US exports over a 10 year period may well have offset most of the original cost of the Marshall Plan to the United States.

The typical conditions under which the transfer of aid to developing countries takes place lie somewhere between these two extremes. There are many instances of foreign exchange bottlenecks whose elimination makes aid more productive, but most of them are the result of mismanagement of trade and investment policies that could well have been avoided. One situation in which the Marshall Plan analogy has some validity was the world oil crisis of 1973, when concessional lending to oil importing developing countries made it possible to sustain higher rates of growth by the lenders as well as the borrowing countries.

The main objective of international aid programmes is long-term economic development, which requires transforming the structure of production and trade (Chenery and Strout, 1966). Recently the distributional aspects of this process have been stressed by both donors and recipients, implying that greater weight should be given to the reduction of poverty than to the mere growth of total income. This shift in emphasis makes the design of aid programmes more complex and increases the need for coordination of donor activities.

Since the time of the Marshall Plan the general concepts of international aid and the effectiveness of specific approaches have been the subject of controversy. The neoclassical critique objects to the implied interference with market forces and the possibility that aid may perpetuate inefficient policies (Johnson, 1967). The radical critique objects to the external control or 'conditionality' that can be exerted by the major donor agencies as a group and their tendency to resist radical change. These criticisms will be addressed in the course of

TABLE 1 *AID BY MAJOR DONORS* (Official Development Assistance)

| | ODA Volume (1983 $ billion) | | | | ODA as Per Cent of Donor GNP | | | |
| | (Share of World Total) | | | | | | | |
Donor	1950–55	1960–61	1970–71	1983–84	1950–55	1960–61	1970–71	1983–84
United States	$4.0	8.7	7.0	8.2	0.32	0.56	0.31	0.24
	(50)	(46)	(31)	(22)				
European	$3.7	6.5	7.1	11.8	0.52	0.64	0.42	0.52
Community	(47)	(35)	(31)	(32)				
Japan	$0.1	0.6	1.6	4.0	0.04	0.22	0.23	0.34
	(1)	(3)	(9)	(11)				
Other OECD	$0.1	0.6	2.1	4.3	—	0.18	0.36	0.43
	(1)	(3)	(9)	(12)				
Total OECD (%)	7.9	16.3	17.8	28.4	0.35	0.52	0.34	0.36
	(100)	(88)	(79)	(77)				
OPEC (%)	—	—	1.1	5.0	—	—	0.78	0.95
			(5)	(13)				
E. Europe (%)	—	1.9	2.6	3.1	—	—	0.15	0.21
		(10)	(11)	(9)				
Other (%)	—	0.4	1.1	0.3	—	—	—	—
		(2)	(5)	(1)				
TOTAL WORLD	7.9	18.8	22.8	37.0	0.30	0.41	0.33	0.37
	(100)	(100)	(100)	(100)				

Source: OECD, *Twenty-five Years of Development Cooperation*, 1985, Table III.

considering the aid process in more detail, first from the recipient and then from the donor point of view.

THE RECIPIENT VIEW. Recipients and donors can agree on the broad objectives of international aid: long-term development and efficient use of resources. These can be achieved more readily if donors and recipients collaborate in the design and execution of aid programmes. Disagreements arise over the magnitude of aid provided, its allocation among countries and the ways in which the resource transfer is made.

In macroeconomic terms foreign aid performs two functions: it adds to the resources available for investment and it augments the supply of foreign exchange to finance imports. Although additional aid serves both these purposes, their relative importance varies according to the economic structure of the recipient. Since many of the goods that are critical to development must be imported – machinery, fuels, and raw materials – a shortage of foreign exchange can become a bottleneck when the cost of imports increases more rapidly than export earnings.

Two major techniques of aid administration have been designed to fit the different conditions under which aid is provided. The first, project aid, is designed to increase output and efficiency in specified productive units. The magnitude of external loans is related to the import of capital goods and other inputs required to carry out the project: disbursement takes place over the period of five years or more required to design and construct the physical plant. In addition, technical assistance to improve the design and operation of projects – or even of whole sectors – has increasingly become an integral part of the project aid package.

The second major technique of aid administration is programme lending, which is designed to support the recipient country's macroeconomic policies and particularly its increased need for foreign exchange during a period of structural change. Although a programme loan may be based on the imports needed by particular sectors, the fungibility of resources requires an assessment of policies related to the balance of payments as a whole. Programme loans are typically disbursed much more rapidly than projects and are preferred by borrowers for this reason.

During periods of stability in the world economy, such as the 1960s, an aid system based largely on project lending can function relatively efficiently since the long-term factors of investment and productivity growth are the dominant problems. However, in times of disequilibrium and rapid change, such as the early 1950s and, again, the 1970s, the need for foreign exchange to maintain imports and support structural adjustment becomes more important. In recent years, this need for 'structural adjustment loans' has been one of the principal issues between borrowers and lenders. Although donors tend to prefer project loans because they are more easily monitored, the recipient's need for imported materials to run existing plants may be greater than its need for additional productive capacity.

THE DONOR VIEW. Although virtually all the objectives of aid recipients can be included under the heading of development, the motives of donors are more complex. To a greater or lesser degree they regard foreign aid as an instrument of foreign policy, which affects both the allocation of bilateral aid among countries and the choice between the more visible project aid and the less visible programme aid. Most donors also use aid as a means of supporting national producers by limiting eligible goods procured with aid funds.

The principal economic distortion in the aid system is the common practice of tying procurement to the donor country. This issue has been attacked with some success by the OECD, most of whose members have agreed to extend procurement to developing countries as well (see OECD, 1985, ch. 10). It has been estimated that the practice of tying bilateral aid reduces its value by 25 per cent or more as compared to the system of competitive bidding that is followed by the World Bank and other multilateral agencies. On the other hand, some donors argue that this support to domestic exporters is a political cost that must be incurred in order to secure parliamentary approval of aid appropriations.

The collective views of the donor countries are expressed in the recommendations of the OECD Development Assistance Committee and in the discussions of the World Bank and International Monetary Fund. In general these agencies have tried to secure larger volumes of aid, more efficient means of transferring resources, and larger allocations to the neediest countries. For example, the Commission on International Development appointed by the World Bank recommended that the share of aid channelled through international institutions be substantially increased from its level of ten per cent in order to secure more equitable use of available funds (Pearson Commission Report, 1969). This shift has been successfully accomplished: in 1985 the multilateral share of OECD aid stands at 30 per cent, of which the World Bank accounts for more than half.

The growth of multilateral lending has made it possible to address more effectively the issue of allocation among countries. Although economists are reluctant to make interpersonal or international comparisons of the marginal utility of income, there is widespread support among the donor countries for allocating aid on the basis of need, which is equated with low per capita income. This principle conflicts with allocation on the basis of efficiency in the use of external resources, which tends to be higher in middle-income countries than in the poorest. A compromise has been reached by reserving the most concessional loan terms for very poor countries while allowing for variation within this category according to the efficiency of use. However, there is still a substantial political element in most bilateral aid programmes.

The growth of multilateral aid has had indirect benefits to both donors and recipients. Perhaps the most important is the extension of arrangements for aid coordination to most aid recipients. The original pattern was established in the early 1960s in the aid consortia set up for India, Pakistan, and Turkey by the World Bank and the OECD. These arrangements included the evaluation of aid needs, government policies and donor pledges of future aid. Subsequent consultative groups under the World Bank have moved toward less formal arrangements aimed at agreement on the diagnosis of country economic needs and the coordination of project and programme lending from many donors.

In the turbulent economic conditions that have prevailed since 1973 the importance of collaboration among donors and recipients has increased. For many countries the traditional form of project lending is no longer the most effective approach to restructuring the economy, and new procedures for structural adjustment lending have had to be developed. These involve greater coordination between the short-run approach of the IMF and the medium- to long-term analysis of the World Bank. The need for new approaches has been particularly acute in sub-Saharan Africa, where the combination of external shocks and inefficient internal policy responses has led to a general decline in per capita incomes (World Bank, 1984, 1985).

PERFORMANCE OF THE INTERNATIONAL AID SYSTEM. Over the past quarter century the transfer of concessional aid from richer to poorer countries has become an established part of the world economy. Although donors and recipients have somewhat different political and social objectives, there is a core of agreement that the main purpose of this transfer is to promote long-term development. Development is now understood to mean not only increasing per capita income but also reducing poverty and the structural changes needed to sustain these processes.

How well does the system work? Since aid is only one of a number of influences on development, the answer can be only partial. However, it is possible to compare the role of aid in the 1960s, when the international economy as a whole was functioning quite well, to the more recent periods in which the strains on the international system were much greater.

In retrospect the period 1960–73 appears as an episode of high and relatively stable growth in both the advanced and developing countries. As shown in Table 1, Official Development Assistance (ODA), as measured by the OECD, rose from 0.35 per cent of the GNP of the advanced countries in the early 1950s to 0.52 per cent in the early 1960s. This level of aid was equivalent to about 20 per cent of the investment of the recipient countries, although increased aid was offset to some extent by increased consumption.

For a low-income country, borrowing of this magnitude would permit a rise in investment – in physical and human capital – of perhaps 15 per cent and a corresponding increase in the rate of growth. While this is consistent with the observed increase in aggregate growth of developing countries from under five per cent in the 1950s to nearly six per cent in the 1960s, it can only be said that aid was one of several important factors contributing to this result.

On the foreign exchange side, aid added a margin of some 20 per cent to export earnings in the 1960s. This was particularly important to countries – perhaps a third of the total – in which the inability to expand exports constituted an important limitation to accelerating growth (see Chenery and Strout, 1966). Aid was a critical factor in countries such as Korea, Taiwan, Greece and Israel in supporting the reorientation of their trade policies from import substitution to export expansion. The importance of this contribution was shown by counterfactual simulations of the development of these countries with less aid than was actually received.

By 1970 the system of international aid was being criticized for being excessively oriented toward growth and not paying sufficient attention to the distribution of its benefits (Faber and Seers, 1972). Poverty alleviation was adopted by most donors and recipients as a primary objective of the aid process. Subsequent studies have shown that while there may be political resistance to this objective, there is no necessary economic conflict between poverty alleviation and growth; in the long run the two are likely to be mutually reinforcing by producing a more productive labour force.

The instability of the world economy following the oil crisis of 1973 has had a large impact on the design and performance of the international aid system. The worsening in terms of trade of oil-importing developing countries in 1973–5 and again in 1980–82 involved larger transfers than the total of concessional aid, and they had to be financed mainly by borrowing on commercial terms. The poorest countries, whose debt servicing capacity is limited, had to restrict their imports to fit the volume of aid available, which in many African countries led to the reduction of economic growth below the rate of population increase. Many middle income countries with more diversified economies – particularly in East Asia – were able to restructure their economies and limit their borrowing to fit the growth of export earnings. The net effect has been to increase the gap between sub-Saharan African and other developing countries.

Although energy markets have now moved closer to equilibrium conditions, the decade of adjustment 1974–84 has had a lasting effect on the foreign aid system. Even though the traditional OECD donors plus the new oil surplus countries financed a modest increase in concessional aid (from 0.33 to 0.37 per cent of their GNP), it fell far short of the amounts needed to sustain the growth of the poorest countries (see World Bank, 1984 and 1985). While borrowing from private banks at low real interest rates offset this shortage for the more creditworthy countries in the 1970s, it also led to the debt crisis of the early 1980s, in which the recession in the advanced countries made economic restructuring and export expansion more difficult.

PROSPECTS FOR FOREIGN AID. In the forty years since the start of the Marshall Plan, the international aid system has performed an increasing variety of functions:

(1) Aid accelerated the structural changes required by postwar reconstruction and performed a similar – but more limited – function in adjusting to the energy crisis.

(2) A cooperative model of international support for development was perfected in the 1960s and functioned fairly effectively at relatively low levels of concessional aid so long as world trade expanded rapidly.

(3) A shift in emphasis from growth to poverty reduction was attempted in the 1970s, but its impact was limited by the overriding requirements to readjust the production and trading structure of the oil-importing countries.

Although the volume of aid has failed to grow rapidly enough to meet the demands of the past decade, there are several aspects of the present system that make it more durable than it appeared to be in the early 1960s. In the first place, while the reduction in the US share from 50 to 22 per cent of the total (shown in Table 1) has lowered the rate of growth, the present system is less vulnerable to changes in American foreign policy and the whims of the US congress. While it would be rash to predict much of an increase in the overall Aid/GNP ratio, it should be easier to maintain the existing share.

Secondly, as per capita incomes of aid recipients have risen, a considerable number of the more successful countries have graduated from concessional to commercial borrowing, particularly in East Asia and Latin America. Continuation of this trend would make it possible to concentrate the aid system increasingly on the more intractable development problems of sub-Saharan Africa and South Asia, which contain most of the absolute poor.

Finally, the most enduring aspect of aid is likely to be the discovery and dissemination of knowledge to fit the development needs of poor countries. A notable success has been the joint sponsorship of agricultural research by multilateral and bilateral aid agencies over the past fifteen years. Knowledge is a classic case of the economist's 'public good', and the expansion of this aspect of the international aid system should command wide support.

HOLLIS CHENERY

See also DEPENDENCY; FISCAL AND MONETARY POLICIES IN DEVELOPING COUNTRIES; INEQUALITY BETWEEN NATIONS; NORTH–SOUTH ECONOMIC RELATIONS; UNEVEN DEVELOPMENT.

BIBLIOGRAPHY
Bhagwati, J.N. 1969. *Amount and Sharing of Aid.* Overseas Development Council Monograph No. 2, Washington, DC.
Chenery, H.B. and Strout, A.M. 1966. Foreign assistance and economic development. *American Economic Review* 56, September, 679–733.
Faber, M., and Seers, D. (eds) 1972. *The Crisis in Planning.* London: Chatto & Windus.
Johnson, H.G. 1967. *Economic Policies Toward Less Developed Countries.* Washington, DC: Brookings.
Little, I.M.D. and Clifford, J.M. 1965. *International Aid.* Chicago: Aldine.
Organization for Economic Cooperation and Development. 1985. *Twenty-Five Years of Development Cooperation.* Paris: OECD.
Pearson, L.B. 1969. *Partners in Development: Report of the Commission on International Development.* New York: Praeger.
Price, H.B. 1955. *The Marshall Plan and its Meaning.* Ithaca: Cornell University Press.
World Bank. 1984. *World Development Report.* New York: Oxford University Press.
World Bank. 1985. *World Development Report.* New York: Oxford University Press.

foreign investment. Defined narrowly, foreign investment is the act of acquiring assets outside one's home country. These assets may be financial, such as bonds, bank deposits and equity shares or they may be so-called direct investment and involve the ownership of means of production such as factories and land. Direct investment is considered to take place also if the ownership of equity shares provides control over the operation of a firm. Johnson (1970) has suggested the expansion of the concept of foreign investment so that it parallels the modern Fisherian approach and distinguishes physical, human and knowledge capital. Accordingly, schooling abroad and technology transfers through the purchase of patents and licences represent foreign investment broadly defined.

In the 19th century, foreign investment involved mostly the ownership of financial assets (Iversen, 1936). After World War II direct foreign investment began to dominate and attract much theoretical and empirical research efforts of economists and the concerns of politicians (Hymer, 1976; MacDougall, 1960; Reddaway, 1967, 1968; Kindleberger, 1968; Johnson, 1970; Caves, 1971; Dunning, 1981; Vernon, 1966). The brain drain, international technology transfers and international bank-lending occupied many researchers after the 1960s.

MOTIVES FOR FOREIGN INVESTMENT. The most fundamental motive for foreign investment is the desire of wealth-holders to maximize the value of their portfolio or net worth. However, this basic motive has been clarified and extended by the inclusion of risk, and analysts now often consider risk-adjusted rated of return to wealth-portfolios as the main motive for foreign investment. Under this approach, foreign investment is possible even if the yield on assets abroad is expected to be lower than that on domestic assets simply because an imperfect correlation of changes in foreign and domestic yields is expected to increase the risk-adjusted rate of return to the entire portfolio (Grubel, 1968). Numerous studies have documented the benefits from the international diversification of portfolios as well as direct investment holdings (Rugman, 1979).

DIRECT FOREIGN INVESTMENT. There are other motives for the purchase of assets abroad. They involve either externalities or market imperfections, which are internalized or eliminated by the multinational enterprise.

Technological externalities arise, for example, from the very high fixed costs in capital-intensive industries. In such industries great efficiency gains can be had by measures which stabilize operations at a high level of output. The ownership or control over suppliers and marketing permits firms in these industries to achieve such stabilization objectives which would be unattainable if separate owners pursued independent profit-maximization strategies. Given that raw materials, energy sources and finished-product markets often are located in different countries, vertically integrated companies in these industries frequently are multinational (Kindleberger, 1968; Caves, 1971).

Imperfections in factor-input markets which give rise to direct foreign investment are due to economies of scale, mainly those arising from the use of knowledge. Such knowledge is especially important in the design, production and marketing of differentiated consumer goods but many also involve management systems and information about customers and sellers. In addition, firms are motivated to own foreign production facilities in order to assure control over the quality of products and the maintenance of commercial secrecy. Furthermore, through direct foreign investment, firms are able to capture the international spillover effects of advertising expenditures.

The final major explanation of direct foreign investment involves distortions introduced by government policies. Tariffs and other protective devices as well as subsidies and taxes can create conditions under which it is more profitable to produce in, rather than export to, a foreign country.

The theory of direct foreign investment has been enriched by the analysis of additional, somewhat less-central issues. These involve the firms' choice of location, the decision to license rather than exploit technological assets through direct foreign investment, the legal forms of foreign ownership and the role of diversification. The usefulness of direct foreign investment as a method for diversification has been questioned in arguments which point to the opportunities of individual stockholders to obtain all the benefits of international diversification in their own portfolios, much like the Miller–Modigliani model questioned the need of individual firms to concern themselves with their capital structure. Some of the most useful insights about the nature of direct foreign investment have been gained by the analysis of reasons for its postwar growth (Kindleberger, 1968).

Attempts have been made to capture most of the motives noted above under the concept of 'internalization' and the 'eclectic theory of direct foreign investment' (Dunning, 1977). These approaches to the explanation of direct foreign investment have not been accepted widely, probably because the phenomenon is too complex to be captured adequately by the theory of internalization. The eclectic theory, on the other hand, is too broad by its inclusion of all of the many driving forces behind foreign direct investment (Black and Dunning, 1982; Buckley and Casson, 1976; Kojima, 1978 – for reviews and marginal extension).

Empirical studies have found support for all of the motives noted above. While none dominates the others clearly, of some special importance appear to be economies of scale due to the ownership of knowledge capital and motives created by government.

WELFARE EFFECTS. During the 1960s concern over the welfare effects of foreign investment centred on its influence on the balance of payments as both the United States and the United Kingdom suffered from large and growing deficits. Two landmark studies (Reddaway, 1967, 1968; Hufbauer and Adler, 1968) did much to sort out the different influences and interdependencies and produced some empirical estimates. Interest in the balance of payments effects of direct foreign investment has disappeared almost totally since the increased flexibility of exchange rates in the early 1970s.

Interest remains strong in the more general welfare effects of international investment, which received an influential early treatment by MacDougall (1960). The knowledge in this field can most easily be discussed with the help of Figure 1, where

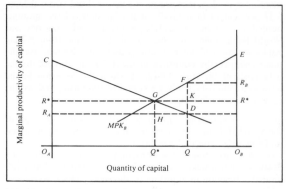

Figure 1

the marginal productivity and the quantity of capital are on the vertical and horizontal axes, respectively. We assume that the world capital stock consists of the quantity $O_A–O_B$, and that there are two countries A and B, the marginal productivity of capital schedules of which are shown originating on the left and right side of Figure 1 respectively. In the initial equilibrium before the opening of capital flows $O_B–Q$ capital is held in country A and has a yield of $O_A–R_A$. Country B holds the rest of the capital with a yield of $O_B–R_B$.

Now assume that capital flows are permitted between the two countries and that as a result the owners of capital in country A invest $Q^*–Q$ in country B. These investments reduce output in A and decrease it in B by the amounts Q^*QDC and Q^*QFG, respectively. As a result, rates of return are equalized in the two countries at Q^*R^*. Most important, the total productivity of the world's capital stock is increased by the area GDF. Such an output gain is the result of all capital movements, regardless of whether they take the form of investment in bonds, common shares, land, factories, human or knowledge capital.

In the new equilibrium the amount Q^*QKG represents the capital yield which accrues to its owners in country A, who therefore enjoy a net gain equal to the triangle GDK. Residents in the host country receive a net gain of GKF. Within country A the lowered capital–labour ratio raises the relative yield on capital and lowers that on labour. In country B the opposite effects take place.

As MacDougall (1960) pointed out, empirically the most important welfare effect of international capital-flows probably arises from taxation of profits and dividends by the country hosting the capital in combination with double taxation agreements which permit the foreign-tax payments to be deducted fully from tax obligations at home. In terms of Figure 1, one half of the area Q^*QDH accrues to the residents of country B at the expense of the residents of country A, under the assumption that the tax rate is 50 per cent. This net gain to the host country is reduced by any subsidies or free services which its government provides to the foreign investment. Empirical studies of this taxation have shown it to involve large welfare effects (Grubel, 1974).

Direct foreign investment often embodies new technology which cannot be acquired separately, and it leads to the net creation of workers' skills. In terms of Figure 1 these effects result in an upward shift of the marginal productivity schedule of investment in the host country B. An additional area of output is created thereby, which accrues to the residents of the host country. Direct foreign investment can lead to increased competition in the host country and, through it, increased efficiency in the use of all domestic resources. Other, more secondary welfare effects arise from changes in the two countries' terms of trade, which could go either way.

In models where there is unemployment equilibrium due to wage and price rigidities or underemployment as in developing countries, direct foreign investment can influence these conditions in both host and recipient countries. The efficiency models typified by Figure 1 cannot deal with these welfare effects, and they have been relatively neglected in the literature under discussion here, even though they are of great concern to politicians.

SOME WELFARE COSTS. The preceding neoclassical model has very little analysis of costs of direct foreign investment except for the tax effects and the usually peripheral issue of dynamic adjustment costs. This model has been attacked for neglecting several important ways in which direct foreign investment can reduce the welfare of the recipient country. Thus the owners of direct foreign investment can make investment, employment and output decisions that maximize rates of return but do not necessarily serve the interests of the host country; they can frustrate the achievement of monetary control as they draw on global capital sources; they can use their large resources to influence public opinion and elections in the interest of a foreign power or ideology; they compete unfairly with domestic producers who do not have access to low-cost capital and technology; they destroy domestic culture and traditions by the introduction of new and cheap goods, entertainment and art; they exploit monopoly and monopsony positions and thus charge too much for their products and pay too little for local inputs; they use transfer-pricing tricks to avoid the payment of host-country taxes; they create dependency on foreign supplies.

The evaluation of the preceding and many other arguments against direct foreign investment is difficult. Many of them are based on analytical paradigms which differ fundamentally from neoclassical economics. Others are based on empirical propositions that are nearly impossible to evaluate with available data. Still others involve value-judgements and implicit views on the relative efficiency of government substitute policies.

In the publications on foreign investment there is little interplay between the standard neoclassical approach to welfare effects and the analysis which stresses the costs. The former is taught and tends to dominate attitudes in industrial countries, while the latter is most popular and often very influential in developing countries and international organizations (Myrdal, 1956; Hymer, 1976; Behrman and Fischer, 1980; Lall and Streeten, 1977; United Nations, 1973, 1978).

POLICY IMPLICATIONS. The central policy issue in the field of international investment is whether or not it should be free, directed to achieve certain policy objectives or prohibited completely. The neoclassical paradigm implies that it should be free and that undesirable consequences accompanying it should be dealt with through policies directed at the problems themselves. As Bhagwati (1971) has shown, this approach permits the correction of market-failures without any sacrifice of the benefits from free trade in assets.

Other paradigms imply controls over foreign investment. At one extreme has been the complete prohibition of foreign investment in the Soviet Union and China after the Communist revolutions. These policies have been abandoned. Most countries of the world have some restrictions on foreign investment. Many insist that foreign investment has to be approved by a government agency, which uses acceptance criteria consistent with political and economic concerns of the time and the ruling party. Some countries restrict foreign ownership to minority holdings, which leave effective control with native entrepreneurs or governments. All of these restrictions involve costs of administration and diminish the level of international capital-flows. Therefore, they reduce the potential welfare gains below those attainable under the policy of dealing with market-failures directly.

HERBERT G. GRUBEL

See also INTERNATIONAL INDEBTEDNESS; PERIPHERY; VENT FOR SURPLUS.

BIBLIOGRAPHY
Behrman, J.H. and Fischer, W.A. 1980. *Oversees RD Activity of Transnational Companies*. Cambridge, Mass.: Oelschlager, Gunn and Hain.
Bhagwati, J. 1971. The general theory of distortions and welfare. In *Trade, Balance of Payments and Growth*, ed. J. Bhagwati et al., Amsterdam: Elsevier, North-Holland.
Black, J. and Dunning, J.H. (eds) 1982. *International Capital Movements*. London: Macmillan.
Buckley, P.J. and Casson, M.C. 1976. *The Future of Multinational Enterprise*. London: Macmillan.
Caves, R.E. 1971. International corporations: the industrial economics of foreign investment. *Economica* 38, February, 1–27.
Caves, R.E. 1982. *Multinational Enterprise and Economic Analysis*. Cambridge and New York: Cambridge University Press.
Dunning, J.H. 1977. Trade, location and economic activity and the multinational enterprise: a search for an eclectic approach. In *The International Allocation of Economic Activity*, ed. B.P. Ohlin, P.O. Hesselborn and P.J. Wiskman, London: Macmillan.
Dunning, J.H. 1981. Explaining the international direct investment position of countries: towards a dynamic or development approach. *Weltwirtschaftliches Archiv* 117(1), 30–64.
Grubel, H.G. 1968. Internationally diversified portfolios: welfare gains and capital flows. *American Economic Review* 58(5), December, 1299–314.
Grubel, H.G. 1974. Taxation and rates of return from some US asset holdings abroad. *Journal of Political Economy* 82(3), August, 469–87.
Hufbauer, G. and Adler, M. 1968. *Overseas Manufacturing Investment and the US Balance of Payments*. Washington, DC: US Treasury Department.
Hymer, S.H. 1976. *The International Operations of National Firms: Study of Direct Foreign Investment*. Cambridge, Mass.: MIT Press. (PhD dissertation, MIT, 1960.)
Iversen, C. 1936. *International Capital Movements*. Oxford: Oxford University Press.
Johnson, H.G. 1970. The efficiency and welfare implications of the international corporation. In *The International Corporation*, ed. C.P. Kindleberger, Cambridge, Mass.: MIT Press.
Kindleberger, C.P. 1968. *American Business Abroad*. New Haven: Yale University Press.
Kojima, K. 1978. *Direct Foreign Investment*. London: Croom Helm.
Lall, S. and Streeten, P. 1977. *Foreign Investment, Transnationals and Developing Countries*. London: Macmillan.
MacDougall, G.D.A. 1960. The benefits and costs of private investment from abroad: a theoretical approach. *Economic Record* 36(73), March, 13–35.
Myrdal, G. 1956. *Development and Underdevelopment*. Cairo. National Bank of Egypt. Reprinted in *Leading Issues in Economic Development*, ed. G. Meier, New York: Oxford University Press, 1970.
Reddaway, W.B. 1968a. *Effects of UK Direct Investment Overseas*. Cambridge: Cambridge University Press.
Reddaway, W.B. 1968b. *Effects of UK Direct Investment Overseas: Final Report*. Cambridge: Cambridge University Press.
Rugman, A.M. 1979. *International Diversification and the Multinational Enterprise*. Lexington, Mass.: D.C. Heath.

Stopford, J.M. and Dunning, J.H. 1983. *Multinationals: Company Performance and Global Trends*. London: Macmillan.

United Nations. 1973. *Multinational Corporations in World Development*. New York: United Nations.

United Nations. 1978. *Transnational Corporations in World Developments: A Reexamination*. New York: United Nations.

foreign trade. The pure theory of trade constitutes, in principle, no more than an application of the general theory of value, distribution and resource allocation. It follows at once, of course, both that each possible approach to general economic theory has its corresponding theory of trade and that any changes or developments in general theory must have implications for the theory of international trade. In particular, this is true of certain debates over value, distribution and capital goods which flourished in the 1960s, following the publication of Piero Sraffa's *Production of Commodities by Means of Commodities* (1960).

It need hardly be said that capital goods – that is, produced inputs, whether they be long-lived or short-lived – are of the very greatest importance in all modern economies. And it is no less true that international trade flows, far from consisting solely of consumption commodities, contain a large and growing volume of producer goods. International trade statistics are not conveniently classified into 'finished consumer goods' and 'other goods' but it appears from the classifications that are available that finished consumer goods probably account for less than some 30 per cent of the value of world trade. Any adequate theory of trade and resource allocation must, then, be able to deal, in a clear and coherent manner, with the important role of produced inputs and it is therefore to be expected that produced inputs would feature prominently in trade theory and that 'capital theory', broadly interpreted, should have significant implications for trade theory. But in fact, when we turn to basic trade theory, we find that capital goods are noticeable only by their absence, all the attention being centred on final consumption commodities.

With respect to capital theory, it is now well known that, in a competitive, constant-returns-to-scale economy using produced inputs (a) relative prices depend on the rate of interest, even for a given technique; (b) capital-intensity depends on the rate of interest, even for a given technique; (c) the choice of technique need not be monotonically related to the rate of interest; and (d) capital-intensity, in a multi-technique economy, need not be inversely related to the rate of interest. (See, for example, the *QJE* Symposium, 1966 and Pasinetti, 1977.) Also well-known are the results that, in an economy experiencing steady growth, there is a 'consumption-growth rate' trade-off which is identical to the 'wage-profit rate' frontier and that only if the growth rate equals the profit rate – the so-called Golden Rule case – is it ensured that the competitive choice of technique will be optimal with respect to the consumption/growth trade-off.

Suppose now that production is carried out using inputs of homogeneous land, as well as homogeneous labour, and produced inputs. Let there be a given, positive rate of interest on the value of capital (the produced inputs); it is then no longer the case that a rising rent/wage ratio must necessarily be associated with a falling land/labour ratio; quite the opposite relationship may hold (Metcalfe and Steedman, 1972; Montet, 1979). It follows that, in the presence of a positive rate of interest, an *increase* in the relative price of the more land-intensive commodity may be associated with a *decrease* in the output of that commodity (and an increase in the output

of the labour-intensive commodity). In other words, there may be a 'perverse' supply response.

In brief, then, capital theory discussions have alerted us (or realerted us, for Wicksell (1901) was well aware of some of these complications) to the distribution-relative nature of relative commodity prices, to the fact that capital-intensity depends on distribution as well as on technical conditions, to the possibility that both capital-intensities and land-labour ratios may respond in 'unexpected' ways to changes in interest, wage and rent rates, to the fact that supply responses can differ from those traditionally supposed and to the possibility that competitive technique choice need not be optimal with respect to the consumption-growth rate trade-off. We now turn to the implications of these findings for the pure theory of trade.

'TEXTBOOK' RICARDIAN THEORY. The reader will be thoroughly familiar with the textbook version of Ricardian trade theory, in which wages are the only kind of income, labour is homogeneous and – as a result of these two assumptions – the autarky price ratios in an economy are exactly proportional to the quantities of labour required to produce the various commodities. Yet when we turn to Ricardo's famous Chapter VII, 'On Foreign Trade' (1817), we see at once that Ricardo supposes there to be a *positive* rate of profit and, indeed, shows how the opening of trade can increase that rate. To this extent, then, 'textbook' Ricardian trade theory is a travesty of Ricardo's theory. Any attempt to excuse this vulgarization of Ricardo would probably appeal to the fact – and it is a fact – that in his Chapter VII Ricardo, whilst acknowledging the presence of both wages and profits, took no account of the influence of distribution on autarky relative prices; he simply identified these latter with relative labour quantities. Yet a large part of Ricardo's Chapter I, 'On Value', is concerned precisely with the fact that, as was noted above, relative commodity prices depend on distribution and not on technical conditions of production alone. The apparent inconsistency is explained by Ricardo's readiness to assume that relative labour costs provide a 'good enough approximation' to relative prices, even though he fully acknowledged that prices really depend on distribution. This explanation, though, is not a justification of Ricardo's procedure in Chapter VII, for he gave quite inadequate grounds for his claim about the 'good enough approximation'. It follows that we should examine carefully what happens to Ricardo's propositions concerning foreign trade when full recognition is given to the distribution-relative nature of autarky prices.

Consider then a two-country, two-consumption commodity model in which, in each country, the autarky price ratio of the two consumption commodities *depends on* the ruling (r, w) under autarky. Such a dependence could arise from the use of (nontradeable) machines in making the consumption commodities; or from the fact that the consumption commodities are *also* capital goods, being used in the production of one another; or from the fact that wages are paid in advance and that the production period over which they have to be advanced differs as between the two consumption commodities. There are many different models which capture the dependence of relative prices on (r, w), all of them providing examples of what Samuelson (1975) has called 'time-phased Ricardian systems'. Now if, in either economy, the autarky rate of interest should happen to be zero, the autarky price ratio of the two consumption commodities will indeed equal the ratio of their total (direct and indirect) labour costs. This must be true when the only form of income payment is that of wages paid to homogeneous labour. But if, as will generally be the case, the autarky interest rate is not zero and fluke technical conditions do not obtain, that autarky price ratio will *not* equal the corresponding labour cost ratio.

Let free trade be opened between our two economies. Will the direction of trade be determined by a comparison of the two countries' autarky price ratios or by a comparison of their labour cost ratios? By the former, of course, since competition works via wages, interest rates and prices. Each country will export that commodity for which it has the lower relative autarky *price*. It may or may not export that commodity for which it has the lower relative labour cost and certainly the pattern of trade is not determined by technical conditions alone but depends also on the autarky (r, w) in each country, simply because autarky relative prices so depend. Notice the corollary that two economies with the *same* technical conditions, for producing commodities by means of homogeneous labour and produced commodities, could enter into free trade if their autarky (r, w) would be different. It is not the case that 'Ricardian' trade models must necessarily suppose different technical conditions in each country – even if it is the case both that Ricardo did make such an assumption and that it is eminently sensible to do so.

Consider now a single, small economy of the kind considered above, which faces given terms of trade for trade in the two consumption commodities. Its pattern of trade will depend on how the given terms of trade compare with its autarky *price ratio*. But whether its fully-specialized, free trade consumption bundle lies outside its autarky consumption-possibility-frontier will depend on that pattern of trade and on how the terms of trade compare with the economy's *labour cost ratio*. Since this latter ratio is not equal, in general, to the autarky price ratio, it is *not* ensured that the with-trade bundle will lie outside the autarky frontier. Consider Figure 1, in which c_1 and c_2 are quantities of the first and second consumption commodities per unit of employment. C_2C_1 is the autarky consumption-possibility-frontier, whose absolute slope is of course equal to the labour cost ratio for the two consumption commodities. P_2P_1 is a line whose absolute slope is equal to the economy's autarky price ratio and T_2T_1 a line with slope equal to the

given terms of trade. Since T_2T_1 is less steep than P_2P_1 the economy will be driven to specialize in commodity 2 – but, since T_2T_1 is steeper than C_2C_1, the economy's free trade consumption bundle, T, which must of course lie on T_2T_1, will be *below* the autarky frontier C_2C_1 (unless at C_2 itself). It will be clear that this result would not obtain if T_2T_1 were either steeper than P_2P_1 (with specialization at C_1) or less steep than C_2C_1 (with specialization at C_2). But the fact remains that Ricardo was able to be 'sure' about the gain from trade only because he illegitimately supposed C_2C_1 and P_2P_1 to have the same slope. This argument can be extended to a steadily growing economy, to show that in the 'Golden Rule' case the with-trade bundle must lie outside the achievable autarky frontier but that if the growth-rate is less than the profit-rate then it may or may not do so (as in Figure 1, which provides simply a special case of this result, with a growth-rate of zero). Since the adoption of a particular specialization can, from a formal point of view, be thought of as a particular choice of technique, the present argument is just an application, to the trade context, of the capital theory result concerning competitive choice of technique and its possible non-optimality in terms of consumption and growth. It is important to notice that this result, concerning the possible (not certain) 'loss from trade', belongs to the class of 'comparative dynamics' results; it is best thought of as providing a *comparison* between a small closed economy and an (otherwise identical) small open economy. It is not a result about the effects on a given economy of the process of opening up to trade, full account being taken of what happens during the transition from the autarky state to the free trading state. But the same is true, it must be noted, of the textbook demonstrations of the gain from trade, in a 'Ricardian' framework, with which the reader is familiar.

While 'factor price equalization' is most often discussed within the Heckscher–Ohlin–Samuelson (HOS) framework, it is of interest to consider whether free trade in all commodities will bring about real wage rate and interest rate equalization in the type of model considered here. If all the freely trading economies have the same available choice of techniques, in a constant-returns-to-scale and homogeneous labour world, then it is certainly true that, if they all have the same rate of interest, they will all have the same set of relative prices. But the converse does *not* hold, when there is a choice of techniques; all the economies could face the same set of relative commodity prices and yet have different interest rates and real wage rates. Hence free trade in all commodities does not entail wage and interest equalization, even when all the economies have the same technical possibilities and are incompletely specialized. (The same negative conclusion holds, even when there is no choice of technique, if there are non-traded commodities.)

(On the pattern of trade and the gain from trade see Mainwaring, 1974; Samuelson, 1975; Steedman and Metcalfe, 1973a, 1979; Steedman, 1979a. On interest rate (non-) equalization see Mainwaring, 1976, 1978; Samuelson, 1975; Steedman and Metcalf, 1973b.)

LAND, LABOUR AND A POSITIVE INTEREST RATE. We now turn to the much-loved HOS model of international trade, in which two countries produce the same two commodities, using the same two primary inputs (which are in fixed supply) and having the same, constant-returns-to-scale technology. The primary inputs are qualitatively the same in both countries, fully mobile within each economy but completely immobile between them. There are no factor-intensity reversals, there is completely free trade and all consumers, in both countries,

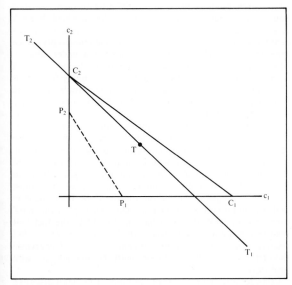

Figure 1

share a common homothetic preference map (so that consumption proportions depend only on the commodity price ratio, being quite independent of income distribution). If the two primary inputs are homogeneous land and homogeneous labour, and if there are no produced inputs (capital goods) of any kind, then the HOS theorem on the pattern of trade (in both its price and quantity forms), the factor price equalization theorem, the Stolper–Samuelson theorem and the Rybczynski theorem are all logically valid theorems.

Suppose now that, retaining all the other assumptions, we allow the two consumption commodities also to be capital goods, being necessary inputs to the various productive processes. What difference does this introduction of produced inputs make to the standard theorems? None whatever! It is now more appropriate to think of land/labour intensities in production in terms of *total* (direct and indirect) uses of land and labour but, since the intensity ranking of commodities in these total terms is necessarily the same as that in direct terms, this introduces no really significant difference from the model without produced inputs. Thus far then, produced inputs make no difference. But the position changes as soon as we allow not only for the presence of such produced inputs but also for a given, *positive* rate of interest on the value of those inputs (circulating capital goods). The presence of a positive interest rate does not alter the fact that the relative price of the land-intensive commodity will be a monotonically increasing function of the rent/wage ratio. But, as was pointed out above, it does mean that an increase in the rent/wage ratio is not necessarily associated with a fall in the land/labour ratio; it then follows that, if land and labour are always fully employed, an increase in the relative price of the land-intensive commodity may be associated with a *fall* in its net output. In Figure 2, which relates to a single economy, y_i is the net product of i, p_i is the price of i, SS is the full employment 'relative supply curve' and DD is the 'relative demand curve' derived from the common homothetic preference map; the figure illustrates the case of a 'perverse' supply response. It will be seen at once that such a supply response immediately gives rise to the possibility of multiple equilibria, the 'first' and 'third' equilibria both being stable.

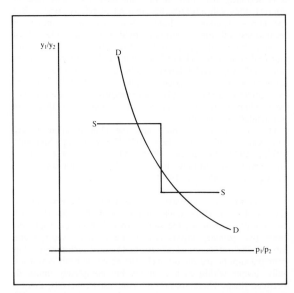

Figure 2

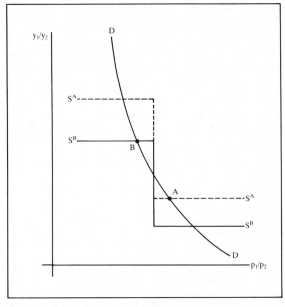

Figure 3

Let two economies, A and B, have the same positive rate of interest; let A be relatively better endowed with land and let commodity 1 be the land-intensive commodity. Figure 3 extends Figure 2 to this case, S^iS^i being the full employment relative supply curve for economy i. Suppose that point A represents A's autarky equilibrium, while point B represents B's. At every (p_1/p_2) lying *between* the autarky price ratios, S^AS^A and S^BS^B both lie on the same side of DD; hence no such price ratio can be an equilibrium terms of trade. The terms of trade lie *outside* the autarky price range. Whether the international equilibrium is found to the left of B or to the right of A, economy A (which is well endowed with land) will be exporting commodity 1 (which is the land-intensive commodity). Thus the HOS quantity theorem holds good. Yet the HOS price theorem, which is sometimes thought rather trivial, is actually *false* here. Since A has the higher autarky (p_1pp_2), it has the higher autarky rent/wage ratio, so that A is exporting the commodity which uses intensively A's relatively *expensive* factor under autarky. Notice also that if international equilibrium is found to the left of B, (p_1/p_2) will have fallen, with trade, in economy A and thus the wage/rent ratio will have *risen*; in fact trade will have *benefited* A's relatively scarce factor (labour), contrary to the usual HOS prediction.

If A and B have the *same* positive interest rate, as above, they have the same relationship between (p_1/p_2) and the rent/wage ratio; it is thus not surprising that free trade will equalize rents and wages (with incomplete specialization) and that the Stolper–Samuelson theorem also holds good. If A and B have *different* positive interest rates, however, almost everything collapses. The exception is the Rybczynski theorem and it is important to understand why. All 'capital theoretic' problems for HOS theory reduce in the end to the fact that relative commodity prices vary with the rate of interest – but relative prices are fixed *by assumption* in the Rybczynski theorem, so that that theorem must be immune to such problems.

Consider now a single, small economy of the kind discussed immediately above. In the presence of a positive interest rate, the price ratio at which a switch of techniques takes place will not be equal, in general, to the physical rate of transformation between the two net outputs. It follows that, when we compare the small open economy with an otherwise identical autarkic economy, we find that the value of consumption in the small open economy, at the given international prices, may be either greater than or less than the corresponding value in the autarkic economy. The 'comparative static' gain from trade may be either positive or negative.

In the land and labour model, then, the presence of produced inputs makes no difference *per se*. But a positive rate of interest on their value does make a difference to some (but not all) HOS theorems, if it is the same in both countries, while a difference in interest rates undermines all the standard HOS results, other than the Rybczynski theorem. For the single, small, open economy the presence of a positive interest rate means that the 'comparative' gain from trade can be positive or negative.

(For the closed economy background see Metcalfe and Steedman, 1972, Montet, 1979; for the trade theory applications Samuelson 1975, Steedman and Metcalfe, 1977; for the gain from trade Metcalfe and Steedman, 1974, Samuelson, 1975.)

LABOUR AND CAPITAL. In the typical textbook presentation of HOS theory the two 'factors' in given supply are not labour and land, as above, but labour and 'capital'. (Although Samuelson (1948, 1949) was careful to stipulate labour and land.) Yet that typical presentation suggests no immediate connection between the two produced commodities and the physical composition of the capital stock, despite the fact that 'capital goods' are, by definition, produced means of production! Indeed, one interpretation of most textbook theory is that 'capital' is simply a misnomer for land, the problems of capital theory being evaded by a simple misuse of terms. Alternatively (and more favourably), the 'given capital supply' can be interpreted to mean that the total *value* of capital goods must always be equal to – or, at least, not greater than – an exogenously given value. An immediate difficulty with this interpretation is that, since relative autarky prices differ between the two economies, the very *ranking* of the two countries' capital/labour endowments ratios may depend on which standard of value is used to measure capital. And what does it *mean* economically to suppose that total capital value is given in terms of one standard and yet, necessarily, is *not* given in terms of all other possible standards (since relative commodity prices are to be determined endogenously)? Even if we ignore these questions – which there is no justification for doing – we know from capital theory that value capital/labour ratios need not be related inversely or, indeed, even monotonically to the rate of interest. This, of course, immediately suggests that some of the HOS theorems may be at risk. Moreover, it can be shown that in a model with many produced inputs, the price ratio between any two particular commodities need not be monotonically related to the rate of interest, *even when* one of the two commodities is always more value capital-intensive than the other. But if neither the capital/labour ratios nor the relative commodity prices need be monotonically related to the rate of interest – even in the absence of factor-intensity reversals – then it will at once be clear that HOS theorems (other than the Rybczynski theorem) cannot be logically valid when one of the two factors is a 'given value of capital'. This stems fundamentally from the simple fact that Wicksell clearly stated many years ago:

> Whereas labour and land are measured each in terms of its own *technical* unit ... capital ... is reckoned, in common

parlance, as a sum of *exchange value* – whether in money or as an average of products. In other words, each particular capital-good is measured by a unit extraneous to itself. [This] is a theoretical anomaly which disturbs the correspondence which would otherwise exist between all the factors of production ([1901] 1967, p. 149).

To illustrate the above negative conclusions, we may use an example in which there are two consumption commodities (two kinds of 'corn'), each producible by means of many alternative types of machine. The consumption commodities are tradeable but the machines are not. Full numerical details of this example can be found in Metcalfe and Steedman (1973); here we confine ourselves to the diagrammatic presentation of Figure 4, in which k_i is the value capital/labour ratio involved, directly and indirectly, in the production of the ith consumption commodity, expressed in terms of the first consumption commodity. It will be seen on the right of figure 4 that neither k_1 nor k_2 is monotonically related to r but that $k_1 > k_2$ at all r; on the left we see that, the absence of factor-intensity reversal notwithstanding, the price ratio (p_1/p_2) is not monotonically related to r. It follows at once that the 'factor price' equalization theorem, the Stolper–Samuelson theorem and the price form of the HOS theorem on the pattern of trade are *not* of general logical validity. But if the pattern of trade theorem is not valid in its price form then it will not be valid in its quantity form either, even if it is the case (which it may not be) that the economy with the higher capital/labour endowment ratio has the lower autarky interest rate.

When produced inputs are introduced into HOS theory in the form that one of the two 'factors' is taken to be a given total value of capital, that theory simply disintegrates. This is so notwithstanding the apparent denial of this negative conclusion by Ethier (1979), who states that 'The central message ... is simple. The four basic theorems of the modern theory of international trade ... are insensitive to the nature of capital' (p. 236). In fact Ethier's paper constitutes a striking confirmation of our negative conclusion, because in order to maintain the *appearance* that capital has no influence on HOS trade theorems, Ethier finds himself compelled to *replace* the familiar theorems, which predict trade outcomes on the basis of exogenous data, by entirely different theorems, which merely describe trade outcomes in terms of trade equilibrium prices, etc.

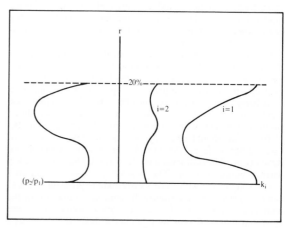

Figure 4

(For the example used in this section, see Metcalfe and Steedman, 1973; on Ethier's conjuring with HOS theorems, see Metcalfe and Steedman, 1981.)

GROWTH, INTERNATIONAL INVESTMENT AND TRANSITIONS. To focus on the role of capital goods in trade and in trade theory is, implicitly, to direct attention also to such matters as growth (capital accumulation), international investment and transitions between steady growth paths. Since the typical trading economy uses many produced inputs – some traded and some not – accumulates capital goods and experiences (often embodied) technical change, along both quantitative and qualitative dimensions, the ideal theory of trade would be able to handle all these closely related issues, in a manner which was both informative and simple. Needless to say, such an ideal theory is not available; international trade theory in these respects can, in the long run, be no more advanced than the general theory of accumulation and technical progress. The preceding discussion can, however, serve to warn us that growth models in which there is a single, physical capital good can almost certainly not be readily generalized to the many capital good case and are thus of *very* limited interest. It is also useful to note, as a simple matter of fact, that while the number of countries in the world is of the order of 200, the number of distinct commodities – when defined at the level of detail relevant to careful value theory – runs into millions. This both tells us that incomplete specialization must be the rule and directs our attention to economic growth models in which the number of commodities can be arbitrarily large; the von Neumann model perhaps deserves to be used more extensively by trade theorists than it has been, its very abstract nature notwithstanding.

When thinking of capital accumulation, the international economist will naturally pay considerable attention to the role of international investment. Here it is most important to recognize that, although they are often connected in practice, there is a perfectly clear – indeed a sharp – distinction between international investment as a flow of finance, on the one hand, and trade in physical capital goods, on the other. This is obvious enough perhaps when stated explicitly but it is to be noted that the idea of a 'factor' capital, conceived of as a sum of value, in fact makes it dangerously easy to confuse financial flows with capital goods flows. The trade theorist would do well to avoid the concept of a 'quantity of capital' altogether, referring only to stocks and flows of specified capital goods, on the one hand, and to international flows of finance, on the other. Such a practice would not only make it easier to avoid capital theory traps but would also facilitate thought about the badly needed integration of pure trade theory with international monetary economics.

We turn now to the question of 'transitions'. Consider first a closed economy whose homogeneous land and homogeneous labour are allocated between strawberry production and raspberry production. No produced inputs are used – not even strawberry and raspberry plants! (Which reminds us, incidentally, of just *how* strained is any picture of direct production of consumption commodities by primary inputs.) If free trade should suddenly become possible, at terms of trade different from the autarky price ratio, there is no difficulty at all in reallocating the land and labour to the newly desired output pattern. The 'transition' from the autarky steady-state to the with-trade steady-state is problem free and can be achieved instantaneously. By contrast, consider now the analogous 'transition' for an economy which does use produced inputs. Except by a complete fluke, the economy's industries will use the various produced inputs in different

proportions from one another and it will now *not* be possible to change to the free trade pattern of output instantaneously. Since the production of the produced inputs takes time, there will have to be a 'transitional' period, during which the physical composition of the economy's aggregate capital stock is adjusted to the new output pattern. Just how long this period will be depends, of course, on how different the input requirements are as between industries, on whether or not some previously used capital goods simply have to be scrapped, on how many of the capital goods are tradeable and how many non-tradeable, etc. Changing the pattern of net output is a far more complicated process in an economy using produced inputs. This issue is avoided in textbook discussions of the gain from trade *and* in the 'comparative dynamics' results given above. Yet it can hardly be denied that the issue is important in many trade policy applications and in many day-to-day debates about trade protection, industries which are under increased international competitive pressure, and so on. It is therefore important that trade theorists should develop *explicit* analyses of transitional processes in the presence of produced inputs. At the same time, however, it would be quite wrong simply to dismiss 'comparative dynamic' results showing that a 'loss from trade' is possible, merely on the grounds that (by definition) they do not take account of transitions. The traditional comparisons of a world of autarky economies with a world of trading economies are designed to show that the with-trade state of the world (which we observe) is preferable to the autarky state (which is purely hypothetical). For the purpose of such an abstract, *hypothetical comparison*, the analysis of transitions would have no significance and it is indeed the purely 'comparative' analysis which is relevant. (There is, of course, no inconsistency in saying also that a transitional analysis is relevant for the study of an actual economy considering the possibility of, say, changing its tariff structure.)

(A trade theory application of the von Neumann model is given in Steedman, 1979c, for a single, small economy; growth in a two country world is discussed by Parrinello, 1979. On transitions see Metcalfe and Steedman, 1974; Smith, 1979.)

CONCLUSION. Sufficient reason has perhaps been given above to justify the rather general conclusion that when one finds trade theorists referring to 'capital' one should immediately be 'on guard'. The presence of produced inputs, with a positive rate of interest on their value, does make a considerable difference to the logical coherence of HOS theory, as has been seen in some detail above. Moreover, 'textbook' Ricardian trade theory, which appears to make no reference to 'capital' at all, ought to make such reference and, if it did, would discover that here again the presence of a positive interest rate makes it far harder to reach any clear cut, logically valid theorems. In seeking to develop a trade theory which *does* give central importance to capital goods and hence to profits, accumulation and technical progress (e.g. Steedman, 1979a) one must expect that simple results may not be abundant. And one must recognize that the assumptions which make growth theory relatively easy, such as constant returns to scale and the absence of land, themselves do violence to the complex realities of international trade. (Its many shortcomings notwithstanding, HOS theory is right to stress the importance of land and labour endowments, even while it is wrong to take them to be qualitatively homogeneous and fully employed.) The role of capital goods is by no means the only important issue in trade theory and recognition of that role certainly makes trade theory more difficult. But can these be good reasons for ignoring capital goods, when that theory is intended to aid our understanding of a world in which produced inputs are, in fact, centrally important?

IAN STEEDMAN

See also HECKSCHER–OHLIN TRADE THEORY; INTERNATIONAL TRADE.

BIBLIOGRAPHY

Ethier, W.J. 1979. The theorems of international trade in time-phased economies. *Journal of International Economics* 9(2), May, 225–38.

Mainwaring, L. 1974. A neo-Ricardian analysis of international trade. *Kyklos* 27(3), 537–53. Reprinted in Steedman (1979b).

Mainwaring, L. 1976. Relative prices and 'factor price' equalisation in a heterogeneous capital goods model. *Australian Economic Papers* 15(26), June, 109–18. Reprinted in Steedman (1979b).

Mainwaring, L. 1978. The interest rate equalisation theorem with nontraded goods. *Journal of International Economics* 8(1), February, 11–19. Reprinted in Steedman (1979b).

Metcalfe, J.S. and Steedman, I. 1972. Reswitching and primary input use. *Economic Journal* 82, March, 140–57. Reprinted, with minor corrections, in Steedman (1979b).

Metcalfe, J.S. and Steedman, I. 1973. Heterogeneous capital and the Heckscher–Ohlin–Samuelson theory of trade. In *Essays in Modern Economics*, ed. J.M. Parkin, London: Longman. Reprinted in Steedman (1979b).

Metcalfe, J.S. and Steedman, I. 1974. A note on the gain from trade. *Economic Record* 50, December, 581–95.

Metcalfe, J.S. and Steedman, I. 1981. On the transformation of theorems. *Journal of International Economics* 11(2), 267–71.

Montet, C. 1979. Reswitching and primary input use: a comment. *Economic Journal* 89, September, 642–47.

Parrinello, S. 1979. Distribution, growth and international trade. In Steedman (1979b).

Pasinetti, L.L. 1977. *Lectures on the Theory of Production*. London: Macmillan.

Quarterly Journal of Economics. 1966. Symposium on paradoxes in capital theory. *Quarterly Journal of Economics* 80(4), November, 503–83.

Ricardo, D. 1817. *Principles of Political Economy and Taxation*. Ed. P. Sraffa. Cambridge: Cambridge University Press, 1951.

Samuelson, P.A. 1948. International trade and the equalization of factor prices. *Economic Journal* 58, June, 163–84.

Samuelson, P.A. 1949. International factor-price equalization once again. *Economic Journal* 59, June, 181–97.

Samuelson, P.A. 1975. Trade pattern reversals in time-phased Ricardian systems and intertemporal efficiency. *Journal of International Economics* 5(4), November, 309–63.

Smith, M.A.M. 1979. Intertemporal gains from trade. *Journal of International Economics* 9(2), May, 239–48.

Sraffa, P. 1960. *Production of Commodities by Means of Commodities*. Cambridge: Cambridge University Press.

Steedman, I. 1979a. *Trade Amongst Growing Economies*. Cambridge: Cambridge University Press.

Steedman, I. (ed.) 1979b. *Fundamental Issues in Trade Theory*. London: Macmillan.

Steedman, I. 1979c. The von Neumann analysis and the small open economy. In Steedman (1979b).

Steedman, I. and Metcalfe, J.S. 1973a. On foreign trade. *Economia Internazionale* 26(3–4) August–November, 516–28. Reprinted in Steedman (1979b).

Steedman, I. and Metcalfe, J.S. 1973b. The non-substitution theorem and international trade theory. *Australian Economic Papers* 12(21), December, 267–69. Reprinted in Steedman (1979b).

Steedman, I. and Metcalfe, J.S. 1977. Reswitching, primary inputs and the Heckscher–Ohlin–Samuelson theory of trade. *Journal of International Economics* 7(2), May, 201–8. Reprinted in Steedman (1979b).

Steedman, I. and Metcalfe, J.S. 1979. The golden rule and the gain from trade. In Steedman (1979b).

Wicksell, K. 1901. *Lectures on Political Economy*. Vol. 1. Trans. E. Classen, ed. Lionel Robbins, London: G. Routledge and Sons, 1934.

foreign trade multiplier. Though its ancestry may be traced to certain ideas of the Mercantilist School, the foreign trade multiplier appears in modern form in a textbook (Harrod, 1933) written three years before Keynes's *General Theory*. There in its simplest form Harrod introduced the familiar equation for the determination of the national income flow consistent with balance in the current account of the balance of payments – that the national income is equal to the product of the volume of national exports and the reciprocal of the average propensity to import. In this form it emphasized the equilibrating role of income over price variations in determining balance of payments adjustment, and the importance that trade performance played in determining the level of activity at which external balance would be achieved.

By the 1939 edition Harrod was expounding his foreign trade multiplier as an application of Keynes's theory of effective demand by incorporating domestic investment and the savings propensity within the analysis.

After World War II great strides were achieved in integrating relative prices and income into the analysis of balance of payments adjustment (Laursen and Metzler, 1950; Harberger, 1950; Meade, 1950). The foreign trade multiplier was subsumed in a synthesis of balance of payments theory which became known as the absorption approach (Alexander, 1952). Theory became organised around the identity for the current account of the balance of payments expressed as the difference between the national income and domestic expenditure or absorption per period.

A further generalization was to analyse the foreign trade multiplier by including the foreign repercussions of changes in spending and income which feed back onto the home country, initiating the change in autonomous expenditure. An essentially identical treatment of the multiplier could be applied either to a multi-sectoral, regional or country analysis of the determination of income by autonomous spending in any single unit of the system. Thus the focus changed from the relationship between the national economy and the rest of the world to the inter-relationships of a multi-country world economy in which trade shares and absorption propensities played an important role in determining the global and distributional generation of income (Goodwin, 1949; Metzler, 1950). Within this framework the post-war problems of international trade and growth could be addressed, where for the world as a whole, the closed economy assumption of theory is literally (and almost uniquely) true.

By the middle of the 1950s the major theoretical developments of multiplier analysis in open-economy macroeconomics had been established.

In contrast with the pure theory of international trade, set in a context ensuring full employment of resources and concerned mainly with factors which determine efficient patterns of trade, the foreign trade multiplier is directly relevant to the determination of the level of output and employment between nations. The extension of effective demand theory into international economics has given greater prominence to income and employment changes rather than relative price changes at full employment in balance of payments adjustment. Whether under fixed or floating exchange rate regimes the foreign trade multiplier indicates how changes in *net* exports, i.e. the excess of export growth over the growth of import penetration, affect both the balance of payments on current account and the degree of domestic capacity utilization.

For all countries which trade significantly with one another, the foreign trade multiplier serves to remind one that the pursuit of full employment and economic growth policies by governments must be accompanied by policies which ensure that they are matched by satisfactory balance of payments performance. This implies that the growth of net exports makes a significant contribution to the growth of national income.

A problem of paramount importance for any system of international settlements adopted between nations is how national fiscal and monetary policies, enacted largely independently of one another, can achieve mutually reconcilable balance of payments outcomes without forcing at least some nations to abandon desirable domestic objectives. The foreign trade multiplier, appropriately more complex than Harrod's original formulation, provides an illuminating framework for addressing this major question.

K.J. COUTTS

See also ABSORPTION APPROACH TO THE BALANCE OF PAYMENTS; INTERNATIONAL TRADE; TRANSFER PROBLEM.

BIBLIOGRAPHY
Alexander, S.S. 1952. Effects of a devaluation on a trade balance. *IMF Staff Papers* 2, April, 263–78.
Goodwin, R.M. 1949. The multiplier as matrix. *Economic Journal* 59, December, 537–55.
Harberger, A.C. 1950. Currency depreciation, income, and the balance of trade. *Journal of Political Economy* 58(1), February, 47–60.
Harrod, R.F. 1933. *International Economics.* Cambridge Economic Handbooks, Cambridge: Cambridge University Press; revised edn, 1939.
Laursen, S. and Metzler, L.A. 1950. Flexible exchange rates and the theory of employment. *Review of Economics and Statistics* 32(4), November, 281–99.
Meade, J.E. 1951. *The Theory of International Economic Policy.* Vol. I: *The Balance of Payments.* London: Oxford University Press.
Metzler, L.A. 1950. A multiple region theory of income and trade. *Econometrica* 18(4), October, 329–54.

forests. Traditional forestry economics has been chiefly concerned with wild or cultured forests as commercial, agricultural, enterprises. For these, net economic benefits stem from the harvested timber and the objective is to calculate the optimal pattern of harvesting over time. While there is a venerable literature on the standing, *in situ*, values of trees (see J. Nisbet's entry in *Palgrave* (1912), Vol. II, pp. 113–18), these have been incorporated only recently in formal optimizing models.

The early work of Martin Faustmann (1849) is noteworthy for its originality and for providing the correct solution to the central problem of the optimal rotation period for a sequence of harvests. For modern interpretations see Gaffney (1960), Pearse (1967) and Gregory (1972).

Stripped to its essentials, the Faustmann problem is to maximize the present value of bare land (V) which will support an indefinite sequence of harvests $(n = 1, 2, \ldots, \infty)$. The trees grow in value (net of harvesting costs) as they mature according to $P(T)$. $P(T)$ is nil for a while, then increases at a decreasing rate, reaching a maximum then falling (as rot sets in). The harvest is instantaneous at T. Regeneration costs are nil. The problem is to

$$\max_{\{T\}} V(T) = \sum_{n=1}^{n=\infty} D(T)^n P(T). \tag{1}$$

The discount factor, applied to the harvest when the trees are T years old, is $D(T) = \exp(-T)$ for continuous compounding and $D(T) = (1 + i)^{-t}$ for annual compounding. For example, the contribution to $V(50)$ from the third harvest is by fifty-year-old trees and is computed as $D(50)^3 P(50)$.

The problem is solved by first recognizing that $D(T)^n P(T)$ in (1) is a declining, infinite, geometric series having a finite sum. Thus

$$V(T) = [D(T)/(1 - D)]P(T) \tag{2}$$

$V(T)$ at first rises, then falls as T is extended beyond the interval while $P(T) = 0$. If T is only a little greater than this, V is still nearly zero since $P(T)$ is small. If T is long, $P(T)$ is large but the expression in square brackets is small. $V(T)$ is a maximum when $V'(T^*) = 0$ or

$$P'(T^*)/P(T^*) = -D'(T^*)/[D(T^*)(1 - D(T^*))] \tag{3}$$

This solves (1) for T^*, the age of *financial maturity* (Duerr, Fedkiw and Guttenberg, 1956). It is the (by now) famous *Faustmann* (1849) *Formula.*

For continuous compounding, $D'(T)/D(T)$ equals $(-r)$ and

$$P'(T^*)/P(T^*) = r(1 - \exp rT^*)^{-1} \tag{3'}$$

Substituting this result back into (2) gives

$$P'(T^*) = r(P(T^*) + V(T^*)) \tag{4}$$

The RHS of (4) contains the maximized present value of the bare land, $V(T^*)$. This is called the *land-expectation* value. Added to this is the optimal *stumpage* value of the trees, $P(T^*)$. The two values together are the value of the land plus the value of the trees, both evaluated at T^*. This sum multiplied by r is the momentary interest payment earned by selling the cut trees *plus* the bare land, then putting the money 'in the bank' earning interest at the rate r. The LHS of (4) is the momentary increase in the value of the trees if left on the stump to grow in the ground. The time to cut, T^*, is when the marginal increment in accrued wealth is the same with the trees 'in the ground' as 'in the bank'.

A special case of the Faustmann result is obtained by supposing that the land has no value other than supporting the first growth of trees. Then, from (4), the *present net worth* of the trees alone is maximized when

$$P'(T^*) = rP(T^*) \tag{5}$$

This result is associated with I. Fisher (1930), Hotelling (1925), von Thünen (1826). It has been called the *Fisher Rule.* When the trees are younger than T^*, they grow in value in the ground faster than their value if cut would increase in the bank. And vice versa after T^*. The moment to cut and sell is the moment of indifference. Since $P'(T^*)$ is greater for smaller T, the Fisher rule signals a later cut than does the Faustmann formula. This is because the Fisher Rule does not capture the impatience of a forester to cut out the old growth to make room for younger and faster growing trees.

Another special case of the Faustmann formula is obtained by supposing that the relevant rate of interest (r) is zero. Then, using l'Hôpital's rule in (3')

$$\lim_{r \to \infty} P'(T^*)/P(T^*) = 1/T^*. \tag{6}$$

Since younger trees grow faster in value, (6) signals a later cut than does (3) with r positive.

The rule in (6) can be obtained directly by maximizing the

mean annual increment (MAI) with respect to the age of trees when harvested.

$$\max_{\{T\}} \text{MAI} = P(T)/T.$$

This objective has been widely accepted by professional foresters since the late 18th century (Osmaston, 1968). The harvesting rule, expressed in (6), is elegant and parsimonious of information requirements. In practice, $P(T)$ is usually approximated by the volume of merchantable timber contained in a tree. Hence, the rule has no economic content except in so far as it may serve as a practical rule-of-thumb approximation of a necessary condition for an economic optimum.

Samuelson (1976) provides a concise and readable comparison of these versions of the Faustmann formula. He concludes that Faustmann's original formulation is the 'correct' one for maximizing the social contribution of forests. It should, however, be interpreted as 'pure theory', subject to the austere conditions imposed to reveal the kernel of the problem.

The Faustmann formula has been enriched to encompass various kinds of silviculture including artificial regeneration, thinning, fertilization, insect and fungus control. Felling, yarding, bucking and transportation costs have also been explicitly included in net revenue functions. Clark (1976), Ledyard and Moses (1976) and Heaps (1981) provide examples. These are exercises in operations research and serve as guides to formulating cost-benefit studies on a case-by-case basis. A general consideration is that costs incurred early in a rotation (artificial regeneration, for example) will be compensated by earlier cuts or enhanced net current values many years later. Calculations made for coastal forest land in northwest United States (southwest Canada) compute the economic value of re-established forests after clear-cutting by approved methods (Smith, 1978). The reference forest is a naturally regenerated wild stand of mixed low and high valued trees harvested after 76 years. The enhanced forest is planted with genetically improved high valued species and the harvest is accelerated after a (non-commercial) thining at 15 years. The additional cost of planting and thinning is $580 per hectare, increasing the *current* value of the harvest by $2794. But the increase in *present* value calculated at a 5 per cent p.a. interest rate is only $78 per hectare. Plantation forests on good, flat land in warmer climates have shorter economic rotations so thinnings and other kinds of silviculture may have net commercial value.

Traditional forest management contemplates *sustained yield* from a *regular*, or *normal*, *forest* which encompasses sufficient geographic scope to sustain a continuous harvest of trees of the desired age. This requires x even-age, equal-area, stands of trees ranging in age from zero to $(x-1)$. A stand is cut when the trees are x years old. Each stand is to be harvested during a year of an x year *rotation*. If the interest rate is zero in the Faustmann formula, the x is T for the maximum MAI to obtain *Maximum Sustained* (in perpetuity) *Yield* (MSY) in terms of timber volume. Rotations are shorter for positive interest rates and fewer even-age stands are required in a management unit.

The regular forest is the forester's ideal. But the initial condition of a forest may be characterized by irregular age distributions. *Ad hoc* formulae for conversion to a regular forest have been proposed by Hanzlik (1922) and others. See Hennes, Irving and Navon (1971). Naudial and Pearse (1967) model conversion as an economic problem. Using a linear programme, they solve for an optimal rotation period (T^*), *given* a time period for conversion. But V is seen to rise as the

period is extended, implying that the regular forest is not optimal. Heaps (1981) and Heaps and Neher (1979) provide the most general treatment of the problem using the Maximum Principle for processes with an endogenous delay (between regeneration and cutting). Variable ('u'-shaped average) costs of harvest are allowed for. The Forestry Maximum Principle leads to a dynamical system of functional differential equations. It is seen that a convergent harvesting policy yields the Faustmann T^* for a regular forest. Global (asymptotic) stability has not been proved but it seems likely that conversion to a regular forest over a period of several rotations is practically optimal.

Other recent contributions allow for stock uncertainty and for standing values of existing tree stocks. Reed (1984) shows that the expected valuation of sustained yield is maximized by planning to cut before the Faustmann T^*. The early cut forestalls the possibility of catastrophic loss. Also see Kao (1984), Martell (1980), and Reed and Errico (1985).

The standing value of trees has long been recognized in the literature (Palgrave, 1912). Indeed, some notable modern forests are extant relics of ancient game preserves (Epping Forest, London, for example). Vast forested areas of the United States are administered by the Forest Service under multiple-use mandates which include preservation of wildlife habitat and recreational values. Recent widespread clearings of tropical rainforest have focused attention on the value of forested areas to sustain (sometimes unique) fauna and flora. Hartman (1976) and Neher (1976) provide early examples of theoretical frameworks for including standing values in intertemporal optimizing models.

The inclusion of standing values presents both interesting theoretical problems and challenging management difficulties. The kernel of the theoretical problem is compactly exposed by absorbing the age-class demographic structure of the forest into a simple, biomass aggregate of trees (B). Let B grow naturally according to $G(B) > 0$ with $G(B)$ a maximum at $B_M > 0$, $G(0) = G(\bar{B}) = 0$ and $G''(B)$. $\bar{B}$ represents the climax forest. Cardinal social benefits depend upon the harvest flow (H) and upon the standing stock (B). In short, $U = H(H, B)$. Let U_H, $U_B > 0$ and U_{HH}, $U_{BB} < 0$. However $U_{BH} = U_{HB}$ is not signed *a priori*. The value of (V) of the harvesting plan is the (undiscounted) sum (integral) of these U's over the planning interval $[0, T]$. V is to be maximized subject to constraints that $B(0) = B_0$ (the original state of the forest is given) and that $\dot{B} = G(B) - H$ (nature's own 'budget constraint') is satisfied. Thus, $U = U(G(B) - \dot{B}, B)$. It is well known that the necessary conditions for an optimal programme are not sufficient if $U()$ is not jointly concave in (B, B). In this case, sufficiency is not guaranteed unless marginal enjoyments of H and B are independent of each other $(U_{HB} = U_{BH} = 0)$. This phenomenon was originally identified by Kurtz (1968) in neoclassical growth where multiple equilibria were identified. Cropper (1976) provides a more recent example in the optimal control of pollution. Additional sources of multiple equilibrium are introduced by considering many (natural and produced) capital stocks (Heal, 1982) and by discounting future benefits (Cass and Shell, 1976). Much work has yet to be done before there is general theoretical understanding of optimal multiple-use forestry.

It is not surprising that practical, implementable, multiple-use models have not been devised. Single-use (timber value) commercial models are conceptually more straightforward, incorporating (if at all) standing values as exogenous constraints on the available commercial area. These are typically linear programming models. See Johnson and Scheurman (1977) for an evaluative survey. The discrete maximum principle (Halkin, 1966) offers a promising alternative (Lyon and Sedjo, 1983). Large-scale programming models have been implemented with arguable success (Timber Resources Allocation Model (Timber RAM); Navon, 1971; FORPLAN, Johnson, Jones and Kent, 1980).

P.A. NEHER

See also FAUSTMANN, MARTIN; NATURAL RESOURCES.

BIBLIOGRAPHY

Cass, D. and Shell, K. 1976. *The Hamilton Approach to Dynamic Economics.* New York: Academic Press.
Clark, C.W. 1976. *Mathematical Bioeconomics.* New York: Wiley.
Cropper, M.L. 1976. Regulating activities with catastrophic environmental effects. *Journal of Environmental Economics and Management* 3, 1–15.
Duerr, W.A., Fedkiw, J. and Guttenburg, S. 1956. Financial maturity. *US Department of Agriculture. Technical Bulletin* No. 1146.
Faustmann, M. 1849. On the determination of the value which forest land and immature stands pose for forestry. English translation in 'Martin Faustmann and the evolution of discounted cash flow', ed. M. Gane, *Oxford Institute Paper* 42, 1968.
Fisher, I. 1930. *The Theory of Interest.* New York: Macmillan.
Gaffney, M.M. 1960. Concepts of financial maturity of timber and other assets. *Agricultural Economics Information Series* No. 62, Raleigh: North Carolina State College.
Gregory, G.R. 1972. *Forest Resource Economics.* New York: Ronald Press.
Halkin, H. 1966. A maximum principle of the Pontryagin type for systems described by non-linear difference equations. *SIAM Journal on Control* 4, 90–111.
Hanzlik, E.J. 1922. Determination of the annual cut on a sustained basis for virgin American forests. *Journal of Forestry* 20, 611–25.
Hartman, R. 1976. The harvesting decision when a standing forest has value. *Economic Inquiry* 16, 52–8.
Heal, G. 1982. The use of common property resources. In *Explorations in Natural Resource Economics*, ed. V.K. Smith and J.V. Krutilla, Baltimore: Johns Hopkins Press.
Heaps, T. 1981. The qualitative theory of optimal rotations. *Canadian Journal of Economics* 14, 686–99.
Heaps, T. and Neher, P.A. 1979. The economics of forestry when the rate of harvest is constrained. *Journal of Environmental Economics and Management* 6, 279–319.
Hennes, L.C., Irving, M.J. and Navon, D.I. 1971. Forest control and regulation. *USDA Forest Service Research Note PSW–231*, Berkeley.
Hotelling, H. 1925. A general mathematical theory of depreciation. *Journal of the American Statistical Association* 20, 340–53.
Johnson, K.N., Jones, D.B. and Kent, B. 1980. A user's guide to the forest planning model (FORPLAN). Land Management Planning, USDA Forest Service, Ft. Collins, Colorado.
Johnson, K.N. and Scheurman, H.L. 1977. Techniques for prescribing optimal timber harvest and investment under different objectives. *Forest Science Monographs* 18.
Kao, C. 1984. Optimal stocking levels and rotation under uncertainty. *Forest Science* 30, 921–7.
Kurz, M. 1968. Optimal economic growth and wealth effects. *International Economic Review* 9, 348–57.
Ledyard, J. and Moses, L.N. 1976. Dynamics and land use: the case of forestry. In *Public and Urban Economics*, ed. R.E. Frieson, Lexington, Mass.: D.C. Heath.
Lyon, K.S. and Sedjo, R.A. 1983. An optimal control theory model to estimate the regional long-term supply of timber. *Forest Science* 29, 798–812.
Martell, D. 1980. The optimal rotation of a flamable forest stand. *Canadian Journal of Forest Research* 10, 30–34.
Navon, D.J. 1971. Timber RAM. *USDA Forest Service Research Paper PNW–70*, Pacific Southwest Forest and Range Experimental Station, Berkeley, California.
Naudial, J.C. and Pearse, P.H. 1967. Optimizing the conversion to a sustained yield. *Forest Science* 13, 131–9.
Neher, P.A. 1976. Democratic exploitation of a replenishable resource. *Journal of Public Economics* 5, 361–71.

Osmaston, F.C. 1968. *The Management of Forests.* London: Allen & Unwin.
Palgrave, R.H.I. (ed.) 1912. *Dictionary of Political Economy*, Vol. II. 2nd edn, London: Macmillan.
Pearse, P.H. 1967. The optimum forest rotation. *Forestry Chronicle* 43, 178–95.
Reed, W.J. 1984. The effect of risk of fire on the optimal rotation of a forest. *Journal of Environmental Economics and Management* 11, 180–90.
Reed, W.J. and Errico, D. 1985. Optimal harvest scheduling at the forest level in the presence of the risk of fire. *Canadian Journal of Forest Research* 15, 680–87.
Samuelson, P. 1976. Economics of forestry in an evolving society. *Economic Inquiry* 14, 466–92.
Smith, J.H.G. 1978. Management of Douglas-fir and other forest types in the Vancouver Public Sustained Yield Unit. In *Forest Management in Canada*, Vol. 11, Ottawa: Environment Canada.
Thünen, J.H. von. 1826. *The Isolated State.* Trans. and ed. Peter Hall, London: Pergamon Press.

forward markets. *See* SPOT AND FORWARD MARKETS.

Fourier, François Marie Charles (1772–1837). According to Fourier, the poverty which accompanies the 'colossal' advance of industry constitutes the main cause of social disorders. Those responsible are the tradesmen; competition has resulted in the creation of a mercantile feudal system.

To set up a new social order, we must use the passions as nature gave them to us. Man is guided by his quest for pleasure, and his passions are always the same, the principal ones being the desire for luxury, the desire to adhere to a group, and that of forming part of a 'series' or a work or play group.

These groupings of individuals must be such that they allow psychological differences to complement each other. Individuals should live together in a 'phalanstery', where, in giving themselves over entirely to their passions, they will form a harmonious and pacific social order. The association will encourage emulation and the disappearance of rivalry. Work will be almost infinitely divided up, its duration will be short, and everyone will be free to choose the work of his choice. Associated property will consist of property brought by the participants who do not keep goods for themselves. Manual work, capital and 'talent' will be remunerated.

The application of Fourier's system has been attempted several times, and has generally ended in failure. However, the experiment led by J.B. Godin at Guise in France can be considered half successful as it was followed up after his death and lasted until 1969.

Fourier was essentially concerned with psychology and social psychology, and can be seen as the precursor of studies conducted from the 1920s onwards on the ways in which work groups function.

J. WOLFF

SELECTED WORKS

1808. *Théorie des quatre mouvements et des destinées générales.*
1822. *Traité de l'association domestique agricole.*
1829. *Le nouveau monde industriel et sociétaire.*
1835–6. *La fausse industrie.*

The complete works of Fourier were republished in eleven volumes by Editions Anthropos, Paris, 1966, a reprint of the third edition of 1846.

BIBLIOGRAPHY

Bourgin, H. 1905. *Fourier.* Paris.
Lehouck, E. 1966. *Fourier aujourd'hui.* Paris: Denoel.
Pinloche, A. 1933. *Fourier et le socialisme.* Paris: F. Alcan.

Foxwell, Herbert Somerton (1849–1936). Born at Shepton Mallet, Somerset, Foxwell received his early education at home and then at schools in Taunton. He matriculated into the University of London in 1866 and received his BA in 1868. He entered St John's College, Cambridge, in 1868. After being placed Senior in the Moral Science Tripos in 1870, he won the Whewell Scholarship in International Law in 1872. He was elected a Fellow of the College in 1874. Under the old Statutes he was forced to vacate his fellowship when he married in 1898, but was able to resume it in 1905 and retained it until his death in 1936. From 1875 until 1905 he served as College Lecturer. At first he taught the whole area of the Moral Sciences, but while Marshall was at Bristol from 1877 until 1884, Foxwell taught courses in economics. When Marshall became Professor of Political Economy at Cambridge in 1885, he quickly overshadowed Foxwell both at St John's and in Cambridge economics generally. Appointed as a University Extension Lecturer in 1874, Foxwell taught widely in the North of England, claiming that it had brought him into close contact 'with the actual conditions of practical life'. In 1876 he became a Lecturer at University College, London, and in 1881 he succeeded W.S. Jevons as its Professor of Political Economy. Despite his frequent travels to London, Foxwell remained firmly committed to Cambridge life. After his retirement from active lecturing at St John's, he served as its Director of Economic Studies until his death. In private life, he lived a few doors from J.N. Keynes in Harvey Road and was well known in Cambridge economic circles.

Foxwell's primary interests were in bibliography, banking and money, and economic history. As a book collector, he assembled several large libraries of economic literature, especially for the period 1740–1848, which became the basis for the Goldsmiths' collection in London and the Kress Library of Business and Economics at Harvard. Along with Jevons, whom he knew well, Foxwell was a severe critic of Ricardo. Indeed, one of the aims of his book-collecting was to demonstrate that England had produced other economic traditions than that of Ricardo. Foxwell held that Ricardo's use of deduction had been excessive and that it had produced a tradition of socialist thought. Foxwell's own conservative position is demonstrated in his historical introduction to Anton Menger's *Right to the Whole Produce of Labour* (1899).

Foxwell attributed his interest in the instability of capitalism to both Jevons and Arnold Toynbee. His major original work in economics, *Irregularity of Employment and Fluctuations of Employment* (1886) held that free competition had produced both wealth and poverty for the workers. Poverty, he argued, was primarily a result of the irregularity of employment due to an unstable level of prices, as well as the persistence of a customary low level of wages for many groups in society. Fearful of social revolution, and building on the work of J.S. Mill, he fashioned a counter-revolutionary programme of state intervention, cooperative schemes, profit sharing, and the benefits which could be derived from regulated monopolies. He especially promoted a system of counter-cyclical state expenditures in such areas as housing, health, education and public works. A strong advocate of bimetallism, he also called for the adoption of a managed system of international payments.

In 1887, Foxwell published his influential 'Economic Movement in England' in the *Quarterly Journal of Economics*, in which he sympathetically chronicled the rise of an ethical and historical economics in England which he claimed as a superior guide to the formulation of public policy than the dominant Ricardian tradition. Subordinate to Marshall at Cambridge, Foxwell increasingly allied himself with the historical criticism of Marshall's economics. In 1903 he even joined the historical economists' attack on free trade. From 1895 he lectured at the London School of Economics and Political Science, which had been established in part in conscious opposition to Marshall's vision of economics, and in 1907 was named Professor of Political Economy at the University of London. His implacable hostility to Ricardo, whom Marshall vigorously defended, his increasing attention to economic history and bibliographical work, his inability to contribute to the development of neoclassical economic theory, his bimetallism, and his difficult personality, made him unacceptable to Marshall as his successor in 1908. Instead, the appointment when to the able and more suitable theorist A.C. Pigou. It was not until 1929, when he was elected President of the Royal Economic Society, that Foxwell reconciled himself to having been passed over by Marshall at Cambridge.

GERARD M. KOOT

See also ECONOMICS LIBRARIES AND DOCUMENTATION; MARSHALL, ALFRED.

SELECTED WORKS

1899. Introduction and Bibliography to Anton Menger, *The Right to the Whole Produce of Labor*. London. Reprinted, New York: Augustus M. Kelley, 1968.
1886. Irregularity of employment and fluctuations of employment. Lecture VI in the *Claims of Labour*, Edinburgh: Cooperative Printing Company.
1887. The economic movement in England. *Quarterly Journal of Economics* 2, October, 84–103.

BIBLIOGRAPHY

Coase, R.H. 1972. The appointment of Pigou as Marshall's successor. *Journal of Law and Economics* 15, October, 473–86.
Coats, A.W. 1972. The appointment of Pigou as Marshall's successor: a comment. *Journal of Law and Economics* 15, October, 487–95.
Foxwell, H.S. 1939. Herbert Somerton Foxwell: a portrait. In *Kress Library of Business and Economics Publication* No. 1, Boston: Harvard Graduate School of Business Administration, 3–30.
Keynes, J.M. 1936. H.S. Foxwell. *Economic Journal* 46, December, 589–614. Reprinted in *The Collected Works of John Maynard Keynes*, Vol. X, *Essays in Biography*, London: Macmillan, 1972.
Koot, G.M. 1977. H.S. Foxwell and English historical economics. *Journal of Economic Issues* 2, September, 561–86.

Franklin, Benjamin (1706–1790). One of the founding fathers of the United States, Franklin is remembered as 'the wisest American' for his many accomplishments as statesman, scientist and writer. As a writer he extolled the virtues of industry and thrift in many memorable phrases, some of which have become household maxims. They lent support to Max Weber's thesis of the Protestant origin of capitalism and were cited by him.

Franklin was a man of wide reading and pronounced intellectual curiosity, whose scientific contributions were mainly in natural science. He has, however, a number of economic writings to his credit. The two most important treat of monetary expansion and population growth. In 1728, at the age of 22, when he was active as a printer, he published *A Modest Inquiry into the Nature and Necessity of a Paper Currency*, in which he made a successful plea for an issue of colonial paper money. If money is tight, Franklin argued, interest rates will be high, prices low, immigration discouraged and imports stimulated. Moneylenders and lawyers may benefit from this, but other groups will suffer. If the paper money is issued on the security of land, the value of money will not decline.

In 1755 Franklin published *Observations Concerning the Increase of Mankind and the Peopling of Countries*. Like the *Modest Inquiry*, it was widely read and influential, and like the other work it shows the influence of Sir William Petty (1623–87). Both Petty and Franklin were convinced of the advantages of a large and swiftly growing population. While the central tendency of Franklin's work runs counter to that of Malthus's later work on population, there are certain notions that can be found in the writings of both men: the idea that population tends to double in 25 years, and the notion of prudence as constituting a check to early marriages and thereby to population growth.

In the *Modest Inquiry* Franklin observed that 'trade in general being nothing else but the exchange of labour for labour, the value of all things is ... most justly measured by labour' (Spiegel, 1960, p. 16). This elicited praise from Marx, who extolled Franklin as 'one of the first economists after William Petty who grasped the nature of value' (*Capital*, vol. 1, ch. 1). Marx also noted Franklin's definition of man as a tool-making animal, a definition he described as characteristic of Franklin's Yankeedom (*ibid.*, ch. 11).

HENRY W. SPIEGEL

BIBLIOGRAPHY

Carey, L.J. 1928. *Franklin's Economic Views.* New York: Doubleday.
Dorfman, J. 1946. *The Economic Mind in American Civilization 1606–1865.* New York: Viking, vol. 1.
Spiegel, H.W. 1960. *The Rise of American Economic Thought.* Philadelphia: Chilton.

fraud. An agent is said to have committed fraud when he misrepresents the information he has at his disposal so as to persuade another individual (principal) to choose a course of action he would not have chosen had he been properly informed. The essential element of this phenomenon is the presence of two individuals both of whom have something to gain from co-operating with each other but who have conflicting interests and differential information. More specifically, it is critical that the agent be both better informed than the principal and in a position to use his superior knowledge to affect the principal's actions so as to increase his own share of the total benefit at the principal's expense. As the choice of terminology indicates, fraud is a special case of a more general class of economic phenomena known as agency relationships. (For a more elaborate discussion and citations see Arrow, 1985.)

Fraud may assume different forms. To focus our discussion, however, we consider the provision by a producer (agent) of misinformation so as to induce customers (principals) to purchase goods or services which, if adequately informed, they would not buy. Our discussion draws heavily upon Darby and Karni (1973), which was the first and so far the most elaborate attempt at an economic analysis of the phenomenon of fraud.

THE PREVALENCE OF FRAUD. Fraud is as prevalent and as persistent as the asymmetrical information necessary to support it. Thus fraud may occur whenever the cost of verification of the producer's claims prior to the actual purchase of the good or service is prohibitively high. For some goods the producer's claims are easily verifiable through their use, for example, the performance of a car, the effectiveness of a painkiller. In these cases, if the population participating in the market is sufficiently stable, the scope for fraud by established firms is limited by the need to maintain their reputations. In such markets fraud may nevertheless be practised by transient firms and fly-by-night operators.

Fraudulent practices of a more persistent nature may occur in service industries where the separation of the diagnosis from provision of the service itself is impractical and where, moreover, the assessment of the quality of service is difficult if not impossible. This is the case when the ultimate performance of the good being serviced depends on several inputs and/or the relation between the service input and the ultimate performance is stochastic. To grasp the point consider a patient who complains of stomach pain. Suppose that the patient is treated with two different medications and undergoes surgery. Should the pain disappear the patient would be unable to determine which, if any, of the possible remedies was responsible for his cure.

THE ECONOMIC CONSEQUENCES OF FRAUD. The opportunities for fraud manifest themselves in voluntary arrangements that define the principal-agent relation, whose purpose is to inhibit the actual perpetration of fraud, and in resource misallocation.

Voluntary arrangements and institutions such as formal warranties and service contracts may be regarded as insurance schemes. However, by placing responsibility for the cost of maintenance on the supplier these contracts eliminate the supplier's incentive to defraud his customers. Thus, in the absence of direct means of verification, extended warranties and service contracts may be regarded as means by which producers authenticate their claims (see Hirshleifer, 1973, for a discussion of authentication as an information-induced behavioural mode). The scope for formal service contracts and warranties is limited by the usual 'moral hazard' problem. In other words, the adverse effect on the owner's incentive to take the necessary care in using the good may undermine these institutions.

A less formal arrangement is the 'client relationship'. This form of principal–agent relation is an implicit agreement that the customer will continue to patronize the service shop as long as he has no reason to suspect fraud. Lacking the means necessary for a direct assessment of the service provided, customers may exploit the opportunity afforded by repeated relations to detect whether a supplier performs at the desired level by using statistical methods. Recognizing this and the need to cultivate a clientele discourages the supplier from defrauding regular customers. This personal relationship replaces the anonymity typical of markets in which information is symmetrically endowed. Obviously this consideration does not apply to transient clientele. Indeed, large parts of the folklore surrounding the tourist industry consist of accounts of flagrant fraudulent practices. (For a more detailed discussion of the client relationship, see Karni and Darby, 1973; Glazer, 1984.)

The profit opportunities made possible by fraud attract resources to industries where such opportunities exist. When barriers to entry do not exist excessive profits are eliminated. The resulting resource allocation, however, is distorted as scarce resources are employed in the provision of unnecessary services.

THE DETERRENCE OF FRAUD. Successful detection and prosecution of fraud have a deterrent effect that benefits society. Thus, a case can be made for social intervention. This may take the form of awarding multiple damages to successful prosecution of fraud that would reflect the full social benefit from its deterrence. Such a policy would have the effect of increasing private vigilance in dealing with fraudulent practices and, with appropriate penalties on the practitioners, reduce the amount of fraud to a socially desirable level. Alternatively, adherence to non-fraudulent practices may be enforced by the law enforcement agencies of the government. (For a detailed discussion, see Darby and Karni, 1973.)

Since the provision of misinformation may just as well be the result of sheer incompetence as of intentional deception, successful fraud-deterring policy will also increase the competence level of the suppliers of services. Unlike the elimination of intentional misrepresentation of information, however, increasing the level of competence involves investment of scarce resources on the part of the suppliers. Therefore, in setting the goals for a policy whose aim is to reduce fraud, the additional gains from the associated increase in the level of competence must be weighed against the corresponding resource cost. The optimal level of fraud may not be zero.

EDI KARNI

See also ASYMMETRIC INFORMATION.

BIBLIOGRAPHY

Arrow, K.J. 1985. The economics of agency. In *Principals and Agents: The Structure of Business*, ed. J.N. Pratt and R. Zeckhauser, Cambridge, Mass.: Harvard Business School Press.

Darby, M.R and Karni, E. 1973. Free competition and the optimal amount of fraud. *Journal of Law and Economics* 16(1), April, 67–88.

Glazer, A. 1984. The client relationship and a 'just' price. *American Economic Review* 74(5), December, 1089–95.

Hirshleifer, J. 1973. Where are we in the theory of information? *American Economic Review* 63(2), May, 31–9.

free banking. Free banking is the term applied in the United States to a system under which (1) banking powers are granted to all applicants under certain prescribed conditions, and (2) bank-notes issued under such authority are protected by a deposit of security held by the government which establishes the system. The earlier banks in the United States, whether established by congress or by the state legislatures, were organized under special charters. Various expedients were resorted to for the prevention of unsound issues, with various degrees of success, but without arriving at any generally acceptable method. The suspension of specie payments in May 1837, and the extraordinary confusion of the paper currency which ensued, finally brought the general discontent to a climax in New York, and the legislature of that state, in June 1838, passed an act for the free organization of banks issuing a secured currency. Under this act, as amended and revised, any group of persons proposing to form a banking association, and contributing a capital in no case less than $25,000, say £5000, can be incorporated with full banking powers, subject to uniform regulations as to the conduct of their business, its supervision by the state, and their corporate liabilities and duties. Individual bankers and firms, who use the name 'bank', are also required to conform to the system, although they may remain unincorporated. The right of issue is given to any association or individual coming under the system. The notes are prepared and registered by a public officer, are delivered to the issuing bank only after the deposit of security of a prescribed kind and amount, and must be signed by the officers of the bank before issue. Banks organized upon such a system are called free banks.

Free banking does not imply, then, an unrestricted management of the business, or complete liberty in the issue of notes. Such a system is called free because the right to organize, upon compliance with fixed conditions, is extended to all, free from any requirement of special legislation. It is not essential that there should be any engagement by the state to make the notes good, if the security, of which the state is trustee, proves insufficient. Neither does the deposit of security for the ultimate payment of the notes answer the question as to proper provision for daily redemption. As the provision for secured notes gave promise of insuring the ultimate solvency of bank notes, it settled the one banking question as to which the public were most sensitive, and enabled the legislature to renounce the task of deciding upon applications for special charters. The system adopted by New York was copied by many other states before the civil war, but in some cases with relaxations which impaired its safety. In 1861 the New York free banks, having on deposit stocks of the United States and other solid securities, met the strain of war with success. In several states, where the law was less rigid, many free banks went down, and their notes, secured in some cases chiefly by bonds of seceded states and others in low credit, caused heavy losses to the holders. Two years later Congress adopted the free banking system on a great scale, by a law providing for national banks, to be organized on application under a general act, and to issue notes with United States bonds as the only admissible security. In 1865 Congress laid a tax of ten per cent on all bank notes other than national, thus excluding from the field all issues authorized by the states. Several of the states, however, still retain their laws as to circulation, although these have been entirely dormant since 1866. Free banking under the national system was for some years seriously limited, by the provision that the aggregate of notes issued by all the national banks should not exceed $300,000,000, afterwards $354,000,000 (say £60,000,000, and £70,800,000) although the organization of banks was still free to all. The act of 1875, for resuming specie payments, removed the limit of aggregate circulation, and thus completely established free banking under the national government. The rapid rise in price of United States bonds, and the low return yielded by an investment in them, have since put a new check upon the system; and if the use of bank-notes is to continue, the alternative may soon be presented, of either finding for deposit by national banks some other security than United States bonds, or removing the prohibitory tax upon issues authorized by the states.

[C.F. DUNBAR]
Reprinted from *Palgrave's Dictionary of Political Economy.*

free disposal.

'I should like to buy an egg, please' she said timidly. 'How do you sell them?'
'Fivepence farthing for one – twopence for two,' the Sheep replied.
'Then two are cheaper than one?' Alice said, taking out her purse.
'Only you must eat them both if you buy two,' said the Sheep.
'Then I'll have one please', said Alice, as she put the money down on the counter. For she thought to herself, 'They mightn't be at all nice, you know.'

Lewis Carroll, *Through the Looking-Glass*

If I dislike a commodity, you may have to pay to get me to accept it. But so long as some otherwise non-sated consumer finds this commodity to be desirable, or at least harmless, it could not have a negative price in competitive equilibrium. Likewise, if some firm can dispose of arbitrary amounts of a commodity without using any other inputs or producing any other (possibly noxious) outputs, its price in competitive equilibrium cannot be negative. Therefore competitive equilibrium analysis can be confined to the case of non-negative prices if every commodity is either harmless to someone or freely disposable.

If a commodity is not freely disposable and is a 'bad' in the sense that everyone prefers less of it to more, it is possible to redefine the 'commodity' as the absence of the bad. The commodity so defined can then be treated as a good with a positive price. More generally, it might be possible to choose some alternative coordinate system in which to measure commodity bundles so that in the new coordinate system either there is free disposability or more is preferred to less. But if people are willing to pay a positive sum for a small amount of a commodity and less for a large amount, then the question of whether that commodity will have a positive or negative price in competitive equilibrium cannot be decided in advance. The sign of the equilibrium price will in general depend on supplies of this and other goods and on the detailed configuration of preferences in the economy.

Sometimes a noxious by-product of production or consumption can be transformed into a useful output if sufficient other resources are used. Then the equilibrium price for the by-product may be either positive or negative, depending on the prices of the other inputs and of the output into which it is transformed. This is particularly evident when commodities are distinguished by location. Garbage located in the centre of a city is undesirable to everyone. To bury or incinerate it is costly and generates no valuable outputs. Therefore, if garbage is disposed of in this way, its equilibrium price must be negative. But the garbage could be transported to the country, boiled and fed to pigs. Depending on the costs of this process and the price of pork, it may turn out that converting garbage to pig feed is profitable even when garbage at the city centre has a zero or positive price. Both the ultimate disposition of garbage and the sign of its price have to be determined endogenously in the competitive process.

Early proofs of the existence of competitive equilibrium (Arrow and Debreu, 1954; Gale, 1955; Debreu, 1959) assumed that all commodities are freely disposable or, equivalently, defined equilibrium so as to allow the possibility that in equilibrium some goods might be in excess supply but have zero price. Debreu (1956) shows how the assumptions of free disposal and monotonicity can be greatly relaxed. McKenzie (1959) and Debreu (1962) present general theorems on the existence of equilibrium in which free disposal is not assumed. Rader (1972), Hart and Kuhn (1975), Bergstrom (1976) and Shafer (1976) suggest further generalizations and simplifications in dealing with negative prices in equilibrium.

The formal treatment of negative prices in existence proofs presents an interesting mathematical problem. Most of the standard existence proofs apply the Kakutani fixed-point theorem to a correspondence that maps the set of possible equilibrium prices into itself in such a way that a fixed point for the mapping is a competitive equilibrium price vector. The Kakutani theorem applies to an upper hemicontinuous mapping from a closed bounded convex set to its compact, convex subsets. If the only prices to be considered are non-negative, then the domain for this correspondence can be chosen to be the unit simplex. If all price vectors, positive and negative, must be considered, then an obvious candidate for the domain of this mapping would be the unit sphere.

$\{p \in R^n | p \cdot p = 1\}$. But this is not a convex set. The closed unit ball $\{p \in R^n | p \cdot p \leqslant 1\}$ is a convex set, but it contains the vector zero, at which point the excess demand mapping is not upper hemicontinuous.

Debreu (1956) solved this problem neatly in a brief, elegant paper that has received less attention than it deserves. The existence proofs that assume free disposability of all goods had shown that there exists a non-negative price vector at which the excess demand vector is either zero or belongs to the negative orthant. Debreu generalized this result to show that if there is free disposability on any convex cone which is not a linear subspace, then a price vector can be found at which excess demand is either zero or belongs to the cone of free disposability. Furthermore, this price vector gives a non-positive value to every activity in the cone of free disposability. In particular, consider the case where one good is assumed to be freely disposable. Then, from Debreu's theorem, it follows that there exists some price vector at which excess demand for all goods other than the freely disposable good is zero, and at which there is either zero or negative excess demand for the freely disposable good. From Walras's Law and the fact that excess demand for all other goods is zero, it follows that the price of the freely disposable good can be positive only if excess demand is zero. Therefore this price vector is a competitive equilibrium. Thus Debreu weakened the free disposability assumption from 'all goods are freely disposable' to 'at least one good is freely disposable'.

We can take Debreu's argument one step further and eliminate the assumption of even one freely disposable good. Nowhere in Debreu's proof is it necessary to assume that the freely disposable good is desirable to anyone. This suggests that the existence of a freely disposable good is not likely to be essential for the existence of equilibrium. For suppose that there is an economy with no freely disposable goods. A fictional good could be introduced which is freely disposable but totally useless and totally harmless to everyone. For the augmented economy found by adding this fictional good to the original economy, by Debreu's theorem there would exist a competitive equilibrium. In this new economy it turns out that the equilibrium price of the useless, freely disposable good must be zero and the vector of equilibrium prices for the other goods can serve as a competitive equilibrium price vector for the original economy.

The approach taken by Bergstrom (1976) is equivalent to introducing a useless and harmless fictional good into Debreu's model. Taking the formal steps of this argument directly without intermediary fictions leads to an upper hemicontinuous mapping from the unit ball into itself for which there is a fixed point on the boundary of the unit ball. This fixed point turns out to be a competitive equilibrium price vector. An interesting alternative approach was taken by Rader (1972) and by Hart and Kuhn (1975). Instead of the Kakutani theorem, they use a theorem about fixed and antipodal points of a continuous mapping from the unit sphere into itself, and are thereby able to deal with all prices on the unit sphere as potential equilibrium prices.

The first and second welfare theorems and the theorems about the equivalence between the core and the set of competitive equilibria apply straightforwardly when there is not free disposal. For example, in order to prove the Pareto optimality of competitive equilibrium in an exchange economy, we simply argue along the usual lines that if any allocation is Pareto superior to a competitive equilibrium, then at the original competitive prices, the aggregate value of consumption in the proposed Pareto superior allocation must exceed the aggregate value of initial endowments. But if the proposed allocation is feasible, then the aggregate consumption vector in the proposed allocation must equal the aggregate initial endowment vector. It follows, whether prices are positive, negative or zero that if the two vectors are equal they must have the same value at the competitive price vector. Therefore there cannot be a feasible allocation which is Pareto superior to a competitive equilibrium. Similar arguments apply to the core theorem. The only matter in which a bit of care must be taken is in defining the activities available to a potential blocking coalition so as to exclude the possibility of dumping undesirable commodities. This simply amounts to the assumption that a blocking coalition must exactly equalize its total consumption of all goods to its total endowment.

TED BERGSTROM

See also EXCESS DEMAND AND SUPPLY; FIXED POINT THEOREMS; FREE GOODS; GENERAL EQUILIBRIUM.

BIBLIOGRAPHY

Arrow, K.J. and Debreu, G. 1954. Existence of an equilibrium for a competitive economy. *Econometrica* 22, July, 265–90.

Bergstrom, T. 1976. How to discard free disposability – at no cost. *Journal of Mathematical Economics* 3(2), 131–4.

Debreu, G. 1956. Market equilibrium. *Proceedings of the National Academy of Sciences of the USA* 42, 876–8.

Debreu, G. 1959. *The Theory of Value*. New York: Wiley.

Debreu, G. 1962. New concepts and techniques for equilibrium analysis. *International Economic Review* 3, 257–73.

Gale, D. 1955. The law of supply and demand. *Mathematica Scandinavica* 3, 155–69.

Hart, O. and Kuhn, H. 1975. A proof of the existence of equilibrium without the free disposal assumption. *Journal of Mathematical Economics* 2(3), 335–43.

McKenzie, L. 1959. On the existence of general equilibrium for a competitive market. *Econometrica* 27(1), 54–71.

Rader, T. 1972. *Theory of General Economic Equilibrium*. New York: Academic Press.

Shafer, W. 1976. Equilibrium in economies without ordered preferences or free disposal. *Journal of Mathematical Economics* 3(2), 135–7.

freedom. *See* ECONOMIC FREEDOM; LIBERTY.

free enterprise. *See* CAPITALISM; ECONOMIC FREEDOM; LAISSEZ FAIRE.

free entry. *See* CONTESTABLE MARKETS; ENTRY AND MARKET STRUCTURE.

free goods. Free goods are 'goods', whether consumer goods or productive inputs, which are useful but not scarce; they are in sufficiently abundant supply that all agents can have as much of them as they wish at zero social opportunity costs (cf. ch. 11, §3, of Carl Menger's *Principles of Economics*, 1871). Goods which have a positive social opportunity cost but a zero price – for example, because there are no property rights in them, or because they are fully subsidized – are *not* free goods. Any 'gift of nature', whether it be a good such as air, or a primary input such as labour or land (in the narrow sense), might be a free good under certain circumstances. But a produced commodity can be a free good, other than in the market period, only if it is a joint product. As is at once obvious from the example of air, the free nature of a good is not an intrinsic property; thus air above the earth's surface is, in most circumstances, a free good but air under water or in

deep mines is not. More abstractly, then, a free good is a good for which supply is not less than demand at a zero price (in the sense of social opportunity cost). But since both supply of and demand for any good depend on the prices of all goods, it is clear that whether a particular good is or is not a free good is a general equilibrium, not a partial equilibrium, issue.

Consider first a Walrasian analysis of general equilibrium. Under the standard assumptions of such an analysis, Walras's Law (or identity) holds, so that $pS \equiv 0$, where p is a row vector of prices and S a column vector of excess supplies. (This is an *identity*, holding at all prices, not only at equilibrium prices.) Now, by definition, in a Walrasian analysis any equilibrium excess supply vector satisfies $S^* \geqslant 0$. Hence if it is ensured that any equilibrium price vector satisfies $p^* \geqslant 0$, it follows – from $pS \equiv 0$ and $S^* \geqslant 0$ – that if $S_j^* > 0$ then $p_j^* = 0$. That is, the Rule of Free Goods holds in such a Walrasian equilibrium, applying to all 'goods', whether produced or non-produced. Two points are to be noted. The less important one is that while $S_j^* > 0$ implies $p_j^* = 0$, $p_j^* = 0$ does not imply $S_j^* > 0$, since $p_j^* = 0 = S_j^*$ is possible. The more important point is that the Rule of Free Goods is not implied by $pS \equiv 0$ and $S^* \geqslant 0$ alone; they must be supported by the condition $p^* \geqslant 0$. This last condition is often underpinned by an assumption of the possibility of *free disposal* (see below). Such an assumption rules out the possibility that any $p_j^* < 0$, for there would be an unlimited demand for a good for which one 'paid' a negative price – that is *received* a positive price – and which one could dispose of at zero cost.

It was noted above that the Rule of Free Goods is applied in Walrasian flex-price analyses to both products and primary inputs. With respect to the latter, it is instructive to consider the linear programming formulation which is sometimes given for the 'supply' side of a general equilibrium existence proof for an economy with linear technical conditions. In the primal problem one is asked to maximize the value of net output, at parametrically given product prices, subject to not using more than the exogenously fixed supply of any primary input. The complementary slackness conditions, corresponding to these last constraints, give immediate expression to the Rule of Free Goods, as applied to the primary inputs. And the non-negativity constraints in the dual problem stipulate, of course, that the solution factor prices cannot be negative. Thus every solution factor price will be non-negative and will be zero if the relevant factor is less than fully utilized.

It is essential to note that not all types of economic analysis impose the Rule of Free Goods with respect to all primary inputs. In the von Neumann model, for example, that rule is certainly imposed with respect to all the produced commodities, but it is not applied to labour, which receives an exogenously given real wage bundle which is independent of the degree of utilization of labour. At most, one could say that a 'Rule of Zero "Excess" Wages' is applied because labour is less than fully employed. Similarly, in Keynes's analysis the presence of involuntarily unemployed labour does not drive the wage to zero but only to an exogenously given minimum (a market level reservation price). Clearly, then, the three assertions $pS \equiv 0$, $S^* \geqslant 0$ and $p^* \geqslant 0$ are not all accepted within Keynes's analysis. But since $S^* \geqslant 0$ and $p^* \geqslant 0$ *are* accepted, it can only be Walras's Law which is being rejected – and this is indeed the case, for in Keynes's analysis we have only the condition that, the elements of S being defined in terms of *desired* supplies and demands, $pS \geqslant 0$. The weaker relation is, of course, perfectly consistent with $S_j^* \geqslant 0$ and $p_j^* \geqslant 0$ (see Morishima, 1976, pp. 203–11).

It was noted above that the 'free-disposal' assumption has the convenient consequence that no equilibrium price can be negative; this means that the search for Walrasian equilibrium price vectors can be confined to the unit simplex. Although this restriction on prices is *not* a necessary ingredient of all general equilibrium existence proofs, the free-disposal assumption is sufficiently widely adopted (it is sometimes even described as obviously reasonable) to merit a close examination of its justification. Consider first the proposal that the commodity to be disposed of in a disposal process is the *only* input to the latter. This means that the only form of 'disposal activity' allowed is that of simply *leaving* the commodity to be disposed of *where* it is and leaving it *as* it is. If it moves or changes its form, that must be the result solely of non-human and non-produced agencies. (Note that one cannot defend the disposal activity assumption by saying that it applies only to the 'last stage' of a real-world-like disposal process, which first uses labour, lorries, etc. to take waste chemicals, for example, to a particular place. This is because the assumption is supposed to apply to *all* commodities, including, for example, the chemical waste *situated at the point of its production*.)

This leads us naturally to a consideration of the second – and even more objectionable – aspect of the disposal activity assumption, the proposal that the activity has *no* outputs. Taken literally, this proposal simply contradicts one of *the* most fundamental laws constituting our conception of the physical universe – the law of conservation of mass-energy. If one takes the conservation law for granted, for the purposes of economic theory, then *either* the zero-output assumption is incomprehensible *or* it means that all the outputs from the disposal process lie outside the commodity set which is taken as the basis for the economic analysis. A defence of the latter interpretation would have to involve both an account of the principles according to which that set is defined on a non-arbitrary basis and an explanation of why the outputs of disposal processes can be supposed – non-arbitrarily – to lie outside that set.

It might be said that the disposal-activity assumption simply provides one interpretation of the basic axiom of free disposal ($x \in X$ and $x' \leqslant x$ implies $x' \in X$, where X is the production set) and that the latter may be acceptable even while the former is not. How then might the axiom be understood in the absence of disposal processes? Suppose that together with all the other inputs and outputs (which will be held constant), a certain input of fertilizer and a certain output of maize define an activity belonging to the production set. It is then proposed that, *ceteris paribus*, the same fertilizer input and a smaller maize output also define a feasible activity. We cannot suppose that some of the fertilizer is simply not used, for then an output (that of fertilizer) would have been increased. Thus all the fertilizer must be used. If it is used in the same way as 'before' then a smaller maize output, *ceteris paribus*, involves different laws of nature. If it is used but used differently from 'before', then some other input has changed, contrary to hypothesis. Hence the presence of a disposal activity is, after all, required.

In the above example, the 'other input' which has changed when fertilizer is used differently is some human agency. For to say that fertilizer is used differently is precisely to say that *someone* has acted differently. Suppose then that we now change the example, replacing the given fertilizer input by a given quantity of a specified type of labour input and including fertilizer amongst the (given) 'other' inputs and outputs. In the absence of disposal processes, does the fact that a certain labour input and a certain maize output define an activity in the production set, mean that the *same* labour input and a *smaller* maize output (perhaps even a *negative* one) also define such an activity? If free disposal is ruled out, the laws of nature are constant and labour is precisely defined, the answer would again seem to be No. Thus, again, the free-disposal axiom does indeed rest on the presence of disposal activities. Objections to the disposal-activity assumption are thus also objections to the axiom of free disposal itself.

It has already been noted that general equilibrium existence proofs can dispense with the free-disposal axiom and that Keynes's theory does not apply the Rule of Free Goods to labour. More generally, the rule of free goods should not simply be assumed to apply to non-produced inputs, for it must always be considered whether their owners place a positive reservation price on them. With respect to produced commodities, free disposal should not be assumed (for the reasons given above), as it commonly is in linear programming models, in studies of balanced growth within closed production models with convex cone production sets, and in proofs of turnpike theorems. In each case, on dispensing with the free-disposal axiom, one must decide how to represent preferences over 'bads'. These apparently abstract issues are, of course, of immediate relevance in the discussion of such policy issues as pollution control, environmental protection and waste disposal. (If there were no joint production, or if free disposal were possible, there could be no problems of pollution control and waste disposal.) When there are disposal activities which involve a negligible private cost but a significant social cost, policy will involve bringing the positive costs of disposal to bear on the individual agents concerned. This may induce them, in turn, to discover or invent new uses for the previously undesired 'commodities'; the costly nature of disposal has spurred changes in technical knowledge.

IAN STEEDMAN

See also FREE DISPOSAL; GENERAL EQUILIBRIUM.

BIBLIOGRAPHY

Debreu, G. 1959. *Theory of Value: an axiomatic analysis of economic equilibrium.* New York: Wiley.

Morishima, M. 1976. *The Economic Theory of Modern Society.* Cambridge: Cambridge University Press.

free lunch. 'There's no such thing as a free lunch' dates back to the 19th century, when saloon and tavern owners advertised 'free' sandwiches and titbits to attract mid-day patrons. Anyone who ate without buying a beverage soon discovered that 'free lunch' wasn't meant to be taken literally; he would be tossed out unceremoniously.

'Free lunch' passed over into political economy during the New Deal era, and is loosely credited to various conservative journalists, including H.L. Mencken, Albert Jay Nock, Henry Hazlitt, Frank Chodorov and Isabel Paterson. (All efforts to identify the true originator proved unavailing.) The phrase signified that the welfare state is an illusion: government possesses no wealth of its own, so it can only redistribute wealth it has seized by taxation.

During the Vietnam war era, 'free lunch' took on a libertarian cast. When defenders of the draft argued that young men *owed* military service because they had accepted free tuition and subsidized school lunches as youngsters, the 'free lunch' expression became a libertarian shorthand to denote that citizens never get something for nothing, that sooner or later they are presented with a bill for all the favours or 'freebies' they accepted from government.

'Free lunch' would have passed into oblivion if it had not

been able to pass a crucial test of its viability in the marketplace of ideas. In the early 1970s, every political or philosophical idea had to be able to fit on a T-shirt or automobile bumpersticker. The new version, TANSTAAFL (there ain't no such thing as a free lunch), was popularized in a science fiction bestseller by Robert Heinlein (*The Moon is a Harsh Mistress*) and in Milton Friedman's widely read columns in *Newsweek* magazine.

ROBERT HESSEN

free rider. *See* INCENTIVE COMPATIBILITY.

free trade and protection. The question of 'free trade *versus* protection' is one of the oldest and most controversial issues in economics. The present article will not attempt to review this controversy from the standpoint of the history of doctrine, nor will it attempt to trace the evolution of trade policy in particular countries. Its focus is on the analytic aspects of the problem as discussed in the modern literature, which can be taken as dating from the seminal investigations of Samuelson (1939, 1962), in the tradition of Paretian welfare economics.

In keeping with this tradition we postulate that the criterion in terms of which any economic situation is to be evaluated, the 'social welfare function', is of the 'individualistic' type, i.e. it depends only upon the well-being of the individual agents themselves in terms of their own preferences, rather than on objectives such as national self-sufficiency, economic growth or some other vaguely defined concept of national interest.

An essential distinction to bear in mind is whether a 'cosmopolitan' or 'nationalist' perspective is adopted, i.e. are all individuals wherever located to count or only those belonging to the 'home' country. The modern view is that free trade is Pareto-optimal in the first case but not in the second, if the home country possesses some degree of monopoly power in foreign trade, which it can then exploit to improve the welfare of its own nationals at the expense of foreigners.

The first result follows from the familiar proposition that a perfectly competitive equilibrium is Pareto-optimal, in the absence of externalities in production and consumption. Marginal rates of substitution in consumption are equal for all individuals everywhere, since they face the same relative prices under free trade. Marginal rates of transformation in production are also everywhere equal, for the same reason, and equal to the corresponding marginal rates of substitution in consumption. The necessary conditions for a Pareto-optimum are therefore satisfied, and sufficiency can be shown to follow from the convexity of preference and production sets.

Since free trade is therefore globally Pareto-optimal, any restriction of trade such as a tariff or quota must be at the expense of someone. The idea of 'letting the foreigner pay the duty' is at the heart of what is known as the 'optimum tariff' argument. If we adopt a purely 'nationalist' perspective, ignoring the effects of our actions on foreign welfare, it is possible to raise our own welfare by a suitable degree of trade restriction to take advantage of our monopoly power in international markets. If the home country is 'small', in the sense that it faces fixed relative prices in the world market for all tradable goods, any tariff or import quota that it adopts will simply reduce its volume of trade without improving its terms of trade, so that free trade is the first-best policy even with a nationalist perspective. If it does have monopoly power,

however, it can balance the improvement in the terms of trade resulting from its restrictive policy against the reduction in the volume of trade that this entails. It can be shown that the formula for this 'optimum tariff' is equal to the reciprocal of the foreign elasticity of demand for imports minus one. The marginal cost of imports, and the marginal revenue for exports, deviate from world prices in the presence of monopoly power by the home country. This is why domestic producers and consumers have to equate their marginal rates of transformation and substitution to tariff-inclusive domestic prices rather than the world prices that would prevail under free trade.

The optimum tariff argument, going back at least to J.S. Mill but re-stated in modern terms of Bickerdike, Kaldor, Samuelson, Graaff and others, is the *only* argument for national trade restriction or 'protection' of import-competing sectors that the modern theory of trade and welfare recognizes. Even then, the argument that a tariff increases national welfare only holds strictly if it is assumed that foreigners do not retaliate, setting off a 'tariff war'. The outcome of such a process, as Johnson (1954) showed, is uncertain, with everybody worse off than under free trade a distinct possibility. Trade policy at the regional or global level thus becomes another example of the familiar 'prisoners' dilemma' situation explored in game theory.

The famous 'infant industry' argument for protection, on the grounds that it takes time for the arts of manufacture to be learnt, thus justifying temporary assistance in the form of tariffs imposed on competing imports, is *not* accepted as a legitimate 'first best' argument for tariffs by the modern theory. The reason is that even if manufacturing production creates externalities in the training of labour and the formation of skills the 'first best' intervention would be an output subsidy rather than a tariff. The reason is that the output subsidy would have the same beneficial effects on learning as the tariff but *without* the restrictive effect on imports and consumption. Welfare would therefore be higher. Similarly, the argument that tariff protection is necessary to tide over initial losses is countered by the contention that this could be accomplished by the capital market, any imperfections of which are best dealt with directly. Other arguments for tariffs, on the ground that urban wages are artificially high compared with rural wages, thus requiring off-setting tariff protection for manufactures, are also countered by the argument that an urban wage subsidy is the best intervention in this case.

All these separate cases are covered by the powerful and elegant theory of optimal intervention, developed by Bhagwati and Ramaswami (1963), and extended by Johnson (1965), Bhagwati (1971) and Corden (1974) with important earlier contributions by Haberler (1950) and Hagen (1958). The basic principle is that if a perfectly competitive equilibrium is not Pareto-optimal from a national perspective, it must be because there is some 'distortion' (see article) in international or domestic product and factor markets. The optimal intervention is to eliminate the distortion 'at the source', rather than to attempt to off-set it by an intervention that creates some other distortion as well. Thus, in keeping with the Lipsey-Lancaster theory of the 'second best', tariffs *may* improve national welfare in all of the above cases, but it is only in the 'optimum tariff' case that they constitute a 'first best' intervention.

The theory of optimal intervention, however, assumes that the subsidies necessary to maximize national welfare can be financed by non-distortionary means, such as lump sum taxes, and that there are no collection and disbursement costs. If these are allowed for, and it is recognized that any means of finance is itself going to be distortionary, the case for tariffs as 'second best' instruments will presumably become stronger, relative to the output and wage subsidies that the theory of optimal intervention blithely dispenses in disregard of any realistic government budget constraint.

R. FINDLAY

See also HECKSCHER–OHLIN TRADE THEORY; INTERNATIONAL TRADE; TARIFFS; TERMS OF TRADE.

BIBLIOGRAPHY

Bhagwati, J.N. 1971. The generalized theory of distortions and welfare. In *Trade, Balance of Payments and Growth*, ed. J.N. Bhagwati et al., Amsterdam: North-Holland.

Bhagwati, J.N. and Ramaswami, V.K. 1963. Domestic distortions, tariffs and the theory of optimum subsidy. *Journal of Political Economy* 71, February, 44–50.

Corden, W.M. 1974. *Trade Policy and Economic Welfare*. Oxford: Oxford University Press.

Haberler, G. 1950. Some problems in the pure theory of international trade. *Economic Journal* 60, June, 223–40.

Hagen, E.E. 1958. An economic justification of protectionism. *Quarterly Journal of Economics* 72, November, 496–514.

Johnson, H.G. 1954. Optimum tariffs and retaliation. *Review of Economic Studies* 21, 142–53.

Johnson, H.G. 1965. Optimal trade intervention in the presence of domestic distortions. In *Trade, Growth and the Balance of Payments*, ed. R.E. Caves et al., Chicago: Rand-McNally.

Samuelson, P.A. 1939. The gains from international trade. *Canadian Journal of Economics and Political Science* 5, 195–205.

Samuelson, P.A. 1962. The gains from international trade once again. *Economic Journal* 72, December, 820–29.

frictional unemployment. See NATURAL RATE OF UNEMPLOYMENT; UNEMPLOYMENT.

frictions and rigidities. See IMPERFECTIONIST MODELS.

Friedman, Milton (born 1912). EARLY YEARS; 1946–1955; THE MONETARY REVOLUTION AND THE RISE OF MONETARISM, 1956–1975; 1975–1985; THE PUBLIC IMAGE OF FRIEDMAN.

EARLY YEARS

Born 31 July 1912 in New York City, Friedman was the son of a poor immigrant dry-goods merchant, who died when Friedman was 15. Friedman was clearly outside the East-coast establishment of the United States, although he did spend a year on graduate studies at the Ivy league school of Columbia. He graduated (BA) at Rutgers University in 1932 and completed his AM at Chicago the following year. After the fellowship at Columbia in 1933–4, he returned to Chicago as a research assistant to Henry Schultz to work on demand analysis until 1935, when he joined the staff of the National Resources Committee. From 1937 he started that long association with the National Bureau of Economic Research which persisted until 1981. From 1938 he began another long association – with Rose Director, as his wife, which produced, *inter alia*, a son and a daughter.

In 1940 there followed a brief period as visiting professor of economics at Wisconsin. Then after a two-year stint (1941–3) in the Treasury in the division of tax research, he became the associate director of the statistical research group in the division of war research at Columbia University until the end of World War II. He then spent a year as associate professor at the University of Minnesota, before returning to Chicago as a professor of economics in 1946, the year in which he received a PhD from Columbia. His teachers at Rutgers were Homer Jones and Arthur Burns; at Chicago, Frank Knight, Lloyd Mints and Jacob Viner; and at Columbia, Harold Hotelling, J.M. Clark and Wesley Mitchell.

Superficially this record does not seem impressive. Yet it encompasses what some scholars, particularly statisticians, would regard as Friedman's most impressive contributions. Inspired by Hotelling's work on the rank correlation coefficient, his first seminal contribution was the development (1937) of the use of rank order statistics to avoid making the assumption of normality in the analysis of variance. After fifty years this article is still regarded as one of the two or three critical papers in the development of non-parametric methods in the analysis of variance, and it was followed by a discussion of the efficiency of tests of significance of ranked data. It is not surprising that these papers have been of considerable practical use, since they were largely a development of Friedman applying his mind to the practical problems he encountered in analysing incomes and consumer expenditure at the National Bureau and in Washington. Even at this early stage Friedman's work bears the imprint that readily identifies all his subsequent work. It is seemingly 'simple', eschewing complexities and complications, concentrating on essentials: all combined into a lucid exposition.

The detailed analysis of data on incomes and expenditures was Friedman's main occupation during these years. With the exception of Kuznets, Mitchell and Burns, it is difficult to find any eminent economist who acquired such a grounding in the basic empirical material of economics. It is characteristic of all his work that the organization of such data would suggest theoretical developments and new ways of arranging the material, and above all new insights into the economic process. His first published article (1934) was on a method of using the separability of the utility function to measure price elasticities from budgetary data. This exploration of new insights into old data was particularly evident in his book (1945; with Kuznets as joint author) on incomes from private professional practice; there one sees the first signs of the permanent income hypothesis and, indeed, the perceptive reader may guess what is likely to follow. In this book, which Friedman submitted as a doctoral thesis, he argued that the process of state licensure enabled the medical profession more effectively to limit entry into their profession and so enabled them to exploit their patients, keeping fees high and competitors out. The fact that the argument was tightly constructed and buttressed with convincing evidence generated the most vehement opposition and animosity from that proud profession, which appears unabated four decades later.

The wartime service in the statistical research group, although an interlude in Friedman's basic work on incomes and expenditures, generated one of the most remarkable advances in statistical theory since the seminal contributions of Sir Ronald Fisher. The group was a galaxy, consisting of Abraham Wald, Allen Wallis, Jacob Wolfowitz, Harold Hotelling and many other distinguished statisticians. The sampling inspection of wartime production of munitions etc was a tedious process of selecting a sample of a given size and testing to see the fraction of good ones in the batch. Friedman together with Allen Wallis and a Captain Schuyler, observed that testing a given size of sample was clearly wasteful. The process of testing itself gave information which enabled one to determine the degree of confidence achieved. Thus instead of continuing to test up to a fixed size of sample, the testing could be halted whenever a predetermined level of confidence in the decision had been reached. Friedman formulated the basic idea of what later came to be called 'sequential sampling' and caught the interest and imagination of Wald, who developed and proved the theorem underlying the probability ratio test and eventually produced the influential book *Sequential Analysis*. The ideas were adapted very rapidly and sequential analysis became the standard method of quality control inspection. Like so many of Friedman's contributions, in retrospect it seems remarkably simple and obvious to apply basic economic ideas to quality control; that however is a measure of his genius.

At the end of World War II, it is clear that Friedman could have continued his work as a statistician. He would have achieved a stature probably as great as that of his most influential teacher, Harold Hotelling. Alternatively he had all the basic qualifications to take the lead in developing the burgeoning field of econometrics, with its great emphasis on the adaptation of statistical theory to modelling economic phenomena. He chose neither. His excursions into statistics were utilitarian rather than speculative, and he could see little to be gained by the endless sharpening of statistical knives which was the stuff of econometrics during those years following World War II. In this decade, his contributions to statistics were even more intimately linked with his strong belief, implanted largely by Mitchell, that economics could acquire plausibility only by being subjected to empirical verification. In spite of the predilections of many economists, Friedman believed that economics should be viewed as an empirical science.

1946–1955

This decade at Chicago, much influenced by the wisdom of Frank Knight, witnessed the rapid development of economics as a positive science with its own methodology. The prevailing view of economic theory, as developed by Lionel (later Lord) Robbins, was that the veracity of theory could be tested primarily by the correspondence between the assumptions and the facts. In his 'Methodology of Positive Economics' (in *Essays in Positive Economics*, 1953), Friedman argued per contra that even if one could specify empirical correlates for the assumptions (and this cannot be done in cases where the assumptions are 'ideal types' such as homo economicus), that is irrelevant for judging the usefulness of the theory. Only by the correspondence of the *predictions* and facts should theories be provisionally accepted or rejected. Results, not assumptions, should be the main focus of our scientific activity in understanding the real world. This approach applied the new philosophy of science, developed by Karl Popper, to economics and by implication to associated social sciences. To countless students, Friedman provided an agenda for what Imre Lakatos later called a progressive research programme. The simplicity of a theory in its ability to explain a lot in exchange for a little input and the degree of 'surprise' in the prediction were the hallmarks of the new approach to theory. But it was in the efficacy and power of the empirical tests that substantial progress was to be made.

In subsequent years the 'Methodology...' has been the subject of enormous controversy. There is general agreement that in applying the theory one cannot dismiss the factual basis of the assumptions in quite such a cavalier manner. Furthermore, no one would be rash enough to declare a (refutable) theory discredited if there were a single or a few counter examples to contradict the predictions. Such absolutism has given way to more subtle interpretations depending, as

Lakatos argued, on the new and surprising insights to be obtained. Most theories coexist with small subsets of anomalous results that strictly should discredit them, and yet they remain useful theories and superior to any suggested alternative. But there is no doubt that the substance of Friedman's 'Methodology ...' has not merely stood the test of time but has also had a profound and lasting effect on the profession.

The application of this methodological approach reached its apotheosis in what most academic economists would regard as Friedman's greatest work *A Theory of the Consumption Function* (1957). The fundamental proposition that emerged from Keynes's *General Theory* was that households expanded their consumption spending by an amount less than the increase in their current income, and that this relationship was sufficiently stable to form the basis for the multiplier through which an increase in autonomous expenditure at the macro level generated a considerably larger increase in real aggregate demand. Since the regularity and predictability of the consumption function was central for the Keynesian control of the economy, it was with trepidation that many observers found that there were considerable inconsistencies between the patterns of household behaviour, particularly from the cross section data of household surveys and the time series of the historical record. Certainly it appeared that the data were quite inconsistent with the Keynesian consumption function. Friedman showed that the Keynesian concept of household behaviour was fundamentally flawed, and that the statistical results suffered from the regression fallacy. People adjusted their consumption with respect to variations in their long term expected (or 'permanent') income, and paid little heed to transitory variations. This basis idea was not new – indeed it can be found in the 18th-century writings of Bernoulli – but Friedman's development showed his genius for simplicity and for the insights of thinking concretely.

But the main quality of *A Theory of the Consumption Function* was the incomparable amassing, organization and interpretation of the evidence. The relatively low propensities to consume evident in the cross section data were shown to be entirely consistent with the much higher propensities that emerged from analyses of aggregate time series, when both sets of figures were interpreted in the form of the permanent income hypothesis. Because of the transitory component in the cross section samples of households, the variance of measured income exceeded the variance of permanent income, and so the slope of the regression of consumer spending on income was much lower than in the aggregate time series regressions, where the transitory component was trivially small. The permanent income hypothesis adequately passed the acid test of using little to explain much.

The integrity of scholarship was demonstrated by the diligent search to find evidence that would discredit the permanent income hypothesis. It was not, and is not, normal practice to scour the literature and statistical evidence for material that might discredit a theory. But Friedman used the hypothesis in the most imaginative way to forecast, for example, the values of regression coefficients for different groups with varying fractions of transitory to permanent income. And he left instructions for other researchers to guide them in tests to be made with further analyses of different data. One of the great contributions of this book was to give a new standard for empirical economics generally. Clearly this was how it *should* be done. The second important effect was the introduction of the concept of permanent income into virtually every field of applied economics, such as monetary economics, housing, transportation and international trade. It was a new way of thinking about chance variations and people's decisions in the real world.

A particularly fruitful theoretical approach to the utility analysis of risk and the measurement of utility, based on the work of von Neumann and Morgenstern, appeared in two papers with L.J. Savage (1948, 1952). Using axioms which most observers would regard as acceptable and reasonable, these papers showed that choice under conditions of uncertainty could be represented as a simple process of maximizing expected utility. Thus the utilities of each of the chance outcomes were weighted by the probability of that outcome, and the sum gave an index of expected utility which, given the axioms, would be maximized by choosing from the alternative uncertain prospects. Again the basic idea was not new (it was developed originally by Bernoulli in solving the St Petersburg Paradox), but Friedman and Savage discovered new insights and implications with wide-ranging applications. Apart from rationalizing the widespread practice of simultaneously gambling and insuring, the hypothesis had a profound effect on the theory and practice of portfolio selection. For the pure economic theorist it offered the attractive proposition that, up to an arbitrary linear transformation of origin and scale, utility should be regarded as a cardinal magnitude.

Subsequent discussion (particularly by Maurice Allais) suggested that one of the axioms (the so-called 'strong independence axiom' which asserted that the preference order would not be affected by mixing these outcomes with equiprobable alternative outcomes) was clearly implausible and violated frequently in practical decisions. Research suggested also that in some fields, for example in air passenger insurance, the expected utility hypothesis was discredited. Nevertheless the hypothesis still forms a cornerstone of all work – and particularly practical work – in choice among risky alternatives. With some minor exceptions, these papers mark the last contributions of Friedman to the pure theory of statistics and decision making. Many statisticians regard the diversion of such a fertile mind from its natural field as a great shame and loss.

The gain to empirical economics – and during these years, particularly to the theory of price – was, one suspects, worth the loss. The reformulation of Marshallian demand theory, as a practical instrument of analysis (1949) was an exercise in meticulous scholarship in the history of thought but one which also argued for approaching demand analysis as a positive rather than a normative discipline, an approach which he attributed to Marshall. But the analysis of economic policy, and particularly a critique of the logical structure of the arguments and the empirical evidence adduced to support proposals on economic policy, became increasingly important. Thus the critique of the arguments showing the inferiority of excise taxes compared with alternative income taxes (1952) exposed basic methodological weaknesses in what were the standard treatments of the day.

The demonstration of the uses, as well as some abuses, of the theory of price was one of the highlights of Friedman's lectures, from 1946 to 1976 (with a gap from 1963 to 1973), at the graduate school of the University of Chicago. The exploitation of demand and supply as an 'engine of discovery' reached out well beyond those conventionally defined limits of the subject. In these lectures Friedman gave full rein to his persistence and determination fearlessly to pursue the argument, with subtlety and imagination, wherever it led. To the students it opened up new vistas – such as the theory of human capital – and exciting ways of unravelling puzzles and resolving problems. In his hands, economics had both power and point, reality and relevance (e.g. 1962). As distinct from

much economic work, where complicated ideas are developed in a simple way, Friedman showed how to interpret simple ideas in a most sophisticated way.

This characterized his work on money which, with the inauguration of his monetary workshop in 1951 began to be a major interest for Friedman himself and the distinguished students and faculty that he inspired. The motivations for studying money were firmly implanted when Friedman was at the Treasury dealing with wartime inflation management, but the immediate incentive was the request from the National Bureau to contribute a study on money for Wesley Mitchell's project on long-term business cycles. Monetary policy as a main tool of macroeconomic management was consistent with a wide degree of free unfettered enterprise and so had an obvious appeal to the liberal (which will be used here in the 19th-century sense) Friedman. The prevailing Keynesian orthodoxy, with its emphasis on expanding the public sector, appeared to threaten the liberal society. The post-Keynesian contempt for money was a tempting target difficult to resist. But undoubtedly Friedman's imagination had been challenged by the Chicago School's preference (particularly by Knight and Simons) for rules rather than authorities in macroeconomic as well as microeconomic policy. The uncertainties of the economic environment would be much reduced if the Federal Reserve Board followed simple rules. Friedman first suggested (1948) a countercyclical rule of financing recession-induced increases in the federal budget deficit by money creation and correspondingly by retiring money during a boom-induced surplus. The empirical evidence that he explored in subsequent years, however, led him to formulate the rule of a fixed and known expansion of the money stock, rather than indulging in countercyclical operations in vain attempts to stabilize the economy. Whatever his motives however (and one should note that motives are quite irrelevant in judging substantive propositions), for the next thirty years Friedman's work was focused on money. At last monetary economics was to be interpreted as part of the central corpus of price theory; it was to be integrated into economics.

THE MONETARY REVOLUTION AND THE RISE OF MONETARISM, 1956–1975

In the late 1950s, to anyone subjected to the Anglo-Saxon schools of economics during the previous two decades any attempt to revive monetary economics appeared to be foolhardy, like flogging a decomposing horse. The Radcliffe Committee, advised by the most eminent economists, had reported in 1959 that the quantity of money was of little or no interest since the velocity of circulation had no limits. The quantity theory of money was subject to particular scorn as a mere identity without content. As Friedman was to point out, however, all theory consists of tautologies; all that theory does is to rearrange the implications of the axioms to produce interesting, even surprising, consequences. But they remain empty and devoid of substantive as distinct from speculative content, until they have been tested against a wide body of facts.

Of course for many years the quantity theory of money had been tested against experience and data and over several critical periods of change. The most distinguished exponents of such tests had included Irving Fisher and Keynes himself, as well as the irrepressible Clark Warburton. Yet the methodology was murky, the statistics slim, and interrelationships between data and theory obscure. In *Studies in the Quantity Theory of Money* (1956), Friedman and his co-authors re-defined the quantity theory in terms of statements specifying a degree of stability in the *demand* for money. It was proposed that the demand for money by the individual household would be a stable function of its money income (later thought to be permanent income or wealth) and the cost of holding money represented by the rate of interest and the expected rate of inflation.

Friedman's presentation of the theory of the demand for money in the first essay in *Studies* is one of his most widely quoted papers, primarily because it is thought to show that in presenting the money demand function as a portfolio decision with respect to alternative assets, rather than a demand related to the flow of transactions and income, Friedman was a closet Keynesian. Substantively this was a side issue; the main point was the stability of demand, particularly with respect to nominal income or wealth. Unfortunately this first essay was not one of Friedman's better expositions. The other essays in *Studies*, particularly that of Cagan on hyperinflations and Selden on velocity, however, established the value of examining nominal income and inflation in the context of the demand for money. The quantity theory in its new reborn Chicago form had passed its first tests.

The unknowns, however, remained legion. The vexed question of the nature of the regime controlling the supply of money, and how to interpret the problem of identifying the demand function in the data were to persist, in the eyes of many critics, as the major weakness in such studies. Was the stock of money reacting passively to changes in nominal income (or wealth) or were prices and output responding to endogenous changes in the supply of money? The Chicago workshop averred that the answer to such questions could be obtained only by painstaking research into the history of the monetary process. Undoubtedly there were occasions when the money stock passively responded to changes in nominal income, but equally obvious were instances where the money supply changed for reasons quite independent of past or contemporaneous movements in money incomes. The role of the balance of payments and the exchange rate regime was clearly recognized, and it is not difficult to discover the genesis of the monetary theory of the balance of payments in 'Real and Pseudo Gold Standards' (1961) and other essays in *Dollars and Deficits* (1968).

Although the detailed development of the history of the money supply process and the relationships with gold and exchange rates were to appear in the monumental *A Monetary History of the United States, 1867–1960* (1963), Friedman had already made it perfectly clear that a stable growth of the money supply was unlikely to be feasible under a regime of fixed exchange rates. His advocacy of flexible exchange rates (in 1953) followed logically on his views of the efficacy of free markets. Friedman was one of the very few economists (Gottfried Haberler and Egon Sohmen were among them) who clearly showed that the ambient dollar shortage was merely a consequence of fixed exchange rates and divergent monetary policies. His analysis was amply justified when by the 1960s, due to the change in monetary policies, the dollar shortage had turned into a dollar glut.

Yet in spite of the increasing attention paid to the balance of payments and the money supply process generally, the prime focus of Friedman's work remained the examination of the effects of monetary variations on nominal income, prices and output. The main questions were: (a) what was the relative importance of monetary compared with fiscal variations (the Keynes vs Monetarist debate); (b) what was the time pattern of adjustment; and (c) could expansionary financial or fiscal policies affect real output in the short or long run? The answers which evolved from Friedman's were: to (a) although an increased fiscal deficit had an impact effect on nominal income this soon disappeared, whereas after a lag the increased rate of money growth permanently augmented the rate of inflation; to (b), the adjustment of nominal income to an increased rate of monetary growth involves lags that are 'long and variable'; and to (c), in the long run additional monetary growth affects only the rate of inflation and has virtually no effect on either the level of output or its growth rate. In essence Friedman found that variations in the rate of growth of the money supply had short run effects, sometimes as in 1931 of a devastating magnitude, on real output as well as on prices; but in the long run (more than three years) the only substantial effect was on prices.

Over the 1960s and 1970s the results of Friedman's research for the long run were widely accepted. The logic as well as the data were appealing: nominal variations (in money) have nominal effects (on prices) and no real effects (on output). But such agreement did not readily extend to his short run claims of first, the impotence of fiscal policy in countering cyclical oscillations and shocks; and secondly the large but unpredictable effects of monetary variation on real output and employment. The claims of Keynesian economists for the stability and size of the fiscal multipliers continued, but it is noteworthy that estimates of the size of the multipliers, except for those produced by the Cambridge (England) school, were substantially reduced in the 1980s. (One is not able to determine whether the economists or the economies have become less Keynesian and more monetarist.)

One of the abiding criticisms of Friedman's work on money (much of it in joint authorship with Anna Schwartz) is that it has no theoretical structure – or more charitably that such theoretical structure as exists is implicit rather than explicit. Processes of monetary transmissions as he describes them are alleged to be 'black boxes' with no precise specification of the way in which money works its magic. Friedman attempted to produce a theoretical underpinning for his approach to research in *Milton Friedman's Monetary Framework* (1974) by producing a seven equation basic model of the (closed) economy. The critical difference between the Keynesian and classical model was the choice of the last equation; the Keynesians chose to specify the price level as fixed by exogenous forces and the level of output as a variable determined by the level of aggregate demand, whereas the classical economists held that the level of real output was fixed by technology, skill, etc. and that the price level was determined by the model. With this simple model, Friedman was able to highlight the differences of method and approach as primarily different views about the size and stability of the coefficients of the system. In principle, at least, such issues could be resolved by appeals to the evidence. But in one respect the *Framework* made little progress – this was in providing a sound basis for the dynamics of the adjustment, through output, price and interest rate effects, to the new long-run equilibrium. The transmission mechanism and dynamics remain enshrouded in the gloom of a black box.

Yet in spite of what many theoretical economists considered to be drastic limitations for sound theoretical developments, in the most important and influential paper in macroeconomics in the post-war years, his presidential address to the American Economic Association, Friedman showed that the view of macroeconomic policy as a trade-off between unemployment and inflation was fundamentally flawed (1968). In the long run there was no such trade-off, while in the short run the trade-off took place only during the adjustment to the new inflationary environment, and then only because people were temporarily surprised by the new environment. The overriding objective of contractual arrangements was to fix *real* wages and prices. Money served as a veil, sometimes seductive but always obscuring underlying reality. The so-called Phillips Curve was a short term temptation rather than a long term choice.

Friedman caught opinion at ebb and turned it into a flood. Throughout the 1960s the trade-off between unemployment and inflation appeared more and more illusory. Unemployment went up but inflation did not go down; it also increased. Into the 1970s and particularly during the great inflationary recession of 1974/75, when both inflation and unemployment reached new highs in most OECD countries, it appeared that only Friedman's view made any sense. Like Keynes's *General Theory*, it was one of the very few contributions that changed both the approach of professional economists and the policies adopted by finance ministers. Sometime during the 1970s most governments recognized that the road to fuller employment did not lie over the high sierra of soaring inflation. Doctrinally, economists took into their toolbox the Friedman concept of a 'natural rate' of unemployment where inflation would neither accelerate nor decelerate. (The word 'natural' usually being considered either normative or even desirable was generally eschewed in favour of the term 'non-accelerating inflation rate of unemployment' or NAIRU).

The natural level of unemployment was held to be determined by the nature of labour markets, such as the conventions of wage contracts, the degree of mobility, the level of unemployment benefits, the marginal utility of income, and many other 'structural' factors which are independent of the rate of inflation. As in the case of the permanent income hypothesis, to which it is distantly related, the concept had applications in fields far from labour markets. At the same time it provided one of the many missing links between the macroeconomics of aggregate output and inflation and the microeconomics of industrial adjustment and resource allocation. Again in retrospect it all seems obvious; but that merely measures the magnitude of the contribution.

By any standards – even those of Keynes and the *General Theory* – Friedman's contribution to monetary analysis and policy must be ranked very high. Every economist, finance minister and banker felt his influence. But, as an accomplishment of the intellect, one suspects that most of Friedman's peers would still regard his work on the consumption function as the maximum maximorum of his contributions to economics. Friedman's monetary analysis did not have that sense of comprehensiveness and structural balance that are the hallmarks of his work on consumer spending. One closed *A Theory of the Consumption Function*, not with the feeling that nothing more need be said, but that whatever was discovered in the future must fit neatly into this superb and satisfying framework. The architecture could accommodate, and indeed so far has shaped and absorbed all new contributions. The *Monetary History* and the *Framework*, however, although probably more influential in doctrine and policy, did not provide the commodious and harmonic form of the *Consumption Function*. A number of awkward corners left one wondering what to do. And since the theoretical plans were left obscure, sometimes there were questions whether the superstructure would really hold up. But this does not belittle the *Monetary History* so much as praise the *Consumption Function*.

1975–1985

The award of the Nobel Prize for Economics, long overdue in 1977, at last recorded that Friedman's great contributions had

even penetrated the Swedish academies. Inevitably Friedman's rise to stardom had given many more opportunities to persuade electorates through the medium of the popular press (primarily in the columns of *Newsweek* from 1966 to 1984) and television (in the popular PBS and BBC series *Free to Choose* in 1980). His contributions to persuasive journalism delighted many, infuriated some, and made all his serious readers, if not wiser, then certainly better informed. In all these popular articles the high professional standards of integrity were maintained. But at the same time Friedman continued with his scholarly work on monetary analysis (again mainly with Anna Schwartz as co-worker). The main output, after more than 20 years of effort, was *Monetary Trends in the United States and the United Kingdom, Their Relation to Income, Prices and Interest Rates, 1867–1975* (1982).

The main methodological decision lying behind this study was that since there was too much inexplicable variation in short run variations in income and money, it was best to ignore these and concentrate on comparing the cyclical phase averages. These would screen out the short term effect and would enable an analysis to be made of the underlying long-term money–income–interest relationships. Even for this team of Friedman and Schwartz, the treatment of the data and the integrity of their analysis reached new heights of meticulous scholarship.

Yet, considering the enormous value of the input of time and energy, the results are, as the authors confess, hardly worth the cost. For the most part the study confirms, and demonstrates with comparative data for the USA and the UK, the basic propositions on velocity, real income, prices and interest rates that had emerged in the *History*.

In his scholarly work in the decade 1976–85, it may be claimed that Friedman has fallen prey to the same temptations that affected Alfred Marshall. For many years of his mature professional career, Marshall spent much of his time revising and refining his great *Principles*. In retrospect it seemed to be a great loss to scholarship that Marshall did not leave the *Principles* well alone and turn to his projected study of the economics of the state. The opportunity was missed. It would be, however, a travesty to draw a close parallel between Friedman and Marshall in their mature years. Perhaps with the example of Marshall in mind, Friedman has generally launched his studies on the profession and then left them largely to fend for themselves. (The only exception is the textbook *Price Theory: A Provisional Text* (1962), which was revised in 1976.) Yet there is a sense in which Friedman, trapped by his immense success in monetary economics, has been prevented from deploying his mind in scholarly work in other fields of economics.

The possibilities are revealed in Friedman's more popular writings on issues such as public spending, price and rent control, taxation, and many issues in microeconomics. Characteristic flashes of insight and phrase, together with the innovations of approach – especially the simplifications – give the professional reader a tantalizing taste of what might have been yet another great contribution to economic science. Many economists have always believed that, in spite of his great strides in money, Friedman's relative advantage was always in the study of price theory and its manifest applications. There is the measure of the man.

THE PUBLIC IMAGE OF FRIEDMAN

The conventional view of Friedman is that he has been one of the most ardent and most effective advocates of free enterprise and monetarist policies over the four decades 1945 to 1985. If far short of his wishes, the success of his advocacy has by any objective standard been enormous. Opinion in Western countries, even among the clerisy, has moved decisively in its preference for those economic freedoms that he has so eloquently advocated.

It is not possible to parcel out any neat attribution of influence on these great changes in attitude and policy. Friedman himself would probably give by far the largest weight to the experience of the 1970s, particularly the disappointments over failure to restrain the growth of government spending and the great inflation from 1965 to 1981. The explanation of these events and the development of an alternative strategy with institutions that would ensure individual economic liberty and freedom from inflation have been, in the public perception, Friedman's great contribution to the reforms. In his appearances in the various media he has been a great persuader, his role being critical in promoting such ideas as an all volunteer army, the voucher schemes for education and health, and indexing income tax. In effectiveness, breadth and scope, his only rival among the economists of the 20th century is Keynes.

ALAN WALTERS

See also MONETARISM; QUANTITY THEORY OF MONEY.

SELECTED WORKS

1934. Professor Pigou's method for measuring elasticities of demand from budgetary data. *Quarterly Journal of Economics* 1, November, 151–63.

1937. The use of ranks to avoid the assumption of normality implicit in the analysis of variance. *Journal of the American Statistical Association* 32, December, 675–701.

1940. A comparison of alternative tests of significance for the problem of m rankings. *Annals of Mathematical Statistics* 11, March, 86–92.

1945. (With Simon Kuznets.) *Income from Independent Professional Practice*. New York: National Bureau of Economic Research.

1948a. (With L.J. Savage.) The utility analysis of choices involving risk. *Journal of Political Economy* 56, August, 279–304.

1948b. (With H.A. Freeman, F. Mosteller, and W. Allen Wallis.) *Sampling Inspection*. New York: McGraw-Hill.

1949. The Marshallian demand curve. *Journal of Political Economy* 57, December, 463–95. Reprinted in (1953).

1952. (With L.J. Savage.) The expected utility hypothesis and the measurability of utility. *Journal of Political Economy* 60, December, 463–74.

1953. *Essays in Positive Economics*. Chicago: University of Chicago Press.

1956. (ed.) *Studies in the Quantity Theory of Money*. Chicago: University of Chicago Press.

1957a. (With Gary S. Becker.) A statistical illusion in judging Keynesian models. *Journal of Political Economy* 65, February, 64–75.

1957b. *A Theory of the Consumption Function*. Princeton: Princeton University Press.

1959. *A Program for Monetary Stability*. New York: Fordham University Press.

1962a. *Capitalism and Freedom*. Chicago: University of Chicago Press.

1962b. *Price Theory: A Provisional Text*. Chicago: Aldine Publishing Co.

1963a. (With Anna J. Schwartz.) *A Monetary History of the United States, 1867–1960*. Princeton: Princeton University Press for the National Bureau of Economic Research.

1963b. (With David Meiselman.) The relative stability of monetary velocity and the investment multiplier in the United States, 1897–1958. In *Stabilization Policies*, a series of studies prepared for the Commission on Money and Credit, Englewood Cliffs, NJ: Prentice-Hall, 165–268.

1967. (With Robert V. Roosa.) *The Balance of Payments: Free versus Fixed Exchange Rates*. AEI Rational Debate Seminar. Washington, DC: American Enterprise Institute.

1968a. The role of monetary policy. (Presidential Address, American Economic Association, 29 December 1967.) *American Economic Review* 58, March, 1–17. Reprinted in (1969).

1968b. *Dollars and Deficits: Inflation, Monetary Policy and the Balance of Payments*. Englewood Cliffs, NJ: Prentice-Hall.

1969. *The Optimum Quantity of Money and Other Essays*. Chicago: Aldine Publishing Co.

1970. (With Anna J. Schwartz.) *Monetary Statistics of the United States*. New York: Columbia University Press for the National Bureau of Economic Research.

1972a. (With Wilbur J. Cohen.) *Social Security: Universal or Selective?* AEI Rational Debate Seminar. Washington, DC: American Enterprise Institute.

1972b. *An Economist's Protest: Columns on Political Economy*. Glen Ridge, NJ: Thomas Horton & Daughters.

1973. *Money and Economic Development*. Horowitz Lectures of 1972. New York: Praeger.

1974. *Milton Friedman's Monetary Framework: A Debate with His Critics*. Edited and with an Introduction by Robert J. Gordon, Chicago: University of Chicago Press.

1976. *Price Theory*. Chicago: Aldine Publishing Co. Revised and enlarged version of the 1962 edition.

1978. *Tax Limitation, Inflation and the Role of Government*. Dallas: Fisher Institute.

1980. (With Rose Friedman.) *Free to Choose*. New York: Harcourt Brace Jovanovich.

1982. (With Anna J. Schwartz.) *Monetary Trends in the United States and the United Kingdom*. Chicago: University of Chicago Press, for the National Bureau of Economic Research.

1984. (With Rose Friedman.) *Tyranny of the Status Quo*. San Diego, New York, and London: Harcourt Brace Jovanovich.

Friend, Irwin (1915–1987). Friend was born in Schenectady, New York and received his PhD from American University in 1953. He then became a professor of finance and economics in the Wharton School, University of Pennsylvania, being president of the American Finance Association in 1972.

His many books and articles deal with securities markets, financial institutions, tax policy, capital asset pricing, consumption and saving functions, econometric models, and the usefulness of expectations and anticipations data. Perhaps Friend's most important contribution to economic theory *per se* is in 'The Demand for Risky Assets', (1975), the first part of which greatly extends the capital asset pricing model in several directions. These include the explanation of the determinants of the basic risk premium between risky assets as a whole and the riskfree rate, incorporating income taxes, allowing the riskfree asset to have a positive supply, and allowing for human capital. The paper demonstrates how the micro theory can be aggregated to obtain a macro model suitable for testing with the time series data, as is done in the latter part of the article.

It is generally recognized that theorists assume away many problems in order to highlight a central issue. The restrictions imposed by theorists often provide strong conclusions, which would not hold without the restrictions. Friend has often tested these restrictions and his results require changes in theory. For example, in the paper just cited, the question is posed of whether typical investors have increasing or proportional risk aversion. The paper presents fairly strong evidence that the appropriate utility function should have proportional risk aversion as part of its properties and provides measures of risk aversion for the market as a whole. Similarly, Friend pushed the permanent income hypothesis to its limits and helped refine it in 'Consumption Patterns and Permanent Income', showing it is not valid to restrict the marginal propensity to consume out of transitory income to be zero. Other areas in which his work has seriously questioned the usefulness of basic assumptions almost universally made by theorists include the supposed irrelevance of unique risks in the pricing of risky assets, the applicability of the customary factor analysis in confirming the arbitrage pricing theory, and the complete faith of many economists in the stock market's efficiency and the undesirability of any form of government intervention (including mandated disclosure).

PAUL TAUBMAN

SELECTED WORKS

1957. (With I. Kravis.) Consumption patterns and permanent income. *American Economic Review* 47(2), May, 536–55.

1964. (With P. Taubman.) A short-term forecasting model. *Review of Economics and Statistics* 46(3), August, 229–36.

1975. (With M.E. Blume.) The demand for risky assets. *American Economic Review* 65(5), December, 900–922.

1981. (With R. Westerfield.) Risk and capital asset prices. *Journal of Banking and Finance* 5(3), September, 291–315.

1983. (With J. Hasbrouck.) Saving and after-tax rates of return. *Review of Economics and Statistics* 65(4), November, 537–43.

1984. Economic and equity aspects of securities regulation. In *Management under Government Intervention: The View from Mt. Scopus*, ed. Lanzillotti and Peles, Greenwich, Conn.: JAI Press.

1985. (With P.J. Dhrymes, M.N. Gultekin and N.B. Gultekin.) New tests of the APT and their implications. *Journal of Finance* 40(3), July, 659–74.

Frisch, Ragnar Anton Kittel (1895–1973). Frisch lived a long, varied and extremely productive life. He graduated in Economics at the University of Oslo in 1919 (although as a son of a goldsmith he 'supplemented' this by finalizing his apprenticeship as a goldsmith in 1920!). He studied in France from 1921 to 1923 and in Britain in 1923; was an associate at the University of Oslo from 1925 and received his doctorate in 1926 in mathematical statistics (Frisch, 1926a). Further studies abroad in the USA, France and Italy (1927–8) were followed by an associate professorship at the University of Oslo (1928) and a full professorship in 1931. Frisch was head of the (newly established) Institute of Economics in Oslo from 1932 to his retirement in 1965. He was also chief editor of *Econometrica* (1933–55), followed by his chairmanship of the editorial board. He was one of the founders (1930) and, in fact, the driving force behind the creation of the Econometric Society. He was a member of a number of national and international expert committees and adviser on several occasions to developing countries (India 1954–5 and Egypt several times over the years 1957–64). He received honorary doctorates from a number of universities (*inter alia* Stockholm, Copenhagen, Cambridge, Birmingham) and was – together with Jan Tinbergen – the first (1969) to receive the Alfred Nobel Memorial Prize in Economics. In addition he received (as the first recipient) the Schumpeter Prize (1955), the Feltrinelli Prize (1956) and three *Festschriften*. He was a visiting professor or guest lecturer to a number of universities – Yale, Minnesota, Paris, Pittsburgh, for example – and he was a very active participant at numerous international meetings of economists, statisticians and mathematicians. In the late 1940s there was a joke among Norwegian students that he was also a 'visiting' professor in Oslo. This was unfair. In particular during the 1930s he put a lot of effort into his teaching and was writing a series of lecture notes, most of them seminal, though many remained unpublished. The impressive list of his publications (cf. Haavelmo, 1973) and activities could be continued because he was a genius, cutting

through problems like a warm knife through butter, and because his working power was extraordinary.

To survey his contributions is not easy for the simple reason that there is scarcely any area of economics Frisch has not been into and left his imprint. To cooperate with him was not always easy. He was too strong, as shown by the fact that he seldom had co-authors. The list of his published and printed works include about 160 items. But to this should be added a long series of mimeographed contributions – many will recall the 'Memoranda fra Socialkonomisk Institutt' – from about 1946 onwards. They amounted altogether to 6500 pages, and most of them are still waiting for publication (though Frisch himself argued that 'for editing it needs a very good man, and if he is good enough, he should write himself').

Frisch began his academic work in the theory of mathematical statistics. This profession today acknowledges his early contributions and regrets his departure from it, though admitting that in terms of the more applied theory of statistics he made noticeable contributions later on. His years in Paris, where he concentrated on mathematics, were not in vain.

It is, however, in economics that Frisch made his name, He was at most of the centres and many of the corners of the subject. One may, however, also argue that his most significant contribution is in economic methodology. This comes out not only in his applications of methods but also in their general presentation. A very good example was written overnight in a hotel room at Colmar, after a day's discussions at a meeting of the Econometric Society. The article (Frisch, 1936b) is a classic, clearing the ground about the very meaning of static versus dynamic analysis. This is by now elementary, but it is elementary because of Frisch. In his principal works on methodology, he used and unified the tools he had mastered so well: economic theory, mathematics and statistics. It is no accident that he invented the word econometrics, for in general he enriched our methodological vocabulary by a number of precise concepts: macro- versus micro-analysis, statics versus dynamics, exogenous versus endogenous variables, the concept of autonomous relations, the problem of identification of relations, confluent relations, decision models, conjectural behaviour (of firms) – a full list would be very long.

Few would hesitate to agree that Frisch 'created' econometrics in the modern sense of the word. It is much more notable that he warned again and again against misuses of the new tools. In the first issue of *Econometrica* he wrote: 'The policy of *Econometrica* will be as heartily to denounce futile playing with mathematical symbols in economics as to encourage their constructive use.' In Frisch (1970), he argued that 'the econometric army has now grown to such proportions that it cannot be beaten by the silly arguments that were used against us previously. This imposes on us a *social and scientific responsibility* of high order in the world of today' (p. 153). But in the very same article he also stressed (p. 163) that 'I have insisted that econometrics must have relevance to concrete realities – otherwise it degenerates into something which is not worthy of the name econometrics, but ought rather to be called playometrics.'

Always underlying Frisch's contributions to methodology were his consistent efforts to turn economics into a precise science, quantifying the variables and the structures. This is different from the traditional 'on the one hand and on the other', where on balance the answer is left in the air. But this 'aggressive' view also presents new challenges. The economist must be prepared for the troublesome work of gathering data, to face the difficulties in estimating structures and in the end to attempt a balanced interpretation of the outcome. Frisch saw this and contributed to this debate throughout his career; illustrations might be Frisch (1933a, 1934b, 1936a and 1939). These and many other contributions had a profound influence and wide applications in pre-war as well as post-war econometrics. Again, and sadly enough, one could also refer to a number of unpublished papers, though these were influential as contributions to scientific gatherings. A supreme example is a paper on *Statistical versus Theoretical Relations in Macrodynamics*, a contribution to a conference sponsored by the League of Nations in 1938 to discuss Jan Tinbergen's work for the League of Nations on the trade cycle.

One may wonder why Frisch left so many pathbreaking contributions without taking the trouble to publish them. I think there is a double answer. On the one hand Frisch was an impatient man: if he had given the gist of the solution to a problem, he tended to go on to new problems. On the other hand, he was extremely careful: a publication going to the printer had to be perfect and finalized – a troublesome process which he often tended to avoid. Haavelmo (1973), reports that Frisch often argued that proofreading is one of the most difficult and important tasks of a scientist.

The general assessment above can be verified by considering Frisch's contributions in the field of demand analysis, the theory of production and the theory of macroeconomics.

In demand analysis he began as early as 1926 (cf. Frisch, 1926b), by formulating a number of basic axioms and from these to deduce a theory of demand. Thus utility functions were not postulated but were derived from more basic axioms, all of these being formulated as being, in principle, open to testing. It may be fair to say that his work in this field culminated in Frisch (1959). It is a tribute to his work that it has in fact been used in practice, for example in Norwegian planning.

In the theory of production Frisch was a forerunner, formulating the theory in a strict mathematical form but also applying it on concrete problems. An example is Frisch (1935). However, most of his works were in the form of mimeographed lecture notes in the 1930s and remained unpublished until 1962 and later (cf. Frisch, 1962, 1963). The main results, however, were internationally known through the works of Schneider and Carlson, who at times were research associates in Oslo and very much influenced by Frisch.

Also in the theory of macroeconomics, Frisch was at the front, even, it can be argued, ahead of Keynes. Anybody reading his booklet, Frisch (1933b), and the subsequent articles in *Econometrica* (1934), might be willing to argue that Frisch made it first. He shows convincingly, how a capitalist economy may go into a deadlock when, to put it in a simple way, the tailor cannot sell to the shoemaker because the shoemaker cannot sell to the tailor:

> ... the cause of great depressions, such as the one we are actually in, is ... fundamentally connected with the fact that modern economic life has been divided into a number of regions or groups.
>
> Under the present system, the blind 'economic laws' will under certain circumstances, create a situation where these groups are forced mutually to undermine each other's position. Each group is forced to curtail the use goods produced and services rendered by the other groups, which, in turn, will cause a still further contraction of the demand for its own products, and so on (Frisch, 1934a, pp. 259f).

He also, in the 1934 articles, outlined (a couple of years before Leontief) an input–output analysis. His contributions in these areas were not appreciated at the time, but in a historical perspective they are pathbreaking. This also holds for his

contribution to the (famous) Cassel *Festschrift*, (Frisch, 1933c), where a dynamic system for the economy as a whole was outlined and where he in a sharp and fruitful way distinguished between the impulses and the propagation mechanism. In this context one may also, as an illustration of his interest in the development of economic theory, make a reference to his excellent analysis of Marshall (Frisch, 1950).

There is a direct line from here to his systems of national accounts (first published in Frisch, 1939, and, later, Frisch, 1973) which had a profound influence on the planning in Norway and elsewhere after the Liberation. In the context of macroeconomics it is illustrative to mention Frisch's discussion with J.M. Clark over the acceleration principle (Frisch, 1931, 1932). In an amazingly simple way Frisch cleared up the issue, that is, the interplay between the pure acceleration principle and the reinvestment cycle, as the following quotation shows:

'Let z be consumer-taking [in present day language this is simply consumption] per unit of time, w capital production per unit of time, and W the capital stock that exists at any moment of time. All the three magnitudes z, w, and W are, of course, functions of time. In practice they would be represented by time series.

Let us, for simplicity, make the following two assumptions: A. Consumer-taking z is the same as the production of the consumer good, and this again is at any time proportional to the existing capital stock W. In other words, we have

$$W = kz, \qquad (1)$$

where k is a constant independent of time. B. The depreciation per unit of time u, that is to say, the capital production that is needed for replacement purposes, is proportional to the existing capital stock. In other words, we have

$$u = hW, \qquad (2)$$

where h is a constant independent of time. Now, the rate of change with respect to time of the capital stock is equal to

$$\dot{W} = w - u. \qquad (3)$$

By virtue of (1) we have, however,

$$\dot{W} = k\dot{z}, \qquad (4)$$

where $\dot{z}$ is the rate of change of consumer-taking. Inserting this into (3), and expressing u in terms of z by (2) and (1), we get

$$k\dot{z} = w - khz.$$

So that we finally have

$$w = k(hz + \dot{z}). \qquad (5)$$

The rate of change with respect to time of capital production is thus equal to

$$\dot{w} = k(h\dot{z} + \ddot{z}). \qquad (6)$$

Formula (5) indicates the two parts of which total capital production is made up. In the first place we have the part khz that represents capital production for replacement purposes. This part is (under our simplified assumption) proportional to the *size* of consumer-taking. In the second place, we have the part $k\dot{z}$ representing capital production for expansion purposes. This part is (under the present simplified assumption) proportional to the *rate of change* of consumer-taking. Thus there are two forces that act upon total capital production. If consumer-taking is increasing, but at a constantly decreasing rate, the first of these two forces tends to

429

increase, and the second tends to slow down capital production. Which one of the two forces shall have the upper hand depends on the *manner* in which the increase in consumer-taking slows down, and it depends also on *the rate of depreciation*' (Frisch, 1931, pp. 647f.)

As will be seen it is all so simple, *provided* the problem is *formulated* clearly. And formulating problems in a fruitful way was one of his secrets.

What is a genius? It might be argued that Frisch up till now was one of the ten in our profession in this century. Not that he cannot be criticized. On occasion he used his brains more or less in vain, for example, on unimportant calculating schemes. Long after electronic calculators were on the market, he used time and effort on inventing schemes for inverting a matrix on a simple desk calculator. His various methods for the solution of programming problems – 'the logarithmic potential method', 'the multiplex method' and 'the nonplex method' (e.g. Frisch, 1956, 1957, 1961a and 1961b) – are still disputable, taking present-day techniques into account. In other words, he might have had a weak point in not always being able to evaluate the importance of a problem, that is, he might now and then have used his immense working power on issues where his opportunity costs were too high.

Even so his life work is impressive. And so was the man himself. His political attitude was rather to the left than to the right – while at the same time he was a devout Christian. He felt a strong social responsibility, as proved through his work on the problems of the 1930s as well as, and perhaps even more so, by his consciousness towards the less developed countries. He could at times be a bit harsh on colleagues who did not live up to his own standards for serious work. At the same time he was extremely kind and helpful to students doing their best. He never failed to encourage. And few will forget when his strong blue eyes were shining with joy.

P. NØRREGAARD RASMUSSEN

SELECTED WORKS

1926a. Sur les semi-invariants et moments employés dans l'étude des distributions statistiques. *Skrifter utgitt av det Norske Videnskaps-Akademi i Oslo II. Hist.-Filos. Klasse 1926*, No. 3.

1926b. Sur un problème d'économie pure. (An attempt at developing an axiomatic foundation of utility as a quantitative notion and at measuring statistically the variation in the marginal utility of money on the basis of data from the Paris Cooperative Society.) *Norsk Matematisk Forenings Skrifter* Series I, No. 16, 1–40.

1931, 1932. The Interrelation between capital production and consumer taking. *Journal of Political Economy* 39(5), October, 1931, 646–54. A Rejoinder, 40(2), April 1932, 253–5. A final word, 40(5), October 1932, 694.

1933a. *Pitfalls in the Statistical Construction of Demand and Supply Curves.* Leipzig: H. Buske.

1933b. *Sparing og Cirkulasjonsregulering* Oslo.

1933c. Propagation problems and impulse problems in dynamic economics. *Economic Essays in Honor of Gustav Cassel.* London: Allen & Unwin. Reprinted in *Readings in Business Cycles*, ed. R.A. Gordon and L.R. Klein, London: Allen & Unwin, 1966.

1934a. Circulation planning. *Econometrica*, Pt I, 2(3), July, 258–336; Pt II, 2(4), October, 422–35.

1934b. *Statistical Confluence Analysis by Means of Complete Regression Systems.* Oslo: University Institute of Economics.

1935. The principle of substitution. An example of its application in the chocolate industry. *Nordisk Tidsskrift for Teknisk Økonomi*, September, 12–27.

1936a. Annual survey of general economic theory: the problem of index numbers. *Econometrica* 4(1), January, 1–38.

1936b. On the notion of equilibrium and disequilibrium. *Review of Economic Studies* 3(2), February, 100–105.

1939. Nasjonalregnskapet. *Transactions of the Nordisk Statistiker Møte*, Oslo.

1943. Økosirk-systemet. 1943. *Ekonomisk Tidskrift* 45, 106–21.

1950. Alfred Marshall's theory of value. *Quarterly Journal of Economics* 64, November, 495–524.

1952. Frisch on Wicksell. In *The Development of Economic Thought*, ed. H.W. Spiegel, New York: John Wiley & Sons.

1956. La résolution des problèmes de programme linéaire par la méthode du potential logarithmique. In *Séminaire économétrique*, ed. René Roy, Paris: CNRS.

1957. The multiplex method for linear programming. *Sankhya*, 18(3–4), September, 329–60.

1959. A complete scheme for computing all direct and cross demand elasticities in a model with many sectors. *Econometrica* 27(2), April, 177–96.

1961a. Mixed linear and quadratic programming by the multiplex method. In *Money, Growth and Methodology, Festskrift til Johan Åkerman*, ed. H. Hegeland, Lund: Gleerup.

1961b. Quadratic programming by the multiplex method in the general cases where the quadratic form may be singular. *Bulletin of the International Statistical Institute* 38(4) (Tokyo), 283–332.

1962. *Innledning til Produksjonsteorien* . Oslo: Universitets Forlaget.

1963. *Lois techniques et économiques de la production.* Paris: Dunod. Trans. as *Theory of Production*, Dordrecht: D. Reidel; Chicago: Rand McNally, 1965.

1970. Econometrics in the world of today. In *Induction, Growth and Trade; Essays in Honour of Sir Roy Harrod*, ed. W.A. Eltis, M.F. Scott and J.N. Wolfe. London: Clarendon Press.

BIBLIOGRAPHY

Arrow, K.J. 1960. The work of Ragnar Frisch, econometrician. *Econometrica* 28, April, 175–92.

Edvardsen, K. 1970. A Survey of Ragnar Frisch's contributions to the science of economics. *Economist* 118(2), March–April, 174–96.

Haavelmo, T. 1973. Minnetale over Professor, dr. philos. Ragnar Frisch. *Årbok det Norske Videnskabs-Akademi i Oslo.*

Johansen, L. 1969. Ragnar Frisch's contributions to economics. *Swedish Journal of Economics* 71(4), December, 302–24. All of these surveys contain extensive bibliographic references, though Haavelmo (1973) gives the most extensive, covering 159 printed works.

Frobenius theorem. *See* PERRON–FROBENIUS THEOREM.

full and limited information methods. Econometricians have developed a number of alternative methods for estimating parameters and testing hypotheses in simultaneous equations models. Some of these are limited information methods that can be applied one equation at a time and require only minimal specification of the other equations in the system. In contrast, the full information methods treat the system as a whole and require a complete specification of all the equations.

The distinction between limited and full information methods is, in part, simply one of statistical efficiency. As is generally true in inference problems, the more that is known about the phenomena being studied, the more precisely the unknown parameters can be estimated with the available data. In an interdependent system of equations, information about the variables appearing in one equation can be used to get better estimates of the coefficients in other equations. Of course, there is a trade-off: full information methods are more efficient, but they are also more sensitive to specification error and more difficult to compute.

Statistical considerations are not, however, the only reason for distinguishing between limited and full information approaches. Models of the world do not come off the shelf. In any application, the choice of which variables to view as endogenous (i.e. explained by the model) and which to view as exogenous (explained outside the model) is up to the analyst. The interpretations given to the equations of the model and the specification of the functional forms are subject to considerable discretion. The limited information and full information distinction can be viewed not simply as one of statistical efficiency but one of modelling strategy.

The simultaneous equations model can be applied to a variety of economic situations. In each case, structural equations are interpreted in light of some hypothetical experiment that is postulated. In considering the logic of econometric model building and inference, it is useful to distinguish between two general classes of applications. On the one hand, there are applications where the basic economic question involves a single hypothetical experiment and the problem is to draw inferences about the parameters of a single autonomous structural equation. Other relationships are considered only as a means for learning about the given equation. On the other hand, there are applications where the basic economic question being asked involves in an essential way an interdependent system of experiments. The goal of the analysis is to understand the interaction of a set of autonomous equations.

An example may clarify the distinction. Consider the standard competitive supply-demand model where price and quantity traded are determined by the interaction of consumer and producer behaviour. One can easily imagine situations where consumers are perfectly-competitive price takers and it would be useful to know the price elasticity of market demand. One might be tempted to use time-series data and regress quantity purchased on price (including perhaps other demand determinants like income and prices of substitutes as additional explanatory variables) and to interpret the estimated equation as a demand function. If it could plausibly be assumed that the omitted demand determinants constituting the error term were uncorrelated over the sample period with each of the included regressors, this interpretation might be justified. If, however, periods where the omitted factors lead to high demand are also the periods where price is high, then there will be simultaneous equations bias. In order to decide whether or not the regression of quantity on price will produce satisfactory estimates of the demand function, the mechanism determining movements in price must be examined. Even though our interest is in the behaviour of consumers, we must consider other agents who influence price. In this case a model of producer behaviour is needed.

This example captures the essence of many econometric problems: we want to learn about a relationship defined in terms of a hypothetical isolated experiment but the data we have available were in fact generated from a more complex experiment. We are not particularly interested in studying the process that actually generated the data, except in so far it helps us to learn about the process we *wish* had generated the data. A simultaneous equations model is postulated simply to help us estimate a single equation of interest.

Some economic problems, however, are of a different sort. Again in the supply-demand set-up, suppose we are interested in learning how a sales tax will affect market price. If tax rates had varied over our sample period, a regression of market price on tax rate might be informative. If, however, there had been little or no tax rate variation, such a regression would be useless. But, in a correctly specified model, the effects of taxes can be deduced from knowledge of the structure of consumer and producer decision making in the absence of taxes. Under competition, for example, one needs only to know the slopes of the demand and supply curves. Thus, in order to predict the effect of a sales tax, one might wish to estimate the system of structural equations describing market equilibrium.

The distinction between these two situations can be summarized as follows: in the one case we are interested in a

structural equation for its own sake; in the other case our interest is in the reduced-form of an interdependent system. If our concern is with a single equation, we might prefer to make few assumptions about the rest of the system and to estimate the needed parameters using limited information methods. If our concern is with improved reduced-form estimates, full-information approaches are natural since specification of the entire system is necessary in any case. A further discussion of these methodological issues can be found in Hood and Koopmans (1953, chs 1 and 6.)

LIMITED INFORMATION METHODS. Consider a single structural equation represented by

$$y = Z\alpha + u \tag{1}$$

where y is a T-dimensional (column) vector of observations on an endogenous variable, Z is a $T \times n$ matrix of observations on n explanatory variables, α is an n-dimensional parameter vector, and u is a T-dimensional vector of random errors. The components of α are given a causal interpretation in terms of some hypothetical experiment suggested by economic theory. For example, the first component might represent the effect on the outcome of the experiment of a unit change in one of the conditions, other things held constant. In our sample, however, other conditions varied across the T observation. The errors represent those conditions which are not accounted for by the explanatory variables and are assumed to have zero mean.

The key assumption underlying limited-information methods of inference is that we have data on K predetermined variables that are unrelated to the errors. That is, the error term for observation t is uncorrelated with each of the predetermined variables for that observation. The $T \times K$ matrix of observations on the predetermined variables is assumed to have rank K and is denoted by X. By assumption, then, $E(X'u)$ is the zero vector. Some of the explanatory variables may be predetermined and hence some columns of Z are also columns of X. The remaining explanatory variables are thought to be correlated with the error term and are considered as endogenous. Implicitly, equation (1) is viewed as part of a larger system explaining all the endogenous variables. The predetermined variables appearing in X but not in Z are assumed to be explanatory variables in some other structural equation. Exact specification of these other equations is not needed for limited information analysis.

In most approaches to estimating α it is assumed that nothing is known about the degree of correlation between u and the endogenous components of Z. Instead, the analysis exploits the zero correlation between u and X. The simplest approach is the method of moments. Since $X'u$ has mean zero, a natural estimate of α is that vector a satisfying the vector equation $X'(y - Za) = 0$. This is a system of K linear equations in n unknowns. If K is less than n, the estimation method fails. If K equals n, the estimate is given by $(X'Z)^{-1}X'y$, as long as the inverse exists. The approach is often referred to as the method of instrumental variables and the columns of X are called instruments.

If K is greater than n, any n independent linear combinations of the columns of X can be used as instruments. For example, for any $n \times K$ matrix D, α can be estimated by

$$(D'X'Z)^{-1}D'X'y, \tag{2}$$

as long as the inverse exists. Often D is chosen to be a selection matrix with each row containing zeros except for one unit element; that is, n out of the K predetermined variables are selected as instruments and the others are discarded. If Z

contains no endogenous variables, it is a submatrix of X; least squares can then be interpreted as instrumental variables using the regressors as instruments.

The estimator (2) will have good sampling properties if the instruments are not only uncorrelated with the errors but also highly correlated with the explanatory variables. To maximize that correlation, a natural choice for D is the coefficient matrix from a linear regression of Z on X. The instruments are then the predicted values from that regression. These predicted values (or projections) can be written as NZ where N is the idempotent projection matrix $X(X'X)^{-1}X'$; the estimator becomes

$$(Z'NZ)^{-1}Z'Ny \qquad (2')$$

Because $N = NN$, the estimator (2) can be obtained by simply regressing y on the predicted values NZ. Hence, this particular instrumental variables estimator is commonly called *two-stage least squares*.

The two-stage least-squares estimator is readily seen to be the solution of the minimization problem

$$\min(y - Za)'N(y - Za). \qquad (3)$$

As an alternative, it has been proposed to minimize the ratio

$$\frac{(y - Za)'N(y - Za)}{(y - Za)'M(y - Za)} \qquad (4)$$

where $M = I - N$ is also an idempotent projection matrix. This yields the *limited-information maximum-likelihood* estimator. That is, if the endogenous variables are assumed to be multivariate normal and independent from observation to observation, and if no variables are excluded a priori from the other equations in the system, maximization of the likelihood function is equivalent to minimizing the ratio (4). This maximum likelihood estimate is also an instrumental variable estimate of the form (2). Indeed, the matrix D turns out to be the maximum likelihood estimate of the population regression coefficients relating Z and X. Thus the solutions of (3) and (4) are both instrumental variable estimates. They differ only in how the reduced-form regression coefficients used for D are estimated.

The sampling distribution of the instrumental variable estimator depends, of course, on the choice of D. The endogenous variables in Z are necessarily random. Hence, the estimator behaves like the ratio of random variables; its moments and exact sampling distribution are difficult to derive even under the assumption of normality. However, large-sample approximations have been developed. The two-stage least-squares estimate and the limited information maximum-likelihood estimate have, to a first order of approximation, the same large-sample probability distribution. To that order of approximation, they are optimal in the sense that any other instrumental variable estimators based on X have asymptotic variances at least as large. The asymptotic approximations tend to be reasonably good when T is large compared with K. When $K - n$ is large, instrumental variable estimates using a subset of the columns of X often outperform two-stage least squares. Further small-sample results are discussed by Fuller (1977).

FULL INFORMATION METHODS. Although limited-information methods like two-stage least squares can be applied to each equation of a simultaneous system, better results can usually be obtained by taking into account the other equations. Suppose the system consists of G linear structural equations in G endogenous variables. These equations contain K distinct predetermined variables which may be exogenous or values of endogenous variables at a previous time period. The crucial

assumption is that each predetermined variable is uncorrelated with each structural error for the same observation.

Let $y_1, \ldots, y_G$ be T-dimensional column vectors of observations on the G endogenous variables. As before, the $T \times K$ matrix of observations on the predetermined variables is denoted by X and assumed to have rank K. The system is written as

$$y_i = Z_i \alpha_i + u_i \qquad (i = 1, \ldots, G) \qquad (5)$$

where Z_i is the $T \times n_i$ matrix of observations on the explanatory variables, u_i is the error vector, and α_i is the parameter vector for equaion i. Some of the columns of Z_i are columns of X; the others are endogenous variables.

Again, estimates can be based on the method of moments. Consider the set of GK equations

$$X'(y_i - Z_i a_i) = 0 \qquad (i = 1, \ldots, G) \qquad (6)$$

If, for any i, K is less than n_i, the corresponding parameter α_i cannot be estimated; we shall suppose that any equation for which this is true has already been deleted from the system so that G is the number of equations whose parameters are estimable. If $n_i = K$ for all i, the solution to (6) is obtained by using limited information instrumental variables on each equation separately. If, for some i, $n_i < K$, the system (6) has more equations than unknowns. Again, linear combinations of the predetermined variables can be used as instruments. The optimal selection of weights, however, is more complicated than in the limited-information case and depends on the pattern of correlation among the structural errors.

If the structural errors are independent from observation to observation but are correlated across equations, we have the specification

$$E(u_i u_j') = \sigma_{ij} I \qquad (i, j = 1, \ldots, G)$$

where the σ's are error covariances and I is a T-dimensional identity matrix. As a generalization of (3), consider the minimization problem

$$\min \sum_i \sum_j (y_i - Z_i a_i)'N(y_j - Z_j a_j)\sigma^{ij} \qquad (7)$$

where the σ^{ij} are elements of the inverse of the matrix $[\sigma_{ij}]$. For given σ's, the first-order conditions are

$$\sum_j Z_i'N(y_j - Z_j a_j)\sigma^{ij} = 0 \qquad (i = 1, \ldots, G) \qquad (8)$$

which are linear combinations of the equations in (6). It can be demonstrated that the solution to (8) is an instrumental variables estimator with asymptotically optimal weights. In practice, the σ's are unknown but can be estimated from the residuals of some preliminary fit. This approach to estimating the α's is called *three-stage least squares* since it involves least-squares calculations at three stages, first to obtain the projections NZ_j, again to obtain two-stage least-squares estimates of the σ's, and finally to solve the minimization problem (7). For details, see Zellner and Theil (1962).

If the structural errors are assumed to be normal, the likelihood function for the complete simultaneous equations system has a relatively simple expression in terms of the reduced-form parameters. However, since the reduced form is nonlinear in the structural parameters, analytic methods for maximizing the likelihood function are not available and iterative techniques are used instead. Just as in the limited-information case, the maximum-likelihood estimator can be interpreted as an instrumental variables estimator. If in (8) the least-squares predicted values NZ_i are replaced by maximum-likelihood predictions and if the σ's are replaced by their maximum-likelihood estimates, the resulting solution is the (full-information) maximum-likelihood estimate of the α's. See Malinvaud (1970, ch. 19) for details.

At one time full-information methods (particularly those using maximum likelihood) were computationally very burdensome. Computer software was almost non-existent, rounding error was hard to control, and computer time was very expensive. Many econometric procedures became popular simply because they avoided these difficulties. Current computer technology is such that computational burden is no longer a practical constraint, at least for moderate-sized models. The more important constraints at the moment are the limited sample sizes compared with the number of parameters to be estimated and limited confidence we have in the orthogonality conditions that must be imposed to get any estimates at all.

THOMAS J. ROTHENBERG

See also ECONOMETRICS; ESTIMATION; IDENTIFICATION; INSTRUMENTAL VARIABLES; MACROECONOMETRIC MODELS; SIMULTANEOUS EQUATIONS MODELS; TWO-STAGE LEAST SQUARES.

BIBLIOGRAPHY

Fuller, W. 1977. Some properties of a modification of the limited information estimator. *Econometrica* 45, May, 939–53.

Hood, W. and Koopmans, T. (eds) 1953. *Studies in Econometric Method.* Cowles Foundation Monograph No. 14, New York: Wiley.

Malinvaud, E. 1970. *Statistical Methods of Econometrics.* 2nd edn, Amsterdam: North-Holland.

Zellner, A. and Theil, H. 1962. Three-stage least squares: simultaneous estimation of simultaneous equations. *Econometrica* 30, January, 54–78.

Fullarton, John (1780?–1849). John Fullarton shared at least one characteristic with his great predecessor, Ricardo: he also seemed, in the words of Lord Brougham, 'as if he had dropped from another planet'. Although Fullarton is described in the *Dictionary of National Biography* as a 'traveller and writer on the currency', travel occupied by far the greater proportion of his life, along with a keen interest in the world of art and literature. Yet the single published work on which his considerable reputation as an economist is based had an impact comparable with that of Ricardo's intervention in the Bullion Controversy at the turn of the century.

In his early twenties, Fullarton became a surgeon in India and found time to edit a Calcutta newspaper. There he subsequently made a fortune in banking and began the first of his extensive tours through 'our eastern possessions', as the *Dictionary of National Biography* endearingly calls them. On this tour, Fullarton collected vast amounts of information and made many notes of his observations, but these were never published. In 1823, having returned to England to live, he contributed articles to the *Quarterly Review* on the reform crisis; however, it was not long before he resumed his travels, this time around Britain and the continent in a coach specially fitted with a library. In 1833, as a Fellow of the Royal Asiatic Society, Fullarton went again to India, and, in the following year, to China; but his zeal evaporated along with his fortune as a result of the failure of his bankers, and he moved back to London permanently.

It was in 1844, during the passage of the Bank Charter Act through the House of Commons, that Fullarton published his major work, *On the Regulation of Currencies*, subtitled 'an examination of the principles on which it is proposed to restrict, within certain fixed limits, the future issues on credit of the Bank of England, and of the other banking establishments throughout the country'. It was immediately hailed as a formidable challenge to the Currency School orthodoxy, whose support for the Bank Charter Act had overwhelmed Tooke's lonely opposition in the opening round of the 'currency–banking debate'. Indeed, according to Gregory, Fullarton's 'penetrating tract' was 'perhaps the most subtle and able production emanating from the Banking School' (introduction to Tooke, 1838/57, p. 81).

Fullarton's aim was a simple one, to bolster Tooke's case against what they both saw as ill-conceived banking legislation; in doing so, however, he not only improved its presentation, but also developed the theoretical basis of the argument in a number of important respects, taking the opportunity to lament the fact that 'Mr. Tooke himself has been exceedingly slow in following out his original conclusions on the subject of price to all their consequences' (1844, p. 18).

The Currency School had asserted that convertibility would not be a sufficient safeguard against the overissue of bank notes and their consequent depreciation; and that the quantity of notes in circulation would have to be regulated in accordance with the movement of bullion across the foreign exchanges. The response of Fullarton and the Banking School took three main lines. First, starting from the assumption that legal convertibility necessarily implied economic convertibility, they pointed out that any discrepancy between the note issue and a purely metallic system arose from the Currency School's erroneous theory of metallic circulation rather than from the supposed autonomy of the notes. Second, any effect on prices attributed to bank notes could not be denied to a range of financial assets excluded by the Currency School from their definition of money. Third, bank notes were in any case not money but credit, and therefore never could be overissued, though the credit structure as a whole might be extended beyond the limits of real accumulation by speculation. It was in this context that Fullarton developed the famous 'law of reflux', which he called 'the great regulating principle of the internal currency' (1844, p. 68).

Tooke, in turn, warmly welcomed Fullarton's analysis in the subsequent volume of his massive *History of Prices*, and gave some indication of the surprise he must have experienced upon its publication:

[L]est his estimate of the value of my contributions to an extension of the knowledge of this subject, should be ascribed to the bias of friendship, I think it right to state that the distinguished author was unknown to me, except by name and reputation, till after the publication of his treatise, and that I had not the slightest knowledge of such a work being in preparation (1838/57, IV, pp. x–xi).

Tooke then paid Fullarton the compliment of quoting extensively from his work, repeatedly praising the 'wonderful clearness and vigour which distinguish his writings' (V, p. 537). Nor was Fullarton above self-promotion: it appears that he had a hand in a *Quarterly Review* article, 'The Financial Pressure', which saw the crisis of 1847 as confirming the warnings of 'Mr Fullarton's masterly treatise' (see Fetter, 1965, p. 212).

It is certainly true that Fullarton's work 'enjoyed, in England and on the Continent, a persistent success such as few contributions to an ephemeral controversy have ever enjoyed' (Schumpeter, 1954, p. 725). Marx, for example, included Fullarton among 'the best writers on money' ([1867], p. 129);

in his view, 'the economic literature worth mentioning since 1830 resolves itself mainly into a literature on currency, credit, and crises' ([1894] 1971, pp. 492–3). Hilferding, too, drew heavily on Fullarton (Hilferding, 1910); and even Keynes was impressed with his 'most interesting' contribution to monetary thought (Keynes, 1936, p. 364 n.). Many of Fullarton's arguments later resurfaced in the Radcliffe Report of 1959, and are still today being 'rediscovered'. As Fullarton himself pointed out (1844, p. 5), 'this is a subject on which there never can be any efficient or immediate appeal to the public at large. It is a subject on which the progress of opinion always has been, and always must be, exceedingly slow'.

ROY GREEN

See also BANKING SCHOOL, CURRENCY SCHOOL AND FREE BANKING SCHOOL.

SELECTED WORKS
1844. *On the Regulation of Currencies.* London: John Murray.

BIBLIOGRAPHY
Fetter, F.W. 1965. *Development of British Monetary Orthodoxy, 1797–1875.* Reprinted, Fairfield: Kelley, 1978.
Hilferding, R. 1910. *Finance Capital: A Study of the Latest Phase of Capitalist Development.* London, Routledge & Kegan Paul, 1981.
Keynes, J.M. 1936. *The General Theory of Employment, Interest and Money.* London: Macmillan.
Marx, K. 1867. *Capital,* Vol. I. Moscow: Progress Publishers, n.d.
Marx, K. 1894. *Capital,* Vol. III. Moscow: Progress Publishers, 1971.
Radcliffe (Lord), (Chairman). 1959. *Committee on the Working of the Monetary System.* London: HMSO.
Schumpeter, J. 1954. *History of Economic Analysis.* London: George Allen & Unwin, 1955.
Tooke, T. 1838–57. *History of Prices, and of the state of the circulation from 1792 to 1856.* London, P.S. King & Son, 1928.

full communism. In Marx and Marxism, Full Communism is that final state of humanity in which productivity is higher than wants and everyone can help himself in the warehouses (not shops!). Since productivity cannot be unlimited, this entails that wants are limited: a direct contradiction to one of the basic propositions of Western economics. This is only possible because *wants have been reduced to needs.* Originally a governmental concept, needs are accepted as valid by each consumer, and internalized to become the new wants.

If wants are to fall below productivity, people must work seriously but voluntarily, that is *work too must become a need and so again a want.* The link between labour and reward is cut, so that everyone gets a 'dividend' and no one gets a wage, however much or little, well or ill, he or she works – and never mind at what job. Moreover that dividend must in total quantity correspond to the individual's consumption needs, so it is *nearly equal* for all people.

Since people would be 'well brought up', they would not help themselves to more than their 'need dividend' should they have the opportunity – for example, in the common mess hall or at the clothing warehouse. In more moderate versions large durables and housing are not offered in profusion without control, but *rationed.* However, the basic principle is not to ration, but to issue on demand, to a body of consumers too idealistic to 'break the bank'. Either way, *no money* is used inside the community. Moreover in the extreme version *nothing is scarce.* The lack of scarcity *removes the optimal allocation problem,* and causes the end of economics (if we accept that definition of it) as an intellectual subject.

Though allocations need no longer be optimal they must still be made, both of goods and of labour. The *state,* however, meaning the coercive organs of the governing class, in this case the proletariat, *has withered away;* so there is a big question-mark over the nature of this allocating authority. At least, since economic scarcity has ceased, its yoke is light. On the other hand this authority must be conducting the propaganda that persuades everyone to internalize the new value system. Short on police power, the authority is long on spiritual power. It might, for instance, well be a Communist party without a security police.

In particular, however, unpopular labour, and labour threatening to convey political power to its performers (notably within the allocating authority), must both be *rotated.* Indeed, in extreme versions, all jobs are rotated, to relieve boredom and broaden human development. This is the (utterly impossible and now very embarrassing to Soviet scholars) abolition of the division of labour. This foolishness stems from Marx and Lenin's notion that advanced technology simplifies all labour.

We have only used the words 'utterly impossible' once, and we have presented the whole concept in ordinary Western language. This is partly because the kibbutz does embody Full Communism in practice, as indeed do most monasteries and nunneries. Elements of it are also included by other organizations such as cities under siege, countries immediately after Communist revolutions, and military forces. Perhaps above all the nuclear family, even the extended family, brings this utopia down to earth.

The kibbutz and the family, the former hardly Marxist, the latter originally scheduled to disappear under Full Communism, both illuminate the Marxist neglect of the *spiritual diseconomies of scale.* The altruism that we feel in not 'breaking the bank' with our consumption need not be very warm, but it must be there, if only as a sense of duty. The larger our community, the less warmth and eventually the less duty we feel. *Homo economicus* simply becomes an empirically more probable mode. But for Full Communism he must be altogether negated, at least on the consumption side. However generous a view we take of needs, only a very 'well-brought-up' population can reduce its wants to that, or indeed to any other than an infinitely high, level. In particular, while we can always want very little more than what we now have, it is almost impossible to want nothing more. So wants always grow, and are fed by *envy* and exceed needs by more and more.

It is a commonplace that the modern kibbutz cannot stop people consuming, but it can make people work. Work, after all, is in part natural. Up to a (very variable) point it is thought of as a duty and a pleasure. Deprivation of it is felt as painful, even when income is constant. *Homo economicus* explains work very badly, however large or small, rich or poor, capitalist or socialist, our community: he is already negated, in all systems.

PLANNING UNDER FULL COMMUNISM. The kibbutz has a labour committee, which has the fairly simple task of drawing up a labour plan each week; and a consumption committee which, in the avowed presence of economic scarcity, adopts a mix of the following allocation instruments:

(i) Free supply; one just takes what one wants. This rule reigns, in respect of quantity but not quality, in the mess hall. Note that if there had been prices demand here would have been inelastic in respect to both price and income. Similarly when Russia went through its Full Communism post-revolutionary fit (June 1918–April 1921) local transport and postage were made uncompromisingly moneyless.

(ii) Rationed supply: housing and all durables, even clothing.

(iii) Pocket-money and actual prices: 'imported' luxuries such as cigarettes and sweets; coin-boxes such as telephones (also 'imported').

The pocket-money is of course divided equally, but the intrusion of money into utopia is viewed with grave misgiving. Not only is it bad in itself, but it leads to 'heterogeneous but equal' consumption. People receive unequal quantities of each thing, and this is supposed to give rise to envy, despite the overall equality of consumption volume. Another intrusion of 'money' is the use of shadow-prices by the labour committee. This is less bad in itself, but leads to narrow rationalistic calculations, whereas Full Communism requires the broad sweep of 'policy' irrespective of mere economics.

Mutatis mutandis Communist governments take the same attitudes as kibbutzim. Of course, after their post-revolutionary fit they recognize that they are only in the 'socialist' transitional phase, in which only the enterprise and not the worker/consumer figures in the command plan; the latter is guided by prices and wage-rates. But they feel they should at least be tending the sprouts of the higher phase to come. To the shadow-price problem described above is added the fact that passive inter-enterprise wholesale prices exist in reality. These must, for accounting and bonus-formation purposes, be actually paid, but have no allocative function (the far smaller kibbutz needs no such thing). It would be convenient and rational to bring the passive prices into line with the shadow-price (which has an allocative function but is never paid). Perhaps such a society, in which there were at least no retail prices and instruments (i) and (ii) of consumption planning were used, could be called Full Communism.

The *official Marxist name* for Full Communism is 'Communism'; we have used the longer phrase for clarity. The first post-revolutionary phase is 'Socialism'. Marx describes this in his Critique of the Gotha Programme in very brief terms that correspond respectably to what the Soviet economy has become. Thus it is false that Marx left no post-revolutionary blueprint, but he certainly had a very foreshortened time path. He called the intermediate phase the 'Dictatorship of the Proletariat', and Full Communism, 'Socialism' or 'Communism' indifferently.

FULL COMMUNISM AND INTERNATIONAL RELATIONS. A kibbutz is, in theoretical economics, a country. Hence our use above of the term 'imports'. People who leave it are 'emigrants', and so on. Like a communist country it uses 'foreign' money for its 'foreign' trade. But it is and is meant to be, even in high ideology, subject to the Israeli state, which is not about to wither away. However the Communist state is supposed to wither away, so who will guard its borders and administer migration and foreign trade? Some of these organs are by definition coercive. They can only wither away in a single world state – an irrefragable conclusion only lightly touched upon in Marxist writings.

P.J.D. WILES

See also ANARCHISM; COMMUNISM; SOCIALISM; UTOPIAS.

full-cost pricing. *See* AVERAGE COST PRICING.

full employment. An expression which came into general use in economics after the Depression of the 1930s, full employment applies to industrially developed economies in which the majority of the economically active are the employees of firms or public authorities as wage and salary earners.

There has always been some unemployment in the course of development of capitalist economies and views have differed as to its causes and as to the extent to which it was a matter of public concern. In the first part of the 20th century three principal strands of thought about unemployment can be distinguished. Firstly, the followers of Marx believed that cycles were an integral part of capitalist development and would lead to ever deepening crisis: the attempt to evade this by colonial expansion would only lead to conflict between imperialist powers. A second group of analysts paid particular attention to the measurement and dating of business cycles, distinguishing cycles of different periodicity, but they did not, as a rule, offer systematic theories. The third strand consisted of those economists who argued that in capitalist economies, if the forces of the market were left to work themselves out, there would always be a tendency towards an equilibrium, in modern parlance towards full employment.

Table 1 shows average rates of unemployment in six developed countries for various periods of the 20th century. National estimates of unemployment are obtained either by sample survey or as the by-product of administration, such as a system of unemployment insurance. There are many problems in counting both the numbers unemployed and the labour force, whose ratio is to constitute the 'rate' of unemployment. There have been attempts to standardize rates obtained in different countries by different methods and over different periods. The figures in Table 1, taken from Maddison (1982) and OECD *Main Economic Indicators* are thought to be reasonably comparable. Only in two cases was it feasible to give estimates before World War I. We have four countries for the interwar years and all six after 1950. It will be seen that in the Depression years 1930–34 the average rates of unemployment were far higher than in any earlier period in the 20th century and that even in the later 1930s the rates remained abnormally high except in Germany.

The time was ripe for a theory which could account for the persistence of large-scale unemployment and it was provided by John Maynard Keynes in *The General Theory of Employment, Interest and Money* (1936), which the author himself said was all about 'my doctrine of full employment'. The self-equilibrating tendencies expounded by those whom

TABLE 1 *Unemployed as a Percentage of the Total Labour Force*

	France	Germany	Japan	Sweden	U.K.	U.S.A.
1900–1913	—	3	—	—	4.3	4.7
1920–1929	—	3.8	—	3.1	7.5	4.8
1930–1934	—	12.7	—	6.3	13.4	16.5
1935–1938	—	3.8	—	5.4	9.2	11.4
1950–1959	1.4	5.0	2.0	1.8	2.5	4.4
1960–1969	1.6	0.7	1.3	1.7	2.7	4.7
1970–1979	3.7	2.8	1.6	2.0	4.3	5.4
1980–1984	7.9	6.1	2.4	2.8	11.8	8.2

Sources: 1900–1979 A. Maddison, *Phases of Capitalist Development*, Oxford University Press, 1982. 1980–1984 OECD. *Main Economic Indicators*, Paris.

(In the overlapping years 1975–1979 there are small discrepancies between Maddison and OECD for Germany and UK. The latest OECD figures were adjusted to be consistent with Maddison.)

Keynes called 'classical' economists did not necessarily function in the manner prescribed for them and capitalist economies could get stuck with persistent unemployment. According to orthodox theory, unemployment should entail falling wages which would eliminate any 'involuntary' unemployment. Similarly, interest rates would fall, bringing about a recovery of investment. Keynes argued that money wages might be 'sticky', and even if they were not, falls in money wages would not entail corresponding falls in real wages, since prices would also fall. As to rates of interest, there was no guarantee that such falls as could occur would give a strong enough impetus to recovery. The analysis points clearly to the idea, which others developed more explicitly, that fiscal policy, that is, the adjustment of the budget balance between revenue and expenditure, could prove a more powerful lever to bring about full employment.

Within less than ten years, the British wartime coalition government, in a famous White Paper, had accepted 'as one of their primary aims and responsibilities' the maintenance of 'a high and stable level of employment', and other governments, in Australia, Canada and Sweden, for instance, made similar affirmations. Article 55 of the United Nations Charter called on members to promote 'higher standards of living, full employment, and conditions of economic and social progress and development'. This remarkable change in public policy cannot be attributed simply to the 'Keynesian Revolution' in economic thought. More powerful was the observation that twice in a generation full employment had only been realized in war. How far the new principles were responsible for the performance of economies in the postwar period is a disputed question. The facts are that for the twenty-five years after 1945 the growth rates of productivity in European countries were much higher, and the average levels of unemployment much lower than they had ever been. Fluctuations in output and employment were smaller than in the past. A group of OECD experts reporting in 1968 said that the results of using fiscal policy to maintain economic balance had been encouraging, though there was room for further improvement. In the United States, the government's attitude towards the new ideas was initially somewhat cooler. By its own past standards, productivity growth was not exceptional, and unemployment, though much lower than in the Depression, was much the same as in the 1920s and before 1914. The Keynesian battle was not truly joined in the USA until the 1960s. In the majority of countries, the era of exceptional growth and full employment came to an end in the early 1970s, since when longer spells of high unemployment have been experienced.

Full employment does not mean zero unemployment. There can be dislocations where large numbers of workers are displaced from their present employment, and time is needed before new workplaces can be created. This can happen at the end of a war, or following some major technological change. Apart from such special cases, regular allowance must be made for frictional and seasonal unemployment. Policy would not aim, therefore, at zero but at the elimination of unemployment attributable to demand deficiency. Governments targeting full employment would like to know the level of measured unemployment to which this corresponds. Three attempts to answer this question deserve mention. (1) The definition given by Beveridge (1944) was that the number of unemployed (U) should equal the number of unfilled vacancies (V). When U is very high, we would expect to find V low, and vice versa. If, over a number of fluctuations, U and V trace out a fairly stable downward sloping curve, we could pick the point on it where U = V as indicating full employment. (2) Phillips (1958) claimed that for Britain there was a good statistical relationship between the level of unemployment and the rate of change of money wages. By choosing the level of unemployment delivering zero wage inflation, or when labour productivity was rising, the slightly higher level delivering zero price inflation, we could pinpoint full employment. (3) Friedman (1968) objected that in the long run there was no trade-off between unemployment and inflation: instead he argued that there was a 'natural' rate of unemployment, such that if the actual level was pushed below this, there would be not only inflation, but accelerating inflation. If this theory could be substantiated, one could choose the 'non-accelerating inflation rate of unemployment' (NAIRU) as the target. It is evident that the usefulness of each of the above approaches turns on the closeness and stability of the statistical relationship actually observed. Experience in different countries has varied, and the British evidence should be regarded as illustrative. For the period from the early 1950s to the later 1960s econometric analysis produced reasonably stable relationships for all three approaches, yielding estimates of the full employment level of unemployment of the order of 2–3 per cent. But in the 1970s any stability of the Phillips curve crumbled, and estimates of NAIRU shot up from below two to over ten per cent, but without any clear indication of the institutional or structural changes which must have occurred to bring about so large a shift in so short a time. The UV relationship did not escape entirely unscathed either, but a plausible story can be told in terms of an outward shift of the UV curve. Brown (1985) reckoned that the United States, the United Kingdom and France suffered increases in the imperfections of the labour market in the period from the early 1960s to 1981 which might account in full employment (U = V) conditions for extra unemployment of two per cent or less. It would seem that the substantial rises in unemployment, especially in Europe, in the 1970s and 1980s can only be accounted in a smaller part by a rise in 'full employment' unemployment and that a greater part denotes a shortfall below it.

If the growth of output of developed economies after 1945 was exceptional, so also was the rate of price increase: in Britain, for example, such a sustained and substantial rise (3–4 per cent a year on average) had not been seen in peacetime for more than two centuries. Some countries had faster rises, but, in most cases, there was no clear sign of acceleration. A marked change of gear in price inflation occurred between the 1960s and the 1970s, precipitated by two large cost impulses. Around 1969 there was in many countries a distinct surge in wage increases which Phelps Brown (1983) has called 'the Hinge' and in 1973 there was the first of the great OPEC oil price rises. Confronted with these spontaneous boosts in costs, the authorities had to choose between allowing their consequences to be worked out within the bounds of the existing monetary and fiscal stance and adjusting that stance to accommodate them, which would mean that final prices would also jump. They began increasingly to opt for the former course. In doing so they received intellectual support from the first wave of the 'monetarist' counter-revolution against the now orthodox Keynesian demand management. Firstly, it was said that to push unemployment below the 'natural rate' would cause accelerating inflation. In any case, too little was known about the structure of the economy, in particular its time lags, for fine tuning to be a sensible policy. Better to adopt simple rules, such as fixed targets for the growth of the supply of money, which would keep inflation under control, and output and employment would adjust to the level indicated by the 'natural rate' of unemployment. Later developments in the new classical economics went further and denied altogether the possibility that governments, by loan financed expenditure, for instance, could effect lasting changes in employment. Instead, it was suggested, the only way to bring down unemployment was to reduce the monopoly power of trade unions, and to take other steps to free labour markets, such as abolishing minimum wage legislation and reducing unemployment benefit. Though not supported by any substantial body of evidence, these new ideas undoubtedly helped to persuade central banks to adopt fixed monetary targets, or rules, and after the second OPEC price rise in 1979, most governments followed restrictive monetary policies with more severe budgets. Calculations of 'constant employment' budget balances show a tightening equivalent to several percentage points of GNP in some cases, especially in Europe where unemployment rose considerably after 1980. On the other hand the United States broke ranks in 1983, allowing both actual and 'constant employment' deficits to rise, and it was the one major economy to experience falling unemployment.

If there is little evidence of a unique 'natural rate' of unemployment, it is nevertheless clear that to bring down a cost-induced inflation by demand restriction may involve high unemployment for a great many years. A wide range of 'income policies' has been attempted, and others canvassed, to secure that firms and workers would settle for lower prices and wages than they would seek if they were acting alone, provided others would do the same. It is unlikely that full employment of the kind experienced in Europe in the 1950s and 1960s could return without the aid of such policies. Throughout the great postwar expansion world trade grew at an unprecedented rate. Fixed exchange rates, with permission to change parities if needed, worked well enough for most countries to maintain their external balance. However, the Bretton Woods system crumbled and was succeeded by generally floating exchange rates, while at the same time controls over capital movements were being dismantled. Exchange rates came to be determined as much by capital movements as by trade, and they can diverge widely and for long periods from any level suggested by purchasing power parity. Thus full employment is also seen to depend increasingly on the joint action of all, or of a large number, of countries.

Employment policy has been linked with the welfare state in contradictory ways. On the one hand, higher unemployment is tolerated on the grounds that welfare provision mitigates the economic hardship involved: on the other hand, higher welfare costs are perceived as a growing burden on economies with high unemployment.

G.D.N. WORSWICK

See also EMPLOYMENT; INVOLUNTARY UNEMPLOYMENT; NATURAL RATE OF UNEMPLOYMENT; STRUCTURAL UNEMPLOYMENT; UNEMPLOYMENT; WAGE FLEXIBILITY.

BIBLIOGRAPHY
Beveridge, W. 1944. *Full Employment in a Free Society.* London: George Allen & Unwin.
Brown, A.J. 1985. *World Inflation since 1950.* Cambridge: Cambridge University Press.
Friedman, M. 1968. The role of monetary policy. *American Economic Review* 58(1), March, 1–17.
Keynes, J.M. 1936. *The General Theory of Employment, Interest and Money.* London: Macmillan.
Maddison, A. 1982. *Phases of Capitalist Development.* Oxford: Oxford University Press.
OECD. 1968. *Fiscal Policy for a Balanced Economy.* Paris: Organization for Economic Cooperation and Development.
Phelps Brown, E.H. 1983. *The Origins of Trade Union Power.* Oxford: Clarendon Press.
Phillips, A.W. 1958. The relation between unemployment and the rate of change of money wage rates in the United Kingdom. *Economica* 25, November, 283–99.

full employment budget surplus. The full or high employment budget surplus is a device for measuring fiscal stance and, specifically, a means of distinguishing the effects of discretionary budgetary policy on the economy from the autonomous effects on the budget of variations in economic activity. In other words, by estimating what public sector outlays, government revenue and, therefore, the budget balance would be, on the basis of current tax rates and expenditure programmes, the implications of policy action can potentially be isolated and the often misleading nature of changes in the actual budget balance kept in perspective.

Its origins lie in the recommendation made by the Committee for Economic Development in the United States that budgetary policy should be designed to 'yield a moderate surplus at high-employment national income' (Committee for Economic Development, 1947, pp. 22–5). The purpose was essentially twofold: to try to make sure that automatic stabilizers – i.e. the tendency for the budget deficit to increase during a recession and to contract during a boom – were allowed to function without being nullified by policy action to bring the budget back to balance; and at the same time to limit the use of discretionary fiscal policy to stimulate economic activity and thereby to cause an unwanted and what was regarded as potentially damaging accumulation of public sector debt. It was a means therefore of keeping Keynesian demand management policies in bounds, which was important in a fiscally conservative country like the United States.

The concept was used most influentially by E. Cary Brown in 1956 in an analysis of the 1930s to demonstrate that federal deficits were caused predominantly by the depth of the recession rather than by lax fiscal policies. It was then taken up by a number of economists, Herbret Stein and Charles Schultze among others (Stein, 1961 and Schultze, 1961) to analyse policy in the economic downturn of 1960–61 and from then on has featured regularly in the US policy debate. Estimates have frequently been presented in the President's Budget documents, in annual reports of the Council of Economic Advisers, in Congressional Budget Office and in academic analyses of policy (such as Schultze (1970–) and Pechman (1978–)).

In practice, the concept has been deployed both in periods of recession, in support of expansionary policies or as a warning against excessively deflationary ones, and in periods of economic upturn, to indicate the unsustainable nature of the budget deficits incurred as a means of shifting the economy out of recession. Given the process of fiscal policy-making in the United States, where any action taken is usually a compromise introduced only after a prolonged battle between the President and Congress, it is understandable that the reliance on fiscal stabilizers should be greater than in other countries and that the focus should be more on the longer term implications of present decisions. Though flawed, the full employment budget surplus plays a useful role in this respect. It is relatively simple and straightforward to estimate – though there is often some disagreement over the rate of unemployment taken to represent full employment and the rate of growth required to maintain such a level – and therefore widely accepted as a meaningful if limited indicator of fiscal stance.

TERRY WARD

See also BUDGETARY POLICY; BUILT-IN STABILIZERS; DEMAND MANAGEMENT; FINE TUNING; STABILIZATION POLICY.

BIBLIOGRAPHY

Brown, E.C. 1956. Fiscal policy in the Thirties: a reappraisal. *American Economic Review* 46(5), December, 857–79.

Committee for Economic Development. 1947. *Taxes and the Budget: A Program for Prosperity in a Free Economy.* Washington, November.

Pechman, J.A. 1978–. *Setting National Priorities.* Washington: Brookings Institution.

Schultze, C.L. et al. 1970–. *Setting National Priorities.* Washington: Brookings Institution.

Schultze, C.L. 1961. In *Current Economic Situation and Short-Run Outlook*, Hearings before the Joint Economic Committee, 86 Cong. 2 sess., Washington.

Stein, H., 1961. In *January 1961 Economic Report of the President and the Economic Situation and Outlook*, Hearings before the Joint Economic Committee, 87 Cong. 1 sess., Washington.

functional analysis. Functional analysis is a branch of mathematics mainly concerned with infinite-dimensional vector spaces and their maps. Elements (points) of certain important specific spaces are functions, hence the term 'functional analysis'.

An important role in the development of functional analysis was played by set theory, abstract algebra and axiomatic geometry. General topology, measure theory, differential equations and some other branches of mathematics evolved in close contact with functional analysis, so that it is difficult to indicate where these disciplines end and functional analysis begins.

The fundamental ideas of functional analysis appeared at the turn of the century; by the 1920s it had already evolved into an autonomous discipline. Among its founders were Banach, Fréchet, Hadamard, Hilbert, von Neumann, Riesz and Volterra.

The creation of functional analysis resulted in basic changes in the approach to many mathematical problems. The study of individual functions and equations was replaced by that of families of such objects. Abstract forms of investigation ensured a unified approach to questions which seemed distant at first glance; they were instrumental in finding more general, yet deeper and more concrete relationships.

From the outset, the development of functional analysis was stimulated by the intrinsic requirements of mathematics, as well as by applications, especially to quantum mechanics. Today the language of functional analysis is actually used in all of continuous mathematics. Its methods have become the foundation of a whole series of new branches of research, both theoretical and applied, such as the theory of random processes, differential topology, dynamic systems, optimal control theory, mathematical programming, and so on. Functional methods penetrate deeper and deeper into theoretical physics and into different engineering disciplines. These methods find more and more widespread applications in mathematical economics.

Spaces studied in functional analysis usually belong to the class of linear (vector) topological spaces, i.e. linear spaces supplied with a topology (a system of open sets and hence a notion of limit), for which the linear operations are continuous. A narrower class of spaces are metric vector spaces, for which distance between points is defined. The distance is given by a function (the metric, assigning a non-negative number to each pair of vectors) which possesses certain specific properties of ordinary distance. The topology in such spaces is naturally induced by the metric.

An important subclass of metric spaces are normed spaces, i.e. linear spaces in which to each element x a non-negative number $\|x\|$, called the norm of x, is assigned, and the following conditions are satisfied:

(1) $\|x\| = 0$ if and only if $x = 0$;

(2) $\|\lambda x\| = |\lambda| \cdot \|x\|$ for any scalar λ (homogeneity);

(3) $\|x + y\| \leqslant \|x\| + \|y\|$ (triangle inequality).

The norm is an abstraction of the notion of 'vector length'. The function $d(x, y) = \|x - y\|$ is the metric in normed spaces. It is said that a sequence x_t of elements converges to the element x in the strong topology, if $\|x_t - x\| \to 0$ as $t \to \infty$. A normed space is said to be a Banach space if it is complete; this means that any of its fundamental sequences (i.e. such that $\|x_t - x_s\| \to 0$ as $t, s \to \infty$) has a limit. Banach spaces often appear in applications.

A Banach space X is said to be a Hilbert space if it is supplied with a numerical function (x, y), called scalar product of vectors $x, y \in X$, related to the norm by the identity $\|x\|^2 = (x, x)$ and satisfying the conditions:

(1) (x, y) and (y, x) are complex conjugates (in particular, for real vector spaces, $(x, y) = (y, x)$);

(2) $(\lambda_1 x_1 + \lambda_2 x_2, y) = \lambda_1 (x_1, y) + \lambda_2 (x_2, y)$;

(3) $(x, x) \geqslant 0$ and $(x, x) = 0$ only if $x = 0$. The scalar product makes it possible to characterize the 'angle between vectors' and, in particular, to introduce the notion of orthogonal vector. As a result, the geometry of Hilbert spaces is close to Euclidean geometry.

Let us present some examples of specific spaces. The space $l_p (1 \leqslant p < \infty)$ of all numerical sequences $x = (\alpha_n)$ with the norm

$$\|x\| = \left(\sum_{n=1}^{\infty} |\alpha_n|^p \right)^{1/p}$$

is a Banach space. For $p = 2$ it is a Hilbert space if the scalar product is defined by the formula

$$(x, y) = \sum_{n=1}^{\infty} \alpha_n \bar{\beta}_n, \quad x = (\alpha_n), \quad y = (\beta_n),$$

where $\bar{\beta}_n$ is the complex number conjugate to β_n. The space $L_2(a, b)$ of all real functions defined on the closed interval $[a, b]$, square integrable in the Lebesgue sense, is a Hilbert space (functions which differ on a set of zero measure are identified) if the scalar product is defined by the formula

$$(x, y) = \int_a^b x(t) y(t) \, dt.$$

$L_2(a, b)$ is a particular case of the Banach spaces $L_p (1 \leqslant p \leqslant \infty)$ of functions defined on so-called measure spaces. The theory of the spaces L_p is part of the foundations of probability theory, where the functions from L_p are interpreted as random variables. For $p \neq 2$ the spaces l_p and L_p are not Hilbert spaces.

Another important example is the Banach space $C(S)$ – the collection of all continuous scalar functions on the compact space S, with the norm

$$\|x\| = \max_{s \in S} \|x(s)\|.$$

All the spaces listed above are infinite dimensional, i.e. contain an infinite subset of linearly independent vectors (the notion of linear independence here is the same as in linear algebra). A finite dimensional vector space may be transformed into a Banach space in many different ways by appropriate choices of norms, but the convergence in any norm will be equivalent to the coordinate one.

Although many facts of classical analysis can be generalized to Banach spaces, the infinite dimensional theory is essentially different from the finite dimensional one in many ways. One of the reasons is that a bounded sequence (with respect to norm) in a Banach space does not necessarily contain any fundamental subsequences and therefore may have no limiting points; such is the sequence $l_n, n = 1, 2, \ldots$ in l_2, whose nth element l_n is the vector all of whose coordinates are zero, except the nth, which equals 1.

A function from one space into another is often said to be an operator. Operators with scalar values are called functionals. The operators most thoroughly studied are the linear ones. An operator T from the vector space X to the vector space Y is called linear if

$$T(\lambda_1 x_1 + \lambda_2 x_2) = \lambda_1 T(x_1) + \lambda_2 T(x_2)$$

for all $x_1, x_2 \in X$ and arbitrary scalars λ_1, λ_2. In particular, the derivation and integration operations determine linear operators for appropriate choices of the spaces X, Y. If X and Y are finite dimensional, linear operators from X to Y are determined by matrices.

The theory of linear operators in Banach spaces is one of the most developed sections of functional analysis. It is a far-reaching generalization of linear algebra and, in particular, of matrix theory. However, the purely algebraic approach is insufficient in the infinite dimensional case. One of the reasons is the necessity of distinguishing continuous and discontinuous linear operators (continuity is not an algebraic notion), while for operators in finite dimensional space linearity implies continuity.

For a linear operator from one Banach space to another to be continuous, it is necessary and sufficient that it be bounded, i.e. that it map bounded sets into bounded sets.

The set $B(x, y)$ of continuous linear operators from X to Y is a linear space with respect to the natural operations of addition and multiplication by scalars. This set becomes a Banach space if the norm $\| T \|$ of the operator T is defined by the formula

$$\| T \| = \sup_{\|x\| \leqslant 1} \| T(x) \|.$$

In the particular case when Y is the set of scalars, we get the Banach space X^* of all linear continuous functionals on X, which is called adjoint to X. The study of adjoint spaces is not only of intrinsic interest but is also needed to obtain deeper results about the initial space X.

The adjoint space of an n-dimensional space is also n-dimensional. The space adjoint to l_p coincides, in a certain sense, with the space l_q, where $1/q + 1/p = 1$ (a similar statement holds for L_p). A complete description of linear continuous functionals has been obtained for many specific spaces. We only mention F. Riesz's famous theorem describing the general form of a linear continuous functional on the space $C(S)$ of continuous functions. In the particular case when S is the closed interval $[a, b]$ on the numerical line, any element $f \in C^*(a, b)$ can be represented in the form

$$f(x) = \int_a^b x(t) \, d\phi(t),$$

where ϕ is a function of bounded variation.

The operation of taking adjoint spaces can be iterated, yielding a sequence of Banach spaces $X, X^*, X^{**}, \ldots$ each of which is adjoint to the previous one. Each vector $x \in X$ can be viewed as an element of the second adjoint space X^{**} by putting $x(f) = f(x)$ for any $f \in X^*$; the functional thus defined is linear, continuous and its norm coincides with $\| x \|$. If all the elements of X^{**} can be represented in this way, the initial Banach space X is called reflexive.

In certain aspects reflexive spaces have more resemblance to finite dimensional ones than do non-reflexive spaces.

A sequence x_n in a Banach space X is said to converge weakly to $x \in X$ if $f(x_n) \to f(x)$ as $n \to \infty$ for any functional $f \in X^*$. This definition implicitly supplies X with the weak topology which differs, as a rule, from the original one. The consideration of different versions of convergence on the same linear space and the study of their relationships is typical of functional analysis.

Among the numerous facts of Banach space it is customary to single out three theorems which, because of their importance and manifold applications, are known as the main principles of linear analysis.

The extension principle (Hahn–Banach Theorem) states that every continuous linear functional defined on a subspace of a normed space can be extended to the entire space, preserving norm. Using this principle it is possible to prove so-called separation theorems, which claim that under appropriate conditions two non-intersecting convex sets in a Banach space may be separated by a hyperplane, i.e. a set of the form $\{x \mid f(x) = \alpha\}$, where f is a non-zero continuous linear functional and α is a scalar. Separation theorems make possible the wide use of geometric ideas in the study of Banach spaces.

The uniform boundedness principle (Banach–Steinhaus Theorem) states that a sequence of linear continuous operators $T_n \in B(X, Y)$ is pointwise convergent, i.e. $T_n(x) \to T(x)$ as $n \to \infty$ for all $x \in X$ if and only if the two following conditions hold:

(1) such a convergence takes place on a set of arguments whose linear envelope is dense in X;
(2) the norms of all the T_n are uniformly bounded with respect to n.

According to the openness principle (Banach Theorem), any continuous linear operator from one Banach space to another sends open sets into open sets.

The development of the theory of linear operators, especially at its initial stage, was stimulated by the problem of solving linear operator equations.

$$T(x) = y \qquad (1)$$

where x, y are elements of infinite dimensional spaces.

The similarity between linear functionals and algebraic equations, previously noted for linear differential equations, turned out to be just as productive for integral equations, whose foundations were laid at the beginning of the century by Fredholm, Hilbert, Noether and Volterra.

An exhaustive theory has only been constructed for certain classes of equations (1). In particular, the case when $T = I + K$ where I is the identity operator and K is compact (i.e. maps bounded sets into sets with compact closure) has been conclusively studied. Compact operators often appear in applications and are very similar to finite dimensional ones.

In the study of operator equations and in many applications of operator theory a leading role is played by the notion of spectrum. The spectrum of a continuous linear operator T defined in a complex Banach space is by definition the set of all scalars λ for which the operator $T - \lambda I$ has no inverse, i.e. $T - \lambda I$ is either not injective (one-to-one) or not surjective (onto). Non-zero solutions of the equation $T(x) = \lambda x$ are called eigen-vectors of the operator T, while the values of λ for which such solutions exist are its eigen-values. All the eigen-values are contained in the spectrum, but, unlike the finite dimensional case, the spectrum may also contain other values. A compact operator has a spectrum containing a finite or countable number of distinct numbers; in the latter case they

439

converge to zero. Spectral analysis – the branch of functional analysis studying the properties of operator spectra – has achieved penetrating advances in the theory of Banach and operator algebras (Gelfand, von Neumann).

A linear operator T in Hilbert space is called self-adjoint if $(T(x), y) = (x, T(y))$ for all x, y. A compact self-adjoint operator has properties similar to that of a symmetric matrix; for example, there exists an orthonormal basis consisting of its eigen-vectors (Hilbert–Schmidt Theorem).

Among the branches of functional analysis beyond the framework of the theory of Banach spaces, the theory of distributions (or 'generalized functions'), initially developed (by Sobolev and Schwartz) as a rigorous foundation for formal operations with δ-functions used in physics, should be mentioned.

In many theoretical and applied problems – in particular, in mathematical economics – it is necessary to consider semi-ordered vector spaces, characterized by the fact that some of their elements are involved in a comparison relation. The most important are those semi-ordered spaces for which every bounded (in the sense of the order relation) subset possesses a least upper bound. The foundations of the theory of such spaces were developed in the 1930s by Kantorovich and are called Kantorovich spaces (K-spaces). For example, the spaces l_p and L_p have a natural partial order relation: one sequence is greater than another, if all the coordinates of the first are greater than the corresponding coordinates of the second; the function x is greater than y if $x(t)$ is greater than $y(t)$ for almost all t. A somewhat wider class is constituted by vector lattices, in which the existence of l.u.b. is guaranteed only for finite sets. In semi-ordered spaces the notion of positive (not necessarily linear) operator can be introduced in a natural way; this notion has been used to generalize the theory of positive matrices.

Positive operators are an important class of maps studied in non-linear functional analysis. Another important class – the monotone operators – includes operators in Hilbert space satisfying the inequality

$$(T(x) - T(y), x - y) \leqslant 0 \qquad \text{for all} \qquad x, y$$

A third example is that of contraction operators, i.e. operators such that

$$\| T(x) - T(y) \| < \alpha \| x - y \| \qquad \text{for some} \qquad \alpha < 1.$$

For those (and some other) classes of non-linear operators, conditions for the existence and uniqueness of operator equation solutions have been obtained in global terms. But, just as in classical analysis, the most universal means of studying non-linear problems is the differential calculus. Many facts of classical differential calculus (in particular, Taylor expansions and the implicit function theorem) have been generalized to Banach spaces.

Among the main instruments of mathematical economics, convex analysis and fixed-point theorems should be noted. Both are in essence branches of functional analysis. The recent extremely rapid advances in convex analysis have been stimulated by the requirements of the theory of extremal problems in abstract spaces (mathematical programming and optimal control). A typical extremal problem is to find the maximum of the functional $f(x)$ defined on the subset G of the space X under the constraints $T(x) \geqslant 0$, $x \in G$ where T is an operator from X to a linear topological space Y supplied with the partial order $\geqslant$. As in the finite-dimensional situation, here the necessary and sufficient conditions for the existence of an extremum (under appropriate assumptions) may be stated in terms of saddle points of the Lagrange function

$$L(x, y^*) = f(x) + y^*(T(x)),$$

where the Lagrange multiplier y^* is an element of the space Y^* adjoint to Y. In deducing this condition, separation theorems, the differential calculus and theorems on the representation of linear functionals play a fundamental role.

In order to solve functional equations and extremal problems in functional spaces, various computational procedures have been developed. In particular, generalizations of gradient methods and Newton's method have been obtained (the first results here are due to Kantorovich); the Newton-Kantorovich method also turned out to be a powerful means of proving existence and uniqueness of solutions. Another approach to computational problems is based on the approximation of the given functional equation by a simpler one. The application of functional analysis methods leads to a general theory of such approximation methods within whose framework the rate of convergence is studied and error estimates are given for a series of computational procedures.

In certain cases approximate solutions may be obtained by computer in analytic rather than numerical form ('deductive computations').

The necessity of considering infinite-dimensional models arises in economics in many problems, among which the following may be distinguished: (1) assessment of random effects in a situation with an infinite number of natural states; (2) study of effects arising from a 'very large' number of participants (competition models); (3) problems of spatial economics; (4) study of economic development in continuous time, in particular, with due regard for lags; (5) economic growth on an infinite time interval; (6) influence of commodity differentiation on exchange processes. This list is not exhaustive.

As a rule, it is possible in principle to use a finite dimensional model and then pass to the limit if necessary. However, the 'infinite dimensional' statement of the problem is often easier to study because a more powerful analytic apparatus may be applied.

The concept of adjoint (dual) spaces mentioned above is of fundamental importance in economics. In a typical case the elements of the given space are interpreted as utilized and produced goods, while elements of the adjoint space (continuous linear functionals) are prices; the value of the functional on the given product vector determine its cost (expenditures, profits, etc.). Then semi-ordered vector spaces, expressing the 'greater than' relationship for certain pairs of expenditure and production vectors and taking into consideration the positivity of prices, turn out to be a natural instrument.

In the use of functional analysis methods, a very delicate question is that of choosing the functional space into which the model should be 'embedded'; it is closely related to the chosen estimate of economic and social values.

As an example let us consider a problem of type (5). In stating dynamical optimal planning problems considerable difficulties are involved in the choice of a plan horizon and objectives for the end of a planning period. However, in many cases the initial interval of the optimal trajectory depends very weakly on these parameters and is close to the corresponding interval of the optimal (in a certain sense) infinite trajectory. This is one of the reasons growth on an infinite time interval is worth studying.

For a wide class of models it is possible to show that any optimal trajectory is the result of maximizing integral profits calculated in appropriately chosen prices. An effective way of studying this question is the following. Let us embed the set of all admissible trajectories of economic growth (i.e. trajectories satisfying technological and resource constraints) in an appropriate Banach space X so that the adjoint space X^* is interpreted as the space of prices; the value of a continuous linear functional on a vector $x \in X$ may be interpreted as the integral of the profits obtained in motion along the trajectory x. The set of trajectories which are better than the optimal one does not intersect the set of admissible trajectories. Under appropriate conditions these two sets may be separated by a hyperplane. The corresponding continuous linear functional will determine the required price trajectory. Using this approach, it is possible to investigate the relationship between competitive equilibrium and optimum for an infinite time interval.

Another example of productive application of functional analysis concerns the influence of commodity differentiation on market processes, a problem occupying an important place in the theory of monopolistic competition. In the simplest case, product differentiation is characterized by a scalar parameter assuming values in the closed interval $[a, b]$. Each consumer may choose any finite number of different goods (i.e. a finite number of points t_i on the interval) and acquire them in arbitrary quantities x_i as long as he satisfies his budget restrictions for the given prices. It is natural to assume that the price $p(t)$ depends continuously on the characteristic of the product $t \in [a, b]$, i.e. $p(t) \in C(a, b)$. The result of a consumer's choice is a finite set of pairs x_i, t_i which determines a continuous linear functional in the price space $C(a, b)$ according to the rule

$$z(p) = \sum_i x_i p(t_i);$$

then $z \in C^*(a, b)$. But $C(a, b)$ can be identified with a subset of its second adjoint space (see above). Thus, as usual, price is a continuous linear functional of the space of collections of goods $C^*(a, b)$. The fact that this space is adjoint to a certain Banach space considerably facilitates its study, since adjoint spaces possess useful topological properties. The analysis of models based on this construction yields conditions under which a market with differentiated commodities and 'small' participants, similar to contemporary competitive markets, ensures an optimal distribution of resources (Mas-Colell, 1975).

The proof of the existence of competitive equilibrium in the finite dimensional case is based on fixed-point theorems. Several such theorems, including the Kakutani Theorem, are also valid for Banach spaces; however, in this case their application becomes more difficult because of the essential trait of infinite dimensional spaces mentioned previously – the non-compactness of the unit sphere. Another trait of infinite dimensional spaces is that special conditions are required for the separability of non-intersecting convex sets. Both of these circumstances considerably complicate the study of economic models.

In discussing the economic applications of functional analysis, two other disciplines closely related to it – measure theory and global analysis – should be mentioned. The first is widely used in the study of probabilistic models, as well as in models with a continuum of participants or products (see Hildenbrand, 1974; Mas-Colell, 1975). Global analysis, introduced into mathematical economics by Debreu and Smale, allowed us to understand the deeper structures of the sets of equilibrium states and to advance to the solution of equilibrium stability problems (see Smale, 1981).

Above we mentioned some applications of functional analysis to economics. In their turn, the problems of economics have influenced the development of mathematics. This is natural since economics is a vast field of research,

differing in principle from those classical physical and mathematical disciplines on the basis of which functional analysis developed. The theory of systems of linear inequalities developed a hundred years later than the theory of linear equations, and precisely because of the needs of economics.

Another interesting and important example is the transportation problem, which was first studied under the name of mass shifting problem by Kantorovich in 1942. The metric introduced in its study (interpreted as the expenditures required to shift a unit mass) has found numerous applications in functional analysis and some other fields. Many mathematical problems from functional analysis originating in economics still await their solution. In particular, the functional equations describing macroeconomic dynamics taking into account the differentiation of funds according to their time of creation have not been exhaustively studied (e.g. see Kantorovich, Zhiyanov and Khovansky, 1978). It can be expected that further advances in the mathematical analysis of economics will become an even more powerful source in the development of mathematical methods, including functional analysis.

LEONID KANTOROVICH AND VICTOR POLTEROVICH

See also CALCULUS OF VARIATIONS; CONTROL AND COORDINATION OF ECONOMIC ACTIVITY; COURT, LOUIS MEHEL; EVANS, GRIFFITH CONRAD; NON-STANDARD ANALYSIS; OPTIMAL CONTROL AND ECONOMIC DYNAMICS; PONTRYAGIN'S PRINCIPLE OF OPTIMALITY; ROOS, CHARLES FREDERICK.

BIBLIOGRAPHY

Dunford, N. and Schwartz, J.T. 1958. *Linear Operators*. New York: Interscience Publishers.

Ekeland, I. and Temam, R. 1976. *Convex Analysis and Variational Problems*. Vol. 1 of *Studies in Mathematics and Its Applications*. Amsterdam: North-Holland.

Hildenbrand, W. 1974. *Core and Equilibria of a Large Economy*. Princeton: Princeton University Press.

Kantorovich, L.V. and Akilov, G.P. 1984. *Functional Analysis*. London: Pergamon Press.

Kantorovich, L.V., Zhiyanov, V.I. and Khovansky, A.G. 1978. The principle of differential optimization as applied to a single-product dynamical economic model. *Sibirski matematicheskii zhurnal*, September–October.

Kutateladze, S.S. 1983. *Foundations of Functional Analysis*. Novosibirsk: Nauka.

Mas-Colell, A. 1975. A model of equilibrium with differentiated commodities. *Journal of Mathematical Economics* 2, June–September, 263–95.

Schaefer, H.H. 1971. *Topological Vector Spaces*. New York: Springer.

Smale, S. 1981. Global analysis and economics. In *Handbook of Mathematical Economics*, ed. K.J. Arrow and M.D. Intriligator, Amsterdam: North-Holland, Vol. 1, ch. 8.

functional finance. In two remarkable papers in 1941 and 1943, Abba Lerner wrote down the rules that he thought should govern macroeconomic policy in a monetary economy where wages and prices are too sticky to keep all markets in the neighbourhood of a (Walrasian) general equilibrium. He called the rules Functional Finance – functional because

> The central idea is that government fiscal policy, its spending and taxing, its borrowing and repayment of loans, its issue of new money and its withdrawal of money, shall all be taken with an eye only to the *results* of these actions on the economy and not to any established traditional doctrine about what is sound or unsound (Lerner, 1943, p. 298).

(Paul Samuelson reports having asked Lerner why he did not call his doctrine 'Lernerism'. Lerner, concerned as ever with function, replied that he would do nothing to limit the doctrine's popularity (Samuelson, 1964).)

The Lerner (macroeconomic) rules were as follows (1941):

1. The government shall maintain a reasonable level of demand at all times. If there is not enough spending so that there is excessive unemployment, the government shall reduce taxes or increase its own spending. If there is too much spending the government shall prevent inflation by reducing its own expenditures or by increasing taxes.

2. By borrowing money when it wishes to raise the rate of interest and by lending money or repaying debt when it wishes to lower the rate of interest, the government shall maintain that rate of interest which induces the optimum amount of investment.

Believing that the 'timidity of the proponents' had served to strengthen resistance to Keynesian doctrines, Lerner made a point of the 'breach with tradition' (1941):

> Spending by the government must be regarded not as something to be done when it can be afforded ... but as a regular and painless way of maintaining prosperity, to be undertaken when the society is poor on account of unemployment ... Taxes must be regarded not as a means to which the government has to resort in order to get money ... but as merely a device for reducing the income and therefore also the expenditures of members of society. The quantity of money must be regarded not as something to be regulated strictly according to the sacred rules of some gold standard ... but as something that is of no account in itself ... completely subservient to the rules for maintaining the right amount of spending and investment. An increase in government debt must be regarded not as a measure of last resort ... but as a matter of very little consequence ... completely subjected to the rules for maintaining prosperity and preventing inflation.

Much of Lerner's fire was aimed at what he thought was a misplaced concern about the (internal) national debt (1943):

> ... there is no reason for assuming that, as a result of the continued application of Functional Finance ... the government must always be borrowing more money ...: First, full employment *can* be maintained by printing the money needed for it ... Second ... the guarantee of permanent full employment will make private investment much more attractive ... Third, as the national debt increases, and with it the sum of private wealth, there will be an increased yield from taxes ... [that] do not represent reductions of spending by the taxpayers ... Fourth, as the national debt increases it acts as a self-equilibrating force ... The greater the national debt the greater is the quantity of private wealth [no foolishness here about Ricardo equivalence] ... the less is the incentive to add [to wealth] by saving out of current income ... This increase in private spending makes it less necessary for the government to undertake deficit financing ... Fifth, if for any reason the government does not wish to see private property grow too much ... it can check this by taxing the rich ... [who] will not reduce their spending significantly ... .

Further,

> Even if the national debt does grow, the interest on it does not have to be raised out of current taxes; Even if the

interest on the debt is raised out of current taxes, these taxes constitute only the interest on only a fraction of the benefit enjoyed from the government spending, and are not lost to the nation but are merely transferred from taxpayers to bond holders; High income taxes need not discourage investment, because appropriate deduction for losses can diminish the capital actually risked by the investor in the same proportion as his net income from the investment is reduced.

(As Scitovsky (1984) points out, Lerner ignored the substitution effect of income taxes on labour supply.)

REMARKS. (1) It is striking how *neo*-Keynesian Lerner's thought was by the early 1940s:

He was not at all preoccupied with under-employment *equilibrium*. Whether money wages are merely sticky, or whether they actually get stuck, the need is for a 'steering wheel' in the form of an active fiscal and monetary policy. ('To seek the alleviation of a depression by reducing money wages, rather than by directly reducing the rate of interest or otherwise encouraging investment or consumption, is to abandon the high road for a devious, dark difficult and unreliable path...' (1936).)

Lerner was not concerned about 'secular stagnation'. Should there be a tendency for private investment to fall short of intended private saving at full employment, even after the government has driven interest rates to feasibly low levels, then, presumably, budget deficits can make up the difference. Moreover, the wealth effect on private spending of the resulting increase in the stock of outside money and government debt would in time automatically correct the condition.

There is hardly a trace of pro-fiscal bias; money, government purchases, transfers and taxes all affect aggregate demand. Even in the 1936 review of the *General Theory*, there is barely a mention of the possibility of a liquidity trap. Choice of a policy mix – from among the many combinations of budget settings and money supply compatible with maintaining full employment (and avoiding inflation) – should reflect preferences with respect to the distribution of income and the division of output between consumption and investment, private and public. (The wording of his first rule suggests that Lerner thought of the budget as the active instrument. But notice that he would have the government issue and withdraw money, not bonds, to offset any resulting budget imbalance. At the same time the government would actively buy and sell bonds to manage the interest rate and thus investment. He does not say whether the optimal quantity of investment would be expected to vary cyclically. In any case, for Lerner the government is the government, whatever may be its institutional subdivisions. He uses the world 'fiscal' to cover monetary as well as budgetary actions.)

(2) There are three gaps in the early 1940s version of Functional Finance:

Lerner pays no attention to feasibility and implementation in a setting of uncertainty, forecast errors, variable lags, and shifting coefficients. He would say that he was concerned with first principles not tactics. (It is unlikely that he would allow the problematic quality of that distinction to worry him much.)

He says next to nothing about the complications posed by international trade, capital movements and exchange rates.

The 1941–43 papers implicitly assume – in hindsight, it is a serious flaw – that the full employment level of output would coincide with the non-inflationary level of output. That Lerner would assume that in the early 1940s is a puzzle only because he told Scitovsky that 'he went to see Keynes in 1935 or 1936 to raise the question whether full employment policies might not start an inflationary process before assuring full employment, but Keynes did not get his point' (Scitovsky, 1984). That in turn is odd, because there are passages in Chapter 21 of the *General Theory* that come very close to identifying the stagflation problem. Keynes called it 'semi-inflation' and thought it had 'a good deal of historical importance'. (Keynes formulation is ambiguous only because most of that chapter concentrates on the level of prices and wages, not on their rates of change.)

By 1946, in his *Encyclopedia Britannica* article on money, Lerner makes a point of 'inflationary tendencies long before unemployment has been reduced to a satisfactory level'. He devoted much of the remainder of his working life to the problem of stagflation.

(3) Lerner's answer to critics concerned about giving politicians financial discretion was implicit in his own allegory of the car without a steering wheel.

'Of course, we have no steering wheel!' says the occupant rather crossly ... 'Suppose we had a steering wheel and somebody held onto it when we reached a curb! He would prevent the automatic turning of the wheel and the car would surely be overturned! And besides we believe in Democracy and cannot give anyone the extreme authority of life and death over all the occupants of the car. That would be Dictatorship.'

It is said that graduate students no longer study Lerner. If true, it is a shame. Extraordinary analytic insight and originality amply make up for the institutional oversimplifications. The clarity and concision of the writing, and the spiritedness of expression, make reading Lerner a special pleasure. (For discussion of the modern debate about functional finance, see the entry on FINE TUNING.)

FRANCIS M. BATOR

See also FINE TUNING; LERNER, ABBA PTACHYA.

BIBLIOGRAPHY

Keynes, J.M. 1936. *The General Theory of Employment, Interest and Money.* London: Macmillan.
Lerner, A.P. 1936. Mr Keynes' 'General Theory of Employment, Interest and Money'. *International Labor Review*, October, 435–54.
Lerner, A.P. 1941. The economic steering wheel. *The University Review* (Kansas City), June, 2–8.
Lerner, A.P. 1943. Functional finance and the Federal Debt. *Social Research* 10, February, 38–51.
Lerner, A.P. 1944. *The Economics of Control: Principles of Welfare Economics.* New York: Macmillan.
Lerner, A.P. 1946. Money. In *Encyclopedia Britannica*, London and Chicago: Encyclopaedia Britannica.
Lerner, A.P. 1951. *Economics of Employment.* New York: McGraw-Hill.
Lerner, A.P. 1983. *Selected Economic Writings of Abba P. Lerner.* Ed. D.C. Colander, New York: New York University Press.
Samuelson, P.A. 1964. A.P. Lerner at sixty. *Review of Economic Studies* 31(3), June, 169–78.
Scitovsky, T. 1984. Lerner's contribution to economics. *Journal of Economic Literature* 22(4), December, 1547–71.

functions. *See* CORRESPONDENCES; FUNCTIONAL ANALYSIS.

fundamental disequilibrium. The Articles of Agreement of the International Monetary Fund stipulate in Article IV(5)a that 'a member shall not propose a change in the par value of its currency except to correct a fundamental disequilibrium'.

The term itself was present from the earliest drafts of the American proposals for a postwar international monetary institution which noted that changes in exchange rates 'shall be made only when essential to correction of a fundamental disequilibrium' (Horsefield, 1969, vol. III, p. 43). The term became part of an agreed Anglo-American text relating to exchange-rate changes on 15 September 1943, when the British suggested the form of words eventually embodied in the Articles of Agreement. At that time there was an attempt to define the considerations which the Fund should or should not take into account in determining whether such a disequilibrium existed. There were also some subsequent discussions of whether it would be possible to devise an 'objective test' by which the appropriateness of an exchange rate might be determined. These attempts to define fundamental disequilibrium were later dropped as impracticable. As Harry White later remarked, 'It was felt ... that the subject matter was so important, and the necessity for a crystallization of a harmonious view so essential, that it was best left for discussion and formulation by the Fund' (Dam, 1982, p. 91).

Since its inauguration the Fund has never attempted to define the term. In 1946, when it was asked by the United Kingdom whether, as the government had committed itself to full employment, steps necessary to protect a member from unemployment of a chronic or persistent character would be considered measures to correct a fundamental disequilibrium, the Fund replied that, yes, such measures were among those necessary to correct a fundamental disequilibrium and that on each occasion when a member proposed a rate change to correct a fundamental disequilibrium the Fund was required to determine in the light of all relevant circumstances whether the change was necessary (Horsefield, 1966, vol. III, p. 227). The matter came up again in 1948, when in connection with a French devaluation it was asked whether the Fund could object to a par-value change if in its opinion the change was insufficient to correct a fundamental disequilibrium. The Fund resolved the question by accepting that it could in principle object, but that in reaching a decision on any proposed exchange-rate change the member country 'should be given the benefit of any reasonable doubt' (ibid.). These matters rested until the redrafting of the Articles associated with the Jamaica Second Amendment of 1976. At that time, a par-value system like that of 1946–73 was only a possible future system, but the notion of fundamental disequilibrium still remained – and remained undefined. Perhaps the last word should lie with the Bank for International Settlements, which noted in 1945 that the likely practical test of the notion would be that 'a disequilibrium which cannot be eliminated by any method other than an alteration of exchange rates must be regarded as fundamental' (1945, p. 109, n. 1).

D.E. MOGGRIDGE

See also INTERNATIONAL MONETARY INSTITUTIONS; INTERNATIONAL MONETARY POLICY.

BIBLIOGRAPHY

Bank for International Settlements. 1945. *Annual Report.* Basle: Bank for International Settlements.
Dam, K.W. 1982. *The Rules of the Game: Reform and Evolution in the International Monetary System.* Chicago: University of Chicago Press.

de Vries, M.G. 1985. *The International Monetary Fund 1972–1978: Cooperation on Trial.* 3 vols, Washington, DC: International Monetary Fund.

Horsefield, J.K. 1969. *The International Monetary Fund 1945–1965: Twenty Years of International Cooperation.* 3 vols, Washington, DC: International Monetary Fund.

fungibility. Fungibility is a central notion in economics, though often unnoticed and unnamed. It means merely 'substitutable', and is in origin a Latin legal term meaning 'such that any unit is substitutable for another' (from *fungor* meaning 'do, discharge'). A debt can be discharged with any money, not merely moneys from a particular account. The task of a low-level administrator is to make accounts fungible with each other, so that pencil money may be spent for office parties when required; the task of a high-level administrator is to prevent this. Mother cannot give money 'for' a new refrigerator: the gift merely raises the recipient's income. Likewise the World Bank rule that the items 'financed by' the Bank must attain a certain level of social return is pointless. The $100 million given to a government will be used anyway for the marginal project in the government's list; the project 'for which the money is given' can be claimed to be any intramarginal one.

Because demands for grain are fungible a cut in Soviet orders for American grain does not cause a one-for-one fall in demands on American suppliers. Because money is fungible the prospect of a government pension will reduce the incentive to save privately. The last, 'winning' points in a football game are in no coherent sense *the* winning points, since points are fungible. On the same grounds 'the reasons' for a decision are meaningless: criteria for the decision are fungible.

Donald N. McCloskey

Fuoco, Francesco (1774–1841). Born in Migano (Naples) on 22 January 1774, Fuoco devoted almost all of his life to the study of political economy and was a member of the Scientific Academies of Naples, Turin and Palermo. He died in Naples on 2 April 1841.

His work can be set within the framework of the development of the contemporary Italian school of thought, and he reflects some of its typical subjectivistic features: the idea of necessity as the basis of the functioning of the economic system; the subjective evaluation of the value of goods; the idea of economic activity as the outcome of natural tendencies; and the idea of the 'public happiness' as a state of equilibrium. At the same time he can be considered atypical of his school in view of several theoretical and methodological contributions which place Fuoco among the followers of David Ricardo, both for his deductive reasoning and for the central role attributed to the theory of rent. The type of society from which he took his inspiration was, after all, that of industrial Lombardy and of its entrepreneurial middle class. Especially famous among his work was *La magia del credito svelata*, elaborated as a consequence of collaboration with the businessman Guiseppe De Welz.

A. Quadrio-Curzio

SELECTED WORKS

1824. *La magia del credito svelata.* 2 vols, Naples.
1825–7. *Saggi economici.* Pisa. Anastatic reprinting, ed. Oscar Nuccio, 2 vols, Rome: Bizzarri, 1969.
1829a. *Introduzione allo studio della economia industriale. Principi di economia civile applicati all'uso della forze.* Naples: Tip. Trani. Reprinted in *Rassegna monetaria,* 1937.
1829b. *Le banche e l'industria.* Naples.

BIBLIOGRAPHY

Anziani, V.M. 1978. La scuola classica in Italia: il caso di Francesco Fuoco. *Richerche economiche* 32(1), January–March, 65–96.
Cossa, L. 1892. *Introduzione allo studio dell'economia politica.* Milan. *Dizionario biografico universale.* 1842. Vol. II, Passigli.

futures markets, hedging and speculation. Futures markets for grain emerged in Chicago in the middle of the 19th century and spread rapidly to other commodities and centres. Forward contracts, in which two agents agree on the details of a transaction for delivery at a specified future date, must date back to the beginnings of commerce itself, but the distinctive feature of a futures market is that the contracts are standardized, transactions costs minimized, and liquidity is high, so that contracts can be, and typically are, bought and sold many times during their lifetime, in contrast to most forward contracts. The standard explanation for the role of futures markets is that they help to spread and hence reduce risks, and to motivate the collection and dissemination of relevant information. Forward markets provide the same risk-sharing opportunities, but the greater transparency and liquidity of futures markets makes the latter far more potent institutions for 'price discovery'.

The question of how well futures markets (and securities markets more generally) perform this role of collecting, aggregating and disseminating information is a large and important topic, best handled under the wider heading of Information. If we assume agents have rational expectations and share common information, then the price-discovery role of futures markets can be ignored and remaining issues of risk-sharing studied in isolation. In this case there is little conceptual difference between futures and forward markets, and we can concentrate attention on the two characteristic modes of behaviour exhibited by these markets – speculation and hedging.

Speculation is the purchase (or temporary sale) of goods for later resale (repurchase), rather than use, in the hope of profiting from the intervening price changes. In principle, any durable good could be the subject of speculative purchase, but if carrying costs are high, or the good is illiquid, then the margin between the buying and selling price will be large, and speculation in that good will be normally be unattractive. Liquidity in this context means that there exists a perfect, or near-perfect, market in which the good can be sold immediately for a well-defined price, and this requirement severely limits the range of assets available for large-scale speculation. There are two types of assets – commodities traded on organized futures markets, and financial assets (bonds, shares) whose properties lend themselves particularly to speculation. Hedging, on the other hand, typically refers to a transaction on a futures markets undertaken to reduce the risks arising from some other risky activity, either producing the commodity, storing it or processing it for final sale.

Thus a risk-averse wheat farmer may hedge his future harvest by selling October wheat futures in January, in which case he is 'long' in actuals and 'short' in futures. A risk-averse miller who anticipates being short of wheat may hedge by buying futures now, in which case he will be a 'long' hedger. Speculators may be on the long or short end of any transaction, but in aggregate their position must offset any net imbalance in the long and short hedgers' positions.

It might appear from this that hedging consists in shifting the price risk onto the speculators in return for a risk premium. This view of speculation, advanced by Keynes (1923) and Hicks (1946), has been challenged by Working

(1953, 1962), who denies any fundamental difference between the motivations of hedgers and speculators. One danger with looking exclusively at the price risk is that it ignores the more fundamental quantity risks that give rise to the price risks. Once this is appreciated, it is possible to formulate a simple theoretical model in which all agents are alike in attempting to maximize their expected utility but differ in the risks to which they are exposed, and these differences motivate trade on futures markets. Whilst the activities of speculators are quite well defined, those of 'hedgers' are in general a mixture of insurance and speculation, as we shall see.

The simplest model of speculation and hedging has just two time periods. In the first period farmers plant their wheat, and the futures market opens. In the second period the wheat is harvested, sold, and the futures contracts expire. There are only three types of agents – farmers, who produce wheat but do not consume it; speculators, who neither produce nor consume wheat; and consumers, who neither produce wheat nor trade on futures markets. All agents are assumed to have beliefs about the relevant variables, which can be described by (subjective) probability distributions, and their behaviour is described by the theory of expected utility maximization. There are n farmers, and for the moment suppose that they have no choice over the amount of wheat to plant, but only over the size of their sales on the futures market. In the first period farmer i believes that his second period output will be $\tilde{q}_i$ (a random variable), and that the market clearing price will be $\tilde{p}^i$, also a random variable. In particular, he believes that $\tilde{q}_i$ and $\tilde{p}^i$ are jointly normally distributed. The price of futures is f, observable now, and he sells z_i futures, so that he believes his second period income will be

$$\tilde{y}_i = \tilde{p}^i \tilde{q}_i + z_i(f - \tilde{p}^i), \qquad (1)$$

a random variable. The farmer's utility function exhibits constant absolute risk aversion, A_i, and takes the form $U^i(y) = -k_i \exp(-A_i \tilde{y})$, where $\tilde{y}$ is the random component of his income. (Any non-random components can be absorbed into the constant, k_i.) This particular form has the property that maximizing expected utility is equivalent to maximizing

$$W = Ey - \tfrac{1}{2}A \operatorname{Var} y, \qquad (2)$$

where Ey is the expected value of income, $\operatorname{Var} y$ is its variance, provided, as in the case here, that y is normally distributed. (These are the standard assumptions of the capital asset pricing model for portfolio choice, and can be viewed as second-order approximations to more general utility functions; see Newbery and Stiglitz, 1981.) If equation (1) is substituted in (2), and if z_i can be positive (futures sales) or negative (purchases) then the value of z_i that maximizes W is

$$z_i = \frac{\operatorname{Cov}(\tilde{p}^i, \tilde{p}^i \tilde{q}_i)}{\operatorname{Var} \tilde{p}^i} - \frac{E\tilde{p}^i - f}{A_i \operatorname{Var} \tilde{p}^i}. \qquad (3)$$

Speculator j has no risky production, so for him $\tilde{q}_j$ is zero, and the first terms in (1) and (3) vanish. Thus the second term in (3) can be identified as the speculative term, and is readily interpreted. The perceived riskiness of the futures contract is measured by $\operatorname{Var} \tilde{p}^i$, and the cost of this risk as $A_i \operatorname{Var} \tilde{p}^i$. The expected return to selling a futures contract is $f - E\tilde{p}^i$. In order to persuade a risk-averse speculator to *buy* futures and accept the risk, the return to *selling* must be negative, hence f must be below the expected spot price, $E\tilde{p}^i$ – a situation of *normal backwardation*. The first term in (3) is the pure hedging term, for if the futures market appears *unbiased* (i.e. $f = E\tilde{p}^i$) then there is no expected speculative profit, and the only motive for trade is the income insurance offered by the price insurance.

The quality of income insurance depends on how well income pq and price risks are correlated; that is, on the ratio of the covariance to the variance. If output is perfectly certain, then income and price are perfectly correlated, the first term will be equal to q_i, and the farmer would sell his entire crop on the futures market if he believed it to be unbiased. In general, though, he will not believe it to be unbiased, and he will wish to speculate in addition to hedging. His net futures trade will reflect the balance of the desire to insure and the returns to speculating.

The futures market clears, so that the sum of z_i across all participants must be zero, and this condition will yield a value for the futures price. What this implies for the value of f and its relation for the subsequent spot price, p, depends on beliefs, as well as preferences. If agents hold *rational expectations*, and have full information about the nature of all production and demand risks, then they will agree on the common values of the expected spot price, Ep, and its variance, $\operatorname{Var} p$. In such a case the only motive for trading on the futures market is to share risk, and speculators will be willing to absorb some of the risk in return, on average, for some profit. If all farmers face perfectly correlated production risk, and if the coefficient of variation of output is σ_q, of price is σ_p, and the correlation coefficient between price and output is r, then market clearing on the futures market gives the bias as

$$\frac{Ep - f}{Ep} = \frac{\bar{Q} \cdot Ep\sigma_p^2(1 + r\sigma_q/\sigma_p)}{\sum 1/A_i}, \qquad (4)$$

and a farmer's futures sales will be

$$\frac{z_i}{Eq_i} = \beta_i(1 + r\sigma_q/\sigma_p), \qquad \beta_i \equiv 1 - \frac{\bar{Q}}{Eq_i A_i \sum_j 1/A_j}, \qquad (5)$$

where $\bar{Q} = \Sigma Eq_i$ is average total output (see Newbery and Stiglitz, 1981, p. 186). Thus β_i is a measure of the extent to which the farmer is more risk-averse than the average (the term in A_i) and more exposed to risk ($\bar{q}_i/\bar{Q}$). If there are n identical farmers and m identical speculators, all with the same coefficient of absolute risk aversion, A, then $\beta = m/(n+m)$. If there is no output risk, so $\sigma_q = 0$, then whilst a farmer would sell his entire crop forward on an unbiased futures market, here he would only sell a fraction β, representing the fraction of the total risk which the speculators are willing to bear. If the only source of risk is supply variability, then $r = -1$, $\sigma_q/\sigma_p = \epsilon$, the elasticity of demand, and the farmer will sell a fraction of his crop $\beta(1 - \epsilon)$ on the futures market, possibly negative.

What lessons can be drawn from this very simplified model? First, futures markets allow speculators to bear some of the farmer's risks. The more highly correlated income and price risks, the better the market is at insuring farmers, but in general it will only provide partial insurance. It is, however, much better suited to providing insurance to stockholders who store the commodity after the harvest until needed for consumption or processing, and it is not surprising that most hedging is done by stockholders rather than farmers. Second, the greater the agreement over the expected spot price, and the less risk-averse are the speculators, the smaller will be the average perceived bias, and the larger will be the fraction of hedging to speculative sales by producers (or stockholders). Third, the greater the degree of agreement on the expected spot price, the more will speculation be a response to the demand for hedging services. The greater the disagreement on the expected spot price, the more likely it is that speculation, in the form of gambling over the expected spot price, will dominate the market. In a masterly series of studies, Holbrook Working showed that most commodity futures markets depend primarily on hedging for their existence, that the size of the open interest follows the demand for hedging of seasonal storage closely, with speculators standing ready to assume the risks offered by the hedgers (Working, 1962). The cost of these hedging services (i.e. the return to the speculators) was quite remarkably small. Thus for cotton traders, the *gross* profit per dollar of sales over a sample of some 3000 trades was 0.023 of 1 per cent with the traders making losses on 15 out of 43 trading days. (Net profits after paying commissions and expenses were substantially less; Working, 1953). The issue of bias turns out to be more complex than the simple Keynes–Hicks risk-premium view, for even in a bilateral market of farmers and speculators, the bias can go either way. Once stockholders and processors are brought into the picture, the relative demands for long and short hedges will change yet again, and in turn influence the direction of speculation (long or short) and hence of the risk premium, or bias.

Several important questions can be asked about the role of speculators. Do they tend to destabilize the spot market and/or the futures market? Do they improve efficiency? Do they have adverse macroeconomic effects? To the layman the association of speculative activity with volatile markets is often taken as proof that speculators are the cause of the instability, though the body of informed opinion is that the volatility creates a demand for hedging or insurance, which is met by the willingness of speculators to bear the risk. It is hard to test the proposition that speculation is stabilizing, for speculative activity (notably, stockholding) can take place without futures markets. In practice, the usual question is, do futures markets, which, by lowering transaction costs, greatly facilitate speculative behaviour, improve the stability of the spot market? Even this question is not straightforward. Futures markets provide an incentive to collect information about the future market-clearing spot price, though, as often with information gathering, there are public good problems associated with its use. Much theoretical effort has been devoted to the question of whether futures prices perfectly reveal the relevant information available to participants, and if so, what incentives would remain for its collection. It now appears that, except in special cases, the information is only partially revealed in the market, leaving incentives for its collection, but nevertheless improving the forecasts of otherwise uninformed traders. If so, and if the spot market is intrinsically volatile (because of variations in supply caused by weather, or demand caused by the trade cycle) then better forecasts of future spot prices will tend to elicit compensating supply responses – if prices are expected to be high tomorrow, then it will pay to produce more, and to carry more stocks forward, tending to reduce, or stabilize, price fluctuations. To the extent that futures markets reduce storage risks, storage becomes cheaper, and this will tend to stabilize supplies and prices directly. On the other hand, anticipated disturbances will have a more immediate effect on current prices, and will tend to make them more responsive to news. A frost in Brazil expected to affect next year's coffee production is likely to have a more rapid effect on current coffee prices in the presence of a futures market than in its absence. Nevertheless, it improves the efficiency of the current market if it does respond to this relevant information.

The clearest example of the stabilizing effect of futures market is provided by cobweb models, in which producers base current production decisions on last year's realized price, with consequent self-sustaining fluctuations in output without any exogenous shocks. If a futures market is set up, then producers initially planning to expand production in response

to last year's high price, and selling futures, would cause the futures price to fall to the predicted spot price, and would lead them to revise their incorrect production plans, hence eliminating the cobweb and stabilizing the market.

Two other factors bear on the question of market stability. It is clear that much hinges on the nature of expectations. Speculation without hedging is a zero-sum game, and if two speculators, each holding different views of the future price, $E\bar{p}^i$, trade with each other, one will gain whilst the other will lose. If they are rational, and risk-averse, they should not be willing to engage in such swaps. On this view, speculators who are more successful at forecasting the future price will make money, and those who are less successful will lose, and be forced to leave the market, until only the good forecasters are left, and they only make money in the course of moving futures prices towards the forecast spot price. However, it is possible that a steady supply of less good speculators, who add noise to the system, lose money and exit to be replaced by others. Their presence may worsen the predictive power of the futures price or, by increasing the returns to information gathering by the informed speculators, may actually improve the predictive power of the futures prices (Anderson; 1984a, Kyle, 1984). Depending on the direction of the net effect of uninformed speculators, the presence of a futures market (which provides them with the opportunity to gamble) may improve or worsen the efficiency of the spot market.

The other possibility is that futures markets will provide opportunities for market manipulation, either by the better informed at the expense of the less well-informed (corners, squeezes) or of the larger at the expense of the smaller. It is easy to show that the futures price has an effect on production decisions by extending the model of equation (1) to allow producers to choose inputs. In the case of pure demand risk (no output uncertainty) it can be shown that the producer will base his production decisions solely on the future price. Large producers (Brazil for coffee, OPEC for oil, etc.) may then find it profitable to intervene in the futures market to influence the production decisions of their competitors in the spot market, and in extreme cases may find it profitable to increase price instability, though the extent to which this is feasible will be limited by the supply of and risk tolerance of other speculators in the futures market (Newbery, 1984). This is true even if all agents hold rational expectations, and share full information (except about the actions of the large producers). If some agents use naive forecasting rules to guide their futures trading, and if these rules are known to other agents who possess market power, then it may pay the large rational agents to destabilize the price and exploit the irrationalities in the forecasting behaviour of the naive agents (Hart, 1977).

Although speculation may stabilize prices, it is quite possible for it to make prices more unstable, even if all agents have equal information and hold rational expectations. Compare two possible arrangements. In the first, futures markets are prohibited, the commodity is perishable, so there is no scope for speculative storage or speculation on the futures market. The commodity can be produced by two methods, one perfectly safe, the other risky, but on average more profitable (e.g. two varieties of irrigated rice, one higher yielding, but susceptible to rust in certain weather conditions). Farmers allocate their land between the two production techniques but, in the absence of the futures market, find the risky technique relatively unattractive and so produce little. In the second arrangement, futures markets are permitted and speculators are willing to trade for a very low risk-premium. Farmers are now able to sell the crop forward, and are therefore more willing to produce the risky crop, whose supply is very variable. Total supply variability increases, and hence the spot price becomes more variable.

It is quite possible that destabilizing speculation of this type yields higher potential social welfare, for yields are higher, if riskier, and the risks are borne at relatively low cost. It is also perfectly possible for speculation on a futures market to be stabilizing (by reducing the costs of storage and therefore improving arbitrage between crop years) and yet make everyone worse off (e.g. see Newbery and Stiglitz, 1981). We now know that if the market structure is incomplete, creating additional markets can make matters worse. Speculation, which creates a market in price risks, does not thereby complete the market structure because quantity risks may remain imperfectly insured. The reason is that the market in price risks causes changes in the market equilibrium which affects the degree to which the other risks (income and quantity risks) are effectively insured. In particular, if prices are stabilized, but quantities remain unstable, incomes may be less stable than if prices were free to move in response to the quantity changes.

Finally, there remains the old Keynesian question of whether speculation which succeeds in stabilizing prices will exacerbate income fluctuations. The argument, due to Kaldor (1939), is straightforward. Speculators undertake or assume the risks for storage, which then responds to mismatches in supply and demand. These stocks, or inventories of goods, will fluctuate markedly and will have the same macroeconomic effect as fluctuations in investment, tending, through the multiplier, to have a magnified effect on national income. Whether these speculative stock movements are stabilizing or destabilizing then turns on whether they offset or amplify the fluctuations in income associated with the mismatch in demand and supply that caused the stock change. Kaldor's view was that stock changes caused by supply shocks would tend to stabilize total income, whilst those caused by demand stocks would be destabilizing, but much will depend on the commodity price elasticities of demand and the nature of the various transmission mechanisms, particularly the lag structure. Nevertheless, the OPEC oil shocks have demonstrated that commodity supply shocks can cause significant macroeconomic disturbances, whilst the increasing ease of currency speculation as restrictions are removed and transaction costs lowered, has reawakened the fear that speculation may, in some cases, destabilize income and impose needless costs.

DAVID M. NEWBERY

See also ARBITRAGE; HEDGING; INTERTEMPORAL PORTFOLIO THEORY AND ASSET PRICING; OPTION PRICING; OPTIONS; PRESENT VALUE.

BIBLIOGRAPHY

Anderson, R.W. 1984a. The industrial organization of futures markets: a survey. Ch. 1 of Anderson (1984b).
Anderson, R.W. (ed.) 1984b. *The Industrial Organization of Futures Markets*. Lexington, Mass.: Lexington Books.
Hart, O.D. 1977. On the profitability of speculation. *Quarterly Journal of Economics* 91(4), November, 579–97.
Hicks, J.R. 1946. *Value and Capital*. 2nd edn, Oxford: Oxford University Press.
Kaldor, N. 1939. Speculation and economic stability. *Review of Economic Studies* 7, October, 1–27. Reprinted in N. Kaldor, *Essays on Economic Stability and Growth*, London: Duckworth, 1960.
Keynes, J.M. 1923. Some aspects of commodity markets. *Manchester Guardian Commercial, Reconstruction Supplement* 29, March.
Kyle, A.S. 1984. A theory of futures market manipulation. Ch. 5 of Anderson (1984b).
Newbery, D.M.G. 1984. The manipulation of futures markets by a dominant producer. Ch. 2 of Anderson (1984b).
Newbery, D.M.G. and Stiglitz, J.E. 1981. *The Theory of Commodity Price Stabilization*. Oxford: Clarendon Press.
Working, H. 1953. Futures trading and hedging. *American Economic Review* 43, June, 314–43.
Working, H. 1962. New concepts concerning futures markets and prices. *American Economic Review* 52, June, 432–59.

futures trading. The object of futures trading is the *futures contract*, which may be defined as a highly standardized forward contract. Although the terms 'forward' and 'futures' are often used interchangeably in the older literature, the distinction is essential to the understanding of futures trading. Forward contracts are widely used; thus an agreement in which an automobile dealer undertakes to deliver a car of a specified make, type and colour to a customer at some later date is a forward contract; so is an employment contract, in which the employee promises to perform specified services during a certain period of time. Because forward contracts are typically quite specific, the employee in the last example cannot substitute another worker for himself without the employer's consent. Futures contracts, by contrast, exist only for a limited number of commodities and financial instruments, and are used only by a relatively small number of firms and individuals.

Futures contracts are of two types. The traditional contract provides for actual delivery of the underlying merchandise or financial instruments. In the early 1980s contracts with 'cash settlement' were introduced; they are settled not by delivery but by calculating traders' gains and losses from a known price, for instance an index of equity prices. Cash settlement is inherently simpler than delivery, but it is of limited application because in most markets there is no single price that could be used for this calculation. The following discussion focuses on futures contracts with delivery, though most of it also applies to cash-settlement contracts.

The standardization characteristic of futures contracts generally involves five elements: (1) *Quantity*: buyers and sellers can deal only in lots of fixed size, for instance 5000 bushels of wheat or bonds with a face value of $100,000; of course they can buy or sell any number of such lots. (2) *Quality*: the commodity or instrument is usually not completely specified but can be anywhere in a range (e.g. all wheat of certain grades, or all government bonds maturing within a certain interval). (3) *Delivery time*: the lot can be delivered at any time within a specified period, say a month. In most markets only contracts for selected delivery months are traded; thus the bond futures market has contracts for March, June, September and December. (4) *Location*: the lot must be delivered in specified places (e.g. warehouses or banks) in one or more specified cities. (5) *Identity of contractors*: after the initial contract is established, the buyer and seller normally have no further dealings with each other, thus eliminating credit risk. The execution is guaranteed by a clearing house, which acts as seller to all buyers and as buyer to all sellers. The clearing house can offer this guarantee by virtue of the security deposits, known as 'margin', it collects from its members.

The immediate purpose of this standardization is to minimize transaction costs and thereby to endow the futures contract with the ready negotiability that forward contracts, heterogeneous as they are, normally lack. Futures contracts are intended to be traded by 'open outcry' on the floor of an organized exchange. Such exchanges are found in a number of commercial centres, especially in Chicago, New York and London.

The overall market for a commodity or financial instrument can be divided into the futures market, which is centralized and trades only standardized contracts, and the cash market, which is dispersed and deals in actual parcels of the commodity or instrument. The cash market can be further divided into the spot market and the forward market.

Traders may have long or short positions in any or all of these three markets; thus a merchant who holds a physical inventory is considered to be long in the spot market. A trader whose net position in the case market is offset by his position in the futures market is called a *hedger*; more particularly he is a 'short hedger' if he is long in the cash market and short in the futures market, and a 'long hedger' if these positions are reversed. Traders who are net long or net short in the overall market (and hence in at least one of its submarkets) are known as *speculators*. In the futures market there also 'spreaders' or 'straddlers', whose long position in one or more futures contracts exactly matches their short position in other futures contracts.

In both the futures and the forward markets the net position of all traders combined must be zero, since there is a sale for every purchase. This is not true in the spot market , where the aggregate net position is positive to the extent of the existing inventories. The total of all long (or short) positions in the futures market is called the 'open interest'.

The prices prevailing in the cash and futures markets at any time are not necessarily equal. However, there are two main links between these markets; one is provided by the delivery mechanism and the other by hedging. As to delivery, when a futures contract reaches maturity (as the May contract does in the month of May) the remaining shorts have to deliver what they have sold, and the remaining longs have to accept and pay for what they have bought. Clearly the shorts will not deliver anything that could be sold at a higher price in the spot market, nor will the longs take delivery of anything that they could buy more cheaply elsewhere. At delivery time, therefore, the futures price must be equal to the spot price of the items that are actually delivered. Since this ultimate equality is widely anticipated, it will also influence futures and spot prices prior to delivery time.

Hedging also serves to relate futures prices and spot prices. As Working (1953) pointed out, it is essentially a form of arbitrage between the two markets. If a futures price is high compared to a spot price, hedgers will buy in the spot market and sell futures. They can do so without risk if the futures price exceeds the spot price by more than the *carrying charge*, which is the cost of holding physical inventories between the present and the maturity of the futures contract. The futures price therefore cannot exceed the current spot price by more than the prevailing carrying charge.

It does not follow, however, that a futures price must always exceed the spot price by the relevant carrying charge. Positive inventories may be held even if the spot price is above the futures price. This is because inventories have what Kaldor (1939) called a 'convenience yield', derived from their availability when buyers need them. The profits of merchants, in fact, depend in large part on their ability to assess and realize the convenience yield. Its size depends primarily on the size of total inventories; if they are small, the marginal convenience yield will be high, but if they are large, it may be zero. Working (1953) described the relationship between the size of inventories and the return of them as the *supply curve of storage*.

The view of hedging expressed above is not necessarily inconsistent with the older interpretation of hedging as an effort to shift the price risk inherent in holding inventories to those (namely the speculators) willing to assume this risk in the hope of profiting from favourable price movements. It should be noted, however, that hedging need not reduce the total risk to which a hedger is exposed. Bankers are generally willing to finance a larger proportion of the value of hedged inventories than of unhedged inventories. By hedging, consequently, a merchant can support a larger inventory with his own capital, thereby giving more scope to the exercise of his merchandising skills. The connection between hedging and risk aversion is not as clear-cut as the older view would suggest.

Regardless of the economic interpretation of hedging, its existence has another important implication discovered by Keynes (1923, 1930) and elaborated by Hicks (1939) and Houthakker (1968). If merchants can increase their profits by hedging, they must be willing to pay a *risk premium* for the opportunity to do so. It is conceivable that short hedging (defined above) exactly offsets long hedging, in which case any premiums paid by hedgers would cancel out. There is considerable evidence, however, that in most markets short hedging exceeds long hedging at most times. The basic reason for this asymmetry is that, as pointed out earlier, the net position in the spot market (and hence in the overall market) is positive. In seasonal commodities an excess of long hedging over short hedging is usually found only towards the end of the crop year, when inventories are small.

Now if the hedgers are net short in futures, the speculators in futures must be net long. Keynes and his followers argued that speculators will only be net long if they expect futures prices to rise. At any particular moment the speculators may of course be wrong, but on the average they are right, and each futures price will tend to rise until, at the maturity of the contract, it equals the relevant spot price. The speculators' gain is the hedgers' loss; thus the speculators receive a risk premium proportionate to the amount of hedging they make possible. This risk premium is implicit in the hedgers' willingness to sell futures contracts that have a tendency to appreciate.

This, in brief, is Keynes' theory of *normal backwardation*. ('Backwardation' designates a situation where the futures price is below the spot price; strictly speaking the term 'normal backwardation' applies only to the nonseasonal markets that Keynes had in mind, but the fundamental idea carries over to markets with seasonality.) The theory anticipated the positive relation between risk and return that is the main result of the Capital Asset Pricing Model developed in the 1960s. Consistency with CAPM also requires, however, that the risk of buying futures cannot be eliminated by diversification, and that has not yet been demonstrated. The theory of normal backwardation can also be summarized as saying that futures prices, when viewed as predictors of the spot price in the future, have a downward bias.

The empirical validity of the theory of normal backwardation remains in dispute. Favourable evidence has been presented by Houthakker (1957, 1961, 1968), Cootner (1960) and Bodie and Rozansky (1980). For adverse evidence see Telser (1958, 1981), Gray (1961), Rockwell (1967) and Dusak (1973). According to the latter group of authors, futures prices are unbiased predictors of spot prices, and no risk premium is paid. The most telling argument of the critics of normal backwardation is that as a body, small speculators appear to lose money rather consistently.

If true, the theory of normal backwardation would also shed light on an observation made earlier, namely the fairly limited scope of futures trading. To be viable, the theory implies, a futures market has to be nourished by the risk premium transferred from the hedgers to the speculators; in its absence the latter would be gradually driven out by the transaction costs they incur. The futures contract must therefore be primarily designed to attract hedging.

It is not a simple matter to design futures contracts that will attract enough hedging to ensure their continued viability. Hedgers need a high correlation between the futures prices and the particular spot prices in which they are interested; consequently the contract should be neither too broad (i.e. include too many deliverable grades) nor too narrow. There must also be enough variability in prices to make hedging and speculation worthwhile.

This is why futures trading was for many years confined to grains, oilseeds, sugar, cotton, non-ferrous metals and a few other staples that can be easily graded and have volatile prices. There is no futures trading in such important commodities as steel, paper and synthetic fibres. In the 1970s, when exchange rates and interest rates became more variable, futures trading was successfully introduced in various financial instruments – first in foreign exchange, then in government securities and similar claims, and most recently in indexes of share prices. Financial futures now account for most of the activity in futures markets. The most important recent addition in the non-financial sector has been futures trading in crude oil and some of its derivatives.

Despite the controversy over normal backwardation it is widely agreed that one of the economic functions of futures trading is risk transfer. Another such function is sometimes called *price discovery*. It consists in the establishment of a competitive reference price for a commodity or financial instrument. Since the cash market is typically heterogeneous, it is convenient to have a single price from which spot and forward prices can be derived as differences. Thus the forward price for a specific transaction may be quoted as a number of cents over or under the May futures price.

Futures trading also facilitates the *allocation of production and consumption over time*, particularly by providing market guidance in the holding of inventories through the supply curve of storage (see above). More generally futures prices provide information relevant to the planning of production and consumption; if the futures prices for distant deliveries are well below those for early delivery, for instance, postponing consumption is more attractive.

The economic functions of futures markets will be performed most effectively when they are highly competitive. If one or more traders are large enough to assert their market power, futures prices (and quite possibly cash prices) may not reflect the underlying supply and demand conditions. The prevention of such distortions – particularly of 'corners', where one or more longs manipulate both the cash and the futures market – is a major concern of futures exchanges and their regulators. In the United States the Commodity Futures Trading Commission supervises the markets with a view to preventing and penalizing these and other abuses, though it has not always succeeded. In Britain the Bank of England has somewhat similar responsibilities.

H.S. HOUTHAKKER

See also BACKWARDATION; FUTURES MARKETS; HEDGING AND SPECULATION; HEDGING.

BIBLIOGRAPHY
Bodie, Z. and Rozansky, V.J. 1980. Risk and return in commodity futures. *Financial Analysts' Journal* 36, May–June, 27–31, 33–39.
Cootner, P. 1960. Returns to speculators: Telser *vs.* Keynes. *Journal of Political Economy* (with reply by Telser and rejoinder by Cootner) 68, August, 396–418.
Dusak, K. 1973. Futures trading and investor returns: an investiga-

tion of commodity market risk premiums. *Journal of Political Economy* 81(6), November–December, 1387–1406.
Gray, R. 1961. The search for a risk premium. *Journal of Political Economy* 69, June, 250–60.
Hicks, J.R. 1939. *Value and Capital.* Oxford: Clarendon Press.
Houthakker, H.S. 1957. Can speculators forecast prices? *Review of Economics and Statistics* 39, May, 143–52.
Houthakker, H.S. 1961. Systematic and random elements in short-term price movements. *American Economic Review, Papers and Proceedings* 51, May, 164–72.
Houthakker, H.S. 1968. Normal backwardation. In *Value, Capital and Growth*, ed. J.N. Wolfe, Edinburgh: Edinburgh University Press.
Kaldor, N. 1939. Speculation and economic stability. *Review of Economic Studies* 7, October, 1–27.
Keynes, J.M. 1923. Some aspects of commodity markets. *Manchester Guardian Commercial, Reconstruction Supplement* 29, March. Reprinted in *The Collected Writings of John Maynard Keynes*, Vol. VII, London: Macmillan, 1973.
Keynes, J.M. 1930. *A Treatise on Money*, Vol. II. London: Macmillan.
Rockwell, C.S. 1967. Normal backwardation, forecasting and the return to commodity futures traders. *Food Research Institute Studies* 7, Supplement, 107–30.
Telser, L.G. 1958. Futures trading and the storage of cotton and wheat. *Journal of Political Economy* 66, June, 233–55.
Telser, L.G. 1981. Why are organized futures markets. *Journal of Law and Economics* 24(1), April, 1–22.
Working, H. 1953. Hedging reconsidered. *Journal of Farm Economics* 35, November, 544–61.

fuzzy sets. The scope of fuzzy economics is to bring into play a new body of concepts in which imprecision (or fuzziness) is accepted as a matter of science. Accurate mathematical methods are used; they are based on the concept of *fuzzy set*. Intuitively, a fuzzy set is compounded of elements which appertain to it *more or less*. The transition from membership to non-membership is soft rather than crisp, as in the case of an ordinary set. In the same manner, *fuzzy logic* handles imprecise truths, and fuzzy connectives and rules of inference, contrary to classical two-valued logic.

The theory of fuzzy sets was initiated by Zadeh (1965). Since then the literature has been plentiful but scattered. Periodically, some handbooks have gathered important results (Kaufmann, 1975; Dubois and Prade, 1980; Zimmermann, 1985).

The word *fuzzy set* is a misuse of language. More exactly, the proper term is *fuzzy subset* because the reference set is not fuzzy. In what follows ordinary (non-fuzzy) concepts are in bold italic, whereas fuzzy concepts are not. For example, $X \subset \mathbf{E}$ is read: X is a fuzzy subset of the ordinary reference set $\mathbf{E}$.

Let $\mathbf{E} = \{x\}$ be a non-empty, finite or not, set and $\mathbf{M}$ a preordered set, with Card $\mathbf{M} \geq 2$. Let $\mathbf{M}^{\mathbf{E}}$ be the set of the mappings from $\mathbf{E}$ into $\mathbf{M}$. By definition a fuzzy subset X of the reference set $\mathbf{E}$ is an element of $\mathbf{M}^{\mathbf{E}}$ such that $X = \{x, \mu_X; \forall x \in \mathbf{E}: \mu_X(x) \in \mathbf{M}\}$, where μ_X is a mapping from $\mathbf{E}$ into $\mathbf{M}$. The mapping $\mu_X(x)$ is called the membership function of x to X and expresses the degree of membership of the element x of $\mathbf{E}$ to the fuzzy subset X of $\mathbf{E}$.

Many particular fuzzy subsets theories can be stated according to the characterization of the membership set $\mathbf{M}$. First, $\mathbf{M}$ is a non-numerical set; its elements are linguistic variables which are applied to approximate reasoning (Zadeh, 1975). Second, $\mathbf{M}$ is a set of ordinary numbers; then different fuzzy subsets can be defined according to the structure of each particular set of numbers which is chosen as membership set. For elaborating theoretical properties and empirical applications, it is convenient to make a distinction depending on whether $\mathbf{M}$ is a lattice or a lattice of intervals. In numerous theoretical statements and in the quasi-totality of applications, $M = [0, 1]$. This characterization was initially proposed by Zadeh (1965). In the most general case, M can be any lattice (Goguen, 1967); fuzzy subsets having more or less general properties are defined, according to the properties of lattices: distributive, complemented, boolean lattices, etc. If M is a lattice of intervals, denoted by $[a_i, a_j] \subseteq [0, 1]$, then still more general fuzzy subsets can be defined. Sambuc (1975) has initially stated the theory, named phi-fuzzy subsets theory. The value of the membership function, denoted by $\Phi_x(x) = [a_i, a_j]$, is equal to the whole interval (a_i, a_j), not to a number included into the interval. Of course, other particular specifications using a set of ordinary numbers as membership set can be stated.

Now M can be a set of fuzzy numbers, whose theory was initiated by Dubois and Prade (1980). A fuzzy number expresses that the value of a variable is not exactly equal to a precise number; the exact value is more or less credible. Consider a fuzzy membership function, denoted by μ_n, from $\mathbb{R}$ into $[0, 1]$ and such that $\forall x \in \mathbb{R}: \mu_n(x) \in [0, 1]$. Thus n is a fuzzy subset of $\mathbb{R}$. If the two following conditions are fulfilled: μ_n has the normality property and is quasi-concave, then the associated fuzzy subset n is called a fuzzy number.

All these specifications must be carefully distinguished because most of the properties of fuzzy subsets are induced by that of the membership set M. Furthermore, in applications, if M is a set of ordinary numbers, the fuzziness which is associated with a datum is expressed in an exact manner, whereas it is expressed in a fuzzy manner when M is a set of fuzzy numbers (Ponsard, 1985b).

Of course, if $M = \{0, 1\}$, ordinary set theory is found again, as a particular case.

The axiomatic framework of fuzzy subsets theory includes that of the theory of measurable sets. A fuzzy measure is defined on a fuzzy σ-algebra over the reference set. A fuzzy σ-algebra differs from a σ-algebra owing to the fact that it does not have the property of complementation. A fuzzy measure on a fuzzy σ-algebra is a mapping with co-domain a preordered and bounded set satisfying some axioms which are less restrictive than the conditions required for an ordinary measure. In particular, a fuzzy measure need not be additive.

So, a careful distinction must be made between the theory of fuzzy subsets and the theory of probability. A probability measure is a mapping from a σ-algebra (with the complementation property) over the reference set into $\mathbb{R}^+$ such that the additive property, among all the axioms, is necessarily verified. Concepts of fuzziness and risk being distinguished, a theory of fuzzy random sets which handles the probabilities of fuzzy events can be stated (Zadeh, 1968).

Finally, in the same manner, the relation between the concepts of fuzziness and uncertainty have to be settled (Zadeh, 1978). A distribution of *possibilities* is a function, denoted by φ, from E into $[0, 1]$ such that

$$\operatorname*{Sup}_{x \in E} \varphi(x) = 1.$$

Possibilities are not additive, contrary to probabilities. Clearly, the theory of risk (or probability) formulates what *must* occur, whereas the theory of uncertainty (or possibility) expresses what *may* happen.

In economics, fuzzy analysis was initiated by Ponsard (1975). Then the Institute of Economic Mathematics (University of Dijon, France) devoted a programme to the field in the framework of spatial economic analysis. Ponsard (1983) specified the place of fuzzy space analysis in the context of modern spatial economic theory.

Many types of fuzzy economic spaces were studied by several contributors: attraction zones for sale-points, areas of fuzzy spatial interactions, fuzzy regional dynamic systems, fuzzy interregional relations, fuzzy urban spaces, mental maps, etc. Indeed, the description of economic spaces has now at its disposal pertinent and sophisticated mathematical tools. For example, in regional analysis, Tranqui (1978) states an automatic classification method which integrates fuzzy data on the observed territories and applies it to the French economy. Then Ponsard and Tranqui (1985) apply the same method to the European economy. More or less fine subdivisions result as a function of the more or less strictness of the chosen degree of similarity and described regions are separated or overlapped. From a complementary point of view, economic regions are analysed as a central places system, where agglomerations are linked together by flows which generate a set of numerous interrelations. The influences exerted by each agglomeration on the others are diffuse and vague by nature. So, the use of several indicators allows us to surround the minimal and maximal bounds of the magnitude of each influence relation in a realistic manner. Ponsard (1977) builds up a phi-fuzzy network such that the arcs which join any pair of agglomerations are valued by an interval which expresses the margin of fuzziness in a given influence relation. In this framework, the fuzzy hierarchical structure of a central places system is revealed.

Besides fuzzy spaces, the analysis of fuzzy spatial behaviours is an important and complementary field whose scope is to state the microeconomic foundations of macroeconomic spaces and the conditions for partial and general equilibria.

In the present state of the art, the locations of economic agents are given, so that partial equilibria are analysed in terms of produced and exchanged quantities of goods, and the general equilibrium in terms of quantities and prices. Three stages have to be distinguished.

First, the economic agent does not generally manifest a perfect aptitude to discriminate clearly, among alternatives between those he prefers and those he does not prefer. It follows that his behaviour does not obey a binary logic of the type preference–non-preference, but a fuzzy logic (Ponsard, 1981a; 1985a). Let $E = \{x_i\}$ be a set of a priori possible alternatives. The behaviour of the economic agent is characterized by a structure $(E, \mathscr{R})$ where $\mathscr{R}$ is a fuzzy binary relation between the elements of E^2. It is such that:

$$x_i \mathscr{R} x_j = \{(x_i, x_j), \mu_{\mathscr{R}}; \forall x_i \in E, \forall x_j \in E: \mu_{\mathscr{R}}(x_i, x_j) \in M\}$$

where M is a preordered and bounded membership set and $\mu_{\mathscr{R}}(x_i, x_j)$ expresses the degree of fuzziness which characterizes the correspondence between two given alternatives. The structure $(E, \mathscr{R})$ has many interesting properties: a strong degree of preference for x_i with respect to x_j can be distinguished from a weak degree of preference for x_j with respect to x_i, fuzzy reflexivity property, Max–Min transitivity property (whose definition is weaker than the classical one), totality property (so that non-comparability does not raise specific problems). Finally, a fuzzy total preorder on E is obtained; the classes of indifference are antisymmetrical and, as such, they form between themselves a fuzzy order relation. Then, under some conditions which assure the existence of a fuzzy topological totally preordered space, a fuzzy continuous utility function, denoted by μ_u, is stated. Now, $M = [0, 1]$ in order to a numerical representation of preference be determined. The utility function is such that, $\forall x_i \in E, \mu_u(x_i) \in [0, 1]$.

Thus the theory of fuzzy spatial preference and utility is neither ordinal nor cardinal. The functions taking their values in any set M or in the interval $[0, 1]$ are fuzzy measures so that ordinal and cardinal theories are particular cases of this *valu-*

ation theory, *valuation* being taken to mean fuzzy measure in short.

Second, the models of fuzzy spatial equilibria of consumer and producer are based on specifications which are peculiar to their respective fields (Ponsard, 1981b; 1982a). They are particular cases of the economic calculation of optimizing a fuzzy objective function under an elastic resource limitation constraint. Again let E be a set of alternatives. A fuzzy decision, denoted by D, in E is by definition the intersection of the fuzzy subset F, $F \subset E$, describing the aimed objective, and the fuzzy subset C, $C \subset E$, describing the constraint. So $D = F \cap C$ with a membership function, denoted by μ_D, such that,

$$\forall x \in E, \mu_D(x) = \mu_F(x) \wedge \mu_C(x), \quad \text{with } \mu_D(x) \simeq 1 \text{ iff } x$$

is good for F *and* C and $\mu_D(x) \simeq 0$ iff x is bad for F *or* C. In fuzzy algebra, the intersection operation makes use of the *Min* operator (denoted by $\wedge$). Then an optimal decision is such that:

$$\operatorname*{Sup}_{x \in E} \mu_D(x) = \operatorname*{Sup}_{x \in E} [\mu_F(x) \wedge \mu_C(x)].$$

This formulation calls on an important remark in the framework of spatial partial equilibria theories: objective and constraint are two fuzzy subsets of the same reference set and have the same role in decision making; their relations are symmetrical since the intersection operation is commutative. Tanaka, Okuda and Asai (1974) have proved that the solution for the problem of finding the best possible decision is to select an element x in E such that:

$$\operatorname*{Sup}_{x \in E} \mu_D(x) = \operatorname*{Sup}_{x \in A} \mu_F(x),$$

with $A \subset E$ and $A = \{x; x \in E: \mu_C(x) \geqslant \mu_F(x)\}$. In clear language, A is a non-fuzzy subset of E such that the value of the constraint membership function is at least equal to the value of the objective membership function. The conditions for the function

$$\operatorname*{Sup}_{x \in A} \mu_F(x)$$

to be continuous are only mildly restrictive. Among them, the condition that the fuzzy subset which describes the objective be strictly convex (in the weaker sense of convexity in fuzzy analysis). Mathematically, it would be indifferent to place the strict convexity condition on the constraint rather than the objective, since they have the same part in the decision making. In economic analysis, it is accurate to place it on the objective. Indeed, in the consumer and producer spatial equilibria theories, it guarantees the continuity property of the fuzzy objective functions. Moreover, in producer equilibrium theory, the awkward situation in which returns are increasing does not pose a specific problem since the strict convexity condition is not placed on the technological constraint. Moreover, the solution is generally not unique, which is an expected result in a fuzzy context. Finally, in the particular case where the objective is precise and the constraint alone is fuzzy, then the fuzzy economic calculation can be solved by a different and much simpler method (Ponsard, 1982b).

In the third stage, a theory of spatial general equilibrium with fuzzy behaviours is stated (Ponsard, 1984). Excess demand, denoted by e, is dependent on a spatial delivered price system, denoted by p (a price vector). So, an excess demand fuzzy point-to-set mapping denoted by φ is defined from $(\boldsymbol{P} \times E)$ to $\mathscr{P}(\hat{\boldsymbol{P}} \times E)$ where $\hat{\boldsymbol{P}}$ designates the set of standard prices and $\mathscr{P}(\hat{\boldsymbol{P}} \times E)$ the fuzzy power-set of $(\hat{\boldsymbol{P}} \times E)$. At the equilibrium, the condition that $e \leqslant 0$ has to be verified. The conditions which ought to be fulfilled by $e(p)$ in order for p to be such

that $e(p) \leqslant 0$ exists, must be stated. The analysis is based on Butnariu's theorems (1982) which extend Brouwer's and Kakutani's theorems to fuzzy functions and fuzzy point-to-set mappings respectively. Economic results are the generalization of Walras's Law to an economic space where behaviours are soft, and the formulation of the following theorem: if the excess demand fuzzy point-to-set mapping is closed and has images which are non-empty, normal and convex, and verifies the generalized Walras's Law, then a competitive equilibrium exists, i.e. there exist a price vector $p^* \in \boldsymbol{P}$ and an excess demand vector $e^* \in e(p^*)$ such that $e^* \leqslant 0$. This theorem is a generalization of a famous result of Debreu (1959) to the case of a spatial economy characterized by fuzzy behaviours of agents. It is true whatever the distribution of locations. Finally, the concept of fuzzy expected utility which brings into play fuzzy random sets and possibility theory is stated by Mathieu-Nicot (1985).

In fact, the chief difficulty is to determine the membership function and the fuzzy measure for the fuzzy subsets of a referential. Of course, there exist no general and unique method. A solution must be found in every case. However this difficulty is not peculiar; in the same manner, the determination of a distribution of probability in stochastic models is often hard.

Finally, it is easy to look forward to further research not only in the field of spatial analysis, but also in general economic theory.

Claude Ponsard

BIBLIOGRAPHY

Butnariu, D. 1982. Fixed points for fuzzy mappings. *Fuzzy Sets and Systems* 7(2), 191–207.

Debreu, G. 1959. *Theory of Value: an Axiomatic Analysis of Economic Equilibrium*. Cowles Foundation Monograph No. 17, New York: John Wiley & Sons.

Dubois, D. and Prade, H. 1980. *Fuzzy Sets and Systems: Theory and Applications*. New York: Academic Press.

Goguen, J.A. 1967. L-fuzzy sets. *Journal of Mathematical Analysis and Applications* 18, 145–74.

Kaufmann, A. 1975. *Introduction to the Theory of Fuzzy Subsets*. Vol. 1: Fundamental Theoretical Elements. New York: Academic Press (trans. of French edn of 1973).

Mathieu-Nicot, B. 1985. *Espérance mathématique de l'utilité floue*. Coll. IME 29, Dijon, Librairie de l'Université.

Ponsard, C. 1975. L'imprécision et son traitement en analyse économique. *Revue d'Economie Politique* 1, 17–37.

Ponsard, C. 1977. Hiérarchie des places centrales et graphes phi-flous. *Environment and Planning A* 9, 1233–52.

Ponsard, C. 1981a. An application of fuzzy subsets theory to the analysis of the consumer's spatial preferences. *Fuzzy Sets and Systems* 5(3), 235–44.

Ponsard, C. 1981b. L'équilibre spatial du consommateur dans un contexte imprécis. *Sistemi Urbani* 3, 107–33.

Ponsard, C. 1982a. Producer's spatial equilibrium with a fuzzy constraint. *European Journal of Operational Research* 10, 302–13.

Ponsard, C. 1982b. Partial spatial equilibria with fuzzy constraints. *Journal of Regional Science* 22, 159–75.

Ponsard, C. 1983. *History of Spatial Economic Theory*. Texts and Monographs in Economics and Mathematical Systems. Berlin: Springer-Verlag.

Ponsard, C. 1984. A theory of spatial general equilibrium in a fuzzy economy. Working Paper No. 65, IME. Revised version in *Fuzzy Economics and Spatial Analysis*, ed. C. Ponsard and B. Fustier, Coll. IME 32, Dijon: Librairie de l'Université, 1986.

Ponsard, C. 1985a. Fuzzy sets in economics: foundation of soft decision theory. In *Management Decision Support Systems using Fuzzy Sets and Possibility Theory*, ed. J. Kacprzyk and R.R. Yager, Coll. Interdisciplinary Systems Research No. 83, 25–37, Cologne: Verlag TUV Rheinland.

Ponsard, C. 1985b. Fuzzy data analysis in a spatial context. In *Measuring the Unmeasurable*, ed. P. Nijkamp, H. Leitner and

N. Wrigley. Series D, no. 22, NATO ASI Series, 487–508. Dordrecht: Martinus Nijhoff.

Ponsard, C. and Tranqui, P. 1985. Fuzzy economic regions in Europe. *Environment and Planning*, Series A 17, 873–87.

Sambuc, R. 1975. Fonctions phi-floues. Application à l'aide au diagnostic en pathologie thyroïdienne. PhD thesis, Université de Marseille.

Tanaka, H., Okuda, T. and Asai, K. 1974. On fuzzy mathematical programming. *Journal of Cybernetics* 3, 37–46.

Tranqui, P. 1978. *Les régions économiques floues: Application au cas de la France*, Coll. IME 16, Dijon: Librairie de l'Université.

Zadeh, L.A. 1965. Fuzzy sets. *Information and Control* 8, 338–53.

Zadeh, L.A. 1968. Probability measures of fuzzy events. *Journal of Mathematical Analysis and Applications* 23, 421–27.

Zadeh, L.A. 1975. The concept of a linguistic variable and its application to approximate reasoning. *Information Sciences*, Part 1: 8, 199–249; Part 2: 8, 301–57; Part 3: 9, 43–80.

Zadeh, L.A. 1978. Fuzzy sets as a basis for a theory of possibility. *Fuzzy Sets and Systems* 1(1), 3–28.

Zimmermann, H.J. 1985. *Fuzzy Set Theory and its Applications*. Dordrecht: Kluwer–Nijhoff.

G

gains from trade. Questions relating to the gainfulness or otherwise of international trade and investment have always interested economists, from Adam Smith to the present day. We now have at our disposal a very large arsenal of propositions concerning the trading gains of single countries and of groups of countries under alternative institutional arrangements. However, most of these propositions relate to the limiting case of small countries. For example, much ingenuity has been expended in tracking the welfare implications of autonomous changes in the world prices faced by a small country or in the vector of tariffs imposed by such a country. Evidently the fruits of such investigations are of only modest general interest. Here we concentrate on two propositions which are valid for economies of any size and which are of considerable historical and intellectual interest. For an accurate summary of small-country results, and for the relevant references to the literature, see Woodland (1982, chs 9 and 11).

THE BENEFITS OF FREE AND COMPETITIVE TRADE. We begin with the oldest and best-known of all propositions in the literature concerning the gains from trade, indeed in the history of economic thought.

Proposition 1: If an initially autarkic or non-trading country *s* is exposed to free commodity trade with one or more other countries, either in the whole set of producible goods or in some subset, and if preferences, technologies and endowments are restricted in the manner of Arrow and Debreu (1954) and if markets are complete, then there is a competitive world trading equilibrium (possibly with lump sum transfers within *s*) such that no individual in *s* is worse off than in autarky.

Proposition 1 is widely accepted. However, it is not immediately plausible. Of course the opening of trade between countries enlarges the set of feasible worldwide consumption vectors. It then follows from the Second Theorem of Welfare Economics that there exists a competitive world equilibrium possibly with lumpsum transfers, such that no individual is worse off than under universal autarky. It might be thought therefore that there is nothing to understand and nothing to prove, that Proposition 1 is embedded in a standard theorem of welfare economics. However, in the statement of the welfare theorem there are no restrictions on the scope of transfers whereas in the statement of Proposition 1 transfers are required to balance within each country. Thus there is indeed something to prove.

Nor is the proof easy. Indeed it was not until 1972, nearly two hundred years after the *Wealth of Nations*, that formal and general statements and proofs became available (see Grandmont and McFadden, 1972; and Kemp and Wan, 1972). One reason for the long lag between conjecture and proof is, undoubtedly, the technical difficulty of establishing the existence of a lump-sum compensated world equilibrium; the appropriate tools for such a demonstration became known to economists only after World War II.

It has been noted that Proposition 1 rests on assumptions of Arrow–Debreu (1954) type. In particular, the number of goods is required to be finite and the set of markets complete. Without both of those assumptions there is no assurance that free trade is gainful to all participating countries. Kemp and Long (1979) have shown that in an infinite-horizon model with overlapping finite generations, and therefore with an infinity of dated goods, trade can be unambiguously harmful to one of the trading partners, even though all countries are competitive and free of conventional distortions, externalities, non-convexities and learning processes. Of course, Malinvaud (1953, 1962) had shown long ago that closed economies of the type studied by Kemp and Long can be inefficient; and it is not surprising perhaps that trade between inefficient economies is not always mutually gainful. Similarly, Newbery and Stiglitz (1981, ch. 23) have shown that if there is an incomplete set of markets in each trading country, so that the several autarkic equilibria are inefficient, then the opening of trade can leave every individual worse off.

Moreover, it is essential to the conclusions of the proposition that compensation be lumpsum. In their recent book, Dixit and Norman (1980) appear to suggest that if trade is strictly gainful with lumpsum compensation then it is strictly gainful with compensation effected by carefully chosen (non-lumpsum) taxes on goods. The suggestion is an interesting one for, if valid, it would imply that any internal misallocation generated by the (carefully chosen) commodity taxes is always more than offset by the possibility of trading at world prices. However, it has been shown by counter example that the suggestion is ill-founded, that non-lumpsum compensation is an adequate substitute for lumpsum compensation only in special cases; see Kemp and Wan (1986a).

Proposition 1 affirms that, for each participating country, free trade is preferable to autarky. It does not state that, for each country, free trade is preferable to all other kinds of trade. Indeed it was recognized quite early, by Sir Robert Torrens (1821, 1844) and John Stuart Mill (1844), that a large trading country, with market power, can improve its position by manipulating its trade with the aid of taxes and subsidies on its exports and imports; indeed, by offering all-or-nothing contracts a large country can do even better than indicated by Torrens and Mill.

THE WELFARE ECONOMICS OF CUSTOMS UNIONS. The interest of economists in customs unions goes back at least to the Prussian Zollverein of 1819–31. For the most part, however, that interest has focused on the trade-distorting effects of unions rather than their welfare-distorting effects. Indeed, it was not until quite recently that a welfare proposition of any generality was established. The following proposition was first stated by Kemp (1964) and later proved under Arrow–Debreu assumptions by Kemp and Wan (1976, 1986b).

Proposition 2: Consider any competitive world trading equilibrium with any number of countries and any finite

453

number of commodities, and with no restrictions on the tariffs and other commodity taxes of individual countries. Let any subset of the countries form a customs union. Then there exists a common tariff vector and a system of lumpsum compensatory payments, involving only members of the union, such that there is an associated competitive equilibrium in which each individual, whether a member of the union or not, is not worse off than before the formation of the union. Proposition 2 has been extended by Grinols (1981) who displayed a particular scheme of compensation based on observable features of the pre-union equilibrium only.

Proposition 2 shows that there is an incentive for trading countries to move towards worldwide free trade, the ultimate customs union. That we do not observe a free-trading world, or even an unmistakable drift to free trade, can be traced to game-theoretical conflicts about the choice of partners, the division of the gains and the enforcement of agreements; to the non-economic objectives of nations; and to the unrealism of some of the Arrow–Debreu assumptions, notably the assumptions that there are no externalities and that production sets and preferences are convex.

MURRAY C. KEMP

See also FOREIGN TRADE; INTERNATIONAL TRADE.

BIBLIOGRAPHY

Arrow, K.J. and Debreu, G. 1954. Existence of an equilibrium for a competitive economy. *Econometrica* 22, July, 265–90.

Dixit, A.K. and Norman, V. 1980. *Theory of International Trade.* Welwyn, Herts.: J. Nisbet and Cambridge University Press.

Grandmont, J.M. and McFadden, D. 1972. A technical note on classical gains from trade. *Journal of International Economics* 2(2), May, 109–25.

Grinols, E.L. 1981. An extension of the Kemp–Wan theorem on the formation of customs unions. *Journal of International Economics* 11(2), May, 259–66.

Kemp, M.C. 1964. *The Pure Theory of International Trade.* Englewood Cliffs, NJ: Prentice-Hall.

Kemp, M.C. and Long, N.V. 1979. The under-exploitation of natural resources: a model with overlapping generations. *Economic Record* 55, September, 214–21.

Kemp, M.C. and Wan, H.Y., Jr. 1972. The gains from free trade. *International Economic Review* 13(3), October, 509–22.

Kemp, M.C. and Wan, H.Y., Jr. 1976. An elementary proposition concerning the formation of customs unions. *Journal of International Economics* 6(1), February, 95–7.

Kemp, M.C. and Wan, H.Y., Jr. 1986a. Gains from trade with and without lumpsum compensation. *Journal of International Economics* 21(1–2), August, 99–110.

Kemp, M.C. and Wan, H.Y., Jr. 1986b. The comparison of second-best equilibria: The case of customs unions. *Zeitschrift für Nationalökonomie.*

Malinvaud, E. 1953. Capital accumulation and efficient allocation of resources. *Econometrica* 21, April, 233–68.

Malinvaud, E. 1962. Efficient capital accumulation: a corrigendum. *Econometrica* 30(3), July, 570–73.

Mill, J.S. 1844. *Essays on Some Unsettled Questions of Political Economy.* London: John W. Parker.

Newbery, D.M.G. and Stiglitz, J.E. 1981. *The Theory of Commodity Price Stabilization: a study in the economics of risk.* Oxford: Oxford University Press.

Torrens, R. 1821. *An Essay on the Production of Wealth.* London: Longman, Hurst, Rees, Orme and Brown.

Torrens, R. 1844. *The Budget. On Commercial and Colonial Policy.* London: Smith, Elder & Co.

Woodland, A.D. 1982. *International Trade and Resource Allocation.* Amsterdam: North-Holland.

Gaitskell, Hugh Todd Naylor (1906–1963).   Hugh Gaitskell was born in 1906 in London and educated at Winchester and New College, Oxford. The General Strike, which occurred mid-way through his undergraduate studies, led to Gaitskell's first active involvement in politics when he assisted local supporters of the Trade Union Council: this experience, and the aftermath of the Strike, began his life-long commitment to the labour movement. Having graduated in 1927 with first class honours in 'Modern Greats' (Politics, Philosophy and Economics) his first job was as Workers' Educational Association lecturer at University College, Nottingham, but after only a year's teaching there he was offered, and accepted, a post as lecturer in economics in the Department of Political Economy at University College, London. The move south did not, however, stem from any desire to pursue a more conventional university career, as he wrote to his mother from Nottingham in the spring of 1928:

> I shall probably not become Academic for (a) I dislike the academics and their attitude and their bourgeoisieness (b) I am likely to continue my association with the Labour movement. I have seen enough of Working Class conditions, industrial war and Class war here to make it probable that on and off through my life ... I shall be taking part in the Working Class movement (quoted by Williams, 1982, p. 36).

Most of Gaitskell's research and writing on economic theory and policy was done during the next eleven years, spent at University College. His academic output was not prolific and most of it was concerned with the 'Austrian' approach to economic theory: he published two highly regarded papers on the period of production (in German); contributed to the translation of Haberler's *Theory of International Trade* (1935); and began, but never completed, the translation of some of Böhm-Bawerk's writings on capital theory. He played a very active role during this period in the formulation, discussion and dissemination of Labour Party economic policy. In particular, he was a leading member of the New Fabian Research Bureau whose activities in the 1930s grouped together a wide circle of the younger socialist-inclined economists. An indication of Gaitskell's views of the appropriate policy response to the problem of mass unemployment is provided in his essay 'Financial Policy in the Transition Period' which appeared in 1935. Most of the paper was concerned with the policies which an incoming Labour government might adopt to counter the 'financial panic' which it was widely believed would accompany their election, but it also contained some more general remarks on the nature of the 'expansionist programme' which a Labour government should pursue:

> The efficacy of monetary policy as a method of curing industrial depression is still a matter of controversy. But that at certain times the banking system as a whole has the power to stimulate industrial expansion can scarcely be questioned,... . There is no doubt, for example, that the very moderate measure of recovery achieved by this country is due in the main to the abandonment of the gold standard and the subsequent policy of the Bank of England. This policy has been of the 'orthodox' character of simply creating and maintaining low rates of interest through the instruments of bank rate and open-market policy... . But although a low long-term rate is certainly essential ... its action is always very slow, and it may by itself be more or less ineffective. What is needed, after all, is not simply an increase in the funds available for secure

investment, but an increase in the money in the hands of industrialists....

Firm control, or even nationalization, of the banking system may therefore be required. He then continued:

The prosperity programme should not consist entirely of monetary measures. The Government should make every effort to expand the demand for, as well as the supply of, credit. This should be done by the orthodox method of a public works programme, the encouraging of Government departments and local authorities to push on with construction and development work....

In 1939 Gaitskell went to work for Hugh Dalton at the Ministry of Economic Warfare, never to return to his University College post or to academic economics. During the course of the war he served in a number of increasingly senior positions in the Civil Service but at the war's end he declined a permanent appointment, choosing instead to continue a political career. Gaitskell was elected as the Labour member for South Leeds in 1945 and after a short time in Parliament he was appointed (in 1950) Chancellor of the Exchequer. The Labour Party was defeated in the election of 1951 and Gaitskell became the 'Shadow' Chancellor until December 1955 when, on the retirement of Attlee, he was elected leader of the Parliamentary Labour Party and became Leader of the Opposition. So he remained until his death in January 1963.

M. ANYADIKE-DANES

SELECTED WORKS
1929. *Chartism*. London: Workers' Educational Association.
1933. Four monetary heretics. In *What Everyone Wants to Know about Money*, ed. G.D.H. Cole, London: Gollancz.
1935. Financial policy in the transition period. In *New Trends in Socialism*, ed. G. Catlin, London: Lovat Dickson and Thompson.
1936, 1938. Notes on the period of production. *Zeitschrift für Nationalökonomie* 7(5), (193)6 and 9(2), (1938).
1940. *Money and Everyday Life*. London: Labour Book Service.

BIBLIOGRAPHY
Williams, P.M. 1982. *Hugh Gaitskell*. Oxford: Oxford University Press.

Galbraith, John Kenneth (born 1908). J.K. Galbraith is a paradox. Born in Canada in 1908, he began his professional career armed with a PhD in agricultural economics from the University of California. During World War II he was in charge of price controls and immediately after was director of the Strategic Bombing Survey. Later he was in charge of economic affairs in the occupied countries and was awarded the Medal of Freedom for his efforts. He became a Professor of Economics at Harvard, a President of the American Economic Association, and an advisor to Presidents and Presidential candidates, the latter leading to his appointment as Ambassador to India during the Kennedy Administration.

Yet throughout this distinguished career the economics profession was moving steadily toward more formal mathematizable models and exhibiting less and less interest in old-fashioned political economy, while Galbraith himself never moved an iota in either direction. In the spirit that one might expect from a former editor of *Fortune* magazine, his books were written always in the form of verbally persuasive economic tracts, without a hint of mathematics. His interests were always those of political economy, with political considerations ranking at least as high and most often higher than those of economics.

Perhaps because of his writings on the causes and consequences of the Great Depression in *The Great Crash*, and his successful experience as a price controller during World War II, he was never a believer in the wisdom of the invisible hand. If there is an essential theme in his economic writings, it is that the government has a role to play in economic planning and economic planning has a role to play in successful economies.

The Affluent Society documents the tendency of the invisible hand to promote private splendour and public squalor. Others have made that case (before and since; analytically and verbally), but no one has ever grabbed the public's attention with vivid examples as he did. I still remember being required to read it during my freshman orientation week at college and the hilariously funny description of a splendid private yacht on a polluted public river. There were other forces also at work, but much of the effort to improve the quality of the public sector during the 1960s can be traced to his writings.

Planning, however, was not just something for the public sector. Planning was essential to the smooth functioning of the private sector. As a result large firms had an important role to play in the private economy. They were not just actual or potential anti-trust threats. In many ways *American Capitalism* and its doctrines of countervailing power have come to be the accepted wisdom. Big is no longer automatically bad. Major new government anti-trust cases have almost disappeared.

In the Galbraith view in *The New Industrial State*, large firms are essential since they finance much of the research and development that leads to the technical innovations that are necessary to secure a rising standard of living. Technical change has always stood outside of economics as an exogenous force. Galbraith placed it where it should be at the centre of his analysis and it led to very different conclusions regarding the role of the large firm. Today it is fashionable to point to the many formerly small firms that have become technological leaders, but Galbraith could reply that most of these firms can be shown to have sprung from the laboratories of some large firm or university. Digital Equipment, for example, sprang from an Air Force laboratory run by MIT.

Given Galbraith's focus on planning it is perhaps not surprising that his books have sold even better in Japan than they have in the United States. The strategic planning that has made Japan so successful on world markets might be a good case example of what he has been talking about. The more-planned Japanese economy beats the less-planned American economy.

The invisible hand systematically leads to too few resources for the public sector, too few resources for research and development, and poor coordination between firms, but it also, in Galbraith's view, leads to too few resources for the poor. In *Economic Development*, *The Nature of Mass Poverty*, and *The Voice of the Poor*, he has systematically argued for public actions to redress the imbalances produced by the market in the distribution of income. He has never been a believer in the virtues of 'trickle down'. And as the percentage of total income going to the bottom 40 per cent of the population falls in the mid-1980s under the impact of America's current experiment with benign neglect, he can claim vindication for his earlier arguments.

The result is an economist out of the mainstream of economic thought, but in the mainstream of economic events.

LESTER C. THUROW

SELECTED WORKS
1952a. *A Theory of Price Control*. Cambridge, Mass.: Harvard University Press.

1952b. *American Capitalism*. Boston: Houghton Mifflin.
1955. *The Affluent Society*. Boston: Houghton Mifflin.
1961. *The Great Crash*. Boston: Houghton Mifflin.
1962. *Economic Development*. Cambridge, Mass.: Harvard University Press.
1967a. *The New Industrial State*. Boston: Houghton Mifflin.
1967b. *How to Get Out of Viet Nam*. New York: New American Library.
1969a. *Ambassador's Journal*. Boston: Houghton Mifflin.
1969b. *How to Control the Military*. New York: Doubleday.
1973a. *A China Passage*. Boston: Houghton Mifflin.
1973b. *Economics and the Public Purpose*. Boston: Houghton Mifflin.
1975. *Money, Whence it Came, Where it Went*. Boston: Houghton Mifflin.
1977. *The Age of Uncertainty*. Boston: Houghton Mifflin.
1979a. *The Nature of Mass Poverty*. Cambridge, Mass.: Harvard University Press.
1979b. *Annals of an Abiding Liberal*. Boston: Houghton Mifflin.
1981. *Life in our Times*. Boston: Houghton Mifflin.
1983. *The Anatomy of Poverty*. Boston: Houghton Mifflin.

Galiani, Ferdinando (1728–1787).   Galiani was born at Chieti, Italy, on 2 December 1728 and died in Naples on 30 October 1787. At the age of seven he was sent to Naples, where he received a classical education under the supervision of his uncle Celestino Galiani, chief almoner to the king. The young Galiani was in close touch with the cultural circles of the time and was soon introduced to the study of economics. In 1744 he translated some of Locke's writings on money. One year later he took religious orders. His extensive monetary studies culminated in the publication of *Della moneta* (1751), his main work. In 1759 he was appointed secretary of the Neapolitan embassy in Paris where he lived, almost without interruptions, for about ten years. At the end of his stay he wrote the *Dialogues sur le commerce des bléds* (1770). After his return to Naples, Galiani held several high positions in the civil service and published other essays on policy issues and in fields outside economics (Galiani 1974, 1975).

Most of Galiani's theoretical work can be found in his *Della moneta* (1751) which appeared when he was twenty-two. Despite the variety of topics addressed in the book, the basic contributions concern value and monetary theory. Having defined value as a relationship of subjective equivalence between a quantity of one commodity and a quantity of another, Galiani argues that value depends on utility (*utilitá*) and scarcity (*raritá*) (1751, pp. 36–56). Utility is the property of commodities to procure welfare or happiness. Man does not wish to satisfy only primary wants – like eating, drinking, and sleeping – because once the latter have been satisfied, several others emerge so that full satisfaction is not attainable. Thus, a non-satiation postulate is assumed to hold. Scarcity refers to the quantity of goods available in the market. Although the interdependence between price and quantity in the determination of market equilibrium is clearly explained by Galiani (1751, pp. 53–4), together with the concept of demand elasticity with respect to wealth, he states that the value of commodities is given by the quantity of labour. Galiani's stress is on value as a relative notion, not related to the intrinsic properties of commodities (1751, p. 119). This theoretical framework allows him to offer a lucid explanation of the so-called paradox of value; according to Schumpeter (1954, p. 300), he 'carried this analysis to its 18th-century peak'.

The main subject of Galiani's 1751 book, however, is money. In order to analyse the properties of a monetary economy, he inquires into the feasibility of dispensing with the use of money altogether, as in religious communities (1751,

pp. 87–91). In a large society, goods could be deposited in public warehouses where each producer would be given a receipt (*bullettino*) stating the quantity of commodities deposited so that he would be entitled to withdraw an equivalent amount of commodities. Relative prices would be fixed by the prince. Yet these receipts are nothing but money; money is the means by which everyone's product is represented. Galiani's analysis foreshadows a basic idea shown by recent research (Ostroy, 1973), that is, that money is a mechanism to avoid inconsistent claims on commodities on the part of individuals who are motivated by self-interest (1751, p. 90). This analysis notwithstanding, he vigorously rejects a fiduciary monetary system, likely under the influence of events related to John Law's experience in France. These results provide the basis for his theory of the origin of money (1751, pp. 74–81). Media of exchange were not deliberately introduced by man but emerged because some goods had properties that let them be used as means of payment. Galiani's important insight – that the commodities performing monetary functions should be of uniform quality and easily recognizable in order to bring about the reduction of transaction costs and the production of information – can be found in recent work on the subject (Jones 1976, p. 775).

The validity of the quantity theory of money is taken for granted by Galiani. There is, however, a dynamic process through which equilibrium is attained and during this adjustment period changes in money supply affect the economy (1751, pp. 187–9). The same argument was advanced by David Hume in a celebrated passage (1752, pp. 37–8), one year after the publication of *Della moneta*. Although the inefficacy of expected inflation is clearly stated by Galiani (1751, p. 189), an unexpected increase in prices is thought to bring about benefits and costs. Both are discussed at length, but the analysis is rather poor and marred by inconsistencies. However, Galiani clearly understands that inflation is a concealed way of levying taxes (1751, pp. 198–9, 203–4, 208) and favours the recourse to such a policy in a critical situation when the benefits will more than offset the eventual costs (Cesarano, 1976, 1983).

As regards the international aspects of monetary economics, several passages in *Della moneta* show the basic principles of the theory of balance of payments adjustment, pointing out that money flows should not be tampered with by laws or regulations. Galiani views the balance of payments as an essentially monetary phenomenon and payments imbalances as a necessary event which should never be meddled with. Finally, the rate of interest is defined as the relative price of goods dated at different points in time (1751, pp. 290–91), stressing the role of different degrees of risk. In this analysis, an anticipation of the time preference theory of interest may be found.

Galiani places full trust on the laws of nature which regulate economic phenomena. These laws have universal validity and, like physical laws, can never be violated. Hence, the implementation of policy actions is constrained by the existence of natural laws (Cesarano, 1976, section 1). The economic process is guided by a 'supreme Hand' (1751, p. 57) which is the religiously biased counterpart (Galiani was an abbot) of Adam Smith's 'invisible hand' a quarter of a century later. This methodological standpoint can also be found in his later book *Dialogues sur le commerce des bléds* (1770), a discussion of the 1764 French law liberalizing corn exports. The theoretical contributions of this work are not as remarkable as those of *Della moneta*. Nevertheless, the *Dialogues* are to be noted for the rather modern treatment of the principles of economic policy. The latter (1770, pp. 319–23) centres upon the fixing of a target and the choice of the means to achieve it. Galiani stresses the need to avoid abrupt changes in policy and to consider the institutional and political setting before following a specific policy. Although natural laws cannot be violated in the long run and so impose a constraint on policy actions, the latter can be effective in the short run.

Galiani's work on economics reveals a large number of contributions putting him far ahead of his time. Concerning the theory of value, Schumpeter stated:

> ... he [Galiani] displayed sure-footed mastery of analytical procedure and, in particular, neatness in his carefully defined conceptual constructions to a degree that would have rendered superfluous all the 19th-century squabbles – and misunderstandings – on the subject of value had the parties to these squabbles first studied his text, *Della moneta*, 1751 (1954, pp. 300–301).

His analysis of the subject of money embodies a rather coherent theoretical structure showing the basic principles upon which classical monetary theory is built.

FILIPPO CESARANO

SELECTED WORKS

1751. *Della moneta*. Introduction by A. Caracciolo, edited by A. Merola, Milan: Feltrinelli, 1963. Reprinted 1780 with the addition of a foreword, 35 notes and an epilogue. Other editions include: P. Custodi (ed.) in *Scrittori classici italiani di economia politica*, Parte Moderna, Vols. 3 and 4, Milan: Destefanis, 1803; F. Nicolini (ed.), Bari: Laterza, 1915. English translation by Peter R. Toscano, as *On Money*, Ann Arbor: University Microfilms International, 1977. A French partial translation has been edited by G.H. Bousquet and J. Crisafulli, *De la monnaie*, Paris: Rivière, 1955. An English translation of the main passages can be found in A.E. Monroe, *Early Economic Thought*, Cambridge, Mass.: Harvard University Press, 1924.

1770. *Dialogues sur le commerce des bléds*. Ed. F. Nicolini, Naples: Ricciardi, 1959.

1974. *Nuovi saggi inediti di economia*. Introduction by G. Demaria, ed. A. Agnati, Padua: Cedam.

1975. *Opere*. Ed. F. Diaz and M. Guerci. Naples: Ricciardi.

BIBLIOGRAPHY

Accademia Nazionale dei Lincei. 1975. *Ferdinando Galiani*. Quaderno N. 211. Rome: Accademia Nazionale dei Lincei.

Cesarano, F. 1976. Monetary theory in Ferdinando Galiani's *Della moneta*. *History of Political Economy* 8(3), Fall, 380–99.

Cesarano, F. 1983. The rational expectations hypothesis in retrospect. *American Economic Review* 73(1), March, 198–203.

Einaudi, L. 1953. Galiani economista. In *Saggi bibliografici e storici intorno alle dottrine economiche*, Rome: Edizioni di Storia e Letteratura.

Hume, D. 1752. Of money. In *Writings on Economics*, ed. E. Rotwein, Madison: University of Wisconsin Press, 1955.

Jones, R.A. 1976. The origin and development of media of exchange. *Journal of Political Economy* 84(4), August, Part 1, 757–75.

Ostroy, J.M. 1973. The information and efficiency of monetary exchange. *American Economic Review* 63(4), September, 597–610.

Schumpeter, J.A. 1954. *History of Economic Analysis*. New York: Oxford University Press.

Tagliacozzo, G. (ed.) 1937. *Economisti napoletani dei sec. XVII e XVIII*. Bologna: Cappelli.

games with incomplete information. Classical economic models almost universally assume that the resources and preferences of individuals (or firms) are known not only to the individuals themselves but also to their competitors. In practice, this assumption is rarely correct. Once the attempt is made to include uncertainty (not just about the environment but also about other strategic actors) within economic models, it becomes necessary to broaden those models substantially, to include considerations about the beliefs of individuals concerning the status of their competitors, as well as about learning as it takes place over time. A standard approach for doing this is to model the situation under investigation as a game with incomplete information, and to study the (Bayesian) equilibrium points of that game.

This approach has been used in recent years to analyse such issues as negotiation, competitive bidding, social choice, limit pricing, the signalling roles of education and advertising, together with a variety of other phenomena which arise under the general heading of industrial organization.

GAMES IN STRATEGIC FORM. Consider first games in strategic form, wherein the competitors each must choose a single action. In principle, any game can be reduced to this form by letting the actions available to the players be sufficiently complex (e.g. poker can be modelled in this manner).

An n-player *game with incomplete information* consists of the following elements: (1) for each player i, a probability space T_i of that player's possible types, a set A_i of actions available to that player, and a pay-off function u_i defined for every combination $(t, a) = (t_1, \ldots, t_n, a_1, \ldots, a_n)$ of player types and actions; and (2) a probability measure μ on the space $T = T_1 \times \cdots \times T_n$. It is assumed that the elements of the game are commonly known to the players. At the start of the game, the n-tuple of player types is determined according to μ. Each player is privately informed of his own type, and then the players simultaneously announce their chosen actions. Each player finally receives the pay-off corresponding to the combination (t, a) of types and announced actions.

For example, assume that each player in a game knows his own preferences but is uncertain about the preferences (and hence, about the strategic motivations) of his competitors. This situation may be modelled as a game in which the pay-off functions have the form $u_i(t_i, a)$. The realization of the random variable $\mathbf{t}_i$ is player i's type, known to him but unknown to the other players.

In contrast, assume that the preferences of the players are known to all, but that the pay-offs are affected by some chance event represented by the random variable $\mathbf{t}_0$; that is, each pay-off function can be written in the form $v_i(\mathbf{t}_0, a)$. The variable $\mathbf{t}_i$ represents a private signal received by player i prior to his choice of an action. Note that a player's signal may be informative about the signals of the others, as well as just about the chance event, through the joint distribution of $(\mathbf{t}_0, \mathbf{t}_1, \ldots, \mathbf{t}_n)$. In this case, the expected pay-off of a player, given that the vector $t = (t_1, \ldots, t_n)$ of signals has arisen and the players have selected the actions $a = (a_1, \ldots, a_n)$, is $u_i(t, a) = E[v_i(\mathbf{t}_0, a) | t_1, \ldots, t_n]$.

THE NOTION OF 'TYPE'. The type-based formulation of a game with incomplete information is due to Harsanyi (1967–8), who proposed it as a way of cutting through the complexities of modelling not only a player's information and preferences but also his beliefs about other players' information and preferences, and his beliefs about their beliefs, and so on.

Mertens and Zamir (1985) subsequently presented a formulation of games with incomplete information which unifies the type-based approach with the beliefs-about-beliefs (and so on) approach to settings of incomplete information. By specifically modelling the iterated sequence of beliefs which determines a player's state of knowledge at the beginning of the game, and then considering 'consistent beliefs-closed subspaces' of the general space of players' beliefs, they were able to show that the original Harsanyi formulation involves no essential loss of generality.

STRATEGIES AND EQUILIBRIA. A *strategy* for a player specifies the action (or randomized choice of action) to be taken by each potential type of that player. The action specified for his actual type can be thought of as his 'private strategy'. In practice, even when a player has already learned his type, in order to decide upon his own appropriate action he must form a hypothesis concerning the strategies to be used by the others. But to analyse their strategic problems, he must ask himself what strategy they will expect him to follow. Therefore, it is necessary for him to consider the strategic choices his other potential types would make, in order to select an appropriate action for his actual type.

A (Bayesian) *equilibrium point* of a game is an n-tuple of strategies, in which the private strategy of each type of each player is a best response for that type of the $(n-1)$-tuple of strategies specified for the other players. This definition directly generalizes that of a Nash equilibrium point for a game with complete information.

As an example, consider two individuals who jointly own a piece of land. They have decided to sever their relationship, and for one of the two to buy the land from the other. Each knows how valuable the land is to himself, but is unsure of its worth to the other. They agree that each will write down a bid; the high bidder will keep the land and will pay the amount of his bid to the other.

Assume that each is equally likely to value the land at any level between $0 and $1200, and that both know this. At the unique Bayesian equilibrium point of the bidding game, each bids one-third of his own valuation. If, for example, one of them values the land at $300 and believes the other to be following the indicated equilibrium strategy, then by bidding $100 he has an expected pay-off of $1/4 \cdot \$200 + 3/4 \cdot \$250$; that is, he expects to win with probability 1/4, and when he loses, he expects the other's (winning) bid to be between $100 and $400. This private strategy is optimal for him, given his belief about the other's behaviour. More generally, given his belief that his partner will bid a third of the partner's valuation, his own expected pay-off, when his valuation is v and he bids b, is $(3b/1200) \cdot (v - b) + (1 - 3b/1200) \cdot (b + 400)/2$. This is maximized by taking $b = v/3$.

DISTRIBUTIONAL STRATEGIES. In order to study the sensitivity of equilibrium results to variations in the informational structure of a game, it is necessary to define topologies on both the spaces of player strategies and the space of games. The first may be done by recasting the definition of a strategy in distributional form:

A *distributional strategy* v for a player is a probability measure on the product of his type and action spaces, with the property that the marginal distribution of v on the player's type space coincides with the original marginal distribution induced by μ. Player i, knowing his type t_i, chooses his action according to the conditional distribution $v(\cdot | t_i)$; an outside observer, seeing the player's action a_i, will revise his beliefs concerning the players type to $v(\cdot | a_i)$. A natural topology on a player's strategy space v is the topology of weak convergence of probability measures.

Taking this distributional perspective, Milgrom and Weber (1985) proved a general equilibrium existence theorem; in particular, it follows from this theorem that any game with compact action spaces, uniformly continuous pay-off functions, and for which the type distribution is absolutely continuous with respect to the product of the marginal type distributions (i.e. for which the joint distribution of types has a corresponding joint probability density function), has an equilibrium point in distributional strategies. They also showed that, with the appropriate topology defined on the space of games, any limit point of equilibria of a sequence of games is an equilibrium point of the limit game. One consequence of the distributional approach is that when the games in a sequence provide a player with private information which disappears in the limit game, a sequence of pure strategies for that player can converge to a randomized strategy in the limit game. This reinforces an observation first offered by Harsanyi (1973) to explain why, in practice, decision-makers are rarely observed to randomize their choices of actions: the existence of a slight amount of private information is sufficient, in most cases, to allow the decision-makers to follow pure strategies which present, to their competitors, the appearance of a randomized choice of actions. In essence, competitors observe the marginal distribution, induced by a player's distributional strategy, on his action space.

INEFFICIENCIES CREATED BY INCENTIVE CONSTRAINTS. In many circumstances, parties holding private information can find it difficult, or even impossible, to arrange efficient trades. A simple example, drawn from a class of problems first discussed by Akerlof (1970), concerns the owner of a car, attempting to arrange the sale of that car to a prospective buyer. Assume that the value of the car to the seller is primarily based on the quality of the car, and that the seller knows this value. Further assume that, whatever the car is worth to the seller, it is worth 50 per cent more to the buyer. And finally, assume that the buyer's only knowledge about the seller's value is that it is uniformly distributed between $0 and $1000.

In this case, it is commonly known to the two parties that a mutually advantageous trade exists. Nevertheless, no sale can be expected to take place, since the seller's willingness to accept any price x signals to the buyer that the seller's valuation lies between $0 and x, and therefore that the expected value of the car to the buyer is most likely no more than $3/2 \cdot (x/2) = 3/4 \cdot x$. As long as the initial uncertainty persists (i.e. as long as no pre-sale verification of the car's quality is possible), and as long as no contingent trade can be arranged (i.e. as long as no warranty can be written), trade is impossible – even if the parties agree to consult an intervenor.

Intervenors in settings of incomplete information typically act as game designers, influencing the flow of information between parties, enforcing agreements and in some cases actually specifying the final resolution of a dispute (e.g. binding arbitration). Essentially, an intervenor creates a game which the parties must play. Any theory of intervention must therefore be tied to the issue of designing games with desirable equilibrium outcomes.

The Akerlof example shows that if intervenors are restricted from playing an auditing role, and if the outcome of the game cannot be made contingent on the parties' true types, then ex-post inefficient outcomes are at times inevitable. This understanding has led to the development of the theory of 'incentive-efficient mechanism design'.

THE REVELATION PRINCIPLE. In the area of game design, a simple, yet conceptually deep, type of analysis has become standard. Consider any equilibrium pair of strategies in a particular two-person game. (The following analysis is equally valid for games involving more than two players.) Each party's strategy can be viewed as a book, with each chapter detailing the private strategy of one that party's types. Given the two actual types, a pairing of the private strategies in the appropriate chapters of the two books will lead to an outcome of the game.

Now, step back from this setting and imagine the two parties in separate rooms, each instructing an agent on how to act on his behalf. Each agent holds in hand the strategy book of his side; all he must be told is which chapter to use. From this new perspective, the original two parties can be thought of as playing an 'agent-instruction' game, in which the strategy books are prespecified and each must merely tell his agent his type (or, equivalently, point to a chapter in his strategy book). An equilibrium point in this new 'type-revelation' game is for each to tell the truth to his agent. Otherwise, the original strategies could not have been in equilibrium in the original game. Consequently, anything which can be accomplished at equilibrium through the use of any particular dispute-resolution procedure can also be accomplished through the use of some other procedure in which the only actions available to the parties are to state their (respective) types, and in which it is in equilibrium for each to reveal his type truthfully.

This observation, known as the 'revelation principle', reduces the problem of game design to the problem of optimizing the designer's objective function, subject to a collection of 'incentive constraints', one for each type of each player. An early application of this approach was to the design of auction procedures which maximize the seller's expected revenue. Myerson (1984) subsequently applied the approach to the problem of bargaining under uncertainty, and provided a generalization of the classical complete-information Nash bargaining solution. A central feature of this generalization is the incorporation of intrapersonal (i.e. intertype) equity considerations.

GAMES IN EXTENSIVE FORM. A game with incomplete information in extensive form begins with a chance move which determines the types of the players, and continues with an information structure which preserves the privacy of each player's information. Many multi-stage bargaining problems can be represented in this form; typically, such games have a large number of equilibria, including equilibria in which one party is completely intransigent and the other concedes immediately, as well as equilibria in which both parties make information-revealing concessions over the series of stages.

A classical approach to the identification of 'plausible' equilibria in games with complete information is to seek equilibria which are subgame-perfect; that is, which specify optimal actions for all parties in all subgames of the original game. For example, Rubinstein (1982) presented a repeated offer-counteroffer game with many equilibria, and demonstrated that the requirement of subgame perfection uniquely identified one of those equilibria. However, subgame perfection is a concept of little use in distinguishing between equilibria of a game with incomplete information, since the privacy of the players' information typically results in the original game having no proper subgames.

Selten (1975), with his notion of 'trembling-hand' perfection, and Kreps and Wilson (1982), with their closely related notion of sequential equilibrium, provided extensions of the concept of subgame perfection which require that players act optimally at positions off the equilibrium path of the game. Central to the Kreps–Wilson approach is the incorporation of players' interim beliefs (about the other players' types, and past and future actions) at all game positions in the specification of an equilibrium point. Subsequent work on equilibrium selection in games with incomplete information has relied heavily on the study of justifiable out-of-equilibrium beliefs.

REPEATED GAMES. A special kind of extensive-form game consists of an initial chance move which determines the players' types, followed by the repeated play of a single game with type-dependent pay-offs. Players are not allowed to observe the actual stage-to-stage pay-offs during play, but are allowed to monitor the stage-to-stage actions of their competitors. The study of such games provides insight into the way players learn about one another over time; that is, insights into the way reputations are developed and maintained or changed.

Beginning in 1965 with research sponsored by the US Arms Control and Disarmament Agency, substantial effort has been focused on the study of infinitely repeated games with incomplete information. A principal result in the two-person, zero-sum case is that optimal strategies typically involve a single initial reference to the information a party holds, followed by period-to-period moves which depend only on the outcome of that single reference. (In an infinitely repeated game, short-term pay-offs are unimportant. Whatever behaviour a player adopts, his opponent's beliefs will converge to some limit; the long-term pay-offs will depend only on the limiting beliefs of the players. Therefore, in a strictly competitive environment it is sufficient for a player to determine at the beginning of the game precisely how much information he will eventually reveal.) Hart (1985) extended this analysis to games with private information on one side, and gains available to the players through cooperative actions. His work demonstrates that, when mutual gains are available, equilibrium behaviour may involve a series of references by the informed player to his information, interspersed with joint randomizing actions between the players which determine what information will next be revealed.

For many years, the finitely repeated Prisoners' Dilemma posed a dilemma for game theorists. Set in the framework of complete information, this game has a unique equilibrium outcome: the players never cooperate with one another. However, experiments repeatedly showed that actual players frequently establish a pattern of cooperation which persists until the game approaches its final stage. Kreps, Milgrom, Roberts and Wilson (1982) finally offered an explanation for this discrepancy, by demonstrating that a slight change in the initial informational framework yields games with equilibrium outcomes similar to the observed experimental outcomes. For example, assume that each player initially assigns a small positive probability to his opponent being the type of individual who (irrationally) will always respond to cooperation in one stage with further cooperation in the next. Then there will be equilibria in which, even when both players are actually rational, they will (with high probability) cooperate until near the end of the game. An interpretation of such equilibrium behaviour is that each finds it to his benefit to build a reputation as the irrational, cooperative type. The incomplete information model is necessary to obtain this behaviour. If the initial uncertainty as to type did not exist in the mind of a player's opponent, such a reputation would be impossible to build. An emerging 'theory of reputation' has its roots in this analysis.

ROBERT J. WEBER

See also GAME THEORY; INCOMPLETE CONTRACTS; OLIGOPOLY AND GAME THEORY.

BIBLIOGRAPHY

Akerlof, G. 1970. The market for lemons: qualitative uncertainty and the market mechanism. *Quarterly Journal of Economics* 84, 488–500.

Harsanyi, J.C. 1967–8. Games with incomplete information played by Bayesian players. *Management Science* 14, 159–82, 320–34, 486–502.

Harsanyi, J.C. 1973. Games with randomly-distributed payoffs: a new rationale for mixed-strategy equilibrium points. *International Journal of Game Theory* 2, 1–23.

Hart, S. 1985. Nonzero-sum two-person repeated games with incomplete information. *Mathematics of Operations Research* 10, 117–53.

Kreps, D.M. and Wilson, R. 1982. Sequential equilibria. *Econometrica* 50, 863–94.

Kreps, D.M., Milgrom, P., Roberts, J. and Wilson, R. 1982. Rational cooperation in the finitely repeated Prisoner's Dilemma. *Journal of Economic Theory* 27, 245–52.

Mertens, J.-F. and Zamir, S. 1985. Formulation of Bayesian analysis for games with incomplete information. *International Journal of Game Theory* 14, 1–29.

Milgrom, P.R. and Weber, R.J. 1985. Distributional strategies for games with incomplete information. *Mathematics of Operations Research* 10, 619–32.

Myerson, R.B. 1984. Two-person bargaining problems with incomplete information. *Econometrica* 52, 461–87.

Rubinstein, A. 1982. Perfect equilibrium in a bargaining model. *Econometrica* 50, 97–109.

Selten, R. 1975. Reexamination of the perfectness concept for equilibriu points in extensive games. *International Journal of Game Theory* 4, 25–55.

game theory. INTRODUCTION; 1910–1930; 1930–1950; 1950–1960; 1960–1970; 1970–1986; CONCLUDING REMARKS.

INTRODUCTION

'Interactive Decision Theory' would perhaps be a more descriptive name for the discipline usually called Game Theory. This discipline concerns the behaviour of decision makers (*players*) whose decisions affect each other. As in non-interactive (one-person) decision theory, the analysis is from a rational, rather than a psychological or sociological viewpoint. The term 'Game Theory' stems from the formal resemblance of interactive decision problems (*games*) to parlour games such as chess, bridge, poker, monopoly, diplomacy or battleship. The term also underscores the rational, 'cold', calculating nature of the analysis.

The major applications of game theory are to economics, political science (on both the national and international levels), tactical and strategic military problems, evolutionary biology, and, most recently, computer science. There are also important connections with accounting, statistics, the foundations of mathematics, social psychology, and branches of philosophy such as epistemology and ethics. Game theory is a sort of umbrella or 'unified field' theory for the rational side of social science, where 'social' is interpreted broadly, to include human as well as non-human players (computers, animals, plants). Unlike other approaches to disciplines like economics or political science, game theory does not use different, ad-hoc constructs to deal with various specific issues, such as perfect competition, monopoly, oligopoly, international trade, taxation, voting, deterrence, and so on. Rather, it develops methodologies that apply in principle to *all* interactive situations, then sees where these methodologies lead in each specific application. Often it turns out that there are close relations between results obtained from the general game-theoretic methods and from the more ad-hoc approaches. In other cases, the game-theoretic approach leads to new insights, not suggested by other approaches.

We use a historical framework for discussing some of the basic ideas of the theory, as well as a few selected applications.

But the viewpoint will be modern; the older ideas will be presented from the perspective of where they have led. Needless to say, we do not even attempt a systematic historical survey.

1910–1930

During these earliest years, game theory was preoccupied with *strictly competitive* games, more commonly known as *two-person zero-sum* games. In these games, there is no point in cooperation or joint action of any kind: if one outcome is preferred to another by one player, then the preference is necessarily reversed for the other. This is the case for most two-person parlour games, such as chess or two-sided poker; but it seems inappropriate for most economic or political applications. Nevertheless, the study of the strictly competitive case has, over the years, turned out remarkably fruitful; many of the concepts and results generated in connection with this case are in fact much more widely applicable, and have become cornerstones of the more general theory. These include the following:

(i) The *extensive* (or *tree*) *form* of a game, consisting of a complete formal description of how the game is played, with a specification of the sequence in which the players move, what they know at the times they must move, how chance occurrences enter the picture, and the *payoff* to each player at the end of play. Introduced by von Neumann (1928), the extensive form was later generalized by Kuhn (1953), and has been enormously influential far beyond zero-sum theory.

(ii) The fundamental concept of *strategy* (or pure strategy) of a player, defined as a complete plan for that player to play the game, as a function of what he observes during the course of play, about the play of others and about chance occurrences affecting the game. Given a strategy for each player, the rules of the game determine a unique outcome of the game and hence a payoff for each player. In the case of two-person zero-sum games, the sum of the two payoffs is zero; this expresses the fact that the preferences of the players over the outcomes are precisely opposed.

(iii) The strategic (or matrix) form of a game. Given strategies $s^1, \ldots, s^n$ for each of the n players, the rules of the game determine a unique payoff $H^i(s^1, \ldots, s^n)$ for each player i. The *strategic* form is simply the function that associates to each profile $s := (s^1, \ldots, s^n)$ of strategies, the *payoff profile*

$$H(s) := (H^1(s), \ldots, H^n(s)).$$

For two-person games, the strategic form often appears as a matrix: the rows and columns represent pure strategies of Players 1 and 2 respectively, whereas the entries are the corresponding payoff profiles. For zero-sum games, of course, it suffices to give the payoff to Player 1. It has been said that the simple idea of thinking of a game in its matrix form is in itself one of the greatest contributions of Game Theory. In facing an interactive situation, there is a great temptation to think only in terms of 'what should I do?'. When one writes down the matrix, one is led to a different viewpoint, one that explicitly takes into account that the other players are also facing a decision problem.

(iv) The concept of *mixed* or *randomized* strategy, indicating that rational play is not in general describable by specifying a single pure strategy. Rather, it is often non-deterministic, with specified probabilities associated with each one of a specified set of pure strategies. When randomized strategies are used, payoff must be replaced by expected payoff. Justifying the use of expected payoff in this context is what led to expected utility theory, whose influence extends far beyond game theory (see *1930–1950*, viii).

(v) The concept of 'individual rationality'. The *security level* of Player i is the amount max min $H^i(s)$ that he can guarantee to himself, independent of what the other players do (here the max is over i's strategies, and the min is over $(n-1)$-tuples of strategies of the players other than i). An outcome is called *individually rational* if it yields each player at least his security level. In the game tic-tac-toe, for example, the only individually rational outcome is a draw; and indeed, it does not take a reasonably bright child very long to learn that 'correct' play in tic-tac-toe always leads to a draw.

Individual rationality may be thought of in terms of pure strategies or, as is more usual, in terms of mixed strategies. In the latter case, what is being 'guaranteed' is not an actual payoff, but an expectation; the word 'guarantee' means that this level of payoff can be attained in the mean, regardless of what the other players do. This 'mixed' security level is always at least as high as the 'pure' one. In the case of tic-tac-toe, each player can guarantee a draw even in the stronger sense of pure strategies. Games like this – i.e. having only one individually rational payoff profile in the 'pure' sense – are called *strictly determined*.

Not all games are strictly determined, not even all two-person zero-sum games. One of the simplest imaginable games is the one that game theorists call 'matching pennies', and children call 'choosing up' ('odds and evens'). Each player privately turns a penny either heads up or tails up. If the choices match, 1 gives 2 his penny; otherwise, 2 gives 1 his penny. In the pure sense, neither player can guarantee more than -1, and hence the game is not strictly determined. But in expectation, each player can guarantee 0, simply by turning the coin heads up or tails up with $1/2 - 1/2$ probabilities. Thus $(0,0)$ is the only payoff profile that is individually rational in the mixed sense. Games like this – i.e. having only one individually rational payoff profile in the 'mixed' sense – are called *determined*. In a determined game, the (mixed) security level is called the *value*, strategies guaranteeing it *optimal*.

(vi) *Zermelo's theorem*. The very first theorem of Game Theory (Zermelo, 1913) asserts that chess is strictly determined. Interestingly, the proof does not construct 'correct' strategies explicitly; and indeed, it is not known to this day whether the 'correct' outcome of chess is a win for white, a win for black, or a draw. The theorem extends easily to a wide class of parlour games, including checkers, go, and chinese checkers, as well as less well-known games such as hex and gnim (Gale, 1979, 1974); the latter two are especially interesting in that one can use Zermelo's theorem to show that Player 1 can force a win, though the proof is non-constructive, and no winning strategy is in fact known. Zermelo's theorem does not extend to card games such as bridge and poker, nor to the variant of chess known as kriegsspiel, where the players cannot observe their opponents' moves directly. The precise condition for the proof to work is that the game be a two-person zero-sum game of *perfect information*. This means that there are no simultaneous moves, and that everything is open and 'above-board': at any given time, all relevant information known to one player is known to all players.

The domain of Zermelo's theorem – two-person zero-sum games of perfect information – seems at first rather limited; but the theorem has reverberated through the decades,

creating one of the main strands of game theoretic thought. To explain some of the developments, we must anticipate the notion of *strategic equilibrium* (Nash, 1951; see *1950–1960*, i). To remove the two-person zero-sum restriction, H.W. Kuhn (1953) replaced the notion of 'correct', individually rational play by that of equilibrium. He then proved that *every n-person game of perfect information has an equilibrium in pure strategies*.

In proving this theorem, Kuhn used the notion of a *subgame* of a game; this turned out crucial in later developments of strategic equilibrium theory, particularly in its economic applications. A subgame relates to the whole game like a subgroup to the whole group or a linear subspace to the whole space; while part of the larger game, it is self-contained, can be played in its own right. More precisely, if at any time, all the players know everything that has happened in the game up to that time, then what happens from then on constitutes a subgame.

From Kuhn's proof it follows that every equilibrium (not necessarily pure) of a subgame can be extended to an equilibrium of the whole game. This, in turn, implies that every game has equilibria that remain equilibria when restricted to any subgame. R. Selten (1965) called such equilibria *subgame perfect*. In games of perfect information, the equilibria that the Zermelo–Kuhn proof yields are all subgame perfect.

But not all equilibria are subgame perfect, even in games of perfect information. Subgame perfection implies that when making choices, a player looks forward and assumes that the choices that will subsequently be made, by himself and by others, will be rational; i.e. in equilibrium. Threats which it would be irrational to carry through are ruled out. And it is precisely this kind of forward-looking rationality that is most suited to economic applications.

Interestingly, it turns out that subgame perfection is not enough to capture the idea of forward-looking rationality. More subtle concepts are needed. We return to this subject below, when we discuss the great flowering of strategic equilibrium theory that has taken place since 1975, and that coincides with an increased preoccupation with its economic applications. The point we wished to make here is that these developments have their roots in Zermelo's theorem.

A second circle of ideas to which Zermelo's theorem led has to do with the foundations of mathematics. The starting point is the idea of a game of perfect information with an infinite sequence of stages. Infinitely long games are important models for interactive situations with an indefinite time horizon – i.e. in which the players act as if there will always be a tomorrow.

To fix ideas, let A be any subset of the unit interval (the set of real numbers between 0 and 1). Suppose two players move alternately, each choosing a digit between 1 and 9 at each stage. The resulting infinite sequence of digits is the decimal expansion of a number in the unit interval. Let G_A be the game in which 1 wins if this number is in A, and 2 wins otherwise. Using Set Theory's 'Axiom of Choice', Gale and Stewart (1953) showed that Zermelo's theorem is false in this situation. One can choose A so that G_A is not strictly determined; that is, against each pure strategy of 1, Player 2 has a winning pure strategy, and against each pure strategy of 2, Player 1 has a winning pure strategy. They also showed that if A is open or closed, then G_A is strictly determined.

Both of these results led to significant developments in foundational mathematics. The axiom of choice had long been suspect in the eyes of mathematicians; the extremely anti-intuitive nature of the Gale–Stewart non-determinateness example was an additional nail in its coffin, and led to an alternative axiom, which asserts that G_A is strictly determined for every set A. This axiom, which contradicts the axiom of choice, has been used to provide an alternative axiomatization for set theory (Mycielski and Steinhaus, 1964), and this in turn has spawned a large literature (see Moschovakis, 1980, 1983). On the other hand, the positive result of Gale and Stewart was successively generalized to wider and wider families of sets A that are 'constructible' in the appropriate sense (Wolfe, 1955; Davis, 1964), culminating in the theorem of Martin (1975), according to which G_A is strictly determined whenever A is a Borel set.

Another kind of perfect information game with infinitely many stages is the *differential game*. Here time is continuous but usually of finite duration; a decision must be made at each instant, so to speak. Typical examples are games of pursuit. The theory of differential games was first developed during the 1950s by Rufus Isaacs at the Rand Corporation; his book on the subject was published in 1965, and since then the theory has proliferated greatly. A differential game need not necessarily be of perfect information, but very little is known about those that are not. Some economic examples may be found in Case (1979).

(vii) *The minimax theorem*. The minimax theorem of von Neumann (1928) asserts that every two-person zero-sum game with finitely many pure strategies for each player is determined; that is, when mixed strategies are admitted, it has precisely one individually rational payoff vector. This had previously been verified by E. Borel (e.g. 1924) for several special cases, but Borel was unable to obtain a general proof. The theorem lies a good deal deeper than Zermelo's, both conceptually and technically.

For many years, minimax was considered the elegant centrepiece of game theory. Books about game theory concentrated on two-person zero-sum games in strategic form, often paying only desultory attention to the non-zero sum theory. Outside references to game theory often gave the impression that non-zero sum games do not exist, or at least play no role in the theory.

The reaction eventually set in, as it was bound to. Game theory came under heavy fire for its allegedly exclusive concern with a special case that has little interest in the applications. Game theorists responded by belittling the importance of the minimax theorem. During the fall semester of 1964, the writer of these lines gave a beginning course in Game Theory at Yale University, without once even mentioning the minimax theorem.

All this is totally unjustified. Except for the period up to 1928 and a short period in the late Forties, game theory was never exclusively or even mainly concerned with the strictly competitive case. The forefront of research was always in n-person or non-zero sum games. The false impression given of the discipline was due to the strictly competitive theory being easier to present in books, more 'elegant' and complete. But for more than half a century, that is not where most of the action has been.

Nevertheless, it is a great mistake to belittle minimax. While not the centrepiece of game theory, it *is* a vital cornerstone. We have already seen how the most fundamental concepts of the general theory – extensive form, pure strategies, strategic form, randomization, utility theory – were spawned in connection with the minimax theorem. But its importance goes considerably beyond this.

The fundamental concept of non-cooperative n-person game theory – the strategic equilibrium of Nash (1951) – is an outgrowth of minimax, and the proof of its existence is

modelled on a previously known proof of the minimax theorem. In cooperative n-person theory, individual rationality is used to define the set of *imputations*, on which much of the cooperative theory is based. In the theory of repeated games, individual rationality also plays a fundamental role.

In many areas of interest – stochastic games, repeated games of incomplete information, continuous games (i.e. with a continuum of pure strategies), differential games, games played by automata, games with vector payoffs – the strictly competitive case already presents a good many of the conceptual and technical difficulties that are present in general. In these areas, the two-person zero-sum theory has become an indispensable spawning and proving ground, where ideas are developed and tested in a relatively familiar, 'friendly' environment. These theories could certainly not have developed as they did without minimax.

Finally, minimax has had considerable influence on several disciplines outside of game theory proper. Two of these are statistical decision theory and the design of distributed computing systems, where minimax is used for 'worst case' analysis. Another is mathematical programming; the minimax theorem is equivalent to the duality theorem of linear programming, which in turn is closely related to the idea of shadow pricing in economics. This circle of ideas has fed back into game theory proper; in its guise as a theorem about linear inequalities, the minimax theorem is used to establish the condition of Bondareva (1963) and Shapley (1967) for the non-emptiness of the core of an n-person game, and the Hart-Schmeidler (1988) elementary proof for the existence of correlated equilibria.

(viii) *Empirics*. The correspondence between theory and observation was discussed already by von Neumann (1928), who observed that the need to randomize arises endogenously out of the theory. Thus the phenomenon of bluffing in poker may be considered a confirmation of the theory. This kind of connection between theory and observation is typical of game theory and indeed of economic theory in general. The 'observations' are often qualitative rather than quantitative; in practice, we do observe bluffing, though not necessarily in the proportions predicted by theory.

As for experimentation, strictly competitive games constitute one of the few areas in game theory, and indeed in social science, where a fairly sharp, unique 'prediction' is made (though even this prediction is in general probabilistic). It thus invites experimental testing. Early experiments failed miserably to confirm the theory; even in strictly determined games, subjects consistently reached individually irrational outcomes. But experimentation in rational social science is subject to peculiar pitfalls, of which early experimenters appeared unaware, and which indeed mar many modern experiments as well. These have to do with the motivation of the subjects, and with their understanding of the situation. A determined effort to design an experimental test of minimax that would avoid these pitfalls was recently made by B. O'Neill (1987); in these experiments, the predictions of theory were confirmed to within less than one per cent.

1930–1950

The outstanding event of this period was the publication, in 1944, of the *Theory of Games and Economic Behavior* by John von Neumann and Oskar Morgenstern. Morgenstern was the first economist clearly and explicitly to recognize that economic agents must take the interactive nature of economics into account when making their decisions. He and von

Neumann met at Princeton in the late Thirties, and started the collaboration that culminated in the *Theory of Games*. With the publication of this book, Game Theory came into its own as a scientific discipline.

In addition to expounding the strictly competitive theory described above, the book broke fundamental new ground in several directions. These include the notion of a cooperative game, its coalitional form, and its von Neumann–Morgenstern stable sets. Though axiomatic expected utility theory had been developed earlier by Ramsey (1931), the account of it given in this book is what made it 'catch on'. Perhaps most important, the book made the first extensive applications of game theory, many to economics.

To put these developments into their modern context, we discuss here certain additional ideas that actually did not emerge until later, such as the core, and the general idea of a solution concept. At the end of this section we also describe some developments of this period not directly related to the book, including games with a continuum of strategies, the computation of minimax strategies, and mathematical advances that were instrumental in later work.

(i) *Cooperative games.* A game is called *cooperative* if commitments – agreements, promises, threats – are fully binding and enforceable (Harsanyi 1966, p. 616). It is called *non-cooperative* if commitments are not enforceable, even if pre-play communication between the players is possible. (For motivation, see *1950–1960*, iv.)

Formally, cooperative games may be considered a special case of non-cooperative games, in the sense that one may build the negotiation and enforcement procedures explicitly into the extensive form of the game. Historically, however, this has not been the mainstream approach. Rather, cooperative theory starts out with a formalization of games (the coalitional form) that abstracts away altogether from procedures and from the question of how each player can best manipulate them for his own benefit; it concentrates, instead, on the possibilities for agreement. The emphasis in the non-cooperative theory is on the individual, on what strategy he should use. In the cooperative theory it is on the group: What coalitions will form? How will they divide the available payoff between their members?

There are several reasons that cooperative games came to be treated separately. One is that when one does build negotiation and enforcement procedures explicitly into the model, then the results of a non-cooperative analysis depend very strongly on the precise form of the procedures, on the order of making offers and counter-offers, and so on. This may be appropriate in voting situations in which precise rules of parliamentary order prevail, where a good strategist can indeed carry the day. But problems of negotiation are usually more amorphous; it is difficult to pin down just what the procedures are. More fundamentally, there is a feeling that procedures are not really all that relevant; that it is the possibilities for coalition forming, promising and threatening that are decisive, rather than whose turn it is to speak.

Another reason is that even when the procedures are specified, non-cooperative analyses of a cooperative game often lead to highly non-unique results, so that they are often quite inconclusive.

Finally, detail distracts attention from essentials. Some things are seen better from a distance; the Roman camps around Metzada are indiscernible when one is in them, but easily visible from the top of the mountain. The coalitional form of a game, by abstracting away from details, yields valuable perspective.

The idea of building non-cooperative models of cooperative games has come to be known as the *Nash program* since it was first proposed by John Nash (1951). In spite of the difficulties just outlined, the programme has had some recent successes (Harsanyi, 1982; Harsanyi and Selten, 1972; Rubinstein, 1982). For the time being, though, these are isolated; there is as yet nothing remotely approaching a general theory of cooperative games based on non-cooperative methodology.

(ii) A *game in coalitional form*, or simply *coalitional game*, is a function v associating a real number $v(S)$ with each subset S of a fixed finite set I, and satisfying $v(\varnothing) = 0$ ($\varnothing$ denotes the empty set). The members of I are called *players*, the subsets S of I *coalitions*, and $v(S)$ is the *worth* of S.

Some notation and terminology: The number of elements in a set S is denoted $|S|$. A *profile* (of strategies, numbers, etc.) is a function on I (whose values are strategies, numbers, etc.). If x is a profile of numbers and S a coalition, we write $x(S) := \Sigma_{i \in S} x^i$.

An example of a coalitional game is the *3-person voting game*; here $|I| = 3$, and $v(S) = 1$ or 0 according as to whether $|S| \geqslant 2$ or not. A coalition S is called *winning* if $v(S) = 1$, *losing* if $v(S) = 0$. More generally, if w is a profile of non-negative numbers (*weights*) and q (the *quota*) is positive, define the *weighted voting game* v by $v(S) = 1$ if $w(S) \geqslant q$, and $v(S) = 0$ otherwise. An example is a parliament with several parties. The players are the parties, rather than the individual members of parliament, w^i is the number of seats held by party i, and q is the number of votes necessary to form a government (usually a simple majority of the parliament). The weighted voting game with quota q and weights w^i is denoted $[q; w]$; e.g., the three-person voting game is $[2; 1, 1, 1]$.

Another example of a coalitional game is a *market game*. Suppose there are l natural resources, and a single consumer product, say 'bread', that may be manufactured from these resources. Let each player i have an endowment e^i of resources (an l-vector with non-negative coordinates), and a concave production function u^i that enables him to produce the amount $u^i(x)$ of bread given the vector $x = (x_1, \ldots, x_l)$ of resources. Let $v(S)$ be the maximum amount of bread that the coalition S can produce; it obtains this by redistributing its resources among its members in a manner that is most efficient for production, i.e.

$$v(S) = \max\left\{ \sum_{i \in S} u^i(x^i) : \sum_{i \in S} x^i = \sum_{i \in S} e^i \right\}$$

where the x^i are restricted to have non-negative coordinates.

These examples illustrate different interpretations of coalitional games. In one interpretation, the payoff is in terms of some single desirable physical commodity, such as bread; $v(S)$ represents the maximum total amount of this commodity that the coalition S can procure for its members, and it may be distributed among the members in any desired way. This is illustrated by the above description of the market game.

Underlying this interpretation are two assumptions. First, that of *transferable utility* (TU): that the payoff is in a form that is freely transferable among the players. Second, that of *fixed threats*: that S can obtain a maximum of $v(S)$ no matter what the players outside of S do.

Another interpretation is that $v(S)$ represents some appropriate index of S's strength (if it forms). This requires neither transferable utility nor fixed threats. In voting games, for example, it is natural to define $v(S) = 1$ if S is a winning coalition (e.g. can form a government or ensure passage of a bill), 0 if not. Of course, in most situations represented by voting games, utility is not transferable.

Another example is a market game in which the x^i are consumption goods rather than resources. Rather than bread, $\Sigma_{i \in S} u^i(x^i)$ may represent a social welfare function such as is often used in growth or taxation theory. While $v(S)$ cannot then be divided in an arbitrary way among the members of S, it still represents a reasonable index of S's strength. This is a situation with fixed threats but without TU.

Von Neumann and Morgenstern considered strategic games with transferable payoffs, which is a situation with TU but without fixed threats. If the profile s of strategies is played, the coalition S may divide the amount $\Sigma_{i \in S} H^i(s)$ among its members in any way it pleases. However, what S gets depends on what players outside S do. Von Neumann and Morgenstern defined $v(S)$ as the maxmin payoff of S in the two-person zero-sum game in which the players are S and $I \backslash S$, and the payoff to S is $\Sigma_{i \in S} H^i(s)$; i.e., as the expected payoff that S can assure itself (in mixed strategies), no matter what the others do. Again, this is a reasonable index of S's strength, but certainly not the only possible one.

We will use the term *TU coalitional game* when referring to coalitional games with the TU interpretation.

In summary, the coalitional form of a game associates with each coalition S a single number $v(S)$, which in some sense represents the total payoff that that coalition can get or may expect. In some contexts, $v(S)$ fully characterizes the possibilities open to S; in others, it is an index that is indicative of S's strength.

(iii) *Solution concepts.* Given a game, what outcome may be expected? Most of game theory is, in one way or another, directed at this question. In the case of two-person zero-sum games, a clear answer is provided: the unique individually rational outcome. But in almost all other cases, there is no unique answer. There are different criteria, approaches, points of view, and they yield different answers.

A *solution concept* is a function (or correspondence) that associates outcomes, or sets of outcomes, with games. Usually an 'outcome' may be identified with the profile of payoffs that outcome yields to the players, though sometimes we may wish to think of it as a strategy profile.

Of course a solution concept is not just any such function or correspondence, but one with a specific rationale; for example, the strategic equilibrium and its variants for strategic form games, and the core, the von Neumann–Morgenstern stable sets, the Shapley value and the nucleolus for coalitional games. Each represents a different approach or point of view.

What will 'really' happen? Which solution concept is 'right'? None of them; they are indicators, not predictions. Different solution concepts are like different indicators of an economy; different methods for calculating a price index; different maps (road, topo, political, geologic, etc., not to speak of scale, projection, etc.); different stock indices (Dow Jones, Standard and Poor's NYSE, etc., composite, industrials, utilities, etc.); different batting statistics (batting average, slugging average, RBI, hits, etc.); different kinds of information about rock climbs (arabic and roman difficulty ratings, route maps, verbal descriptions of the climb, etc.); accounts of the same event by different people or different media; different projections of the same three-dimensional object (as in architecture or engineering). They depict or illuminate the situation from different angles; each one stresses certain aspects at the expense of others.

Moreover, solution concepts necessarily leave out altogether some of the most vital information, namely that not entering the formal description of the game. When applied to a voting game, for example, no solution concept can take into account

matters of custom, political ideology, or personal relations, since they don't enter the coalitional form. That does not make the solution useless. When planning a rock climb, you certainly want to take into account a whole lot of factors other than the physical characteristics of the rock, such as the season, the weather, your ability and condition, and with whom your are going. But you also do want to know about the ratings.

A good analogy is to distributions (probability, frequency, population, etc.). Like a game, a distribution contains a lot of information; one is overwhelmed by all the numbers. The median and the mean summarize the information in different ways; though other than by simply stating the definitions, it is not easy to say how. The definitions themselves do have a certain fairly clear intuitive content; more important, we gain a feeling for the relation between a distribution and its median and mean from experience, from working with various specific examples and classes of examples over the course of time.

The relationship of solution concepts to games is similar. Like the median and the mean, they in some sense summarize the large amount of information present in the formal description of a game. The definitions themselves have a certain fairly clear intuitive content, though they are not predictions of what will happen. Finally, the relations between a game and its core, value, stable sets, nucleolus, and so on is best revealed by seeing where these solution concepts lead in specific games and classes of games.

(iv) *Domination, the core and imputations.* Continuing to identify 'outcome' with 'payoff profile', we call an outcome y of a game *feasible* if the all-player set I can achieve it. An outcome x dominates y if there exists a coalition S that can achieve at least its part of x, and each of whose members prefers x to y; in that case we also say that S can *improve upon* y. The *core* of a game is the set of all feasible outcomes that are not dominated.

In a TU coalitional game v, feasibility of x means $x(I) \leq v(I)$, and x dominating y via S means that $x(S) \leq v(S)$ and $x^i > y^i$ for all i in S. The core of v is the set of all feasible y with $y(S) \geq v(S)$ for all S.

At first, the core sounds quite compelling; why should the players be satisfied with an outcome that some coalition can improve upon? It becomes rather less compelling when one realizes that many perfectly ordinary games have empty cores, i.e. every feasible outcome can be improved upon. Indeed, this is so even in as simple a game as the 3-person voting game.

For a coalition S to improve upon an outcome, players in S must trust each other; they must have faith that their comrades inside S will not desert them to make a coalition with other players outside S. In a TU 3-person voting game, $y := (1/3, 1/3, 1/3)$ is dominated via $\{1, 2\}$ by $x := (1/2, 1/2, 0)$. But 1 and 2 would be wise to view a suggested move from y to x with caution. What guarantee does 1 have that 2 will really stick with him and not accept offers from 3 to improve upon x with, say, $(0, 2/3, 1/3)$? For this he must depend on 2's good faith, and similarly 2 must depend on 1's.

There are two exceptions to this argument, two cases in which domination does not require mutual trust. One is when S consists of a single player. The other is when $S = I$, so that there is no one outside S to lure one's partners away.

The requirement that a feasible outcome y be undominated via one-person coalitions (*individual rationality*) and via the all-person coalition (*efficiency* or *Pareto optimality*) is thus quite compelling, much more so than that it be in the core. Such outcomes are called *imputations*. For TU coalitional games, individual rationality means that $y^i \geq v(i)$ for all i (we do not distinguish between i and $\{i\}$), and efficiency means that $y(I) = v(I)$. The outcomes associated with most cooperative solution concepts are imputations; the imputations constitute the stage on which most of cooperative game theory is played out.

The notion of core does not appear explicitly in von Neumann and Morgenstern, but it is implicit in some of the discussions of stable sets there. In specific economic contexts, it is implicit in the work of Edgeworth (1881) and Ransmeier (1942). As a general solution concept in its own right, it was developed by Shapley and Gillies in the early Fifties. Early references include Luce and Raiffa (1957) and Gillies (1959).

(v) *Stable sets.* The discomfort with the definition of core expressed above may be stated more sharply as follows. Suppose we think of an outcome in the core as 'stable'. Then we should not exclude an outcome y just because it is dominated by *some* other outcome x; we should demand that x itself be stable. If x is not itself stable, then the argument for excluding y is rather weak; proponents of y can argue with justice that replacing it with x would not lead to a more stable situation, so we may as well stay where we are. If the core were the set of all outcomes not dominated by any element of the core, there would be no difficulty; but this is not so.

Von Neumann and Morgenstern were thus led to the following definition: A set K of imputations is called *stable* if it is the set of all imputations not dominated by any element of K.

This definition guarantees neither existence nor uniqueness. On the face of it, a game may have many stable sets, or it may have none. Most games do, in fact, have many stable sets; but the problem of existence was open for many years. It was solved by Lucas (1969), who constructed a ten-person TU coalitional game without any stable set. Later, Lucas and Rabie (1982) constructed a fourteen-person TU coalitional game without any stable set and with an empty core to boot.

Much of the *Theory of Games* is devoted to exploring the stable sets of various classes of TU coalitional games, such as 3- and 4-person games, voting games, market games, compositions of games, and so on. (If v and w have disjoint player sets I and J, their *composition* u is given by $u(S): = v(S \cap I) + w(S \cap J)$.) During the 1950s many researchers carried forward with great vigour the work of investigating various classes of games and describing their stable sets. Since then work on stable sets has continued unabated, though it is no longer as much in the forefront of game-theoretic research as it was then. All in all, more than 200 articles have been published on stable sets, some 80 per cent of them since 1960. Much of the recent activity in this area has taken place in the Soviet Union.

It is impossible here even to begin to review this large and varied literature. But we do note one characteristic qualitative feature. By definition, a stable set is simply a set of imputations; there is nothing explicit in it about social structure. Yet the mathematical description of a given stable set can often best be understood in terms of an implicit social structure or form of organization of the players. Cartels, systematic discrimination, groups within groups, all kinds of subtle organizational forms spring to one's attention. These forms are endogenous, they are not imposed by definition, they emerge from the analysis. It is a mystery that just the stable set concept, and it only, is so closely allied with endogenous notions of social structure.

We adduce just one, comparatively simple example. The TU 3-person voting game has a stable set consisting of the three imputations (1/2, 1/2, 0), (1/2, 0, 1/2), (0, 1/2, 1/2). The social structure implicit in this is that all three players will *not* compromise by dividing the payoff equally. Rather, one of the three 2-person coalitions will form and divide the payoff equally, with the remaining player being left 'in the cold'. Because any of these three coalitions can form, competition drives them to divide the payoff equally, so that no player will prefer any one coalition to any other.

Another stable set is the interval $\{(\alpha, 1 - \alpha, 0)\}$, where α ranges from 0 to 1. Here Player 3 is permanently excluded from all negotiations; he is 'discriminated against'. Players 1 and 2 divide the payoff in some arbitrary way, not necessarily equally; this is because a coalition with 3 is out of the question, and so competition no longer constrains 1 and 2 in bargaining with each other.

(vi) *Transferable utility.* Though it no longer enjoys the centrality that it did up to about 1960, the assumption of transferable utility has played and continues to play a major role in the development of cooperative game theory. Some economists have questioned the appropriateness of the TU assumption, especially in connection with market models; it has been castigated as excessively strong and unrealistic.

This situation is somewhat analogous to that of strictly competitive games, which as we pointed out above (*1930–1950*, vii), constitute a proving ground for developing and testing ideas that apply also to more general, non-strictly competitive games. The theory of NTU (non-transferable utility) coalitional games is now highly developed (see *1960–1970*, i), but it is an order of magnitude more complex than that of TU games. The TU theory is an excellent laboratory or model for working out ideas that are later applied to the more general NTU case.

Moreover, TU games are both conceptually and technically much closer to NTU games than strictly competitive games are to non-strictly competitive games. A very large part of the important issues arising in connection with non-strictly competitive games do not have any counterpart at all in strictly competitive games, and so simply cannot be addressed in that context. But by far the largest part of the issues and questions arising in the NTU theory do have counterparts in the TU theory; they can at least be addressed and dealt with there.

Almost every major advance in the NTU theory – and many a minor advance as well – has had its way paved by a corresponding advance in the TU theory. Stable sets, core, value, and bargaining set were all defined first for TU games, then for NTU. The enormous literature on the core of a market and the equivalence between it and competitive equilibrium (c.e.) in large markets was started by Martin Shubik (1959a) in an article on TU markets. The relation between the value and c.e. in large markets was also explored first for the TU case (Shapley, 1964; Shapley and Shubik, 1969b; Aumann and Shapley, 1974; Hart, 1977a), then for NTU (Champsaur, 1975, but written and circulated circa 1970; Aumann, 1975; Mas-Colell, 1977; Hart, 1977b). The same holds for the bargaining set; first TU (Shapley, 1984), then NTU (Mas-Colell, 1988). The connection between balanced collections of coalitions and the non-emptiness of the core (*1960–1970*, viii) was studied first for TU (Bondareva, 1963; Shapley, 1967), then for NTU (Scarf, 1967; Billera, 1970b; Shapley 1973a); this development led to the whole subject of Scarf's algorithm for finding points in the core, which he and others later extended to algorithms for finding market equilibria and fixed points of mappings in general. Games arising from markets were first abstractly characterized in the TU case (Shapley and Shubik, 1969a), then in the NTU case

(Billera and Bixby, 1973; Mas-Colell, 1975). Games with a continuum of players were conceived first in a TU application (Milnor and Shapley, 1979, but written and circulated in 1960), then NTU (Aumann, 1964). Strategic models of bargaining where time is of the essence were first treated for TU (Rubinstein, 1982), then NTU (Binmore, 1982). One could go on and on.

In each of these cases, the TU development led organically to the NTU development; it isn't just that the one came before the other. TU is to cooperative game theory what *Drosophila* is to genetics. Even if it had no direct economic interest at all, the study of TU coalitional games would be justified solely by their role as an outstandingly suggestive research tool.

(vii) *Single play*. Von Neumann and Morgenstern emphasize that their analysis refers to 'one-shot' games, games that are played just once, after which the players disperse, never to interact again. When this is not the case, one must view the whole situation – including expected future interactions of the same players – as a single larger game, and it, too, is to be played just once.

To some extent this doctrine appears unreasonable. If one were to take it literally, there would be only one game to analyse, namely the one whose players include all persons ever born and to be born. Every human being is linked to every other through some chain of interactions; no person or group is isolated from any other.

Savage (1954) has discussed this in the context of one-person decisions. In principle, he writes, one should 'envisage every conceivable policy for the government of his whole life in its most minute details, and decide here and now on one policy. This is utterly ridiculous ...' (p. 16). He goes on to discuss the *small worlds* doctrine, 'the practical necessity of confining attention to, or isolating, relatively simple situations ...' (p. 82).

To a large extent, this doctrine applies to interactive decisions too. But one must be careful, because here 'large worlds' have qualitative features totally absent from 'small worlds'. We return to this below (*1950–1960, ii, iii*).

(viii) *Expected utility*. When randomized strategies are used in a strategic game, payoff must be replaced by expected payoff (*1910–1930, iv*). Since the game is played only once, the law of large numbers does not apply, so it is not clear why a player would be interested specifically in the mathematical expectation of his payoff.

There is no problem when for each player there are just two possible outcomes, which we may call 'winning' and 'losing', and denominate 1 and 0 respectively. (This involves no zero-sum assumption; e.g. all players could win simultaneously.) In that case the expected payoff is simply the probability of winning. Of course each player wants to maximize this probability, so in that case use of the expectation is justified.

Suppose now that the values of i's payoff function H^i are numbers between 0 and 1, representing win probabilities. Thus, for the 'final' outcome there are still only two possibilities; each pure strategy profile s induces a random process that generates a win for i with probability $H^i(s)$. Then the payoff expectation when randomized strategies are used still represents i's overall win probability.

Now in any game, each player has a most preferred and a least preferred outcome, which we take as a win and a loss. For each payoff h, there is some probability p such that i would as soon get h with certainty as winning with probability p and losing with probability $1 - p$. If we replace all the h's by

the corresponding p's in the payoff matrix, then we are in the case of the previous paragraph, so use of the expected payoff is justified.

The probability p is a function of h, denoted $u^i(h)$, and called i's von Neumann–Morgenstern *utility*. Thus, to justify the use of expectations, each player's payoff must be replaced by its utility.

The key property of the function u^i is that if h and g are random payoffs, then i prefers h to g iff $Eu^i(h) > Eu^i(g)$, where E denotes expectation. This property continues to hold when we replace u^i by a linear transform of the form $\alpha u^i + \beta$, where α and β are constants with $\alpha > 0$. All these transforms are also called utility functions for i, and any one of them may be used rather than u^i in the payoff matrix.

Recall that a strictly competitive game is defined as a two-person game in which if one outcome is preferred to another by one player, the preference is reversed for the other. Since randomized strategies are admitted, this condition applies also to 'mixed outcomes' (probability mixtures of pure outcomes). From this it may be seen that a two-person game is strictly competitive if and only if, for an appropriate choice of utility functions, the utility payoffs of the players sum to zero in each square of the matrix.

The case of TU coalitional games deserves particular attention. There is no problem if we assume fixed threats and continue to denominate the payoff in bread (see ii). But without fixed threats, the total amount of bread obtainable by a coalition S is a random variable depending on what players outside S do; since this is not denominated in utility, there is no justification for replacing it by its expectation. But if we do denominate payoffs in utility terms, then they cannot be directly transferred. The only way out of this quandary is to assume that the utility of bread is linear in the amount of bread (Aumann, 1960). We stress again that no such assumption is required in the fixed threat case.

(ix) *Applications*. The very name of the book, *Theory of Games and Economic Behavior*, indicates its underlying preoccupation with the applications. Von Neumann had already mentioned *Homo Economicus* in his 1928 paper, but there were no specific economic applications there.

The method of von Neumann and Morgenstern has become the archetype of later applications of game theory. One takes an economic problem, formulates it as a game, finds the game-theoretic solution, then translates the solution back into economic terms. This is to be distinguished from the more usual methodology of economics and other social sciences, where the building of a formal model and a solution concept, and the application of the solution concept to the model, are all rolled into one.

Among the applications extensively treated in the book is voting. A qualitative feature that emerges is that many different weight-quota configurations have the same coalitional form; [5; 2, 3, 4] is the same as [2; 1, 1, 1]. Though obvious to the sophisticated observer when pointed out, this is not widely recognized; most people think that the player with weight 4 is considerably stronger than the others (Vinacke and Arkoff, 1957). The Board of Supervisors of Nassau County operates by weighted voting; in 1964 there were six members, with weights of 31, 31, 28, 21, 2, 2, and a simple majority quota of 58 (Lucas, 1983, p. 188). Nobody realized that three members were totally without influence, that [58; 31, 31, 28, 21, 2, 2] = [2; 1, 1, 1, 0, 0, 0].

In a voting game, a winning coalition with no proper winning subsets is called *minimal winning* (mw). The game [q; w] is *homogeneous* if $w(S) = q$ for all minimal winning S; thus

[3; 2, 1, 1, 1] is homogeneous, but [5; 2, 2, 2, 1, 1, 1] is not. A *decisive* voting game is one in which a coalition wins if and only if its complement loses; both the above games are decisive, but [3; 1, 1, 1, 1] is not. TU decisive homogeneous voting games have a stable set in which some mw coalition forms and divides the payoff in proportion to the weights of its members, leaving nothing for those outside. This is reminiscent of some parliamentary democracies, where parties in a coalition government get cabinet seats roughly in proportion to the seats they hold in parliament. But this fails to take into account that the actual number of seats held by a party may well be quite disproportional to its weight in a homogeneous representation of the game (when there is such a representation).

The book also considers issues of monopoly (or monopsony) and oligopoly. We have already pointed out that stable set theory concerns the endogenous emergence of social structure. In a market with one buyer (monopsonist) and two sellers (duopolists) where supply exceeds demand, the theory predicts that the duopolists will form a cartel to bargain with the monopsonist. The core, on the other hand, predicts cut-throat competition; the duopolists end up by selling their goods for nothing, with the entire consumer surplus going to the buyer.

This is a good place to point out a fundamental difference between the game-theoretic and other approaches to social science. The more conventional approaches take institutions as given, and ask where they lead. The game theoretic approach asks how the institutions came about, what led to them? Thus general equilibrium theory takes the idea of market prices for granted; it concerns itself with their existence and properties, calculating them, and so on. Game Theory asks, *why* are there market prices? How did they come about? Under what conditions will all traders trade at given prices?

Conventional economic theory has several approaches to oligopoly, including competition and cartelization. Starting with any particular one of these, it calculates what is implied in specific applications. Game Theory proceeds differently. It starts with the physical description of the situation only, making no institutional or doctrinal assumptions, then applies a solution concept and sees where it leads.

In a sense, of course, the doctrine is built into the solution concept; as we have seen, the core implies competition, the stable set cartelization. It is not that game theory makes no assumptions, but that the assumptions are of a more general, fundamental nature. The difference is like that between deriving the motion of the planets from Kepler's laws or from Newton's laws. Like Kepler's laws, which apply to the planets only, oligopoly theory applies to oligopolistic markets only. Newton's laws apply to the planets and also to apples falling from trees; stable sets apply to markets and also to voting.

To be sure, conventional economics is also concerned with the genesis of institutions, but on an informal, verbal, ad-hoc level. In Game Theory, institutions like prices or cartels are outcomes of the formal analysis.

(x) Games with a *continuum of pure strategies* were first considered by Ville (1938), who proved the minimax theorem for them, using an appropriate continuity condition. To guarantee the minimax (security) level, one may need to use a continuum of pure strategies, each with probability zero. An example due to Kuhn (1952) shows that in general one cannot guarantee anything even close to minimax using strategies with finite support. Ville's theorem was extended in the fifties to strategic equilibrium in non-strictly competitive games.

(xi) *Computing* security levels, and strategies that will

guarantee them, is highly non-trivial. The problem is equivalent to that of linear programming, and thus succumbed to the simplex method of George Dantzig (1951a, 1951b).

(xii) The major advance in relevant mathematical methods during this period was *Kakutani's fixed point theorem* (1941). An abstract expression of the existence of equilibrium, it is the vital active ingredient of countless proofs in economics and game theory. Also instrumental in later work were Lyapunov's theorem on the range of a vector measure (1940) and von Neumann's selection theorem (1949).

1950–1960

The 1950s were a period of excitement in game theory. The discipline had broken out of its cocoon, and was testing its wings. Giants walked the earth. At Princeton, John Nash laid the groundwork for the general non-cooperative theory, and for cooperative bargaining theory; Lloyd Shapley defined the value for coalitional games, initiated the theory of stochastic games, co-invented the core with D.B. Gillies, and, together with John Milnor, developed the first game models with continua of players; Harold Kuhn worked on behaviour strategies and perfect recall; Al Tucker discovered the prisoner's dilemma; the Office of Naval Research was unstinting in its support. Three Game Theory conferences were held at Princeton, with the active participation of von Neumann and Morgenstern themselves. Princeton University Press published the four classic volumes of *Contributions to the Theory of Games*. The Rand Corporation, for many years to be a major centre of game theoretic research, had just opened its doors in Santa Monica. R. Luce and H. Raiffa (1957) published their enormously influential *Games and Decisions*. Near the end of the decade came the first studies of repeated games.

The major applications at the beginning of the decade were to tactical military problems: defense from missiles, Colonel Blotto, fighter-fighter duels, etc. Later the emphasis shifted to deterrence and cold war strategy, with contributions by political scientists like Kahn, Kissinger, and Schelling. In 1954, Shapley and Shubik published their seminal paper on the value of a voting game as an index of power. And in 1959 came Shubik's spectacular rediscovery of the core of a market in the writings of F.Y. Edgeworth (1881). From that time on, economics has remained by far the largest area of application of game theory.

(i) An *equilibrium* (Nash, 1951) of a strategic game is a (pure or mixed) strategy profile in which each player's strategy maximizes his payoff given that the others are using their strategies. See the entry on NASH EQUILIBRIUM.

Strategic equilibrium is without doubt the single game theoretic solution concept that is most frequently applied in economics. Economic applications include oligopoly, entry and exit, market equilibrium, search, location, bargaining, product quality, auctions, insurance, principal-agent, higher education, discrimination, public goods, what have you. On the political front, applications include voting, arms control, and inspection, as well as most international political models (deterrence, etc.) Biological applications of game theory all deal with forms of strategic equilibrium; they suggest a simple interpretation of equilibrium quite different from the usual overt rationalism (see *1970–1986*, i). We cannot even begin to survey all this literature here.

(ii) *Stochastic and other dynamic games*. Games played in stages, with some kind of stationary time structure, are called *dynamic*. They include stochastic games, repeated games with or without complete information, games of survival (Milnor and Shapley, 1957; Luce and Raiffa, 1957; Shubik, 1959) or ruin (Rosenthal and Rubinstein, 1984), recursive games (Everett, 1957), games with varying opponents (Rosenthal, 1979), and similar models.

This kind of model addresses the concerns we expressed above (*1930–1950*, vii) about the single play assumption. The present can only be understood in the context of the past and the future: 'Know whence you came and where you are going' (Ethics of the Fathers III:1). Physically, current actions affect not only current payoff but also opportunities and payoffs in the future. Psychologically, too, we learn: past experience affects our current expectations of what others will do, and therefore our own actions. We also teach: our current actions affect others' future expectations, and therefore their future actions.

Two dynamic models – stochastic and repeated games – have been especially 'successful'. *Stochastic* games address the physical point, that current actions affect future opportunities. A strategic game is played at each stage; the profile of strategies determines both the payoff at that stage and the game to be played at the next stage (or a probability distribution over such games). In the strictly competitive case, with future payoff discounted at a fixed rate, Shapley (1953a) showed that stochastic games are determined; also, that they have optimal strategies that are stationary, in the sense that they depend only on the game being played (not on the history or even on the date). Bewley and Kohlberg (1976) showed that as the discount rate tends to 0 the value tends to a limit; this limit is the same as the limit, as $k \to \infty$, of the values of the k-stage games, in each of which the payoff is the mean payoff for the k stages. Mertens and Neyman (1981) showed that the value exists also in the undiscounted infinite stage game, when payoff is defined by the Cesaro limit (limit, as $k \to \infty$, of the average payoff in the first k stages). For an understanding of some of the intuitive issues in this work, see Blackwell and Ferguson (1968), which was extremely influential in the modern development of stochastic games.

The methods of Shapley, and of Bewley and Kohlberg, can be used to show that non-strictly competitive stochastic games with fixed discounts have equilibria in stationary strategies, and that when the discount tends to 0, these equilibria converge to a limit (Mertens, 1982). But unlike in the strictly competitive case, the payoff to this limit need not correspond to an equilibrium of the undiscounted game (Sorin, 1986b). It is not known whether undiscounted non-strictly competitive stochastic games need at all have strategic equilibria.

(iii) *Repeated* games model the psychological, informational side of ongoing relationships. Phenomena like cooperation, altruism, trust, punishment, and revenge are predicted by the theory. These may be called 'subjective informational' phenomena, since what is at issue is information about the behaviour of the players. Repeated games of incomplete information (*1960–1970*, ii) also predict 'objective informational' phenomena such as secrecy, and signalling of substantive information. Both kinds of informational issue are quite different from the 'physical' issues addressed by stochastic games.

Given a strategic game G, consider the game G^∞ each play of which consists of an infinite sequence of repetitions of G. At each stage, all players know the actions taken by all players at all previous stages. The payoff in G^∞ is some kind of average of the stage payoffs; we will not worry about exact definitions here.

The reader is referred to the entry on REPEATED GAMES. Here we state only one basic result, known as the *Folk Theorem*. Call an outcome (payoff profile) x *feasible* in G if it is achievable by the all-player set when using a correlated randomizing device; i.e. is in the convex hull of the 'pure' outcomes. Call it *strongly individually rational* if no player i can be prevented from achieving x^i by the other players, when they are randomizing independently; i.e. if $x^i \geqslant \min\max H^i(s)$, where the max is over i's strategies, and the min is over $(n-1)$-tuples of mixed strategies of the others. The Folk Theorem then says that the equilibrium outcomes in the repetition G^∞ coincide with the feasible and strongly individually rational outcomes in the one-shot game G.

The authorship of the Folk Theorem, which surfaced in the late Fifties, is obscure. Intuitively, the feasible and strongly individually rational outcomes are the outcomes that could arise in cooperative play. Thus the Folk Theorem points to a strong relationship between repeated and cooperative games. Repetition is a kind of enforcement mechanism; agreements are enforced by 'punishing' deviators in subsequent stages.

(iv) The *Prisoner's Dilemma* is a two-person non-zero sum strategic game with payoff matrix as depicted in Figure 1. Attributed to A.W. Tucker, it has deservedly attracted enormous attention; it is said that in the social psychology literature alone, over a thousand papers have been devoted to it.

One may think of the game as follows: Each player decides whether he will receive $1000 or the other will receive $3000. The decisions are simultaneous and independent, though the players may consult with each other before deciding.

The point is that ordinary rationality leads each player to choose the $1000 for himself, since he is thereby better off *no matter what the other player does*. But the two players thereby get only $1000 each, whereas they could have gotten $3000 each if both had been 'friendly' rather than 'greedy'.

The universal fascination with this game is due to its representing, in very stark and transparent form, the bitter fact that when individuals act for their own benefit, the result may well be disaster for all. This principle has dozens of applications, great and small, in everyday life. *People who fail to cooperate for their own mutual benefit are not necessarily foolish or irrational*; they may be acting perfectly rationally. The sooner we accept this, the sooner we can take steps to design the terms of social intercourse so as to encourage cooperation.

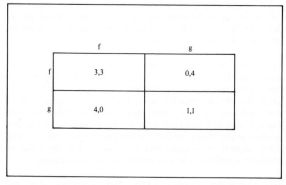

	f	g
f	3,3	0,4
g	4,0	1,1

Figure 1

One such step, of very wide applicability, is to make available a mechanism for the enforcement of voluntary agreements. 'Pray for the welfare of government, without whose authority, man would swallow man alive' (Ethics of the Fathers III:2). The availability of the mechanism is itself sufficient; once it is there, the players are naturally motivated to use it. If they can make an *enforceable* agreement yielding (3, 3), they would indeed be foolish to end up with (1, 1). It is this that motivates the definition of a cooperative game (*1930–1950*, i).

The above discussion implies that (*g*, *g*) is the unique strategic equilibrium of the prisoner's dilemma. It may also be shown that in any finite repetition of the game, all strategic equilibria lead to a constant stream of 'greedy' choices by each player; but this is a subtler matter than the simple domination argument used for the one-shot case. In the infinite repetition, the Folk Theorem (iii) shows that (3, 3) is an equilibrium outcome; and indeed, there are equilibria that lead to a constant stream of 'friendly' choices by each player. The same holds if we discount future payoff in the repeated game, as long as the discount rate is not too large (Sorin, 1986a).

R. Axelrod (1984) has carried out an experimental study of the repeated prisoner's dilemma. Experts were asked to write computer programmes for playing the game, which were matched against each other in a 'tournament'. At each stage, the game ended with a fixed (small) probability; this is like discounting. The most successful program in the tournament turned out to be a 'cooperative' one: Matched against itself, it yields a constant stream of 'friendly' choices; matched against others, it 'punishes' greedy choices. The results of this experiment thus fit in well with received theoretical doctrine.

The design of this experiment is noteworthy because it avoids the pitfalls so often found in game experiments: lack of sufficient motivation and understanding. The experts chosen by Axelrod understood the game as well as anybody. Motivation was provided by the investment of their time, which was much more considerable than that of the average subject, and by the glory of a possible win over distinguished colleagues. Using computer programmes for strategies presaged important later developments (*1970–1986*, iv).

Much that is fallacious has been written on the one-shot prisoner's dilemma. It has been said that for the reasoning to work, pre-play communication between the players must be forbidden. This is incorrect. The players can communicate until they are blue in the face, and agree solemnly on (*f*, *f*); when faced with the actual decision, rational players will still choose *g*. It has been said that the argument depends on the notion of strategic equilibrium, which is open to discussion. This too is incorrect; the argument depends only on strong domination, i.e. on the simple proposition that people always prefer to get another $1000. 'Resolutions' of the 'paradox' have been put forward, suggesting that rational players will play *f* after all; that my choosing *f* has some kind of 'mirror' effect that makes you choose it also. Worse than just nonsense, this is actually vicious, since it suggests that the prisoner's dilemma does not represent a real social problem that must be dealt with.

Finally, it has been said that the experimental evidence – Axelrod's and that of others – contradicts theory. This too is incorrect, since most of the experimental evidence relates to repeated games, where the friendly outcome is perfectly consonant with theory; and what evidence there is in one-shot games does point to a preponderance of 'greedy' choices. It is true that in long finite repetitions, where the only equilibria are greedy, most experiments nevertheless point to the friendly outcome; but fixed finite repetitions are somewhat artificial,

and besides, this finding, too, can be explained by theory (Neyman, 1985; see *1970–1986*, iv).

(v) We turn now to cooperative issues. A model of fundamental importance is the *bargaining problem* of Nash (1950). Formally, it is defined as a convex set C in the Euclidean plane, containing the origin in its interior. Intuitively, two players bargain; they may reach any agreement whose payoff profile is in C; if they disagree, they get nothing. Nash listed four *axioms* – conditions that a reasonable compromise solution might be expected to satisfy – such as symmetry and efficiency. He then showed that there is one and only one solution satisfying them, namely the point *x* in the non-negative part of C that maximizes the product $x^1 x^2$. An appealing economic interpretation of this solution was given by Harsanyi (1956).

By varying the axioms, other authors have obtained different solutions to the bargaining problem, notably Kalai–Smorodinski (1975) and Maschler–Perles (1980). Like Nash's solution, each of these is characterized by a formula with an intuitively appealing interpretation.

Following work of A. Rubinstein (1982), K. Binmore (1982) constructed an explicit bargaining model which, when analyzed as a non-cooperative strategic game, leads to Nash's solution of the bargaining problem. This is an instance of a successful application of the 'Nash program' (see *1930–1950*, vi). Similar constructions have been made for other solutions of the bargaining problem

An interesting qualitative feature of the Nash solution is that it is very sensitive to risk aversion. A risk loving or risk neutral bargainer will get a better deal than a risk averse one; this is so even when there are no overt elements of risk in the situation, nothing random. The very willingness to take risks confers an advantage, though in the end no risks are actually taken.

Suppose, for example, that two people may divide $600 in any way they wish; if they fail to agree, neither gets anything. Let their utility functions be $u^1(\$x) = x$ and $u^2(\$x) = \sqrt{x}$, so that 1 is risk neutral, 2 risk averse. Denominating the payoffs in utilities rather than dollars, we find that the Nash solution corresponds to a dollar split of $400–$200 in favour of the risk neutral bargainer.

This corresponds well with our intuitions. A fearful, risk averse person will not bargain well. Though there are no overt elements of risk, no random elements in the problem description, the bargaining itself constitutes a risk. A risk averse person is willing to pay, in terms of a less favourable settlement, to avoid the risk of the other side's being adamant, walking away, and so on.

(vi) The *value* (Shapley, 1953b) is a solution concept that associates with each coalitional game *v* a unique outcome *φv*. Fully characterized by a set of axioms, it may be thought of as a reasonable compromise or arbitrated outcome, given the power of the players. Best, perhaps, is to think of it simply as an index of power, or what comes to the same thing, of social productivity (*see* SHAPLEY VALUE).

It may be shown that Player *i*'s value is given by

$$\phi^i v = (1/n!) \sum v^i(S_R^i),$$

where Σ ranges over all *n*! orders on the set I of all players, S_R^i is the set of players up to and including *i* in the order *R*, and $v^i(S)$ is the *contribution* $v(S) - v(S \setminus i)$ of *i* to the coalition S; note that this implies linearity of *φv* in *v*. In words, $\phi^i v$ is *i*'s mean contribution when the players are ordered at random; this suggests the social productivity interpretation, an interpretation that is reinforced by the following remarkable theorem (Young

1985): Let ψ be a mapping from games v to efficient outcomes ψv, that is symmetric among the players in the appropriate sense. Suppose $\psi' v$ depends only on the 2^{n-1} contributions $v^i(S)$, and monotonically so. Then ψ must be the value ϕ. In brief, if it depends on the contributions only, it's got to be the value, even though we don't assume linearity to start with.

An intuitive feel for the value may be gained from examples. The value of the 3-person voting game is $(1/3, 1/3, 1/3)$, as is suggested by symmetry. This is not in the core, because $\{1, 2\}$ can improve upon it. But so can $\{1, 3\}$ and $\{2, 3\}$; starting from $(1/3, 1/3, 1/3)$, the players might be well advised to leave things as they are (see *1930–1950*, iv). Differently viewed, the symmetric stable set predicts one of the three outcomes $(1/2, 1/2, 0)$, $(1/2, 0, 1/2)$, $(0, 1/2, 1/2)$. Before the beginning of bargaining, each player may figure that his chances of getting into a ruling coalition are 2/3, and conditional on this, his payoff is 1/2. Thus his 'expected outcome' is the value, though in itself, this outcome has no stability.

In the homogenous weighted voting game $[3; 2, 1, 1, 1]$, the value is $(1/2, 1/6, 1/6, 1/6)$; the large player gets a disproportionate share, which accords with intuition: 'l'union fait la force.'

Turning to games of economic interest, we model the market with two sellers and one buyer discussed above (*1930–1950*, ix) by the TU weighted voting game $[3; 2, 1, 1]$. The core consists of the unique point $(1, 0, 0)$, which means that the sellers must give their merchandise, for nothing, to the buyer. While this has clear economic meaning – cutthroat competition – it does not seem very reasonable as a compromise or an index of power. After all, the sellers do contribute something; without them, the buyer could get nothing. If one could be sure that the sellers will form a cartel to bargain with the buyer, a reasonable compromise would be $(1/2, 1/4, 1/4)$. In fact, the value is $(2/3, 1/6, 1/6)$, representing something between the cartel solution and the competitive one; a cartel is possible, but is not a certainty.

Consider next a market in two perfectly divisible and completely complementary goods, which we may call right and left gloves. There are four players; initially 1 and 2 hold one and two left gloves respectively, 3 and 4 hold one right glove each. In coalitional form, $v(1234) = v(234) = 2$, $v(ij) = v(12j) = v(134) = 1$, $v(S) = 0$ otherwise, where $i = 1$, 2, and $j = 3$, 4. The core consists of $(0, 0, 1, 1)$ only; that is, the owners of the left gloves must simply give away their merchandise, for nothing. This in itself seems strange enough. It becomes even stranger when one realizes that Player 2 could make the situation entirely symmetric (as between 1, 2 and 3, 4) simply by burning one glove, an action that he can take alone, without consulting anybody.

The value can never suffer from this kind of pathological breakdown in monotonicity. Here $\phi v = (1/4, 7/12, 7/12, 7/12)$, which nicely reflects the features of the situation. There *is* an oversupply of left gloves, and 3 and 4 do benefit from it. Also 2 benefits from it; he always has the option of nullifying it, but he can also use it (when he has an opportunity to strike a deal with both 3 and 4). The brunt of the oversupply is thus born by 1 who, unlike 2, cannot take measures to correct it.

Finally, consider a market with 2,000,001 players, 1,000,000 holding one right glove each, and 1,000,001 holding one left glove each. Again, the core stipulates that the holders of the left gloves must all give away their merchandise, for nothing. True, there *is* a slight oversupply of left gloves; but one would hardly have imagined so drastic an effect from one single glove out of millions. The value, too, takes the oversupply into account, but not in such an extreme form; altogether, the left-glove holders get about 499,557 pairs, the right about

500,443 (Shapley and Shubik, 1969b). This is much more reasonable, though the effect is still surprisingly large: The short side gains an advantage that amounts to almost a thousand pairs.

The value has many different characterizations, all of them intuitively meaningful and interesting. We have already mentioned Shapley's original axioms, the value formula, and Young's characterization. To them must be added Harsanyi's (1959) dividend characterization, Owen's (1972) fuzzy coalition formula, Myerson's (1976) graph approach, Dubey's (1980) diagonal formula, the potential of Hart and Mas-Colell (1986), the reduced game axiomatization by the same authors, and Roth's (1977) formalization of Shapley's (1953b) idea that the value represents the utility to the players of playing a game. Moreover, because of its mathematical tractability, the value lends itself to a far greater range of applications than any other cooperative solution concept. And in terms of general theorems and characterizations for wide classes of games and economies, the value has a greater range than *any* other solution concept, bar none.

Previously (*1930–1950*, iii), we compared solution concepts of games to indicators of distributions, like mean and median. In fact the value is in many ways analogous to the mean, whereas the median corresponds to something like the core, or to core-like concepts such as the nucleolus (*1960–1970*, iv). Like the core, the median has an intuitively transparent and compelling definition (the point that cuts the distribution exactly in half), but lacks an algebraically neat formula; and like the value, the mean has a neat formula whose intuitive significance is not entirely transparent (though through much experience from childhood on, many people have acquired an intuitive feel for it). Like the value, the mean is linear in its data; the core, nucleolus, and median are not. Both the mean and the value are very sensitive to their data: change one datum by a little, and the mean (or value) will respond in the appropriate direction; neither the median nor the core is sensitive in this way: one can change the data in wide ranges without affecting the median (or core) at all. On the other hand, the median can suddenly jump because of a moderate change in just one datum; thus the median of 1,000,001 zeros and 1,000,000 ones is 0, but jumps to 1 if we change just one datum from 0 to 1. We have already seen that the core may behave similarly, but the mean and the value cannot. Both the mean and the value are mathematically very tractable, resulting in a wide range of applications, both theoretical and practical; the median and core are less tractable, resulting in a narrower (though still considerable) range of applications.

The first extensive applications of the value were to various voting games (Shapley and Shubik, 1954). The key observation in this seminal paper was that the value of a player equals his probability of *pivoting* – turning a coalition from losing to winning – when the players are ordered at random. From this there has grown a very large literature on voting games. Other important classes of applications are to market games (*1960–1970*, v) and political-economic games (e.g. Aumann and Kurz, 1977; Neyman, 1985b).

(vii) *Axiomatics*. The Shapley value and Nash's solution to the bargaining problem are examples of the axiomatic approach. Rather than defining a solution concept directly, one writes down a set of conditions for it to satisfy, then sees where they lead. In many contexts, even a relatively small set of fairly reasonable conditions turn out to be self-contradictory; there is no concept satisfying all of them. The most famous instance of this is Arrow's (1951) impossibility theorem for social welfare functions, which is one of the earliest applications of axiomatics in the social sciences.

It is not easy to pin down precisely what is meant by 'the axiomatic method'. Sometimes the term is used for any formal deductive system, with undefined terms, assumptions, and conclusions. As understood today, all of game theory and mathematical economics fits that definition. More narrowly construed, an axiom system is a small set of individually transparent conditions, set in a fairly general and abstract framework, which when taken together have far-reaching implications. Examples are Euclid's axioms for geometry, the Zermelo–Fraenkel axioms for set theory, the conditions on multiplication that define a group, the conditions on open sets that define a topological space, and the conditions on preferences that define utility and/or subjective probability.

Game theoretic solution concepts often have both direct and axiomatic characterizations. The direct definition applies to each game separately, whereas most axioms deal with relationships between games. Thus the formula for the Shapley value ϕv enables one to calculate it without referring to any game other than v. But the axioms for ϕ concern relationships between games; they say that if the values of certain games are so and so, then the values of certain other, related games must be such and such. For example, the additivity axiom is $\phi(v+w) = \phi v + \phi w$. This is analogous to direct vs. axiomatic approaches to integration. Direct approaches such as limit of sum work on a single function; axiomatic approaches characterize the integral as a linear operator on a *space* of functions. (Harking back to the discussion at (vi), we note that the axioms for the value are quite similar to those for the integral, which in turn is closely related to the mean of a distribution.)

Shapley's value and the solutions to the bargaining problem due to Nash (1950), Kalai–Smorodinski (1975) and Maschler–Perles (1980) were originally conceived axiomatically, with the direct characterization coming afterwards. In other cases the process was reversed; for example, the nucleolus, NTU Shapley value, and NTU Harsanyi value were all axiomatized only years after their original direct definition (see *1960–1970*). Recently the core, too, has been axiomatized (Peleg, 1985, 1986).

Since axiomatizations concern relations between different games, one may ask why the players of a given game should be concerned with other games, which they are not playing. This has several answers. Viewed as an indicator, a solution of a game doesn't tell us much unless it stands in some kind of coherent relationship to the solutions of other games. The ratings for a rock climb tell you something if you have climbed other rocks whose ratings you know; topographic maps enable you to take in a situation at a glance if you have used them before, in different areas. If we view a solution as an arbitrated or imposed outcome, it is natural to expect some measure of consistency from an arbitrator or judge. Indeed, much of the law is based on precedent, which means relating the solution of the given 'game' to those of others with known solutions. Even when viewing a solution concept as a norm of actual behaviour, the very word 'norm' implies that we are thinking of a function on classes of games rather than of a single game; outcomes are largely based on mutual expectations, which are determined by previous experience with other games, by 'norms'.

Axiomatizations serve a number of useful purposes. First, like any other alternative characterization, they shed additional light on a concept, enable us to 'understand' it better. Second, they underscore and clarify important similarities between concepts, as well as differences between them. One

example of this is the remarkable 'reduced game property' or 'consistency principle', which is associated in various different forms with just about every solution concept, and plays a key role in many of the axiomatizations (see *1970–1986*, vi). Another example consists of the axiomatizations of the Shapley and Harsanyi NTU values. Here the axioms are exact analogues, except that in the Shapley case they refer to payoff profiles, and in the Harsanyi case to 2^n-tuples of payoff profiles, one for each of the 2^n coalitions (Hart, 1985a). This underscores the basic difference in outlook between those two concepts: The Shapley value assumes that the all-player coalition eventually forms, the intermediate coalitions being important only for bargaining chips and threats, whereas the Harsanyi value takes into account a real possibility of the intermediate coalitions actually forming.

Last, an important function of axiomatics relates to 'counter-intuitive examples', in which a solution concept yields outcomes that seem bizarre; e.g. the cores of some of the games discussed above in (vi). Most axioms appearing in axiomatizations do seem reasonable on the face of it, and many of them are in fact quite compelling. The fact that a relatively small selection of such axioms is often categoric (determines a unique solution concept), and that different such selections yield different answers, implies that all together, these reasonable-sounding axioms are contradictory. This, in turn, implies that any one solution concept will necessarily violate at least some of the axioms that are associated with other solution concepts; thus if the axioms are meant to represent intuition, counter-intuitive examples are inevitable.

In brief, axiomatics underscores the fact that a 'perfect' solution concept is an unattainable goal, a *fata morgana*; there is something 'wrong', some quirk with every one. Any given kind of counterintuitive example can be eliminated by an appropriate choice of solution concept, but only at the cost of another quirk turning up. Different solution concepts can therefore be thought of as results of choosing not only which properties one likes, but also which examples one wishes to avoid.

1960–1970

The Sixties were a decade of growth. Extensions such as games of incomplete information and NTU coalitional games made the theory much more widely applicable. The fundamental underlying concept of common knowledge was formulated and clarified. Core theory was extensively developed and applied to market economies; the bargaining set and related concepts such as the nucleolus were defined and investigated; games with many players were studied in depth. The discipline expanded geographically, outgrowing the confines of Princeton and Rand; important centres of research were established in Israel, Germany, Belgium and the Soviet Union. Perhaps most important was the forging of a strong, lasting relationship with mathematical economics and economic theory.

(i) *NTU coalitional games and NTU value*. Properly interpreted, the coalitional form (*1930–1950*, ii) applies both to TU and to NTU games; nevertheless, for many NTU applications one would like to describe the opportunities available to each coalition more faithfully than can be done with a single number. Accordingly, define a game in *NTU coalitional form* as a function that associates with each coalition S a set V(S) of S-tuples of real numbers (functions from S to $\mathbb{R}$). Intuitively, V(S) represents the set of payoff S-tuples that S can achieve. For example, in an exchange economy, V(S) is the set of utility S-tuples that S can achieve when its members trade among themselves only, without recourse to agents outside of S. Another example of an NTU coalitional game is Nash's bargaining problem (*1950–1960*, iii), where one can take $V(\{1, 2\}) = C$, $V(1) = \{0\}$, $V(2) = \{0\}$.

The definitions of stable set and core extend straightforwardly to NTU coalitional games, and these solution concepts were among the first to be investigated in that context (Aumann and Peleg, 1960; Peleg, 1963a; Aumann, 1961). The first definitions of NTU value were proposed by Harsanyi (1959, 1963), but they proved difficult to apply. Building on Harsanyi's work, Shapley (1969) defined a value for NTU games that has proved widely applicable and intuitively appealing.

For each profile λ of non-negative numbers and each outcome x, define the *weighted outcome* λx by $(\lambda x)^i = \lambda^i x^i$. Let $v_\lambda(S)$ be the maximum total weight that the coalition S can achieve,

$$v_\lambda(S) := \max \left\{ \sum_{i \in S} \lambda^i x^i, x \in V(S) \right\}.$$

Call an outcome x an *NTU value* of V if $x \in V(N)$ and there exists a weight profile λ with $\lambda x = \phi v_\lambda$; in words, if x is feasible and corresponds to the value of one of the coalitional games v_λ.

Intuitively, $v_\lambda(S)$ is a numerical measure of S's total worth and hence $\phi^i v_\lambda$ measures i's social productivity. The weights λ^i are chosen so that the resulting value is feasible; an infeasible result would indicate that some people are overrated (or underrated), much like an imbalance between supply and demand indicates that some goods are overpriced (or underpriced).

The NTU value of a game need not be unique. This may at first sound strange, since unlike stability concepts such as the core, one might expect an 'index of social productivity' to be unique. But perhaps it is not so strange when one reflects that even a person's net worth depends on the prevailing (equilibrium) prices, which are not uniquely determined by the exogenous description of the economy.

The Shapley NTU value has been used in a very wide variety of economic and political-economic applications. To cite just one example, the Nash bargaining problem has a unique NTU value, which coincides with Nash's solution. For a partial bibliography of applications, see the references of Aumann (1985).

We have discussed the historical importance of TU as pointing the way for NTU results (*1930–1950*, vi). There is one piquant case in the reverse direction. Just as positive results are easier to obtain for TU, negative results are easier for NTU. Non-existence of stable sets was first discovered in NTU games (Stearns 1967), and this eventually led to Lucas's famous example (1969) of non-existence for TU.

(ii) *Incomplete information.* In 1957, Luce and Raiffa wrote that a fundamental assumption of game theory is that 'each player ... is fully aware of the rules of the game and the utility functions of each of the players ... this is a serious idealization which only rarely is met in actual situations' (p. 49). To deal with this problem, John Harsanyi (1967) constructed the theory of games of incomplete information (sometimes called differential or asymmetric information). This major conceptual breakthrough laid the theoretical groundwork for the great blooming of information economics that got under way soon thereafter, and that has become one of the major themes of modern economics and game theory.

For simplicity, we confine attention to strategic form games in which each player has a fixed, known set of strategies, and the only uncertainty is about the utility functions of the other players; these assumptions are removable. Bayesian rationality in the tradition of Savage (1954) dictates that all uncertainty

can be made explicit; in particular, each player has a personal probability distribution on the possible utility (payoff) functions of the other player. But these distributions are not sufficient to describe the situation. It is not enough to specify what each player thinks about the other's payoffs; one must also know what he thinks they think about his (and each others') payoffs, what he thinks they think he thinks about their payoffs, and so on. This complicated infinite regress would appear to make useful analysis very difficult.

To cut this Gordian knot, Harsanyi postulated that each player may be one of several *types*, where a type determines both a player's own utility function and his personal probability distribution on the types of the other players. Each player is postulated to know his own type only. This enables him to calculate what he thinks the other players' types – and therefore their utilities – are. Moreover, his personal distribution on their types also enables him to calculate what he thinks they think about his type, and therefore about his utility. The reasoning extends indefinitely, and yields the infinite regress discussed above *as an outcome*.

Intuitively, one may think of a player's type as a possible state of mind, which would determine his utility as well as his distribution over others' states of mind. One need not assume that the number of states of mind (types) is finite; the theory works as well for, say, a continuum of types. But even with just two players and two types for each player, one gets a non-trivial infinite string of beliefs about utilities, beliefs about beliefs, and so on.

A model of this kind – with players, strategies, types, utilities, and personal probability distributions – is called an *I-game* (incomplete information game). A *strategic equilibrium* in an I-game consists of a strategy for each *type* of each player, which maximizes that type's expected payoff given the strategies of the other players' types.

Harsanyi's formulation of I-games is primarily a device for thinking about incomplete information in an orderly fashion, bringing that wild, bucking infinite regress under conceptual control. An (incomplete) analogy is to the strategic form of a game, a conceptual simplification without which it is unlikely that game theory would have gotten very far. Practically speaking, the strategic form of a particular game such as chess is totally unmanageable, one can't even begin to write it down. The advantage of the strategic form is that it is a comparatively simple formulation, mathematically much simpler than the extensive form; it enables one to formulate and calculate examples, which suggest principles that can be formulated and proved as general theorems. All this would be much more difficult – probably unachievable – with the extensive form; one would be unable to see the forest for the trees. A similar relationship holds between Harsanyi's I-game formulation and direct formulations in terms of beliefs about beliefs. (Compare the discussion of perspective made in connection with the coalitional form (*1930–1950*, i). That situation is somewhat different, though, since in going to the coalitional form, substantive information is lost. Harsanyi's formulation of I-games loses no information; it is a more abstract and simple – and hence transparent and workable – formulation of the same data as would be contained in an explicit description of the infinite regress.)

Harsanyi called an I-game *consistent* if all the personal probability distributions of all the types are derivable as posteriors from a single prior distribution p on all n-tuples of types. Most applications of the theory have assumed consistency. A consistent I-game is closely related to the ordinary strategic game (C-*game*) obtained from it by allowing 'nature' to choose an n-tuple of types at random according to the distribution p, then informing each player of his type, and then playing the I-game as before. In particular, the strategic equilibria of a consistent I-game are essentially the same as the strategic equilibria of the related C-game. In the cooperative theory, however, an I-game is rather different from the related C-game, since binding agreements can only be made after the players know their types. Bargaining and other cooperative models have been treated in the incomplete information context by Harsanyi and Selten (1972), Wilson (1978), Myerson (1979, 1984), and others.

In a repeated game of incomplete information, the same game is played again and again, but the players do not have full information about it; for example, they may not know the others' utility functions. The actions of the players may implicitly reveal private information, e.g. about preferences; this may or may not be advantageous for them. We have seen (*1950–1960*, iii) that repetition may be viewed as a paradigm for cooperation. Strategic equilibria of repeated games of incomplete information may be interpreted as a subtle bargaining process, in which the players gradually reach wider and wider agreement, developing trust for each other while slowly revealing more and more information (Hart 1985b).

(iii) *Common knowledge*. Luce and Raiffa, in the statement quoted at the beginning of (ii), missed a subtle but important point. It is not enough that each player be fully aware of the rules of the game and the utility functions of the players. Each player must also be aware of this fact, i.e. of the awareness of all the players; moreover, each player must be aware that each player is aware that each player is aware, and so on ad infinitum. In brief, the awareness of the description of the game by all players must be a part of the description itself.

There is evidence that game theorists had been vaguely cognizant of the need for some such requirement ever since the late Fifties or early Sixties; but the first to give a clear, sharp formulation was the philosopher D.K. Lewis (1969). Lewis defined an event as *common knowledge* among a set of agents if all know it, all know that all know it, and so on ad infinitum.

The common knowledge assumption underlies all of game theory and much of economic theory. Whatever be the model under discussion, whether complete or incomplete information, consistent or inconsistent, repeated or one-shot, cooperative or non-cooperative, the model itself must be assumed common knowledge; otherwise the model is insufficiently specified, and the analysis incoherent.

(iv) *Bargaining set, kernel, nucleolus*. The core excludes the unique symmetric outcome (1/3, 1/3, 1/3) of the three-person voting game, because any two-person coalition can improve upon it. Stable sets (*1930–1950*, v) may be seen as a way of expressing our intuitive discomfort with this exclusion. Another way is the bargaining set (Davis and Maschler 1967). If, say, 1 suggests (1/2, 1/2, 0) to replace (1/3, 1/3, 1/3), then 3 can suggest to 2 that he is as good a partner as 1; indeed, 3 can even offer 2/3 to 2, still leaving himself with the 1/3 he was originally assigned. Formally, if we call (1/2, 1/2, 0) an *objection* to (1/3, 1/3, 1/3), then (0, 2/3, 1/3) is a *counter objection*, since it yields to 3 at least as much as he was originally assigned, and yields to 3's partners in the counter-objection at least as much as they were assigned either originally or in the objection. In brief, the counter-objecting player tells the objecting one, 'I can maintain my level of payoff and that of my partners, while matching your offers to players we both need.' An imputation is in the core iff there is

no objection to it. It is in the *bargaining set* iff there is no *justified* objection to it, i.e. one that has no counter-objection.

Like the stable sets, the bargaining set includes the core (dominating and objecting are essentially the same). Unlike the core and the set of stable sets, the bargaining set is for TU games never empty (Peleg, 1967). For NTU it may be empty (Peleg, 1963b); but Asscher (1976) has defined a non-empty variant; see also Billera (1970a).

Crucial parameters in calculating whether an imputation x is in the bargaining set of v are the *excesses* $v(S) - x(S)$ of coalitions S w.r.t. x, which measure the ability of members of S to use x in an objection (or counter-objection). Not, as is often wrongly assumed, because the initiator of the objection can assign the excess to himself while keeping his partners at their original level, but for precisely the opposite reason: because he can parcel out the excess to his partners, which makes counterobjecting more difficult.

The excess is so ubiquitous in bargaining set calculations that it eventually took on intuitive significance on its own. This led to the formulation of two additional solution concepts: the *kernel* (Davis and Maschler, 1965), which is always included in the bargaining set but is often much smaller, and the *nucleolus* (Schmeidler, 1969), which always consists of a single point in the kernel.

To define the nucleolus, choose first all those imputations x whose maximum excess (among the 2^n excesses $v(S) - x(S)$) is minimum (among all imputations). Among the resulting imputations, choose next those whose second largest excess is minimum, and so on. Schmeidler's theorem asserts that by the time we have gone through this procedure 2^n times, there is just one imputation left.

We have seen that the excess is a measure of a coalition's 'manoeuvring ability'; in these terms the greatest measure of stability, as expressed by the nucleolus, is reached when all coalitions have manoeuvring ability as nearly alike as possible. An alternative interpretation of the excess is as a measure of S's total dissatisfaction with x, the volume of the cry that S might raise against x. In these terms, the nucleolus suggests that the final accommodation is determined by the loudest cry against it. Note that the *total* cry is determining, not the average cry; a large number of moderately unhappy citizens can be as potent a force for change as a moderate number of very unhappy ones. Variants of the nucleolus that use the average excess miss this point.

When the core is non-empty, the nucleolus is always in it. The nucleolus has been given several alternative characterizations, direct (Kohlberg, 1971, 1972) as well as axiomatic (Sobolev, 1975). The kernel was axiomatically characterized by Peleg (1986), and many interesting relationships have been found between the bargaining set, core, kernel, and nucleolus (e.g. Maschler, Peleg and Shapley, 1979). There is a large body of applications, of which we here cite just one: In a decisive weighted voting game, the nucleolus constitutes a set of weights (Peleg 1968). Thus the nucleolus may be thought of as a natural generalization of 'voting weights' to arbitrary games. (We have already seen that value and weights are quite different: see *1950–1960*, vi.)

(v) *The Equivalence Principle*. Perhaps the most remarkable single phenomenon in game and economic theory is the relationship between the price equilibria of a competitive market economy, and all but one of the major solution concepts for the corresponding game (the one exception is the stable set, about which more below). By a 'market economy' we here mean a pure exchange economy, or a production economy with constant returns.

We call an economy 'competitive' if it has many agents, each individual one of whom has too small an endowment to have a significant effect. This has been modelled by three approaches. In the *asymptotic approach*, one lets the number of agents tend to infinity, and shows that in an appropriate sense, the solution concept in question – core, value, bargaining set, or strategic equilibrium – tends to the set of competitive allocations (those corresponding to price equilibria). In the *continuum approach*, the agents constitute a (non-atomic) continuum, and one shows that the solution concept in question actually equals the set of competitive allocations (see the entry on LARGE ECONOMIES). In the *non-standard* approach, the agents constitute a non-standard model of the integers in the sense of Robinson (1970), and again one gets equality. Both the continuum and the non-standard approaches require extensions of the theory to games with infinitely many players; see vi.

Intuitively, the equivalence principle says that the institution of market prices arises naturally from the basic forces at work in a market, (almost) no matter what we assume about the way in which these forces work. Compare (*1930–1950*, ix).

For simplicity in this section, unless otherwise indicated, the terms 'core', 'value', etc., refer to the limiting case. Thus 'core' means the limit of the cores of the finite economies, or the core of the continuum economy, or of the non-standard economy.

For the core, the asymptotic approach was pioneered by Edgeworth (1881), Shubik (1959) and Debreu and Scarf (1963). Anderson (1986) is an excellent survey of the large literature that ensued. Early writers on the continuum approach included Aumann (1964) and Vind (1964); the non-standard approach was developed by Brown and Robinson (1975). Except for Shubik's, all these contributions were NTU. See the entry on CORE. After a twenty-year courtship, this was the honeymoon of game theory and mathematical economics, and it is difficult to convey the palpable excitement of those early years of intimacy between the two disciplines.

Some early references for the value equivalence principle, covering both the asymptotic and continuum approaches, were listed above (*1930–1950*, vi). For the non-standard approach, see Brown and Loeb (1976). Whereas the core of a competitive economy equals the set of *all* competitive allocations, this holds for the value only when preferences are smooth (Shapley, 1964; Aumann and Shapley, 1974; Aumann 1975; Mas-Colell, 1977). Without smoothness, every value allocation is competitive, but not every competitive allocation need be a value allocation. When preferences are kinky (non-differentiable utilities), the core is often quite large, and then the value is usually a very small subset of the core; it gives much more information. In the TU case, for example, the value is always a single point, even when the core is very large. Moreover, it occupies a central position in the core (Hart, 1980; Tauman, 1981; Mertens, 1987); in particular, when the core has a centre of symmetry, the value is that centre of symmetry (Hart, 1977a).

For example, suppose that in a glove market (*1950–1960*, vi), the number (or measure) of left-glove holders equals that of right-glove holders. Then at a price equilibrium, the price ratio between left and right gloves may be anything between 0 and ∞ (inclusive!). Thus the left-glove holders may end up giving away their merchandise for nothing to the right-glove holders, or the other way around, or anything inbetween. The same, of course, holds for the core. But the value prescribes precisely equal prices for right and left gloves.

It should be noted that in a finite market, the core contains the competitive allocations, but usually also much more. As the number of agents increases, the core 'shrinks', in the limit leaving only the competitive allocations. This is not so for the value; in finite markets, the value allocations may be disjoint from the core, and a fortiori from the competitive allocations (*1950–1960*, vi).

We have seen (*1930–1950*, iv) that the core represents a very strong and indeed not quite reasonable notion of stability. It might therefore seem perhaps not so terribly surprising that it shrinks to the competitive allocations. What happens, one may ask, when one considers one of the more reasonable stability concepts that are based on domination, such as the bargaining set or the stable sets?

For the bargaining set of TU markets, an asymptotic equivalence theorem was established by Shapley and Shubik in the mid-Seventies, though it was not published until 1984. Extending this result to NTU, to the continuum, or to both seemed difficult. The problems were conceptual as well as mathematical; it was difficult to give a coherent formulation. In 1986, Shapley presented the TU proof at a conference on the equivalence principle that took place at Stony Brook. A. Mas-Colell, who was in the audience, recognized the relevance of several results that he had obtained in other connections; within a day or two he was able to formulate and prove the equivalence principle for the bargaining set in NTU continuum economies (Mas-Colell, 1988). In particular, this implies the core equivalence principle; but it is a much stronger and more satisfying result.

For the strategic equilibrium the situation had long been less satisfactory, though there were results (Shubik, 1973; Dubey and Shapley, 1980). The difficulty was in constructing a satisfactory strategic (or extensive) model of exchange. Very recently Douglas Gale (1986) provided such a model and used it to prove a remarkable equivalence theorem for strategic equilibria in the continuum mode.

The one notable exception to the equivalence principle is the case of stable sets, which predict the formation of cartels even in fully competitive economies (Hart, 1974). For example, suppose half the agents in a continuum initially hold 2 units of bread each, half initially hold 2 units of cheese, and the utility functions are concave, differentiable, and symmetric (e.g., $u(x, y) = \sqrt{x} + \sqrt{y}$). There is then a unique price equilibrium, with equal prices for bread and cheese. Thus each agent ends up with one piece of bread and one piece of cheese; this is also the unique point in the core and in the bargaining set, and the unique NTU value. But stable set theory predicts that the cheese holders will form a cartel, the bread holders will form a cartel, and these two cartels will bargain with each other as if they were individuals. The upshot will depend on the bargaining, and may yield an outcome that is much better for one side than for the other. Thus at each point of the unique stable set with the full symmetry of the game, each agent on each side gets as much as each other agent on that side; but these two amounts depend on the bargaining, and may be quite different from each other.

In a sense, the failure of stable set theory to fall into line makes the other results even more impressive. It shows that there isn't some implicit tautology lurking in the background, that the equivalence principle makes a substantive assertion.

In the *Theory of Games*, von Neumann and Morgenstern (1944) wrote that

> when the number of participants becomes really great, some hope emerges that the influence of every particular participant will become negligible ... These are, of course, the classical conditions of 'free competition' ... The current assertions concerning free competition appear to be very

valuable surmises and inspiring anticipations of results. But they are not results, and it is scientifically unsound to treat them as such.

One may take the theorems constituting the equivalence principle as embodying precisely this kind of 'result'. Yet it is interesting that Morgenstern himself, who died in 1977, never became convinced of the validity of the equivalence principle; he thought of it as mathematically correct but economically wrongheaded. It was his firm opinion that economic agents organize themselves into coalitions, that perfect competition is a fiction, and that stable sets explain it all. The greatness of the man is attested to by the fact that though scientifically opposed to the equivalence principle, he gave generous support, both financial and moral, to workers in this area.

(vi) *Many players*. The preface to *Contributions to the Theory of Games* I (Kuhn and Tucker, 1950) contains an agenda for future research that is remarkable in that so many of its items – computation of minimax, existence of stable sets, *n*-person value, NTU games, dynamic games – did in fact become central in subsequent work. Item 11 in this agenda reads, 'establish significant asymptotic properties of *n*-person games, for large *n*'. We have seen ((v)) how this was realized in the equivalence principle for large economies. But actually, political game models with many players are at least as old as economic ones, and may be older. During the early Sixties, L.S. Shapley, working alone and with various collaborators, wrote a series of seven memoranda at the Rand Corporation under the generic title 'Values of Large Games', several of which explored models of large elections, using the asymptotic and the continuum approaches. Among these were models which had both 'atoms' – players who are significant as individuals – and an 'ocean' of individually insignificant players. On example of this is a corporation with many small stockholders and a few large stockholders; see also Milnor and Shapley (1978). 'Mixed' models of this kind – i.e. with an ocean as well as atoms – have been explored in economic as well as political contexts using various solution notions, and a large literature has developed. The core of mixed markets has been studied by Drèze, Gabszewicz and Gepts (1969), Gabszewicz and Mertens (1972), Shitovitz (1973) and many others. For the nucleolus of 'mixed' voting games, see Galil (1974). Among the studies of values of mixed games are Hart (1973), Fogelman and Quinzii (1980), and Neyman (1987).

Large games in which *all* the players are individually insignificant – *non-atomic* games – have also been studied extensively. Among the early contributions to value theory in this connection are Kannai (1965), Riker and Shapley (1968), and Aumann and Shapley (1974). The subject has proliferated greatly, with well over a hundred contributions since 1974, including theoretical contributions as well as economic and political applications.

There are also games with infinitely many players in which *all* the players are atoms, namely games with a denumerable infinity of players. Again, values and voting games loom large in this literature. See, e.g., Shapley (1962), Artstein (1972) and Berbee (1981).

(vii) *Cores of finite games and markets*. Though the core was defined as an independent solution concept by Gillies and Shapley already in the early Fifties, it was not until the Sixties that a significant body of theory was developed around it. The major developments centre around conditions for the core to be non-empty; gradually it came to be realized that such conditions hold most naturally and fully when the game has

an 'economic' rather than a 'political' flavour, when it may be thought of as arising from a market economy.

The landmark contributions in this area were the following: the Gale–Shapley 1962 paper on the core of a marriage market; the work of Bondareva (1963) and Shapley (1967) on the balancedness condition for the non-emptiness of the core of a TU game; Scarf's 1967 work on balancedness in NTU games; the work of Shapley and Shubik (1969a) characterizing TU market games in terms of non-emptiness of the core; and subsequent work, mainly associated with the names of Billera and Bixby (1972), that extended the Shapley–Shubik condition to NTU games. Each of these contributions was truly seminal, in that it inspired a large body of subsequent work.

Gale and Shapley (1962) asked whether it is possible to match *m* women with *m* men so that there is no pair consisting of an unmatched woman and man who prefer each other to the partners with whom they were matched. The corresponding question for homosexuals has a negative answer: the preferences of four homosexuals may be such that no matter how they are paired off, there is always an unmatched pair of people who prefer each other to the person with whom they were matched. This is so, for example, if the preferences of *a*, *b*, and *c* are cyclic, whereas *d* is lowest in all the others' scales. But for the heterosexual problem, Gale and Shapley showed that the answer is positive.

This may be stated by saying that the appropriately defined NTU coalitional game has a non-empty core. Gale and Shapley proved not only the non-emptiness but also provided a simple algorithm for finding a point in it.

This work has spawned a large literature on the cores of discrete market games. One fairly general recent result is Kaneko and Wooders (1982), but there are many others. A fascinating application to the assignment of interns to hospitals has been documented by Roth (1984). It turns out that American hospitals, after fifty years of turmoil, finally developed in 1950 a method of assignment that is precisely a point in the core.

We come now to general conditions for the core to be non-empty. Call a TU game v *superadditive* at a coalition U if $v(U) \geq \Sigma_j v(S_j)$ for any partition of U into disjoint coalition S_j. This may be strengthened by allowing partitions of U into disjoint 'part-time' coalitions θS, interpreted as coalitions S operating during a proportion θ of the time ($0 \leq \theta \leq 1$). Such a partition is therefore a family $\{\theta_j S_j\}$, where the total amount of time that each player in U is employed is exactly 1; i.e., where $\Sigma_j \theta_j \chi_{S_j} = \chi_U$, where χ_S is the indicator function of S. If we think of $v(S)$ as the revenue that S can generate when operating full-time, then the part-time coalition θS generates $\theta v(S)$. Superadditivity at U for part-time coalitions thus means that

$$\sum_j \theta_j \chi_{S_j} = \chi_U \text{ implies } v(U) \geq \sum_j \theta_j v(S_j).$$

A TU game v obeying this condition for $U = I$ is called *balanced*; for all U, *totally balanced*.

Intuitively, it is obvious that a game with a non-empty core must be superadditive at I; and once we have the notion of part-time coalitions, it is only slightly less obvious that it must be balanced. The converse was established (independently) by Bondareva (1963) and Shapley (1967). Thus *a TU game has a non-empty core if and only if it is balanced.*

The connection between the core and balancedness (generalized superadditivity) led to several lines of research. Scarf (1967) extended the notion of balancedness to NTU games, then showed that every balanced NTU game has a non-empty core. Unlike the Bondareva–Shapley proof, which is based on linear programming methods, Scarf's proof was more closely related to fixed-point ideas. Eventually, Scarf realized that his methods could be used actually to prove Brouwer's fixed-point theorem, and moreover, to develop effective algorithms for approximating fixed points. This, in turn, led to the development of algorithms for approximating competitive equilibria of economies (Scarf, 1973), and to a whole area of numerical analysis dealing with the approximation of fixed points (*see* COMPUTATION OF GENERAL EQUILIBRIA).

An extension of the Bondareva–Shapley result to the NTU case that is different from Scarf's was obtained by Billera (1970a).

Another line of research that grew out of balancedness deals with characterizing markets in purely game-theoretic terms. When can a given coalitional game v be expressed as a market game (*1930–1950*, ii)? The Bondareva–Shapley theorem implies that market games have non-empty cores, and this also follows from the fact that outcomes corresponding to competitive equilibria are always in the core. Since a subgame of a market game is itself a market game, it follows that *for v to be a market game, it is necessary that it and all its subgames have non-empty cores*, i.e., that the game be totally balanced. (A *subgame* of a coalitional game v is defined by restricting its domain to subcoalitions of a given coalition U.) Shapley and Shubik (1969a) showed that *this necessary condition is also sufficient*. Balancedness itself is not sufficient, since there exist games with non-empty cores having subgames with empty cores (e.g., $|I| = 4$, $v(S) := 0, 0, 1, 1, 2$ when $|S| = 0, 1, 2, 3, 4$, respectively).

For the NTU case, characterizations of market games have been obtained by Billera and Bixby (1973), Mas-Colell (1975), and others.

Though the subject of this section is finite markets, it is nevertheless worthwhile to relate the results to non-atomic games (where the players constitute a non-atomic continuum, an 'ocean'). The total balancedness condition then takes on a particularly simple form. Suppose, for simplicity, that v is a function of finitely many measures, i.e., $v(S) = f(\mu(S))$, where $\mu = (\mu_1, \ldots, \mu_n)$, and the μ_j are non-atomic measures. Then v is a market game iff f is concave and 1-homogeneous ($f(\theta x) = \theta(fx)$) when $\theta \geqslant 0$). This is equivalent to saying that v is superadditive (at all coalitions), and f is 1-homogeneous (Aumann and Shapley, 1974).

Perhaps the most remarkable expression of the connection between superadditivity and the core has been obtained by Wooders (1983). Consider coalitional games with a fixed finite number k of 'types' of players, the coalitional form being given by $v(S) = f(\mu(S))$, where $\mu(S)$ is the profile of type sizes in S, i.e. it is a vector whose i'th coordinate represents the number of type i players in S. (To specify the game, $\mu(I)$ must also be specified.) Assume that f is superadditive, i.e. $f(x + y) \geqslant f(x) + f(y)$ for all x and y with non-negative integer coordinates; this assures the superadditivity of v. Moreover, assume that f obeys a 'Lipschitz' condition, namely that $|f(x) - f(y)|/\|x - y\|$ is uniformly bounded for all $x \neq y$, where $\|x\| := \max_j |x_j|$. Then for each $\epsilon > 0$, when the number of players is sufficiently large, the ϵ-core is non-empty. (The ϵ-core is defined as the set of all outcomes x such that $x(S) \geqslant v(S) - \epsilon |S|$ for all S.) Roughly, the result says that the core is 'almost' non-empty for sufficiently large games that are superadditive and obey the Lipschitz condition. Intuitively, the superadditivity together with the Lipschitz condition yield 'approximate' 1-homogeneity, and in the presence of 1-homogeneity, superadditivity is equivalent to concavity. Thus f is approximately a 1-homogeneous concave function, so that we are back in a situation similar to that treated in the previous paragraph. What makes this result so remarkable is that other than the Lipschitz condition, the only substantive assumption is superadditivity.

Wooders (1983) also obtained a similar theorem for NTU; Wooders and Zame (1984) obtained a formulation that does away with the finite type assumption.

1970–1986

We do not yet have sufficient distance to see the developments of this period in proper perspective. Political and political economic models were studied in depth. Non-cooperative game theory was applied to a large variety of particular economic models, and this led to the study of important variants on the refinements of the equilibrium concept. Great strides forward were made in almost all the areas that had been initiated in previous decades, such as repeated games (both of complete and of incomplete information), stochastic games, value, core, nucleolus, bargaining theory, games with many players, and so on (many of these developments have been mentioned above). Game Theory was applied to biology, computer science, moral philosophy, cost allocation. New light was shed on old concepts such as randomized strategies.

Sociologically, the discipline proliferated greatly. Some 16 or 17 people participated in the first international workshop on game theory held in Jerusalem in 1965; the fourth one, held in Cornell in 1978, attracted close to 100, and the discipline is now too large to make such workshops useful. An international workshop in the relatively restricted area of repeated games, held in Jerusalem in 1985, attracted over fifty participants. The *International Journal of Game Theory* was founded in 1972; *Mathematics of Operations Research*, founded in 1975, was organized into three major 'areas', one of them Game Theory. Economic theory journals, such as the *Journal of Mathematical Economics*, the *Journal of Economic Theory*, *Econometrica*, and others devoted increasing proportions of their space to game theory. Important centres of research, in addition to the existing ones, sprang up in France, Holland, Japan, England, and India, and at many Universities in the United States.

Gradually, game theory also became less personal, less the exclusive concern of a small 'in' group whose members all know each other. For years, it had been a tradition in game theory to publish only a fraction of what one had found, and then only after great delays, and not always what is most important. Many results were passed on by word of mouth, or remained hidden in ill-circulated research memoranda. The 'Folk Theorem' to which we alluded above (*1950–1960*, iii) is an example. This tradition had both beneficial and deleterious effects. On the one hand, people did not rush into print with trivia, and the slow cooking of results improved their flavour. As a result, phenomena were sometimes rediscovered several times, which is perhaps not entirely bad, since you understand something best when you discover it yourself. On the other hand, it was difficult for outsiders to break in; non-publication caused less interest to be generated than would otherwise have been, and significantly impeded progress.

Be that as it may, those days are over. There are now hundreds of practitioners, they do not all know each other, and sometimes have never even heard of one another. It is no longer possible to communicate in the old way, and as a result, people are publishing more quickly. As in other disciplines, it is becoming difficult to keep abreast of the important developments. Game theory has matured.

(i) *Applications to biology*. A development of outstanding importance, whose implications are not yet fully appreciated, is the application of game thory to evolutionary *biology*. The high priest of this subject is John Maynard Smith (1982), a biologist whose concept of *evolutionarily stable' strategy*, a variant of strategic equilibrium, caught the imagination both of biologists and of game theorists. On the game theoretic side, the theme was taken up by Reinhard Selten (1980, 1983) and his school; a conference on 'Evolutionary theory in biology and economics', organized by Selten in Bielefeld in 1985, was enormously successful in bringing field biologists together with theorists of games to discuss these issues. A typical paper was tit for tat in the great tit (Regelmann and Curio, 1986); using actual field observations, complete with photographs, it describes how the celebrated 'tit for tat' strategy in the repeated prisoners' dilemma (Axelrod, 1984) accurately describes the behaviour of males and females of a rather common species of bird called the great tit, when protecting their young from predators.

It turns out that ordinary, utility maximizing rationality is much more easily observed in animals and even plants than it is in human beings. There are even situations where rats do significantly better than human beings. Consider, for example, the famous probability matching experiment, where the subject must predict the values of a sequence of i.i.d. random variables taking the values L and R with probabilities 3/4 and 1/4 respectively; each correct prediction is rewarded. It is of course optimal always to predict L; but human subjects tend to match the probabilities, i.e. to predict L about 3/4 of the time. On the other hand, while rats are not perfect (i.e. do not predict L *all* the time), they do predict L significantly more often than human beings.

Several explanations have been suggested. One is that in human experimentation, the subjects try subconsciously to 'guess right', i.e. to guess what the experimenter 'wants' them to do, rather than maximizing utility. Another is simply that the rats are more highly motivated. They are brought down to 80 per cent of their normal body weight, are literally starving; it is much more important for them to behave optimally than it is for human subjects.

Returning to theory, though the notion of strategic equilibrium seems on the face of it simple and natural enough, a careful examination of the definition leads to some doubts and questions as to why and under what conditions the players in a game might be expected to play a strategic equilibrium. See the entry on NASH EQUILIBRIUM. Evolutionary theory suggests a simple rationale for strategic equilibrium, in which there is no conscious or overt decision making at all. For definiteness, we confine attention to two-person games, though the same ideas apply to the general case. We think of each of the two players as a whole species rather than an individual; reproduction is assumed asexual. The set of pure strategies of each player is interpreted as the locus of some gene (examples of a locus are eye colour, degree of aggressiveness, etc.); individual pure strategies are interpreted as alleles (blue or green or brown eyes, aggressive or timid behaviour, etc.). A given individual of each species possesses just one allele at the given locus; he interacts with precisely one individual in the other species, who also has just one allele at the locus of interest. The result of the interaction is a definite increment or decrement in the fitness of each of the two individuals, i.e. the number (or expected number) of his offspring; thus the payoff in the game is denominated in terms of fitness.

In these terms, a mixed strategy is a distribution of alleles throughout the population of the species (e.g., 40% aggressive, 60% timid). If each individual of each species is just as likely to meet any one individual of the other species as any other one, then the probability distribution of alleles that each individual faces is precisely given by the original mixed strategy. It then follows that a given pair of mixed strategies is a strategic equilibrium if and only if it represents a population equilibrium, i.e. a pair of distributions of characteristics (alleles) that does not tend to change.

Unfortunately, sexual reproduction screws up this story, and indeed the entire Maynard Smith approach has been criticized for this reason. But to be useful, the story does not have to be taken entirely literally. For example, it applies to evolution that is cultural rather than biological. In this approach, a 'game' is interpreted as a *kind* of confrontational situation (like shopping for a car) rather than a specific instance of such a situation; a 'player' is a role ('buyer' or 'salesman'), not an individual human being; a pure strategy is a possible kind of behaviour in this role ('hard sell' or 'soft sell'). Up to now this is indeed not very different from traditional game theoretic usage. What is different in the evolutionary interpretation is that pure or mixed strategic equilibria do not represent conscious rational choices of the players, but rather a population equilibrium which evolves as the result of how successful certain behaviour is in certain roles.

(ii) *Randomization as ignorance*. In the traditional view of strategy randomization, the players use a randomizing device, such as a coin flip, to decide on their actions. This view has always had difficulties. Practically speaking, the idea that serious people would base important decisions on the flip of a coin is difficult to swallow. Conceptually, too, there are problems. The reason a player must randomize in equilibrium is only to keep others from deviating. For himself, randomizing is unnecessary; he will do as well by choosing any pure strategy that appears with positive probability in his equilibrium mixed strategy.

Of course, there is no problem if we adopt the evolutionary model described above in (i); mixed strategies appear as population distributions, and there is no explicit randomization at all. But what is one to make of randomization within the more usual paradigm of conscious, rational choice?

According to Savage (1954), randomness is not physical, but represents the ignorance of the decision maker. You associate a probability with every event about which you are ignorant, whether this event is a coin flip or a strategic choice by another player. The important thing in strategy randomization is that the *other* players be ignorant of what you are doing, and that they ascribe the appropriate probabilities to each of your pure strategies. It is not necessary for you actually to flip a coin.

The first to break away from the idea of explicit randomization was J. Harsanyi (1973). He showed that if the payoffs to each player *i* in a game are subjected to small independent random perturbations, known to *i* but not to the other players, then the resulting game of incomplete information has *pure* strategy equilibria that correspond to the mixed strategy equilibria of the original game. In plain words, nobody really randomizes. The appearance of randomization is due to the payoffs not being exactly known to all; each player, who does know his own payoff exactly, has a unique optimal action against his estimate of what the others will do.

This reasoning may be taken one step further. Even without perturbed payoffs, the players simply do not know which strategies will be chosen by the other players. At an equilibrium of 'matching pennies', each player knows very well what he himself will do, but ascribes $1/2 - 1/2$ probabilities to the other's actions; he also knows that the other ascribes those probabilities to his own actions, though it is admittedly not quite obvious that this is necessarily the case. In the case of a general n-person game, the situation is essentially similar; the mixed strategies of i can always be understood as describing the uncertainty of players other than i about what i will do (Aumann, 1987).

(iii) *Refinements of strategic equilibrium.* In analysing specific economic models using the strategic equilibrium – an activity carried forward with great vigour since about 1975 – it was found that Nash's definition does not provide adequately for rational choices given one's information at each stage of an extensive game. Very roughly, the reason is that Nash's definition ignores contingencies 'off the equilibrium path'. To remedy this, various "refinements' of strategic equilibrium have been defined, starting with Selten's (1975) 'trembling hand' equilibrium. Please refer to our discussion of Zermelo's theorem (*1930–1950*, vi), and to Section IV of the entry on NASH EQUILIBRIUM.

The interesting aspect of these refinements is that they use *irrationality* to arrive at a strong form of rationality. In one way or another, all of them work by assuming that irrationality cannot be ruled out, that the players ascribe irrationality to each other with a small probability. True rationality requires 'noise'; it cannot grow in sterile ground, it cannot feed on itself only.

(iv) *Bounded rationality.* For a long time it has been felt that both game and economic theory assume too much rationality. For example, the hundred-times repeated prisoner's dilemma has some $2^{2^{100}}$ pure strategies; all the books in the world are not large enough to write this number even once in decimal notation. There is no practical way in which all these strategies can be considered truly available to the players. On the face of it, this would seem to render statements about the equilibrium points of such games (*1950–1960*, iv) less compelling, since it is quite possible that if the sets of strategies were suitably restricted, the equilibria would change drastically.

For many years, little on the formal level was done about these problems. Recently the theory of automata has been used for formulations of bounded rationality in repeated games. Neyman (1985a) assumes that only strategies that are programmable on an automaton of exogenously fixed size can be considered 'available' to the players. He then shows that even when the size is very large, one obtains results that are qualitatively different from those when all strategies are permitted. Thus in the n-times repeated prisoner's dilemma, only the greedy-greedy outcome can occur in equilibrium; but if one restricts the players to using automata with as many as $e^{o(n)}$ states, then for sufficiently large n, one can approximate in equilibrium any feasible individually rational outcome, and in particular the friendly–friendly outcome. For example, this is the case if the number of states is bounded by any fixed polynomial in n. In unpublished work, Neyman has generalized this result from the prisoner's dilemma to arbitrary games; specifically, he shows that a result similar to the Folk Theorem holds in any long finitely repeated game, when the automaton size is limited as above to subexponential.

Another approach has been used by Rubinstein (1986), with dramatically different results. In this work, the automaton itself is endogenous; all states of the automaton must actually be used on the equilibrium path. Applied to the prisoner's dilemma, this assumption leads to the conclusion that in equilibrium, one cannot get anywhere near the friendly-friendly outcome. Intuitively, the requirement that all states be used in equilibrium rules out strategies that punish deviations from equilibrium, and these are essential to the implicit

enforcement mechanism that underlies the folk theorem. See the discussion at (*1950–1960*, iii) above.

(v) *Distributed computing.* In the previous subsection (iv) we discussed applications of computer science to game theory. There are also applications in the opposite direction; with the advent of distributed computing, game theory has become of interest in computer science. Different units of a distributed computing system are viewed as different players, who must communicate and coordinate. Breakdowns and failures of one unit are often modelled as malevolent, so as to get an idea as to how bad the worst case can be. From the point of view of computer tampering and crime, the model of the malevolent player is not merely a fiction; similar remarks hold for cryptography, where the system must be made proof against purposeful attempts to 'break in'. Finally, multi-user systems come close to being games in the ordinary sense of the word.

(vi) *Consistency* is a remarkable property which, in one form or another, is common to just about all game-theoretic solution concepts. Let us be given a game, which for definiteness we denote v, though it may be NTU or even non-cooperative. Let x be an outcome that 'solves' the game in some sense, like the value or nucleolus or a point in the core. Suppose now that some coalition S wishes to view the situation as if the players outside S get their components of x so to speak exogenously, without participating in the play. That means that the players in S are playing the 'reduced game' v_x^S, whose all-player set is S. It is not always easy to say just how v_x^S should be defined, but let's leave that aside for the moment. Suppose we apply to v_x^S the same solution concept that when applied to v yields x. Then the consistency property is that $x|S$ (x restricted to S) is the resulting solution. For example, if x is the nucleolus of v, then for each v, the restriction $x|S$ is the nucleolus of v_x^S.

Consistency implies that it is not too important how the player set is chosen. One can confine attention to a 'small world', and the outcome for the denizens of this world will be the same as if we had looked at them in a 'big world'.

In a game theoretic context, consistency was first noticed by J. Harsanyi (1959) for the Nash solution to the *n*-person bargaining game. This is simply an NTU game V in which the only significant coalitions are the single players and the all-player coalition, and the single players are normalized to get 0. The Nash solution, axiomatized by Harsanyi (1959), is the outcome x that maximizes the product $x^1 x^2 \ldots x^n$. To explain the consistency condition, let us look at the case $n = 3$, in which case $V(\{1, 2, 3\})$ is a subset of 3-space. If we let $S = \{1, 2\}$, and if x_0 is the Nash solution, then 3 should get x_0^3. That means that 1 and 2 are confined to bargaining within that slice of $V\{(1, 2, 3)\}$ that is determined by the plane $x^3 = x_0^3$. According to the Nash solution for the two-person case, they should maximize $x^1 x^2$ over this slice; it is not difficult to see that this maximum is attained at (x_0^1, x_0^2), which is exactly what consistency requires.

Davis and Maschler (1965) proved that the kernel satisfies a consistency condition; so do the bargaining set, core, stable set, and nucleolus, using the same definition of the reduced game v_x^S as for the kernel (Aumann and Drèze, 1974). Using a somewhat different definition of v_x^S, consistency can be established for the value (Hart and Mas-Colell, 1986). Note that strategic equilibria, too, are consistent; if the players outside S play their equilibrium strategies, an equilibrium of the resulting game on S is given by having the players in S play the same strategies that they were playing in the equilibrium of the large game.

Consistency often plays a key role in axiomatizations. Strategic equilibrium is axiomatized by consistency, together with the requirement that in one-person maximization problems, the maximum be chosen. A remarkable axiomatization of the Nash solution to the bargaining problem (including the 2-person case discussed at *1950–1960*, v), in which the key role is played by consistency, has been provided by T. Lensberg (1981). Axiomatizations in which consistency plays the key role have been provided for the nucleolus (Sobolev, 1975), core (Peleg, 1985, 1986), kernel (Peleg, 1986), and value (Hart and Mas-Colell, 1986). Consistency-like conditions have also been used in contexts that are not strictly game-theoretic, e.g. by Balinski and Young (1982), W. Thomson, J. Roemer, H. Moulin, H.P. Young and others.

In law, the consistency criterion goes back at least to the 2000-year old Babylonian Talmud (Aumann and Maschler, 1985). Though it is indeed a very natural condition, its huge scope is still somewhat startling.

(vii) The fascination of *cost allocation* is that it retains the formal structure of cooperative game theory in a totally different interpretation. The question is how to allocate joint costs among users. For example, the cost of a water supply or sewage disposal system serving several municipalities (e.g. Bogardi and Szidarovsky, 1976); the cost of telephone calls in an organization such as a university or corporation (Billera, Heath and Raana, 1978); or the cost of an airport (Littlechild and Owen, 1973, 1976). In the airport case, for example, each 'player' is one landing of one airplane, and $v(S)$ is the cost of building and running an airport large enough to accommodate the set S of landings. Note that $v(S)$ depends not only on the number of landings in S but also on its composition; one would not charge the same for a landing of a 747 as for a Piper, for example because the 747 requires a longer runway. The allocation of cost would depend on the solution concept; for example, if we are using the Shapley value ϕ, then the fee for each landing i would be $\phi^i v$.

The axiomatic method is particularly attractive here, since in this application the axioms often have rather transparent meaning. Most frequently used has been the Shapley value, whose axiomatic characterization (*see* SHAPLEY VALUE) is particularly transparent (Billera and Heath, 1982).

The literature on the game theoretic approach to cost allocation is quite large, probably several hundred items, many of them in the accounting literature (e.g. Roth and Verrecchia, 1979).

CONCLUDING REMARKS

(i) *Ethics*. While game theory does have intellectual ties to ethics, it is important to realize that in itself, it has no moral content, makes no moral recommendations, is ethically neutral. Strategic equilibrium does not tell us to maximize utility, it explores what happens when we do. The Shapley value does not recommend dividing payoff according to power, it simply measures the power. Game Theory is a tool for telling us where incentives will lead. History and experience teach us that if we want to achieve certain goals, including moral and ethical ones, we had better see to the incentive effects of what we are doing; and if we do not want people to usurp power for themselves, we had better build institutions that spread power as thinly and evenly as possible. Blaming game theory – or, for that matter, economic theory – for selfishness is like blaming bacteriology for disease. Game theory studies selfishness, it does not recommend it.

(ii) *Mathematical methods*. We have had very little to say about mathematical methods in the foregoing, because we wished to stress the conceptual side. Worth noting, though, is that mathematically, game theoretic results developed in one context often have important implications in completely different contexts. We have already mentioned the implications of two-person zero-sum theory for the theory of the core and for correlated equilibria (*1910–1930*, vii). The first proofs of the existence of competitive equilibrium (Arrow and Debreu, 1954) used the existence of strategic equilibrium in a generalized game (Debreu, 1952). Blackwell's 1956 theory of two-person zero-sum games with vector payoffs is of fundamental importance for n-person repeated games of complete information (Aumann, 1961) and for repeated games of incomplete information (e.g. Mertens, 1982; Hart, 1985b). The Lemke–Howson algorithm (1962) for finding equilibria of 2-person non-zero sum non-cooperative games is seminal in the development of the algorithms of Scarf (1967, 1973) for finding points in the core and finding economic equilibria.

(iii) *Terminology*. Game Theory has sometimes been plagued by haphazard, inappropriate terminology. Some workers, notably L.S. Shapley (1973b), have tried to introduce more appropriate terminology, and we have here followed their lead. What follows is a brief glossary to aid the reader in making the proper associations.

Used here	*Older term*
Strategic form	Normal form
Strategic equilibrium	Nash equilibrium
Coalitional form	Characteristic function
Transferable utility	Side payment
Decisive voting game	Strong voting game
Improve upon	Block
Worth	Characteristic function value
Profile	*n*-tuple
1-homogeneous	Homogeneous of degree 1

R.J. AUMANN

See also DECISION THEORY; EXCHANGE; NASH EQUILIBRIUM; OLIGOPOLY AND GAME THEORY; PRISONER'S DILEMMA; SHAPLEY VALUE.

BIBLIOGRAPHY

Anderson, R.M. 1986. Notions of core convergence. In Hildenbrand and Mas-Colell (1986), 25–46.

Arrow, K.J. 1951. *Social Choice and Individual Values*. New York: John Wiley.

Arrow, K.J. and Debreu, G. 1954. Existence of an equilibrium for a competitive economy. *Econometrica* 22, 265–90.

Artstein, Z. 1972. Values of games with denumerably many players. *International Journal of Game Theory* 3, 129–40.

Asscher, N. 1976. An ordinal bargaining set for games without side payments. *Mathematics of Operations Research* 1, 381–9.

Aumann, R.J. 1960. Linearity of unrestrictedly transferable utilities. *Naval Research Logistics Quarterly* 7, 281–4.

Aumann, R.J. 1961. The core of a cooperative game without side payments. *Transactions of the American Mathematical Society* 98, 539–52.

Aumann, R.J. 1964. Markets with a continuum of traders. *Econometrica* 32, 39–50.

Aumann, R.J. 1975. Values of markets with a continuum of traders. *Econometrica* 43, 611–46.

Aumann, R.J. 1985. On the non-transferable utility value: a comment on the Roth–Shafer examples. *Econometrica* 53, 667–77.

Aumann, R.J. 1987. Correlated equilibrium as an expression of Bayesian rationality. *Econometrica* 55, 1–18.

Aumann, R.J. and Drèze, J.H. 1974. Cooperative games with coalition structures. *International Journal of Game Theory* 3, 217–38.

Aumann, R.J. and Kurz, M. 1977. Power and taxes. *Econometrica* 45, 1137–61.

Aumann, R.J. and Maschler, M. 1985. Game theoretic analysis of a bankruptcy problem from the Talmud. *Journal of Economic Theory* 36, 195–213.

Aumann, R.J. and Peleg, B. 1960. Von Neumann–Morgenstern solutions to cooperative games without side payments. *Bulletin of the American Mathematical Society* 66, 173–9.

Aumann, R.J. and Shapley, L.S. 1974. *Values of Non-Atomic Games*. Princeton: Princeton University Press.

Axelrod, R. 1984. *The Evolution of Cooperation*. New York: Basic Books.

Balinski, M.L. and Young, H.P. 1982. *Fair Representation*. New Haven: Yale University Press.

Berbee, H. 1981. On covering single points by randomly ordered intervals. *Annals of Probability* 9, 520–28.

Bewley, T. and Kohlberg, E. 1976. The asymptotic theory of stochastic games. *Mathematics of Operations Research* 1, 197–208.

Billera, L.J. 1970a. Existence of general bargaining sets for cooperative games without side payments. *Bulletin of the American Mathematical Society* 76, 375–9.

Billera, L.J. 1970b. Some theorems on the core of an n-person game without side payments. *SIAM Journal of Applied Mathematics* 18, 567–79.

Billera, L.J. and Bixby, R. 1973. A characterization of polyhedral market games. *International Journal of Game Theory* 2, 253–61.

Billera, L.J. and Heath, D.C. 1982. Allocation of shared costs: a set of axioms yielding a unique procedure. *Mathematics of Operations Research* 7, 32–9.

Billera, L.J., Heath, D.C. and Raanan, J. 1978. Internal telephone billing rates – a novel application of non-atomic game theory. *Operations Research* 26, 956–65.

Binmore, K. 1982. Perfect equilibria in bargaining models. *ICERD Discussion Paper* No. 58, London School of Economics.

Blackwell, D. 1956. An analogue of the minimax theorem for vector payoffs. *Pacific Journal of Mathematics* 6, 1–8.

Blackwell, D. and Ferguson, T.S. 1968. The big match. *Annals of Mathematical Statistics* 39, 159–63.

Bogardi, I. and Szidarovsky, F. 1976. Application of game theory in water management. *Applied Mathematical Modelling* 1, 11–20.

Bondareva, O.N. 1963. Some applications of linear programming methods to the theory of cooperative games (in Russian). *Problemy kibernetiki* 10, 119–39.

Borel, E. 1924. Sur les jeux où interviennent l'hasard et l'habilité des joueurs. In *Eléments de la theorie des probabilités*, ed. J. Hermann, Paris; Librairie Scientifique, 204–24.

Braithwaite, R.B. (ed.) 1950. F.P. Ramsey, *The Foundations of Mathematics and Other Logical Essays*. New York: Humanities Press.

Brams, S.J., Lucas, W.F. and Straffin, P.D., Jr. (eds) 1983. *Political and Related Models*. New York: Springer.

Brown, D.J. and Loeb, P. 1976. The values of non-standard exchange economies. *Israel Journal of Mathematics* 25, 71–86.

Brown, D.J. and Robinson, A. 1975. Non standard exchange economies. *Econometrica* 43, 41–55.

Case, J.H. 1979. *Economics and the Competitive Process*. New York: New York University Press.

Champsaur, P. 1975. Cooperation vs. competition. *Journal of Economic Theory* 11, 394–417.

Dantzig, G.B. 1951a. A proof of the equivalence of the programming problem and the game problem. In Koopmans (1951), 330–38.

Dantzig, G.B. 1951b. Maximization of a linear function of variables subject to linear inequalities. In Koopmans (1951), 339–47.

Davis, M. 1964. Infinite games with perfect information. In Dresher, Shapley and Tucker (1964), 85–101.

Davis, M. 1967. Existence of stable payoff configurations for cooperative games. In Shubik (1967), 39–62.

Davis, M. and Maschler, M. 1965. The kernel of a cooperative game. *Naval Research Logistics Quarterly* 12, 223–59.

Debreu, G. 1952. A social equilibrium existence theorem. *Proceedings of the National Academy of Sciences of the United States* 38, 886–93.

Debreu, G. and Scarf, H. 1963. A limit theorem on the core of an economy. *International Economic Review* 4, 236–46.

Dresher, M.A. and Shapley, L.S. and Tucker, A.W. (eds) 1964.

Advances in Game Theory. Annals of Mathematics Studies Series 52, Princeton: Princeton University Press.

Dresher, M.A., Tucker, A.W. and Wolfe, P. (eds) 1957. *Contributions to the Theory of Games III*. Annals of Mathematics Studies Series 39, Princeton: Princeton University Press.

Drèze, J.H., Gabszewicz, J. and Gepts, S. 1969. On cores and competitive equilibria. In Guilbaud (1969), 91–114.

Dubey, P. 1980. Asymptotic semivalues and a short proof of Kannai's theorem. *Mathematics of Operations Research* 5, 267–70.

Dubey, P. and Shapley, L.S. 1980. Non cooperative exchange with a continuum of traders: two models. *Technical Report of the Institute for Advanced Studies*, the Hebrew University of Jerusalem.

Edgeworth, F.Y. 1881. *Mathematical Psychics*. London: Kegan Paul.

Everett, H. 1957. Recursive games. In Dresher, Tucker and Wolfe (1957), 47–78.

Fogelman, F. and Quinzii, M. 1980. Asymptotic values of mixed games. *Mathematics of Operations Research* 5, 86–93.

Gabszewicz, J.J. and Mertens, J.F. 1971. An equivalence theorem for the core of an economy whose atoms are not 'too' big. *Econometrica* 39, 713–21.

Gale, D. 1974. A curious nim-type game. *American Mathematical Monthly* 81, 876–79.

Gale, D. 1979. The game of hex and the Brouwer fixed-point theorem. *American Mathematical Monthly* 86, 818–27.

Gale, D. 1986. Bargaining and competition, Part I: Characterization; Part II: Existence. *Econometrica* 54, 785–806; 807–18.

Gale, D. and Shapley, L.S. 1962. College admissions and the stability of marriage. *American Mathematical Monthly* 69, 9–15.

Gale, D. and Stewart, F.H. 1953. Infinite games with perfect information. In Kuhn and Tucker (1953), 245–66.

Galil, Z. 1974. The nucleolus in games with major and minor players. *International Journal of Game Theory* 3, 129–40.

Gillies, D.B. 1959. Solutions to general non-zero-sum games. In Luce and Tucker (1959), 47–85.

Guilbaud, G.T. (ed.) 1969. *La décision: aggrégation et dynamique des ordres de préférence*. Paris: Editions du CNRS.

Harsanyi, J.C. 1956. Approaches to the bargaining problem before and after the theory of games: a critical discussion of Zeuthen's, Hicks' and Nash's theories. *Econometrica* 24, 144–57.

Harsanyi, J.C. 1959. A bargaining model for the cooperative n-person game. In Tucker and Luce (1959), 325–56.

Harsanyi, J.C. 1963. A simplified bargaining model for the n-person cooperative game. *International Economic Review* 4, 194–220.

Harsanyi, J.C. 1966. A general theory of rational behavior in game situations. *Econometrica* 34, 613–34.

Harsanyi, J.C. 1967–8. Games with incomplete information played by 'Bayesian' players, parts I, II and III. *Management Science* 14, 159–82, 320–34, 486–502.

Harsanyi, J.C. 1973. Games with randomly disturbed payoffs: a new rationale for mixed strategy equilibrium points. *International Journal of Game Theory* 2, 1–23.

Harsanyi, J.C. 1982. Solutions for some bargaining games under the Harsanyi–Selten solution theory I: Theoretical preliminaries; II: Analysis of specific games. *Mathematical Social Sciences* 3, 179–91; 259–79.

Harsanyi, J.C. and Selten, R. 1987. *A General Theory of Equilibrium Selection in Games*. Cambridge, Mass.: MIT Press.

Harsanyi, J.C. and Selten, R. 1972. A generalized Nash solution for two-person bargaining games with incomplete information. *Management Science* 18, 80–106.

Hart, S. 1973. Values of mixed games. *International Journal of Game Theory* 2, 69–86.

Hart, S. 1974. Formation of cartels in large markets. *Journal of Economic Theory* 7, 453–66.

Hart, S. 1977a. Asymptotic values of games with a continuum of players. *Journal of Mathematical Economics* 4, 57–80.

Hart, S. 1977b. Values of non-differentiable markets with a continuum of traders. *Journal of Mathematical Economics* 4, 103–16.

Hart, S. 1980. Measure-based values of market games. *Mathematics of Operations Research* 5, 197–228.

Hart, S. 1985a. An axiomatization of Harsanyi's nontransferable utility solution. *Econometrica* 53, 1295–314.

Hart, S. 1985b. Non zero-sum two-person repeated games with

incomplete information. *Mathematics of Operations Research* 10, 117–53.

Hart, S. and Mas-Colell, A. 1986. The potential: a new approach to the value in multi-person allocation problems. Harvard University Discussion Paper 1157.

Hart, S. and Schmeidler, D. 1988. Correlated equilibria: an elementary existence proof. *Mathematics of Operations Research*.

Hildenbrand, W. (ed.) 1982. *Advances in Economic Theory*. Cambridge: Cambridge University Press.

Hildenbrand, W. and Mas-Colell, A. 1986. *Contributions to Mathematical Economics in Honor of G. Debreu*. Amsterdam: North-Holland.

Hu, T.C. and Robinson, S.M. (eds) 1973. *Mathematical Programming*. New York: Academic Press.

Isaacs, R. 1965. *Differential Games: A Mathematical Theory with Applications to Warfare and Pursuit, Control and Optimization*. New York: John Wiley.

Kakutani, S. 1941. A generalization of Brouwer's fixed point theorem. *Duke Mathematical Journal* 8, 457–9.

Kalai, E. and Smorodinsky, M. 1975. Other solutions to Nash's bargaining problem. *Econometrica* 43, 513–18.

Kaneko, M. and Wooders, M. 1982. Cores of partitioning games. *Mathematical Social Sciences* 3, 313–27.

Kannai, Y. 1966. Values of games with a continuum of players. *Israel Journal of Mathematics* 4, 54–8.

Kohlberg, E. 1971. On the nucleolus of a characteristic function game. *SIAM Journal of Applied Mathematics* 20, 62–6.

Kohlberg, E. 1972. The nucleolus as a solution to a minimization problem. *SIAM Journal of Applied Mathematics* 23, 34–49.

Koopmans, T.C. (ed.) 1951. *Activity Analysis of Production and Allocation*. New York: Wiley.

Kuhn, H.W. 1952. *Lectures on the Theory of Games*. Issued as a report of the Logistics Research Project, Office of Naval Research, Princeton University.

Kuhn, H.W. 1953. Extensive games and the problem of information. In Kuhn and Tucker (1953), 193–216.

Kuhn, H.W. and Tucker, A.W. (eds) 1950. *Contributions to the Theory of Games I*. Annals of Mathematics Studies Series 24, Princeton: Princeton University Press.

Kuhn, H.W. and Tucker, A.W. (eds) 1953. *Contributions to the Theory of Games II*. Annals of Mathematics Studies Series 28, Princeton: Princeton University Press.

Lemke, L.E. and Howson, J.T. 1962. Equilibrium points of bimatrix games. *SIAM Journal of Applied Mathematics* 12, 413–23.

Lensberg, T. 1981. The stability of the Nash solution. Unpublished.

Lewis, D.K. 1969. *Convention*. Cambridge, Mass.: Harvard University Press.

Littlechild, S.C. 1976. A further note on the nucleolus of the 'airport game'. *International Journal of Game Theory* 5, 91–5.

Littlechild, S.C. and Owen, G. 1973. A simple expression for the Shapley value in a special case. *Management Science* 20, 370–72.

Lucas, W.F. 1969. The proof that a game may not have a solution. *Transactions of the American Mathematical Society* 137, 219–29.

Lucas, W.F. 1983. Measuring power in weighted voting systems. In Brams, Lucas and Straffin (1983), ch. 9.

Lucas, W.F. and Rabie, M. 1982. Games with no solutions and empty core. *Mathematics of Operations Research* 7, 491–500.

Luce, R.D. and Raiffa, H. 1957. *Games and Decisions, Introduction and Critical Survey*. New York: John Wiley.

Luce, R.D. and Tucker, A.W. (eds) 1959. *Contributions to the Theory of Games IV*. Annals of Mathematics Studies Series 40, Princeton: Princeton University Press.

Lyapounov, A.A. 1940. On completely additive vector-functions (in Russian, abstract in French). *Akademiia Nauk USSR Izvestiia Seriia Mathematicheskaia* 4, 465–78.

Martin, D.A. 1975. Borel determinacy. *Annals of Mathematics* 102, 363–71.

Maschler, M. (ed.) 1962. *Recent Advances in Game Theory*. Proceedings of a Conference, privately printed for members of the conference, Princeton: Princeton University Conferences.

Maschler, M., Peleg, B. and Shapley, L.S. 1979. Geometric properties of the kernel, nucleolus, and related solution concepts. *Mathematics of Operations Research* 4, 303–38.

Maschler, M. and Perles, M. 1981. The superadditive solution for the Nash bargaining game. *International Journal of Game Theory* 10, 163–93.

Mas-Colell, A. 1975. A further result on the representation of games by markets. *Journal of Economic Theory* 10, 117–22.

Mas-Colell, A. 1977. Competitive and value allocations of large exchange economies. *Journal of Economic Theory* 14, 419–38.

Mas-Colell, A. 1988. An equivalence theorem for a bargaining set. *Journal of Mathematical Economics*.

Maynard Smith, J. 1982. *Evolution and the Theory of Games*. Cambridge: Cambridge University Press.

Mertens, J.F. 1982. Repeated games: an overview of the zero-sum case. In Hildenbrand (1982), 175–82.

Mertens, J.F. 1987. The Shapley value in the non-differentiable case. *International Journal of Game Theory*.

Mertens, J.F. and Neyman, A. 1981. Stochastic games. *International Journal of Game Theory* 10, 53–66.

Milnor, J.W. 1978. Values of large games II: Oceanic games. *Mathematics of Operations Research* 3, 290–307.

Milnor, J.W. and Shapley, L.S. 1957. On games of survival. In Dresher, Tucker and Wolfe (1957), 15–45.

Moschovakis, Y.N. 1980. *Descriptive Set Theory*. New York: North-Holland.

Moschovakis, Y.N. (ed.) 1983. *Cabal Seminar 79–81: Proceedings, Caltech-UCLA Logic Seminar 1979–81*. Lecture Notes in Mathematics 1019, New York: Springer-Verlag.

Mycielski, J. and Steinhaus, H. 1964. On the axiom of determinateness. *Fundamenta Mathematicae* 53, 205–24.

Myerson, R.B. 1977. Graphs and cooperation in games. *Mathematics of Operations Research* 2, 225–9.

Myerson, R.B. 1979. Incentive compatibility and the bargaining problem. *Econometrica* 47, 61–74.

Myerson, R.B. 1984. Cooperative games with incomplete information. *International Journal of Game Theory* 13, 69–96.

Nash, J.F., Jr. 1950. The bargaining problem. *Econometrica* 18, 155–62.

Nash, J.F., Jr. 1951. Non-cooperative games. *Annals of Mathematics* 54, 289–95.

Neyman, A. 1985a. Bounded complexity justifies cooperation in the finitely repeated prisoner's dilemma. *Economics Letters* 19, 227–30.

Neyman, A. 1985b. Semivalues of political economic games. *Mathematics of Operations Research* 10, 390–402.

Neyman, A. 1987. Weighted majority games have an asymptotic value. *Mathematics of Operations Research*.

O'Neill, B. 1987. Non-metric test of the minimax theory of two-person zero-sum games. *Proceedings of the National Academy of Sciences of the United States* 84, 2106–9.

Owen, G. 1972. Multilinear extensions of games. *Management Science* 18, 64–79.

Peleg, B. 1963a. Solutions to cooperative games without side payments. *Transactions of the American Mathematical Society* 106, 280–92.

Peleg, B. 1963b. Bargaining sets of cooperative games without side payments. *Israel Journal of Mathematics* 1, 197–200.

Peleg, B. 1967. Existence theorem for the bargaining set $M_1^{(i)}$. In Shubik (1967), 53–6.

Peleg, B. 1968. On weights of constant-sum majority games. *SIAM Journal of Applied Mathematics* 16, 527–32.

Peleg, B. 1985. An axiomatization of the core of cooperative games without side payments. *Journal of Mathematical Economics* 14, 203–14.

Peleg, B. 1986. On the reduced game property and its converse. *International Journal of Game Theory* 15, 187–200.

Pennock, J.R. and Chapman, J.W. (eds) 1968. *Representation*. New York: Atherton.

Ramsey, F.P. 1931. Truth and probability. In Braithwaite (1950).

Ransmeier, J.S. 1942. *The Tennesee Valley Authority: A Case Study in the Economics of Multiple Purpose Stream Planning*, Nashville: Vanderbilt University Press.

Regelmann, K. and Curio, E. 1986. How do great tit (Parus Major) pair mates cooperate in broad defence? *Behavior* 97, 10–36.

Riker, W.H., and Shapley, L.S. 1968. Weighted voting: a mathematical analysis for instrumental judgements. In Pennock and Chapman (1968), 199–216.

Robinson, A. 1974. *Non-Standard Analysis*. Amsterdam: North-Holland.

Rosenthal, R.W. 1979. Sequences of games with varying opponents. *Econometrica* 47, 1353–66.

Rosenthal, R.W. and Rubinstein, A. 1984. Repeated two player games with ruin. *International Journal of Game Theory* 13, 155–77.

Roth, A.E. 1977. The Shapley value as a von Neumann–Morgenstern utility. *Econometrica* 45, 657–64.

Roth, A.E. 1984. The evolution of the labor market for medical interns and residents: a case study in game theory. *Journal of Political Economy* 92, 991–1016.

Roth, A.E. and Verrecchia, R.E. 1979. The Shapley value as applied to cost allocation: a reinterpretation. *Journal of Accounting Research* 17, 295–303.

Rubinstein, A. 1982. Perfect equilibrium in a bargaining model. *Econometrica* 50, 97–109.

Rubinstein, A. 1986. Finite automata play the repeated prisoner's dilemma. *Journal of Economic Theory* 39, 83–96.

Savage, L.J. 1954. *The Foundations of Statistics*. New York: John Wiley.

Scarf, H.E. 1967. The core of an n-person game. *Econometrica* 35, 50–69.

Scarf, H.E. 1973. *The Computation of Economic Equilibria*. New Haven: Yale University Press.

Schelling, T.C. 1960. *The Strategy of Conflict*. Cambridge, Mass.: Harvard University Press.

Schmeidler, D. 1969. The nucleolus of a characteristic function game. *SIAM Journal of Applied Mathematics* 17, 1163–70.

Selten, R.C. 1965. Spieltheoretische Behandlung eines Oligopolmodells mit Nachfragetragheit. *Zeitschrift für die gesammte Staatswissenschaft* 121, 301–24; 667–89.

Selten, R.C. 1975. Reexamination of the perfectness concept for equilibrium points in extensive games. *International Journal of Game Theory* 4, 25–55.

Selten, R.C. 1980. A note on evolutionary stable strategies in asymmetric animal conflicts. *Journal of Theoretical Biology* 84, 101–101.

Selten, R.C. 1983. Evolutionary stability in extensive two-part games. *Mathematical Social Sciences* 5, 269–363.

Shapley, L.S. 1953a. Stochastic games. *Proceedings of the National Academy of Sciences of the United States* 39, 1095–100.

Shapley, L.S. 1953b. A value for n-person games. In Kuhn and Tucker (1953), 305–17.

Shapley, L.S. 1962. Values of games with infinitely many players. In Maschler (1962), 113–18.

Shapley, L.S. 1964. Values of large games, VII: a general exchange economy with money. *RAND Publication* RM–4248, Santa Monica, California.

Shapley, L.S. 1967. On balanced sets and cores. *Naval Research Logistics Quarterly* 14, 453–60.

Shapley, L.S. 1969a. On market games. *Journal of Economic Theory* 1, 9–25.

Shapley, L.S. 1969b. Pure competition, coalitional power and fair division. *International Economic Review* 10, 337–62.

Shapley, L.S. 1969c. Utility comparison and the theory of games. In Guilbaud (1969), 251–63.

Shapley, L.S. 1973a. On balanced games without side payments. In Hu and Robinson (1973), 261–90.

Shapley, L.S. 1973b. Let's block 'block'. *Econometrica* 41, 1201–2.

Shapley, L.S. 1984. Convergence of the bargaining set for differentiable market games. Appendix B in Shubik (1984), 683–92.

Shapley, L.S. and Shubik, M. 1954. A method for evaluating the distribution of power in a committee system. *American Political Science Review* 48, 787–92.

Shitovitz, B. 1973. Oligopoly in markets with a continuum of traders. *Econometrica* 41, 467–501.

Shubik, M. 1959a. Edgeworth market games. In Luce and Tucker (1959), 267–78.

Shubik, M. 1959b. *Strategy and Market Structure*. New York: John Wiley.

Shubik, M. (ed.) 1967. *Essays in Mathematical Economics in Honor of Oskar Morgenstern*. Princeton: Princeton University Press.

Shubik, M. 1973. Commodity, money, oligopoly, credit and bankruptcy in a general equilibrium model. *Western Economic Journal* 11, 24–36.

Shubik, M. 1982. *Game Theory in the Social Sciences, Concepts and Solutions*. Cambridge, Mass.: MIT Press.

Shubik, M. 1984. *A Game Theoretic Approach to Political Economy*. Cambridge, Mass.: MIT Press.

Sobolev, A.I. 1975. Characterization of the principle of optimality for cooperative games through functional equations (in Russian). In Vorobiev (1975), 94–151.

Sorin, S. 1986a. On repeated games of complete information. *Mathematics of Operations Research* 11, 147–60.

Sorin, S. 1986b. An asymptotic property of non-zero sum stochastic games. *International Journal of Game Theory* 15(2), 101–7.

Tauman, Y. 1981. Value on a class of non-differentiable market games. *International Journal of Game Theory* 10, 155–62.

Ville, J.A. 1938. Sur le théorie générale des jeux où intervient l'habilité des joueurs. In *Traité du calcul des probabilités et de ses applications*, Vol. 4, ed. E. Borel, Paris: Gauthier-Villars, 105–13.

Vinacke, W.E. and Arkoff, A. 1957. An experimental study of coalitions in the triad. *American Sociological Review* 22, 406–15.

Vind, K. 1965. A theorem on the core of an economy. *Review of Economic Studies* 32, 47–8.

von Neumann, J. 1928. Zur Theorie der Gesellschaftsspiele. *Mathematische Annalen* 100, 295–320.

von Neumann, J. 1949. On rings of operators. Reduction theory. *Annals of Mathematics* 50, 401–85.

von Neumann, J. and Morgenstern, O. 1944. *Theory of Games and Economic Behavior*. Princeton: Princeton University Press.

Vorobiev, N.N. (ed.) 1975. *Mathematical Methods in Social Science* (in Russian). Vipusk 6, Vilnius.

Wilson, R. 1978. Information, efficiency, and the core of an economy. *Econometrica* 46, 807–16.

Wolfe, P. 1955. The strict determinateness of certain infinite games. *Pacific Journal of Mathematics* 5, 841–7.

Wooders, M.H. 1983. The epsilon core of a large replica game. *Journal of Mathematical Economics* 11, 277–300.

Wooders, M.H. and Zame, W.R. 1984. Approximate cores of large games. *Econometrica* 52, 1327–50.

Young, H.P. 1985. Monotonic solutions of cooperative games. *International Journal of Game Theory* 14, 65–72.

Zermelo, E. 1913. Über eine Anwendung der Mengenlehre auf die theorie des Schachspiels. *Proceedings of the Fifth International Congress of Mathematicians* 2, 501–4.

gaming contracts. A gaming or wagering contract is one by which two persons professing to hold opposite views touching a future uncertain event mutually agree that, dependent upon that event, one shall receive from the other, and the other shall pay or hand over to him, a sum of money or other stake; neither of the contracting parties having any other interest in that contract than the sum or stake he will so win or lose, there being no other real consideration for the making of such contract by either of the parties (Justice Hawkins in *Carlill* v. *Carbolic Smoke Ball Company* (1892), Queen's Bench 484).

Contracts of this description are declared to be null and void by an act passed during the present reign [of Queen Victoria] (8 & 9 Vict. c. 109. p. 18). Notwithstanding this act it was held that a betting agent who had paid the amount due by his principal on the loss of a bet was entitled to recover the same from the latter (*Read* v. *Anderson*, 10 Queen's Bench Division 100; 13 Queen's Bench Division 779), but this indirect recognition of betting transactions has lately been set aside by the Gaming Act of 1892, which enacts that no action shall be brought to recover any sum of money paid in respect of any gaming or wagering contract. The subject of gaming contracts has recently been discussed with reference to the 'missing word competitions' organized by certain newspapers, which were held to be illegal, as the result did not depend on skill and judgement but upon mere chance (*Barclay* v. *Pearson* (1893), 2

Chancery 154). The principle of the statute against gaming and wagering contracts was, to a certain extent, already recognized by the statute of 14 Geo. III. c. 48, which forbids the insurance of a life in which the insurer has no interest, and which is still in force.

Much discussion has taken place both in England and abroad on the question whether certain time bargains on the stock exchange and in the produce markets are to be considered as partaking of the nature of wagers, and the result of the decisions seems to be that a contract is not enforceable where it can be proved that it was not the intention of the parties to deliver or receive a certain quantity of securities or produce at a certain price, but that the payment of the difference between the price at which the bargain was made and the market price at the time fixed for the completion of the bargain was the sole object of the transaction; in the absence of such proof the parties must be presumed to have intended a real sale.

[E. Schuster]
Reprinted from *Palgrave's Dictionary of Political Economy.*

Ganilh, Charles (1785–1836).   Ganilh was born in Allanches (Cantal) on 6 January 1785, and died in Paris in 1836. The main claims to (modest) fame of this barrister-turned-politician-turned-economist are his *Systèmes d'économie politique* (1809) and his *Dictionnaire d'économie politique* (1826). These two works are respectively the first systematic history of thought and the first theoretically oriented dictionary of economics ever published. Unfortunately, save this claim for priority, these two prestigious titles barely conceal the analytical poverty of their contents.

Ganilh's two main analytical books can best be seen as potted (though not uncritical) introductions of the main Smithian theses to the French educated layman. In the *Systèmes*, leaving aside all anecdotal reference to economists and their intellectual environment, Ganilh concentrates exclusively on a history of economic theory. Centred around main concepts (wealth, labour, value, capital, money etc.), Ganilh's systematic treatment ranges from the Greek philosophers down the centuries to the Mercantilist and Physiocratic schools which, together with the *Wealth of Nations*, are given the pride of place. In a similar way, in the *Dictionnaire*, what Ganilh considers to be the main theoretical concepts (excluding any factual or biographical entries) are not only individually discussed in alphabetic order but also logically connected by means of a cross-reference system.

Free trade and the notion of productive labour are the two themes around which Ganilh's resistance to Adam Smith are articulated. Without reverting to the Mercantilist doctrine that dominated French regulatory practice right up to the fall of the *ancien régime*, Ganilh thinks fit (for practical reasons) to bring some restriction to the then newly discovered free trade doctrine. Anticipating in some ways Frederic List's *National System of Political Economy* (1841), Ganilh strongly advocates trade barriers in favour of France's nascent industries (1826, pp. 142–150). For his critical remarks on Smith's concept of *productive labour* (i.e. labour supported by capital to produce material goods), Ganilh was caught in a heated argument with *inter alia* Malthus, Ricardo, Buchanan and Lauderdale (1826, pp. 102–4 and 413–28). With his opinion that 'any labour the exchange of which gives rise to a value' (1826, p. 415) is productive, Ganilh espoused the then dominant utilitarian doctrine of Say; and for that, he was later to be disparagingly condemned by Marx.

In addition to these two analytical works, Ganilh wrote extensively in the field of public finance (1815).

P. Bridel

SELECTED WORKS
1809. *Des systèmes d'économie politique.* 2 vols, Paris: Déterville.
1815. *Théorie de l'économie politique.* 2 vols, Paris: Déterville.
1826. *Dictionnaire analytique de l'économie politique.* Paris: Ladvocat.

Garnier, Clément Joseph (1813–1881).   French economist, born at Beuil (Alpes Maritimes) on 3 October 1813; died at Paris on 25 September 1881. Joseph Garnier (not to be confused with the translator of Adam Smith) came from a family of prosperous farmers, but showed no inclination to follow his ancestral heritage. He made his way to Paris in 1829, when only 16. Poised to join the banking firm of Lafitte, he was induced instead to enter the Ecole Supérieure de Commerce by a family friend, Adolphe Blanqui. Later following in Blanqui's footsteps, Garnier became both teacher and principal at the Ecole.

Garnier remained in the mainstream of French economics throughout the mid-19th century. He was one of the founders of the Société d'Economie Politique and its permanent secretary from 1842 until his death in 1881. A tireless teacher, he held professorships at five different schools. In 1843 he began a series of lectures at the Athenée, which eventually evolved into his *Eléments de l'économie politique*, a popular and encyclopedic treatise that went through multiple editions from 1846 to 1907. The work was retitled in its fourth edition to the *Traité de l'économie politique*, and in this form it was eventually translated into Spanish, Italian and Russian. Already on the strength of the first edition, Garnier had been named in 1846 to the newly created chair of political economy at the École des Ponts et Chaussées (Dupuit's *alma mater*), a position he held for 35 years. He also spent a quarter of a century as the chief editor of the *Journal des économistes*, the leading French economics journal of the period.

His fortuitous placement at the nerve centres of French economics (the faculty of a *grande école*, the leading journal and professional society) gave Garnier a measure of influence beyond what could be expected from the originality of his ideas, which in any case was minimal. He was chiefly an exponent of classical economics, occupying the middle ground between the virulent strain of liberalism that afflicted Molinari and the less strident version associated with Blanqui and Dunoyer. Although he produced a popular, annotated French edition of Malthus's *Essay on Population*, Garnier chose in his own works to expand the optimistic doctrines of Smith and Say rather than the underside of Ricardo and Malthus. His *Traité* is a good example of French economics before Mill. Its popularity must be attributed in part to this orthodoxy, but also to Garnier's depth of knowledge and his orderly presentation.

Garnier was named a member of the French Institute in 1873. Three years later he was elected to the French Senate by voters in his home district, thus capping a long academic and literary career with a final dimension of public service.

R.F. Hébert

SELECTED WORKS
1846. *Eléments de l'économie politique, exposé des notions
     fondamentales de cette science.* Paris: Guillaumin.
1857. *Du principe de population.* Paris: Garnier frères.
1858a. *Premières notions d'économie politique, sociale ou industrielle.*
     Paris: Guillaumin.
1858b. *Eléments de finances suivis de éléments de statistique, de la
     misère, l'association et l'économie politique.* Paris: Garnier frères.

gauge functions. Consider the standard two-product diagram which depicts an opportunity set P with production frontier fr(P). For any point x^1 inside P it would be useful to have a measure of just how inefficient it is, i.e. to gauge how far it is from the frontier. A simple way of doing this is, first to find that point $\bar{x} \in$ fr(P) which is just a scale change of x^1, so that $x^1 = \lambda_1 \bar{x}$ for some $\lambda_1 \in [0, 1)$. Then a function $J(.|P)$ that calibrates any such point with respect to P is defined by putting $J(x^1|P) = \lambda_1$. For this to be a sensible measure of efficiency, it should obviously have the property that $J(x|P) = 1$ if and only if (iff) $x \in$ fr(P).

Similarly, for any point x^2 outside P it can be asked: How much would productive capacity (i.e. P) have to grow in order that x^2 be producible? Again, a simple measure of this would be to find $\ddot{x} \in$ fr(P) such that $x^2 = \lambda_2 \ddot{x}$ for some $\lambda_2 > 1$, and put $J(x^2|P) = \lambda_2$. Thus $J(x|P)$ becomes a general measure of the producibility of x with respect to P.

In the same way, in the theory of consumer's preferences consider a better set $B^t = \{x \in R^n : x \gtrsim x^t\}$ for some 'target' bundle x^t, and let x^3 be any bundle that lies above the indifference surface I^t which bounds B^t from below. An obvious measure of how much x^3 could be reduced and the resulting bundle still remain in B^t is given by finding $\bar{x} \in I^t$ such that $x^3 = \mu_3 \bar{x}$ for some $\mu_3 > 1$. Provided $j(x|B^t) = 1$ iff $x \in I^t$, μ_3 is then a measure of the redundancy of x^3 in achieving the target level of satisfaction represented by x^t. Again, if x^4 lies below I^t and $x^4 = \mu_4 \ddot{x}$ for some $\ddot{x} \in I^t$ and $\mu_4 \in [0, 1)$, then μ_4 is a measure of the shortfall of x^4 in achieving x^t. In each case, putting $j(x|B^t) = \mu$ defines a function that gauges the performance of the actual bundle x with respect to the set B^t, hence to the target bundle x^t.

These two functions $J(.|P)$ and $j(.|B^t)$ are examples of what in this essay will be called *gauge* and *s-gauge* functions, respectively. Such functions form the basis of one of the two main duality schemes in economics, the other scheme being that of Fenchel transforms (for which see DUALITY).

1. HISTORY

As already indicated, gauge functions are of two types. The first, simply called *gauges*, are a direct generalization of Minkowski's Distanzfunktion (e.g. (1911)) and so are sometimes referred to as *Minkowski functionals*. Functions of this type are often used in mathematics but are as yet rarely employed (at least explicitly) in economics. They are best suited for bounded sets lying near the origin, such as P above and the trading sets X_i of McKenzie (1981, p. 820).

The second type of gauge function is almost unknown in mathematics but often used in economics, chiefly for unbounded sets that do not contain the origin, such as B^t above and analogous sets in the theory of production. Economists have given these functions many names, among them 'distance', 'transformation' and 'deflation' function. The name *s-gauges* used here both pays homage to Shephard (1953) and emphasizes their affinity with gauges.

S-gauges were introduced more or less simultaneously by Debreu (1951), Malmquist (1953) and Shephard (1953), each in his own way. Debreu, concerned with general equilibrium, defined his function 'the coefficient of resource utilization' for the better set in commodity space whose lower boundary is a Scitovsky community indifference surface; it was obtained as the solution to an optimization problem and is actually the inverse of an s-gauge. Malmquist, concerned with index numbers and hampered by misprints, defined a 'quantity index' (pp. 230–32) as s-gauges of better sets B^t, and a 'price index' (pp. 213–15) as s-gauges of the sets B_0^t (see Section 3(b)) that are dual to B^t.

Shephard, concerned with production functions, defined s-gauges for surfaces in the strictly positive orthant and showed that cost functions could be regarded as s-gauges of dual surfaces in the price space. Oddly, he appeared to regard s-gauges as an example of Minkowski's Distanzfunktion (1953, pp. 6), even though the latter is a convex function defined only for convex compact sets with the origin as interior point, while Shephard's function is concave and defined only for convex unbounded sets that clearly do not contain the origin. Later he implicitly recognized this anomaly by referring to s-gauges as 'an adaptation ... of the Minkowski distance function' (1970, p. 66).

Notwithstanding, his discussion was by far the most comprehensive of the three and fully warrants Diewert's judgement (1982, p. 551): 'the first modern, rigorous treatment of duality theory' – at least in economics.

After this pioneering work there was a long gap, until s-gauges reappeared in the 1970s with the work (in alphabetical order) of Blackorby, Primont and Russell (e.g. 1978), Deaton (1979), Deaton and Muellbauer (1980), Diewert (e.g. 1982, which contains a full bibliography), Gorman (1970, 1976), Hanoch (1978), Jacobsen (1972), McFadden (1978), Ruys (1972), Ruys and Weddepohl (1979), Shephard (1970) and Weddepohl (1972). It is sometimes argued that Wold (1943) and Uzawa (1964) were early contributors to this literature, but Wold used only a Euclidean norm and explicitly rejected a Minkowskian approach (1943, pp. 231–2), while Uzawa's paper was actually a nice formulation of the 1–1 correspondence between closed convex sets and their support functions.

Although these later investigations showed a wide range of application for s-gauges, apart from Weddepohl's contribution their formal theory remains more or less as Shephard left it in 1953, still seriously incomplete. For example, economists usually define s-gauges with respect to utility and production *functions* rather than sets (which was the original Minkowskian tradition) and partly for this reason place severe and unexplained restrictions on the relevant domains. Moreover, their discussions are typically confined to convex preferences and technologies and to finite-dimensional spaces.

This essay sets out a formal and coherent account of gauges and s-gauges taken together, although no proofs are given. The primal (e.g. quantity) and dual (e.g. price) spaces will be denoted X and Y, respectively. Though the non-specialist reader need only consider the familiar space $X = R^n = Y$, the natural frame of reference of this theory is where $\langle X, Y \rangle$ are a pair of (in general) infinite-dimensional Hausdorff topological vector spaces 'in duality' (see e.g. Robertson and Robertson, 1964), a class which for example includes Banach spaces and their duals.

Bonnesen and Fenchel (1934) give a detailed discussion of and references to Minkowski's original Distanzfunktion on R^n; Cassels (1959) is also useful. Discussions of gauges (Minkowski functionals) in more general spaces are to be found in works on functional analysis, e.g. Bourbaki (1953), Kothe (1969) and Moreau (1967). For mathematical concepts not explained here, see the entries on CONVEX PROGRAMMING and DUALITY.

2. DEFINITIONS AND SIMPLE PROPERTIES

(a) Gauges. A set $C \subset X$ is a *cone with vertex 0* if $x \in C \Rightarrow \lambda x \in C$ for all $\lambda > 0$. Deconstructing this idea into two others, *starred* and *haloed* sets, the *starred hull* $e(a)$ of any set $A \subset X$ is given by $e(A) = \{v \in X : v = \lambda x, \ x \in A \ 0 \leqslant \lambda \leqslant 1\}$ and the *haloed hull* $h(A)$ by $h(A) = \{w \in X : w = \lambda x, \ x \in A, \ \lambda \geqslant 1\}$.

Then A is *starred* iff $A = e(A)$ and *haloed* iff $A = h(A)$. The *conical hull* $C(A)$ is $e(A) \cup h(A)$, and A is a cone with vertex 0 iff $A = C(A)$.

For any $A \subset X$ its *gauge* is the numerical function $J(.\,|A): X \to [0, \infty]$ defined by

$$J(x|A) = \inf\{\lambda > 0: x \in A\} \quad \text{if } x \neq 0$$
$$= 0 \quad \text{if } x = 0 \quad (1)$$

In 'seeing' this definition, think A as rather like the unit sphere, or at least as being near the origin and possibly bounded. If $x \notin A$ the set is to be enlarged by the magnification factor $\lambda > 1$ until it just engulfs x, the corresponding value of λ then being $J(x|A)$, while if $x \in A$ then A is to be uniformly shrunk by the contraction factor $\lambda \leqslant 1$ until it is on the verge of parting company with x.

It follows from (1) that $J(.\,|A)$ is always proper, and positively homogeneous (ph), (i.e. $J(\lambda x|A)) = \lambda J(x|A)$ for all $\lambda \geqslant 0$. More surprisingly, $J(.\,|e(A)) \equiv J(.\,|A)$, so that it is natural for the theory of gauges to deal only with starred sets. Notice that if $A = C(A)$ then $J(.\,|A) \equiv \delta(.\,|A)$, where the latter function is the *indicator* of A, i.e. $\delta(x|A) = 0$ if $x \in A$, and $= \infty$ otherwise.

If A is starred and (topologically) closed, then $J(.\,|A)$ is lower semicontinuous (lsc) on the whole of X, and decreasing in A (i.e. $A^1 \subset A^2$ iff $J(.\,|A^1) \geqslant J(.\,|A^2)$), while $A = \{x \in X: J(x|A) \leqslant 1\}$. Convex sets that contain the origin (and so are starred) have gauges that are not only ph but also *subadditive* ($J(v + w|A) \leqslant J(v|A) + J(w|A)$ for all $v, w \in X$), and so they are convex functions. Finally, if A is convex, closed and such that 0 is in the topological interior (int) of A, then the (topological) boundary or *frontier* of A is $\{x \in X: J(x|A) = 1\}$.

As already noted, Minkowski's Distanzfunktion was defined only for convex compact $A \subset R^n$ such that $0 \in \text{int } A$, whereas its generalization $J(.\,|A)$ can be used for much wider classes of both sets and spaces. However, in mathematics gauges are in fact typically limited to sets that are convex, closed, *balanced* (if $|\lambda| \leqslant 1$ and $x \in A$ then $\lambda x \in A$) and such that $0 \in \text{int } A$, i.e. sets that are very much like the unit sphere.

(b) *S-gauges.* The formal treatment of s-gauges in mathematics seems confined to Phelps (1963), although economists have made some valuable contributions, in particular Shephard (1953) and Weddepohl (1972). The next definition is best 'seen' if B is taken to be haloed (hence unbounded) and *not* containing the origin.

For any $B \subset X$ its *s-gauge* is the numerical function $j(.\,|B): X \to \{-\infty\} \cup [0, \infty]$ defined by

$$j(x|B) = \sup\{\mu > 0: x \in \lambda B\} \quad \text{if } x \neq 0$$
$$= 0 \quad \text{if } x = 0 \quad (2)$$

If $x \notin B$, $j(x|B)$ is found by pulling B uniformly towards the origin via the contraction factor $\mu \in (0, 1]$ until μB just reaches x. If $x \in B$, then $j(x|B)$ is found by making B recede radially away from the origin through multiplication by the expansion factor $\mu \geqslant 1$ until μB is on the verge of leaving x behind. Notice that $j(x|B) = 0$ iff $x = 0$, unlike the situation with gauges.

A point $v \in B$ is *internal to* B if for any other $x \in X$ there exists $\epsilon > 0$ such that $(v + \lambda x) \in B$ for all λ with $|\lambda| < \epsilon$; this is an algebraic rather than topological idea of what it means to be inside a set. If the origin 0 is one of B's internal points, so that every point x in X can as it were be drawn into B by scaling that point down by a suitable contraction factor λ, then B is *absorbent*. Denote by $X \backslash B$ the set-theoretic difference X *less* B, and put $c(B) = C(B) \backslash \{0\}$.

It follows from (2) that $j(.\,|B)$ is ph and that it is n-proper if $X \backslash B$ is absorbent. Phelps (1963, Prop, 2iii) proved that $j(.\,|h(B)) \equiv j(.\,|B)$, so it is natural to assume that B is always haloed. If B is haloed and closed, then $j(.\,|B)$ is increasing in B, i.e. $B^1 \subset B^2$ iff $j(.\,|B^1) \leqslant j(.\,|B^2)$, and $B = \{x \in X: j(x|B) \geqslant 1\}$. If B is closed and $0 \notin B$, then $j(.\,|B)$ is upper semicontinuous (usc) on $C(B)$. If B is convex, haloed and such that X/B is absorbent, then $j(.\,|B)$ is superadditive ($j(v + w|B) \geqslant j(v|B) + j(w|B)$ for all $v, w \in X$), and so concave. Finally, if B is haloed, convex, closed and such that int $B \neq \varnothing$, then the frontier of B is $\{x \in X: j(x|B) = 1\}$.

These properties of s-gauges should be compared with the corresponding properties of gauges.

3. DUAL SETS

(a) *Polar sets.* For any $A \subset X$ its *polar set* $A^0 \subset Y$ is defined by

$$\mathbf{A}^0 = \{y \in Y: \langle x, y \rangle \leqslant 1 \text{ for all } x \in A\} \quad (3)$$

where $\langle x, y \rangle$ denotes the value of the linear functional y at x (if $X = R^n = Y$ then it is the inner product of x and y). If $A = C(A)$ then it is easy to check that A^0 coincides with the *polar cone* of A ($= \{y \in Y: \langle x, y \rangle \leqslant 0 \text{ for all } x \in A\}$), whence the name of this generalization.

Let $\Gamma^0(Y)$ denote the set of all those subsets of Y that are convex, closed and contain the origin (hence are starred); similarly for $\Gamma^0(X)$. From (3) each y in A^0 satisfies a family of weak linear inequalities (one for each x in A), so that A^0 is convex and closed; moreover, obviously $0 \in A^0$ for any A. Hence $A^0 \in \Gamma^0(Y)$.

The *bipolar set* $A^{00} \subset X$ of any $A \subset X$ is given by

$$A^{00} = \{x \in X: \langle x, y \rangle \leqslant 1 \quad \text{for all} \quad y \in A^0\}. \quad (4)$$

One can define A^{000}, $A^{0000}, \ldots$ in ways analogous to (3) and (4) respectively, but it is easily shown that $A^{000} = A^0$, $A^{0000} = A^{00} \ldots$ etc. Other simple properties are that $A \subset A^{00}$, and that $A_1 \subset A_2$ implies $A_2^0 \subset A_1^0$ and $A_1^{00} \subset A_2^{00}$.

Since $A^{00} \in \Gamma^0(X)$ it is clearly necessary for $A = A^{00}$ that A itself be in $\Gamma^0(X)$. The next result, saying that this condition is also sufficient, is the fundamental theorem of the theory of gauges. Discovered by Dieudonné and Schwartz (1950), it is equivalent to the Hahn–Banach Theorem, to 'the' Theorem of the Separating Hyperplane, and to the Fenchel–Moreau theorem (for which see DUALITY).

Theorem 1. (Bipolar Theorem). For any $A \subset X$, $A^{00} = K(e(A))$.

Here, for any set $M \subset X$, $K(M)$ denotes its *closed convex hull*, i.e. the intersection of all the convex closed sets that contain M. An equivalent version of Theorem 1 is that $A^{00} = K(A \cup \{0\})$, so that for example McKenzie's $\bar{X}_i$ (1981, p. 825) is simply X_i^{00}.

(b) *S-polar sets.* For any $B \subset X$ its *s-polar set* $B_0 \subset Y$ is given by

$$B_0 = \{y \in Y: \langle x, y \rangle \geqslant 1 \quad \text{for all} \quad x \in B\} \quad (5)$$

and its *s-bipolar set* $B_{00} \subset X$ by

$$B_{00} = \{x \in X: \langle x, y \rangle \geqslant 1 \quad \text{for all} \quad y \in B_0\} \quad (6)$$

As before, it can be shown that $B \subset B_{00}$, that $B_{000} = B_0$ and $B_{0000} = B_{00}$ etc., and that $B^1 \subset B^2$ implies $B_0^2 \subset B_0^1$ and $B_{00}^1 \subset B_{00}^2$. However, it is obvious from (5) that if $0 \in B$ then $B_0 = \varnothing$, which indicates a major asymmetry between polar and s-polar sets.

Denote the class of all non-empty convex closed haloed subsets of Y that do *not* contain the origin by $\Gamma_0(Y)$, and similarly for $\Gamma_0(X)$; since these sets are haloed, they are all unbounded. Although $A^0 \in \Gamma^0(Y)$ and $A^{00} \in \Gamma^0(X)$ for each

$A \subset X$, it is *not* true that $B_0 \in \Gamma_0(Y)$ and $B_{00} \in \Gamma_0(X)$ for each $B \in X$. For example, suppose $X = R$ and that $B = \{-b, b\}$ for some number $b \neq 0$. Then $B_0 = \varnothing$ and $B_{00} = R$, though neither $\varnothing$ nor R is in $\Gamma_0(R)$.

Fortunately, by using a separating hyperplane theorem one can obtain precise information about when B_0 exists.

Theorem 2. For any $B \subset X$, $B_0 = \varnothing$ iff $0 \in K(B)$.

Corollary 1. $0 \notin K(B)$ iff $[B_0 \in \Gamma_0(Y)$ and $B_{00} \in \Gamma_0(X)]$.

A basic theorem corresponding to Theorem 1 is

Theorem 3. For any $B \subset X$ such that $0 \notin K(B)$, $B_{00} = K(h(B))$.

A partial converse to this is

Proposition 1. If $K(h(B)) \neq X$, then $B_{00} = K(h(B))$ implies $0 \notin K(B))$.

That the conditional in this result is essential follows from the previous example of $B = \{-b, b\}$, since there $R = B_{00} = K(h(B))$ yet $0 \in K(B)$.

4. TRANSFORMS

(*a*) *Polar transforms.* For any $A \subset X$, the *polar transform* of its gauge $J(.\,|A)$ is the gauge $J(.\,|A^0)$ of its polar set A^0, and the *bipolar transform* of $J(.\,|A)$ is the gauge $J(.\,|A^{00})$ of its bipolar set A^{00} (the term is due to Young (1969, p. 108)). Since each of these sets is convex, closed and contains the origin, it follows from earlier results that each of the transforms is convex, ph and lsc on X.

Define the *support function* $S(.\,|A): Y \to [-\infty, \infty]$ of A by

$$S(y\,|A) = \sup\{\langle x, y \rangle : x \in A\} \tag{7}$$

These functions are ph, lsc iff A is closed, and convex iff A is convex.

The next result is simple to prove but very important.
Theorem 4. For any $A \subset X$,

$$J(.\,|A^0) \equiv S(.\,|A^{00}) \quad \text{and} \quad J(.\,|A^{00}) \equiv S(.\,|A^0).$$

While each transform maps to the interval $[0, \infty]$, for several reasons (such as the problem of invertibility) it is important to know when it is actually positive and finite, i.e. maps to $(0, \infty)$.
Proposition 2. For any $A \subset X$,

(a) $[\{y \in Y : 0 < J(y\,|A^0) < \infty\} = c(A^0)]$

   iff $[y \neq 0 \to S(y\,|A^{00}) \neq 0]$

(b) $[\{x \in X : 0 < J(x\,|A^{00}) < \infty\} = c(A^{00})]$

   iff $[x \neq 0 \to S(x\,|A^0) \neq 0]$.

These conditions are precise but restrictive. Moreover, it is not easy to 'see' what A should look like in order that they be satisfied. So it is useful to have

Proposition 3. For any $A \subset X$,

(a) (i) $[\{y \in Y : 0 < J(y\,|A^0) < \infty\} = c(A^0)] \to [A^{00} \text{ absorbent}]$

   (ii) $[0 \in \text{int } A^{00}] \to [\{y \in Y : 0 < J(y\,|A^0) < \infty\} = c(A^{00})]$

(b) $[\{x \in X : 0 < J(x\,|A^{00}) < \infty\} = c(A^{00})]$ iff $[A \text{ bounded}]$

(c) If $X = R^n = Y$,

   $[\{y \in Y : 0 < J(y\,|A^0) < \infty\} = c(A^0)]$ iff $0 \in \text{int } A^{00}$.

The last of these results can be deduced from Rådström (1949–50, para. 4, p. 28) or Rockafellar (1970, Cor. 14.5.1, p. 125).

(*b*) *s-polar transforms.* For any $B \subset X$, the *s-polar transform* of $j(.\,|B)$ is the s-gauge $j(.\,|B_0)$ of B_0, and similarly the

s-bipolar transform of $j(.\,|B)$ is the s-gauge $j(.\,|B_{00})$ of B_{00}. Provided B_0 exists, then from earlier results each s-polar transform is concave, ph and usc on $C(B_0)$ (or $C(B_{00})$, as the case may be). Unlike the case of polar transforms, the positivity and finiteness of s-polar transforms offers no difficulty, as shown by

Proposition 4. For any $B \subset X$,

(a) $\{y \in Y : 0 < j(y\,|B_0) < \infty\} = c(B_0)$

(b) $[\{x \in X : 0 < j(x\,|B_{00}) < \infty\} = c(B_{00})]$ iff $B_0 \neq \varnothing$

As an application of (b), suppose that B is a better set for some target x^t and that $B \in \Gamma_0(X)$, so that $B_0 \neq \varnothing$ and $B_{00} = B$. If the indifference surface I^t is asymptotic to each axis then $j(x\,|B)$ is positive and finite for each strictly positive bundle x. This is a rationale for the common requirement that 'distance functions' be defined only on the strictly positive orthant.

For s-polar transforms it is not support functions that are relevant but *concave support functions* $s(.\,|B): Y \to [-\infty, \infty]$ of B, defined by

$$s(y\,|B) = \inf\{\langle x, y \rangle : x \in B\} \tag{8}$$

Such functions are ph, usc iff B is closed and concave iff B is convex. Unfortunately, the simple universality of Theorem 4 is not available for s-polar transforms. Without further assumption, the best that can be done is

Proposition 5. For any $B \subset X$,

(a) $y \in c(B_0) \to j(y\,|B_0) = s(y\,|B_{00})$

(b) $[x \in c(B_{00}) \to j(x\,|B_{00}) = s(x\,|B_0)]$ iff $B_0 \neq \varnothing$.

In looking for conditions to strengthen these results a hint is provided by the fact, noted earlier, that s-gauges vanish *only* at the origin. It follows that to have identity between s-polar transforms and concave support functions the latter must have that property as well. Now Arrow's famous 'exceptional case' (1951, pp. 527–8) referred precisely to a concave support function (i.e. the cost function for Individual 2's better set) which vanished for some nonzero price vector. This motivates the following

Definition. A set $M \subset X$ such that $y \neq 0 \to s(y\,|M) \neq 0$ is called *Arrovian*, and similarly for any $N \subset Y$ such that $x \neq 0 \to s(x\,|N) \neq 0$.

Geometrically, a non-empty $M \subset X$ is Arrovian iff none of the affine hyperplanes that support it from below passes through the origin. Economically, the idea is closely related to the existence of the 'locally cheaper points' of McKenzie (1957). That it is just the strengthening condition required is shown by

Theorem 5. For any $B \subset X$:

(a) $j(.\,|B_0) \equiv s(.\,|B_{00})$ iff B_{00} is Arrovian.

(b) $[j(.\,|B_{00}) \equiv s(.\,|B_0)$ iff B_0 is Arrovian] iff $B_0 \neq \varnothing$.

(*c*) *Comparison between polar and s-gauge transforms.* The assumption that a set be Arrovian is clearly an s-gauge version of the corresponding assumptions for $S(.\,|A^{00})$ and $S(.\,|A^0)$ that were prescribed in Propositions 2(a) and 2(b), respectively. However, Proposition 3 showed the latter to be closely related (if not equivalent) to more intuitively understandable assumptions, namely, absorbency and boundedness. Hence there was no need for a separate labelling of 'gauge-Arrovian' sets. For s-gauges, however, there appear to be no such simple characterizations of the Arrovian property.

It follows from these considerations that, in any given application, properties which at first sight look very different may simply be 'gauge/s-gauge' versions of the same idea. Suppose $A \in \Gamma^0(R^n)$ and $B \in \Gamma_0(R^n)$ so that $A^{00} = A$, $B_0 \neq \varnothing$ and $B_{00} = B$. Then from Propositions 2 and 3 the gauge version of the s-gauge condition that B be Arrovian (e.g. that it always has locally cheaper points) is that $0 \in \text{int } A$, while the gauge version of B_0 being Arrovian is that A be bounded. Thus different appearances may mask similar roles, forming disguises which are not easily penetrated without the help of duality theory.

5. MAHLER'S INEQUALITY

(*a*) *Gauges.* Polar and bipolar transforms of gauges satisfy a fundamental inequality, which plays the same role in the theory of gauges that (W.H.) Young's Inequality plays in the theory of Fenchel transforms (see DUALITY). L.C. Young (1939) attributed the statement and proof (for R^n) of this inequality of Mahler (1939), though saying also (p. 569) that 'it is implicitly contained in the work of Minkowski'. An inkling of its importance is that the celebrated Cauchy-Schwarz Inequality is a special case (see INEQUALITIES).

Theorem 6. (Mahler's Inequality). For any $A \subset X$.

$$\forall x \in X, \quad \forall y \in Y \quad J(x|A^{00})J(y|A^0) \geqslant \langle x, y \rangle \quad (9)$$

Those pairs $(x\,y)$ such that equality holds in (9) may be called *polar* to each other. Notice that if $A \in \Gamma^0(X)$ then, from Theorems 1 and 4, (9) is equivalent to

$$\forall x \in X, \quad \forall y \in Y \quad J(x|A)S(y|A) \geqslant \langle x, y \rangle \quad (10)$$

In economic applications, if x is a quantity vector and y a price vector then in (10) $J(.|A)$ is dimensionless and $S(.|A)$ has dimension 'price × quantity'. New characterizations of the polar and bipolar transforms are provided by

Proposition 6. For any $A \subset X$:

(a) If $0 \in \text{int } A^{00}$, $\forall x \in X$

$$J(x|A^{00}) = \sup\{\langle x, y \rangle : y \in Y, J(y|A^0) = 1\} \quad (11)$$

(b) If A is bounded, $\forall y \in Y$

$$J(y|A^0) = \sup\{\langle x, y \rangle : x \in X, J(x|A^{00}) = 1\} \quad (12)$$

If $A \in \Gamma^0(X)$ then (11) and (12) simplify considerably, just as (10) simplifies (9).

(*b*) *S-gauges.* The s-gauge version of Mahler's Inequality, though still remarkable enough, is not so unrestricted as that for gauges.

Theorem 7. For any $B \subset X$ such that $0 \notin K(B)$:

(a) If B_{00} is Arrovian, $\forall x \in X$, $\forall y \in C(B_0)$

$$j(x|B_{00})j(y|B_0) \leqslant \langle x, y \rangle \quad (13)$$

(b) If B_0 is Arrovian, $\forall x \in C(B_{00})$, $\forall y \in Y$

$$j(x|B_{00})j(y|B_0) \leqslant \langle x, y \rangle \quad (14)$$

The characterizations of the s-polar and s-bipolar transforms which are analogous to (11) and (12) are similarly restricted.

Proposition 7. For any $B \subset X$ such that $0 \notin K(B)$:

(a) If B_{00} is Arrovian, $\forall x \in c(B_{00})$

$$j(x|B_{00}) = \inf\{\langle x, y \rangle : y \in Y, j(y|B_0) = 1\} \quad (15)$$

(b) If B_0 is Arrovian, $\forall y \in c(B_0)$

$$j(y|B_0) = \inf\{\langle x, y \rangle : x \in X, j(x|B_{00}) = 1\} \quad (16)$$

A close reading of Gorman (1976) and Deaton (1979) will show that Theorem 7 and Proposition 7 formalize and generalize several of the results for s-gauges ('distance functions') obtained in those papers.

6. SUBDIFFERENTIALS

Sub- and superdifferentials generalize differentials in a way especially appropriate to convex and concave functions, respectively; for a fairly detailed discussion see DUALITY (Section II). In particular, conditions of optimality (e.g. minimization of cost, maximization of profit) are expressible naturally by the non-emptiness of sub- and super differentials. Denote the subdifferential (resp., superdifferential) of the gauge (resp., s-gauge) of any set M by $\partial J(\cdot|M)$ (resp., $\Delta j(\cdot|M)$). Then a comprehensive result for subdifferentials of polar and bipolar transforms is *Theorem 8.* For any $A \subset X$:

(a) $y^1 \in \partial J(x^1|A^{00})$ iff $[y^1 \in A^0$ and y^1 achieves $S(x^1|A^0)]$

(b) $x^1 \in \partial J(y^1|A^0)$ iff $[x^1 \in A^{00}$ and x^1 achieves $S(y^1|A^{00})]$

(c) If $y^1 \in c(A^0)$ and $0 \in \text{int } A^{00}$,

then

$$(y^1/J(y^1|A^0)) \in \partial J(x^1|A^{00})$$
$$\text{iff } [J(x^1|A^{00})J(y^1|A^0) = \langle x^1, y^1 \rangle]$$

(d) If $x^1 \in c(A^{00})$

and A is bounded, then

$$(x^1/J(x^1|A^{00})) \in \partial J(y^1|A^0)$$
$$\text{iff } [J(x^1|A^{00})J(y^1|A^0) = \langle x^1, y^1 \rangle].$$

The combination of (c) and (d) obviously gives conditions under which the subdifferentials of the polar and bipolar transforms are *inverse* to each other in the sense of set-valued mappings, i.e. when

$$(y^1/J(y^1|A^0)) \in \partial J(x^1|A^{00}) \quad \text{iff } (x^1/J(x^1|A^{00})) \in \partial J(y^1|A^0).$$

As one might expect by now, the corresponding results for superdifferentials of s-gauges are slightly more restrictive.

Theorem 9. For any $B \subset X$ such that $0 \notin K(B)$:

(a) If B_0 is Arrovian,

$$y^1 \in \Delta j(x^1|B_{00}) \quad \text{iff } [y^1 \in B_0 \text{ and } y^1 \text{ achieves } s(x^1|B_0)]$$

(b) If B_{00} is Arrovian,

$$x^1 \in \Delta j(y^1|B_0) \quad \text{iff } [x^1 \in B_{00} \text{ and } x^1 \text{ achieves } s(y^1|B_{00})]$$

(c) If $y^1 \in c(B_0)$ and B_0 is Arrovian, then

$$(y^1/j(y^1|B_0)) \in \Delta j(x^1|B_{00})$$
$$\text{iff } [j(x^1|B_{00})(y^1|B_0) = \langle x^1, y^1 \rangle]$$

(d) If $x^1 \in c(B_{00})$ and B_{00} is Arrovian, then

$$(x^1/j(x^1|B_{00})) \in \Delta j(y^1|B_0)$$
$$\text{iff } [j(x^1|B_{00})j(y^1|B_0) = \langle x^1, y^1 \rangle]$$

Notice the subtle changes between this theorem and Theorem 8. As already noted, the condition $0 \in \text{int } A^{00}$ corresponds to B_{00} being Arrovian. In Theorem 8(c) the first of these conditions is used to assure the meaningfulness of the ratio $y^1/J(y^1|B^0)$, while in Theorem 9(d) the second condition is used, not to validate the ratio $x^1/j(x^1|B_{00})$ (this follows from Proposition 4(b)), but to assure that $j(y|B_{00}) = s(y|B_0)$ for all $y \in Y$ (Theorem 5(a)). A similar analysis holds for the conditions that A be bounded (Theorem 8(d)) and that β_0 be Arrovian (Theorem 9(c)).

Observe that Theorem 9(c) and (d) provide conditions under which the superdifferential mappings associated with s-polar and s-bipolar transforms are inverse to each other in the sense of set-valued mappings. Indeed, since superdifferentiability generalizes the differentiability of concave functions, Theorem

9 formalizes and generalizes most of the optimality conditions in the literature on s-gauges, e.g. Properties 5–8 in Deaton (1979, pp. 394–6).

7. CONCLUSION

Gauges and their polar transforms, together with s-gauges and their s-polar transforms, constitute a powerful and general duality scheme for economic theory. This scheme has already had many applications to the theory of production and consumption, to index number theory, to optimal taxation, and to the theory of income distribution, to name just a few. Many more can be expected, including applications to general equilibrium theory. The scheme is especially well suited to situations where the appropriate normalization of prices is by total expenditure or income, just as the other main duality scheme – Fenchel transforms – is suitable when the appropriate normalization is by a single numeraire good.

Finally, the close parallelism between the theory of gauges and that of s-gauges suggests the existence of some transformation of variables under which the two theories would be seen simply as two aspects of one unified theory.

PETER NEWMAN

See also COST MINIMIZATION AND UTILITY MAXIMIZATION; DUALITY; HOMOGENEOUS AND HOMOTHETIC FUNCTIONS; RATIONING.

BIBLIOGRAPHY

Arrow, K.J. 1951. An extension of the basic theorems of classical welfare economics. In *Proceedings of the Second Berkeley Symposium on Mathematical Statistics and Probability*, ed. J. Neyman, Berkeley: University of California Press, 507–532.

Blackorby, C., Primont, D. and Russell, R.R. 1978. *Duality, Separability and Functional Structure: Theory and Economic Applications*. New York: Elsevier/North-Holland.

Bonnesen, T. and Fenchel, W. 1934. *Theorie der Konvexen Körper*. Berlin: Springer. Reprinted, New York: Chelsea, 1948.

Bourbaki, N. 1953. *Espaces vectoriels topologiques, Chapitres I–II*. Actualités Scientifiques et Industrielles No. 1189. Paris: Hermann.

Cassels, J.W.S. 1959. *An Introduction to the Geometry of Numbers*. Berlin: Springer-Verlag.

Deaton, A. 1979. The distance function in consumer behaviour with applications to index numbers and optimal taxation. *Review of Economic Studies* 46, 391–406.

Deaton, A. and Muellbauer, J. 1980. *Economics and Consumer Behaviour*. Cambridge: Cambridge University Press.

Debreu, G. 1951. The coefficient of resource utilization. *Econometrica* 19, 273–92.

Dieudonné, J. and Schwartz, L. 1950. La dualité dans les espaces ($\mathscr{F}$) et ($\mathscr{LF}$). *Annales de L'Institut Fourier*, Université de Grenoble, Vol. 1, 61–101.

Diewert, W.E. 1982. Duality approaches to microeconomic theory. In *Handbook of Mathematical Economics*, Vol. II, ed. K.J. Arrow and M.D. Intriligator, Amsterdam: North-Holland, 535–99.

Fuss, M. and McFadden, D. (eds) 1978. *Production Economics: A Dual Approach to Theory and Applications*, Vol. 1, *The Theory of Production*. Amsterdam: North-Holland.

Gorman, W.M. 1970. Quasi-separable preferences, costs and technologies. Mimeo, Department of Economics, University of North Carolina, Chapel Hill, NC.

Gorman, W.M. 1976. Tricks with utility functions. In *Essays in Economic Analysis*, ed. M.J. Artis and A.R. Nobay, Cambridge: Cambridge University Press.

Hanoch, G. 1978. Symmetric duality and polar production functions, Chapter 1.2, 111–131, in Fuss and McFadden (1978).

Jacobsen, S.E. 1972. On Shephard's duality theorem. *Journal of Economic Theory* 4, 458–64.

Köthe, G. 1969. *Topological Vector Spaces I*. New York: Springer-Verlag.

McFadden, D. 1978. Cost, revenue and profit functions. Chapter 1.1, 3–109, in Fuss and McFadden (1978).

McKenzie, L.W. 1957. Demand theory without a utility index. *Review of Economic Studies* 24, 185–9.

McKenzie, L.W. 1981. The classical theorems on existence of competitive equilibrium. *Econometrica* 49, 819–41.

Mahler, K. 1939. Ein Übertragungsprinzip für konvexe Körper. *Casopis pro Pěstováni Matematiky a Fysiky* 63, 93–102.

Malmquist, S. 1953. Index numbers and indifference surfaces. *Trabajos de Estadistica* 4, 209–241.

Minkowski, H. 1911. Theorie der konvexen Körper. *Gesammelte Abhandlungen*. Leipzig–Berlin: Teubner II, 131–229.

Moreau, J.-J. 1967. Fonctionelles convexes. Séminaire sur les equations aux derivèes partielles, II. College de France, Mimeo.

Phelps, R.R. 1963. Support cones and their generalizations. In *Convexity*, ed. V.L. Klee, Proceedings of Symposia in Pure Mathematics, Volume VII, Providence, RI: American Mathematical Society.

Rådström, H. 1949–50. II. Polar reciprocity. Seminar on Convex Sets, Princeton, NJ: Institute for Advanced Study, mimeo.

Robertson, A.P. and Robertson, W. 1964. *Topological Vector Spaces*. Cambridge Tracts in Mathematics and Mathematical Physics, No. 53, Cambridge: Cambridge University Press.

Rockafellar, R.T. 1970. *Convex Analysis*. Princeton, Princeton University Press.

Ruys, P.H.M. 1972. On the existence of an equilibrium for an economy with public goods only. *Zeitschrift für Nationalökonomie* 32, 189–202.

Ruys, P.H.M. and Weddepohl, H.N. 1979. Economic theory and duality. In *Convex Analysis and Mathematical Economics*, ed. J. Kriens, Lecture Notes in Economics and Mathematical Systems. No. 168, New York: Springer-Verlag, 1–72.

Shephard, R.W. 1953. *Cost and Production Functions*. Princeton: Princeton University Press.

Shepard, R.W. 1970. *Theory of Cost and Production Functions*. Princeton: Princeton University Press.

Uzawa, H. 1964. Duality principles in the theory of cost and production. *International Economic Review* 5, 216–20.

Weddepohl, H.N. 1972. Duality and equilibrium. *Zeitschrift für Nationalökonomie* 32, 163–87.

Wold, H. 1943. A synthesis of pure demand analysis: Part II. *Skandinavisk Aktuarietidskrift* 26, 220–63.

Young, L.C. 1939. On an inequality of Marcel Riesz. *Annals of Mathematics* 40, 567–74.

Young, L.C. 1969. *Lectures on the Calculus of Variations and Optimal Control Theory*. Philadelphia: W.B. Saunders Company.

Gayer, Arthur David (1903–51). Arthur (Archie) Gayer was a rising star in economic research in the United States during the 1930s but his promise was not sustained in later years of his foreshortened life. Born of English parentage at Poona in British India on 19 March 1903, he was educated at St Paul's School, London, 1916–21, and Lincoln College, Oxford (BA, 1925; D. Phil., 1930). On arriving in the United States in 1927 as a Rockefeller Foundation fellow, he did graduate work at Columbia University, where he taught economics at Barnard College from 1931 to 1940. From 1940 until his death on 17 November 1951 as the result of an automobile accident, he taught at Queens College of the City University of New York.

Gayer's research covered three main subjects: public works in the United States; monetary policy and economic stabilization; and United Kingdom growth and cyclical experience, 1790–1850. He had begun work in 1929 as an assistant to Leo Wolman at the National Bureau of Economic Research on a statistical analysis of the volume, distribution, and fluctuations of local and federal construction expenditures. Gayer's study, covering the period 1919–34, was generally regarded as a thorough examination of the empirical record as well as a careful summary of the theory of public works as a stabilizing device. In his study of monetary policy, in which he reviewed the traditional and interwar gold standard, Gayer advocated British–US cooperation in establishing a satisfactory international monetary system over the dollar-sterling area. In honour of Irving Fisher's seventieth birthday, he edited a volume of essays on the lessons of monetary experience. From 1936 to 1941, Gayer directed a study designed to expand his doctoral thesis, 'Industrial Fluctuation and Unemployment in England, 1815–1850'. The product of this research, in which he collaborated with a team of young scholars, was published in two volumes after Gayer's death. Conceptually, the study was heavily influenced by ideas in the *General Theory*, although the cyclical analysis was based on measures pioneered by the National Bureau. By the time the study appeared, the National Bureau's method of analysis commanded attention neither in the United States nor in Britain.

ANNA J. SCHWARTZ

SELECTED WORKS

1935a. *Public Works in Prosperity and Depression*. New York: National Bureau of Economic Research.

1935b. *Economic Stabilization: a Study of the Gold Standard*. London: A. & C. Black.

1937 (ed.) *The Lessons of Monetary Experience: Essays in Honor of Irving Fisher*. New York: Farrar & Rinehart.

1953. (With W.W. Rostow and Anna J. Schwartz.) *The Growth and Fluctuation of the British Economy, 1790–1850*. 2 Vols, Oxford: Clarendon Press. Reprinted, Hassocks: Harvester, 1975.

gearing. The question whether a company's choice of the proportion of debt to equity finance in its capital structure matters has involved a great deal of controversy. This choice, known as the gearing decision in the UK and the leverage decision in the USA, is widely regarded by corporate finance directors, investors, stock market participants and many others as an issue of considerable importance, yet the basic result of conventional economic theory applied to this question is that the gearing decision is irrelevant – there is no advantage to a firm in choosing one debt–equity ratio rather than another. This striking contrast between theory and practice has, of course, led to much critical examination of the assumptions of the theory, and some progress has been made in identifying ways in which gearing may matter. However it remains true that the determinants of a firm's gearing decision, and its importance, are not yet fully understood.

The argument that the gearing decision is a matter of irrelevance, affecting neither the firm's value nor its cost of capital (and hence its investment decision), is due to Modigliani and Miller in a celebrated article (Modigliani and Miller, 1958). Their fundamental insight was that, in a world of perfect and complete capital markets in which taxation and asymmetric information are absent, individual investors can create any particular pattern of returns from holdings of securities by borrowing on their own account. This ability of investors to engage in 'home-made leverage' means that there is no reason for firms to concern themselves about the amount of debt in their capital structure: investors can create for themselves any pattern of returns which would be given by a share in a firm with a particular gearing ratio, so firms cannot gain by offering one such pattern rather than another.

To see this, consider the following simple illustration of the Modigliani–Miller argument (based on Nickell, 1978). Suppose that a firm possesses assets which yield 1,000 $\tilde{\theta}$ per annum in perpetuity, where θ is a random variable, and that this firm has 1000 equity shares outstanding, but no debt in its capital structure. The price of a claim to an income stream yielding $\tilde{\theta}$ in perpetuity is determined by a perfect capital market to be 1, so the value of the firm's equity is 1000. If the firm borrows,

say, 200 at a rate of 10% per annum with no risk of default each share will now yield $\tilde{\theta}$–0.02 per annum in perpetuity, because there is a certain interest payment to be made from the returns on the firm's assets in each year before shareholders receive anything. If individual investors can borrow at the same interest rate as the firm the price of a share will now be 0.8, since an investor could have created a $\tilde{\theta}$–0.02 income stream in the original situation by borrowing 0.2 (on which the annual interest payment is 0.02) and using 0.8 of his own to buy one share for 1. Thus if the firm does borrow 200 the value of its equity will fall to 800, and its overall value (the sum of the values of outstanding debt and equity) remains constant at 1,000. Home-made leverage enables an individual investor to create any combination of $\tilde{\theta}$ and a certain return whatever combination of $\tilde{\theta}$ and a certain return the firm offers, so that there is no reason for the firm to concern itself with the choice of a particular combination of the two.

Modigliani–Miller's original argument rested on a number of restrictive assumptions, such as the existence of risk classes within which firm's operating earnings were perfectly correlated, which were relaxed in subsequent work. One of the most general proofs of the Modigliani–Miller theorem was that by Stiglitz (1974), which did not need to make any assumptions about the existence of risk classes, the source of uncertainty, individuals having the same expectations, or the interest rate paid by a firm being unaffected by the amount of capital it raises. Stiglitz noted three critical limitations to his proof however, and it is these which begin to suggest ways in which gearing may be a relevant decision for firms. One is that individuals' expectations about future prices and firm valuations must not be affected by changes in companies' financial policy: this in effect rules out the possibility of financial policy acting as a signal in a world of asymmetric information. The second limitation is that individual borrowing must be a perfect substitute for firm borrowing, while the third is that there must be no bankruptcy.

At first sight these last two limitations would appear to indicate clearly why gearing is important in practice. But matters are not so straightforward. Companies may be able to borrow on better terms than individuals, but this may be because they are better risks: the Modigliani–Miller theorem only requires that individuals and firms borrow on the same terms for debt of equivalent risk. It is certainly not true that companies can always borrow on better terms than individuals: mortgages for house purchase, for example, are sometimes available at rates below those charged on corporate borrowing. Even if it is true that firms can borrow on better terms than individuals for equivalent risk loans, so that they can gain in value by offering this service to individuals, firms can compete by so doing and this may eliminate the gain in value: the supply of corporate debt expands until the Modigliani–Miller proposition is re-established.

A similar argument applies in the case of bankruptcy. When a firm issues risky debt it creates a security which individuals, lacking limited liability, cannot replicate by borrowing on their own account. This expands the range of portfolio opportunities open to investors, which they should in principle be willing to pay for thus enabling the firm to increase its value by use of some debt finance. However the extent to which a particular firm can gain by issuing risky debt depends on whether it can offer something special to investors that is not already available; it is difficult to believe that one more firm's risky debt significantly expands the set of portfolio opportunities available to investors.

A more promising approach to understanding the importance attached to the gearing decision in practice would appear to be the relaxation of the assumption that there is no taxation. Most corporation tax systems allow interest payments on debt finance to be deductible against corporation tax, and this has been widely argued to provide the obvious explanation for the use of some debt in a firm's capital structure (by Modigliani–Miller among many others). Tax advantages to debt might imply that firms should use all-debt finance, but this unsatisfactory conclusion has been avoided by introducing costs of bankruptcy and financial distress, which reduce the size of the total payout to investors in certain contingencies that are more likely the larger the firm's gearing ratio. These include costs of reorganization and liquidation associated with bankruptcy, together with costs of financial distress, such as the foregoing of profitable investment opportunities which may be necessitated by bankruptcy, and the making of suboptimal investment decisions in an attempt to forestall bankruptcy. These costs will result in the firm's market value beginning to decline beyond some level of gearing, so that together with the corporation tax advantage of debt a theory of optimal gearing ratios seems to result.

This theory is perhaps the most commonly accepted one for explaining the importance of gearing, but it is by no means uncontroversial. One reason for its lack of universal acceptance is that evidence on the size of the costs of bankruptcy and financial distress is limited and, where available (Warner, 1977), it does not suggest that they are large. Another reason is that this theory only takes account of corporate taxes in arguing that there is a tax advantage to the use of debt finance. Investors are subject to personal taxes on interest and dividend income and capital gains, and these tax rates differ usually between income and capital gains, and certainly between different investors. This causes a number of problems. If personal tax is higher on debt interest than on equity income (taking account of both dividends and capital gains) the use of debt finance may reduce the firm's value. The variation of personal tax rates across investors means that it is likely that some would prefer debt on tax grounds while others would prefer equity. Indeed differences in personal tax rates were used by Miller (1977) to reintroduce an irrelevance result (in which bankruptcy costs were ignored): he argued that investors would specialize their holdings in debt or equity according to whether their after-all-tax income from a unit of pre-tax debt cash flow (1 − personal tax rate on debt income) was greater or less than their after-tax income from a unit of pre-tax equity cash flow ((1 − personal tax rate on equity income) times (1 − corporation tax rate)). There would be a determinate aggregate debt–equity ratio, at which point marginal investors would be indifferent between holding debt or equity on tax grounds, but the gearing decision would be irrelevant for individual firms.

Miller's argument shows that when heterogeneous personal tax rates are considered, as they must be, it is not obvious that there is a tax advantage to corporate borrowing. But there are problems with this argument too. Auerbach and King (1983) show that the Miller equilibrium requires the existence of certain constraints on investors: without such constraints (on, for example, borrowing and short-selling) questions arise concerning the existence of an equilibrium, for with perfect capital markets realistic tax systems provide opportunities for unlimited arbitrage at government expense between investors and firms in different tax positions. Auerbach and King also show that the combined effect of taxation and risk is to produce a situation in which gearing is relevant. With individual investors facing different tax rates and wishing to hold diversified portfolios the Miller equilibrium can no longer be sustained: investors who on tax grounds alone would hold

only equity may nevertheless hold some debt because an equity-only portfolio would be too risky. Miller's argument also does not take account of the implications of uncertainty and the asymmetric treatment of profits and losses which is a feature of most corporation tax systems. De Angelo and Masulis (1980) argued that the probability of interest tax shields being lost or deferred precluded the Miller equilibrium, and suggested that an optimal gearing ratio existed for a firm where the cost of debt finance, taking account of the probability of being unable to offset interest fully against corporation tax, equalled the cost of equity.

Miller's argument should thus be seen as one which raises important questions about whether taxation really does give incentives for individual firms to use debt finance, but does not clearly establish that there are no such incentives. It therefore weakens the theory based on trading off tax advantages of debt against costs of bankruptcy and financial distress, but does not destroy it. Another way in which this theory has been weakened is as a result of work on capital structure and financial policy which drops the assumption that the probability distribution of a firm's profits is common knowledge and independent of the firm's financial structure (this is essentially the first of the three critical limitations to Stiglitz' proof of the Modigliani–Miller theorem discussed above). There are a number of models based on asymmetric information of one sort or another in which a firm's gearing decision is not irrelevant. One type of model is where the firm's managers know more about the firm's possible returns than do outside investors. Ross (1977) assumes that managerial rewards depend on the current value of the firm and its future returns, and managers know the distribution of future returns while outside investors do not. The amount of debt chosen acts as a signal: managers of firms with higher expected future returns choose larger amounts of debt because only the managers of the better firms are willing to incur the increased risk of bankruptcy and its related costs associated with higher debt. Another type of model is based on principal–agent considerations: firms are run by managers (agents) on the behalf of shareholders (principals), but managers have some scope for pursuing their own interests at shareholders' expense because of asymmetric information. This is, however, recognized by the shareholders. The general form of models of this type is that managers choose a financial structure of the firm which determines managerial incentives: the capital market understands the incentives implied by a particular financial structure, and values the firm accordingly: this evaluation is taken into account by the managers in choosing financial structure. It is clear how a determinate capital structure can emerge from this framework, and Jensen and Meckling (1976) and Grossman and Hart (1982) are two examples of papers where gearing is important because of these reasons.

This work on asymmetric information and capital structure is highly suggestive of factors which may make gearing important, but as yet all we have in this area are insights rather than a complete and coherent theory. In particular there has been little integration of the traditional taxation arguments into the asymmetric information approach. Hence economists' understanding of firms' gearing decisions is still imperfect.

J.S.S. EDWARDS

See also DIVIDEND POLICY; FINANCE; RETENTION RATIO.

BIBLIOGRAPHY

Auerbach, A.J. and King, M.A. 1983. Taxation, portfolio choice and debt-equity ratios: a general equilibrium model. *Quarterly Journal of Economics* 98(4), November, 587–609.

De Angelo, H. and Masulis, R.W. 1980. Optimal capital structure under corporate and personal taxation. *Journal of Financial Economics* 8(1), March, 3–29.

Grossman, S.J. and Hart, O.D. 1982. Corporate financial structure and managerial incentives. In *Economics of Information and Uncertainty*, ed. J. McCall, Chicago: University of Chicago Press.

Jensen, M.C. and Meckling, W.H. 1976. Theory of the firm: managerial behavior, agency costs and ownership structure. *Journal of Financial Economics* 3(4), October, 305–60.

Miller, M.H. 1977. Debt and taxes. *Journal of Finance* 32(2), May, 261–75.

Modigliani, F. and Miller, M.H. 1958. The cost of capital, corporation finance, and the theory of investment. *American Economic Review* 48, June, 261–97.

Nickell, S.J. 1978. *The Investment Decisions of Firms*. Cambridge: Cambridge University Press.

Ross, S.A. 1977. The determination of financial structure: the incentive-signalling approach. *Bell Journal of Economics* 8(1), Spring, 23–40.

Stiglitz, J.E. 1974. On the irrelevance of corporate financial policy. *American Economic Review* 64(6), December, 851–66.

Warner, J.B. 1977. Bankruptcy costs: some evidence. *Journal of Finance* 32(2), May, 337–47.

Geary, Robert Charles (1896–1983).   Geary was born on 11 April 1896 in Dublin, and died on 8 February 1983, also in Dublin. He was educated at University College, Dublin, from 1913 to 1918, and at the Sorbonne from 1919 to 1921. In 1922 he became assistant lecturer in mathematics at University College, Southampton. He held the post of statistician in the Department of Industry and Commerce in Dublin from 1923 to 1949. From 1946 to 1947 he was a Senior Research Fellow in the Department of Applied Economics in Cambridge. He held the position of First Director in the Central Statistical Office of Eire from 1949 to 1957. In 1957 he was appointed Head of the National Accounts Branch of the United Nations Statistical Office, which post he held until 1960. From 1960 to 1966 he was director of the newly founded Economic (and Social) Research Institute in, Dublin, remaining attached to it as consultant from 1966 until his death.

Geary was a mathematical statistician of international standing. His statistical writings cover a wide range of topics, among them testing for normality, the distribution of ratios, parameter estimation etc., some of which are relevant to econometric methodology. Indeed, much of his work has an explicitly economic content: he was probably responsible for the excellent report prepared by the Department of Industry and Commerce containing the first national accounts of Eire (Eire, Minister for Finance, 1946); he wrote many papers on the determination of relationships between variables, notably Geary (1948 and 1949), in the second of which he uses instrumental variables; he derived the form of the utility function underlying the linear expenditure system (Geary, 1950–51); he was part-author of a monograph on linear programming applied to economics (Geary and McCarthy, 1964); and he built a model of the Irish economy based on an accounting framework (Geary, 1963–4). His published output numbers 112 titles, more than half of which appeared after his 65th birthday. A full bibliography is appended to Spencer (1976, 1983).

J.R.N. STONE

SELECTED WORKS

1946. Eire, Minister for Finance. *National Income and Expenditure 1938–1944*. Dublin: Stationery Office.

1948. Studies in relations between economic time series. *Journal of the Royal Statistical Society*, Series B 10(1), 140–58.

1949. Determination of linear relations between systematic parts of variables with errors of observation the variances of which are unknown. *Econometrica* 17, January, 30–58.

1950–51. A note on 'A constant utility index of the cost of living'. *Review of Economic Studies* 18(1), 65–6.

1964. (With M.D. McCarthy.) *Elements of Linear Programming with Economic Applications*. London: Griffin. 2nd edn (with J.E. Spencer), 1973.

1963–4. Towards an input–output decision model for Ireland. *Journal of the Statistical and Social Inquiry Society of Ireland* 21, Pt 2, 67–119.

BIBLIOGRAPHY

Spencer, J.E. 1976. The scientific work of Robert Charles Geary. *Economic and Social Review* 7(3), April, 233–47.

Spencer, J.E. 1983. Robert Charles Geary – an appreciation. *Economic and Social Review* 14(3), April, 161–4.

Gee, Joshua (*fl.*1725–1750).   Joshua Gee's place in the history of economics rests on his contributions to the protectionist literature during the early decades of the 18th century. He collaborated with Henry Martin among others in publishing the *British Merchant* that argued the protectionist case in 1713 and 1714 against the Treaty of Commerce proposed at Utrecht, and he published an extensive discussion of England's foreign trade, together with strong protectionist sentiments, in *The Trade and Navigation of Great Britain Considered* in 1729.

Addressing the current decline in the English export trades, the high level of imports of certain commodities, the demand for which, particularly in the case of French fashion goods, could be met by home produced import substitutes, the declining health of the woollen industry, and the currently widespread unemployment, Gee made a number of proposals for government regulation of trade and manufacturing. These proposals were directed principally to the need for 'finding effectual ways for employing the poor', thereby aligning his work with the widespread employment argument of the time. To the same end he advocated also a wider development of workhouses. Following Josiah Child, Gee proposed that trade with the colonial plantations should be regulated in such a way as not only to encourage their production of the materials needed for English manufacturing industries, thereby 'employing all the poor', but also to facilitate 'supplying our plantations with everything they want and all manufactured within ourselves'.

Gee's descriptive essays exhibited an understanding of the interdependence of economic activities and processes, 'one employment depending on another', and of 'the circulation of commerce that must infuse riches into every part'. He argued that higher domestic commodity prices would induce workers to increase their supply of labour, leading to higher incomes and also higher discretionary consumption expenditures. But such potentially important notions were not accorded any systematic or analytical development.

DOUGLAS VICKERS

See also MERCANTILISM.

SELECTED WORKS

1729. *The Trade and Navigation of Great Britain Considered*. London.

1742. *An Impartial Enquiry into the Importance and Present State of the Woollen Manufactures of Great Britain*. London.

1742. *The Grazier's Advocate, or Free Thoughts of Wool and the Woollen Trade*. London.

gender. The term gender has traditionally referred, as has sex, to the biological differences between men and women. More recently a movement has arisen both in social science writings and in public discourse to expand this definition to encompass also the distinctions which society has erected on this biological base, and further to use the word gender in preference to sex to refer to this broader definition. In this essay, we describe the relationship of this expanded concept of gender to economic theory.

Historically, gender has not been perceived to be a central concept in economic analysis, either among the classical and neoclassical schools or among Marxist economists. However, as the force of current events has thrust gender-related issues to the fore, economists have responded by seeking to analyse these issues. The outcome of this process has been not only a better understanding of the nature of gender differences in economic behaviour and outcomes, but also an enrichment of the discipline itself.

While, as noted above, the mainstream of economic analysis paid scant attention to gender-related issues, the 19th century campaign for female suffrage did focus some attention on gender inequality. Among classical economists, J.S. Mill (1878) eloquently argued for the 'principle of perfect equality' (p. 91) between men and women. Not only did he favour equality of the sexes within the family, but also women's 'admissibility to all the functions and occupations hitherto retained as the monopoly of the stronger sex'. He also expressed the belief 'that their disabilities elsewhere are only clung to in order to maintain their subordination in domestic life' (p. 94). In the Marxist school, Engels (1884) tied the subjection of women to the development of capitalism and argued that women's participation in wage labour outside the home, as well as the advent of socialism, was required for their liberation. The belief in the emancipating effects of a fuller participation in employments outside the home was shared not only by Mill and Engels, but also by such contemporary feminist writers as Gilman (1898).

The passage of time has proved these views oversimplified. As Engels and Gilman correctly foresaw, there has been an increase in the labour force participation of women, particularly of married women, in most of the advanced industrialized countries. This has undoubtedly altered both the relationship between men and women and the very organization of society in many ways. However, while women's labour force participation has in many instances risen dramatically, it nonetheless remains the case that the types of jobs held by men and women as well as the earnings they receive continue to differ markedly.

The contribution of modern neoclassical analysis, which comprises the main focus of this essay, has been to subject to greater scrutiny and more rigorous analysis both women's economic roles within the family and the causes of gender inequality in economic outcomes. We examine each of these areas below. However, the interrelationships between the family and the labour market, most importantly the consequences of labour market discrimination against women for their roles and status in the family, have tended to be neglected. Nonetheless, the possible existence of such feedback effects is an important issue which is also considered here.

TIME ALLOCATION IN THE FAMILY CONTEXT

Prompted in part by their desire to understand the causes of the rising labour force participation of married women in the post-World War II period, economists extended the traditional theory of labour supply to consider household production

more fully. The consequence was not only a better understanding of the labour supply decision, but also the development of economic analyses of the related phenomena of marriage, divorce and fertility.

The traditional theory of labour supply. The traditional theory of labour supply, also known as the labour–leisure dichotomy, was a simple extension of consumer theory. In this model, individuals maximize their utility, which is derived from market goods and leisure, subject to budget and time constraints. Where an interior solution exists, utility is maximized when the individual's marginal rate of substitution of income for leisure is set equal to the market wage.

Since in this model all time not spent in leisure is spent working, a labour supply (leisure demand) function may be derived with the wage, non–labour income, and tastes as its arguments. The well-known results of consumer theory are readily obtained. An increase in non–labour income, all else equal, increases the demand for all normal goods including leisure, inducing the individual to consume more leisure and to work fewer hours (the income effect). An increase in the wage, ceteris paribus, has an ambiguous effect on work hours due to two opposing effects. On the one hand, the increase in the wage is like an increase in income and in this respect tends to lower work hours due to the income effect. On the other hand, the increase in the wage raises the price (opportunity cost) of leisure inducing the individual to want to consume less of it, i.e., a positive substitution effect on work hours.

The theory sheds light on the labour force participation decision when it is realized that a corner solution will arise if the marginal rate of substitution of income for leisure at zero work hours is greater than the market wage. In this case, the individual maximizes utility by remaining out of the labour force. The impact of an increase in the wage is unambiguously to raise the probability of labour force participation, since, at zero work hours, there is no off-setting income effect of a wage increase.

Household production and the allocation of time. While the simple theory is sufficient for some purposes, it has limited usefulness for understanding the determinants of the gender division of labour in family and the factors influencing women's labour force participation, both at a point in time and trends over time. The key to addressing these issues is a fuller understanding and analysis of the household production process.

The first step in this direction was taken by Mincer (1962) who pointed out the importance, especially for women, of the three-way decision among market work, non-market work and leisure. He argued that the growth in married women's labour force participation was due to their rising real wages which increased the opportunity cost of time spent in non-market activities. But since, during the same period, the real wages of married men were also increasing, this must mean that the substitution effect associated with women's own real wage increases dominated the income effect associated with the growth in their husbands' real wages. While this part of the analysis could be accommodated in the framework of the traditional model, the next question Mincer raised could not. Why should the substitution effect dominate the income effect for women when such time series evidence as the declining work week suggested a dominance of the income effect over the substitution effect for men? The answer, according to Mincer, lay in women's responsibility for non-market production. The opportunities for substituting market time (through the purchase of market goods and services) are greater for time spent in home work than for time spent in leisure. Thus, since married women spend most of their non-market time on household production while men spend most of theirs on leisure, the substitution effect of a wage increase would be larger for married women than for men.

Becker (1965) advanced this process considerably by proposing a general theory of the allocation of time to replace the traditional theory of labour supply. In this and other work (summarized in Becker, 1981), he laid the foundations of what has become known as the 'new home economics', and spearheaded the development of economic analyses of time allocation, marriage, divorce and fertility. Interestingly, while Mincer opened a window on household production by distinguishing non-market work from leisure where the traditional labour supply theory had not done so, Becker was able to provide a further advance by again eliminating the distinction. However, while in the traditional labour supply model all non-market time is spent in leisure, in Becker's model all non-market time is spent in household production.

Specifically, Becker assumes that households derive utility from 'commodities' which are in turn produced by inputs of market goods and non-market time. It is interesting to note that Becker's 'commodities', produced and consumed entirely in the home, are the polar opposite of Marx's (1867) 'commodities', produced and exchanged in the market. Examples of Becker's commodities range from sleeping, which is produced with inputs of non-market time and of market goods like a bed, sheets, a pillow and a blanket (and in some cases, perhaps, a sleeping pill); to a tennis game that is produced by inputs of non-market time combined with tennis balls, a racquet, an appropriate costume, and court time; to a clean house produced with inputs of non-market time and a vacuum cleaner, a bucket and a mop, and various cleaning products.

In this model the production functions for the commodities are added to the constraints of the utility maximization problem. Utility can still be expressed as a function of the quantities of market goods and non-market time consumed; however, market goods and non-market time now produce utility only indirectly through their use in the production of commodities. Relative preferences for market goods versus home time depend on the ease with which the household can substitute market goods for non-market time in consumption and production. Substitution in consumption depends on their preferences for 'goods intensive' commodities – those produced using relatively large inputs of market goods in comparison to non-market time – relative to 'time intensive' commodities – those produced using relatively large inputs of non-market time in comparison to market goods. Substitution in production depends on the availability of more goods-intensive production techniques for producing the same commodity.

The usefulness of these ideas may be illustrated by considering the relationship of children to women's labour force participation. Children (especially when they are small) may be viewed as a time-intensive 'commodity'. Traditionally, it has been the mother who has been the primary care-giver. Moreover, while it is possible to substitute market goods and services for home time in caring for children (in the form of babysitters, day care centres, etc.), these alternative production techniques tend to be costly and it is sometimes difficult to make suitable alternative arrangements (in terms of quality, scheduling, etc.). Thus, at a point in time, the probability that a woman will participate in the labour force is expected to be inversely related to the number of small children present. Over time, the increase in women's participation rates has been associated with decreases in birthrates, as well as increases in the availability of various types of child care facilities, formal and informal. Changes in social norms (Brown, 1984) making it more acceptable to substitute for the time of parents in the care of young children may also have been a factor, although it is difficult to know, in this case as in others, the extent to which attitude change precedes or follows change in the relevant behaviour.

The relationship between labour force participation and fertility is reinforced by the impact of the potential market wage on women's fertility decisions. Greater market opportunities for women have increased the opportunity cost of children (in terms of their mothers' time inputs) and induced families to have fewer of them. Similarly, the greater demand for alternative child care arrangements (also due to the increased value of women's market time) has made it profitable for more producers to enter this sector.

The gender division of labour. In our discussion of children, we simply assumed that women tend to bear the primary responsibility for child care. However, the gender division of labour in the family is also an issue which the new home economics addresses. According to Becker (1981), the division of labour will be dictated by comparative advantage. To the extent that women have a comparative advantage in household and men in market production, it will be efficient for women to specialize to some extent in the former while men specialize in the latter. In this view, the increased output corresponding to this arrangement constitutes one of the primary benefits to marriage. Thus, women's increasing labour force participation is seen to have reduced the gains from marriage thereby contributing to the trend towards higher divorce and lower marriage rates.

The notion that, where families are formed, it is generally efficient and thus optimal for one member, usually the wife, to specialize to some extent in household production, while the husband specializes entirely in market work, has important consequences for women's status in the labour market. As we shall see in greater detail below, human capital theorists expect such a division of labour to lower the earnings of women relative to men, due to work force interruptions and smaller investments in market-oriented human capital. For this and other reasons, it is important to consider in greater detail whether such specialization is indeed as desirable for the family as the model suggests, and, by implication, whether it is apt to continue into the future. There are three points to be made in this regard.

First, such a division of labour may not be as advantageous for women as it is for men (Ferber and Birnbaum, 1977; Blau and Ferber, 1986). Thus, even if such a specialization is efficient in many respects, it may not maximize the family's utility. Indeed, when there are conflicts of interest or even pronounced differences in tastes between the husband and wife, the concept of the family utility function itself becomes less meaningful, since the way in which the preferences of family members can meaningfully be aggregated to form such a utility function has not been satisfactorily specified.

What are the disadvantages to women of their partial specialization in household production? First, in a market economy, such an arrangement makes them to a greater or lesser degree economically dependent on their husbands (see also Hartmann, 1976). This is likely to reduce their bargaining power relative to their husbands' in family decision-making, as well as to increase the negative economic consequences for them (and frequently for their children) of a marital break-up. In the face of recent increases in the divorce rate, such specialization has become a particularly risky undertaking. Second, as more women come to value their careers in much the same way as men do, both in terms of achievement and earnings, their specialization in housework to the point where it is detrimental to their labour market success is not apt to be viewed with favour by them. The utility-maximizing family will take these disadvantages into account in conjunction with the efficiency gains of specialization in allocating the time of family members.

If specialization is indeed considerably more productive than sharing of household responsibilities, it may be possible for this higher output to be used in part to compensate women for the disadvantages detailed above. However, it is likely that the gains to such specialization will shrink over time relative to the disadvantages of such an arrangement. As women anticipate spending increasingly more of their working lives in the labour market, their investments in market-oriented capital may be expected to continue to grow and their comparative advantage in home work relative to men to decline. Moreover, as the quality of the opportunities open to women in the labour market continues to improve, the disadvantages of specialization in home work in the form of foregone earnings and possibilities for career advancement will also rise. Thus, greater sharing of household responsibilities between men and women is likely to become increasingly prevalent, even if women in general retain a degree of comparative advantage in household production for some time to come.

A second point to be made with regard to women's specialization in household production is that comparative advantage does not comprise the only economic benefit to family or household formation (Ferber and Birnbaum, 1977; Blau and Ferber, 1986). Families and households also enjoy the benefits of economies of scale in the production of some commodities, as well as the gains associated with the joint consumption of 'public' goods. These benefits of collaboration would be unaffected by a reduction in specialization, even if those based on comparative advantage would be diminished. Other benefits of marriage or household formation may actually be increased by a more egalitarian division of household responsibilities. For example, two-earner families are in a sense more diversified and thus enjoy greater income security than families which depend on only one income. It may also be the case that the enjoyment derived from joint consumption is enhanced when the members of a couple have more in common, as when both participate in market and home activities. Thus, the incentives of couples to adhere to the traditional division of labour in order to enjoy the economic benefits of marriage may not be as strong as suggested when only the gains to comparative advantage are considered.

Finally, it is important to point out that women's comparative advantage for household production may stem not only from the impacts of biology and gender differences in upbringing and tastes, but also from the effect of labour market discrimination in lowering women's earnings relative to men's. Decisions based to some extent on such market distortions are not optimal from the perspective of social welfare even though they may be rational from the perspective of the family. The importance of such feedback effects are considered in greater detail below.

GENDER DIFFERENCES IN LABOUR MARKET OUTCOMES

We turn now to the contribution of economic analysis to an understanding of the causes of gender inequality in economic outcomes. Here, the consideration of gender issues has been accommodated principally through the development of new and interesting applications of existing theoretical approaches. The particular challenge posed to the theories by women's economic status is the existence of occupational segregation as well as earnings differentials by sex. Occupational segregation refers to the concentration of women in one set of predominantly female jobs and of men in another set of predominantly male jobs. The reasons for such segregation and its relationship to the male–female pay differential are two key questions to be addressed.

As in the case of the analysis of women's roles in the family, the catalyst for the development of these approaches was provided by external events. Some moderate degree of interest in this issue was generated in England by the World War I experience. Pursuant to the war effort, there was some substitution of women into traditionally male civilian jobs, although not nearly to the degree that there would be during World War II. Questions of the appropriate pay for women under these circumstances arose and stimulated some economic analyses of the gender pay differential – all of which gave a prominent causal role to occupational segregation. These included the work of Fawcett (1918) and Edgeworth (1922) (which provided the antecedents for Bergmann's (1974) overcrowding model, discussed below) and Webb (1919).

The analysis of gender differentials in the labour market received another impetus in the early 1960s, this time in the United States, with the development of the women's liberation movement and the passage of equal employment opportunity legislation. Two broad approaches to the issue have since evolved. First is the human capital view which lays primary emphasis on women's own voluntary choices in explaining occupation and pay differences. Second are a variety of models of labour market discrimination which share the common characteristic of placing the onus for the unequal outcomes on differential treatment of equally (or potentially equally) qualified men and women in the labour market. While these two approaches may be viewed as alternatives, it is important to point out that they are in fact not mutually exclusive. Both may play a part in explaining sex differences in earnings and occupations and the empirical evidence suggests that this is the case (see, e.g., Treiman and Hartmann, 1981). Indeed, as we shall see, their effects are quite likely to reinforce each other. We now consider each of these approaches in turn.

The human capital explanation. The human capital explanation for gender differences in occupations and earnings, developed by Mincer and Polachek (1974), Polachek (1981) and others, follows directly from the analysis of the family described above. It is assumed that the division of labour in the family will result in women placing greater emphasis that men on family responsibilities over their life cycle. Anticipating shorter and more discontinuous work lives as a consequence of this, women will have lower incentives to invest in market-oriented formal education and on-the-job training than men. Their resulting smaller human capital investments will lower their earnings relative to those of men.

These considerations are also expected to produce gender differences in occupational distribution. It is argued that women will choose occupations for which such investments are less important and in which the wage penalties associated with work force interruptions (due to the skill depreciation that occurs during time spent out of the labour force) are minimized. Due to their expected discontinuity of employment, women will avoid especially those jobs requiring large investments in firm-specific skills (i.e. skills which are unique to a particular enterprise), because the returns to such investments are reaped only as long as one remains with the firm. The shorter expected job tenure of women in comparison with that of men is also expected to make employers reluctant to hire women for such jobs in that employers bear some of the costs of such training. Thus, to the extent that it is difficult to distinguish more from less career-oriented women, the former may be negatively affected (see the discussion of statistical discrimination below).

More recently, Becker (1985) has further argued that, even when men and women spend the same amount of time on market jobs, women's homemaking responsibilities can still adversely affect their earnings and occupations. Specifically, he reasons that since child care and housework are more effort intensive than are leisure and other household activities, married women will spend less effort than married men on each hour of market work. The result will be lower hourly earnings for married women and, to the extent that they seek less demanding jobs, gender differences in occupations.

Thus, the human capital analysis provides a logically consistent explanation for gender differences in market outcomes on the basis of the traditional division of labour by gender in the family. An implication generally not noted by those who have developed this approach is that, to the extent that the human capital explanation is an accurate description of reality, it serves to illustrate graphically the disadvantages for women of responsibility for (specialization in) housework which we discussed above. To the extent that gender differences in economic rewards are not fully explained by productivity differences, we must turn to models of labour market discrimination to explain the remainder of the difference.

Models of labour market discrimination. As noted earlier, models of discrimination were developed to understand better the consequences of differences in the labour market treatment of two groups for their relative economic success. The starting point for models of labour market discrimination is the assumption that members of the two groups are equally or potentially equally productive. That is, except for any direct effects of the discrimination itself, male and female labour (in this case) are perfect substitutes in production. This assumption is made not because it is necessarily considered an accurate description of reality, but rather because of the question which discrimination models specifically address: why do equally qualified male and female workers receive unequal rewards? Such models may then be used to explain how discrimination can produce pay differentials between men and women in excess of what could be expected on the basis of productivity differences.

Theoretical work in this area was initiated by Becker's (1957) model of racial discrimination. Becker conceptualized discrimination as a taste or personal prejudice. He analysed three cases, those in which the tastes for discrimination were located in employers, co-workers and customers, respectively. As Becker pointed out, for such tastes to affect the economic status of a particular group adversely, they must actually affect the behaviour of the discriminators.

One may at first question whether such a model is as applicable to sex as to race discrimination in that, unlike the case of racial discrimination, men and women are generally in close contact within families. However, the notion of socially appropriate roles, not explicitly considered by Becker, both sheds light on this question and establishes a link between his theory and occupational segregation. Thus, employers may be quite willing to hire women as secretaries, receptionists or nursery school teachers but may be reluctant to employ them

as lawyers, college professors or electricians. Co-workers may be quite comfortable working with women as subordinates or in complementary positions, but feel it is demeaning or inappropriate to have women as supervisors or as peers. Customers may be happy to have female waitresses at a coffee shop, but expect to be served by male waiters at an elegant restaurant. They may be delighted to purchase women's blouses or even men's ties from female clerks, but prefer their appliance salesperson, lawyer or doctor to be a man. Such notions of socially appropriate roles are quite likely a factor in racial discrimination as well.

Employers with tastes for discrimination against women in particular jobs will be utility rather than profit maximizers. They will see the full costs of employing a woman to include not only her wages but also a discrimination coefficient ($d_r \geqslant 0$) reflecting the pecuniary value of the disutility caused them by her presence. Thus, they will be willing to hire women only at lower wages than men ($w_f = w_m - d_r$). If men are paid their marginal products, employer discrimination will result in women receiving less than theirs. When employers differ in their tastes for discrimination, the market-wide discrimination coefficient will be established at a level which equates supply and demand for female labour at the going wage. Thus the size of the male-female pay gap will depend on the number of women seeking work, as well as on the number of discriminatory employers and on the magnitude of their discrimination coefficients.

One of the particularly interesting insights of Becker's (1957) analysis is that profit-maximizing employers who do not themselves have tastes for discrimination against women will nonetheless discriminate against them if their employees or customers have such prejudices. Male employees with tastes for discrimination against women will act as though their wage is reduced by $d_e(\geqslant 0)$, their discrimination coefficient, when they are required to work with women. Thus, they will consent to be employed with women only if they receive a higher wage – in effect a compensating wage differential for this unpleasant working condition.

The obvious solution to this problem from the employer's point of view is to hire a single-sex work force. If all employers followed such a strategy, male and female workers would be segregated by firm, but there would be no pay differential. Yet, as Arrow (1973) has noted, employers who have made a personnel investment in their male workers, in the form of recruiting, hiring or training costs, may not find it profitable to discharge all their male employees and replace them with women, even if the latter become available at a lower wage. While such considerations cannot explain how occupations initially become predominantly male, it can shed light on one factor – the necessity of paying a premium to discriminatory male workers to induce them to work with women – contributing to the perpetuation of that situation. Further, where women do work with discriminating male workers, a pay differential will result.

Some extensions of Becker's (1957) analysis of employee discrimination are also of interest. Bergmann and Darity (1981) point out that employers may be reluctant to hire women into traditionally male jobs because of adverse effects on the morale and productivity of the existing male work force. Given the replacement costs discussed above this would be an important consideration. As Blau and Ferber (1986) note, employee discrimination may also directly lower women's productivity relative to that of men. For example, since much on-the-job training is informal, if male supervisors or coworkers refuse or simply neglect to instruct female workers in these job skills, women will be less productive than men workers. Similarly, the exclusion of women from informal networks and mentor-protégé relationships in traditionally male occupations can diminish their access to training experiences and even to the information flows needed to do their jobs well.

Customer discrimination can also reduce the productivity of female relative to male employees. Customers with tastes for discrimination against women will act as if the price of a good or service provided by a woman were increased by their discrimination coefficient, $d_c(\geqslant 0)$. Thus, at any given selling price, a female employee will bring in less revenue than a male employee. Women either will not be hired for such jobs or will be paid less. The potential applicability of this model is not only to conventional sales jobs. In our 'service economy', a large and growing number of jobs entail personal contact between workers and customers/clients.

Models based on the notion of tastes for discrimination are consistent with occupational segregation, but do not necessarily predict it. If wages are flexible, it is altogether possible that such discrimination will result in lower pay for women, but little or no segregation. However, if discriminatory tastes against women in traditionally male pursuits (on the part of employers, employees and/or customers) are both strong and prevalent, women may tend to be excluded from these areas. On the other hand, even if such segregation occurs, it may or may not be associated with gender pay differentials. In the presence of sufficient employment opportunities in the female sector, equally qualified women may earn no less than men.

The relationship between occupational segregation and earnings differentials is further clarified in Bergmann's (1974) overcrowding model. If for whatever reason – labour market discrimination or their own choices – potentially equally qualified men and women are segregated by occupation, the wages in male and female jobs will be determined by the supply and demand for labour in each sector. Workers in male jobs will enjoy a relative wage advantage if the supply of labour is more abundant relative to demand for female than for male occupations. Such 'crowding' of female occupations can also widen differentials between male and female jobs that would exist in any case due to women's smaller human capital investments or to employers' reluctance to invest in their human capital.

Perhaps the most serious question that has been raised about the Becker analysis, particularly of the case of employer discrimination, is its inability to explain the persistence of discrimination in the long run. Assuming that tastes for discrimination vary, the least discriminatory firms would employ the highest proportion of lower-priced female labour. They would thus have lower costs of production and, under constant returns to scale, could in the long run expand and drive the more discriminatory firms out of business (Arrow, 1973).

This issue has provided the rationale, at least in part, for the elaboration of alternative models of discrimination, including the statistical discrimination model discussed below. Others, not considered here, have emphasized non-competitive aspects of labour markets (e.g. Madden, 1973). However, this criticism of the Becker model is a double-edged sword in that it has led some economists to doubt that labour market discrimination is responsible, in whole or part, for gender inequality in economic rewards. Yet it is important to recognize that the phenomenon which we seek to understand is intrinsically complex. From this perspective it is not surprising that no easy solution has been found to the question of why discrimination has persisted. Similarly, the various models of discrimination, each emphasizing different motivations and different sources

of this behaviour, need not be viewed as alternatives. Rather, each may serve to illuminate different aspects of this complex reality.

As noted above, models of statistical discrimination were developed by Phelps (1972) and others to shed light on the persistence of discrimination. They do so by imputing a motive for employer discrimination which, in an environment of imperfect information, is consistent with profit maximization. Statistical discrimination occurs when employers believe that, all else equal, women are on average less productive or less stable workers than men. The common perception that women are more likely to quit their jobs than men would be an example of this.

As in the employer taste for discrimination model, statistical discrimination would cause employers to prefer male workers and to be willing to hire women only at a wage discount. A difference is, however, that in this case male and female workers are not perceived to be perfect substitutes. Further, if women are viewed as less stable workers, there will tend to be substantive differences between male and female jobs, with the former emphasizing firm-specific skills to a greater extent. This is essentially the picture painted by the dual market model (Piore, 1971; Doeringer and Piore, 1971). In this view, women tend to be excluded from the 'primary sector', jobs requiring firm-specific skills and thus characterized by relatively high wages, good promotion opportunities and low turnover rates, and to find employment in the 'secondary sector', comprised of low paying, dead-end jobs in which there tends to be considerable turnover.

Like the human capital model, the notion of statistical discrimination provides a link between women's roles in the family and gender differences in market outcomes. However, the connection is in terms of differences in the treatment of men and women, rather than differences in the choices they make.

One crucial issue is of course whether employers' perceptions are indeed correct. If they are, as Aigner and Cain (1977) have pointed out, then in some sense labour market discrimination as conventionally defined does not exist: women's lower wages are due to their lower productivity. Nonetheless, the employer's inability to distinguish between more and less career-oriented women certainly creates an inequity for the former vis-à-vis their male counterparts.

On the other hand, employer perceptions may be incorrect or exaggerated. Differentials based on such erroneous views undoubtedly constitute discrimination as economists have defined it. However, as Aigner and Cain (1977) have persuasively argued, gender differentials based on employers' mistaken beliefs are even less likely to persist in the long run than those based on employers' tastes for discrimination. Nonetheless, in times of rapid changes in gender roles, there may be considerable lags in employers' perceptions. Employers' incorrect views could also magnify the impact of employee or customer discrimination, as when such discrimination is either less extensive or more susceptible to change than employers believe.

A potentially more powerful role for statistical discrimination is provided in models which allow for feedback effects, for example Arrow's (1973) model of perceptual equilibrium. In this case, men and women are assumed to be potentially perfect substitutes in production, but employers believe that, for example, women are less stable workers (Arrow, 1976). They thus allocate women to jobs where the cost of turnover is minimized and women respond by exhibiting the unstable behaviour employers expect. The employers' assessments are correct *ex post*, but are in fact due to their own discriminatory actions. This equilibrium will be stable even though an alternative equilibrium is potentially available in which women are hired for jobs which are sufficiently rewarding to inhibit instability. More generally, any form of discrimination can adversely affect women's human capital investments and labour force attachment by lowering the market rewards to this behaviour (see also, Blau, 1984; Blau and Ferber, 1986; Ferber and Lowry, 1976; and Weiss and Gronau, 1981).

CONCLUSION

We have considered the contributions of neoclassical economic theory to our understanding of women's labour supply decisions, the gender division of labour within the family, and male–female differences in labour market outcomes. With the introduction of feedback effects, the separate strands of neoclassical theory analysing women's economic roles in the family and their labour market outcomes may be more tightly woven together. The causation runs not only from women's roles within the family to their resulting economic success, as human capital theorists emphasize, but also from their treatment in the labour market to their incentives to invest in market-oriented human capital and to participate in the labour force continuously. Thus, even a small amount of discrimination at an early stage of the career can have greatly magnified effects over the work life. While it is unlikely that labour market discrimination created the traditional division of labour between men and women in the family, it could certainly help to perpetuate it.

However, it is also the case that increasing opportunities for women in the labour market create powerful incentives to reduce gender differences in family roles and labour market behaviour. At the same time, women's increased attachment to the labour force, due not only to these increased opportunities but also to changes in household technology and in tastes, may be expected to increase their market productivity and hence their earnings directly, and also to reduce statistical discrimination against them. Similarly, the movement of women into traditionally male jobs has the potential not only to increase the wages of those who become so employed, but to reduce overcrowding and increase wages in female jobs as well. Thus, just as a fuller understanding of the interrelationships between women's roles in the family and their status in the labour market helps us to understand the persistence of gender inequality in economic outcomes, it also enables us to appreciate how changes in either one of these spheres, or both, can induce a mutually reinforcing process of cumulative change. Recent signs of progress in reducing the pay gap in many of the advanced industrialized countries may well signal the beginnings of such a process.

In our emphasis upon the interdependence of women's status within the family and the labour market, we have in some respects returned to our starting point, for this conclusion bears a close resemblance to the views of the 19th-century observers which we reviewed at the outset. However, it is also clear that neoclassical economic theory has enhanced our understanding of the causes of gender differences in both the family and the labour market, as well as allowing us to comprehend better the links between the two sectors.

FRANCINE D. BLAU

See also DISCRIMINATION; FAMILY; HOUSEHOLD PRODUCTION; HOUSEWORK; INEQUALITY BETWEEN THE SEXES; LABOUR MARKET DISCRIMINATION; LABOUR SUPPLY OF WOMEN; OCCUPATIONAL SEGREGATION; WOMEN AND WORK; WOMEN'S WAGES.

BIBLIOGRAPHY

Aigner, D. and Cain, G. 1977. Statistical theories of discrimination in labor markets. *Industrial and Labor Relations Review* 30(2), January, 175–87.

Arrow, K. 1973. The theory of discrimination. In *Discrimination in Labor Markets*, ed. O. Ashenfelter and A. Rees, Princeton: Princeton University Press.

Arrow, K. 1976. Economic dimensions of occupational segregation: comment I. *Signs* 1(3), Part II, 233–7.

Becker, G. 1957. *The Economics of Discrimination.* 2nd edn, Chicago: University of Chicago Press, 1971.

Becker, G. 1965. A theory of the allocation of time. *Economic Journal* 75, September, 493–517.

Becker, G. 1981. *A Treatise on the Family.* Cambridge, Mass.: Harvard University Press.

Becker, G. 1985. Human capital, effort, and the sexual division of labor. *Journal of Labor Economics* 3(1), January, 533–58.

Bergmann, B. 1974. Occupational segregation, wages and profits when employers discriminate by race or sex. *Eastern Economic Journal* 1, April/July, 103–10.

Bergmann, B. and Darity, W., Jr. 1981. Social relations in the workplace and employer discrimination. *Proceedings of the Thirty-Third Annual Meeting of the Industrial Relations Research Association.* ed. B.D. Dennis, New York: Industrial Relations Research Association, 155–62.

Blau, F. 1984. Discrimination against women: theory and evidence. In *Labor Economics: Modern Views*, ed. W. Darity, Boston: Kluwer-Nijhoff Publishing.

Blau, F. and Ferber, M. 1986. *The Economics of Women, Men, and Work.* Englewood Cliffs, NJ: Prentice-Hall.

Brown, C. 1984. Consumption norms, work roles, and economic growth. Paper presented at the conference on Gender in the Workplace, Washington, DC: Brookings Institution, November.

Doeringer, P. and Piore, M. 1971. *Internal Labor Markets and Manpower Analysis.* Lexington, Mass.: D.C. Heath and Co.

Edgeworth, F. 1922. Equal pay to men and women for equal work. *Economic Journal* 32, December, 431–57.

Engels, F. 1884. *The Origin of the Family, Private Property and The State.* New York: International Publishers, 1972.

Fawcett, M.G. 1918. Equal pay for equal work. *Economic Journal* 28, March, 1–6.

Ferber, M. and Birnbaum, B. 1977. The 'new home economics': retrospects and prospects. *Journal of Consumer Research* 4(1), June, 19–28.

Ferber, M. and Lowry, H. 1976. The sex differential in earnings: a reappraisal. *Industrial and Labor Relations Review* 29(3), Apr. 377–87.

Gilman, C. 1898. *Women and Economics: a study of the economic relation between men and women as a factor of social evolution.* New York: Harper & Row, 1966.

Hartmann, H. 1976. Capitalism, partriarchy and job segregation by sex. *Signs* 1(3), Part II, 137–69.

Madden, J. 1973. *The Economics of Sex Discrimination.* Lexington, Mass.: D.C. Heath and Co.

Marx, K. 1867. *Capital: A Critique of Political Economy*, Vol. I. New York: International Publishers, 1967.

Mill, J.S. 1869. *The Subjection of Women.* 4th edn, London: Longmans, Green, Reader & Dyer, 1878.

Mincer, J. 1962. Labor force participation of married women. In *Aspects of Labor Economics*, National Bureau of Economic Research, Princeton: Princeton University Press.

Mincer, J. and Polachek, S. 1974. Family investments in human capital: earnings of women. *Journal of Political Economy* 82(2), Part II, S76–S108.

Phelps, E. 1972. The statistical theory of racism and sexism. *American Economic Review* 62(4), September, 659–61.

Piore, M. 1971. The dual labor market: theory and implications. In *Problems in Political Economy: an urban perspective*, ed. D. Gordon, Lexington, Mass.: D.C. Heath and Co.

Polachek, S. 1981. Occupational self-selection: a human capital approach to sex differences in occupational structure. *Review of Economics and Statistics* 63(1), February, 60–69.

Treiman, D. and Hartmann, H. (eds) 1981. *Women, Work, and Wages: equal pay for jobs of equal value.* Washington, DC: National Academy Press.

Webb, B. 1919. *The Wages of Men and Women: should they be equal?* London: Fabian Bookshop.

Weiss, Y. and Gronau, R. 1981. Expected interruptions in labour force participation and sex-related differences in earnings growth. *Review of Economic Studies* 48(4), October, 607–19.

general equilibria, computation of. *See* COMPUTATION OF GENERAL EQUILIBRIA.

general equilibrium. General equilibrium theory is in contrast with partial equilibrium theory where some specified part of an economy is analysed while the influences impinging on this sector from the rest of the economy are held constant. In general equilibrium the influences which are treated as constant are those which are considered to be noneconomic and thus beyond the range of economic analysis. Of course, this does not guarantee that these influences will in fact remain constant when the economic factors change, and the usefulness of economic analysis for predictive purposes may depend on to what degree influences treated as noneconomic are really independent of the economic variables.

The institution whose phenomena are the primary subject matter of economic analysis is the market, made up of a group of economic agents who buy and sell goods and services to one another. In partial equilibrium theory the group of agents may be confined to those who are involved in one industry, either buying or selling its product or buying or selling the materials and productive services used in making its product. However, in general equilibrium theory all the agents involved in exchanges with each other should ideally be included and all their sales and purchases should be allowed for. However, it may happen that the activities of many agents are only treated in the aggregate and the list of goods and services may be reduced by aggregation. The aggregation of agents and commodities into a few categories is especially important when general equilibrium theory is applied to special areas of public policy such as the government budget, money and banking, or foreign trade. Much of the theory developed for these subjects is general equilibrium theory in aggregated form.

The general equilibrium implies that all subsets of agents are in equilibrium and in particular that all individual agents are in equilibrium. The conscious development of a formal general equilibrium theory stated in mathematical terms seems to have been inspired by a formal theory of the equilibrium of the individual consumer faced with a given set of trading opportunities or prices. This theory was developed by the marginal utility, or neo-classical, school of economists in the third quarter of the 19th century, independently, by Gossen (1854), Jevons (1871), and Walras (1874–7), who used mathematical notations, and by Menger (1871) who did not. The step was taken in the most effective way by Walras.

THE EQUILIBRIUM OF AN EXCHANGE ECONOMY. Walras assumed that the utility derived from the consumption of a good was given as a function of the amount of that good alone that was consumed and independent of the amounts consumed of other goods. He also assumed that the first derivative of the utility function was positive and decreasing up to a point of satiation when one exists. He then gave a rigorous derivation of the demand for a good by a consumer from the maximization of utility subject to a budget constraint. The demand functions give the equilibrium quantities traded by the consumer as a function of market prices. As Walras saw, this is a crucial step in the development of a general equilibrium theory for an economy. It has remained in a generalized form the cornerstone of general equilibrium theory since Walras.

The simplest problem of general equilibrium arises in the theory of the exchange economy without production. In this economy the budget constraint of the trader is established by his initial stocks and the list of prices. Then the individual demand function represents the equilibrium of the single trader in face of a given price system. The market demand function is the sum of the individual demand functions, and the equilibrium of the market occurs at a price for which the sum of demands, including offers as negative demands, is equal to 0 for each good, or, if free disposal is allowed, is not positive for any good. This idea was expressed in classical economic theory by the equality of supply and demand in each market, but its expression in a set of equations to be satisfied by the list of equilibrium prices was due to Walras, although Cournot (1838) had foreshadowed the Walrasian analysis in his discussion of the international flow of money and Mill (1848) in his discussion of foreign trade.

Suppose there are n goods to be traded and there are m traders. Let w_i^h be the quantity of the ith good held initially by the hth trader. Let $u^h(x)$ where $x = (x_1, ..., x_n)$ be the utility to the hth trader of possessing the quantities $x_1, ..., x_n$ of the n goods traded. Then the hth trader is in equilibrium at the prices $p = (p_1, ..., p_n)$ and the quantities x^h if $u^h(x)$ is a maximum at x^h over all values of x which satisfy $\Sigma_1^n p_i x_i \leqslant \Sigma_1^n p_i w_i^h$. If smoothness and concavity conditions are met by the utility function, and the goods are divisible, the maximizing x will be unique and will define a function $f^h(p)$ over an appropriate price domain. Since the set of commodity bundles x at which the utility function is maximized does not change when the prices p are multiplied by a positive scalar, this function will satisfy $f^h(p) = f^h(\alpha p)$ for $\alpha > 0$.

The market demand function is $f(p) = \Sigma_1^m f^h(p)$. Then the market equilibrium for a trading economy is given by a price vector p and an allocation of goods $(x^1, ..., x^m)$ such that $x^h = f^h(p)$ and $\Sigma_1^m x^h = \Sigma_1^m w^h$, or, assuming free disposal, $\Sigma_1^m x^h \leqslant \Sigma_1^m w^h$. The first condition expresses the equilibrium of the individual trader and the second condition is the equality of supply and demand. Thus there are n scalar equations $\Sigma_{h=1}^m f_i^h(p = \Sigma_{h=1}^m w_i^h$ to determine the n equilibrium prices p_i. The given data are the consumer tastes, expressed in the utility functions u^h, and the initial stocks of goods w^h.

It is clear that the market demand function satisfies the homogeneity condition $f(\alpha p) = f(p)$ for $\alpha \geqslant 0$. Thus equilibrium prices are only determined up to multiplication by a positive number. This reflects the fact that the equilibrium of the consumer is not affected if prices are multiplied by α and market equilibrium is the simultaneous equilibrium of all consumers at the same prices. It is often convenient to adopt some normalization of prices. Walras chooses a good whose price is known to be positive in equilibrium and gives this good, which he calls the numeraire, the price 1. Another convention which is useful when free disposal is assumed, so that prices are necessarily non-negative, is to choose p such that $\Sigma_1^n p_i = 1$. Then the domain of definition for the demand functions may be taken to be all p such that $p_i \geqslant 0$ and $\Sigma_1^n p_i = 1$.

There is an analogy between the equilibrium of the trading economy and the equilibrium of mechanical forces. Indeed, one of the inspirations for the theory of Walras appear to have been a treatise on statics by Poinsot (1803, 1842). According to the principle of virtual work an infinitesimal displacement of a mechanical system, which is at equilibrium under the stress of forces and subject to constraints, does no work. In the economy at equilibrium an infinitesimal displacement of the allocation of goods $(x^1, ..., x^m)$ cannot increase the utility of one trader unless it reduces the utility of another. This is an easy implication of the fact that utility is maximized over the budget constraint, provided no one is saturated. This means that a new allocation to a trader cannot preserve his utility level if its value at the equilibrium prices falls. On the other hand, the utility level of a trader cannot increase unless his allocation becomes more valuable at the equilibrium prices. But then the new allocations $x^{h'}$ would satisfy

$$\sum_{h=1}^{m} \sum_{i=1}^{n} p_i x_i^{h'} > \sum_{h=1}^{m} \sum_{i=1}^{n} p_i w_i^h$$

which is impossible since the total allocation cannot exceed the total supply of goods. Indeed, if each trader holds all goods in his equilibrium allocation and the utility functions are differentiable, which implies that goods are divisible, an infinitesimal reallocation would have no effect on utility levels if it has no effect on the levels of individual budgets. This property of market equilibrium was first recognized by Pareto (1909), and an allocation of goods with the property that no displacement of it can benefit one consumer unless it harms another is said to be Pareto optimal. The implication from competitive equilibrium to Pareto optimality requires that no consumer be locally satiated. It is also true that a Pareto optimal allocation may be realized as a competitive equilibrium given an appropriate distribution of initial stocks but the conditions are more severe. The first general theorems were proved by Arrow (1951).

EQUILIBRIUM WITH PRODUCTION. The next step in developing the general equilibrium of an economy is to introduce production under the condition that the output matures without a lapse of time. This step was taken by Walras who introduced linear activities which list the quantities of productive services required to produce on unit of a good. There may be many alternative activities for the production of any given good and a choice is made among them in order to minimize the cost of production at given market prices. Let $z = (z_1, \ldots, z_r)$ be a list of quantities of productive services and let $g^i(z)$, $i = 1, \ldots, n$, be production functions for the n goods. Since linear activities are assumed, the production functions will satisfy $\alpha g^i(z) = g^i(\alpha z)$. In particular, we may consider the unit isoquant or the set A_i such that $g^i(z) = 1$ for z in A_i. Then the activities which minimize cost at given prices q are represented by production coefficients $a^i(q)$, contained in A_i, where $q'a^i(q) \leqslant q'z$ for z in A_i. Equilibrium in the production sector is given by price vectors p and q and activity vectors $a^i(q)$ where $p_i \leqslant \sum_{j=1}^{r} q_j a_j^i(q)$ for all i and equality holds if the ith good is produced.

In an equilibrium of the production sector any quantities y of outputs may be produced provided quantities z of productive services are available where $z_j = \sum_{i=1}^{n} y_i a_j^i(q)$. In order to include the productive sector in a market equilibrium the utility functions of consumers must be extended to include productive services among their arguments. They may be written $u^h(x, z)$. If we reinterpret x_i as the quantity of a good traded rather than the quantity consumed, the initial stocks may be suppressed. This is convenient since it is not clear how initial stocks of labour services can be specified. Then the individual consumer is in equilibrium given prices p and q for goods and productive services when the quantities traded (x^h, z^h) maximize $u^h(x, z)$ over all (x, z) such that $\sum_i^n p_i x_i - \sum_1^r q_i z_i \leqslant 0$. The maximizing quantities need not be unique in general, so it is necessary to represent demand by a correspondence that takes a set of trades as its value and write $(x^h, z^h) \in f^h(p, q)$ when (x^h, z^h) is a maximizer given prices p and q.

As before market equilibrium is achieved when all economic agents are in equilibrium at the same prices and supply is equal to demand. Since risk is not present in this economy, the productive services involved in organizing production need not be given a distinguished role. Activities may be treated as conducted by the whole set of owners of the productive services involved in them. Then if it should happen that $p_i > \sum_{j=1}^{r} q_j a_j^i(q)$ for the ith good, there will be an opportunity for some owners of productive services to earn larger returns producing the ith good than those prevailing generally as given by q. Thus productive services will leave other activities and flow to this activity, so equilibrium does not obtain for owners of productive services. This equilibrium now requires, on the one hand, equilibrium of each economic agent as consumer of goods and provider of productive services, that is, $(x^h, z^h) \in f^h(p, q)$, and, on the other hand, equilibrium of each economic agent as a participant in production, that is, $p_i \leqslant \sum_{j=1}^{r} q_j a_j^i(q)$, with equality if the ith good is produced. However, market equilibrium also requires that $\sum_{h=1}^{m} z_j^h = \sum_{h=1}^{m} \sum_{i=1}^{r} x^h a_j^i(q)$, that is, the supply of productive services must equal the quantities needed to produce the quantities of goods demanded. As before, if surplus productive services may be freely disposed of, the equality in the last equation may be replaced by an inequality.

The demand functions $f_i^n(p, q)$ and the supply functions $f_{n+j}^h(p, q)$ express the equilibrium of the household sector. Therefore, the relation $\sum_{i=1}^{n} p_i f_i^h(p, q) = \sum_{j=1}^{r} q_j f_{n+j}^h(p, q)$ holds for all values of p and q in the price domain. Let x_i be the amount of the ith good produced and let z_j be the amount of the jth factor used in production. Then equilibrium in the production sector implies that

$$\sum_{i=1}^{n} p_i x_i = \sum_{j=1}^{r} \sum_{i=1}^{n} q_j x_i a_j^i = \sum_{j=1}^{r} q_j z_j.$$

Let $f(p, q) = \sum_{h=1}^{m} f^h(p, q)$. Then household equilibrium implies $\sum_{i=1}^{n} p_i f_i(p, q) = \sum_{j=1}^{r} q_j f_{n+j}(p, q)$. Let excess demand for a good be $e_i(p, q) = f_i(p, q) - x_i$, and excess demand for a productive service be $e_{n+j}(p, q) = z_j - f_{n+j}(p, q)$. Then equilibrium in the production and household sectors together implies that $\sum_{i=1}^{n} p_i e_i(p, q) + \sum_{j=1}^{n} q_j e_{n+j}(p, q) = 0$, or the value of excess demand is zero whatever price system is set. This relation is referred to as Walras's Law.

If there is free disposal, prices must be non-negative. Otherwise, disposal would be profitable. Also with free disposal the condition for equilibrium of the market is $e(p, q) \leqslant 0$. Then Walras's Law immediately implies $p_i e_i(p, q) = q_j e_{n+j}(p, q) = 0$, and if any good or productive service is in excess supply in equilibrium, its equilibrium price must be 0. This might be termed Wald's Law, since he made crucial use of it in the first rigorous proof that equilibrium exists in a competitive economy (Wald, 1935, 1936a).

A production sector composed of activities with single outputs is the model used by Walras who was responsible for the first fully developed general equilibrium theory. The natural generalization of this model is to introduce more than one output. Then the kth activity is represented by an output vector $b^k = (b_1^k, \ldots, b_n^k)$ and an input vector for productive services $a^k = (a_1^k, \ldots, a_r^k)$. Assume that activities may be replicated and are independent of each other. Then if (a^k, b^k) is a possible input–output combination for the kth activity, so is $(\alpha a^k, \alpha b^k)$ where α is any non-negative integer. Indeed if all inputs and outputs are divisible it is possible for α to take as its value any real number.

This model of the production sector which embraces the transformation of productive services into goods and services is due to Walras in the context of a theory of general equilibrium. It is convenient to think of the market as held periodically to arrange for the delivery of goods and services over a certain basic period of time. This view of the market, which is also a device of Walras, leads to a theory of temporary equilibrium. The theory was further elaborated by Hicks (1939) and in recent years by other authors. In order to explain the demand and supply of products and productive services in the periodic market it is necessary to introduce some assumptions on the formation of expectations for the prices which will prevail in future markets. The simplest assumption is that the prices arrived at in one market are expected to prevail in future markets. This type of expectation formation is sometimes referred to as static expectations. Walras usually appears to assume static expectations. Hicks introduced a notion of elasticity of expectations to allow expectations of future prices to depend on the change of prices from one temporary equilibrium to another. In recent work analysis has proceeded upon more general assumptions, using various formal properties of dependencies between past prices and expected prices. A quite different approach to expectations which enjoys much current popularity is to assume that expectations are correct, at least in a stochastic sense. The rationale of this approach is that any persistent bias in forecasts of future prices implies that there are unexploited opportunities for profit from further trading which eventually should be recognized.

The model of the production sector as a set of potential linear activities was subsequently used by Cassel (1918) in a simplified Walrasian model which preserved the demand functions and the production coefficients but which did not deduce the demand functions from utility functions or preferences. The model was generalized to allow joint production in a special context by von Neumann (1937). It was given a thorough elaboration and analysis in a model where intermediate products are introduced explicitly by Koopmans (1951). In the Walrasian picture intermediate products were eliminated through the combination of activities so that activities were described as transforming productive services directly into final products whether consumer goods and services or capital goods. However, such a description of the economy depends for its relevance on prices which do not change from one temporary equilibrium to another, so that the choice of activities is not changing.

In the general linear model of production it is no longer adequate to treat the choice of activities as a process of cost minimization given the price vectors p and q. Cost minimization must be replaced by the condition that no activity may offer a profit and no activity which is used in competitive equilibrium may suffer a loss. This is exactly the condition '*ni benefice ni perte*' which Walras used to define equilibrium in production, initially in a model with fixed coefficients of production. However, this condition was first used in a general production model by von Neumann, so it might be termed von Neumann's Law for an activities model of production. Koopmans explored the relation between efficient production and von Neumann's Law. He established an equivalence between the proposition that an output is efficient and the proposition that prices exist such that von Neumann's Law is satisfied when the activities used are those needed to produce this output. Moreover, if each good or service is either desired in unlimited quantities or freely disposable the prices must be non-negative. Thus under these demand and supply conditions any competitive equilibrium must include an efficient output from the production sector. The activities approach to the production sector of a competitive economy was used by Wald and then by McKenzie (1954) in proofs of existence for competitive

equilibrium. It was also used by Scarf (1973) in an algorithm for finding a competitive equilibrium given the technology, the resources, and the demand functions.

An alternative model of the production sector emphasizes the productive organization or firm rather than the activities or technology. A set of actual or potential firms is given and each firm is endowed with its own set of possible input–output combinations. The set of possible input–output combinations achievable by the economy, independently of resource availabilities, is the sum of the sets of input–output combinations achievable by the firms. The condition for equilibrium in the production sector is that each firm maximizes its profits, that is, the value of the input–output combination over its production possibility set, given the prices of inputs and outputs. This view of production was explicit in a partial equilibrium context in Cournot. It was at least implicit in the work of Marshall (1890) and Pareto, and became quite explicit in a general equilibrium context in the work of Hicks (1939) and Arrow and Debreu (1954).

In the Hicksian model a firm is associated with each economic agent who is a consumer and who may be a worker and owner of resources, but who also may be an entrepreneur. As an entrepreneur he owns a possible production set based on his personal characteristics and perhaps some other non-marketed resources. Of course, most of these individual enterprises will be inactive. A difficulty with this model is that it seems unrealistic to treat the entrepreneur as a profit maximizer unless all the resources which he himself supplies have market prices so that they could equally well be bought by him from the market or sold by him to the market. But if that is the case we are back to the concept of the entrepreneur used by Walras and it seems more realistic to refer to activities, which are impersonal, rather than to individual enterprises.

In the model of Arrow and Debreu, which is the first complete general equilibrium model in which the existence of equilibrium was rigorously proved, the production sector is made up of firms which are described as joint stock companies. Each firm has a production possibility set based on resources which it owns and the ownership of the firm is spread in a prescribed way over a set of consumers. The production sector is in equilibrium when each firm has chosen an input–output combination from its production possibility set which maximizes profit at the market prices. Since the outputs of one firm may be inputs of another and the resort to integrated activities which convert productive services directly into products is not available in a model based on firms, it is convenient to distinguish inputs from outputs by signs rather than be lists. Let Y_j denote the production possibility set of the jth firm, and let $y = (y_1, \ldots, y_n)$ denote an element of this set. There are n goods and services in the economy, and $y_i < 0$ denotes an input, while $y_1 > 0$ denotes an output. Let y^j be the input–output vector of the jth firm. Then equilibrium in the production sector requires that the condition $p \cdot y^j \geqslant p \cdot y$ for all $y \in Y_j$ hold for all j, where j indexes the set of firms, and p is the market price vector.

The Arrow–Debreu approach to the production sector involves a major difficulty. It is not well adapted to handle the formation of new firms and the dissolution of old ones. If firms are based on the assembly of a set of resources jointly owned by the shareholders, it becomes critical to give the principle which underlies such an assembly. If the firm's resources are priced and traded, so the firm's production may be treated like an activity, there is no difficulty since von Neumann's Law may be applied. Otherwise, the rules governing the entry and exit of firms are unclear. The problem is similar to the general problem of coalition formation in the theory of cooperative games.

A FORMAL MODEL. A formal model of the competitive economy, presented in the form of a series of axioms, was developed in the 1950s. It was intended that the axioms should be interpretable to apply to real economic systems, albeit in some approximate sense. However, as a formal mathematical model the implications of the axioms could be developed independently of the applications. The selection of axioms was influenced by the possibility of making useful interpretations, but also by the facility with which results can be derived.

Two closely related sets of assumptions were developed. One, developed primarily by McKenzie (1959), is a formalization of the Walrasian theory and uses a linear model of production. The other, developed primarily by Arrow and Debreu, is a formalization of the Hicksian theory where the production sector is described as an assembly of firms. On the side of consumers and the market there are no significant differences at a fundamental level, although there are sometimes differences of approach. A history of the problem of existence of equilibrium for the formal models may be found in Weintraub (1983).

In the fully developed McKenzie model (see McKenzie, 1981) two assumptions are made for the consumption sector, two for the production sector, and two assumptions relate the consumption and production sectors. On the consumption side there is a finite number m of consumers indexed by h, and each consumer has a set X_h of trades which are feasible for him. There are n goods and the sets X_h are contained in R^n, the n dimensional Euclidean space. The convention is used that quantities supplied by consumers are negative and quantities received by consumers are positive. The consumer has preferences defined on X_h by a correspondence P_h. The preference correspondence P_h takes as its value at $x \in X_h$ the subset of X_h each of whose members is preferred to x. This subset may be empty. The assumptions on the consumers which hold for all h, are

(1) X_h is convex, closed and bounded below.
(2) P_h is open valued relative to X_h and lower semicontinuous. Also x is not in convex hull $P_h(x)$.

Convexity of X_h implies that a good is divisible if someone can consume it in more than one quantity. X_h bounded below means that the consumer is not able to supply an indefinite quantity of any good. Closedness and boundedness are needed to provide compact feasible sets.

On the production side there is an activities model with no limitation on the number of activities. The activities are linear and give rise to a possible production set Y contained in R^n. if $y \in Y$, the negative components of y denote quantities of inputs and the positive components denote quantities of outputs. The assumptions on Y are

(3) Y is a closed convex cone.
(4) $Y \cap R^n_+ = \{0\}$. R^n_+ is the set of non-negative vectors in R^n.

That Y is a convex cone is equivalent to the production set being generated by linear activities. It means that if y and y' are producible, that is, elements of Y, then $\alpha y + \beta y'$ is also producible, that is, an element of Y, for any non-negative numbers α and β. Thus producible goods are divisible. Closedness is needed for the compactness of the feasible set. Assumption (4) is not restrictive. It is a recognition that goods which are never scarce are irrelevant to problems of economizing.

Finally two assumptions relate the consumption sector and the production sector. Let X be the total possible consumption set, that is, $X = \Sigma^m_{h=1} X_h$. The first relation is

(5) relative interior $X \cap$ relative interior $Y \neq \phi$.

Here the relative interior of a set is relative to the smallest linear subspace that contains it. This assumption insures that someone has income at any price vector which is consistent with equilibrium in the production sector, that is, satisfies von Neumann's Law. The second relation is an assumption that the economy is irreducible. Let I_1 and I_2 refer to nonempty subsets of consumers such that $I_1 \cup I_2$ includes all consumers and $I_1 \cap I_2 = \phi$. Let $X^1 = \Sigma X_h$ for $h \in I_1$, and similarly for I_2. Let $\bar{X}_h$ be the convex hull of X_h and the origin of R^n. The irreducibility assumption is

(6) However I_1 and I_2 may be selected, if $x^1 = y - x^2$ with $x^1 \in X^1$, $y \in Y$, and $x^2 \in \bar{X}^2$, then there is also $\tilde{y} \in Y$ and $w \in \bar{X}^2$, such that $\tilde{x}^1 = \tilde{y} - x^2 - w$ and $\tilde{x} \in P(x^h)$ for all $h \in I_1$.

Assumption (6) guarantees that everyone has income if anyone has income. The meaning of having income is that the consumer is able to reduce his spending at the market price vector below the cost of his allocation and remain within his possible consumption set X_h.

Competitive equilibrium is defined by a price vector p, an output vector y, and vectors $x^1, \ldots, x^m$ of consumer trades. There is equilibrium in the production sector if von Neumann's Law holds, that is.

(I) $y \in Y$ and $p \cdot y = 0$, and for any $y' \in Y, p \cdot y' \leqslant 0$.

When y satisfies (I) it is not possible for the owners of inputs to withdraw them from activities where they are being used and employ them in other activities, whether in use or not, so that the receipts from the resulting outputs allow some inputs to earn larger returns while none of them earns less. This is the same condition for equilibrium in production that was given by Walras, or, for that matter, by Adam Smith (1776).

There is equilibrium in the consumer sector if the x^h satisfy

(II) $x^h \in X_h$ and $p \cdot x^h \leqslant 0$, and $p \cdot z > 0$ for any $z \in P_h(x^h)$, $h = 1, \ldots, m$.

When x^h satisfies condition (II), there is no preferred bundle of goods, including goods or services that are supplied by the consumer, which is available to him under his budget constraint. This is essentially the same condition used by Walras, except that he assumed that preferences could be represented by a strictly concave utility function. Thus he is able to refer to maximization of the utility function over the budget set uniquely at x^h.

Finally, there is market equilibrium when

(III) $\Sigma^m_{h=1} x^h = y$.

This is the condition that markets clear which was used by Walras.

If there is free disposal. Wald's Law may be derived directly from equilibrium in the production sector. The possibility of free disposal is recognized by the inclusion of disposal activities in the production cone, that is, an activity y^i for $i = 1, \ldots, n$ which has $y^i_i = -1$ and $y^i_j = 0$ for $j \neq i$. The condition $p \cdot y^i \leqslant 0$ implies that $p_i \geqslant 0$ must hold. Then if disposal occurs the condition $p \cdot y^i = 0$ implies that $p_i = 0$.

On the basis of Assumptions 1 through 6 it is possible to prove that a competitive equilibrium exists. This was first achieved in a model with assumptions for the demand sector put directly on preferences, in the manner of Walras, by Arrow and Debreu. At the same time McKenzie proved existence for a model with assumptions put on the demand functions rather than directly on preferences. Also McKenzie assumed a linear technology rather than a set of firms. This was a generalization of a model of Wald in which joint production was absent and the very special assumption was made that the market demand functions satisfied the weak axiom of revealed preference. The weak axiom says if x is demanded at p and x' at p', then $p \cdot x' \leqslant p \cdot x$ implies that $p' \cdot x' < p' \cdot x$. This is a consistency requirement on choice under budget constraints. Wald's assumption was a deep insight. He anticipated the statement of this principle by Samuelson (e.g. 1947) who applied it to the demand of the individual consumer to derive most of the propositions of demand theory. Wald showed that the weak axiom assumed for the market leads to uniqueness of equilibrium. Subsequently it was shown by Arrow and Hurwicz (1958) that the weak axiom is implied by the assumption that all goods are gross substitutes. They also proved that the weak axiom confined to a comparison of choices between the equilibrium prices and other prices implies the global stability of a process of price adjustment in which the prices of goods are increased if excess demand exists and lowered if excess supply exists. Wald (1936b) wrote another paper on equilibrium in an exchange market which used assumptions closer to those of Arrow and Debreu, but this paper unfortunately was lost.

The only important distinction between the approach of Arrow and Debreu (see Debreu, 1962) and the approach expressed in Assumptions 1 through 6 is the use of a set of firms rather than a set of activities to generate the production set. Mathematically, through the introduction of entrepreneurial factors the approaches can be reconciled. However, the intentions of the two approaches are quite different. The linear model is intended to represent free entry into any line of production by cooperating factors, however organized in a legal sense, where economies of scale are sufficiently small to allow approximate linearity to be achieved by the multiplication of producing units. The lumpiness which is present is compared to that resulting from goods which are in fact indivisible, although they are treated as divisible. This leads to a reasonable approximation to real markets only if units are small compared with the levels of trade. This view of the competitive economy is consistent with the analysis of Marshall as well as Walras. Of course, it has to be recognized that in real economies some sectors cannot be approximated in this way. However, when linearity becomes a bad approximation to the production sector, convexity has in all likelihood become an equally bad approximation to the production sets of firms.

Recently an explicit modelling of the approach of the firms economy to the activities economy has been given by Novshek and Sonnenschein (1980). They use the model of quantity adjusting firms developed in a partial equilibrium context by Cournot to find an equilibrium for the firms economy. Then they let the firm size shrink and show that the Walrasian equilibrium of an activities economy is approached in the limit.

TWO INTERPRETATIONS OF THE FORMAL MODEL. Two basic interpretations of the general equilibrium model were described by Hicks and referred to as the spot economy and the futures economy. The spot economy is a market held on 'Monday' at which all transactions are arranged that involve delivery during the 'week'. This is the economy described by Walras. The equilibrium of the spot economy is called temporary equilibrium in the modern literature. Some effort has been devoted to an analysis of the path followed by such an economy through a succession of temporary equilibria. The role of expectations in the spot economy is critical, as Hicks recognized.

The futures economy on the other hand has a single market in which all future transactions are negotiated at once. Hicks does not treat this economy in detail, but turns to a sequence of spot markets with trading that is guided by expectations. In the futures economy goods available in different periods would be treated as different goods, so that the number of goods would be finite only if the economic horizon is finite. If there is perfect foresight the futures economy is a reasonable alternative and there is not reason why markets should reopen. However, when the future is uncertain and the available futures contracts are for sure delivery, or at least do not exist in sufficient variety to take account of all contingencies, there is no assurance that the contracts entered into will remain desirable or indeed can be executed. For this reason Hicks chose to do a dynamic analysis of a sequence of temporary equilibria in the main body of his work.

In order to avoid the problem of the feasibility of plans and the need to reopen markets, Debreu (1959) following a lead of Arrow (1953) introduced a specification of goods by the event in which they are made available. The set of events would have to discriminate all the circumstances that might make delivery impossible or undesirable, so there would be no motive for traders to reopen markets. Despite this complexity, it is a consistent model which may have relevance to the real world. In order to keep the set of goods finite they assume a finite horizon and a finite set of events, in addition to assuming a finite list of goods in terms of location and physical characteristics.

With this interpretation of the formal model there is no room for borrowing and lending since payments are cleared only once, at the beginning of time. Uncertainty is present since there is no assumption that the event realized at any future time is known. Rather it will be revealed when the time arrives. There is no reason for spot markets to arise since the transactions which have been made for the future event that is revealed are the ones each trader desired at the prices paid in those circumstances. Thus if a spot market were opened no transactions would take place.

Of course it is idealization to suppose that all relevant events could be described in advance, or, if they could, that it would be feasible to establish markets discriminating between them. An alternative is to use a succession of markets in which temporary equilibria are established while some trading in futures contracts takes place. However, the limiting cases of the pure spot economy or the pure futures economy have an analytical tractability that the mixed cases lack and for this reason they remain of great importance.

TEMPORARY EQUILIBRIUM. Once a sequence of markets is contemplated, rather than a single comprehensive market, plans for future trades become relevant and, therefore, expectations of the prices at which they can be made. Also money stocks and loans become useful in making financial preparations for the trading that is planned. Also, if there may be forward trading as well as spot trading, arbitrage is possible, and speculative trading arises which expresses disagreement among consumers about probable price levels on future spot and forward markets.

These complications were handled by Walras without an explicit analysis of demand by consumers for goods in the future using utility functions in which these goods appear. Rather he reduces the demand for future goods to a demand for assets in general which would provide the means for future purchases. On the other hand, he carefully distinguishes between stocks of goods and their services, and the investments of the consumer are treated as if they were made directly in the stocks of goods whose services are sold to the entrepreneurs, or directly to consumers in the case of services of consumer goods.

The spirit of this analysis is to choose a period short enough that it is not too great a distortion of reality to suppose that all trades for this period can be concluded in advance as in the Arrow–Debreu model for the entire horizon, but the forms of industrial organization are abstracted from, so that attention may be concentrated on the productive activities and the ultimate beneficial owners of the resources whose services are used in them. Also to give the future some role in the decisions of the consumers but not a role requiring detailed analysis, Walras assumed that present market prices are expected to persist. In contrast, Hicks and Arrow–Debreu deal explicitly with intertemporal planning by firms and consumers. In a succession of markets this allows Hicks to analyse the effects of changes in expectations on the present market prices and the plans of agents.

The theory of Walras provides the most complete and detailed model of temporary general equilibrium that has ever been given, an impressive performance since it was also the first formal model of general equilibrium. He was able to deal with money, production, lending, and capital accumulation, and in his model an interest rate, price levels, and prices of capital goods and their services are all determined. He showed that the system was not overdetermined, and probably not underdetermined either, in that the number of independent functional relationships and the number of economic quantities to be determined are equal. He was not able to give a proof that an equilibrium in non-negative real variables exists for his model. However, proofs have since been given for simplified versions of it.

A fundamental difference between temporary equilibrium and equilibrium over a horizon is that part of the consumers' demand for goods in the temporary equilibrium is intended for investment rather than for consumption within the period while in the economy of the classical existence theorem consumers' demand is entirely aimed at consumption within the horizon. This raises two problems. One is to distinguish between resources devoted to this period's consumption and resources reserved for the support of consumption in future periods. The other is to explain how the decision to reserve a certain quantity of resources for future use is made. Walras went further to make the distinction between current and future use than any of his successors. They, on the other hand, have done much more analysis of the relation between investment and expectations. The Walrasian assumption on expectations was usually to project the prices arrived at in the current market into the future. This assumption is only appropriate for a stationary, or a steadily progressive, state of the economy. Of course, it has often been remarked that it is only in these conditions that expectations are likely to be correct.

Walras distinguished between consumption goods and services which are consumed in one use and consumption goods which are in effect capital goods providing consumer services, that is, having more than one use. Among the consumption goods which serve as capital goods he included consumption goods which are held in stocks to provide, as Walras put it, services of availability. Thus part of a person's income for a period may be invested in new stocks of consumer goods as well as in capital goods which are intended for use in productive activities. By the same token some of the productive activities which occur may occur in the household rather than in the factory, and these should satisfy the same profit conditions as the productive activities that occur in the firms.

The Walrasian approach to temporary equilibrium is entirely appropriate only to steady states where underlying circumstances, technology, tastes, and resources are constant, perhaps with capital stocks and population expanding at uniform rates. Then the comparative statics that can be done is a comparison of different steady states. On the other hand, in the Hicksian model where expectations of price changes are allowed, it is possible to consider the effect on the temporary equilibrium of changes in price expectations which need not duplicate changes in current prices. However, the approach of Walras allows him to ignore the consumer's portfolio problem and treat the consumer as only making a saving decision, since all assets of equal value are treated as indifferent with equal rates of return after allowing for depreciation and insurance costs. When there is uncertainty, the treatment of all assets as indifferent in this fashion is not justified even by the mean-variance theory of portfolio selection. The variances and covariances of asset returns must be taken into account. Thus Walras's theory of investment requires that expectations be held with certainty, although he only explicitly assumes certainty within the horizon of a single period, after allowing for fully insurable risks.

There are two features of the Walrasian theory of investment which are quite effective, even by modern standards. One is the analysis of the demand for money. Money is needed during the period to make payments which are planned in advance and the cost of this money service is simply the interest on a loan of that amount for the period. This is very close to the treatment of the demand for money for transactions purposes in modern theory. The demand for money as an asset is merged with the general demand for assets, since any net money balance at the end of the period will be expected to be lent at the current interest rate for the next period, either to others or implicitly to oneself. This represents a cash balance approach to monetary theory where cash balances are only wanted for transactions purposes. It leads to a strict quantity theory of the price of money in terms of other goods in comparisons between steady states.

The second effective feature of Walras's theory of investment is the recognition that the cost of investment goods will depend on the level of investment, since in the general equilibrium high levels of investment will raise the prices of the productive services needed to produce investment goods and thus the prices of the investment goods themselves. In this way the Walrasian theory takes account of the distinction between the marginal efficiency of investment and the marginal efficiency of capital familiar in the Keynesian literature, as well as the modern notion of the cost of adjustment resulting from an increase in the level of investment.

The two main deficiencies of the Walrasian theory of temporary equilibrium are its lack of an analysis of the demand for assets in general in terms of the future consumption streams that the assets are expected to support and the expected utility they promise to yield, and its lack of an analysis of the demand for particular assets in terms of the distribution of their expected returns.

The neglect of future plans for consumption in determining current demand was addressed by Hicks. He did not suppose that consumers make detailed plans but that they form vague plans and expectations of future prices, which still allow some comparative statics methods to be applied in estimating the effect on current demand of changes in current or future expected prices.

Since firms are recognized explicitly in Hicks's model, they are also represented as making plans for future inputs and outputs in the light of price expectations, which in his case can be identified with the expectations of individuals who become entrepreneurs. The equilibrium of such a model in one period is a set of prices for all the goods and services traded in the market of that period such that the demand for each good or service, including any contract for future delivery that happens to be traded, equals the supply.

Hicks assumes that each consumer and each firm in its planning applies actual or expected interest rates to discount expected future prices to the present so that the problem of maximizing utility for the consumer, or present value for the firm, does not differ, in principle, from the static problem. However, he must assume that agents are risk neutral or in any case that distributions of prices may be replaced by single prices, or certainty equivalents. Thus he is no more able than Walras to analyse how the value of an asset is influenced by the distribution of its returns. But he is able to consider how changes in current prices influence expected future prices, when expected future prices do not necessarily change by the same amounts. This may be the most significant advance made by Hicks beyond Walras, together with the corollary of planning by firms and consumers for a future that involves expected prices changes.

EXPECTATIONS IN TEMPORARY EQUILIBRIUM. A natural way to generalize the Hicksian model and one which has been followed in recent years, for example, by Grandmont, is to impute to each trader an expectation function which gives a probability distribution over future prices, and perhaps over other relevant variables, both market and environmental, as functions of previous values taken by the same variables. Then assuming that each trader has a criterion by which he can choose an optimal trade plan given his expectations, he will determine an excess demand as a function of current prices. Then equilibrium is achieved if there is market clearing at the current prices. Since in the Walrasian or Hicksian model there are two kinds of traders, consumers and entrepreneurs or firms, criteria must be found for each kind of trader.

The criterion for the consumer is rather easily arrived at. It is assumed that each consumer has a von Neumann–Morgenstern utility function, so that any current trade can be evaluated in terms of the expected utility which it makes possible. The utility in turn is derived from the utility of the various possible consumption streams multiplied by their probabilities of occurrence. Of course, these consumption streams and their probabilities logically underlie the expected utilities but they cannot be known to the consumer in detail. The probability distribution on consumption streams is induced by the probability distribution on prices and environmental variables, together with the current trade of the consumer and his plans for future trades, which are in turn contingent on the prices and environmental variables realized in the future. As Hicks points out the consumer may only try to plan levels of spending and certain large expenditures for the future. Particular price expectations will affect these plans and current spending, in total as well as on specific items. What is needed for the theory is to express consumer's demand finally as a function of current prices so that the condition of market clearing will characterize equilibrium prices. The logic of this analysis is entirely compatible with the methods of Walras, given stationary conditions for tastes, technology, and resources. In simple models it can be spelled out in detail.

On the other hand, there is little agreement on an appropriate criterion for the firm. The difficulty arises that the firm is usually owned by many consumers whose preferences and probability beliefs differ. The consumer does not own capital goods directly but only stock in firms. Moreover, the

firms make investment plans and plan their dividend streams in considerable independence of their owners. Walras abstracts from these difficulties in his formal development by two means. First, he treats the consumer as the owner of capital goods which are rented to the entrepreneur. Second, he values the capital goods on the assumption that prices of productive services, interest rates, depreciation rates, and insurance rates will be constant in the future. Given the prices of the productive services arbitrage in the market for capital goods results in a uniform ratio between the net rental of the capital goods, or the prices of their productive services less depreciation and insurance charges, and the prices of the capital goods. In Walras's notation $P_k = p_k/(i + \mu_k + v_k)$ where k indexes capital goods, P_k is the price of the capital good, p_k is the price of its service, i is the interest rate per period, $\mu_k P_k$ is the depreciation change per period, and $v_k P_k$ is the insurance charge per period. In equilibrium the consumer will be indifferent between capital goods in making investments since they all promise equally attractive returns. This also applies in a similar way to investments in circulating capital or in loans.

Hicks adapts the Walrasian viewpoint to a model in which expectations are point valued but not static by imputing to the entrepreneur, who now owns the capital goods, a plan of inputs, including initial stocks, and outputs, including terminal stocks, whose values are discounted back to the present. Then the entrepreneur chooses a plan with the largest discounted value. In this case the firm achieves maximum value in the eyes of its owner. Radner (1972) adapts the Hicksian viewpoint to a model in which point estimates of future prices are not a sufficient basis for decisions. In a temporary equilibrium model his approach imputes to each firm a Von Neumann–Morgenstern utility function over alternative dividend streams. This would imply an expected utility for alternative investments in the current period in the same way that the utility of alternative consumption plans implies expected utility for current spending by the consumer.

On the other hand, by use of the stock market it is possible to bring consumers into the decision-making of firms. The firm's criterion is then to choose a plan of production and investment which leads to a maximum value for its shares on the stock market. It can be argued that if the firm chooses a plan which fails to maximize its value in the stock market the stock market will not be in equilibrium, since there is a profitable arbitrage opportunity for someone to buy controlling interest in the firm and revise its planning.

Existence theorems for temporary equilibrium have been proved in many special cases, particularly for trading economies where production does not enter and the number of periods is taken to be finite. Typically the method of proof parallels a method of proof for the model with complete markets, that is, appropriate continuity properties for individual, and thus market, excess demand functions are proved for the goods and services, and the futures contracts, if any, which are traded in the current period. The application of a fixed point theorem completes the proof that a price system exists which results in market clearing, that is, puts each excess demand function equal to zero. However, some special problems do arise.

Consider a market at the start of period 1 when there are two periods and a second market will be held at the start of period 2. There is uncertainty about the endowment of period 2 and about the spot prices of the second market. All goods are perishable. Suppose there is trading in contracts for current delivery and in forward contracts for delivery in the second period. Let x_1^h, x_2^h be the vectors of goods and services delivered to the hth consumer in periods 1 and 2 respectively. Denote by w_1^h and w_2^h the vectors of endowments for the hth consumer in periods 1 and 2 respectively. Let $\psi^h(p_1, q_1)$ be the expectation function for the hth consumer, that is, the value of ψ^h is a probability distribution of (w_2^h, p_2), where p_1 and p_2 are the vectors of spot prices in periods 1 and 2, while q_1 is the vector of forward prices in period 1 for sure delivery in period 2. There is a finite set of goods and services in each period and a finite number of consumers each of whom holds positive initial stocks in the first period. The possible consumption sets are $X_1^h = R_+^{n_1}$ and $X_2^h = R_+^{n_2}$, the positive orthants of the respective commodity spaces.

The following assumptions are made for the consumer.

(1) There is a concave and monotone utility function u^h of von Neumann–Morgenstern type, that is, preferences over trades in the first period may be determined by taking the sum of the utilities of the resulting consumption vectors weighted by these probabilities of occurrence.

(2) The expectation function $\psi^h(p_1, q_1)$ is continuous in an appropriate sense.

(3) For every (p_1, q_1), $\psi^h(p_1, q_1)$ gives probability 1 to the set of (w_2^h, p_2) for which p_2 is positive.

(4) The support of ψ^h is independent of (p_1, q_1). The convex hull of the projection of the support of ψ^h on the second period price space has a non-empty interior Π^h.

With these assumptions a necessary and sufficient condition for the existence of competitive equilibrium is that the intersection Π of the Π^h not be empty. In other words there must be an open set of spot prices in the second period which all traders believe to have a positive probability of occurrence. Then, if the forward prices q_1 lie in Π and p_1 and q_1 are positive, excess demand is well defined. Let D be the set of (p_1, q_1) satisfying these conditions. As (p_1, q_1) converges to the boundary of D, excess demand diverges to ∞. This happens because preferences are monotone and for q_1 outside Π unlimited arbitrage becomes profitable to some trader. These results were reached by Green (1973). It should be noted that point expectations are not consistent with the assumption that Π is not empty, unless all traders expect the same prices next period. However, Π might not be needed to bound short sales if other considerations limit the commitments that will be accepted in view of the likelihood that they can be fulfilled.

MONEY IN TEMPORARY EQUILIBRIUM. There is little difficulty in introducing money into the temporary equilibrium model. It must be recognized that money serves in at least two capacities, to facilitate exchange, and as an asset with its own prospects for losing or gaining value relative to other goods. In addition it may serve as a numeraire, in terms of which prices are stated. In its capacity as an asset in a market with uncertainty, money may contribute to a diversified portfolio. On the other hand, in its capacity to facilitate exchange money balances will affect the cost of making transactions and thus the stream of consumption which is realizable from given resources. Given his context, where risks are assumed to be insurable, Walras is particularly clear in his treatment of money. If some good other than money serves as numeraire, the price of the service of availability of money is written by Walras as p_m, and the price of money itself as P_m. Then as for any asset the ratio of the net rental to the asset price is equal to the interest rate or $p_m/P_m = i$. Thus if money serves as the numeraire, $P_m = 1$ and $p_m = i$. Although his analysis seems somewhat artificial because uninsurable risks are absent, Walras indicates clearly how cash balances may contribute to productive efficiency and to consumer utility.

If attention is concentrated on the asset role of money, so

that the transaction role is neglected, it may be shown that the assumption of static expectations may lead to the absence of equilibrium for the current period. Static expectations imply that the relative prices of present and future goods cannot be changed. Therefore, price changes leading to intertemporal substitution are prevented. Only the wealth effects of price changes have free play since price level decreases raise the value of the money stock and conversely for increases. However, as Grandmont (1983) has demonstrated, these real balance effects may be insufficient to equate supply and demand. For example, if there is excess demand for current goods, this excess demand may not be eliminated by increases in the current price level which are accompanied by equally large increases in the future price level. In a trading economy the effect of the price increases is to reduce the wealth of the traders toward the endowment point $(w(1), w(2))$ in a two period model. Suppose there is only one good, which is perishable, and money is the only store of value. Then if the marginal utility of the current endowment exceeds the marginal utility of the second period endowment for all traders, the price of the good cannot rise high enough to reduce current demand to the current endowment. The same dilemma may arise when the Hicksian elasticity of expectations is equal to one, even though expected prices do not equal current prices.

Grandmont considers a model of this type where trading in futures contracts is excluded so that point expectations do not cause difficulties. It is a trading economy in which consumers receive an endowment of perishable goods in each period of their lives and an initial money stock in the first period. In the current period they maximize a utility function of consumption over the remaining periods of life (assuming the life span to be known) subject to budget constraints of the form $p_t x_t + m_t = p_t w_t + m_{t-1}$, where future prices p_t are equal to functions ψ_t of present prices p_1. He assumes

(1) The utility function $u^h(x_1, \ldots, x_{n(h)},)$ is continuous, increasing, and strictly quasi-concave for every h.

(2) The endowments w_t^h are positive for all h and t, $1 \leqslant t \leqslant n(h)$.

(3) Total money stock $M = \Sigma_h m_h$ is positive.

He then proves that the temporary monetary equilibrium exists, that is, money prices are well defined, if every agent's price expectations ψ_t^h are continuous and, for at least one agent, who will be living in the next period and who has a positive money stock, price expectations are bounded away from 0 and ∞. In Grandmont's opinion this result leaves the existence of temporary equilibrium 'somewhat problematic'.

However, it seems quite inappropriate to deal with a money which has no role to play in facilitating transactions. Grandmont and Younes (1972) have studied general equilibrium in a model similar to the model just described except that lifetimes are taken to be infinite and utility functions are separable by time period, that is $u^h(x_1, \ldots) = \Sigma_{t=1}^\infty \delta^t u^h(t)$ for $0 < \delta < 1$. Also money is now assigned a role in transactions, that is, only part of the proceeds of sales in the current period can be used to finance purchases in this period. Thus in each period there is both a budgetary constraint as before and, in addition, a liquidity constraint, which may be written in simplest form as $p_t(x_t - w_t)^+ \leqslant m_t + k p_t(x_t - w_t)^-$, where for any vector z we write $z_i^+ = \max(z_i, 0)$ and $z_i^- = \max(-z_i, 0)$, and $0 < k < 1$. Thus the fraction k of receipts from sales can be used to buy goods in the current period. This fraction could be allowed to vary by consumer and by good. The constraint on purchases is entirely in the spirit of Walras. It is an explicit modelling of a need for liquidity that he left implicit in his account.

In order to prove that a monetary equilibrium exists an assumption to bound expected prices is made which is very similar to the previous assumption for this purpose, and also very similar to the assumption made by Green to obtain existence of temporary equilibrium in a non-monetary economy with futures trading. The assumption is that the set of expected prices, over a finite planning horizon, that result from all possible choices of current prices, which are assumed positive, lie in a compact subset of the set of positive future prices. Then if all consumers have continuous expectations which satisfy this assumption, and the assumptions of the previous model are also met, there will exist a temporary equilibrium in this case also. Indeed, the case $k = 1$, where the liquidity motive is lacking, can be allowed.

In the second model where money has a transactions role expectations are described as depending on past prices as well as current prices, which leads inelastic price expectations to be more plausible. It also gives plausibility to correct foresight in states of stationary equilibrium over sequences of periods. Grandmont and Younes (1973) prove that the stationary equilibria of the model are not Pareto optimal. However, they can be made Pareto optimal by use of a lump sum tax to reduce the quantity of money by a factor equal to the discount factor for utility. It is then proved that a continuum of such equilibria exist to sustain any Pareto optimal allocation, since the price level falls by the same factor, and it is not worthwhile to reduce a money stock, even if it is in excess of transaction requirements. Moreover, if the tax rate is set slightly too high, the consumer will always wish to increase his real balances and no stationary equilibrium will exist. Grandmont and Younes are not able to prove that an exact stationary equilibrium exists for a fixed money stock, although a near equilibrium exists of the discount factor is near 1.

In addition to proofs of existence and non-optimality for monetary equilibria, Grandmont and Younes show that the quantity theory holds between stationary equilibria, that is, if p and m_h, $h = 1, \ldots, m$, provide a stationary equilibrium, then λp and λm_h also provide one. This is the conclusion of Walras as well. On the other hand, the stationary equilibria of a monetary economy will differ from the stationary equilibria of a barter economy unless $\delta = 1$. This is apparent from the fact that the barter economy's equilibria are Pareto optimal and the monetary economy's equilibria are not, unless $\delta = 1$. Thus the simple 'classical dichotomy' does not hold.

EQUILIBRIUM OVER TIME. In addition to temporary equilibrium Hicks considered the possibility of equilibrium over time, in the sense that the expectations held by traders in one market about prices on future markets are realized when those future markets are held. However, when there is uncertainty it is not clear what is meant by the realization of expectations. If expectations take the form of a non-atomic probability measure over future prices, any vector of prices within the support of the measure is as likely as any other, that is, it has zero probability. Nor does the Hicksian trick of replacing the probability distribution by a representative price, depending on the trader, avoid the difficulty, since the representative price is not typically a statistic of the price distribution, such as the mean or the mode. Thus even if all traders held the same expectations in the sense of a probability distribution for prices, they would not have the same representative prices except by the chance that their circumstances and their risk preferences also coincide.

A way to resolve this dilemma was provided by Radner (1972). His solution is a type of perfect foresight. All traders hold the same point expectations for prices with certainty, contingent on the event in which the market is held. Only a finite number of dates are allowed and only a finite number of events may occur in each. From the viewpoint of a given market the relevant elementary events are the possible sequences of states of nature that may occur up to the horizon. For any such sequence the traders expect correctly a corresponding sequence of prices. This does not lead to a grand initial market in which all future exchanges are arranged because the set of forward commitments which are actually available in the market is a small subset of all those associated with future events. For example, it may be that most commodities are traded for sure delivery and only one commodity (money or the numeraire) is traded on a contingent basis (insurance). It should be noted that this construction does not depend on any agreement between traders on the probabilities of the alternative events. Thus the expectation functions which were introduced in the discussion of temporary equilibrium would not be likely to be the same for different traders.

In this setting the trader plans a sequence of consumptions contingent on the events in which they occur and also a sequence of trades on the markets which are open. Spot markets are open for all commodities at all dates but only a small subset of the possible markets in forward contracts may be open at any particular date. In any case since the number of dates and states of nature and thus of elementary events is finite, only finitely many prices will arise.

Let X_h be the consumption set of the hth consumer. Let M be the set of elementary date-events pairs. A consumption-trade plan for the hth consumer is a pair (x^h, z^h) where x_m^h is the consumption planned for $m \in M$ and z_m^h is the trade planned for $m \in M$. Let $\Gamma_h(p)$ be the set of feasible plans for h, given prices p. In particular, (x^h, z^h) in $\Gamma_h(p)$ implies that consumption x_m^h plus net deliveries $\tilde{z}_m^h$ due at m are not greater than resource endowments w_m^h for each m and the budget constraint $p_m z_m^h$ holds at each $m \in M$.

Let $\gamma^h(p)$ be the set of plans in $\Gamma_h(p)$ which arae optimal for h. An equilibrium of plans and price expectations (including current prices which are known) is given by plans (x^h, z^h) and expected prices p such that (x^h, z^h) is in $\gamma^h(p)$ for each h, that is, the plans are preferred at the expected prices, and the sum $\Sigma_h \Sigma_m^h$ of commitments at each m is non-negative, and the value of commitments $p_m \Sigma_h z_m^h = 0$ at each m, that is, Walras's Law holds. In such a purely trading economy for perishable goods with a finite set of dates and events and under assumptions of the usual kind on preferences, and positive endowments which lie in the interior of consumption sets, Radner proves that an equilibrium exists.

It is not difficult to bring production into this setting if firms are introduced with fixed production plans and with shares which are traded on a stock exchange. The ownership of a share of a firm can be equated to the ownership of a share of its output, including the end of the period capital stock. The output of a firm at any date would depend on the event, and the function relating this output to the events would be known by traders, just as future prices of goods are known, contingent on events. Now, in addition to goods prices, share prices are foreseen in each event at each date with certainty. As before the number of dates and events is finite.

A feature of this model not present in the trading model is that consumers do not own the resources of the firms as individual goods but as proportions of the batch of goods that firms hold. The consumer can buy and sell goods forward by means of long and short positions in the stock market but the trade he arranges by these means for one event at the next date determines his trade for all other events at that date. Thus spot markets still may offer useful alternatives, quite aside from the practical difficulties of physically dissolving the firm. Of course, given the presence of spot markets, dissolution of the firms is not needed if the value of the firm equals or exceeds the value of its resources.

If one tries to go further to specify how the production and trading plans are arrived at, a major problem arises of setting the objectives of the firm. Hicks solves this problem by assuming that the production plan chosen would have the maximum discounted value among those available. This value could be calculated since expectations were single-valued and interest rate, actual or expected, could be used in arriving at present values. Moreover, firms were treated like single proprietorships. In the modern literature firms have sometimes been assigned utility functions defined on the streams of profits. Another suggestion is to suppose that the firm adopts the plan that maximizes the value of its shares on the stock market. This would seem to be the approach most in accord with other parts of general equilibrium theory. However, it encounters the difficulty that the judgement of the management and the judgement of the market on the probability of different events may not coincide. If this difference of judgement exists, the market solution would be for the firm to be purchased through a takeover by those who value its potential most highly and the management displaced. Markets which work in this way would correspond quite well to the original Walrasian model.

Various results on the existence of a general equilibrium have been reached with special models of production by firms. One theorem of Radner extends the existence of an equilibrium of plans and price expectations to this context. His assumptions are:

(1) Consumers satisfy the usual conditions on convexity, non-satiation, and positive endowments.

(2) Consumers own the shares of firms and each consumer owns shares in every firm.

(3) Producers have closed, convex production sets with free disposal. The total production set satisfies the condition that the negative of a producible vector of commodities is not producible.

(4) Each firm has a continuous, strictly concave utility function on profit streams.

With these assumptions he does not achieve a full existence theorem because the model is not well adapted to handle the entry and exit of firms. What may happen is that some firms show an excess supply of shares in some events and dates. Then since the firms are treated like partnerships with unlimited liability, negative share prices might be justified at this point. In any case the questions of entry and exit of firms is one that the Arrow–Debreu model also fails to deal with. The theorem proved by Radner only finds a 'pseudo-equilibrium' where the value of total excess supply (of shares) is minimized.

In the foregoing discussion it has been assumed that only a subset, possibly small, of the potential Arrow–Debreu markets is open. It is possible to justify the selection of markets which are open by postulating costs for carrying out transactions. If the markets which are open are given, the previous equilibria may be supported by assigning infinite transactions costs to the lost markets and zero costs to the open ones. Otherwise the open markets will be endogenous to the general equilibrium. In the analysis of markets with transaction activities which consume resources the same convexity or linearity assumptions have been used as for the production technology. Then it is not difficult to prove existence of equilibrium under assumptions of the usual sort.

RATIONAL EXPECTATIONS. It has been implicitly assumed in the preceding discussion of temporary equilibrium that the traders have the same information available. If this is not the case the complication arises that the equilibrium price may convey information. For example, in the market for umbrellas if some traders have the benefit of weather forecasts and some do not, a high price based on the demand of informed traders will signal to uninformed traders that rain is expected. Then all traders are informed and an equilibrium price must be consistent with fully informed demand.

A difficulty arises if it happens that the utilities of consumers depend on events in contrary ways, that is, uninformed consumers use umbrellas to ward off sun and informed traders to ward off rain. Then price will be higher if rainy weather is expected by informed traders but if uniformed traders perceive this and become informed, the high price may not appear and a fully informed market may not show a price difference depending on the weather forecast. But then no information is transmitted so the weather forecast cannot be read out of equilibrium prices. The conclusion is that no equilibrium is possible. However, the result requires an exact balance in the effects of rain and sun on the two sets of traders, so it is unlikely to hold. More robust examples of nonexistence were given by Green (1977) and Kreps (1977). The idea of the discontinuity was first proposed by Radner (1967).

A rational expectations equilibrium is said to exist if there is a function ϕ mapping states of the world into equilibrium prices which is invertible, that is, ϕ^{-1} exists, mapping prices, from a normalized set, into states of the world. It is clear that such a function will exist if the equilibrium price which appears when all traders are fully informed is uniquely determined by the elementary event, and the relation is one to one. It is also clear, given a finite set of elementary events, that the correspondence of prices to elementary events will be one to one in all but exceptional cases. Then the equilibrium is said to be revealing. But the price function of a revealing full information equilibrium is a price function that provides a rational expectations equilibrium. This observation is due to Grossman (1981).

The situation is more complicated when the possibility is recognized that spending resources will allow more information to be gathered. The information that is disseminated free of charge by prices will discourage the use of resources to gather information and thus prevent the attainment of a Pareto optimum. In welfare terms a suboptimal amount of resources will be devoted to information activities.

AN INFINITE HORIZON. In the Arrow–Debreu model of general equilibrium there are a finite number of periods, a finite number of locations, a finite number of events, and a finite number of commodity types, so the number of distinct goods when all these grounds for distinguishing goods have been recognized is still finite. The principal objection to the restriction to a finite number of goods is that it requires a finite horizon and there is no natural way to choose the final period. Moreover, since there will be terminal stocks in the final period there is no natural way to value them without contemplating future periods in which they will be used. The finiteness of the number of locations and commodity types is achieved by making a discrete approximation to a continuum, and perhaps the finiteness of the number of states of nature can also be viewed in this light. But in the case of time, a discrete approximation by periods still leaves a denumerable infinity of dates.

There are two principal models in which an infinite number of goods appear. In one model there is a finite number of infinitely lived consumers. Such a consumer may be considered to represent a series of descendants stretching into the indefinite future, so that consumers alive in the present period have an interest in the goods of all periods. The other model has an infinite number of consumers, but only a finite number of them are alive in any period. This model is called the overlapping generations model. It was first proposed and explicitly analysed by Samuelson (1958).

A model of general competitive equilibrium with a finite number of consumers and an infinite number of commodities was first presented in rigorous form by Peleg and Yaari (1970). They assumed the number of commodities to be denumerable. This is a basic case since a noncompact but separable commodity space can be approximated arbitrarily closely with a denumerable set of commodities in the same sense that a compact commodity space can be approximated by a finite set of commodities. This assumes that a sensible neighbourhood system can be defined in the commodity space, as Debreu does for the dimensions of location and time with places and periods.

Peleg and Yaari present a trading model without production. The commodity space s is the space of all real sequences. In order to discuss continuity the space must be given a topology, in this case, the product topology. Thus a sequence of points converges if it converges in every coordinate, that is, $x^s \to x$, $s = 1, 2, \ldots$, if $x^s(i) \to x(i)$ for $i = 0, 1, \ldots$. The space is presented as a sequence of real numbers but by grouping terms it may equally well represent a sequence of vectors, for example, commodity bundles occurring in successive time periods. The hth trader has an initial stock w_h, where $w_h \in s$ and a preference relation $\succsim_h$, which is reflexive, transitive, and complete on s_+, the set of non-negative sequences. Strict preferences $\succ_h$ is defined by $x \succ_h y$ if $x \succsim_h y$ and not $y \succsim_h x$.

Peleg and Yaari prove an existence theorem for this economy on the following assumptions.

(1) Desirability. If $x \geq y$, then $x \succsim_h y$.

(2) Strong convexity. If $x \neq y$ and $x \succsim_h y$, then $\alpha x + (1 - \alpha)y \succ_h y$ for $0 < \alpha < 1$.

(3) Continuity. The two sets $\{y \mid y \succsim_h x\}$ and $\{y \mid x \succsim_h y\}$ are closed.

(4) Positivity of total supply. Let $w = \sum_{h=1}^m w_h$. Then $w > 0$.

A price system is a real sequence $\pi > 0$ which satisfies $\sum_{i=0}^\infty \pi(i)w_h(i) < \infty$, the value of the initial bundles is finite. This implies that $\pi(i)w_h(i)$ converges to 0 as $i \to \infty$. A competitive equilibrium is given by $(x_1, \ldots, x_m; \pi)$ such that π is a price system, $\sum_{i=0}^\infty \pi(i)x_h(i) \leq \sum_{i=0}^\infty \pi(i)w_h(i)$, for each h, and $\sum_{i=0}^\infty \pi(i)x(i) \leq \sum_{i=0}^\infty \pi(i)w_h(i)$ implies $x_h \succsim_h x$. Peleg and Yaari prove that a competitive equilibrium exists.

It is clear from their discussions, and it has become even clearer in subsequent work, that the use of a topology such that, in the context of an infinite horizon interpretation of the model, impatience is implied by continuity of preferences is the crucial assumption for a proof of existence. That the product topology implies impatience may be seen in the following way. If $x \succ_h y$ then by continuity there is a neighbourhood U of x such that $z \in U$ implies $z \succ_h y$. However, a neighbourhood U is defined by $|z(i) - x(i)| < \epsilon > 0$ for a finite number of coordinates where the remaining coordinates are free. Thus given $y \succ_h x$, there must exist $N > 0$ such that $z(i) = x(i)$ for $i \leq N$ and $z(i) = 0$ for $i > N$, and $z \succ_h y$. These conditions are met if the preference order is representable by a separable utility function which is the sum of periodwise utilities discounted back to the present at a constant rate per period, and these utilities are continuous and uniformly bounded. Such a utility function is a common way of expressing impatience.

A model of general competitive equilibrium which allows for

production where there is an infinite number of commodities was first presented in a rigorous form by Bewley (1972). A preference relation $\succsim_h$ is assumed for each consumer as in Peleg and Yaari. We will describe Bewley's model for the case of a sequence of periods with an infinite horizon where N_1 is the finite set of commodities available in the tth period. Then the set of all commodities is

$$M = \bigcup_{t=1}^\infty N_t.$$

It is assumed that $M = M_c \cup M_p$ where M_c and M_p are disjoint and M_c contains the consumption goods. Bewley confines attention to the commodity space l_∞ of bounded sequences of real numbers. Let $K_c = \{x \in l_\infty \mid x(i) = 0 \text{ for not } i \in M_c, x(i) \geq 0 \text{ for } i \in M_c\}$, and similarly for K_p. Let $\tilde{K}_c$ have the same definition as K_c except that $x(i) > \epsilon > 0$ for all $i \in M_c$ for some given ϵ. Bewley's existence theorem holds for a weaker notion of continuity than that of componentwise convergence, but we will stay with the definition used by Peleg and Yaari for the sake of simplicity. Then the assumptions on the consumer sector are

(1) The consumption sets $X_h = K_c - w_h$ where w_h is the endowment of the hth consumer.

(2) The sets $\{y \mid y \succsim_h x\}$ and $\{y \mid x \succsim_h y\}$ are closed. Also $\{y \mid y \succsim_h x\}$ is convex.

(3) M_c is not empty and for each h, if $x \in X_h$ and $y \in \tilde{K}_c$, then $x + y \succ_h x$.

The production sector is defined by means of production sets Y_t which convert inputs belonging to N_{t-1} into outputs belong to N_t. Then $Y = \sum_{t=1}^\infty Y_t$. The assumptions on the production sector are

(4) Y is a convex closed cone with vertex at 0.

(5) If $w \in l_\infty$, then $Y + w \cap l_\infty$ is bounded.

(6) If $y \in Y$, then $y^n \in Y$ where $y_t^n = y_t$ for $t = 0, \ldots, n$, and $y_t^n = 0$ for $t > n$.

(7) $-K_p \subset Y$.

Assumption (4) means that each Y_t is a linear activities model as Walras assumed. Assumption (5) excludes unbounded production from given inputs. Assumption (6) allows production to end at any time with free disposal of the final outputs. Assumption (7) allows free disposal of all goods other than consumption goods.

In addition there is one assumption which relates the consumption sector and the production sector.

(8) For each consumer h, there exists $\bar{x}_h \in X_h$ and $\bar{y}_h \in Y$ such that $\bar{y}_h(i) - \bar{x}_h(i) > \epsilon > 0$ for all i and some $\epsilon > 0$.

Assumption (8) protects consumer income in the sense that the consumer is not reduced to the subsistence level in equilibrium. That is to say, there are cheaper consumption bundles within this consumption set at equilibrium prices. An equilibrium is an allocation $(x_1, \ldots, x_m, y)$ and a price sequence $\pi = (\pi(0), \pi(1), \ldots)$ where $\pi(i)$ is non-negative for all i but different from zero for some i, which satisfy the conditions:

(I) $y \in Y$ and $\pi y = 0$, $\pi z \leq 0$ for all $z \in Y$. The profit condition.

(II) $x_h \in X_h$ and $\pi x_h = 0$, all h, and $z \succ_h x_h$ implies that $\pi z > \pi x_h$. The demand condition.

(III) $\sum_{h=1}^m x_h = y$. The balance condition.

On the basis of the assumptions Bewley is able to prove that an equilibrium exists where the price system $\pi \in l_1$, that is, $\sum_{i=0}^\infty \pi(i) < \infty$. This represents a generalization of the classical existence theorem in the form given by McKenzie to the case of denumerably many commodities, retaining the assumption of a finite number of consumers. The argument is stated in terms of an infinite horizon and a finite number of goods in each period, but the original theorem is more general and applies to the case of uncertainty with an infinite number of events as well as to models with a continuum of commodities. The continuum of commodities may arise from a variation in the physical properties of the goods and services.

OVERLAPPING GENERATIONS. In the overlapping generations model of general equilibrium the number of consumers as well as the number of commodities is infinite. However, at any given time the number of both is finite. While the model with a finite number of infinitely lived consumers treats the consumers who are living as if their lives were extended into the indefinite future by the lives of their descendants, in the classical overlapping generations model bequests are neglected and each generation is assumed to be interested only in its own consumption.

The first rigorous analyses of an overlapping generations model in a general equilibrium setting were done by Balasko, Cass and Shell (1980) and by Wilson (1981). They treat an exchange model in which all goods perish in each period, and each consumer receives an endowment in each period. They assume that each consumer lives for two periods. However, this assumption is not essential. What is essential is that lifetimes are finite in length and some of the people alive at any date have lifetimes which overlap the lifetimes of some people who are born later than they.

The formal model makes these assumptions.

(1) In each period $t(t = 1, 2, \ldots)$ there is an arbitrary, finite number of perishable commodities $n' \geqslant 1$.

(2) Each consumer $h = 1, 2, \ldots$ lives for two periods. At the start of period t an arbitrary but finite number of consumers is born with indices $h \in G^t$.

(3) Consumption sets $X_h = R_+^{n(0)}$ for $h \in G^0$ the consumers alive when the economy begins and $X_h = R_+^{n(t)} \times R_+^{n(t+1)}$ for $h \in G^t, t \geqslant 1$. Write $x_h = x_h'$ for $h \in G^0$ and $x_h = (x_h(t), x_h(t+1))$ for $h \in G^t$.

(4) Each consumer has a utility function, $u_h(x(1))$ for $h \in G^0$ and $u_h(x(t), x(t+1))$ for $h \in G^t$. Utility functions u_h are continuous, quasi-concave, and without local maxima.

(5) Each consumer receives an endowment, $w_h = w_h(1)$ for $h \in G^0$ and $w_h = (w_h(t), w_h(t+1))$ for $h \in G'$. For each h, $w_h \geqslant 0$ and $w_h \neq 0$.

(6) The economy is intertemporally irreducible. Let $I(t) = \{h \mid h \in G^s \text{ for } 0 \leqslant s \leqslant t\}$. Then there exists a sequence $t_\mu \to \infty$ with the following property. Given any allocation $x = (x_1, x_2, \ldots)$ and $I_1(t_\mu)$ and $I_2(t_\mu) \neq \phi$, with $I_1(t_\mu) \cap I_2(t_\mu) = \phi$, and $I_1(t_\mu) \cup I_2(t_\mu) = I(t_\mu)$, there exist $y_h \geqslant 0$ for $h \in I_1(t_\mu)$ and $x_h' \geqslant 0$ for $h \in I_2(t_\mu)$ such that $\Sigma_{h \in I_1(t_\mu)} y_i(t) = 0$ when $\Sigma_{h \in I_1(t_\mu)} w_{hi}(t) = 0$, for $1 \leqslant i \leqslant n^t, 1 \leqslant t \leqslant t_\mu + 1$, and

$$\sum_{h \in I_2(t_\mu)} x_h' \leqslant \sum_{h \in I_1(t_\mu)} (w_h + y_h) + \sum_{h \in I_2(t_\mu)} w_h.$$

Moreover, $u_h(x_h') \geqslant u_h(x_h)$ for all $h \in I_2(t_\mu)$ with the strict inequality for some h.

Assumption (6) is the irreducibility assumption of McKenzie adapted to economies made up of the consumers born by the period t_μ. It says that it is always possible to increase the welfare of the second subgroup if the scale of the endowment of the first subgroup is increased.

Let $p = (p(1), p(2), \ldots)$ where $p(t) \in R^{n(t)}$. Then the pair (x, p) is a competitive equilibrium if

(I) For all $h, u_h(x_h)$ is maximal over all z_h such that $p(t)z_h(t) + p(t+1)z_h(t+1) \leqslant p(t)w_h(t) + p(t+1)w_h(t+1)$ if $h \in G^t, t \geqslant 1$, and $p(1)z_h(1) \leqslant p(1)w_h(1)$ if $h \in G^0$, where $z_h \geqslant 0$.

(II) $\Sigma_h x_{hi}(t) \leqslant \Sigma_h w_{hi}(t)$ with equality if $p_i(t) > 0$, where the summation is over $h \in G^{t-1} \cup G^t, 1 \leqslant i \leqslant M(t)$ and $t \geqslant 1$.

Condition (I) is the usual demand condition and condition (II) is the balance condition. Balasko, Cass and Shell (1980) prove that the six assumptions listed imply the existence of a competitive equilibrium. They show that the artificial assumptions on birthdates and lifetimes are irrelevant by a redefinition of the period. They also conjecture that the introduction of production and consumption sets of the usual classical type, which are closed, convex, and bounded below, would cause no major difficulties.

Wilson (1981) treats an economy which may contain both finite lived and infinite lived consumers and which may be specialized to either. He also allows intransitive preferences. He uses a somewhat simpler version of irreducibility and proves existence in an exchange economy where the number of goods in each period is finite in two circumstances (1) when the consumers are all finite lived and (2) when a finite subset of infinite lived consumers own a positive fraction of the endowment in all but a finite number of periods. If preferences are transitive and strictly convex, the competitive equilibrium is also Pareto optimal. Thus Wilson's results contain the theorems on existence of Bewley and Balasko, Shell, and Cass as special cases while also providing conditions in the model sufficient for Pareto optimality.

A striking difference between the competitive equilibria of economies where the number of consumers is finite, and the competitive equilibria of economies with overlapping generations and an infinite horizon, where the number of consumers is infinite, is that with perfect foresight the former equilibria are also Pareto optima while the latter need not be. This is the major point emphasized by Samuelson in his initial paper. The most general theorem proving that competitive equilibria are Pareto optima even when the number of commodities is infinite provided that the number of consumers is finite is due to Debreu (1954). Under some additional smoothness conditions on utility and boundedness conditions on prices and allocations Balasko and Shell (1980) prove that the allocation x of a competitive equilibrium is Pareto optimal if and only if $\Sigma_t(1/\|p_t\|) = \infty$. This is a condition which had already been shown to characterize efficiency in neo-classical production economies by Cass. It is clear that $\lim \inf(\|p_{t+1}\| / \|p_t\|) = r \leqslant 1$ implies that the condition for Pareto optimality is satisfied since the sums dominate $\Sigma_t(1/r^t)$ which diverges. Intuitively, for a stationary economy if the interest rates are asymptotically non-negative, the competitive equilibria will be Pareto optimal, or if the economy is growing, if the interest rates exceed the growth rate, Pareto optimality follows.

LIMITATIONS OF THE ANALYSIS. As mentioned in the beginning the claim of the theories described as general equilibrium theories to be 'general' is qualified by the set of conditions considered to be constant. Walras as well as most subsequent theorists classified the constant factors as tastes, technology, and resources, including population. However, all three of these categories have been treated by some economists as responding, in ways amenable to analysis, to market variables. These studies have usually been confined to a few variables and have usually been partial equilibrium in character, although the classical school of economists included population as a major variable in models of economic development. Their models are comprehensive but lack the market equilibrium analysis of the general equilibrium theories, whose inspiration appears to have been found in the marginal utility theory of consumer demand. Similarly, tastes have sometimes been modelled to depend on past consumption or advertising, and technology has been modelled to depend on research and development spending and on the rewards to innovation. Also natural resources, in terms of resources known to exist, are often treated as responding to prices.

From this perspective general equilibrium theory is a partial theory of economic affairs with a special set of *ceteris paribus* assumptions. The variables which are left free are chosen because they lend themselves to a particularly elegant theory in terms of consumer demand under budget constraints and producer supplies with profit conditions where these constraints and conditions are established by prices equating demand and supply. This was the vision of Walras, perhaps guided by the theory of static equilibrium of mechanical forces which he found in Poinsot.

Another direction of abstraction in general equilibrium theory in its classic expressions has been to ignore the effects of processes which do not pass through the market. In particular each consuming unit is described as interested only in its own consumption in the theory of Pareto optimality and as uninfluenced in its choices by the choices made by other households. Similarly, the production possibilities of one firm or process are treated as independent of the productive activities of other firms. Some attempts have been made to incorporate these effects in the general equilibrium models but not with complete success. In particular there is not a good theory of existence when consumer possibility sets or production sets are affected by levels of consumption and production.

The convexity assumptions which have appeared in general equilibrium models from the time of Walras are often not good approximations of reality though they are depended on for many of theorems of the subject, such as the theorems on existence and Pareto optimality. However, there is a theory of approximate equilibria and of limiting results as the size of the market increases relative to the participants which does something to bridge the gap between theory and fact.

Finally, the assumption that the market participants take prices as independent of their actions fails to describe many markets, and describes very few exactly. Nonetheless, this assumption may be useful for a theory that embraces all markets, whose special features cannot be described in detail. It may, that is, give a good approximation to the working of the economy as a whole. Also it is useful for its implications for optimality, a point which was perceived, albeit through a glass darkly, by Walras. The proper notion was later found by Pareto.

Just as the model does not accommodate monopoly easily, government does not fit in well. A chief difficulty arises from its compulsory features which allow it to extract resources by force rather than by voluntary agreement. Government is not easily described either as a producer selling services, or as a voluntary organization performing acts of collective consumption, though in ways it resembles both. Voluntary societies also do not fit perfectly in the scheme of producers and households though the disparity is less, since they must meet their expenses from contributions by the membership who will not contribute unless the services of the society to them are worth the dues they pay.

PROPERTIES OF GENERAL EQUILIBRIUM. Walras set the major objectives of general equilibrium theory as they have remained ever since. First, it was necessary to prove in any model of general equilibrium that the equilibrium exists. Then its optimality properties should be demonstrated. Next it should be shown how the equilibrium would be attained, that is, the stability of the equilibrium and its uniqueness should be studied. Finally, it should be shown how the equilibrium will

change when conditions of demand, technology, or resources are varied, the subject now called comparative statics. He contributed to all these lines of research.

Walras's arguments for existence are not conclusive but he did contribute a basic principle, that the model should be neither underdetermined nor overdetermined. That is, the number of independent equations to be satisfied and the number of variables to be determined should be equal. Some critics saw right away that this equality did not ensure a meaningful solution to the equation system, for example, that the solution to such an equation system is not guaranteed to be real. The question was not taken up seriously until the 1930s and the first rigorous treatment was given by Wald (1935, 1936b). Then in the 1950s more complete solutions on neo-classical assumptions were found by Arrow and Debreu (1954), McKenzie (1954) and Nikaido (1956).

In the discussion of models of general equilibrium that have been given above, the first requirement has been a set of assumptions from which existence could be inferred. This approach to the subject was begun in the papers of Wald and von Neumann, presented to the colloquium of Karl Menger (mathematician and son of Carl Menger, the neoclassical economist) in Vienna in the 1930s.

The optimality that Walras claimed for competitive equilibrium, under conditions of certainty, except for insurable risk, did not seem to go beyond individual maximization of utility in face of an equilibrium price system. However, Pareto gave a genuinely social definition that the allocation of goods and services in a competitive equilibrium is such that no reallocation is possible with some consumer better off unless some consumer is made worse off. In fact, Walras seemed to be groping for the same definition and his arguments may be slightly extended to establish Pareto's proposition.

As noticed in the earlier discussion of markets with certainty, Pareto optimality is implied by maximization of preference under budget constraints and von Neumann's law, or maximization of profit given the technology. The former implies that an allocation which improves one consumer's position and harms none must, given local non-satiation for all consumers, be more valuable at equilibrium prices while the latter implies that no more valuable allocation is achievable. This argument depends on the finiteness of the value of the goods in the economy. Otherwise the impossibility of a more valuable allocation is not meaningful. Thus when the horizon is infinite and the discount factor is too large, for example, equal to 1 if the economy is stationary, or in general greater than or equal to the reciprocal of the growth rate, Pareto optimality may fail in competitive equilibrium, as Samuelson showed. Also there is no reason to expect Pareto optimality, in an exact sense, when some markets are missing, a very likely eventuality when there is uncertainty and goods must be traded on every possible contingency to provide complete markets.

A second theorem on Pareto optimality asserts that any Pareto optimum can be realized as a competitive equilibrium. This theorem requires assumptions which are similar to those leading to existence, in particular, assumptions providing local non-satiation for some consumers and convexity of the preferred sets and the feasible set. Moreover, when the number of goods is infinite as in the case of an infinite horizon an additional condition is needed to give the existence of the prices. This condition may be that the sum of consumers' preferred sets has an interior or that the production set has an interior. In the case of the product topology and free disposal by consumers the preferred sets will have interiors if the periodwise utility functions are continuous and bounded (see

Debreu, 1954). Finally it was shown by Arrow (1953) that in order for the Pareto optimal allocation to maximize preference over the budget set rather than only to minimize the cost of achieving a given preference level, it is useful to assume that x_i, the consumption set of the ith consumer, contains a point which is cheaper than the allocation he receives, for $i = 1, \ldots m$.

The stability theory for general equilibrium has been largely devoted to the stability of the Walrasian tâtonnement, or process of groping for equilibrium prices through a process of price revision according to excess demand. That is, prices rise or fall depending on whether excess demand is positive or negative. In the tâtonnement there is no trading until equilibrium prices have been reached. The most convincing theorems concern local stability and the dominant assumption leading to local stability is that the market excess demand function satisfies the weak axiom of revealed preference between the equilibrium price and any other price in a sufficiently small neighbourhood of the equilibrium price. That is, if $\bar{p}$ is an equilibrium price and e is the excess demand function, $p \cdot e(\bar{p}) - p \cdot e(p) \leqslant 0$ implies $\bar{p} \cdot e(\bar{p}) - \bar{p} \cdot e(p) < 0$. Since $\bar{p}$ is an equilibrium price, $e(\bar{p}) = 0$, and $p \cdot e(p) = 0$ by Walras's Law. Therefore, the condition holds and we may conclude that $\bar{p} \cdot e(p) > 0$. The weak axiom for the market may be expected to hold if the net income effect of price changes is small.

Consider the price revision process given by $\mathrm{d}p_i/\mathrm{d}t = \dot{p}_i = e_i(p)$, $i = 1, \ldots, n-1$, where the nth good is numeraire so $\dot{p}_n \equiv 0$. Then consider the function $|p(t) - \bar{p}|^2$, the square of the distance from the equilibrium price vector to the price vector at time t. We derive

$$\mathrm{d}/\mathrm{d}t(|p(t) - \bar{p}|)^2 = 2\sum_1^n (p_i - \bar{p}_i)\dot{p}_i = 2\sum_1^n (p_i - \bar{p}_i)e_i < 0,$$

using the weak axiom of revealed preference and Walras's Law. Thus the distance of $p(t)$ from $\bar{p}$ constantly falls, or $p(t) \to \bar{p}$ as $t \to \infty$. Since locally the rate of price change can be equated to excess demand for any continuous tâtonnement by choice of units, this is a general argument. Since the assumption of gross substitutes ($e_{ij} < 0$ for $i \neq j$ and $e_{ij} = \partial e_i(p)/\partial p_j$) implies the weak axiom, and the assumption of a negative definite Jacobian $[e_{ij}]$, $i, j = 1, \ldots, n-1$, at equilibrium is equivalent to the weak axiom locally, the weak axiom is a dominant condition for local stability. All global stability results are very special and relatively unconvincing.

A rigorous treatment of the stability problem for the tâtonnement was given by Arrow and Hurwicz (1958) and Arrow, Block and Hurwicz (1959). A stability theory which allows for trading was given by Hahn and Negishi (1962). These theories do not allow for speculative trading although profitable arbitrage opportunities would be likely to exist for any speculator who correctly inferred what the price revision process was. The stability of the tâtonnement was conjectured by Walras, to be the normal case for economies with many goods and essentially correct arguments were given by Walras for the case of exchange economies with two goods. He recognized and illustrated the case of locally unstable equilibria in the two good case.

Finally, as Walras saw, it may be possible through a general equilibrium analysis to determine the effect of changes in the exogenous factors, resources, technology, or tastes, on the economic variables in equilibrium. This is analogous to the effect of a change in the constraints on the equilibrium of mechanical forces, an analogy with which Walras would have been familiar from the book of Poinsot. In the case of the exchange of two commodities Walras derives some simple and correct results for comparative statics just as he does for stability. He observes that an increase in the marginal utility of

a good or a reduction in its supply will raise its price. In drawing this inference form his demand and offer curves he confines himself to stable equilibria as the only equilibria of interest.

Hicks used the comparative static result of Walras in a market with many goods to define stability of equilibrium. Samuelson (1947) pointed out that stability of equilibrium, where stability is given a dynamic interpretation as in a continuous tâtonnement, may imply comparative static results as a general principle. However, the straightforward generalization of Walras is the use of conditions which are sufficient to imply stability as a basis for deriving theorems on comparative statics. The most interesting theorem may be that derived from the revealed preference assumption at equilibrium.

Suppose that $e(\bar{p}) = 0$ but excess deman changes so that the new excess demand function $e_i'(\bar{p}) = e_i(\bar{p})$ for $i \neq 1$ or n and $e_1'(\bar{p}) = \delta_1 < 0$ while $e_n'(\bar{p}) = \delta_n > 0$. Let n be numeraire. This change can be arranged by taking δ_n of the nth good from some holder and compensating him with $\delta_1 = \delta_n/\bar{p}_1$ of the first good. Suppose that the new equilibrium price is p, or $e'(p) = 0$. By Walras's Law $\bar{p} \cdot e(\bar{p}) = 0$ and by the assumption of revealed preference $p \cdot e'(\bar{p}) > 0$. Thus $(\bar{p} - p) \cdot e'(\bar{p}) < 0$, or $(\bar{p}_1 - p_1)\delta_1 < 0$, or $\bar{p}_1 - p_1 > 0$. Any good falls in price when the excess demand for the numeraire rises at the expense of that good (see Allingham, 1975).

A type of stability has been proved for competitive equilibrium over time which concerns the path of equilibrium prices over real time rather than the path of disequilibrium prices over virtual time, that is, the time of the tâtonnement. It was shown by Negishi (1960) that there is a social welfare function associated with a competitive equilibrium which is maximized in the equilibrium over feasible allocations. Suppose each consumer has a concave utility function which is given by a discounted sum of periodwise utilities. Then the social welfare function which is maximized is also a discounted sum of periodwise utilities equal to a weighted sum of the individual utilities. Then using results from turnpike theory for optimal capital accumulation it has been shown by Bewley (1982) that the competitive equilibrium allocations converge over time to the allocations of a stationary competitive equilibrium whose capital stocks and allocations are the same as those of the unique optimal stationary path of capital accumulation given the social welfare function. The utility functions and the production functions are assumed to be strictly concave and the discount factors are the same for all consumers and sufficiently near 1. However, these conditions may be relaxed.

Comparative static and comparative dynamic results have been derived from stability conditions in the context of optimal capital accumulation, which is equivalent to competitive equilibrium over time with a representative consumer. We may say that an optimal stationary path of capital is regular if an increase in the discount factor implies an increase in the value of capital stocks at initial prices. Then there are sufficient conditions for local stability of the optimal stationary path which imply that the path is regular. Similar dynamic results may be achieved for non-stationary paths as well (see Araujo and Scheinkman, 1979). It may be possible to extend these results to Bewley type economies.

Lionel W. McKenzie

See also ARROW–DEBREU MODEL; EXISTENCE OF GENERAL EQUILIBRIUM; MATHEMATICAL ECONOMICS; OVERLAPPING GENERATIONS MODEL; UNVERTAINTY AND GENERAL EQUILIBRIUM.

BIBLIOGRAPHY
Allingham, M. 1975. *General Equilibrium*. New York: Wiley.

Araujo, A.P. de and Scheinkman, J.A. 1979. Notes on comparative dynamics. In *General Equilibrium, Growth, and Trade*, ed. J.R. Green and J.A. Scheinkman, New York: Academic Press.

Arrow, K.J. 1951. An extension of the basic theorems of classical welfare economics. In *Proceedings of the Second Berkeley Symposium*, ed. J. Neyman, Berkeley: University of California Press.

Arrow, K.J. 1953. Le rôle des valeurs boursières pour la répartition la meilleure des risques. *Économétrie*, Paris: Centre National de la Recherche Scientifique. Trans. as 'The role of securities in the optimal allocation of risk-bearing', *Review of Economic Studies* 31, (1964), 91–6.

Arrow, K.J. and Debreu, G. 1954. Existence of an equilibrium for a competitive economy. *Econometrica* 22, 265–90.

Arrow, K.J. and Hurwicz, L. 1958. On the stability of the competitive equilibrium I. *Econometrica* 26, 522–52.

Arrow, K.J., Block, H.D. and Hurwicz, L. 1959. On the stability of the competitive equilibrium II. *Econometrica* 27, 82–109.

Balasko, Y. and Shell, K. 1980. The overlapping generations model, I: the case of pure exchange without money. *Journal of Economic Theory* 23, 281–306.

Balasko, Y., Cass, D. and Shell, K. 1980. Existence of competitive equilibrium in a general overlapping generations model. *Journal of Economic Theory* 23, 307–22.

Bewley, T.F. 1972. Existence of equilibria in economies with infinitely many commodities. *Journal of Economic Theory* 4, 514–40.

Bewley, T.F. 1982. An integration of equilibrium theory and turnpike theory. *Journal of Mathematical Economics* 10, 233–68.

Cassel, G. 1918. *Theoretische Sozialökonomie*. 5th German edn, trans. as *The Theory of Social Economy*, New York: Harcourt Brace, 1932.

Cournot, A. 1838. *Recherches sur les principes mathématiques de la théorie des richesses*. Paris: Hachette. Trans. as *Researches into the Mathematical Principles of the Theory of Wealth*, New York: Kelley, 1960.

Debreu, G. 1954. Valuation equilibrium and Pareto optimum. *Proceedings of the National Academy of Sciences* 40, 588–92.

Debreu, G. 1959. *Theory of Value*. New York: Wiley.

Debreu, G. 1962. New concepts and techniques for equilibrium analysis. *International Economic Review* 3, 257–73.

Gossen, H. 1854. *Entwicklung der Gesetze des menschlichen Verkehrs*. 3rd edn, Berlin: Prager, 1927.

Grandmont, J.M. 1973. On the efficiency of a monetary equilibrium. *Review of Economic Studies* 40, 149–65.

Grandmont, J.M. 1977. Temporary general equilibrium theory. *Econometrica* 45, 535–72.

Grandmont, J.M. 1983. *Money and Value*. New York: Cambridge University Press.

Grandmont, J.M. and Younes, Y. 1972. On the role of money and the existence of a monetary equilibrium. *Review of Economic Studies* 39, 355–72.

Green, J.R. 1973. Temporary general equilibrium in a sequential trading model with spot and future transactions. *Econometrica* 41, 1103–23.

Green, J.R. 1977. The nonexistence of informational equilibria. *Review of Economic Studies* 44, 451–63.

Grossman, S.J. 1981. An introduction to the theory of rational expectations under asymmetric information. *Review of Economic Studies* 48, 541–60.

Hahn, F.H. and Negishi, T. 1962. A theorem on non-tâtonnement stability. *Econometrica* 30, 463–9.

Hicks, J.R. 1939. *Value and Capital*, Oxford: Clarendon Press.

Jevons, W.S. 1871. *The Theory of Political Economy*. London: Macmillan; 5th edn, New York: Kelley and Millman, 1957.

Koopmans, T.C. 1951. Analysis of production as an efficient combination of activities. In *Activity Analysis of Production and Allocation*, ed. T.C. Koopmans, Wiley: New York.

Kreps, D.M. 1977. A note on fulfilled expectations equilibria. *Journal of Economic Theory* 14, 32–43.

McKenzie, L.W. 1954. On equilibrium in Graham's model of world trade and other competitive systems. *Econometrica* 22, 147–61.

McKenzie, L.W. 1959. On the existence of general equilibrium for a competitive market. *Econometrica* 27, 54–71.

McKenzie, L.W. 1981. The classical theorem on existence of competitive equilibrium. *Econometrica* 49, 819–41.

Marshall, A. 1890. *Principles of Economics*. 8th edn, London: Macmillan, 1920.

Menger, C. 1871. *Grundsätze der Volkswirtschaftslehre*. Vienna. Trans. as *Principles of Economics*, Glencoe, Ill.: Free Press, 1950.

Mill, J.S. 1848. *Principles of Political Economy*. London: Parker. New edn, London: Longmans, 1909.

Negishi, T. 1960. Welfare economics and existence of an equilibrium for a competitive economy. *Metroeconomica* 12, 92–7.

Neumann, J. von. 1937. Über ein Ökonomisches Gleichungssystem und eine Verallgemeinerung des Brouwerschen Fixpunktsatzes. *Ergebnisse eines mathematischen Kolloquiums* 8, 73–83. Trans. in *Review of Economic Studies* 13, (1945), 1–9.

Nikaido, H. 1956. On the classical multilateral exchange problem. *Metroeconomica* 8, 135–45.

Novshek, W. and Sonnenschein, H. 1980. Small efficient scale as a foundation for Walrasian equilibrium. *Journal of Economic Theory* 22, 243–55.

Pareto, V. 1909. *Manuel d'économie politique. Paris*. Trans. from 1927 edn as *Manual of Political Economy*, New York: Kelley, 1971.

Peleg, B. and Yaari, M.E. 1970. Markets with countably many commodities. *International Economic Review* 11, 369–77.

Poinsot, L. 1803. *Eléments de statique*. 8th edn, Paris, 1842.

Radner, R. 1967. Equilibre des marchés à terme et au comptant en cas d'incertitude. *Cahiers d'Économétrie*, Paris: CNRS, 4, 35–52.

Radner, R. 1972. Existence of equilibrium of plans, prices and price expectations in a sequence of markets. *Econometrica* 40, 289–303.

Samuelson, P.A. 1947. *Foundations of Economic Analysis*. Cambridge, Mass.: Harvard University Press.

Samuelson, P.A. 1958. An exact consumption – loan model of interest with or without the social contrivance of money. *Journal of Political Economy* 66, 467–82.

Scarf, H.F. (With T. Hansen) 1973. *The Computation of Economic Equilibria*. New Haven: Yale University Press.

Smith, A. 1776. *An Inquiry into the Nature and Causes of the Wealth of Nations*. 5th edn, ed. E. Cannan, London: Methuen, 1906.

Wald, A. 1935. Über die eindeutige positive Losbarkeit der neuen Produktionsgleichungen. *Ergebnisse eines mathematischen Kolloquiums* 6, 12–20.

Wald, A. 1936a. Über die Produkionsgleichungen der Ökonomischen Wertlehre. *Ergebnisse eines Mathematischen Kolloquiums* 7, 1–6.

Wald, A. 1936b. Über einige Gleichungssysteme der mathematischen Ökonomie. *Zeitschrift für Nationalökonomie* 7, 637–70. Trans. as 'On some systems of equations of mathematical economics', *Econometrica* 19, 1951, 368–403.

Walras, L. 1874–7. *Elements d'économie politique pure*. Lausanne: Corbaz. Trans. by W. Jaffé as *Elements of Pure Economics*, London: George Allen & Unwin, from the 1926 definitive edition, 1954.

Weintraub, E. R. 1983. On the existence of a competitive equilibrium: 1930–1954. *Journal of Economic Literature* 21, 1–39.

Wilson, C.A. 1981. Equilibrium in dynamic models with an infinity of agents. *Journal of Economic Theory* 24, 95–111.

general equilibrium, existence of. *See* EXISTENCE OF GENERAL EQUILIBRIUM.

general systems theory. The term 'general systems' refers to a movement among a wide variety of scholars to overcome the barriers of communication which divide the established disciplines, by developing theoretical concepts and systems which are common to the different disciplines. Biologist Ludwig von Bertalanffy originated the movement with his concept of 'open systems'. The Society for General Systems Research, originally called the Society for the Advancement of General Systems, was founded at a meeting at the American Association for the Advancement of Science in Berkeley, California, in December 1954. The economist Kenneth E. Boulding was the first president. The Society issues the *General Systems Yearbook*, partly of reprinted, partly of original articles, of which the first editor was Anatol Rapoport, a mathematician and game theorist. The yearbooks are still published, and a number of journals now contribute to the field.

In Europe general systems has frequently been identified with 'cybernetics', originated by Norbert Wiener of MIT in 1948, which is the study of both the equilibrium and disequilibrium systems which involve feedback. A thermostat is a good example of an equilibrium system with negative feedback. The equilibrium is manipulable. It is the temperature at which the thermostat is set. If the temperature rises above this, the thermostat turns the furnace off; if it falls below it, the thermostat turns the furnace on. All such cybernetic systems, of which there are many, such as homoeostatic mechanisms of the body, exhibit cycles, the period and magnitude of which depend mainly on the time of response of the feedback. The tendency of the market price system of relative prices to fluctuate around an equilibrium and the tendency of competitive markets in commodities and securities to fluctuate in aggregate or average prices is a good example of negative feedback provided by the behavioural reactions to price above or below what is regarded as normal. Inflation is frequently an example of positive feedback, especially hyperinflation, where a rise in the price level produces both expectations which lead to a continued rise and also a partial collapse of the tax system and budget deficits, which likewise feed the continuing rise. Deflation, such as occurred during the Great Depression of 1929–33, is also a positive-feedback process, in which, for instance, declining profits produce declining investment, which produces further declining profits, further declining investments, and so on.

Another important line of development of general systems has been the development of a general theory of the ontogeny, structure and behaviour of organisms, ranging from the cell, the organ, the living organism, the group, the social organization, the nation-state, and so on. James Grier Miller has made important contributions to this, particularly in regard to the taxonomy of organisms, and has identified at least 19 necessary components of such structures, common to all of them.

The structure of organisms and organizations is also influenced by the principle of allometry, developed especially by von Bertalanffy, but going back to D'Arcy Thompson in his work *On Growth and Form*. This is the principle that an increase in the linear dimensions of any structure, keeping the same proportions, will increase the areas as a square and the volumes as the cube of the linear increase. Thus a two-inch cube has four times the area and eight times the volume of a one-inch cube. This explains why structure is a function of scale. Some properties depend on the linear dimensions, like the transmission of information or fluids. Some depend on the areas, like chemical exchange and structural strength; some depend on the volumes, like weight or mass. This goes a long way to explaining why structure changes with size and must change with growth. This principle also applies to social organizations and firms in terms of hierarchy, specialization, diffusion of responsibility, and so on. In economics it is responsible for such phenomena as diseconomies of scale unless there is structural change and also for the less well recognized phenomenon of organizational failure when growth takes place without adequate structural change.

Another aspect of general systems is the development of more general ecological and evolutionary theory. There are many parallels between ecological theory in biology and the theory of a general equilibrium and development of commodities. One may claim Adam Smith, indeed, as perhaps the first ecological and evolutionary theorist, perceiving the economic system to be an ecosystem of commodities with equilibrium populations at which births (production) and deaths (consumption) are equal, with the equilibrium population of each commodity being a function of the population of all others. This gives us a system of n equations and n unknowns, as developed, for instance, in economics by Walras. Adam Smith further recognizes what today would be called 'mutation'; that is, changes in the parameters of the system, leading to new equilibrium positions continually as changes take place in the genetic factors in the production of commodities. Adam Smith also recognized that these genetic factors primarily involved changes in human knowledge as a result of a learning process. The classical economist also had something like a food chain theory of the ecosystem of commodities, with the input of food into the food-producer producing a surplus of food which could then feed the producers of other commodities. The main difference between biological and economic systems is that the genetic structure for biological products is contained in the products themselves, whereas in the case of commodities the genetic structure is contained in many other human minds and human artifacts. The economic system is multi-parental. These ideas are not very widely accepted by economists, who still cling to a somewhat Newtonian view of the system, perhaps because they do not conform easily to quantification and mathematization, and emphasize that the real world consists of structure rather than number.

General systems has not established itself well in the role structure of universities; very few have formal programmes in it. The Society for General Systems Research, however, is still very much alive. Other aspects of general systems, such as theories of autopoiesis – that is, the instability of chaos and the spontaneous formation of structures – may turn out to have considerable relevance to economic problems like entrepreneurship and innovation. It cannot be claimed that general systems has had much impact on economics to date, but it is not much more than 30 years old.

KENNETH E. BOULDING

see also BERTALANFFY, LUDWIG VON, BOULDING, KENNETH EWART; STABILITY.

Genovesi, Antonio (1712–1769). Genovesi was born near Salerno and died at Naples: he took holy orders in 1736. In 1741 he taught metaphysics at the University of Naples. He was intimately acquainted with Bartolomeo Intieri, who induced him to follow Broggia and Galiani in the study of economics; and when, in 1754, by the advice of Intieri and with funds liberally supplied by him, the teaching of economics, then termed mechanics and commerce, was established at Naples, Genovesi was called to the chair. He was 'the most distinguished and the most moderate of all Italian mercantilists... . Commerce was for him not an end only, but also a means by which the products of industry at large were brought to the right market. He, moreover, distinguished between useful commerce which exported manufactured goods and brought back in return raw material, and harmful commerce which exported raw material and imported foreign goods; he also insisted that useful commerce calls rather for liberty than for protection, while upon harmful commerce the strictest embargo should be laid, or at least it should as far as possible be bound hand and foor' (Cossa, *Introduction to Political Economy*, translation, p. 235).

These ideas, neither new nor original even in his time, were maintained by Genovesi in many of his works, and brought together, but without any systematic order, in his *Lezioni di Commercio ossia di Economia Civile* (Napoli, 1765, e. ii. ediz. 1768–70, 2 vols). Though the *Lezioni* do not form a regular treatise, they contain the author's opinions on the mercantilist system and the most important principles of economics, which he terms *Civile 'la scienza che abbraccia le regole per rendere la sotto-posta nazione popolata, potente, saggia, polita'* (the science which embraces the laws which make a nation populous, powerful, wise, and cultured), limiting thus the science to the increase of population and the production of wealth.

As to population, Genovesi follows the mistaken principle of his times, exaggerating the advantage of a large population, proposing that government should encourage marriages by granting privileges and honours. He says that the population ought not only to be numerous but supplied with comforts, and he sees the relation between population and means of subsistence or production of wealth.

As a writer he is a mercantilist, though he does not regard money as the only form of riches; he says that the wealth of a nation is quite apart from the quantity of money treasured up.

He derives the idea of value from demand, distinguishing different degrees of demand according to their abstract importance in several categories, maintaining that a thing which satisfies a want repeatedly has a higher value than what satisfies only a few wants or the same only sometimes (*puo soddisfare ad un bisogno più volte, ha maggior prezzo che non quella, la quale o non puo soddisfare che pochi bisogni o al medesimo qualche volta*). What is able to satisfy a great want is of more value than what satisfies a small want (*una cosa fatta a soddisfare il maggior bisogno si apprezza più che quella la quale non è fatta che a soddisfare ad un minore*); and further he asserts that the quality of things influences the value. Graziani (*Storia della teoria del valore in Italia*, Milano, 1889, p. 108) justly remarks that in this Genovesi approaches the important question which Galiani answered: namely, why do luxuries generally cost more than necessaries? In this he is obliged to have recourse to the element of scarcity, a line of argument which he does not know how to reconcile with those previously mentioned. Genovesi's want of originality is obvious, as F. Ferrara has shown (*Bibl. dell' Econo.*, 1ª. S. vol. iii. Introduz.) in contradistinction to the exaggerated opinion which Bianchini held respecting him (*La scienza del ben vivere sociale*), since the Socialists of the Chair persist, erroneously, in considering him as a precursor of their opinions. This tendency is also attributed to Genovesi, as well as to Beccaria, Verri, and Romagnosi by the French socialist B. Malon; which is a further example of the errors of the socialists in their historical criticism of political economy.

[ANGELO BERTOLINI]
Reprinted from *Palgrave's Dictionary of Political Economy*.

See also MERCANTILISM.

George, Henry

George, Henry (1839–1897). Henry George was by turns sailor, prospector, printer, reporter, San Francisco newspaper editor and publisher, orator and political activist before closeting himself to write on political economy. His *Progress and Poverty* (1879) electrified reformers, catapulted him to fame and began a worldwide movement for land reform and taxation, opening to George an extraordinary career in radical politics. Returning from Ireland as reporter for *The Irish World* of New York he was lionized by Irish-New Yorkers for his stand on the Irish land question. With ethnic, union and socialist backing he formed the United Labour Party and ran for mayor of New York in 1886, nearly winning.

He toured Britain and won over the Radical-Liberals; then toured Australia as a folk-hero. At home he was courted by Democrat and later by Populist leaders. He died in 1897 while running again for New York mayor, but his followers rose in and helped shape the Progressive movement which dominated the next twenty years. His name has become a byword for ideas and policies he espoused.

George is best known today for *Progress and Poverty* (1879). Eloquent, timely and challenging, it soon became and remains the all-time best-seller on economic theory and policy.

George defines 'The Problem' as increase of want with increase of wealth. Refuting Malthusian fatalism as merely a device to rationalize privilege, George attributes low wages and unemployment rather to artificial scarcity of land, and barriers to free exchange. Artificial scarcity results from unequal dispensation of public lands, concentration and 'speculation'. George's speculation is pervasive market failure endemic to land, which failure he attributes to holding for the unearned increment.

George proposed to raise the *ad valorem* property tax rate on bare land (broadly defined as all natural opportunities), thus socializing rent without excess burden. He would remove other taxes, calling them barriers to commerce, employment and capital formation. The cash drain of the *ad valorem* tax, while neutral at the margin, would move and lubricate the land market as a whole, forcing land into full use. Observation persuaded him that otherwise speculation overrode the incentives to use land fully.

Release of hoarded lands would open wider opportunities for both labour and man-made capital. His overriding concern was for labour, but he saw capital mainly as a form of labour, produced by labour, complementing labour. So in an era before payroll taxes it was actually capital and commerce he sought to untax for the benefit of labour – a preview of 'business Keynesianism' in W. Heller's production of Camelot.

George did not see investment employing labour, but labour producing capital, a difference that to him was more than a nuance. While admiring Quesnay he never absorbed the Physiocratic idea of '*avances*'. Instead he attacked its English derivative the wages-fund theory with its advances of subsistence, a concept he rejected as condescending to labour. He developed no concept of economic circulation, either of capital or spending. He lacked a good capital theory, belittling Austrian interest theory and botching his own. These faults narrowed the effective scope of his otherwise seminal work and ultimately limited its influence, which is still wide and sustained but mainly outside the macroeconomic field it addressed.

His programme would level barriers to exchange and specialization and production and synergy. These include spatial barriers forced by land speculation (e.g. scattered settlement and urban sprawl); fiscal barriers like excise and wage taxes; and social barriers from unequal wealth and contempt for workmanship, which he (like Veblen) traced to the influence of privilege and unearned wealth. This 'true free trade' would unleash technological, scientific, cultural and spiritual development in a more egalitarian and moral society organized around a perfected market mechanism.

George drew on earlier thinkers: Quesnay, Smith, Ricardo, Spencer, and Mill. And he contributed much to later thinking.

George was system-minded and sought to unify the laws of production and distribution in a coordinated harmonious system. His theoretical framework is an early adumbration of the marginal productivity theory of wages, which he integrates with Ricardo's rent law. J.B. Clark was a nemesis, and P. Wicksteed a friend, but both were formalizing insights from George.

Although best known as a deductive thinker, the journalist was also an observer with statistical intuition. In debate with Francis Walker on 'The March of Concentration' in farming, George anticipated Lorenz's method of analysing size distributions and goaded the US Census into publishing farm size data in that form.

George wanted radical redistribution but without revolution. He pioneered the idea that taxation, properly crafted, can redistribute wealth without damage to the market. His influence on Fabianism was early and wide; also on American reformers like Tom L. Johnson, Upton Sinclair, John R. Commons and Norman Thomas. The modern 'mixed economy' is in the Georgist spirit of reform within traditional forms.

Continued heavy reliance on real estate taxation in Canada and the United States, with separate assessment of land value, reflects George's influence, as do the inclusion of land rents and gains in the income tax base, and the efforts of Lloyd George, Asquith and Snowden to introduce national land taxes in Britain.

Free provision of public goods, social dividends, and marginal-cost pricing for urban mass transit and utilities are vintage Henry George. H. Hotelling and W. Vickrey have acknowledged their debt.

The optimistic 'economics of abundance' idea owes much to George. The prevailing 'dismal' economics was a science of choice where all the choices were bad and leaders could only call for more sacrifices. George promised full employment at higher wages by unlocking natural opportunities now held in speculation. Needed capital would be formed in the very process of making jobs, an idea pervading Keynes. Social synergy would produce a surplus that spills over into higher land rents, a 'free lunch' that government may tap in lieu of taxes that penalize and abort useful activity.

George lives too in Urban Economics and City Planning. George's emphasis on the synergistic gains from urban linkages, and the wastes of sprawl caused by failure of the land market anticipates much of planning doctrine. Ebenezer Howard is an obvious link: his 'Garden City' presupposed Georgist taxation to move the land market.

The idea that environment is a common heritage for future generations in pure Georgism. 'Spaceship Earth', common property, and rights of the unborn are his very phrases.

As to Economic Development the economists are legion who have recommended a 'dose of Henry George' to help LDCs take off, and some like Taiwan, belatedly following the counsel of the Georgist Dr Sun Yat-sen, have taken the dose with good results.

On the conservative side, George was a pioneer of tax limitation, insisting that land rent set an upper limit on government spending. The resurgence of libertarianism and supply-side economics may set a new stage for George, whose programme was mainly oriented to increasing production in the private sector. Religion in politics should not threaten George, who unabashedly presented economic policy as an implementation of religious ideals.

George's blend of radicalism and conservatism can puzzle one until it is seen as a reconciliation of the two. The system is internally consistent but defies conventional stereotypes.

MASON GAFFNEY

See also LAND TAX.

SELECTED WORKS
Complete Works. Garden City, NY: The Fels Fund; also other publishers.

BIBLIOGRAPHY
Andelson, R. (ed.) *Critics of Henry George*. London and Cranbury, NJ: Associated University Presses.
Barker, C.A. 1955. *Henry George*. New York: Oxford University Press.
Cord, S.B. 1965. *Henry George: Dreamer or Realist?* Philadelphia: University of Pennsylvania Press.
de Mille, A.G. 1950. *Henry George: Citizen of the World*. Chapel Hill: University of North Carolina Press.
Geiger, G.R. 1933. *The Philosophy of Henry George*. New York: Macmillan Co.
George, H., Jr. 1900. *The Life of Henry George*. Reprinted, New York: Robert Schalkenbach Foundation, 1943.
Lawrence, E.P. 1957. *Henry George in the British Isles*. East Lansing: University of Michigan Press and Vanguard Press.
Post, L.F. 1930. *The Prophet of New York*. New York: Vanguard Press.
Sawyer, R.A. 1926. *Henry George and the Single Tax, a Catalogue of the Collections in the New York Public Library*. New York: New York Public Library.
The Standard. 1887–92. (weekly) New York.

Georgescu-Roegen, Nicholas (born 1906). Born in Constanza, Rumania, on 4 February 1906, Georgescu-Roegen obtained his first degree in mathematics in 1926 from the University of Bucharest. He then went to Paris where, under the supervision of E. Borel and G. Darmois, he received in 1930 the doctorate in mathematical statistics. In October of the same year he moved to London to pursue further research with K. Pearson. By 1932 Georgescu was Professor of Statistics at the University of Bucharest. His life was inextricably bound up with the social and political events of his country, which explains the emergence of his interest in economics and his consequent decision to spend a two-year 'apprenticeship' (1934–6) at Harvard where he was able to work closely with Schumpeter. In 1937 he returned to Rumania, where he combined an active academic career with increasing responsibilities in public institutions. In February 1948 he fled from his country and, after a short stay at Harvard, was appointed professor at Vanderbilt University, where he remained until his retirement in 1976.

Georgescu-Roegen's scientific work is notable for an early phase centred around consumer theory, input–output analysis and production theory at large, and a later phase mainly devoted to growth modelling, methodological issues and the ambitious attempt to develop a 'bioeconomic' approach to economic thinking. The early phase is well represented by his 1936 classic article on consumer theory and his 1954 famous paper on 'Choice, Expectations and Measurability'. In the former article, which deals with the 'mysterious' problem of integrability in the theory of demand, one finds two major results: the demonstration that the integral varieties do not necessarily coincide with the indifference varieties – whence the distinction between mathematical integrability and economic integrability – and the demonstration that the two kinds of varieties come to the same thing in the presence of the postulate of transitivity of preferences. The latter essay, focusing on the non-existence of the indifference map of the consumer as a consequence of the pervasiveness of lexicographic ordering of preferences, allowed him to prove what he called the 'ordinalist fallacy' and to inquire about the origin and implications of probabilistic preferences, a subject that is at the very frontiers of economics even today.

On the other front, three contributions are particularly noteworthy. In Georgescu (1951a) we find the first and the most general statement of the celebrated non-substitution theorem: justifying the separation of scale and composition in linear multisectoral models, the theorem provides a theoretical underpinning and analytic rationale for the consistency of input–output analysis. The (1951b) paper offers the first 'geometric' proof of the existence of a von Neumann's equilibrium by using the separating hyperplane theorem – a theorem that was to enter the tool-box of the economist. In his (1951c) essay, Georgescu challenged the two most intractable problems in macrodynamics – non-linearities and discontinuities – providing, on the basis of an innovative application of the theory of relaxation oscillations, a fundamental result for investigations of regime switching.

The later phase begins with the 1966 famous methodological essay containing Georgescu-Roegen's critique of standard economics for having reduced the economic process to a mechanical analogue and a proposal of a new alliance between economic activity and the natural environment – what later would become his 'bioeconomic programme'. The key to such a project is found in the entropy law ('the most economical of physical laws'), which brought Georgescu to inquiry on the fundamental relation between mankind's existence and its environmental dowry. This problem prompted him to step over the fence of economics into thermodynamics, where he formulated a new law (the 'fourth law'): the impossibility of the perpetual motion of the third kind defined as a *closed* system that could perform work at a constant rate indefinitely. The implications for economics of this line of thinking and in particular of his strong rebuttal of the 'energetic dogma' ('only energy matters') are nicely developed in his 1971 and 1976 books. In this last book, Georgescu lays the foundations of a new approach to production theory: the 'flow-fund' model as a radical alternative to both the production function model and the activity analysis model, models whose main drawback lies in their inability to tackle properly the time element in the productive process.

The long introductory essay (145 pages) that Georgescu wrote in 1983 for the English edition of Gossen's *The Law of Human Relations* is not simply a splendidly written intellectual biography, showing the depth and breadth of his economic culture, but it contains also a restatement in modern analytical terms and an expansion of Gossen's theory of economic behaviour. Georgescu-Roegen is one of those rare scientists able to couple a remarkable expertise in their specific field with a philosophical bent of mind. In this sense he is a true renaissance man, which perhaps helps to explain the generalized *fin de non recevoir* of the profession with respect to his critical message, the message of a scholar who cannot be identified with any single school of economic thought and whose intellectual endeavour is best seen as a major contribution to the shifting of the frontiers of economic theory and methodology.

STEFANO ZAMAGNI

SELECTED WORKS

1936. The pure theory of consumers' behavior. *Quarterly Journal of Economics* 50, August, 545–93.
1951a. Some properties of a generalized Leontief model. In *Activity Analysis of Production and Allocation*, ed. T.C. Koopmans et al., New York: Wiley.
1951b. Relaxation phenomena in linear dynamic models. In *Activity Analysis of Production and Allocation*, ed. T.C. Koopmans et al., New York: Wiley.
1951c. The aggregate line as production function and its applications to Von Neumann's economic model. In *Activity Analysis of Production and Allocation*, ed. T.C. Koopmans et al., New York: Wiley.
1954. Choice, expectations and measurability. *Quarterly Journal of Economics* 68, November, 503–34.
1966. Some orientation issues in economics. In N. Georgescu-Roegen, *Analytical Economic Issues and Problems*. Cambridge, Mass.: Harvard University Press.
1971. *The Entropy Law and the Economic Process*. Cambridge, Mass.: Harvard University Press.
1976. *Energy and Economic Myths: Institutional and Analytical Economic Essays*. Oxford: Pergamon Press.
1983. Hermann Heinrich Gossen: his life and work in historical perspective. In H.H. Gossen, *The Laws of Human Relation and the Rules of Human Action Derived Therefrom*. Trans. from the original German edn of 1854 by R.C. Blitz, Cambridge, Mass.: MIT Press.

German historical school. The German historical school is very closely connected to Romanticism and the rise of nationalism in Germany; it is considered a reaction to English enlightenment and classical economics. This reaction to English classical economics manifested itself in two different ways; by developing different methods and by seeking alternative aims in economic research.

The classical school's deductive method is criticized as being too abstract. The German historical school puts the emphasis on the inductive method. Historians point out that economic development is unique, so there can be no 'natural laws' in economics. The economist can only try to show patterns of development common to different economies. Instead of searching for generally applicable laws, the historical school therefore tried to describe the particulars of each era, society and economy. A rational approach to human behaviour is criticized as being unable to show correctly the amplitude of human motives – these being influenced by non-economic principles, even where economics are concerned.

The aims of economic research were put differently: research for research's sake must be abandoned, it must be seen as a means of achieving sensible economic policy, useful for society. This leads to another aspect of the German historical school: ethics. One of the reasons for the rise of the historical school was the social question, namely the problems arising in Germany in the middle of the 19th century. These led to the belief that free trade was unable to solve problems of industrialization in a country totally different from England. From the ethical point of view the German historians demanded that the state had an important role to play in economic affairs. The historical school can be considered as the beginning of the end of liberal economic policy in Germany.

Friedrich List, considered as a forerunner of the historical school, criticized 'free trade' and put forward the idea that it was the duty of the state to protect the still young German industry from the competition presented by a much further developed English industry. He also suggested that the state should protect the socially weak sections of the population. These ideas arose from the phenomenon of 'Pauperismus' in Germany in the 1830s and 1840s – the poverty of millions of people who were no longer able to find work in agriculture nor in slowly developing industry.

The ever-widening rift between economic theories and experienced reality set off a new direction in economic research. With industrialization progressing in countries, whose social conditions and economic basis were totally different from those in 18th-century England, it seemed necessary to adapt economic research to changing reality.

An attempt to bridge this rift was made in two ways: on the one hand there was the attempt to find a totally new theory which would be more comprehensive than classical theory; on the other hand there was a tendency to dismiss theory and try to see the depiction of reality, in a historical perspective, as the only sensible aim of economic research. For these reasons the historical school is characterized by the development of statistics and economic history.

It is difficult to discover the general opinions of all the economists of the historical school. Few of their ideas were formulated in a clear, non-ambiguous way. The general ideas common to all of them must be filtered out from their works, and this leads to a subjective interpretation. Generally it can be said that all the economists of the German historical school put forward criticisms of the methods of classical economy, especially of deductive methods – even if some of them used such methods in their own works.

Another point common to them all is their criticism of the classical belief in harmony that results from the individual's knowledge and rational following of his economic advantages. German historians emphasized the non-rational influences which lead to human actions, and they also stressed the fact that the individual is part of a socially unique context, which differs in time and space (e.g. differences between the industrialization of England and Germany in the 19th century).

The German historical school has been divided into two epochs, the older and the younger. The older historical school can be attributed to the 1840s–1870s.

The beginning of the historical school is dated 1843 because the first representative of the school, Wilhelm Roscher, then published his book *Grundriss zu Vorlesungen über die Staatswirtschaft nach geschichtlicher Methode*. In view of this he is seen as the founder of the Historical School. He tried to illustrate classical theory with historical examples and his goal was to use the classical theory as a basis for practical economic policy. He confronted the universal claim of the classical theory with the individuality of each single national economy. Economics as a science should try to find out the interactions between ethical, political and economic phenomena. The most important result of Roscher's work was to put forward the non-economic factors which influence economic life. He tried to find laws of development in economies using the method of comparative induction and comparing different times, peoples, countries and cultures.

The second representative of the German historical school is Bruno Hildebrand. He had a more ambitious programme of research than Roscher. His main, uncompleted work is *Die Nationalökonomie der Gegenwart und Zukunft* (1848). He stresses much more sharply than Roscher the differences between the German historical school and classical economics. For Hildebrand, history is a means of renewing economic research and thought. He tried to show the differences between the economies of different times, people and states. He especially tried to find out the laws of economic development (*Lehre der Entwicklungsgesetze der Völker*) with the help of statistical data. In order to help this research he founded the journal *Jahrbücher für Nationalökonomie und Statistik*, which still exists.

The new method of the historical school is theoretically best illustrated by Karl Knies. His book *Die politische Ökonomie vom Standpunkt der geschichtlichen Methode* (1853) is, from a theoretical point of view, more refined than the books of Hildebrand and Roscher. He also accentuates the need to find a new method in economic research. This new method is somewhat different from what Hildebrand and Roscher advocated. Knies was sceptical about the laws of economic development which Hildebrand tried to discover. For Knies, there are only analogies and not 'laws' of economic development in different peoples; economic thought develops alongside economic conditions.

Deciding which economists to attribute to the younger historical school is a point of controversy, since every German economist at the end of the 19th century was formed by this school. The head of the younger school was surely Gustav Schmoller, who dominated German economics from the 1870s to the end of the 19th century.

Characteristic of Schmoller and his school is the fact that they do not specifically deny that 'laws' and regularities exist in economic and social life – in some ways they are themselves deterministic when they try to find out these regularities. They wrote a large number of monographs, which can be considered works of economic history. As well as this they found another area of research, the solving of practical problems of the day, especially in the social field.

In economic policies the work of the younger historical school can be characterized by its desire to eliminate the negative results of economic liberalism (especially after the 'Gründerkrise' of 1873), by demanding that the state intervene. Schmoller states that the classical theory is unable to solve the problems of the working classes. The discussion now arises around the question of *how* the state should intervene.

In the field of economic policy the younger historical school had its greatest practical success. The historians were called 'Kathedersozialisten' because most of them were professors. They asked for social laws, insurance against illness, accident, old age and unemployment and founded the 'Verein für Socialpolitik', a forum where these demands were put forward and discussed. The practical result of these demands were the social laws of the 1880s which gave German workers insurance against illness, accident and old age – then unique in Europe.

The younger historical school has found fame through a discussion of methods between Gustav Schmoller and Carl Menger. Menger published in 1883 *Untersuchungen über die Methoden der Sozialwissenschaften und der Politischen Ökonomie insbesondere*, to which Schmoller answered with his article *Zur Methodologie der Staats- and Sozialwissenschaften*.

The books of Menger and Schmoller gave rise to a very polemical discussion about the methods of economic research. Menger defended the deductive method against the historical research work of the historical school. In this fight over method all those aspects which had been brought forward in the discussion of the older historical school arose again – although in a more refined way.

It is difficult to say which of the writers at the end of the 19th century can be counted among the economists of the younger historical school: it has been said that Albert Schäffle belonged to it. Schäffle believed in the compatibility of planned production with individual liberty to consume. These ideas were opposed by Lujo Brentano, also attributed to the younger historical school, who pointed out that it was impossible to have individual consumer freedom while there was a central production plan, because consumer demand was mostly irrational. Adolph Wagner has also been counted among the representatives of the younger historical school. His main works dealt with public finance and he gave the state an important role in directing the course of economy. Karl Bücher put forward the idea of stages of economic evolution, which had been discussed since the first half of the 19th century. Werner Sombart, whose major work *Der moderne*

Kapitalismus describes the history of capitalism, was influenced by the younger historical school, but cannot be attributed to it, because he later put the accent on very different problems.

The German historical school cannot be understood without knowledge of the economic history of Germany in the 19th century. It is mostly the result of social problems arising from population growth at this time and those emerging with industrialization in Germany. It is also the result of increasing nationalistic feelings in a country divided into more than 39 sovereign states. For the younger historical school the economic crisis of the 1870s was an important departure point in demanding state intervention in economics.

The historical background leads to the fact that apart from many different ways in reacting to classical economics the economists of the historical school had many things in common, which justify their incorporation under the same heading. The main idea is that each economic phenomenon is a product of its social context, having grown historically as the result of a long process.

The historical school was typical for Germany in the 19th century, having little influence elsewhere. Its view of human behaviour asked for research in the field of social psychology. In France this led to the development of sociology and social history. The younger historical school had some influence in the United States, where institutionalism can be seen as an epoch of American economic thought.

F. SCHINZINGER

See also SCHMOLLER, GUSTAV.

BIBLIOGRAPHY

Böhm-Bawerk, E. von. 1924. *Historische und theoretische Nationalökonomie.* Jena: Fischer.
Brentano, L. 1888. *Die klassische Nationalökonomie.* Leipzig.
Brinkmann, C. 1937. *Gustav Schmoller und die Volkswirtschaftslehre.* Stuttgart: Kohlhammer.
Bücher, K. 1893, 1918. *Die Entstehung der Volkswirtschaft. 6 Vorträge.* Pts I and II. Tübingen, Laupp.
Cunningham, W. 1894–5. Why had Roscher so little influence in England? *Annals of the American Academy* 5.
Diehl, K. 1941. *Die sozialrechtliche Richtung in der Nationalökonomie.* Jena.
Eisermann, G. 1956. *Die Grundlagen des Historismus in der deutschen Nationalökonomie.* Stuttgart.
Hildebrand, B. 1848. *Die Nationalökonomie der Gegenwart und Zukunft.* Vol. 1, Frankfurt am Main.
Keynes, J.N. 1891. *The Scope and Method of Political Economy.* London: Macmillan.
Knies, K.G.A. 1850. *Die Statistik als selbständige Wissenschaft.* Kassel: Verlag der J. Luckhardt'schen Buchhandlung.
Knies, K.G.A. 1853. *Die politische Ökonomie vom Standpunkte der geschichtlichen Methode.* Braunschweig: Schwetschke.
List, F. 1925. *Das nationale System der politischen Ökonomie.* 8th edn, Stuttgart and Berlin.
Marshall, A. 1897. The older generation of economists and the new. *Quarterly Journal of Economics* 11.
Menger, C. 1883. *Untersuchungen über die Methode der Socialwissenschaften, und der politischen Ökonomie insbesondere.* Leipzig: Duncker & Humblot.
Montaner, A. 1948. *Der Institutionalismus als Epoche amerikanischer Geistesgeschichte.* Tübingen: Mohr.
Roscher, W.G.F. 1843. *Grundriß zu Vorlesungen über die Staatswirtschaft nach geschichtlicher Methode.* Göttingen.
Roscher, W.G.F. 1854. *System der Volkswirtschaft. Ein Hand- und Lesebuch für Geschäftsmänner und Studierende.* Vol. 1: *Die Grundlagen der Nationalökonomie.* Stuttgart: Cotta.
Schäffle, A.E.F. 1878. *Enzyklopädie der Staatslehre.* Tübingen.
Schäffle, A.E.F. 1873. *Das gesellschaftliche System der menschlichen Wirtschaft.* 2nd edn, Tübingen.
Schmoller, G. 1888. *Die Schriften von Menger und W. Dilthey zur Methodologie der Staats- und Sozialwissenschaften (1883).* In G. Schmoller, *Zur Litteraturgeschichte der Staats- und Sozialwissenschaften,* Leipzig: Duncker & Humblot, 275–304.
Schmoller, G. 1897. *Wechselnde Theorien und feststehende Wahrheiten im Gebiete der Staats- und Sozialwissenschaften und die heutige deutsche Volkswirtschaftslehre.* Berlin.
Schmoller, G. 1890. *Zur Sozial- und Gewerbepolitik der Gegenwart.* Leipzig.
Sombart, W. 1930. *Die drei Nationalökonomien. Geschichte und System der Lehre von der Wirtschaft.* Munich.
Sombart, W. 1925. *Die Ordnung des Wirtschaftslebens.* Berlin.
Sombart, W. 1902. *Der moderne Kapitalismus.* 2 vols, Leipzig: Duncker & Humblot.
Sombart, W. 1903. *Die deutsche Volkswirtschaft im 19. Jahrhundert.* Berlin: Georg Bondi.
Spiethoff, A. (ed.) 1938. *Gustav von Schmoller und die deutsche geschichtliche Volkswirtschaftslehre, Festgabe zur 100. Wiederkehr seines Geburtstages.* Berlin.
Veblen, T. 1919. *The Place of Science in Modern Civilization.* New York: Huebsch.
Wagner, A. 1895. *Sozialismus, Sozialdemokratie, Katheder– und Staatssozialismus.* Berlin.
Wagner, A. 1899. *Finanzwissenschaft.* Leipzig.
Wagner, A. 1912. *Die Strömungen in der Sozialpolitik und der Katheder- und Staatssozialismus.* Berlin: Volkstümliche Bücherei.
Wagner, A.D.H. 1912. *Finanzwissenschaft, Britische Besteuerung im 19. Jahrhundert und bis zur Gegenwart (1815–1910).* Vol. 2(3), 2nd edn, Leipzig.
Weber, M. 1922. *Roscher und Knies und die logischen Probleme der historischen Nationalökonomie, gesammelte Aufsätze zur Wissenschaftslehre.* Tübingen: Mohr.

Gerschenkron, Alexander (1904–1978). Gerschenkron was born in Odessa in 1904 and died in Cambridge, Massachusetts, in 1978. He left Russia in 1920 and settled in Austria. In 1938, a decade after receiving the degree of *doctor rerum politicarum* from the University of Vienna, he emigrated to the United States and spent the next six years at Berkeley. After a short period at the Federal Reserve Board, he went to Harvard in 1948 to teach both economic history and Soviet Studies. His passion for the former dominated, and he flourished there as the *doyen* of economic history in the United States. He influenced a generation of Harvard economists through his required graduate course in economic history and attracted several to his seminar and the field. His erudition and breadth were legendary, and defined an indelible, if unattainable, standard of scholarship for his colleagues and students.

Gerschenkron's principal contribution to economics was the elaboration of a model of late-comer economic development. Its central hypothesis is the positive role of relative economic backwardness in inducing systematic substitution for supposed prerequisites for industrial growth. State intervention could, and did, compensate for the inadequate supplies of capital, skilled labour, entrepreneurship and technological capacity found in follower countries. Thus the German institutional innovation of the 'great banks' provided access to needed capital for industrialization, even while greater Russian backwardness required a larger and more direct state role.

Gerschenkron's analysis is consciously anti-Marxian: it rejected the English Industrial Revolution as the normal pattern of economic development and deprived the original accumulation of capital of much of its conceptual force. Elements of modernity and backwardness could survive side by side, and did in a systematic way. Apparent disadvantageous initial conditions of access to capital could be overcome. Success was rewarded with proportionately more rapid growth, signalled by a decisive spurt in industrial expansion.

This model, first presented in 1952 in an essay entitled 'Economic Backwardness in Historical Perspective' (reprinted in 1962), underlay Gerschenkron's extensive research into the specific developmental experiences of Russia, Germany, France, Italy, Austria and Bulgaria. Out of those historical studies emerged a comparative, all-encompassing European picture. 'In this fashion, the industrial history of Europe is conceived as a unified and yet graduated pattern' (Gerschenkron, 1962, p. 1). In turn, his hypotheses became progressively more precise. They may be summarized as follows:

(1) Relative backwardness creates a tension between the promise of economic development, as achieved elsewhere, and the reality of stagnancy. Such a tension motivates institutional innovation and promotes locally appropriate substitution for the absent preconditions of growth.

(2) The greater the degree of backwardness, the more interventionist was the successful channelling of capital and entrepreneurial guidance to nascent industries. Also, the more coercive and comprehensive were the measures to reduce domestic consumption.

(3) The more backward the economy, the more likely were: an emphasis upon producers' goods rather than consumers' goods; use of capital intensive rather than labour intensive methods of production; emergence of larger rather than smaller units of both plant and enterprise; dependence upon borrowed, advanced technology rather than indigenous techniques.

(4) The more backward the country, the less likely was the agricultural sector to provide a growing market to industry through rising productivity, and the more unbalanced the resulting productive structure of the economy.

The considerable and continuing appeal of the Gerschenkron model derives from its logical and consistent ordering of the process of European development, the conditional nature of its predictions and its generalizability to the experience of the late late-comers of the present Third World. His formulation rises above other theories which emphasize stages of growth both because of its attention to historical detail and its insistence upon the special attributes of late-comer development that cause differential evolution. In Gerschenkron's own hands, his propositions afforded an opportunity to blend ideology, institutions and the historical experience of industrialization, especially that of Russia, in a dazzling fashion. For others, his approach has proved a useful starting point for the discussion of non-European late-comers, including Japan and the newly industrializing countries.

The model is, of course, not without its limitations. History, even of Europe alone, does not in every detail bear easily the weight of such a grand design. In other parts of the world, as might be expected from a concept rooted in the special features of the historical European experience, larger amendments are frequently required. And somewhat surprisingly, in view of Gerschenkron's own pathbreaking essay in political economy, *Bread and Democracy in Germany*, there is too little attention to the domestic classes and groups whose interests the interventionist state must adequately incorporate if it is to play the central role required. Backwardness too easily becomes an alternative, technologically rooted explanation, distracting attention from the state rather than focusing upon its opportunities and constraints.

Still, the concept of relative backwardness, and Gerschenkron's always insightful and rich elaborations in so many national contexts, represent a brilliant and original contribution to economic history for which he is justly celebrated. It is not the only one. The 'Gerschenkron effect',

arising from the difference between calculated Paasche and Laspeyre indexes of Soviet machinery output (1951), also commemorates him. Current price weights will tend to underestimate the extent of growth because prices and quantities are negatively correlated, just as base year weights exaggerate it. The larger is the difference between the alternatively constructed quantity indexes, the greater is the degree of structural change. Again, divergence rather than uniformity is the source of useful information about historical processes.

Alexander Gerschenkron has few peers, past or present, in his command of comparative economic history. Scholarly interest in contemporary economic development has brought him an increasing following. His insights thus continue to influence a new generation of scholars and guarantee him a central place in any assessment of the evolution of the discipline of economic history.

ALBERT FISHLOW

See also BACKWARDNESS.

SELECTED WORKS

1943. *Bread and Democracy in Germany*. Berkeley: University of California Press.
1951. *A Dollar Index of Soviet Machinery Output*. Santa Monica: Rand Corporation.
1962. *Economic Backwardness in Historical Perspective*. Cambridge, Mass.: Harvard University Press.
1968. *Continuity in History and Other Essays*. Cambridge, Mass.: Harvard University Press.
1970. *Europe in the Russian Mirror*. London: Cambridge University Press.
1977. *An Economic Spurt That Failed*. Princeton: Princeton University Press.

Gervaise, Isaac (*fl.* 1680–1720). Merchant and economist of French Huguenot extraction. Gervaise was born in the second half of the 17th century, probably in Paris, and migrated with his family to London in 1681. With his father he was associated with the Royal Lustring Company (1688–1720) engaged in the manufacture of a fine, light, black, glossy silk under patent granted by parliament. Ironically, the year the company lost its charter saw the publication of Gervaise's 34-page pamphlet, *The System or Theory of the Trade of the World* with its attack on exclusive companies. Foxwell (1940, p. 167) described it as 'one of the earliest formal systems of political economy ... stating one of the most forcible practical arguments for free trade'. Quite unlike much contemporary writing on trade, Gervaise's pamphlet is tersely written and especially noted for its peculiar terminology and highly abstract argument. Gervaise is presumed to have died in London by 1739.

The real discoverer of Gervaise's work, Viner (1937, pp. 79–80), has described it as 'an elaborate and close reasoned exposition of the nature of international equilibrium and of the self-regulating mechanism whereby specie obtained its 'natural' or proper international distribution'. The novelty of Gervaise's treatment of the specie mechanism is his emphasis on the role of income rather than prices in strong contrast with subsequent treatments by Cantillon, Vanderlint and Hume. The starting point for the analysis is the proposition that the equilibrium bullion stock of any nation is proportioned to its output in terms of labour and that such a stock also maintains the balance between consumption and production, exports and imports. Excess bullion breaks these balances by raising consumption and reducing production, thereby lowering exports and raising imports, hence bullion will be exported and the balances will be resorted. An inadequate bullion stock leads to specie inflow by raising production relative to consumption, and exports to imports. Gervaise treats credit as if it were bullion; oversupply or deficiency is self-correcting via the balance of trade, though the adjustment process with credit is more rapid through its additional income effects of interest payment to suppliers of credit whom he sees as consumers rather than producers. Hence 'credit is of pernicious consequences to that Nation that uses or encourages it beyond nature' (Gervaise, 1720, p. 14) – a comment perhaps not unrelated to contemporary developments with Law's system in France. War, capital consumption or export, and restrictions on trade may prevent or postpone attainment of monetary equilibrium. For this reason, and for the resource misallocation potential flowing from encouragement of specific manufactures through companies, laws or taxes on imports, Gervaise (1720, pp. 17–18) concludes that 'Trade is never in a better condition, than when it's natural and free.' Gervaise also pointed out that the 'natural proportion' of bullion for specific countries is influenced by their situation, particularly as regards proximity to water transport and that implementing policies of debasing the currency had similar effects on trade and the balance of consumption and production as credit oversupply. Although less elegantly written than Hume's later account of the specie mechanism, the emphasis on adjustment through income rather than price effects, though not always clearly explained, makes Gervaise's short and penetrating contribution to the subject more modern than most of its successors in the century and a half which followed.

PETER GROENEWEGEN

SELECTED WORKS

1720. *The System or Theory of the Trade of the World.* Reprinted with a biographical introduction by J.M. Letiche and a foreword by Jacob Viner, Baltimore: Johns Hopkins Press, 1954.

BIBLIOGRAPHY

Foxwell, H.S. 1940. Comment reproduced in *Catalogue of the Kress Library of Business and Economics*, Baker Library, Harvard Graduate School of Business Administration, Boston.
Viner, J. 1937. *Studies in the Theory of International Trade.* New York: Harper and Brothers.

Gesell, Silvio (1862–1930). Gesell was born in Germany but emigrated to the Argentine in 1886, where he was so successful as an importer that he retired to Switzerland in 1900 to farm and to continue to write. The 'retirement' included a return to the Argentine to manage his late brother's business and an involvement in Bavarian politics at their most chaotic. As a deposed Minister of Finance he was tried for high treason, and acquitted.

His prolific writing began in Argentina, provoked by the economic chaos of the late 1880s there. But his fame rests on *The Natural Economic Order*, originally published in two parts in 1906 and 1911. It was translated into English in 1929. Rent-free land and interest-free money characterize that Order. Land would be nationalized, its owners compensated by the issue of state bonds. Through the device of stamped money, which would remain current only if a stamp, obtained at a cost set by government, was regularly affixed, the rate of interest on these bonds and other lending instruments would eventually be driven to zero. With no income diverted to rent or interest the worker would receive the full value of his

output. Mothers were to receive income from annuities based on the nationalized land, since their 'output', the population, was the source of demand for land and hence rent.

Gesell attributed depressions to inadequate investment and the latter to the fall in the expected rate of return as investment continued, coupled with a money rate of interest which was prevented from falling by the alternative opportunity of hoarding. This analysis substantially anticipates Keynes's (1936), as Keynes amply acknowledges (pp. 353–8). Gesell suggested adjusting the stamp duty on money to force down the rate of interest.

The stamped money principle was three times applied on a local scale in the 1930s: in Bavaria, in the Austrian Tyrol and in Alberta, Canada. In each case the scheme successfully raised demand and employment, but the money was soon banned by the authorities.

Though theoretical inadequacies and practical difficulties are claimed against Gesell's theory, its *aim* is probably more responsible for its eclipse. But it lives on furtively, below the surface, in the underworlds of Keynes's *General Theory* and Fisher's *Booms and Depressions*.

VICTORIA CHICK

SELECTED WORKS

1891a. *Currency reform as a bridge to the social state.* Buenos Aires. Trans. by Philip Pye, typescript, 1951.
1891b. *Nervus rerum: continuation of 'currency reform as a bridge to the social state'.* Buenos Aires. Trans. by Philip Pye, typescript, 1951.
1929. *The Natural Economic Order: a plan to secure an uninterrupted exchange of the products of labour.* Trans. by Philip Pye, Berlin: Neo-Verlag.

BIBLIOGRAPHY

Fisher, I. 1933. *Booms and Depressions.* London: George Allen & Unwin, Appendix VII.
Gaitskell, H. 1969. *Four Monetary Heretics.* Christchurch: Lyn Christie & Son.
Keynes, J.M. 1936. *The General Theory of Employment, Interest and Money.* London: Macmillan.
Wise, L. (n.d.) *Silvio Gesell.* London: Holborn Publishing Co.

Giblin, Lyndhurst Falkiner (1872–1951). Giblin was born and died in Hobart, Tasmania. Trained in mathematics and statistics at King's College, Cambridge, he became teacher, gold-miner, fruit-grower, Labor politician in Tasmania, and soldier before beginning his career as statistician/economist in 1919. In his official positions as Tasmanian Government Statistician (1919–28), Acting Commonwealth Statistician (1931–2), Ritchie Research Professor of Economics at the University of Melbourne (1929–39), member of the Committee of Enquiry into the Australian Tariff (1927–9), of the Commonwealth Grants Commission (1933–6), the Commonwealth Bank Board (1935–42) and chairman of the Finance and Economic Policy Committee (1939–46), he exercised significant influence on economic policy-making in Australia. He shepherded the small band of Australian economists to preserve cohesion within the profession, provided links with governments, and endeavoured to raise public awareness of the nature and dimensions of economic problems.

In the course of his quantitative work, Giblin dealt with the economic choices faced by a politically federated nation, intent on economic development, but with an open economy subject to the vicissitudes of international trade and capital movements. As a basis for Federal financial relations, he pioneered the measurable concepts of relative taxable capacity and severity of taxation. He attempted to measure the 'excess costs' of protection and their redistributive effects. In 1929, while demonstrating the repercussive effects on Australian incomes of adverse movements in the terms of trade, he formulated a first version of the foreign trade multiplier. During the 1940s he devised the money control tool of requiring the trading banks to hold special deposits with the Commonwealth Bank.

M. HARPER

SELECTED WORKS

1929. (With J.B. Brigden, D.B. Copland, E.C. Dyason, and C.H. Wickens.) *The Australian Tariff: An Economic Enquiry.* Melbourne: Melbourne University Press (Economic Series No. 6).
1930a. State disabilities – with special reference to Tasmania: a memorandum submitted to the Committee of Public Accounts as evidence of Tasmanian disabilities. Appendix J, *The Case for Tasmania, 1930,* Hobart: Government Printer.
1930b. *Australia, 1930.* Melbourne: Melbourne University Press, (Economic Series No. 8).
1951. *The Growth of a Central Bank: The Development of the Commonwealth Bank of Australia 1924–1945.* Melbourne: Melbourne University Press.

BIBLIOGRAPHY

For a full bibliography see
Copland, D. (ed.) 1960. *Giblin: the Scholar and the Man.* Melbourne: F.W. Cheshire.

Gibrat, Robert Pierre Louis (1904–1980). Gibrat was born on 23 March 1904 in Lorient, France, and died on 13 May 1980 in Paris. He studied at Saint Louis de Paris, as well as in Rennes, Lorient and Brest, and in 1922 he entered the Ecole Polytechnique to become a mining engineer. He received a bachelor's degree in science and a doctorate in law from the University of Paris. He was a technical consultant in private firms before being named director of electricity in the Ministry of Public Works, 1940–42. He became Secretary of State for Communications under the Laval government but resigned after the Allied invasion of North Africa. After the Liberation he was chief engineer of mines. He was consulting engineer for French Electric on tidal energy (1945–68) and served as Director General for atomic energy (Indatom, 1955–74), president of the scientific and technical committee of Euratom (1962), and as a consulting engineer for Central Thermique, 1942–80. He taught at Ecole des Mines from 1936 to 1968. He served as President of the French Society of Electricians, Vice President and President of the Civil Engineers of France, President of the Statistical Society of Paris (1966), President of the French Statistical Society, President and Honorary President of the Technical Committee for the Hydrotechnical Society of France, President of the French Meteorological Society (1969), Honorary President of the World Federation of Organizations of Engineers, and President of the French Section of the American Nuclear Society. Gibrat was the author of reports to the Academy of Sciences, some 100 professional articles and two books (on economics and tidal energy), and a Knight of the Legion of Honor.

His major contribution to economics is known as Gibrat's Law. This states that the expected growth rate for a firm is independent of its size. Gibrat's Law has been successfully tested by French and American investigators, among others. His famous economics work, *Les inégalités économiques* was published in 1931. For his contributions to economics and mathematical economics he was elected a Fellow of the Econometric Society in 1948.

DAVID E.R. GAY

SELECTED WORKS

1930. Une loi des répartitions économiques: l'effet proportionnel. *Bulletin de la Statistique générale de la France et du Service d'observations des prix,* July–September.
1931a. *Les inégalités économiques.* Paris: Sirey.
1931b. Les inégalités économiques. *Revue de l'industrie minérale,* July.
1935. La science économique. Méthodes et philosophie. *Actualités scientifiques et industrielles,* Paris.

Gibrat's Law. Gibrat's Law is a proposition concerning the process of growth in size of firm. According to this law, the probability of a given proportional change in size (during a particular period) is the same for all firms in a given industry, regardless of their size at the beginning of the time period (see Gibrat, 1931). Thus, a firm with sales of $1 billion is as likely to double in size during a given period as a firm with sales of $10 million. Put differently Gibrat's Law states that

$$X_{ij}^{t+\Delta} = U_{ij}(t, \Delta)X_{ij}^t, \tag{1}$$

where X_{ij}^t is the size of the i_{th} firm in the jth industry at time t, $X_{ij}^{t+\Delta}$ is its size at time $t + \Delta$, and $U_{ij}(t, \Delta)$ is a random variable distributed independently of S_{ij}^t.

This law is a basic component of many mathematical models designed to explain the shape of the size distribution of firms. As is well known, this distribution ordinarily is highly skewed, and Herbert Simon and others have used Gibrat's Law to explain this fact (see, for example, Simon and Bonini, 1958). Without question, Gibrat's Law is, for many purposes, a useful first approximation, although econometric studies indicate that it should be used with caution (see Mansfield, 1962).

Gibrat's Law can be formulated in at least three ways, depending on the treatment of the death of firms and the comprehensiveness claimed for the law. First, one can postulate it holds for all firms, including those that exit the industry during the period. If the end-of-period size of each departing firm is regarded as zero, this version of the law does not fare very well, because the probability that a firm will leave is not independent of its size. Econometric studies indicate that smaller firms are more likely than larger firms to leave an industry.

Second, one can postulate that the law holds for all firms other than those that leave the industry. This version of the law also seems to run into difficulties. Equation (1) implies that

$$\ln X_{ij}^{t+\Delta} = V_i(t, \Delta) + \ln X_{ij}^t + e_{ij}(t, \Delta), \tag{2}$$

where $V_i(t, \Delta)$ is the mean of $\ln U_{ij}(t, \Delta)$ and $e_{ij}(t, \Delta)$ is a homoscedastic random variable with zero mean. Consequently, if $\ln X_{ij}^{t+\Delta}$ is plotted against $\ln X_{ij}^t$, the data should be scattered with constant variance about a line with slope of one, if this version of the law holds. In fact, studies indicate that the slope tends to be less than one, and the variance of $X_{ij}^{t+\Delta}/X_{ij}^t$ tends to be inversely related to X_{ij}^t.

Third, one can postulate that the law holds only for firms exceeding the minimum efficient size in the industry. This is the version put forth by Simon and Bonini (1958), although it seems to be a stronger assumption than they require. Excluding firms that exit the industry, the available evidence often is quite consistent with the hypothesis that the expected value of $X_{ij}^{t+\Delta}/X_{ij}^t$ does not vary with X_{ij}^t. However, the variance of $X_{ij}^{t+\Delta}/X_{ij}^t$ often tends to be inversely related to X_{ij}^t.

EDWIN MANSFIELD

See also LOGNORMAL DISTRIBUTION.

BIBLIOGRAPHY

Gibrat, R. 1931. *Les inegalités économiques*. Paris: Recueil Sirey.

Mansfield, E. 1962. Entry, Gibrat's law, innovation, and the growth of firms. *American Economic Review* 52, December, 1023–51.

Simon, H. and Bonini, C. 1958. The size distribution of business firms. *American Economic Review* 48, September, 607–17.

Gide, Charles (1847–1932). Gide was born at Uzès (Gard) and died in Paris. A strong Huguenot family tradition marked his austere and moralizing personality. His doctoral thesis was on *Le droit d'association en matière religieuse*. He was a founder of the *Revue d'Economie Politique* (1887). He taught at the Faculty of Law in Bordeaux (1874), in Montepellier (1880), in Paris (1898) and at the Collège de France (1919).

Best known for the *History of Economic Doctrines* he wrote with Charles Rist, his *Principes* and his *Cours d'économie politique*, Gide's claim to creativity lies in the field of social ethics. In reaction to Bastiat, whose works introduced him to economics, Charles Gide drew inspiration from Fourier and Robert Owen to become a moral critic of the competitive system. Against the self-seeking competitive spirit, he opposed the values of solidarity and cooperation. Like J.S. Mill, Gide drew a distinction between the realm of natural necessity and that of human volition. The principle of solidarity and cooperation, as a moral duty, should ultimately supersede the struggle of man against man. As the founder of the Ecole de Nîmes, a Huguenot intellectual centre, Gide remained all his life a preacher of the cooperative gospel. Like Fourier, he was a visionary of a better world to come; like J.S. Mill, he was aware of the historical relativity of human institutions. His artist's mind was unwilling to bend to the canons of any rigid science. The purpose of economics was to enlighten the road and inspire the endeavour to a better world, though his message of hope was tempered by a measure of scepticism. In international relations he was a pacifist.

ROGER DEHEM

SELECTED WORKS

1872. Le droit d'association en matière religieuse. Paris, Faculté de droit, thesis.

1884. *Principes d'économie politique*. Paris: Larose et Forcel. 26th edn, 1931.

1887. La notion de valeur dans Bastiat au point de vue de la justice distributive. *Revue d'économie politique* 1(3).

1894–9. Proudhon (III, 237–8); J.B. Say (III, 357–8); Saint-Simon (III, 346–7); Solidarity (III, 444–5); Sully (III, 486–7). In *Dictionary of Political Economy*, ed. R.H. Inglis Palgrave, London: Macmillan.

1904. *Les sociétés coopératives de consommation*. Paris: Colin.

1907. Economic literature in France at the beginning of the twentieth century. *Economic Journal* 17, June, 192–212.

1905. *Les institutions du progrès social au début du XXe siècle*. Paris: Larose et Tenin.

1915. (With Ch. Rist.) *A History of Economic Doctrines*. London: G. Harrap. 2nd edn, 1948.

1920. *Des institutions en vue de la transformation ou de l'abolition du salariat*. Paris: Giard.

1924. *Fourier, précurseur de la coopération*. Paris: Association pour l'Enseignement de la Coopération.

1928. *L'école de Nîmes*. Paris: Association pour l'Enseignement de la Coopération.

1930. *Les colonies communistes et coopératives*. Paris: Association pour l'Enseignement de la Coopération.

1930–35. Boyve (II, 670); Cooperation (IV, 376–81). In *Encyclopaedia of the Social Sciences*, London: Macmillan & Co.

1931. (With W. Oualid.) Le bilan de la guerre pour la France. In *Histoire économique et sociale de la guerre mondiale*, Paris: Presses Universitaires de France.

1932. *La solidarité*. Paris: Presses Universitaires de France.

Giffen, Robert (1837–1910). Robert Giffen's name seems likely to be known by students of economics for generations to come in relation to the famous result in the theory of consumer demand which bears his name but about which, so far as can be determined, he had nothing to say. Marshall originated the tradition when he associated the result with Giffen's name in the third edition of his *Principles* in 1895 (p. 208).

Giffen was born in Lanarkshire. At the age of 13 he was apprenticed to a solicitor in Strathaven, and continued in the same vocation until 1860 (though during the last seven years of this period he resided and worked in Glasgow). Still only twenty-three years old, Giffen struck out on a career in journalism – in which he was to be successful in establishing his reputation in economic circles of the day. He begun as a sub-editor for the *Stirling Journal*, moved to London in 1862 to work at the *Globe*, transferred to the *Fortnightly Review* in 1866, and in 1868 became assistant editor at the *Economist* – a post at which he remained until his next change of vocation in 1876. He was also city editor at the *Daily News* between 1873 and 1876. Giffen's third and final career was as a professional civil servant, first as chief of the statistical department at the Board of Trade, and then in 1882 as its Assistant Secretary. He retired from the civil service at the age of sixty. Giffen served on numerous Royal Commissions (including the Gold and Silver Commission of 1886–8); he was editor of the *Journal of the Royal Statistical Society* (1876–91), President of that Society (1882–4), twice presided over the economics section of the British Association (1887 and 1901), and was one of the founders of the Royal Economic Society. In short, he was one of those figures encountered frequently in British economics whose not inconsiderable power and prestige appears to be disproportionate to their actual contribution to economic science.

In so far as he was primarily a statistician, Giffen's work did attempt to alert economists to the dangers of theory without measurement. His presidential address to the Royal Statistical Society in 1882 was devoted to the subject, and in 1901 as president of Section F of the British Association (his second term in that office) he returned to the same theme (see 1904, II, chs 13 and 28). Indeed, according to Higgs in his edition of this *Dictionary* (1925), Giffen's statistical prowess was one of the factors which helped to secure the respect of theorists. His article on international statistical comparisons in the *Economic Journal* for 1892, for example, can be singled out for special mention since it treats for the first time a problem which has still not been adequately resolved. Of course, it was not always the case that Giffen's careful mustering of the statistical evidence allowed him, any more than the theorists, to avoid the pitfalls of making predictions which subsequent experience has proven to be silly – witness his claim that the whole protectionist school would die out within a decade (1898, p. 16).

However, in the final analysis it is in Giffen's attempts to provide reasonably accurate measurements of indicators like wage rates, economic growth (see 1884), and national product (1889), that one should isolate his main contribution. While it is true that subsequent work in this field has advanced well beyond Giffen's early efforts, he remains one of the pioneers of applied economics in its modern sense.

It seems that Giffen was also a strong supporter of a Channel tunnel: not for one between England and France, but between Ireland and England. He died on 12 April 1910 and is buried in Strathaven.

MURRAY MILGATE

SELECTED WORKS

1872. (With B. Cracroft.) *American Railways as Investment*. London.

1873. The production and movement of gold since 1848. In Giffen (1880).

1877. *Stock Exchange Securities*. London: G. Bell & Sons.

1880. *Essays in Finance*. First Series, Vol. I. London: G. Bell & Sons.

1884. *The Progress of the Working Class in the Last Half Century*. London: G. Bell & Sons.

1886. *Essays in Finance*. Second Series, Vol. II. London: G. Bell & Sons.

1887. The recent rate of material progress in England. Address as president of Section F of the British Association. *Journal of the Royal Statistical Society* 50, December, 615–47. Reprinted in Giffen (1904), Vol. II.

1889. *The Growth of Capital*. London: G. Bell & Sons.

1892a. *The Case Against Bimetallism*. London: G. Bell & Sons.

1892b. On international statistical comparisons. *Economic Journal* 2, June, 209–38.

1898. Protection for manufactures in new countries. *Economic Journal* 8, March, 3–16.

1904. *Economic Inquiries and Studies*. 2 vols, London: G. Bell & Sons.

Giffen's paradox. This is the occurrence of a positively-inclined demand curve. Henry Beeke remarked upon the paradox in 1800 (Rashid, 1979), but a far more clear and complete statement of it was given by Simon Gray in 1804 (Gray, 1815). 'Part of the population lives so much on bread', he wrote, because 'their incomes are not sufficient to enable them to buy meat, &c. even when bread is cheap. All the surplus after buying bread, and some necessary articles of clothing, is laid out, indeed, on these more desirable species of food.' The result is that 'to raise the price of corn [grain] in any great degree, tends directly to increase the general consumption of that necessary', because some of the surplus has to be used to buy more bread; and 'in proportion as bread rises in price, the surplus decreases, till at length there is no surplus at all' (ibid., pp. 505–6; and see Powell, 1896; Masuda and Newman, 1981). The most famous statement of the paradox was given by Alfred Marshall:

As Mr. Giffen has pointed out, a rise in the price of bread makes so large a drain on the resources of the poorer labouring families and raises so much the marginal utility of money to them, that they are forced to curtail their consumption of meat and the more expensive farinaceous foods: and, bread being still the cheapest food which they can get and will take, they consume more, and not less of it (Marshall, 1895, p. 208; 1920, p. 132).

It has not been verified that Robert Giffen, an English statistician who was Marshall's contemporary, mentioned the paradox (Stigler, 1947; but see Giffen, 1909, p. 334).

The modern explanation of it identifies two consequences of a rise in the price of a commodity on the quantity of it demanded by a consumer. First, the consumer's purchasing power falls; he is poorer because he has the same money income but the price of a commodity has risen. Consequently he purchases less of the more expensive commodities and to sustain himself has to buy additional amounts of commodities that are cheap but that he regards as inferior because they are not very appetizing or are boring, including more of the inferior good whose price has risen. That is, the income effect on the quantity demanded of that commodity is positive. For the income effect to be large, a considerable amount of purchasing power must be lost by the rise in the price, and therefore a large proportion of the consumer's income must be spent on the inferior good, as is true for some people in some cultures for commodities like beans or bread or rice. Second, because the price of the inferior good has risen relative to the prices of its substitutes, the consumer tends to displace it to some degree by purchasing more of them. That is, the substitution effect on the quantity demanded of the inferior commodity is negative. If the positive income effect is sufficiently large to more than offset the negative substitution effect, the paradoxical result occurs that the quantity demanded rises when the price rises. The commodity is then not merely inferior but a Giffen good to the individual. If it is a Giffen good over a particular range of prices to all or many of the consumers, then the market demand curve for the commodity will be positively sloped.

Giffen's paradox has generated an extensive literature, for three major reasons.

First, economists have wished to explore the theoretical conditions under which a Giffen market demand curve occurs in order to determine whether or not it is likely to occur very frequently. The issue is important because if the paradox occurs frequently, much of economic theory and many policy prescriptions would need to be reformulated, for they involve the assumption that price and quantity demanded are inversely related.

Second, a number of interesting problems arise in trying to show how the paradox does or does not fit in with Marshall's general theory of demand. The frequently contradictory arguments on these problems depend upon debatable characterizations of both Marshall's general theory of demand and of the relation of the paradox to it. One argument is that Giffen's paradox is an exceptional empirical case that lies outside Marshall's general demand theory because the latter depends upon assumptions with which the paradox is inconsistent, namely, that the purchasing power of money is constant, that a small part of the budget is spent on any commodity, and that the marginal utility of money is constant (Gramm, 1970). Another argument is that Marshall was inconsistent because his general definition of an individual demand curve requires that changes in money income or purchasing power offset the effects of changes in the price of the commodity so as to hold the consumer's level of welfare constant, whereas along the demand curve for a Giffen good changes of purchasing power occur and so therefore do changes of the level of welfare (Friedman, 1949, pp. 83–5). Another is that the paradox is inconsistent with Marshall's assumption of additive utilities (Stigler, 1950, p. 327). Yet another is that the paradox is consistent with Marshall's general demand theory because the level of welfare changes along all Marshallian individual demand curves and because he did not assume that utilities are additive (Mayston, 1976, p. 504). The points that have been settled are that Marshall did assume additively separable utilities, and that Giffen's paradox is consistent with that assumption (Silberberg and Walker, 1984). It has also been shown that, contrary to Marshall's belief, the marginal utility of income falls when the price of the Giffen good rises (ibid.).

Third, a number of theoretical and practical problems have confounded attempts to make econometric tests of the paradox and led to renewed efforts to explore its mysteries (Boland, 1977). The underlying presumed structure of market supply and Giffen demand curves can generate equilibria that do not materialize because they are unstable (Dougan, 1982). Intractable problems arise in trying to identify an upward-sloping demand curve as distinct from an upward-sloping locus of points of equilibrium on different negatively inclined demand curves. Instances that fit the conditions of the paradox are hard to find. For example, the Irish potato famine of 1845–8, which has frequently been cited as an example, does not fit. For if the market supply and demand curves are both positively inclined, if price is the independent variable, and if equilibrium is stable, the increase in price required for the paradox can come about only as a result of a decrease in supply or an increase in demand, or both of those changes simultaneously. In each of those cases there would be an increased equilibrium quantity supplied and consumed, which obviously did not happen in Ireland (Dwyer and Lindsay, 1984). Similarly, the situation in which a commodity absorbs a large part of consumers' expenditures ordinarily occurs when it is a kind of food, and often the consumers are producers of much or all of that food. In each of the foregoing cases the producer's income would be increased, which, if the consumers are the producers, would offset the decreased purchasing power caused by the increase of the price, and that would mask the Giffen effect.

As yet there has been no empirical verification of the existence of a Giffen individual demand curve using data on household behaviour, nor of a Giffen market demand curve. Indeed, the probability of discovering a Giffen market demand curve that has actually existed is slight (Dougan, 1982). Giffen's paradox therefore remains an elusive possibility, sufficiently remote in likelihood of either occurrence or detection as to fail to give rise to apprehension for the foundations of economic science.

DONALD A. WALKER

BIBLIOGRAPHY

Boland, L.A. 1977. Giffen goods, market prices and testability. *Australian Economic Papers* 16(28), June, 72–85.

Dougan, W.R. 1982. Giffen goods and the law of demand. *Journal of Political Economy* 90(4), August, 809–15.

Dwyer, G.P., Jr., and Lindsay, C.M. 1984. Robert Giffen and the Irish potato. *American Economic Review* 74(1), March, 188–92.

Friedman, M. 1949. The Marshallian demand curve. *Journal of Political Economy* 57(6), December, 463–95.

Giffen, R.G. 1909. City notes. *Economic Journal* 19(74), June, 332–5.

Gramm, W.P. 1970. Giffen's paradox and the Marshallian demand curve. *Manchester School of Economics and Social Studies* 38(1), March, 65–71.

Gray, S. 1815. *The Happiness of States: Or An Inquiry Concerning Population, The Modes of Subsisting and Employing It, and the Effects of All on Human Happiness.* London: Hatchard.

Marshall, A. 1895. *Principles of Economics.* 3rd edn, London: Macmillan.

Marshall, A. 1920. *Principles of Economics.* 8th edn, London: Macmillan.

Masuda, E. and Newman, P. 1981. Gray and Giffen goods. *Economic Journal* 91, December, 1011–14.

Mayston, D.J. 1976. On the nature of marginal utility – a neo-Marshallian theory of demand. *Economic Journal* 86, September, 493–507.

Powell, E. 1896. Simon Gray. In *Dictionary of Political Economy*, ed. R.H.I. Palgrave, Vol. 2, London: Macmillan.

Rashid, S. 1979. The Beeke good: a note on the origins of the 'Giffen good'. *History of Political Economy* 11(4), Winter, 606–7.

Silberberg, E. and Walker, D.A. 1984. A modern analysis of Giffen's Paradox. *International Economic Review* 25(3), October, 687–94.

Stigler, G.J. 1947. Notes on the history of the Giffen Paradox. *Journal of Political Economy* 55(2), April, 152-6.

Stigler, G.J. 1950. The development of utility theory. Pts I and II. *Journal of Political Economy* 58(4), August, 307–27; 58(5), October, 373–96.

gifts. A gift, according to the *Concise Oxford Dictionary*, is a 'voluntary transference of property; thing given, present, donation'. For most economists, especially those familiar only with industrial capitalist economies, this is all that need be said on the matter: it is obvious what gift exchange is and there is nothing to be explained. The only problem the phenomenon of exchange poses for the economist is that of 'value' and this arises in the context of commodity exchange.

For the anthropologist, however, the phenomenon of exchange poses questions about the nature of gift exchange. These lie at the centre of the discipline and the topic has been the subject of much theoretical debate. Anthropologists stress that while gifts appear to be voluntary, disinterested and spontaneous, they are in fact obligatory and interested. It is this underlying obligation that anthropologists seek to understand: What is the principle whereby the gift received has to be repaid? What is there in the thing given that compels the recipient to make a return?

It is clear that the one economic category – exchange – means fundamentally different things to different people and that these contrary perceptions of the exchange process have given rise to quite distinct theoretical traditions. The reasons for this are to be found in the historical conditions which gave rise to the development of the academic disciplines of economics and anthropology. The history of economic thought must be understood with reference to the development of mercantile and industrial capitalism in Europe; the development of anthropological theorizing, on the other hand, must be situated in the context of the imperialist expansion of European capitalism and especially the colonial conquest of Africa and the Pacific towards the end of the 19th century. The fact that economists have been preoccupied with commodity exchange whilst anthropologists have been primarily concerned with gift exchange simply reflects the fact that the modern European economy is organized along very different lines from the indigenous economies of Africa, the Pacific and elsewhere. The data anthropologists have collected from these countries over the past one hundred years has revolutionized our understanding of tribal economy and the theory of the gift; their theoretical reflections on this data constitutes a major contribution to the theory of comparative economic systems and also to the theory of development and underdevelopment. This anthropological literature takes us far beyond the superficial dictionary definition of the gift and raises important questions about the seemingly unrelated issue of shell money; interestingly it also brings us back to the original meaning of the word as 'payment for a wife' and 'wedding' found in *The Oxford Dictionary of English Etymology*.

Anthropological accounts of gift giving first began appearing towards the end of the 19th century; by the end of World War I a large quantity of data had been collected. The most spectacular accounts came from the Kwakiutl Indians of the northwest coast of America and from the Melanesian Islanders of the Milne Bay District of Papua New Guinea. Among the Kwakiutl vast amounts of valuable property (mainly blankets) are ceremonially destroyed in a system called 'potlatch' (Boas, 1897). In the potlatch system the prestige of an individual is closely bound up with giving: a would-be 'big-man' or 'chief' is constrained to give away or to destroy everything he possesses. The principles of rivalry and antagonism are basic to the system and people compete with one another, each trying to outgive the other in order to gain prestige. The status and rank of individuals and clans is determined by this war of property. In Papua New Guinea, the classic home of competitive gift exchange, the instruments of 'gift warfare' are food (Young, 1971) and shells of various shapes and sizes (Leach and Leach, 1983). These are not destroyed but transacted by status seekers according to complicated sets of rules which we are only now beginning to understand. Most Papua New Guinean societies are without any form of ascribed status and the egalitarian ideology of these societies means that competitive gift giving is primarily concerned with the maintenance of equal status rather than dominance. Staying equal is, as Forge (1972) has pointed out, an extremely onerous task requiring continual vigilance and effort: perfect balance is impossible to achieve as the temporal dimension of gift exchange necessarily introduces status inequalities. Perhaps the most complicated gift exchange system in Melanesia is the Rossel Island 'monetary system'. This was first described by Armstrong (1924) and has recently been restudied by Liep (1983). On Rossel Island there are two kinds of 'shell money' *ndap* and *ko*. A single unit of *ndap* is a polished piece of *spondylus* shell a few millimetres thick, having an area varying from 2 to 20 square centimetres and roughly triangular in shape. A single unit of *ko* consists of ten pieces of *chama* shell of roughly the same size and thickness with a small hole in the centre for binding them together. Each shell group contains some forty-odd hierarchical divisions. What is unusual about these divisions is that they have rank rather than value, that is, they are ordinally related rather than cardinally related. For example, the relationship of a big *ndap* shell to a small one is analogous to that between an ace of hearts and a two of hearts rather than that between a dollar and a cent.

The publication of Armstrong's (1928) ethnography of Rossel Island and Malinowski's (1922) now classic description of an inter-island gift exchange system called *kula* sparked off a debate about the nature of 'shell money' which still rages today. This debate is kept alive not from an antiquarian interest in 'archaic' money systems but because these gift exchange systems are still flourishing despite their incorporation into the world capitalist economy (MacIntyre and Young, 1982; Gregory, 1980, 1982). On Rossel Island, for example, not only are the *ndap* and *ko* shells still transacted as gifts according to the complicated rules of old, but the demand for Rossel Island *chama* shells for use in the flourishing *kula* gift exchange system of neighbouring islands has transformed Rossel Island into a major commodity producer and exporter of *chama* shells (Liep, 1981; 1983).

These facts raise conceptual questions about the difference between gift exchange and commodity exchange, and theoretical and empirical questions about the nature of the interaction between them. Neoclassical economics answers these questions within a framework that employs the universalist and subjectivist concept 'goods', a category which, by definition, cannot explain the particularist and objective nature of gift and commodity exchange (Gregory, 1982). A 'gift' therefore becomes a 'traditional good' and highly questionable psychological criteria are used to distinguish this from a 'modern good'. For example, Einzig argues that 'the intellectual standard' of people in tribal societies 'is inferior and their mentality totally different from ours' (1948, p. 16); Stent and Webb (1975, p. 524) argue that 'traditional' consumers in Papua New Guinea are on the bliss point of their indifference curves. A further difficulty economists have with the problem of contrasting economic systems – and this is not restricted to neoclassical economic thought – is the habit of beginning an argument with an analysis of barter in an 'early and rude state of society'. The barter economies of these theories are figments of a Eurocentric imagination that bear no resemblance at all to actual tribal economies. Economic anthropologists have been making this point for over fifty years but without much success (Malinowski, 1922, pp. 60–61; Polanyi, 1944, pp. 44–5). What is needed, then, is an

empirically based theory of comparative economy. The foundations of such a theory were laid by Marx (1867) but the rise to dominance of neoclassical theory precluded any further development of the theory of comparative economy within the economics discipline. The theoretical advances have come from without and have been made by anthropologists, sociologists and economic historians.

The outstanding contribution to the 20th-century literature is undoubtedly Mauss's *The Gift: Forms and Functions of Exchange in Archaic Societies*, first published in French in 1925 as 'Essai sur le don, forme archaïque de l'echange' in Durkheim's journal, *L'Année Sociologique*. Mauss (1872–1950) was Durkheim's nephew and became a leading figure in French sociology after his uncle's death. His essay on the gift is a remarkable piece of scholarship. Not only did he survey all extant ethnographic data on gift giving from Melanesia, Polynesia, northwest America and elsewhere, he also examined the early literature from Ancient Rome, the Hindu classical period and the Germanic societies. His essays conclude with a critique of western capitalist society by drawing out the moral, political, economic and ethical implications of his analysis.

The key to understanding gift giving is apprehension of the fact that things in tribal economies are produced by non-alienated labour. This creates a special bond between a producer and his/her product, a bond that is broken in a capitalist society based on alienated wage-labour. Mauss's analysis focused on the 'indissoluble bond' between things and persons in gift economies and argued that 'to give something is to give a part of oneself' (1925, p. 10). Gifts therefore become embodied with the 'spirit' of the giver and this 'force' in the thing given compels the recipient to make a return. This does not exist in our system of property and exchange which is based on a sharp distinction between things and persons, that is, alienation (1925, p. 56). The wage-labourer in a capitalist society gives a 'gift' which is not returned (1925, p. 75). Capitalism for Mauss then was a system of non-reciprocal gift exchange; a system where the recipients of a gift were under no obligation to make a return gift.

This analysis of the wage-labour contract under capitalism has a Marxian ring about it. However, Mauss was no revolutionary and he drew very different policy conclusions from his analysis of the wage-labour relation. He argued for a welfare capitalism where the state, through its social legislation, provided recompense to the workers for their gifts.

A feature of Mauss's work, and indeed a feature of much early theorizing about the gift, was the evolutionary framework within which the ethnographic data was analysed. The tribal economies studied by anthropologists were seen as living fossils from European pre-history, hence the use of terms such as 'archaic' and 'primitive'. These early theorists, then, were only concerned with the intellectual contribution this data could make to the study of comparative economy. To the extent that they were concerned with the welfare of living people it was the welfare of their European countrymen and women; they were not concerned with policies for the development of tribal peoples.

The other outstanding theorist in this evolutionary tradition was another Frenchman, Claude Lévi-Strauss. His theory of the gift is contained in his *The Elementary Structures of Kinship* (1949). Like Mauss's *The Gift*, Lévi-Strauss's book is an encyclopaedic survey of the ethnographic literature. Its central focus is marriage. In line with a long tradition in anthropology he conceptualizes this as an exchange of women. However, Lévi-Strauss's innovation is to argue that women are the 'supreme gift' and that the incest taboo is the key to understanding gift exchange. The virtual universal prohibition on marriage between close kin, he argues, is the basis of the obligation to give, the obligation to receive, and the obligation to repay.

Lévi-Strauss's theory is an analytical synthesis of literally thousands of ethnographic accounts from the Australian Aborigines, the Pacific and Asia. The original or most elementary form of gift exchange, according to Lévi-Strauss, is 'restricted' exchange where the moieties of a population exchanged sisters at marriage; the second form is 'delayed' exchange where a woman is given this generation and her daughter returned the next; the most advanced form is 'generalized' exchange where one clan gives women to another clan but never receives any in return, the closure of the system being brought about by a circle of giving. In the movement from one stage to another, extra spheres of gift exchange are developed as symbolic substitutes for women. These are needed to maintain the ever widening marriage alliances brought about by the shift from restricted to generalized exchange. This movement from marriage to exchange is an aspect of an opposing movement from exchange to marriage. Lévi-Strauss sees a continuous transition from war to exchange, and from exchange to intermarriage as effecting a transition from hostility to alliance, and from fear to friendship.

Lévi-Strauss's theory has attracted considerable critical attention and has been described by his principal opponent as 'in large measure fallacious' (Leach, 1970, p. 111). Whatever its shortcomings his theory nevertheless manages to establish the important link between gift giving and the social organization of kinship and marriage. In other words, he has established a relationship between the obligation to give and receive gifts and the biological and social basis of human reproduction.

While Lévi-Strauss was developing his theory of the gift, an economic historian, Karl Polanyi, was approaching the problem from an altogether different perspective in his classic study, *The Great Transformation* (1944). His problem was the analysis of the emergence of the 'self-regulating market' and in order to grasp the 'extraordinary assumptions' underlying such a system he developed a theory of comparative economy based on ethnographic and historical evidence.

Polanyi correctly identified the Smithian 'paradigm of the bartering savage', which is accepted as axiomatic by many social scientists, as a barrier to an adequate understanding of non-market economy. In a tribal economy, notes Polanyi, the propensity to truck, barter and exchange does not appear: there is no principle of labouring for remuneration, the idea of profit is banned and giving freely is acclaimed a virtue. How, then, is production and distribution ensured, he asks. Polanyi devoted only ten pages of his book to answering this question but his insights have had a significant impact on anthropological thought (see e.g. Dalton and Kocke, 1983). Tribal economy, he argued, is organized in the main by two principles; *reciprocity* and *redistribution*. Reciprocity works mainly in regard to the sexual organizations of society, that is, family and kinship, and it is that broad principle which helps to safeguard both production and family sustenance. Redistribution refers to the process whereby a substantial part of all the produce of the society is delivered to the chief who keeps it in storage. This is redistributed at communal feasts and dances when the villagers entertain one another as well as neighbours from other districts.

Reciprocity and redistribution are able to work because of the institutional patterns of *symmetry* and *centricity*. Tribes, says Polanyi, are subdivided along a symmetrical pattern and this duality of social organization forms the 'pendant' on which the system of reciprocity rests. (Lévi-Strauss's restricted exchange model of gift exchange also presupposes dual social organization.) The institution of territorial centricity forms the basis of redistribution.

To these two principles, Polanyi adds a third – *householding*, production for use with *autarky* as its basis – and argues that all economic systems known to us up to the end of feudalism were organized on either the principle of reciprocity, or redistribution, or householding, or some combination of the three. These made use of the patterns of symmetry, centricity and autarky, with custom, law, magic and religion cooperating to induce the individual to comply with the rules of behaviour.

Capitalism, in Polanyi's view, implies the wholesale destruction of these principles and the establishment of free markets in land, money and labour run according to the profit principle. Like Marx, Polanyi sees the emergence of free wage-labour as a commodity as the crucial defining characteristic of capitalism. Labour was the last of the markets to be organized in England and both Marx and Polanyi saw the enclosure movements, especially those at the time of the industrial revolution, as central to this process. Polanyi is more precise in his historiography however. He sees the Poor Law Reform of 1834, which did away with the final obstruction to the functioning of a free labour market, as the beginning of the era of the self-regulating market.

Postwar developments in the theory of the gift have built on the foundations laid by Mauss, Lévi-Strauss and Polanyi. The influential contributions of Godelier (1966, 1973), Meillassoux (1960, 1975) and Sahlins (1972) in particular are heavily indebted to these theorists whose ideas they attempt to develop in the light of Marx's theory of comparative economy. Recent empirical research (e.g. Strathern, 1971; Young, 1971; Leach and Leach, 1983) has provided, and will continue to provide, the basis for new comparative insights into the theory of the gift (Forge, 1972).

An important postwar development in the theory of the gift has been the analysis of the impact of colonization and capitalist imperialism on tribal societies.

For the early contributors to this literature the problem was how to explain the process of destruction brought about by capitalism. Paul Bohannan (1959), an American anthropologist with fieldwork experience in West Africa, developed a theory of the impact of money on a tribal economy based on Polanyi's ideas. Commodity exchange, according to Polanyi, is a 'uni-centric economy' because of the nature of 'general purpose money' which reduces all commodities to a common scale. In a tribal society, by way of contrast, the economy is 'multi-centric': there are multiple spheres of exchange, each with 'special purpose' money that could only circulate within that sphere. Among the Tiv of West Africa, for example, there were three spheres of exchange. The first sphere contained locally produced foodstuffs, tools and raw materials; the second sphere contained non-market 'prestige' goods such as slaves, cattle, horses, prestige cloth (*tuguda*) and brass rods; the third sphere contained the 'supreme gift', women. Bohannan's argument was that the general purpose money introduced by the colonial powers reduced all the various spheres to a single sphere thereby destroying them.

Bohannan's theory was applied to the analysis of the impact of colonization in other parts of the world, Papua New Guinea among others (e.g. Meggitt, 1971). While Bohannan's theory makes an important conceptual advance in comparative economy it is now recognized that his theory of the impact of colonization has a number of shortcomings as a description of what happened in West Africa (see Dorward, 1976); furthermore, it does not pose the problem to be explained.

Today, it is now realized, the problem is not, 'How was the tribal gift economy destroyed?' but rather, 'Why has it flourished under the impact of colonization?'

Take the famous potlatch system, for example. The establishment of a canning industry in the area in 1882 led to a rapid increase in the per capita income of the Kwakiutl, a rapid increase in the number of blankets that could be purchased, and hence a rapid increase in the number of blankets given away in potlatch ceremonies. Before the canning industry was established the largest potlatch consisted of 320 blankets, but during the period 1930–1949 potlatch ceremonies involving as many as 33,000 blankets were recorded (Codere, 1950, p. 94). This rapid growth in potlatch occurred despite the institution in 1885 of a law prohibiting the ceremonies. The system has not retained its pristine form, however. Legal and other influences have brought about a variety of outward changes in form but the original purpose of the system still persists: the presentation of a claim to a specific social status (Drucker and Heizer, 1967. pp. 47–52).

In Papua New Guinea, to take another example, the establishment of one of the world's largest copper mines in Bougainville has stimulated a flourishing import of shells into the island. The shells are manufactured by the Langalanga people of western Malaita in the Solomon Islands some 1550 kilometres away. The mine has given the people of Bougainville income earning opportunities unavailable to other islanders and they are able to outbid other purchasers for the Langalangan shells. The Langalangans, for their part, have oriented all their production away from local purchasers to the Bougainville market. In Bougainville the shells are used mostly by the Siwai people who give them as marriage gifts and traditional gift exchanges involving land and pigs; they are also used as ornaments (see Connell, 1977).

This symbiosis between commercialization and gift exchange is found elsewhere in Papua New Guinea. The famous *kula* gift exchange system in the Milne Bay District still persists despite more than one hundred years of colonization (Leach and Leach, 1983). Milne Bay is now something of an economic backwater, its heyday of commercial development being the gold mining era early in this century. Labour is probably one of the area's most important exports today. These migrants maintain close contact with their villages and often send home money, some of which is channelled into *kula* transactions. The migrants, who are senior public servants, entrepreneurs, and politicians, also take their culture with them to the urban areas. The result is that the *kula* ring now extends to Port Moresby, where Mercedes cars and telephones have replaced outrigger canoes and conch shell horns as the principal means of communication.

There is some empirical evidence that appears to contradict the theses that gift exchange has effloresced under the impact of colonization. Prior to the European colonization of West Africa and India these countries were part of a flourishing international cowrie-shell economy. The shells (*cyprae moneta*) were produced in the Maldive Islands of the Indian Ocean and were shipped to West Africa and India where they were used primarily as instruments of exchange but also for religious and ornamental purposes (Heimann, 1980). The cowrie shells were an important and profitable item of international trade in the mercantile era. They were purchased very cheaply in the Maldives – where they grow in great profusion – and exported to India or Europe. The merchants of Europe re-exported them to West Africa where they used them to purchase slaves.

This international shell economy, which had persisted for many centuries, began to collapse around the middle of the 18th century. The supply of shells began to increase rapidly and their price began to fall. For example, in 1865, 1636 tons of cowries were imported into Lagos; by 1878 imports totalled 4472 tons, which was the peak; ten years later imports had fallen to a mere ten tons. Cowrie shell prices (measured in pounds sterling) collapsed over this period. In 1851 two thousand cowries cost 4s. 9d. but by 1876–79 the price had fallen to 1s. 0d. (Hopkins, 1966; Johnson, 1970). By the beginning of the 20th century cowrie shells were no longer current; their place had been taken by the fiat money of the respective colonial government.

This evidence of the destructive impact of colonization only appears to contradict the 'efflorescence of gift exchange' thesis however. The reality is otherwise and the evidence demonstrates the point that exchange is a social relationship which varies depending upon the political and historical context. Objects, such as shells, have many uses, and the historical fact that they have been used as instruments of gift exchange here, as objects of commodity exchange there, and as currency in other places has caused great confusion in the literature. The issue is further confused by the fact that in contemporary Papua New Guinea for example, a shell may be used in all three roles during the same day. The issue can be clarified somewhat by inquiring into the primary role of an exchange object and situating this historically and comparatively in terms of the mode of reproduction of a society. The uniqueness of a place such as Papua New Guinea becomes apparent from this perspective. Papua New Guinea, unlike West Africa or India, was not part of an international mercantile economy prior to European colonization, and as a result commercial exchange transactions were a subordinate and insignificant part of total exchange. Pre-colonial India and West Africa, on the other hand, were highly commercialized: land and labour were freely transacted as commodities with gold and silver commodity monies being used as the principal instruments of exchange. The colonization of West Africa transformed it from being a stateless commodity economy to a state controlled one. This involved a suppression of the stateless commodity monies and their substitution by state fiat money. In India a similar process occurred as the British Government established strong centralized administrative control over numerous weak, corrupt princely states. The destruction of the cowrie shell economy must be seen as part of this process of transition from stateless commodity money to state fiat money. Cowries were the small change of gold and silver. The relationship of cowries to gold and silver, then, finds its counterpart in the relationship of pennies to shillings and pounds. However, whereas the relationship between gold and cowries is determined by production conditions and changes from day to day, the relationship between pounds and pennies is set by government decree and never changes. Where a stable government exists, and the value of money remains constant, it is obvious that a merchant or consumer will prefer to use the latter.

The shells used in West Africa and India, then, were used primarily as instruments of commodity exchange and the term 'shell money' is correct in this context. However, the shells used in the exchange systems of Melanesia and elsewhere were not used as the small change of commodity monies in pre-colonial times. They were used primarily as instruments of gift-exchange and the term 'shell gifts' is more appropriate in this context. Colonization has resulted in the efflorescence of gift-exchange in Melanesia because the colonial state brought an end to tribal warfare and facilitated a transition from fighting with weapons to fighting with gifts. These gifts take the form of women, shells, food and even money nowadays. These gifts do not involve a 'voluntary transference of property' as the *Oxford English Dictionary* would have it. They are the results of obligations imposed on people struggling to achieve status and wealth in a situation where indigenous systems of land tenure, kinship and marriage are being incorporated into an international economic and political order, over which tribespeople and peasants have little control.

C.A. GREGORY

See also EXCHANGE; ECONOMIC ANTHROPOLOGY.

BIBLIOGRAPHY

Armstrong, W.E. 1924. Rossel Island money: a unique monetary system. *Economic Journal* 34, 423–9.
Armstrong, W.E. 1928. *Rossel Island: An Ethnological Study*. Cambridge: Cambridge University Press.
Boas, F. 1897. *Kwakiutl Ethnography*. Ed. H. Codere, Chicago: University of Chicago Press, 1966.
Bohannan, P. 1959. The impact of money on an African subsistence economy. *Journal of Economic History* 19(4), 491–503.
Codere, H. 1950. *Fighting with Property*. New York: Augustin.
Connell, J. 1977. The Bougainville connection: changes in the economic context of shell money production in Malaita. *Oceania* 48(2), December, 81–101.
Dalton, G. and Köcke, J. 1983. The work of the Polanyi group: past, present and future. In S. Ortiz (ed.), 1983.
Dorward, D.C. 1976. Precolonial Tiv trade and cloth currency. *The International Journal of African Historical Studies* 9(4), 576–91.
Drucker, P. and Heizer, R.F. 1967. *To Make My Name Good: A Reexamination of the Southern Kwakiutl Potlatch*. Los Angeles: UCLA Press.
Einzig, P. 1948. *Primitive Money*. London: Eyre and Spottiswoode.
Forge, A. 1972. The Golden Fleece. *Man* 7(4), 527–40.
Godelier, M. 1966. *Rationality and Irrationality in Economics*. London: New Left Books, 1972.
Godelier, M. 1973. *Perspectives in Marxist Anthropology*. Cambridge: Cambridge University Press, 1977.
Gregory, C.A. 1980. Gifts to men and gifts to god: gift exchange and capital accumulation in contemporary Papua. *Man* 15(4), 626–52.
Gregory, C.A. 1982. *Gifts and Commodities*. London: Academic Press.
Heimann, J. 1980. Small change and ballast: cowry trade and usage as an example of Indian Ocean economic history. *South Asia* 3(1), 48–69.
Hopkins, A.G. 1966. The currency revolution in south-west Nigeria in the late nineteenth century. *Journal of the Historical Society of Nigeria* 3(3), 471–83.
Johnson, M. 1970. The cowrie currencies of West Africa. *Journal of African History* 11(1), 17–49; 11(3), 331–53.
Leach, E.R. 1970. *Lévi-Strauss*. London: Fontana.
Leach, J.W. and Leach E. (eds) 1983. *The Kula*. Cambridge: Cambridge University Press.
Lévi-Strauss, C. 1949. *The Elementary Structures of Kinship*. Trans., London: Eyre and Spottiswoode, 1969.
Liep, J. 1981. The workshop of the Kula: production and trade of shell necklaces in the Louisade Archipelago. *Folk og Kultur* 23, 297–309.
Liep. J. 1983. Ranked exchange in Yela (Rossel Island). In J.W. Leach and E. Leach (eds), *The Kula*, Cambridge: Cambridge University Press.
MacIntyre, M. and Young, M. 1982. The persistence of traditional trade and ceremonial exchange in the Massim. In *Melanesia: Beyond Diversity*, ed. R.J. May and Hank Nelson, Canberra: Australian National University.
Malinowski, B. 1922. *Argonauts of the Western Pacific*. New York: E.P. Dutton, 1961.
Marx, K. 1867. *Capital*. Vol. 1: *A Critical Analysis of Capitalist Production*. Moscow: Progress Publishers, n.d.
Mauss, M. 1925. *The Gift*. London: Routledge and Kegan Paul, 1974.
Meggitt, M.J. 1971. From tribesman to peasants: the case of the Mae-Enga of New Guinea. In *Anthropology in Oceania*, ed. L.R. Hiatt and C.J. Jayawardena, Sydney: Angus and Robertson.

Meillassoux, C. 1960. Essai d'interprétation du phénomène économique dans les sociétés traditionelles d'auto-subsistance. *Cahiers d'Etudes Africaines* 4, 38–67.
Meillassoux, C. 1975. *Maidens, Meal and Money*. Cambridge: Cambridge University Press, 1981.
Ortiz, S. (ed.) 1983. *Economic Anthropology: Topics and Theories*. New York: University Press of America.
Polanyi, K. 1944. *The Great Transformation*. New York: Rinehart.
Sahlins, M. 1972. *Stone Age Economics*. Chicago: Aldine.
Stent, W.R. and Webb, L.R. 1975. Subsistence, affluence and market economy in Papua New Guinea. *Economic Record* 51, 522–38.
Strathern, A.J. 1971. *The Rope of Moka*. Cambridge: Cambridge University Press.
Young, M.W. 1971. *Fighting with Food: Leadership, Values and Social Control in a Massim Society*. Cambridge: Cambridge University Press.

Gilbert, Milton

Gilbert, Milton (1909–1979). Gilbert was born in Philadelphia and educated in that city, receiving his doctorate from the University of Pennsylvania. His contributions to economics were both in the statistical and substantive domains.

His work as an economic statistician occupied the early and middle years of his career and was centred on the development of national accounts. His first major position was in the US Department of Commerce as Editor of the *Survey of Current Business*. He became chief of the national income division in the Commerce Department in 1941 and for the next 10 years presided over the development of the US system of national income and product accounts. In 1951 he went to the Organization for European Economic Cooperation (OEEC) first as head of the Statistics and National Accounts Division and then as head of the combined Economics and Statistics Division. At the OEEC he initiated and co-authored the first systematic international comparison of national products and currency purchasing powers. His growing attention to substantive economic issues was reflected in his leading role in an influential OEEC report on the wage–price spiral.

In 1960 Gilbert went to the Bank for International Settlements (BIS) as Economic Adviser and later headed the Bank's Monetary and Economic Department. In addition to his analyses of world monetary problems, much of it focused on the workings and the breakdown of the Bretton Woods system, Gilbert greatly strengthened the statistical work of the BIS especially with regard to the collection and analysis of data on the Euro-currency market.

IRVING B. KRAVIS

See also INTERNATIONAL INCOME COMPARISONS; SOCIAL ACCOUNTING.

SELECTED WORKS

1947. (With others.) *U.S. National Income Supplements*. Washington, DC: US Government Printing Office.
1954. (With I.B. Kravis.) *An International Comparison of National Products and the Purchasing Power of Currencies: A study of the United States, the United Kingdom, France, Germany and Italy*. Paris: Organization for European Economic Co-operation.
1968. *The Gold–Dollar System: Conditions of Equilibrium and the Price of Gold*. Princeton: International Finance Section, Princeton University.
1980. *Quest for World Monetary Order: the Gold–Dollar System and its Aftermath*. Posthumous. Edited by P. Oppenheimer and M.G. Dealtry, New York: Wiley.

Gilman, Charlotte Perkins

Gilman, Charlotte Perkins (1860–1935). Gilman was born on 3 July 1860 in Hartford, Connecticut, and died on 17 August 1935 in Pasadena, California. Known world-wide as a feminist theorist and a generally iconoclastic social critic, Gilman was a major intellectual force in turn-of-the-century America. Largely self-educated, problems with her first marriage led her to separate from her first husband and begin an unconventional freelance life based in California, earning her living from her lecturing and writing. *Women and Economics* (1898) was her first book-length exposition of her theory of the evolution of gender relations. Influenced by the ideas of Edward Bellamy, Lester Frank Ward, Darwin, the Webbs and G. Bernard Shaw, she explained that human institutions (like the species itself) has evolved over time, favouring the survival of the best adapted. A major exception, however, was the definition of 'women's place'. Here social development had been frozen by Tradition. Women were confined to households which were no longer the locus of any socially productive activity, since now the factory produced the needed consumption goods, and children were better raised in schools, by professionals. The role of full-time housewife and mother had become anachronistic, reducing women to the state of social parasites. As she also argued in her 1903 classic *The Home*, for their own progress and for the progress of human civilization overall, women would have to leave these domestic prisons and take up socially useful work in the larger world of production. In the articles and didactic fiction that she wrote for her monthly magazine, the *Forerunner*, she developed a wide range of startlingly rational ideas for social reorganization.

B. BERCH

SELECTED WORKS

1892. The yellow wall-paper. *New England Magazine*, January. Reprinted in *The Charlotte Perkins Gilman Reader*, ed. A. Lane, New York: Pantheon, 1980.

1898. *Women and Economics*. Ed. C. Degler, New York: Harper & Row, 1966.

1903. *The Home*. Introduction by W. O'Neill, Urbana: University of Illinois Press, 1972.

1915. Herland. *The Forerunner*, Volume 6. Reprinted, with introduction by A. Lane, as *Herland: A Lost Feminist Utopian Novel*, New York: Pantheon, 1978.

Gini, Corrado (1884–1965). Gini, perhaps best known to economists because of the Gini Coefficient, was born in Motta di Livenza, Italy and died in Rome. He studied at the University of Bologna; his doctoral thesis 'Il sesso dal punto di vista statistico' (1908), defended in 1905, was awarded the Vittorio Emanuele prize for social sciences. Gini distinguished himself as a teacher and a researcher. In 1909 he was appointed an assistant professor of the University of Cagliari, becoming full professor a year later. Gini won a chair at the University of Padova in 1913, then joined the University of Rome in 1925, where in 1955 he was awarded the distinction of emeritus professor. Social scientist and statistician, Gini taught economics, statistics, sociology and demography, making pathbreaking contributions to these highly related disciplines. Among them we mention the neo-organicist theory (Gini, 1909, 1924a) that presents a dynamic theory of society in which demographic factors (differential birth rates among social classes and social mobility) play a basic role. In this theory, Gini introduced and analysed self-conservation, self-regulative and self-reequilibrating mechanisms, thus offering a well-structured anticipation of Wiener's cybernetics, von Bertalanffy's general system theory and modern disequilibrium economics. He provided new insights to the analysis of inter- and intra-national migrations (Gini, 1948) and demographic dynamics (Gini, 1908, 1909, 1912a, 1931).

He developed a methodology to evaluate the income and wealth of nations (Gini, 1914a, 1959) including a discussion of human capital, already present in his research on the causes and consequences of international migrations. In this context he specified a model of income and wealth distributions and a measure of income and wealth inequalities (Gini, 1909, 1912b, 1914b, 1955). Gini's research interests motivated important contributions to statistics and economics, such as the Gini identity (1921, 1924b) on price index numbers, the Gini mean difference (1912b), the transvariation theory (Gini, 1916, 1960), the index of dissimilarity (Gini, 1914) and the Gini Coefficient. Gini founded several scientific journals, such as *Metron* and *Genus*; academic institutions, such as the Institute and Faculty of Statistics, Demography and Actuarial Sciences of the University of Rome; and was the organizer and first president (1926–32) of the Istituto Centrale di Statistica. An extraordinarily prolific writer and thinker, endowed with powerful new ideas that he developed in more than 70 books and 700 articles, Gini was in the 20th century a true renaissance man.

CAMILO DAGUM

SELECTED WORKS

1908. *Il sesso dal punto di vista statistico*. Milan: Sandrom.

1909. Il diverso accrescimento delle classi sociali e la concentrazione della ricchezza. *Giornale degli Economisti*, January.

1912a. *I fattori demografici dell'evoluzione delle nazioni*. Turin: Bocca.

1912b. Variabilità e mutabilità. Reprinted in Gini (1955).

1914a. *L'ammontare e la composizione della ricchezza delle nazioni*. 2nd edn, Turin: UTET, 1962.

1914b. Sulla misura della concentrazione e della variabilità dei caratteri. Reprinted in Gini (1955).

1914c. Di una misura della dissomiglianza tra due gruppi di quantità e delle sue applicazioni allo studio delle relazioni statistiche. *Atti del R. Istituto Veneto di Scienze, Lettere ed Arti*.

1916. Il concetto di transvarizione e le sue prime applicazioni. Reprinted in Gini (1960).

1921. Sull'interpolazione di una retta quando i valori della variabile indipendente sono affetti da errori accidentali. *Metron*, January–April.

1924a. *Patologia economica*. 5th edn, Turin: UTET, 1954.

1924b. Quelques considérations au sujet de la construction des nombres indices des prix et des questions analogues. *Metron*, July–September.

1931. *Le basi scientifiche della politica della popolazione*. Catania: Studio Editoriale Moderno.

1948. Apparent and real causes of American prosperity. *Banca Nazionale del Lavoro Quarterly Review*, July–September.

1955. *Memorie di metodologia statistica*. Vol. I: *Variabilità e concentrazione*. Ed. A. Pizzetti and T. Salvemini, Rome: Veschi.

1959. *Ricchezza e reddito*. Turin: UTET.

1960. *Transvariazione*. Ed. G. Ottaviani, Rome: Libreria Goliardica.

Gini ratio. In a national economy, the price system determines both resource allocation and the income distribution. The imputation to the factors of production of the mass of income associated with an economy's output determines its distribution by factor shares, or *functional* income distribution. This mainstream of research follows Ricardo's (1817) contribution. Another mainstream of research was initiated by Pareto (1895, 1897), and deals with the distribution of a mass of income among the members of a set of economic units (family, household, individual), considering either the total income of each economic unit or its disaggregation by source of income, such as wages and salaries, property income, self-employment income, transfers, etc. This type of inquiry deals with distribution by size of income, or *personal* income distribution, and the quantitative assessment of the relative degree of income inequality among the members of a given set of economic units. Such inquiries provide basic quantitative information in support of a comprehensive research strategy on income distributions, including causal explanations for social welfare and policy.

It is of interest to remark that Pareto's research on income distribution was motivated by the polemic he engaged in with French and Italian socialists concerning the ways and means of achieving a less unequal distribution. Thus, the actual *measurement* of inequality was brought to the fore, with its main purposes the assessment of (i) the evolution of inequality in a given country or region, and (ii) the relative degree of inequality between countries or regions.

In a series of methodological and applied contributions Corrado Gini (1955) enriched this field of research. In 1910 he corrected the interpretation of Pareto's inequality parameter and, in 1912, proposed a new measure of income inequality, the Gini ratio.

Pareto (1896, 1897) specified three versions of this model of income distribution. The most widely used model is Pareto Type I

$$S(x) = 1 - F(x) = (x/x_0)^{-\alpha}, \quad 0 < x_0 < x, \quad \alpha > 1, \quad (1)$$

where $S(x) = P(X > x)$ is the survival distribution function (SDF) of the income variable X, $F(x)$ is the cumulative distribution function (CDF), x_0 is the minimum value of X, α is a scale-free inequality parameter, and the mathematical expectation of income is

$$\mu = E(X) = \alpha x_0^\alpha \int_{x_0}^\infty x^{-\alpha}\, dx = \alpha x_0/(\alpha - 1). \quad (2)$$

Pareto seems to have assumed that income growth implies less income inequality. This assumption, together with eqn (2), led him to the conclusion that income inequality is an increasing function of α. Gini (1910) reversed this interpretation, proving that, given model (1), income inequality is a decreasing function of α. Gini's rationale was as follows: given n units with incomes $x_1 \leqslant x_2 \leqslant x_3, \ldots, \leqslant x_n$, the average of the last $m\,(m \leqslant n)$ income units $\Sigma_{i=0}^{m-1} x_{n-i}/m$ is greater than or equal to the average income $\mu = \Sigma_{i=0}^n x_i/n$ of the population, hence, there exists a $\delta \geqslant 1$ such that

$$\left(\sum_{i=0}^{m-1} x_{n-i} \bigg/ \sum_{i=1}^n x_i \right)^\delta = m/n, \quad \delta \geqslant 1, \quad (3)$$

Equation (3) is known as the Gini model. Gini (1910) interpreted the scale-free parameter δ as a measure of income inequality and called it a *concentration ratio* because it is an increasing function of the concentration of income in the upper income groups. For this reason, Gini called eqn (3) a *concentration curve*, where the abscissa represents the CDF $F(x_m) = m/n$ and the ordinate the income share $\Sigma_{i=1}^m x_i/\Sigma_{i=1}^n x_i, m = 1, 2, \ldots, n, \delta$ being an unknown parameter that has to be estimated.

Using the CDF $F(x)$ and the Lorenz curve $L(x)$ (also called the Lorenz–Gini curve since it was independently introduced by both authors), eqn (3) takes the form

$$1 - F(x) = [1 - L(x)]^\delta, \quad \delta \geqslant 1, \quad (4)$$

where

$$L(y) = (1/\mu) \int_0^y x\, dF(x). \quad (5)$$

Replacing $F(x)$ from model (1) into eqns (4) and (5), Gini (1910) proved that $\delta = \alpha/(\alpha - 1)$ and thus reversed Pareto's interpretation of α. In fact, when $\alpha \to \infty$, $\delta \to 1$ and $F(x) = L(x)$, and the mass of income is equally distributed.

Gini (1912) specified the Gini mean difference with and without replacement. The latter is by definition

$$\Delta = \sum_{j=1}^n \sum_{i=1}^n |x_j - x_i|/n(n-1), \quad 0 \leqslant \Delta \leqslant 2\mu, \quad (6)$$

and using the Riemann–Stieltjes integral, which covers, as particular cases, both discrete and continuous distributions, we have

$$\Delta = \int_0^\infty \int_0^\infty |y - x|\, dF(x)\, dF(y), \quad (7)$$

where X and Y are identically and independently distributed variables. When $x_1 = x_2 = \cdots = x_n$, $\Delta = 0$, and when $x_1 = x_2 \cdots = x_{n-1} = 0$ and $x_n = n\mu$ (the total income), $\Delta = 2\mu$.

Since Δ is a monotonic increasing function of the degree of income inequality, Gini (1912) specified

$$G = \Delta/2\mu, \quad 0 \leqslant G \leqslant 1 \quad (8)$$

as an income inequality measure. Equation (8) is known as the Gini ratio or Gini index and it is widely used in theoretical and applied research on income and wealth distributions.

Gini (1914) proved the important theorem that $G = \Delta/2\mu$ is equal to twice the area between the equidistribution line $F(x) = L(x)$ and the Lorenz curve $L(x)$ (see fig. 1). Moreover,

$$G = \Delta/2\mu = 2 \int_0^1 (F - L)\, dF$$

$$= (2/\mu) \int_0^\infty x[F(x) - \tfrac{1}{2}]\, dF(x)$$

$$= (2/\mu) \int_0^\infty x[\tfrac{1}{2} - S(x)]\, dF(x). \quad (9)$$

For the discrete case, it follows from eqns (6) and (8), that

$$G = [2/n(n-1)\mu] \sum_{k=1}^n kx_k - (n+1)/(n-1)$$

$$= (n+1)/(n-1) - [2/n(n-1)\mu] \sum_{k=1}^n (n-k+1)x_k, \quad (10)$$

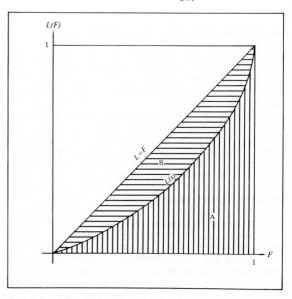

Figure 1 Lorenz Curve $L(x)$ and Gini ratio G
$L(F(x)) = (1/\mu)\int_0^y y\, dF(y)$
$G = 2B = 1 - 2A = 1 - 2\int_0^1 L\, dF$

showing that the welfare function underlying the Gini ratio is a rank-order-weighted sum of the economic units' income shares.

The properties that an income inequality measure must fulfil were first discussed by Dalton (1920). It can be shown (Dagum, 1983, pp. 34–5) that G fulfils the properties of (i) transfer, (ii) proportional addition to incomes, (iii) equal addition to incomes, (iv) proportional addition to persons, (v) symmetry, (vi) normalization, and (vii) operationality.

The Gini ratio is sensitive to transfers to all income levels. In fact, it follows from eqn (10), that a transfer of h dollars from the richer jth to the poorer ith, without modifying their income ranks, is

$$\Delta G(j, i; h) = -2(j - i)h/n(n - 1)\mu > 0, \qquad j > i, \quad (11)$$

therefore $-\Delta G$ is an increasing function of $j - i = F(x_j) - F(x_i)$ and a decreasing function of both n and μ. The maximum reduction of G is achieved when $h = (x_j - x_i)/2$, and is not necessarily given by eqn (11) unless the transfer fulfils certain conditions with respect to the original income ranking of the population.

Often, the Gini ratio is misinterpreted when it is incorrectly claimed that it attaches more weight to transfers to income near the mode of the distribution than at the tails. In particular, the misinterpretation arises when eqn (11) instead of eqn (9) is applied to unimodal distributions when assessing the relative sensitivity of G to income transfers. Consequently, the assumptions supporting the mathematical structure of eqn (11) are ignored.

It follows from eqn (10) that the Gini ratio fulfils the duality principle between the representation of an inequality measure (I) satisfying the principle of transfer, i.e. $I = E[V(x)]$, and that of a social welfare (SW) function, i.e. $SW = E[-V(x)]$, where $-V(x)$ is concave, or more generally, S-concave (Berge, 1966). It follows from eqn (9), that two equivalent forms of $V(x)$ in $G = E[V(x)]$ are

$$V(x) = 2xF(x)/\mu - 1, \quad \text{and} \quad V(x) = x[2F(x) - 1]/\mu. \quad (12)$$

Sen (1974) introduced an axiomatic system for the SW interpretation of the Gini ratio based on the individual income ranking of the population suggested by the structure of eqn (10). Following Sen's ideas, Kakwani (1980, pp. 77–9) presented a SW interpretation of the Gini ratio as a function of income. Both approaches can be presented in a compact form by making use of the SDF $S(x) = 1 - F(x)$ and the first moment survival distribution function $S_1(x) = 1 - L(x)$. In fact, specifying the SW function

$$SW(X) = E[Xv(X)] \quad (13)$$

where $v(X)$ is a decreasing and differentiable function of X, and making use of $v(X) = 2S(X) = 2[1 - F(X)]$, i.e., twice the frequency of economic units with income greater than X, we deduce

$$SW(X) = 2 \int_0^\infty xS(x) \, dF(x) = \mu(1 - G), \quad (14)$$

which proves Sen's (1974, p. 410) theorem that the SW function (14) ranks a set of distributions of a constant total income and population in precisely the same way as the negative of the Gini ratio of the respective distributions, i.e. in reverse order from that by the cardinal value of the Gini ratio. On the other hand, making $v(X) = bS_1(X) = b[1 - L(X)]$, $b > 0$ and $\int_0^\infty v(x) \, dF(x) = 1$, where $S_1(x)$ is the income share of the economic units with income greater than x, we deduce

$b \int_0^\infty [1 - L(x)] \, dF(x) = b(1 + G)/2 = 1$, and

$$SW(X) = [2/(1 + G)] \int_0^\infty x[1 - L(x)] \, dF(x)$$
$$= \mu/(1 + G), \quad (15)$$

which also states that the SW function (15) is a decreasing function of the Gini ratio. The result obtained in eqn (14) supports Sen's (1976, p. 384) cogent statement that 'one might wonder about the significance of the debate on the non-existence of any additive utility function which ranks income distributions in the same order as the Gini ratio'.

The Gini ratio stimulated important contributions such as:

(i) The construction of a confidence interval for G. Given a random sample of size n, eqn (10) is an unbiased estimator of G. However, income distribution data are presented by class intervals, hence Gini (1914) proposed the formula $G_L = 1 - 2A$, where A is the area under the Lorenz curve (Figure 1) estimated by application of the trapezoidal approximation to

$$\int_0^1 L \, dF,$$

thus underestimating G because the trapezoidal rule implies that within each interval, income is equally distributed. Gastwirth (1972) derived an upper bound G_u by maximizing the spread within each income interval, and proposed (G_L, G_u) as a confidence interval within which a parametric estimate of G should fall. Dagum (1980a) proved that his confidence interval is a necessary but not sufficient condition to assess a model goodness of fit.

(ii) The Gini ratio gives a welfare ranking (weak ordering) of a set of income distributions of a constant mass of income and over a constant population, and a strict partial ordering among the subset of income distributions with non-intersecting Lorenz curves. This conclusion is further supported by eqns (14) and (15).

(iii) The welfare ranking of income distributions with equal and different means can be obtained via a decision function $R(G, D)$, where the ratio G states the preference for less inequality (inequality aversion) regardless of the mean income (so that the partial derivative $R_G < 0$), and the relative economic affluence D (Dagum, 1980b, 1987) states the preference for more income (poverty aversion), so that the partial derivative $R_D > 0$.

(iv) Research on the economics of poverty led Sen (1976) to the specification of an axiomatic structure of a new poverty measure as a function of (a) the relative frequency of the poor members of the population, (b) a weighted average of the poverty gap, i.e. the aggregate shortfall from the poverty line of the poor population, and (c) the Gini ratio of the income distribution of the subpopulation with incomes below the poverty line.

(v) Gini (1932) introduced a new coordinate system taking as the abscissa the egalitarian line $F = L$ and as the ordinate the distance between the Lorenz curve and the egalitarian line. Gini thoroughly analysed this new coordinate system and its relation to the G ratio. Kakwani (1980, ch. 7) worked with a similar transformation.

(vi) Analysing consumer behaviour in India, Mahalanobis (1960) extended and generalized the Lorenz curve and the Gini ratio with the introduction of the concentration curve and ratio, respectively. Other authors such as Kakwani (1980, chs 8–14) made further contributions and dealt with the relationships among the distribution of several economic variables such as expenditures and income after tax, and investigated the degree of tax and public expenditure progressivity or regressivity. If $y = g(x)$ is the function of income that is the object of inquiry, $g(x)$ must be non-negative. For the particular case of $g(x) = x$, the concentration curve and ratio are identical to the Lorenz curve and Gini ratio, respectively. Moreover, if $g(x)$ is an increasing and differentiable function of x, i.e. $g'(x) > 0$, then the concentration ratio is equal to the Gini ratio for the function $g(x)$.

(vii) The decomposition approach disaggregates a population according to some relevant socio-economic attributes and analyses the equality within each subpopulation and between them, and assesses the contribution of each subpopulation to overall inequality. This approach also disaggregates the income variable by source of income such as wages and salaries, self-employment, pension and government transfers. Bhattacharya and Mahalanobis (1967) were the first to deal with the decomposition of the Gini ratio. Several authors made further contributions to this topic, among them Pyatt (1976) and Shorrocks (1983).

CAMILO DAGUM

BIBLIOGRAPHY

Berge, C. 1966. *Espaces topologiques, fonctions multivoques.* 2nd edn, Paris: Dunod.

Bhattacharya, N. and Mahalanobis, P.C. 1967. Regional disparities in household consumption in India. *Journal of the American Statistical Association* 62, 143–61.

Dagum, C. 1980a. The generation and distribution of income, the Lorenz curve and the Gini ratio. *Economie Appliquée* 33(2), 327–67.

Dagum, C. 1980b. Inequality measures between income distributions with applications. *Econometrica* 48(7), November, 1791–803.

Dagum, C. 1983. Income inequality measures. In *Encyclopedia of Statistical Sciences,* Vol. IV, ed. S. Kotz and N.L. Johnson, New York: John Wiley & Sons, 34–40.

Dagum, C. 1987. Measuring the economic affluence between populations of income receivers. *Journal of Business and Economic Statistics* 5(1), 5–12.

Dalton, H. 1920. The measurement of the inequality of incomes. *Economic Journal* 30, September, 348–61.

Gastwirth, J.L. 1972. The estimation of the Lorenz curve and Gini index. *Review of Economics and Statistics* 54, 306–16.

Gini, C. 1910. Indici di concentrazione e di dependenza. *Atti della III Riunione della Società Italiana per il Progresso delle Sciencze,* in Gini (1955), 3–120.

Gini, C. 1912. Variabilità e mutabilità. *Studi economico-giuridici, Università di Cagliari* III, 2a, in Gini (1955), 211–382.

Gini, C. 1914. Sulla misura della concentrazione e della variabilità dei caratteri. *Atti del R. Istituto Veneto di Scienze, Lettere ed Arti,* in Gini (1955), 411–459.

Gini, C. 1932. Intorno alle curve di concentrazione. *Metron* 9(3–4), in Gini (1955), 651–724.

Gini, C. 1955. *Memorie di metodologia statistica.* Vol. 1: *Variabilità e concentrazione.* Ed. E. Pizetti and T. Salvemini, Rome: Libreria Eredi Virgilio Veschi.

Kakwani, N. 1980. *Inequality and Poverty: methods of estimation and policy applications.* Oxford: Oxford University Press.

Mahalanobis, P.C. 1960. A method of fractile graphical analysis. *Econometrica* 28(2), April, 325–351.

Pareto, V. 1895. La legge della domanda. *Giornale degli Economisti,* 59–68.

Pareto, V. 1896. *Ecrits sur la courbe de la répartition de la richesse.* In *Oeuvres complètes de Vilfredo Pareto,* ed. Giovanni Busino, Geneva: Librairie Droz, 1965.

Pareto, V. 1897. *Cours d'économie politique.* Ed. G.H. Bousquet and G. Busino, Geneva: Librairie Droz 1964.

Pyatt, G. 1976. On the interpretation and disaggregation of Gini coefficients. *Economic Journal* 86, June, 243–55.

Ricardo, D. 1817. *Principles of Political Economy.* In *Works and Correspondence of David Ricardo,* Vol. I, ed. Piero Sraffa, Cambridge: Cambridge University Press, 1951.

Sen, A.K. 1974. Information bases of alternative welfare approaches. *Journal of Public Economics* 3, 387–403.

Sen, A.K. 1976. Poverty: an ordinal approach to measurement. *Econometrica* 44(2), March, 219–31.

Shorrocks, A.F. 1983. The impact of income components on the distribution of family income. *Quarterly Journal of Economics* 98(2), May, 311–26.

Gioia, Melchiorre (1767–1829). Born in Piacenza, Italy, Melchiorre Gioia actively participated in the turbulent political life of his time, ending up in prison more than once.

Gioia became a Catholic priest; however, his gospel was that man should obtain the 'maximum product with the minimum expenditure of effort'. This 'principle' inspires his main contribution to the development of economic analysis: the principle of the association and of division of work. The principle of the association of work states that the cooperation of qualitatively equal kinds of labour increases efficiency in production. Cooperation has considerable advantages even if it is not associated with the division of labour (i.e. the association of qualitatively different kinds of labour). As to the nature and the advantages of the division, Gioia differentiates himself from Smith on two points. Firstly, he maintains that the division of labour (as well as undifferentiated cooperation) can and does exist in the animal and human world independently of exchange. Secondly, Gioia relates the advantages of the division of labour more to the saving of natural or acquired human skills than to the creation of job specific skills. Natural skills are saved if the individuals possessing them dedicate themselves only to those occupations which require them while other people perform the remaining activities. As to the scope of acquired skills, specialization saves training time. The more pronounced the specialization of the members of an association, the narrower the set of tasks which they have to learn.

Melchiorre Gioia criticized the laissez-faire policies advocated by the English school of political economy. He anticipated some aspects of the theories of market failures due to externalities and monopolies and favoured state intervention for correcting them. He also advocated protectionistic policies aimed at the development of infant industries.

Gioia's greatest work is *Il Nuovo Prospetto delle Scienze Sociali* (1815), the first volume of which contains his principle of the association and division of labour. His views on state intervention are well expressed in *Discorso popolare sulle manifatture nazionali e tariffe daziarie dei commestibili ed il caro prezzo del vitto* (1802). His other works include: *Del merito e delle ricompense,* etc. (1818), *Filosofia della Statistica* (1826), *Indole, estensione e vantaggi della Statistica,* (1809), *Logica Statistica* (1808) and *Tavole Statistiche* etc. (1808).

An exhaustive examination of the economic theory of Melchiorre Gioia can be found in Barucci (1965).

U. PAGANO

SELECTED WORKS

1815. *Il nuovo prospetto delle scienze sociali.* Lugano.

BIBLIOGRAPHY

Barucci, P. 1965. *Il pensiero economico di Melchiorre Gioia.* Milan.

global analysis in economic theory. The goal here is to illustrate 'global analysis in economics' by putting the main results of classical equilibrium theory into a global calculus context. The advantages of this approach are fourfold:

(1) The proofs of existence of equilibrium are simpler. Kakutani's fixed point theorem is not used, the main tool being the calculus of several variables.

(2) Comparative statics is integrated into the model in a natural way, the first derivatives playing a fundamental role.

(3) The calculus approach is closer to the older traditions of the subject.

(4) In so far as possible the proofs of equilibrium are constructive. These proofs may be implemented by a speedy algorithm, which is Newton's method modified to give global convergence. On the other hand, the existence proofs are sufficiently powerful to yield the generality of the Arrow–Debreu theory.

Only two references are given at the end of this entry, each containing an extensive bibliography with historical notes. The two references themselves give detailed, expanded accounts of the subject of global analysis in economic theory.

Let us proceed to an account of this model. The basic equation of equilibrium theory is 'supply equals demand', or in symbols, $S(p) = D(p)$. Since we are in a situation of several markets, there are several variables in this equation. Equilibrium prices are obtained by setting the excess demand $z = D - S$ equal to zero and solving. Consider this function z on a more abstract level.

Suppose that given an economy of l markets, or of l commodities with corresponding prices written as $p_1, \ldots, p_l$, the excess demand for the ith commodity is a real valued function $z_i = z_i(p_1, \ldots, p_l)$ $p_j \geq 0$ and we form the vector $z = (z_1, \ldots, z_l)$. Thus the excess demand can be interpreted as a map, which we take to be sufficiently differentiable, from $\mathbb{R}^l_+$ to $\mathbb{R}^l$, where $\mathbb{R}^l$ is Cartesian l-space and $\mathbb{R}^l_+ = \{p \in \mathbb{R}^l / p_i \geq 0\}$. An economic equilibrium is a set of prices $p = (p_1, \ldots, p_l)$ for which excess demand is zero, that is, $z(p) = 0$.

Economic theory imposes some conditions on the function z which go as follows.

First and foremost is Walras's Law, which is expressed simply by $p \cdot z(p) = 0$ (inner product). Written out, this is

$$\sum_{i=1}^{l} p_i \cdot z_i(p_1, \ldots, p_l) = 0$$

and states that the value of the excess demand is zero. This is a budget constraint which asserts that the excess demand is consistent with the total assets of the economy. It can be proved from a reasonable microeconomic foundation, as can be seen below.

Second is the homogeneity condition $z(\lambda p) = z(p)$ for all $\lambda > 0$. Changing all prices by the same factor does not affect excess demand. This condition reflects the fact that the economy is self-contained; prices are not based on anything lying outside the model.

The final condition is the boundary condition that $z_i(p) \geq 0$ if $p_i = 0$. This may be interpreted as: if the ith good is free, then there will be a non-negative excess demand for it.

The following result and its generalizations and ramifications lie at the heart of economic theory.

Existence Theorem. Suppose that an excess demand z satisfies Walras's Law, homogeneity, and the boundary condition. Then there is a price equilibrium.

We will give the proof under the additional mild non-degeneracy condition that the derivative of z is non-singular somewhere on the boundary of $\mathbb{R}^l_+$. This proof is based on

Sard's theorem and the inverse function theorem, two basic theorems of global analysis.

Consider a differentiable map f from a set U contained in $\mathbb{R}^k$ to $\mathbb{R}^n$. A vector $y \in \mathbb{R}^n$ is said to be a *regular* value if at each point $x \in U$ with $f(x) = y$, the derivative $Df(x)$: $\mathbb{R}^k \to \mathbb{R}^n$ is surjective. A subset of $\mathbb{R}^n$ is of *full measure* if its complement has measure zero.

Sard's Theorem. If a map f: $U \to \mathbb{R}^n$ is of sufficient differentiability class (C^r, $r > k - n, 0$) then the set of regular values of f has full measure.

A subset V of $\mathbb{R}^n$ is called a k-dimensional *submanifold* if for each point, there is a neighbourhood U in V and a change of coordinates of $\mathbb{R}^n$ which throws U into a coordinate subspace of dimension k.

Inverse Function Theorem. If $y \in \mathbb{R}^n$ is a regular value of a smooth map f: $U \to \mathbb{R}^n$, $U \subset \mathbb{R}^k$, then either $f^{-1}(y)$ is empty or it is a submanifold of dimension $k - n$.

Let us sketch out the proof of the Existence Theorem. Define the space of normalized prices by

$$\Delta_1 = \left\{ p \in \mathbb{R}^l_+ \Big/ \sum p_i = 1 \right\}$$

A space auxiliary to the commodity space is defined by

$$\Delta_0 = \left\{ z \in \mathbb{R}^l \Big/ \sum z_i = 0 \right\}$$

From the excess demand map z: $\mathbb{R}^l_+ - 0 \to \mathbb{R}^l$, define an associated map ϕ: $\Delta_1 \to \Delta_0$ by $\phi(p) = Z(p) - \Sigma z^i(p)p$. Note that $\phi(p)$ is well-defined (i.e. $\phi(p) \in \Delta_0$) and also smooth. Note moreover that if $\phi(p) = 0$, then $z(p) = 0$, and that p is a price equilibrium. This follows from Walras's Law as follows. If $\phi(p) = 0$, then $z(p) = \Sigma z^i(p)p$ and so $p \cdot z(p) = \Sigma z^i(p)$ $p \cdot p = 0$. Therefore $\Sigma z^i(p) = 0$, since $p \cdot p \neq 0$. By the previous equation $z(p)$ must be zero.

The boundary condition on z implies that ϕ satisfies a similar boundary condition. That is, if $p_i = 0$ then $\phi_i(p) = z_i(p) \geq 0$.

It is now sufficient to show that $\phi(p) = 0$ for some $p \in \Delta_1$. The argument for this proceeds by defining yet another map $\hat{\phi}$ by

$$\hat{\phi}(p) = \frac{\phi(p)}{\|\phi(p)\|}$$

where $E = \phi^{-1}(0)$ and S^{l-2} is the set of unit vectors in Δ_0.

By definition the set E is the set of price equilibria, which is to be shown not empty.

Let p_0 be a price vector on the boundary of Δ_1 where the derivative $D\phi(p_0)$ is non-degenerate (our special hypothesis implies the existence of this p_0). One applies Sard's theorem to obtain a regular value y of $\hat{\phi}$ in S^{l-2} near $\hat{\phi}(p_0)$, where $\hat{\phi}^{-1}(y)$ is non-empty.

From the inverse function theorem it follows that $\hat{\phi}^{-1}(y)$ is a smooth curve in Δ_1 (a 1-dimensional submanifold). From the boundary condition and a short argument which we omit, it follows that this curve cannot leave Δ_1.

Since the curve $\hat{\phi}^{-1}(y)$ is a closed set in $\Delta_1 - E$ and has no end points (the inverse function theorem implies that) it must tend to E. In particular, E is not empty and therefore the existence theorem is proved.

The above proof is 'geometrically'' constructive in that a curve $\gamma = \hat{\phi}^{-1}(y)$ is constructed which leads to a price equilibrium. This picture can be made analytic by showing that γ is a solution of the ordinary differential equation 'Global Newton', $d\phi/dt = \lambda D\phi(p)^{-1}\phi(p)$, where λ is $+1$ or -1 determined by the sign of the determinant of $D\phi(p)$. As a consequence the Euler method of approximating the solution of an ordinary differential equation can be used to obtain a discrete algorithm for locating a price equilibrium. By an appropriate choice of steps, ± 1, this discrete algorithm near that equilibrium is Newton's Method; thus the appellation 'Global Newton' for the differential equation.

One would like to understand the process of convergence to equilibrium in terms of decentralized mechanics of price adjustment. Unfortunately the situation in this respect is unclear.

Next we give a brief picture of how global analysis relates to a pure exchange economy. This will allow a microeconomic derivation of the excess demand function discussed above, so that the existence theorem just proved will imply an existence theorem for a price equilibrium of a pure exchange economy. Continuing in this framework one can prove Debreu's theorem on generic finiteness of price equilibria, by putting the structure of a differentiable manifold on the big set of price equilibria. The equilibrium manifold is a natural setting for comparative statics.

A trader's preferences will be supposed to be represented by a smooth utility function $u: P \to \mathbb{R}$, where $p = \{x \in \mathbb{R}^l, x_i > 0\}$ is commodity space. The indifference surfaces are those $u^{-1}(c) \subset P$. We make strong versions of classical hypotheses on this function.

Monotonicity. The gradient, grad $u(x)$, has positive coordinates.

Convexity. The second derivative $D^2 u(x)$ is negative difinite on the tangent space at x of the corresponding indifference surface.

Boundary condition. The indifference surfaces are closed sets in $\mathbb{R}^l$ (not just P).

From the utility function, one defines for the individual trader a demand function. $f: \mathbb{R}^l_+ \times \mathbb{R}_+ \to P$ of prices $p \in \mathbb{R}^l_+$ and wealth $w > 0$. For this, consider the budget set $B_{p,w} = \{x \in P \,|\, p \cdot x = w\}$. Then $f(p, w)$ is the maximum of f on $B_{p,w}$.

One can prove:

Proposition. The demand function f satisfies

(a) grad $u(f(p, w)) = \lambda p$, for some $\lambda > 0$

(b) $p \cdot f(p, w) = w$

(c) $f(\lambda p, \lambda w) = f(p, w)$ any $\lambda > 0$

(d) f is smooth.

A pure exchange economy will be a set of m traders, each with preferences as discussed above, associated to utility functions u_i, $i = 1, \ldots m$ defined on the same commodity space P. Also associated to the ith trader is an endowment vector $e_i \in P$. At prices p, this trader's wealth is the value of his endowment $p \cdot e_i = w_i$. A *state* is an allocation $(x_1, \ldots, x_m)$, $x_i \in P$ and a price system $p \in \mathbb{R}_+$.

Feasibility is the condition:

(F) $\sum x_i = \sum e_i$

A kind of satisfaction condition of a state is

(S) For each i, x_i maximizes u_i on the budget set

$$B = \{y \in P \,|\, p \cdot y = p \cdot e_i\}$$

An economic equilibrium of a pure exchange economy $(e_1, \ldots, e_m, u_1, \ldots, u_m)$ is a state $[(x_1, \ldots, x_m), p]$ satisfying (F) and (S).

Theorem. There exists a price equilibrium of every pure exchange economy.

The proof goes by applying the previous existence theorem above. Define the excess demand $Z = D - S$ as follows:

$$S(p) = \sum e_i, \qquad D(p) = \sum f_i(p, p \cdot e_i),$$

where f_i is the above defined demand of the ith trader. One then shows Walras's Law:

$$p \cdot Z(p) = p \cdot D(p) - p \cdot S(p) = \sum p \cdot f_i(p, p \cdot e_i) - p \cdot \sum e_i = 0$$

using (b) of the proposition above.

Use (c) of the proposition to confirm the homogeneity of z. The use of the boundary condition is more technical. But under the rather strong hypotheses, this gives a fairly complete existence proof for a price equilibium of a pure exchange economy.

This existence proof extends to prove the Arrow–Debreu theorem in the generality of the latter's *Theory of Value.*

STEVE SMALE

See also CATASTROPHE THEORY; GENERAL EQUILIBRIUM; MATHEMATICAL ECONOMICS; REGULAR ECONOMIES.

BIBLIOGRAPHY

Mas-Colell, A. 1985. *The Theory of General Economic Equilibrium, a Differentiable Approach.* Cambridge: Cambridge University Press.

Smale, S. 1981. Global analysis and economics. In *Handbook of Mathematical Economics, Volume 1*, ed. K. J. Arrow and M. D. Intriligator. Amsterdam: North-Holland.

global stability. *See* ADJUSTMENT PROCESSES AND STABILITY; STABILITY; TATONNEMENT AND RECONTRACTING.

gluts. *See* AGGREGATE DEMAND AND SUPPLY ANALYSIS; DÉBOUCHÉS, THÉORIE DES; OVERPRODUCTION.

Godwin, William (1756–1836). The first and greatest exponent of philosophical anarchism, Godwin was born in Wisbech, Cambridgeshire in 1756. He was brought up in the Dissenting tradition in Norfolk, attended Hoxton Academy, and became a candidate minister. He gradually lost his faith, and in his late twenties turned to political journalism for the Whig cause. Inspired by the French Revolution, he wrote *An Enquiry concerning Political Justice* (1793). It earned him immediate recognition: 'no work in our time', Hazlitt wrote, 'gave such a blow to the philosophic mind of the country'. His novel *Caleb Williams* (1794) was considered no less of a masterpiece. But as the reaction to the French Revolution grew, so Godwin's reputation waned. Despite a long series of novels, histories, plays, essays and children's books, he was unable to recapture the public imagination. He died in 1836, and was buried beside the feminist Mary Wollstonecraft, who had died in childbirth. Their daughter Mary eloped with Godwin's greatest disciple, Percy Bysshe Shelley.

Godwin's economics, like his politics, are an extension of his ethics. His starting point is a belief in the perfectibility of man: since man is a rational and voluntary being, education will suffice to make him enlightened, generous and free. In his ethics, he is a thoroughgoing and consistent utilitarian, defining good as pleasure and arguing that 'I should contribute everything in my power to the general good'. Indeed, his bold application of the principle of utility, coupled with the principle of impartiality, led him to condemn private affections, positive rights, promises, gratitude and patriotism. In his politics, he concluded that government and law are unnecessary evils and in their place proposed a decentralized and simplified society of autonomous communities.

Godwin considers the subject of property (or economics) as the keystone that completes the fabric of political justice. His treatment has a critical and a constructive phase. He sees a close link between property and power: the rich are always 'directly or indirectly the legislators of the state'. Moreover, accumulated property has disastrous effects on rich and poor alike: it creates a 'servile and truckling spirit', makes wealth the universal passion, and reduces society to the narrowest selfishness.

But since we have a common nature, it follows from the principle of impartial justice that the good things of the world are a 'common stock', upon which one man is as entitled as another to draw what he wants. Property therefore should be considered a trust to be employed in the best possible way in order to promote liberty, knowledge and virtue. Just as every man has a duty to help his neighbour, so his neighbour has a claim to assistance.

Developing the labour theory of value, Godwin further argues that money is only a means of exchange and that there is no wealth except the labour of man. The producer should therefore retain what is necessary for his subsistence from the produce of his labour and then distribute the surplus to the most needy. Godwin also distinguishes between four classes of things: the means of subsistence, the means of intellectual and moral improvement, inexpensive pleasures, and luxuries. It is the last which is the chief obstacle to the just distribution of the previous three.

In place of the capitalist economic system, Godwin looks to small-scale production for the local market in a decentralized society. Production would be organized voluntarily with the producers controlling distribution. There would be a voluntary sharing of material goods, without barter or exchange. Anticipating the liberating effects of new technology, Godwin suggests that if all able-bodied people worked, production time could be reduced drastically, thereby giving people the leisure to develop their intellectual and moral potential.

When Malthus asserted that such a scheme would result in over-population, Godwin replied with his doctrine of moral restraint or prudence as a check. In his *Of Population* (1820), he went on to question the validity of Malthus's ratios, and argued that people would tend to reproduce less as their living standards improved.

Godwin's economic theory is clearly both profound and original. He was the first to write systematically about the competing claims of capital, need and production. Marx and Engels recognized his importance in developing the theory of exploitation. He not only strongly influenced the early socialist thinkers Robert Owen, William Thompson and Thomas Hodgskin, but the Owenites and Chartists took note of what he had to say. While Malthus has been most remembered, it is arguable that Godwin will be proved right in the long run. His scheme of voluntary communism remains moreover a thoughtful and persuasive ideal.

PETER MARSHALL

SELECTED WORKS

1793. *An Enquiry concerning Political Justice and its Influence on General Virtue and Happiness.* 2 vols, London: Robinson.
1794. *Things as They Are; or, The Adventures of Caleb Williams.* 3 vols, London: B. Crosby.
1797. *The Enquirer: Reflections on Education, Manners and Literature.* London: Robinson.
1801. *Thoughts occasioned by the perusal of Dr. Parr's Spital Sermon.* London: Robinson.
1820. *Of Population. An Enquiry concerning the Power of Increase in the Numbers of Mankind.* London: Longman, Hurst, Rees, Orme & Brown.
1831. *Thoughts on Man, his Nature, Productions and Discoveries.* London: Wilson.

BIBLIOGRAPHY

Brailsford, H.N. 1913. *Shelley, Godwin and their Circle.* Oxford: Oxford University Press.
Clark, J.P. 1977. *The Philosophical Anarchism of William Godwin.* Princeton: Princeton University Press.
Hazlitt, W. 1820. William Godwin. In *The Spirit of the Age*, Oxford: Oxford University Press.
Locke, D. 1980. *A Fantasy of Reason: The Life and Thought of William Godwin.* London, Boston and Henley: Routledge & Kegan Paul.
Marshall, P.H. 1984. *William Godwin.* New Haven and London: Yale University Press.
Monro, D.H. 1953. *Godwin's Moral Philosophy.* Oxford: Oxford University Press.
Paul, C.K. 1876. *William Godwin: his Friends and Contemporaries.* 2 vols, London: H.S. King.
Priestley, F.E.L. 1946. Introduction to *Enquiry concerning Political Justice,* Vol. III. Toronto: University of Toronto Press.

golden age. The notion of equilibrium or steady state in economics differs from that in mechanics because, unlike particles and molecules, economic agents are guided in their action by expectations of the future. Moreover, expectations are often being falsified by actual events forcing the economic actors to revise continuously and adapt their expectations in the light of experience. However, in an economic equilibrium expectations are *never* falsified; what happens must be compatible with what people expected to happen in every period of time and all expectations are being continuously fulfilled. Since all expectations cannot always be fulfilled in actual history, this in itself should be an adequate warning about the utter implausibility of the notion of equilibrium in economics. Economic equilibrium is something that we can never observe in reality; at best, it has to be recognized as a 'thought experiment' designed to facilitate analysis (Robinson, 1980).

An economically valid 'thought experiment' on the properties of steady state (or equilibrium) growth in a capitalist economy must then entail an analysis of how the various economic agents guided by their expectations, behave along such a path. This, in essence, would distinguish an economically meaningful description of steady state growth from its mere mechanical analogy in terms of a set of balance equations. Joan Robinson coined the term 'golden age' to demonstrate that, once expectations are explicitly dealt with, steady state growth represents an almost 'mythical state of affairs not likely to obtain in any actual economy'. (Robinson, 1956, p. 99).

The mechanical balance condition for the 'golden age' is the equality over time between the natural, the warranted and the actual rate of growth of national income ($g_n = g_w = g_a$) in the sense of Harrod (1948). This, in itself, represents a highly stringent set of conditions that are satisfied only accidentally. For instance, unless labour productivity rises at a uniform rate in all the sectors (in a multisectoral context) and the real (and individual product) wage rate(s) also rises at that same rate, technical progress would tend to upset the system of relative prices as well as the class distribution of income between wage and profit ruling along any steady growth path. Thus, technical progress, unless it is 'neutral' in the above sense, would not be compatible with golden age growth. But this could only be *accidentally* true, if one accepts that labour productivity growth through technical progress has a largely autonomous character.

Along the golden age, a constant set of relative prices rule and there is a uniform rate of profit (r) in all sectors, given by the Cambridge formula $r = g_n/s_p$, where s_p is the propensity to save out of profit and, g_n is the natural rate of growth (Robinson, 1956; Pasinetti, 1962). However, this *realized* profit rate must also be the *expected* profit rate ($\hat{\gamma}$) of the capitalists. Given that investment decisions are *autonomously* taken by the capitalists in the light of their profit expectations, at that expected rate of profit ($\hat{\gamma}$), the capitalists must autonomously decide to carry out just sufficient investment required at the golden age of growth (hence, $g_w = g_a$; Robinson, 1962; Marglin, 1984). Similarly, one can also presume that, with powerful trade unions, their *expected* real wage rate must always correspond to its actual path along the golden age. Thus, unless real wage is expected to rise at the same rate as labour productivity under neutral technical progress, collective wage bargain may divert the economy from the mythically tranquil conditions of social harmony represented along the golden age.

But even all this may be inadequate for maintaining the golden age. By its very nature, the equality between the natural, warranted and actual growth ($g_n = g_w = g_a$) ensures only a *flow equilibrium*, for example on this growth path the *additional* demand created by the multiplier through rising investment equals *additional* supply generated through net investment in each period. However, it does *not* ensure that any under- or over-utilization of the historically inherited capacities of the initial period would be alleviated. An *arbitrary* initial condition, say in terms of under-utilization of initial capacities or large initial unemployment, may therefore jeopardize the possibility of the golden age. This leads to its most paradoxical aspect: the initial conditions have to be suited exactly to the requirements of equilibrium growth rather than being arbitrary to sustain a golden age, that is, the economy must already be in equilibrium to continue along it!

AMIT BHADURI

See also ROBINSON, JOAN VIOLET; STEADY STATE.

BIBLIOGRAPHY

Harrod, R. 1948. *Towards a Dynamic Economics.* London: Macmillan.
Marglin, S. 1984. Growth, distribution and inflation: a centennial synthesis. *Cambridge Journal of Economics* 8(2), June, 115–44.
Pasinetti, L.L. 1962. Rate of profit and income distribution in relation to the rate of economic growth. *Review of Economic Studies* 29, October, 267–79.
Robinson, J. 1956. *The Accumulation of Capital.* London: Macmillan.
Robinson, J. 1962. A simple model of accumulation. In J. Robinson, *Essays in the Theory of Economic Growth,* London: Macmillan.
Robinson, J. 1979. History vs. equilibrium. In J. Robinson, *Collected Economic Papers,* Vol. 5, Oxford: Basil Blackwell.

golden rule. The so-called golden rule, or golden rule of capital accumulation, is a proposition about the consequences for national consumption – more broadly, for national welfare possibilities – of alternative paths of national wealth, and hence of national saving, in a closed economy. It developed out of the dynamic models of capital accumulation and output growth, generally in a setting of steady technical progress and demographic increase, begun in 1956 by R.M. Solow and T.W. Swan after some early explorations by Harrod, Domar and Robinson. Solow and Swan had shown that, provided diminishing returns to capital set in strongly enough (whether or not smoothly), there exists a state of steady growth corresponding to each possible value of the saving–output ratio, and, interestingly enough, this steady-state growth rate is independent of the value of the saving ratio – and often called the natural rate of growth. Hence, an upward shift of saving at each level of output, through increased private thrift or else higher taxes or lower spending by the government, cannot have a permanent, or non-vanishing, effect on the growth rate of output, only a transient effect.

If this theorem was the first law of the new 'growth economics', the golden rule of accumulation was the second law of growth economics. It states that the steady-growth state that gives the maximum path of consumption – the path layered on top of all the other steady-growth consumption tracks – is the one along which national consumption equals the national wage bill and thus national saving equals 'profits' (gross of interest in the present use of the term). Equivalently, the consumption-maximizing steady-growth path is the steady state along which the competitive rate of interest, which is the social rate of return to investment and to saving, is equal to the natural rate of growth. (To see the equivalence, divide profits and saving by capital.) Hence, a country (with any given history) that now plans forever to equate saving to profits could not hope to achieve a sustainable increase in the consumption path by some date in the future through a shift of policy towards increased saving; very possibly, even a temporary increase of consumption would not result. The reason is that despite a boost to future output brought by greater accumulation, the increase in future investment would eat up the increase in future output – and then some.

The arrival *circa* 1960 of the golden rule result was a classic case of multiple discoverers. And discovery seems the apt word, since the golden rule theorem was just a simple insight about a set or sets of equations in existence for several years that was waiting to be noticed, not a creative vision of the world springing from an independent empirical sense; accordingly, many or most of the discoverers were fledgling, pre-flight theorists still working on the ground of existing models. The earliest publishers of the result were Phelps (1961), Robinson (1962), and Swan (1963). However, it quickly became apparent that there were also discoveries on the Continent by von Weizsäcker and Allais, and even within the tiny space of the Cowles Foundation at Yale there were additional independent discoveries by Beckmann and Srinivasan. Robinson coined the proposition 'the neoclassical theorem', but eventually Phelps' coinage, the golden rule, became the standard. This was not a case of bad money driving out good, as will be explained.

The term 'golden rule' was something of a play on words. Mrs Robinson had dubbed states of steady growth as 'golden ages', so a proposition (if not exactly a maxim) about choosing among golden ages was natural to call a golden rule. In addition, there was also an allusion in the term to the biblical golden rule, do unto others as you would have them do unto you. The sense of that maxim, presumably, is that if one asserts a right to a certain policy, or treatment, from others, then in one's own treatment of others one must accord them the right to the same policy; so the choice of the rights to assert is subject to a reciprocity or cost constraint, which is a useful thing, for otherwise one would demand the most extreme sacrifices of others. Of course, this precept – the national saving policy, or national consumption function, that a future generation would have preceding ones follow, in view of its self-interests, it must likewise adopt on behalf of succeeding generations – does not by itself determine the just policy of national saving. Yet the golden rule perspective serves to alert us that there will be a limit to the austerity that future generations would ask of the present generation if they are obliged to practice the same austerity that they choose to preach. To make this effective, it should be noted, the saving policies from which society is to choose must be linear-homogeneous, and thus expressed in terms of saving as a ratio to output or profits or some related variable. (Otherwise, a future generation could piously call for lower consumption only at the present, comparatively low level of national income – and thus travel as a free rider.)

The meaning and indeed the significance of the golden rule becomes quite transparent in the special case of an economy in which the technology and the (working-age) population are constant, so that the natural rate of growth of the economy is zero. In this case the golden rule state is the Schumpeterian zero-interest stationary state; and since the net rate of return to investment is zero, gross profits are simply depreciation allowances and equal to gross investment, which is entirely replacement investment in a stationary state. Here it is abundantly clear why an alternative stationary state with a constant negative rate of interest would actually yield a lower path of consumption: the extra replacement investment would more than eat up the extra gross output, leaving an actual diminution of the net national product and consumption (to which NNP is equal). It is also perfectly clear that a society is not required as a matter of efficiency to aim for the Schumpeterian state; if the initial rate of return to investment is positive, it would take *lower* consumption in the present than would otherwise be possible on a sustainable basis (simply by consuming income) in order to move to the Schumpeterian state so that in the future a higher consumption level could be sustained than would otherwise be possible. Neither is such a move required as a matter of justice. From the utilitarian side there are economists who cheerfully discount the utilities of future people, and from the 'maximin' perspective it is obvious that present people would not optimally sacrifice to make better-off those who were not worse-off than they to begin with. The basic significance of the golden rule, then, is as a warning against national policies of over-saving, or counterproductive austerity. The golden rule theorem is simply a generalization to a growing economy of these observations.

Further results on the inefficiency entailed by exceeding, so to speak, the golden rule in certain respects were later obtained. It was shown with the help of T.C. Koopmans that keeping the capital–output ratio indefinitely in excess (by a non-vanishing amount) of the golden rule level would be dominated in terms of consumption, and thus utility, by another path, feasible from the same initial conditions, along which the capital–output ratio is always 'epsilon' smaller (Phelps, 1965, 1966). A much more general analysis came later from D. Cass in 1972 in which the borderline between efficient and dynamically inefficient paths is systematically examined.

'But the golden rule path could be the social optimum, couldn't it? Certainly it is very beautiful, and not obviously unjust!' There was a tendency among some to regard it as the optimum at least provisionally, for working purposes. However, any budding claims that may have existed for the 'optimality' of the golden rule path met with an objection by I.F. Pearce (1962). Start there at T_1, Pearce said, and end there at T_2. Then, if there is steady population growth, so the golden rule interest rate is positive (being equal to the population growth rate), it will increase the integral of total utility to save more now and less later, causing the capital stock to arch over its golden-rule track, since more saving will increase output. There could be no denying this, although some utilitarians prefer to sum the *per capita* utility of people

over time (or the utility of per capita consumption), which suggests there is a maximin impulse in their otherwise utilitarian hearts; from this angle it would not be preferable to deviate along Pearce's arching detour. Then, using the per capita utility version of utilitarianism, P.A. Samuelson (1967) took up the cudgels with a revision of the Pearce argument: if there is steady *technical progress*, so the golden rule interest rate exceeds the population growth rate, it will increase the integral of *per capita* utility to cause the capital stock to arch above its golden rule track since more saving will increase per capita output – as long as the interest rate remains above Samuelson's 'biological' level, which is the population growth rate. Again, it is *nolo contendere* from the golden rule side. Yet maximin advocates might object that if the per capita utility received by succeeding generations is rising, and unavoidably so, so that the oldest generation extant is always the worst-off, the detour from the golden rule path proposed by Samuelson would presumably entail some belt-tightening by the oldest generation along with the others in order to produce the consumption splurge for the benefit of some younger or future generations – hence a reduction in *minimum* per capita utility across generations. That cannot be a maximin improvement and is indeed a maximin worsening. Thus goes the maximin rejoinder to the turnpike 'refutation' of the golden rule.

EDMUND S. PHELPS

See also NEOCLASSICAL GROWTH MODELS; OPTIMAL CONTROL AND ECONOMIC DYNAMICS; OPTIMAL SAVINGS; RAMSEY MODEL; ROBINSON, JOAN VIOLET.

BIBLIOGRAPHY
Cass, D. 1972. On capital overaccumulation in the aggregative neoclassical model of economic growth: a complete characterization. *Journal of Economic Theory* 4(2), April, 200–23.
Pearce, I.F. 1962. The end of the golden age in Solovia. *American Economic Review* 52, December, 1088–97.
Phelps, E.S. 1961. The golden rule of accumulation: a fable for growthmen. *American Economic Review* 51, September, 638–43.
Phelps, E.S. 1965. Second essay on the golden rule of accumulation. *American Economic Review* 55, September, 793–814.
Phelps, E.S. 1966. *Golden Rules of Economic Growth*. New York: Norton; Amsterdam: North-Holland.
Robinson, J. 1962. A neo-classical theorem. *Review of Economic Studies* 29, June, 219–26.
Samuelson, P.A. 1967. A turnpike refutation of the golden rule in a welfare-maximizing many-year plan. In *Essays on the Theory of Optimal Economic Growth*, ed. K. Shell, Cambridge, Mass.: MIT Press.
Solow, R.M. 1956. A contribution to the theory of economic growth. *Quarterly Journal of Economics* 70, February, 65–94.
Swan, T.W. 1956. Economic growth and capital accumulation. *Economic Record* 32, November, 334–61.
Swan, T.W. 1963. Of golden ages and production functions. In *Economic Development with Special Reference to East Asia*, ed. K.E. Berrill, London: Macmillan.

Goldsmith, Raymond William (born 1904). Born in Brussels, Goldsmith studied in Berlin (PhD, 1927). He emigrated to the United States and served on the US Securities and Exchange Commission (1934–41), the War Production Board (1942–7) and as a member of the Senior Staff, National Bureau of Economic Research (1953–78). He was a Professor at New York University (1956–61) and at Yale University (1962–73; Emeritus, 1973 onwards). He pioneered the measurement of national and sector saving, investment, wealth, and balance sheets.

In the 1930s Goldsmith initiated Securities and Exchange Commission estimates of saving, obtained by a new method: subtracting additions to liabilities from additions to assets. During 1947–68 the Commerce Department national income accounts reported personal saving derived from the Securities and Exchange Commission series as an alternative to estimates obtained as income less consumption and as investment less corporate and government saving. The Federal Reserve Board then absorbed the Securities and Exchange Commission work into its flow-of-funds accounts. These had been started by Copeland in research partly paralleling Goldsmith's (Copeland, 1952, p. xvi).

A still more important breakthrough, in 1950, was Goldsmith's perpetual inventory method of estimating the stock and consumption of durable capital. The United States and other countries use it to obtain official national income and capital stock series. Goldsmith's monumental *Study of Saving* followed, providing balance sheets, saving, and wealth, for the nation and sectors. It featured broad scope, great detail, and series running back to the 1890s (to 1805 for reproducible wealth; in Goldsmith, 1951b). Goldsmith subsequently updated these series and introduced similar ones for other countries. Lifelong adherence to the principle of reproducibility, which calls for sufficient description of all estimates to enable the reader to reproduce them, increases the usefulness of Goldsmith's data.

Goldsmith has written extensively about financial institutions and capital markets. He introduced the financial interrelations and financial intermediation ratios (Goldsmith, 1966, 1969). In the 1980s Goldsmith was estimating ancient and medieval national products.

E. DENISON

See also SOCIAL ACCOUNTING.

SELECTED WORKS

1939. (Assisted by Walter Salant.) The volume and components of saving in the United States 1933–1937. *Studies in Income and Wealth*, Vol. 3, New York: National Bureau of Economic Research, 217–93.

1951a. A perpetual inventory of national wealth. *Studies in Income and Wealth*, Vol. 14, 5–61.

1951b. The growth of reproducible wealth of the United States of America from 1805 to 1950. In *Income and Wealth of the United States: Trends and Structure*, ed. S. Kuznets, Income and Wealth, Series II, Cambridge: Bowes and Bowes.

1955–6. *A Study of Saving in the United States*. 3 vols, Princeton: Princeton University Press.

1958. *Financial Intermediaries in the American Economy Since 1900*. Princeton: Princeton University Press.

1962. *The National Wealth of the United States in the Postwar Period*. Princeton: Princeton University Press.

1963. (With R. Lipsey and M. Mendelson.) *Studies in the National Balance Sheet of the United States*. 2 vols, Princeton: Princeton University Press.

1966. *The Determinants of Financial Structure*. Paris: Organization for Economic Cooperation and Development.

1969. *Financial Structure and Development*. New Haven: Yale University Press.

1982. *The National Balance Sheet of the United States, 1953-1980*. Chicago: University of Chicago Press.

1984. An estimate of the size and structure of the national product of the Early Roman Empire. *Review of Income and Wealth*, September, 263–88.

1985. *Comparative National Balance Sheets: A Study of Twenty Countries, 1688–1978*. Chicago: University of Chicago Press.

BIBLIOGRAPHY

Copeland, M. 1952. *A Study of Moneyflows in the United States*. New York: National Bureau of Economic Research.

Goldsmith, Selma (née Selma Evelyn Fine) (1912–1962). A government economic statistician specializing in distributions of consumer units by size of income, Selma Goldsmith studied at Cornell (BA, 1932) and Radcliffe (PhD, 1936), and was the wife of the economist Raymond W. Goldsmith.

After participating in income estimation at the National Resources Committee and the Agriculture Department, Goldsmith initiated the Office of Business Economics size distributions. Distributions were prepared annually as an adjunct to the office's national income accounts, and older distributions were adjusted for comparability. Before this work was suspended, Goldsmith and her successors published before-tax distributions of personal income in current and constant dollars for 20 of the years from 1929 through 1963 and after-tax distributions for 1950–62 (Fitzwilliam, 1964).

Estimators elsewhere had supplemented surveys of consumer units (which contain large reporting errors) with information for upper income groups from tax returns, and adjusted resulting distributions to conform to independent totals for income and units. Goldsmith's innovations included starting with earnings from tax returns for the whole nonfarm distribution, using survey information only for farm units, to fill gaps, and to combine tax payers into family groups; adopting OBE's personal income concept; and constructing time series. Ingenious techniques, often devised by H. Kaitz, and meticulous use of all information characterized the distributions. However, data grouped by class intervals, rather than microfiles based on statistical or exact matching of tax and survey data, usually had to be used in combining sources or adjusting to control totals.

Goldsmith also compiled capital gains statistics (Seltzer, 1951). At her death she was Chief of the Income and Statistics Branch of the Bureau of the Census.

E. DENISON

See also SOCIAL ACCOUNTING.

SELECTED WORKS

1939. Fine, S. (With E. Baird.) The use of income tax data in the National Resources Committee estimate of the distribution of income by size. *Studies in Income and Wealth*, Vol. 3, New York: National Bureau of Economic Research, 149–203.

1951. Appraisal of basic data available for constructing income size distributions. *Studies in Income and Wealth*, Vol. 13, New York: National Bureau of Economic Research, 266–373.

1951. Seltzer, L., assisted by Goldsmith, S., and Kendrick, M. *The Nature and Tax Treatment of Capital Gains and Losses*. New York: National Bureau of Economic Research.

1954. (With G. Jaszi, H. Kaitz and M. Liebenberg.) Size distribution of income since the mid-thirties. *Review of Economics and Statistics* 36, February, 1–32.

1955. Income distribution in the United States, 1950–53. *Survey of Current Business* 35(3), March, 14–27.

1955. (With G. Jaszi, assisted by H. Kaitz and M. Liebenberg.) *Income Distribution in the United States, A Supplement to the Survey of Current Business*.

1958. The relation of census income distribution statistics to other income data. (With comment by J. Pechman.) *Studies in Income and Wealth*, Vol. 23, 65–113.

1959. Income distribution by size – 1955–58. *Survey of Current Business* 39(4), April, 9-16.

1960. Size distribution of personal income, 1956–59. *Survey of Current Business* 40(4), April, 8–15.

BIBLIOGRAPHY

Fitzwilliam, J. 1964. Size distribution of income in 1963. *Survey of Current Business* 44(4), April, 3–11.

gold standard. For nearly three thousand years coined weights of metal have been used as money and for just as long, gold, silver and copper have been the preferred metals for minting coins. Currencies consisting of coins whose value is expressed by the weight of the metal contained in them at market prices, however, have seldom been used.

Until the inception of metallist reforms, the necessary amounts of the metals required for coinage were brought to the Mint by the sovereign or by the public. The sovereign's monetary prerogative consisted in fixing the Mint price of the metal, i.e. how many coins of a certain denomination could be coined from a given weight of metal. The Mint price was established with reference to a money of account, and it could diverge, and very often did, from the market price of the metal. When the sovereign deemed it necessary to devalue his coinage, he could change the parity between the coins and an 'ideal' currency, usually one which had circulated in the past. This made it unnecessary to resort to recoinage operations. The latter were also available but were considered as more radical policies, while changing the relative value of coins in terms of the 'ideal' currency was a more gradual instrument, which could be used almost daily if need be. But it was only very rarely that the sovereign would renounce his prerogative of giving coins a face value by law. Indeed, for many students of money this sovereign prerogative was what transformed a coined metal into money. Theodor Mommsen wrote some unforgettable passages (*Histoire de la Monnaie Romaine*, vol. 3, p. 157) on Emperor Constantine's decision, in the 4th century AD, to resort to free minting of gold coins whose value was given by their weight at market prices. To Mommsen this was the lowest point reached in the degradation of Roman monetary sovereignty. A Roman coin had always been taken at its face value, whatever its weight. Indeed, this had been the chief testimony of the credibility of the Roman State.

After the fall of the Roman Empire, none of the successor states can be said to have enjoyed, in the following centuries, an equivalent degree of monetary sovereignty. The plurality of successor states implied a plurality of currencies, with ample possibility of speculating and arbitraging between currencies. Citizens learned to defend themselves when the sovereign's privilege became exorbitant. Seignorage tended to be inversely correlated with the commercial openness of States. The less truck subjects had with foreigners, the greater the divergence between the legal and intrinsic value of coins could be. A trading nation very soon found that it had to have a currency whose value corresponded to its metallic content. With the rise of the absolute state, the sovereign's monetary privilege tended again to become exorbitant. The ratio of internal to international trade increased, in spite of the rise of mercantilism, and the fiscal use of money by the State became greater. It is obvious that this sovereign prerogative, when it became exorbitant, made life very difficult for the sovereign's subjects, who could only defend themselves by changing prices, provided that had not been declared illegal by the sovereign. Metallist reforms were an expression of the new power acquired by the subjects vis-à-vis the State. At the core of these reforms was the actual coinage of the 'ideal' money as a real full-bodied coin, whose weight and fineness was decreed, and this coin became the 'standard' of the national monetary system. The sovereign, by these reforms, saw his monetary prerogatives diminished to that of a keeper of weights and measures. In the intentions of the reform's advocates, it was a way of constitutionalizing the sovereign, so that he would be compelled to resort openly to his fiscal powers, which had been constitutionalized long before.

The Gold Standard was just one of the possible metallic standards. It was adopted in England, while the French preferred to choose a silver standard. In the course of the following centuries intellectual debate centred around the choice of metal, and economists, statesmen, intellectuals, declared themselves in favour of or against silver or gold, in favour of or against bimetallism or monometallism. But the basic choice in favour of a pure metallic standard, where an actual coin whose value as given by the weight of its metal content at market prices was the only money of account available, was not seriously discussed again for a long time, until the development of banking and the integration of world commodity and financial markets gave reason to challenge existing institutions.

The great metallist reforms were the outcome of the intellectual movement which would later take the name of 'political economy'. This is now used to define an academic subject, taught in universities, but between the second half of the 17th century and the first decades of the 19th century, it became an intellectual, almost a political movement. It was composed of men who, in many countries, believed that human society was organized according to natural principles, which could be studied by the same methods used to inquire into the world of nature. By scientific inquiry the laws which governed society could be discovered and the action of the state could be made to agree with them. In particular, the laws governing the production and distribution of commodities could be discovered, and the principles according to which value was conferred upon goods and services. The political economists soon found themselves considerably disturbed by the existence of a human institution, Money, which continually interfered with the progress of commodity valuation. As hinted above, they tried to devise a solution which would allow society to enjoy the advantages of using money while being spared the problems which the creation and use of money entailed. The solution was a commodity money, a monetary regime whose standard would be a coin made of metal of fixed weight and fineness. They hoped that a commodity money would free the economic world from the uncertainties induced by the raids of those who exercised or usurped monetary sovereignty. A pure metallic money would be subject to the same laws of value to which other commodities were subject; its demand and supply would be determined strictly by the needs of trade.

By advocating the adoption of a pure metallic standard, the political economists were thus killing two birds with one stone. They were putting a stop to the exorbitant privilege of the State, which used its monetary prerogative to tax people without asking for the powers to do so, and were also recommending a type of money which would not disturb the functioning of economic laws, since it obeyed those laws. By the adoption of a pure metallic standard, a truly neutral money could be relied upon.

If the desires of political economists were important in actually pushing forward the adoption of metallic standards, however, it was more because both governing circles and public opinions were anxious to put an end to the previous system, which was based on uncertainty and sovereign privilege, than because of a widely felt need to put economic theory on a sounder theoretical footing.

In this respect British experience is different from the French. In England, a metallic standard had been in use since soon after the great recoinage of the end of the 17th century. At the turn of the next century, Sir Isaac Newton, the Master of the Mint, had established the canonical weight for the Pound sterling in gold, at 123.274 grains of gold at 22/24 carats (corresponding to 7.988 grammes at the title of 0.916). Free minting remained possible but, as very little silver was coined, silver coins were soon demoted by the public to the role of subsidiary currency, as they were still of the old sort, without milled edges, and were badly worn, because of the repeated clipping. Thus, early in the 18th century, England went on the Gold Standard.

In France, metallist reform had to wait until the Revolution. After an early attempt to introduce the Gold Standard, and a gigantic outflow of gold under the Terror government, in the year XI of the Revolution the free minting of both gold and silver was declared. One franc was given a weight of 5 grammes of silver at 9/10 title. A fixed parity was also established between gold and silver, although the French legislators, in the Report of the Comité de Monnayes in 1790, had declared that a permanently fixed parity between the metals was impossible, and had quoted Newton and Locke to corroborate their declaration. Bimetallism was thus instituted in France, and would last almost as long as the Gold Standard in England, but from the beginning it was understood that the parity between gold and silver would have to be changed when necessary, even if it was to be done by law each time. The French lawmakers gave life, therefore, to a system which we would call today of fixed but adjustable parity between gold and silver.

Contemporary literature devoted much attention to the relative virtues of mono- and bimetallism. Modern economic literature, however, starting from the end of World War I, has almost exclusively focused on Gold Monometallism. From the point of view of monetary history this is a pity, because what commonly goes under the name of the International Gold Standard was, on the contrary, a complex system composed of a monometallist and a bimetallist part, where the importance of the former was not greater, for the functioning of the whole system, than that of the latter. We shall see, in what follows, how the smooth functioning of the Gold Standard essentially required the existence of a bimetallist periphery which surrounded the monometallist centre.

Let us first concentrate on the British Gold Standard. After it had been in existence for close to a century it had to be suspended in 1797, because of the difficulties which the Napoleonic wars entailed for monetary management. In the period of over twenty-five years in which cash payments remained suspended, a very lively debate took place among political economists, politicians, bankers, and industrialists on how suspension affected internal and international economic relations. Some of the best pages in the history of political economy were written as contributions to that debate.

Specie payments had been suspended by an Order in Council in February 1797. The same decree had undertaken that they be resumed, at par, six months after a definitive peace treaty had been signed. In the intervening period of open hostilities, the currency had depreciated, the Government had incurred a huge debt which was largely in the hands of City financiers, and war demand for all sorts of commodities had favoured the amassing of great fortunes by a bevy of 'homines novi'. As peace approached, it was found that a resumption at the old parity would enhance the post-war slump which already appeared after Waterloo. This prospect united landowners and industrialists, who had been natural enemies heretofore, against creditors, Government Debt holders and, in general, people with fixed incomes. Because of its huge debt, the Government ought to have been on the same side as the debtors. It had, however, muddled through the war by putting up a system by which it held bond prices up and kept the financial market favourable to new debt issues. The system

consisted of redeeming old long term debt and replacing it by floating debt. Pascoe Grenfell and David Ricardo were quick to chastise the Government's debt management policy. In 1816 and 1817 the Government's balancing act was successful but in 1818 it came unstuck, as the Government had to buy stock dear and sell it cheap. Meanwhile, the ratio of funded to floating debt had fallen, and this precluded the possibility of reducing the main debt.

Resumption was as highly political a measure as Restriction had been. The Whig opposition railed against Restriction, calling the Government a committee of the Bank of England. And, indeed, the Bank did its best to make the accusation credible. It tried to blackmail the Government into a continuation of Restriction by threatening to stop its support of the Government's debt management policy. It also threatened to stop accommodating Meyer Nathan Rothschild, who was the principal holder of Government Stock. But Resumption had also its advocates within the Cabinet; Huskisson, for instance, who with Parnell, Henry Thornton, and Francis Horner, had drafted the Bullion Report in 1810 and had thus permanently alienated traditional City interests. He had advocated a prompt resumption in a memorandum he submitted in 1816, and again, early in 1819, he submitted a memorandum calling for prompt resumption accompanied by fiscal deflation. The Government then appointed a Secret Committee to consider resumption, which soon became dominated by opinion in favour. When the Committee's Report was discussed in Parliament, Ricardo's vehement advocacy of resumption definitively swung parliamentary opinion. Payments were resumed, at the old parity, in May 1819. Ricardo called the decision to resume 'a triumph of science, and truth, over prejudice, and error'. It certainly was a triumph of new City blood over old financial interests, who had thrived in the easy days of inflationary finance, lending to financially weak Governments, at rates they themselves pushed up by manipulating the money market.

After resumption at the old parity, a shock wave went through all British economic circles. The Gold Standard did seem to have no advocates left among manufacturers and financiers. The Bank of England had been against it all along, and so menacing had been its representations that the Government had been driven by such impudence to breaking its useful wartime alliance with it, which had rested, it now appeared, on easy money. Landowners, on the contrary, were pleased. A measure that made them poorer in capital values, gave them, at the same time, a greater real value for their rents. It also represented a restoration of old values against the encroachment of industry and its social evils, which had occurred during Restriction. If the Gold Standard was bad for industry, which had flourished under the Paper Pound, then the relative power of the Old Order, which Agriculture represented, would grow again.

The Bullionists, who had campaigned for a resumption at the old parity, believed that a deflation would purge society of the most glaring speculators, of unsound industrialists, and, more generally, of upstarts who had grown rich on easy money. At the same time they believed that the Gold Standard would transform Britain – and we have Huskisson's testimonial to this belief – into the chief bullion market of the world. London would become the 'settling house of the money transactions of the world'. The intention was thus to favour the New City, to be 'Mart and Banker' to the world, rather than its workshop.

Finally, Resumption was seen as an instrument of social justice. Deflation would give back to creditors, who had lent their money to their country in wartime, the full value of what they had lent. To politicians, an automatic Gold Standard looked like a relief from the heavy responsibilities of managing the economy. It would restore them to true political activity, and mark the final transition to Peace.

The French monetary reform was very different. It was aimed directly against the Ancien Régime, seen however in its fiscal capacity, and not as an unholy alliance of politicians and financiers. It ended up by establishing a long lasting bimetallic system, which did not overlook the interests of those whose prices were fixed in silver, like wage earners and petty traders. The Reform thus did not represent a clearly determined social choice, like the Resumption in Britain. A more neutral system was devised, which tried to accommodate both the third and the fourth estate. The revolutionary experience was too recent to invite, by a deflationist monetary regime, new social disorders. It is somewhat ironic to see how the country, on the verge of defeat, opted for a regime much less radically deflationist than the one the victorious country would choose. In both countries money was constitutionalized, but in Britain, the coalescence of interests of the New City and the landowners made the country the world pivot of monetary radicalism. It is fair to say that the expectations of those who had favoured the Gold Standard in Britain were not fulfilled. Deflation brought in its wake unemployment and social disorder. It also induced British industrial producers to invade world markets, as home demand shrank. The benefits the New City interests had expected did indeed materialize, but only a few decades later. The Gold Standard induced export-led growth in Britain, and to become 'mart and banker' to the world she first had to become the workshop of the world. The mechanisms the bullionists had set working thus functioned in reverse. But it would be unfair to say that Huskisson's expectations were representative of those of all bullionists. David Ricardo, for one, expected the Gold Standard to bring about industrial expansion. And he wanted industrial growth to employ the labour made available by the working of the Law of Population, in which he firmly believed.

The Gold Standard was supposed to check the power of the Bank of England, which seemed to have become so great under the Paper Pound as to represent a threat to a truly constitutional institution like Parliament. But its actual functioning enhanced that power even further. As Britain became the workshop of the world and Sterling was more and more widely used as an international currency, the importance of London as a financial centre grew apace. The Bank of England thus became pivot of an international payments system founded on Britain's industrial and financial supremacy. The Bank of England's importance as a commercial bank was enhanced by its monopoly position as a joint stock bank. Just as it had flourished as the chief source of Government finance under restriction, the Bank flourished as a commercial bank as a result of Britain's ascent to industrial and commercial leadership.

Its international pre-eminence was always dependent on its domestic primacy. Under the auspices of the Bank's monopoly, the centralized reserve system, which remained for a long time unique to Britain, was developed. It was a very lean and efficient system, which minimized the amount of cash needed to oil the wheels of the domestic payments network. But it was also highly unstable, since its leanness did not tolerate any serious obstacle which might appear in the national and international flow of cash and capital. The fact that it could carry on for such a long time, until the First World War, is explained by a series of fortunate circumstances which occurred in succession. We shall examine them in some detail.

We must, however, strongly underline the fact that, under the Gold Standard, Britain experienced very strong cyclical swings. The hundred years after resumption were marked by commercial and financial crises which recurred about every ten years, even if the last part of the period saw crises appear at longer intervals than before. The regularity of crises gave rise to much monocausal theorizing, and the Gold Standard was often indicted as one of the chief culprits. It was, by contemporary opinion, accused of being a monetary regime too inflexible to allow for the smooth growth of the economy. Critics invariably quoted the French and then the German monetary systems as preferable, since they were supposed to possess a greater degree of flexibility and made possible better management of the economy.

Yet, in spite of a very lean centralized reserve, and of recurrent financial crises, Britain never abandoned the Gold Standard. One of the most important reasons why she was not compelled to do so under the pressure of crises must be found, as we noted earlier, in the peculiar features the international financial system possessed in the combination of a monometallic and a bimetallic part. Since oldest antiquity, silver was the metal preferred by the Far East for coinage. And for almost as long as history goes, the Western trade balance with the Far East has shown a deficit. A structural trade imbalance with the Far East meant a continuous support of silver towards the East. Around the middle of the 19th century, this structural trend combined with gold discoveries to depress the price of gold. In the last thirty years of the century, however, the trend was reversed, as silver started to be abandoned by most developed countries as a monetary standard. The gold–silver parity rose accordingly.

Throughout the century, London retained a quasi-monopoly of gold and silver transactions. And it maintained, without any interruption, a free gold market. It is certain that it could not have afforded to do so, had not first France and, later on, the Indian Empire come to the rescue.

The Anglo-French financial connection is one of the most fascinating, and least researched, features of the 19th century international payments system. From what we know, however, it appears that the much greater liquidity the French monetary system retained throughout the 19th century was skilfully exploited by Britain. Bank rate would be raised when pressure was felt on the Bank of England's reserve, but the expectation was that gold would flow mainly from Paris. Why did it flow? First of all, because there was a lot there, because of both the wealth of the French economy and of the underdevelopment of the French banking system, which rendered the use of gold coins for large transactions necessary (whereas in Britain cheques were commonly used). We must not forget, however, the essential role played by the House of Rothschild in connecting the French and British money markets. The archival evidence available shows that, in most British financial crises, the reserves of the Bank of England were refurbished with gold procured by Rothschild from France. The House of Rothschild intermediated between the gold and silver sides of the international monetary system. They were the super arbitrageurs who had the huge reserves and prestige necessary to play successfully a role which remains to be described in full detail, but whose importance it is possible to detect even in the present state of research. They were the 'protectors' of Bank Rate. It is not without importance that a Rothschild sat in the Court of Directors of the Bank of England and a French Rothschild occupied an equivalent position in the Directorate of the Banque de France.

Towards the end of the century, however, the precipitous fall of silver, induced by and in turn determining the abandonment of silver as a monetary standard in the whole developed world, reduced the role played by the French monetary system as a stabilizer of the Gold Standard. France had herself to close the mints to silver, to avoid being flooded by a metal nobody seemed to want any more. In the remaining period, commonly referred to as the 'heyday of the Gold Standard', the Bank of England's balancing act could continue with the help of two other shock absorbers, the Indian monetary system, still based on silver, and South African gold production. The Empire of India was kept by the British on a silver standard even when silver was fast depreciating against gold. This made exports of primary commodities and raw materials easy and was undoubtedly responsible to a large extent for the large export surplus India earned in the last part of the pre-war period. It is in the management of this surplus in a way conducive to the stability of the Gold Standard that the British financial elite proved most imaginative and successful. The Indian surplus was invested in London, in Government bonds or in deposits with the banking system. The 'Council Bills' system, which had been devised to effect financial transfers between India and the Metropolis, was managed so as to keep the Rupee's value stable. The whole system, called the 'Gold Exchange Standard', was extolled as a paragon of skill and efficiency by J.M. Keynes, in the book that first gave him notoriety, *Indian Currency and Finance* (1913). Indeed, the young Keynes was right, as far as the functioning of the Gold Standard was concerned. Whether it was also efficient from the point of view of promoting Indian economic development, is entirely another matter, and one with which Indian economic historians have seldom concerned themselves.

South African gold production also helped to stabilize the Gold Standard. All the gold mined there was commercialized in London, and the proceeds invested there, at least in the short run. It is easy to imagine how important the British monetary authorities considered the control of that huge flow. This became evident, after World War I, when an attempt was made to revive the Gold Standard. Following Professor Kemmerer's advice, South Africa decided in favour of the Gold Standard, and against pegging its currency to Sterling. The connection with London was cut, to the great discomfort of Montagu Norman, who saw one of the main props of Sterling suddenly disappear.

If France, India and South Africa contributed to making the Gold Standard stable, the United States represented, throughout the century in which the Gold Standard lasted, one of the great, perhaps the single greatest, disturbing elements to its smooth functioning. After the political and economic forces which stood for an orderly financial development of the Republic had been routed in the first decades of the 19th century, the growth of the American economy took the spasmodic features it would keep until the Second World War. The United States was deliberately deprived of the Central Bank that Alexander Hamilton, imitating the Bank of England, had designed. Banks proliferated everywhere, following a model of wildcat finance which, if it promoted the phenomenal growth of the US economy, also gave it a very strong cyclical pattern. For the whole Gold Standard period, the Bank of England was called to play the difficult role of being the lender of last resort to the American financial system. The growth of American farm exports, coupled to local industrial growth and the peculiar development-underdevelopment of the US banking system, gave rise to a notorious seasonal pattern of financial difficulties, which was called the 'autumn drain'. This recurred every year, when American crops were sold on world markets and the proceeds disappeared into the entrails of the completely decentralized American banking system, and, more generally, into the hands of American farmers. A gold drain was felt first in New York, the main US financial centre. Interest rates rose violently, as there was no centralized banking reserve in New York, and the US Treasury, which kept a very large gold reserve, knew only very imperfectly how to use it for stabilization purposes. The rise of New York rates would thus be transmitted to London, which kept the only free gold market. Gold would thus flow to New York and it would be months before it could be seen again, as farmers spent the proceeds of crop sales and US local banks recycled the money back to New York.

To this seasonal drain, to which the Bank of England was never able to find a remedy, other sudden drains would be added, when the peculiar American banking system went into one of its recurrent panics. After the most violent of them had, in 1907, brought chaos to the whole international economy, the US Congress decided to move and the Federal Reserve System was established, in 1913. But it took another twenty years, and another huge crisis, that of 1929–33, before it really began to work as a central bank.

We have dedicated considerable space to a summary of US financial history because it must be fully appreciated what the peculiar structure of American finance meant for the world financial system in the age of the Gold Standard. What by the end of the century had become both the largest industrial producer and largest agricultural exporter, was still importing huge financial resources from the rest of the world. It lacked a central bank and had developed a thoroughly decentralized banking system which, if it was functional to rapid economic growth, had also a strong vocation for recurrent instability. The US Congress and Government also did their part to enhance instability, by unwise and partisan policies concerning, for instance, silver prices, and the management of fiscal revenues.

In the last decade of the 19th century the crisis of silver induced a veritable stampede by Governments and Parliaments, in most countries, to adopt the Gold Standard. More than the unidirectional movement of silver, it was its wild oscillations, made deeper by the inconsiderate silver policies of the United States, that convinced most interested parties to opt for gold. Even European farmers, who were fighting a desperate war against cheap New World imports, were reduced to favouring the Gold Standard by the impossibility of forecasting a price for their harvest at sowing time. Industrialists in developing countries who had started import substitution activities were in favour of a strong currency, to repay foreign loans without problems, and preferred protection as a means of keeping out foreign industrial products. Most countries, when they went on the Gold Standard, also started a centralized gold reserve, which they intended as an exchange stabilization fund. Very often they surrounded this reserve by an outward layer of foreign currency reserves, which they called upon under pressure in order to keep their gold reserve intact.

Contrary to what British monetary authorities thought and did after the First World War, their pre-war predecessors were extremely worried by the universal trend in favour of Central Banking and of the Gold Standard. They very correctly understood that Britain had succeeded in staying at the centre of the system as long as it remained a free flow system, where the only stock of gold was the one kept by the Bank of England. French gold accumulation had been seen favourably, as it enhanced the *masse de manoeuvre* of the Bank of England at almost no cost. But already German gold accumulation was a threat, as Germany did not believe in a free gold market and

Bank Rate found obstacles in attracting gold from there. The German pattern was, unfortunately for Britain, the one that found the largest number of followers among countries that established a Central Bank, and a central reserve to manage the Gold Standard. The result was the increasing seclusion of previously free-flowing gold into large stocks, over which the British traditional control instrument Bank Rate scarcely exercised any leverage and which dwarfed in size the reserves of the Bank of England.

To these external difficulties with which British monetary authorities were greatly concerned, others of a more domestic nature had to be added.

The British financial system had emerged from the turmoil of the Napoleonic wars apparently unscathed. It was formed by a cluster of merchant banks and other financial institutions, like the discount house, and by the great commodity and service exchanges, and it had the Bank of England at its centre. The composition of the governing body of the Bank of England ensured that most City voices would get a fair hearing. It is impossible to exaggerate its internal homogeneity and cohesion (especially at a time like the present, when the system is definitively being demolished). A good study of the City in the years of the Gold Standard ought to be conducted by structural anthropologists, rather than by economists. A very serious threat to this semi-tribal system, which had succeeded in controlling world trade and payments for many decades, was developing fast in the late years of the Gold Standard. It was represented by the rapid concentration of British deposit banking, which resulted in the survival of only a handful of giant joint-stock banks. The Clearing Banks – as they came to be called – provided the City with a large part of the short term funds which were used as raw material to finance world commodity trade. They had huge branch networks which channelled savings from the remotest corners of Britain to London, and thence to all parts of the world through City intermediation. Thus the Clearing Banks provided the base for the whole British financial system. But their power was not constitutionalized by any matching responsibility. They had no say in the conduct of monetary policy. They were not represented in the Court of Directors of the Bank of England. Moreover, as concentration increased, the Clearing Banks thought they might as well invade some of the markets traditionally reserved to merchant banks, and in particular, they started invading the field of commodity trade financing. Finally, they began to lay the foundations for their own centralized gold reserve, alternative to that kept by the Bank of England.

Speaking more generally, a trend can be noticed in the last 25 years of the pre-war Gold Standard, away from homogeneity and towards decentralization, in the British financial system. The Clearing Banks increasingly baulked at being disciplined by the Bank of England. Often, especially in times of crisis, they pulled the rug from under the financial establishment, by withdrawing their short term deposits with City houses. By this behaviour they showed their muscle and demanded recognition. This pattern is clearly detectable in the 1890, 1907 and 1914 crises. It was a trend that greatly disturbed the financial elite and contributed, with the exogenous factors we have mentioned before, to making the Gold Standard more unstable. It could even be said that the loss of cohesiveness and homogeneity of the British financial system brought the Gold Standard to its demise, in July 1914. The system collapsed long before Britain entered the conflict.

THE GOLD STANDARD AND THE ECONOMISTS. The development of

Gold Standard theory coincides with the development of economic theory. We have already mentioned the role played by commodity money in the theoretical apparatus of the classical economists. A commodity money would obey the rules dictated by Nature (of which even human behaviour was part) as far as its supply, demand, and price were concerned. Thus a monetary economy based on a pure metallic standard would enjoy all the advantages afforded by the presence of money, without being subject to the many disadvantages induced by a man-made currency not tied to a metal. For David Ricardo, the recommendation to adopt the Gold Standard meant not only preventing the Bank of England from usurping monetary sovereignty, which he recognized as a Parliamentary prerogative, it also meant giving the economic system a standard, like gold, which had the virtue of being a good approximation to his invariant measure of value. He wanted to see the price system uninfluenced by political power, so that Nature would be free to play her game and gold would be distributed among the 'different civilized nations of the earth, according to the state of their commerce and wealth, and therefore according to the number and frequency of the payments which they had to perform' (Ricardo [1811], 1951, p. 52). If freedom of gold movements existed, this redistribution would soon bring about a state of rest, when gold had been allotted to each nation according to its needs and would not move again. If all countries promoted metallist reforms, fixing a gold weight for their currencies, arbitrageurs would operate, within the gold points, to keep gold prices uniform. Gold would function as the numeraire of the world economic system and it would be enough to ensure gold arbitrage to guarantee uniformity of all the world price systems. There would be no need for arbitrage to involve other, bulkier, commodities, whose transportation would imply greater costs. This, of course, did not mean that international trade would not take place. Commodities would move across countries according to the Law of Comparative Advantage, and the Gold Standard would make sure that this law did not suffer perturbations because of 'unregulated' money supplies. 'Regulation', of course, meant that fiduciary money would depend, for its supply, on the dynamics of the gold reserve of the issuing agency. Ricardo's view of how the world economy worked, based on his analysis of commodity currency systems, rapidly conquered not only the economics profession but also politicians and intellectuals. It was a scientific system of political economy, whose core was the Gold Standard. John Stuart Mill and Alfred Marshall were to refine and qualify that world view.

Mill analysed with great care the implications of a commodity money, whose exchange value would be equal to its cost of production. He did, however, clearly point out the importance of the existing stock of gold relative to its current or even potential flow. The gold stock/flow ratio made full adjustment a lengthy process, so that, in the short run, the price level would be determined by the demand for, and supply of, money. He never doubted, however, that a commodity money would not be able to change the international production relations as they existed under barter. To him money was, like oil in the wheels of moving mechanisms, 'a contrivance to reduce friction'. He fully trusted that David Hume's adjustment mechanism would have only nominal consequences in the case of a discovery of a hoard of treasure in one country. This would raise prices there, discourage exports and induce imports. The resulting balance of payments deficit would redistribute the hoard to the rest of the world and lower prices in the original country to their previous level. In Mill's opinion, real effects would, however, result in the case of a loan from one country to another. Then a real transfer would have to be effected.

Neither Mill nor Marshall considered the Gold Standard a perfect system. Both of them opposed bimetallism at fixed rates. They believed that relative changes in the costs of production of the two metals would be likely, and that would involve a scarcity of the dearer metal and a shift in favour of the monetary use of the cheaper one. Instability was therefore built into the bimetallist system. Mill preferred a 'limping' gold standard, where gold would be the only legal tender, and silver would be coined at market prices. John Locke's tradition obviously lived on.

Marshall's creative thinking in the field of monetary standards included 'symetallism' and the 'Tabular standard'. According to the first scheme, vaguely reminiscent of the oldest currency, the Lydian 'Elektron', if the public wanted to give paper currency and receive metals it could only get gold and silver together, in bars of fixed proportions. Marshall thought this would link the paper currency to the mean of the values of the two metals and make possible, by this more stable currency, a world monetary area including both the gold and silver countries. It is easy to recognise in Marshall's scheme a forerunner of the contemporary European Monetary System's ECU.

Marshall's Tabular standard, on the other hand, reintroduced the concept of a money of account separate from the medium of exchange. The money of account would serve for long term contracts and would be tied to an 'official index number, representing average movements of the prices of important commodities' (Marshall, 1935, p. 36).

As far as the adjustment mechanism under a gold standard regime was concerned, Marshall clearly saw the growing integration of capital markets replacing traded goods, arbitrage and gold movements as the chief instrument of adjustment. This of course meant recognizing the importance of interest rate differentials and interest arbitrage. And, in turn, giving a great role to play to banks and Central Banks.

With J.S. Mill, Marshall, and, in particular, Irving Fisher, we begin to get out of the 'naturalist' world view which permeates the writings of Ricardo, his inspirers, and his followers. The world is not run solely according to forces of nature, which it is the economist's role to discover and which cannot be violated without meeting an inevitable punishment. The Gold Standard is not a 'scientific method' of organizing a monetary regime. Like Marshall, Irving Fisher thinks of it more in historical rather than scientific terms. It is something the world embraced by historical accident. Supply and demand conditions for gold and silver are unstable. The system is not perfect and is perfectible, justifying proposals to make it work better.

As we advance toward what has been called the 'heyday' of the gold standard, in the eyes of contemporary economists its virtues seem to pale and its vices to come into relief. To Knut Wicksell, under a commodity standard there is no guarantee that a causal link will be able to exist between money supply and price level movements. Such a link can be seen to exist only if we take a very long view. Like the practitioner-theorists who staffed the British Treasury before 1914, Wicksell noticed that central banks, by keeping large gold reserves, had interposed themselves between gold supply and price movements. The price stabilization function of central banks is recognized and the new institutional set-up is in any case superior to a pure metallic standard, which in Wicksell's eyes would be totally at the mercy of the vagaries of demand for and supply of gold.

The 'heyday of the gold standard' which (as we hope to have shown above) was in historic reality the beginning of its decline, were thus also days of decline as far as Gold Standard theory was concerned. A growing scepticism begins to engulf Hume's price–specie flow mechanism. Commodity arbitrage is seen as prevailing over gold arbitrage. Adjustment must involve real, not just nominal changes. International capital movements are brought increasingly into the picture. Stock adjustment in all sorts of markets is a phenomenon which fascinates the economic theorists of this age. From recognizing stock adjustments to advocating stock management is a short intellectual distance and most of these theorists cover it at great speed.

Ironically, in the theoretical cycle the pendulum had swung in the 25 years before World War I, away from Ricardo and Locke and towards Lowndes and Thornton. The pure metallic standard has lasted only *l'espace d'un matin* both in theory and in practice. Economists had not been able to ignore the giant strides of banking and of world economic and financial integration. From Ricardo's golden rules, simple and infallible, we move to Mill, Marshall, Wicksell, and Fisher and their inventive recipes for national and international monetary management. Doubts prevail over certainties. We cannot accept J.M. Keynes's post-war strictures about the pre-war perception of the Gold Standard. It was *not* seen as immutable, frictionless, and automatic by its contemporaries. The seeds of post-war criticism and disenchantment were firmly sown before the war. In fact, we might go as far as to say that pre-war learned opinion was much less apologetic of the pure metallic standard than would be the post-war economists and politicians. Pre-war observers had realized that the Gold Standard was a game which had become increasingly hard to play, precisely because everybody had learned – and wanted – to play it.

MARCELLO DE CECCO

See also BULLIONIST CONTROVERSY; CENTRAL BANKING; INTERNATIONAL FINANCE; INTERNATIONAL LIQUIDITY; SPECIE-FLOW MECHANISM.

BIBLIOGRAPHY

Ashton, T.S. and Sayers, R.S. 1953. *Papers in English Monetary History*. Oxford: Clarendon Press.

Bagehot, W. 1873. *Lombard Street*. Reprint of the 1915 edn, New York: Arno Press, 1969.

Bloomfield, A.I. 1959. *Monetary Policy under the International Gold Standard*. New York: Federal Reserve Bank of New York.

Bordo, M. and Schwartz, A.J. (eds) 1984. *A Retrospective of the Classical Gold Standard 1821–1931*. Chicago: University of Chicago Press.

Clapham, J.H. 1944. *A History of the Bank of England*. 2 vols, Cambridge: Cambridge University Press.

Cottrell, P.L. 1980. *Industrial Finance 1830–1914*. London: Methuen.

De Cecco, M. 1984. *The International Gold Standard: Money and Empire*. 2nd edn, London: Frances Pinter.

Fanno, M. 1912. *Le banche e il mercato monetario*. Roma: Loescher.

Feaveryear, A. 1963. *The Pound Sterling*. 2nd edn, Oxford: Clarendon Press.

Fetter, F.W. 1965. *Development of British Monetary Orthodoxy 1717–1875*. Cambridge, Mass.: Harvard University Press.

Ford, A.G. 1962. *The Gold Standard 1880–1914: Britain and Argentina*. Oxford: Clarendon Press.

Goodhart, C.A.E. 1972. *The Business of Banking 1891–1914*. London: Weidenfeld & Nicholson.

Hilton, B. 1977. *Cash, Corn and Commerce: the Economic Policies of the Tory Governments 1815–1830*. Oxford: Oxford University Press.

Ingham, G. 1984. *Capitalism Divided? The City and Industry in British Social Development*. London: Macmillan.

Keynes, J.M. 1913. *Indian Currency and Finance*. In Vol. 1 of *The Collected Writings of J.M. Keynes*, London: Macmillan, 1971.

Lindert, P.H. 1969. *Key Currencies and Gold 1890–1913*. Princeton Studies in International Finance, No. 24, Princeton: Princeton University Press.

Marshall, A. 1923. *Money, Credit and Commerce*. London: Macmillan.

McCloskey, D.N. and Zecher, J.R. 1976. How the Gold Standard worked. In *The Monetary Approach to Balance of Payments Theory*, ed. J. Frenkel and H.G. Johnson, Toronto: University of Toronto Press.

Mommsen, T. 1865. *Histoire de la monnaie romaine*. Paris: Rollin et Feuardent.

Morgan, E.V. 1965. *The Theory and Practice of Central Banking 1797–1913*. London: Frank Cass.

Nogaro, B. 1908. L'expérience bimétalliste du XIXᵉ siècle. *Revue d'Economie Politique* 22(10), October, 641–721. *Report from the Select Committee on the High Price of Bullion* 1810. New York: Arno Press, 1978.

Ricardo, D. 1811. *The High Price of Bullion: A Proof of the Depreciation of Bank Notes*. In *The Works and Correspondence of David Ricardo*, ed. P. Sraffa, Vol. 3, Cambridge: Cambridge University Press, 1951.

Sayers, R.S. 1936. *Bank of England Operations 1890–1914*. London: P.S. King & Son.

Supino, C. 1910. *Il mercato monetario internazionale*. Milano: Hoepli.

Thornton, N. 1802. *An Inquiry into the Nature and Effects of the Paper Credit of Great Britain*. New York: Augustus Kelley, 1978.

Triffin, R. 1964. *The Evolution of the International Monetary System*. Princeton Essays in International Finance No. 12, Princeton: Princeton University Press.

Williams, D. 1968. The evolution of the sterling system. In *Essays in Honour of R.S. Sayers*, ed. C.R. Wittlesley and J.S.G. Wilson, Oxford: Clarendon Press.

Gonner, Edward Carter Kersey (1862–1922). The character, causes and consequences of that long historical process which witnessed the transformation of British agriculture from a system of open field cultivation based on entitlements and obligations fixed by the custom of the manor, to a system of large-scale enclosed farming characterized by modern relations between wage-labour and capital with the ownership of the land vested in private hands, constitutes perhaps the most difficult, controversial and fascinating set of historical questions facing the economist. From the time of the agrarian disturbances of the 16th century, popular opinion has associated the process with physical deprivation among the agricultural population, the depopulation of the countryside and, later, with the throwing of whole populations into urban centres where they were forced to subsist under conditions of severe economic hardship. While scholarly opinion on the subject has not proved to be so single-minded, there are a few contributions to it that seem likely to have a long life; Gonner's *Common Land and Inclosure* (1912a) is one of them.

Published in the same year as Tawney's *Agrarian Problem of the Sixteenth Century* and less than a year after the Hammond's *Village Labourer*, both of which dealt with the same subject, Gonner sets out to avoid 'general approval or condemnation' of varying views (p. v) by addressing the question using a descriptive and statistical account of actual movements and their effects. Of course, the basic scholarly positions had been set down in the preceding century in the writing of Marx, on the one hand, and that of authors like Cunningham on the other. The former supported a conflictual interpretation: Marx saw the movement as one of 'bloody legislation' which tore peasants from the soil and created a propertyless proletariat. Cunningham, however, rejected this view and saw the process as a more voluntaristic one essential to economic progress – enclosure was in a sense in everyone's interest. He also invoked the demographic transition to account for the origins of the industrial workforce. In an

important way, the work of the Hammonds and Tawney represents an attempt to re-state the conflictual position using more detailed historical analysis than Marx (or, for that matter, Cunningham) had invoked. Though they qualified the orthodox Marxian vision in crucial ways, not least in terms of their rejection of the simplistic explanation which entailed the idea of an immanent contradiction between the social relations and the forces of production which dominated Marxist doctrine, they called into question the more optimistic position taken by Cunningham and others. Gonner's account seems to fall somewhere in between. He distinguishes between necessary and unnecessary effects and is on the whole convinced that the dislocation that did occur was an unnecessary consequence. He finds no evidence of any radical decline in employment in agriculture, but like Tawney he emphasizes the importance of the penetration of market relations into medieval agriculture over a very long period of time.

Gonner's early works include a tract against socialism (1895) and a study of the social philosophy of Rodbertus (1899). With his *Interest and Saving* of 1906, however, he turned his attention to a topic more susceptible to empirical analysis, and found there a mode of analysis more suited to his talents.

In addition to this, Gonner edited two volumes of Ricardo's works which are still quite well known: the *Principles* in 1891, and the *Economic Essays* published posthumously in 1923. To judge from his introductions to these volumes, Gonner seems to have had very little deep regard for the work of Ricardo. His verdict on the *Principles* is harsh, holding that 'not only is it remarkable for infelicity of language, with all its fatal consequences of exaggeration and obscurity, but the grammar itself is halting and the accuracy often apparent, fallaciously apparent rather than real' (p. xxiv). In the much later volume of Ricardo's *Economic Essays*, while apparently more generous in finding Ricardo to be 'in reality far less abstract and far more inductive than many of his critics have allowed' (p. xviii), Gonner adds immediately a rough jibe at Ricardo's 'overstrained and incorrect' assumptions. He also attacks Ricardo's analysis of the relationship between the rate of profit and accumulation, and repeats the charge originally levelled at Ricardo by Malthus that the exceptions to which Ricardo's theory is subject may be encountered frequently (pp. xxxiv–xxxvi).

Even in his defence of Ricardo (1890) against the attacks of Jevons, Ingram, and the almost hysterical Adolph Held, Gonner's purpose is not so much to absolve Ricardo of either theoretical error or of making extreme assumptions (though he does reject the still widely held misapprehension that Ricardo adopted a wages-fund theory), as it is to claim that the critics exaggerated their case in the cause of polemics. His defence of Ricardo against the strictures of Jevons, for example, seems to consist not in arguing that Ricardo's economics was on the correct lines, but rather that Ricardo himself could not have 'shunted the car of economic science onto the wrong lines' because he was no more than following a tradition established by others before him. It is also interesting to note that in this article, Gonner issues the claim (using it as a defence of Ricardo) that Ricardo never intended his *Principles* to be a complete coverage of the subject. This, of course, was the famous excuse for Marshall's wholesale re-interpretation of Ricardo, which has since been deprived of its veracity with the discovery and publication of Mill's correspondence with Ricardo over the book.

So far as can be determined, Gonner's life was uneventful. After graduating from Oxford, he took up his first academic post at Bristol in 1885. From there he moved in 1888 to the newly founded University College of Liverpool, eventually occupying (in 1891) the Brunner Chair of Economic Science in that College's institutional successor, the University of Liverpool, a post in which he remained until his death on 25 February 1922. Like so many British academics, he was seconded to government service during World War I, serving in the Ministry of Food – first as economic adviser and later as its Director of Statistics. From time to time he served as official arbitrator in industrial disputes for the Ministry of Labour.

MURRAY MILGATE AND ALASTAIR LEVY

SELECTED WORKS
1890. Ricardo and his critics. *Quarterly Journal of Economics* 4, 276–90.
1891. (ed.) D. Ricardo, *Principles of Political Economy and Taxation.* London: G. Bell & Sons.
1893. The survival of domestic industries. *Economic Journal* 3, 23–32.
1895. *The Socialist State: Its Nature, Aims and Conditions.* London: W. Scott.
1899. *The Social Philosophy of Rodbertus.* London: Macmillan.
1906. *Interest and Saving.* London: Macmillan.
1912a. *Common Land and Inclosure.* London: Macmillan.
1912b. The economic history. In J.H. Rose, C.H. Herford, E.C.K. Gonner and M.E. Sadler, *Germany in the Nineteenth Century*, Manchester: Manchester University Press.
1923. (ed.) D. Ricardo, *Economic Essays.* London: G. Bell & Sons.

goods and commodities. Towards the end of the 1950s, Harry Johnson produced the following theorem on value theory:

> Define a good as an object or service of which the consumer would choose to have more. Then the collection of goods he chooses when he has more money to spend (prices being constant) must represent more goods than that he chooses when he has less money to spend (since he could have had more of each separate good).
>
>   i. If his income rises, he buys more goods; this implies a presumption that normally the income effect is positive.
>
>   ii. If he chooses collection B when he could have had collection A for the same money (i.e. $\Sigma p_b q_b = \Sigma p_b q_a$), he does not choose A if he could have had B for less money, because that would mean collection B represented less goods than collection A, and conflict with the definition of goods. Hence, when A is chosen B must be at least as expensive (i.e. $\Sigma p_a q_b \geqslant \Sigma p_a q_a$). This establishes that the substitution effect is non-negative (by subtraction, $\Sigma(p_b - p_a)(q_b - q_a) \leqslant 0$).
>
>   Hence we derive both parts of the law of demand from the definition of goods. The hypothesis from which we have deduced it is that goods are goods (1958, p. 149).

The idea that the definition of goods carried with it the whole of the theory of demand, that the explanation of the determination of the exchangeable value of 'things' was intimately bound up with the definition of the 'things' themselves, probably struck many of Harry Johnson's readers as amusing. Indeed, Johnson may even have had this end in mind – no doubt many in the profession were beginning to wonder whether the frequency with which reconsiderations of something apparently so obvious as the theory of demand were being undertaken was entirely necessary. However, it did not strike at least one of his readers as being just another amusing aphorism, Kelvin Lancaster, upon whose review of Hicks's *Revision of Demand Theory* Johnson was commenting when he produced his theorem, took it seriously. Eight years later in the *American Economic Review* Lancaster advanced the so-called characteristics theory of demand. The argument was a simple corollary of the Johnson theorem: if it is the aim of the theory of demand to determine the prices of goods, then one ought to specify as clearly as possible the goods which are being demanded. After all, on this line of reasoning one demands not just physical objects, but the qualities with which they are endowed; it is to their *characteristics* that the potential purchaser first turns his attention.

The interesting features of this little episode, however, are not exhausted in a consideration of the ideas to which it gave rise. Quite as important are the implications which follow upon the recognition of the fact that the kinds of questions which lie behind Johnson's theorem had been debated before in contexts where certain useful results were generated. At least since the time of Adam Smith, economists have struggled to be clear about what it is in the nature of the things which are daily exchanged on markets that gives rise to exchangeable value. When Smith discussed the famous water–diamonds paradox, and drew from it (however perilously) the conclusion that the theory of exchangeable value should focus upon what may be called the objective conditions of production of things, rather than upon the subjective conditions of their consumption, he was engaged in just such an endeavour.

Smith was followed in this project by Ricardo. In the opening passages of the *Principles*, by establishing a clear line of demarcation between scarce and reproducible commodities, Ricardo reached Smith's conclusion by a different route. Marx praised this passage from Ricardo and focused his attention exclusively upon what he termed the commodity form. Moreover, this was not exclusively a classical preoccupation. Later writers, to whom modern economics seems to owe much more, also took the question very seriously indeed. Having returned to Smith's original paradox, they applied the distinction between total and marginal utility and to their satisfaction resolved it. This deprived Smith's original conclusion of its validity and allowed neoclassical writers to rebuild the theory of exchangeable value upon the basis of the subjective conditions of consumption of goods. Marshall was very clear about this at the beginning of the second chapter of Book II of his *Principles*.

The questions that Johnson's theorem prompts, therefore, include also those which were raised in these widely publicized and not insignificant debates in the theory of value over the distinction between those physical objects whose main characteristic is that they can be said to be in short supply, and those whose quantity may be increased by reproduction on an extended scale. To what extent, if at all, the choice of terminology by earlier writers reflects these differences is the subject matter of an investigation into goods and commodities.

ETYMOLOGICAL PRELIMINARIES. In English the word *good* derives from the Old English word *gōd*. It is related also to the Old Frisian *gōd*, the Old High German *guot*, the Old Saxon *gōd*, and the Old Norse *gòdr*. The word is defined in the *Oxford Dictionary of English Etymology* as 'the most general adjective of commendation'. The substantive plural form, *goods*, while sharing the origins, seems not to have appeared in English until the 13th century with a meaning much as it has today: objects or things which confer some advantage or produce some desirable effect upon their owner. Two further points may be noted. The first is that although there exists a genitive singular in Old Norse, no Teutonic language seems to have possessed a substantive plural form. Its usage in this manner probably derives from the Latin *bona*. The second is that despite the standard *O.E.D.* classification of *goods* as indeclinable, a substantive singular form has become common among economists.

In both modern French and German, the adjectives *bien* and *gut* share the meaning and sense as *good* in English (the German sharing the same Old Teutonic origins). The substantive plurals *biens* and *Güter* likewise share meaning and sense, together with the partial Latin origin of the English.

James Bonar's definition of the term *goods* in the original edition of this *Dictionary* – that 'by the plural (Goods) is denoted concrete embodiments of usefulness' – suggests that nothing of substance had been altered in the definition of the word even after it had been co-opted into the formal terminology of economic theory. Furthermore, Bonar's statement that *goods* are the physical embodiment of the metaphysical quality of *good*, seems to apply across all three languages. Of course, given that a substantive singular form is now in common usage, that part of Bonar's definition which went on to argue that the substantive singular *commodity* 'is employed by economists to represent the missing singular of goods', must be abandoned.

The word *commodity* is of entirely different origin and meaning. Its roots are in Latin and it is defined in the *Oxford English Dictionary* as 'a thing produced for use or sale, an article of commerce, an object of trade'.

COMMODITIES. Questions as to the essential properties of the things exchanged in a market economy, though they had arisen in the work of earlier economists, took on an entirely new dimension with the commencement of the systematic study of exchangeable value in the last half of the 18th century. Following immediately upon the definition of wealth as the the 'annual production' of the system, and the analysis of the effects of progress in the division of labour on the 'proportion between annual production and consumption', Adam Smith had confronted the issue of the valuation of this 'quantity of commodities annually circulated'. The problem was to establish, in the first instance, the sphere within which exchangeable value was to be examined. His answer, though failing to take into account conditions of relative scarcity, illustrates just as clearly as Johnson's theorem how an apparently neutral choice of language may be the bearer of certain theoretical precepts upon which an entire argument rests.

Perhaps even more importantly, Smith seems to have established not only the formal framework for the theory of value, but also the very language in which it was transmitted in orthodox circles right down to the time of Ricardo. That argument is sufficiently familiar not to have to be rehearsed here – the essential ingredient that is relevant to us is its rejection of the notion that the 'utility of some particular object' has anything to do with determining exchangeable value and that, instead, the exchangeable value of an object is to be explained in terms of what Smith variously called the 'toil and trouble of obtaining it' or its 'difficulty and facility of production'.

These objects are quite consistently called by Smith *commodities* and not *goods*. The terminology and the theoretical construct seem to match quite well. If one is to follow Smith into an investigation of the relationship between conditions of production and relative prices, the usage of the term *goods* would be less than apposite. This, of course, is not to say that the familiar word *goods* does not crop up from time to time in the *Wealth of Nations* (opening the book at random would quickly disprove such a strong assertion). Nor is it to claim that Smith even bothered to take the time to explain his pattern of usage. But a determinate pattern there surely is.

Consider, for example, the discussion of the water–diamonds paradox, a passage where *goods* appears twice. This very short passage is followed by a carefully constructed paragraph setting out in a quite formal and purposeful way the project for the remainder of Book One. In that particular place, the term *commodities* is used exclusively. Indeed, the following three chapters, on real and nominal price, the component parts of price, and natural and market price, adhere fairly rigidly to this pattern of formal usage. The index, which was added to the original in its third edition of 1784, contains a lengthy entry under *commodities* but not one for *goods*. What is also interesting is that as between the two words, *goods* usually appears in those more discursive passages of the *Wealth of Nations*, whereas *commodities* is reserved for passages of a more formal, theoretical kind.

A remarkable parallel is to be found in the third edition of Ricardo's *Principles*. There, in the first paragraphs of the chapter on value, Ricardo makes a significant attempt to define just what it is that is important in the nature of those objects whose prices are determined on markets. At the same time, it should be noted, he replaces Smith's argument as to why exchangeable value is to be investigated in the sphere of production. The argument is pure Ricardo. Utility is 'essential to exchangeable value', objects which contribute in no way towards 'gratification' would be 'destitute of exchangeable value', but it does not determine it. Two conditions then remain to determine exchangeable value – Smith's difficulty and facility of production and, what Smith had passed over, scarcity. There follows Ricardo's famous twofold classification of commodities: those which are currently reproduced (produced commodities) and those which are fixed in quantity (scarce commodities). The exchangeable value of the former, when competition operates without restraint, is to be investigated in terms of the available methods of production. Relative prices of the latter depend upon the 'wealth and inclinations of those who are desirous of possessing them'. Ricardo restricts the investigation to produced commodities and his use of terms resembles the pattern one discerns in the *Wealth of Nations*.

This particular argument was taken up subsequently by two writers who stand in contrasting positions with respect to this classical conception of the framework for the analysis of exchangeable value – John Stuart Mill and Karl Marx. Both built quite self-consciously on the work of Smith and Ricardo. But as it happens, while Mill was effectively to put in place ideas (which admittedly had been in the air for some time) that were quite dramatically to modify the classical position, Marx was to revivify it. How closely terminological conventions reflect these factors is a question of some importance in the present context.

To begin with, let us turn to the German language, and to Marx, who claimed that Ricardo's argument for restricting the domain of the theory of value and distribution to the sphere of produced commodities had been 'formulated and expounded in the clearest possible manner' (1859, p. 60). Quite unlike Ricardo, Marx not only consistently avoided the use of term *Güter* (goods) – it is hardly possible to forget that the first chapter of *Capital* bears the title *Wares* – but actually considered the theoretical consequences of these terminological conventions in the *Contribution to the Critique of Political Economy*. Though not especially satisfying in itself, the theoretical argumentation of the *Critique* is simple enough: 'use-value as such, since it is independent of the determinate economic form, lies outside the sphere of political economy' (1859, p. 28).

GOODS. As the basis of the theory of value shifted away from the old classical idea of production as a circular process, towards the newer and different idea of an economic process resembling a one-way street – from 'factors of production' to 'goods' – there began simultaneously a retreat from the examination of exchangeable value in terms of the objective conditions surrounding the production of commodities, and an advance towards a theory of value grounded in the subjective conditions surrounding the consumption of goods. Of course, this was not entirely an unprecedented idea (the work of Lauderdale and Bailey comes to mind), what is different is the fact that these notions are now placed on a firmer theoretical footing than had hitherto been the case and that they come to form the mainstream of the discipline.

The orientation thereby imparted to the theory of exchangeable value by the economists in the vanguard of this change took as its starting point precisely those passages of the *Wealth of Nations* which had been so important in establishing the conceptual apparatus of the earlier classical economists. But the lesson that was drawn from them was not that which had been drawn by Smith. They, too, were keenly interested in the properties of the actual objects of market exchange, but from Smith's water–diamonds paradox they did not reach the classical conclusion, but rather one that held that the joint conditions of scarcity and utility would act to determine relative prices. As Pareto was eloquently to put it, economics became the study of equilibrium between man's tastes and the obstacles to satisfying them. Exchangeable value, to borrow Jevons's terminology, would be determined by the final degree of utility.

The 1870s were, of course, the years in which the basic provisions of the new constitution of economics were laid down almost simultaneously in Britain, France and Germany. Precursors had been sought out and honoured by those in the vanguard of the new theory, and the battle against the 'noxious influence of authority', as Jevons put it, already promised success – even the sterner opposition of the historical school was beginning to seem less formidable. Yet despite these quite rapid developments, in the initial years of the marginal revolution the language and usage in English economics seems to have remained essentially as it had been in the classical period. An example may serve to highlight the point.

William Stanley Jevons, who by the second edition of his *Theory of Political Economy* in 1879 had succeeded in substituting 'economics' for the older term 'political economy' in everything but the title of his book, retained the substantive *commodities* even though it was to their want-satisfying qualities that he wished to defer in his explanation of exchangeable value. Usage of the term *goods*, which conveys with greater accuracy the theoretical conceptions at the base of this new approach to the theory of value, appears to have been consciously avoided by Jevons. There is a particularly interesting passage from the *Theory of Political Economy* that illustrates the degree to which Jevons grappled with the language in which to express his theory;

> It will be allowable ... to appropriate the good English word *discommodity*, to signify any substance or action which is the opposite of *commodity*, that is to say, *anything which we desire to get rid of* ... . Discommodity is, indeed, properly an abstract form signifying inconvenience or disadvantage (1879), p. 114, italics in original).

It is impossible to resist the temptation to add that 'the good English word' *discommodity* is of Latin origin, and that the formal introduction of the simple Old English word *goods* at this juncture would have relieved Jevons of the need to conduct such linguistic exercises (see also, Jevons, 1882,

p. 11). Nevertheless, the example is sufficient to indicate that in the English language at least, *goods* was not at this time in formal use in theoretical economics.

Despite this, however, the appearance of the term goods in the formal literature of economics is inextricably linked with the rise of the neo-classical theory of exchange and demand. But to see how this is so, it is necessary to turn to the writings of German economists of the new school.

In the German neoclassical literature, quite precise definitions were given for the formal usage of the substantives *Gut* and *Güter*. Carl Menger's *Grundsätze der Volkswirtschaftslehre* (1871) provides a particularly striking example of this:

> Diejenigen Dinge, welche die Tauglichkeit haben in Causalzusammenhang mit der Befriedigung menschlicher Bedürfnisse gesetzt zu werden, nennen wir Nützlichkeiten, wofern wir diesen Causalzusammenhang aber erkennen und es zugleich in unserer Macht haben, die in Rede stehenden Dinge zur Befriedigung unserer Bedürfnisse tatsächlich heranzuziehen, nennen wir sie Güter (1871, pp. 1–2).

Menger, in fact, went so far as to devote an exceedingly long footnote (printed as an appendix to the *Werke* edition of the *Grundsätze*) to the history of the usage of this term in this sense.

How and when the equivalent term entered the formal language of English economics – and when it might be said to have established itself – is not a difficult question to answer. Alfred Marshall's *Principles* (1890) seems to be the innovator. In a passage from the second edition of the *Principles* dating from 1891, Marshall remarked that lacking any short term in common use to represent all desirable things, that is 'things that satisfy human wants', he proposed 'to use the term *Goods* for that purpose' (1961, I, p. 54, italics in original). In the second edition Marshall appended a footnote to the effect that he intended to replace the singular *commodity* with the term *good*, and gives as explicit justification for this the correspondence between his usage and that of the German economists (see 1961, II, p. 185e). This appears to be the first systematic application of the term *goods* in the formal terminology of economic theory – what is more, its usage is derived from the German. Note that a substantive singular form also appears.

It would seem reasonable to conclude, therefore, that it is from this source (that is, from Marshall's *Principles*) that the term *goods* gained wide circulation in economic theory. The date by which it might be reckoned to have established itself would appear to be around the mid-1890s, as it was in the fourth edition of the *Principles* in 1898 that Marshall chose to delete the footnote alluded to in the previous paragraph. This would accord broadly with the date at which the original edition of this *Dictionary* appeared containing James Bonar's entry under the heading *goods*. It is interesting to note that by the end of the 1920s the term had been so fully absorbed into the language of economic theory that Robbins chose to omit from the English edition of Wicksell's *Lectures* a paragraph where the question of this terminology is discussed. According to Robbins's editorial note, this paragraph was 'of no interest to English readers' (Wicksell, 1934, I, p. 15 n.1).

The originators of the modern theory of exchange and demand, nevertheless, had established a terminology through which to convey one of the basic tenets of their argument. Not only did they appreciate that goods are goods, but they expended a considerable amount of time and energy establishing that the subjective conditions of consumption were the appropriate place to locate the analysis of exchangeable value.

CHARACTERISTICS. This brings us back to Harry Johnson's theorem. The grounding of a theory of exchangeable value upon the notion of *goods* was taken in certain circles to require a closer specification of the want-satisfying qualities of the 'things' which are daily exchanged on markets – since these are, in the final analysis, the *goods* which form the subject of the examination. When one contemplates the kinds of developments in the theory of exchange which might contribute towards the fulfilment of this requirement, nearly all of them seem to entail a widening of the gap between the actual 'things' which are exchanged on markets, and their want-satisfying characteristics which are the real subjects of demand. This, of course, is the direction in which the characteristics theory of demand has already taken us. Its problematic, of course, is to establish a transformation from characteristics to the actual objects through which these characteristics are transmitted. In the language of Lancaster, what is required is a well-defined mapping from the characteristics space to the goods space – since in the end the prices that are thrown up on markets are attached to actual objects and not to their characteristics. As the entry on characteristics in this Dictionary illustrates, the theory of separable utility functions has been of immense assistance in this regard.

However, if the classification of these characteristics could be rendered sufficiently fine, then a concomitant implication would seem to be that the idea of securing a theory of the prices attached to actual objects exchanged on markets would need to be sought in some other direction. Otherwise, we should be left with a theory of exchange and demand which made no contact, even at an abstract theoretical level, with the material realities of market exchange in modern economies. We should certainly have a theory of *goods*, but to what form of economic organization such a theory could be held to apply, if any, is not at all obvious.

So that such speculations should not be thought to be idle, it is interesting to note that language is not only a vehicle for the transmission of theoretical conceptions; it is often the vehicle through which a whole array of structural and cultural data about social interaction is conveyed. Exchange in different kinds of societies frequently embodies these complex social relations – so much so that the familiar idea of economists of the modern school that a universally applicable analysis of exchange is somehow desirable, or even possible, would seem to be fraught with pitfalls.

MURRAY MILGATE

See also CHARACTERISTICS; EXCHANGE; GIFTS.

BIBLIOGRAPHY
Jevons, W.S. 1871. *The Theory of Political Economy*. Ed. R.D.C. Black, Harmondsworth: Penguin, 1970.
Jevons, W.S. 1882. *The Principles of Economics*. Ed. H. Higgs. Reprinted, New York: Kelley, 1965.
Johnson, H. 1958. Demand theory further revisited or goods are goods. *Economica* 25, May.
Lancaster, K.J. 1966. Change and innovation in the technology of consumption. *American Economic Review, Papers and Proceedings* 56, May, 14–23.
Maitland, J. (Earl of Lauderdale) 1804. *An Inquiry into the Nature and Origin of Public Wealth*. Reprinted, New York: Kelley, 1962.
Malthus, T.R. 1827. *Definitions in Political Economy*. Reprinted, New York: Kelley, 1963.
Marshall, A. 1961. *Principles of Economics*. 9th variorum edition, London: Macmillan. (1st edn, 1890.)
Marx, K. [1857]. *Grundrisse*. Harmondsworth: Penguin, 1973.
Marx, K. 1859. *A Contribution to the Critique of Political Economy*. London: Lawrence & Wishart, 1971.
Marx, K. 1867. *Capital*. 4th edn, 1909. 3 vols, New York: International Publishers, 1967.
Mauss, M. 1925. *The Gift*. Ed. E.E. Evans-Pritchard, London: Routledge & Kegan Paul, 1970.
Menger, C, 1871. *Grundsätze der Volkswirtschaftslehre*. Reprinted in *The Collected Works of Carl Menger*, Vol. I, London: LSE Reprints No. 17, 1934.
Mill, J.S. 1848. *Principles of Political Economy*. Peoples edn, London: Longmans, 1873.
Palgrave, R.H.I. (ed.) 1894–9. *Dictionary of Political Economy*. 3 vols, London: Macmillan.
Roscher, W. 1854. *Principles of Political Economy*. Translated from the 13th German edition, Chicago: Callaghan and Coy, 1882.
Say, J-B. 1803. *A Treatise on Political Economy*. Translated by C.R. Prinsep. New York: Kelley. Reprinted, 1971.
Smith, A. 1776. *An Inquiry into the Nature and Causes of the Wealth of Nations*. 2 vols, ed. E. Cannan, London: Methuen, 1961.
Sraffa, P. (ed. with the collaboration of M.H. Dobb) 1951–73. *The Works and Correspondence of David Ricardo*. Cambridge: Cambridge University Press.
Sraffa, P. 1960. *Production of Commodities by Means of Commodities*. Cambridge, Cambridge University Press.
Wicksell, K. 1934. *Lectures on Political Economy*. 2 vols, ed. L. Robbins, London, Routledge & Kegan Paul.

Gordon, Robert Aaron (1908–1978). Gordon was born 26 July 1908 in Washington, D.C., and died 7 April 1978 in Berkeley, California. He was a policy-oriented economist whose research style was quantitative but not econometric and whose influence was felt not only through his basic research but also as a dedicated and tireless public servant. After his formative years as graduate student and instructor at Harvard University, 1929–38, he accepted a position at the University of California (Berkeley), where he remained until his retirement from teaching in 1976.

His early work in industrial organization culminated in the influential volume on *Business Leadership in the Large Corporation* (1944), which is noteworthy for its pioneering use of empirical data in the field. Another major strand is his work on unemployment, its structural and cyclical causes, and the goal of full employment (1962, 1967).

His lifelong research interests were primarily focused, however, on business cycles and their causes. He championed the quantitative-historical method of cycle analysis over the National Bureau of Economic Research (Burns–Mitchell) and econometric modelling (Cowles Commission) approaches (1949), and he devoted much of his career to implementing his approach in studies of the business fluctuations of the interwar period (1951) and before and after World War II (1955a, 1969, 1974). His eclectic analysis of the causes of business cycles emphasized the Schumpeter–Hansen distinction between major and minor business cycles (1956) and attributed the major cycles to the appearance and exhaustion of investment opportunities in particular industries (1955b).

He combined a formidable talent for economic analysis with a sense of historical and institutional relevance and was impatient with the tendency he discerned in the profession at large to favour rigour over relevance in economic theorizing. This impatience was expressed at an early stage with regard to price theory (1948), reiterated in his Presidential Address to the American Economic Association on 'Rigor and Relevance in a Changing Institutional Environment' (1976), and repeated in his last piece on 'A Skeptical Look at the "Natural Rate" Hypothesis' (1978).

His contributions to the public weal were many and lasting, but two in particular may be cited. In 1956–1959 Gordon undertook a massive study of business education jointly with James E. Howell for the Ford Foundation, and their 1959 report provided the stimulus for a radical reorientation of MBA programmes in graduate business schools toward the use of analytical methods drawn from economics, statistics, and the behavioural sciences. He also served as chair of the President's Committee to Appraise Statistics on Employment and Unemployment in 1961–62, which led to important reforms in the nation's statistics in this vital area.

B.G. Hickman

SELECTED WORKS

1944. *Business Leadership in the Large Corporation.* Washington, DC: Brookings Institution.

1948. Short-period price determination in theory and practice. *American Economic Review* 38, June, 265–88.

1949. Business cycles in the interwar period: the 'quantitative-historical' approach. *American Economic Association, Papers and Proceedings* 39, May, 47–63.

1951. Cyclical experience in the interwar period: the investment boom of the twenties. In Universities–National Bureau Committee, *Conference on Business Cycles,* New York: National Bureau of Economic Research.

1955a. Investment opportunities in the US before and after World War II. In *The Business Cycle in the Post-War World,* ed. Erik Lundberg, New York: Macmillan.

1955b. Investment behavior and business cycles. *Review of Economics and Statistics* 37, February, 23–34.

1956. Types of depression and programs to combat them. In Universities–National Bureau Committee, *Policies to Combat Depression,* Princeton: Princeton University Press.

1959. (With James E. Howell.) *Higher Education for Business.* New York: Columbia University Press.

1962. (Co-author.) *Measuring Employment and Unemployment.* Report of the President's Committee on Employment and Unemployment Statistics, Washington, DC: Government Printing Office.

1967. *The Goal of Full Employment.* New York: Wiley.

1969. The stability of the U.S. economy. In *Is the Business Cycle Obsolete?,* ed. Martin Bronfenbrenner, New York: Wiley.

1974. *Economic Growth and Instability: The American Record.* New York: Harper.

1976. Rigor and relevance in a changing institutional setting. *American Economic Review* 66(1), March, 1–14.

1978. A skeptical look at the 'natural rate' hypothesis. In *Economic Theory for Economic Efficiency: Essays in Honor of Abba P. Lerner,* Boston: MIT Press.

Goschen, George Joachim, Viscount (1831–1907). British statesman and financier of German origin, born in London on 10 August 1831; died at Seacox Heath, Surrey, on 7 February 1907. Goschen joined his father's firm of merchant bankers in London on leaving Oxford University. He became a director of the Bank of England in 1858 and an MP in 1863. He was given his first cabinet appointment in 1866. As Chancellor of the Exchequer under Lord Salisbury, Goschen's brilliant political career is chiefly remembered for his conversion and consolidation of the greater part of the National Debt and his reform of the gold coinage. He also set up two important Royal Commissions – that on the Depression of Trade and Industry in 1886, and the Gold and Silver Commission of the following year. Alfred Marshall's written Memoranda and Oral Evidence for both Commissions remained for nearly 40 years the half-forgotten source from which grew the Cambridge 'oral tradition' in monetary theory.

Goschen did not claim to be a professional economist though he wrote a number of essays and addresses on economic and monetary subjects. His famous *Theory of the Foreign Exchanges* (1861) remained for decades the standard work of reference on the subject. Although rather new in its blend of theory and facts, Goschen's book is extremely traditional in its analysis of the mechanism whereby international price adjustments are brought about. Building on Mill's version of the Hume–Ricardo quantity theory-cum-specie-flow doctrine, Goschen offers a clear presentation of how a country's exchange rate is determined by the amount of its short term indebtedness, the size of its monetary stock and the domestic price level. His discussion of the working of the Gold Exchange Standard (in particular the gold points mechanism) and an explicit statement of the purchasing power parity theory still repay study. However, his free trade convictions and his convinced support for all *laissez faire* policies led him to some debatable propositions in his policy chapters. In particular, his strong desire to avoid all interference with what he called the 'natural' workings of the price system made him argue that the exchange rate market is self-adjusting and that, in altering its discount rate, the Central Bank is *following* rather than controlling market conditions. Bearing in mind the extensive use of the discount rate policy made at the time by the Bank of England this argument – neither grounded in facts nor in theory – gave rise to much debate and criticism in Goschen's own time.

In 1890 the Royal Economic Society was founded, Goschen holding the office of president from that year until his death.

P. Bridel

SELECTED WORKS

1861. *The Theory of Foreign Exchanges.* London: Effingham Wilson, 1890.

1905. *Essays and Addresses on Economic Questions.* London.

Gossen, Hermann Heinrich (1810–1858). Gossen was born in Düren (between Aachen and Cologne) on 7 September 1810; he died in Cologne on 13 February 1858. Little is known about his life, partly because the inconspicuous bachelor did not attract attention, partly because most of those who had known him were dead by the time he became famous, partly also because his literary remains, scant as they must have been, are lost. The principal biographical source is the essay by Walras (1885). The available facts are admirably surveyed by Georgescu-Roegen (1983), on whose masterly introduction to the English translation of Gossen's book the following life sketch is mostly based.

Gossen's father was a tax collector under Napoleon and subsequently the Prussian administration; later he managed his wife's estate near Godesberg. Hermann obtained a good high school education, showing ability in 'elementary mathematics', but his mathematical training never went beyond that level. Since his father insisted on a government career in the tradition of his forebears, his university studies in Bonn and Berlin concentrated on law and government.

In 1834, Gossen entered the civil service as a 'Referendar' (junior law clerk) in Cologne. While he seems to have been a well-mannered young man, the performance of his duties left much to be desired. He simply had no interest in a government career and loved the good things in life. There were complaints and reprimands, and the promotion to the rank of 'Regierungsassessor' came rather later than usual. Finally, in 1847, though his superiors seem to have shown considerable sympathy, he had no choice but to resign.

The transition to a new career was perhaps eased by his father's death, which spared him recriminations about his failure and provided him with the means for a new start. Gossen went to Berlin, where he seems to have sympathized with the liberal revolution, and then returned to Cologne as a partner in a new accident insurance firm. He soon withdrew from the firm, but continued to devise grandiose insurance projects.

Living with his two sisters, Gossen now devoted most of his energies to developing the unorthodox ideas he had expressed in his civil service examination papers into his *magnum opus*. The preface suggests that he hoped this would not only make him the Copernicus of the social universe but also open the door to an academic career. In 1853 an attack of typhoid fever undermined his health, and the disappointment about the fate of his book depressed him. Death came from pulmonary tuberculosis. He seems to have been an amiable, sincere and idealistic human being with broad interests, including music and painting. Brought up a Catholic, he developed into an enthusiastic hedonist. Dreaming of reforming the world, he lacked the force to conquer it.

The *Entwickelung der Gesetze des menschlichen Verkehrs* was published in 1854 at Gossen's expense by the publisher Vieweg in Brunswick. Very few copies were sold and the book remained unnoticed for years. Shortly before his death, Gossen withdrew it from circulation and the unsold copies were returned to him. After the author had become famous, Vieweg's successor, Prager, bought this stock from Gossen's nephew, a professor of mathematics by the name of Hermann Kortum, and put it on the market again with a new title page, as a 'second edition', in 1889. There is an Italian translation by Tullio Bagiotti and there is now, since 1983, a careful English translation by Rudolph C. Blitz, nicely divided into chapters. The manuscript of a French translation by Walras was apparently lost.

The first known references to Gossen's book were by Julius Kautz (1858/60), but they only show that their author did not understand the problems Gossen had solved. Slightly more understanding was shown by F.A. Lange, but again in no more than a footnote. Fortunately, Kautz's reference was seen by Robert Adamson, who was able to get hold of a copy and reported its content to Jevons. In the second edition of *The Theory of Political Economy* Jevons included a generous acknowledgement of Gossen's priority 'as regards the general principles and method of the theory of Economics', which became the ignition point of Gossen's posthumous fame.

Though Gossen's name became famous, his book remains largely unread to this day. Pantaleoni (1889/98) was the only notable economist who based his own work on it. This inspired many disparaging remarks about the 'immaturity' of economic science and the like. The simple fact is that the book, even for a German-speaking economic theorist, is very hard to read. It is true that the reasoning is precise and the material reasonably well organized, but there are no chapter headings, the style is involved, the copious algebra is inelegant, the numerical examples are tedious, and the sermonizing is often disconcerting. Gossen had brilliant thoughts, but he never learnt to communicate them effectively. Sombart's often-quoted description of him reflects sadly on the style of academic discourse in Imperial Germany, but if it is modified to 'idiosyncratic genius' it becomes a fitting epitome of Gossen's tragedy.

At the level of individual behaviour, Gossen's basic theoretical problem concerns optimization with limited resources (references are to the 1889 edition; they are followed by the corresponding references to the English translation,

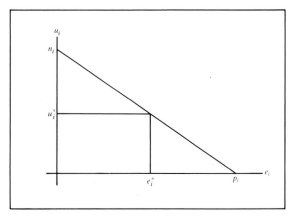

Figure 1

marked T). Resources are first visualized as time (p. 1 f.; T ch. 1), thus foreshadowing an approach recently developed by Gary Becker. The given life-time, $\bar{E}$, has to be allocated to enjoyable activities $I \ldots i \ldots m$ in such a way that life-time enjoyment or, in modern terminology, utility, U, is maximized. In symbols,

$$\max U = \Sigma U_i(e_i), \qquad \text{s.t. } \Sigma e_i = \bar{E}.$$

For a given activity, marginal utility, u_i, is assumed to be a declining function of the time spent on it, e_i. In Gossen's words, 'The magnitude of a given pleasure decreases continuously if we continue to satisfy this pleasure without interruption until satiety is ultimately reached' (p. 4f.; T p. 6). This is the postulate Wilhelm Lexis (1895) christened 'Gossen's First Law'. *In itself*, it was neither new nor profound. W.F. Lloyd had expressed it twenty years earlier just as clearly, it had a long ancestry reaching back to Bentham, the French 'subjectivists', Daniel Bernoulli, and the scholastics, and it is essentially commonplace. To simplify, Gossen assumes the marginal utility curves to be linear, as drawn in Figure 1. It is important to note that Gossen's curves do *not* describe the decline in the marginal utility of a good as its quantity increases, but the decline in the utility from the marginal unit of resources as the quantity of resources is increased. While this facilitated the analysis in some respects, it became a crucial handicap in others.

Gossen realized that each of these marginal utility functions must be thought of as being derived by solving a suboptimization problem, inasmuch as time allocated to activity i must be spent in the most enjoyable way, probably with interruptions. However, his analysis of this difficult subproblem, though original and suggestive, remained incomplete and unsatisfactory, leaving much to do for future research on the allocation of time.

Gossen recognized at once that a necessary condition for the optimal allocation of resources is the equality of the marginal utilities in different activities. This is 'Gossen's Second Law', which he had printed in heavy type: 'The magnitude of each single pleasure at the moment when its enjoyment is broken off shall be the same for all pleasures' (p. 12; T p. 14). This theorem is Gossen's principal claim to fame. In it he had no forerunners. It was the key that opened the door to a fruitful analytical use of the First Law and thus initiated the 'marginal revolution' in the theory of value.

The resulting allocation of resources was summarized in a brilliantly constructed graph by the horizontal addition of the marginal-utility curves (Figure 2). The resulting solid line

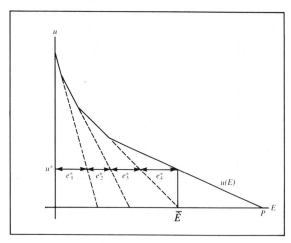

Figure 2

represents the marginal utility of resources, u, as a function of total resources. Its level at the point $E = \bar{E}$, u^*, is the marginal utility of resources in the optimal plan. The resources allocated to each activity, e_i^*, can be simply read off the horizontal distances between the individual marginal-utility curves at the level of u^*.

Gossen also succeeds in determining u^* algebraically by starting from a simple tautology,

$$u^* = u^* \frac{\sum p_i - \sum e_i^*}{\sum (p_i - e_i^*)} = \frac{P - \bar{E}}{\sum \dfrac{p_i - e_i^*}{u^*}} = \frac{P - \bar{E}}{\sum \dfrac{p_i}{n_i}},$$

where P is the sum of the horizontal intercepts. The non-tautological step is the last one, where Gossen recognizes that the ratio $(p_i - e_i^*)/u^*$ for each activity corresponds to the slope of the respective marginal utility curve.

Once u^* is known, it is easy to determine total utility as the sum of the triangular areas under the marginal utility curves minus the sum of the triangles with bases $p_i - e_i^*$ and height u^*. These results are derived without the use of calculus. The latter is used, however, to show that u^* corresponds to the increase in total utility obtainable from a relaxation of the resources constraint, namely to $dU/d\bar{E}$.

In a second step, the model is extended to include production (p. 34f.; T ch. 2). This is achieved by reinterpreting activities as products and resources as 'effort', where total effort can be varied. This requires a transformation of utility curves, each of which now shows the marginal utility resulting from the effort spent on a certain product, taking into account the required amount of effort per unit of product. The time constraint is now replaced by a linear function representing the marginal utility of effort. A small amount of effort is assumed to be pleasurable, but after a point diminishing marginal utility changes into growing marginal disutility. As Georgescu-Roegen has noted, an analysis of leisure is lacking. In Figure 3, the marginal utility of effort, v, is measured downward, positive values thus expressing marginal disutilities. The remainder of the graph corresponds to Figure 2.

The optimal input of effort, E^*, is characterized by the equality between the utility of the marginal effort spent on each product and the marginal disutility of effort. Total utility is then described by the curved triangle between the two marginal curves. Again Gossen in able to describe the optimal solution algebraically in terms of the intercepts of the

individual marginal curves and he also determines the comparative-static effects on this solution of various changes in the underlying parameters (p. 48f.; T chs 4–6).

The third stage is reached with the introduction of exchange (p. 80f.; T ch. 7). Gossen begins with the bilateral case. He immediately perceives that there are many different opportunities for mutually beneficial exchange, but his discussion of these possibilities is, understandably, inconclusive. As a necessary condition for optimal exchange he postulates that the marginal utilities must be equalized between individuals for each product. While this formulation requires both cardinality and interpersonal comparability of utility, its economic substance, since it can be expressed in terms of marginal rates of substitution, is independent of these assumptions. The concept of a 'contract curve', however, is not used. The statement that each individual would usually be willing to forego a portion of what he receives suggests some notion of consumer's surplus.

The analysis is then extended to market exchange, where each individual can exchange goods and effort at parametrically given prices, expressed in a common numéraire called money. This means that E is again reinterpreted, this time in the sense of expenditure or income. The product curves now relate to marginal utility per dollar spent on a given product, the solid convex line relates to the marginal utility of income and the rising line expresses the disutility from earning a marginal dollar of income at the going prices. We thus end up with the optimization problem that became the banner of the 'marginal revolution'. The 'Second Law' can then be expressed by the condition that 'the last atom of money creates the same pleasure in each pleasurable use' (p. 93f.; T p. 109).

The solution to this problem determines the individual's market demand and supply for each product and effort. Gossen also shows how the value of intermediate products can sometimes be derived from that of the final goods, thereby foreshadowing Menger's theory of 'imputation', but he is careful to note that the market mechanism works even where imputation fails (p. 24f.; T p. 28f.). If prices are specified at random, aggregate demand and supply will generally differ. Gossen explains how this exerts pressure on prices until all markets are cleared. Prices are thus endogenously determined by general equilibrium. This argument, though concise, is presented in verbal form only. The mathematical formulation of general equilibrium, foreshadowed by Cournot, had to wait for Walras.

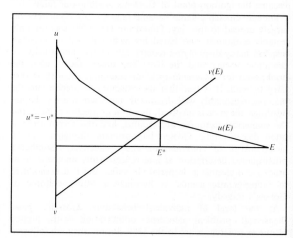

Figure 3

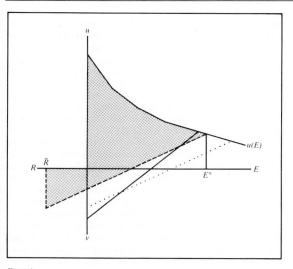

Figure 4

In the fourth stage, Gossen introduces rent (p. 102f; T chs 8–12). If the profundity of an economist can be gauged by his treatment of rent, he comes out near the top. The worker described by Figure 3 is assumed to own a specific piece of land. Suppose he is now offered the use of land at a superior location, owned by another individual. This does not affect his utility curves, but for the amount of effort for which the marginal utility of effort is just zero he can now earn a higher income. At the same time the marginal disutility curve becomes flatter because the same total enjoyment is now spread over a higher income. In Figure 4 this change is expressed by the dotted line.

In the absence of rent, the superior location would, of course, promise higher income. However, moves to superior locations are not free, but cost rent. This means that at the new location the individual has to earn a certain amount, $\bar{R}$, before he can even begin to buy commodities. The origin of the marginal disutility curve is thus shifted to the left as indicated by the broken line.

What is the maximum rent an individual is willing to pay for a superior location? This 'warranted rent' is reached at the point where total utility at the superior location, measured by the shaded area, is equal to the total utility at the original location, described by the solid triangle. Gossen shows algebraically that with rent at the warranted level superior locations are associated with higher earned income and higher consumption.

A corresponding experiment can be conducted for a move to an inferior location. In this case, the marginal disutility curve is steeper than the solid line and its origin is shifted to the right, the individual now receiving rent and thus spending more than he earns. Competition in the market for land will see to it that in full equilibrium all rents are at the warranted level.

While Gossen developed a novel and fruitful way to incorporate rent into a general equilibrium framework, his theory of rent is less rich than von Thünen's, published 28 years before. Gossen was no better in reading his predecessors than the later 'marginalists' were in reading Gossen.

The fifth stage introduces capital and interest (p. 114; T ch. 13). The basic question concerns the highest amount of present utility that could be sacrificed for a piece of land with a given annual rent, continuing into the distant future. Gossen finds the answer by discounting the utility of each future rent payment at the appropriate rate of psychological time preference (as we would call it), reflecting uncertainty of expectations (pp. 30, 115; T pp. 35, 134). This promising idea is not successfully exploited, however, and the adaptation of the land paradigm to capital goods remains sketchy. Gossen thinks in terms of land and labour, while capital goods are played down (p. 172; T p. 194). He also makes an effort to determine the optimal amount of saving by the condition that the highest price the individual is willing to pay for a source of rent should be equal to the market price, but he seems to confuse average and marginal concepts and the sense of his argument remains obscure.

In an effort to interpret everyday observations in the light of his theory, Gossen offers an elaborate discussion of the effect of price changes on demand and expenditure. This discussion anticipates a lot of later work on demand elasticities, but it is also cumbersome. The reason is that Gossen's analytical engine, while permitting a brilliantly simple determination of the optimal budget at given prices, is ill suited for the analysis of price change. Since Gossen's curves, as observed above, relate the marginal utility of expenditure to expenditure, they have to be redrawn after each price change. The insights which Marshall's apparatus made so easy to communicate, remained virtually incommunicable for Gossen. This may be one of the main reasons why his achievement, though at the highest intellectual level, remained sterile. If he had read Cournot, his fate might have been different.

The second part of Gossen's book is largely devoted to social philosophy and policy. It shows its author as a passionate libertarian. Through free markets, mankind would succeed without effort where all socialist planning must fail, namely in reaching the highest possible happiness. Abhorring all forms of protection, Gossen was in favour of free trade, the protection of property rights and a liberal education for both sexes. To prevent fluctuations in the value of money, he advocated a metallic currency and the abolition of paper money. That he also asked for restrictions on child labour and government sponsorship of credit unions seems to indicate that he knew externalities and market imperfections when he saw them. Competitive equilibrium was for him much more than an economic theory or an ideology; it was the gospel, revealing the perfection of a benevolent creator. For him, the 'invisible hand' was not a didactic metaphor, but religion itself. Today, this apotheosis of competition, in language closer to a revival meeting than to scientific discourse, strikes one as bizarre.

Major sources of inefficiency, Gossen thought, were distortions in the allocation of land, preventing land from actually being used by the potentially most efficient user. To correct this defect, he proposed that the government use borrowed money to buy land on the free market and then lease it to the highest bidder (p. 250f.; T ch. 23). Since governments differ from individuals by (1) being immortal, (2) having a higher credit rating, and (3) a lower time preference, such a scheme, he argued, would actually improve government wealth, and the initial debt could eventually be repaid out of rising rent income. For a given year, the scheme would be viable if the price paid by government for a piece of land, A, did not exceed the sum of the rent, a, and the annual increase in the value of land, capitalized at the market rate of interest, z. This led him to the condition

$$A \leqslant \frac{a + \dfrac{z'}{z} a}{z},$$

553

where z' is the annual rate of rent increase. It is understandable that Walras was attracted to this scheme. It is also evident, however, that Gossen was not a 'land socialist'; he was not concerned about 'land monopoly' and the 'socialization of rent'. His objective was the correction of a market imperfection without any limitation of property rights.

Gossen, though perhaps not quite a genius, had a brilliant, original and precise mind. With his one book, he moved constrained optimization into the centre of the theory of value and allocation, where it has since remained. With respect to economic content, his was probably the greatest single contribution to this theory in the 19th century. He failed, however, to develop the basic principle into a usable analytical engine. As a consequence, the so-called 'founders' of the modern theory of value had to rediscover those principles before they could proceed with their engineering work.

JÜRG NIEHANS

SELECTED WORKS

1854. *Entwickelung der Gesetze des menschlichen Verkehrs, und der daraus fliessenden Regeln für menschliches Handeln.* Braunschweig: Vieweg. Reprinted, Amsterdam: Liberac, 1967. 2nd edn, Berlin: Prager, 1889. 3rd edn (with introduction by F.A. Hayek), Berlin: Prager, 1927. Italian translation by T. Bagiotti as *Ermanno Enrico Gossen, Sviluppo delle leggi del commercio umano.* Padua: CEDAM, 1950. English translation by R.C. Blitz (with introductory essay by N. Georgescu-Roegen) as *The Laws of Human Relations and the Rules of Human Action Derived Therefrom*, Cambridge, Mass.: MIT Press, 1983.

BIBLIOGRAPHY

Bagiotti, T. 1957. Reminiszenzen anlässlich des hundertsten Jahrestages des Erscheinens des Buches von Gossen. *Zeitschrift für Nationalökonomie*, 17.
Bousquet, G.-H. 1958. Un centenaire: l'oeuvre de H.H. Gossen (1810–1858) et sa véritable structure. *Revue d'économie politique* 68.
Edgeworth, F.Y. 1896. Gossen, Hermann Heinrich. In *Dictionary of Political Economy*, ed. R.H. Inglis Palgrave, Vol. 2, London: Macmillan.
Georgescu-Roegen, N. 1983. Introduction to H.H. Gossen, *The Laws of Human Relations and the Rules of Human Action Derived Therefrom*, trans. R.C. Blitz, Cambridge, Mass.: MIT Press.
Jevons, W.S. 1879. *The Theory of Political Economy.* 2nd edn, London: Macmillan.
Kautz, J. 1858, 1860. *Theorie und Geschichte der National-Oekonomik.* 2 vols, Vienna: Gerold.
Krauss, O. 1910. Gossen, Hermann Heinrich. In *Allgemeine Deutsche Biographie*, Vol. 55, Leipzig: Duncker & Humblot.
Lange, F.A. 1875. *Die Arbeiterfrage. Ihre Bedeutung für Gegenwart und Zukunft.* 3rd edn, Winterthur: Bleuler–Hausheer.
Lexis, W. 1895. Art. Grenznutzen. *Handwörterbuch der Staatswissenschaften*, Vol. 1 (Supplement), Jena: Fischer.
Liefmann, R. 1910. Hermann Heinrich Gossen und seine Lehre. Zur hundertsten Wiederkehr seines Geburtstages am 7. September 1910. *Jahrbücher für Nationalökonomie und Statistik* 40.
Neubauer, J. 1931. Die Gossenschen Gesetze. *Zeitschrift für Nationalökonomie* 2.
Pantaleoni, M. 1889. *Principii di economia pura.* Florence: G. Barbèra. Trans. by T.B. Bruce as *Pure Economics*, London: Macmillan, 1898.
Riedle, H. 1953. *Hermann Heinrich Gossen 1810–1858. Ein Wegbereiter der modernen ökonomischen Theorie.* Winterthur: Keller.
Walras, L. 1885. Un économiste inconnu: Hermann-Henri Gossen. *Journal des Economistes.* Reprinted in L. Walras, *Etudes d'économie sociale*, Lausanne: Rouge, 1896, 351–74.

Gournay, Jacques Claude Marie Vincent, Marquis de (1712–1759). French economist, merchant and government official, Gournay was born at St. Malo in 1712. After a long career as merchant, spent largely in Cadiz (1729–44), his partner's death in 1746 permitted his retirement two years later from active trade and his entry into public life and more serious research into economics. Gournay has been traditionally associated with the propagation in France of free trade ideas such as deregulation of colonial trade, abolition of the guilds and of the system of government inspection of manufactures, aspects of his work illustrated by the important place generally assigned to him in the history of the phrase, *laissez faire, laissez passer* (Schelle, 1897, pp. 214–17). Turgot (1759, pp. 30–32) has noted, however, that his free trade position should be qualified and in addition that, unlike the Physiocrats, he accorded an important role in economic development to industry and trade as well as agriculture. He has therefore sometimes been described as the founder of a separate non-Physiocratic free trade school, whose members, among others, included Turgot, Morellet and Trudaine. Apart from *Observations sur l'agriculture, le commerce et les arts de Bretagne* (1757), only his notes accompanying the translation of Child (1754), now edited by Tsuda (1983), appear to have survived. His long friendship with Turgot exerted some influence on the latter's economics, partly because Turgot accompanied Gournay on his tours of inspection of industry between 1753 and 1756. Gournay's most important contribution to French economics seems to have been the encouragement he gave to the study of English economics literature. With Butel-Dumont he had himself translated Child and Culpeper (1754), he encouraged Forbonnais to abridge King's *The British Merchant*, Turgot to translate one of Tucker's pamphlets and, most importantly, may have been responsible for the publication of Cantillon's *Essay* in 1755 (Morellet, 1821, pp. 36–7). His death in 1759 provided the occasion for Turgot's eulogy on which much of the information about his life and work is based, though as Ashley (1900, p. 306) warns, there are reasons for being hesitant in accepting Turgot's eulogy (1759) 'as evidence of Gournay's opinions'.

PETER GROENEWEGEN

BIBLIOGRAPHY

Ashley, W.J. 1900. Gournay. In W.J. Ashley, *Surveys, Historic and Economic*, London: Longman & Co.
Child, J. 1754. *Traités sur le commerce et sur les avantages qui résultent de la reduction de l'intérêt de l'argent par Josias Child ... avec un petit traité contre l'usure par Thomas Culpeper.* Traduit de l'Anglois [sic] [by Gournay and Butel-Dumont], Amsterdam and Berlin.
Morellet, l'Abbé. 1821. *Mémoires de l'Abbé Morellet.* Paris: Librairie Française.
Schelle, G. 1897. *Vincent de Gournay.* Paris: Guillaumin.
Tsuda, T. 1983. *Josiah Child, Traités sur le commerce de Josiah Child, avec les remarques inédites de Vincent de Gournay.* Tokyo: Kinokuniya Company.
Turgot, A.R.J. 1759. In praise of Gournay. In *The Economics of A.R.J. Turgot*, ed. and trans. P.D. Groenewegen, The Hague: Nijhoff, 1977.

government borrowing. *See* BUDGETARY POLICY; DEFICIT FINANCING; PUBLIC SECTOR BORROWING.

government budget restraint. Macroeconomic policy analysis has changed since the government budget restraint (GBR) was incorporated into macroeconomic models. The GBR is the requirement that the total of government expenditure for all purposes – interest, other transfer payments, and goods and services – must equal the total of government financing from all sources – taxes, borrowing from the central bank (i.e. printing money), borrowing from others, and net reduction in reserves of assets such as gold, foreign currency or minerals. For simplicity we ignore changes in reserves.

The GBR implies that the authorities cannot exogenously fix the paths of all the macroeconomic policy variables. They can fix all but one, whereupon the economy and the GBR will endogenously determine the path of the remaining one. For example, if paths of tax rates and spending and borrowing are fixed exogenously, then the change in the monetary base cannot be freely chosen: it must equal spending minus taxes minus borrowing. The authorities can decide which policy variable is to be endogenous and can then fix the paths of the rest. When an exogenous change is made in just one policy variable, the endogenous policy variable must change in response, as must other endogenous variables such as income, prices and interest rates.

The most striking consequence of the GBR is that the economy's path of response to an exogenous change in one policy variable depends on which of the other policy variables is chosen to adjust endogenously, both at impact and during the subsequent adjustment period. Consider a balanced-budget equilibrium that is disturbed by an exogenous step-decrease in tax rates. The economy's response path will be different, depending upon whether the endogenous policy variable is government purchases or transfer payments (either of these will continuously balance the budget) or the monetary base or private holdings of government debt (either of these will at least temporarily involve a budget deficit). Not only does the new equilibrium of the system depend on the choice of the policy variable that is to be endogenous; whether the equilibrium is stable or unstable may also depend on that choice. For example, several authors have concluded that the system is likely to be unstable under a particular form of the monetarist rule; namely, when the monetary base, tax rates and government purchases are fixed exogenously and government debt is endogenous (see below).

Though the facts of government financing and of money creation have long been known, they were not incorporated into mathematical models for some time. Modigliani's (1944) influential model and many of its successors adopted Keynes's (1936) liquidity preference equation and did not include bonds, though non-mathematical accounts often treated liquidity preference as describing the substitution between money and bonds in portfolios. Metzler (1951) broke new ground by introducing an explicit variable to represent assets (equities) that could be exchanged for money in open-market operations, and Patinkin (1956) explicitly included bonds in his models. Shortly thereafter, the GBR began to appear in macroeconomic analysis.

The flavour of GBR analysis is conveyed by a simple model of a closed economy with no real growth, adapted from Christ (1978, 1979) by assuming that prices adjust instantly so as to maintain real income at full capacity. The model contains the GBR of the consolidated central government sector including the central bank, excluding local governments since they cannot print money.

Symbols are as follows. B = number of privately held perpetual government bonds each of which pays $1 a year nominal interest, g = real government purchases of goods and services, H = monetary base (high powered money), P = price level, π = expected inflation rate, r = nominal interest rate, u = marginal tax rate, V = autonomous nominal taxes, and y = real income (both the actual and the capacity level). V is negative; that is, the tax system is progressive. Transfer payments other than interest are assumed zero, but they could easily be handled by defining u and V to be taxes net of transfers.

Consider the model after it has been reduced to three equations in three endogenous variables: P, r and the endogenous policy variable which may be either B, g, H, u or V. All other variables are exogenous, including y and π. One equation is a financial-assets equilibrium condition, similar to the familiar LM equation except that it allows the demand for the real monetary base to depend on private holdings of government bonds as well as on income and the interest rate. When solved for the latter it becomes:

$$1/r = \lambda(y, H/P, B/P) \tag{1}$$

The second equation is the aggregate demand function:

$$y = \phi(P, \pi; B, g, H, u, V) \tag{2}$$

It was obtained by substituting (1) into a familiar IS equation. The partial derivatives ϕ_P, ϕ_u, and ϕ_V are negative; ϕ_g, ϕ_H and ϕ_π are positive; ϕ_B is of uncertain sign, but it must be less than ϕ_H/r since an open-market purchase increases aggregate demand. The third equation is the GBR:

$$g + B/P = uy + uB/P + V/P + \dot{H}/P + \dot{B}/rP \tag{3}$$

where a dot above a symbol denotes its derivative with respect to time. The left side of (3) is real government expenditures for goods and services g and for interest B/P. The right side is real government finance; that is, real taxes at the marginal rate u on real income y and on real interest B/P, plus real autonomous taxes V/P, plus $\dot{H}/P$ (the real value obtained from the issue of base money) plus $\dot{B}/rP$ (the real value obtained from the issue of bonds).

In order to permit a balanced equilibrium to exist, consider the case where the growth rates of all the nominal exogenous policy variables are the same. Assume that the exogenous expected inflation rate π is equal to that same growth rate (any plausible expectations-formation process will have this property in equilibrium). Then the system can have a dynamic equilibrium path with steady inflation at the rate π.

For simplicity, consider the case where π and these growth rates are zero. Then, at the static equilibrium, $\dot{H}$ and $\dot{B}$ and r drop out of (3), which takes the following static form:

$$P = [(1-u)B - V]/(uy - g) \tag{4}$$

Hence from (2) and (4) one can find the static equilibrium values of P and the endogenous policy variable, and the comparative static effect of any exogenous variable upon P. The effect depends on the choice of which policy variable is to be endogenous. For example, the comparative static effect of the monetary base on the price level, $\partial P/\partial H$, is:

$-\phi_H(1 - u)/P\Delta_B$

if B (bonds) is endogenous $\qquad$ (5)

$-\phi_H/\Delta_g$

if g (government purchases) is endogenous $\qquad$ (6)

$\phi_H/P\Delta_V$

if V (autonomous nominal taxes) is endogenous $\qquad$ (7)

where Δ with a subscript stands for the determinant of the linearized system (2) and (4) when the variable in the subscript of Δ is endogenous. In particular:

$$\Delta_B = [(1-u)B - V]\phi_B/P^2 + (1-u)\phi_P/P \tag{8}$$

$$\Delta_g = [(1-u)B - V]\phi_g/P^2 + \phi_P \tag{9}$$

$$\Delta_H = [(1-u)B - V]\phi_H/P^2 > 0 \tag{10}$$

$$\Delta_V = [(1-u)B - V]\phi_V/P^2 - \phi_P/P > 0 \tag{11}$$

The nature and stability of the dynamic path also depend on which policy variable is endogenous. From the GBR (3) it can be seen that if the monetary base H and private holdings of government bonds B are exogenously held fixed during the adjustment period, the system is not dynamic at all: following any exogenous disturbance, the endogenous response is to balance the budget instantaneously. However, if the endogenous policy variable is either B or H, then the GBR (3) is a dynamic equation, and so the system is dynamic. Then its stability depends on whether B or H is the endogenous variable, as follows.

Suppose H is the endogenous variable. Then the GBR (3) shows that the dynamic path of H is given by:

$$\dot{H} = P \cdot [g - uy - V/P + (1-u)B/P] \qquad (12)$$

This is stable iff $\partial \dot{H}/\partial H < 0$; that is,

$$\text{iff } [(1-u)B - V]P^{-1} \phi_H/\phi_P < 0.$$

Since $\phi_H > 0$ and $\phi_P < 0$, the system is stable when H is the endogenous variable.

Now suppose B is the endogenous variable. Then the GBR (3) shows that the dynamic path of B is given by:

$$\dot{B} = rP \cdot [g - uy - V/P + (1-u)B/P] \qquad (13)$$

This is stable iff $\partial \dot{B}/\partial B < 0$; that is [using (8)], iff $rP\Delta_B/\phi_P < 0$.

Since the sign of ϕ_B is uncertain, the sign of Δ_B in (8) seems uncertain. However, note from (5) that if the effect of the monetary base upon the equilibrium price level is positive when B is endogenous, as is plausible, then the positive sign of ϕ_H implies that $\Delta_B < 0$, and hence the system is unstable when B is the endogenous variable. Similar results have been obtained by several others; for example, Tobin and Buiter (1976).

There is a large literature on the GBR, in which the foregoing analysis has been extended in several directions. Steady-inflation equilibrium paths have been considered. So have steady-real-growth equilibrium paths. (In either case, the equilibrium path has a constant real budget deficit. A steady-inflation equilibrium is possible only if the real deficit is not too large to be financed by the inflation tax.) Deviations of output from the capacity level (as in business cycles) have been introduced. The foreign sector has been included. Further, two or more policy variables can be endogenous at the same time, if there is a policy rule governing their joint responses.

Does the Ricardian equivalence theorem of Barro (1974) obviate the need for the GBR? The theorem says that under suitable conditions (including the dubious assumption that the interest on government debt will certainly be covered by future taxes, without inflation or default) bond finance is equivalent to tax finance. It implies that any change in the timing of tax payments will have no effect on private behaviour as long as the present value of tax payments is not altered. As Barro recognizes, this is not true for persons who are at a corner solution, consuming less than they would if they could shift some purchasing power from the future to the present.

But suppose there are never any such persons. The Equivalence Theorem implies that government bonds are not net wealth. According to Sargent (1979, ch. 4) and McCallum (1978), this means that models like (1)–(2) above should be modified in such a way that, once the time-paths of the monetary base and government purchases are fixed exogenously, the GBR and the proportion of bond finance to tax finance are irrelevant for private behaviour. If this were correct, a government could set its taxes permanently at zero and finance its expenditures, including interest, solely by issuing new debt forever, without affecting the interest rate, prices or output.

Suppose that the interest rate exceeds the economy's growth rate. This is one of the premises of the Equivalence Theorem. It is plausible, since the golden rule of economic growth in Phelps (1965) shows that the economy cannot be on an optimum steady state path if the reverse inequality holds. Then, as Barro (1974, 1976), Sargent and Wallace (1981), and McCallum (1984) recognize, it is not possible to pursue forever a policy of continually borrowing to pay the debt interest. It would eventually make the debt interest exceed the revenue capacity of the tax system. Hence debt interest could not be covered by taxes, and the Equivalence Theorem would fail. The result would be inflation or outright default. Thus the omission of the GBR and government debt can lead to incorrect conclusions.

CARL F. CHRIST

See also PUBLIC DEBT; RICARDIAN EQUIVALENCE THEOREM.

BIBLIOGRAPHY

Barro, R. 1974. Are government bonds net wealth? *Journal of Political Economy* 82, December, 1095–117.

Barro, R. 1976. Reply to Feldstein and Buchanan. *Journal of Political Economy* 84, April, 343–9.

Christ, C.F. 1978. Some dynamic theory of macroeconomic policy effects on income and prices under the government budget restraint. *Journal of Monetary Economics* 4, January, 45–70.

Christ, C.F. 1979. On fiscal and monetary policies and the government budget restraint. *American Economic Review* 69, September, 526–38.

Keynes, J.M. 1936. *The General Theory of Employment Interest and Money.* New York: Harcourt Brace.

McCallum, B.T. 1978. On macroeconomic instability from a monetarist policy rule. *Economics Letters* 1, 121–4.

McCallum, B.T. 1984. Are bond-financed deficits inflationary? A Ricardian analysis. *Journal of Political Economy* 92, February, 123–35.

Metzler, L.A. 1951. Wealth, saving and the rate of interest. *Journal of Political Economy* 59, April, 93–116.

Modigliani, F. 1944. Liquidity preference and the theory of interest and money. *Econometrica* 12, January, 45–88.

Patinkin, D. 1956. *Money, Interest, and Prices.* Evanston: Row Peterson.

Phelps, E.S. 1965. Second essay on the golden rule of accumulation. *American Economic Review* 55, September, 793–814.

Sargent, T. 1979. *Macroeconomic Theory.* New York: Academic Press.

Sargent, T. and Wallace, N. 1981. Some unpleasant monetarist arithmetic. *Federal Reserve Bank of Minneapolis Quarterly Review,* Fall, 1–17.

Tobin, J. and Buiter, W. 1976. Long-run effects of fiscal and monetary policy on aggregate demand. In *Monetarism,* ed. Jerome L. Stein, Amsterdam: North-Holland.

government ownership. *See* NATIONALIZATION.

government regulation. *See* REGULATION AND DEREGULATION.

Graham, Frank Dunstone (1890–1949). Graham was born in Halifax, Nova Scotia, and died in Princeton, New Jersey. He is known mainly for his work in the theory of international trade, and especially for his attack on classical and neoclassical trade theory. He received his doctorate from Harvard, where he came under the influence of Taussig. After teaching at Rutgers and Dartmouth, he joined the Princeton faculty in 1921, becoming a full professor in 1930. In addition to undergraduate teaching, he taught the Princeton graduate courses in international trade and in monetary theory.

In a path-breaking article (1923b), Graham argued that J.S. Mill, by using a two-country, two-commodity model, had reached erroneous conclusions concerning the effect of changes in international demand on the commodity terms of trade. Mill had reasoned that, within the limits set by comparative cost (limits which he – but not Graham – regarded as improbable cases), an increase in a country's demand for imports would worsen the country's terms of trade. Retaining Mill's assumptions of free trade, costless transportation and constant cost per unit of output, Graham concluded that when a given commodity is produced by more than one country, the cost structures of the affected countries are locked together and that, therefore, changes in international demand do not affect the equilibrium terms of trade so long as the same commodities continue to be produced by the same countries; instead, within possibly wide limits, international adjustment takes place through shifts in output, the limits occurring when commodities disappear from, or are added to, national production schedules.

To illustrate these points, Graham devised a multi-country, multi-commodity model, with all variables expressed in real terms. Operating with assumed national opportunity-cost ratios, national productive capacities and national demand functions, he was able to derive, by a trial-and-error process using simple arithmetic, an equilibrium solution specifying the commodity terms of trade and each country's consumption, production (if any) and exports or imports of each commodity. In his final work, *The Theory of International Values* (1949), he developed these ideas at length, using illustrations with as many as ten countries and ten commodities. Because of the assumption of costless transportation, domestic (non-traded) goods do not appear in the trade model, but Graham examined their role in international adjustment in his earliest article (1922) and in his 1949 treatise.

Although the Graham model, which assumes full employment, can be used to demonstrate that national and world real output are maximized under free trade, Graham was not a doctrinaire free-trader. In an early article (1923a), he made a case for permanent protection for decreasing-cost industries. The article was attacked on various grounds by Knight (1924) and others, but Graham retained the argument in his book, *Protective Tariffs* (1934), which, while critical of most arguments for tariffs, included a chapter on 'Rational Protection'.

In the field of money, Graham's major work was his treatise (1930) on the German hyperinflation after World War I. Perhaps his most significant conclusion was the concept of 'ceiling velocity'. He found that, in the German case, monetary velocity reached an upper limit which was about 25 times the prewar normal; thereafter, the German price level rose at approximately the same rate as the German money supply.

Graham had a passionate interest in economic policy. He was an early advocate of flexible exchange rates (on a managed basis), and during the Great Depression he devised various plans to promote recovery. Later, he advocated a commodity-reserve monetary standard as a means of achieving price-level stability and full employment.

An iconoclast with a caustic wit, Graham was an unusually stimulating teacher and had a profound influence on his students, two of whom – T.M. Whitin and L.W. McKenzie – extended his work on the trade model. In a 1953 article which illustrated the model geometrically, Whitin concluded that Graham's work 'anticipated linear programming models by many years', and McKenzie, in a powerful 1954 article employing a theorem from topology, demonstrated what Graham firmly believed but was never able to prove: that his

trade model yields an equilibrium for any continuous demand functions and that this solution is unique for the demand functions which Graham actually used.

RANDALL HINSHAW

SELECTED WORKS

1922. International trade under depreciated paper: the United States, 1862–79. *Quarterly Journal of Economics* 36, February, 220–73.
1923a. Some aspects of protection further considered. *Quarterly Journal of Economics* 37, February, 199–227.
1923b. The theory of international values re-examined. *Quarterly Journal of Economics* 38, November, 54–86.
1930. *Exchange, Prices, and Production in Hyperinflation: Germany, 1920–1923.* Princeton: Princeton University Press.
1932a. The theory of international values. *Quarterly Journal of Economics* 46, August, 581–616.
1932b. *The Abolition of Unemployment.* Princeton: Princeton University Press.
1934. *Protective Tariffs.* New York: Harper & Bros. Reprint edn, Princeton: Princeton University Press, 1942.
1942. *Social Goals and Economic Institutions.* Princeton: Princeton University Press.
1949. *The Theory of International Values.* Princeton: Princeton University Press.

BIBLIOGRAPHY

Knight, F.H. 1924. Some fallacies in the interpretation of social cost. *Quarterly Journal of Economics* 38, August, 582–606.
McKenzie, L.W. 1954. On equilibrium in Graham's model of world trade and other competitive systems. *Econometrica* 22, April, 147–61.
Metzler, L.A. 1950. Graham's theory of international values. *American Economic Review* 40, June, 301–22.
Whitin, T.M. 1953. Classical theory, Graham's theory, and linear programming in international trade. *Quarterly Journal of Economics* 67, November, 520–44.
Whittlesey, C.R. 1952. Frank Dunstone Graham, 1890–1949. *Economic Journal* 62, June, 440–45.

Gramsci, Antonio (1891–1937). Italian communist and Marxist theorist, Gramsci was born in 1891 in Ales (Sardinia), and died in Rome in 1937.

Gramsci's work acquired national importance in Italy when in 1919, together with A. Tasca, U. Terracini and P. Togliatti, he founded the weekly magazine *Ordine nuovo*. The aim of this publication, under the influence of the Russian revolution, was to disseminate the idea of a proletarian dictatorship based on workers' 'councils' and on the alliance between the workers of northern Italy and the poor peasants of the south. Gramsci was elected member of Parliament (1924–6) and became secretary of the Italian Communist Party (PCI) in 1924. In spite of the fact that he had opposed the left in the CPSU in 1926, Gramsci, in bitter opposition to Togliatti, warned of the danger that Bukharin and Stalin's aim was to crush their opponents and subject the International to Russian national interests. Gramsci was arrested in November 1926 and condemned by the special fascist Tribunal to twenty years' imprisonment. During his time in prison he made notes and kept records which were collected in *Quaderni del carcere* (first published between 1948 and 1951). In 1930 he rejected the theory of 'social fascism' supported by the third International. He later became seriously ill and died in a clinic in Rome in 1937.

While in jail Gramsci was aided in many ways by his friend the eminent economist Piero Sraffa, who had moved to Cambridge, England, in 1927.

Gramsci's Marxist beliefs developed into an anti-positivistic attitude. He was affected by the idealism of Gentile and Croce, and also by the thought of Sorel and Bergson. It seemed to him that Lenin and the Bolshevik Party were the ideological incarnation of the new Marxism, organized into an active political force, in direct contrast to the old deterministic Marxism of the social democratic parties of the Second International.

In 1919–20 Gramsci had believed in the supremacy of the revolutionary intitiative of the 'Workers' Councils'. After 1921, as a result of a deeper understanding of Leninism, he changed his perspective and underlined instead the primacy of the party as interpreter of the revolutionary process.

During the time that he was in prison, he reflected on the causes of the defeat of the Revolution in the West. He wrote in the *Quaderni* that the social, political and cultural differences between the East and West were such that the Russian Revolution could not be adopted as a model to be copied automatically. In the West the accession to power would have to be preceded by a period of intense political struggle ('war of position') during which the Communist Party (the 'Modern Prince') and the proletariat would have to form a broad front of social alliances and win a wide political and cultural 'consensus' (the theory of 'hegemony').

Gramsci believed that Italy had missed out on the opportunity of producing a national bourgeoisie capable of ensuring the development of a modern society. Italy's inability to solve the problems of the South ('the southern question') bore witness to this. Gramsci believed that it was up to the PCI to change Italian society and, by creating a new socialist order, to accomplish the difficult task of 'national' unification.

Gramsci's beliefs exerted a wide influence on the left, first in Italy, and then in Western Europe. The PCI, which had at the beginning judged him to be a great 'orthodox Leninist', later used Gramsci's 'theory of hegemony' as its main theoretical inspiration for 'Eurocommunism', thus forming a political strategy aimed at surmounting the limits of Leninism.

Gramsci never paid any systematic attention to economic theory. Nevertheless he wrote on it, especially in the *Quaderni* which includes many methodological notes. He was against using the concept of 'laws' according a deterministic pattern both in economics and sociology. In his opinion, only Marxism was able to establish a 'critical' conception of economics. The 'value' – he stated – is the very core of Marxist economic theory, as far as it explains the 'relationship between the worker and the industrial forces of production'. And whereas the bourgeois idea of 'market' is an 'abstract' one, the Marxist idea is related to 'historicism', that is it is based on the consciousness of the social and historical conditions of the market itself, which have to be changed in consequence of the revolutionary process.

Massimo L. Salvadori

SELECTED WORKS

1947. *Lettere dal carcere*. Turin: Einaudi. Trans. by Lynne Lawner as *Letters from Prison*, New York: Harper & Row, 1973.
1948–51. *Quaderni del carcere*. 6 vols, Turin: Einaudi. New edn, 4 vols, Turin: Einaudi, 1975. Selections trans. by Quintin Hoare and Geoffrey Nowell Smith as *Selections from the Prison Note-books*, London: Lawrence & Wishart, 1971.
1954. *L'ordine nuovo*. Turin: Einaudi. Another edn, 1975.
1971. *La costruzione del Partito Comunista*. Turin: Einaudi.

BIBLIOGRAPHY

Adamson, W.L. 1980. *Hegemony and Revolution. A study of Antonio Gramsci's political and cultural theory*. Berkeley: University of California Press.
Buci-Glucksmann, C. 1975. *Gramsci et l'état*. Paris: Fayard.
Clark, M. 1977. *Antonio Gramsci and the Revolution that Failed*. New Haven: Yale University Press.
Fiori, G. 1966. *Vita di Antonio Gramsci*. Bari: Laterza.

Romeo, R. *Risorgimento e capitalismo*. Bari: Laterza.
Salvadori, M.L. 1970. *Gramsci e il problema storico della democrazia*. Turin: Einaudi.
Spriano, P. 1967. *Storia del Partito Comunista italiano*, Vol. 1. Turin: Einaudi.
Spriano, P. 1977. *Gramsci in carcere e il Partito*. Rome: Editori Riuniti. Trans. by John Fraser as *Antonio Gramsci and the Party: the Prison Years*, London: Lawrence & Wishart, 1979.
Togliatti, P. 1967. *Gramsci*. Rome: Editori Riuniti.

graph theory. Graph theory is a part of that field of mathematics referred to as combinatorics. Although the basic concepts have a simple intuitive interpretation and correspond to many features of social and economic organization this tool has been very little exploited in economics. The reasons for this are the same as those which explain economists' lack of interest in combinatorics in general (see COMBINATORICS).

The reader will be familiar with the diagrammatic representation of what he thinks of as a graph. It consists of a collection of points some of which are linked by lines or 'edges'. The edges may be directed, in the sense that there may be a link from point *a* to point *b* but not vice versa, or they may be undirected, in which case a link from a to b implies a link from *b* to *a*. The first person to resolve a problem specifically formulated in terms of a graph was Euler who proved that the famous 'Konigsberg Bridge problem' had no solution. Graph theory has since been used to study electric networks, enumeration problems in organic chemistry, interpersonal relations in psychology, interaction problems in statistical mechanics, Markov chains in probability theory and simplicial complexes in combinatorial topology. This last application is closely related to the problem of developing algorithms for finding the equilibria of economic systems.

However the first direct application to an economic problem was to the study of optimal flows in networks. (Standard references are Ford and Fulkerson, 1962 and Berge and Ghouila Houri, 1965.) In this type of problem the points of a graph correspond to the physical locations of goods and a directed edge from one place to another and an associated number to a channel and its maximal capacity. The sort of question involving the optimal way in which to move different goods to different locations has been regarded as belonging to the province of operational research rather than to that of economics though for no good reason. Such questions have received some attention in 'locational economics', in 'regional economics' and in 'transport economics'. Nevertheless they have remained very marginal.

In order to understand why the problems which graph theory typically addresses are largely ignored by economic theorists it suffices to look at the way in which economic theory has developed. A graph typically describes the structure of communication between the points which may be identified with individual traders, with firms or with locations. Yet it is precisely this organizational structure which is lacking in standard economic models. If any description of the communication between individuals is given it is usually of a trivial nature. Consider the Walrasian model of general equilibrium, the only communication that takes place is between the traders and the fictitious central 'auctioneer'. The graph corresponding to this rudimentary organization is a 'star' as in Figure 1. In those rare cases where genuinely decentralized trading is considered as for example in discussing under what conditions Pareto optima can be achieved by a sequence of bilateral trades, (see e.g. Feldman, 1973), it is assumed that all pairs of individuals can trade. If the

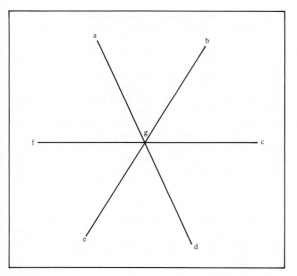

Figure 1

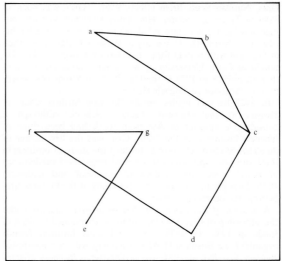

Figure 3

possibility of trade is represented by an edge then we have the 'completely connected graph' in Figure 2 for example. Think of an edge between two points or individuals as representing the relation 'are in contact with' then in studying the cooperative game theoretic approach to economics we might wish to restrict the formation of coalitions to only those where all the members are in contact with each other. Thus, in Figure 3, for example, the coalition containing *a*, *b*, *c* can form whilst that containing *b*, *c*, *d* cannot. In this case to assume, as is typically done, that all coalitions can form, is to assume again the structure given by Figure 2. The problem of an incomplete coalition structure can be formulated in graph theoretic terms and the resultant solutions examined. Kirman (1983) and Kirman et al. (1986) considered the case in which the communication structure is stochastic, that is individuals only contact each other with a certain probability. In this case, imposing the same condition as before, the set of admissible

coalitions is now a random variable. However, in this case even if we allow the probability that individuals are in contact with each other to go to zero as the economy becomes large, the standard result that the core 'shrinks' to the set of competitive equilibria, is still true with probability one. These results use stochastic graph theory in an essential way. Ioannides (1986) has studied the problem of dispersed trading using a similar approach.

What is curious nevertheless is that in a world where the importance of the organizational structure of an economy is being more and more explicitly recognized, where external effects and who causes what to whom are more frequently discussed, where the question of the search for opportunities in a world of uncertainty receives considerable attention, and where the question as to who learns what from whom is increasingly posed, in such a world a mathematical tool, graph theory, particularly apt to handle such problems remains largely unused.

A.P. KIRMAN

See also CAUSALITY IN ECONOMIC MODELS; COMBINATORICS; COMPUTATION OF GENERAL EQUILIBRIUM; CRITICAL PATH ANALYSIS; INTEGER PROGRAMMING; OPERATIONS RESEARCH; QUALITATIVE ECONOMICS; RANK.

BIBLIOGRAPHY Good introductions to Graph Theory are given for example by Berge (1962) and Harary (1969).

Berge, C. 1962. *The Theory of Graphs*. London: Methuen.

Berge, C. and Ghouila-Houri, A. 1965. *Programming, Games and Transportation Networks*. London: Methuen.

Erdos, P. and Renyi, A. 1960. On the evolution of random graphs. *Publications of the Mathematical Institute of the Hungarian Academy of Sciences* 5, 17–61.

Feldman, A. 1973. Bilateral trading processes, pairwise optimality and Pareto optimality. *Review of Economic Studies* 40, October, 463–79.

Ford, L.R. and Fulkerson, D.R. 1962. *Flows in Networks*. Princeton: Princeton University Press.

Harary, F. 1969. *Graph Theory*. Reading, Mass.: Addison-Wesley.

Ioannides, Y. 1986. Dispersed trading. Discussion Paper, Athens Business School.

Kirman, A.P. 1983. Communication in markets: a suggested approach. *Economic Letters* 12, 101–8.

Kirman, A.P., Oddou, C. and Weber, S. 1986. Stochastic communication and coalition formation. *Econometrica* 54, 129–38.

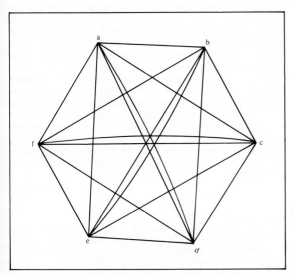

Figure 2

Gras, Norman Scott Brien (1884–1956). Gras was born in 1884 in Toronto, Canada. His family were not well off, but high intelligence won him scholarships and made possible his education, first at the University of Western Ontario, and then at Harvard with a PhD thesis directed by Edwin F. Gay. He taught at Clark University in Worcester, Massachusetts, and then from 1918 to 1927 at the University of Minnesota, where he was Professor of English History.

In 1927 he was invited to fill the new Strauss Chair in History at the Harvard Business School. Although a considerable amount of American work had been done in business history since the 1880s, Gras was the first holder of an endowed chair. He took as his field the study of business in the history of capitalism and adopted time-based subdivisions of 'petty', 'mercantile', 'industrial', 'financial' and 'national'. In the later periods, particularly, the study of the business firm became the dominant theme.

A *Journal of Economic and Business History* started under the editorship of Gay in 1928, had to be abandoned for lack of funds in 1932. In 1938, Gras and local business friends organized the Business Historical Society with a semi-annual *Bulletin* that led in 1954 to the Quarterly *Business History Review*. The depression was probably responsible to some degree for a narrowing of Gras's own emphasis and that of his students to the history of the firm, usually a self-sustaining type of publication.

In 1939 Gras published his *Business and Capitalism*: An *Introduction to Business History*, which illustrates his historical rather than theoretical emphasis. Two years before his retirement in 1950s, Gras and Professor Henrietta M. Larson of the Business School founded the Business History Foundation to write the history of Standard Oil and other companies that might apply. Gras died in 1956 at Cambridge, Massachusetts.

Thomas C. Cochran

SELECTED WORKS

1915. Evolution of *the English Corn Market from the Twelfth to the Eighteenth Century*. Harvard Economic Studies Vol. 13, Cambridge, Mass.: Harvard University Press.

1918. *The Early English Customs System: A Documentary Study of the Institutional and Economic History of the Customs From the Thirteenth to the Sixteenth Century*. Harvard Economic Studies Vol. 18, Cambridge, Mass.: Harvard University Press.

1922. *An Introduction to Economic History*. New York: Harper.

1925. *A History of Agriculture in Europe and America*. 2nd edn, New York: Crofts, 1940.

1930a. *Industrial Evolution*. Cambridge, Mass.: Harvard University Press.

1930b. (With Ethel C. Gras.) *The Economic and Social History of an English Village (Crawley, Hampshire), AD 909–1928*. Harvard Economic Studies Vol. 34, Cambridge, Mass.: Harvard University Press.

1937. The Massachusetts First National Bank of Boston: 1784–1934. *Harvard Studies in Business History* No. 4, Cambridge, Mass.: Harvard University Press.

1939a. *Business and Capitalism: An Introduction to Business History*. New York: Crofts.

1939b. (With Henrietta M. Larson.) *Casebook in American Business History*. New York: Crofts.

1942. *Harvard Cooperative Society Past and Present: 1882–1942*. Cambridge, Mass: Harvard University Press.

1962. *Development of Business History up to 1950*. Ed. Ethel C. Gras, Lincoln Educational Foundation. (Published posthumously.)

Graunt, John (1620–1674). Graunt was born on 24 April 1620 in Hampshire and died on 18 April 1674 in London. At the age of 16 he was apprenticed to his father as a haberdasher of small wares, but remarkably little is known of his life before he published his *Natural and political observations made upon the Bills of Mortality* in 1662. Graunt had formed a friendship with William Petty, who came from a social and economic background similar to his own and who may have drawn Graunt's attention to the data in the London Bills of Mortality. Six months after the publication of the *Natural and political observations* Graunt was made a fellow of the Royal Society, in whose foundation Petty had been greatly involved. The publication of this volume had an immediate impact; a second edition was published the same year and two others in 1665. Graunt subsequently fell into disgrace following his conversion to Catholicism and he died in poverty despite generous help from Petty. The latter's own work, especially that on urban growth, owed much to methodologies initiated by Graunt.

It has been claimed that the *Natural and political observations* 'created the subject of demography' (Glass, 1963) as it involved the first truly analytic study of births and deaths within a population precisely situated in space and time. To do this Graunt employed the Bills of Mortality (Greenwood, 1948) and the records of christenings in 17th-century London in order to investigate mortality and population growth in the city. The study's most outstanding qualities are revealed by the search for regularities and configurations in mortality and fertility along with a critical and very insightful appreciation of the quality of the data. He was greatly concerned to establish mortality rates by age and through this interest he came remarkably close to constructing the first formal life table. Using the cause of death evidence in the Bills of Mortality, Graunt without any information whatsoever on ages proceeded to estimate the extent of mortality in infancy and childhood by selecting those causes of death which he guessed would only affect children 'under four or five years old'. To these he added half of the deaths from smallpox and measles which he thought would fall upon children under six, along with slightly less than one third of the plague victims. From these assumptions he derived an estimate suggesting that 36 per cent of all deaths occurred to children under six years. Similar principles were employed to estimate that 7 per cent of deaths through 'ageing' could be attributed to those over 70 years. He then proceeded to take these not implausible estimates of mortality at young and old ages to construct an elementary life table which unfortunately was flawed by the unrealistic assumption that above age six the deaths in each specified age period amounted to about three-eighths of the survivors at the beginning of the period. Nonetheless, what Graunt had evolved was an outstanding innovation and was very soon taken up and developed by others. Graunt's interests in the *Natural and political observations* were innovative in other respects; he attempted to calculate the size and rate of growth of 17th-century London, the incidence of plague mortality (Sutherland 1963, 1972), sex ratios at birth and levels of maternal mortality, and he made some highly believable estimates of the levels of immigration to London that were needed to sustain the city's remarkable growth in the 17th century.

R.M. Smith

SELECTED WORKS

1662. *Natural and political observations ... upon the Bills of Mortality*. There was a further edition later in 1662, and the plague of 1665 called forth a third edition in the early summer and a fourth edition in November (printed in Oxford) of that year.

1938. *Natural and political observations*. Ed. Walter F. Wilcox, Baltimore: Johns Hopkins University Press.

1973. Natural and Political observations made upon the Bills of Mortality. In *The Earliest Classics*, ed. P. Laslett, Farnborough, Hampshire: Gregg International Publishers.

BIBLIOGRAPHY

Glass, D.V. 1963. John Graunt and his 'Natural and political observations'. *Royal Society of London Notes and Records* 19, 63–100.

Greenwood, M. 1948. *Medical Statistics from Graunt to Farr*. Cambridge: Cambridge University Press.

Sutherland, I. 1963. John Graunt: a tercentenary tribute. *Proceedings of the Royal Society*, Series A 126, 537–56.

Sutherland, I. 1972. When was the Great Plague? In *Population and Social Change*, ed. D.V. Glass and R. Revelle, London: Edward Arnold.

gravitation. *See* CENTRE OF GRAVITATION; COMPETITION; CLASSICAL CONCEPTIONS.

gravity models. This name is often used generically for a large family of quantitative models or 'laws' which seek to describe observed spatial regularities by means of simple equations. Strictly speaking, however, the gravity model is that developed in the 1940s by James Q. Stewart (1947, 1948), a Princeton astronomer who thought the social sciences could profitably use the physical sciences' early strategy of looking for simple mathematical relations among variables. Analogizing from the Newtonian gravity formula (hence his name for the model, demographic gravitation), he found many strong correlations of the form $I_{ij} = k(P_i^* P_j)/(d_{ij})^2$, where I_{ij} is some form of interaction between places i and j (such as traffic, migration or communications), P_i and P_j are the populations of those two places, d_{ij} the distance between them, and k a constant. Others since have elaborated mightily upon this simple formulation, as will be discussed below.

The gravity model proper is but one in the generic family to which that name is commonly applied. The tradition of this search for regularities goes back at least to E.G. Ravenstein's laws of migration (1885 and 1889) and in this century extends from the commercial rule of thumb of W.J. Reilly's law for defining retail-market areas by 'retail gravitation' (1931) to G.K. Zipf's rank-size rule of city sizes (1949), which states that if a region's or a nation's cities are ranked in order of population size, the size of the nth city will be approximately that of the largest city divided by n. Zipf's interests were broad, and his theorizing imaginative, although it has not attracted followers. He was a lecturer in linguistics at Harvard and, like Stewart, an intellectual eccentric with a passion for numerical regularities. According to legend, this got him into trouble in the late 1930s, when he supported Hitler's annexation of Austria and the Sudetenland on the grounds that Germany's urban system did not conform to the rank-size rule, but that the resulting pan-German system did. Be that as it may, the rank-size rule has held more or less true for the United States from 1790 to the present, if one squints and makes allowance for changing definitions of cities and metropolitan areas (interestingly, the relation deteriorated in the decades preceding the Civil War and recovered afterwards) and holds, in modified forms, for other countries and regions. Later workers have added exponents (e.g. $P_n = P_1/n^\alpha$) to make the relation more flexible, and have suggested that the magnitude of α is linked to economic development: the higher α, the greater the level of urban primacy, and so (ambiguously) the lower the level or the higher the pace of development. Zipf (1946) also worked on something like gravity models, calling them 'the $P_1 P_2/D$ hypothesis' and slightly anticipating Stewart;

he applied them to the movements of people, goods and information. Stewart himself made another important contribution, the population potential model. By analogy to physics again, the potential, V_i, is the local scalar value of the system's distance-discounted population distribution [$\Sigma(P_j/d_{ij})$], and is found to correlate with the local intensity of such things as per-capita income, the intensity of economic activity, and land values. See Carrothers (1956) for a useful if technically dated review of the early literature.

Many mainstream economists of today, trained in the deductive mode, regard such models as unconventional and to a degree bizarre, like sun-spots theories of the market. To be sure, they stress empirical regularities and are not based on well-specified theory; yet their general forms are not surprising. Population potential, for instance, may be viewed as a measure of access from a locality to the national population when the cost of distance is taken into account. It is akin to the US Bureau of the Census's 'center of population', which now lies just west of the Mississippi; the Bureau's measure is calculated as the (minimum) of the function $V_i = \Sigma(P_j^* d_{ij}^\beta)$, where β is 1, whereas β in the case of potential is -1, which makes more sense for the purpose. If distance were of no consequence, β would be zero and every locality would have equal access to the national population. Similarly, in the case of gravity models, one would expect that the interaction between two localities would be proportional to their sizes, and negatively related to the ease of access from one to the other. The question is not so much one of the *direction* of effect of the variables as one of the particular functional forms. Note, for instance, that a doubling of populations in the gravity model would imply a quadrupling of interactions.

Before returning to the standing of these models in theory, it is important to note that they have been used widely in applied work. The potential model has been used in many contexts, from industrial-location and market-areas analysis (Reilly's law is a form of potential), to theoretical two-dimensional function-smoothing in geography, to a recent test of F. Braudel's intuition of the shifting geography of European centres of power from 1500 to 1800 (de Vries, 1984). But it is the gravity model, in its original form and in modern variants, which has been the workhorse in applications, most notably as an important component in simulation models of urban traffic and land-use and in interregional models of development. Some of these models, which began to be developed in the 1950s, are big by any standard in the social sciences, involving several years' work, thousands of equations, and budgets in the millions.

In modern work the gravity model is often considerably modified, and estimated by the latest econometric techniques. The population variables may be replaced by complex functions with multiple arguments, such as wages, unemployment, measures of amenity, and so forth. To the degree possible, some functional rather than geographic measure of distance is used and, because they are more tractable, exponential rather than power functions are preferred. Normalizing terms are often employed to control, say, for the competition among alternative destinations. Thus a modern variant of the gravity model may be an econometrically estimated equation of the form:

$$M_{ij} = k^* F_i^* (G_j^* \exp\{-cd_{ij}\}) \bigg/ \left[\sum_k (G_k^* \exp\{-cd_{ik}\}) \right],$$

where M_{ij} is the movement from i to j, k and c are constants, d_{ij} and d_{ik} are functional distances between i and j and k, respectively, and F_i, G_j and G_k are functions with multiple arguments estimated at i, j and k, respectively. The line has thus become blurred between gravity models and workaday econometric models of a multiplicative or exponential sort. Additionally, there are variants which use optimization techniques, probability and information theories. Hua and Porell (1979) and Harris (1985a, b) provide excellent reviews of this rich literature.

Although there have been several attempts to account for the empirical power of these models using neoclassical theory, the results have not been satisfactory, either requiring overly-restrictive assumptions or obtaining too-narrow results. This may be because the geographic regularities which gravity models describe are statistical by nature. To make yet another analogy with physics, the neoclassical approach may be compared with Newtonian mechanics, best suited to the analysis of the behaviour of comparatively few and well-defined bodies, while these observed geographic regularities may require an approach more like that of statistical mechanics. Some recent literature moves in this direction, trying to place gravity models in the context of statistical processes and the mathematics of their formulation (Alonso, 1978; Willekens, 1980), and begins to shed light on their relation to other models and approaches such as input–output, Markov transitions, logit models, logarithmic contingency tables, and entropy-based analysis.

Over the years gravity models and their kin have generated a fair amount of nonsense, but perhaps no more than any other branch of the social sciences, while they have been very valuable in applied work of various sorts. Their worth in the long term may lie in the question they pose in various forms: what causes the patterns we observe? They are not just statistical artifacts (although other functional forms may be better descriptors), and yet our current kit of theory cannot account for them. Their value, then, may lie in their forcing us to think hard and in new directions.

WILLIAM ALONSO

See also ADJUSTMENT PROCESSES AND STABILITY; CENTRAL PLACE THEORY; CHRISTALLER, WALTER; INTERNAL MIGRATION; LOCATION OF ECONOMIC ACTIVITY; STABILITY.

BIBLIOGRAPHY

Alonso, W. 1978. A theory of movement. In *Human Settlement Systems*, ed. N. Hansen, Cambridge, Mass.: Ballinger.

Carrothers, G.A.P. 1956. A historical review of the gravity and potential concept of human interaction. *Journal of the American Institute of Planners* 22, 94–102.

De Vries, J. 1984. *European Urbanization, 1500–1800*. Cambridge, Mass.: Harvard University Press.

Harris, B. 1985a. Synthetic geography: the nature of our understanding of cities. *Environment and Planning* A 17, 443–64.

Harris, B. 1985b. Urban simulation models in regional science. *Journal of Regional Science* 25, 545–68.

Hua, C.-I. and Porell, F. 1979. A critical review of the development of the gravity model. *International Regional Science Review* 4(2), 97–125.

Ravenstein, E.G. 1885, 1889. The law of migration. *Journal of the Royal Statistical Society* pt 1, 48, 167–235; pt 2, 52, 241–305.

Reilly, W.J. 1931. *The Law of Retail Gravitation*. New York: W.J. Reilly Co.

Stewart, J.Q. 1947. Suggested principles of 'social physics'. *Science* 106, 179–80.

Stewart, J.Q. 1948. Demograpic gravitation: evidence and application. *Sociometry* 1, 31–58.

Willekens, F. 1980. Entropy, multiproportionality adjustments and analysis of contingency tables. *Sistemi Urbani* 2, 171–201.

Zipf, G.K. 1941. *National Unity and Disunity*. Bloomington: Principia Press.

Zipf, G.K. 1946. The PP/D hypothesis: on the intercity movement of persons. *American Sociological Review* 11, 677–86.

Zipf, G.K. 1949. *Human Behavior and the Principle of Least Effort: An Introduction to Human Ecology*. Cambridge, Mass.: Addison-Wesley.

Gray, Alexander (1882–1968).   Gray was a civil servant (1905–21), professor of political economy (Aberdeen University, 1921–35; University of Edinburgh, 1935–56) and, throughout his career, a public servant, poet and translator. Gray combined an early training in mathematics with Scottish pragmatism and an international outlook developed as a graduate student at Göttingen and Paris. He was not interested in esoteric economic theory, but in applications, in public policy and in the historical development of ideas. His early writings (1923, 1927) drew on his civil service experience in establishing the welfare state. Later, in his lectures, he became a leading analyst of the growth of the nationalized industries in Britain. His writings in the history of economic thought (1931, 1946) were remarkable for their range, dissecting ideas from Ancient Greece to modern times and showing a profound knowledge of the literature in many languages. He did not espouse any particular thesis about the development of ideas, but was temperamentally critical of socialism and of the growth of state intervention. He was an inspiring teacher, setting economics in context, historically and politically. Outside economics his translations into broad Scots of European ballads and of Heine were sensitive and much admired.

T. JOHNSTON

SELECTED WORKS

1923. *Some Aspects of National Health Insurance*. London: P.S. King & Son.

1929. *Family Endowment*. London: Ernest Benn.

1931. *The Development of Economic Doctrine*. London: Longmans, Green & Co.

1946. *The Socialist Tradition. Moses to Lenin*. London: Longmans, Green & Co.

Gray, John (1799–1883).   From the time that he went to work in a London manufacturing and wholesale house at the age of 14, Gray was, apparently, interested in social reform and, after attending the London tavern debates of 1817, his thinking began to assume an Owenite socialist complexion. It was this interest in cooperative socialism that led him to visit the community established by Abram Combe at Orbiston in 1825, though his reaction as expressed in *A Word of Advice to the Orbistonians* (1826) was critical. Nevertheless his first major work, *A Lecture on Human Happiness* (1825), did embrace the communitarian ideal. In it he argued that under competitive capitalism the real income of the country, which consisted of the quantity of wealth annually produced by the labour of the people, was taken from its producers through the rent, interest and profits of those who bought labour at one price and sold it at another. This failure to exchange equivalents would be eliminated through the formation of cooperative communities and the abolition of the competitive system of exchange itself.

Yet if it was the *Lecture* which secured him a measure of contemporary notoriety, *The Social System: A Treatise on the Principle of Exchange* (1831) is Gray's most interesting work. Here his explanation of the impoverishment of labour is similar to that of the earlier book. Exploitation was deemed to occur in the sphere of exchange, producing poverty, distress, underconsumption and thence arrested economic development. Gray's solution, however, represented a significant move away from his earlier communitarianism in the direction of a centrally controlled, technocratically run economy, more Saint-Simonian than Owenite in character.

As in the *Lecture*, the competitive market economy was to be abolished but in *The Social System* its pricing, allocative, distributive and equilibrating functions were to be performed by a central authority denominated by Gray the National Chamber of Commerce. Thus

the Social System recognises as useful, but one controlling and directing power, but one judge of what is prudent and proper to bring into the market, either as respects kind or quantity – the Chamber of Commerce – who, having the means of ascertaining, at all times, the actual stock of any kinds of goods in hand would always be able to say at once where production should proceed more rapidly, where at its usual pace and where also it should be retarded.

Further the Chamber would employ 'agents' to organize the 'cultivation of the land and the management of all those trades and manufactures which had been brought together by a voluntary association of landowners, capitalists and traders'. It would also be responsible for ensuring that demand was commensurate with supply, for making, as Gray phrased it, 'Production the uniform and never failing cause of demand.'

The central themes of *The Social System* were to be reiterated in a work entitled *An Efficient Remedy for the Distress of Nations* (1842), two works which taken together represent the first significant attempt, however inadequate, in the history of British socialist thought to discuss how central control or authority might be applied to a modern, complex, interdependent, industrial economy.

Yet by 1848, and the publication of his *Lectures on the Nature and Use of Money*, Gray had abandoned both communitarianism and central economic planning, arriving instead at the conclusion that the system of exchange might be rationalized without any transcendence or abolition of the market and without the institutional paraphernalia of communities or chambers of commerce. There was now no problem as there had been in the *Lecture*, in reconciling free and equitable exchange with free competition. 'The great principle of individual competition should be left free and unfettered as the air we breathe.' All that was required to make competitive capitalism work, all that was needed to ensure generalized prosperity was a medium of exchange that could be expanded *pari passu* with the level of output. This would eliminate the incidence of general economic depressions and lay the basis for rising living standards and social harmony. As Gray saw it 'A few salutary money laws are all that are wanted' because 'our false monetary position is the one and only cause of our misfortunes'.

By his death in 1883, after a successful career in publishing, Gray had become reconciled to market capitalism at a practical as well as a theoretical level.

N.W. THOMPSON

SELECTED WORKS

1825. *A Lecture on Human Happiness*. London.
1826. *A Word of Advice to the Orbistonians on the principles which ought to regulate their present proceedings, 29 June 1826*. Edinburgh.
1831. *The Social System: A Treatise on the Principle of Exchange*. Edinburgh.
1842. *An Efficient Remedy for the Distress of Nations*. Edinburgh.
1848. *Lectures on the Nature and Use of Money*. Edinburgh.

BIBLIOGRAPHY

Beales, H.L. 1933. *The Early English Socialists*. London: Hamish Hamilton.
Beer, M. 1953. *A History of British Socialism*, 2 vols. London: Allen & Unwin.
Cole, G.D.H. 1977. *A History of Socialist Thought*. 5 vols. Vol. 1: *Socialist Thought: The Forerunners, 1789–1850*, London: Macmillan.
Foxwell, H.S. 1899. Introduction to the English translation of A. Menger, *The Right to the Whole Produce of Labour*. London: Macmillan.
Gray, A. 1967. *The Socialist Tradition, Moses to Lenin*. London: Longman.
Hunt, E.K. 1980. The relation of the Ricardian socialists to Ricardo and Marx. *Science and Society* 44, 177–98.
Kimball, J. 1946. *The Economic Doctrines of John Gray 1799–1883*. Washington, DC: Catholic University of America Press.
King, J.E. 1981. Perish Commerce! Free trade and underconsumption in early British radical economics. *Australian Economic Papers* 20, 235–57.
King, J.E. 1983. Utopian or scientific? A reconsideration of the Ricardian socialists. *History of Political Economy*, 15, 345–73.
Lowenthal, E. 1911. *The Ricardian Socialists*. New York: Longman.
Martin, D. 1982. John Gray 1799–1883. In *Dictionary of Labour Biography*, ed. J. Saville and J. Bellamy, London: Macmillan, Vol. 6.
Thompson, N.W. 1984. *The People's Science: The Popular Political Economy of Exploitation and Crisis, 1816–34*. Cambridge: Cambridge University Press.

Gray, Simon (alias George Purves, LL.D.) (*fl.* 1795–1840). Little is known about Gray except that he worked at the War Office in the late 18th and early 19th centuries. His major work (if that is the right phrase) was *The Happiness of States*, a book which 'he meant to have published in 1804 but was prevented', with the result that it did not appear until 1815. A large and cranky work, it is at once anti-Physiocrat, anti-Smith (the similarity of its title to *The Wealth of Nations* is no accident) and anti-Malthus.

Its reception was so poor that in 1817 and 1818, under the pseudonym of George Purves LL.D., he published two other works whose main purpose was to praise his own *Happiness of States* (see Sraffa, 1952, p. 38n). Such self-puffery was of course not unknown, even for works of the first magnitude. (When his *Treatise of Human Nature* 'fell "dead-born from the press" ' in 1739, the great but dejected Hume responded by publishing anonymously a year later *An Abstract of a Book lately Published; Entituled, A Treatise of Human Nature, &c.*, in which he praised his otherwise ignored work (Keynes and Sraffa, 1938).) Whilst Hume's puff was unsuccessful (no second edition of the *Treatise* appearing in his lifetime) Gray's was not, at least in his own judgement, since in 1819 he brought out a second edition of his original work.

Gray could be left in well-deserved obscurity were it not for one thing. Chapter V of Book VII of *Happiness*, entitled 'Of Scarcities', contains a most detailed account of what economists usually call Giffen goods. 'To raise the price of corn in any great degree, tends directly to increase the general consumption of that necessary' (p. 505). 'There is no paradox here', since bread is both an inferior good and a large item in the budgets of the poor. Thus 'By raising the price of bread corn,... we force them to live more on it; ... However, paradoxical, therefore, it may be in seeming, it is a plain substantial fact, that the higher the price of corn and potatoes, the greater is the consumption ...' (pp. 509–10, original punctuation).

Perhaps we should call them not Giffen but Gray goods.

PETER NEWMAN

SELECTED WORKS

1797. *The Essential Principles of the Wealth of Nations, illustrated, in opposition to some false doctrines of Dr. Adam Smith, and others ....* London: T. Becket.

1815. *The Happiness of States: Or An Inquiry Concerning Population, The Modes of Subsisting and Employing It, and the Effects of All on Human Happiness.* London: Hatchard. 2nd edn, 1819.

1817. [As George Purves.] *All Classes Productive of National Wealth,... .* London: Longman. 2dn edn, 1840.

1818. [As George Purves.] *Gray versus Malthus, The Principles of Population and Production Investigated,... .* London: Longman etc.

1820. *Remarks on the production of wealth ... in a letter to the Rev. T.R. Malthus,... .* London.

1823. [As George Purves] *The Grazier's Ready Reckoner, being a complete set of tables, shewing the weight of cattle, calves and pigs by admeasurement, etc.* Warrington.

1839. *The Spaniard; or, Relvindez and Elzora, a tragedy [in five acts and in verse], and The Young Country Widow [in five acts and in prose],... .* London.

1842. *The Messiad; or, the life, death, resurrection, and exaltation of Messiah, the Prophet of the Nations.* London.

BIBLIOGRAPHY

Keynes, J.M. and Sraffa, P. (eds) 1938. *An Abstract of A Treatise of Human Nature 1740, by David Hume.* Cambridge: Cambridge University Press.

Masuda, E. and Newman, P. 1981. Gray and Giffen goods. *Economic Journal* 91, 1011–14.

Powell, E.G. 1896. Gray, Simon. In *Dictionary of Political Economy,* ed R.H.I. Palgrave, London: Macmillan, Vol. II, 257–8.

Sraffa, P. (ed. with the collaboration of M.H. Dobb.) 1952. *The Works and Correspondence of David Ricardo,* Vol. VIII. Cambridge: Cambridge University Press for the Royal Economic Society.

Gregory, Theodore Emanuel Gugenheim (1890–1970). Gregory was born in London and after an education at St Owen's School in Islington, at Stuttgart and at the London School of Economics, he became an assistant lecturer at the LSE from 1913 to 1919. Appointed Cassel Reader in International Trade in 1920, he became Dean of the Faculty of Economics in London (1927–1930) and was Sir Ernest Cassel Professor of Economics from 1927 to 1937.

Robbins refers to him as 'one of the last of the generation of gifted teachers who, in the twenties, contributed so much to the international standing of the London School of Economics' (*The Times*, 3 February 1970).

Gregory sat on various official commissions including the Macmillan Committee on Industry and Finance (1929–31), the Irish Free State Banking Commission (1934–37), and acted as economic adviser to the Indian government from 1938 to 1946. He was knighted in 1942.

Gregory was 'acutely conscious ... of the disintegrating elements at large in the world' in the interwar period (Robbins, 1970). His interwar books on currency reflect this outlook, supporting a return to gold as less liable to inflationary abuse, despite some flaws, rather than a managed system. His analysis was meticulous, examining the problems of timing, parities and the long-term supply and demand outlook for gold. He argued that exchange rate stability reflects the stability of relative prices between countries; thus the best way of achieving this was to link prices to the same standard, i.e. gold. He argued strongly against the inflation of paper money, and although gold itself has no intrinsic stability of value he argued that experience had shown it to be more stable than managed currencies. He further suggested that the improved stability of some paper currencies was due to exchange rate targets based on the US dollar which had remained tied to gold (an effective Gold Exchange Standard).

His most significant work was concerned with the history of banking and currency culminating in his introduction to Tooke and Newmarch's *History of Prices* and a history of the Westminster Bank that was a valuable contribution to the study of 19th-century monetary history. However, Gregory's interests were not restricted to the monetary field and Schumpeter, alongside his commendation of the introduction to Tooke, praises Gregory's article on 'The Economics of Employment in England, 1660–1713' (*Economica* 1921) and says 'There is no equally valuable survey for any other country' (p. 272).

R.J. BIGG

SELECTED WORKS

1921a. *Foreign Exchange before, during and after the war.* World of Today Series, Oxford: Oxford University Press.

1921b. The economics of employment in England, 1660–1713. *Economica* 1, January, 37–51.

1925. *The Return to Gold.* London: Ernest Benn.

1928. *Introduction to Tooke and Newmarch's History of Prices.* London: P.S. King & Son. Reprinted, London: LSE, 1962.

1929. *Select Statutes, Documents and Reports relating to British Banking 1832–1928.* 2 vols, London: Oxford University Press.

1936. *The Westminster Bank Through a Century.* 2 vols, London: Oxford University Press.

Gresham, Thomas (c1519–1579). The second son of Sir Richard Gresham, merchant, Sir Thomas Gresham was educated at Gonville Hall, Cambridge, apprenticed to his uncle Sir John Gresham, also a merchant, and admitted a member of the Mercers' company in 1543. In 1551 or 1552 he became royal agent or king's factor at Antwerp, in which post he received twenty shillings a day, and which he retained with few intervals during three reigns until 1574, employed in spite of his Protestant views even by Mary. His business was to negotiate royal loans with Flemish merchants, to buy arms and military stores, and to smuggle into England as much bullion as possible. He succeeded in raising the rate of exchange from 16s. to 22s. in the £, and is said to have saved in this way 100,000 marks to the crown and 300,000 to the nation. His operations greatly benefited English trade and credit, though the government could not be induced to pay its debts as punctually as Gresham would have liked. He did not hesitate to remonstrate with and advise Elizabeth and Cecil; but he was so useful and trustworthy that he was never seriously out of favour, except just after Mary's accession. On Mary's death he advised Elizabeth to restore the base money, to contract little foreign debt, and to keep up her credit, especially with English merchants. Later he taught her how to make use of these English merchants when political troubles in the Netherlands curtailed her foreign resources; at his suggestion the Merchant Adventurers and Staplers were forced by detention of their fleets to advance money to the state; but as they obtained interest at 12 per cent instead of the legal maximum of 10, and the interest no longer went abroad, the transaction proved advantageous to all parties and increased Gresham's favour. His journeys to and from Antwerp were very frequent, but in his later years he entrusted most of his public work to his agent, and not known to have been at Antwerp after 1567. In 1554 he was sent to Spain to procure bullion, a very difficult task in which he was only partially successful; and in 1559 he was employed as ambassador to the Duchess of Parma, regent of the Netherlands; it was on this occasion that he was knighted.

In addition to his public services he continued throughout his life to do the work of 'the greatest merchant in London'. He was, in the language of the day, a banker and goldsmith, with a shop in Lombard Street, as well as a mercer; but he was a considerable country gentleman besides, with estates, chiefly in Norfolk, where his father held considerable property, and with several country houses besides the house in Bishopsgate which he built and bequeathed to London as Gresham College. He twice entertained Queen Elizabeth as his guest. His wealth was mainly earned by his private business, but he cannot be acquitted of enriching himself at the public expense by at least one dishonourable manœuvre; and he habitually forwarded his schemes by bribery. The money so gained he applied to public uses, his only son having died young: the foundation of the royal exchange, of Gresham College, and of eight almshouses, and the establishment of the earliest English paper-mills on his estate at Osterley, show the breadth of his interests, his liberality, his charity, his culture, and his commercial enterprise.

(ELEANOR G. POWELL]
Reprinted from *Palgrave's Dictionary of Political Economy*.

BIBLIOGRAPHY

Anon. 1883. *A Brief Memoir of Sir Thomas Gresham, with an abstract of his will, and of the act of parliament for the foundation of Gresham's College.* London.
Bindoff, S.T. 1973. *The Fame of Sir Thomas Gresham.* London: Jonathan Cape.
Boorstin, D.J. 1980. *Gresham's Law: Knowledge or Information?* Washington, DC: Library of Congress.
Burgon, J.M. 1839. *Life and Times of Sir Thomas Gresham.* 2 vols. London.
de Roover, R.A. 1949. *Gresham on Foreign Exchange.* Cambridge, Mass.: Harvard University Press.
Fox Bourne, H.R. 1866. *English Merchants.* London.
Fuller, T. 1662. *Worthies of England.* London.
Hall, H. 1886. *Society in the Elizabethan Age.* London.
Holinshed, R. 1807–8. *Chronicles of England, Scotland and Ireland.* 6 vols. London.
Salter, R.F. 1925. *Sir Thomas Gresham.* London: Parsons.
Ward, J. 1740. *Life of Sir Thomas Gresham.* London.

Gresham's Law.   This familiar term was introduced by MacLeod in 1858 (*Elements of Political Economy*, p. 477), and has since been generally accepted by economists.

It denotes that well-ascertained principle of currency which is forcibly though not quite adequately expressed in the dictum – 'bad money drives out good'. It has also not infrequently been explained by the statement that where two media of exchange come into circulation together the more valuable will tend to disappear. The principle in its broadest form may be stated as follows: Where by legal enactment a government assigns the same nominal value to two or more forms of circulatory medium whose intrinsic values differ, payments will always, as far as possible, be made in that medium of which the cost of production is least, and the more valuable medium will tend to disappear from circulation; in the case where the combined amount in circulation is not sufficient to satisfy the demand for currency, the more valuable medium will simply run to a premium.

This is a principle which obviously has its roots in the ordinary instincts of commercial life, and the cases in which it has asserted itself may be divided into those where (1) of two media intrinsically good, one is by error under-valued; where (2) it is sought to keep a debased metallic currency in circulation on a par with that of a better metal; where (3) an inconvertible paper has been made to run by the side of a metallic currency. The reference to paper suggests the observation that the circulation of convertible paper side by side with, and as 'the shadow of' gold, is explained by reference to a different principle.

The dictum above quoted appears to have been used first in the proclamation of 1560 respecting the decrial of the base silver coin; and we know that Sir Thomas Gresham took a prominent part in advising Queen Elizabeth and Cecil on the reform of the currency. (See Burgon's *Life and Times of Sir Thomas Gresham*.) We do not, however, find the principle stated in his own handwriting.

The actual instances of the operation of the law are numerous, and we shall cite only a few. MacLeod quotes Aristophanes, *Frogs*, as the earliest instance of its recognition. Amongst many in the history of the United Kingdom, besides the prevalence of debased silver which formed the occasion of Gresham's dictum, we may cite the following from Lord Liverpool, referring to the over-valuation of silver prior to the reign of James I.

It is certain that the rise in the value of Gold made by James I in the 2nd and 3rd year of his reign was rendered necessary by the exportation of the Gold Coin, which had for some time been experienced, and by the very small quantity of it that was then left in circulation. This rise ... produced a partial and temporary relief ... but [Stowe] confesses that this plenty of Gold coin did not continue in circulation for any length of time, and that it afterwards began to be exported. It soon became evident that the last-mentioned rise of the value of Gold in the Coins of this Kingdom was not sufficient to make it equal to the relative value of Gold to Silver at the market.

In Mr Chalmers's recent work on the *History of Colonial Currency*, we have without difficulty noted eighteen instances, and there are many more. In a pamphlet of 1740 on the *Currencies of the British Plantations in America*, we have the following:

In sundry of our Colonies were enacted laws against passing of light Pieces of Eight. These laws not being put into execution, heavy and light Pieces of Eight passed promiscuously; and, as it always happens, a bad currency drove away the good currency; heavy Pieces of Eight were shipped off.

And on this Mr Chalmers comments as follows:

Imitating the practices familiar to them in England, dishonest persons traded on the desire of the young communities for a metallic currency, by circulating clipped money at the full rate; and this malpractice was condoned by the Colonies when it was found that the light money was more apt to stay with them than 'broad' pieces.

So that throughout the 17th and 18th centuries in the American and West Indian colonies we constantly find a debased currency in possession of the field. Again, when the Treasury tried to introduce sterling into the colonies in 1838, and wrongly valued the dollar for concurrent circulation with the shilling, 'by a familiar law, the over-rating of the dollar sufficed to drive out the shillings' which had been shipped from Great Britain.

In the case of the United States we have excellent instances both in the effect of the legislation of 1837 as driving out silver dollars, and in the way in which the inconvertible 'greenbacks' formed the chief part of the currency from 1864 to 1879.

[C. ALEXANDER HARRIS]
Reprinted from *Palgrave's Dictionary of Political Economy*.

See also BIMETALLISM.

Grossmann, Henryk (1881–1950). Grossmann was born on 14 April 1881 in Cracow and died on 24 November 1950 in Leipzig. He studied in Cracow and lived from 1908 to 1918 in Vienna (collaborating with Carl Grünberg) and 1918 to 1925 in Warsaw (at the Central Statistical office and the Free University). From 1925 to 1933 he was a political refugee in Germany (University of Frankfurt/Main) and later in France, England and USA. Grossmann spent his last years in the German Democratic Republic, at the University of Leipzig.

His main work (1929), based on Marx, deals with the inevitability of the breakdown of capitalism. The method of Marx, in his view, is a step-wise approximation to reality which starts from a simplified abstract model of the accumulation process (based on the reproduction schema and assuming a closed system, two classes only, no credit, commodities sold at their values, constant value of money) and proceeds by gradually adding realistic details of secondary importance ('surface phenomena') among which he counts monopoly, money and credit, capital exports and the struggle for raw materials. In following this method Grossmann demonstrates the inevitability of breakdown and then deals with the factors which counteract and therefore delay the breakdown.

The starting point of his theory is an arithmetic example of Otto Bauer based on the reproduction schema of Marx which was intended by Bauer (in a polemic against Rosa Luxemburg) to demonstrate that realization under extended reproduction was perfectly possible. Bauer worked out his example only for four years, but Grossmann extended it to 35 years in order to demonstrate that the accumulation process could not proceed without limit. Following Bauer, he made the following assumptions: 5 per cent growth of variable capital (determined exogenously by the growth of population) while the constant capital was to grow by 10 per cent; surplus value was to be constant at 100 per cent.

Since the organic composition of capital was continuously increasing, the Marxian conclusion of a declining rate of profit held good, as also shown in Bauer's example. This implied, with constant growth of capital, that the share of consumption in surplus value had to decrease. Grossmann sees no difficulty in this so long as the absolute amounts of profit and consumption increase (owing to the increase in capital). For this reason he considers Marx's theory as incomplete. His own contribution is to show that the *absolute* amounts also have to decrease. The step taken from Marx seems to be fairly simple: if the declining rate of profit is combined with an exogenously given constant rate of growth of capital, then the share of consumption in surplus must ultimately go down to zero and below. This marks the point of breakdown.

Grossmann deals extensively with counteracting tendencies such as new colonial markets and real wage cuts. These can only delay and not avoid the breakdown. Their effect will appear in the form of cyclical crises which Grossmann expected would become more and more serious.

Grossmann strongly criticizes all Marxist writers before him (in particular, Hilferding and Luxemburg) for having distorted the content of Marx's teaching. His aim is to restore the orthodoxy of the true Marx which in his view is embodied in the breakdown thesis based on the increase in the organic composition of capital. Other aspects of Marx he plays down (historical materialism) or ignores (the realization problem). Since his book largely takes the form of a polemic it may be used as a source of information on Marxist and other literature, but he does scant justice to the ideas of some of these writers.

Grossmann was, however, a man of culture and learning, with considerable knowledge of the economic doctrines of the 18th and early 19th centuries, and a highly esteemed historian who wrote a pioneering study on the Principality of Warsaw (a short-lived state created by Napoleon) based on census material. His surviving papers are in the archives of the Polish Academy of Sciences.

JOSEF STEINDL

SELECTED WORKS

1914. *Österreichs Handelspolitik mit Bezug auf Galizien in der Reformperiode 1772–1790.* Vienna: Konegen.
1924. *Simonde de Sismondi et ses théories économiques.* Warsaw: Bibliothèque Universitaire Libre Polonaise.
1925. Struktura spoleczna i gospodarcza Księstwa Warszawskiego na podstawie spisów ludności 1808–1810 (Social and economic structure of the Warsaw Principality on the basis of the census of population 1808 to 1810). *Kwartalnik Statystyczny,* Warsaw.
1929. *Das Akkumulations- und Zusammenbruchsgesetz des kapitalistischen Systems.* Leipzig: C.L. Hirschfeld.
1975. *Marx, l'économie politique classique et le problème de la dynamique.* Preface by Paul Mattick. Paris: Editions Champ Libre.

gross revenue. *See* REVENUE, GROSS AND NET.

gross substitutes. The assumption that goods are gross substitutes is applied to a set of excess demand functions $e_i(p_1, \ldots, p_n)$, $i = 1, \ldots, n$, where p_i is the price of the ith good and e_i is the excess demand for the ith good. The concept was introduced by Mosak (1944) in the context of a pure trading model. However, his definition required that $\partial e_i / \partial p_j$ have the same sign as the substitution term in the Slutsky equation as well as be of positive sign. That is, the income effect should not overbalance the substitution effect. At about the same time Metzler (1945) said simply that the jth good is a gross substitute for the ith good if $e_{ij} = \partial e_i(p)/\partial p_j > 0$ holds, and this has been the meaning used in later papers when the functions e_i have been assumed to be differentiable.

A definition with wider application is the one used by Morishima (1964). By this definition the jth good is a gross substitute for the ith good if $e_i(p) < e_i(p')$ whenever $p \leqslant p'$, $p_j < p'_j$, $p'_k = p_k$ for $k \neq j$. We will say that the assumption of gross substitutes holds if $e_i(p) < e_i(p')$, $i = 1, \ldots, n$, for all p and p' such that $p \leqslant p'$, $p \neq p'$, and $p_i = p'_i$. If $e(p)$ is differentiable, this assumption implies that the $n \times n$ matrix $[e_{ij}(p)]$ has positive off-diagonal elements for all p in the interior of the domain of $e(p)$. A matrix with this property and a negative diagonal is often referred to as a Metzler matrix. We will say that the assumption of weak gross substitutes holds if the condition $e_i(p) < e_i(p')$ is replaced by the weak inequality $e_i(p) \leqslant e_i(p')$. In the case of differentiable $e(p)$ it is implied by the assumption of weak gross substitutes that $[e_{ij}(p)]$ has non-negative off-diagonal elements. That these assumptions are not empty is shown by the case of excess demands defined by Cobb–Douglas utility functions of the form $U(x) = \Pi_{i=1}^n x_i^{\alpha_i}$, where $a_i > 0$ and $\Sigma \alpha_i = 1$. The excess demand function for a consumer in a pure exchange economy, holding initial stocks $\bar{x}$, is $e_i(p) = (\alpha_i \Sigma_{k=1}^n p_k \bar{x}_k / p_i) - \bar{x}_i$, so $e_{ij}(p) = \alpha_i \bar{x}_j / p_i \geqslant 0$ for $i \neq j$. If the initial stock of every good is positive for the whole market, the assumption of gross substitutes is satisfied.

A price vector p is said to be an equilibrium of the set of demand functions if $e(p) \leqslant 0$. The gross substitute assumption is used to establish the existence and uniqueness of an equilibrium and to prove the equilibrium to be stable for a dynamic adjustment system for prices. The gross substitute assumption also implies results of comparative statics, that is, results on the displacement of equilibrium that follows from shifts in demand or changes in initial stocks.

The following assumptions will be made on $e(p)$.

(B) $e(p)$ is defined for all $p > 0$, and $p^s \to p$ where $p_i = 0$ for $i \in I$, $p \neq 0$ implies $\Sigma_{i \in I}|e_i(p)| \to \infty$. Also $e(p)$ is bounded below.

(C) $e(p)$ is single-valued and continuous for $p > 0$.

(H) $e(p) = e(\lambda p)$ for any $\lambda > 0$, that is $e(p)$ is positively homogeneous of degree 0.

(W) $\Sigma_{i=1}^n p_i e_i(p) = 0$, that is, $e(p)$ satisfies Walras's Law.

The example of Arrow–Hahn (1971), pp. 29–30, where demand is derived from the utility function $u(x) = x_1^{1/2} + x_2^{1/2} + x_3^{1/2}$ where x is the consumption bundle, shows that assumption B cannot easily be improved.

UNIQUENESS. The existence of a positive equilibrium under assumptions B, C, H, and W does not require an assumption of gross substitutes (see Debreu, 1970). However, if there should exist an equilibrium price p when gross substitutes is assumed, $p > 0$ must hold. This follows from the fact that $p_i = 0$ for some i, and $p \neq 0$, is inconsistent with assumption H in that case, since it must at the same time be true, for $\lambda > 1$, that $e_i(\lambda p) = e_i(p)$ and $e_i(\lambda p) > e_i(p)$. Thus $e(p)$ cannot be defined at such a p.

Make the assumption that all goods are gross substitutes. Assume that two equilibria exist, say p and p' where $p' \neq \lambda p$ for any $\lambda > 0$. By assumption H we may choose p and p' so that $p_i = p_i'$ for some i and $p_j \leqslant p_j'$ for $j \neq i$. Then by gross substitutes $e_i(p) < e_i(p')$. But p and $p' > 0$ and assumption W implies that $e_i(p) = e_i(p') = 0$, which is a contradiction. Thus an equilibrium price vector is unique up to multiplication by a positive number.

Consider a partition of the goods into two non-empty subsets I and J. Say that the excess demand functions are connected if for any such partition $p_i = p_i'$ for all $i \in I$ and $p_j < p_j'$ for all $j \in J$ implies that $e(p) \neq e(p')$. Then by a similar argument, uniqueness of equilibrium may be seen to hold when weak gross substitutes and connectedness are assumed. Strictly speaking, connectedness need only hold at an equilibrium point.

If weak gross substitutes is assumed without connectedness, uniqueness may fail. However, it may be shown that the set of equilibrium price vectors is convex (McKenzie, 1960). Arrow and Hurwicz (1960) proved that the weak axiom of Revealed Preference holds between any equilibrium price vector p and any non-equilibrium price vector p' when weak gross substitutes is assumed. This means that $0 = p'e(p) = p'e(p')$ implies $pe(p') > pe(p) = 0$. In other words, $pe(p') > 0$. Suppose p and p' are both equilibria and consider $p'' = \alpha p + (1 - \alpha)p'$ for $0 < \alpha < 1$. If p'' is not an equilibrium $p''e(p'') = \alpha pe(p'') + (1 - \alpha)p'e(p'') > 0$ which contradicts assumption W, Walras's Law. Thus p'' is an equilibrium and the set of equilibria is convex.

COMPARATIVE STATICS. The modern approach to the comparison of equilibria after a shift of demand was begun by Hicks (1939). The fact that the Hicksian theorems hold locally when the excess demand functions satisfy the gross substitute assumption was proved by Mosak (1944). A global treatment of comparative statics in this context was given by Morishima (1964). Assume weak gross substitutes. Let demand shift from the ith good to the jth good. Let p and p' be the old and new equilibrium prices, e and e' the old and new excess

demand functions. Then

$$p_i e_i'(p) + p_j e_j'(p) = \sum_{i=1}^n p_k' e_k'(p) > 0,$$

by the Weak Axiom. (1)

On the other hand,

$$p_i e_i'(p) + p_j e_j'(p) = \sum_{i=1}^n p_k e'_k(p) = 0, \quad \text{by Walras' Law.} \quad (2)$$

Multiply (1) by p_i and (2) by p_i' and subtract to obtain $p_i p_j' - p_i' p_j > 0$, or $p_j'/p_j > p_i'/p_i$. Thus the price of the jth good increases relative to the price of the ith good.

Assume that all goods are gross substitutes. Then a shift in demand from the ith good to the jth good raises the price of the jth good relative to all other goods and lowers the price of the ith good relative to all other goods. These results are immediate from the fact that the good that rises in price relative to some goods and falls relative to none must experience a fall in demand and the good that falls in price relative to some goods and rises relative to none must experience a rise in demand. But only the jth good can absorb a rise and only the ith good can absorb a fall and still have demand equal to 0 after the shift has occurred. All other goods have zero excess demands at the old equilibrium prices after the shift, and the excess demands at the new equilibrium prices must also be zero. The same results follow from weak gross substitutes if any subset of the excess demand functions with $n - 1$ members is connected.

If the $e_i(p)$ are assumed to be continuously differentiable, the local theory of comparative statics is equivalent to determining the sign pattern of the inverses of the principal submatrices of order $n - 1$ of the Jacobian $[e_{ij}i, j = 1, \ldots, n$. The gross substitutes assumption implies that $e_{ij} > 0$ for $i \neq j$. Then either assumption H or W implies that the inverses of these submatrices have all elements negative. Choose the nth good as numeraire, and choose units so equilibrium prices $p_i = 1$, all i. Then $dp_i/d\alpha = -([e_{ij}]_{nn})_{ih}^{-1}$ if $\partial e_h(p, \alpha)/\partial \alpha = 1$. This is minus the ith element of the hth column of the inverse matrix of the submatrix where the nth row and column are omitted. Thus $dp_i/d\alpha > 0$, and it may be shown that $dp_h/d\alpha > dp_i/d\alpha$ for $i \neq h$ or n. If weak gross substitutes is assumed but the Jacobian and its $n - 1$ principal minors are indecomposable the same conclusions follow.

The local results may be extended to the case where there is a numeraire and the goods other than the numeraire may be partitioned into two non-empty subsets with indices in I and J, such that $e_{ij} > 0$ for $i \neq j$ and i and j in the same subset, while $e_{ij} < 0$ for i and j in different subsets. If the principal minors of $[e_{ij}]$ of order $n - 1$ have dominant diagonals, at equilibrium with the equilibrium prices as multipliers, the shifts of demand raise the price of the good to which demand has shifted and also the prices of all other goods in the subset of the partition to which it belongs, while lowering the prices of the goods in the other subset. Also the beneficiary of the demand shift has the largest change in equilibrium price in absolute value. These results are seen to follow from those for gross substitutes by considering the matrix formed by pre- and post-multiplying a principal minor of $[e_{ij}]$ by a diagonal matrix D with $d_{ii} = 1$ for $i \in I$ and $d_{jj} = -1$ for $j \in J$. This case was first analysed by Morishima (1952). The gross substitute case and the Morishima case may be shown to be the only sign patterns for a Jacobian matrix of the demand functions with all elements non-zero which allow the inverse matrix to be signed without quantitative information (see Bassett, Habibagahi and Quirk, 1967).

STABILITY. Since weak gross substitutes implies the weak axiom of revealed preference which in turn implies local stability of the tâtonnement for both the usual price adjustment models, with and without a numeraire (Arrow and Hurwicz, 1958), there is no special advantage for gross substitutes in local analysis of stability. However, for global results on stability the gross substitute assumptions are the only ones known with much plausibility. In order to use the weak axiom the adjustment must be, after proper choice of units,

$$\text{(I)} \quad \dot{p}_i = e_i(p),$$

for all goods other than the numeraire, if any, where $\dot{p}_i = \partial p_i / \partial t$, and (I) must hold globally. Excess demand is now assumed to be continuously differentiable. In order to use the other major possibility, a dominant diagonal for the matrix of demand elasticities, assuming a numeraire, the adjustment process

$$\text{(II)} \quad \dot{p}_i = p_i e_i(p),$$

is used for the non-numeraire goods (see Arrow and Hahn, 1971, p. 293). While (II) may be more reasonable than (I) for a global adjustment rule it is also very special.

On the other hand, when weak gross substitutes is assumed, stability may be proved for the adjustment rule

$$\text{(III)} \quad \dot{p}_i = h_i(p),$$

for all goods other than the numeraire, if any. The only special requirements placed on $h_i(p)$ are that it should be continuously differentiable and have the sign of $e_i(p)$. The adjustment rule III was proposed by McKenzie (1960) and global stability was proved using as a Lyapunov function the value of positive excess demand, $V[p(t)] = \Sigma_{i \in P} p_i e_i(p)$, $P = \{i | e_i(p) \geq 0\}$. The tâtonnement is shown to converge to the convex set of equilibrium prices. If the excess demand functions are connected, the equilibrium is unique.

Arrow and Hurwicz (1962) proved that the convergence of process III with weak gross substitutes is, in fact, to a particular equilibrium price vector, which will depend on the initial prices. This is clear once it is recognized that the goods whose prices are highest relative to some equilibrium price cannot be in excess demand and their prices cannot rise, and *mutatis mutandis* for the prices which are lowest. Thus the prices during tâtonnement cannot retreat from any equilibrium price vector. In the case of gross substitutes this line of argument is very simple and effective, since the prices actually fall and rise respectively.

In the theory of tâtonnement prices are revised according to excess demand but no trading occurs until equilibrium is reached. However, the stability of the adjustment process for a pure exchange economy is not lost if trading occurs at market prices, so long as the excess demand that drives the tâtonnement is determined by the maximization of utility by each trader under a budget equal to the value of the stocks he currently holds (Negishi, 1961). The crucial fact is that trading at market prices has only second order effects on excess demand. However, the price to which the process converges now depends on initial conditions and the course of trading.

It was pointed out by Rader (1972) that the production sector of the economy is unlikely to satisfy the gross substitute assumption in the demand for factors. As a consequence it seemed that the range of application of the gross substitutes assumption was effectively confined to pure trading economies. However, Rader was able to prove a local stability theorem assuming gross substitutes only for households. The production sector is made up of a finite number of firms with strictly

convex production possibility sets. The key to the argument is that $p\{e_{ij}^H\} = 0$ at equilibrium, where $e^H(p)$ is household excess demand. This is established by differentiating Walras's Law $pe^F(p) + pe^H(p) = 0$ to give

$$e^F(p) + e^H(p) + p[e_{ij}^F] + p[e_{ij}^H] = 0. \tag{3}$$

Since the first two terms sum to 0 at equilibrium, and the third term is 0 by profit maximization, (3) implies $p[e_{ij}^H] = 0$. Then the Jacobian of the adjustment system I with a numeraire is negative definite at equilibrium and local stability follows.

GENERALIZATIONS. Mukherji (1972) pointed out that some gross substitute theorems carry over if the weak gross substitute pattern is established in a transformed goods space. In particular, if there exists a matrix S such that $S^{-1}[e_{ij} + e_{ji}]S$ is indecomposable with off-diagonal elements non-negative, the tâtonnement is locally stable for the process I, with or without a numeraire. Also a rise in demand for the ith good causes the ith equilibrium price to rise. Ohyama (1972) shows that similar results follow if there is a stochastic matrix G which is positive definite and $G[e_{ij}]$ or $G[e_{ij} + e_{ji}]$ satisfies the conditions above. Uzawa (1960) proved that a discrete tâtonnement defined by $P_i(t + 1) = \max\{0, p_i(t) + f_i(t)\}$ where $f_i(t) = \beta_i\{e_i[p(t)]\}$ is globally stable when the Weak Axiom holds and $\beta_i > 0$ is sufficiently small. He assumes a numeraire and some additional differentiality and nonsingularity conditions.

Howitt (1980) defines a generalized gross substitute notion where $e(p)$ is allowed to be a convex valued correspondence rather than a function. This assumption of generalized gross substitutes holds if

(a) For all $p, p' > 0$, if there is a partition of indices for goods into non-empty subsets I and J where $p_i' = p_i$ for all $i \in I$ and $p_j' > p_j$ for all $j \in J$, then $\Sigma_{i \in I} p_i x_i' \geq \Sigma_{i \in I} p_i x_i$ for all $x \in e(p)$, $x' \in e(p')$.

(b) Strict inequality holds in (a) if p is an equilibrium.

Howitt proves that the equilibrium price vector is unique and globally stable for the price adjustment process

$$\text{(IV)} \quad \dot{p}_i \in e(p),$$

under generalized gross substitutes, when assumptions C, W, and H hold and an equilibrium exists. Excess demand $e(p)$ is assumed to be upper semi-continuous. He applies his result to the linear economy described by Gale (1976) and shown to satisfy gross substitutes by Cheng (1979).

Arrow and Hurwicz (1962) extended the adjustment process III to include expected prices. Let q represent expected prices. Then their process with adaptive expectations is

$$\dot{p}_i = h_i(p, q), \quad \text{if} \quad p_i > 0 \quad \text{and} \quad h_i(p, q) > 0, = 0, \quad \text{otherwise}$$

$$\text{(V)} \quad \text{sign } h_i(p, q) = \text{sign } e_i(p, q), \text{or} = 0 \text{ if } i \text{ is a numeraire}$$

$$\dot{q}_i = a_i(q_i - p_i)$$

They prove global stability for this process under weak gross substitutes with some auxiliary assumptions. Arrow and Hahn (1971) give a proof of global stability with an assumption of gross substitutes and assumptions B, C, H, and W. By the gross substitute assumption in this context is meant that $\partial e_i(p, q)/\partial p_j > 0$ for $i \neq j$, and $\partial e_i(p, q)/\partial q_j > 0$ for all j. The adjustment function h_i is assumed to be continuously differentiable.

Arrow and Hahn also prove a global stability theorem for a model in which expected prices q_i are given as functions $q_i(p)$ of current prices. This is in accord with models of temporary equilibrium. They prove global stability for adjustment process II with a numeraire, assuming gross substitutes for $e(p, q)$ and the Hicksian elasticity of substitution $\epsilon_j = \mathrm{d}\log q_j/\mathrm{d}\log p_j \leqslant 1$, which is consistent with Hicks's presumption when the strict inequality holds (see Hicks, 1939, p. 251).

In this model of the tâtonnement for temporary equilibrium $q(p)$ is presumably the expected price on the assumption that p is an equilibrium price. It is not so clear how to justify adaptive expectations in the tâtonnement setting.

LIONEL W. MCKENZIE

See also ADJUSTMENT PROCESSES AND STABILITY; SUBSTITUTES AND COMPLEMENTS; STABILITY.

BIBLIOGRAPHY

Arrow, K.J. and Hahn, F.H. 1971. *General Competitive Analysis.* San Francisco: Holden-Day.

Arrow, K.J. and Hurwicz, L. 1958. On the stability of the competitive equilibrium, I. *Econometrica* 26, 522–52.

Arrow, K.J. and Hurwicz, L. 1960. Competitive stability under weak gross substitutability: the 'Euclidean distance approach'. *International Economic Review* 1, 38–49.

Arrow, K.J. and Hurwicz, L. 1962. Competitive stability under weak gross substitutability: nonlinear price adjustment and adaptive expectations. *International Economic Review* 3, 233–55.

Bassett, L. Habibagahi, H. and Quirk, J. 1967. Qualitative economics and Morishima matrices. *Econometrica* 35, 221–33.

Cheng, H.-C. 1979. Linear economies are 'gross substitute' systems. *Journal of Economic Theory* 20, 110–17.

Debreu, G. 1970. Economies with a finite set of equilibria. *Econometrica* 38, 387–92.

Gale, D. 1976. The linear exchange model. *Journal of Mathematical Economics* 3, 205–59.

Hicks, J.R. 1939. *Value and Capital.* Oxford: Clarendon Press.

Howitt, P. 1980. Gross substitutability with multi-valued excess demand functions. *Econometrica* 48, 1567–75.

McKenzie, L.W. 1960. Stability of equilibrium and the value of positive excess demand. *Econometrica* 28, 606–17.

Metzler, L. 1945. Stability of multiple markets: the Hicks conditions. *Econometrica* 13, 277–92.

Morishima, M. 1952. On the laws of change of price-system in an economy which contains complementary commodities. *Osaka Economic Papers* 1, 101–13.

Morishima, M. 1964. *Equilibrium, Stability and Growth.* Oxford: Clarendon Press.

Mosak, J.L. 1944. *General Equilibrium Theory in International Trade.* Bloomington, Indiana: Principia Press.

Mukherji, A. 1972. On complementarity and stability. *Journal of Economic Theory* 4, 442–57.

Negishi, T. 1961. On the formation of prices. *International Economic Review* 2, 122–6.

Ohyama, M. 1972. On the stability of generalized Metzlerian systems. *Review of Economic Studies* 39, 193–204.

Rader, T. 1972. General stability theory with complementary factors. *Journal of Economic Theory* 4, 372–80.

Uzawa, H. 1960. Walras' tâtonnement in the theory of exchange. *Review of Economic Studies* 27, 182–94.

Grotius (de Groot), Hugo

Grotius (de Groot), Hugo (1583–1645). Legal theorist, philosopher and theologian, Grotius was born in Delft on 10 April 1583 and died in Rostock on 28 August 1645. An infant prodigy, Grotius entered the University of Leyden at the age of eleven, and at fifteen was hailed by Henry IV of France as 'the miracle of Holland'. Deciding on a legal career, he had become Advocate General of Holland, Zealand and West Friesland by the age of 24. In this period he wrote a treatise on the Law of Prize, of which the part dealing with freedom of the seas (*Mare Liberum*) was published in 1609. Because of his support for the moderate Arminians against the Calvinists, he was in 1619 imprisoned in Loevestein castle, and while there wrote an introduction to the law of Holland (*Inleidinge tot de Hollandsche Rechtsgeleertheyd*, published in 1631) and a tract on the truth of the Christian religion, the first of many theological writings. After two years, his wife arranged his escape in a chest ostensibly holding books, and thereafter he lived mainly in France, where he served for ten years as ambassador of Sweden.

Grotius' greatest work is *De iure belli ac pacis* (On the law of war and peace), published in 1625 and widely translated (six editions appeared in English before 1750). Written during the upheavals of the Thirty Years War, it laid down certain fundamental principles of law which purported to have the certainty of mathematics and absolute validity in all times and in all places. These principles both provided a standard for measuring the validity of the positive law of any state and also formed the basis for governing the relations between states. The work had enormous influence on the ethical and legal thought of the 17th and 18th centuries and is regarded as the beginning of 'the law of nature and of nations', the forerunner of modern international law.

Grotius built on the learning of late scholastic writers on natural law, such as Suarez, but he tried to make it independent of theological doctrine, so that amid the factionalism of the Reformation its principles would be unaffected by conflicting religious views. For him these principles could be proved in two ways, *a priori*, by logically deducing them from the rational and social nature shared by all mankind, and *a posteriori*, by showing that they were generally accepted by the consensus of writers – at least in more civilized nations – through the ages. For when many writers 'at different times and different places affirm the same thing as certain, that ought to be referred to a universal cause', which must be either correct conclusion drawn from the principles of nature or common consent (Prolegomena, sec. 40). Grotius concentrated on the latter approach, and dealt particularly with property and contract, the area of law of most concern to market societies and to nation states dealing with each other at arm's length.

Relying on the Bible narrative and on accounts of American Indians, he envisaged a primitive state of nature in which everything was held in common. When primitive simplicity gave way to specialization in agriculture and cattle raising, the conflicts that arose led first to division of lands among nations and then to division among families; thus community of property was replaced by private property. 'This happened not by a mere act of will ... but rather by a kind of agreement, either expressed, as by a division, or implied, as by occupation' (*De iure belli* II.2,21–5).

His doctrine of contracts was loosely based on Aristotle. He tolerated monopolies but condemned combinations to raise prices or to prevent the movement of goods by fraud or force. Although the law of nature did not forbid usury, divine positive law forbade it for Christians. However, Grotius adopted the canonist distinction between usury, which was forbidden, and receiving interest, which was permissible, if the rate was reasonable as in the positive law of Holland.

P.G. STEIN

BIBLIOGRAPHY

The most convenient modern edition of *De iure belli ac pacis* is in Classics of International Law Series, Oxford, 1925 (Vol. 1: the Latin text of 1646; Vol. 2: English translation by F.W. Kelsey). General surveys are in W.S.M. Knight, *Life and Works of Hugo Grotius*, London, 1925, and E. Dumbauld, *Life and Writings of Hugo Grotius*, Norman, Oklahoma, 1969. Full bibliographies are in the annual volumes of *Grotiana* (New Series), Assen, Netherlands, 1980 onwards. See also P. Haggenmacher, *Grotius et la doctrine de la guerre juste*, Paris, 1983.

group decisions. *See* INTERDEPENDENT PREFERENCES; SOCIAL CHOICE; VOTING.

group (Lie group) theory. Although it was nearly a century ago that a Norwegian mathematician by the name of Sophus Lie developed his theory of transformation groups, economists have only recently discovered that his group theory can be productively applied to such areas of economic inquiry as the theory of technical change, the theory of duality, dynamic symmetries, economic conservation laws and the theory of invariant index numbers, to name a few (see, e.g., Sato, 1981). The main feature of Lie's work on transformation groups (see Lie, 1888–1893 and 1891; Lie and Scheffers, 1893) is the study of the relationship between groups and differential equations. A survey of this particular aspect of Lie's theory is contained in the Appendix to Sato (1981).

The adoption of Lie group theory follows a long-standing tradition where economists have adapted such mathematical tools as calculus, matrix algebra, set theory, topology, probability theory, optimal control theory, game theory, etc. to their study of economic behaviour. Of course, with such a large inventory of mathematical tools, the question naturally arises: Why add group theory to the tool-kit? The answer to that question is that the application of Lie group theory is the most powerful and most systematic way of analysing 'invariant' relationships among economic variables, where often the relationships are represented by (partial) differential equation systems.

LIE GROUP CONCEPTION OF TECHNICAL CHANGE. To illustrate what is meant by a Lie group, consider a situation where the technical progress taking place in some production process is *a priori* known to have the simple 'neutral', or uniform factor augmenting form:

$$T_t: \bar{K} = e^{\alpha t}K, \qquad \bar{L} = e^{\alpha t}L,$$

where K represents capital, L represents labour, $\alpha(\alpha \geqslant 0)$ depicts the rate of technical progress, $\bar{K}$ represents 'effective' capital, $\bar{L}$ represents 'effective' labour, and t serves as the index of technical change. The equations characterizing $\bar{K}$ and $\bar{L}$ may be called the technical progress functions for capital and labour respectively.

Let the index of technical progress change from t_0 to t_1. Then, effective capital, $\bar{K}$, and effective labour, $\bar{L}$, is transformed from

$$T_{t_0}: \bar{K}_0 = e^{\alpha t_0}K, \qquad \bar{L}_0 = e^{\alpha t_0}L,$$

to

$$T_{t_1}: \bar{K}_1 = e^{\alpha t_1}K, \qquad \bar{L}_1 = e^{\alpha t_1}L.$$

The transformation of $\bar{K}$ and $\bar{L}$ satisfy the following conditions: (i) (*Composition*) The result of the successive performance of T_{t_0} and T_{t_1} is the same as that of the single transformation

$$T_{t_2}: \bar{K}_2 = \exp(\alpha(t_0 + t_1))K, \qquad \bar{L}_2 = \exp(\alpha(t_0 + t_1))L.$$

(ii) (*Identity*) When there is no technical change, i.e. $t = 0$, then $\bar{K} = K$ and $\bar{L} = L$. (iii) (*Inverse*) The inverse functions of T_t are also members of T when t is replaced by $-t$, i.e.

$$T_t^{-1} = T_{-t}: K = e^{-\alpha t}\bar{K}, \qquad L = e^{-\alpha t}\bar{L}.$$

Since the transformations governing $\bar{K}$ and $\bar{L}$ satisfy the above-mentioned conditions, the technical progress functions for $\bar{K}$ and $\bar{L}$ constitute a Lie group. More specifically, they constitute a one-parameter Lie group of continuous transformations. For a more formal definition of Lie group, refer to the Appendix of Sato (1981).

HOMOTHETIC PRODUCTION FUNCTIONS AND HICKS NEUTRAL TECHNICAL CHANGE. Suppose the production function characterizing the production process mentioned in the previous section was a homothetic one:

$$Y = F[f(K, L)], \tag{1}$$

where Y = output, K = capital, L = labour, f is a continuously differentiable function homogeneous of degree one with respect to K and L, with $f_K, f_L > 0$ and $f_{KK}, f_{LL} < 0$, and F any strictly monotone increasing (or homothetic) function of f. Under the uniform factor augmenting type of technical progress mentioned in the previous section, the impact of technical progress can always be represented by another member of the class of homothetic production functions:

$$\bar{Y} = F[f(\bar{K}, \bar{L})] = F[f(e^{\alpha t}K, e^{\alpha t}L)]$$
$$= F[e^{\alpha t}f(K, L)] = G_{(t)}[f(K, L)]. \tag{2}$$

The impact of this type of technical change on the technology of production is 'neutral' in the sense of Hicks. After technical change has occurred, the underlying isoquant map is *invariant*, with the exception of a change in the output levels associated with each isoquant.

Since the homotheticity property of production functions is associated with the notion of scale effects in production, the result mentioned above is disconcerting from an empirical perspective. The result implies that if time series data is used to estimate the (homothetic) form of the production function, or the rate of Hicks neutral technical change, one would not be able to distinguish between the effects of Hicks neutral technical change and returns to scale.

HOLOTHETICITY OF A TECHNOLOGY. Consider the general technical progress functions:

$$T_t: \bar{K} = \Phi(K, L, t), \qquad \bar{L} = \Psi(K, L, t). \tag{3}$$

Assume that the technical progress functions Φ and Ψ satisfy the conditions of Lie transformation groups. When technical progress occurs, it will in general affect the manner in which nominal K and L are combined in production. This in turn results in an efficiency improvement leading to effective $\bar{K}$ and effective $\bar{L}$. While there is no compelling reason for technical change to affect technology in such a way as *not* to alter the underlying isoquant map, let us confine ourselves only to the study of 'isoquant invariant' technical change.

For this kind of analysis, we require an appropriate definition.

Definition: (Holotheticity). When the complete effect of technical progress T, working through the technical progress functions Φ and Ψ within a production function $f(K, L)$, is represented by some strictly monotone transformation F, then the production function is said to be 'holothetic' (complete-transformation type) under a given T, i.e.:

$$\bar{Y} = f(K, L, t) = f(\Phi, \Psi, 0) = f(\Phi, \Psi) = f(\bar{K}, \bar{L})$$
$$= f[\Phi(K, L, t), \Psi(K, L, t)] = g[h(K, L), t]$$
$$= F_{(t)}[f(K, L)] = F_{(t)}(Y), \tag{4}$$

where $h(K, L) = f(K, L, 0) = f(K, L)$.

The holotheticity condition (4) can be thought of as the condition for 'generalized' Hicks neutrality, since it encompasses all technical change (of the Lie group type), working through some production function form (technology) which would leave the underlying isoquant map invariant. Thus, we can alternatively think of the holotheticity condition as the condition for 'isoquant invariant' technical change. The impact of this kind of technical change on the underlying production function is completely of a scale effect nature (homotheticity).

One practical value of this kind of analytical framework is that in doing empirical estimation, the researcher can determine if the hypothesized production function is holothetic under the hypothesized technical progress functions. If it is, then there is no way scale effects can be separated from the effects of technical change. If it is *not*, then scale effects can in principle be separated from the effects of technical progress. The holotheticity concept can be extended to r-parameter transformations via the concept of 'G(group) neutrality'.

INFINITESIMAL OPERATOR (LIE DIFFERENTIATION). The study of production function forms (technology) which are holothetic under specific type(s) of technical progress functions is facilitated by the use of the concept of the *infinitesimal* transformation (Lie differentiation) of the technical progress functions which satisfy the Lie group properties. The infinitesimal transformations of the general technical progress functions Φ and Ψ are defined as:

$$\left.\begin{array}{l} \left(\dfrac{\partial \bar{K}}{\partial t}\right)_{t=0} = \left(\dfrac{\partial \Phi}{\partial t}\right)_{t=0} = \xi(K, L) \\[3mm] \left(\dfrac{\partial \bar{L}}{\partial t}\right)_{t=0} = \left(\dfrac{\partial \Psi}{\partial t}\right)_{t=0} = \eta(K, L) \end{array}\right\} \qquad (5)$$

which are components of the infinitesimal operator U:

$$U = \xi(K, L)\frac{\partial}{\partial K} + \eta(K, L)\frac{\partial}{\partial L}. \qquad (6)$$

(For a more extended discussion of *Lie derivative*, see Lovelock and Ruud, 1975.) By making use of (6) we can express the holotheticity condition (4) as:

$$\left(\frac{\partial Y}{\partial t}\right)_{t=0} = Uf = \xi(K, L)\frac{\partial f}{\partial K} + \eta(K, L)\frac{\partial f}{\partial L} = G(f). \qquad (4')$$

Equation (4') points out that the holotheticity condition (4) is equivalent to the *invariance* condition of Lie group transformations.

The expression for the infinitesimal operator stated as holotheticity condition (4') can be given the following economic interpretation:

$$\begin{pmatrix} \text{The measure of} \\ \text{technical change} \end{pmatrix} = \begin{pmatrix} \text{infinitesimal} \\ \text{transformation} \\ \text{of capital} \end{pmatrix} \times \begin{pmatrix} \text{marginal product} \\ \text{of capital} \end{pmatrix}$$

$$+ \begin{pmatrix} \text{infinitesimal} \\ \text{transformation} \\ \text{of labour} \end{pmatrix} \times \begin{pmatrix} \text{marginal} \\ \text{product} \\ \text{of labour} \end{pmatrix} = \begin{pmatrix} \text{some transformation} \\ \text{of the production} \\ \text{function itself} \end{pmatrix}$$

OTHER APPLICATIONS OF LIE GROUP THEORY. In the above discussion, it is demonstrated that Hicks neutral technical change (isoquant invariant) can be 'generalized' to the notion of holothetic technology.

Other areas of economic analysis that can benefit from the application of Lie group theory are comparative statics, implicit economic functions and the duality of consumer preferences and producer technologies. The fundamental theorem of comparative statics can be expressed in terms of the infinitesimal operator. It can be demonstrated that the integrability conditions of demand and utility analysis are simultaneous invariance conditions of Lie groups. Implicit

economic functions and the duality and self-duality of economic functions are also areas amenable to Lie group application. For example, Lie group theory helps one to formulate exact conditions for duality and self-duality.

A theorem by Noether (1918) is most useful in the study of the invariance properties of optimal dynamic economic models, such as the Ramsey and general von Neumann types. Noether's theorem implies that if the fundamental integral of the calculus of variations is *invariant* under the r-parameter (Lie) group of transformations, then there are r conservation laws.

The theory of index numbers will also benefit from the application of Lie group theory. The Fisher–Frisch criteria of index numbers, as well as Divisia index analysis, can be represented in terms of Lie group transformations. Noether's theorem can also be applied in this area of analysis. Unquestionably, there will be many other areas of economic analysis that will benefit from the application of Lie group theory. Those readers who would like to explore the possibilities are referred to Sato (1981).

RYUZO SATO

See also CES PRODUCTION FUNCTION; MEANINGFULNESS AND INVARIANCE; MEASUREMENT, THEORY OF; TRANSFORMATIONS AND INVARIANCE.

BIBLIOGRAPHY
Lie, S. 1888–93. *Theorie der Transformationsgruppen.* Ed. F. Engel, 3 vols, Leipzig: Teubner.
Lie, S. 1891. *Vorlesungen über Differentialgleichungen, mit bekannten infinitesimalen Transformationen.* Ed. G. Scheffers, Leipzig: Teubner.
Lie, S. and Scheffers, G. 1893. *Vorlesungen über continuierliche Gruppen mit geometrischen und anderen Anwendungen.* Leipzig: Trubner.
Lovelock, D. and Ruud, H. 1975. *Tensors, Differential Forms, and Variational Principles.* New York: Wiley.
Noether, E. 1918. Invariante variationsprobleme. *Nachrichten Akademie Wissenschaft Gottingen, Mathematisch-Physischen Kl.II.* Trans. M.A. Tavel as 'Invariant variational problems', *Transport Theory and Statistical Physics* 1(3).
Sato, R. 1975. The impact of technical change on the holotheticity of production functions. Paper presented at the World Congress of the Econometric Society, Toronto. Published as Sato (1980).
Sato, R. 1980. The impact of technical progress on the holotheticity of production functions. *Review of Economic Studies* 47, 767–76.
Sato, R. 1981. *Theory of Technical Change and Economic Invariance: Application of Lie Groups.* New York: Academic Press.

Groves–Ledyard mechanism. *See* INCENTIVE COMPATIBILITY.

growth, immiserizing. *See* IMMISERIZING GROWTH.

growth, measurement of. *See* MEASUREMENT OF ECONOMIC GROWTH.

growth accounting. Growth accounting refers to allocation of growth rates of national output or output per person employed among the determinants of output that changed and caused growth; to allocation of international differences in output levels among the determinants responsible for level differences; to estimation of the size of the effect upon output of a given change in each output determinant; and to related estimates. It is a way to organize quantitative information in a convenient and systematic way. Growth accounting studies usually measure output net rather than gross of capital consumption because, insofar as a large output is a proper goal of society, it is net product that measures the degree of success in achieving this goal. There is no reason to wish to maximize the quantity of capital goods used up in production.

Growth accounting stems from an investigation by Edward F. Denison of the sources of growth in the United States from 1909 to 1958, of the probable future growth rate (obtained by adding the expected contributions of these sources), and of the amounts by which the future growth rate could be altered by each of a 'menu' of possibilities (Denison, 1962a, 1962b; Stein and Denison, 1960). Among the output determinants examined were characteristics of labour that affect its knowledge, skill, and energy. This met the criticism levelled against then-conventional economics by T.W. Schultz: ' ... treating a count [of employed persons] as a measure of the quantity of an economic factor is no more meaningful than it would be to count the number of all manner of machines to determine their economic importance either as a stock of capital or as a flow of productive services' (Schultz, 1961, p. 3).

Denison found that the largest positive sources of growth were increases in employment, in education held by employed persons, in capital stock, and in size of markets; improved resource allocation; and advances in the state of knowledge relevant to production. The study's most important lesson was that large and costly changes would be required if deliberate policy was to raise the high-employment growth rate appreciably above what it would be otherwise. 'The tale of a kingdom lost for want of a nail appears in poetry, not economic history'. This finding contrasted with a common view that it would be easy to add whole percentage points to the growth rate.

In *Why Growth Rates Differ,* Denison assisted by Poullier (1967) estimated the sources of differences among the postwar growth rates of the United States, Belgium, Denmark, France, West Germany, Italy, the Netherlands, Norway and the United Kingdom, and sources of differences among the 1960 levels of output per person employed in these countries, then explored the relationship between growth and level of output. Walters (1968, 1970) provided a companion study for Canada. Denison found that the growth rate of the United States was below rates in the continental countries because of differences in opportunities for growth (including, notably, opportunities for various kinds of 'catching up' such as elimination of inefficient farms and small non-farm enterprises), not of better policy or greater efforts to achieve growth. Contrary to frequent assertions, investment in physical capital was found not to be a uniquely important determinant of differences in growth rates but only one of several important determinants.

Similar investigations of postwar growth cover Japan (Kanamori, 1972; Denison and Chung, 1976), India (Dholakia, 1974), and South Korea (Kim and Park, 1985). Denison and Chung found that the very high 1953–71 Japanese growth rate was not ascribable to any single determinant. Rather, with one minor exception Japan received a greater contribution from all of five major categories of growth sources – the increase in labour, the increase in capital, the reallocation of labour from agriculture and self-employment, the incorporation of improved technology into production, and economies of scale – than did any of ten Western countries with which it was compared. Excellent studies of very long term growth in Japan, France, and the United Kingdom conducted for the Social Science Research Council include full or partial growth accounting estimates (Ohkawa and Rosovsky, 1973; Carré, Dubois and Malinvaud, 1975; Matthews, Feinstein and Odling-Smee, 1982).

In *Accounting for United States Economic Growth* (1974),

Denison for the first time introduced separate estimates for non-residential business and three other sectors, and extended growth accounting to include potential output. Annual estimates of potential output were derived from actual output by adjusting each output determinant to its value under specified potential economic conditions and then adjusting output for the difference between the actual and potential values of all determinants. The procedure yields annual estimates of the sources of the difference between actual and potential output, and of the sources of growth of potential output. Changes in actual output and each determinant can be divided between changes in the economy's capacity to produce and changes in the utilization of that capacity.

All the studies cited used sources-of-growth estimates to explain changes in growth rates from period to period. Growth accounting investigations conducted after 1974, when growth of productivity declined abruptly, tended to focus on this slowdown (e.g. Denison, 1979, 1985; Kendrick, 1979). These studies identified and quantified the contributions of many determinants that contributed to the slowdown. Equally important, they rejected many 'explanations' that had been advanced as possibilities. However, much of the slowdown could not be definitively assigned to particular determinants.

Growth accounting starts by recognizing that many different determinants govern the size of a country's output at any given time. It deals in the first instance with direct determinants of output such as the number, hours, demographic composition, and education of employed persons; quantities of land and capital; the stock of knowledge; the size of markets; the extent to which actual practice departs from lowest cost practice; the amount by which resource allocation departs from the output – maximizing allocation; and the intensity with which factor inputs are used. Changes in these determinants cause changes in output, or growth. Sources-of-growth tables are obtained by measuring changes in each such determinant and the effect that this change had upon output. Each determinant's contribution is the amount by which the growth rate of output would have been reduced (or increased, when the contribution is negative) if that determinant had not changed while other determinants had changed as they actually did. Its size depends upon the determinant's importance and how much it changes.

Direct determinants of output are, of course, influenced by a host of indirect determinants such as tax structure, attitudes toward work, inflation, deaths in war, or birth control information. Growth accounting studies do not ignore such indirect determinants of output, but recognize that their effects can best be appraised by first judging the amounts by which a change in any of them (or a difference between two situations, such as two tax structures) alters all of the direct determinants it influences and then calculating the effect of these alterations upon output.

Four characteristics of a desirable classification of direct growth sources have been suggested (Denison, 1972). (1) It should identify effect with cause (e.g. the effect on productivity of Edison's inventions should appear as a contribution of advances in knowledge – not of capital, which should be related to saving). (2) Categories should be those that economists are accustomed to use and which facilitate application of their tools. (3) The classification must have the characteristics that an output determinant that does not change between two dates contributes nothing to growth and that the contribution of every determinant is measured in comparison with a no-change situation. (4) A practical classification must take into account the possibilities of estimation.

Various techniques are used to estimate the contributions of determinants, and their reliability varies. Denison (1972) distinguishes five types. The following paragraphs provide the percentages of the 1929–82 growth rate of the total actual national income of the United States that represents contributions estimated in each way, calculated from Denison (1985). Because contributions can be negative as well as positive, net contributions are only an imperfect guide to importance, which in any case varies among periods and places.

(1) The contributions of residential capital, international assets, and labour employed by general government, households, and institutions can – and, to avoid inconsistencies, must – be obtained directly from the details of the output estimates themselves. They contributed 26 per cent of the specified growth rate.

(2) The contributions made to the growth rate of national income in non-residential business by changes in labour, capital, and land input in that sector, and by reduction of overallocation of labour to farms and to non-farm self-employment, are calculated by a common technique. It derives from the principle of proportionality: if economic units combine factors of production in such a way as to minimize their costs – as they have every incentive to seek to do and as competition tends to compel – then the marginal products of labour, land, fixed capital, and inventories must be proportional to their earnings (Denison, 1967, 1974; Daly, 1972). Unless this condition is satisfied, enterprises could reduce costs by substituting one factor for another. Departures from this situation are assumed to be small or offsetting.

It follows from the principle of proportionality that if an increase of, say, one per cent in the quantity of all inputs in non-residential business would raise the sector's output by x per cent, then an increase of one per cent in the quantity of only one input (such as all types of labour) would raise sector output by a percentage of x equal to that factor's share of non-residential business national income. The proposition holds not only for total labour, capital and land but also for detailed categories of inputs. Suppose that, on the average, individuals with 12 years of education earn 22 per cent more than otherwise similar individuals with 8 years of education. It is inferred that the marginal products of the first group are 22 per cent higher – and that, on the average, an additional worker with 12 years of education adds 22 per cent more to output than an otherwise similar additional worker with 8 years of education. As a matter of classification, contributions of all determinants except economies of scale are calculated as if a one per cent increase in all inputs in non-residential business raises sector output by one per cent, even though it would actually raise sector output by a larger percentage. The difference is a contribution of economies of scale.

The contribution of an input to the non-residential business sector growth rate is the product of the growth rate of the input and its income share, plus a trifling interaction term. Results are appropriately reduced to obtain contributions of the sector's inputs to the growth rate of output in the economy as a whole. To obtain series for inputs inevitably requires judgements for which information is inadequate – for example, as to the effect of shortening work hours upon output per hour. To enrich the results, contributions of employment and labour characteristics such as hours, age–sex composition, and education, are measured separately, as are contributions of fixed capital and inventories.

Estimates based on income shares and, in some cases, earnings differentials accounted for 47 per cent of the specified growth rate.

(3) An investigator often concludes that an output determinant did not change in a particular period. He can then put its contribution at zero even though he would have no method to calculate the contribution if the determinant had changed. Obviously, none of the specified growth rate stems from such determinants.

(4) Special procedures, ranging from satisfactory to inadequate, are developed for other determinants – such as the diversion of resources to satisfy environmental regulation, economies of scale, and the effect of fluctuations in demand upon output per unit of input. Determinants in this category contributed 4 per cent of the specified growth rate, but this results from netting minus 5 per cent against plus 9 per cent.

(5) No way has been found to estimate directly the contributions of advances in knowledge or certain miscellaneous determinants, so all growth accounting tables contain a residual. It contributed 23 per cent of the specified growth rate. In Denison's opinion the net contribution of the miscellaneous determinants probably was small in the United States in 1948–73 so the residual provided a tolerable estimate of the contribution of advances in knowledge – but this was not so thereafter nor in some other periods and places.

The heavy reliance of growth accounting studies on the second method has caused all investigations that use observed income shares and earnings differentials to be described as using 'growth accounting techniques', even though both the use of income shares to combine labour, capital and land and the use of earnings differentials to equate types of labour (Strumilin, 1925) antedate growth accounting. The intent is to distinguish these investigations from studies that estimate elasticities of output with respect to individual inputs by correlation analysis, a method those using growth accounting techniques consider inferior (Denison, 1984a, pp. 15–17). Studies continue to be made that calculate factor input by use of share weights but do not attempt a fuller breakdown of sources of growth. The Bureau of Labor Statistics (1983) introduced a continuing series that does not take account of changes in the composition of labour while others do (e.g. Christensen, Cummings and Jorgenson, 1980).

Most criticisms of growth accounting have referred to the estimated contribution of an individual output determinant. When the criticism is based on new information or analysis and is found persuasive, the original estimate is changed. For example, Denison changed his estimates for the education of employed persons when discussants persuaded him that an adjustment of data based on highest grade completed to allow for changes in school days per year should be omitted for persons who reached college and reduced for others (Denison, 1967, pp. 380–83); he modified his estimates of the reduction in output per unit of input that was caused by safety legislation for coal mining in response to comments by mining experts (Denison, 1985, pp. 66–8). Growth accounting is open to such piece-by-piece improvement because it can accommodate estimates prepared by any method. Frequently, however, assertions that particular estimates are too small reflect differences in classification rather than substantive disagreement. Proponents of the dominance of knowledge, capital, education, resource reallocation and market size all tend to claim the same contributions to growth, usually without providing a complete classification of growth sources that would make their differences apparent (Denison, 1972).

Nelson has a broader commentary, in its latest version not so much a criticism of growth accounting as a stressing of its (and orthodox theory's) limitations (Nelson, 1984, pp. 403–09). Two may be mentioned. First, as Nelson sees it, time series for output and input are presumed to have been generated by a moving competitive equilibrium, and while the analyst on occasion can get outside of that framework and consider certain disequilibrium aspects and market imperfections, 'there is not much room for serious analysis' of their effects. Second, interaction among determinants is important and not revealed by growth accounting. Nelson's first point, however, is inaccurate: any change in a disequilibrium condition or market imperfection whose effect on growth can be quantified by any technique available is easily incorporated into the estimates, and growth account studies have attempted more such quantifications than other branches of economics. As to the second, economic interaction is a fact of life that poses no more difficulty for growth accounting than for any theory that relates cause to effect by assuming 'other conditions' to be unchanged. Because growth accounting requires a consistent non-overlapping classification, it has had to pay especially careful attention to the ways in which determinants interact. To eschew analysing individual output determinants would be to abandon quantitative analysis of growth and growth policy.

E. DENISON

See also MEASUREMENT PROBLEMS; PRODUCTION FUNCTIONS; PRODUCTIVITY; TOTAL FACTOR PRODUCTIVITY.

BIBLIOGRAPHY

Bureau of Labor Statistics. 1983. *Trends in Multifactor Productivity, 1948–81*. Bulletin 2178. Washington, DC: US Government Printing Office.

Carré, J., Dubois, P. and Malinvaud, E. 1975. *French Economic Growth*. Stanford: Stanford University Press.

Christensen, L., Cummings, D. and Jorgenson, D. 1980. Economic growth, 1947-73; an international comparison. With 'Comment' by D. Daly and 'Reply'. In *New Developments in Productivity Measurement and Analysis*, ed. J. Kendrick and B. Vaccara, Studies in Income and Wealth, Vol. 44, Chicago: University of Chicago Press, 595–698.

Daly, D. 1972. Combining inputs to secure a measure of total factor input. *Review of Income and Wealth*, March, 27–53.

Denison, E. 1962a. *The Sources of Economic Growth in the United States and the Alternatives Before Us*. New York: Committee for Economic Development.

Denison, E. 1962b. How to raise the high-employment growth rate by one percentage point. *American Economic Review, Papers and Proceedings*, May, 67–75.

Denison, E. 1964. Measuring the contribution of education (and the residual) to economic growth. With comments by F. Edding, E. Malinvaud, E. Lundberg and J. Sandee, and reply. In Study Group in the Economics of Education, *The Residual Factor and Economic Growth*, Paris: Organization for Economic Co-operation and Development, 11–100.

Denison, E. (Assisted by J. Poullier.) 1967. *Why Growth Rates Differ: Postwar Experience in Nine Western Countries*. Washington, DC: Brookings.

Denison, E. 1972. Classification of sources of growth. *Review of Income and Wealth*, March, 1–25.

Denison, E. 1974. *Accounting for United States Economic Growth, 1929–1969*. Washington, DC: Brookings.

Denison, E. 1979. *Accounting for Slower Economic Growth: The United States in the 1970s*. Washington, DC: Brookings.

Denison. E. 1984a. Accounting for slower economic growth: an update. In *International Comparisons of Productivity and Causes of the Slowdown*, ed. J. Kendrick, Cambridge, Mass.: Ballinger Publishing Co., 1-45.

Denison, E. 1984b. Productivity analysis through growth accounting. In *Productivity Research in the Behavioural and Social Sciences*, ed. A. Brief, New York: Prager Publishers, 7–55.

Denison, E. 1985. *Trends in American Economic Growth, 1929–1982*. Washington, DC: Brookings.

Denison, E. and Chung, W. 1976. *How Japan's Economy Grew So Fast: The Sources of Postwar Expansion*. Washington, DC: Brookings.

Dholakia, B. 1974. *The Sources of Economic Growth in India*. Baroda, India: Good Companions.

Gollop, M. and Jorgenson, D. 1980. US productivity growth by industry, 1947–73. With 'Comment' by E. Berndt. In *New Developments in Productivity Measurement and Analysis*, ed. J. Kendrick and B. Vaccara, Studies in Income and Wealth, Vol. 44, Chicago: University of Chicago Press, 17–136.

Jorgenson, D., Griliches, Z. and Denison, E. 1972. The measurement of productivity [an exchange of views]. *Survey of Current Business*, May, Pt II. (Includes reprinted material with corrections.)

Kanamori, H. 1972. What accounts for Japan's high rate of growth. *Review of Income and Wealth*, June, 155–171.

Kendrick, J. 1961. *Productivity Trends in the United States*. Princeton: Princeton University Press.

Kendrick, J. 1979. Productivity trends and the recent slowdown: historical perspective, causal factors, and policy options. In W. Fellner, (Project Director), *Contemporary Economic Problems*, Washington, DC: American Enterprise Institute.

Kim, K. and Park, J. 1985. *Sources of Economic Growth in Korea: 1963–1982*. Seoul: Korea Development Institute.

Matthews, R., Feinstein, C. and Odling-Smee, J. 1982. *British Economic Growth 1856–1973*. Stanford: Stanford University Press.

Nelson, R. 1984. Where are we in the discussion? Retrospect and prospect. In *International Comparisons of Productivity and Causes of the Slowdown*, ed. J. Kendrick, Cambridge, Mass.: Ballinger Publishing Co.

Ohkawa, K. and Rosovsky, H. 1973. *Japanese Economic Growth: Trend Acceleration in the Twentieth Century*. Stanford: Stanford University Press.

Román, Z. 1982. *Productivity and Economic Growth*. Budapest: Akadémiai Kiadó.

Schultz, T. 1961. Investment in human capital. *American Economic Review* 51, March, 1–17.

Stein, H. and Denison, E. 1960. High employment and growth in the American economy. In President's Commission on National Goals, *Goals for Americans*, New York: Prentice-Hall.

Strumilin, S. 1925. The economic significance of national education. Trans. B. Jeffrey, in *The Economics of Education*, ed. E. Robinson and J. Vaizey, London: Macmillan, 1966.

Walters, D. 1968. *Canadian Income Levels and Growth: An International Perspective*. Economic Council of Canada, Ottawa: Queen's Printer.

Walters, D. 1970. *Canadian Growth Revisited, 1950–1967*. Economic Council of Canada, Ottawa: Queen's Printer.

growth and cycles. Cycles can exist without growth and, conversely, growth can exist without cycles. The special problem posed by an economy is that they are so closely interconnected that it is quite improper to analyse one without the other. Thus both the upper turning point and the lower turning point in economic cycles are normally higher each time, signalling both growth and an altered productive structure. In dealing with fluctuations it is not too difficult to explain the upper turning point but the lower one has eluded any simple, compelling explanation: there is no systematic reason for the renewal of investment, and hence demand and output, when there is general excess capacity. On the other hand there are good simple theories of steady-state growth, but, being by their very nature dynamically unstable, such a system can never be observed on such a path, since any deviation from it will increase with time.

Perceptive analysis begins with Marx, who, taking a Darwinian view of society and the economy, investigated the reasons for change and growth, and reached the conclusion that they could only proceed irregularly, a conclusion amply confirmed by the statistics of economic history. The first step is to acknowledge that for at least two centuries industrialized economies have experienced involuntary unemployment more often than not. This fact, which any usable economic theory must explain, has to be emphasized since, strangely enough, the dominant economic paradigm, competitive laissez faire, does not do so. Deriving from Adam Smith's benediction of capitalism with the remarkable proposition that uncontrolled pursuit of private gain would lead to the general good, economists have tended to embrace a theory which assumed that competition would clear all markets in an optimal way. Walras convinced himself and others that this was indeed the case. This involved the demonstrably false proposition that not only markets for produced goods but also for unproduced goods, or 'factors of production', would all be cleared. Wicksell, more aware than most, stated explicitly that in his analysis he assumed full employment and hence elaborated an analysis of disequilibrium in terms of banking misbehaviour, the 'output' of banks being essentially different from other production. This type of theory became understandably popular since it enabled economists to reconcile a theory which denied persistent unemployment with the fact of its evident existence. It was only Keynes, thoroughly conversant with the monetary mechanism, who unmasked orthodoxy, by saying not only that unemployment must be explained as a part of the normal operation of the economy, but also that it could be, in some sense, an equilibrium position.

To understand this it is helpful to go back to Marx, who said capitalism constituted a new mode of production in which the essential decisions about both production and employment were made by the capitalists not the workers. It follows quite simply from this that employment is determined by producers in the context of market demand, not by individuals offering labour. It is therefore not surprising that Marx described unemployment, in the form of the Industrial Reserve Army, as endemic in capitalism. Say's Law in its original formulation offers no salvation, since it only yields a neutral equilibrium of aggregate demand and output at any level, including any degree of unemployment.

To see the nature and necessity of fluctuations in output, it is helpful to consider first the inventory cycle, which is both well substantiated statistically and easily explicable. The more important effects of durable goods cycles will then only reinforce the analysis and conclusions. Suppose the rate of production is determined by the aim of maintaining a desired level of stocks of finished goods, that is, an example of simple negative feedback error control. If stocks, s_t, are greater than desired stocks, s^*, output q, is reduced:

$$\dot{q} = -\beta(s_t - s^*), \qquad \beta > 0.$$

The rate of change of stocks is the difference between demand and output. Measuring output in deviations from the equilibrium level determined by exogenous demand, then for a linear savings function,

$$\dot{s} = q - \alpha q, \qquad 1 > \alpha > 0.$$

Hence

$$\ddot{q} = -\beta\dot{s} = -\beta(1 - \alpha)q,$$

which produces simple harmonic motion and implies oscillating unemployment, since employment can never be greater than the available work force.

A fuller analysis requires consideration of investment in durable goods, plant, equipment and construction. This complicates and enriches the analysis since investment in durable goods occurs only when there is growth and net growth can only occur when there has been a prior accumulation of durable goods. A basic difference between the behaviour of stocks of durables as compared with non-durables is that the latter can be reduced relatively rapidly, whereas durables, once produced, can only be reduced slowly after a much longer time. Consequently durable goods may be substantially increased during an upswing, but can only be reduced to a negligible extent during a contraction. For this reason it is just as necessary to consider growth in explaining the cycle, as it is to include the cycle in any explanation of the irregularity of growth.

When output increases above previous levels it becomes necessary to invest first in order to expand capacity. The demand resulting from the investment raises further the need for added capacity, so that the economy becomes unstable upward. The demand effect occurs rapidly whereas the rise in capacity is necessarily slower. Consequently, depending on the structural parameters, the expansion may come to an end endogenously, as the capacity finally catches up with output, thus ending the necessity for investment and precipitating a rapid decline in demand and output. If this endogenous effect does not arrest the boom, the constraint of existing resources, particularly available trained labour, will bring about a deceleration and consequent downturn. The decline will end most investment and the economy will be reduced to the level determined by exogenous expenditure and the resulting demand. The fact that capacity can only somewhat slowly be increased and cannot be decreased at all significantly, means that the major business cycle lasts considerably longer than the stocks cycle. Since the depression phase is characterized by excess capacity, there is no automatic revival of investment, even when the decline has ceased.

Therefore it is necessary to appeal to Schumpeter's conception of innovational investment, which is independent of the existence of excess capacity and, indeed is encouraged by the existing low level of profits resulting from excess capacity. The fact that real wages fall little if any in the depression, means that the new investment will be heavily oriented towards labour saving. In consequence of innovational technical progress each succeeding upswing can proceed beyond the preceding peak both because of increased labour productivity and any increased available labour supply. The upward instability of the economy means that each expansion is likely to carry the economy to a new higher level, thus yielding growth out of the cycle. Viewed in this light one sees that growth is likely to occur in spurts, in waves. The economy first grows too fast, exhausting the essential 'reserve army' of the unemployed: it then grows too slowly, or even declines, with falling investment leading to falling demand and output. Just as the upper turning point is normally higher than the previous one, so also is the lower turning point, by virtue of exogenous expenditure which is expanded by the long term growth. The depression generates excess capacity and regenerates the army of the unemployed, creating the necessary condition for rapid growth and the renewed pressure for innovative technology to restore profits, investment and growing demand. Hence it is the relentless search for profit which leads to the growth and the upward instability, driving the economy repeatedly back towards the ceiling of full employment. Because of unemployment, wage rates tend to rise less than productivity, thus giving rising profits both per unit of output and because of increased output. Rising profits provide both the incentive and the monetary means for further investments, in both new technology and accumulation of additions to existing capacity. Beginning with unemployment, the economy can expand at a rate which cannot be continued as unemployment tends towards zero. Gradually labour scarcities will bring rising real wages, moderating the growth rate. The durability of capital goods means that investment becomes highly sensitive to growth rates as well as levels of output. Since it takes significant time to expand capacity, and since, once expanded, it will last well into the future, the expected growth rate becomes of great importance. Any deceleration reduces the urge to invest and any reduction in investment reduces the profits available for investment.

With some loss of realism, a model of the system can be formulated, applying the same type of adjustment mechanism to capacity, k, as to stocks;

$$k = -\gamma(k - q).$$

In the same way let output adjust to demand:

$$\dot{q} = -\delta[q - (\alpha q + \varkappa k)]$$

where $\varkappa$ is the 'capital-output ratio', i.e. the amount of current flow of output necessary per unit of the stock of capacity. Since q is a 'fast' variable and k a 'slow' one, $1 > \delta > \gamma > 0$, yielding very different magnitudes of the distributed lags. The resulting system in matrix form is:

$$\begin{Bmatrix} \dot{q} \\ \dot{k} \end{Bmatrix} = \begin{Bmatrix} +\delta[\varkappa\gamma - (1-\alpha)] & -\delta\varkappa\gamma \\ +\gamma & -\gamma \end{Bmatrix} \begin{Bmatrix} q \\ k \end{Bmatrix}$$

Since the off-diagonal terms are of opposite sign, it must oscillate and since the sum of the diagonal terms is positive, it must do so with ever increasing amplitude. Given the fact that the economy cannot grow beyond full employment, the result is a limit cycle, thus explaining its persistence. This, combined with the reality of increased productivity and increased labour force over each cycle, means that successive peaks are higher. Furthermore the fact of exogenous innovational investment both removes the necessity for lengthy disinvestment in the depression and explains the successive higher peaks. Such a model is obviously only a skeleton, which has to be fleshed out with realistic detail.

The timing of actually realized increases in productivity is difficult to specify; most innovations, whether in processes or in new goods, usually require many improvements and adaptations to diverse uses. Though somewhat simplistic, perhaps the best assumption is a steady rate of growth of labour productivity, for which there is some statistical support. This is especially true for aggregative analysis, where productivity is the result of a very large number of independent events. There are, of course, great differences between different sectors of the economy. The more progressive sectors find it easy to grant wage increases and are vulnerable to pressure for such increases. These rising wage rates then exercise varying degrees of pressure to increase wages in the less progressive sectors, entailing rising costs and rates of inflation. The over-all result of this complex process is a homeodynamic system which determines a rising structure of real wage rates appropriate to the economy: appropriate in this context means appropriate given the degree of technological progress and labour supply growth; these vary markedly between economies. Industrialized economies have consistently generated wage and profit shares which vary over the cycle in such a way that they drive the economy towards full employment and maintain it, in the long run, in the general neighbourhood of full employment – a level which has shown large growth over the last two centuries. If wages remain too low, the economy will only reach full employment more rapidly with ensuing rapidly rising wages. If the wages are too high, there will be slow growth and little tendency to further increase in wages.

A growth path is inherently unstable, so that a slight deceleration leads to further deceleration and and a strong tendency to cessation of both innovational and accelerational investment. When this deceleration ceases, there may be a resumption of investment, depending on the technological situation. As Schumpeter pointed out long ago, the great innovations in energy, for example, steam, electricity, oil, required a long time to be fully integrated into the whole productive structure. Therefore such innovations are by no means completed in one fluctuation, so that their expansion and adaptation will be quickly and vigorously renewed after a decline. In this way two important aspects of fluctuating growth can be explained. Each succeeding wave will be different in magnitude and in duration. Furthermore in this way it is possible to explain the so-called long waves. Since the implementation of a major innovation may take up to nearly a century, there will be a long period of vigorous expansions and short, sharp depressions. In this sense formal theory cannot dispense with the accidents of economic history.

R.M. GOODWIN

See also ACCELERATION PRINCIPLE; AGGREGATE DEMAND AND SUPPLY ANALYSIS; HARROD–DOMAR MODEL; NEOCLASSICAL GROWTH MODELS; SADDLEPOINTS.

BIBLIOGRAPHY For a more complete listing, see *Nonlinear Models of Fluctuating Growth*, ed. R.M. Goodwin, M. Krüger and V. Vercelli, New York Springer, 1984.

Balducci, R., Candela, G. and Ricci, G. 1984. A generalization of R. Goodwin's model with rational behaviour of economic agents. In *Nonlinear Models*, ed. R.M. Goodwin et al., New York: Springer, 1984.

Desai, M. 1973. Growth cycles and inflation in a model of the class struggle. *Journal of Economic Theory* 6(6), December, 527–45.

Goodwin, R.M. 1955. A model of cyclical growth. In *The Business Cycle in the Post-War World*, ed. E. Lundberg, London: Macmillan.

Goodwin, R.M. 1967. A growth cycle. In R.M. Goodwin, *Essays in Economic Dynamics*, London: Macmillan, 1982, 165–70.

Kalecki, M. 1954. *Theory of Economic Dynamics*. London: Unwin, 1965.

Rose, H. 1967. On the non-linear theory of the employment cycle. *Review of Economic Studies* 34, 153–73.

Schumpeter, J.A. 1934. *The Theory of Economic Development*. New York: Oxford University Press, 1961.

growth and inflation. See INFLATION AND GROWTH.

growth and international trade. The relationship between international trade and economic development is at the root of the opposition to mercantilism by classical political economy and of some of the most important and long-lasting economic debates at least since the late 18th century. In spite of the importance of the subject at the level of economic doctrine and policy, the theory of international trade during the 19th and most of the present century has mainly developed in a static context while the development of growth theory, both in classical political economy and in its revival after World War II, took place to a large extent in a closed-economy framework. The theory of international trade and growth is, paradoxically, a relatively new field of enquiry, approached from a diversity of points of view and it is in constant and rapid evolution.

Classical economists did not develop the more formal and theoretical aspects of the relationship between growth and international trade. The analysis of the effects of trade on economic growth was confined mainly to the links that could be established in Ricardo's theory between the gains from trade according to comparative advantage specialization, the rate of profit and the rate of capital accumulation, wherein foreign trade played a role analogous to technical progress or, more precisely, to a once and for all improvement in the methods of production. The classical views on the effects of growth on international trade were mostly limited to the concern for the deterioration of the terms of trade between agriculture and manufacturing due to the presence of diminishing returns in agriculture (cf. Torrens, 1821), and to John Stuart Mill's analysis in his *Principles of Political Economy* of the effects of technical progress in the export industries.

The predominantly static nature of international trade theory was accentuated in neoclassical economics up to the 1950s. Neoclassical theory concentrated on the analysis of the effects of trade on the allocation of given resources and of the static gains arising from specialization according to comparative advantage, paying little attention to the long-term effects that this would have on the rate and the pattern of economic growth. An interesting exception is Edgeworth (1894), who, following John Stuart Mill, examined the conditions under which an increase in exports would have a negative effect on a country's national income. This analysis is an antecedent of modern 'immiserizing growth' models (see below).

The modern neoclassical theory of growth and trade originates in a now classic paper by Hicks (1953) focusing on the long-term development of the trade balance and the terms of trade in a two-country model, inspired by the experience of persistent deficit in Britain's balance of trade in the late 1940s and early 1950s. The analysis of the effects of growth on trade volumes, the terms of trade and national income became the main object of later developments taking into account the different possible sources of growth, changes in factor endowments and technical progress (cf. Johnson, 1958; Findlay and Grubert, 1959; Kemp, 1964; Södersten, 1964).

In spite of these theoretical developments, the analysis of the effect of international trade on the allocation of resources has continued to receive disproportionate attention in a static framework, and the normative conclusions of traditional comparative advantage have remained, to a large extent, unaffected. However, when all the complexities of real economies – differential rates of technical progress among industries, variable returns to scale and different income elasticities of demand for individual commodities – are considered, it can be argued that the analysis of the effects of trade and its normative implications may be considerably altered in a dynamic context.

Let us examine the interactions between trade and economic growth by considering, successively, the cases of small and large resource constrained countries and, later, the case of a balance of payments constrained economy. Particular attention will be given to the implications of the pattern of specialization induced by international trade on the growth path of the economy.

Take, first, the simplest case of a resource-constrained small economy, an economy facing exogenous terms of trade and no demand constraints, since imports and exports adjust respectively to the excess demand for importable goods and the excess supply for exportable goods at the given terms of trade. Output growth is determined by changes in factor supplies and exogeneous technical progress. When opened up to trade we shall assume, first, that the economy considered faces constant terms of trade through time. In this process, the economy will experience a reallocation of resources and approach a pattern of specialization biased towards those

industries possessing a comparative advantage in international trade. The additional production for exports in the expanding industries, resulting from the absorption of resources from the contracting industries, will be able to purchase, through trade, a larger quantity of those commodities that were previously produced under autarchy, due to the lower relative price of these commodities under free trade. In the initial period, real income will thus be larger, in general, under free trade than under autarchy.

If technical progress and returns to scale were uniform among industries, this static gain from trade will be the only effect on international trade. However, under more general and realistic assumptions, this effect may be outweighed in the long term if international trade and comparative advantage have led the economy to specialize in the technologically less progressive industries, this having the effect or retarding, by comparison with the autarchic situation, the overall rate of technical progress. Having started from an initially higher level, real incomes in the trading economy will, after a certain period, fall below the level that they could have attained in the autarchic economy. This phenomenon we may term dynamic or long-run losses from trade arising from the pattern of specialization adopted.

Similar outcomes can occur when the terms of trade are changing through time. Assume that, for the small open economy, the rate of change of the terms of trade reflects the difference between productivity growth rates in the production of the exported and imported commodities in the rest of the world. It can be shown then (Pasinetti, 1981, ch. X) that the pattern of trade which is in the long-term interest of the economy to adopt is to specialize in the production of those commodities for which the country can achieve the highest comparative rates of growth of productivity. This pattern of specialization, however, may or may not coincide with that induced by static comparative advantage arising from free trade.

When demand constraints are introduced, the best pattern of production and trade will depend also crucially on the price and income elasticities of domestic and foreign demand for the commodities on which the country specializes. Consider, now, a resource-constrained economy which is, however, large enough to affect the terms of trade. Output growth is, again, determined by changes in factor supplies and exogenous technical progress, but changes in the terms of trade are now the equilibrating mechanism in balancing the growth of demand and productive capacity. When opened up to trade the economy considered has a comparative advantage and specializes in the production and export of commodities with low income and price elasticities of demand, importing commodities with a high income elasticity of demand. Assume, also, that the economy has a relatively high rate of output growth due to demographic factors, leading to a rapid growth of exportable surpluses. Under these circumstances, the three sets of forces determining the evolution of the terms of trade – rates of growth, income and price elasticities – will combine to generate a downward trend in the relative price of exported commodities. It is possible, then, that when compared to autarchy or to alternative patterns of specialization, the economy will be suffering long-term or dynamic losses resulting from the pattern of specialization induced by free trade.

The example above is reminiscent of the export-led growth experience of the primary producing countries of Latin America during the first decades of the present century, an experience which gave rise, in the 1940s and 1950s, to the structuralist school of Latin American economists led by

Prebisch at the United Nations Economic Commission for Latin America. This school developed an original analysis of economic growth and international trade between industrial or 'central' countries and primary producing or 'peripheral' countries (cf. UN Economic Commission for Latin America, 1950; Prebisch, 1959). The international division of labour between industrial and primary producing countries, together with a number of differences concerning market conditions and the adaptability of their productive structures, were seen as the causes of an unequal diffusion of the gains from technical progress in the international economy. This led to an unfavourable long-term evolution of the terms of trade against the 'peripheral' countries. Thus, it was argued, that pattern of trade according to static comparative advantage had led primary producing countries to specialize in the production and exportation of commodities with a low price elasticity of demand, in highly competitive and slowly growing markets. Under these circumstances, economic growth based on primary exports was bound to produce a deterioration in the terms of trade, the 'peripheral' countries exporting to the 'centre' the productivity gains achieved in their exporting sectors, thus perpetuating their state of underdevelopment. It is on these grounds that the Latin American structuralist school became a strong advocate of the deliberate promotion of industrialization in underdeveloped countries through the protection of manufacturing industries.

An extreme case of the previous example occurs when the deterioration of the terms of trade is so intense that the expansion of domestic productive capacity leads to a reduction of national welfare because the loss due to deteriorating terms of trade outweighs the gain due to increased production. This is the case of 'immiserizing growth' (cf. Bhagwati, 1958), which has received considerable attention in modern literature. The phenomenon can occur under different sets of circumstances. Using a two-country model of complete specialization where each of the two economies produces only one good, all of which is exported, the condition for immiserizing growth for a country, when the other is stationary, turns out to be that foreign-demand elasticity is less than unity.

Effective demand and income elasticities become even more important when we turn to the case of balance of payments constrained economies. In contrast to the previous cases, economic expansion is now limited by the growth of effective demand which, in turn, is determined by a balance of payments constraint. The terms of trade do not play, in this view, an equilibrating role in international trade, but are rather internally determined by domestic market structures and income distribution. The dynamic balance between productive capacity and demand, as in several post-Keynesian growth models, is achieved then through the long-run adjustment of capacity to the expansion of demand, the presence of domestic or foreign labour reserves along with the long-term flexibility of capacity utilization and the propensity to save accounting for the non-limiting role of the factor endowments under normal circumstances. In a simple version, assuming a conventional import demand function (dependent only on domestic income and relative prices), a balanced trade account and constant terms of trade through time, this approach can be summarized in the formula determining the rate of output growth by the growth rate of exports divided by the income elasticity of the demand for imports. Now the presence of dynamic gains or losses from trade will depend crucially on the income elasticities of foreign and domestic demand of the commodities on which the economy specializes. The long-term advantage of the trading economy will be to specialize in those commodities having the highest income

elasticity of domestic demand and the highest potential for export growth.

The determination of output growth by a balance of payments constraint can be combined with the presence of dynamic economies of scale as expressed in the relationship known as Verdoorn's Law, linking the rates of growth of output and productivity (Verdoorn, 1949). Following Kaldor (1966) and considering technical progress as a phenomenon essentially endogeneous to the economic growth process, we can interpret Verdoorn's Law as a causal relationship determining the growth of productivity by the rate of output growth. This interpretation is based on the presence of increasing returns to scale in manufacturing industries and on the effects of economic growth on investment expansion incorporating new technological developments.

The combination of the balance of payments constraint and a technical progress function of the Verdoorn type provides the means to substantiate the view that economic development, in a system open to international trade, can assume an unequal and divergent character. Starting from an initial competitive advantage (or disadvantage), and through a chain of causal relationships involving export demand, output growth and productivity changes, a circular and cumulative process is set in motion giving rise to an increase of the initial advantage (or disadvantage). Thus, for example, a country starting from a weak competitive position and facing a loss of foreign markets may be forced, through the balance of payments constraint, to retard capital accumulation, thereby depressing productivity growth, and end up falling further behind in the competitive race. This circular and cumulative process can take place either through the dynamics of relative prices and price competitiveness, or through the links between market growth and non-price competitiveness – such as the development of new or differentiated products. The emphasis of recent models on income elasticities of demand and on non-price factors, rather than on price elasticities, seems to reflect the growing awareness that those factors play in international trade, after the major changes in exchange rates and relative prices, since the late 1960s, were found to be less effective than expected in promoting equilibrium in international trade flows.

Theories of economic growth constrained by the balance of payments have had several contributors since the early 1960s (Beckerman, 1962; Kaldor, 1966 and 1970; Thirlwall, 1979), but their origins are to be found in the classical works of Harrod (1933), on the foreign trade multiplier, and of Prebisch on the causes of balance of payments disequilibria in underdeveloped countries. Myrdal's analysis of circular and cumulative causation (1957) has also provided elements taken up in that approach. Its recent revival and development seems to be related to the process of economic development in industrialized countries during the postwar period of increasing economic integration and rapid development of world trade. This experience showed that while some countries, such as Japan and Germany, were clearly benefiting from the rapid expansion in international trade, other countries were suffering negative consequences. A typical case was Great Britain, for which the unfavourable trends in the balance of payments were acting as a brake on economic growth.

Summing up the discussion presented, a full analysis of the interactions between international and economic growth needs to take into consideration the dynamic gains or losses from trade occurring in complex processes of economic development. Then, even when there exists a pattern of specialization and trade which yields more desirable results than autarky and is in the best long-term interest of the economy, this pattern need not coincide with that induced by free international trade. That pattern appears to be determined much less by static comparative advantage than by factors such as the comparative potential for technical progress, the type of returns to scale among industries, and the income and price elasticities of domestic and foreign demand among commodities.

The lack of a full analytical consideration, in traditional theory, of the relationship between trade and growth and the ease with which the normative conclusions of static comparative advantage theory have been extended to a dynamic world probably has its origins in the beneficial effects that trade liberalization had on Britain's industrial development during the 19th century and the first decades of the present century, and to the strong impetus for economic growth experienced, following the expansion of Britain's imports by Scandinavian countries and the 'regions of recent settlement' such as Canada, Australia and New Zealand. However, other historical experiences – such as the slow and highly vulnerable pace of development in Latin America in the first decades of the present century or the unequal and divergent character of economic growth during the postwar period – suggest that the links between economic growth and international trade can be much more complex than traditional theory assumes. Once a dynamic framework is brought into consideration, the theory of comparative advantage may retain a descriptive value but will, indeed, fail to carry the normative conclusions that it has in a static world.

JAIME ROS

BIBLIOGRAPHY

Beckerman, W. 1962. Projecting Europe's growth. *Economic Journal* 72, December, 912–25.
Bhagwati, J. 1958. Immiserizing growth: a geometrical note. *Review of Economic Studies* 25, June, 201–5.
Edgeworth, F.Y. 1894. The theory of international values. Pts I–III. *Economic Journal* 4, Pt I, March, 35–50; Pt II, September, 424–43; Pt III, December, 606–38.
Findlay, R. and Grubert, H. 1959. Factor intensities, technological progress, and the terms of trade. *Oxford Economic Papers* 11, February, 111–21.

Harrod, R.F. 1933. *International Economics*. New York: Harcourt, Brace.
Hicks, J.R. 1953. An inaugural lecture. *Oxford Economic Papers*, NS 5, June, 117–35.
Johnson, H. 1958. *International Trade and Economic Growth*. London: George Allen and Unwin.
Kaldor, N. 1966. *Causes of the Slow Rate of Economic Growth of the United Kingdom: An Inaugural Lecture*. Cambridge: Cambridge University Press.
Kaldor, N. 1970. The case for regional policies. *Scottish Journal of Political Economy* 17(3), November, 337–48.
Kemp, M.C. 1964. *The Pure Theory of International Trade*. Englewood Cliffs, NJ: Prentice-Hall.
Myrdal, G. 1957. *Economic Theory and Underdeveloped Regions*. London: Duckworth.
Pasinetti, L. 1981. *Structural Change and Economic Growth*. Cambridge: Cambridge University Press.
Prebisch, R. 1959. Commercial policy in the underdeveloped countries. *American Economic Review, Papers and Proceedings* 49, May, 251–73.
Södersten, B. 1964. *A Study of Economic Growth and International Trade*. Stockholm: Almqvist and Wiksell.
Thirlwall, A.P. 1979. The balance of payments constraint as an explanation of international growth rate differences. *Banca Nazionale del Lavoro Quarterly Review* 128, March, 45–53.
Torrens, R. 1821. *An Essay on the Production of Wealth*. London: Longman, Hurst, Rees, Orme & Brown.
United Nations Economic Commission for Latin America. 1950. *The Economic Development of Latin America and Its Principal Problems*. New York: United Nations.
Verdoorn, P.J. 1949. Fattori che regolano lo sviluppo della produttività del lavoro. *L'Industria* 1, 45–53.

growth models, classical. *See* CLASSICAL GROWTH MODELS.

growth models, neoclassical. *See* NEOCLASSICAL GROWTH THEORY.

growth theory. *See* CLASSICAL GROWTH MODELS; HARROD–DOMAR MODEL; MULTISECTOR GROWTH MODELS; NATURAL AND WARRANTED RATES OF GROWTH; NEOCLASSICAL GROWTH THEORY.

H

Haavelmo, Trygve (born 1911). Haavelmo was born in Skedsmo, Norway. He graduated from the University of Oslo in 1933 and joined Ragnar Frisch's newly created Institute of Economics as a research assistant. He spent the war years working for the Norwegian government in the United States. After a year's stay at the Cowles Commission at the University of Chicago, he returned to Norway in 1947, becoming professor of economics at the University of Oslo in 1948. He retired from his chair in 1979.

Haavelmo first made his name by a series of pathbreaking contributions to the theory of econometrics, most of which were written during his years in the United States. His 1943 article in *Econometrica* was the first to consider the statistical implications of simultaneity in economic models. This paper was one of the main sources of inspiration for the extensive work carried out in this area over the next decade, particularly at the Cowles Commission. Haavelmo developed his ideas further in the famous 1944 supplement to *Econometrica*; the main general contribution of this work was to base econometrics more firmly on the foundations of probability theory.

After his return to Norway, Haavelmo turned away from econometrics to economic theory as his main field of interest. In his 1957 presidential address to the Econometric Society (published the next year) he emphasized the need for a more solid theoretical foundation for empirical work as well as the need for theory to be inspired by empirical research.

Haavelmo's *Study in the Theory of Economic Evolution* (1954), is a broad exploration of the contributions that analytical economics can make to the understanding of global economic inequality. As an early contribution to growth theory it is less notable for simple models and precise theorems than for its imaginative and experimental attitude towards hypotheses concerning population growth, education, migration and the international struggle for redistribution. The open-mindedness of the approach is very characteristic of the author.

Similar remarks apply to his 1960 book, *A Study in the Theory of Investment*. Its main objective is to provide a firmer microeconomic foundation for the macroeconomic theory of investment demand. To this end Haavelmo probes deeply into capital theory, emphasizing strongly, however, that a theory of optimum capital use does not in itself provide a theory of investment. This insight, and his clear statement of what has since been known as the neoclassical theory of capital accumulation, has been a major influence on late work in this area, both theoretical and applied.

Of Haavelmo's other contributions to economic theory, special mention should be made of his 1945 analysis of the balanced budget multiplier. The expansionary effect in a Keynesian unemployment situation of a balanced increase of public expenditure and taxes had been pointed out before, but Haavelmo was the first to provide a rigorous theoretical analysis of it.

Haavelmo has also been very active as a teacher. His lecture notes on a wide range of topics in economic theory have exerted a formative influence on generations of Norwegian economists.

AGNAR SANDMO

SELECTED WORKS

1943. The statistical implications of a system of simultaneous equations. *Econometrica* 11, January, 1–12.
1944. The probability approach in econometrics. Supplement to *Econometrica* 12, July, S1–115.
1945. Multiplier effects of a balanced budget. *Econometrica* 13, October, 311–18.
1947a. Methods of measuring the marginal propensity to consume. *Journal of the American Statistical Society* 42(237), March, 105–22.
1947b. (With M.A. Girshick.) Statistical analysis of the demand for food: examples of simultaneous estimation of structural equations. *Econometrica* 15, April, 79–110.
1954. *A Study in the Theory of Economic Evolution*. Amsterdam: North-Holland.
1958. The role of the econometrician in the advancement of economic theory. *Econometrica* 26, July, 351–7.
1960. *A Study in the Theory of Investment*. Chicago: University of Chicago Press.
1970. Some observations on welfare and economic growth. In *Induction, Growth and Trade: Essays in Honour of Sir Roy Harrod*, ed. W.A. Eltis, M.F.G. Scott and J.N. Wolfe, Oxford: Clarendon Press.

Habakkuk, John Hrothgar (born 1915). Born in Wales in 1915, Habakkuk graduated from Cambridge in 1936, where he was a Fellow of Pembroke College from 1938 until 1950. He held the Oxford chair of economic history from 1950 to 1967, when he became Principal of Jesus College, Oxford. He retired in 1984. As a member of the Advisory Council on Public Records (1958–70), the Royal Commission on Historic Manuscripts, and the British Library Organizing Committee, amongst other bodies, he has been active in the field of public records; he was knighted in 1974.

His major contribution was to the study of the rates of technological change in Britain and America in the 19th century and the reasons for the much more rapid development and use of manufacturing technology in the States. In his book, *American and British Technology in the Nineteenth Century*, American industrial development is roughly divided into two important stages, the period before the first wave of immigration in the 1840s, which laid the ground for future development, and the period after 1870 when abundant natural resources and rapid growth of market demand provided the stimulus for growth.

Habakkuk argues that American technological development in the early period, by contrast with Britain, was stimulated by the high cost of labour relative to capital and the relative inelasticity of labour supply. The expanding manufacturer, to avoid a falling marginal rate of profit, was more likely than his British counterpart to look to capital-intensive and labour-saving technology. Though Habakkuk was also keen to stress the importance of social factors, the suspicion of British employers and the hostility of British labour to new techniques, his explanation of the disparity is grounded in economic relationships.

Habakkuk's thesis has come under considerable scrutiny; recent research has tended to suggest that there was considerable diversity, both on a regional basis and between different industries, in development on both sides of the Atlantic. Economic historians have also questioned the timing of significant development in the States and chosen to put greater stress on non-economic explanations.

Habakkuk also made notable contributions to the debates on British population growth in the late 18th century and on the changing pattern of land-holding as smaller holdings gave way to larger units in the same period.

RAVI MIRCHANDANI

SELECTED WORKS

1953. English population in the eighteenth century. *English Historical Review* 6(2).

1962. *American and British Technology in the Nineteenth Century: the search for labour-saving inventions.* Cambridge: Cambridge University Press.

1968. *Industrial Organisation since the Industrial Revolution.* Southampton: University of Southampton Press.

1971. *Population Growth and Economic Development since 1750.* Leicester: Leicester University Press.

1979–81. The rise and fall of English landed families, 1600–1800. *Royal Historical Society Transactions* 29–31.

Haberler, Gottfried (born 1900). Gottfried Haberler was born on 20 July 1900 in Purkersdorf, near Vienna. He studied economics at the University of Vienna under Friedrich von Wieser and Ludwig von Mises, where he received doctorates in law (1923) and economics (1925). After two years in the United States and Britain he returned to Vienna, received his habilitation in 1928, and was appointed lecturer, later Professor of Economics, at the University of Vienna, from 1928 to 1936. He was appointed professor at Harvard University in 1936 where he remained until his retirement in 1971. Since that time he has been a resident scholar at the American Enterprise Institute, Washington, DC. He was President of the International Economic Association (1950–51), the National Bureau of Economic Research (1955), and the American Economic Association (1963). In 1980 he was awarded the Antonio Feltrinelli prize.

Haberler's first major work was his habilitation thesis (1927), *The Meaning of Index Numbers*, summarized in Koo (1985, pp. 546–9). This work stimulated a great deal of subsequent research on the theory of the price or cost-of-living index. Haberler defined the 'true change in the price level' as 'the ratio of the money income in the first period to the money income in the second period that would leave the individual indifferent' (Koo, 1985, p. 547). Haberler's main concern was to find conditions under which this 'true price index' would be bounded by the Laspeyres and Paasche price indices. Some of the difficulties with this approach (and with the similar, earlier approach of Konüs, 1924) were discussed by Bortkiewicz (1928), Neisser (1929), Staehle (1935), and Frisch (1938). Frisch remarked (p. 25) that Haberler's definition of the 'true change of the price level' involved an implicit assumption of expenditure proportionality (homothetic preferences), and attributed this point (but apparently without justification) to Bortkiewicz; he also interpreted Haberler (1929) in his reply to Neisser and Bortkiewicz as accepting this point. In terms of contemporary concepts we may say that homothetic preferences characterize indirect utility functions of the form $Y/C(p)$

where Y is income and $C(p)$ is a homogeneous-of-degree-1 function of prices.

Undoubtedly Haberler's most significant contribution was his reformulation of the theory of comparative costs (Haberler, 1930a), which revolutionized the theory of international trade. Prior to this paper, the Ricardian theory still held sway, but had been so amended with ill-defined concepts such as 'real cost' and 'units of productive power' taking the place of labour allocation that it had lost all its simplicity and elegance. Haberler introduced the production 'substitution curve' (now usually known as the production-possibility frontier), allowing for several factors of production, and taken to be concave to the origin as a result of diminishing returns. This laid the foundations for Ohlin's theory, as well as Lerner's and Samuelson's. True, as recently brought to light by Maneschi and Thweatt (1987), a footnote contained in the posthumous edition of Barone's *Principi* (1936, pp. 170–73), depicting a (non-concave) production-possibility frontier and a community indifference curve, was actually present in the first (1908) edition – but not subsequent ones; hence Barone must be accorded priority. But Haberler's independent discovery – and the use to which he put it – is what transformed the theory of international trade. Haberler also introduced the concept of a 'specific factor' – one that is completely immobile among industries – and used this concept once again with great effect in Haberler (1950) to illustrate the proposition that the gains from trade do not depend on the assumption of factor mobility.

Haberler has made numerous other contributions to international economics, including (1) his synthesis and clarification of the Keynes-Ohlin debate on the transfer problem (Haberler, 1930b); (2) his judicious use of purchasing-power-parity calculations to set exchange rates (Haberler, 1945); (3) his introduction of the concept of supply and demand schedules for foreign exchange (1936) and his subsequent use of them (Haberler, 1949) in qualified support of the proposition that a devaluation in a pegged-rate regime could improve a country's balance of payments – but subject to the important proviso (1949, p. 213) that it would, through monetary expansion, likely shift these schedules; (4) his advocacy of free trade as the best policy for developing countries (Haberler, 1959); (5) numerous contributions to past and current history of international economic relations (cf. Koo, 1985).

The third area in which Haberler has made major contributions is business-cycle theory (Haberler, 1937, 1942). His classic synthesis, notably in the third edition of *Prosperity and Depression* (1941), introduced the important 'real-balance effect', initially called the 'Pigou effect' by Patinkin (1948), although Patinkin in his 1951 revision acknowledged Haberler's priority over Pigou (1943). In the 1970s and 1980s Haberler furnished trenchant analyses of the phenomenon of worldwide inflation and the political economy of stagflation (cf. Koo, 1985), displaying the unique combination of clarity and wisdom that are characteristic of his writings.

Information on Haberler's life and work may be found in Schuster (1979), Chipman (1982), Baldwin (1982), Officer (1982), and Willett (1982). A complete bibliography of his writings is contained in Koo (1985).

JOHN S. CHIPMAN

SELECTED WORKS

1927. *Der Sinn der Indexzahlen.* Tübingen: J.C.B. Mohr (Paul Siebeck).

1929. Der volkswirtschaftliche Geldwert und die Preisindexziffern. *Weltwirtschaftliches Archiv* 30, July, 6**–14**.

1930a. Die Theorie der komparativen Kosten und ihre Auswertung für die Begründung des Freihandels. *Weltwirtschaftliches Archiv* 32, July, 350–370. Trans. as 'The theory of comparative costs and its use in the defense of free trade' in Koo (1985), 3–19.

1930b Transfer und Preisbewegung. *Zeitschrift für Nationalökonomie* 1(4), 547–54; 2(1), 100–102. Trans. as 'Transfer and price movements' in Koo (1985), 133–42.

1933. *Der internationale Handel. Theorie der weltwirtschaftlichen Zusammenhänge sowie Darstellung und Analyse der Aussenhandelspolitik.* Berlin: Julius Springer. Translated (revised by the author) as *The Theory of International Trade with its Applications to Commercial Policy,* London: William Hodge & Co. 1936.

1937. *Prosperity and Depression.* Geneva: League of Nations. 3rd edition enlarged by Part III, 1941. 5th and 6th editions, Cambridge, Mass.: Harvard University Press, 1958, 1964.

1942. *Consumer Instalment Credit and Economic Fluctuations.* New York: National Bureau of Economic Research.

1945. The choice of exchange rates after the war. *American Economic Review* 35, June, 308–318.

1949. The market for foreign exchange and the stability of the balance of payments. *Kyklos* 3 (3), 193–218. Reprinted in Koo (1985), 143–165.

1950. Some problems in the pure theory of international trade. *Economic Journal* 60, June, 223–240. Reprinted in Koo (1985), 37–54.

1955. *A Survey of International Trade Theory.* Special Papers in Economics No. 1, International Finance Section, Princeton University. Revised and enlarged edition, 1961. Reprinted in Koo (1985), 55–108.

1959. *International Trade and Economic Development.* Cairo: National Bank of Egypt, Reprinted in Koo (1985), 495–527.

BIBLIOGRAPHY

Baldwin, R.E. 1982. Gottfried Haberler's contributions to international trade theory and policy. *Quarterly Journal of Economics* 97, February, 141–59.

Barone, E. 1908. *Principi di economia politica.* Rome: Tipografia Nazionale di Giovanni Bertero e C. German translation from the third (1913) edition, *Grundzüge der theoretische Nationalökonomie,* Bonn: Kurt Schroeder, 1927. Posthumous edition, *Le opere economiche,* Vol. II, Bologna: Nicola Zanichelli Editore, 1936.

Bortkiewicz, L. von. 1928. Review of *Der Sinn der Indexzahlen* by Gottfried Haberler. *Magazin der Wirtschaft* 4(11), 15 March, 427–9.

Chipman, J.S. 1982. Salute to Gottfried Haberler on the occasion of his 80th birthday. *Journal of International Economics* [Supplement], January, 25–30.

Frisch, R. 1936. Annual survey of general economic theory: the problem of index numbers. *Econometrica* 4, January, 1–38.

Konüs, A.A. 1924. The problem of the true index of the cost of living. *The Economic Bulletin of the Institute of Economic Conjuncture* 9–10, October, 64–71. Translated from the Russian in *Econometrica* 7, January 1939, 10–29,

Koo, A.Y.C. 1985. *Selected Essays of Gottfried Haberler.* Cambridge, Mass.: MIT Press.

Maneschi, A. and Thweatt, W.O. 1987. Barone's 1908 representation of an economy's trade equilibrium and the gains from trade. *Journal of International Economics.*

Neisser, H. 1929. Der volkswirtschaftliche Geldwert und die Preisindexziffern, *Weltwirtschaftliches Archiv* 29, Part I, 6**–18**. Schlusswort, Part II, July, 14**–17**.

Officer, L.H. 1982. Prosperity and depression – and beyond. *Quarterly Journal of Economics* 97, February, 149–59.

Patinkin, D. 1948. Price flexibility and full employment. *American Economic Review* 38, September, 543–64. Revised version in *Readings in Monetary Theory,* ed. F.A. Lutz and L.W. Mints, Philadelphia: Blakiston, 1951, 252–83.

Pigou, A.C. 1943. The classical stationary state. *Economic Journal* 53, December, 343–51.

Schuster, H. (ed.) 1979. Univ.-Prof. Dr. Gottfried Haberler. In *Österreicher, die der Welt gehören.* Vienna: Mobil Oil Austria AG, 34–43.

Staehle, H. 1935. A development of the economic theory of price index numbers. *Review of Economic Studies* 2, June, 163–88.

Willett, T.D. 1982. Gottfried Haberler on inflation, unemployment, and international monetary economics: an appreciation. *Quarterly Journal of Economics* 97, February, 161–9.

habit persistence. *See* RELATIVE INCOME HYPOTHESIS.

Hadley, Arthur Twining (1856–1930). American economist, educator and public servant, Hadley was educated at Yale and at the University of Berlin, where he studied under German historicists. In a remarkable career, Hadley was, in turn, a freelance writer and lecturer on railway economics, a professor of political economy at Yale (1891–9), president of the American Economic Association, president of Yale University (1899–1921), chairman of the Railroad Securities Commission providing the Hadley Report on Railway finances in 1911, and was widely sought after as a political candidate for high political office in the United States. An inveterate traveller, Hadley died aboard ship in Kobe harbour in 1930.

Hadley was an extremely prolific and eclectic writer, but the bulk of his important work in economics was completed before the turn of the century. His reputation rests essentially on two works, *Railway transportation* (1885) and a basic text, *Economics: An Account of the Relations between Private Property and Public Welfare* (1896), which received high praise from his friend and colleague, Irving Fisher.

In *Railway Transportation* Hadley revealed himself as the most creative railway economist of the day through an integration of sophisticated (certainly for the time) economic analysis with the problems of railway organization. Among other theoretical insights Hadley formalized a theory of monopoly and price discrimination; developed, in the mathematical terms of Cournot, a marginal rule for profit maximization; and anticipated the period analysis of Marshall's *Principles.* More importantly, perhaps, he developed a modern and complete theory of cartels, showing that, in the presence of open competition, such unsanctioned behaviour on the part of railroads would lead to the benefits of competition without the attendant disadvantages. In another perspicacious insight Hadley correctly characterized railway regulation as resulting from the capture, by the industry, of legal sanctions to obtain rate stability. In the main, Hadley viewed regulation as representing a low-cost cartel enforcement device.

In *Economics* Hadley went further than Marshall by explicitly developing the interrelations between property rights, economic evolution and economic efficiency. Hadley utilized the real world examples of the fisheries and mining to demonstrate the impact of ill-defined property rights on depletable resources, emphasizing the necessity of altered systems to obtain optimal resource use and allocation. This contribution, along with his prophetic analyses of transport market structure, establishes Hadley as one of the most inventive pre-20th-century American economists.

ROBERT B. EKELUND, JR.

SELECTED WORKS

1885. *Railway Transportation: Its History and its Laws.* New York.

1890. The prohibition of railroad pools. *Quarterly Journal of Economics* 4, January, 158–71.

1896. *Economics, An Account of the Relations between Private Property and Public Welfare.* New York.

BIBLIOGRAPHY

Cross, M.L. and Ekelund, R.B., Jr. 1980. A.T. Hadley on monopoly theory and railway regulation: an American contribution to economic analysis and policy. *History of Political Economy* 12(2), Summer, 214–33.

Cross, M.L. and Ekelund, R.B., Jr. 1981. A.T. Hadley: the American invention of the economics of property rights and public goods. *Review of Social Economy* 39(1), April, 37–50.

Fisher, I. 1930. Obituary: Arthur Twining Hadley. *Economic Journal* 40, September, 526–33.

Locklin, D.P. 1933. The literature on railway rate theory. *Quarterly Journal of Economics* 47, February, 167–230.

Hagen, Everett Einar (born 1906). Hagen was born in Holloway, Minnesota. He graduated from St Olaf College (BA, 1927) and the University of Wisconsin (MA, 1932; PhD, 1941). After a short period at the University of Illinois (1948–51) he became Professor of Economics at the Massachusetts Institute of Technology (1953–72); from 1970 to 1972 he was Director of the Center for International Studies at MIT.

Since World War II developing nations have received unprecedented attention from economists and large financial resources from the industrialized world. Dr Hagen has been an important contributor to analysing key problems and processes of economic development.

Before concentrating on economic development, Hagen served in the Bureau of the Budget as a close associate of Gerhard Colm in the application of Keynesian principles to US fiscal policies. His firm commitment to Keynes's concepts was a factor in his transfer to the MIT from the University of Illinois, where more traditionalist faculty and top officialdom were hostile to the views of Keynes and of the New Deal.

In his book *On the Theory of Social Change* (1962), Hagen correctly concluded that economics alone could not provide the theoretical or policy directions for economic development. He studied deeply the role of human behaviour based on studies of anthropologists, sociologists and political scientists. Hagen's multidisciplinary approach provided invaluable insights for formulating development plans and policies.

In his fourth edition of *The Economics of Development* (1986), Hagen continued to elaborate on theoretical aspects as well as policies and implementation processes essential to development progress. Hagen up-dates the most promising lessons from successful nations replicable in the lagging nations.

Hagen disputes the common view that high population growth rates are a major deterrent to development. He also documents the thesis that protectionism is helpful to the developing world. He sets forth a strong case for attributing considerable unemployment to technological change. These somewhat unorthodox views are persuasively articulated and documented.

Of major importance are Hagen's conceptual formulations, his analyses based on personal experiences, and his challenges to economists and members of other disciplines to work jointly to overcome the persistent barriers to significant progress in the lagging nations.

ROBERT R. NATHAN

SELECTED WORKS

1962. *On the Theory of Social Change*. Homewood, Ill.:Dorsey Press.
1963. (ed.) *Planning Economic Development*. Homewood, Ill.: Richard D. Irwin.
1968. *The Economics of Development*. Homewood, Ill.: Richard D. Irwin. Revised, 1980, 1986.

'Hahn problem'. Harrod (1939), who inaugurated the postwar concern with growth theory, distinguished between three growth rates: the natural, the warranted and the actual. True to his Keynesian heritage he argued that there were circumstances in which the warranted rate of growth permanently exceeds the natural rate. More importantly from the point of view of this essay he claimed that the warranted growth path was highly unstable – he called it a 'knife-edge'. By this he meant that small disturbances of the warranted growth path would lead to a cumulative divergence of actual from warranted growth. The argument was simple. Suppose, for instance, that for some exogenous reason the actual growth rate fell a little below the warranted rate. By virtue of the accelerator mechanism, savings would exceed investment (ex ante) and income would be given a further impulse taking it below its warranted level. This leads to further reductions in investment and to further downward displacement of the actual path. This process continues. Hicks (1950) quickly saw that this theory could easily serve as an explanation of cycles.

Many economists, however, took the view that Harrod had underestimated the prevalence of stabilizers in a market economy. In particular his theory had little to say about the behaviour of relative prices and had ruled out substitution possibilities by assuming fixed coefficients of production. Not only did he thereby overdetermine the long run equilibrium system (the equation: natural rate = warranted rate had only exogenously given variables on both sides) but he allowed no scope to the price mechanism to stabilize the economy against small shocks. This argument found its clearest expression in a famous article by Solow (1956).

For a fuller discussion of Solow's work the reader should consult the entry on Neoclassical Growth Theory, here it is very briefly summarized. Let y = output per man and k = capital per man and let

$$y = f(k)$$

be the production function which is concave and has the property

$$f'(0) = \infty, \quad f'(k) > 0, \quad \text{all } k \in (0, \infty).$$

Let n be the rate of population growth and s the propensity to save. For an equilibrium, saving per man must equal investment per man, write it as i. But

$$i = \dot{k} + nk$$

so we require

$$\dot{k} + nk = sy. \tag{1}$$

In steady state $\dot{k} = 0$ and we must solve

$$nk = sf(k) \quad \text{or} \quad n = s\frac{f(k)}{k}$$

which is Harrod's equation. Given the assumptions on $f(k)$ there always exists k^* which solves the equation. This then answers one of Harrod's arguments to the effect that it may not be possible to bring the natural rate (n) into equality with the warranted rate $[sf(k)/k]$.

Now divide both sides of (1) by k and rearrange to give

$$\frac{\dot{k}}{k} = s\frac{f(k)}{k} - n. \tag{2}$$

By the concavity of $f(k)$, $f(k)/k$ is a diminishing function of k. Hence starting at any $k(0) \neq k^*$ and following a path for which (a) employment grows at the rate n and (b) savings are always equal to investment (call this a 'warranted' path), the economy will be driven to the steady state k^*, (where $\dot{k} = 0$). This was the gist of Solow's argument.

It will be noticed straight away that this argument has no bearing on Harrod's knife-edge claim. Harrod had not proposed that warranted paths diverge from the steady state but that actual paths did. The latter are neither characterized by a continual equality of ex ante investment and savings nor by continual equilibrium in the market for labour. Thus although Solow thought that he was controverting the knife-edge argument he had only succeeded in establishing the convergence of warranted paths to the steady state.

However, even here it was not at all clear how robust *that* conclusion was to a relaxation of some of its rather strong assumptions. In particular it was widely agreed that the aggregate production function in terms of an aggregate capital input was a 'fable' (Samuelson, 1962). The question was whether this fable was instructive or misleading. An attempted answer which was closely related to the pioneering work on turnpikes by Dorfman, Samuelson and Solow (1958) was christened the 'Hahn problem', although it was not really a problem nor was Hahn's analysis of startling novelty.

Before giving a precise account it will be helpful to have a bird's eye view.

Suppose that there are many different capital goods used in their own production as well as in the production of a single consumption good. Let $t = 0$ be the initial date at which we take the capital stock as determined by past history up to that date. (For simplicity capital goods are assumed to be infinitely durable.) Let agents have expectations concerning the change in relative prices between $t = 0$ and $t = 0 + \epsilon$. These expectations together with the technological conditions of production will determine investment in the various capital goods. This will have the property that everyone is, at the margin, indifferent between investing in one good rather than another. Once that has been determined the economy is, as it were, on rails from which it cannot deviate if we require expectations to be correct and production to be intertemporarily efficient. For the correctness of the price expectations for $t = 0 + \epsilon$ imply what prices must be in all subsequent time periods. However, the 'rails' which the economy gets onto depend on the arbitrarily postulated expectations at $t = 0$. There are in fact an infinity of such rails depending on initial expectations. Most of these, however, lead away from the steady state and not to it (in the example of Hahn, 1966, all of them except one lead the economy away from the steady state). There are thus many warranted paths and they do not conform to the Solow proposition for the single capital good. There seems to be both indeterminacy and instability of the steady state under warranted paths deviations. However, it may be that the rails which lead the economy away from the steady state are also leading it into an abyss. That is, the paths may eventually become infeasible because some capital good needed in production has disappeared. However, if we postulate some form of myopia in expectations, by which is meant no more than that agents cannot predict prices into the infinitely distant future, there is nothing to prevent the economy following such errant warranted paths for a 'long time'. However, we return to this matter below after the technical discussion.

The story which has just been told informally exemplifies the difficulties which arise in an economy which does not have a full set of Arrow–Debreu markets. Such an economy must act on the basis of price expectations and these in turn open up the possibility of 'bootstrap' warranted paths: the economy evolves the way in which it does because expectations are what they are and not for any 'real' reason. In the conclusion we return to these intuitive explanations. But first we must demonstrate the existence of many warranted paths which do not seek the steady state.

Let there be m capital goods whose quantities *per man* are denoted by the vector $k = (k_1, \dots, k_m)$ and let $y = (y_1, \dots, y_m)$ be the output vector (per man) of the capital goods. The output of consumption good per man is written as y_0. Let p_0 be the price of the consumption good and $p = (p_1, \dots, p_m)$ the price vector of capital goods. All prices are reckoned in unit of account. There are constant returns to scale and one defines

$$A(k) = \{(y, y_0) \mid F(y_0, y \cdot k) \geqslant 0\}$$

as the production possibility set of the economy given k. In this definition $F(\cdot)$ is assumed C^2, strictly concave function with the property:

$$\frac{\partial F}{\partial k_i} > 0, \qquad \text{for } k_i < \infty,$$

$$\frac{\partial F}{\partial k_i} < +\infty, \qquad \text{for } k_i = 0 \text{ all } i.$$

A competitive economy in equilibrium will at all dates behave as if it solved the problem:

$$\max_{A(k)} (p \cdot y + p_0 y_0).$$

Let R be this maximized sum. Then we can write

$$R = R(p_0, p, k).$$

Classical duality theory gives

$$R_i(p, p, k) = y, \qquad i = 0, \dots, m.$$

where $R_i = \partial B / \partial p_i$. Moreover we know that R is convex in (p_0, p) and concave in k. If we suppose that population is growing at the geometric rate n then the evolution of the capital stock per man is given by the differential equation

$$R_i(p_0, p, k) - nk_i = \dot{k}_i, \qquad i = 1, \dots, m. \tag{3}$$

But if the economy has perfect foresight so that the expected rate and actual rate of all price changes coincide then it must satisfy arbitrage equations which ensure that investment in all directions is equally profitable. If we let $R_{m+i} = \partial R / \partial k_i$ this means that there is at each date a scalar, r, such that

$$R_{m+i}(p_0, p, k) + \dot{p}_i = rp_i, \qquad i = 1, \dots, m. \tag{4}$$

Since we can choose one good as numeraire (say the consumption good) we need only one more equation to be able to trace the evolution of all variables from given initial conditions. That equation must refer to the common rate of return r. This will depend on the savings decisions of agents and on technology and so on (p_0, p, k, r). Write

$$\dot{r} = c(p_0, p, k)$$

In steady state: $\dot{r} = \dot{k} = \dot{p} = 0$. Let r^*, p^*, k^* be the solution of (3), (4), (5) in such a steady state. (On present assumptions such a solution exists.) To study the warranted growth path of the economy near the steady state we take a first order Taylor expansion of these three equations at (r^*, p^*, k^*). We write: $\tilde{p} = p - p^*$, $\tilde{k} = k - k^*$, $\tilde{r} = r - r^*$ and set p_0^* identically equal to unity. Also

$$R_{ij} = \frac{\partial R_i(p^*, k^*)}{\partial P_j}, \qquad R_{im+j} = \frac{\partial R_i(p^*, k^*)}{\partial k_j} \text{ etc.}$$

and

$$c_r = \frac{\partial c(r^*, p^*, k^*)}{\partial r}.$$

We obtain

$$\sum_j R_{im+j}\bar{k}_j - n\bar{k}_j + \sum_j R_{ij}\bar{p}_j = \dot{\bar{k}}_i, \quad i = 1, \ldots, n \quad (6)$$

$$\bar{r}p_i^* - \sum_j R_{m+im+j}\bar{k}_j - \sum_j R_{m+ij}\bar{p}_j + r^*\bar{p}_i = \dot{\bar{p}}_i, \quad i = 1, \ldots, n \quad (7)$$

$$\bar{r}c_r + \sum_j c_{m+j}\bar{k}_j + \sum_j c_p\bar{p}_j = \dot{\bar{r}} \quad (8)$$

Let R_{pk} be the $n \times n$ matrix of elements $[R_{im+j}]$, R_{pp} the $n \times n$ matrix of elements $[R_{ij}]$, R_{kk} the $n \times n$ matrix of elements $[R_{m+i, m+j}]$ and I the $n \times n$ identity matrix. Then the above equations can be written compactly as

$$\begin{bmatrix} c_r & \{c_{m+j}\} & \{c_j\} \\ \{0\} & R_{pk} - nI & R_{pp} \\ p^* & -R_{kk} & -R_{kp} + r^*I \end{bmatrix} \begin{bmatrix} \bar{r} \\ \bar{k} \\ \bar{p} \end{bmatrix} = \begin{bmatrix} \dot{\bar{r}} \\ \dot{\bar{k}} \\ \dot{\bar{p}} \end{bmatrix} \quad (9)$$

Note that $R_{pk} = R'_{kp}$.

Let us consider the unbordered matrix:

$$A = \begin{bmatrix} R_{pk} - nI & R_{pp} \\ -R_{kk} & -R_{kp} + p^*I \end{bmatrix}$$

If we make the assumption that profits are all saved and wages are all spent then $r^* = n$. Make this assumption: Let

$$T = \begin{bmatrix} 0 & -I \\ I & 0 \end{bmatrix}$$

so that $T' = -T$. Then

$$TA = \begin{bmatrix} R_{kk} & R_{kp} - nI \\ R_{pk} - nI & R_{pp} \end{bmatrix} det\ B$$

and B is a symmetric matrix. Now let $Ax = \lambda x$ be the characteristic equation for A with eigenvalue λ. Then

$$\lambda Tx = Bx = B'x \quad (10)$$

But $TA = A'T' = B'$ and $A'T' = A'(-T)$. Let $Tx = y$. Then from (10)

$$\lambda y = -A'y \quad \text{or} \quad -\lambda y = A'y.$$

Hence if λ is a root of A so is $-\lambda$. One says that A has the saddle point property. The phase diagram for p and k in two dimensions is given in Figure 1.

If $\dot{r}$ remained constant at its steady state value r^* then Figure 1 would show all the warranted paths of the economy. It will be seen that only one of these approaches the steady state. On the other hand all the other paths may eventually become infeasible – they lead to one of the axes. Infinite perfect foresight would rule all these paths out of consideration. However, the postulate of such foresight seems farfetched.

When the whole system of equation (9) is considered matters are more complicated. One way out of the complication is to suppose that the economy behaves as if it were solving an infinite 'Ramsey problem'. The behaviour of r would then be fully determined by the Euler–Lagrange equations for this problem. But once again, in the absence of discounting, all paths but the convergent one would be ruled out and the 'Hahn problem' would disappear. But also once again the realism and relevance of such a postulate must be in doubt (see Hahn, 1968; Kurz, 1968).

The alternative is to proceed by way of a model of overlapping generations or simply by a descriptive savings function. Work along these lines (and also with more than one consumption good), has been undertaken by a number of economists. Shell and Cass (1976) have provided a good

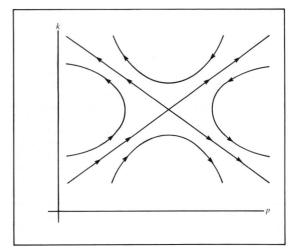

Figure 1

general treatment of systems such as (9). The main conclusion is that in addition to the divergent paths there is also a manifold of paths which converge. This is interesting since now even with infinite perfect foresight and convergence, there is nothing to tell us which of the convergent paths the economy will choose. The same difficulty has been encountered in the overlapping generations' literature which has been very ably summarized by Woodford (1984). However in both approaches divergent warranted paths remain and some of these (in the case of overlapping generations) are viable over infinite time.

There are a number of technical matters which have not been considered in the above account – for instance the relation of the problem to turnpike theory (Samuelson 1960) and the role of intertemporal efficiency conditions which have been encapsulated in the dual formulation here adopted. But enough has been provided to allow an intuitive summary.

The requirement of perfect foresight equilibrium is certainly too weak to determine the path of an economy if perfect foresight is not over the infinite future. Moreover there are then many paths which do not converge to the steady state. This is due to the circumstances that arbitrage, given initial expectations, imposes a particular path on the economy if expectations are to be fulfilled in the future, and if the arbitrage equations are to hold. Thus the invisible hand may for a long time provide coherence in the economy even while it is guiding it to eventual disaster. But even when there is perfect foresight over the infinite future and that future is discounted, there will be many paths that do not converge to the steady state. (Kurz, 1968, gives a case where (9) has been converted into a Ramsey problem with discounting and where all paths diverge from the steady state.) It would seem that in general the price-mechanism even with correct expectations will not bear out the rather optimistic conclusion of Solow with which this essay started.

Two matters remain to be mentioned. Warranted paths which do not converge may yet be Pareto-efficient (see Cass, 1972), provided of course they are feasible over infinite time. However this does not mean that such paths do not provide an occasion for policy since they may be associated with very undesirable inter-temporal distributions of welfare between generations. (This applies to models in which agents are not infinitely lived and in which agents do not value their descendants' utility as they do their own.)

The second matter is this: one may ask whether the steady state would not be stable if one allowed for false expectations, that is if one considered actual and not warranted paths. This question was posed by Shell and Stiglitz (1967) and is also discussed in Hahn (1969). Although Shell and Stiglitz did indeed find that with relatively inelastic expectations the steady state was stable their model was very special, particularly in the manner in which it incorporates the heterogeneity of capital goods. In a more general model Hahn (1969) found no general presumption that the steady state was stable unless expectations were completely inelastic, as was postulated by Morishima (1964). In that latter case there are no expected capital gains and losses and the arbitrage equation takes on a degenerate form. Nonetheless it remains an interesting question which set of circumstances leads to false expectations being stabilizing. No general answers are now available. But the 'Hahn problem' was concerned with correct (albeit myopic) expectations.

F.H. HAHN

See also HAMILTONIANS; SUNSPOT EQUILIBRIA; TULIPMANIAS; TURNPIKE THEOREMS.

BIBLIOGRAPHY

Cass, D. 1972. On capital overaccumulation in the aggregative, neoclassical model of economic growth. *Journal of Economy Theory* 4, 200–23.

Dorfman, R. Samuelson, P.A. and Solow, R.M. 1958. *Linear Programming and Economic Analysis*. New York: McGraw-Hill.

Hahn, F.H. 1966. Equilibrium dynamics with heterogeneous capital goods. *Quarterly Journal of Economics* 80, 633–45.

Hahn, F.H. 1968. On warranted growth paths. *Review of Economic Studies* 35, 175–84.

Hahn, F.H. 1969. On some equilibrium paths. In *Models of Economic Growth*, ed. J. Mirrlees and N.H. Stern, London: Macmillan, 1973.

Harrod, R.F. 1939. An essay in dynamic theory. *Economic Journal* 49, 14–33.

Hicks, J.R. 1950. *A Contribution to the Theory of the Trade Cycle*. Oxford: Clarendon Press.

Kurz, M. 1968. The general instability of a class of competitive growth processes. *Review of Economic Studies* 35, 155–74.

Morishima, M. 1964. *Equilibrium, Stability and Growth*. Oxford: Clarendon Press.

Samuelson, P.A. 1960. Efficient paths of capital accumulation in terms of the calculus of variations. *Stanford Symposium on Mathematical Methods in the Social Sciences*, ed. K.J. Arrow, S. Karlin and P. Suppes, Stanford: Stanford University Press.

Samuelson, P.A. 1962. Parable and realism in capital theory: the surrogate production function. *Review of Economic Studies* 29, 193–206.

Shell, K. and Cass, D. 1976. The structure and stability of competitive dynamical systems. *Journal of Economic Theory* 12, 31–70.

Shell, K. and Stiglitz, J. 1967. The allocation of investment in a dynamic economy. *Quarterly Journal of Economics* 81, 592–610.

Solow, R.M. 1956. A contribution to the theory of economic growth. *Quarterly Journal of Economics* 70, 65–94.

Woodford, M. 1984. Indeterminancy of equilibrium in the overlapping generation model: a survey. Mimeo, Columbia University.

Halévy, Elie (1870–1937). Elie Halévy was one of the foremost historians of 19th-century English thought and politics. He was born at Etretat, France, and educated in Paris at the Lycée Condorcet and the Ecole Normale. His early training was philosophical, and he remained throughout his life associated with the *Revue de métaphysique et de morale*. He passed his *agrégation* in 1892, and was invited to lecture at the Ecole des Sciences Politiques on the evolution of political ideas in England; this was to establish the course of his career. In 1900–1903 he published his first major work, *La formation du radicalisme philosophique en Angleterre*, a study tracing the development of the utilitarian doctrine from 1776 to 1832. As an offshoot of this project he also published a short study, *Thomas Hodgskin* (1903), which presents Hodgskin as a precursor of Marx. Halévy's major historical writings were the volumes of his *Histoire du peuple anglais au XIXe siècle* (1912–32), most notably vol. I, *England in 1815*, and vol. V, *Imperialism and the Rise of Labour*.

La formation du radicalisme philosophique is, among other things, a signal contribution to the history of economic thought. Halévy's subject is less utilitarianism in general than the application of utilitarian principles to criticize the established order and to justify grand proposals of reform: in sum, Philosophic Radicalism, or what Bentham referred to as the exposure of 'political fallacies'. The book offers a detailed exposition, at once historical and analytical, of works by Bentham and James Mill, and to a somewhat lesser extent the classical economists (Smith, Malthus, Ricardo). Halévy is at pains to demonstrate the connection between utilitarianism as a moral and political doctrine, and classical political economy. Indeed he summarizes his argument in the formula. 'The morality of the Utilitarians is their economic psychology put into the imperative' (Halévy, 1928, p. 478), a formula which nicely captures utilitarianism's debt to economics as well as the ambiguities inherent in the doctrine.

Bentham held that only the principle of utility – the principle of promoting the greatest happiness for the greatest number – can offer a satisfactory criterion for evaluating action. Not only is this principle commonly acceptable to reasonable men and women, it is moreover grounded in and reinforced by human psychology. Human beings are creatures who cannot but pursue pleasure and avoid pain. The difficulty in the argument arises in the comparison and summing up of individual pleasures. Is the greatest happiness of the greatest number simply a summation of individual happinesses egoistically pursued? Or does it require that an individual's pursuit of his private pleasure coincide with a pursuit of the greatest happiness of the greatest number? For his own part Bentham was ambiguous on this point: on the one hand, he acknowledged the (potential and actual) conflict between private interests and the public interest, and hence the need for molding or transforming human nature. This is the sphere of the 'artificial identification of interests', where as Halévy puts it, 'the science of the legislator must intervene to identify interests which are naturally divergent' (p. 508). On the other hand, Bentham argued, more optimistically, that there is social order 'realised spontaneously, by the harmony of egoisms' (p. 508). This is the part of the argument utilitarianism shares most closely with, or borrows from, classical economics. It provides the climax of Halévy's history:

insensibly, the progress of the new political economy had determined the preponderance within [Utilitarianism] of another principle, the principle according to which egoisms harmonise of themselves in a society which is in conformity to nature. From this new point of view, the fundamental moral notion for the theorists of Utilitarianism is no longer that of obligation, but that of exchange ... . The Utilitarian moralist dispenses the legislator from intervening just in so far as, by his advice and by his example, he tends, in conformity with the hypothesis of the political economists, to realise in society the harmony of egoisms (p. 478).

In the event, as Halévy shows in conclusion, this synthesis was precarious. The harmony of egoisms was too tenuous a factual basis for utilitarian morality, and the ambiguities of Philosophic Radicalism were supplanted by new and simplified versions of utilitarianism, like the 'Manchester philosophy'.

M. DONNELLY

SELECTED WORKS

1912–32. *History of the English People in the Nineteenth Century.* 6 vols, trans. E.I. Watkin and D.A. Barker, London: Benn, 1924–49.

1928. *The Growth of Philosophic Radicalism.* Trans. Mary Morris, London: Faber.

Hamilton, Alexander (1755–1804).

One of the founding fathers of the United States and Secretary of the Treasury in President Washington's cabinet, in which Thomas Jefferson served as Secretary of State. The two great men differed widely in their views about the destiny of the young nation. Jefferson wanted to preserve the position of the states and assign to the national government not much more than authority over foreign affairs. Hamilton favoured a strong and active central government. Jefferson was eager to preserve the rural economy in which he had grown up in Virginia. Hamilton proposed to promote economic development, especially manufacture, and vest in the national government the function of actively fostering such development. Jefferson took a dim view of public debts, paper money and financial institutions. Hamilton favoured them all. Jefferson was more of an egalitarian and had greater faith in the common man than Hamilton, who placed his trust in an alliance of government and the aristocracy of wealth: neither could flourish without the support of the other. Hamilton died in a duel with a political adversary during Jefferson's presidency, but his ideas were strong enough to survive him. The exigencies of the time caused Jefferson himself to adopt a number of Hamiltonian policies.

Thus Hamilton became the architect of what in *The Federalist* (1787, No. XI) he had called 'the great American system', later to be buttressed by such economic writers as the Careys, Daniel Raymond and Frederick List, and by Henry Clay in politics. He set forth his economic ideas in a series of state papers, published under his name when serving as Secretary to the Treasury. These papers are the first and second *Report on the Public Credit* (1790 and 1795), the *Report on a National Bank* (1790), and the *Report on Manufactures* (1791). The state papers are justly famous, not as repositories of economic analysis, but as a masterly presentation of a case of which Hamilton, who had been trained in the law, was an eloquent advocate.

The apotheosis of credit found in Hamilton's reports refers to public and private credit as well. According to Hamilton, credit is a substitute of capital almost as useful as gold and silver. It has, and Hamilton has no doubt about this, a tendency to lower the interest rate. If the public credit is in a bad state, it can only have deleterious effects on private credit. The preservation of a healthy credit system is thus an important task. Foreign creditors should enjoy the same protection as domestic ones. Domestic holders of the public debt should be protected against the imposition of taxes on the public funds, as foreign creditors should be protected from repudiation or expropriation.

With the help of the public debt it will be possible to promote the economic development of the country. Scrupulous attention must be paid to the rights of the creditors, both for the sake of public expediency and as a moral obligation. The public debt of the United States should be funded, that is, arrangements should be made for the service of the debt by putting aside funds for the payment of interest and principal. A funded debt has great benefits. It will facilitate the use of instruments of debt as money and bring about lower interest rates, and will result in an increase in land values, which have declined in consequence of the scarcity of money.

Hamilton also proposed that the Union assume responsibility for the debts of the states, and that the funding of the debt should be financed in part from new duties on imported spirits and taxes on domestic ones and on stills. These proposals met considerable opposition because of the windfall gains that would accrue to speculators who had purchased instruments of the debt at low prices. To obtain Jefferson's support for this measure, Hamilton had to agree that the future capital of the nation would be located in the South, that is, in what is now Washington, DC.

The national bank, which Hamilton proposed to establish, was designed to aid in the expansion of the money supply, thereby facilitating the payment of taxes, the reduction of interest rates, the fulfilment of public functions, and the development of the national economy. The bank was to be under private rather than public direction, with the government playing the role of a minor shareholder. When the question was raised whether the Constitution granted the Federal Government the authority to establish a bank, Hamilton resolved it by referring to 'implied powers', that is, the power to employ suitable means to pursue constitutional ends. This solution was to have far-reaching consequences for the future development of constitutional law.

The *Report on Manufactures* goes into considerable detail examining the relative merits of agriculture and industry. Hamilton underlines the merits of both and the benefits which each derives from the other. He stresses that both are productive, a point that had to be made, and made forcefully, in view of the teachings of the physiocrats. Hamilton demonstrates great ingenuity in enumerating the factors that are responsible for favourable effects of industrial development on the national income. Among these factors he mentions the division of labour, the more extensive use of machinery, the utilization of manpower that is not suited for agricultural pursuits, the promotion of immigration, the widening of opportunities for the exercise of entrepreneurial talent, and the strengthening of demand for agricultural products.

As far as international trade is concerned, Hamilton holds that the benefits from free trade are more imaginary than real because of the obstacles which foreign countries place in the way of United States exports. Moreover, foreign governments support domestic industries in various ways, and the United States should adopt similar policies by imposing protective and prohibitive duties, granting subsidies to domestic industries, and promoting internal improvements that facilitate the flow of commerce. Subsidies are liable to be abused, but their advantages outweigh the disadvantages. Lastly, Hamilton proposes that a board be established to promote economic development by bringing in skilled workers from abroad, rewarding useful improvements and inventions, paying premiums to importers of machinery, and similar means.

HENRY W. SPIEGEL

SELECTED WORKS

1790a. *Report on the Public Credit.* Reprinted, Washington: Government Printing Office, 1908.

1790b. *Report on a National Bank.* Reprinted, Washington: Government Printing Office, 1908.

1791. *Report on Manufactures*. Reprinted, Washington: Government Printing Office, 1913.
1795. *Report on the Public Credit*.
1934. *Papers on Public Credit, Commerce and Finance*. Ed. Samuel McKee, Jr. New York: Columbia University Press.

BIBLIOGRAPHY

Dorfman, J. 1946. *The Economic Mind in American Civilization 1606–1865*. Vol. 1, New York: Viking.
Spiegel, H.W. 1960. *The Rise of American Economic Thought*. Philadelphia: Chilton.

Hamilton, Earl Jefferson (born 1899). Hamilton was born on 17 May 1899 in Houlka, Mississippi. After graduating from the University of Mississippi (and coaching football and playing minor league baseball), he received in 1929 his doctorate in economics at Harvard. He taught at Duke, Northwestern, and finally for twenty years at the University of Chicago, during the banishment and eventual rehabilitation of the quantity theory of money.

Though he has worked on several topics in the early history of the Atlantic economy, his main contribution to economic science is the documentation of the dependence of the price level on the quantity of precious metals, 1351–1800. Spain through its centuries of prosperity and decline was his field of study, and he is accounted the major historian of that country (hon. Dr University of Madrid, 1967). Involved nearly from its beginnings in the 1920s with the International Committee on Price History, Hamilton constructed indexes of prices, wages and money from primary sources. The historical weight and economic ingenuity of his volumes, 1934, 1936, 1947, made them central to the modern quantity theory. Various attempts to revise his history of prices (attaching it to population, for instance) have had difficulties with the sheer mass of evidence that Hamilton accumulated, Kepler-like. Hamilton, further, is prominent in the thin, bright stream of historical economists *avant la lettre*. His combination of economic and historical erudition is a model of cliometrics, exhibited best in his lucid reply to his revisers (1960).

DONALD N. MCCLOSKEY

SELECTED WORKS

1934. *American Treasure and the Price Revolution in Spain, 1501–1650*. Cambridge, Mass.: Harvard University Press.
1936. *Money, Prices, and Wages in Valencia, Aragon, and Navarre, 1351–1500*. Cambridge, Mass.: Harvard University Press.
1947. *War and Prices in Spain, 1651–1800*. Cambridge, Mass.: Harvard University Press.
1960. The history of prices before 1750. In *XIe Congrès International des Sciences Historiques*, Stockholm.
1968. John Law. In *International Encyclopedia of the Social Sciences*, Vol. 9, New York: Macmillan.

Hamiltonians. The laws of motion for a perfect-foresight economy, whether centrally planned or competitive, can be described by a Hamiltonian dynamical system or by a simple perturbation thereof. The Hamiltonian dynamical system and the Hamiltonian function which generates it are named for their inventor, the great Irish mathematician William Rowan Hamilton (1805–1865).

Hamilton's differential equations serve as the basic mathematical tool of classical particle mechanics (including celestial mechanics). Let $x(t) = (x_1(t), \ldots, x_i(t), \ldots, x_m(t))$ and $y(t) = (y_1(t), \ldots, y_i(t), \ldots, y_m(t))$ be m-vectors dependent on time t. Let H be a continuous, differentiable function of x, y, and t, H:

$R^m \times R^m \times R \to R$. Think of H as the Hamiltonian function (HF) which generates Hamilton's differential equations,

$$dx_i(t)/dt = -\partial H(x(t), y(t), t)/\partial y_i(t)$$

and

$$dy_i(t)/dt = \partial H(x(t), y(t), t)/\partial x_i(t)$$

for $i = 1, \ldots, m$. If the Hamiltonian function H depends on time only through the variables $x(t)$ and $y(t)$, i.e., $\partial H/\partial t \equiv 0$, then the corresponding Hamiltonian dynamical system (HDS) is said to be *autonomous*. These differential equations are frequently interpreted in physics as solutions to some extremization problem. In mechanics for example, HDS is implied by the principle of least action. Since economic planning and many other economic problems involve maximization or minimization over time, it is unsurprising that the Hamiltonian formalism has substantial application in economics. Its appeal to economists goes much further than this. There is a duality (conjugacy, in the language of mechanics) between $x_i(t)$ and $y_i(t)$ which allows us to interpret one as a (primal) economic flow and the other as a (dual) economic price. Given this point of view, the Hamiltonian function (HF) itself has important economic interpretations. Hamiltonian dynamics not only arises in economic optimization problems but it also arises in descriptive economic models in which there is perfect foresight about asset prices. Hamiltonian dynamics applies in discrete time as well as in continuous time. In discrete time, the system of differential equations is replaced by a closely related system of difference equations. The right side of the equations describing Hamilton's law of motion need not be single-valued. The theory accommodates differential correspondences or difference correspondences, which naturally arise in economics.

Consider first the application of Hamiltonian approach to the theory of economic growth; see, e.g. the Cass–Shell (1976a) volume. A large class of economic growth models can be described by simple laws of motion based on the instantaneous production set T, with feasible production satisfying.

$$(c, z, -k, -l) \in T \subset \{(c, z, -k, -l) | (c, k, l) \geqslant 0\},$$

where c denotes the vector of consumption-goods outputs, z the vector of net investment-goods outputs, k the vector of capital-goods inputs, and l the vector of primary-goods inputs. There is an equivalent representation of static technological opportunities that is better suited to dynamic analysis: the representation of the static technology by its Hamiltonian function H.

Let p be the vector of consumption-goods prices and q be the vector of investment-goods prices. Define the Hamiltonian function $H(p, q, k, l)$ by

$$H(p, q, k, l) = \max_{(c', z')} \{pc' + qz' | (c', z', -k, -l) \in T\},$$

H is defined on the non-negative orthant and can be interpreted as the maximized value of net national product at the output prices (p, q) given input endowments (k, l).

Obviously, if we know the set T, then we know precisely the function H. If T is closed, convex, and permits free disposal, then H is continuous, convex and homogeneous of degree one in the output prices (p, q), and concave in the input stocks (k, l). If H is a function of (p, q, k, l) which is continuous, convex and homogeneous of degree one in (p, q), and concave in (k, l), then H corresponds to a unique T among closed, convex technologies permitting free disposal. In many dynamic applications, it is only the H representation which matters. Relax, for example, the free-disposal assumptions on T. For a given function H, the set T might not be unique, but the dynamics would be independent of the particular set T which generated the function H. Relax, as another example, the assumption that T is convex. Given an H which is convex in (p, q) and concave in (k, l), the set T will not be unique, but the continuous dynamics (HDS) will not be altered in an essential way.

Representation of the static technology by the Hamiltonian function permits one to describe the economic laws of motion as a Hamiltonian Dynamical System. In continuous time, the motion is described by

$$\dot{k}(t) \in \partial H(p(t), q(t), k(t), l(t))/\partial q(t)$$

(HDS)

$$\dot{q}(t) \in -\partial H(p(t), q(t), k(t), l(t))/\partial k(t)$$

where $\dot{k}(t)$ and $\dot{q}(t)$ are vectors of time derivatives and $(\partial H/\partial q)$ and $(\partial H/\partial k)$ are gradients (derivatives when H is differentiable). The first line of (HDS) is immediate from the definition of net investment since it reduces to $\dot{k}(t) = z(t)$, where $z(t)$ is the vector of net investment. The second line is an equal-asset-return condition which reduces to $\dot{q}(t) + r(t) = 0$, where $r(t)$ is the dual vector of shadow rental rates.

For discrete time, the Hamiltonian dynamical system is

$$k_{t+1} \in k_t + \partial H(p_t, q_t, k_t, l_t)/\partial q_t$$

(HDS)′

$$q_{t+1} \in q_t - \partial H(p_{t+1}, q_{t+1}, k_{t+1}, l_{t+1})/\partial k_{t+1}.$$

Line 1 is equivalent to $k_{t+1} = k_t + z_t$, and line 2 is equivalent to $q_{t+1} - q_t - r_{t+1} = 0$, where z_t is the time-t gross investment vector and r_{t+1} is the dual vector of shadow capital-goods rental rates in period $(t + 1)$.

For openers, let us analyse the case where H is autonomous. This occurs if $p(t) = \bar{p}$ and $l(t) = \bar{l}$ for (HDS) or $p_t = \bar{p}$ and $l_t = \bar{l}$ for (HDS)′. Let (q^*, k^*) be a rest point to (HDS) or (HDS)′. Hence, we have

$$0 \in \partial H(\bar{p}, q^*, k^*, \bar{l})/\partial q,$$

$$0 \in \partial H(\bar{p}, q^*, k^*, \bar{l})/\partial k.$$

Consider the linear approximations about (q^*, k^*) of (HDS) and (HDS)′ (taken, for example, as if H were quadratic). Study the characteristic roots of the linearized systems. A simple but remarkable theorem due to Poincaré tells us that if λ is a root for the linearized, *autonomous* version of (HDS) then so is $-\lambda$. For the linearized, *autonomous* version of (HDS)′, we have if λ is a root, then so also is $1/\lambda$. If for (HDS), we could rule out pure imaginaries (Re $\lambda \neq 0$), then we would have: The dimension of the manifold in (q, k) – space of solutions tending to (q^*, k^*) as $t \to \infty$ is equal to the dimension of the manifold of solutions tending to (q^*, k^*) as $t \to -\infty$. This is the *saddle-point property*, where the manifold of forward solutions and the manifold of backward solutions each have dimension equal to half the total dimension of the space. Similarly, we would have the saddle-point property for (HDS)′, if the modulus $|\lambda|$ is unequal to unity.

Poincaré's result nearly gives us the saddle-point property. In the autonomous cases, the saddle-point property can be assured if the geometry of the Hamilton function is correct. We need to add very little to the convexity-concavity assumption (see Cass and Shell (1976b) and Rockafellar (1976)). Strict convexity in q and strict concavity in k will do the trick. So will a weaker uniform Hamiltonian steepness condition, which reduces to a value-loss condition; see, e.g., McKenzie (1968), and Cass–Shell (1976b).

What about non-autonomous systems, such as optimal economic growth with the constant, positive discount rate ρ? Here $c(t)$ or c_t is a scalar called felicity and usually denoted in optimal-growth problems by $u(t)$ or u_t. In this case, present prices must satisfy

$$-\dot{p}(t)/p(t) = \rho$$

or

$$-(p_t - p_{t-1})/p_t = \rho.$$

For simplicity, allow only for a single fixed factor and adopt the convention $l(t) = 1$, or $l_t = 1$.

It is natural then to re-express the systems (HDS) and (HDS)′ in terms of current prices $Q \equiv q/p$, rather than in terms of present prices q. We then have

$$\dot{k} \in \partial H(Q, k)/\partial Q$$

(PHDS)

$$\dot{Q} \in -\partial H(Q, k)/\partial k + \rho Q$$

and

$$k_{t+1} \in k_t + \partial H(Q_t, k_t)/\partial Q_t$$

(PHDS)′

$$Q_{t+1} \in Q_t - \partial H(Q_{t+1}, k_{t+1})\partial k_{t+1} + \rho Q_t.$$

The systems (PHDS) and (PHDS)′ are *perturbed* Hamiltonian dynamical systems. We no longer have Poincaré's root-splitting theorems in pure form: the roots split but not about 0 for (HDS) nor 1 for (HDS)′. The trick here is to strengthen the geometry of H to give a saddle-point property or something like it.

This is the basics of the approach taken by Cass and Shell (1976b), Rockafellar (1976) and Brock and Scheinkman (1976). Conditions are found on H which assure that either (PHDS) or (PHDS)′ along with transversality conditions defines a globally stable system. It suffices to strengthen the convexity-concavity of H by an amount dependent on ρ or (weaker) to strengthen the steepness of H by an amount dependent on ρ. (The Lyapunov function which does the trick is $V = (Q - Q^*)$ $(k - k^*)$ in the continuous-time model.)

The Hamiltonian approach *through the Hamiltonian function* has proved remarkably successful in establishing sufficient conditions for the saddle-point property and related stability questions in a class of optimal economic growth models. The parallel programme of using the Hamiltonian formalism in optimal-growth theory to yield sufficient conditions for cycling or other dynamic configurations has not yet been pursued in a systematic fashion but should prove equally successful when applied. The success of the Hamiltonian approach in decentralized and descriptive growth theory has so far been very limited; see, Cass and Shell (1976b, Section 4). This has been disappointing. I still hope to see the Hamiltonian approach playing a pivotal technical role in, say, the dynamical analysis of overlapping-generations models, but there has not been much tangible encouragement for this hope.

Many of us first met Hamiltonian dynamical systems as necessary conditions for intertemporal maximization in the form of Pontryagin's Maximum Principle; see Pontryagin et al. (1962). See Shell (1967) for applications to economics and references.

Samuelson and Solow (1956) were probably the first in economics to mention the Hamiltonian formalism. For some of the history of Hamiltonian dynamics in economics, mathematics, and physics, and for some of the classical references see Magill (1970).

KARL SHELL

See also OPTIMAL CONTROL AND ECONOMIC DYNAMICS; SADDLEPOINTS.

BIBLIOGRAPHY

Brock, W.A. and Scheinkman, 1976. Global asymptotic stability of optimal economic systems with applications to the theory of economic growth. In Cass and Shell (1976a).

Cass, D. and Shell, K. 1976a. *The Hamiltonian Approach to Dynamic Economics.* New York: Academic Press. [Reprinted from the *Journal of Economic Theory* 12, February 1976, Symposium: 'Hamiltonian dynamics in economics'.]

Cass, D. and Shell, K. 1976b. The structure and stability of competitive dynamical systems. In Cass and Shell (1976a).

Magill, M.J.P. 1970. *On a General Economic Theory of Motion.* Berlin: Springer-Verlag.

McKenzie, L.W. 1968. Accumulation programs of maximum utility and the von Neumann facet. In *Value, Capital and Growth*, ed. J.N. Wolfe, Edinburgh: Edinburgh University Press.

Pontryagin, L.S. 1962. *The Mathematical Theory of Optimal Processes.* New York: Interscience.

Rockafellar, R.T. 1976. Saddlepoints of Hamiltonian systems in convex Lagrange problems having nonzero discount rate. In Cass and Shell (1976a).

Samuelson, P.A. and Solow, R.M. 1956. A complete model involving heterogeneous capital goods. *Quarterly Journal of Economics* 70(4), November, 537–62.

Shell, K. (ed.) 1967. *Essays on the Theory of Optimal Economic Growth.* Cambridge, Mass.: MIT Press.

Shell, K. 1969. Applications of Pontryagin's maximum principle to economics. In *Mathematical Systems Theory and Economics, I*, ed. H.W. Kuhn and G.P. Szegö, Berlin: Springer-Verlag, 241–92.

Hammarskjöld, Dag (1905–1961). Hammarskjöld was born in Jönköping, Sweden, and died in an aircrash near Ndola in Zambia. He came from a family (which was knighted early in the 17th century) with a long tradition of public service; his father Hjalmar Hammarskjöld was Prime Minister of Sweden in 1914–17. Hammarskjöld's achievements were many and varied: a PhD at the University of Stockholm, 1933; Docent at the University of Stockholm, 1933–41; Under Secretary of the Ministry of Finance, 1936–45; Chairman of the Board of Governors of the Bank of Sweden, 1941–8; Under Secretary of the Foreign Office, 1946–51; Vice-chairman of the executive committee of Organization for European Economic Cooperation, 1948–9; an expert non-party member of the Swedish Cabinet as Minister without portfolio in charge of foreign economic relations, 1951–3; Vice-chairman of Sweden's delegation to the UN General Assembly in 1952 and chairman in 1953; elected Secretary General of the UN for 1953–8 and re-elected for a five-year term in 1958; Fellow of the Swedish Academy of Letters, 1954; posthumously awarded the Nobel Peace Prize, 1961.

While Hammarskjöld is mainly known for his outstanding career as a civil servant and international statesman, he also made a contribution to economics. After his MA in 1928 he became secretary to the Royal Commission on Employment (1927–35). Hammarskjöld's dissertation (1933) was published as a report to the committee and he also wrote the theoretical introduction to its final report. The aim of the dissertation was to show the determinants of the price level of consumer goods for an expired period, which implies an ex post analysis. Hammarskjöld went beyond the earlier formulas of Keynes and Lindahl, since his construction had profit as the mechanism by which given changes in prices and purchasing power are transmitted to the next period, i.e. a form of disequilibrium process. In this context, he was the first among Swedish economists actually to define the length of a period, namely, the duration of time for which plans are unchanged determines the length of the unit period. However, his work was not particularly influential among his colleagues, since his exposition was extremely complicated. During his career as a civil servant he published very little in the economic field.

BJÖRN HANSSON

See also STOCKHOLM SCHOOL.

SELECTED WORKS

1932. Utkast till en algebraisk metod för dynamisk prisanalys. (A sketch of an algebraic method for a dynamic analysis of prices.) *Ekonomisk Tidskrift* 34, 157–76.

1933. *Konjunkturspridningen. En teoretisk och historisk undersökning.* (The propagation of business cycles. A theoretical and historical investigation.) Stockholm: P.A. Norstedt.

Hammond, John Lawrence le Breton (1872–1949) and **Lucy Barbara** (1873–1961). Lawrence Hammond was born in Yorkshire in 1872. He married Barbara Bradby in 1901; they had no children. Both the Hammonds received a classical education, which they drew on in their literary work. At Oxford, Lawrence was a Scholar of St John's College and Barbara of Lady Margaret Hall, where she made a striking impression as an early feminist. She became active in social work in London at the turn of the century, while Lawrence was making his career as a Liberal journalist and later as Secretary of the Civil Service Commission (1907–13). But their increasingly precarious health (hers tubercular, his mainly coronary) led to a steady withdrawal to a life of authorship in the country, punctuated by Lawrence's intermittent work for the *Manchester Guardian* in later years. It is as pioneer social historians that they are remembered, especially for their 'labourer' trilogy. Their account of how agricultural workers fared under the enclosure measures of the period 1760–1830 opened up a far-reaching debate. They did not deny the economic rationality of the process, but pointed to the way in which its costs were borne by the rural poor (Hammond and Hammond, 1911). Their work was given contemporary salience by the inception of Lloyd George's Land Campaign in 1913. In turning their attention to the urban working class, the Hammonds helped establish the 'pessimistic' view of the Industrial Revolution. Again, they did not disparage industrialization itself but focused on its exploitative effects, given the prevailing ideologies of an age which took social inequality for granted (Hammond and Hammond, 1917). Published amid wartime planning for reconstruction, their findings once more fed current political debate. Finally, the Hammonds analysed the impact of technological change in making skilled craftsmen redundant in the early 19th century, and offered a new understanding of the Luddite movement (Hammond and Hammond, 1919).

The Hammonds' view of the Industrial Revolution was countered in the 1920s by that of J.H. Clapham, with all his authority as Professor of Economic History at Cambridge. He mounted an 'optimistic' case on the standard of living by constructing a real wage index which showed substantial gains by industrial workers (Clapham, 1926). The Hammonds publicly bowed on this point in face of the apparent weight of the new statistical evidence (Hammond, 1930), though subsequent research has shown that Clapham's claims themselves depended upon a flawed price index. Insofar as the Hammonds rested their interpretation upon a quantitative assessment, therefore, it was by no means overturned; and its main thrust, in fact, was qualitative in its concern for the impact of economic change upon the lives of ordinary people. While they depicted the 'bleak age' of early industrialization, they also pointed to the civilizing process which urban life underwent from the middle of the 19th century (Hammond and Hammond, 1930). Though often identified as socialists, the Hammonds remained liberal reformists in their outlook. Their work came to serve as a straw man, the object of ritual slights from a new generation of 'optimists' among professional economic historians; but its scholarly credentials have survived with remarkable resilience.

Peter Clarke

SELECTED WORKS

1911. *The Village Labourer, 1760–1832*. London: Longman, Green.
1917. *The Town Labourer, 1760–1832*. London: Longman, Green.
1919. *The Skilled Labourer, 1760–1932*. London: Longman, Green.
1925. *The Rise of Modern Industry*. London: Methuen.
1930. *The Age of the Chartists*. Revised as *The Bleak Age*, Harmondsworth: Penguin, 1947.
[Hammond, J.L.] 1930. The Industrial Revolution and discontent. *Economic History Review* 2, January, 215–28.

BIBLIOGRAPHY

Clapham, J.H. 1926. *An Economic History of Modern Britain*. Vol. 1: The Early Railway Age, 1820–1850. Cambridge: Cambridge University Press.

Hansen, Alvin (1887–1975).   Alvin Hansen grew up in Viborg, South Dakota, a small rural community with a one-room school house and traditional values. Preferring academic pursuits to farm work, he proceeded to Sioux Falls for his high school education and then to Yankton College for his BA degree. Several years of high school teaching followed, with rapid advancement to principal and superintendent. The financial basis for his graduate work thus laid, Hansen entered the University of Wisconsin in 1914, where John R. Commons and R.T. Ely were to impress him with the importance of data and their institutional setting. In 1919 he moved to Brown University as Assistant Professor. There he completed his dissertation, later published as *Cycles of Prosperity and Depression* (1921). He then accepted a position at the University of Minnesota, where he remained for nearly twenty years. His major works during the 1920s included a solid *Principles* text, co-authored with F.B. Garver (1928) and an historical study of *Business Cycle Theory* (1927). Ranging from Malthus to Spiethoff and Hawtrey, stress was on structural shifts in investment rather than on monetary factors, and special attention was given to the interaction of short cycles with longer waves of economic development.

A Guggenheim fellowship in 1928 permitted extensive travel abroad, an experience that he continued to cherish and renew in later years. The early thirties also brought a growing policy involvement outside the campus, activities that in subsequent years were to claim an increasing share of his time. Such early activities included that of Director of Research for the Committee of Inquiry on International Economic Relations (1933–34) and service as adviser on Trade Agreements to Secretary Cordell Hull.

In 1936 Harvard University had received a grant to establish the Littauer School of Public Administration and Hansen was appointed as its first Lucius S. Littauer Professor of Political Economy. As fortune had it, his arrival at Harvard in the fall of 1937 closely followed the appearance of Keynes's *General Theory*. Hansen, distressed by the wastes of the Great Depression soon (though with some initial hesitation) adopted the Keynesian approach. With Harvard's Fiscal Policy Seminar as his base, Hansen became the leading analyst and expositor of Keynesian economics in the United States. Driven by his enthusiasm for new ideas, his determination to find policy solutions, and his eagerness to learn as well as to teach, the seminar left a deep impact on the course of macroeconomics. The still obscure components of the Keynesian system had to be sorted out and new tools, such as the concept of governmental net contribution, the multiplier-accelerator model and the balanced budget theorem, were forged. With the application of these new tools to the setting of the US economy as its challenge, the seminar thus became the training ground for a generation of US policy economists.

The output of these years may be traced in Hansen's writings, beginning with the two key volumes of *Full Recovery and Stagnation* (1938) and *Fiscal Policy and Business Cycles* (1941). Other volumes followed, including *Business Cycles and National Income* (1951) and his widely used *A Guide to Keynes* (1953). The persistent theme was that of unemployment, caused by a failure of private investment to match the level of saving at a full employment income. With the effectiveness of monetary policy reduced by inelastic investment and high liquidity preference in a sluggish economy, the required level of aggregate demand would have to be provided by fiscal expansion, responded to in the private sector by a multiplier-accelerator process. The need for expansionary fiscal policy, however, would not be one of pump-priming only. Linking back to his earlier interest in the long waves of the cycle, the weakness of the economy was seen as the downward phase of a long wave, with the declining population growth the most depressing factor. The stagnation thesis, offered in his presidential address before the American Economic Association (1937) placed the Keynesian model in a historical perspective and once more emphasized the strategic role of expansionary budget policy. Events, to be sure, proved different. World War II generated massive budgetary expansion and a strengthened postwar economy called for a correction, combining the traditional role of monetary policy with that of fiscal controls. Hansen, the pragmatist welcomed the neoclassical synthesis of the mid-sixties.

While macro policy and stabilization remained his major concern, his activities during the Harvard years covered a much wider range. As a member of the Advisory Council on Social Security in 1937–38, he helped to shape the Social Security System. During 1941–43 he served as Chairman of the US–Canadian Joint Economic Commission, and from 1940 to 1945 he acted as Economic Advisor to the Federal Reserve Board. At the close of the war he participated in the monetary reconstruction of Bretton Woods and the birth of IMF. At the same time, he played a strategic role in the creation of the Full Employment Act of 1946 and the Council of Economic Advisers. After retiring from Harvard in 1956, Hansen remained in Belmont, Massachusetts, until 1972, when he joined his daughter in Virginia. He died there in 1975.

Throughout Hansen's work, the goal of full employment was central, as was the need for fiscal action to achieve it. His social philosophy was expressed 'in the democratic ideal of providing for all individuals a reasonable approach to equality of opportunity'. Beyond this, he was pragmatic and non-ideological in approach. For him, economics – as James Tobin put it when presenting him with the Walker Medal at his 80th birthday – was a science for the service of mankind.

Richard A. Musgrave

SELECTED WORKS

1921. *Cycles of Prosperity and Depression in the United States, Great Britain and Germany: A Study of Monthly Data 1902–1908.* Madison: University of Wisconsin Press.
1927. *Business-Cycle Theory: Its Development and Present Status.* Boston: Ginn.

1928. (With F.B. Garver.) *Principles of Economics*. Boston: Ginn.
1938. *Full Recovery or Stagnation?* New York: Norton.
1941. *Fiscal Policy and Business Cycles*. New York: Norton.
1951. *Business Cycles and National Income*. New York: Norton.
1953. *A Guide to Keynes*. New York: McGraw-Hill.

Hardenberg, Georg Friedrich Philipp von. *See* NOVALIS.

Hardy, Charles Oscar (1884–1948). Financial economist; born Island City, Missouri, 2 May 1884, died 30 November 1948. Hardy held posts at the University of Kansas and between 1918 and 1922 was a lecturer at the University of Chicago, where he had received the PhD in 1916. Hardy was also vice-president of the Federal Reserve Bank of Kansas City and was long associated with the Brookings Institution and with monetary policy debates of his time. As a Brookings scholar Hardy authored a number of books dealing with currency problems, focusing especially on the functioning of the gold standard. Hardy argued (1936) that the increase in the world's monetary gold stock (since 1929) led to undesirable expansions in floating credit and the potential for monetary instability. Further he thought that balance of trade shifts along with changes in long-term investments created havoc in the central bank's ability to have an impact upon domestic stability. He therefore argued for large-scale modifications in the gold standard as it was then practised. Hardy later advocated an activist fiscal policy, coordinated with monetary policy, to promote economic stabilization.

Hardy's most original and important contribution was to the theory of risk. In a 1923 paper (co-authored with Leverett S. Lyon, 1923b) Hardy analysed the functioning of futures markets in detail, carefully and correctly explaining why hedging contracts cannot be expected to provide complete protection to the user against the risk of adverse price changes. In the same year Hardy authored a pre-Knightian textbook on risk (1923a). In it Hardy features uncertainty as well as risk as elements in production and investment, crediting his colleague Frank Knight for access to preliminary versions of Knight's *Risk, Uncertainty, and Profit*.

ROBERT B. EKELUND, JR.

SELECTED WORKS
1923a. *Risk and Risk-Bearing*. Chicago: University of Chicago Press.
1923b. (With L.S. Lyon.) The theory of hedging. *Journal of Political Economy* 31, 276–87.
1932. *Credit Policies of the Federal Reserve System*. Washington: The Brookings Institution.
1936. *Is There Enough Gold?* Washington: The Brookings Institution.

harmony. *See* ECONOMIC HARMONY.

Harris, Seymour Edwin (1897–1975). Harris was born in Brooklyn, New York, and graduated from Harvard University, where he also took his doctorate. His career, apart from the World War II period and a few post-retirement years at the University of California at La Jolla, was spent at Harvard. In World War II, he was in charge of the pricing of exported and imported products and various liaison tasks for the Office of Price Aministration. Throughout his life he undertook numerous regional and developmental assignments in New England and was one of the founders of the highly successful Massachusetts community college system.

Harris's early academic work, including a major history of the Federal Reserve System, was competent, orthodox and, as

he would later view it, uninspired. Upward progress in his academic career at Harvard was also gradual and unspectacular, a circumstance related at the time to his Jewish origins. In later years he emerged as one of the most highly regarded members of the Cambridge (USA) economic and university community. He became a highly respected Chairman of the Harvard Economics Department and was the Editor of the *Review of Economics and Statistics* and of numerous essay collections by fellow economists. He did not entirely escape criticism from his more relaxed colleagues for his prodigious work and publication schedule. President John F. Kennedy, shortly before he was killed, told of his intention of making Harris his next appointment to the Board of Governors of the Federal Reserve System.

From his earlier orthodox, even conservative tendencies Harris was released by Keynes and the New Deal. His work came to reflect a strong commitment to Keynesian economics and policy and to the broad welfare measures of the Roosevelt, Kennedy and Johnson years. He was not a compelling writer; in his books, however, this was more than compensated for by the solid competence of his research and preparation, his strongly compassionate views on welfare issues and his very evident desire to extend knowledge on a great range of subject matter. On the economics of health care, education, social security, international monetary policy, central-bank policy, monetary history and literally a dozen other topics, he provided the basic source material from which legislators learned what could be done, what should be done and how it might be done. A full listing of his works would be among the longest in this Dictionary. Among the prominent later examples are those listed below.

J.K. GALBRAITH

SELECTED WORKS
1947. *The New Economics: Keynes' Influence on Theory and Public Policy*. New York: A.A. Knopf.
1948a. *How Shall We Pay For Education? Approaches to the Economics of Education*. New York: Harper.
1948b. *Saving American Capitalism; A Liberal Economic Program*. New York: A.A. Knopf.
1949. *The Market for College Graduates, and Related Aspects of Education and Income*. Cambridge, Mass.: Harvard University Press.
1952. *The Economics of New England; Case Study of an Older Area*. Cambridge, Mass.: Harvard University Press.
1953. *National Health Insurance, and Alternative Plans for Financing Health*. Foreword by Alfred Baker Lewis. New York: League for Industrial Democracy.
1955. *John Maynard Keynes, Economist and Policy Maker*. New York: Scribner.
1962. *Higher Education: Resources and Finance*. New York: McGraw-Hill.
1964. *The Economics of American Medicine*. New York: Macmillan.

Harris–Todaro model. The replacement of the equality of wages by the equality of expected wages as the basic equilibrium condition in a segmented, but homogeneous, labour market has proved to be an idea of seminal importance in development economics. Attributed originally to Todaro (1968, 1969) and Harris–Todaro (1970), and commonly referred to as the Harris–Todaro hypothesis, the idea was very much in the air around the late 1960s as can be seen from the contemporaneous writings of Akerlof–Stiglitz (1969), Blaug et al. (1969) and Harberger (1971), among others.

The motivation for the Harris–Todaro hypothesis lies in an attempt to explain the persistence of rural to urban migration in the presence of widespread urban unemployment, a pervasive phenomenon in many less developed countries. It is natural to ask why such unemployment does not act as a deterrent to further migration. According to the Harris–Todaro hypothesis, the answer lies in the migrant leaving a secure rural wage W_r for a higher expected urban wage W_u^e even though the latter carried with it the possibility of urban unemployment. The expected wage is computed by using the rate of urban employment as an index for the probability of finding a job. Thus

$$W_u^e = W_u(L_u/(L_u + U)) + 0(U/(L_u + U))$$
$$= W_u(1/(1 + U/L_u)) = W_u(1/(1 + \lambda)) \qquad (1)$$

where W_u is the urban wage, L_u is the number of urban employed, U the number of urban unemployed and λ the rate of urban unemployment. Thus, the Harris–Todaro hypothesis is precisely formulated by the equilibrium condition

$$W_r = W_u^e \Leftrightarrow W_u = W_r(1 + \lambda). \qquad (2)$$

Once the Harris–Todaro equilibrium condition is embedded in a two-sector, so-called general equilibrium model (see Johnson, 1971), we obtain the Harris–Todaro model. However, it should be noted that the Harris–Todaro hypothesis introduces a further unknown, namely the equilibrium rate of unemployment, and thus, in contrast to the standard two-sector model, the Harris–Todaro model must be buttressed by a theory of urban wage determination. The simplest setting is the one originally adopted by Harris–Todaro and subsequently by Bhagwati–Srinivasan (1973, 1974, 1977), and others. This work assumes the urban wage to be an exogenously given constant and typically rationalizes it as a consequence of government fiat.

In the 1970s, however, several theories of endogenous urban wage determination were proposed. Foremost among these is the work of Joseph Stiglitz who provides a microfoundation for the urban wage in terms of labour-turnover (Stiglitz, 1974) or in terms of biological efficiency considerations (Stiglitz, 1976). One may also mention in this context the work of Calvo (1978), who sees the equilibrium urban wage as an outcome of trade union behaviour; and of Calvo–Wellisz (1978), who see a higher urban wage as a consequence of costly supervision. At this stage of the development of the literature, each theory of urban wage determination led to a particular version of the Harris–Todaro model and the common structural similarities were obscured.

In Khan (1980a), the elementary observation is made that all these variants of the Harris–Todaro model could be studied under one rubric if the determination of urban wages is seen in a somewhat more abstract way, i.e.,

$$W_u = \Omega(W_r, \lambda, R, \tau) \qquad (3)$$

where R is the rental on capital and τ a shift parameter. This led to the, so-called, generalized Harris–Todaro (GHT) model whose importance lay not so much in synthesizing the several variants of the Harris–Todaro model but in emphasizing the points of contact of this literature with the trade theory literature. In particular, when (3) collapses to

$$W_u = W_r \qquad (4)$$

that is, when the elasticities of the omega function with respect to λ, R and τ are all zero, we obtain the Heckscher–Ohlin–Samuelson (HOS) model in the case when capital is intersectorally mobile or the Ricardo–Viner model in the case when it is not (for details on these basic constructions of trade theory, see, for example, Caves–Jones, 1985).

Let us now consider in some detail the GHT model in the case when capital is intersectorally mobile. Let a country consist of an urban and a rural sector, indexed by u and r respectively, and be endowed with positive amounts of labour L and capital K. Let the ith sector produce a commodity i in amount X_i in accordance with a production function

$$X_i = F_i(L_i, K_i) \qquad i = u, r, \qquad (5)$$

which is assumed to exhibit constant returns to scale and is twice continuously differentiable and concave. The allocation of labour and capital, L_i and K_i, is determined through marginal productivity pricing. Thus, we have

$$P_r F_r^K = R = P_u F_u^K \qquad (6)$$

$$P_r F_r^L = W_r, \quad \text{and} \quad P_u F_u^L = W_u \qquad (7)$$

where F_i^j is the derivative of $F_i(i = u, r)$ with respect to j ($j = L, K$). The country is considered too small to influence the positive international prices of the two commodities, P_u and P_r. On rewriting the equilibrium condition (2) in a slightly more general form,

$$W_u = \rho W_r(1 + \lambda); \qquad \rho \text{ a shift parameter,} \qquad (8)$$

(5), (6), (7) and (8), along with the material balance equations below, complete the specification of the model.

$$K_r + K_u = K \quad \text{and} \quad L_r + L_u(1 + \lambda) = L. \qquad (9)$$

The first point to be noticed about this model is a *decomposability* property whereby the factor prices, W_u, W_r and R and the unemployment rate λ are independent of the endowments of labour and capital and depend solely on P_u, P_r and the shift parameters τ and ρ. This can be seen most easily if we subsume the marginal productivity conditions (6) and (7) into price equal unit cost equations

$$P_i = C_i(W_i, R) \qquad i = u, r. \qquad (10)$$

This allows one to decompose the model into a subsystem comprising equations (8) and (4) along with (10).

This basic observation leads to several interesting characteristics of the equilibria of the GHT model. First, the market rural wage and market rental correctly measure the social opportunity cost of labour and capital if we use the international value of GNP as the relevant measure of social welfare. Second, despite the presence of a distorted labour market, there is no possibility of immiserizing growth. Third, an increase in capital (labour) increases the output of the capital (labour) intensive commodity and decreases the output of the labour (capital) intensive commodity provided the intensities are measured in *employment adjusted* terms, that is

$$\frac{k_u}{1 + \lambda} = \frac{K_u}{L_u(1 + \lambda)} \gtreqless \frac{K_r}{L_r} = k_r. \qquad (11)$$

This third property is an analogue of the Rybczynski property of the HOS model. Not surprisingly, we also obtain an analogue of the Stolper–Samuelson property whereby the effect of changes in international prices on factor returns depends on factor intensities, provided these are now measured in *elasticity adjusted* terms. The urban sector is said to be capital intensive in elasticity adjusted terms if

$$\theta_{rL}(\theta_{uK}(1 - e_\lambda) + \theta_{uL}e_R) > \theta_{uL}\theta_{rK}(e_W - e_\lambda) \qquad (12)$$

where θ_{ij} is the share of the jth factor ($j = K, L$) in the ith sector ($i = u, r$), and e_i is the elasticity of the $\Omega(\cdot)$ function with respect to the relevant variable.

In the setting where e_W equals unity and e_λ, e_R and e_τ are all zero, (11) and (12) collapse to the conventional physical and value intensities of Magee (1976) and Jones (1971) for the HOS model with proportional wage differentials. Under the further specialization that ρ in (8) equals unity, there is no difference between these two kinds of intensities and a perfect correspondence between the Rybczynski and Stolper–Samuelson theorems.

The appearance of the physical and value intensities of the wage-differential model leads us to inquire into the possibility of downward-sloping supply curves of X_r and X_u. This is indeed possible and a sharp generalization is available in the result that there are perverse price-output responses in the GHT model if and only if the employment adjusted factor intensities do not conflict with the elasticity adjusted intensities; see Khan (1980b) for details.

Another direct consequence of the decomposability property of the model is a generalization of the Bhagwati (1968), Johnson (1971), Brecher–Alejandro (1977) paradox. This states that capital inflow in the presence of a tariff and with full repatriation of its earnings is immiserizing if and only if the imported commodity is capital intensive in employment adjusted terms. This result is independent of the various mechanisms for the determination of urban wages. For details, see Khan (1982a).

The results that we have presented so far have a trade-theoretic flavour but one question that has remained in the forefront of analytical work on the Harris–Todaro model relates to the effect of urban wage subsidies on urban unemployment and urban output. A seminal result here is the Corden–Findlay (1975) paradox which draws attention to the fact that urban employment and urban output could rise if the urban wage is increased. This question has now been resolved by Khan (1980b) and Neary (1981).

So far we have focused on the comparative-static properties of the Harris–Todaro equilibrium. It is also worth emphasizing that the actual existence of the Harris–Todaro equilibrium cannot be taken for granted and must be proved. In the original Harris–Todaro model with an exogenously given rigid wage, equilibrium exists if and only if the rural sector is more capital intensive in employment adjusted terms; see Khan (1980a). Furthermore, following Neary's (1978) lead for the wage differential model, one can present 'reasonable' adjustment processes of the Marshallian type under which the Harris–Todaro equilibrium is locally asymptotically stable if and only if the employment adjusted factor intensities do not conflict with the elasticity adjusted intensities. Since the elasticity adjusted intensities of (12) collapse to $\theta_{rL}\theta_{uK}$ in the Harris–Todaro model with a rigid wage, we have the satisfying result that the criteria for the existence of equilibrium and its stability coincide.

There are several aspects of the Harris–Todaro model that we have not covered. In particular, we have confined our attention to the version with intersectorally mobile capital. In many ways, the case of sector-specific capital is more difficult and also more interesting. In addition, even for the version that we have discussed, several substantive questions of interest have not been touched on. These relate to gains from trade which now depend on the asymmetric nature of the model and as to whether the rural or the urban commodity is being exported as in Khan–Lin (1982); to the possibility of underemployment as in Bhagwati–Srinivasan (1977); to the possibility of educated unemployment as in Chaudhuri–Khan (1984); to the consequences of a distorted capital market as in Khan–Naqvi (1983); to the interaction of ethnic groups as in Khan (1979) and Khan-Chaudhuri (1985); to the introduction

of an informal sector as in Fields (1975) and Stiglitz (1982); to cost–benefit analyses as in Svinivasan–Bhagwati (1975) and Stiglitz (1977, 1982). However, what we have hoped to show is that the GHT model is a versatile and useful analytical tool for a variety of questions arising in international and development economics.

M. Ali Khan

See also DISGUISED UNEMPLOYMENT; DUAL ECONOMIES; INTERNATIONAL MIGRATION; LABOUR SURPLUS ECONOMIES; SEGMENTED LABOUR MARKETS.

BIBLIOGRAPHY
Akerlof, G. and Stiglitz, J.E. 1969. Capital, wages and structural employment. *Economic Journal* 79, 269–81.
Bhagwati, J.N. 1968. Distortions and immiserizing growth. *Review of Economic Studies* 35, 481–5.
Bhagwati, J.N. and Srinivasan, T.N. 1971. The theory of wage differentials: production response and factor price equalization. *Journal of International Economics* 1, 19–35.
Bhagwati, J.N. and Srinivasan, T.N. 1973. The ranking of policy interventions under factor market imperfections: the case of sector-specific sticky wages and unemployment. *Sankhya*, Series B, 35(4), December, 405–20.
Bhagwati, J.N. and Srinivasan, T.N. 1974. On reanalyzing the Harris–Todaro model: policy rankings in the case of sector-specific sticky wages. *American Economic Review* 64, 502–8.
Bhagwati, J.N. and Srinivasan, T.N. 1977. Education in a job ladder model and the fairness-in-hiring rule. *Journal of Public Economics* 7(1), 1–22.
Blaug, M., Layard, P.R.G. and Woodhall, M. 1969. *The Causes of Graduate Unemployment in India*. London: Allen Lane.
Brecher, R.A. and Diaz-Alejandro, C.F. 1977. Tarrifs, foreign capital and immiserizing growth. *Journal of International Economics* 7, 317–22.
Calvo, G.A. 1978. Urban unemployment and wage determination in LDC's: trade unions in the Harris–Todaro model. *International Economic Review* 19, 65–81.
Calvo, G.A. and Wellisz, S. 1978. Supervision, loss of control and the optimum size of the firm. *Journal of Political Economy* 86, 943–52.
Caves, R.E. and Jones, R.W. 1985. *World Trade and Payments*. 4th edn, Boston: Little, Brown & Co.
Chaudhuri, T.D. and Khan, M. Ali. 1984. Educated unemployment, educational subsidies and growth. *Pakistan Development Review* 23, 395–409.
Corden, W.M. and Findlay, R. 1975. Urban unemployment, intersectional capital mobility and development policy. *Economica* 42, 59–78.
Fields, G.S. 1975. Rural-urban migration, urban unemployment and job-search activity in LDCs. *Journal of Development Economics* 2, 165–87.
Harberger, A.C. 1971. On measuring the social opportunity cost of labour. *International Labour Review* 103, 559–79.
Harris, J.R. and Todaro, M. 1970. Migration, unemployment and development: a two sector analysis. *American Economic Review* 40, 126–42.
Johnson, H.G. 1971. *The Two-Sector Model of General Equilibrium*. Yrjö Jahnsson Lectures, Chicago: Aldine-Atherton.
Jones, R.W. 1971. Distortions in factor markets and the general equilibrium model of production. *Journal of Political Economy* 79, 437–59.
Khan, M. Ali. 1979. A multisectoral model of a small open economy with non-shiftable capital and imperfect labor mobility. *Economic Letters* 2, 369–75.
Khan, M. Ali. 1980a. The Harris–Todaro hypothesis and the Heckscher-Ohlin-Samuelson trade model: a synthesis. *Journal of International Economics* 10, 527–47.
Khan, M. Ali. 1980b. Dynamic stability, wage subsidies and the generalized Harris–Todaro model. *Pakistan Development Review* 19, 1–24.
Khan, M. Ali. 1982a. Social opportunity costs and immiserizing growth: some observations on the long run returns versus and short. *Quarterly Journal of Economics* 96, 353–62.

Khan, M. Ali. 1982b. Tariffs, foreign capital and immiserizing growth with urban unemployment and specific factors of production. *Journal of Development Economics* 10, 245–56.

Khan, M. Ali and Chaudhuri, T.D. 1985. Development policies in LDCs with several ethnic groups – a theoretical analysis. *Zeitschrift für Nationalökonomie* 45, 1–19.

Khan, M. Ali and Lin, P. 1982. Sub-optimal tariff policy and gains from trade with urban unemployment. *Pakistan Development Review* 21, 105–26.

Khan, M. Ali and Naqvi, S.N.H. 1983. Capital markets and urban unemployment. *Journal of International Economics* 15(3–4), 367–85.

Magee, S.P. 1976. *International Trade and Distortions in Factor Markets*. New York and Basle: Marcel-Dekker.

Neary, J.P. 1978. Dynamic stability and the theory of factor market distortions. *American Economic Review* 68, 672–82.

Neary, J.P. 1981. On the Harris–Todaro model with intersectoral capital mobility. *Economica* 48, 219–34.

Srinivasan, T.N. and Bhagwati, J. 1975. Alternative policy rankings in a large open economy with sector-specific minimum wages. *Journal of Economic Theory* 11, 356–71.

Srinivasan, T.N. and Bhagwati, J. 1978. Shadow prices for project selection in the presences of distortions: effective rates of protection and domestic resource costs. *Journal of Political Economy* 86, 91–116.

Stiglitz, J.E. 1974. Alternative theories of wage determination and unemployment in LDC's: the labor-turnover model. *Quarterly Journal of Economics* 88, 194–227.

Stiglitz, J.E. 1976. The efficiency wage hypothesis, surplus labor, and the distribution of income in the LDCs. *Oxford Economic Papers* 28, 185–207.

Stiglitz, J.E. 1977. Some further remarks on cost–benefit analysis. In *Project Evaluation*, ed. H. Schwartz and R. Berney, Washington, DC: Inter-American Development Bank.

Stiglitz, J.E. 1982. The structure of labor markets and shadow prices in LDCs. In *Migration and the Labor Market in Developing Countries*, ed. R.H. Sabot, Boulder: Westview Press.

Todaro, M.P. 1968. An analysis of industrialization: employment and unemployment in LDCs. *Yale Economic Essays* 8, 329–492.

Todaro, M.P. 1969. A model of labor migration and urban unemployment in less developed countries. *American Economic Review* 59, 138–48.

Harrod, Roy Forbes (1900–1978). Roy Harrod was born in February 1900 and died in 1978. His father, Henry Dawes Harrod, was a businessman and author of two historical monographs. His mother, Frances (née Forbes-Robertson) was a novelist, and sister of the notable Shakespearean actor-manager, Sir Johnston Forbes-Robertson. Henry Harrod's business failed in 1907, but Roy won a scholarship to St Paul's School in 1911 and a King's Scholarship to Westminster in 1913. He became Head of his House, and in 1918 won a Scholarship in History to New College, Oxford, his father's College. He enlisted in September 1918 and was commissioned in the Royal Field Artillery, but the war ended before his training was completed.

He went up to Oxford in early 1919 and first read Literae Humaniores (Classical Literature, Ancient History and Philosophy). He might well have devoted his career to academic philosophy and he valued his publications in that subject more highly than his seminal contributions to economics. He has remarked that significant economic problems have only attracted the attention of profound thinkers for about two hundred years and interest in them might well disappear in another two hundred. In contrast deep thought has been devoted to the great philosophical problems (such as the validity of inductive methods of thought) for more than two thousand years and new contributions will be read

for so long as civilized life remains. But his philosophy tutor at New College, H.W.B. Joseph, deterred him from devoting his life to that subject, by reacting extremely negatively to his essays. Harrod has left an account of a seminar on Einstein's theory of relativity in Oxford in 1922 where Joseph drew attention to a few terminological problems and believed this had undermined the theory. Einstein's theory of relativity survived, but Harrod was persuaded not to pursue a career in academic philosophy. In later years he published in the distinguished philosophical journal, *Mind*, and his *Foundations of Inductive Logic* (1956) has received serious critical attention from philosophers as distinguished as A.J. Ayer (1970), but his main scholarly work was not to be in Philosophy.

He followed his First Class Honours in Literae Humaniores in 1922 with a First Class in Modern History just one year later, and in 1923, Christ Church, Oxford elected him to a Tutorial Fellowship (confusingly described as a Studentship in that College) to teach the novel subject, Economics, which was to be part of Oxford's new Honour School of Politics, Philosophy and Economics.

Harrod was allowed two terms away from Oxford so that he could learn enough economics to teach it, and it was suggested that he might spend this time in Europe, but he first went to Cambridge where he attended a wide range of lectures and wrote weekly essays on Money and International Trade for John Maynard Keynes. He was equally fortunate when he returned to Oxford, for while he was critically discussing the economics essays of Christ Church's undergraduates he was himself writing weekly microeconomic essays for the Drummond Professor of Political Economy, Francis Ysidro Edgeworth.

In addition to his new academic work Harrod took a notable part in the administration of his College (where he was Senior Censor in 1929–31, the most responsible office a Student of Christ Church can be called upon to discharge), and also the University where he was elected to Oxford's Governing Body (the Hebdomadal Council) in 1929 before he was thirty. In the University and in Christ Church, he fought powerful campaigns on behalf of Professor Lindemann (subsequently Lord Cherwell) who held Oxford's Chair of Experimental Philosophy (Physics), and became principal scientific adviser to Winston Churchill's wartime government and a member of his postwar cabinet.

By 1930 his economics had developed to the point where he was able to publish his first important and original contribution, 'Notes on Supply', in which he was the first 20th-century economist to derive the marginal revenue curve. This should have appeared in 1928 to produce a claim for international priority, but Keynes, the editor of the *Economic Journal*, sent the article to Frank Ramsey who first believed there were difficulties with the argument. He subsequently appreciated that his objections rested on a misunderstanding, but Harrod's new contribution was less startling in 1930 than it would have been in 1928. He followed this initial contribution to the imperfect competition literature with an important article, 'Doctrines of Imperfect Competition' (1934), in which he summarized the essential elements of the new theories of Edward Chamberlin and Joan Robinson.

During the 1930s Harrod frequently stayed with Keynes and he was increasingly drawn into the group of brilliant young economists which included Richard Kahn and Joan Robinson who were helping him develop the new theories which culminated in *The General Theory of Employment, Interest and Money*. Harrod had written a number of important and influential articles in the press advocating new reflationary policies in the early 1930s and these together with his extension of Kahn's employment multiplier to international trade in his *International Economics* (1933) prompted Joseph A. Schumpeter to write in 1946 in his obituary article on Keynes, 'Mr Harrod may have been moving independently toward a goal not far from that of Keynes, though he unselfishly joined the latter's standard after it had been raised.'

Shortly after the *General Theory* appeared, Harrod published *The Trade Cycle* (1936) in which he developed some of the dynamic implications of the new theory of effective demand. The conditions where output would grow were a central theme in Adam Smith's, *The Nature and Causes of the Wealth of Nations*, and it had been much analysed in the great 19th-century contributions of Malthus, Ricardo, Mill and Marx, but the long-term dynamic implications of immediate changes to particular economic variables received virtually no attention in the neoclassical work that followed the marginal revolution. In the *General Theory* Keynes mostly went no further than to work through completely the immediate effects *on a formerly stationary economy* of a variety of disturbances such as an excess of the saving which would occur at full employment over the investment businessmen considered it prudent to undertake. Harrod went a vital step further and showed what could be expected to occur if saving was *permanently high* in relation to *the long-term opportunity to invest*. In 1939 he followed *The Trade Cycle* with 'An Essay in Dynamic Theory', and after the war he developed his growth theory further in the book, *Towards a Dynamic Economics* (1948). Important articles followed including a 'Second Essay in Dynamic Theory' (1960), and 'Are Monetary and Fiscal Politics Enough?' (1964). It is almost certainly because of Harrod's rediscovery of growth theory in the 1930s and his notable contributions to it that Assar Lindbeck, the Chairman of the Nobel Prize Committee, chose to state that he was among those who would have been awarded a Nobel Prize in Economics if he had lived a little longer. The nature of Harrod's original contribution, and the gradual evolution of his theory from 1939 to 1964 is set out in the second part of this article. The detailed technical characteristics of Harrod's growth model are the subject of a separate article, *The Harrod–Domar Growth Model*.

In the Second World War Harrod's friendship with Lindemann and his increasing distinction as an economist led to an invitation to join the Statistical Department of the Admiralty (S Branch) which Churchill set up when he again became First Lord in 1939. This moved to Downing Street when Churchill became Prime Minister in 1940, but Harrod did not have a particular talent for detailed statistical work and he developed an increasing interest in the international financial institutions, the International Monetary Fund and the World Bank, which would need to be set up as soon as the war was won, and from 1942 onwards he pursued this work in Christ Church. In the immediate postwar years he took a strong interest in national politics, and stood for Parliament unsuccessfully as a Liberal in the General Election of 1945 and for a time he was a member of that Party's Shadow Cabinet. He had served on Labour Party committees before the war, and in the 1950s with Churchill's support he unsuccessfully sought adoption as a Conservative parliamentary candidate: his economic advice was warmly welcomed by Harold Macmillan, Conservative Prime Minister in 1957–63. Harrod received the honour of knighthood in 1959 in recognition of his public standing and his notable academic achievements in the prewar and postwar decades.

He had succeeded Keynes as editor of the *Economic Journal* in 1945, and in partnership with Austin Robinson (who looked after the book reviews) he sustained its reputation and quality until his retirement from the editorship in 1966.

His own postwar academic work included important contributions in three areas. In addition to the continuing development and refinement of his prewar work on dynamic theory, he published extensively on the theory of the firm and on international monetary theory which had been his particular concern during the war.

The Oxford Economists' Research Group had begun to meet prominent British industrialists before the war. A group of Oxford economists which generally included Harrod invited individual industrialists to dine in Oxford, and after dinner they were questioned extensively on the considerations which actually influenced their decisions. This led to the publication of a number of much cited articles and the book, *Oxford Studies in the Price Mechanism* (1951) to which Harrod himself did not contribute. Propositions which emanated from these dinners included the notion that businessmen took little account of the rate of interest in their investment decisions, and that they did not seek to profit maximize, but priced instead by adding a margin they considered satisfactory to their average or 'full' costs of production. In his important articles, 'Price and Cost in Entrepreneurs' Policy' (1939) and 'Theories of Imperfect Competition Revised' (1952), Harrod set out a theoretical account of how firms price in which industrialists follow something like these procedures. Their object is especially to achieve a high market share and by setting prices low enough to deter new entry, they actually succeed in maximizing their long-run profits and avoid the excess capacity that Chamberlin and Joan Robinson had considered an inevitable consequence of monopolistic or imperfect competition. This attempt to reconcile the 'rules of thumb' that the businessmen revealed with the propositions of traditional theory was more highly regarded outside Oxford than some of the books and articles in the new tradition.

His work on the world's international monetary problems occupied a good deal of his time and attention in the postwar decades. Keynes himself had considered the breakdown in international monetary relations a crucial element in the collapse of effective demand in so many countries in the 1930s, and he devoted much of the last years of his life to the creation of new institutions which would avoid a repetition of these disasters. Harrod believed he was continuing this vital work when he devoted much thought and energy to these questions. He arrived at the conclusion that there was bound to be some inflation in a world which was successfully pursuing Keynesian policies, and that the liquidity base of the world's financial system was bound to become inadequate if the price of gold failed to rise with other prices. He believed that underlying world liquidity which rested on gold in the last resort must be allowed to rise in line with the international demand for money. He therefore came to focus on the price of gold, and in his book, *Reforming the World's Money* (1965), he proposed that a substantial increase in the price of gold would be needed if subsequent international monetary crises were to be avoided. Harry Johnson (1970) has summarized his contribution to this debate.

Harrod took a great interest in actual developments in the United Kingdom economy, and published seven books and collections of articles in the first two postwar decades which were directly concerned with the policies Britain should follow. There was in addition an immense range of articles in the academic journals, the bank reviews and the press on these questions, not to mention monthly stockbrokers letters for Phillips and Drew. Harrod argued strongly and powerfully that nothing was to be gained by running the economy below full employment, which meant an unemployment rate of less than 2 per cent in the 1950s and the 1960s. In the late 1950s he was deeply concerned that the removal of import controls would render it increasingly difficult for Britain to pursue such Keynesian policies, and he was a vigorous opponent of European Common Market entry. He attached more significance than some distinguished Keynesians to holding down inflation but he published statistics in *Towards a New Economic Policy* (1967) to show that in Britain, this had tended to be faster when the economy was in recession than when output was allowed to expand. He argued therefore that deflationary policies could play no useful role in policies to control the rate of cost inflation, which he considered the essential element in inflation in Britain. Policy swung sharply away from this Keynesian tradition in the last years of his life, and he wrote a final letter to *The Times* on 21 July 1976 in which he praised the economics of Tony Benn and Peter Shore for their opposition to the Labour government's public expenditure cuts, for, 'To cut public spending when there is an undesirably high rate of unemployment is crazy.'

His advocacy of import controls and his adverse reaction to deflationary policies at all times might suggest that he was an economist of the Left, but his willingness to support each of the British political parties at various times underlines how his approach to economic and social problems cannot be typecast. The lines of policy he supported always followed directly from his understanding of the significance of the major interrelationships, and it was his belief that Keynesian theory (which he had so notably helped to refine and develop) provided the appropriate tools for the analysis of Britain's economic problems that led him towards the expansionist policies he so consistently advocated. But further theoretical and empirical relationships which he believed were equally well founded led him to advocate a series of social policies to which very Right wing labels can be attached.

Just before the 1959 election his article, 'Why I Shall Vote Conservative', in *The Sunday Times*, put forward the startlingly unfashionable argument that only the Conservatives would allow more money to go to the better off who had most to contribute to the future of Britain. Harrod's strong belief in the importance of the *quality* of the country's population stock (which, he held, mattered no less than the physical capital stock) lay behind this article. Harrod thought the quality of the population would be bound to deteriorate if the middle classes continued to have fewer children than the poor. He was a strong believer in the inheritance of every kind of ability, and a provocative conversational conclusion he drew was that in an ideal world, one-third of Christ Church's much sought after undergraduate places should be sold to the rich. Their children often had insufficient academic ability to perform well in examinations, but they had inherited abilities of other kinds which would take them to the highest positions, so they should go to Oxford first. Harrod's reasoning on the inheritence of ability and its implications is set out in detail in the Memorandum he submitted to the Royal Commission on Population in 1944. There he suggested that a difficulty in finding servants was one reason why the middle classes had fewer children. Among his suggestions to remedy this state of affairs was that Diplomas in Domestic Service should be established, and that it should become common practice for servants to have latch-keys and the same rights as their mistresses to enjoy social lives with no questions asked. His Memorandum reads strangely in the 1980s when it is widely regarded as unacceptable that any practical conclusions may be drawn from the proposition that human abilities are inherited. Harrod never hesitated to carry his arguments to

their limits, and he always went where his reasoning took him, irrespective of the predictable reactions of others.

The unselfconsciousness of both his academic and his public writing comes out especially in his two biographical volumes, the official life of Keynes (commissioned by the executors) which he published in 1951 and *The Prof* (1959), his personal sketch of Lord Cherwell. As well as providing magnificent accounts of their subjects from the standpoint of one who had known them intimately (and who profoundly understood the economic problems Keynes wrestled with), these books contain extensive autobiographical passages which will enable later generations to know more of Harrod than any biographer can begin to convey.

He ceased to lecture in Oxford in 1967 upon reaching the statutory retirement age of 67, but as a Visiting Professor he continued to teach in several distinguished North American Universities. He died in his Norfolk home in 1978 eleven years after his Oxford work came to an end.

HARROD'S REVIVAL OF GROWTH THEORY AND HIS CONTRIBUTION TO KEYNESIAN MACROECONOMICS. Harrod was intimately involved in the origins and development of Keynesian economics. As the galley proofs of the *General Theory* emerged from the printers from June 1935 onwards, copies were sent to Harrod, to Kahn and to Joan Robinson and with their assistance, Keynes rewrote extensively for final publication. Harrod helped to clarify the relationship between Keynes's new theory of the rate of interest and the then ruling neoclassical theory where this depended upon the intersection of ex-ante saving and investment schedules. In the course of their correspondence, Harrod showed Keynes how well he understood the essence of the *General Theory* by setting out its novelty and its principal elements in ten lines on 30 August 1935:

Your view, as I understand it is broadly this:-

Volume of investment determined by	marginal efficiency of capital schedule rate of interest
Rate of interest determined by	liquidity preference schedule quantity of money
Volume of employment determined by	volume of investment multiplier
Value of multiplier determined by	propensity to save

Keynes responded, 'I absolve you completely of misunderstanding my theory. It could not be stated better than on the first page of your letter.'

Almost immediately after the appearance of the *General Theory*, Harrod published *The Trade Cycle* which contained for the first time in the Keynesian literature the concept of an economy growing at a steady rate. Keynes wrote of it to Joan Robinson on 25 March 1937, 'I think he has got hold of some good and important ideas. But, if I am right, there is one fatal mistake', and to Harrod himself on March 31, 'I think that your theory in the form in which you finally enunciate it is not correct, being fatally affected by a logical slip in the argument.' Harrod replied devastatingly on April 6th, 'There is no slip ... The fact is that you in your criticism are still thinking of once over changes and that is what I regard as a static problem. My technique relates to steady growth.' Harrod's slip was in fact the first step towards the reinstatement of growth theory into mainstream economic analysis.

597

Harrod convinced Keynes, who, on 12 April congratulated him for 'having invented so interesting a theory', but with the reservation, 'I should doubt whether any reader who has not talked or corresponded with you could be aware that the whole of the last half of the book was intended to be in relation to a moving base of steady progress.' Keynes added that it was vital that Harrod carry his ideas further and restate them more comprehensibly.

Harrod made important progress in the next fifteen months, and on 3 August 1938 he sent Keynes a preliminary draft of the article, 'An Essay in Dynamic Theory', and wrote in his accompanying letter,

> my re-statement of the dynamic theory ... is, I think, a great improvement on my book ... I have been throwing out hints in a number of places of the possibility of formulating a simple law of growth and I want to substantiate the claim. It is largely based on the ideas of the general theory of employment; but I think it gets us a step forward.

A lengthy correspondence then developed between Harrod and Keynes in which the two most original elements in Harrod's contribution which later excited much interest and controversy in the economics profession were extensively discussed.

Harrod's principal innovation was the invention of a *moving equilibrium growth path* for the economy, and he described this as the 'warranted' line of growth. Harrod had perceived before he wrote *The Trade Cycle* that there was a fundamental contradiction between the assumptions prevalent in the microeconomic theory of the firm and industry, to which he had made notable contributions, and the new Keynesian macroeconomics. In the theory of the firm, long-term investment was zero, for firms had no motivation to undertake further investment once they were in long period equilibrium. But the new Keynesian macroeconomics required that there be net investment by firms or the government whenever there was any net saving in the macroeconomy. A theory compatible with both macro and microeconomic equilibrium therefore required that firms invest all the time, so that they can continually absorb total net saving. Harrod's formulation of the warranted rate of growth, his novel discovery, was an attempt to set out this necessary equilibrium growth path that industrial and commercial investment decisions must all the time follow in order to achieve a complete economic equilibrium.

Harrod's moving equilibrium or warranted growth path required that saving (of s per cent of the national income) be continually absorbed into investment, so he asked the question: at what rate of growth will firms all the time choose to invest the s per cent of the national income, which equilibrium growth requires? To answer this question, he made use of the acceleration principle or 'the relation' as he called it, that firms need say C_r units of additional capital to produce an extra unit of output. It follows from these premises that the warranted rate of growth of output will be s/C_r per cent per annum. Since each rise in output by 1 unit entails that C_r extra units be invested, a rise in output by s/C_r per cent of the national income will call for an equilibrium investment of C_r times this which is precisely s per cent of the national income, the ratio of ex-ante saving in the national income. In Harrod's examples at this time, he suggested a typical s of 10 per cent of the national income and a C_r of 4, to produce a warranted rate of growth of $2\frac{1}{2}$ per cent.

This idea that if there is continual saving, then equilibrium entails a continual geometric growth in production came as a considerable surprise to Keynes and the other members of the 'circus'. As Harrod had already explained in April 1937,

> The static system provides an analysis of what happens where there is no increase [in output] which entails (as in Joan Robinson's long-period analysis) that saving = 0. Now I was on the lookout for a steady rate of advance, in which the rates of increase would be mutually consistent.

But Harrod's second discovery had equally radical implications. Suppose the actual growth of output is marginally above the equilibrium or warranted rate of growth. In Harrod's numerical example with s 10 per cent and C_r 4, it can be supposed that output actually grows 0.1 per cent faster than the warranted rate, that is by 2.6 per cent instead of 2.5 per cent. Then with 2.6 per cent output growth, the acceleration principle or relation will entail that 4 times 2.6 per cent be added to the capital stock, so that ex-ante investment is 10.4 per cent of the national income. With ex-ante saving limited to 10.0 per cent, the 0.1 per cent excess of actual growth over warranted growth then produces an excess in ex-ante investment over ex-ante saving of 0.4 per cent of the national income. Any excess in ex-ante investment over ex-ante saving will be associated with extra expansion of the national income according to the economics of the *General Theory*. Thus if the actual rate of growth exceeds the warranted rate of s/C_r per cent, the tendency will be for actual growth to rise and rise, for as soon as actual growth rises from 2.6 to say 3 per cent, required investment will rise further to 4 times 3 per cent which equals 12 per cent and so exceed the 10 per cent savings ratio by a still greater margin. Conversely, when actual growth comes out at a rate just short of the warranted 2.5 per cent, ex-ante investment will be below the 10 per cent savings ratio, which will cause the rate of growth to decline. This second discovery, which became known as Harrod's knife-edge, was therefore that any rate of growth in excess of the equilibrium or warranted path he had discovered would set off a continual acceleration of growth, while any shortfall would set off deceleration. He wrote to Keynes of this discovery on 7 September 1938:

> If in static theory producers produce too little, they will be well satisfied with the price they get and feel happy; but this is not taken to be the *right* amount of output; they will be stimulated to produce more. The equilibrium output is taken to be that which *just* satisfies them and induces them to go on as before. Similarly the warranted rate [of growth] is that which just satisfies them and leaves them going on as before. The difference between the warranted rate and the old equilibrium (i.e. the difference between dynamic and static theory) is, on my view, that if they produce above the warranted rate, they will be more than satisfied and be stimulated, and conversely, while in the case of equilibrium in static conditions the opposite happens. The 'field' round the [static] equilibrium contains centripetal, that round the warranted centrifugal forces.

It took Keynes time to absorb Harrod's startling discovery. On September 19th he proposed a counterexample in which C_r was merely one-tenth, while s was also one-tenth. With this counterexample, a deviation of output by a small amount from the warranted path, say by δx, which would raise planned investment above the level at which it would otherwise be by $C_r \delta x$ would merely raise this by $0.10 \delta x$, which would equal the rise in planned saving of $s\delta x$, which would also come to $0.10 \delta x$, so there would be no tendency towards an explosive growth in effective demand. This would grow explosively if C_r was one-ninth (in which case planned investment would rise by

0.11 δx and saving by only 0.10 δx) but the further growth of output would be damped if C_r was merely one-eleventh, so, Keynes insisted, 'neutral, stable or unstable equilibrium' are equally likely.

Harrod protested on 22 September, 'it is absurd to suppose extra capital required [C_r] only $\frac{1}{10}$ of annual output, when the capital required in association with the pre-existent level of incomes in England today is 4 or 5 times annual output.' The probability that C_r would exceed s so that ex-ante investment would rise by more than ex-ante saving in order to produce instability was therefore overwhelming.

But several qualifications emerged. In comparing the increase in ex-ante investment to the increase in ex-ante saving following a small deviation of output from the warranted rate:

(1) The relevant marginal capital coefficient (C_r) which determines how much planned investment will rise as the net new requirement of *induced* investment. In so far as investment decisions are autonomous of short-term fluctuations in output, the relevant C_r will be lower than the economy's overall capital output ratio.

(2) The relevant coefficient which determines the increase in planned saving is the *marginal* and not the average propensity to save. Planned saving will rise more where output deviates upward from the warranted rate, the greater is the marginal propensity to save in relation to the average propensity.

The circumstances that could produce a stable upward deviation of growth from the warranted rate and the avoidance of Harrod's knife-edge are therefore a very high marginal propensity to save in combination with a situation where most investment is autonomous so that the induced investment coefficient, C_r, is considerably less than 1. In 'An Essay in Dynamic Theory', Harrod covered this possibility with the caveat, 'when long-range capital outlay is taken into account ... the attainment of a neutral or stable equilibrium of advance may not be altogether improbable in certain phases of the cycle.' The possibility he had in mind here is that in the early stages of a cyclical recovery there may be so much excess industrial capacity that C_r will be quite low for a time, and therefore quite possibly lower than the marginal propensity to save. But in general any deviation of growth from the warranted line of advance would raise ex-ante investment by a greater margin than ex-ante saving with the result that the rate of growth would deviate further.

In addition to establishing the existence of the warranted line of advance and its instability, Harrod had to define the equilibrium investment behaviour by businesses which would actually lead to expansion at the requisite rate. In his 1939 article he omitted to offer any behavioural rule but simply asserted that the warranted rate was, 'that rate of growth which, if it occurs, will leave all parties satisfied that they have produced neither more nor less than the right amount'. That is no more than a description of equilibrium growth, and much the same can be said of his definition of the warranted rate in *Towards a Dynamic Economics* (1948) as, 'that over-all rate of advance which, if executed, will leave entrepreneurs in a state of mind in which they are prepared to carry on a similar advance'. It was only in the article, 'Supplement on Dynamic Theory' (1952) that Harrod arrived at a behavioural assumption that matched his algebraic formulation of the warranted rate:

Let the representative entrepreneur on each occasion of giving an order repeat the amount contained in his order for the last equivalent period, adding thereto an order for an amount by which he judges his existing stock to be deficient, if he judges it to be deficient, or subtracting

therefrom the amount by which he judges his stock to be redundant, if he does so judge it.

With that assumption an economy which once achieves growth at the warranted rate will sustain it, while any upward or downward deviations will lead to still greater deviations wherever C_r exceeds the marginal propensity to save.

But it emerged by 1964 when Harrod published, 'Are Monetary and Fiscal Policies Enough?', that even that assumption fails to define growth at the warranted rate, for it must also be assumed that the representative entrepreneur will expand at a rate of precisely s/C_r when he judges his capital to be neither deficient nor redundant. This requires an expectation by the representative entrepreneur that his market will grow at a rate of precisely s/C_r. Hence the full requirement for growth along Harrod's warranted equilibrium path is that entrepreneurs expect growth at this rate and expand and continue to expand at that rate so long as their capital stock continues to grow in line with their market so that it is neither deficient nor redundant. They will of course increase their rate of expansion if their capital should prove deficient, and curtail it if part of their stock becomes redundant.

The warranted rate of growth and its instability were Harrod's great innovations. From 1939 onwards he contrasted this equilibrium rate with the natural rate of growth, 'the rate of advance which the increase of population and technological improvements allow', which was entirely independent of the warranted rate. Harrod defined the rate of technical progress more precisely in 1948 as the increase in labour productivity 'which, at a constant rate of interest, does not disturb the value of the capital coefficient'. This then entered the language of economics as Harrod-neutral technical progress, which, together with growth in the labour force, determines the natural rate of growth, that is the rate at which output can actually be increased in the long run. This raised few theoretical problems in 1939, and there was nothing novel in the proposition that long-term growth must depend on the rate of increase of the labour force and technical progress. Keynes himself had said as much several years earlier in, 'Economic Possibilities for our Grandchildren' (1930). But the contrast between this natural rate, and Harrod's innovatory warranted rate offered entirely new insights.

If the warranted rate exceeds the feasible natural rate, the achievement of equilibrium growth must be impractical because the economy cannot continue to grow faster than the natural rate. It must deviate downwards from the warranted rate towards the natural rate far more than it deviates upwards with the result that 'we must expect the economy to be prevailingly depressed'. If the natural rate is greater, output will tend to deviate upwards towards the natural rate with the result that the economy should enjoy 'a recurrent tendency towards boom conditions'.

Keynes's own reaction to the dichotomy between the warranted and natural rates was characteristically (his letter to Harrod on 26 September 1938) that the warranted rate always exceeded the natural:

In actual conditions ... I suspect the difficulty is, not that a rate in excess of the warranted is unstable, but that the warranted rate itself is so high that with private risk-taking no one dares to attain it ...

I doubt if, in fact, the warranted rate – let alone an unstable excess beyond the warranted – has ever been reached in USA and UK since the war, except perhaps in 1920 in UK and 1928 in USA. With a stationary population, peace and unequal incomes, the warranted rate sets a pace which a private risk-taking economy cannot normally reach and can never maintain.

That is characteristic Keynes, but Harrod had persuaded him to express his familiar analysis in the language of his new theory of growth. In the immediate postwar decades when full employment and creeping inflation prevailed, it was widely argued that the natural rate had come to exceed the warranted. The richness of Harrod's model is demonstrated by its ability to illuminate both kinds of situation.

Evsey Domar's growth model which has a good deal in common with Harrod's was published seven years after 'An Essay in Dynamic Theory', and a considerable literature emerged in the next fifteen years on the stability conditions and other important features of what came to be known as the Harrod–Domar growth model. This is elegantly summarized by Frank Hahn and Robin Matthews in their celebrated 1964 survey article.

The development of neoclassical growth theory in the 1950s led to an increasing realization that the warranted and natural growth rates could be equated by an appropriate rate of interest. If the warranted rate was excessive so that oversaving led to slump conditions, a lower interest rate which raised C_r sufficiently would bring it down to the natural rate. Conversely the inflationary pressures that resulted from an insufficient warranted rate would be eliminated if higher interest rates reduced C_r sufficiently. If the real rate of interest and C_r responded in this helpful way, s/C_r, the warranted rate could always be brought into equality with the natural rate.

Harrod's response included his 'Second Essay in Dynamic Theory' (1960), a title which underlines its significance. He proposed that there was an optimum real rate of interest r_n which would maximize utility, with a value of G_p/e, G_p being the economy's long-term rate of growth of labour productivity and e the elasticity of the total utility derived from real per capita incomes with respect to increases in these. If a 1 per cent increase in real per capita incomes raises per capita utility $\frac{1}{2}$ per cent, e will be 0.5, and r_n the optimum rate of interest which maximizes utility will be $G_p/0.5$, viz. twice the rate of growth of labour productivity. If the marginal utility of income does not fall at all as real per capita incomes rise, per capita utility will grow 1 per cent when incomes rise 1 per cent so that e is unity, and r_n equals G_p. The more steeply the marginal utility of incomes fall, the more e will fall below unity, and the more the optimum real rate of interest, G_p/e, will exceed the rate of growth of labour productivity.

If a society actually seeks to establish the optimum rate of interest determined in this kind of way, the value of C_r will depend upon this optimum rate of interest, so it will not also be possible to use the rate of interest to equate the natural and warranted rates of growth in the manner the neoclassical growth models of, for instance, Robert Solow (1956) and Trevor Swan (1956) propose. There will therefore still be difficulties because the warranted rate of growth with real interest rates at their optimum level will not in general be equal to the natural rate. Therefore as Harrod suggested in the final articles he published in 1960 and 1964, governments will have to run persistent budget deficits or surpluses if they are to avoid the difficulties inherent in discrepancies between the natural and the warranted rates of growth.

So Harrod remained a convinced Keynesian who continued to believe that a long-term imbalance between saving, the main determinant of the warranted rate, and investment opportunity would call for persistent government intervention. When that approach to economic policy again becomes fashionable, economists may learn a good deal from Harrod's later articles which have not yet received the same attention from the economics profession as his seminal work in the 1930s and the 1940s.

WALTER ELTIS

SELECTED WORKS

The 'Bibliography of the works of Sir Roy Harrod', in *Induction, Growth and Trade*: *Essays in Honour of Sir Roy Harrod*, ed. W.A. Eltis, M.FG. Scott and J.N. Wolfe, Oxford: Oxford University Press, 1970, pp. 361–76, includes all the articles he published in books, journals and magazines from 1928 to 1969, and some of his most influential newspaper articles. The present Bibliography is confined to his books and academic articles in books, and academic journals.

BOOKS

1933. *International Economics*. Cambridge: Cambridge University Press. 1st revised edn, 1939; 2nd revised edn, 1957; 3rd revised edn, mainly rewritten, 1974.

1936. *The Trade Cycle: An Essay*. Oxford: Oxford University Press.

1946. *A Page of British Folly*. London: Macmillan.

1947. *Are These Hardships Necessary?* London: Rupert Hart-Davis.

1948. *Towards a Dynamic Economics: Some Recent Developments of Economic Theory and Their Application to Policy*. London and New York: Macmillan.

1951a. *The Life of John Maynard Keynes*. London and New York: Macmillan.

1951b. *And so it goes on: Further Thoughts on Present Mismanagement*. London: Rupert Hart-Davis.

1952a. *Economic Essays*. London and New York: Macmillan.

1952b. *The Pound Sterling*. Princeton Essays in International Finance No. 13, Princeton: Princeton University Press.

1953. *The Dollar*. London and New York: Macmillan. 2nd edn with new introduction, New York: The Norton Library, 1963.

1956. *Foundations of Inductive Logic*. London and New York: Macmillan.

1958a. *The Pound Sterling, 1951–58*. Princeton Essays in International Finance, Princeton: Princeton University Press.

1958b. *Policy against Inflation*. London and New York: Macmillan.

1959. *The Prof: A Personal Memoir of Lord Cherwell*. London: Macmillan.

1961. *Topical Comment: Essays in Dynamic Economics Applied*. London: Macmillan; New York: St Martin's Press.

1963. *The British Economy*. New York: McGraw Hill.

1964. *Plan to Increase International Monetary Liquidity*. Brussels and London: European League for Economic Co-operation.

1965. *Reforming the World's Money*. London: Macmillan. New York: St Martin's Press.

1967. *Towards a New Economic Policy*. Manchester: Manchester University Press.

1969. *Money*. London: Macmillan; New York: St Martin's Press.

1970. *Sociology, Morals and Mystery*. Chichele Lectures, All Souls College, Oxford. London: Macmillan.

1973. *Economic Dynamics*. London: Macmillan; New York: St Martin's Press.

ARTICLES AND OTHER CONTRIBUTIONS PUBLISHED IN BOOKS

1945. Memorandum to the *Royal Commission on Equal Pay for Men and Women*. Appendix IX in the Fourth Volume of Memoranda of Evidence. London: HMSO.

1948. The economic consequences of atomic energy. In *The Atomic Age*, Sir Halley Stewart Lectures, London: Allen & Unwin.

1950. Memoranda (Submitted in August and December 1944). *Papers of the Royal Commission on Population*, Vol. 5. London: HMSO.

1952a. Theory of imperfect competition revised. In R.F. Harrod, *Economic Essays*, London and New York: Macmillan.

1952b. Supplement on dynamic theory. In R.F. Harrod, *Economic Essays*, London and New York: Macmillan.

1959. Inflation and investment in underdeveloped countries. In *Economi, Politik, Samhälle: en bok Tillagnad Bertil Ohlin*, ed. John Bergvall, Stockholm: Bokförlaget Folk och Samhälle.

1960. Evidence submitted to the Radcliffe Committee on the Working of the Monetary System, May 1958. *Principal Memoranda of Evidence*, Vol. 3. London: HMSO.

1961. The dollar problem and the gold question. In *The Dollar in Crisis*, ed. S.E. Harris, New York: Harcourt, Brace and World.

1963a. Desirable international movements of capital in relation to growth of borrowers and lenders and growth of markets. In *International Trade Theory in a Developing World*, ed. R.F. Harrod and D.C. Hague, London and New York: Macmillan.

1963b. Liquidity. In *World Monetary Reform*, ed. H.C. Grubel, Stanford: University Press.

1964a. Comparative analysis of policy instruments. In *Inflation and Growth in Latin America*, ed. Werner Baer and Issac Kerstenetzky, Homewood, Ill.: Richard Irvin.

1964b. Retrospect on Keynes. In *Keynes' General Theory*, ed. R. Lekachman, New York and London: Macmillan.

1966. Optimum investment for growth. In *Problems of Economic Dynamics and Planning: Essays in Honour of Michael Kalecki*, Oxford: Pergamon Press.

1967. Increasing returns. In *Monopolistic Competition Theory: Studies in Impact: Essays in Honour of Edward H. Chamberlin*, ed. Robert E. Kuenne, New York: John Wiley.

1968. What is a model? In *Value, Capital and Growth: Papers in Honour of Sir John Hicks*, ed. J.N. Wolfe, Edinburgh: Edinburgh University Press.

ARTICLES IN ACADEMIC JOURNALS

1930a. Notes on supply. *Economic Journal* 40, June, 232–41.

1930b. Progressive taxation and equal sacrifice. *Economic Journal* 40, June, 704–7.

1931. The law of decreasing costs. *Economic Journal* 41, December, 566–76. Addendum: September 1932, 490–92.

1933. A further note on decreasing costs. *Economic Journal* 43, June, 337–41.

1934a. Professor Pigou's theory of unemployment. *Economic Journal* 44, March, 19–32.

1934b. Doctrines of imperfect competition. *Quarterly Journal of Economics* 48, May, 442–70.

1934c. The equilibrium of duopoly. *Economic Journal* 44, June, 335–7.

1934d. The expansion of credit in an advancing economy. *Economica* NS 1, August, 287–99. Rejoinders: November 1934, 476–8; and February 1935, 82–4.

1936a. Utilitarianism revised. *Mind* 45, April, 137–56.

1936b. Imperfect competition and the trade cycle. *Review of Economics and Statistics* 18, May, 84–8.

1937. Mr Keynes and traditional theory. *Econometrica* 5, January, 74–86.

1938. Scope and method of economics. *Economic Journal* 48, September, 383–412.

1939a. Modern population trends. *Manchester School of Economics and Social Studies* 10(1), 1–20. Rejoinder: April 1940, 47–58.

1939b. Price and cost in entrepreneurs' policy. *Oxford Economic Papers* 2, May, 1–11.

1939c. An essay in dynamic theory. *Economic Journal* 49, March, 14–33. Errata, June 1939, 377.

1939d. Value and Capital by J.R. Hicks. *Economic Journal* 49, June, 294–300.

1942. Memory. *Mind* 51, January, 47–68.

1943. Full employment and security of livelihood. *Economic Journal* 53, December, 321–42.

1946a. *Price Flexibility and Employment*. By Oscar Lange. *Economic Journal* 56, March, 102–7.

1946b. Professor Hayek on individualism. *Economic Journal* 56, September, 435–42.

1947. A comment on R. Triffin's 'National Central Banking and The International Economy'. *Review of Economic Studies* 14(2), 95–7.

1948. The fall in consumption. *Bulletin of the Oxford University Institute of Statistics* 10, May, 162–7. Rejoinder: July–August, 235–44; September, 290–93.

1951. Notes on trade cycle theory. *Economic Journal* 61, June, 261–75.

1952. Currency appreciation as an anti-inflationary device: comment. *Quarterly Journal of Economics* 66, February, 102–16.

1953a. Imbalance of international payments. *International Monetary Fund Staff Papers* 3, April, 1–46.

1953b. Foreign exchange rates: a comment. *Economic Journal* 63, June 294–8.

1953c. Sir Hubert Henderson, 1890–1952. *Oxford Economic Papers* NS 5, supplement, June, 59–64.

1953d. Full capacity vs. full employment growth: a comment on Pilvin. *Quarterly Journal of Economics* 67, November, 553–9.

1955. Investment and population. *Revue Economique*, May, 356–67.

1956a. The British boom, 1954–55. *Economic Journal* 66, March, 1–16.

1956b. Walras: a re-appraisal. *Economic Journal* 66, June, 307–16.

1957a. The Common Market in perspective. *Bulletin of the Oxford Institute of Statistics* 19, February, 51–5.

1957b. Review of *International Economic Policy* by J.E. Meade. *Economic Journal* 67, June, 290–95.

1957c. Clive Bell on Keynes. *Economic Journal* 67, December, 692–9.

1958a. The role of gold today. *South African Journal of Economics* 26, March 1958, 3–13. Rejoinder: March 1959, 16–22.

1958b. Questions for a stabilization policy in primary producing countries. *Kyklos* 11(2), 207–11.

1958c. Factor-price relations under free trade. *Economic Journal* 68, June, 245–55.

1959. Domar and dynamic economics. *Economic Journal* 69, September, 451–64.

1960a. New arguments for induction: reply to Professor Popper. *British Journal for the Philosophy of Science* 10(40), February, 309–12.

1960b. Keynes' attitude to compulsory military service. *Economic Journal* 70, March, 166–7.

1960c. Second essay in dynamic theory. *Economic Journal* 70, June, 277–93. Comment, December 1960, 851. Rejoinder: December 1962, 1009–10.

1961a. The general structure of inductive argument. *Proceedings of the Aristotelian Society, 1960–61* 61, 41–56.

1961b. Real balances: a further comment. *Economic Journal* 71, March, 165–6.

1961c. A plan for increasing liquidity: a critique. *Economica* NS 28, May, 195–202.

1961d. The 'neutrality' of improvements. *Economic Journal* 71, June, 300–304.

1961e. Review of Sraffa's *Production of Commodities by Means of Commodities*. *Economic Journal* 71, December, 783–7.

1962a. Economic development and Asian regional cooperation. *Pakistan Development Review* 2, 1–22.

1962b. Dynamic theory and planning. *Kyklos* 15(3), February, 68–79.

1963. Themes in dynamic theory. *Economic Journal* 73, September, 401–21. Corrigendum: December 1963, 792.

1964. Are monetary and fiscal policies enough? *Economic Journal* 74, December, 903–15.

1966. International liquidity. *Scottish Journal of Political Economy* 13, June, 189–204.

1967a. Methods of securing equilibrium. *Kyklos* 20(1), February, 24–33.

1967b. World reserves and international liquidity. *South African Journal of Economics* 35, June, 91–103.

1967c. Assessing the trade returns. *Economic Journal* 77, September, 499–511.

1970a. Reassessment of Keynes's views on money. *Journal of Political Economy* 78(4), July–August, 617–25.

1970b. Replacements, net investment, amortisation funds. *Economic Journal* 80, December, 24–31.

1972. Imperfect competition, aggregate demand and inflation. *Economic Journal* 82, March, 392–401.

BIBLIOGRAPHY

Ayer, A.J. 1970. Has Harrod answered Hume? In *Induction, Growth and Trade: Essays in Honour of Sir Roy Harrod*, ed. W.A. Eltis, M.FG. Scott and J.N. Wolfe, Oxford: Oxford University Press.

Blake, R. 1970. A personal memoir. In *Induction, Growth and Trade: Essays in Honour of Sir Roy Harrod*, ed. W.A. Eltis et al., Oxford: Oxford University Press.

Domar, E. 1946. Capital expansion, rate of growth, and employment. *Econometrica* 14, April, 137–47.

Domar, E. 1947. Expansion and employment. *American Economic Review* 37, March, 34–55.

Hahn, F.H. and Matthews, R.C.O. 1964. The theory of economic growth: a survey. *Economic Journal* 74, December, 779–902.

Johnson, H.G. 1970. Roy Harrod on the price of gold. In *Induction, Growth and Trade: Essays in Honour of Sir Roy Harrod.*

Keynes, J.M. 1930. Economic possibilities for our grandchildren. In *The Collected Writings of John Maynard Keynes*, Vol. IX: *Essays in Persuasion*, London: Macmillan, 1972.

Keynes, J.M. 1973. *The General Theory and After* (Correspondence and Articles). Vols XIII and XIV of *The Collected Writings of John Maynard Keynes*, London: Macmillan.

Lindbeck, A. 1985. The Prize in Economic Science in memory of Alfred Nobel. *Journal of Economic Literature* 23(1), March, 37–56.

Phelps-Brown, H. 1980. Sir Roy Harrod: a biographical memoir. *Economic Journal* 90, March, 1–33.

Schumpeter, J.A. 1946. John Maynard Keynes 1883–1946. *American Economic Review* 36, September, 495–518.

Solow, R.M. 1956. A contribution to the theory of economic growth. *Quarterly Journal of Economics* 70, February, 65–94.

Swan, T.W. 1956. Economic growth and capital accumulation. *Economic Record* 32, November, 334–61.

Wilson, T. and Andrews, P.W.S. 1951. *Oxford Studies in the Price Mechanism*. Oxford: Oxford University Press.

Harrod–Domar growth model. The Keynesian revolution led Roy Harrod (1939) and Evsey Domar (1946 and 1947) to work out the implications of permanent full employment. In *The General Theory of Employment, Interest and Money* (1936) Keynes himself showed how full employment could be reached, but he made no attempt to work out the long-term conditions which must be satisfied before an economy can continue to produce at that level. Harrod's and Domar's analyses of this problem show that long-term full employment requires that two fundamental conditions be satisfied.

First, the economy must invest full employment saving every year. If saving is s_f per cent of the full employment national income, and investment falls short of this, then as Keynes showed, effective demand is bound to be insufficient for full employment.

Second, for continuous full employment, the rate of growth of output must equal the growth of the physical labour force, plus the rate of increase in labour productivity. If there are n per cent more workers every year, and each produces a per cent more output, then continuous full employment requires that production grow $(n+a)$ per cent a year. There will be no need to make use of n per cent more workers if output grows less than this, so all the extra workers who wish to join the labour force will not find employment.

Harrod and Domar both discovered a truism which allows formulae for g, the rate of growth, to be derived from these fundamental conditions. g can be defined as $\delta Y/Y$, where δY is 'increase in output' and Y the level of output. $\delta Y/Y$ is identically equal to $\delta K/Y$ divided by $\delta K/\delta Y$, where $\delta K/Y$ is 'increase in capital/output', that is, 'investment/output', while $\delta K/\delta Y$ is 'increase in capital/increase in output' or the *marginal* capital-output ratio. There is therefore the truism that:

$$g \equiv \text{Investment/output } (I/Y) \div \text{ the capital-output ratio } (C).$$

This can be combined with two basic full employment conditions. The result is presented first in the manner suggested by Harrod (whose model was published seven years prior to Domar's).

The condition that for full employment the share of investment must equal the full employment savings ratio, s_f, means that in the above formula, it is necessary that:

$$g = s_f \text{ (which has to equal } I/Y) \text{ divided by } C.$$

There will be one particular level of C, the marginal capital–output ratio, which profit maximizing entrepreneurs consider ideal, for which Harrod used the symbol, C_r, and when this is substituted for C in the above expression, one necessary condition for continuous equilibrium growth at full employment is arrived at:

$$g = s_f/C_r$$

A second condition which needs to be satisfied if there is to be continuous full employment is that the economy's rate of growth must equal $(n+a)$, the rate of growth of the physical labour force plus labour productivity. Hence, if there is to be continuous full employment growth, it is necessary that:

$$g = s_f/C_r = n + a$$

So growth has to equal both s_f/C_r and $(n+a)$. Harrod called the first of these the 'warranted' rate of growth for which he used the symbol g_w and the second the 'natural' rate for which he wrote g_n. An economy will only be able to achieve continuous full employment if its rate of growth is equal to both g_w and g_n. Since in Harrod's account, s_f and C_r which determine the 'warranted' rate, and $(n+a)$ which determines the natural rate, are exogenously given and independent, g_w and g_n will only be equal by chance. It follows that actual economies will find it virtually impossible to achieve continuous full employment, a Keynesian result which follows naturally from Harrod's Keynesian assumptions.

In the version Domar published in 1946 and 1947 which he sent to the printers before he was aware of Harrod's 1939 article, 'the rate of growth required for a full employment equilibrium' (Harrod's g_n) is described as r, the economy's long-term saving ratio (s_f) is α, and the annual output produced by a unit of capital in the long term $(1/C_r)$ is σ. Domar's equivalent to Harrod's condition for long term full employment equilibrium that g_n must equal s_f/C_r is (Harrod, 1959) the identical proposition that r must equal $\alpha\sigma$. Harrod's symbols are more often used than Domar's because g, s, and C are more readily thought of as the growth rate, the savings ratio and the capital-output ratio than, r, α and $1/\sigma$.

Harrod and Domar were both then unaware of the work of Fel'dman, who had produced a growth model quite similar to theirs in the Soviet Union in 1928. Domar published an account of Fel'dman's model, 'A Soviet Model of Growth', in his *Essays in the Theory of Economic Growth* (1957), a collection of papers in which his own model of growth and its implications for public policy are fully developed.

The consequences of the all but inevitable failure to achieve Harrod's and Domar's conditions provide illuminating insights into the long term development of real economies which often fail to achieve full employment over considerable periods. Harrod's first condition is that g, the economy's actual rate of growth must equal the 'warranted' rate, s_f/C_r. The meaning of this condition is that equilibrium growth entails that full employment saving be continuously invested, as in table 1, where a full employment savings ratio (s_f) of 12 per cent, and a required capital-output ratio (C_r) of 4 are assumed, so that the warranted rate is exactly 3 per cent. The real national income is 100 in the first year, and the initial capital stock is exactly the one required, namely four times this or 400.

TABLE 1. A Table to Illustrate Growth at the Warranted Rate $s_f = 12$ and $C_r = 4$

Year	Capital Stock	National Income	Desired Capital	Investment
	$K = K_{-1} + I_{-1}$	Y	$C_r . Y$	$I = s . Y$
1	400.00	100.00	400.00	12.00
2	412.00	103.00	412.00	12.36
3	424.36	106.09	424.36	12.73

Investment which is always 12 per cent of the national income is added to the capital stock of the previous year, and the national income (which grows at exactly the warranted rate of 3 per cent) is always exactly one-quarter the capital stock, so the 'desired capital stock' (which is C_r times the national income) is always in line with the actual stock. This means that if the economy grows at precisely the 'warranted' rate (3 per cent), entrepreneurs will be satisfied that they have undertaken the commercially correct rate of investment. In 1939 Harrod defined the 'warranted' rate of growth as 'that rate of growth which, if it occurs, will leave all parties satisfied that they have produced neither more nor less than the right amount', which is precisely the situation in the table where the actual capital stock always equals the desired stock.

Table 2 illustrates what goes wrong when g, the actual rate of growth is less than g_w. It is assumed that g is only 2 per cent, while with s_f 12 per cent and C_f 4 as before, g_w is still 3 per cent.

TABLE 2. Growth where the Actual Rate (g) is 1 per cent less than the Warranted Rate (g_w)

Year	Capital Stock	National Income	Desired Capital	Investment
	$K = K_{-1} + I_{-1}$	Y	$C_r . Y$	$I = s . Y$
1	400.00	100.00	400.00	12.00
2	412.00	102.00	408.00	12.24
3	424.24	104.04	416.16	12.48
4	436.72	106.12	424.48	12.73

Here, where the rate of growth is slightly less than the warranted rate, the capital stock actually increases *faster* than the one entrepreneurs consider ideal. This margin of excess capital grows continuously, year after year, so the time is bound to come where entrepreneurs will respond by cutting investment. According to Harrod (1952) the rate at which firms invest to expand will be determined as follows:

Let the representative entrepreneur on each occasion of giving an order repeat the amount contained in his order for the last equivalent period, adding thereto an order for an amount by which he judges his existing stock to be deficient, if he judges it to be deficient, or subtracting therefrom the amount by which he judges his stock to be redundant, if he does so judge it (p. 284).

In the conditions set out in Table 2 where g_w exceeds g, part of the capital stock of the representative entrepreneur gradually becomes redundant, so investment and therefore effective demand and growth will begin to fall. Thus Harrod arrived at the extremely uncomfortable conclusion that if actual growth is less than the 'warranted' rate, it will come to fall still further below this. It can be shown similarly that if g exceeds g_w for any reason, the economy will become increasingly short of capital with the result that g will rise further and further above g_w.

There are propositions in microeconomic theory which claim to demonstrate that if there is a surplus of any particular commodity, then the rate at which it is supplied will fall off with the result that market forces respond in the direction required to remove the surplus. The economy is therefore expected to respond to a shortage or surplus of an individual commodity in the manner required to remove it; but according to Harrod's instability theorem, at the macroeconomic level, any chance deviation of actual growth below the warranted rate will lead to excess capacity, and as this grows, investment and hence effective demand will be curtailed, which will lead to the creation of still more excess capacity. The response of the macro-economy to excess capital will therefore be the opposite of that required to remove the excess, with the result that economies are inherently unstable at the macro level.

Domar arrived at a similar result by directly contrasting the rate of growth of effective demand to the growth of productive capacity. In his formulation (but using Harrod's symbols) the growth in demand equals the increase in investment (δI) times the multiplier ($1/s$) while the growth of productive capacity equals total investment (I) divided by the long term capital-output ratio (C_r), with the result that where the growth of demand equals the growth of capacity:

$$\delta I / I = s / C_r.$$

A slight upward deviation of investment from this critical rate of growth (which corresponds to Harrod's 'warranted' rate) will raise $\delta I / I$ (which equals the growth of demand) relative to s/C_r, the growth of capacity, and this can be expected to lead to further increases in investment. Thus as in Harrod's argument, any chance deviation in *the rate of growth of investment* from the critical s/C_r growth rate of productive capacity can be expected to lead to further deviations in the same direction.

The difficulties capitalist economies must overcome to achieve continuous expansion at full employment are still greater because in order to grow all the time at the 'warranted' rate and so escape the instability inherent in any departure of g from s_f/C_r, the 'warranted' rate itself must equal the natural rate, but there is no reason why s_f/C_r should equal $(n + a)$.

Suppose the conditions assumed in the above tables ($s_f = 12$ per cent and $C_r = 4$ so that $g_w = 3$ per cent) but that the labour force grows at only 0.5 per cent and productivity at 1.5 per cent so that g_n is just 2 per cent. Then the economy's full employment output can grow no more than 2 per cent a year, so it will be possible for the economy to achieve the 3 per cent growth rate required to prevent the emergence of continual excess capacity for a few years at most. Its actual long term growth rate is likely to approximate to the 2 per cent 'natural' rate with the result that g, the actual rate will fall short of g_w most of the time. Then years with excess capacity leading to economic depression will predominate over periods of expansion. The continual tendency towards depression will reduce average actual saving (s) below full employment saving (s_f). Then via unemployment and underproduction, the economy's actual long term savings ratio will come into line with the lower investment ratio (C_r times g_n) which physical conditions actually allow the economy to sustain.

Conversely, where g_n exceeds g_w, market forces will all the time attempt to push actual growth above the 'warranted' rate, with the result that conditions where capital is scarce and saving inadequate will predominant. In the first instance this will lead to excess demand for capital and therefore to a predominance of inflation over deflation which is what Harrod emphasized in 1948: 'we may have plenty of booms and a frequent tendency to approach full employment, the high employment will be of an inflationary and therefore unhealthy character' (p. 88). However, if investment of less than $C_r(n+a)$ causes the rate of growth of productive capacity to fall short of $(n+a)$, then there will be insufficient growth of the real capital stock to provide enough physical capital equipment to raise employment at the rate at which the physical labour force is growing (n), with the result that the economy will suffer from growing *structural* unemployment.

Harrod's theory therefore predicts that incompatibilities between long term saving and investment opportunity are all but certain to cause prolonged unemployment (which will be structural where g_n exceeds g_w and demand deficient where g_w exceeds g_n) with persistent inflation in addition wherever long term saving is inadequate for the natural rate of growth. This raises fundamental problems for public policy, and Harrod argued in 1939 that 'the difficulties may be too great to be dealt with by a mere anti-cycle policy'. He suggested that where an economy suffers from a long term tendency to over saving with the result that the 'warranted' rate exceeds the 'natural' rate, then a generous attitude to public investment is appropriate so that more will be undertaken than commercial and social considerations call for. Conversely governments should seek to generate more long term saving and to curtail long range and social investment where the 'natural' rate exceeds the 'warranted' rate.

By the later 1950s the United States and several West European economies were achieving full employment and negligible inflation which led a number of distinguished economists to develop models of economic growth which were less prone to predict secular unemployment or inflation. Robert Solow (1956) and Trevor Swan (1956) produced neoclassical growth models where market forces adjust the equilibrium capital–output ratio (C_r) so that this automatically equates g_w to g_n (which is achieved when $C_r = (n+a)/s_f$). Nicholas Kaldor (1955–6 and 1957) evolved a Keynesian model of growth and income distribution where shifts between wages and profits will adjust the savings ratio until this becomes the one required $(C_r(n+a))$ to equate g_w and g_n. A few years earlier, Alexander (1950) had questioned the inevitability of Harrod's knife-edge which sent an economy soaring upwards or downwards wherever g diverged from g_w.

The unemployment and stagflation of the 1970s and the 1980s has surprisingly failed to restore some of the former prestige of the Harrod–Domar model. In the 20th century in the leading Western economies there have been prolonged periods when more saving would have been beneficial, and others with every appearance of inadequate effective demand. The Harrod–Domar growth model is one of the few which actually predicts this, so it still deserves serious attention.

WALTER ELTIS

See also AGGREGATE DEMAND AND SUPPLY ANALYSIS; NATURAL AND WARRANTED RATES OF GROWTH.

BIBLIOGRAPHY
Alexander, S.S. 1950. Mr Harrod's dynamic model. *Economic Journal* 60, December, 724–39.
Domar, E. 1946. Capital expansion, rate of growth, and employment. *Econometrica* 14, April, 137–47.
Domar, E. 1947. Expansion and employment. *American Economic Review* 37, March, 34–55.
Domar, E. 1957. *Essays in the Theory of Economic Growth.* New York: Oxford University Press.
Harrod, R.F. 1939. An essay in dynamic theory. *Economic Journal* 49, March, 14–33.
Harrod, R.F. 1948. *Towards a Dynamic Economics.* London: Macmillan.
Harrod, R.F. 1952. Supplement on dynamic theory. In R.F. Harrod, *Economic Essays,* London: Macmillan.
Harrod, R.F. 1959. Domar and dynamic economics. *Economic Journal* 69, September, 451–64.
Kaldor, N. 1955–6. Alternative theories of distribution. *Review of Economic Studies* 23(2), 83–100.
Kaldor, N. 1957. A model of economic growth. *Economic Journal* 67, December, 591–624.
Keynes, J.M. 1936. *The General Theory of Employment, Interest and Money.* London: Macmillan.
Solow, R.M. 1956. A contribution to the theory of economic growth. *Quarterly Journal of Economics* 70, February, 65–94.
Swan, T.W. 1956. Economic growth and capital accumulation. *Economic Record* 32, November, 334–61.

Hart, Albert Gailord (born 1909). Born in Oak Park, Illinois, Hart received his BA from Harvard in 1930 and his PhD from the University of Chicago in 1936. Most of his career – from 1946 until his retirement in 1979 – was spent as Professor of Economics at Columbia University. Much of his noteworthy work concerned the implications of uncertainty for policy makers, but he should also be remembered as having worked with Kaldor and Tinbergen (1964) to produce an ingenious proposal for a commodity reserve currency: this would serve to improve international liquidity simultaneously with providing a means of protecting incomes of primary producers against shrinkage in times of depression.

Hart's work on uncertainty included a monograph (1940), one notable feature of which was an attempt to analyse how decision makers can judge their success or failure, and thence reformulate their expectations, in the light of partial knowledge of performance distributions. From 1936 onwards, he emphasized the rationality, in situations of uncertainty, of choosing flexible production technologies which, though they might not be perfectly adapted to any specific output rate, would not be disastrously expensive to run over a range of outputs. This idea, which was also promoted by his Chicago contemporary Stigler (1939), led Hart to be critical of much writing on decision theory. He felt it misleading to theorize as if firms assign probabilities to rival hypothetical outputs, aggregate these weighted values and then build their plans around the weighted average of probable output rates (1942). Hart was also irritated by Keynes's tendency to speak of expectations in terms of certainty equivalents, and he warned that, 'generally speaking, the business policy appropriate to a complex of uncertain anticipations is different in kind from that appropriate for any set of certain expectations' (1947, p. 422).

Hart carried this theme into work critical of deterministic macroeconomic model-building and fiscal policy formulation (1945), and into a distinctive approach to monetary theory (1948, especially part II). In the latter, he introduced the 'margin of safety' motive for holding liquid assets, arguing that the structure of economic affairs is such that risks are usually linked: a single disappointment is prone to cause many other things to go wrong in consequence. Hart's concern with surprise, flexibility, and structural linkages in many ways foreshadows themes that emerged in the 1980s in the business policy literature on scenario planning and strategic choices. However, he is not usually credited as the pioneer of this kind of thinking: having been largely ignored by mainstream writers, his ideas were sufficiently poorly known to end up being reinvented.

PETER EARL

SELECTED WORKS
1940. *Anticipations, Uncertainty and Dynamic Planning.* Chicago: University of Chicago Press.

1942. Risk, uncertainty and the unprofitability of compounding probabilities. In *Studies in Mathematical Economics and Econometrics*, ed. O. Lange, F. McIntyre and T.O. Yntema, Chicago: University of Chicago Press.

1945. 'Model-building' and fiscal policy. *American Economic Review* 35, September, 531–58.

1947. Keynes's analysis of expectations and uncertainty. In *The New Economics: Keynes's Influence on Theory and Public Policy*, ed. S.E. Harris, New York: Knopf.

1948. *Money, Debt and Economic Activity*. New York: Prentice-Hall.

1964. (With N. Kaldor and J. Tinbergen.) The case for an international commodity reserve currency. (Paper submitted to UNCTAD, Geneva, March–June 1964.) In *Essays on Economic Policy*, ed. N. Kaldor, Vol. II, London: Duckworth.

BIBLIOGRAPHY

Stigler, G.J. 1939. Production and distribution in the short run. *Journal of Political Economy* 47, June, 305–27.

Hawkins–Simon conditions. In a Leontief system of interindustrial input–output relationships consisting of n sectors of industry, each of which produces a single good, without joint products, under constant returns to scale, and using n goods as input in fixed proportions, the balance of demand for and supply of goods is represented by a system of linear equations

$$x_i = \sum_{j=1}^{n} a_{ij}x_j + c_i, \qquad (i = 1, 2, \ldots, n),$$

where a_{ij} are non-negative input coefficients of the jth sector, x_j is the level of output of the jth sector and c_i is the level of final demand for the ith good $(i, j = 1, \ldots, n)$.

With the input coefficient matrix A having a_{ij} in the ith row and the jth column, the output vector x having x_j in the jth component, and the final demand vector c having c_i in the ith component the system is represented in matrix form by the equation

$$x = Ax + c.$$

The system is productive enough to give positive net output over input, if x_j non-negative units of output of the jth sector $(j = 1, \ldots, n)$ are produced to meet a bill of positive final demand $c_i (i = 1, \ldots, n)$.

The *productivity* of the system, which is equivalent to the condition that the n-dimensional square matrix $I - A$, where I is the identity matrix, have an inverse matrix $(I - A)^{-1}$ having all the elements non-negative, hinges on and is completely determined by the magnitudes of the input coefficients. A necessary and sufficient condition for such productivity, stated in terms of inequalities constraining the magnitudes of the input coefficients and referred to as the Hawkins-Simon conditions, after the names of its discoverers (Hawkins and Simon, 1949), is that all the principal minor determinants of the matrix I-A be positive. This is equivalent to the seemingly weaker conditions that the n principal minor determinants located in the ascending order on the upper left corner of the matrix I-A be positive

$$\Delta_k = \begin{vmatrix} 1 - a_{11} & -a_{12} & \ldots & -a_{1k} \\ -a_{21} & 1 - a_{22} & \ldots & -a_{2k} \\ \ldots & & & \\ -a_{k1} & -a_{k2} & \ldots & 1 - a_{kk} \end{vmatrix} > 0,$$

$$(k = 1, \ldots, n).$$

As a mathematical result the equivalence of the Hawkins–Simon conditions to productivity is very easy to prove, as can readily be shown by transforming the equation $(I - A)x = c$ through Gaussian elimination to a triangular form

$$b_{11}x_1 + b_{12}x_2 + \cdots + b_{1n}x_n = d_1$$
$$b_{22}x_2 + \cdots + b_{2n}x_n = d_2$$
$$\ldots$$
$$b_{nn}x_n = d_n,$$

where

$$b_{ij} \leqq 0 \ (i < j), \qquad d_i \geqq 0 \ (i = 1, \ldots, n)$$

and

$$\Delta_k = b_{11}b_{22} \ldots b_{kk} \ (k = 1, \ldots, n).$$

Since the Hawkins–Simon conditions ensure the productivity of the system, they are a primary prerequisite for the Leontief system, and enlarged systems involving it as a built-in subsystem, to be well-behaved. They also make the Leontief system dynamically well-behaved. In the multiplier process over discrete time,

$$x_i(t + 1) = \sum_{j=1}^{n} a_{ij}x_j(t) + c_i, \qquad (i = 1, \ldots, n)$$

the solution converges to the equilibrium output levels supplying net output equal to the final demand $c_i \ (i = 1, \ldots, n)$, if and only if the Hawkins–Simon conditions are satisfied. This stability is equivalent to the convergence of the matrix geometric progression

$$I + A + A^2 + \cdots + A^t + \cdots$$

to the inverse matrix $(I - A)^{-1}$. In the multiplier process over continuous time,

$$dx_i/dt = \alpha_i \left(\sum_{j=1}^{n} a_{ij}x_j + c_i - x_i \right), \qquad (i = 1, \ldots, n)$$

the Hawkins–Simon conditions are necessary and sufficient as well for the convergence of the solution to the same equilibrium output levels, which is equivalent to the condition that the real parts of all the eigenvalues of the matrix A-I be negative.

HUKUKANE NIKAIDO

See also LINEAR MODELS; PERRON–FROBENIUS THEOREM.

BIBLIOGRAPHY

Hawkins, D. and Simon, H.A. 1949. Note: some conditions of macroeconomic stability. *Econometrica* 17, July–October, 245–8.

Hawtrey, Ralph George (1879–1975). Hawtrey was born in Slough, near London, and went up to Trinity College, Cambridge, from Eton in 1898. Three years later he graduated 19th Wrangler in the Mathematical Tripos. Hawtrey remained at Cambridge for a further period to read for the Civil Service examinations, as was quite common at that time. This latter study included some economics with lectures largely by G.P. Moriarty and J.H. Clapham. In 1903 he entered the Admiralty, but in 1904 he transferred to the Treasury, where he was to remain until retirement in 1947 (his official retirement at 65 was in 1944). Hawtrey's only academic appointments in economics were in 1928–9, when he was given special leave from the Treasury to lecture at Harvard (as a visiting professor) and after his retirement, when he was elected Price Professor of International Economics at the

Royal Institute of International Affairs (1947–52). Hawtrey served as President of the Royal Economic Society between 1946 and 1948.

Hawtrey was not, therefore, directly a part of the 'Cambridge School' of economics. Marshall took no immediate part in Hawtrey's economic education which was, for the most part, acquired in the Treasury. Nonetheless he had close contacts with the Cambridge economists. Away from economics he was involved with both the Apostles and with Bloomsbury, whilst within the subject he was a visitor to Keynes's Political Economy Club at Cambridge and his major work, *Currency and Credit* (1919) became a standard work in Cambridge in the 1920s. Furthermore, although there were differences in approach between Hawtrey and the Cambridge School in some areas, Keynes himself noted in reviewing *Currency and Credit* the similarities between Hawtrey's approach to the theory of money and that of the Cambridge School – though Keynes remarked that Hawtrey had reached his results independently (Keynes, 1920).

I. Hawtrey was primarily a monetary economist; his major contributions related to the Quantity Theory and the trade cycle. He was one of the first English economists to stress the primacy of credit-money rather than metallic legal tender. Furthermore his income-based approach, like that of the Cambridge School, led to a closer integration of the theories of money and output. For Hawtrey, money income determines expenditure, expenditure determines demand and demand determines prices.

Hawtrey summarized his aims in monetary theory in the preface to *Currency and Credit*:

> Scientific treatment of the subject of currency is impossible without some form of the quantity theory ... but the quantity theory by itself is inadequate, and it leads up to the method of treatment based on what I have called the consumers' income and the consumers' outlay – that is to say, simply the aggregates of individual incomes and individual expenditures (1919, p. v).

Investment (the result of Saving) is included in consumers' outlays, since it is spent on fixed capital. Consumers' balances are then the difference between outlays and income and thus consist only of accumulated cash balances (including money held in bank accounts). In addition there is a similar demand for money balances by traders related to their turnover. Of course individual agents may hold both consumers' and traders' balances – Hawtrey notes that the true income of traders is the profits of the business and that this is included in consumers' income.

The 'unspent margin', or total money balances, consists of the consumers' and traders' balances taken together. From this Hawtrey derives a form of the quantity theory. Hawtrey argues that traders' balances are relatively stable, and thus the operational relationships are concerned with the supply of money (in a wide sense taken to include credit) and consumers' income and outlay. It is worth noting that compared to the Cambridge income-based approach Hawtrey's places greater emphasis on the demand for nominal balances rather than real balances. It is also interesting to note that Keynes used a similar balances approach to the quantity theory in the period after 1925 leading up to the theory presented in the *Treatise on Money* (1930), where he distinguishes first between investment and cash deposits and later between income, business and savings deposits.

The demand for money is also analysed in terms of motives. Hawtrey identifies a transaction demand, a precautionary demand, and a residual demand which reflects a gradual accumulation of savings balances or what Joan Robinson has called short-hoards (Robinson, 1938). Hawtrey envisages agents as saving gradually but investing only larger sums periodically. In the meantime these short-hoards act as a buffer stock. The main cost of holding money balances is the interest foregone, and thus Hawtrey points to a balancing process between costs and advantages in determining desired balances. The introduction of a banking system into the model allows agents to substitute borrowing power for money balances (Hawtrey, 1919, pp. 36–7).

II. Hawtrey also introduces a concept of effective demand:

> The total effective demand for commodities in the market is limited to the number of units of money of account that dealers are prepared to offer, and the number they are prepared to offer over any period of time is limited according to the number they hope to receive (1919, p. 3).

Later, in *Trade and Credit* (1928) Hawtrey points to a flaw in the theory of an elastic supply of labour based on marginal utilities (or disutilities) of product and effort. He argues that whilst a difference between the marginal utility of the product and the disutility of effort may prompt an additional supply of labour 'in the simple case of a man working on his own account' (1928, p. 148), this is not the general case since: 'the decision as to the output to be undertaken is in the hands of a limited numbers of employers, and the workmen in the industry are passively employed by them for the customary hours at the prevailing rates of wages' (1928, p. 149). In this case output decisions are based not on the gross proceeds, but on the net profit margin.

The factor of expectations is also present in Hawtrey's analysis of fluctuations. Hawtrey suggests that during a downturn in activity money balances will be reduced more quickly than they are replenished in an upswing. This is because as income drops initially consumers will draw on their balances to maintain their outlay.

There is then a further level of adjustment as changes in consumers' outlays impacts on traders. Consider an upswing: the increase in consumers' outlays will increase the nominal receipts of traders and reduce their physical stocks. Traders, finding their balances have increased can either order more stock from manufacturers or reduce their bank indebtedness. Prices will tend to rise as traders find they are unable to replenish their stocks fast enough. For Hawtrey quantity adjustments occur *before* price adjustments, indeed often the price movements result from the quantity movements. Thus 'the rise of prices, when it occurs, is caused by the activity; it is a sign that production cannot keep pace with demand' (1928, p. 156). The role of stocks in Hawtrey's theory is pivotal, in general it is quantity signals rather than price signals which are the more effective. The existence of traders' stocks means that it is nearly always possible to meet the demand for increased consumption in the short term, which implies that at least in the short term a naive proportionality between increases in the money supply and prices does not hold. Furthermore the model opens the possibility of short run quantity adjustments in disequilibrium. Thus, argues Hawtrey:

> It is only in times of equilibrium, when the quantity of credit and money in circulation is neither increasing nor decreasing, that the relation of prices and money values to that quantity of credit and money is determined by the individual's considered choice of the balance of purchasing power appropriate to his income. ... In practice it seldom,

perhaps never, happens that a state of equilibrium is actually reached (1919, p. 46).

Nonetheless Hawtrey's theory of the trade cycle is money-driven. It is the fluctuations in money and credit which stimulate and support the price and quantity movements. Hawtrey argued that the periodic nature of the trade cycle was solely due to monetary factors. Traders stocks are viewed as being highly interest elastic since they are held on borrowed funds, investment in fixed capital is also interest elastic (based on a marginal efficiency of capital analysis).

Thus an increase in the rate of interest will tend to reduce the demand for credit due to a lower demand for stocks and a reduced level of new investment. If the increased rate of interest is itself the result of a decreased supply of credit then there may also be some quantitative restrictions of borrowing. To reduce their stocks traders will stop giving new orders to manufacturers, leading to a drop in the level of output which will further diminish the demand for credit, as well as the level of income and demand. Traders may reduce prices to stimulate sales to accelerate the destocking process. There is thus a tendency to a cumulative decline in output, credit and prices until the banks find themselves with excess reserves and believe it to be profitable to reduce the interest rate and expand credit. For Hawtrey, macro-economic disequilibrium was defined in terms of monetary disequilibrium.

The solution was also therefore monetary, and in particular the short-term rate of interest (the long-term rate of interest was seen as relatively ineffective as a means of control because of its relatively slow impact on investment). Hawtrey viewed the psychological factors in the trade cycle as secondary, arguing that no amount of good news or bad could seriously affect the cycle if monetary factors were not accommodating. He also opposed the public works solution to a slump in output along similar lines – and in this respect is associated with the 'Treasury View' (see Hawtrey, 1925). In later life Hawtrey did acknowledge that public works could play a role in severe depressions, but as Haberler (1939, p. 23) points out Hawtrey viewed those occasions when cheap money would fail to stimulate a revival as generally very rare – although he accepts that this was the case in the 1930s.

III. For Hawtrey, investment decisions were made on a Marshallian marginal productivity of capital basis. In a perfectly competitive market, the marginal return on capital employed would be equalized across every industry. In these circumstances Hawtrey identifies the 'ratio of labour saved per annum to the labour expended on first cost' as 'a physical property of the capital in use' (1913, p. 66) and as a 'natural rate' of interest. Under stable monetary conditions and in the absence of a banking system this natural rate is equal to the market rate of interest or the profit rate, as in the standard marginal efficiency of capital analysis. But changes in monetary conditions will generate changes in prices and thus profits; hence the market rate will diverge from the natural rate in the same direction as the movement in prices.

With the addition of a banking system, the actual rate of interest will depend on the behaviour of the banks, and in particular their reserve position. Thus the interest rate will diverge from the profit rate. There is a three-way equilibrium condition, relating the physical return on capital, the profit rate and the balance position of banks, i.e. $N = p = r$ where N is the Natural rate p is the profit rate and r is the interest rate. An increase in the supply of money will cause a rise in prices and the availability of credit; thus $N < p$ at the same time the banks will find themselves with excess reserves and

thus interest rates will tend to be lower than otherwise to stimulate borrowing, i.e. $p > r$. This will be generally expansive, demand, investment and output will all tend to rise—but the seeds of the eventual slump are already present. The rising prices and relatively low rate of interest will encourage firms to over-invest, expecting returns greater than those actually accruing. On the downward cycle $N > p$ and $p < r$.

It is worth briefly considering the relationship between Hawtrey's natural rate and that associated with Wicksell. In his early work Wicksell took the natural rate as that prevailing if loan transactions were made in kind, but he later revised this to equate the natural rate with the rate of profits received in the form of money (see Lindahl, 1939, p. 261; Lindahl also discusses a physical return on capital 'natural rate' similar to Hawtrey's). Thus Wicksell's natural rate can be seen as closer to Hawtrey's profit rate. Wicksell, like Hawtrey, also associates the natural rate with an equilibrium between savings and investment and stability in the price level.

Hawtrey does not place great stress on this natural rate analysis, concentrating more on the relationship of the profit rate and the interest rate. There are also considerable practical problems in determining Hawtrey's natural rate, particularly in imperfect capital markets (see the discussion in Lindahl, 1939, and Haberler, 1939).

IV. Hawtrey, like most of the inter-war Cambridge economists, had a fundamental belief in the self-adjusting nature of the economic system, even though much of the analysis of the period would suggest otherwise. Hawtrey believed that the system was continually approaching or seeking an equilibrium, though in practice the next shock would come before the adjustment process was complete. However, Hawtrey's theoretical approach was to concentrate on the processes of adjustment to monetary disequilibrium.

The income/inventories approach to the trade cycle is mirrored in Hawtrey's analysis of Savings and Investment. For Hawtrey, Savings were directed into Investment opportunities by securities dealers who acted like traders, holding stocks of securities financed by bank borrowing, intermediating between the savers and investors. In the early 1930s Hawtrey developed this analysis into a model where an imbalance between Savings and Investment results in an unanticipated change in physical stocks of goods as a result of changes in consumers' incomes (and outlays).

Savings are the excess of consumers income over desired consumption and are represented by investment; an increase in money balances; or purchases of goods. Net investment is defined as the total of securities sold less those bought by securities dealers. Clearly the price of securities (and by implication the long term rate of interest) will move to achieve an equilibrium between the net amount of investment and capital raised, but planned savings can exceed the resources seeking investment, in which case the excess must flow into additional money balances or additional consumption – or vice versa. In either case an expansion or contraction of demand is set in motion. Both Saulnier (1938) and Haberler (1939) note the similarity of this analysis with that of D.H. Robertson. This aspect of Hawtrey's theory is also reviewed by Davis (1981).

Hawtrey's disequilibrium analysis where unintended changes in stocks bring about an equality of actual savings and investment, but a further chain of adjustment if intended savings and investment are not equal is remarkably close to the modern textbook presentation of the Keynesian equilibrium adjustment process. It is interesting therefore to briefly examine the discussions between Keynes and Hawtrey leading up to the *General Theory*. Indeed in Hawtrey's comments on the drafts of the *Treatise* he is often more 'Keynesian' than Keynes himself! (see *CW* XIII, pp. 138–69). At this stage Keynes envisages:

(1) A decline in fixed investment relatively to saving.

(2) A fall of prices ...

(3) A fall of output, as a result of the effect of falling prices and accumulating stocks on the minds of entrepreneurs (Letter to R.G. Hawtrey, 28 November 1930, *CW* XIII, p. 143).

The fall in output leads to a disinvestment in working capital, and eventually to a situation where total investment and prices fall too far. Once output stops declining this leads to a slight rise in prices, and given the low level of stocks at this point, so starts the upturn. Hawtrey, on the other hand, sees a direct effect on output from the contraction in demand at unchanged prices, and criticizes Keynes for only taking account of the reduction in prices relative to costs in his fundamental equations (*CW* XIII, pp. 151–2). Hawtrey argues that 'the change in prices when it does occur is not by itself an adequate measure of the departure from equilibrium' (*CW* XIII, p. 151). And later comments that: 'A manufacturer restricts output, not because he believes that prices are about to fall, but because he cannot secure sufficient sales at the existing price' (Letter from R.G. Hawtrey, 6 December 1930, *CW* XIII, p. 165).

Prices are reduced only gradually in an attempt to boost orders, but Hawtrey also points out that it is the level of retail prices which will determine the ultimate level of sales – and this will depend on how quickly retailers pass on the manufacturers' reductions. Both Hawtrey and Keynes realize that the decline in output will rebound on savings, but do not appear to treat this as the main equilibrating factor (as in the later Keynesian theory).

V. The high point of Hawtrey's official career came with the Genoa International Financial Conference in 1922. The conference was concerned with the problems relating to a general return to the international Gold Standard after World War I. In particular there was concern that the quantity of gold might be insufficient for a return to the system at the old pre-war parities, other concerns centred on problems relating to fluctuations in demand for monetary gold. The result was greater interest in a joint Sterling–Gold Standard along the lines of the Gold Exchange Standard operated earlier by India and other countries.

Hawtrey's main suggestions adopted by the Genoa conference related to greater cooperation between central banks to manage the demand for monetary gold and to regulate credit so as to stabilize the purchasing power of gold. However, the Genoa Resolutions were never acted on, largely as a result of US scepticism, and the failure of other central banks to participate in the planned follow-up conference (see Davis, 1981).

At the Treasury Hawtrey had argued that there were two primary considerations for monetary policy: the stabilization of internal prices and the stabilization of the foreign exchanges. Given the UK's status as a financial centre he argued that exchange instability was particularly damaging and would make the covering of trade finance offered through London increasingly difficult. This predisposed him towards the Gold Standard as the *de facto* most practical means of achieving exchange stability.

Though Hawtrey was aware of possible deflationary problems associated with the return to Gold, he appears to have believed that the exchange rate would return to par naturally, and that the necessary adjustments would come from American inflation rather than UK deflation (see the discussion in Moggridge, 1972, pp. 71–2, 91).

VI. Despite a long and active life, Hawtrey's main theoretical contributions to Economics came largely in the interwar period. His first book, *Good and Bad Trade*, was published in 1913 and sets out a view of the trade cycle which received a more rigorous theoretical treatment in *Currency and Credit* (1919), but which remained little changed thereafter, although the debates surrounding Keynes's *Treatise* prompted some refinements and revisions, as did the experience of the 1930s depression. The last major contemporary studies of his work were Saulnier (1938), which also reviewed the theories of D.H. Robertson, F.A. von Hayek and J.M. Keynes, and Haberler (1937, 1939). Interest in Hawtrey revived in the later 1970s following his death (see for example Davis, 1977 and 1981).

In the 1920s innovative monetary theory in England was largely associated with the Cambridge School and in particular D.H. Robertson and Keynes. Hawtrey with his close Cambridge contacts contributed to this work, as the correspondence with Keynes now reprinted in the *Collected Works* shows. The three were often working along similar lines in this period and their work reflects (to varying degrees) an increasing failure of conventional theory to match the problems of the age.

R.J. BIGG

SELECTED WORKS
1913. *Good and Bad Trade*. London: Constable.
1919a. *Currency and Credit*. London: Longmans.
1919b. The Gold Standard. *Economic Journal* 29, December, 428–42.
1921. *The Exchequer and the Control of Expenditure*. London: World of Today.
1922. The Genoa Resolutions on currency. *Economic Journal* 32, September, 290–304.
1923. *Monetary Reconstruction*. London: Longmans.
1924. Discussion on monetary reform. *Economic Journal* 34, June, 155–76.
1925. Public expenditure and the demand for labour. *Economica* 5, March, 38–48.
1926. *The Economic Problem*. London: Longmans.
1927. *The Gold Standard in Theory and Practice*. London: Longmans.
1928. *Trade and Credit*. London: Longmans.
1930. *Economic Aspects of Sovereignty*. London: Longmans.
1931. *Trade Depression and the Way Out*. London: Longmans.
1932. *The Art of Central Banking*. London: Longmans.
1933. Saving and hoarding. *Economic Journal* 43, December, 701–8.
1934. Monetary analysis and the investment market. *Economic Journal* 44, December, 631–49.
1937. Alternative theories of the rate of interest: three rejoinders. *Economic Journal* 47, September, 436–43.
1938. *A Century of Bank Rate*. London: Longmans.
1939. *Capital and Employment*. London: Longmans.
1944. *Economic Destiny*. London: Longmans.
1946. *Economic Rebirth*. London: Longmans.
1946. *Bretton Woods: for better or worse*. London: Longmans.
1949. *Western European Union. Implications for the UK*. London: Royal Institute of International Affairs.
1950. *Balance of Payments and the Standard of Living*. London: Royal Institute of International Affairs.
1954. *Towards the Rescue of Sterling*. London: Longmans.
1955. *Cross Purposes in Wage Policy*. London: Longmans.
1961. *The Pound at Home and Abroad*. London: Longmans.
1965. *An Incomes Policy*. London: Woolwich Economic Papers No. 4.
1967. *Incomes and Money*. London: Longmans.

BIBLIOGRAPHY

Cambridge University, 1917. *Historical Register of the University of Cambridge to 1910*. Cambridge: Cambridge University Press.

Davis, E.G. 1977. The economics of R.G. Hawtrey. *Carleton Economic Papers*, 77(12).

Davis, E.G. 1981. R.G. Hawtrey, 1879–1975. In *Pioneers of Modern Economics in Britain*, ed. D.P. O'Brien and J.R. Presley, London: Macmillan.

Haberler, G. 1937. *Prosperity and Depression*. Geneva: League of Nations, Economic & Financial 1936, II.A.24.

Haberler, G. 1939. *Prosperity and Depression*. 2nd edn, Geneva: League of Nations, Economic & Financial 1939, II.A.4.

Hutchison, T.W. 1953. *A Review of Economic Doctrines 1870–1929*. Oxford: Oxford University Press.

Keynes, J.M. 1920. Review of Hawtrey's *Currency and Credit*. *Economic Journal* 30, September, 362–5.

Keynes, J.M. 1923. *A Tract on Monetary Reform*. London: Macmillan, Reprinted as *Collected Writings*, Vol. IV, London: Macmillan, 1971.

Keynes, J.M. 1930. *A Treatise on Money*, 2 vols. London: Macmillan. Reprinted as *Collected Writings*, Vols. V and VI, London: Macmillan, 1971.

Keynes, J.M. 1973. *Collected Writings*, Vol. XIII. London: Macmillan.

Lindahl, E. 1939. *Studies in the Theory of Money and Capital*. London: George Allen & Unwin. Reprinted, New York: Augustus M. Kelley, 1970.

Moggridge, D.E. 1972. *British Monetary Policy 1924–1931*. Cambridge: Cambridge University Press.

Moggridge, D.E. (ed.) 1973. *The General Theory and After*. Part I: Preperation. Vol. XIII in the *Collected Writings of John Maynard Keynes*, London: Macmillan.

Robinson, J.V. 1938. The concept of hoarding. *Economic Journal* 48, June, 231–6.

Rouse-Ball, W.W. and Venn, J.A. (eds) 1913. *Admissions to Trinity College, Cambridge*. Vol. 5: *1851 to 1900*. London: Macmillan.

Saulnier, R.J. 1938. *Contemporary Monetary Theory*. Columbia University Studies in the Social Sciences No. 443, New York: Columbia University Press.

The Times. 1975. Sir Ralph Hawtrey CB (Obituary). 22 March.

Hayek, Friedrich August von (born 1899). Friedrich August von Hayek, a central figure in 20th-century economics and foremost representative of the Austrian tradition, 1974 Nobel laureate in economics, a prolific author not only in the field of economics but also in the fields of political philosophy, psychology and epistemology, was born in Vienna on 8 May 1899. Following military service as an artillery officer in World War I, Hayek entered the University of Vienna, where he attended the lectures of Friedrich von Wieser and Othmar Spann and obtained doctorates in law and political science. After spending a year in New York (1923–4), Hayek returned to Vienna where he joined the famous *Privatseminar* conducted by Ludwig von Mises. In 1927 Hayek became the first director of the Austrian Institute for Business Cycle Research. On an invitation from Lionel Robbins, he lectured at the London School of Economics in 1931 and subsequently accepted the Tooke Chair. Hayek soon came to be a vigorous participant in the debates that raged in England during the 1930s concerning monetary, capital, and business-cycle theories and was a major figure in the celebrated controversies with John Maynard Keynes, Piero Sraffa and Frank H. Knight.

During the late 1930s and early 1940s Hayek's research focused on the role of knowledge and discovery in market processes, and on the methodological underpinnings of the Austrian tradition, particularly subjectivism and methodological individualism. His contributions in these areas were an outgrowth of his participation in the debate over the possibility of economic calculation under socialism.

In 1950 Hayek moved to the United States, joining the Committee on Social Thought at the University of Chicago. His research there engaged the broader concerns of social, political and legal philosophy. He returned to Europe in 1962 with appointments at the University of Freiburg, West Germany, and then (1969) at the University of Salzburg, Austria. Since 1977 Hayek has resided in Freiburg.

Hayek's scholarly output spans more than six decades. Still growing in the mid-1980s, his bibliography (Gray, 1984) includes 18 books, 25 pamphlets, 16 books edited or introduced, and 235 articles. Although these publications have brought Hayek international renown and honours in several disciplines, his contributions to other social sciences emerged, to a significant degree, as extensions of his scholarship in the field of economics and its methodological foundations. The following survey refers rather narrowly to the career and contributions of Hayek the economist.

ECONOMICS AS A COORDINATION PROBLEM. Throughout all of Hayek's writings, both the questions asked and the answers given reflect his general conception of economics as a coordination problem (O'Driscoll, 1977). Thoughtful observation of market economies suggests that they are characterized by order more complex and intricate than can be explained in terms of deliberate efforts to achieve coordination among individual activities. According to Hayek (1952, p. 39), it is precisely the existence of this 'spontaneous order' that provides the subject matter for the science of economics.

While market economies are better coordinated than can be accounted for by references to deliberate planning, they are always less than fully coordinated, hence the coordination *problem*. In one important sense, coordination failures are an integral part of an ongoing market process that iterates towards a greater degree of coordination. An oversupply or undersupply of some particular good, for instance, is evidence that the plans of producers and consumers of that good are not well coordinated one with the other. But the discoordination itself provides both an indication of the inconsistency in plans and the incentive for producers and consumers to make the appropriate adjustments.

But market economies do occasionally experience profound economy-wide coordination failures. Much of Hayek's research has been aimed, either directly or indirectly, toward discovering the set of circumstances or, more appropriately, the sequence of events that could cause such failures, i.e. that could cause an economy to collapse into economic depression. The focus of his research is *intertemporal* discoordination. The coordination of activities over time is inherently more difficult, more problematic, than the coordination of activities in a given period. Producers must make decisions now in anticipation of decisions that other producers and, ultimately, consumers will make sometime in the future. The fact that production is time consuming, the more so the more well developed the economy, figures importantly in Hayek's theorizing. This essential time element increases the likelihood of erroneous investment decisions and gives scope for cumulative investment errors. A spate of intertemporally discoordinated investments, whether triggered by a real or a monetary disturbance, can increase employment opportunities producing an artificial boom. But the eventual realization of the discoordination will necessitate a partial liquidation, which constitutes a bust. In this context, the Austrian theory is differentiated from other macroeconomic theories by its attention to the problem of intertemporal coordination *within* the investment sector. The more conventional treatments of macroeconomic coordination problems focus on the general *level* of investment in comparison with the level of saving or the size of the labour force.

Hayek adopted a two-tier approach to the study of business cycles. Prerequisite to the question of how an economywide coordination failure could occur is the question of how any degree of intertemporal coordination can be achieved at all in market economies. In Hayek's words, 'before we explain why people commit mistakes, we must first explain why they should ever be right' (1937, p. 34). His account first of how a market economy works to coordinate activities over time and then of what can go wrong draws from several different fields of study within the science of economics. In particular, it draws in fundamental ways from price theory, capital theory and monetary theory.

Each of these fields required further development before becoming part of Hayek's account. Price theory had to be recast so as to emphasize the role of the price system as a communication network and as the most efficient means of making use of economic information. Capital theory had to be detailed so as to give play to the individual elements of the capital structure, which is made up of heterogeneous pieces of capital of various degrees of specificity and durability and related to one another by various degrees of intertemporal substitutability and complementarity. And monetary theory had to be extended in scope so as to allow the identification of systematic relative price effects associated with the process of monetary expansion or contraction.

While Hayek contributed importantly to each of these fields of study, his ultimate achievement consists in the integration of price theory, capital theory and monetary theory. Hayek integrated his own developments in these fields into a cohesive account of a market process that tends towards intertemporal coordination and of central-bank policies that can interfere with that process in such a way as to cause artificial economic booms which are inevitably followed by economic busts. Hayek's business cycle theory provided a basis for interpreting much of 19th- and 20th-century economic history, for evaluating alternative macroeconomic theories – especially those of John Maynard Keynes, and for promoting institutional reform of the kind that will prevent or minimize intertemporal discoordination.

SUBJECTIVISM AND METHODOLOGICAL INDIVIDUALISM. The methodological norms adopted by Hayek are a direct reflection of his perception of the subject matter: economic phenomena as *spontaneous order*. Fundamental institutions in society owe their existence to no identifiable creator. They are the 'results of human action but not of human design'. The most obvious examples of spontaneous order are the use of language and, among economic phenomena, the use of money. Money, the most commonly accepted medium of exchange, came to be accepted, commonly accepted, and then most commonly accepted as a result of a long sequence of actions on the part of a multitude of individual traders none of whom *intended* to create the institution of money. Other economic phenomena – from the simple division of labour to the more broadly conceived organization of industry – are to be understood as instances of spontaneous order.

If there were no order in society except for what was consciously designed, Hayek argued, there would be no scope and no need for the social sciences. The task of these sciences in a world characterized by spontaneous order is precisely to account for those aspects of social order that were not consciously designed.

A central methodological theme that has consistently pervaded Hayek's investigation of spontaneous order stems from his insistence that it is inappropriate to apply uncritically the methods of the physical sciences to the phenomena of the social sciences. Hayek used the term *scientism* to refer to the slavish imitation of the methods of the physical sciences without regard for the innate differences between physical and non-physical reality. Scientism, which unavoidably overlooks crucial aspects of social reality, such as perception, intent and anticipation, was the focus of two long and critical articles published by Hayek during World War II. In these articles, which constitute the central core of his 1952 book, *The Counter-Revolution of Science: Studies on the Abuse of Reason*, Hayek spelled out the case for subjectivism and methodological individualism in the social sciences. 'It is probably no exaggeration,' according to Hayek, 'to say that every important advance in economic theory during the last hundred years was a further step in the consistent application of subjectivism' (1952, p. 31).

Classical economists had focused their attention on the *objects* being valued and had looked for common denominators of value in terms of labour input or costs of production. The Austrian economists, particularly Menger, Mises and Hayek, are to be credited with shifting attention from the objects being valued to the subjects engaged in valuation. The value attributed to the various objects of economic actions, Hayek emphasized, can be accounted for only with reference to human purposes and in terms of the views that people hold about those objects.

Hayek's thoroughly subjectivist outlook and his adherence to the strictures of methodological individualism were mutually reinforcing. Methodological individualism is not a prescription of how to engage in economic research but rather a recognition of what counts as an economic explanation. To explain the undesigned aspects of a spontaneous order is to trace those aspects to the consciously taken individual actions that gave rise to that order. In Hayek's own words, 'it is the concepts and the views held by individuals which are directly known to us and which form the elements from which we must build up, as it were, the more complex phenomena ...' (Hayek, 1952, p. 38).

The contention that Hayek's crusade against scientism has consistently informed his substantive work is at least partly in conflict with a recent argument by T.W. Hutchison, who has sought to establish that Hayek's 1937 article 'Economics and Knowledge' marked a sharp change in his methodology towards a 'falsificationist' approach to economic science (Hutchison, 1981). This argument has been effectively disputed by John Gray (1984, pp. 16–21), who recognizes that the 1937 article was intended to persuade Mises that, contrary to Mises' own 'praxeology', there is an essential empirical element in our understanding of economic phenomena. Further, Hayek's (1952) commitment to subjectivism and methodological individualism, and his emphasis on the fallacies of scientism suggest in fact a deepening, rather than an erosion, of his recognition of the extent to which economic theory is independent of – in fact a prerequisite for – empirical economic observation.

THE PRICE SYSTEM AS A COMMUNICATION NETWORK. It is a short step from Hayek's appreciation of the phenomenon of spontaneous order to his understanding of the price system as a communication network. The key contribution of the price system to social well-being consists, Hayek demonstrated, in the system's capacity to transmit information from one part of the market to another. In the event of a natural disaster which has curtailed the availability of a specific raw material, for example, the fact of a reduced supply will be effectively communicated to potential users through the medium of a higher price – which also provides the incentive for the socially desirable economizing of the particular raw material (Hayek, 1945, p. 85–6). The need for such a communication network arises out of the fact that the information to be communicated is dispersed throughout the society. This insight into the nature of prices as *signals* has, during the past decade and a half, come to be fairly widely recognized and expounded in modern textbooks.

In his treatment of the use of knowledge in society, Hayek made a sharp distinction between two kinds of knowledge: (1) scientific, or theoretical, knowledge and (2) the knowledge of the particular circumstances of time and place. The first-mentioned category is the proper concern of the economist; the second-mentioned category is the proper concern of the market participant. Failure to recognize this 'division of knowledge' can lead to one of two serious errors. The assumption that *economists* can assimilate both kinds of knowledge leads to the conclusion that 'rational planning' can outperform – or at least duplicate – the market itself; the assumption that *market participants* can assimilate both kinds of knowledge leads to the conclusion that 'rational expectations' can nullify the systematic effects of monetary manipulation.

Hayek recognized and emphasized that if a fully adjusted system of prices – one corresponding to attained equilibrium – can be held to offer a system of coordinated and mutually reinforcing signals, such a system must depend on some prior groping process of market *discovery*. Hayek saw this process as consisting of market *competition* – which meant for him not the state of affairs consistent with the conditions for so-called perfect competition, but rather the rough-and-tumble process of market agitation kept in motion by complete freedom for competitive entrepreneurial entry. What such a competitive process can accomplish, Hayek argued, is the discovery of possibilities and preferences that no one had realized hitherto (Hayek, 1968).

These insights concerning knowledge and discovery articulated by Hayek in a number of profound papers from the late 1930s to the mid-1940s (Hayek, 1948) were partly responsible for, and partly emergent from, Hayek's participation in the celebrated interwar debate over the possibility of economic calculation under a socialist system. In deepening and widening the case originally presented by Mises in 1920, which challenged the feasibility of such calculation in the absence of market prices for factors of production, Hayek came to perceive the market process itself as crucial for the generation of that very knowledge which it would be necessary for a central planning authority to possess *before* it could hope to achieve a successful and efficient allocation of societal resources.

It was especially this Hayekian appreciation for the market as a discovery process that has significantly contributed to the contemporary revival of interest in the Austrian paradigm. In this context the Austrian contribution is to be distinguished from the more formal, or mathematically tractable, theories by its emphasis on the role of the entrepreneurial discovery in those systematic market processes upon which we must depend, in a world of ignorance and disequilibrium, for any possible tendency toward mutual coordination among the market participants. What Hayek showed was that much modern economics misconstrues the nature of the economic problem facing society by assuming away the problems raised by the fact of dispersed information. To imagine (as earlier critics of Mises and Hayek had proposed) that it would be possible to run a socialist system by simulating the market and promulgating non-market 'prices' for the guidance of socialist managers is to ignore the extent to which market prices – both of consumer goods and of the capital goods that constitute the economy's capital structure – *already* express the outcome of an entrepreneurial discovery procedure that draws upon scattered existing knowledge.

THE INTERTEMPORAL STRUCTURE OF CAPITAL. Hayek's contribution to the development of capital theory is commonly regarded as his most fundamental and pathbreaking achievement (Machlup, 1976). His early attention (1928) to 'Intertemporal Price Equilibrium and Movements in the Value of Money' (English translation in Hayek, 1984) provided both the basis and inspiration for many subsequent contributions in this area, most notably for those of John Hicks. The widely recognized but rarely understood Hayekian triangles, introduced in his *Prices and Production* (1935), provided a convenient but highly stylized way of describing the changes in the intertemporal pattern of the capital structure. The formal and comprehensive analysis in *The Pure Theory of Capital* (1941) fleshed out the earlier formulations and established the centrality of the 'capital problem' in questions about the market's ability to coordinate economic activities over time.

The essential element of time in the economy's production process coupled with the inherent complexities of the capital structure gives special significance to the problem of intertemporal coordination. Individual producers must commit resources in the present on the basis of some production plan. Intertemporal coordination in the strictest sense requires that all such plans be mutually compatible and that they be jointly consistent with resource availabilities. The extent to which such compatibility and consistency actually exists is determined only through the market process in which each producer attempts to carry out his own plan. The individual production plans take shape as non-specific capital (e.g. raw material) is committed to a specific use (e.g. a particular tool or machine); the passage of time and the efforts of each producer to secure the additional capital needed to complete his own production plans reveal the extent to which the capital structure is intertemporally coordinated or discoordinated. The actual availability of some raw material complementary to already-committed capital may be less, for instance, than the amount needed for each producer to carry out his plan. As such discoordination is revealed (by an increase in the price of the raw material), production plans are revised. In Hayek's formulation, the capital goods that make up the production process are neither so specific that such plan revision is impossible nor so non-specific that it is costless.

In his *Pure Theory of Capital* (1941), Hayek provides a detailed treatment of capital goods in terms of reproducibility, durability, specificity, substitutability and complementarity. These multifaceted characteristics of various capital goods and of relationships among them cause the structure of production, taken as a whole, to be characterized by a longer or shorter 'period of production', a greater or lesser degree of 'roundaboutness'. The degree of roundaboutness, the extent to which the production process ties up resources over time, is determined by the market rate of interest – with the 'market rate' broadly conceived as the terms of trade between goods available in the present and goods available in the future. The market process works to translate intertemporal preferences into production plans. For instance, a fall in the rate of interest reflecting an increased willingness to forgo present goods for future goods creates incentives for engaging in production processes of greater degrees of roundaboutness. The characteristics, mentioned above, of the individual capital goods and of the relationships among them determine the extent to which the existing capital structure is actually adaptable to changes in intertemporal preferences.

MONEY AND ITS EFFECTS ON PRICES. Hayek's contribution to monetary theory and to trade cycle theory are intertwined, a circumstance that reflects the nature of his contribution in both areas. In summary terms, Hayek's monetary theory consists of integrating the idea of money as a medium of exchange with the idea of the price system as a communication network. His trade-cycle theory consists of integrating monetary theory and capital theory – in which a particular aspect of the price system, namely the system of intertemporal prices, is emphasized.

Both in his *Monetary Theory and the Trade Cycle* (1933) and his *Prices and Production* (1935), Hayek argued against the then-dominant (and still-prevalent) idea that the appropriate focus of monetary theory is on the relationship between the quantity of money and the general level of prices. The kernel of truth in the quantity theory of money was not to be denied, but progress in monetary economics was to be made by moving beyond the simple proportionalities implied by a relatively stable velocity of circulation. According to Hayek (1935, p. 127), the proper task of monetary theory requires a thorough reconsideration of the pure theory of price determination, which is based on the assumption of barter, and a determination of what changes in the conclusions are made necessary by the introduction of indirect exchange.

Hayek introduced the concept of 'neutral money' in part as a means to contrast his own view of money with the more aggregative views. By definition, neutral money characterizes a monetary system in which money, while facilitating the coordination of economic activities, is itself never a source of discoordination. According to the aggregative views, money is neutral so long as the value of money (as measured by the general level of prices) remains unchanged. Thus, increases in economic activity require proportionate increases in the quantity of money in circulation. According to Hayek, monetary neutrality requires the absence of 'injection effects'. When the quantity of money is increased, the new money is injected in some particular way, which temporarily distorts relative prices causing the price system to communicate false information about consumer preferences and resource availabilities.

The contrasting views on the requirements for monetary neutrality had important implications for US monetary policy during the prosperous decade of the 1920s. The rate of monetary growth during that period was roughly equivalent to the rate of real economic growth, a circumstance which resulted in a near-constant price level. The absence of price inflation was taken by most monetary economists to be a sign of monetary stability. Hayek's contrary assessment (1925) that the injection of money through credit markets must result in a misallocation of resources despite the price-level stability was the basis for his prediction that the money-induced boom would eventually lead to a bust.

It should be noted that in other writings, both early and late in his career (e.g. 1933 and 1984), Hayek was ambivalent about the choice between a monetary policy that avoids injection effects (a constant money supply despite a positive real growth rate) and a monetary policy that avoids price deflation (a money growth rate that 'accommodates' real growth).

THE TRADE CYCLE AS INTERTEMPORAL DISCOORDINATION. Hayek's contribution to the theory of the trade cycle consists in his developing the idea that monetary injections can have a systematic effect on the intertemporal pattern of prices. The Austrian theory of the trade cycle was first formulated by Mises (1912), who showed that money-induced movements in the interest rate (as identified by Knut Wicksell) have identifiable effects on the capital structure (as conceived by Eugen von Böhm-Bawerk). Hayek's major contribution to the theory (1935), as well as many subsequent developments of it, was based on an extremely stylized portrayal of the economy's time-consuming production process. The relevant characteristics of the 'structure of production' were identified with the dimensions of a right triangle. One leg of the triangle represents the time dimension of the structure of production, the degree of roundaboutness; the other leg represents the money value of the consumer goods yielded up by the production process. Slices of the triangle perpendicular to the time leg represent stages of production; the heights of individual slices represent the money value of the yet-to-be-completed production process.

Resources are allocated among the different stages of production as a result of entrepreneurial actions guided by price signals. But because of the distinct temporal dimension of the structure of production, the supplies and demands for resources associated with the different stages are differentially sensitive to changes in the rate of interest: the demand for the output of extraction industries, for example, is more interest-elastic than the demand for the output of service industries. Changes in the rate of interest will have a systematic effect on the pattern of prices that allocates resources among the different stages of production. A fall in the rate of interest, for instance, will strengthen the relatively interest-elastic demands drawing resources into the early stages of production. This modification is represented by a relative lengthening of the temporal dimension of the Hayekian triangle.

A crucial distinction is made between interest-rate changes attributable to changes in the intertemporal preferences of consumers and interest-rate changes attributable to central-bank policy. In the first instance (Hayek, 1935, pp. 49–54), entrepreneurial actions and resulting changes in the pattern of prices allow the structure of production to be modified in accordance with the changed consumer preferences; in the second instance (Hayek, 1935, pp. 54–62), similar changes in the pattern of prices induced by the injecting of new money through credit markets constitute 'false signals', which result in a misallocation of resources among the stages of production. The artificially low rate of interest can trigger an unsustainable boom in which too many resources are committed to the early stages of production. The market process triggered by the injection of money through credit markets, Hayek showed, is a self-reversing process. More production projects are initiated than can possibly be completed. Subsequent resource scarcities turn the artificial boom into a bust. Economic recovery must consist of liquidating the 'malinvestments' and reallocating resources in accordance with actual intertemporal preferences and resource availabilities.

Hayek (1939) recognized that expectations about future movements in the rate of interest and entrepreneurial interpretations of intertemporal price movements can have an important effect on the course of the trade cycle. That is, prices are signals, not marching orders. But Hayek did not assume, as some modern economists do, that falsified price signals plus 'rational' expectations are equivalent to unfalsified price signals. Such an equivalence would require that market participants make use of knowledge of the kind that they cannot plausibly possess; it would require that they have knowledge of the 'real' factors independent of the price system that supposedly communicates that knowledge.

CRITIQUE OF KEYNESIANISM. Hayek's critique of Keynesian theory and policy followed directly from his own theories of capital and of money. Hayek argued that by ignoring the intertemporal structure of production and particularly the intertemporal complementarity of the stages of production, Keynes failed to identify the market process that could achieve intertemporal coordination: 'Mr Keynes's aggregates conceal the most fundamental mechanisms of change' (Hayek, 1931, p. 227). And by shifting the focus of analysis from money as a medium of exchange to money as a liquid asset, Keynes failed to see the harm caused by policies of injecting newly created money through credit markets or of spending it directly on public projects.

Hayek had emphasized that in functioning as a medium of exchange, money 'constitutes a kind of loose joint in a self-equilibrating apparatus of the price mechanism which is bound to impede its working – the more so the greater the play in the loose joint'. Keynesian theory and policy were the specific targets of Hayek's criticism when he warned that

the existence of such a loose joint is no justification for concentrating attention on that loose joint and disregarding the rest of the mechanism, and still less for making the greatest possible use of the short-lived freedom from economic necessity which the existence of this loose joint permits (Hayek, 1941, p. 408).

In the decades that followed the debate between Hayek and Keynes, economic theory was dominated by Keynesianism, and the corresponding macroeconomic policies consisted precisely of those measures that Hayek had warned against: monetary manipulation for political advantage. Monetary injections during the Great Depression, conceived as 'pump priming', soon gave way to a more broadly conceived policy of 'demand management'. The short-run trade-off between inflation and unemployment were treated in the political arena – and in some academic circles – as a societal menu from which elected officials, and hence voters, could choose; deviations of the economy from some conception of full employment or from some long-run growth path were taken as mandates for macroeconomic 'fine tuning' to be implemented by the central bank in cooperation with the fiscal authority.

As Hayek clearly recognized in his critique of Keynes's theories and his analysis of the actual effects of Keynesian policies, the political exploitation of the monetary loose joint contains an inherent inflationary bias. Newly created money can be used to hire the unemployed and to finance politically popular spending programmes. Monetary injections through the commercial banking system can stimulate the economy by triggering an artificial economic boom. The undesirable effects of inflating the money supply, the eventual collapse of the artificial boom and the general increase in the level of prices, are removed in time from the initial, politically desirable effects and are less conspicuously identified with the elected officials who engineered the monetary expansion (Hayek, 1960, pp. 324–39). As the political process continues, elected officials face the choice of monetary passivity which would permit the market to undergo the painful adjustments to earlier monetary injections or further monetary injections which would reproduce the desirable effects in the short run

while staving-off the eventual adjustment. The cumulative effects of the play-off between political advantage and economic necessity is the theme of Hayek's critique of Keynesianism. Excerpts of 'a forty years' running commentary on Keynesianism by Hayek, compiled by Sudha Shenoy, is appropriately entitled *A Tiger by the Tail* (1972).

DENATIONALIZATION OF MONEY. Hayek as a monetary reformer is interested in minimizing the potential for discoordination that is inherent in monetary mechanisms and precluding the manipulation of money for political advantage. He has long doubted that the government has either the will or the ability to manipulate the money supply in the public interest.

In his early writings Hayek took for granted the existence of a central bank and focused his analysis on the consequences of different policy goals, for example, the goal of stimulating economic growth or the goal of stabilizing the general price level. In his later writings, he began to see the monopolization of the money supply as the ultimate cause of monetary disturbances. As early as 1960, though still

convinced that modern credit banking as it has developed requires some public institutions such as central banks, [he was] doubtful whether it is necessary or desirable that they (or the government) should have the monopoly of the issue of all kinds of money (1960, p. 520, n.2).

In the mid-1970s Hayek's interest in the denationalization of money (1976) was renewed. Having lost all hope of achieving monetary stability through the instruments of highly politicized monetary institutions, Hayek suggested – by his own account, almost as a 'bitter joke' – that the business of issuing money be turned over to private enterprise. Soon taking this suggestion seriously, he began to explore the feasibility and the consequences of competing currencies.

Hayek's proposal for competition in the issue of money is not subject to the standard objection based on the so-called common-pool problem. The proposal is not that private issuers should compete by issuing some generic currency. Clearly, competition on this basis would produce an explosive inflation. The proposal, rather, is that each competitor issue his own trade-marked currency. Under this arrangement, each issuer would have an incentive to maintain a stable value of his own currency and to minimize the difficulties of using this currency in an environment where other currencies are used as well.

In spelling out just how such a system of competing currencies would or could work, Hayek has had to walk the fine line between constructivism on the one hand and blind faith in the market process on the other. His discussions of possible outcomes of the market process should not be taken as prescriptions for the provision of competing currencies, but rather as a basis for believing that competition between private issuers is feasible. Individuals may choose one currency over another on the basis of the issuer's demonstrated ability to achieve purchasing-power stability for that currency. Their choice may be influenced, Hayek has suggested, by what particular price level serves as the issuer's guide for managing the currency. Or it may be that public confidence can be maintained only by a currency that is convertible at a fixed rate into some stipulated commodity or basket of commodities. Hayek does doubt that a gold standard would re-emerge as a result of the competitive process, largely because the confidence and stability of gold was based upon beliefs and attitudes on the part of the public that no longer exist and cannot easily be recreated. But if gold did prevail in a competitive environment, there would be no basis for objection.

More importantly, Hayek's proposal for monetary reform should be seen not as an aberration from but as thoroughly consistent with his view of economics as a spontaneous order. Markets serve to coordinate the activities of individual market participants. The use of money, while greatly facilitating economic coordination, contains an inherent potential for discoordination. Competition in the market for money holds that potential in check and allows market participants to take the fullest advantage of the remaining elements of the spontaneous order.

ROGER W. GARRISON AND ISRAEL M. KIRZNER

SELECTED WORKS

1925. (In German.) The monetary policy of the United States after the recovery from the 1920 crisis. In F.A. Hayek, *Money, Capital, and Fluctuations: Early Essays*, ed. R. McCloughry, Chicago: University of Chicago Press, 1984.

1928. (In German.) Intertemporal price equilibrium and movements in the value of money. In F.A. Hayek, *Money, Capital, and Fluctuations: Early Essays*, ed. R. McCloughry, Chicago: University of Chicago Press, 1984.

1931–2. Reflections on the pure theory of money of Mr J.M. Keynes I–II. *Economica*, Pt I, 11, August 1931, 270–95; Pt II, 12, February 1932, 22–44.

1933. (In German.) On 'neutral money'. In F.A. Hayek, *Money, Capital, and Fluctuations: Early Essays*, ed. R. McCloughry, Chicago: University of Chicago Press, 1984.

1933. *Monetary Theory and the Trade Cycle*. New York: Augustus M. Kelley, 1975.

1935. *Prices and Production*. 2nd edn, New York: Augustus M. Kelley, 1967.

1937. Economics and knowledge. *Economica* NS 4, February, 33–54.

1939. Price expectations, monetary disturbances, and malinvestments. In F.A. Hayek, *Profits, Interest, and Investment*, Clifton, NJ: Augustus M. Kelley, 1975.

1941. *The Pure Theory of Capital*. Chicago: University of Chicago Press.

1945. The use of knowledge in society. *American Economic Review* 35, September, 519–30. Reprinted in F.A. von Hayek, *Individualism and Economic Order*, London: Routledge & Kegan Paul, 1949.

1949. *Individualism and Economic Order*. London: Routledge & Kegan Paul.

1952. *The Counter-Revolution of Science: Studies on the Abuse of Reason*. Glencoe, Ill.: Free Press.

1960. *The Constitution of Liberty*. Chicago: Henry Regnery & Co., 1972.

1968. Competition as a discovery procedure. In *New Studies in Philosophy, Politics, Economics and the History of Ideas*, Chicago: University of Chicago Press.

1972. *A Tiger by the Tail*. Ed S. Shenoy, London: Institute for Economic Affairs.

1975. *Full Employment at Any Price*. Occasional Paper No. 45, London: Institute of Economic Affairs.

1976. *Denationalization of Money*. London: Institute of Economic Affairs.

1984. The future monetary unit of value. In *Money in Crisis: The Federal Reserve, the Economy, and Monetary Reform*, ed. B. Siegel, Cambridge, Mass.: Ballinger.

BIBLIOGRAPHY

Gray, J. 1984. *Hayek on Liberty*. Oxford: Basil Blackwell.

Hutchison, T.W. 1981. *The Politics and Philosophy of Economics: Marxians, Keynesians, and Austrians*. New York: New York University Press, ch. 7.

Keynes, J.M. 1936. *The General Theory of Employment, Interest, and Money*. New York: Harcourt, Brace.

Machlup, F. 1976. Hayek's contribution to economics. In *Essays on Hayek*, ed. F. Machlup, Hillsdale, Mich.: Hillsdale College Press.

Mises, L. 1912. *The Theory of Money and Credit*. New Haven: Yale University Press, 1953.

O'Driscoll, G. 1977. *Economics as a Coordination Problem: The Contribution of Friedrich A. Hayek*. Kansas City: Sheed Andrews & McMeel.

Hayek effect. *See* RICARDO–HAYEK EFFECT.

health economics. Health economics is an applied field in which empirical research predominates. It draws its theoretical inspiration principally from four traditional areas of economics: finance and insurance, industrial organization, labour, and public finance. Some of the most useful work employs only elementary economic concepts but requires detailed knowledge of health technology and institutions. Policy-oriented research plays a major role, and many important policy-relevant articles are published in journals read by physicians and others with direct involvement in health (e.g. Enthoven, 1978).

The systematic application of economic concepts and methods to the health field is relatively recent. In a comprehensive bibliography of health economics based on English language sources through 1974 (Culyer, Wiseman and Walker, 1977), fewer than 10 per cent of the entries are dated prior to 1963. In that year a seminal article by Arrow (1963) discussed many of the central theoretical problems, and a few years later a major monograph based on modern econometric methods appeared (M.S. Feldstein, 1967).

The literature prior to 1963, thoroughly reviewed by Klarman (1965), was primarily institutional and descriptive. Significant contributions include discussions of US medical care institutions (Davis and Rorem, 1932; Ginzberg, 1954; Somers and Somers, 1961), mental illness (Fein, 1958), public health (Weisbrod, 1961), and the British National Health Service (Lees, 1961). The first US conference on health economics was held in 1962 (Mushkin, 1964), the first international conference in 1973 (Perlman, 1974), and the first widely adopted textbook did not appear until 1979 (P. Feldstein, 1983).

The field divides naturally into two distinct, albeit related, subjects: the economics of health *per se*, and the economics of medical care. The latter has received much more attention from economists than the former, but it is useful to consider health first because the demand for medical care is, in part, derived from the demand for health, and because many of the theoretical and empirical problems in the economics of medical care arise because of difficulties in measuring, valuing, and analysing health.

HEALTH

Concepts, measures, and valuation. Health is multidimensional. With the exception of the dichotomy between life and death, there is no completely objective, invariant ordering across individuals or populations with respect to health. Health can be defined according to criteria such as life expectancy, capacity for work, need for medical care, or ability to perform a variety of personal and social functions. Economists' attempts to measure and analyse differences in health across individuals and populations have typically focused on mortality (especially age-specific and age-adjusted death rates), morbidity (as evidenced by symptoms or diagnosed illnesses), or self-evaluations of health status. There have also been several attempts to take account simultaneously of mortality, morbidity, and health-related limitations by weighting years of life according to illness and disability.

Despite claims that health is more important than any other goal and that human life is priceless, economists note that individuals make tradeoffs between health and other goals and that the valuation of health (including life itself) is necessary for the rational allocation of scarce resources. The two leading approaches to the valuation of human life are 'discounted future earnings' and 'willingness to pay'. Rice (1966) estimated the costs of various illnesses as the sum of direct expenditures for medical care, the foregone earnings attributable to morbidity, plus the cost of premature death, which is assumed to be equal to the present value of future earnings. Willingness to pay is usually defined as the amount of money an individual would require (pay) in exchange for an increase (decrease) in the risk of death. This approach is preferred on theoretical grounds (Schelling, 1968; Mishan, 1971; Jones-Lee, 1974), but difficult to estimate empirically. Two oft-quoted studies that infer the value of life from risk-related wage differentials differ five-fold in their estimates (Thaler and Rosen, 1976; Viscusi, 1978).

The determinants and consequences of variations in health. Health is sometimes modelled as a dependent and sometimes as an independent variable, although frequently causality runs in both directions. Health has been studied as a function of medical care, income, education, age, sex, race, marital status, environmental pollution, and personal behaviours such as cigarette smoking, diet, and exercise. Grossman (1972) developed a model of the *demand* for health – both as a consumption commodity that enters directly into utility and as an investment commodity that contributes to the production of other goods and services. In his model, variables such as age and schooling affect the optimal level of health by changing its shadow price. Most studies that make health the dependent variable take a *production function* approach (Auster, Leveson and Sarachek, 1969) with health depending on income, medical care, education, and other inputs.

There is a strong positive correlation between income and health among less developed countries, but, *ceteris paribus*, the relationship tends to disappear at higher levels of income. This may reflect a high income elasticity of demand for other goods and services that adversely affect health, or may result from stress or other harmful side effects of earning more money. Also, as the average level of income rises, those diseases that stem from poverty tend to disappear and those that are not related to income form an increasing share of the burden of illness.

Advances in medical care such as the introduction of antibiotics have had significant effects on mortality and morbidity, but holding constant the state of medical science, the marginal effect of an increase in the quantity of medical care on health appears to be small in developed countries. Elimination of financial barriers to care, as in the British National Health Service, has not been accompanied by a reduction in the traditional mortality differentials across social classes. In all countries health is strongly correlated with years of schooling, but the explanation is not firmly established. Education may increase the efficiency with which individuals produce health. Alternatively, some third variable, such as time preference, may simultaneously affect schooling and health (Farrell and Fuchs, 1982). Marital status and health are strongly correlated (more so for men than for women), but the causality probably runs both ways. Other interesting correlations include those between wife's education and husband's health, and between parents' education and children's health.

Health has frequently been used as an independent variable to explain labour force participation, particularly at older ages. Not only do retired persons frequently cite poor health as the reason for retirement, but current workers who report a health limitation are more likely to withdraw from work in subsequent years. Health status has also been used to explain wages, productivity, school performance, fertility, and the demand for medical care. The results are often sensitive to the particular measure of health that is used, but the direction of effect generally confirms *a priori* predictions.

Health as a commodity. Health is both an intermediate commodity that affects production and a final commodity that affects utility directly. When health is included in the utility function several questions arise. Do standard theories regarding risk aversion and time discounting with respect to income apply equally to health? How is the marginal utility of income affected by changes in health levels? Both income and health are, in part, exogenously determined by initial endowments, and these endowments are likely to be positively correlated. The endogenous aspects of income and health are subject to many forces, some that may produce positive correlation and some the reverse. Although utility is a function of health, there can be a considerable difference between maximizing health (as measured, say, by life expectancy) and maximizing utility. The value of a given reduction in the probability of death (as evidenced by willingness to pay) is higher when the probability is high than when it is low. Thus programmes to treat the seriously ill at high cost per death averted are often preferred to preventive programmes that avert deaths at lower costs.

Many health problems have a significant genetic component while others are attributable to unfavourable experiences during the foetal period, at delivery, or in childhood. The resulting heterogeneity among individuals poses a variety of problems for analysis and policy. Unobserved heterogeneity can bias inferences about the effects of health interventions, especially in studies based on non-experimental data (Rosenzweig and Schultz, 1982). For instance, if people born with weak hearts are less inclined to exercise vigorously, the true effect of exercise on heart disease may be less than that inferred from observational studies. When heterogeneity is observable, the problem becomes primarily one of incorporating both efficiency and distributional considerations into policy analysis. Whose health should be considered in setting standards for air pollution or occupational safety? Under what circumstances should persons in poor health be required to pay actuarially fair health-insurance premiums?

The externalities associated with health have attracted considerable discussion. Analysis of the benefits of vaccination or the costs of pollution is fairly straightforward, but other externalities are less conventional. Individuals may derive utility from knowing that the poor sick among them are receiving medical care. They could attempt to achieve this through voluntary philanthropy, but the amount purchased is likely to be less than socially optimal because each individual's contribution would maximize private utility, ignoring the effects on others (Pauly, 1971). These health-centred philanthropic externalities are sometimes invoked to explain the widespread subsidization of medical care for the poor through national health insurance or other institutional arrangements in preference to general income redistribution.

MEDICAL CARE

Medical care accounts for more than 10 per cent of gross national product in the United States and approximately that much in several other developed countries. By contrast, restraints on input prices and quantities result in about a 6 per cent share in the United Kingdom. The effect on health of such wide variation in medical care expenditures is not clear. Governments in all countries play a large role as regulators, subsidizers, direct buyers, or producers of medical care. Economists have paid considerable attention to the reasons for and consequences of these governmental interventions. One useful way of categorizing economic research on medical care is to relate it to the older, better established areas of specialization that furnish most of the concepts used in health economics. Some studies, to be sure, draw their inspiration from and enrich more than one area.

Finance and insurance. Risk aversion and uncertainty about future health create a demand for health insurance (Arrow, 1963; Phelps, 1973). Once insurance is in place, moral hazard leads to over-utilization of medical care (Pauly, 1968). These two observations have generated a huge amount of research on the role of health insurance in health services (Rosett, 1976). The effect of insurance on the demand for care is better understood than is the demand for insurance itself. The usual risk aversion story, for example, cannot explain why many people purchase policies that cover the first dollar of expenditure but have a ceiling beyond which expenditures are not covered.

Asymmetry in information about potential demand for medical care creates another analytical and policy problem for insurance markets. When the consumer knows a great deal more about his health status and preferences than does the insurance company, adverse selection can lead to a breakdown in the free market in insurance. Group insurance is a typical solution, with participation achieved by compulsion, direct subsidies, or indirect subsidies via the tax system. Compulsory health insurance is also advocated to deal with the free rider problem. The possibility of free riders implies that even countries without explicit national insurance have public or private programmes that provide some kind of implicit universal coverage.

Many research methods, including a large-scale prospective controlled experiment (Newhouse et al., 1981) have been used to study the effect of insurance on the demand for care. As a result, few empirical propositions in economics have been as well established as the downward slope of the demand curve for medical care. Nevertheless, precise estimates of price elasticity (net of insurance) are difficult to obtain, in part because the features of some insurance policies – deductibles, varying coinsurance rates, and limits on indemnity payments – imply that the consumer faces a variable price under uncertainty (Keeler, Newhouse et al., 1977).

One solution to the risk aversion/moral hazard dilemma is for insurance to be provided in the form of contingent claims. For each identifiable condition the insured would be covered for care up to the point where the marginal benefit equals the marginal cost. Consumers, when well, may prefer that type of contract, but once sick they will want any care that has positive marginal benefit. Furthermore, the physician may want to provide care up to the point where marginal benefit is zero. The insurer's task is to enforce the tighter standard on the patient and the physician. This will often require some deception or nonprice rationing which patients can try to offset by strategic behaviour. The resulting loss of trust and less than candid exchange of information between patient and physician can adversely affect the production of care.

Industrial organization. Probably the largest range of problems and the largest volume of research in health economics falls in the area of industrial organization. There are many different, though related, industries to study: for example, physicians' services, hospitals, drugs, nursing homes, dental care. The topics covered range from licensure and regulation (Peltzman, 1973) to price discrimination (Kessel, 1958) and nonprice rationing (Friedman, 1978) to technological innovation and diffusion (Russell, 1979). Organization in the narrow sense of the term is of considerable interest because of the admixture of public, private nonprofit, and for-profit hospitals, and because the modes of physician practice range from solo fee-for-service to huge groups of salaried physicians. Behaviour inside the organization is particularly important in analysing nonprofit hospitals (the dominant form of organization in the United States) because the trustees, the administrator, the attending physicians, and the house staff all have considerable power and frequently have different objectives.

Medical care is, in many respects, the quintessential service industry. First, it is extremely difficult to measure output. As a result, standard economic accounts show little or no gain in productivity over time despite large expenditures for research and development and rapid technological change. Second, the consumer frequently plays a major role in the production process. This means not only that the value of the patient's time is part of the cost of care, but that the patient's knowledge, skill, and motivation, and the level of trust between patient and physician can affect the outcome of the care process. Third, the physician often knows a great deal more than the patient does about the patient's need for various types of care. This asymmetry of information has been used to explain the unwillingness of most societies to rely solely on competitive market forces to insure appropriate behaviour by physicians. Fourth, because output cannot be stored and short-run supply is relatively inelastic, productivity is sensitive to changes in demand. The stochastic nature of the demand for hospital care results in excess capacity and the problem is exacerbated by systematic variation in demand according to day of week and month of year.

Despite the difficulty of measuring output, numerous estimates of hospital short-run and long-run cost functions have been made (Lave and Lave, 1970). In the short run, marginal cost is below average cost in most hospitals. Long-run average cost tends to fall with increasing size until about 200 beds and then tends to be constant with a possible rise after a size of about 500 beds. Most researchers have defined output as a day of care or as a hospital admission. Standardization for patient mix can only be done incompletely, and failure to measure the effects of care on health is a serious limitation. To be sure, changes in health are only one aspect of output. A considerable fraction of resources in hospitals and nursing homes is devoted to *caring* for people who are in pain or who are disabled, regardless of whether their health status is (or even can be) improved. Also, a significant fraction of physicians' time is devoted to providing information and validation independently of any intended or actual effect on health status.

The problem of measuring output also increases the difficulty of analysing the demand for care. Additional complications result from the possibility that physicians may shift the patient's demand (Sloan and Feldman, 1978; Evans, 1974). Some economists concede that physicians have the power to shift demand, but believe that it is sufficiently and uniformly exploited so that further shifting can be ignored. Others argue that the amount of shifting varies with exogenous changes in the physician/population ratio. This follows from a model in which physicians maximize utility as a function of income, leisure, and 'correct practice'. The empirical evidence is consistent with this model (Fuchs, 1978), but it is consistent with other explanations as well.

Perhaps less controversial is the proposition that physicians can and do change their patients' demand for hospital care. One example of this in the United States is the lower hospital utilization by patients enrolled in prepaid group practice plans such as Kaiser Permanente (Luft, 1981). Patients pay a single annual premium for total care regardless of the quantity of services used; physicians typically receive a salary or a share of the net income after the hospital and other costs have been paid. An even more spectacular change in hospital utilization emerged when Medicare (the US publicly funded insurance for the elderly) changed from retrospective cost-reimbursement to prospective payment per admission in a particular diagnosis-related group. In just two years the average length of stay of patients 65 years of age and older in short-term general hospitals fell 12 per cent without any change in conventional demand variables.

The demand for care usually increases as health worsens, but not always. For instance, an elderly person who is in good general health may demand a variety of surgical interventions for specific problems such as hip replacement or lens implantation; a person of the same age who is in bad health may not.

Problems in measuring output imply problems in measuring price. An alternative approach to the estimation of price change is to measure change in the total cost of treating a defined illness or medical condition. A price index calculated by this method was found to rise more rapidly than the conventional medical care price index (Scitovsky, 1967), possibly as a result of unmeasured changes in output. Technological advances may have increased the cost of care by making it possible for the physician to do more things for the patient.

Economic research on drugs falls into two main categories that reflect differing policy concerns in the United States before and after the 1962 Kefauver–Harris Amendments to the Pure Food and Drug Act. Prior to the amendments, attention was focused on price fixing, price discrimination (manufacturers' prices vary in several ways, depending upon the type of customer), and on alleged socially wasteful expenditures on product differentiation. Since 1962 economic research has shifted to the volume and character of innovation. A decrease in the flow of new drugs has been attributed to the increased cost of satisfying regulatory requirements (Grabowski et al., 1978). Despite the questions raised by economists regarding the net benefit of tighter controls, other countries have tended to follow the policy direction set by the United States in 1962.

In recent years in the United States and many other countries, the major health policy questions have revolved around efforts to contain the cost of care. Most economists have argued for greater reliance on market mechanisms and less on regulation (Zeckhauser and Zook, 1981), but an intricate web of social, political, and economic considerations seems to preclude a pure laissez-faire approach to health (Fuchs, 1974).

Labour economics. Much of the research on the economics of *health* comes out of labour economics, especially the human capital branch. Concerning medical care, labour economists have been primarily interested in the demand and supply of various health occupations. Numerous studies of the earnings of physicians have mostly confirmed the results of a pioneering study (Friedman and Kuznets, 1945) that physicians, on average, realize an excellent return on their investment in medical education. Research on choice of specialty and

location, however, does not support any simple model of physicians as income maximizers. Some specialties appear to have more intrinsic appeal than others, and the wide geographical variation in the physician/population ratio in the United States is not primarily the result of variation in fees or income. Changes in physician distribution by specialty or location, however, do conform closely to predictions based on standard utility maximization (Newhouse et al., 1982).

Research on nurses in the United States has focused heavily on an alleged persistent "shortage". One explanation is that the principal employers of nurses – hospitals – have monopsony power (Yett, 1975). Faced with rising supply curves, hospitals equate the marginal cost of nurses with their marginal revenue product and set the wage on the supply curve. The 'shortage' simply reflects the fact that the hospital administrator would like to hire more nurses at the going wage, but has no incentive to raise the wage.

Public finance. Ever since Bismarck introduced compulsory health insurance to Germany in 1883, the financing of health care has been of increasing concern to governments and to economists who specialize in public finance. Even in the United States, the last major holdout against national health insurance, government pays directly for about 42 per cent of all health-care expenditures, and pays indirectly for an appreciable additional share through tax exemptions and allowances. Numerous reasons have been offered as to why governments pay for health care, but each has its shortcomings.

All explanations that are health related (e.g., 'health is a right', 'the government has an obligation to reduce or eliminate class differentials in mortality') are suspect because in Bismarck's time there was virtually no connection between medical care and health, and even today the connection at the margin is highly circumscribed. The explanation that medical care is a 'merit' good seems circular. Governments are said to subsidize 'merit' goods, but the only way to identify them is by the presence of government subsidies. The standard externality-public good explanation applies to the prevention and treatment of communicable diseases, but this accounts for only a small fraction of health care. Subsidies in other industries such as agriculture or the merchant marine can frequently be explained by pressure from producers, but government subsidies for health care have usually been opposed by the producers of care.

Subsidies for health care are frequently defended on the grounds that it is unfair to allow the distribution of health care to be determined by the distribution of income. Many economists counter by saying it would be more efficient to redistribute income and then let the poor decide how much of the increase they want to devote to health care and how much to other goods and services. The crux of the problem seems to be that the amount of redistribution that society wants to make to an individual may depend on the individual's need for care. The greater the need, the greater society's willingness to redistribute. It may be more efficient to combine the determination of need with the redistribution via the delivery of care than to separate the functions.

One special problem arises in this field from the proposition that in order to reduce inequality, governments should limit the amount of care that individuals can obtain. This view is virtually inescapable once a government is committed to equality in health care and constrained to keep the health budget within limits set by general budgetary considerations. No country can afford to provide health care for all its citizens up to the point where the marginal benefit is zero.

Economists spend a great deal of time deploring the fact that no country shows much interest in evaluating the outcome of medical care. It might be more fruitful to try to explain why this is so. No doubt part of the answer is that evaluation is very difficult, but part may be related to the symbolic and political role that medical care plays in modern society. When governments, insurance companies, or employers promise to finance all necessary and appropriate care, they typically have to introduce implicit constraints to keep from going bankrupt. A thorough evaluation would make these constraints explicit and could create a great deal of dissatisfaction.

Future research. Because health economics is predominantly applied and policy-oriented, future research will undoubtedly be influenced by the changing nature of health problems and by developments in medical science. Thus, it is reasonable to expect to see more attention to the health problems of the elderly – chronic illness and the need for long-term care. Among the non-elderly, health problems stemming from substance abuse are large and growing in importance. The new understanding of the role of genetic factors in disease creates dramatic opportunities for screening and intervention, but these opportunities will pose problems of enormous complexity for analysis and policy. The gap between what is technically possible and what is economically feasible will probably widen; thus, the demand for guidance concerning the efficiency and equity implications of alternative health policies is likely to grow. Further development of health economics as an established field of inquiry will help to meet that demand.

VICTOR R. FUCHS

See also NON-PROFIT ORGANIZATIONS; PUBLIC HEALTH; STATE PROVISION OF MEDICAL SERVICES; VALUE OF LIFE.

BIBLIOGRAPHY

Arrow, K.J. 1963. Uncertainty and the welfare economics of medical care. *American Economic Review* 53(5), December, 941–73.

Auster, R., Leveson, I. and Sarachek, D. 1969. The production of health, an exploratory study. *Journal of Human Resources* 4, Fall, 412–36.

Culyer, A.J., Wiseman, J. and Walker, A. 1977. *An Annotated Bibliography of Health Economics.* London: Martin Robertson.

Davis, M.M. and Rorem, C.R. 1932. *The Crisis in Hospital Finance.* Chicago: University of Chicago Press.

Enthoven, A.E. 1978. Consumer-choice health plan. *New England Journal of Medicine* 298, 23 March and 30 March.

Evans, R.G. 1974. Supplier-induced demand: some empirical evidence and implications. In *The Economics of Health and Medical Care,* ed. M. Perlman, London: Macmillan.

Farrell, P. and Fuchs, V.R. 1982. Schooling and health: the cigarette connection. *Journal of Health Economics* 1(3), December, 217–30.

Fein, R. 1958. *Economics of Mental Illness.* New York: Basic Books.

Feldstein, M.S. 1967. *Economic Analysis for Health Service Efficiency.* Amsterdam: North-Holland.

Feldstein, P. 1983. *Health Care Economics.* 2nd edn, New York: John Wiley & Sons.

Friedman, B. 1978. On the rationing of health services and resource availability. *Journal of Human Resources* 13 (Supplement), 57–75.

Friedman, M. and Kuznets, S. 1945. *Income from Independent Professional Practice,* General Series No. 45, New York: National Bureau of Economic Research.

Fuchs, V.R. 1974. *Who Shall Live? Health, Economics, and Social Choice.* New York: Basic Books.

Fuchs, V.R. 1978. The supply of surgeons and the demand for operations. *Journal of Human Resources* 13 (Supplement), 35–56.

Ginzberg, E. 1954. What every economist should know about health and medicine. *American Economic Review* 44(1), March, 104–19.

Grabowski, H.G., Vernon, J.M. and Thomas, L.G. 1978. Estimating the effects of regulation on innovation: an international compara-

tive analysis of the pharmaceutical industry. *Journal of Law and Economics* 21(1), April, 133–63.

Grossman, M. 1972. *The Demand for Health: A Theoretical and Empirical Investigation.* New York: National Bureau of Economic Research.

Jones-Lee, M. 1974. The value of changes in the probability of death or injury. *Journal of Political Economy* 82(4), July–August, 835–49.

Keeler, E.B., Newhouse, J.P. et al. 1977. Deductibles and demand: a theory of the consumer facing a variable price schedule under uncertainty. *Econometrica* 45(3), April, 641–55.

Kessel, R.A. 1958. Price discrimination in medicine. *Journal of Law and Economics* 1(2), October, 20–53.

Klarman, H.E. 1965. *The Economics of Health.* New York: Columbia University Press.

Lave, J.R. and Lave, L.B. 1970. Hospital cost functions: estimating cost functions for multi-product firms. *American Economic Review* 60(3), June, 379–95.

Lees, D.S. 1961. *Health Through Choice.* Hobart Paper No. 14, London: Institute of Economic Affairs.

Luft, H.S. 1981. *Health Maintenance Organizations: Dimensions of Performance.* New York: John Wiley & Sons.

Mishan, E.J. 1971. Evaluation of life and limb: a theoretical approach. *Journal of Political Economy* 79(4), 687–705.

Mushkin, S.J. (ed.) 1964. *The Economics of Health and Medical Care.* Ann Arbor: University of Michigan Press.

Newhouse, J.P. et al. 1981. Some interim results from a controlled trial of cost sharing in health insurance. *New England Journal of Medicine* 305, 17 December, 1501–7.

Newhouse, J.P. et al. 1982. Does the geographical distribution of physicians reflect market failure? *Bell Journal of Economics* 13(2), Autumn, 493–505.

Pauly, M.V. 1968. The economics of moral hazard: comment. *American Economic Review* 58(3), June, 531–6.

Pauly, M.V. 1971. *Medical Care at Public Expense, A Study in Applied Welfare Economics.* New York: Praeger.

Peltzman, S. 1973. An evaluation of consumer protection legislation: the 1962 drug amendments. *Journal of Political Economy* 81(5), September–October, 1049–91.

Perlman, M. (ed.) 1974. *The Economics of Health and Medical Care.* London: Macmillan.

Phelps, C. 1973. *The Demand for Health Insurance: A Theoretical and Empirical Investigation.* Santa Monica: The Rand Corporation, No. R-1054–OEO.

Rice, D.P. 1966. *Estimating the Cost of Illness.* Washington, DC: USDHEW, Public Health Service Publication, 947–6.

Rosenzweig, M.R. and Schultz, T.P. 1982. The behavior of mothers as inputs to child health: the determinants of birth weight, gestation, and rate of fetal growth. In *Economic Aspects of Health,* ed. V.R. Fuchs, Chicago: University of Chicago Press.

Rosett, R.N. (ed.) 1976. *The Role of Health Insurance in the Health Services Sector.* New York: National Bureau of Economic Research.

Russell, L.B. 1979. *Technology in Hospitals.* Washington, DC: Brookings.

Schelling, T.C. 1968. The life you save may be your own. In *Problems in Public Expenditure Analysis,* ed. S.B. Chase, Washington, DC: Brookings.

Scitovsky, A.A. 1967. Changes in the costs of treatment of selected illnesses, 1951–65. *American Economic Review* 57(4), December, 1182–95.

Sloan, F., and Feldman, R. 1978. Competition among physicians. In *Competition in the Health Care Sector: Past, Present, and Future,* ed. W. Greenberg, Washington, DC: Federal Trade Commission.

Somers, H.M. and Somers, A.R. 1961. *Doctors, Patients, and Health Insurance.* Washington, DC: Brookings.

Thaler, R. and Rosen, S. 1976. The value of saving a life: evidence from the labor market. In *Household Production and Consumption,* ed. N.E. Terleckyj, New York: National Bureau of Economic Research.

Viscusi, K.W. 1978. Labor market valuations of life and limb: empirical evidence and policy implications. *Public Policy* 26(3), Summer, 359–86.

Weisbrod, B.A. 1961. *Economics of Public Health*. Philadelphia: University of Pennsylvania Press.

Yett, D.E. 1975. *An Economic Analysis of the Nurse Shortage*. Lexington, Mass.: Heath.

Zeckhauser, R. and Zook, C. 1981. Failures to control health costs: departures from first principles. In *A New Approach to the Economics of Health Care*, ed. Mancur Olson, Washington, DC: American Enterprise Institute for Public Policy Research.

Hearn, William Edward (1826–1888).

Hearn was born in County Cavan, Ireland and died in Melbourne, Australia. Educated at Trinity College Dublin, he was appointed professor of political economy and other subjects at the University of Melbourne in 1854. Subsequently a member of the Legislative Council of the State of Victoria and contributor to the local press, Hearn is known to economists principally for his *Plutology* (1863).

Plutology explains increasing wealth as a result of the competitive exchange of services. The analysis owes a good deal to Herbert Spencer and Frederic Bastiat. Competition is held to have three general results. It is: beneficent, since prices reflect the minimum cost of procuring a service; just, because recompense is in proportion to merit; and equalizing, since no recompense permanently reflects the effects of chance. As an 'unfailing rule', the pursuit of self interest means services are produced in 'order of their social importance'. Competition results in a natural social order, ordained by Providence, in which the principles of Darwin's natural selection are applied to industry (ch. 19).

The price of any service, determined by the extent of demand and supply, oscillates towards the minimum cost of production. The upper price limit is set where the purchaser equates desire for a service with the sacrifice necessary to either directly produce or obtain it from another source. The minimum price must cover any outlays and provide the 'average' reward for the vendor's type of service (ch. 14). The discussion of price formation is not conducted in marginalist terms and owes a good deal, via J.S. Mill, to De Quincey.

The distribution of income is explained according to the general principles of exchange. The manager of an enterprise, for example, contracts with the vendors of labour and 'capital' for their services at a fixed price. Discounting all costs and gross returns, the manager then has full title to the output, assuming responsibility for losses and receiving net gains. If capital is supplied in 'commodity' form (machinery, buildings) rent is paid; if it is supplied in money form (loans, insurance) interest accrues. Directly following Bastiat's *Harmonies*, Hearn argues ground rent cannot be a gratuitous gift of nature as land has a price only if labour is bestowed on it (ch. 18).

The role of a central government in an 'advanced' nation is thus basically a nightwatchman, although it may undertake some limited regulation. It is acknowledged, however, that the accumulation path will be impeded to some extent. The most serious problems result from enterprises mistaking market demand and engaging in speculative ventures. Still, fluctuations in output and investment have relatively little importance. 'Failures, poverty, suffering and privation' are not part of the 'ordinary course of events', any 'ravages' are soon repaired and objects destroyed in commercial fluctuations would have mainly been consumed rather than invested. Any 'disturbances' are thus 'incidental' to the natural laws of economic organization (ch. 24).

Marshall and Edgeworth bestowed high praise on *Plutology*, while Jevons considered its arguments were 'nearly identical' to those in his *Theory*. Subsequent commentary has noted Hearn's dogmatism and plagiarism, especially from John Rae and Bastiat.

M. WHITE

SELECTED WORKS

1851. *The Cassell Prize Essay on the Condition of Ireland*. London.

1863. *Plutology: or the Theory of the Efforts to Satisfy Human Wants*. Melbourne: Robertson; London: Macmillan, 1864.

1867. *The Government of England, its Structure and its Development*. Melbourne: Robertson; London: Longmans Green, Reader & Dyer.

1878. *The Aryan Household, its Structure and its Development*. Melbourne: Robertson; London: Longmans, Green & Co., 1879.

1883. *The Theory of Legal Duties and Rights*. Melbourne: Government Printer; London: Trubner & Co.

BIBLIOGRAPHY

Copland, D.A. 1935. *W.E. Hearn: First Australian Economist*. Melbourne: Melbourne University Press.

La Nauze, J.A. 1949. *Political Economy in Australia*. Melbourne: Melbourne University Press.

Heckscher, Eli Filip (1879–1952).

Born into a Jewish family in Stockholm, Heckscher studied history under Hjärne and economics under Davidson at Uppsala University from 1897. In 1907 he became a docent at Stockholm University College of Commerce, and from 1909 to 1929 he was professor of economics and statistics. Then, because of his great research productivity, the college authorities changed his position to research professor, lightened his teaching duties and made him director of the newly established Institute of Economic History. Heckscher continued in this position until he retired in 1945. He succeeded in establishing economic history as a subject of graduate study in Sweden's universities.

In 1950, the Ekonomisk-historiska Institutet, Stockholm, through Bonniers Co., published the *Eli F. Heckscher bibliografi 1897–1979* (123 pp.). It contains 1148 entries for his 36 books, 174 articles in professional journals, his chapters in government reports, and the more than 700 short articles he wrote for the weekend issues of Stockholm's leading newspapers. Only a few of his books and articles have been translated and will be referred to by their English titles; other works will be mentioned only by the English translation of their original titles and identified by their numbers as entries in the Heckscher bibliography.

By 1929, when he was able to specialize in economic history, Heckscher had already written a dozen books on such diverse subjects as *Economic Principles* (1910, No. 158), *The Continental System* (1918, No. 443, later republished, Oxford, 1922) and *Economics and History* (1922, No. 478). As a result of his teaching, his contributions to economics are a blend of innovations in economic theory and a new methodology for economic history research, an approach to quantitative research very different from that used by leaders in his field such as Schmoller, Cunningham and Sombart.

Heckscher's most significant contributions to economic theory may be found in two articles. 'Effects of Foreign Trade on Distribution of Income' (1919) is the origin of the modern Heckscher–Ohlin factor proportions theory of international trade, developed further in Ohlin (1933).

'Intermittently Free Goods' presents a theory of imperfect competition nine years ahead of that by Joan Robinson and Edward Chamberlin, and a discussion of collective goods not priced by the market. Heckscher observed that significant new products are introduced by firms with investment in plants that have a capacity which far exceeds initial demand for their products. The latter are sold at prices which barely cover unit variable costs, and so, for a time, the services of the fixed investment are provided as 'free goods'. Then, as weaker firms are eliminated, demand shifts to the remaining larger firms who use up their production capacity. By and by these firms expand, enjoy economies of scale, differentiate their products and become prosperous oligopolies dividing a mass market into more or less definite shares.

A situation of the opposite kind arises when the smallest feasible production facility has a production capacity which suffices for a growing and indefinitely large demand without affecting the costs and service life of the production unit. This is the case with many so called 'pure' public or collective goods. Heckscher used street illumination as an example of a collective good, which can be used simultaneously by few or many persons, a service that cannot be priced per unit of individual use. The costs of providing this service, then, are usually met by an increase in local government taxes. In that case, and in contrast to that of intermittently free goods, the citizens pay the full-cost price of the service from the outset in their current taxes. Then, as activities in and use of lighted streets increase over time, the citizens derive increased utility per tax dollar spent for street lighting.

At the Institute of Economic History Heckscher's first work, one of his major and most widely known treatises, was *Mercantilism* (1931). His other major work, the fruit of many years of pioneering research devoted to his own country, was *Sweden's Economic History from the Reign of Gustav Vasa* (vols. I-II, 1935–6, No. 878; vols. III-IV, 1950, No. 1146). He also wrote a popular version of this work, *Life and Work in Sweden from Medieval Times to the Present* (1941), No. 1014, republished as *An Economic History of Sweden* (1954). Among his other books of particular interest are *Materialist and Other Interpretations of History* (1944, No. 1052), *Industrialism, Its Development from 1750 to 1914* (1946, No. 1123) and *Studies of Economic History* (1936, No. 918). It was in this work he presented a new methodology he proposed for economic history research in his essay 'Aspects of Economic History', pp. 9–69. This was reinforced in his articles, 'A Plea for Theory in Economic History' (1929) and 'Quantitative Measures in Economic History' (1939).

For the analysis of any epoch of economic history – as distinct from factual description of a chronologically arranged body of heterogeneous source materials – Heckscher proposed consideration of a succession of its 'economic aspects', to introduce order and inject economic theory into the interpretation of that epoch. Unlike 'periods', 'aspects' are *not necessarily* time dependent. They are theoretical and imply hypotheses that are, within limits, testable against the given data. A series of aspects, for instance of (i) the exchange processes; (ii) natural resources and technologies; (iii) labour force and capital; (iv) forms of enterprise organization; and (v) extent and composition of demand, form an economic model of the epoch. This done, the function of the economic historian is to provide a synthetic overview and explanation of the relations between the aspects of the model.

Thus Heckscher bridged the gap between economic history and theory by addressing broad questions or hypotheses to the source materials for intensive and critical study. He always preferred to present his finding supported by statistical data. That done, he was not satisfied until he had explained and illuminated these by economic analysis, that is by applying cognate principles of economic theory to their interpretation.

CARL G. UHR

See also HECKSCHER–OHLIN TRADE THEORY; STOCKHOLM SCHOOL.

SELECTED WORKS

1918. *Kontinentalsystemet*. Stockholm: P.A. Norstedt & Söner. Trans. as *The Continental System*, ed. Harald Westergaard, Oxford: Clarendon Press, 1922.
1919. Effects of foreign trade on distribution of income. *Ekonomisk Tidskrift*. Reprinted in AEA, *Readings in the Theory of International Trade*, ed. H.S. Ellis and L. Metzler, Philadelphia: Blakiston, 1949.
1924. Intermittently free goods. *Ekonomisk Tidskrift* 553. Reprinted in German in *Ein Beitrag zur Sozialwissenschaft und Sozialpolitik* 59, 1928.
1929. A plea for theory in economic history. *Economic Journal*, Historical Supplement No. 4, 523–54.
1931. *Merkatilismen*. Vols. I and II, Stockholm: P.A. Norstedt & Söner. Authorized trans. by Mendel Shapiro as *Mercantilism*, London: G. Allen & Unwin, 1935. Revised edn, New York: Macmillan; London: G. Allen & Unwin, 1955.
1935–6, 1949. *Sveriges Ekonomiska Historia från Gustav Vasa*. Vols. I and II, 1935–6; Vols. III and IV, 1949, Stockholm: Bonnier.
1939. Quantitative measures in economic history. *Quarterly Journal of Economics* 53, 167–93.
1941. *Svenskt arbete och liv från medeltiden till nutiden*. Stockholm. Trans. by G. Ohlin as *An Economic History of Sweden*, Cambridge, Mass.: Harvard University Press, 1954.
1950. *Eli F. Heckscher bibliografi 1897–1949*. Stockholm: Bonnier.

BIBLIOGRAPHY

Ohlin, B. 1933. *Interregional and International Trade*. Cambridge, Mass.: Harvard University Press.

Heckscher–Ohlin trade theory. Eli Heckscher (1919) and Bertil Ohlin (1933) laid the groundwork for substantial developments in the theory of international trade by focusing on the relationships between the composition of countries' factor endowments and commodity trade patterns as well as the consequences of free trade for the functional distribution of income within countries. From the outset general equilibrium forms of analysis were utilized in these developments, which gradually came to be sorted out into four 'Core Propositions' (Ethier, 1974) in the pure theory of international trade.

I. THE FOUR THEOREMS. Although all four of the propositions to be discussed are an outgrowth of the seminal work of Heckscher and Ohlin, only one of these propositions bears their name explicitly. The *Heckscher–Ohlin theorem* states that countries export those commodities which require, for their production, relatively intensive use of those productive factors found locally in relative abundance. The twin concepts of relative factor intensity and relative factor abundance are most easily defined in the small dimensional context in which the basic theory is usually developed. Two countries are engaged in free trade with each producing the same pair of commodities in a purely competitive setting, supported by constant returns to scale technology that is shared by both countries. Each commodity is produced separately with inputs of two factors of production which, in each country, are supplied perfectly inelastically. Following the Ricardian distinction, commodities are freely traded but productive factors are internationally immobile.

Although one country may possess a larger endowment of each factor than another, the presumed absence of returns to scale guarantees that only relative factor endowments are important. The home country is said to be relatively labour abundant if the ratio of its endowment of labour to that (say) of land exceeds the corresponding proportion abroad. This is known as the physical version of relative factor abundance. An alternative involves a comparison of autarky relative factor prices in the two countries: the home country can be defined to be relatively labour abundant if its wage rate (compared with land rentals) is lower before trade than is the foreign wage (relative to foreign land rentals). Since autarky factor prices are determined by demand as well as supply conditions, these two versions need not correspond. In particular, if the home country is, in the physical sense, relatively labour abundant it might nonetheless have its autarky wage rate relatively high if taste patterns at home are strongly biased towards labour-intensive commodities compared with tastes abroad. In such a case the trade pattern reflects the autarky factor–price comparison: the home country exports the physically land-intensive commodity. As discussed below, the link between commodity price ratios (the proximate determinant of trade flows) and factor price ratios is more direct than that between commodity price ratios and physical factor endowments. Thus the Heckscher–Ohlin theorem is more likely to hold if relative factor abundance is defined in terms of relative factor prices prevailing before trade. The procedure typically followed in the literature is to assume that both countries share identical and homothetic taste patterns. Such an assumption, in conjunction with the presumed identity of technology at home and abroad (with an even stronger version of homotheticity–linear homogeneity) helps to isolate the separate influence of physical factor supplies and makes the validity of the Heckscher–Ohlin theorem with the physical definition of factor abundance as likely as with the autarky factor price definition.

These assumptions are less than sufficient to guarantee the Heckscher–Ohlin theorem, even in the simple context of two-country, two-factor, two-commodity trade. The potential stumbling block is the fact that even though countries share the same technology, the commodity that is produced by relatively labour-intensive techniques at home may be produced by relatively land-intensive techniques abroad. This is the phenomenon of factor-intensity reversal. If production processes are independent of each other, there is nothing (other than bald assumption) to rule out its appearance. The bald assumption would assert that regardless of factor endowments one industry always employs a relatively higher ratio of labour to land, where techniques are chosen with reference to the wage/rental ratio common to both industries. If this is not the case, and if the commodity that is relatively labour-intensive at home is produced by relatively land-intensive techniques abroad, the phrasing of the Heckscher–Ohlin theorem that explicitly states 'each country exports the commodity that is produced in that country making relatively intensive use of the factor found in relative abundance in that country' is fatally flawed. If the relatively labour abundant country exports its labour-intensive commodity, it must do so in exchange for the commodity which, in the relatively land abundant foreign country, is produced by labour-intensive techniques. Thus if one country satisfies the theorem, the other country cannot (Jones, 1956).

In the event of factor-intensity reversal, it must be the case that whichever commodity is exported by the labour-abundant home country, the ratio of labour to land employed in its production must exceed the labour/land intensity adopted in foreign exports. However, this observation is of little value if one wishes to infer from an intensity comparison between exportables and import-competing goods within a given country whether that country is relatively labour abundant compared with some foreign country. Such an inference lay behind the celebrated study of Leontief (1953) on United States trade patterns. This research, the conclusions of which came to be known as the Leontief Paradox (American

exportables are produced by labour-intensive techniques compared with import-competing goods) provided the major stimulus to developing and defining the meaning and conditions supporting the Heckscher–Ohlin theorem.

Earlier work in Heckscher–Ohlin trade models was focused on the pricing relationships embodied in Heckscher–Ohlin theory. Ohlin (1933) stressed the effect which free trade would tend to have on the distribution of income within countries; relative factor prices would move in the direction of equality between trading countries which share the same technology. Ohlin's mentor, Heckscher, went even further in his pioneering 1919 article. *Absolute* factor price equalization was purported to be 'an inescapable consequence of trade'. Nonetheless, Ohlin's view of partial equalization seems to have dominated, with the exception of Lerner's unpublished 1933 manuscript (which surfaced after Samuelson's articles), until the statement of the *Factor Price Equalization theorem* in articles by Samuelson in 1948 and 1949. Rejecting his earlier tacit acceptance of the Ohlin thesis of partial equalization (in the Stolper–Samuelson article, which appeared in 1941), Samuelson proved that within the traditional confines of the $2 \times 2 \times 2$ model (with no factor-intensity reversals and each country incompletely specialized), free trade would drive wage rates to absolute equality in the two countries (and, as well, would equate land rentals) despite the assumption that labour (and land) are immobile between countries.

The logic of the argument for the simple 2×2 case can be stated briefly. In a competitive equilibrium unit cost equals price if the commodity is produced. Thus let A represent the matrix of input–output coefficients, a_{ij}, w the vector (pair) of factor prices, and p the vector (pair) of commodity prices. Techniques need not be constant; in general they depend upon prevailing factor prices so that $A = A(w)$. Therefore the competitive profit conditions if both goods are actually produced dictate that:

$$A(w) \cdot w = p \qquad (1)$$

Assuming no factor-intensity reversals, $A(w)$ is non-singular. Therefore if countries share the same technology and face the same pair of free-trade commodity prices, they must face exactly the same set of factor prices if each country produces both goods.

This approach may suggest that the crucial issue in the factor price equalization argument is the unique dependence of factor price vector w on commodity price vector p, and an extensive literature has developed which focuses on this issue. In the 2×2 case uniqueness is a simple question – it depends on factor intensities differing between sectors and not reversing. But from the outset Samuelson pointed out that this was not the only issue. The question of uniqueness involves properties of technology alone, whereas under appropriate circumstances two countries in free trade will have factor prices equalized only if factor endowments are reasonably similar. For, if factor endowments are too dissimilar, it will be impossible for both countries to produce both commodities, in which case the equalities in (1) cannot universally hold.

These ideas can be made more precise by considering a concept due to McKenzie (1955), which Chipman (1966) called the 'cone of diversification'. For any factor price vector, w, there is determined a pair of techniques (labour/land ratios) for the two commodities. Both factors can be fully employed only if the country's endowment vector is contained within the cone spanned by these techniques. Suppose two countries face a common free-trade commodity price vector, p, and that the commonly shared technology associates a unique factor price w corresponding to this p. Then if the endowment vectors of both countries lie within the cone of diversification, their factor prices must be equalized (McKenzie, 1955).

More recently, Dixit and Norman (1980) and Helpman and Krugman (1985) have utilized box-diagram techniques suggested by Lancaster (1957) and Travis (1964) to build upon Samuelson's 1949 'parable' of an integrated world economy in which all factors are initially internationally mobile. Suppose all tastes are identical and homothetic. Then for this world economy an equilibrium commodity price vector p and factor price vector w exist. In a box diagram whose dimensions reflect world supplies of land and labour there can be inscribed a diamond-shaped region formed from the two cones of diversification from each origin. The diagonal of the box splits this region into symmetrical halves. Any division of the world into two nations such that the point representing the allocation of endowments lies within the diamond-shaped region must reproduce exactly the same world production and prices as in the integrated world economy. Given the technology, the factor price equalization theorem holds if and only if factor endowments are sufficiently similar to lie in this region.

Some seven years prior to Samuelson's first factor price equalization essay there appeared the article by Stolper and Samuelson (1941), which must be ranked a classic not only for its discussion of what became known as the *Stolper–Samuelson theorem*, but because it is one of the first concrete developments of the ideas of Heckscher and Ohlin in the explicit format of a two-factor, two-commodity, general equilibrium model. Their argument supposedly concerns the effect of protection on real wages, and in the course of the argument they assume that a tariff does not change the terms of trade so that locally import prices rise. Subsequently, in what has become known as the 'Metzler tariff paradox', Metzler (1949) showed that with sufficiently inelastic demand a tariff might so improve a country's terms of trade that the relative domestic price of imports falls. If so, the Stolper–Samuelson contention that a tariff yields an increase in the real return to a country's relatively scarce factor would be reversed. However, it is now commonly agreed that the Stolper–Samuelson theorem refers to the general phenomenon whereby an increase in the relative domestic price of a commodity (whether brought about by a tariff increase, decrease, or some other reason) must unambiguously raise the real return to the factor of production used relatively intensively in the production of that commodity.

Introducing the production-box diagram technique (for a single country), Stolper and Samuelson illustrate how a rise in the relative price of labour-intensive watches attracts resources from land-intensive wheat. To clear factor markets, both sectors must then use labour more sparingly. That is, the ratio of land to labour utilized in each sector rises, which implies an unambiguous increase in labour's marginal productivity measured either in watches or in wheat. Thus regardless of workers' taste pattern, protection has increased the *real* wage.

The logic of the Stolper–Samuelson argument rests heavily upon the presumed lack of joint production. It takes labour and land to produce watches and, in a separate activity, a higher land/labour ratio is used to produce wheat. In competitive settings any change in a commodity's price must reflect an average of factor price changes so that unit costs change as much as price. Therefore one factor price must rise relatively more than either commodity price. Which factor gains depends only upon the factor-intensity ranking. If the price of watches rises, and wheat does not, the wage rate must rise by more. And this result follows even if techniques are frozen so that no resources can be transferred between sectors

(as in the Stolper–Samuelson discussion) and marginal products are not well defined (Jones, 1965).

To round out the quartet of theorems, the *Rybczynski theorem* (1955) deals with the same model but focuses on the relationship between factor endowments and commodity outputs. Suppose commodity prices are kept fixed in the 2×2 setting and an economy is incompletely specialized. Then by the factor price equalization theorem, factor prices are determined and fixed as well, which implies also that techniques of production remain constant. If the economy's endowment of one factor rises, while its endowment of the other factor remains constant, the economy must in some sense grow (the transformation schedule shifts out). However, this growth is strongly asymmetric: one output actually falls. The factor-intensity ranking selects the loser – the commodity which uses intensively the factor which is fixed in overall supply must decline. The reasoning is simple. As one factor expands, it must be absorbed in producing the commodity using it intensively. But with techniques frozen (since prices are assumed fixed), the expanding sector must be supplied with doses of the non-expanding factor as well. The only source for this factor is the other industry which must, perforce, contract.

II. RELATIONSHIPS AMONG THE THEOREMS. All four propositions are based on the same 'mini-Walrasian' general equilibrium model of trade and there are some interesting relationships and distinctions among them. Perhaps most importantly, both the Heckscher–Ohlin theorem and the factor price equalization theorem refer explicitly to a comparison between two countries, whereas the Stolper–Samuelson and Rybczynski propositions are involved with relationships within a single country. This distinction implies that the assumption that countries share an identical technology is not necessary for the latter two propositions. Thus, for example, a country could protect the factor used intensively in its import-competing sector in real terms (according to Stolper and Samuelson) regardless of the level or type of technology adopted by other countries.

The factor price equalization theorem is a razor's-edge type of result. Should the technology available to two countries differ only slightly, any presumption of exact factor-price equalization in the absence of explicit international factor markets disappears. The Heckscher–Ohlin theorem is a little more robust in this regard. In general, trade patterns depend on all those variables which influence prices: tastes, technology, and factor endowments (not to mention taxes or other distortions). If tastes are identical (and homothetic) but factor endowments are not, the latter difference will tend to dominate the trading pattern even if technologies differ as long as these differences are 'less important'. At issue is a weighing of endowment differences with the Ricardian emphasis on technology differences.

Two versions of the Heckscher–Ohlin theorem have been cited, depending on which definition of relative factor abundance is selected. If physical factor proportions are chosen as the criterion, the basis for the Heckscher–Ohlin theorem resides in the kind of link between endowment patterns and outputs for a single economy exemplified by the Rybczynski theorem. An extension of this theorem allows a comparison of the transformation schedules for two economies with similar technologies. The relatively labour abundant (physical definition) country will produce relatively more of the labour-intensive commodity at common commodity prices (Jones, 1956). Therefore, unless taste differences are sufficiently biased to counter this effect, the labour-intensive good will, in autarky, be cheaper in the labour-abundant country and, with trade, will be exported. The Stolper–Samuelson theorem is closely linked to the alternative form of the Heckscher–Ohlin theorem. Suppose there are no factor-intensity reversals. Then if both goods are produced there is a monotonic relationship between the wage/rent ratio and the relative price of the labour-intensive good such that a rise in the latter is associated with a greater than proportionate rise in the former. Thus the relatively low wage country must, in autarky, have been the relatively cheap producer of the labour-intensive commodity. As mentioned earlier, no caveat must be added about tastes, since these are already incorporated in the autarky factor-price comparison.

Although a comparison of factor endowments between countries is crucial in considering both the Heckscher–Ohlin theorem and the factor price equalization theorem, such a comparison works in opposite directions for these two propositions. Thus if factor endowment proportions are sufficiently *dissimilar*, trade patterns suggested by the Heckscher–Ohlin theorem *must* hold (aside from the possibility of factor-intensity reversals) whereas free trade *cannot* bring about factor price equalization. Sufficiently different factor endowments entail one country's transformation schedule being everywhere flatter than the other country's. At least one country must be specialized with trade. By contrast, the factor price equalization result holds if factor endowments are close enough so that international differences in the composition of outputs are capable of absorbing these endowment differences at the same set of techniques (and factor prices). If endowments are this close, it would always be possible for demand differences to be so biased that the physically labour-abundant country exports the land-intensive commodity. Indeed, if such a demand reversal of the Heckscher–Ohlin theorem takes place, free trade *must* result in factor-price equalization (Minabe, 1966).

Samuelson's name occurs so frequently in the literature on Heckscher–Ohlin trade theory it is often appended to the other two names. One of his results not cited heretofore is the *reciprocity relationship* (Samuelson, 1953). This states that in any general equilibrium model the effect of an increase in a commodity price (say p_j) on a factor return (say w_i) is the same as the effect of an increase in the corresponding factor endowment (V_i) on the output of commodity j. Of course, in each case some other set of variables is being held constant. Thus:

$$\frac{\partial w_i}{\partial p_j} = \frac{\partial x_j}{\partial V_i}, \tag{2}$$

with all other commodity prices and all endowments held constant in the left-hand derivative and all other endowments and all commodity prices held constant in the right-hand derivative. This relationship is easy to prove (see, for example, Jones and Scheinkman, 1977). It also reveals the *dual* nature of the Stolper–Samuelson and Rybczynski theorems. If an increase in the price of watches lowers land rentals, then an increase in the endowment of land (at constant prices) would lower the output of watches. In each case it is the presumed labour-intensity of watches that is operative.

In the 2×2 setting both the Stolper–Samuelson and Rybczynski theorems reflect the 'magnification effects' (Jones, 1965) that stem directly from the assumed lack of joint production. Letting a ' ^ ' over a variable designate relative changes, if watches are labour intensive and wheat land intensive and if the relative price of watches rises,

$$\hat{w} > \hat{p}_{wa} > \hat{p}_{wh} > \hat{r}. \tag{3}$$

In addition, should an economy grow, but with labour (L) growing more rapidly than land (T),

$$\hat{x}_{wa} > \hat{L} > \hat{T} > \hat{x}_{wh}. \tag{4}$$

Inequality ranking (3) shows commodity price changes trapped between factor-price changes (since two factors are required to make a single good), while inequality (4) shows that in order to absorb endowment changes, the composition of outputs (each of which uses both factors) must change more drastically. Stolper and Samuelson stressed the first inequality in (3), while Rybczynski focused on the last inequality in (4), assuming $\hat{T}$ equals zero.

III. EXTENSIONS: HIGHER DIMENSIONS AND JOINT PRODUCTION. International trade theory generally, and Heckscher–Ohlin trade theory in particular, has frequently been criticized for its restriction to the low dimensionality represented by two commodities, two factors, and two countries. In fairness to both Heckscher and Ohlin it should be stressed that their discussions typically were not so confined. But neither were their conclusions as precise as those subsequently developed by Samuelson and others in the 2×2 versions of the four core propositions. And in the years following Samuelson's pioneering work on factor price equalization, scores of articles have indeed appeared dedicated to the question of robustness of these results in higher-dimensional contexts. A highly detailed and thorough discussion of the issue of dimensionality is provided in Ethier (1984), and an earlier critique of the limitations imposed by small numbers of goods and factors is found in Ethier (1974) and Jones and Scheinkman (1977).

Part of the difficulty embedded in the move to higher dimensions lies in the ambiguity involved in what the propositions should state for cases beyond 2×2. The one proposition for which this is not the case is the factor price equalization theorem. Consider the case of equal numbers of factors and produced commodities, with all goods traded and factors immobile internationally. The uniqueness of a factor price vector, w, corresponding to a given commodity price vector, p, is not guaranteed even in the 2×2 case; a factor-intensity reversal could lead to two (or more) values of w consistent with a given p. For the $n \times n$ case Gale and Nikaido (1965) provided conditions sufficient to guarantee global univalence of the factor price, commodity price relationship: the $A(w)$ matrix of input-output coefficients should be a 'P-matrix', that is a matrix with all positive principal minors. These conditions have been slightly weakened by Mas-Colell (1979), and earlier a fundamental interpretation of the conditions was supplied by Uekawa (1971). It remains the case, however, that this condition on technology alone is somewhat remote from the issue of factor-price equalization. Just as in the 2×2 case, two countries sharing a common technology and each capable of producing the same set of n commodities (at the same traded-goods prices) with n productive factors, will, if techniques of production are independent, have their factor prices driven to equality if their factor endowments are sufficiently close. The concept of the 'cone of diversification' within which both endowment vectors must lie for factor price equalization is as meaningful and relevant in n dimensions as it is in two.

Although the factor price equalization theorem has an unambiguous meaning in higher dimensions, it is a theorem that cannot be expected to hold if the number of productive factors exceeds the number of freely traded commodities. The reasoning is basic, and can be linked to equations (1). These competitive profit conditions supply n links between factor prices and traded commodity prices, where n is the number of traded commodities. If r, the number of factors, should exceed n, equations (1) are insufficient in number to provide a solution for the vector w for given p. Other conditions are required, and these are provided by the full employment conditions, one for each productive factor. Thus a nation's endowment bundle, V, becomes a determinant of factor prices that is additional to the commodity price vector, p. For example, in the simple three-factor, two-commodity 'specific-factor' model (Jones, 1971; Samuelson, 1971), suppose a country faces a given world price vector, p, and experiences a slight increase in its endowment of a factor 'specific' in its use in the first industry. The intensity with which factors are utilized depends upon factor prices, and if these do not change, there is no way in which outputs can adjust to clear all factor markets. The return to the factor specific to the first industry must fall so as to encourage the further use of that factor. Two countries of this type with different endowments will generally have different sets of factor prices with trade, even if they share a common technology.

The case in which a country is capable of producing a greater number of traded commodities than it has productive factors provides a useful context in which to re-examine the Heckscher–Ohlin theorem concerning the pattern of trade. To simplify, suppose two countries share a given technology whereby a number of commodities might be produced and that there are only two factors (e.g. labour and land) and that with trade factor prices are equalized. The patterns of production, and therefore trade, become indeterminate. The transformation surface for each country has 'flats', regions in which it is possible to alter outputs without changing techniques, factor prices, or unit costs while still maintaining full employment. (The Ricardian example of a linear transformation schedule between two commodities when there is only one productive factor is a more primitive version of these 'flats'.) Thus the relatively labour-abundant country might export some of the commodity that is most land-intensive in its use of factors. This observation has led some (Melvin, 1968; Vanek, 1968) to enquire as to the factor content of trade. This is a more simple matter. Following the procedure adopted in the 2×2 case to neutralize the role of demand, suppose both countries share identical, homothetic taste patterns. Then the pattern of demands is identical up to a scale factor (given by relative size). But net exports equal production minus demand, so that regardless of the commodity composition of trade the exports of the labour-abundant country must embody a higher labour/land content than do the exports of the land/abundant country.

A variation in this scenario dispenses with the problem of indeterminacy and reveals the strong influence of factor endowments on production and trading patterns. Suppose commodity prices for traded goods are determined in a world market composed of a number of different countries with potentially a wide variation in technologies. Against this background consider a pair or more of countries which do share the same technology. Given world prices, such countries share a Hicksian composite unit-value isoquant for all traded goods (Jones, 1974), made up of strictly bowed-in sections (where only one commodity is produced) alternating with flats (where a pair of commodities is produced). Countries differ in factor endowments, and a ranking from most to least labour abundant can be made for countries sharing the common technology. Regardless of the number of commodities, each country engaged in trade need produce only one or two (in the two-factor case), and these commodities will be the ones requiring factors in proportions close to that country's

endowment ratio. In this setting the spirit of the Heckscher–Ohlin theorem is that each country concentrates its resources on a small range of commodities whose factor requirements mirror closely that country's endowment base; the country exports some or all commodities in this set and imports commodities that are more labour-intensive than these good as well as those that are more capital-intensive. Two countries whose endowments are fairly similar may or may not produce the same pair of goods and thus achieve factor price equalization with trade. Countries further apart in endowment composition will have disparate sets of factor prices and may produce completely different bundles of commodities (see also Krueger, 1977).

With many factors and many commodities a different approach can be taken. The ability of autarky commodity price comparisons to predict trade patterns item by item is severely questioned, so that little hope remains of linking endowment differences to the detailed composition of trade. But statements about aggregates or 'correlations' between trade patterns and autarky prices can be made (Deardorff, 1980; Dixit and Norman, 1980). A nation's net imports, M, are positively correlated with the comparison of its autarky commodity price vector, p^A, and the vector of free-trade commodity prices, p^T. Thus:

$$(p^A - p^T)M \geqslant 0 \qquad (5)$$

(see Ethier, 1984, p. 139). This idea can be extended to the further relationship between autarky commodity prices and the vector of autarky factor prices (as in (1)) to establish that countries possess a comparative advantage, on average, in commodities using intensively factors that are relatively cheap in autarky. (See Deardorff, 1982, and Ethier, 1984, for more details.)

The reciprocity relationship expressed in (2) is quite general in terms of dimensionality and thus serves to link the Rybczynski theorem in a dual relationship to the Stolper–Samuelson theorem. However, when the number of factors exceeds the number of produced commodities, differences between the two types of theorems do appear. This basic asymmetry is linked to the failure of the factor price equalization theorem when factors exceed commodities in number.

The condition that all factors (V_i) be fully employed is shown in (6):

$$\sum_j a_{ij} x_j = V_i, \qquad i = 1, \ldots, r. \qquad (6)$$

If endowments are slightly perturbed, and commodity prices are held constant (as in the original 2×2 context for the Rybczynski theorem), equation (7) shows the balance between relative changes in outputs, endowments and techniques:

$$\sum_j \lambda_{ij} \hat{x}_j = \hat{V}_i - \left\{ \sum_j \lambda_{ij} \hat{a}_{ij} \right\}, \qquad i = 1, \ldots, r. \qquad (7)$$

(The λ_{ij} refer to the fraction of V_i that is allocated to output x_j.) If the number of factors, r, exceeds the number of produced commodities, a non-uniform change in endowments at constant commodity prices changes factor prices and thus techniques. By contrast, if factors and commodities balance in number, the last bracketed expression in (7) vanishes since factor prices do not change if commodity prices are kept constant. Every endowment change is thus a weighted average of output changes and magnification effects such as that described in (4) can be expected to hold: for any single $\hat{V}_i > 0$, there would exist some output that would fall and some output that would rise by a greater proportion than V_i.

This statement represents one generalized form of the Rybczynski theorem for the $n \times n$ case. The specific-factors 3×2 model illustrates how this form of the theorem can fail when factors exceed commodities in number. If two sectors each use a different factor exclusively, as well as a (mobile) factor in common, a rise in the endowment of the latter causes all outputs to grow at constant commodity prices. Alternatively, if the endowment of a specific factor rises, the output in which it is used does not expand by relatively as much (the other output falls).

Commodity price-factor price relationships are in general obtained by differentiating the competitive profit conditions (1):

$$\sum_i \theta_{ij} \hat{w}_i = \hat{p}_j - \left\{ \sum_i \theta_{ij} \hat{a}_{ij} \right\}, \qquad j = 1, \ldots, n \qquad (8)$$

(The θ_{ij} refer to the distributive share of factor i in the jth industry.) Although the structure of (8) parallels that of (7), the condition that firms choose techniques so that unit costs are minjmized entails that $\Sigma \theta_{ij} \hat{a}_{ij}$ always vanishes, regardless of the number of factors and commodities. Thus for a change in the price of any produced commodity, there exists at least one factor whose return rises relative to that commodity price and at least one other factor whose return is lowered (relative to any commodity price change). This represents one generalized form of the Stolper–Samuelson theorem, and in contrast to the fate of the Rybczynski theorem, the validity of this form is not restricted to the $n \times n$ case.

Much of the effort devoted to extending the Stolper–Samuelson and Rybczynski theorems to the $n \times n$ case has considered more detailed structure. In particular, is it possible that each factor can be associated with a particular, unique, commodity, such that a rise in any single commodity price results in a relatively greater increase in the return to the associated factor and a reduction in the return to all other factors? Technically, if $[\theta]$ represents the matrix of distributive factor shares, where commodity j is 'intensively' associated with factor j, the Stolper-Samuelson theorem is said to generalize if the inverse of the $[\theta]$ matrix has diagonal elements all exceeding unity and negative off-diagonal elements (Chipman, 1969; Kemp and Wegge, 1969; Uekawa, 1971). If the technology possesses such a structure it is also the case that the inverse of $[\lambda]$, the matrix of factor allocation fractions, is characterized by diagonal elements greater than unity and negative off-diagonal elements. (Matrices of this type are called Minkowski matrices.) That is, a strong generalization of the Rybczynski theorem would hold: the increase in any single factor endowment would, at constant commodity prices, encourage a greater than proportional expansion in the output of the commodity intensively associated with that factor and an absolute reduction in all other outputs.

These are indeed strong generalizations of the two propositions. For example, if there is any industry that does not use all factors in positive amounts, these strong generalizations must fail. Suppose a_{ik} is zero and p_i rises, all other prices constant. If the strong form of Stolper–Samuelson holds, w_i would have to rise and all other factor returns fall. But p_k is assumed constant and k does not use any of the single factor (i) which has gone up in price. Therefore the cost of producing k would have to fall, which violates the competitive profit conditions since k's price is assumed constant.

These remarks should serve to illustrate the ambiguity surrounding the question of whether the Stolper–Samuelson theorem survives into the $n \times n$ or $r \times n$ higher dimensional contexts. Indeed, Ethier (1984) poses an alternative possible generalization. Since in the 2×2 context *any* relative price change results in one factor unambiguously gaining and the other losing (as in magnification effect (3)), under what conditions is it the case that *any* relative price changes in the $r \times n$ case leave only two classes of factor returns – unambiguous winners or unambiguous losers. There are no such conditions: with at least three commodities it is always possible to trap any factor price change between commodity price changes.

Although these strong interpretations of the Stolper–Samuelson theorem either cannot hold or, in the case in which θ^{-1} must exhibit the Minkowski sign pattern, impose much additional structure on the technology, it would be misleading to conclude that the Stolper–Samuelson theorem rarely generalizes to higher dimensions. Stolper and Samuelson pointed out how a particular policy aimed at changing relative commodity prices (a tariff policy in their case) could unambiguously raise the real return to a particular productive factor. There is a sense in which the move to the higher-dimensional $n \times n$ case does not, by itself, alter this conclusion. Assume away degenerate cases so that n independent non-joint productive activities are undertaken, each requiring inputs of at least two factors, with each factor employed in at least two sectors. Then for *any* factor i it is possible to find some proper subset of commodities such that if all their prices are raised in proportion relative to fixed prices for the remaining commodities, the return to factor i rises by a greater relative amount. Thus a simple relative commodity price change suffices in the $n \times n$ case to raise the real return to any factor, irrespective of any other form of aid (such a redistribution of tariff proceeds). It may not be possible to accomplish this rise in factor i's real reward merely by increasing a single commodity price if factor i bulks large in the national income (Jones and Scheinkman, 1977). But there always exists some group of commodities which is sufficient for the purpose (Jones, 1985).

The specific-factors model illustrates that such a result is limited to the 'even' $n \times n$ case. Holding endowments fixed, the change in the mobile factor return is a *positive* weighted average of all commodity price changes, and so cannot exceed them all. Thus this generalization of the Stolper–Samuelson theorem suggests that it is not higher dimensionality *per se* that destroys the results of 2×2 theory, but rather the existence of more factors than goods. And this is precisely the setting in which the factor-price-equalization result no longer holds.

Much the same kind of result serves to generalize the Rybczynski theorem in the $n \times n$ case. Any single endowment increase at constant commodity prices must cause at least one output to fall and at least some other output to rise by a greater proportion. If a particular output, j, bulks large in the national income, there may be no single endowment increase which would serve to raise output j by a greater relative amount. But some proper subset of endowments exists such that a uniform increase in the supply of these endowments is sufficient to raise x_j by a greater proportional amount.

These results also serve to emphasize that both the Stolper–Samuelson theorem and the Rybczynski theorem are essentially reflections of the asymmetry between factors and commodities. This asymmetry is characterized by the assumption that productive activities are non-joint: in the non-degenerate cases more than one input is required to produce, separately, each output. Thus each commodity price change is a positive weighted average of the changes in rewards to factors used to produce that commodity. This implies that regardless of the ranking of commodity price changes, there is some factor reward that would rise relative to any commodity price rise and at least one factor reward which would rise by less (or fall by more) than any commodity price change. Allowing joint production potentially destroys this asymmetry and thus the basis for the magnification effects.

There is a small literature dealing with this issue (Jones and Scheinkman, 1977; Chang et al., 1980; Uekawa, 1984). Much depends on the range of output proportions in any productive activity compared with the range of input proportions. For example, in the 2×2 case suppose one activity produces primarily the first commodity, but also a small amount of the second, while the other activity reverses these proportions. Furthermore, suppose this 'output' cone of diversification contains the standard 'input' cone of diversification. In this case traditional magnification effects underlying the Stolper–Samuelson and Rybczynski theorems remain valid. New results emerge if these cones intersect or the input cone contains the output cone. (Cones can be made comparable by using distributive shares of inputs and outputs in activities.)

Joint production does not, by itself, interfere with the status of the factor-price-equalization theorem, but does suggest an alteration in the Heckscher–Ohlin theorem. Instead of concentrating on the link between factor endowments and the location of commodity outputs (and therefore trading patterns), the focus is on the location of productive activities. Each activity requires, as before, an array of inputs, and the allocation of endowment bundles among countries helps determine where these activities are located. The pattern of commodity trade must then reflect, as well, the output composition of these activities.

Although the existence of joint production potentially destroys the asymmetry between the ranking of commodity prices and factor prices, the concern with the 'real' reward of a factor can be accommodated by considering the relation between the nominal change in w_i and the change in some price index. In a closed economy an index using overall production weights would match one using consumption weights. In such a case if all factors shared similar homothetic taste patterns, the change in the real reward of a factor would be captured by a comparison between $\hat{w}_i$ and the average of all factor price changes. Such a comparison is simpler than one that involves, as well, a comparison with individual commodity price changes. For an open economy consumption and production patterns differ so that even if, say, factor i's return changes by an amount equal to the average for all factors (which is a production price index), factor i might nonetheless gain if the price changes reflect an improvement in the country's terms of trade.

IV. CONCLUDING REMARKS. The theory of international trade that has developed from the seminal writings of Heckscher and Ohlin is fundamentally based on the twin observations that countries differ from each other in the composition of their factor endowments and productive activities are distinguished by the different relative intensities with which factors are required. As this theory has been developed four core propositions have served to summarize its content. The strict validity of each of these propositions has been seen to depend upon further specification of the technology (e.g. ruling out factor intensity reversals, joint production, and non-constant returns to scale), demand (e.g. requiring all individuals to possess identical homothetic taste patterns), or dimensionality

(e.g. requiring a small number of factors and commodities, or a matching number of both). To conclude this discussion of the core propositions it is possible to point out the less precise, broad message of each.

(1) *The Heckscher–Ohlin theorem.* Production patterns reflect different compositions of endowments and, unless demand differences are significant, so will patterns of trade. The possibility introduced by trade of a greater concentration of resources encourages specialization in production of those activities requiring factors in proportions similar to the endowment bundle and allows a country to import commodities whose factor requirements are far from proportions found at home.

(2) *The factor price equalization theorem.* Even if the mobility of factors of production is limited by national frontiers, free trade in commodities helps to even out disparities in demand relative to supply of factors and to diminish the discrepancy between factor returns among countries. Two or more countries sharing the same technology will find that free trade brings factor returns to absolute equality if their endowments are sufficiently similar and they produce in common a sufficient number of commodities (at least equal to the number of distinct productive factors).

(3) The *Stolper–Samuelson theorem.* Changes in relative commodity prices, such as those brought about by trade or interferences in trade, have strong asymmetric effects on factor rewards. If no joint production prevails some factors find their real rewards unambiguously raised and other rewards are unambiguously lowered by relative price changes. If, further, the number of factors equals the number of produced commodities, as in the original 2×2 setting, relative commodity price changes can be constructed which, without the aid of any direct subsidies, will raise the real reward of any particular factor.

(4) *The Rybczynski theorem.* Unbalanced growth in factor supplies tends, at given commodity prices, to lead to stronger asymmetric changes in outputs. If the numbers of factors and commodities are evenly matched and production is non-joint, this asymmetry entails that growth in some, but not all, factors serves to force an actual reduction in one or more outputs. By similar reasoning, differences in the composition of endowments among countries with similar technologies results in stronger asymmetries in production patterns when all face free trade commodity prices. If tastes are somewhat similar, these endowments differences are apt to support the trading patterns described by the Heckscher–Ohlin theorem.

RONALD W. JONES

See also FOREIGN TRADE; INTERNATIONAL TRADE; LEONTIEF PARADOX.

BIBLIOGRAPHY

Chang, W., Ethier, W. and Kemp, M. 1980. The theorem of international trade with joint production. *Journal of International Economics* 10, 377–94.

Chipman, J. 1966. A survey of the theory of international trade, Part 3. *Econometrica* 34, January, 18–76.

Chipman, J. 1969. Factor price equalization and the Stolper–Samuelson Theorem. *International Economic Review* 10 399–406.

Deardorff, A. 1980. The general validity of the law of comparative advantage. *Journal of Political Economy* 88, 941–57.

Deardorff, A. 1982. The general validity of the Heckscher–Ohlin theorem. *American Economic Review* 72, 683–94.

Dixit, A. and Norman, V. 1980. *Theory of International Trade.* Cambridge: Cambridge University Press.

Ethier, W. 1974. Some of the theorems of international trade with many goods and factors. *Journal of International Economics*, 199–206.

Ethier, W. 1984. Higher dimensional issues in trade theory. Ch. 3 in *Handbook of International Economics*, Vol. 1, ed. R. Jones and P. Kenen, Amsterdam: North-Holland.

Gale, D. and Nikaido, H. 1965. The Jacobian matrix and the global univalence of mappings. *Mathematische Annalen* 159, 81–93.

Heckscher, E. 1919. The effect of foreign trade on the distribution of income. *Ekonomisk Tidskriff*, 497–512. Translated as chapter 13 in American Economic Association, *Readings in the Theory of International Trade*, Philadelphia: Blakiston, 1949, 272–300.

Helpman, E. and Krugman, P. 1985. *Market Structure and Foreign Trade*. Cambridge, Mass.: MIT Press.

Jones, R. 1956. Factor proportions and the Heckscher–Ohlin theorem. *Review of Economic Studies* 24, 1–10.

Jones, R. 1965. The structure of simple general equilibrium models. *Journal of Political Economy* 73, 557–72.

Jones, R. 1971. A three-factor model in theory, trade, and history. In *Trade, Balance of Payments, and Growth*, ed. J.N. Bhagwati, R.W. Jones, R.A. Mundell and J. Vanek, Amsterdam: North-Holland.

Jones, R. 1974. The small country in a many-commodity world. *Australian Economic Papers* 13, 225–36.

Jones, R. 1985. Relative prices and real factor rewards: a reinterpretation. *Economic Letters 19(1)*, 47–9.

Jones, R. and Scheinkman, J. 1977. The relevance of the two-sector production model in trade theory. *Journal of Political Economy* 85, 909–35.

Kemp, M. and Wegge, L. 1969. On the relation between commodity prices and factor rewards. *International Economic Review* 10, 407–13.

Krueger, A. 1977. Growth, distortion, and patterns of trade among many countries. Princeton Studies in International Finance, No. 40, Princeton: Princeton University Press.

Lancaster, K. 1957. The Heckscher–Ohlin trade model: a geometric treatment. *Economica* 24, 19–39.

Leontief, W. 1953. Domestic production and foreign trade: the American capital position re-examined. *Proceedings of the American Philosophical Society* 97, 332–49.

Lerner, A. 1933. Factor prices and international trade. Mimeo. Published in *Economica* 19, (1952), 1–15.

Mas-Colell, A. 1979. Two propositions on the global univalence of systems of cost functions. In *General Equilibrium, Growth, and Trade*, ed. J. Green and J. Scheinkman, New York: Academic Press.

McKenzie, L. 1955. Equality of factor prices in world trade. *Econometrica* 23, 239–57.

Melvin, J. 1968. Production and trade with two factors and three goods. *American Economic Review* 58, 1248–68.

Metzler, L. 1949. Tariffs, the terms of trade, and the distribution of national income. *Journal of Political Economy* 57, 1–29.

Minabe, N. 1966. The Heckscher–Ohlin theorem, the Leontief paradox, and patterns of economic growth. *American Economic Review* 56, 1193–211.

Ohlin, B. 1933. *Interregional and International Trade*. Cambridge, Mass.: Harvard University Press, 1966.

Rybczynski, T.M. 1955. Factor endowments and relative commodity prices. *Economica* 22, 336–41.

Samuelson, P. 1948. International trade and the equalization of factor prices. *Economic Journal* 58, 163–84.

Samuelson, P. 1949. International factor-price equalization once again. *Economic Journal* 59, 181–97.

Samuelson, P. 1953. Prices of factors and goods in general equilibrium. *Review of Economic Studies* 21, 1–20.

Samuelson, P. 1971. Ohlin was right. *Swedish Journal of Economics* 73, 365–84.

Stolper, W. and Samuelson, P. 1941. Protection and real wages. *Review of Economic Studies* 9, 58–73.

Travis, W. 1964. *The Theory of Trade and Protection*. Cambridge, Mass.: Harvard University Press.

Uekawa, Y. 1971. Generalization of the Stolper–Samuelson theorem. *Econometrica* 39, 197–213.

Uekawa, Y. 1984. Some theorems of trade with joint production. *Journal of International Economics* 16, 319–33.

Vanek, J. 1968. The factor-proportions theory: the n-factor case. *Kyklos* 28, 749–55.

hedging. Hedging is the purchasing of an asset or portfolio of assets in order to insure against wealth fluctuations from other sources. A hedge portfolio is any asset or collection of assets purchased by one or more agents for hedging. A grain dealer may hedge against losses on an inventory of grain by selling grain futures; a Middle Eastern businessman may hedge against political turmoil (and the resulting losses) by buying gold; a pension fund may hedge against capital losses on its equity portfolio by buying stock index put options.

1. A COMPETITIVE EQUILIBRIUM MODEL OF HEDGING. The fundamental concepts of hedging can best be described in a state space model. Consider a one period economy with M agents and one end-of-period consumption good. For simplicity, assume that there is no consumption at the beginning of the period. Each agent possesses a real asset which produces a random amount of the consumption good at the end of the period. Agents have homogeneous beliefs. There are N possible states of nature, with probabilities $\Pr(1), \ldots, \Pr(N)$. Agents have concave, possibly state-dependent utility functions and wish to maximize the expected utility of end-of-period consumption. Let $U_j(C_j, \theta_i)$ denote the end-of-period utility of agent j given that his consumption is C_j and the state is θ_i.

A *financial asset* is a claim to a random amount of end-of-period output, which is traded between agents at the beginning of the period. A *hedge portfolio* is a particular type of financial asset or collection of financial assets which protects an agent against some particular risky outcome(s).

The analysis is simplest if we assume that the hedge portfolio consists of a mixed asset/liability with positive payoffs in some states and negative payoffs in other states, balanced so as to give a competitive equilibrium price of zero. Under this formulation, a hedge portfolio is a portfolio which pays off positively in states where the agent would otherwise have a high marginal utility of consumption (i.e. in 'bad' states) and negatively in states where he would otherwise have a low marginal utility of consumption. If the agent's 'expected' marginal utility (marginal utility times the probability of the state) is equalized across the relevant states after purchasing the hedge portfolio, then he is fully hedged; if the hedge position lowers but does not eliminate the disparity, then he is partially hedged.

Who takes the other side of the hedging transaction? There are three possibilities. First, if there exist two agents who have real asset cash flows which vary inversely, then they can trade in a way which allows both to hedge simultaneously. For example, a grain dealer who holds an inventory of grain may be able to sell a futures contract to a bread producer who has committed himself to using grain at a later stage of his production process. Both parties consider themselves as hedging. Second, one agent may be less risk-averse toward certain states of nature than another. The less risk-averse may be willing to sell the hedge asset to the more risk-averse at a price which produces mutual gains in expected utility. Third, the hedging agent may be able to trade small quantities of the hedge asset with many agents, who can then eliminate all or most of the risk of the trade by combining the asset with many others (i.e. by diversifying away the risk). For example, insurance companies can sell fire insurance policies to many individuals and leave very little risk to be absorbed by the company's shareholders.

Let the number of distinct types of assets be K and let Y denote the $N \times K$ matrix of their payoffs in the N possible states of nature. The set of available trades is span(Y) where span(.) denotes the subspace spanned by the matrix. In an economy without frictions, agents will create new financial assets until all mutually beneficial trade opportunities are in span(Y). All mutually beneficial trades have been consummated if there exist positive scalars $\lambda_1, \ldots, \lambda_M$ such that

$$\lambda_j U_j'(C_j, \theta_i) = \lambda_h U_h'(C_h, \theta_i) \qquad i = 1, \ldots, N;$$
$$j, h = 1, \ldots, M \quad (1)$$

The invisible hand drives agents toward creating all the types of financial assets which can lead to mutually beneficial trades. However, there are many external factors which can offset this tendency. If agents have some control over outcomes, then moral hazard problems may limit hedging opportunities. For example, agents may not be able to hedge against changes in labour income if work requires imperfectly observable effort. If agents have special knowledge, then adverse selection can similarly limit trade. If a car owner knows more about its quality than a prospective buyer, then an agent cannot sell his car at a reasonable price when he experiences financial distress. The administrative costs of trade can also limit hedging before the full efficiency condition (1) is fulfilled.

The model described above is static. In an intertemporal model, dynamic strategies increase the set of hedging opportunities beyond the linear span of the matrix of asset payoffs. Agents can create a rich set of payoff claims by dynamically varying the proportions invested in the individual assets. With continuous trading, this process reaches its natural limit: if an asset price follows Brownian Motion, then a continuously adjusted portfolio consisting of only this risky asset and a riskless asset can be constructed which replicates the payoff to any put or call option on the risky asset.

The proliferation of complex financial assets, such as options on futures and interest rate and currency options, and the increased sophistication of traders, has led to a bewildering array of dynamic hedging strategies, especially by large institutional investors. *Portfolio insurance* provides a good example of the kind of sophisticated new hedging instrument which can be created with a dynamic trading strategy. Consider a pension fund with a large equity portfolio and an aversion to large capital losses on this portfolio. A portfolio insurance strategy can put a floor on the random rate of return to the pension fund's portfolio. The return floor can be any rate lower than the available riskless rate (it can be a negative net return, so that the fund bounds its losses rather than assuring itself of a small gain). The strategy works as follows. At the starting date of the insurance strategy, the fund has most of its money invested in equities and a small proportion in a riskless asset (i.e. government notes). If the equities fall in price, the fund sells some of the equities and places the cash in the riskless asset. If equity prices continue to fall, the fund increases the proportion of investment in the riskless asset. If there is a sustained fall in equity prices, the fund will end the insurance programme invested entirely in the riskless asset. It will have earned a rate equal to the pre-chosen minimally acceptable return. The fund makes a 'soft landing' at this minimal value: the proportion of money invested in the riskless asset approaches one as the value of the portfolio approaches the minimally acceptable level.

Portfolio insurance is not a free lunch. In exchange for the return floor, the pension fund sacrifices some of the upside potential of pure equity investment. For example, if the equity market declines sharply and then rises, the fund will miss the upturn, since it will have defensively decreased its position in equities before the upturn.

There are numerous other dynamic hedging strategies, not only in equity markets but in fixed income, currency and options markets. In terms of the volume of trade, hedging in financial markets now greatly outpaces the activity in commodities futures markets, which is the original and classic example of a market often used for hedging.

2. RISK PREMIA AND HEDGING. An economically interesting question is whether agents 'pay a premium' to hedge. Assume again that the current price of the hedge portfolio is set to zero by appropriate balancing of the asset and liability sides of the hedge (a futures contract is a natural example). If the expected cash flow is negative (positive) next period, then the hedge portfolio carries a positive (negative) implicit risk premium. If the expected cash flow is zero, then the implicit risk premium is zero.

Much of the early literature on hedging was centred on hedging in commodity futures contracts. One of the key questions was whether agents who sold futures pay a positive risk premium. Keynes (1930) considers this problem for the case of commodity futures contracts. He argues that the natural supply of short hedgers (sellers of futures contracts) outnumber long hedgers (buyers) in this market. Therefore, the implicit risk premia for holding a futures contract should be negative, in order to induce other agents (henceforth called 'speculators') to absorb the excess hedging demand of short hedgers. This will be true if the futures price increases on average over the life of the contract, so that the expected cash flow from holding the contract is positive. The empirical evidence for this positive drift (sometimes called *normal backwardation*) in commodity futures prices is weak at best.

Keynes's analysis implicitly assumes that the commodity futures market is isolated from other asset markets so that hedgers must pay other agents (the speculators) a premium to induce them to take a position in the market. In an integrated set of asset markets, hedgers need not pay any premium to induce other agents to trade. Rather, the existence of a risk premium depends upon the covariance between the payoffs to the hedge asset and the economy-wide risks faced by all the agents. If the hedge asset is uncorrelated with market-wide risks, then it will carry no risk premium, even though it may have a high value to a particular hedger due to his specific income stream. A hedge asset which protects against market risk will carry a risk premium.

There is another source of return to speculators, which is not captured in the competitive pricing model. Speculators may charge an explicit or implicit bid–ask spread when trading with hedgers. If hedgers buy and hold for a long period, then this is equivalent to the return premium described above. However, if hedgers trade frequently then a bid–ask spread can lower their realized returns, and raise the realized returns to speculators, without affecting the observed long-run return premium of the hedge asset as reflected in transactions prices. This may explain the lack of empirical evidence for normal backwardation in commodities futures markets. This effect of a bid–ask spread was not recognized in most of the early literature on commodity futures markets.

The bid–ask spread need not be explicit. Even open-floor markets will contain a set of implicit bid–ask spreads, to the extent that trader strategies reflect a greater willingness to sell at higher prices and to buy at lower ones. One can view the feverish activity which is common in floor trading as speculators searching for transactions at the outer edges of an implicit bid–ask spread. Hedgers, who are off the floor and are more anxious to complete a particular trade, take the losing side of the implicit spread.

3. THE ROLE OF HEDGERS IN A MARKET WITH HETEROGENEOUS INFORMATION. The model in sections 1 and 2 assumes

homogeneous information across agents. If agents have differential information about the payoffs to assets then the trading strategies of rational agents cannot have the simple competitive form. Agents must treat trade opportunities as signals of the information of other agents about the value of the trade.

The presence of differential information can lead to fewer hedging opportunities and/or raise the expected cost of hedging. Milgrom and Stokey (1982) show that rational agents will not trade solely because they have different beliefs about the value of an asset. If agents are distinguished merely by their differential information, then they will refuse all trades, since the willingness of the other agent to trade signals that the terms are unfavourable. This means that a financial asset market will fail to open in the absence of other motives for trade. This is a market failure due to adverse selection.

The needs of some agents for hedging can provide an additional reason for trade which overcomes the adverse selection problem and eliminates the market failure. Hedgers will be willing to trade even if they suspect that the other party to the transaction has superior information. Informed agents will be gaining at the expense of hedgers, but they will also be providing an insurance/liquidity service to hedgers, and so hedgers may be willing to trade with them despite their informational disadvantage. This in turn has the side-effect of permitting superior information to be reflected in market prices.

I will follow Glosten and Milgrom (1985) and assume that there exist costless, competitive, risk-neutral market makers who intervene in all trades. This is for analytical convenience and is not necessary to the basic model. Suppose that certain agents ('hedgers') have a strong preference for a given hedge asset, i.e., their preference is such that they will buy (and sell) some non-zero amount at a price higher (lower) than the market-clearing equilibrium price. This implies that they are willing to trade even if they must pay a bid–ask spread around the equilibrium price. Informed agents (henceforth 'speculators') will also trade despite a bid–ask spread as long as the expected profit from their superior information is larger than the bid–ask spread. The market makers can set an equilibrium bid–ask spread which allows the hedgers to trade at an expected loss and the speculators at an expected gain, leaving the market makers with an expected profit of zero (their equilibrium condition). The market markers will respond to the net demands of all traders (which partially reveals the net demand of speculators) to adjust the bid and ask prices, and so (partially) capture in market price the superior information of speculators.

One intriguing feature of this model is the symbiotic roles of speculators and hedgers. Without speculators (informed traders), the hedgers would lose liquidity; without hedgers, speculators would lose the opportunities to gain on their superior information. Without both speculators and hedgers, the price in the market would no longer provide a useful signal for agents making production and consumption decisions. Kyle (1984) develops a model in which this symbiotic relationship is made clear and describes the effects of more or fewer speculators or hedgers on the informational efficiency and liquidity of the market. Some of the results are counter-intuitive: for instance, increasing the number of hedgers, who are uninformed, can *increase* the informational efficiency of prices.

4. RISK PREMIA ON HEDGE PORTFOLIOS AND GENERAL EQUILIB-RIUM PRICING. In section 2, I described two types of hedge portfolios – those with and without risk premia – and how the distinction between them depends on the covariance between the hedge portfolio's returns and the market-wide risks in the economy. In this section, I will describe a relationship between hedge portfolios which protect against market-wide risks and the general equilibrium pricing of assets.

Let Q_t denote the discounted expected utility of lifetime consumption for some agent at time t:

$$Q_t = E_t \left[\sum_{\tau=t+1}^{\infty} \rho^\tau U(C_\tau, \theta_\tau) \right],$$

where ρ is the agent's discounted factor and $U(,)$ is his utility function. Let J_t denote the change in discounted expected utility given a change in the agent's time t wealth:

$$J_t = \partial Q_t / \partial W_t,$$

where W_t is his wealth at time t. Note that at time $t-1$, J_t is a random variable. Let r_{it} denote the return from time $t-1$ to t of the ith financial asset. If the agent holds an equilibrium amount of this asset then the following first order condition is satisfied:

$$E_{t-1}[r_{it} J_t] = \partial U(C_{t-1}, \theta)/\partial C,$$

which can be re-written (using $E[ab] = E[a]E[b] + \text{cov}[a, b]$) as:

$$E_{t-1}[r_{it}] = r_{0t} + (1/\gamma)\text{cov}_{t-1}[r_{it}, J_t], \qquad (2)$$

where $\gamma = \partial U(C_{t-1}, \theta)/\partial C$ and r_{0t} is the expected return on an asset with a riskless payoff at time t. Suppose that, at time $t-1$, J_t equals a sum of a set of K uncorrelated random variables $Z_{1t}, Z_{2t}, \ldots, Z_{Kt}$:

$$J_t = Z_{1t} + \cdots + Z_{Kt}. \qquad (3)$$

The variables $Z_1, \ldots, Z_K$ describe the K random shocks which affect the agent's marginal utility. They could be interest rate movements, output shocks, inflation shocks, etc. Assume that there exists a set of K portfolios with returns $r_{1t}^*, \ldots, r_{Kt}^*$ such that the jth portfolio has perfect negative correlation with Z_{jt}:

$$\text{cov}_{t-1}[r_{jt}^*, z_{jt}] = (\text{var}_{t-1}[r_{jt}^*]\text{var}_{t-1}[Z_{jt}])^{1/2}. \qquad (4)$$

These portfolios are potential hedges against the K types of risk which affect the investor. (The agent would short-sell the portfolio to hedge since the portfolio return varies inversely with marginal utility.) I will call $r_{1t}^*, \ldots, r_{Kt}^*$ an *indexing set of hedge portfolios* since the portfolios index the random shocks to the agent's marginal utility. Using (3) and (4) we can re-write (2) as:

$$E_{t-1}[R_{it}] = r_{0t} + \beta_{i1t}\pi_{1t} + \cdots + \beta_{iKt}\pi_{Kt}, \qquad (5)$$

where

$$\beta_{ijt} = \text{cov}_{t-1}[r_{it}, r_{jt}^*]/\text{var}_{t-1}[r_{jt-1}^*]$$

and

$$\pi_{jt} = E_{t-1}[r_{jt}^* - r_{0t}].$$

Equation (5) is an asset pricing relationship: it says that the expected return on any asset equals the riskless return plus a linear combination of the covariances of the asset's return with an indexing set of hedge portfolios.

In general, there will not exist a finite set of portfolios fulfilling (4). Merton (1973) develops a continuous time model in which there does exist an indexing set at each instant of time. Breeden (1984) shows that Merton's model with $K > 1$ can be simplified to a $K = 1$ model without loss of generality.

One limitation of this model is that the beta coefficients (β_{ijt}) and risk-premia (π_{jt}) in (5) are only defined at a single point in time. In empirical applications of the model, one must use time series of asset returns and other variables to estimate these parameters. Time-varying parameters are not estimable from a time series sample unless their time-series behaviour is specified. A typical assumption, which is not always consistent with the theoretical model, is that the betas and risk premia are constant through time. Despite the limitations on estimatibility, this model is important and illustrates the role of hedging in asset pricing theory.

GREGORY CONNOR

See also BACKWARDATION; FINANCE; FUTURES MARKETS, SPECULATION AND HEDGING; FUTURES TRADING; OPTION PRICING THEORY; OPTIONS.

BIBLIOGRAPHY

Arrow, K.J. 1964. The role of securities in the optimal allocation of risk-bearing. *Review of Economic Studies* 31, 91–6.

Breeden, D.T. 1979. An intertemporal asset pricing model with stochastic consumption and investment opportunities. *Journal of Financial Economics* 7(3), 265–96.

Breeden, D.T. 1984. Futures markets and commodity options: hedging and optimality in incomplete markets. *Journal of Economic Theory* 32, 275–300.

Brennan, M.J. 1958. The supply of storage. *American Economic Review* 48, 50–72.

Cootner, P.H. 1967. Speculation and hedging. *Food Research Institute Studies*, Supplement to Volume 7, 65–105.

Duffie, D. and Huang, C. 1985. Implementing Arrow–Debreu equilibria by continuous trading of a few long-lived securities. *Econometrica* 53, 1337–56.

Glosten, L. and Milgrom, P.R. 1985. Bid, ask, and transaction prices in a specialist market with heterogeneously informed traders. *Journal of Financial Economics* 14, 71–100.

Grauer, F.L. and Litzenberger, R.H. 1979. The pricing of nominal bonds and commodity futures contracts under uncertainty. *Journal of Finance* 34, 69–84.

Gray, R.W. 1961. The search for a risk premium. *Journal of Political Economy* 69, 250–60.

Grossman, S. 1976. On the efficiency of competitive stock markets when traders have diverse information. *Journal of Finance* 31, 573–85.

Hoffman, G.W. 1954. Past and present theory regarding futures trading. *Journal of Farm Economics* 19, 1–11.

Houthakker, H.S. 1957. Can speculators forecast prices? *Review of Economics and Statistics* 39, 143–52.

Keynes, J.M. 1930. *A Treatise on Money. Volume II: The Applied Theory of Money.* London: Macmillan.

Kyle, A.S. 1984. Market structure, information, futures markets, and price formation. In *International Agricultural Trade: Advanced Readings in Price Formation, Market Structure, and Price Instability,* ed. G. Storey, A. Schmitz and A. Sarris, London: Westview.

Leland, H. 1980. Who should buy portfolio insurance. *Journal of Finance* 35, 581–94.

Malinvaud, E. 1972. The allocation of individual risk in large markets. *Journal of Economic Theory* 4, 312–28.

Merton, R.C. 1973. An intertemporal capital asset pricing model. *Econometrica* 41, 867–87.

Milgrom, P. and Stokey, N. 1982. Information, trade and common knowledge. *Journal of Economic Theory* 26, 17–27.

Working, H. 1953a. Futures trading and hedging. *American Economic Review* 43, 314–43.

Working, H. 1953b. Hedging reconsidered. *Journal of Farm Economics* 35, 544–61.

Working, H. 1967. Tests of a theory concerning floor trading on commodity exchanges. *Food Research Institute Studies*, Supplement to volume 7, 5–48.

hedonic functions and hedonic indexes.

I. HEDONIC FUNCTIONS

A hedonic function is a relation between prices of varieties or models of heterogeneous goods – or services – and the quantities of characteristics contained in them:

$$P = h(c) \tag{1}$$

where P is an n-element vector of prices of varieties, and (c) is a $k \times n$ matrix of characteristics. The theory providing its economic interpretation rests on the *hedonic hypothesis* – heterogeneous goods are aggregations of characteristics, and economic behaviour relates to the characteristics.

The hedonic hypothesis implies that a transaction is a tied sale of a bundle of characteristics, so the price of a variety is interpreted as itself an aggregation of lower-order prices and quantities. Characteristics are assumed the true arguments of utility functions; they are the inputs to the production process, in the case of heterogeneous materials, capital goods, or labour services. Hence:

$$Q = Q(c, M) \tag{2}$$

where Q is utility (output), M is a vector of other, homogeneous consumption goods (productive inputs), and for expositional simplicity we specify only one heterogeneous good in the system, with characteristics (c). For a heterogeneous labour type, productive characteristics are typically assumed to have been acquired through investment in human capital, so that (1) is a hedonic wage equation, and human capital characteristics appear in (2). It is common to assume, for durable goods and for labour, that services of characteristics are proportional to their stocks, through characteristics may decay at varying rates. Analysis of consumer behaviour toward characteristics of goods is frequently linked to the literature on household production, but the two subjects are conceptually distinct, and the latter is ignored here, in the interest of brevity.

Production of the heterogeneous good is the joint production of a bundle of characteristics:

$$t(c, K, L) = 0 \tag{3}$$

Characteristics may be attached to goods through externalities (air quality as a housing characteristic) or by an act of nature (risk as an attribute of jobs) as well as by explicit production decisions of producers.

Equations (2)–(3) exhibit the extreme form of the hedonic hypothesis: *only* the characteristics of heterogeneous goods enter behavioural relations. Plausible cases exist where both quantities of goods and of their characteristics matter, particularly where there are complementarities in (2) between characteristics and other inputs or outputs (two small cars are not necessarily equivalent to a large one with the same total quantities of characteristics because consumption also requires input of driving time), or when conventional scale economies are present in (3). For present purposes, such additional structure is dispensed with because it complicates the exposition, and because it is more relevant to investigating the demand and supply of characteristics than for explaining hedonic functions. For the same reasons, we cannot explore interesting cases of production where both inputs and outputs are heterogeneous.

It is well-established – but still not widely understood – that the form of $h(\cdot)$ cannot be derived from the form of $Q(\cdot)$ or of $t(\cdot)$, nor does $h(\cdot)$ represent a 'reduced form' of supply and demand functions derived from $Q(\cdot)$ and $t(\cdot)$. Establishing this

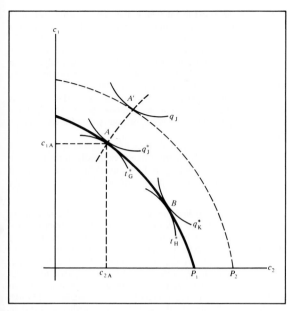

Figure 1

result requires consideration of buyer and seller behaviour toward characteristics.

The buyer, or user, side. It is expositionally convenient to represent the user's choice of characteristics as a two-stage budgeting process. Suppose that (2) can be written

$$Q = Q(q(c), M) \qquad (4)$$

where $q(\cdot)$ is an aggregator over the characteristics (c). Then conditional on M and a utility (output) level Q^*, the allocation of characteristics (choice of variety) can be determined by minimizing the cost of attaining the sub-aggregate $q(c)$. Thus, if q^* is a value of $q(\cdot)$ such that $Q^* = Q(q^*, M)$, the optimal choice of (c) is the solution to:

$$\min_{c} h(c), \quad \text{s.t. } q(c) = q^* \qquad (5)$$

Marginal conditions for an optimum are (where the subscript shows partial derivative with respect to c_i or c_j):

$$q_i/q_j = h_i/h_j \qquad (6)$$

The ratio of marginal 'sub-utilities' of c_i and c_j must equal the ratio of acquisition costs for incremental units of c_i and c_j. Note that the ratio h_i/h_j is the slope of $h(\cdot)$ in the c_i/c_j plane, variety price held constant.

Suppose for illustration a non-linear, two-characteristic, continuous hedonic function such that, for any fixed price P^*, the graph of

$$P^* = h(c_1, c_2) \qquad (7)$$

has the form of the contours P_1 and P_2 in Figure 1. The locus P_1 connects all varieties selling for the price P_1 – point A designates a variety described by the vector $[P_1, c_{1A}, c_{2A}]$, point B by $[P_1, c_{1B}, c_{2B}]$, and so forth. The slope of P_1 at any point gives relative marginal acquisition costs for characteristics c_1 and c_2. The solution to the choice of variety problem is shown in Figure 1 by the tangency of q_J^* – a partial or conditional indifference curve (isoquant) for user J – and P_1.

A quantal choice problem is contained in this optimization: The buyer selects the variety whose embodied characteristics are closest to the optimal ones. When the spectrum of varieties is continuous in c_1, c_2, the quantal choice is trivial so long as only one unit of the good is bought; Lancaster (1971), following Gorman (1980, but written in 1956), models the non-continuous case by specifying the P_1, P_2,... contours as piece-wise linear, and permitting the buyer to obtain an optimal set of characteristics by combining two varieties.

The remainder of the user optimization problem proceeds as in other two-stage allocations. Total expenditure on characteristics (the level of 'quality' when only one unit is bought) is determined by:

$$\max Q(q(c), M)$$

$$\text{subject to: } q(c) \cdot v(c) + P_M M = y \qquad (8)$$

where $v(c)$, the price of the composite commodity q – or alternatively, the 'price of quality' – is the slope of the hedonic surface above an expansion path such as AA′ in figure 1. With respect to any good i in M, the solution entails:

$$Q_q/Q_i = v(c)/p_i \qquad (9)$$

and the set of such conditions determines total expenditure on characteristics (equals the price of the model chosen).

The characteristics–space problem has many similarities to normal 'goods space' problems. The hedonic frontiers P_1, P_2,... provide analogues to conventional budget constraints (isocost lines) and serve to constrain the agent's optimization problem in characteristics space. These are the constraints themselves, *not* the cost functions of conventional duality theory. The constraints may be non-linear; if so, characteristics prices are not fixed, but are uniquely determined for each buyer by the buyer's location on the hedonic surface (compare the slope of P_1 at A and B). It is observed that varieties having differing characteristics are available at the same price and are chosen by different buyers (in figure 1, model B is chosen by buyer K). This suggests that divergence of tastes and technologies is an essential part of the theory for hedonic functions, and that 'representative consumer' (firm) models do not describe market outcomes.

If there are a large number of buyers, Rosen (1974) shows that each frontier $P_1, P_2, \dots$, will trace out an envelope of tangencies with relations such as $q_J^*, q_K^*, \dots$. As with any envelope, the form of $h(\cdot)$ is independent of the form of $Q(\cdot)$ – except for special cases – and is determined on the demand side by the distribution of buyers across characteristics space. This is an important result for the interpretation of hedonic functions

Forming measures of 'quality'. It is well known that (4) implies weak separability of Q on (c), which permits consistent aggregation over the characteristics in (c) – see Blackorby, Primont and Russell (1978). It is natural to take such an aggregate as a measure of 'quality'.

One can thus use weak separability on characteristics to rationalize the common practice of writing scalar 'quality' in the utility or production function, as for example Houthakker's (1952) model of quantity and quality consumed – a model that has many empirical progeny, and much appeal for its simplicity. Weak separability on characteristics also provides the analytic bridge between characteristics–space models and Fisher and Shell's (1972) notion of 'repackaging', in which quality change enters the utility function by scalar multiplication of the good whose quality changed. Because hedonic functions have mostly been *used* for purposes (like

631

constructing a 'quality-adjusted' price index for automobiles: Griliches, 1971) for which separability was assumed (usually implicitly), separability assumptions on characteristics are thus a common thread through most analysis of 'quality'.

Obviously, when Q is not separable on (c), no consistent scalar measure of 'quality' can be formed. It is not hard to think of cases where characteristics separability is not realistic (are refrigerator characteristics separable from what is stored in them, or transportation equipment characteristics separable from energy consumption?). One should note that characteristics–space approaches could be adapted to certain non-separable cases (computing the cost per mile of constant-quality transportation services), where scalar approaches may be more problematic. Moreover, since weights for the aggregator are the marginal subtilities q_1 and q_2, the quality measure will depend on relative characteristics prices – properly, on the position of the P-contours in Figure 1 – whenever substitution among characteristics quantities is possible; a scalar quality measure is therefore not in general unique, even when consistent. These points suggest that a major advantage of hedonic, or characteristics–space, methods is their potential for dealing with non-separable cases and with changing relative characteristics costs regimes, though there is little demonstration of this potential in existing empirical work.

The production side. A comparable theory shows how a price-taking producer selects the optimal variety or varieties to sell, given (1) and (3). For a particular level of input usage or production cost, a two-characteristic form of $t(\cdot)$ yields transformation surfaces, for producers or production processes G and H, like t_G^* and t_H^* in Figure 1. Revenue from increments of characteristics added to the design can be computed from partial derivatives of the hedonic function. Optimal product design is determined by:

$$t_1/t_2 = h_1/h_2 \qquad (10)$$

The quantity produced of the optimal design is determined in the usual way by setting the marginal cost of quantities of the optimal design (equation omitted here for brevity) equal to the variety price (given by the hedonic function, $h(\cdot)$).

The production-side theory is problematic, compared with the user case, because in the absence of scale economies in the production of varieties, producers would build 'custom products', offering all product designs on the hedonic surface where variety price exceeds cost, rather than specializing in the most profitable variety. The competitive, large numbers case is thus not an appealing one, unless product design is to an extent fixed by sellers' endowments at least in the short run (the normal assumption for labour markets, and for land).

If there are a large number of sellers, Rosen (1974) shows that, except for special cases, each hedonic frontier $P_1, P_2, \ldots$ will trace out an envelope of tangencies with relations such as t_G^* and t_H^*. As in the user case, the form of $h(\cdot)$ is therefore influenced on the supply side by the distribution of sellers across characteristics space and by their output scales, but the form of $h(\cdot)$ cannot in general be derived from the form of $t(\cdot)$.

Special cases. If $q(\cdot)$ is identical for all users, then only a single set of q^* contours appears in Figure 1, and each hedonic frontier $P_1, P_2, \ldots$ traces out the associated q^* contour. In this case, the form of $h(\cdot)$ is determined by the form of $q(\cdot)$, up to a monotonic transformation, and should conform to the principles of classical utility theory (which means that each hedonic frontier, P, bows inward, toward the origin, rather than as drawn in Figure 1).

If $t(\cdot)$ is identical fo all sellers, then only a single set of t^* contours appears in Figure 1, and each hedonic frontier $P_1, P_2, \ldots$ traces out the associated t^* contour. In this case, the form of $h(\cdot)$ is determined by the form of $t(\cdot)$, and the usual reasons for assuming convexity of production sets apply, so that the P-frontiers should bow outward from the origin, in the manner of a normal production transformation curve.

If there is no diversity on either side of the market, only one design will be available at each model price. The hedonic frontiers degenerate into a series of points, one for each model price.

Of these possibilities, uniformity of $t(\cdot)$ across sellers (except for labour services) is the most likely, especially in the long run when access to technology is freely available. Uniformity of $q(\cdot)$ is improbable, and appears inconsistent with available evidence.

Functional forms for hedonic functions. Neither classical utility nor production theory can specify the functional form of $h(\cdot)$. The P-frontiers can bow in, bow out, or take the form of straight lines (or even irregular shapes). In particular, and contrary to assertions that have appeared in the literature, nothing in the theory rules out the semi-logarithmic form (which has often emerged as best in goodness-of-fit tests in both labour and product market hedonic studies). Though non-linear in P and (c), the semi-log is nevertheless linear in the $[c_i, c_j]$ plane and thus even has some 'nice' properties (because all buyers and sellers face the same characteristics prices, for equal expenditure on, or revenue from, characteristics).

Hedonic 'demand' studies. Hedonic functions have sometimes been used to generate demand or 'willingness to pay' estimates (particularly, of the value of air quality or neighbourhood amenities in land and housing prices, and of risk in labour markets). However, as Figure 1 shows, buyers J and K, though located on the same hedonic price surface, may face different characteristics prices as a consequence of their preference functions; the slopes of the P-function at A and B do not represent exogenous price variance that determines characteristics allocations.

Unless one is willing to assume that all buyers have identical tastes, cross-section characteristics demand studies founder for the same reason as cross-section 'goods' demand studies: Variations in quantities reflect taste differences and not shifts in the slope of the budget constraint. Moreover, the situation depicted in figure 1 cannot be reduced to a demand estimation problem by treating it as some variant of an econometric demand–supply identification problem, despite some attempts to do so in the literature.

II. HEDONIC INDEXES

A hedonic price index is one that makes use of information from the hedonic function. Adding time dummy variables to a multi-period regression on (1) is a favourite empirical procedure, but is by no means the only way to compute a hedonic price index. A characteristics price index is any index that is defined on the characteristics of goods, or on behavioural functions in which characteristics are arguments. A hedonic price index is thus a particular implementation of a characteristics price index. Almost any empirical application of hedonic functions (e.g. use of hedonic wage regressions to estimate race or sex discrimination in labour markets) can be interpreted as an index number, so the theory of characteristics–space indexes has wide applicability. To conserve space, the following is couched in terms of a cost-of-living index but application to other contexts can be made by suitable extensions (Triplett, 1983).

The exact characteristics–space index. A cost-of-living (COL) index shows the minimum change in cost between two periods that leaves utility unchanged. Using (1) and (2), the minimum cost of attaining utility level Q^* in any period is:

$$C^* = C(P_M, h(\cdot), Q^*) = \min_{c, M} [P_M M + h(c): Q(c, M) = Q^*]$$

(11)

The form of the cost functional, C, depends on the form of $Q(\cdot)$ and the budget constraint; the hedonic function makes up that portion of the budget constraint that pertains to the acquisition of characteristics. The cost-of-living index between periods r and s is then:

$$\underset{r, s}{COL} = C(P_{Mr}, h(\cdot)_r, Q^*)/C(P_{Ms}, h(\cdot)_s, Q^*)$$

(12)

Generally, the full index is intractable, and there is need to consider a less comprehensive measure that is more nearly congruent with the problem at hand. For the separable utility function (4) an exact 'subindex' (Pollak, 1975) can be computed that involves only the heterogeneous good. Define the cost functional d by:

$$d = d(h(\cdot), q^*) = \min_c [h(c): q(c) = q^*]$$

(13)

Then the *characteristics price index* is:

$$I_{r, s} = [d(h(\cdot)_r, q^*]/[d(h(\cdot)_s, q^*]$$

(14)

where the subscripts designate characteristics costs in period r and s, respectively. Expression (14) is the ratio of the costs, under two characteristics price regimes, of a constant-utility collection of characteristics.

Note that (14) does *not* hold characteristics constant – it is not the price of the same, or 'matched', variety in two periods. Rather, (14) permits substitution among characteristics as relative characteristics costs change, in a manner analogous to the normal COL defined on goods – (14) would be implemented by finding a variety (bundle of characteristics) in period s that was *equivalent* in utility to the one chosen in period r, but which minimized consumption costs in the relative price regime of period s.

Information requirements. The normal 'goods' COL index requires knowledge of the utility function. The form of the characteristics price index (14) depends on the form of the utility function (or the 'branch' utility function, $q(\cdot)$) *and* the form of the hedonic function, $h(\cdot)$. Both are unobservable or must be estimated. The reason (14) requires more information than the analogous 'goods-space' COL index is that in general the hedonic function is non-linear and therefore its form enters into $d(\cdot)$. In contrast, 'goods' COL indexes assume a bounding hyperplane, whose linearity implies a mirror-image duality between the utility function and the consumption cost function. Use in characteristics space of the demand-systems approaches that have been used to estimate goods–space COL indexes (Braithwait, 1980) is complicated by the non-linearity of the hedonic function and by the necessity to estimate both the demand equations and the budget constraint.

Note that, contrary to assertions that have appeared in the literature, imposing 'nice' functional forms (that is, those with properties of classical utility theory) on the hedonic function does nothing to identify index (14) – unless the special case of uniform preferences, where the hedonic function sketches out the characteristics–space preference map, obtains.

Bounds and approximations: empirical hedonic price indexes. It is evident that (14) is not an index number that can be computed from the hedonic function alone, so it is not an empirical hedonic price index. It is important to specify the relation of empirical hedonic indexes to (14).

In the usual goods case the budget constraint is assumed a hyperplane. Accordingly, bounds on goods–space COL indexes are fixed-weight (Laspeyres or Paasche) indexes – the denominator of the Laspeyres index, for example, is the equation for the reference period budget constraint, and the numerator is the equation for another budget constraint. Fixed-weight indexes are also convenient approximations to COL indexes, since they require only knowledge of one actual budget constraint and two price regimes, and one knows that the fixed-weight index differs from the true index only by the expenditure saving from substitution.

For characteristics–space price indexes, it is natural to follow an analogous procedure and construct approximations to (14) from the characteristics–space budget constraint, when it is known. The characteristics–space budget constraint is precisely the information provided by the hedonic function. Accordingly, hedonic price index numbers – those computed from hedonic functions – can be interpreted as approximations to the true characteristics–space indexes (14) in the same sense that fixed weight Laspeyres and Paasche price indexes approximate goods COL indexes: The approximations are, in each case, based solely on the budget surface, where the true indexes, in each case, require knowledge of the utility function.

Hedonic indexes differ from goods–space approximating indexes in two major respects. In the characteristics–space case, the form of the budget surface must be estimated empirically. When the hedonic function is linear or is semi-log, the P-contours of Figure 1 are linear – each budget constraint is a hyperplane. Otherwise, the constraints are non-linear. Secondly, and as a corollary, the form of the approximating hedonic index depends on the form of the hedonic function. A third, subsidiary, point is that with usual procedures, the hedonic index records the shift in the whole hedonic surface, rather than, as goods–space fixed-weight indexes are usually calculated, a shift in a single selected budget hyperplane.

Hedonic indexes may also be bounds on the true index, though this interpretation requires more careful empirical specification of the hedonic function than has often been the case, and it is not clear whether they are the *best* bounds. The theory of bounds for characteristics–space indexes is not well worked out.

JACK E. TRIPLETT

See also CHARACTERISTICS; INDEX NUMBERS; SEPARABILITY.

BIBLIOGRAPHY

Blackorby, C., Primont, D. and Russell, R.R. 1978. *Duality, Separability and Functional Structure: Theory and Economic Applications.* New York: North Holland.
Braithwait, S.D. 1980. The substitution bias of the Laspeyres price index: an analysis using estimated cost-of-living indexes. *American Economic Review* 70, 64–77.
Fisher, F. and Shell, K. 1972. *The Economic Theory of Price Indices: Two Essays on the Effects of Taste, Quality, and Technological Change.* New York: Academic Press.
Gorman, W.M. 1980. A possible procedure for analysing quality differentials in the egg market. *Review of Economic Studies* 47, 843–56.
Griliches, Z. (ed.) 1971. *Price Indexes and Quality Change: Studies in New Methods of Measurement.* Cambridge, Mass.: Harvard University Press.

Houthakker, H.S. 1952. Compensated changes in quantities and qualities consumed. *Review of Economic Studies* 19, 155–64.

Lancaster, K. 1971. *Consumer Demand: A New Approach*. New York: Columbia University Press.

Pollak, R.A. 1975. Subindexes of the cost of living. *International Economic Review* 16, 135–50.

Rosen, S. 1974. Hedonic prices and implicit markets: product differentiation in pure competition. *Journal of Political Economy* 92, 34–55.

Triplett, J.E. 1983. Concepts of quality in input and output price measures: a resolution of the user value–resource cost debate. In *The U.S. National Income and Product Accounts: Selected Topics*, ed. Murray F. Foss, Conference on Research in Income and Wealth, Volume 47, University of Chicago Press for the National Bureau of Economic Research.

hedonism. From the Greek *hedone*, 'pleasure', this term is used of two different theses, one a psychological thesis about motivation (psychological hedonism), the other a thesis about what is intrinsically valuable in a person's life (ethical hedonism).

Psychological hedonism refers to the claim that a person acts solely to promote his own pleasure. (It is usual to limit the scope of the claim to acts that meet some minimal standards of rationality, that is, excluding confused, involuntary or habitual acts).

Ethical hedonism refers to the claim that only pleasure is intrinsically valuable, that all other things that are valuable are so only instrumentally as means to pleasure. This root form of ethical hedonism is not strictly an ethical view at all (so the name is something of a misnomer); it is rather a view about what constitutes the quality of an individual life. This root form is often combined with a view about action, namely that all action should aim at maximizing pleasure. This combination can easily turn into a view about rationality (e.g. that a rational agent acts to maximize his own pleasure – what could be called egoistic hedonism), or into a view about morality (e.g. that each person should act to maximize pleasure for persons generally – universalististic hedonism).

It is hard to supply a satisfactory analysis of the concept of 'pleasure' or to understand its relation to 'happiness'. Pleasure is not a physical sensation (think of the pleasure of country walks), nor a psychological one (think of the pleasure from one's work). What we enjoy is so heterogeneous that no unified account of 'pleasure' may be possible. Sidgwick (1907), aware of how different our states of minds can be when we enjoy ourselves, thought that the unifying feature was desire: 'pleasure', he proposed, is 'desirable consciousness'. J.S. Mill thought that the relation between 'pleasure' and 'happiness' was relatively simple: he defined 'happiness' as 'pleasure and the absence of pain'. But the terms mark different features of life: a martyr might go to the stake happily, but is unlikely to do so with pleasure. The lack of breadth of physical and psychological explanations of 'pleasure' has led to more behavioural ones: for example, what we find pleasant is what we do, or would, give ourselves to eagerly. Similarly, 'pain' applies to more than just physical pains, and the painfulness even of physical pains is a matter not only of our sensations but also of how we react to them. So the need to get a sufficiently broad analysis of 'pain' has also led to behavioural explanations; for example what we find painful is what we wish to avoid, have alleviated, etc. The terms 'pleasure' and 'pain' sometimes become so broad in the course of the statement or defence of either psychological or ethical hedonism that they become, in effect, technical terms. When that happens, we have to ask what their technical sense is. For example, near the end of his life Sigmund Freud refused strong pain-killing drugs, preferring, he said, to think in torment than to be confused in comfort. We might wish to use 'pleasure' in such a way that we should say that Freud found clear thought in pain more 'pleasant' than confused pleasure. But we might also think that the technical sense of the word would now be so far removed from the ordinary sense that it would be better to find another way of speaking altogether.

Psychological hedonism is an empirical thesis, and widely thought to be false. As Butler, Hume and others point out, one's actions are often explained by desires for things other than one's own pleasure or avoidance of pain (e.g. the desire to eat is often more effective than the desire for pleasure from eating). Perhaps the clearest counter-examples are the desires that many people have about what will happen after their deaths, when (they assume) they will not exist. A psychological hedonist might reply that what really explains such actions are desires for the pleasure of knowing how things will go after one's death. However, a simple thought experiment would often show this to be false. Suppose that a father working to provide for his family after his death is told that he can choose between (1) his family's being provided for though he will never know it, and (2) his family's not being provided for though he will think that they will be, and that, after he makes his choice, he will be made to forget it. Many persons would choose (1) rather than (2), thereby showing that their action is not prompted by any concern for their own future mental states. (The falsity of psychological hedonism leads to the *paradox of hedonism*, namely, that typically one cannot promote one's own pleasure by aiming to promote it, that pleasure is usually an unintended accompaniment of action aimed at another goal.)

The root form of ethical hedonism is a thesis about what affects the quality of a single life. It too is widely disputed. Even on Sidgwick's broad account of 'pleasure' as 'desirable consciousness', anything that affects the quality of life must enter consciousness. But we desire, when fully informed, things other than states of consciousness and, moreover, we seem to regard them as making our lives better. For instance, we may desire a good reputation among people we do not know, or posthumous fame, or to do something important with our lives. An ethical hedonist may object that these desires are irrational, but this cannot be plausibly claimed about all of these desires. And we have many desires of this kind. This point is made forcefully by Robert Nozick with a piece of science fiction, his *Experience Machine* (a variant on earlier Pleasure Machines). Suppose we could plug into a machine that would give us any state of consciousness we desired. Would we plug in? What could matter except how life feels from the inside? Most people would respond that they want not just the experience of helping their children or doing something important, but also actually to do these things; most people also want to be in touch with reality, even at the cost of some desirable consciousness. This suggests that the notions of 'quality of life' or 'well-being' cannot be understood entirely in hedonistic terms, even when 'hedonism' is generously defined.

In the last two centuries ethical hedonism has figured most prominently as the value theory of classical utilitarianism. However, it is not essential to utilitarianism; other value theories can be, and are, substituted for it. Modern economists, having borrowed the conceptual framework of utility maximization from classical utilitarians, largely ignored hedonism in favour of a more neutral value theory: what is valuable is the fulfilment of desire, where it is left open what the objects of desire may be (pleasure, no doubt, being one but not necessarily the only one). Modern philosophers in this tradition are divided; some wish to stay in the hedonist tradition, at least broadly interpreted, and insist that only states of consciousness affect the quality of a life, while others drop that requirement and prefer to develop the notion of the fulfilment of informed desire.

JAMES GRIFFIN AND DEREK PARFIT

BIBLIOGRAPHY

A classic discussion of these issues is Henry Sidgwick, *The Methods of Ethics*, 7th edn, Bk I, chs 4 and 9; Bk II, chs 1–4; Bk III, ch. 14; Bk IV, ch. 1. For a good discussion of Sidgwick's views, see Schneewind (1977), ch. 11. For modern discussions, see Brandt (1979), ch. 13; Edwards (1979); Parfit (1984), Appendix I; Griffin (1986), Pt I.

Brandt, R.B. 1979. *A Theory of the Good and the Right*. Oxford: Clarendon Press.

Edwards, R.B. 1979. *Pleasures and Pains*. Ithaca: Cornell University Press.

Griffin, J. 1986. *Well-Being*. Oxford: Clarendon Press.

Parfit, D. 1984. *Reasons and Persons*. Oxford: Clarendon Press.

Schneewind, J.B. 1977. *Sidgwick's Ethics and Victorian Moral Philosophy*. Oxford: Clarendon Press.

Sidgwick, H. 1874. *The Methods of Ethics*. 7th edn, London: Macmillan, 1907.

Hegelianism. The origins and concerns of the political ideas of the German philosopher G.W.F. Hegel (1770–1831) are traditionally thought to be economic rather than political. However, a preoccupation with issues of political economy is present in his earliest theological writings and lies at the centre of his wider philosophical project (Hegel, 1793–1800). Broadly speaking, Hegel wished to construct an ethical theory appropriate for the specific problems of the modern world. He believed ancient and medieval societies had been bound together by a communal code of behaviour, with social roles mirroring a putative natural or divine order. The harmony of the natural macrocosm and the social microcosm had been sundered in modern societies by a growing awareness of individuality on the part of their members. Hegel traced this development to two sources: the primacy accorded to the individual conscience within Christianity, especially the Lutheranism he personally espoused, and the individualism encouraged by the capitalist mode of production. Contrary to recent influential critics (e.g. Popper, 1945), Hegel did not wish to stifle individual liberty by returning to the organic community theorized by Plato. Instead, he sought to describe the conditions necessary for the freedom of each person to be compatible with the freedom of all.

Hegel traces the development of this consciousness of subjective freedom in the *Phenomenology* (1807) and the *Lectures on World History* (1822–30). He regards the symbol of Christ, of the divine present within humankind, as emblematic of this sense of personal freedom and simultaneously the death of any notion of a transcendent God standing outside of human existence. The individual becomes the fount and locus of all value, confronting a material world with judgements he or she has chosen and endowing it with meaning. This process is given substance through human labour and the physical transformation of nature to suit human purposes, an idea Hegel borrows from Locke. Drawing on the stadial model of economic development advocated by the political economists of the Scottish Enlightenment, particularly Sir James Steuart and Adam Smith, he went on to elaborate how this new ethic had spawned a completely new civilization.

Commercial society broke the old ties of dependence of agrarianism and feudalism by freeing humanity from a subordination to nature. Humans no longer live in a created world, but create their own environment. However, the exchange economy generates new social bonds by involving individual producers within mutual service relationships. 'Civil Society' (*burgerliche Gesellschaft*), according to Hegel, is united by a 'system of needs' (Hegel, 1821, para. 189). The division of labour reduces our self-sufficiency and makes us dependent on others for the provision of our wants. Production too becomes a cooperative venture, both in the interests of efficiency and because more specialized skills are required. As our technical ability to create new commodities increases, so does the complexity of our needs. The labour process becomes ever more subdivided and the interrelationships deriving from mutual services more intricate. Hegel regards these developments as double-edged. On the one hand, he fully embraces the classical liberals' praise of market society as increasing individual liberty. To a certain extent he endorses their claim that the interrelatedness of the system of needs makes it self-regulating. Such duties as the obligation to obey promises, notions of fair exchange, bans on stealing etc. ... emerge within civil society itself, and he agrees with Hume that certain criteria of justice derive from the mutual self-interest of property owners in conditions of scarcity. He appropriately locates police functions within civil society. On the other hand, he does not believe that the needs of the market alone can lead to a well-ordered community. Hegel points to two potential sources of instability. First, he expands the insights of Smith, Adam Ferguson and Schiller's *Letters on the Aesthetic Education of Man* into the ennervating and alienating effects of modern industrial labour. With Smith's famous pin-factory example in mind, he notes the stupefying and mechanical nature of factory work, predicting that it will ultimately be taken over by machines. Second, he foresaw capitalism's propensity for periodic crises of overproduction. The business classes' uncontrolled pursuit of conspicuous consumption leads them to produce more goods than there are consumers. The bottom falls out of the market and workers, who, because of the extent of the division of labour, rely entirely upon this single commodity for their employment, will lose their livelihood. This group becomes 'a rabble of paupers' (Hegel, 1821, para. 244) outside of society and unprovided for by Humean economic justice.

Hegel gave the problem of poverty considerable thought and he dwells on it in a number of writings. He suggests two solutions, state charity funded by taxation and the direct creation of employment by state interference in the economy. He rejected the latter as merely exacerbating the problem, since overproduction was its root cause. The former, whilst more appealing, is equally inadequate. He notes that poverty is relative as well as absolute, and that charity can therefore create a stigma which increases the inferiority of the recipients and undermines their self-respect. Hegel's solution was to introduce a political dimension into social decision making. He parts company with classical political economists here, maintaining that our understanding of the true nature of society is incomplete as long as we remain within the restricted perspective of the market mentality. Like the more mercantilist Steuart, he contends that an awareness of our mutual obligations can grow through membership of occupational associations (*Korporation*) and social groups (*Stände*). He advocated a system of indirect democracy, whereby representatives from these bodies are sent to a national assembly which can enact social legislation. Hegel maintained that participation within these institutions would moderate the individualist self-seeking which led to economic crises. People would appreciate their mutual debts, implicit in capitalist production, and alter their behaviour accordingly to further the common good of the whole community.

Some commentators have regarded this solution as a sleight of hand (Avineri, 1972 Plant, 1977). Following Marx, they regard Hegel as having correctly expounded the contradictions of capitalist society, but assert he has merely carried them into the political sphere by enfranchising functional groups rather than individuals. Ending poverty requires the radical restructuring of productive relations demanded by communism (Marx, 1845). Hegel failed to make this step because he limited the philosopher's task to understanding society rather than changing it. However, Hegel's purpose was to preserve modern individuality. For him Marxism would have represented an unacceptably anachronistic return to the organic communities of the past.

Liberals also dispute Hegel's political response. They accuse him of subverting liberty by imposing a corporate mentality upon the free transactions of individuals within society. This misunderstanding of Hegel's intentions stems from their view of the relation of society to the state. Whereas liberals regard the state as merely providing the minimal means necessary for our pursuit of our private projects, without fear of undue hindrance from others, Hegel defines it in terms of certain shared ethical norms presupposed by all our activities. The public sphere is not the outcome of individual choices but what is presumed by them, the medium within which they are formed. This is the ethical life or *Sittlichkeit* of a community, which the state represents and upholds.

Whilst the corporatist policies of fascist authors, such as Giovanni Gentile (1875–1944), seem to justify the fears of both liberal and Marxist writers, others have understood him better. The British idealists in particular, such as T.H. Green (1836–82) and Bernard Bosanquet (1840–1923), shared his concern with poverty and suggested schemes for the state regulation of industry, education and poor relief which provided the intellectual origins for later proposals for the welfare state. Like Hegel, they regarded social and political institutions as instrumental in fostering an awareness of the complex of mutual rights and duties necessary for the adoption of such policies. They were similarly ambivalent about the degree to which poverty arose from a weakness of will on the part of the poor or social conditions. Nevertheless, an unresolved paradox persists in Hegel's theory. He claims community is an unconscious presupposition of maximizing individuals in commercial society, but it is not at all obvious how the market would operate once people become conscious of this fact and adopt the community-minded behaviour Hegel believed they would. Clearly civil society would then be thoroughly politicized; whether or not with the dire consequences liberals fear, or in a self-contradictory manner as Marx opined, is beyond the competence of this article to judge.

R.P. Bellamy

See also DIALECTICAL MATERIALISM.

BIBLIOGRAPHY
Avineri, S. 1972. *Hegel's Theory of the Modern State*. Cambridge: Cambridge University Press.
Bosanquet, B. 1899. *The Philosophical Theory of the State*. London: Macmillan.
Chamley, P. 1963. *Economie politique et philosophie chez Steuart et Hegel*. Paris: Presses Universitaires de France.

Gentile, G. 1946. *Genesis and Structure of Society*. Trans H.S. Harris, Urbana: University of Illinois Press, 1960.
Green, T.H. [1878–80.] *Lectures on the Principles of Political Obligation*. Ed. Paul Harris and John Morrow, Cambridge: Cambridge University Press, 1986.
Hegel, G.W.F. 1793–1800. *Early Theological Writings*. Trans. T.M. Knox, Chicago: Chicago University Press, 1948.
Hegel, G.W.F. 1807. *The Phenomenology of Spirit*. Trans. A.V. Miller, Oxford: Oxford University Press, 1977.
Hegel, G.W.F. 1817–19. *Die Philosophie des Rechts: Die Mitschriften Wannenman (Heidelberge 1817/18) und Homeyer (Berlin 1818/19)*. Ed. K.-H. Itling, Stuttgart: Keltt-Cotta, 1983.
Hegel, G.W.F. 1818–31. *Vorlesungen über Rechtsphilosophie, 1818–31*. 4 vols, ed. K.-H. Itling, Stuttgart-Bad Cannstatt: Frommann-Holzboog, 1973–4.
Hegel, G.W.F. 1821. *Philosophy of Right*. Trans. T.M. Knox, Oxford: Oxford University Press, 1958.
Hegel, G.W.F. 1822–30. *Lectures on the Philosophy of World History: Introduction*. Trans. H.B. Nisbet, Cambridge: Cambridge University Press, 1975.
Marx, K. 1843. *Critique of Hegel's Philosophy of Right*. Trans. J. O'Malley, Cambridge: Cambridge University Press, 1970.
Pelczynski, Z.A. (ed.) 1984. *The State and Civil Society: Studies in Hegel's Political Philosophy*. Cambridge: Cambridge University Press.
Plant, R. 1977. Hegel and political economy. *New Left Review* No. 103, 79–92; No. 104, 103–13.
Popper, K. 1945. *The Open Society and its Enemies*, Vol. 2. London: Routledge & Kegan Paul.
Vincent, A. and Plant, R. 1984. *Philosophy, Politics and Citizenship: The Life and Thought of the British Idealists*. Oxford: Blackwell.

Helfferich, Karl (1872–1924). Helfferich was an economist with a particular expertise in currency problems; at times, he was also a civil servant, a banker and a politician. He was born in Neustadt/Palatinate and died in a railway accident in Bellinzona, Switzerland. Helfferich studied in Munich, Berlin and Strasbourg, where he took his PhD (1894). In the heated discussion during the years between 1895 and 1901 over whether or not Germany should stay with the gold standard or move to bimetallism, he fought vigorously for the former position. In 1899 he became a lecturer at the University of Berlin. From 1901 to 1906 he was in the Colonial Department of the Foreign Office in charge of currency and transport matters in the German colonies of that time. He then joined the Deutsche Bank, first in a high position in Istanbul and later as director in Berlin. Early in 1915 Helfferich became the secretary of state in the German Treasury Office. In financing the war, he made recourse far less to additional taxes than to borrowing, including borrowing from the Reichsbank – a method which was strongly criticized later on because of its inflationary consequences. In the following year Helfferich took the same post in the Office of the Interior, from which he resigned one year later. From 1920 until his death, Helfferich was a member of the German Reichstag and strongly influenced the policy of the Deutschnationale Volkspartei.

As a scholar he taught and wrote primarily on monetary and currency matters. But he also did some substantial work in other economic fields, such as trade policy, national income and wealth, and in politics. His most important scientific publication is the book *Das Geld*, which between 1903 and 1923 went into six editions (English edn, 1927). It was one of the best textbooks of its time covering in a very systematic way historical, theoretical, organizational and political issues.

Most important was Helfferich's role in the German currency reform of 1923. It was he who invented the idea by introducing an auxiliary currency (originally the *Roggenmark*, finally the *Rentenmark*) to provide for a stable-value legal tender as well as the stabilization of the Mark. This combination of aims and the restoration of confidence in the new currency by making the *Rentenmark* redeemable in *Rentenbriefe* (which were issued on the basis of the agricultural and industrial property) were the decisive conditions for the success of the currency reform of 1923. In 1923 Helfferich was recommended for the presidency of the Reichsbank, but for political reasons the opportunistic Dr Schacht was given the position.

K. SCHMIDT

SELECTED WORKS

1898. *Die Reform des deutschen Geldwesens nach der Gründung des Reiches*. 2 vols, Leipzig: Duncker & Humblot.
1901. *Handelspolitik*. Leipzig: Duncker & Humblot.
1903. *Des Geld*. 6th edn, Leipzig: Hirschfeld, 1923. English edition as *Money*; trans. L. Infield and ed. with Introduction by T.E. Gregory, London: Benn, 1927; New York: Adelphi, 1927.
1913. *Deutschlands Volkswohlstand 1888/1913*. 5th edn, Berlin: Stilke, 1923.

BIBLIOGRAPHY

Lumm, K. von. 1926. *Karl Helfferich als Währungspolitiker und Gelehrter*. Leipzig: Hirschfeld.
Reichert, J.W. 1929. Helfferich, Karl. In *Handwörterbuch der Staatswissenschaften*, Supplement to 4th edn, Jena: Fischer.

Heller, Walter Wolfgang (1915–1987). Heller was born in Buffalo on 27 August 1915. He grew up in Seattle and Milwaukee, and graduated from Oberlin College. He received a doctorate in economics from the University of Wisconsin, where he studied with Harold M. Groves, who greatly influenced a generation of public finance scholars. He has spent his entire academic career as professor of economics at the University of Minnesota.

Heller has made important scholarly contributions to the study of public finance, but his major claim to fame was his highly successful term as chairman of the Council of Economic Advisers under Presidents John F. Kennedy and Lyndon B. Johnson from 1961 to 1964. Since leaving the government, he has been influential as a consultant and adviser to Presidents, Congress, and business. He has written widely on current economic developments, tax policy, and state-local finance, and is also known as a stimulating lecturer and commentator on economic policy issues. In 1974, he served as president of the American Economic Association.

Heller began his professional career as an expert on state and local taxation. He wrote his doctoral dissertation on the administration of state income taxes, and later originated the idea of federal revenue sharing with the states and local governments. The details of revenue sharing were developed by a task force appointed by President Johnson, but it was enacted by Congress only after it was recommended by President Richard M. Nixon in 1972. The revenue sharing legislation was extended until the end of September 1986.

During World War II, Heller moved to the Treasury Department, where he contributed to the development of tax policy to finance the war. In 1947–8, he was tax adviser to the US Military Government in Germany, where he played an important role in designing the currency and fiscal reforms that helped launch the postwar German economic revival. He also served as a consultant to the Treasury Department during the late 1940s and early 1950s. He has been a strong advocate of progressive taxation and was one of the first to recognize that unnecessary deductions and tax preferences narrow the income tax base, require higher marginal tax rates to raise the necessary revenues, and distort economic decisions.

As chairman of the Council of Economic Advisers, Heller supported innovative macroeconomic policies to promote economic growth and stability. He persuaded President Kennedy to propose a major tax cut to stimulate demand, advocated the enactment of an investment tax credit and liberalized depreciation allowances to increase investment incentives. His Council developed the first, and most successful, voluntary wage-price guidelines to help contain inflationary pressures as the economy moved to full employment.

Heller's Council pioneered fiscal analysis based on the concepts of potential gross national product – the output the economy would produce at full employment – and the full-employment surplus. It is also noted for its advocacy of the neoclassical Keynesian synthesis of fiscal and monetary policies required to achieve full employment and increase economic growth. To reach full employment, it proposed the use of stimulating budget and monetary policies. To increase growth at full employment, it stressed the need for a full-employment surplus and monetary ease to support private investment in plant and equipment, combined with public investments in education, research, and development. It also urged the dismantling of barriers to free trade among nations to achieve the benefits of international specialization and exchange.

As a result of the policies pursued by the Kennedy and Johnson administrations, the nation enjoyed a long period of economic growth and prosperity without inflation. From the fourth quarter of 1960 to the fourth quarter of 1964 (when Heller left his CEA post), US real GNP grew at an average annual rate of 4.9 per cent, consumer prices rose 1.2 per cent a year, and long-term federal bond yields never exceeded 4.2 per cent.

Heller combined his advocacy of sound economic policies with an understanding of the need to help the disadvantaged and underprivileged. He helped to persuade President Johnson to design and implement an antipoverty programme to provide economic opportunities for low-skilled workers and a decent income for those who cannot earn their own livelihood. 'We cannot relax our efforts to increase the technical efficiency of economic policy', he wrote in 1966. 'But it is also clear that its promise will not be fulfilled unless we couple with improved techniques of economic management a determination to convert good economics and a great prosperity into a good life and a great society.'

JOSEPH A. PECHMAN

SELECTED WORKS

1952. Limitations of the federal individual income tax. *Journal of Finance* 7(2), May, 185–202.
1959. (With Clara Penniman.) *State Income Tax Administration*. Madison: University of Wisconsin Press.
1966. *New Dimensions of Political Economy*. Cambridge, Mass.: Harvard University Press.
1968. A sympathetic reappraisal of revenue sharing. In *Revenue sharing and the City*. ed. H.S. Perloff and R.P. Nathan, Baltimore: Johns Hopkins University Press.
1969. *Monetary vs. Fiscal Policy*, a dialogue with Milton Friedman. New York: W.W. Norton.
1975. What's right with economics? *American Economic Review* 65(1), March, 1-26.
1976. *The Economy: Old Myths and New Realities*. New York: W.W. Norton.
1982. Kennedy economics revisited. In *Economics in the Public Service* (papers in honor of Walter W. Heller), ed. J.A. Pechman and N.J. Smiler, New York: W.W. Norton.

Helvetius, Claude Adrien (1715–1771). There are, for Helvetius, a certain number of fundamental points: the individual is led, spontaneously, to seek pleasure and to avoid pain, and this engenders self-esteem; having realized what his needs are, the search for objects able to satisfy them determines his behaviour; personal interest governs his decisions, and these vary as interests do according to individuals, the social environment and the era; education, custom and environment form the whole man; men would be equally happy if they could fill all the different moments of their lives agreeably.

The question to ask is if, and how, one can guarantee general happiness. To maintain universal contentment there has to be a reciprocal dependence between all the members of society; that is to say, they should all be 'equally' occupied, or work should be 'equally' divided amongst them. For this to be the case, there must not be too large an inequality of wealth, condemning some to deprivation and excessive work whilst others are corrupted by luxury. This is all the more true today, as people almost everywhere are divided into two classes, one of which lacks necessities whilst the other has too much and consequently grows bored.

The only way to proceed is greatly to increase the number of landowners and therefore to redistribute land. This is always a difficult step to take as it constitutes a violation of a sacred right, the right of ownership.

It is the government which, to a large extent, is responsible for the happiness of the individual. It can and must 'mould' men and take every possible measure to secure for them the equality of happiness which is their right. It must endeavour to reduce the wealth of some and increase that of others, ensure that the poor have property and combat concentrations of wealth by means of taxation and laws of succession. This would only be possible by making very gradual changes. Moreover, the legislator could, by means of a wise education, show men that they can be happy without being equally rich.

Helvetius was Farmer-General from 1738 to 1751, that is, one of the financiers entrusted by the monarchy with the task of collecting tax by means of outright payment. He was also one of the Encyclopédists, a group which included Diderot, d'Alembert and d'Holbach.

He was influenced by Locke. He preached the right to rational criticism in all matters. For him nothing is innate, everything is acquired. The individual is the integral product of his environment and circumstances, which is a sort of rudimentary materialism.

His first book, *De l'ésprit*, was condemned and burnt in 1759. Thereafter it was re-edited several times in London and Amsterdam, and was illegally brought into France, where it was widely read. Helvetius' ideas had an extremely important influence on Bentham and on the formation of utilitarianism; he was also to influence J.S. Mill and Beccaria in Italy. He was translated into German and read in Russia, and praised by Marx for having emphasized the determining role of social conditions in the development of humanity. Curiously, it could be said that his thought has been forgotten during the last sixty years.

J. WOLFF

SELECTED WORKS

1758. *De l'ésprit*. Paris.
1795. *De l'homme, de ses facultés intellectuelles et de son éducation.* (Posthumous.)
The complete works of Helvetius were published by Editions Didot, Paris, in 1795 and by Editions Lepetit, Paris, in 1818.

BIBLIOGRAPHY
Garaudy, R. 1948. *Les sources françaises du socialisme scientifique.* Paris: Editions Sociales.
Horowitz, I.L. 1954. *C. Helvetius.* New York: Paine, Whitman.
Keim, H. 1907. *Helvetius – sa vie et son oeuvre.* Paris: Alcan.
Lichtenberger, H. 1895. *Le socialisme au XVIIIème siècle.* Paris: Alcan.

Henderson, Alexander (1914–1954). Described by J.R. Hicks as one of the most brilliant students in Cambridge in the 1930s, Henderson's professional career was cut short by a long spell of war service (1940–45) and by his early death at the age of 39. He held professorial appointments at the University of Manchester (1949–50) and Carnegie Institute of Technology, Pittsburgh (1951–4). His major journal articles were in the field of microeconomics, the best known being a note in the *Review of Economic Studies* (1941) which markedly influenced Hicks's well known exposition of the meaning and measurement of consumer's surplus. Of more lasting interest, perhaps, is his development of public utility pricing theory in respect of the case where marginal cost pricing theory would require a public enterprise to make a loss. He argued that as a loss would have to be covered by a tax, the problem became one of choosing the 'best' tax or combination of taxes. Taxes were labelled 'good' if (a) they ensured that once an investment in a public enterprise had taken place it would be used by all who would be willing to pay the marginal cost; (b) they ensured that an investment would not be undertaken if its cost exceeded consumers' surplus; and (c) they would place the burden where political preferences would wish it to be put, meaning that if the distribution of income was optimal before the investment were undertaken, any tax should be levied on the users of the product of the investment. Applying these criteria he was able to make some trenchant criticisms of established views of the financing of public enterprise investment, notably concerning the two-part tariff system. He wrote on population problems, international trade and took part in the somewhat arid debate on the welfare effects of direct versus indirect taxes during the 1940s. He also co-authored with Charnes and Cooper (1953) one of the best known earlier texts on the application of linear programming to economic problems.

ALAN PEACOCK

SELECTED WORKS
1941. Consumer's surplus and the compensating variation. *Review of Economic Studies* 8, February, 117–21.
1947. The pricing of public utility undertakings. *Manchester School of Economics and Social Studies* 15, September, 223–50.

BIBLIOGRAPHY
Charnes, A., Cooper, W.W. and Henderson, A. 1953. *An Introduction to Linear Programming*, New York: Wiley.

Henderson, Hubert Douglas (later Sir Hubert) (1890–1952). Henderson was born of a Scottish family. Educated at Rugby School and Cambridge, he began his university studies as a not very successful mathematician but then changed over to economics and at once found his metier. He was placed in the first class with Dennis Robertson and two others in 1912, at a time when Cambridge economics had become a very lively school, very much in the hands of a younger generation, with Pigou as a very young professor and Maynard Keynes, Walter Layton and Ryle Fay as active young lecturers.

Like most of them, Henderson was drawn off into wartime activities. Unfit for military service, he was first in the Board of Trade and subsequently in the Cotton Control Board, whose history he later wrote. After the war he retired to Cambridge with a fellowship at Clare College, lecturing ostensibly on monetary problems but in practice, to the enjoyment of my own generation of undergraduates, on the economic problems of the moment. In this period he wrote the small book *Supply and Demand*, which for thousands of English students during the following 30 years was their first introduction to economics. But he was never by choice an economic theorist and in later life apt to be out of touch with the latest theoretical developments.

In the Cambridge of the 1920s Henderson, with Keynes and others, was in the thick of the re-thinking of Liberal economic policies with Lloyd George as figurehead. When a group of Liberals acquired in 1923 the weekly *Nation and Athenaeum* Henderson, with Keynes as his chairman, became its editor. For the next seven years the *Nation* under his editorship was compulsive reading for every political economist. He might discuss with Keynes, but it was always Henderson who wrote. This, it seems clear in retrospect, was the peak of his career and the job he did best.

In 1930 Henderson was persuaded to give up the *Nation* to become the chief economist of the Economic Advisory Council, then newly created by Ramsay MacDonald's Labour government. He was faced by an impossible task at an impossible time. Britain, saddled by Winston Churchill's decision when Chancellor of the Exchequer to return to the prewar gold standard, was struggling with the hopelessly inconsistent tasks of deflating to achieve that and simultaneously expanding to overcome a mountain of unemployment. It was not Henderson's fault that despite the ingenuities of Keynes and the debates of countless committees they failed to do so. It was the fault of a generation of politicians who could not be persuaded to grasp the nettle. But these years of frustration left Henderson a different man. He was no longer the crusading optimist. He had become the eternal critic, with a duty to ensure that no one should ever overlook any possible difficulties of any proposed source of action.

The outbreak of war in 1939 found him a member with Lord Stamp and Henry Clay of a committee to examine the war plans of government departments and more generally the problems of the war economy. When, soon after Churchill became Prime Minister in 1940 this came to an end, Henderson was absorbed into the Treasury with no very specific responsibility. For his period there he was the arch-critic, always engaged, as has been said, in detecting difficulties, and something of a discouragement to those who were trying to design policies for a better world. Administration, the achievement of consensus around the best practicable answer, was not his role.

In 1944 he was offered and gladly accepted a special research fellowship at All Souls College, Oxford; a year later he was elected to the long-established Drummond Professorship of Political Economy in the University of Oxford. He was back in the atmosphere in which he was completely happy. He could forget the problems of consensus. He could be right in a minority of one. He had enthusiastic undergraduate audiences to hear his views on the interwar years. He was by now out of touch with the theories, not only of his very able younger Oxford colleagues but also of Keynes and his own contemporaries. But in the vigorous argument of an Oxford common room he had few equals. Shortly before his death early in 1952 he had been elected Warden of All Souls. He did not live to take up the office.

E.A.G. Robinson

heredity. *See* NATURAL SELECTION AND EVOLUTION.

Herfindahl index. This is one form within the species of 'comprehensive' indices of market structure, and is used in industrial economics to suggest the degree of monopoly power. The Hirschman–Herfindahl Index (HHI) is the sum of the squared values of all firms' market shares in a given market. If shares are measured from 0 to 1.0, the HHI ranges from minimal to 1. If the shares are taken as per cent values from 0 to 100, then the HHI ranges from minimal to 10,000.

The index first acquired the name of Orris C. Herfindahl (an energy economist) in the 1950s, but Albert O. Hirschman used the index earlier in assessing foreign trade patterns, hence the dual name. Its users note that it is comprehensive, while the standard concentration ratio covers only the leading firms. The ratio gained a certain technical vogue in the 1980s, but has not displaced concentration ratios as the mainstream basis for estimating the degree of market power.

The HHI presents three problems. First, as a pure number it lacks content. Users must translate it into equivalent 'real' concentration ratios, in order to convey its possible meaning. Thus a 1000–2000 HHI range has no intrinsic meaning. It is (*very*) roughly comparable to four-firm concentration ratios of 50 to 80, and that is the way in which the ratios have come to be evaluated. Second, the HHI's data requirements are heavy. If market shares are known for individual firms, those details are the key facts to use, rather than to submerge them in a single index. Finally, the weighting of shares by an exponent of 2 (or any other specific value) has no basis in theory or empirical patterns. As one result, the upper ranges of market shares give very high HHI values (e.g. a firm with 70 per cent of the market has, by itself, an HHI of 4900). Such high numbers may correctly reflect an extreme degree of monopoly power held by dominant firms, but the issue has not been researched. HHI users have preferred to look only at oligopoly patterns in the lower HHI ranges of 1000 and 2500.

Other comprehensive indexes ('entropy', 'numbers equivalent', etc.) offer variations on the HHI in the hopeful quest for a single 'best' index. All of these technical variations suffer from problems of lack of content, burdensome data needs, and debatable weighting. None of them is likely to displace the standard concentration ratio for mainstream analytical purposes.

WILLIAM G. SHEPHERD

See also CONCENTRATION RATIOS; INDEX NUMBERS; MARKET SHARE.

BIBLIOGRAPHY
Hirschman, A.O. 1964. The paternity of an index. *American Economic Review* 54, September, 761.
Scherer, F.M. 1970. *Industrial Market Structure and Economic Performance*. 2nd edn, Chicago: Rand-McNally, 1980.

Hermann, Friedrich Benedict Wilhelm von (1795–1868). Hermann was born in Dinkelsbuhl, Germany. his career spanned the half-century or more in which German economics came to terms with English classical political economy, first welcoming it and then rejecting it, particularly in its Ricardian variety. After teaching mathematics in a secondary school, Hermann was appointed to the chair in what was still called *Kameralwissenschaften* (Cameralism) – an old title soon to be discarded – at the University of Munich in 1827. He made his reputation with *Staatswirthschaftliche Untersuchungen* (Investigations into Political Economy) (1832), a book which owed much to *The Wealth of Nations* but little to the writings of either Malthus or Ricardo. The book was organized around the simple but appealing idea that all economic variables are the outcome of the forces of demand and supply, so that economic analysis consists essentially of an investigation of the factors lying behind demand and supply. The book revelled in endless definitions and classifications of types of goods, wants, costs, capitals, and so on, but did not clutter the analysis with endless attacks on the deductive method of the English school. Together with Rau (1792–1870), Hermann thereby laid the foundations on which Mangoldt (1824–68) and Thünen (1783–1850) were soon to build a German brand of classical economics. No wonder Marshall much admired 'Hermann's brilliant genius' and frequently quoted Hermann's treatise in his own *Principles of Economics* (1890).

Hermann became a Director of the Bavarian Statistical Bureau in 1839 and organized the first official life table covering an entire German state. As a member of the Frankfurt Parliament in 1848, he advocated the unification of all German states.

MARK BLAUG

Herskovits, Melville Jean (1895–1963). Herskovits was born in Bellefontaine, Ohio, and died in Evanston, Illinois. He studied history at the University of Chicago (BA, 1920) and anthropology at Columbia University (PhD, 1923) as a student of Franz Boas. He taught at Columbia and Howard universities before going to Northwestern in 1927, where he spent the rest of his academic career. Herskovits did anthropological fieldwork in West Africa, the Caribbean and Brazil, and was among the first American anthropologists to specialize in African societies as well as blacks in the Caribbean and the US. He started the first Program of African Studies in the US, at Northwestern.

Herskovits was an early contributor to the field of study now established as economic anthropology. The first edition of his book on this topic was called *The Economic Life of Primitive Peoples* (1940), the revised edition being *Economic Anthropology* (1952).

Herskovits is best remembered by economic anthropologists for his views on a theoretical issue of importance that arose in his controversy with Frank Knight, who reviewed the 1940 edition of Herskovits's book. In the 1940 edition, Herskovits criticized the conventional economics of Marshallian microtheory for its uselessness to anthropologists trying to understand the underlying principles which explain the working of primitive economies – such as African tribal economies not yet changed by European colonial rule – primitive economies lacking capitalism's core attributes of machine technology, modern money, and market organization for the transaction of inputs and outputs. In his book review, Frank Knight criticized Herskovits for misunderstanding the 'abstract' and 'intuitive' nature of economic theory. (I doubt that Knight's portrayal of economics, as stated there, would be shared today by many economists.) Knight's review, together with a rejoinder by Herskovits, are reprinted in *Economic Anthropology* (1952).

The relevance of conventional economic theory to the analysis of pre-industrial, non-capitalist economies remains an unresolved issue to this day. It is an issue much more important today because of the much greater interest now in the study of early and primitive economies, and in the study of the large, diverse set of developing economies in the Third World. This inability to agree on the relevance of conventional economics to the analysis of non-market economies finds expression in economic anthropology's literature of acrimoni-

ous theoretical dispute and in the existence side by side of three radically different theoretical systems all employed by archaeologists, anthropologists and historians to analyse non-capitalist economies: Formalism (i.e. conventional microeconomic theory); Marxism; and Substantivism (i.e. Karl Polanyi's system of analysis described in his *Trade and Market in the Early Empires*, 1957, and his *Primitive, Archaic, and Modern Economies*, 1971).

GEORGE DALTON

See also ECONOMIC ANTHROPOLOGY; KNIGHT, FRANK HYNEMAN; POLANYI, KARL.

SELECTED WORKS

1940. Anthropology and economics. In *The Economic Life of Primitive Peoples*, New York: Knopf.
1941. Economics and anthropology: a rejoinder. *Journal of Political Economy* 49, 269–78.
1952. *Economic Anthropology*. New York: Knopf.

BIBLIOGRAPHY

Dalton, G. 1961. Economic theory and primitive society. *American Anthropologist* 63, February, 1–25.
Dalton, G. 1969. Theoretical issues in economic anthropology. *Current Anthropology* 10, February, 63–102.
Knight, F. 1941. Anthropology and economics. *Journal of Political Economy* 49, 247–68.
Martin, J. and Knapp, K. (eds) 1967. *The Teaching of Development Economics*. Chicago: Aldine Press.
Simpson, G.E. 1973. *Melville J. Herskovits*. New York: Columbia University Press.

heterogeneous capital models. *See* 'HAHN PROBLEM'.

heteroskedasticity. One of the basic assumptions of the classical regression model

$$Y_i = \beta_1 + \beta_2 X_{i2} + \cdots + \beta_K X_{iK} + \epsilon_i \qquad (i = 1, 2, \ldots, n)$$

is that the variance of the regression disturbance ϵ_i is constant for all observations, that is, that $\mathrm{Var}(\epsilon_i) = \sigma^2$ for all i. This feature of ϵ_i is known as *homoskedasticity* and its absence is called *heteroskedasticity*. The homoskedasticity assumption is quite reasonable for observations on aggregates over time, since the values are of a similar order of magnitude for all observations. It is, however, implausible with respect to observations on microeconomic units such as households or firms included in a survey, since there are likely to be substantial differences in magnitude of the observed values. For example, in the case of survey data on household income and consumption, we would expect less variation in consumption of low-income households, whose average level of consumption is low, than in consumption of high-income households, whose average level of consumption is high. Empirical evidence suggests that this expectation is in accord with actual behaviour. Heteroskedasticity also arises when the data are in the form of group averages and the groups are of unequal size.

Heteroskedasticity has two important consequences for estimation: (1) The least squares estimators of the regression coefficients are no longer efficient or asymptotically efficient. (2) The estimated variances of the least squares estimators are, in general, biased, and the conventionally calculated confidence intervals and tests of significance are invalid. The second of these consequences is more serious than the first since inefficiency of estimation can be compensated for by a large number of observations.

The deficiencies of the least squares estimation can be remedied by adopting a *weighted* (or *generalized*) least squares procedure. This method involves weighting each observation by the reciprocal of the respective standard deviation of the disturbance, and then applying the least squares method to the transformed equation

$$(Y_i/\sigma_i) = \beta_1(1/\sigma_i) + \beta_2(X_{i2}/\sigma_i) + \cdots + \beta_K(X_{iK}/\sigma_i) + u_i,$$

where

$$\sigma_i = \sqrt{\mathrm{Var}(\epsilon_i)} \quad \text{and} \quad u_i = \epsilon_i/\sigma_i.$$

The difficulty with the weighted least squares method is that its implementation requires knowledge of σ_i, which is rarely available. This difficulty is usually overcome by making certain assumptions about σ_i or, when possible, by estimating σ_i. The assumptions typically involve associating σ_i with some variable Z_i, normally represented by one of the explanatory variables of the regression equation. For instance, in a microconsumption function the variance of the disturbance is frequently positively associated with income. In general, two forms of association between σ and Z have been proposed in the literature and applied in practice: a *multiplicative* and an *additive* form. Multiplicative heteroskedasticity – which is more common – can be described as

$$\sigma_i^2 = \sigma^2 Z_i^\delta,$$

where σ and δ are parameters to be estimated. A frequent representation of additive heteroskedasticity is

$$\sigma_i^2 = a + bZ_i + cZ_i^2,$$

where a, b, and c are parameters to be estimated. Estimation of the parameters involved in the specification of σ_i can be carried out simultaneously with the estimation of the regression coefficients by using the method of maximum likelihood. No assumptions about the form of heteroskedasticity are necessary where σ_i can be estimated from replicated data which, unfortunately, are rather rare in applied economic research.

The presence or absence of heteroskedasticity may be subjected to a test. Several suitable tests, some developed only recently, are available and are described in recent econometric texts.

The problem of heteroskedasticity and its consequences was brought to the attention of applied economists by two seminal research monographs, Stone (1954) and Prais and Houthakker (1955). The subject has been further developed by a number of econometricians and is now standard fare in all introductory courses of econometrics; see, for example Kmenta (1986).

J. KMENTA

See also LEAST SQUARES; REGRESSION AND CORRELATION ANALYSIS; RESIDUALS.

BIBLIOGRAPHY

Kmenta, J. 1986. *Elements of Econometrics*. 2nd edn, New York: Macmillan Publishing Co.

Prais, S.J. and Houthakker, H.S. 1955. *The Analysis of Family Budgets*. Cambridge: Cambridge University Press.

Stone, J.R.N. 1954. *The Measurements of Consumers' Expenditure and Behaviour in the United Kingdom, 1920–1938*. Vol. 1, Cambridge: Cambridge University Press.

Hicks, John Richard (1904–1989)

1 BIOGRAPHY AND INTELLECTUAL DEVELOPMENT. Hicks was born in Warwick. He studied at Oxford (1922–6) and taught at the London School of Economics (1926–35). He was Professor at Manchester University (1935–46) from where he moved to Oxford, first as Fellow of Nuffield College, and from 1952 until he retired, from teaching but not from writing, as Drummond Professor of Political Economy and Fellow of All Souls College. In 1935 he married Ursula Webb, a distinguished public finance specialist, and he collaborated with her in the preparation of numerous works on public finance, its theory and its application to various countries. Ursula Hicks, as she was subsequently known, died in 1985. John Hicks was a member of the Royal Commission on the Taxation of Profits and Income in 1951. He became a Fellow of the British Academy in 1942, a Knight in 1964 and was awarded the Nobel Prize in Economics (jointly with Kenneth J. Arrow) in 1972. He died in 1989.

Hicks was the product of a generation which was the last to produce in abundance all round economic theorists – economists who could turn their minds to almost any theoretical problem. Its leading lights, among whom Hicks is certainly to be counted, left their marks on most of the major new branches and issues of economics as these in turn attracted the interest of themselves and their contemporaries. Hicks's powerful and original mind first made itself felt in what is now called microeconomics, particularly in *The Theory of Wages* (1932, 2nd edition 1963) and with R.G.D. Allen, 'A reconsideration of the theory of value' (*Economica*, 1934) and in welfare economics. However his best-known work, *Value and Capital* (1939), goes beyond microeconomics to offer an economic dynamics and discussion of monetary theory which reaches into the new macroeconomics.

Before Keynes's *General Theory* fundamentally altered the way in which economists viewed their subject, the theory of value, including the theory of the firm, shared the field with monetary theory. Hicks was first a value theorist but he never neglected monetary theory, and it was an area to which he was frequently to return. It was a value theorist with an interest in monetary economics who provided in 'Mr Keynes and the classics' (*Econometrica*, 1937) an exposition of Keynes's *General Theory* that was probably more directly influential than the original. There followed work on the trade cycle, *A Contribution to the Theory of the Trade Cycle* (1950); on growth, *Capital and Growth* (1965); and an unusual approach to capital theory, *Capital and Time: a New Austrian Theory* (1973).

Each decade of Hicks's life seemed to find him more eclectic and innovative than the last. Indeed, his willingness to speculate about and write on areas in which he had not seeped himself as a specialist was a notable feature of his later writing. Striking examples are *A Theory of Economic History* (1969), in which Hicks undertook the risks inherent in proposing a grand theory of economic history, and *Causality in Economics* (1979), in which he entered ground normally reserved for philosophers and statisticians. These works can be criticized but as their author always commands a well-provisioned base camp in the economics which is his own, they are never merely amateurish. Hicks is an economist of outstanding breadth and erudition.

With hindsight it is remarkable that the author of such a formidable theoretical corpus should write ('Commentary' in the 1963 edition of *The Theory of Wages*, p. 306): '... at first I regarded myself as a labour economist, not a theoretical economist at all'. Lionel Robbins is given the credit for interesting Hicks in theory: '... he moved me from Cassel to Walras and Pareto, to Edgeworth and Taussig to Wicksell and the Austrians – with all of whom I was more at home at that stage than I was with Marshall and Pigou' (ibid.). It would be foolish to attempt to explain why Hicks became the distinctive economist that he was to become. However the above snatches of autobiography probably go some way to explaining why *Value and Capital* turned out to be a book like no other that an English economist had written before.

Hicks's huge output (for the papers see the three-volume *Collected Essays on Economic Theory*, 1981–3) is all the more remarkable when one considers that he seldom simply reacted to the work of others. There are no papers by Hicks pointing out mistakes by other writers and none which embody minor changes to or extensions of existing models. Naturally Hicks produced work which follows paths opened up by others. However when he did so, as in *A Revision of Demand Theory* (1956), or with the famous IS/LM model, his approach was so distinctive that the commentary is recognizably a contribution of Hicks. Other writers feature mainly in footnotes and even such a powerful contribution as Samuelson's treatment of Walrasian stability earns no more than two pages in the Second Edition of *Value and Capital*. There is a streak of self-centredness and parochialism in Hicks which mirrors that to be found in other English economists of his generation and those before. It would be insufferable in an economist less gifted and genuinely self-critical.

2 THE THEORY OF WAGES. Writing later (1963) of the First Edition of *The Theory of Wages* its author remarks that '... there has been no date this century to which the theory that I was putting out could have been more inappropriate.' However, Hicks was careful not to attribute the shortcomings of his first book to the misfortune of publishing in the worst year of the depression and a few years ahead of the reassessment of the theory of the firm brought about by the writings of Chamberlin and Joan Robinson and, worse fortune still, ahead of the *General Theory*. In this he was right. *The Theory of Wages* set out to examine the determination of wages under supply and demand in a competitive market. This admittedly limited task is not without importance and had it been perfectly accomplished it would not be sensible to criticize the resulting work for not solving other problems, such as wages under imperfect competition or the consequences of nominal wage bargaining, important though those problems might be. However the truth is that there were shortcomings in Hicks's treatment even given its chosen emphasis. It was not as good a book as Hicks was later to show that he could write, though it was surely a better book than the later Hicks's embarrassment at its shortcomings allowed him to admit.

G.F. Shove (whose fairly hostile review Hicks reprinted in the Second Edition) identified a number of the shortcomings. Notable among these is the relatively weak treatment of the supply side of labour markets and the consequently limited ability to treat unemployment. Shove also seems to accuse Hicks of failing to provide a treatment of the general equilibrium of many labour markets, which must be counted a rather common failing among labour economists. Shove, not surprisingly, was clear on minimum cost and the adding-up problem where Hicks's account needed improvement – it was after all Shove's bread and butter at the time. A point which Shove missed is that Hicks always discussed differences in the productivity of different workers as equivalent to differences in the quantity of effective labour provided per hour of work. In other words, like Marx before him, he fudged the problem of aggregating different types of labour.

These legitimate criticisms apart, there were very considerable merits. By concentrating on the long-run determinants of wage rates Hicks was able to examine some of the most interesting influences at work. He saw changes in the demand for labour as consisting of two components quite analogous to the income and substitution effects in demand that he was to investigate later. A lower wage rate leads to an expansion of output, because the cost curve has fallen, which induces a higher demand for labour. In addition a lower wage rate induces the adoption of more labour intensive methods of production, which increases the demand for labour for a given output. The analysis of this last effect lead to the discovery of the new concept of the elasticity of substitution, not quite as neat in Hicks's formulation as in Joan Robinson's later presentation, but this was the original. In general, Hicks's definition of the elasticity of substitution is different from Joan Robinson's, but the two are equivalent in the two-factor case.

Many topics discussed only briefly and not deeply analysed were far ahead of their time. There is the idea that because capital tends to accumulate faster than labour, technical progress tends to be labour saving – the induced bias of technical progress as we would now say. There is the first ever attempt to model a labour dispute which may culminate in a strike, and more besides.

3 VALUE THEORY. This area and welfare economics are fields to which Hicks contributed the writings that would have made him a great economist if he had done nothing else. In making the 1972 Nobel Prize award to Hicks jointly with Arrow the Committee mentioned 'general equilibrium and welfare economics'. The reference in Hicks's case was clearly to *Value and Capital* on the one hand, and to the various papers which established the Kaldor–Hicks criterion in welfare economics on the other.

Hicks's paper with R.G.D. Allen, 'A reconsideration of the theory of value' (1934) was written when both authors were at the London School of Economics, but its pedigree goes back to Slutsky, who had discovered the income and substitution effects in demand as early as 1915. However Slutsky's work was almost entirely unknown to economists in the West, and this included, as Hicks informs us, himself and Allen ('... I never saw Slutsky's work until my own was very far advanced, and some time after the substance of these chapters had been published in *Economica* by R.G.D, Allen and myself' (1939, p. 19).

Value and Capital is a work so rich in ideas that a short account of it cannot hope to do it justice. It showed that the basic results of consumer theory could be obtained from ordinal utility; it expounded what became known as the 'Hicksian substitution effect', obtained by varying income as relative prices changed so as to maintain an index of utility constant; it developed the parallel results for production theory; and it popularized among English speaking economists the notion of a general equilibrium of markets. Unlike Arrow, his fellow Nobel laureate, Hicks did not take the existence argument beyond equation and variable counting. There was about the Walrasian approach, Hicks concluded, '... a certain sterility' (1939, , p. 60). The way to overcome this was to consider the 'laws of change' of a general equilibrium system. This lead Hicks to the first ever attempt to analyse the stability of a system of multiple exchange.

This is the same question as was examined by Paul Samuelson in various papers of the 1940s and in his *Foundations of Economic Analysis* (1947). However Hicks's method of tackling the problem was importantly different from the one later adopted by Samuelson. Consider a system of M markets with prices $p_1, p_2, \ldots, p_M$ and excess demands for the goods $X_1, X_2, \ldots, X_M$. Making the dependence of excess demands on all prices explicit, this system can be written as:

$$X_1(p_1, p_2, \ldots, p_M)$$
$$X_2(p_1, p_2, \ldots, p_M)$$
$$\ldots\ldots\ldots\ldots \quad (3.1)$$
$$X_M(p_1, p_2, \ldots, p_M)$$

In equilibrium prices are such that all excess demands are zero. Now consider one good, which may be taken without loss of generality to be good 1. Select any value for p_1 and suppose that there are unique values of the remaining prices such that the excess demands for goods 2 to M are zero. If the excess demands for the other goods are always maintained at zero by changes in their prices, all other prices become implicit functions of p_1. The Hicks stability condition is then the one that would be required of a single market – X_1 should decrease with p_1. Full stability requires that this condition should be satisfied for each good in turn.

At first sight the condition appears to be asymmetrical but as the condition must be satisfied by all goods, there is no genuine asymmetry involved. However each test does involve a certain kind of asymmetry, and this was what Samuelson objected to. When we look at good 1 we implicitly assume that prices in other markets react more rapidly to disequilibrium than does the price of good 1. When we look at good 2 we make the same implicit assumption for the price of good 2, and so on. What Samuelson did was to make the time rate of change of each price a function of the excess demand in its own market hence arriving at the system of simultaneous differential equations:

$$\dot{p}_1 = a_1 \cdot X_1(p_1, p_2, \ldots, p_M)$$
$$\dot{p}_2 = a_2 \cdot X_2(p_1, p_2, \ldots, p_M)$$
$$\ldots\ldots\ldots\ldots \quad (3.2)$$
$$\dot{p}_M = a_M \cdot X_M(p_1, p_2, \ldots, p_M)$$

The Hicksian stability condition can be shown to be neither necessary nor sufficient for the stability of (3.2). Hicks however defended his own approach, on the ground that it answers a different but interesting question, in the Second Edition of *Value and Capital* (Additional note C).

It was in Parts III and IV of *Value and Capital* that Hicks showed the full extent of his originality. In these Parts he adapted the static theory of the earlier parts to create an economic dynamics which borrowed equally from the Marshallian–Keynesian tradition of the short period and the Walras–Wicksell tradition of long-period equilibrium. The key idea was the concept of temporary equilibrium – an equilibrium of current markets in which future markets make their influence felt indirectly, through the expectations held by agents, which influence their behaviour in current markets. From this emerged the concept of the elasticity of expectations, an idea which proved to be crucial in much later work on macroeconomic theory.

4 WELFARE ECONOMICS. Hicks's writings on welfare economics are largely accounted for by work on four closely connected fields of interest: the foundations of welfare economics, including the famous compensation test; the valuation of social income; the definition and measurement of consumer's surplus; and, lastly, the measurement of capital.

Hicks was one of the pioneers of the 'new welfare economics', an approach which owed its inception to Kaldor's 'Welfare propositions in economics' (*Economic Journal*, 1939). The problem at issue is inescapable and fundamental to the justification of recommendations when these are advanced by economists. By the time the debate arose cardinal utility was

no longer generally accepted and the need was felt to differentiate between 'scientific' propositions and 'value judgements'. The notion of a 'Pareto improvement' – a change that would make no individual worse off and at least one better off – was familiar but was seen to be limited as a basis for recommendations, as nearly all actual changes made at least one person or group worse off. In Robbins's telling example, economists could not state scientifically that the abolition of the Corn Laws was a good thing because this reform made landlords worse off.

Hicks's suggested solution to the difficulty was the same as that proposed by Kaldor – a compensation test. A reform should be counted an improvement if the gainers could afford to compensate the losers and still be better off. In 'The foundations of welfare economics' (*Economic Journal*, 1939), Hicks discussed the question of whether compensation must be paid for the improvement to count without a sense of how crucial this question was to prove to be. It was of course central to the issue posed by the Scitovsky example, which showed that the Kaldor–Hicks rule could lead to contradictory recommendations if compensation were not paid. A well-argued solution to this problem was proposed by I.M.D. Little (1950), but this required explicit value judgements concerning whether income distribution had improved or not in a movement from one position to another, hence negating the original intention of the exercise, which had been to remove value judgements from welfare economics.

Hicks seemed to see these developments as fairly unimportant qualifications to the original idea. In 'The measurement of real income' (1958), he writes of the 'new welfare economists'; 'They were indeed over-confident in their belief that they had found a means of direct comparison which will always work. But I still maintain that they did find a means of direct comparison which will often work' (reprinted in *Collected Essays*, Vol. I, p. 168). For a statement of Hicks's mature views on these questions see 'The scope and status of welfare economics' (1975). Perhaps the most interesting thing to notice about Hicks's long involvement with the foundations of welfare economics is that he seems never to have wholly accepted the conclusion upon which the majority of economists have been willing to settle. Briefly put, this view says that value judgements are an inescapable element in welfare evaluations and this should be accepted and the judgements made explicit. Hence the design of policy by the means of the maximization of an explicit social welfare function – the welfare weights of cost-benefit analysis – never engaged Hicks's interest.

It is evident that the problem of the measurement of income is closely allied to the issue of welfare improvements and Hicks, as would be expected, contributed to this area as well. Hicks discussed social accounting in his text book *The Social Framework* (1942), and the valuation of social income in a paper of that title in *Economica* (1940).

Hicks concluded that the measurement of income could mean measurement in terms of utility or measurement in terms of cost, and that the two measures were in general different. The most interesting issue to which this gave rise was the problem of how to treat indirect taxation and government expenditure on goods and services in the valuation of social income. This led Hicks into controversy with Kuznets (*Economica*, 1948; see also Essay 7 in Volume I of the *Collected Essays*). The usual practice is to measure prices at factor cost and to value public services at cost. Hicks's original position may be briefly summarized as follows:

(i) As there is no market test where public goods are concerned the taxation which pays for them is not a reliable measure of their value to the consumer; and (ii) even if consumers were to be regarded as implicitly choosing public expenditure exactly as they choose private expenditure, the appropriate price weights would not be average costs but marginal costs. For a mature statement, see the Addendum to Essay 7 in Volume I of the *Collected Essays*.

Between 1941 and 1946 Hicks published a number of papers on consumers' surplus in the *Review of Economic Studies* that did much to revive interest in a concept which had seemed to lose its validity when measurable utility went out of fashion. His most important contribution to the controversial question of the measurement of capital, significantly entitled 'Measurement of capital in relation to the measurement of other economic aggregates', is in Lutz and D.C. Hague (1961).

5 THE KEYNESIAN REVOLUTION AND THE THEORY OF MONEY. Hicks's first response to the *General Theory* is described in detail in 'Recollections and documents' (*Economica*, 1973, included in *Economic Perspectives*, 1977). However the response for which he is best known was an expository piece 'Mr Keynes and the "classics" ' (1937) that perfectly fulfilled the innate demand for a more readily accessible account of the essentials of Keynes's argument. It is important to make clear that what was provided was more than an *haut vulgarisation* of Keynes, because the paper has been widely criticized for vulgarization and still more for seriously misrepresenting what the *General Theory* is about. This case has never been rigorously argued and it is hard to see how it could succeed. Hicks reproduced rather faithfully Keynes's various specifications, but by working with a two-sector model produced a framework which resulted in a simple diagram – the IS/LM diagram – which became to macroeconomic textbooks what the benzene ring diagram is to textbooks of organic chemistry. It is no surprise therefore that Keynes on reading the paper wrote to Hicks that he had '... next to nothing to offer by way of criticism'.

In fact Hicks's way of presenting the argument is in some ways superior to that adopted in the *General Theory* because the original IS/LM model brings out very clearly how the relative price of capital and consumption goods enters into the determination of the solution – a point which is somewhat obscure in Keynes. How ironic therefore that one of the arguments later advanced against the IS/LM model, admittedly with simpler versions than Hicks's in mind, was that it omitted an essential feature of Keynes – relative prices of capital and other goods.

Hicks's IS curve is based on the striking observation that if the capital stocks in the two sectors of the economy are given, and if the money wage is known, then outputs in the two sectors depend on the nominal prices of their products through short-term profit maximization conditions. Given these outputs and prices, the value of nominal total income follows. The output of the investment sector depends on the rate of interest through the marginal efficiency of capital relation, hence, given the rate of interest, the nominal price of the investment good follows and the part of income generated in that sector. Now choose an arbitrary value, which can be thought of as a guess at the level of total nominal income. As the part of nominal income generated in the investment goods sector is known, given the rate of interest, the guess implies a certain level of nominal income to be generated in the consumption good sector. We now have a value of total income and a value of total consumption, both in nominal terms. If these values are consistent with the consumption function our guess for the value of total income was correct

and we have discovered the level of income on the IS curve for the rate of interest with which we were working.

We have discussed only the IS curve but the LM curve is relatively uncomplicated – there is less going on behind it. The beauty of this elegant and lucid way of expounding Keynes's model is that it brings out clearly the vital role played in the model by aggregation assumptions which have the effect that the model decomposes, so that parts of it can be dealt with in partial isolation from the complete system. The simple specifications of the determinants of investment and the consumption function produce this result. The role played by income and working in terms of nominal values – which are equivalent to wage units, as the nominal wage has been taken as given – are all brought out clearly.

In the hands of others the IS/LM model often became merely a model of an economy with all prices fixed and was often misused, as when it was applied to long-run questions for which it is not suitable. However it made the *General Theory* intelligible to a whole generation, not because it left out the subtleties, it was never intended to substitute for the text, but because it perfectly captured the part of Keynes's message which is amenable to formalization.

A Contribution to the Theory of the Trade Cycle (1950) provides an example of the type of model that explains cycles as the outcome of the interaction between the multiplier and the accelerator. These systems are linear in their simplest formulations when they lead to cycles which are almost certainly either damped or anti-damped. Three different ideas have been proposed to yield an outcome in conformity to the stylized model of a capitalist economy with regular cycles of constant amplitude. The underlying solution may be anti-damped and buffers, in the form of a floor on or a ceiling to the level of economic activity, may be added to keep the solution within bounds. The system may be made non-linear, which is equivalent to buffers which make their influence felt continuously rather than abruptly. Finally, the underlying solution may be damped, in which case the cycle will have to be kept alive by the frequent intervention of random shocks. Hicks's main model embodies the last type of approach

From 1937 Hicks continued to write regularly on questions of macroeconomics. Volume II of his *Collected Essays* contains a selection of his best work in this vein. Essay 18, 'Methods of dynamic analysis', proposes the distinction between the fixprice and the flexiprice economy which was to be developed in *Capital and Growth*. In his Yrjö Jahnsson lectures, *The Crisis in Keynesian Economics* (1974), Hicks offers reflections on the Keynesian theory and particularly on the impact of inflation on a Keynesian model.

Hicks never remained far from monetary theory. *Critical Essays in Monetary Theory* (1967), shows the richness of his early writings on monetary economics, while Essay 19 in Volume II of the *Collected Essays* gives a good indication of his later work. It is tempting to say that if Hicks had written nothing but his work on monetary economics he would be counted a considerable economist. However the truth is that he could not have written on monetary economics as he did write had he not been the broad economic theorist that he was. Hicks always placed monetary theory centrally in equilibrium theory. This was the distinctive idea of his first paper on the subject, 'A suggestion for simplifying the theory of money' (*Economica*, 1935), and it is a theme which he was to carry through all his later work.

6 GROWTH AND CAPITAL THEORY. Hicks's two other books with 'Capital' in their titles, *Capital and Growth* (1965) and *Capital and Time: a New Austrian theory* (1973), have little else in

common. *Capital and Growth* was Hicks's response to the frantic interest in growth theory which infected the 1960s. It was a characteristically personal response in which Hicks tried to apply the framework for dynamic analysis that he had developed in *Value and Capital* to the construction of a growth model.

The analogue of the static problem of Part I of *Value and Capital* was now the steady state growth path, but once again Hicks found the most interesting question to be the dynamic adjustment to equilibrium, and once again he attacked this problem with an approach which was all his own. The 'traverse' was the history of the movement of an economy from one steady state to another. This approach to growth theory was not very influential and the reason was not so much that the new interest in growth had extinguished interest in equilibrium theory. Rather it was that equilibrium theory and its sister economic dynamics had moved on a great deal since *Value and Capital*. Hicks, who had taught a generation how to do general equilibrium economics, was no longer talking a language that most economic theorists found congenial.

Capital and Time was not the product of the latest fashion in economic theory but was surely the result of long meditation starting from that wonderfully fruitful comparative ignorance of Marshall and Pigou as against the Austrians and other continentals, which was noted above. Hicks always conceded a place to the old classical idea that capital accumulation means more 'waiting'. In *Value and Capital* (pp. 197–8) however, he pointed out that the conclusion that the rate of interest is the marginal product of waiting is a special case of more general rules which apply to an intertemporal equilibrium. This conclusion, that Austrian models of capital are special cases of the more general von Neumann model of capital accumulation, remains valid. However special cases permit of special results, and Hicks's analysis of the Austrian model was remarkably successful in showing how that framework permits some strong and definite conclusions to be drawn.

7 OTHER TOPICS. We consider only *A Theory of Economic History* (1969) and *Causality in Economics* (1979), as these constitute the most audacious of Hicks's expeditions far from the mainstream of economic theory. A longer review of Hicks's work would have to find space to discuss his writings on economic policy (for a sample of which see *Essays in World Economics*, 1959) and on the history of economic thought (for some of which see Volume III of the *Collected Essays*), but here we merely note that these are serious omissions from the present survey.

We begin with *Causality in Economics*. This book was not the eventual product of long years of mental rumination but the result, its author tells us frankly, of dissatisfaction with the 1974 International Economic Association conference on 'The Micro-foundations of Macroeconomics' which Hicks attended. It is book of interesting ideas on economics which are reluctantly regimented by a Sergeant-Major called 'causality'. This gentleman turns out to be only remotely related to the 'causation' of for example Aristotle or Kant. Hicks's definition of causality is reminiscent of Hume, but without the idea that the validity of induction is importantly involved. Causation is seen as conjunction of events, possibly in a complex form. This idea is an old one and was very effectively criticized by the Cartesians but their contribution is not considered. As an essay in philosophy *Causality in Economics* cannot be taken seriously. The economics of course is of a higher standard. The last chapter provides a statement of Hicks's views on the

meaning of probability and on econometric methodology. These are *obiter dicta* not the fruits of profound investigation.

A Theory of Economic History is as ambitious a sortie into foreign territory as *Causality in Economics*, but is the product of more thought and reading and must be regarded as much more successful. The main idea, that economic history is tied up with the development of the market, is one that few would question. However most historians would be tempted to take cover behind a safe position according to which developments of ideas, knowledge, social institutions, etc., would all be seen as progressing in parallel with the development of the market, which consequently would enjoy no special status as a motive force. Put simply, Hicks's account gives a much more leading role to the market, although he does not of course go so far as to argue that the market drives history.

Such a strong argument could not fail to attract criticism, particularly from professional historians. A long book would have done the same but a very short book was a particularly provocative target. As the argument gave a lot of attention to the ancient world this proved to be a contested area. However while *A Theory of Economic History* was criticized it received respectful criticism. It may be only a way of looking at economic history but it was generally judged to be a good way. Hicks's reply to his critics may be found in *Economic Perspectives* (1977, pp. 181–4).

8 RETROSPECT. Schumpeter argued that the ideas of a great economist are more or less in place by the age of 40 – the rest is nurturing and polishing. At first glance Hicks appears to be an exception. He was 65, for example, when his theory of economic history was announced to the world. Yet probably on closer examination he will be seen to conform to the Schumpeter pattern. In the case of the *A Theory of Economic History* he tells us in the foreword that he had nursed the idea for years. There is indeed a powerful sense of direction to Hicks's intellectual journey. He often returns to old themes and new themes are examined from older perspectives. Probably after 40 Hicks was only nurturing and polishing, but it is no contradiction of that claim to say that the second half of his life produced some of his most creative work.

It remains to mention two particular qualities of Hicks. First, he wrote beautifully, in a style that is very correct from the formal point of view yet almost conversational in its flow and ease. Secondly he was the most approachable of men, never pulling rank or flaunting his formidable distinction.

CHRISTOPHER BLISS

SELECTED WORKS

1932. *The Theory of Wages*. London: Macmillan. 2nd edn, London: Macmillan, 1963.
1934. (With R.G.D. Allen.) A reconsideration of the theory of value. Pts I–II. *Economica* 1, Pt I, February, 52–76; Pt II, May, 196–219.
1935. A suggestion for simplifying the theory of money. *Economica* 2, February, 1–19.
1937. Mr Keynes and the 'classics'. *Econometrica* 5, April, 147–59.
1939a. *Value and Capital*. Oxford: Clarendon Press.
1939b. The foundations of welfare economics. *Economic Journal* 49, December, 696–712.
1940. The valuation of the social income. *Economica* 7, May, 105–24.
1942. *The Social Framework*. Oxford: Clarendon Press.
1948. The valuation of the social income: a comment on Professor Kuznets' reflections. *Economica* 15, August, 163–72.
1950. *A Contribution to the Theory of the Trade Cycle*. Oxford: Clarendon Press.
1956. *A Revision of Demand Theory*. Oxford: Clarendon Press.
1958. The measurement of real income. *Oxford Economic Papers* 10, June, 125–62.
1959. *Essays in World Economics*. Oxford: Clarendon Press.

1961. Measurement of capital in relation to the measurement of other economic aggregates. In *The Theory of Capital*, ed. F.A. Lutz and D.C. Hague, London: Macmillan; New York: St Martin's Press.

1965. *Capital and Growth*. Oxford: Clarendon Press.

1967. *Critical Essays in Monetary Theory*. Oxford: Clarendon Press.

1969. *A Theory of Economic History*. Oxford: Clarendon Press.

1973a. *Capital and Time: a neo-Austrian theory*. Oxford: Clarendon Press.

1973b. Recollections and documents. *Economica* 40, February, 2–11.

1974. *The Crisis in Keynesian Economics*. Oxford: Basil Blackwell.

1975. The scope and status of welfare economics. *Oxford Economic Papers* 27(3), November, 307–26.

1977. *Economic Perspectives*. Oxford: Clarendon Press.

1979. *Causality in Economics*. Oxford: Basil Blackwell.

1981–3. *Collected Essays on Economic Theory*. Oxford: Basil Blackwell.

Hicks, Ursula Kathleen (neé Webb) (1896–1985). An Irish-born economist specializing in public finance, Lady Hicks's long career spanned teaching and research at the London School of Economics and Political Science, University of Liverpool, and the University of Oxford (latterly as Foundation Fellow of Linacre College), as well as the holding of many visiting posts in foreign universities and service as a member of advisory missions on fiscal matters, notably in the Caribbean, India and Africa.

She made three significant contributions to her specialism, the theory and practice of public finance. Her paper 'The Terminology of Tax Analysis' (1946) questioned the usefulness of the distinction between direct and indirect taxes and argued persuasively for distinguishing between taxes on income and taxes on expenditure (outlay), the dichotomy now used in national accounting. She also explored the difference between the formal incidence of taxes (the liability to pay taxes) and the effective incidence (the determination of tax burdens). Second, in collaboration with her husband, Sir John Hicks, she endeavoured to produce coherence between what the aims of government should be and how fiscal institutions should be organized to achieve them (cf. 1947, Part III). Third, she applied a unique knowledge of fiscal systems to the study of federal and local finance particularly in developing countries (cf., e.g., 1961).

No account of her contribution would be complete without mentioning her immense influence as a teacher of students of public finance from all parts of the world and her part in the foundation of the *Review of Economic Studies* (together with Abba Lerner and Paul Sweezy), of which she was Managing Editor from 1933 to 1961.

ALAN PEACOCK

SELECTED WORKS

1946. The terminology of tax analysis. *Economic Journal* 56, March, 38–50.

1947. *Public Finance*. London: Nisbet & Co.; Cambridge: Cambridge University Press.

1961. *Development From Below*. Oxford: Clarendon Press.

BIBLIOGRAPHY

David W.L. (ed.) 1973. Introduction to *Public Finance, Planning and Economic Development Essays in Honour of Ursula Hicks*. London: Macmillan.

hidden actions, moral hazard and contract theory. 'Moral hazard' in the literal sense refers to the adverse effects, from the insurance company's point of view, that insurance may have on the insuree's behaviour. As an extreme but standard example, a fire insurance holder may burn the property in order to obtain the insured sums. Although the expression can be found in earlier literature, its extensive use in economics can be dated from Arrow's *Essays in the Theory of Risk-bearing* (1971), which had a decisive influence in popularizing the term as well as in stimulating a systematic study both of the subject itself and of related phenomena. Arrow stresses that the complete set of markets required for first best efficiency often cannot be organized. The (so-called) Arrow–Debreu contracts which are needed would have to be contingent on states of nature. This term, 'states of nature', has to be taken in its meaning in decision theory where it refers to random events whose realization reflects an exogenous choice by 'Nature', and not an endogenous choice by agents. However, states of nature may not be observable either directly or indirectly, so that real contracts have to rely upon imperfect proxies. Take the overly simple fire insurance example: Arrow–Debreu contingent contracts would make indemnification conditional only on the occurrence of those natural events that can cause fire, such as thunderstorms, whereas actual real-world contracts make it dependent upon the occurrence of fire itself, whether due to an unusual exogenous event, or to a more normal exogenous event coupled with insufficient care.

Following Arrow, modern economic terminology has come to use 'moral hazard' to mean the unobservability of contingencies, about which information is needed in order to design first-best efficient contracts. Considering now a general framework of contracts, it is normally the case that the relevant contractual information can be obtained through observation of actions and outcomes, the latter themselves dependent on states of nature. Assuming that outcomes are always observed, moral hazard is therefore restricted to mean that some actions of one or more of the parties are not publicly observable (i.e. by all parties to the contract). With the more suggestive terminology of Arrow, moral hazard is thus associated with the existence of *hidden actions* in a contractual relationship.

This definition deserves three comments.

(i) 'Moral hazard' has unfortunate ethical connotations. Given that parties to contracts are usually modelled as standard maximizers of utility, it seems preferable to employ the term *hidden actions*.

(ii) Recent literature on contracts distinguishes between the observability and the verifiability of actions. A variable can be observable by all the parties to a contract, but not to outsiders to the contract. In particular, it may provide no evidence for a court of law. It is then said to be non-verifiable. Then *hidden actions* conveys the right idea but not the right nuance, and we should rather speak of *unverifiable actions*.

(iii) Difficulties in organizing a contractual relationship arise not only from actions that some parties can hide but also from the limited accessibility of the information that some parties use before taking actions. This may be private information of one party about itself (an agent usually knows his own characteristics better than do his partners in the contract), or information on some relevant states of nature which can influence the outcome of the relationship. Such difficulties are thus due to *hidden knowledge* as well as to hidden actions. Consider again an example drawn from insurance. Insurance companies (life insurance, car insurance) face both good risks and bad risks, i.e. agents who for a given level of care or prevention have to be assigned different probabilities of injury. This distinction thus refers to privately known characteristics of the insurees themselves rather than to the actions they take.

Hidden knowledge generates opportunism. Faced with a set of contracts, high risk and low risk people will select different contracts; this is self-selection or, from the company's point of view, adverse selection. The distinction between hidden actions and hidden knowledge seems more suggestive than the more usual distinction in contract theory between moral hazard and adverse selection. Although we will examine some problems in which hidden actions and hidden knowledge are mixed, the main subject of this article is the analysis of contractual problems raised by hidden actions. Attention will be focused primarily on an abstract hidden action model, rather on the subject-specific discussions which generated the main building blocks of that model.

I. THE BASIC HIDDEN ACTION MODEL OF A BILATERAL RELATIONSHIP

The prototype model considered in this section owes much to the pioneering work of Ross (1973), Mirrlees (1974 and 1976) and Holmstrom (1979), and its presentation here draws heavily on the syntheses of Grossman and Hart (1983a). It is a principal–agent model with one principal and one agent. The agent chooses from among available actions one which together with random events (states of nature) determines a measurable result, which most of the time is a money payment to the principal. The principal is interested in the results as well as in the money remuneration he gives to the agent. The agent has a utility function depending upon the action taken and on the money transfer he receives from the principal. Some actions are more costly or involve higher 'effort'. Indeed, in many specific models the action variable is a loosely defined effort level: effort of the manager when the principal consists of shareholders, effort of firms when the principal is a bank.

In the simpler version of the model considered here, each utility function is separable, and risk-neutrality vis-à-vis income obtains, with a utility linear in income. It is assumed that the agents' actions are not observable but that the results are verifiable. A contract between the principal and the agent then consists of a reward schedule which associates a money transfer to any possible result. Analytically an optimal contract for the principal is a solution (if any) to a programme which maximizes over the set of all reward schedules, under a constraint of individual rationality for the agent.

The solution just sketched calls for preliminary comments:

(i) In the degenerate case where there is no choice of action – the principal–agent problem reduces to a pure risk-sharing question, whose solution depends on the risk-aversion of the parties concerned. In particular, a risk-neutral agent bears all income fluctuations and provides full insurance to his risk-averse partners. An optimal contract between a risk-neutral firm and risk-averse workers leads to utility profiles of workers constant across states of nature and a constant wage in states where workers are employed. This latter remark is at the core of the theory of implicit contracts initiated by Azariadis (1975), Bailey (1974) and Gordon (1974).

(ii) With a non-degenerate set of actions, but with an observability assumption, the optimal contract trades off between efficiency and risk-sharing considerations. Following the usual terminology, the corresponding contract is referred to as first best. When actions are not observable, the reward scheme has to be based on results only. It is generally impossible to reward actions indirectly in a way which mimics the first best contract. We then have to determine a second best contract.

First insights into the model are obtained when the reward schedule is restricted to be an affine function of the money outcome. Then, when the principal is risk-neutral and the agent is risk-averse, the optimal contract trades off between incentives and risk-sharing requirements in a way which confirms intuition. A positive fixed fee has to be combined with a linear schedule the slope of which is, however, smaller than the marginal value of the performance for the principal.

The derivation of the optimal non-linear second best contract leads to a serious analytical difficulty, which has been of primary concern to analysts. In the context of moral hazard this difficulty was initially stressed by Mirrlees (1975), and was independently discovered and analysed in the context of a general equilibrium second best problem by Guesnerie and Laffont (1978). It can be described as follows: For a given reward schedule, the agent's utility as an indirect function of actions is not generally quasi-concave. Hence, when the parameters of the reward schedule are modified the optimal response of the agent may jump. Although this jump only occurs for exceptional values of the parameters, it may still be the case that the optimal contract systematically selects such exceptional values (this is really the essence of the point made by Mirrlees and Guesnerie–Laffont). Then the local description of the agent's local behaviour from the first order conditions of utility maximization – which is analytically very convenient – becomes invalid. This failure of the so-called 'first-order approach' has generated contributions which are decisive for a rigorous analysis of the problem (see e.g. Rogerson, 1985).

The research has led us to a much more thorough understanding of an optimal schedule. In particular, it has made clear that the reward associated to a given result reflects the Bayesian statistical inference made by the principal from this result, although this convenient interpretation should not hide the fact that the principal does not ignore the agent's action! However, the results on the shape of the optimal schedule are somewhat deceptive. As the statistical inference argument suggests, few restrictions on it can be deduced from general theory. Even monotonicity – higher rewards for higher results – cannot be guaranteed, without strong assumptions on the distribution of results conditional on actions. For example, monotonicity obtains with the monotone likelihood ratio property introduced in this problem by Milgrom (1981) and the concavity of distribution function condition (see Grossman and Hart, 1983a). Non-monotonicity is hardly surprising; imagine that the most desirable actions from the principal's point of view give rise to high results and to low results with smaller probability but never to intermediate results. Conceivably, intermediate results will thus be less rewarded than low results.

In this rather disappointing picture, a result of general relevance does emerge. Although weak, it is remarkably robust. All variables that are correlated with the noise carry useful information for the design of optimal contracts. New information is redundant only when existing variables are sufficient statistics (see Holmstrom, 1979; Gjesdal, 1982).

To complete the picture, cases where the first best is implementable have to be stressed.

(i) If the agent is risk-neutral, a reward schedule which gives him the money result up to some constant provides correct incentives (such a reward schedule is reminiscent of the Groves scheme in an adverse selection problem). The agent then acts as a residual claimant and chooses the first best action.

(ii) Suppose that one result signals for sure that some non-optimal action has been chosen (i.e. this result has probability zero when the optimal action is taken). Then, if a high penalty is associated with this result, the agent will be deterred from choosing any action for which this result can occur with non-zero probability. It follows that the first-best action will be chosen if there is a subset of highly penalized results that are reached with probability zero when the optimal action is taken, and with positive probability when any non-optimal action is taken. In particular, if the result is a noisy estimate of the action, the first best is implementable when the noise is additive and has compact support. The power of high penalties, at least in some contexts, is a striking feature of moral hazard problems. We will come back to this point later.

II. FURTHER GENERAL CONSIDERATIONS ON HIDDEN ACTION MODELS

We will examine briefly four directions of development for the basic hidden action model described in section I.

II.1 The complexity of the optimal reward schedule. The results described in the previous section suggest contractual arrangements which are more complex than those observed in real situations. Several explanations have been suggested: for example, bounded rationality of the parties is a plausible argument for the use of unsophisticated reward schemes. Another possible explanation might be found in the inadequacy of the modelling options described in section I. This is a subject of current research and an interesting point has recently been made by Holmstrom and Milgrom (1985). They modify the basic model by assuming that the agent has progressive information on the occurrence of the outcome so that he can continuously adapt his action (here, his effort) in the time interval where the relationship takes place. They show that the optimal reward schedule, which in the standard version of these problems is highly non-linear, becomes linear. Although this conclusion relies on special assumptions concerning the agent's utility and the noise, it suggests that the enrichment of the action space of the agent leads to simpler reward schedules.

II.2 Mixing adverse selection and moral hazard. It has been argued above that hidden action and hidden knowledge determine two polar cases in the theory of contracts – in fact many contract problems involve both hidden action and hidden knowledge. In the mixed case the non-linear reward scheme thus has three different roles. It should provide correct incentives by limiting the distortion between the value of outcome for the principal and the agent's reward, and should induce adequate risk-sharing; these two functions are already central to the hidden action model. In addition, it should keep control of the self-selection process by inducing satisfactory choices of agents of different characteristics. The determination of the optimal contract in the mixed case then assimilates the analytical difficulties of each of the polar cases (each of these polar cases is reasonably well understood, and for a synthesis on an adverse selection principal-agent problem, in a spirit similar to Grossman and Hart's article on moral hazard, see Guesnerie and Laffont (1984). The understanding of the intricacies of the general case requires further investigation. The analysis of an intermediate case provides a useful benchmark. It is presented now.

Consider a pure hidden knowledge problem when actions of the agent are observable although characteristics are not. Let us introduce the moral hazard ingredient that actions are no longer perfectly observable. Their observation is affected by noise. The new problem calls for two immediate remarks: first, if the parties are risk-averse, the introduction of noise will reduce social welfare (when compared to its pure adverse

selection maximum level); second, if the adverse selection problem is degenerate, i.e. the agent's characteristics are known, there is no welfare loss when agents are risk-neutral. This absence of welfare loss can be shown to extend to a non-degenerate hidden knowledge model. For a large class of noises, with risk-neutral agents, the maximum adverse selection welfare, can be at least approximately reached when the observation of actions becomes noisy. In other words, the second best adverse selection welfare can still be implemented with noisy observations. This (quasi) implementation obtains either by using a family of quadratic schedules (see Picard, 1987) or by using a single schedule, different from the adverse selection optimal schedule, but obtained from it as the solution of a convolution equation when the noise is additive (see Caillaud, Guesnerie and Rey, 1986) or a Fredholm equation for non-additive noise (see Melumad and Reichelstein, 1986). Furthermore, when one of the action variables can be observed, a family of linear schedules may serve for implementation whatever the distribution of noise (it is then a universal family of schedules) or a family of truncated linear schedules may serve for implementation when the noise is small. However, these appealing properties are likely to hold in circumstances which are rather special (see Laffont and Tirole (1986) for one of these special cases, and Caillaud, Guesnerie and Rey (1986) for a comprehensive analysis of this problem).

II.3 Monitoring devices and high penalties. We have provided an interpretation of the basic hidden action model where 'results' are an intrinsic and unavoidable outcome of the relationship. There are cases, however, where the principal is only interested in the actions taken by the agents and where the inference on the action is made from observations which are obtained from a special device: examples of such monitoring devices which allow more precise inference of the behaviour of an agent are audits.

If the basic frame is easily adapted to the study of such a situation when the monitoring device is given, or even if there are several possible monitoring devices, a basic difficulty occurs when the frequency of use of such a monitoring device is not fixed. The nature of the difficulty is the following: the use of the monitoring device being costly, the principal can economize on expected costs by writing a contract which stipulates that the control device will only be used with probability smaller than one, rather than for sure. But whatever the probability chosen, it is often the case that the principal can reduce it further and modify penalties and rewards accordingly in such a way that the choice of action is unchanged. This argument was made in particular by Polinsky and Shavell (1979; see also Rubinstein, 1979) who were considering the substitutability between the probability and the magnitude of legal fines. The fact is that the expected value of the fine may be held constant when the probability of control is decreased and the magnitude of the fine is increased. This argument has proved to be remarkably robust (for extensions, see Nalebuff and Scharfstein, 1985). In particular, it does not depend upon the risk-aversion of agents, at least in a bilateral relationship. In our framework, it suggests that the optimal contract, when the use of monitoring devices is costly, may be stochastic and may involve a low probability of control, together with (possibly) high penalties and rewards. Again, real contractual arrangements do suggest neither the use of high penalties, nor the substitution of penalties to control frequencies at least to the extent predicted by the above theory. A more careful analysis of the problems suggests at least three reasons for the first noted discrepancies (and at the same time three directions of improvements for the basic model).

(i) Our argument holds in the special case of a hidden action relationship in which all the elements of the problem are in the language of the theory, 'common knowledge'. For example, it assumes that the agent's risk-aversion is exactly known by the principal or that the distribution functions of the random variables is common knowledge. Giving up one of these assumptions amounts to introducing hidden knowledge into the relationship. The efficiency of high penalties does not seem to be robust to the introduction of these considerations.

(ii) The credibility of the principal's commitment to some probability of control is problematic. It would require the implementation of some kind of public lottery.

(iii) The outcome of control via a monitoring device should be verifiable. If not, the principal would have an incentive to announce results which highly penalize the agent. Some neutral third party is required. But the danger of collusion between this third party and another party increases with the amount of penalties (or reward). (For an analysis of collusion in a three parties relationship, see Tirole, 1986.)

II.4 The dynamics of moral hazard contracts. Assume that the basic principal–agent relationship is repeated. The one-period game described above is extended to a large number of periods (assuming for simplicity separability between periods). It is intuitively clear from the law of large numbers that time filters out uncertainty and allows a more and more accurate knowledge of the mean action taken by the agent. Repetition should thus alleviate moral hazard problems. The formal analysis confirms and makes precise these findings, at least when parties to the contract put enough weight on the future. If agents are interested only in the average pay-off over an infinity of periods or if both have a (common) discount rate close enough to one, there exist dynamic contracts which allow one to approximate the first best welfare level (see Radner, 1981, 1985). It would, however, be premature to conclude from this neat result that moral hazard problems disappear within a long enough relationship. Let us make clear the limits of this result.

(i) The result only holds for discount rates close to one. Even then, it does not provide a characterization of the truly optimal policy (it uses an a priori policy which is shown to be quasi-optimal). *A fortiori*, the characteristics of the optimal policy for lower values of the discount rates are not well understood. The study of simple cases such as the one considered by Henriet and Rochet (1984) suggests that the present reward at any period should put more weight on observed performances which are more recent. This is in sharp contrast with what happens in an adverse selection problem, where the observation serves to estimate the value of unobservable characteristics, a case in which the Henriet–Rochet model leads one to base the reward on the mean of observation before the present period.

(ii) The model supposes both that the principal can commit himself to the announced strategy and that the agent is locked-in in the relationship, but is not necessarily needed for the conclusion. In addition, the principal's policy relies on the threat of high penalties, a feature of contracts the adequacy of which has been questioned in the previous subsection. The commitment assumption is subject to the usual objections. The lock-in assumption for the agent is also much debatable. The agent should at least be allowed to smooth his income through time by access to financial markets. Exit of the agent via financial markets is a subject of present research.

(iii) As in the static case, many interesting dynamic problems

mix hidden action and hidden knowledge. This leads to more intricate phenomena as demonstrated by the models of Holmstrom (1982b) or Harris and Holmstrom (1982). Assume as in these models, the output of a worker is the product of an unobservable characteristic (say skill) and of an action (say effort). The firms' inference from the sequence of outputs aims at determining both effort and skill. In their turn, workers are induced to over-invest in effort in the first periods to signal high skill and to under-invest when their position has been established. This has some resemblance to real academic life rather than to a pure hidden action dynamic model.

II.5 Tournaments and moral hazard in a group. The so-called tournament model focuses attention on a relationship involving one principal and several agents. With several agents, the contracts are not necessarily independent: the reward of one agent can be based not only upon his performance but also upon the performance of the other. One polar case of interdependent contract is the contract associated with a rank-order tournament where actual outputs are ranked and the reward jumps with the rank. With two agents the winner has the highest prize (R&D competition for patents induces a similar structure of rewards: see Guesnerie and Tirole, 1985). Let us briefly mention the main direction explored by the tournament literature.

(i) Lazear and Rosen (1981), in a two-agent model, compare the rank-order tournament with special independent contracts, i.e. linear contracts, and discuss the relative merits of both.

(ii) Independent non-linear contracts are dominated by dependent contracts only when the principal can infer more information on the variables faced by the agent (before his decision was made) from the whole set of outcomes than he can infer from any single outcome. In such circumstances Green and Stokey (1983), Holmstrom (1982a), and Nalebuff and Stiglitz (1983) focus attention on situations in which the mean of outcomes is a sufficient statistic for the variables unknown to the principal. Thus, the optimal contract is only dependent upon the mean of outcomes and the individual outcome.

(iii) First best can be approximately implemented from rank order tournaments with high penalties when the number of participants is large enough (see Holmstrom, 1982; Nalebuff and Stiglitz, 1983). In the different but related context of moral hazard in teams, Holmstrom (1982) has stressed that a team can behave poorly in the solution of moral hazard problems when no agent in the group can act as residual claimant. The group thus cannot commit itself credibly to use a sharing rule which induces efficient effort. The existence of a residual claimant is essential for making credible the threat of destruction.

CONCLUSION

One of the two obvious omissions in the present review has already been stressed. Applications of the basic ideas to different subjects have not been reviewed. The 'horizontal' presentation adopted here should be complemented by 'vertical' readings which describe the implications of the basic ideas in different fields. The second omission is the fact that the work reviewed is only of partial equilibrium nature. However, the hidden action model is part of the contractual approach to economics which has developed since the 1970s from a recognition of the failure of the impersonal market hypothesis to explain certain phenomena. The corresponding literature had the more or less explicit ambition of assessing the aggregate implications of the existence of contractual arrangements at the micro level. In particular, the study of the general equilibrium implications of moral hazard is an important topic. It has not been presented here, partly from lack of space, and partly because a coherent presentation of existing work is more difficult.

In conclusion, let us briefly mention a number of directions of present research.

First, the nature of competition is affected by the presence of moral hazard at the micro level. Helpman and Laffont (1975), Arnott and Stiglitz (1985), and Hellwig (1987) analyse this problem.

Second, normative economics should take into account the specification of contractual relationships. In particular, one can expect that the contractual approach will favour a better assessment of the informational constraints faced by government action. Also, moral hazard at the micro level is responsible for externalities, the particular features of which are analysed in the case of the labour market by Arnott and Stiglitz (1985).

Finally, the examination of the aggregate consequences of contractual arrangements in the labour market is a subject of intensive research – Shapiro and Stiglitz (1984) argue that in the absence of direct penalties (for reasons discussed above) for breach of labour contracts, unemployment serves as a 'discipline device'. Other work on the general equilibrium consequences of the contractual labour conditions – in case of hidden action – include Malcomson and MacLeod (1986).

ROGER GUESNERIE

See also ADVERSE SELECTION; ASYMMETRIC INFORMATION; EXTERNALITIES; INCOMPLETE CONTRACTS; MORAL HAZARD; PRINCIPAL AND AGENT; SIGNALLING.

BIBLIOGRAPHY

Arnott, R. and Stiglitz, J. 1985. Labor turnover, wage structures, and moral hazard: the inefficiency of competitive markets. *Journal of Labor Economics* 3, 434–62.
Arrow, K.J. 1964. *Essays on the Theory of Risk-Bearing*. Chicago: Aldine.
Arrow, K.J. 1985. The Economics of Agency. In *Principals and Agents: The Structure of Business* ed. J. Pratt and R. Zeckhauser, Boston: Harvard Business School Press, 37–51.
Azariadis, C. 1975. Implicit contracts and underemployment equilibria. *Journal of Political Economy* 83, 1183–202.
Bailey, M. 1974. Wages and employment under uncertain demand. *Review of Economic Studies* 41, 37–50.
Bester, H. 1985. Screening versus rationing in credit markets with imperfect information. *American Economic Review* 75(4), 850–55.
Bhattacharya, S. 1983. Tournaments and incentives: heterogeneity and essentiality. Research Paper no. 695, Graduate School of Business, Stanford University.
Caillaud, B., Guesnerie, R. and Rey, P. 1986. Contracts with adverse selection and moral hazard: the case of risk neutral partners. Mimeo.
Calvo, G. and Wellicz, S. 1978. Supervision, loss of control and the optimal size of the firm. *Journal of Political Economy* 86(5), 943–52.
Diamond, D. 1984. Financial intermediation and delegated monitoring. *Review of Economic Studies* 51(3), 393–414.
Fama, E. 1980. Agency problems and the theory of the firm. *Journal of Political Economy* 88, 268–307.
Gibbons, R. 1985. Essays on labor markets and internal organization. Unpublished dissertation, Stanford University, July.
Gjesdal, F. 1982. Information and incentives: the agency information problem. *Review of Economic Studies* 49, 373–90.
Gordon, D. 1974. A neo-classical theory of Keynesian unemployment. *Economic Inquiry* 12, 431–59.
Green, J. and Stokey, N. 1983. A comparison of tournaments and contracts. *Journal of Political Economy* 91, 349–64.

Grossman, S. and Hart, O. 1983a. An analysis of the principal–agent problem. *Econometrica* 51, 7–45.
Grossman, S. and Hart, O. 1983b. Implicit contracts under asymmetric information. *Quarterly Journal of Economics*, Supplement, 71, 123–57.
Guesnerie, R. and Laffont, J.J. 1978. Taxing price makers. *Journal of Economic Theory* 19(2), 423–55.
Guesnerie, R. and Laffont, J.J. 1984. A complete solution to a class of principal–agent problem with an application to a self managed firm. *Journal of Public Economics* 25(3), 329–69.
Guesnerie, R. and Tirole, J. 1985. L'économie de la recherche développement. *Revue économique* 5, 843–71.
Harris, M. and Holmstrom, B. 1982. A theory of wage dynamics. *Review of Economic Studies* 49, 315–33.
Hellwig, M. 1987. Some recent developments in the theory of competition in markets with adverse selection. *European Economic Review* 31(1/2), 319–25.
Helpman, E. and Laffont, J.J. 1975. On moral hazard in general equilibrium theory. *Journal of Economic Theory* 10(1), 8–23.
Henriet, D. and Rochet, J.C. 1984. The logic of bonus-penalty systems in automobile insurance. Working Paper No. A273 0784, Ecole Polytechnique.
Holmstrom, B. 1979. Moral hazard and observability. *Bell Journal of Economics* 10, 74–91.
Holmstrom, B. 1982a. Moral hazard in teams. *Bell Journal of Economics* 13, 324–40.
Holmstrom, B. 1982b. Managerial incentive problems – a dynamic perspective. In *Essays in Economics and Management in Honor of Lars Wahlbeck*, Helsinki: Swedish School of Economics.
Holmstrom, B. and Milgrom, P. 1985. Aggregation and linearity in the provision of intertemporal incentives. Cowles Discussion Paper no. 742, April.
Holmstrom, B. and Ricart-Costa, J. 1984. Managerial incentives and capital management. Cowles Discussion Paper no. 729, November.
Holmstrom, B. 1983. Equilibrium long-term contracts. *Quarterly Journal of Economics*, Supplement, 98, 23–54.
Joskow, P. 1985. Vertical integration and long-term contracts. *Journal of Law, Economics and Organization* 1, Spring, 33–80.
Laffont, J.J. and Tirole, J. 1986. Using cost observation to regulate firms. *Journal of Political Economy* 94(3), Pt I, 614–41.
Lambert, R. 1986. Executive effort and selection of Risky projects. *Rand Journal of Economics* 16, 77–88.
Lazear, E. and Rosen, S. 1981. Rank order tournaments on optimum labour contracts. *Journal of Political Economy* 89(5), 841–64.
Malcomson, J. 1984. Work incentives, hierarchy, and internal labor markets. *Journal of Political Economy* 92(3), 486–507.
Melumad, N. and Reichelstein, S. 1985. Value of communication in agencies. Mimeo.
Milgrom, P. 1981. Good news and bad news: representation theorems and applications. *Bell Journal of Economics* 12, 380–91.
Mirrlees, J. 1974. Notes on welfare economics, information and uncertainty. In *Essays in Economic Behavior Under Uncertainty*, ed. M. Balch, D. McFadden and S. Wu, Amsterdam: North Holland, 243–258.
Mirrlees, J. 1975. The theory of moral hazard and unobservable behavior – Part I. Mimeo, Nuffield College, Oxford.
Mirrlees, J. 1976. The optimal structure of authority and incentives within an organization. *Bell Journal of Economics* 7, 105–31.
Mookherjee, D. 1984. Optimal incentives schemes with many agents. *Review of Economic Studies* 51(3), 433–46.
Nalebuff, B. and Stiglitz, J. 1983. Prizes and incentives: towards a general theory of compensation and competition. *Bell Journal of Economics* 13, 21–43.
Newbery, D. and Stiglitz, J. 1983. Wage rigidity, implicit contracts and economic efficiency: are market wages too flexible? Economic Theory Discussion Paper 68, Cambridge University.
Picard, P. 1987. On the design of incentives schemes under moral hazard and adverse selection. *Journal of Public Economics*.
Polinsky, A. and Shavell, S. 1979. The optimal tradeoff between the probability and the magnitude of fines. *American Economic Review* 69(5), 880–89.
Radner, R. 1981. Monitoring cooperative agreements in a repeated principal–agent relationship. *Econometrica* 49, September, 1127–48.

Radner, R. 1985. Repeated principal–agent games with discounting. *Econometrica* 53, 1173–98.

Rogerson, W. 1985. 'The First-Order Approach to Principal-Agent Problems', *Econometrica*, 53, 1357–68.

Ross, S. 1973. The economic theory of agency: the principal's problem. *American Economic Review* 63, 134–9.

Rubinstein, A. 1979. Offenses that may have been committed by accident – an optimal policy of retribution. In *Applied Game Theory*, ed. S. Brahms, A. Shotter and G. Schwödiauer, Würtzburg: Physica-Verlag, 406–413.

Shapiro, C. and Stiglitz, J. 1984. Equilibrium unemployment as a worker incentive device. *American Economic Review* 74, 433–44.

Shavell, S. 1979. Risk sharing and incentives in the principal and agent relationship. *Bell Journal of Economics* 10, 55–73.

Stiglitz, J. and Weiss, A. 1985. Credit rationing in markets with imperfect information. *American Economic Review* 71(3), 393–410.

Stiglitz, J. 1974. Incentives and risk sharing in sharecropping. *Review of Economic Studies* 41(2), 219–55.

Tirole, J. 1986. Hierarchies and bureaucracies: on the role of collusion in organizations. *Journal of Law Economics and Organization* 2(2), 181–214.

Williamson, O. 1985. *The Economic Institutions of Capitalism.* New York: Free Press.

Yaari, M. 1976. A law of large numbers in the theory of consumer's choice under uncertainty. *Journal of Economic Theory* 12, April, 202–17.

hidden hand. *See* INVISIBLE HAND.

hierarchy. The economic theory of hierarchies attempts to explain the typical organization chart of firms, with lines of command clearly defined in a top-to-bottom fashion. Modern approaches are able to rationalize some of the observed wage-scale peculiarities of organizations, particularly the tendency of wages to escalate more than in proportion to 'natural' abilities as one moves to the top of the organization ladder.

There are essentially two type of approaches to the theory of hierarchies. the first, pioneered by Mayer (1960), looks at hierarchies as production structures that contribute to the enhancement of labour productivity by facilitating the transmission of talents from the more to the less gifted workers. This approach does not signify a radical departure from the textbook theory of the firm, since the theories developed along these lines involve, essentially, just a richer specification of the former. An example following Mayer (1960) could be illuminating.

Imagine that there is only one type of homogeneous output which is equal to the input of 'effective' labour services of production workers (to be defined below), and that there is a given distribution of individuals according to their labour endowment; for the sake of simplicity, we will assume that there are only two types of individuals, and indicate their labour endowments by α and β, where $\alpha > \beta$. If individuals work independently from one another (if, in other words, they were to set up their own firms), then each individual would be a production worker; thus, if we assume that under those circumstances effective labour services are equal to the individual's labour endowment, it follows that, in a competitive equilibrium, we will have firms of different sizes (α and β), and that the net income of an individual is just proportional to his/her labour endowment. In practice, however, one finds that individuals of greater ability or experience are able to teach or train some of those who are less endowed. To capture this phenomenon, let us assume (with

Mayer, 1960) that if an individual of type α forms a firm with n ($n \geqslant 1$) individuals of type β, then the effective labour services of each one of the latter becomes α. An interpretation of this is that the type-α individual supervises (or trains, or co-ordinates) the n type-β workers, in which case output of the firm would be $n\alpha$. For the sake of simplicity, we further assume that, in equilibrium, some type-β individuals have their own firm (we will later make the necessary assumptions for this to hold in general equilibrium). Since in that case, the opportunity cost of a type-β worker is just β, his/her competitive wage will be β, and, thus, a firm headed by a type-α manager will make a net profit equal to

$$n(\alpha - \beta) \tag{1}$$

Thus, a type α individual has an incentive to form a firm if the expression in (1) exceeds his net income if he worked alone ($= \alpha$), or, equivalently, if

$$(n-1)\alpha - \beta > 0 \tag{2}$$

Under our assumptions, (2) will hold if n is sufficiently large. Thus, letting the upper bound on n be denoted by $\bar{n}$ (usually called the 'span of control'), the above reasoning implies that if $\bar{n}$ is sufficiently large, then two-level firms will be formed in equilibrium. Further assuming that the ratio of type-β to type-α individuals exceeds $\bar{n}$ allows us to ensure that some type-β individuals will form their own firms, thus validating the assumption under which the firm optimization exercise was performed.

The example shows that (a) firms of different size can exist in equilibrium, and (b) that the most able individuals may get an income which is larger than what they would obtain if they did not team up with those of lower ability; as a consequence, the model rationalizes the existence of hierarchies, and the empirical finding that within an organization the distribution of income appears to be more skewed to the right than the distribution of abilities (see Simon, 1957; Lydall, 1968). This type of model has been extended by several authors (see Rosen, 1982) to account for more general talent-transmission technologies. An important characteristic of the present approach is that the equilibrium allocation is a Pareto optimum since firms end up being devices that allow a full internalization (by managers or owners) of the externalities involved in the talent-transmission mechanism.

The second type of approach has quite different efficiency implications. The basic assumption of this approach is that supervision is necessary for ensuring performance (see Calvo and Wellisz, 1979). To illustrate how this assumption can give rise to a hierarchy, let us assume that there exists only one competitive firm headed by one manager/owner who, as in the previous example, has access to a labour-only technology; except for the manager/owner all the other individuals in the economy have identical qualities, and, in particular, they are each endowed with one unit of labour services. We assume that the labour contract specifies a wage w if, when supervised, the worker is not idle (assume, for simplicity, that a worker either works full-time or stays idle full-time); otherwise, that is, if the worker is supervised and he/she is idle, the worker is fired and becomes self-employed with income h. Following Becker and Stigler (1974), we assume that, given w, the utility of staying idle (or shirking) and not being caught by the supervisor is $w + k$, where $k > 0$, while the utility of working is just w; k can be interpreted as the utility derived from on-the-job leisure, stealing, etc. Thus, if the probability of detection is P, the expected utility of shirking is $Ph + (1 - P)(w + k)$; on the other hand, the expected utility of working full-time is, obviously, just w. Thus, assuming that in case of indifference the worker does not stay idle, the result is the minimum wage that can be offered in order to be able to elicit labour services from the workers that are employed is

$$h + (1/P - 1)k \qquad (3)$$

which is larger than h if $P < 1$ (imperfect supervision). Thus, if in equilibrium some workers remain self-employed, the latter will receive an income smaller than that of their (otherwise) equals who are employed at the firm. Secondly, in order for this to be an equilibrium the technology employed at the firm has to be superior to that of self-employment (in order to be able to 'support' a wage larger than h). If, in addition, we assume that P is inversely related to the level of employment, there will normally be an optimal level for the latter, N; thus, if N is smaller than total (active) population, some workers will remain in the self-employment sector. These are some of the basic reasons why the equilibrium solutions to this kind of model tend to be Pareto inefficient: the need for supervision in the organized sector contributes to the survival of inferior technologies (in our example, self-employment). A proof is given in Calvo (1985).

Despite its inefficiency, it is possible to show that it may be profitable for the firm to add a layer of pure supervisors. This is so because if a supervisor can be induced to work full-time, the workers under him will be in the same situation as in the earlier owner-workers' case; thus, if workers have access to the same technology as before, revenue (not subtracting the supervisors' wages) will be multiplied by the number of supervisors. This, of course, has to be balanced against the additional supervision cost, but in the present and more general examples one can find plausible conditions where more supervision layers are always profitable (see Williamson, 1967; Calvo and Wellisz, 1978, 1979). In addition, in this example, one can show that if adding a layer of supervisors is profitable, then the optimal wage of a supervisor is larger than that of a production worker, *even when they are both assumed to have the same quality*. The reason for this is closely related to the imperfection detected in (3), and to the fact that the cost of an idle supervisor is magnified by the resulting inducement to shirk given to every member of his team.

The above framework can easily accommodate different labour qualities. Calvo and Wellisz (1979) have examples where a labour quality improves as one moves towards the top of the hierarchical ladder, but as in the Mayer-type examples the wage is shown to increase faster than it is granted by quality alone.

The Pareto inefficiency implied by the supervision approach is interesting, because it suggest that wage-cum-supervision and hierarchical structures may be improved upon by other distribution schemes which could be made conditional on, for instance, total or team output. One would hope that that kind of consideration could help explain, for instance, why profit-related bonuses are more prevalent in the upper rather than lower layers of an organization.

GUILLERMO A. CALVO

See also INCENTIVE CONTRACTS; RANK.

BIBLIOGRAPHY

Becker, G.S. and Stigler, G.J. 1974. Law enforcement, malfeasance and compensation of enforcers. *Journal of Legal Studies* 3(1), January, 1–18.

Calvo, G.A. 1985. The inefficiency of unemployment: the supervision perspective. *Quarterly Journal of Economics* 100(2), May, 373–88.

Calvo, G.A. and Wellisz, S. 1978. Supervision, loss of control and the optimum size of the firm. *Journal of Political Economy* 86(5), October, 943–52.

Calvo, G.A. and Wellisz, S. 1979. Hierarchy, ability and income distribution. *Journal of Political Economy* 87(5), October, 991–1010.

Lydall, H.F. 1968. *The Structure of Earnings.* Oxford: Clarendon Press.

Mayer, T. 1960. The distribution of abilities and earnings. *Review of Economics and Statistics* 42(2), May, 189–95.

Rosen, S. 1982. Authority, control and the distribution of earnings. *Bell Journal of Economics* 13(2), Autumn, 311–23.

Simon, H.A. 1957. The compensation of executives. *Sociometry* 20(1), March, 32–5.

Williamson, O.E. 1967. Hierarchical control and optimum firm size. *Journal of Political Economy* 75(2), April, 123–38.

higgling. Higgling of the market is described by Adam Smith as a process by which 'exchangeable value' is adjusted to its measure 'quantity of labour':

> It is often difficult to ascertain the proportion between two different quantities of labour ... it is not easy to find any accurate measure either of hardship or ingenuity. In exchanging, indeed, the different productions of different sorts of labour for one another, some allowance is commonly made for both. It is adjusted, however, not by any accurate measure, but by the higgling and bargaining of the market, according to that rough equality which, though not exact, is sufficient for carrying on the business of common life (*Wealth of Nations*, bk. i, ch. v).

Compare Fleeming Jenkin:

> The higgling of the market, ascertaining the result of the relative demand and supply in that market, does not in the long run determine the price of either eggs or tea; it simply finds out the price which had been already determined by quite different means ('Time-Labour System', *Papers, Literary, Scientific*, etc., p. 139).

It is possible to accept the writer's account of the *market process* (ibid. p. 123) without contrasting so strongly the determination of price by demand and supply and by cost of production (cf. Marshall's *Principles*, Preface to 1st edn, p. xi.). Prof. Marshall at the beginning, when treating of the theory of the equilibrium of demand and supply, gives an excellent type of the action of a market (ibid, 5th edn, bk. v, ch. ii, § 2). The subject can hardly be apprehended without mathematical conceptions. Thus Mill, in his description of the play of demand and supply (*Political Economy*, bk. iii, ch. ii, § 4), in the absence of the idea of a demand-curve or function, may seem to use the phrases 'demand increases', 'demand diminishes', loosely. A more distinct idea is thus expressed by Fleeming Jenkin in his *Graphic Representations*: 'If every man were openly to write down beforehand exactly what he would sell or buy at each price, the market price might be computed immediately.' A similar idea is presented by Prof. Walras (*Éléments d'économie pure*, article 50). In some later passages he has formulated the higgling of the market more elaborately. The present writer, criticizing these passages (*Revue d'économie politique*, January 1891), has maintained that even if the dispositions of all the parties were known beforehand, there could be predicted only the position of equilibrium, not the particular course by which it is reached. Of course special observation may supply the defects of theory. For instance there may be evidence of the incident which Cantillon attributes to the 'altercation' of a market, namely the predominant influence of a few buyers or sellers; 'le prix réglé par quelques uns est ordinairement suivi par les autres' (*Essai*, part ii, ch. ii. Des prix des marchés). Compare Condillac:

'Aussitôt que quelues uns seront d'accord sur la proportion à suivre dans leurs échanges les autres prendront cette proportion pour règle' (*Le Commerce et le Gouvernement*, ch. iv: *Des marchés*).

'Higgling' is not always qualified as 'of a market'. The term may be used in much the same sense as the 'art of bargaining' is used by Jevons, with reference to a transaction between two individuals, in the absence of competition (*Theory*, p. 124, 3rd edn). Thus Professor Marshall, in an important passage relating to the case in which agents of production are held by two monopolists, says that there is 'nothing but "higgling and bargaining" ' to settle the proportions in which a certain surplus will be divided between the two (*Principles of Economics*, bk. v, ch. xi). Moses, in the *Vicar of Wakefield*, did not require a fair for the exercise of the skill which is thus attributed to him: 'He always stands out and higgles and actually tires them till he gets a bargain.'

[F.Y. EDGEWORTH]
Reprinted from *Palgrave's Dictionary of Political Economy*.

BIBLIOGRAPHY
Cantillon, R. 1755. *Essai sur la nature du commerce en général*. Paris. Ed. H. Higgs, London: Macmillan.

Condillac, E.B.A.M. 1776. *Le commerce et le gouvernement considerés relativement l'un à l'autre*. In *Oeuvres complètes de Condillac*, Vol. 4, Paris: Briére, 1821.

Jenkin, H.C.F. 1887. *Graphic Representation of the Laws of Supply and Demand*. London. Reprinted in his *Papers, Literary, Scientific & c.*, London: Longmans, Green & Co.

Jenkin, H.C.F. 1887. Time-labour system. In his *Papers, Literary, Scientific & c.*, ed. S.C. Colvin and J.A. Ewing, London: Longmans, Green & Co.

Marhsall, A. 1890. Preface to his *Principles of Economics*. London: Macmillan. 5th edn., 1907.

Mill, J.S. 1848. *Principles of Political Economy*. London: J.W. Parker.

Smith, A. 1776. *An Inquiry into the Nature and Causes of the Wealth of Nations*. London: W. Strahan & T. Cadell.

Walras, L. 1874–7. *Eléments d'économie pure*. Lausanne: Corbaz.

Higgs, Henry (1864–1940). The original edition of this *Dictionary* was reprinted (with revisions) by Inglis Palgrave on a number of occasions during his lifetime. The only edition compiled by someone other than Palgrave himself was that which Henry Higgs published between 1923 and 1926. That revised edition, which for the first time incorporated Palgrave's name into the title, important as it was, had to be compiled under the severe restriction of having to use the original plates for the bulk of the text. This permitted Higgs only two avenues for bringing the original up to date. The first was to add, in an appendix to each volume, biographical notices of economists who had died since the compilation of the original and, in a few cases, continuations of articles already to be found in the body of the text. The second was even more inhibiting and involved replacing sections of the text of the original with new material of exactly the same number of words. One senses behind Higgs's bland explanation of this course of action – that to reset the whole would 'have necessitated a prohibitive price for the volume' – more than a note of regret.

Henry Higgs was born on 4 March 1864, the eleventh of thirteen children of a Cornish landowner. At the age of eighteen he entered the civil service as a Lower Division Clerk in the War Office, moving to the Postmaster General's department in 1884. In 1899 he was transferred to the Treasury, and when Sir Henry Campbell-Bannerman took office as Prime Minister in December 1905 Higgs was appointed his Private Secretary. Upon Campbell-Bannerman's death in 1908 he returned to the Treasury. Higgs remained a civil servant until his retirement in 1921. Like his friend James Bonar, Higgs seems to have found a career in the civil service sufficiently flexible to admit of active research into the history of economics and a close involvement with teaching and the professional associations of British economists. In this latter context, particularly to be noted is the instrumental part he played in securing for the British Economic Association (of which he was a founding member in 1890) its Royal Charter in 1902 when it changed its name to the Royal Economic Society. From 1892 until 1905 Higgs served as Secretary of the RES, and from 1896 until 1905 he was the assistant editor (to Edgeworth) of the *Economic Journal*.

In 1884 Higgs began attending lectures on jurisprudence and Roman law at University College, London, finally securing (after having first to matriculate) his LL.B in 1890. His only formal instruction in economics seems to have come from Foxwell, whose lectures at University College Higgs attended in 1885–6 and 1886–7, and it seems likely that it was from this source that his interest in the work of Richard Cantillon derived. (Higgs's book on Physiocracy is dedicated to Foxwell: 'my master and friend'.) Cantillon was probably the perfect subject for Higgs – he had been hailed by Jevons as having established the 'nationality of political economy', but his work and name were scarcely known; he had a mysterious personal history; and the book to which so much credit was being given apparently existed only in a French translation of the missing English original. Research in the Bibliothéque Nationale during his annual vacations provided the material upon which his article on Cantillon for the first volume of the *Economic Journal* in 1891 was based. On the same subject there followed his entry for Palgrave's *Dictionary* in 1894 and, after his retirement, his now standard edition of Cantillon's *Essai* was published under the auspices of the Royal Economic Society (1931).

From these researches undoubtedly sprang Higgs's other great interest in the area: the economics of Physiocracy. In May and June of 1896 Higgs gave a series of six lectures on the Physiocrats at the London School of Economics. These were published in the following year as *The Physiocrats*. In 1894 he had written the entry on the *Economistes* for Palgrave, but that piece ran to only two short paragraphs (one of which was little more than a list of names) and referred the reader to the much longer entry on the Physiocrats which was written by Gustave Schelle. His stamp was thus more permanently impressed on the study of Cantillon than it was upon that of Physiocracy.

Of Higgs's other contributions to the history of economics, only two need to be noted. First, and not surprisingly, Higgs was among the most stalwart of supporters of Palgrave's *Dictionary*, contributing nineteen entries to its original edition (including those on Cantillon, Mirabeau, Turgot, and the *Economistes*), and forty more to his own edition of 1923-6. His entry on 'Débouchés' has been retained in the present work. It is clear that the *Dictionary* is the vehicle which will perpetuate his name. Secondly, in the later years of his life he undertook to edit and produce for the Royal Economic Society bibliographical volumes on the literature of economics. The idea was to capitalize upon, and to record for posterity, the legacy of Foxwell's activities as a scholar and book collector extraordinary. Unfortunately, only one volume appeared (1935) – as Keynes remarked, it may have been that at so late a stage of his life the task no longer suited his gifts (1940, p. 556).

It might also be noted that Higgs's one disservice to the history of economics was his edition of an unfinished manuscript by Jevons which was published in 1905 under the title *Principles of Economics*. So fragmented is this, that it is very difficult to imagine just how Higgs could have been persuaded to print it. Appearing as it did when the climate of opinion about Jevons (largely due to Marshall's efforts) was somewhat less than enthusiastic, it numbers as one of those unfortunate incidents which have combined to diminish the reputation of Jevons in a way that is entirely unwarranted.

In addition to these works, Higgs published two books on the financial system of Britain. The first, *The Financial System of the United Kingdom*, appeared in 1914 was an attempt to provide a connected account of governmental financial procedure. The second, *Financial Reform*, appeared in 1924 and was more an account of the conduct of government policies as Higgs had directly experienced it. He also delivered the Newmarch Lectures at University College in 1892 and 1893 on household budgets, and was the editor of the Centenary Volume of the Political Economy Club.

An exemplary description of Higgs in later life was written by Keynes for the December number of the *Economic Journal* for 1940. So improbable is it that this will ever be bettered, it is reproduced below with the permission of the Society which Higgs helped to found (in the text, Keynes is referring to his recollections of meetings of the council of the Royal Economic Society):

> Becoming, at the last, extremely deaf and quite unable to hear the comments of others present, in which indeed he seemed to take no interest, his argument would continue as an entirely solo performance, frequently on some other item of the agenda than that under discussion; the only Chairman, in my experience, who was able to make him desist until his oration was really finished, being Edwin Cannan, who used to take him almost by the throat, shouting down his ear that we were not discussing that matter, and putting his hand over his mouth until he gave up. Or on other occasions when he had more curiosity as to what was going on, he would push towards whomever was speaking his highly unreliable electrical machine, which would proceed to deliver a thunder-and-lightning storm above which nothing could be heard. I wish I could give some slight indication of Higgs's very individual and oratorical manner of address. It could be a bore and a hindrance if one was in a hurry in this modern age, – or in such circumstances as the above! But if only one could be patient, it had in truth extraordinary finish and a sort of beauty of its own; unquestionably great style in it. These orations to our Council, delivered on the wrong items of the agenda, were often delightful in themselves, elaborately prepared beforehand, I sometimes thought, really remarkable in their own way and the best and most characteristic product of his personality (Keynes, 1940, pp. 557–8).

MURRAY MILGATE

See also PALGRAVE'S DICTIONARY OF POLITICAL ECONOMY.

SELECTED WORKS
1891. Richard Cantillon. *Economic Journal* 1, June, 262–91.
1897. *The Physiocrats*. London: Macmillan & Co.
1905. W.S. Jevons: *Principles of Economics*. Edited with Harriet Jevons, London: Macmillan & Co.
1914. *The Financial System of the United Kingdom*. London: Macmillan & Co.
1923-6. (ed.) *Palgrave's Dictionary of Political Economy*. 3 vols, London: Macmillan & Co.
1924. *Financial Reform*. London: Macmillan & Co.
1931. (ed.) R. Cantillon, *Essai sur la nature du commerce en général*. London: Macmillan & Co. for the Royal Economic Society.
1935. *A Bibliography of Economics down to 1700*. Cambridge: Cambridge University Press.

BIBLIOGRAPHY
Keynes, J.M. 1940. Obituary: Henry Higgs. *Economic Journal* 50, December, 555–8.

high-powered money and the monetary base. The concept of high-powered money or a monetary base appears as an important term in any analysis addressing the determinants of a nation's money stock in regimes exhibiting financial intermediation. Two types of money can be distinguished in such institutional contexts. One type only occurs as a 'monetary liability' of financial intermediaries. It characteristically offers a potential claim on another type of money. The contractual situation between customers and intermediaries reveals that this potential claim, to be exercised any time at the option of the owner, forms a crucial condition for the marketability of the intermediaries' monetary liabilities. This second type offers in contrast no such potential claim. While it is exchangeable for other objects, it is a sort of 'ultimate money' without regress to other types of money.

This characterization differs from the widely used classification 'outside-inside' money. 'Inside money' matches in a consolidated balance sheet of 'money producers' a corresponding amount of private debt. Money which cannot be matched in this way forms the outside money. But outside money does not necessarily coincide with the monetary base. The latter magnitude exceeds the volume of outside money by the amount of private debt acquired by the Central Bank in fiat regimes. The two concepts refer, however, to the same magnitude in pure commodity regimes and even in some possible Central Bank regimes with specific arrangements. It follows that the monetary base covers a somewhat wider range than outside money. This difference corresponds to the different analytic purposes of the two concepts. The 'monetary' base is designed for explanations of the behaviour of a nation's money stock, whereas 'outside money' was advanced to express the monetary system's contributions to the economy's net wealth.

The distinction between monetary base and the nation's money stock is hardly informative or relevant for pure commodity money regimes. The distinction becomes important with the emergence of intermediation. Financial intermediation inserts a wedge between the monetary base and the money stock (see article on money supply). But regimes with intermediation cover a wide range of arrangements bearing on the nature of the monetary base. High-powered money may consist of commodity money with or without fiat component or of pure fiat money. These differences are characteristically associated with significant differences in the supply conditions of high-powered money.

The measurement of the monetary base for any country involves, at this stage of monetary evolution, the consolidated balance sheet of the Central Bank system. But the Central Bank is usually not the only producer of 'ultimate money'. The balance sheet of other agencies may also have to be considered. This extension covers in the USA a special Treasury monetary account summarizing the Treasury's money creating activity. In other cases, a balance sheet of the mint or an exchange equalization account may have to be added. But whatever the range of ultimate money producers may be, we need to consolidate their respective balance sheets into a single statement. The monetary 'liabilities' of this consolidated statement, i.e., all items listed on the right-side of the consolidated statement which are money, constitute the monetary base.

The consolidated statement determines that the monetary base can be expressed in two distinct ways. It can be exhibited as the sum of its uses by banks and public. The 'uses statement' thus presents the monetary base as the sum of bank reserves in form of base money and currency held by the public. A 'source statement' complements the uses statement. The sources statement can be immediately read from the balance sheet. The monetary base appears thus as the sum of all assets listed on the left-side of the consolidated statement minus the sum of all non-monetary liabilities. Both statements can be easily derived from the published data in the USA. More difficulties may be encountered for other countries.

The comparatively simple case of the USA may be used to exemplify the sources statement needed for the subsequent discussion. We can write the following expression:

Monetary Base = Federal Reserve Credit

(i.e., earning assets of Central Bank consisting of government securities and advances to banks) + gold stock (including SDR's) minus treasury cash (i.e. free gold) + treasury currency (mostly coin) + a mixture of other assets minus other liabilities (including net worth).

Both uses and sources statement refer to important aspects of the money supply process. The uses statement refers in particular to the allocation of base money, determined by the public's and the bank's behaviour, between bank reserves and currency held by the public. This allocation contributes to shape the link between monetary base and money stock. The sources statement on the other hand directs our attention to an examination of possible (or relevant) supply conditions of base money.

The measurement, but not the definition, of the base clearly depends on prevailing institutions. One particular institution, viz. the imposition of variable reserve requirements on financial intermediaries, suggests a useful extension of the money base. Changes in reserve requirements release or absorb reserves similar to transactions between banks and Central Bank, e.g., an open-market operation. Similar consequences follow with respect to both money stock and 'bank credit'. Thus appeared an extension of the monetary base beyond the 'sources base' (or the volume of high-powered money) defined by the sources statement. The monetary base is understood as the sum of the 'sources base' and a reserve adjustment magnitude (RAM). This magnitude is the cumulated sum of all past releases and absorption of reserves due to changes in reserve requirements. This practice has become the standard procedure in the reports published by the Federal Reserve Bank of St Louis. The extended concept of the base offers the further advantage that the resulting magnitude only reflects actions of the monetary authorities and also reflects all the most important actions proceeding within a given institutional framework.

The sources statement offers a useful starting point for an analysis of the supply conditions of the monetary base. The study of these conditions is motivated by the systematic relation between base and money supply. Changes in the monetary base are a necessary condition for persistently large or substantially accelerated monetary growth in most countries for most of the time. Substantial changes in the monetary base

are frequently also a sufficient condition for corresponding changes in the money supply.

The sources statement yields a means to examine the sources of all changes in the base. We can thus investigate which of the sources dominate the trend, the variance of cyclical movements and the variances of middle range or very short-run movements. The patterns shift over time with the monetary regime and vary substantially between countries. Trend and longer-term variance in the USA are dominated, for instance, by the behaviour of the Federal Reserve Credit (i.e., the earning assets of the Central Bank system). We find in contrast for the Swiss case that trend and variance of the base are dominated by the behaviour of the gold stock and foreign exchange holdings. The portfolio of government securities play a comparatively small role. Such examination can also be exploited in order to judge whether movements in the base are essentially temporary or can reasonably be expected to persist with a longer duration.

The stochastic structure of the major and minor source components constitute the supply conditions of the monetary base. These conditions are sensitively associated with a variety of institutional arrangements under the control of legislative bodies or policymakers. The procedures instituted, for instance, by the Federal Reserve system to offer check collection services to banks contribute to the shortest run variance of the monetary base. Reserve requirements imposed on the liabilities of financial institutions offer policy-makers an opportunity to raise the proportion of outstanding government debt held by the Central Bank. Higher reserve requirements raise the level of the monetary base required to produce a given money supply. Correspondingly a larger volume of government securities can be held by the Central Bank.

The supply conditions may disconnect the behaviour of the base from the economy. This will happen whenever the processes governing the source components operate essentially independently of the economy's movements. In general some dependence may be produced by the prevailing institutions and policies. Such a feedback creates a role for the interaction within asset markets, and also between asset markets and output markets in the determination of the monetary base. The supply conditions of the monetary base acquire thus a central role in our monetary affairs. This is most particularly the case as these conditions emerge from legislative decisions and policy strategies. They fully characterize under the circumstances an important component of a monetary regime. Different monetary regimes are reflected by variations in the supply conditions. The growing dissatisfaction with the discretionary regime, which produced the Great Depression and the inflation of the 1970s, initiated in recent years much public debate about the nature of an adequate monetary regime. A rational examination requires in this case an evaluation of the consequences associated with alternative supply conditions governing the monetary base. This programme still needs some attention by the professions and ultimately (and very hopefully) even by politicians.

KARL BRUNNER

See also MONETARY POLICY; MONEY SUPPLY.

Hildebrand, Bruno (1812-1878). Hildebrand was born in Naumburg (Thuringia), the son of a clerk to the court. He studied in Leipzig and Breslau. In 1841 he was promoted full professor of Staatswissenschaften (of government, which included political economy) at the University of Marburg.

Hildebrand had always been an activist in the liberal and patriotic movement. He faced political persecution before the 1848 revolution, during which he was elected deputy of the Frankfurt National Assembly. In the subsequent period of restoration he was forced to emigrate to Switzerland, where he became not only a professor but also the director of a railway company, and founded the first Swiss statistical office (at Berne). In 1861 he was appointed professor at the University of Jena. He was founder (in 1862) and editor of the *Jahrbücher für Nationalökonomie und Statistik* and contributed to the establishment of the statistical office of the United Thuringian States (in 1864).

Hildebrand is considered as one of the founders of the German historical school. He was opposed to the deductive method of the classicals and denied the existence of 'natural laws' in economic life (Hildebrand, 1863). His most important work was *Die Nationalökonomie der Gegenwart und Zukunft*, where he discussed the theories of Friedrich List, Adam Müller, and especially those of Adam Smith. With his sharp criticism of self-interest and egoism as the central determinant of Smith's economic system – and the emphasis on ethical principles and the historically changing patterns of economic development – Hildebrand launched the attacks on Smith and the classical economists that were subsequently continued by many German historical economists.

The largest part of his main work was devoted to a discussion of socialism and communism, which he sharply rejected. Hildebrand focused his attention on the then little known Friedrich Engels and his recently published *Conditions of the Working Class in England* (Hildebrand [1848], 1922, pp. 125–90). He particularly criticized Engel's euphemistic description of pre-industrial conditions and contrasted it with empirical data that showed quite a different picture.

While being aware of current social problems Hildebrand perceived capitalist development most optimistically and envisioned as its last stage of development – the so-called 'credit economy' – a society where an advanced banking system would provide credit to a worker according to his morals and character and where thereby the monopoly of the capitalist class on capital would be broken (Hildebrand [1864], 1922, pp. 354–5). This theory of stages has to be regarded as Hildebrand's capitalist utopia, his liberal answer to socialism and communism.

Hildebrand's importance and his influence on the German historical school has generally been underestimated; after all Hildebrand was – as Max Weber remarked – the only one really to work with the historical method. He undertook statistical studies – he regarded statistics as an important tool for detailed historical and empirical research (Hildebrand, 1865) – and wrote historical monographs (Hildebrand, 1866). He thus anticipated much of the research programme of the 'younger historical school' and the Verein für Socialpolitik, which he joined – as the only economist of the 'older historical school' – as a charter member in 1873.

Hildebrand stood for a kind of progressive liberalism that intended to reshape Germany along the lines of England, which he admired.

HERMANN REICH

SELECTED WORKS

1848. *Die Nationalökonomie der Gegenwart und Zukunft und andere gesammelte Schriften*. Ed. H. Gehrig, Jena: Gustav Fischer, 1922. It contains the articles *Die gegenwärtige Aufgabe der Wissenschaft der Nationalökonomie* (1863), *Die wissenschaftliche Aufgabe der Statistik* (1865), and *Natural-, Geld- und Kreditwirtschaft* (1864).
1866. Zur Geschichte der deutschen Wollenindustrie. *Jahrbücher für Nationalökonomie und Statistik*, VI and VII.

Hilferding, Rudolf (1877–1941). Hilferding blended Marxist economics and Social Democratic politics in a career cut tragically short by the rise of fascism in Germany. He studied medicine at the University of Vienna, but soon showed more interest in organizing the student socialist society. After graduating in 1901, he helped Max Adler to found the *Marx-Studien* (1904–23), a series which was to become the theoretical flagship of 'Austro-Marxism'. The first volume contained a vigorous defence of the labour theory of value by Hilferding himself against Böhm-Bawerk's marginalist critique, *Zum Abschluss des Marxschen Systems* (1896). It earned him his intellectual spurs in the German-speaking socialist movement.

At the same time, Hilferding was already contributing to debate within the German Social Democratic Party (SPD) through its journal, *Die Neue Zeit*. There, on the controversial 'mass strike' issue, he steered a course for the party leadership between Eduard Bernstein's 'revisionist' abandonment of the socialist goal and Rosa Luxemburg's revolutionary commitment to it (1903/4, 1904/5). He was rewarded with an appointment in 1906 as economics lecturer at the party school in Berlin, and then as foreign editor of the party newspaper, *Vorwärts*. From 1907, he also wrote regularly for the newly established journal of the Austrian Social Democrats, *Der Kampf*.

Hilferding published his major work, *Das Finanzkapital*, in 1910; it was immediately hailed by such diverse figures as Kautsky (1911), Lenin (1916) and Bukharin (1917), as a path-breaking development of Marxist economic analysis. Essentially, Hilferding argued that the concentration and centralization of capital had led to the domination of industry and commerce by the large banks, which were transformed into 'finance capital' (1910, p. 225). The socialization of production effected by finance capital required a correspondingly increased economic role for the state. Society could therefore plan production by using the state to control the banking system:

> The socializing function of finance capital facilitates enormously the task of overcoming capitalism. Once finance capital has brought the most important branches of production under its control, it is enough for society, through its conscious executive organ – the state conquered by the working class – to seize finance capital in order to gain immediate control of these branches of production ... . Even today, taking possession of six large Berlin banks would mean taking possession of the most important spheres of large-scale industry ... (ibid., pp. 367–8).

This chain of reasoning, however, tended to exaggerate not only the leverage of the banks over industry, but also the role of the state in the organization of production. While it convinced Hilferding that socialism could be introduced by a determined majority in parliament, it demonstrated to Lenin that socialism would not be possible unless the state was 'overthrown' by a determined minority outside parliament. Their common point of reference was the centrality of the state – rather than society – in the 'latest phase of capitalist development'. It forced socialists to make a choice between parliamentarism and insurrection, the very nature of which contributed to the defeat of the labour movement in Germany and the rise of party dictatorship in Russia (Neumann, 1942, pp. 13–38). Although theory cannot be held responsible for the course of history, it may influence political judgements which tip the balance at decisive moments. Hilferding's generation lived through many such moments.

When war broke out in 1914, Hilferding associated himself with the SPD minority which voted against war credits and which later formed the Independent Social Democrats (USPD). He spent most of the war on the Italian front, having been drafted into the Austrian army as a doctor, and returned to Berlin as editor of the USPD journal, *Freiheit*. Hilferding successfully opposed USPD affiliation to the Third International; his speech against Zinoviev at the Halle conference of 1920 – published under the title, 'Revolutionäre Politik oder Machtillusionen?' – was a decisive turning point. Once the embryonic Communist Party (KPD) forced a split on the issue, however, he saw no alternative to reunification with the remnants of the SPD.

During the 1920s, Hilferding turned his attention almost entirely to the political and economic problems facing the new German republic. He was a leading member of the Reich Economic Council, twice Minister of Finance and an active participant in the discussions on 'workers' councils' and the government's 'socialization' programme. Hilferding's first stint as Minister of Finance lasted only seven weeks in the Stresemann government of 1923. Although he had no opportunity to implement his proposals, he devised a plan for currency reform involving the introduction of a *Rentenmark* backed by gold as part of an anti-inflation package. By the time Hilferding returned to the same post in the Müller government of 1928/9, economic conditions had worsened; his predicament was appreciated by Schumpeter who wrote, 'we now have a socialist minister who faces the exceptionally difficult task of curing or improving a situation bequeathed by non-socialist financial policies' (quoted in Gottschlacht, 1962, p. 24). A less sympathetic observer, however, portrayed Hilferding at this time as 'the theorist of coalition politics in the period of capitalist stabilisation' (see Gottschlacht, 1962, p. 204), blinded by theory to the imminent fascist danger.

Pursuing the logic of *Das Finanzkapital*, Hilferding had developed a theory of 'organized capitalism', a term he first used in 1915 in *Der Kampf*, and then explained more fully in 1924 in *Die Gesellschaft*. He summarized the approach at the SPD's Kiel conference in 1927: 'Organized capitalism means replacing free competition by the social principle of planned production. The task of the present Social Democratic generation is to invoke state aid in translating this economy, organized and directed by the capitalists, into an economy directed by the democratic state' (see Neumann, 1942, p. 23). Ironically, this was the very position of an earlier Social Democratic leadership which Marx had singled out for criticism. Commenting on the demand for a 'free state' in the 1875 Gotha programme, Marx wrote:

It is by no means the goal of workers who have discarded the mentality of humble subjects to make the state 'free'. In the German Reich the 'state' has almost as much 'freedom' as in Russia. Freedom consists in converting the state from an organ superimposed on society into one thoroughly subordinate to it; and even today state forms are more or less free depending on the degree to which they restrict the 'freedom of the state' (Marx, 1891, p. 354).

While Hilferding understood that in capitalist society power lay with capital and was exercised by the representatives of capital in the management structure of the great corporations, he failed to see that democratic control over the productive forces would require a change in the relationship of power *within* the corporation itself. Organized labour could use the state to accelerate this process of social transformation and to create the centralized institutional machinery necessary for the

'associated producers' to plan directly the whole economy; but the notion that the state itself could perform this task rested upon an illusion. In attempting to replace the domination of capitalist employers with the domination of a 'democratic state', Hilferding and the party leadership achieved only one practical result: 'Unwittingly, they strengthened the monopolistic trends in German industry' (Neumann, 1942, p. 21). The state domination which followed was far from democratic.

Hilferding, a Jew, was forced into exile after 1933, first in Switzerland via Denmark and then in France. In an unfinished manuscript, *Das historische Problem*, he set about revising his whole conception of the state. The problem was now said to consist 'in the change in the relation of the state to society, brought about by the *subordination of the economy* to the coercive power of the state ...' (quoted by Bottomore, Introduction to Hilferding, 1981, p. 16, emphasis in original). Hilferding briefly presented his new approach in the New York *Socialist Courier* in 1940; there, like Marx, he drew a rueful comparison between Germany and Russia. The state had not 'withered away' under Soviet communism:

History, that 'best of all Marxists', has taught us another lesson. It has taught us that, in spite of Engels' expectations, the 'administration of things' may become an unlimited 'domination over men', and thus lead not only to the emancipation of the state from the economy but even to the subjection of the economy by the holders of state power (Hilferding, 1981, p. 376 fn.).

It was too late for Hilferding's brave reassessment to influence the course of events. In 1941, he died in the hands of the Gestapo.

ROY GREEN

SELECTED WORKS Books
1904. *Böhm-Bawerk's Criticism of Marx*. Ed. P. Sweezy, London: Merlin Press, 1975.
1910. *Finance Capital: A Study of the Latest Phase of Capitalist Development*. London: Routledge & Kegan Paul, 1981. Articles
1902/3. Der Funktionswechsel des Schutzzolles. Tendenz der modernen Handelspolitik. *Die Neue Zeit* XXI, 2.
1903/4. Zur Frage des Generalstreik. *Die Neue Zeit* XXII, 1.
1904/5 Parliamentarismus und Massenstreik. *Die Neue Zeit* XXIII, 2.
1915. Historische Notwendigkeit und notwendige Politik. *Der Kampf* VIII.
1915. Arbeitsgemeinschaft der Klassen? *Der Kampf* VIII.
1924. Probleme der Zeit. *Die Gesellschaft* I, 1.
1924. Realistischer Pazifismus. *Die Gesellschaft* I, 2.
1933. Zwischen den Entscheidungen. *Die Gesellschaft* X.
1933/4. Revolutionärer Sozialismus. *Zeitschrift für Sozialismus* I.
1934/5. Macht ohne Diplomatie – Diplomatie ohne Macht. *Zeitschrift fur Sozialismus* II.
1940. State capitalism or totalitarian state economy. *Socialist Courier*, New York; reprinted in *Modern Review* I, (1947). Published Speeches
1919. Zur Sozialisierungsfrage. 10th Congress of the German trade unions, Nuremberg, 30 June – 5 July 1919. Berlin.
1920. Revolutionäre Politik oder Machtillusionen?. Speech against Zinoviev at the annual conference of the USPD in Halle, 1920. Berlin.
1920. Die Sozialisierung und die Machtverhältnisse der Klassen. 1st Congress of Works Councils, 5 October 1920. Berlin.
1927. Die Aufgaben der Sozialdemokratie in der Republik. Annual conference of the SPD in Kiel, 1927. Berlin.
1931. Gesellschaftsmacht oder Privatmacht über die Wirtschaft. 4th AFA (Allgemeiner freier Angestelltenbund) trade union congress in Leipzig, 1931. Berlin.

BIBLIOGRAPHY
Böhm-Bawerk, E. von. 1886. *Karl Marx and the Close of his System*. Ed. P. Sweezy, London: Merlin Press, 1975.

Bukharin, N. 1914. *Imperialism and World Economy*. London: Merlin Press, 1972.

Gottschlacht, W. 1962. *Struktur veränderungen der Gesellschaft und politisches Handeln in der Lehre von Rudolf Hilferding*. Berlin: Duncker & Humblot.

Kautsky, K. 1911. Finanzkapital und Krisen. *Die Neue Zeit* 29, 764–72, 797–803, 838–46, 874–83.

Lenin, V.I. 1916. *Imperialism, the Highest Stage of Capitalism*. Moscow: Foreign Languages Publishing House, 1947.

Marx, K. 1891. Critique of the Gotha Programme. In *The First International and After*, ed. D. Fernbach, Harmondsworth: Penguin, 1974.

Neumann, F. 1942. *Behemoth*. London: Victor Gollancz.

Hill, Polly (born 1914). Polly Hill was born on 10 June 1914 into a remarkable Cambridge family that includes Nobel Prize winning physiologist A.V. Hill (her father) and J.M. Keynes (her mother's brother) among its many distinguished members. She graduated from Cambridge in 1936 with a degree in economics.

Her first job upon leaving university was with the Royal Economic Society as an editorial assistant, a position she held for two years (1936–8). Her next appointment was a one year (1938–9) research position with the New Fabian Research Bureau (which almost immediately re-amalgamated with the Fabian Society) where she wrote her first book *The Unemployment Services* (1940). This book was concerned to expose the inefficiency and inhumanity of the system of unemployment relief and to make constructive proposals. Polly Hill's commitment to social justice has not waned: economic inequality is the central theme of all her books.

At the outbreak of the war she was obliged, as an unmarried young woman, to become a temporary civil servant. She worked first, briefly, in the Treasury, then for a long time in the Board of Trade and finally in the Colonial Office. She resigned in 1951. After a period of unemployment she became a journalist for the weekly *West Africa*. She married in 1953 and moved to Ghana with her husband where, at the age of forty, she began her academic career. The academic posts she held there involved no teaching and she was able to become, as she has put it, 'a pupil of the migrant cocoa farmers of southern Ghana'. She began her fieldwork as an economist and collected data using the questionnaire method, producing her second book, *The Gold Coast Cocoa Farmer: A Preliminary Survey* (1956) with characteristic speed and efficiency. The prevailing orthodoxy had it that sedentary food farmers in southern Ghana had suddenly taken up cocoa farming at the end of the 19th-century with such a degree of success that cocoa exports had risen from nil to over 50,000 tons by 1914 – the largest quantity for any country. Polly Hill had uncritically accepted this orthodoxy and her subsequent realization that most farmers appeared to be migrants who had bought their land was to have a profound effect upon her intellectual methods. She abandoned the questionnaire method of data collection in favour of one that sought to develop generalizations on the basis of: (1) detailed fieldwork in one village; (2) fieldwork done by others elsewhere; (3) archival sources. She also began a lifelong struggle with development economists and other purveyors of orthodoxies based on casual empirical observation and 'commonsense'. She drifted towards anthropology and history where the qualities of her empirical findings were recognized for what they were: revolutionary. She spent three and a half years collecting detailed evidence to substantiate her claim that the cocoa farmers were migrants and made many fascinating discoveries

in the process. For example, she found that the matrilineal farmers adopted an entirely different mode of migration from patrilineal farmers: the former bought family lands with the aid of their kin, and were prepared to grant usufructural rights to their male and female kinsfolk; the latter clubbed together in so-called 'companies', groups of non-kin, the land being divided into strips from a base line, according to the contribution each had made, with subsequent division in inheritance always being longitudinal. Upon hearing of this Professor Meyer Fortes, then Professor of Social Anthropology at Cambridge, encouraged her to apply for a Smuts Visiting Fellowship. This enabled her to write *The Migrant Cocoa-Farmers of Southern Ghana: A Study in Rural Capitalism* (1963) which is now widely regarded as a classic. (She was awarded a PhD in social anthropology from Cambridge under new special regulations in 1966 on the basis of it.) Mainstream writers on development have by and large ignored the book even though it contains telling criticisms of aspects of W.A. Lewis's work.

Following more fieldwork in Ghana, Nigeria and India she produced a further stream of books (1970a, 1970b, 1972, 1977, 1982, 1985, 1986) and many articles of outstanding quality which established her reputation as the world's foremost economic anthropologist. She was appointed a Fellow of Clare Hall in Cambridge in 1965 (a position she still holds) and subsequently to the prestigious Smuts Readership in Commonwealth Studies (1973–9). Her publications document in painstaking detail the complexity of agrarian relations in the tropical regions of the world in which she has worked. The books as a whole constitute an encyclopaedia of knowledge on the socio-economic conditions of poverty and economic inequality and her work ranges in scope from 'agrestic servitude' to 'zamindars'. Her oeuvre is much more than a compilation of facts though. Her own data and that of others is presented in a theoretical context which has broadened as her own field experience has widened. She has been unrelenting in her empirically based critiques of development economists and her latest book (1986) is a concerted attempt to make them see the error of their ways.

C.A. GREGORY

SELECTED WORKS

1940. *The Unemployment Services*. London: Routledge.

1956. *The Gold Coast Cocoa Farmer: A Preliminary Survey*. Oxford: Oxford University Press.

1963. *The Migrant Cocoa-Farmers of Southern Ghana: A Study in Rural Capitalism*. Cambridge: Cambridge University Press.

1970a. *The Occupations of Migrants in Ghana*. Ann Arbor: University of Michigan Press.

1970b. *Studies in Rural Capitalism in West Africa*. Cambridge: Cambridge University Press.

1972. *Rural Hausa: A Village and a Setting*. Cambridge: Cambridge University Press.

1977. *Population, Prosperity and Poverty: Rural Kano 1900 and 1970*. Cambridge: Cambridge University Press.

1982. *Dry Grain Farming Families: Hausaland (Nigeria) and Karnataka (India) Compared*. Cambridge: Cambridge University Press.

1985. *Indigenous Trade and Market Places in Ghana, 1962–64*. Jos Oral History and Literature Texts, University of Jos, Nigeria.

1986. *Development Economics on Trial: The Anthropological Case for a Prosecution*. Cambridge: Cambridge University Press.

Hirschman, Albert Otto (born 1915). Hirschman was born on 7 April 1915 in Berlin. After attending the Sorbonne and the London School of Economics he obtained a doctorate in economic science from the University of Trieste in 1938. His early career was dominated by the struggle against fascism in Europe (Coser, 1984). He actively supported the underground opposition to Mussolini while in Italy in the mid-1930s, fought with the Spanish Republican Army in 1936 and later with the French Army until its defeat in June 1940. He stayed on in Marseilles six months more, engaging in clandestine operations to rescue political and intellectual refugees from Nazi-occupied Europe. He avoided arrest by leaving France for the United States in January 1941. There he produced his first book, *National Power and the Structure of Foreign Trade* (1945), which introduced some of the main themes of what is now called 'dependency theory'.

After the war he served as an economist in the Federal Reserve Board until 1952, when he left for Colombia where he stayed four years. Beginning in 1956 he held professorships successively at Yale, Columbia and Harvard, and in 1974 was appointed professor at the Institute for Advanced Study in Princeton.

Hirschman has been a leading figure in economic development since the publication in 1958 of his second book, *The Strategy of Economic Development*. Hirschman's analysis grew out of extensive practical experience in Colombia as an adviser both to its government and to private firms. Characteristically, Hirschman dissented from orthodox views of both right and left, arguing that neither *laissez faire* nor 'rational' economy-wide planning made sense for poor countries. Government needed to encourage 'unbalanced growth', deploying its scarce decision-making capacities strategically to set up disequilibria that would stimulate effort and mobilize hidden and underutilized resources. Targeting development efforts on key industries with strong 'linkages' to other parts of the economy could stimulate a favourable dynamic.

Hirschman later provided the label 'possibilism' (1971) for the outlook that shaped much of his thought on development and on which he elaborated in many further books and articles. When social science focuses exclusively on the search for general laws, it obscures the irreducible role of the unique and the unpredictable in human affairs. This causes progress to be viewed either as ensured by the application of general rules or thwarted by the presence of inescapable obstacles. But history reveals that actual social change often follows paths that are *a priori* quite unlikely, turning obstacles into opportunities and confounding rules with unanticipated consequences. From this starting point, Hirschman has cultivated an approach to development problems which embodies respect for complexity and openness to the possibility of genuine novelty – what he once called 'the discovery of 'an entirely new way of turning a historical corner' (1971, p. 27).

Since 1970, Hirschman has been bringing his possibilist approach to bear on broader problems of social theory. His slim volume, *Exit, Voice, and Loyalty* (1970) revealed the unexpected richness to be found in comparing the implications of dissatisfied clients alternatively *exiting* from an organization or giving *voice* to their complaints. This volume, like Hirschman's more recent work (1982b) on the forces that propel individuals and societies into and out of periods of intense political involvement, explores issues on the borderline between economics and politics. But unlike most economists with an interest in 'public choice', Hirschman shows no inclination to reduce politics to economics. Indeed, both works stress that standard models of economic behaviour fail to make sense of familiar forms of 'public-minded' behaviour such as voicing one's convictions on public matters,

participating in demonstrations or working to support candidates for office.

Hirschman's propensity to devise analytical formulations that express rather than conceal the complexities of human motivations and institutions is evident also in his studies of historical views of capitalism (1977, 1982a). Hirschman shows that capitalism has been seen as a powerful civilizing influence and alternatively as a destroyer of the moral and social fabric; still other views have portrayed capitalism, for better or worse, as too feeble to overcome the restraints of preceding social forms. These competing ideological views have evolved, Hirschman notes, in total isolation from one another. A fuller view would recognize that all these contradictory tendencies are present at once, but to recognize this truth would be highly inconvenient, making it 'much more difficult for the social observer, critic, or "scientist" to impress the general public by proclaiming some inevitable outcome of current processes' (1982a).

'But', Hirschman concludes, in a question that captures well his own unique stance in modern social science, 'after so many failed prophecies, is it not in the interest of social science to embrace complexity, be it at some sacrifice of its claim to predictive power?' (1982a, p. 1483).

M.S. McPHERSON

See also EXIT AND VOICE; LINKAGES.

SELECTED WORKS

1945. *National Power and the Structure of Foreign Trade*. Berkeley and Los Angeles: Bureau of Business and Economic Research, University of California.
1958. *The Strategy of Economic Development*. New Haven: Yale University Press.
1970. *Exit, Voice, and Loyalty: Responses to Decline in Firms, Organizations, and States*. Cambridge, Mass: Harvard University Press.
1971. Introduction: political economics and possibilism. In A.O. Hirschman (ed.), *A Bias for Hope: Essays on Development and Latin America*, New Haven: Yale University Press.
1977. *The Passions and the Interests: Political Arguments for Capitalism Before Its Triumph*. Princeton: Princeton University Press.
1982a. Rival interpretations of market society: civilizing, destructive, or feeble? *Journal of Economic Literature* 20, December, 1463–84.
1982b. *Shifting Involvements: Private Interest and Public Action*. Princeton: Princeton University Press.
1986. *Rival Views of Market Society and Other Essays*. New York: Viking–Penguin International.

BIBLIOGRAPHY

Coser, A. 1984. *Refugee Scholars in America: Their Impact and Their Experiences*. New Haven: Yale University Press.

historical cost accounting. Historical cost is the cost at which an asset was actually purchased. This is the value traditionally imputed to assets in accounts. Valuation at historical cost was a natural process in the early days of accrual accounting. Historical cost represented money which had been paid out (or a liability created) which was not to be charged against profit because it represented the creation of an asset, rather than an expense. Thus, the logic of double entry suggested that assets should, initially at least, be valued at historical cost.

However, the survival to the present day of historical cost as a valuation basis in accounts is not due merely to its easy assimilation into double entry book-keeping. For accountants, historical cost has at least two attractions relative to current valuation bases, such as current replacement cost or realizable value. Firstly, it is relatively objective, having been established by a verifiable transaction on which two independent accountants would be likely to take the same view, whereas current values involve estimating what would happen if a transaction (replacement or sale) were to occur. Secondly, it is conservative, insofar as it does not recognize gains in value which have taken place since the asset was acquired.

It is alleged (e.g. by Ijiri, 1971) that these properties help historical cost accounts to fulfil the stewardship function of providing users of accounts with a relatively objective statement of the financial transactions of those responsible for managing the assets of the business. Accounts which were surrounded by greater uncertainty, due to the subjectivity of the valuation base, might not fulfil this function so well. Furthermore, the conservative practice of not showing any gains in the value of assets due to price rises since the acquisition date is a protection against the manipulation of accounts by unduly optimistic or unscrupulous managers.

On the other hand, the principle of conservatism has been applied so strongly that it has been allowed to modify historical cost in certain cases where current market value is lower than cost. Thus, in the United Kingdom, the valuation rule for current assets such as stocks and work in progress is, in conventional accounts, 'cost or current market value, whichever is the lower'. For fixed assets of limited life, depreciation is traditionally written off the historical cost of the asset over its lifetime. Written down historical cost does not claim to be a close approximation to current market value, but it is less likely to exceed market value than unadjusted historical cost. The estimation of depreciation reduces the objectivity of historical cost valuation, as does the introduction of lower market values, thus diminishing one of the important advantages claimed for historical cost.

Another common breach of the historical cost system in conventional accounting practice is the periodic revaluation of fixed assets in the balance sheet. This has become accepted practice in the United Kingdom, as a response to the pressure for more relevant information in a period of rising prices. The integrity of historical cost profit is usually preserved by not passing the revaluation through the profit and loss account, that is, the increased value of the assets is regarded as a capital gain, giving rise to an increase in undistributable reserves rather than in profit. On the other hand, the principle of conservatism is applied so that future depreciation charges against profit are based on the revalued amount, so that the charges are higher, and profits lower, than if the revaluation had not taken place. Thus, the effect of the revaluation is to depress the future accounting rate of return by increasing the numerator (profit, after charging depreciation) and increasing the denominator (net assets).

The above description applies to current conventional accounting practice in the United Kingdom. However, historical cost is currently the basis of conventional financial accounts in all major capitalist economies, and in each case there are departures from strict historical cost to meet difficulties which have been encountered in practice; for example, certain Latin American countries which have suffered very high inflation rates have requirements for applying indexation to historical cost. The widespread survival of historical cost accounting can be attributed to two factors. Firstly, the firm transactions base of historical cost accounting gives it a degree of objectivity which, although not as great as might appear at first sight, is not matched by alternative systems. Secondly, vast experience of implementing historical cost has accumulated. Thus, accountants are better equipped to implement it rather than alternative systems, such as current cost accounting.

Accounting practice has evolved as a pragmatic response to practical difficulties, and most accountants think of it in this way rather than as the rational application of theoretical principles. Thus, it seems likely that there will be powerful support from the accounting profession for the continued evolution of generally accepted accounting principles (known in the United States as GAAP), based on historical cost but with an increasing degree of modification, rather than its revolutionary replacement by a different valuation base, such as current cost accounting (as proposed by the Sandilands Report (1975) in the United Kingdom).

G. Whittington

See also ACCOUNTING AND ECONOMICS; INFLATION ACCOUNTING.

BIBLIOGRAPHY

Ijiri, Y. 1971. A defence for historical cost accounting. In *Asset Valuation and Income Determination, A Consideration of the Alternatives,* ed. R.R. Sterling, Houston, Texas: Scholars Book Co., 1975.

Sandilands Report. 1975. *Inflation Accounting: Report of the Inflation Accounting Committee under the Chairmanship of F.E.P. Sandilands,* Cmnd. 6225, London: HMSO, September 1975.

historical demography. What is historical demography? Some writers define it by its object, i.e. by the populations of the past. Most define it by its sources and methods. Since ancient statistics are imperfect, incomplete or even non-existent, new means had to be found to investigate past populations; in particular, material could be extracted from historical sources which had been compiled not for scientific purposes but to allow civil and religious authorities to control populations. These included baptism, marriage and burial registers, tax rolls, cadastral surveys, or family papers such as contracts, inventories and genealogies.

The historical demographers of our time are the distant successors to the 17th century political arithmeticians. Graunt, Petty and King attempted to calculate the size of the English population in the absence of a census, with the help of incomplete non-demographic documents (London mortality reports, chimney taxes, and so on). Today Mols, Goubert, Henry, Laslett, Wrigley and others seek not only to determine the size and distribution of past populations, but also to measure their structure and behaviour with the help of an even broader variety of sources and more refined techniques.

It is possible to classify their methods under two major rubrics: *microanalysis* and *aggregative analysis.*

Microanalysis consists in deploying the maximum amount of information about a given group of individuals, to create a representative sample of the population to study. Genealogy is the most ancient and the most typical example of microanalysis. Genealogy provides the skeleton of most cultures. It did not, however, become precise and complete until the Renaissance. Moreover, it did not, until recently, become the object of truly scientific analysis. Examples of such analysis are Henry's study of the bourgeoisie of Geneva (1956) and Hollingsworth's study of the British ducal families (1957).

Ascendant genealogies provide all the ancestors of an individual along the paternal as well as the maternal lines. From the point of view of social history, ascendant genealogies have the serious fault of being non-representative. They are limited by the upper-class origin of their authors, by the narrow and select range of groups under study. From the historical demographic point of view, they suffer from a constitutional vice: since not all social groups have reproduced themselves in similar proportions, ascendant genealogies create an illusion of low death rates but of high fertility rates in past populations. In particular, such genealogies never contain dead branches.

Descendant genealogies can escape such criticism. However, the founders of a line have mostly been chosen for their past glory (for example Charlemagne). In order to get a genuine image of past populations, such ancestors should be chosen according to scientific criteria, for only then would it be possible to constitute a representative sample. But this in turn leads to a technical problem. Due to geographical mobility, it is difficult to follow the destiny of all descendants; many individuals are known only by their birth certificates.

The *family reconstruction* method apparently provides a means of avoiding such difficulties. It consists in following individuals within a territorial framework – a parish or a group of parishes – counting migrants as 'profits and losses'. This method was in the first instance applied to the population of the small town of Durlach by the German historian Roller (1907). It was then perfected, first by the geneticist Scheidt (1928), then by the Swedish demographer Hyrenius (1942), and finally formalized by the French historian Louis Henry (Fleury and Henry, 1965; Gautier and Henry, 1968). According to this method, parish births (or baptisms), marriages and deaths (or burials) are all noted down on cards in chronological order. They are then regrouped on large size cards, called family cards (one per couple). These cards contain boxes for calculating age and intervals (age at marriage and at death, age of the mother at the birth of each child, interval between marriage and the first born, between successive births, between widowhood and remarriage). This makes it possible to form a picture of marriage rates, death rates and, what is most important, of fertility rates, i.e. of female fertility according to age group and age at the time of marriage, fertility according to the length of marriage and age at the time of marriage, late fertility, age at the time of the last born, ultimate descendancy and so on.

The investigation is complete only in the case of families where the female age, date of marriage and the end of the observation are precisely known. Other types of records are much less thorough. Henry has however devised a technique for measuring and correcting any under-registration of infant births and deaths.

The invention of the family reconstitution method has had important repercussions in demography, in history and in the social sciences in general. In demography, this method has led to progress in longitudinal analysis. Here events are related to a prior event and not to the actual population as a whole. For example, remarriage is not considered in terms of the number of widowers in the population, but in terms of the age at the wife's death and the interval between remarriage and the wife's death. In history, this method marked the high point of 'cliometry' – at no time thus far has it reached such a degree of formalization and abstraction. Moreover, Henry's studies have led to greater sensitivity among historians towards the notion of representativeness, stimulating them to reach conclusions that can be generalized. Indeed, in social science, the invention of the family reconstitution method has been equal to the invention of the microscope in natural science. It has revealed all that we can expect from micro-observation, provided that it is used with rigour and discernment.

Aggregative analysis has never been as well formalized, and its methods have never been the object of systematic study. In general, these methods are related to those of classical demography, with the use of statistical tools such as the calculation of central parameters (median and average) and dispersion parameters (variance), correlation and dispersion coefficients, significance tests and factorial analysis. For historians, progress in information technology has eased the manipulation of these techniques. The construction of models has been more beneficial in demography than in economics, for the number of parameters has been more limited. With the theory of stable populations, it is relatively easy to link the distribution of a population according to age, fertility, mortality and growth, as demonstrated already in 1760 by Euler and then since 1907 by Lotka. In 1966, Coale and Demeny constructed a network of mortality tables of stable populations on this basis. This has been most useful to demography in general and to historical demography in particular. The Princeton school has launched a systematic investigation into demographic transition in Europe. The most remarkable book to emerge from this investigation contains a complete reconstruction of the 19th century French female population (Van de Walle, 1974). More recently, E.A. Wrigley and R.S. Schofield have been able to regularly evaluate the total population, migratory records and even fertility and mortality records (Wrigley and Schofield, 1981) by reconstituting statistics of baptism and death in England since 1540. Their research was based on the back-projection method, invented by R. Lee.

For most populations of the past, the construction of models tends to produce very precise results. For in the long term, birth and death rates are liable to reach an equilibrium, a characteristic of stationary populations. It is possible to construct a mortality table and an age pyramid based on the distribution of death according to age; in this case life expectancy varies inversely with the mortality rate.

The constitution of historical demography as an autonomous discipline has had important repercussions on the history of populations. It is now applied in all countries that possess satisfactory records of births, marriages and deaths (the whole of Europe, except those countries in which the Orthodox Church remained predominant, the two Americas, the older colonies) making it possible to identify the demographic state of peasant populations and even of a few urban populations (Bardet, 1983; Perrenoud, 1979), in particular, important characteristics such as late marriage, the rather important frequency of celibacy, high legitimate fertility, low illegitimate fertility, very high death rates which vary from year to year (Flinn, 1981). Despite frequent crisis, these populations were kept in balance by a self-regulating system based on late marriage. Young people were allowed to procreate only when they had the means to settle down. But the number of settlements was limited by the set of techniques used, the stagnant character of the economy, and the system of distribution of landownership (Dupaquier, 1979).

The family reconstitution method has generally produced parish monographs such as that for Crulai (Gautier and Henry, 1958), most of which concern the 17th and 18th centuries. In Italy and in France, interesting debates have focused on the extent to which these monographs are representative (Dupaquier, 1984). In Canada, the department of demography at the University of Montreal has started an integral reconstitution of French Canadian families since their arrival in Quebec. In France, INED has reconstituted the families of a representative sample of forty villages during the period 1670–1829 which resulted in important conclusions concerning the evolution of peasant fertility.

Aggregative analysis is much less efficient in measuring behaviour. But it throws light on the size and structure of populations. It has been applied with growing rigour and supported by the construction of models. This method made it possible to correct 19th century statistics (Van der Walle, 1974), to study demographic transition in Europe (Knodel, 1974; Livi-Bacci, 1977), to reconstitute the history of the French population since 1740 (*Population*, special issue, November 1975), and that of the English population since 1540 (Wrigley and Schofield, 1981). These last two studies have yielded particularly interesting results. The reconstitution of the French population was carried out by INED on the basis of a sample (at the rate of 1/500) in state registers of a group of communes which formed a representative sample. This made it possible to calculate not only the movement of the population, but its distribution according to age groups, with five year intervals, on the basis of statistics. These provide age at death and are classified according to year and generation. Mortality tables are also used for the same purpose. According to this investigation, France contained at least 24,600,000 inhabitants in 1740 (within its present borders), 28,100,000 in 1790, and 30 million in 1810. So demographic growth preceded economic growth. This in itself threw into doubt the old theory that attributed changed to an 'agricultural revolution'. Finally, the work of Wrigley and Schofield is based on an aggregate of 530 parish series which have been calculated by many correspondents of the Cambridge Group. Since they knew the flows (baptism = entry flow; death = exit flow), the authors proceeded to look for the basic characteristics of the population which produced these flows. They concluded that the growth of the 18th-century English population was three-quarters accounted for by an increase in the number of marriages (in particular by an increase in early marriage) and only one-quarter by a fall in the death rate. For Wrigley and Schofield, the high death rates of the past must be attributed first of all to exogenous factors (bad harvests, epidemics). They cannot be attributed to fluctuations in wages, nor to a deterioration in living standards.

The success of historical demography has had its most spectacular repercussions in the area of historical anthropology, in particular in the history of sexuality and the family. For the first demographic historians, one of the most striking discoveries concerned the low frequency of illegitimate births, at least in the French countryside. From the historical and literary evidence, it had previously appeared that traditional societies were overflowing with bastards. However, the proportion of illegitimate infant baptisms was actually contained within the 0.5 to 2 per cent mark, that is, four or five times less than in contemporary Western societies – even before the spread of reliable methods of contraception. In the towns, the frequency of illegitimacy was higher, partly because of the flight towards the towns of pregnant peasant girls. But it remained lower than that of the 19th and 20th centuries, despite the relatively advanced age of marriage and the non-negligible incidence of celibacy.

These observations led historians to consider the history of sexuality. Most historians see the 17th century as a puritan time following a long period of sexual freedom (the end of the Middle Ages and the Renaissance) and preceding a slackening in morals at the beginning of the 18th century. Some historians have even attributed the creativity of the classical period to the sublimation of the sexual instinct (Chaunu, 1966). This view has been contested by J.L. Flandrin, who has questioned its statistical basis. But analysis of the interval between marriage and the appearance of the first-born demonstrates the equally low frequency of pre-marital conceptions, except in countries like England where marriage

remained a private affair. It is in this atmosphere of controversy that the history of sexuality began and resulted in such works as those of Shorter (Canada), Stone and Laslett (Britain).

The history of the family is deeply indebted to Laslett, the founder of the Cambridge Group. Laslett discovered that it was difficult to apply Henry's family reconstitution method in England due to the small content of title deeds. He had recourse to other sources, in particular ancient lists of names. This led him to new research perspectives such as the relationship between intermixing within the population and the structure of households. Laslett challenged the theory of 'contraction' in sociology, after observing the overwhelming predominance of the nuclear family in modern England, even before industrialization. According to the 'contraction' theory, there is a gradual evolution from the 'patriarchal' model to the present small family. In September 1969, in order to broaden his conclusions, Laslett organized an international colloquium at Cambridge on the structure of households. This led to the no doubt exaggerated impression of the general predominance of the mononuclear model in traditional societies (Laslett and Wall, 1972). This idea was in turn quickly challenged by the American L.K. Berkner, who demonstrated that the 'social family' could not appear with its true frequency on the basis of name selection. For even when it constituted the predominant model, it could reach full development only during one phase in the family cycle. This study and all those that followed led historians to reject the idea of a universal model. Nevertheless, all these studies had the merit of turning the history of the family into a fashionable subject, mainly in England and in the United States. They also opened up broad inter-disciplinary perspectives.

In 1960, the French historian Philippe Aries published his famous book 'The Child and Family Life under the Ancien Regime'. Aries has revived the history of 'mentalités' and the analysis of the emotional life of the family. Unfortunately he often does not see the wood for the trees and seems to have ignored a fundamental requirement of science, that of being representative. So these subjects have remained outside the scope of historical demography.

In social history, historical demography at first achieved a high level of success. It created the opportunity to investigate and analyse the behaviour of the lower classes, in particular that of peasants – while former studies had been limited to the nobility and the upper middle classes. Nevertheless, historical demography has as yet far from exhausted itself. In fact, methods of micro-observation should make it possible to analyse not only the specific behaviour of social classes, but should also give scope for scientific measurement of geographic and social mobility. For a long time, historian-demographers have considered completing the family cards with facts derived from sources other than state registers. These would include tax rolls, contracts of marriage, post-mortem inventories. However, the limited scope of the majority of such documents means that part of the population – usually the poorest section – is excluded from the research. Moreover, due to technical exigencies underlying the case of the family cards, this research is limited to stable families, i.e. to those that spend all their lives within the framework of the village. There has been, and there remains, much discussion concerning the demographic representativeness of such stable families, whose fertility rate may not have been very different from that of mobile families. Yet it is doubtful that their respective marriage and death rates were equal. In any case, these stable families do not convey a true image of social phenomena. For the study of such families conveys the false impression of a static society. Social mobility is closely linked to geographic mobility. With the help of micro-observation methods, any social change can be traced only by following the fate both of mobile individuals and of stable families. This implies that the area of study should not be limited by territory; that is should not be limited to parish monographs. In other words, for historical demography to become legitimate social history, the territorial perspective will have to make way for the genealogical perspective. As we have seen above, the technical difficulties involved are considerable, and the problem of how representative selected families are is acute. Only when such problems have been overcome can historical demography make a fresh start and conquer new fields.

J. DUPÂQUIER

BIBLIOGRAPHY

Journals
Annales de démographie historique. 1965 onwards (annual). Paris: Ecole des Hautes Etudes en Sciences Sociales.
Population. 1946 onwards (6 issues per year). Paris: Institut National d'Etudes Démographiques.
Population Studies. 1947 onwards (4 issues per year).

Books
Aries, P. 1960. L'enfant et la vie familiale sous l'ancien régime. *Année sociologique* 60, 253–7.
Bardet, J.P. 1983. *Rouen aux XVIIe et XVIIIe siècles. Les mutations d'un espace social.* Paris: SEDES.
Charbonneau, H. 1970. *Tourouvre-au-Perche aux XVIIe et XVIIIe siècles.* INED, Travaux et Documents Vol. 55, Paris: PUF.
Chaunu, P. 1966. *La civilisation de l'Europe classique.* Collection 'Les grandes civilisations', Paris: Arthaud.
Coale, A. and Demeny, P. 1966. *Regional Model Life Tables and Stable Populations.* Princeton: Princeton University Press.
Cook, S. and Borah, W. 1971. *Essays in Population History: Mexico and the Caribbean.* Berkeley: University of California Press.
Dupaquier, J. 1979. *La population rurale du Bassin parisien à l'époque de Louis XIV.* Paris and Lille: EHESS.
Dupaquier, J. 1984. *Pour la démographie historique.* Collection 'Histoires', Paris: PUF.
Dupaquier, J. and M. 1985. *Histoire de la démographie.* Collection 'Pour l'Histoire', Paris: Librairie Académique Perrin.
Eversley, D.E.C., Laslett, P. and Wrigley, E.A. 1966. *An Introduction to English Historical Demography.* Ed. E.A. Wright, London: Weidenfeld & Nicolson.
Fleury, M. and Henry, L. 1965. *Nouveau manuel de dépouillement et d'exploitation de l'état civil ancien.* Paris: INED.
Flinn, M.W. 1981. *The European Demographic System.* Brighton: Harvester Press.
Gautier, E. and Henry, L. 1958. *La population de Crulai paroisse normande. Etude historique.* INED, Travaux et Documents Vol. 33, Paris: PUF.
Glass, D.V. and Eversley, D.E.C. (eds) 1965. *Population in History. Essays in Historical Demography.* London: Edward Arnold.
Glass, D.V. and Revelle, E. (eds) 1972. *Population and Social Change.* London: Edward Arnold.
Goubert, P. 1960. *Beauvais et le Beauvaisis de 1600 à 1730.* Paris: SEVPEN.
Henry, L. 1980. *Techniques d'analyse en démographie historique.* Paris. INED.
Hollingworth, T.H. 1969. *Historical Demography.* Ithaca: Cornell University Press.
Imhof, A. 1976. *Aspekte der Bovolkerungs Entwicklung in den Nordischen Ländern, 1729–1750.* 2 vols, Berne, Francke Verlag.
Imhof, A. 1977. *Einführung in die Historische Demographie.* Munich: C.H. Beck.
Knodel, J. 1974. *The Decline of Fertility in Germany 1871–1939.* Princeton: Princeton University Press.
Laslett, P. 1971. *The World we have Lost: England before the Industrial Age.* 2nd edn, New York: Charles Scribner's Sons.

Laslett, P. and Wall, R. (eds) 1972. *Household and Family in Past Times.* Cambridge: Cambridge University Press.
Lebrun, F. 1971. *La mort en Anjou au XVIIe et XVIIIe siècles – Essai de démographie et de psychologie historiques.* Paris: La Haye, Mouton.
Le roy Ladurie, E. 1966. *Paysans de Languedoc.* 2 vols, Paris: SEVPEN.
Livi-Bacci, M. 1977. *A History of Italian Fertility during the last two Centuries.* Princeton: Princeton University Press.
Marcilio, M.L. and Charbonneau, H. (eds) 1979. *Démographie Historique.* Paris: PUF.
Mols, R. 1954–6. *Introduction à la démographie historique des villes d'Europe du XIVe au XVIIIe siècle.* Louvain: Duculot.
Perrenoud, A. 1979. *La population de Genève du XVIe au début du XIXe siècle: étude démographique,* Vol. I. Geneva: Société d'Histoire et d'Archéologie de Genève.
Reinhard, M., Armengaud, A. and Dupaquier, J. 1978. *Histoire générale de la population mondiale.* Paris: Montchrestien.
Van de Walle, E. 1974. *The Female Population of France in the 19th Century. A Reconstitution of 82 Departments.* Princeton: Princeton University Press.
Willigan, J.D. and Lynch, K.A. 1982. *Sources and Methods of Historical Demography.* New York, London: Academic Press.
Wrigley, E.A. (ed.) 1972. *Nineteenth Century Society. Essays in the Use of Quantitative Methods for the Study of Social Data.* Cambridge: Cambridge University Press.
Wrigley, E.A. and Schofield, R.S. 1981. *The Population History of England 1541–1871. A Reconstruction.* London: Edward Arnold.

historical school. *See* GERMAN HISTORICAL SCHOOL; ENGLISH HISTORICAL SCHOOL.

hoarding. *See* FINANCE AND SAVING; ROBERTSON, DENNIS; SAVINGS EQUALS INVESTMENT.

Hobbes, Thomas (1588–1679). The greatest English political theorist and philosopher, Hobbes was born at Malmesbury and died at Hardwick, the seat of the Earl of Devonshire, who had been Hobbes's patron for many years. After attending Magdalen Hall, Oxford (BA 1608), Hobbes entered the Devonshire household as tutor to the son, and made several trips to the Continent, on one of which (in 1636) he conversed with Galileo, whose resolutive-compositive method Hobbes took over, and whose laws of motion he later carried over and applied to the motions, internal and external, of men. In 1640, fearing that his earliest work would offend the Long Parliament, he went into voluntary exile in Paris, where for a time (1646–8) he tutored the future Charles II in mathematics. He returned to England in 1651 and from then on lived as inconspicuously as he could.

Economic insights are to be found in his three main works of political theory, *The Elements of Law, Natural and Politic* (1640); *De Cive* (1642), translated (by Hobbes) as *Philosophical Rudiments Concerning Government and Society* (1651); *Leviathan* (1651); and in his history of the Long Parliament and the Civil War, *Behemoth* (1682). Hobbes's great work was his political science, of which his economic ideas seem to be only an incidental part. Yet we may notice that his political edifice rested on economic assumptions, in that his model of society was the atomistic bourgeois market society whose seismic rise in England in his own time he had certainly noticed. However, he did not attempt anything along the lines of the classical political economy of the 18th century, or even of the political arithmetic of his own century: he offered neither a general theory of exchange value nor a theory of distribution, that is, of the determinants of rent, interest, profits and wages, nor even a theory of the balance of trade or of foreign exchange. But he did set down a few general economic principles. One is a supply and demand theory of exchange value, as in: 'The value of all things contracted for, is measured by the Appetite of the Contractors' (*Leviathan*, ch. 15, p. 208) and in his more striking statement

> The *Value* or WORTH of a man, is as of all other things, his Price; that is to say, so much as would be given for the use of his Power: and therefore is not absolute; but a thing dependent on the need and judgement of another ... . And as in other things, so in men, not the seller, but the buyer determines the Price ... (ibid., ch. 10, pp. 151–2).

The two statements are consistent only on the assumption of an endemic surplus of wage-labourers, an assumption which Hobbes did explicitly make. The able-bodied poor, who were expected to increase indefinitely,

> are to be forced to work: and to avoyed the excuse of not finding employment, there ought to be such Lawes, as may encourage all manner of Arts; as Navigation, Agriculture, Fishing, and all manner of Manifacture that requires labour. The multitude of poor, and yet strong people still encreasing, they are to be transplanted into Countries not sufficiently inhabited: where neverthelesse, they are not to exterminate those they find there; but constrain them to inhabit closer together, and not range a great deal of ground, to snatch what they find; but to court each little Plot with art and labour, to give them their sustenance in due season (*Leviathan*, ch. 30, p. 387; cf. *Behemoth*, p. 126).

Another general proposition is that 'a mans Labour also, is a commodity exchangeable for benefit, as well as any other thing' (*Leviathan*, ch. 24, p. 295).

More important than such general principles are his many policy recommendations to the Sovereign, all of which are designed to increase the wealth of the nation by promoting the accumulation of capital by private enterprisers seeking their own enrichment. Typical are his recommendations about taxation. Taxes are justified only because they provide the income which enables the sovereign power to maintain the conditions for private enterprise: 'the Impositions that are layd on the People by the Sovereign Power, are nothing else but the Wages, due to them that hold the publique Sword, to defend private men in the exercise of severall Trades, and Callings' (ibid., ch. 30, p. 386). Taxes on wealth are bad, for they discourage accumulation. The best taxes are those on consumption, which discourage 'the luxurious waste of private men' (p. 387). Hobbes's recommendations to the Sovereign all follow from his most general rule, as set out in the opening paragraph of chapter 30:

> The office of the Soveraign, (be it a Monarch, or an Assembly,) consisteth in the end, for which he was trusted with the Soveraign Power, namely the procuration of *the safety of the people* ... But by Safety here, is not meant a bare Preservation but also all other Contentments of life, which every man by lawfull Industry, without danger, or hurt to the Common-wealth, shall acquire to himself (p. 376).

Most important of all was his insistence that the sovereign was above the law and could not be limited by any of the traditional rights of leasehold or copyhold tenants, or by any traditional limits on market transactions, or traditional protections of the poor: 'it belongeth to the Common-wealth, (that is to say, to the Sovereign), to appoint in what manner, all kinds of contract between Subjects, (as buying, selling,

exchanging, borrowing, lending, letting, and taking to hire), are to bee made; and by what words and signes they shall be understood for valid' (ibid., ch. 24, p. 299).

In short, the job of the state was to clear the way for capitalism. It is evident that Hobbes's doctrine was particularly appropriate to the period of primary capital accumulation. It is scarcely too much to say that it was his perception of the needs of such a period which determined the main lines of his political theory. What was needed was a sovereign powerful enough to override all the protections of the common law, and, to justify such a power, a new, untraditional basis for political obligation. That is what Hobbes's doctrine provided. In effect, it is the legitimation of the early capitalist state.

C.B. MACPHERSON

SELECTED WORKS

1650. *Elements of Law Natural and Politic*. Ed. F. Tonnies, Cambridge: Cambridge University Press, 1928.
1651. *Philosophic Rudiments Concerning Government and Society*. Published as *De Cive or The Citizen*. Ed. S.P. Lamprecht, New York: Appleton-Century-Crofts, 1949.
1651. *Leviathan*. Ed. C.B. Macpherson, Harmondsworth: Penguin, 1968.
1682. *Behemoth; or the Long Parliament*. Ed. F. Tonnies, London: Simpkin, Marshall & Co., 1889.

BIBLIOGRAPHY

Macpherson, C.B. 1983. Hobbes's political economy. *Philosophical Forum* 14 (3–4).

Hobson, John Atkinson (1858–1940). Hobson was born in Derby in 1858 and died at home in Hampstead in 1940. He was educated at Derby School and Lincoln College, Oxford, where he read Greats, 1876–80, but only gained a Third. He taught Classics at Faversham and Exeter, 1880–81, before moving to London, where he supplemented his private income (from the Derby newspaper which his father had owned) with intermittent earnings from journalism, lecturing and his books. It is significant that his work as a university extension lecturer in economics was thwarted by Professor F.Y. Edgeworth, who considered his views disturbingly heterodox. Hobson was never offered an academic post in an English university nor invited to contribute to the *Economic Journal*. He remained a publicist, propagating his economic views through more than fifty books and a series of organs of radical Liberal and socialist leanings: the *Progressive Review*, run by the Rainbow Circle in the 1890s; C.P. Scott's *Manchester Guardian*, especially at the time of the Boer War; the *Nation*, under the editorship of H.W. Massingham from 1907, and, in later years, its successor, the *New Statesman & Nation*.

Hobson's subsequent reputation has been coloured by his supposed role as a predecessor, not only of Lenin and his theory of imperialism, but also of Keynes and his concept of effective demand. Neither connection is wholly factitious but both have been open to an unhistorical distortion of Hobson's own concerns. Temperamentally he was both iconoclastic and idiosyncratic; he had few intellectual debts and acknowledged fewer. After early professional rebuffs, he proclaimed himself an economic heretic with a characteristic mixture of defiance and irony. Working against the grain of orthodox economic opinion, cut off from academic colleagues, writing to immediate deadlines for a wide range of publications, Hobson left an *oeuvre* which is not easy to assess and in which formal inconsistencies are not difficult to find. He conveys, nonetheless, a general vision of the scope and nature of economics which is both distinctive and coherent.

Hobson has long been best known as an underconsumptionist. His first book (Mummery and Hobson, 1889) was written in collaboration with A.F. Mummery, a businessman, who seems to have been the senior partner. The book set out to expose fallacies in classical political economy as expounded by J.S. Mill. Its central proposition was that trade depression was caused by a deficiency in effective demand since it was the level of consumption in the immediate future which limited profitable production. It followed that there was a limit to the amount of useful savings which a community could make. Each individual could save with advantage to himself, but the overall result might be a position of underconsumption, for which oversaving was another name. Hobson was to seize on this self-defeating process as an example of what he called the protean fallacy of individualism – an idea that pervades his work in a far more general way than the particular concept of underconsumption. The polemical thrust of this early book was thus against the tendency of economists to extol thrift insofar as this neglected the crucial importance of maintaining sufficient demand. This statement of the underconsumptionist case, reiterated in two further books (Hobson, 1894 and 1896), identified Hobson as an economic heretic, a role which he took up by broadening rather than narrowing his dissent from neo-classical analysis. His distrust of marginalism, on the ground that it rested upon an unreal process of abstraction, marked a further breach with Marshallian orthodoxy (Hobson, 1901b and 1926a).

Hobson was to supplement his account of underconsumption with a theory of distribution (Hobson, 1900a) which drew heavily upon the Fabian theory of rent. He distinguished the costs of subsistence for any factor of production from its rent element, and argued that in principle surplus value might accrue to land, labour or capital. He further introduced the idea of 'forced gains' as an assertion of superior bargaining power in this process, with the result that 'unearned income' accrued to certain individuals and classes. Hobson further assumed that the proportion of income which was in this sense economically functionless varied directly with the absolute level of income received. It followed that progressive taxation would not in practice impair any necessary incentive to production.

This analysis was later elaborated (Hobson, 1909b) to distinguish a 'productive surplus' which covered the costs of growth from an 'unproductive surplus', distributed according to no functional principle. Morally this was the property of the community which had created it. If redistributive taxation could restore it to its rightful possessors, oversaving by the rich would be curtailed and underconsumption by the poor rectified. This functional view of the proper working of the economic system, with effort matched to reward by rooting out parasitism, reappears constantly as a paradigm in Hobson's writings. He dignified it with the name 'the organic law' and often suggested an evolutionary provenance for it. But he also claimed the authority of John Ruskin, of whom he wrote an admiring study (Hobson, 1898), for seeing consumption, not production, as the qualitative end of economic activity. He sought to unite these ideas in one of the most frequently reprinted of his books (Hobson, 1894) by adopting the formula: 'From each according to his powers, to each according to his needs.'

Hobson's theory of imperialism, advanced as a response to the Boer War (1899–1902), built upon these foundations. As a 'pro-Boer', Hobson sought to analyse the reasons for the annexation by Britain of the two South African republics. After visiting South Africa for the *Manchester Guardian*, he pointed to the influence of powerful interests who controlled the goldfields and the press, and in this context referred to 'the class of financial capitalists of which the foreign Jew must be taken as the leading type' (Hobson, 1900b, p. 189). Whether such references amount to anti-semitism is a controversial point. At any rate, Hobson countered the claim that a clear national interest was at stake by stressing its cosmopolitan character. He identified speculative investment in undeveloped territories as a cause of imperialism and claimed that it arose from oversaving by a parasitic class at home. In this sense underconsumption was the economic taproot of imperialism (Hobson, 1902). What he vigorously rejected was the proposition that there was sufficient profit to the country as a whole from trade and investment in Africa to counterbalance the costs of aggression. In contrast to Lenin, therefore, Hobson denied that imperialism was a structural necessity of the metropolitan economy. It could and should be checked at home by a policy of redistributive taxation, which would have the reciprocal effect of cutting the taproot (ending oversaving) and stimulating domestic demand (ending underconsumption).

The economic implication was that Britain could easily make up any loss on foreign trade by generating wealth at home – an argument for protection which Hobson chose to suppress in the context of the time. For it was the Liberal and Labour parties, with their commitment to Free Trade, to which Hobson looked for reformist amelioration. He was confident that imperialism could be beaten by democratic means precisely because it did not serve the interests of the majority but only of a privileged section of the nation. In his most famous book, therefore, Hobson devotes more than twice as much space to the politics than to the economics of imperialism (Hobson, 1902). He needed to do so because the puzzle was how a policy that was bad business for the nation as a whole had come to be adopted. The answer was that finance was the governor of an engine whose motor power came from elsewhere. Much of Hobson's work was thus concerned with the forces of nationalism and social psychology which fuelled the politics of self-assertion (Hobson, 1901a). His view on the utility of overseas investment was strongly coloured by the context in which he saw it. In at least one book (Hobson, 1911b) he commended cosmopolitan finance as a force for peace and development. To a surprising extent he remained a lifelong Cobdenite (Hobson, 1919a), and he finally left the Liberal party during World War I when he thought it had betrayed the cause of Free Trade.

It will be apparent that Hobson was no single-minded underconsumptionist. In the early 1900s his energies were directed towards permeating the Liberal Party with a broad-based conception of economics which would justify it in rejecting the classical nostrums of laissez faire in favour of interventionist policies designed to further social justice (Hobson 1909). This was the New Liberalism, of which Hobson and his friend L.T. Hobhouse became the leading intellectual exponents. The publication of Hobson's *The Industrial System*, which consolidated much of his previous work, opportunely coincided with Lloyd George's People's Budget in 1909 and offered a defence of the policy of redistributive taxation via the concept of the surplus. This aspect overshadowed the restatement of Hobson's underconsumptionist position; though he now went further than before in analysing the dynamic process by which oversaving reduced all real incomes in the economy until automatic checks came into play (Hobson, 1909b, ch. 18). One might call this Hobson's most accomplished exercise in macroeconomics.

It was in the context of the depression after World War I

that Hobson once more returned to this theme (Hobson, 1922 and 1929), and it was in this period that his economic views enjoyed greatest publicity. He was now loosely identified with the Labour Party and found a natural application for his ideas in mounting an economic case for a 'living wage' (Hobson, 1926b). His central contentions on oversaving had not significantly changed but when he reiterated them, amid widespread unemployment, he found a more sympathetic response, even among professional economists who had previously accepted a full-employment assumption. In particular, by 1930 Hobson was on cordial terms with J.M. Keynes, who had in earlier years scorned his work. But Keynes was still anxious to keep his distance, as he made clear (Keynes, 1930, pp. 160–61). The reason was that when Keynes wrote of oversaving he meant under-investment; whereas for Hobson saving and investment were two names for the same thing, and by over-saving he had always meant underspending. It followed also that Keynes had more interest in policies of public works as a means of promoting investment, whereas Hobson concentrated on the case for redistribution as a means of stimulating consumption. It was not until Keynes had virtually finished the *General Theory* that he fully realized how far his new concept of effective demand has been foreshadowed in Hobson's work, to which he paid a handsome if belated tribute (Keynes, 1936, pp. 364–71).

PETER CLARKE

SELECTED WORKS

1889. (With A.F. Mummery.) *The Physiology of Industry*. London: Murray.
1894. *The Evolution of Modern Capitalism*. London: Walter Scott.
1896. *The Problem of the Unemployed*. London: Methuen.
1898. *John Ruskin, Social Reformer*. London: Nisbet.
1900a. *The Economics of Distribution*. New York.: Macmillan.
1900b. *The War in South Africa: its causes and effects*. London: Nisbet.
1901a. *The Psychology of Jingoism*. London: Nisbet.
1901b. *The Social Problem*. London: Nisbet.
1902. *Imperialism: A Study*. London: Nisbet.
1909a. *The Crisis of Liberalism*. Ed. P.F. Clarke, Brighton: Harvester Press, 1974.
1909b. *The Industrial System*. London: Longman.
1911a. *The Science of Wealth*. 4th edn, with preface by R.F. Harrod, Oxford: Home University Library, 1950.
1911b. *An Economic Interpretation of Investment*. London: Financial Review of Reviews.
1914. *Work and Wealth, a human valuation*. London: Macmillan.
1919a. *Taxation in the New State*. London: Methuen.
1919b. *Richard Cobden: The International Man*. London: J.M. Dent.
1922. *The Economics of Unemployment*. London: Macmillan.
1926a. *Free Thought in the Social Sciences*. London: Allen & Unwin.
1926b. (With H.N. Brailsford, A. Creech Jones and E.F. Wise.) *The Living Wage*. London: Independent Labour Party.
1929. *Wealth and Life*. London: Macmillan.
1930. *Rationalisation and Unemployment*. London: Allen and Unwin.
1938. *Confessions of an Economic Heretic*. With an introduction by M. Freeden, Brighton: Harvester Press, 1976.

BIBLIOGRAPHY

Keynes, J.M. 1930. *A Treatise on Money*. Vol. 1: The Pure Theory of Money. London: Macmillan for the Royal Economic Society, 1971.
Keynes, J.M. 1936. *The General Theory of Employment, Interest and Money*. London: Macmillan for the Royal Economic Society, 1973.

Hodgskin, Thomas (1787–1869). Thomas Hodgskin joined the navy at the age of twelve, rose to the rank of lieutenant and was then forcibly retired on half pay after a contretemps with his authoritarian captain. On the advice of Francis Place he subsequently embarked, in 1815, upon a continental tour with the object of collecting material on the social and economic conditions of post-Napoleonic war Europe. It was this that formed the basis of his *Travels in the North of Germany*, which was published in 1820.

It was in the early 1820s that his sympathy for the working classes was aroused through his involvement in the struggle for the repeal of the Combination Acts and the attempts to establish a London Mechanics' Institute. The former led to the publication of Hodgskin's most famous work, *Labour Defended against the Claims of Capital* (1825), while the latter led to a series of lectures, given at the Institute, which formed the basis of *Popular Political Economy* (1827). It was these two works which, together with his *Natural and Artificial Right of Property Contrasted* (1832), established his reputation as one of the major 19th-century anti-capitalist writers. After 1832 financial necessity forced Hodgskin to abandon his more serious intellectual labours and to concentrate upon a journalistic career that had begun in the early 1820s and which lasted until the early 1850s when he worked for the *Economist*.

While generally dubbed a 'Ricardian socialist', the single most important influence upon the thought of Thomas Hodgskin was Adam Smith. From Smith he believed he had derived the central tenet of his social and economic philosophy, namely that the material world was shaped by natural laws emanating from an omniscient and beneficent Providence, all interference with which was either superfluous or pernicious. Such views are apparent in Hodgskin's *Travels in the North of Germany*, where he attacked the malign interference of government with the natural laws that should be left to regulate trade and industry, and they were to provide the philosophical underpinning of all his major works.

Thus in *Labour Defended*, written in defence of trade union activity, Hodgskin attacked profit, the reward of the capitalist, as a violation of the natural laws of value and distribution. Here Hodgskin confronted the classical argument that the capitalist derived his entitlement to a share in labour's product from his ownership of fixed and circulating capital which he provided for the use of his workforce. The idea of a fund of circulating capital Hodgskin dismissed as a fiction. What labour depended upon during the period necessary to make and bring a commodity to market was co-existing labour, while fixed capital was simply the result of past exertions, utilized, maintained and ultimately replaced by present labour. Thus the capitalist's reward derived not from his exertions but from the economic power which allowed him to transform 'natural' into 'social' price through the addition of profit to natural value. Here Hodgskin distinguished Smith's additive explanation of the determination of exchange value under capitalism from Ricardo's labour embodied theory. For Hodgskin, while Ricardo had explained what *should* determine natural value, Smith had made clear what *were* the actual determinants of prices under capitalism namely wages plus profits plus rents. Thus

> Mr. R. appears to me to have confounded in the whole of his speculations real natural price with exchangeable value. The former is accurately measured by the quantity of *labour* necessary to obtain any commodity from nature, the latter on the contrary is the quantity of labour augmented by the amount of rent and profits.

It was, therefore, in Smithian rather than Ricardian terms that Hodgskin formulated the profit-upon-alienation theory of labour exploitation to which he adhered in his major works.

As a good Smithian it became a central concern of Hodgskin not only to ensure that the natural laws of economic life prevailed but also to demolish the arguments of those who would impugn their beneficence. Thus Hodgskin's *Popular Political Economy* challenged in particular Malthusian population theory, which seemed to suggest that there existed insuperable natural obstacles to material prosperity in the form of Nature's parsimony and Man's sexual incontinence. For Hodgskin, the obstacles which existed were not natural but the consequence of the coercive exercise of power bolstered by social regulations and artificial laws. Contrary to Malthus, he saw population increase not as a cause of indigence but of material improvement, with demographic pressure creating new demands and needs, stimulating the inventive faculties of Mankind and so enhancing society's capacity to produce.

Hodgskin was equally critical of the slur upon Nature cast by Ricardian rent theory. This he saw as implying the necessary advent of a poverty stricken stationary state due to the finite nature of land resources. For Hodgskin poverty was not a dictate of niggardly nature. On the contrary, it resulted from the unwarranted exactions of capitalist, landowner, State and Church; in particular it was 'the overwhelming nature of the demands of capital, sanctioned by the customs of men, enforced by the legislature ... which keep the labourer in poverty and misery'.

In his last major work, *The Natural and Artificial Right of Property Contrasted* (1832), Hodgskin went on to consider the nature of the rights conferred by these positive laws as against those granted by Nature herself. As he saw it positive law generally legitimized the gains reaped by the exercise of coercive force, while at best it did no more than mirror the dictates of natural law, imposing upon the present a conception of rights which historical progress would rapidly render redundant.

Hodgskin's achievement was to integrate a teleological optimism based upon an anti-Malthusian conception of the consequences of population increase, with a penetrating critique of contemporary capitalism which turned to critical use the tools, concepts and analytical constructs of political economy. It is this critique which explains his categorization as a 'socialist'.

Yet, while their critical analyses are similar in many respects, Hodgskin's vision of the future never permitted any flirtation with the Owenites or the principles of Owenite socialism. His was an individualistic utopia and he never doubted that private property was a *sine qua non* of material progress. His just and equitable society was essentially atomistic, an unplanned consequence of the spontaneous, unrestricted actions of individuals. It is, therefore, in the company of William Godwin and Herbert Spencer, rather than Owen, Bray and Marx that Hodgskin should be placed.

Marx did, of course, see much of worth in Hodgskin's political economy, in particular, his theory of capital, but it would be entirely wrong to see Hodgskin as his intellectual precursor. For, while the Ricardianism of Marx led him to locate exploitation at the point of production, the Smithianism of Hodgskin led him to place it in the sphere of circulation or exchange. Thus while Marx preached the working-class seizure of the means of production, Hodgskin advocated the creation of equitable exchange relations through the liberation of market forces. This is the essence of Hodgskin's libertarian political economy which was eventually to evolve from the acerbic anti-capitalism of the 1820s to the 1840s when, writing for the *Economist* (1848–53), he began to deny the existence of any necessary antagonism between capital and labour. In the final analysis, Hodgskin wished to purify and generalize capitalism rather than to destroy it.

N.W. THOMPSON

SELECTED WORKS

1820. *Travels in the North of Germany, describing the Present State of the Social and Political Institutions, the Agriculture, Manufactures, Commerce, Education, Arts and Manners in that Country, particularly in the Kingdom of Hannover.* 2 vols, Edinburgh.
1825. *Labour Defended against the Claims of Capital; or, the unproductiveness of capital proved with reference to the present combinations amongst journeymen, by a Labourer.* London.
1827. *Popular Political Economy, Four Lectures delivered at the London Mechanics' Institution.* London.
1832. *The Natural and Artificial Right of Property Contrasted, a series of letters addressed without permission to H. Brougham, Esq.* London.

BIBLIOGRAPHY

Beales, H.L. 1933. *The Early English Socialists.* London: Hamish Hamilton.
Beer, M. 1953. *A History of British Socialism.* 2 vols, London: Allen & Unwin.
Cole, G.D.H. 1977. *A History of Socialist Thought.* 5 vols. Vol. 1: *Socialist Thought, The Forerunners, 1789–1850.* London: Macmillan.
Foxwell, H.S. 1899. Introduction to the English translation of A. Menger, *The Right to the Whole Produce of Labour.* London: Macmillan.
Gray, A. 1967. *The Socialist Tradition, Moses to Lenin.* London: Longman.
Halévy, E. 1903. *Thomas Hodgskin.* Trans. with an introduction by A.J. Taylor, London: Benn, 1956.
Hollander, S.G. 1980. The post-Ricardian dissension: a case study of economics as ideology. *Oxford Economic Papers* 32, 370–410.
Hunt, E.K. 1977. Value theory in the writings of the classical economists. *History of Political Economy* 9, 322–45.
Hunt, E.K. 1980. The relation of the Ricardian socialists to Ricardo and Marx. *Science and Society* 44, 177–98.
King, J.E. 1981. Perish Commerce! Free trade and underconsumption in early British radical economics. *Australian Economic Papers* 20, 235–57.
King, J.E. 1983. Utopian or scientific? A reconsideration of the Ricardian socialists. *History of Political Economy* 15, 345–73.
Lowenthal, E. 1911. *The Ricardian Socialists.* New York: Longman.
Marx, K. 1969. *Theories of Surplus Value.* 3 vols, Moscow: Progress.
Thompson, N.W. 1984. *The People's Science: The Popular Political Economy of Exploitation and Crisis, 1816–34.* Cambridge: Cambridge University Press.

Hollander, Jacob Harry (1871–1940). Born in Baltimore, Maryland on 23 July 1871, Hollander spent his entire career at the Johns Hopkins University, studying under R.T. Ely and J.B. Clark, graduating AB in 1891, PhD in 1894, and joining the faculty immediately thereafter. A versatile scholar, his special fields were labour economics, the history of economic thought, and public finance. In the first of these he ran a notable seminar for several decades with his colleague George Barnett, and both were elected President of the American Economic Association, Hollander in 1921, Barnett in 1932. As a doctrinal historian Hollander is especially remembered for his discovery and editing of Ricardo's letters, and the latter's important *Notes on Malthus*. He also collected a major library of works on economics. As a tax and financial expert Hollander held numerous local, state, federal and international posts, especially in Puerto Rico (1900–1901) and in the Dominican Republic (1905–7), where he continued to serve as financial adviser up to 1910. He was a pacifist, opposing US membership of the League of Nations, and a defender of Prohibition.

A.W. COATS

SELECTED WORKS

1895. *Letters of David Ricardo to John Ramsay McCulloch 1816–1823*. New York: Macmillan Co.

1899a. *The Financial History of Baltimore*. Baltimore: Johns Hopkins Press.

1899b. (ed., with J. Bonar.) *Letters of David Ricardo to Hutches Trower*. Oxford: Clarendon Press.

1905a. *Debt of Santo Domingo ... . Report on the debt of Santo Domingo submitted to the President of the United States*. Washington, DC: Government Printing Office.

1905b. (ed., with G. Barnett.) *Studies in American Trade Unionism*. New York: Holt and Co. Reprinted, 1912.

1910. *David Ricardo, A Centenary Estimate*. Baltimore: Johns Hopkins Press.

1914. *The Abolition of Poverty*. Boston and New York: Houghton Mifflin Co.

1919. *War Borrowing: A Study of Treasury Certificates of Indebtedness of the United States*. New York: Macmillan Co.

1928. (ed., with T.E. Gregory.) *Notes on Malthus' 'Principles of Political Economy' by David Ricardo*. Baltimore: Johns Hopkins Press; London: Oxford University Press.

Homan, Paul Thomas (1893–1969). Homan was born on 12 April 1893 in Indianola, Iowa, and died on 3 July 1969 in Washington, DC. Educated at Williamette University, Oxford University and the Brookings Institution (then Graduate School of Economics and Government) (PhD, 1926), he taught at Cornell (1919–47), the University of California at Los Angeles (1950–69), and Southern Methodist University (1953–63). He was managing editor of the *American Economic Review*, 1941–52. He served with the War Production Board, UNRRA, UNESCO, and the Council of Economic Advisers. He was on the staff of the Brookings Institution and also was associated with Resources for the Future.

An expert on the National Recovery Administration, Homan later wrote on oil conservation regulation, estimating oil and gas reserves, costing in the petroleum industry, and, several years before the OPEC oil embargo, problems of Middle Eastern oil for the western world.

Homan's *Contemporary Economic Thought* (1928) was an influential interpretation of the state of the discipline at the time. He emphasized its enormous diversity, treating the heterodox work of John A. Hobson, Thorstein Veblen and Wesley C. Mitchell alongside the more orthodox doctrines of John Bates Clark and Alfred Marshall, all as serious inquiry within general economic theory. Although admiring economics as a science, he recognized that economics has the quality of a system of beliefs, both influenced by and influencing general philosophical and ideological points of view in society. He found the principal axis of diversity to lie between those who emphasized the static, deductive, mathematical individualist approach, and those who pursued realism, empiricism and holistic evolutionism. His personal view was complex. He clearly thought that value theory, logical deduction, mechanical analogy, and the study of the price system were central to economics, and that a framework of thought commanding more general assent might be desirable; but that diversity was not objectionable *per se*, there being room for the study of the price system, institutions, and the meaning of economic life.

WARREN J. SAMUELS

SELECTED WORKS

1928. *Contemporary Economic Thought*. New York: Harper.

1934. (With others.) *The ABC of the NRA*. Washington, DC: Brookings Institution.

1935. (With others.) *The National Recovery Administration: An Analysis and Appraisal*. Washington, DC: Brookings Institution.

1945. (ed., with F. Machlup.) *Financing American Prosperity*. New York: Twentieth Century Fund.

1964. (With W.F. Lovejoy.) *Problems of Cost Analysis in the Petroleum Industry*. Dallas: Southern Methodist University Press.

1965. (With W.F. Lovejoy.) *Methods of Estimating Reserves of Crude Oil, Natural Gas, and Natural Gas Liquids*. Baltimore: Johns Hopkins University Press for Resources for the Future.

1967. (With W.F. Lovejoy.) *Economic Aspects of Oil Conservation Regulation*. Baltimore: Johns Hopkins University Press for Resources for the Future.

1971. (With S.H. Schurr.) *Middle Eastern Oil and the Western World; Prospects and Problems*. New York: Elsevier.

homogeneous and homothetic functions.

HOMOTHETIC ORDERINGS. Given a cone E in the Euclidean space $\mathbb{R}^n$ and an ordering $\leqslant$ on E (i.e. a reflexive and transitive binary relation on E), the ordering is said to be homothetic if for all pairs x, y, $\in E$

$$x \leqslant y \Rightarrow \lambda x \leqslant \lambda y \quad \text{for all} \quad \lambda > 0.$$

For each $x \in E$, denote by $L(x)$ the indifference surface

$$L(x) = \{ y \in E : y \leqslant x \quad \text{and} \quad x \leqslant y \}.$$

Hence, geometrically, if the ordering is homothetic, then for all $x \in E$ and $\lambda > 0$

$$L(\lambda x) = \{ \lambda y : y \in L(x) \}.$$

HOMOTHETIC FUNCTIONS. Recall that a real function f on a set E defines a complete (or total) ordering on E via the relation

$$x \leqslant y \quad \text{if and only if} \quad f(x) \leqslant f(y).$$

By definition, f is said to be homothetic if the ordering is homothetic (implying that the domain E of f is a cone). Thus utility functions which represent a homothetic ordering are homothetic.

Assume, now, that f is a homothetic and differentiable function on an open cone E of $\mathbb{R}^n$. Assume also that $\nabla f(x) \neq 0$ for all $x \in E$. Hence for all $\lambda > 0$ and all $x \in E$ there exists $k > 0$ such that

$$\frac{\partial f}{\partial x_i}(\lambda x) = k \frac{\partial f}{\partial x_i}(x), \quad \text{for } i = 1, 2, \ldots, n.$$

In economic terms, this property means that the marginal rate of substitution remains constant along any ray from the origin. In fact, under some suitable assumptions, this property characterizes homothety of functions.

POSITIVELY HOMOGENEOUS FUNCTIONS. A real function f defined on a cone E of $\mathbb{R}^n$ is said to be positively homogeneous of order p if for all $x \in E$

$$f(\lambda x) = \lambda^p f(x) \quad \text{for all} \quad \lambda > 0.$$

If $p = 1$, the function is said to be positively homogeneous or linearly homogeneous.

If $p = 0$, then the definition becomes

$$f(\lambda x) = f(x) \quad \text{for all} \quad \lambda > 0 \quad \text{and} \quad x \in E.$$

Clearly, positively homogeneous functions of any order are homoethetic. Conversely, under some suitable assumptions on E and f (for instance E is the positive orthant in $\mathbb{R}^n$ and f is increasing on E) then, if f is homothetic there exist a positively homogeneous function g of order 1 on E and an increasing function k on $\mathbb{R}$ such that

$$f(x) = k[g(x)] \qquad \text{for all} \quad x \in E.$$

(This property is sometimes used as an alternative definition of homothety for functions.) As a consequence, under reasonable economic assumptions, a homothetic preference ordering can be represented by a linearly homogeneous utility function.

Production functions are often assumed to be positively homogeneous of order p. For example, the so-called Cobb–Douglas function

$$f(x_1, x_2, \ldots, x_n) = K x_1^{\alpha_1} x_2^{\alpha_2} \ldots x_n^{\alpha_n} \qquad x_i > 0,$$

where $K, \alpha_1, \alpha_2, \ldots, \alpha_n$ are positive constants, is homogeneous of order $p = \alpha_1 + \alpha_2 + \cdots + \alpha_n$.

In consumer theory, demand functions are positively homogeneous of order zero in prices and wealth.

POSITIVELY HOMOGENEOUS CONVEX (OR CONCAVE) FUNCTIONS. Since convexity is a fundamental concept in economics, special attention should be paid to positively homogeneous functions which are convex or concave.

Let E be a convex cone and f a real function on E. Then a necessary and sufficient condition for f to be convex (concave) and positively homogeneous of order 1 on E is that for all $x \in E$ and $\lambda \geqslant 0$

$$f(\lambda x) = \lambda f(x)$$

and for all pairs $x, y \in E$

$$f(x + y) \leqslant (\geqslant) f(x) + f(y).$$

The producer's cost function illustrates a concave positively homogeneous function: assuming that only one output is produced using n inputs, the cost function is given by

$$c(y, p) = \operatorname*{Min}_{x} [p^t x : F(x) \geqslant y]$$

where p_i, $i = 1, 2, \ldots, n$, is the unit price of input i and $F(x)$, the production function, is the maximal amount of output which can be produced with the input vector $x = (x_1, x_2, \ldots, x_n)$. Then, for a fixed price vector p, $c(y, p)$ is the minimal cost of producing y units of the output. For y fixed, $c(y, p)$ is concave and positively homogeneous of order 1 in p. Similarly, in consumer theory, if F now denotes the consumer's utility function, the $c(y, p)$ represents the minimal price for the consumer to obtain the utility level y when p is the vector of utility prices.

A fundamental property is as follows. Let f be a real continuous function on a closed convex cone of $\mathbb{R}^n$. Then f is convex and positively homogeneous of order 1 if and only if there exists a closed convex set S of $\mathbb{R}^n$ such that

$$f(x) = \operatorname{Sup}[y^t x / y \in S].$$

This set S is unique and the function is called the support function of S (by symmetry, the same result holds when replacing convex by concave and Sup by Inf). Duality in consumer's (as well as in producer's) theory is based on this property.

We conclude with three examples of functions widely used in mathematics. A *semi-norm* on $\mathbb{R}^n$ is a convex positively homogeneous function f of order one on $\mathbb{R}^n$ such that $f(x) = f(-x)$ for all x (then $f(x) \geqslant 0$ for all x). A *norm* is a semi-norm for which $x = 0$ whenever $f(x) = 0$. Finally, given a convex set C which contains the origin, the *gauge* of C is the function f defined by

$$f(x) = \operatorname{Inf}[\lambda \geqslant 0 / x \in \lambda C].$$

A gauge function is convex and positively homogeneous of order one. Moreover, if the origin belongs to the interior of C and C is balanced (i.e. $x \in C$ implies that $x \in -C$), then the gauge is a norm.

POSITIVELY HOMOGENEOUS QUASI-CONCAVE (QUASI-CONVEX) FUNCTIONS. Let $\geqslant$ be a preference ordering on a set E. In view of economic considerations, a common and reasonable assumption is the convexity of the ordering (i.e. for all $x \in E$, the set $\{y \in E / y \geqslant x\}$ is convex). Then the utility functions which represent the ordering are quasi-concave but in general, a concave representation does not exist. However, in the case where the ordering is homothetic, it does. Indeed, a quasi-concave linearly homogeneous function which takes only positive (negative) values on the interior of its domain is concave [Newman] (by symmetry the same result holds for quasi-convex functions). It follows that a representable preference ordering which is homothetic and convex admits a representation by a concave linearly homogeneous utility function.

J.-P. CROUZEIX

See also AGGREGATION OF DEMAND; COBB–DOUGLAS FUNCTIONS; EULER'S THEOREM; GAUGE FUNCTIONS; QUASI-CONCAVITY; SEPARABILITY; TRANSFORMATIONS AND INVARIANCE.

BIBLIOGRAPHY

Barten, A.P. and Bohm, V. 1981. Consumer theory. In *Handbook of Mathematical Economics*, Vol. 2, ed. K.J. Arrow and M.D. Intriligator, New York: North-Holland Publishing Company, 381–429.

Diewert, W.E. 1981. Duality approaches to microeconomic theory. In *Handbook of Mathematical Economics*, Vol. 2, ed. K.J. Arrow and M.D. Intriligator, New York: North-Holland Publishing Company, 353–99.

Green, J. and Heller, W.P. 1981. Mathematical analysis and convexity with application to economics. In *Handbook of Mathematical Economics*, Vol. 1, ed. K.J. Arrow and M.D. Intriligator, New York: North-Holland Publishing Company, 1–52.

Katzner, D.W. 1970. *Static Demand Theory*. New York: Macmillan.

Newman, P. 1969. Some properties of concave functions. *Journal of Economic Theory* 1, 291–314.

homoskedasticity. *See* HETEROSKEDASTICITY.

Horner, Francis (1778–1817). Horner was born in Edinburgh in 1778, the son of a merchant. He was educated at the Royal High School and the local university and was a member of the group of former students of Dugald Stewart who in 1802 founded the *Edinburgh Review* – a Whig quarterly which became the main reviewing periodical in political economy during the first third of the 19th century. Horner was the expert on political economy within the founding group, and his advice on books and reviewers was often crucial to the editor, Francis Jeffrey. Horner's other early claim to be of note is that he was probably one of the world's first *students* of political economy, having attended Stewart's pioneering course of lectures on the subject on no less than three occasions. The record of his studies during this period reveals him to have been a close but by no means uncritical student of the *Wealth of Nations*, and an admirer of the work of Turgot, whose writings he hoped to translate, having already translated Euler's *Elements of Algebra* from the French in 1797.

After graduation Horner joined the Scottish Bar, but in 1803 decided to move to the English Bar. He entered Parliament in 1806 under the patronage of a Whig magnate, but did not follow the party line on all matters, especially on foreign policy, where, for example, he opposed any attempt to restore the monarchy in France after Napoleon's defeat. He was one of the prime movers in calling for the establishment of the Bullion Committee in 1810, and his reputation as one of the leading parliamentary experts on political economy made him the obvious candidate for its chairmanship. Although Ricardo was a member of this committee and was later to become an outspoken advocate of its main recommendation in favour of resumption of cash payments by the Bank of England, the report was chiefly written by Horner, Huskisson and Thornton. Horner's efforts in 1816 to gain acceptance of the report's views by means of a commitment to return to convertibility in two years' time were unsuccessful; but he had already played a major part in the process which led to acceptance by the Bank of England of its public responsibilities as lender of last resort and guardian of monetary orthodoxy.

Horner's best-known article for the *Edinburgh Review* was the generally appreciative one he wrote on Henry Thornton's *Inquiry into the Nature and Effects of the Paper Credit of Great Britain* (1802), a review that is credited with being more systematic than the book itself. It corrects Thornton's erroneous opinion that domestic inflation consequent upon the over-issue of inconvertible paper money would make goods dearer abroad rather than generate a gold outflow when the market price of gold rose above the mint price. He contributed to the debate on the corn bounty in 1804 in an article which upholds Smith's conclusions against the bounty system, but not the theory on which they were based. Horner has also received attention for the article which he persistently *failed* to write, namely a review of Malthus's *Essay on Population* – possibly because he found himself in disagreement with someone with whom he had become friendly as a result of common interests and Whig sympathies. Horner's letters to Malthus on the Corn Laws reveal a proto-Ricardian response to Malthus's heresy in giving his support to the retention of a measure of protection.

Horner died of consumption in Pisa in 1817, aged 38. It was widely thought that he had good chances of becoming Chancellor of the Exchequer in a future Whig ministry; and one of the arguments used by James Mill to convince Ricardo that his services were needed in Parliament was that he would replace Horner as the spokesman for 'correct principles'.

DONALD WINCH

SELECTED WORKS
1853. *Memoirs and Correspondence of Francis Horner, M.P.* Ed. L.J. Horner. 2 vols. Boston.
The Economic Writings of Francis Horner in the Edinburgh Review, 1802-6. Ed. F.W. Fetter, London: LSE Selected Reprints, 1957.

Hotelling, Harold (1895–1973). Harold Hotelling, a creative thinker in both mathematical statistics and economics, was born in Fulda, Minnesota, on 29 September 1895 and died in Chapel Hill, North Carolina, on 26 December 1973. His influence on the development of economic theory was deep, though it occupied a relatively small part of a highly productive scientific life devoted primarily to mathematical statistics; only ten of some 87 published papers were devoted to economics, but of these six are landmarks which continue to this day to lead to further developments. His major research, on mathematical statistics, had, further, a generally stimulating effect on the use of statistical methods in different specific fields of application, including econometrics.

His early interests were in journalism; he received his BA in that field from the University of Washington in 1919. Later in classes, he would illustrate the use of dummy variables in regression analysis by a study (apparently never published) of the effect of the opinions of different Seattle newspapers on the outcome of elections and referenda. The mathematician and biographer of mathematicians, Eric T. Bell, discerned talent in Hotelling and encouraged him to switch his field. He received an MA in mathematics at Washington in 1921 and a PhD in the same field from Princeton in 1924; he worked under the topologist, Oswald Veblen (Thorstein Veblen's nephew), and two of his early papers dealt with manifolds of states of motion.

The year of completing his PhD, he joined the staff of the Food Research Institute at Stanford University with the title of Junior Associate. In 1925 he published his first three papers, one on manifolds, one on a derivation of the F-distribution, and one on the theory of depreciation. Here, apparently for the first time, he stated the now generally accepted definition of depreciation as the decrease in the discounted value of future returns. This paper was a turning-point both in capital theory proper and in the reorientation of accounting towards more economically meaningful magnitudes.

In subsequent years at Stanford he became Research Associate of the Food Research Institute and Associate Professor of Mathematics, teaching courses in mathematical statistics and probability (including an examination of Keynes's *Treatise on Probability*) along with others in differential geometry and topology. In 1927, he showed that trend projections of population were statistically inappropriate and introduced the estimation of differential equations subject to error; he returned to the statistical interpretation of trends in a notable joint paper (1929a) with Holbrook Working, largely under the inspiration of the needs of economic analysis. The same year he published the famous paper on stability in competition (1929b), in which he introduced the notions of locational equilibrium in duopoly. This paper is still anthologized and familiar to every theoretical economist. As part of the paper, he noted that the model could be given a political interpretation, that competing parties will tend to have very similar programmes. Although it took a long time for subsequent models to arise, these few pages have become the source for a large and fruitful literature.

The paper was in fact a study in game theory. In the first stage of the game, the two players each chose a location on a line. In the second, they each chose a price. Hotelling sought what would now be called a subgame perfect equilibrium point. However, there was a subtle error in his analysis of the second stage, as first shown by d'Aspremont, Gabszewicz, and Thisse (1979). Hotelling indeed found a local equilibrium, but the payoff functions are not concave; if the locations are sufficiently close to each other, the Hotelling solution is not a global equilibrium. Unfortunately, this is the interesting case, since Hotelling concluded that the locations chosen in the first stage would be arbitrarily close in equilibrium. In fact, the optimal strategies must be mixed (Dasgupta and Maskin, 1986, pp. 30–32).

His paper on the economics of exhaustible resources (1931a) applied the calculus of variations to the problem of allocation of a fixed stock over time. All of the recent literature, inspired by the growing sense of scarcity (natural and artificial), is essentially based on Hotelling's paper. Interestingly enough, according to his later accounts, the *Economic Journal* rejected the paper because its mathematics was too difficult (although it had published Ramsey's papers earlier); it was finally published in the *Journal of Political Economy*.

In 1931, he was appointed Professor of Economics at Columbia University, where he was to remain until 1946. There he began the organization of a systematic curriculum in theoretical statistics, which eventually attained the dignity of a separate listing in the catalogue, though not the desired end of a department or degree-granting entity. Toward the end of the 1930s, he attracted a legendary set of students who represented the bulk of the next generation of theoretical statisticians. His care for and encouragement of his students were extraordinary: the encouragement of the self-doubtful, the quick recognition of talent, the tactfully-made research suggestion at crucial moments created a rare human and scholarly community. He was as proud of his students as he was modest about his own work.

He also gave a course in mathematical economics. The general environment was not too fortunate. The predominant interests of the Columbia Department of Economics were actively anti-theoretical, to the point where no systematic course in neoclassical price theory was even offered, let alone prescribed for the general student. Nevertheless, several current leaders in economic theory had the benefit of his teaching. But his influence was spread more through his papers, particularly those (1932, 1935) on the full development of the second-order implications for optimization by firms and households (contemporaneous with Hicks and Allen) and above all by his classic presidential address (1938) before the Econometric Society on welfare economics. Here we have the first clear understanding of the basic propositions (Hotelling, as always, was meticulous in acknowledging earlier work back to Dupuit), as well as the introduction of extensions from the two-dimensional plane of the typical graphical presentation to the calculation of benefits with many related commodities. He argued that marginal-cost pricing was necessary for Pareto optimality even for decreasing-cost industries, used the concept of potential Pareto improvement, and showed that suitable line integrals were a generalization of consumers' and producers' surplus for many commodities. Here also we have the clearest expression in print of Hotelling's strong social interests which motivated his technical economics. His position was undogmatic but in general it was one of market socialism. He had no respect for acceptance of the *status quo* as such, and the legitimacy of altering property rights to benefit the deprived was axiomatic with him; but at the same time he was keenly aware of the limitations on resources and the importance in any human society of the avoidance of waste.

Important as was his contribution to economics, most of his effort and his influence were felt in the field of mathematical statistics, particularly in the development of multivariate analysis. In a fundamental paper (1931b), he generalized Student's test to the simultaneous test of hypotheses about the means of many variables with a joint normal distribution. In the course of this paper, he gave a correct statement of what were later termed 'confidence intervals'. In two subsequent papers (1933, 1936) he developed the analysis of many statistical variables into their principal components and developed a general approach to the analysis of relations between two sets of variates. The statistical methodologies of these papers and in particular the last contributed significantly to the later development of methods for estimating simultaneous equations in economics.

In 1946, he finally had the long-desired opportunity of creating a department of mathematical statistics, at the University of North Carolina, where he remained until retirement. He continued his active interest in economics there.

Space forbids more than the brief mention of his important work in the foundation of two learned societies, the Econometric Society and the Institute of Mathematical Statistics, both of which he served as President at a formative stage. He received many formal honours during his lifetime, including honorary degrees from Chicago and Rochester; he was the first Distinguished Fellow of the American Economic Association when that honour was created, as well as member of the National Academy of Sciences and the Accademia Nazionale dei Lincei, Honorary Fellow of the Royal Statistical Society and Fellow of the Royal Economic Society.

KENNETH J. ARROW

SELECTED WORKS

1925. A general mathematical theory of depreciation. *Journal of the American Statistical Association* 20, 340–53.
1927. Differential equations subject to error. *Journal of the American Statistical Association* 22, 283–314.
1929a. (With H. Working). Applications of the theory of error to the interpretation of trends. *Journal of the American Statistical Association* 24, 73–85.
1929b. Stability in competition. *Economic Journal* 39, 41–57.
1931a. The economics of exhaustible resources. *Journal of Political Economy* 39, 137–75.
1931b. The generalization of Student's ratio. *Annals of Mathematical Statistics* 21, 360–78.
1932. Edgeworth's taxation paradox and the nature of supply and demand functions. *Journal of Political Economy* 40, 577–616.
1933. Analysis of a complex of statistical variables with principal components. *Journal of Educational Psychology* 24, 417–41, 498–520.
1935. Demand functions with limited budgets. *Econometrica* 3, 66–78.
1936. Relation between two sets of variates. *Biometrika* 28, 321–77.
1938. The general welfare in relation to problems of taxation and of railway and utility rates. *Econometrica* 6, 242–69.

BIBLIOGRAPHY

Dasgupta, P. and Maskin, E. 1986. The existence of equilibrium in discontinuous economic games, II: Applications. *Review of Economic Studies* 53, 27–41.
d'Aspremont, C., Gabszewicz, J.-J. and Thisse, J. 1979. On Hotelling's 'Stability in Competition'. *Econometrica* 47, 1145–50.

hot money. Hot money describes large-scale international movements of short-term capital under a fixed exchange rate system driven either by speculation on an imminent devaluation (or revaluation) or by interest rate differentials apparently greater than exchange risk. In the two decades prior to World War I, hot money flows were rare – so great was the level of confidence in the maintenance of the gold standard in the major countries. It was quite different in the interwar years. Major episodes of hot money flows included the flood of foreign funds into France in 1926–8 on speculation that the franc would be revalued. Then in the mid-1930s, there were huge outflows of hot money from the gold bloc currencies into London and New York.

Hot money flows reached a new crescendo in the final years of the Bretton Woods system. The Sterling devaluation of 1967, the devaluation of the French franc and revaluation of the Deutsche mark in 1969, and the floating of the mark in 1971 were all preceded by huge speculative flows of capital. The biggest ever movement of hot money was in the first quarter of 1973. Speculation was rife that the Smithsonian Agreement would break down. The Nixon Administration was pursuing a prices and wages policy whilst US interest rate were held at low levels. In contrast, the Bundesbank was seeking to combat inflationary pressures by instituting a monetary squeeze and pushing interest rates to much higher levels. A general devaluation of the dollar in mid-February failed to arrest the hot money flows out of the dollar (principally into the mark). Finally, on 12 March, the EEC currencies were jointly floated. In the era of floating exchange rates, the main examples of hot money flows have been within the Snake (and its successor, the EMS) and into or out of the British pound during periods when its rate has been temporarily stabilized (either against the dollar or some weighted basket).

Hot money flows are usually a source of instability in the domestic economy, in that they induce sudden and occasionally perverse changes in monetary conditions. The country losing funds suffers deflation, sometimes intensely, as interest rates are pushed to high levels in defence of the currency. The deflationary cost of sticking to a parity in defiance of market pressure often proves unacceptable politically, even when policy-makers strongly believe that the present parity is consistent with 'fundamental equilibrium'. Thus hot money flows may produce self-fulfilling prophecies. The same is true in the opposite direction. Hot money inflows into a country on speculation of a revaluation may force such action, or else the continued swelling of the domestic money supply would threaten an outbreak of inflation.

Governments have turned to a variety of weapons to combat hot money flows in order not to be deflected from their chosen policy course. One has been direct controls on capital movements. For example, banks may be restricted in their covered interest arbitrage operations. In consequence, some speculative pressure would be absorbed by the differential between the forward exchange rate and its interest rate parity level, and less pressure would fall directly on interest rates. An alternative option is the introduction of a dual exchange market, whereby capital flows are channelled through a financial tier, in which the rate floats freely. Then speculation on a change in the official rate gives rise to a change in the free rate rather than to a loss or gain of reserves together with interest rate changes. In practice, though, it is difficult to prevent leaks between the two tiers. Central banks subject to large-scale money inflows from abroad may impose raised reserve requirements on domestic banks' external liabilities and on domestic corporations' borrowing from abroad.

An alternative to direct controls as a method of insulating domestic monetary conditions from speculative pressures are policies of sterilization, where the central bank seeks to offset the effect of foreign reserve changes on the money supply by undertaking open market or swap operations. In practice, sterilization policies have rarely been applied forcefully – mainly because they tend to aggravate the flow of hot money and the amount of foreign exchange intervention necessary to support the parity. A central bank which desists from raising interest rates when its currency is under attack not only fails to increase the cost of speculation but also confirms suspicions that it is set on an easy money policy, inconsistent with exchange rate stability.

BRENDAN BROWN

See also CAPITAL FLIGHT; EXCHANGE CONTROLS; INTERNATIONAL CAPITAL FLOWS.

BIBLIOGRAPHY
Emminger, O. 1977. The D-Mark in the conflict between internal and external equilibrium 1948–75. *Essays in International Finance* No. 122, Princeton, June.
McKinnon, R. 1974. Sterilization in four dimensions: major trading countries, Euro-currencies and the United States. In *National Monetary Policies and the International Financial System*, ed. R.Z. Aliber, Chicago: University of Chicago Press.

Swoboda, A.K. 1974. The dual exchange-rate system and monetary independence. In *National Monetary Policies and the International Financial System*, ed. R.Z. Aliber, Chicago: University of Chicago Press.

hours of labour. *See* EXPLOITATION.

household budgets. The earliest known example of systematically collected household budgets can be found in *The State of the Poor* by Eden (1797). To assess the living conditions of the lower classes Eden wanted to know, in addition to other matters, the 'Earnings and expenses of a labourer's family for a year: distinguishing the number and ages of the family; and the price and quantity of their articles of consumption' (Preface, p. iv). He obtained this information for households from some 50 parishes in England. Eden reports for these families their earnings by type of income (mostly wages) and income earner, and their expenses by type of expenditure (food, rent, fuel, clothing). Prices and quantities are only rarely given but the composition of the family and the occupation of its head are usually precisely described. Another well known early example is the collection of 199 budgets for Belgian labouring class families in 1853, published by Ducpétiaux (1855), which provided the statistical material for the formulation of Engel's Law (Engel, 1857). Ducpétiaux used a uniform classification of expenditures to facilitate comparison of consumption patterns across families. The 19th century has seen a gradual extension of such household budget surveys mostly conducted by private (groups of) persons on an incidental basis. In more recent times official institutions organize these surveys more or less regularly as part of their normal operations. They may cover thousands of families.

DESCRIPTIVE ASPECTS

In current usage a household budget is a summary of how a particular household allocates its expenditure over well defined items or groups of items during a given period (month, year). Usually, the items are grouped according to a uniform system of classes for all households participating in the same survey.

The emphasis is on expenditure rather than on earnings, contrary to the early examples. Information on earnings is more sensitive than that on expenses and is in general less accurately reported by the participants. The breakdown of expenditure into a quantity and a price component is whenever possible desirable but not always realized.

The unit, the household, consists of members of a family and others sharing living and eating arrangements. There are basically two ways in which the information on its expenditure is collected. One is to ask the household to record all expenditure as soon as it is made in a specially provided notebook. The other is to ask the household to recall its expenditure over a given period of time in the past. The first method is more demanding on the household and excludes those that have not enough literacy or discipline. The second one is clearly less reliable.

The type of household considered depends on the purpose of the survey. Many of the early surveys were conducted to obtain information about poverty and concentrated therefore on low income households. Surveys are sometimes used to obtain appropriate weights for cost of living indexes. If these indexes are used to gauge the real income of wage earners, the collected budgets are those of families with a wage earner as head. If a household budget survey is meant to provide detailed information on consumer habits in general one will try to have a more or less representative sample from the population. Because certain types of consumer units like one-member households, collective units (for example, boarding schools), illiterate and irregular families, families changing residence and composition tend to be excluded from the sample one should not expect full representativity.

Still, as a source of detailed information on consumer behaviour with respect to very finely detailed commodities and services there is no alternative of the same quality to the budget survey. It can provide useful information about the extent of the market for a certain product or about the type of families that have special interest in certain expenditures.

The degree of detail has a limit, however. To keep track of all available shades of quality is virtually infeasible. Some aggregation over qualities is unavoidable, which might cause apparent differences in unit prices owing to differences in the quality composition of the aggregate.

Other problems are the value of gifts and of the consumed own production of farmers. They are usually solved in a pragmatic way.

As such a budget survey gives synchronic information. It observes a group of families during the same, rather short, period of time. It provides no information about the changes in behaviour over time. For example, usually prices do virtually not change during the period of observation, which rules out the possibility to study responses to price changes. Clearly, repetition of a budget survey adds a time dimension and opens the possibility to analyse time dependent changes. So-called *panel studies* are such repeated surveys where in each new round a part of the participants is replaced by other households with similar characteristics. The combination of diachronic and synchronic information such a panel offers is of great value.

The value of budget surveys is also increased if the expenditure behaviour of a household can be related to its various characteristics like residence, race, degree and type of labour participation, composition of the family (sex, age), education, owned or rented housing, ownership of durables, hobbies, pets, and so on. Such additional information is not always fully collected, or frequently not made available in detail to the public in order to avoid identification of the participants by outsiders.

Budget surveys have their limits. They are usually costly. The participants will not (accurately) respond to certain questions for various reasons. As already mentioned some of the collected information is not published to protect the participants.

Normative budgets. The early interest on household budgets had a humanitarian motive. The actual expenditure of a family was compared with what a family of that type needs. These needs were determined in the form of a set of minimal quantities of various items (usually foodstuffs). Given corresponding actual prices the total means to purchase these quantities can be calculated and constitutes a normative budget. A household with an income below the norm qualifies for support. The selection of the minimal quantities is not without ambiguity. A norm very close to a physical survival level leaves little choice, but norms corresponding to social viability in a modern society are difficult to define in an indisputable way.

ANALYSIS OF HOUSEHOLD BUDGETS

Differences in the expenditure patterns across families can be attributed to three factors: (i) variation in available means, (ii) variation in relative prices, (iii) differences in other family characteristics. These three topics will be taken up one by one in what follows. Among differences in family characteristics differences in size and composition of the household have historically played an important role. This justifies their discussion in a special section.

Engel curves. The relation between demand for (or expenditure on) a good and the means of a consumer unit is frequently named *Engel curve* after Ernst Engel (1857) who on the basis of his analysis of the budget data of Ducpétiaux stated his law that the share of food in total expenditure is a decreasing function of the level of prosperity of the family. Engel's law appears to hold almost universally – see, for example, Houthakker (1957). One would like to amend it somewhat in the sense that it pertains to the share of staple food items in the budget rather than to that of all types of food. This distinction is less relevant for families in the lowest income groups than for more well-to-do ones.

A generalization of Engel's law states that with increasing prosperity the budget share of any good initially increases (except for some basic subsistence good) and later on decreases. Increasing budget shares correspond with the 'luxury' status of a good. Its budget or income elasticity of demand is larger than one. A 'necessity' has a decreasing budget share. Its budget or income elasticity is smaller than one. Note that decreasing budget share does not imply decreasing quantity, although this might occur. A necessity with a diminishing quantity is an 'inferior' good. One with a constant or growing quantity as prosperity increases is a 'normal' or 'superior' good. A commodity may go through a prosperity cycle, being a luxury and normal commodity for the very poor, a necessity and normal commodity for the better-off and perhaps a necessity and inferior commodity for the very rich.

For empirical research the issue of the measurement of 'available means' or 'level of prosperity' arises. Total wealth of a family, defined as the market value of its real and financial assets plus the present value of expected future income from other sources, might be the most appropriate concept. It escapes direct measurement, however, because it is based on subjective expectations while also the ownership of assets is not well observed. The same holds for the concept of permanent income which is the amount of money which may be consumed leaving total wealth unchanged. Current income is apt to include transitory components and is presumably anyway not faithfully recorded. The amount of total expenditure is usually readily available and might be closely related to total wealth or permanent income. This makes it an attractive proxy for available means or prosperity when explaining the pattern of expenditure.

The explanation of expenditure patterns as a function of total expenditure is a typical allocation model. Let q_i denote the quantity bought of item i ($i = 1, \ldots, n$) and p_i its price. Then

$$e_i = p_i q_i \qquad (1)$$

is the amount paid. By definition total expenditure m is given by

$$\sum_{i=1} e_i = m \qquad (2)$$

The Engel curves $E_i(m)$ should satisfy the following adding-up condition

$$\sum_{i=1} E_i(m) = m \qquad (3)$$

673

Condition (2) is automatically satisfied by the data. Property (3) depends in part on the functional form of the Engel curves used in the explanation. Linear Engel 'curves' easily satisfy (3) but cannot deal with the possibility that goods change from superior to inferior commodities over the prosperity cycle. They can also not guarantee non-negative consumption. An Engel curve system satisfying (3), allowing for a prosperity cycle and excluding non-negative consumption is still not available.

Zero consumption for certain items, a common phenomenon, provides another complication of Engel curve analysis. Of the full range of commodities on the market only a limited number will be bought by a given family. The statistical distribution of e_i conditional on m is then a bimodal one with zero for the smaller mode. Least-squares regressions using all data will not estimate either of these modes. Leaving out observations with a zero consumption level estimates correctly the non-zero mode. This leads to problems, however, if one wants to estimate a full system of Engel curves simultaneously, because only a very few families will report non-zero expenditures for all items.

In the case when zero consumption is owing to the fact that the amount of money needed to acquire the smallest available quantity of a desirable item is more than the household can afford there is a link with prosperity, because for a sufficiently high level of prosperity the household will buy the commodity. Still there is a statistical complication due to the mixture of discreteness (zero versus non-zero) and continuity (if non-zero how much?) of the relationship. Tobin (1958) gives an elegant statistical solution to this problem using a combination of qualitative and quantitative response modelling – see also Maddala (1983).

Price responses. Differences in expenditure pattern may be also due to differences in prices paid by the households. As already mentioned budget surveys do not usually provide a good data base to observe and analyse price responses. Sometimes, however, there is price variation owing to geographical distance or other supply factors. If the prices are reported their effects can be analysed. The same is true for panel data with price variations over time (and space).

If price information is available it can be employed to explain variation in expenditure patterns. One may write

$$q_i = f_i(m, p_1, \ldots, p_n) \qquad (4)$$

to express that the quantity purchased of commodity i does depend on total means m or total expenditure (see above) and the prices (p_i) of all goods in the budget. On the assumption that given his budget m the consumer will select the set of quantities that satisfies him best economic theory has specified several properties of price responses – see, for example, Deaton and Muellbauer (1980). These can be taken into account in the estimation or can be tested on their empirical validity.

Effects of household characteristics. Variation in available means and relative prices account for a relatively small part of the variation in expenditures over various items across families. The remaining variation is to be attributed to differences in preferences, tastes. Such differences are not necessarily random. Engel (1857) previously pointed out the importance of differences in climate. Later (1883, 1895), he elaborated the effects of household composition. Next to such physiological factors there may be cultural ones like race and religion. The profession of the head of the household appears to have explanatory power – see Prais and Houthakker (1955). It is a mixture of a physiological (physical effort) and a sociological (reference group) effect. The urban/rural difference

matters too. In part this difference can reflect differences in price structure, in part differences in proximity of shops and availability of public transportation. There may be also a sociological aspect to this difference. More economic in nature are differences in ownership of household appliances and in the extent to which the mother participates in the labour market. A variable like years of school education overlaps largely with the factors already mentioned.

As far as these factors are purely qualitative they can be taken into account in two ways. One is to split the sample into cells which are qualitatively homogeneous and estimate for each cell the effect of quantitative determinants. Some of these cells might be sparsely populated. Another possibility is to treat the qualitative factors as dummies (covariance analysis). A formal test can then supply the answer to the question whether a certain quality makes a significant difference for the expenditure pattern.

The effects of household characteristics on expenditure patterns can be analysed in economic terms as follows. Consider two households with the same m and the same price system. They belong to a different social class and have a different consumption behaviour. They can afford each others' life style but clearly prefer their own. By way of a system of positive and negative subsidies one can induce one household to purchase the same set of quantities as the other household. This change involves in part a reduction in real income because it cannot any more afford the originally preferred set of purchases and in part a change in relative prices. Since the argument is symmetric in both families one cannot say that one is better off than the other. For welfare comparison one needs other information than that on purchasing behaviour supplied by a budget survey.

Household size and composition. The treatment of household size and composition is a subject of long standing in household budget analysis. As a first approximation the effect of differences in family size can be taken into account by considering average expenditure per member as the variable to be explained and average total expenditure per member as the appropriate explanatory variable. Obviously, this approach ignores the possibility that members of the same household have different basic needs. To handle this issue one has experimented with a rescaling of the number of members into a number of equivalent members. Engel (1883) proposed as unit the 'quet', which corresponded with a newborn baby. The normal weight of a person of given age and sex divided by that of the infant defined the number of quets for that person. Family size was measured by the sum of quets of the members.

More recent approaches, however, take the male adult as the unit and the members of the household are converted into an equivalent male adult. The Amsterdam scale for example, assigns a factor 0.9 to a female adult and a factor of 0.1 to a one-year-old child.

The early equivalent adult scales reflected mostly physiological differences. They were usually meant to correct for household size effects on consumption of food and were established *a priori*. Many different scales have been introduced. Sydenstricker and King (1921) introduced commodity specific equivalent adult scales together with an overall scale. This can be formalized by writing the Engel curve as

$$e_{ih}/s_{ih} = E_i(m_h/s_h) \qquad (5)$$

where i denotes commodity, h household h, s_{ih} is the size of household h using weights specific for i and s_h is the size of household h using overall weights. According to (5) the addition of a member to the household will have a direct, usually positive, effect on the expenditure on i by way of its impact on s_{ih} and an indirect, usually negative, effect by way of its reduction of total means per equivalent adult. Sydenstricker and King also suggested the estimation of the weights of the scales along with the other parameters of the Engel curves. These contributions went largely unnoticed until reintroduced by Prais and Houthakker (1955).

There are at least two problems with formulation (5). The first one is that it is nonlinear in the size variables causing estimation problems. The second problem is the one of (in)compatibility of the overall size variable s_h with the commodity specific scales. On the basis of (3) and (5) one has that

$$m_h = \sum_i s_{ih} E_i(m_h/s_h) \qquad (6)$$

which may be seen as an implicit definition of s_h involving m_h, the total means of the household. To put this another way, given the specific weights the overall weights are determined and they are generally not independent of m_h. Estimating the overall weights independently of the specific ones leads to problems.

Another approach using commodity specific scales is to estimate

$$e_{ih} = \pi_{ih} f_i(m_h, \pi_{1h}, \ldots, \pi_{nh}) \qquad (7)$$

where $f_i(\ )$ is a demand function with as prices $\pi_{jh} = p_j s_{jh}$. The family size effect is in this way assimilated with a price effect.

Also here, an increase in the family results in a direct, positive, effect via the π_{ih} factor right after the equality sign in (7) and an indirect effect which the changes in the relative prices exert on demand. This latter effect takes the place of the overall effect in (5). This reformulation is formally justified by redefining the utility function of the consumer unit in terms of $x_{ih} = q_{ih}/s_{ih}$, that is in quantities per equivalent adult. This function is then maximized subject to the budget definition written as

$$\sum_i \pi_{ih} x_{ih} = m_h \qquad (8)$$

The optimal x_{ih} are given by $f_i(\ )$. Multiplying by π_{ih} yields (7). This approach, proposed by Barten (1964), avoids incompatibility problems.

There is an identification issue, however – see Muellbauer (1975, 1980). One needs additional information on the weights of the specific scales in order to identify them from the observations. This prior information can take the form of assigning for each age–sex class the value of the weight in one of the n specific scales (for example 1 for male adult in the scale for tobacco; 1 for infants in the scale of babyfood; 1 for female adults in the scale for cosmetics). One can also formulate a restriction, again for each age–sex class, involving weights in more than one specific scale (for example, equality of the weights of teenagers in the scales for bread and for meat).

The specification of the scales deserves some further discussion. Define

$$s_{ih} = \sum_j b_{ij} c_{jh} \qquad (9)$$

with c_{jh} being the number of members of household h in age–sex class j, while the b_{ij} are the corresponding weights. The linear discrete specification treats the b_{ij} as constants which are either estimated or fixed extraneously. The continuous scale approach of Friedman (1952) makes the b_{ij} a continuous function of age and sex: hm_i (age j) for male members and hf_i (age j) for female members. Various restrictions on these functions result in scales which are smooth at the end points and parsimonious in parameters. There may be a problem, however, in obtaining a proper monotone behaviour.

A related issue is that of incorporating scale effects into the family size measure. Kapteyn and van Praag (1976) let the weight depend on the age rank (r) of the member of class j in the family. Following this approach one could specify as weight

$$b_{ijr} = b_{ij0} + b_{ij1}(r - 1) + b_{ij2}(r - 1)^2$$

The measurement of family size effects has sometimes been motivated by the desire to obtain a more objective, empirical basis for family allowance schemes. The welfare implications of varying family composition are not unambiguous, however. As already stated above when discussing the impact of differences of family characteristics in general one cannot conclude directly from observable behaviour that such differences imply being better or worse off. What holds in general is then also true for differences in family composition.

A.P. BARTEN

See also CHARACTERISTICS; CONSUMERS' EXPENDITURE; DEMAND THEORY; ENGEL'S LAW; HEDONIC FUNCTIONS AND HEDONIC INDEXES; SEPARABILITY.

BIBLIOGRAPHY

Barten, A.P. 1964. Family composition, prices and expenditure patterns. In *Econometric Analysis for National Economic Planning*, eds. P.E. Hart, G. Mills and J.K. Whitaker, Butterworths: London.

Deaton, A. and Muellbauer, J. 1980. *Economics and Consumer Behavior*. Cambridge: Cambridge University Press.

Ducpétiaux, A.E. 1855. *Budgets économiques des classes ouvrières en Belgique*. Brussel: Hayez.

Eden, F.M. 1797. *The State of the Poor*. 3 vols. London: J. Davis. Facsimile edn, London: Frank Cass & Co., 1966.

Engel, E. 1857. Die Productions- und Consumtions verhältnisse des Königreichs Sachsen. *Zeitschrift des Statistischen Büreaus des Königlich Sächsischen Ministerium des Innern*, No. 8(9), 22 November. Reprinted in *Bulletin de l'Institut International de la Statistique* 9, (1895), 1–54.

Engel, E. 1883. Der Werth des Menschen; I. Teil: Der Kostenwerth des Menschen. *Volkswirtschaftliche Zeitfragen*, Vols 37–38, Berlin: L. Simion, 1–74.

Engel, E. 1895. Die Lebenskosten belgischer Arbeiter-Familien früher und jetzt. *Bulletin de l'Institut International de Statistique* 9, 1–124.

Friedman, M. 1952. A method of comparing incomes of families differing in composition. *Studies in Income and Wealth* 15, 9–24.

Houthakker, H.S. 1957. An international comparison of household expenditure patterns, commemorating the centenary of Engel's law. *Econometrica* 25, 532–51.

Kapteyn, A. and Van Praag, B. 1976. A new approach to the construction of family equivalence scales. *European Economic Review* 7, 315–35.

Maddala, G.S. 1983. *Limited-dependent and Qualitative Variables in Econometrics*. Cambridge: Cambridge University Press.

Muellbauer, J. 1975. Identification and consumer unit scales. *Econometrica* 43, 807–9.

Muellbauer, J. 1980. The estimation of the Prais–Houthakker model of equivalence scales. *Econometrica* 48, 153–76.

Prais, S.J. and Houthakker, H.S. 1955. *The Analysis of Family Budgets*. Cambridge: Cambridge University Press.

Sydenstricker, E. and King, W.I. 1921. The measurement of the relative economic status of families. *Journal of the American Statistical Association* 17, 842–57.

Tobin, J. 1958. Estimation of relationships for limited dependent variables. *Econometrica* 26, 24–36.

household production. Even a casual survey of recent developments in neoclassical economics will reveal a self-conscious intellectual imperialism. Substantive areas traditionally the private preserve of other social science disciplines have experienced significant incursions: fertility, voting behaviour, crime, education, and others. But perhaps the most visible and influential expansion of neoclassical economics has been into the formation, functioning, and dissolution of families. Under the banner of the 'new home economics', conventional utility maximization with fixed preferences claims to provide explanations for an enormous variety of decisions made by households and their members.

There is little doubt that these developments have gained a number of adherents. Among economists, much of this success can be attributed to the obvious appeal of creatively moving a well-known theoretical apparatus to a novel setting. But, there have also been converts from outside economics. For them, perhaps more important than the merits of a new home economics is the absence of persuasive alternatives; the new home economics is effectively directed at the soft underbellies of other social-science disciplines.

In particular, family sociology has conventionally applied conceptual frameworks placing instrumental activities in the market and expressive activities in the home (e.g. Blood and Wolf, 1960). It does not occur to most sociologists, therefore, to think about households as 'productive', nor to see household activities as the concrete manifestation of production functions; home life is about affect. In addition, many sociologists are inductively inclined, preferring to work up from data not down from theory. Their literature, as a result, is rich in facts that are not easily placed under a single theoretical rubric; perhaps knowing more has meant knowing less.

DO FAMILIES REALLY OPTIMIZE? General criticisms of utility maximization are well known and need not be reviewed here (e.g. Hollis and Nell, 1975; Leibenstein, 1976; Lesourne, 1977; Simon, 1978). For at least two reasons, however, utility maximization may be especially problematic within the family setting.

First, all neoclassical economic perspectives on households require that households optimally allocate their resources. This assumption has been supported in part with the argument that inefficient households either will not form or will not survive (Becker, 1981, pp. 40–2, 66–82, 219–36); households form and dissolve within a 'marriage market', which performs the same functions as any other free market.

However, as Blaug (1980, p. 119) has observed in a somewhat different context,

> to survive, it is only necessary to be better adapted to the environment than one's rivals, and we can no more establish from natural selection that surviving species are perfect than we can establish from economic selection that surviving firms are profit maximizers.

At best, therefore, only family partnerships better suited to the environment need survive; the survivors are not required to be optimally adapted. In other words, the assumption of optimization cannot be justified by recourse to market forces.

Second, there is some scepticism about whether families can adjust quickly to a changing environment. Schultz (1974, p. 6) observes,

> The typical family that we observe, especially in rich countries, lives in an economy in which economic conditions are and have been changing substantially over time. As these changes occur, thinking in terms of economics, there are presumably responses – responses in the age at which marriage occurs, responses in spacing and number of children, and responses in the amount of family resources devoted to investment in children. Furthermore,

before these families have fully adjusted and have arrived at an equilibrium with respect to any given economic change, additional and unexpected changes will have occurred. Thus, the families we observe are seldom, if ever, in a state of economic equilibrium.

WHOSE WELLBEING IS BEING MAXIMIZED? Almost all neoclassical perspectives on the family assume that the decision-making unit is the family as a whole, and that there is a single household utility function. In the face of considerable skepticism (e.g. Nerlove, 1974; Mancer and Brown, 1980; McElroy and Horney, 1981; Witte, Tauchen and Long, 1984), Becker's use of 'altruism' (1981, pp. 172–201) is perhaps the best justification.

However, according to Ben-Porath (1982, p. 54), Becker's formulation requires some very strong assumptions, such as perfect information, despite powerful incentives for household members not to reveal accurately how well off they are. In a similar manner, Pollak (1985, p. 599) argues that Becker's results do not depend on altruism per se, but on 'implicit assumptions about power, or equivalently, about the structure of the bargaining game'. Perhaps most important, there is lots of evidence that *ongoing* conflict and coercion characterize a significant number of households. For example, one is a very long way from a single utility function when a recent report from the United States Attorney General's Office asserts (Hart, 1984, p. 11),

> Battery is a major cause of injury to women in America. Nearly a third of female homicide victims are killed by their husbands or boyfriends. Almost 20 percent of all murders involve family relationships. Ascertainable reported cases of child abuse and neglect have doubled from 1976 to 1981. In addition to one million reported cases of child maltreatment, there may be another million unreported cases. Untold numbers of children are victims of sexual abuse, and uncounted older persons suffer abuse.

WHAT ABOUT JOINT PRODUCTION? Given the linear budget constraint, Pollak and Wachter (1975) point out that joint production is effectively excluded from the recent neoclassical approaches to the family. Thus, it is impossible to obtain psychic gratification and a concrete household commodity from the same household activity (e.g. cooking a meal). Berk and Berk (1983, p. 388) observe that joint production could be incorporated with a nonlinear budget constraint, but *additional assumptions* would have to be made. For example, one would need to specify through the appropriate elasticities how responsive to changes in family money income each of the joint products happened to be. It is very unlikely that data could be found to inform meaningfully such an exercise.

The key question, therefore, is whether joint production is common, and what little research that exists (e.g. Berk and Berk, 1979, pp. 237–250), coupled with everyday experience, suggests that it is widespread. One has only to introspect a bit about the nature of child care.

ARE THERE CONSTANT RETURNS TO SCALE? The assumption of constant returns to scale also creates difficulties. In recent statements (e.g. Becker, 1981), household commodities are rather general entities such as prestige, health, esteem and the like. There is no reason to assume that for these outputs, constant returns to scale hold. Indeed, common experience suggests quite the opposite.

For example, doubling the amount of food one ingests will affect one's health in rather different ways depending on how much food one ordinarily ingests. For malnourished individuals, a rather dramatic improvement in health will probably be seen. For well-fed individuals, little improvement will result, and depending on the kind of food eaten, health could actually decline. In short, the linear budget constraint is once again inappropriate so that the usual formulations of the household production function no longer yields signed results. And again, the use of a nonlinear budget constraint requires new assumptions that are very unlikely to have any meaningful justification.

WHAT ABOUT TRANSACTION COSTS? Despite an explicit interest in household production, the recent neoclassical economics of the household so abstracts the production process that it become difficult to recognize the daily activities in which we all engage. Berk (1980, p. 136) has observed, 'One of the ironies of the New Home Economics is that with all the talk about the household production function, scant attention is paid to the actual production processes implied.' More recently, Pollak (1985, p. 582), has noted that

> Since neoclassical economics identifies firms with their technologies and assumes that firms operate efficiently and frictionlessly, it precludes any serious interest in the economizing properties and internal structure and organization of firms. The new home economics, by carrying over this narrow neoclassical view from firms to households, thus fails to exploit fully the insight of the household production approach.

Pollak goes on to propose a transactions cost approach to households in which the family is conceptualized as a governance structure rather than a preference ordering. Special emphasis is placed on how families are able to provide incentives to their members and monitor their performance. For example, because important instrumental and expressive activities are carried out in the same setting, families are able to apply rewards and punishments not readily available to other institutions. Yet at the same time, the intermingling of economic and personal relationships means that quarrels initiated in one sphere may carry over into another. Whatever the merits of Pollak's perspectives, they emphasis how much of family life has been lost in the neoclassical abstraction.

MODEL SPECIFICATION IN EMPIRICAL WORK. The ultimate validation of any theory must come from how it performs in the empirical world. By and large, the empirical work done to date within the new home economics has been roughly consistent with theoretical predictions. However, the effects of key variables are often very small and/or statistically indistinguishable from zero (e.g. Layard and Mincer, 1985). More important, as Pollak asserts (1985, p. 584), 'because of the central role of unobservable variables (e.g., preferences, household technology, genetic endowments), the new home economics view of the family does not lead simply or directly to a model capable of empirical implementation'.

For example, it is one thing to 'hold constant' the role of a priori preferences when extracting the essentials for theory development, but quite another to omit sound measures of tastes from one's econometric models (Berk and Berk, 1983, pp. 380–1). Unless the omitted taste variables are uncorrelated with either the outcome variable or the explanatory variables that are included, biased estimates will result. Hence, even when statistical results appear consistent with economic theory, it is not clear what has been demonstrated. And to date, the empirical literature has typically failed to introduce reasonable measures of family members' preferences.

CONCLUSIONS. Given the current state-of-the-art, economists probably ask far too much of their theories. Nowhere is this more true than in the recent applications of neoclassical microeconomics to families. In the search for signed results, enormous simplifications and abstractions have been introduced. One is left with a perspective that if taken literally will probably fail.

First, the requisite assumptions, if accepted at face value, make the theory of dubious relevance for most households. Consequently, one is in practice reduced to arguing about how closely the theory approximates reality, and almost any empirical findings may be dismissed. If, for example, in certain developing countries women's labour force participation does not respond in expected ways to increases in market wages, one may simply claim that the market economy is insufficiently mature.

Second, many of the theory's key concepts are typically unobserved in practice and perhaps even unobservable in principle. This means that all empirical efforts are undermined by errors in variables and model misspecification. Once again, therefore, virtually any empirical finding may be discarded. For example, if women with more education spend fewer hours caring for their children than women with less education, it may be that one is witnessing the substitution effects (via greater market wages) predicted by economic theory. Alternatively, with greater education comes a preference for market activities. Or, women who already prefer market activities to home activities obtain more education. However, *all* of these interpretations may be easily dismissed. Neither the going, occupationally specific wage nor preferences for market activities are directly measured.

In contrast, the sensitizing role of recent efforts by neoclassical economists to understand family life has been extraordinarily useful. The new home economics force one to address seriously the nature of household production and the degree to which concepts from neoclassical economics can be instructive. In other words, we are told where to look and given some initial tools to aid in that process. These are major accomplishments.

RICHARD A. BERK

See also FAMILY; GENDER; HOUSEWORK; WOMEN AND WORK.

BIBLIOGRAPHY

Becker, G.S. 1981. *A Treatise on the Family*. Cambridge: Harvard University Press.

Ben-Porath, Y. 1982. Economics and the family – match or mismatch? *Journal of Economic Literature* 20(1), March, 52–64.

Berk, R.A. 1980. The new home economics: an agenda for sociological research. In *Women and Household Labor*, ed. S.F. Berk, Beverly Hills, California: Sage.

Berk, R.A. and Berk, S.F. 1979. *Labor and Leisure at Home*. Beverly Hills: Sage Publications.

Berk, R.A. and Berk, S.F. 1983. Supply-side sociology of the family: the challenge of the New Home Economics. In *Annual Review of Sociology*, Vol. 9, Palo Alto, California: Annual Reviews Inc.

Blaug, M. 1980. *The Methodology of Economics*. Cambridge: Cambridge University Press.

Blood, R.O. and Wolf, D.W. 1960. *Husbands and Wives: The Dynamics of Married Living*. New York: Macmillan.

Hannan, M.T. 1982. Families, markets and social structure. *Journal of Economic Literature* 20(1), March, 65–72.

Hart, W.L. 1984. *Attorney General's Task Force on Family Violence*. Washington, DC: US Department of Justice.

Henderson, J.M. and Quandt, R.E. 1980. *Micro-economic Theory: A Mathematical Approach*. 3rd edn, New York: McGraw Hill.

Hollis, M. and Nell, E. 1975. *Rational Economic Mann*. Cambridge: Cambridge University Press.

Layard, R. and Mincer, J. (guest eds.) 1985. Trends in women's work, education and family building. *Journal of Labor Economics* 3(1), January, i–iii.

Leibenstein, H. 1976. *Beyond Economic Man.* Cambridge: Harvard University Press.

Lesourne, J. 1977. *A Theory of the Individual for Economic Analysis.* New York: North Holland.

McElroy, M.B. and Horney, M.J. 1981. Nash – bargained household decisions: toward a generalization of the theory of demand. *International Economic Review* 22, June, 333–49.

Mancer, M. and Brown, M. 1980. Marriage and household decision-making. *International Economic Review* 21, February, 31–44.

Nerlove, M. 1974. Toward a new theory of population and economic growth. In *Economics of the Family,* ed. T.W. Schultz, Chicago: University of Chicago Press.

Pollak, R.A. 1985. A transaction cost approach to families and households. *Journal of Economic Literature* 23(2), June, 581–608.

Pollak, R.A. and Wachter, M.L. 1975. The relevance of the household production function and its implications for the allocation of time. *Journal of Political Economy* 83(2), April, 255–77.

Schultz, T.W. 1974. Fertility and economic values. In *Economics of the Family,* ed. T.W. Schultz, Chicago: University of Chicago Press.

Simon, H.A. 1978. Rationality as process and as product of thought. *American Economic Review* 68(2), May, 1–16.

Witte, A.D., Tauchen, H.V. and Long, S.K. 1984. Violence in the family: a non-random affair. Working paper no. 89, Department of Economics, Wellesley College, October.

housework. Housework consists of childrearing and the satisfaction of basic human needs through the provision of meals, clothing and shelter within the home. The functioning of the home economy ensures reproduction while it maintains adults so that they can engage in paid work outside the home. Housework is an essential part of an economic and social system because it not only provides essential services but also helps to maintain the unequal class structure. Within the home, children learn their place in the social structure while they prepare for their place in the labour market.

Social rules and customs (i.e. institutions) govern housework – the required amounts, the way it is done and who does it. Housework will therefore vary across culture, over time, and by class within each culture. Within a class, however, what housework is done and who does it will be socially determined. The gender division of labour that has occurred in most industrial societies, with men engaged in production in return for wages and women engaged in family reproduction in return for sharing income, relegated women to the private sphere of the home. Although all adults engage in some housework, wives are primarily responsible for housework and engage in 30 to 50 hours of housework weekly in the US, depending primarily on the number and ages of their children (Walker and Woods, 1976). Since women have primary responsibility for the home, they can engage in paid labour only after making sure that the socially required housework activities are done.

With very limited access to money-producing activities and with their services primarily rendered to their children and husband, women do not acquire the power that is associated with exchange in public life (Friedl, 1975). Production of status through housework rather than commodities in the marketplace ensured women's inferior position (Benston, 1969; Papanek, 1979).

This institutional analysis of housework assumes that people's needs and desires are formed by the social structure. Alternatively, neoclassical models assume that people's preferences are idiosyncratic and that the marketplace responds to people's desires. In a neoclassical world, housework is abstracted from the social structure, and the household is analysed as a small firm that produces commodities with time and market inputs (Becker, 1965). Systematic substitution between time and market goods occurs by choosing among different consumption bundles and by varying the production process. In this model, women's work decisions are made on efficiency grounds, so that the wife equalizes the marginal return on unpaid (i.e. homework) and paid (i.e. market) work. Specifically, the wife is viewed as having flexibility in deciding how to combine her time with market goods in producing family meals, a pleasant home, presentable clothing and well-behaved children. However, empirical studies of the home-making process and family budgets in the US have shown that very little substitution occurs between the home-maker's time and market goods in housework. Empirically, after standardizing by family income, the employed wife uses few market goods and services outside of childcare to substitute for her own time, and both employed and full-time home-markers use the same techniques in performing housework (Brown, 1979; Strober, 1980; Berk and Berk, 1979). The main substitution tends to be between the wife's market time and her leisure time.

The lack of substitution between housework and purchased goods and services reflects the social norms governing activities that provide family life. In addition, the services provided by the home-maker and the goods and services purchased in the marketplace are generally not comparable. The home economy specializes in producing mothering and the nurturing of family members, along with personalized care in providing food, clothing and shelter. The marketplace produces sophisticated medical care, advanced education, the means of transportation and communication, urban housing and the ability to pool risks through insurance, as well as mass-produced food, clothing, cars and other consumer durables. The family's evaluation of these dissimilar home-produced and market-produced goods and services will be a major determinant of whether the wife works exclusively in the home or also has a job. The family's evaluation will vary with social position and experiences over time.

Although a great deal of attention has been paid to estimating the market value of housework, the lack of comparability between housework and market substitutes makes these estimates problematic. The full-time home-maker's provision of round-the-clock care of family members' needs makes it impossible to equate the value of her time with her permanent replacement cost (i.e. the wage rate such services would command in the marketplace). The personalized and on-call nature of her work prevents us from evaluating the services of the housewife as a combination of so many hours of chauffeur, cook, baby-sitter and laundress per day. In the real world, the household cannot contract to buy these services – more impersonalized than the housewife's – in the small amounts of time and at the random hours that the housewife actually performs these duties. The purchased services usually are not equivalent to the service which the housewife provides because she knows intimately the family members she is serving and takes responsibility for the organizing and providing of care as it is needed. Even in societies where a servant class provides cheap domestic labour, the servant must be directed and supervised by someone, usually the home-maker, and this affects the experience of family life.

Housework has evolved historically as the economy has developed and as social needs have changed (Reid, 1934; Gilman, 1910). Two distinct stages characterize the interaction between the home and the industrialized market. Early industrialization began the process of transferring some production processes (e.g. cloth-making, sewing, ready-made crackers) from the home to the marketplace. Although the home economy could still produce these goods, the processes were arduous and the market economy was more efficient. The more important second stage was evident in the early part of the 20th century as the marketplace began producing goods and services that had never been produced by the home economy, and the home economy was unable to produce them (e.g. electricity and electrical appliances, the automobile, telephone, television, advanced education, sophisticated medical care). In the second stage, the question of whether the home economy was less efficient in producing these new goods and services was irrelevant; if the family were to enjoy these fruits of industrialization, they would have to be procured in the marketplace. The traditional ways of taking care of these needs in the home, such as nursing the sick, became socially unacceptable (and, in most serious cases, probably less successful). Just as the advent of the automobile made the use of the horse-drawn carriage illegal and then impractical, and the advent of television changed the radio from a major source of news and entertainment to background music, so most fruits of economic growth did not increase the flexibility for the home economy in producing these goods and services in modern capitalist economies. Growth brought with it new requirements, such as more mobility in urban areas and increased diversity in consumption goods, along with increased consumer reliance on the marketplace. In order to consume these goods and services, the family had to enter the marketplace as wage-earners and consumers.

Meanwhile, the primary housework activities of meal preparation and clothing care used a declining share of the family's budget. A housewife's efforts to decrease expenditures in these areas by direct work activities (e.g. baking from scratch) or more careful shopping had less impact on the family's budget. Purchases of food and clothing – 56 per cent of the average wage-earner family's disposable income in 1918 and 40 per cent in 1950 – accounted for only 26 per cent in 1972 (Brown, 1986). Thirty per cent of the family's budget that had previously been spent on food and clothing now became freed for other kinds of expenditures, primarily transportation, insurance and retirement, and home ownership. Although the output of the home economy has declined in relative importance as a determinant of the family's total consumption standard, privatized housework is still an essential part of the prevailing social and economic structure.

The demand by the women's movement for economic independence and the equalization of sex roles has brought into clear relief the contrast and contradictions between housework and paid market work. The differences between the two economies have helped perpetuate sexual inequality. Because women have been prepared to run the home economy when they assume their roles as wives and mothers, their sense of identity and personal power is grounded in this economy. The market economy and home economy have their own value structures, work structures and reward structures, which can be contrasted by five major characteristics:

(1) Supervision. The housewife is her own supervisor, while most workers have a formal supervisor, who decides what work needs to be done in what manner.

(2) Pay. No systematic relationship exists between the output of the housewife's effort and the family income, while a paid worker has a rate of pay for a job performed, with rules governing behaviour on the job, sick leave, vacation days and hours of work.

(3) Mobility. 'Changing jobs' for the housewife usually means continuing to work (i.e. caring for the children) without a guarantee of pay, while the worker can usually find another suitable job and has unemployment insurance during job search.

(4) Measure of value. Housework is socially required and does not carry a money value since the market has not provided a permanent replacement, while workers have the exchange value (i.e. wage) as the measure of value for work performed.

(5) Personal behaviour. The home economy is based on the concept of mutual aid and service to others, with cooperative rather than competitive behaviour, while the competitive market economy rewards the individual.

Since money and individual advancement are not part of the reward structure of the home economy, a woman who takes the cooperative and service values of the home economy with her into the market economy will be at a disadvantage in demanding equitable compensation for her work according the values of the market economy.

Besides these conflicts between home and market economies, conflict over production and redistribution issues occurs between family members within the household and between the household and public bodies, such as the state and the workplace. The division of labour by gender that creates the basis of the conflict within the home, especially around housework, also creates interdependence among family members (Hartmann, 1981).

How one views the structure of housework and its role in the economy and society determines how one evaluates female unemployment and the income distribution. In a neoclassical world, wives' unemployment results in only a small loss for both the individual and society, because the wife can substitute her own time in place of the market goods and services purchased with her earnings. Her family's income falls only to the extent that the market is more efficient in providing these goods and services. From an institutional perspective, the wife's unemployment results in an economic loss for both the individual and society that approximates her pay cheque because she cannot use her time to produce the market goods and services purchased with her pay cheque without a major change in the family's life-style. The decline in money income determines how short of expected social standards the family falls. Since income and housework time are both required and are not interchangeable, they cannot be aggregated into a single measure of 'full income' as an indicator of economic well-being. Provided that the family's required housework is being done, measured income determines a family economic well-being.

The neoclassical and institutional models of housework also suggest different strategies for women to use in their struggle for equality in the workplace. Neoclassicists can ignore the burden imposed on employed women by their housework time, since it assumes that this time can be bought off as desired. Institutionalists recognize that for equality to prevail in the labour market, either both spouses must share equally the housework hours required to sustain family life, or childcare and meal preparation must be transferred outside the home to the community or to the marketplace. However, industrializing housework within a private market economy will not necessarily fulfil the basic human needs now served by personalized housework. Such changes in housework will require a fundamental social restructing that will radically alter family life as the norms governing everyday life are transformed.

CLAIR BROWN AND AMELIA PREECE

See also GENDER; HOUSEHOLD PRODUCTION.

BIBLIOGRAPHY

Becker, G.S. 1965. A theory of the allocation of time. *Economic Journal* 75, September, 493–517.

Benston, M. 1969. The political economy of women's liberation. *Monthly Review* 21(4), September, 13–27.

Berk, R.A. and Berk, S.F. 1979. *Labor and Leisure at Home: Content and Organization of the Household Day*. Beverly Hills: Sage.

Brown, (Vickery) C. 1979. Women's economic contribution to the family. In *The Subtle Revolution: Women at Work*, ed. R. Smith, Washington, DC: Urban Institute.

Brown, C. 1986. *Consumption Norms, Work Roles and Economic Growth in Urban America, 1918–1980*. Washington, DC: Brookings Institution.

Friedl, E. 1975. *Women and Men: An Anthropologist's View*. New York: Holt, Rinehart & Winston.

Gilman, C.P. 1910. *The Home: Its Work and Influence*. New York: Charlton Co.

Hartmann, H.I. 1981. The family as the locus of gender, class and political struggle: the example of housework. *Signs* 6(3), Spring, 366–94.

Papanek, H. 1979. Family status production: the 'work' and 'non-work' of women. *Signs* 4(4), Summer, 775–81.

Reid, M.G. 1934. *Economics of Household Production*. New York: Wiley.

Strober, M.H. and Weinberg, C.B. 1980. Strategies used by working and nonworking wives to reduce time pressures. *Journal of Consumer Research*, March, 338–48.

Walker, K.E. and Woods, M.E. 1976. *Time Use: A Measure of Household Production of Family Goods and Services*. Washington, DC: American Home Economics Association.

housing markets. The principal features that distinguish housing from other goods in the economy are its relatively high cost of supply, its durability, its heterogeneity, and its locational fixity. Of course, many other commodities exhibit one of these features. However, the interaction of these distinguishing characteristics complicates theoretical and empirical analyses of the housing market.

Durability, heterogeneity and fixity together indicate that the housing market is really a collection of loosely related but segmented markets for particular packages of underlying commodities differentiated by size, physical arrangement, quality and location. These sub-markets are connected in a predictable way. At neighbouring locations, differences in prices between submarkets cannot exceed the cost of converting a housing unit from one sub-market to another. At different sites, variations in prices within any sub-market cannot exceed the transport cost differentials for the marginal consumer. However, a price-inelastic demand for some of the attributes jointly purchased, combined with inelastic supply in the short-run, can make the pattern of housing prices rather complex, even in a market in temporary equilibrium.

Analyses of the supply and demand for housing are complicated by these somewhat peculiar characteristics. Consider the demand side of the market; take the case of renters. Presumably, quantity demanded depends upon price and income. The 'quantity' in this case consists of a vector of attributes. This quantity can, of course, be summarized by its market rent, but the rent of a dwelling unit is neither a price nor a quantity. Rent is measured in the units of price-times-quantity, and it is a formidable task to disentangle the two for statistical purposes.

The third variable included in the demand relationship, income, is equally difficult to measure in the housing market. Given the high costs of transforming residential capital and the high costs of moving, it follows that housing decisions are based upon some long-run or 'permanent' notion of income (Friedman, 1957), a concept which has proved difficult to specify without ambiguity.

The relevant notion of the price of housing for the decision of owner occupants is even more elusive. By observing transactions in the market, the value (V) of an owner-occupied dwelling can be ascertained. Under familiar but quite restrictive competitive conditions (infinite durability, no depreciation or maintenance, capital gains or taxes), the annual rent (R) for this dwelling is:

$$R = iV \qquad (1)$$

where i is the rate of interest, assumed equal to the mortgage interest rates. Under more realistic conditions the annual cost of a dollar of residential capital, the so-called 'user cost' of capital (Jorgenson, 1971), can be estimated. It varies with four broad classes of circumstance: (a) the expected rate of increase in housing equity, in alternative investments (including rents) and the fraction of the purchase financed by borrowing; (b) the type of mortgage and the holding period; (c) the rate of depreciation (δ) gross of maintenance expenditures, and the fixed cost of buying and selling the residence; and (d) the marginal tax rates for income (T_y), property (T_p), and capital gains (T_g), and the rules for tax liability. Assume four classes of simplified market conditions; (a) the rate of increase of rents equals the rate of increase in housing values (γ); (b) the net mortgage rate $i(1 - T_y)$ equals the net rate of return on alternative investment, for a fixed rate mortgage with an infinite holding period; (c) the buying and selling costs are zero; and (d) interest and property taxes are deductible from taxable income and the imputed return from living in a dwelling is not taxed. Under these simplified conditions (Rosen, 1985), the annual cost of housing capital may be represented as:

$$R = [(1 - T_y)i - (1 - T_g)\gamma + \delta + (1 - T_y)T_p]V \qquad (2)$$

Equation (2) emphasizes the importance of taxes and capital gains, as well as interest rates, in defining the effective price of housing services to an owner-occupant. For example, the expectation of capital gains (γ) decreases the effective cost of housing. This may be partly offset by capital gains taxation (T_g), but in many countries housing transactions are essentially free of this tax. As long as capital gains tax rates are less than marginal income tax rates, general inflation (i.e. increases in interest rates and capital gains) reduces the cost of home ownership more for higher-income households. Tax provisions, especially the tax-free nature of imputed rent, reduce the relative cost of home ownership at higher-income levels.

Together, these price and income concepts have been used to estimate the parameters governing housing demand and tenure choice. There seems to be some general agreement that: the elasticity of demand for the composite housing good is low for annual income, but much higher, approaching one, for average (one 'permanent') income; and that housing demand is price-inelastic. Evidence also suggests that tenure choice is rather insensitive to the relative prices of owning and renting dwellings.

The spatial pattern of housing and households defines the economic geography of urban life and the development of metropolitan regions. Modern economic theory which explains these spatial patterns (e.g. Muth, 1969) owes much to the German economic geographers of the 19th century. In particular, the seminal work of von Thünen (1826) considers the question of agricultural production on an isolated plain relative to a central market place. The modern treatment considers the residential locations of workers employed at a central worksite. Workers (or farmers) are willing to pay a premium for central locations to reduce transport costs, so housing (or agricultural land) must become cheaper at more distant locations. To illustrate, assume consumers derive utility $U(h, x)$ from housing (h) and other goods (x). They confront a budget constraint which requires them to allocate exogenous income Y between housing consumption, whose price $P(t)$ varies with distance t, other goods (at a price of one), and transportation costs $k(t, y)$, which vary with distance and income; i.e., $y = x + P(t)h + k(t, y)$. Maximizing utility subject to this constraint yields:

$$hP' = -\partial k / \partial t. \qquad (3)$$

The consumer chooses to locate at that point where the marginal savings from cheaper housing exactly offset the marginal costs of additional commuting. Clearly, the location chosen depends upon the household's preferred amount of housing. It can be shown (by differentiating (3) with respect to income) that higher-income households will choose less accessible ('suburban') locations under reasonable conditions (i.e., as long as the income elasticity of housing demand exceeds the income elasticity of marginal transport costs). The theory thus provides an explanation for the central location of the poor, an explanation which is quite distinct from competing theories based upon the prior location of the oldest housing stock (e.g. Burgess, 1925).

The concentrations of low-income households and the existence of slum housing raise several questions about the operation of housing markets, the 'filtering' of dwellings and the role of externalities in housing. The concept of filtering arises from the observation that 'most households live in second-hand housing, even the Queen of England' (Grigsby, 1963). If a middle-income household is induced to move to a newly built dwelling, it sets off a chain of moves, as the rent which can be charged for a vacated dwelling declines, making each one available to households of lower income. Under what circumstances does the filtering process make lower-income households better off? If the quality of housing were truly exogenous (for example, if it were only related to the vintage of the dwelling), then low-income households could benefit directly from the filtering process as higher-quality housing became available. On the other hand, if housing quality is sufficiently responsive to landlord maintenance decisions, then demand price declines may be matched by quality declines. It thus requires a very special view of housing to conclude that the 'filtering' process will lead to improved housing for the poor, even under static conditions.

Externalities in the housing market may arise from physical, 'social', or pecuniary conditions. The propinquity of dwellings does suggest that the maintenance decisions of landlords may be subject to a kind of 'prisoners' dilemma' in low-income neighbourhoods. Because the rent of a unit reflects the quality of adjacent dwellings, owners of neighbouring properties may maximize returns if 'the other guy' invests. Thus, housing or rehabilitation investment which is jointly profitable may not be undertaken at all.

The policy prescription for economic efficiency in this case is joint ownership or decision-making (or public renewal). But suppose the externality is of a social or demographic character. For example, suppose members of each of two races can only tolerate neighbourhoods in which they constitute at least x per cent of all households. Under such circumstances, Thomas Schelling (1978) has shown that integrated neighbourhoods will result, at least for some distributions of x. But he has also shown that this result may be highly unstable; the integrated outcome could easily unravel in response to an exogenous movement of a few households. Either segregated or integrated solutions may be 'efficient' in

some very narrow sense. Suppose instead that members of one group have a uniform aversion to living with members of another group (or 'a taste for discrimination', in the terminology of Becker, 1957). In this case a segregated pattern of occupancy may satisfy narrow allocative efficiency principles, since those who discriminate will be required to 'pay for their prejudices'. These different economic models of discrimination are disturbing, but they have only limited application to the housing market, since most empirical evidence suggests a different pattern of prices. Minority households pay higher prices for otherwise comparable housing, at least in North American markets (e.g. Kain and Quigley, 1975).

As noted, the level of new construction is subject to great fluctuation: long term, in response to immigration and population readjustment (Kuznets, 1952), as well as short term, in response to interest rates and credit availability. To some extent the organization of the industry may be a reflection of this cyclicity. The industry is still dominated by small firms, often undercapitalized, producing low levels of output. The relatively low rate of productivity growth in housebuilding may thus reflect an adjustment to cyclicity in demand as well as inherent technological considerations.

JOHN M. QUIGLEY

See also MONOCENTRIC MODELS; PROPERTY TAXATION; TIEBOUT HYPOTHESIS; URBAN ECONOMICS; URBAN HOUSING.

BIBLIOGRAPHY

Becker, G.S. 1957. *The Economics of Discrimination*. Chicago: University of Chicago Press.

Burgess, E.W. 1925. The growth of the city. In *The City*, ed. R.E. Park, E.W. Burgess and R.D. McKenzie, Chicago: University of Chicago Press.

Friedman, M. 1957. *A Theory of the Consumption Function*. Princeton: Princeton University Press.

Grigsby, W. 1963. *Housing Markets and Public Policy*. Philadelphia: University of Pennsylvania Press.

Jorgenson, D.W. 1971. Econometric studies of investment behavior: a survey. *Journal of Economic Literature* 9, December, 1111–47.

Kain, J.F. and Quigley, J.M. 1975. *Housing Markets and Racial Discrimination: A Microeconomic Analysis*. New York: Columbia University Press.

Kuznets, S. 1952. Long term changes in national income of the United States since 1870. In *Income and Wealth*, Series II, ed. S. Kuznets, London: Cambridge University Press.

Muth, R.F. 1969. *Cities and Housing*. Chicago: University of Chicago Press.

Quigley, J.M. 1979. What have we learned about housing markets. In *Current Issues in Urban Economics*, ed. P. Mieszkowski and M. Straszheim, Baltimore: Johns Hopkins Press.

Rosen, H.S. 1985. Housing subsidies: effects on housing decisions, efficiency, and equity. In *Handbook of Public Economics*, Vol. I, ed. A.J. Auerbach and M. Feldstein, Amsterdam: North-Holland.

Schelling, T. 1978. *Micromotives and Macrobehavior*. New York: W.W. Norton.

von Thünen, J.H. 1826. *Der isolierte Staat in Beziehung auf Landwirtschaft und Nationalökonomie* . Hamburg.

human capital. Human capital refers to the productive capacities of human beings as income producing agents in the economy. The concept is an ancient one, but the use of the term in professional discourse has gained currency only in the past twenty-five years. During that period much progress has been made in extending the principles of capital theory to human agents of production. Capital is a stock which has value as a source of current and future flows of output and income. Human capital is the stock of skills and productive knowledge embodied in people. The yield or return on human capital investments lies in enhancing a person's skills and earning power, and in increasing the efficiency of economic decision-making both within and without the market economy. This account sketches the main ideas, and the bibliography is necessarily restrictive. For additional detail and alternative interpretations, the reader should consult the surveys by Blaug, Rosen, Sahota and Willis, which also present complete bibliographies.

Differences in form between human and non-human capital are of less import for analysis than are differences in the nature of property rights between them. Ownership of human capital in a free society is restricted to the person in whom it is embodied. By and large a person cannot, even voluntarily, sell a legally binding claim on future earning power. For this reason the exchange of human capital services is best analysed as a rental market transaction. Quantitative analysis is restricted to the income and output flows that result from human capital investments: wage payments and earnings flows are viewed as the equivalent of rentals of human capital value, because a person cannot sell asset claims in himself. Even the long-term commitments found in enduring employment relationships are best viewed as a sequence of short-term, renewable rental contracts. By contrast, the legal system places many fewer restrictions on the sale and voluntary transfer of title to non-human capital. In fact, substantial activity on non-human capital asset markets is a hallmark of an enterprise system of organization.

Flexibility must be maintained, however, in these distinctions, which are not always hard and fast. The institution of slavery was the primary example of a transferable property right in human capital. To be sure, the involuntary elements of slavery are essential, but even voluntary systems have not been unknown. Similarly, indentured servitude was an example of a legally enforceable long-term contractual claim on the human capital services of others. And in many societies today there are severe legal restrictions on transfer of title to non-human capital: the chief example is collective and state ownership of non-human capital in planned economies.

BACKGROUND. Classical economics maintained a tripartite distinction among the factors of production, Land, Labour and Capital; whereas modern economics is much less rigid in these divisions. Viewed from the perspective of supply, factors of production, whatever their form, can be increased and improved at some cost. To the extent that these improvements involve weighing future benefits against current costs, the principles of capital theory are applicable.

William Petty, the early actuary and national income accountant, is generally credited with the firt serious application of the concept of human capital, when in 1676 he compared the loss of armaments, machinery and other instruments of warfare with the loss of human life. Elements of such comparisons survive to the present day. However, Adam Smith set the subject on its main course. *The Wealth of Nations* identified the improvement of workers' skills as a fundamental source of economic progress and increasing economic welfare. It also contained the first demonstration of how investments in human capital and labour market skills affect personal incomes and the structure of wages. Alfred Marshall stressed the long-term nature of human capital investments, and the role of the family in undertaking them. He also pointed out that non-monetary considerations would play a unique role in these decisions because of the dual nature of workers as

factors of production and as consumers of their work environments. The distinguished actuary and scientist Alfred Lotka provided the first quantitative application of human capital in collaboration with Dublin, calculating the present value of a person's earnings to serve as guidelines for the rational purchase of life insurance. J.R. Walsh made the first cost imputation of human capital value. Frank Knight focused upon the role of improvements in society's stock of productive knowledge in overcoming the law of diminishing returns in a growing economy.

These early contributions stand as landmarks. However, the impetus for rapid progress in this area came from the quantitative revolution in economics after World War II, when extensive data sources revealed certain systematic regularities. The first of these stems from economists' interest in understanding the nature and sources of economic growth and development in the 1950s and 1960s. Detailed calculations by national income accountants showed that conventional aggregate output measures grow at a more rapid pace than aggregate measures of factor inputs. A fundamental conservation law in economics would be violated unless the unexplained 'residual' was identified with (unexplained) technical change. Research associated with T.W. Schultz and Edward Denison attributed much of the measured residual to improvements in factor inputs. Schultz adopted an all-inclusive concept of human capital. At its heart lay secular improvements in workers' skills based on education, training and literacy; but he also pointed to sources of progress in improved health and longevity, the reduction in child mortality and greater resources devoted to children in the home, and the capacity of a more educated population to make more intelligent and efficient economic calculations. John Kendrick systematically pursued the empirical implications of these ideas and demonstrated that the rate of return on these inclusive human capital investments is of comparable magnitude to yields on non-human capital. This line of research as a whole proves that an investment framework is of substantial practical value in accounting for many of the sources of secular economic growth.

Another parallel strand of development arose from professional interest in the nature and determinants of the personal distribution of income and earnings. This problem was propelled, in addition, by substantial public interest in the problem of poverty and prospects for redistributing resources to the poor. Empirical bases for this inquiry were, and continue to be, supported by extensive personal survey instruments (such as Census and allied records) that have become widely available in the post-war period. Much of this work has focused on the role of education and training as important determinants of personal wealth and income. Herman Miller's updating and elaboration of Dublin and Lotka's calculation found a strong and systematic relationship between education and personal economic success, a finding that has been replicated many times in virtually every country where data are available to make the calculations.

The fundamental conceptual framework of analysis for virtually all subsequent work in this area was provided by Gary Becker, who not only organized the emerging empirical observations but also provided a systematic method for seeking new results and implications of the theory. Practically every idea in his book has been pursued at length in the research of the past two decades. Following Schultz's lead, Becker organized his theoretical development around the rate of return on investment, as calculated by comparing the earnings streams in discounted present value on alternative courses of actions. Rational agents pursue investments up to the point where the marginal rate of return equals the opportunity cost of funds. Hence, conditional on the sources of financing investments through the market and family resources, there is a tendency for rates of return to be equated at the margin. This theory of *supply* of human capital implies empirically refutable restrictions on intertemporal and interpersonal differences in the patterns of earnings and other aspects of productivity. In focusing on the development of a person's skills and earning capacity over the life cycle, human capital theory has evolved as a theory of 'permanent income' and wealth.

Becker also made a distinction between human capital that is specific to its current employment in a firm, and that which has more general value over a broader set of employments. The concept of firm-specific capital is closely allied with organizational capital, a person's contribution to a specific organization, the value of which is lost and must be reproduced by costly investment when the employment relationship is terminated. General human capital represents skills that are not specifically tied to a single firm and whose employment can be transferred from one firm to another without significant loss of value. This distinction has proved valuable for analysing the determinants of turnover and firm-worker attachments and its ramifications are still being pursued. For example, the concept of firm-specific capital underlies the transactions cost basis for recent research on labour market and other contracts.

THE RATE OF RETURN. The connection between the rate of return on investment in human capital and observable earnings is illustrated by Smith's discussion of the relative earnings of physicians and other professional workers. A person who contemplates entering one of these fields must look forward to a long period of training and costly personal investment before any income is forthcoming. Furthermore, the long training period cuts into the period of actual practice and reduces the period of positive earnings. Consequently earnings must *compensate* for the cost and effort required to practice the trade: if they did not, fewer people would find it attractive to enter.

The compensatory nature of earnings on prior investments, equivalent to a rate of return, is the fundamental insight of human capital theory. First, it points to the opportunities foregone by an action as a fundamental cost of undertaking it. Thus the direct tuition and other costs of education are only one component of the true cost. The fact that the person defers entering the market and gives up a current source of earnings is also properly counted as a cost. Second, the focus on the intertemporal and life-cycle nature of these decisions leads to a much different concept of income and inequality than simply examining current earnings. Human capital theory suggests that the distributions of *lifetime earnings* and human capital *wealth* are the keys to analysing the distribution of economic welfare, because earnings are the result of prior investments.

Two methods are widely used to calculate the return on human capital investments. Consider one alternative, call it the null alternative, which yields an earnings flow of $x_0(t)$. Consider another alternative, call it the investment alternative, which yields an earnings flow of $x_1(t)$. For example, in the leading case $x_0(t)$ is the expected flow of earnings in year t if one terminates education after high school graduation and $x_1(t)$ is the earnings that can be expected if one continues on to college. The time index t commences as of high school graduation, so $x_1(t)$ will typically show a phase (during the period of college attendance) of much smaller values than does $x_0(t)$. However, in later life $x_1(t)$ is generally larger than $x_0(t)$. This is precisely the investment content of the decision to continue school: there is a current cost in terms of income foregone, but a deferred benefit in terms of greater earnings prospects in the future. Write the difference $z(t) = x_1(t) - x_0(t)$. Then $z(t)$ shows a systematic pattern of negative values when t is small and positive values when t is large; $z(t)$ is increasing from negative to positive in between. Observed earnings in the two choices allows calculation of the internal rate of return, defined as the rate of interest which equates the present discounted value of the two earnings streams. If i is the internal rate, then $\Sigma z(t)/(1+i)^t = 0$.

Of course, it is not possible to observe earnings in the path not taken. A person either stops school or continues on to the next level. In practice, the calculation is made by using observed average earnings of college graduates at different ages as an estimate of $x_1(t)$ and using the observed average earnings of high school graduates as an estimate of $x_0(t)$. The typical calculation produces an estimate of i in the neighbourhood of 10 per cent, comparable to the rate of return on investment in physical capital. Hanoch presents the most complete treatment of this problem. Remarkably, rates of return on education in the vicinity of 10 per cent are found in a wide variety of countries and economic institutions.

Another method of calculation, first presented by Jacob Mincer, brings out the economic aspects of these estimates more clearly. Suppose a person contemplates a level income in amount $y(s)$ over the life work-life cycle if s years of schooling are undertaken. If schooling is productive we must have that $y'(s) = dy/ds$ is positive, that is, anticipated earnings must be increasing in years of schooling. The present discounted value of wealth associated with some choice s, from the point of view of the present time, is simply

$$W(s) = y(s) \int_s^n e^{-rt} \, dt,$$

where the index of integration runs from s, the time the person completes school and enters the market, to n, the time the person retires. Since n is large, we may take the approximation

$$W(s) = y(s) \int_s^\infty e^{-rt} \, dt = y(s) e^{-rs}/r.$$

Assume that the schooling decision is made to maximize human capital wealth $W(s)$. Then differentiating with respect to s, the first order condition is $[y'(s) - ry(s)]e^{-rs} = 0$, or $y'(s)/y(s) = r$. y'/y is nothing other than the marginal internal rate of return on investment in schooling, so schooling is chosen such that its marginal internal rate equals the rate of interest. This rule, similar to the economic problem of when to cut a tree or uncork the wine, is one that maximizes lifetime consumption prospects for the person.

Now extend this argument to many people. In an economy with many similar individuals making schooling choices, all would choose the same value of s, satisfying $d \log y(s)/d \log s = r$. Since there would be no differences in schooling choices among them occupations and jobs that required either more or less education would go unfilled, and the labour market would not clear. Yet, if we observe that in the market equilibrium different people choose different amounts of schooling, with some actually choosing more education and some actually choosing less, then the market earnings on jobs with different schooling requirements must adjust so that the marginal condition is an identity for all possible values of s. That is, people must be indifferent as to how much education they choose. Viewing the marginal condition as a differential equation in y and s and integrating yields the *restriction* $y(s) = y_0 e^{rs}$, where y_0 is the earnings of a person without any schooling. Substituting this back into the definition of $W(s)$, we have

$$W(s) = y_0 e^{rs} \int_s^{\infty} e^{-rt} \, dt = y_0/r$$

is *independent of* s. Writing $W(s) = W$ to reflect this fact, we have $y(s) = (rW) e^{rs}$, and $\log y(s) = \log(rW) + rs$. Think of this last expression as a regression equation. Then after adjusting the income data for age and experience, a regression of the log of income on years of school yields an estimate of the marginal internal rate of return to education (r) as the regression coefficient on schooling. The constant term in the regression estimates 'earning capacity' $\log(rW)$.

The economic logic underlying this development clearly shows the compensatory nature of the returns to schooling and its relationship to the theory of supply. The equilibrium earnings–schooling function is an equalizing difference on the foregone opportunity and other costs of attending school. If people are alike, earnings must rise with schooling to cover the direct and interest costs. Otherwise no one would be inclined to undertake these investments. Notice that in this example, income differences are equalized on cost at every point and that the human wealth (W) is the same for all. Thus there is inequality of earnings, but complete equality of human capital wealth or life cycle earnings. Restricting attention to inequality in the observed distribution of earnings would give a highly misleading indication of inequality in the true distribution of economic welfare in this case.

This simple decision problem provides a convenient and powerful conceptual framework around which much of the research in this area has been organized. The value of this framework was first demonstrated by Becker, who expanded it to include interpersonal differences in abilities and talents and in family circumstances. Interpersonal differences in the rate of interest r, are identified with financial constraints on human capital investments associated with family background and related factors. A person confronting a higher rate of interest would be unable to finance human capital investments on favourable terms and would therefore rationally choose to invest less than a person who was able to borrow at lower rates. Similarly, there may be interpersonal differences in talents among people. Some may be more skilled in learning, which makes schooling effectively cheaper for them, or they may have natural talents which either complement or substitute for schooling in producing earning capacity.

Considerations such as these lead to an identification problem in the schooling–earnings relationship observed *across* different individuals (see Rosen, 1977, for elaboration; also Willis). To begin, let us isolate the effects of family background and financial constraints by restricting attention to a subset of individuals with the same natural talents and abilities. Then differences in school choices within this group would be provoked by corresponding differences in family backgrounds and financial constraints. The reason for this goes back to the institutional feature of human capital assets noted above, that a person cannot sell an asset claim to future earning power. Thus human capital does not serve as collateral for investments in anywhere near the same way as title to physical capital does for non-human investment. A

house, for example, serves as collateral for a mortgage. If the purchaser defaults on the mortgage then the creditor gains title to the house, which can then be sold to settle the debt. Non-transferable titles to human capital make this kind of arrangement impossible for personal investments. Relaxing these kinds of constraints is, of course, the fundamental economic logic behind the public provision of education in most countries throughout the world. But since direct tuition and related costs are only a part of the true costs of schooling, the importance of foregone earnings costs suggests that financial constraints would still remain a factor in educational decision-making. As Marshall noted, the social and economic status of the family play an important role in educational choices.

From the point of view of econometric estimation, observing a subset of the population where abilities are roughly constant, but where financial constraints dictate different schooling choices allows identification of the schooling–earnings relationship for that ability level. This in turn enables the analyst to calculate the social rate of return on investment, and to determine empirically the effect on personal and aggregate wealth of social policies that relax the financial constraints. Earnings of otherwise similar people who were less constrained serve as excellent estimates of the true earnings prospects for more constrained individuals.

Extensive empirical investigation of the connection between schooling, earnings, and family background shows a very strong and systematic relationship between parents' socioeconomic status and background and the school quality and completion levels of their children (e.g. Griliches, 1970, 1977). This is prima facie evidence of financial constraints on educational choices, though it does not rule out other routes by which family background affects a person's economic success, such as complementary investments in the home in child care and quality. These studies also indicate a direct connection between family background and earnings given the schooling choices of children. The causal link between these direct effects of family background and earnings remain to be established. It could reflect common but unobserved variance components across generations within families, such as unobserved ability; and also unmeasured factors, such as school quality and the quality of parental inputs, that are correlated with family background. Whatever their source, these direct linkages are numerically small compared with the effect of schooling itself on earnings. Most of the effect of family background on economic success works through its effects on the educational decisions of children and through that to economic success as measured by income and earnings. The direct effect on income, while persistent and significant, is quantitatively small.

SOME APPLICATIONS. Perhaps the main policy area where these ideas on financial constraints are important is in public provision of training and 'manpower' development programmes for the poor. The logic of these policies rests on the proposition that a person's income in a market economy reflects the quantity of resources that the person controls and the value of these resources. People who are permanently poor have less skills and also less valuable skills then the non-poor. So an attractive policy to help eliminate poverty is to give them more and better resources through education and training. The rate of return has been widely used for programme evaluation. For if the social return to investment in subsidized training is less than the rate of return on other forms of social investment, then programmes emphasizing direct monetary and other transfers to the poor are better bets for society overall than devoting resources to skill enhancement. There now exists a voluminous literature on manpower programme evaluation along these lines, largely stemming out of the social programmes what were instituted in the 1960s and 1970s in the United States. The evidence is mixed. While many examples of successful programmes can be found, the prevailing assessment among experts is that the average programme has not been clearly successful (Ashenfelter, 1978). This empirically based conclusion suggests that the underlying causes of poverty are more complicated than simple family constraints on resources which thwart human capital investments. Lack of motivation, discrimination, ability, low quality prior education and insufficient investments in children in the home, as well as constraints on financing are among many of the possibilities that present themselves as causal factors in reducing personal investments in human capital.

The changing role of women in the workplace and in the home has refocused current professional interest on the role of families in determining economic success of children. While these intergenerational connections between the wealth and economic status of parents and their children have long been recognized as a key element in the question of poverty and the size distribution of income, these aspects have only been linked to human capital theory in very recent years. Again, the impetus for this interest lines in the empirical findings summarized above, and also in some that have come from unexpected quarters, namely the economic success of immigrants and their children.

Recent work by Barry Chiswick (1978) has established a systematic empirical pattern for many immigrant groups into the United States. Chiswick finds that members of the first generation of immigrants earn less than comparable native born citizens in the first two decades of their life in the US. At that point their incomes reach parity with native born citizens and beyond it actually surpass the incomes of the native population. More remarkably, the sons of these immigrants – the members of the second generation – earn incomes which exceed those of the sons of native born workers. However, by the third generation there is parity, and the effects of foreign-born status wash out. While certain aspects of Chiswick's findings remain controversial and are being studied at length, they support the 'melting pot' view of economic life in the US. There is obviously substantial interest and importance in examining similar phenomena in other countries.

The chief theoretical work in the intergenerational transmission of wealth and economic status through families is contained in the research of Becker and Tomes. This work directly addresses intergenerational linkages through preferences and attitudes of parents toward their children, through natural hereditary transfers of ability and through discretionary transfers of resources through the generations. This work is the most complete theoretical description of the intergeneration distribution of wealth available so far. Inheritability of abilities is known from statistical theory to imply a regression-toward-the-mean phenomenon. Thus the fortunes of one generation are not only linked by direct transfers of non-human wealth and human capital investments, but also by inherited traits. These two forces interact in the intergenerational transmission mechanism. The economic fortunes of generations are more closely linked the greater the degree of inheritability of ability and the greater the propensity of parents to invest in their children's human capital. The effects of good fortune in one generation spills over to the next through the transfer mechanism. Interestingly, it may spill over to several subsequent generations. Thus regression toward the mean may occur only after several generations

rather than after only one. When borrowing constraints are imposed on this structure even more persistence is implied because low income families do not have sufficient resources to invest in their children, whose incomes as parents are smaller than they would otherwise be. These issues are important for understanding social and economic mobility, and only recently have data become available to study them empirically. In the end this may be one of the most important developments in human capital theory.

ABILITY BIAS. The other major area where considerable research progress has been made is the role of ability in determining economic success. In terms of the decision model above, interpersonal differences in ability shift the earnings–schooling relationship. More able persons earn more at a given level of schooling than the less able, so the observed income-schooling relationship does not necessarily represent the returns available to a given person. Thus consider a group of individuals who have the same financial resources (the same value of r in the term discussion above). If ability is complementary with schooling then the rate of return to schooling will be larger for the more able and they will choose to invest more. A person observed choosing less education rationally does so because the personal return is relatively small under these circumstances. Comparing the earnings of persons who choose less education with those of persons choosing more education leads to a biased assessment of the returns due to differences in their abilities. This 'ability bias' issue has been examined in much detail.

The basic issue was originally posed by Becker, using the discounted earning stream comparisons presented above. If $x_0(t)$ is the earnings stream of people who stop school after high school completion and $x_1(t)$ is the earning stream of those who continue to college, then $x_1(t)$ is likely to be a biased estimate of the earnings prospects of high school graduates had they continued on to college. In so far as their average ability is lower than college graduates, their earnings had they chosen to continue on to college are likely to be smaller than $x_1(t)$. Similarly, the higher average abilities of college going persons makes it probable that $x_0(t)$ is a downward biased measure of what they would have earned had they stopped their education after high school graduation. Thus comparing $x_1(t)$ with $x_0(t)$ yields an upward biased estimate of the rate of return to education for either group.

In order to correct this bias it is necessary to purge the earnings data of the direct effects of ability. Several methods have been proposed, and most find that the effect of ability biases in rate of return calculations is positive but relatively small (Griliches). The fundamental reason for this is due to a finding of Welch, that while the direct effect of measured ability on earnings is positive (given schooling), its numerical effect is quite small. Even a person whose measured ability is one standard deviation above the mean receives, on average, an income that is only a few percentage points above average.

Most of the research in this area has concentrated on indexes of ability associated with IQ and other measures meant to predict school performance. However, predictors of school performance and grades are not necessarily good predictors of economic success. The most sophisticated studies employ factor analytic statistical models, in which measured abilities embodied in IQ scores and the like serve only as indicators of underlying and unobserved 'true' abilities. These studies show that 'raw' rate of return estimates unadjusted for ability differences overstate 'true' rate of return calculations by only a few percentage points. The rate of return to schooling remains substantial, and of comparable magnitude to that on other forms of investment even after ability adjustments have been made.

Most of this ability-bias research assumes that ability can be captured statistically as a single factor (in the statistical sense). However, some recent work is based on a multiple-factor view of ability in which there are different dimensions and components (Willis and Rosen, 1978). This multi-factor framework is familiar from the theory of comparative advantage in economics. A unidimensional specification of ability only allows for absolute advantage, where a person who is more able in one thing is necessarily more able in everything else. By contrast, a comparative advantage specification allows for both absolute and relative advantages. A person may be very talented in all things (absolute advantage), but may also be relatively more talented in some things than others. Furthermore, absolute advantage may not be so important. A great musician is not necessarily adept at non-musical activities such as accounting; and the typical accountant may well have no more than the average musical ability in the entire population. An extension of the model above shows that people would naturally select themselves into those occupations and educational categories that exploit their comparative advantage. Thus those who choose to specialize their human capital investments in musical activities would be likely to have more natural talent for it than the population at large. Similarly, those who learn the plumbing trade would be likely to have more mechanical ability than those who make some other choice. These types of selection problems gain research interest because educational and occupational choices are closely linked. While much important work remains to be done in this area, available evidence is at least consistent with the existence of comparative advantage and occupational selection. If so, the overall ability bias in simple rate of return calculations is likely to be relatively small.

The question of ability bias and selection comes up in a quite different manner in the literature on educational screening and signalling (Spence, 1973). In its most extreme form, the signalling literature maintains the hypothesis that education has no direct effect on improving a person's skills, but rather serves as an informational device for identifying more and less talented people. This model rests on a unidimensional view of ability and also on the suppositions that direct observation of a person's ability and productivity is very costly and that a person knows much more about his own abilities than other persons do. In these circumstances, education serves as a signal of ability if the more able can purchase the educational signal on more favourable terms athan the less able. For then education and ability are highly correlated, and the higher income earned by those with more schooling is supported in equilibrium by their higher ability-productivity.

Several points must be made in this connection. The first is, that taken on its own terms, the signalling and human capital models have very similar implications for the rational choice of schooling. In fact they appear to be econometrically indistinguishable on the basis of income and schooling data alone. The chief difference is a normative one, that schooling has little social value when it serves as a signal, and has much social value when it produces real human capital. Second, the data reveal considerable 'noise' in the schooling–earnings relationship. An investigator does very well when a third of the total variance in earnings can be 'explained' in the analysis of variance sense by observable personal factors such as education, experience, ability measures, family background and other factors. The

schooling–earnings relationship is very strong in the sense of population averages, but the error in prediction is very large for any given person. Large personal prediction errors dull the value of education as a signal. This fact also suggests that education is a personally risky investment. Third, when the signalling model is expanded, it does not necessarily imply that educational signals are socially unproductive. Education may have significant social value in identifying naturally talented people if there is social value in classification and sorting. For example, there may be significant interactions among workers in an organization. If so, then the organization must be structured to choose the optimal *distribution* of talent within it; for example, it may be socially beneficial for the most talented people to work together. In so far as the educational system serves to classify people for these purposes, it is producing a form of human capital (information in this case) which has both private and social value. Finally, the value of education in assisting persons to find their niche in the overall scheme of the economy, precisely because they do not know so much about themselves, has never been quantified.

SIGNALLING AND INFORMATION. A definitive empirical study capable of distinguishing signalling and human capital views of investment in education is yet to be produced in spite of many attempts to do so. Most work in this area has floundered on the fact that the two views imply very similar equilibrium implications about the observed relationship between earnings and schooling, so that if any real progress is to be made, future investigations will have to look elsewhere. A promising area is to examine the direct effects of education on productivity (and not on income alone). Much research has been done on educational production functions, which have an obvious bearing on these linkages and how a different form of education might affect them. For example, some evidence suggests that preschool training can overcome the adverse effects of a poor home environment in educational success. Hanushek (1977) reviews the literature on educational production.

Surprisingly few studies have attempted to examine the schooling–productivity linkage directly, probably because data on personal productivity measures are hard to find, but those few that have managed to do so have found some very impressive results. Griliches reviews the issues at the aggregate level. However, the sharpest results have arisen in agriculture, a sector which has shown an enormous and sustained growth in productivity for at least five decades. The rate of return to education among farmers is substantial. Since most of these persons are self-employed and sell their produce in impersonal, competitive markets, it is difficult to make an *a priori* case that signalling plays any significant role in their educational decisions. Moreover, detailed study shows how these returns come about. More educated farmers control larger resources in the form of larger farms. It is possible that there is a common connection with family background and wealth. However, available evidence suggests that these farmers are also much more efficient in their techniques of production, and that their education is used primarily to keep them informed of recent technological changes in agricultural production, which they adopt with greater frequency and with quicker response. The case that education makes farmers more efficient processors of new information is very well made in the work of Welch (1976, 1978). Schultz indicates that similar findings would apply to much of agricultural production throughout the world, and broadens the argument to make it more generally applicable to all walks of life.

NON-MONETARY CONSIDERATIONS. Another potential source of bias in rate of return calculations arises from the limitations of earnings data. Using expected discounted earnings as the choice criterion is a first order approximation to a more complete formulation. Discounted expected *utility* is the ideal choice index, because an employment relationship is a tie-in between the productive services rendered by human capital skills on the one hand, and the consumption of non-pecuniary aspects of the work environment on the other. The imputed monetary equivalent value of these job-consumption items should be added to earnings in a complete calculation. The same is true of the skills that are utilized outside of the market sector, such as in home production (see Michael, 1982).

That individuals may differ in their tastes for employment of alternative forms of human capital leads to the existence of rents in human capital valuations. Furthermore, the evidence suggests that on-the-job consumption values increase with education and skill. Jobs which require more schooling are likely to be more desirable on *both* monetary and non-monetary grounds (this evidence is reviewed in Rosen, 1986). Economic theory suggests that some portion of earning capacity would be 'spent' on more desirable and more amenable jobs. To the extent that the value of work amenities increase with schooling, observed earnings are a downward biased estimate of total earnings for the more educated, and measured rates of return are downward biased.

These issues are most sharply drawn in the treatment of hours worked in rate of return calculations. For example, if observed earnings alone are used in the calculations, groups such as physicians are found to exhibit large rates of return on their medical education, whereas groups such as teachers are found to earn much lower returns. But physicians work very long hours, perhaps as much as 40 per cent more than the typical worker, whereas teachers work far fewer hours than most other workers; they do not work in the summer, for instance. It is necessary to make judgements about the imputed value of leisure to deal adequately with these differences. If leisure is valued at the wage rate, the proper calculation refers to 'full' income at a common hours-worked standard. Similar considerations apply to growth accounting calculations: The secular increase in embodied skills and human capital has been accompanied by a secular decrease in working hours among the employed population. The imputed value of the quantity and quality of increased 'leisure' should be counted in a measure of welfare. Also, using only market transactions as a basis for calculation conceals the significant value of human capital in home production among those groups, especially women, whose activities have shifted between the non-market and market sectors.

OCCUPATIONAL CHOICE. The discussion so far has concentrated on the role of formal schooling in human capital production. A small but important literature has used these ideas to analyse occupational choice, especially among the professions. The first, and still significant work in this area is due to Friedman and Kuznets, who set the general framework in terms of wealth maximization and rate of return calculations on entry into law, medicine and dentistry. Subsequent literature, of which the work of Freeman is especially notable, has applied modern time-series statistical methods to these problems, concentrating especially on the role of income prospects in attracting or repelling new entrants into a profession.

The human capital perspective suggests that longer term income prospects should play an important role in occupational decisions of the young and that short-term and transitory fluctuations should be of lesser consequence because they have small impact on expected lifetime wealth. Nevertheless, a central finding in this literature is that current market conditions have large effects on occupational choice, and that supply to a specific occupation is relatively elastic with respect to current wages. The effects of long-term prospects have been much more difficult to isolate empirically, depending as they do on specific formulations of expectations and the connections between future earnings expectations and current and past realizations. In so far as a person is 'locked in' to a profession after choosing it, economic theory suggests that long-term expectations should be the primary determinant of choice. The finding that current prospects are highly significant in these choices suggests considerable mobility and recalibration of choices after training. For example, many lawyers use their skills outside the formal practice of law and in complementary ways in the business sector more generally. However, the nature and extent of ex-post mobility possibilities remains to be thoroughly examined.

LEARNING FROM EXPERIENCE. From the theoretical point of view, formal schooling decisions are only half the story in human capital accumulation and skill development. Investment does not cease after schooling: there is another sense in which it just begins. Formal schooling sets the stage for accumulation of specific skills and learning in concrete work situations, through on-the-job training. The human capital literature interprets the term 'on-the-job training' very broadly. Only a small part of the overall concept is included in formal training programmes, apprenticeships and the like. The greater part is associated with learning from experience. This broad and inclusive interpretation is supported by persistent empirical observations on the evolution of earnings over the life-cycle. The age structure of earnings shows remarkably systematic patterns. Earnings rise rapidly in the first several years of working life, but the rate of growth falls toward mid-career and tends to turn negative toward retirement. In panel data, wage rates rise throughout the life cycle, with the greatest rate of increase in the early years. An attractive interpretation of these observations is that the increase in earnings with work experience is due to increasing productivity and human capital accumulation over the entire life cycle.

A fruitful empirical approach for studying these patterns has been developed by Jacob Mincer (1974). The conception of the problem extends the education model above. A person is viewed as making human capital investment choices at each point in the life cycle. Workers who choose to invest more pay for their choice by accepting lower earnings when young and earn returns on their prior investments in the form of larger earnings when they are older. This is essentially a choice between a level experience–earnings pattern (if investments are small) and a 'tilted' one, starting at a lower point and rising to a higher one if investments are large. Mincer develops the concept of 'overtaking' to impute the total return to human capital. The basic idea extends the Smithian principle of compensation to on-the-job training investments. Suppose a person has a large variety of possible investment opportunities after completing school. If no further investments are made, the experience earnings profile is relatively flat. The slope of the earnings–experience profile is increasing and the intercept of the profile is decreasing with the magnitude of investment. Hence the investment level defines an entire family of age earnings profiles, which are spun out around a roughly common crossing point, labelled the 'overtaking' point, if in market equilibrium wealth is approximately independent of investment.

The model has a very sharp empirical prediction that in a cohort of individuals with the same schooling level and different post-school investments, the interpersonal variance of earnings should be decreasing with experience up to the overtaking point and increasing thereafter. These systematic variance patterns have been found by many investigators in a variety of data sources. The assumptions that on-the-job investments are completely equalizing and that human wealth is the same for all investment paths makes it possible to decompose total investments into formal education and on-the-job components. Mincer reports that the on-the-job components are substantial, of the order of a third or more of the total.

The complete education–experience human capital model has important implications for the analysis of poverty and income distributions. In a nutshell, human capital theory suggests that life-time earnings is the appropriate construct for understanding inequality. To the extent that age-earnings patterns are the result of rational investments in human capital, it is misleading to use unadjusted cross-section annual earnings data for inequality analysis. For those young persons who are intensively engaged in investment activities and whose current income is therefore small at present may be classified erroneously as poor even though they are not poor in the lifetime sense. These life cycle issues have not been given sufficient attention in the extensive literature on the social welfare consequences of inequality, in spite of the fact that Paglin (1975) conclusively shows that they have large consequences for the measurement of inequality. Taking the life cycle view yields Gini coefficient estimates of real inequality that are smaller than when only current incomes are used in the calculations.

More detailed econometric work on the dynamic structure of individual earnings based on panel data helps resolve questions of the extent to which poverty status is permanent or transitory over the life cycle. The most sophisticated study so far (Lillard and Willis, 1978) decomposes earnings into several components. One is measureable characteristics of persons, such as education and experience, which reflect human capital and other considerations. Another is a 'person effect' capturing unmeasured components of ability, health, and related factors which permanently affect a person's earning power relative to his cohort. Finally, the third component reflects more transient variations, reflecting such factors as luck and other random events which may persist for a time but which eventually die out. Each component explains about one-third of the total variance of earnings. Since the measurable factors are, by human capital theory, largely equalizing on prior investments and the transitory effects have only small effects on life cycle wealth, this leaves about one-third of the total variance of life cycle earnings as attributable to permanent differences among persons or to 'pure' inequality. Certainly this is quite a different picture than emerges from examining the cross-section distribution of current earnings.

Other approaches to understanding age–earnings profiles in the human capital framework have used a more formal capital–theoretic structure. Here human capital is associated with the latent stock of embodied skills and investment with skill acquisition and learning. A person must give up current income to learn more and increase the stock of skills available for rental at a later date. The optimal investment programme maximizes the present value of lifetime earnings. This basic set-up of the problem was first formulated in an important paper by Ben-Porath (1967), who structured the investment control as choice of the division of a person's time between working and investing. An extension by Rosen (1972) structures it as choice among a spectrum of jobs which offer different learning environments and opportunities. The wage on a job that offers more learning possibilities is lower and the programme is implemented by a 'stepping stone' progression of positions.

This capital theoretic formulation of the problem has virtues in demonstrating the conceptual commonalities between capital and growth theory and human capital theory. However, its generality comes at the cost of providing less robust predictions. Thus it seems fair to say that extensive work attempting to implement these rigorous ideas empirically has not met with overwhelming success in extracting information from observed age–experience trajectories. It appears that other important forces also affect these patterns. Several possibilities have been suggested. One relates to investments in information and search for enduring long job attachments. Job turnover is much larger among young workers than older ones. While this is a form of human capital accumulation and much recent work has been devoted to these issues, it has so far proven difficult to link this class of problems with the ideas reviewed here. Nor has human capital theory yet adequately come to terms with the fact that job patterns typically exhibit discrete jumps and 'promotions', where the character of human capital services rendered changes at each step. Competition for higher ranking positions is properly considered within the human capital framework, but little analysis is available so far.

Any review of human capital would be remiss in not calling attention to parallel developments and important applications in economic historians' interpretation of slavery. The work of Fogel and Engerman (1974) stands out as the primary example of the approach. Here the empirical work focuses on direct human capital valuations rather than on earnings. The principles of capital valuation are used to examine such issues as the long-term economic viability of slavery as an economic institution in the absence of intervention. In addition, some important and fascinating agency problems must be confronted because of an inherent conflict in the master–slave relationship. The conflict arises because the owner naturally desires more effort than the slave prefers to put forth. Various institutions, involving *both* punishments and rewards, were structured to help resolve these conflicts. Mention also should be made of research on indentured servitude by economic historians (Galenson, 1981), which is analysed as a response to a capital market imperfection. A person voluntarily indentured himself for a period of years as payment for a loan to provide transportation and connections in the New World. Repayment was guaranteed by a legally binding claim on the person's services for the period of the contract.

DEMOGRAPHIC EFFECTS. Over the years there has been increasing recognition of the relationship between human capital and economic demography. This is inherent in the role of families as both producers and financiers of human capital investments. Two important recent developments strongly rest on these connections.

The first one is related to large demographic changes in the age structure of the population in the post-war period (the 'baby boom') in the United States. Rates of return on education had remained remarkably constant for a thirty-year period. This in spite of the fact that there had been an enormous increase in education over that period. However, Freeman identified a decline in the rate of return commencing in the late 1960s. The evidence currently available suggests that the rate fell by several percentage points for a 10–12 year period throughout the 1970s, but had gradually returned to its prior level. The leading explanation for this has been provided by Welch (1978) and relates to increased competition for jobs within cohorts as a function of their size.

A stable age distribution of the working population provides a naturally stable progression of work and job opportunities over a person's working life. Not only the level, but also the nature and productive role of human capital changes over the life cycle. Young workers perform different tasks and have different responsibilities than do older workers. Therefore competition and supply of human capital of various types in the labour market is strongly age related. Thus as the large birth cohorts of the 1950s began to enter the market in the late 1960s and 1970s, the increased supply of educated young workers lowered their wage rates and reduced the rate of return. These effects are diffused as the large cohort ages and works its way through the age distribution, and as the structure of work is altered to accommodate their large numbers. The weight of extensive research in this area has shown that returns and wage rates are affected by cohort size. The consequences of this research for the future development of human capital theory will be important, because it requires considering heterogeneous human capital investments and the evolution and development of different types of skills over working life. It may ultimately require analysing how work itself is organized and structured.

HUMAN CAPITAL AND DISCRIMINATION. A final important recent development proceeds on somewhat more conventional theoretical grounds. It addresses the role of human capital in observed wage differences between men and women, and is ultimately related to questions of labour market discrimination. The work in this area is firmly based on empirical calculations. The main fact to be explained is that women earn less than men, even after adjusting for differences in occupational status and hours worked. Labour market discrimination against women is one possible interpretation. However, there may be more subtle forces at work. Mincer and Polachek (1974) build an alternative interpretation on the observation that earnings–experience profiles of women are flatter and exhibit much less life-cycle growth than that of men, and tied it to the well known fact that women traditionally have exhibited less stronger labour force attachments than men due to the sexual division of labour in the home and the bearing and raising of children.

The value of an investment increases with its rate of utilization. Compare two persons: one who expects to utilize an acquired skill very intensively and one who expects to utilize it less intensively. Suppose further that the costs of acquiring the skill are approximately independent of its subsequent utilization. Then the rate of return on investment is larger for the intensive user and that person will tend to invest more. The application to male–female wage differential is apparent upon connecting intensity of utilization with labour force attachments and hours worked. In so far as married women play dual roles in the market and in the household, there is a tendency to invest less in labour market skills and more in non-market skills. The opposite is true of men, given prevailing marriage institutions. These differential incentives can account for differences in age earnings patterns between men and women as well as the larger average wages of men. Research on female labour supply supports the point by showing overwhelming evidence that labour force activities of married women are severely constrained by the presence of children in the home. Mincer and Polachek provided direct empirical support by demonstrating that earnings of never-married women closely approximate those of men.

Considerable research is in progress on these ideas (see, for example, *Journal of Labor Economics*, 1985). At a minimum, the human capital perspective shows that these issues are more complicated than appears on the surface. Yet there are some unresolved puzzles. In spite of the vast increase in female labour force participation in the past two decades, the relative wages of men and women have not changed very much in the United States, though they have come closer to parity in a number of other countries. Part of this may be due to differences in the importance of the government sector as employers of women, as well as differences in compliance with equal pay legislation. A definitive answer is not yet on the horizon.

This essay started by noting the twin origins of developments of the theory of human capital in understanding the sources of economic growth on the one hand and the distribution of economic rewards on the other. Much progress has been made on both counts. However, these two branches have not yet been clearly joined. Future progress will have to come to terms with the issue of how private incentives to acquire human capital affect the available social stock of productive knowledge and how changes in social knowledge become embodied in the skills of subsequent generations.

SHERWIN ROSEN

See also DISCRIMINATION; FAMILY; GENDER; HOUSEHOLD PRODUCTION; INTELLIGENCE, MEASUREMENT OF; SIGNALLING; VALUE OF TIME.

BIBLIOGRAPHY
Ashenfelter, O. 1978. Estimating the effect of training programs on earnings. *Review of Economics and Statistics* 60(1), February, 47–57.
Becker, G. 1964. *Human Capital*. 2nd edn, New York: Columbia University Press, 1975.
Becker, G. and Tomes, N. 1978. An equilibrium theory of the distribution of income and intergenerational mobility. *Journal of Political Economy* 87(6), December, 1153–89.
Ben-Porath, Y. 1967. The production of human capital and the life cycle of earnings. *Journal of Political Economy* 75(4), Pt 1, August, 352–65.
Blaug, M. 1976. The empirical status of human capital theory: a slightly jaundiced survey. *Journal of Economic Literature* 14(3), September, 827–55.
Chiswick, B. 1978. The effect of Americanization on the earnings of foreign-born men. *Journal of Political Economy* 86(5), October, 897–921.
Denison, E. 1962. *The Sources of Economic Growth in the United States and the Alternatives Before Us*. New York: Committee for Economic Development.
Dublin, L. and Lotka, A. 1930. *The Monetary Value of a Man*. New York: Ronald Press.
Fogel, R. and Engerman, S. 1974. *Time on the Cross*. New York: Little, Brown.
Freeman, R. 1971. *The Market for College-Trained Manpower*. Cambridge, Mass.: Harvard University Press.
Freeman, R. 1976. *The Overeducated American*. New York: Academic Press.
Friedman, M. and Kuznets, S. 1954. *Income from Independent Professional Practice*. Princeton: Princeton University Press.
Galenson, D. 1981. *White Servitude in Colonial America*. Cambridge: Cambridge University Press.
Griliches, Z. 1970. Notes on the role of education in production functions and growth accounting. In *Education, Income and Human Capital*, ed. L. Hansen, New York: National Bureau of Economic Research.
Griliches, Z. 1977. Estimating the returns to schooling: some economic problems. *Econometrica* 45(1), January, 1–22.
Hanushek, E. 1977. *A Reader's Guide to Educational Production Functions*. Institution for Social Policy Studies, Yale University.
Kendrick, J. 1976. *The Formation and Stocks of Total Capital*. New York: National Bureau of Economic Research.

Knight, F. 1944. Diminishing returns from investment. *Journal of Political Economy* 52, March, 26–47.

Lillard, L. and Willis, R.J. 1978. Dynamic aspects of earnings mobility. *Econometrica* 46(5), September, 985–1012.

Marshall, A. 1920. *Principles of Economics*. 8th edn. London: Macmillan, 1930.

Michael, R. 1982. Measuring non-monetary benefits of education: a survey. In *Financing Education: Overcoming Inefficiency and Inequity*, ed. W. McMahon and T. Geske, Urbana: University of Illinois Press.

Miller, H. 1960. Annual and lifetime income in relation to education, 1929–1959. *American Economic Review* 50, December, 962–86.

Mincer, J. 1958. Investment in human capital and personal income distribution. *Journal of Political Economy* 66, August, 281–302.

Mincer, J. 1974. *Schooling, Experience and Earnings*. New York: Columbia University Press.

Mincer, J. and Polachek, S. 1974. Family investment in human capital: earnings of women. *Journal of Political Economy* 82(2), Pt II, March–April, S76–S108.

Paglin, M. 1975. The measurement and trend of inequality: a basic revision. *American Economic Review* 65(4), September, 589–609.

Petty, W. 1676. *Political Arithmetic*. In *The Economic Writings of Sir William Petty*, ed. C. Hull, Vol. 1, Cambridge: Cambridge University Press, 1899.

Rosen, S. 1986. The theory of equalizing differences. In *Handbook of Labour Economics*, ed. O. Ashenfelter and R. Layard.

Rosen, S. 1977. Human capital: a survey of empirical research. In *Research in Labor Economics*, Vol. 1, ed. R. Ehrenberg, Greenwich, Conn.: JAI Press.

Rosen, S. 1985. The theory of equalizing differences. In *Handbook of Labour Economics*, ed. O. Ashenfelter and R. Layard, Amsterdam: North-Holland.

Sahota, G. 1978. Theories of personal income distribution: a survey. *Journal of Economic Literature* 16(1), March, 1–55.

Schultz, T. 1961. Investment in human capital. *American Economic Review* 51, March, 1–17.

Schultz, T. 1975. The value of the ability to deal with disequilibria. *Journal of Economic Literature* 13(3), September, 827–46.

Smith, A. 1776. *An Inquiry into the Nature and Causes of the Wealth of Nations*. Modern Library Edition, New York: Random House, 1947.

Spence, M. 1973. Job market signaling. *Quarterly Journal of Economics* 87(3), August, 355–74.

Walsh, J. 1935. Capital concept applied to man. *Quarterly Journal of Economics* 49, February, 255–85.

Welch, F. 1970. Education in production. *Journal of Political Economy* 78(1), January–February, 35–59.

Welch, F. 1976. Ability tests and measures of differences between black and white Americans. Rand Corporation.

Welch, F. 1979. Effects of cohort size on earnings: the baby boom babies' financial bust. *Journal of Political Economy* 87(5), Pt II, October, S65–97.

Willis, R. 1986. Wage determinants: a survey and reinterpretation of human capital earnings functions. In *Handbook of Labour Economics*, ed. O. Ashenfelter and R. Layard, Amsterdam: North-Holland.

Willis, R. and Rosen, S. 1978. Education and self-selection. *Journal of Political Economy* 87(5), Pt II, October, S65–S97.

human nature. *See* ECONOMIC MAN; INDIVIDUALISM; SELF-INTEREST.

humbug production function. Neoclassical economics has always tried to portray wages and profits as mere technical variables. At an aggregate level, this is accomplished by connecting labour and capital to output through a 'well-behaved' aggregate production function, with the marginal products of labour and capital equal to the wage rate and profit rate, respectively. Thus in competitive equilibrium each social class is pictured as receiving the equivalent of the marginal product of the factor(s) it owns (Shaikh, 1980).

The original optimism that aggregate production functions and their corresponding marginal productivity rules could be derived from more detailed general equilibrium models eventually gave way to the sobering realization that the conditions for any such a derivation were 'far too stringent to be believable' (Fisher, 1971). Yet neoclassical economists continue to use aggregate production functions, apparently because they seem to fit the data well and their estimated marginal products closely approximate the observed wage and profit rates (so-called factor prices).

This apparent empirical strength of aggregate production functions is often interpreted as support for neoclassical theory. *But there is neither theoretical nor empirical basis for this conclusion.* We already know that such functions cannot be derived theoretically, except under conditions which neoclassical theory itself rejects (e.g. the simple labour theory of value) (Garegnani, 1970). Moreover, Fisher (1971) discovered through simulation studies that the aggregate data generated by microeconomic production functions were not generally well fitted by aggregate production functions; that the functions which did best fit this data are not neoclassical in nature (this is a common finding, e.g. Walters, 1963); and that in simulation runs where the wage share happened to be roughly constant and aggregate Cobb–Douglas production functions happened to work well, this goodness of fit was puzzling because it held even when the theoretical conditions for aggregate production functions were flagrantly violated.

Shaikh (1974, 1980) has shown that this last result is simply an artifact of the constancy of the wage share. To see this, let r_t represent the rate of profit, and q_t, w_t, k_t the per worker net output, wages and capital, respectively, all at time t. Then the national accounting identity $q_t = w_t + r_t k_t$ can be differentiated to yield percentage rates of change q', w', etc., weighted by the profit share $s_t = r_t k_t / q_t$ and the labour share $1 - s_t = w_t / q_t$:

$$q'_t = B'_t + s_t k'_t, \quad \text{where} \quad B'_t = (1 - s_t) w'_t + s_t r'_t. \quad (1)$$

The preceding relation says nothing about the nature of the underlying economic processes, since it is derived from an identity. But if social forces happen to produce a stable profit (and hence wage) share, so that $s_t = s$ (a constant), we can immediately integrate both sides of (1) to get

$$q_t = A_t k_t^s, \quad \text{where} \quad A_t = C\, e^{\int B'_t\, dt}, \ C = a \text{ contants.} \quad (2)$$

Equation (2) looks like an aggregate Cobb–Douglas production function with constant returns to scale, marginal products equal to factor prices, and a technical change shift parameter A_t. It will even seemingly reflect neutral technical change if the rate of change B'_t can be expressed as a function of time. And yet *it is not a production function at all, but rather merely the algebraic expression of any social forces resulting in a constant share – even when the underlying processes are definitely not neoclassical in nature.* To illustrate this, we will now demonstrate that even a very simple 'anti-neoclassical' (Robinsonian) economy will fit such a function.

Consider an economy at time t_0, in which all possible techniques of production are dominated by a *single* linear technique (linear because capital–labour ratios are equal across all sectors). With one dominant technique, there is no neoclassical substitutability among techniques, and the linear wage–profit curve of the dominant technique is also the wage–profit frontier for the whole economy (the line $q_0 R$ in Figure 1, for the given time period). Because q, k and R (net output/capital) are all *constant* along the wage–profit frontier, the marginal products of labour and capital therefore cannot

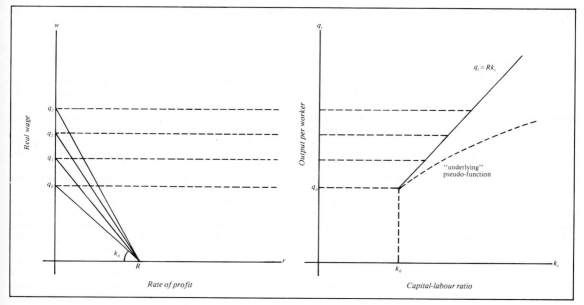

Figure 1   Figure 2

even be defined. The determination of the so-called factor prices w and r cannot possibly be tied to some corresponding marginal products. Lastly, because q and k are constant for any given frontier, a frontier such as $q_0 R$ in Figure 1 contributes only a single point q_0, k_0 to the q_t, k_t space in Figure 2.

Now consider Harrod-neutral technical change, in which both output per worker q_t and the capital–labour ratio k_t rise at the same rate, so that the output–capital ratio R remains constant:

$$q_t/q_0 = k_t/k_0 = e^{at}, \quad \text{and since} \quad q_0/k_0 = R, q_t/k_t = R \quad (3)$$

This is depicted in Figure 1 by the successive wage–profit frontiers and in Figure 2 by the corresponding (solid) straight line q_t of slope R.

If we were simply concerned with the best relation between inputs and output, then the *true* relation $q_t = Rk_t$ would be the correct one. But within neoclassical theory, such a fitted function would imply a constant marginal product of capital, a zero marginal product of labour (Allen, 1968, pp. 45–6), and no technical change (since the 'shift parameter' R is constant). A good neoclassical would therefore have to reject this best (and true) fitted function in favour of some more 'appropriate' functional form (Fisher, 1971, pp. 312–13). How then might an aggregate production function fare in our anti-neoclassical world?

We have already assumed a constant profit share $r_t k_t/q_t = s$, and since the output–capital ratio $q_t/k_t = R$ is constant (equation (3), it follows that the rate of profit $r_t = sR$ is constant. Similarly, the assumption of a constant wage share $w_t/q_t = 1 - s$ and a steadily growing output per worker $q_t = q_0 e^{at}$ (equation (3)), implies a steadily growing real wage $w_t = (1 - s) q_0 e^{at}$. All this allows us to solve explicitly for B_t' and A_t in equations (1)–(2):

$$B_t' = (1 - s) w_t' + s r_t' = (1 - s)a \quad (4)$$

$$q_t = C\, e^{(1 - s)at} k_t^s, \quad \text{since} \quad A_t = C\, e^{(1 - s)at} \quad (5)$$

Thus when the wage share is constant, *even a fixed proportion technology undergoing Harrod-neutral technical change is perfectly consistent with an aggregate pseudo-production function* (equation (5)). This is, however, a law of algebra, not a law of production. The above reasoning has been shown to have grave implications for production function studies (Shaikh, 1980). For instance, Solow's (1957) so-called seminal technique for assessing technical change amounts to decomposing the true production relation into an 'underlying' pseudo-production function and a residual A_t whose rate of change is then taken to measure technical progress (figure 2). But this measures nothing more than distributional changes, since B_t is simply the weighted average of the rates of change of observed wage and profit rates (equations (1)–(2)). Similarly, Fisher's previously mentioned puzzle concerning the empirical strength of aggregate Cobb–Douglas production functions can be shown to be an artifact of the stability of the wage share over those particular simulation runs. Last, and perhaps most strikingly, it is interesting to note that even data points which spell out the word 'HUMBUG' can be well fitted by a Cobb–Douglas production function apparently undergoing neutral technical change and possessing marginal products equal to the corresponding 'factor prices'! Surely there is a message in this somewhere?

ANWAR SHAIKH

See also COBB–DOUGLAS FUNCTIONS.

BIBLIOGRAPHY
Allen, R.G.D. 1968. *Macro-Economic Theory: A Mathematical Treatment.* London: Macmillan.
Fisher, F. 1971. Aggregate production functions and the explanation of wages: a simulation experiment. *Review of Economics and Statistics* 53(4), November, 305–25.
Garegnani, P. 1970. Heterogeneous capital, the production function, and the theory of distribution. *Review of Economic Studies* 37(3), 407–36.
Shaikh, A. 1974. Laws of algebra and laws of production: the humbug production function. *Review of Economics and Statistics* 51(1), 115–20.

Shaikh, A. 1980. Laws of algebra and laws of production: the humbug production function II. In *Growth, Profits and Property: Essays on the Revival of Political Economy*, ed. E.J. Nell, Cambridge: Cambridge University Press.

Solow, R. 1957. Technical change and the aggregate production function. *Review of Economics and Statistics* 39, 312–20.

Walters, A.A. 1963. Production functions and cost functions: an econometric survey. *Econometrica* 31(1), 1–66.

Hume, David (1711–1776). David Hume's economic essays (which originally appeared in 1752 in a volume entitled *Political Discourses*) comprise a small portion of his writings. The scope of Hume's thought was vast. He wrote extensively in philosophy (the area in which his reputation primarily lies), explored several of the social sciences and the humanities, and was deeply interested in history. His multi-volume *History of England* (1754–61) was a pathbreaking work in the field. Nonetheless, in the literature Hume's economic writings have typically been treated as an entirely self-contained aspect of his work. This is not surprising, since in his economic essays he does not allude to his other writings, and subsequent disciplinary specialization has not encouraged consideration of any interrelationships between the two. For their part, philosophers have often treated Hume's philosophical writings in isolation from his other work.

For Hume, however, there was no such sharp disjunction. In the Advertisement prefixed to his first and major philosophical work – *A Treatise of Human Nature* (1739) – he states that he expects his philosophy to serve as the 'capital or centre' of all the 'moral' (i.e. psychological and social) sciences and that he hopes to expand the *Treatise* to accommodate a study of these areas. Owing perhaps to the poor reception accorded his *Treatise*, Hume did not carry out his original intention. His treatment of the moral sciences was left mainly to his essays. But there are many links between Hume's philosophical thought and his essays, and this is true with respect to his economic essays. Indeed, in light of the importance of these links, Hume may be regarded as the outstanding philosopher-economist of the 18th century.

Viewed in most general form, what is the nature of the relationship between Hume's economic and philosophical thought? Hume regarded the foundation of his entire philosophical system – its 'capital or centre' – as a body of 'principles of human mature', or elements and relations concerning human understanding and human passions that he believed to be irreducible and universal. These principles, which constitute the analytical phase of Hume's system of thought, are treated in Books I and II of the *Treatise*. In the second and synthetic phase Hume then relates various aspects of 'human nature' to environmental forces in seeking to frame laws of human behaviour, or generalizations indicating how man may be expected to behave under different specific conditions. These generalizations comprise the substance of the 'moral sciences' with which, as indicated, Hume dealt principally in the essays. An explicit and deep interest in psychology is thus a salient characteristic of Hume's treatment of the 'moral sciences' in general, and this is conspicuously evident in his economic analysis.

What were Hume's views concerning the prospects of developing reliable generalizations in the 'moral sciences?' That Hume should have distinct views on this issue is scarcely surprising in light of the depth of his interest, as a philosopher, in the epistemological basis of science. As he had argued, the contrary of any generalization concerning relations between matters of fact is always conceivable and hence always possible. Consequently, the only way of developing an understanding of these relations, he contended, is through empirical observation; and this can only yield probabilities, never certainty. With respect to his own principles of human nature, Hume believed that his propositions carried the highest order of probability because of the abundance of evidence on which they rested.

On the other hand, recognizing the complexity of the interrelationships between man's 'nature' and his environment, he stressed the difficulty in framing valid laws of human behaviour. He calls attention to the effect on human behaviour of imperceptible influences, emphasizes the extent to which it could be altered by changing conditions and notes the impracticality of conducting controlled experiments in the realm of psychological phenomena. He thus warns that in the social sciences 'all general maxims ... ought to be established with the greatest caution' and states that 'I am apt ... to entertain a suspicion that the world is still too young to fix many general truths in [the area of the social sciences] which will remain true to the latest posterity' (Hume, *Philosophical Works*, ed. Green and Grose, vol. III, pp. 156–7). Of all the social fields, however, he believed that a field such as economics lent itself especially well to scientific study, and here he was cautiously optimistic concerning the possibility of developing reliable generalizations through direct observation of man in the course of his day-to-day affairs. As he argued, behaviour here was governed by mass passions, which were 'gross' or 'stubborn', or were not as affected by imperceptible influences as passions governing the behaviour of small numbers of individuals. Uniformities in behaviour therefore could here be more readily discerned (*Philosophical Works*, p. 176). It should be noted that, in accord with this view, Hume introduces his economic essays by contrasting the potential for scientific analysis in economics with the very limited prospects for such analysis in a field such as foreign diplomacy, where events are controlled by the behaviour of a small number of individuals (*Writings on Economics*, ed. Rotwein, pp. 3–4).

To return to the substance of Hume's economic thought, in addition to emphasizing psychological considerations Hume's analysis displays a deep interest in historical sequence. Hume's interest in history developed at a very early age, even before he undertook his *Treatise*. As it appears in his essays, however, his treatment of history differs from conventional historiography (with its concern with unique particulars) which predominates in his *History of England*. For, writing as a 'moral scientist', Hume sought to reduce historical sequence to generalizations which explain how transformations in human behaviour result from the impact of changing historical circumstance on 'human nature'. This type of study (which bore a relationship to the 'conjectural history' and the French '*histoire raisonée*' of the period) Hume termed 'natural history' – the term 'natural' here denoting the recurrent or probable, or the substance of laws of human behaviour. There are clusters of what Hume regards as historical laws of human behaviour in several of the essays. One essay bears the title 'The Natural History of Religion'. And in the economic essays the approach of 'natural history' is of fundamental importance.

This can be seen when Hume's economic essays are viewed on three different levels of analysis. The first is economic psychology, where Hume deals with economic motivation, or what he terms the 'causes of labour'. This is the most basic level of his economic analysis in the sense that here one finds the links between his economic thought and his treatment of 'human nature' in the *Treatise*. On this level the analysis takes the form of a natural history of 'the rise and progress of commerce'. In a word, Hume introduces the question of economic motivation in seeking to explain how changing environmental influences stimulated the economic growth of his general period through their impact on various human passions. Here Hume observes that there are four 'causes of labour' – the desire for consumption, the desire for action, the desire for liveliness and the desire for gain.

The first of these, which is commonly stressed by economists, simply denotes all the wants that may be gratified by consumption. The desire for action refers to a desire for challenging activity as such. However, its full effectuation, as Hume stressed, requires activity whose end or objective has independent value. Like hunting and gaming, economic pursuits (and especially the activities of the merchant and, more generally, the 'industrious professions') are seen as meeting these conditions. By the desire for liveliness Hume meant the desire for the experience of active passion as such (which he contrasts with a state of no passion, or in effect a state of waking sleep). This is not a completely independent cause of labour but is an important ingredient common to both consumption and interesting activity. The last cause of labour is the desire for monetary gain, which is a desire to accumulate the tokens of success in the economic 'game'.

Hume argues that all these motives play a role in a nation's economic growth – the initial stimulus to which he finds in the expansion of international trade. As compared with the treatments of economic motivation by economists (which commonly accord exclusive or overshadowing emphasis to the desire for consumption), a striking characteristic of Hume's treatment lies in its multi-dimensionality. This multi-dimensionality is also found in Hume's criticism of the doctrine of psychological hedonism. Here he argues that, in addition to seeking pleasure, man is driven by a variety of 'instincts' which lead him to do things for their own sake, and therefore will not automatically lead him to act in his own best interests. Hume's position thus precludes any simple identification of wealth with welfare.

The second level of Hume's economic analysis is his political economy, or his treatment of market relations. It is this which makes up the bulk of his economic essays. Here Hume considers several of the major economic issues of his own period, including monetary theory, interest theory, the question of free versus regulated trade, the shifting and incidence of taxes, and fiscal policy. In this context the natural history of 'the rise and progress of commerce' plays a dominant role. For repeatedly in his critical treatment of the economic doctrines of his period Hume seeks to show that their major deficiency lies in a failure to give proper attention to the importance of economic growth and to the underlying psychological and other factors associated with this growth process.

Let us consider first Hume's quantity theory specie flow doctrine, which he presents (in the essay 'Of the Balance of Trade') in criticism of the mercantilist view that without restraints on international trade a nation would suffer losses in its money supply. Hume's position, which has been recognized as an early anticipation of the classical view, is that, owing to the effects of specie flows on price levels in trading nations, the amount of specie in each automatically tends towards an equilibrium at which its exports and imports are in balance. Any attempt through restraints on trade to increase the amount of specie beyond this equilibrium level, as Hume argues, is destined to fail (assuming the money circulates domestically) because the specie movement from abroad will raise the nation's prices relative to those abroad, reduce exports and increase imports, and generate a return outflow of specie.

The relationship of this analysis to Hume's historical perspective is evident in the purpose with which he introduces this doctrine. For in employing the quantity theory of money he is here arguing that the extent to which a specie inflow into a nation affects its prices depends on its total output. Consequently, as he is seeking to show, it is the level of a nation's economic development, or its productive capacity as determined by its population and the spirit of industry of its people, that controls the amount of specie a nation can attract and retain. As he states, 'I should as soon dread that all our springs and rivers should be exhausted as that money should abandon a kingdom where there are people and industry' (Hume, *Writings on Economics*, p. 61).

To consider another of Hume's anticipations of the classical position – his interest theory presented in his essay 'Of Interest' – here he attacks the mercantilist view that the rate of interest is determined by the money supply. On quantity theory grounds he argues that an increased money supply will simply raise all prices and, necessitating an offsetting increased demand for loans to finance expenditures, will leave interest rates unaffected. It is therefore the supply of real capital that determines interest rates. The bulk of Hume's discussion, however, is concerned with the factors affecting the supply of real capital itself; and here he turns to a historical analysis in which he considers the effect of economic growth on the class structure of society and, through this, on economic incentives. In this context every 'cause of labour' considered in the natural history of 'the rise and progress of commerce' is brought into his treatment. In a feudal society, he points out, the supply of capital is low because there are only two classes – the peasants and the landed aristocracy. The peasants cannot save since they are poor. On the other hand, the landed aristocracy tend to be heavy borrowers. For, as they are idle and lack the sense of liveliness that interesting activity affords, they seek liveliness wholly through extravagant consumption expenditures. Capital is therefore scarce and interest rates are high. Economic development, however, spawns the growth of the merchant class and the industrious professions. These groups derive a sense of liveliness from economic activity. Consumption expenditure drops for this reason and also because the pursuit of profit nourishes a desire to accumulate gain as a token of success in the economic game. As the new industrious classes earn a substantial share of the growing national income, their disposition to save thus results in a significant increase in the capital supply and a decline in interest rates.

As noted, Hume employs the quantity theory of money in criticizing the mercantilist position. But Hume's monetary theory also exhibits a similarity to the mercantilist view. However, his treatment here too springs from an attempt to call attention to the importance of economic growth. Thus (in his essay 'Of Money') Hume – assuming a condition of less than full employment – grants that an increase in the quantity of money (as against a greater absolute quantity of money as such) need not simply raise prices but can stimulate economic activity. Here, in tracing the impact of the increased money supply as it courses through the economy, he presents a lucid description of the multiplier process. He denies, however, that the stimulating effect on industry – when resulting from a short-run increase in the money supply – can prove anything more than ephemeral. No justification for this view is given. But it serves to underscore the conclusion of his analysis. For he goes on to argue that if the increase in the money supply is gradual and continues over a long period of time, its stimulating effects on output will prove enduring because it will nourish the 'spirit of industry' and therefore economic growth itself. Similarly, although Hume argued that an increase in the money supply does not affect interest rates, near the conclusion of his essay 'Of Interest' he points out that a long-run increase in the supply of money, by stimulating economic growth and inducing a change in spending and saving patterns, can increase the supply of capital and lower the interest rate.

Another noteworthy area of Hume's analysis is his treatment of the issue of free versus regulated markets. Since the relevant comments are not found in his economic essays but rather lie scattered through his *History of England*, the full extent to which Hume anticipated Adam Smith's 'invisible hand' argument has not been generally recognized. These comments make clear that Hume understood the role of a free price mechanism is governing the allocation of resources (*Writings on Economics*, pp. lxxviii–lxxx).

In applying the argument for free markets to the case of international trade, Hume emphasizes that free trade makes it possible for nations to enjoy the gains from an exchange of the products of their different resource endowments. However, in his most thorough treatment of the issue of international free trade (in his essay 'Of the Jealousy of Trade') it is not this static approach to the question that predominates. Rather, once again, it is economic growth considerations that receive primary emphasis. For here, where Hume seeks to meet the mercantilist argument that foreign economic development adversely affects home industry and employment, he takes the position that expansion abroad, on the contrary, commonly promotes economic development at home. By increasing foreign income, he argues, economic growth abroad not only leads to an expansion of foreign demand for domestic output but, through an emulation of foreign technological innovations, promotes the advance of technology at home. Hume goes on to argue that even when foreign expansion competes with domestic output, there is no need for concern provided the nation's 'spirit of industry' – which is itself nourished by foreign trade – is preserved. For as long as a nation remains industrious it need not fear that other nations will encroach on the market for its staple and, even in the unlikely event that this does occur, an industrious nation can readily divert its resources to other uses. Moreover, in stimulating the spirit of industry, foreign trade also promotes the diversification of a nation's resource use, and so reduces the impact of any shrinkage of demand that may occur from time to time in particular markets.

There are indications that Hume was more fully aware of the possible costs of free trade than one would gather from the main argument in the essay 'Of the Jealousy of Trade'. Elsewhere he treats the interests of poor and rich countries as incompatible, and in one place he also justifies the use of a tariff in specific cases (*Writings on Economics*, pp. 34–5, 76, 199–205). In the essay 'Of the Jealousy of Trade' itself he recognizes, in a modification of his main argument, that there are circumstances in which a nation facing a loss of markets to foreign countries may find resource diversion difficult (p. 81). The character of this essay as a whole (which appeared six years after the other economic essays) suggests, however, that after much reflection and groping Hume had concluded that free trade would have a markedly favourable effect on long-term economic growth for all nations, and that, with this end in view, any associated costs – which would be of a shorter-term nature – would be well worth sustaining.

A further illustration of the role of natural history in Hume's political economy is found in his treatment of the shifting and incidence of taxes (in his essay 'Of Taxes'), where he considers the view that an expansion of taxes creates an expanded ability to pay the levies by increasing 'proportionably the industry of the people'. This view was commonly held by the mercantilists and, in what came to be known as 'the utility of poverty' doctrine, was employed to justify the imposition of excises on goods consumed by the poor. Hume's position here is twofold. He points out that history shows that natural burdens, such as relatively infertile soil, often stimulate industry, and he argues that artificial burdens such as taxes may have the same effect. This position springs from Hume's view concerning the importance of a desire for interesting action as a 'cause of labour' since he here emphasizes that in order to prove interesting the activity must be difficult and challenging. On the other hand, he emphasizes that since economic activity is also motivated by a desire for consumption, increasing difficulty beyond a certain level in achieving consumption ends will lead to despair. From the viewpoint of its stimulating effect on industry there is thus an optimum tax level, and Hume takes the view that taxes on the poor throughout Europe have already so substantially exceeded that optimum that they are threatening to 'crush all art and industry'. Considered as a whole, Hume's position represents an amalgam of both the mercantilist and the later classical view. He rejects the mercantilist 'utility of poverty' doctrine with its unqualified endorsement of higher taxes on goods consumed by the poor, but also would reject the view (which is based on the subsistence or accustomed standard of living theory of wages found in the writings of Smith and Ricardo) that any tax on labour would inevitably result in a reduction in its supply.

Hume's treatment of fiscal policy – the last major aspect of his political economy – does not reveal significant relationships to his natural history of the rise and progress of commerce. Owing to space limitations, his analysis – contained in the long essay 'Of Public Credit' – cannot here be considered in detail. It should be observed, however, that this essay, which deals specifically with the question of large and continually mounting public debt, constitutes in all essential respects a 'natural history of the rise and collapse of public credit'. Particularly noteworthy in this analysis are the extensive relationships Hume draws between economic and other social developments, especially of a political and sociological character. Of all aspects of his political economy, this essay most fully exhibits Hume's awareness, as a moral scientist, of significant interrelations between different realms of social experience.

The third and last level of Hume's economic thought in his economic philosophy, which is his appraisal, on ultimate moral grounds, of the desirability of a commercial and industrial society. In light of his general concern, as a philosopher, with moral questions, it is hardly surprising to find that the question of the moral aspects of commercial and industrial growth was of basic importance for Hume. Appearing in the second of the economic essays – 'Of Refinement in the Arts' – he considers this question before turning to an analysis of market problems. Although the essay is brief, its scope is broad; for Hume discusses the impact of the development of an advanced economy both on the individual and on society as a whole.

The standard for moral judgement Hume employs is drawn from the utilitarian ethic – a position which he himself had expounded and defended in his philosophical analysis. And here the role played by his natural history of the rise and progress of commerce is fundamental. As observed, in this natural history Hume dealt with various 'causes of labour'. In his economic philosophy three of these motives – the desires for consumption, for interesting activity and for liveliness – are now treated as ends which are regarded as major ingredients of the happiness of the individual. Here he argues that – by providing new consumption experiences, enlarging the scope for the enjoyment of economic activity as a form of interesting action and (through both the latter) enhancing a sense of liveliness – economic growth advances the fulfilment of all these ends. Economic growth, he contends, contributes to the fulfilment of a fourth end of importance to human welfare – a sense of peace and tranquillity or a state of no passion – which he argues is enjoyable only in 'recruiting the spirits' after intensive indulgence in lively experiences. It is noteworthy that Hume's treatment of these ingredients of human happiness bears a direct relationship to the principal conceptions of the good life as Hume construes these in an earlier series of essays entitled 'The Epicurean', 'The Stoic' and 'The Platonist'. Further, the pluralism reflected in his multi-dimensional prescription for human happiness springs from the position taken in a fourth essay on the good life entitled 'The Sceptic' (Hume, *Writings on Economics*, pp. xcv–xcix).

Turning to a treatment of the effect of economic development on major aspects of social relations, Hume now expands the 'natural history' to encompass non-economic considerations. He argues that economic growth contributes to the growth of knowledge in the liberal as well as the mechanical arts, nurtures a sense of humanity and fellow-feeling, enhances a nation's spiritual as well as its economic ability to defend itself and, through its impact on the growth of knowledge and fellow-feeling, advances an understanding of the art of government and political harmony. A final political consideration, to which Hume gives special attention, is the charge (drawn from the experience of Rome) that luxury is corrupting and debasing and therefore is inimical to liberty. Hume argues that history shows that precisely the opposite is true. For the growth of commerce brings the expansion of the merchant class – the 'middling rank of men' who above all are interested in uniform laws protecting their property; and it is this development, he emphasizes, which has led to the growth of parliamentary government and the associated respect for individual liberty. Hume thus perceived the link between the growth of economic individualism and political liberty that has drawn so much attention since his time. Although Hume recognized that the development of commerce and industry could produce evils of its own, he argued that these were outweighed by its benefits. Owing apparently to an overzealous desire to counter the common religious objections to luxury, Hume over-extends himself and leaves some of his arguments in support of economic growth open to criticism (Hume, *Writings on Economics*, pp. cii–civ). His treatment nonetheless stands as an unusually broad and penetrating appraisal of a wealth-orientated individualistic society. In light of this it deserves recognition as an early classic.

Throughout our discussion, attention has been given to Hume's interest in the psychological and historical aspects of economic activity. A similar interest – pursued in varying degree – is found among other writings of Hume's own period. However, owing to his own searching analysis as a philosopher and historian, Hume's treatment was of a particularly high order; equally extraordinary was the extent to which he employed the method of 'natural history' in the treatment of a wide range of issues of economic theory and policy.

Comparing Hume with Adam Smith (his close friend), one is struck by the brevity of Hume's economic writings. Hume wrote a series of relatively short 'discourses' on selected topics. Smith's *Wealth of Nations* (1776) is a general economic treatise. In contrast to Smith, Hume moreover gives little systematic attention to price and distribution theory, which

was to become the major concern of classical and neoclassical economics. In point of the general analysis of psychological and historical influences on economic activity, however, Hume's work is more comprehensive, more highly organized and more penetrating than Smith's. When dealing with the subjective aspects of human behaviour, Smith not infrequently regards them as universals (e.g. his assertion that there is an innate disposition among men to 'truck and barter'), where Hume treats them as historical variables and himself seeks to explain the nature of the specific historical influences at work (Hume, *Writings on Economics*, pp. cvii–cx). In this Hume did not foreshadow the mainstream of subsequent economic thought; it was Adam Smith's tendency in his economic theory to abstract from history that was to become the dominant characteristic of later economic analysis. In point of general perspective (though often not its conceptual framework) Hume's economic thought bears a relation to other subsequent lines of development – to the historical and institutional schools of economics, to the more current revived analytical interest in economic growth along with its associated cultural aspects, to the concern with psychological factors in dealing both with macroeconomics and the economics of non-competitive markets, and to the normative appraisals of economic systems in their fuller social settings.

In the standard histories of economic thought Hume has been accorded relatively little attention. He is often ignored altogether or treated cursorily as a predecessor of Adam Smith. Various studies of the technical aspects of economic analysis have called attention to several of Hume's contributions. These aspects of Hume's analysis are noteworthy in their own right. Their significance deepens and broadens when they are related to Hume's work as a philosopher and historian and are seen to take form within the context of 'natural history'.

EUGENE ROTWEIN

See also BULLIONIST CONTROVERSY; GOLD STANDARD; SPECIE-FLOW MECHANISM.

SELECTED WORKS

1752. *Political Discourses.* Edinburgh: A. Kincaid & A. Donaldson.
1875. *The Philosophical Works of David Hume.* Ed. with notes by T.H. Green and T.H. Grose, 4 vols, London: Longmans, Green & Co.
1955. *Writings on Economics.* Ed. Eugene Rotwein, London: Nelson.

hunger. *See* FAMINE; NUTRITION; POVERTY.

hunting and gathering economies. Men and women (*Homo erectus*) who were culturally and biologically distinguishable from other hominoids have lived on the planet Earth for about 1.6 million years (Pilbeam, 1984). It is likely that the biological changes since that time form a microevolutionary continuum: archaic *H. sapiens*, including the Neanderthal, appeared 125,000 years ago and anatomically modern *H. sapiens* appeared about 45,000 years ago. The record suggests that *H. erectus* fabricated and used tools, and his use of fire may have begun by 700,000 years ago. The changes identified in the prehistoric period appear only to distinguish less advanced from more advanced stone age technology. Consequently, the dominating message seems to be that over almost the whole of man's epoch on earth he lived successfully as an exceptionally well-adapted hunter. It is only recently, in the last 8–10,000 years (less than one per cent of his time on Earth), that man abandoned the nomadic life of the hunter to begin growing crops, husbanding domesticated animals, and living in villages. It is difficult to exaggerate the importance of this agricultural or first economic revolution (North and Thomas, 1977) in understanding who we are, and what we have become. Once man opted for the farmer–herder way of life it was but a short step to mankind's much more sophisticated development of specialization and exchange, greatly enlarged production surpluses, the emergence of the State, and finally the industrial revolution. Our direct knowledge of early man is confined to the record of the durables he left behind. Yet when combined with anthropological evidence from the study of recent hunter–gatherer economies the evidence can be interpreted as demonstrating that all the ingredients associated with the modern wealth of nations – investment in human capital, specialization and exchange, the development of property right or contracting institutions, even environmental 'damage' – had their development in the course of that vast prehistorical, pre-agricultural, period.

What accounts for this sudden abandonment of the nomadic hunting life? We do not know for we have no direct observations on the transformation from hunting to agriculture. This transformation is perhaps the pre-eminent scientific mystery, since all of that which we have called civilization, all the great achievements of industry, science, art and literature stem from that momentous event within the last few minutes of man's day on Earth. Yet there are common factors that dominated the evolution of man from his earliest form to modern *H. sapiens*, and his primary intellectual and social development, which suggest an underground continuity between the pre-agricultural, Paleolithic hunting period, and the agricultural and subsequent periods.

MAN THE HUNTER-GATHERER. There are many widely held beliefs concerning the characteristic features of the hunter–gatherer way of life that stretch back several hundred years in academic writings, and persist as part of the folklore of contemporary man's misperception of his own prehistoric past; until recently these beliefs dominated even the anthropological view of hunter–gatherer 'subsistence'. These beliefs tend to obscure the striking continuity in man's ability to respond to changes in his environment by substituting new inputs (labour, capital and knowledge) for old, and develop new products to replace the old when effort prices were altered by the environment.

Ever since Hobbes there has prevailed the perception that life in the state of nature was 'solitary, poor, nasty, brutish and short'. A more accurate representation (if not strictly correct in all aboriginal societies) would argue that the hunter culture was the original affluent society (Lee and DeVore, 1968). Extensive earlier data on extant hunter–gatherers show that with rare exceptions (such as the Netsilik Eskimos) their food base was at minimum reliable, at best very abundant. The African Kung Bushman inhabited the semi-arid northwest region of the Kalahari Desert, an inhospitable environment, characterized by drought every second or third year. These conditions had served more to isolate the Kung from their agricultural neighbours than to condemn them to a brutish existence. Adults typically worked 12–19 hours per week in getting food. As with all such societies for the most part the women gathered, the men hunted. The caloric-protein returns exceeded several measures of nutritional adequacy. Gathering was the more reliable and productive activity with women producing over twice as many edible calories per hour as men. Both men and women bought leisure with this work schedule – resting, visiting, entertaining and (for the men) trance dancing. About 40 per cent of the population were children,

unmarried young adults (15–25 years of age) or elderly (over 60 years of age), who did not contribute to the food supply and were not pressured to contribute.

A comparable macroeconomic picture applied to the Hazda in Tanzania. Large and small animals were numerous and all – with the exception of the elephant – were hunted and eaten by Hazda. Hunting was the speciality of men and boys, conducted as an individual pursuit that relied primarily on poisoned arrows. The Hazda spent on average no more than two hours a day hunting. The principal leisure activity of the men was gambling which consumed more time than hunting.

Other hunting (or fishing) peoples of Africa, Australia, the Pacific Northwest, Alaska, Malaya and Canada have shown comparably effective adaptation to this form of livelihood. Malnutrition, starvation and chronic diseases were rare or infrequent, although accidental death was high in certain cases such as the Eskimo.

The argument that life in the Paleolithic must have been intolerably harsh is simply not borne out by the many ethnographic studies of extant hunting societies in the past century. With few exceptions such societies have fared well, and did not leap to embrace the agricultural or pastoral pursuits of their neighbours. Whether life in the Paleolithic mirrored this modern experience cannot be known with any assurance, but certainly there is no support for the proposition that hunting, *per se*, means an intolerably harsh existence. In fact the Paleolithic hunting economy had demonstrably high survival value in a world far more plentifully endowed with game than has existed since the great megafaunal extinctions of the late Pleistocene, and therefore a world which might indeed have been marked by numerous original affluent economies.

Although it is natural to suppose that man's uniqueness derived from his intellectual superiority, what is more likely is that man's physical superiority was also important in giving him a superpredator's advantage over other species. His endowment of physical human capital would probably have been of significance even in the absence of his investment in tools and the human capital required to produce and use tools. As noted by J.B.S. Haldane, only man can swim a mile, walk twenty, and then climb a tree. Add to this observation the four-minute mile, unsurpassed long-distance endurance running, the ability to carry loads in excess of body weight, high altitude performance, American Indian capacity literally to run down a horse or deer by pacing the animal, the incredible accomplishments of acrobats and gymnasts, and finally the finger agility and coordination required to milk a cow, and you are left with the physical portrait of an astonishingly superior species. It appears that man's basic foundation of physical superiority was laid by his upright stance, to which of course the addition of knowledge made him truly formidable, even in the presence of the various giant proboscidea (mastodon, mammoth, elephant) which early man did not hesitate to hunt and to kill on three continents.

The idea that primitive man was too puny and too few in number to have had a significant influence on his environment underestimates man's uniqueness as a tool using, fire using, highly mobile species who, with minor exceptions (Madagascar, New Zealand and Antarctica), had populated the world by 8000 BC. The archaeological record suggests that man was a big game hunter *par excellence*. He hunted mammoth, mastodon, horse, bison, camel, sloth, reindeer, shrub oxen, red deer, aurochs (wild cattle), and other large mammals, for perhaps a minimum of 30–40,000 years, ceasing only with the great megafaunal extinctions throughout much of the world some 8–12,000 years ago. Paul Martin (1967) has argued the case for the overkill hypothesis that man was a significant causative factor in these extinctions. Essentially, the argument is that the alternatives to overkill, principally the climate hypothesis, fail to account for the worldwide pattern of these extinctions which appear to have begun in Africa and perhaps southeast Asia 40–50,000 years ago, spread north through Eurasia 11–13,000 years ago, jumped to Australia perhaps 13,000 years ago, and entered North America in the last 11,000 years, followed by South America 10,000 years before the present. The most recent extinctions are in New Zealand (numerous species of flightless moa birds) 900 years ago and in Madagascar 800 years ago, shortly after the remarkably late migration of man to those islands.

Man's use of fire as a tool in the management and control of natural resources must be counted as having a profound effect on his ecological environment. Numerous authors who have studied patterns of land burning by primitive peoples have concluded that most of the greatest grasslands of the world represent fire-vegetation that is manmade (see Heizer, 1955, for a summary). Where tree growth is strongly favoured by climatic conditions, regular burning will select for certain species of tree such as the pine stands of southern New York and to the West, which have been attributed to Indian burning. Contemporary man's attempts to prevent fires, which today are almost entirely caused by lightning, has probably produced far more ecological damage than the controlled use of fire that has characterized aboriginal cultures. Recurrent fire prevents the accumulation of brush which then fuels the holocaust wildfire that destroys all forest vegetation.

A third source of ecological change produced by primitive peoples was their transportation of seed, in their migrations as hunter–gatherers, which introduced numerous botanical exotics into new regions. Archaeologists have frequently observed the association of various plants with ancient campsites and dwellings. For example, the wide distribution of wild squash, gathered for its seed, appears to be associated with man. The introduction of exotics can and has produced significant environmental changes in modern times, but the phenomenon has ancient origins and may have been considerably more disruptive as the first men moved from one 'pristine natural' region to another.

Success as a hunter–gatherer requires human capital usually associated only with agricultural and industrial man: learning, knowledge transfer, tool development and social organization. Comprehensive studies of the aboriginal use of fire for game and plant management show clearly that primitive men demonstrated extensive knowledge of the reproductive cycles of shrubs and herbaceous plants, and used fire to encourage the growth and flowering of the plants used in gathering, and to discourage the growth of undesirable plants (Lewis, 1973). This required one to know when, where, how and with what frequency to apply the important tool of controlled burning for managing the resources that allow gathering to make an efficient, productive and sustainable contribution to living. Primitive men knew that the growing season can be advanced by spring burns designed to warm the earth, that in dry weather fires should be set at the top of hills to prevent wild fires, but in damp air they should be set in depressions to avoid being extinguished, that the burning of underbrush aided the growth of the oak whose acorns were eaten and attracted moose who avoid underbrush, and that deer and other animals congregate to feed on the proliferation of tender new plants that sprout following a fall burn.

To live by hunting is to be committed to an intellectually and physically demanding activity that requires technology, skill, social organization, some division of labour, knowledge of animal behaviour, the habit of close observation, inventiveness, problem solving, risk bearing, and high motivation, since the rewards are great and the penalties severe. Such exceptional demands could have been highly selective in man's long evolution, and disciplined the development of the intellectual and genetic equipment that facilitated his subsequent rapid creation of modern civilization. This natural selection could have been intensified by the widespread practice among aboriginals of rewarding superior hunters with many wives.

It was as a hunter that man learned to learn. In particular he understood that young boys must be imbued with the habit of goal-oriented observations, and with knowledge of animal behaviour and anatomy. To know that many ungulates travel in an arc meant that tracking success could be improved by transversing the chord. Knowledge of animal behaviour was a substitute for weapon development. Even the weapons of the later pre-agricultural period (spears, bow and arrow, harpoon) required the hunter to approach the prey within ten yards for a best shot. This might require hours crouched on the ground waiting for a shift in the wind, for just the right change in the animal's position, or for the mammoth to get deeper into the bog in a watering hole. The weapons changed with shifts to new prey. Thus the Clovis fluted point, widely distributed throughout North America, was used to kill mammoth and mastodon 11–12,000 years ago. The Folsom point was then developed and used to kill the large, now extinct *Bison antiquus*, which then gave way to the Scottsbluff point associated with the killing of the slightly smaller, now extinct *Bison occidentalis* (Haynes, 1964; Wheat, 1967). These observations suggest high specialization which required new forms of human and physical capital to meet the specialized demands of new prey.

The organizational requirements of the hunt are illustrated at the Olsen–Chubbuck site in Colorado, where the excavated remains of bones and projectile points of the Scottsbluff design show that about 8500 years ago some two hundred *Bison occidentalis* were stampeded into an arroyro 5–7 feet deep. Armed hunters in the arroyo on each side of the stampede then slaughtered the injured or escaping animals with their weapons (Wheat, 1967).

Primitive man has often been modelled as 'cultural' not 'economic' man, but the power and importance of the opportunity cost principle in conditioning the choice of all peoples was perceptively stated by the Kung Bushman, who, when asked why he had not turned to agriculture, replied, 'Why should we plant, when there are so many mongongo nuts in the world?' (Lee and DeVore, 1968, p. 33). This Bushman, I would hypothesize, stated the answer to the scientific question: why did man the hunter tend to abandon that which appeared to serve him so well for 1.6 million years and to which he seems to have adapted ever more successfully, as indicated by the growing complexity of his tools and weapons as he evolved from *H. erectus* to anatomically modern *H. sapiens*? Man would not have given up the hunter–gatherer life had there not been a change in the terms of trade between man and nature that made the hunting way of life more costly relative to agriculture. This *hypothesis* does not leave 'culture' out of the equation. Thus to describe hunter–gatherers as directly seeking the cultural goal of prestige does not contradict the hypothesis that man, like nature, ever economizes. Attaching prestige to the hunt may simply be an astute means of advertising, teaching and propagating the discovery that hunting and its attendant technology is the best means of livelihood, with the result that each new generation does not have to rediscover this knowledge. Myths of the great hunter, of great rewards, of great penalties for lost technique, of killing the goose that lays golden eggs are part of the oral tradition by which the economy preserves this human capital.

The hypothesis that the agricultural revolution was due to a major decrease in the productivity of labour in hunting–gathering relative to agriculture (Smith, 1975; North and Thomas, 1977) is consistent with the observations that this cultural shift (a) occurred at different times in different parts of the world, with small aboriginal hunting enclaves still in existence, and (b) did not occur once and for all in every such tribe. With respect to (a), the great wave of terrestrial animal extinctions occurred over a period of several thousand years, and therefore the relative increase in the cost of hunting struck different regions at different times. Also different peoples in different environments with different opportunity costs would be expected to provide different mechanisms of adaptation, with some persisting as gatherers and small game hunters, and others turning to or perhaps persisting as fisherman (e.g. the Aleutian Eskimos and the Pacific Northwest Indians) in regions unsuitable for agriculture. With respect to (b) the reintroduction of the horse in North America by the Spanish (in the hardy form of *Equus caballus* just 8000 years after other members of the genus became extinct in the Americas) had a major modifying impact on the economy of the plains Indians. In the northern plains the 'fighting' Cheyenne, as they were later to be termed by the Europeans, and the Arapahoe quickly abandoned their villages along with their pottery arts and horticulture to become nomadic Bison hunters (see the references in Smith, 1975). Apparently, agricultural productivity was dominated by the enormous increase in the bison harvest made possible by a technological change that combined the horse with the bow and arrow. To the south, where the growing season was longer and the climate more favourable, the Pawnee preserved their maize agriculture when they turned to Bison hunting, creating a mixed agricultural–hunting economy. The southwestern Apache, reported by Coronado in 1541 to be subsisting as bison hunters, simply adapted the horse to their pre-existing hunter culture. The vast bison-hide tepee encampments witnessed by the first Europeans to cross the plains were already the product of a technologically transformed native American, many of whom had only recently abandoned their agricultural economies.

PLEISTOCENE EXTINCTIONS AND THE RISE OF AGRICULTURE. Here then is a model of the epoch of man: he arrives 1.6 million years ago as a hunter among hunters, but distinguishable in terms of his human capital endowment and his ability to invest in the development of human and physical capital. His tools become more complex and knowledge of the use of fire, perhaps his most significant tool, is added to his stock of human capital. There is a gradual improvement in weapons technology – clubs, stones, stone axes, spears, stone projectile points, the atlatl (which applies the leverage principle) and, in the late pre-agricultural period, the bow (which combines the leverage principle with temporary storage of energy for increased mechanical advantage). The combination of his physical superiority, tools and fire make him a superpredator without equal. At some unknown point this success brings relative affluence, and the important commodity 'leisure', which might have contributed to the development of language and other forms of investment in human and physical capital.

Although *H. erectus* and archaic *H. sapiens* were advanced hunters who apparently spread from Africa to Eurasia and Asia, it remained for modern *H. sapiens* to establish himself as a big game hunter *par excellence*, who populated most of the world by 8000 BC. Associated with this radiation is recorded a wave of extinction that was largely confined to the large terrestrial herbivores and their dependent carnivores and scavengers. (Other extinction episodes in the Earth's history had affected plants and marine life, as well as animals.) There appear to be no continents or islands where these accelerated late Pleistocene extinctions precede man's invasion (Martin, 1967). Whether men caused these extinctions cannot be known with any certainty, but Martin's overkill hypothesis is clearly consistent with a common property resource model of the economics of megaherbivore hunting (Smith, 1975). Thus the large gregarious animals that suffered extinction provided low search cost and high kill value. The lack of appropriation (branding or domestication) provided disincentives for conservation and sustained yield harvesting. There are numerous stampede kill sites (pitfalls and cliffs) in Russia, Europe and North America that indicate wastage killing in excess of immediate butchering requirements. Considering the complex of suitabilities necessary for the remains of such a site to have been preserved, it is likely that only the tip of such phenomena has been observed. Finally, the slow growth, long lives and long maturation of the megafauna made them more vulnerable than other animals to extinction by hunting pressure.

But our model of economizing man need not sustain such a controversial hypothesis as overkill. It is sufficient that the easy, valuable prey disappeared, precipitating a decline in the productivity of hunting. Substitution is to be expected, given a change in relative effort 'prices'. Hence, it is in this late pre-agricultural period that the archaeological record shows the appearance of bows and arrows, seed grinding stones, boiling vessels, boats, more advanced houses, even 'villages' (probably clan group abodes), animal drawn sledges and the dog (almost certainly derived from domesticating the wolf). These developments strongly suggest the substitution of new tools and techniques for the old, which allowed new products to substitute for the loss of big game that could be harvested by stampeding and/or dispatch with thrusting or throwing weapons. Now the bow and arrow becomes adaptive, and gathering becomes more crucial to maintaining overall food productivity. Whereas formerly, gathering emphasized seeds and plants that could be eaten on the run, now some of the seeds gathered were inedible without grinding, soaking, boiling. All this paraphernalia implies more sedentary, less nomadic, hunting and gathering.

Hence the incentive to invest in facilities such as utensils, sledges and houses. The boat allows fishing, sealing and whaling. The wolf, also characterized by its capacity to apply organization to the hunt, is now enlisted with man in the hunting of the game still available. Perhaps more important, the wolf may have been the model for domesticating other animals since the dog was a companion and pet that enabled children to learn about domesticated animal behaviour. With a more sedentary life, and the accumulation of personal property and real estate, would come more complex property right and contracting arrangements. The study of precolonial aboriginal societies in Northwest America and Melanesia reveals the existence of elaborate *multilateral contracting* arrangements in the form of 'ceremonial exchanges' such as the potlatch, kula, moka and abutu (Dalton, 1977). The use of valuables or commodity money (bracelets, pearl shells, cowries, young women) in these primitive societies was more complex than that of cash used in nation states with well-defined legal bases for exchange. These valuables not only bought other valuables in ordinary internal or external market exchange, they bought kinship ties with the exchange of women, military assistance when attacked, the right of refuge if invasion required the abandonment of homes, and emergency aid in times of poor harvest, hunting or fishing. In short they bought political stability, and a property right environment that made ordinary exchange and specialization possible. Property was owned by corporate descent lineages and included land, fishing sites, cemetery plots and livestock, but, interestingly, also public goods like crests, names, dances, rituals and trade routes, that could be assigned to many groups or individuals. These practices, which characterize stateless hunter–gatherer aboriginals, demonstrate that the phenomenon of multilateral contracting (Williamson, 1982), so common to the market economy in nation states, has ancient origins which antedate the State and the agricultural revolution.

Man's long existence as a hunter had brought knowledge of animals; extinction brought a change in relative costs; gathering brought knowledge of seeds and eggs; life became more sedentary, with property, contracting and exchange becoming more important. Under these more stable conditions it was a short step for mankind to plant for harvest, and/or to husband some of the more docile game that had been hunted previously. With agriculture and herding came a more sophisticated development of the earlier hunter–gatherer institutions of contract, property, exchange and specialization; and ultimately the continuing industrial–communication revolution. But long before these sweeping changes can be seen the dim outline of continuity in the development of man's capacity to adapt by creating cheaper products and techniques to substitute for dearer ones.

VERNON L. SMITH

See also AGRICULTURAL GROWTH AND POPULATION CHANGE; ECONOMIC ANTHROPOLOGY.

BIBLIOGRAPHY

Dalton, G. 1977. Aboriginal economies in stateless societies: interaction spheres. In *Exchange Systems in Pre-History*, ed. J. Erickson and T. Earle, New York: Academic Press.

Haynes, C.V. 1964. Fluted projectile points: their age and dispersion. *Science* 145, 25 September, 1408–13.

Heizer, R. 1955. Primitive man as an ecological factor. *Krober Anthropological Society Papers* No. 13.

Lee, R.B. and DeVore, I. 1968. *Man the Hunter*. Chicago: Aldine.

Lewis, H. 1973. *Patterns of Indian Burning in California: Ecology and Ethnohistory*. Ballena Press, Anthropological Papers, No. 1.

Martin, P. 1967. Prehistoric overkill. In *Pleistocene Extinctions*, ed. P.S. Martin and H.E. Wright, Jr., New Haven: Yale University.

North, D.C. and Thomas, R.P. 1977. The first economic revolution. *Economic History Review* 30(2), May, 229–41.

Pilbeam, D. 1984. The descent of hominoids and hominids. *Scientific American* 250(3), 60–69.

Smith, V.L. 1975. The primitive hunter culture, pleistocene extinction, and the rise of agriculture. *Journal of Political Economy* 83(4), August, 727–55.

Wheat, J.B. 1967. A Paleo-Indian bison kill. *Scientific American* 216(1), January, 44–52.

Williamson, O. 1984. Credible commitments: using hostages to support exchange. *American Economic Review* 74(3), September, 488–90.

Huskisson, William (1770–1830). Huskisson is better remembered for the manner of his death than for his not inconsiderable achievements as a statesman and economist. While it is true that he enjoyed 'little success in public life

compared with that which his rare abilities should have commanded' (*Dictionary of National Biography*), there were few major debates which were not enhanced by his contribution. Huskisson first entered Parliament in 1796 and remained a member, with only one short break, for over thirty years. He served in the cabinet from 1823, and held a number of key Government posts, including Secretary of the Treasury, President of the Board of Trade and Secretary of State for War and the Colonies. He figured prominently in the Bullion Controversy and the subsequent discussion on the resumption of cash payments; and he initiated the process of tariff reform which was to culminate in the repeal of the Corn Laws. His abilities may be gauged by the tributes paid by his contemporaries. It was said that 'there is no man in Parliament, or perhaps out of it, so well versed in finance, commerce, trade or colonial matters' (Charles Greville, in Melville, 1931, p. viii); and that 'the knowledge of theory and practice were never possessed by any one in so high a degree' (Kirkman Finlay, in Huskisson, 1831, I, p. 161; also Alexander Baring and Henry Brougham, ibid., pp. 120–21). Indeed, according to some observers, Huskisson might easily have become Chancellor of the Exchequer, but for his almost disingenuous loyalty to George Canning and the offence which he regularly caused to traditional Tory interests. These 'failings' earned him a remarkably fulsome tribute from J.S. Mill: 'With the exception of Turgot, the history of the world does not perhaps afford another example of a minister steadfastly adhering to general principles in defiance of the clamours of the timid and interested of all parties...' (*Westminster Review*, 1826, cit. Tucker introduction to Huskisson, 1830, p. xv). Even his closest supporters, however, could not pretend that Huskisson was an eloquent speaker; to his everlasting shame, he was born and brought up outside London and the Home Counties. As a consequence, no doubt, he was 'a wretched speaker with no command of words, with awkward motions, and a most vulgar, uneducated accent' (Sir Egerton Brydges, cit. *Dictionary of National Biography*). Huskisson's interest in political economy began in Paris, where, as a young man, he moved in French liberal circles, and is said to have met Franklin and Jefferson. There, in 1790, he presented a paper on the currency to the monarchist 'Club of 1789'; once the French Government started issuing assignats, however, he resigned from the club and, shortly afterwards, returned to Britain. In 1810, Huskisson had an opportunity to make his mark on British financial policy; he did so in conjunction with Henry Thornton and Francis Horner in the Bullion Report, and then on his own in a pamphlet defending the report against its 'anti-bullionist' critics. This pamphlet, *The Question concerning the Depreciation of our Currency* (1810), ran to several editions and drew praise not only from Ricardo, as might be expected, but also from the more critical Thomas Tooke (1838/57, IV, p. 98); its main target was the 'real bills doctrine' pleaded by the Bank of England directors as an adequate principle of limitation even when the currency was inconvertible. In the Parliamentary debates on the Bullion Report, Huskisson likened the views of the Bank directors to those of John Law, and made a strong case for the resumption of cash payments (Fetter, 1965, p. 43). After the passage of resumption legislation in 1819, however, Huskisson confessed to private doubts: 'The wheel of depreciation producing high prices, etc., was turning one way whereby many interests suffered and were ruined; to attempt to turn the wheel back, without some equitable adjustment ... has always appeared to me madness' (Letter to J.C. Herries, 20 December 1829, cit. Melville, 1932, p. 312). The sharp decline in prices which followed resumption particularly affected agricultural products. A Committee on Agriculture was formed in 1821 whose report – drafted mainly by Huskisson and Ricardo – accepted many of the arguments against the Act of 1819 but came down in favour of its retention. Thomas Attwood, after giving evidence to the Committee, wrote: 'The stupid landowners ... are all as dull as beetles, whilst Huskisson and Ricardo are as sharp as *needles* and as active as bees' (*cit.*, Ricardo, 1951/52, VIII, p. 370). A year later, Huskisson headed off Western's motion to reopen the issue with an amendment in the same terms as Montague's resolution of 1696, 'That this House will not alter the Standard of Gold or Silver, in fineness, weight, or denomination'. During the 1820s, Huskisson became an effective spokesman for the manufacturing interest, defending, 'with singular success and ability, the general principles of commercial freedom' (Tooke, 1838/57, V, p. 414). He took part in debates on the silk trade, agricultural protection, tax reform, shipping and the repeal of the Combination Acts; and he was almost alone in foreseeing the crisis of 1825, expressing concern as early as March 1822, 'that this universal Jobbery in Foreign Stock will turn out the most tremendous Bubble ever known' (Hudson Gurney, *cit.* Fetter, 1965, pp. 111–12). Having disregarded his warnings, the Bank of England directors sought to blame Huskisson for promoting the crisis:

> Such is the detestation in which he is held in the City that Ld L[iverpool] & Mr. Canning did not think it prudent to summon him to London till all the Cabinet were sent for &, in the discussions with the Bank, he is kept out of sight. He repays them with equal hatred ... (Mrs. Arbuthnot, 17 December 1825, *cit.* Fetter, 1965, p. 117).

In June 1827, Huskisson, responding to a memorandum circulated by James Pennington, wrote of the need to 'prevent ... those alterations of excitement and depression which have been attended with such alarming consequences to this country'; he went on:

> This, for a long time, has appeared to me one of the most important matters which can engage the attention of the Legislature and the Councils of this country. The subject is certainly intricate and complicated; but the too great facility of expansion at one time, and the too rapid contraction of paper credit (I speak of it in the largest sense) at another, is unquestionably an evil of the greatest magnitude (*cit.* Fetter, 1965, p. 131; also Viner, 1937, p. 224).

Huskisson asked Pennington for suggestions as to how these fluctuations could be minimised, and Pennington submitted a second memorandum which was to form the basis of the 'currency principle'.

Huskisson resigned from the Government in 1828 over a seemingly trivial but symbolic issue – the allocation of a Parliamentary seat to a sparsely populated rural hundred, instead of a manufacturing town. He died soon afterwards in unusual, not to say bizarre, circumstances. On 15 September 1830, he attended the opening ceremony of the Manchester and Liverpool railway:

> At that moment several engines were seen approaching along the rails between which Huskisson was standing. Everybody made for the carriages on the other line. Huskisson, by nature uncouth and hesitating in his motions, had a peculiar aptitude for accident... . On this occasion he lost his balance in clambering into the carriage and fell back upon the rails in front of the Dart, the advancing engine. It ran over his leg ... He lingered in great agony for nine hours, but gave his last directions

calmly and with care, expiring at 9 P.M. (*Dictionary of National Biography*).

That would be the end of the story but for a fine piece of detective work by G.S.L. Tucker and his assistant, Helen Bridge, who, in 1976, established beyond reasonable doubt that the author of an anonymously published 1830 tract, *Essays on Political Economy*, was none other than William Huskisson. In addition to the circumstantial evidence of style and argument, the publisher's Commission Ledger was signed by a certain 'George Robertson', a name unknown to political economy at that time. It was then demonstrated by Detective Sergeant D.G. Stuckey of the Document Examination Unit, New South Wales Police, that the signature belonged not to 'George Robertson' at all but to Huskisson's half-brother, Thomas, with whom he was on close terms (Fay, 1951, pp. 300–301). Thomas Huskisson was a captain the Royal Navy; and there is evidence that in return for career advancement (William Huskisson was treasurer of the Navy from 1823 to 1827), he would perform errands of this kind (ibid.).

Although the *Essays* had a poorer reception than if they had appeared under Huskisson's own name, he presumably felt that he could not take the risk of further embarrassing the Government with his forthright views. The *Essays* are basically Smithian in approach, and, in most respects, were already superseded by Ricardo's *Principles*. They do, however, propose some important financial reforms (Huskisson, 1830, pp. 149–151 and 152–153), repudiate the landowners' monopoly (*ibid.*, p. 255) and, most notably, anticipate J.S. Mill's concept of a 'general glut' (ibid., pp. 448–52 and 454–5). Overall, they epitomize Huskisson's economic philosophy and were even cited approvingly by Marx ([1867], p. 495 n.); this philosophy was reflected clearly and consistently in a life of ceaseless activity: 'Whatever ridicule might be attempted to be thrown on the science of political economy', he said, 'that science could not be discredited. It was the result of general principles warranted by observation, and constituted the guide in the regulation of political measures' (Huskisson, 1831, II, p. 128).

ROY GREEN

SELECTED WORKS
1830. *Essays on Political Economy*. Canberra: Australia National University, 1976.
1831. *The Speeches of the Right Honourable William Huskisson*. London: John Murray.

BIBLIOGRAPHY
Fay, C.R. 1951. *Huskisson and his Age*. London: Longmans.
Fetter, F.W. 1965. *Development of British Monetary Orthodoxy, 1797–1875*. Reprinted, Fairfield: Kelley, 1978.
Marx, K. 1867. *Capital*, Vol. I. Moscow: Progress Publishers, n.d.
Melville, L. 1931. *The Huskisson Papers*. London: Constable.
Ricardo, D. 1951–73. *The Works and Correspondence of David Ricardo*. Ed. P. Sraffa, Cambridge: Cambridge University Press.
Tooke, T. 1838–57. *A History of Prices, and of the State of Circulation, from 1792 to 1856*. London: P.S. King, 1928.
Viner, J. 1937. *Studies in the Theory of International Trade*. London: George Allen & Unwin.

Hutcheson, Francis (1694–1746). Hutcheson was born on 8 August 1694. His father, John, was a Presbyterian minister in Armagh, Ireland, and Francis spent his early years at nearby Ballyrea. In 1702 Francis and his elder brother Hans went to live with their grandfather, Alexander Hutcheson, at Drumalig in order to attend school near Saintfield. At the age of 14 he moved to a small denominational college or academy at Killyleagh, County Down.

In 1711 Hutcheson matriculated at Glasgow University where he was particularly influenced by Robert Simson (Mathematics), Gerschom Carmichael (Moral Philosophy), Alexander 'old' Dunlop (Greek) and John Simpson the 'heretical' divine. Hutcheson graduated in 1713 and embarked on a course of study in theology under Simpson's guidance. Hutcheson was back in Ireland by 1719, when he was licensed as a probationary minister, but moved to Dublin, where he set up an academy of which he remained head until 1730. It was during his period in Dublin that he was closely associated with Lord Molesworth, Edward Synge and James Arbuckle, all of whom had been influenced by the work of Shaftesbury.

His reputation established, Hutcheson was elected to the Chair of Moral Philosophy in Glasgow in 1730, when he delivered an inaugural lecture defending his principle of benevolence. It was as a lecturer that he made his mark on the University: brilliant and stylish, using English rather than Latin, Hutcheson's career amply confirms the accuracy of Adam Smith's reference to 'the abilities and virtues of the never to be forgotten' Francis Hutcheson (*Correspondence*, 1977, letter 274, p. 309).

Hutcheson lectured five days a week on Natural Religion, Morals, Jurisprudence, and Government (Leechman, 1755, p. xxxvi). On three days, he lectured on classical theories of morality thus contributing (with Dunlop) to a revival of classical learning in Glasgow and forming an important channel for Stoic philosophy (Scott, 1937, pp. 31–2); a branch of philosophy which was to have a profound influence on his pupil, Adam Smith. Hutcheson died on 8 August (his birthday) 1746 and was buried in St Mary's Churchyard, Dublin.

ETHICS AND JURISPRUDENCE. Hutcheson shared with his correspondent David Hume a concern with empirical study and an interest in elucidating the principles of human nature (1755, Book I, chapter 1). He noted that men are capable of acts of will which may be 'calm' or 'turbulent' where both may be selfish or benevolent, thus introducing the need for some means of control.

He argued that man is equipped with the faculty of reason and has powers of perception which 'introduce into the mind all its materials of knowledge' and which are associated with 'acts of the understanding' (1755, i. 7). This argument in turn led to the discussion of the senses of beauty, harmony and design. He added: 'Another important determination or sense of the soul we may call the *sympathetick*', which differs from *external* senses such as sight, sound, or taste and 'by which, when we apprehend the state of others, our hearts naturally have a fellow-feeling with them' (i. 19).

But in practice Hutcheson placed most emphasis on the *moral sense* representing a capacity for moral judgement, whose deployment (reinforced by the senses of honour or of shame) encourages the individual to virtuous actions. Hutcheson went on to distinguish the moral sense from the senses of decency and dignity. He added, in an important passage, that 'A penetrating genius, capacity for business, patience of application and labour ... are naturally admirable, and relished by all observers; but with quite a different feeling from moral approbation' (i. 28).

The position is admirably summarized in the introduction to Smith's *Theory of Moral Sentiments*, where the editors note that Hutcheson 'held (against egoism) that moral action and moral judgement are disinterested and (against rationalism) that they both depend on natural feeling. Moral action is

motivated by the disinterested feeling of benevolence', and since benevolence aims at producing happiness, 'the morally best action is that which procures the greatest happiness for the greatest number' (1976, p. 12).

It was Hutcheson's contention that men were inclined to, and fitted for, society: 'their curiosity, communicativeness, desire of action; their sense of honour, their compassion, benevolence, gaiety and the moral faculty, could have little or no exercise in solitude' (1755, i. 34).

This discussion was to lead to Hutcheson's treatment of natural rights and of the state of nature in a manner which is reminiscent of Locke. He also advanced the Lockean claim that the state of nature is a state not of war but of inconvenience which can only be resolved by the establishment of government in terms of a complex double contract:

> Civil power is most naturally founded by these three different acts of a whole people. 1. An agreement or contract of each one with all the rest, that they will unite into one society or body, to be governed in all their common interests by one council. 2. A decree or designation, made by the whole people, of the form or plan of power, and of the persons to be intrusted with it. 3. A mutual agreement or contract between the governors thus constituted and the people, the former obliging themselves to the faithful administration of the powers vested in them for the common interest, and the latter obliging themselves to obedience' (ii. 227).

This has been described as the 'Real Whig position' (Winch, 1978, p. 46; Robbins, 1968) and may explain the considerable influence of Hutcheson's political ideas in the American Colonies (Norton, 1976). Hutcheson's 'warm love of liberty' was attested by Principal Leechman in his introduction to the *System* (1755, pp. xxxv–xxxvi); a sentiment which was echoed by Hugh Blair (Winch, 1978, pp. 47–8) in a contemporary review of the book.

The argument was developed in a number of directions which include the analysis of the 'Several Forms of Polity, with their Principal Advantages and Disadvantages' (Book III, ch. 6).

POLITICAL ECONOMY. The economic analysis, which is an important feature of the *System*, is woven into the broader fabric of the argument. It is concentrated primarily in Book 2, chapters 4, 6, 7, 12 and 13. Economic policy of a broadly 'mercantilist' tenor is discussed in Book 3, chapter 9, section 4 and taxation (including a discussion of the canons of taxation) is considered briefly in section 16 of the same chapter.

The argument is interesting in that it illustrates Hutcheson's concern with the advantages which accrue from the organised social state and further illustrates his grasp of the role of self-interest. The order of the argument is equally noteworthy.

Hutcheson effectively began his treatment of economic topics with a discussion of the division of labour (i. 287–9):

> Nay, tis well known that the product of the labour of any given number, twenty for instance ... shall be much greater by assigning to one, a certain sort of work of one kind, in which he will soon acquire skill and dexterity, and to another assigning work of a different kind, than if each one of the twenty were obliged to employ himself, by turns, in all the different sorts of labour requisite for his subsistence, without sufficient dexterity in any. In the former method each procures a great quantity of goods of one kind, and can exchange a part of it for such goods obtained by the labours of others as he shall stand in need of (i. 288–9).

The discussion of the division of labour led in turn to Hutcheson's treatment of property and further illustration of the claim that men have a right to property which is important of itself and which, where respected, also provides a significant stimulus to economic activity (i. 320–24).

The longest continuous discussion is found in chapter 12 of the second Book (ii. 53–64) where Hutcheson, following the lead of Gershom Carmichael, introduced the discussion of value. Starting from the proposition that the 'natural ground of all value or price is some sort of use which goods afford in life' (ii. 53), 'usefulness' is then defined to include 'any tendency to give any satisfaction, by prevailing custom or fancy' (ii. 54). He concluded that:

> When some aptitude to human use is pre-supposed, we shall find that the prices of the goods depend on these two jointly, the *demand* on account of some use or other which many desire, and the *difficulty* of acquiring or cultivating for human use (ibid).

On the supply side, Hutcheson argued that 'value' is affected by the labour involved, the materials used, the ingenuity of the artist, the remuneration of the employer, rent and interest. He added that 'value is also raised, by the dignity of the station in which, according to the custom of a country, the men must live who provide us with certain goods or works of art' (ii. 55).

Hutcheson proceeded from this point to the discussion of the means of exchange (money) and to comment on the advantages of the metals and the need for coinage (ii. 56–8). The remainder of the chapter is largely concerned with the issue of debasement, and makes the point that money is not of itself an adequate measure of value.

The analytical section of the work is concluded in the following chapter where Hutcheson demonstrates the need for *interest*, since were it prohibited 'none would lend' (ii. 72). He argued that the rate would be determined by 'the state of trade and the quantity of coin' recognizing that 'as men can be supported by smaller gains in proportion upon their large stocks, the profit made upon any given sum employed is smaller, and the interest the trader can afford must be less' (ibid.) Hutcheson was well aware of the relationship between the rate of interest and other forms of return such as rent, and also introduced an allowance for risk. In sum, an interesting and often sophisticated analysis which is likely to have made an important general impression on Adam Smith, while influencing his treatment of particular topics, such as the division of labour.

ADAM SMITH. There can be little doubt that Smith owed much to his teacher. There are marked parallels in the treatment of beauty (Scott, 1900, 188, 284). In the ethics, there is the same emphasis on the role of immediate sense and feeling, on the point that moral judgement is disinterested, and further elaboration of the concept of the 'spectator'. It has also been noted that there is a marked parallel in the treatment of jurisprudence (see for example, Meek, 1976; Pesciarelli, 1986). The similarity in respect of the treatment of economic topics in Smith's lectures and in Hutcheson's work has been elaborated most notably by W.R. Scott (1900, 1937) and by W.L. Taylor (1965). Edwin Cannan went so far as to claim with respect to Hutcheson's treatment of price, that 'Probably it is in this chapter that the germ of the *Wealth of Nations* is to be found' (1896, p. xxvi).

At the same time there are differences of emphasis which are analytically interesting. While Smith admired benevolence, he rejected Hutcheson's claim that self-love cannot be 'in any case

a motive for virtuous actions' (1759, VII, ii. 3.13), together with the view that moral judgement depended upon a special sense (1759, VII, iii. 3). If the economic analysis found in Smith's lectures follows the order of his master's treatment, the discourse is not only more elaborate, but also presented as a complete and self-contained account. But the most remarkable contrast is to be found in Hutcheson's acceptance, and Smith's rejection, of the contract theory of government. Hutcheson's radicalism has been contrasted with Smith's more conservative position (Winch, 1978, pp. 52–4). The same writer has also drawn attention to the associated point that Smith's treatment of jurisprudence was more explicitly historical (ibid., p. 65).

Yet a *System of Moral Philosophy* would have been as good a title for Smith's more ambitious programme, as it was for Hutcheson's summary of his lecture course; a course which embraced ethics, jurisprudence, government and political economy and which treated these different disciplines as the parts of a single, organic, whole (Scott, 1900, p. 227).

ANDREW SKINNER

SELECTED WORKS All references to Hutcheson (and Leechman) in the text are to the *System of Moral Philosophy* (1755).

1725. *Inquiry into the Original of our Ideas of Beauty and Virtue.* London. 2nd edn, 1726.

1725–6. *Reflections upon Laughter and Remarks on the Fable of the Bees. Dublin Journal* No. 11 (5 June 1725); No. 12 (12 June); No. 13 (19 June) and No. 45 (5 February 1726); No. 46 (12 February); No. 47 (19 February). Also in *Hibernicus's Letters* (1729; 2nd edn, 1734).

1728a. *Essay on the Nature and Conduct of the Passions, with Illustrations upon the Moral Sense.* London and Dublin.

1728b. *Letters between the late Mr G. Burnet and Mr Hutcheson, London Journal.*

1730. *Hutchesoni Oratio Inauguralis.*

1735. *Considerations on Patronage addressed to the Gentlemen of Scotland.*

1742a. *The Meditations of M. Aurelius Antoninus. Newly Translated from the Greek, with Notes and an Account of His Life.* Glasgow.

1742b. *Metaphysicae Synopsis Ontologiam et Pneumatologiam complectens.* 2nd edn, 1744.

1742c. *Philosophiae Moralis Institutio Compendiaria, Ethices et Jurisprudentiae Naturalis Elementa continens, Libri Tres.* Glasgow. 2nd edn, 1745. Published as *A Short Introduction to Moral Philosophy in Three Books, containing the Elements of Ethics and the Law of Nature,* Glasgow, 1747.

1755. *A System of Moral Philosophy in Three Books, written by the late Francis Hutcheson, LL.D., Professor of Moral Philosophy in the University of Glasgow. Published from the original MS. by his son Francis Hutcheson, M.D. to which is prefixed Some Account of the Life, Writings, and Character of the Author, By The Reverend William Leechman, D.D., Professor of Divinity in the same University.* Glasgow.

1756. *Logicae Compendium, &c.* Glasgow. *Illustrations on the Moral Sense* was edited by Bernard Peach in 1971. Hutcheson's *Collected Works* are available in the Georg Olms Verlagsbuchhandlung edition (Hildesheim, 1969).

BIBLIOGRAPHY

Blackstone, W.T. 1975. *Francis Hutcheson and Contemporary Ethical Theory.* Athens, Georgia: University of Georgia Press.

Campbell, T.D. 1982. Francis Hutcheson. In *The Origins and Nature of the Scottish Enlightenment,* ed. R.H. Campbell and A.S. Skinner, Edinburgh: Edinburgh University Press.

Cannan, E. (ed.) 1896. *Adam Smith's Lectures on Justice, Police, Revenue and Arms.* Oxford: Clarendon Press.

McCosh, J. 1875. *The Scottish Philosophy From Hutcheson to Hamilton.* Princeton: Princeton University Press.

Meek, R.L. 1976. New light on Adam Smith's Lectures on Jurisprudence. *History of Political Economy* 8(4), Winter, 439–77.

Norton, D.F. 1976. Francis Hutcheson in America. In *Studies on Voltaire and the Eighteenth Century*, ed. T. Besterman, Vol. CLIV, Oxford: Voltaire Foundation.

Pesciarelli, E. 1986. On Adam Smith's Glasgow Lectures on Jurisprudence. *Scottish Journal of Political Economy* 33(1), February, 74–85.

Raphael, D.D. 1947. *The Moral Sense*. London: Oxford University Press.

Raphael, D.D. 1969. *British Moralists 1650–1800*. Oxford: Clarendon Press.

Robbins, C. 1954. When is it that colonies may turn independent: an analysis of the environment and politics of Francis Hutcheson. *William and Mary Quarterly* 11(2), 214–51.

Robbins, C. 1968. *The Eighteenth-Century Commonwealth Man*. New York: Atheneum.

Scott, W.R. 1900. *Francis Hutcheson, His Life, Teaching and Position in the History of Philosophy*. Cambridge: Cambridge University Press.

Scott, W.R. 1932. Hutcheson, Francis In *Encyclopedia of the Social Sciences*, Vol. 7, ed. E.R.A. Seligman, New York: Macmillan.

Scott, W.R. 1937. *Adam Smith as Student and Professor*. Glasgow: Jackson, Son & Co.

Smith, A. 1759. *The Theory of Moral Sentiments*. Ed. D.D. Raphael and A.L. Macfie, Oxford: Clarendon Press, 1976.

Smith, A. 1977. *The Correspondence of Adam Smith*. Ed. E. Mossner and I.S. Ross, Oxford: Clarendon Press.

Taylor, W.L. 1965. *Francis Hutcheson and David Hume as Predecessors of Adam Smith*. Durham, North Carolina: University of North Carolina Press.

Teichgraeber, R.E. 1986. *Free Trade and Moral Philosophy*. Durham, North Carolina: University of North Carolina Press.

Winch, D. 1978. *Adam Smith's Politics: An Essay in Historiographic Revision*. Cambridge: Cambridge University Press.

Hutchison, Terence Wilmot

Hutchison, Terence Wilmot (born 1912). Hutchison was born in Bournemoth and took his BA at the University of Cambridge in 1934. He began teaching at the London School of Economics after World War II but moved to the University of Birmingham in the 1950s, serving as Mitsui Professor of Economics at Birmingham from 1956 to 1978, the year of his retirement. A prolific essayist in the methodology and history of economics, he established an early reputation with *The Significance and Basic Postulates of Economic Theory* (1938), a book which in time acquired the status of a classic: it was written in reaction to *The Nature and Significance of Economic Science* (1932) by Lionel Robbins and the pamphlet, *Economics is a Serious Subject* (1932) by Joan Robinson, his Cambridge tutor, and was the first work to apply Popper's falsificationist philosophy of science to economics (Coats, 1983).

The years at the London School of Economics resulted in a major contribution to the history of economic thought: *A Review of Economic Doctrines 1870–1929* (1953). This work was remarkable for two reasons: it displayed an unusual knowledge (unusual among Anglo-American writers) of the continental literature of economics, and it was deeply coloured by the Keynesian interpretation of economic problems; thus, the book gave almost as much attention to macroeconomics, business cycles and monetary theory as to the theory of value and welfare economics. Hutchison went on subsequently in a number of essays on 18th-century economists to underline the Keynesian thesis that English classical political economy had been responsible for burying concern among 17th- and 18th-century writers with the question of full employment.

Misgivings about the Keynesian message, however, made its appearance with *Economics and Economic Policy 1946–1966* (1968), a devastating study of the post-war policy writings of some of Keynes's leading British followers. The book was a sequel to Hutchison's earlier study of the relationship between *'Positive' Economics and Policy Objectives* (1961), but the style of the new book was more polemical and the tone was both angry and bitter. Other writings over the years were brought together in three more books (Hutchison, 1977, 1978, 1981). The early emphasis on falsificationism in Hutchison's writings is qualified but retained in these later works, which also contain a number of passionate and lively diatribes against what he regards as the distortions and double-think of much neo-Marxist and post-Keynesian historiography in economics.

MARK BLAUG

SELECTED WORKS

1938. *The Significance and Basic Postulates of Economic Theory*. London: Macmillan. 2nd edn, New York: Augustus Kelley, 1960.

1953. *A Review of Economic Doctrines, 1870–1929*. Oxford: Oxford University Press.

1961. *'Positive' Economics and Policy Objectives*. London: Allen & Unwin.

1968. *Economics and Economic Policy in Britain, 1946–1966: Some Aspects of Their Interrelations*. London: Allen & Unwin.

1977. *Knowledge and Ignorance in Economics*. Oxford: Blackwell.

1978. *On Revolutions and Progress in Economic Knowledge*. Cambridge: Cambridge University Press.

1981. *The Politics and Philosophy of Economics, Marxists, Keynesians, and Austrians*. Oxford: Blackwell.

BIBLIOGRAPHY

Coats, A.W. 1983. Half a century of methodological controversy in economics: as reflected in the writings of T.W. Hutchison. In *Methodological Controversy in Economics: Historical Essays in Honor of T.W. Hutchison*, ed. A.W. Coats, Greenwich, Conn.: JAI Press.

Hymer, Steven Herbert

Hymer, Steven Herbert (1934–1974). Hymer was born on 15 November 1934 in Montreal, Canada, and died tragically at the age of 39 in a car accident, returning from a winter holiday, on a New York State thruway in February 1974.

Hymer began his study of economics as an undergraduate at McGill University and then received his PhD in economics from MIT in 1960. He worked in Ghana for several years in the early 1960s and then returned to the United States to teach at Yale from 1964 to 1970. He moved increasingly in radical and then Marxian directions in the late 1960s. Having been denied tenure by Yale – a common fate at elite US graduate schools for leftists of his generation – he moved to the Graduate Faculty of the New School for Social Research, where he helped found and then foster a political economy programme until his sudden death in 1974.

Hymer's main analytic contributions flowed from his analyses of foreign direct investment by multinational corporations. As early as his seminal dissertation (1960), Hymer broke away from international trade theory, viewing foreign direct investment as a consequence of the particular internal contradictions of multinational enterprises and their drive to extend territorial control. Despite his short productive working life, Hymer's work in this area had wide-ranging influence in both the advanced and developing worlds in shaping both analysis and policy discussions.

Though less widely known for this work, Hymer was also making important contributions in his last several years to the articulation of a modern, complex, analytically rigorous Marxian political economy. Some of his most original and provocative papers in this effort, along with his best essays on multinationals and the global economy, were posthumously collected and published in *The Multinational Corporation* (1979).

DAVID M. GORDON

SELECTED WORKS
1960. *The International Operations of National Firms: a study of direct foreign investment*. PhD dissertation; Cambridge, Mass.: MIT Press, 1976.
1979. *The Multinational Corporation: A Radical Approach*. Ed. R.B. Cohen et al., New York: Cambridge University Press.

Hyndman, Henry Mayers (1842–1921). A British Marxist theorist and politician, Hyndman was born in London to a prosperous merchant family of staunchly Conservative politics; he was educated at Trinity College, Cambridge. He became in turn a journalist, imperial traveller and financial adventurer. An enthusiast for Empire, he stood for Parliament in 1880 as an independent on a Tory radical programme but withdrew from the contest. Increasingly acquainted with continental socialism, he read Marx's *Capital* in French in 1880 and became personally acquainted with Marx in London. This began the process whereby Hyndman, during the 1880s, emerged as the pioneer of British Marxism, the founder and leader of a Marxist party (the Social Democratic Federation) and the leading theorist and propagandist of Marxism in Britain. In essentials, this was the role he continued to play for the rest of his life.

Hyndman has had a bad press. In part this may be attributed to the easy caricature of him as the Marxist cricketer, stockbroker and national chauvinist, armed with top hat and frock coat. In part, too, it derives from his overbearing personality and sectarian political leadership. However, it is also directly related to the nature of his presentation of Marx's economic theory. He set himself the task of explaining this theory to the British public, relating it to British conditions, and drawing the appropriate political lessons from it. His *England for All* (1881), with its indirect tribute to Marx's work but omission of his name (thereby beginning the personal breach with Marx and Engels), began this task, which was then taken further in his best book, *The Historical Basis of Socialism in England* (1883), with its application of Marxist economic theory to the economic history of England since the 15th century. Its preface recorded his 'indebtedness to the famous German historical school of political economy headed by Karl Marx, with Friedrich Engels and Rodbertus immediately following'.

Hyndman's presentation of Marxist economics was narrowly literal and inflexible, which meant that he could neither develop it creatively nor defend it against its critics with sufficient rigour. When he departed from Marx's own position this was not because of any intention to do so but because he had either failed to understand Marx on the point, or had access only to a limited range of Marx's work, or because when he cited other economic authorities (such as Rodbertus and Lassalle) he was unaware of Marx's disagreements with them. Hence his exposition of the Lassallean 'iron law of wages' as Marxist orthodoxy. In the 1880s, on the basis of Marx's work then available, it was certainly possible to present this as Marx's own position, but Hyndman's later and most developed discussion of economics in his *Economics of Socialism* (1896) showed him still substantially attached to a theory by then repudiated in Marx's mature work. It was on the basis of this doctrinal position that Hyndman poured scorn on the trade unions for the futility of their economic activities.

Similar limitations prevented Hyndman from defending a tenable version of Marx's theory of value when this came under criticism and discussion in the 1880s, especially in Wicksteed's critique of it in terms of Jevonian marginal utility

theory. If Fabian intellectuals like Shaw and Webb could respond to this critique by restating the economic case against capitalism in terms of a theory of economic rent rather than of Marxist surplus value, Hyndman lacked the equipment to mount an effective counter-offensive of his own. He continued to be a vigorous propagandist for what he understood as Marxist economic orthodoxy, but the intellectual battle was lost and it was left to a later generation of British Marxist economists to take the argument further.

ANTHONY WRIGHT

SELECTED WORKS
1883. *The Historical Basis of Socialism in England*. London: Kegan Paul.
1896. *The Economics of Socialism*. London, Boston: Small, Maynard and Co.

BIBLIOGRAPHY
Collins, H. 1971. The Marxism of the Social Democratic Federation. In *Essays in Labour History 1886–1823*, ed. A. Briggs and J. Saville, London: Macmillan.
Hobsbawm, E. 1964. Hyndman and the SDF. In E. Hobsbawm, *Labouring Men*, London: Weidenfeld & Nicolson.
Tsuzuki, C. 1961. *H.M. Hyndman and British Socialism*. Oxford: Oxford University Press.

hyperinflation. Hyperinflation is an extremely rapid rise in the general level of prices of goods and services. It typically lasts a few years or in the most extreme cases much less before moderating or ending. There is no well-defined threshold. It is best described by a listing of cases, which vary enormously. The numerous cases have provided a testing ground for theories of monetary dynamics reported in a vast literature.

HISTORICAL SURVEY. The world's record occurred in Hungary after World War II when an index of prices rose an average 19,800 per cent per month from August 1945 to July 1946 and $4 \cdot 2 \times 10^{16}$ per cent in the peak month of July. Also in the aftermath of World War II extreme price increases occurred in China, Greece, and Taiwan. Hyperinflations followed World War I in Austria, Germany, Hungary, Poland and Russia. If we measure the total increase in prices from the first to last month in which the monthly increase exceeded 50 per cent and afterwards stayed below that rate for a year or more, a price index rose from 1 to $3 \cdot 8 \times 10^{27}$ in the record Hungarian episode, 10^{11} in China, 10^{10} in Germany, and ranged down to 70 in Austria and 44 in the first Hungarian episode after World War I. In the last, the mildest of those cited above, the rise in prices averaged 46 per cent per month.

Prior to World War I extreme inflations were rare. A price index rose from 1 to about 18 from mid-1795 to mid-1796 at the height of the *assignats* inflation in France, from 1778 to 1780 in the American War of Independence, to 12 from 1863 to 1865 in the Confederacy during the American Civil War, and comparable inflation rates were reported for Columbia in 1902. The oft-cited currency depreciations of the ancient and medieval world and of Europe in the 17th century from the influx of precious metals were mild by modern experience. Earlier extreme inflations were rare because of the prevalence of commodity monies and convertibility. Only inconvertible paper currencies can be expanded rapidly without limit to generate hyperinflation.

Although the greatest hyperinflations have occurred in countries devasted by war, non-war-related inflation rates of several hundred per cent per year were reached briefly in 1926 in Belgium and France. Since World War II to the time of writing (1985) the frequency of both mild and extreme inflations unrelated to war has increased throughout the world. While rates of several hundred per cent per year or more for short periods have become common since World War II, few cases have exceeded 1000 per cent per year for even a few months (Meiselman, 1970), and hence they fall far short of the great hyperinflations. The rate of over 10,000 per cent per year in Bolivia in 1985 is a major exception.

MONETARY CHARACTERISTICS. Extreme increases in the price level cannot occur without commensurate increases in the money stock, which are usually less than proportionate because of decreases in the demand for real money balances. Governments resort to issuing money rapidly when they are unable to contain expanding budget expenditures and to raise sufficient funds by conventional taxation and borrowing from the public. Money creation is a special form of taxation which is levied on the public's holdings of money. It is administratively easy to impose and collect. Excessive money issues to finance the government budget add to aggregate spending and raise prices; the resulting depreciation in the purchasing power of outstanding real money balances imposes the tax. Bailey (1956) finds the social costs of this tax to be high compared with other forms of taxation.

Escalations of inflation at any level tend to stimulate economic activity temporarily. Since high rates of inflation tend to distort relative prices, however, much of the economic activity is socially wasteful. Many businesses and workers are dependent on prices and wages that lag behind the general inflation, and thus suffer severe declines in real income. In addition, unanticipated depreciations in financial and monetary assets in real terms produce major redistributions of wealth. These effects are socially and politically disruptive (Bresciani-Turroni, 1931). Yet the de-escalation of inflation temporarily contracts aggregate demand, which is also disruptive and therefore politically difficult to undertake.

THEORETICAL ISSUES. The depreciation of money during inflation greatly increases the cost of holding it. Although depreciating currencies are not abandoned completely, testifying to the great benefits of a common medium of exchange, the public undertakes costly efforts to reduce holdings of a rapidly depreciating money, including barter arrangements and the use of more stable substitutes such as foreign currencies (Barro, 1970). These efforts result in a large reduction in money balances in real terms and a large rise in monetary velocity.

A study of this result by Cagan (1956) estimated the demand for real money balances in hyperinflation as inversely dependent on the *expected* rate of inflation. Expectations about future developments can differ from concurrent conditions and determine the public's response to inflation. Cagan hypothesized that expectations are formed adaptively, whereby expected values are adjusted in proportion to their discrepancy from actual values. The theoretical implication is that expected inflation can be estimated as an exponentially weighted average of past inflation rates.

Such adaptive expectations lag behind the changes in actual values, which can explain why hyperinflations characteristically tend to escalate. As the inflation tax extracts revenue from real money balances, the expected inflation rate increases to match the higher actual rate and the revenue declines in real terms, but with a lag. The real revenue can be increased by speeding up money creation, but only until expectations adjust to the higher inflation rate. If the inflation rate were to remain constant so that the expected rate eventually matched it, the real revenue from money creation would be sustainable at a constant level. Among such constant levels a maximum real revenue is obtainable by a particular constant inflation rate, which depends on the elasticity of demand for real money balances with respect to the inflation rate. This revenue can be raised further by continually increasing monetary growth and the inflation rate. The hyperinflations kept escalating well beyond the maximizing constant rate to obtain more revenue.

Inflation also usually reduces the real value of other tax revenues because of lags between the imposition of a tax and its collection. A tax on money balances must exceed the reduced real collections of other taxes in order to prevent a decline in total government real revenue. This is often true initially under civil disorder, but hyperinflation reduces all taxes in real terms and in a short time largely destroys its revenue justification.

Adaptive expectations can be a 'rational' way for people to distinguish between transitory and one-time permanent changes in a variable (Muth, 1960). But if the inflation rate is continually rising, adaptive expectations as a weighted average of past rates are always too low, and such a series of correlated expectational errors is inconsistent with rational behaviour. The theory of rational expectations argues that the public uses all available information in predicting the inflation rate, including economic models of the process. This implies in particular that expectations of inflation, taking into account the importance of money, will focus on the money creating policies of the monetary authorities. If the government is after a certain amount of revenue, the public may be able to estimate the rate of money creation, which can be translated into a path for prices. Usually, however, the amount of money issued may change unpredictably or otherwise not be knowable with much precision.

Rational expectations have two important empirical implications for hyperinflation. First, if money is consistently issued to raise a certain revenue in real terms, monetary growth will depend on the inflation rate (Webb, 1984). The money stock is then statistically endogenous to the inflationary process. Sargent and Wallace (1973) and Frenkel (1977) presented statistical evidence that money depends on prices in the German hyperinflation, though such evidence has been contested (Protopapadakis, 1983).

To find that the money supply is endogenous does not mean that money demand is no longer dependent on the expected rate of inflation. But the finding discredits econometric regressions of real money balances on inflation rates. Other variables are needed to measure the expected cost of holding money. Frenkel (1977) used the forward premium on foreign exchange in the German hyperinflation, which reflected the market's estimate of future depreciation of the foreign exchange rate and presumably was dominated by expectations of inflation. (This also helps avoid possible spurious correlation when a price series for calculating real money balances is used in the same regression to derive the rate of price change.) The forward premium does explain movements in real money balances and confirms as a proxy the effect of the expected inflation rate. The forward premium in Germany was also found to be uncorrelated with past inflation, which satisfies the rational expectations requirement that the premium should not depend on past information and not involve lagged adjustments. This leaves unexplained, however, why the German inflation rate escalated beyond the revenue-maximizing constant rate, since rational expectations prevent anticipated escalation from increasing the revenue. Sargent (1977) suggests that the revenue-maximizing rate, when properly estimated under the hypothesis of rational expectations, was not exceeded by actual inflation rates.

Another possibility is that public behaviour did not fully anticipate the successive increases in inflation rates, which thus temporarily added to the government's inflation revenue.

STABILIZATION REFORM. Hyperinflation, if driven by rising expectations of inflation rather than rising money growth, can become a self-generating process. This has never occurred, however, except conceivably for very brief periods. Hyperinflations can always be stopped, therefore, by ending the monetary support. But the revenue from money creation is often difficult for governments to replace or survive without, explaining why some countries are subjected to high inflation for long periods.

Yet many hyperinflations have been stopped all at once with a programme of reform and without prolonged economic disruption. After a short period the economy usually recovers and prospers. These stabilization programmes have been studied to determine the necessary conditions for success (Sargent, 1982; Bomberger and Makinen, 1983). Some attempted reforms have failed, notably twice in Greece after World War II (Makinen, 1984). First of all, it is critical to gain control over monetary growth, and this requires an end to the government's dependence on money creation to finance its budget. Successful reforms involve a reorganization of government finances, both to cut expenditures and to raise taxes, and legal authority for the central bank to refuse to create money to lend to the government. Although a new currency unit is often issued to replace the depreciated one, this is symbolic only. Foreign loans or financial aid to bolster foreign exchange reserves and to finance government deficits for a while help to inspire confidence in the success of stabilization, but have not always been necessary. Convertibility of the new currency into gold or a key foreign currency assures the reform but is not always introduced immediately. Such convertibility to end a severe inflation has proven difficult in the post-Bretton Woods environment in which the key foreign currency (usually the dollar) floats in value. Fixing the foreign exchange rate can then produce massive trade deficits (if the key currency appreciates) which are impossible to maintain. Chile in the early 1980s is a notable example (Edwards, 1985).

Most reforms are initially popular, promising to bring back the benefits of a well-functioning monetary system. A resurgence of public confidence in the currency usually occurs, which produces a substantial increase in money demand from low hyperinflation levels. This allows a one-time increase in the money supply without raising prices. The monetary expansion must not continue beyond the demand increase, however, or it will set off a new round of inflation, and the stabilization will fail. Many reforms that eventually failed gained credibility initially and had an increase in money demand, but then subsequently over issued money and returned to high inflation rates. To avoid this outcome there mut be a commitment to maintain a stable price index or convertibility.

Stabilization reforms that have achieved an immediate end to hyperinflations contrast with the protracted efforts to subdue many moderate inflations. One difference is that in hyperinflations long-term contracts specifying prices or interest rates and wage agreements are no longer entered into because of great uncertainty over the inflation rate. Consequently, few parties are injured by such contracts when hyperinflation suddenly ends, and inflexibilities in the price system do not impede the required substantial readjustment of relative prices. Contracts that index financial and wage contracts to previous price movements impart a momentum to inflation that makes it more disruptive to end the process. The wide use of indexing, as in Brazil and Israel in the early 1980s, reduces differential price and wage movements but creates an obstacle to successful reform.

Hyperinflations on the order of those following the two World Wars remain rare and, when they occur, soon escalate to levels that necessitate an ending by drastic measures. Inflations on the order of 50 to several hundred per cent a year have been difficult to end, though they often subside for varying periods. These inflations despite their serious economic consequences give no indication of disappearing.

PHILLIP CAGAN

See also INFLATION; INFLATIONARY EXPECTATIONS.

BIBLIOGRAPHY
Bailey, M. 1956. The welfare cost of inflationary finance. *Journal of Political Economy* 64, April, 93–110.
Barro, R.J. 1970. Inflation, the payments period and the demand for money. *Journal of Political Economy* 78, November/December, 1228–63.
Bomberger, W.A. and Makinen, G.E. 1983. The Hungarian hyperinflation and stabilization of 1945–46. *Journal of Political Economy* 91, October, 801–24.
Bresciani-Turroni, C. 1931. *The Economics of Inflation: A Study of Currency Depreciation in Post-War Germany: 1914–1923.* Trans., London: Allen & Unwin, 1937.
Cagan, P. 1956. The monetary dynamics of hyperinflation. In *Studies in the Quantity Theory of Money*, ed. M. Friedman, Chicago: University of Chicago Press.
Edwards, S. 1985. Stabilization with liberalization: an evaluation of ten years of Chile's experiment with free-market policies, 1973–1983. *Economic Development and Cultural Change* 33, January, 223–54.
Frenkel, J.A. 1977. The forward exchange rate, expectations, and the demand for money: the German hyperinflation. *American Economic Review* 67, September, 653–70.
Makinen, G.E. 1984. The Greek stabilization of 1944–46. *American Economic Review* 74, December, 1067–74.
Meiselman, D. (ed.) 1970. *Varieties of Monetary Experience.* Chicago: University of Chicago Press.
Muth, J. 1960. Optimal properties of exponentially weighted forecasts. *Journal of the American Statistical Association* 55, June, 299–306.
Protopapadakis, A. 1983. The endogeneity of money during the German hyperinflation: a reappraisal. *Economic Inquiry* 21, January, 72–92.
Sargent, T.J. 1977. The demand for money during hyperinflation under rational expectations I. *International Economic Review* 18, February, 59–82.
Sargent, T.J. 1982. The ends of four big inflations. In *Inflation: Causes and Effects*, ed. R. Hall, Chicago: University of Chicago Press.
Sargent, T.J. and Wallace, N. 1973. 'Rational' expectations and the dynamics of hyperinflation. *International Economic Review* 14, June, 328–50.
Webb, S.B. 1984. The supply of money and Reichsbank financing of government and corporate debt in Germany, 1919–1923. *Journal of Economic History* 44, June, 499–507.

hypothesis testing.

1. TESTING RESTRICTIONS ON PARAMETERS. For those who believe that economic hypotheses have to be confirmed by empirical observations, hypothesis testing is an important subject in economics. As a classical example, when an economic relation is represented by a linear regression model:

$$Y = X\beta + \epsilon \qquad (1)$$

where Y is a column vector of n observations on the dependent variable y, X is an $n \times k$ matrix with each column giving the corresponding n observations on each of k explanatory variables (which typically include a column of ones), β is a column of k regression coefficients and ϵ is a vector of n independent and identically distributed residuals with mean zero and variance σ^2, it is of interest to test a hypothesis consisting of m linear restrictions on β:

$$R\beta = r \qquad (2)$$

where R is $m \times k$ and r is $m \times 1$. A most common case occurs when there is only one restriction ($m = 1$) and (2) is reduced to $\beta_i = 0$, the hypothesis being that the ith explanatory variable has no effect on y.

Among the statistical tests often employed in economic research are the likelihood ratio (LR) test, the Wald test and the Lagrangian multiplier (LM) test. The LR test, due to Neyman and Pearson (1928), uses as the test statistic the likelihood ratio:

$$\mu = \frac{L(Y, \hat{\theta}^*)}{L(Y, \hat{\theta})}$$

where L is the likelihood function, $\hat{\theta}^*$ is the maximum-likelihood estimator of a parameter vector θ under the null hypothesis to be tested, or subject to a vector $h(\theta) = 0$ of m restrictions such as (2), and $\hat{\theta}$ is the ML estimator of θ without imposing the restrictions. A high value of the likelihood ratio μ favours the null hypothesis. The Wald test, proposed by Wald (1943), uses the test statistic:

$$W = h(\hat{\theta})'[\text{Cov } h(\hat{\theta})]^{-1} h(\hat{\theta}) \qquad (4)$$

where Cov denotes covariance matrix. The null hypothesis $h(\theta) = 0$ will be accepted if the vector $h(\hat{\theta})$ is sufficiently close to zero, or if the statistic W is sufficiently small. Wald (1943) has shown that under general conditions, the statistics W and $-2 \ln \mu$ have the same asymptotic distribution.

The LM test, suggested by Silvey (1959), uses the Lagrangian multiplier $\hat{\lambda}$ obtained by maximizing the Lagrangian expression:

$$n^{-1} \ln L(Y, \theta) + \lambda' h(\theta) \qquad (5)$$

or by solving the associated first-order conditions for $\hat{\theta}^*$ and $\hat{\lambda}$:

$$n^{-1} \frac{\partial \ln L(Y, \hat{\theta}^*)}{\partial \theta} + H_\theta \hat{\lambda} = 0$$

$$h(\hat{\theta}^*) = 0 \qquad (6)$$

where H_θ denotes the $k \times m$ matrix $\partial h'(\theta)/\partial \theta$. The solution of (6) gives the maximum-likelihood estimator $\hat{\theta}^*$ subject to the restriction $h(\theta) = 0$ and the associated Lagrangian multiplier $\hat{\lambda}$. Under the null hypothesis $h(\theta) = 0$, $\sqrt{n}\hat{\lambda}$ has a normal limiting distribution with mean zero and a certain covariance matrix $-R$. Hence the statistic $-\hat{\lambda}' R^{-1} \hat{\lambda}$ is distributed asymptotically as $\chi^2(m)$. This statistic can be rewritten as a score statistic (see Chow, 1983, pp. 286–9):

$$-n\hat{\lambda}' \hat{R}^{-1} \hat{\lambda} = [\partial \ln L(Y, \hat{\theta}^*)/\partial \theta']$$
$$\times [-\partial^2 \ln L(Y, \hat{\theta}^*)/\partial\theta\partial\theta']^{-1} [\partial \ln L(Y, \hat{\theta}^*)/\partial \theta] \qquad (7)$$

As is well known, under the null hypothesis $\partial \ln L(Y, \theta)/\partial \theta$ has mean zero and covariance matrix $-E\partial^2 \ln L/\partial\theta\partial\theta'$. If the vector $\partial \ln L(Y, \hat{\theta}^*)/\partial \theta'$ is very different from zero, as measured by the statistic (7), one would be inclined to reject the null hypothesis. Silvey (1959) has shown that under fairly general assumptions:

$$-p \lim(2 \log \mu) = p \lim W = -p \lim n\hat{\lambda}' R^{-1} \hat{\lambda} \qquad (8)$$

and that the LR test, the Wald test and the LM test are asymptotically equivalent in the sense that their test statistics have the same asymptotic distribution. The equivalence for testing the hypothesis (2) in the linear regression case with normal residuals is shown in Chow (1983, pp. 290–1).

An example of (2) often encountered in practice is the hypothesis that certain subsets of coefficients in two linear regressions are equal. The test serves to detect whether certain economic parameters have changed from one sample period to another or whether they are different in two different situations (see Chow, 1960). Let the two samples of n_1 and n_2 observations be represented by:

$$Y_i = X_i \beta_i + \epsilon_i = Z_i \gamma_i + W_i \delta_i + \epsilon_i, \qquad (i = 1, 2) \qquad (9)$$

We wish to test H_0: $\gamma_1 = \gamma_2$, each with k_1 elements. A linear regression model for both samples can be written as:

$$\begin{bmatrix} Y_1 \\ Y_2 \end{bmatrix} = \begin{bmatrix} Z_1 & 0 & W_1 & 0 \\ 0 & Z_2 & 0 & W_2 \end{bmatrix} \begin{bmatrix} \gamma_1 \\ \gamma_2 \\ \delta_1 \\ \delta_2 \end{bmatrix} + \begin{bmatrix} \epsilon_1 \\ \epsilon_2 \end{bmatrix} \qquad (10)$$

The null hypothesis $\gamma_1 = \gamma_2$ can be written as a set of k_1 linear restrictions:

$$R\beta = [I \quad -I \quad 0 \quad 0] \begin{bmatrix} \gamma_1 \\ \gamma_2 \\ \delta_1 \\ \delta_2 \end{bmatrix} = 0 \qquad (11)$$

When the elements of ϵ_1 and ϵ_2 are normal, the test statistic is:

$$\frac{(A - B)/k_1}{B/(n_1 + n_2 - 2k)} \qquad (12)$$

where A is the sum of squared residuals of (10) estimated by imposing the k_1 restrictions (11) and B is the sum of squared residuals estimated without imposing the restrictions. Under H_0, the statistic (12) has an $F(k_1, n_1 + n_2 - 2k)$ distribution.

Much useful information concerning economic relations can be ascertained by testing hypotheses about the parameters of economic models. For example, one question in applying the regression model (1) to time-series data is whether the elements ϵ_t are serially correlated. One may postulate a first-order autoregressive model $\epsilon_t = \rho\epsilon_{t-1} + \eta_t$ for the residuals, where η_t is assumed to be independent and identically distributed. The hypothesis of interest is $\rho = 0$. As another example, one may ask whether the relation between y and a certain explanatory variable x_j is linear. A partial answer is given by introducing powers of x_j in the regression and testing whether their coefficients are significantly different from zero.

2. TESTING NON-NESTED HYPOTHESES. In the last section, the hypothesis to be tested consists of a set of restrictions $g(\theta) = 0$ on the parameter vector θ. Since the null hypothesis states that the parameter θ lies in a subspace of a parameter space, it is nested within a more general hypothesis. Comparing a more general alternative hypothesis with a more restrictive null hypothesis nested within the former is to test a nested hypothesis. When the two hypotheses to be compared are not nested, we are testing *non-nested* hypotheses. One important example of non-nested hypothesis consists of two regression models, (1) and:

$$Y = Z\gamma + u \qquad (13)$$

where Z is an $n \times p$ matrix including a different set of explanatory variables from those included in X of model (1). X and Z may have some variables in common, but neither hypothesis can be derived from restricting the values of the parameter vector permitted by the other hypothesis. In general, one may wish to choose between two non-nested hypotheses represented by two density functions $f_1(y, \theta_1)$ and $f_2(y, \theta_2)$ for generating y.

For the purpose of choosing between two competing density functions, Cox (1961, 1962) suggests combining them in the model:

$$h(y; \theta_1, \theta_2, \lambda) = k f_1(y, \theta_1)^\lambda f_2(y, \theta_2)^{1-\lambda} \quad (14)$$

If the maximum-likelihood estimate of λ is close to 1, choose f_1; if it is close to zero, choose f_2; if neither, the result is inconclusive. Quandt (1974) proposes an alternative way of combining the two density functions, namely:

$$h(y; \theta_1, \theta_2, \lambda) = \lambda f_1(y, \theta_1) + (1 - \lambda) f_2(y, \theta_2) \quad (15)$$

For choosing between two normal linear regression models (1) and (13), all parameters in (15) are identifiable, whereas for (14) one cannot separately identify $\lambda, \beta, \gamma, \sigma_1^2 = E\epsilon_i^2$ and $\sigma_2^2 = Eu_i^2$.

A common approach to choosing between non-nested models is to formulate a more general model nesting them and reduce the problem to one of testing a nested hypothesis, as exemplified by the methods just described. As another example, to choose between linear regression models (1) and (13), one may formulate a more general linear regression model including both sets of explanatory variables X and Z. If this general model is assumed to be the true model, then both (1) and (13) may be false. Nevertheless, one may still ask which has a smaller error in predicting y by testing the null hypothesis that the residual variances of these models are equal. The residual variance of the regression of Y on X is:

$$n^{-1}[E(Y'Y) - (EY)'X(X'X)^{-1}X'(EY)]$$

and similarly for the regression of Y on Z. In the general model, let $EY = [X \ Z]\alpha$. The equality of these two residual variances means:

$$\alpha'[X \ Z]'[X(X'X)^{-1}X'$$
$$- Z(Z'Z)^{-1}Z'][X \ Z]\alpha \equiv \alpha'H\alpha = 0 \quad (16)$$

This is a quadratic restriction on the coefficient vector α in a linear regression model. It can be tested by the methods of (3), (4) and (5). See Chow (1980; 1983, pp. 278–84). Some other works on testing non-nested hypotheses are cited in Chow (1983, pp. 284–6).

3. TESTING MODEL SPECIFICATIONS. When an economist wishes to find out whether a certain model is correctly specified, tests of model specification can be used. The situation here differs from that of section 2 in having no specific model to compete with the model in question. It differs from that of section 1 in not singling out, at least in the first instance, certain parameters as the likely sources of model misspecifications. If one believes that an omitted variable in a regression model may be the culprit, one would test whether its coefficient is significantly different from zero. If one believes that the residuals may be serially correlated, one might add an autoregressive structure to the residual and test the significance of its coefficients. Likewise, one may drop certain explanatory variables by testing the significance of their coefficients. In tests of model specifications, the alternatives are less specific. The tests aim at detecting misspecifications of a model against a variety of alternatives.

One approach to specification testing, initiated by Wu (1973)

and studied by Hausman (1978), is based on comparing two estimators of a parameter vector which are both consistent and asymptotically normal if the model is correctly specified. One estimator $\hat{\gamma}^0$ is asymptotically efficient if the model is correctly specified but is inconsistent if the model is incorrectly specified. The second estimator $\hat{\gamma}$ is consistent even if the model is incorrectly specified. If the difference $\hat{q} = \hat{\gamma} - \hat{\gamma}^0$ is large, one tends to reject the null hypothesis that the model is correctly specified. Let $V(\hat{q})$ be the covariance matrix of the asymptotic distribution of $\sqrt{n}\hat{q}$ and $\hat{V}(\hat{q})$ be a consistent estimate of $V(\hat{q})$. Then under the null hypothesis, which implies $p \lim q = 0$:

$$n\hat{q}'\hat{V}(\hat{q})^{-1}\hat{q} \quad (17)$$

will have $\chi^2(k)$ as its asymptotic distribution, k being the number of elements of $\hat{q}$. As an example, consider testing whether X is correlated with ϵ in model (1). Under the null hypothesis $p \lim n^{-1}X'\epsilon = 0$, an asymptotically efficient estimator is the least-squares estimator β^0. Even if the null hypothesis does not hold a consistent estimator is the instrumental variable estimator $\beta = (W'X)^{-1}Y$ where we assume $p \lim n^{-1}W'X$ to be a nonsingular matrix and $p \lim n^{-1/2}W'\epsilon$ to converge in distribution to k-variate normal with zero mean. A $\chi^2(k)$ statistic can be constructed to test the null hypothesis, using the difference $\hat{q} = \beta - \beta^0$ and its covariance matrix. Another example is to test the correct specification of simultaneous equations by comparing a three-stage least-squares estimator $\hat{\gamma}^0$ and a two-stage least-squares estimator $\hat{\gamma}$.

A convenient framework of Newey (1985) views specification testing as choosing some function $m(y, \theta)$ which satisfies the moment condition:

$$E[m(y, \theta_0)] = 0 \quad (18)$$

if the model $f(y, \theta)$ is correctly specified, and testing this condition by using the sample moment $\Sigma_{t=1}^n m(y_t, \hat{\theta})/n$. For example, the information matrix text of White (1982) compares two estimates of the information matrix and uses as elements of the vector function $m(y, \theta)$:

$$m_h(y, \theta) = \frac{\partial \ln f(y, \theta)}{\partial \theta_i} \cdot \frac{\partial \ln f(y, \theta)}{\partial \theta_j} + \frac{\partial^2 \ln f(y, \theta)}{\partial \theta_i \, \partial \theta_j} \quad (19)$$
$$(h = i + j - 1; \quad i = 1, \ldots, j; \quad j = 1, \ldots, k)$$

where k is the number of parameters. The Hausman test using (17) is shown by Newey (1985) to be asymptotically equivalent to a particular moment-condition test.

Economists using various specification tests should be reminded that these tests serve the same purpose as the many diagnostic checks for statistical models used in the literature. Examples are the diagnostic checks of Box and Jenkins (1970) for time-series models and those of Belsley, Kuh and Welsch (1980) for regression models.

4. MODEL SELECTION CRITERIA. The statistical tests presented so far are based on the notion that if a model is true (an assumption to be tested), it will be chosen. This nation might be questioned because the true model can be very complicated and in practice one may prefer to use a simpler model for estimation or prediction purposes. Consider the choice between model (1), with $X\beta = X_1\beta_1 + X_2\beta_2$ and normal ϵ, and the smaller linear model using X_1 alone as explanatory variables, where X_1 is $n \times k_1$ and X_2 is $n \times k_2$. The standard treatment using the methods of section 1 is to test the null hypothesis $\beta_2 = 0$, but a question remains as to what level of significance to use. An alternative viewpoint is to choose the model which is estimated to have smaller prediction errors. Specifically, let n future, out-of-sample, observations be:

$$\tilde{Y} = \tilde{X}\beta + \tilde{\epsilon} \quad (20)$$

under the assumption that the larger model (1) is the true model. Let the model be selected which has a smaller expected sum of squared prediction errors.

Using the small model with X_1 alone and denoting the corresponding maximum-likelihood estimate of β by $\hat{\beta}_1$ [consisting of $(X_1' X_1)^{-1} X_1' Y$ and 0], one easily evaluates $E(\hat{\beta}_1 - \beta)(\hat{\beta}_1 - \beta)'$. Then using the estimated small model and the predictor $\tilde{X}\beta_1$ for $\tilde{Y}$, one finds the expected sum of squared prediction errors to be:

$$E(\tilde{X}\hat{\beta}_1 - \tilde{Y})'(\tilde{X}\hat{\beta}_1 - \tilde{Y})$$
$$= E(\hat{\beta}_1 - \beta)'\tilde{X}'\tilde{X}(\hat{\beta}_1 - \beta) + E\tilde{\epsilon}'\tilde{\epsilon}$$
$$= k_1\sigma^2 + \beta_2' X_2'[I - X_1(X_1' X_1)^{-1} X_1']X_2\beta_2 + n\sigma^2 \quad (21)$$

Using the large model (1) and letting $\hat{\beta} = (X'X)^{-1}X'$, we have:

$$E(X\hat{\beta} - \tilde{Y})'(\tilde{X}\hat{\beta} - \tilde{Y}) = (k_1 + k_2)\sigma^2 + n\sigma^2 \quad (22)$$

Comparing (21) and (22), we find that the small model, though not being the true model, should be used if and only if:

$$\beta_2' X_2'[I - X_1(X_1' X_1)^{-1} X_1']X_2\beta_2 \equiv \beta_2' X_{2\cdot 1}' X_{2\cdot 1}\beta_2 < k_2\sigma^2 \quad (23)$$

where $X_{2\cdot 1}$ is the matrix of residuals of the regression of X_2 on X_1. To apply the criterion (23), one may replace $\beta_2' X_{2\cdot 1}' X_{2\cdot 1}\beta_2$ by its unbiased estimate $\hat{\beta}_2' X_{2\cdot 1}' X_{2\cdot 1}\hat{\beta}_2 - k_2\sigma^2$, and replace σ^2 in the resulting inequality by the unbiased estimate s^2 to yield:

$$\hat{\beta}_2' X_{2\cdot 1}' X_{2\cdot 1}\hat{\beta}_2 < 2k_2 s^2$$
$$\equiv 2k_2(Y - X\hat{\beta})'(Y - X\hat{\beta})/(n - k_1 - k_2) \quad (24)$$

as the condition for selecting the small model. This criterion amounts to setting the critical value of the F ratio $\hat{\beta}_2' X_{2\cdot 1}' X_{2\cdot 1}\hat{\beta}_2/k_2 s^2$ for testing the null hypothesis $\beta_2 = 0$ equal to 2. It is the C_p criterion of Mallows (1973) and is motivated by the desire for more accurate prediction. Comparing (21) and (22) we observe that omitting the variables X_2 might yield a better model for prediction even when (1) is the true model and $\beta_2 \neq 0$.

The information criterion of Akaike (1973, 1974) is also motivated by the desire for more accurate prediction. However, instead of using the expected squared prediction errors, one uses the following expected information:

$$E[\ln g(\tilde{y}, \theta_0) - \ln f(\tilde{y}, \theta)] \quad (25)$$

to measure how good the density function $f(\cdot)$ of the model used for predicting a future observation y is, as compared with the true model $g(\cdot)$. Akaike has implemented this criterion by estimating (25), suggesting the criterion for selecting a model if its maximum log likelihood minus the number of estimated parameters is the highest among the competing models. A model having more parameters will tend to have a higher value for its maximum log likelihood, but this value has to be reduced by the number of parameters estimated. Sawa (1978) has provided a better estimate of (25) for linear regression models while Chow (1981a, b) has provided better estimates of (25) for general statistical models and simultaneous-equation models.

5. THE POSTERIOR-PROBABILITY CRITERION. Another criterion for selecting models is the Jeffrey–Bayes posterior-probability criterion. Let $p(M_j)$ be the prior probability for model M_j to be correct and $p(\theta | M_j)$ be the prior density for the k_j-dimensional parameter vector θ_j conditioned on M_j being correct. Assume that a random sample of n observations $(y_1, y_2, \ldots, y_n) = Y$ is available. By Bayes's theorem the posterior probability of the jth model being correct is:

$$p(M_j | Y) = \frac{p(M_j)p(Y | M_j)}{p(Y)} = \frac{p(M_j)p(Y | M_j)}{\sum_j p(M_j)p(Y | M_j)} \quad (26)$$

where

$$p(Y | M_j) = \int L_j(Y, \theta)p(\theta | M_j)\,d\theta \quad (27)$$

with $L_j(Y, \theta_j)$ denoting the likelihood function for the jth model. Since $p(Y)$ is a common factor for all models, the model with the highest posterior probability of being correct is the one with the maximum value for:

$$p(M_j)p(Y | M_j) = p(M_j)\int L_j(Y, \theta)p(\theta | M_j)\,d\theta$$

If the prior probabilities $p(M_j)$ are equal for the models, the one with the highest $p(Y | M_j)$ will be selected.

To evaluate $p(Y | M_j)$ for large samples we apply a theorem of Jeffreys (1961, pp. 193ff.) on the posterior density $p(\theta | Y, M_j)$ of θ_j given model M_j:

$$p(\theta | Y, M_j) = \frac{L_j(Y, \theta)p(\theta | M_j)}{p(Y | M_j)}$$
$$= (2\pi)^{-k_j/2}|S|^{1/2}\exp[-\tfrac{1}{2}(\theta - \hat{\theta}_j)'S(\theta - \hat{\theta}_j)]$$
$$\times [1 + 0(n^{-1/2})] \quad (28)$$

where $\hat{\theta}_j$ is the maximum-likelihood estimate of θ_j and the inverse covariance matrix is $S = -[(\partial^2 \ln L_j)/(\partial\theta\,\partial\theta')]_{\hat{\theta}} \equiv nR_j$. $0(n^{-1/2})$ is a function of order $n^{-1/2}$. Thus, for large samples, the posterior density of a parameter vector θ in model j is asymptotically normal with mean equal to the maximum-likelihood estimate $\hat{\theta}_j$ and covariance matrix which can be approximated by the inverse of S. Evaluating both sides of (28) at $\theta = \hat{\theta}_j$ and taking natural logarithms, we obtain, noting $|S| = |nR_j| = n^{k_j}|R_j|$,

$$\ln p(Y | M_j) = \ln L_j(Y, \hat{\theta}_j) - \frac{k_j}{2}\ln n - \tfrac{1}{2}\log|R_j|$$
$$+ \frac{k_j}{2}\ln 2\pi + \ln p(\hat{\theta}_j | M_j) + 0(n^{-1/2}) \quad (29)$$

If we retain only the first two terms $\ln L_j(Y, \hat{\theta}_j)$ and $-k_j(\tfrac{1}{2}\ln n)$ in (29), we obtain the formula of Schwarz (1978) for approximating $\log p(Y | M_j)$.

In practice $\ln p(Y | M_j)$ may not be well approximated by using only the first two terms of (29), as it will depend on the prior density $p(\theta | M_j)$ of the parameter vector chosen for each model M_j. Bayesian statisticians, including Jeffreys (1961), Pratt (1975), and Leamer (1978), among others, have recognized the difficult problem of choosing a prior distribution $p(\theta | M_j)$ for the parameters of each model to be used to compute $p(Y | M_j)$. Unlike the estimation of parameters by Bayesian methods, even for large samples the choice of models by the posterior-probability criterion is very sensitive to the prior distribution $p(\theta | M_j)$ assumed for each model.

In this essay I have summarized some of the important ideas and methods employed in hypothesis testing and model selection in econometrics. The choice of an econometric model is a complicated subject. Many approaches have to be explored in practice for choosing and evaluating econometric models. Some of these approaches are discussed in Chow and Corsi (1982) and in Belsley and Kuh (1986).

GREGORY C. CHOW

See also ECONOMETRICS; INFORMATION THEORY; LIKELIHOOD; NON-NESTED HYPOTHESES; REGRESSION AND CORRELATION ANALYSIS; STATISTICAL INFERENCE.

BIBLIOGRAPHY

Akaike, H. 1973. Information theory and an extension of the maximum likelihood principle. In *Proceedings of the 2nd International Symposium for Information Theory*, ed. B. Petrov and F. Cśaki, Budapest: Akademiai Kiadó.

Akaike, H. 1974. A new look at the statistical model identification, *IEEE Transactions on Automatic Control* AC-19, 716–23.

Belsley, D. and Kuh, E. 1986. *Model Reliability*. Cambridge, Mass.: MIT Press.

Belsley, D., Kuh, E. and Welsch, R. 1980. *Regression Diagnostics*. New York: Wiley.

Box, G. and Jenkins, G.M. 1970. *Time-Series Analysis: Forecasting and Control*. San Francisco: Holden-Day.

Chow, G. 1960. Tests of equality between sets of coefficients in two linear regressions. *Econometrica* 28, 591–605.

Chow, G. 1980. The selection of variates for use in prediction: a generalization of Hotelling's solution. In *Quantitative Econometrics and Development*, ed. L. Klein, M. Nerlove and S.C. Tsiang, New York: Academic Press.

Chow, G. 1981a. A comparison of the information and posterior probability criteria for model selection. *Journal of Econometrics* 16, 21–33.

Chow, G. 1981b. Evaluation of econometric models by decomposition and aggregation. In *Methodology of Macro-Economic Models*, ed. J. Kmenta and J. Ramsey, Amsterdam: North-Holland.

Chow, G. 1983. *Econometrics*. New York: McGraw-Hill.

Chow, G. and Corsi, P. (eds) 1982. *Evaluating the Reliability of Macro-Economic Models*. London: Wiley.

Cox, D. 1961. Tests of separate families of hypotheses. In *Proceedings of the 4th Berkeley Symposium on Mathematical Statistics and Probability*, Berkeley: University of California Press.

Cox, D. 1962. Further results on tests of separate families of hypotheses. *Journal of the Royal Statistical Society* Series B24, 406–24.

Hausman, J. 1978. Specification tests in econometrics. *Econometrica* 46, 1251–72.

Jeffreys, H. 1961. *Theory of Probability*. 3rd edn, Oxford: Clarendon Press.

Leamer, E. 1978. *Specification Searches*. New York: Wiley.

Mallows, C. 1973. Some comments on C_p. *Technometrics* 15, 661–75.

Newey, W. 1985. Maximum likelihood specification testing and conditional moment tests. *Econometrica* 53, 1047–70.

Neyman, J. and Pearson, E. 1928. On the use of interpretation of certain test criteria for the purpose of statistical inference. *Biometrika* 20A, Part I, 175–240; Part II, 263–294.

Pratt, J. 1975. Comments. In *Studies in Bayesian Econometrics and Statistics*, ed. S. Fienberg and A. Zellner, Amsterdam: North-Holland.

Quandt, R. 1974. A comparison of methods for testing nonnested hypotheses. *Review of Economics and Statistics* 56, 92–9.

Sawa, T. 1978. Information critiera for discriminating among alternative regression models. *Econometrica* 46, 1273–92.

Schwarz, G. 1978. Estimating the dimension of a model. *Annals of Statistics* 6, 461–4.

Silvey, S. 1959. The Lagrangian multiplier test. *Annals of Mathematical Statistics* 30, 389–407.

Wald, A. 1943. Tests of statistical hypotheses concerning several parameters when the number of observations is large. *Transactions of the American Mathematical Society* 54, 426–82.

White, H. 1982. Maximum likelihood estimation of misspecified models. *Econometrica* 50, 1–25.

Wu, D. 1973. Alternative tests of independence between stochastic regressors and disturbances. *Econometrica* 41, 733–50.

I

ideal indexes. Among many index numbers, the two most favoured because of algebraic simplicity and ease of computation are those advocated by E. Laspeyres in 1864 and by H. Paasche in 1874. There are n commodities, indexed from 1 to n. At time point t, the price vector is $p_t = \{p_{1t}, \ldots, p_{nt}\}$ and the quantity vector $q_t = \{q_{1t}, \ldots, q_{nt}\}$. $p_s q_t$ denotes $\Sigma_{i=1}^{n} p_{is} q_{it}$. Let P_{st} and Q_{st} be the price and quantity indexes from time s to t. Then, these two indexes are

$$\text{Laspeyres } P_{st}^{L} = p_t q_s / p_s q_s, \qquad Q_{st}^{L} = p_s q_t / p_s q_s$$

$$\text{Paasche } P_{st}^{P} = p_t q_t / p_s q_t, \qquad Q_{st}^{P} = p_t q_t / p_t q_s$$

There are several desirable properties that an index ought to satisfy (Samuelson and Swamy, 1974; Allen, 1975, pp. 40–47). Three basic tests (stated for the price index) which any reasonable index must meet are:

(1) Identity test: $P_{tt} = 1$.
(2) Proportionality test: $P_{st'} = kP_{st}$, when $p_{it'} = kp_{it}$, $q_{it'} = q_{it}$ for all i.
(3) Dimensional test: changes of units do not affect the index value.

The next three are not always satisfied:

(4) Time-reversal test: $P_{st} P_{ts} = 1$.
(5) Circular test: $P_{rs} P_{st} = P_{rt}$.
(6) Factor-reversal test: $P_{st} Q_{st} = E_{st}$, where $E_{st} = p_t q_t / p_s q_s$ is the expenditure index and P and Q are matching indexes in the sense that they share a common form except that p and q are interchanged between them.

Irving Fisher (1922), who most energetically pursued the topic of index numbers, emphasized the factor-reversal test and regarded PQ/E (where P and Q are matching indexes) as the bias of an index. Very few indexes satisfy (6). For the Laspeyres and Paasche indexes, the following identities are seen to hold:

$$P_{st}^{L} Q_{st}^{P} = P_{st}^{P} Q_{st}^{L} = E_{st},$$

i.e., the Laspeyres and Paasche indexes are 'factor antitheses'. Then, their geometric averages

$$P_{st}^{F} = \sqrt{P_{st}^{L} P_{st}^{P}}, \qquad Q_{st}^{F} = \sqrt{Q_{st}^{L} Q_{st}^{P}}$$

satisfy (6). Fisher regarded this index to be the best or 'ideal' among 134 indexes he compared. This index has been known as Fisher's ideal index even though he was not the only one who discussed this index at the time.

A log-change index has also been popular. It is given the form

$$\ln P_{st} = \sum_{i} s_i (\ln p_{it} - \ln p_{is}),$$

$$\ln Q_{st} = \sum_{i} s_i (\ln q_{it} - \ln q_{is})$$

where $s_i \geq 0$, $\Sigma s_i - 1$. Expenditure shares are used for weights. Let w_{it} be the share of good i in total expenditure at time t. Loglinear analogues of the Laspeyres, Paasche, and Fisher indexes are obtained by setting (i) $s_i = w_{is}$, (ii) $s_i = w_{it}$, and (iii) $s_i = \frac{1}{2}(w_{is} + w_{it})$. The last one, which is attributed to Törnqvist, does not satisfy the factor-reversal test.

Log-change indexes may be considered as discrete approximations to the continuous Divisia index obtained by integrating

$$d \ln P = \sum_{i} w_i d \ln p_i, \quad d \ln Q = \sum_{i} w_i d \ln q_i$$

from s to t.

Suppose that (p, q) represents the behaviour of a consumer maximizing utility. Assume that the consumer's utility is represented by a preference function of a certain homogeneous form. It can then be shown that the Divisia index also assumes a certain form. This index is said to be 'exact' with the preference function (Diewert, 1976). (The Laspeyres is exact with a linear utility function, the Paasche with a Leontief-type utility function, and the Törnqvist with a translog utility function.)

The correspondence between a preference function and an index can be given the following heuristic argument: The preference ordering can be represented either in a direct form $[U(q)]$ or in an indirect form $[V(E/p)]$. Interpreting the quantity index as a constant-utility index, we have $Q_{st} = U(q_t)/U(q_s)$. By the same token, the price index is associated with the indirect utility function so that $P_{st} = V(E/p_t)/V(E/p_s)$. When U and V are alternative representations of a preference function, they form a dual pair. P and Q which are exact with them are factor antitheses, namely, $P_{st} Q_{st} = E_{st}$. As P and Q are not in general matching indexes, they do not meet the factor reversal test.

When the duel pair, U and V, share a common functional form, they are called 'self-dual' (Houthakker, 1965). It then follows that Q and P which are exact with the dual pair must also share a common form, i.e., they are matching indexes. Thus, an important proposition holds: there are as many ideal index numbers as there are self-dual preference functions as they are equivalent to each other. There are only three known self-dual preference functions: (a) Cobb–Douglas, (b) quadratic, and (c) constant-elasticity-of substitution (CES). Ideal indexes which correspond to these are as follows:

(a) A log-change index with fixed weights. The weights are exponents of the Cobb–Douglas. Since expenditure shares do not remain constant over time, this index violates reality.

(b) Fisher's ideal index. This correspondence was noted by Konüs and Byushgens already in the 1920s (Afriat, 1977).

(c) A log-change index with variable weights where s_i is given by $(w_{it} - w_{is})/(\ln w_{it} - \ln w_{is})$, divided by its sum over i. Though complicated in form, these weights are seen to be in the nature of geometric averages. This index was discovered independently by Sato (1976) and Vartia (1976).

No other self-dual preferences not ideal indexes have been discovered since.

Kazuo Sato

See also INDEX NUMBERS.

BIBLIOGRAPHY

Afriat, S.N. 1977. *The Price Index.* London: Cambridge University Press.

Allen, R.G.D. 1975. *Index Numbers in Theory and Practice*. Chicago: Aldine.

Diewert, W.E. 1976. Exact and superlative index numbers. *Journal of Econometrics* 4, 115–45.

Fisher, I. 1922. *The Making of Index Numbers*. Boston: Houghton Mifflin.

Houthakker, H.S. 1965. A note on self-dual preferences. *Econometrica* 33, October, 797–801.

Samuelson, P.A. and Swamy, S. 1974. Invariant economic index numbers and canonical duality: survey and synthesis. *American Economic Review* 64, September, 566–93.

Sato, K. 1976. The ideal log-change index number. *Review of Economics and Statistics* 58, May, 223–8.

Vartia, Y.O. 1976. Ideal log-change index numbers. *Scandinavian Journal of Statistics* 3, 121–6.

ideal output. Pigou, writing in *The Economics of Welfare*, calls 'the output in any industry which maximizes the national dividend, and, apart from the differences in the marginal utility of money to different people, also maximises satisfaction, the ideal output'. He goes on to argue that 'this output is attained – the possibility of multiple maximum positions being ignored – when the value of the marginal social net product of each sort of resource invested in the industry under review is equal to the value of the marginal social net product of resources in general'. And, finally, it 'will be that output which makes the demand price of the output equal to the money value of the resources engaged in producing a marginal unit of output' (1932, pp. 802, 803).

The line of argument that comes through so clearly in these quotations can be traced back to Pigou's earlier *Wealth and Welfare* (1912) and indeed to Marshall; but in the last fifty years it has been overtaken by the development of a more powerful strand of analysis that stems from Pareto (1897) and Barone (1908) and has culminated in the theory of the General Optimum of Production and Exchange. In it one maximizes in turn the welfare of each member of the community, subject to the constraint of the social production function and to holding on each occasion the welfare of each other member constant. The resulting first order conditions include the marginal equivalences enumerated in the theory of ideal output (Graaff, 1957). Any modern discussion of the theory must therefore be set against the background of the one that has incorporated and replaced it.

The more modern theory has the virtues of elegance, simplicity and generality. It embraces exchange as well as production. It deals with commodities and firms (or event plants) instead of industries. It does not need the doctrine of maximum satisfaction, or any assumption about inter-personal comparisons of utility. But at the end of the day it does not reach any substantial conclusion that the theory of ideal output, correctly employed, would not itself have reached.

The problem, especially in the early development of the theory, was that it was not all that easy to apply it correctly. It was not originally recognized that (at least in a closed economy) the correct way to reckon the value of a marginal social net product is at *constant* prices. The same remark applies to the calculation of marginal social cost. If higher prices have to be offered to factors of production to attract them to an industry undergoing expansion, the element of the cost of the expansion caused by the higher prices represents a transfer payment to the factors (in the form of a rent or quasi-rent), not a cost to society. The cost to society is the value of the output sacrificed when the factors are withdrawn from their previous use. That value was reckoned at the original prices of the factors. Those prices must therefore be used in reckoning their cost to society in their new use.

Clarification of this issue was the result of a famous debate of the 1920s – much of it reprinted in *Readings in Price Theory* (Stigler (ed.), 1953) – on the desirability of taxing industries subject to diminishing returns and paying bounties to those subject to increasing returns, a result to which the theory of ideal output at one stage seemed to point. As competitive conditions were meant to be prevailing, the industries enjoying increasing returns had to be assumed to comprise firms whose unit costs were falling because of *external* economies; and as external economies were themselves recognized as possible reasons for a divergence between private and social net products, the opportunities for getting muddled were legion. It is to the credit of the participants – among them D.H. Robertson, G.F. Shove, F.H. Knight and J. Viner – that these dangers were largely avoided.

Much of the motivation for the theory of ideal output seems to have been a desire to see when competitive output was ideal, and when interference in a competitive economy would be justified. Today we ask, rather more formally (cf. Debreu, 1959), when a competitive equilibrium would also be a General Optimum. The answer, very briefly, is when the technology is convex, there are no external effects in production or consumption, no public goods and no foreign trade.

Apart from the fact that the existence of public goods was glossed over, ideal output theory would not have given a very different answer. The importance of the foreign trade exception was recognized. (The marginal social cost of importing goods subject to rising supply price is higher than the marginal private cost. The rents that accrue to *foreigners* are not mere transfers within the domestic community, but a part of social cost). Divergences between private and social costs due to external economies and diseconomies in production, and between private and social benefits due to external economies and diseconomies in consumption, were fully discussed. The counterpart of the modern insistence on a convex technology was the painstaking treatment of increasing returns. The conditions under which competitive output would approach the ideal were pretty clearly defined.

Pigou also discussed the deviation from the ideal of the outputs of discriminating monopolists. (Not surprisingly, they fell short.) R.F. Kahn (1935) extended the analysis to imperfect competition. He argued that (taking diseconomies as negative economies) all industries could be arranged in descending order on a scale according to the extent of the external economies they generated and the degree of monopoly (measured by the gap between price and marginal cost) they enjoyed and that at a certain point on the scale there would be an average industry. Above this point all should expand to produce ideal outputs; below it all should contract. Adjustment could be achieved by a set of taxes and bounties. When all industries had expanded or contracted to conform to the average degree of monopoly and the average capacity to create external economies, their marginal social products would diverge from their marginal private products to the same extent and ideal output would be attained.

Note that this treatment avoids the error of making 'piecemeal' recommendations of the sort so often found in partial analysis. All industries must move to the average. It may not help if one or two do. That may just increase the gap between those that conform and those that do not. (In technical terms, the first order conditions for a maximum must be satisfied simultaneously.)

In this sense Kahn's treatment is very general. In another it is not general enough. Proportionality of marginal products is not sufficient. For a full optimum, equality is essential (Lerner, 1944, ch. 9). This may require an adjustment in the number of hours worked, and an expansion or contraction in the level of output as a whole.

The view that suitable corrective taxes and bounties can and should be used to bring marginal private products into line with marginal social products, when they diverge, was once very popular. On the whole it has weathered less well than ideal output theory itself, although the latter is by now no more than an episode in the history of economic thought.

J. DE V. GRAAFF

See also PARETO EFFICIENCY; TAXES AND SUBSIDIES.

BIBLIOGRAPHY
Barone, E. 1908. Il Ministerio della Produzione nello stato colletivista. *Giornale degli Economisti*. Trans. in *Collectivist Economic Planning*, ed. F.A. Hayek, London: Routledge, 1935.
Debreu, G. 1959. *Theory of Value*. New York: Wiley.
Graaff, J. de V. 1957. *Theoretical Welfare Economics*. Cambridge: Cambridge University Press.
Kahn, R.F. 1935. Some notes on ideal output. *Economic Journal* 45, March, 1–35.
Lerner, A.P. 1944. *The Economics of Control*. New York: Macmillan.
Pareto, V. 1897. *Cours d'économie politique*. Lausanne: Rouge.
Pigou, A.C. 1912. *Wealth and Welfare*. London: Macmillan.
Pigou, A.C. 1932. *The Economics of Welfare*. 4th edn, London: Macmillan.
Stigler, G.J. (ed., with K.E. Boulding) 1953. *Readings in Price Theory*. London: Allen & Unwin.

ideal type. This is the term used by Max Weber to describe the distinctive concepts and models developed by economic and social theorists, and employed in the activity of empirical analysis. The term also defines the characteristic method which Weber saw as distinctive of the social sciences. Social life is infinitely complex and can never be exhaustively described or explained. In order to make sense of it, the social scientist uses artificially pure concepts, e.g. 'natural economy', 'handicraft', 'capitalism', which are intellectual constructs involving a high degree of abstraction from the actual world. They comprise the most typical elements which have been isolated from a historically repeated pattern of action, relationship or institution, as seen from a partial point of view (economic, political, etc.), and combined into an internally consistent and inherently intelligible unity. With the help of such constructs the social scientist is able to characterize a particular object of study, and make its complexity intelligible according to its degree of conformity to the stipulations of the relevant concept or model. Often a particular social complex will require a combination of such concepts for its elucidation, as for example the class structure of a given society can be understood as a combination of the analytically separable elements of property ownership ('class'), social esteem ('status') and authority position ('power'). Ideal types have nothing to do with ideals (though there can be ideal-type of ideals, e.g. 'individualism') and could perhaps less confusingly be called 'pure types'.

Weber's characterization of the ideal-type method is best understood in the context of the 'Methodenstreit' between the historical and theoretical schools of German political economy. He developed it to rebut what he saw as a mistaken understanding of theory on the part of certain members of the historical school. In their view economic theory should involve the quest for universal laws of a natural-scientific kind, arrived at inductively on the basis of exhaustive empirical studies of economic phenomena. They saw the work of the historical school as the necessary preliminary stage to the discovery of such laws. Measured against this conception, the theoretical work of Carl Menger and the marginal utility school was judged to be excessively abstract, one-sided in its assumptions about human nature, and above all premature.

Weber's ideal-type method provided a critique of this 'scientistic' understanding of economic and social theory. The focus of interest of the social scientist, he argued, lay in the historically specific, not the most general, aspects of phenomena. The latter were both the most banal and the least useful for explanatory purposes. The distinctive method of social-scientific abstraction involved not a quest for universal laws, but a process of isolating what was most typical and essential to a pattern of action or social relation, and rendering it intelligible as an internally coherent whole. It was by the same method of abstraction that the typical historical preconditions and consequences of a given social institution were to be elucidated. Weber argued that this was in practice the method adopted by economic theoreticians, Marxists and marginalists alike, though they did not always recognize its implications. The error of the former was to treat their theoretical deductions about the typical consequences of capitalist competition as actual historical tendencies or laws, in advance of any empirical confirmation. The error of the latter was their failure to recognize the actual historical preconditions for the rigorous calculation and maximization of economic interests, which made their theoretical models historically specific rather than applicable to all times and places (Weber, 1903, 1904a).

In this manner Weber's account of the ideal-type method offered a resolution of the controversy between the historical and theoretical schools. On the one hand it demonstrated the historical specificity of even the most abstract theorizing. On the other hand it revealed the irreducibly theoretical character of the concepts used in historical economics, which was anything but a merely descriptive activity on whose successful 'completion' the construction of theory was itself supposedly dependent. Properly understood, the respective emphases of theory and history were mutually complementary, a conjunction which Weber's own work such as *The Protestant Ethic and the Spirit of Capitalism* (Weber, 1904b) or the more theoretical formulations of *Economy and Society* (Weber, 1921) amply demonstrated.

Subsequent discussion of the ideal-type method has taken place within sociology and political science, rather than within the discipline of economics, which provided its original intellectual location. Most social scientists would accept the necessity for typological construction, but disagree over both its manner and the criteria for its assessment. Weber's approach has been criticized for its inherent subjectivity, in two quite different senses. First, his method of 'Verstehen' or 'understanding', which is necessary for assessing the internal coherence of ideal-type constructs, has been seen as unavoidably arbitrary. To this it can be simply replied that the criteria for the intelligibility of social action are interpersonal, not private, despite the obvious difficulties in respect of alien cultures.

Secondly, following the neo-Kantianism of Heinrich Rickert, Weber argued that the objects of study and hence the concepts used in the social sciences are determined according to their 'value-relevance', i.e. their significance for our values. Unlike Rickert, however, he did not believe that these value standpoints could be objectively grounded in human reason.

Some commentators have therefore concluded that it is impossible to rescue Weberian concept formation from the subjectivity of the investigator's own values. Such a conclusion overlooks Weber's insistence that ideal-type constructs must satisfy the criterion of explanatory power as well as of significance, and thus 'be valid for all who seek the truth'. The ultimate test for ideal-type construction must be an objective one: its fruitfulness in identifying and resolving explanatory problems.

DAVID BEETHAM

See also WEBER, MAX.

BIBLIOGRAPHY

Weber, M. 1903. *Roscher and Knies: The Logical Problems of Historical Economics.* Ed. G. Oakes, New York: Free Press, 1975.

Weber, M. 1904a. 'Objectivity' in social science. In *The Methodology of the Social Sciences*, ed. E.A. Shils and H.A. Finch, New York: Free Press, 1959.

Weber, M. 1904b. *The Protestant Ethic and the Spirit of Capitalism.* London: Allen & Unwin, 1930.

Weber, M. 1921. *Economy and Society.* New York: Bedminster Press, 1968.

identification. In economic analysis we often assume that there exists an underlying structure which generated the observations of real-world data. However, statistical inference can relate only to characteristics of the distribution of the observed variables. A meaningful statistical interpretation of the real world through this structure can be achieved only if there is no other structure which is also capable of generating the observed data.

To illustrate, consider X as being normally distributed with mean $E(X) = \mu_1 - \mu_2$. Then $\mu_1 - \mu_2$ can be estimated using observed X. But the parameters μ_1 and μ_2 are not uniquely estimable. In fact, one can think of an infinite number of pairs (μ_i, μ_j), $i, j = 1, 2, \ldots, (i \neq j)$ such that $\mu_i - \mu_j = \mu_1 - u_2$. In order to determine $\mu_1 - \mu_2$ uniquely, we need additional prior information, such as $\mu_2 = 3\mu_1$ or some other assumption.

The problem of whether it is possible to draw inferences from the probability distribution of the observed variables to an underlying theoretical structure is the concern of econometric literature on identification. The first economist to raise this issue was Working (1925; 1927). The general formulations of the identification problems were made by Frisch (1934), Marschak (1942), Haavelmo (1944), Hurwicz (1950), Koopmans and Reiersøl (1950), Koopmans, Rubin and Leipnik (1950), Wald (1950) and many others. An extensive treatment of the theory of identification in simultaneous equation systems was provided by Fisher (1966). A survey of recent advances in the subject can be found in Hsiao (1983).

1 DEFINITIONS

It is generally assumed in econometrics that economic variables whose formation an economic theory is designed to explain have the characteristics of random variables. Let $\mathbf{y}$ be a set of such observations. A structure S is a complete specification of the probability distribution function of $\mathbf{y}$. The set of all a priori possible structures, T, is called a model. In most applications, $\mathbf{y}$ is assumed to be generated by a parametric probability distribution function $F(\mathbf{y}, \boldsymbol{\theta})$, where the probability distribution function F is assumed known, but the $m \times 1$ parameter vector $\boldsymbol{\theta}$ is unknown. Hence, a structure is

described by a parametric point $\boldsymbol{\theta}$, and a model is a set of points $A \subset R^m$.

Definition 1: Two structures, $S^0 = F(\mathbf{y}, \boldsymbol{\theta}^0)$ and $S^* = F(\mathbf{y}, \boldsymbol{\theta}^*)$ are said to be *observationally equivalent* if $F(\mathbf{y}, \boldsymbol{\theta}^0) = F(\mathbf{y}, \boldsymbol{\theta}^*)$ for ('almost') all possible $\mathbf{y}$. A model is identifiable if A contains no two distinct structures which are observationally equivalent. A function of $\boldsymbol{\theta}, g(\boldsymbol{\theta})$, is identifiable if all observationally equivalent structures have the same value for $g(\boldsymbol{\theta})$.

Sometimes a weaker concept of identifiability is useful.

Definition 2: A structure with parameter value $\boldsymbol{\theta}^0$ is said to be *locally identified* if there exists an open neighborhood of $\boldsymbol{\theta}^0, \Omega$, such that no other $\boldsymbol{\theta}$ in Ω is observationally equivalent.

2 GENERAL RESULTS

Lack of identification is a reflection that a random variable has the same distribution for all values of the parameter. R. A. Fisher's information matrix provides a measure of sensitivity of the distribution of a random variable due to small changes in the value of the parameter point (Rao, 1962). It can, therefore, be shown that subject to the regularity conditions, $\boldsymbol{\theta}^0$ is locally identified if and only if the information matrix evaluated at $\boldsymbol{\theta}^0$ is nonsingular (Rothenberg, 1971).

It is clear that unidentified parameters cannot be consistently estimated. There are also pathological cases where identified models fail to possess consistent estimators (e.g. Gabrielson, 1978). However, for most practical cases we may treat identifiability and the existence of a consistent estimator as equivalent (for precise conditions, see Le Cam, 1956; Deistler and Seifert, 1978).

3 SPECIFIC MODELS

The choice of model structure is one of the basic ingredients in the formulation of the identification problem. In this section we briefly discuss some identification conditions for different types of models in order to demonstrate the kind of prior restrictions required.

3.1 Linear models. Consider a theory which predicts a relationship among the variables as

$$B\mathbf{y}_t + \Gamma \mathbf{x}_t = \mathbf{u}_t, \qquad (1)$$

where $\mathbf{y}_t$ and $\mathbf{u}_t$ are $G \times 1$ vectors of observed and unobserved random variables, respectively, $\mathbf{x}_t$ is a $K \times 1$ vector of observed non-stochastic variables, B and Γ are $G \times G$ and $G \times K$ matrices of coefficients, with B being nonsingular. We assume that $\mathbf{u}_t$ is independently normally distributed with mean $\mathbf{O}$ and variance-covariance matrix Σ. Equations (1) are called *structural equations*. Solving the endogenous variables, $\mathbf{y}$, as a function of the exogenous variables, $\mathbf{x}$, and the disturbance $\mathbf{u}$, we obtain

$$\mathbf{y}_t = -B^{-1}\Gamma\mathbf{x}_t + B^{-1}\mathbf{u}_t$$
$$= \Pi\mathbf{x}_t + v_t, \qquad (2)$$

with $E v_t = 0$, $E v_t v_t' = V = B^{-1}\Sigma B^{-1'}$. Equations (2) are called the *reduced form* equations derived from (1) and give the conditional likelihood of $\mathbf{y}_t$ for given $\mathbf{x}_t$.

Premultiplying (1) by a $G \times G$ nonsingular matrix D, we get a second structural equation

$$B^*\mathbf{y}_t + \Gamma^*\mathbf{x}_t = \mathbf{u}_t^*, \qquad (3)$$

where $B^* = DB$, $\Gamma^* = D\Gamma$, and $\mathbf{u}_t^* = D\mathbf{u}_t$. It can be readily seen that (3) has the same reduced form (2) as (1). Therefore, the two structures are observationally equivalent and the model is non-identifiable.

To make the model identifiable, additional prior restrictions have to be imposed on the matrices B, Γ and/or Σ. Consider the problem of estimating the parameters of the first equation in (1), out of a system of G equations. If the parameters cannot be estimated, the first equation is called *unidentified* or *underidentified*. If given the prior information, there is a unique way of estimating the unknown parameters, the equation is called *just identified*. If the prior information allows the parameters to be estimated in two or more linearly independent ways, it is called *overidentified*. A necessary condition for the first equation to be identified is the number of restrictions on this equation be no less than $G - 1$ (order condition). A necessary and sufficient condition is that a specified submatrix of B, Γ and Σ be of rank $G - 1$ (rank condition) (Fisher, 1966; Hausman and Taylor, 1983). For instance, suppose the restrictions on the first equation are in the form that certain variables do not appear. Then this rank condition says that the first equation is identified if and only if the submatrix obtained by taking the columns of B and Γ with prescribed zeros in the first row is of rank $G - 1$ (Koopmans and Reiersøl, 1950).

3.2 Dynamic models. When both lagged endogenous variables and serial correlation in the disturbance term appear, we need to impose additional conditions to identify a model. For instance, consider the following two equation system (Koopmans, Rubin and Leipnik, 1950)

$$y_{1t} + \beta_{11}y_{1,t-1} + \beta_{12}y_{2,t-1} = u_{1t},$$
$$\beta_{21}y_{1t} + y_{2t} = u_{2t}. \tag{4}$$

If (u_{1t}, u_{2t}) are serially uncorrelated, (4) is identified. If serial correlation in (u_{1t}, u_{2t}) is allowed, then

$$y_{1t} + \beta_{11}^{*}y_{1,t-1} + \beta_{12}^{*}y_{2,t-1} = u_{1t}^{*},$$
$$\beta_{21}y_{1t} + y_{2t} = u_{2t}, \tag{5}$$

is observationally equivalent to (4), where $\beta_{11}^{*} = \beta_{11} + d\beta_{21}$, $\beta_{12}^{*} = \beta_{12} + d$, and $u_{1t}^{*} = u_{1t} + du_{2t}$.

Hannan (1971) derives generalized rank conditions for the identification of this type of model by first assuming that the maximum orders of lagged endogenous and exogenous variables are known, then imposing restrictions to eliminate redundancy in the specification and to exclude transformations of the equations that involve shifts in time. Hatanaka (1975), on the other hand, assumes that the prior information takes only the form of excluding certain variables from an equation, and derives a rank condition which allows common roots to appear in each equation.

3.3 Non-linear models. For linear models we have either global identification or else an infinite number of observationally equivalent structures. For models linear in the variables but nonlinear in the parameters the state of the mathematical art is such that we can only talk about local properties. That is, we cannot tell the true structure from any other substitute; however, we may be able to distinguish it from other structures which are close to it. A sufficient condition for local identification is that the Jacobian matrix formed by taking the first partial derivatives of

$$\omega_i = \Psi_i(\theta), \qquad i = 1, \ldots, n$$
$$0 = \phi_j(\theta), \qquad j = 1, \ldots, R \tag{6}$$

with respect to θ be of full column rank, where ω_i are the n population moments of $\mathbf{y}$ and ϕ_j are the R *a priori* restrictions on θ (Fisher, 1966).

When the Jacobian matrix of (6) has less than full column rank, the model may still be locally identifiable via conditions implied by the higher order derivatives. However, the estimator of a model suffering from first order lack of identification will in finite samples behave in a way that is difficult to distinguish from the behaviour of an unidentified model (Sargan, 1983).

3.4 Bayesian analysis. In Bayesian analysis all quantities, including the parameters, are random variables. Thus, a model is said to be identified in probability if the posterior distribution for θ is proper. When the prior distribution for θ is proper, so is the posterior, regardless of the likelihood function of $\mathbf{y}$. In this sense unidentifiability causes no real difficulty in the Bayesian approach. However, basic to the Bayesian argument is that all probability statements are conditional. That is, it consists essentially in revising the probability of a fixed event in the light of various conditioning events, the revision being accomplished by Bayes' theorem. Therefore, in order for an experiment to be informative with regard to unknown parameters (i.e. the prior to be different from the posterior) the parameter must be identified or estimable in the classical sense and identification remains as a property of the likelihood function (Kadane, 1975).

Drèze (1975) has commented that exact restrictions are unlikely to hold with probability one and has suggested using probabilistic prior information. In order to incorporate a stochastic prior, he has derived necessary rank conditions for the identification of a linear simultaneous equation model.

4 CONCLUDING REMARKS

The study of identifiability is undertaken in order to explore the limitations of statistical inference (when working with economic data) or to specify what sort of a priori information is needed to make model parameters estimable. It is a fundamental problem concomitant with the existence of a structure. Logically it precedes all problems of estimation or of testing hypotheses.

An important point that arises in the study of identification is that without a priori restrictions imposed by economic theory it would be almost impossible to estimate economic relationships. In fact, Liu (1960) and Sims (1980) have argued that economic relations are not identifiable because the world is so interdependent as to have almost all variables appearing in every equation, thus violating the necessary condition for identification. However, almost all the models we discuss in econometrics are only approximate. We use convenient formulations which behave in a general way that corresponds to our economic theories and intuitions, and which cannot be rejected by the available data. In this sense, identification is a property of the model but not necessarily of the real world.

The problem of identification arises in a number of different fields such as automatic control, biomedical engineering, psychology, systems science, etc., where the underlying physical structure may be deterministic (e.g. see Aström and Eykhoff, 1971). It is also aptly linked to the design of experiments (e.g. Kempthorne, 1947; Bailey, Gilchrist and Patterson, 1977). Here, we restrict our discussion to economic applications of statistical identifiability involving random variables.

CHENG HSIAO

See also ECONOMETRICS; ENDOGENEITY AND EXOGENEITY; ESTIMATION; SIMULTANEOUS EQUATIONS MODELS.

BIBLIOGRAPHY

Aström, K.J. and Eykhoff, P. 1971. System identification – a survey. *Automatica* 7, 123–62.

Bailey, R.A., Gilchrist, F.H.L. and Patterson, H.D. 1977. Identification of effects and confounding patterns in factorial designs. *Biometrika* 64, 347–54.

Deistler, M. and Seifert, H. 1978. Identifiability and consistent estimability in econometric models. *Econometrica* 46, 969–80.

Drèze, J. 1975. Bayesian theory of identification in simultaneous equations models. In *Studies in Bayesian Econometrics and Statistics*, ed. S.E. Fienberg and A. Zellner, Amsterdam: North-Holland.

Fisher, F.M. 1966. *The Identification Problem in Econometrics*. New York: McGraw-Hill.

Frisch, R. 1934. *Statistical Confluence Analysis by Means of Complete Regression Systems*. Publication No. 5, Oslo: Universitetes Økonomiske Institutt.

Gabrielson, A. 1978. Consistency and identifiability. *Journal of Econometrics* 8, 261–63.

Haavelmo, T. 1944. The probability approach in econometrics. *Econometrica* 12, Supplement, 1–115.

Hannan, E.J. 1971. The identification problem for multiple equation systems with moving average errors. *Econometrica* 39, 751–65.

Hatanaka, M. 1975. On the global identification of the dynamic simultaneous equations model with stationary disturbances. *International Economic Review* 16, 545–54.

Hausman, J.A. and Taylor, W.E. 1983. Identification, estimation and testing in simultaneous equations models with disturbance covariance restrictions. *Econometrica* 51, 1527–49.

Hsiao, C. 1983. Identification. In *Handbook of Econometrics*, Vol. I, ed. Z. Griliches and M. Intriligator, Amsterdam: North-Holland.

Hurwicz, L. 1950. Generalization of the concept of identification. In *Statistical Inference in Dynamic Economic Models*, Cowles Commission Monograph no. 10, New York: John Wiley.

Kadane, J.B. 1975. The role of identification in Bayesian theory. In *Studies in Bayesian Econometrics and Statistics*, ed. S.E. Fienberg and A. Zellner, Amsterdam: North-Holland.

Kempthorne, O. 1947. A simple approach to confounding and factorial replication in factorial experiments. *Biometrika* 34, 255–72.

Koopmans, T.C. and Reiersøl, O. 1950. The identification of structural characteristics. *Annals of Mathematical Statistics* 21, 165–81.

Koopmans, T.C., Rubin, H. and Leipnik, R.B. 1950. Measuring the equation systems of dynamic economics. In *Statistical Inference in Dynamic Economic Models*, Cowles Commission Monograph No. 10, New York: John Wiley.

Le Cam, L. 1956. On the asymptotic theory of estimation and testing hypotheses. In *Proceedings of the Third Berkeley Symposium on Mathematical Statistics and Probability*, Vol. 1, Berkeley: University of California Press.

Liu, T.C. 1960. Underidentification, structural estimation, and forecasting. *Econometrica* 28, 855–65.

Marschak, J. 1942. Economic interdependence and statistical analysis. In *Studies in Mathematical Economics and Econometrics*, Chicago: University of Chicago Press, 135–50.

Rao, C.R. 1962. Problems of selection with restriction. *Journal of the Royal Statistical Society*, Series B 24, 401–5.

Rothenberg, T.J. 1971. Identification in parametric models. *Econometrica* 39, 577–92.

Sargan, J.D. 1983. Identification and lack of identification. *Econometrica* 51, 1605–33.

Sims, C.A. 1980. Macroeconomics and reality. *Econometrica* 48, 1–48.

Wald, A. 1950. Note on the identification of economic relations. In *Statistical Inference in Dynamic Economic Models*, Cowles Commission Monograph No. 10, New York: John Wiley.

Working, E.J. 1925. The statistical determination of demand curves. *Quarterly Journal of Economics* 39, 503–43.

Working, E.J. 1927. What do statistical 'demand curves' show? *Quarterly Journal of Economics* 41, 212–35.

ideology. Now and then one comes across the claim that, unlike, for example, physics, 'economics is thoroughly permeated by ideology ...' (Ward, 1979, p. viii). The exact import of this claim regarding the epistemological status of economics is not clear, since the noun 'ideology' is employed in a variety of senses. However, it should be stressed at once that, despite occasional criticisms (e.g. McCloskey, 1983, p. 334), most economists long ago accepted Hume's insistence that policy proposals cannot be deduced from descriptive statements alone (Klappholz, 1964) and have therefore stressed the distinction between positive and normative economics. The claim discussed in this essay appears to be directed at both the positive, as well as the normative, parts of economics, but we shall be concerned mainly with its import for positive economics. In section I we interpret the claim that economics is ideological as the view that economic theories can be explained by the social position and attitudes of those who put them forward, that is, by the Sociology of Knowledge (discussed critically in Popper, 1957, chs 23 and 24). In section II we consider the suggestion that ideology is pseudo-science. In section III we consider it as consisting of non-scientific views. Finally, in section IV, we draw on the preceding discussion to appraise the claim that economists' policy proposals are ideological.

I. The pursuit of scientific research is a social activity, and thus must have a sociological dimension. In an epistemological and methodological context, however, interest centres, not on the sociological aspects, but on how to appraise scientific theories. In that context any explanation, even a successful one, of how people's social position causes them to hold certain views and beliefs does not imply anything about the truth of those beliefs (Popper, 1959, pp. 31–2). To see this, consider the proposition, sometimes called 'the principle of sociologism', that *all* theories are ideological. It is sometimes argued (e.g. Popper, 1957, notes 7 and 8 to ch. 24, pp. 353–6) that this proposition implies the contradictory view 'all statements are false', but this may not be the case. It is sufficient to make the more modest inference that all theories are equally arbitrary. But if all theories are ideological, then so is this claim about all theories. Hence this particular theory of ideology is arbitrary, and must be rejected if the idea of objective truth is to be retained. Indeed, this is implicitly conceded when physics is deemed not to be ideological. It then follows that the socio-psychological motives which may induce people to advance certain factual views cannot imply anything about the truth of those views. To suppose the contrary is to commit the genetic fallacy, the fallacy that the truth of statements is decidable on the basis of their originators' motives in uttering them and, perhaps, believing them to be true (Rosenberg, 1976, pp. 202–3). Of course, if it could be shown that economists' adherence to particular theories is conditioned by their social position, or other extraneous factors, and is unrelated to logical and empirical considerations, their methods would indeed be unscientific. Attempts to show this can be found (e.g. Wiles, 1979–80), but they cannot be appraised here.

Mention must be made of an idea related to, but not identical with, the view that all theories are ideological. This view asserts that people can communicate successfully, even within a given subject such as economics, only if they share a common intellectual framework. Sir Karl Popper styled this view 'The Myth of the Framework' (1976). If this view were true, it would imply, for example, that supporters of the rational expectations, market-clearing paradigm of the functioning of a market economy could not communicate successfully with those economists who do not work within that paradigm. A glance at the professional literature shows that the view is false.

II. We saw that, if we use 'ideological' in the sense of section I, we must reject the statement, 'all statements are ideological'. This nevertheless leaves open the possibility that economics itself consists of statements which express 'biased' (i.e. false) views, although whether they *are* false is not decidable on the grounds of their originators' psychological motives, or social position. Without committing the genetic fallacy, writers who think of economics as 'impregnated with ideology' have suggested that ideological utterances be regarded as pseudo-scientific.

One suggestion is that 'ideological statements ... be ... defined as value judgments parading as statements of facts' (reported by Blaug, 1980, p. 138), i.e. as *covert* prescriptions, all the more suspect, since they are supposedly motivated by attempts to promote some 'class interest' (Rosenberg, 1976, pp. 203-4, examines this claim). It has been suggested that economics does, or must, consist only of such ideological pseudo-statements, and therefore cannot be scientific (a suggestion criticized in Klappholz, 1964). No doubt a careless reader could mistake disguised value judgements for factual statements, but this possibility is a subject for psychological, rather than methodological, consideration, despite occasional suggestions to the contrary (e.g. Blaug, 1980, p. 138).

Turning to statements which are descriptive, i.e. have a truth value, the following are among other suggested jointly sufficient conditions for economic statements to be ideological, i.e. pseudo-scientific: (a) that they be false; (b) that they support a given political philosophy, or be convenient for those with an interest in perpetuating some political or social order; (c) that the given political philosophy, or the convenience of the belief, be the cause of the false statements being believed (Mingat, Salmon and Wolfelsperger, 1985, pp. 353-5 and Rosenberg, 1976, pp. 204-9, critically discuss these characterizations).

The philosophic problem of demarcating scientific from other kinds of discourse cannot be discussed here. It must suffice to point out that the above characterizations would render (a set of) statements pseudo-scientific if one subscribed to the epistemological view that 'true science' consists of statements known to be true by being logically derived from facts (Lakatos, 1978, ch. 1). Few, if any, philosophers subscribe to this infallibilistic view of science and, in its absence, the above characterizations do not render statements pseudo-scientific (although, as noted above, (c) alone would not be a methologically satisfactory reason for an economist to support a theory).

Thus, if statements are judged ideological, not because they are false, but because they are possibly false, then one could not say they are pseudo-scientific, since all scientific theories are possibly false. Again, if a universal theory is viewed as ideological because it is regarded as false, for example, as is Newton's theory, but at the same time is accepted for certain technological purposes (Klappholz and Agassi, 1959, pp. 31-3), it is still not pseudo-scientific. Indeed, if such theories are regarded as pseudo-scientific, the view of ideology considered here leads to the no doubt unintended, but nevertheless absurd consequence that the available stock of pseudo-science increases with scientific progress. References to the 'convenience' of certain views, i.e. to (b), as alleged explanations of why supposedly false theories are believed, i.e. to (c), direct criticism towards individuals' conscious or unconscious motives, in the spirit of the Sociology of Knowledge, rather than to the objective scientific issues.

So far we have discussed the possible or actual falsity of theories. Theories are falsified if observations come to light which are in conflict with them. These observations are reported in what have been called basic statements (Popper, 1959, chs IV, V), i.e. statements the acceptance of which does not give rise to controversy.

For example, economists advance theories about the determinants of unemployment. These theories might be thought to be testable with the help of observations of unemployment, which, for example, lead to observation reports such as 'the level of unemployment in the UK in March 1985 was 13.3 per cent or 3.2 million people'. This is not an explanatory statement and therefore, presumably, not pseudo-scientific. However, as is well known, it is also not a basic statement, since it is controversial. Controversy is aroused, not only because the statement raises problems of statistical interpretation, but also theoretical problems, such as the observations which would be needed to measure the extent of involuntary unemployment (although, given the way unemployment is measured, large changes in the measured figures have led economists to reconsider their theories of unemployment). This is merely an example of some of the well-known problems encountered in attempts to test economic theories. Therefore, these theories are not obviously false, as seems to be required of a theory if it is to be ideological in the sense of the present Section. However, this discussion suggests that economics contains factual theories which may not be scientific, i.e. testable.

III. Factual theories which are not scientific – rather than pseudo-scientific – have been called ideological (e.g. Schumpeter, 1949; Robinson, 1962.) If one does not view scientific theories as consisting of statements known infallibly to be true, but rather as tentative hypotheses, which can be revised in the light of new evidence, then one can easily think of statements which are not scientific, but which nevertheless play a role in discussions of economic theories. Here we are referring to metaphysical statements, as well as to expressions of belief regarding the truth of competing theories among which do decisions can be made on the basis of tests.

Some economic theories may be testable (for example, the appearance of stagflation must be regarded as an anomaly for *all* previous economic theories that are relevant to the subject). However, many appear not to be testable. For instance, it has been held that general equilibrium theories are not testable (e.g. Hausman, 1981). This consideration may account for Friedman's well-known remark that reports of the corroboration of some economic theories he endorsed are 'hard to document' (Friedman, 1953, p. 22). Indeed, it has been argued that, since economic data are derived from situations which cannot be controlled for disturbing factors, statistical inference is possible only on the basis of *prior* beliefs, the differences among which cannot be objectively justified (Leamer, 1983).

Where theoretical conflicts of views cannot be resolved by available evidence, it is possible to suspend judgement. However, those engaged in research need to choose a programme, that is, to judge which theory is most likely to offer the best prospects for scientific progress. This choice may be influenced by people's Weltanschauung and preferences, in short, their ideologies. In this respect the situation in economics does not seem to differ from that in other sciences, and the mere fact that ideology, in the sense of the present section, may play a part in discussions of economic theory need not give rise to 'concern for [its] conceptual status' (Rosenberg, 1976, p. 202). Concern may be expressed with good reason if and when unwarranted claims to scientific knowledge are made.

IV. Historically, the charge of 'ideological bias' has been directed especially at economists' views on desirable economic policies, as suggested by remarks that economists have tended to 'justify the ways of Mammon to men' (Robinson, 1962, p. 25). We now consider this issue in the light of the preceding discussion.

It was noted above that policy recommendations cannot be deduced solely from economic theory: in addition, some value, i.e. non-scientific, premises are required. The Paretian value premise, widely adopted by economists, reflects an individualistic political philosophy and may be regarded as ideological (Klappholz, 1968).

Apart from adopting the Paretian value premise, economists have advocated policies which show a preference for organizing economic activities through markets (Kearl et al., 1979). However, it is difficult to take seriously the view, referred to in sections I and II, that this stance is to be explained by those economists' 'position in the social structure' or by their 'interest in perpetuating the system'. If the preference for market-organized economic activity is less marked among, for example, sociologists, wherein lies the difference in their social position, or their interest in perpetuating the system, compared to that of economists? Thus, it is more plausible to suppose that economists' preference for markets has been shaped by the dominant paradigm of the invisible hand, and by the fact that there is the most widespread professional consensus on the consequences of overriding markets, by, for example, such policies as rent control.

It was noted that, where theoretical differences cannot be resolved, judgement may be suspended. In the case of policy, policy makers and their advisers cannot suspend judgement, since decisions cannot be avoided, even if the implicit decision is to take no action. Assuming no well-grounded consensus regarding the consequences of alternative courses of action, it seems plausible that ideological views will influence judgements on the most likely consequences, thus influencing decisions, quite apart from the value premises which are logically indispensable for reaching them. In general, there seems to be less consensus regarding the effects of policies in the area of economics than in policies based mainly on the natural sciences, although lack of consensus in the latter case is not unknown. Thus, not surprisingly, there is more scope for ideological influence in decisions about economic policy.

However, given the absence of consensus, and the relevance of economics to public policy, differences in ideological views, (be they differences in value judgements or differences in beliefs about the outcome of policies) can be viewed as part of the mechanism of the public aspect of scientific activity which promotes criticism and, through it, may help us to learn more about the issues at hand. Those for, and those against, a given policy *all* may have an ideologically based incentive to try to show, as objectively as possible, the practical consequences any given policy will have. This view is opposed to the conventional wisdom, according to which ideology is a 'Weltanschauung felt passionately *and defended unscrupulously*' (Wiles, 1978-80, p. 61, italics added). Ideological views need not lead to dogmatism, or to lack of scruples, and there is, in any case, no way of ensuring the absence of dogmatic people. All one can do is to shun discussion with them.

KURT KLAPPHOLZ

See also PHILOSOPHY AND ECONOMICS; RHETORIC; VALUE JUDGEMENTS.

BIBLIOGRAPHY

Blaug, M. 1980. *The Methodology of Economics*. Cambridge: Cambridge University Press.

Friedman, M. 1953. *Essays in Positive Economics*. Chicago: University of Chicago Press.

Hausman, D.M. 1981. Are general equilibrium theories explanatory? In *Philosophy in Economics*, ed. J. Pitt, Dordrecht: D. Reidel. Reprinted in *The Philosophy of Economics*, ed. D.M. Hausman, Cambridge: Cambridge University Press, 1984.

Kearl, J.R., Pope, C.L., Whiting, G.T. and Wimmer, L.T. 1979. A confusion of economists. *American Economic Review* 69(2), May, 28–37.

Klappholz, K. 1964. Value judgments and economics. *British Journal for the Philosophy of Science*, August. Reprinted in *The Philosophy of Economics*, ed. D.M. Hausman, Cambridge: Cambridge University Press, 1984.

Klappholz, K. 1968. What redistribution may economists discuss? *Economica* 35, May, 194–7.

Klappholz, K. and Agassi, J. 1959. Methodological prescriptions in economics. *Economica*, February Reprinted in *Readings in Microeconomics*, ed. D.R. Kamerschen, New York: 1967.

Lakatos, I. 1978. Introduction to *The Methodology of Scientific Research Programmes, Philosophical Papers*. Vol. I, ed. G. Currie and J. Worral. Cambridge: Cambridge University Press.

Leamer, E.E. 1983. Let's take the con out of econometrics. *American Economic Review*, March. Reprinted in *Appraisal and Criticism in Economics*, ed. B. Caldwell, London: Allen & Unwin, 1984.

McCloskey, D.N. 1983. The rhetoric of economics. *Journal of Economic Literature*, June. Reprinted in *Appraisal and Criticism in Economics*, ed. B. Caldwell, London: Allen & Unwin, 1984.

Mingat, A., Salmon, P. and Wolfelsperger, A. 1985. *Méthodologie économique*. Paris: Presses Universitaires de France.

Popper, K.R. 1957. *The Open Society and its Enemies*. 3rd edn, Vol. II, London: Routledge & Kegan Paul.

Popper, K.R. 1959. *The Logic of Scientific Discovery*. London: Hutchinson.

Popper, K.R. 1976. The myth of the framework. In *The Abdication of Philosophy: Philosophy and the Public Good*, ed. E. Freeman, La Salle, Ill.: Open Court.

Robinson, J. 1962. *Economic Philosophy*. London: Watts. Reprinted London: Pelican Books, 1964.

Rosenberg, A. 1976. *Micro-Economic Laws: A Philosophical Analysis*. Pittsburgh: University of Pittsburgh Press.

Schumpeter, J. 1949. Science and ideology. *American Economic Review*, March. Reprinted in *The Philosophy of Economics*, ed. D.M. Hausman, Cambridge: Cambridge University Press, 1984.

Ward, B. 1979. *The Ideal World of Economics: Liberal Radical and Conservative Economic World Views*. London: Macmillan.

Wiles, P. 1979–80. Ideology, methodology, and neoclassical economics. *Journal of Post Keynesian Economics*, Winter. Reprinted in *Why Economics is not a Science*, ed. A.S. Eichner, New York: M.E. Sharpe, 1983.

immigration. *See* INTERNATIONAL MIGRATION.

immiserizing growth.   The theory of immiserizing growth has been developed by theorists of international trade, though it has recently been the focal point of research also by mathematical economists. It is central to understanding several important paradoxes in economic theory and has significant policy implications.

That growth in a country could immiserize it is a paradox that was first noted by trade theorists such as Bhagwati (1958) and Johnson (1955) in the context of the postwar discussions of dollar shortage. They established conditions under which, in a two-country, two-traded-goods framework of conventional theory, the growth-induced deterioration in the terms of trade would outweigh the primary gain from growth. It was shown that this paradox, unlike the paradox of donor-enriching and recipient-immiserizing transfers, was compatible with Walras-stability.

The phrase 'immiserizing growth' was invented by Bhagwati (1958) and has now been widely accepted (including by literary editors who have long ceased to insist on changing it to the correct English versions such as 'immiserating'), the theory itself being generally attributed (e.g. Johnson, 1967) to this 1958 article. Interestingly, as often in economics, Bhagwati happened to chance upon an early contribution by Edgeworth (1894), where Edgeworth developed an example of what he called 'indamnifying' growth; and the controversy surrounding this result at the time and its relationship to the Bhagwati–Johnson analyses of the 1950s was reviewed in Bhagwati and Johnson (1960).

Later, Johnson (1967) demonstrated another paradox of immiserizing growth. If a small country had a distortionary tariff in place, and then exogenously it experienced growth, the result again could be to immiserize the country. Later, Bertrand and Flatters (1971) and Martin (1977) established formally the conditions under which this new paradox of immiserizing growth could arise.

Bhagwati (1968) got to the bottom of these paradoxes and produced the central insight that explains why these, and other immiserizing-growth paradoxes, can readily arise. He showed that, if an economy was suboptimally organized, the primary gain from growth, measured hypothetically as if the economy had an optimal policy in place before and after the growth, could be outweighed by accentuation of the loss from the distortion-induced suboptimality when growth occurred. In the original Bhagwati (1958) example, since the terms of trade could deteriorate, the economy had monopoly power in trade but was following free trade policy which is evidently suboptimal. In the Johnson (1967) example, the tariff was being used by a small country with given terms of trade and was therefore also a suboptimal policy. In both cases the suboptimal policy produced losses which were accentuated by the growth and then managed to outweigh the primary gains from growth that would have occurred if optimal policies were in place. The result was a powerful generalization that placed the theory of immiserizing growth squarely into the central theory of distortions and policy intervention (Srinivasan, 1987) that lies at the core of the modern theory of trade and welfare. Evidently, immiserizing-growth paradoxes could arise only if there was a distortion present.

This central result has immediate implications. If an economy has a suboptimal money supply, growth could be immiserizing. If trade policy is highly distorted, growth could be immiserizing. The well-known results of trade theory, which show that free trade need not be welfare-improving relative to autarky (e.g. Haberler, 1950) under distortions are also seen as instances of immiserizing-growth theory; free trade augments the availability set relative to autarky, implying 'as-if' growth, and if distortions are present, there is no surprise to the immiseration that free trade brings. Again, if a country uses tariffs to induce foreign investment (the so-called tariff-jumping investment that developing countries often used in the postwar period), such investment could immiserize the host country: this being a simple extension of the Johnson (1967) demonstration, argued to be relevant to analysis of developing countries in Bhagwati (1978), and analysed extensively in Bhagwati (1973), Brecher and Alejandro (1977), Hamada (1974), Minabe (1974), Uzawa (1969) and Brecher and Findlay (1983). Yet another important insight from the immiserizing-growth theory is that, in the new and growing theory of DUP (directly-unproductive profit-seeking) activities, which incorporates several quasi-political activities essentially into the corpus of economic theory, a DUP activity that wastes resources directly need not cause ultimate loss of welfare. This is because the waste may occur from a suboptimal situation, thus resulting in welfare-improvement paradoxically. This is

the obverse of immiserizing growth: in one case, growth immiserizes; in the other, throwing away or wasting resources enriches. This is at the heart of the contention in Bhagwati (1980) that an exogenous tariff at t per cent may be welfare-superior to an endogenous tariff, procured by tariff-seeking lobbies that have diverted uses to such DUP activity, also at t per cent. Several such implications of the theory of immiserizing growth are discussed in Bhagwati and Srinivasan (1983, ch. 25).

Two further developments need to be cited. First, the dual of immiserizing growth, when such growth is due to factor accumulation, clearly yields negative shadow factor prices. This aspect is relevant to certain formulations in cost–benefit analysis; see, in particular, Findlay and Wellisz (1976), Diamond and Mirrlees (1976), Srinivasan and Bhagwati (1978), Bhagwati, Srinivasan and Wan (1978) and Mussa (1979).

Next, mathematical economists such as Aumann and Peleg (1974), and then Mas-Colell (1976) and Mantel (1984) among others, have rediscovered the original immiserizing-growth paradox, illustrating how economists working apart or in different traditions may rediscover one another's findings, often decades apart. A synthesis of the two literatures has been provided in Bhagwati, Brecher and Hatta (1984). A complete and formal reconciliation of the conditions established in Bhagwati (1958) and in Mas-Colell (1976) and Mantel (1984) for the original immiserizing-growth paradox is provided by Hatta (1984).

JAGDISH N. BHAGWATI

See also DISTORTIONS; STRATEGIC REALLOCATIONS OF ENDOWMENTS; TERMS OF TRADE.

BIBLIOGRAPHY
Aumann, R.J. and Peleg, B. 1974. A note on Gale's example. *Journal of Mathematical Economics* 1, 209–11.
Bertrand, T. and Flatters, F. 1971. Tariffs, capital accumulation and immiserizing growth. *Journal of International Economics* 1(4), 453–60.
Bhagwati, J. 1958. Immiserizing growth: a geometrical note. *Review of Economic Studies* 25, June, 201–5. Reprinted in *International Trade: Selected Readings*, ed. J. Bhagwati, Cambridge, Mass.: MIT Press, 1981.
Bhagwati, J. 1968. Distortions and immiserizing growth: a generalization. *Review of Economic Studies* 35, October. Reprinted in J. Bhagwati, *The Theory of Commercial Policy*, Vol. I, Cambridge, Mass.: MIT Press, 1983.
Bhagwati, J. 1973. The theory of immiserizing growth: further applications. In *International Trade and Money*, ed. M. Connolly and A. Swoboda, Toronto: University of Toronto Press.
Bhagwati, J. 1978. *Foreign Trade Regimes and Economic Development: The Anatomy and Consequences of Exchange Control.* Cambridge, Mass.: Ballinger.
Bhagwati, J. 1980. Lobbying and welfare. *Journal of Public Economics* 14, December, 355–63.
Bhagwati, J., Brecher, R. and Hatta, T. 1984. The paradoxes of immiserizing growth and donor-enriching 'recipient-immiserizing' transfers: a tale of two literatures. *Weltwirtschaftliches Archiv* 120(4), 228–43.
Bhagwati, J. and Johnson, H.G. 1960. Notes on some controversies in the theory of international trade. *Economic Journal* 60, 74–93.
Bhagwati, J. and Srinivasan, T.N. 1983. *Lectures on International Trade.* Cambridge, Mass.: MIT Press.
Bhagwati, J., Srinivasan, T.N. and Wan, H., Jr. 1978. Value subtracted, negative shadow prices of factors in project evaluation, and immiserizing growth: three paradoxes in the presence of trade distortions. *Economic Journal* 88, 121–5.
Brecher, R. and Diaz-Alejandro, C. 1977. Tariffs, foreign capital and immiserizing growth. *Journal of International Economics* 7, 317–22. Reprinted in *International Trade: Selected Readings*, ed. J. Bhagwati, Cambridge, Mass.: MIT Press, 1981.

Brecher, R. and Findlay, R. 1983. Tariffs, foreign capital and national welfare with sector-specific factors. *Journal of International Economics* 14, 277–88.

Diamond, P. and Mirrlees, J. 1976. Private constant returns and public shadow prices. *Review of Economic Studies* 43, 41–8.

Edgeworth, F.Y. 1894. The theory of international values. *Economic Journal* 4, 35–50, 424–43, 606–38.

Findlay, R. and Wellisz, S. 1976. Project evaluation, shadow prices and trade policy. *Journal of Political Economy* 84(3), 543–52.

Haberler, G. 1950. Some problems in the pure theory of international trade. *Economic Journal* 60, 223–40.

Hamada, K. 1974. An economic analysis of the duty-free zone. *Journal of International Economics* 4(3), 225–41.

Hatta, T. 1984. Immiserizing growth in a many-economy setting. *Journal of International Economics* 17, 335–45.

Johnson, H.G. 1955. Economic expansion and international trade. *Manchester School of Economic and Social Studies* 23(2), 95–112.

Johnson, H.G. 1967. The possibility of income losses from increased efficiency or factor accumulation in the presence of tariffs. *Economic Journal* 77, 151–4. Reprinted in *International Trade: Selected Readings*, ed. J. Bhagwati, Cambridge, Mass.: MIT Press, 1981.

Mantel, R. 1984. Substitutability and the welfare effects of endowment increases. *Journal of International Economics* 17, 325–34.

Martin, R. 1977. Immiserizing growth for a tariff-distorted, small economy. *Journal of International Economics* 3(4), 323–6.

Mas-Colell, A. 1976. En torno a una propiedad poco atractiva del equilibrio competitivo. *Moneda y Credito* 136, 11–27.

Minabe, N. 1974. Capital and technology movements and economic welfare. *American Economic Review* 64, 1088–100.

Mussa, M. 1979. The two-sector model in terms of its dual: a geometric exposition. *Journal of International Economics* 9(4), 513–26. Reprinted in *International Trade: Selected Readings*, ed. J. Bhagwati, Cambridge, Mass.: MIT Press, 1981.

Srinivasan, T.N. and Bhagwati, J. 1978. Shadow prices for project selection in the presence of distortions: effective rates of protection and domestic resource costs. *Journal of Political Economy* 86(1). Reprinted in *International Trade: Selected Readings*, ed. J. Bhagwati, Cambridge, Mass.: MIT Press, 1981.

Uzawa, H. 1969. Shinon jiyutato kokumin keizai (Liberalization of foreign investments and the national economy). *Ekonomisuto* 23, December, 106–22 (in Japanese).

impatience. Impatience refers to the preference for earlier rather than later consumption an idea which stems from Böhm-Bawerk (1912) and Fisher (1930), among others. Preference orderings that exhibit impatience are also described as being myopic or as embodying discounting. Because in many contexts the future has no natural termination date, an infinite horizon framework is most appropriate and convenient for the analysis of many problems in intertemporal economics. The open-endedness of the future raises several issues surrounding impatience (its presence, degree, and the precise form it takes) which do not arise in finite horizon models.

Consider a world with a countable infinity of time periods or generations, $t = 0, 1, \ldots, T, \ldots$, where there is a single good which can be consumed or accumulated. Let $x = (x_0, \ldots, x_t, \ldots)$ represent a consumption programme where x_t denotes the consumption of the representative consumer for the tth generation. Given an initial (capital) stock k_0 of the good, and a technology that transforms capital into a flow output, the set of feasible consumption programmes, denoted $S(k_0)$, is determined.

At issue is the optimal programme of consumption and accumulation. Suppose it is determined by a central planner who ranks programmes in $S(k_0)$ according to the utility functional

$$U(x) = \sum_0^\infty (1 + \rho)^{-t} u(x_t). \tag{1}$$

This is a common specification. For $\rho = 0$ it dates from Ramsey (1928); for the general case see Koopmans (1966). The instantaneous utility function $u(.)$ is increasing and concave (diminishing marginal utility).

The parameter ρ equals the rate of time preference. Impatience (in the sense of any of the precise definitions given below) is present if (and only if) $\rho > 0$. There is a preliminary technical problem with certain values of ρ. When $\rho = 0$, for example, the infinite sum in (1) diverges for many of the paths to be compared. Ramsey provides one device for getting around this difficulty. Another device is von Weizsäcker's (1965) overtaking criterion, according to which x^* is *optimal* in $S(k_0)$ if it is feasible and if for any other feasible path x,

$$\sum_0^T (1 + \rho)^{-t} u(x_t^*) \geq \sum_0^T (1 + \rho)^{-t} u(x_t),$$

for all sufficiently large T. This notion of optimality is well-defined for any value of ρ, even for negative values; and an optimal x^* maximizes U on $S(k_0)$ if $U(x^*)$ is finite.

The specification of ρ is crucial and presumably reflects the ethical principles of the planner. Ramsey (1928) objects to discounting on ethical grounds and thus assumes $\rho = 0$. But Koopmans (1966, 1967) argues that there are technical limitations on the specification of ρ which are imposed by the requirement that an optimal plan x^* exist for a range of choice environments. The potential difficulty is readily understood: a positive return to saving provides an incentive to postpone consumption. Positive (negative) discounting provides an offsetting (reinforcing) incentive. Finally, diminishing marginal utility and diminishing marginal productivity in production induce a smoothing of consumption over time. For many specifications, the net incentive is to postpone and to do so idefinitely, which is clearly not optimal. Consequently an optimal programme fails to exist. The existence problem is mitigated the larger is ρ, in the sense that if $\rho_1 < \rho_2$ and if an optimum in $S(k_0)$ exists when $\rho = \rho_1$, then it exists also when $\rho = \rho_2$. In particular, in order that an optimum exist in several simplified but commonly specified choice environments, it is necessary that $\rho > 0$ and hence that the future be discounted. (See also von Weizsäcker, 1965.)

The existence of solutions to optimization problems is a basic question in mathematical programming which is most commonly resolved by application of the Weierstrass Theorem (or its many extensions). The Theorem guarantees existence of a solution if the objective function is continuous and the constraint set is compact. It is valid in general topological spaces and so is applicable also to the present setting where the choice variable x lies in an infinite dimensional space. The Theorem is the basis for the proof by Magill (1981) of the existence of an optimum to infinite horizon optimization problems. When specialized to the constant discount rate functional (1), his analysis confirms the consequences for existence of large ρ. Moreover, it shows 'why' a large ρ is beneficial – the larger is ρ, the more stringent the form of continuity satisfied by the utility functional and hence the broader the class of constraint sets to which the Weierstrass Theorem is applicable.

To pursue the link between impatience and continuity, it is necessary to consider the latter more carefully. First, however, restrict attention to bounded consumption profiles, that is, to the set

$$L_+^\infty = \{x = (x_0, \ldots, x_t, \ldots): \quad x_t \geq 0 \\ \text{for all } t \quad \text{and sup } x_t < \infty\}.$$

Secondly, the existence of a utility function is an unnecessarily restrictive assumption. Thus consider preference relations $\succeq$ on L_+^∞, with strict preference denoted by $\succ$.

To discuss continuity, we need to specify a topology for L_+^∞; that is, we need to define what it means for two consumption paths to be 'close' to one another. This is most simply done by specifying when a sequence of consumption paths $\{x^n = (x_0^n, x_1^n, \ldots, x_t^n, \ldots)\}_{n=1}^\infty$ *converges* to a path x in L_+^∞. (Strictly speaking, generalized sequences called nets should be used, but the use of sequences is adequate for this informal discussion.) For many topologies that are of interest in economics 'closeness' can be measured by a *metric* or distance function d such that $d(x, y)$ measures the 'distance' between x and y. When such a metric exists, convergence of $\{x^n\}$ to x means simply that $d(x^n, x)$ approaches 0 as $n \to \infty$, in which case we refer to the d-*convergence* of the sequence.

Table 1 defines four topologies by specifying the conditions for convergence imposed by each. When a metric exists, it is also specified. Of course many other plausible topologies could be considered.

Continuity of a preference relation means roughly that consumption paths that are close to one another are ranked similarly vis-à-vis other paths. More formally, say that the relation $\succeq$ is continuous in the topology Γ (or Γ-continuous) if for each x and y in L_+^∞, and for any sequences $\{x^n\}$ and $\{y^n\}$ that converge to x and y respectively according to Γ, it is the case that

$$x \succ y \Rightarrow x \succ y^n \quad \text{and} \quad x^n \succ y$$

for all sufficiently large values of n.

Which topology should be adopted? The question does not arise in finite dimensional contexts. The reason is simply that all 'natural' topologies on finite dimensional Euclidean spaces are *equivalent* in the sense that the corresponding convergence definitions are logically equivalent to one another. This is the case, for example, with the four topologies in the table if they are adapted in the obvious way to a finite horizon context. In all cases, convergence is identical to the usual notion based on the Euclidean metric. Thus the corresponding notions of continuity are also identical.

In contrast, in the infinite horizon model, the noted equivalence fails. It is easily shown that

$$d_s\text{-convergence} \Rightarrow d_\infty\text{-convergence}$$
$$\Rightarrow \text{Mackey-convergence}$$
$$\Rightarrow d_p\text{-convergence.} \qquad (2)$$

But none of the reverse implications is true. For example, define the sequences $\{x^n\}$, $\{y^n\}$ and $\{z^n\}$ as follows:

$$x_t^n = 0 \quad \text{if} \quad 0 \le t \le n \quad \text{and} \quad = n \quad \text{if} \quad t > n$$
$$y_t^n = 0 \quad \text{if} \quad 0 \le t \le n \quad \text{and} \quad = 1 \quad \text{if} \quad t > n$$
$$z^n = (n^{-1}, n^{-1}, n^{-1}, \ldots).$$

Then $\{x^n\}$ converges to $(0, 0, \ldots)$ in the product topology but not in the Mackey topology. In the former case x^n is viewed as being close to the zero consumption path for large n, because the first n generations all have zero consumption. Thus the product topology discounts the fact that in x^n infinitely many generations enjoy large consumption levels which are unbounded as n grows. It is the latter feature which explains why x^n and $(0, 0 \ldots)$ are not viewed as being close to one another by the Mackey topology. (Take $a_t = t^{-1/2}$ in the definition of Mackey convergence.) Thus, for example, in the case of $\{y^n\}$ where the consumption of future generations is bounded in n, the sequence is Mackey-convergent to the zero consumption path. The sequence $\{y^n\}$ is not d_∞-convergent since not all generations have consumption near 0. Finally, $\{z^n\}$ converges to $(0, 0 \ldots)$ in the sup topology, but it is not d-convergent since the 'aggregate' deviation of consumption levels between the two paths is large (indeed $\Sigma_0^\infty |z_t^n| = \infty$).

When topologies are not equivalent continuity of a preference relation has different meaning depending upon which topology is adopted. Thus (2) implies immediately that

$$d_p\text{-continuity} \Rightarrow \text{Mackey-continuity}$$
$$\Rightarrow d_\infty\text{-continuity}$$
$$\Rightarrow d_s\text{-continuity}, \qquad (3)$$

and none of the reverse implications is valid. In finite dimensional analysis continuity is a purely technical assumption which is innocuous from an economist's point of view. But the discussion of convergence in the above four topologies strongly suggests that in infinite horizon models the specification of a topology and the assumption of continuity can have economic content. Indeed, continuity in some topologies can imply impatience.

One demonstration of the crucial role played by a topology is provided by Diamond (1965) and Svensson (1980). Call a preference relation *equitable* if it provides equal treatment for all generations in the sense that for all x and y in L_+^∞, $x \succeq y \Leftrightarrow \pi x \succeq \pi y$, where πx (or πy) is obtained from x (or y) by permuting finitely many of its components. A preference relation is weakly monotonic if $x_t > y_t$ for all $t \Rightarrow x \succ y$. Diamond shows that there does not exist an equitable and weakly

Table 1

Topology	Definition of convergence of $\{x^n\}$ to x	Metric
product	$x_t^n \xrightarrow[n \to \infty]{} x_t$ for all t	$d_p(x, y) = \sup_t \dfrac{2^{-t}|x_t - y_t|}{\{1 + |x_t - y_t|\}}$
Mackey	$\sup_t |a_t \cdot (x_t^n - x_t)| \xrightarrow[n \to \infty]{} 0$ for all sequences of real numbers $\{a_t\}_0^\infty$ that converge to 0	—
supremum	$\sup_t |x_t^n - x_t| \xrightarrow[n \to \infty]{} 0$	$d_\infty(x, y) = \sup_t |x_t - y_t|$
Svensson	$\sum_0^\infty |x_t^n - x_t| \xrightarrow[n \to \infty]{} 0$	$d_s(x, y) = \min\left(1, \sum_0^\infty |x_t - y_t|\right)$

monotonic preference relation that is also continuous in the product metric. This preclusion of equity is perhaps not surprising given the discounting of the future that is built into the definition of d_p. But even given the apparently 'time neutral' metric d_∞, the scope for equity is limited. Diamond proves that equity and d_∞-continuity are incompatible given strong monotonicity ($x_t \geqslant y_t$ for all t and $x_\tau > y_\tau$ for some $\tau \Rightarrow x > y$.) If only weak monotonicity is imposed, then all postulates are satisfied by the maximin ordering, whereby

$$x \geqslant y \Leftrightarrow \inf x_t \geqslant \inf y_t. \tag{4}$$

The view, based on finite dimensional analysis, that continuity is an innocuous technical assumption, would lead one to interpret Diamond's results as demonstrating the non-existence of equitable orderings that satisfy minimal additional regularity conditions. But, the correct interpretation is the Diamond's theorems demonstrate the strong ethical content of d_p-continuity and d_∞-continuity. The latter view is fortified by Svensson (1980). He shows that if the d_s metric is adopted, then there exist equitable and strongly monotonic orderings which are d_s-continuous. Since d_s is a priori plausible, the onus is clearly shifted to the metric. At the extreme, continuity can be imposed with total impunity if the metric d_0 is adopted, where

$$d_0(x, y) = \begin{bmatrix} 0, & \text{if} & x = y \\ 1, & \text{if} & x \neq y \end{bmatrix}.$$

The topology corresponding to d_0 is called the discrete topology. According to this metric distinct consumption paths cannot be close to one another, so continuity is automatic. A natural open question is the characterization of metrics d (and more general topologies) such that d-continuous, equitable and (weakly or strongly) monotonic preference relations exist.

At this point it is worth recalling a principal reason that continuity is of interest – namely that by (an extension of) the Weierstrass Theorem, it will guarantee the existence of optimal elements in compact sets. Given a topology Γ on L_+^∞, a set $K \subset L_+^\infty$ is Γ-compact if every (generalized) sequence of points in K has a (generalized) subsequence that converges according to Γ to a point in K. As the topology changes in such a way as to permit more continuous functions, the family of compact sets shrinks (see (2) and (3)). Thus as continuity becomes easier to achieve it also becomes less significant. (For example, K is d_0-compact only if it consists of finitely many points; and there exist many economically relevant sets K that are compact in the product topology but not in the sup topology. One example arises in an exhaustible resource model where feasible consumptions plans satisfy $\Sigma_0^\infty x_t \leqslant w$, and w is the initial stock of the good.) If there is a class of constraint sets where the existence of optimal elements is desired, then the 'useful' topologies are those that make each of the constraint sets compact. This approach (emphasized by Campbell, 1985) would remove some of the arbitrariness from the choice of a topology.

Diamond's results suggest that continuity may imply 'some form of impatience', since equity can be viewed as the lack of impatience. A more precise definition of impatience is required for a clearer demonstration of the link between the latter and continuity. For example, impatience could be taken to mean that interchanging the consumption levels of generations 1 and t results in a strictly preferred plan if period t consumption was initially larger. If the preceding statement is valid only for t sufficiently far into the future, then *eventual impatience* could be said to prevail. This latter notion captures not only a preference for the advancement of the timing of satisfaction,

but also the idea that the taste for future consumption diminishes as the time of consumption recedes into the future. These and related definitions appear in Koopmans (1960), Koopmans et al. (1964) and Diamond (1965). Their proofs that appropriate continuity implies (eventual) impatience depend, with the single exception of Diamond (p. 174), on maintained separability assumptions on the preference relation. The separability assumptions can be deleted if the existence of a differentiable utility function is assumed (Burness, 1973).

Brown and Lewis (1981) define some notions of asymptotic impatience. For example, they call a preference relation *strongly myopic* if for all x, y and z in L_+^∞, $x > y \Rightarrow x > y + {}_nz$ for all sufficiently large n, where ${}_nz = (0, \ldots, 0, z_{n+1}, z_{n+2}, \ldots)$. In other words, the preference for x over y is unchanged by an increase in the latter programme in the consumption of infinitely many generations, as long as the increase occurs only for generations that are situated sufficiently far into the future.

Interpret a preference relation as belonging to a consumer rather than to a central planner. Consumption programmes in L^∞ describe the consumption of that consumer and his descendants; the latters' consumption levels matter because of intergenerational altruism. This is a common framework in the capital theory literature where the behavioural assumption of impatience is often maintained. This suggests that from the perspective of capital theory, economically interesting topologies are those which (through continuity) imply myopia. For example, any preference relation which is d_p-continuous is necessarily strongly myopic. But the implication is false if the product metric is replaced by d_∞ or d_s. Brown and Lewis show that the Mackey topology bears a special relationship to strong myopia. Mackey-continuity is the weakest continuity requirement (corresponding to topologies in a broad and convenient class) that can be imposed on a preference relation in order that strong myopia be implied. Thus it is a 'natural' topology if strong myopia is desirable.

There is an important link between the Mackey topology and strong myopia on the one hand and general equilibrium analysis in the framework of 'infinitely lived' agents on the other. Bewley (1972) points out that the Mackey topology is particularly appropriate for general equilibrium analysis because continuity requirements weaker than Mackey-continuity do not guarantee the existence of equilibria with price systems that can be represented by absolutely summable sequences ($p_0, \ldots, p_t, \ldots$), rather than merely for more general mathematical constructs that have no economic interpretation. In light of the relationship between Mackey-continuity and strong myopia, the latter seems necessary for meaningful general equilibrium analysis.

Brown and Lewis sharpen the link between impatience and general equilibrium analysis. They prove that if individual preferences are suitably monotonic, then Mackey-continuity and strong myopia are unnecessarily strong assumptions. But a form of asymptotic impatience is still relevant. Call a preference ordering *weakly myopic* if the implication defining strong myopia is valid for all constant programmes z. Then even if individual preferences are weakly monotonic, the existence of economically interpretable equilibrium price systems as above can be guaranteed only by continuity requirements which imply weak myopia.

Suppose that we are willing to accept more general constructs (linear functionals on L_+^∞) as price systems. Can we then dispense with impatience? Araujo (1985) provides a negative partial answer. He restricts attention to a well-defined subset of those continuity conditions which lie 'between' d_∞-continuity and d_p-continuity. Then he shows that the existence of such general price systems can be guaranteed only if continuity requirements are imposed which imply strong myopia, or, when suitable monotonicity is maintained for preferences, weak myopia. Existence of equilibria cannot be guaranteed in such cases as the maximin ordering (4) which exhibits no impatience.

We offer one final comment. In a planning context, continuity of the social preference relation may be desirable not necessarily for its own sake nor because it may imply myopia, but primarily to guarantee that the preference relation be *effective*, that is, that optimal consumption paths exist. From this perspective, it seems more pertinent to investigate the link between effectiveness and impatience directly, without involving continuity which is, after all, at best sufficient and definitely not necessary for the existence of optimal paths. Thus, for example, a pertinent question is whether impatience (in some precise sense) is necessary for effectiveness in a relevant set of choice environments. While this question has been addressed to some extent in the growth theory literature cited earlier based on the additive utility functional (1), an analysis comparable in generality to that of Brown and Lewis or Araujo has yet to be performed.

LARRY G. EPSTEIN

See also FISHER, IRVING; PRESENT VALUE; TIME PREFERENCE.

BIBLIOGRAPHY

Araujo, A. 1985. Lack of Pareto optimal allocations in economies with infinitely many commodities: the need for impatience. *Econometrica* 53(2), March, 455–61.

Bewley, T. 1972. Existence of equilibria in economies with infinitely many commodities. *Journal of Economic Theory* 4(3), June, 514–40.

Böhm-Bawerk, E. von 1912. *Positive Theory of Capital.* South Holland, Ill.: Libertarian Press, 1959.

Brown, D.J. and Lewis, L.M. 1981. Myopic economic agents. *Econometrica* 49(2), March, 359–68.

Burness, H.S. 1973. Impatience and the preference for advancement in the timing of satisfactions. *Journal of Economic Theory* 6(5), October, 495–507.

Diamond, P.A. 1965. The evaluation of infinite utility streams. *Econometrica* 33, January, 170–77.

Fisher, I. 1930. *The Theory of Interest.* New York: Macmillan.

Koopmans, T.C. 1960. Stationary ordinal utility and impatience. *Econometrica* 28, April, 287–309.

Koopmans, T.C. 1966. On the concept of optimal economic growth. In *The Econometric Approach to Development Planning*, Amsterdam: North-Holland.

Koopmans, T.C. 1967. Objectives, constraints, and outcomes in optimal growth models. *Econometrica* 35, January, 1–15.

Koopmans, T.C., Diamond, P.A. and Williamson, R.E. 1964. Stationary utility and time perspective. *Econometrica* 32, January–April, 82–100.

Magill, M.J.P. 1981. Infinite horizon programs. *Econometrica* 49(3), May, 679–711.

Ramsey, F.P. 1928. A mathematical theory of saving. *Economic Journal* 38, December, 543–59.

Svensson, L.G. 1980. Equity among generations. *Econometrica* 48(5), 1251–56.

von Weizsäcker, C.C. 1965. Existence of optimal programs of accumulation for an infinite time horizon. *Review of Economic Studies* 32, April, 85–104.

imperfect competition. Imperfect competitors are individuals or firms who face downward-sloping demand curves or upward-sloping supply curves for some product(s). This is to be contrasted with perfect competitors who, by definition, face perfectly elastic demand and supply curves for all products. Notice we define perfect competitors not just as price-takers, but as rational price-takers: perfect competitors cannot influence the levels of market clearing prices. By contrast imperfect competitors, by their presence, can influence some equilibrium prices. As simple as these definitions sound, they hold within themselves a world of meaning that we will explore a little in this entry.

Since the early days of economics as a science, the importance of the force of competition has been stressed. Adam Smith viewed the force of competition as a central benefactor of society, which both (a) guards people against the possibility of monopolistic exploitation by insuring that the long run price will not exceed the cost of production; and (b) automatically provides for long-run progress by firing entrepreneurs' restless search for new profit potentials.

In contrast to Smith, modern-day economists are becoming increasingly uncertain whether the force of competition is entirely beneficent. The image of wasteful competition between individuals and between firms is gaining repute. Theories of imperfect competition are becoming increasingly popular, reflecting a dissatisfaction with the predictive power of the perfectly competitive model of economic reality.

The insight that competition can be wasteful, not necessarily beneficent, was popularized by Edward Chamberlin, who along with Joan Robinson is typically credited with renewing economists' interest in imperfect competition beginning in the 1930s (Chamberlin, 1933; Robinson, 1933). As a contender to the perfectly competitive image of economic reality, Chamberlin offered his image of many firms selling differentiated products, contending with one another, but nevertheless each facing a downward-sloping demand curve. His famous 'excess capacity theorem' was the caricature he offered of wasteful competition.

As in Chamberlin, the current modelling of imperfect competition tends to be partial equilibrium. A popular practice is to make the assumption that firms will interact in a Cournot–Nash fashion. Perhaps more ambitious and interesting are current explorations at the interface of game theory and industrial organization theory. Many (small group) models of imperfectly competitive interactions are available, each with its own idiosyncratic, stylized features. These models are a beginning toward analysing imperfect competition between individuals and between firms as an active process. But there does not currently exist a standard paradigm of imperfect competition (either partial equilibrium or general equilibrium). This contrasts sharply with the case of perfect competition, which is typically idealized using a Walrasian general equilibrium model. Perhaps models of imperfect competition must necessarily be legion and case-specific?

We will not try to survey existent models of imperfect competition here. Rather, we will try to offer some overview in terms of a unifying principle. In particular, we will argue that *increasing returns* is the usual source of imperfect competition. Knowledge of such a unifying source will hopefully help the reader make sense of the plethora of available idiosyncratic models. It should also help the reader understand why imperfect competition, in contrast to perfect competition, may be wasteful.

1. THE MEANING OF INCREASING RETURNS: FROM PERFECT COMPETITION TO IMPERFECT COMPETITION. To understand the concept of increasing returns, as applied to the economy as a whole (rather than to a particular firm), it is useful to first understand how economists usually ensure that a model of the economy as a whole will be *perfectly competitive*. This will provide us with a benchmark from which to proceed since, as we shall see, a perfectly competitive economy typically exhibits constant returns, in contrast to increasing returns. (The

observation that constant returns typifies perfect competition is also central to Samuelson (1967). He proceeds in a somewhat different fashion, but his article may be read as a useful complement to this one.)

To ensure that an economy will be perfectly competitive economists typically assume a finite number of homogeneous private goods. Then, keeping the set of goods fixed, they *replicate* the economy by increasing the number of buyers and sellers of each commodity indefinitely. The resulting, limiting economy will be perfectly competitive in the sense that the force of competition between the many alternative sellers of each commodity and the many alternative buyers will be sufficient to ensure that no one individual will possess any monopoly or monopsony power. That is, no one individual will be able to influence the levels of the prices that equilibrate supply and demand. For example, if some seller tries to exploit some buyer there will be plenty of perfect substitute sellers available ready to take the buyer away from him.

Notice that the image of 'thick markets', i.e. homogeneous private goods with many small sellers and buyers of each good, is central to economists' image of perfect competition. It is this image of thick markets that Chamberlin found to be a grotesque caricature of our economic reality.

It is easy to see that a large, replicated private-good economy exhibits constant returns to scale in the sense that a small subset of its participants could do as well on their own as they could participating in the economy as a whole. The economy can be 'disintegrated' without loss of consumers' surplus or gains from trade.

The analogy to ordinary production theory can be made more precise in an idealized special case, that of transferable utility – where utility can be regarded as cardinal and additive over individuals. (Notice this is essentially equivalent to assuming that everyone always enjoys constant marginal utility from income.) In this case, one can construct an analogy to an ordinary firm production function for the economy as a whole (a sort of 'aggregate production function'), and one can show that in the limit, replication will result in this function exhibiting constant returns. Further, in a perfectly competitive equilibrium all individuals will be rewarded with their marginal products to the economy as a whole, calculated from this 'aggregate production function'.

This idealized special case is useful for gaining parable-like insights into the nature of not only perfect competition, but also imperfect competition. So we shall first sketch some of the claims made for it above (for further details, see Makowski and Ostroy, 1987). The basis for its usefulness is that, if we assume utility is cardinal and additive over individuals, then we can formalize the idea that the economy as a whole is in the business of producing *utility* for its participants. In particular, with this assumption we can let $g(S)$ equal the total potential gains from trade possible in a subeconomy consisting only of the set of individuals S; i.e. $g(S)$ equals the maximum total utility achievable by S when it can only trade within itself. Then we can regard g, the total potential gains from trade *function* (defined over all possible subeconomies S) as the economy's 'aggregate production function'. Notice that the range of g is defined in utility space: the economy as a whole produces utility as its output. And the domain of g is subsets of individuals: individuals are the 'inputs' used to produce utility, by exploiting the gains from trade. (In cooperative game theory, the g function would be called a 'characteristic function'. But we shall restrict our attention to non-cooperative, bilateral interactions; this may be rationalized by assuming that multilateral coalition formation is prohibitively costly.)

Just as with any production function, we can define the marginal product of each factor of production – now each individual rather than each commodity since the domain of g is subsets of individuals. In particular, it is natural to define the potential marginal product of individual i to the economy as a whole, MP_i, as the difference between the potential gains from trade in the economy as a whole, $g(A)$ (where A is the set of all individuals) and the potential gains from trade in the absence of individual i, $g(A^i)$ (where A^i is the set of all individuals in the economy except i); i.e. $MP_i \equiv g(A) - g(A^i)$. Notice MP_i just equals individual i's contribution to the total potential gains from trade in the economy.

It can be shown that in any perfectly competitive economy, each individual's final utility level (say u_i) just equals his potential marginal product to the economy as a whole. That is, the total gains from trade are distributed under perfect competition so that $u_i = MP_i$ for each individual i. Thus the analogy to ordinary production theory under perfect competition, where each factor earns its MP, is complete. Since any perfectly competitive equilibrium is efficient (i.e. the actual gains from trade equal the maximum potential gains), this implies there must be 'adding-up' in any perfectly competitive economy: the sum of all individuals' MP's to the economy as a whole must equal the total potential product of the economy, $g(A)$.

Constant returns and adding-up are intimately related. Both are achieved by replication as follows. Typically the above g function will initially exhibit increasing returns in the sense that the sum of all individuals' MP's will exceed the total potential 'output'. But, for larger and large economies this sum approaches $g(A)$. The process is idealized in the limit – when we can regard individuals as infinitesimal, i.e. points on a line. In this limiting, continuum-of-individuals case the g function will be homogeneous: multiplying all 'inputs' by any factor will just multiply the total achievable gains from trade by the same factor. Hence, 'adding-up' in the limit is ensured by Euler's Theorem. (Individual i's potential marginal product in the limiting, continuum economy just equals the partial derivative of g with respect to that individual, evaluated at A, rather than the finite different $g(A) - g(A^i)$.)

Thus, the connection between replication, constant returns, and the nature of perfectly competitive economies is clarified. In particular, we now see that such economies exhibit, in the limit, constant returns *over* (the 'inputs') *individuals*. One deeper result from perfect competition theory will also be useful, before we leave this benchmark case for the domain of imperfect competition. It can be shown that not only does perfect competition imply

(i) $u_i = MP_i$ for each individual i; and (ii) $\Sigma MP_i = g(A)$, but conversely, (i) and (ii) also imply perfect competition. Thus, perfectly competitive economies are essentially *equivalent* to ones in which constant returns over individuals prevails. In the absence of such constant returns, we could not rely on Euler's Theorem to ensure adding-up, (ii); consequently, it would be a mere accident if one could reward everyone with their MP's to the economy as a whole.

This last, equivalence observation provides us with a key for transiting into the realm of imperfect competition. Since the presence of constant returns over individuals essentially characterizes perfectly competitive economies, its absence essentially characterizes economies without perfect competition, i.e. economies in which competition must necessarily be *imperfect*. But under what circumstances will competition necessarily be imperfect? Or, expressed in terms of our idealized special case, under what circumstances will the g function not exhibit constant returns over individuals?

The replication image of perfect competition gives us our

first insight into such imperfectly competitive economies. They are economies in which there are not sufficient perfect substitute sellers or buyers for the force of competition to ensure that no individual can influence the levels of market clearing prices. But what does this mean in terms of our gains from trade function?

As noted above, in the absence of perfect competition (e.g., in small economies) the g function will typically exhibit *increasing returns* over individuals, in the sense that the sum of all individuals' MP's will typically exceed the total potential gains from trade. To illustrate with a paradigmatic example of imperfect competition – bilateral monopoly – consider an economy with just one buyer and one seller, and with potential gains from trade between them. Then each individual is *crucial* to realizing the gains from trade. In particular, without either there would be zero gains from trade, so the MP of *each* equals the total potential gains from trade, $g(A)$. But then the sum of their MP's exceeds the total potential gains from trade since $\Sigma MP_i = g(A) + g(A) = 2g(A)$. So, there are increasing returns over individuals in bilateral monopoly situations. Obviously each person cannot appropriate all of $g(A)$.

That the sum of the two individuals' MP's exceeds the potential gains from trade between them has the following interpretive significance. Imagine the buyer and seller contending with one another over their respective shares of the total economic pie, $g(A)$. Each might insist on receiving his full potential contribution to the size of the pie, his MP. But in cases of imperfect competition, this is impossible to achieve. (Note that, by contrast, under perfect competition each seller (respectively, buyer) receiving his full MP would be the *inevitable outcome of competition* between alternative competing buyers of the seller's output (respectively, alternative competing sellers to the buyer). The consequence in terms of prices is that under perfect competition no one buyer (respectively, seller) can influence the level of market clearing prices.) We might next imagine each individual engaging in devious bargaining tactics to win at least as much of the pie for himself as he can. Such manoeuvrings are generally resource costly, hence the whole size of the pie may well diminish in the process of bargaining for shares of it. This is the image of wasteful competition! Our story indicates how increasing returns over individuals, and the consequent failure of adding-up of individuals' MP's to the economy as a whole, can give rise to wasteful competition. That the potential economic pie cannot be naturally imputed to individuals, via their contributions to the size of the pie (their MP's), makes the potential gains from trade a common property resource to be contended over wastefully.

2. AN EXAMPLE OF WASTEFUL COMPETITION. To make the discussion more concrete, we now present a more explicit example involving bilateral monopoly. Imagine an economy with just one barber B and one customer, C. B can cut hair costlessly, and C is willing to pay up to w dollars for one haircut (he does not want more than one); hence $g(A) = w$ which, recall, also equals each individual's MP. Will the full potential gains from trade be realized?

Suppose at the beginning of the world nature picks C's willingness to pay for a haircut from a distribution between 0 and 10, so that any w in this interval is an equally likely choice by nature. Suppose further that Mr C knows his actual type, w, but Mr B only knows the distribution from which nature has picked C's type. Then bilateral bargaining will not generally result in all the potential gains from trade being realized. To see why suppose B is a tough bargainer and can commit himself to a take-it-or-leave-it price for a haircut.

Then, given his incomplete information about C's type, it is easy to see he will commit himself to a price of \$5/haircut; this maximizes his expected profits. But then, whenever C's true willingness to pay is less than \$5, he will not get a haircut although it is efficient for him to do so given B's cost of haircuts is zero; $g(A)$ will not be realized. For example, suppose $w = \$4$, then although C may go to B and say 'I am willing to get a haircut if you will lower the price to something less than \$4,' B will rationally not believe him and change his price, since if he believed C in this case then C would rationally pretend to have a w less than \$5 even when his true w is greater than \$5.

Notice that the basic source of the inefficiency when $w < \$5$ is the potential deviousness by C about his true willingness-to-pay – in an effort to induce a lower price and hence a bigger share of his full potential marginal product, w – coupled with B's contrary effort to extract the biggest possible share of *his* potential marginal product, w, by making a price commitment that reflects his ignorance about w. Summarizing, (wasteful) competition between B and C over the potential gains from trade results in the actual gains, zero, falling short of the potential gains, w, whenever $w < \$5$. Wasteful competition is reflected in the *underproduction* of haircuts.

In contrast, notice that in a replicated economy with many identical B's and C's, (perfect) competition between barbers for customers would force the price of haircuts down to their true cost, zero. Hence, the full potential gains from trade would be realized without devious, wasteful competition. (The reader can check that in this replicated case the MP of any one barber equals zero while that of any one customer equals w; hence there is 'adding-up' in this case.)

The fact that imperfect competition is generally inefficient – it frustrates Adam Smith's Invisible Hand – is so central to our understanding of the economic import of imperfect competition that it is perhaps useful to re-phrase the source of market failure under imperfect competition in terms of 'externalities' since it is well-understood that externalities give rise to market failures. Under perfect competition each individual appropriates his full potential contribution to society, his MP. Consequently, he creates *no externalities*, beneficial or harmful, to others. By contrast, under imperfect competition not everyone can appropriate his full potential contribution to society, his MP. Consequently, if an equilibrium allocation with imperfect competition is to be efficient, some individual(s) must create external benefits *for others* (since some individuals must receive less than their marginal products). But no one cares about *external* benefits, only about the benefits he can internalize (i.e. appropriate). Consequently, in trying to internalize as much of his contribution as possible, an imperfect competitor will engage in wasteful market tactics most of whose harmful consequences others must bear.

Multilateral examples of imperfect competition, more in the spirit of Chamberlin, can also be constructed. In such examples, increasing returns lead to the gains from trade between producers and consumers being a common property resource that cannot be naturally imputed to agents using the MP reward principle. This can lead to 'excess capacity' as some industries' potential profits become a common property resource to be contended over wastefully via over-entry. In contrast, under perfect competition entry is efficiently guided since each firm's profits just reflect *its* MP; not any share of some other firm's potential MP that it can steal away by entering the industry.

Notice that throughout this article we are supposing there do not exist any non-market external effects between economic agents. So, all interactions are voluntary and involve exchange.

But this does not exclude the possibility of external effects between economic agents in their trade relationships, so called 'pecuniary externalities'. Indeed, the possibility of such trade-related externalities is the essence of imperfectly competitive interactions and the source of the Invisible Hand's failure to achieve Pareto efficient outcomes under imperfect competition. (A terminological note: We refer to imperfect competition as 'wasteful' relative to the benchmark of achieving pure Pareto efficiency. A related question that we do not address in this entry is: Can one find institutions that could improve on the market outcome in the presence of imperfect competition? Some economists would argue that the answer is 'no'; hence that the market outcome provides the best *realistic* benchmark even in the presence of imperfect competition, for example see Demsetz, 1959.)

3. INDIVISIBILITIES, COMPLEMENTARITIES AND INCREASING RETURNS. There is a tradition in economic theory that views some sort of indivisibility as the main source of increasing returns. In this tradition, if just doubling the amounts of all factors results in more than double the output, the source of increasing returns is interpreted in terms of indivisibilities in some specialized functions of factors.

That indivisibilities are the usual source of increasing returns was disputed by Chamberlin in a famous controversy with Kaldor; the latter subsequently recanted his position (see Kaldor, 1972). Without clouding ourselves in the smoke raised by this issue, we can shed some light on the central substantive aspect. At the heart of the dispute is the question, will sufficiently large economies necessarily be perfectly competitive? (Notice that the idea of indivisibilities suggests that at some sufficiently large level of production all scale economies will be exhausted.) Thus it is interesting to observe that increasing returns can exist even in large economies.

In particular, *how* one replicates an economy is crucial to whether a replicated economy will become closer and closer to a perfectly competitive one. For perfect competition to result in the limit, (1) it is essential that one only allows private goods, not collective goods: replicating an economy with collective goods generally does not diminish the presence of monopsony on the buyers' side since each buyer never competes with other buyers for units of a *collective* good. This monopsony power gives rise to manifestations of wasteful competition by each buyer – to try to appropriate the biggest possible share of his contribution to the gains from trade, his *MP* – such as 'free rider problems'.

(A bibliographical note: Samuelson introduced the concept of collective goods to Anglo-American economists in a series of articles (Samuelson 1954, 1955 and 1958). He forcefully argues that public goods differ fundamentally from private goods insofar as the ability of the Invisible Hand to allocate them efficiently is concerned. One can detect, in reading his three articles chronologically, a maturing in Samuelson's appreciation of the source of market failure as being due to some sort of increasing returns in public good economies. This point is made in Head (1962), whose article may be read as a useful complement to this one. Head stresses difficulties in appropriation as the source of market failure with collective goods, without explicitly using the MP concept.)

More in the spirit of Chamberlin, (2) it is also essential to keep the set of private commodities relatively fixed while one replicates: if the set of commodities expands at the same rate as the set of buyers and sellers, then perfect competition need not emerge even in the limit. Some sellers may still be 'special' as far as some buyers are concerned; thus a seller may still face a downward sloping demand curve reflecting the tastes of

buyers who regard the seller's product as special (e.g., see Hart, 1985). In this context, the right image of a large economy is an ever-expanding nexus of complementarities between individuals, that never becomes large enough to be 'disintegrated' without loss in potential gains from trade (Kaldor, 1972, and Allyn Young's classic 1928 paper may be usefully read on this point). In this image the possibilities for increasing returns are never exhausted since essential complementarities between individuals are never exhausted. Notice that the reason for increasing returns here is more easily explained in terms of the existence of *complementarities* between individuals, rather than indivisibilities. Expressed in terms of our idealized special case, as long as there exist essential complementarities between individuals, the gains from trade function will continue to exhibit increasing returns *over individuals*. In this common case, the force of competition will not be sufficient to guarantee that everyone has perfect substitutes. Thus competition between individuals may remain imperfect – and wasteful – even in large economies.

LOUIS MAKOWSKI

See also COMPETITION; ENTRY AND MARKET STRUCTURE; MONOPOLISTIC COMPETITION; PERFECTLY AND IMPERFECTLY COMPETITIVE MARKETS.

BIBLIOGRAPHY
Chamberlin, E.H. 1933. *The Theory of Monopolistic Competition.* Cambridge, Mass.: Harvard University Press.
Demsetz, H. 1959. The nature of equilibrium in monopolistic competition. *Journal of Political Economy* 67, February, 21–30.
Hart, O.D. 1985. Monopolistic competition in the spirit of Chamberlin. *Review of Economic Studies* 52, 529–46.
Head, J.G. 1962. Public goods and public policy. *Public Finance/Finances Publiques* 17(3), January, 197–219.
Kaldor, N. 1972. The irrelevance of equilibrium economics. *Economic Journal* 82, December, 1237–55.
Makowski, L. and Ostroy, J.M. 1987. Vickrey–Clarke–Groves mechanisms and perfect competition. *Journal of Economic Theory*, June.
Robinson, J. 1933. *Economics of Imperfect Competition.* London: Macmillan.
Samuelson, P.A. 1954. The pure theory of public expenditure. *Review of Economics and Statistics* 36, November, 387–9.
Samuelson, P.A. 1955. Diagrammatic exposition of a theory of public expenditure. *Review of Economics and Statistics* 37, November, 350–56.
Samuelson, P.A. 1958. Aspects of public expenditure theories. *Review of Economics and Statistics* 40, November, 332–8.
Samuelson, P.A. 1967. The monopolistic competition revolution. In *Monopolistic Competition Theory: Studies in Impact. Essays in Honor of Edward H. Chamberlin*, ed. R.E. Kuenne, New York: Wiley.
Young, A. 1928. Increasing returns and economic progress. *Economic Journal* 38, December, 527–42.

imperfectionist models. The term 'imperfectionist' was applied by Eatwell and Milgate (1983) to those models which rely on imperfections or arbitrary constraints in order to analyse the phenomenon under consideration. In other words, an imperfectionist analysis involves the construction of a model which, when innocent of those arbitrary constraints, does not display the phenomenon. The leading species of this genus to be found in economics today are models of unemployment in which imperfections such as sticky prices, or the effects of uncertainty, are imposed on a Walrasian model, thus disrupting the Walrasian relationship between price formation and the determination of levels of output which implies clearing of the markets for endowments of factor services.

The key issues in any consideration of the relationship between the theory of output and the theory of value and distribution can be revealed by the answers given to two questions:

(1) Does the determination of relative prices in a market economy also involve the determination of the size and composition of output and, in particular, is the level of output such that labour is fully employed (in the sense that at the going wage all workers willing to offer labour would be able to find employment)?

(2) Are variations in relative prices associated with variations in output such that the economy tends towards a level of output compatible with the full employment of labour?

Each of these questions can be supplemented with a further question: if not, why not?

The significance of these questions can be illustrated in terms of the most elementary piece of orthodox neoclassical analysis. According to this account, 'equilibrium' is determined at the point of intersection of a function relating price to quantity demanded and another relating price to quantity supplied. When this view of price determination is extended to the economic system as a whole, the equilibrium position of the economy is characterized by a set of market-clearing prices, with associated quantities (levels of commodity output and levels of 'factor' utilization), such that the markets for all commodities and all 'factors of production' clear. In particular, the labour market clears at the equilibrium level of the wage (relative to the associated set of equilibrium prices).

In terms of this familiar approach to the analysis of price formation the answer to the first question is obvious. Equilibrium prices and equilibrium quantities are determined simultaneously. The theory of value, based on demand and supply, is one and the same thing as the theory of output. If there exists an equilibrium set of prices then there exists an equilibrium set of outputs – equilibrium in the sense of market clearing, including the full employment of labour, as defined above. Furthermore, this theory of the simultaneous determination of prices and quantities is typically presented in such a way – by juxtaposing demand and supply functions – that the idea that prices adjust automatically so as to clear markets, thus tending to push the economic system towards a full-employment level of output, seems to follow as a self-evident corollary of the theory. (It does not in fact follow as readily as might appear at first sight, since the stability of an equilibrium is far more difficult to demonstrate than its existence.)

Here, then, one has the demand-and-supply (neoclassical) analysis of prices and quantities in a nutshell: the equilibrium set of outputs (and levels of 'factor' utilization) is determined simultaneously with the equilibrium set of prices (of commodities and 'factors of production'); variations in relative prices sparked off by an imbalance between demand and supply, will be associated with variations in quantities in a direction which ensures that both prices and quantities tend towards their equilibrium levels. Neoclassical analysis, therefore, answers the first two questions posed above in the affirmative.

An analysis of unemployment may then be derived directly from these relationships between prices and quantities. Any *inhibition* to the tendency of prices and quantities to find their equilibrium (market-clearing) levels will leave the economic system in disequilibrium with, perhaps, either an excess demand for labour or an excess supply of labour (ie unemployment). An enormous variety of analyses of unemployment are constructed in this way.

The general tenor of the neoclassical analysis of the causes of unemployment is that while the economy would be self-regulating in the best of all possible worlds (ie the implicit tendency towards the full employment of labour would be realized) – the market is *inhibited* from fulfilling this task by the presence of certain 'frictions' or 'rigidities'. In the literature on the problem of unemployment, examples of such inhibitions are legion. They include: 'sticky' prices (particularly 'sticky' or even rigidly fixed wages and/or 'sticky' interest rates); institutional barriers to the efficacy of the price mechanism, such as monopoly pricing (by firms or individual groups of workers); inefficiencies introduced into the working of the 'real' economy by the operations of the monetary system; the failure of individual agents to respond appropriately to price signals because of disbelief in those signals, the disbelief being derived from uncertainty about the current or future state of the market, or from incorrect expectations concerning future movements in relative prices, or from false 'conjectures' about the actual state of the market.

Indeed, examples of 'frictions' and 'rigidities' can be multiplied at will – any factor which causes the market to work *imperfectly* will do. It will be convenient, therefore, to group all the authors of the myriad of arguments of this kind together under the general heading of 'imperfectionists'. (It should be noted that by referring to this kind of analysis as 'imperfectionist' I do not intend to imply that the envisaged failure of the market mechanism to operate in the way depicted by the underlying demand-and-supply theory *necessarily* derives from imperfections of competition.)

Underlying them all is a fundamental similarity: that if the particular aspect (or aspects) of the economic system which gives rise to the breakdown of the market mechanism were to be absent, then the system would tend towards the full employment of labour (and other 'factors of production'). Thus, in all cases, the analysis of unemployment is viewed as no more than an aspect of the neoclassical theory of value and distribution. According to this approach, whether a relatively 'optimistic' or 'pessimistic' stance is taken with respect to the efficacy of the market mechanism in promoting full employment, the analysis of output and employment is part and parcel of the theory of relative price determination. This is so even in the case of those imperfectionists who feel that the essential workings of the theory are distorted gravely in the real world.

In marked contrast to the analysis outlined above are those theories of employment which propose no particular functional relationship between prices and quantities. The central proposition of neoclassical analysis, that the theory of value and distribution is also the theory of output, is rejected, together with the connected notion that appropriate variations in relative prices will promote variations in quantities, so moving the economic system in the direction of a full-employment equilibrium.

Unfortunately, this rejection of the neoclassical theory of value and distribution – of the entire apparatus of demand-and-supply analysis – has not always been backed up by rigorous analytical argument; so much so that it has sometimes been confused with an imperfectionist position. A striking example of this is the rejection by a number of writers of the neoclassical theory of value, and their advocacy of the idea that relative prices, far from being determined by demand and supply, are determined by a mark-up over normal prime cost where this mark-up is insensitive to variations in the conditions of demand (see, for example, Kalecki, 1939; Neild, 1963; Godley and Nordhaus, 1972). Quite apart from the obvious shortcomings of 'mark-up' analysis as a theory of price formation – it is in essence a proposition about the stability of the ratio between prices and costs rather than a theory about the determination of either of those magnitudes, or even of the size of the ratio – this attempt to separate the study of relative price determination from the analysis of output may readily be confused with an imperfectionist argument based on 'sticky' prices arising from the presence of monopolistic or oligopolistic influences in commodity markets. (Thus Malinvaud (1977) cites the results of Godley and Nordhaus (1972) in support of his orthodox imperfectionist position.) Moreover, the bald assertion that prices and quantities do not bear the well-defined functional relationship to one another that is postulated in neoclassical theory does not provide a satisfactory analytical basis upon which to build up a critique of the neoclassical position.

Yet the requisite critique does exist, and is to be found in the outcome of the debate over the neoclassical theory of distribution and, in particular, over its treatment of 'capital' as a 'factor of production' on a par, so to speak, with land and labour. While this debate is seen by many as a rather esoteric controversy in the more abstract realms of economic theory, its implications are more far-reaching than has hitherto been appreciated. The central conclusion of the debate may be summed up, in broad terms, as follows: when applied to the analysis of a capitalistic economy (that is, an economic system where some of the means of production are reproducible), the neoclassical theory is logically incapable of determining the long-run equilibrium of the economy and the associated general rate of profit whenever capital consists of more than one reproducible commodity. Since, in equilibrium, relative prices may be expressed as functions of the general rate of profit, the neoclassical proposition that equilibrium prices are determined by demand and supply (or, more generally, by the competitive resolution of individual utility maximization subject to constraint) is also deprived of its logical foundation.

The relevance of this critique of the neoclassical theory of value and distribution to the problem of the missing critique of the neoclassical theory of output and employment should be apparent from what has already been said. Because the neoclassical analysis of the determination of prices and the determination of quantities is one and the same theory (that of the mutual interaction of demand and supply), the critique of the neoclassical theory of value is simultaneously a critique of the neoclassical theory of output and employment. Therefore, the first of the two questions that were posed at the very outset of this discussion must, on the grounds of the requirement of logical consistency alone, be answered in the negative. The second question, from which neoclassical theory derives the idea that under the operation of the market mechanism there is a long-run tendency towards a determinate full-employment equilibrium, is rendered superfluous.

But this is not all. If the general (or long-run) case of the neoclassical model has been shown to be logically deficient, then all imperfectionist arguments of the introduction of particular (or short-run) modifications into the general case – are incapable of providing a satisfactory analysis of the problem of unemployment. This is not to say that many of the features of the economic system cited by the imperfectionists will have no role to play in a theory of employment based on quite different foundations to those adopted by the neoclassicals. After all, much of the credibility of imperfectionist arguments derives from their pragmatic objections to the direct applicability of the assumptions of the more abstract versions of demand-and-supply theory. But pragmatism is not enough. The implications of more realistic hypotheses must be explored in the context of a general

theoretical framework within which they are integral parts, not imperfections.

JOHN EATWELL

See also KEYNESIANISM.

BIBLIOGRAPHY
Eatwell, J. and Milgate, M. (eds) 1983. *Keynes's Economics and the Theory of Value and Distribution.* London: Duckworth.
Godley, W.A.H. and Nordhaus, W.D. 1972. Pricing in the trade cycle. *Economic Journal* 82, September, 853–82.
Kalecki, M. 1939. *Essays in the Theory of Economic Fluctuations.* London: Allen & Unwin.
Malinvaud, E. 1977. *The Theory of Unemployment Reconsidered.* Oxford: Basil Blackwell.
Neild, R.R. 1963. *Pricing and Employment in the Trade Cycle.* Cambridge: Cambridge University Press.

imperialism. Few subjects of such conspicuous historical importance have so consistently escaped lucid theoretical exposition as imperialism. The neoclassical economists have made no theoretical gains whatsoever in the field, having chosen to ignore the subject altogether. Their starting and ending point is a short essay borrowed from Schumpeter in which imperialism in the 19th and 20th centuries is attributed to the atavism of states, acting on feudal and absolutist impulses from an earlier precapitalist era. The field, therefore, has been dominated by Marxists. 'To write about theories of imperialism is already to have a theory,' states Barratt Brown (1972). In modern times, just to use the word is to label what is said as Marxist. The word – like capitalism itself – also implies a theory of broadly construed economic systems and long historical epochs. The sweep of the subject matter is reflected in the breadth of the two major propositions that Marxists have posed: that imperialism and monopoly capitalism are synonymous; and that capitalism underdevelops the third world. The sweep of the subject matter has lent itself to meaningless generalizations and reductionist arguments. But to ignore imperialism altogether on the ground that it is a political phenomenon is to abrogate a responsibility to study a major dimension of economic life, in particular the relationship between the operations of the market and coercive mechanisms.

Part of the problem lies in the ambiguity of the term. Since there is no agreement on the referent of imperialism, there is none on the meaning of the word itself. Marx and Engels did not discuss imperialism as such so they bequeathed no definition. To one of their followers, Rosa Luxemburg (1913), it was the political expression of the accumulation of capital in its competitive struggle for what is still left of the non-capitalist regions of the world. To another, Nikolai Bukharin (1917), it was a policy of conquest by finance capital that is characteristic of one stage of capitalist development. To a follower of a later generation, Samir Amin (1976), it was the perpetuation and expansion of capitalist relations abroad by force or without the willing consent of the affected people. Schumpeter (1919) defined it as the objectless disposition on the part of a state to unlimited forcible expansion.

While no consensus exists, most definitions share an idea that interactions between two social formations are in some sense imperialist if they depend upon force. And the use of force is all the more likely if the two entities are of unequal strength. This is not to say that only military domination qualifies as imperialism. Or that any exchange, commercial or financial, between two parties of unequal strength is imperialism. Rather, even if the use of force is only implicit, perpetrated by the fountain pen, it qualifies as imperialist if the weaker collectivity is subjected to some sort of control by the stronger. So defined, and such is the definition followed below, imperialism is ultimately a political phenomenon, whatever its underlying tap-root.

There appear to be as many explanations for the motivations underlying imperialism as there have been wars. Yet the economic explanations are qualitatively distinct from the rest – geopolitical, psychological – because they reflect the fact that different economic systems reproduce themselves differently. In societies where reproduction was constrained by the availability of land, territorial expansion was the impetus. In societies dependent upon slavery, there was warring for slaves. To buy cheap and sell dear in the age of mercantilism, there was resort to plunder. Come the capitalist system, imperialism evolved into something more complex than theft. It was embodied in exchange relationships. And since exchange could occur peacefully, without the use of force, some, like Schumpeter, presumed that capitalism and imperialism were antithetical. Yet force has been used to accelerate the onset of exchange relationships, to preserve them, and to improve the terms of exchange. Imperialism under centralized planning involves still another dynamic, since the driving imperative for markets (for economic surplus) is absent. It has been attributed by Ota Sik, the Czechoslovak planner, to the requirement of reducing uncertainty through the control of inputs and outputs (Owens and Sutcliffe, 1972). A complex of causes, however, is evident even for an imperialism defined sensibly for a specific historical period. The so-called 'new imperialism', which is the concern here and which dates from the 1870s–80s and onwards, is attributed to economic factors by, say, Hobson (1902) and Hilferding (1910); to European diplomatic rivalries by Fieldhouse (1966) and Langer (1935); and to extreme nationalism by Hayes (1941) and Mommsen (1980).

Precisely where to draw the dividing line between imperialist episodes, however, is contentious; and more than a mere theoretical quibble in the case of the 'new imperialism'. Robinson and Gallagher (1953) argue that there is little that distinguishes the allegedly 'indifferent' mid-Victorian imperialism, when free-trade beliefs were at their height, from the 'enthusiastic' late-Victorian imperialism, when such beliefs were in decline, along with British competitiveness. According to the authors, the indifference–enthusiasm polarization leaves out too many of the facts. There were numerous additions to empire, both formal and informal, in the indifferent decades. Between 1841 and 1851 Great Britain occupied or annexed New Zealand, the Gold Coast, Labuan, Natal, the Punjab, Sind and Hong Kong. In the next twenty years British control was asserted over Berar, Oudh, Lower Burma and Kowloon, over Lagos and the neighbourhood of Sierra Leone, over Basutoland, Griqualand and the Transvaal; and new colonies were established in Queensland and British Columbia. What is more, in the supposedly laissez-faire period, before the 1870s, the economy of India was managed along the best mercantilist lines. Such continuity in 19th century imperialism contradicts 'those who have seen imperialism as the high stage of capitalism and the inevitable result of foreign investment ... [in] ... the period after the 1880s', Lenin included.

Lenin's towering influence on Marxist theorists derives from his pamphlet, *Imperialism, the Highest Stage of Capitalism*, written in 1916 in response to the outbreak of war. The academic establishment in Europe attributed the First World War mostly to the official mind. Lenin ascribed it to monopoly capitalism, the economic mainspring of imperialist rivalry:

Railways are a summation of the basic capitalist industries: coal, iron and steel; ... The uneven distribution of the railways, their uneven development – sums up, as it were, modern monopolist capitalism on a world-wide scale. And this summary proves that imperialist wars are absolutely inevitable under such an economic system ... (Preface, pp. 4–5).

The economic system of monopoly capitalism is first portrayed by Lenin as being highly productive. According to a US Commission that he cites, the trusts expand their market share on the basis of scale economies and superior technology: 'Their superiority over competitors is due to the magnitude of ... [their] ... enterprises and their excellent technical equipment.' This leads Lenin to state: 'Competition becomes transformed into monopoly. The result is immense progress ... . In particular, the process of technical invention and improvement becomes socialized' (p. 24). He goes on to argue, however, that industrial capital falls prey to finance capital. He also embraces the prevailing academic view of monopoly, that it is unproductive, although he is far more cautious about this than his followers were to be:

Certainly, the possibility of reducing cost of production and increasing profits by introducing technical improvements operates in the direction of change. But the *tendency* to stagnation and decay, which is characteristic of monopoly, continues to operate, and in certain branches of industry, in certain countries, for certain periods of time, it gains the upper hand (p. 119).

Stagnation, in turn, leads to the export of capital, but Lenin is vague in his explanation for why this should be so:

The necessity for exporting capital arises from the fact that in a few countries capitalism has become 'overripe' and (owing to the backward stage of agriculture and the impoverished state of the masses) capital cannot find a field for 'profitable' investment (p. 74).

The direction of capital exports is to the backward countries:

... surplus capital will be utilized ... for the purpose of increasing profits by exporting capital abroad to the backward countries. In these backward countries profits are usually high, for capital is scarce, the price of land is relatively low, wages are low, raw materials are cheap (p. 73).

For Lenin, therefore, imperialism becomes organically inseparable from monopoly capitalism. Whereas in common usage imperialism means forced economic gain on a global scale, to Lenin it means much more. The most concise definition he gives is 'imperialism is the monopoly stage of capitalism', uniquely characterized, it should be added, by capital export.

Capital exports rose dramatically after the turn of the twentieth century. Yet neither underconsumption, as expounded by Hobson, nor a superabundance of capital, as Lenin suggested, nor a declining profit rate, a conceivable consequence of rising capital investments at home, provide particularly good explanations. Instead, Magdoff (1972) argues that in addition to the immediate causes of the sudden upsurge of capital exports (more competitors, more exporters; more tariff walls, more foreign investment to jump them), '[t]he desire and need to operate on a world scale is built into the economics of capitalism' (p. 148). Competition creates pressures for the expansion of markets. The emergence of a significant degree of concentration does not mean the end of competition. 'It does mean that competition has been raised to a new level ... . Since capital operates on a world scale,... the competitive struggle among the giants for markets stretches over large sections of the globe' (p. 157). Although the scramble for colonies preceded rather than followed the rise of monopoly and capital exports, annexation was not what Lenin meant by imperialism. On the contrary, Sutcliffe states, in response to Robinson and Gallagher, 'it was a prelude to imperialism ... . The system changed its character at the end of the century because from then on both expansion and rivalry between the major capitalist powers would have to take new forms since the chances of territorial expansion had been exhausted' (Sutcliffe, 1972, p. 314).

Lenin based his analysis of imperialism on the stranglehold of finance capital, by which he meant the leading role that banks came to play in economic decision making. The financiers were perceived to have the biggest stake in imperialism and their hunger for quick returns led to economic chaos. Yet in fact after World War I finance capital decidedly took a back seat as the multinational firm grew in the US, Europe and, belatedly, England. As evidence for this, there was a shift over time away from indirect foreign investment, that is, portfolio or debt capital, to direct foreign investment, or equity capital. Roughly two-thirds of foreign investment took the form of debt capital before World War I. Thereafter, direct foreign investment became predominant, although a new type of portfolio investment rose again sharply in the late 1970s–early 1980s.

Chandler (1980) writes about the *form* that the growth of large-scale firms assumed:

> ... modern industrial enterprise ... grew by adding new units of production and distribution, by adding sales and purchasing offices, by adding facilities for producing raw and semi-finished materials, by obtaining ... transportation units, and even by building research laboratories ... (p. 397).

These new specializations of large business enterprises are the crux of Hymer's (1976) explanation for why capital exports were increasingly direct rather than indirect. According to him, the specializations that Chandler mentions – management expertise, capability in manufacturing, technology, distribution – constituted firm-specific monopolistic assets. To take full monetary advantage of them, firms exerted direct control over overseas operations, through equity ownership.

Yet foreign investment, whether direct or indirect, did not flow preponderantly to backward regions. In the interwar period and even before 1914, the main destination for overseas funds was Europe and North America. British colonies, including India, accounted for only about 20 per cent and South America, for another 20 per cent (Barratt Brown, 1972). After 1929, the share of the advanced countries in the inflow of direct foreign investment rose even further, reaching around 75 per cent of the total in the mid-1970s. The share was higher still for direct foreign investment in the manufacturing sector (USDC, various years). Thus, while the locus of socialist revolutions was backward regions, not advanced ones, capital exports flowed increasingly to advanced regions, not backward ones. The direction of foreign investment is significant because it suggests an altogether different centre of gravity in economic activity under monopoly capitalism from the one Lenin's followers entertained.

Beginning at the turn of the century, the principal orientation of the economic activity of advanced countries was, in general, toward each other, not the backward regions. Like foreign investment, foreign trade in manufactures largely engaged the advanced countries. Their competitive struggle involved mainly invasions of each other's markets. The major contest in economic strength after World War II, between the US and Japan, barely stretched to third world shores.

Explanations other than international differences in gross profit rates must be sought for the geographical distribution of foreign investment. No definitive data exist to compare profit rates across countries. Yet profit rates are likely to have been relatively higher in backward countries, as Lenin suggests, because rates of surplus value, in the Marxist accounting sense, were higher there, at least in the 1970s in the manufacturing sector (Amsden, 1981). One reason why foreign investment and trade primarily occupied the richer countries is that their per capita incomes were growing faster than the poorer regions; the newly industrializing countries excluded. The higher *level* of income in advanced countries also made them better markets. In turn, high income markets complemented the type of competition that became characteristic of monopoly capitalism. The monopolistic assets of large business enterprises were the competitive weapons. The coming of age of industrial capital witnessed an intensification of competition on the technology front. New products, new processes, new production systems constituted the razor's edge of the competitive battle, moderating the demand for protection and price-fixing cartels. Such technology was not designed with third world domination in mind. The location of industry in the course of a product cycle from the 1950s at least through the 1980s progressed from the innovating country, to other advanced countries and only belatedly to backward regions (Vernon, 1966); and then only if new discoveries did not short circuit the cycle such that production returned to the innovator's country of origin.

The monopolistic assets of large business enterprises did not all work productively, and Marxists pointed to the wasteful effects of advertising and to the ruinous effects of financial manipulation in the form of takeover waves at home and periodic, aggressive bouts of lending to the backward regions. But technological competition was the stuff out of which monopoly capitalism was made after World War II. So to equate monopoly capitalism and imperialism robs both terms of much of their meaning. The two cannot be reduced to one another.

Even if, following Stokes (1969), one attributes to Lenin what has come to be a non-'Leninist' view, that the contestation of imperialist rivalries occurs not in the third world but in the monopolized countries themselves, then the conflation of monopoly capitalism and imperialism is still obfuscating. Whereas such rivalries engaged Europe in war at the time Lenin was writing, they were mediated peacefully there for at least forty years after World War II.

Nor is Lenin especially illuminating on why capital exports are the *specifica differentia* of monopoly capitalism. Was foreign investment more likely to precipitate the use of force than foreign trade? No, because trade in raw materials in the 19th century presupposed foreign investment. And what is the significance of the shift from indirect to direct foreign investment? Marxists have not systematically explored the answer. History teaches us that finance capital increasingly falls under the control of a few large banks, but it comprises much less differentiable products than industrial capital and, therefore, is more at the mercy of the laws of supply and demand. To prevent interest rates from falling, the banks look overseas for profitable investment outlets, and when they compete on the basis of price, they look in particular to the backward regions. The upsurge of portfolio investment in the late 1970s–early 1980s was accounted for overwhelmingly by the third world. Presumably the backward regions will become a more important locale for industrial capital as technological competition among advanced countries grows more even and product differentiation converges. Then manufacturers may be expected to locate their production facilities in lower wage, higher profit countries in order to compete better on the basis of price. That they did not do so to any significant extent before the 1980s suggests not a shortage of profitable investment outlets in advanced countries, supposedly a hallmark of monopoly, but a surplus of such outlets. Even though profit rates in the manufacturing sector were lower in the advanced countries, assuming the numbers are correct, marginal profit rates are likely to have been equal or higher, due to an outpouring of innovations.

The backward regions, however, were hardly inconsequential, to either industrial or finance capital. Certain third world raw materials, not least of all petroleum, remained critical business cost factors. The third world's debt crises undermined global monetary stability. The 'defection' of third world countries to socialism precipitated armed intervention. And, while the capital that flowed from the advanced countries to the third world throughout most of the tenure of the 'new imperialism' amounted to a mere trickle, there was a massive net transfer of surplus from the third world to the advanced countries (Bagchi, 1982). Capitalism, after all, had become a world system. The relationship between imperialism and the economic development of the backward regions was the subject of as much literature as the relationship between imperialism and monopoly capitalism. Indeed, more was written on the former, because the neoclassical economists contributed; discreetly, the term imperialism never being mentioned.

Imperialism before and after World War II was quite distinct, as formal colonialism ended and large portions of Asia and Africa gained independence. One would expect economic growth in the backward regions to be quite distinct in each period as well, as a consequence of such political change. Yet, curiously, both Marxist and neoclassical economists saw continuity. In the neoclassical view, the backward regions had as good a chance to develop under colonialism as under independent rule so long as they organized their economies in the pursuit of comparative advantage. For the Marxists, underdevelopment was the expected outcome whatever the political regime, so long as the economic mechanisms of imperialism were fundamentally unaltered.

But how did these mechanisms operate? And did they remain unaltered amidst shifts in political circumstance? Schumpeter's argument, that imperialism under capitalism was a throwback to precapitalist impulses, was based on the premise that peaceful exchange was preferable to the use of force for all self-interested parties, and that ultimately reason would prevail over atavism. Yet, at minimum, force might be rational for one party to hasten another's *entry* into capitalist exchange relationships or to prevent another's *exit* into an altogether different economic system. The latter appears to have driven a good deal of US imperialism after World War II, notwithstanding the fact that the US had no precapitalist history. In the war's aftermath, American aid to Greece and Turkey limited leftist activity and the US government helped opponents of socialist and communist candidates for office in France and Italy. Vietnam apart, the US intervened either directly through the military or covertly through the Central Intelligence Agency to halt what was perceived as socialist aggression in Greece, Iran, Guatemala, Indonesia, Lebanon, Laos, Cuba, the Congo, British Guiana, the Dominican Republic, Chile and possibly Brazil.

The onset of capitalist relations in the third world was also replete with the use of force. In many colonies where foreign enclaves were established in the 19th century for the purpose of producing primary products for export, population was scarce, so in retrospect 'overpopulation' cannot be held responsible for the underdevelopment that ensued. Indeed, one would have expected not underdevelopment but the onset of a 'high wage economy', given a scarcity of labour and a growing demand for labour's services in the mines and on the plantations. But wages did not rise (Myint, 1964). For the neoclassical paradigm of peaceful market exchange, this constitutes a paradox. For more institutionally oriented economists, this seeming paradox was resolved with the artifice of the 'backward bending labor supply curve'. It was imagined that self-sufficient peasants who migrated to the mines and plantations offered their services with the limited purpose of obtaining only a 'target' income. If higher wages were paid, their objective would be met all the faster, with the consequence of a smaller, not larger labour supply. In fact, foreign firms in the mining and plantation sectors were faced with a decision – of whether to pay in excess of labour productivity in the short run or to coerce an adequate labour supply at a low wage rate equal to (or below) the prevailing level of productivity – and they opted for force. Colonial authorities passed legislation that indirectly compelled natives to work: but taxes were imposed that had to be paid in cash, not kind and alternative income-earning opportunities were limited through encroachments on land and restrictions on the cultivation of cash crops. The result was the onset of a 'low wage economy', that effectively channelled the 'secondary multiplier effects', of enclave production to the advanced countries and doomed the backward regions to a 'vicious circle' of poverty (Myrdal, 1957; Nurkse, 1953; Singer, 1950).

Outright appropriation of land and labour was more blatant in the earlier than the later phases of imperialism and in some backward regions (Indonesia, the Congo) than in others (India, Latin America). But it was often possible to extract more surplus through indirect taxation and through purchase of commodities and sale of manufactures from and to the peasants. 'The British', writes Bagchi, 'may indeed be regarded as the real founders of modern neocolonialism, for both in Latin America and in India in the late nineteenth century they depended more on economic power and political influence than on direct use of political power at every stage for obtaining the lion's share of the surplus of the dominated economies' (1982, p. 78). Land taxes, payable in cash, either reduced the peasants to landless proletarians or required them to produce export crops, with little surplus to diversify in the event of unfavourable terms of trade. Free trade itself destroyed domestic manufactures, made it unprofitable to invest in anything other than export crops and impeded the growth of capitalist classes that could have challenged foreign domination. Even in the bottom of the barrel, backward regions characterized by peasant export economies with little to offer foreigners in the way of raw materials or markets (say, West Africa, Burma, Thailand and Vietnam), the functioning of the market mechanism was not devoid of coercive elements. Peasants who entered the money economy became vulnerable to international commodity price fluctuations. Foreigners, acting as monopolistic middlemen, gained the upper hand and reinvested the surplus elsewhere. Local moneylenders, who controlled credit, foreclosed on indebted peasants where land had become alienable. Railways and other infrastructure supported external rather than internal exchange, thereby discouraging domestic manufactures.

In reality, therefore, no ideal, pure market exchange between rich and poor countries existed that could be delinked neatly from imperialism. Mechanisms of coercion and mechanisms of exchange operated hand-in-hand. From the Marxist perspective it followed that imperialism was neither atavistic nor limited merely to entry and exit to and from capitalist exchange. Rather, force was pervasive and imperialism was business as usual.

If, to varying degrees, force was pervasive in market relationships, then as force changed its colours in tandem with political change, one would expect some change in market relationships as well. Imperialism, after all, is a political phenomenon. Yet in the post-World War II period, no attempt was made by Marxists to distinguish the intrinsic from the historical effects of different economic practices on growth prospects. Instead, all intercourse with advanced countries was condemned as leading to underdevelopment, in sharp contradistinction to Marx, Engels, and even Lenin. The economic practices singled out for special opprobrium were those in which intercourse between the advanced countries and backward regions was most direct – foreign trade, foreign investment and even foreign aid. As Brenner (1977) put it, Adam Smith was turned on his head.

Yet the effect of any given economic practice on economic development clearly depended on the political setting. Aid helped Europe after World War II but seemingly hurt Bangladesh. Whereas export-led growth based on a primary product or 'staple' led to underdevelopment in the backward regions, it led to prosperity in the regions of recent settlement (Canada, Australia, New Zealand, white South Africa, white Rhodesia, etc.). Evidently there was nothing inherent in exporting that led irrevocably to either development or underdevelopment. Rather, what happened depended on local conditions. Unlike the backward regions, the regions of recent settlement retained the surplus by dint of their 'high wage economies' and reared a manufacturing sector by erecting protective tariffs. In the case of direct foreign investment, the expected gains to the 'host' country were *a priori* indeterminate. On the one hand, direct foreign investment promised a transfer of modern management techniques to backward regions. On the other hand, motivated by a wish to make use of monopolistic assets, there was nothing to insure that the multinationals would share their know-how with local managers. In fact, the outcome depended on the political conditions imposed on foreign capital; so Canada benefited far more than say, Chile, from overseas investment.

If Marxists saw foreign trade and foreign investment as dooming the third world to underdevelopment, neoclassical economists followed the same logic but arrived at an opposite conclusion: that foreign trade and foreign investment were the key to third world prosperity (Little, 1982). Now this flew in the face of reality. The economies of the backward regions had long been oriented to foreign trade and foreign investment but were hardly prosperous. Two different tacks were taken to reconcile any seeming inconsistency between theory and practice. One, it was argued that the backward regions had not been sufficiently singleminded in their pursuit of free trade. They had broken faith after World War II in particular, by embracing the 'dogma of dirigisme' (whereupon, it may be added, they grew the fastest ever; Lal, 1984). Two, it was argued that, in fact, the backward regions had long been growing at a fairly rapid clip, although to be sure, there were exceptions to the rule. According to Reynolds (1985): '... against the view that "life began in 1950," ... the third world has a rich record of prior growth, beginning for most countries in the 1850–1914 era' (p. 4). In anticipation of the obvious objection, that developing countries are still desperately poor, Reynolds writes:

Certainly people in Western Europe and the United States

are much better off than people in Sri Lanka [the example he uses], though not as much better off as the World Bank tables suggest ... conversion from local currencies to U.S. dollars at official exchange rates exaggerates the actual difference in consumption levels (p. 40).

Both Marxist and neoclassical analysis suffered from a failure to look beyond either the historical specificities of 'export-led exploitation' (the term is Bagchi's) or the formalism of export-led growth, as the case may be, to the underlying power structures in the backward regions. Beginning with Baran (1957), Marxists portrayed political and social life in the third world simplistically. The state and whatever local capitalists existed were seen as corrupt puppets of advanced country powers. No scope was given to the possibility of local initiatives to mediate foreign trade, foreign investment and foreign aid to advantage. It is fair to say the neoclassical economists largely ignored local conditions in developing countries, even economic ones. When Jacob Viner (1953) delivered a lecture series in Brazil in 1950, he expressed confidence in a growth strategy based on agricultural exports. As evidence, he pointed to the correlation between high per capita incomes and agricultural exports in the regions of recent settlement, overlooking any other factors in these regions that may also have contributed to growth. The result was an inability to grasp what came to constitute a serious challenge to both theories: the economic development along capitalist lines after World War II of a handful of nations (or nation states) in East Asia, South Korea and Taiwan in particular.

The development of these countries posed a challenge to neoclassical theory because, while all the countries in question were highly oriented to trade, they were by no means committed to laissez-faire (Amsden, 1985). They exerted strong centralized control over their economies. They flouted static comparative advantage and were protectionist. Their large private or public conglomerates were a mirror image of concentrations of economic power under monopoly capitalism in advanced countries. They fought force with force, as it were, in dealing with foreign capital. To say that these countries could have grown even faster had they adopted laissez-faire policies is beside the point. The development of these countries posed a challenge to Marxist theory because it wasn't supposed to happen. Such development, therefore, was preemptively dismissed. It was attributed either to a fluke – geopolitics and a superabundance of foreign aid [sic] – or repression of workers, although Engels (1878) cautions against the view that it is possible to industrialize by the gun.

The one dissenting voice among Marxists against the notion that capitalism underdevelops the third world missed the point. For Warren (1980), the problem of underdevelopment was not too much foreign capital but too little. Yet, however great the flow of foreign capital to South Korea and Taiwan (mostly, it may be noted, in the form of finance rather than industrial capital), much more accounted for development in these countries than capital per se.

The intellectual antecedents of Warren's view are traceable directly to Marx, so to suggest that Warren missed the point about economic development is also to suggest that Marx himself missed the point. Marx's point is that colonies like India were destined to develop because the capitalist system was compelled to replicate itself around the globe. With the destruction of the Asiatic mode of production, with the imposition of market relationships and with the arrival of the railroad, India would become another England (Marx and Engels, 1960). Yet markets and technology alone do not make for economic development. What appears to be critical are the power relationships and institutions that unfold on their own terms to guide the accumulation process. But Marx is silent about these.

The dirigiste state stands at the opposite extreme of Marx's liberal view of the market as the engine of growth. But neither is a dirigiste regime a sufficient condition for economic development. Dirigisme and underdevelopment are both rampant in the third world. Instead, what Japan and a few South Koreans suggest is that economic development in the 20th century hinges on a delicate relationship between the operations of the market and coercive mechanisms.

Marxists have focused on this relationship in the general case, which is the starting point for any theory of imperialism, and presuppose that markets and force are impenetrable. Yet their equation of imperialism and monopoly capitalism led them to misjudge the relationship after World War II, because imperialism was not the key to the rapid growth of the advanced countries. And their second *idée fixe*, that capitalism underdevelops the third world, led again to the relationship's misjudgement, because proof of economic development in even a handful of third world countries deprived their theory of analytical clarity. Nonetheless, to operate with the world view of the neoclassicists – of a separation between markets and power – is to deny the very existence of imperialism and to forego the conceptual tools to analyse it.

ALICE H. AMSDEN

See also COLONIALISM; COLONIES; HOBSON, JOHN ATKINSON; LENIN, VLADIMIR ILYICH; NATIONALISM; PERIPHERY; UNEQUAL EXCHANGE.

BIBLIOGRAPHY
Amin, S. 1976. *Unequal Development*. New York: Monthly Review Press.
Amsden, A.H. 1981. An international comparison of the rate of surplus value in manufacturing industries. *Cambridge Journal of Economics* 5(3), September, 229–49.
Amsden, A.H. 1985. The state and Taiwan's economic development. In *Bringing the State Back In*, ed. P. Evans et al., Cambridge: Cambridge University Press.
Bagchi, A. 1982. *The Political Economy of Underdevelopment*. Cambridge: Cambridge University Press.
Baran, P.A. 1957. *The Political Economy of Growth*. New York: Monthly Review Press.
Barratt Brown, M. 1963. *After Imperialism*. London: Merlin Press, 1970.
Barratt Brown, M. 1972. A critique of Marxist theories of imperialism. In Owen and Sutcliffe (1972).
Brenner, R. 1977. The origins of capitalist development: a critique of neo-Smithian Marxism. *New Left Review* No. 104, July–August, 25–92.
Bukharin, N. 1914. *Imperialism and World Economy*. New York: Monthly Review Press. 1972.
Chandler, A. 1980. The growth of the transnational industrial firm in the United States and the United Kingdom: a comparative analysis. *Economic History Review* 33, August, 396–410.
Engels, F. 1878. *Anti-Dühring*. New York: International Publishers.
Fieldhouse, K.D. 1966. *The Colonial Empires: A Comparative Study from the Eighteenth Century*. London: Weidenfeld & Nicolson.
Hayes, C.J.H. 1941. *A Generation of Materialism 1871–1900*. New York: Harper.
Hilferding, R. 1910. *Finance Capital: A Study of the Latest Phase of Capitalist Development*. London: Routledge & Kegan Paul, 1981.
Hobson, J.A. 1902. *Imperialism: A Study*. London: Allen & Unwin, 1938.
Hymer, S. 1976. *The International Operations of National Firms: A Study of Direct Foreign Investment*. Cambridge, Mass.: MIT Press.
Lal, D. 1984. *The Poverty of 'Development of Economics'*. London: Institute of Economic Affairs.

Langer, W.L. 1935. *The Diplomacy of Imperialism 1890–1902*. 2nd edn, New York: Knopf.
Lenin, V.I. 1973. *Imperialism, The Highest Stage of Capitalism*. Peking: Foreign Language Press.
Little, I. 1982. *Economic Development: Theory, Policy, and International Relations*. New York: Basic Books.
Luxemburg, R. 1913. *The Accumulation of Capital*. Trans. Agnes Schwarzchild, London: Routledge & Kegan Paul, 1951.
Magdoff, H. 1972. Imperialism without colonies. In Owen and Sutcliffe (1972).
Marx, K. and Engels, F. 1960. *On Colonialism*. London: Lawrence & Wishart.
Mommsen, W. 1980. *Theories of Imperialism*. Trans. P.S. Falla, New York: Random House.
Myint, H. 1964. *The Economics of the Developing Countries*. New York: Praeger.
Myrdal, G. 1957. *Rich Lands and Poor: The Road to World Prosperity*. New York: Harper.
Nurkse, R. 1953. *Problems of Capital Formation in Underdeveloped Countries*. New York: Oxford University Press.
Owen, R. and Sutcliffe, B. (eds) 1972. *Studies in the Theory of Imperialism*. New York: Longmans.
Reynolds, L.G. 1985. *Economic Growth in the Third World, 1850–1980*. New Haven: Yale University Press.
Robinson, R. and Gallagher, R. 1953. The imperialism of free trade. *Economic History Review*, 2nd Series 6(1), 1–15.
Schumpeter, J. 1919. *Imperialism and Social Classes*. Trans., New York: Augustus M. Kelley.
Singer, H.W. 1950. The distribution of gains between investing and borrowing countries. *American Economic Review, Papers and Proceedings* 40, May, 473–85.
Sutcliffe, B. 1972. Conclusion. In Owen and Sutcliffe (1972).
Stokes, E. 1969. Late nineteenth-century colonial expansion and the attack on the theory of economic imperialism: a case of mistaken identity? *Historical Journal* 12, 285–301.
United States Department of Commerce (USDC). (Various years.) *Survey of Current Business*.
Vernon, R. 1966. International investment and international trade in the product cycle. *Quarterly Journal of Economics* 80, May, 190–207.
Viner, J. 1953. *International Trade and Economic Development*. Oxford: Clarendon Press.
Warren, B. 1980. *Imperialism: Pioneer of Capitalism*. London: Verso.

implicit contracts. An implicit contract is a theoretical construct meant to describe complex agreements, written and tacit, between employers and employees, which govern the exchange of labour services when various types of job-specific investments inhibit labour mobility and opportunities to shed risk are limited by imperfectly developed markets for contingent claims. This construct differs from the more familiar one of a neoclassical labour exchange in emphasizing a trading process, frequently over a long period of time, between two *specific* economic units (say a worker and a firm, union and management, etc.) rather than the impersonal, and often instantaneous, market process in which wages decentralize and coordinate the actions of labour suppliers and labour demanders.

Adam Smith's exposition of occupational wage differentials (1776, book I, ch. 10) recognized very early the idiosyncratic nature of the labour market and, in particular, that employment risk affected wages in various occupations. Since then economists have accumulated many facts, raw or stylized, which are best understood if one abandons the traditional view that the shadow price of labour is simply the wage rate. Prominent among explananda are the widespread use of temporary layoffs as a means of regulating the volume of employment (Feldstein, 1975); the continuity of jobs by many primary wage earners (Hall, 1982); the collective bargaining

733

tradition of leaving the volume of employment at the discretion of management while predetermining money wage rates two or three years in advance.

To these, one must add certain 'impressions' or softer facts about the labour market which arise from the central role labour services possess in macroeconomic models. There is, indeed, among macroeconomists a shared impression (Hall, 1980) that, over a typical business cycle, average real compensation per hour fluctuates considerably less than does the marginal revenue-product of labour or, for that matter, the total volume of employment.

One consequence is that wage and price rigidity are among the key assumptions of Keynesian macroeconomics, both in the Hicksian IS/LM framework and in the concept of quantity-constrained equilibrium originally developed by Clower (1965) and formalized by Bénassy (1975) and Drèze (1975). Another is the overwhelming importance of words like 'jobs' and 'unemployment', both in our colloquial vocabulary and in the specialized lexicon of economics. In particular, 'involuntary unemployment' is for many academic economists the *sine qua non* of modern macroeconomics.

The technically minded reader will find many of these issues surveyed in a number of specialized papers of which the most recent are Hart (1983) and Rosen (1985).

WAGES AND EMPLOYMENT. The earliest literature on implicit contracts exploits an insight of Frank Knight (1921), who argued that inherently 'confident and venturesome' entrepreneurs will offer to relieve their employees of some market risks in return for the right to make allocative decisions. The formal development of this idea began with three independently written papers by Baily (1974), Gordon (1974), and Azariadis (1975), motivated by the seeming puzzle of layoffs. In an unusual coincidence, all three authors took the employment relation not simply as a sequential spot exchange of labour services for money, but as a more complicated long-term attachment; labour services are traded in part for an insurance contract that protects workers from random, publicly observed fluctuations in their marginal revenue-product. The idea was that workers could purchase insurance only from their employers, not from third parties.

Risk-averse workers deal with risk-neutral entrepreneurs who head firms consisting of three departments: a production department that purchases labour services and credits each worker with his marginal revenue-product (MRPL); an insurance department that sells actuarially fair policies and, depending on the state of nature, credits the worker with a net insurance indemnity (NII) or debits him with a net insurance premium; and an accounting department that pays each employed worker a wage, w, with the property that $w = MRPL + NII$ in every state of nature.

Favourable states of nature are associated with high values of MRPL; in these the net indemnity is negative and wage falls short of the MRPL. Adverse states of nature correspond to low values of MRPL, to positive net insurance indemnities, and to wages in excess of MRPL. An implicit contract is then a complete description, made before the state of nature becomes known, of the labour services to be rendered unto the firm in each state of nature, and of the corresponding payments to be delivered to the worker. The contract is implementable if we assume the state of nature is as easily verifiable as events are in a normal insurance contract.

An immediate consequence of this framework is that wages are disengaged from the marginal revenue-product of labour. In fact, if the amount of labour performed by employed workers per unit time is fixed institutionally, then each

worker's consumption is proportional to the wage rate; an actuarially fair insurance policy should make this consumption independent of the MRPL by stabilizing the purchasing power of wages over states of nature. Therefore, the real wage rate is rigid.

In traditional macroeconomic models of course, wage rigidity by itself is sufficient to cause unemployment: if wages do not adjust for some reason, then neither does the demand for labour. The argument does not carry over to implicit contracts because of the very separation between wages and the marginal revenue product of labour. A complete theory of unemployment must explain why layoffs are preferred to work-sharing in adverse states of nature, and why laid-off workers are worse off than their employed colleagues.

This is not a simple task if one thinks of implicit contracts as ordinary, explicit, timeless insurance contracts between risk-averse workers and risk-neutral entrepreneurs. All contracts of this type would share a basic property of optimum insurance schemes; namely, keeping the worker's marginal utility of consumption independent of all random, publicly observed events – including such events as 'employment' or 'unemployment'.

To explain layoff unemployment, we need to distort or complicate the insurance contract in some significant way. A distortion that was noted early in the implicit contract literature is the dole. In an extremely adverse state of nature, the flow of insurance indemnities to workers can become a substantial drain on profit; one way to staunch losses is to place the burden of insurance on an outside party, the dole.

The practice of layoffs is simply the administrative counterpart of this insurance-shifting manoeuvre; workers consent in advance that some of them may be separated from their jobs in order to become eligible for unemployment insurance (UI) payments from an outside public agency. Furthermore, no worker will contract his labour unless the expected value (utility) of the total package, taken over all possible states of nature, exceeds the value of being on the dole in every state. This means, in turn that employed workers receive a wage in excess of UI payments and are therefore to be envied by their laid-off colleagues – a situation that many economists would call 'involuntary unemployment'.

The fact that laid-off workers would gladly exchange places with their employed colleagues is not in itself sufficient to establish a misallocation of resources. After all, accident victims may very well envy more fortunate individuals without any implication that the insurance industry works poorly. Layoffs, by themselves, could be no more than the luck of the draw unless we can demonstrate that they constitute, in some sense, socially inefficient underemployment. This is clearly impossible within the Walras–Arrow–Debreu model; and it is for this reason that the early literature on contracts turned to institutions like the dole in order to explain layoff unemployment.

PRIVATE INFORMATION. One fundamental departure from the Walrasian paradigm that received much attention in the early 1980s was a weakening of the information assumptions: information becomes 'private' or 'asymmetric', which simply means that not everyone is equally informed about the relevant state of nature. This is a perfectly sensible observation, for what justifies the trading of implicit contracts in the first place is that third parties simply are not as well informed about someone's income or employment status as is his employer; the employer, in turn, may be less informed about an employee's non-labour income and job opportunities than is the worker himself.

The thread was picked up by a number of authors who studied the properties of wages and employment for two main cases: in the first, entrepreneurs possess superior information about labour demand (Hall and Lilien, 1979; Grossman and Hart, 1981; Azariadis, 1983; Farmer, 1984); in the second case, workers possess superior information about labour supply, as in Cooper (1983). Suppose, for instance, that wages and employment do not depend on the unobservable true state of nature but on what the better informed contractant (say, the employer) *announces* that state to be. The question now becomes how to design contracts that reward entrepreneurs who tell the truth and punish those who lie.

One desirable property of contracts is that the truth should be the value-maximizing strategy for firms: truth-telling ought to be consistent with equality between the marginal cost and the marginal revenue-product of labour. Furthermore, entrepreneurs who misrepresent actual conditions should be punished, say, for knowingly under-reporting demand.

Under-reporting demand does turn out to be a problem in contracts that permit employers to slash both workforce and the wage bill when demand is slack, and do it in such a manner as to reduce cost more than revenue. To avoid this temptation, a properly designed contract specifies a highly variable pattern of employment over states of nature; that is, one in which employment is below what is socially optimal and the marginal product of labour is correspondingly above the marginal rate of substitution between consumption and leisure. It is in this sense that asymmetric information is said to result in socially inefficient underemployment or unemployment.

What relation is there between the layoffs we all know and the inefficient underemployment of a model economy that suffers from asymmetric information? To go from the latter to the former, one must understand first why layoffs are a more common means of reducing employment than is work-sharing. Second, a general equilibrium picture of underemployment would require an explanation of why underemployed (or unemployed) individuals are not hired by other employers. Third, and most important, the unemployment found in this private-information story is a response to private, firm-specific risk; most economists, however, consider the unemployment observed in market economies to be a reaction to social risks, especially to business cycles set in motion by aggregate demand disturbances. Unless one intends to make the far-fetched claim that the general public is unaware of, or cannot observe, whatever disturbances set off business cycles (e.g. changes in government consumption, money supply or consumer confidence), does it not appear that information-based unemployment simply describes the behaviour of an isolated firm?

The answer is not obvious. Note, however, that in order to have an inefficient volume of equilibrium employment, it is sufficient that *some but not all* information be private. In fact, it is not difficult to imagine general equilibrium extensions of the work we are discussing that would include both public and private information. Such extensions will be useful, especially if they manage to establish a firm link between inefficient underemployment and extreme values of some publicly observed aggregate disturbance.

EMPIRICAL IMPLICATIONS. Whether information is publicly shared or in the private domain, wages in implicit contracts do not merely reflect the marginal product of labour or the workers' marginal rate of substitution between consumption and leisure, as they might in more conventional theories. The empirical implications of this insight are just being worked

out, and they seem to be quite considerable. At the most aggregative level, one can make sense of the oft-verified fact (Neftci, 1978) that hourly wages in manufacturing show little cyclical variability and are best described as a random walk.

In fact, it seems preferable to have empirical investigations of this sort at a less aggregated level. Aggregate studies are victims of selection bias: they fail to capture changes in the composition of output or of the labour force, which are themselves sufficient to induce substantial cyclical movement in economy-wide wages even if the business cycle does not affect the real wage of any skill grade in any industry.

Consider, for instance, a fictitious economy with homogeneous labour in which almost all industries experience little cyclical fluctuation except one, the quad industry, which is thoroughly buffeted by the business cycle. If labour mobility is good across industries, quad workers will suffer more layoffs and enjoy a wage higher than elsewhere whenever they are employed. The economy-wide average wage will vary procyclically.

Another phenomenon accounted for naturally by implicit contracts is the behaviour of occupational wage differentials (i.e. of the unskilled-to-skilled wage ratio). These have shown a definite countercyclical tendency, widening in contractions and narrowing in booms, both in the US and in the UK.

To see why, suppose that we drop the postulate of labour homogeneity in the economy just described and admit two skill grades. For simplicity, assume that the cycle is of such amplitude that there is no unemployment outside the quad industry, while unemployment in the quad industry falls solely on common labourers. These workers are thus the only group in the economy to suffer layoffs; in return they receive a wage above that of common workers outside the quad industry and below that of skilled workers – in the quad industry or out. As the cycle unfolds, then, the economy-wide wage average for craftsmen remains unaltered, the one for labourers changes procyclically, and occupational wage differentials follow a countercyclical pattern.

Intertemporal labour supply models of the type pioneered by Lucas and Rapping (1969) are another area that may in the future make fruitful use of implicit contracts. Econometric work on intertemporal labour substitution identifies the preferences of a 'typical' working household from time-series data on wages and salaries. The outcome is invariably an estimate of the wage-elasticity of labour supply that is so low as to be inconsistent with time-series data on employment (Kydland and Prescott, 1982). In other words, someone who believes that the wage rate represents an important conditioning factor for labour supply and demand will find that wage rates do not vary sufficiently over the business cycle to account for observed fluctuations in employment.

Employment in an implicit contract, however, reflects the underlying value of labour's marginal revenue-product, whereas wages are smoothed averages of the MRPL over time or states of nature. Small fluctuations in contract wages are in principle consistent with substantial variations in contract employment; whether these are mutually consistent *in practice* remains to be seen from empirical work.

MACROECONOMIC ASPECTS. From empirical labour economics we turn to the macroeconomic issues that provided the original impetus for the development of implicit contracts. Unemployment, says this theory, is the result of differential information: a credible signal from employers to employees that product demand is slackening, or one from employees to employers that job opportunities are really better elsewhere.

Newer ideas that seem to be building on this basic piece of

intuition are outlined later in this article. But whatever progress we have made towards understanding fluctuations in employment has not dispelled the dense fog that still shrouds the issue for wage rigidity. All we have to go on is the early result of Martin Baily that insurance makes the wage rate less variable than it otherwise might be. This stickiness, however, is a property of the *real* rather than the nominal wage rate, and it is the latter that is assumed to be rigid in Keynesian macroeconomics.

Rigidity, of course, does not necessarily imply complete time-invariance, nor does it require money wages to change less frequently than other prices; it is simply an information-processing failure. The standard procedure in collective bargains, for instance, is to predetermine money wages several years in advance; more often than not those wages are invariant to any information that may accumulate over the duration of the contract. Only in exceptional circumstances are money wages in the United States allowed to reflect *any* contemporaneous developments in the cost of living (indexation) or in the profitability of the employer (bankruptcy).

The mystery of wage rigidity is then the failure of contracts to set money wages as *functions* of publicly available information that is obviously relevant to the welfare of all parties. Why does the wage-setting process choose to ignore this information? One answer is transaction costs and/or bounded rationality: contracts are cheaper to evaluate and implement when they are defined by a few simple numbers rather than by complicated rules that condition employment or wages on contingent events. Another possibility is to exploit the great multiplicity of equilibria that is typical of economies with missing securities markets (Azariadis and Cooper, 1985). One of these equilibria features predetermined prices and wages, while employment and other quantity variables adjust fully to short-term disturbances. Wage rigidity here is like a Nash equilibrium: it is the best response of a firm in a labour market in which the wages paid by all other firms fail to reflect new information instantaneously.

IMPLEMENTATION. An implicit contract is formally defined as a collection of schedules describing how the terms of employment for one person or group of persons change in response to unexpected changes in the economic environment. What brings contractants together? How detailed are their agreements? And what mechanisms are there to enforce such agreements once they are reached? After an initial stage of fairly rapid development, research is returning to these elementary questions as if trying to clarify the axiomatic basis of the underlying theory.

What brings potential contractants together is the opportunity jointly to reap substantial returns on investments peculiar to their relationship. The idea is apparent in Becker's theory of specific human capital (1964) and in Williamson's hypothesis (1979) of physical assets that are specific to a given supplier–customer pair. To reap any returns, contractants must wed themselves to one partner, forsaking all others, for some period of time. Maintaining such a special relationship involves the transactions costs of creating an idiosyncratic asset, as well as an implicit contract; that is, a number of rules that define how the partners have decided to share the returns in various possible future circumstances.

There are, of course, circumstances that are not explicitly covered, either because they are not observable at reasonable cost or because contractants think of them as unlikely or unworthy of note. Irrespective of the possible events that are covered and of the prior rules that govern the distribution of returns to shared investments, all contractants are required to

bear risk and to subordinate their short-term interest to longer-term considerations.

Workers, for instance, suffer layoffs in recessions while firms hoard labour in order to preserve a long-term relationship. What mechanisms keep contractants together in adverse circumstances?

One mechanism – studied extensively by Radner (1981), Townsend (1982) and others – is reputation: if somebody deviates from the terms of the contract, the deviation becomes widely known, and the deviant finds it difficult to locate trading partners in the future. That works well if the time horizon is fairly long or the future is fairly important relative to the present; reputations are likely to be important for firms, less so for workers.

Another method of enforcement is by a third party: a monitor, arbitrator or court of law. In order for a third party to enforce a contract, it has to be able to observe all the prices and all the quantities specified in it – the employment status, hours worked and wage rate of every worker. That is an unreasonably large informational burden to place on someone who is outside the special relationship called a contract. Outsiders can be expected to observe at low cost only certain aggregates or averages, but not very much in the way of idiosyncratic detail.

How does one design and enforce contracts when outsiders are poorly informed about the trades among contractants? According to Hölmstrom (1983) and Bull (1986), self-interest will enforce contracts that third parties are not sufficiently informed to implement.

In particular, workers will put in the required amount of effort on the job, not because effort can be ascertained easily by an outside arbitrator but rather because they know that their wages and speed of promotion depend on performance. And employers will be careful not to break even the most implicit of their commitments if doing so will compromise their ability to attract workers in the future. As of this writing, the design of self-enforcing contracts seems to be the central theoretical problem in the field of implicit contracts.

COSTAS AZARIADIS

See also ASYMMETRIC INFORMATION; INCENTIVE CONTRACTS; LABOUR ECONOMICS; LAYOFFS.

BIBLIOGRAPHY

Azariadis, C. 1975. Implicit contracts and underemployment equilibria. *Journal of Political Economy* 83, 1183–202.

Azariadis, C. 1983. Employment with asymmetric information. *Quarterly Journal of Economics* 98, Supplement, 157–72.

Azariadis, C. and Cooper, R. 1985. Nominal wage-price rigidity as a rational expectations equilibrium. *American Economic Review, Papers and Proceedings* 75, 31–5.

Baily, M. 1974. Wages and employment under uncertain demand. *Review of Economic Studies* 41, 37–50.

Becker, G. 1964. *Human Capital.* New York: Columbia University Press.

Benassy, J.-P. 1975. Neo-Keynesian disequilibrium in a monetary economy. *Review of Economic Studies* 42, 502–23.

Bull, C. 1986. The existence of self-enforcing implicit contracts. *Quarterly Journal of Economics.*

Clower, R. 1965. The Keynesian counter-revolution: a theoretical appraisal. In *The Theory of Interest Rates,* ed. F. Hahn and F. Brechling, London: Macmillan.

Cooper, R. 1983. A note on overemployment/underemployment in labor contracts under asymmetric information. *Economics Letters* 12, 81–7.

Drèze, J. 1975. Existence of an equilibrium under price rigidity and quantity rationing. *International Economic Review* 16, 301–20.

Farmer, R. 1984. A new theory of aggregate supply. *American Economic Review* 74, 920–30.

Feldstein, M. 1975. The importance of temporary layoffs: an empirical analysis. *Brookings Papers on Economic Activity* 3, 725–44.

Gordon, D.F. 1974. A neoclassical theory of Keynesian unemployment. *Economic Inquiry* 12, 431–49.

Grossman, S. and Hart, O. 1981. Implicit contracts, moral hazard and unemployment. *American Economic Review, Papers and Proceedings* 71, 301–7.

Hall, R. 1980. Employment fluctuations and wage rigidity. *Brookings Papers on Economic Activity* 8, 91–124.

Hall, R. 1982. The importance of lifetime jobs in the US economy. *American Economic Review, Papers and Proceedings* 72, 716–24.

Hall, R. and Lilien, D. 1979. Efficient wage bargains under uncertain supply and demand. *American Economic Review* 69, 868–79.

Hart, O. 1983. Optimal labour contracts under asymmetric information: an introduction. *Review of Economic Studies* 50, 3–35.

Hölmstrom, B. 1983. Equilibrium long-term labor contracts. *Quarterly Journal of Economics* 98, Supplement, 23–54.

Knight, F. 1921. *Risk, Uncertainty and Profit.* Boston: Houghton Mifflin.

Kydland, F. and Prescott, E. 1982. Time to build and aggregate fluctuations. *Econometrica* 50, 1345–70.

Lucas, R. and Rapping, L. 1969. Real wages, employment and inflation. *Journal of Political Economy* 77, 721–54.

Neftci, S. 1978. A time-series analysis of the real wages–employment relationship. *Journal of Political Economy* 86, 281–91.

Radner, R. 1981. Monitoring cooperative agreements in a repeated principal–agent relationship. *Econometrica* 49, 1127–48.

Rosen, S. 1985. Implicit contracts: a survey. *Journal of Economic Literature* 23, 1144–75.

Smith, A. 1776. *The Wealth of Nations.* London: Pelican Books, 1970.

Townsend, R. 1982. Optimal multiperiod contracts and the gain from enduring relationships under private information. *Journal of Political Economy* 90, 1166–86.

Williamson, O. 1979. Transaction-cost economics: the governance of contractual relations. *Journal of Law and Economics* 22(2), October, 233–61.

import duties. *See* OPTIMAL TARIFFS; TARIFFS.

import substitution and export-led growth. In an economy in which expansion is limited by a balance of payments constraint, action must be taken either to boost exports or to limit imports. This truism takes on an added dimension when the trade strategy adopted is part of a general development strategy. In these circumstances the evaluation of any particular trade strategy must include not only the implications for the allocation of resources, but also the consequences for the rate of accumulation and of technological progress.

In the 1950s and early 1960s, the years of the dollar shortage, the balance of payments constrained industrial countries adopted quite different trade and industrial strategies. West Germany pursued a strategy of export expansion by means of an undervalued Deutschmark and subsidies to export industries. With world trade in manufactures growing rapidly, and West Germany's share of that trade growing too, the rapid growth of manufactured exports provided the foundation for domestic expansion (Shonfield, 1962). Italy pursued a similar strategy by means of regular devaluation of the lira, devaluations often being associated with a surplus on Italy's current account.

These two examples of export-led growth contrast markedly with the strategies adopted by France and by Japan. Both countries vigorously protected their home markets, using industrial expansion within the home market as a springboard for the capture of export markets. The rationale behind this policy of import substitution was spelt out by Vice-Minister

Ojimi, of the Japanese Ministry of International Trade and Industry:

> After the war, Japan's first exports consisted of such things as toys or other miscellaneous merchandise and low-quality textile products. Should Japan have entrusted its future, according to the theory of comparative advantage, to these industries characterized by intensive use of labour? That would perhaps be rational advice for a country with a small population of 5 or 10 million. But Japan has a large population. If the Japanese economy had adopted the simple doctrine of free trade and had chosen to specialise in this kind of industry, it would almost permanently have been unable to break away from the Asian pattern of stagnation and poverty ...
>
> The Ministry of International Trade and Industry decided to establish in Japan industries which require intensive employment of capital and technology, industries that in consideration of comparative cost should be the most inappropriate for Japan, industries such as steel, oil refining, petro-chemicals, automobiles, industrial machinery of all sorts, and electronics, including electronic computers. From a short-run, static viewpoint, encouragement of such industries would seem to be in conflict with economic rationalism. But from a long-range viewpoint, these are precisely in industries where income elasticity of demand is high, technological progress is rapid, and labour productivity rises fast ... (Ojimi, 1970).

Ojimi's argument encapsulates the dispute over import substitution or export-led growth as development strategies. The orthodox theory of international trade suggests that resources are most efficiently allocated in a regime of free trade. Efficient development would therefore require the adoption of free trade, with variation in exchange rates being used as the means of balancing trade.

This argument rests on a number of strong assumptions, in particular the assumptions that all countries have access to the same technologies, that factor markets clear (labour is fully employed), and that all countries have equal access to all markets – including equal access to all financial markets. If these, and other well-known assumptions, are not fulfilled, then the argument for free trade *on these grounds* no longer stands, and is superseded by the uncertainties of the second best.

Rejection of arguments for the efficiency of the price mechanism, for example on Keynesian grounds, also lead to the rejection of the efficiency of free trade. It was Keynesian arguments that underpinned the so-called ECLA strategy for structural change in Latin America. If expansion of domestic demand could be prevented, by protective measures, from leaking abroad then savings and fiscal revenues at home would finance domestic investment and government expenditure. Moreover, the profitability of protected domestic production would encourage further investment. The process of expansion would be self-sustaining.

The application of import-substitution strategies in Latin America in the 1950s met initially with considerable success. Output of domestically produced manufactured goods grew rapidly, as did industrial employment. Later the policy fell into disrepute. It was argued that import substitution took place primarily in 'soft' consumer goods industries, whereas investment goods continued to be imported. Hence after the early growth associated with import substitution in consumer goods, growth was once again constrained by the necessity of importing machinery. Moreover, it was argued that protected domestic industry was relatively inefficient, and unable to compete on world markets. These matters are the subject of considerable dispute, particularly as they involve not only questions of economic efficiency, but also issues of national sovereignty, since the IMF has responded to the difficulties in which some Latin American countries have found themselves by demanding the removal of the trade protection on which the earlier development strategy was based.

These criticisms of import substitution extend beyond the traditional case for free trade, to consideration of the implication of different trade strategies for structural development and technological change. It was on exactly these grounds that Ojimi sought to justify Japan's strategy of import substitution. The Japanese case suggests that the traditional dichotomy between import substitution and export-led growth is invalid. Whilst Japanese industry was developed within a rapidly growing and protected home market, that growth proved to be springboard for expansion into world markets. Exports were domestic-growth led.

The performance of the successful Japanese (and French) examples of import substitution, and the problems encountered in Latin America, cannot be evaluated using static conceptions of allocative efficiency. Success (and lack of it) have clearly been associated with technological progress and industrial modernization. The case for free trade must be made on the ground that it encourages the most rapid adoption of the new techniques which determine competitive advantage.

Nicholas Kaldor's version of Verdoorn's Law (Kaldor, 1966), whereby it is argued that the rate of productivity growth in manufacturing industry is a function of the rate of growth of demand for manufactured products, provides a framework within which trade strategies may be evaluated (see, for example, Brailovsky, 1981).

The growth of demand for a country's manufactures is a function of the rate of growth of its home market, the rate of growth of its export markets, and the rate of change of its share of those markets. Changing market shares is a slow and uncertain business. It is growth of markets which is the major determinant of growth of demand. Since all countries are competing for shares of (roughly) the same export market, it is growth of the home market which typically differentiates the growth of demand for the manufactures of one country from those of another. This would suggest that manipulation of growth of the home market, using whatever means are necessary to relax the balance of payments constraint, is the most efficient development strategy.

However, the Verdoorn argument does not encompass the scale of productivity response to any given growth of demand. The implementation of industrial policies which both ensure that the expansion of industrial structure is 'balanced', and hence not overly dependent on imports, and directs demand toward those sectors which have both greatest competitive potential and which have the highest ratio of domestic value-added to import content, are more likely produce a greater response than if these issues are neglected.

The efficiency of any given trade strategy is not independent of the performance of the world economy as a whole. All countries cannot achieve export-led growth at once. Moreover, the success of the West Germany recovery strategy was undoubtedly enhanced by the fact that it was implemented in a period of rapid growth in world trade. In an era in which world trade is expanding relatively slowly, reliance on export demand is unlikely to prove a successful foundation for rapid growth of demand and hence for rapid technological progress.

JOHN EATWELL

See also AUTARKY; EFFECTIVE PROTECTION; FREE TRADE AND PROTECTION; IMMISERIZING GROWTH; INFANT INDUSTRY; QUOTAS AND TARIFFS; VENT FOR SURPLUS.

BIBLIOGRAPHY
Brailovsky, V. 1981. Industrialisation and oil in Mexico: a long-term perspective. In *Oil or Industry?*, ed. T. Barker and V. Brailovsky, London: Academic Press.
Kaldor, N. 1966. *The Causes of the Slow Rate of Growth of the UK.* Cambridge: Cambridge University Press.
Ojimi, V. 1970. Japan's industrialisation strategy. In OECD, *Japanese Industrial Policy*, Paris: OECD.
Shonfield, A. 1963. *Modern Capitalism.* Oxford: Oxford University Press.

impossibility theorem. *See* ARROW'S THEOREM.

imprisonment. *See* CRIME AND PUNISHMENT.

imputation. This is a term introduced into economics as *Zurechnung* by the Austrian School economist Friedrich Freiherr von Wieser (Wieser, 1889). The term was a legal one, and the analogy was based on the legal method by which the jurist *imputes* guilt or liability to one or another criminal or tortfeasor. Imputation was a central concern of the Austrian School, since its analysis centred on the nature of the means–ends relationship (Mises, 1949), and on the process by which the subjective valuations and value-preferences of individual consumers 'impute' value to the goods being produced. As Carl Menger, founder of the Austrian School, pointed out, the valuations by consumers of their satisfactions, or ends, impute values to the consumer goods, the means, that are expected to satisfy those wants (Menger, 1871). And since producers' goods are only means to the production and sale of consumer goods, the values of the factors of production will in turn be determined by and be equal to the expected values of the consumer goods to the consumers. In short, values are 'imputed' back to the prices of the factors of production; the rents of Champagne land are high because the consumers value the champagne highly, and not the other way round. 'Costs' of resources are reflections of the value of products foregone.

While this process was clear in principle, there were considerable difficulties in working out the specifics. Essentially, Menger and his student Böhm-Bawerk stuck close to the realities of the market process, and focused on value imputation as a process of estimating how much of a product would be lost if the producer were deprived of one unit of a factor. Wieser, on the other hand, presumed that the marginal value of each factor could be found with great precision; in doing so, he assumed illegitimately that subjective values can be added and multiplied to arrive at the total value of a quantity of goods. But by its nature subjective value is an expression of ordinal preferences and therefore can neither be added nor measured.

The modern theory of marginal productivity has essentially solved these problems and shown how values of products can be imputed back to productive factors. One exception is the current assumption that the existence of variable proportions solves the problem of pricing factors and leaves no theoretical room for arbitrary bargaining between factor owners. But the more important solution depends on whether factors are purely specific to one line of production or are relatively non-specific, that is, can be employed in the production of more than one good. If two factors are each purely specific to a given product, then even if their proportions are variable, there is still no principle by which the market can determine their relative prices except by arbitrary bargaining (Mises, 1949, p. 336). In the real world, of course, the existence of such purely specific factors, and hence the scope for such bargaining, will be extremely limited.

The other important point is that values cannot be added or divided, and that the imputation process takes place, not automatically or precisely in an abstract realm of 'values', but only concretely and by trial and error, in the realistic market process of changing prices. In other words, although consumers can evaluate consumer goods and determine their prices directly by valuation, the prices of productive factors are only determined indirectly through market prices and entrepreneurial trial and error. There is no direct, abstract or pure process of imputing values.

This problem became strikingly relevant during the well-known debate over the Mises–Hayek demonstration that socialist governments cannot calculate economically. Joseph Schumpeter brusquely dismissed this contention with the statement that economic calculation under socialism follows 'from the elementary proposition that consumers in evaluating ("demanding") consumers' goods *ipso facto* also evaluate the means of production which enter into the production of these goods' (Schumpeter, 1942, p. 175). Hayek's perceptive reply points out that the '*ipso facto*' assumes complete knowledge of values, demands, scarcities, etc. to be 'given' to everyone, thereby ignoring the reality of the universal lack of complete knowledge, as well as the necessary function of the market economy, and the market price system, in conveying knowledge to all its participants (Hayek, 1945).

The analysis of imputation began in a neglected work of Aristotle, the *Topics*. Here, Aristotle analysed the ends–means relationship, and pointed out that the means, or 'instruments of production', necessarily derive their value from the ends, the final products useful to man, 'the instruments of action'. The more desirable the final good, the more valuable will be the means to arrive at the product. Aristotle introduced the theme of marginality by stating that if the addition of a good *A* to an already desirable good *C* yields a more desirable result than the addition of good *B*, then *A* will be more highly valued than *B*. Indeed, he also added a pre-Böhm-Bawerkian note by stressing the differential value of the loss rather than the addition of a good. Good *A* will be more valuable than *B*, if the loss of *A* is considered to be worse than the loss of *B*. While critics have noted that Aristotle only slightly applied his analysis to the economic realm, his imputation theory was still an important contribution to the general theory of action of which economic theory is a highly developed part (Spengler, 1955).

MURRAY N. ROTHBARD

See also ADDING-UP PROBLEM; AUSTRIAN SCHOOL OF ECONOMICS; MARGINAL PRODUCTIVITY THEORY.

BIBLIOGRAPHY

Aristotle. *Topica*. Trans. into English by W.A. Pickard-Cambridge, and included in Vol. I of *The Works of Aristotle*, ed. W.D. Ross, Oxford: Clarendon Press, 1928.

Hayek, F.A. 1926. Some remarks on the problem of imputation. In F.A. Hayek, *Money, Capital, and Fluctuations: early Essays*, Chicago: University of Chicago Press, 1984.

Hayek, F.A. 1945. The use of knowledge in society. In F.A. Hayek, *Individualism and Economic Order*, Chicago: University of Chicago Press, 1948.

Kauder, E. 1965. *A History of Marginal Utility Theory*. Princeton: Princeton University Press.

Menger, C. 1871. *Principles of Economics*. Glencoe, Ill.: Free Press, 1950.

Mises, L. von. 1949. *Human Action: a treatise on economics*. 3rd edn, Chicago: Regnery, 1966.

Schumpeter, J.A. 1942. *Capitalism, Socialism, and Democracy*. New York: Harper.

Spengler, J. 1955. Aristotle on economic imputation and related matters. *Southern Economic Journal* 21, April, 371–89.

Stigler, G. 1941. *Production and Distribution Theories: The Formative Period*. New York: Macmillan.

Wieser, F. von. 1889. *Natural Value*. New York: Kelley & Millman, 1956.

incentive compatibility. Allocation mechanisms, organizations, voting procedures, regulatory bodies, and many other institutions are designed to accomplish certain ends such as the Pareto-efficient allocation of resources or the equitable resolution of disputes. In many situations it is relatively easy to conceive of feasible processes; processes which will accomplish the goals if all participants follow the rules and are capable of handling the informational requirements. Examples of such mechanisms include marginal cost pricing, designed to attain efficiency, and equal division, designed to attain equity. Of course once a feasible mechanism is found, the important question then becomes whether such a mechanism is also informationally feasible and compatible with 'natural' incentives of the participants. Incentive compatibility is the concept introduced by Hurwicz (1972, p. 320) to characterize those mechanisms for which participants in the process would not find it advantageous to violate the rules of the process.

The historical roots of the idea of incentive compatibility are many and deep. As was pointed out in one of a number of recent surveys,

> the concept of incentive compatibility may be traced to the 'invisible hand' of Adam Smith who claimed that in following individual self-interest the interests of society might be served. Related issues were a central concern in the 'Socialist Controversy' which arose over the viability of a decentralized socialist society. It was argued by some that such societies would have to rely on individuals to follow the rules of the system. Some believed this reliance was naive; others did not. (Groves and Ledyard, 1986, p. 1).

Further, the same issues have arisen in the design of voting procedures. Concepts and problems related to incentives were already identified and documented in the 18th century in discussions of proposals by Borda to provide alternatives to majority rule committee decisions. (See STRATEGY-PROOF ALLOCATION MECHANISMS for further information on voting procedures.)

Incentive compatibility is both desirable and elusive. The desirability of incentive compatibility can be easily illustrated by considering public goods, goods such that one consumer's consumption of them does not detract from another consumer's simultaneous consumption of that good. The existence of these collective consumption commodities creates a classic situation of *market failure*; the inability of markets to arrive at a Pareto-optimal allocation. It was commonly believed, prior to Groves and Ledyard (1977), that in economies with public goods it would be impossible to devise a decentralized process that would allocate resources efficiently since agents would have an incentive to 'free ride' on others' provision of those goods in order to reduce their own share of providing them. Of course Lindahl (1919) had proposed a feasible process which mimicked markets by creating a separate price for each individual's consumption of the public good. This designed process was, however, rejected as unrealistic by those who recognized that these 'synthetic markets' would be shallow (essentially monopsonistic) and therefore buyers would have no incentive to treat prices as fixed and invariant to their demands. The classic quotation is '... it is in the selfish interest of each person to give *false* signals, to pretend to have less interest in a given collective consumption activity than he really has...' (Samuelson, 1954, pp. 388–9). Allocating public goods efficiently through Lindahl pricing would be feasible and successful if consumers followed the rules; but, it would not be successful since the mechanism is not incentive compatible. If buyers do not follow the rules, efficient resource allocation will not be achieved and the goals of the design will be subverted because of the motivations of the participants. Any institution or rule, designed to accomplish group goals, must be incentive compatible if it is to perform as desired.

The elusiveness of incentive compatibility can be most easily illustrated by considering a situation with only private goods. Economists generally model behaviour in private goods markets by assuming that buyers and sellers 'follow the rules' and take prices as given. It is now known, however, that as long as the number of agents is finite then any one of them can still gain by misbehaving and, furthermore, can do so in a way which can not be detected by anyone else. The explanation is provided in two steps. First, if there are a finite number of traders, and none have a perfectly elastic offer curve (which will be true if preferences are non-linear) then one trader can gain by being able to control prices. For example, a buyer would want to set price where his marginal benefit equalled his marginal outlay and thereby gain monopsonistic benefits. Of course, if the others know that buyer's demand curve (either directly or through inferences based on revealed preference) then they would know that the buyer was not 'taking prices as given' and could respond with a suitable punishment against him. This brings us to our second step. Even though others can monitor and prohibit price setting behaviour, our benefit-seeking monopsonist has another strategy which can circumvent this supervision. He calculates a (false) demand curve which, when added to the others' offer curves, produces an equilibrium price equal to that which he would have set if he had direct control. He then calculates a set of preferences which yields that demand curve and participates in the process *as if he had these (false) preferences*. Usually this involves simply acting as if one has a slightly lower demand curve than one really does. Since preferences are not able to be observed by others, he can follow this behaviour which looks like it is price-taking, and therefore 'legal', and can do individually better. The unfortunate implication of such concealed misbehaviour is that the mechanism performs other than as intended. In this case, resources are artificially limited and too little is traded to attain efficiency.

In 1972 Hurwicz established the validity of the above intuition. His theorem can be precisely stated after the introduction of some notation and a framework for further discussion.

THE IMPOSSIBILITY THEOREM. The key concepts include economic environments, allocation mechanisms, incentive compatibility, the no-trade option, and Pareto-efficiency. We take up each in turn.

An *economic environment*, those features of an economy which are to be taken as given throughout the analysis, includes a description of the agents, the feasible allocations they have available and their preferences for those allocations. While many variations are possible, I concentrate here on a simple model. Agents (consumers, producers, politicians, etc.) are indexed by $i = 1, \ldots, n$. X is the set of feasible allocations where $x = (x^i, \ldots, x^n)$ is a typical element of X. (An exchange environment is one in which X is the set of all $x = (x^1, \ldots, x^n)$ such that $x^i \geqslant 0$ and $\Sigma x^i = \Sigma w^i$, where w^i is i's initial endowment of commodities.) Each agent has a selfish utility function $u^i(x^i)$. The environment is $e = [I, X, u^1, \ldots, u^n]$. A crucial fact is that initially *information is dispersed* since i, and only i, knows u^i. We identify the specific knowledge i initially has as i's *characteristic*, e^i. In our model, $e^i = u^i$.

Although there are many variations in models of allocation mechanisms, I begin with the one introduced by Hurwicz (1960). An *allocation mechanism* requests information from the agents and then computes a feasible allocation. It requests information in the form of messages m^i from agent i through a *response function* $f^i(m^i, \ldots, m^n)$. Agent i is told to report $f^i(m, e^i)$ if others have reported m and i's characteristic is e^i. An equilibrium of these response rules, for the environment e, is a joint message m such that $m^i = f^i(m, e^i)$ for all i. Let $\mu(e, f)$ be the set of equilibrium messages for the response functions f in the environment e. The allocation mechanism computes a feasible allocation x by using an *outcome function* $g(m)$ on equilibrium messages. The net result of all of this in the environment e is the allocation $g[\mu(e, f)] = x$ *if all i follow the rules, f*. Thus, for example, the *competitive mechanism* requests agents to send their demands as a function of prices which are in turn computed on the basis of the aggregate demands reported by the consumers. In equilibrium, each agent is simply allocated their stated demand. (An alternative mechanism, yielding exactly the same allocation in one iteration, would request the demand *function* and then compute the equilibrium price and allocation for the reported demand functions.) It is well known, for exchange economies with only private goods, that if agents report their true demands then the allocations computed by the competitive mechanism will be Pareto-optimal.

It is obviously important to be able to identify those mechanisms, those rules of communication, that have the property that they are self-enforcing. We do that by focusing on a class of mechanisms in which each agent gains nothing, and perhaps even loses, by misbehaving. While a multitude of misbehaviours could be considered it is sufficient for our purposes to consider a slightly restricted range. In particular we can concentrate on undetectable behaviour, behaviour which no outside agent can distinguish from that prescribed by the mechanism. We model this limitation on behaviour by requiring the agent to restrict his misrepresentations to those which are consistent with some characteristic he might have. An allocation mechanism is said to be *incentive compatible* for all environments in the class E if there is no agent i and no environment e in E and no characteristic e^{*i} such that (e/e^{*i}) is in E (where (e/e^{*i}) is the environment derived from e by replacing e^i with e^{*i}) and such that

$$u^i\{g[\mu(e, f)], e^i\} > u^i\{g[\mu(e/e^{*i}, f], e^i\}$$

where $u^i(x^*, e^i)$ is i's utility function in the environment e. That is, no agent can manipulate the mechanism by pretending to have a characteristic different from the true one and do better than acting according to the truth. The agent has an incentive to follow the rules and the rules are compatible with his motivations.

Incentive compatibility is at the foundation of the modern *theory of implementation*. In that theory, one tries to identify conditions under which a particular social choice rule or performance standard, $P : E \to X$, can be recreated by an allocation mechanism under the hypothesis that individuals will follow their self-interest when they participate in the implementation process. In our language, the rule P is implementable if and only if there is an incentive compatible mechanism (f, g) such that $g[\mu(e, f)] = P(e)$ for all e in E. The theory of implementation seeks to answer the question 'which P are implementable?' We will see some of the answers below for P which select from the set of Pareto-efficient allocations. Those interested in more general goals and performance standards should consult Dasgupta, Hammond and Maskin (1979) or Postlewaite and Schmeidler (1986).

An allocation mechanism is said to have the *no trade-option* if there is an allocation θ at which each participant may remain. In exchange environments the initial endowment is usually such an allocation. Mechanisms with a no-trade option are non-coercive in a limited sense. If an allocation mechanism possesses the no-trade option then the allocation it computes for an environment e, if agents follow the rules, must leave everyone at least as well off, using the utility functions for e, as they are at θ. That is, for all i and all e in E

$$u^i\{g[\mu(e, f)], e^i\} > u^i(\theta, e^i).$$

An allocation mechanism is said to be *Pareto-efficient* in E if the allocations selected by the mechanism, when agents follow the rules, are Pareto-optimal in e. That is, for each e in E there is no allocation x^* in X such that, for all i,

$$u^i(x^*, e^i) \geqslant u^i\{g[\mu(e, f)], e^i\}$$

with strict inequality for some i.

With this language and notation, Hurwicz's theorem on the elusive nature of incentive compatibility in private markets, subsequently expanded by Ledyard and Roberts (1974) to include public goods environments, can now be easily stated. *Theorem*: In classical (public or private) economic environments with a finite number of agents, there is no incentive compatible allocation mechanism which possesses the no-trade option and is Pareto-efficient. (Classical environments include pure exchange environments with Cobb–Douglas utility functions.)

A more general version of this theorem, in the context of social choice theory, has been proven by Gibbard (1973) and Satterthwaite (1975) with the concept of a 'non-dictatorial social choice function' replacing that of a 'mechanism with the no-trade option'. (See STRATEGY-PROOF ALLOCATION MECHANISMS.)

There are a variety of possible reactions to this theorem. One is simply to give up the search for solutions to market failure since the theorem seems to imply that one should not waste any effort trying to create institutions to allocate resources efficiently. A second is to notice that, at least in private markets, if there are a very large number of individuals in each market then efficiency is 'almost' attainable (see Roberts and Postlewaite, 1976). A third is to recognize that the behaviour of individuals will generally be different from that implicitly assumed in the definition of incentive compatibility. A fourth is to accept the inevitable, lower one's sights, and look for the 'most efficient' mechanism among those which are incentive compatible and satisfy a voluntary participation constraint. We consider the last two options in more detail.

OTHER BEHAVIOUR: NASH EQUILIBRIUM. If a mechanism is incentive compatible, then each agent knows that his best strategy is to follow the rules according to his true characteristic, *no matter what the other agents will do*. Such a strategic structure is referred to as a dominant strategy game

and has the property that no agent need know or predict anything about the others' behaviour. In mechanisms which are not incentive compatible, each agent must predict what others are going to do in order to decide what is best. In this situation agents' behaviour will not be as assumed in the definition of incentive compatibility. What it will be continues to be an active research topic and many models have been proposed. Since most of these are covered in Groves and Ledyard (1986), I will concentrate on the two which seem most sensible. Both rely on game-theoretic analyses of the strategic possibilities. The first concentrates on the outcome rule, g, and postulates that agents will not choose messages to follow the specifications of the response functions but to do the best they can against the messages sent by others. Implicitly this assumes that there is some type of iterative process (embodied in the response rules) which allows revision of one's message in light of the responses of others. We can formalize this presumed strategic behaviour in a new concept of incentive compatibility. An allocation mechanism (f, g) is called *Nash incentive compatible* for all environments in E if there is no environment e, no agent i, and no message m^{*i} which i can send such that

$$u^i(g[\mu(e, f)/m^{*i}, e^i]) > u^i(g[\mu(e, f), e^i])$$

where $\mu(e, f)$ is the 'equilibrium' message of the response rules f in the environment e, $g(m)$ is the outcome rule, and $[m/m^{*i}]$ is the vector m where m^{*i} replaces m^i. In effect this requires the equilibrium messages of the response rules to be Nash equilibria in the game in which messages are strategies and payoffs are given by $u[g(m)]$. It was shown in a sequence of papers written in the late 1970s, including those by Groves and Ledyard (1977), Hurwicz (1979), Schmeidler (1980), and Walker (1981), that Nash incentive compatibility is not elusive. The effective output of that work was to establish the following. *Theorem*: In classical (public or private) economic environments with a finite number of agents, there are many Nash incentive compatible mechanisms which possess the no-trade option and are Pareto-efficient.

With a change in the predicted behaviour of the participants in the mechanism, in recognition of the fact that in the absence of dominant strategies agents must follow some other self-interested strategies, the pessimism of the Hurwicz theorem is replaced by the optimistic prediction of a plethora of possibilities. (See Dasgupta, Hammond and Maskin, 1979, Postlewaite and Schmeidler (1986) and Groves and Ledyard (1986) for comprehensive surveys of these results including many for more general social choice environments.) Although it remains an unsettled empirical question whether participants will indeed behave this way, there is a growing body of experimental evidence that seems to me to support the behavioural hypotheses underpinning Nash incentive compatibility, especially in iterative tâtonnement processes.

OTHER BEHAVIOUR: BAYES' EQUILIBRIUM. The second approach to modelling strategic behaviour of agents in mechanisms, when dominant strategies are not available, is based on Bayesian decision theory. These models, called *games of incomplete information* (see Myerson, 1985), concentrate on the beliefs of the players about the situation in which they find themselves. In the simplest form, it is postulated that there is a common knowledge (everyone knows that everyone knows that ...) probability function, $\pi(e)$, which describes everyone's prior beliefs. Each agent is then assumed to choose that message which is best against the expected behaviour of the other agents. The expected behaviour of the other agents is also constrained to be 'rational' in the sense that it should be best against the behaviour of others. This presumed strategic behaviour is embodied in a third type of incentive compatibility. (It could be argued that the concept of incentive compatibility remains the same, based on non-cooperative behaviour in the game induced by the mechanism, while only the presumed information structure and sequence of moves required to implement the allocation mechanism are changed. Such a view is not inconsistent with that which follows.) An allocation mechanism (f, g) is called *Bayes incentive compatible* for all environments in E given π on E if there is no environment $e*$, no agent i, and no message m^{*i} which i can send such that

$$\int u^i\{g[\mu(e, f)/m^{*i}], e^{*i}\}\,d\pi(e\,|\,e^{*i})$$
$$> \int u^i\{g[\mu(e, f), e^{*i}]\,d\pi(e,|\,e^{*i})\}$$

where, as before, μ is the equilibrium message vector and g is the outcome rule. Further, $\pi(e\,|\,e^{*i})$ is the conditional probability measure on e given e^{*i}, and u^i is a von Neumann–Morgenstern utility function. In effect, this requires the equilibrium messages of the response rules to be Bayes equilibrium outcomes of the incomplete information game with messages as strategies, payoffs $u[g(m)]$ and common knowledge prior π.

There are two types of results which deal with the possibilities for Bayes incentive compatible design of allocation mechanisms, neither of which is particularly encouraging. The first type deals with the possibilities for incentive compatible design which is independent of the beliefs. The typical theorem is illustrated by the following result proven by Ledyard (1978). *Theorem*: In classical economic environments with a finite number of agents, there is no Bayes incentive compatible mechanism which possesses the no-trade option and is Pareto-efficient *for all π on E*. Understanding this result is easy when one realizes that any mechanism (f, g) is Bayes incentive compatible for all π for all e in E if and only if it is (Hurwicz) incentive compatible for all e in E. Thus the Hurwicz impossibility theorem again applies.

The second type of result is directed towards the possibilities for a specific prior π; that is, towards what can be done if the mechanism can depend on the common knowledge beliefs. The most general characterizations of the possibilities for Bayes incentive compatible design can be found in Palfrey and Srivastava (1987) and Postlewaite and Schmeidler (1986). They have shown that two conditions, called monotonicity and self-selection, are necessary and sufficient for a social choice correspondence to be implementable in the sense that there is a Bayes incentive compatible mechanism that reproduces that correspondence. The details of these conditions are not important. What is important is that many correspondences do not satisfy them. In particular, there appear to be many priors π and many sets of environments E for which there is no mechanism which is Bayes incentive compatible, provides a no-trade option and is Pareto-efficient. Thus, impossibility still usually occurs even if one allows the mechanism to depend on the prior.

One recent avenue of research which promises some optimistic counterweight to these negative results can be found in Palfrey and Srivastava (1987). In much the same way that the natural move from Hurwicz incentive compatibility to Nash incentive compatibility created opportunities for incentive compatible design, these authors have shown that a move back towards dominant strategies may also open up possibilities. Refinements arise by varying the equilibrium

concept in a way that reduces the number of (Bayes or Nash) equilibria for a given e or π. Moore and Repullo use subgame perfect Nash equilibria. Palfrey and Srivastava eliminate weakly dominated strategies from the set of Nash equilibria. They have discovered that, in pure exchange environments, virtually all performance correspondences are implementable if behaviour satisfies these refinements. In particular, any selection from the Pareto-correspondence is implementable for these refinements, and so there are many refined-Nash incentive compatible mechanisms which are Pareto-efficient and allow a no-trade option. It is believed that these results will transfer naturally to refinements of Bayes equilibria, but the research remains to be done.

INCENTIVE COMPATIBILITY AS A CONSTRAINT. Another of the reactions to the Hurwicz impossibility result is to accept the inevitable, to view incentive compatibility as a constraint, and to design mechanisms to attain the best level of efficiency one can. If full efficiency is possible, it will occur as the solution. If not, then one will at least find the second-best allocation mechanism. Examples of this rapidly expanding research literature include work on optimal auctions (Harris and Raviv, 1981; Matthews, 1983; Myerson, 1981), the design of optimal contracts for the principle-agent problem, and the theory of optimal regulation (Baron and Myerson, 1982). As originally posed by Hurwicz (1972, pp. 299–301), the idea is to adopt a social welfare function $W(x, e)$, a measure of the social welfare attained from the allocation x if the environment is e and then to choose the mechanism (f, g) to maximize the (expected) value of W subject to the 'incentive compatibility constraints', the constraint that the rules (f, g) be consistent with the motivations of the participants. One chooses (f, g) to

$$\text{maximize } \int W\{g[\mu(e, f)], e\}\,d\pi(e)$$

subject to, for every i, every e, and every e^{*i},

$$\int u^i\{g[\mu(e/e^{*i}, f)], e^i\}\,d\pi(e\,|\,e^i) \leqslant \int u^i(g[\mu(e, f), e^i])\,d\pi(e\,|\,e^i).$$

As formalized here the incentive compatibility constraints embody the concept of Bayes incentive compatibility. Of course, other behavioural models could be substitued as appropriate.

Sometimes a voluntary participation constraint, related to the no-trade option of Hurwicz, is added to the optimal design problem. One form of this constraint requires that (f, g) also satisfy, for every i and every e,

$$\int u^i\{g[\mu(e)], e^i\}\,d\pi(e\,|\,e^i) \geqslant \int u^i(\theta[e], e^i)\,d\pi(e\,|\,e^i).$$

In practice this optimization can be a difficult problem since there are a large number of possible mechanisms (f, g). However, an insight due to Gibbard (1973) can be employed to reduce the range of alternatives and simplify the analysis. Now called the *revelation principle*, the observation he made was that, to find the maximum, it is sufficient to consider only mechanisms, called direct revelation mechanisms, in which agents are asked to report their own characteristics. The reason is easy to see. Suppose that (f^*, g^*) solves the maximum problem. Let (F^*, G^*) be a new (direct revelation) mechanism defined by $F^{*i}(m, e^i) = e^i$ and $G^*(m) = g[\mu(m, f)]$. Each i is told to report his characteristic and then G^* computes the allocation by computing that which would have been chosen if the original mechanism (f, G^*) had been used honestly in the reported environment. (F^*, G^*) yields the same allocation as (f^*, g^*), *if each agent reports the truth*. But the incentive compatibility constraints, which (f*, g*) satisfied, ensure that each agent will want to report truthfully. Thus, whatever can be done, by any arbitrary mechanism subject to the Bayes incentive compatibility constraints, can be done with direct revelation mechanisms subject to the constraint that each agent wants to report their true characteristic. One need only choose a function $G : E \to X$ to

$$\text{maximize} \int W[G(e), e] \, d\pi(e)$$

subject to, for every i, e and e^i,

$$\int u^i[G(e/e^{*i}), e^i] \, d\pi(e \mid e^i) \leqslant \int u^i[G(e), e^i] \, d\pi(e \mid e^i),$$

and

$$\int u^i[G(e), e^i] \, d\pi(e \mid e) \geqslant \int u^i(\theta[e], e^i) \, d\pi(e \mid e^i).$$

There are at least two problems with this approach to organizational design. The first is that the choice of mechanism depends crucially on the prior beliefs, π. This is a direct result of the use of Bayes incentive compatibility in the constraints. Since the debate is still open let me simply summarize some of the arguments. One is that if the mechanism chosen for a given situation does not depend on common knowledge beliefs then we would not be using all the information at our disposal to pursue the desired goals and would do less than is possible. Further, since the beliefs are common knowledge we can all agree as to their validity (misrepresentation is not an issue) and therefore to their legitimate inclusion in the calculations. An argument is made against this on the practical grounds that one need only consider actual situations, such as the introduction of new technology by a regulated utility or the acquisition of a major new weapons system by the government, to understand the difficulties involved in arriving at agreements about the particulars of common knowledge. Another argument against is based on the feeling that mechanisms should be robust. A 'good' mechanism should be able to be described in terms of its mechanics and, while it probably should have the capacity to incorporate the common knowledge relevant to the current situation, it should be capable of being used in many situations. How to capture these criteria in the constraints or the objective function of the designer remains an open research question.

The second problem with the optimal auction approach to organizational design is the reliance on the revelation principle. Restricting attention to direct revelation mechanisms, in which an agent reports his entire characteristic, is an efficient way to prove theorems, but it provides little guidance for those interested in actual organization design. For example it completely ignores the informational requirements of the process and any limitations, if any, in the information processing capabilities of the agents or the mechanism. Writing down one's preferences for all possible consumption patterns is probably harder than writing down one's entire demand surface which is certainly harder than simply reacting to a single price vector and reporting only the quantities demanded at that price. A failure to recognize the information processing constraints in the optimization problem is undoubtedly one of the reasons there has been limited success in using the theory of optimal auctions to explain the existence of pervasive institutions, such as the first-price sealed-bid auction used in competitive contracting or the posted price institution used in retailing.

SUMMARY. Incentive compatibility captures the fundamental positivist notion of self-interested behaviour that underlies almost all economic theory and application. It has proven to be an organizing principle of great scope and power. Combined with the modern theory of mechanism design, it provides a framework in which to analyse such diverse topics as auctions, central planning, regulation of monopoly, transfer pricing, capital budgeting, and public enterprise management. Incentive compatibility provides a basic constraint on the possibilities for normative analysis. As such it serves as the fundamental interface between what is desirable and what is possible in a theory of organizations.

JOHN O. LEDYARD

See also BIDDING; EFFICIENT ALLOCATION; EXTERNALITY; LINDAHL EQUILIBRIUM; ORGANIZATION THEORY; PUBLIC GOODS; REVELATION OF PREFERENCES.

BIBLIOGRAPHY
Baron, D. and Myerson, R. 1982. Regulating a monopolist with unknown costs. *Econometrica* 50, 911–30.
Dasgupta, P., Hammond, P. and Maskin, E. 1979. The implementation of social choice rules: some general results on incentive compatibility. *Review of Economic Studies* 46, 185–216.
Gibbard, A. 1973. Manipulation of voting schemes: a general result. *Econometrica* 41, 587–602.
Groves, T. and Ledyard, J. 1977. Optimal allocation of public goods: a solution to the 'free rider' problem. *Econometrica* 45, 783–809.
Groves, T. and Ledyard, J. 1986. Incentive compatibility ten years later. In *Information, Incentives, and Economic Mechanisms*, ed. T. Groves, R. Radner and S. Reiter. Minneapolis: University of Minnesota Press.
Harris, M. and Raviv, A. 1981. Allocation mechanisms and the design of auctions. *Econometrica* 49, 1477–99.
Hurwicz, L. 1960. Optimality and informational efficiency in resource allocation processes. In *Mathematical Methods in the Social Sciences*, ed. K. Arrow, S. Karlin and P. Suppes, Stanford: Stanford University Press, 27–46.
Hurwicz, L. 1972. On informationally decentralized systems. In *Decision and Organization: A Volume in Honor of Jacob Marschak*, ed. R. Radner and C.B. McGuire, Amsterdam: North-Holland, 297–336.
Hurwicz, L. 1979. Outcome functions yielding Walrasian and Lindahl allocations at Nash equilibrium points. *Review of Economic Studies* 46, 217–25.
Ledyard, J. 1978. Incomplete information and incentive compatibility. *Journal of Economic Theory* 18, 171–89.
Ledyard, J. and Roberts, J. 1974. On the incentive problem with public goods. Discussion Paper No. 116, Center for Mathematical Studies in Economics and Management Science, Northwestern University.
Lindahl, E. 1919. *Die Gerechtigkeit der Besteuerung*. Lund. Partial translation in *Classics in the Theory of Public finance*, ed. R.A. Musgrave and A.T. Peacock, London: Macmillan, 1958.
Matthews, S. 1983. Selling to risk averse buyers with unobservable tastes. *Journal of Economic Theory* 30, 370–400.
Moore, J. and Repullo, R. 1986. Subgame perfect implementation. London School of Economics, Working Paper.
Myerson, R.B. 1981. Optimal auction design. *Mathematics of Operations Research* 6, 58–73.
Myerson, R.B. 1985. Bayesian equilibrium and incentive compatibility: an introduction. In *Social Goals and Social Organization: Essays in Memory of Elisha Pazner*, ed. L. Hurwicz, D. Schmeidler and H. Sonnenschein, Cambridge: Cambridge University Press.
Palfrey, T. and Srivastava, S. 1986. Implementation in exchange economies using refinements of Nash equilibrium. Graduate School of Industrial Administration, Carnegie-Mellon University, July 1986.
Palfrey, T. and Srivastava, S. 1987. On Bayesian implementable allocations. *Review of Economic Studies* 54(2), 193–208.
Postlewaite, A. and Schmeidler, D. 1986. Implementation in differential information economics. *Journal of Economic Theory* 39(1), June, 14–33.

Roberts, J. and Postlewaite, A. 1976. The incentives for price-taking behavior in large economies. *Econometrica* 44, 115–28.

Samuelson, P. 1954. The pure theory of public expenditure. *Review of Economics and Statistics* 36, 387–9.

Satterthwaite, M. 1975. Strategy-proofness and Arrow's conditions: existence and correspondence theorems for voting procedures and social welfare functions. *Journal of Economic Theory* 10, 187–217.

Schmeidler, D. 1980. Walrasian analysis via strategic outcome functions. *Econometrica* 48, 1585–93.

Walker, M. 1981. A simple incentive compatible scheme for attaining Lindahl allocations. *Econometrica* 49, 65–71.

incentive contracts. Incentives are the essence of economics. The most basic concept, demand, considers how to induce a consumer to buy more of a particular good; that is, how to give him an incentive to purchase. Similarly, supply relationships are descriptions of how agents respond with more output or labour to additional compensation.

Incentive contracts arise because individuals love leisure. In order to induce them to forgo some leisure, or put alternatively, to put forth effort, some form of compensation must be offered. The theme of this essay is that different forms of incentive contracts deal with some aspects of the problems better than others. The strength of one type of contract is the weakness of another. The labour market trades off these strengths and weaknesses and thereby selects a set of institutions. In what follows, the development of the literature on incentive contracts is briefly discussed. The emphasis is on concepts rather than specific papers or authors, so the bibliography is far from exhaustive.

To discuss incentive contracts, the most general concepts must be narrowed. This essay does that in two ways. First, attention here is restricted to the labour market. At a more general level, incentive contracts can relate to other areas as well. For example, the government may want to have a space satellite built at the lowest possible cost. To do so, incentives must be set appropriately or the producer may charge too much or fail to meet desired quality standards. This problem is analogous to those that arise in the labour context, but for the most part they are ignored, except when isomorphic with the labour market paradigm. Similarly, the law and economics literature is another area where incentive problems are studied, usually in the context of accident liability (see, for example, Green, 1976; Polinsky, 1980; Shavell, 1980). These specific questions are ignored as well, except as they border on the labour market context. Second, the focus is on observability problems. Standard labour supply functions, where hours of work can be observed and paid, are incentive contracts. However, standard labour supply issues are eliminated from consideration since they are dealt with in other essays in *The New Palgrave*.

GENERAL FRAMEWORK

An employer in a competitive environment must induce a worker to perform at the efficient level of effort or face extinction. The reason is simple: if one employer can, through clever use of an incentive contract, get a worker to perform at a more efficient level, that firm's cost will be lower. Lower costs imply that higher wages can be paid to workers and all workers will be stolen from inefficient firms. As a result, the objective function that is taken as standard for the firm is:

$$\text{Max } F(Q, E) - C(E), \qquad (1)$$
$$\scriptstyle F$$

where Q is output and E is worker effort. Thus $F(Q, E)$ is the compensation schedule that the firm announces to the worker; $C(E)$ is the worker's cost of effort function, to be thought of as the dollar cost associated with supplying effort level E.

The competitive nature of the firm in factor and product markets implies that the firm must maximize worker net wealth as in (1) subject to the zero profit constraint:

$$Q = F(Q, E). \qquad (2)$$

Output is defined so that each unit sells for $1 (the numeraire). Thus (2) merely says that output, Q, must be paid entirely to the worker otherwise another firm could steal the worker away by paying more.

The incentive problem arises because the worker takes the compensation scheme $F(Q, E)$ as given and chooses effort to maximize expected utility. Once the worker has accepted the job, his problem is:

$$\text{Max } F(Q, E) - C(E). \qquad (3)$$
$$\scriptstyle E$$

The worker's effort supply function comes from solving the first-order condition associated with (3) or

$$C'(E) = \frac{\partial F}{\partial Q} \cdot \frac{\partial Q}{\partial E} + \frac{\partial F}{\partial E}, \qquad (4)$$

which says that the worker sets the marginal cost of effort equal to its marginal return to him. The transformation of effort into output, (i.e. $\partial Q/\partial E$) depends on the production function. A convenient specification is

$$Q = E + v, \qquad (5)$$

so that output is the sum of effort, E, and luck, v.

An incentive contract selects $F(Q, E)$ subject to the zero-profit constraint, (2), taking into account that the worker behaves according to (4). There are an infinite variety of incentive contracts that are subsumed by $F(Q, E)$. To make things clear, we consider two polar extremes – the salary and the piece rate (for a more detailed treatment, see Lazear, 1986).

Let us define a salary as compensation that depends only on input so that $F(Q, E)$ takes the form $S(E)$. An hourly wage is an example. Irrespective of the amount that is produced during the hour, the worker receives a fixed amount that depends only on the fact that he supplies E of effort for the hour. (Of course, difficulty in measuring E may be a compelling reason to avoid this form of incentive contract.) At the other extreme is a piece rate where compensation depends only on output so that $F(Q, E)$ takes the form of $R(Q)$. There, no matter how much or how little effort the worker exerts, his compensation depends only on the number of units produced. Both salaries and piece rates are incentive contracts; the first provides incentives by paying workers on the basis of input. The second provides incentives by paying on the basis of output. More sophisticated incentive contracts, which blend the two or use multiperiod approaches are discussed later.

THE PRINCIPAL–AGENT PROBLEM

At the centre of the incentive contract literature is the 'principal–agent' problem. The principal, say, an employer, wants to induce its agent, say, a worker, to behave in a way that is beneficial to the employer. The problem is that the principal's knowledge is imperfect; either he cannot see what the agent does (as in the case of a taxi driver who can sleep on the job) or he cannot interpret the actions (as in the case of an auto mechanic who replaces a number of parts to correct a perhaps simple malfunction). The incentive contracts that can be used to address the problem were discussed early by Ross (1973), Mirrlees (1976), Calvo and Wellisz (1978) and by Becker and Stigler (1974). The last in particular, uses a sampling approach. For example, a politician can be required to post a large bond on taking office. If he is caught engaging in some malfeasant behaviour, he forfeits the bond. This contract is based on output, which is observed infrequently or imperfectly. Other kinds of incentive contracts are discussed in the following sections.

PAYMENT BY OUTPUT

Sharecropping. One of the earliest examples of incentive contracts that is based on output is sharecropping. In sharecropping, the owner contracts to split the output of the land in some proportion with the individual who farms and lives on it. It was also one of the first incentive schemes that was clearly analysed (see Johnson, 1950, and later Cheung, 1969, and Stiglitz, 1974). The original problem as formulated in sharecropping can be seen as follows.

Payment is conditional only on Q and by some fixed proportion so that the worker receives γQ. Using (4) and (5), compensation of this sort implies that the worker's first-order condition is

$$C'(E) = \gamma,$$

so that the worker sets the marginal cost of effort equal to γ. But (5) implies that the marginal value of effort is \$1, which exceeds γ so that the worker puts forth too little effort. This is inefficient. Additionally, if the farmer can obtain land without limit, he pushes his sharecropping acreage to the point where the next unit of land has zero marginal product. This is clearly inefficient but can be remedied if landowners can select sharecroppers and terms according to the amount of land each works. Both the owner and worker could be made better off if the worker could be induced, by another incentive contract, to produce where $C'(E) = 1$.

Renting the land to the farmer and allowing the farmer to keep all of the output accomplishes this. Under rental, the worker's compensation is [Q-Rent]. By (4) and (5), the worker is induced to set $C'(E) = 1$; the marginal cost and marginal value of output are equated. Of course, rental does not solve all of the problems. Absent in the production function in (5) is that maintenance may be required. For example, if the farmer does not fertilize the land, it may not produce as well in the future. A renter, who can move on to the next plot after the soil is drained of minerals, has little incentive to put resources into the land. Thus the solution is to sell the land to the farmer. Then the individual who works the land has the correct incentives, either because he will continue to use it in the future or because the sale price will reflect the quality of the land. But sale of the land begs most of the questions. The sale may not come about because of the farmer's capital constraints, because of his lack of entrepreneurial skill, or because of his distaste for risk. (Note that risk is shifted from owners to farmers even in sharecropping and renting. Only labour contracts based exclusively on effort shift the risk entirely to the owner.)

The sharecropping paradigm applies to industrial production as well. Profit-sharing arrangements are, in many respects, like sharecropping. This is especially true when there is only one worker. Partnerships are similar. The same incentive problems arise. A worker who can quit and move on to another firm without penalty does not have the same desire to maintain the equipment as the firm's owner. Again the solution is to sell the capital to the worker, but this simply redefines the owner.

Then there is no principal–agent problem because there is no agent. This can be considered in more detail in the next section.

Piece rates. Piece-rate compensation is not much different from sharecropping, the latter being a special case of the former (see Stiglitz, 1975). The owner allows the worker (or farmer) to use his capital (or land) and pays the worker according to some function of output. In the simplest scheme, a linear piece rate is used and the worker is paid rate R per unit Q so that compensation is RQ. The worker's maximization problem (3) and (4) implies that the worker sets $C'(E) = R$. The firm's zero-profit constraint in (2) implies that $Q = RQ$ or that $R = 1$. Thus the piece rate is efficient because the worker sets the marginal cost of effort equal to its marginal social value, \$1.

The issue is only slightly more complicated if capital is involved. A linear piece rate with an intercept (i.e. compensation equal to $A + RQ$) will do the job. This incentive contract achieves first-best efficiency. The worker's first-order condition, (4), still guarantees that he sets $C'(E) = R$. The intercept drops out. But the zero-profit constraint now becomes:

$$Q - \text{rental cost of capital} = A + RQ.$$

The firm must 'charge' the worker for the cost of using the capital, but how should this be done? R can be reduced below 1 or A can be set to a negative number. The answer is that $A = -(\text{rental cost of capital})$ and $R = 1$. Since (4) does not contain A, the worker does not respond to changes in A. However, reducing R below 1 causes the worker to reduce effort. Thus the efficient incentive contract, which also maximizes worker wealth subject to the firm's zero-profit constraint, requires that $R = 1$. Zero profit requires that $A = -(\text{rental cost of capital})$.

A major advantage to the use of piece rates as an incentive contract is that it tolerates heterogeneity of worker ability. More able – that is, lower effort cost – workers choose higher levels of effort but are paid more. There is no inefficiency involved in having workers of both types in the firm. Of course, if capital is important so that the worker is 'charged' A for the right to work on a machine, only workers above some threshold ability level will choose to work. But workers self-sort. There is no need for the firm to do anything other than pay the efficient piece rate, in this case $R = 1$.

Linear piece rates are no longer appropriate incentive contracts if workers are risk-averse. In general, a non-linear scheme will do better but will fail to achieve first-best solutions. As long as asymmetric information exists, so that individual actions cannot be observed and contracted upon, Pareto optimal risk-sharing is precluded (see Hölmstrom, 1979; Harris and Raviv, 1979.)

Payment of relative output. The study of relative compensation has become increasingly important. There are two approaches in this literature. The first, from Lazear and Rosen (1981), characterizes the labour market as a tournament, where one worker is pitted against another. The one with the highest level of output receives the winning prize (i.e. the high-wage job) while the other gets the losing prize (i.e. the low-wage job). By increasing the spread between the winning and losing prizes, incentives are provided to work hard. The optimum spread induces workers to move to the point where the marginal cost of effort exactly equals the marginal (social) return to it. The major advantages to payment by tournament method are twofold. First, tournaments require only that relative comparisons be made. It may be cheaper to observe that one worker produces more than another than to determine the actual amount that each produces. Second, compensation by rank 'differences out' common noise. For example, sales may be low because the economy is in a slump, which has nothing to do with worker effort. Risk aversion operates against penalizing or rewarding workers for factors over which they have no control. But since the slump affects both workers equally, relative comparisons are unaffected. The best worker still produces more, even though both produce small amounts.

Tournament-type incentive contracts induce workers to behave efficiently if they are risk neutral. They are easy to use but carry one major disadvantage. Workers increase the probability of winning, not only by doing well themselves but also by causing the opponent to do poorly. Thus tournaments discourage cooperation. This results in wage compression, which works to discourage the aggressive behaviour of workers who are competing for the same job. Other work in the area of tournament-type incentive contracts includes Nalebuff and Stiglitz (1983), Green and Stokey (1983) and Carmichael (1983).

The second approach, from Hölmstrom (1982), suggests that if levels of output can be observed, then payments can be based, at least in part, on a team average. As Hölmstrom points out, a tournament is not a sufficient statistic, so that using a team average allows the firm to better address risk aversion. This incentive device also takes out common noise. A peer average picks up disturbances that are common to the industry and allows the firm to cater to the tastes of risk-averse workers.

PAYMENT BY INPUT

Observability of effort. It is commonly alleged that payment of a salary or hourly wage does not provide workers with the appropriate incentives. Whether or not this is true depends on the connection between the measurement of time and measurement of effort. To see this, suppose that effort can be observed perfectly, but that output cannot be observed at all. For example, suppose that it is easy to measure the number of calories burned up by a worker during his work day, but it is impossible to separate his output from that of his peers. Payment by effort is a first-best incentive contract. The compensation scheme that pays the worker $1 per unit of effort exerted induces him to set $C'(E) = 1$, which, as we have seen, is first best. Note further that this is first best even for risk-averse workers since compensation does not vary with random productivity shocks, v (see Hall and Lilien, 1979).

The allegation that effort pay does not provide incentives is based on the difference between hours of work and effort. If hours were a perfect proxy for effort, then payment of an hourly wage would be an optimal incentive contract. But because workers can vary work per hour, the connection breaks down. Payment per hour provides appropriate incentives for choice of the number of hours, but does not deal with what is done within the hour.

Payment by effort and worker sorting. Piece rates induce workers to sort appropriately. Above, it was argued that workers who cannot produce a sufficiently high level of output will not come to a firm that 'charges' for use of capital. Salaries (or hourly wages) that pay on the basis of an imperfect measure of effort encourage the lower-quality workers to come to the firm. Lazear (1986) demonstrates that a separating equilibrium (see, e.g., Rothschild and Stiglitz, 1976; Salop and Salop, 1976) exists where high-quality workers choose to work at firms that pay piece rates and low-quality ones choose salaries. The difference in quality across firms might lead one to conclude that movement to output-based incentive contracts increases total output. In fact, the reverse may well be true. In the same sense that screening in Spence (1973) is socially unproductive, forcing salary firms to adopt piece-rate incentive contracts wastes resources on a potentially useless signal.

Incentive contracts and product quality. Sometimes quantity is easier to observe than quality. The problem with incentive contracts that are based on output quantity is that they induce the worker to go for speed and to ignore quality. If quality can be observed, then the worker can be compensated appropriately for quantity and quality. The appropriate compensation function is essentially the consumer's demand for the product as it varies with quality and quantity. But if quality cannot be observed, payment by input 'solves' the quantity/quality problem. If the worker is paid, say, by hour, and is merely instructed to produce goods of a given quality, he has no incentive to deviate from that instruction. Compensation is based only on input, so there is no desire to rush the job. Of course, this requires a method of monitoring effort cheaply (see Lazear, 1986, for a full discussion of the trade-offs).

OTHER ISSUES IN INCENTIVE CONTRACTING

Efficient separation and long-term investments. A properly structured incentive contract must induce the correct amount of long-term investment. The problem is most clearly seen in the context of specific human capital, as in Becker (1962, 1975). Specific human capital is only valuable when the worker is employed at the current firm. As such, workers are reluctant to invest in specific capital because the firm may capriciously fire the worker, in which case the investment is lost. Similarly, firms are reluctant to invest because the worker may capriciously quit. The incentive contract that Becker suggests is a sharing of investment costs and returns by both workers and firms (Hashimoto and Yu, 1980, model this more precisely.) Kennan (1979) points out that a particular kind of severance pay solves the investment problem. It is akin to the liability rules that are efficient in auto accident problems. But as Hall and Lazear (1984) argue, these rules may actually induce too much investment. Since a worker is compensated for the full investment whether work occurs or not, he has no incentive to account for situations that make a separation optimal. For example, if it were optimal to sever the work relationship 25 per cent of the time, the worker should behave as if a specific investment that yields $1 return only yields $0.75. A full-reimbursement severance pay arrangement ensures a full $1, irrespective of the status of work, and induces too much investment.

More general issues of efficient separation arise in the labour market context, and incentive contracts must be structured to deal with these problems. Hall and Lazear (1984) consider a variety of different incentive contracts and conclude that none generally achieves first best. One that comes close to doing so is Vickrey's (1961) bilateral auction approach. There, compensation and work are separated so that the worker and firm have incentives to reveal the true relevant values. Another scheme is coordinated severance pay, suggested by d'Aspremont and Gerard-Varet (1979). Sufficiently high penalties on the firm associated with a worker's refusal to work induces the firm to behave in a manner that is apparently first best.

Intertemporal incentive contracts. Sometimes, the fact that workers live for more than one period allows contracts to be structured in a way that solves incentive problems. This is the subject of Lazear (1979, 1981). The problem is that as a worker approaches the end of his career, he has an incentive to shirk because the costs, even of being fired, are reduced as his retirement date draws near. A way to discourage shirking is to tilt the age-earnings profile and couple it with a contingent pension. Young workers are paid less than their marginal products; old workers are paid more. In equilibrium, shirking is discouraged and workers receive exactly their lifetime marginal products. The distortion in the timing of the payments implies that workers do not voluntarily choose to work the correct number of hours. Thus hours constraints are required, an extreme form of which is mandatory retirement. Other work that has refined or provided empirical support for that concept is Kuhn (1986) and Hutchens (1986a, b).

There are other papers that focus on the intertemporal aspects of incentive contracts. The first, Fama (1980) argues that the market provides a discipline on workers. In a spot market, the wage that another firm is willing to offer a worker next period depends on how well he did last period. Fama shows that this can act as a perfect incentive device. Of course, no end-game problems are addressed by this mechanism, but it does demonstrate the possibility of incentive provision even without explicit or implicit contracts. The second idea is attributable to Rogerson (1985). The emphasis here is on risk-sharing, but the work has some features in common with Fama (1980). In particular, memory plays a strong role in these incentive contracts, so that an outcome that affects the current wage also affects the future wage.

Intertemporal strategic behaviour by firms. Once intertemporal contracts are considered, it is necessary to examine the issue of opportunistic behaviour by firms. It may be that a firm does not know a worker's cost of effort function, $C(E)$. Actions that the worker takes may reveal information about that function. The firm can use that information in subsequent periods against the worker. As a result, the worker attempts to disguise $C(E)$, leading to inefficiencies. Such is the case of salesmen, whose next period quota depends on this period's performance. In Lazear (1986) it is shown that a properly structured contract in a competitive labour market can undo the effects of this kind of strategic behaviour. This is a specific example of the general theorem on revelation presented in Harris and Townsend (1981). It is also related to the literature on planned economies, since bureaucrats tend to make things look worse than they are to lessen next period's requirements or to increase next period's budget allocation (see, e.g., Weitzman, 1976, 1980; Fan, 1975).

Insurance. Finally, there is a closely related literature that examines insurance contracts. That literature focuses, for the most part, on the trade-off between insurance and efficiency in the labour market. Some of the more important papers in that literature include Harris and Hölmstrom (1982), Grossman and Hart (1983) and Green and Kahn (1983).

CONCLUSION

Although incentive problems are pervasive, the market has found a number of solutions. These involve payment by output of the piece rate or sharecropping variety; payment by relative output, exemplified by labour market tournaments; payment by measured input, such as hours of work; and multi-period incentive contracts. The contracts do not always achieve the first best, especially when risk aversion is an issue.

Still, the rich variety of institutions that address incentive problems and the large amount of literature devoted to study attest to the problem's importance in the labour market context.

EDWARD P. LAZEAR

See also IMPLICIT CONTRACTS; INCOMPLETE CONTRACTS; LAYOFFS; PRINCIPAL AND AGENT.

BIBLIOGRAPHY

Becker, G.S. 1962. Investment in human capital: a theoretical analysis. *Journal of Political Economy* 70, October, 9–49.

Becker, G.S. 1975. *Human Capital: A Theoretical and Empirical Analysis, with Special Reference to Education.* 2nd edn, New York: Columbia University Press for the National Bureau of Economic Research.

Becker, G.S. and Stigler, G.J. 1974. Law enforcement, malfeasance, and compensation of enforcers. *Journal of Legal Studies* 3, January, 1–18.

Calvo, G. and Wellisz, S. 1978. Supervision, loss of control and optimum size of the firm. *Journal of Political Economy* 86, October, 943–52.

Carmichael, H.L. 1983. The agent–agents problem: payment by relative output. *Journal of Labor Economics* 1, January, 50–65.

Cheung, S.N.S. 1969. *The Theory of Share Tenancy: With Special Application to Asian Agriculture and the First Phase of Taiwan Land Reform.* Chicago: University of Chicago Press.

d'Aspremont, C. and Gerard-Varet, L.A. 1979. Incentives and incomplete information. *Journal of Public Economics* 11, February, 25–45.

Fama, E. 1980. Agency problems and the theory of the firm. *Journal of Political Economy* 88, April, 288–307.

Fan, L.-S. 1975. On the reward system. *American Economic Review* 65, March, 226–9.

Green, J.R. 1976. On the optimal structure of liability laws. *Bell Journal of Economics* 7, Autumn, 553–74.

Green, J.R. and Kahn, C. 1983. Wage employment contracts. *Quarterly Journal of Economics* 98, 173–88.

Green, J.R. and Stokey, N.L. 1983. A comparison of tournaments and contracts. *Journal of Political Economy* 91, June, 349–64.

Grossman, S. and Hart, O. 1983. Implicit contracts under asymmetric information. *Quarterly Journal of Economics* 71, 123–57.

Hall, R.E. and Lazear, E.P. 1984. The excess sensitivity of layoffs and quits to demand. *Journal of Labor Economics* 2, April, 233–58.

Hall, R.E. and Lilien, D. 1979. Efficient wage bargains under uncertain supply and demand. *American Economic Review* 69, December, 868–79.

Harris, M. and Hölmstrom, B. 1982. A theory of wage dynamics. *Review of Economic Studies* 49, July, 315–33.

Harris, M. and Raviv, A. 1979. Optimal incentive contracts with imperfect information. *Journal of Economic Theory* 20(2), April, 231–59.

Harris, M. and Townsend, R. 1981. Resource allocation under asymmetric information. *Econometrica* 49, January, 33–64.

Hashimoto, M. and Yu, B. 1980. Specific capital, employment contracts, and wage rigidity. *Bell Journal of Economics*, Autumn, 536–49.

Hölmstrom, B. 1979. Moral hazard and observability. *Bell Journal of Economics* 10, Spring, 74–91.

Hölmstrom, B. 1982. Moral hazard in teams. *Bell Journal of Economics* 13, Autumn, 324–40.

Hutchens, R. 1986a. Delayed payment contracts and a firm's propensity to hire older workers. *Journal of Labor Economics* 4(4), October, 439–57.

Hutchens, R. 1986b. An empirical test of Lazear's theory of delayed payment contracts. Working paper, Cornell University Institute for Labor and Industrial Relations.

Johnson, D.G. 1950. Resource allocation under share contracts. *Journal of Political Economy* 58, 111–23.

Kennan, J. 1979. Bonding and the enforcement of labor contracts. *Economic Letters* 3, 61–6.

Kuhn, P.J. 1986. Wages, effort, and incentive compatibility in life-cycle employment contracts. *Journal of Labor Economics* 4, January, 28–49.

Lazear, E.P. 1979. Why is there mandatory retirement? *Journal of Political Economy* 87, December, 1261–84.

Lazear, E.P. 1981. Agency, earnings profiles, productivity and hours restrictions. *American Economic Review* 71, September, 606–20.

Lazear, E.P. 1986. Salaries and piece rates. *Journal of Business* 59(3), July, 405–31.

Lazear, E.P. and Rosen, S. 1981. Rank order tournaments as optimum labor contracts. *Journal of Political Economy* 89, 841–64.

Mirrlees, J.A. 1976. The optimal structure of incentives with authority within an organization. *Bell Journal of Economics* 7, Spring, 105–31.

Nalebuff, B.J. and Stiglitz, J.E. 1983. Prizes and incentives: toward a general theory of compensation and competition. *Bell Journal of Economics* 14, Spring, 21–43.

Polinsky, A.M. 1980. Strict liability vs. negligence in a market setting. *American Economic Review* 70(2), May, 363–7.

Rogerson, W.P. 1985. Repeated moral hazard. *Econometrica* 53, January, 69–76.

Ross, S.A. 1973. The economic theory of agency: the principal's problem. *American Economic Review* 63, May, 134–9.

Rothschild, M. and Stiglitz, J.E. 1976. Equilibrium in competitive insurance markets: an essay on the economics of imperfect information. *Quarterly Journal of Economics* 90, November, 629–49.

Salop, J. and Salop, S. 1976. Self-selection and turnover in the labor market. *Quarterly Journal of Economics* 90, November, 619–27.

Shavell, S. 1980. Strict liability versus negligence. *Journal of Legal Studies*, January, 1–25.

Spence, A.M. 1973. Job market signalling. *Quarterly Journal of Economics* 87, August, 355–74.

Stiglitz, J.E. 1974. Incentive and risk sharing in sharecropping. *Review of Economic Studies* 41, April, 219–55.

Stiglitz, J.E. 1975. Incentives, risk, and information: notes toward a theory of hierarchy. *Bell Journal of Economics and Management Science* 6, Autumn, 552–79.

Vickrey, W. 1961. Counterspeculation, auctions, and competitive sealed tenders. *Journal of Finance* 16, 8–37.

Weitzman, M. 1976. The new Soviet incentive model. *Bell Journal of Economics* 7(1), Spring, 251–7.

Weitzman, M. 1980. The 'Ratchet Principle' and performance incentives. *Bell Journal of Economics* 11(1), Spring, 302–8.

incidence of taxation. *See* TAX INCIDENCE.

income. Like 'supply', 'demand', 'rent', 'welfare' and 'utility', the word 'income' is a part of common speech that has entered economics as a technical term. *The Concise Oxford Dictionary* defines income as 'receipts from one's lands, work, investment etc'. That meaning carries over into economic theory, where, for instance, a consumer may be said to maximize utility subject to an income constraint or a firm may be said to maximize income accruing to its stockholders. The meaning of income is somewhat modified in the construction of income statistics. These are employed in two quite distinct contexts: as the basis for income taxation and as generalized to national income.

In each context, the definition of income is governed by the purpose of the statistics. Personal and corporate income are defined to serve as criteria for taxing people and corporations, the main principles behind the definitions being equity or fairness among people with different sources of income and efficiency in the economy as a whole. The purpose of income

within the national accounts is less easily defined. The national accounts are an intricate set of statistics intended to describe the economy as a whole, primarily but not exclusively to facilitate counter-cyclical policy. The simple concept of income as applied to a person is extended to the entire nation in several ways: national income is the sum of the earnings of all factors of production; national product is the value of output of all goods and services; national expenditure is the sum of each person's expenditure on goods and services. All three would be equal in a world without depreciation, indirect taxation or subsidies. Income statistics also serve as the basis for comparisons among regions, products and occupations; converted to real income, they become the basis for the measurement of the rate of economic growth.

As a first approximation, we may say that national income is the sum of all personal incomes, but the definition of income for tax purposes differs in several important respects from the definition in the national accounts. The major differences can be classified under the headings of scope, intermediation and timing.

The scope of income is a trade-off between two objectives, to include all benefits to consumers as part of income, even benefits arising from non-market activity, and to construct statistics that are reasonably precise and beyond dispute. The latter consideration is relatively more important for tax purposes. Thus money values of the services of owner-occupied housing, food grown and consumed on farms, and direct provision of food and lodging for the armed services, are usually included in national income but almost never as part of the tax base. Housework, on the other hand, is included in neither definition, giving rise to the old paradox that the national income falls if a man marries his housekeeper. There is an ongoing debate in public finance as to whether the base for personal taxation should be income as a whole (consumption plus investment) or just consumption.

The major issue in the timing of income concerns capital gains which are sometimes included in income for tax purposes but are never part of the national income. They are excluded from the national income so that beneficial changes in technology and other aspects of the economy appear as part of income in the years when they materialize as goods and services rather than when they are first anticipated. Personal income is another matter. A person becomes wealthy in the year his assets appreciate, regardless of the dates of the increases in the marginal products of the corresponding capital goods. The usual justification for including capital gains as part of taxable income is that a person should be taxed when he becomes wealthy, just as he is taxed when he earns ordinary income.

The inclusion of capital gains in the tax base creates several problems: assets may appreciate as part of a general inflation; taxation of capital gains may only be feasible when gains are realized rather than when they accrue; gain on human capital is automatically exempt from the tax; taxation of capital gain may be double taxation of the return to capital because future tax on the earnings to capital goods is discounted back in the price at which the goods are sold. However, the exclusion of capital gains from the tax base creates an incentive for firms to seek ways to disburse money to people as non-taxable capital gains rather that as taxable income.

Both personal and national income are defined net of the cost of intermediate products. The income of the travelling salesman is net of the cost of his car, and the national income is net of the aggregate cost of transport for business purposes – of the cost of haulage of goods, business travel, and so on. The boundary between final and intermediate products is sometimes problematic. For example, it is not always clear how to classify the business lunch, especially for participants who would rather diet. Current expenditure of government is classified in the national accounts as consumption, yet it is arguable that most such expenditure is either intermediate product (social overhead for the economy as a whole) or intangible investment (the obvious instance being current expenditure on research).

Depreciation is like an intermediate product. An input to production that is used up in the course of the year or incorporated into output is unambiguously intermediate, and its cost is excluded on that account from the measure of income. An input to production that lasts more than a year but depreciates somewhat during the first year is financially equivalent to the sum of an ordinary intermediate product and an input that only becomes available at the end of the first year. To deduct depreciation from income is to treat the intermediate component of investment like an ordinary intermediate good. On the other hand, it is often difficult in practice to determine what depreciation ought to be. The tax code specifies rates of allowable depreciation for each type of capital equipment, for it is more important in this context to be precise and predictable than to be right. There is more flexibility in the national accounts and an attempt is made to measure the true loss over the year in the value of capital goods. Loss of value may be deterioration or obsolescence. Both belong as part of depreciation, especially when obsolescence is anticipated, for it makes no difference at the time a machine is purchased whether it is destined to deteriorate through use or to become worthless as better machines are developed. Unanticipated obsolescence is more problematic, for there is something anomalous in reducing this year's national income for the fall in the value of capital goods brought about by expected technical change next year when the benefit of the change itself is excluded. Income would fall though there would be no reduction of output in the current year and people would be better off in the long run.

D. USHER

See also CAPITAL GAINS AND LOSSES; DEPRECIATION; NATIONAL INCOME; REAL INCOME; SOCIAL ACCOUNTING.

BIBLIOGRAPHY

On the definition of personal income, see Simons (1938). The classic discussion of the concept of national income, related conceptual problems and the history of the development of the national accounts is Studenski (1961). Most countries produce national accounts and publish details of how they are compiled. For a proposed standardization, see *A System of National Accounts*, United Nations, 1968. Current research in national accounting is likely to appear in *The Review of Income and Wealth*.

Simons, H. 1938. *Personal Income Taxation: The Definition of Income as a Problem of Fiscal Policy.* Chicago: University of Chicago Press.
Studenski, P. 1961. *The Income of Nations.* New York: New York University Press.
United Nations. 1968. *A System of National Accounts.* New York: United Nations, Department of Economic and Social Affairs. Studies in Methods, series F, no. 2.

income distribution. *See* DISTRIBUTION THEORIES: CLASSICAL; DISTRIBUTION THEORIES: KEYNESIAN; DISTRIBUTION THEORIES: MARXIAN; INEQUALITY BETWEEN PERSONS; MARGINAL PRODUCTIVITY THEORY; PROFIT AND PROFIT THEORY; RENT; WAGES, REAL AND MONEY.

income–expenditure analysis. The term 'income–expenditure analysis' serves as a short-hand expression for the dominant type of conceptual framework for macroeconomic analysis to emerge from the debate which crystallized around Keynes's *General Theory* (1936). As Coddington (1976) notes, income-expenditure analysis was not the only thing to be learned from the *General Theory*, but it has certainly been the dominant one, forming the central message of Keynesian economics as generally understood. Although the term does not appear to have been used by Keynes himself it is to be found, freely used, in the early works of exposition of the *General Theory* and the Keynesian Revolution. At the formal and simplest level it can be taken to refer to the 45° 'Keynesian Cross' diagram, at a more sophisticated level to the IS/LM analysis.

At the outset of the *General Theory*, Keynes noted the inability of traditional theory to explain the Great Depression. His analysis was evolved to make good this deficiency and does so by sidestepping the concerns of that theory, in that the power of the price system to ensure the equilibration of the economy is simply denied. Relative prices being set on one side, the analysis focuses on the interaction of flows of expenditure to explain economic fluctuations and the determination of output. In the more familiar embodiments of income–expenditure analysis, the abstraction is clearer and proceeds further than it does in the *General Theory* itself. In particular, both wages and prices are taken as fixed, whilst the stock of productive capital, wealth and the 'state of expectations' are also taken as given. Capital markets are imperfect. Keynes's concept of the propensity to consume is central and leads to the key result of the multiplier. Because expenditures are cash-constrained and output is demand-constrained, an autonomous increment in demand relaxes the constraint on spending and so on output, raising incomes and so stimulating further 'rounds' of expenditures. The ultimate increase in demand may exceed the initial stimulus. In the simplest version of this analysis, exemplified in the 'Keynesian Cross' diagram, asset prices are taken as fixed along with wages and commodity prices. In the IS/LM version this restriction is relaxed and demand increases may then spend themselves at least partly in changes in asset prices which in turn affect the desire to acquire capital goods. In this way the multiplier process may be attenuated if monetary conditions are tight rather than passive.

The framework readily accommodates policy, inviting the tools of fiscal and monetary policy to be used in the management of demand with the objective of output stabilization. Further, the analysis lends itself to quantification; with the addition of a modelling of the wage-price sector and of the foreign exchanges (or capital flows), income–expenditure analysis was the initial basis for the construction of macroeconometric models of the kind used by Finance Ministries and Central Banks and indeed remains the basic foundation for such models today.

The abstraction from relative prices of this approach, its highly aggregative and often heavily quantified nature has exposed income-expenditure analysis to the criticism that it is excessively mechanical (Coddington (1976) describes it as 'hydraulic' Keynesianism in reference to this criticism and perhaps to the fact that some early teaching machines of this model were literally hydraulic); it was said to 'lack microfoundations'. Modern temporary equilibrium theory of the kind pioneered by, for example, Barro and Grossman (1976) has supplied such foundations – or, to be more accurate, has shown how a general equilibrium model with fixed prices generates properties which are very similar to those to be found in income–expenditure analysis –

involuntary unemployment may exist, the multiplier process is replicated, and fiscal and monetary policy can be given a demand management rationale, for example. The condition on which these results are generated is that wages and prices fail to equilibrate the model and this condition is imposed as a stylized fact rather than explained by the model itself, a weakness in the eyes of critics who would argue that the existence of such rigidities implies unexploited opportunities for profitable trade to (well-informed) rational agents. On the other hand, there appear to be a number of reasons why wages and prices are in fact sticky, supporting the principal strategic simplification of income–expenditure analysis, whilst modern exploration of the properties of rational expectations models has confirmed the wisdom of Keynes's tactic of treating expectations as parametric in view of the problem of multiple equilibria. Certainly, the resultant mode of analysis has been very successful and still retains a dominant place in macroeconomics, despite the flourishing new classical school.

MICHAEL ARTIS

See also AUTONOMOUS EXPENDITURES; INFLATIONARY GAP; IS–LM ANALYSIS; NEOCLASSICAL SYNTHESIS.

BIBLIOGRAPHY
Barro, R. and Grossman, H.I. 1976. *Money, Employment and Inflation.* London and New York: Cambridge University Press.
Coddington, A. 1976. Keynesian economics: the search for first principles? *Journal of Economic Literature*, December. Reprinted in A. Coddington, *Keynesian Economics: The Search for First Principles*, London: George Allen & Unwin, 1983.
Keynes, J.M. 1936. *The General Theory of Employment, Interest and Money.* Reprinted in *The Collected Writings of John Maynard Keynes*, Vol. VII, London: Macmillan, 1971.

incomes policies. Keynes's *General Theory* attacked the foundation of the quantity theory of money – the proposition that the level of activity is determined by real forces. But though Keynes provided a new theory of output, he offered no systematic explanation of the price level. He simply took money wages as given, and argued that their level was the key determinant of all nominal magnitudes.

In *How to Pay for the War* Keynes took the analysis further, presenting two complementary theories of the determination of the *rate of change* of money wages and prices. The rate of inflation was affected on the one hand, by the pressure of demand; on the other, by the attempt of workers to maintain the real value of their incomes in a recession or when the terms of trade deteriorated.

These two explanations reinforce one another in Keynes's exposition, but they are formally distinct. The pressure of demand may explain the rate of change of wages independently of any predetermined real wage; this was one central implication of Phillips's (1958) study of the relationship between money wage inflation and the level of unemployment, and of later refinements of the Phillips curve. However, the efforts of workers to maintain real incomes may determine the rate of change of money wages and prices relatively *independently* of the pressure of demand (though they may be affected by rapid changes in the pressure of demand).

If the two processes are combined, the result may be formulated as a relationship between the pressure of demand and the rate of change of the rate of inflation. In some writers' view, the pressure of demand leads to real wage bids which cannot be satisfied and hence to cumulative rises in wages and prices, as workers and employers bid for shares of real income which total more than one (Rowthorn, 1977).

When the Phillips curve analysis was presented, it appeared to carry an important message for macroeconomic policy: that there is a trade-off between unemployment and the rate of inflation (or between unemployment and the rate of change of the rate of inflation). The inflation which was believed to be associated with macroeconomic expansion provided a constraint on such expansion. If inflationary pressures could be reduced, it was argued, economies could produce more. From that perspective, the purpose of an incomes policy was to reduce the inflationary pressure associated with any given level of demand.

Discussion of incomes policy often fails to make it clear whether the policy is intended to operate solely on nominal wages, or on real wages. In fact, the distinction is important for both analytical and practical reasons. If the level of money wages is given and unrelated to real variables (as Keynes appeared to suggest in the *General Theory*), then it may be argued that the role of incomes policy is to moderate the rate of change of nominal magnitudes, with no particular implications for real wages, or for the distribution of income between wages and profits. In which case, the pressure to disrupt an income policy will logically come from groups intent on changing the distribution of real income in their favour.

But if the level of money wages is the outcome of bargaining over real incomes, then the purpose of an incomes policy will usually be to persuade workers to achieve lower real wages and thus change the distribution of income. In which case, the policy must either deal directly with the forces which determine real incomes in the first place; or it will be placed under considerable strain – perhaps breaking down as those forces reassert themselves (Tarling and Wilkinson, 1977).

In reality, incomes policies have often been justified as a way of reducing inflation in nominal magnitudes; but they have also changed the distribution of income, usually by squeezing real wages.

In the view of those who favour incomes policy, successful policies – those that have lasted the longest and been associated with relatively low rates of inflation – have been those which (a) recognize explicitly that the distribution of real income is at stake, and plan the rate of change of money wages as part of a socio-political 'deal' (war-time policies are good examples); and/or (b) are implemented when real national income is growing rapidly, so that all real wages can increase even as the distribution of real income is changed.

Although labelled 'incomes policy', most such policies are concerned only with wages. They may sometimes be linked to controls on dividends to provide an aura of 'fairness'; or to a prices policy, in which case they are overtly a policy for real incomes.

In the OECD countries five main types of incomes policies have been tried. One version relies on *exhortatory guidelines*. It does little more than encourage employers and employees to settle for lower increases in nominal wages and salaries. As one author sympathetic to incomes policy has argued, 'a principal objective of incomes policy must be to inform public opinion and develop a consensus on the appropriate rate of increase for most wages and salaries ...' (Braun, 1986). To this end, a government may set an example itself in the sectors in which it is the direct employer (such as the civil service) or the indirect financier (as may be the case in some nationalized industries). And the government may also argue that monetary and fiscal expansion will be restrained if inflation is 'too high'.

If exhortation fails, a government may resort to *temporary measures*, such as a wage freeze. For example, it may be argued that a freeze is a temporary measure designed to adjust inflationary expectations in such a way as to minimize the consequences of deflationary policies. The most dramatic examples of the use of temporary measures have occurred in countries suffering from very rapid inflation (Argentina, Bolivia and Israel in 1985, and Brazil in 1986), and have often been accompanied by price freezes and currency reform.

A *statutory norm* involves a government laying down limits for the increase of nominal wages allowed in any one year. The increase has usually been couched in percentage terms, though sometimes this has been combined with an absolute limit. The rate is set for the economy as a whole, with increases beyond the norm being sanctioned by some form of arbitration tribunal – usually on the grounds of exceptional productivity growth. This approach has been widely used in Britain, and, from time to time, in the United States. The relative success or failure of the statutory norm has been determined by its consequences for the rate of increase of real wages, and by the impact on wage differentials of productivity-based 'exceptions'.

Recognition that incomes policies are concerned with the distribution of real income has led to the suggestion that nominal magnitudes should be *indexed*. This usually means that wage agreements contain automatic escalator clauses. Such agreements are undermined when the increase in real incomes implicit in nominal wage negotiations cannot be sustained – as, for example, when the quadrupling of oil prices in 1973–4 resulted in a transfer of real income from the OECD countries to the oil producers. In such circumstances, indexation can become a source of explosive inflation, at least in the short term.

Milton Friedman has claimed (1974) that

> widespread escalator clauses would make it easier for the public to recognize changes in the rate of inflation, would thereby reduce the time-lag in adapting to such changes, and thus make the nominal price level more sensitive and variable ... . But, if so, the *real variables* would be *less* sensitive and *more* stable – a highly beneficial trade-off.

In short, indexation would allow governments to disinflate (by appropriate monetary measures) with the least harmful consequences for the real economy. Wage earners would find that their wages adjusted quickly to disinflationary policies. Without that prompt adjustment, they would have obtained unjustifiably large increases, which would cause bankruptcies and unemployment.

Indexation may be favoured on other grounds. For example, it may give confidence to particular bargaining groups, who will therefore have no incentive to bid for large increases in nominal wages as a means of protecting real wages.

As a policy device, indexation was adopted by several countries in the late 1940s, among them Belgium, Luxembourg, Italy, Denmark and Norway. During the 1970s indexation was introduced in Britain, the Netherlands, Ireland, Switzerland and Australia. Indexation had been common in France until 1958, when it was abolished at the same time as the franc was devalued. This proved a pointer to the subsequent abandonment of indexation in many other countries. Governments found that indexation prevented them from making desired changes in relative prices (such as changes in exchange rates) which had implications for the distribution of income. In such circumstances indexation is a device for institutionalizing inflation.

A different automatic device for regulating wage increases is the *tax-based incomes policy* (TIP). In this scheme the government sets a norm for wage increases. Firms that pay less than the norm receive a reward, typically in the form of lower corporate taxes. Those that pay more face a tax penalty. A variant would provide employees with tax incentives to settle for wage increases below the norm.

Schemes of this type have been tried only in the most general form. In Britain in 1977–8, for example, the Labour government promised to reduce income tax if the national increase in wages was moderated. In Austria in 1967–8, the government achieved a wage-tax bargain. When it tried again, in 1974, its proposal was rejected by the unions.

In practice, incomes policies have tended to follow a sequence of initial acceptance and effectiveness, followed by growing opposition and circumvention, and then breakdown. Explanations of this phenomenon vary according to each author's view of the price mechanism. For example, some economists argue that incomes policies fail because they seek to over-ride market forces. An incomes policy unsupported by suitably anti-inflationary macroeconomic policies is bound to fail; but one that is so supported is superfluous. Indeed it may do microeconomic harm because it slows down and distorts market adjustments: 'the theory of incomes policy, as opposed to the desperate "ad-hocery" of practice, has not come to grips with resolving some form of wage, dividend and price control with the resource-allocating function that both goods and factor prices are held to play' (Ball and Doyle, 1969; see also Paish, 1968).

A different explanation of the failure of incomes policy comes from those who take a somewhat jaundiced view of the efficiency of markets, particularly of the labour market. They attribute failure to inefficient implementation – imposing an incomes policy as part of a deflationary package, rather than using it as part of an expansionary programme. Other analysts point to particular groups of workers who break norms. Still others single out the impact of external shocks, which destroy the assumptions on which the policy had been framed.

RUPERT PENNANT-REA

See also DEMAND MANAGEMENT; FULL EMPLOYMENT; INFLATION; STABILIZATION POLICY; WAGE INDEXATION.

BIBLIOGRAPHY

Ball, R.J. and Doyle, P. (eds) 1969. *Inflation.* London: Penguin.

Braun, A. R. 1986. *Wage Determination and Incomes Policy in Open Economies.* Washington: International Monetary Fund.

Friedman, M. 1974, Monetary correction. In M. Friedman, *Essays on Inflation and Indexation,* Washington DC: American Enterprise Institute for Public Policy Research.

Keynes, J.M. 1936. *The General Theory of Employment, Interest and Money.* London: Macmillan.

Keynes, J.M. 1940. *How to Pay for the War.* London: Macmillan.

Paish, F.W. 1986. The limits of incomes policies. In F.W. Paish and J. Hennessy, *Policy for Incomes,* 4th edn, Hobart Paper 29, London: Institute of Economic Affairs.

Phillips, A.W.H. 1958. The relation between unemployment and money wage rates in the United Kingdom, 1861–1957. *Economica* 25, November, 283–99.

Rowthorn, R. 1977. Conflict, inflation and money. *Cambridge Journal of Economics* 1, 215–39.

Tarling, R. and Wilkinson, S.F. 1977. The social contract – post-war incomes policies and their inflationary impact. *Cambridge Journal of Economics* 1(4), December, 395–414.

income taxes. *See* TAXATION OF INCOME.

incomplete contracts. The past decade has witnessed a growing interest in contract theories of various kinds. This development is partly a reaction to our rather thorough understanding of the standard theory of perfect competition under complete markets, but more importantly to the resulting realization that this paradigm is insufficient to accommodate a number of important economic phenomena. Studying in more detail the process of contracting – particularly its hazards and imperfections – is a natural way to enrich and amend the idealized competitive model in an attempt to fit the evidence better.

In one sense, contracts provide the foundation for a large part of economic analysis. Any trade – as a quid pro quo – must be mediated by some form of contract, whether it be explicit or implicit. In the case of spot trades, however, where the two sides of the transaction occur almost simultaneously, the contractual element is usually down-played, presumably because it is regarded as trivial (although we will argue below that this need not be the case). In recent years, economists have become much more interested in long-term relationships where a considerable amount of time may elapse between the quid and the quo. In these circumstances, a contract becomes an essential part of the trading relationship.

Research on contracts has progressed along several different lines. Two prominent areas of work are principal-agent theory and implicit labour contract theory. In these literatures, the focus is on risk-sharing or income-smoothing as the motivation for a contract; that is, on the gains the parties receive from transferring income from one state of the world or one period to another. For example, in implicit contract theory, it is supposed that workers are constrained in their ability to get insurance or to borrow on the open market and that employers therefore offer these services as part of an employment contract.

While 'income-smoothing' is undoubtedly important, there are arguably more fundamental factors underlying the existence of long-term contracts. A basic reason for long-term relationships is the existence of investments which are to some extent party specific; that is, once made, they have a higher value inside the relationship than outside. Given this 'lock-in' effect, each party will have some monopoly power ex-post, although there may be plenty of competition ex-ante, before investments are sunk. Since the parties cannot rely on the market once their relationship is underway, a long-term contract is an important way for them to regulate, and divide up the gains from, their trade. This will be the case even if the parties are risk neutral and have access to perfect capital markets, that is, even if the income-smoothing role is completely inessential. Moreover, in the case, say, of supply contracts involving large firms, risk neutrality and perfect capital markets may be reasonable approximations in view of the many outside insurance and borrowing/lending opportunities available to such parties.

In spite of their importance, contracts whose *raison d'être* is the regulation of specific relationships have been the subject of little analysis. A notable early reference is Becker's (1964) analysis of worker training. More recently, Williamson (1985) and Klein et al. (1978) have emphasized the difficulty of writing contracts which induce efficient relationship-specific investments as an important factor in explaining vertical integration.

In this entry I will try to summarize what is known theoretically about contracts of this type. I will focus particularly on the problems which arise when the parties write a contract which is incomplete in some respects. Given the rudimentary state of our knowledge of the area, the entry is inevitably quite speculative in nature. The reader who is interested in an elaboration of some of the ideas presented here, and how they fit into the rest of contract theory, might want to consult Hart and Holmstrom (1987).

1. THE BENEFITS OF WRITING LONG-TERM CONTRACTS GIVEN RELATIONSHIP-SPECIFIC INVESTMENTS. The role of a long-term contract when there are relationship-specific investments can be seen from the following example (based on Grout, 1984). Let B, S be, respectively, the buyer and seller of (one unit of) an input. Suppose that in order to realize the benefits of the input, B must make an investment, a, which is specific to S; for example, B might have to build a plant next to S. Assume that there are just two periods; the investment is made at date 0, while the input is supplied and the benefits are received at date 1. S's supply cost at date 1 is c, while B's benefit function is $b(a)$ (all costs and benefits are measured in date 1 dollars).

If no long-term contract is written at date 0, the parties will determine the terms of trade from scratch at date 1. If we assume that neither party has alternative trading partners at date 1, there is, given B's sunk investment cost a, a surplus of $b(a) - c$ to be divided up. A simple assumption to make is that the parties split this 50:50 (this is the Nash bargaining solution). That is, the input price p will satisfy $b(a) - p = p - c$. This means that the buyer's overall payoff, net of his investment cost, is

$$b(a) - p - a = \frac{b(a) - c}{2} - a. \qquad (1)$$

The buyer, anticipating this payoff, will choose a to maximize (1), i.e. to maximize $1/2\, b(a) - a$.

This is to be contrasted with the efficient outcome, where a is chosen to maximize total surplus, $b(a) - c - a$. Maximizing (1) will lead to underinvestment; in fact, in extreme cases, a will equal zero and trade will not occur at all. The inefficiency arises because the buyer does not receive the full return from his investment – some of this return is appropriated by the seller in the date 1 bargaining. Note that an upfront payment from S to B at date 0 (to compensate for the share of the surplus S will later receive) will not help here, since it will only change B's objective function by a constant (it is like a lump-sum transfer). That is, it redistributes income without affecting real decisions.

Efficiency can be achieved if a long-term contract is written at date 0 specifying the input price p^* in advance. Then B will maximize $b(a) - p^* - a$, yielding the efficient investment level, a^*. An alternative method is to specify that the buyer must choose $a = a^*$ (if not he pays large damages to S) – the choice of p can then be left until date 1, with an upfront payment by S being used to compensate B for his investment. The second method presupposes that investment decisions are publicly observable, and so in practice may be more complicated than the first (see below).

We see then that a long-term contract can be useful in encouraging relationship-specific investments. The word 'investment' should be interpreted broadly here; the same factors will apply whenever one party is forced to pass up an opportunity as a result of a relationship with another party (e.g., A's 'investment' in the relationship with B may be not to lock into C). That is, the crucial element is a sunk cost (direct or opportunity) of some sort (an effort decision is one example of a sunk cost). Note that the income-transfer motive for a long-term contract is completely absent here; there is no uncertainty and everything is in present value terms.

Given the advantages of long-term contracts in specific relationships, the question that obviously arises is why we do not see more of them, and why those we do see seem often to be limited in scope. To this question we now turn.

2. THE COSTS OF WRITING LONG-TERM CONTRACTS. Contract theory is sometimes dismissed because 'we don't see the long-term contingent contracts that the theory predicts'. In fact, there is no shortage of complex long-term contracts in the world. Joskow (1985), for example, in his recent study of transactions between electricity generating plants and mine-mouth coal suppliers finds that some contracts between the parties extend for fifty years, and a large majority for over ten years. The contractual terms include quality provisions, formulae linking coal prices to costs and prices of substitutes, indexation clauses, and so on. The contracts are both complicated and sophisticated. Similar findings are contained in Goldberg and Erickson's (1982) study of petroleum coke.

At a much more basic level, a typical contract for personal insurance, with its many conditions and exemption clauses, is not exactly a simple document. Nor for that matter is a typical house rental agreement. On the other hand, labour contracts are often surprisingly rudimentary, at least in certain respects (for example, there is little indexation of wages to retail prices or to firm employment or sales; layoff pay is limited, etc.).

Given that complex long-term contracts are found in some situations but not others, it is natural to explain any observed contract as an outcome of an optimization process in which the relative benefits and costs of additional length and complexity are traded off at the margin. In the last section, we indicated some of the benefits of a long-term contract. (The example considered was sufficiently straightforward that the ideal long-term contract was a simple noncontingent one; however, with the inclusion of such factors as uncertainty about payoffs and variable quality of the input, the optimal contract would be a (possibly much more complex) contingent one.) But what about the costs? These are much harder to pin down since they fall under the general heading of 'transaction costs', a notoriously vague and slippery category. Of these, the following seem to be important: (1) the cost to each party of anticipating the various eventualities that may occur during the life of the relationship: (2) the cost of deciding, and reaching an agreement about, how to deal with such eventualities; (3) the cost of writing the contract in a sufficiently clear and unambiguous way that the terms of the contract can be enforced; and (4) the legal cost of enforcement.

One point to note is that *all* these costs are present also in the case of *short-term* contracts, although presumably they are usually smaller. In particular, since the short-term future is more predictable, the first cost is likely to be much reduced, and so possibly is the third. However, it certainly is not the case that there is a sharp division between short-term contracts and long-term contracts, with, as is sometimes supposed, the former being costless and the latter being infinitely costly.

It is also worth emphasizing that, when we talk about the cost of a long-term contract, we are presumably referring to the cost of a 'good' long-term contract. There is rarely significant cost or difficulty in writing *some* long-term contract. For example, the parties to an input supply contract could agree on a fixed price and level of supply for the next fifty years. They do not presumably because such a rigid arrangement would be very inefficient. (In some cases the courts will not enforce such an agreement, taking the point of view that the parties could not really have intended it to apply unchanged for such a long time. A clause to the effect that the parties really do mean what they say should be enough to overcome this difficulty, however. In other cases, it may be impossible to write a binding long-term contract because the

identities of some of the parties involved may change. For example, one party may be a government that is in office for a fixed period, and it may be impossible for it to bind its successors. This latter idea underlies the work of Kydland and Prescott (1977) and Freixas et al. (1985).)

Due to the presence of transaction costs, the contracts people write will be *incomplete* in important respects. The parties will quite rationally leave out many contingencies, taking the point of view that it is better to 'wait and see what happens' than to try to cover a large number of individually unlikely eventualities. Less rationally, the parties will leave out other contingencies that they simply do not anticipate. Instead of writing very long-term contracts the parties will write limited term contracts, with the intention of renegotiating these when they come to an end. (A paper which explores the implications of this is Crawford, 1986.) Contracts will often contain clauses which are vague or ambiguous, sometimes fatally so.

Anyone familiar with the legal literature on contracts will be aware that almost every contractual dispute that comes before the courts concerns a matter of incompleteness. In fact, incompleteness is probably at least as important empirically as asymmetric information as an explanation for departures from 'ideal' Arrow–Debreu contingent contracts. In spite of this, relatively little work has been done on this topic, the reason presumably being that an analysis of transaction costs is so complicated. One problem is that the first two transaction costs referred to above are intimately connected to the idea of bounded rationality (as in Simon, 1982), a successful formalization of which does not yet exist. As a result, perhaps, the few attempts that have been made to analyse incompleteness have concentrated on the third cost, the cost of writing the contract.

One approach, due to Dye (1985), can be described as follows. Suppose that the amount of input, q, traded between a buyer and seller should be a function of the product price, p, faced by the buyer: $q = f(p)$. Writing down this function is likely to be costly. Dye measures the costs in terms of how many different values q takes on as p varies; in particular, if $\#\{q \mid q = f(p) \text{ for some } p\} = n$, the cost of the contract is $(n - 1)c$, where $c > 0$. This means that a noncontingent statement '$q = 5$ for all p' has zero cost, the statement '$q = 5$ for $p \leqslant 8$, $q = 10$ for $p > 8$' has cost c, and so on. The costs Dye is trying to capture are real enough, but the measure used has some drawbacks. It implies for example, that the statement '$q = p^{1/2}$ for all p' has infinite cost if p has infinite domain, and does not distinguish between the cost of a simple function like this and the cost of a much more complicated function. As another example, a simple indexation clause to the effect that the real wage should be constant (i.e. the money wage $= \lambda p$ for some λ) would never be observed since, according to Dye's measure, it too has infinite cost. In addition, the approach does not tell us how to assess the cost of indirect ways of making q contingent; for example, the contract could specify that the buyer, having observed p, can choose any amount of input q he likes, subject to paying the seller σ for each unit.

There is another way of getting at the cost of including contingent statements. This is to suppose that what is costly is describing the state of the world ω rather than writing a statement per se. That is, suppose that ω cannot be represented simply by a product price, but is very complex and of high dimension – e.g., it includes the state of demand, what other firms in the industry are doing, the state of technology, etc. Many of these components may be quite nebulous. To describe the state ex-ante in sufficient detail that an outsider, e.g. the courts, can verify whether a particular state $\omega = \hat{\omega}$ has occurred, and so enforce the contract, may be prohibitively costly. Under these conditions, the contract will have to omit some (in extreme cases, all) references to the underlying state.

Similar to this is the case where what is costly is describing the *characteristics* of what is traded or the *actions* (e.g. investments) the parties must take. For example, suppose that there is only one state of the world, but that q now represents the quality of the item traded rather than the quantity. An ideal contract would give a precise description of q. However, quality may be multidimensional and very difficult to describe unambiguously (and vague statements to the effect that quality should be 'good' may be almost meaningless). The result may be that the contract will have to be silent on many aspects of quality and/or actions.

Models of this sort of incompleteness have been investigated by Grossman and Hart (1987) and Hart and Moore (1985) for the case where the state of the world cannot be described and by Bull (1985) and Grossman and Hart (1986, 1987) for the case where quality and/or actions cannot be specified. These models do not rely on any asymmetry of information between the parties. Both parties may recognize that the state of the world is such that the buyer's benefit is high or the seller's cost is low, or that the quality of an item is good or bad or that an investment decision is appropriate or not. The difficulty is conveying this information to others. *That is, it is the asymmetry of information between the parties on the one hand, and outsiders, such as the courts, on the other, which is the root of the problem.*

To use the jargon, incompleteness arises because states of the world, quality and actions are *observable* (to the contractual parties) but not *verifiable* (to outsiders).

We describe an example of an incomplete contract along these lines in the next section.

3. INCOMPLETE CONTRACTS: AN EXAMPLE. We will give an example of an incomplete contract for the case where it is prohibitively costly to specify the quality characteristics of the item to be exchanged or the parties' investment decisions. Similar problems arise when the state of the world cannot be described. The example is a variant of the models in Grossman and Hart (1986, 1987), Hart and Moore (1985).

Consider a buyer B who wishes to purchase a unit of input from a seller S. B and S each make a (simultaneous) specific investment at date 0 and trade occurs at date 1. Let I_B, I_S denote, respectively, the investments of B and S, and to simplify assume that each can take on only two values, H or L (high or low). These investments are observable to B and S, but are not verifiable (they are complex and multidimensional, or represent effort decisions) and hence are noncontractible. We assume that at date 1 the seller can supply either 'satisfactory' input or 'unsatisfactory' input. 'Unsatisfactory' input has zero benefit for the buyer and zero cost for the seller (so it is like not supplying at all). 'Satisfactory' input yields benefits and costs which depend on ex-ante investments. These are indicated in Figure 1.

The first component refers to the buyer's benefit, v, and the second to the seller's cost, c. So when $I_S = H$, $I_B = H$, $v = 10$ and $c = 6$ (if input is 'satisfactory'). From these gross benefits and costs must be subtracted investment costs, which we assume to be 1.9 if investment is high and zero if it is low (for each party). (All benefits and costs are in date 1 dollars.) Note that there is no uncertainty and so attitudes to risk are irrelevant.

Our assumption is that the characteristics of the input (e.g. whether it is 'satisfactory') are observable to both parties, but

	$I_B = H$	$I_B = L$
$I_S = H$	(10, 6)	(9, 7)
$I_S = L$	(9, 7)	(6, 10)

Figure 1

are too complicated to be specified in a contract. The fact that they are observable means that the buyer can be given the option to reject the input at date 1 if he does not like it. This will be important in what follows.

An important feature of the example is that the seller's investment affects not only the seller's costs but also the buyer's benefit and the buyer's investment affects not only the buyer's benefit but also the seller's costs. The idea here is that a better investment by the seller increases the quality of 'satisfactory' input; and a better investment by the buyer reduces the cost of producing 'satisfactory' input, that is input that can be used by the buyer.

For instance, one can imagine that B is an electricity generating plant and S a coal mine that the plant is sited next to. I_B might refer to the type of coal-burning boiler that the plant installs and I_S to the way the coal supplier develops the mine. By investing in a better boiler, the power plant may be able to burn lower quality coal, thus reducing the seller's costs, while still increasing its gross (of investment) profit. On the other hand, by developing a good seam, the coal supplier may raise the quality of coal supplied while reducing its variable cost.

The first-best has $I_B = I_S = H$, with total surplus equal to $(10 - 6) - 3.8 = 0.2$ (if $I_B = H$ and $I_S = L$, or vice versa, surplus $= 0.1$ and if $I_B = I_S = L$, no trade occurs and surplus is zero). This could be achieved if *either* investment *or* quality were contractible as follows. If investment is contractible, an optimal contract would specify that the buyer must set $I_B = H$ and the seller $I_S = H$ and give the buyer the right to accept the input at date 1 at price p_1 or reject it at price p_0. If $10 > p_1 - p_0 > 6$, the seller will be induced to supply satisfactory input (the gain, $p_1 - p_0$, from having the input accepted exceeds the seller's supply cost) and the buyer to accept it (the buyer's benefit exceeds the increment price $p_1 - p_0$). If, on the other hand, quality is contractible, the contract could specify that the seller must supply input with the precise characteristics which make it satisfactory when $I_B = I_S = H$. Each party would then have the socially correct investment incentives since, with specific performance, neither party's investment affects the other's payoff (there is no externality).

We now show that the first-best cannot be achieved if investment and quality are both noncontractible. A second-best contract can make price a function of any variable that is verifiable. Investment and quality are not verifiable (nor is v or c), but we shall suppose that whether the item is accepted or rejected by the buyer is, so the contract can specify an acceptance price, p_1, and a rejection price, p_0. In fact, p_0, p_1 can also be made functions of (verifiable) messages that the buyer and seller send each other, reflecting the investment decisions that both have made (as in Hart and Moore, 1985). The following argument is unaffected by such messages and so, for simplicity, we ignore them (the interested reader is referred to Hart and Holmstrom, 1987).

Can we sustain the first-best by an appropriate choice of p_0, p_1? The seller always has the option of choosing $I_S = L$ and producing an item of unsatisfactory quality, which yields him a net payoff of p_0. In order to induce him not to do this, we must have

$$p_1 - 6 - 1.9 \geqslant p_0, \quad \text{i.e.} \quad p_1 - p_0 \geqslant 7.9. \qquad (2)$$

Similarly the buyer's net payoff must be no less than $-p_0$ since he always has the option of choosing $I_B = L$ and rejecting the input. That is,

$$10 - p_1 - 1.9 \geqslant -p_0, \quad \text{i.e.} \quad p_1 - p_0 \leqslant 8.1. \qquad (3)$$

So $(p - p_0)$ must lie between 7.9 and 8.1.

Now the seller has an additional option. If he expects the buyer to set $I_B = H$, he can choose $I_S = L$ and, given that $8.1 \geqslant p_1 - p_0 \geqslant 7.9$, still be confident that trade of 'satisfactory' input will occur under the original contract at date 1 (the buyer will accept satisfactory input since $v = 9 > p_1 - p_0$, while the seller will supply it since $p_1 - p_0 > 7 = c$). But if the seller deviates, his payoff rises from $p_1 - 6 - 1.9$ to $p_1 - 7$. (The example is symmetric and so a similar deviation is also profitable for the buyer.) Hence the $I_B = I_S = H$ equilibrium will be disrupted.

We see, then, that the first-best cannot be sustained if investment and quality are both noncontractible. The reason is that it will be in the interest of the seller (or the buyer) to reduce investment since, although this reduces social benefit by lowering the buyer's (or seller's) benefit, it increases the seller's (or buyer's) own profit. The optimal second-best contract will instead have $I_B = H$, $I_S = L$ (or vice versa), which will be sustained by a pair of prices p_0, p_1 such that $9 > p_1 - p_0 > 7$. Total surplus will be 0.1 instead of the first-best level of 0.2. (Note the importance of the assumption that both the buyer and seller can choose $I = H$ or L. If only the buyer (or the seller) can choose $I = L$, the first-best can be achieved by choosing $p_1 - p_0$ between 6 and 7 (or 9 and 10): any deviation by the buyer (or the seller) will then be unprofitable since it will lead to no trade.)

The conclusion is that inefficiencies can arise in incomplete contracts even though the parties have common information (both observe investments and both observe quality). The particular inefficiency that occurs in the model analysed is in ex-ante investments. Ex-post trade is always efficient relative to these investments since p_1, p_0 can and will be chosen such that $v > p_1 - p_0 > c$, i.e. the seller wants to supply and the buyer to receive satisfactory input. The example can be regarded as formalizing the intuition of Williamson (1985) and Klein et al. (1978) that relationship-specific investments will be distorted due to the impossibility of writing complete contingent contracts – note that this result is achieved without imposing arbitrary restrictions on the form of the permissible contract (e.g. we have not ruled out the existence of long-term contracts from the start). (There is one exception to this statement – we have excluded the participation of a third party to the contract; for a discussion and justification of this, see Hart and Holmstrom, 1987.)

The example may be used to illustrate a theory of ownership presented in Grossman and Hart (1986, 1987). It is sometimes suggested that when transaction costs prevent the writing of a complete contract, there may be a reason for firm integration (see Williamson, 1985). Consider the payoffs of Figure 1 and suppose that B takes over S. The control that B thereby gains over S's assets may allow B to affect S's costs in various ways, and this may reduce the possibility of opportunistic behaviour by S. To take a very simple (and contrived) example, suppose that if S chooses $I_S = L$, B can take some action, α with respect to S's assets at date 1 so as to make S's cost of supplying either satisfactory or unsatisfactory input equal to 9 (in the coal–electricity example, α might refer to the part of the mine's seam the coal is taken out of; note that we now drop the assumption that the cost of supplying unsatisfactory input is zero). Imagine furthermore that this action increases B's benefit, so that B will indeed take it at date 1 if S chooses L. Then with this extra degree of freedom, the first-best can be achieved. In particular, if $p_1 = p_0 + 6.1$, $I_S = I_L = H$ is a Nash equilibrium since, by the above reasoning, any deviation by the seller will be punished, while if the buyer deviates, the seller will supply unsatisfactory input given that $p_1 < p_0 + 7$.

Note that if action α could be specified in the initial contract, there would be no need for integration: the initial contract would simply say that B has the right to choose α at date 1. Ownership becomes important, however, if (i) α is too complicated to be specified in the date 0 contract and therefore qualifies as a residual right of control; and (ii) residual rights of control over an asset are in the hands of whomever owns that asset. The point is that under incompleteness the allocation of residual decision rights matters since the contract cannot specify precisely what each party's obligations are in every state of the world. To the extent that ownership of an asset guarantees residual rights of control over that asset, vertical and lateral integration can be seen as ways of ensuring particular – and presumably efficient – allocations of residual decision rights. (While in the above example, integration increases efficiency, this is in no way a general conclusion. In Grossman and Hart (1986, 1987), examples are presented where integration reduces efficiency.)

Before concluding this section, we should emphasize that for reasons of tractability we have confined our attention to incompleteness due to a very particular sort of transaction cost. In practice, some of the other transactions costs we have alluded to are likely to be at least as important, if not more so. For example, in the type of model we have analysed, although the parties cannot describe the state of the world or quality characteristics, they are still supposed to be able to write a contract which is unambiguous and which anticipates all eventualities. This is very unrealistic. In practice, a contract might, say, have B agreeing to rent S's concert hall for a particular price. But suppose S's hall then burns down. The contract will usually be silent about what is meant to happen under these conditions (there is no hall to rent, but should S pay B damages and if so how much?), and so, in the event of a dispute, the courts will have to fill in the 'missing provision'. (A situation where it becomes impossible or extremely costly to supply a contracted for good is known as one of 'impossibility' or 'frustration' in the legal literature.) An analysis of this sort of incompleteness, although extremely hard, is a very important topic for future research. It is likely to yield a much richer and more realistic view of the way contracts are written and throw light on how courts should assess damages (this latter issue has begun to be analysed in the law and economics literature; see, e.g., Shavell, 1980).

4. SELF-ENFORCING CONTRACTS. The previous discussion has been concerned with explicit binding contracts that are enforced by outsiders, such as the courts. Even the most casual empiricism tells us that many agreements are not of this type. Although the courts may be there as a last resort (the shadow of the law may therefore be important), these agreements are enforced on a day to day basis by custom, good faith, reputation, etc. Even in the case of a serious dispute, the

parties may take great pains to resolve matters themselves rather than go to court. This leads to the notion of a self-enforcing or implicit contract (the importance of informal arrangements like this in business has been stressed by Macaulay (1963) and Ben-Porath (1980) among others).

People often by-pass the legal process presumably because of the transaction costs of using it. The costs of writing a 'good' long-term contract discussed in Section 2 are relevant here. So also is the skill with which the courts resolve contractual disputes. If contracts are incomplete and contain missing provisions as well as vague and ambiguous statements, appropriate enforcement may require abilities and knowledge (what was in the parties' minds?) that many judges and juries do not possess. This means that going to court may be a considerable gamble – and an expensive one at that. (This is an example of the fourth transaction cost noted in Section 2.)

Although the notion of implicit or self-enforcing contracts is often invoked, a formal study of such agreements has begun only recently (see, e.g. Bull, 1985), with a considerable stimulus coming from the theory of repeated games. This literature has stressed the role of *reputation* in 'completing' a contract. That is, the idea is that a party may behave 'reasonably' even if he is not obliged to do so in order to develop a reputation as a decent and reliable trader. In some instances such reputational effects will operate only within the group of contractual parties – this is sometimes called *internal* enforcement of the contract – while in others the effects will be more pervasive. The latter will be the case when some outsiders to the contract, for example other firms in the industry or potential workers for a firm, observe unreasonable behaviour by one party, and as a result are more reluctant to deal with it in the future. In this case the enforcement is said to be *external* or *market-based*. Note that there may be a tension between this external enforcement and the reasons for the absence of a legally binding contract in the first place – the more people can observe the behaviour, the more likely it is to be verifiable.

The distinction between an incomplete contract and a standard asymmetric information contract should be emphasized here. It is the former that allows reputation to operate since the parties have the same information and can observe whether reasonable behaviour is being maintained. In the latter case, it is unclear how reputation can overcome the asymmetry of information between the parties that is the reason for the departure from an Arrow–Debreu contract.

The role of reputation in sustaining a contract can be illustrated using the following model (based on Bull (1985) and Kreps (1984); this is an even simpler model of incomplete contracts than that of the last section). Assume that a buyer, B, and a seller, S, wish to trade an item at date 1 which has value v to the buyer and cost c to the seller, where $v > c$. There are no ex-ante investments and the good is homogeneous, so quality is not an issue. Suppose, however, that it is not verifiable whether trade actually occurs. Then a legally binding contract which specifies that the seller must deliver the item and the buyer must pay p, where $v > p > c$, cannot be enforced. The reason is that, assuming (as we shall) that simultaneous delivery and payment are infeasible, if the seller has to deliver first, the buyer can always deny that delivery occurred and refuse payment, while if the buyer has to pay first, the seller can always claim later that he did deliver even though he did not. As a result, if the parties must rely on the courts, a gainful trading opportunity will be missed.

The idea that not even the level of trade is verifiable is extreme, and Bull (1985) in fact makes the more defensible assumption that it is the quality of the good that cannot be

verified (in Bull's model, S is a worker and quality refers to his performance). Bull supposes that quality is observable to the buyer only with a lag, so that take it or leave it offers of the type considered in the last section are not feasible. As a result the seller always has an incentive to produce minimum quality (which corresponds in the above model to zero output). Making quantity nonverifiable is a cruder but simpler way of capturing the same idea (this is the approach taken in Kreps, 1984).

Note that in the above model incompleteness of the contract arises entirely from transaction cost (3), the difficulty of writing and enforcing the contract.

To introduce reputational effects one supposes that this trading relationship is repeated. Bull (1985) and Kreps (1984) follow the supergame literature and assume infinite repetition in order to avoid unravelling problems. This approach, as is well known, suffers from a number of difficulties. First, the assumption of infinite (or in some versions, *potentially* infinite) life is hard to swallow. Secondly, 'reasonable' behaviour, i.e. trade, is sustained by the threat that if one party behaves unreasonably so will the other party from then on. While this threat is 'credible' (more precisely, subgame perfect), it is unclear why the parties could not decide to continue to trade after a deviation, i.e. to 'let bygones be bygones' (see Farrell, 1984.

It would seem that a preferable approach is to assume that the relationship has finite length, but introduce asymmetric information, as in Kreps and Wilson (1982) and Milgrom and Roberts (1982). The following is based on some very preliminary work that Bengt Holmstrom and I have undertaken along these lines.

Suppose that there are two types of buyers in the population, honest and dishonest. Honest buyers will always honour any agreement or promise that they have made while dishonest ones will do so only if this is profitable. A buyer knows his own type, but others do not. It is common knowledge that the fraction of honest buyers in the population is π, $0 < \pi < 1$. In contrast, all sellers are known to be dishonest. All agents are risk neutral.

Assume for simplicity that a single buyer and seller are matched at date 0 with neither having any alternative trading partners at this date or in the future. Consider first the one-period case. Then a date 0 agreement can be represented as follows.

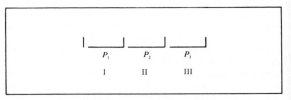

Figure 2

The interpretation is that the buyer promises to pay the seller p_1 before date 1 (stage I); in return, the seller promises to supply the item at date 1 (stage II); and in return for this, the buyer promises to make a further payment of p_2 (stage III).

We should mention one further assumption. Honest buyers, although they never breach an agreement first, are supposed to feel under no obligation to fulfil the terms of an agreement that has already been broken by a seller (interestingly, although this is a theory of buyer psychology, it has parallels in the common law). Note that if a buyer ever breaks an agreement first, he reveals himself to be dishonest, with the consequence that no further self-enforcing agreement with the seller is possible and hence trade ceases.

What is an optimal agreement? Consider Figure 2. The seller knows that he will receive p_2 only with probability π since a dishonest buyer will default at the last stage. Since the seller is himself dishonest, he will supply at Stage II only if it is profitable for him to do so, i.e. only if

$$\pi p_2 - c \geqslant 0. \tag{4}$$

Assume for simplicity that the seller has all the bargaining power at date 0 (nothing that follows depends on this). Then the seller will wish to maximize his overall payoff

$$p_1 + \pi p_2 - c, \tag{5}$$

subject to (4) which makes it credible that he will supply at stage II and also the constraint that he does not discourage an honest buyer from participating in the agreement at date 0. Since with (4) satisfied, buyers know that they will receive the item for sure, this last condition is

$$v - p_1 - p_2 \geqslant 0. \tag{6}$$

Note that a dishonest buyer's payoff $v - p_1$ is always higher than an honest buyer's payoff given in (6), so there is no way to screen out dishonest buyers. In the language of asymmetric information models, the equilibrium is a pooling one.

Since the seller's payoff is increasing in p_1, (6) will hold with equality (the buyer gets no surplus). (More generally, changes in p_1 simply redistribute surplus between the two parties without changing either's incentive to breach.) If we substitute for p_1 in (5), the seller's payoff becomes $v - p_2(1-\pi) - c$, which, when maximized subject to (4), yields the solution $p_2 = c/\pi$. The maximized net payoff is

$$v - c/\pi, \tag{7}$$

which is less than the first-best level, $v - c$.

We see then that the conditions for trade are more stringent in the absence of a binding contract. If $c/(\pi) > v > c$, there are gains from trade which would not be realized in a one-period relationship.

Suppose now that the relationship is repeated. Consider a two-period version of the above and assume no discounting. Now the diagram shown in Figure 3 applies.

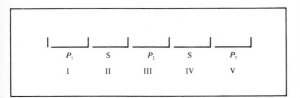

Figure 3

That is, the agreement says that the buyer pays, the seller supplies the first time, the buyer pays more, the seller supplies a second time, and the buyer makes a final payment. Rather than solving for the optimal arrangement, we shall simply show that the seller can do better than in the one period case. Let $p_3 = c/\pi$, $p_2 = c$ and $p_1 = 2v - c - (c)/\pi$. Then (i) the seller will supply at Stage IV (if matters have got that far), knowing that he will receive p_3 with probability π; (ii) both honest and dishonest buyers will pay p_2 at Stage III, the latter because, at

a cost of c, they thereby ensure supply worth $v > c$ at Stage IV; (iii) the seller will supply at stage II because this gives him a net payoff of $p_2 + \pi p_3 - 2c \geqslant 0$, while if he does not the arrangement is over and his payoff is zero; (iv) an honest buyer is prepared to participate since his surplus is non-negative (actually zero).

The seller's overall expected net payoff is

$$p_1 + \pi_2 + \pi p_3 - 2c = 2v - c - \frac{c}{\pi}, \tag{8}$$

which *exceeds* twice the one-period payoff. Hence trade is more likely to take place in a two-period relationship than in a one-period one. In fact it can be shown that the above is an *optimal* two-period agreement.

Repetition improves things by allowing the honest buyer to pay less second time round (Stage III) than third time round (Stage V). That is, the arrangement *back-loads* payments. This is acceptable to the seller because he knows that even a dishonest buyer will not default at Stage III since he has a large stake in the arrangement continuing. To put it another way, the dishonest buyer does not want to reveal his dishonesty at too early a stage.

The same arrangement can be used when there are more than two periods: the buyer promises to pay c at every stage except the last, when he pays (c/π). In fact the per period surplus of the seller from such an arrangement converges to the first-best level $(v - c)$ as the number of periods tends to ∞ (assuming no discounting, of course).

Although the above analysis is extremely provisional and sketchy, we can draw some tentative conclusions about the role of reputation and indicate some directions for further research. First, the notion of a psychic cost of breaking an agreement seems to be a useful – as well as a not unrealistic – basis for a theory of self-enforcing contracts. It is obviously desirable to drop the assumption that some agents are completely honest and others completely dishonest, and assume instead that the typical trader has a finite psychic cost of breaking an agreement, where this cost is distributed in the population in a known way. In other words, everybody 'has their price', but this price varies. Preliminary work along these lines suggests that the above results generalize; in particular, repetition makes it easier to sustain a self-enforcing contract.

Of course, asymmetries of information about psychic costs are not the only possible basis for a theory of reputation. For example, the buyer and seller could have private information about v and c, and might choose their trading strategies to influence perceptions about the values of these variables (the role of uncertainty about v and c in determining reputation has been investigated by Thomas and Worrall, 1984). A theory of self-enforcing contracts should ideally generate results which are not that sensitive to where the asymmetry of information is placed. The work of Fudenberg and Maskin (1986) in a related context, however, suggests that this may be a difficult goal to achieve.

There are a number of other natural directions in which to take the model. One is to introduce trade with other parties. For example, the seller may trade with a succession of buyers rather than a single one. The extent to which repetition increases per period surplus in this case depends on whether new buyers observe the past broken promises of the seller. (This determines the degree to which external enforcement operates; more generally, a new buyer may observe that default occurred in the past, but be unsure about who was responsible for it.) If new buyers do not observe past broken promises, repetition achieves nothing, which gives a very strong prediction of the possible benefits of a long-term relationship between a fixed buyer and seller. Even if past broken promises are observed perfectly, it appears that, *ceteris paribus*, a single long-term agreement may be superior to a succession of short-term ones. The reason is that in the latter case the constraint is imposed that each party must receive non-negative surplus over *their* term of the relationship whereas in the former case there is only the single constraint that surplus must be non-negative over the whole term (see Bull, 1985; Kreps, 1984).

Probably the most important extension is to introduce incompleteness due to other sorts of transaction costs, e.g. the 'bounded rationality' costs (1) and (2) discussed in Section 2. The problem is that the same factors which make it difficult to anticipate and plan for eventualities in a formal contract apply also to informal arrangements. That is, an informal arrangement is also likely to contain many 'missing provisions'. But then the question arises, what constitutes 'reasonable' or 'desirable' behaviour (in terms of building a reputation) with regard to states or actions that were not discussed ex-ante? Custom, among other things, is likely to be important under these conditions: behaviour will be 'reasonable' or 'desirable' to the extent that it is generally regarded as such (for a good discussion of this, see Kreps, 1984). This raises many new and interesting (as well as extremely difficult) questions.

5. SUMMARY AND CONCLUSIONS. The vast majority of the theoretical work on contracts to date has been concerned with what might be called 'complete' contracts. In this context, a complete contract means one that specifies each party's obligations in every conceivable eventuality, rather than a contract that is fully contingent in the Arrow–Debreu sense. In particular, according to this terminology, the typical asymmetric information contract found in the principal-agent or implicit contract literatures (see Hart and Holmstrom, 1987) is complete.

In reality it is usually impossible to lay down each party's obligations completely and unambiguously in advance, and so most actual contracts are seriously incomplete. In this entry, we have tried to indicate some of the implications of such incompleteness. Among other things, we have seen that incompleteness can lead to departures from the first-best even when there are no asymmetries of information among the contracting parties (and, moreover, the parties are risk neutral).

More important perhaps than this is the fact that incompleteness raises new and difficult questions about how the behaviour of the contracting parties is determined. To the extent that incomplete contracts do not specify the parties' actions fully, i.e. they contain 'gaps', additional theories are required to tell us how these gaps are filled in. Among other things, outside influences such as custom or reputation may become important under these conditions. In addition, outsiders, such as the courts (or arbitrators), may have a role to play in filling in missing provisions of the contract and resolving ambiguities rather than in simply enforcing an existing agreement. Incompleteness can also throw light on the importance of the allocation of decision rights or rights of control. If it is too costly to state precisely how a particular asset is to be used in every state of the world, it may be efficient simply to give one party 'control' of the asset, in the sense that he is entitled to do what he likes with it, subject perhaps to some explicit (contractible) limitations.

While the importance of incompleteness is very well recognized by lawyers, as well as by those working in law and economics, it is only beginning to be appreciated by economic theorists. It is to be hoped that work in the next few years will lead to significant advances in our formal understanding of this phenomenon. Unfortunately, progress is unlikely to be easy since many aspects of incompleteness are intimately connected to the notion of bounded rationality, a satisfactory formalization of which does not yet exist.

As a final illustration of the importance of incompleteness, consider the following question. Why do parties frequently write a limited term contract, with the intention of renegotiating this when it comes to an end, rather than writing a single contract that extends over the whole length of their relationship? In a complete contract framework such behaviour cannot be advantageous since the parties could just as well calculate what will happen when the contract expires and include this as part of the original contract. It is to be hoped that future work on incomplete contracts will allow this very basic question to be answered.

OLIVER HART

See also ADVERSE SELECTION; ASYMMETRIC INFORMATION; BOUNDED RATIONALITY; EXCHANGE; HIDDEN ACTIONS, MORAL HAZARD AND CONTRACT THEORY; IMPLICIT CONTRACTS.

BIBLIOGRAPHY

Becker, G. 1964. *Human Capital*. New York: Columbia University Press.

Ben-Porath, Y. 1980. The F-connection: families, friends, and firms and the organization of exchange. *Population and Development Review* 6, March, 1–30.

Bull, C. 1985. The existence of self-enforcing implicit contracts. C.V. Starr Center, New York University.

Crawford, V. 1986. Long-term relationships governed by short-term contracts. Princeton University.

Dye, R. 1985. Costly contract contingencies. *International Economic Review* 26, 1, 233–50.

Freixas, X., Guesnerie, R. and Tirole, J. 1985. Planning under incomplete information and the ratchet effect. *Review of Economic Studies* 52(2), 169, 173–92.

Fudenberg, D. and Maskin, E. 1986. The Folk Theorem in repeated games with discounting and with incomplete information. *Econometrica* 54(3), 533–54.

Goldberg, V. and Erickson, J. 1982. Long-term contracts for petroleum coke. Department of Economics Working Paper Series No. 206, University of California, Davis, September.

Grossman, S. and Hart, O. 1986. The costs and benefits of ownership: a theory of vertical and lateral integration. *Journal of Political Economy* 94(4), August, 691–719.

Grossman, S. and Hart, O. 1987. Vertical integration and the distribution of property rights. In *Economic Policy in Theory and Practice*, Sapir Conference Volume, London: Macmillan Press.

Grout, P. 1984. Investment and wages in the absence of binding contracts: a Nash bargaining approach. *Econometrica* 52(2), March, 449–60.

Hart, O. and Holmstrom, B. 1987. The theory of contracts. In *Advances in Economic Theory, Fifth World Congress*, ed. T. Bewley, Cambridge: Cambridge University Press.

Hart, O. and Moore, J. 1985. Incomplete contracts and renegotiation. London School of Economics, Working Paper.

Joskow, P. 1985. Vertical integration and long-term contracts. *Journal of Law, Economics and Organization* 1, Spring.

Klein, B., Crawford, R. and Alchian, A. 1978. Vertical integration, appropriable rents and the competitive contracting process. *Journal of Law and Economics* 21, 297–326.

Kreps, D. 1984. Corporate culture and economic theory. Mimeo, Stanford University, May.

Kreps, D. and Wilson, R. 1982. Reputation and imperfect information. *Journal of Economic Theory* 27, 253–79.

Kydland, F. and Prescott, E. 1977. Rules rather than discretion: the inconsistency of optimal plans. *Journal of Political Economy* 85(3), 473–92.

Macaulay, S. 1963. Non-contractual relations in business: a preliminary study. *American Sociological Review* 28, February, 55–67.

Milgrom, P. and Roberts, D.J. 1982. Predation, reputation and entry deterrence. *Journal of Economic Theory* 27, 280–312.

Shavell, S. 1980. Damage measures for breach of contract. *Bell Journal of Economics* 11(2), Autumn, 466–90.

Simon, H. 1982. *Models of Bounded Rationality*. Cambridge, Mass.: MIT Press.

Thomas, J. and Worrall, T. 1984. Self-enforcing wage contracts. Mimeo, University of Cambridge.

Williamson, O. 1985. *The Economic Institutions of Capitalism*. New York: Free Press.

incomplete information, games with. *See* GAMES WITH INCOMPLETE INFORMATION.

incomplete markets. Markets are complete when every agent is able to exchange every good either directly or indirectly with every other agent. When markets are not complete, two fundamental properties of a competitive equilibrium with complete markets may no longer be satisfied. First, stockholders may not agree on the optimal production plan for a firm. Secondly, even in a model of pure exchange, a competitive allocation may not be Pareto optimal even when we restrict attention to allocations which are 'consistent' with the market structure. It is with the failure of this optimality property that this essay will be concerned.

Most of the literature on incomplete markets has been motivated by the analysis of competitive markets in the context of uncertainty. A good starting point for a discussion of this literature is the work of Arrow (1971). He demonstrated that extending the static analysis of competitive markets under certainty to the case of uncertainty requires the introduction of a contingent claims market for each good for every possible state of the world. He went on to examine the extent to which the same result could be achieved with a smaller number of markets. He showed that if spot markets are complete and their prices are perfectly forecast, then a complete set of markets can be attained with only one contingent claim (security) for each state. Before the state is realized, agents trade the state contingent securities. Once the state realized, a spot market opens and the consumer adjusts his consumption bundle taking into account any transfers of purchasing power resulting from his earlier exchange of state contingent securities. This market structure reduces the number of securities required to achieve any Arrow-Debreu allocation by a factor equal to the number of goods in each state.

Arrow's argument can be generalized to allow for an arbitrary set of securities. In general, a typical security may specify a different payment of goods in any number of states of the world. Nevertheless, if the set of securities is to duplicate the effect of Arrow's state contingent securities for every possible price vector, the number of securities must always be at least as large as the number of states (see e.g. Townsend, 1978). A moment of reflection is enough to convince almost anyone that this is still an unreasonably large number. It is probably impossible even to list all of the contingencies one can imagine, much less set up an explicit market for each one. Evidently, a realistic analysis of competitive markets under uncertainty must allow for markets which are incomplete.

Once we consider incomplete markets, the efficiency properties of the competitive allocation must be reexamined. If, for instance, we use the Pareto criteria and ignore any limitations on the feasible allocations implied by the market structure, then we should generally not expect a competitive allocation to be Pareto optimal. The reason is that, when markets are incomplete, there is no guarantee that any Pareto optimal allocations can be attained simply by reallocating available securities and then letting agents trade on the spot market.

To illustrate the point with the simplest possible example, consider a pure exchange economy consisting of two agents and two states of nature with a single good in each state. Without a securities market to allow the transfer of purchasing power between states, no trade between the agents is possible. In this example, therefore, the competitive allocation is Pareto optimal if and only if the initial endowment is Pareto optimal.

If the model is to provide the foundation for the analysis of the effect of government intervention, however, the unrestricted Pareto criterion may be of little use. A more appropriate welfare criterion should compare the market allocation to the set of allocations which can be achieved by a feasible institution. If the absence of certain security markets reflects some fundamental technological constraints, then those same constraints need to be reflected in the analysis of the performance of any other institution. On the other hand, one of the instruments that an institution might use is the assignment of ownership rights to existing securities. At the very least, therefore, we should test the efficiency of the market allocation relative to the set of allocations which can be implemented simply by redistributing the existing set of securities.

In order to formulate such a definition we must be precise about what we mean by the set of allocations which are consistent with a given allocation of securities. Once a state is realized, there must be a restriction on the allocations which can be implemented for any given distribution of securities. Otherwise, a mechanism in the spot markets can be introduced to undo any restrictions imposed by the allocation of securities. One possibility is to suppose that final allocations must be attainable as competitive equilibria in the spot markets, given the distribution of income implied the allocation of securities. This approach leads to the following definition suggested by Hart (1975). An allocation x is *constrained Pareto optimal* for a set of securities Q if (i) it is a competitive equilibrium for some allocation of securities in Q, and (ii) there is no Pareto superior allocation $\hat{x}$ which is a competitive equilibrium allocation for some other allocation of securities in Q. (This is a slight restatement of Hart's definition. It corresponds to the definition of *constrained Pareto suboptimal* allocation given by Geanakoplos and Polemarchakis, 1985.)

With a complete set of securities, the first welfare theorem implies that any competitive equilibrium is constrained Pareto optimal. (It is also an implication of the second welfare theorem that the set of constrained Pareto optimal allocations corresponds to the set of unconstrained Pareto optima.) A similar result can be established as long as we restrict ourselves to pure exchange economies with only one good in each state. For these economies, the allocation of securities determines the final allocation of goods in each state independently of the preferences of the individual agents. Consequently, the economy is equivalent to an economy in which preferences are defined directly on the allocation of securities. It then follows again from the first welfare theorem that any competitive allocation is constrained Pareto optimal.

Once we allow for more than one good in each state,

however, this conclusion changes. The reason is that a redistribution of the ownership of securities may lead to a change in the spot market prices which in turn implies redistribution of real income which cannot be duplicated by the exchange of securities at the existing prices. (In contrast, when markets are complete, any allocation of real income can be achieved by redistributing securities.) Consequently, it is possible to increase the welfare of every agent in the economy by increasing the purchasing power of some agents at the expense of others through a redistribution of the ownership of securities. The idea can be illustrated with a simple example.

Consider an economy with three agents, a, b and c, and two states of the world, 1, 2. In each state i there are two goods, labelled X and Y (Table 1). Suppose the preferences and endowments of the agents are given by the following table where x_i^α and y_i^α is the consumption of goods X and Y respectively by agent α in state i.

TABLE 1

Agent	Endowments		Utility
	(X_1, Y_1)	(X_2, Y_2)	
a	(0, 2)	(2, 0)	$x_1^a + \epsilon \min\{x_2^a, y_2^a\}$
b	(2, 0)	(0, 2)	$\epsilon \min\{x_1^b, y_1^b\} + x_2^b$
c	(1, 1)	(1, 1)	$y_1^c + y_2^c$

In this economy, agent a is endowed with two units of good Y in state 1 and two units of good X in state 2. He consumes only good X in state 1 and always consumes an equal amount of both goods in state 2. For each pair of units of the two goods he consumes in state 2 he is willing to give up ϵ units of his consumption of good X in state 1. The endowment and preferences of agent b are the same except that the role of the two states is reversed. Agent c is endowed with one unit of good X in both states but consumes only good Y. His marginal rate of substitution between consumption in the two states is unity.

Suppose there is a single security which promises to deliver one unit of good X in each state. Since there is nothing for which to exchange this security, the equilibrium income and spot prices in each state will be determined solely by the endowments of the agents in that state. It is easy to check that the relative price of the two goods is unity in both states. Agent a consumes two units of good X in state 1 and one unit of each good in state 2. Agent b consumes one unit of each good in state 1 and two units of good X in state 2. Agent c consumes two units of good Y in both states.

Even though the security will never be traded in the market, it can be used by the government to redistribute purchasing power in the two states and thereby change the spot prices. Suppose, for instance, that agents a and b must each supply agent c with two units of the security. Then the effect is the same as if the endowments were changed as shown in Table 2.

TABLE 2

Agent	Endowments	
	(X_1, Y_1)	(X_2, Y_2)
a	(−2, 2)	(0, 0)
b	(0, 0)	(−2, 2)
c	(5, 1)	(5, 1)

For this economy the equilibrium price of good Y in terms of good X in each state is 5/2. Agent a consumes the three units of good X in state 1 and nothing in state 2. Agent b consumes nothing in state 1 and all three units of good X in state 2. Agent c consumes the three units of good Y in both states.

Now compare the welfare of the two agents in the two economies. Without the transfer payments, agents a and b attain an expected utility of $2 + \epsilon$ while agents c attains an expected utility of 4. With the transfer payments, agent a and b both attain an expected utility of 3 while agent c attains a utility of 6. Consequently, for $0 < \epsilon < 1$, the equilibrium with transfer payments Pareto dominates the equilibrium without transfer payments. By transferring purchasing power to agent c in both states, the economy has made the price of the goods demanded by agents a and b cheaper in those states where they value their increased welfare the most. Notice that with a complete set of markets, this Pareto superior allocation would also have been achieved without any transfer payments. In this case, the equilibrium price of good Y in terms of good X is 1/2 in each state with a unitary relative price of the goods across states.

The possibility that securities can be reallocated to attain a Pareto superior allocation when markets are incomplete was first illustrated by Hart (1975). He provided an example in which removing securities and hence decreasing the possibilities for trade actually resulted in a Pareto superior allocation. The intuition is similar to that provided in the example above. If markets are not complete, the introduction of a new security may change the spot market prices in such a way that utility of agents decreases unless they can make trades which are not available with the existing set of securities.

Hart also provided an example to illustrate another important point. If a typical security promises to pay (or receive) goods in more than one state, then the set of trades spanned by the market may depend on the spot prices. He used this insight to construct an example in which there are two competitive equilibria, one of which Pareto dominates the other. Hart's work has been generalized in a number of directions by Geanakoplos and Polemarchakis (1985). They have established that when markets are incomplete and there are at least two goods in each state, then the allocation is 'generically' constrained suboptimal; this implies in particular that Hart's example of Pareto dominance is not generic.

Recently, Cass (1985) and Geanakoplos and Mas-Colell (1985) have investigated the implications for equilibrium when some of the assets are 'nominal', i.e. assets in which the returns in any state are denominated in some unit of account. Subject to a few qualifications, they have demonstrated that, so long as markets are incomplete, the dimension of indeterminancy is generally equal to the number of states minus one.

To understand the logic of this result, suppose that the prices in each spot market $s \in \{1, \ldots, S\}$, are normalized so that they sum to q_s. Then for each vector, $q = (q_1, \ldots, q_s)$, any given nominal asset corresponds to a unique real asset which pays q_s units of each good in each state s. Since markets are incomplete, any nonproportional change in q will generically change the set of (state contingent) commodity bundles spanned by the set of securities (together with the spot markets). Consequently, the equilibrium allocation associated with any two (non-proportional) vectors q will generally be different. We conclude that generically the dimension of equilibrium allocations should be $S - 1$.

CHARLES WILSON

See also ARROW–DEBREU MODEL; PERFECTLY AND IMPERFECTLY COMPETITIVE MARKETS; OVERLAPPING GENERATIONS MODEL.

BIBLIOGRAPHY

Arrow, K. 1953. The role of securities in the optimal allocation of risk-bearing. Reprinted in *Essays in the Theory of Risk-Bearing*, Chicago: Markham Publishing Co., 1973

Cass, D. 1985. On the 'number' of equilibrium allocations with incomplete financial markets. CARESS Working Paper No. 85–16, University of Pennsylvania, 1985.

Geanakoplos, J. and Mas-Colell, A. 1985. Real indeterminancy with financial assets. Yale University, October.

Geanakoplos, J. and Polemarchakis, H. 1985. Existence, regularity, and constrained suboptimality of competitive portfolio allocations when the asset market is incomplete. Yale University, October.

Hart, O. 1975. On the optimality of equilibrium when the market structure is incomplete. *Journal of Economic Theory* 11(3), December, 418–43.

Townsend, R. 1978. On the optimality of forward markets. *American Economic Review* 68(1), March, 54–66.

increasing returns to scale. The focus of this essay is the set of positive propositions that can be obtained when technology exhibits increasing returns to scale. The basic incompatibility of perfect competition and increasing returns to scale is examined separately in a section on existence of equilibria, in which we discuss how one should model economies exhibiting such technologies, i.e. essentially how to modify the Walrasian equilibrium concept in order to guarantee existence of equilibria. Welfare and purely empirical problems are not considered. *Definitions*: A technology exhibits increasing returns to scale if a proportionate increase in all inputs allows for a more than proportionate increase in outputs; in the single-output case, this implies a decreasing average cost curve.

I. DIVISION OF LABOUR AND INCREASING RETURNS TO SCALE. Adam Smith (1776), Babbage (1832), Marshall (1890, 1919) and Young (1928) considered the process of division of labour as the main reason why we observe technologies that exhibit increasing returns to scale. A version of their arguments runs as follows: Let A be the set of tasks to be executed in order to produce good x; a partition $A_1, \ldots, A_n$ is called a first-stage division of labour. Each sub-task A_i, $i = 1, \ldots, n$ is executed by (potentially but not necessarily) different kinds of machinery and primary factors, to be called first-stage intermediate goods. The set of tasks to be executed in order to produce each first-stage intermediate good is also subject to division of labour, to be called second-stage division of labour. Each subtask generated by a second-stage division of labour is executed by intermediate goods, to be called second-stage intermediate goods. Clearly, this process can go on indefinitely. We say that the process of division of labour stops at the nth stage if the n-stage intermediate products are all primary factors; a process is feasible if it stops after a finite number of stages and if the demand for primary factors that it generates does not exceed supply. Suppose, now, that production processes are indivisible, i.e. that when an intermediate good is utilized in the production of some other good, its quantity cannot fall short of a minimum irreducible amount, to be called a fixed cost. An increase in the degree of division of labour is defined as either a finer partition of the set of tasks to be executed in order to produce some good, with the number of stages fixed; or an increase in the number of stages. Clearly, then, an increase in the degree of division of labour implies an increase in fixed costs; discarding inferior divisions of labour, therefore, means that an increase in the degree of division of labour has to imply a decrease in variable cost coefficients.

Smith (1776, p. 7) gave three reasons for such a decrease:

first, ... the increase of dexterity in every particular workman; secondly ..., the saving of the time which is commonly lost in passing from one species of work to another; and lastly ... the invention of a great number of machines which facilitate and abridge labour, and enable one man to do the work of many.

(See also Babbage, 1832, ch. xix.) From now on, the (degree of) division of labour and the degree of increasing returns are used as synonyms.

II. ADAM SMITH. Adam Smith formulated the following propositions:

(1) The division of labour is limited by the extent of the market (Book 1, ch. 3).

(2) The extent of the market is positively related to population size and density, the amount of natural resources and accumulated capital available, and the ease of transportation (Book 1, ch. 3; Book 2, pp. 259–61).

(3) Small economies devote most of their resources to agriculture, while large economies specialize in industry, because the latter affords a greater degree of division of labour. For exactly the same reason, increases in market size decrease the price of industrial products relative to primary products, and as a consequence the profit rate in industry declines (Book I, ch. XI, pp. 242–7; Book III, ch. I).

(4) Trade increases market size and allows each trader (country, region, individual) to specialize and reap the benefits of increased division of labour. Trade is therefore beneficial to all parties involved, it increases real income of all classes, and therefore should not be restricted by governments (Book IV, ch. II; Book I, ch. II).

(5) Economic activity is located in areas in which transportation is least costly, and therefore in areas with the largest potential for division of labour and trade (Book I, pp. 18–21).

(6) The division of labour is limited by the stability of the market (this is not explicitly stated by Smith, but a number of passages indicated that he was aware of it: Book I, p. 21; Book IV, p. 430).

Notice that (1), (2) and (6) are general propositions, while (3), (4) and (5) are applications.

Proposition (1) has generated many important subsidiary propositions, to be described below. Smith used it to derive (3), (4) and (5) without paying attention to the fact that he never demonstrated how the division of labour is determined (as opposed to limited) by the extent of the market.

Marx (1867, Vol. I, Part IV, section 4), Young (1928), Coase (1937) and Stigler (1951) utilized Proposition (1), often unwittingly, to provide the rudiments of a theory of vertical integration and production roundaboutness. Marx (1867, Vol. I) considered the two as different aspects of the same problem, i.e. vertical (dis)integration is 'division of labour in the society' and production roundaboutness is 'division of labour in the workshop'. The following quotation is from Book I, ch. XIV, section 4, p. 355:

But what is it that forms the bond between the independent labours of the cattle-breeder, the tanner, and the shoemaker? It is the fact that their respective products are commodities. What, on the other hand, characterizes division of labour in manufactures? The fact that the detail labourer produces no commodities. It is only the common product of all the detail labourers that becomes a commodity. Division of labour in society is brought about by the purchase and sale of the products of different branches of industry, while the connexion between the detail operations in a workshop is due to the sale of labour-power of several workmen in one capitalist, who applies it as combined labour-power. The division of labour in the workshop implies concentration of the means of production in the hands of one capitalist; the division of labour in society implies their dispersion among many independent producers of commodities. While within the workshop, the iron law of proportionality subjects definite numbers of workers to definite functions, in the society outside the workshop, chance and caprice have full play in distributing the producers and their means of production among the various branches of industry.

Marx also saw that the degree of vertical integration is higher the higher the degree of market imperfection:

> the distinction between division of labour in society and in manufacture was practically illustrated to the Yankees. One of the new taxes devised at Washington during the Civil War, was the duty of 6% 'on all industrial products'. Question: What is an industrial product? Answer of the legislature: A thing is produced 'when it is made', and it is made when it is ready for sale ... . The New York and Philadelphia manufacturers had previously been in the habit of 'making' umbrellas with all their belongings. But since an umbrella is a mixtum compositum of very heterogeneous parts, by degrees these parts became the products of various separate industries, carried on independently in different places. They entered as separate commodities into the umbrella manufactory, where they were fitted together. The Yankees have given to articles thus fitted together the name of 'assembled articles', a name they deserve, for being an assemblage of taxes. Thus the umbrella 'assembles' first, 6% on the price of each of its elements, and a further 6% on its own total price' (ibid., p. 355, footnote 2).

Coase (1937) rediscovered and generalized these observations of Marx and constructed a theory of the firm out of them: price-mediated transactions are costly, and firms exist in order to economize on these costs by organizing transactions in a different, non-price mediated way. At this level of generality, the theory is tautological. Stigler (1951) was the first to try to make it operational: he assumes that a single-output firm executes a set of functions, some of them subject to diminishing and others to increasing average cost. The reason why the firm does not become a monopoly is that the increasing cost functions eventually prevail over the decreasing cost ones, so that the firm's average cost curve is U-shaped. (This is clearly not in the spirit of the classical economists, who assumed global increasing returns.) The reason why with small market size, a firm performs the increasing returns functions itself, instead of abandoning them to specialized firms and so sharing fixed costs with other buyers, is that the fixed cost of these functions is too high relative to market size to allow for the survival of even one specialized firm. This argument is based on the implicit assumption that it is profitable for an integrated firm to perform the increasing returns to scale function, while a specialized firm would make a loss because it would not be able to capture all the surplus of the downstream firms, i.e. it would be able to practise only a sufficiently imperfect degree of price discrimination. As market size increases, though, the position of the specialized firm is strengthened, and eventually it can extract enough surplus from the downstream firms to make positive profit; at this point integrated firms abandon the increasing returns function and become downstream firms (buyers) as far as this function is concerned. Spence and Porter (1977) have provided a formal, partial-equilibrium model along these lines.

The nature of the trade-off is different in Vassilakis (1986b): there are global increasing returns to scale, and firms can choose both the degree of division of labour in the production of the final good (i.e. production roundaboutness) and the extent to which they will make their own intermediate goods (vertical integration). Integrated firms do not buy their intermediate goods, and so they avoid monopolistic exploitation associated with the non-price taking behaviour of intermediate goods sellers; on the other hand, they have to pay the fixed cost of producing intermediate goods. For specialist firms, the trade-off is reversed. Also, a firm that adopts a high degree of division of labour has to pay higher fixed cost, but lower variable cost, than a firm that produces the same product with a lower degree of division of labour. In equilibrium, the ratio of specialist to integrated firms (the degree of vertical disintegration), and the degree of division of labour within each firm (production roundaboutness), are such that the costs and benefits of marginal changes cancel out. Increases in market size (the number of agents) increase vertical disintegration and production roundaboutness for the same reason: it pays to exploit economies of scale more fully now, both by sharing fixed costs with other buyers instead of bearing them unilaterally, and by reducing variable cost through increases in fixed cost. In this sense, market size determines the degree of division of labour. Very clear anticipations of these views on vertical integration are to be found in Austin Robinson (1931, pp. 19, 65, 96, 110).

Proposition (3), another application of Proposition (1), has not been subject to equally intensive theoretical investigation. Kaldor (1978, Essay 9) and Negishi (1986, ch. 3) provide some clarifications. Proposition (4) reappears in Ohlin (1933, ch. 3). For the empirical puzzles that led to the reintroduction of increasing returns to scale in formal trade theory, see Helpman and Krugman (1985, pp. 2–4).

Proposition (5) can be found in Ohlin (1933, pp. 200–211), who generalizes it considerably; increasing returns to scale in production and transportation favour concentration of economic activity in as few points as possible, while the dispersion of natural resources and the fact that certain economic activities are resource-intensive favour decentralization. It is also important whether raw materials or final products are cheaper to transport, with the obvious implications for localization of activities. The result of these considerations is a generalization of Proposition (5).

> [5'] Districts with good transport relations tend to attract plenty of labour and capital and become important market; consequently they tend to specialize in industries which (1) are market-localized and show important advantages from large-scale production; and (2) produce goods which are difficult to transport. On the other hand, districts with poor transport relations become scantily populated and tend to specialize in goods which are easy to transport and can be advantageously produced on a small scale (Ohlin, 1933, p. 208).

Implicit in Ohlin (1976, pp. 48–50) is the proposition that increases in market size increase geographical concentration of economic activity; the reason seems to be that with increased size there is more to be gained by fuller exploitation of scale economies, i.e. by higher concentration of economic activity, and this gain more than compensates for loss due to increased transportation costs.

Proposition (6) has been exploited by Piore and Sabel (1984). Given that a reduction in demand uncertainty is equivalent to an increase in market size, reductions in uncertainty will increase the degree of division of labour. Piore and Sabel view the coexistence of large and small firms, inventory holding, long-term contracts tying buyers to sellers and vertical integration as uncertainty-reducing devices that allow for a higher degree of division of labour. Also, collective wage bargaining and government stabilization policies are attempts to control that part of uncertainty that cannot be affected by individual firms. Weitzman (1982) and then Kaldor (1983) went even further and argued that a necessary condition for involuntary unemployment, and therefore for Keynesian economics, is the presence of increasing returns to scale, otherwise the unemployed can 'produce themselves out of unemployment', since non-increasing returns to scale imply that small-scale production is at least as efficient as large-scale. (see also the Symposium on Increasing Returns and Unemployment Theory, 1985).

III. MILL AND MARX. Mill and Marx gave two closely interrelated propositions:

(7) Increases in market size result in increased concentration of economic activity, in the sense that a higher percentage of the population earn income by selling labour (and not by producing). See Mill (1848, Book I, ch. IX, p. 3) and Marx (1867, ch. XXXII).

(8a) Increases in market size, and the resulting concentration and increase in the scale of production of each firm, is an unqualified benefit from the efficiency point of view, but not necessarily from the equity point of view (Mill, ibid.; Marx, ibid. and ch. XXV).

Both (7) and (8a) are derived as a consequence of the fact that concentration allows for fuller exploitation of scale economies; Marx added another reason, i.e. that the skills of small-scale producers are 'rendered worthless' by division of labour, which subdivides and simplifies the tasks to be executed in order to produce a commodity (Marx, 1848, in McLellan, 1977, p. 227).

Another proposition of Marx on the same subject is:

(8b) Increases in market size increase the distance between the economy's actual and potential performance (Marx, 1867, ch. XXXII); Elster (1985, ch. 5), provides a rather exhaustive discussion of the exact meaning of this proposition.

We now make (8b) more precise by thinking of increases in market size as generating two contradictory forces: on the one hand, efficiency increases because the increase in market size and the resulting increase in concentration (Proposition 7) allow for fuller exploitation of scale economies; on the other hand, this very increase in concentration that results in fuller exploitation of scale economies, hampers efficiency by increasing the distortionary effects associated with non-price taking behaviour. In other words, economies of scale are created faster than they are exploited. Finally, we can safely attribute to Marx the following proposition, a variant of his law of the falling rate of profit (Marx, 1894, Vol. III, Part III).

(9) Increases in market size reduce the profit rate.

Proposition (9) differs from Proposition (3) of Smith (and Ricardo), because it does not rely on the law of diminishing returns due to land scarcity. (In Marx's words, Ricardo 'fled from economics to seek refuge in organic chemistry' in order to generate a falling profit rate). It is formulated in this particular way, because it has been shown that under constant returns to scale and a constant real wage, the law does not hold, while with a rising real wage it holds only under very restrictive assumptions that turn the law into an improbable special case (Roemer, 1981, chs 4, 5 and 6). On the other hand, Negishi (1985, ch. 4) has provided some textual evidence to support the view that Marx had in mind an

economy with increasing returns to scale technology and producers facing downward sloping demand, so that (9) is the only version of the law that might be sustainable. Indeed, increasing concentration and a falling profit rate have been obtained in Vassilakis (1986a) as a result of increases in market size; the profit rate, though, falls because both the real wage and the proportion of workers in the population rise in a full employment model, so this version of the law is not entirely in the Marxian spirit. As for Proposition (8a), the formal literature supports the view that with increasing returns to scale only in the neighbourhood of the origin, increases in market size reduce Pareto inefficiency and in the limit they eliminate it (Novshek and Sonnenschein, 1978; Hart, 1979). On the other hand, Hart and Guesnerie (1985) have found that with global increasing returns, Pareto inefficiency does not disappear in the limit, although per capita inefficiency does; Vassilakis (1986a) finds that even per capita welfare loss can be positive in the limit, for a particular choice of technology; the difference in the result is due to the fact that the latter reference assumes that the alternative to producing is being a worker and earning wage income, while Hart and Guesnerie assume the opportunity cost of a producer to be zero. So, it is fair to say that there is some support for Proposition (8a), while Proposition (8b) remains untested.

IV. MARSHALL. Marshall (1890, p. 318; 1919, pp. 186–9) believed that all industries exhibit global increasing returns to scale, checked only by short-run fixities or land scarcities; in this case he agreed with the classical economists. As Stigler (1941, p. 78) remarked, though, 'if the economies of large scale production are so important ..., how do small concerns manage to exist at all?' and 'either the division of labour is limited by the extent of the market, and, characteristically, industries are monopolized; or industries are characteristically competitive, and the theorem is false or of little significance' (Stigler, 1951). Marshall tried to reconcile economies of scale and perfect competition in three different ways, namely:

(a) (Some) economies of scale are external to the firm.

(b) Increasing returns to scale is a dynamic phenomenon, and its full effects take so long to manifest themselves that 'the guidance of the business falls into the hands of people with less energy and less creative genius' (Marshall, 1890, p. 316).

(c) Transportation costs rise so fast in some industries as to restrict the market area of each firm.

Clearly, (a) assumes the problem away; in Marshall's own words,

> ... with the growth of capital, the development of machinery, and the improvement of the means of communication the importance of internal economies has increased steadily and fast, while some of the old external economies have declined in importance (Marshall, 1919, p. 167).

But even if we assume that most economies of scale are external to the firm, competition is not the most likely outcome; one still has to explain why firms do not merge to internalize external economies, in which case oligopoly is the most likely outcome, or why markets for external effects do not emerge, in which case again, external economies become internal, and we are back to square one. (Starret and Heller (1976) analyse external effects as absence of markets; Makowski (1980) analyses mergers as a way to internalize external effects.)

Explanation (b) is at best of limited importance, unless one can show that expansion by merger is impossible or that the market for managers is so imperfect that a long-lived firm is doomed to fall in the hands of the inept. Also, in Stigler's words, 'if Marshall's discussion of economies is correct and approximately complete, it would not require an extraordinarily high calibre of entrepreneurship to secure a monopoly, or at least a dominant position, in almost any industry' (Stigler, 1941, p. 81). Finally, explanation (c) is of limited applicability because it ignores increasing returns to scale in transportation. Marshall himself thought that it cannot be elevated to a general explanation of the coexistence of competition and increasing returns, so he had to invent explanation (b); (Marshall, 1919, pp. 315–16). We have to conclude that (not only) in 'competitive, stationary economies, Marshall clearly fails to provide the conditions of stable equilibrium' (Stigler, 1941, p. 81). Downward sloping demand and non-price taking behaviour cannot be avoided, therefore; based on Marshall's cues (Marshall, 1890, pp. 286–7, 453–8), Sraffa (1926), Robinson (1933) and Chamberlin (1933) reintroduced downward sloping demand almost one hundred years after Cournot.

Despite the fact that Marshall did not have a formal theory of increasing returns economies, he relentlessly applied 'the principle of Increasing Return' to generate propositions. He is the only one after Smith, Marx and Mill to propose a new general proposition (not an application), namely:

(10) '... almost every kind of horizontal extension tends to increase the internal economies of production on a large scale, but as rule, an increase in the variety of output lessens the gain in this direction' (Marshall, 1919, p. 216).

In other words, increasing variety reduces efficiency. Proposition (10) is then utilized by Marshall to explain the coexistence of large and small firms, and to determine the range of products of a multiproduct firm. Large firms produce those goods that are most in demand and/or afford the greatest degree of division of labour; their product range is determined by the condition that the addition of one more product would increase cost (due to lost scale economies) by more than it would increase revenue (due to increased market area). Small firms produce goods whose demand is so low, and/or afford so small a degree of division of labour, that large firms do not want to produce, because they can be better off devoting their resources to increase production of the commodities they already produce. As market size increases, there is more to be gained by concentration, so firm size tends to increase. On the other hand, though, small firms will survive at all market sizes, because of three factors (Marshall, 1919, ch. III and IV).

(a) The increased income generated by increased market size allows consumers to demand goods closer to their ideal specifications, so the variety of goods demanded increases.

(b) Increased market size increases household specialization, i.e. goods previously produced within the household become commodities.

(c) Increased size increases vertical disintegration.

An obvious implication of this theory is that increases in market size will have different effects, depending on the degree of demand homogeneity and on whether demand is concentrated on goods that afford considerable division of labour. Marshall (1919, Book I) attributes the different growth patterns of industrial economies to differences in the size stability and perfection of their respective markets.

V. EXISTENCE OF EQUILIBRIUM. The incompatibility of price-taking behaviour and increasing returns to scale was first noticed by Cournot (1838, pp. 59–60), but rigorous examination of the issue has been taken up only very recently.

No general existence theorems are available because there is no generally accepted model to imperfect competition. What is available, though, points to the importance of the following three factors: (i) downward sloping demand; (ii) a variable number of firms; (iii) a large number of agents relative to the degree of increasing returns. Downward sloping demand is clearly a necessary condition for existence in economies with global increasing returns to scale, for otherwise firms would have an incentive to produce an unlimited amount of some output. The number of firms should be variable for three reasons: first, because of fixed costs, the number of firms cannot be too large for otherwise profit would be negative; secondly, the number of firms should be sufficiently large to ensure that the demand price faced by each firm is lower than average costs for large enough output levels, otherwise firms would produce arbitrarily large amounts; thirdly, the number of firms should be sufficiently large to discourage entry, so as to ensure that if one more agent sets up a firm, he will earn less than his earnings in the best alternative occupation. Finally, one needs a large number of agents relative to the degree of increasing returns to ensure that the number of firms is large enough to satisfy the conditions above, and in order to convexify reaction correspondences, so that fixed-point theorems can be applied (see Roberts and Sonnenschein, 1976, and Novshek and Sonnenschein, 1978). All models in the literature on increasing returns to scale base their existence results on (i), (ii) and (iii) above, although they differ in specifics. Thus, we have Bertrand models, in which the agents' strategic variable is price, and Cournot models, in which agents compete in quantities. Also, we have symmetric models, in which all agents are allowed the same strategic possibilities; and non-symmetric models, in which the set of agents is, *a priori* and once and for all, divided into two disjoint sets: the set of consumers–factor suppliers–price takers, and the set of producers–factor demanders–price makers (or quantity setters).

Existence in non-symmetric Cournot models with increasing returns only in a small neighbourhood of the origin is proved in Novshek and Sonnenschein (1978), and with global increasing returns in Hart and Guesnerie (1985); existence in symmetric Cournot models with global increasing returns is proved in Vassilakis (1986a). All proofs with global increasing returns refer to a single-input, single-output economy. For non-symmetric Bertrand games, with a single input, see Hart (1985) and Economides (1982; 1983); for the single-output many-inputs case see Sharkey (1982, ch. 8).

SPYROS VASSILAKIS

See also COMPETITION; DIVISION OF LABOUR; LEARNING BY DOING.

BIBLIOGRAPHY

Babbage, C. 1832. *On the Economy of Machinery and Manufactures.* London: C. Knight.

Chamberlin, E. 1933. *The Theory of Monopolistic Competition.* Cambridge, Mass.: Harvard University Press.

Coase, R. 1937. The nature of the firm. *Economica* 4, November, 386–405.

Cournot, A. 1838. *Recherches sur les principes mathématiques de la théorie des richeses.* Paris: M. Rivière.

Economides, N. 1982. Oligopoly in differentiated products with three or more competitors. Columbia University Discussion Paper No. 153.

Economides, N. 1983. Symmetric equilibrium existence and optimality in differentiated products markets. Columbia University Discussion Paper No. 197.

Elster, J. 1985. *Making Sense of Marx*. Cambridge: Cambridge University Press.

Hart, O. 1979. Monopolistic competition in a large economy with differentiated commodities. *Review of Economic Studies* 46, January, 1–30.

Hart, O. 1985. Monopolistic competition in the spirit of Chamberlin: a general model. *Review of Economic Studies* 52(4), October, 529–46.

Hart, O. and Guesnerie, R. 1985. Welfare loss due to imperfect competition: asymptotic results for Cournot–Nash equilibria with and without free entry. *International Economic Review* 26(3), October, 525–45.

Heller, W. and Starret, D. 1976. On the nature of externalities. In *Theory and Measurement of Economic Externalities*, ed. S. Lin, New York: Academic Press.

Helpman, E. and Krugman, P. 1985. *Market Structure and Foreign Trade*. Cambridge, Mass.: MIT Press.

Kaldor, N. 1978. *Further Essays on Economic Theory*. London: Duckworth.

Kaldor, N. 1983. Keynesian economics after fifty years. In *Keynes and the Modern World*, ed. D. Worswick and J. Trevithick, Cambridge: Cambridge University Press.

Makowski, L. 1980. Perfect competition, the profit criterion, and the organization of economic activity. *Journal of Economic Theory* 22(2), April, 222–42.

Marshall, A. 1890. *Principles of Economics*. London: Macmillan.

Marshall, A. 1919. *Industry and Trade*. London: Macmillan.

Marx, K. 1867–94. *Capital*. 3 vols, Harmondsworth: Penguin Books, 1976.

McLellan, D. 1977. *Karl Marx, Selected Writings*. Oxford: Oxford University Press.

Mill, J.S. 1848. *Principles of Political Economy*. London: J.W. Parker.

Negishi, T. 1985. *Economic Theories in a non-Walrasian Tradition*. Cambridge: Cambridge University Press.

Novshek, W. and Sonnenschein, H. 1978. Cournot and Walras equilibrium. *Journal of Economic Theory* 19, 223–66.

Ohlin, B. 1933. *Interregional and International Trade*. Cambridge, Mass.: Harvard University Press.

Ohlin, B. et al. 1976. *The International Allocation of Economic Activity*. New York: Holmes & Meier.

Piore, M. and Sabel, C. 1984. *The Second Industrial Divide*. New York: Basic Books.

Roberts, J. and Sonnenschein, H. 1977. On the foundations of the theory of monopolistic competition. *Econometrica* 45, 101–13.

Robinson, E.A.G. 1931. *The Structure of Competitive Industry*. Cambridge: Cambridge University Press.

Robinson, J. 1933. *The Economics of Imperfect Competition*. London: Macmillan.

Roemer, J. 1981. *Analytical Foundations of Marxian Economic Theory*. Cambridge: Cambridge University Press.

Sharkey, W.W. 1982. *The Theory of Natural Monopoly*. Cambridge: Cambridge University Press.

Smith, A. 1776. *An Inquiry into the Nature and the Courses of the Wealth of Nations*. Ed. E. Cannan, London: Methuen, 1961.

Spence, M. and Porter, M. 1977. Vertical integration and differentiated inputs. Harvard Discussion Paper No. 576.

Sraffa, P. 1926. The laws of returns under competitive conditions. *Economic Journal* 36, December, 535–50.

Stigler, G. 1941. *Production and Distribution Theories*. New York: Macmillan.

Stigler, G. 1951. The division of labor is limited by the extent of the market. *Journal of Political Economy* 59, June, 185–93.

Symposium on Increasing Returns and Unemployment Theory. 1985. *Journal of Post Keynesian Economics* 7(3), Spring, 350–409.

Vassilakis, S. 1986a. Increasing returns and strategic behavior I: the worker–employer ratio. Johns Hopkins Working Paper, No. 168.

Vassilakis, S. 1986b. Increasing returns and strategic behavior II: the division of labor. Johns Hopkins Working Paper, No. 169.

Weitzman, M. 1982. Increasing returns and the foundations of unemployment theory. *Economic Journal* 92, December, 787–804.

Young, A. 1928. Increasing returns and economic progress. *Economic Journal* 38, December, 527–42.

indentured labour. *See* SLAVERY.

indexed securities. Conventional securities are generally offered at a fixed coupon rate that incorporates the underlying expected real rate of return in the economy, the market's expectation at the time the security is issued of inflation over the duration of the instrument, a premium to compensate for the fact that future rates of inflation are uncertain, and an adjustment reflecting the tax treatment of interest on behalf of both the lender and the borrower. For simplicity, it is useful to abstract temporarily from the inflation risk premium and taxes, although both these factors will be discussed later.

With these simplifying assumptions, if the real rate of return in the economy is 3 per cent, and inflation is expected to remain constant at 4 per cent annually, the nominal return will be 7 per cent. If expectations should prove incorrect and inflation turns out to be lower than anticipated, say 2 per cent, investors will receive more income in present value terms than they expected and experience an increase in their real rate of return, reflecting the unanticipated decline in the inflation rate. On the other hand, if inflation turns out to be 6 per cent, then investors will receive less in real terms than expected and their real return will fall below the rate initially negotiated. If they attempt to sell the security in the higher inflationary environment, they will experience a capital loss.

Index bonds are financial instruments designed to protect investors fully against the erosion of principal and interest due to inflation. This protection is accomplished in one of two ways. Under the first option, the bond is issued at a specified real coupon rate and both coupon payment and repayment of principal are scaled up or down by the change in prices that occurs between the time that the money is borrowed and the time the payments are made. For example, if inflation is 4 per cent annually, the coupon on a five-year $1000 bond issued at a real interest rate of 3 per cent would increase from $30 in the first year to $35.10 in the fifth year. At maturity, the government then would adjust the principal for inflation over the life of the bond; thus, in the above example, the government would repay $1217 at the end of the five-year period. This approach is similar to the index bonds that have been sold in Great Britain.

Under the second approach the entire inflation adjustment is made through the coupon payment and the bondholder is repaid his original principal at maturity. For example, if the real rate is set at 3 per cent and inflation averages 4 per cent, the total annual interest cost would be 7 per cent. This approach mimics the current method of compensating the lender for inflation, except that instead of trying to predict inflation at the time of the loan and incorporating this expectation into the stated nominal interest rate, actual observations on price are used to determine annual interest payments.

Either of these two approaches will protect the investor against the risks associated with unanticipated price changes if the index bond is held to maturity; however, it is important to emphasize that neither produces a risk-free investment. As with any long-term security, bondholders selling an index bond before maturity would take a capital loss if the underlying expected real rates have increased since the date of purchase. The result is that these bonds would probably not be the ideal assets for individuals to purchase directly unless they were certain that they could hold them to maturity. For index bonds to serve as risk-free inflation-protected investments, financial intermediaries are required which will hold the bonds to maturity and offer repackaged investments free of the real-return risk.

IMPACT ON THE GOVERNMENT BUDGET. Arguments about the potential impact of index bonds on the government budget have figured prominently in debates about this form of financing and the range of opinion has been extraordinary. In Great Britain, some opponents argued that index bonds would cost the government more than fixed-interest securities to service, since they would have to be issued at positive real interest rates as opposed to the negative real returns received by investors on nominal debt during the period 1973–8 (Rutherford, 1983). On the other hand, in hearings before the Joint Economic Committee in May 1985, a major proponent of index bonds projected that, because excessive inflation premiums were incorporated in current yields, the US government could save $9 billion in the first year and $135 billion over a five-year period by issuing indexed rather than conventional long-term debt (Joint Economic Committee, 1985).

These conflicting statements are based on opposite assumptions about people's ability to project future inflation. The contention that index bonds will cost money assumes that individuals will continually underestimate future inflation and always end up with lower than anticipated or negative returns; the argument that the Treasury can reduce costs with indexed debt assumes individuals will consistently overestimate inflation and demand excessive inflation premiums. It is unclear why, over the long run, individuals should systematically err on one side or another in their inflation projections.

In the last 15 years, the relationship between expected and actual inflation has varied over time; during the 1970s average expectations about near-term inflation tended to prove too low, while since 1981 inflation has generally fallen one or two percentage points below projections. Although no evidence is available on investors' ability to forecast inflation over longer periods, of say 20 or 30 years, the same pattern is likely to emerge as swings in short-run expectations affect the longer-run outlook. Hence, the most reasonable conclusion is that in the long run forecasting errors will cancel out, and have little impact on the relative costs of indexed versus unindexed debt.

On the other hand, the uncertainty surrounding future rates of inflation means that investors demand an inflation-risk premium before they are willing to take on fixed-coupon debt. In this case, the guarantee of a real return provided by indexed securities, which eliminates the risk of reduced real returns and capital losses caused by unanticipated inflation, would lower the yield that lenders will require in order to provide their funds. In other words, the lender would be willing to accept a somewhat lower rate in return for the privilege of having the government guarantee the real return on the loan.

Little evidence exists about the size of the inflation risk premium (an exception is Bodie, Kane and McDonald, 1986). As long as the outlook for price increases is moderate, the premium is probably relatively small; at higher and more volatile rates of inflation, the importance of risk protection would increase. Even if this premium proved to be quite small, however, its elimination could produce substantial savings in view of the enormous magnitude of government debt. The problem is that, in the short run at least, the risk premium effect is likely to be dominated by the difference between expected and actual real returns caused by errors in investors' expectations. Hence, for any defined period of time, it would be impossible to predict whether substituting index bonds for traditional government securities would cost or save the Treasury money. In the long run, however, if errors in inflation forecasts cancel out, index bonds should save the government the inflation-risk premium on long-term securities.

While the net interest saving to the government is difficult to predict, the pattern of government borrowing would certainly be altered if the British indexing option were adopted. Even in an environment where inflationary expectations always prove correct and the inflation premium is zero, an index bond that defers the principal adjustment for inflation until maturity reduces the Treasury's borrowing in the intervening years.

TAX POLICY AND INDEX BONDS. Uncertainty about how index bonds would be taxed has been viewed as a major impediment to their introduction. The tax questions are indeed critical, because they determine not only how index bonds would affect revenues but also who might be the likely buyers of these securities and the potential yields.

If the tax code does not distinguish between real and inflationary returns, then the most likely results would be to tax both the real component of interest and the inflation adjustment as ordinary income. This would be quite straightforward in the case of the indexing method that incorporates the inflation adjustment in the interest rate, but some complexities arise in the case of the British approach. In order to make the treatment of bonds indexed in this fashion analogous to that accorded conventional and zero-coupon bonds, the annual appreciation of principal due to inflation would have to be taxed as it accrued.

Taxing the principal adjustment as if it were received each year would make index bonds less attractive than their unindexed counterparts. Not only would owners of securities have to pay taxes on illusory gains, which they do in the case of conventional bonds, but they would also have to pay the tax before they received their inflation compensation. On the other hand, deferring the tax on the adjustment of principal until the bond is redeemed at maturity would favour the indexed over the unindexed security and result in a loss of revenue for the Treasury.

The second problem with applying current tax law to index bonds is that it would no longer be possible to guarantee a constant real after-tax rate of return. Under the current system, taxes would rise with inflation and the real after-tax return would decline. For example, if the tax rate were 30 per cent, the real return on a bond with a 3 per cent coupon would be 2.1 per cent in an environment of no inflation. If inflation should rise to 4 per cent and the nominal coupon rises to only 7 per cent, the after-tax yield is 4.9 per cent or 0.9 per cent real. The only way to avoid this problem is to exempt the nominal adjustments for inflation from taxation. This approach, however, would introduce a type of inflation indexing not found elsewhere in the tax system.

PRIVATE ISSUES OF INDEX BONDS. Some sceptics charge that if index bonds were such a great idea, they would have been offered by the private sector. Indeed, theoretical work by Stanley Fischer leads to the conclusion that firms should be equally willing to issue index bonds as conventional nominal bonds (Fischer, 1982). Fischer offers two possible reasons for the lack of the private sector innovation: the relatively stable rates of inflation traditionally experienced in the United States and the possibility that borrowers' expectations about inflation have been systematically higher than those of lenders. Others contend that the issuance of index-linked debt may actually have been illegal in the United States until 1977 (McCulloch, 1980). Another problem is that an aggregate price index may not correlate with prices received by an individual firm. The most persuasive reason, however, relates to the lack of indexation in the corporate income tax, which causes the effective tax rate to increase with inflation. If firms were to issue index bonds, this inverse relation between inflation and profitability would worsen, since corporations would forfeit the mitigating effect of the decline in the value of outstanding liabilities as inflation increased. Hence, the non-issuance of index bonds by the corporate sector may be one of the major casualties of an unindexed tax structure.

The only serious objection ever levelled against index bonds is that protecting bondholders from inflation might reduce public pressure to maintain price stability. If part of the pain of inflation is removed, this reasoning goes, the public's resolve to control inflation will weaken, and inflation will ultimately get worse. On the other hand, one could argue in economic terms that index bonds might help in the fight against inflation by providing an attractive investment vehicle that would encourage saving and, as argued by Tobin, by offering the monetary authorities a tool that would strengthen their control of the economy (Tobin, 1971). In political terms, it would seem that the issuance of index bonds would eliminate one of the main incentives for the government to inflate the economy. With indexed debt the government can no longer reduce the real value of its outstanding liabilities by allowing prices to rise; instead, inflation will produce an immediate increase in required expenditures. Finally, index bonds do not appear to have encouraged inflation in Great Britain; the inflation rate has declined from 15 to 5 per cent since 1981, the year the bonds were introduced.

ALICIA H. MUNNELL AND JOSEPH B. GROLNIC

See also INFLATION; MONETARY POLICY; WAGE INDEXATION.

BIBLIOGRAPHY

This essay is abstracted from Munnell and Grolnic (1986).

Bodie, Z., Kane, A. and McDonald, R. 1986. Risk and required returns on debt and equity. In *Financing Corporate Capital Formation*, ed. B.M. Friedman, New York: National Bureau of Economic Research, 51–66.
Fischer, S. 1982. On the nonexistence of privately issued index bonds in the US capital market. In *Inflation, Debt, and Indexation*, ed. R. Dornbusch and M.H. Simonsen, Cambridge, Mass.: MIT Press, 247–66.
Joint Economic Committee. 1985. Inflation Indexing of Government Securities. Hearing before the Subcommittee on Trade, Productivity, and Economic Growth, 99 Congress, 1 session, 14 May.
McCulloch, J.H. 1980. The ban on indexed bonds, 1933–77. *American Economic Review*, December, 1081–21.
Munnell, A.H. and Grolnic, J.B. 1986. Should the US Government issue index bonds? Federal Reserve Bank of Boston, *New England Economic Review*, September/October, 3–21.
Rutherford, J. 1983. Index-linked gilts. *National Westminster Review*, November, 2–17.
Tobin, J. 1971. An essay on the principles of debt management. In J. Tobin, *Essays in Economics, Volume 1: Macroeconomics*, Chicago: Markham, 439–47.

indexing. See INDEXED SECURITIES; WAGE INDEXATION.

index numbers. The index number problem may be phrased as follows. Suppose we have price data $p^i \equiv (p_1^i, \ldots, p_N^i)$ and quantity data $x^i \equiv (x_1^i, \ldots, x_N^i)$ on N commodities that pertain to economic unit i or that pertain to the same economic unit at time period i for $i = 1, 2, \ldots, I$. The *index number problem* is to find I numbers P^i and I numbers X^i such that

$$P^i X^i = p^i \cdot x^i \equiv \sum_{n=1}^{N} p_n^i x_n^i \quad \text{for} \quad i = 1, \ldots, I. \tag{1}$$

P^i is the *price index* for period i (or unit i) and X^i is the corresponding *quantity index*. P^i is supposed to be representa-

tive of all of the prices $p_n^i, n = 1, \ldots, N$ in some sense while X^i is to be similarly representative of the quantities $x_n^i, n = 1, \ldots, N$. In what precise sense P^i and X^i represent the individual prices and quantities is not immediately evident and it is this ambiguity which leads to different approaches to index number theory. Note that we require that the product of the price and quantity indexes, $P^i X^i$, equal the actual period (or unit) i net expenditures on the N commodities, $p^i \cdot x^i$. Thus if the P^i are determined, then the X^i may be implicitly determined using equations (1), or vice versa.

Each individual consumes the services of thousands of commodities over a year and most producers utilize and/or produce thousands of individual products and services. Index numbers are used to reduce and summarize this overwhelming abundance of microeconomic information. Hence index numbers intrude themselves on virtually every empirical investigation in economics.

Index number theory splits naturally into two divisions, depending on the size of I. If $I = 2$, so that there are data for only two time periods or two economic units, then we are in the realm of *bilateral index number theory* while if $I > 2$, then we are in the realm of *multilateral indexes*. Bilateral approaches are considered in sections 1–5 below and multilateral approaches are considered in sections 6–10.

The four main approaches to index number theory are: (i) *statistical* (section 1), (ii) *test or axiomatic* (sections 2 and 9), (iii) *microeconomic* which relies on the assumption of maximizing or minimizing behaviour (sections 3, 4 and 5), and (iv) *neostatistical* (section 10).

1. STATISTICAL APPROACHES

Let $I = 2$ and consider the following formula for p^2/p^1 due originally to Dutot (1738):

$$P^2/P^1 = \left(\sum_{n=1}^{N} p_n^2/N \right) \bigg/ \left(\sum_{n=1}^{N} p_n^1/N \right). \tag{2}$$

Thus the average level of prices in say period 2 relative to period 1 is set equal to the arithmetic average of the period 2 prices divided by the arithmetic average of the period 1 prices. The right-hand side of (2) is called an *index number formula*.

Given an index number formula, we may solve the aggregation problem (1) as follows: set p^2 equal to the index number formula and determine p^1, X^i and X^2 by:

$$P^1 = 1, \quad X^1 = p^1 \cdot x^1 \quad \text{and} \quad X^2 = p^2 \cdot x^2/P^2. \tag{3}$$

Setting $P^1 = 1$ is regarded as an arbitrary normalization; any other convenient normalization such as $P^1 = 100$ could be chosen, in which case $X^1 \equiv p^1 \cdot x^1/P^1$, $P^2 \equiv (P^2/P^1)P^1$ and $X^2 \equiv P^2 \cdot x^2/P^2$, where P^2/P^1 is the index number formula.

Rather than taking P^2/P^1 to be a ratio of average prices, Carli (1764) suggested taking an average of the price ratios as follows:

$$P^2/P^1 = \sum_{n=1}^{N} (p_n^2/p_n^1)/N. \tag{4}$$

The average of the price ratios in (4) is an arithmetic average. Jevons (1865) suggested using a geometric average:

$$P^2/P^1 = \prod_{n=1}^{N} (p_n^2/p_n^1)^{1/N}. \tag{5}$$

Once the ratio P^2/P^1 has been determined by (4) or (5), P^1, X^1 and X^2 may be determined using (3). But which of the three alternative formulae for P^2/P^1 should we use?

Walsh (1901) criticized the use of the Dutot formula (2) on the grounds that the index was not invariant to changes in the units of measurement. This criticism was a telling one, and virtually nobody uses formula (2) at present. However, formulae (4) and (5) are invariant to changes in the units of measurement, so we must still discriminate between them.

Jevons argued that changes in the quantity of money between the two periods would lead to proportional changes in all prices except for random errors. In particular, Jevons argued that the price ratios, p_n^2/p_n^1, would be independently and symmetrically distributed around a common mean. If this distribution happened to be the normal distribution, then the maximum likelihood estimator for the common mean leads to the index number formula (4). If the ratios p_n^2/p_n^1 happened to be lognormally distributed, then statistical considerations would lead us to the index number formulae (5).

Bowley (1928) attacked the use of both (4) and (5) on two grounds. First, from an empirical point of view, he showed that price ratios were not symmetrically distributed about a common mean and their logarithms also failed to be symmetrically distributed. Secondly, from a theoretical point of view, he argued that it was unlikely that prices or price ratios were independently distributed. Keynes (1930) developed Bowley's second objection in more detail; he argued that changes in the money supply would not affect all prices at the same time. Moreover, real disturbances in the economy could cause one set of prices to differ in a systematic way from other prices, depending on various elasticities of substitution and complementarity. In other words, prices are not randomly distributed, but are systematically related to each other through the general equilibrium of the economy.

The above criticisms led to a movement away from the use of unweighted averages of price ratios to represent price movements independently of quantity movements. Walsh (1901) and others suggested that the quantity observations x_n^i that were associated with the individual price observations p_n^i should be used as weights in the price index formula.

Scrope (1833) suggested the following formula:

$$P^2/P^1 = p^2 \cdot x / p^1 \cdot x \tag{6}$$

where $x \equiv (x_1, \ldots, x_N)$ was a somewhat vaguely specified quantity vector which was used to weight the price vectors p^1 and p^2 as in (6).

Laspeyres (1871) recommended that x be set equal to x^1, the period 1 quantity vector, while Paasche (1874) suggested that x be set equal to x^2, the period 2 quantity vector. This led to the following formulae:

$$P^2/P^1 = p^2 \cdot x^1 / p^1 \cdot x^1 \equiv P_L(p^1, p^2, x^1, x^2); \tag{7}$$

$$P^2/P^1 = p^2 \cdot x^2 / p^1 \cdot x^2 \equiv P_P(p^1, p^2, x^1, x^2). \tag{8}$$

Note that we are now following custom in the index number literature by defining an index number formula to be a function $P(p^1, p^2, x^1, x^2)$ of the price and quantity vectors that pertain to the two observations or time periods under consideration: P_L defines the Laspeyres price index while P_P defines the Paasche price index.

Pigou (1920) and Irving Fisher (1922) advocated taking a geometric mean of the Paasche and Laspeyres indexes and the resulting formula has come to be known as the Fisher ideal price index P_F:

$$P^2/P^1 = [P_L P_P]^{1/2} \equiv P_F(p^1, p^2, x^1, x^2). \tag{9}$$

Rather than taking a geometric average of (7) and (8), Walsh (1901) (1921) advocated using formula (6) where the weight

vector x was chosen to be the vector of geometric means of the two quantity vectors:

$$P^2/P^1 = \sum_{n=1}^{N} p_n^2 (x_n^1 n_n^2)^{1/2} \Big/ \sum_{n=1}^{N} p_n^1 (x_n^1 x_n^1)^{1/2}$$
$$\equiv P_W(p^1, p^2, x^1, x^2). \tag{10}$$

Törnqvist (1936) advocated a weighted geometric mean of the price ratios of the following form:

$$P^2/P^1 = \prod_{n=1}^{N} (p_n^2/p_n^1)^{s_n} \equiv P_T(p^1, p^2, x^1, x^2) \tag{11}$$

where $s_n \equiv (1/2)(p_n^1 x_n^1 / p^1 \cdot x^1) + (1/2)(p_n^2 x_n^2 / p^2 \cdot x^2)$ is the average expenditure share on good n for $n = 1, \ldots, N$.

It turns out that formulae (7), (8), (9), and (11) are the most widely used formulae for a price index. However, at this point, we have no way of justifying their popularity. Walsh (1901) and Fisher (1922) present scores of functional forms for price indexes–on what basis are we to choose one as being better than the others?

This question leads us to discuss the test or axiomatic approach to index number theory.

2. THE TEST APPROACH TO BILATERAL INDEXES

Consider $P(p^1, p^2, x^1, x^2)$, a function of the N period i prices, $p^i \equiv (p_1^i, \ldots, p_N^i)$, and the N period i quantities, $x^i \equiv (x_1^i, \ldots, x_N^i)$ for $i = 1, 2$. The price index P is supposed to represent the level of prices in period 2 relative to period 1. What properties or *tests* should such an index number formula satisfy? The following nine tests (or closely related variants) have been considered in the literature.

BT1: *Identity Test*:
  $P(p^1, p^2, \alpha x^1, \beta x^2) = 1$ for all numbers $\alpha > 0$,
    $\beta > 0$ if $p^1 = p^2$ and $x^1 = x^2$.

BT2: *Proportionality Test*:
  $P(p^1, \alpha p^2, x^1, x^2) = \alpha P(p^1, p^2, x^1, x^2)$ for $\alpha > 0$.

BT3: *Invariance to Changes in Scale Test*:
  $P(\alpha p^1, \alpha p^2, \beta x^1, \gamma x^2) = P(p^1, p^2, x^1, x^2)$
    for all $\alpha > 0, \beta > 0$ and $\gamma > 0$.

BT4: *Invariance to Changes in Units (Commensurability) Test*:
  $P(\alpha_1 p_1^1, \ldots, \alpha_N p_N^1; \alpha_1 p_1^2, \ldots, \alpha_N p_N^2;$
    $\alpha_1^{-1} x_1^1, \ldots, \alpha_N^{-1} x_N^1; \alpha_1^{-1} x_1^2, \ldots, \alpha_N^{-1} x_N^2)$
    $= P(p^1, p^2, x^1, x^2)$ for $\alpha_1 > 0, \ldots, \alpha_N > 0$.

BT5: *Symmetric Treatment of Countries or Time (Country or Time Reversal Test)*:
  $P(p^2, p^1, x^2, x^1) = 1/P(p^1, p^2, x^1, x^2)$.

BT6: *Symmetric Treatment of Commodities (Commodity Reversal) Test*:
  $P(\tilde{p}^1, \tilde{p}^2, \tilde{x}^1, \tilde{x}^2) = P(p^1, p^2, x^1, x^2)$
    where $\tilde{p}^i$ denotes a permutation of the elements of the vector p^i and $\tilde{x}^i$ denotes the same permutation of the elements of x^i, $i = 1, 2$.

BT7: *Monotonicity*:
  $P(p^1, p^2, x^1, x^2) \leqslant P(p^1, p^3, x^1, x^2)$
    if $p^2 \leqslant p^3$; i.e., if $p_n^2 \leqslant p_n^3$ for $n = 1, \ldots, N$.

BT8: *Mean Value Test*:
  $\min_n \{p_n^2/p_n^1\} \leqslant P(p^1, p^2, x^1, x^2) \leqslant \max_n \{p_n^2/p_n^1\}$.

BT9: *Circularity*:
  $P(p^1, p^2, x^1, x^2) P(p^2, p^3, x^2, x^3) = P(p^1, p^3, x^1, x^3)$.

Tests BT1 and BT3 may be found in Vartia (1985) who calls BTE the *Strong Monetary Unit Test*. Test BT2 may be found in Walsh (1901), tests BT2 and BT4 are in Fisher (1911), tests BT5 and BT6 are in Fisher (1922), tests BT8 and a stronger version of BT9 are in Eichhorn and Voeller (1976) and test BT9 may be traced back to Westergaard (1890).

BT1 may be interpreted as follows: if prices and quantities are all equal in the two periods (or for the two regions under consideration), then the price index should be unity. This equality should still hold even if all quantities in period 1 are multiplied by the same number α and all quantities in period 2 are multiplied by the same β.

BT2 means if all period 2 prices are multiplied by α, then the new price index should equal α times the old price index.

Tests BT3–BT6 are invariance or symmetry tests. BT3 says that the price index should remain unchanged when each price in both periods is multiplied by the same number α and when quantities in period 1(2) are all multiplied by $\alpha(\beta)$. BT4 says that the index should remain unchanged if each good is measured in different units. BT5 says that if we interchange the role of periods 1 and 2 in our price index, then the new price index should equal the reciprocal of the original index. BT6 says that the index number formula should treat all commodities in an evenhanded way: no commodity can be singled out to play an asymmetric role. For example, suppose $P(p^1, p^2, x^1, x^2) \equiv p_1^2/p_1^1$. Then for $N \geq 2$, this formula, which equals the price ratio for commodity 1 only, fails BT6.

BT7 says that if period 2 prices increase in any manner, then the price index cannot decrease.

BT8 says that the price index should lie between the smallest and largest price ratio over all commodities.

BT9 is a transitivity test which looks beyond the case of only 2 periods of countries. BT9 says that if we have price and quantity data for 3 time periods, then the product of the price index going from period 1 to period 2 times the price index going from period 2 to 3 should equal the price index going from period 1 to 3 directly.

All of the above tests seem to be reasonable and desirable.

If $N = 1$ so that there is only one commodity, then BT1 and BT2 imply that $P(p^1, p^2, x^1, x^2)$ must equal p_1^2/p_1^1 and of course, this index formula will satisfy all of the remaining tests.

In the general N commodity case (assuming that all prices and quantities are positive), which tests are satisfied by the index number formulae defined in the previous section?

It can be shown that the Dutot index (2) satisfies all tests except BT4 (but this is a fatal flaw), the Carli index (4) fails only BT5 and BT9, the Jevons geometric index (5) satisfies all tests, the Laspeyres and Paasche indexes defined by (7) and (8) fail BT5 and BT9, the Fisher and Walsh indexes (9) and (10) fail only BT9, and the Törnqvist index fails BT7 and BT9. Thus from the viewpoint of the test approach, it would appear that the geometric index is best.

The above conclusion is not warranted since our list of desirable tests is incomplete. We shall consider an additional two tests where the geometric index receives a failing grade.

Consider an index number formula that utilizes positive price and quantity information for N commodities, $P^N(p^1, p^2, x^1, x^2)$ say. Now consider the same functional form that uses information on only the first $N - 1$ commodities, P^{N-1} say. Then we may want the index number formula to satisfy the following property:

$$\lim_{x_N^1 \to 0, \, x_N^2 \to 0} P^N(p^1, p^2, x^1, x^2)$$
$$= P^{N-1}(p_1^1, \ldots, p_{N-1}^1, \, p_1^2, \ldots, p_{N-1}^2,$$
$$x_1^1, \ldots, x_{N-1}^1, \, x_1^2, \ldots, x_{N-1}^2). \qquad (12)$$

Thus as the quantity of commodity N tends to 0 in both periods (and thus commodity N becomes irrelevant), the N commodity price index tends to the $N - 1$ commodity price index which has deleted the prices and quantities of good N

from the formula. This might be called the *irrelevance of tiny commodities test*, test BT10. Obviously, this test only makes sense if the basic index number formula p satisfies BT6, so that all commodities are treated symmetrically. The geometric price index (5) fails test BT10 as do the other unweighted formulae (2) and (4). However, the quantity weighted price indexes, (7)–(11), all pass this test. Thus from the viewpoint of passing tests, the Fisher and Walsh indexes, (9) and (10), now look just as good as the geometric index (5).

There is another reason for not preferring the geometric price index. Recall our basic aggregation problem (1). It is clear that we can interchange the role of prices and quantities in the two periods and define a *quantity index* $Q(p^1, p^2, x^1, x^2)$ in much the same way that we defined the price index $P(p^1, p^2, x^1, x^2)$. We set $P^2/P^1 = P(p^1, p^2, x^1, x^2)$ and we may set $X^2/X^1 = Q(p^1, p^2, x^1, x^2)$. From (1), we deduce that the product of the price and quantity indexes should equal the value ratio for the two periods; i.e., P and Q should satisfy the following *product test* due to Frisch (1930):

$$P(p^1, p^2, x^1, x^2) Q(p^1, p^2, x^1, x^2) = p^2 \cdot x^2 / p^1 \cdot x^1. \qquad (13)$$

However, rather than defining Q independently of P, (13) may be rearranged to yield a *definition* of Q in terms of P. The resulting quantity index is called the *implicit quantity index* that corresponds to P. If we define the implicit quantity index, Q_G say, which corresponds to the geometric price index (5) and consider the quantity counterparts to Tests BT1–BT9 above (BT10 does not have a sensible quantity counterpart), we find that Q_G fails the quantity counterpart to BT8, the mean value test. Hence we have another reason for failing the geometric price index and its corresponding implicit quantity index. On the other hand, the implicit Fisher quantity index satisfies tests BT1–BT8 adapted for the quantity context.

The implicit Fisher quantity index Q_F is:

$$Q_F(p^1, p^2, x^1, x^2)$$
$$\equiv p^2 \cdot x^2 / p^1 \cdot x^1 P_F(p^1, p^2, x^1, x^2)$$
$$= (p^1 \cdot x^2 p^2 \cdot x^2 / p^1 \cdot x^1 p^2 \cdot x^1)^{1/2} \quad \text{using (9)}$$
$$= P_F(x^1, x^2, p^1, p^2). \qquad (14)$$

Thus Q_F has the same functional form as P_F except that the role of prices and quantities has been interchanged. We have shown that P_F and Q_F satisfy Fisher's (1922) *factor reversal test* which may be stated as follows:

$$P(p^1, p^2, x^1, x^2) P(x^1, x^2, p^1, p^2) = p^2 \cdot x^2 / p^1 \cdot x^1. \qquad (15)$$

Of the price indexes defined in the previous section, only the Fisher price index satisfies (15). However, the Fisher price index is by no means the only index number formula that satisfies the factor reversal test. Walsh (1921) showed how to generate hundreds of formulae that would satisfy the test: take a price index $P(p^1, p^2, x^1, x^2)$, and define its factor antithesis by $p^2 \cdot x^2 / p^1 \cdot x^1 P(x^1, x^2, p^1, p^2)$. Define a new price index by taking the square root of the product of the original index and its factor antithesis. This new index will automatically satisfy the factor reversal test.

The consistency and independence of various bilateral index number tests was studied in some detail by Eichhorn and Voeller (1976). Our conclusion at this point echoes that of Frisch (1936): the test approach to index number theory, while extremely useful, does not lead to a single unique index number formula. Thus we turn to economic approaches to index number theory to see if we are led to a more definite conclusion.

3. MICROECONOMIC APPROACHES TO PRICE INDEXES

Before a definition of a microeconomic price index is presented, it is necessary to make a few preliminary definitions.

Let $F(x)$ be a function of N variables, $x \equiv (x_1, \ldots, x_N)$. In the consumer context, F represents a consumer's preferences; i.e., if $F(x^2) > F(x^1)$, then the consumer prefers the commodity vector x^2 over x^1. In this context, F is called a *utility function*. In the producer context, $F(x)$ might represent the output that could be produced using the input vector x. In this context, F is called a *production function*. In order to cover both contexts, we follow the example of Diewert (1976) and call F an *aggregator function*.

Suppose the consumer or producer faces prices $p \equiv (p_1, \ldots, p_N)$ for the N commodities. Then the economic agent will generally find its useful to minimize the cost of achieving at least a given utility or output level u; we define the *cost function* or *expenditure function* C as the solution to this minimization problem:

$$C(u, p) \equiv \underset{x}{\text{minimum}} \{p \cdot x : F(x) \geqslant u\} \qquad (16)$$

where $p \cdot x \equiv \Sigma_{n=1}^{N} p_n x_n$ is the inner product of the price vector p and quantity vector x.

Note that the cost function depends on $1 + N$ variables: the utility or output level u and the N commodity prices in the vector p. Moreover, the functional form for the aggregator function F completely determines the functional form for C.

We say that an aggregator function is *neoclassical* if F is: (i) continuous, (ii) positive; i.e. $F(x) > 0$ if $x \gg 0_N$ (which means each component of x is positive), and (iii) linearly homogeneous; i.e., $F(\lambda x) = \lambda F(x)$ if $\lambda > 0$. If F is neoclassical, then the corresponding cost function $C(u, p)$ equals u times the unit cost function, $c(p) \equiv C(1, p)$, where $c(p)$ is the minimum cost of producing one unit of utility or output; i.e.,

$$C(u, p) = uC(1, p) \equiv uc(p). \qquad (17)$$

Shephard (1953) formally defined an aggregator function F to be *homothetic* if there exists an increasing continuous function of one variable g such that $g[F(x)]$ is neoclassical. However, the concept of homotheticity was well known to Frisch (1936) who termed it expenditure proportionality. If F is homothetic, then its cost function C has the following decomposition:

$$C(u, p) \equiv \min_x \{p \cdot x : F(x) \geqslant u\}$$
$$= \min_x \{p \cdot x : g[F(x)] \geqslant g(u)\}$$
$$= g(u)c(p) \qquad (18)$$

where $c(p)$ is the unit cost function that corresponds to $g[F(x)]$.

Let $p^1 \gg 0_N$ and $p^2 \gg 0_N$ be positive price vectors pertaining to periods or observations 1 and 2. Let $x > 0_N$ be a non-negative, non-zero reference quantity vector. Then the Konüs (1924) *price index* or *cost of living index* is defined as:

$$P_K(p^1, p^2, x) \equiv C(F(x), p^2) / C(F(x^1), p^1). \qquad (19)$$

In the consumer (producer) context, P_K may be interpreted as follows. Pick a reference utility (output) level $u \equiv F(x)$. Then $P_K(p^1, p^2, x)$ is the minimum cost of achieving the utility (output) level u when the economic agent faces prices p^2 relative to the minimum cost of achieving the same u when the agent faces prices p^1. If $N = 1$ so that there is only one consumer good (or input), then it is easy to show that $P_K(p_1^1, p_1^2, x_1) = p_1^2 x_1 / p_1^1 x_1 = p_1^2 / p_1^1$.

Using the fact that a cost function is linearly homogeneous in its price arguments, it can be shown that P_K has the following homogeneity property; $P_K(p^1, \lambda p^2, x) = \lambda P_K(p^1, p^2, x)$ for $\lambda > 0$ which is analogous to the proportionality test BT2 in the previous section. P_K also satisfies $P_K(p^2, p^1, x) = 1 / P_K(p^1, p^2, x)$ which is analogous to the time reversal test, BT5.

Note that the functional form for P_K is completely determined by the functional form for the aggregator function F which determines the functional form for the cost function C.

In general, P_K depends not only on the two price vectors p^1 and p^2, but also on the reference vector x. Malmquist (1953), Pollak (1971) and Samuelson and Swamy (1974) have shown that P_K is independent of x and is equal to a ratio of unit cost functions, $c(p^2)/c(p^1)$, if and only if the aggregator function F is homothetic.

If we knew the consumer's preferences or the producer's technology, then we would know F and we could construct the cost function C and the Konüs price index P_K. However, we generally do not know F or C and thus it is useful to develop *bounds* that depend on observable price and quantity data but do not depend on the specific functional form for F or C.

Samuelson (1947) and Pollak (1971) established the following bounds on P_K. Let $p^1 \gg 0_N$ and $p^2 \gg 0_N$. Then for every reference quantity vector $x > 0_N$, we have:

$$\min_n \{p_n^2 / p_n^1\} \leqslant P_K(p^1, p^2, x) \leqslant \max_n \{p_n^2 / p_n^1\}; \qquad (20)$$

i.e., P_K lies between the smallest and largest price ratios. Unfortunately, these bounds are usually too wide to be of much practical use.

To obtain closer bounds, we now assume that the observed quantity vectors for the two periods, $x^i \equiv (x_1^i, \ldots, x_N^i)$, $i = 1, 2$, are solutions to the producer's or consumer's cost minimization problems; i.e., we assume:

$$p^i \cdot x^i = C(F(x^i), p^i), \qquad p^i \gg 0_N, x^i > 0_N, \qquad i = 1, 2. \quad (21)$$

Given the above assumptions, we now have two natural choices for the reference quantity vector x that occurs in the definition of $P_K(p^1, p^2, x)$: x^1 or x^2. The *Laspeyres–Konüs price index* is defined as $P_K(p^1, p^2, x^1)$ and the *Paasche–Konüs price index* is defined as $P_K(p^1, p^2, x^2)$.

Under the assumption of cost minimizing behaviour (21), Konüs (1924) established the following bounds:

$$P_K(p^1, p^2, x^1) \leqslant p^2 \cdot x^1 / p^1 \cdot x^1 \equiv P_L(p^1, p^2, x^1, x^2); \qquad (22)$$

$$P_K(p^1, p^2, x^2) \geqslant p^2 \cdot x^2 / p^1 \cdot x^2 \equiv P_P(p^1, p^2, x^1, x^2) \qquad (23)$$

where P_L and P_P are the Laspeyres and Paasche price indexes defined earlier by (7) and (8). If in addition, the aggregator function is homothetic, then Frisch (1936) showed that for any reference vector $x > 0_N$,

$$P_P \equiv p^2 \cdot x^2 / p^1 \cdot x^2 \leqslant P_K(p^1, p^2, x) \leqslant p^2 \cdot x^1 / p^1 \cdot x^1 \equiv P_L. \quad (24)$$

In the consumer context, it is unlikely that preferences will be homothetic; hence the bounds (24) cannot be justified in general. However, Konüs (1924) showed that bounds similar to (24) would hold even in the general nonhomothetic case, provided that we choose a reference vector $x \equiv \lambda x^1 + (1 - \lambda)x^2$ which is a $\lambda, (1 - \lambda)$ weighted average of the two observed quantity points. Specifically, Konüs showed that there exists a λ between 0 and 1 such that if $P_P \leqslant P_L$, then

$$P_P \leqslant P_K[p^1, p^2, \lambda x^1 + (1 - \lambda)x^2] \leqslant P_L \qquad (25)$$

or if $P_P > P_L$, then

$$P_L \leqslant P_K[p^1, p^2, \lambda x^1 + (1 - \lambda)x^2] \leqslant P_P. \quad (26)$$

The bounds on the microeconomic price index P_K given by (20) and (22)–(26) are the best bounds that we can obtain without making further assumptions on F. In the time series context, the bounds given by (25) or (26) are quite satisfactory: the Paasche and Laspeyres price indexes for consecutive time periods will usually differ by less than 1 per cent. However, in the cross-section context where the observations represent, for example, production data for two producers in the same industry but in different regions, the bounds are often not very useful since P_L and P_P can differ by 50 per cent or more in the cross-sectional context; see Ruggles (1967).

In section 5 below, we will make additional assumptions on the aggregator function F or its cost function dual C that will enable us to determine P_K exactly. Before we do this, in the next section, we will define various quantity indexes that have their origins in microeconomic theory.

4. MICROECONOMIC APPROACHES TO QUANTITY INDEXES

In the one commodity case, a natural definition for a quantity index is x_1^2/x_1^1, the ratio of the single quantity in period 2 to the corresponding quantity in period 1. This ratio is also equal to the expenditure ratio, $p_1^2 x_1^2/p_1^1 x_1^1$, divided by the price ratio p_1^2/p_1^1. This suggests that in the N commodity case, a reasonable definition for a quantity index would be the expenditure ratio divided by the Konüs price index, P_K. This course of action was suggested by Pollak (1971). Thus we define the *Konüs–Pollak quantity index*, Q_K, by:

$$Q_K(p^1, p^2, x^1, x^2, x)$$

$$\equiv p^2 \cdot x^2/p^1 \cdot x^1 P_K(p^1, p^2, x)$$

$$\equiv p^2 \cdot x^2/p^1 \cdot x^1 P_K(p^1, p^2, x)$$

$$= [C(F(x^2), p^2)/C(F(x), p^2)]/[C(F(x^1), p^1)/C(F(x), p^1)]$$

$$(27)$$

where the second line follows from the definition of P_K, (19), and the assumption of cost minimizing behaviour in the two periods, (20).

The definition of Q_K depends on the reference vector x which appears in the definition of P_K. The general definition of Q_K simplifies considerably if we choose x to be x^1 or x^2. Thus define the *Laspeyres–Konüs quantity index* as

$$Q_K(p^1, p^2, x^1, x^2, x^1) \equiv C[F(x^2), p^2]/C[F(x^1), p^2] \quad (28)$$

and the *Paasche–Konüs quantity index* as

$$Q_K(p^1, p^2, x^1, x^2, x^2) \equiv C[F(x^2), p^1]/C[F(x^1), p^1]. \quad (29)$$

It turns out that the indexes defined by (28) and (29) are special cases of another class of quantity indexes. For any reference price vector $p \gg 0_N$, define the *Allen (1949) quantity index* by

$$Q_A(x^1, x^2, p) \equiv C[F(x^2), p]/C[F(x^1), p]. \quad (30)$$

If p is chosen to be p^1, (30) becomes (29) and if $p = p^2$, then (30) becomes (28).

Using the properties of cost functions, it can be shown that if $F(x^2) \geqslant F(x^1)$, then $Q_A(x^1, x^2, p) \geqslant 1$ while if $F(x^2) \leqslant F(x^1)$, then $Q_A(x^1, x^2, p) \leqslant 1$. Thus the Allen quantity index correctly indicates whether the commodity vector x^2 is larger or smaller than x^1. It can also be seen that Q_A satisfies a counterpart to the time reversal test; i.e., $Q_A(x^2, x^1, p) = 1/Q_A(x^1, x^2, p)$.

Just as the price index P_K depended on the unobservable aggregator function, so also do the quantity indexes Q_K and Q_A.

Thus it is useul to develop bounds for the quantity indexes that do not depend on the particular function form for F.

Samuelson (1947) and Allen (1949) established the following bounds for (28) and (29):

$$Q_A(x^1, x^2, p^1)$$

$$= Q_K(p^1, p^2, x^1, x^2, x^2) \leqslant p^1 \cdot x^2/p^1 \cdot x^1 \equiv Q_L; \quad (31)$$

$$Q_A(x^1, x^2, p^2)$$

$$= Q_K(p^1, p^2, x^1, x^2, x^1) \geqslant p^2 \cdot x^2/p^2 \cdot x^1 \equiv Q_P. \quad (32)$$

Note that the observable *Laspeyres* and *Paasche quantity indexes*, Q_L and Q_P, appear on the right-hand sides of (31) and (32).

Diewert (1981), utilizing some results of Pollak (1971) and Samuelson and Swamy (1974), established the following results: if the underlying aggregator function F is neoclassical, then for all $p \gg 0_N$ and $x \gg 0_N$,

$$Q_P \leqslant Q_A(x^1, x^2, p)$$

$$= Q_K(p^1, p^2, x^1, x^2, x) = F(x^2)/F(x^1) \leqslant Q_L. \quad (33)$$

Thus if the aggregator function F is neoclassical, then the Allen quantity index for all reference vectors p equals the Konüs quantity index for all reference quantity vectors x which in turn equals the ratio of aggregates $(F(x^2)/F(x^1))$. Moreover, Q_A and Q_K are bounded from below by the Paasche quantity index Q_P and bounded from above by the Laspeyres quantity index Q_L in the neoclassical case.

In the general nonhomothetic case, Diewert (1981) showed that there exists a λ between 0 and 1 such that $Q_K[p^1, p^2, x^1, x^2, \lambda x^1 + (1 - \lambda)x^2]$ lies between Q_P and Q_L and there exists a λ^* between 0 and 1 such that $Q_A[x^1, x^2, \lambda^* p^1 + (1 - \lambda^*)p^2]$ also lies between Q_P and Q_L. Thus the observable Paasche and Laspeyres quantity indexes bound both the Konüs quantity index and the Allen quantity index, provided that we choose appropriate reference vectors between x^1 and x^2 and p^1 and p^2 respectively.

Using the linear homogeneity property of the cost function in its price arguments, we can show that the Konüs quantity index has the desirable homogeneity property, $P_K(p^1, \lambda p^1, x) = \lambda$ for all $\lambda > 0$; i.e., if period 2 prices are proportional to period 1 prices, then P_K equals this common proportionality factor. It would be desirable for an analogous homogeneity property to hold for quantity indexes. Unfortunately, it is not in general true that $Q_K(x^1, \lambda x^1, p^1, p^2, x) = \lambda$ or that $Q_A(x^1, \lambda x^1, p) = \lambda$. Thus we turn to a third microeconomic approach to defining a quantity index which does have the desirable quantity proportionality property.

Let x^1 and x^2 be the observable quantity vectors in the two situations as usual, let $F(x)$ be an increasing, continuous aggregator function, and let $x \gg 0$ be a reference quantity vector. Then the *Malmquist (1953) quantity index* Q_M is defined as:

$$Q_M(x^1, x^2, x) \equiv D[F(x), x^2]/D[F(x), x^1] \quad (34)$$

where $D(u, x^i) \equiv \max_k \{k : F(x^i/k) \geqslant u, k > 0\}$ is the *deflation* or *distance function* which corresponds to F. Thus $D[F(x), x^2]$ is the biggest number which will just deflate the quantity vector x^2 onto the boundary of the utility (or production) possibilities set $\{z : F(z) \geqslant F(x)\}$ indexed by the reference quantity vector x while $D[F(x), x^1]$ is the biggest number which will just deflate the quantity vector x^1 onto the set $\{z : F(z) \geqslant F(x)\}$ and Q_M is the ratio of these two deflation factors.

Q_M depends on the unobservable aggregator function F and as usual, we are interested in bounds for Q_M.

Diewert (1981) showed that Q_M satisfied bounds analogous to (20); i.e.,

$$\min_n \{x_n^2/x_n^1\} \leqslant Q_M(x^1, x^2, x) \leqslant \max_n \{x_n^2/x_n^1\}. \tag{35}$$

It should be noted that we do not require the assumption of cost minimizing behaviour in order to define the Malmquist quantity index or to establish the bounds (35). However, in order to establish the following bounds due to Malmquist (1953) for Q_M, we need the assumption of cost minimizing behaviour (21) and we require the reference vector x to be x^1 or x^2:

$$Q_M(x^1, x^2, x^1) \leqslant p^1 \cdot x^2/p^1 \cdot x^1 \equiv Q_L; \tag{36}$$

$$Q_M(x^1, x^2, x^2) \geqslant p^2 \cdot x^2/p^2 \cdot x^1 \equiv Q_P. \tag{37}$$

Diewert (1981) showed that under the hypothesis of cost minimizing behaviour, there exists a λ between 0 and 1 such that $Q_M[x^1, x^2, \lambda x^1 + (1 - \lambda)x^2]$ lies between Q_P and Q_L. Thus the Paasche and Laspeyres quantity indexes provide bounds for a Malmquist quantity index for some reference indifference or product surface indexed by a quantity vector which is a λ, $(1 - \lambda)$ weighted average of the two observable quantity vectors, x^1 and x^2.

Pollak (1971) showed that if F is neoclassical, then we can extend the string of equalities in (33) to include the Malmquist quantity index $Q_M(x^1, x^2, x)$, for any reference quantity vector x. Thus in the case of a linearly homogeneous aggregator function, all three theoretical quantity indexes coincide and this common theoretical index is bound from below by the Paasche quantity index Q_P and bounded from above by the Laspeyres quantity index Q_L.

In the general case of a nonhomothetic aggregator function, our best theoretical quantity index, the Malmquist index, is also bounded by the Paasche and Laspeyres indexes, provided that we choose a suitable reference quantity vector.

We noted in the price index context that the Paasche and Laspeyres price indexes were usually quite close in the time series context. A similar remark also applies to the Paasche and Laspeyres quantity indexes. Thus taking an average of the Paasche and Laspeyres indexes, such as the Fisher price and quantity indexes, will generally approximate underlying microeconomic price and quantity indexes sufficiently accurately for most practical purposes. However, this observation does not apply to the cross-sectional context, where the Paasche and Laspeyres indexes can differ widely. In the following section, we offer another microeconomic justification for using the Fisher indexes that also applies in the context of making interregional and cross-country comparisons.

5. EXACT AND SUPERLATIVE INDEXES

Assume that the producer or consumer is maximizing a neoclassical aggregator function f subject to a budget constraint during the two periods. Under these conditions, it can be shown that the economic agent is also minimizing cost subject to a utility or output constraint. Moreover, the cost function C that corresponds to f can be written as $C[f(x), p] = f(x)c(p)$ where c is the unit cost function (recall (17) above).

Suppose a price index $P(p^1, p^2, x^1, x^2)$ and a quantity index $Q(p^1, p^2, x^1, x^2)$ of the type considered in sections 1 and 2 are given. The quantity index Q is defined to be *exact* for a neoclassical aggregator function f with unit cost dual c, if for every $p^1 \gg 0_N, p^2 \gg 0_N$ and $x^1 \gg 0_N$ a solution to the aggregator maximization problem $\max_x \{f(x): p^i \cdot x \leqslant p^i \cdot x^i\} = f(x^i) > 0$ for $i = 1, 2$, we have

$$Q(p^1, p^2, x^1, x^2) = f(x^2)/f(x^1). \tag{38}$$

Under the same hypothesis, the price index P is *exact* for f and c if we have

$$P(p^1, p^2, x^1, x^2) = c(p^2)/c(p^1). \tag{39}$$

In (38) and (39), the price and quantity vectors are not regarded as being independent. The p^i can be independent, but the x^i are solutions to the corresponding aggregator maximization problem involving p^i, for $i = 1, 2$. Note that if Q is exact for a neoclassical f, then Q can be interpreted as a Konüs, Allen or Malmquist quantity index and the corresponding P defined implicitly by (13) can be interpreted as a Konüs price index.

The concept of exactness is due to Konüs and Byushgens (1926). Below, we shall give some examples of exact index number formulae. Additional examples may be found in Afriat (1972), Pollak (1971) and Samuelson and Swamy (1974).

Konüs and Byushgens (1926) showed that Irving Fisher's ideal quantity index Q_F defined by (14) and the corresponding price index P_F defined by (9) are exact for the homogeneous quadratic aggregator function f defined by

$$f(x_1, \ldots, x_N) \equiv \left(\sum_{n=1}^{N} \sum_{m=1}^{N} a_{nm} x_n x_m\right)^{1/2} \equiv (x^T A x)^{1/2} \tag{40}$$

where $A \equiv [a_{nm}]$ is a symmetric N by N matrix of constants. Thus under the assumption of maximizing behaviour, we can calculate $f(x^2)/f(x^1) = Q_F$ and $c(p^2)/c(p^1) = P_F$ where f is defined by (40) and c is the unit cost function that corresponds to f. The important thing to note is that f depends on $N(N + 1)/2$ unknown a_{nm} parameters but we do not need to know these parameters in order to evaluate $f(x^2)/f(x^1)$ and $c(p^2)/c(p^1)$.

Diewert (1976) showed that the Törnqvist price index P_T defined by (11) is exact for the unit cost function $c(p)$ defined by:

$$\ln c(p) \equiv \alpha_0 + \sum_{n=1}^{N} \alpha_n \ln p_n + (1/2) \sum_{m=1}^{N} \sum_{n=1}^{N} \alpha_{mn} \ln p_m \ln p_n \tag{41}$$

where the parameters α_n and α_{mn} satisfy the following restrictions:

$$\sum_{n=1}^{N} \alpha_n = 1, \sum_{n=1}^{N} \alpha_{mn} = 0 \quad \text{for} \quad m = 1, \ldots, N \quad \text{and}$$

$$\alpha_{mn} = \alpha_{nm} \quad \text{for all} \quad m, n. \tag{42}$$

Thus we may calculate $c(p^2)/c(p^1) = P_T$ and $f(x^2)/f(x^1) = p^2 \cdot x^2/p^1 \cdot x^1 P_T \equiv \tilde{Q}_T$ where c is the unit cost function defined by (41), f is the aggregator function which corresponds to this c, and $\tilde{Q}_T$ is the implicit Törnqvist quantity index. Note that we do not have to know the parameters α_n and α_{mn} in order to evaluate $c(p^2)/c(p^1)$ and $f(x^2)/f(x^1)$.

The unit cost function defined by (41) is the *translog* unit cost function defined by Christensen, Jorgenson and Lau (1971). Since P_T is exact for this translog functional form, P_T is sometimes called the *translog price index*.

Before we present our final example of an exact index number formula, we need to define a family of quantity indexes Q_r that depend on a number, $r \neq 0$:

$$Q_r(p^1, p^2, x^1, x^2) \equiv \left[\sum_{n=1}^{N} s_n^1 (x_n^2/x_n^1)^{r/2}\right]^{1/r}$$

$$\times \left[\sum_{m=1}^{N} s_m^2 (x_m^2/x_m^1)^{-r/2}\right]^{-1/r} \tag{43}$$

where $s_n^i \equiv p_n^i x_n^i/p^i \cdot x^i$ is the period i expenditure share for good n. For each $r \neq 0$, define the corresponding implicit price index by:

$$\tilde{P}_r(p^1, p^2, x^1, x^2) \equiv p^2 \cdot x^2/p^1 \cdot x^1 Q_r(p^1, p^2, x^1, x^2). \tag{44}$$

A bit of algebra will show that when $r = 2$, $\tilde{P}_2 = P_F$, the Fisher price index defined by (9) and when $r = 1$, $\tilde{P}_1 = P_W$, the Walsh price index defined by (10).

Diewert (1976) showed that Q_r and $\tilde{P}_r$ are exact for the *quadratic mean of order r aggregator function* f_r defined by:

$$f_r(x_1, \ldots, x_N) \equiv \left(\sum_{m=1}^{N} \sum_{n=1}^{N} a_{mn} x_m^{r/2} x_n^{r/2} \right)^{(1/r)} \quad (45)$$

where $A \equiv [a_{mn}]$ is a symmetric matrix of constants. Thus the Walsh price index P_W is exact for $f_1(x)$ defined by (45) when $r = 1$.

Diewert (1974) defined a linearly homogeneous function of N variables f to be *flexible* if it could provide a second order approximation to an arbitrary twice continuously differentiable linearly homogeneous function. It can be shown that f defined by (40), c defined by (41) and (42) and f_r defined by (45) for each $r \neq 0$ are all examples of flexible functional forms.

Let the price and quantity indexes P and Q satisfy the product test equality, (13). Then Diewert (1976) defined P and Q to be *superlative indexes* if either P is exact for a flexible unit cost function c or Q is exact for a flexible aggregator function f. Thus P_F, P_W, P_T and $\tilde{P}_r$ are all superlative price indexes.

At this point, it may seem that we are in the same position that we were at the end of section 2 where we could not find an index number formula that satisfied all reasonable tests; hence we could not single out any formula as being *the* best. In the present context, we have a similar problem: how are we to discriminate between P_F, P_W and P_T? Fortunately, it does not matter very much which of these formulae we choose to use in applications: they will all give the same answer to a reasonably high degree of approximation. Diewert (1978) showed that all known superlative index number formulae approximate each other to the second order when each index is evaluated at an equal price and quantity point. This means the P_F, P_W, P_T and each $\tilde{P}_r$ have the same first and second order partial derivatives with respect to all $4N$ arguments when the derivatives are evaluated at a point where $p^1 = p^2$ and $x^1 = x^2$. A similar string of equalities also holds for the corresponding implicit quantity indexes defined using the product test (13). Empirically, it has been found that superlative index number formulae typically approximate each other to something less than 0.2 per cent in the time series context and to about 2 per cent in the cross-section context; see Fisher (1922) and Ruggles (1967).

Diewert (1978) also showed that the Paasche and Laspeyres indexes approximate the superlative indexes to the first order at an equal price and quantity point. In the time series context, for adjacent periods, the Paasche and Laspeyres price indexes typically differ by less than 0.5 per cent; hence these indexes may also provide acceptable approximations to a superlative index.

Having considered the case of two observations at great length, we now turn our attention to the I observation case. In the next section, we consider the case of I consecutive time series observations of prices and quantities on the same producer or consumer. We shall consider the general case of I observations separated by space (and possibly by time) in section 9 below.

6. THE FIXED BASE VERSUS THE CHAIN PRINCIPLE

Consider the case of I consecutive observations through time on the prices and quantities of N goods utilized by an economic unit, $p^i \equiv (p_1^i, \ldots, p_N^i)$ and $x^i \equiv (x_1^i, \ldots, x_N^i)$, $i = 1, \ldots, I$.

Suppose that we have decided on a price index P and a quantity index Q where P and Q satisfy (13). This allows us to compare prices and quantities for any two observations. Based on the test approach, we would probably choose P and Q to be

the Fisher indexes, P_F and Q_F, since they satisfy the most tests. Since the Fisher indexes also lie between the Paasche and Laspeyres bounds derived in sections 3 and 4 and moreover are superlative indexes, the choice of the Fisher indexes also fits in well with the microeconomic approach to index number theory.

However, if we choose any superlative price index P in order to make bilateral comparisons of prices, we are faced with a problem when $I \geqslant 3$: the circular test BT9 will not be satisfied. Thus choose observation 1 as the base and make all price comparisons relative to period 1, so that the relative price level in period i is $P(p^1, p^i, x^1, x^i)$. Now choose another observation, say I, as the base so that the relative price level in period i relative to I is $P(p^I, p^i, x^I, x^i)$. In order to make the price level in period 1 equal to unit, divide each $P(p^I, p^i, x^I, x^i)$ by $P(p^I, p^1, x^I, x^1)$. If the circular test held, then the two price series would coincide; i.e., we would have

$$P(p^1, p^i, x^1, x^i)$$
$$= P(p^I, p^i, x^I, x^i)/P(p^I, p^1, x^I, x^1), \quad i = 1, \ldots, I. \quad (46)$$

However, since the circular test does not hold in general, we are faced with a problem: which period should we choose as the base?

There are a number of alternative strategies that could be followed, including: (i) choose the first observation as the base, (ii) take an average over all possible choices of the base period, (iii) abandon the use of a bilateral formula and develop an entirely new multilateral approach, or (iv) use the *chain principle*, which will be explained below.

Alternative (i) seems rather arbitrary but it has simplicity to recommend it. Virtually all official time series indexes are constructed using fixed base Paasche or Laspeyres indexes with the base year being changed every 5 to 15 years.

Alternative (ii) seems attractive at first glance since it treats each period in a symmetric fashion. The problem with the method is that economic history has to be rewritten every time a new observation is added to the initial I observations.

Alternative (iii) may also seem attractive. For example, in making a price comparison between i and j, we may want to utilize the quantity information for all periods, so that the bilateral index number formula $P(p^i, p^j, x^i, x^j)$ could be replaced by $P^*(p^i, p^j, x^1, \ldots, x^I)$. For example, we could use the Scrope index (6) where x could be taken to be the average quantities over all periods, $\Sigma_{i=1}^I x^i/I$, or we could use the Törnqvist index (11) where s_n could be set equal to the average over all periods commodity n expenditure share, $\Sigma_{i=1}^I (1/I) p_n^i x_n^i / p^i \cdot x^i$. Both of these examples lead to indexes that satisfy the circularity property for the original I observations. However, as was the case with alternative (ii), these new multilateral indexes would have to be recomputed as new time series observations become available.

Alternative (iv) is to use the *chain principle*, originally suggested by Marshall (1887), although the term is due to Fisher (1911). This principle makes use of the natural order provided by the march of time. One first chooses a bilateral index number formula P. The period 1 price level is set equal to unity and the period 2 price level is set equal to $P(p^1, p^2, x^1, x^2)$. The period 3 price level is set equal to $P(p^1, p^2, x^1, x^2) \times P(p^2, p^3, x^2, x^3)$. The period 4 price level is set equal to the period 3 price level times $P(p^3, p^4, x^3, x^4)$, and so on. Thus the period i price level is not obtained by the direct comparison of period i prices with period 1 prices, $P(p^1, p^i, x^1, x^i)$, but rather as the product of the period by period relative price levels; i.e., by travelling along the links of a chain.

The chain principle has one substantial disadvantage; if $p^i = p^j$ and $x^i = x^j$ and periods i and j are not adjacent, then it is not necessarily the case that the price level in period i will coincide with the price level in period j.

However, the chain principle has a number of advantages: (i) no single period is singled out to play an asymmetric role, (ii) the price levels for I periods are not changed as additional periods are added to the data set, (iii) if a good disappears or a new good is introduced so that N, the number of commodities, changes, then the chain price indexes (and the corresponding implicit quantity indexes) will still be comparable for all periods before and after the change in N and (iv) all superlative indexes will closely approximate each other if the chain principle is used, since changes in prices and quantities tend to be small for adjacent time periods.

The above-mentioned disadvantage of the chain method is due to the lack of circularity in the bilateral formula $P(p^1, p^2, x^1, x^2)$. However, experience has shown (e.g., see Fisher, 1922) that deviations from circularity for superlative index number formulae are small in the time series context. In fact, one can show that if the bilateral index P is superlative (or equal to the Paasche or Laspeyres index), then P satisfies the circular test to the first order; i.e., the first order derivatives of $P(p^1, p^3, x^1, x^3)$ and of $P(p^1, p^2, x^1, x^2) P(p^2, p^3, x^2, x^3)$ with respect to the components of p^1, p^2, p^3, x^1, x^2 and x^3 coincide when evaluated at an equal price and quantity point where $p^1 = p^2 = p^3$ and $x^1 = x^2 = x^3$. (This is a new result.)

Our conclusion at this point is that alternatives (ii) and (iii) do not look very attractive in the time series context: the use of a fixed base or the chain principle seems preferable. However, in the context of cross-section data where there is no natural way of ordering the data sequentially, alternatives (ii) and (iii) become much more attractive. We consider genuine multilateral index number formulae in sections 9 and 10 below.

7. AGGREGATION OVER CONSUMERS

Thus far, our discussion of the microeconomic approach to indexes has been limited to the one consumer or producer case (or the case of two consumers or producers who have identical preferences or technologies). In this section, we relax these restrictions and discuss aggregate consumer or household indexes. In the following section, we discuss aggregate output price and quantity indexes.

Consider the case of two countries or regions (or two time periods for the same country). We suppose that there are H_i households in country i and the preferences of household h in country i are represented by a continuous utility function, $F^{ih}(x, z^{ih})$, where $x \equiv (x_1, \ldots, x_N) \geqslant 0_N$ is a non-negative vector of market goods, $p^i \equiv (p_1^i, \ldots, p_N^i) \gg 0_N$ is the corresponding vector of positive prices that each household in region i faces and z^{ih} is a vector of demographic variables or consumption of public goods by household h in country i.

The *restricted cost or expenditure function* of household h in country i is defined as:

$$C^{ih}(u^{ih}, p^i, z^{ih}) \equiv \min_x \{p^i \cdot x : F^{ih}(x, z^{ih}) \geqslant u^{ih}\},$$
$$i = 1, 2; h = 1, \ldots, H_i, \quad (47)$$

where u^{ih} is a utility or welfare level. We assume that $x^{ih} > 0_N$, the household h in the country i observable consumption vector, solves (47) with $u^{ih} = F^{ih}(x^{ih}, z^{ih})$ for $i = 1, 2$ and $h = 1, \ldots, H_i$.

We may use the preferences and observed choices of household h in country i to define a Konüs price index P^{ih} for the level of prices in country 2 relative to country 1:

$$P^{ih}(p^1, p^2) \equiv C^{ih}(u^{ih}, p^2, z^{ih})/C^{ih}(u^{ih}, p^1, z^{ih}),$$
$$i = 1, 2; h = 1, \ldots, H_i. \quad (48)$$

Suppose that each expenditure function C^{ih} has a translog functional form; i.e., $\ln C^{ih}(u, p, z)$ is a quadratic function in the logarithms of its variables, similar to the right-hand side of (41) except that now (u, p, z) replaces p. We note that we can approximate arbitrary preferences to the second order using this functional form; i.e., the translog restricted expenditure function is a flexible functional form. Caves, Christensen and Diewert (1982b) establish the following result: if household h in country 1 and household k in country 2 have translog expenditure functions with identical coefficients on the second order terms in commodity prices (this forces some similarity in preferences), then the geometric mean of the two theoretical Konüs price indexes, P^{1h} and P^{2k}, equals the observable Törnqvist or translog price index $P_T(p^1, p^2, x^{1h}, x^{2k})$; i.e., we have

$$[P^{1h}(p^1, p^2) P^{2k}(p^1, p^2)]^{1/2} = P_T(p^1, p^2, x^{1h}, x^{2k}),$$
$$h = 1, \ldots, H_1; k = 1, \ldots, H_2. \quad (49)$$

Rather than deal with the individual household indexes p^{ih} defined by (48), one can define an average index, where the average is taken over all households in the country. Thus for a non-negative weights vector $\alpha^i \equiv (\alpha_1^i, \ldots, \alpha_{Hi}^i)$ such that $\Sigma_{h=1}^{Hi} \alpha_h^i$, define the *country i average price index* P^i as an α^i weighted geometric mean of the individual indexes:

$$P^i(p^1, p^2, \alpha^i) \equiv \prod_{h=1}^{Hi} P^{ih}(p^1, p^2)^{\alpha_h^i}, \quad i = 1, 2. \quad (50)$$

The theoretical index P^i defined in (50) utilizes the price vectors p^1 and p^2 in both countries but utilizes only the preferences of households in country i. As a final bit of averaging, we take the geometric mean of P^1 and P^2 to obtain a final theoretical price index that treats each household in each country in a symmetric fashion:

$$P(p^1, p^2, \alpha^1, \alpha^2) \equiv [P^1(p^1, p^2, \alpha^1) P^2(p^1, p^2, \alpha^2)]^{1/2}. \quad (51)$$

The natural choices for the household weighting vectors α^1 and α^2 are: (i) *democratic weights* and (ii) *plutocratic weights*. In case (i), each household in each country is given an equal weight; i.e., $\alpha_h^i \equiv 1/H_i, i = 1, 2$ and $h = 1, \ldots, H_i$. In case (ii), each household gets a weight that is proportional to its share of consumption in its own country; i.e., $\alpha_h^i \equiv p^1 \cdot x^{ih}/p^i \cdot x^i \equiv s_h^i$ for $i = 1, 2$ and $h = 1, \ldots, H_i$ where $x_i \equiv \Sigma_{h=1}^{Hi} x^{ih}$ is country i's aggregate consumption vector. If each household has a translog restricted expenditure function with identical coefficients on the second order terms in commodity prices, then making repeated use of (49), we can deduce that the aggregate price index (51) in case (i) is

$$P(p^1, p^2, 1/H_1, \ldots, 1/H_1, 1/H_2, \ldots, 1/H_2)$$
$$= \prod_{h=1}^{H_1} \prod_{k=1}^{H_2} P_T(p^1, p^2, x^{1h}, x^{2k})^{1/H_1 H_2} \quad (52)$$

while in case (ii), the aggregate price index is

$$P(p^1, p^2, s^1, s^2) = P_T(p^1, p^2, x^1, x^2) \quad (53)$$

where $s^1 \equiv [s_1^1, \ldots, s_{H_1}^1]$ is the country i expenditure share vector for households in country i and $x^i \equiv \Sigma_{h=1}^{H_i} x^{ih}$ is the aggregate country i consumption vector. Thus if individual household consumption data x^{ih} are available, then the *democratic aggregate price index* defined by (52) can be evaluated as a geometric mean of individual household translog prices indexes. If only aggregate country consumption data x^i are available, then the *plutocratic aggregate price index* defined by (53) can be evaluated as a translog price index (recall (11)), using the aggregate consumption vectors x^1 and x^2 as quantity weights.

The terms democratic and plutocratic are due to Prais (1959). For other approaches to aggregate consumer price indexes, see Prais (1959), Pollak (1981), and Diewert (1983a).

We turn now to the construction of an aggregate quantity index for the two periods. Let the preferences of households h in country i be represented by the continuous, increasing utility function $F^{ih}(x)$ where we have absorbed the vector of demographic variables into the function F^{ih}. For $i = 1, 2$ and $h = 1, \ldots, H_i$, define the household h in country i *deflation function* $D^{ih}(u, x)$ for $x > 0_N$ and u in the range of F by

$$D^{ih}(u, x) \equiv \max_{\delta} \{\delta : F^{ih}(x/\delta) \geq u, \delta \geq 0\}. \qquad (54)$$

As in section 5, define the Malmquist quantity index for $x^* > 0_N$ relative to $x > 0_N$ using the preferences of household h in country i by:

$$Q^{ih}(x, x^*, u) \equiv D^{ih}(u, x^*)/D^{ih}(u, x),$$

$$i = 1, 2; h = 1, \ldots, H. \quad (55)$$

Let the observed consumption vector of household h in country i be $x^{ih} > 0_N$ and define the corresponding utility level by $u^{ih} \equiv F^{ih}(x^{ih})$.

Define the *index of average household consumption* in country j relative to i by:

$$Q^{ij} \equiv \sum_{k=1}^{H_j} Q^{jk} \left(\sum_{h=1}^{H_i} x^{ih}/H_i, x^{jk}, u^{jk} \right) \Big/ H; i, j = 1, 2, i \neq \mathrm{j}. \quad (56)$$

To explain the meaning of (56), define the country i average or per capita consumption vector by

$$\bar{x}^i \equiv \sum_{h=1}^{H_i} x^{ih}/H_i, \qquad i = 1, 2. \qquad (57)$$

Then Q^{12} is an average of the individual country 2 Malmquist indexes $Q^{2k}(\bar{x}^1, x^{2k}, u^{2k})$ which is turn compares the observed consumption vector of household k in country 2 with the average consumption vector $\bar{x}^1$ for country 1, using the household k in country 2 indifference surface through x^{2k} as the reference indifference surface. Similarly, Q^{21} is an average of the individual country 1 Malmquist indexes $Q^{1k}(\bar{x}^2, x^{1k}, u^{1k})$.

Diewert (1986) showed that under the assumption of expenditure minimizing behaviour on the part of consumers in both countries, Q^{12} has the lower bound $p^2 \cdot \bar{x}^2/p^2 \cdot \bar{x}^1$, a Paasche quantity index in per capita quantities, and $[Q^{21}]^{-1}$ has the upper bound $p^1 \cdot \bar{x}^2/p^1 \cdot \bar{x}^1$, a Laspeyres quantity index in per capita quantities. Diewert also shows that there exists a $0 \leq \lambda \leq 1$ such that $\lambda Q^{12} + (1 - \lambda)[Q^{21}]^{-1}$ lies between these per capita Paasche and Laspeyres quantity indexes. This suggests that we can approximate the theoretical index $\lambda Q^{12} + (1 - \lambda)[Q^{21}]^{-1}$ by an average of these Paasche and Laspeyres indexes such as the Fisher index $Q_F(p^1, p^2, \bar{x}^1, \bar{x}^2) \equiv [p^2 \cdot \bar{x}^2/p^2 \cdot \bar{x}^1]^{1/2} [p^1 \cdot \bar{x}^2/p^1 \cdot \bar{x}^1]^{1/2}$ where the per capita quantity vector $\bar{x}^i$ are defined by (57).

To summarize this section: we have shown that the translog price index P_T and the Fisher quantity index Q_F, which had very satisfactory economic interpretations in the case of one consumer, also have reasonable economic interpretations in the many consumer case.

8. AGGREGATION OVER PRODUCERS

We assume that there are two regions or countries that are to be compared. The two countries could represent the same country at different time periods. The private production sector in country i uses a vector of primary inputs $v^i \equiv (v_1^i, \ldots, v_M^i)$, where v_M^i is the amount of input m used in country i. These inputs are different types of labour, capital, land and other natural resources. There are N net outputs that can be produced by the private production sector in each country. These goods are different types of consumer and investment goods, exports and imports. The net output vector for country i is $y^i \equiv (y_1^i, \ldots, y_N^i)$ and the corresponding price vector is the positive vector $w^i \equiv (w_1^i, \ldots, w_N^i) \gg 0_N$. We assume $w^i \cdot y^i \equiv \sum_{n=1}^N w_n^i y_n^i > 0$ for $i = 1, 2$. If $y_n^i < 0$, then good n is utilized as an input in country i; this good could be an imported good or it could be an intermediate input that is produced by the government sector in country i. Note that in contrast to the consumer case, we no longer assume that quantity vectors are non-negative.

The technology set for the private production sector in country i is the set $S^i \equiv \{y^i, v^i\}$, a feasible set of net output vectors y^i and primary input vectors v^i. If knowledge is freely transferable across countries, then $S^i = S$ for $i = 1, 2$ so that there is a common technology set across countries. However, we do not require this assumption in what follows.

Define country i's *private national product function* g^i by

$$g^i(w, v) \equiv \max_{y} \{w \cdot y = (y, v) \quad \text{belongs to} \quad S^i\},$$

$$i = 1, 2. \quad (58)$$

The number $g^i(w, v)$ is the maximum value of outputs (less the value of imports) that the private production sector of country i can produce, given that each producer in the country faces the price vector w and the aggregate private economy has at its disposal the primary input vector v. If each producer faces the same price vector w and behaves competitively, then we do not have to concern ourselves with individual producer output vectors: all that matters is the aggregate net output vector. The national product function was introduced into the economics literature by Samuelson (1953); it is sometimes called a variable or restricted profit function or a net revenue function.

The Fisher–Shell (1972) *output price index* of country 2 relative to 1 using the country i technology set S^i and primary input vector v^i is defined as

$$P^i(w^1, w^2) \equiv g^i(w^2, v^i)/g^i(w^1, v^i), \qquad i = 1, 2 \quad (59)$$

We assume optimizing behaviour on the part of producers in each country so that the observed country i price and net output vectors, w^i and y^i, satisfy $w^i \cdot y^i = g^i(w^i, v^i) > 0$ for $i = 1, 2$.

Assume that the private national product function $g^i(w, v)$ has a translog functional form for each i (recall (41) except that now $\ln g^i(w,v)$ is a quadratic form in the logarithms of w_n and v_m) and further assume that the coefficients for the quadratic terms in the logarithms of output prices are the same across the two countries. Then Diewert (1986) showed that the geometric mean of the two theoretical output price indexes defined by (59) is exactly equal to the observable translog price index $P_T(w^1, w^2, y^1, y^2)$, where P_T is defined by (11); i.e.,

$$[P^1(w^1, w^2) P^2(w^1, w^2)]^{1/2} = P_T(w^1, w^2, y^1, y^2). \quad (60)$$

Thus the translog price index P_T again turns out to have a strong microeconomic justification.

The Malmquist quantity index has been applied to the problem of constructing output indexes by Caves, Christensen and Diewert (1982b), and Diewert (1986). It is first necessary to define the country i output deflation function d^i: for a primary input vector v and a net output vector y, define

$$d^i(y, v) \equiv \min_{\delta} \{\delta : (y/\delta, v) \quad \text{belongs to}$$

$$S^i, \delta^i \geq 0\}, \qquad i = 1, 2 \quad (61)$$

where S^i is the technology set for country i. Thus $d^i(y, v)$ denotes the amount the net output vector y must be deflated so that the deflated output vector and the reference input vector v are just on the frontier of the country i production possibilities set S^i.

The Malmquist output index of country 2 relative to country 1 using the country i technology and primary input vector v^i is:

$$Q^i(y^1, y^2) \equiv d^i(y^2, v^i)/d^i(y^1, v^i), \qquad i = 1, 2. \quad (62)$$

Assume the observed country i net output vector y^i solves the country i private product maximization problem, $\max_y \{w^i \cdot y : (y, v^i) \text{ belongs to } S^i\}$ for $i = 1, 2$. Then Diewert (1986) established the following bounds for the Malmquist indexes Q^i defined by (62):

$$Q^1(y^1, y^2) \geqslant w^1 \cdot y^2 / w^1 \cdot y^1 \equiv Q_L(w^1, w^2, y^1, y^2); \quad (63)$$

$$Q^2(y^1, y^2) \leqslant w^2 \cdot y^2 / w^2 \cdot y^1 \equiv Q_P(w^1, w^2, y^1, y^2) \quad (64)$$

where Q_L and Q_P are the Laspeyres and Paasche quantity indexes.

It is also possible to define a deflation function d that uses a convex combination of the input vectors, $\lambda v^1 + (1 - \lambda)v^2$, and a convex combination of the technology sets for the two countries, $\lambda S^1 + (1 - \lambda)S^2$:

$$d(y, \lambda) \equiv \min_{\delta > 0} \{\delta : (y/\delta), [\lambda v^1 + (1 - \lambda)v^2] \text{ belongs to } \lambda S^1 + (1 - \lambda)S^2\}. \quad (65)$$

The *Malmquist λ weighted average output index* for country 2 relative to 1 may be defined as:

$$Q^\lambda(y^1, y^2) \equiv d(y^2, \lambda)/d(y^1, \lambda). \quad (66)$$

Assuming maximizing behaviour, Diewert (1986) shows that there exists a λ such that $0 \leqslant \lambda \leqslant 1$ and $Q^\lambda(y^2, y^2)$ lies between the Laspeyres and Paasche quantity indexes Q_L and Q_P defined in (63) and (64). Thus the Fisher quantity index, $Q_F \equiv [Q_L Q_P]^{1/2}$, should provide an adequate approximation to the theoretical output index $Q\lambda(y^1, y^2)$.

Our conclusion is that the translog price index P_T and the Fisher quantity index Q_F have reasonably strong justifications in the bilateral aggregate private production context.

We now turn our attention to the multilateral case.

9. MULTILATERAL TEST APPROACHES

We are finally in a position to study the multilateral index number problem which was set out in the introduction. To review the notation, there are I positive price vectors $p^i \equiv [p_1^i, \ldots, p_N^i]$ and I quantity vectors $x^i \equiv [x_1^i, \ldots, x_N^i]$ with $p^i \cdot x^i > 0$ for $i = 1, \ldots, I$. We wish to find $2I$ positive numbers p^i (price indexes) and X^i (quantity indexes) such that $P^i X^i = p^i \cdot x^i$ for $i = 1, \ldots, I$. The I data points (p^i, x^i) will typically be observations on production or consumption units that are separated spatially but yet are still comparable. We have argued that the chain principle should be used if the I observations are time series observations on the same economic unit, but the reader is free to interpret the I data points as observations on the same economic unit over time if this seems desirable. For the sake of definiteness, we shall refer to the I data points as countries. Each commodity n is supposed to be the same across all countries. This can always be done by a suitable extension of the list of commodities.

Our first approach to the construction of a system of multilateral price and quantity indexes is based on the use of a bilateral quantity index Q. In this method, the first step is to pick the 'best' bilateral index number formula; e.g., the Fisher index Q_F defined by (14) or the implicit translog quantity index

defined by $\tilde{Q}_T(p^1, p^2, x^1, x^2) \equiv p^2 \cdot x^2 / p^1 \cdot x^1 P_T(p^1, p^2, x^1, x^2)$ where P_T is defined by (11). Secondly, pick a numeraire country, say country 1, and then calculate the aggregate quantity for each country i relative to country 1 by evaluating the quantity index $Q(p^1, p^i, x^1, x^i)$. In order to put these relative quantity measures on a symmetric footing, we convert each relative to country 1 quantity measure into a share of world quantity by dividing through by $\Sigma_{k=1}^I Q(p^1, p^k, x^1, x^k)$. For a general numeraire country j, define the *share of world quantity for country i, using country j as the numeraire country* by:

$$\sigma_i^j(p, x) \equiv Q(p^j, p^i, x^j, x^i) \bigg/ \sum_{k=1}^I Q(p^j, p^k, x^j, x^k),$$
$$i = 1, \ldots I \quad (67)$$

where $p \equiv [p^1, \ldots, p^I]$ is the N by I matrix of price data and $x \equiv [x^1, \ldots, x^I]$ is the N by I matrix of quantity data. Once the numeraire country j has been chosen and the country i shares σ_i^j calculated, we may set $X^i \equiv \sigma_i^j$ and $p^i \equiv p^i \cdot x^i / X^i$ for $i = 1, \ldots, I$. Thus we have provided a solution to the multilateral index number problem (1). Of course, one is free to renormalize the resulting p^i and X^i if desired; i.e., all X^i can be multiplied by a number provided all p^i are divided by this same number. Kravis (1984) calls this method the *star system*, since the numeraire country plays a starring role: all countries are compared with it and it alone.

We shall assume throughout this section that the index number formula Q satisfies the quantity counterparts to the bilateral price tests BT1 through BT6. (The quantity counterpart to BT1 is $Q(\alpha p^1, \beta p^2, x^1, x^2) = 1$ if $p^1 = p^2$ and $x^1 = x^2$, to BT2 is $Q(p^1, p^2, x^1, \alpha x^2) = \alpha Q(p^1, p^2, x^1, x^2)$ for $\alpha > 0$ and so on.) This assumption is not restrictive since our best bilateral formulae, Q_F and $\tilde{Q}_T$, both satisfy these tests.

Of course, the problem with the star system for making multilateral comparisons is its lack of invariance to the choice of the numeraire or star country. Different choices for the base country will in general give rise to different indexes P^i and X^i. This problem can be traced to the lack of circularity of the bilateral formula Q: if Q satisfies the time reversal test BT5 and the circular test BT9 for quantity indexes, then $\sigma_i^j = \sigma_i^k$ for all i, j and k; i.e., the shares σ_i^j defined by (67) do not depend on the choice of the numeraire country j. However, given that the bilateral formula Q does not satisfy the circularity test (as is the case with Q_F and $\tilde{Q}_T$), how can we generate multilateral indexes that treat each country symmetrically?

Fisher (1922) recognized that the simplest way of achieving symmetry was to average base specific index numbers over all possible bases. Thus define country i's share of world output $S_i(p, y)$ by

$$S_i(p, x) \equiv \sum_{j=1}^I \sigma_i^j(p, x)/I, \qquad I = 1, \ldots, I \quad (68)$$

where the σ_i^j are defined by (67). We can now define country i quantities and prices by

$$X^i \equiv S_i(p, x), \ P^i \equiv p^i \cdot x^i / X^i, \qquad i = 1, \ldots, I. \quad (69)$$

Fisher (1922) called this method of constructing multilateral indexes the *blend method* while Diewert (1986) called it the *democratic weights method*, since each share of world output using each country as the base is given an equal weight in the formation of the average.

Of course, there is no need to use an arithmetic average of the σ_i^j as in (68); one can use a geometric average:

$$\sigma_i(p, x) \equiv \left[\prod_{j=1}^I \sigma_i^j(p, x)\right]^{1/I} \qquad i = 1, \ldots, I. \quad (70)$$

Of course, the resulting shares no longer sum to one in general, so country i's share of world output is now defined as:

$$S_i(p, x) \equiv \sigma_i(p, x) / \sum_{k=1}^{I} \sigma_k(p, x), \qquad i = 1, \ldots, I. \quad (71)$$

If the Fisher index Q_F is used in the definition of the σ_i^j, then

$$S_i(p, x)/S_j(p, x) = \left[\prod_{k=1}^{I} Q_F(p^k, p^i, x^k, x^i) \middle/ \prod_{m=1}^{I} Q_F(p^m, p^j, x^m, x^j) \right]^{1/I} \quad (72)$$

and in this case, the multilateral method defined by (70) reduces to a method recommended by Eltetö and Köves (1964) and Szulc (1964), the *EKS method*. Instead of using the Fisher formula in (72), Caves, Christensen and Diewert (1982a) advocated the use of the translog quantity index Q_T while Diewert (1986) suggested the use of the implicit translog quantity index $\tilde{Q}_T$, since $\tilde{Q}_T$ is well defined even in the case where some quantities x_n^i are negative (whereas Q_T is not). We call the indexes generated by (69) and (71) for a general bilateral index Q *generalized EKS* indexes.

When forming averages of the σ_i^j as in (68) or (70), there is no necessity to use equal weights: one can define country j's value share of world output as $\beta_j \equiv p^j \cdot x^j / \Sigma_{k=1}^{I} p^k \cdot x^k$ (this requires all prices to be measured in units of a common currency) and then we may define a plutocratic share weighted average of the σ_i^j:

$$S_i(p, x) \equiv \sum_{j=1}^{I} \beta_j(p, x) \, \sigma_i^j(p, x). \quad (73)$$

Diewert (1986) called this method of constructing multilateral indexes the *plutocratic weights method*.

Another multilateral method that is based on a bilateral index Q may be described as follows. Define

$$\alpha_i(p, x) \equiv \sum_{j=1}^{I} [Q(p^j, p^i, x^j, x^i)^{-1}]^{-1}, \qquad i = 1, \ldots, I. \quad (74)$$

If there is only one commodity so that $N = 1$ and the bilateral index Q satisfies BT1, BT2 and BT3, then $\alpha_i = [\Sigma_{j=1}^{I} [x^i/x^j]^{-1}]^{-1} = [\Sigma_{j=1}^{I} x^j/x^i]^{-1} = x^i/\Sigma_{j=1}^{I} x^j$ which is country i's share of world product. In the general case where $N > 1$, the 'shares' α_i do not necessarily sum up to unity, so it is necessary to normalize them:

$$S_i(p, x) \equiv \alpha_i(p, y) \middle/ \sum_{k=1}^{I} \alpha_k(p, y), \qquad i = 1, \ldots, I. \quad (75)$$

Diewert (1986) called this the *own share method* for making multilateral comparisons.

The above methods for achieving consistency and symmetry rely on averaging over various bilateral index number comparisons. Fisher (1922) realized that symmetry could be achieved by making comparisons with an average; he called this broadening the base. Thus the *basket method* (which corresponds to Fisher's (1922) formula 6053 and to method 8 described in Ruggles (1967)) may be described as follows. The price level of country i relative to country j is set equal to $p^i \cdot (\Sigma_{k=1}^{I} x^k / I) / p^j \cdot (\Sigma_{k=1}^{I} x^k / I)$. This index number formula is a Scrope index (6), where the reference quantity vector is chosen to be the average market basket, $\Sigma x^k / I$. The same result could be achieved if we chose x to be the total market basket, $\Sigma_k x^k$. Now define $Q^{ji} \equiv p^i \cdot x^i / p^j \cdot x^j [p^i \cdot (\Sigma_k x^k) / p^j \cdot (\Sigma_k x^k)]$ to be the implicit output of country i relative to j. Choose a j as a numeraire country and calculate country i's share of world output as:

$$S_i(p, y) \equiv Q^{ji} \middle/ \sum_{k=1}^{I} Q^{jk} = \left(p^i \cdot x^i / p^i \cdot \sum_k x^k \right) \times \middle/ \sum_{m=1}^{I} \left(p^m \cdot x^m / p^m \sum_k x^k \right), \qquad i = 1, \ldots, I. \quad (76)$$

Note that the final expression for S_i does not depend on the choice of the numeraire country j. As usual, once the share functions S_i have been defined, the aggregate X^i and P^i may be defined by (69).

A variation on the basket method due to Geary (1958) and Khamis (1972) is defined by (77)–(79) below:

$$\pi_n \equiv \sum_{i=1}^{I} p_n^i x_n^i \middle/ p^i \sum_{k=1}^{I} x_n^k, \qquad n = 1, \ldots, N; \quad (77)$$

$$P^i \equiv \sum_{n=1}^{N} p_n^i x_n^i \middle/ \sum_{m=1}^{N} \pi_m x_m^i, \qquad i = 1, \ldots, I; \quad (78)$$

$$X^i \equiv p^i \cdot x^i / P^i, \qquad i = 1, \ldots, I. \quad (79)$$

π_n is interpreted as an average international price for good n. From (78), it can be seen that P^i, the price level of purchasing power parity for country i, is a Paasche-like price index for country i except that the base prices are chosen to be the international prices π_n. The π_n and $(P^i)^{-1}$ can be solved for as a system of simultaneous linear equations (up to a scalar normalization) or the $(P^i)^{-1}$ may be determined as the components of the eigenvector that corresponds to the maximal positive eigenvalue of a certain matrix. The P^i can be normalized so that the quantities X^i defined by (79) sum up to unity. This *GK method* for making multilateral comparisons has been widely used in empirical applications: see Kravis et al. (1975).

We have defined seven methods for making multilateral comparisons: the star method (67), the democratic (68) and plutocratic (73) weights methods, the generalized EKS method (71), the own share method (74), the basket method (76) and the GK method (79). How can we discriminate between them?

One helpful approach would be to define a system of *multilateral tests* and then evaluate how the above methods satisfy these tests.

In the bilateral situation, it was natural to phrase the tests in terms of the price index P or the quantity index Q, since if either of these functions were given (along with a single normalization such as $P^1 = 1$), then the aggregates P^1, P^2, X^1 and X^2 were all determined.

In the multilateral situation, it seems natural to phase the tests in terms of the properties of the system of world output share functions, $S(p, y) \equiv [S_1(p, y), \ldots, S_I(p, y)]$, since given these share functions, we may set $X^i \equiv S_i(p, y)$ and $P^i = p^i \cdot x^i / X^i$ for $i = 1, \ldots, I$.

The tests MT1 to MT6 listed below are multilateral counterparts to the bilateral tests BT1 to BT6 applied to quantity indexes rather than price indexes. MT0 is a preliminary test that does not have a bilateral counterpart. Recall that $p \equiv [p^1, \ldots, p^I]$ and $x \equiv [x^1, \ldots, x^I]$.

MT0: *Share Test*:
$\Sigma_{i=1}^{I} S_i(p, x) = 1$.

MT1: *Multilateral Identity Test*:
$S^i(\alpha_1 p^1, \ldots, \alpha_I p^I, \beta_1 x^1, \ldots, \beta_1 x^I) = \beta_i$ for $i = 1, \ldots, I$ for all $\alpha_i > 0$, $\beta_i > 0$ if $p^1 = \cdots = p^I$ and $x^1 = \cdots = x^I$.

MT2: *Proportionality Test*:
For $i = 1, \ldots I$ and $\lambda_i > 0$, $S_i(p, x^1, \ldots, x^{i-1}, \lambda_i x^i, x^{i+1}, \ldots, x^I)/S_j(p, x^1, \ldots, x^{i-1}, \lambda_i x^i, x^{i+1}, \ldots, x^I) = \lambda_i S_i(p, x)/S_j(p, x)$ for $j = 1, \ldots, i-1, i+1, \ldots, I$.

MT3: *Invariance to Changes in Scale Test*:
$S_i(\alpha_1 p^1, \ldots, \alpha_I p^I, \beta x^1, \ldots, \beta x^I) = S_i(p, x)$ for   all
$\alpha_i > 0, \beta > 0, i = 1, \ldots, I.$

MT4: *Invariance to Changes in Units Test*:
$S_i(\alpha_1 p_1^1, \ldots, \alpha_N p_N^1; \ldots; \alpha_1 p^I, \ldots, \alpha_N p_N^I; \alpha_1^{-1} x_1^1, \ldots, \alpha_N^{-1}$
$\times x_N^1; \ldots, \alpha_1^{-1} x_1^I, \ldots, \alpha_N^{-1} x_N^I) = S_i(p, x)$ for $i = 1, \ldots, I$
and $\alpha_1 > 0, \ldots, \alpha_N > 0.$

MT5: *Symmetric Treatment of Countries Test*:
Let $\hat{p}$ denote a permutation of the I columns of the N by I matrix p, let $\hat{x}$ denote the same permutation of the I columns of x, and let $\hat{S}(p, x)$ denote the same permutation of the I columns of the row vector $S(p, x)$. Then $\hat{S}(p, x) = S(\hat{p}, \hat{x})$.

MT6: *Symmetric Treatment of Commodities Test*:
Let $\tilde{p}$ denote a permutation of the N rows of p and let $\tilde{x}$ denote the same permutation of the N rows of x. Then $S_i(\tilde{p}, \tilde{x}) = S_i(p, x)$ for $i = 1, \ldots, I$.

MT7: *Country Partitioning Test*:
Let $S_j^I \equiv S_j(p^1, \ldots, p^I; x^1, \ldots, x^I)$ for $j = 1, \ldots, I$ and let $0 < \lambda_i < 1$. Define $S_j^{I+1} \equiv S_j(p^1, \ldots, p^I, p^i; x^1, \ldots, x^{i-1}, \lambda_i x^i, x^{i+1}, \ldots, x^I, (1 - \lambda_i) x^i)$ for $j = 1, \ldots, I, I + 1$. Then $S_j^I = S_j^{I+1}$ for $j = 1, \ldots, i - 1, i + 1, \ldots, I,$ $S_i^{I+1} = \lambda_i S_i^I$ and $S_{I+1}^{I+1} = (1 - \lambda_i) S_i^I$. This property is to hold no matter which country i is partitioned.

The functions S_j^I are the share functions for the initial world economy that consists of I countries. The functions S_j^{I+1} are the share functions for a new world economy, where the original country i with price vector p^i and quantity vector x^i, has been partitioned into two countries with price vectors p^i and p^i and quantity vectors $\lambda_i x^i$ and $(1 - \lambda_i) x^i$. MT7 says that under these conditions, the original country i share S_i^I splits into $\lambda_i S_i^I$ and $(1 - \lambda_i) S_i^I$ and the remaining shares are unaffected.

MT8: *Irrelevance of Tiny Countries Test*:
Let $\lambda_i > 0$ and define $S_j^I(\lambda_i) \equiv S_j(p, x^1, \ldots, x^{i-1}, \lambda_i x^i, x^{i+1}, \ldots, x^I)$ for $j = 1, \ldots I$. Define $S_j^{I-1} \equiv s_j$ $(p^1, \ldots, p^{i-1}, p^{i+1}, \ldots, p^I; x^1, \ldots, x^{i-1}, x^{i+1}, \ldots, x^I)$ for $j = 1, \ldots, I - 1$. Then $\lim_{\lambda_i \to 0} S_j^I(\lambda_i) = S_j^{I-1}$ for $j = 1, \ldots, i - 1$ and $\lim_{\lambda_i \to 0} S_j^I(\lambda_i^I) = S_{j-1}^{I-1}$ for $j = 1 + 1, i + 2, \ldots, I$. This property is to hold for all choices of the disappearing country i.

In the above test, the quantity vector for country i is deflated down to a zero vector. Consider the resulting system of limiting share functions. For all countries except i, the limiting share is equal to the share we would get if we simply deleted the data for country i and defined a system of share functions for a world economy consisting of only $I - 1$ countries. MT8 is a country counterpart to the irrelevance of tiny commodities test, (12).

The above multilateral tests were proposed by Diewert (1986). Our final multilateral test is a new one.

MT9: *Bilateral Properties Test*:
When $I = 2$, $S_2(p^1, p^2, x^1, x^2)/S_1(p^1, p^2, x^1, x^2)$ satisfies the bilateral tests BT1–BT6 for quantity indexes.

When $I = 2$, the multilateral system collapses down to a bilateral system. Hence it seems perfectly sensible to demand that S^2/S^1 should satisfy tests BT1–BT6 at least, since our best bilateral indexes, Q_F and $\tilde{Q}_T$, satisfied these tests.

The seven multilateral methods mentioned above satisfy most of the multilateral tests, assuming that the bilateral index Q satisfies BT1–BT6 (recall that all of the multilateral methods utilize a bilateral index Q except the basket and GK methods)

The star system fails MT5 (it is obviously not symmetric)

and it fails MT8 when the tiny country is chosen to be the numeraire country.

The plutocratic weights method fails MT2 and MT3; thus the resulting quantity indexes are not invariant to country inflation rates, a very severe defect.

The democratic weights method fails the multilateral proportionality test MT2 and the two consistency in aggregation tests, MT7 and MT8. This method is dominated by the generalized EKS method which fails only MT7 and MT8.

The basket and GK methods fail MT2 and MT9: both methods fail BT2 when $I = 2$. When $I = 2$, $S_2/S_1 = p^2 \cdot x^2 p^1 \cdot (x^1 + x^2)/p^1 \cdot x^1 p^2 \cdot (x^1 + x^2)$ for the basket method, and $S_2/S_1 = p^2 \cdot x^2/p^1 \cdot x^1 P_{GK}(p^1, p^2, x^1, x^2)$ for the GK method where the GK bilateral price index is defined as

$$P_{GK}(p^1, p^2, x^1, x^2)$$
$$\equiv \left[\sum_{n=1}^N p_n^2 x_n^1 x_n^2/(x_n^1 + x_n^2)\right] \Big/ \left[\sum_{k=1}^N p_k^1 x_k^1 x_k^2/(x_k^1 + x_k^2)\right]. \quad (80)$$

The five multilateral methods that use a bilateral index Q as a building block all have the property that $S_2/S_1 = Q$ when $I = 2$, which explains why these methods pass MT9.

Unfortunately, none of the above multilateral methods satisfies all nine multilateral tests. Our tentative conclusion is that if a symmetric multilateral method is desired, then the choice seems to be between the EKS (which fails MT7 and MT8, the consistency in aggregation properties) and the own share method (which fails the multilateral proportionality test, MT2). However, the systematic study of multilateral methods has only begun, so it may well be that better methods will be discovered in the future.

We turn now to another class of methods for constructing multilateral indexes.

10. NEOSTATISTICAL APPROACHES TO MULTILATERAL INDEXES

In one of the early statistical approaches to the construction of a price index, P^2/P^1, the index can be found by minimizing the sum of squared residuals, $\sum_{n=1}^N (p_n^2/p_n^1 - P^2/P^1)^2$ with respect to P^2/P^1. The resulting price index turns out to equal (4). Note that this price index was defined independently of quantities.

Theil (1960) initiated a *neostatistical approach* where the price and quantity indexes, P^i and X^i, are simultaneously determined. Theil's *best linear price and quantity indexes* may be found by solving the following constrained minimization problem:

$$\min_{P^1, \ldots, P^I, X^1, \ldots, X^I} \sum_{i=1}^I \sum_{j=1}^I (p^i \cdot x^j - P^i X^j)^2 \quad (81)$$

subject to a normalization on the P^i or X^i such as:

$$\sum_{i=1}^I X^i = 1 \quad (82)$$

which means that we can interpret the Theil X^i as shares. We may define $e_{ij} \equiv p^i \cdot x^j - P^i X^j$ as an error and then the interpretation of (81) becomes straightforward: we choose the P^i and X^i to minimize the sum of squared errors subject to (82).

The Theil index numbers have not been widely used, since they do not satisfy the product test equalities:

$$P^i X^i = p^i \cdot x^i, \quad i = 1, \ldots, I. \quad (83)$$

Kloek and De Wit (1961) suggested a number of variants of the Theil indexes including one where the constraints (83) were imposed. Thus define the Kloek and de Wit multilateral indexes as the P^i and X^i which solve (81) subject to (82) and (83). Unfortunately, the resulting indexes do not have very satisfactory properties: they fail the multilateral tests MT2,

MT3, MT7, MT8 and MT9 (when $I = 2$, the resulting bilateral quantity index fails BT2 and BT3 for quantity indexes).

Another neostatistical approach has been suggested by Van Yzeren (1957) which he called the *balanced method*. The price indexes P^i are determined (up to a normalization or factor of proportionality) by solving the following minimization problem:

$$\min_{P1,\ldots,\,PI} \sum_{i=1}^{I} \sum_{j=1}^{I} (P^i)^{-1}(p^i \cdot x^j / p^j \cdot x^j)P^j. \tag{84}$$

The X^i are then determined by (83) and the price normalization may be chosen so that (82) is satisfied. In this case, the errors e_{ij} may be defined by $(P^i X^j)^{-1}(p^i \cdot x^j / p^j \cdot x^j)P^j X^j = 1 + e_{ij}$ and the P^i may be found by minimizing

$$\sum_{i=1}^{I} \sum_{j=1}^{I} e_{ij}$$

subject to a normalization.

The Van Yzeren system of share functions does rather well in the multilateral test examination: they pass all tests except the consistency in aggregation tests MT7 and MT8. When $I = 2$, X^2 / X^1 turns out to equal the Fisher quantity index, Q_F. Thus the balanced method satisfies the same tests as the EKS method. However, the EKS method has a more satisfactory economic interpretation and is easier to construct numerically.

Many other neostatistical approaches to the construction of multilateral indexes could be explored. However, the resulting methods seem to be rather arbitrary and moreover, they lack economic interpretations.

11. OTHER ASPECTS

There are many aspects of index number theory that we cannot cover in this brief survey, such as: (i) *sampling problems* (see Fisher, 1922; and Allen, 1975), (ii) the treatment of *seasonality* (see Turvey, 1979; Balk, 1980; and Diewert, 1983c), (iii) consistency in aggregation and the theory of *subindexes* (see Vartia, 1976; Pollak, 1975; and Diewert, 1983a), (iv) productivity indexes (see Jorgenson and Griliches, 1967; Caves, Christensen and Diewert, 1982b; and Denny and Fuss, 1983), and (v) *econometric approaches* to cost of living indexes (see Jorgenson and Slesnick, 1983).

However, one area of concern that must be discussed is the *new goods problem*. Suppose that we are in the time series context and we have price and quantity data for $N - 1$ commodities in periods 1 and 2, p_n^t and x_n^t for $t = 1, 2$ and $n = 1, \ldots, N - 1$. Suppose in addition, that x_N^2 units of a new good are sold at the price p_N^2 during period 2. How are we to compute the bilateral price index, $P(p^1, p^2, x^1, x^2)$, when we do not know p_N^1, the price of the new good in period 1? Of course, we can assume $x_N^1 = 0$, so determining the quantity of the new good in period 1 is no problem.

From the viewpoint of the microeconomic approach to index number theory, Hicks (1940) provided a formal solution to this new good problem: if we are in the consumer context, p_N^1 should be the price which would just make the consumer's demand for good N in period 1 equal to zero. The practical problem is that this shadow price is not observable: we require a knowledge of the consumer's indifference surfaces to calculate it. Of course, econometric techniques could be used to estimate these shadow prices (see Diewert (1980) for an example of such a technique in the producer context), but most index number practitioners will find it inconvenient to resort to econometrics. In practice, most official indexes ignore the existence of new goods.

In order to illustrate the price index bias that can result from the omission of new goods, we shall present a hypothetical

example. Suppose that there are 3 periods, one 'old' good with constant price and quantity, $p_1^t = x_1^t = 1$ for $t = 1, 2, 3$ and one new good which appears in period 2, so that $x_2^1 = 0$. Typically, new goods follow a product cycle: they are introduced at a relatively high price and then the price declines over time. Thus we assume that $p_2^2 = 2$ and $p_2^3 = 1$, so that the period 2 price for the new good is twice as high as the period 3 price. We assume that the quantity purchased of the new good in period 2 is $f > 0$, where f is a fraction which represents the period 2 proportion of new goods to old goods. We assume that the quantity purchased of the new good in period 3 is $2f$ and that the shadow price p_2^1 in period 1 that would make the demand for the new good equal to zero is 4.

If the new good is ignored, we find that $P^t = 1$ for $t = 1, 2, 3$ for any reasonable index number formula. The true chain Laspeyres price indexes which do not ignore the new good are: $P^1 = 1$, $P^2 = 1$ and $P^3 = (1 + f)/(1 + 2f)$. The reader can verify that the same indexes result no matter what shadow price p_2^1 is chosen. Evaluating P^3 for various reasonable values of f yields the following period 3 price indexes: if $f = 0.01$, then $P^3 = 0.9902$; if $f = 0.02$, then $P^3 = 0.9808$ and if $f = 0.05$, then $P^3 = 0.9545$. Thus the conventional Laspeyres price index which ignores the existence of new goods will have an *upward bias* of about 1 to 4.5 per cent compared with the true Laspeyres index.

In order to evaluate the bias in the conventional chained Paasche and Fisher price indexes, we have to use our assumption that $p_2^1 = 4$. Under our assumptions, we obtain the following values for the true chained Paasche price indexes in period 3: if $f = 0.01$, then $P^3 = 0.9619$; if $f = 0.02$, then $P^3 = 0.9273$ and if $f = 0.05$, then $P^3 = 0.8403$. We also obtain the following values for the true chained Fisher price indexes in period 3: if $f = 0.01$, the $P^3 = 0.9759$; if $f = 0.02$, then $P^3 = 0.9537$ and if $f = 0.05$, then $P^3 = 0.8956$. Thus the conventional Fisher ideal price index will have an upward bias of about 2.5 to 10.5 per cent in period 3, depending on the fraction f of new goods introduced in period 2.

The above analysis of bias is only illustrative but it does indicate that ignoring new goods could lead to a substantial overestimation of price inflation and a corresponding underestimation of real growth rates, especially in advanced market economies where millions of new goods are introduced each year.

W. E. DIEWERT

See also DIVISIA INDEX; GAUGE FUNCTIONS; HEDONIC FUNCTIONS AND HEDONIC INDEXES; HERFINDAHL INDEX; IDEAL INDEXES; TRANSFORMATIONS AND INVARIANCE.

BIBLIOGRAPHY

Afriat, S.N. 1972. The theory of international comparisons of real income and prices. In *International Comparisons of Prices and Outputs*, ed. D.J. Daley, New York: Columbia University Press, 13–69.

Allen, R.G.D. 1949. The economic theory of index numbers. *Economica*, NS 16, 197–203.

Allen, R.G.D. 1975. *Index Numbers in Theory and Practice*. London: Macmillan.

Balk, B.M. 1980. A method for constructing price indices for seasonal commodities. *Journal of the Royal Statistical Society*, Series A 143, 68–75.

Bowley, A.L. 1928. Notes on index numbers. *Economic Journal* 38, 216–37.

Carli, G.R. 1764. Del valore e della proporzione de'metalli monetati con i generi in Italia prima delle scoperte dell'Indie colonfronto del valore e della proporzione de'tempi nostri. In *Opere scelte di Carli*, Vol. 1, Milan, 299–366.

Caves, D.W., Christensen, L.R. and Diewert, W.E. 1982a. Multilateral comparisons of output, input and productivity using superlative index numbers. *Economic Journal* 92, 73–86.

Caves, D.W., Christensen, L.R. and Diewert, W.E. 1982b. The economic theory of index numbers and the measurement of input, output and productivity. *Econometrica* 50, 1393–414.

Christensen, L.R., Jorgenson, D.W. and Lau, L.J. 1971. Conjugate duality and the transcendental logarithmic production function. *Econometrica* 39, 255–6.

Denny, M. and Fuss, M. 1983. A general approach to intertemporal and interspatial productivity comparisons. *Journal of Econometrics* 23, 315–30.

Diewert, W.E. 1974. Applications of duality theory. In *Frontiers of Quantitative Economics*, ed. M.D. Intriligator and D.A. Kendrick, Vol. II, Amsterdam: North-Holland, 106–71.

Diewert, W.E. 1976. Exact and superlative index numbers. *Journal of Econometrics* 4, 115–45.

Diewert, W.E. 1978. Superlative index numbers and consistency in aggregation. *Econometrica* 46, 883–900.

Diewert, W.E. 1980. Aggregation problems in the measurement of capital. In *The Measurement of Capital*, ed. Dan Usher, Chicago: University of Chicago Press, 433–528.

Diewert, W.E. 1981. The economic theory of index numbers: a survey. In *Essays in the Theory and Measurement of Consumer Behaviour in Honour of Sir Richard Stone*, ed. A. Deaton, London: Cambridge University Press, 163–208.

Diewert, W.E. 1983a. The theory of the cost-of-living index and the measurement of welfare change. In *Price Level Measurement*, ed. W.E. Diewert, and C. Montmarquette, Ottawa: Statistics Canada, 163–233.

Diewert, W.E. 1983b. The theory of the output price index and the measurement of real output change. In *Price Level Measurement*, ed. W.E. Diewert and C. Montmarquette, Ottawa: Statistics Canada, 1049–113.

Diewert, W.E. 1983c. The treatment of seasonality in a cost-of-living index. In *Price Level Measurement*, ed. W.E. Diewert and C. Montmarquette, Ottawa: Statistics Canada, 1019–45.

Diewert, W.E. 1986. Microeconomic approaches to the theory of international comparisons. Technical Working Paper No. 53, National Bureau of Economic Research, Cambridge, Mass.

Dutot, Charles de Ferrare. 1738. *Réflexions politiques sur les finances et le commerce*. The Hague.

Eichhorn, W. and Voeller, J. 1976. *Theory of the Price Index*. Berlin: Springer-Verlag.

Eltetö, O. and Köves, P. 1964. On a problem of index number computation relating to international comparison. *Statisztikai Szemle* 42, 507–18.

Fisher, F.M. and Shell, K. 1972. The pure theory of the national output deflator. In F.M. Fisher and K. Shell, *The Economic Theory of Price Indices*, New York: Academic Press, 49–113.

Fisher, I. 1911. *The Purchasing Power of Money*. London: Macmillan.

Fisher, I. 1922. *The Making of Index Numbers*. Boston: Houghton Mifflin.

Frisch, R. 1930. Necessary and sufficient conditions regarding the form of an index number which shall meet certain of Fisher's tests. *American Statistical Association Journal* 25, 397–406.

Frisch, R. 1936. Annual survey of economic theory: the problem of index numbers. *Econometrica* 4, 1–39.

Geary, R.G. 1958. A note on comparisons of exchange rates and purchasing power between countries. *Journal of the Royal Statistical Society*, Series A 121, 97–9.

Hicks, J.R. 1940. The valuation of the social income. *Economica* 7, 105–24.

Jevons, W.S. 1865. Variations of prices and the value of currency since 1762. *Journal of the Royal Statistical Society* 28, 294–325.

Jorgenson, D.W., and Griliches, Z. 1967. The explanation of productivity change. *Review of Economic Studies* 34, 249–83.

Jorgenson, D.W. and Slesnick, D.T. 1983. Individual and social cost-of-living indexes. In *Price Level Measurement*, ed. W.E. Diewert and C. Montmarquette, Ottawa: Statistics Canada, 241–323.

Keynes, J.M. 1930. *Treatise on Money*, Vol. 1. London: Macmillan.

Khamis, S.H. 1972. A new system of index numbers for national and international purposes. *Journal of the Royal Statistical Society*, Series A 135, 96–121.

Kloek, T. and de Wit, G.M. 1961. Best linear unbiased index numbers. *Econometrica* 29, 602–16.

Kravis, I.B. 1984. Comparative studies of national incomes and prices. *Journal of Economic Literature* 22, 1–39.

Kravis, I.B., Kenessey, Z., Heston, A. and Summers, R. 1975. *A System of International Comparisons of Cross Product and Purchasing Power*. Baltimore: Johns Hopkins University Press.

Konüs, A.A. 1924. The problem of the true index of the cost of living, Trans. in *Econometrica* 7, (1939), 10–29.

Konüs, A.A., and Byushgens, S.S. 1926. K probleme pokupatelnoi cili deneg. *Voprosi konyunkturi* 2(1), 151–72.

Laspeyres, E. 1871. Die Berechnung einer mittleren Warenpreissteigerung. *Jahrbücher für Nationalökonomie und Statistik* 16, 296–314.

Malmquist, S. 1953. Index numbers and indifference surfaces. *Trabajos de estadistica* 4, 209–42.

Marshall, A. 1887. Remedies for fluctuations of prices. *Contemporary Review*, March. Reprinted as ch.8 in *Memorials of Alfred Marshall*, ed. A.C. Pigou, London: Macmillan, 1925.

Paasche, H. 1874. Über die Preisentwicklung der letzten Jahre nach den Hamburger Börsennotirungen. *Jahrbücher für Nationalökonomie und Statistik* 23, 168–78.

Pigou, A.C. 1920. *The Economics of Welfare*. London: Macmillan.

Pollak, R.A. 1971. The theory of the cost of living index. Research Discussion Paper 11, Bureau of Labor Statistics, Washington, DC. Published in *Price Level Measurement*, ed. W.E. Diewert and C. Montmarquette. Ottawa: Statistics Canada, 1983, 87–161.

Pollak, R.A. 1975. Subindexes of the cost of living. *International Economic Review* 16, 135–50.

Pollak, R.A. 1981. The social cost of living index. *Journal of Public Economics* 15, 311–36.

Prais, S. 1959. Whose cost of living? *Review of Economic Studies* 26, 126–34.

Ruggles, R. 1967. Price indexes and international price comparisons. In *Ten Economic Studies in the Tradition of Irving Fisher*, ed. W. Fellner et al., New York: John Wiley, 171–205.

Samuelson, P.A. 1947. *Foundations of Economic Analysis*. Cambridge, Mass.: Harvard University Press.

Samuelson, P.A. 1953. Prices of factors and goods in general equilibrium. *Review of Economic Studies* 21, 1–20.

Samuelson, P.A. and Swamy, S. 1974. Invariant economic index numbers and canonical duality: survey and synthesis. *American Economic Review* 64, 566–93.

Scrope, G.P. 1833. *Principles of Political Economy*. London: Longman, Rees, Orme, Brown, Green and Longman.

Shephard, R.W. 1953. *Cost and Production Functions*. Princeton: Princeton University Press.

Szulc, B. 1964. Indices for multiregional comparisons. *Przeglad statystyczny* 3, 239–54.

Theil, H. 1960. Best linear index numbers of prices and quantities. *Econometrica* 28, 464–80.

Törnqvist, L. 1936. The Bank of Finland's consumption price index. *Bank of Finland Monthly Bulletin* 10, 1–8.

Turvey, R. 1979. The treatment of seasonal items in consumer price indices. *Bulletin of Labour Statistics*, 4th quarter, Geneva: International Labour Office, 13–33.

Van Yzeren, J. 1957. *Three Methods of Comparing the Purchasing Power of Currencies*. Statistical Studies No. 7, The Hague: Centraal Bureau Voor de Statistiek.

Vartia, Y.O. 1976. Ideal log-change index numbers. *Scandinavian Journal of Statistics* 3, 121–6.

Vartia, Y.O. 1985. Defining descriptive price and quantity index numbers: an axiomatic approach. Paper presented at the Fourth Karlsruhe Symposium on Measurement in Economics, University of Karlsruhe, July.

Walsh, C.M. 1901. *The Measurement of General Exchange Value*. New York: Macmillan.

Walsh, C.M. 1921. The best form of index number: discussion. *Quarterly Publication of the American Statistical Association* 17, 537–44.

Westergaard, H. 1890. *Die Grundzüge der Theorie der Statistik*. Jena: Fischer.

indicative planning.

1. THE PRINCIPLE OF INDICATIVE PLANNING. The aim of pure indicative planning is to improve the performance of an economy by the provision of better economic information; forecasts or targets are published but compliance with them is purely voluntary. The underlying logic is that the plan, via collective action, can supply economically valuable information which, as a public good, the market mechanism does not disseminate efficiently. If the plan is correctly drawn up, it will indicate an optimal path for the economy which, it is hoped, would then be spontaneously followed by everyone without compulsion. Planning without coercion marries well with Keynesian ideas, and was indeed foreshadowed in Keynes's own writings. But the idea first reached the UK via France in the 1960s.

France had been using a form of indicative planning since 1946, and in a number of English-language presentations in the early 1960s. Pierre Massé, the French economist and Planning Commissioner argued that their indicative plan acted as an informational substitute for forward markets that did not exist. Anyone can observe the prices for goods and services to be sold in the present but not for those to be sold at future dates. Such market uncertainty would inhibit investment. Buyers might well have firm plans to purchase and suppliers be willing to sell, but transactions might be frustrated and growth slowed by economic actors' ignorance of one another's intentions. Market forecasts in the plan would give prospective buyers and sellers the confidence that both markets and suppliers would be available in due course. Potential overcapacity and shortage could be anticipated and avoided. Without this 'collective market research' the economy risked being in state of disequilibrium with expectations unfulfilled, thus jeopardizing the supposed optimality properties of the market process. The text of the French Fourth Plan (1961–65) claimed to have calculated a 'general equilibrium' for the economy in 1965; this, it said, should be substantially self-implementing, for if it had been correctly worked out it represented the optimal allocation of resources for the economy which was in everyone's interest, adding, however, that policy measures might however be added to supplement this spontaneous implementation.

Formalizing the idea, J.E. Meade linked Massé's theory in the Arrow–Debreu general equilibrium model of welfare maximizing perfect competition with 'forward markets' extending over time and to cover risk. Meade showed that optimality properties could be obtained even without a full set of forward contracts provided that economic agents made non-binding declarations of how they intended to act in the future. Suppose that there is complete knowledge about exogenous or environmental factors. (including the weather, technology and foreign trade), then all the uncertainty in the economy is endogenous, i.e. stemming from each economic agent's ignorance about the demand and supply intentions of others. If every economic actor made known what their prospective demand and supply intentions were for every future date, equilibrium prices and quantities could be calculated to make an indicative plan. The predictions of the indicative plan would necessarily be realized since they correspond to optimal behaviour by buyers and sellers. If exogenous shocks are ruled out there is nothing to prevent the initially declared intentions from being carried out, and so the plan will implement itself.

This assumes that agents have declared their true intentions. If individual agents are large enough to influence prices, there are incentives against revealing true preferences. For example, by exaggerating their potential demand, component users can induce an artificial glut, of which they can take advantage. Such problems of course also beset long-term contractual arrangements if one party has an incentive to cheat and the other is unable to monitor or to prevent this.

But even with honest participants, a single-variant plan would not make any sense when, as in reality, the economy is subject to many external shocks. So a useful indicative plan would in fact have to contain not one path over time, but as many as there were possible scenarios for the exogenous environment. The multi-scenario plan could not eliminate exogenous uncertainty but it could eliminate endogenous or market uncertainty. One would not know what the weather would be like but one would know what the consequences of any given environment would be. Risk could only be totally eliminated through insurance contracts. All this would of course require a vast number of scenarios: if the plan contained only ten exogenous variables capable of taking on three values each in any one year, the total number scenarios to be investigated is about 24 million.

Massé's original formulation had brushed aside this difficulty and had in fact stressed the positive virtues of having 'a common view of the future'. This feature of his approach was seized on by pragmatic proponents of indicative planning as well as subsequently by critics. There can be substantial gains to the economy if all actors work on the basis of identical expectations, but *if and only if these expectations are in fact correct*. The economy will then be in equilibrium, probably a superior equilibrium to any achievable without the plan. This is most evident in macroeconomic terms. Following Massé, many UK economists in the 1960s argued that if only agents could be persuaded to believe that economic growth would accelerate and they were shown the detailed implications for their sectors of the fast growth scenario, then they would know what to do to implement this plan. Sir Roy Harrod advocated the use of indicative planning in the context of his model of economic growth. By focusing expectations on the underlying natural rate of growth, indicative planning could stabilize the warranted and actual growth rates on this figure, and even raise it by improvements in the effectiveness of investment. Indicative planning thus came to be viewed as a way to pull economic growth up by the bootstraps of macroeconomic expectations. A detailed examination of French experience suggests that the main effect of indicative planning there was on macro rather than micro expectations, but the formal theory emphasized the micro side and failed to come to terms with the fact that there are a great many factors constraining the growth of a national economy other than the expected rate of growth of aggregate demand.

2. DISAPPOINTING RESULTS AND CRITICAL REACTIONS. In 1965 the British government attempted to put this logic into practice. A National Plan was constructed which calculate sectoral implications of an increase in the growth rate of GNP to 4% p.a. It was hoped that if everyone committed themselves to these targets they would come about. More emphasis was placed on the effects on expectations of publishing the plan than the potentially more useful task the planners also tried to undertake, namely using the projections to identify what constraints might occur to obstruct the fulfillment of the plan. Within a year or so of its promulgation, a balance of payments crisis occurred, in principle forseeable but unanticipated in the plan, and deflation became the central aim of government policy.

In the same period in France, there was also a growing disillusionment with detailed indicative planning. Indeed, many years later Pierre Massé commented when asked if the theory he had outlined genuinely corresponded to the reality of the plan:

> As Planning Commissioner, when you address the general public you have to speak about 'Planning Growth' but when you are addressing a conference of sceptical Anglo-Saxon economists, you have to come up with a justification in terms that will impress them.

French practice in the 1960s differed markedly from the model that was refined elsewhere. There was much sporadic state interventionism, but after the late 1960s there was no attempt to make the Plan itself anything more than a modest forecasting exercise, which lost much of its value as the overwhelming source of uncertainty was external. By the 1980s political economic and technical developments led to the use of multi-scenario projections with no targets at all.

Indicative planning was also attacked as not only ineffective but harmful. 'Austrian' economists such as Lutz denounced any collective forecasting as perniciously undermining competition. At a macro level, single variant indicative planning can lead too many eggs being put into one, potentially wrong, basket. When planners claim that individuals are ill-equipped to forecast exogenous variables, and that the assembling of all information by a public agency could lead to a better result, allowing economic agents to take better-informed decisions, the anti-planners reply that official forecasters have often been notoriously misleading. For Austrians, the essence of free competition is that there must be pluralism in economic information. Competition means that the entrepreneur who makes the best guesses will win. This should create an incentive to explore a wide 'portfolio' of economic hypotheses, in the knowledge that there are substantial monopoly profits to be made if you alone are right; someone should continue to explore for oil during a glut. Society thus spreads its risks. This criticism does not apply to the full-blown multi-scenario indicative plan as imagined by Meade, but it does apply to any realistic model of indicative planning, since even if there is a multiplicity of scenarios, there is only a very limited scope for varying key parameters such as time lags and micro-economic substitution elasticities.

Furthermore, even if the forecasters have correctly forecast a particular market, conditionally or otherwise, the sales of individual firms remain uncertain, unless market shares are pre-determined, e.g. by collusion. If the economy is generally competitive, no problem arises in Meade's theoretical model where the plan was forecasting *prices*, and the supply and demand decisions of firms are implicit. But any real plan makes quantity forecasts for sectors. Consider a real example: in the UK since the late 1970s there has been substantial new investment in consumer electronics. Manufacturers apparently meet regularly to inform one another of their prospective production and investment plans. The total of these is then compared with independent forecasts of the scale of the total market.

Producers then readjust their own plans. We cannot say if this exercise improves efficiency. However, the problem still exists in the orthodox model of market behaviour when demand rises and permits supranormal profits. Entry is clearly indicated but not by whom or on what scale. The optimum size of the firm under constant returns to scale has always been a puzzle for the traditional model of markets. We must always choose between the devil of collusion and the deep blue sea of uncoordinated investment. Collusion may lead to the maintenance of too many sub-optimal plants, while totally independent action may lead to alternate waves of over- and

under-investment, with shortages then excess capacity and scrapping. New developments in industrial economics remind us that the full benefits of competition require not merely that entry to a sector should be easy but that there should be no problems of recovering capital costs if the investor wants to disinvest.

Our example illustrates other aspects of micro-level indicative planning. As Massé noted, the evolution of sectoral demand depends on factors probably more easily visible at a macro-level than by individual firms, e.g. growth of overall consumer spending, and a variety of public sector actions. The failure to foresee these led to problems in the early 1970s. Collective action to forecast the market has a real role, though, it need not necessarily be done by the state. At the same time, when investments are especially lumpy, and product lead-times long, knowledge of other firms' intentions becomes crucial. However, without some form of pressure to coordinate, whether by collusion or state intervention, it is not at all clear that merely having the information about others' stated intentions is enough to lead investors to converge voluntarily on an optimal path.

3. AN ALTERNATIVE APPROACH. In my view, the theory of the 1960s was deeply misleading. By trying to imitate the optimal equilibrium of the Arrow–Debreu model, it obscured the fact that the extent and nature of economic uncertainty and the bounds on human rationality mean that economic systems are doomed to lurch around in a permanent state of disequilibrium. Expectations can never be fully realized. As Keynes observed, people prefer to hold money instead of making detailed and complex futures and contingent contracts. Most of the barriers to the existence of full futures markets would make the writing of a serious comprehensive indicative plan impossible too. However, this does not mean we must accept the Austrian conclusion that nothing can be done to improve the situation. We may not be able to achieve economic nirvana by indicative planning, but we may be able to reach a better disequilibrium. Economic information will never be allocated efficiently by the market, as compared with what is actually achievable and not merely by utopian standards. The profitability of consultancies and financial services testify to the value of market information that is not directly acquired through experience. Private production and market sale of information are driven by the search for monopoly profits. Thus information once generated will not be given away free, even though its marginal cost of distribution is zero. Add to this the observation that even with modern micro-computers there are major economies of scale in information processing, and we have a modest but firm case for some form of public indicative planning. It would probably have an open-ended 'strategic' quality, alerting decision-makers to contingencies they might not have thought of; the process of exchanging information would be as important as any figures finally published. An actual plan might not be able to produce totally coherent forecasts for the economy if what is known when it is drawn up is merely that present trends are unsustainable. This knowledge is of value to agents, even if it is accompanied only by limited indications as to which ways the disequilibrium can be resolved.

The experience of the last decades suggests some areas where indicative planing can be of value. At a micro economic level the Austrian critics are surely right to suppose that individual entrepreneurs can analyse their own markets better than the state; but in order to do so they require information that they will not automatically acquire in the course of their business. They will normally need macroeconomic information but in some cases where investments are lumpy and irreversible there is a case for checking the consistency of behaviour among actors in a given sector.

At a macroeconomic level it is arguable that the government is a particularly well-informed actor. A public agency can usefully check the consistency of the intentions of state and other actors, for the present and over time. If the planned investments by big firms in the economy are inconsistent with estimates of aggregate investment, or plausible future patterns of demand and trade, then something will have to give, though the first iteration of an indicative plan cannot determine what it will be. At a broader macroeconomic level, expectations, whilst not as malleable as the early indicative planners hoped, are surely not as rigid as in some modern views, especially Rational Expectations. Credible promises by governments to act to support full employment must have sustained economic stability before 1973, even though actual interventions were sparing. Half-hearted promises after 1973 may account for the half-hearted world recovery in the late 1970s; deflationary policies after 1979 must have affected growth expectations at least as much as inflation expectations. Indeed failure to check the mutual inconsistency underlying national economic policies has been a major problem for the world economy in the 1980s. Looking at the world economy highlights the potential utility, if insufficiency, of some form of indicative planning. Major governments have simultaneously adopted deflationary strategies while assuming constant global aggregate demand! Creditor nations have demanded repayment of debts while restricting imports from their debtors. Agencies exist, such as the OECD, whose job it is to spot these inconsistencies, but then only moral pressure can force governments to amend their policies.

Fortunately, within a national economy, the government could decide to adopt counter-balancing policies faced with indicative plan forecasts showing a prospective disequilibrium, if private agents declined to alter their behaviour. The government is also able to ensure compliance with voluntary agreements made among national economic actors. There are obvious implications for an incomes policy. The mere provision of forecasts showing that real income gains expected by the totality of economic agents are unsustainable may have to be reinforced by some form of contractual agreement in order to secure a stabilizing response. (This problem is not unique to indicative planning however: monetarist policies aim to reduce inflationary expectations purely by the supposed connection between announcements of monetary policy and prices.)

The public character of a plan may be important in legitimating the forecasts in it. One factor inhibiting social actors from making contracts, implicit or explicit, of macroeconomic significance is lack of reliable non-partisan information. French experience suggests that expectations about the prospects for growth or the need for austerity measures can indeed be influenced by the existence of a consensual planning process, and also that this function is undermined if the state tries to manipulate forecasts opportunistically. We should also remember that the state itself is a cluster of often imperfectly coordinated agencies who can benefit from indicative planning as well for the purposes of their own strategic planning. For these reasons there is a case for saying that indicative planning activities may be best carried out by an agency in the public sector but with no operational responsibilities of its own, thus able to be impartial towards other public as well as private agents.

4. CONCLUSION. If the economic system were constrained to a uniquely determined equilibrium, indicative planning could achieve nothing; but the real world is a messy and uncertain place. Indicative planning cannot produce a perfect Arrow–Debreu equilibrium, but by promoting a better exchange of even imperfect information it may still be able to improve the quality of decision-making by public and private economic actors and the interaction between them.

PETER HOLMES

See also PLANNING.

BIBLIOGRAPHY

Beckerman, W. (ed.) 1965. *The British Economy in 1975*. Cambridge: Cambridge University Press.

Cave, M. and Hare, P. 1981. *Alternative Approaches to Economic Planning*. London: Macmillan.

Estrin, S. and Holmes, P. 1983. *French Planning in Theory and Practice*. London: George Allen & Unwin.

Estrin, S. and Holmes, P. 1985. Uncertainty, deficient foresight and economic planning in Keynesian Economics. *Journal of Post-Keynesian Economics*, Summer, 463–73.

Grossman, S.J. and Stiglitz, J.E. 1980. On the impossibility of informationally efficient markets. *American Economic Review* 70, 393–408.

Harrod, R.F. 1973. *Economic Dynamics*. London: Macmillan.

Johansen, L. 1978. *Lectures on Macro-Economic Planning*, Vol. 2. Amsterdam: North-Holland.

Keynes, J.M. 1931. The end of laissez-faire. In J.M. Keynes, *Essays in Persuasion*, London: Macmillan.

Kirzner, I. 1979. *Perception Opportunity and Profit*. Chicago: University of Chicago Press.

Lutz, V. 1969. *Central Planning for the Market Economy*. London: Longman.

Massé, P. 1965. French planning and economic theory. *Econometrica* 33, 265–76.

Massé, P. 1968. *Le plan ou l'anti-hasard*. Paris: Gallimard.

Meade, J.E. 1970. *The Theory of Indicative Planning*. Manchester: Manchester University Press.

Meade, J.E. 1971. *The Controlled Economy*. London: George Allen & Unwin.

Miller, J.B. 1979. Meade on indicative planning. *Journal of Comparative Economics* 3, 27–40.

Quinet, E. and Touzéry, L. 1986. *Le plan français*. Paris: Economica.

indicators.

TYPES AND STRUCTURE. Economic indicators, as a general category, are descriptive and anticipatory data used as tools for the analysis of business conditions and forecasting. There are potentially as many subsets of indicators in this sense as there are different targets at which they can be directed. For example, some indicators may relate to employment, others to inflation.

This brings to mind the uses of such time series as lagged explanatory variables in econometric models and regression equations. But there is a different, established meaning to what is often called the 'indicator approach'. This is a system of data and procedures designed to monitor, signal and confirm cyclical changes, especially turning points, in the economy at large. The series that serve this purpose are selected for being comprehensive and systematically related to business cycles and are known as *cyclical indicators*.

Business cycles are recurrent sequences of alternating phases of expansion and contraction that involve a great number of diverse economic processes. These movements are both sufficiently diffused and sufficiently synchronized to show up as distinct fluctuations in comprehensive series that measure production, employment, income and trade-aspects of aggregate economic activity. The end of each expansion is marked by a cluster of peaks in such series, the end of each contraction by a cluster of troughs. Analysts at the National Bureau of Economic Research (NBER) base the dating of business cycle peaks and troughs on the identification and analysis of such clusters, that is, the consensus of the corresponding turning points in the principal *coincident indicators*. This is done because (1) the co-movement of the indicators is itself an essential characteristic of the business cycle; (2) no single adequate measure of aggregate economic activity is available in a consistent form for a long historical period; and (3) economic statistics generally are subject to error, so that the evidence from a number of independently compiled indicators tends to be more reliable than the evidence from any individual series. The NBER reference chronologies of business cycle peaks and troughs (Burns and Mitchell, 1946, ch. 4; Moore, 1961, chs 5 and 6; Zarnowitz and Moore, 1977, 1981) are widely used in academic as well as current business research.

The specific cycles observed across a wide spectrum of variables differ greatly and in part systematically. Thus many economic time series called the *leading indicators* tend to reach their turning points *before* the corresponding business cycle turns. There are also many series that tend to reach their turning points *after* the peaks and troughs in the business cycle, and they are the *lagging indicators*. The leading series represent largely flow and price variables that are highly sensitive to the overall cyclical influences but also to shorter random disturbances; hence they show large cyclical rises and declines but also high volatility. Coincident series have generally smaller cyclical movements and are at the same time much smoother. Lagging indicators include some massive stock variables which have modest cyclical functions yet are extremely smooth.

Most indicators display, in addition to the cyclical fluctuations that dominate the developments over spans of several years, trends that prevail across decades and reflect largely economic growth and, for nominal variables, inflation. Seasonal variations are likewise widespread but these stable or evolving patterns of intra-year change show much diversity, hence are often weakened by aggregation, unlike the longer movements which are in general positively cross-correlated. It is a common practice to assume that the seasonal movements are exogenous and separable, and cyclical indicators are used predominantly in 'seasonally adjusted' form. The reason is to show the trends and cycles in monthly or quarterly data more

clearly, but control against significant errors from faulty seasonal adjustments is also necessary, and often neglected.

The indicators, then, are viewed as composites of trends, cyclical and 'irregular' movements. The latter are generally small and of stable random appearance, apart from occasional outliers due to some particular disturbances such as major strikes or unseasonable weather. The 'classical' decomposition approach does not rule out some interactions among the component movements. What is alien to the indicator analysis, however, is the more recent notion that the trends and cycles themselves are purely stochastic phenomena, essentially random walks or results of the cumulation of random changes.

In a growing economy business expansions must be on the average larger than contractions in terms of output, employment, etc., and they are also likely to be longer. The individual cycles and their phases, however, vary greatly in duration and amplitude. These differences are systematically related to the scope or diffusion of the cyclical movements among different units of observation (e.g. activities, regions, industries). Vigorous expansions are generally more widespread than weak expansions; severe contractions are more widespread then mild contractions. But the timing sequences and amplitude differences among the indicators are observed during long and short, strong and weak cycles.

Diffusion indexes are time series showing the percentage of items in a given population that are rising over a specified unit period. Information about the direction of the change can often be obtained much more readily than information about the size of the change, hence surveys designed to produce timely diffusion measures on actual or expected sales, prices, profits, etc., are popular in many countries. Moreover, diffusion indices are correlated with rates of change in the corresponding aggregates and tend to lead the levels of these aggregates.

SIGNIFICANCE IN BUSINESS CYCLE THEORIES. The indicators in current use play important roles in many areas viewed as critical in business cycle theories. This is illustrated by the summary in Table 1, based on a long series of studies (for references, see Zarnowitz, 1972, 1985; Moore, 1983, pp. 347–51).

The literature on business cycles, though rich in ingenious hypotheses of varying plausibility and compatibility, produced no unified theory (Haberler, 1964; Zarnowitz, 1985). There is evidence in support of a number of different models that focus on period-specific or sector-specific aspects of the economy's

TABLE 1

Theories or models	Some of the main factors	Evidence from time series
Accelerator-multiplier models; hypotheses on autonomous investment, innovations, and gestation lags.	Interaction between investment, final demand and savings.	Large cyclical movements in business investment commitments (orders, contracts) lead total output and employment; smaller movements in investment realizations (shipments, outlays) coincide or lag
Inventory investment models.	Stock adjustments in response to sales changes and their effects on production.	Inventory investment tends to lead; its declines during mild recessions are large relative to those in final sales.
Old monetary over-investment and current monetarist theories.	Changes in the supply of money, bank credit, interest rates, and the burden of private debt.	Money and credit flows (rates of change) are highly sensitive, early leaders; velocity, market rates of interest, credit outstanding coincide or lag.
Hypotheses of cost-price imbalances, volatility of prospective rates of return, and expectational errors.	Changes in costs and prices, in the diffusion, margins, and totals of profits, and in business expectations.	Profit variables and stock price indexes are sensitive early leaders. Unit labor costs lag.

motion. Monocausal theories may help explain some episodes but are invalidated by long experience. The regularities noted above are complementary in the interdependent economic system but some of them may be more important under certain temporarily prevailing conditions, others under different conditions. Thus, for business cycles analysis and forecasting, groups of leading, coincident and lagging indicators representing a whole set of these relationships are expected to outperform any individual indicators or subsets representing fewer regularities. This insight provides a general rationale for the line of research summarized below.

SELECTING AND EXPLAINING THE PRINCIPAL INDICATORS. Cyclical indicators have been selected and analysed in a series of studies by the NBER and most recently by the Bureau of Economic Analysis (BEA) in the US Department of Commerce (Mitchell and Burns, 1938; Moore, 1950, 1961; Moore and Shiskin, 1967; Zarnowitz and Boschan, 1975a, 1975b). The results include a cross-classification of over 100 series by several broad 'economic-process' groups (e.g. production and income, fixed capital investment, money and credit) and typical timing at business cycle peaks and troughs. The data are regularly presented in a monthly report of BEA, *Business Conditions Digest* (BCD). A detailed weighting scheme was developed to score each of these series by seven major criteria: economic significance, statistical adquacy, consistency of cyclical timing, conformity to business expansions and contractions, smoothness, prompt availability or currency, and reliability of preliminary as compared with revised data. As far as possible, the assessments were based on statistical measures to ensure their consistency and replicability.

The information thus collected served as a basis for the construction of *composite indexes* of leading, coincident and lagging indicators. These indexes incorporate the best-scoring series from the different economic-process categories and combine those with similar cyclical timing, using their overall performance scores as weights. The series are all monthly; all but a few, as noted below, represent real rather than nominal variables.

The coincident index comprises non-farm employment, industrial production, real personal income less transfer payments, and real manufacturing and trade sales. Repeated tests showed this index to have a better record of conformity, timing and currency than alternative indexes including real GNP and the unemployment rate.

There are good reasons to expect the sequences of the leading, coincident and lagging indexes to persist, as indeed they do. Several of the component leaders represent early stages of production and investment processes – commitments that precede the later stages of outlays, construction put in place and deliveries. This subset includes new business formation, contracts and orders for plant and equipment, new orders for consumer goods and materials, and permits for new housing.

The timing relations depend not only on technology but also on the state of the economy. Thus delivery periods get progressively longer just before and during recoveries and especially in booms when orders back up and strain the capacity to produce; and they get progressively shorter when an expansion slows down and a contraction develops. This explains the leads of vendor performance, percent of companies receiving slower deliveries and also, in part, the fact that the leads of the indicators tend to be considerably longer, but also more variable, at peaks than at troughs.

The change in manufacturing and trade inventories on hand and on order tends to turn before sales to which the desired level of the stocks is adjusted (a type of accelerator relationship). This series, a volatile mixture of intended and unintended investment, requires some smoothing. Total inventories move sluggishly; the ratio of inventories to sales is a component of the lagging index.

Sensitive prices of industrial materials are related to new orders, vendor performance and inventory investment. The leading composite now includes the rate of change in an index of these prices but this is a very volatile series, even in somewhat smoothed form. In times of low inflation, the index itself (i.e. the level of such prices) would probably make a better indicator.

Another nominal indicator, the rate of change in business and consumer credit outstanding, leads because the new loans principally serve to finance investment in processes that are themselves leading (in inventories, housing and consumer durables; also in plant and equipment, where the loans are largely taken out early in the process). Here too, there are timing sequences that reflect stock-flow relationships: new increments lead, totals lag. The stock of commercial and industrial loans outstanding (deflated) is a component of the lagging index, and so is the ratio of consumer installment credit outstanding to personal income.

Compared with the overall credit flows, rates of growth in monetary aggregates show in general lower cyclical conformities and amplitudes and more random variations. They have historically led at business cycles turns by highly variable but mostly long intervals. The aggregates themselves are dominated by strong upward trends and show persistent declines only in cycles with severe contractions. However, a measure of 'real balances', the broadly defined money supply M2 deflated by a consumer price index, anticipated most of the recent business turns and is included in the current leading index. In late stages of expansion (contraction) money increased less (more) than prices.

The Standard & Poor's price index of 500 common stocks is included in the leading composite without adjustment for inflation. The market apparently tracks or anticipates well the movement of corporate earnings which is itself characterized by early timing. Money wages often rise less than prices in recoveries and more than prices late in expansion, while output per hour of labour fluctuates procyclically around a rising trend, generally with leads. Labour costs per unit of output, therefore, also move procyclically relative to their upward trends but with lags (they are a component of the lagging index). As a result of these tendencies connected with cyclical changes in sales and the rates of utilization of labour and capital, profit margins and totals swing widely in each cycle with sizeable leads.

Stock prices also tend to react inversely to changes in market interest rates. It is when an expansion (contraction) is well advanced and sufficiently strong that bank rates and bond yields tend to rise (decline) substantially, that is, interest rates generally lag. The average prime rate charged by banks is included in the lagging index.

Finally, there are the labour market indicators. Changes in hours are less binding than changes in the number employed, so the average workweek in manufacturing leads because it is altered early in response to uncertain signs of shifts in the demand for output. Initial claims for unemployment insurance lead the unemployment rate by short intervals. The average duration of unemployment lags the unemployment rate and is a component of the lagging index. These series, of course, show strong countercyclical movements, so they are used in inverted form.

FUNCTIONS. When used collectively, the indicators provide over the course of business cycles a revolving flow of signals. Shallow and spotty declines in the leading series provide only weak and uncertain warnings; a run of several large declines increases a risk of a general and serious slow-down or recession. The latter may suggest some stabilizing policy actions which, if effective, could falsify the warning. The coincident indicators confirm or invalidate the expectations based on the behaviour of the leaders and any related policy decisions.

The lagging indicators provide further checks on the previously derived inferences, in particular on any early designation of the timing of a business cycle turn. Moreover, they also act as predictors. The turning points in the lagging index systematically precede the opposite turns in the leading index. Unit labour costs, interest rates, outstanding debt, and inventories measure or reflect the costs of doing business. For this reason, these series, when inverted, show very long leads. For example, declines in inventories and interest rates during a recession pave the way for an upturn in new orders and then output of materials and finished goods (Zarnowitz and Boschan, 1975b, and Moore, 1983, ch. 23).

CRITIQUE AND EVIDENCE. Enough has been said above on the reasons for the observed behaviour of indicators and their links to business cycle theories to weaken if not disprove the charge of 'measurement without theory'. If the reasons are simple so much the better. Macroeconomic forecasting, which the indicator system is designed to aid, must be essentially consistent with the ascertained regularities of business fluctuations, however difficult it may be to reconcile these 'stylized facts' with the preconceptions of general equilibrium theory.

The real problems with the indicators are mainly practical. Large amounts of random noise, large revisions of originally published figures, and short lead times (which occur mostly at troughs of short recessions) detract from the usefulness of some leading series. Those irregular variations and data errors in its components that are independent tend to cancel out in the leading index, which is therefore relatively smooth. This reduces but does not eliminate the problem of extra turns or false warnings. The index signalled each of the eight recessions but also each of the four major slow-downs (phases of below-average but still positive growth) in 1948–85. In sum, the leading indicators predict best the 'growth cycles', that is, fluctuations in trend-adjusted aggregates of output, employment, etc. This was found to be true as well for Japan, Canada and the major countries of Western Europe (Moore, 1983, chs 5 and 6). A sequential signalling system designed to safeguard against false signals and discriminate in a timely fashion between recessions and slow-downs has been devised and tested with promising results (Zarnowitz and Moore, 1982).

Forecasting with leading indicators has a long history of applications, elaborations, and revisions occasioned by new data and research findings, and changes in the workings of the economy (Burns, 1950; Moore, 1983, ch. 24). Repeated tests were made of both the turning-point predictions and forecasts of series such as real GNP and industrial production (Hymans, 1973; Neftci, 1979; Auerbach, 1982). Tests have also been made by duplicating the US indicator test using data for other countries (Klein and Moore, 1985). The most demanding, correctly performed tests produced generally positive results (see Auerbach, 1982; Moore, 1983, chs 24 and 25).

V. ZARNOWITZ

See also BUSINESS CYCLES; DEMAND MANAGEMENT; STABILIZATION POLICY.

BIBLIOGRAPHY
Auerbach, A. 1982. The index of leading economic indicators: 'Measurement without theory' thirty-five years later. *Review of Economics and Statistics* 64(4), November, 589–95.
Burns, A.F. 1950. *New Facts on Business Cycles.* New York: National Bureau of Economic Research.
Burns, A.F. and Mitchell, W.C. 1946. *Measuring Business Cycles.* New York: Columbia University Press for the National Bureau of Economic Research.
Haberler, G. 1964. *Prosperity and Depression.* New edn. Cambridge, Mass.: Harvard University Press.
Hymans, S. 1973. On the use of leading indicators to predict cyclical turning points. *Brookings Papers on Economic Activity* 2, Washington, DC: Bookings Institution.
Klein, P.A. and Moore, G.H. 1985. *Monitoring Business Cycles in Market-Oriented Countries.* Cambridge, Mass.: Ballinger for the National Bureau of Economic Research.
Mitchell, W.C. and Burns, A.F. 1938. Statistical Indicators of Cyclical Revivals. *Bulletin 89* New York: National Bureau of Economic Research.
Moore, G.H. 1950. *Statistical Indicators of Cyclical Revivals and Recessions.* Occasional Paper No. 31, New York: National Bureau of Economic Research.
Moore, G.H. 1961. *Business Cycles Indicators.* New York: National Bureau of Economic Research.
Moore, G.H. 1983. *Business Cycles, Inflation, and Forecasting.* 2nd edn, Cambridge, Mass.: Ballinger for the National Bureau of Economic Research.
Moore, G.H. and Shiskin, J. 1967. *Indicators of Business Expansions and Contractions.* New York: National Bureau of Economic Research.
Neftci, S. 1979. Leading-lag relations, exogeneity, and prediction of economic time series. *Econometrica* 47(1), January, 101–13.
Zarnowitz, V. (ed.) 1972. *The Business Cycle Today.* New York: National Bureau of Economic Research.
Zarnowitz, V. 1985. Recent work on business cycles in historical perspective: a review of theories and evidence. *Journal of Economic Literature* 23(2), June, 523–80.
Zarnowitz, V. and Boschan, C. 1975a. Cyclical indicators: an evaluation and new leading index. *Business Conditions Digest*, May, v–xxii.
Zarnowitz, V. and Boschan, C. 1975b. New composite indexes of coincident and lagging indicators. *Business Conditions Digest*, November, v–xxiv.
Zarnowitz, V. and Moore, G.H. 1977. The recession and recovery of 1973–1976. *Explorations in Economic Research* 4(4), Fall, 471–557.
Zarnowitz, V. and Moore, G.H. 1981. The timing and severity of the 1980 recession. *NBER Reporter*, Spring, 19–21.
Zarnowitz, V. and Moore, G.H. 1982. Sequential signals of recession and recovery. *Journal of Business* 55(1), January, 57–85.

indifference, law of. A designation applied by Jevons to the following fundamental proposition: 'In the same open market, at any one moment, there cannot be two prices for the same kind of article.'

This proposition, which is at the foundation of a large part of economic science, itself rests on certain ulterior grounds: namely, certain conditions of a perfect market. One is that monopolies should not exist, or at least should not exert that power in virtue of which a proprietor of a theatre, in Germany for instance, can make a different charge for the admission of soldiers and civilians, of men and women. The indivisibility of the articles dealt in appears to be another circumstance which may counteract the law of indifference in some kinds of market, where price is not regulated by cost of production. [Jevons, *Theory of Exchange*, 2nd edn, p. 99 (statement of the law). Walker, *Political Economy*, art. 132 (a restatement). Mill, *Political Economy*, bk. ii. ch. iv. § 3 (imperfections of actual markets). Edgeworth, *Mathematical Psychics*, pp. 19, 46 (possible exceptions to the law of indifference).]

[F.Y. EDGEWORTH]
Reprinted from *Palgrave's Dictionary of Political Economy*.

BIBLIOGRAPHY
Edgeworth, F.Y. 1881. *Mathematical Psychics* London: Kegan Paul.
Jevons, W.S. 1875. *Money, the Mechanism of Exchange*. London.
Mill, J.S. 1848. *Principles of Political Economy*. London: J.W. Parker.
Walker, F.A. 1886. *A Brief Textbook of Political Economy*. London.

indifference curves. *See* EDGEWORTH, FRANCIS YSIDRO; PARETO, VILFREDO.

indirect taxes. It is conventional to describe direct taxes as taxes where the person legally liable to pay the tax is also the person whose income or welfare is reduced as a result of its imposition: while indirect taxes are those where liability can be shifted to someone else. This distinction is essentially an arbitrary one. All taxes can be shifted to some degree: only in exceptional circumstances can any agent shift a tax completely. In common usage, indirect taxes are those which are paid by retailers, wholesalers or manufacturers, but believed to be shifted toward final consumers. In this entry indirect taxation is regarded as synonymous with commodity taxation.

There are three main categories of indirect tax. Excise duties fall on particular commodities, especially those goods which are traditionally subject to particularly heavy taxation, such as tobacco products and alcoholic drinks. Such taxes are often specific – charged per unit of the commodity concerned – but may be ad valorem – assessed as a percentage of the retail price. In many countries these specific tax rates have failed to keep pace with inflation, and this has tended to reduce the incidence of these taxes in relation to total revenue and to the price of the commodities concerned.

More broadly based indirect taxes are usually set at ad valorem rates. These may be single stage taxes, collected at wholesale or retail level. Alternatively, multi-stage taxes may be imposed at each part of the production process. Turnover or cascade taxes are of this kind, but the most common multi-stage tax is a value added tax. This has been obligatory for member states of the EEC since directives based on the Neumark Report of 1963, and is now used in around forty countries. This tax is payable at all stages of production, but recoverable by all business purchasers, who are however obliged to charge tax on their own output. The consequence is that net tax is payable only on sales to final consumers.

The value added tax is consistent with one important result from the theory of commodity taxation. This is that commodity taxes should be levied only on purchases of goods by final consumers, and not on intermediate transactions between producers (Diamond and Mirrlees, 1971). The reason for this is that the distributional or other objectives of commodity taxation can in all cases be achieved equally well by the taxation of final commodities; the taxation of intermediate goods achieves no advantage in this but additionally distorts the choices of inputs made by producers. It therefore imposes the avoidable distortion of production inefficiency on top of the inevitable deadweight loss in consumption which is common to any system of commodity taxation.

Commodity taxes impose deadweight losses on consumers. These arise from the distortions of consumer choice which create costs over and above the tax revenue derived by governments. Traditionally these have been expressed in terms of consumer surplus triangles but are now more effectively expressed using the dual formulation of demand theory implied by the expenditure function (Diamond and McFadden, 1974). This gives deadweight loss as

$$L = E(p^+ t, u) - E(p, u) - tx$$

where p is a vector of producer prices, t is the tax vector, x the purchased commodity vector and u the reference utility level for evaluation of the expenditure function.

The optimal structure of commodity taxation may be derived from the minimization of L, and this leads to two schools of thought on the appropriate structure of commodity taxes. By choosing a vector t to minimize L we derive the Ramsey (1927) rules for optimal commodity tax rates in the implicit form

$$\sum_j t_j \left(\frac{\partial x_i}{\partial t_j} \right)_u = \lambda x_i$$

for some λ increasing in tax revenue. This can be interpreted as requiring that the compensated demand for all goods should be reduced in the same proportion. If there is no net complementarity or substitutability, this condition reduces to

$$\frac{t_i}{p_i} = \lambda \frac{x_i}{p_i} \left(\frac{\partial x_i}{\partial t_i} \right)_u$$

which yields an inverse elasticity rule: tax should bear most heavily on commodities in inelastic demand.

A weakness of this analysis is that L is still more effectively minimized – indeed reduced to zero – by the imposition of a lump sum tax. While lump sum taxes varying across individuals and independent of their economic behaviour are generally reckoned to be impracticable, a uniform lump sum tax is feasible. The primary reasons for rejecting a poll tax – concern for distributional effects among households with different tastes and endowments – are abstracted from in the Ramsey formulation. A more direct way of reaching the same conclusion is to observe that life is the most inelastically demanded commodity of all. It follows that any set of Ramsey taxes will always be dominated by a poll tax.

If, however, a distributional objective is introduced then any set of commodity taxes will generally be inferior to a tax related to the total income of the consumer (or his total consumption, since there is no difference at the present level of abstraction). Thus there is a role for commodity taxes only if they are related to other household characteristics which cannot be observed and taxed directly, such as household skill levels. The argument illustrates a general feature of recent optimal tax theory, which shows that efficient tax structures are often very sensitive to the assumptions made about the other policy instruments available.

These results direct attention towards a different tradition (see, for example Hotelling, 1938a), which favours uniformity of rates of commodity taxes, on the grounds that this leaves relative commodity prices equal to relative marginal costs. This would be appropriate if all commodities were taxable, but there is at least one important good – leisure – which cannot be subjected to taxation. This takes the problem of optimal commodity taxes into the realm of the second best and suggests relatively high rates of taxation on those goods which are complementary with leisure and lower rates on those which are substitutes for it.

There are other reasons for departures from uniformity. Merit goods (Musgrave, 1959) are commodities, such as education, whose consumption is thought to have some value, either social or for the individual concerned, beyond his own personal assessment. This may be a reason for specially low rates of tax on particular commodities or, more commonly, for specially high excise taxes. Corrective taxes are also a means by which market outcomes can be induced to reflect the externalities – good or bad – which are associated with particular kinds of production or consumption.

With these exceptions, the theoretical arguments for extensive departures from a general principle of uniformity in commodity taxation do not seem strong. In the main, most objectives which governments seek through elaborately differentiated rate structures can be more effectively achieved in other ways. Since this uniformity has considerable administrative advantages, both analytical and practical considerations point in a similar direction. The widespread move throughout the world to broadly based value-added taxes as a primary instrument of indirect taxation reflects the application of these principles.

JOHN KAY

See also PUBLIC FINANCE; TAX INCIDENCE; VALUE-ADDED TAX.

BIBLIOGRAPHY
Atkinson, A.B. and Stiglitz, J.E. 1980. *Lectures on Public Economics.* New York: McGraw-Hill.
Diamond, P.A. and McFadden, D. 1974. Some uses of the expenditure function in public finance. *Journal of Public Economics* 3(1), February, 3–21.
Diamond, P.A. and Mirrlees, J.A. 1971. Optimal taxation and public production, Parts I and II. *American Economic Review* 61, June, 8–27 and 261–78.
Hotelling, H. 1938. The general welfare in relation to problems of taxation and of railway and utility rates. *Econometrica* 6, July, 242–69.
Musgrave, R.A. 1959. *The Theory of Public Finance.* New York: McGraw-Hill.
Ramsey, F.P. 1927. A contribution to the theory of taxation. *Economic Journal* 37, March, 47–61.

indirect utility function. After many independent discoveries that were widely separated in time and space, the indirect utility function has in the last 35 years gradually become a standard part of demand theory. Its first discovery was made as early as 1886 by Antonelli in Italy, who also derived what has come to be known as Roy's Identity (see Chipman's introduction to the translation of Antonelli (1886) in Chipman et al., 1971). Later contributions came from Konyus (1924, 1926) and Byushgens in Russia, from Hotelling (1932) and Court (1941, pp. 284–97) in the United States, from Roy (1942, 1947) and Ville (1946) in France, and from Wold (1943–4) and Malmquist (1953) in Sweden; a good brief history may be found in Diewert (1982, pp. 547–50).

But it was not until the early 1950s and the contributions of Houthakker (1951–2, 1960) that the indirect utility function became an integral part of the theory of consumer's behaviour. Indeed, the very names in standard use appear to be due to him, 'indirect utility function' in (1951–2, p. 157) and 'Roy's Identity' in (1960, p. 250).

1. Definition and simple properties. Suppose that the consumer has completely preordered preferences defined over the commodity space R^{n+} of non-negative bundles $x = (x_1, x_2, \ldots, x_n)$, that those preferences are representable by a real-valued utility

function u, that he (or she) faces competitively determined positive money prices $(p_1, p_2, \ldots, p_n) = p$ for the n goods, and has exogenously determined monetary wealth $\omega > 0$. It is standard in demand theory to assume that the consumer chooses a bundle x^* by solving the optimization problem:

Max(p, ω) Find $x \in R^{n+}$ to max $u(x)$

$$\text{subject to } \langle p, x \rangle \leqslant \omega \quad (1)$$

where the notation $\langle \cdot, \cdot \rangle$ means the inner product of the two vectors concerned.

Assume that Max(p, ω) has a unique solution x^*, for which it suffices that preferences be monotonically increasing and strictly convex. Then the number

$$\tau^* = u(x^*) \quad (2)$$

is the *value* of Max(p, ω). This joint determination of solution and value once p and ω are known implies the existence of two functions of the price-wealth pair (p, ω), called respectively the *ordinary* (or Marshallian) *demand function* f: $R^{n++} \times R^{++} \to R^{n++}$, defined by

$$x^* = f(p, \omega) \quad (3)$$

and the *indirect utility function* v: $R^{n++} \times R^{++} \to R$, defined by

$$\tau^* = v(p, \omega) \quad (4)$$

Define the attainable (or budget) set $A(p, \omega)$ by

$$A(p, \omega) = \{x \in R^{n+} : \langle p, x \rangle \leqslant \omega\}$$

From (1) it follows that for any $\lambda > 0$, $A(\lambda p, \lambda \omega) = A(p, \omega)$, so that both f and v are positively homogeneous of degree zero in (p, ω). Next, if $(p^1 - p^2) \in R^{n+}$ and $p^1 \neq p^2$ then $A(p^1, \omega) \subset A(p^2, \omega)$, from which $v(p^1, \omega) \leqslant v(p^2, \omega)$; for similar reasons, $v(p, \cdot)$ is nondecreasing. It can be shown further that if u is continuous then so is v (see e.g. Varian, 1984, pp. 121, 326–7).

A useful result is that $v(\cdot, \omega)$ is quasi-convex. To prove this let $p^t = tp^1 + (1 - t)p^2$, where $t \in [0, 1]$. Then for any $x \in A(p^t, \omega)$,

$$t\langle p^1, x \rangle + (1 - t)\langle p^2, x \rangle \leqslant \omega. \quad (5)$$

If $t = 0$, $x \in A(p^2, \omega)$, while if $t = 1$, $x \in A(p^1, \omega)$. Otherwise, suppose that x is in neither $A(p^1, \omega)$ nor $A(p^2, \omega)$. Then $t\langle p^1, x \rangle > t\omega$ and $(1 - t)\langle p^2, x \rangle > (1 - t)\omega$, which on addition yield a contradiction to (5). So x is in either $A(p^1, \omega)$ or $A(p^2, \omega)$. Hence $v(p^t, \omega)$, which is the sup of $u(\cdot)$ on $A(p^t, \omega)$, can be no larger than $\max[v(p^1, \omega), v(p^2, \omega)]$, which are themselves the sups of $u(\cdot)$ on $A(p^1, \omega)$ and $A(p^2, \omega)$, respectively. But the condition $v(p^t, \omega) \leqslant \max[v(p^1, \omega), v(p^2, \omega)]$ is the original definition of the quasi-convexity of the function $v(\cdot, \omega)$ (see Fenchel, 1953, p. 117).

2. Relations between the ordinary demand functions and the indirect utility function. For simplicity, the following assumptions are made: (a) x^* is a strictly positive vector. (b) Each function involved is as differentiable as required. (c) At any $x \in R^{n+}$ there is at least one commodity in which u is strictly increasing (this implies local non-satiation of preferences).

Suppose that at x^* the constraint (1) is 'slack', i.e. $\omega - \langle p, x^* \rangle = \delta > 0$. Let k be any good with property (c) at x^*, and define a new bundle x^1 by putting $x_i^1 = x_i^*$ for $i \neq k$, and $x_k^1 = x_k^* + (\delta/p_k)$. Then by construction $\langle p, x^1 \rangle = \omega$, while from (c) $u(x^1) > u(x^*)$, contradicting the hypothesis that x^* solves max(p, ω). So

$$\langle p, x^* \rangle = \omega \quad (6)$$

Next, define $L: R^{n+} \times R^{n++} \to R$ by

$$L(x^1, p^1) = v(p^1, \langle p^1, x^1 \rangle) - u(x^1) \quad (7)$$

where x^1 and p^1 are arbitrary. From (2) and (4), for any x^1 the value $v(p^1, \langle p^1, x^1 \rangle)$ is the *maximized* level of utility when prices are p^1 and wealth $\langle p^1, x^1 \rangle > 0$. Hence, $L(\cdot, p^1)$ is positive semi-definite, i.e. $x^1 \in R^{n+}$ implies $L(x^1, p^1) \geqslant 0$ for any p^1. Putting $p^1 = p$, the actual prices, if x^* solves Max(p, ω) it follows from (2), (4) and (6) that

$$L(x^*, p) = v(p, \langle p, x^* \rangle) - u(x^*) = 0 \quad (8)$$

Hence x^* attains the infimum of $L(\cdot, p)$. So from (6), (8) and the Chain Rule,

$$\forall i = 1, 2, \ldots, n \qquad v_\omega(p, \omega) p_i = u_i(x^*) \quad (9)$$

From (c), $u_i(x^*) > 0$ for at least one i. Since $p_i > 0$ this implies the simple but important result

$$v_\omega(p, \omega) > 0 \quad (10)$$

i.e. the marginal utility of wealth is positive.

From (2), (3) and (4) the equation

$$v(p, \omega) = u(f(p, \omega))$$

is an identity in (p, ω). So differentiating each of the individual demand functions f_i with respect to (wrt) each p_j and ω yields,

$$\forall j = 1, 2, \ldots, n \qquad v_j(p, \omega) = \sum u_i(x^*) f_{ij}(p, \omega)$$
$$v_\omega(p, \omega) = \sum u_i(x^*) f_{i\omega}(p, \omega) \quad (11)$$

From (6) and (3),

$$\langle p, f(p, \omega) \rangle = \omega$$

This is another identity in (p, ω), and differentiating it wrt each p_j and ω results in

$$\forall j = 1, 2, \ldots, n \qquad f_j(p, \omega) + \sum p_i f_{ij}(p, \omega) = 0 \quad (12)$$
$$\sum p_i f_{i\omega}(p, \omega) = 1$$

From (11) and (9),

$$\forall j = 1, 2, \ldots, n \qquad v_j(p, \omega) = v_\omega(p, \omega) \sum p_i f_{ij}(p, \omega)$$

and from this and (12),

$$\forall j = 1, 2, \ldots, n \qquad v_j(p, \omega) = -f_j(p, \omega) v_\omega(p\omega) \quad (13)$$

Equation (13) is the main result connecting v with f. From (3), (a), (10) and (13) there follow

$$\forall j = 1, 2, \ldots, n \qquad v_j(p, \omega) < 0 \quad (14)$$

and Roy's Identity (1942, p. 24; 1947, p. 217),

$$\forall j = 1, 2, \ldots, n \qquad x_j^* = -v_j(p, \omega)/v_\omega(p, \omega) \quad (15)$$

As deservedly famous as is (15), its equivalent version (13) reveals the structures involved more clearly, since it focuses sharply on the relations between the functions v and f rather than the particular quantities x_j^*. In each of Roy's contributions the identity is first given in the form $v_j(p, \omega)/x_j^* = -v_\omega(p, \omega)$, and is used primarily to prove (14); later, in Roy (1947, p. 220), the identity takes the more usual form (15).

Since (13) is an identity, differentiating it wrt any p_i yields $\forall i, j = 1, 2, \ldots, n$

$$-v_{ji}(p, \omega) = f_{ij}(p, \omega) v_\omega(p, \omega) + f_j(p, \omega) v_{\omega i}(p, \omega)$$

Applying Young's Theorem to these equations, by symmetry, $\forall i, j = 1, 2, \ldots, n$

$$f_{ij}(p, \omega) v_\omega(p, \omega) + f_i(p, \omega) v_{\omega j}(p, \omega) = f_{ji}(p, \omega) v_{\omega i}(p, \omega) \quad (16)$$

Now make the quite restrictive assumption that for each $p_i, v_{\omega i}(p, \omega) = 0$; this requires in effect that each good have unitary elasticity of demand (see Samuelson, 1942, pp. 80–81). Then from (10) and (16),

$$\forall i, j = 1, 2, \ldots, n \qquad f_{ij}(p, \omega) = f_{ji}(p, \omega)$$

which are Slutsky-like equations that apply not to compensated but to ordinary demand functions.

3. Relations with the cost function and the compensated demand functions. Suppose now that a *target* level τ of utility is specified and the following new optimization problem posed:

Min(p, τ): Find $x \in R^{n+}$ to min$\langle p, x \rangle$

$$\text{subject to } u(x) \geqslant \tau \quad (17)$$

Assume that a unique solution x^{**} to this problem exists, yielding a value $\langle p, x^{**} \rangle$. This implies the existence of two functions of the price-target pair (p, τ), called the *compensated (or Hicksian) demand function* $h: R^{n++} \times R \to R^{n+}$, defined by

$$x^{**} = h(p, \tau) \quad (18)$$

and the *cost (or expenditure) function* $\gamma: R^{n++} \times R \to R^+$, given by

$$\langle p, x^{**} \rangle = \gamma(p, \tau) \quad (19)$$

Retain assumptions (a)–(c), replacing x^* by x^{**}. Define $M: R^{n+} \times R^{n++} \to R$ by putting

$$M(x^1, p^1) = \gamma(p^1, u(x^1)) - \langle p^1, x^1 \rangle \quad (20)$$

where x^1 and p^1 are arbitrary, as before. It follows that $M(\cdot, p^1)$ is negative semi-definite. Putting $p^1 = p$, the actual prices, it follows that if x^{**} solves Max(p, ω) then

$$M(x^{**}, p) = \gamma(p, u(x^{**})) - \langle p, x^{**} \rangle = 0 \quad (21)$$

so that x^{**} maximizes $M(\cdot, p)$. Then a development *exactly* like that of the last section leads to a simple but basic result on the interrelations betwen h and γ, namely:

$$\forall j = 1, 2, \ldots, n \qquad \gamma_j(p, \tau) = h_j(p, \tau) \quad (22)$$

where h_j is the compensated demand function for the jth good. From (22) and (a),

$$\forall j = 1, 2, \ldots, n \qquad \gamma_j(p, \tau) > 0 \quad (23)$$

From (18), (22) can be rewritten in the more customary version that has come to be called *Shephard's Lemma* (Shephard, 1953), although it dates back at least to Hotelling (1932).

$$\forall j = 1, 2, \ldots, n \qquad x_j^{**} = \gamma_j(p, \tau) \quad (24)$$

Thus (22) (or the Lemma) plays a role in the analysis of this problem which is symmetrical to that played by (13) (or Roy's Identity) in the analysis of Max(p, ω).

However, there are two important structural asymmetries between the problems max(p, ω) and min(p, τ). First, suppose that for some reason (such as incompleteness of preferences) the utility function u does not exist, so that v does not exist either. Clearly, since Max(p, ω) requires a scalar measure of utility it cannot be defined in this new situation. However, by replacing the target level τ of utility by a target bundle x^τ, one can still define a perfectly sensible minimum problem Min(p, x^τ).

The second asymmetry is that while $v(\cdot, \omega)$ is only quasi-convex, $\gamma(\cdot, \tau)$ is actually concave, and this without any assumptions on preferences. Since (full) concavity imposes sharper restrictions on any function than does quasi-convexity, the analysis of Min(p, τ) (or of Min(p, x^τ)) yields easier proofs of basic results than does that of Max(p, ω). For example, from (22) $h_{jj}(p, \tau) = \gamma_{jj}(p, \tau)$, and since $\gamma(\cdot, \tau)$ is concave $\gamma_{jj}(p, \tau) \leqslant 0$, proving that the substitution effect is non-positive.

4. Duality. It is not productive to oppose the virtues of minimum problems to those of maximum problems. Indeed, the most efficient path of the derivation of such propositions as the 'Fundamental Equation or Value Theory' (Hicks, 1939, p. 309) is by a judicious mixture of the two, i.e. by first solving $\max(p, \omega)$ to obtain $\tau^* = v(p, \omega)$ and $x^* = f(p, \omega)$, and then showing that x^* also solves $\min(p, \tau^*)$. One interesting result that one can reach by this route relates all four functions v, f, γ and h in one equation:

$$\forall i, j = 1, 2, \ldots, n$$
$$\gamma_i(p, \tau^*) f_{j\omega}(p, \omega) = -v_i(p, \omega) h_{jt}(p, \tau^*) \quad (25)$$

Since from Shephard's Lemma the left-hand side of (25) is the Hicksian income effect of a change in p_i on the demand for good j, so is the right-hand side (RHS). Notice that although each of the components of the RHS is affected by choice of the utility index u, their product is not.

Revert now to the assumptions of section 1. The problems $\text{Max}(p, \omega)$ and $\text{Min}(p, \tau^*)$ are often referred to in the literature as dual to each other. For reasons given in detail in the entry on COST MINIMIZATION AND UTILITY MAXIMIZATION, this usage seems inappropriate. However, as pointed out by Konyus and Byushgens (1926, p. 159) and Houthakker (1951–2, pp. 157–8), there *is* an interesting duality between the functions u and v. To show this, first rewrite the given prices and income (p, ω) as (p^*, ω^*), where $\omega^* > 0$ will be kept constant throughout. Next, define new income-normalized prices $q \in R^{n++}$ for any p by

$$q = (\omega^*)^{-1} p$$

Then use the homogeneity of f and v in (p, ω) to put them in the normalized forms $F: R^{n++} \rightarrow R^{n+}$ and $w: R^{n++} \rightarrow R$, defined by

$$F(q^*) \equiv f(p^*, \omega^*) \quad (26)$$

and

$$w(q^*) \equiv v(p^*, \omega^*) \quad (27)$$

Let

$$A(q^*) = \{x \in R^{n+} : \langle q^*, x \rangle \leqslant 1\} = A(p, \omega).$$

Then $\text{Max}(p^*, \omega^*)$ can also be written in a new form:

$$\text{Max}(q^*): \text{Find } x \in A(q^*) \text{ to max } u(x).$$

The data of $\text{Max}(q^*)$ are q^* and u. In the same way, the chosen bundle x^* and w are the data for a problem dual to $\text{Max}(q^*)$. Let $B(x^*) = \{q \in R^{n++} : \langle x^*, q \rangle \leqslant 1\}$. Then the dual problem, situated in the space of normalized prices q, is

$$\text{Min}(x^*): \text{Find } q \in B(x^*) \text{ to min } w(q).$$

A unique solution q^{**} to $\text{Min}(x^*)$ (for which the strict quasi-convexity of w would suffice) implies the existence of two functions $\Phi: R^{n++} \rightarrow R^{n+}$ and $U: R^{n++} \rightarrow R$, defined analogously to (3) and to (2)-cum-(4) by

$$q^{**} = \Phi(x^*) \quad (28)$$

$$U(x^*) = w(q^{**}). \quad (29)$$

By the construction of $\text{Min}(x^*)$, $x^* \in A(q)$ for any q. So $w(q)$ must be at least as large as the utility level at x^*. But since x^* is bought at q^*, that utility level is $w(q^*)$. Thus

$$\forall q \in B(x^*) \qquad w(q) \geqslant w(q^*) \quad (30)$$

Since $\min(x^*)$ is assumed to have a unique solution, (30) says that it must be q^*. If follows from this, (26) and (28) that Φ is actually the *inverse* demand function F^{-1}. Moreover, $U(x^*) = w(q^*)$. So from this, (27), (4) and (2),

$$U(x^*) = u(x^*) \quad (31)$$

However, it cannot be concluded from (31) that $U \equiv u$ unless every bundle x in the domain of u is bought at some price-income pair (p, ω) and so can be an optimizing bundle such as x^*. This property requires that u be strictly quasi-concave. Granted that, (31) shows that the direct utility function u is recoverable from the indirect utility function w, just as w is obtainable from u.

PETER NEWMAN

See also DEMAND THEORY; INDEX NUMBERS; ROY, RENÉ.

BIBLIOGRAPHY

Antonelli, G.B. 1886. *Sulla teoria matematica della economia politica.* Pisa: Tipografia del Folchetto.

Chipman, J.S., Hurwicz, L. Richter, M.K. and Sonnenschein, H.F. (eds) 1971. *Preferences, Utility and Demand.* New York: Harcourt, Brace, Jovanovich.

Court, L.M. 1941. Entrepreneurial and consumer demand theories for commodity spectra. *Econometrica* 9, 135–62, 241–97.

Diewert, W.E. 1982. Duality approaches to microeconomic theory. In *Handbook of Mathematical Economics,* Vol. II, ed. K.J. Arrow and M.D. Intriligator, Amsterdam: North-Holland, ch. 12, 535–99.

Fenchel, W. 1953. Convex cones, sets and functions. Department of Mathematics, Princeton University, mimeo.

Hicks, J.R. 1939. *Value and Capital.* Oxford: Clarendon Press.

Hotelling, H. 1932. Edgeworth's taxation paradox and the nature of demand and supply functions. *Journal of Political Economy* 40, 577–616.

Houthakker, H.S. 1951–2. Compensated changes in quantities and qualities consumed. *Review of Economic Studies* 19, 155–64.

Houthakker, H.S. 1960. Additive preferences. *Econometrica* 28, 244–57.

Konyus, A.A. 1924. The problem of the true index of the cost of living. *Economic Bulletin of the Institute of Economic Conjuncture* 9–10, 64–71. Trans. in *Econometrica* 7, 1939, 10–29.

Konyus, A.A. and Byushgens, S.S. 1926. K probleme pokupatelnoi cili deneg. *Voprosi Konyunkturi* 2(1), 151–72.

Malmquist, S. 1953. Index numbers and indifference surfaces. *Trabajos de estadistica* 4, 209–41.

Roy, R. 1942. *De l'utilité: contribution à la théorie des choix.* Paris: Hermann.

Roy, R. 1947. La distribution du revenu entre les divers biens. *Econometrica* 15, 205–25.

Samuelson, P.A. 1942. Constancy of the marginal utility of income. In *Studies in Mathematical Economics and Econometrics: In Memory of Henry Schultz,* ed. O. Lange, F. McIntyre and T.O. Yntema, Chicago: University of Chicago Press, 75–91.

Shephard, R.W. 1953. *Cost and Production functions.* Princeton: Princeton University Press.

Varian, H.R. 1984. *Microeconomic Analysis.* 2nd edn, New York: Norton.

Ville, J. 1946. Sur les conditions d'existence d'une ophelimité totale et d'un indice du niveau des prix. *Annales de l'Université de Lyon* 9, 32–9. Trans. in *Review of Economic Studies* 19, 1951, 123–8.

Wold, H.O.A. 1943–4. A synthesis of pure demand analysis. *Skandinavisk Aktuarietidskrift* 26, 85–144, 220–75; 27, 69–120.

individualism. Individualism is social theory or ideology which assigns a higher moral value to the individual than to the community or society, and which consequently advocates leaving individuals free to act as they think most conducive to their self-interest. The term was also, as noted below, sometimes used in the 19th century as a name for an actual economic system. When so used, the term denoted the competitive market system which lets the direction of the economy be the unintended outcome of the decisions made by myriad individuals about the uses to which they will put their own labour and resources.

The first edition (1896) of Palgrave's *Dictionary of Political Economy* defined individualism in the latter, narrower sense. The article entitled Individualism began by reporting that John Stuart Mill had applied the term to 'that system of industrial organisation in which all initiative is due to private individuals, and all organisation to their voluntary agreement'. The article then remarked: 'The natural antithesis to individualism is COLLECTIVISM or we may say SOCIAL-ISM, a system under which industry is directly organized by the state, which owns all means of production and manages all processes by appointed officers.' The author defined the fundamentals of the system of individualism quite precisely:

> The essential features of individualism are, (1) private property in capital, to which are added almost of necessity the rights of bequest and inheritance, thus permitting unlimited transfer and accumulation. (2) competition, a rivalry between individuals in the acquisition of wealth, a struggle for existence in which the fittest survive.

There could hardly be a better definition of capitalism, at least of the neo-classical economists' model of capitalism. John Stuart Mill's *Socialism* is cited as authority for such a use of 'individualism', properly enough: his *Chapters on Socialism* (1879) does describe 'the principle of individualism' as 'competition, each one for himself and against all the rest. It is grounded on the opposition of interests, not the harmony of interests, and under it everyone is required to find his place by struggle, by pushing others back or being pushed back by them'; and later in the same work individualism is equated with 'quarrelling about material interests'. One might also cite Mill's earlier (1851) 'Newman's Political Economy', where 'the existing individualism', described as 'arming one human being against another, making the good of each depend upon evil to others', is said to be so morally inferior to socialism that socialism is 'easily triumphant' over it.

It may be thought that the Palgrave definition of individualism is unduly narrow: a modern scholar (Lukes, 1973) has distinguished no less than eleven meanings the term may have, ranging from respect for human dignity, autonomy, privacy, and self-development, to epistemological and method-ological individualism. Most of these meanings are indeed not considered in Palgrave's *Dictionary*, but since it is a dictionary *of political economy*, only meanings with an economic connotation can be expected to be treated. However, although that charge of undue narrowness may be dismissed, it may still appear that, considered historically, his usage is too narrow to be accurate for the whole modern Western tradition down to his own time.

The idea that the individual is morally more important than society goes back of course, in modern times, to the Renaissance. The same view, in religious terms, emerged at the Reformation, which made each individual, rather than the Church, the guardian of his own salvation; and this view got wider currency in 17th-century Puritanism. Neither the Renaissance nor the Reformation and the subsequent Puritanism reduced individuals to atoms of matter in motion, each seeking power and wealth at the expense of every other one. That step was taken by Hobbes in the mid-17th century: in his view, society was simply a congeries of colliding atoms in unceasing motion. That puts Hobbes's individualism close to, but leaves it broader than, Palgrave's concept.

In the 18th century Adam Smith gave full market individualism a more pleasant face, arguing that the most beneficent possible social result would be attained by leaving individuals free to make self-interested bargains in a competitive market: that was the doctrine of *laissez-faire*. And the market economy was solidly enough established in England by Smith's time that it could be accepted as a part of the natural order by that venerator of the traditional hierarchy of ranks, Smith's contemporary, Edmund Burke, though in Burke's hands the market economy became a much less pleasant affair. His *Thoughts and Details on Scarcity* (1795) was an unqualified endorsement of *laissez-faire*: it issued a shrill warning against 'breaking the laws of commerce, which are the laws of nature, and consequently the laws of God'. Governments must not interfere with 'the great wheel of circulation' even though it dooms 'so many wretches' to 'innumerable servile, degrading, unseemly, unmanly, and often most unwholesome and pestiferous occupations'.

In the 19th century, Bentham relentlessly restated and elaborated Hobbes's atomic individualism, and Benthamism became the dominant ideology. Its doctrine of human nature was summed up in its crudest form in James Mill's article *Government* (1820): 'The desire ... of that power which is necessary to render the persons and properties of human beings subservient to our pleasures in a grand governing law of human nature.'

So we may say that historically, at least down to 1820 or so, the Palgrave definition is not at all too narrow. But it is too narrow for the latter part of the century, for it leaves out a quite different idea of individualism, one which John Stuart Mill promoted implicitly in his *Principles of Political Economy* (1848) and explicitly in his *On Liberty* (1859), with its opening laudatory quotation from Wilhelm von Humboldt: 'The grand, leading principle, towards which every argument unfolded in these pages directly converges, is the absolute and essential importance of human development in its richest diversity.'

Let us call this *developmental individualism*. It is the antithesis of *possessive individualism*, which assumes that the human being is essentially a striver for, and a receptacle for the acquisition of, material goods. The whole doctrine of *On Liberty* puts Mill squarely in the camp of developmental individualism. And the famous chapter 'Of the Stationary State' (*Principles*, Bk. IV, ch. 6) is eloquent testimony to the depth of his revulsion from the existing acquisitive individual-ism of the competitive market economy. So, although the developmental ideal of individualism is not found as positively in Mill's *Political Economy* as it is in his *On Liberty*, we may treat the former text also as being on the developmental side. If we do so, however, we must add that Mill was himself so confused a political economist that he did not see that the acquisitive behaviour he denounced was entailed in the capitalist structure he accepted: he did not see that it was that structure which effectively denied a developmental life to the bulk of the wage earners.

A greater political economist than Mill, namely Marx, saw through this confusion and took the logical way out. Marx may be classified as the ultra-collectivist but it is important to see that for him the collective control of the economy was simply a necessary means to an end which was ultra-individualistic, that is, to a flowering of individuality which would be possible when capitalism with its alienation of labour had been surpassed. Marx condemned capitalism morally because it denied any such flowering.

> In bourgeois society ... the past dominates the present; in Communist society, the present dominates the past. In bourgeois society capital is independent and has individu-ality, while the living person is dependent and has no individuality. And the abolition of this state of things is

called by the bourgeois, abolition of individuality and freedom! And rightly so. The abolition of bourgeois individuality, bourgeois independence, and bourgeois freedom is undoubtedly aimed at (*Communist Manifesto*, 1848, sect.2).

And the final outcome of the communist revolution was to be 'an association, in which the free development of each is the condition for the free development of all' (ibid.).

Similarly:

> In a higher phase of communist society ... after labour has become not only a means of life but life's prime want; after the productive forces have also increased with the all-round development of the individual, and all the springs of co-operative wealth flow more abundantly – only then can the narrow horizon of bourgeois right be crossed in its entirety and society inscribe on its banner: From each according to his ability, to each according to his needs! (*Critique of the Gotha Programme*, 1875, I, 3).

The *Manifesto's* vision of a fully developed individual as the highest human attainment, echoed in the *Critique of the Gotha Programme*, puts Marx as firmly as Mill in the developmental camp. And just as Mill is there not only by virtue of his *On Liberty* but also by virtue of his *Political Economy*, so Marx is there not only by virtue of the *Manifesto* but also of the *Critique*. And we may add that Marx is there just as firmly in Volume I of *Capital* (1867), where he refers scornfully to the capitalist mode of production as that 'in which the labourer exists to satisfy the needs of self-expansion of existing values instead of, on the contrary, material wealth existing to satisfy the needs of *development on the part of the labourer*' (emphasis added).

There is no warrant in any of this for trying, as some commentators used to do, to drive a wedge between the young 'humanist' Marx and the 'mature' political economist. And, of course, Marx had a strongly developmental vision in his earliest work, the *Economic-Philosophic Manuscripts of 1844*. Thus from his earliest to his latest economic writings there is this development vision. Development individualism is at the very heart of his political economy.

We find, then, that by the time of Mill and Marx developmental individualism is well established: in the liberal tradition it takes place alongside the continuing possessive individualism; in Marx's theory it was inherent from the beginning.

What of the late 20th century? The liberal tradition still contains the two strands of individualism. On the one hand, two of the most esteemed liberal individualists of our time – Isaiah Berlin and John Rawls – are clearly developmental individualists. And on the other hand, the two most noted economic individualists – Friedrich Hayek and Milton Friedman – are equally clearly possessive individualists. Friedman, who would dismantle the welfare state and leave the distribution of economic benefits to an unrestrained competitive market, may be cited as the very model of a possessive individualist. Hayek, whose economic philosophy was set out succinctly in his 1945 lecture *Individualism, True and False*, tries to give market individualism a more agreeable image. He does this by claiming as 'true' individualists the great names in one line of the British tradition, a line from Locke through Mandeville, Hume, Tucker, Ferguson, Smith and Burke, down to Lord Acton, and by categorizing as false individualists the 19th-century Benthamists and Philosophical Radicals, and, on the continent, those infected by Cartesian

rationalism, notably the French Encyclopaedists, Rousseau and the Physiocrats. True individualism, he says,

> affirms the value of the family and all the common efforts of the small community and group,... believes in local autonomy and voluntary associations ..., and ... its case rests largely on the contention that much for which the coercive action of the state is usually invoked can be done better by voluntary collaboration.

In sharp contrast, false individualism 'wants to dissolve all these smaller groups into atoms which have no cohesion other than the coercive rules imposed by the state ...'. But Hayek's attempt to humanize market individualism cannot hide the fact that his 'true' individualism, being tied to the free market economy, compels everyone to compete atomistically. Both his kinds of individualism must be graded possessive. Market freedom, the individual freedom to choose between different uses of one's abilities and resources, is, he recognizes, 'incompatible with a full satisfaction of our individual views of distributive justice'. And the individual's freedom is limited by 'the hard discipline of the market'. Hayek's 'true' individualism, for all its smoothness, in the end comes down to the atomistic 'rugged individualism' of Calvin Coolidge and Herbet Hoover: it is rugged individualism with a smooth false front.

It is clear, then, that the liberal tradition in the late 20th century, including within itself both the developmental individualism of Berlin and Rawls and the possessive individualism of Hayek and Friedman, does contain two antithetical positions and cannot be reduced to either one.

We have said that the old Palgrave definition of individualism was an accurate enough description of the prevailing ideology in the earlier part of the 19th century but was too narrow for the latter part of the century, when the view we have called developmental individualism emerged alongside of the earlier purely possessive individualism. We may go on to ask, what brought about this change? What brought developmental individualism into the picture?

Clues are to be found in John Stuart Mill's own writings. In the first place is his perception that the unrestrained market economy had produced a kind of society which would no longer be tolerated by the working class it had produced. In his 1845 article 'The Claims of Labour' he took the rise of the Chartist movement, with its threat of physical force, to be evidence that the British working classes would no longer put up with things as they were, and he believed that 'the more fortunate classes' must see the writing on the wall: 'While some, by the physical and moral circumstances which they saw around them, were made to feel that the condition of the labouring classes *ought* to be attended to, others were made to see that it *would* be attended to, whether they wished to be blind to it or not.'

In the second place, perhaps partly because of this apprehension of class violence, Mill became a more sensitive and humane liberal than his father or Bentham, denouncing as utterly unjust the existing relation of effort and reward, by which the produce of labour was apportioned 'almost in an inverse ratio to the labour' (*Principles of Political Economy*, Bk. II, ch. 1, sect.3), and deploring the fiercely competitive character of the market-dominated society of his day, 'the trampling, crushing, elbowing, and treading on each other's heels, which form the existing type of social life' (*Principles*, Bk. IV, ch. 6, sect.2). In reacting as early as 1848 against this kind of society, Mill was a harbinger of the more humane social conscience which became noticeable in early 20th-century liberal thinking and which in mid-20th century brought the welfare state.

We conclude that the old Palgrave definition of individualism, already too narrow when it was promulgated, became increasingly inadequate in the subsequent decades. It was made inadequate by the rise and growth of developmental individualism, which in turn was the result of two distinct but related phenomena – the apprehension by middle-class thinkers of a danger of working-class violence, and the somewhat delayed reaction of those same minds to the shocking brutality of the industrial *laissez-faire* society. The two factors together ensured that developmental individualism would coexist with possessive individualism in the heyday of free capitalist enterprise.

How much longer they will coexist is not readily predictable. The danger of class violence now within advanced capitalist welfare states is less than Mill thought it to be in the society of his time, but what may well be called class violence as between undeveloped (or misdeveloped) and developed states is not far to seek in our time. And the working and living conditions of wage-earners in developed countries are less savage now than they were in Mill's day, but the increasing speed and tension of much of the work presses heavily on them. All we can say is that the probability of our advanced societies continuing to afford any substantial measure of developmental individualism varies inversely with the degree of industrial speed-up and the amount of class violence, national and international.

C.B. MACPHERSON

See also ALTRUISM; ECONOMIC MAN; SELF-INTEREST.

BIBLIOGRAPHY

Burke, E. 1795. *Thoughts and Details on Scarcity, originally presented to the Right Hon. William Pitt, in the month of November, 1795.* London, 1800.

Hayek, F.A. 1946. *Individualism: True and False. The twelfth Finlay Lecture. ... 1945.* Dublin: Hodges, Figgis & Co.; Oxford: B.H. Blackwell.

Lukes, S. 1973. *Individualism.* Oxford: Blackwell.

Macpherson, C.B. 1962. *The Political Theory of Possessive Individualism.* Oxford: Oxford University Press.

Marx, K. and Engels, F. 1848. *The Communist Manifesto.* London.

Marx, K. 1867. *Capital.* London: Lawrence & Wishart, 1970.

Marx, K. 1891. *Critique of the Gotha Programme.* London: Lawrence & Wishart, 1938.

Marx, K. 1959. *Economic and Philosophic Manuscripts of 1844.* Moscow: Foreign Languages Publishing House.

Mill, J. 1820. *Government.* (Originally written for the supplement to the fifth edn of the *Encyclopaedia Britannica* which was completed in 1824.)

Mill, J.S. 1845. The claims of labour. *Edinburgh Review.* In *Collected Works of John Stuart Mill*, Vol. IV, ed. J.M. Robson, Toronto: University of Toronto Press, 1967, 363–89.

Mill, J.S. 1848. *Principles of Political Economy.* 2 vols, London: J.W. Parker.

Mill, J.S. 1851. Newman's political economy. *Westminster Review.* In *Collected Works of John Stuart Mill*, Vol. V, ed. J.M. Robson, Toronto: University of Toronto Press, 1967, 439–57.

Mill, J.S. 1859. *On Liberty.* London: J.W. Parker.

Mill, J.S. 1879. Chapters on socialism. *Fortnightly Review.* In *Collected Works of John Stuart Mill*, Vol. V, ed. J.M. Robson, Toronto: University of Toronto Press, 1967, 703–53.

Palgrave, R.H.I. (ed.) 1894–9. *Dictionary of Political Economy.* London: Macmillan & Co.

indivisibilities. A commodity is indivisible if it has a minimum size below which it is unavailable, at least without significant qualitative change. Most commodities are indivisible but this is often unimportant. Half a chair has little use, but this makes little difference for analysis of market demand because so many are sold that there is little inaccuracy in treating an increase in sales of chairs from 10 million to 10,000,001 as a change in a continuous variable. In other cases, minimum size is so large relative to usage that it requires special analytic approaches and has substantial behavioural consequences: a Boeing 747 passenger aircraft is a large outlay for any airline; to carry *any* freight from New York to Chicago a railroad must lay at least two rails, each about 1000 miles long.

FIXED COST AND SUNK COST. The *fixed cost* of a firm is defined as the minimum outlay it must incur to carry out any activity. If we write (assuming input prices fixed) the long run cost function as $C(y) = k + f(y)$, where $k = $ constant, $f(0) = 0$, and $y = $ the vector of output quantities, then k is the fixed cost. As in the railroad example, the need for indivisible equipment is the normal source of fixed costs.

Fixed costs are important in economics as a source of economies of scale, of impediments to the workings of the price mechanism, of breakdown in the convexity conditions usually relied upon in optimization calculations and in the uniqueness of solutions.

Fixed costs are often confused with *sunk costs*, which are also related to indivisibilities. A sunk cost may or may not be larger than the minimum outlay a firm needs to operate but, once incurred, it cannot be withdrawn for some substantial period without significant loss. An automobile producer may build a plant much larger than the minimum needed to turn out one car, and once the capital is sunk it may only be possible to retrieve it gradually as vehicles are sold. Thus, sunk costs (like the car plant) need not be fixed and fixed costs (like an aircraft) need not be sunk.

ECONOMIES OF SCALE AND SCOPE. Indivisible inputs by their nature yield economies of scale and scope. An indivisibility requires a producer of even a small output volume to acquire relatively large capacity, part of which must be unused. The firm can then increase its outputs without increasing costs proportionately (economies of scale). Formally, strict economies of scale are defined to be present at output vector y if $C(ay)/a > C(y)$, where $0 < a < 1$, that is, if average cost is declining along the ray ay. With fixed costs this becomes $[k + f(ay)]/a > k + f(y)$, $k > 0$. Then, assuming that $f(y)$ is bounded from both above and below, say, $0 \leqslant f(y) \leqslant M < \infty$, the scale economies criterion must clearly be satisfied as a approaches zero. Thus, the presence of fixed costs always introduces scale economies (so defined), at least in any neighbourhood of the origin.

If the indivisible item is not too specialized, the firm can add commodities to its product line without the combined costs equalling the sum of those of several more specialized enterprises which together produce the same output vector as our firm. The latter attribute is referred to as *economies of scope*. Formally, using the three product case $y = (y_1, y_2, y_3)$ for simplicity, strict economies of scope are defined by $C(y) < C(y_1, 0, 0) + C(0, y_2, 0) + C(0, 0, y_3)$.

Together, economies of scale and scope are what underlie the phenomenon of natural monopoly. An industry is said to be a *natural monopoly* at output y if one single firm can produce y more cheaply than can be done by *any* combination of two or more firms. Formally, if y^i is the output vector of firm i, then the industry is a natural monopoly at y if $C(y) < \Sigma C(y^i)$ for each and every set of y^i such that $\Sigma y^i = y$.

Scale economies lead to natural monopoly because in their

absence it may be possible to save resources by dividing the industry's output among several firms, each providing similar proportions of the industry's output vector. Specifically, the absence of (weak) scale economies at y means that for some values of a, $C(ay)/a < C(y)$, $0 < a < 1$. Suppose there exists such a value of a at which $b = 1/a$ is an integer. Then the industry can reduce cost by dividing output among b firms each producing $y^i = ay$, at total cost

$$\Sigma C(y^i) = bC(ay) = C(ay)/a < C(y),$$

thus violating the criterion of natural monopoly. Economies of scope are relevant because in their absence it may be possible to save resources by dividing up the industry's products among specialized enterprises. Specifically, for example in the two product-case, absence of weak economies of scope means $C(y_1,0) + C(0, y_2) < C(y) = C(y_1,y_2)$, also violating the natural monopoly requirement.

It can also be shown that scale economies together with an attribute closely related to economies of scope are sufficient (but not necessary) for an industry to be a natural monopoly (see Baumol, Panzar and Willig (1982), pp. 178, 187–8).

INDIVISIBILITIES, SUNK COSTS AND BARRIERS TO ENTRY. The literature offers various definitions of 'barriers to entry', some mutually inconsistent. If one defines them as impediments to the invisible hand mechanism, then sunk costs are entry barriers while fixed costs are not.

The need to sink capital into an enterprise constitutes a risk which obviously can deter a potential entrant and thus can protect incumbents from potential competition. So, in an industry with relatively large sunk costs, monopoly profits and inefficiencies become possible.

On the other hand, even where indivisibilities impose large fixed costs, if they are not sunk, potential competition can impose behaviour upon incumbents that is consistent with economic efficiency. Where the fixed capital is highly mobile and there is an active market on which it can readily be sold (as with, for example, ocean cargo vessels) then the fixed capital constitutes no special risk and is no impediment to entry. Even if the indivisibilities make the industry a natural monopoly it will be unable to earn excess profits, operate inefficiently or behave like a protected monopolist in other ways, because this will attract entry that – with no sunk costs – incurs little risk and punishes the misbehaving monopolist.

INDIVISIBILITIES AS IMPEDIMENT TO EFFICIENT PRICING. Perhaps the most significant of the ways in which indivisibilities can impede efficiency in pricing is the existence of indivisible input–output vectors that are efficient but which are not profit maximizing at any positive scalar prices. This is best shown diagrammatically. In Figure 1 (Frank 1969. pp. 5, 42–3) $y_1 \leqslant 0$ and $y_2 \geqslant 0$ are the input and ouput quantities respectively. With both of them indivisible, the dots, or *lattice points*, represent the only feasible input-output combinations. Point $A = (-2, 1)$ is efficient since no feasible lattice point lies to its northeast. However, A lies inside the convex hull of the (nonconvex) feasible region whose northeast boundary is ray OR. Hence, any line given by $p_1y_1 + p_2y_2 =$ profit, through point A must lie below at least one lattice point on OR (here, either B or 0). Thus, at any non-negative prices efficient point A must be less profitable than O or B – no simple prices can lead profit maximizing firms to produce A. Only a set of 'nonlinear prices' (e.g., two-part tariffs), which lead to a curved isoprofit locus such as PP, can induce production of A.

The diagram demonstrates how indivisibilities lead to nonconvexity. For example, a line segment connecting points A and B in Figure 1 clearly is not composed entirely of lattice points, that is, it is not entirely contained in the feasible set of lattice points, and so that set is not convex.

The graph also shows in another way how indivisibilities introduce scale economies. Consider D, a nonlattice point on OR to the right of A. Let c be the smallest integer for which $c(\text{distance } AD) \geqslant 1$ (in the graph $c = 2$). Then cA will be a feasible lattice point (point E), but there will be a point (B) between E and $E - c(AD)$ which is a feasible lattice point, with the same output and a smaller input quantity than those at E. Since E is an integer multiple of efficient point A, one can multiply output by $c > 1$ while multiplying input by a smaller amount, that is, there must be scale economies.

Indivisibilities impede the price system in yet another way. By creating scale economies they make marginal cost pricing unprofitable. Specifically, let y be an output vector at which there are scale economies so that $C(ay) = a^bC(y)$ in the neighbourhood of y, with $b < 1$. Then, the function is locally (approximately) homogeneous of degree b and by Euler's theorem $\Sigma y_i \partial C/\partial y_i = bC < C$. Hence, if prices are set equal to marginal costs the supplier must lose money. In that case, financial feasibility requires the substitution of Ramsey prices (see RAMSEY PRICING) for marginal cost prices to achieve a second best optimal resource allocation. This is true not only for the individual firm – the entire economy may have no parametric price option that is superior to Ramsey prices. For all outputs must be sold to suppliers of inputs and the receipts from output sales are paid out as wages, profits, etc. to the input suppliers. This imposes (in the absence of lump sum payments with parametric prices of inputs and outputs) the economy's circular flow requirement $\Sigma p_iy_i = 0$, again taking input quantities to be negative. Now, a set of Pareto optimal prices p_i^* will, in general, not satisfy this constraint. The second best prices, p_i, which are constrained to satisfy this requirement are by definition the Ramsey prices and the differences $t = p_i - p_i^*$ between Ramsey prices and optimal prices may be interpreted as the optimal vector of taxes needed for compliance with the economy's circular flow constraint.

That is the form in which Frank Ramsey's original treatment is expressed. As we have just seen from the Euler's theorem argument, where costs are differentiable with respect to outputs, the first best prices of the outputs, which are their

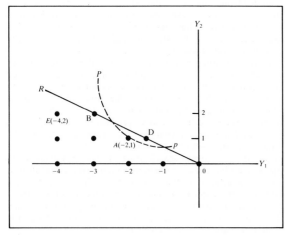

Figure 1

marginal costs, will not satisfy the circular flow constraint when there are scale economies. This shows that in general, where indivisibilities create scale economies, optimality in pricing cannot avoid the complications of Ramsey theory.

There is a third way in which indivisibilities complicate the optimization process. As is well known, where the feasible set is not convex, as must be true when there are indivisibilities, a multiplicity of local maxima is likely to be present and an iterative solution process that always follows a direction in which profit (or the value of the social objective function) is increasing may well lead toward a local optimum rather than one which is global.

INTEGER PROGRAMMING AND THE ANALYSIS OF INDIVISIBILITIES. Integer programming is the mathematical technique that is naturally suited to optimality analysis involving indivisibilities. An integer programme is a mathematical programme in which only integer values are admissible for some or all of the variables. The constraint requiring $x =$ number of locomotives to be an integer is what keeps the solution from including the absurd recommendation that 1.783 locomotives be produced.

Integer programming also permits the solution of more subtle indivisibility problems, such as those involving scale economies or either/or choices, which have resisted other analytical techniques. As an example, consider a firm required to produce y units of output using either a machine of type 1 or a machine of type 2, where x is the vector of other inputs, and x_1 and x_2 are the respective numbers of the two types of machines purchased, $\Pi(y, x, x_1, x_2)$ is the profit function and $y \leqslant f(x, x_1, x_2)$ is the production constraint. Then the firm must

$$\text{maximize } \Pi(y, x, x_1, x_2)$$

subject to the constraints

$$y \leqslant f(x, x_1, x_2)$$
$$y, x, x_1, x_2 \geqslant 0$$
$$x_1 + x_2 \leqslant 1$$
$$x_1, x_2 \text{ integer.}$$

The last two constraints guarantee that x_1 will take either the value zero or unity and that (at least) one of them will be zero, as an either/or decision requires.

Economies of scale and scope raise related issues. Such cases tend to yield corner rather than interior solutions. If there are n firms, each with different attributes, which are candidate producers of industry output vector y, it is likely to be most economical for just one of them to produce all of y. But which one of the n firms should do the job? That is obviously an extended either/or issue whose formal statement is perfectly analogous to that just described.

Indivisibilities give rise to other complex combinatorial problems. The choice among m machines may, for example be constrained by the fact that a machine of type A will work only if a machine of type B is also purchased. This is dealt with via the constraints $x_a \leqslant x_b$, x_a, x_b integer. In such problems the indivisibility feature is fundamental and cannot be avoided by non-integer approximation. In sum, indivisibilities raise basic issues for theory and for methods of analysis which bear little resemblance to those pertinent to cases of divisibility.

WILLIAM J. BAUMOL

See also ASSIGNMENT PROBLEM; CONTESTABLE MARKETS; INCREASING RETURNS TO SCALE; INTEGER PROGRAMMING; NATURAL MONOPOLY; NON-CONVEXITY.

BIBLIOGRAPHY

Baumol, W.J., Panzar, J.C. and Willig, R.D. 1982. *Contestable Markets and the Theory of Industry Structure.* San Diego: Harcourt Brace, Jovanovich.

Dupuit, J. 1844. De la mésure de l'utilité des travaux publiques. *Annales des Ponts et Chaussées*, 2nd Series, Vol. 8. Reprinted in *International Economic Papers* No. 2, 1952, London: Macmillan, 83–110.

Frank, C.R., Jr. 1969. *Production Theory and Indivisible Commodities.* Princeton: Princeton University Press.

Gomory, R.E. 1965. On the relation between integer and non-integer solutions to linear programs. *Proceedings of the National Academy of Sciences* 53, February, 260–65.

Gomory, R.E. and Baumol, W.J. 1960. Integer programming and pricing. *Econometrica* 28, July, 521–50.

Koopmans, T.C. 1957. *Three Essays on the State of Economic Science.* New York: McGraw-Hill.

Lewis, W. 1949. *Overhead Costs.* London: George Allen & Unwin.

induction. Induction, in its most general form, is the making of inferences from the observed to the unobserved. Thus, inferences from the past to the future, from a sample to the population, from data to an hypothesis, and from observed effects to unobserved causes are all aspects of induction, as are arguments from analogy. A successful account of induction is required for a satisfactory theory of causality, scientific laws, and predictive applications of economic theory. But induction is a dangerous thing, and especially so for those who lean towards empiricism, the view that only experience can serve as the grounds for genuine knowledge. Because induction, by its very nature, goes beyond the observed, its use is inevitably difficult to justify for the empiricist. In addition, inductive inferences differ from deductive inferences in three crucial respects. First, the conclusion of an inductive inference does not follow with certainty from the premises, but only with some degree of probability. Second, whereas valid deductive inferences retain their validity when extra information is added to the premises, inductive inferences may be seriously weakened. Third, whereas there is widespread agreement upon the correct characterization of deductive validity, there is widespread disagreement about what constitutes a correct inductive argument, and indeed whether induction is a legitimate part of science at all.

Approaches to these issues generally fall into two categories. The older, more philosophical approaches attempt to provide an extremely general justification for the use of inductive methods and to isolate the universal characteristics which make for a correct inductive inference. The second kind of approach focuses on what are called 'local inductions' – analyses of very specific kinds of inferences, applicable in precisely detailed circumstances. With the enormously increased power of statistical methods which is characteristic of this century, the second approach has become more and more the province of theoretical and applied statistics. It would be inappropriate to discuss specifically statistical issues here and the reader is referred to the excellent Barnett (1982). However, it should be recognized that although these detailed mathematical techniques have increased our understanding of induction immensely, they do not by themselves answer all questions about the soundness of inductive procedures. In collecting data, for example, judgements must be made about which situations are similar to one another, and hence a combination of analogical principles and judgements of causal relevance will be needed. The principles of experimental design, not only for field data but also in the growing subject of experimental economics, generally require such judgements.

Bayesian statistical methods need principles on which to attribute prior probabilities, and there is an extensive philosophical literature on the acceptability of such principles. Finally, it should be emphasized that most statistical techniques were developed within a climate of extreme empiricism or positivism, and that the application and integration of statistical models to economic systems requires a delicate inductive sensibility that cannot be reduced to algorithmic procedures.

PHILOSOPHICAL APPROACHES. The use of induction as the basis of a general scientific method was first systematically advocated by Francis Bacon. His suggested methods will seem queer to the modern reader, but the importance of his break with the deductive traditions of Greek and medieval thought should not be undervalued. He himself realized this in entitling his principal work *Novum Organum* to mirror Aristotle's *Organum* of logic, and his methods partially anticipate the eliminative methods later championed by J.S. Mill and Popper. Important as Bacon's work was, all modern work on induction lives under the shadow of the later 'problem of induction'.

'The problem of induction' is to state conditions under which an inductive inference can be rationally justified. Ever since its statement in his *Treatise of Human Nature* (1739), it has been associated with the name of the Scots philosopher David Hume. It can be broadly stated in this way: when one infers from the observed O to the unobserved U, O and U are always logically distinct, at least in the sense that one can conceive of O holding, yet U not. So there is no logical necessity for U to follow from O. What then could form the grounds for asserting U, given O? For an empiricist (such as Hume) there was nothing that one could observe which would fit the bill – possible stopgaps such as natural necessity or causal powers were simply metaphysical fictions. There was, in short, merely a succession of events, and nothing we can observe guarantees that the unobserved will continue the pattern of the observed. Furthermore, any attempt to justify induction by a deductive argument would be inappropriate, for induction is essentially ampliative, in that the conclusion goes beyond what is contained in the premises, whereas deductive inferences are always conservative. Conversely, an inductive justification of induction, on the grounds that it has worked well so far, would appear to be circular.

The philosophical responses to this problem can be of two kinds. One response is to acknowledge that inductive inferences are unjustifiable, and that consequently they should play no role in a rational enterprise such as science. Thus many authors have placed great emphasis on eliminative methods, whereby various potential explanations of the inductive evidence are eliminated as impossible or highly improbable, using primarily deductive methods. The best-known modern advocate of this view is Karl Popper, whose *Logic of Scientific Discovery* (1959) is, in part, a sustained defence of a purely deductive scientific methodology. Mill's famous methods of experimental inquiry (1843, Book III) are eliminative, as is part of Keynes's (1921) theory of induction, and a large portion of Bacon's approach. It is also possible to view in this way the objectivist statistical methodology of hypothesis testing, where the emphasis is on the rejection of statistical hypotheses. There is serious doubt with all of these approaches, however, as to whether they can function properly without tacitly employing inductive methods at some stage.

The second, and more common response is to provide some reasons why inductive inferences are indeed rationally justifiable. The 'missing premise' approach, for example,

suggests that we view inductive arguments as incomplete deductive arguments, or enthymemes. By adding some extra assumption, usually a variant of a uniformity of nature principle, one can convert inductive arguments into deductive. Holders of this view have often felt compelled to adopt such a uniformity of nature principle as an a priori truth, one without which science would be impossible. The problems with this approach are many, primarily: what exact form should the uniformity of nature principle take, and how is it to be justified? 'The future resembles the past' is too vague, and almost certainly false, for dissimilarities are at least as common as similarities. 'Every event falls under a law of nature' may be true, but which law for which event? Mill tried to solve the problem by claiming that all inductions were inferences from particulars to particulars, although also asserting that a general uniformity of nature principle could be established inductively using the success of many more specific inductive generalizations.

The pragmatic approach to induction, credited to Hans Reichenbach (1949, pp. 469–82), argues that while induction cannot be guaranteed to work, if any method succeeds then induction will do just as well. Hence one may as well employ what Reichenbach called the 'straight rule', – infer that the relative frequency of observed positive instances of an effect will continue in the future. There is, however, an infinite number of alternative rules that are consistent with Reichenbach's procedure, and hence the vagueness problem is still with him.

Much philosophical work was done in the middle part of this century to construct systems of inductive logic using a logical probability function i.e. a numerical function which attributes a degree of inductive confirmation to an hypothesis, given certain evidence statements. This work, the most developed of which was carried out by Rudolf Carnap (1950), is generally regarded as having failed to achieve its aims. It did, however, produce a number of useful insights into the nature of inductive inferences, among which was the principle of total evidence, which asserts that in applications of inductive logic, no relevant evidence should be omitted.

CONTRIBUTIONS OF ECONOMISTS. Among those who have made important contributions to both economics and the study of induction, we may count primarily J.S. Mill (1843), W.S. Jevons (1874), J.M. Keynes (1921), and R.F. Harrod (1956). Economists who have also written explicitly on induction include A.A. Cournot, F.Y. Edgeworth, F.P. Ramsey, John Hicks, Herbert Simon, and F.A. Hayek. (It is worth mentioning that Hume himself made a seminal contribution to economics with his theory of gold-flow equilibrium and defence of free trade.) Mill's views have been described earlier. Jevons's principal work is *The Principles of Science* (1874), within which the use of the hypothetico-deductive method is heavily stressed, as well as the allocation of subjective probabilities to those hypotheses by means of inverse probability methods. Jevons's inductive views are now for the most part regarded as combining exceptional insight with generally fallacious reasoning.

Keynes's only philosophical book, *A Treatise on Probability* (1921), is, like most of his work, of great originality. Here one can find one of the first systematic expositions of logical probability. Keynes is also perhaps the first to have insisted that logical probabilities are relative to evidence and cannot be separated from such. Hence there is no rule of detachment for probabilistic inductive logic, in the sense that evidence premises cannot be detached from the inductively supported conclusion, as is possible in deductive logic. This work on inductive logic was an important precursor of Rudolf Carnap's contributions in this area. Keynes also introduced the Principle of Limited Independent Variety, which essentially asserts that all inductive inferences concern objects with a finite number of independent properties, or, that there cannot be an infinite plurality of causes for an effect. This principle was necessary in order to attribute finite prior probabilities to the hypotheses under consideration. Harrod's (1956) theory cannot be swiftly stated: suffice it to say that he argues for the intrinsic acceptability of certain inductive arguments based on probability without supplementation by additional assumptions. His work has not attracted wide support.

PROSPECTS. Is there a solution to Hume's problem? A characteristic of both kinds of approach discussed above has been their tendency towards an increased level of abstraction, symbolized by increasingly powerful mathematical and logical techniques. Useful as these techniques are, inductive inferences can rarely be made confidently without careful attention to causal relationships. Hume's problem itself arose directly from his argument that there is nothing more to causal connections than the regular succession of temporally ordered contiguous events. Mill gave careful attention to the causal foundations of inductions, but many empiricists are uneasy with causal talk, and the 20th century has largely eschewed causes in favour of mathematics. Because induction, causality and probability are so intimately connected, one may be able to rectify this neglect by making use of a specifically causal concept of probability (e.g. Humphreys, 1985). That is, rather than construing probabilities as logical relations, subjective degrees of belief, or relative frequencies, one may take them to be propensities, i.e. probabilistic dispositions whose concrete structural basis is the economic system under investigation. Indeed, much of the work by Marschak, Hurwicz and by Simon (1977) on identifiability of structural parameters within causally isolated systems lends itself to this kind of approach. Those theories are ultimately reliant upon an understanding of causation which comes from experimental interventions, and since we are undoubtedly acquainted with primitive causal relations in that way, using such relations to justify others will not result in circularity. By localizing such inferences, there need be no vagueness about the inductive claims made. This approach does suffer from the extreme difficulty of identifying causal relationships within complex economic systems, and this difficulty is, of course, why one often must replace the experimental controls of simpler physical sciences by statistical surrogates for economic purposes. Complete certainty about inductive inferences is impossible, but the clear and discoverable differences between stable and unstable systems, equilibrium and disequilibrium, and isolated and non-isolated systems lie at the heart of the difference between secure and insecure inductive inferences from the past to the future, and a judicious mixture of statistical techniques with causal models seems to offer a promising alternative to the acausal inductive heritage of Hume.

A comprehensive bibliography up to 1921 may be found in Keynes (1921). More recent work is cited in Swinburne (1974). The best survey is still Kneale (1949) and an elementary source is Skyrms (1986).

PAUL W. HUMPHREYS

See also ANALOGY; HUME, DAVID.

BIBLIOGRAPHY
Bacon, F. 1620. *Novum Organum.* Reprinted as *The New Organon,* ed. F.H. Anderson. Indianapolis: Bobbs-Merrill, 1960.

Barnett, V. 1982. *Comparative Statistical Inference.* 2nd edn, Chichester: John Wiley.
Carnap, R. 1950. *Logical Foundations of Probability.* Chicago: University of Chicago Press.
Harrod, R. 1956. *Foundations of Inductive Logic.* New York: Harcourt, Brace.
Hume, D. 1739. *A Treatise of Human Nature.* Ed. L.A. Selby-Bigge, Oxford: Oxford University Press, 1888.
Humphreys, P. 1985. Why propensities cannot be probabilities. *Philosophical Review* 94, 557–70.
Jevons, W.S. 1874. *The Principles of Science.* London: Macmillan.
Keynes, J.M. 1921. *A Treatise on Probability.* London: Macmillan.
Kneale, W. 1949. *Probability and Induction.* Oxford: Clarendon Press.
Mill, J.S. 1843. *A System of Logic.* London: J.W. Parker.
Popper, K. 1959. *The Logic of Scientific Discovery.* London: Hutchinson.
Reichenbach, H. 1949. *The Theory of Probability.* Berkeley: University of California Press.
Simon, H. 1977. Causal ordering and identifiability. In H. Simon, *Models of Discovery,* Dordrecht: D. Reidel.
Skyrms, B. 1986. *Choice and Chance.* 3rd edn, Belmont: Wadsworth.
Swinburne, R. 1974. *The Justification of Induction.* Oxford: Oxford University Press.

industrialization. Industrialization is a process. The following are essential characteristics of an unambiguous industrialization process. First, the proportion of the national (or territorial) income derived from manufacturing activities and from secondary industry in general goes up, except perhaps for cyclical interruptions. Secondly, the proportion of the working population engaged in manufacturing and secondary industry in general also shows a rising trend. While these two ratios are increasing, the income per head of the population also goes up except again for temporary interruptions (Datta, 1952; Kuznets, 1966, 1971; Sutcliffe, 1971). There are cases in which the per capita income goes up, income derived from secondary industry per head of the population also goes up, but there may be little growth either in the proportion of income derived from the secondary sector or in the ratio of the working force engaged in that sector. Such cases, except when they are observed for a highly developed country, not only make the unambiguous labelling of the process of development as industrialization difficult; they also pose questions regarding the sustainability of the process that has been observed.

Other characteristics are also often associated with industrialization or a more general process of what Kuznets has called 'modern economic growth' (Kuznets, 1966, ch. 1). These include a narrowing and ultimate closing of the gap between productivity per head in the secondary sector and in the primary sector (that is, agriculture, forestry and fishing), continual changes in the methods of production, the fashioning of new products, rise in the proportion of population living in towns, changes in the relative ratios of expenditures on capital formation and consumption and so on.

Most of these associated characteristics were derived from the experience of Great Britain, or more narrowly, England and Wales, which was the first country to industrialize. That experience has remained unique in many ways. But since England was the original centre for diffusion of the economic and technical changes associated with the industrialization process, it is important to understand what happened in that country.

At least since the days of Karl Marx, England has been known as the first country in which feudalism broke down, and capitalism brought the economy under its sway (Dobb, 1946). This meant that all means of production came to be owned by a small group of property-owners called capitalists, and the rest of the working people became free wage-workers who earned their livelihood by selling their labour power to the capitalists (Dobb, 1946). It has been claimed that while serfdom broke down all over western Europe, England was the only country where a group of landlords managed to concentrate most of the land in their hands and prevent the consolidation of a free peasantry which could be used by an absolutist state to defeat the rise of capitalist agriculture (Moore, 1967, ch. 1; Brenner, 1976). It has been further claimed that economic individualism which has been taken as the hallmark of the motivation of an entrepreneur in capitalist society, goes back in England to the 12th–13th centuries, so that capitalism went through a long process of birth in the first industrializing nation (Macfarlane, 1978, chs 5–7). By the time of the first industrial revolution, England was a society in which the abiding interest of the rulers was to make money from agriculture, trade and industry, and in which the rulers were prepared rationally to order the affairs of the state so as to enable the entrepreneurs to conquer foreign countries and markets, by the force of arms if need be and had the financial and military might to carry out those plans. England had also become the leader in trade and finance among the countries of western Europe, after the decline of Amsterdam (Braudel, 1984).

The English industrial revolution is traditionally associated with the rise of machine-based industry powered by steam (Marx, 1887, chs XIV and XV; Mantoux, 1928). Certainly the classic age of British dominance of world industry, which is roughly the period from the end of the Napoleonic wars up to 1870, was characterized by the conquest of production methods by machines with moving parts of iron and steel, powered by steam, and operated by scores or even hundreds of operatives concentrated in single factories. However, what is becoming apparent is that for practically the whole of the eighteenth century, traditional techniques and materials (such as wood), and traditional sources of power such as muscles of men, women and children, animals, and water and wind, were responsible for the growth and spread of factory industry (Musson, 1972; Von Tunzelmann, 1978; Crafts, 1985).

The experience of England lends credence to the postulation of a stage of 'industrialization before industrialization' or 'proto-industrialization' (Mendels, 1972). This has been defined as 'the development of rural regions in which a large part of the population lived entirely or to a considerable extent from industrial mass production for inter-regional and international markets' (Kriedte, Medick and Schlumbohm, 1981, p. 6). The growth of industry in England was spearheaded by an explosion in the development of cotton spinning; and the cotton mills which utilized the new spinning machines sought out suitable sources of water power and labour – mostly in the rural areas or small towns. Steam engines were an element in the industrial revolution, but they did not come into their own as the major prime movers in manufacturing industry until perhaps the second quarter of the nineteenth century.

In England, cotton textiles were a relatively new industry; and they grew at first by redressing the balance of labour power needed in traditional spinning methods, so that no major displacement of labour took place within the system of proto-industrialization, in the 18th century. But machine spinning stimulated handloom production, and handloom weavers were pauperized even in England when powerlooms displaced handlooms (Bythell, 1969). In other countries, where traditional handicrafts were displaced by the new machine-made fabrics, and because of political or internal social factors

they were not replaced, or not replaced quickly enough, by any considerable growth of machine industry, pauperization and de-industrialization were widespread and in some cases they became endemic phenomena (Bagchi, 1976; Kriedte, 1981). So Ricardo's worries about the possible employment-displacing effects of machinery were justified after all (Ricardo, 1821, ch. 31; see also Hicks, 1969). But in Britain, continued growth of external trade and the coming of the railway age helped in overcompensating the labour-displacing effects. Not all countries had the same advantages.

The proto-industrialized order was succeeded in England by the system of machine manufacture perhaps because the former faced its severest crisis there: social relations there had already been transformed in a fully capitalist mould by the time smallscale manufacture reached its fullest development. Developments in science, technology and statecraft almost certainly helped resolve the crisis in favour of a higher stage of industrial development.

The fact that England had a decisive lead in the use of machine manufacture and steam power, and had formal or informal colonies where she could ignore barriers erected by the USA or Continental European countries made her the supreme industrial nation of the world for almost three-quarters of a century (cf. Robinson, 1954).

Once the revolution in textiles and steam power had been pioneered in England it could, however, be diffused to other countries, provided the latter possessed suitable political and social conditions. It is on the basis of the timing, speed and social mechanism of diffusion of the industrial revolution that we can distinguish three clusters of countries which have gone through an unambiguous process of industrialization. The first is the cluster of countries on both sides of the North Atlantic seaboard and overseas colonies with populations of predominantly European origin; the second consists of Japan and the four islands of industrialization in the Far East, viz., South Korea, Taiwan, Hong Kong and Singapore, and the third is the cluster of socialist countries led by the Soviet Union. The rest of the world are still struggling, with only varying degrees of success, to get a sustained process of industrialization going (Bagchi, 1982).

The English industrial revolution was, to start with, very much a matter of textiles; it was only in the 19th century that it affected other industries, especially iron and steel and mechanical engineering in general, on a large scale. The uniqueness of England, with all the advantages of a first start (Robinson, 1954) allowed her to expand her markets overseas in an almost unrestrained manner until the USA and other western European countries expanded their home production, not only of textiles, but also of other manufactures, often behind walls of protection against the English manufactures. The west European industrialization was helped very much by the nearness of England: from Britain flowed information about the new inventions, machines, men and capital, although there was for a time an attempt to restrict the exports of new machinery from England (Landes, 1965). Capital flows from England and to a lesser extent from France, were particularly important in supporting the movement of European populations to the USA, Canada, Australia, South Africa, New Zealand and Argentina (Kuznets, 1971; Bagchi, 1972; Edelstein, 1982).

Yet despite more active support by politically independent governments, the spread of industrialization to western Europe took a surprisingly long time to get going (Lewis, 1978, chs 7 and 8; Crafts, 1985, ch. 3). One set of reasons had to do with political, social and structural factors. The French needed a major revolution before the bourgeoisie could take possession of the state apparatus. Even then, the entrenchment of peasant agriculture in the countryside probably delayed the full conversion of the primary sector to capitalist relations. In other countries, even the 1848 revolution did not complete the process of capitalist take-over. Associated with these lags went the fact that by English standards, too high a proportion of population continued to depend on agriculture and a large gap between agricultural and industrial productivity continued to persist down to the eve of World War I. Such political and social lags, of course, even more impeded the process of industrialization in the countries of central, southern and eastern Europe down to the period between the two world wars. We will have to pay separate attention to the case where the logjam in the process of industrialization was only broken with the Bolshevik Revolution, with most other countries of eastern Europe following after World War II (Berend and Ranki, 1982).

As the process of industrialization spread, the supply of importable technologies and the financial requirements for implementing such technologies both increased. According to one estimate, gross domestic investment as a proportion of GDP in Great Britain increased from around 4 per cent in 1700 to 5.7 per cent in 1760, 7.9 per cent in 1801 and 11.7 per cent in 1831 (Crafts, 1983), and remained between 10 and 12 per cent between 1831 and 1860 (Feinstein, 1978, p. 91). By contrast in countries such as Germany, Sweden or Denmark the rate of investment in their phase of industrialization (after 1860) often reached 15 per cent and more of GDP. In the USA the social preconditions for industrialization were much more favourable than in most European countries, and the export of capital from Europe considerably aided her industrialization process until she in turn became a creditor nation around the turn of the 19th century.

The latecomers among the western European countries, and Japan on the other side of the world, used state intervention on a much wider scale and much more purposively than Britain did. This intervention did not take the same form in all countries: in a country such as Germany, financing of industry was far more widely supported by the state and by new instruments of finance created for the purpose than, say, in Italy. It is doubtful whether a general pattern of successful state intervention to overcome economic backwardness can be discerned from the historical experience as has sometimes been claimed (Gerschenkron, 1962). What can be asserted is that state intervention in industry was much more likely to succeed in countries where capitalist relations had advanced far than where intervention from the top was used as a substitute for social change which might upset the balance of class forces among the rulers (cf. Berend and Ranki, 1982).

The example of Russia is especially instructive in showing the limits of state action in a society where capitalist relations had taken root only to an imperfect degree. In Russia serfdom had been consciously introduced in the 17th century, and the system bore particularly heavily on regions producing grain which was a major export of eastern European lands. The so-called village communes (*obschina*) produced both agricultural products and handicrafts. Beginning around the 1830s modern machinery was employed in the processing of beet sugar and in the spinning of cotton yarn. Even after the abolition of serfdom in 1861, handicrafts remained predominant (Crisp, 1978), but later on, the system of domestic production and production by handicrafts became more and more integrated into the system of capitalist production (Lenin, 1898). It was only in the 1880s that large-scale industry employing modern machines experienced an accelerated growth in Russia (Lyaschenko, 1949; Crisp, 1978).

The development of modern industry and capitalism in general in Russia gave rise to a vigorous debate which still has contemporary relevance in many countries of the third world. Some of the Russian Populists (*Narodniks*) contended that the development of capitalist industry was impossible in a backward country such as Russia. They argued that modern machine-based industry destroys handicrafts and small peasant agriculture, the incomes of people dependent on them consequently shrink, and thus modern industry faces a severe – indeed insurmountable – realization problem. Countering this argument, Lenin pointed out that capitalism created its own markets by converting goods produced within a household or barter economy into tradable commodities and by continually generating new methods of production. The latter in their turn create demands for new equipment and materials (Lenin, 1897, 1899). Lenin did not deny that capitalism needed foreign markets. But at that stage he attributed the need not to the impossibility of realizing the surplus value but to intercapitalist competition and the continuous drive of capital towards expansion. In the process of discussing the analytical issues involved Lenin enunciated a law of development of capital, namely, that 'constant capital grows faster than variable capital, that is to say, an ever larger share of newly-formed capital is turned into that department of the social economy which produces means of production' (Lenin, 1897, pp. 155–6).

While markets expanded in Russia with state support for development of railways and war-related industries, the process of industrialization before the Revolution of 1917 remained ridden with numerous contradictions. Before the Stolypin reforms (which were initiated after the abortive revolution of 1905) the spread of individual ownership in agriculture was held up by numerous restrictions on peasant mobility and on the transferability of land. Even after the Stolypin reforms (or reaction) landlords' social and economic power continued to limit the development of capitalism in agriculture (see, e.g., Lenin, 1912). A substantial proportion of growth in the industrial capital stock was financed by foreign banks and foreign entrepreneurs (McKay, 1970). Large-scale industry was regionally and sectorally concentrated (Portal, 1965) and the proportion of the working force engaged in industry (including construction) was only 9 per cent in 1913; it was only after the Bolshevik Revolution and the implementation of the two Five Year Plans that there was a decisive change in the occupational structure. The proportion of the working force engaged in industry and construction climbed to 23 per cent in 1940 and 39 per cent in 1979; correspondingly the proportion engaged in the agriculture and forestry declined from 75 per cent in 1913 to 54 per cent in 1940 and 21 per cent in 1979 (Sarkisyants, 1977, p. 180; see also Kuznets, 1966, p. 107).

In Japan, the course and the pattern of industrialization differed considerably from the sequence witnessed in western Europe and the USA, and also from that followed in socialist countries. Under Tokugawa rule, Japan was characterized by what has been called 'centralized feudalism' (Ohkawa, 1978, p. 140) with the *shogun* exercising supreme power through the *daimyos* and a rigid hierarchy going down to the village level. But the increasing use of money for the payment of taxes, countrywide transactions in money required to support the *daimyos*' and their retainers' expenditures in their travels to the capital and back, and the increasing indebtedness of many *daimyos* to merchants enhanced the power of the latter . The merchants' ambitions, the peasants' discontent and the frustrations of many of the feudal lords in the face of the increasing threat posed by the military and technological advance of the Western powers ultimately led to the end of the shogunate and the restoration of the Meiji emperor. The fierce nationalism bred among the nobility under the isolation enforced on the country earlier by the shogunate led them to define their objectives in the image of the activities of the Western imperialist powers (Beasley, 1963; Norman, 1943; Smith, 1961).

While abolishing many of the privileges of the warrior class the new Japanese rulers held on to the rigid rules of hierarchal control descending from the emperor through the nobility and the higher ranks of merchants to the village headmen, and down to the peasants working in the fields. The rigid subjugation of family members, especially of women, to the patriarch and the use of communal ties to enforce authoritarian rule continued unabated, and was adapted to the requirements of modern industry (Morishima, 1982). A high level of land taxes imposed on the peasantry financed much of the economic growth in Japan which accelerated from the 1880s. Young women, more or less bonded to the factories by their fathers or other family heads provided cheap labour. The first steps in the industrialization process were taken under the guidance of the state which built or financed shipyards, telegraph lines, railways and armament works (Lockwood, 1968). The actual pace-setter in the industrialization process, in Japan as in Britain, was textiles, and for a long time, handicraft methods continued to be used alongside of machine methods in producing Japan's industrial goods. Silk, indemnities from foreign conquest, and exports of cotton yarn and cotton goods allowed Japan to do without much foreign investment in her drive towards industrialization. As in Britain, so in Japan, external markets and imperial conquest played an important role in the rise of modern industry (Lockwood, 1968).

Japan's industrial growth was already impressive in its diversity and sophistication during the interwar years. But it is since World War II that her growth has surpassed earlier historical standards (Ohkawa, 1978; Armstrong, Glyn and Harrison, 1984). The reserve army of labour in agriculture was finally exhausted there under the dual impact of land reforms imposed by the American occupation authorities and rates of industrial growth that often exceeded 15 per cent per year. Accompanying the Japanese growth was domination of trade and industry by a handful of giant conglomerates, the *zaibatsu*, giant firms and general trading corporations or *soga soshas*, acting in close collaboration with the Ministry of International Trade and Industry, and the subjugation of the labour movement to company objectives. It is these characteristics combined with a systematic exclusion of foreign capital from practically all fields that led observers to use the phrase 'Japan Inc.' to characterize the Japanese system of management. Japan eventually surpassed all capitalist countries except the USA in the value of her industrial production and in her technological advance.

While the countries on the two sides of the north Atlantic seaboard were industrializing and Japan was slowly emerging as a challenger to the industrial and political supremacy of the Western powers in the Far East, the majority of the people living in Asia, Africa and Latin America hardly experienced any positive process of industrialization. The movement of neither capital nor labour favoured such a process in China, India, Egypt, Peru, Brazil or Mexico, even in the exceptional days of massive British investments overseas that enriched the USA, Australia or Canada with men or materials (Edelstein, 1982; Davis and Huttenback, 1985). Only a small fraction of foreign investment made by Britain and France went to the non-white, dependent colonies, or formally independent, but

effectively dependent countries peopled by non-white populations. These investments went generally into plantations, mines, railways rather than manufacturing industries. While the British dominions such as Canada or Australia pursued their economic policies largely independently of metropolitan control and protected their nascent industries, India, Egypt or even China and Turkey were forced to pursue *laissez faire* policies under the pressure of the metropolitan powers. The small flows of foreign investment into the colonies were swamped in the case of India, West Indies or even Brazil by the outflow of capital to the metropolitan countries (and thence to their colonies of settlement) as political tribute, and profit on external trade, foreign exchange transactions or plantation and railway enterprises (Bagchi, 1982, chs 3 and 4).

Policies of free trade or state intervention in favour of metropolitan trade and industry generally led to a decline in handicrafts and domestic industry on a large scale in such countries as India, China and Turkey. This erosion of proto-industrial output and employment was only very inadequately compensated by the rise of modern industry. Colonial rule also led in many cases to the strengthening of ties of bondage of various kinds in the rural areas. When migration occurred on a large scale from these countries, it was often organized by the merchants from the metropolitan countries, and the migrants often entered into a semiservile condition in the plantations of Assam (India), Trinidad, Guiana, or mines of South Africa. The effective control of modern plantation and mining enterprises and many areas of trade, especially wholesale internal and external trade by merchants from metropolitan countries, policies of free trade and processes of de-industrialization retarded the development of an indigenous mercantile community in most of the dependent colonies and often delayed the onset of any process of industrialization until the 1950s.

In many Latin American countries industrialization was quickened in the 1930s as a result of import restriction policies forced on the governments by the deep depression, especially in primary commodity exports, and attendant balance of payments crises. Following on from this experience, many of them adopted industrialization as a strategy of development and the basic objective of planning. The Prebisch–Singer thesis of a secular decline in terms of trade of primary products provided the rationale for such a strategy in Latin America (Prebisch, 1950; Singer, 1950; Spraos, 1982). Elsewhere, the success of the Soviet experiment provided an inspiration for planning.

However, after some initial successes, in most countries of the third world, the process of industrialization was caught up in multiple contradictions. In few countries were there land reforms conferring the right of ownership and control on the cultivating peasantry. This failure rendered the supply of food grains and other farm products inelastic and enabled the entrenched landlords and traders to speculate in these commodities. As a result, any stepping up of investment through governmental efforts soon met inflation barriers and balance of payments crises. The latter were aggravated by a tendency to import newer and newer consumer goods for the upper and the middle classes, by an inability to bargain from a position of strength with the suppliers of technology, and by the oversell practised by many of the aid-givers wanting to tie the loans or grants to purchases of goods from the donor country. In many Latin American countries, threats of social revolution were met by imposition of authoritarian regimes, generally with US connivance or assistance. The case of Chile where a popular government under the presidentship of Salvador Allende was replaced by a brutal military dictatorship is perhaps the most glaring example of this tendency, but Argentina, Brazil and Uruguay all fitted the same pattern. The primary commodities boom in the early 1970s, rise in oil prices in 1973–4 and 1978–9 along with the attraction provided to transnational corporations by explicit policies of wage repression and labour regimentation boosted the rate of industrial growth in countries as widely dispersed as Brazil and Iran. However, in most of these countries, including some oil-exporters such as Mexico and Nigeria, astronomically large external debts and debt servicing charges put a stop to most development efforts by the early 1980s.

A few economies in east Asia, more specifically the two enclaves of Hong Kong and Singapore, and the two medium-sized economies of South Korea and Taiwan went through a process of successful industrialization. In South Korea and Taiwan, radical land reforms, partly brought about through the defeat of the Japanese in 1945 who had been major land-holders in these two provinces, and partly imposed by the US authorities fearing a Communist revolution in emulation of the People's Republic of China, enormously speeded up the movement of trading capital into industry, increased the elasticity of supply of farm products and widened the market for basic consumer and producer goods. Chinese overseas capital had for a long time dominated trade and money-lending in many countries of east and south-east Asia. Communist take-over of mainland China drove out a sizeable section of big mercantile capital. The newly migrating and old Chinese overseas capital then turned to industrial investment in many of these countries. Increased US military activities in the region, attempted economic blockade of Communist China by the Western capitalist countries and the large expenditures attending US military aggression in Vietnam provided multiple opportunities to the traders and industrialists in the region for capital accumulation and expansion. Many Japanese, American and western European transnational corporations found Singapore, Hong Kong, Taiwan and South Korea useful as export platforms since these four economies provided the attraction of low wages, a disciplined (and regimented) labour force and privileged access to US, EEC and Japanese markets.

However, despite the fact that many Asian economies have continued to experience positive growth in a period of global recession, it cannot be said yet that the east Asian experience is catching or easily diffusible. Most of Africa is experiencing negative growth, and large clusters of population are caught there in the clutches of famine. Most Latin American economies are yet to get out of the debt trap. The only other countries which are still experiencing a positive process of industrialization to a greater or lesser extent are the socialist countries which have embraced some variant of Marxism as their guiding ideology. The share of industry in GNP rose steeply in most of these countries and often exceeded 40 per cent. The high rate of economic growth in these countries was financed by the confiscation of rent incomes from land, by the channelling of all surpluses into investment and by allowing only a moderate rise in real wages until an acceptable level of GNP was reached (cf. Ellman, 1975; Lippit, 1974). One feature that has distinguished socialist industrialization is that usually the share of services in GNP and employment has been lower than in most non-socialist economies. In a country such as China, the abnormally low share of services has been seen as a defect associated with the phase of extensive growth.

Most of the socialist countries are also now grappling with problems of lower productivity growth. Effective decentralization of planning processes, increased responsiveness to changes in relative scarcity and signalling of such changes through changes in relative prices and provision of adequate incentives to managers and workers have been generally seen as the answer to these problems. Increased imports of technology from the OECD countries and their effective absorption are also seen as part of the answer, but the successful pursuit of such strategies is intertwined with the issue of economic reforms on the one hand and geopolitical manoeuvring between the two blocks on the other.

One problem that will continue to bedevil industrialization strategies in most large third world countries for a long time is the very high ratio of the working population engaged in agriculture to the total working force. Even in a country such as China, which has experienced a trend rate of industrial growth of more than 10 per cent over the years since the Communist revolution in 1949, and where the share of industry in national income went up to 42.2 per cent in 1982, agriculture and forestry continued to employ 71.6 per cent of the labour force in the same year (China, 1983, pp. 24, 121). It is only some medium-sized economies with a high rate of industrial growth such as South Korea and Taiwan that have experienced any major shift in the population balance towards industrial employment.

The experience of the structural changes within the east Asian group of capitalist economies shows that under favourable circumstances, it is possible for the less industrialized economies to grow at high rates if there is a sustained shedding off of the lower-productivity sectors by the more advanced regions and the grafting of the shedded output on to the structures of the less developed economies (Yamazawa, Taniguchi and Hirata, 1983). The process is very similar to that observed in western Europe in the early part of the 19th century, except that the role of migration of population to countries outside the region (such as the USA in the case of western Europe) in easing population pressure has been minimal. But the roles of direct investment by Japanese and other OECD firms and of privileged access to extra-regional markets have been more significant than in the case of western Europe. (It could, of course, be argued that western European countries had a privileged access to markets in their dependent colonies.)

The general developments in the advanced capitalist bloc of countries (within which Japan occupies a unique position because of her maintenance of moderate to high rates of growth and near full employment and her large trade surpluses with most other countries), however, preclude the replication of the east Asian pattern in the rest of the third world. Most of them are afflicted by high rates of unemployment – exceeding levels witnessed since the end of the 1930s. Some countries such as the UK experienced an absolute decline in manufacturing (Singh, 1977). These developments aggravated protectionism in these countries, thus creating barriers against the expansion of exports from the third world countries, while the OECD group of countries continued to constitute the biggest market for manufactured goods in the world. Developments in microelectronic technology posed major threats to the further expansion of labour-intensive textile products and clothing exports from the third world to the OECD countries (UNCTAD, 1981). More generally, the spread of microelectronic technologies embracing whole branches of manufacture are threatening to remove many assembly operations which the OECD-based transnational corporations had earlier found it profitable to subcontract to the favoured export enclaves including the east Asian group of newly industrializing countries (Kaplinsky, 1984).

Within the OECD group, the USA has become the biggest magnet for capital flows from all over the world. The high interest rate and large budget deficits maintained by the US government have forced most other OECD governments to pursue deflationary policies within their borders. There is little sign as yet that such trends will be reversed. The Japanese, who have run up large trade surpluses (exceeding US $40 billion) with the USA have proceeded to invest most of their export surplus in the US. Thus the diffusion that is borne on the backs of foreign investment within the order of capitalism has been severely hampered by these developments.

The only alternative that is left for most third world countries is to rely on building industries on the basis of domestic resources and domestic markets. But the guidelines laid down by the International Monetary Fund seeking to impose severely deflationary policies on most countries applying for its assistance in meeting their debt problems, the power exerted by OECD-based transnational corporations in effectively restricting the flow of technology, and the internal social structures in most of these countries blocking the spread of literacy and accrual of purchasing power to common people are likely to hamper the feeble efforts at industrialization on a self-reliant basis. The other path of industrialization, building on growing exports and expanding international investment flows, would appear also to be beset with dangerous pitfalls for most of the poor countries of the world. Thus the spread of industrialization in the near term to the poorer countries is likely to be very slow compared with the speed witnessed between, say, 1950 and 1978. At the other end of the spectrum, in countries such as the USA and UK, services and finance have gained tremendously at the expense of manufacturing industry, and it is through the use of financial instruments as much as advanced technology in manufacturing (including armaments production) and services that the USA dominates the economies of most of the capitalist countries. But as Japan continues to forge ahead even in frontier technologies such as the mass production of semiconductor chips for use in the most advanced microelectronic processes (cf. Gregory, 1985), a change in the balance within the capitalist order is very likely. In the meanwhile continued growth in the socialist world will also affect the global balance in manufacturing and economic power.

AMIYA KUMAR BAGCHI

See also BACKWARDNESS; DUAL ECONOMIES; GERSCHENKRON, ALEXANDER; INDUSTRIAL REVOLUTION; LABOUR SURPLUS ECONOMIES; MANUFACTURING AND DEINDUSTRIALIZATION; MODES OF PRODUCTION.

BIBLIOGRAPHY

Armstrong, P., Glyn, A. and Harrison, J. 1984. *Capitalism since World War II: The Making and Breaking of the Great Boom*. London: Fontana.

Bagchi, A.K. 1972. Some international foundations of capitalist growth and underdevelopment. *Economic and Political Weekly* 7(31–33), Special Number, August, 1559–70.

Bagchi, A.K. 1976. De-industrialization in India in the nineteenth century: some theoretical implications. *Journal of Development Studies* 12(2) January, 135–64.

Bagchi, A.K. 1982. *The Political Economy of Underdevelopment*. Cambridge: Cambridge University Press.

Bairoch, P. 1975. *Economic Development of the Third World since 1900*. London: Methuen.

Beasley, W.G. 1963. *The Modern History of Japan*. New York: Praeger.

Berend, I.T. and Ranki, G. 1982. *The European Periphery and Industrialization 1780–1914*. Cambridge: Cambridge University Press.

Blackaby, F. (ed.) 1979. *De-industrialization*. London: Heinemann.

Braudel, F. 1984. *The Perspective of the World: Civilization & Capitalism, 15th-18th Century*. London: Collins.

Brenner, R. 1976. Agrarian class structure and economic development in pre-industrial Europe. *Past and Present* 70, February, 30–75.

Bythell, D. 1969. *The Handloom Weavers: A Study in the English Cotton Industry during the Industrial Revolution*. Cambridge: Cambridge University Press.

China. 1983. *Statistical Yearbook of China 1983*. Hong Kong: State Statistical Bureau PRC and Economic Information & Agency.

Crafts, N.F.R. 1983. British economic growth, 1700–1831: A review of the evidence. *Economic History Review* 36(2), May, 177–99.

Crafts, N.F.R. 1985. *British Economic Growth during the Industrial Revolution*. Oxford: Clarendon Press.

Crisp, O. 1978. Labour and industrialization in Russia. In Mathias and Postan (1978).

Datta, B. 1952. *The Economics of Industrialization*. Calcutta: World Press.

Davis, L. and Huttenback, R.A. 1985. The export of British finance, 1865–1914. *Journal of Imperial and Commonwealth History* 13(3), May, 28–76.

Dobb, M. 1946. *Studies in the Development of Capitalism*. London: Routledge & Kegan Paul.

Edelstein, M. 1982. *Overseas Investment in the Age of High Imperialism; The United Kingdom 1850–1914*. London: Methuen.

Ellman, M. 1975. Did the agricultural surplus provide the resources for the increase in investment in the USSR during the first five year plan? *Economic Journal* 85, December, 844–63.

Feinstein, C.H. 1978. Capital formation in Great Britain. In Mathias and Postan (1978).

Gerschenkron, A. 1962. *Economic Backwardness in Historical Perspective*. Cambridge, Mass.: Harvard University Press.

Gregory, G. 1985. Chip shop of the world. *New Scientist*, 15 August, 28–31.

Habakkuk, H.J. and Postan, M. (eds) 1965. *The Cambridge Economic History of Europe*. Vol. VI: *The Industrial Revolution and After*, Pts 1 and 2, Cambridge: Cambridge University Press.

Hicks, J. 1969. *A Theory of Economic History*. Oxford: Clarendon Press.

Hilton, R. (ed.) 1976. *The Transition from Feudalism to Capitalism*. London: New Left Books.

Kaplinsky, R. 1984. The international context for industrialization in the coming decade. *Journal of Development Studies* 21(1), October, 75–96.

Kriedte, P. 1981. The origins, the agrarian context, and the conditions in the world market. In Kriedte, Medick and Schlumbohm (1981).

Kriedte, P., Medick, H. and Schlumbohm, J. 1981. *Industrialization before Industrialization: Rural Industry before the Genesis of Capitalism*. Cambridge: Cambridge University Press.

Kuznets, S. 1966. *Modern Economic Growth: Rate, Structure and Spread*. New Haven: Yale University Press.

Kuznets, S. 1971. *Economic Growth of Nations: Total Output and Production Structure*. Cambridge, Mass.: Harvard University Press.

Landes, D. 1965. Technological change and development in Western Europe 1750–1914. In Habakkuk and Postan (1965).

Lenin, V.I. 1897. A characterisation of economic romanticism (Sismondi, and our native Sismondists). Trans. from Russian in Lenin, *Collected Works*, Vol. 2, Moscow: Foreign Languages Publishing House, 1963.

Lenin, V.I. 1898. The handicraft census of 1894–95 in Perm Gubernia and general problems of handicraft industry. Trans. from Russian in Lenin, *Collected Works*, Vol. 2, Moscow: Foreign Languages Publishing House, 1963.

Lenin, V.I. 1899. *The Development of Capitalism in Russia*. Text of the 2nd edn of 1908, trans. from Russian in Lenin, *Collected Works*, Vol. 3, Moscow: Progress Publishers, 1964.

Lenin, V.I. 1912. The last valve. Trans. from Russian in Lenin, *Collected Works*, Vol. 18, Moscow: Progress Publishers, 1968.

Lewis, W.A. 1978. *Growth and Fluctuations 1870–1913*. London: Allen & Unwin.

Lippit, V.D. 1974. Land reform and economic development in China. *Chinese Economic Studies* 7(4), Summer, 3–181.

Lockwood, W.W. 1968. *The Economic Development of Japan*. Princeton: Princeton University Press.

Lyaschenko, P.T. 1949. *History of the National Economy of Russia to the 1917 Revolution*. London: Macmillan.

Macfarlane, A. 1978. *The Origins of English Individualism*. Oxford: Basil Blackwell.

McKay, J.P. 1970. *Pioneers for Profit: Foreign Entrepreneurship and Russian Industrialization*. Chicago: University of Chicago Press.

Mantoux, P. 1928. *The Industrial Revolution in the Eighteenth Century*. London: Jonathan Cape.

Marx, K. 1867–94. *Das Kapital*. Trans. by S. Moore and E. Aveling as *Capital: A Critical Analysis of Capitalist Production*, Vol. 1. Reprinted, Moscow: Foreign Languages Publishing House, n.d.

Mathias, P. and Postan, M.M. (eds) 1978. *The Cambridge Economic History of Europe*. Vol. VII, *The Industrial Economies: Capital, Labour and Enterprise*, Pts 1 and 2, Cambridge: Cambridge University Press.

Mendels, F. 1972. Proto-industrialization: the first phase of the industrialization process. *Journal of Economic History* 32(1), March, 241–61.

Moore, B. Jr., 1967. *Social Origins of Dictatorship and Democracy: Land and Peasant in the Making of the Modern World*. London: Allen Lane.

Morishima, M. 1982. *Why has Japan 'Succeeded'?* Cambridge: Cambridge University Press.

Musson, A.E. (ed.) 1972. *Science, Technology and Economic Growth in the Eighteenth Century*. London: Methuen.

Norman, E.H. 1943. *Soldier and Peasant in Japan*. New York: Institute of Pacific Relations.

Ohkawa, K. 1978. Capital formation in Japan. In Mathias and Postan (1978).

Portal, R. 1965. The industrialization of Russia. In Habakkuk and Postan (1965), Pt 2.

Prebisch, R. 1950. *The Economic Development of Latin America and its Principal Problems*. United Nations, New York: Reprinted in *Economic Bulletin for Latin America* 7(1), February 1962, 1–22.

Ricardo, D. 1821. *On the Principles of Political Economy and Taxation*. 3rd edn, reprinted in *The Works and Correspondence of David Ricardo*, Vol. I, ed. P. Sraffa with the collaboration of M.H. Dobb, Cambridge: Cambridge University Press, 1951.

Robinson, E.A.G. 1954. The changing structure of the British economy. *Economic Journal* 64, September, 443–61.

Sarkisyants, G.S. (ed.) 1977. *Soviet Economy: Results and Prospects*. Moscow: Progress Publishers.

Singer, H. 1950. The distribution of gains between investing and borrowing countries. *American Economic Review* 40, May, 473–85.

Singh, A. 1977. UK industry and the world economy: a case of de-industrialisation? *Cambridge Journal of Economics* 1(2), June, 113–36.

Smith, T.C. 1961. Japan's aristocratic revolution. *Yale Review* 50(3), Spring, 370–83.

Spraos, J. 1982. Deteriorating terms of trade and beyond. *Trade and Development. An UNCTAD Review*, No. 4, Paris: UNCTAD.

Sutcliffe, R.B. 1971. *Industry and Underdevelopment*, London: Addison-Wesley.

UNCTAD. 1981. *Fibres and Textiles: Dimensions of Corporate Marketing Structure*. Geneva: United Nations.

Von Tunzelmann, G.N. 1978. *Steam Power and British Industrialization to 1860*. Oxford: Clarendon Press.

Yamazawa, I., Taniguchi, K. and Hirata, A. 1983. Trade and industrial adjustment in Pacific Asian countries. *Developing Economies* 21(4), December, 281–312.

industrial organization. Based on the activities of those who consider themselves in the field, industrial organization (or industrial economics) today may be broadly defined as the field of economics concerned with markets that cannot easily be analysed using the standard textbook competitive model. In such markets the positive and normative implications of models of imperfect competition are generally of interest, as are the design and effects of government antitrust and regulatory policies aimed at improving market performance. Because there are many models of imperfect competition, and because general policies must be applied to particular cases, much of the research in industrial organization has been and continues to be empirical.

Historically, industrial organization emerged as a distinct field after the rise of the modern manufacturing enterprise around the turn of the century (compare Chandler, 1977 and Hay and Morris, 1979, ch. 1). Early writers largely equated 'industrial' with 'manufacturing' and focused on markets for manufactured products. Students of industrial organization today do not limit themselves exclusively to the manufacturing sector, but, in part because of the availability of data, departures from that sector are selective.

Thus securities markets, which seem to approximate perfect competition well, are not studied in industrial organization, but competition among financial institutions and regulation of their behaviour have been investigated. Studies of transportation and traditional public utilities are common, in part because of the important role played by government policy in these sectors.

Industrial organization has also retained a strong focus on the firm as an object of study. In microeconomic theory, the firm is a given cost or production function assumed to be operated to maximize profits; in industrial organization the structure and behaviour of firms are objects of study. In contrast, relatively little attention is devoted to household behaviour.

The national markets for manufactured goods that were created early in this century have two important and apparently novel characteristics, stressed in Chamberlin's (1933) seminal and controversial analysis. First, products in many manufacturing markets are *differentiated*; that is, buyers do not view them as perfect substitutes. In such markets, non-price competition, involving product design, advertising, and other selling expenses, is often important. The sources and consequences of product differentiation and non-price competition have been intensively studied.

Second, some industrial markets came to be dominated by a relatively small number of firms. A good deal of work in industrial organization has attempted to explain differences in the 'organization' of markets, focusing on *seller concentration*, the extent to which sales are concentrated in the hands of a small number of firms. The consequences of seller concentration have also been intensively studied, and the analysis of oligopoly behaviour has accordingly played a central role in industrial organization.

The main objective of the field has been to develop tools to analyse market processes and their consequences for economic performance. Since Bain (1959), it has been customary to work with the concepts of structure, conduct, and performance. Market *structure* refers to a set of variables that are relatively stable over time, observable (at least in principle), and that are important determinants of buyer or seller behaviour. All scholars in the field agree implicitly or explicitly that there exists such a set of variables; otherwise market behaviour is in principle unpredictable.

Intrinsic market structure variables (termed *basic conditions* by Scherer, 1980, ch. 1) are essentially completely determined by the nature of the product and the available technology: all modern steel industries are capital-intensive, for instance. Other elements of market structure are *derived* in that they may reflect government policy, corporate strategies, or accidents of history: the concentration that was created by the US Steel merger in the USA in 1901 is an obvious example (see Chandler, 1977, ch. 11 and Stigler, 1968, ch. 9). Intrinsic structural variables also affect derived variables to some extent: even if the US Steel merger was not inevitable, it is difficult to imagine an atomistic steel industry. The strength of these effects is perhaps inevitably controversial, since the stronger they are, the less scope there is for governments to enhance efficiency by changing market structures.

In any complete market model, market structure determines the *conduct* of buyers and sellers. Compare, for instance, structure and conduct (rules for output choice) in pure monopoly and in perfect competition. An important objective of industrial organization is to describe and predict the conduct of actual industries in terms of a continuum joining these two polar cases: as one moves toward the competitive end, the intensity of *rivalry* increases and profits fall accordingly. But a single dimension cannot describe conduct fully; market behaviour typically involves choosing which products to produce, the corresponding vector of prices or outputs, distribution and advertising strategies, and levels and directions of research and development activity.

Market *performance* is assessed by comparing the results of market behaviour in efficiency terms to first-best ideals or feasible alternatives. One might compare prices with marginal costs, for instance, or the array of products produced with some ideal array. Performance is determined by all aspects of conduct, along with the intrinsic elements of market structure.

To this relatively static framework, one must add dynamic effects of buyer and seller behaviour on market structure. Intrinsic structural variables can be changed by innovation, for instance, and seller concentration can be changed by mergers. Established sellers may be able to take actions to inhibit the entry of new rivals.

Much early work in industrial organization eschewed formal theory, in part because there did not exist an adequate general theory of behaviour in oligopolistic markets. Many scholars concentrated on induction from case studies of particular markets. At the same time, others sought to reinterpret and extend the standard competitive and monopoly models to enhance their explanatory power. Beginning in the 1950s, cross-section statistical work on samples of manufacturing industries became common. In the 1960s, particularly in the USA, industrial organization economists began to look beyond antitrust policy and to examine systematically the effects of government regulatory programmes.

Recently, formal models of imperfect competition have been studied intensively, and the tools of noncooperative, extensive form game theory have assumed central importance in this work (see Schmalensee, 1982; Waterson, 1984; and Roberts, 1985). Laboratory experiments are being performed more frequently, as are case studies relying heavily on formal models and econometric analysis of firm behaviour. Implications of imperfect competition for international trade and for national welfare in a world economy are receiving increased attention.

In the remainder of this essay I discuss briefly some of the questions that have been studied intensively by industrial organization economists. For more detailed treatments of many of these issues, see Bain ([1959], 1968), Hay and Morris (1979), Scherer (1980), Schmalensee and Willig (1988), Stigler (1968), and Waterson (1984). Space constraints preclude an explicit discussion of antitrust or regulatory policy.

Are firms managed so as to maximize profit (or, more generally, the wealth of their owners), as microeconomic theory assumes? This question is most important in imperfectly competitive markets, since non-maximizing firms cannot survive in the long run under perfect competition. Two alternatives to profit-maximization have been advanced. Neither has yet proved to be more generally useful, though both yield valuable insights in some situations.

First, some scholars have argued that firms' problems are so complex and their information so imperfect that maximization is effectively impossible. They stress the importance of routines, rules of thumb, experimentation, and learning in actual business behaviour.

Second, others note that the many shareholders who nominally own large corporations cannot effectively review managers' decision-making. These scholars model managers as pursuing a variety of their own objectives (such as firm size or growth) subject to constraints (often relating to profitability) imposed by owners. Recent work using agency theory to model the manager–owner relation shows considerable promise here.

In an ideal world, shareholders would always use the *market for corporate control* to replace managers who did not effectively pursue the owners' interests. And shareholders in fact often force mergers and sometimes elect boards of directors opposed by incumbent managers. But ours is not an ideal world, and the importance of frictions and imperfections in the market for corporate control is widely debated.

What determines the boundaries between firms and markets? A number of authors have focused on the implications of cost minimization for firm structure. Oliver Williamson (1975) has stressed the difficulty of writing long-term contracts that allow for all possible contingencies. If efficient production requires making investments that cannot be easily shifted to alternative uses, this difficulty may make it more efficient to integrate related activities within a single firm, rather than to attempt to coordinate them by contract and risk the effects of contractual breakdown.

Stigler (1968, ch. 12) stressed the importance of *economies of scale*. He argued that, as markets grew, specialized firms would arise to perform functions in which economies of scale were important. More recently, Baumol, Panzar and Willig (1982) have argued that multi-product firms may arise to take advantage of *economies of scope*, which lower cost when the production of multiple products is carefully coordinated. Scope economies arise when assets can be readily shared among processes producing several outputs.

Another even more diverse body of theoretical literature argues that imperfections in competition may produce other incentives for firms to expand their activities. Relatively few empirical tests of any of these models have been performed, however.

What is a market? In microeconomic theory, a market is the locus of trades in a single, perfectly homogeneous product. This definition would make almost all real firms monopolists. In practice, markets must be defined by aggregating products that are relatively close substitutes in demand or supply. For most purposes, particularly in the context of antitrust policy, it is useful to define a market as the smallest aggregate that could profitably be monopolized. It is rarely easy to implement this definition in a fully satisfactory way, however, and government data-collection agencies rarely try. This poses real problems for empirical work.

What are the key elements of market structure? In the seminal work on this point, Bain (1959) argued that there were four such elements, all of which he seemed to treat as derived: the extent of seller concentration, the extent of buyer concentration, the importance of product differentiation, and the conditions of entry. Seller concentration was held to facilitate non-competitive behaviour; buyer concentration was held to make such behaviour harder to sustain. Bain argued that product differentiation insulated sellers from each others' actions and changed the focus of rivalry from price to non-price competition. He also argued that the easier it was for new competitors to enter an industry, the more difficult it would be for established firms to maintain prices above costs and earn supra-normal profits. (Baumol, Panzar and Willig (1982) have coined the term *contestable* to describe markets in which entry is so easy that potential competition alone suffices to eliminate excess profits.)

Bain (1956) went on to identify four sources of *barriers to entry*, four reasons why established firms might be able to earn excess profits without facing the threat of entry. (See Stigler, 1968, ch. 6 for an important alternative definition of this term.) First, substantial economies of scale might make potential entrants reluctant to enter at efficient scale for fear of depressing prices below costs. Second, established firms might have cost advantages over potential entrants, perhaps because of proprietary production processes. Third, established firms might have demand-side or product differentiation advantages over potential entrants, perhaps because of patented products or buyers' reluctance to switch brands. Finally, Bain felt it was possible that imperfections in capital markets would inhibit entry when large initial investments were required. This last possibility remains controversial.

While this framework remains influential, it is increasingly under attack. Bain and many of his followers assigned what now seems to be excessive importance to seller concentration. Bain's framework neglects firm structure, but firms that operate in many markets may behave differently from single-market enterprises, since actions taken in one market may affect costs or strategic opportunities in others. Caves and Porter (1977) have argued that the notion of entry barriers must be generalized to include *mobility barriers*, which impede entry into *strategic groups* of sellers with similar capabilities and strategic objectives. They and others contend that the structure of strategic groups within industries can materially affect conduct.

Bain's framework now seems to many scholars to omit a number of critical structural variables. Recent work attaches particular importance to cost conditions and information. Baumol, Panzar and Willig (1982) have stressed the impact of *sunk costs*, costs required to enter a market that cannot be recovered if the market is later abandoned. The more important sunk costs are, the greater the risk of entry, and the more important scale economies are as a barrier to entry, all else being equal.

Stigler (1968, ch. 5) pointed out that sellers are more likely to be able to sustain non-competitive behaviour the better their information on each others' actions. Recent game-theoretic work has expanded on this insight and stressed the importance of information about rivals' capabilities and objectives as well. Buyers' information about prices and qualities may also play a central role in determining marketing and distribution arrangements and the form and intensity of rivalrous behaviour. If buyers must spend time to learn the prices of competing sellers, for instance, each seller has some monopoly power even if there are many firms marketing identical products.

Bain ([1959] 1968, p. 9) argued that one ought to restrict attention to a small number of structural characteristics because 'meaningful intermarket comparisons and meaningful generalizations about the influence of structure on behaviour are effectively forestalled if the content of "structure" is made so comprehensive that no two markets could be viewed as structurally alike'. But many scholars now feel that simple generalizations of the type that Bain sought may not have much predictive power.

How are the derived elements of market structure determined? Most work has focused on the determinants of seller concentration, with special attention given to the hypothesis that concentration is determined by economies of scale. This hypothesis is supported by the observation that the same industries tend to be concentrated in all developed economies, despite different histories and government policies.

Scale economies in manufacturing at the plant and firm levels have been measured by statistical methods, by interview studies (the 'engineering' approach), and by comparing the sizes of units that prosper and decline (the 'supervisorship' approach). These studies have been criticized because of the inherent difficulty of measuring non-production scale economies (those in marketing and distribution, for instance) that occur at the firm level. In general this work suggests that concentration in most US manufacturing industries is higher than required for the exploitation of economies of scale in production. This is consistent with the observation that among large industrialized economies, absolute levels of concentration in particular industries are not sensitive to differences in the size of the national market.

A number of other potential sources of concentration have been identified. Spence (1981) has shown that *economies of learning*, which cause unit cost to decline with cumulative production, can mandate high concentration even when scale economies are absent. Demsetz (1973) has argued that persistent efficiency differences, along with the tendency for efficient firms to expand at the expense of their rivals, are an important source of concentration. Many others have studied the impacts of random variations in firm growth rates and of mergers on concentration. Mergers are an important source of concentration in some countries but not in others; the empirical importance of the other factors remains controversial.

What determines the intensity of rivalry in oligopolies? We still have no fully satisfactory, general model of oligopoly, but theoretical work has yielded a number of valuable insights.

The basic problem faced by any set of sellers is that posed by the classic prisoners' dilemma game. In a static setting, all sellers do well if prices are kept high. If all other sellers set high prices, however, any single firm can usually increase its profits by charging a lower price (or producing more than its assigned quota). If all behave selfishly in this fashion, all will charge low prices and receive low profits. That is, *non-cooperative*, selfish behaviour in this setting tends to produce competitive, low-profit outcomes.

Interest thus attaches to the possibility of *cooperative* or *collusive* behaviour that can produce monopolistic performance, with high prices and high profits. In principle, collusive behaviour can be overt, with firms explicitly agreeing on strategies, or tacit, with firms reaching an unspoken understanding about acceptable policies. It is more difficult to reach agreement tacitly than overtly; agreement is also more difficult the more firms there are and the greater the differences among them. Collusion can either be total, covering all decisions, or partial, covering only some variables under firms' control. It may be easier to collude on price than on advertising, for instance.

Collusive agreements are inherently unstable, since individual sellers can usually increase their profits, as in the prisoners' dilemma game, by departing unilaterally from the agreement. Stability requires the ability to detect cheating and to make a *credible threat* (one that it would actually be rational to carry out) to impose a sufficiently severe penalty to render cheating unprofitable. The game-theoretic notion of *perfect* (Nash) *equilibrium*, in which players noncooperatively pursue their own interests but noncredible threats are ruled out, has been used heavily in recent work on oligopoly theory and entry deterrence (discussed below).

Stigler (1968, ch.5) argued that cheating can be more reliably detected in concentrated markets. A number of authors have recently built on his work and devised multi-period game-theoretic models in which firms announce credible threats that make cheating irrational, even when cheating can only be imperfectly detected. In these models threats are sometimes carried out (price wars occur) even though nobody ever cheats.

A number of econometric time-series studies of individual oligopolistic markets have been undertaken in recent years. These often employ the non-game-theoretic formalism of *conjectural variations*. In a market in which products are undifferentiated and firms set outputs, a firm's conjectural variation is its expectation of the derivative of all other firms' output with respect to its own. Estimates of the conjectural variations consistent with observed market outcomes provide a summary description of the intensity or rivalry. If all conjectural variations equal minus one, behaviour is perfectly competitive; larger values imply departures from the competitive ideal. Data limitations make it hard to apply this approach to many industries.

Can the conduct of established sellers discourage the entry of new rivals? In the classical limit-pricing model of Bain (1956), an established monopoly in an industry with significant scale economies could discourage entry by raising output above the monopoly level and threatening not to reduce production if entry occurred. The incumbent would optimally select its output so that entry at efficient scale would raise total output so much as to depress price (just) below cost.

Unfortunately, the threat in this model is not credible (i.e., the no-entry equilibrium is not perfect). If entry did occur, the incumbent could generally increase its own profits by reducing its output, and a potential entrant has no reason to believe that the incumbent would forego such an opportunity. Recent work (see, especially, Roberts, 1987) imposes the requirement of credibility. There are three strands to this literature.

First, when information is imperfect, incumbent firms may attempt to use pre-entry price to deceive potential entrants. If potential entrants don't know the incumbent's costs, for instance, the incumbent may lower its pre-entry price below the monopoly level in order to persuade potential rivals that its costs are too low to permit viable entry. This resembles classic limit-pricing, but it turns out on average not to deter sophisticated entrants, who understand the incumbent's incentive to attempt deception.

Second, an incumbent may be able to make credible threats by making *commitments* in advance of entry. That is, it may be able to take actions before entry that alter its post-entry incentives in a way that makes a hostile response to entry more attractive. Spence (1977), who began this line of work, considered investment in production capacity as a vehicle for commitment.

Third, if entrants are uncertain about an incumbent's objectives, it may be rational for the incumbent to take predatory actions designed to eliminate entrants when they appear. Such a policy may give it a *reputation* for aggressive (or irrational) behaviour, which may serve to deter even sophisticated potential entrants.

Do statistical analyses of data from multiple industries shed light on the validity of the hypotheses discussed above? Many cross-section studies have been performed by students of industrial organization, but the interpretation of many of their statistical findings is controversial, and the intertemporal stability of some key relationships has recently been questioned.

All cross-section studies employ accounting data, and most focus on determinants of profitability. But accounting data do not provide exact measures of real, economic profitability, in part because of differences in riskiness, the way in which long-lived investments are depreciated, and the possible ability of labour unions to capture rents generated by collusive behaviour. It has proven difficult to obtain good measures of other theoretical constructs as well, particularly product differentiation and barriers to entry. The growing importance of firms operating in several markets poses yet another measurement problem.

Many cross-section studies find a weak but statistically significant positive relation between seller concentration and industry profitability. Until recently this was generally interpreted as supporting the Bainian hypothesis that concentration facilitates collusion. But Demsetz (1973) has offered an alternative explanation: in a world without collusion, substantial efficiency differences among rival sellers are likely to produce both concentration, as noted above, and high industry-level profits, because efficient firms earn rents. Where efficiency differences are unimportant, one would expect both concentration and profits to be low. This hypothesis is consistent with the strong positive correlation between market share and profitability in some industries, but not many industries follow this pattern. It has proven difficult to discriminate between these two hypotheses empirically.

Similarly, numerous studies have found a strong positive correlation between advertising/sales ratios and profitability. This has often been taken to support Bain's (1956, 1959) hypotheses about the effects of product differentiation and product differentiation advantages of established firms. But advertising is logically only one input affecting those structural variables, and it is an endogenous variable, determined by profit-seeking sellers. Unfortunately, it has been difficult to specify good simultaneous equations models in this area. Finally, the effects of advertising on demand probably persist over time, so that advertising should be treated as an investment, not a current expense. If advertising's effects are assumed generally to decay slowly enough, the correlation between advertising intensity and (corrected) profitability measures disappears, but the appropriate decay rate assumption remains controversial.

Because of these and other problems of measurement and interpretation, inter-industry empirical work seem to have lost the central place it formerly held in industrial organization. Still, such work remains an important source of the general stylized facts needed to guide the construction of useful theoretical tools.

What sorts of price structures are imposed by firms with market power? What are the welfare implications of price discrimination? If a seller has some control over its price (i.e., it is not a perfect competitor), can identify (even imperfectly) customers with different demand characteristics, and can prevent (or at least inhibit) trade among its customers, it will generally pay it to practise price discrimination. That is, it will adopt a price policy in which different customers pay different marginal or average prices depending on their demand characteristics. The ability to earn excess profits is not required; price discrimination can persist in a free-entry equilibrium of the Chamberlin (1933) type.

In practice, sellers employ many devices for identifying customers of different types. Bulk discounts, in which the average price paid falls with volume, provide one method. (This is a special case of *non-linear pricing*, in which the buyer's bill is a nonlinear function of the quantity he purchases.) Delivered pricing, in which a buyer's price depends on his location, provides another. Discrimination may also be effected by bundling or tying arrangements, which require buyers to purchase related products from a single seller. And there are a host of market-specific devices: US airlines, for instance, charge a much lower fare for trips that involve spending a Saturday away from home, thus generally charging lower prices to tourists than to business travellers.

The large theoretical literature on the consequences of such practices contains few sharp results. Prohibiting price discrimination in most cases produces both gainers and losers; the net welfare effect is usually ambiguous.

How are product quality and variety determined in imperfect markets? Are the outcomes likely to be optimal in any sense? If 'quality' is simply inserted as an additional variable in a standard monopoly model, one can show that 'quality' may be either too high or too low in equilibrium, depending on the details of the demand function. If there are many sellers and buyers rely on firms' reputations for quality in making decisions, firms producing high quality (and thus high cost) products must be able to charge prices above marginal cost in equilibrium. If not, they would have an incentive to lower quality and exploit their reputations until buyers caught on.

Models in which variety is determined usually assume economies of scale in the form of brand-specific fixed costs; otherwise it would generally be socially efficient and privately optimal to produce all possible brands. This assumption rules out purely competitive equilibria and forces second-best welfare comparisons. It also provides a rationale for intra-industry international trade: such trade expands the market and makes greater variety economically feasible.

In some models of variety determination, the demand side of the market is a single representative consumer who desires variety. In others, consumers have different ideal brands, and each desires, all else equal, to consume the brand that is 'closest' to his ideal in the space of all possible brands. In a third class of models, consumers agree on the ranking of all possible brands but differ in their willingness to pay for quality. Market equilibria in all these models generally involve a non-optimal set of brands, but the nature of deviations from optimality depends on the details of the model. Chamberlin's (1933) view that monopolistic competition implies excessive variety is not generally valid.

Are market-determined levels of advertising excessive? Do they increase barriers to entry? Neither of these traditional questions has yet been answered, and neither may in fact have a general answer.

In order to assess the optimality of any level of advertising, one must make some assumption about how advertising affects consumer behaviour. One extreme assumption is that advertising simply provides consumers with information; the other is that it simply changes their tastes. Under the first assumption, market-determined advertising levels are optimal only under very special conditions; under the second assumption the optimal level of advertising depends on what tastes are used as a yardstick. In fact, neither extreme assumption is likely to be generally correct.

The theoretical effect of advertising on conditions of entry depends, again, on the way advertising affects consumers, and this is likely to differ among markets. In some markets restrictions on advertising are observed to increase prices; in others, heavily advertised brands sell at substantially higher prices than apparently physically identical brands that are not advertised. Advertising may be less important in markets in which retailers are an important source of information. Product differentiation advantages of established brands may depend more on satisfied buyers' rational reluctance to experiment with new brands than on the effects of advertising.

Are large firms in concentrated markets the major sources of technical progress, as Schumpeter (1942) argued? It is difficult to measure any firm's contribution to technical progress; counts of patents or significant innovations are frequently used but obviously imperfect indicators. Most studies have found that, in most industries, large firms are not disproportionate sources of innovations, especially not significant innovations, but there are exceptions. It is more difficult to assess the impact of market structure on technical progress, since one must control for differences in the opportunities for innovation across markets. The available evidence provides at most weak support for Schumpeter's view of the effects of concentration.

On the theoretical side, a number of authors have recently modelled research and development rivalry in game-theoretic terms. In many of these models, firms spend money (perhaps over time) to increase their chances of winning a single prize, usually interpreted as a patent. Under some conditions, it may be rational for an incumbent monopolist to outspend potential entrants in order to prevent their entry. In general, theoretical work indicates that market-determined levels of research and development spending may be excessive or inadequate. This work also makes clear that concentration and other structural variables are in the long run determined by the intrinsic opportunities for innovation. Concentration and innovative activity are thus both endogenous variables.

How important are departures from competitive performance? Early studies of this question, which associated differences in profit rates with departures from the competitive ideal, concluded that monopoly power imposed relatively small costs on society. It is now clear that a proper general equilibrium analysis of this issue may imply much larger or even smaller effects, depending on the values of unknown parameters.

Some authors have argued that differences in observed profit rates understate the actual effects of monopoly power because monopoly profits are to some extent dissipated in actions taken to achieve or protect monopoly positions, captured by labour unions, or simply foregone by lazy or inept managers not subject to market discipline. On the other hand, if Schumpeter (1942) was right, short-run measures of the cost of monopoly omit important long-run benefits. Like so much else in this field, the actual importance of departures from pure competition in modern economies remains controversial.

RICHARD SCHMALENSEE

See also ADVERTISING; MARKET STRUCTURE; SELLING COSTS.

BIBLIOGRAPHY
Bain, J.S. 1956. *Barriers to New Competition.* Cambridge, Mass.: Harvard University Press.
Bain, J.S. 1959. *Industrial Organization.* New York: John Wiley. 2nd edn, 1968.
Baumol, W.J., Panzar, J.C. and Willig, R.D. 1982. *Contestable Markets and the Theory of Industrial Structure.* New York: Harcourt, Brace, Jovanovich.
Caves, R.E. and Porter, M.E. 1977. From entry barriers to mobility barriers. *Quarterly Journal of Economics* 91(2), May, 241–61.
Chamberlin, E.H. 1933. *The Theory of Monopolistic Competition.* Cambridge, Mass.: Harvard University Press.
Chandler, A.D. 1977. *The Visible Hand: The Managerial Revolution in American Business.* Cambridge, Mass.: Harvard University Press.
Demsetz, H. 1973. Industry structure, market rivalry, and public policy. *Journal of Law and Economics* 16(1), April, 1–9.
Hay, D.A. and Morris, D.J. 1979. *Industrial Economics: Theory and Evidence.* Oxford: Oxford University Press.
Roberts, D.J. 1987. Battles for market share: incomplete information, aggressive strategic pricing, and competitive dynamics. In *Advances in Economic Theory II*, ed. T. Bewley, Cambridge: Cambridge University Press.

Scherer, F.M. 1980. *Industrial Market Structure and Economic Performance*. 2nd edn, Chicago: Rand-McNally; 1st edn, 1970.

Schmalensee, R. 1982. The new industrial organization and the economic analysis of modern markets. In *Advances in Economic Theory*, ed. W. Hildenbrand, Cambridge: Cambridge University Press.

Schmalensee, R. and Willig, R.D. (eds) 1988. *Handbook of Industrial Organization*. Amsterdam: North-Holland.

Schumpeter, J.A. 1942. *Capitalism, Socialism, and Democracy*. New York: Harper.

Spence, A.M. 1977. Entry, capacity, investment and oligopolistic pricing. *Bell Journal of Economics* 8(2), Autumn, 533–44.

Spence, A.M. 1981. The learning curve and competition. *Bell Journal of Economics* 12(1) Spring, 49–70.

Stigler, G.J. 1968. *The Organization of Industry*. Homewood, Illinois: Irwin.

Waterson, M. 1984. *Economic Theory of the Industry*. Cambridge: Cambridge University Press.

Williamson, O.E. 1975. *Markets and Hierarchies: Analysis and Antitrust Implications*. New York: Free Press.

industrial partnership. *See* CODETERMINATION AND PROFIT-SHARING.

industrial relations.   Two great historical developments accounted for the emergence of industrial relations as a locus of social and economic tension in the 19th century: the Industrial Revolution and the extension of political democracy and public education. The factory system resulted in the assembly of large groups of workers in large-scale establishments where they were often subjected to machine-paced and authoritarian discipline and to increased economic insecurity, but the extension of political democracy and public education heightened the expectations of citizens as members of the dependent labour force. For Marx, writing in the 19th century, the relationship between labour and management under capitalism was inherently one of exploitation. But to observers in the 20th century, this relationship came to be characterized by the power of wage earners to resist downward pressure on money wages during slumps, which became a central point in the continuing debates over cause and cure of cyclical unemployment.

On the other hand, competitive market analysis has generally maintained that there is nothing necessarily one-sided (in either direction) or economically inefficient in relationships between firms and workers. Equilibrium divergences between marginal productivities and wages are acknowledged to exist, but they might be reconciled with competitive equilibrium if interpreted as evidence of mutually profitable investments in human capital rather than as Marxist exploitation (if positive), or of union monopoly power over wages and work practices (if negative). Such specialized functions of personnel (or 'human resources') management as employee recruitment, testing and selection, job classification and evaluation, training, and wage determination might thus be classified with the firm's other investment activities.

Another major role of the firm's personnel policy has been to provide incentives to the individual employees by policy-making and administration in such functional areas as employee discipline and morale, promotional sequences, performance standards, and incentive pay systems. Indeed, management policy governing average levels of pay in the short term can also affect employee morale and unit labour costs; and this aspect of industrial relations has formed the basis of some postwar attempts to explain cyclical wage rigidity and unemployment in the postwar period. A strong version of implicit contract theory holds that a policy of wage rigidity during downswings reduces the risk of temporary reductions in income, to which workers are allegedly more averse than capitalists. Efficiency–wage theories argue that savings from downward wage flexibility would be offset by losses from the induced reductions in employee efficiency. But the former theory fails to explain why management should choose to insure its employees against the lesser risk of a decline in the rate of pay while, in so doing, it increases the greater risk of unemployment. And the latter theory largely ignores recourse to the negative incentives provided by *exemplary discipline* as an alternative to the positive incentive provided by maintaining wage levels during downswings in demand.

CONCERTED BEHAVIOUR. Such models become more plausible if the usual assumption of atomistic behaviour is relaxed sufficiently to take into account the potential for concerted behaviour by wage earners that is afforded by what Hicks (1932) referred to as the 'social' nature of work in modern establishments. Students of industrial relations and of organizational behaviour have long reported on the propensity of small groups of non-union employees to form and to develop informal restraints on conduct governing work and pay, which often work at cross purposes with formal regulations posted by management. Such activity has tended to reflect interdependence of workers' individual preference systems, shared notions of equitable standards of earnings and effort, and often adherence to a 'lump of labour' view of demand elasticity. Hence threatened or actual deterioration of established terms or conditions of employment has often been regarded as breach of implicit contract by unorganized groups of workers, among whom it has tended to arouse common feelings of inequity and to elicit a collective response. The latter has frequently been reflected in a slowdown in productivity which, because of its collective nature, is resistant, although not impervious, to managerial attempts to impose exemplary discipline on individual offenders. Concerted behaviour would therefore tend to raise the disciplinary cost of a desired reduction in pay; it would include the cost of multiple replacement investments in specific human capital that would be necessitated by wholesale dismissals. Such disciplinary costs, however, vary inversely with unemployment as well as with the strength of concerted employee resistance. Therefore, higher levels of unemployment would be required to hold the disciplinary cost to levels at which desired reductions in pay would be profitable.

UNIONS AND COLLECTIVE BARGAINING. Unions, which were described as 'permanent associations' by Marx (1848) and as 'continuous associations' by the Webbs (1894), have generally been in a better position than informal and often ad hoc workplace groups to resist employer efforts to reduce pay during downswings in business activity. As more broadly based organizations, national or regional unions have frequently revealed a bargaining preference for directly defending levels of pay when the greater risk of unemployment was confined to a relatively small minority within their wide jurisdictions; hence risk aversion becomes a more plausible explanation of wage rigidity when considered in the context of explicit and collective rather than implicit and individual contracts.

National unions have attempted to organize or coordinate activity across plants within the firm and/or across firms within an industry. As multiplant organizations, they could resist employer efforts to 'whipsaw', or transfer production from one plant in a firm to another, by striking on a company-wide basis. National unions might even gain a whipsaw advantage for their own side if they could and wished to conduct selective strikes against competing firms on a company-by-company basis. Even when competing firms confronted national unions with industry-wide associations of their own, the union could deploy the strike more effectively than isolated workplace groups. Moreover, these unions were in a better position to 'take wages out of competition', either by raising wages in low-wage sectors of an industry or by resisting competitive erosion of average wage levels during cyclical downswings (Ulman, 1955).

Unions, of course, have always sought to be offensive as well as defensive organizations. According to the Webbs' (1894) classic definition, they are designed to 'improve' as well as 'maintain' the condition of their members' working lives. Formal recognition by employers of unions as representatives of their employees, joint determination of conditions governing employee effort and security as well as pay, substitution of explicit for implicit contracts, and establishment of joint machinery for the disposition of employee grievances and of disputes arising over the terms of the contract, have been characterized as extensions of the procedure of democratic government and judicial processes to the conduct of industrial relations (S. and B. Webb, 1897; Slichter, 1941; Slichter, Healy and Livernash, 1960). This has been the most compelling case for the establishment and support of collective bargaining in democratic societies.

The case for collective bargaining, however, must assume the existence of sufficient economic growth to enable that institution to satisfy historically rising levels of worker expectations reasonably well. It would be difficult for a continuing industrial relationship that is alternately adversarial and reconciliatory in nature to persist without the lubricant provided by growth to minimize conflict over the distribution of income.

The case against collective bargaining from the left proceeded from the belief (a) that capitalist innovation and investment must produce declining (rather than rising) profitability, employment and real wages; (b) that, as a result, distributional conflict was indeed inevitable; (c) that collective bargaining was a blind alley for unions and workers; and (d) that revolutionary methods would be required in its place. The syndicalist tendency was to reject collective bargaining as an instrument of class collaboration and to rely instead on such ad hoc tactics as local strikes or sabotage and on general strikes. Socialists have preferred parliamentary political activity, although they have also been willing to support and participate in collective bargaining, if only as a transitional device. Communists have sought rigorously to subordinate unions to party discipline and to adapt bargaining activity to their contemporary political requirements (which could call for either greater militancy or greater restraint than would seem optimal from the viewpoint of 'business unionism').

WORKER IDEOLOGIES, EMPLOYER REACTIONS, AND BARGAINING SYSTEMS. The case against collective bargaining from the right has been that at best it will not exert an independent effect on economic outcomes and that at worst it is itself destructive of efficiency and growth. Some employers have been ideologically opposed to collective bargaining (notably in the United States), but most have viewed it pragmatically in terms of the costs which it imposes, the costs of averting or eliminating it, and the costs of likely alternatives. The last has depended on currently dominant ideological preferences and prevailing degrees of militancy among the wage earners. When the most

likely alternative was simply an unorganized labour market, the employer's decision was typically to select the lowest cost combination of incentives and discipline which could preclude collective bargaining (provided that it was less than combined strike and settlement costs under collective bargaining). When the most likely alternative consisted of a serious revolutionary threat to the social and economic order, the decision – taken collectively – could be to accept or even encourage the method of collective bargaining, although in the least costly form possible.

BARGAINING SYSTEMS IN EUROPE. As a result, the characteristic scope and structure of a given Western nation's 'system' of industrial relations have been significantly influenced not only by market configurations (Dunlop, 1958), but also by the ideological orientation that prevailed within its (major) labour movement in the late 19th and early 20th centuries. In France and Italy, where anarcho-syndicalist influences have been strong, bargaining institutions have tended to remain relatively weak and underdeveloped. And, after World War II, the largest union movements in both countries came under the control of strong Communist parties which usually did not favour the emergence of independent bargaining at plant, enterprise or industry levels.

Elsewhere in Western Europe, including Germany and Britain, the main challenge from the left took the form of parliamentary political activity by socialist (or labour) parties whose affiliated union movements favoured collective bargaining. When general strikes and social unrest occurred on the continent (in the first decade of the 20th century and during and after World War I), large-scale employers tended to respond by accepting the enactment and extension of public systems of social welfare and, in some instances, the establishment of labour court systems and legal and de facto restraints on employee dismissals. In Germany and Scandinavia, employers resisted extensive strikes by forming strong employer associations, which were armed with strike insurance funds and authority to order lockouts; but they also entered into continuing bargaining relationships with the unions at centralized (mainly industry-wide) levels. These political and industrial developments, however, meant that the domain of collective bargaining was limited in two respects: it was largely confined to the determination of industry-wide levels of wage and hours; and it did not control industrial relations at the workplace, as a result of which important determinants of productivity were left, at least proximately, to the unilateral determination of the employer. (After World War I, left-wing agitation resulted in the establishment of works councils, but they were mostly confined by both management and union efforts to non-adversary roles.).

In Great Britain, recognition of unions by employers appears to have resulted less from class behaviour on their part or from political considerations than on the continent (especially in Germany). British employer associations, moreover, were valued less for the bargaining resistance and cost containment that they offered than for 'taking wages out of competition'; and they generally lacked the resources and authority found in Germany and elsewhere (Phelps Brown, 1983). British unionists, on the other hand, were militant and tenacious bargainers, sometimes tending to sympathetic strike action and generating a class of highly militant and autonomous shop stewards. As a result, employers were obliged to bargain with workplace groups over work rules and technical change at the shop level, as well as with national unions over wages at industry levels.

Both the British and the continental models were revived

after World War II. Relatively high rates of economic growth in the Fifties and Sixties were conducive to the pursuit of collective bargaining. In some instances, however, centralized bargaining arrangements were weakened by the tendency of firms, either unilaterally or in response to pressure from local groups of workers, to grant 'wage drift' (or payments in excess of centrally negotiated increases) during inflationary periods. Under the impetus provided by international competition, large-scale firms expanded at the expense of smaller and less efficient competitors; and they also embarked upon processes of 'rationalization' which often required increased internalization of wage policy in the interest of greater worker efficiency. These developments, however, were paralleled by a tendency for wage earners to react against downward pressures on changes in real and effort wages (also, in the case of more highly skilled and white-collar groups, against reduction in relative pay), and ultimately against the emergence of lay offs, dismissals, and plant closings (as unemployment spread from the non-union periphery of the labour force to the organized core sectors). In the late Sixties and early Seventies worker reactions took the form of unofficial strikes, and in the Seventies, seizure of plants that were scheduled to be shut down (especially in France). At the same time there occurred a revival of interest in the old syndicalist objective of worker control of industry. This took a variety of forms – for example, 'co-determination', with the inclusion of employee elected representatives on company boards of directors (especially in Germany since the end of World War II); worker 'participation' in management (in France); profit-sharing or capital-sharing schemes (notably in Sweden). In some instances larger firms encouraged active employee participation in shop-floor management, along with major redesigning of production jobs, as non-pecuniary incentives to efficiency; in others it was viewed as a way to reinforce the employee's primary allegiance to the company. Whatever the motive, such managerial reaction to employee assertiveness may have ushered in a more bipartisan approach to industrial relations at company level on the continent (Flanagan, Soskice and Ulman, 1983). In the second half of the Seventies and the early Eighties, on the other hand, unions and bargaining institutions on the continent and in Britain weakened under pressures created by the emergence of dramatically higher levels of unemployment in the wake of the oil price shocks, international recession, accelerated technological change, and the loss of international market shares by heavily organized manufacturing industries.

BARGAINING AND NON-UNIONISM IN THE UNITED STATES. In the United States, a pragmatic and implicitly optimistic belief in the efficacy of collective bargaining within the domain of capitalism predominated among unionists over the radical alternatives that were offered to them before World War I. For employers, therefore, the most likely alternative to collective bargaining was the unorganized labour market rather than radical social change; and large-scale corporations could oppose collective bargaining (along with social insurance) instead of accepting it as the lesser of two evils, as some of their foreign counterparts had been doing. The opposition of American employers to unionism and collective bargaining took two forms: direct and often ruthless measures to break up unions (including liberal recourse to strikebreaking, blacklisting, and the labour injunction) and, in the early 1920s, a variety of paternalistic measures designed to forestall unionism among the employees in various large corporations. Such 'welfare capitalism' schemes included the maintenance of relatively high wages, a variety of insurances,

pensions, and other benefit plans, career opportunities through promotion ladders, and the replacement of the pre-World War I 'drive system' of management with an 'enlightened' approach that emphasized consultation and 'human relations in industry'.

Hence, while in Europe industrial unionism in the manufacturing sectors antedated World War I and generally survived the interwar period, it did not take hold in the United States until the occurrence of an extraordinary sequence of events in the 1930s and 1940s. First came the wage cuts, speedups, and layoffs associated with deep depression in the 1930s, which generated widespread labour unrest and militancy. Next came a sharp change in the prevailing political climate ushered in by the New Deal and marked by the passage of the National Labor Relations Act of 1935, which forbade employers to interfere with their employees' efforts to organize and required them to bargain in good faith with union representatives selected by majority vote in government-held elections. Finally, during World War II, economic conditions were created which were to the unions' bargaining advantage; and the unions were compensated for cooperating with wartime wage controls by the widespread adoption of a variety of non-wage benefits and devices which enhanced their 'security' vis-à-vis both employers and the membership in the plants.

There emerged from these developments a set of bargaining relationships in many industries which were distinguished from prevailing European arrangements in the following respects: bargaining between individual firms and national unions in some large-scale industries; plant-level bargaining which was integrated with company-wide bargaining; negotiation of a wide range of non-wage benefits (which in Europe have been provided primarily through government social welfare programmes); long-term contracts (usually of three years' duration), with interim protection for employees provided by cost-of-living escalator clauses and for management via no-strike clauses, and grievance procedures, typically with provision for impartial arbitration. The latter have tended to strengthen the authority of the national union at plant level through inclusion of the latter in appellate stages, and they have permitted management to proceed with the introduction of 'new or changed jobs' subject to subsequent grievance, rather than to prior negotiation.

This system was credited with contributing both to a relative absence of wage drift (due to strong national union authority at local levels) and to greater wage flexibility than prevailed in Europe. During the second half of the Seventies, long-term contracts contributed to money wages lagging behind prices, and in the early Eighties many firms in actual or threatened financial distress prevailed on unions to reopen their long-term contracts and negotiate significant 'give-backs' in wages and benefits.

On the other hand, collective bargaining had opened up large differences between union and non-union levels of compensation during the stagflationary Seventies; and the flexibility in real wages during that period also reflected the low levels and sharp decline of union organization which characterized postwar industrial relations experience in the United States. Union membership fell from a relatively low initial rate of about 35 per cent of the private non-agricultural labour force in the mid-1950s to under 20 per cent in the mid-1980s. The success of union negotiators in raising nominal wages and benefits, taken in conjunction with shifts in product demand, increased competition in product markets and technological change, helped to shrink the historical jurisdictions of unionism. Meanwhile, the unions were unable to prevent the growth of non-union sectors within manufacturing and mining and in the growing services sectors and white-collar labour force.

Their failure was reflected in a resumption by non-union employers of a postwar carrot and stick strategy similar to that employed by their predecessors in the Twenties. The strategy included the adoption of deterrent wage-setting and in some cases of schemes for worker participation in management decision-making at plant level (which elsewhere were integrated into established union–employer relationships). It also included a variety of tactics which, by weakening the effectiveness of protection extended to employees by the Wagner Act of 1935, and by exploiting some of the protections that had been extended to employers by the Taft–Hartley Act of 1947, helped to account for a steady decline in union success rates in representation elections conducted by the National Labor Relations Board. But (as union leaders themselves have implied), a decline in the desire of workers themselves – at least non-union workers – to organize and bargain collectively cannot be ruled out as a cause of organizational failure. Protection of the opportunity to organize certainly remained greater than it had been in the pre-Wagner era. But postwar experience revealed to workers the economic limitations (notably in the area of employment protection), as well as the potential, of traditional collective bargaining as an instrument for continuing the historic advance of industrial democracy into the 21st century.

LLOYD ULMAN

See also IMPLICIT CONTRACTS; TRADE UNIONS; WAGE FLEXIBILITY.

BIBLIOGRAPHY

Azariadis, C. and Stiglitz, J. 1983. Implicit contracts and fixed price equilibria. *Quarterly Journal of Economics*, Supplement 98(3), 1–22.
Baily, M. 1974. Wages and employment under uncertain demand. *Review of Economic Studies* 41(1), January, 37–50.
Doeringer, P. and Piore, M. 1971. *Internal Labor Markets and Manpower Analysis*. Lexington, Mass.: D.C. Heath.
Dunlop, J. 1958. *Industrial Relations Systems*. New York: Holt.
Flanagan, R., Soskice, D. and Ulman, L. 1983. *Unionism, Economic Stabilization, and Incomes Policies: European Experience*. Washington, DC: Brookings.
Freeman, R. and Medoff, J. 1984. *What Do Unions Do?* New York: Basic Books.
Hicks, J. 1932. *The Theory of Wages*. 2nd edn, London: Macmillan, 1963.
Kerr, C., Dunlop, J., Harbison, F. and Myers, C. 1960. *Industrialism and Industrial Man*. Cambridge, Mass.: Harvard University Press.
Marx, K. 1848. *The Communist Manifesto*. London.
Mathewson, S. 1931. *Restriction of Output Among Unorganized Workers*. New York: Viking.
Mayo, E. 1946. *The Human Problems of an Industrial Civilization*. New York: Macmillan.
Mitchell, D. 1980. *Unions, Wages, and Inflation*. Washington, DC: Brookings.
Perlman, S. and Taft, P. 1935. *Labor Movements*. Vol. IV in *History of Labor in the United States, 1896–1932*, ed. J. Commons, New York: Macmillan.
Phelps Brown, E.H. 1983. *The Origins of Trade Union Power*. Oxford: Clarendon.
Roethlisberger, F. and Dickson, W. 1939. *Management and the Worker*. Cambridge, Mass.: Harvard University Press.
Shapiro, C. and Stiglitz, J. 1984. Equilibrium unemployment as a worker discipline device. *American Economic Review* 74(3), June, 433–44.
Slichter, S. 1929. The current labor policies of American industries. *Quarterly Journal of Economics* 43, May, 393–435.
Slichter, S. 1941. *Union Policies and Industrial Management*. Washington, DC: Brookings.
Slichter, S., Healy, J. and Livernash, E. 1960. *The Impact of Collective Bargaining on Management*. Washington, DC: Brookings.
Strauss, G. 1974. Job satisfaction, motivation, and job redesign. In *Organizational Behavior: Research and Issues*, ed. G. Strauss, R.E. Miles, C.C. Snow and A.S. Tannenbaum, Madison: Industrial Relations Research Association.
Ulman, L. 1955. *The Rise of the National Trade Union*. Cambridge, Mass.: Harvard University Press.
Webb, S. and Webb, B. 1894. *The History of Trade Unionism*. London: Longmans, Green.
Webb, S. and Webb, B. 1897. *Industrial Democracy*. London: Longmans, Green.

industrial revolution. Industrial Revolution is one of those pervasive phrases with which we cannot now dispense, but which has assumed so protean a character as to become almost formless. The term originated in France: in 1837 Jérôme-Adolphe Blanqui wrote that 'la révolution industrielle se met en possession de l'Angleterre.' But it was Arnold Toynbee with his *Lectures on the Industrial Revolution* of 1884 who gave the term general currency.

Usage of the term spread from 'the' Industrial Revolution (that of Great Britain, roughly between 1760 and 1830), to a general expression, to be applied to any great acceleration of industrial output wherever it may occur. It has also, along the way, become a concept, though an ill-defined one. Moreover, its scope has varied greatly between concentration on particular countries, and the industrial world as a whole seen as a self-unifying collectivity. This latter mode of thought involves an attempt to understand both the eight- to ten-year cyclical fluctuations and the long waves of economic development (of some fifty years), operating over the world generally, powered by groups or 'bunches' of technological advances.

In short, the challenges associated with an Industrial Revolution are (1) to understand the primary and classic case, that of Britain; (2) to assess the possibility of prescribing in general terms what the minimal critical conditions of such a dramatic and irreversible growth of output might be; (3) to scrutinize the performance of a wide range of industrialized economies in these terms over a wide range of time; and (4) to assess the notion of Industrial Revolution in global terms, in the long or even in the very long run. Finally, there are the questions of (5) the responses of economists to Industrial Revolution thinking, together with (6) the utility of the notion of the Industrial Revolution to the policy-maker.

THE FIRST INDUSTRIAL REVOLUTION. Serious attempts to explain the British Industrial Revolution did not really begin until the 1880s or 1890s, when the British economy had passed its high Victorian greatness and had begun to slip back down the league of industrialized countries. The early attempts at explanation were of the itemizing kind, various observers choosing particular aspects of the process. Against this agenda might be placed the organic or integrated approach, coming a good deal later.

The itemized list is certainly useful, containing as it does the necessary conditions for the classic Industrial Revolution of Britain, and many of those necessary generally. Each item in turn, however, requires a background discussion of its own, of considerable depth, which should eventually cause it to meet all or most of the others. High on the list is the political aspect, requiring that the state protect private property (not least against itself in terms of taxation or confiscation), and that the state refrain from impeding trade. The state must also provide internal conditions of stability and meet the external needs of defence, while hindering to a minimal degree the processes of economic growth. By the late 18th century it was Holland and Britain that came closest to meeting these conditions.

Then the mode of production involved in the new phase of industrial progress must be such as to permit ease of entry on a relatively general scale, allowing entrepreneurs to proliferate free of the monopolistic tendency toward restricted entry. Certain industries met this condition in Britain, cotton spinning being the most dramatic. For an aggressive race of entrepreneurs to appear there must be a supply of innovative men, and the system must permit of the release of their energies. Much debate has gone into the question of the supply of entrepreneurs, leading deep into the sociology of Britain, including the notion of the Protestant ethic. The treatment of the business man has ranged from regarding him as the classical economists did as a mere automatic responder to market stimuli, to seeing him as the hero of Industrial Revolution, the masterly agent of change. The former view has been reinforced by the argument that these men were lucky enough to find themselves in a uniquely favourable economic environment, so that much of their achievement was luck. There is also the thesis that they were impelled to achievement by their value system (in which Calvinism played an important part), or that as religious dissenters they had suffered from status deprivation and so were impelled to assert themselves. But the followers of the heroic or semi-heroic approach have not been routed. The entrepreneur, however, can make little advance (except perhaps through better organization and efficiency), without the inventor.

Here is another set of problems: under what conditions will men start to tinker with the mechanical world and perfect inventions which are profit-generating on a large scale? Certainly this happened in Britain, giving rise among scholars to a range of monographs, both biographical and general.

Similarly with agriculture: if the economy is to progress, the yield to effort on the land must rise and the proportion of the labour force in agriculture must fall: both of these happened in Britain through the advent of high farming from the later 18th century. Both the entrepreneur and the improving farmer required credit on reasonable terms: the development of the London capital market, together with local markets and the banking system, helped by lowering the rate of interest in the longer run. Perhaps even more important, the appearance of encouraging profits made possible the ploughing back of such gains into the new enterprises, assisting in the further growth. The infrastructure, in terms of canals and roads, also became attractive to promoters and investors so that an all-round reduction in costs were achieved, together with a widening of markets. As a general background the population trend was upward. In all previous cases this increase had been choked off on Malthusian lines, the yield of the economy being incapable of sustaining it. But in Britain the convergence of the factors listed above made it possible for the first time to breach the Malthusian barrier and sustain the growing numbers. At the same time the labour supply was increased, making for cheap labour, with the formation of trade unions forbidden by law.

Against the above listing one can place the conditions obtaining among Britain's potential rivals. Among these was the fact that France for most of the period was distracted by revolution and the Napoleonic wars; the USA had still to find its manifest destiny as a political and geographical unit; Germany and Italy were collections of fragments, and Japan continued feudal for two more generations. So it was that

between 1820 and 1850 Britain dominated the industrial world to an extraordinary degree, producing two-thirds of the world's coal, half of its iron and more than half of its steel, half of cotton cloth and forty per cent of hardware, as well as being the leading exponent of engineering in all its forms. Britain was thus placed in a dominant world position, causing an imperial outthrust (though much of this was lagged until after 1850).

By these arguments the British Industrial Revolution has been given a highly dramatic content. It is presented as representing the conditions under which the industrial world 'went critical', carrying other nations into a new and irreversible era by forcing them into the somewhat humble role of trying to catch up with the British achievement, in particular drawing upon the newly formed bank of technological know-how.

There has, however, been a considerable scholarly effort, not least among the British, to de-dramatize the British Industrial Revolution. As the British empire and the *Pax Britannica* have receded into the past, and as an overwhelming wave of world technology based upon advanced science has swept over us since World War II, the British Industrial Revolution has lost something of its miraculous nature, being no longer seen by many as the birth of the modern world.

One form that de-dramatization has taken has been the adoption of a long-term organic approach to the British Industrial Revolution, treating it not in terms of a specifically British experience over a narrow range of years, with emphasis on the 'items' and their interplay, but as a very long-term European, and indeed world phenomenon, bringing in the contributions of ancient China.

This perspective involves a consideration of the condition of Europe from the Middle Ages onward. It requires a sense of geo-economies, namely a realization that Europe, because of its geographical diversity, with its many distinct cultures, was never reduced to a single rule by Charlemagne, Napoleon or anyone else, but yet shared a European culture stemming from Christendom. This meant that a complex interplay of regional achievement and inter-regional borrowing could take place over the continent, the process merely culminating in Britain. Moreover, the organic approach leads to a consideration of how 'backward' other countries really were in terms of the necessary conditions of Industrial Revolution: recent scholarship tends to close the gap between Britain and the rest. De-dramatization goes a good deal further when the approach is via the generalized growth rate of the British economy as calculated in recent years, compares with the rates that have been generated since 1945 in other countries. Britain's great achievement rested on an annual average growth rate of about 1.5 per cent. On the other hand five Asian countries have, between 1960 and 1980, achieved a growth rate of almost seven per cent. Further drama has been emptied from the British Industrial Revolution by looking in depth at what actually happened in particular situations. The impact of the factory system, so striking to contemporaries and so dominant in the minds of Engels and Marx, has been, in quantitative terms at least, exaggerated: there were still more people making boots and shoes in Britain in 1850 than worked in factories. Weaving had yet to be mechanized on a general scale. The railway, regarded by many as an essential vehicle of Industrial Revolution, had scarcely been launched by 1830. The steam engine, it has been claimed, had not really proliferated before 1850. Britain was till in 1830 under an inhibiting tariff system: free trade was an affair of the second half of the century. Finally, in the longer perspective of demography and family structure, the nuclear family, so long regarded as an effect of the British Industrial Revolution, has been shown long to pre-date it, and indeed to have been a contributing cause of Industrial Revolution. The question as to how far the Industrial Revolution was science-based is a vexed one among historians: it would seem that, on balance, the view predominates that the inventions that powered the Industrial Revolution were largely the intuitive perceptions of practical men, though scientific progress since the 17th century may well have had the effect of liberating inquiring minds.

THE MINIMAL CRITICAL CONDITIONS OF THE INDUSTRIAL REVOLUTION. Is it possible to escape from the baffling particularism of the British case into a generalized specification of the conditions of an Industrial Revolution? The most noteworthy and persistent of such attempts comes from W.W. Rostow. After a deep immersion in the British case, followed by others, he came to the conclusion that four great governing conditions were required for an Industrial Revolution, namely that there be a rise in the rate of productive investment from, say, five per cent or less of national income to more than ten per cent; that there appear in industry one or more leading sectors; that the productive process in some sense draw upon the potential of post-Newtonian science (thus invigorating invention and innovation); and that the political, social and institutional framework be appropriate. These were the conditions under which an economy would 'take off' into sustained growth. There has been much debate around them. The third and fourth are the most troublesome, appearing to critics to be little more than self-evident, requiring much more definition: this Rostow in his later writings has sought to provide. The concept of leading sectors had proved fruitful in inspiring progress in industrial and business history, and a search for such sectors in particular economies. The investment requirement has also provoked productive debate, though it may have contributed to an over-facile view of what can be achieved by investment in less developed countries. The Rostow attempt to model Industrial Revolution through the concept of take-off, though stimulating and provocative, has achieved only limited success.

Rostow under critical comment amplified his attempt at a theory of immanent market emergence. He found that take-off at a moment of time, did require qualifications that reached far back in time and so deprived it of much of its immediacy. In a sense, though Rostow called his approach a 'Non-Communist Manifesto', he was pushed in part of his thinking away from immanence toward much larger configurations of societies and cultures and much longer periods of time. He has in a sense provided for this wider perspective with the notion of 'traditional society' preceding the meeting of the conditions of take-off, but the idea of self-equilibration as applied to pre-industrial societies is, to the historical mind, unhistorical.

The Marxist approach seeks to understand change in terms of a transition from feudalism to capitalism, from one grand conceptual framework to another. Such a perspective runs in terms of the long-run evolution of entire societal structures and their associated mentalities. It does not preclude critical phases, and indeed in some forms demands them, as with the notion that European society in some sense underwent general crisis in the 17th century, presumably as a condition of the release of the conditions of capitalism. Such grand transitional thinking encounters difficulties which are the obverse of the critical phase approach to the Industrial Revolution; they have little explanatory power in terms of what was in fact, in spite of many qualifications, a dramatic increase in output and environmental exploitation over a relatively narrow band of decades.

THE INDUSTRIAL REVOLUTION IN DIFFERENT CONTEXTS. As between countries there have been and are many paths to industrialization. One would expect this from their historical and geographical diversity, with associated differences in the gestation period involved. It is these variations that militate against a non-country specific theory of the Industrial Revolution. It is possible to order industrializing countries, as Rostow has done, on a time scale, beginning with Britain, but the timing and nature of the experiences of 'following' countries is open to much debate. Moreover the notion of 'second' Industrial Revolutions further complicates matters: for example, the performance of the West German and Japanese economies since the virtual destruction of their productive apparatus in World War II has produced results that are more striking than the 19th-century achievements of these countries, notable though these were.

The catching-up thesis as affecting those countries which the British Industrial Revolution had left behind led to a search for reasons for such relative failure in various national contexts: for example, in the French case a good deal has been made of the absence of a real entrepreneurial class and the concentration of finance on the needs of the state. This negative approach has had some utility by calling attention to missing or weak industrializing factors. But it has been succeeded by an emphasis on the more positive aspects of the French performance, for example the scientific and engineering expertise of the Grande Ecoles. In Europe of the greater nations, only Spain seems to have had no redeeming Industrial Revolution features, a failure rooted in her imperial past both in Europe and the Americas.

THE INDUSTRIAL REVOLUTION IN GLOBAL TERMS. If we adopt a global approach to the industrializing process, seeing it as a phenomenon generated by the world economy as a whole, rather than by individual countries, two main lines of thought are encountered. One has to do with the longer-term stability of the world system, an aspect of thought associated with the long cycles linked with the name of Kondratieff, of some fifty years. The second is concerned with the possibility that successive world Industrial Revolutions, powered by the bunched technologies of Kondratieff-type thinking, are seriously exploitative of those countries which have not achieved an Industrial Revolution and so constitute peripheral societies, largely dependent on the export of primary products and raw materials, the value added element being expropriated by the industrialized countries. The idea of the Industrial Revolution being based upon western exploitation has a long pedigree, including the belief held by some that the first Industrial Revolution rested on British exploitation of slavery in the West Indies and on the building of fortunes by Nabobs in India, repatriating the proceeds to Britain.

Kondratieff thinking, like that of Rostow, is market based, seeing market phenomena as the principal determinant; core-periphery thinking starts from the idea of monopolistic power as between countries. The exploitation theory is important to Marxian thinking because of the prophecy by Marx that proletarian revolution was imminent in industrializing countries, but was perplexingly delayed: the answer was seen to lie in the idea that the contradictions of capitalism had been postponed in maturing economies by the exploitation of non-Industrial Revolution countries. The extent to which any Industrial Revolution depends upon the exploitation of other countries is difficult to resolve. That there has been such exploitation there can be no doubt: yet the power of the Industrial Revolution to generate a larger product by internal operation cannot be denied.

The long cycle attracted Kondratieff's attention in terms of prices and rates of interest: more recently analysis has been focussed on fluctuations in long-term production and investment. This is consistent with Schumpeter's view that the long cycle was real, and was due to the implementation by entrepreneurs, on a world scale, of successive 'bunches' of investment in newly available technologies. Such a 'bunch' is associated with the first Industrial Revolution (the cotton inventions, the steam engine, machine making and engineering, high farming and heavy chemicals), another with the later 19th century (electricity, organic chemicals, the automobile), and yet another with the post-1945 expansion (electronics, the computer, synthetic materials, drugs, oil and petrochemicals, consumer durables, the aeroplane and vehicles). There does indeed seem to be a case for long-term waves of innovation: if so the idea of successive Industrial Revolutions impacting on a world scale seems realistic. But simple extrapolation is not possible, either in terms of timing or technological content.

THE INDUSTRIAL REVOLUTION AND THE ECONOMISTS. Until Rostow's time, the economists were happy to leave the conditions of industrializing to the historians. Technology and the supply of entrepreneurship were regarded as too impalpable to be made part of economic thinking. The attempt to make economics 'scientific', that is, capable of yielding abstract models, meant that these factors were neutralized by *ceteris paribus* assumptions. Toynbee was not however an economist, but started rather from social premises: for him the Industrial Revolution had generated fearful social problems, with its benefits muted. Toynbee placed great stress on the inventions (the technology) that lay behind the vast changes. The pessimistic view of great economic change was of course much older than Toynbee. Adam Smith considered the possibility that a society that had achieved 'commercial' status (that of his own time) could be worse off than previously, partly from the alienation caused by the division of labour as in his pin factory. Ricardo and Malthus too carried a pessimistic tone, taking the view that diminishing returns to effort in agriculture must in the longer term lead to the stationary state with zero net investment. The notion that a strategic bottleneck would impose an inherent limit on the benefits of Industrial Revolution was deeply embedded in people's minds. Jevons feared an energy crisis through coal exhaustion; in more modern times there has been the fear of educational failure, bureaucratic strangulation, political perversity, resource exhaustion and waste disposal. John Stuart Mill, on the other hand, welcomed the idea of the stationary state as providing a respite during which society could find its bearings.

Equilibristic thinking, whether in terms of marginalism from the later 19th century, or the trade cycle in the interwar years, could be inhibitive upon the development of thinking of the Industrial Revolution kind. But after 1945 there was a rebirth of theories of economic development in an atmosphere of European reconstruction and decolonization. Even more recently economists have made strenuous efforts to introduce into their thinking the impalpables that had formerly been excluded, especially technology.

But a price has to be paid for this widening of horizons. Rostow, for example, started from neoclassical economic theory with the market as his operative mechanism, but by trying to incorporate the supply of entrepreneurial and inventive initiative, technology and science and political systems, what was intended to be an articulated system of reasoning has become disarticulated. The sociologists, on the other hand, from their models, using the thought forms of Marx, Weber and others in their search for a theory of social rather than economic change, never really articulate their thinking except in a very loose way. These two social science disciplines have thus tended to pursue their own perspectives, generating discrepant conceptual frameworks, with associated data, that are very difficult of mutual assimilation.

THE INDUSTRIAL REVOLUTION AND THE POLICY-MAKERS. Of what use is the concept of Industrial Revolution as a guide to national policy-makers? It is a value only if it can be so stated as to fit the requirements of the country whose economic performance is to be improved. In developing economics, with little modern history, not greatly removed from traditional society, and based on agriculture, the concept of Industrial Revolution, in the sense of the possibility of inducing rapid, dramatic and self-sustaining change, is of very limited practical application. More generally any attempt to induce Industrial Revolution from above is a subtle and perhaps dangerous business, disrupting the existing value system, social structure and political framework. And yet there is a sense in which Industrial Revolution can be useful for policy, both to inform and to warn. The possibility of it doing so depends upon reconciling the need to see examples of the Industrial Revolution in particular contexts in their own full historical terms, while being able at the same time to derive such sense of universality concerning industrialization as is really justified, and which may usefully be transferrable to the country that is to be manipulated. It might be argued that the pace of technical change since World War II has been so revolutionary that looking to history in terms of an Industrial Revolution is now irrelevant. This however is not so. The conditions of industrial success have not changed to the degree that history is meaningless, for in spite of the power of new technologies, human structures, needs and responses are not so different from those of the past.

There are two general views of the role of the state in past Industrial Revolutions. Students of the British Industrial Revolution for long held that success depended on the removal of the state from the economy, or at least minimizing its presence. This was the basis of the liberal economic philosophy. Students of the German and Japanese industrial revolutions are inclined to emphasize the directive role of the state. This line of argument was used first in terms of the need to 'catch up' with Britain, but in more recent times has been seen more positively as essential in the present state of world industrialization. It embodies the view that the demands for modern technology are such in terms of R and D, financial provision, the educational system and the need for the setting of long-term priorities, that they can only be met under the constructive aegis of the state. The rise of the Asian economies, going through their recent Industrial Revolutions under strong state surveillance, is a challenge to the British historical model of a spontaneous release of atomistic entrepreneurship operating in untrammelled markets.

The Industrial Revolution, with its focus on the minimal conditions of critical, irreversible change, remains a useful intellectual and teaching tool, one of those devices which helps us to make sense of periods and places in which configurations of circumstance, whether spontaneous or contrived, have carried society to a new plateau of wealth.

S.G. CHECKLAND

See also CONTINUITY IN ECONOMIC HISTORY; ECONOMIC HISTORY; ECONOMIC INTERPRETATION OF HISTORY; INDUSTRIALIZATION.

BIBLIOGRAPHY

Ashton, T.S. 1948. *The Industrial Revolution, 1760–1830.* London: Oxford University Press.

Bezanson, A. 1922. The early use of the term Industrial Revolution. *Quarterly Journal of Economics* 36, 343–9.

Deane, P. 1965. *The First Industrial Revolution.* Cambridge: Cambridge University Press.

Flinn, M.W. 1966. *The Origins of the Industrial Revolution.* London: Longmans.

Foster, J. 1974. *Class Struggle and the Industrial Revolution.* London: Wiedenfeld.

Habakkuk, H.I. and Postan, M. (eds) 1965. *Cambridge Economic History of Europe.* Vol. VI: *The Industrial Revolutions and After: incomes, population and technological change.* Cambridge: Cambridge University Press.

Landes, D. 1969. *The Unbound Prometheus: technological change and industrial development in western Europe from 1750 to the present.* Cambridge: Cambridge University Press.

McClelland, D.C. 1961. *The Achieving Society.* Toronto: Van Nostrand.

Mathias, P. 1983. *The First Industrial Nation: An Economic History of Britain, 1700–1914.* 2nd edn, London: Methuen.

Mitchell, B.R. and Deane, P. 1962. *Abstract of British Historical Statistics.* Cambridge: Cambridge University Press.

Mokyr, J. (ed.) 1985. *The Economics of the Industrial Revolution.* Totowa, NJ: Rowmann & Allanheld.

North, D.C. and Thomas, R.P. 1973. *The Rise of the Westen World.* Cambridge: Cambridge University Press.

Rostow, W.W. 1960. *The Stages of Economic Growth.* Cambridge: Cambridge University Press.

Rostow, W.W. (ed.) 1963. *The Economies of Take-off into Sustained Growth.* London: Macmillan.

Rostow, W.W. 1975. *How it all Began. Origins of the Modern Economy.* New York: McGraw-Hill.

Smith, A. 1776. *An Inquiry into the Nature and Causes of the Wealth of Nations.* Oxford: Oxford University Press, 1976.

Toynbee, A. 1908. *Lectures on the Industrial Revolution of the 18th century in England.* 2nd edn, London: Longmans.

inequalities. Mathematical inequalities are pervasive in economic theory, just as economic inequalities are pervasive in social life. The insistence that quantities (always) and prices (usually) be non-negative, the constraint that expenditure not exceed wealth, the necessity in proving existence of competitive equilibrium that each agent's resources have positive value, are so familiar that we scarcely think of them as requirements of inequality, though that is what they are.

Many of the basic results of economic theory (such as the non-positivity of the substitution effect) take the form of inequalities. These in turn often arise from the definiteness or semidefinitenes of certain matrices, such definiteness being again expressed by inequalities. Yet further along the chain of reasoning, those matrices usually derive such properties from their origin in the convexity or concavity of various functions. For real-valued functions, convexity is defined by *Jensen's Inequality* (1906): The function $f: X \subset R^n \rightarrow R$ is *convex* if

$$\forall x^1, x^2 \in X, \quad \forall \alpha \in [0, 1]$$

$$f(\alpha x^1 + (1 - \alpha)x^2) \leqslant \alpha f(x^1) + (1 - \alpha)f(x^2) \quad (1)$$

(A function g is concave if $-g$ is convex).

There are close connections between convex functions and inequalities in general. Indeed, 'The classical inequalities are ... obtained by verifying that a certain function is convex and by calculating its transforms.' (Young, 1969, p. 112). To illustrate this general proposition by an important special case, consider the *gauge* $J(\cdot|C)$ of any set $C \subset R^n$, together with its *polar transform* $J^0(\cdot|C)$, which is the gauge of the polar set C^0 of C (see GAUGE FUNCTIONS). When C is convex

and closed and contains the origin, $J^0(\cdot|C)$ becomes the *support function* $S(\cdot|C)$ of C. A fundamental inequality of convexity for gauges and their polar transforms is *Mahler's Inequality* (1939), which applied to the present situation reads:

$$\forall x \in R^n, \quad \forall y \in R^n \qquad \sum x_i y_i \leqslant J(x|C)S(y|C) \quad (2)$$

Consider now R^n with its standard Euclidean norm $\|x\|_2 = (\Sigma\, x_i^2)^{1/2}$, and suppose that C is the closed unit sphere $S_c = \{x \in R^n : \|x\|_2 \leqslant 1\}$ of R^n. In this special case it happens that

$$J(\cdot|S_c) = \|\cdot\|_2 = S(\cdot|S_c) \quad (3)$$

(see e.g. Rockafellar, 1970, p. 130). So from (2) and (3),

$$\forall x \in R^n, \quad \forall y \in R^n \qquad \sum x_i y_i = \left(\sum x_i^2\right)^{1/2} \left(\sum y_i^2\right)^{1/2} \quad (4)$$

Since (4) is the famous *Cauchy–Buniakowski–Schwarz Inequality*, this illustrates Young's general proposition above. Young (1969, pp. 112–113) gives further examples (with proofs) of the connections between convexity and the classical inequalities, such as that relating the arithmetic and geometric means, and *Holder's Inequality* (1889):

$$\forall x \in R^n, \quad \forall y \in R^n \qquad \sum x_i y_i \leqslant \left(\sum |x_i|^p\right)^{1/p} \left(\sum |y_i|^q\right)^{1/q} \quad (5)$$

(where $p > 0$, $q > 0$, and $p^{-1} + q^{-1} = 1$), of which (4) is the special case $p = 2 = q$.

It is not surprising then that the classic work on inequalities, the delightful and indispensable book by Hardy, Littlewood and Polya (1934), contains one of the earliest systematic treatments of convex functions in English. A later survey is Beckenbach and Bellman (1961).

PETER NEWMAN

See also CONVEX PROGRAMMING; GAUGE FUNCTIONS.

BIBLIOGRAPHY

Beckenbach, E.F. and Bellman, R. 1961. *Inequalities.* Berlin: Springer-Verlag.

Hardy, G.H., Littlewood, J.E. and Polya, G. 1934. *Inequalities.* Cambridge: Cambridge University Press. 2nd edn, 1952.

Hölder, O. 1889. Über einen Mittelwertsatz. *Göttinger Nachrichten,* 38–47.

Jensen, J.L.W.V. 1906. Sur les fonctions convexes et les inégalités entre les valeurs moyennes. *Acta Mathematica* 30, 175–93.

Mahler, K. 1939. Ein Übertragungsprinzip für konvexe Körper. *Časopis pro Pěstování Matematiky a Fysiky* 63, 93–102.

Rockafellar, R.T. 1970. *Convex Analysis.* Princeton: Princeton University Press.

Young, L.C. 1969. *Lectures on the Calculus of Variations and Optimal Control Theory.* Philadelphia: W.B. Saunders Company.

inequality. To speak of a social inequality is to describe some valued attribute which can be distributed across the relevant units of a society in different quantities, where 'inequality' therefore implies that different units possess different amounts of this attribute. The units can be individuals, families, social groups, communities, nations; the attributes include such things as income, wealth, status, knowledge, power. The study of inequality then consists of explaining the determinants and consequences of the distribution of these attributes across the appropriate units.

This essay on Inequality has four principal objectives. First, I will propose a general typology of *forms of inequality*. This typology will help to map out the conceptual terrain of the discussion. Second, I will examine debates on the conceptual status of one particular type of inequality within this typology, inequality in material welfare. In particular, I will examine the debate over whether or not material inequalities in contemporary societies should be viewed as rooted in *exploitation*. Third, I will examine the implications of these contending views of material inequality for strategies for empirical research on income inequality. Finally, I will discuss the relationship between contending accounts of income inequality and the analysis of social classes.

1. A TYPOLOGY OF INEQUALITIES

Social inequalities can be distinguished along two dimensions: first, whether the unequally distributed attribute in question is a *monadic* attribute or a *relational* attribute; and second, whether the process of acquisition of a particular magnitude of this attribute by the individual can be considered a monadic or relational *process*.

Monadic and relational attributes. A monadic attribute is any property of a given unit (individual, family, community, etc.) whose magnitude can be defined without any reference to other units. Material consumption is a good example: one can assess how much an individual unit consumes in either real terms or monetary terms without knowing how much any other unit consumes. This does not mean that the attribute in question has no social content to it. Monetary income, for example, is certainly a social category: having an annual income of a $30,000 only represents a source of inequality given that other people are willing to exchange commodities for that income, and this implies that the income has an irreducibly social content to it. Nevertheless, income is a monadic attribute in the present sense in so far as one can measure its magnitude without knowing the income of other units. Of course, we would not know whether this magnitude was high or low – that requires comparisons with other units. But the magnitude of any given unit is measurable independently of any other unit.

Relational attributes, in contrast, cannot be defined independently of other units. 'Power' is a good example. As Jon Elster (1985, p. 94) writes, 'In one simple conceptualization of power, my amount of power is defined by the number of people *over whom* I have control, so the relational character of power appears explicitly.' To be powerless is to be controlled by others; to be powerful is to control others. It is impossible to measure the power of any unit without reference to the power of others.

Monadic and relational processes. Certain unequally distributed attributes are acquired through what can be called a monadic process. To describe the distribution process (as opposed to the attribute itself) as monadic is to say that the immediate mechanisms which cause the magnitude in question are attached to the individual units and generate their effects autonomously from other units.

A simple example of a monadic process that generates inequalities is the distribution of body weight in a population. The distribution of weight in a population of adults is certainly unequal – some people weigh three times the average weight of the population, some people weigh half as much as the average. An individual's weight is a monadic attribute – it can be measured independently of the weight of any other individual. And the weight acquisition process is also essentially monadic: it is the result of mechanisms (genes, eating habits, etc.) directly attached to the individual. This is not to say, of course, that these mechanisms are not themselves shaped by social (relational) causes: social causes may influence genetic endowments (through marriage patterns – e.g. norms governing skinny people marrying fat people) and social causes may shape eating habits. Such social explanations of body weight distributions, however, would still generally be part of a monadic process in the following sense: social causes may help to explain why individuals have the weight-regulating mechanisms they have (genes, habits), but the actual weight of any given individual results from these individual weight-regulating mechanisms acting in isolation from the weight-regulating mechanisms of other individuals. The empirical distribution of weights in the population is therefore simply the sum of these monadic processes of the individuals within the distribution.

Now, we can imagine a social process through which weight was determined in which this description would be radically unsatisfactory. Imagine a society in which there was insufficient food for every member of the society to be adequately nourished, and further, that social power among individuals determined how much food each individual consumed. Under these conditions there is a *causal* relation between how much food a fat (powerful) person eats and how little is consumed by a skinny (powerless) person. In such a situation, the immediate explanation of any given individual's consumption of food depends upon the social *relations* that link that individual to others, not simply on monadic mechanisms. Such an inequality generating process, therefore, would be described as relational rather than a monadic process. More generally, to describe the process by which inequalities are generated as relational, therefore, is to say that the mechanisms which determine the magnitude of the unequally distributed attribute for each individual unit causally depends upon the mechanisms generating the magnitude for other individuals.

Taking these two dimensions of inequality together, we can generate the following typology of ideal-typical forms of inequality. This typology (Table 1) is deliberately a simplification: the causal processes underlying the distribution of most inequalities will involve both monadic and relational mechanisms. Nevertheless, the simplification will help to clarify the conceptual map of inequalities which we have been discussing.

'Power' is perhaps the paradigmatic example of a relationally determined relational inequality. Not only is power measurable only relationally, but power is acquired and distributed through a relational process of competition and conflict between contending individuals, groups, nations, etc. (For discussions of power as form of inequality see Lenski, 1966; Lukes, 1974.)

TABLE 1. Typology of forms of inequality

| | | Form of the unequal attribute | |
		Relational	Monadic
Form of the Process of	Relational	Power, status	Income
Distribution of Attributes	Monadic	Talent	Health, weight

Power is not, however, the only example. Social status is also generally an example of a relationally determined relational attribute. Status is intrinsically a relational attribute in that 'high' status only has meaning relative to lower statuses; there is no absolute metric of status. The process of acquisition of such high status is also generally a relational process of exclusion of rival contenders for status through competitive and coercive means. (Under special circumstances status-acquisition may be a largely monadic process. In artistic production, for example, one could imagine a situation in which each individual simply does the best he or she can and achieves a certain level of performance. There is nothing in one person's achievement of a given level of performance that precludes anyone else achieving a similar level. The status that results from that achievement, however, is still relational: if many people achieve the highest possible level of performance, then this level accords them less status than if few do, but the acquisition process would not itself be a relational one. In general, however, since the process by which the level of performance itself is achieved is a competitive one in which people are excluded from facilities for learning and enhancing performance, status acquisition is itself a relational process.)

The distribution of health is largely a monadic process for the distribution of a monadic attribute. In general, as in the weight acquisition case, the mechanisms which determine an individual's health – genetic dispositions, personal habits, etc. – do not causally affect the health of anyone else. There are, however, two important kinds of exceptions to this monadic causal process, both of which imply a relational process for the distribution of health as a monadic inequality. First, infectious diseases are clearly an example of a process through which the mechanisms affecting health in one person causally affect the health of another. More significantly for social theory, where the distribution of health in a population is shaped by the distribution of medical services, and medical services are relatively fixed in quantity and unequally distributed, then the causal mechanism producing health in one person may well affect the health of another in a relational manner.

Talent is an example of a relational attribute that is unequally distributed through a monadic process. A 'talent' can be viewed as a particular kind of genetic endowment – one that enhances the individual's ability to acquire various skills. To be musically talented means to be able to learn to play and compose music easily, not actually to play and compose music well (a potential prodigy who has never seen a piano cannot play it well). Talents are caused through a monadic process since the causal mechanism which determines one person's latent capacities to acquire skills does not affect anyone else's. (Obviously, parents' talent-generating mechanisms – genes – can affect their children's through inheritance. This is identical to the effect of parents' genes in the weight example. The point is that the effectiveness of one person's genes is independent of anyone else's.) The attribute so produced, however, is clearly relational: a talent is only a talent by virtue of being a deviation from the norm. If everyone had the same capacity to write music as Mozart, he would not have been considered talented.

Income inequality, at least according to certain theories of income determination (see below), could be viewed as an example of a relational process for distributing a monadic attribute. Income is a monadic attribute in so far as one individual's income is definable independently of the income of anyone else. But the process of acquisition of income is plausibly a relational one: the mechanisms by which one person acquires an income causally affects the income of others.

2. INEQUALITIES IN MATERIAL WELFARE: ACHIEVEMENT VERSUS EXPLOITATION

More than any other single kind of inequality, inequality of material welfare has been the object of study by social scientists. Broadly speaking, there are two distinct conceptualizations which have dominated the analysis of this kind of inequality in market societies. These I will call the achievement and exploitation perspectives.

Achievement models. The achievement model of income determination fundamentally views income acquisition as a process of individuals acquiring income as a return for their own efforts, past and present. The paradigm case would be two farmers on adjacent plots of land: one works hard and conscientiously, the other is lazy and irresponsible. Assuming no externalities, at the end of a production cycle one has twice the income of the other. This is clearly a monadic process producing a distribution of a monadic outcome.

The story then continues: the conscientious farmer saves and reinvests part of the income earned during the first cycle and thus expands production; the lazy farmer does not have anything left over to invest and thus continues production at the same level. The result is that over time the inequalities between the two farmers increases, but still through a strictly monadic process.

Eventually, because of a continually expanding scale of production, the conscientious farmer is unable to farm his/her entire assets through his/her own work. Meanwhile the lazy farmer has wasted his/her resources and is unable to support him/herself adequately on his/her land. The lazy farmer therefore goes to work as a wage-earner for the conscientious farmer. Now, clearly, a relational mechanism enters the analysis, since the farm labourer acquires income in a wage paid by the farmer-employer. However, in the theory of wage-determination adopted in these kinds of models in which the labourer is paid exactly the marginal product of labour, this wage is exactly equivalent to the income the labourer would have received simply by producing the same commodities on his/her own account for the market. The relational mechanism, therefore, simply mirrors the initial monadic process.

In such achievement models of income acquisition genuinely relational processes may exist, but generally speaking these have the conceptual status of deviations from the pure model reflecting various kinds of disequilibria. In the sociological versions of achievement models – typically referred to as 'status attainment' models of stratification – these deviations are treated as effects of various kinds of ascriptive factors (race, sex, ethnicity) which act as obstacles to 'equal opportunity'. (The best example of status attainment models of inequality is Sewell and Hauser, 1975.) Similarly, in the economic versions of such models – generally referred to as 'human capital' models – the deviations either reflect transitory market disequilibria or the effects of various kinds of extra-economic discrimination. (The classic account of human capital theory is given by Becker, 1975. For his analysis of discrimination see Becker, 1971.) In both the sociological and economic versions, these relational mechanisms of income determination that produce deviations from the pure achievement models mean that certain kinds of people are prevented from getting full income pay-offs from their individual efforts. The inner logic of the process, in short, is monadic with contingent relational disturbances.

Exploitation models. Exploitation models of income inequality regard the income distribution process as fundamentally relational. The basic argument is as follows: In order to obtain income, people enter into a variety of different kinds of social relations. These will vary historically and can be broadly classified as based in different 'modes of production'. Through a variety of different mechanisms, these relations enable one group of people to appropriate the fruits of labour of another group (Cohen, 1979). This appropriation is called exploitation. Exploitation implies that the income of the exploiting group at least in part depends on the efforts of the exploited group rather than simply their own effort. It is in this sense that income inequality generated within exploitative modes of production is intrinsically relational.

There are a variety of different concepts of exploitation contending in current debates. The most promising, in my judgement, is based on the work of Roemer (1983). (For a debate over Roemer's formulation, see *Politics & Society*, 11(2), 1982.) In Roemer's account, different forms of exploitation are rooted in different forms of property relations, based on the ownership of different kinds of productive assets. Roemer emphasizes two types of property in his analysis: property in the means of production (or alienable assets) and property in skills (or inalienable assets). Unequal distribution of the first of these constitutes the basis for capitalist exploitation; unequal distribution of the second constitutes the basis, in his analysis, for socialist exploitation.

While Roemer criticizes the labour theory of value as a technical basis for analysing capitalist exploitation, nevertheless his basic defence of the logic of capitalist exploitation is quite in tune with traditional Marxist intuitions: capitalists appropriate part of the surplus produced by workers by virtue of having exclusive ownership of the means of production. Socialist or skill exploitation is a less familiar notion. Such exploitation is reflected in income returns to skills which is out of proportion to the costs of acquiring the skills. Typically this disproportion – or 'rent' component of the wage – will be reproduced through the institutionalization of credentials. Credentials, therefore, constitute the legal form of property that typically underwrites exploitation based in skills.

Two additional assets can be added to Roemer's analysis. Unequal distribution of *labour power* assets can be seen as the basis for feudal exploitation, and unequal distribution of *organization* assets can be viewed as the basis for state bureaucratic exploitation (i.e. the distinctive form of exploitation in 'actually existing socialism'). The argument for feudalism is basically as follows: in feudal society, individual serfs own less than one unit of labour power (i.e. they do not fully own their own labour power) while the lord owns part of the labour power of each of his serfs. The property right in the serf's labour power is the basis for the lord forcing the serf to work on the manorial land in the case of corvée labour, or paying feudal rents in cases where corvée labour has been converted into other forms of payment. The flight of peasants to the cities, in these terms, is a form of theft from the lord: the theft of the lord's labour power assets. The argument for state bureaucratic societies is based on the claim that control over the organizational resources of production – basically control over the planning and coordination of the division of labour – is the material basis for appropriation of the surplus by state bureaucrats. (For a detailed discussion of these additional types of assets and their relationship to exploitation, see Wright, 1985.) In all of these cases, the ownership and/or control of particular types of productive assets enables one class to appropriate part of the social surplus produced by other classes.

In exploitation models of income distribution, monadic processes can have some effects. Some income differences, for example, may simply reflect different preferences of individuals for work and leisure (or other trade-offs). Some of the income difference across skills may simply reflect different costs of acquiring the skills and therefore have nothing to do with exploitation. Such monadic process of income determination, however, are secondary to the more fundamental relational mechanisms.

3. IMPLICATIONS FOR EMPIRICAL RESEARCH STRATEGIES

As one would suspect, rather different empirical research strategies follow from monadic versus relational conceptions of the process of generating income inequality. In a strictly monadic approach, a full account of the individual (non-relational) determinants of individual income is sufficient to explain the overall distribution of income. This suggests that the central empirical task is first, to assemble an inventory of all of the individual attributes that influence the income of individuals, and second, to evaluate their relative contributions to explaining variance across individuals in income attainment. In the case of the example of the two farmers discussed above this would mean examining the relative influence of family background, personalities, education and other individual attributes in accounting for their different performances. The sum of such explanations of autonomously determined individual outcomes would constitute the basic explanation of the aggregate income distribution.

It follows from this that the heart of statistical studies of income inequality within an achievement perspective would be multivariate micro-analyses of variations in income across individuals. The study of overall income distributions as such would have a strictly secondary role.

In exploitation models of income distribution, the central empirical problem is to investigate the relationship between the variability in the form and degree of exploitation and income inequality. This implies a variety of specific research tasks, including such things as studying the relationship between the overall distribution of exploitation-generating assets in a society and its overall distribution of income, the different processes of income determination within different relationally defined class positions (see Wright, 1979), and the effects of various forms of collective struggle which potentially can counteract (or intensify) the effects of exploitation-mechanisms on income inequalities.

This does not imply, of course, that achievement models of income inequality have no interest in macro-studies of income distribution, nor that exploitation models have no interest in micro-studies of individual income determination. But it does mean that the core empirical agendas of each model of income inequality will generally be quite different.

4. MATERIAL INEQUALITY AND CLASS ANALYSIS

Sociologists are interested in inequalities of material welfare not simply for their own sake, but because such inequality is thought to be consequential for various other social phenomena. Above all, material inequality is one of the central factors underlying the formation of social classes and class conflict.

The two models of income inequality we have been discussing have radically different implications for class analysis. In achievement models of income distribution, there is nothing intrinsically antagonistic about the interests implicated in the income determination process. In the example we discussed, the material interests of the lazy farmer are in no sense intrinsically opposed to those of the industrious farmer. The strictly economic logic of the system, therefore, generates autonomous interests of different economic actors, not conflictual ones.

Contingently, of course, there may be conflicts of interest in the income determination process. This is particularly the case where discrimination of various sorts creates noncompetitive privileges based on ascriptive characteristics such as sex and race. These conflicts, however, are not fundamental to the logic of market economies and they do not constitute the basis for conflicts between economic classes as such.

Conflicts between classes in capitalist societies, therefore, basically reflect either cognitive distortions on the part of economic actors (e.g. misperceptions of the causes of inequality) or irrational motivations (e.g. envy). Conflicts do not grow out of any objective antagonism of interests rooted in the very relations through which income inequalities are generated.

Exploitation models of income inequality, in contrast, see class conflict as structured by the inherently antagonistic logic of the relational process of income determination. Workers and capitalists have fundamentally opposed interests in so far as the income of capitalists depends upon the exploitation of workers. Conflict, therefore, is not a contingent fact of particular market situations, nor does it reflect ideological mystifications of economic actors; conflict is organic to the structure of the inequality-generating mechanisms themselves.

These different stances towards the relationship between interests and inequality in the two approaches means that for each perspective different social facts are treated as theoretically problematic, requiring special explanations: conflict for achievement theories, consensus for exploitation theories. Both models, however, tend to explain their respective problematic facts through the same kinds of factors, namely combinations of ideology and deviations from the pure logic of the competitive market. Exploitation theories typically explain cooperation between antagonistic class actors on the basis 'false consciousness' and various types of 'class compromises' between capitalists and workers, typically institutionalized through the state, which modify the operation of the market (see Przeworski, 1985). Achievement theories, on the other hand, use discriminatory preferences and market imperfections to explain conflict.

ERIK OLIN WRIGHT

See also CAPITAL AS A SOCIAL RELATION; CLASS; DISTRIBUTIVE JUSTICE; ECONOMIC FREEDOM; EQUALITY; JUSTICE; POVERTY; PROPERTY; SOCIAL JUSTICE.

BIBLIOGRAPHY
Becker, G.S. 1971. *The Economics of Discrimination.* 2nd edn, Chicago: University of Chicago Press.
Becker, G.S. 1975. *Human Capital.* 2nd edn, New York: National Bureau of Economic Research.
Cohen, G.A. 1979. The labor theory of value and the concept of exploitation. *Philosophy and Public Affairs* 8(4), 338–60.
Elster, J. 1985. *Making Sense of Marx.* Cambridge: Cambridge University Press.
Lenski, G. 1966. *Power and Privilege.* New York: McGraw-Hill.
Politics & Society. 1982. 11(3). Special issue on John Roemer's theory of class and exploitation.
Lukes, S. 1974. *Power: a Radical View.* London: Macmillan.
Przeworski, A. 1985. *Capitalism and Social Democracy.* Cambridge: Cambridge University Press.
Reich, M. 1981. *Racial Inequality.* Princeton: Princeton University Press.
Roemer, J. 1983. *A General Theory of Exploitation and Class.* Cambridge, Mass.: Harvard University Press.
Sewell, W. and Hauser, R. 1975. *Education, Occupation and Earnings.* New York: Academic Press.
Szymanski, A. 1976. Racial discrimination and white gain. *American Sociological Review* 41, 403–14.
Wright, E.O. 1979. *Class Structure and Income Determination.* New York: Academic Press.
Wright, E.O. 1985. *Classes.* London: New Left Books/Verso.

inequality between nations. From the origins of systematic economic analysis, differences in national prosperity have been remarked upon. Indeed, Adam Smith's *Wealth of Nations* elaborated the theoretical reasons supporting the contemporary observation that income in Europe varied inversely with the extent of state intervention. But it is to the beginning of sustained economic growth in the 18th century and the wonders of compound interest that we trace the present large differences among national per capita incomes. Countries that followed early in the van of England's industrial revolution could multiply their initial levels of per capita income by factors of ten to twenty times over the course of a century and a half. Those that did not faced ever larger disparities. The World Bank's *World Development Report* in 1985 recorded a range of per capita income from $120 in Ethiopia to $22,870 in the United Arab Emirates and $14,110 in the United States.

Conventional measures of inequality applied to such a distribution of income levels, like the Gini coefficient, show values among nations in excess of 0·7, and thus notably greater than inside even the most unequal of countries. On a global level, half the world population commands little more than 5 per cent of total income, while at the other end, 15 per cent of the population receives something like three-fourths of world income.

These comparisons overstate inequality by reason of the use of market exchange rates to convert national currencies to US dollars. In poorer countries, non-traded goods and some wage goods sell for much less than their exchange rate derived equivalent. Systematic collection and analysis of domestic prices in a number of countries in a study conducted under the direction of Irving Kravis (1982) have provided a more accurate indication of the extent of bias introduced by the use of exchange rates to convert national values to dollars. It is considerable. For the poorest countries, incomes relative to those in the United States are understated by a factor of 2·5; in the richest, overstated by 7 per cent. Thus Malawi's 1975 gross product per capita rises from $138 to $352, while France's falls from $6428 to $5977. Even as adjusted for this bias, international inequality remains extreme: the Gini coefficient exceeds 0·6, and the poorest half of the world's population has access to only 12 per cent of the world's product.

This relative gap between the rich and poor has not narrowed in the post-World War II period, despite the much improved performance of developing countries. Since 1950, although developing country annual per capita income growth has accelerated to almost 3 per cent, the industrialized countries have more than kept pace. Indeed, it is only during the disturbed international economic environment of the 1970s that faltering growth in the industrialized economics enabled developing countries to move ahead more rapidly, and there are signs of reversal in the 1980s. Most troubling is the plight of the low income countries (excluding China). Africa and South Asia have fallen further behind, even as the new industrializing countries of East Asia and some Latin American countries have improved their positions. It has become more common to speak of Third and Fourth Worlds, as increased differentiation has emerged among developing countries. Signs of improved performance in South Asia and China in recent years remain unmatched in Africa.

Attention to this widening relative gap obscures the positive attainments of the developing countries over the last 35 years. Their economic performance, even while many were burdened simultaneously with the task of nation-building as artificial colonial boundaries and status were swept away, exceeds that attained by the present industrialized countries in the past. It is a still greater improvement over the progress of the developing countries before 1950, when their growth was much slower and more erratic.

So long as the differences between industrialized and developing countries are the central issue, the perspective cannot help but be pessimistic. Improvements in relative position still imply a continuous increase in the size of the absolute gap until the time when the ratio of per capita incomes equals the inverse of the ratio of growth rates. For middle-income developing countries, whose income per capita is only about a fourth of the industrialized country level after correction for exchange rates, the absolute difference will continue to widen, because growth rates do not come close to being four times greater. Those select developing countries attaining sustained rates of growth greater than the expansion of the industrialized countries nonetheless confront an absolute gap whose elimination is to be reckoned in units of centuries, and not even decades (Morawetz, 1977, pp. 26–30).

Such arithmetic raises the question of the usefulness of gaps, relative or absolute. Why not opt instead for an absolute criterion analogous to the focus upon poverty within countries? As W.A. Lewis has said: 'What will happen to the gap between the rich and the poor countries? ... Since I think what matters is the absolute progress of the LDCs and not the size of the gap, I do not care' (Ranis, 1972, p. 420). With this shift of emphasis, the development problem continues to remain immense. By a World Bank criterion of minimal standard of living, some 40 per cent of the population of developing countries is in a destitute state. Rapid and consistent economic growth, and redistributionist policies, will be required to make a significant dent in the present dismal situation.

But even were such efforts to yield results, concern about inequality between nations will not go away readily. Comparisons will influence economic analysis as well as international politics. The existence of the gap has contributed to new ways of thinking about the process of integration of developing countries into the world economy, as well as inspiring practical efforts to reduce it.

The conceptual issue posed by increasing inequality over the 19th and 20th centuries, despite a more inclusive international economy, is its inconsistency with predictions of convergence inherent in a competitive market framework. Trade, capital flows and migration are all supposed to equalize returns to factors of production, and thereby to diminish national differences. The apparent failure to do so on a global scale has led to a group of theories emphasizing the opposite: the adverse effects of developing country integration in the world economy through declining terms of trade, backwash and demonstration effects, unequal exchange, technological dependence and domestic disintegration. These theories, and the experience of the Great Depression, reinforced an inward, interventionist and industrialization orientation to development efforts in most developing countries in the postwar period.

That point of view, despite its critical reception in industrialized countries, continues to be influential in developing countries, not least among policy makers. From intellectual origins at the Economic Commission for Latin America in the 1940s, this criticism of an asymmetrical international order led to the first UNCTAD conference in 1964 and its continuing service as a permanent forum for the discussion of relationships between developing countries. On that first occasion and at subsequent meetings, the South, united politically, sought to extract aid concessions and preferential trade access from the North. The principal harvest was the eventual acceptance of a Generalized System of Preferences, and thus a modest discrimination in favour of developing countries. The Third World was very much in the role of supplicant in behalf of a limited agenda.

It was not until after the oil crisis and price rise in 1973 and an enhanced sense of commodity power, that developing countries, backed by OPEC, extended their demands and elevated their voices on behalf of a more comprehensive New International Economic Order. The UN General Assembly, in a Sixth Special Session convened in April 1974, concluded its deliberations by committing itself 'to work urgently for the establishment of a new international economic order [NIEO] ... which shall correct inequalities and redress existing injustices, make it possible to eliminate the widening gap between the developed and developing countries'. The range of concrete proposals included stabilization of commodity prices at more favourable levels; enhanced regulation of foreign investment and technology transfer; new rules for the international monetary system, including a linkage of IMF special drawing rights to development finance; enhanced official aid and a debt moratorium for some hard-pressed developing countries; and greater participation in the decision structure of the multilateral economic institutions, the IMF, World Bank and GATT (Fishlow, 1978).

For a brief moment, North–South economic relations stood centre stage in the traditional pride of place of East–West strategic concerns. New international negotiating structures were created. More conciliatory stands on some issues were taken by the industrialized countries, including agreement in principle on a modest commodity stabilization effort. A new international commission under the leadership of Willy Brandt submitted an action report (Independent Commission, 1980). Even a summit meeting was eventually held in 1981. But the sense of urgency and the faith in negotiation of extensive international economic reforms had already receded by that time.

The limited progress did not owe itself simply to the radicalism of the demands. Despite occasional rhetorical flourishes, the platform of the NIEO was reformist, and accepting of the importance of the international economy for the prospects of developing countries. Indeed, emphasis was placed upon a competitive international order, in which developing countries could exploit their new-found comparative advantage. The self-reliance, or delinking, advocated by some Third World spokesmen was implicitly rejected. Many of the specific proposals had been aired before. Yet in their totality, there was also a clear commitment to discrimination and non-reciprocity, a tendency toward intervention in market processes, and demands for a redistribution of resources and power that industrialized countries found objectionable. The more conservative political tide of the late 1970s and early 1980s in the industrialized countries reinforced this opposition even as it became clearer that the feared capacity of developing countries to restrict supply to needed raw materials had been overstated.

But the NIEO agenda also lapsed because of the relative

gains experienced by many developing countries in the 1970s as international trade expanded and capital flows increased in unanticipated fashion. Oil exporters clearly gained from their improved terms of trade. The East Asian newly industrializing countries did likewise by exporting manufactured products at a rate that compensated for their worsened terms of trade. Even under the old order it was possible for some countries to make it. This differentiation among developing countries reduced their political coherence and unity. Commodity agreements might help some, but they also could harm others increasingly dependent upon primary imports. The emergence of the newly industrializing countries and their penetration of developed country markets, defined another path to success.

Nor did the slowing of growth of many of the heavily indebted developing countries in the early 1980s owing to higher oil prices, interest rates and industrialized country recession stimulate a renewed cry for a new order, although some leaders tried. Debt problems, and their remedies, tended to be specific to particular countries. The critical element of Third World unity was largely absent, despite some common origins in external economic circumstances.

Although Latin America was especially afflicted by a debt problem, it involved oil exporters as well as importers. Africa also faced a problem, but of a different kind. Mounting debt service payments on official debt, rather than private, were the proximate source of import compression. Each region sought to mobilize relief for its difficulties, but in different ways. At the same time, a large East Asian debtor like Korea managed to avert reduced growth in part because its low debt-export ratio permitted continuing creditworthiness. Low income China and the countries of the Indian sub-continant, virtually unencumbered by debt, experienced much improved economic performance in the early 1980s.

Yet it would be a mistake to dismiss the efforts of the developing countries in the name of the NIEO as of no practical value. Even in the absence of fully effective political pressure, they have kept alive a concern with the consequences of industrialized country policies. Present opposition to protectionism owes itself in part to the constant vigilance of developing countries. They have also kept in the forefront the need for greater coordination of macroeconomic policy among the developed countries themselves, since they bear many of the consequences of its inadequacy. And they have been instrumental in seeking to increase the role for official flows to finance medium term, rather than immediate, adjustment to changes in the world economy. In regional forums, developing countries have sometimes been more successful in achieving practical cooperation.

Although the focus on the gap between rich and poor nations has faded in recent years compared to its ascendance in the 1970s, the great and persistent inequality between developing and developed countries remains. It is difficult to imagine that there will not be a resurgence in political attention when circumstances again seem more propitious. Redistributionist reforms on a global scale – even when efficiency enhancing and mutually beneficial – will not prove easy, however. In the last analysis, the moral and ethical sense inherent in community that informs national policies to deal with internal inequality is less evident at the international level. Vague, and sometimes even concrete, threats to international order do not fully substitute as a compelling and continuing impulse to action. In a world of sovereign states, the primary burden for relative and eventual absolute reduction of the gap falls upon national development strategies.

ALBERT FISHLOW

See also IMPERIALISM; NORTH–SOUTH ECONOMIC RELATIONS; PERIPHERY; UNEQUAL EXCHANGE.

BIBLIOGRAPHY

Fishlow, A. et al. 1978. *Rich and Poor Nations in the World Economy*. New York: McGraw-Hill.

Independent Commission on International Development Issues. 1980. *North–South: A Program for Survival*. Cambridge, Mass.: MIT Press.

Kravis, I. et al. 1982. *World Product and Income. International Comparisons of Real Gross Product*. Baltimore: Johns Hopkins Press.

Morawetz, D. 1977. *Twenty-five Years of Economic Development, 1950 to 1975*. Washington, DC: The World Bank.

Ranis, G. (ed.) 1972. *The Gap Between Rich and Poor Nations*. London: Macmillan.

inequality between persons. Although inequality between persons can refer to a great variety of issues concerned with the disparate treatment and circumstances of individuals, economic discussion has focused on those aspects that relate to the acquisition and expenditure of income. As a consequence, the study of personal inequality has become largely synonymous with the distribution of income among individuals or households. Early contributors to this subject tended to provide an overall perspective on personal income distribution. In recent years, however, more attention has been paid to the particular dimension of inequality under investigation. Consideration has also been given to the precise way in which 'inequality' should be interpreted and measured, a trend most evident in the adjustments applied to observed incomes in order that the 'true' degree of inequality is revealed.

An initial distinction may be made between those studies which examine the origins of income dispersion and those which are interested in its consequences. The principal concern of the latter is inequality in living standards, or levels of well-being, and here the appropriate methodology is well established. For each household we require a measure of the level of its resources relative to its needs. The resource variable is typically identified with income, so that income distribution is the traditional point of departure in the study of unequal living standards. Ideally, however, income should be interpreted in a broad sense to include not only monetary receipts, but also unrealised capital gains, non-pecuniary benefits and household production which is not marketed. In addition, a long run income concept such as 'permanent income' or 'lifetime income' is preferred to the short run concept (weekly, monthly or annual) which applies to most of the readily available data. These incomes should then be adjusted to allow for different household circumstances. One type of adjustment concerns family characteristics, such as the number and ages of family members, and is accomplished by the use of household equivalence scales. A second type of adjustment relates to the environment in which the household operates and covers such factors as the prevailing level of commodity prices; the shelter, heating and transportation requirements associated with household location; and the level of provision of public goods and services. The aim, as before, is to achieve comparability between households in different circumstances. Needless to say, while this programme of adjustments may be generally accepted as the ideal, most empirical studies of living standards fall a long way short of the target.

A much larger body of literature is concerned with the causes of income inequality. This work focuses on the experience of individuals in factor markets and covers a wide range of issues on which opinions seldom agree. It will be helpful to begin by splitting the income y of an individual into components and writing

$$y = y_1 + y_2 + \cdots + y_n \tag{1}$$

$$y = r_1 x_1 + r_2 x_2 + \cdots + r_n x_n, \tag{2}$$

where r_i and x_i are the 'price' and 'quantity' associated with the ith component of income. A decomposition of this form would be appropriate if the x_i denoted the individual's endowments of productive factors, such as labour, capital and land, and the r_i represented the corresponding factor prices. This immediately suggests two principal causes of income inequality: differences in the endowments of productive resources which individuals own; and the structure of factor prices determined by the combination of institutional, market and social forces which we will call the *common environment* of individuals. Further refinement of this line of reasoning can be achieved by extending the coverage of the x_i to include a variety of other characteristics, and by looking back towards the source of the characteristics: to inheritance, in the form of genetic traits, material wealth and family advantage; to innate and acquired skills; and to the choices individuals make in respect of occupation, location and workhours. There is also, inevitably, a portion of income, often attributed to 'chance' or 'luck', which is not systematically related to any of these factors.

Theories of income distribution differ not only in the particular influences and mechanisms that are stressed, but also in their view of what a theory of income distribution should set out to accomplish. One aim is to account for the overall degree of inequality at any date, and the pattern of changes in aggregate inequality that take place over time. Another topic of interest is the characteristic shape of the frequency distribution which incomes tend to follow. A third objective is to explain why different individuals happen to have different incomes. All three of these issues are valid concerns, and all would be addressed in a satisfactory general theory. On the whole, however, a distinct literature has developed on each of the questions. These are reviewed in turn below.

1. CHANGES IN INCOME DISPERSION OVER TIME. Explanations of changes in income inequality over time have typically drawn attention to the features of the common environment and to the consequent pattern of factor prices. Factor prices play a significant role in most of the discussion of income distribution prior to 1900, and even up to the middle of this century, as indicated by the selection of papers published by the American Economic Association in 1946. This is perhaps a reflection of the rigid social structure in the 19th century, which made it natural to assume that the resource endowments of individuals remained relatively constant over time. In those circumstances the first priority was to account for the level of factor prices or factor shares. Once this was done, aggregate factor payments could be allocated among individuals according to a prearranged pattern of entitlement. A theory of factor prices was, in effect, a sufficient explanation for the distribution of personal income.

The tendency to submerge the theory of personal income distribution within the grander themes of Labour, Capital and Land was not without its critics. Cannan (1905) was prompted to suggest that a student seeking an explanation for the riches and poverty surrounding him would return home in disgust from a typical lecture on the subject. He argued that more attention should be paid to the way that aggregate factor payments were shared between individuals, a suggestion taken up with enthusiasm by Dalton (1920) in one of the earliest and most outstanding volumes devoted to personal income distribution. The ideas of Cannan and Dalton had little immediate impact. But the importance they attributed to inheritance has certainly been echoed in subsequent research, most notably in connection with the sources of wealth inequality.

More recently, the explanation of long-run trends in income dispersion has been particularly associated with the work of Kuznets and Tinbergen, and again regards the common environment as the ultimate source of change. The programme of research that has developed from Kuznets (1955) is concerned with the relationship between economic growth and the distribution of income within countries, and sees demographic movements as a major influence on inequality. Countries begin with a fairly homogeneous population, largely employed in the traditional sector. In the course of development, individuals transfer into the modern sector causing income inequality to first rise and then fall, as the modern sector becomes dominant. Inequality within both the traditional and modern sectors may, however, remain constant. This suggests that observed variations in inequality could be spurious, reflecting the way in which data is recorded rather than a real change in the relative income positions of individuals. Other demographic factors, such as shifts in the age structure and household composition have also been cited as having a similar impact.

Tinbergen (1975) appeals to the common environment influences that determine the relative earnings of skilled and unskilled workers. In his view, the observed trend in income inequality is the outcome of a race between technology and education. Technological progress creates additional demand for skilled workers, while improved educational opportunities increases the supply. The direction of movement of the skilled–unskilled wage differential depends on the relative strength of these two forces. Over the course of this century, education has advanced faster than technology, driving down the relative earnings in professional and skilled occupations with notable consequences for income dispersion. The distinguishing characteristic of Tinbergen's argument is the attention given to the operation of factor markets. Many other studies have been concerned with the role of education and training, but they tend to emphasize the process of skill acquisition and treat factor prices as exogenous data.

2. THE PATTERN OF INCOME FREQUENCIES. It has long been recognized that the density function for incomes has a characteristic shape, sufficiently regular and well documented to merit special attention. The major early contribution to this line of enquiry was undoubtedly Pareto, who, in a series of publications, including his *Cours d'économie politique* (1896), assembled evidence on personal incomes spanning more than four centuries and expounded his universal law of income distribution. Pareto noted that the data were closely matched by the formula

$$\ln N = \ln A - \alpha \ln y, \tag{3}$$

where y is a given level of income and N is the number of people with incomes above y. Furthermore, the slope coefficient α was always approximately 1.5. This, he argued, could not be a coincidence. The statistical regularity must indicate a natural state of affairs which would tend to reassert itself if, for any reason, the income distribution departed temporarily from its stable equilibrium.

Pareto's results, and the inefficacy of redistributive policies which they seemed to imply, soon attracted both dedicated support and hostile opposition. The increasing availability of data eventually undermined the strong version of Pareto's law. However the tendency for the upper tail of incomes to follow the Pareto curve (3) is well established, and remains one of the 'stylized facts' concerning the distribution of both income and wealth. Pareto also had a profound influence on many of the methodological developments that have subsequently taken place. His interest in the collection and summary of data, the parametric description of income frequencies, and the identification and investigation of observed statistical regularities, are all strongly echoed in later research.

Although high incomes tend to follow the Pareto relationship the Lognormal distribution is a better representation over the whole income range. This provides a clue as to how the observed pattern of incomes could arise. For just as normal distributions result from a large number of small and statistically independent effects which combine additively, so lognormal distributions emerge when the effects combine multiplicatively. More formally, if we replace income in (1) with the logarithm of income to obtain

$$\ln y = y_1 + y_2 + \cdots + y_n, \qquad (4)$$

and assume, say, that the y_i are identically and independently distributed variables, then the income pattern will be approximately lognormal when n is sufficiently large. In this argument it is the statistical properties, rather than the origins, of the components y_i that are significant. They may, therefore, be treated as unspecified random effects.

The notion that income distribution can be viewed as the outcome of a process governed by a large number of small random influences was developed by a number of authors, most notably Gibrat (1931) who formulated his 'law of proportionate effect', and Champernowne (1953) who demonstrated how the Pareto upper tail could emerge as a feature of the equilibrium distribution of a Markov Chain. Further modifications to these models were later shown to be capable of generating, as the limit of a stochastic process, a variety of other functional forms which have features in common with the lognormal and Pareto distributions, and which may be regarded as reasonable descriptions of the frequency distribution of income. As explanations of income inequality they have been criticized on the grounds that chance or luck, rather than systematic personal or market forces, appears to be the principal determinant of income differences. But the significance attached to this complaint depends on the question that is being addressed. If we are primarily interested in accounting for the overall features of the density function of incomes, it may be appropriate from an aggregate perspective to treat the incidence of personal success and failure as a random event, while simultaneously accepting that success and failure may be rationalized at the level of specific individuals.

3. INCOME DIFFERENTIALS. The dominant theme of recent research on personal income inequality is the explanation of income differentials – the reasons why particular individuals have different incomes. This work has several distinctive features. One is an increasing concern with empirical issues, and with the empirical evaluation of competing theories, facilitated by the availability of large-scale survey data and the means to process the information. Nowadays the question is not so often which factors have an impact on income distribution, but which factors have the most quantitative significance as explanations of observed income differences. Another distinctive feature is the emphasis placed on the distribution of earnings, again partly a result of the quantity and quality of earnings data. Earnings inequality is important,

not only because wages and salaries form such a large proportion of income, but also because it reflects the extent to which labour markets operate fairly. As a consequence the explanation of the inequality of pay has become the main battleground for opposing views on the origins of personal inequality.

One method of approaching the question of earnings differentials is to regard the labour market as being composed of a set of individuals with different personal traits P, and a set of job opportunities offering various combinations of characteristics J. The process of matching people to jobs then generates a relation between personal traits, job characteristics and pay which is typically captured in an earnings equation of the form

$$\ln w = \alpha_1 P + \alpha_2 J + R, \qquad (5)$$

where w denotes earnings or hourly wage rates, and R is a non-systematic or random effect. In this formulation, the coefficients α_1 and α_2 may be interpreted as the 'prices' which the market imputes to the various characteristics. Their values are often estimated from empirical data. Notice that equation (5) has similarities with (2) and, more especially, with (4). This indicates that the separate influences combine multiplicatively, rather than additively, as Gibrat had suggested earlier.

Many different views on the determinants of earnings can be accommodated within equation (5). One common argument claims that earnings are related to the ability or productivity embodied in individuals. Certain personal traits may therefore be valued because they indicate the actual productive performance that derives from either natural ability or the skills acquired as a result of education, training and work experience. Furthermore since perceived, rather than actual, productivity is rewarded in the market place, other characteristics may also be valued if firms believe them to be correlated with relevant variables that cannot be observed. This can account for the 'prices' imputed to educational credentials, family background, gender and race. There are, however, alternative explanations for sex and race discrimination: consumer and employer prejudice towards the goods and services provided by disadvantaged groups; and exploitation by firms of the different supply elasticities that arise from personal circumstances and role specialization within the family.

The structure of job opportunities and the prices imputed to job characteristics lead to a different set of considerations. One line of argument, associated with segmented labour market models and the Job Competition model (Thurow, 1976), stresses the significance of the distribution of jobs across occupations and industries, which depends on the state of technology, the structure of markets and the other social and institutional factors contained in the common environment. It is this job distribution, together with customary wage and salary differentials, which is the principal determinant of earnings inequality. Personal characteristics appear to be important only because they are used by firms to ration entry into the more attractive jobs.

This argument suggests that the desirability of any given occupation is directly related to its imputed price. Exactly the opposite conclusion applies if the prices of the characteristics are interpreted in terms of Adam Smith's (1776, Book 1, ch. X) concept of compensating differentials. For if individuals can choose to trade-off income against job characteristics such as occupation, location and working conditions, it is precisely those jobs with the least desirable features that need to pay more in order to attract an adequate workforce.

The notion that some part of earnings may compensate for

other job features, and that the process of choice may help to explain observed income variations, has special significance in the analysis of personal inequality. This may be seen by considering a situation in which people are faced with a set of employment prospects each of which offers a level of income and a combination of other characteristics. If all individuals select from the same set of options, there is clearly no 'true' inequality in the sense of unequal opportunity. Yet people with different tastes will choose different alternatives, so observed incomes will typically vary. It follows that observed income dispersion may well exaggerate the true degree of inequality if some income differences are attributable to choice.

Individuals do not, of course, all face the same set of options. So different opportunities, as well as different choices, contribute to observed inequality. Much of the discussion of the determinants of earnings can be viewed in the context of the distinction between these two factors. Indeed, the most controversial aspects of the study of income distribution often reflect conflicting opinions on the relative importance of the 'true' component of inequality arising from different opportunities and the 'spurious' element of inequality that results from choice. Notice that choice is not the only mechanism that separates opportunities from outcomes. Chance also has a role to play if uncertain prospects are among the set of available options. It is therefore most appropriate to decompose inequality into three components: choice, chance and unequal opportunity. The influence of chance will, however, tend to disappear when we examine the typical experience of a group of individuals.

Those who stress the importance of unequal opportunity tend to focus attention on the contribution of natural ability, family background and discrimination. Here we might, perhaps, distinguish between two aspects of unequal opportunity: the unequal inherited endowments associated with natural ability and family background; and the unequal market treatment associated with discrimination. In contrast, the impact of choice is most clearly seen in the decisions relating to hours of work and geographical location. Individual preferences can also explain the choice of occupation and length of training. Thus, for example, a naive version of the Human Capital model suggests that the level of acquired ability is freely chosen under conditions of equal opportunity, so that the earnings differentials corresponding to education and training are purely compensatory. Refinements of the model, however, allow training opportunities to be influenced by ability and family background, and these factors, together with discrimination, are important elements of those arguments which emphasize the lack of equal and open access to training programmes. The impact of choice on skill levels depends, therefore, on the precise process by which skills are acquired and augmented.

The debate on the relative importance of unequal opportunity and choice for earnings inequality has its counterpart in the study of investment income, via the determinants of wealth distribution. Here, individual preferences are captured in the motives for saving: a desire to provide for retirement, to make bequests, or simply to practice thrift. Choices based on these preferences then determine savings behaviour and can account for some wealth differences in terms of past accumulation. Unequal opportunity, on the other hand, appears principally in the guise of material inheritance, but may also arise from the differences in incomes and family circumstances that affect the opportunities for saving.

ANTHONY F. SHORROCKS

See also DISCRIMINATION; LABOUR MARKET DISCRIMINATION.

BIBLIOGRAPHY

American Economic Association. 1946. *Readings in the Theory of Income Distribution*. Philadelphia: Blakiston.
Atkinson, A.B. 1983. *The Economics of Inequality*. 2nd edn, Oxford: Clarendon Press.
Cannan, E. 1905. The division of income. *Quarterly Journal of Economics* 19, May, 341–69.
Champernowne, D.G. 1953. A model of income distribution. *Economic Journal* 63, June, 318–51.
Dalton, H. 1920. *The Inequality of Incomes*. London: Routledge.
Friedman, M. 1953. Choice, chance, and the personal distribution of income. *Journal of Political Economy* 61, August, 277–90.
Gibrat, R. 1931. *Les inégalités économiques*. Paris: Sirey.
Kuznets, S. 1955. Economic growth and income inequality. *American Economic Review* 45, March, 1–28.
Mincer, J. 1958. Investment in human capital and personal income distribution. *Journal of Political Economy* 66, August, 281–302.
Pareto, V. 1896. *Cours d'économie politique*. Lausanne: F. Rouge.
Smith, A. 1776. *An Inquiry into the Nature and Causes of the Wealth of Nations*. Ed. E.R.A. Seligman, London: J.M. Dent, 1910.
Thurow, L. 1976. *Generating Inequality*. London: Macmillan.
Tinbergen, J. 1975. *Income Distribution, Analysis and Policies*. Amsterdam: North-Holland.

inequality between the sexes. Economic theory concerning inequality between the sexes focuses upon inequality in wages, job recruitment, promotion and dismissal, for women and men with similar qualifications and availability. Neoclassical theory explains these inequalities as a result of free and rational choice, based upon the biological differences between the sexes. According to Becker (1981), women's role in reproduction makes it rational for women to specialize more in family skills, and men more in labour market skills, and parents make a rational choice for their children by preparing them for different careers. When women's reproductive role is reduced due to the decline of birth rates, women's availability for the labour market increases, and they begin to invest more in labour market skills than is the case in countries with continued high fertility. So sex-related differences in level and types of human investment and availability provide the explanation for the differences in wages, types of work and promotion.

By focusing upon the biological differences between the sexes, neoclassical theory selects the features which distinguish inequality between sexes from inequality between other discriminated groups, that is the young versus the old, or foreigners versus members of the dominant ethnic or national group. All these inequalities have been characteristic features of human societies since prehistoric times. The basic principle in the organization of societies is that only members of the superior group have adult status or civic rights, while the members of the inferior groups depend upon the benevolence of the 'adults'. In most societies, economic and social development have reduced the inequalities, but nowhere have they been completely eliminated, and the traditional power of the superior male group over the inferior female group cannot be ignored in the economic analysis of inequality between sexes. The power of the male group over the female one is supported by access to the best technology and a monopoly in learning how to use it (Boserup, 1970). Men's monopoly in the use of weapons, superior hunting equipment, and animal-drawn agricultural equipment, is of ancient origin. But even in societies where men have shifted to tractors and other industrial inputs, women often continue to use primitive hand tools for the operations assigned to them, and even in modern mechanized industries, men distribute the tasks and assign the unskilled, routine operations to the female workers.

In primitive subsistence economies, woman's reproductive role does not prevent her being assigned the most onerous tasks with incessant daily toil, and if the mother's work prevents her from taking care of young children, these are cared for by older sisters or other members of the group. At a later stage of development, when specialization of labour leads to the transfer of an increasing share of the labour power of the family to outside work, the reproductive role of women contributes to explaining why more women than men continue to work in the family and for the family, either as unpaid family members or as domestic servants. However, due to their superior status, men have the right to dispose of money incomes earned by female family members within or outside the family enterprise. There may be regional and local differences in women's status, but in most traditional societies women cannot dispose of money or undertake monetary transactions, accept employment or move away from the locality where they live and work, without the permission of a male guardian who decides all these matters, as well as family matters, like marriage, divorce and the fate of the children. The right to take part in decisions on public matters is reserved for members of the male sex.

Gradually, as technological development transfers an increasing number of products and services from family production to production in specialized enterprises and institutions private or public, there is no need for the full labour power of all female family members in the household. Through the same process the family economy becomes more and more dependent upon money income to purchase the products and services which the family no longer produces, and to pay the taxes which finance the growing public sector. As a result, increasing numbers of women become money earners. At this stage, women's ability to engage independently in economic and other transactions, and their lack of responsibility, becomes a handicap not only to themselves but also to their employers, creditors, customers and guardians, as well as to public authorities and male family members who must support them, if they are unable to support themselves and their dependents because of economic disabilities.

In some European countries, 'market women', who were often middle-aged women with dependents, attained adult status many centuries ago. Later, when it became customary for young girls to work for wages before their marriage, and for other single, divorced and widowed women to support themselves and their dependents by wage labour, or by self employment, these categories of women were granted 'adult' status in economic affairs; but married women continued to be denied adult status. In most industrialized countries, married women first attained adult status when further reduction of the domestic sector, together with the decline of birth rates, radically increased their participation in the labour market and made their work in the labour market an important part of the national economy. In most developing countries, women, whether married or not, are still denied adult status in economic affairs; in some countries it severely limits their labour market participation, in other cases it limits the business activities they are able to accomplish.

Human capital investment in 'market skills' becomes more and more important with economic and social development, while investment in family skills loses in importance when more and more activities are transferred from the family setting to private enterprises or public institutions. When the responsibility for physical protection is transferred from the family to the government, and formal education is introduced, educational level may replace the ability to use weapons as a status symbol for male youth. The priority given to boys over girls in formal education is not only a result of their larger labour market participation, as suggested by Becker, but also a means to preserve a higher male status, by letting men reach higher educational levels than women.

The status of parents may require that their daughters be educated as well, but that boys should not lose status by receiving less schooling than their sisters, while to preserve the superior status of the husband, the wife must not be more educated than he is. Universities were long closed to women, and in many countries the difficulties of obtaining marriage partners for educated women make both parents and daughters afraid of continuing their education. The low marriage age for girls, another means to preserve male status in the family, may also prevent continuation of the education of girls. The differences between the sexes in educational levels serve to reinforce inequality not only in the family, but also in the labour market. With economic development the difference becomes limited to the highest educational levels but it has not disappeared, even in countries with very high and uninterrupted female labour force participation.

Usually, differences in access to technical training for girls are much larger than differences in access to formal education. From the day women began to work for wages in urban activities, men have insisted on their priority right to skilled, supervisory, and other better paid work. Both in guilds, and later in industries and public service, men became apprentices and skilled workers while women remained assistants to the male workers, unskilled or semiskilled, working under male supervision. In most cases, male trade unions continued the fight of the guild members against rights for women to training, and even to membership of the organization and right to work in the trade. The inferior position of women was defended by the short stay in the labour market of young girls before they married, with no account taken of the large number of spinsters, poor married women and female heads of households, who were permanent members of the labour force both in European and in many non-European countries.

In addition to the lower position of women in the job hierarchy, female wage rates are usually much lower than male wage rates for similar work. Only in periods of great shortage of labour, for instance in wartime or in agricultural peak seasons, may female wages temporarily rise to the level of male wages. The fact that these wage differences are related to sex, and not to the burden of dependency, belies the usual explanation for them. They are a result of the principle of male superiority, and neoclassical theory has helped to make the principle acceptable. Since the theory assumes that differentials in wages equal differentials in marginal productivity of labour, the lower wage rates for women could be taken as a confirmation of the general assumption of female inferiority, which also applied to women as workers.

The superior status of men is supported when women doing similar work get lower wages; when a wife is prevented from earning as much as her husband, he preserves his superior status as principal breadwinner, even if he is too poor to enjoy the even higher status of being the only breadwinner in the family. Training girls in low-wage occupations and discriminating against women in recruitment for 'on the job training' or access to 'learning by doing' supervisory work, reduces the risk that male staff will lose status by being supervised by women.

When employers in private enterprises and public service pay males higher wages than females for similar work, they include the higher male wages in their production costs, even if that reduces the demand for products made primarily by male labour. If an enterprise or a trade has difficulties in competing, due to the payment of high male wages, employers will not reduce the wage differential, but will instead try to get the workers and the trade unions to accept the recruitment, or additional recruitment, of women. If they succeed, the trade will become less attractive to men, and the labour force will gradually become female as has happened to many trades in which trade unions were weak. The separation of the labour market into masculine and feminine trades and jobs becomes even more pronounced if the principle of equal pay for equal work is introduced by law or labour contract, since sex specialization makes it more difficult to prove that the work paid at different rates is 'equal'.

Inequalities between men and women in the labour market and in the family reinforce each other. While Becker assumes a harmony of interests between the marriage partners and an equal distribution of consumption and leisure between them, Sen (1985) uses bargaining theory to explain the observed inequalities in consumption and leisure, which in some countries include differences in coverage of calorie requirements and in access to health care between husband and wife, and between boys and girls. The wife's bargaining position is directly related to her access to the labour market and position in it, but her bargaining position is also weakened because women are likely to perceive inequalities as natural, and make no objections against them. This feature is due to the family socialization of girls from a young age. In many societies, girls are taught that they are less valuable human beings than their brothers, and virtually everywhere girls must help their mothers to provide domestic and personal services for their brothers, who are allowed much more freedom and leisure.

Even in countries with high and perpetual labour force participation by women, girls' education and training within the family focus on child care and domestic activities, and on beautifying themselves to be able to make a good match and reduce the risk of divorce and abandonment, while boys' interests are stimulated in all other fields. Usually, girls are taught to be obedient, to be modest and to do routine jobs without protest, while boys are encouraged to be enterprising, even aggressive, and more self-confident. The inferiority feelings of the girls may induce them to invest less in education and training than boys, as suggested by Arrow (1973), but even if they have the same formal education and training as male competitors, women are likely to lose in competition with males in the labour market. Girls, who are socialized to accept routine jobs and to be modest and obedient, are unlikely to demand good jobs and advancement, or in other ways to fight actively for their interests in the labour market, even when there are few prejudices against them. Much female aptitude for routine and precision work, unsuitability for leadership and unwillingness to take responsibility results from family socialization in the first years of life. Most often, the schools continue in the same vein, but even when schools aim at abolishing inequality between the sexes, the teachers may be powerless, due to family socialization of pupils of both sexes.

In industrialized countries, the last few decades have seen an acceleration of related and mutually reinforcing changes in technology, labour participation by married women with small children, and birth rates. Decline of birth rates to below replacement level, and increasing female labour force participation provide an inducement to the improvement of household technologies, and the introduction of new products and services as substitutes for women's traditional activities and child care. These technological and social changes further induce increasing female labour force participation. A rapidly increasing proportion of married women continue their money-earning activities without reducing work hours during the period when they have small children. But the traditional sex hierarchy is dying very slowly, and although birth rates are low, female levels of education and professional training fairly high, and labour market participation high and continuous, reductions in sex differentials in earnings have been moderate, if any. Earnings in female occupations, including those requiring professional training, are lower than in male occupations with similar requirements. Except for a small female elite, women continue to occupy the positions at the bottom of the labour market within each occupation, as assistants to men, and often supervised by men even in otherwise female occupations.

Married women with full-time work and young children have much longer working hours than men and little leisure because male patterns of work have changed very little, in spite of reduced working hours and the increasing amount of money wives contribute to family expenditure, and also because of the lack of child-care facilities in many countries. However, in spite of the differences between male and female earnings, most women in the industrialized countries have become less dependent upon male support because of the general increase of all wages and the reduction of working hours in the labour market. Therefore, women can support themselves by work in the labour market, if they choose to, and with the aid of obligatory contributions from the father, and public support to female-headed households, they can support children, although the living standards of female-headed households are usually much lower than those of male-headed households. Consequently, many young women react against unequal work burdens by demanding divorce or leaving the home, or by not entering into a formal marriage or cohabitation. Others react by reducing birth rates even further. Contrary to earlier patterns, female applications for divorce have become more numerous than male ones in some industrialized countries. These social and demographic changes serve to make young men, and public opinion in general, more inclined to consider women's demands for more equality.

In many developing countries, economic and social development are producing changes in female labour force participation and birth rates which resemble earlier changes in industrialized countries. Family legislation has been modernized, there is legal equality or less legal inequality between the sexes, access to divorce has become less easy for men and easier for women, and better access to the labour market provides women with some possibilities for self-support in case of divorce and widowhood. Age differences between the spouses are declining due to higher female marriage age, birth rates are declining, and women's position is gradually improving.

But in many other developing countries, either economic changes are few, or male resistance to changes in the traditional status of women is strong. Except for voting rights to parliaments with little influence, women continue to be legally minor, and in many cases their situation has deteriorated because technological changes, or changes in land tenure, have deprived them of traditional means of self-support. In some countries, the labour market continues to be closed not only to married women, but also to deserted women, divorcees and widows, and if labour market shortages occur, they are met by large scale imports of male labour. In these countries birth rates remain high in spite of economic development. For women, economic support from sons is the only alternative to destitution, when the husband dies or ceases to support his wife, and women also desire to have many sons as a means to reduce the risk of abandonment and divorce.

E. BOSERUP

See also DISCRIMINATION; GENDER; LABOUR MARKET DISCRIMINATION; WOMEN'S WAGES.

BIBLIOGRAPHY

Arrow, K. 1973. The theory of discrimination. In *Discrimination in Labour Markets*, ed. O. Ashenfelter and A. Rees, Princeton: Princeton University Press.

Becker, G. 1981. *A Treatise on the Family*. Cambridge, Mass.: Harvard University Press.

Boserup, E. 1970. *Woman's Role in Economic Development*. London: Allen & Unwin.

Buvinic, M., Lycette, M.A. and McGreevey, W.P. 1983. *Women and Poverty in the Third World*. Baltimore: Johns Hopkins Press.

Cain, G. 1985. Welfare economies of policies towards women. *Journal of Labour Economics* 3(1), special issue, January, 375–96.

Sen, A. 1985. *Women, Technology and Sexual Divisions*. New York: United Nations.

Tilly, L. and Scott, J. 1978. *Women, Work and Family*. New York: Holt, Rinehart & Winston.

inequality of pay. The difference between the hourly rates of pay for two jobs might be seen in the same way as that between the prices of a ton of copper and a ton of steel: there is a common unit of quantity, but the conditions of supply and demand for the two articles are largely independent, and the difference between their prices is only an arithmetic by-product. But people do attach significance to the differences between rates of pay, and their ideas about what these should be help to fix particular rates. The relations between the rates of pay for different jobs are termed *differentials* when the jobs compared lie at different grades within the same occupation or industry, and *relativities* when they are in different ones. Both sorts of comparison are possible because jobs are defined by their requirements – what physical and mental ability, length of training, experience, tolerance of adverse working conditions, and the like, they demand from anyone who is to do them adequately; and these requirements are regarded as being common to jobs of all kinds, but present in different amounts and proportions. The requirements of given jobs are assessed intuitively by those who make practical judgements about the fairness of differentials and relativities, and are set out explicitly in the procedure known as job evaluation. The rate for a job is regarded as made up of the shadow prices of the capacities to meet the several job requirements, together with the extent of each requirement in the given job.

The question then arises, how those shadow prices are determined: how comes it about that skill commands a higher price than muscle? Two answers have been given – convention, and supply and demand. The case for convention opens with an appeal to everyday knowledge of how people insist on the maintenance of customary relations. They require that relative pay conform with status. That the labourer's rate stood at two-thirds of the craftsman's in the building industry of Southern England over more than six centuries (Phelps Brown and Hopkins, 1981) can be attributed only to convention. If women's rates had been simply proportioned to productivity, it is hard to account for their relative rise when the rule of 'equal pay for equal work' was enforced in the UK and the Netherlands. Here convention had been keeping pay down, but John Stuart Mill remarked long ago (1848, II, xiv, 2) that

it was keeping the relative pay of clerks up, after the increase of supply of clerical capacity through extended education had tended to lower it. These observations are all consistent with differences of pay being set to match accepted gradations of status. But on the other hand there are all the instances of market forces moving a particular rate, and so changing a relativity, when there has been no question of an antecedent change in status. Differentials that have long remained constant have changed when an upheaval has loosened the hold of custom, and the way in which they have changed can be explained by market forces. The Soviet-type economies use the differentials and relativities in their pay structure as incentives to attract and deflect the supply of labour to particular employments. In the West, competition of employers for a new skill such as computer technology opens up a differential over the pay commanded by other qualifications at the same level. A mental experiment indicates that if those who need a long and costly training before they can meet a job's requirements – say surgeons – were to be paid no more than the unskilled, then though a certain number would still enter the profession out of interest in the work or a sense of social responsibility, it would not be possible to maintain the numbers for which consumers have shown themselves willing to pay at actual rates.

We conclude that the basic reason for the inequality of pay is that, on the side of demand, users are willing to pay different amounts for the capacity to meet the requirements of different jobs; and on the side of supply, that unless a certain rate of pay is provided, labour capable of doing the job will not eventually be forthcoming in the amount that users wish to employ at that rate. If status and pay commonly agree, that is because the personal capacity that confers status also commands a higher price because of its productivity. But there are zones of tolerance within which market forces do not fix rates closely, and here convention may prevail.

On this view, the major obstacle to the reduction of inequality lies in the limitations of the supply of labour to the better-paid jobs. These are to be found in the genetic distribution of personal potential, and next in the distribution of the factors moulding capacity in early childhood, from the homes that foster development to those that thwart it. The quality and extent of education are then limited by the availability of institutions, and the cost of maintenance of students that falls on their families. Similar limitations restrict the numbers obtaining training following education. Readier access to education and training should lower the supply price of labour to the jobs with more exacting requirements and higher pay; but the limitations imposed by heredity and early upbringing will remain (Rutter and Madge, 1976; Phelps Brown, 1977, chs 6, 7 and 9).

So far the discussion has concerned the pay of different occupations, but the differences between the earnings of individuals in the same occupation also demand consideration. These are commonly wide. Some of the range arises from short-period fluctuations in bonus and overtime. Earnings also vary with age, and the inequality of lifetime pay is much less than that of pay at any one time. But a great part of the range is due to differences of individual performance. There are also some systematic forms of differentiation. Regional differences in the rate of pay for the same work are often substantial. In local labour markets quite large differences are found between the pay for a given occupation in different firms. It is generally higher, the bigger the firm. There may be discrimination against ethnic minorities or women. Discrimination 'before the market' occurs when the victims are denied equal access to the means of acquiring capacity. Discrimination 'within the market' occurs when some persons receive lower pay than others by reason of their ascriptive characteristics and not because of, or in proportion to, their lower capacity. It appears likely that discrimination 'within the market' is much less considerable than discrimination 'before the market'.

The combined outcome of differences of pay between jobs and among individuals in the same job is a distribution of individual earnings. The form of this distribution confronts us with a striking social regularity. When Lydall (1968) brought together comparable data of earnings from more than thirty countries, he found a common form of distribution. This was unimodal, with a long upper tail. The central part was closer to a log-normal than a normal distribution, but both tails were thicker or longer than in the log-normal, and in particular the upper tail was fitted closely by Pareto's formula. Generally, distributions of earnings are now taken to be log-normal but with a Paretean upper tail. Soviet planners are understood to treat the distribution of earnings as log-normal.

The common form does not imply an equal measure of inequality, though there are some striking instances of this. The distribution of the earnings of manual men found in the British wage census of 1886 agrees closely with that found in the 1970s. Bergson (1984) found 'a rather striking similarity in equality, as measured, between the USSR and Western countries,' though the USSR distribution lacks the Paretean tail. But differences appear among European countries. 'Britain, France and Italy show the greatest inequality, and the West German structure is the most egalitarian; Belgium and the Netherlands lie between', and the relative pay of women varies greatly among these countries (Saunders and Marsden, 1981, pp. 61, 238). Dispersion has also varied over time. Kuznets (1963) found a systematic relation between the extent of dispersion and the stage of development of the economy, dispersion increasing in the earlier stages of growth and then decreasing in the developed economy. Williamson and Lindert (1980) found that the course of change in the American pay structure agreed with this. The differential for skill was small before 1816, and then came a rapid widening, down to 1856. A further surge from 1899 brought the differential to its peak in 1916; but after 1929 a process of contraction set in and was maintained until the Korean war. The authors ascribed the change to three major factors. A high rate of investment displaced unskilled more than skilled labour, and was linked with a movement of labour out of agriculture. The uneven advance of productivity in different sectors affected the demand for skilled and unskilled labour differentially. Variations in immigration and fertility affected the relative supply of the unskilled. Elsewhere it has been pointed out that it is differences of pay between occupations that make up the greater part of dispersion, except in Great Britain, and attention has been directed to the impact on these differences of changes in demand and supply (Douty, 1980, ch. 5, for USA). In particular, the extension of education has increased the relative supply of professional and technical qualifications. Trade union policy has taken effect, in Great Britain to reduce or eliminate the formerly very wide regional differences (Hunt, 1973) and in Sweden, in pursuance of solidarity, to reduce the lead of builders' pay, raise the relative pay of women, and reduce the gap between white-collar and manual rates. Government policy has endeavoured to raise the lowest paid, as by the national minimum rate in France, Wages Councils in Great Britain, and the Fair Labor Standards Act in the USA, though the long-run effect on dispersion is uncertain. Incomes policies have affected differentials markedly in the short run. Other short-run changes arise from the trade cycle: rising activity has raised the lower rates relatively, and conversely.

Despite the variability of the pay structure in these ways, the regularity of its main features over space and time remains outstanding. It challenges explanation, but remains a matter of discussion. Theories that have been advanced to account for the distribution of income are relevant here (Sahota, 1978), as is the analysis of the log-normal form (Aitchison and Brown, 1957). We may take it that pay is based on capacity. That the distribution of some measures of capacity such as IQ is normal and not log-normal is no bar to believing this, for it is understandable that as we go up the scale of capacity, what users are willing to pay rises more than proportionately, until we reach the vast earnings of those 'at the top of their profession'. We have then to explain why the distribution of capacity should take a common form in diverse societies. It seems likely that the explanation lies in the life-chances of the individual, and the impact of the myriad forces, beginning with conception, that shape body, mind, personality, training and experience. Though these forces have many different features in different societies, they share a stochastic property that gives a common form to the distribution of capacity, and hence to the inequality of pay.

HENRY PHELPS BROWN

See also DISCRIMINATION; LABOUR ECONOMICS; LABOUR MARKETS; SEGMENTED LABOUR MARKETS.

BIBLIOGRAPHY

Aitchison, J. and Brown, J.A.C. 1957. *The Lognormal Distribution.* Cambridge: Cambridge University Press.

Bergson, A. 1984. Income inequality under Soviet socialism. *Journal of Economic Literature* 22(3), September, 1052–99.

Douty, H.M. 1980. *The Wage Bargain and the Labor Market.* Baltimore and London: Johns Hopkins University Press.

Hunt, E.H. 1973. *Regional Wage Variations in Britain 1850–1914.* Oxford: Clarendon Press.

Kuznets, S. 1963. Quantitative aspects of the economic growth of nations. VIII: Distribution of income by size. *Economic Development & Cultural Change* 9(2), Pt II, January, 1–79.

Lydall, H.F. 1968. *The Structure of Earnings.* Oxford: Oxford University Press.

Mill, J.S. 1848. *Principles of Political Economy.* London: Parker.

Phelps Brown, E.H. 1977. *The Inequality of Pay.* Oxford: Oxford University Press.

Phelps Brown, E.H. and Hopkins, S.V. 1981. *A Perspective of Wages and Prices.* London and New York: Methuen.

Rutter, M. and Madge, N. 1976. *Cycles of Disadvantage.* London: Heinemann.

Sahota, G.S. 1978. Theories of personal income distribution: a survey. *Journal of Economic Literature* 16(1), March, 1–55.

Saunders, C. and Marsden, D. 1981. *Pay Inequalities in the European Community.* London: Butterworth.

Williamson, J.G. and Lindert, P.H. 1980. *American Inequality, a Macroeconomic History.* New York and London: Academic Press.

infant industry. Opposing arguments for free trade and protection constitute the longest-standing policy debate in the history of economic thought. In this debate the infant-industry argument has acquired pride of place as an exception to free trade – especially as trade theory now gives more attention to explicitly dynamic analysis instead of being confined to comparative statics. But the argument must be carefully stated, and when expressed in its precise modern form its applicability is narrowly limited.

During the period of mercantilism the argument was used to justify the granting of trade monopolies in new and hazardous trades and to inventions (Viner, 1937, p. 71). Alexander Hamilton (1791), Friedrich List (1841) and J.S. Mill (1848) were also early prominent exponents of the argument. Since World War II it has acquired increasing emphasis for less developed countries.

The nature and scope of the infant-industry argument has been refined in modern times by the theory of domestic distortions and the application of welfare economics, with their concern for conditions of Pareto efficiency and determination of the cost of protection. It has also been delimited by considering the benefits and costs of alternative policy instruments – a subsidy, tariff or quantitative restriction – in the context of the hierarchy of policy making (Bhagwati and Ramaswami, 1963; Johnson, 1965; Bhagwati, 1971; Corden, 1974).

The essence of the infant-industry argument rests on 'dynamic learning effects', so that the economy's transformation curve shifts outwards over time, and an industry that is not currently competitive may achieve comparative advantage after a temporary period of protection. Properly stated, the conditions necessary for infant-industry protection are: (1) irreversible technological external economies are generated that cannot be captured by the protected industry; (2) the protection is limited in time; and (3) the protection allows the industry to generate a sufficient decrease in economic costs such that the initial excess costs of the industry will be repaid with an economic rate of return equal to that earned on other investments.

If condition (1) is not fulfilled, the private market should be able to yield an efficient allocation unless capital markets are imperfect or there is imperfect information, so that risks are overestimated. Infant-industry protection is justified not by the fact that there are losses until the infant grows up – but by the fact of external economies associated with the learning process, so that there is underproduction from the social point of view. Condition (2) guarantees that the industry is not protected from infancy to geriatric or even senile stages. And condition (3) guarantees that the expected benefit must be sufficiently great to offset, in present value terms, the current costs of the policy required to produce the benefit (Kemp, 1960).

If free trade is not optimal because of the presence of externalities and the possibility of lower costs over time, what then are the optimal policy instruments for protecting the infant industry? The normative theory of international trade policy has established that the first-best policy would be a production subsidy aimed at the source of the distortion (Corden, 1974, pp. 28–31). This would be preferable to a tariff, which would lead to a by-product, consumption distortion. Although the tariff could restore equality between the marginal rate of domestic transformation and the marginal rate of transformation through foreign trade, it also would drive a wedge between the marginal rate of substitution in consumption and that of transformation. A tariff in turn would be preferable to a quantitative restriction, which would yield quota profits instead of customs revenue and would entail the cost of rent-seeking if there are import licences (Krueger, 1974).

Although it is a domestic market failure that justifies the protection, nonetheless under certain types of market failure the first-best policy may not be a production subsidy (Corden, 1984, pp. 91–2). If the learning experience results in dynamic internal economies in which the learning benefits remain wholly within the firm, the market failure may be in the imperfection of the capital market that makes the financing of such investment difficult or too expensive because the capital market is biased against this type of 'invisible' investment in human capital, or because the rate of interest for all long-term investment is too high owing to private myopia. In this case the first-best policy is to improve the capital market directly; a subsidy to that element of factor input or output that gives rise to the learning benefits would be second best, while further down the hierarchy there would be a general output subsidy to the industry, and then a tariff (Corden, 1984, pp. 91–2).

Another case might involve dynamic external economies created by the labour training of a firm, but the firm is not able to retain the workers it has trained. In a perfect market situation the learning effects would be internalized: the workers would accept low wages during the learning stage, financing themselves by borrowing, with recoupment through subsequent mobility. But if the capital market is imperfect, or if there are rigidities in wage determination, this may not be possible. Again, the first-best policy is to improve the capital market; the second-best policy is to provide financing for, or subsidization to, the labour training; while subsidization of the firm's output would be further down the policy hierarchy.

Baldwin (1969) has indicated that a protective duty is no guarantee that individual entrepreneurs will undertake greater investments in acquiring technological knowledge. As long as the learning-by-experience costs are higher than those which other firms must pay to acquire the knowledge, it cannot be assumed that firms will generally be prepared to incur the initial direct learning costs, even if the government imposes a tariff on the product. The duty will tend merely to encourage socially inefficient production as long as the state is willing to provide protection. A production subsidy on an industry-wide basis will have the same effect. What is needed is a direct and selective policy of subsidies to the initial entrants into the industry for discovering better productive techniques.

The infant-industry argument is also sometimes generalized to an 'infant economy' argument in which it is claimed that the entire industrial sector must go through an infancy stage, that the learning by each firm generates benefits for the whole sector and that by their mutual expansion all firms will enjoy a reduction in their production costs. Such a belief may underlie a broad import-substitution strategy with a uniform rate of effective protection to all manufacturing activities (Krueger, 1984, p. 525). But import-substitution strategies beyond the first easy stage have proved excessively costly in developing countries, and their adverse effects on agricultural development and on export promotion have limited the rates of development in countries that have practised import-substitution protection (Balassa, 1980).

In contrast to import substitution, it should be recognized that an export industry may also be an infant industry. Free trade may fail to bring about socially optimal levels of knowledge and factor endowment in new export industries. Policy interventions are then justified. Another possibility is that actual consumption experiences may be required to learn about an export commodity's qualities, but each firm's efforts at overcoming foreign-buyer resistance benefit not only itself but also all other firms that try to sell the same product in the same new market. The social returns of investments in market cultivation exceed the private returns, and subsidization is then justified (Mayer, 1984). The higher rates of economic growth enjoyed by many countries that have promoted exports suggest that it is possible that the infant-industry proponents are correct in their basic argument that there is a period of learning and of relatively high costs, and that an export-promotion strategy is a more efficient way of developing an efficient, low-cost industrial structure (Krueger, 1981, p. 16; Westphal, 1981, p. 22).

Empirical evidence on infant-industry protection, however, is not as extensive as theoretical developments. Taussig (1888) concluded that there was legitimate application of protection 'for young industries' in the United States during the early period of 1789 to 1838. Marshall (1919), however, saw no clear evidence in support of intervention by the state in favour of nascent manufactures. For contemporary economies the empirical evidence with respect to infant industry protection is not definitive on its costs, benefits and duration of protection over time. Krueger and Tuncer (1982) showed that in Turkey there was no evidence to suggest that more protected industries experienced a higher rate of declining costs than less protected industries. The industries did not pass the necessary condition for an economic justification of protection, namely that they experienced more rapid gains in efficiency as judged by comparing domestic resource costs against foreign-exchange savings at shadow prices that properly reflect relative scarcities. Even though a protected industry may grow, the question remains whether it would not have grown in the absence of intervention. And the empirical question of potential benefits being greater than earlier costs must also be examined.

A major study concluded that productivity growth in infant industries appears to be highly variable and that few of the infant enterprises studied in less developed economies have demonstrated the high and continual productivity growth needed to achieve and maintain international competitiveness (Bell et al., 1984, p. 114). Moreover, high levels of protection have also tended to persist beyond a temporary learning period (ibid. p. 117), and there is little evidence that higher rates of protection have been given to industries with greater externalities.

Finally, regarding the degree of protection, Westphal (1981, p. 12) has suggested that – even for an 'efficient' infant industry, and evaluated at prices that properly reflect relative scarcities – the domestic resource costs might initially be as much as twice the value of the foreign exchange saved or earned, with up to a decade being required to bring costs down to competitive levels. If production subsidies are given, the implied starting rate of subsidy in relation to value added is as much as 50 per cent. If, however, tariff protection is utilized, the rate of effective protection implied at the start of production is as high as 100 per cent.

Clearly the empirical justification for infant-industry protection will remain ambiguous until more research is done in quantifying the costs and benefits of protection, and its magnitude and duration. Such research is still in its own infancy.

GERALD M. MEIER

See also COMPARATIVE ADVANTAGE; EFFECTIVE PROTECTION; INDUSTRIALIZATION; NATIONAL SYSTEM; PROJECT EVALUATION; TARIFFS.

BIBLIOGRAPHY

Balassa, B. 1980. *The Process of Industrial Development and Alternative Development Strategies.* Princeton University, Essays in International Finance No. 141, December.

Baldwin, R.E. 1969. The case against infant-industry tariff protection. *Journal of Political Economy* 77(3), May/June, 295–305.

Bardhan, P.K. 1970. *Economic Growth, Development and Foreign Trade.* New York: Wiley.

Bell, M., Ross-Larson, B. and Westphal, L.E. 1984. Assessing the performance of infant industries. *Journal of Development Economics* 16, 101–28.

Bhagwati J.N. 1971. The generalized theory of distortions and welfare. In *Trade, Balance of Payments and Growth*, ed. J.N. Bhagwati, et al., Amsterdam: North-Holland, ch. 12.

Bhagwati, J.N. 1978. *Foreign Trade Regimes and Economic Development: Anatomy and Consequences of Exchange Control Regimes.* New York: National Bureau of Economic Research.

Bhagwati, J.N. and Ramaswami, V.K. 1963. Domestic distortions, tariffs and the theory of optimum subsidy. *Journal of Political Economy* 71, 44–50.

Clemhout, S. and Wan. H.Y. 1970. Learning-by-doing and infant industry protection. *Review of Economic Studies* 37, 33–56.

Corden, W.M. 1971. *The Theory of Protection.* Oxford: Clarendon Press.

Corden, W.M. 1974. *Trade Policy and Economic Welfare.* Oxford: Clarendon Press.

Corden, W.M. 1984. Normative theory of international trade. In *Handbook of International Economics*, ed. R.N. Jones and P.B. Kenen, Vol. 1, Amsterdam: North-Holland.

Hamilton, A. 1791. *Report on Manufactures.* Reprinted in US Senate Documents XXII/172, Washington, DC: Congress, 1913.

Johnson, H.G. 1965. Optimal trade intervention in the presence of domestic distortions. In *Trade, Growth, and the Balance of Payments*, ed. R. Caves, H.G. Johnson and P.B. Kenen, New York: Rand McNally.

Johnson, H.G. 1970. A new view of the infant industry argument. In *Studies in International Economics: Monash Conference Papers*, ed. A. McDougall and R.H. Snape, Amsterdam: North-Holland.

Kemp, M.C. 1960. The Mill–Bastable infant-industry dogma. *Journal of Political Economy* 68, February, 65–7.

Krueger, A.O. 1974. The political economy of the rent-seeking society. *American Economic Review* 64, 291–303.

Krueger, A.O. 1978. *Foreign Trade Regimes and Economic Development: Liberalization Attempts and Consequences.* Cambridge, Mass.: Ballinger for the National Bureau of Economic Research.

Krueger, A.O. 1981. Export led industrial growth. In *Trade and Growth of the Advanced Developing Countries in the Pacific Basin*, ed. W. Hong and L. B. Krause, Seoul: Korea Development Institute.

Krueger, A.O. 1984. Trade policies in developing countries. In *Handbook of International Economics*, Vol. I, ed. R.W. Jones and P.B. Kenen, Amsterdam: North-Holland.

Krueger, A.O. and Tuncer, B. 1982. An empirical test of the infant industry argument. *American Economic Review* 72(5), December, 1142–52.

List, F. 1841. *Das nationale System der Politischen Oekonomie.* Jena: Gustav Fischer, 1920. Trans. by G.P.A. Matile as *National System of Political Economy*, Philadelphia: Lippincott, 1856.

Little, I.M.D., Scitovsky, T. and Scott, M.FG. 1970. *Industry and Trade in Some Developing Countries: A Comparative Study.* London: Oxford University Press.

Marshall, A. 1919. *Industry and Trade.* London: Macmillan, Appendix G, 2.

Mayer, W. 1984. The infant-export industry argument. *Canadian Journal of Economics*, May, 249–69.

Meade, J.E. 1955. *Trade and Welfare.* New York: Oxford University Press.

Mill, J.S. 1848. *Principles of Political Economy.* Ed. W.J. Ashley, London: Longmans Green, 1909.

Taussig, F.W. 1888. *The Tariff History of the United States.* New York: G.P. Putnam's Sons.

Viner, J. 1937. *Studies in the Theory of International Trade.* New York: Harper & Bros.

Westphal, L.E. 1981. Empirical justification for infant industry protection. World Bank Staff Working Paper No.445, March, 1981.

infant mortality. There is extensive variation in the level of infant mortality (deaths under one year of age) across countries, over time within countries, and across subgroups within countries or regions. Social scientists, and demographers in particular, have devoted a great deal of research effort towards identifying the underlying sources of variation, biological and genetic, environmental and behavioural, and their relative importance.

To gain perspective on the degree of cross-sectional and temporal variation in infant mortality rates, the following data are useful. According to World Fertility Survey statistics obtained in the mid-1970s, in Bangladesh 13.5 per cent of all children ever born died before the age of one, in Mexico 7.2 per cent of infants died, and in Malaysia 3.6 per cent. Comparable figures were approximately 1.9 per cent for the USA, 1.3 per cent for Japan, and 1.2 per cent for Sweden. Around 1900 the USA had an infant mortality rate equal to that cited for present-day Bangladesh; around 1925 it was equal to that of present-day Mexico; and directly after World War II to that in present-day Malaysia. Further, within the region of West Africa, a high mortality area, the infant mortality rate in 1972 varied from 12.2 per cent in Ghana to 21.6 per cent in Guinea.

It is not my intention to review the evidence on the relative importance of the various factors thought to influence the level of infant mortality, for in my view that literature has serious methodological flaws stemming from the lack of a consistent theoretical paradigm. Suffice it to say that there is still much debate even among those who would not hold this view. I wish instead to present an economic perspective which has a critical bearing on the methodological approach used to resolve this debate. This perspective draws heavily on the notion of household production (Becker, 1965).

The essence of this approach is that infant health, survival being one albeit very important aspect of health, is produced according to some technological function which includes as inputs the resources devoted to the child (during pregnancy and after birth), such as prenatal care, breastfeeding, parental time devoted to child care, vaccinations, etc.; environmental conditions, such as sanitation and weather; biological conditions, such as the interval between births, the age of the mother at birth, and the genetic endowment of the child. More precisely, these inputs yield a probability distribution over health outcomes. Infant mortality differs systematically among individuals within a society, across societies, and over time within societies because of the differences in the levels and mix of these inputs and because of differences in the characteristics of the technology. In order to estimate the production function from data of any kind, it is necessary to postulate a mechanism which accounts for the input variation in the data.

From the economist's perspective it is natural to think of individuals as optimizing subject to constraints. Inputs have prices, monetary and/or psychic; faced with a given health technology, input levels are chosen depending upon the array of input prices both currently and the distribution expected to prevail in the future, wage rates (current and future), other income sources (current and future), preferences, and family and child endowments not subject to choice to the extent that they are known by the household. Much of the empirical literature can be seen as attempting to estimate technology, although in most of that literature technology is confounded with preferences through the introduction of income or prices. Further, few studies have estimated technology, accounting for the fact that, within an optimization framework, input choices would be conditioned on family or child endowments and on exogenous environmental factors. For example, because an infant's intake of breast milk depends on its ability to suckle, immature or ill infants may thus be breastfed less or not at all, leading to an upward bias in the estimation of the effect of breastfeeding on infant health or survival. A number of recent papers which adopt this household production framework using both micro-level data from developed (Rosenzweig and Schultz, 1983) and less developed (Olsen and Wolpin, 1983) countries have demonstrated the importance of the assumptions about the process generating input variation in estimating input effects.

It is plausible that infant mortality is linked behaviourally to other demographic decisions, such as the number, timing, and spacing of children. Indeed, there is a large literature which has posed the question about the impact of infant deaths on fertility, presuming the variation in infant mortality at the individual level to arise solely or mostly from stochastic events not subject to control. Two distinct fertility strategies have been discussed – replacement and hoarding. Replacement refers to the fertility reaction to a realized death, while hoarding refers to a strategy of acquiring an inventory of children in anticipation of future deaths (Ben-Porath, 1976; Schultz, 1976). Replacement behaviour would arise in the simplest of dynamic models with infant survival uncertainty because an infant death must increase the marginal benefits of an additional child. Hoarding behaviour, which is a response to the *ex ante* survival uncertainty, will arise only if mortality of older children is significant and/or surviving children are desired early in the life cycle. Although it is recognized that this behaviour, if rigorously modelled, would require solving a complicated dynamic stochastic optimization problem, most attempts to estimate these effects have been statistical in nature (Olsen, 1980) and only loosely based on theory. Some estimates of replacement based explicitly on a behavioural formulation have been obtained (Wolpin, 1984), but solving and estimating such models is computationally burdensome. Formulating and estimating a dynamic model with hoarding is a more ambitious undertaking than has yet been accomplished. Incorporating health investment decisions in children in a dynamic choice setting has yet to be implemented, although that is where this literature is and ought to be moving.

But, what does the household choice theoretical framework have to do with the enormous differentials in infant mortality we observe between countries and the historically extraordinary decline in infant mortality throughout the world? Surely the individuals in Bangladesh cannot choose to have an infant mortality rate equal to that in the USA. One can view this question in two ways. At a superficial level, it is clear from the figures cited above that infant mortality is inversely related to per-capita income in the cross-section and the time series. For example, using data from the World Fertility Survey countries, the per-capita income for countries with an infant mortality rate above 10 per cent was around $350, for countries with an infant mortality of between 7.5 and 10 per cent, it was $600, for countries with an infant mortality rate between 5.0 and 7.5 per cent, $1,302, and for countries with infant mortality rates between 2.5 and 5.0 per cent, $2,168. Of course, it is not income per se which causes reductions in infant mortality, but the improved preventive medicine and eradication of disease, the introduction of modern sanitation, and the improved food distribution, which come with economic development. Even ignoring the fact that the relationship between infant mortality and income is far from perfect – for example, Turkey had in the 1970s an infant mortality rate similar to that in Bangladesh but a per-capita income level ten times as large – the relationship between infant mortality and income is not fundamental. To the extent that innovations in medicine and the like parallel changes in economic circumstances of the population as a whole and are intertwined with the fundamental desires of the population, they too require explanation in the context of overall economic and social development. What one learns from the household choice framework is that to understand the cross-section and time-series aggregate data, what is needed is a model of economic growth which incorporates endogenous demographics, by which I mean conscious choice about investments in human capital (infant health, for example) and family size, in addition to investments in physical capital. Introducing modern sanitation, for example, requires real resources in research and implementation and some decision process must be responsible for the diversion of resources to that use.

To Malthus (1798), mortality played a crucial role as a positive check on population growth. One can always argue about what Malthus really meant, but stripped to essentials the argument was that the fixed capacity of land coupled with exogenous population growth causes consumption per capita to fall to the subsistence level. Although Malthus recognized that fertility might respond to falling living standards, this preventive check was in his view weak. The equilibrium mortality rate is that rate which constrains population size to remain at the level consistent with steady state subsistence consumption.

The standard neoclassical growth model, e.g. Solow (1956), takes net population growth (fertility net of mortality) as exogenous. Per-capita consumption is maximized when the marginal product of capital is equal to the net reproduction rate. This model leads to the result that the 'optimal' mortality rate is equal to the excess of the fertility rate over the replacement fertility rate. Samuelson (1975) noticed that in an overlapping generations growth model, the optimal rate of growth of population is infinite because welfare rises continuously the more young there are to support the old. Samuelson conjectured that there would be a deterministic optimum for population in a model which combined the neoclassical and overlapping generations approaches. Unfortunately, an overlapping generations model with capital (Diamond, 1965) does not yield an interior solution for the optimal population growth rate for any unbounded production function (Deardorf, 1976); again, population growth should optimally be zero. Because the overlapping generations framework is based on individual optimization, it allows for a broadening of the choice set. It does not make a great deal of sense to discuss optimal population issues in models where there exists no mechanism to achieve the optimum. There have been some attempts to allow for endogenous population in such models by adopting the notion from the microeconomic literature on fertility that households can choose their fertility, where children are consumption goods requiring expenditures (Eckstein and Wolpin, 1985; Nerlove, Razin and Sadka, 1984). These models are capable of describing the time path of output, consumption and population, and can yield insights as to the impact of technology, taste and endowments on those time paths. Incorporating infant mortality into overlapping generations growth models as a choice outcome, in the sense of allowing for investment in the human and physical capital necessary to affect it, would seem to be a logical and important step forward in understanding the economic and demographic development process. In particular, in many now developed countries, infant mortality decline preceded the decline in fertility so that population growth increased during the transition. Can an economic growth model in which fertility and investments in child survival are explicitly chosen by economic agents account for such a demographic pattern, and can this be fit into the observed timing of economic growth?

The notion that infant mortality, more broadly infant health and even more broadly child human capital, is and has always been an economic decision in a fundamental sense forces one to think more carefully about the determinants of aggregate cross-sectional and time-series demographic and economic variables. Much of this work is itself in an infant stage; its health and survival is also subject to choice.

K. WOLPIN

See also DEMOGRAPHIC TRANSITION; FECUNDITY; FERTILITY; NUTRITION; MORTALITY; PUBLIC HEALTH.

BIBLIOGRAPHY

Becker, G. 1965. A theory of the allocation of time. *Economic Journal* 75, September, 493–517.

Ben-Porath, Y. 1976. Fertility response to child mortality: micro data from Israel. *Journal of Political Economy* 84, August, S163–S178.

Deardorff, A. 1976. The optimum growth rate for population: comment. *International Economic Review* 17(2), June, 510–15.

Diamond, P. 1965. National debt in a neoclassical growth model. *American Economic Review* 55, December, 1126–50.

Eckstein, Z. and Wolpin, K. 1985. Endogenous fertility and optimum population size. *Journal of Public Economics* 27, June, 93–106.

Malthus, T. 1798. *An Essay on the Principle of Population*. Ed. A. Flew, Harmondsworth: Penguin, 1970.

Nerlove, M., Razin, A. and Sadka, E. 1984. Income distribution policies with endogenous fertility. *Journal of Public Economics* 24, March, 221–30.

Olsen, R. 1980. Estimating the effects of child mortality on the number of births. *Demography* 17, November, 429–43.

Olsen, R. and Wolpin, K. 1983. The impact of exogenous child mortality on fertility: a waiting time regression with dynamic regressors. *Econometrica* 51, May, 731–49.

Rosenzweig, M. and Schultz, T.P. 1983. Demand for health inputs, and their effects on birth weight. *Journal of Political Economy* 91, October, 723–46.

Samuelson, P. 1975. The optimum growth rate of populations. *International Economic Review* 16, October, 531–38.

Schultz, T.P. 1976. Interrelationships between mortality and fertility. In *Population and Development: The Search for Selective Interventions*, ed. R. Ridker, Baltimore: Johns Hopkins Press.

Solow, R. 1956. A contribution to the theory of economic growth. *Quarterly Journal of Economics* 70, February, 65–94.

Wolpin, K. 1984. An estimable dynamic stochastic model of fertility and child mortality. *Journal of Political Economy* 92, October, 852–74.

inflation. 'Inflation is a process of continuously rising prices, or equivalently, of a continuously falling value of money' (Laidler and Parkin, 1975, p. 741). Since there are many different ways of measuring prices, there are also many different measures of inflation. The most commonly used measures in the modern world are the percentage rate of change in a country's Consumer Price Index or in its Gross National Product deflator. Measures of inflation in earlier periods are based on fragmentary samples of prices, such as those of corn and other staple commodities, or of labour.

Inflation has been a feature of human history for as long as money has been used as a means of exchange, and a compact account of the history of the phenomenon from antiquity through to modern times is provided in Schwartz (1973). One of the earliest documented inflations in the ancient world occurred following Alexander the Great's conquest of the Persian Kingdom (330 BC); the Roman Empire experienced rapid inflation under Diocletian at the end of the 3rd century AD. Our knowledge about inflation for the thousand years that followed that is nonexistent. Price statistics for that period are not available. We do have data, however, from the Middle Ages onwards. Inflations occurred in the Middle Ages but were modest. Also, there was a tendency for periods of rising prices to be interspersed by periods of falling prices. This pattern of intermittent inflation and deflation persisted all the way through to the Great Depression of the 1930s. Since the Great Depression, there has been a general tendency for prices to rise every year (with trivial exceptions). By the 1970s and early 1980s, serious inflations – of more than 10 per cent per annum – gripped almost all the industrial countries. The middle 1980s saw inflation rates return, however, to the more modest levels experienced in the late 1960s. Individual inflations of spectacular dimensions have been experienced in interwar Europe, the fall of Nationalist China (1948–9), and in modern times in Latin America and Israel. Some of these were hyperinflations–episodes in which the rate of price increase exceeded 50 per cent per month.

It is the fact that inflation has been so variable over time and across countries that gives rise to the question: what are the causes and the consequences of inflation? It is the enormously rich variation in inflationary experience that also provides the data which makes progress in answering those questions possible.

The literature on inflation is large, and several comprehensive, if slightly dated, surveys of it are available (see Bronfenbrenner and Holzman, 1963, Johnson, 1963, and Laidler and Parkin, 1975). Unfortunately, no truly up-to-date survey of the literature on inflation is available.

Attempts to understand inflation have been greatly aided by the insight that anticipated inflation will have very different effects from unanticipated inflation. It will be convenient to use that distinction in organizing this essay. It needs to be borne in mind, however, that the distinction between anticipated and unanticipated inflation is analytical. It is not a distinction that has an immediate or direct correspondence with actual historical inflations.

ANTICIPATED INFLATION. Anticipated inflation is an idealized situation in which prices are rising at a rate at which all economic agents expect them to rise. No one is caught by surprise. What are the effects of a fully anticipated inflation?

There is little disagreement on the answer to this question concerning the effects on nominal variables – on such things as nominal interest rates, wages and foreign-exchange rates. Other things equal, the higher the expected rate of inflation, the higher the *level* of market rates of interest, the higher the rate at which wages rise and the faster the rate of currency depreciation. Furthermore, these effects will all be one for one. An x per cent higher anticipated inflation will be associated with x per cent higher nominal interest rates, with wages rising x per cent faster and with the currency depreciating x per cent faster.

Where there is less than complete agreement concerns the effects of anticipated inflation on real economic variables. It is a feature of all economic theories that, abstracting from transitory adjustment paths, money is neutral in the sense that a one-shot change in the quantity of money will lead to a proportionate change in the levels of all prices (and wages) and will have no real effects. It is not the case, however, that all economic theories predict that money is superneutral – that changes in the growth rate of the quantity of money will be neutral with respect to real variables.

There are three alternative views in the theoretical literature concerning money's superneutrality. One view is that a higher anticipated rate of inflation will be associated with a higher level of output (and economic welfare). A second view is that a higher anticipated rate of inflation will be associated with a lower level of output (and economic welfare). Yet a third view is that money is superneutral so that a higher anticipated rate of inflation will have no effects on output (or on economic welfare).

The first view stems from the so-called Mundell–Tobin effect (Mundell, 1963, 1965; Tobin, 1965). A higher anticipated rate of inflation results in an increase in the opportunity cost of holding real money balances. According to the Mundell–Tobin

view, this higher opportunity cost of holding money results in a portfolio reallocation away from money and towards physical capital. The higher holdings of physical capital result in a higher stock of capital and therefore in a higher capital–labour ratio, which in turn leads to a higher level of output. A rise in the anticipated rate of inflation would put the economy on an adjustment path towards the new higher capital stock that would be associated with a transitory rise in the growth rate and a permanent rise in the level of output. A recent restatement of the Mundell–Tobin position couched in a modern rational expectations terms has been provided by Fischer (1979).

The second view that a higher anticipated rate of inflation lowers output arises in two classes of models. First in an overlapping-generations framework (Samuelson, 1958; Wallace, 1980) a rise in the anticipated rate of inflation leads agents to economize on their holdings of money which, in turn, leads them to save less and transact on a lower scale with the succeeding generation. The other class of models are those that stem from Clower's (1967) suggested technological basis for money – the cash-in-advance constraint. Using Clower's assumption, Stockman (1981) shows that because a higher anticipated inflation rate raises the opportunity cost of holding money, this, in effect, raises the opportunity cost of undertaking all transactions and, therefore, in equilibrium lowers the scale of transactions undertaken. In Stockman's model, this results in a lower investment rate and lower capital stock. Thus a higher expected inflation rate leads to a lower level of output. A rise in the anticipated inflation rate will place the economy on an adjustment path that would result in a lower transitory growth rate and a lower permanent level of income.

The superneutrality result has been most elegantly and clearly stated by Sidrauski (1967). The result also is present in some modern theories of money that pay detailed attention to the physical environment in which monetary exchange arises (see, for example, Townsend, 1980). The essential feature of models that generate superneutrality is that the real rate of interest is imposed by the structure of preferences (intertemporally additive with a constant rate of time preference). The marginal product of capital is set equal to this fixed rate of time preference so that, regardless of what happens to money, the capital stock and output rate are unaffected.

All of the above results can be thought of in terms of the substitute/complement relation between money and capital. If money and capital are substitutes in portfolios, then the Mundell–Tobin result arises. If money and capital are complements, as they implicitly are in the overlapping generations and cash-in-advance models, then higher anticipated inflation leads to lower output.

A further variant of the superneutrality proposition is the natural rate hypothesis. This hypothesis, advanced by Friedman (1968) and Phelps (1968) states that money is superneutral in the particular sense that there is a unique 'natural rate' of unemployment that is independent of the anticipated rate of inflation. Any trade-off between inflation and unemployment is temporary and best thought of as a trade-off between unanticipated inflation and unemployment.

There is an abundance of empirical evidence on the alternative hypotheses about the effects of fully anticipated inflation. The evidence is not, however, entirely unambiguous. Since the very concept of anticipated inflation is analytical and not historical, in examining actual inflationary experience assumptions have to be made concerning the extent to which inflations have been anticipated. The only comprehensive and systematic attempt that has addressed the question in the context of economic growth is that by Kormendi and Meguire (1985). Studying postwar data for 47 countries, Kormendi and Meguire analysed the effects of a change in the anticipated rate of inflation on output growth in a multivariate regression framework. Anticipated inflation was measured as simply the mean growth rate of inflation over the sample period (which went from the late 1940s to 1977). The finding of that study solidly rejects the Tobin–Mundell hypothesis and, in some formulations, fails to reject the opposite view. Yet other formulations, however, are consistent with superneutrality.

Investigations of the neutrality of unemployment with respect to anticipated inflation has been the subject of innumerable studies. Most of these studies up to the mid-1970s are reviewed in Laidler and Parkin (1975). The broad and clear conclusion that has emerged from this work is that the unemployment rate is indeed neutral with respect to anticipated inflation.

The literature just reviewed deals with the consequences of anticipated inflation and not its causes. Questions concerning causality are much more naturally addressed in the context of an investigation of unanticipated inflation.

UNANTICIPATED INFLATION. It is not possible to analyse unanticipated inflation in isolation, independently of other aspects of aggregate economic performance. Fluctuations in the general level of economic activity and in inflation, though far from perfectly correlated, share some common features. There is, for example, a general positive correlation between inflation and real income (or equivalently, a negative correlation between inflation and unemployment). There is also a positive correlation between money and income as well as between the velocity of circulation of money and income.

These 'stylized facts' about the business cycle (shared by all economies) raise difficult questions about cause and effect. Of the four variables – the price level, real output, the money supply and the velocity of circulation – which, if any, is the prime mover? Do fluctuations in the growth rate of the money supply cause fluctuations in the other variables? Do autonomous movements in the price level, perhaps stemming from wage-push pressure, initiate the fluctuations in money, velocity and output? Does the business cycle have its origin in real factors that initiate fluctuations in output, which in turn lead to induced fluctuations in money supply growth, inflation and velocity?

At one level questions such as these are statistical and are capable of being investigated using econometric methods that detect causality, such as those proposed by Granger (1969). Studies based on such methods have not, however, delivered decisive results.

Most investigations of the possible causes of inflation have sought to understand the phenomenon by identifying the sources of inflation and studying the transmission mechanism whereby those sources are translated into variations in the rate of inflation and in other economic aggregates. This approach is one which seeks to understand both inflation and the business cycle as an integrated phenomenon.

There are three broad classes of theories that have been proposed for understanding the unanticipated and cyclical aspects of inflation. The first of these stems from the work of Keynes (1936) and emphasizes both price stickiness and the potential for autonomous movements in prices. On this view, the normal state of affairs would be one in which wages and prices are relatively sticky, responding only gradually to aggregate demand shocks. Shocks to aggregate demand arise from a variety of sources. One possibility is that autonomous

fluctuations in investment produce fluctuations in aggregate demand. Another possible source of aggregate demand fluctuations are fluctuations in wealth and interest rates which in turn are induced by fluctuations in the growth rate of the money supply. Fluctuations in wealth and interest rates can induce fluctuations in investment and consumption. All of these potential sources of variation in aggregate demand lead to cycles in both output and the price level. Initially, a change in demand will have bigger output effects than price-level effects but eventually prices and wages will adjust to reflect fully the change in aggregate demand. The resulting co-movements in output and prices will be positively, though not strongly, correlated.

From time to time this normal state of affairs will be disturbed by autonomous price shocks. The most commonly hypothesized source of price shocks is wage-push. It is suggested that at times of substantial industrial or social unrest, movements in the level of money wages will act as a type of social safety mechanism. The idea that wage-push results from sociological phenomena was particularly popular amongst English economists in the early 1970s (see, in particular, Balogh, 1970; Jones, 1972; Wiles, 1973; Hicks, 1974. By the time the first oil shock occurred (late 1973), 'wage-push' gave way to 'oil-push' as the most commonly identified source of autonomous movement in inflation.

When autonomous movements in the price level occur the phenomenon that came to be known as 'stagflation' quickly follows. The autonomous price rise raises the inflation rate and lowers output (raising unemployment). If the higher unemployment and lower output induces an increase in the growth rate of the money supply then even further price-level rises occur.

This traditional version of the Keynesian theory of inflation and the business cycle, together with some of the sociological embellishments that have been briefly reviewed above, is very thoroughly explained and elaborated in Laidler and Parkin (1975).

More recent and sophisticated versions of the Keynesian theory of cycles and inflation may be found in papers by Fischer (1977), Phelps and Taylor (1977) and Taylor (1979, 1980). The essence of what are sometimes called 'New Keynesian' theories is the existence of long-term contractual arrangements in labour markets. Such arrangements result in wages, the major element of costs, being predetermined. This stickiness of wages and costs results in a stickiness of prices, even if the expectations of prices that form the basis for the long-term labour market contracts are formed rationally.

A second approach to understanding cyclical fluctuations is one based on incomplete contemporaneous information about aggregate demand. This approach, sometimes called the 'New Classical Theory', was first suggested in the early 1970s by Lucas (1972, 1973). The approach is broadly consistent with the Keynesian mechanism of aggregate demand determination but proposes an alternative theory of aggregate supply. Individual economic agents are assumed to operate in informationally isolated 'islands' and to be incapable of distinguishing relative from absolute price level changes. The resulting confusion causes them to respond to absolute price changes as if they were relative price changes. This response results in positive co-movements in output and the price level.

In both the Keynesian and New Classical approaches, the key driving variable generating the cycle – fluctuations in both real output and the inflation rate – is a fluctuating growth rate in the money supply. This is not to deny that other things might, from time to time, shock the economy. Rather, it is a proposition about the major ongoing source of cyclical variation. Within both of the theories, positive co-movements of velocity are explained by appealing to the idea that to some degree the cycle itself is forecastable. To the extent that it is higher rates of inflation at the cyclical peak will in part be anticipated and, therefore, reacted to. It is always efficient to reduce money holdings when the opportunity cost of holding money increases. Higher expected inflation rates, leading to higher nominal interest rates, induce such economizing and are, therefore, the major source of procyclical fluctuations in velocity.

A third approach to understanding aggregate fluctuations denies the primacy of variations in the money supply growth rate, or in any other sources of aggregate demand fluctuation in generating the cycle. This approach, known as 'Real Business Cycle Theory', has yet to gain a major following but has, in recent years, begun to spawn a growing and important literature (see, in particular, King and Plosser, 1984; Kydland and Prescott, 1982; Long and Plosser, 1983; Nelson and Plosser, 1982. Though differing in some details, the essential proposition of the new real business cycle theories is that aggregate fluctuations emanate from technological shocks to the aggregate production function or, in some versions, from sector-specific shocks and from the interactions between sectors of the economy.

Technological shocks that generate fluctuations in full-employment output would, other things equal, generate negative co-movements in prices and, presumably, to the extent that such movements were forecastable, countercyclical movements in velocity. Since such co-movements do not occur, it seems as if the real cycle theories are in substantial trouble. King and Plosser (1984) address this problem directly by proposing that technological shocks which affect real output induce responses in money and credit that accommodate – indeed over-accommodate – the real fluctuations. Thus when there is a positive shock to aggregate supply, this induces an even bigger rise in the total volume of money and credit and, therefore, induces procyclical co-movements in money, prices and output. To the extent that these are forecastable, economizing on real balances generates procyclical velocity.

There is not, at the present time, any definitive and systematic evidence capable of disposing convincingly of any of these three alternative approaches; nor is there any overwhelming evidence suggesting that any of them is clearly in the lead.

INFLATION IN OPEN ECONOMIES. The alternative approaches to understanding inflation that have been reviewed so far have (implicitly) examined inflation in a closed economy. Most practical concerns about inflation arise in individual countries which are open economies. The international trade and international capital market transactions undertaken by such countries have an important bearing on their inflation performance. Also, the foreign-exchange rate regime – fixed or flexible – has an important influence upon a country's inflation performance. It was during the period of rapidly accelerating inflation in the 1970s that open economy theories and the international transmission mechanism gained in prominence (see Parkin and Zis, 1976a, 1976b).

The main feature of the analysis of inflation in an open economy is the emphasis on the limited potency of domestic monetary policy under fixed exchange rates. In a country, or more interestingly in a world, operating an fixed exchange rates, individual countries' monetary policies have no effect on the country's rate of inflation. Instead, monetary policy influences the country's balance of payments. In such a world, inflation is not a national but rather a world-wide phenomenon. It is the growth rate of the world money supply that determines the world average rate of inflation. Theorizing along this line had, in fact, made good progress even as early as the middle of the 18th century at the hands of David Hume (1752). It was rediscovered and popularized in the 1960s and early 1970s by Mundell (1971) and Johnson (1973).

The rediscovery of David Hume's analysis provided interesting insights into the resurgence of world inflation at the end of the 1960s. An attempt on the part of the United States to finance its Great Society programme and the Vietnam War with limited tax increases and with an increase in the growth rate of the money supply – with an increase in the inflation tax – became the engine of an inflation that engulfed the entire fixed exchange rate world.

Understanding the international generation and transmission of inflation in a flexible exchange rate world, such as that which had emerged by the mid-1970s, is still far from settled. At the centre of the problem of understanding inflation is the problem of understanding the determination of foreign exchange rates. Large and rapid movements in foreign-exchange rates are seen as having a potentially powerful and rapid effect upon domestic price levels. The forces which determine the exchange rate are still, however, far from well understood. Viewing the foreign exchange rate as following a random walk is as precise as any structural theories of the exchange rate that have so far been proposed and tested.

POSITIVE THEORIES OF CENTRAL BANK BEHAVIOUR. Recent developments in understanding inflation have been dominated by the rational expectations revolution in macroeconomic analysis. Some of the implications of that revolution have been discussed above and have been to strengthen and refine the theories of inflation that emphasize fluctuations in the growth rate of the money supply as the principal source of fluctuations in inflation and other economic aggregates.

The rational expectations hypothesis holds that expectations are formed by making predictions of future inflation on the basis of the mechanisms that generate actual inflation. If inflation is indeed caused by rapid monetary expansion, then forecasting future inflation is the same thing as forecasting future monetary growth. But monetary growth itself emerges from some ill-understood political process. The direct manipulator of the stock of money is the central bank. In determining the quantity of money and the rate at which to expand it, the central bank is heavily influenced by the economic and political environment in which it is operating and must also take account of the consequences of its actions for the behaviour of the economy as a whole.

In order to understand the inflationary process, with people forming expectations rationally, it becomes necessary to understand the policy-making mechanisms and the forces that generate varying monetary growth rates. The first serious modern analysis of this problem was that by Kydland and Prescott (1977) and the problem has been investigated more recently by Barro and Gordon (1983). In the models proposed by these writers, a central bank's goal is to achieve an optimal combination of inflation and unemployment. Lower inflation and lower unemployment are seen by the central bank as desirable objectives. The bank is constrained, however, by a short-run trade-off between inflation and unemployment – a trade-off arising from the considerations described above. A surprise rise in inflation would produce a cut in unemployment while a surprise drop in inflation would produce a rise in unemployment. The precise way in which the short-run trade-off between inflation and unemployment constrains the central bank depends on the expectations of private agents concerning the bank's behaviour. A central bank that can credibly precommit to a particular rule about inflation – perhaps a zero-inflation rule – would be a bank that could engender rational expectations of zero inflation. It would be optimal for such a bank to in fact precommit to a zero rate of inflation and then deliver that rate.

The ability to precommit and with credibility seems to require some mechanism for binding the central bank that does not have a readily identifiable counterpart in the real world. Central banks are, in fact, free to pursue whatever policies they wish at their discretion. Since this fact is known to all private economic agents it will be rational for them to take it into account when forming expectations about central bank behaviour. The equilibrium that results in this case will be such as to ensure that the actual inflation rate chosen by the bank is one that removes any temptation for the bank to depart from that rate and further exploit the short-run trade-off. Put differently, the inflation rate chosen will be the best available at the natural rate of unemployment. Only in such a situation would the central bank have no further temptation to attempt to exploit the short-run trade-off. Thus without the ability to precommit to a fixed (and presumably zero) rate of inflation, a central bank will end up delivering a higher rate of inflation than that which is socially desirable.

One feature of the positive theories of inflation developed by Kydland–Prescott and Barro–Gordon that some people find disquieting is the time inconsistency. (In game theory language, the equilibrium concept is Nash rather than sub-game perfection). Attempts to develop positive analyses that do not have this feature have been based on reputation. One such approach, in Barro and Gordon (1983), uses the so-called 'trigger strategy' model of reputation suggested by James Friedman (1971). A model is proposed in which the central bank would be punished if it delivered too high a rate of inflation and in which it takes time to restore the bank's reputation. In equilibrium, the bank never does inflate at a rate that requires the punishment to be inflicted.

An alternative approach by Barro (1986) uses the reputation analysis developed by Kreps and Wilson (1982). In this model there are two potential 'types' of central banker, one that likes inflation and one that dislikes it. The inflationary central banker has an incentive to masquerade as a non-inflationary type in order to induce low inflation expectations. By inducing low inflation expectations, the inflationary central bank will, at some point, be able to exploit those low expectations and produce a surprise inflation. it will do this by following initially a strategy of inflating at exactly the same rate as would be chosen by a non-inflationary central bank. At some later point it will pursue a mixed strategy – a strategy analogous to choosing an inflation rate by drawing numbers from an urn. Once this mixed strategy has resulted in a high rate of inflation, the inflationary central banker is revealed and expectations about inflation as well as actual inflation will rise.

Work along similar lines by Backus and Driffill (1985), analyses possible interactions between unified wage determination institutions and central banks in determining inflation. In their analysis inflation (and money supply growth) are determined as the outcome of a game between the central bank and a central wage-setting authority (such as an economy-wide labour union).

INFLATION AND FISCAL POLICY. A further consequence of the rational expectations revolution has been to force attention back to the connection between fiscal and monetary policy. The simple accounting fact that government expenditure has to be financed, either by taxation, by borrowing or by money

creation, implies that any analysis of the determination of the money supply growth must at the same time, make consistent propositions about fiscal policy and deficit financing. It is well understood, of course, that variations in the growth rate of interest-bearing debt can provide a good deal of insulation of money growth from the deficit. Large and persistent deficits, however, may give rise to rational expectations of future money growth, even in the face of currently firm monetary policies. Sargent and Wallace (1981) have shown that if the fiscal authority is the prime mover and follows taxation and spending policies that are independent of monetary policy, then, essentially, inflation and, ultimately, money growth are fiscal phenomena. Whether these findings are of practical importance is a matter of some controversy. Sargent (1982), studying the ends of four big inflations, has argued that adjustments in fiscal policy have been crucial to ending inflation. By implication, the emergence of a large and apparently uncontrolled deficit would be seen as the origin of serious inflation. Recent work by Dornbusch and Fisher (1986) offers a different interpretation, however, placing major importance on the behaviour of the foreign exchange rate.

POLICY TOWARDS INFLATION. Analyses of policies towards inflation have changed over the years. Advocacy of gradually slowing down the growth rate of the money supply and advocacy of controls on wages and prices were the most commonly heard policy suggestions for controlling inflation in the 1960s and early 1970s. Those who saw autonomous wage and price movements as the principal source of inflation saw prices and incomes policies as the major weapon to control it. Those who saw money growth as the source of inflation embraced monetary gradualism as the most obvious cure. A prodigious amount of work attempting to evaluate alternative policies was undertaken, much of which is surveyed by Laidler and Parkin (1975).

As a consequence of the rational expectations revolution, the focus of the policy debate has shifted markedly from that of seeking to manipulate variables such as key wage settlements (prices-and-incomes policy solution) or the growth rate of the money supply (monetarist solution). Instead, attention has turned to thinking about the way in which different institutional arrangements interact to produce different rates of inflation.

One of these developments has been the analysis of the consequences of alternative monetary systems, including the adoption of alternative forms of commodity money or, alternatively, some form of monetary rule that forces the money supply to target on a stable price level (see, in particular, 'Conference on Alternative Monetary Standards', 1983. A further recently advanced idea is that of targeting nominal income growth as a means of conquering and avoiding inflation (Tobin, 1983; Taylor, 1985).

CONCLUSION. Macroeconomics in general, and the theory of inflation in particular, is in a fluid state. The foregoing has attempted to review that state and provide a picture of the path that we have taken in getting to it. We have broad agreement on the facts to be explained and broad agreement on the behaviour of nominal variables (for given real variables) in an inflationary economy in which the path of inflation is anticipated. We also have broad agreement that fully anticipated inflations, though in many theoretical models capable of generating non-neutralities, are nevertheless to a good approximation neutral. Beyond that there is little in the way of firm knowledge. We have a variety of models of macroeconomics and inflation and many clear theoretical results. We do not have much, however, in the way of solidly based rejections of any of the available models. Uncertainty surrounds both the issue of the impulse (or impulses) that generate inflation and other fluctuations and on the propagation mechanisms that translate those impulses into movements in output and the price level.

MICHAEL PARKIN

See also COST-PUSH INFLATION; DEMAND-PULL INFLATION; HYPERINFLATION; INFLATION AND GROWTH; INFLATIONARY EXPECTATIONS; INFLATIONARY GAP; PHILLIPS CURVE; PRICE LEVEL; QUANTITY THEORY OF MONEY; RATIONAL EXPECTATIONS; SUPPLY SHOCKS.

BIBLIOGRAPHY

Backus, D. and Driffill, J. 1985. Rational expectations and policy credibility following a change in regime. *Review of Economic Studies* 3, April, 211–21.

Balogh, T. 1970. *Labour and Inflation*. London: Fabian Society.

Barro, R.J. 1986. Reputation in a model of monetary policy with incomplete information. NBER Working Paper No. 1794.

Barro, R.J. and Gordon, D.B. 1983. Rules, discretion and reputation. A model of monetary policy. *Journal of Monetary Economics* 12, 101–21.

Bronfenbrenner, M. and Holzman, F.D. 1963. A survey of inflation theory. *American Economic Review* 53(4), 593–661.

Clower, R.W. 1967. A reconsideration of the microfoundations of monetary theory. *Western Economic Journal* 6, December, 1–8.

Conference on alternative monetary standards. *Journal of Monetary Economics* 12/1.

Dornbusch, R. and Fischer, S. 1986. Stopping hyperinflations past and present. NBER Working Paper No. 1810.

Fischer, S. 1977. Long-term contracts, rational expectations, and the optimal money supply rule. *Journal of Political Economy* 85(1), 191–205.

Fischer, S. 1979. Anticipations and the non-neutrality of money. *Journal of Political Economy* 87(2), April, 228–52.

Friedman, J.W. 1971. A non-cooperative equilibrium for supergames. *Review of Economic Studies* 38, January, 1–12.

Friedman, M. 1968. The role of monetary policy. *American Economic Review* 58(1), 1–17.

Granger, C.W.J. 1969. Investigating causal relations by econometric models and cross-spectral methods. *Econometrica* 37, 424–38.

Hicks, J.R. 1974. *The Crisis in Keynesian Economics*. Oxford: Blackwell.

Hume, D. 1752. 'Of Money'; 'Of Interest'; 'Of the Balance of Trade'. First published in *Political Discourses*; reprinted in *Three Essays: Moral, Political and Literary*, London: Oxford University Press, 1963.

Johnson, H.G. 1963. A survey of theories of inflation. *Indian Economic Review* 6(4).

Johnson, H.G. 1973. Secular inflation and the international monetary system. *Journal of Money, Credit, and Banking* 5(1), Pt II, 509–20.

Jones, A. 1972. *The New Inflation: The Politics of Prices and Incomes*. London: Penguin Books and André Deutsch.

Keynes, J.M. 1936. *The General Theory of Employment, Interest and Money*. London: Macmillan.

King, R.G. and Plosser, C.I. 1984. Money, credit, and prices in a real business cycle model. *American Economic Review* 74(3), 363–80.

Kormendi, R.C. and Meguire, P.G. 1985. Macroeconomic determinants of growth: cross-country evidence. *Journal of Monetary Economics* 16(2), September, 141–63.

Kreps, D. and Wilson, R. 1982. Reputation and imperfect information. *Journal of Economic Theory* 27, 253–79.

Kydland, F.E. and Prescott, E.C. 1977. Rules rather than discretion: the inconsistency of optimal plans. *Journal of Political Economy* 85, April, 473–91.

Kydland, F.E. and Prescott, E.C. 1982. Time to build and aggregate fluctuations. *Econometrica* 50(6), 1345–70.

Laidler, D. and Parkin, M. 1975. Inflation: a survey. *Economic Journal* 85, December, 741–809.

Long, J.B. and Plosser, C.I. 1983. Real business cycles. *Journal of Political Economy* 91(1), 39–69.

Lucas, R.E., Jr. 1972. Expectations and the neutrality of money. *Journal of Economic Theory* 4(2), 103–24.

Lucas, R.E., Jr. 1973. Some international evidence on output–inflation tradeoffs. *American Economic Review* 63(3), 326–34.

Mundell, R.A. 1963. Inflation and real interest. *Journal of Political Economy* 71, June, 280–83.

Mundell, R.A. 1965. Growth, stability and inflationary finance. *Journal of Political Economy* 73, April, 97–109.

Mundell, R.A. 1971. *Monetary Theory: Inflation, Interest and Growth in the World Economy*. Pacific Palisades, California: Goodyear Publishing Co.

Nelson, C.R. and Plosser, C.I. 1982. Trends and random walks in macroeconomic time series. *Journal of Monetary Economics* 10, 139–62.

Parkin, M. and Zis, G. 1976a. *Inflation in Open Economies*. Manchester: Manchester University Press.

Parkin, M. and Zis, G. 1976b. *Inflation in the World Economy*. Manchester: Manchester University Press.

Phelps, E.S. 1968. Money, wage dynamics, and labor market equilibrium. *Journal of Political Economy* 76(4), Part II, 678–711.

Phelps, E.S. and Taylor, J.B. 1977. Stabilizing powers of monetary policy under rational expectations. *Journal of Political Economy* 85(1), 163–90.

Samuelson, P.A. 1958. An exact consumption-loan model of interest with or without the social contrivance of money. *Journal of Political Economy* 66(6), December, 467–82.

Sargent, T.J. 1982. The ends of four big inflations. In *Inflation: Causes and Effects*, ed. R.E. Hall, Chicago: University of Chicago Press.

Sargent, T.J. and Wallace, N. 1981. Some unpleasant monetarist arithmetic. *Federal Reserve Bank of Minneapolis, Quarterly Review* 5(3), Fall, 1–17.

Schwartz, A.J. 1973. Secular price change in historical perspective. *Journal of Money, Credit, and Banking* 5(1), Pt II, 243–69.

Sidrauski, M. 1967. Inflation and economic growth. *Journal of Political Economy* 75, December, 796–810.

Stockman, A.C. 1981. Anticipated inflation and the capital stock in a cash-in-advance economy. *Journal of Monetary Economics* 8(3), November, 387–93.

Taylor, J.B. 1979. Staggered wage setting in a macro model. *American Economic Review, Papers and Proceedings* 69(2), 108–13.

Taylor, J.B. 1980. Aggregate dynamics and staggered contracts. *Journal of Political Economy* 88(1), 1–23.

Taylor, J.B. 1985. What would nominal GNP targeting do to the business cycle? *Carnegie-Rochester Conference Series on Public Policy* 22, 61–84.

Tobin, J. 1965. Money and economic growth. *Econometrica* 33(4), October, 671–84.

Tobin, J. 1983. Monetary policy: rules, targets and shocks. *Journal of Money, Credit, and Banking* 15, 506–18.

Townsend, R. 1980. Models of money with spatially separated agents. In *Models of Monetary Economies*, ed. J.H. Kareken and N. Wallace, Minneapolis: Federal Reserve Bank of Minneapolis, 265–303.

Wallace, N. 1980. The overlapping generations model of fiat money. In *Models of Monetary Economies*, ed. J.H. Kareken and N. Wallace, Minneapolis: Federal Reserve Bank of Minneapolis, 49–82.

Wiles, P. 1973. Cost inflation and the state of economic theory. *Economic Journal* 83, 377–98.

inflation accounting. The adjustment of the accounts of business enterprises to reflect the consequences of inflation has been the subject of considerable theoretical controversy and practical experimentation in recent years, as a result of historically high inflation rates. The traditional valuation basis of accounts is historical cost, and the gap between historical values and current values tends to be widened by inflation. Furthermore, historical cost accounting does not reflect the gain on borrowing, which arises from having a liability which is fixed in nominal monetary units, in a period of inflation, or its counterpart, the loss on holding money, or assets denominated in nominal monetary units (such as trade debts).

Two apparently competing systems have been proposed to deal with the inflation accounting problem. The first is Current Purchasing Power Accounting, CPP. This applies general price indices to historical cost in order to allow for the decline in the value of money due to inflation. The second system is current value accounting, which revalues assets and liabilities at their current values, or, alternatively, restates historical cost by reference to a specific price index relating to the specific asset type, rather than a general price index. The current value accounting system found most commonly in practice, and featuring most prominently in the theoretical debate, is current cost accounting, CCA, which revalues assets at current cost, that is, replacement cost (typically) or recoverable amount (the higher of net selling price or net present value of future services to its present owners), whichever is the lower. A third system, which combines current valuation with CPP adjustments, and therefore eliminates the need to choose between CPP and CCA, is the real terms system, RT.

The principles of the three systems may be illustrated by a simple numerical example. Suppose an asset is bought at time O for £10,000 when the retail price index is 100. By time 1, the current cost of the asset is £15,000, by which time, the retail price index is 120. CPP accounting would restate the historical cost (£10,000 at time O) as £12,000 ($=£10,000 \times 120/100$) at time 1, because it would take 1.2 time 1 £s to buy the equivalent of 1 time 0 £. On the other hand, the proprietor's capital also needs to be restated in an exactly equivalent manner, so no gain or loss is shown by CPP, unless the asset was financed by borrowing in nominal money units (which would not require restatement, leading to a gain on borrowing) or had a value fixed in nominal money units (which would also not require restatement, leading to a loss on holding monetary assets). Current value accounting, on the other hand, would restate the asset at its current market value, £15,000 at time 1. If we retained the unindexed money capital maintenance convention, this would lead to the recognition of a gain of £5000 ($=£15,000-£10,000$). Alternatively, if we used a form of physical capital maintenance convention, such as is commonly found in CCA systems, we would restate capital also by reference to the specific price change of the physical assets, so no gain would be shown. Real Terms, RT, accounting, on the other hand, would restate the asset at current value (£15,000 at time 1) but would restate the proprietor's capital by reference to the retail price index (£12,000 at time 1), showing a 'real gain' of £3000 ($=£15,000-£12,000$), which is the amount by which the asset has appreciated in excess of the fictitious element due to inflation. There are many detailed variations on the simple systems illustrated here. These are explained and illustrated in Whittington (1983).

The CPP system deals only with the effects of general price level changes and ignores relative changes in the prices of specific assets. It is thus a pure inflation accounting system which would be adequate if all prices changed in the same proportion. The system was developed in Germany during the hyperinflation of the early 1920s. It was developed further and introduced into the North American literature by H.W. Sweeney (1936). During the past two decades, similar systems have been adopted by leading Latin American countries (notably Brazil, Chile and Argentina) under the pressure of very high inflation rates. In practice, CPP seems to be adopted only when pure inflation is seen as an important and urgent problem.

The CCA system owes its origins to the replacement cost accounting systems proposed by American, Dutch and German writers in the first three decades of the 20th century. CCA became prominent in the inflation accounting debate in the English-speaking world in the mid-1970s, with the aid of support from government agencies (such as the Securities and Exchange Commission in the USA and the Sandilands Committee in the UK). An account of this 'CCA counter-revolution' (and other aspects of the history of inflation accounting) will be found in Tweedie and Whittington (1984). The probable motivation for government support for CCA was that this system avoids the use of general indices and that any form of general indexation was regarded, at the time, as reinforcing the process of inflation. The obvious strength of CCA is that it attempts to record the assets held and used by the business at their specific current prices (although the precise definition of current cost is a controversial issue), thus measuring the current performance (in the profit and loss account) and state of the business (in the balance sheet) in terms of economic opportunities currently available in the market place. The weakness of the system is in its treatment of assets and liabilities which are of fixed nominal money value. The two capital maintenance concepts which are naturally associated with CCA are physical capital maintenance (as in the Sandilands Committee's operating profit measure) and nominal money capital maintenance (as in the Sandilands Committee's statement of total gains). Neither of these is capable, in its simplest form, of reflecting the gain on borrowing or loss on holding money which occurs in a period of inflation: indeed, some supporters of CCA would deny that such gains and losses occur. In order to capture these effects, recent British CCA systems (as in the Accounting Standard SSAP16, 1980) have adopted the gearing adjustment and the monetary working capital adjustment. These attempt to capture the gain on borrowing and loss on holding money, respectively, by using specific rather than general price indices. They have proved difficult to implement in practice as well as being difficult to justify theoretically, and are currently under review. They have been proposed but not implemented in a number of other countries (Australia, Canada, Germany, New Zealand, and Sweden). A clear account of the debate surrounding the introduction of CCA in the United Kingdom will be found in Kennedy (1978).

The RT system owes its origins to the work of Sweeney (1936), who pointed out that stabilization of the monetary unit by the use of general indices could be applied to any valuation base. The term CPP is normally restricted to stabilization of a historical cost base, so we use the term RT for stabilization of a current cost or other current value base. Since a CCA system already records assets at their current values, which are denominated in current currency units, the stabilization of such a system to convert it to an RT system requires no further indexation of asset values. The proprietor's capital is, however, adjusted by a general index, so that initial capital is maintained in terms of real purchasing power before a profit or gain is recognized. Thus, the RT system recognizes only real gains on assets (as illustrated earlier): it also recognizes the real gain on borrowing and loss on holding monetary assets in a period of inflation. This system was developed in considerable detail by Edwards and Bell (1961), who showed that it was possible, within the RT framework, to calculate CCA operating profit and then, by adding real gains and losses on holding assets and liabilities (which Edwards and Bell describe as 'real holding gains' or losses) to derive a final total of real profit or gain (which they describe as 'real business profit').

In many ways, the RT system seems to be a logical and

consistent means of recognizing the effects of general inflation in eroding the purchasing power of proprietor's capital, while also recognizing the effects of changing individual prices on the value or cost of the specific assets held and used by the business. It thus deals with the problems dealt with by both CPP and CCA, while avoiding the weaknesses of these two systems. There are strong elements of the RT system in the current USA standard on accounting for changing prices (FAS33), and it may be the system which will ultimately prevail in practice. Its slow emergence has much to do with the fact that CPP and CCA have been espoused by groups which see their particular interests being served by one of these systems; for example, professional accountants tend to be attracted by the relative objectivity (and therefore lower risk of professional liability for error) of CPP, which avoids subjective estimates of current values.

Finally, it should be noted that this essay has dealt only with business accounting. Inflation accounting is also an important problem in national accounts. The traditional adjustments for price changes are replacement cost of capital consumed and the elimination of stock appreciation (see Stone and Stone, 1977). This is analogous to CCA adjustment of business accounts. National accounts are also often restated in constant price terms, but this is a crude transformation rather than CPP or RT adjustment which would require restatement of opening capital figures to reveal real holding gains and losses (including those on monetary items) in various sectors. This issue is explored in Godley and Cripps (1983).

G. WHITTINGTON

See also ACCOUNTING AND ECONOMICS; HISTORICAL COST ACCOUNTING.

BIBLIOGRAPHY

Edwards, E.O. and Bell, P.W. 1961. *The Theory and Measurement of Business Income.* Berkeley: University of California Press.
FAS33. 1979. *Statement of Financial Accounting Standards No.33, Financial Reporting and Changing Prices.* Stamford, Conn.: Financial Accounting Standards Board.
Godley, W. and Cripps, F. 1983. *Macroeconomics.* Oxford: Oxford University Press.
Kennedy, C. 1978. Inflation accounting: retrospect and prospect. *Cambridge Economic Policy Review,* no. 4, March, 58-64.
Sandilands Report. 1975. *Inflation Accounting: Report of the Inflation Accounting Committee under the chairmanship of F.E.P. Sandilands.* Cmnd. 6225 London: HMSO, September 1975.
SSAP16. 1980. *Statement of Standard Accounting Practice No.16, Current Cost Accounting.* London Accounting Standards Committee.
Stone, J.R.N. and Stone, G. 1977. *National Income and Expenditure.* 10th edn. Cambridge: Bowes & Bowes.
Sweeney, H.W. 1936. *Stabilized Accounting.* New York: Harper.
Tweedie, D.P. and Whittington, G. 1984. *The Debate on Inflation Accounting.* Cambridge: Cambridge University Press.
Whittington, G. 1983. *Inflation Accounting: An introduction to the Debate.* Cambridge: Cambridge University Press.

inflation and growth. Although the relationship between inflation and economic growth has interested economists for some time, the nature of this association is still not well understood. Early discussions deliberated on the relative merits of a rising compared to a falling price level on profits, confidence, investment and other macro variables as these affected the growth of the economy, especially productivity. No noticeable consensus emerged from these deliberations.

In more recent times, in particular the post World War II period up until the early 1970s, the historical record gives ambiguous if not misleading clues. For example, cross-country comparisons of rates of inflation and productivity growth in the developed capitalist economies reveal virtually no association between the two. And if the period of rapid growth of productivity of the 1950s and 1960s is compared with the period of stagnation since the early 1970s, over time a negative correlation between inflation and productivity growth is found in each of the economies. The rise in inflation rates is associated with a slowdown in productivity growth.

THE POLITICAL ECONOMY OF INFLATION AND GROWTH. Conventional economic theory is equally inconclusive on the causal connection between the two. However, since the early 1970s, activist government intervention in the various economies has introduced a connecting link, resulting in a definite causal connection between inflation and growth that is likely to persist for some time to come. As a result, a correct understanding of this relationship involves a conceptual framework that is broader than that assumed by conventional economic theory. The causal relationship between inflation and growth must be seen as a problem in political economy, for it is the response of governments to inflation, both actual and predicted, that has and will determine the ultimate impact of inflation on the growth of productivity and output.

To put the matter in its simplest form, inflation leads to slow growth or stagnation because in those countries in which inflation cannot be brought under control by other means (e.g. an incomes policy) governments respond by implementing restrictive aggregate demand policies. Such responses lead, as they have since the early 1970s, to high rates of unemployment and low rates of capacity utilization, investment and productivity growth.

Taking the analysis one step further, in studying the mechanism by which inflation leads to stagnation under existing institutions it is useful to divide the developed capitalist economies into two groups. First, there are economies that experience accelerating rates of inflation under conditions of sustained high employment. To put the matter differently, there are countries in which the non-accelerating inflation rate of unemployment (NAIRU) is greater than the rate of unemployment at which all unemployment is voluntary.

No successful incomes policy can be implemented that would allow involuntary unemployment to be reduced to a minimum without the strong demand conditions leading to accelerating rates of inflation. As a result these countries will adopt restrictive aggregate demand policies in order to increase unemployment enough to reduce the rate of inflation. The fear that stimulative fiscal policies will lead to greater budget deficits and the fear of increased power of labour under full employment conditions, partly because it is believed that each causes inflation rates to accelerate, will reinforce this trend towards restrictive aggregate demand policies.

These economies can be said to suffer from an inflationary bias (Cornwall, 1983, ch. 6). Because of the policy response to this bias, inflation (or even the fear of inflation) will lead to high rates of unemployment and low rates of growth of productivity, that is, economic stagnation.

In contrast there is a second group of economies which, because of favourable institutional arrangements, could achieve full employment without accelerating rates of inflation *if restrictive aggregate demand policies were not adopted by the first group of economies.* These countries would be likely to adopt full employment policies if restrictive policies were not in effect elsewhere. But when they are, this group of economies is also forced to pursue restrictive aggregate demand policies but for quite different reasons than the first group. However, the effect of policy on the growth of output and productivity is the same; it will be greatly reduced.

PLURALIST ECONOMIES. Thus the first step in understanding the relation between inflation and productivity growth is a recognition that the simultaneous achievement of full employment and non-accelerating rates of inflation is not an automatic feature of capitalist economies. Moreover any failure to achieve these goals is not to be attributed to a failure of the authorities to follow some monetary or fiscal rule. Instead, the failure of an economy to handle inflationary pressures while maintaining full employment must be attributed to existing institutional and political arrangements. These make the coordination of wage and price settings in individual markets with the national goal of overall wage and price stability impossible.

These institutions can be said to act as constraints limiting the number and kinds of policy instruments available to the authorities for combating inflation. Going further, since the authorities in these economies respond to accelerating inflation by creating whatever unemployment is politically tolerable in an attempt to reduce inflation, the use of aggregate demand policies as an instrument for realizing desirable employment goals is, therefore, severely constrained by an inflationary bias. The authorities in these economies can be expected to pursue stagnationist policies under existing institutional and political arrangements.

The relation between inflation, the political response to inflation, and growth just described is similar to that seen by Kalecki (1977). However, it is more accurate to limit the kind of 'political theory of the business cycle' foreseen by Kalecki to a special group of capitalist economies which will be referred to as 'pluralist' economies. Pluralist economies share certain features in common. Governments play an essentially passive role in governing, primarily reacting to demands by special interest groups; there is a widespread belief among the powerful economic and political groups that an invisible hand or system of countervailing power exists that guarantees some kind of social optimum; the industrial relations system can be characterized as adversarial; and decision-making within the trade-union movement is decentralized.

The countries today that suffer from an inflationary bias and whose institutional features most clearly coincide with those just mentioned are the developed English-speaking countries, particularly Canada, the United Kingdom and the United States. Very likely, other countries with somewhat different institutions suffer from an inflationary bias, for example, France and Italy, and for the purposes at hand could be included in this group (Barber and McCallum, 1982; Crouch, 1984; McCallum, 1983).

CORPORATIST ECONOMIES. There is a second group of economies which will be referred to as 'corporatist economies'.

Corporatist economies are characterized by a tradition of state intervention in the economy, a high degree of cooperation and collaboration between the major economic groups in policy formation, a disbelief in invisible hands, and a system of industrial relations that can be described as cooperative. Primarily because of these institutions, these economies have been able in one form or another to implement relatively successful voluntary incomes policies in the past. Inflation has not been eliminated to be sure but has been kept at rates lower than would likely have resulted had union and management been unwilling to cooperate with government in the interests of achieving wage and price stability. More certainly, as Table 1 reveals, economies with these characteristics and with powerful trade union movements as well, for example, Austria and Sweden, have been able to reduce unemployment to extremely low levels without experiencing inflation rates much higher (if higher) than those in the pluralist economies.

Unfortunately given the high degree of economic interdependence between economies, most of those economies best able to contain inflation at full employment can no longer do so when the pluralist economies adopt restrictive policies. The economic importance of the pluralist bloc in the world economy forces restrictive policies on the second group of countries. Their importance guarantees that by restricting aggregate demand in their own countries, depressed conditions in the pluralist countries will be exported to the others in the form of a decrease in demand for their exports. Furthermore any attempt by any of the corporatist economies to offset declining exports through stimulative demand policies will lead to current account deficit that cannot be sustained through continuous borrowing (Thirlwall and Hussain, 1982). As a result, the full employment goal must be sacrificed.

Basic to this argument is the assumption that in the face of depressed demand conditions in the pluralist bloc, any corporatist economy acting on its own is not able to offset the adverse effects of a full employment policy on its payments position through an exchange rate policy. Unfortunately changes in the exchange rate are not sufficient to induce the kind of expenditure switching needed to bring the current account of the corporatist economies more or less into balance at full employment. These economies can be said to be limited or constrained in their use of aggregate demand policies for attaining full employment because of a payment constraint.

It is useful for pedagogical reasons to divide the capitalist world into two mutually exclusive groups, pluralist and corporatist. With this simplification in mind, the stagnating capitalist economies fall into one or the other of two groups: those in which restrictive demand policies are employed out of a fear of inflation and those in which a fear of payments problems at full employment caused by the restrictive policies of the first group leads to the same policies. The causal mechanism at work today, whereby inflation (or merely the fear of inflation) in one group of countries leads to worldwide

TABLE 1 *Annual Average Rates of Unemployment (U) and Inflation (ṗ), 1963–73 for Selected Capitalist Economies*

	U^a	$\dot{p}$		U^a	$\dot{p}$		U^a	$\dot{p}$
Austria	1.7%	4.2%	Italy	5.2%	4.0%	Switzerland	0.0^b	4.5%
Canada	4.8	4.6	Japan	1.2	6.2	Sweden	2.0	4.9
France	2.0	4.7	Netherlands	1.2	5.5	United Kingdom	3.0	5.3
Germany	0.8	3.6	Norway	1.7	5.3	United States	4.5	3.6

[a]1965–1973. [b]National definition.

Source: OECD, *Economic Outlook*, Paris, various issues; and OECD, *Labour Force Statistics*, Paris, various issues.

stagnation, now becomes clear. As long as the pluralist group restricts aggregate demand because of an inflationary bias, less than full employment conditions are forced upon the rest of the world. As a result, an inflationary bias in the pluralist group, that is, a tendency for inflation rates to accelerate at or before full employment, leads not just to breakdown in those countries but to worldwide stagnation.

THE FAILURE OF CONVENTIONAL POLICIES. As just argued, worldwide stagnation can be attributed to an inflationary bias in a group of key countries. Seen in another way, the difficulties or sources of stagnation can be traced to a failure of the traditional policy instruments, that is, monetary, fiscal and exchange rate policies, to work successfully in realizing full employment, price stability and external balance. Underlying this failure are certain structural and institutional changes that develop over a prolonged period of full employment such as the quarter of a century following World War II. Simply put, in democratic capitalist societies the rising affluence attributable to a long period of full employment is accompanied by the extension of the welfare state. This greatly increases the relative power of labour. As a result wages (and prices) are no longer determined primarily by the traditional market forces of demand and supply (Hicks, 1974; Okun, 1981; Scitovsky, 1978). This makes aggregate demand policy a highly inefficient means of fighting inflation. While wages and prices may respond eventually if restrictive policies are pushed far enough, the quantity effects on output and employment are substantial and immediate and persist while the policy is in effect. Futhermore, any implementation of an expansionary demand policy following a 'successful' restrictive policy that has brought down inflation rates will merely bring back the inflation in the pluralist economies. Restrictive policies whose aim is to permanently reduce inflation will fail in these countries because they fail to attack the sources of the inflationary bias.

Increased affluence and greater labour power also contribute to the ineffectiveness of exchange rate policy. First, consider that the trend in international trade has been increasingly towards the more highly fabricated goods that are desired for their non-price qualities, for example, design, durability, reliability, delivery dates, etc. This trend can, to a large extent, be attributed to affluence. It results in a downward trend in price elasticities of traded goods making it increasingly unlikely that the Marshall–Lerner conditions will be satisfied. When they are not, devaluation leads to a worsening of the trade deficit, other things being equal.

Second, the successful use of the exchange rate as an instrument for relieving a payments constraint is severely compromised by the existence of real wage resistance. A cheapening of exports relative to imports following devaluation likely leads to a decline in real wages. Under full employment conditions labour will have a strong incentive to press for higher money wage increases in an effort to protect their real wages. The resulting wage–price spiral can lead to the real exchange rate returning to its previous level. Like the inflationary bias, this difficulty arises out of the increased power of labour under full employment conditions and the affluence full employment brings.

Real wage resistance can be a real problem even in corporatist economies that may have had success in the use of income policies in the past. In earlier times the incentive for acceptance by labour of a voluntary incomes policy was provided by a promise of full employment and the rising real wages that full employment encourages. Unfortunately a devaluation of the currency, forced upon the authorities in

their pursuit of full employment by restrictive policies in the pluralist bloc, may lead to non-compliance with the incomes policy.

If the reduction in real wages can be limited, real wage resistance might be avoided. However, the international interdependence of capital markets can and has lead to situations in which the local authorities have little control over the magnitude of the actual depreciation of the exchange rate. A deliberate devaluation generates fears of accelerated inflation in the minds of managers of exceedingly large and mobile capital funds. This leads to large withdrawal of funds from the country, a further depreciation of the currency, greater fears of inflation, etc. As the experience of several countries in the recent past make clear, governments are soon forced to reverse their employment policies in order to protect the exchange rate.

CONCLUSIONS. In order to break the causal chain leading from inflation to restrictive policy to stagnation that prevails under modern capitalist conditions, major structural-institutional changes are required. Most important are changes that would relieve the pluralist economies of their inflationary bias. A different conception of the role of the state in the economy, the development of a cooperative industrial relations system and possibly of centralized collective bargaining would be extremely helpful because these changes increase the possibility that a successful incomes policy could be implemented. The benefits of its success for the rest of the world are apparent.

A second programme for recovery is more limited in that it is restricted to the corporatist bloc. This would take the form of the corporatist economies adopting coordinated trade and lending policies that discriminate against the pluralist group, in order to ease possible payments difficulties from full employment policies.

There are other possibilities involving one or more countries, but, however much they differ in detail, they share one thing in common: all require radical and basic structural changes in key economic and political institutions. Without these adaptations, the present political economy of inflation and stagnation will continue indefinitely and will be worldwide.

JOHN CORNWALL

See also FORCED SAVING; INFLATION; STAGFLATION; SUPPLY SHOCKS.

BIBLIOGRAPHY
Barber, C. and McCallum, J. 1982. *Controlling Inflation; learning from experience in Canada, Europe and Japan*. Ottawa: Canadian Institute for Economic Policy.
Cornwall, J. 1983. *The Conditions for Economic Recovery: a Post-Keynesian Analysis*. Oxford: Blackwell.
Crouch, C. 1984. The conditions for trade-union wage restraint. In *The Politics of Inflation and Economic Stagnation*, ed. L. Lindberg and C. Maier, Washington, DC: Brookings.
Hicks, J. 1974. *The Crisis in Keynesian Economics*. New York: Basic Books.
Kalecki, M. 1977. Political aspects of full employment. In *Selected Essays on the Dynamics of the Capitalist Economy 1933–1970*, ed. M. Kalecki, Cambridge: Cambridge University Press.
McCallum, J. 1983. Inflation and social consensus in the seventies. *Economic Journal* 93, December, 784–805.
Okun, A. 1981. *Prices and Quantities: A Macroeconomic Analysis*. Washington, DC: Brookings.
Scitovsky, T. 1978. Market power and inflation. *Economica* 45, August, 221–33.
Thirlwall, A.P. and Hussain, M. 1982. The balance of payments constraint, capital flows and growth rate differences between developing countries. *Oxford Economic Papers* 34(3), November, 198–510.

inflationary expectations. In a monetary economy, the expectations held by individual agents concerning future inflation rates will influence their current demand and supply decisions in various ways. Differences in expectations across agents will clearly be of relevance for their individual choices and for the extent of asset exchanges occurring at any point of time. In addition, characteristics other than the mean of an agent's subjective distribution of a future inflation rate may be significant. Nevertheless, even if all agents have similar expectations and these are in each case representable in certainty-equivalence fashion as a single number, the magnitude of this number will be of considerable importance from a macroeconomic perspective. The following discussion, consequently, will proceed by assuming certainty equivalence and ignoring heterogeneity across individuals (except for possible informational heterogeneity). In that sense, it will focus on macroeconomic issues. Organizationally, the first main section will discuss the importance of inflationary expectations in macroeconomics and the second will consider alternative theories pertaining to the formation of these expectations. Historical perspective will be provided in the third, and concluding, section.

MACROECONOMIC EFFECTS OF EXPECTED INFLATION

There are two distinct ways in which inflationary expectations enter importantly into macroeconomic analysis. The first of these has to do with period-to-period fluctuations in expected inflation rates, fluctuations that are intimately involved with business cycle analysis. The second, by contrast, pertains to effects on steady-state paths of inflationary expectations maintained at essentially constant levels over long spans of time.

The main involvement of inflationary expectations with business-cycle fluctuations occurs by way of some type of Phillips curve or aggregate supply relationship. There currently exists considerable disagreement over the proper specification of this relationship, which is a crucial component of any macroeconomic model, as is illustrated by the significantly different specifications utilized in notable contributions by Lucas (1972a, 1973), Fischer (1977), Taylor (1980) and Barro (1980). Most current theories imply, nevertheless, that the magnitude of any real stimulus resulting from inflation will be smaller the greater is the extent to which this inflation was previously expected. Indeed, a prominent and important line of thought originated by Friedman (1966) and Phelps (1967) contends that inflation will provide a stimulus to output and employment (via the Phillips relation) *only* to the extent that it is unexpected. As the validity of that viewpoint – often termed the 'natural rate hypothesis' – is highly relevant for stabilization policy, many attempts have been made to conduct statistical tests. The appropriate design of such tests will of course depend significantly on the way in which expectations are formed, a matter that will be discussed below.

The second main category of macroeconomic effects arises essentially because expected inflation constitutes the difference between nominal and real interest rates. Since real rates are relevant for saving and investment decisions while a nominal rate reflects the opportunity cost of holding wealth in the form of money, the expected inflation rate enters into the general equilibrium determination of various macroeconomic magnitudes. Two especially important potential influences are on money demand and capital accumulation, which we now consider in that order.

Even in an economy in which the real rate of interest is invariant to expected inflation, the nominal rate – and thus the quantity of real money balances held – will be influenced by these expectations. In particular, a high expected inflation rate will induce individual agents to hold a relatively small share of their wealth in the form of money. Consequently, since reduced real money balances yield reduced quantities of the transaction-facilitating services that are provided by the medium of exchange, agents are required to devote more time and/or resources to the activity of 'shopping', that is, effecting transactions. In addition, the real rate at which transactions are conducted may also fall. A reduced level of utility is then the consequence for each individual agent of an increased rate of expected inflation. Friedman (1969) has argued that, since there are virtually no resource costs associated with the creation of fiat money, overall efficiency requires a rate of expected inflation that drives the opportunity cost of holding money to zero and thereby satiates agents with real money balances.

In most well-articulated models, expected inflation also has effects on other variables – that is, money is not 'superneutral' (Barro and Fischer, 1976). In models with finite-lived agents the steady-state real rate of interest will be affected by inflationary expectations and, consequently, so will the per capita stock of capital and rate of consumption. Even if individuals are modelled as having infinite time horizons and a fixed rate of time preference – features which together fully determine the steady-state real rate of interest – capital and consumption per capita will under most specifications depend (perhaps weakly) on the expected inflation rate. (The celebrated model of Sidrauski (1967) provides an exception, but only because it ignores individuals' desire for leisure.) In sum, the magnitude of inflationary expectations may have significant allocative consequences, even if one neglects the practically important effects of tax schedules that are set in nominal terms. These allocative effects are in principle operative also at business-cycle frequencies, but the magnitude of the capital stock/investment ratio in developed countries leads to the presumption that the effects are of quantitative significance only over longer spans of time.

Expectation formation. In the consideration of alternative steady states, it is natural to presume that expected inflation rates will match those actually realized, and virtually all contemporary theorizing about steady states proceeds under that assumption. Analysis of quarter-to-quarter or year-to-year variations requires, however, some more consciously adopted hypothesis concerning expectational behaviour. From the time of Cagan's (1956) study of hyperinflations until the middle 1970s, the most widely used hypothesis was that of adaptive expectations – which makes each period's change in the expectational variable proportional to the most recent expectational error – with other autoregressive representations also used to some extent. During the 1970s it became clear, however, that adaptive and other fixed autoregressive specifications permit the occurrence of repeated, systematic expectational errors. But, since such errors are costly to the individual agents who make them, standard neoclassical reasoning suggests that it would be analytically fruitful to assume that agents typically eliminate any *systematic* source of expectational error, subject to available information. This hypothesis of *rational expectations* was introduced by Muth (1961) and developed in a macroeconomic context by Lucas (1972a) and Sargent (1973). It met with some initial resistance, perhaps because of a mistaken impression that it implies homogeneity of information and expectations across agents and/or that activist macroeconomic stabilization policy must necessarily be ineffective. But by the end of the 1970s the rational expectations hypothesis – that an agent's expectational errors are uncorrelated over time with all elements of his information set – had become dominant in both theoretical and applied macroeconomics.

The early development of techniques for the econometric implementation of rational expectations involved attempts to test the Friedman–Phelps natural rate hypothesis mentioned above. Various estimates of the crucial slope parameter attached to the expected-inflation variable in a Phillips-type relationship had been obtained, during the late 1960s and early 1970s, with procedures relying upon the assumption of adaptive expectations (or fixed autoregressive expectations with lag weights summing to 1.0). Typical estimates of the slope parameter obtained in these studies were in the vicinity of 0.4–0.6, well below the value of unity implied by the Friedman–Phelps theory (e.g., Solow, 1969). It was shown analytically by Sargent (1971) and Lucas (1972b), however, that the test strategy utilized would not identify the parameter at issue if expectations are in fact formed rationally. Instead, the estimate would tend to equal this parameter value times the sum of lag coefficients in a univariate forecasting equation for the inflation rate, a sum that need not equal the value of 1.0 presumed by the procedure in question. Estimates using similar (quarterly, US) data sets but taking account of this insight were then found to yield values close to unity (McCallum, 1976). The resulting interpretation, that the true parameter value is approximately unity and that expectations are at least partly rational (see below), was subsequently given indirect support by additional estimates presuming fixed autoregressive expectations, for the values obtained rose over time during the 1970s (Gordon, 1976). As the univariate autoregressive representation of actual inflation was also changing during this period, with the sum of lag coefficients rising from around 0.5 to nearly 1.0, these findings accorded well with the Sargent–Lucas interpretation.

An important implication of the Sargent–Lucas analysis is that, if expectations are in fact rational, one cannot generally measure the 'long-run' effect (i.e., comparative steady-state effect) of one variable on another by the sum of coefficients in a distributed-lag relationship. For example, since expected inflation affects interest rates to a different extent than does unexpected inflation, the sum of coefficients in a distributed-lag regression of interest on inflation will depend on the stochastic properties of inflation (the variable being forecast) as well as the slope coefficient measuring the effect of expected inflation on interest. To test hypotheses about the latter effect, it is necessary to take some account of the type of process generating the variable being forecast. That this principle continues to obtain when frequency-domain statistical techniques are employed has been emphasized by McCallum (1984).

The foregoing principle may also remain true, it should be added, if expectations are not strictly rational. In particular, it will apply if expectations are formed in a manner that reflects full but delayed responsiveness to the properties of the generating system. Expectational behaviour of that type, which might be termed 'asymptotic rationality', can be expressed analytically by the condition that the unconditional mean of the expectational error process be zero, a weaker requirement than the error being uncorrelated with all information variables available at the time of expectation formation. This less stringent type of partial rationality has not been prominent to date, but may become important eventually. It is not the same, it should be noticed, as hypotheses involving *learning* – that is, changing perceptions over time regarding the structure of the system. Some economists anticipate that the latter class of expectational hypotheses will prove fruitful; others doubt that the implied non-stationarity will permit much analytical progress.

Historical considerations. The most celebrated discussion of inflationary expectations in the literature predating World War II (the 'prewar' literature) is that of Irving Fisher. In *Appreciation and Interest* (1896), Fisher emphasizes the real vs. nominal interest rate distinction that is often associated with his name, and in *The Theory of Interest* (1930) he estimates a distributed-lag regression relating interest to current and past inflation rates (interpreting the long lags as due to 'delayed adjustment'). In addition, several other economists (e.g. Marshall, 1890) devoted some attention to the effects of expected inflation, the contribution of Henry Thornton (1802) being perhaps the most prescient (on this topic, see Humphrey, 1983). All in all, however, the subject attracted rather little attention in the prewar literature. Even in Knut Wicksell's (1898) famous analysis of the 'cumulative process' of inflation, there is only brief passing mention (pp. 96, 148) of the possibility that the inflation will be anticipated. Discussion of the effects of sustained inflationary expectations on capital formation seem to be entirely absent. This neglect may perhaps be satisfactorily explained by first noting that it is *sustained* inflation that is relevant and then recalling that the world's major economies normally adhered to some commodity-money standard during this earlier era.

BENNETT T. MCCALLUM

See also ADAPTIVE EXPECTATIONS; EXPECTATIONS; INFLATION; RATIONAL EXPECTATIONS.

BIBLIOGRAPHY

Barro, R.J. 1980. A capital market in an equilibrium business cycle model. *Econometrica* 48, September, 1393–417.

Barro, R.J. and Fischer, S. 1976. Recent developments in monetary theory. *Journal of Monetary Economics* 2, April, 133–67.

Cagan, P. 1956. The monetary dynamics of hyperinflation. In *Studies in the Quantity Theory of Money*, ed. M. Friedman, Chicago: University of Chicago Press.

Fischer, S. 1977. Long-term contracts, rational expectations, and the optimal money supply rule. *Journal of Political Economy*. 85, February, 191–205.

Fisher, I. 1896. Appreciation and interest. *AEA Publications* 3(11), August, 331–442.

Fisher, I. 1930. *The Theory of Interest*. New York: Macmillan.

Friedman, M. 1966. Comments. In *Guidelines, Informal Controls, and the Market Place*, ed. G.P. Shultz and R.Z. Aliber, Chicago: University of Chicago Press.

Friedman, M. 1969. *The Optimum Quantity of Money and Other Essays*. Chicago: Aldine.

Gordon, R.J. 1976. Recent developments in the theory of inflation and unemployment. *Journal of Monetary Economics* 2, April, 185–219.

Humphrey, T.M. 1983. The early history of the real/nominal interest rate relationship. *Federal Reserve Bank of Richmond Economic Review*, May/June, 2–10.

Lucas, R.E., Jr. 1972a. Expectations and the neutrality of money. *Journal of Economic Theory* 4, April, 103–24.

Lucas, R.E., Jr. 1972b. Econometric testing of the natural rate hypothesis. In *The Econometrics of Price Determination Conference*, ed. O. Eckstein, Washington, DC: Board of Governors of the Federal Reserve System.

Lucas, R.E., Jr. 1973. Some international evidence on output–inflation tradeoffs. *American Economic Review* 63, June, 326–34.

McCallum, B.T. 1976. Rational expectations and the natural rate hypothesis: some consistent estimates, *Econometrica* 44, January, 43–52.

McCallum, B.T. 1984. On low-frequency estimates of long-run relationships in macroeconomics. *Journal of Monetary Economics* 14, July, 3–14.

Marshall, A. 1890. *Principles of Economics*. London: Macmillan.

Muth, J.F. 1961. Rational expectations and the theory of price movements. *Econometrica* 29, July, 315–35.

Phelps, E.S. 1967. Phillips curves, expectations of inflation, and optimal unemployment over time. *Economica* 34, August, 254–81.

Sargent, T.J. 1971. A note on the accelerationist controversy. *Journal of Money, Credit and Banking* 3, August, 50–60.

Sargent, T.J. 1973. Rational expectations, the real rate of interest, and the natural rate of unemployment. *Brookings Papers on Economic Activity* No. 2, 429–72.

Sidrauski, M. 1967. Rational choice and patterns of growth in a monetary economy. *American Economic Review* 57, May, 534–44.

Solow, R.M. 1969. *Price Expectations and the Behaviour of the Price Level*. Manchester: Manchester University Press.

Taylor, J.B. 1980. Aggregate dynamics and staggered contracts. *Journal of Political Economy* 88, February, 1–23.

Thornton, H. 1802. *An Enquiry into the Nature and Effects of the Paper Credit of Great Britain*. Ed. F.A. von Hayek, Fairfield: Kelley, 1978.

Wicksell, K. 1898. *Interest and Prices*. Trans. R.F. Kahn, London: Macmillan, 1936.

inflationary gap. This term originates from the analysis of inflation put forward by Keynes in *How to Pay for the War* (1940). If there is a gap between the level of aggregate demand for goods and services and the quantity of available supply, then this will cause inflation.

The 'inflation gap' has been used as a basis for straightforward demand pull theories of inflation. But in *How to Pay for the War* Keynes used his concept of the inflation gap to build a strikingly novel theory of the inflationary process in the UK during World War I, which foreshadowed both the demand pull and the cost push concepts developed later, and which also made a particular use of the effects of inflation on income distribution. First of all, Keynes embodied the assumption of flexible rather than administered prices of produced goods and services; any excess demand associated with the inflation gap causes the prices of goods to rise relative to their costs of production, to the extent necessary to choke off the excess demand. The second distinctive part of the theory is as follows. The rise in prices of goods will be effective in choking off the excess demand – i.e. it will close the inflation gap – for the following two reasons. First, if there is a tax on profits bigger than any tax on wages then profit incomes will leak out of circulation to the government thereby diminishing the consumption of rentiers. Second, if the propensity to consume out of profit incomes is lower than that out of wage incomes – as assumed by Keynes – then a redistribution away from wages will, of itself, lower the propensity to consume. Notice that, the fall in real wages required to close the inflation gap may be large, since it depends upon the existence of the profits tax or upon the difference between the two consumption propensities. The third distinctive element in Keynes's theory is that the resulting reduction of real wages will cause pressure for an increase in money-wages. But if the inflation gap is not to reappear, then the prices of goods must run ahead of the increases in money-wages again. The speed of the inflation depends upon the lag with which wages chase prices, and inflation accelerates if this lag shortens.

The novelty of Keynes theory may be apparent when it is realised that the dominant inflation theory of the time derived from the quantity theory of money, in which the rate of inflation is determined by the rate of monetary growth (Keynes, 1940). Keynes's own theory has remained popular with Latin American structuralist writers; the inflation gap originates both because of chronic supply side ('structural') problems and because of a failure of policy to control demand, and changes in the distribution of income away from wages are thought to play an important role in stabilizing the overall process, as in Keynes's theory. (See Kaldor, 1964; Cardoso, 1980, presents a formal model but without the stabilizing role of lower real wages on demand.)

Orthodox demand pull theories of inflation see an inflation gap as leading to a rather different kind of inflationary process. Again they begin with an excess of aggregate demand over supply. In one version this excess demand will increase the prices of goods relative to wages, since firms are assumed to be price taking profit maximizers, who only increase supplies in the face of an extra demand if the profit margin per unit rises at the same time (Friedman, 1968, 1975; Phelps, 1970). In this case, wages again chase prices as wage earners begin to learn that real wages have fallen, and the process so far is very similar to that presented by Keynes. Or, in another version, the excess demand may not of itself raise prices until it percolates directly through to the labour market and causes money-wages to rise, subsequently inducing increases in prices, and then in turn inducing wages to chase the higher prices (Laidler and Parkin, 1975). But in both of these more conventional demand pull theories, the extent of the inflation which ultimately emerges depends on how demand management policy (i.e. fiscal and monetary policy) responds to dampen down the inflation gap, by curtailing economic activity. Indeed, in the hands of monetarist economists, under the assumption of a constant velocity of circulation of money, the extent of the inflation comes to depend directly on the rate of growth of the money supply, the very view which Keynes had earlier criticized. Crucially, these orthodox theories do not, like Keynes, see changes in the distribution of income away from wages as serving the function of regulating the excess demand, and of thereby determining the severity of the inflationary process which emerges.

Keynes himself did not foreshadow another possibility, almost the reverse of what he analysed. This is that a cost push inflation process could be engendered by a social conflict over the distribution of income, quite independently of any inflationary stimulus coming from the inflation gap (see Rowthorn, 1977; Meade, 1982; and Marglin, 1984).

DAVID VINES

See also DEMAND MANAGEMENT; DEMAND-PULL INFLATION; FISCAL STANCE.

BIBLIOGRAPHY

Cardoso, E. 1980. Food supply and inflation. *Journal of Development Economics* 8, 269–84.

Friedman, M. 1968. The role of monetary policy. *American Economic Review* 58, 1–17.

Friedman, M. 1975. Unemployment versus inflation? Occasional Paper No. 44, Institute of Economic Affairs, London. Reprinted in M. Friedman, *Lectures in Price Theory*, Chicago: Aldine, 1976.

Kaldor, N. 1964. Economic problems of Chile. Chapter 21 of N. Kaldor, *Essays on Economic Policy*, Vol. II, London: Duckworth.

Keynes, J.M. 1940. *How to Pay for the War*. London: Macmillan. Reprinted in *Essays in Persuasion; The Collected Writings of John Maynard Keynes*, Vol. IX, London: Macmillan, 1972.

Laidler, D. and Parkin, M. 1975. Inflation: a survey. *Economic Journal* 85, 741–809.

Marglin, S.A. 1984. Growth distribution, and inflation: a centennial synthesis. *Cambridge Journal of Economics* 8, 115–44.

Meade, J.E. 1982. *Stagflation*, Vol. 1: *Wage Fixing*. London: George Allen & Unwin.

Phelps, E. (ed.) 1970. *Microeconomic Foundations of Employment and Inflation Theory*. London: Macmillan.

Rowthorn, R.E. 1977. Conflict, inflation and money. *Cambridge Journal of Economics* 1, 215–39.

informal economy. The term 'informal economy' became current in the 1970s as a label for economic activities which take place outside the framework of corporate public and private sector establishments. It arose at first in response to the proliferation of self-employment and casual labour in Third World cities; but later the expression came to be used with reference to societies like Britain, where it competed with other adjectives describing deindustrialization – the 'hidden', 'underground', 'black' economy, and so on.

The social phenomenon is real enough and of some antiquity. London's East End in the mid-19th century is a stark example of informal economic organization which rivals in scale any of today's tropical slum areas. Nevertheless, the empirical referents of the 'informal economy' remain elusive, ranging as they do between the extremes of corrupt public finance in Zaire and do-it-yourself in a London suburb. The intellectual history of the concept is clearer. It was provoked by the failure of prevalent economic models to address a large part of the world that they claimed to describe. Sociologists, anthropologists, geographers and historians have grasped the opportunity to embarrass economists by pointing out this deficiency. More remarkably, many economists, including employees of such established bureaucracies as the World Bank and the International Labour Organization (ILO), have identified the 'informal sector' as something they must deal with, at least as a feature of life among the Third World's urban poor.

Some notable attempts have been made to document the economy of the streets. Henry Mayhew's investigations for the *Morning Chronicle* in the 1850s, published as *London Labour and the London Poor*, are a classic source, as are W.F. Whyte's American study *Street Corner Society* (1943) and Oscar Lewis's several accounts of the 'culture of poverty', e.g. *La Vida* (1964). Very little of all this impinged on the world of development economists. The dualistic models of economic development which prevailed in the 1960s took their lead from W. Arthur Lewis's theory of development with unlimited supplies of labour (1954), which postulated a modern sector of capitalist accumulation drawing on a traditional sector composed mainly of subsistence farmers. An influential variant of this approach was put forward by Harris and Todaro (1968). Here it was suggested that rural–urban migration in Africa could be modelled by focusing on the discrepancy between urban unskilled wage rates and marginal returns to agricultural labour, allowing for subsidized urban living standards and the urban unemployment rate.

Hart (1973) argued that the masses who were surplus to the requirements for wage labour in African cities were not 'unemployed', but rather were positively employed, even if often for erratic and low returns. He proposed that these activities be contrasted with the 'formal' economy of government and organized capitalism as 'informal income opportunities'. Moreover, he suggested that the aggregate intersectoral relationship between the two sources of employment might be of some significance for models of economic development in the long run. In particular, the informal economy might be a passive adjunct of growth originating elsewhere or its dynamism might be a crucial ingredient of economic transformation in some cases.

The dualism (formal/informal) and some of the thinking behind it received immediate publicity through its adoption in an influential ILO report on incomes and employment in Kenya (1972), which elevated the 'informal sector' to the status of a major source for national development by the bootstraps, as it were. This was enough to encourage legions of researchers to adopt the term in the 1970s. Before long a substantial critique of the 'informal sector' concept had emerged. Marxists claimed that its proponents mystified the essentially regressive and exploitative nature of this economic zone, which they preferred to call 'petty commodity production'. The study of Third World urban poverty rapidly became a new segment of the academic division of labour; as a key term in its discourse, the informal economy attracted an unusual volume of debate (Bromley, 1978). In recent years, British sociologists have applied the term to their own economy, whose formal institutions now employ a smaller proportion of the active labour force than at any time since the 1930s (Pahl, 1984).

Popularity as a jargon word has not helped the informal economy/sector to acquire a measure of analytical precision. For many the term is a convenient name for an unambiguous empirical phenomenon – what you find in the slums of Manila. Others pay more attention to the logic of conceptual dualism, but vary greatly in their definition of its essence. Thus the distinction is commonly taken to refer to size (large-scale/small-scale), productivity (high/low), visibility (enumerated/unenumerated), pattern of rewards (wages/self-employment), market conditions (monopoly/competitive) and much else. In all this, insufficient attention has been paid to the intellectual origins of the expression and to its usage in standard English. Hart (1973) explicitly derived his analysis from Weber's theory of rationalization, which refers to the growing scope for bureaucratic organization and calculation of rewards in the history of Western economic institutions. Weber believed that economic progress was inhibited by irregularity and unpredictability in social life and he saw the rational/legal state as guarantor of an emergent corporate capitalism. This process was in part one of increased formality in economic organization, as manifested in the planning of concrete enterprises and in an increasingly coherent body of economic theory. In this point of view there is a highly formalized part of all Third World economies today, where states, owing their existence in large degree to international institutions and forces, seek with variable effectiveness to establish their writ over economically backward populations. Equally, much that goes on in these economies is only marginally the product of state regulation: it is by that fact alone 'informal' relative to the forms of publicly organized economic life. This is a qualitative distinction, so that questions of size or productivity cannot be intrinsic to its definition. Informality in this context is a matter of degrees of social organization.

If we consider normal English usage, 'informal' refers to behaviour which relatively speaking lacks *form*. We all know the difference between formal and informal dancing or dress. But what is form? It is the presumptively invariant in the variable – 'presumptively' because what is held to be invariant (the rule) is rarely so in practice. Form is thus what is regular, predictable, reproducible, recognizable; and it is intrinsic to all social behaviour in some degree. When we identify something as informal, it is because it fails to reproduce the pattern of some established form. The consequence for economic analysis is obvious. The 'formal' economy is the epitome of whatever passes for regularity in our contemporary understanding, here the institutions of modern nation states, the more corporate levels of capitalist organization and the intellectual procedures devised by economists to represent and manipulate the world.

The 'informal' economy is anything which is not entailed directly in these definitions of reality. From the standpoint of high civilization, whatever it cannot control or comprehend is 'informal' – that is, irregular, unpredictable, unstable, even invisible. Of course, the people whose activities appear in this light believe that they have social forms which help them to live from day to day; but these forms are usually less powerful and less rigid than those underwritten by state law and immense wealth.

It follows from this that informality is in the eye of the beholder. The informal economy does not exist in any empirical sense: it is a way of contrasting some phenomena with what we imagine constitutes the orthodox core of our economy. Providing that it is self-conscious, such an exercise is almost always beneficial. Without it we remain trapped in the secular theology of a myopic elite. Economic theory proceeds by means of abstraction; but it is as well to consider from time to time what it has left out. The International Monetary Fund imposes its traditional recipe for formal incorporation of insolvent governments into the official economy without regard for the informal pressures to which they are subject; and the international economic order staggers towards its next crisis. More insidiously, the media (especially the television news) reproduce daily the outward signs of the economy – unemployment figures, the exchange rate, share indexes – and our collective understanding fixes on forms without substance.

The formal/informal dualism can have at least three constructions. First, the informal may be the variable *content* of the form; thus street pedlars of cigarettes invisibly complete the chain linking large foreign firms to consumers. Second, it may be the *negation* of formal institutions, whether tax evasion, shop-floor resistance or the world traffic in drugs. Third, it may be the *residue* of what is formal, that is more or less independent of it, not predicated on it, simply other: much of the Third World's countryside is so alien to the urban-based, state-made economy that it would be nonsensical to suggest a dialectical relationship between the two. If we are to restrict the definition of 'informal economy' at all, it seems reasonable to concentrate on the first two constructions – the relatively unspecified content of an economic form and subversion of such a form, its negation. The informal economy can then be taken to be an economic variant of the general theory of formal organizations.

It is nominalism of the most haphazard sort to claim that the urban poor have an informal economy but their rich masters do not; or that the Third World has an informal sector but not the industrialized West. As long as there is formal economic analysis and the *partial* institutionalization of economies around the globe along capitalist or socialist lines, there will be a need for some such remedial concept as the informal economy. Its application to concrete conditions is likely to be stimulated by palpable discrepancies between prevalent intellectual models and observed realities. Such a discrepancy provoked the emergence of the concept in the 1970s, when Third World economies bore the brunt of the depression which marked the end of the West's postwar miracle. Later the accelerating decline of the British economy encouraged some social scientists to adopt the term there. The common strand is the growing gap between modern states and the wider economic environments that sustain them. It is from this contrast that the need for a dualistic analysis, such as that offered by the 'informal economy' concept, derives its impetus.

KEITH HART

See also DEVELOPMENT ECONOMIES; DUAL ECONOMIES; ECONOMIC ANTHROPOLOGY; EXTENDED FAMILY; LABOUR SURPLUS ECONOMIES.

BIBLIOGRAPHY

Bromley, R. (ed.) 1978. The urban informal sector: critical perspectives. *World Development* 6, nos. 9–10.

Harris, J.R. and Todaro, M.P. 1968. Urban unemployment in East Africa: an economic analysis of policy alternatives. *East African Economic Review* 4(2), December, 17–36.

Hart, K. 1973. Informal income opportunities and urban employment in Ghana. *Journal of Modern African Studies* 11, 61–89.

International Labour Office. 1982. *The Urban Informal Sector in Developing Countries: Employment, Poverty and Environment*. Ed. S. Sethuraman, Geneva: ILO.

Lewis, O. 1964. *La Vida: A Puerto Rican family in the culture of poverty – San Juan and New York*. New York: Random House.

Lewis, W.A. 1954. Economic development with unlimited supplies of labour. *Manchester School of Economics and Social Studies* 22(2), 139–91.

Mayhew, H. 1861–2. *London Labour and the London Poor*. 4 vols, London.

Pahl, R. 1984. *Divisions of Labour*. Oxford: Basil Blackwell.

Whyte, W.F. 1943. *Street Corner Society: the social structure of an Italian slum*. Chicago: University of Chicago Press.

information theory. Information theory is a branch of mathematical statistics and probability theory. Thus, it can and has been applied in many fields, including economics, that rely on statistical analysis. As we are concerned with it, the technical concept of 'information' must be distinguished from the semantic concept in common parlance. The simplest and still the most widely used technical definitions of information were first introduced (independently) by Shannon and Wiener in 1948 in connection with communication theory. Though decisively and directly related, these definitions must also be distinguished from the definition of 'information' introduced by R.A. Fisher in 1925 for estimation theory.

Whenever observations are made or experiments conducted, we seek information about the underlying populations. Information theory provides concepts and definitions by means of which we may measure formally what can be inferred from the sampled data. More narrowly interpreted, we may view these concepts and definitions as summary statistics representing the associated (empirical) distributions, much like the moments of distributions. These concepts, however, also admit the intuitive meanings not unlike the notions of 'information' and 'uncertainty' in common parlance. One such concept, the *entropy* of a distribution, is central to information theory and is a measure of disorder or uncertainty. It was the definition of entropy that first caught the attention of Henri Theil and led to the use of information theory in *economics* in the 1960s. Following this discovery two measures of income inequality, measures of divergence and/or concentration in trade and industry, and many other economic applications were introduced in Theil (1967). Further applications in economic theory and other social sciences are discussed in Theil (1972, 1980).

A somewhat different but equally important set of developments of information theory have taken place in *econometrics*. Information criteria exist which measure the 'divergence' between populations. The use of such criteria helps to discriminate statistically between hypotheses, select models and evaluate their forecasting performance. These are essential steps in model evaluation and inference in econometrics.

Given a sample and (possibly) some prior information, a so-called 'maximum entropy (ME)' distribution of the underlying continuous process may be derived. This distribution and its quantiles and moments can be used in order to resolve a number of important problems including that of undersized samples. Theil and Laitinen (1980) formulated the ME distribution, and Theil and Fiebig (1984), give a complete coverage of this topic.

More general concepts of entropy and the associated measures of divergence can be used to develop a family of Generalized Entropy (GE) measures of inequality and in the area of 'multidimensional' welfare analysis. A by-product of this is the development of a 'summary welfare attribute' which serves as an index of several miscellaneous attributes such as income, wealth and physical quality of life indices. Another by-product is the development of 'information efficient' functional forms when the regression function is unknown *a priori*.

Applications to the classical theory of consumer and producer demand equations is noteworthy. The feature of this theory that provides the connection with information theory is the concept of 'value shares', the expenditure on a given commodity (input factor) divided by total expenditure. Thus the value shares have the same mathematical properties as probabilities ($0 \leqslant p_i \leqslant 1$). Theil (1967) noted some of the first attempts by Lisman (1949) and Pikler (1951) to draw analogies between econometric and information theory concepts. Emphasizing a point elaborated upon by De Jongh (1952), Theil wrote: 'the reason information theory is nevertheless important in economics is that it is more than a theory dealing with information concepts. It is actually a general partitioning theory in the sense that it presents measures for the way in which some set is divided into subsets ....' And then, 'it may amount to dividing certainty (probability 1) into various possibilities none of which is certain, but it may also be an allocation problem in economics.'

Our account begins with the introduction of some basic concepts.

1 CENTRAL DEFINITIONS AND CONCEPTS

An extensive literature exists which treats information theory in an axiomatic manner, much of it stimulated by the work of Shannon (1948) and Wiener's suggestive remarks in Wiener (1948). Kullback (1959) provides a comprehensive bibliography. From Stumpers (1953) it is clear, however, that some important contributions had appeared prior to 1948. For our purposes it will suffice to note that the occurrence of an event contains (or conveys) information. Thus one needs 'information functions' that measure the amount of information conveyed. If, as is usual, we are concerned with random events, there usually exist prior probabilities of occurrence of events and posterior probabilities. Hence, given an information function, we may measure the 'information gain' as between the prior and the posterior probabilities. Further, given an experiment and a *set* of observations, we may measure the 'expected information gain'.

The most widely used information function is $-\log p_i$, where p_i denotes the probability of a random variable X taking a value x_i. This function is non-negative and satisfies the 'additivity axiom' for *independent events*. That is, the information that both of two independent events have occurred is equal to the sum of the information in the occurrence of each of the events. More general information functions will be discussed later. The base of the logarithm (usually e or 2) determines the desired 'information unit', 'bits' when 2 is the base, and 'nits' when the natural base is employed. It may be useful to list some of the axioms that are employed to restrict the form of information functions:

Let such functions be denoted by $h(\cdot)$.

Axiom 1 – $h(\cdot)$ is a function only of the probability of events (p, say).

Axiom 2 – $h(p)$ is continuous in $p, 0 < p \leqslant 1$.

Axiom 3 – $h(p)$ is a monotonically decreasing function of p. This last axiom is quite intuitive. For instance, when we receive a definite message (observation) that a most unlikely event (p close to zero) has occurred, we are more highly surprised (informed) than if the event had a high probability (p close to 1) of occurrence. Thus, we may impose the following restrictions:

$$h(0) = \infty, \qquad h(1) = 0 \quad \text{and} \quad h(p_1) > h(p_2),$$
$$\text{if} \quad 0 \leqslant p_1 < p_2 \leqslant 1.$$

There are many functions that satisfy these axioms. $\log 1/p = -\log p$ is uniquely identified (apart from its base) on the basis of a further axiom, the *additivity* of $h(\cdot)$ in the case of independent events.

A generalization of the above definition is needed for general situations in which the messages received are not completely reliable. An important example is the case of forecasts of weather or economic conditions. Suppose p_1 is the probability of occurrence of an event (perhaps calculated on the basis of frequency of occurrence in the past), and p_2 the probability of the same event *given that it had been predicted to occur* (calculated in the same manner). We may then enquire as to the merit or the 'information gain' of such predictions. Given the observation that the predicted event has occurred, this information gain is defined as follows:

$$h(p_1) - h(p_2) = \log p_2/p_1$$

More generally, the information content of data which may be used to 'update' prior probabilities is defined by:

Information gain (IG) = log (posterior probability/prior probability). Since typically, data are subject to sampling variability, the 'expected information gain' is the appropriate measure of information gain. Thus, the expected value of IG defined above may be obtained with respect to either the prior or the posterior distributions over the range of the possible values for the random event. This gain is the difference between the 'expected information' in the two distributions. Consider a set of n mutually exclusive and exhaustive events $E_1, \ldots, E_n$ with corresponding probabilities $p_1, \ldots, p_n$ ($\Sigma p_n = 1$, $p_i \geqslant 0$). Then occurrence of an event contains $h(p_i)$ information with probability p_i. *Before* any observation is made, the 'expected information' in the distribution $p = (p_1, \ldots, p_n)$ is given by:

$$0 \leqslant \sum_{i=1}^{n} p_i h(p_i) = H(p) \leqslant \log n. \tag{1}$$

When $h(p_i) = -\log p_i$, the convention: $p_i \log p_i = 0$ if $p_i = 0$, is used. The maximum, $\log n$, occurs when $p_i = 1/n$ for all $i = 1, \ldots, n$. This is a situation of maximum prior 'uncertainty' or 'disorder' in the distribution p. Contrast this with $p_1 = 1, p_i = 0 \forall i \neq 1$, in which case $H(p) = 0$.

A dual concept to 'expected information' may be defined. This is the 'Entropy' which measures the 'uncertainty' or 'disorder' in a distribution p as defined by $H(p)$ in equation (1). Entropy is a central concept in information theory and its applications. It is also considered as an index of how close a distribution is to the uniform distribution. Note that an otherwise unknown distribution (p) may be determined by maximizing its entropy subject to any available restrictions. For instance the first few moments of the distribution may be specified *a priori*. The distribution so obtained is called the 'Maximum Entropy' (ME) distribution. In Theil and Laitinen (1980), *continuity* as well as the first moment of the observed data are utilized in order to obtain the ME distribution of the data and its higher moments.

The concept of 'divergence' or 'distance' between distributions naturally follows those of expected information gain and entropy. Instead of the prior and posterior distributions referred to earlier, it may be more suggestive to consider competing distributions (perhaps resulting from competing hypotheses), $f(x)$ and $g(x)$, which may generate a random variable x with the range denoted by R. There are two *directional* measures of divergence between $f(\cdot)$ and $g(\cdot)$. These are:

$$I(2,1) = \int_R g(x) \log \frac{g(x)}{f(x)} \, dx$$

and

$$I(1,2) = \int_R f(x) \log \frac{f(x)}{g(x)} \, dx.$$

A *non-directional* measure in the same context may be an average of the two directional criteria given above. For instance:

$$J(1,2) = \int_R [\log f(x) - \log g(x)][f(x) - g(x)] \, dx$$
$$= I(1,2) + I(2,1).$$

This is the well known Kullback–Leibler information criterion used extensively in many applications.

Various generalizations of these criteria are obtained either by generalizations of the form of the information function (which typically include the logarithmic form as a special case), or by generalizations of the *metrics* that include $J(1,2)$ above.

Some properties of these central concepts and their generalizations will be discussed in the following sections. It will be illuminating, however, to close this section by demonstrating an interesting connection between the information concepts defined so far and an important definition given by R.A. Fisher (1925):

Let x be a random variable taking values in the space S with the p.d.f. $f(\cdot, \theta)$ with respect to a σ-finite measure v. Assume $f(\cdot, \theta)$ differentiable w.r.t. θ and:

$$\frac{d}{d\theta} \int_c f(x, \theta) \, dv = \int_c \frac{d f(x, \theta)}{d\theta} \, dv$$

for any measurable set $c \in S$. Fisher defined the following measure of information on θ contained in x:

$$\phi(\theta) = E\left(\frac{d \log f}{d\theta}\right)^2 = V\left(\frac{d \log f}{d\theta}\right)$$

where V denotes the variance when $E(d \log f/d\theta) = 0$.

If there is a unique observation (of x) with probability 1 corresponding to *each value of* θ, then the random variable (i.e. its distribution) has the maximum information. The least information exists if the random variable has the *same distribution* for all θ. Thus, one might measure the sensitiveness of x with respect to θ by the extent to which its distribution changes in response to (infinitesimal) changes in θ. If θ and $\theta' = \theta + \delta\theta$ are two values of θ, a suitable measure of 'distance' or 'divergence', $D[f(\theta), f(\theta')]$, is required. An example of $D[\cdot]$ was given earlier, and many more criteria have been proposed. It may be shown that many such criteria are *increasing* functions of Fisher's information $[\phi(\theta) \geq 0]$. To give an example, consider the Hellinger distance:

$$\cos^{-1} \int [f(x, \theta) \cdot f(x, \theta')]^{1/2} \, dv.$$

Using a Taylor expansion of $f(x, \theta')$ and neglecting terms of power 3 or more in $\delta\theta$, we find:

$$\cos^{-1} \int f(\theta) \left\{ 1 - \frac{1}{8} \left[\frac{f'(\theta)}{f(\theta)} \right]^2 (\delta\theta)^2 \right\} dv$$
$$= \cos^{-1}[1 - \tfrac{1}{8}\phi(\theta)(\delta\theta)^2]$$

where $\phi(\theta)$ is indeed Fisher's information. Thus, such divergence criteria and $\phi(\theta)$ are equivalent measures of the sensitivity of the random variables (p.d.f., s) with respect to small changes in the parameter values. These observations provide a basis for measuring the distance between competing hypotheses (on θ) and model selection techniques in econometrics.

2 APPLICATIONS IN ECONOMICS AND ECONOMETRICS

2.1 Measurement of economic inequality. Theil (1967) observed that the entropy, $H(y)$, was a remarkably useful measure of 'equality'. If $y = (y_1, \ldots, y_N)$ denotes the non-negative income *shares* of N individuals, the entropy of y, by definition, measures its distance from the rectangular distribution, $y_i = 1/N$, which is the case of complete equality. Thus, the difference between $H(y)$ and its maximum value, $\log N$, may be used as a measure of *inequality*. This measure satisfies the 'three fundamental welfare requirements', namely symmetry (S), Homogeneity (H), and the Pigou–Dalton principle of transfers (PT). In addition, Theil (1967) demonstrated extremely useful *additive decomposability* properties of this measure. In recent axiomatic treatments by Bourguignon (1979), Shorrocks (1980), Cowell and Kuga (1981) and Foster (1983), the following question is posed: what is a suitable measure of inequality among general classes of functional forms which are restricted to satisfy the above three requirements in addition to Theil's decomposability? The last condition identifies Theil's first measure, $T1 = \log N - H(y)$, and a second information measure proposed in Theil (1967). The latter is defined as follows:

$$T2 = -\log N - \frac{1}{N} \sum \log y_i.$$

The choice between these two measures implies preferences that may be formulated by Social Welfare Functions (SWF). From a practical viewpoint, however, the choice between T1 and T2 may also be made on the basis of their decomposability properties. These may be briefly described as follows:

Let there by G exclusive sets (groups) of individuals, $S_1, \ldots, S_G$, with N_g denoting the number of individuals in $S_g, g = 1, \ldots, G$ ($\Sigma N_g = N$). Let y_g be the *share* of S_g in total income. Then we have:

$$T1(y) = \sum_{g=1}^{G} y_g \log \frac{y_g}{N_g/N} + \sum_g y_g[\log N_g - H_g(y)].$$

Here, $H_g(y)$ denotes the entropy of group g calculated from (y_i/y_g) for all $i \in S_g$. The first term above is the 'between-group' inequality, and the second term is a weighted average of the 'within-group' inequalities (the term in the square brackets). This decomposition is essential in analysing the incidence of inequality amongst the population subgroups (e.g. defined by age, race, region, etc.)

A similar decomposition formula holds for the second measure T2. The major difference is that the 'within-group' inequalities in T2 are weighted by the groups' *population shares* (N_g/N) rather than their income shares (y_g). The decomposition for T2 is somewhat preferable to that for T1 (see Shorrocks, 1980) as it permits a less ambiguous discussion of such questions as: what is the contribution of the inequality in the gth group to total inequality? This is partly because y_g are sensitive to redistributions (distributional changes) whereas population shares (N_g/N) are not so by design.

A generalization of the concept of information functions and the entropy has been employed by Toyoda, Cowell and Kuga to define the family of Generalized Entropy (GE) inequality indices. The GE is defined as follows:

$$\mu_\gamma(y) = \frac{1}{N} \sum_i [(Ny_i)^{1+\gamma} - 1]/\gamma(\gamma + 1)$$

where $\mu_0()$ and $\mu_{-1}()$ are, respectively, the T1 and T2 defined earlier. $\gamma \leqslant 0$ ensures the convexity of GE members, and for values of $\gamma \ll 0$ the GE is ordinally equivalent to the class of measures proposed earlier by Atkinson (1970) by direct reference to the SWFs. $-\gamma = v \geqslant 0$ is referred to as the 'degree of inequality aversion' exhibited by the underlying SWF. The information function underlying the GE indices is $-1/\gamma(y_i^\gamma - 1)$ which includes $-\log y_i$ as a special case. The generalized entropy corresponding to this function is given by:

$$H_\gamma(y) = \sum_i \frac{1}{\gamma(\gamma + 1)} [(Ny_i)^\gamma - 1] \cdot y_i.$$

The GE measures are also decomposable in the manner described above.

2.2 Multi-dimensional welfare analysis. The recognition that welfare depends on more than any single attribute (e.g. income) has lead to analyses of welfare functions and inequality in the multi-dimensioned space of several attributes, such as incomes (and its factor components), wealth, quality of life and basic needs indices. Pioneering work that deals directly with individual utilities (as functions of these attributes) and Social Welfare Functions is primarily due to Kolm (1977) and Atkinson and Bourguignon (1982). The measurement approach due to Maasoumi (1986a) poses the same question statistically and, surprisingly, provides a pure measurement (index number) interpretation of the SWF approach with equivalent solutions. The measurement approach seeks a 'summary share' or an index which may be employed to represent the miscellaneous attributes of interest. Information theory is utilized to obtain 'summary share' distributions without explicitly imposing any restrictive structure on preferences and behaviour. Noting that a measure of inequality is a summary statistic for a distribution (much as the moments of a p.d.f.). Maasoumi (1986a) poses the following question: Which summary distribution (index) is the 'closest' to the distribution of the welfare attributes of interest? Given suitable information criteria for measuring the 'distance' between distributions, one can find a summary (or representative) *share vector* (distribution) that minimizes this distance. Briefly: let y_{if} be the share of the ith individual, $i \in [1, N]$, from the fth attribute, $f \in [1, M]$. There are M distributions, $y_f = (y_{1f}, \ldots, y_{Nf})$, and we seek a distribution, $S = (S_1, \ldots, S_N)$, which is closest to these M distributions as measured by the following generalized information measure of divergence:

$$D(\gamma) = \frac{1}{\gamma(\gamma + 1)} \sum_{f=1}^{M} \alpha_f \sum_{i=1}^{N} S_i[(S_i/y_{if})^\gamma - 1]$$

where α_f is the weight given to the fth attribute. Minimizing $D(\gamma)$ with respect to S_i subject to $\Sigma S_i = 1$, we find:

$$S_i \propto \left(\sum_f \delta_f y_{if}^{-\gamma}\right)^{-1/\gamma}, \quad \sum_f \delta_f = 1.$$

This is a functional of the CES variety which includes the Cobb–Douglas ($\gamma = 0$) and the linear ($\gamma = -1$) forms as special cases. The S_i may be regarded as individual utility functions with the optimal distributional characteristics implied from $D(\gamma)$.

Once $S = (S_1, \ldots, S_N)$ is so determined, any of the existing measures of inequality may be used to measure multivariate inequality represented in S. Members of GE have been used for this purpose in Maasoumi (1986a). It may be shown that only two members of GE, Theil's T1 and T2, and some values of γ, provide fully decomposable measures of multidimensional inequality. Full decomposition refers to decomposition by population subgroups as well as by the inequality in individual welfare attributes. An interesting alternative method, Principal Components, is seen to be a special case of the above approach. It corresponds to the case $\gamma = -1$, with α_f being the elements of the first characteristic vector of $y'y$, $y = (y_{if})$ being the distribution matrix. This feature of information theory is not surprising. As S. Kullback and others have noted, one of its great advantages is the generality that it affords in the analysis of statistical issues, with the suggestion of new solutions and useful interpretations of the old.

2.3 'Information efficient' functional forms. Economic theory is generally silent on the specific form of functional relationships between variables of interest. Certain restrictions on the general characteristics of such relations are typically available, but ideally one must use the available data to determine the appropriate functional form as well as its parameters. The common practice in econometrics is either to specify flexible functional families, or to test specific functional forms for statistical adequacy (e.g. see Judge et al, 1980, ch. 11). The criteria of section 2.2 may be used, however, to obtain functions of the data with distributions which most closely resemble the empirical distribution of the observations. For instance, using $D(\gamma)$ from the previous section, let $S_i = f(x_i)$ be the indeterminate functional relationship between the variables $x_i = (x_{1i}, \ldots, x_{ki})$ at each observation point $i = 1, \ldots, T$. The variable set x_i may or may not include the endogenous (dependent) variable in an explicit regression context. Maasoumi (1986b) shows that, in the latter case, the CES functional form is 'ideal' according to $D(\gamma)$, and in the former the usual Box–Cox transformation is obtained. According to $D(\gamma)$, any other functional forms will distort the distributional information in the sample. The value of $D(\gamma)$ for any approximate regression function, less its minimum, is an interesting measure of the informational inefficiency of that regression function.

2.4 Tests of hypotheses and model selection. Sample estimates of the measures of divergence $I(1, 2)$, $I(2, 1)$ and $J(1, 2)$ defined above may be used to test hypotheses or to choose the 'best' models. For instance, the minimum value of $I(1, 2)$, denoted by $I(*:2)$ and called 'the minimum discrimination information', is obtained for a given distribution $f_2(x)$ with respect to all $f_1(x)$, such that

$$\int T(x)f_1(x)\, dx = \Theta$$

where Θ are constants and $T(x)$ are measureable statistics. For example, $T(x) = x$ and $\Theta = \mu$, restricts the mean of possible distributions $f_1(x)$. General solutions for $f_1(x)$ [denoted $f^*(x)$] and $I(*:2)$ are given in Kullback (1959, ch. 3) and elsewhere. $I(*:2)$, and the corresponding values for $I(1:*)$ and $J(*)$, may be estimated by replacing Θ (and the other unknown parameters) by their sample estimates ($\hat{\Theta}$) when $f_2(x)$ is the generalized density of n independent observations. Given a sample 0_n, we denote this estimate by $\hat{I}(*:2;0_n)$. This statistic measures the minimum discrimination information between a population with density $f^*(x)$ (with $\Theta = \hat{\Theta}$ etc.), and the population with the density $f_2(x)$. The justification for its use in tests of hypotheses and model selection is that, the non-negative statistic $\hat{I}(\cdot)$ is zero when $\hat{\Theta}$ is equal to Θ of the population with density $f_2(x)$, becoming larger the worse is the resemblance between the sample and the hypothesized population $f_2(x)$.

To illustrate, consider the linear regression model

$$y = X\beta + U$$

where $U \sim N(0, \Sigma)$ and $X \sim T \times K$ and of rank K.

Consider two competing hypotheses:

$$H_1: \beta = \beta \text{ (no restriction)}, \qquad H_2: \beta = \beta^2.$$

Then, it may be verified that:

$$J(1, 2) = 2I(1, 2) = (X\beta - X\beta^2)'\Sigma^{-1}(X\beta - X\beta^2)$$
$$= \frac{1}{\sigma^2}(X\beta - X\beta^2)'(X\beta - X\beta^2), \quad \text{if } \Sigma = \sigma^2 I$$
$$= (\beta - \beta^2)'(X'X)(\beta - \beta^2)/\sigma^2.$$

Replacing the unknown parameters, $\Theta = (\beta, \sigma^2)$, with their respective unbiased OLS estimates, we find:

$$\hat{J}(H_1; H_2) = (\hat{\beta} - \beta^2)'(X'X)(\hat{\beta} - \beta^2)/\hat{\sigma}^2.$$

And if $\beta^2 = 0$:

$$\hat{J}(H_1, H_2) = \hat{\beta}'(X'X)\hat{\beta}/\hat{\sigma}^2$$

which may be recognized as proportional to Hotelling's T^2 statistic with an F distribution with K and $T - K$ degrees of freedom.

The above example produced a statistic with a known finite sample distribution. In this situation, we are in effect rejecting H_2 if:

$$\text{Prob}\{\hat{I}(*:H_2) - \hat{I}(*H_1) \geqslant C | H_2\} \leqslant \alpha$$

where C is chosen to control the size (α) and the power of the test.

More generally, when the exact distributions are not known, asymptotic procedures may be employed. Suppose, for instance, that the competing populations (hypotheses), H, are members of the exponential family [which includes $f^*(x)$]. Let the admissible range of parameter values be denoted by Ω, and the range (value) specified by $H_i \in H$ denoted by ω_i. It may be shown that (see Kullback, 1959):

$$\hat{I}(*:H) = \log\left[\max_{\Omega} f^*(x) \bigg/ \max_{\omega_i} f^*(x)\right] = -\log \lambda_i$$

where λ_i is the Neyman–Pearson (likelihood-ratio) statistic. Under certain regularity conditions, the statistic $-2 \log \lambda_i$ is asymptotically distributed as X^2. Also, in the same situation, for two competing hypotheses $H_1: \Theta \in \omega_1$, $H_2: \Theta \in \omega_2$ it may be shown that:

$$\hat{I}(*:H_2) - \hat{I}(*:H_1) = -\log \lambda *$$

where $\lambda * = \max_{\Theta \in \omega_2} f^*(x)/\max_{\Theta \in \omega_1} f^*(x)$. Variants of such statistics are also useful for tests of 'non-nested' hypotheses. For the distribution of $\lambda *$ and its extensions see (e.g.) Chernoff (1954) and Cox (1961). Finally, we note that the above test reduces to $\hat{I}(*:H_2)$ when $H_2: \Theta \in \omega$ and $H_1: \Theta \in \Omega - \omega$. In such cases $\hat{I}(*:H_1) = 0$.

Variations to the information criteria described above have been proposed for 'model selection' in econometrics. Akaike (1973) proposed a measure based on the Kullback–Leibler

criterion. We give this criterion (AIC) for the problem of choosing an optimal set of regressors in the standard regression model. As before, let $y = X\beta + U$, $U \sim N(0, \sigma_u^2 I)$, be the 'comprehensive' model, with $X = [X_1, X_2]$ and $\beta = (\beta_1', \beta_2')'$ representing a full rank partition $K_1 + K_2 = K$. Under the null hypothesis $R\beta = [0, I_{K_2}]\beta = 0$ ($\beta_2 = 0$), the AIC is as follows:

$$\text{AIC} = -\frac{2}{T} \log l(y, \beta_1) + 2K_1/T$$
$$= \log(y'M_1 y/T) + 2K_1/T$$

where $l()$ is the likelihood function and $M_1 = I - X_1(X_1'X_1)^{-1}X_1'$. One proceeds to choose K_1 (and hence M_1) so as to minimize AIC. The first term decreases with K_1, thus the above criterion incorporates the trade-off between parsimony and 'fit' of a model. If one proceeds as though σ^2 were known in the likelihood function. Amemiya (1976) shows that:

$$\text{AIC} (\sigma^2 \text{ known}) = y'M_1 y/T + \sigma^2(2K_1)/T.$$

An estimate of this last criterion may be based on the unbiased OLS estimates of σ^2 with or without the restrictions ($\beta_2 = 0$). The latter estimate is equivalent to the so-called 'Cp criterion', and the former is equivalent to the 'Prediction Criterion' for model selection. For other variations to AIC (e.g. Sawa's BIC criterion) see Judge et al. (1980, Section 11.5).

CONCLUSIONS

The above examples do not do justice to a remarkable range of currently available applications of the information criteria in economics and econometric inferences. We hope, however, that they suffice to show: (1) the usefulness of the general approach in encompassing many different and often ad hoc procedures in econometric inference; and (2) how new methods with plausible and intuitive appeal may be derived in order to resolve many hitherto unresolved problems. The full potential for further applications of information theory and its formal discipline in economic and econometric theory is great. The current level of interest in this potential is extremely promising.

ESFANDIAR MAASOUMI

See also ENTROPY; HYPOTHESIS TESTING; PREDICTION; SIGNALLING.

BIBLIOGRAPHY

Akaike, H. 1973. Information theory and the extension of the maximum likelihood principle. In *International Symposium on Information Theory*, ed. B.N. Petrov and F. Csaki, Budapest: Akadémiaikiado.

Amemiya, T. 1976. Selection of regressors. Technical Report No. 225, Stanford: Stanford University.

Atkinson, A.B. 1970. On the measurement of inequality. *Journal of Economic Theory* 2, 244–63.

Atkinson, A.B. and Bourguignon, F. 1982. The comparison of multi-dimensional distributions of economic status. *Review of Economic Studies* 49, 183–201.

Bourguignon, F. 1979. Decomposable income inequality measures. *Econometrica* 47, 901–20.

Chernoff, H. 1954. On the distribution of the likelihood ratio. *Annals of Mathematical Statistics* 25, 573–8.

Cowell, F.A. and Kuga, K. 1981. Inequality measurement: an axiomatic approach. *European Economic Review* 13, 147–59.

Cox, D.R. 1961. Test of separate families of hypotheses. In *Proceedings of the Fourth Berkeley Symposium on Mathematical Statistics and Probability*, Vol. 1, Berkeley: University of California Press.

De Jongh, B.H. 1952. *Egalisation, Disparity and Entropy*. Utrecht: A.W. Bruna en Zoons-Uitgevers Maatschapij.

Fisher, R.A. 1925. *Statistical Methods for Research Workers*. London: Oliver & Boyd.

Foster, J.E. 1983. An axiomatic characterization of the Theil measure of income inequality. *Journal of Economic Theory* 31(1), October, 105–21.

Judge, G.G., Griffiths, W.E., Hill, R.C. and Lee, T.C. 1980. *The Theory and Practice of Econometrics*. New York: Wiley.

Kolm, S.-Ch. 1977. Multi-dimensional egalitarianism. *Quarterly Journal of Economics* 91, 1–13.

Kullback, S. 1959. *Information Theory and Statistics*. New York: Wiley.

Lisman, J.H.C. 1949. Econometrics and thermodynamics: a remark on Davis's theory of budgets. *Econometrica* 17, 59–62.

Maasoumi, E. 1986a. The measurement and decomposition of multi-dimensional inequality. *Econometrica* 54(5), September.

Maasoumi, E. 1986b. Unknown regression functions and information efficient functional forms: an interpretation. In *Innovations in Quantitative Economics*, ed. D. Slottje, Greenwich, Conn.: JAI Press.

Pikler, A. 1951. Optimum allocation in econometrics and physics. *Welwirtschaftliches Archiv* 66, 97–132.

Shannon, C.E. 1948. A mathematical theory of communication. *Bell System Technical Journal* 27, 379–423, 623–56.

Shorrocks, A.F. 1980. The class of additively decomposable inequality measures. *Econometrica* 48, 613–25.

Stumpers, F.L.H.M. 1953. A bibliography of information theory; communication theory–cybernetics. *IRE Transactions*, PGIT-2.

Theil, H. 1967. *Economics and Information Theory*. Amsterdam: North-Holland.

Theil, H. 1972. *Statistical Decomposition Analysis With Applications in the Social and Administrative Sciences*. Amsterdam: North-Holland.

Theil, H. 1980. The increased use of statistical concepts in economic analysis. In *Developments in Statistics*, Vol. 3, ed. P.R. Krishnaiah, New York: Academic Press.

Theil, H. and Fiebig, D. 1984. *Exploiting Continuity: Maximum Entropy Estimation of Continuous Distributions*. Cambridge, Mass: Ballinger.

Theil, H. and Laitinen, K. 1980. Singular moment matrices in applied econometrics. In *Multivariate Analysis V*, ed. P.R. Krishnaiah, Amsterdam: North-Holland.

Wiener, N. 1948. *Cybernetics*. New York: John Wiley.

Ingram, John Kells (1823–1907). Ingram's whole professional career was spent at Trinity College, Dublin, of which he became a Fellow in 1846. He subsequently held a remarkable variety of offices there – Professor of Oratory (1852) and English Literature (1855), Regius Professor of Greek (1866), Librarian (1879) and Vice-Provost (1898) – but was never a professional teacher of political economy.

Nevertheless Ingram played a notable part in the debates of the 1870s on the future of political economy and became one of the leading advocates in English of the use of the historical method in that science. Ingram's views were initially stated in his Presidential Address to Section F of the British Association in 1878. Here he attacked the 'vicious abstraction' and attachment to the deductive method of the classical economists, blaming this for the low repute into which political economy had fallen. He advocated the replacement of the deductive by the historical method and that 'the study of the economic phenomena of society ... be systematically combined with that of other aspects of social existence'. In adopting this approach Ingram was influenced partly by his contemporary T.E. Cliffe Leslie (1826–1882) but chiefly by the Positivist philosophy of Auguste Comte, of whom he was an active and lifelong disciple. Of his later economic writings, the best known was, and still is, his *History of Political Economy* which was for long the fullest account in English of the work of the historical school in Germany, France and Belgium. All

Ingram's economic work displayed the holistic and normative outlook which he derived from Comte, but did not go far towards fulfilling the programme of historical and comparative studies to which his earlier critique of classical economics pointed.

R.D. COLLISON BLACK

SELECTED WORKS

1878. The present position and prospects of Political Economy. *Report of the British Association for the Advancement of Science*, 641–58. Reprinted in *Essays in Economic Method*, ed. R.L. Smyth, London: Duckworth, 1962.

1880. *Work and the Workman*. Address to the Trade Union Congress, Dublin, September 1880, Dublin: E. Ponsonby.

1888. *A History of Political Economy*. Edinburgh: A. & C. Black.

1895. *A History of Slavery and Serfdom*. London: A. & C. Black.

1901. *Human Nature and Morals according to Auguste Comte*. London: A. & C. Black.

inheritance. Inheritance, in the strict sense, is the transmission of relatively exclusive rights at death. Such transmission is part of the wider process of the *devolution* of rights between or within the generations (eventually always between), and particularly between persons regarded as holders and heirs. Devolution continues throughout an individual's life, involving him both as giver and as receiver, and entailing transfers *inter vivos*, between the living for education, marriage, house-purchase, etc. as well as the residuum at death. The connection between inheritance and earlier transfers is given explicit recognition in some customary systems of endowment of sons and daughters where what has already been received is deducted from the final share of the parental estate (as in the revision clause of the Paris–Orleans region from the 16th century). In the same way the trend in European and American tax laws, epitomized in the British Capital Transfer Tax, is to treat as a whole the transfers of property from an estate (in the case of an estate tax) or to one individual or donee (in the case of an inheritance tax).

In a society where production is based upon the household and where rights (whether of ownership, tenancy or use) are vested in the domestic group, then the central importance of the devolution of such rights is clear. This is the case in most sectors of pre-industrial economies, but especially in agriculture and crafts. Where individuals have no such rights in the basic means of production, being employed as wage-labourers or as salaried employees, then the productive system is involved in interpersonal transfers only through share ownership, the transmission of managerial functions having been 'bureaucratized'; the handing over of such functions takes place at retirement rather than death and involves succession to 'office' rather than inheritance to property.

Thus in industrial societies of whatever political complexion, inheritance is of less significance for individuals and for society (except as windfall income in the first instance and windfall revenue in the second) than in earlier times when, except for the landless, it involved the transfer of rights in the means of livelihood. Even in industrial societies, the state may make special provision for family farms or firms to ensure continuity of the working group.

A radical instance of such a law was enacted in Germany by the National Socialists in 1933, providing for undivided inheritance and forbidding partition by will, the sale of the land or its encumbrance with a mortgage. The law was repealed after World War II, but in Germany as in France and other European countries, the transmission of farm property within the family is protected with the primary aim of ensuring continuity and providing an incentive to work and improve the farm for the next generation.

THE ARGUMENT ABOUT THE INHERITANCE OF WEALTH. Two divergent views on inheritance are current. On an ideological level, 'socialist' societies, parties and individuals regard inheritance as a way of transmitting inequalities and are therefore in favour of its restriction by taxation or even expropriation. Those espousing 'capitalist' theories look upon the right to transmit acquired wealth to one's offspring as part of the incentive necessary for accumulation, saving and investment. The extreme 'socialist' position is not simply a matter of recent theories of society. At the end of the 18th century the Abbé Raynal declared that at an individual's death, any land he possessed should become a free good. The theme has played a subdominant role in Christian thought over a long period. In 5th-century Gaul, the priest Salvian maintained that since all property came from God, at death it should be returned to his representatives on earth, the Church, for distribution to the poor as well as for its own purposes. Such assumptions left no room for inheritance, so the argument for the social uses of wealth depends not only on the negative case for reducing inequalities but on the positive one for assisting charities. Both positions involve an 'individualistic' view of property, 'freedom' to testate on the one hand and the reduction of the share or relatives (especially of collateral kin) on the other.

At an implicit level, we find a similar spread of ideas in simpler societies. In Africa a distinction is often made between self-acquired property, over which an individual may have a measure of freedom of disposal, and inherited property, especially land, which has come down from his forefathers. In the second case alienation is impossible because an individual is only a temporary custodian, having an obligation (as in some earlier European Laws) to hand the property down in the same line from whence it came. If it had been inherited in the agnatic line, then only agnatic relatives could benefit. In other words the property was 'corporate', or at least 'ancestral'. This notion of an heirloom runs quite contrary to the idea firstly that an individual's wealth should be confiscated in the wider interest either of the government or one of the 'great organizations'; and secondly that he should have completely free disposition over all he has accumulated. Clearly the case for inheritance is more tenable in traditional societies like those in Africa where differences in wealth were small, so that the case for redistribution (motivated either by a positive notion of distributive 'justice' or a negative resentment of inequality) was hardly relevant and where the 'poor' were the responsibility of their kin group. It becomes less tenable with the greater differentiation of capital and income, especially where individuals no longer own the means of production because they are working either for an industrial corporation or for a socialized enterprise, and where it is unnecessary, and often thought undesirable from the bureaucratic standpoint, to attach the next generation to the parental enterprise.

The stress on either the 'socialist' or 'capitalist' pole obviously has some relation to the nature of the ideology and to the organization of productive enterprise. But it is also true that domestic accumulation and extra-domestic redistribution are aspects of all contemporary social systems. In 1926 the Soviet Union reversed its early position which limited inheritance to small amounts passed on to close relatives or to the surviving spouse, providing there were in need. Later property could be left to anyone and its inheritance, listed as one of the rights of citizens in the constitution of 1936, was seen as a useful incentive to productivity. On the other hand every major 'capitalist' country levies some kind of tax at death, the proceeds of which are destined for the public purse rather than for private enjoyment.

Theories of income distribution often start by assuming normal distribution. This assumption is not adequate for all groups because of the inheritance of property at the death of the parents or other kin. But the main reason lies in the differential interest and capacity of parents to 'invest in' and encourage the abilities of their children. Such encouragement is a kind of transfer *inter vivos*, though it is part of the very process of socialization itself, one in which material gifts may play a major part in helping to provide both shelter in the form of house or apartment and more especially training or capital to generate income.

The ability to transfer privilege from one generation to another is, in the end, intrinsic to family life and to the reproduction process itself with its particularistic interests. Many utopian communities and ideologically based communes attempt to equalize opportunity through early 'schooling' or joint upbringing as well as wealth sharing; in the extreme case parenthood has only a physiological function, upbringing being left to the group. In the extreme case the contradictions become apparent in the longer-term development of communities like the Israeli *kibbutz* where family ties, and hence intra-familial differences, begin to manifest themselves, in limited but significant ways, after the initial period of open recruitment.

TAXATION. Whatever their ideological position, all modern societies place a progressive tax on inherited property, a tax that socialist countries see as a means of equalizing advantage, just as attempts are made to counter other benefits derived from family through a national system of education. However, many earlier and some modern taxes were primarily visualized as methods of raising income for the ruler rather than equalizing income among the citizens.

From either standpoint death provides the best moment to raise money, since future beneficiaries are unlikely to offer many objections if they have not yet taken possession. Moreover the property of the deceased often had to be listed, especially under the notarial systems of Europe, as part of the process of handing over: consequently the basis for an assessment already existed. Such forms of taxation have a long history, appearing in Rome as the *vicessima hereditatum*, the twentieth penny of inheritance. In feudal Europe the heriot, payable at the transfer of an estate, accrued to the lord; but already in 1694 a central death tax was introduced in England, taking its modern form in 1779–80. Taxes on inherited wealth thus long preceded taxes on income.

Ways of avoiding tax also had a long history. Under English law, trusts and life-estates could be set up in order to skip a generation in the transfer of property. Discretionary trusts were not available in continental Europe. But other avoidance measures included the handing over of property *inter vivos*, a long-established tradition (even extending to the farm itself), although such a practice now runs up against taxes on gifts or capital transfers: a modern alternative, that of changing one's country of residence, is more difficult to control, except by controlling the total outflow of capital from the country.

Today these possibilities remain relatively little used, and yet the revenue role of death taxes is not great. No taxes, remarks Shoup (1968, p. 559), have had a better reputation to less effect. This is partly because of avoidance and high rates of exemption, but mainly because individuals divest themselves of property to their children or use it for support in their old age, which is especially easy when personal property relates to consumption rather than to production. However, in the USA where charitable gifts are exempt, private foundations reap important benefits, while in the UK major contributions to national collections of art, buildings or land result directly from such taxes.

THE HISTORY OF INHERITANCE. The system of inheritance, involving the transmission of relatively exclusive rights over material objects, clearly varies with the mode of production. In hunting and gathering societies, rights of this kind are minimal; much of a man's property may be destroyed at his death, each individual fashioning himself or acquiring from others the tools he requires for his own use. The destruction is an aspect of the close identification of a person with the property he has created or used that is characteristic of such societies.

In simple agricultural economies, rights in land become elaborated, although with shifting cultivation access is more important than ownership. Access to the basic means of production is likely to be achieved through membership of a kin group (by descent or affiliation) rather than through an inheritance transaction. But other types of property, livestock, houses and exchange items, are transmitted in the course of long funeral ceremonies.

Where animal or other forms of non-human energy can be harnessed in the process of production, land becomes a scarcer resource, more differentiated in its distribution, with a greater complexity of rights, 'ownership' tending to be the prerogative of the dominant groups, and tenancy (or even labouring) the prerogative of others. In the case of tenancy it is landlords that tend to make the rules for the transfer of property, insisting for example on indivision (keeping the holdings intact) or on redistribution (keeping the holdings equal). The system of inheritance itself is influenced by the existence of stratified access to land, each group attempting to employ strategies of heirship to maintain or improve their position. The situation with regard to stratified access to livestock for pastoral peoples is different in certain important respects (the herd is more easily divided, increased and consumed) but tends to produce broadly similar strategies as are found in plough agriculture.

In industrial societies the situation is radically different because the vast majority of the population labour for wages rather than owning rights in the means of production. As we have seen, inheritance consequently plays a very different and more peripheral role.

Everywhere inheritance is basically a kinship transaction. While other persons may be involved, the core relationships are close 'familial' ones. In the simpler societies eligible kin are rarely, if ever, lacking since virtually all relationships are between kin. In complex ones, the definition of eligibility tends to be narrower, friendship supplements kinship, the percentage of unmarried tends to be higher and in any case other institutions, the 'great organizations' of church, state, as well as the charitable foundations, make their own demands; nevertheless the 'family' continues to dominate the process of the transfer of wealth between generations.

In kinship terms one can transmit laterally to spouses or siblings, or lineally to children or to siblings' children: the choice is one of priority since all ultimately has to go to the next generation. Downwards transmission for men can be to the sister's child (uterine inheritance) or to own children (agnatic inheritance). In simple hoe agricultural societies inheritance between spouses is rare, transmission tending to be homoparental, that is, male to male, female to female. Such is the case in much of Africa where economic differentiation is

relatively small and access to land available to all or most free individuals. Historically, these forms of inheritance were usually associated with the presence of unilineal descent groups (clans or lineages) in which property is transmitted between its members of the same sex.

Patrilineal and matrilineal clans with agnatic and uterine systems of inheritance are found in all types of pre-industrial society but matrilineality is more frequent with tropical hoe agriculture (in which women often do much of the farming, continuing their role as food gatherers in hunting societies) while patrilineality predominates when agriculture is combined with the herding of large livestock (whether or not these are used for plough traction).

The alternative form of inheritance, dominant in one form or other since the advent of plough agriculture, is diverging, or bisexual, that is, with children inheriting from both parents, and parents transferring wealth to daughters as well as to sons, but not necessarily at death. One form of early transfer is the direct dowry whereby daughters are 'endowed' when they depart at marriage. While the man-to-man (homoparental) transmission of Africa excludes inheritance by spouses, diverging devolution in Eurasia tends to exclude uterine inheritance by sister's children, concentrating on passing property, after the surviving spouse has been taken care of, directly to one's own 'natural' children, and even encouraging the adoption of outside heirs before allowing property to go to collaterals. The elementary family takes precedence.

Differences in stratification associated with advanced agriculture work in favour of the identification of conjugal statuses. Transmission in such societies is usually bisexual. At marriage some kind of conjugal fund (or identity of interest) is established and the property is transmitted, though not in equal proportions, to the children of both sexes, with certain types tending to be sex-linked; for example, land may be passed down to males alone where it is associated with male status among the nobility, as under the law of the Salian Franks. Since handing down occurs not only at death but on earlier occasions, especially at marriage, questions concerning the equality of 'inheritance' have to be looked at in terms of the total process of devolving property between holder and heir throughout their lifetime. For example, a woman may receive less at death because she has received a larger dowry at the time of her marriage, even as the promise of dower to maintain her as a widow.

The different treatment of siblings depending on birth order takes the form of primogeniture, ultimogeniture, or partition, known to earlier English law as Borough French, Borough English and gavelkind respectively. Rarely if ever does one find the transmission of the entire conjugal estate to a single sibling but rather the preferential treatment of one at the expense of the others. Such a preference may be tied to particular obligations, as when the inheriting child is expected to stay with the parents in their old age. In other cases the preference for one child (unigeniture) is related to the desire to keep the family estate intact, either because it will only support one family (among the poor) or because it is tied in with status consideration (among the rich). The first situation applied to pre-Revolutionary China where the poor tended to live in stem households (containing one member of each generation) while the richer lived in larger, extended ones. The poor either had less children *in toto* or the additional offspring migrated elsewhere or worked locally as landless labourers. The second situation was found in parts of 'feudal' Europe where title and position were linked to estate and income; just as one child succeeded to the title or office, so he had to inherit the bulk of the estate to which it was attached. Younger sons sought their fortunes elsewhere in the great organizations of the church or the army, to which they had access as a consequence of the political power assured by the parental estate.

DEVOLUTION, RETIREMENT AND INTERPERSONAL CONFLICT. Until the end of the 19th century (and still today in some areas of rural Europe), propertied classes endowed their daughters at marriage (and sometimes on entering a convent, on becoming 'a bride of Christ') with part of the 'portion' they would otherwise have inherited at the death of the parents. In some farming communities the parents would hand over the farm to their son or daughter on the occasion of their marriage, reserving for themselves certain rights to bed and board which were sometimes embodied in specific retirement contracts. One of the penalties of such an early handing over of property is that parents are placed in a King Lear situation, overly dependent upon the succeeding generation and running the danger of neglect (or 'ingratitude'). On the other hand to hang on till the end to property that is critical to status or survival leads to the opposite kind of tension characterizing the Prince Hal situation, where the son attempted to grasp his father's crown while he was still alive. These problems are of less significance in wage-earning societies where individuals are more dependent on income than on capital, and it is into training children for future employment that parents invest their time and wealth, rather than devolving property at marriage or even at death. While the state provides a minimum level of support, wealth may be invested in a pension or retained by the elderly for their support, possibly disappearing with their death in the form of annuity. Little conflict arises between holder and heir, who rarely continue to reside together (except in the case of spouses); inheritance tends to come late and to be seen as a 'windfall'. Its distribution may still give rise to conflicts within the group of potential beneficiaries, while even the prospect of a windfall produces enough underlying tension to fill the pages of many a piece of detective fiction (and 19th-century classics like *Middlemarch*), although significantly the plots are frequently located in the past when greater weight attached to rentier income.

RIGHTS TRANSFERRED. The rights transferred from the dead to the living are largely those in material property, houses, land, money, heirlooms, but they may also include rights to receive rent, interest, dividends from shares. In earlier societies they included rights to the services of other humans (of serfs and slaves), even to women as wives and to men as husbands, as in the Jewish practice of leviratic inheritance, taking on the widow of a childless brother with a view to breeding offspring to his name. Any semblance of the inheritance of widows was rigidly excluded from the law of England at the time of the Reformation since it was on grounds of the invalidity of such marriages that Henry VIII set aside his first wife (Catherine of Aragon, the widow of his dead brother, Arthur) in his search for a successor and an heir. Inheritance may also involve other types of right, those of a non-corporeal kind, right to songs and stories (copyright), rights to armorial bearings, titles, etc., although here we touch upon the field of succession to social position, to office, to nobility or to similar benefits.

Inheritance is not only concerned with rights; duties too are involved; debts have to be paid from the estate; the acceptance of an inheritance may involve a change of name, of residence and even, especially in societies that resolve disputes by means of feud, of the specific obligation to settle a score.

A broad distinction is made in Anglo-American law between real and personal property, roughly between land a chattels, between movables and immovables. A similar distinction is found in many other cultures and is related to the special position of land within the general category of property, since it acts both as a factor in production and as a locus for all social activity. Hence different rules and practices are applied to these two categories; in England after the Norman conquest it was the royal courts that dealt with real, and ecclesiastical courts with personal property, the former emphasizing indivision, the latter allowing more testamentary freedom and alienation. Land was subject to different rules until 1926 and its transfer is still hedged about with formalities that mark it off from all other forms of property. For a hierarchy of rights is always involved; these may refer to usufruct, tenancy, mortgage, metayage, and a host of arrangements (including sovereignty itself) to which other property is not subject and some of which need to be acknowledged in the deed of transfer itself.

TESTATE AND INTESTATE. There are two types of inheritance in literate cultures, testate and intestate. The former involves making (writing) a will or testament, the latter describes what happens when there is no such statement. In nonliterate societies inheritance is automatically intestate, and 'custom' lays down how property should be distributed. There is little 'freedom' to alienate goods from the recognized heirs so that even gifts *inter vivos* have to be monitored.

The written will or testament introduces a measure of certainty in situations, reducing possible conflict or indeterminacy; where this is available, nuncupative (that is, oral) wills are considered valid only under exceptional circumstances. But in early times one of the main functions of the testament was to certify that any alienation from customary heirs was according to the wishes of the deceased. In other words its very existence assumed a degree of 'freedom', by which is meant freedom of choice for the holder, limiting the right of 'society' to say who should be the recipient. It is not surprising that the will with its corresponding freedom of testation was encouraged by the early Christian Church as a way of acquiring property to be put to divine purposes. And in more recent times it has been a central instrument for the transfer of wealth to charitable foundations. Without its intervention, inheritance goes to the family.

Testamentary inheritance occurs by means of the written will, although initially the latter term applied only to real property (land), the testament to personal property. Literacy is thus essential either on the part of the testator or on the part of the notary, lawyer or priest who draws up the will. The fact that the document has to be proven in court means that professionals tend to be employed for the purpose. Hence the whole industry of literate legal specialists involved in writing the will, in helping it to come into effect and in administering the estate, the last being currently the most profitable part of the enterprise in anticipation of which other charges may be scaled down. It is these specialists who help to ensure that the formalities are observed, not only in the words but in the witnesses, and that the will is not invalidated for other reasons. All of which tends to take the mechanics of transfer out of the hands of kin, and places the process firmly under the charge of those who engage in it for their livelihood and who tend to create their own specialist language, codes and organization.

TESTAMENTARY FREEDOM. In oral societies, little scope existed for alienation from the heirs who were regarded as the proper recipients. On the other hand testamentary freedom has disinheritance as its corollary. The problem of exclusion became acute in the early days of Christendom since some religious advisors encouraged the old to leave all to the

Church and nothing to their kin. The Church itself, and later 'hell fire' and 'charity-begins-at-home' statutes, legislated against such forms of disinheritance and indeed most contemporary systems reserve a 'legitimate' part for the spouse and the children. In this way testamentary freedom is limited by law so that close kin benefit from a portion of the estate, although not to the same extent as under intestacy. This limitation holds even in ancient Roman and modern Anglo-American law where freedom to disinherit was greater than in the civil law regimes of the continent. In England, the obligation to leave a minimum share of personal property to close kin disappeared in the course of the 17th and 18th centuries, while in 1833 the widow lost her right to a dower. In Scotland, on the other hand, rights such as the bairn's part continued. Indeed in wealthier families in England these rights were always maintained by entails, by the strict family settlement of the 18th century and by earlier devices which prevented the splitting of the estate, while parallel practices deriving from late Roman law existed on the continent. Such arrangements were the subject of objections by some because they kept land from the market and made it impossible to raise a mortgage to effect improvements. In Europe the system collapsed with the French Revolution, following which the Napoleonic Code tried to ensure partition. But in England it persisted until the Settled Land Act of 1882. More recently Family Provision Acts have restored some of the protection given to the surviving spouse and to the children, and in the Soviet Union to anyone previously dependent upon the deceased.

In fact, the beneficiaries of inheritance under a will do not turn out to be greatly different from 'intestate' inheritance, not only because of legal restrictions but because the contents of written wills follow the general sentiments of donors. Indeed because of its flexibility the pattern of testamentary inheritance may be closer to the moral climate of opinion, as in the preference it gives to the 'spouse-all' provisions of modern Anglo-American law, whereas division with the children obtains in the more conservative case of intestacy. The legal formalism connected with literacy 'tends to generalize rules that have originated in connection with special situations into applications beyond their initial scope' (Rheinstein, 1974, p. 590). At the same time the written rule tends to preserve past situations so that intestacy laws have 'frequently looked obsolete, confused, or arbitrary'.

INHERITANCE UNDER CURRENT ANGLO-AMERICAN LAW. Intestate rules in Anglo-American law usually split the property between the surviving spouse and the children. When people make wills, on the other hand, they use the testamentary freedom to leave all to the spouse, usually the wife as she is often younger and lives longer than the husband. In general it is women as widows that benefit most from inheritance. Only after the widow's death does the property drop a generation. The one exception is in the case of a remarriage where specific provision is often made in advance for the children of 'the first bed' in whose welfare the surviving spouse may have less interest.

Even here, despite the potential difference between the outcome of testate and intestate inheritance, the results are very similar since children normally hand over the portion to which they are legally entitled to their parent so that he or she can continue to lead an independent life. When the next generation eventually inherits, the property is usually split equally between children regardless of sex. However, there is one major exception to equality of partition. When one of the siblings has looked after the parents in their old age,

testamentary 'freedom' or intestate adjustment is used to allocate that person a preferential share. This was one of the roles of preferential primogeniture or ultimogeniture in early English law, the last-born son being known in some parts as the *astrier*, the one who remains by the hearth. Otherwise equality is the norm both in law and in practice. Whatever discrimination operates against women in other sections of the society, little is now manifest in testamentary matters, either as spouses or as daughters (however, the 'poor' widow who did not produce a dowry can be helped from the estate, both in Justinian's law and in modern Lousiana). A wife tends to regard an inheritance as her personal peculium, a nest-egg. Given the relatively late age that most people receive legacies, these may make little difference to the lifestyles of the recipients, who sometimes use them as gifts *inter vivos* to assist their own children rather than themselves.

JACK GOODY

See also ECONOMIC ANTHROPOLOGY; FAMILY; INHERITANCE TAXES; RICARDIAN EQUIVALENCE THEOREM.

BIBLIOGRAPHY

Goody, J. 1962. *Death, Property and the Ancestors*. Stanford: Stanford University Press.

Goody, J., Thirsk, J. and Thompson, E.P. (eds) 1976. *Family and Inheritance: rural society in Western Europe 1200–1800*. Cambridge: Cambridge University Press.

Renner, K. 1929. *The Institutions of Private Law and their Social Functions*. Trans. from German, London: Routledge & Kegan Paul, 1949.

Rheinstein, M.Y. 1974. Inheritance. In *Encyclopaedia Britannica*, Vol. 19, Chicago: Encylopaedia Britannica.

Shoup, C. 1966. *Federal Estate and Gift Taxes*. Washington, DC: Brookings Institution.

Shoup, C. 1968. Taxation: death and gift taxes. *International Encyclopedia of the Social Sciences*, Vol. 15, New York: Macmillan.

Sussman, M.B., Cates, J.N. and Smith, D.T. 1970. *The Family and Inheritance*. New York: Russell Sage Foundation.

Wedgwood, J. 1929. *The Economics of Inheritance*. London: G. Routledge & Sons.

inheritance taxes. Taxes on property left by individuals to their heirs are among the oldest forms of taxation. In societies in which property is privately owned, the state protects the property rights of the individual and supervises the transfer from one generation to the next. Consequently, the state has always regarded property transfers as appropriate objects of taxation. However, taxes on bequests and gifts raise very little revenue in modern tax systems.

FORMS OF DEATH TAXES. Taxation of property transfers can take several forms, depending on when the transfers are made. *Estate* taxes are taxes on the privilege of transferring property to one's heirs at death. *Inheritance* taxes are levied on the privilege of inheriting property. Most estate and inheritance taxes are levied at graduated rates (sometimes reaching high levels), with high exemptions.

Taxes at death could be avoided simply by transferring property by gifts *inter vivos* (between living persons). Accordingly, estate taxes are usually associated with a *gift* tax on the donor, and inheritance taxes are associated with a gift tax on the donee. In the United States, the United Kingdom, and other countries, the estate and gift taxes have been unified into one tax. In such cases, the tax is levied on the accumulated bequests and gifts, with instalments paid on the incremental gifts as they are made and the bequest treated as the final gift.

State governments of the United States and many countries levy taxes on inheritances separately from gift taxes or without gift taxes. A unified tax on gifts and inheritances, called *accessions tax*, is not used anywhere.

Some people regard an inheritance or accessions tax as more equitable than an estate tax, because taxes are graduated according to the total wealth received by any one person. On the other hand, most countries use the estate tax because it is easier to administer.

Bequests and gifts, like income from work or investments, are a source of ability to pay. In theory, therefore, they should be taxable to the recipient as income when received. However, bequests and gifts are taxed separately from income in all countries.

HISTORY. Death taxes pre-date both income and sales taxes as a source of government revenue. The first inheritance tax was levied in the Roman Empire beginning in AD 6. During the Middle Ages, various feudal taxes resembled inheritance taxes. Such taxes were in use in several Italian commercial cities by the end of the 14th century and in England, France, Germany, the Netherlands, and Portugal by the end of the 17th century. Estate or inheritance taxes are now levied in practically all industrial countries and in many developing countries, but the type of tax and the degree of progression differs greatly among them.

England's death tax, which dates from the year 1694, took its modern form in 1779, when a flat duty was replaced essentially by a proportional tax; graduation was introduced in 1894. France adopted its inheritance tax in 1796 and introduced graduation in 1902. Italy modelled its tax after the French system in 1862 and made it progressive in 1902. Germany's tax, which was based on the Prussian inheritance tax of 1873, was graduated in 1905. The federal government of the United States levied temporary inheritance taxes during the Civil War and the Spanish–American War, but the tax was already in use in many of the states before the modern, graduated estate tax was adopted in 1916.

RATIONALE. Adam Smith was ambivalent about death taxes, mainly because they bear heavily on families having deaths spaced at short intervals. (This problem is now handled by providing a credit or rebate for taxes on an estate which was recently subject to tax, say, within the last five or ten years.) Smith and David Ricardo believed that death taxes would reduce the funds available for investment. Jeremy Bentham and John Stuart Mill attacked the ethical justification of the institution of inheritance, and supported a limitation on the amount any one person could acquire from others without working. Henry Sidgwick, Alfred Marshall, A.C. Pigou, and many other economists in the classical tradition supported the taxation of wealth transfers to promote greater equality of opportunity, even though some were concerned about the effect on capital accumulation. Josiah Wedgwood (1929), Hugh Dalton, James Meade (e.g. 1976) and others emphasized the objective of reducing the concentration of wealth as a justification for state or inheritance taxation. Henry Simons, who argued strongly for the taxation of what he called 'gratuitous receipts' as income, also supported a supplementary tax on wealth transfers to control the size of inheritances.

Some Keynesian economists in Britain have supported death taxes in order to raise the propensity to consume, but this rationale had no influence either on the development of these taxes or on the public's attitude toward them. Keynes himself

mentioned that death taxes would probably increase the propensity to consume more than other taxes of equal yield, but did not recommend that such taxes be enacted for this reason.

Death taxes have been supported by people in all wealth classes. One of their strongest supporters was the American steel magnate, Andrew Carnegie, who had doubts about the institution of inheritance (because it impairs children's incentives) and felt that wealthy persons are morally obliged to use their fortunes for social purposes. While this view is not widely held, many people believe that the existing distribution of wealth and control of business enterprise should not be perpetuated through succeeding generations and that taxation of bequests and gifts is the most effective method of achieving this objective. Although data are scarce, the available information suggests that inherited wealth is a major reason why the distribution of wealth is highly unequal in market economies.

Opinions about the impact of death taxes on private incentives vary. Some believe that these taxes reduce saving and undermine the economic system. But even they might concede that death taxes have less adverse effects on incentives than do income taxes of equal yield. Income taxes reduce the return from effort and risk-taking as income is earned, whereas death taxes are paid only after a lifetime of work and accumulation and are likely to be given less weight by individuals in their work, saving, and investment decisions. This distinction was emphasized by many economists in the classical tradition (including Mill, Sidgwick, Marshall, and Pigou).

Proposals have been made from time to time to tax inherited wealth more heavily than wealth accumulated out of an individual's own saving. In some plans (for example, a plan proposed by Eugenio Rignano, 1924), inherited wealth would be taxed at progressively higher rates in succeeding generations. Such proposals have never been given serious consideration because of the difficulties of tracing inherited wealth, the harshness of the tax when there are quick successions, and the problems of record-keeping and administration. It is possible to accomplish Rignano's objective by varying the tax on the basis of the number of years during which donors hold their wealth. Such a plan, which was devised by William Vickrey (1947) and later proposed in modified form by a commission headed by James E. Meade (1976), tends to be complicated (because it requires interest adjustments to equate the taxes on bequests and gifts over a given number of years) and no country has shown any interest in this type of tax.

STRUCTURAL PROBLEMS. Since wealth transfers take many forms, estate and gift taxation is inherently complicated. The bases of the estate and gift taxes are intended to consist of all property transferred by gift or death, but only a small fraction of total property transfers is subject to tax. In the United States, less than one per cent of all the estates of those who die in any one year is subject to estate or gift taxes.

The estate and gift tax exemptions tend to be relatively high in most countries and concessions for transfers to children and grandchildren (consanguinity rates) are frequently excessive. The tax base is also eroded by undervaluations of farm and small business properties. Works of art and other personal property often escape taxation. In the United States, family foundations may be used to remove wealth from the estate tax base without relinquishing control of the business enterprise.

The most difficult problems in estate and gift taxation have arisen from the use of trusts to transfer wealth to future generations. Children and grandchildren may receive the income from a trust while they are living, but no tax is due on the property when the trust terminates at their death, thus avoiding tax for one or more generations. In the United States and the United Kingdom, the trust property is treated as if it is owned by the income beneficiaries, but significant tax avoidance possibilities through the use of trusts remain.

Even if the estate and gift taxes are unified, wealthy people may reduce their taxes by transferring property by gifts during their lifetimes. Usually, an annual exclusion is allowed to avoid the need to account for small gifts, but such exclusions permit large amounts of property to be transferred free of tax over a period of years. In addition, the gift tax itself is not included in the tax base, whereas the estate tax is computed on the basis of the donor's entire property, including the tax. Despite the advantage of making gifts, data for the United States suggest that wealthy people prefer to retain the bulk of their property until death.

Another device used by wealthy people to avoid estate taxes is to manipulate the ownership of various classes of stock in a corporation so as to funnel increasing equity values to children or grandchildren. For example, by recapitalizing the equity structure of the business, the owner of a successful closely-held corporation might give his children all the common stock, which is initially given a low value, and retain the preferred stock, which is given a high value. As the corporation prospers, the common stock rises in value to reflect the increased earnings of the corporation, but this increase in wealth never shows up in the estate tax base.

Because of the practical problems of taxes on wealth transfers, some have proposed the enactment of an annual wealth tax to reach property that is not now subject to death (or income) taxes. Annual taxes on net wealth have been enacted in a number of European countries, but these taxes raise very little revenue and are regarded as supplements to estate or inheritance taxes, rather than as substitutes.

REVENUE YIELD. Despite the appeal of estate and gift taxes on social and economic grounds and despite the use of relatively high rates, taxes on property transfers have never provided significant revenues anywhere and have had only modest effects on the distribution of wealth. In 1983, estate and gift tax collections amounted to 0.3 per cent of gross domestic product in France, 0.2 per cent in the United States and the United Kingdom, and 0.1 per cent in Germany.

One can only guess why heavier reliance has not been placed on estate and gift taxes. One explanation is that people resent paying taxes on such wealth as the family home or business, works of art, and other personal property. The public is not aware that the major part of the estate and gift tax bases consists of stocks, bonds, and real estate, and that the exemptions remove the wealth of most people from the base. Another explanation is that greater equality in the distribution of wealth is not generally accepted as an objective of tax policy.

More intensive use of estate and gift taxes would add progressivity to tax systems with less impairment of economic incentives than many other taxes. Major obstacles to increased use of these taxes are public apathy and the lack of understanding of their major features and how they apply in individual circumstances. Resistance to higher death duties by wealthy people is also a factor. The merits of wealth transfer taxes will have to be more widely understood and accepted before they can become effective revenue sources.

JOSEPH A. PECHMAN

See also REDISTRIBUTION OF INCOME AND WEALTH; TAXATION OF CAPITAL; TAXATION OF WEALTH.

BIBLIOGRAPHY
Cooper, G. 1979. *A Voluntary Tax? New Perspectives on Sophisticated Estate Tax Avoidance.* Washington, DC: Brookings Institution.
Meade, J.E. 1976. *The Structure and Reform of Direct Taxation.* Report of a Committee Chaired by Professor J.E. Meade, London: Institute for Fiscal Studies; Allen & Unwin.
Rignano, E. 1924. *The Social Significance of the Inheritance Tax.* Trans. W.J. Schultz, New York: Knopf.
Sandford, C.T., Willis, J.R.M. and Ironside, D.J. 1973. *An Accessions Tax.* London: Institute for Fiscal Studies.
Shoup, C. 1966. *Federal Estate and Gift Taxes.* Washington, DC: The Brookings Institution.
Tait, A.A. 1967. The Taxation of Personal Wealth. Urbana: University of Illinois Press.
Vickrey, W. 1947. *Agenda for Progressive Taxation.* New York: The Ronald Press.
Wedgwood, J. 1929. *The Economics of Inheritance.* London: G. Routledge & Sons.

Innis, Harold Adams (1894–1952). Canadian economist, historian and university administrator. Born in rural Ontario, Innis was educated at McMaster University, Toronto (BA, MA), and at the University of Chicago (PhD). Having served in the Great War, he joined the faculty of the Department of Political Economy, University of Toronto, in 1920. From 1937 until his death he was Head of the Department, and from 1947 he was also Dean of the Graduate School; at his death he was President of the American Economic Association.

A prolific and thoughtful scholar, Innis began by scrutinizing Canadian economic history, both in shorter writings and in such major works as *The Fur Trade in Canada* (1930) and *The Cod Fisheries* (1940), where he concentrated his attention on such great 'staple products' as codfish, fur, wheat and timber. In these works, which have been read with interest in other lands whose economic structures appear to be similar, such as Australia, Innis developed a vision of Canadian economic history that centred on the successive development of natural-resource-based industries. The physical characteristics of these industries' products, Innis believed, had shaped not only the economic but the political and cultural history of Canada. Few would now accept Innis's interpretation of Canadian history as the mere reflection of the 'staple products'. Yet for forty years that interpretation shaped the teaching and writing of economy history in English-speaking Canada, and it affected political historiography as well. Innis's undergraduate education was aimed at the Baptist ministry, and perhaps it was a misfortune that he turned to economics; if he had followed some more speculative vocation, the particular powers of his intellect might have developed more widely and less eccentrically, although Canadian economic history would have been deprived of its most creative practitioner. The broadening of Innis's interests beyond economic history can be detected in his early writings on what he called 'the penetrative power of the price system' – the ability of market mechanisms to reshape social relationships. Uncertain in his grasp of modern economics, Innis ignored the Keynesian revolution, and he was profoundly sceptical about the potential contribution to rational national policy-making which might come not only from economists but from other scholars; better, he thought, for university folk to concentrate upon the safeguarding of the Western cultural tradition. In his later years Innis wrote almost exclusively about very large questions – the interconnections, over very long periods, among imperial structures and means of communication.

These works – *The Bias of Communications, Changing Concepts of Time, Empire and Communications, Minerva's Owl* – have had little impact on economists or economic historians, although they have influenced some students of the humanities – most notably the Canadian literary scholar Marshall McLuhan. Also, during the 1970s and 1980s Innis's writings attracted attention from Canadian nationalists, more or less regardless of discipline; furthermore, in these decades efforts were made to find and explicate new profundities in his writings, or to reinterpret this pessimistic and conservative thinker as an unconscious proto-Marxist. Few economists and fewer historians have found these efforts persuasive.

Ian M. Drummond

See also LINKAGES.

SELECTED WORKS

1930. *The Fur Trade in Canada.* New Haven: Yale University Press; revised edn, Toronto: University of Toronto Press, 1956.
1940. *The Cod Fisheries.* New Haven: Yale University Press; revised edn, Toronto: University of Toronto Press, 1954.
1950. *Empire and Communications.* London: Oxford University Press; revised edn, Toronto: University of Toronto Press, 1972.
1951. *The Bias of Communications.* Toronto: University of Toronto Press.
1956. *Essays in Canadian Economic History.* Toronto: University of Toronto Press.

BIBLIOGRAPHY

Neill, R. 1972. *A New Theory of Value: The Canadian Economics of H.A. Innis.* Toronto: University of Toronto Press.

innovation. Economists of all descriptions have accepted that new products and new processes are the main source of dynamism in capitalist development. But relatively few have stopped to examine in depth the origins of such innovations or the consequences of their adoption. Most have preferred, in Rosenberg's (1982) apt description, not to look 'inside the black box', but to leave that task to technologists and historians, preferring to concentrate their own efforts on '*ceteris paribus*' models, which relegate technical and institutional change to the role of exogenous variables.

The classical economists were generally more ready to look inside the black box; Adam Smith and Marx in particular both showed a deep interest in the relationship between scientific research, technical innovation and the market. Smith (1776), pointed already in the 18th century to the growth of specialization in scientific research and to the links between innovation in the machine-building industries and scientists ('philosophers' or men of 'speculation' whose task is 'to observe everything'). Marx (1848) probably more than any other economist assigned to technical innovation the driving force in economic development and competition – 'the bourgeoisie cannot exist without constantly revolutionizing the means of production.'

But in the first half of the 20th century Schumpeter was almost alone among leading economists in following and developing this classical tradition. Consequently those economists such as Nelson (1977 and 1982) and Rosenberg (1976 and 1982) who have concentrated much of their attention on the economics of innovation are often referred to as 'Schumpeterian' or 'neo-Schumpetarian', even though their ideas may considerably diverge on many topics.

It is to Schumpeter that we owe the threefold distinction between invention, innovation and diffusion of innovations, which has now become the generally accepted convention in analysis of technical change. *Invention* is generally defined as a novel idea, sketch or model for a new or improved product, process or system. It need not necessarily imply any empirical test of feasibility or prototype experience, but as Jewkes (1958) suggests, it usually does convey the first belief that something should work and often the first rough test that it will in fact work.

Nevertheless, Schumpeter was right to stress the distinction between invention and *innovation*. There is an enormous difference between 'working' under laboratory conditions and working under commercial conditions. Schumpeter used the expression 'innovation' to connote the first introduction of a new product, process, method or system into the *economy*. (This is generally taken to include military or health care applications as well as the more purely commercial innovations.) As Schumpeter pointed out, there is many a slip between cup and lip in the development of an invention to the point of commercial introduction. Problems in scaling up from laboratory scale to works scale lead to the demise of many apparently sound ideas and unanticipated 'bugs' are the rule rather than the exception in the exploitation of inventions. Many (perhaps most) inventions are patented, but most patents are never actually used commercially except perhaps as bargaining counters.

Some ambiguity still surrounds the definition of 'innovation', since the word is used both to indicate the date of *first* introduction of a new product or process (e.g. the float glass process was innovated in 1958) and to describe the whole process of taking an invention or set of inventions to the point of commercial introduction, as in 'management of innovation' – a process which may take many years of development work, trial production and marketing.

In fact the date of launch of an innovation is seldom as precise as might appear at first glance, since false starts and modifications to the design of a radical new product or process are commonplace. Thus many different dates can be found for the innovation of well-known products, such as the radio or the electronic computer. National bias plays a part too, as well as definitional problems.

This point is an important one when we come to consider the third aspect of technical change in the Schumpeterian framework – the *diffusion* of innovations. Although almost all economists would agree that the diffusion of innovations through a population of potential adopters is crucial to the achievement of productivity gains and successful competitive performance more generally, they would also agree with Rosenberg (1976), that the product or process which is being diffused is itself usually subject to further change *during* the diffusion process. Indeed, this has been one of the main criticisms of some studies of diffusion in the 1960s and 1970s (Metcalfe, 1981) which tended to make the static assumptions of an unchanged product diffusing through an unchanged environment. Nevertheless, this does not invalidate Schumpeter's analytical distinction, which has proved extremely fruitful both in theoretical and empirical work, as shown notably in the major international conference on diffusion of innovations in Venice in 1986.

When Jewkes and his colleagues (1958) made their original study of the sources of invention, they rightly complained that economists had made very little contribution to the study of invention and innovation, and Rogers (1962) could legitimately make a very similar complaint, in relation to the study of *diffusion* of innovations. However, in the next quarter of a century the picture changed considerably. Following the impetus given especially by Mansfield (1968, 1977) numerous empirical studies in Europe, America and Japan covered much of the territory which Schumpeter sketched out in a preliminary way. Unknown and uncharted territory still remains, however, and its exploration is by no means straightforward (Dosi, 1985).

Thus, for example, we now know a good deal about the conditions surrounding success and failure in the competitive struggle of private firms to innovate, but far less is known about the types of government policies which are most likely to encourage innovators and promote their success. The study of the latter is inhibited by the difficulty of isolating any specific single measure, such as a tax incentive, development subsidy or procurement initiative from other more general influences on the behaviour of the firm and numerous factors specific to individual firms (Rothwell and Zegveld, 1981).

In the analysis of competitive attempts by individual firms to innovate the problems of multiple causality has been partly overcome by the use of statistical techniques in paired comparisons of success and failure, as for example in project SAPPHO (Freeman, 1982; Rothwell et al., 1974) and similar studies in several countries (e.g. Szakasits, 1974). By and large these studies agree in highlighting the main factors leading to successful innovation performance: the depth of understanding of the needs of potential users of the innovations and the steps taken to obtain this knowledge (external communications network): the research and development capability to eliminate or minimize 'bugs' prior to launch of the innovation; internal communications adequate to ensure effective links between those responsible for R&D, marketing and production within the firm; entrepreneurs or 'business innovators' with the status and experience to ensure the necessary mobilisation and coordination of resources within the firm. Studies of failure have been particularly illuminating in demonstrating the tendency of some technical innovators to neglect user needs and the lack of communication between various departments in some large firms (Burns and Stalker, 1961). However, they also show that even in cases, when firms appear to follow all the 'rules' and 'best practices' which lead to good innovation performance, technical and market uncertainties may frustrate their best efforts.

Indeed, the empirical studies of the management of innovation and firm behaviour have undermined the traditional neoclassical theory of the firm. Imperfect information, uncertainty, complex institutional linkages, cumulative in-house technology, and searching modes of behaviour are characteristic of innovation, rather than the tidy, rational, optimizing calculations and perfect foresight postulated by neoclassical theory (Dosi, 1984, 1985). For this reason contributors to innovation studies have also made major new contributions to a revised theory of firm behaviour, which take into account the findings of the stream of empirical research (Nelson and Winter, 1982; Dosi, 1984).

Less clear-cut conclusions have emerged with respect to the influence of size and concentration on innovative performance. Schumpeter (1928 and 1942) is often known for his emphasis on the advantages of large size and monopoly on innovative performance, whilst traditional theory has continued to stress the advantages of competitive market structures. Clearly large size can facilitate innovative efforts in areas where development costs are unavoidably high because of number and complexity of components, as for example in spacecraft, nuclear reactors, or electronic telephone exchanges. The R&D threshold entry barriers in such areas can sometimes be so high as to limit effective competition to only a few large organizations throughout the world; and often innovation costs are partly met by state subsidies.

Even those economists, such as Jewkes et al. (1958), who have stressed the role of individual inventors and small firms

at the stage of *invention*, have accepted that often *development* costs are so high that large firms tend to predominate when it comes to *innovation*. Many of the case studies described by Jewkes et al. illustrate this point, since the small firms or individuals who initiated the inventive work were often obliged to seek the help of larger organizations or were taken over by them before they could launch the new product or process on the market.

However, revolutionary advances in technology, for example the micro-chip, can sometimes lower entry barriers dramatically. In those areas where smaller firms can afford the entry costs they appear to perform relatively well in competition with larger firms. Thus the SAPPHO project did not show size as a variable which discriminated systematically between success and failure.

Schumpeter (1912 and 1928) had himself recognized the advantages of new small innovator–entrepreneurial firms, but believed that the general trend of capitalist development and the rising costs of in-house R&D would lead increasingly to the management of innovation by larger bureaucratic organizations. Galbraith (1972) developed this notion of the 'techno-structure' in large firms in his 'New Industrial State'. However, empirical evidence suggest that small firms have continued to maintain, or increase their share of innovations, even though large firms do indeed now account for more than two-thirds of R&D and of all innovations (Townsend et al., 1982). The share of small firms in innovations is apparently greater than their share of R&D expenditures, and this phenomenon has been explained partly in terms of motivation and good internal communications leading to greater efficiency in the conduct of R&D, and partly in terms of the 'spin-off' of technical innovators who have left large government, industrial or academic laboratories with the idea for an already partly developed product.

The debate continues but with increasingly general acceptance that both very large and new entrepreneurial firms enjoy advantages in distinct types of invention and at different stages of the evolution of new technologies. The previously observed tendency for R&D intensity to decline in the largest firms has been denied by Soete (1979), who maintains that more recent evidence supports the Schumpeterian hypothesis.

Schumpeter's contention that technological competition was more important than price competition with invariant conditions of production has also found increasing confirmation from empirical and theoretical work in the sphere of international trade. Since Hufbauer's (1965) original demonstration of the role of technical innovation in the explanation of patterns of international trade in synthetic materials, evidence has accumulated to confirm that 'neo-technology' theories have greater explanatory power in relation to international trade performance generally than the Heckscher–Ohlin factor proportions theory (Soete, 1981).

The notion that cumulative patterns of advantage in know-how, skills and innovative capability may underlie some of the persistent differences in comparative international trade and productivity performance has also found confirmation in many national studies of innovation and economic development (e.g. Pavitt, 1980). These suggest that institutional innovations in education and training systems, as well as in research institutes and organizations have historically played an important part in building up cumulative technological capability. Thus, for example, German strength since the late 19th century in the chemical and engineering industries has been related to the establishment of the 'Technische Hochschulen' and other new developments in German universities, as well as the establishment of in-house R&D in the leading German chemical and electrical firms. Similar arguments have been advanced with respect to Japanese industrial training and technological innovation systems and the more recent outstanding successes of the Japanese economy (Freeman, 1983).

To sum up, empirical studies of innovations and their diffusion have provided mounting evidence that mainstream neoclassical theories of firm behaviour, competition, international trade and consumer behaviour are seriously deficient in their assumptions and conclusions. However, the 'neo-Schumpeterian' tradition in economics has only begun the task of substituting a more satisfactory theoretical foundation which would take both technical innovation and institutional factors fully into account (Dosi, 1985).

C. FREEMAN

See also SCHUMPETER, JOSEPH ALOIS; TECHNICAL CHANGE.

BIBLIOGRAPHY

Burns, T. and Stalker, G.M. 1961. *The Management of Innovation.* London: Tavistock.

Dosi, G. 1984. *Technical Change and Industrial Transformation – the Theory and an Application to the Semi-conductor Industry.* London: Macmillan.

Dosi, G. 1985. The micro-economic sources and effects of innovation; an assessment of some recent findings. Paper given to Conference on 'Distribution, Growth and Technical Progress', Rome; (mimeo) DAEST, University of Venice.

Freeman, C. 1982. *The Economics of Industrial Innovation.* 2nd edn, Cambridge, Mass.: MIT Press; London: Frances Pinter.

Freeman, C. 1983. *Design and British Economic Performance.* London: Royal College of Art.

Galbraith, J.K. 1972. *The New Industrial State.* 2nd edn, London: André Deutsch.

Hufbauer, G. 1966. *Synthetic Materials and the Theory of International Trade.* London: Duckworth.

Jewkes, J., Sawers, D. and Stillerman, R. 1958. *The Sources of Invention.* London: Macmillan.

Mansfield, E. 1968. *Industrial Research and Technological Innovation: an econometric analysis.* New York: W.W. Norton.

Mansfield, E. et al. 1972. *Research and Innovation in the Modern Corporation.* London: Macmillan.

Marx, K. and Engels, F. 1848. *The Communist Manifesto.* London.

Metcalfe, J.S. 1981. Impulse and diffusion in the study of technical change. *Futures* 13(5), 347–59.

Nelson, R.R. and Winter, S.G. 1977. In search of a useful theory of innovation. *Research Policy* 6(1), 36–76.

Nelson, R.R. and Winter, S.G. 1982. *An Evolutionary Theory of Economic Change.* Cambridge, Mass.: Belknap and Harvard University Press.

Pavitt, K.L.R. 1980. *Technical Innovation and British Economic Performance.* London: Macmillan.

Rogers, E.M. 1962. *The Diffusion of Innovations.* New York: Free Press.

Rosenberg, N. 1976. *Perspectives on Technology.* Cambridge: Cambridge University Press.

Rosenberg, N. 1982. *Inside the Black Box.* Cambridge: Cambridge University Press.

Rothwell, R.R. and Zegveld, W. 1981. *Industrial Innovation and Public Policy.* London: Frances Pinter.

Schumpeter, J.A. 1912. *The Theory of Economic Development.* Leipzig. Trans, Cambridge, Mass.: Harvard University Press, 1934.

Schumpeter, J.A. 1928. The instability of capitalism. *Economic Journal* 38, September, 366–86.

Schumpeter, J.A. 1942. *Capitalism, Socialism and Democracy.* New York: Harper.

Smith, A. 1776. *An Inquiry into the Nature and Causes of the Wealth of Nations.* London: Dent, 1910.

Soete, L.L.G. 1979. Firm size and inventive activity: the evidence reconsidered. *European Economic Review* 12, 319–40.

Soete, L.L.G. 1981. A general test of technological gap trade theory. *Weltwirtschaftliches Archiv* 117(4), 638–66.

Szakasits, G. 1974. The adoption of the SAPPHO method in the Hungarian electronics industry. *Research Policy* 3(1), 18–28.

Townsend, J. et al. 1982. Innovations in Britain since 1945. Social Policy Research Unit, Occasional Paper No. 16, University of Sussex.

input–output analysis. Input–output analysis is a practical extension of the classical theory of general interdependence which views the whole economy of a region, a country and even of the entire world as a single system and sets out to describe and to interpret its operation in terms of directly observable basic structural relationships.

Wassily Leontief, a Russian-born American economist, started the construction of the first input–output tables of the American economy when he joined the faculty at Harvard University in 1932. These tables, for the years 1919 and 1929, were published together with the formulation of a corresponding mathematical model and numerical computation based on it in 1936 and 1937. Thus from the very outset the new methodology – for the development of which Leontief was awarded forty years later a Nobel prize – emphasized the importance of close mutual alignment of systematic fact finding and theoretical formulation.

In the late Twenties Leontief spent three years at the Institute for the World Economy at the University of Kiel (Germany) on derivation of statistical supply and demand curves. That early experience with curve fitting taught him not to rely on indirect statistical inference as a substitute for painstaking direct factual inquiry.

With its emphasis on disaggregation permitting detailed quantitative description of the structural properties of all component parts of a given economic system the input–output analysis moved in a direction directly opposite to that of the highly aggregative approach that began, approximately at the same time, to dominate fundamental economic research under the powerful influence of the Keynesian paradigm presented in Keynes's *General Theory*. Hand-in-hand with a disaggregated data base went an equally disaggregated theoretical model, the empirical implementation of which involved numerical computations exceeding in their complexity and scale anything that had been carried out up to that time along these lines in economics or any other social science.

The limited capabilities of the Wilbur linear analog computer used in the first large scale computation forced Leontief to scale down his problem by neglecting some of the detail contained in the disaggregated data base. Subsequent rounds of computation were carried out at first on Howard Aiken's, Mark I and Mark II computers, and later on the early electronic machines. Thirty years later the race between the economists and statisticians compiling more and more detailed factual information, and engineers constructing more and more powerful machines, was won hands down by the latter.

A standard input–output table contains square arrays of figures arranged in chess-board fashion. Each row and the corresponding column bears the name of one particular sector, say, steel industry, automobile industry, electric power utilities, advertising services, and so on. Each individual entry represents the amount (which can, of course, be zero) of the commodity or service produced by the sector – identified by the name of the row in which it appears – that has been delivered to the sector named at the head of the column in which that entry is placed. The small schematic input–output table presented below (Table 1) describes intersectoral

TABLE 1

	Agriculture	Manufacturing	Households	Total
Agriculture	25	20	55	100 bushels
Manufacturing	14	6	30	50 yards of cloth
Households	80	180	—	260 man-years

transactions between the three sectors of the elementary economy described by it.

Examining these figures, one finds that to produce one bushel of wheat, agriculture requires 0.25 bushels of wheat (seed), 0.14 tons of steel and 0.80 man years of labour. A similar set of technical coefficients – 0.40 units of agricultural and 0.12 of manufactured products – describe the input requirements for production of one yard of cloth. Listed column by column these sets of technical input coefficients represent the structural matrix at the producing part of the given economy.

TABLE 2

	Sector 1	Sector 2
Sector 1	0.25	0.40
Sector 2	0.14	0.12
Household	0.80	3.60

While the figures in Table 2 were derived from the input–output table (Table 1), estimates of the magnitudes of the technical coefficients could be, and in some instances actually are obtained directly from technical, engineering data sources.

The structural matrix of an economy provides a basis for determination of total sectoral output as well as magnitude of inter-sectoral transactions that would enable the producing sectors to deliver to households and to other so-called final users a specified 'bill of goods'. Considering the vector of final demand, consisting of 55 bushels of wheat and 30 yards of cloth, as given, the following set of balanced equations can be used to determine the total amounts of wheat (x_1), of cloth (x_2), as well as the total amount of labour (L) needed to balance under these particular technological conditions the outputs and inputs of both producing sectors,

$$(1 - 0.25)x_1 - 0.14x_2 = y_1 \qquad (1)$$
$$-0.40x_1 + (1 - 0.12)x_2 = y_2$$

The general solution of these two equations:

$$1.457y_1 + 0.662y_2 = x_1 \qquad (2)$$
$$0.232y_1 + 1.242y_2 = x_2$$

permits us to compute the total levels of output of wheat, x_1 and cloth, x_2 required directly and indirectly to satisfy any given vector (y_1, y_2) of 'final demand'.

An increase in the final deliveries of agricultural products, y_1 by one unit would for instance require a rise of total agricultural output, x_1, by 1.1457 units, 0.1457 of which will have to be used to satisfy the additional input requirements of the agricultural and manufacturing sectors.

Formulated in short-hand matrix notation, the balance equations (1), describing the relationship between the column vector of final demand, y, and the column vector, x, of total outputs of all producing sectors can be written as:

$$(I - A)x = y \qquad (3)$$

where A represents the upper, square part of the structural matrix (fig. 2) describing the material input requirements of all producing sectors, x is the column vector of total outputs and y, the column vector of final deliveries of both goods. The general solution of that linear equation is,

$$x = (I - A)^{-1}y \qquad (4)$$

where $(I - A)^{-1}$ represents the so-called inverse of matrix $(I - A)$.

Total labour requirement can be computed in a separate step,

$$L = l'x = l'(I - A)^{-1}y \qquad (5)$$

where l' is a row vector of technical labour coefficient representing the technologically determined amounts of labour that each industry employs per unit of its total output.

The same set, A, of structural coefficients that controls the physical flows, determines also the relationship between the prices of goods and services produced by different industries and the 'value added' payments (expressed in the monetary units) made by each industry per unit of its output. These include wages, profits, taxes, etc. In short, all payments other than those made for goods and services purchased from other producing sectors.

This set of value added–price equations, (often referred to as a 'dual' to set (3) of physical input–output relationships) can be formulated as follows,

$$(I - A')P = V \qquad (6)$$

and its solution for the unknown prices as,

$$P = (I - A')^{-1}V \qquad (7)$$

where P is the column vector of prices of all sectoral outputs and V is the given column vector of values added (per unit of their respective outputs), in different sectors.

In the schematic input–output table (fig. 1) considered above all amounts entered along a particular row are measured in the same appropriately selected physical unit, for instance, wheat – in bushels; cloth – in yards; labour – in man years. No column totals are entered, since adding amounts measured in incomparable physical units would make no sense. In most published input–output tables, all transactions are measured however in value terms – usually in 'base year' prices. Since these are assumed to satisfy the price-value added equations described above – each column total, including the value added per unit of total output, must naturally be equal to the total output figures entered at the end of the corresponding row.

Value figures entered along a particular row can however also be interpreted as representing physical amounts of the good in question, provided the physical unit in which they are measured is implicitly defined as the quantity of that good purchasable for, say, one dollar.

In the case of a table some rows of which are presented in conventional physical amounts, say kwh of electric power, or tons of copper, while some other rows – in monetary units, appropriate 'equilibrium prices' can be computed through solution of the corresponding 'dual' equation (7).

To do so it would suffice to re-define the physical unit of the products of each sector as the amount purchasable for, say, one dollar, or some other monetary unit, at the price actually used in determination of the value figures entered on the base year table. These prices might of course be different from the equilibrium prices.

From the outset the development of input–output analysis was marked by a succession of empirical applications. In Leontief's early volume, *The Structure of American Economy, 1919–1929* (1941), this was the computation of the effects of changes in the input structure of different industries on levels of output and prices of their products, and in particular on the 'standard of living' of households.

With the onset of World War II, attention was centred on the transition from peacetime to a war economy. In particular, on the effects of changes in the level and composition of final demand on the intersectoral distribution of output and employment. The first official US input–output table – for the year 1939, compiled for the US Bureau of Labor Statistics, provided a basis for preparation of a detailed multisectoral projection of postwar production and employment levels. Correctly predicting serious steel shortages, instead of large surpluses anticipated by leading economic and industry experts, that report gained wider interest in the new approach not only in government circles, but among large industrial corporations as well. The Western Electric Company (the manufacturing arm of A.T.&T.) having successfully employed input–output analysis to anticipate impending shortages of lead, one of its principal raw materials, even produced an educational film describing the methodology used.

In one of the early applications of the same modelling technique to what later on became known as operations research the small input–output team organized – under the name Project Scoop – by the US Air Force constructed a detailed structural matrix of its far-flung material procurement and training operations. It was not a square, but rather a rectangular matrix showing for some sectors not one but several input vectors corresponding to two or more alternative technologies that could be used to produce a particular weapon or to provide a particular type of pilot training. Confronted with the problem of optimal choice between alternative 'cooking recipes', Dr George Dantzig, a young mathematician on the Project's staff, invented the still very widely used Simplex method of linear programming, which consists of a series of inversion of structural input–output matrices with sequential substitution at alternative vectors of technical coefficients.

Not unlike research conducted in modern natural sciences, input–output analysis was from the outset most successfully conducted by closely coordinated teams rather than individual investigators. The first of such academic research groups was the Harvard Economic Research Project directed by Leontief over a period of nearly thirty years. Another centre was organized by Richard Stone in the Department of Applied Economics at the University of Cambridge. He was responsible for formal incorporation of input–output tables in the United Nations system of national accounts designed by him.

Many of the young foreign economists who came for completion of their graduate or postgraduate studies to the United States spent from a few months up to several years at the HERP and after returning home introduced input–output analysis not only as a subject of academic instruction and research but also as a new field of governmental statistics.

In Norway, Canada, Japan and in many other countries governmental planning agencies and central statistical offices compile national input–output tables and carry out practical applications of input–output analysis, but also engage in fundamental methodological research. In Soviet Russia this was the first non-marxist, mathematical approach to economics adapted, on the recommendation of Oscar Lange, after World War II as a subject of academic instruction and as a tool of economic planning.

The first International Conference on Input–Output Analysis organized by Professor Tinbergen was held in Dreibergen, Holland in 1950; the eighth has been held in Japan in 1986. Proceedings of these and of other similar scientific meetings published in book form provide a good account of the current state of the art in the general field of input–output analysis and its various applications.

One of the fundamental theoretical questions that came up in connection with the early input–output computations concerned the conditions under which none of the elements of the inverse $(I-A)^{-1}$ can be negative. The answer to it was provided by Herbert Simon – the future Nobel prizewinner – and David Hawkins, a philosopher, in the form of the following theorem:

The necessary and sufficient conditions for some of the elements of $(I-A)^{-1}$ to be positive, and all to be non-negative, are:

$$|\ 1 - a_{11}\ | > 0,$$

$$\begin{vmatrix} (1-a_{11}) & -a_{12} \\ -a_{21} & (1-a_{22}) \end{vmatrix} > 0, \ldots$$

$$\begin{vmatrix} (1-a_{11}) & -a_{12}\ldots & -a_{1n} \\ -a_{21} & (1-a_{22})\ldots & -a_{2n} \\ \vdots & \vdots & \vdots \\ -a_{n1} & -a_{n2}\ldots & (1-a_{nn}) \end{vmatrix} > 0. \quad (8)$$

If these conditions are satisfied for any particular numbering of sectors it will necessarily be satisfied for any other numbering sequence too. The economic interpretation of this theorem is that for a system, in which each sector functions by absorbing directly or indirectly outputs of some other sectors, to be able not only to sustain itself but also to make some positive deliveries to final demand, each one of the smaller and smaller sub-systems contained within it has to be capable of sustaining itself and yielding a surplus deliverable to outside users as well.

An example of a system unable to sustain itself in this sense could be an economy so badly damaged by some natural catastrophe or war that only external assistance, taking the form of an import surplus, could prevent it from complete collapse. Exports are entered in a standard input–output table and in the corresponding set of balance equations, as positive and exports as negative components of the final bill of goods. The negative elements of the inverse $(I-A)^{-1}$ multiplied into such negative components of the vector y of final demand would yield in this case positive total outputs x.

In an attempt to reconcile at least to some extent the so-called fixed coefficient assumption of linear input–output models with the neoclassical production functions allowing for input substitution, Kenneth Arrow, Tjalling Koopmans and Paul Samuelson provided independently from each other three different proofs of the 'non-substitution theorem'. They considered a multisectoral economy in which each productive sector operates on the basis of a neoclassical production function and all sectors use the same single primary factors of production, say labour. The input combinations used by different sectors are chosen so as to minimize the total amount of labour that has to be employed by that economy in order to enable it to deliver to final users an exogenously specified bill of goods. The non-substitution theorem states that the combination of the relative amounts of different inputs chosen in each sector will be independent of the composition of the final bill of goods. That means that even if the structure of final demand changes all producing sectors will behave as if they were operating on the basis of fixed coefficients of production.

Restrictive assumptions – particularly those postulating invariability of production functions that control the operations of all sectors – deprive the non-substitution theorem of much of its practical significance. However, it calls attention to the difference between the ways in which the terms technology, and technological change, are used in neoclassical and in input–output theory. In input–output modelling the technology used in any particular sector is described as a given column vector of coefficients, and a change in any element of that vector is called technological change. In neoclassical modelling the state of the technology employed by a particular sector is described by a much more general – and because of that much more complex – kind of functional relationship that in input–output analysis would have to be viewed as a set of many (strictly speaking, infinitely many) different technologies, each described by a different column vector of input coefficients. While providing a convenient basis for deductive reasoning the neoclassical terminology makes the task of actual observation of the technological structure of a particular economy and empirical description of processes of technological change extremely, not to say, prohibitively difficult.

Since direct observation of a set of isoquants is hardly ever possible, empirical implementation of standard neoclassical models involves nearly exclusive reliance on more and more sophisticated methods of indirect statistical inference.

Neither of the two definitions of technology and technological change can be said to be more correct than the other. The employment of the simpler definition however permitted input–output analysis to advance in the direction of systematic detailed factual inquiry, while reliance on a definition, much less serviceable for purposes of empirical description but much richer in its theoretical implications, propelled neoclassical economics towards construction of elaborate theoretical models erected on a narrow, fragile data base or even on quite arbitrary, purely theoretical assumptions.

In static input–output models, additions to the stocks of building, machinery, and other kinds of productive stocks are treated as a component part of the final demand vector, entered in the right-hand side of the balance equation (6). In the following formulation of a simple dynamic model these terms are transferred to its left-hand sides and described explicitly as serving technologically determined capacity expansion required for a rise in the level of output.

$$(I - A)X_t - B(X_{t+1} - X_t) = Y_t \qquad (9)$$

B is a square matrix of technical capital coefficients each column of which consists of stock-flow ratios, describing the stocks of products of different industries which the sector in question must have on hand per unit of its capacity output.

If the time unit in terms of which the process is observed and described is relatively long, say, covering a five or even ten year period, the stocks might be engaged in production in the same time period during which they have been produced. In this case, the second term on the left-hand side would be $B(X_t - X_{t-1})$. Current inputs required for maintenance of the existing capital stock have of course to be accounted for by the appropriate elements of the A matrix.

While bringing to the fore the crucial role that a complete set of capital coefficients has to play – in addition to a complete set of current input coefficients – in the detailed description of the structural framework of a given economy, such a set of difference equations is too rigid a tool to be used to describe and project the actual process of economic development and change.

More effective, because more flexible, is an approach which takes the form of a step-by-step construction of complete input–output tables of the economy for successive periods of time, each based on the knowledge of its state in the previous period, of anticipated changes in the final bill of goods and expected technological changes.

In more general terms, the input–output relationship between goods produced and consumed over a sequence of successive years can be formally described exactly in the same terms as relationships between different sectors are presented in an ordinary 'static' input–output table for a single year. The solution of a time-phased system of linear equations describing the intertemporal balances of inputs and outputs of goods and services produced and consumed over a long stretch of successive periods of time can be interpreted as inversion of a large triangular matrix; triangular because outputs of one year can become inputs in later years, but not vice versa. The results of this operation describing the direct and indirect relationships between all appropriately timed inputs and outputs has been called the 'dynamic inverse'. Since the sets of flow and capital coefficients controlling the input–output balances in successive stretches of such an historical process do not have to remain the same, both that dynamic matrix and its inverse can accurately represent all kinds of structural change, including elimination of old and introduction of entirely new goods.

Introduction of capital coefficients permits subdivision of the value-added term, V, on the right-hand side of the dual system (8) into its two parts – the returns on capital and wage income:

$$(I - A')P = \lambda B'P + 1w \qquad (10)$$

or, solving for P:

$$(I - A')P - \lambda B'P = 1w$$

λ represents the rate of return on invested capital and w, the wage rate. These equations can be used for calculating the 'trade-off curve' between real wages (i.e. money wage rate divided by a price index) and the rate of return on capital for any given state of technology. Comparison of such curves, each reflecting a different combination of alternative technologies available in different sectors, provides a base for numerical assessment of the influence of the distribution of income between the return on capital and wages upon technological choice.

Practical concerns led quite early to construction of regional input–output tables. The municipal government of the city of Stockholm was the first to compile a detailed metropolitan table. The complex fact-finding task of putting together a detailed input–output map of a particular region seemed to have been inspired sometimes by the desire to assert distinct identity. In Canada, French-speaking economists were the first to construct a regional table, that of Quebec. In Belgium one was compiled for the autonomy-seeking Flemish provinces. In addition to pressing needs of developmental planning, similar considerations seem to have prompted early compilation of input–output tables of many less developed countries.

The next step was construction of multiregional input–output tables and models in which intraregional transactions were linked with each other by interregional flows of goods and services. While comparison of labour, capital and natural resource 'contents' was the object of some of the earliest input–output studies of domestic and internationally traded goods, neither the theoretical formulation nor the available data base are yet sufficiently advanced to permit input–output modelling of international economic transactional trade to be solidly based on direct empirical implementation of the comparative cost theory. In most multiregional input–output models the structure of international transactions is controlled by sets of empirically determined export and import coefficients. A large multiregional input–output model of the world economy constructed under the auspices of the United Nations was published in 1977. Originally intended to provide a basis for a set of alternative projections of the future growth of eight groups of developed and seven groups of less developed countries, this large, highly disaggregated model was used in a series of other studies such as the analysis of economic effects of international arms trade, detailed long-run projections of the production and consumption of non-ferrous metals in the United States and construction of alternative multiregional scenarios of future exploration of agricultural and energy resources.

As the range of its practical applications widened, the scope of input–output modelling had to be broadened, along with the contents of the requisite data bases.

Analysis of the petroleum refining industry in the early Fifties required modelling of multiproduct processes. Thirty years later a similar approach was employed to describe within the framework of a national input–output table the generation and elimination of various polluting substances. Modelling devices adapted in description of the allocation of the output of transportation and trade sectors have later on been adapted in modelling the activities of all service industries. Separation of the description of the physical from the price and costing aspects of government operations proved to be useful in construction and theoretical interpretation of input–output tables of simple, not yet fully monetized economies of the less developed economies. Richard Stone offered the conceptual framework of input-output analysis for the formal description of demographic processes.

To the extent to which it can provide a bridge between aggregate analysis and detailed description of production and consumption of specific goods and services input–output analysis has been incorporated into most of the well-known forecasting econometric models.

The general nature of the approach has made the development of input–output analysis a cumulative process. Each refinement in theoretical structure and each addition to or improvement in the accuracy of factual information incorporated in its data base potentially improved the performance of the general model in application to all special problems.

WASSILY LEONTIEF

See also HAWKINS–SIMON CONDITIONS; LEONTIEF PARADOX.

BIBLIOGRAPHY

Brody, A. 1970. *Proportions, Prices and Planning: A Mathematical Restatement of the Labor Theory of Value*. Amsterdam: North-Holland.

Brody, A. and Carter, A.P. (eds) 1970. *Applications of Input-Output Analysis*. Proceedings of the Fourth International Conference on Input-Output Techniques, Geneva, 8–12 January, 1968, Vol. 2, Amsterdam: North-Holland.

Brody, A. and Carter, A.P. (eds) 1972. *Input-Output Techniques*. Proceedings of the Fifth International Conference on Input-Output Techniques, Geneva, January, 1971. Amsterdam: North-Holland.

Bulmer-Thomas, V. 1982. *Input-Output Analysis in Developing Countries: Sources, Methods and Applications*. New York: John Wiley & Sons.

Carter, A.P. 1970. *Structural Change in the American Economy*. Cambridge, Mass.: Harvard University Press.

Leontief, W. 1941. *The Structure of American Economy, 1919–1939: An Empirical Application of Equilibrium Analysis*. 2nd edn, enlarged, White Plains, NY: International Arts and Sciences Press, 1951.

Leontief, W., et al. 1953. *Studies in the Structure of the American Economy: Theoretical and Empirical Explorations in Input-Output Analysis*. White Plains, NY: International Arts and Sciences Press.

Leontief, W. 1966. *Input-Output Economics*. 2nd edn, New York: Oxford University Press, 1986.

Leontief, W., Carter, A.P. and Petri, P.A. 1977. *The Future of the World Economy*. A United Nations Study. New York: Oxford University Press.

Leontief, W. and Duchin, F. 1985. *The Future Impact of Automation on Workers*. New York: Oxford University Press.

Meyer, U. 1980. *Dynamische Input-Output-Modelle*. Königstein: Athenäum Ökonomie Verlag.

Miller, R.E. and Blair, P.D. 1985. *Input-Output Analysis: Foundations and Extensions*. Englewood Cliffs, NJ: Prentice-Hall.

Polenske, R. and Skolka, Jiři V. (eds) 1976. *Advances in Input-Output Analysis*. Proceedings of the Sixth International Conference on Input-Output Techniques, Vienna, 22–26 April, 1974, Cambridge, Mass.: Ballinger.

Schumann, J. 1968. *Input-Output-Analyse*. Berlin and Heidelberg: Springer-Verlag.

Smyshlyaev, A. (ed.) 1983. *Proceedings of the Fourth IIASA Task Force Meeting On Input-Output Modeling 29 September-1 October 1983*. Laxenburg, Austria: International Institute for Applied Systems Analysis.

institutional economics. Apart from Marxism, which has also the character of a social movement, institutional economics has been the principal school of heterodox thought in economics. Originating in and still concentrated largely, but by no means exclusively, within the United States, institutionalism has served the dual functions of providing critiques of mainstream neoclassical (and Marxian) economics and producing an alternative conception of the economy and of doing economic research and analysis. In so doing, it has represented in part a continuation of the German and English historical traditions, including Max Weber, as well as other writers, such as John Hobson.

The precise relationship of heterodox institutional economics to orthodox neoclassical economics is complicated by several considerations: the awkward sociological status of heterodoxy within the discipline; the ambivalence within institutionalism as to the relationship, some institutionalists feeling that the two are complementary and others that the two are mutually exclusive; and the presence within institutionalism of two different and to some extent conflicting traditions, one emanating from Thorstein Veblen and continuing through Clarence Ayres, the other starting with John R. Commons. The Veblen–Ayres tradition focuses on the progressive role of technology and the inhibitive role of institutions; the Commons tradition is less enamoured of the imperatives of technology and approaches institutions, as modes of collective action, more neutrally; both accept that actual economic performance is a function, *inter alia*, of both technology and institutions. Notwithstanding their differences, there is a common core of institutional analysis of perhaps no greater variety of formulation than within neoclassicism or Marxism.

In contrast with mainstream economics, which maintains that the central economic problems are the allocation of resources, the distribution of income, and the determination of the levels of income, output and prices, institutional economists assert the primacy of the problem of the organization and control of the economic system, that is, its structure of power. Thus, whereas orthodox economists tend strongly to identify the economy solely with the market, institutional economists argue that the market is itself an institution, comprised of a host of subsidiary institutions, and interactive with other institutional complexes in society. In short, the economy is more than the market mechanism: it includes the institutions which form, structure, and operate through, or channel the operation of, the market. The fundamental institutionalist position is that it is not the market but the organizational structure of the larger economy which effectively allocates resources.

To the extent, then, that institutional and neoclassical economists study the same questions, for example, resource allocation, the institutionalists generally encompass a broader or deeper set of explanatory variables: instead of having price, and resource allocation, be a function of demand and supply in a purely conceptual market, these latter are in turn related to the structure of power (wealth, institutions) which help form them. Power structure in turn is related to legal rights, thence to the use of government in forming legal rights of economic significance and thereby influencing the allocation of resources, level of income, and distribution of wealth.

But institutionalists are generally less concerned with price and resource allocation *per se* and more with the problem of the organization and control of the economy, that is, with performance seen as specific to power (rights) structure, as well as to technology. Institutionalists thus are interested, for example, in the formation and role of institutions, and the interrelations between economic and legal systems and between power and belief system.

If institutionalists insist that the economy comprises more than the market mechanism, they also object to the equilibrium and presumptive optimality modes of analysis of neoclassical economics. The search for the deterministic technical conditions of stable equilibrium, it is felt, obscures the fundamental power and choice aspects of the economy. The search for optimality, or for optimal solutions, it is also felt, is either formally empty or can be given substance only by the introduction, typically implicitly, of antecedent normative assumptions as to whose interests count, whereas in the real world such questions have to be worked out both within institutions and through contests over institutional adjustment and reformation.

The central features of institutional thought are its holism and evolutionism. Thus the further principal themes of institutional economics include the following:

(1) A theory of social change, an activist orientation towards social institutions, through focusing on both the substantive impact of institutions on economic performance and the processes of institutional change, treating institutions not as something to be taken as given but as man-made and changeable, both deliberatively and nondeliberatively.

(2) A theory of social control and collective choice, or a theory of institutions, a focus on the formation and operation of institutions as both cause and consequence of the power structure and societized behaviour of individuals and subgroups, and as the mode through which economies are organized and controlled. Instead of focusing on the mechanics of choice from within opportunity sets, a focus on the formation of opportunity sets; instead of a focus on unfettered market freedom, a focus on the total, complex pattern of freedom and control, that is, on the formation and

operation of the system of control through which both actual opportunity sets and freedom are formed.

(3) A theory of the economic role of government, as a principal social process through which both itself and other institutions of economic significance are in part formed and revised. Instead of treating government, law, and the system of rights as either given and/or exogenous, these are treated as both dependent and independent, and always critical, not merely aberrational, economic variables.

(4) A theory of technology, as defining and determining the relative scarcity of all resources, as a principal force in the evolution of economic structure (including the operation of institutions) and performance, and as the basis of the logic of industrialization marking the mentality as well as the practices of modern economies.

(5) The fundamental principle that the real determinant of resource allocation is not the market but the organizational – institutional, power – structure of society.

(6) An emphasis on facets of the value conception which transcend price, on the values represented in and given effect by the habits and customs of social life, on the pragmatic, instrumental values ensconced in the transcendental notion of the life process of man and society, and on the constructive values latent within and given effect by the working rules of law which are both the foundation and the product of the power structure of society. Included are attempts to understand the process by which values are changed, in contrast to the orthodox assumption of given values; that is, to consider within economics such questions as where the values come from, how they are tested, and how they are changed.

In amplification of these themes one finds, for example, Veblen's emphasis on status emulation as a principal force in the formation of economic behaviour, including (through conspicuous consumption and the making of invidious comparisons) the formation of consumer demands; Commons's analysis of the evolution of the fundamental legal foundations of the modern economy; John Dewey's theory of instrumental logic and social value; John Maurice Clark's analysis of the social control of business; Wesley Mitchell's emphasis on the economy as a pecuniary phenomenon; Commons's and Selig Perlman's analyses of labour unions as a mode of representing worker interests and of generating institutional change; Edwin E. Witte's, and Commons's, efforts at creating new institutions for the embodiment and protection of rising interests and for the creative resolution of social conflict and the development of a body of analysis of institutional genesis and adjustment; and, *inter alia*, Veblen's and Ayres's analyses of the formation of the human belief system, including that of economists, under the impact of the contest between traditional and new ways of doing things.

Apropos of the last point, institutionalists have freely pointed to the selectivity and typically implicit nature of the operative assumptions of neoclassical analysis. They insist that by its taking institutional or power structure as given or, more typically, by its selective specification of institutions and power structure, there is a strong tendency towards selective apologetics in orthodox economics, especially in that work which is directed to the identification of 'optimal' solutions. The institutionalist solution to such problems is that of Gunnar Myrdal: to avoid the pretence of value-free economics by making all, or substantially all and certainly the operative, value premises explicit and by generating appraisals thereof.

Accordingly, institutional economists have tended to avoid recourse to methodological individualism and to abstain from puzzle-solving research in the context of models devoid of institutional embodiment and stressing equilibrium, optimality, and purely competitive markets. They have rather attended to theoretical and empirical analyses of real-world problems, such as the operation of particular institutions, business–government relations, and the conditions of economic development. Insofar as they have dealt with economic variables at fundamental conceptual levels, such as government and rights, they have at least tried to do so in both analytically credible and nonpresumptive ways.

The best-known contemporary version of the institutionalist conception of the economy has been that of John Kenneth Galbraith. Following the course laid down by Veblen, and grafting it on to a version of Keynesian economics, Galbraith has explored the corporate nature and planning modes of the business system and the impact of what he considers to be technological imperatives, the social formation of individual preferences underlying demand functions, the power and continuous interaction of the state and the corporate core of the economy, and the factors and forces which influence the formation of opinion and policy in the public sector.

In such fields as labour economics, industrial organization, economic development, law and economics, agricultural and natural resource economics, and macroeconomics, institutionalists, through their primary attention to power structure and belief system, in the context of their overriding concerns with social change and social control, have produced understandings or pictures of economic reality quite different from those of neoclassical economists. These contributions have come through the recent work, in addition to Galbraith, of John Adams, Jack Barbash, Kenneth E. Boulding, Daniel R. Fusfeld, Wendell C. Gordon, Allan G. Gruchy, David B. Hamilton, Gardiner C. Means, Walter C. Neale, Kenneth Parsons, Wallace Peterson, A. Allan Schmid, Robert Solo, Paul Strassmann, Marc Tool and Harry M. Trebing, among others. Some of this work appears in the *Journal of Economic Issues*, published by the Association for Evolutionary Economics. Altogether this work has constituted an alternative analysis of the economic system, especially of capitalism but also of socialism, and a critique of both existing economic systems and orthodox schools of economics.

WARREN J. SAMUELS

BIBLIOGRAPHY

Canterbery, E.R. et al. 1984. Galbraith Symposium. *Journal of Post-Keynesian Economics*, Autumn.

Dorfman, J. et al. 1963. *Institutional Economics*. Berkeley: University of California Press.

Gruchy, A.G. 1947. *Modern Economic Thought*. New York: Prentice-Hall.

Gruchy, A.G. 1972. *Contemporary Economic Thought*. Clifton, NJ: Kelley.

Sharpe, M.E. 1974. *John Kenneth Galbraith and the Lower Economics*. 2nd edn, White Plains, NY: International Arts and Sciences Press.

Thompson, C.C. (ed.) 1967. *Institutional Adjustment*. Austin: University of Texas Press.

Tool, M. 1979. *The Discretionary Economy*. Santa Monica: Goodyear.

instrumental variables. In one of its simplest formulations the problem of estimating the parameters of a system of simultaneous equations with unknown random errors reduces to finding a way of estimating the parameters of a single linear equation of the form $Y = X\beta_0 + \epsilon$, where β_0 is unknown, Y and X are vectors of data on relevant economic variables and ϵ is the vector of unknown random errors. The most common method of estimating β_0 is the method of least squares:

$\hat{\beta}_{0LS} \equiv \text{argmin } \epsilon(\beta)'\epsilon(\beta)$, where $\epsilon(\beta) \equiv Y - X\beta$. Under fairly general assumptions $\hat{\beta}_{0LS}$ is an unbiased estimator of β_0 provided $E(\epsilon_t | X) = 0$ for all t, where ϵ_t is the tth-coordinate of ϵ.

Unfortunately for the empirical economist, it is often the case that the basic orthogonality condition between the errors and the explanatory variables is not satisfied by economic models, due to correlation between the errors and the explanatory variables. Particularly relevant examples of this situation include (1) any case where the data contain errors introduced by the process of collection (errors in variables problem); (2) the inclusion of a dependent variable of one equation in a system of simultaneous equations as an explanatory variable in another equation in the system (simultaneous equations bias); and (3) the inclusion of a lagged dependent variable as an explanatory variable in the presence of serial correlation. For all of these cases,

$$E(\hat{\beta}_{0LS}) = E[(X'X)^{-1}X'Y]$$
$$= \beta_0 + E\left[(X'X)^{-1}\sum_{t=1}^{n} X_t' E(\epsilon_t | X)\right] \neq \beta_0$$

in general, and the bias introduced cannot be determined because the errors ϵ are unknown. Furthermore, in every case the bias fails to go to zero as the sample size increases. Clearly the method of least squares is unsatisfactory for many situations of relevance to economists.

In 1925 the US Department of Agricultural published a study by the zoologist Sewall Wright where the parameters of a system of six equations in thirteen unknown variables were estimated using a method he referred to as 'path analysis'. In essence his approach exploited zero correlations between variables within his system of equations to construct a sufficient number of equations to estimate the unknown parameters. The idea which underlies this approach is that if two variables are uncorrelated, then the average of the product of repeated observations of these variables will approach zero as the number of observations is increased without bound except for a negligible number of times. Thus if we know that a variable of the system Z_i is uncorrelated with the errors ϵ, we can exploit the fact that $n^{-1}\sum_{t=1}^{n} Z_{ti}\epsilon_t \equiv n^{-1}\sum_{t=1}^{n} Z_{ti}(Y_t - X_t\beta_0)$ approaches zero to construct a useful relationship between parameters of the system by setting such averages equal to zero. Provided a sufficient number of such relationships can be constructed which are independent, this provides a method for estimating the parameters of a system of simultaneous equations which should become more accurate as the number of observations increases.

Since the 1940s, when Reiersøl (1941, 1945) and Geary (1949) presented the formal development of this procedure, the variables Z which are instrumental in the estimation of the parameters β_0 have been called 'instrumental variables'. Associated with each instrumental variable Z_i is an equation formed as described in the previous paragraph, called a normal equation, which can be used to form the estimates of the unknown parameters. Frequently there are more instrumental variables than parameters to be estimated. As the equations are formed from relationships between random variables, generally no solution will exist to a system of estimating equations formed in this manner using all possible instrumental variables. As each estimating equation contains relevant information about the parameters to be estimated, it is undesirable just to ignore some of them. Thus we can define a fundamental problem in the application of this method: how can we make effective use of all the information available from the instrumental variables? This problem will occupy the rest of this entry.

Let $\epsilon_t(\theta) \equiv F_t(X_t, Y_t, \theta)$ be a $p \times 1$ vector-valued function defined on a domain of possible parameter values $\Theta \subseteq \mathbb{R}^k$ which represents a system of p simultaneous equations with dependent variables Y_t, a $p \times 1$ random vector, and an $m \times s$ random matrix of explanatory variables X_t for all $t = 1, 2, \ldots, n$. Standard formulations of $F_t(X_t, Y_t, \theta)$ are the linear model $\epsilon_t(\theta) \equiv Y_t - X_t\theta$ and the nonlinear model $\epsilon_t(\theta) \equiv Y_t - f_t(X_t, \theta)$. Let $W_t(\theta)$ be a $p \times r$ random valued matrix defined on Θ for all $t = 1, 2, \ldots, n$. Assume that there exists a unique value θ_0 in Θ such that $E(\epsilon_t^0|W_t^0) = 0$ for all $t = 1, 2, \ldots, n$, where $\epsilon_t^0 \equiv F_t(X_t, Y_t, \theta_0)$ and $W_t^0 \equiv W_t(\theta_0)$. Finally, let $Z_t(\theta)$ be a $p \times 1$ random matrix such that $E(Z_t(\theta)|W_t(\theta)) = Z_t(\theta)$ for all θ in Θ. Any such variables $Z_t(\theta_0)$ may serve as instrumental variables for the estimation of the unknown parameters θ_0 since

$$E(Z_t^{0\prime}\epsilon_t^0) = E(E(Z_t^{0\prime}\epsilon_t|W_t^0)) = E(Z_t^0 E(\epsilon_t^0|W_t^0)) = 0$$

for all $t = 1, 2, \ldots, n$, as long as the functions F_t and W_t and the data generating process satisfy sufficiently strong regularity assumptions to ensure that the uniform law of large numbers is satisfied, i.e.

$$n^{-1}\sum_{t=1}^{n} Z_t(\theta)'\epsilon_t(\theta) \xrightarrow{P} n^{-1}\sum_{t=1}^{n} E[Z_t(\theta)'\epsilon_t(\theta)]$$

uniformly in θ on Θ.

Identification of the unknown parameters θ_0 requires that there be at least as many instrumental variables as there are parameters to be estimated, i.e. $l \geqslant k$. On the other hand, if there are more instrumental variables than parameters to be estimated, there will be no solution to $n^{-1}\sum_{t=1}^{n} Z_t(\theta)'\epsilon_t(\theta) = 0$ in general for finite n as indicated above. One possible solution to this problem is simply to use k of the instrumental variables in the estimation of θ_0. The omitted instrumental variables may then be used to construct statistical tests of the $l - k$ overidentifying restrictions of the unknown parameter vector. A drawback of this approach is that not all of the information available to us is used in the estimation of the unknown parameters and hence, the estimates will not be as precise as they should be. An alternative approach which effectively uses all of the available instrumental variables is to be preferred.

Even though in general the moment function $n^{-1}\sum_{t=1}^{n} Z_t'(\theta)\epsilon_t(\theta) \neq 0$ for any value of θ, its limiting function $n^{-1}\sum_{t=1}^{n} E[Z_t'(\theta)\epsilon_t(\theta)]$ does vanish when $\theta = \theta_0$. This suggests estimating θ_0 with that value of Θ which makes $n^{-1}\sum_{t=1}^{n} Z_t'(\theta)\epsilon_t(\theta)$ as close to zero as possible. The criterion of closeness is of some interest to the econometrician. It affects the size of the confidence ellipsoids of the estimator about θ_0 and hence the precision of the estimate. The Nonlinear Instrumental Variables Estimator (NLIV), $\hat{\theta}_{n,\text{NLIV}} = \text{argmin}_{\theta \in \Theta}$

$$\left[\sum_{t=1}^{n} Z_t'(\theta)\epsilon_t(\theta)\right]' \cdot \left[\text{Var}\sum_{t=1}^{n} Z_t'(\theta_0)\epsilon_t(\theta_0)\right]^{-1} \cdot \left[\sum_{t=1}^{n} Z_t'(\theta)\epsilon_t(\theta)\right]$$

is the optimal instrumental variables estimator in this respect (Bates and White, 1986a).

The NLIV estimator simplifies to well-known econometric estimators in a variety of alternative specifications of the underlying probability model which generated the variables. When the data generating process is independent and identically distributed, $\hat{\theta}_{n,\text{NLIV}}$ is the Nonlinear Three-stage Least Squares estimator of Jorgenson and Laffont (1974). The additional restriction of consideration to a single equation ($p = 1$) results in the Nonlinear Two-stage Least Squares estimator of Amemiya (1974). Furthermore, if the model $\epsilon(\theta)$ is linear in θ, $\hat{\theta}_{n,\text{NLIV}}$ then simplifies to the Three-stage Least Squares estimator of Zellner and Theil (1962) for a system of simultaneous equations and to the Two-stage Least Squares Estimator of

Theil (1953), Basmann (1957) and Sargan (1958) for the estimation of the parameters of a single equation. On the other hand, if we allow for heterogeneity by restricting the data generating process only to be independent, $\hat{\theta}_{n,\text{NLIV}}$ simplifies to White's (1982) Two-stage Instrumental Variables estimator of the parameters of a single linear equation.

As indicated above, it is desirable from consideration of asymptotic precision to include as many instrumental variables as are available for the estimation of the unknown parameters θ_0. This raises the question of the existence of a set of instrumental variables $\{Z^*\} \in \Gamma$ that renders the inclusion of any further instrumental variables redundant, where Γ is the set of all sequences of instrumental variables such that $\hat{\theta}_{n,\text{NLIV}}$ is a consistent estimator of θ_0 with an asymptotic covariance matrix. Bates and White (1986b) provide conditions which imply that such instrumental variables exist, though it may not be possible to obtain them in practice. Suppose there exists a sequence of k instrumental variables $\{\mathscr{Z}\}$ such that for all $\{Z\}$ in Γ

$$E[\mathscr{Z}(\theta_0)'\nabla_\theta\epsilon(\theta_0)] = E[\mathscr{Z}(\theta_0)'\epsilon(\theta_0)\epsilon(\theta_0)'Z(\theta_0)].$$

Then $\{\mathscr{Z}(\theta_0)\}$ is optimal in Γ in the sense of asymptotic precision. Suppose it is also the case that Σ an $np \times np$ matrix with representative element $\sigma_{th\tau g} \equiv E(\epsilon_{th}(\theta_0) \cdot \epsilon_{\tau g}(\theta_0)|W_{th}(\theta_0),$ $W_{\tau g}(\theta_0))$, is nonsingular a.s. and that

$$E[E(\nabla_\theta\epsilon_{th}(\theta_0)|W_{th}(\theta_0))|W_{\tau g}] = E(\nabla_\theta\epsilon_{th}(\theta_0)|W_{th}(\theta_0))$$

for all $t, \tau = 1, 2, \ldots, n$ and $h, g = 1, 2, \ldots, p$ such that $\sigma^{th\tau g} \neq 0$, where $\sigma^{th\tau g}$ is a representative element of Σ^{-1}. Let Z^* be an $np \times k$ matrix with rows

$$Z_{th}^* \equiv \sum_{\tau=1}^{n} \sum_{g=1}^{p} \sigma^{th\tau g} E[\nabla_\theta\epsilon_{\tau g}(\theta_0)|W_{\tau g}(\theta_0)].$$

If $\{Z^*\}$ is in Γ then $\{Z^*\}$ is optimal in Γ.

In many situations it will not be possible to make use of such instrumental variables in practice. However, for some important situations optimal instrumental variables are available. Suppose that $\epsilon(\theta) \equiv Y - X\theta$ and the explanatory variables X are independent of the errors $\epsilon(\theta_0)$. If the errors are independent and identically distributed for all $t = 1, 2, \ldots, n$ and $h = 1, 2, \ldots, p$, then $Z^* = X$. Thus the optimal instrumental variables estimator is given by

$$\hat{\theta}_{n,\text{NLIV}} \equiv \underset{\theta \in \Theta}{\text{argmin}}\ \epsilon(\theta)'X[\sigma^2 E(X'X)]^{-1}X'\epsilon(\theta),$$

where $\sigma^2 = \text{var}[\epsilon_{th}(\theta_0)]$ is a real, nonstochastic scalar for all t and h. If it is also the case that $n^{-1}E(X'X) - n^{-1}X'X^p \to 0$ as $n \to \infty$, $\hat{\theta}_{n,\text{NLIV}}$ is asymptotically equivalent to $\text{argmin}_{\theta \in \Theta}\epsilon(\theta)'X-(X'X)^{-1}X'\epsilon(\theta)$, that is Ordinary Least Squares is the optimal instrumental variables estimator. If there is contemporaneous correlation only, i.e. $\text{var}(\epsilon_t(\theta_0)) = \Omega$, a $p \times p$ nonstochastic matrix, then Zellner's (1962) Seemingly Unrelated Regression Estimator (SURE), is the optimal instrumental variables estimator. If we further relax these assumptions so that $\text{var}(\epsilon(\theta_0))$ is an arbitrary positive definite $np \times np$ matrix, the Generalized Least Squares (Aitken, 1935) is the optimal instrumental variables estimator.

Since the development of the Two-stage Least Squares estimator in the mid-1950s, the method of instrumental variables has come to play a prominent role in the estimation of economic relationships. Without this method much of economic theory cannot be tested in any concrete way. But it is inherently a large sample estimation method based as it is on the law of large numbers and the central limit theorem. Since in general it is not possible to know how much data are required to arrive at acceptable estimates, all conclusions derived from instrumental variables estimates should be tempered with a healthy dose of scepticism.

CHARLES E. BATES

See also ERRORS IN VARIABLES; ESTIMATION; LATENT VARIABLES.

BIBLIOGRAPHY

Aitken, A.C. 1935. On least squares and linear combinations of observations. *Proceedings of the Royal Society of Edinburgh* 55, 42–8.

Amemiya, T. 1974. The nonlinear two-stage least-squares estimator. *Journal of Econometrics* 2, 105–110.

Basmann, R.L. 1957. A generalized classical method of linear estimation of coefficients in a structural equation. *Econometrica* 25, 77–83.

Bates, C.E. and White, H. 1986a. Efficient estimation of parametric models. Johns Hopkins University, Department of Political Economy, Working Paper No. 166.

Bates, C.E. and White, H. 1986b. An asymptotic theory of estimation and inference for dynamic models. Johns Hopkins University, Department of Political Economy.

Geary, R.C. 1949. Determination of linear relations between systematic parts of variables with errors in observation, the variances of which are unknown. *Econometrica* 17, 30–58.

Goldberger, A.S. 1972. Structural equation methods in the social sciences. *Econometrica* 40, 979–1001.

Hausman, J.A. 1983. Specification and estimation of simultaneous equation models. In *Handbook of Econometrics*, Amsterdam: North-Holland, ch. 7.

Jorgenson, D.W. and Laffont, J. 1974. Efficient estimation of nonlinear simultaneous equations with additive disturbances. *Annals of Economic and Social Measurement* 3(4), 615–40.

Reiersøl, O. 1941. Confluence analysis by means of lag moments and other methods of confluence analysis. *Econometrica* 9, 1–24.

Reiersøl, O. 1945. Confluence analysis by means of instrumental sets of variables, *Arkiv for Mathematik, Astronomi och Fysik*, 32A.

Sargan, J.D. 1958. The estimation of economic relationships using instrumental variables. *Econometrica* 26, 393–415.

Theil, H. 1953. Estimation and simultaneous correlation in complete equation systems. The Hague: Centraal Planbureau.

White, H. 1982. Instrumental variables regression with independent observations. *Econometrica* 50, 483–500.

White, H. 1984. *Asymptotic Theory for Econometricians*. Orlando: Academic Press.

White, H. 1985. Instrumental variables analogs of generalized least squares estimators. *Journal of Advances in Statistical Computing and Statistical Analysis* 1.

Wright, S. 1925. Corn and Hog Correlations. Washington, DC: US Department of Agriculture, Bulletin 1300.

Zellner, A. 1962. An efficient method of estimating seemingly unrelated regressions and tests for aggregation bias. *Journal of the American Statistical Association* 57, 348–68.

Zellner, A. and Theil, H. 1962. Three-stage least squares: simultaneous estimation of simultaneous equations. *Econometrica* 30, 54–78.

instruments. *See* TARGETS AND INSTRUMENTS.

insurance. Insurance is an ancient institution. It is impossible to reflect on evolutionary processes without recognizing the intrinsic role of insurance. Any species that relied on nature's harmony and regularity and ignored its stochastic whims was soon extinct. The position adopted here is that uncertainty is one of the decisive determinants of individual behaviour. 'Individual' includes not only early and modern man, but also plants and animals. Furthermore, the response to uncertainty is both adaptive and dynamic. Martingale models are ideally suited to portray these responses. The goal of the individual is to maximize expected utility and for early man this was roughly equivalent to maximizing the probability of survival.

In order to achieve this goal, he devised a variety of insurance mechanisms he established institutions that included a large element of flexibility so they could readily adjust to nature's stochastic quirks. Insurance and the ensuing flexibility were key components of his decisionmaking. Aggregating these individual responses over the entire group reveals one critical aspect of the society's culture. The individual quickly perceived that some of the most effective devices for mitigating uncertainty entailed *cooperative arrangements*.

At the same time that individuals and bands are designing mechanisms for coping with a fluctuating environment, nature is inexorably monitoring these activities and eliminating those individuals and bands who respond too slowly to adversity or are unlucky, and despite their best efforts, are overcome by misfortune. These Job-like extinctions are more likely in harsh and highly variable environments. We expect that those who survive in these environments are quite distinct from survivors where nature is relatively benevolent. The excellent study by Minnis (1985) indicates how persistent misfortune affects existing social institutions and induces significant modifications. Minnis examined three periods of environmental stress and shows the enhanced survival value of innovation and flexibility.

Insurance also manifests itself in unconscious natural selection. Lowell (1985) has shown that as the predictability of the environment decreases, the safety factory of the biological structure increases. Roughly speaking, the organism adapts to the distribution of the maximum stress, rather than to the distribution of the actual stress. In other words, the difference between the stress-capability of the organism and the actual stress encountered, measures the 'slack'. The presence of 'slack' or flexibility enhances the survivability of the organism. Thus a fairly conservative way of handling unanticipated shocks is to imitate nature and use the distributions of the extremes (maximum period of drought or minimum density of prey). Leadbetter et al. (1983) is a fine, comprehensive treatment of extreme value distributions. Flocking (Morse, 1970) is another of the almost endless variety of insurance mechanisms that evolved in response to the threat of extinction. The study of extinction is a discipline in itself. A good sampling of recent research is the book edited by Nitecki (1984) with Diamond's paper on isolated populations an outstanding contribution. Martin's piece on catastrophic extinctions is the most pertinent. Also see Mangel and Ludwig (1977) for a stochastic analysis of extinction in competitive struggles among species. The classic studies by Slobodkin (1961) and MacArthur (1955) are illuminating. An elegant model of extinction could be devised using the contact process methods in Liggett (1985).

THE BASIC ELEMENTS OF INSURANCE. We have already described the essential aspects of insurance. The remaining task is to translate them into the modern language of economics and probability. The original research on the economics of insurance was conducted by Arrow (1971) and Borch (1968). Hirshleifer and Riley (1979) contains a fine survey of insurance; also see Lippman and McCall (1982). The neatest presentation of stochastic dominance in an insurance setting is the paper by Lippman (1972) that has gone unnoticed in spite of its excellence.

Individuals in modern societies are unable to predict the time and magnitude of events that profoundly affect their well-being. Insurance, in all its guises, is the institution that mitigates the influence of uncertainty. The individual invests in a host of activities *now* to insure that the timing and magnitude of unfortunate future events will be less harmful.

These activities enable firms and individuals to trade risks among themselves. The most familiar of these transfers is the ordinary insurance contract. The essence of this contract is the payment of a fee by the insuree in exchange for the insurer's promise to pay a certain sum of money provided a stipulated event occurs.

One of the simplest insurance contracts has the following structure. The insured pays a fixed premium y to avoid the small probability p of incurring L, a large loss. For simplicity, ignore loading charges (i.e. the charge to cover the administrative costs associated with writing and overseeing the insurance contract), and set this premium equal to the actuarial value of the loss plus an amount c, the compensation to the insurer for assuming the risk:

$$y = pL + (1-p)0 + c = pL + c. \tag{1}$$

Such a policy is advantageous to both the insured and insurer. The insured possesses a concave utility function and is therefore eager to pay y to dispense with this risk. The insurance company is able to pool independent risks and via the law of large numbers converts risky contracts into almost 'sure' things.

Because of its fundamental importance we state a simple version of the law of large numbers, namely, the weak law of large numbers for a fair Bernoulli random variable. (This version applies when $L = 1$ and $p = 1/2$.) A fair coin is flipped n times. Let the random variable X_n be given by

$$X_n = \begin{cases} 1, & \text{if a head occurs on } n\text{th flip} \\ 0, & \text{if a tail occurs on the } n\text{th flip,} \end{cases}$$

and S_n is defined by

$$S_n = \sum_{i=1}^{n} X_i.$$

The proportion of heads in the first n trials is simply S_n/n.

At first one might think that the probability of exactly n heads in $2n$ trials should be high because $p = 1/2$, but

$$P(S_{2n} = n) = \binom{2n}{n} 2^{-2n} \simeq 1/\sqrt{\pi n}. \tag{2}$$

Thus, the probability of exactly n heads in $2n$ trials goes to zero as n gets large.

The weak law of large numbers states that the probability of S_n/n deviating from $\frac{1}{2}$ by any fixed positive quantity goes to zero as n goes to infinity. More precisely,

$$\lim_{n \to \infty} P(|S_n/n - \tfrac{1}{2}| > \epsilon) = 0, \quad \text{for any } \epsilon > 0. \tag{3}$$

Proof: Letting $\hat{\Sigma}$ designate the sum over the set of integers k such that $|k/n - \tfrac{1}{2}| > \epsilon$, we have

$$P(|S_n/n - \tfrac{1}{2}| > \epsilon) < \hat{\Sigma}[(k/n - \tfrac{1}{2})/\epsilon]^2 \binom{n}{k} 2^{-n}$$

$$\leq \epsilon^{-2} \sum_{k=0}^{n} (k/n - \tfrac{1}{2})^2 \binom{n}{k} 2^{-n}$$

$$= \epsilon^{-2} \mathrm{Var}(S_n/n) = p(1-p)/n\epsilon^2 \leq 1/4n\epsilon^2,$$

where the first inequality follows from the fact that $(k/n - \tfrac{1}{2})^2 > \epsilon^2$, the second inequality from the fact that additional non-negative terms are being summed, and the final inequality from the fact that $x(1-x)$ is maximized at $x = \tfrac{1}{2}$. Q.E.D.

Note that the law of large numbers does not say that if you and I play a game of chance with a fair coin, I will lead approximately half the time. In fact, if I win the game, I am likely to have led for most of the game.

By this law S_n/n converges in probability to $\tfrac{1}{2}$. Thus, an insurance company has considerable control over its average loss.

When contracting (risk transfer) transpires in an uncertain environment, two basic problems present themselves: *moral hazard* and *adverse selection*. Both are founded on imperfect information. The prototypical example of these problems is the insurance contract between an insurance company (the principal) and the insured (the agent). By paying a premium the agent transfers the risk associated with a particular activity to the principal. This risk transfer affects the incentives and behaviour of the agent. It is these incentive effects that are commonly referred to as the moral hazard problem. It has its roots in the inability of the principal to costlessly observe the actions of the agent. Hence when the untoward event occurs, the principal is not sure whether it was caused by the agent's carelessness or by chance. Moral hazard can be reduced by requiring the agent to bear some of the costs of the contingency and/or by monitoring the agent's behaviour. Adverse selection is similar to moral hazard in that the problem arises because the principal does not have costless access to information possessed by agents and vice versa. For example, some purchasers of health insurance have more information about their health status than insurance companies. Because the insurance company cannot discriminate perfectly between healthy and sickly agents, the latter will pose as healthy agents and be 'adversely selected' (insured) by the principal. Insurance companies can and do cope with these informational asymmetries by (a) experience rating, that is, continually adjusting premiums to reflect the size and incidence of each agent's claims and (b) designing policies that elicit the information necessary for partitioning agents into distinct categories.

These problems have received an enormous amount of study by economic theorists. Much of the analysis is static and, of course, gives rise to many perplexes. When properly formulated as dynamic stochastic control problems the perplexities diminish and the solutions accord with those that have been used for centuries by business firms.

ALTERNATIVE INSURANCE MECHANISMS. The insurance contract is only one of a multitude of devices that have been created for coping with the risks that afflict any economic system. These risks include not only fire, theft, sickness, and death but also fluctuating prices, equipment malfunctions, zero inventory levels causing unsatisfied demands, and failure of basic research ranging from falsely 'proved' theorems to unisolated viruses. The existence of futures contracts permits the farmer or food processor to specialize in production, while the speculator specializes in risk-bearing. The risks of equipment failure can be reduced by improved design and maintenance procedures like redundancy and frequent inspection. The probability of an unfulfilled demand can be diminished by maintenance of larger inventories. The costs of research failure are frequently insured against by initiation of a large number of relatively independent projects (self-insurance), or, where the costs are large and uncertain, by adoption of inefficient contractual procedures like the cost-plus, fixed-fee contract (government insurance).

The basic institution for shifting the risks of business from entrepreneurs to the general public is the securities market. Individuals can diversify their portfolio of stocks to achieve an acceptable level of expected return for a given level of risk. This ability of individuals to spread risks thereby permits firms to engage in projects which otherwise would be unacceptable. Consequently, society is better off.

These insurance arrangements are, however, far from ideal. It is usually impossible for a firm to transfer *only* rights to the outcomes of its highly risky ventures. In contrast with the futures market, the stock market is usually incapable of separating production and risk, leaving the former to the entrepreneur and transferring the latter to the general public. Instead, the stock certificate is a relatively blunt instrument for disentangling risk and production. The fact that society has not created a sharper instrument attests to the refractory nature of this problem.

THE PROBABILITY OF BANKRUPTCY. The goal of many firms and indeed of many organisms is to maximize the probability of survival.

Consider the following simple version of the bankruptcy problem. An entrepreneur begins with wealth $W_o \equiv 0$ and acquires \$1 with probability p and loses \$1 with probability $1 - p$. If he reaches \$$-b$ before he reaches a wealth of \$$a$, he is bankrupt; whereas if he reaches a before $-b$, he can issue stock and essentially reduce his ruin probability to zero. Let z be the probability of hitting a before $-b$, and Y_n the position of the entrepreneur after n periods (trials). Then

$$X_n = \left(\frac{1-p}{p}\right)^{Y_n}$$

is a martingale with $E(X_n) = 1$, all n. Hence, the probability of survival, z, is the solution to:

$$z\left(\frac{1-p}{p}\right)^a + (1-z)\left(\frac{1-p}{p}\right)^{-b} = 1,$$

which implies that

$$z = \frac{b}{a+b}.$$

When survival is the criterion function in a finite horizon problem, it is easy to show that the optimal policy is conservative (timid) for the first few trials (the entrepreneur has control over the size of the bet at each trial), but becomes bold as the horizon approaches and success has not yet been achieved. Using standard terminology, the stochastic behaviour of the entrepreneur would switch from risk aversion to risk preference at some critical point. This problem is solved in Lippman and McCall (1980) and Houston and McNamara (1985).

J.J. McCall

See also ADVERSE SELECTION; LIFE INSURANCE; MORAL HAZARD; RISK; UNCERTAINTY.

BIBLIOGRAPHY

Arrow, K.J. 1964. The role of securities in the optimal allocation of risk-bearing. *Review of Economic Studies* 31, April, 91–6.

Arrow, K.J. 1971. *Essays in the Theory of Risk-Bearing*. Chicago: Markham.

Borch, K.H. 1968. *The Economics of Uncertainty*. Princeton: Princeton University Press.

Bühlmann, H. 1970. *Mathematical Methods in Risk Theory*. Berlin, Heidelberg and New York: Springer-Verlag.

Dubins, L. and Savage, L.J. 1965. *How to Gamble If You Must*. New York: McGraw-Hill.

Ehrlich, I. and Becker, G.S. 1972. Market insurance, self-insurance, and self-protection. *Journal of Political Economy* 80(4), July–August, 623–48.

Hirshleifer, J. and Riley, J.G. 1979. The analytics of uncertainty and information – an expository survey. *Journal of Economic Literature* 17(4), December, 1375–421.

Houston, A. and McNamara, J. 1985. The choice of prey types that minimizes the probability of starvation. *Behavioral Ecology and Sociology* 17, 135–41.

Leadbetter, M.R., Lindgren, G. and Rootzen, H. 1983. *Extremes and Related Properties of Random Sequences and Processes*. New York: Springer-Verlag.

Liggett, T.M. 1985. *Interacting Particle Systems*. New York: Springer-Verlag.

Lippman, S.A. 1972. Optimal insurance. *Journal of Financial Quantitative Analysis* 7, 2151–5.

Lippman, S.A. and McCall, J.J. 1980. Constant absolute risk aversion, bankruptcy, and wealth-dependent decisions. *Journal of Business* 53, 285–96.

Lippman, S.A. and McCall, J.J. 1982. The economics of uncertainty: selected topics and probabilistic methods. In *Handbook of Mathematical Economics*, Vol. 1, ed. K.J. Arrow and M. Intriligator, Amsterdam and New York: North-Holland.

Lowell, R.B. 1985. Selection for increased safety factors of biological structures as environmental unpredictability increases. 1985. *Science* 228, 1009–11.

MacArthur, R.H. 1972. *Geographical Ecology*. Princeton: Princeton University Press. Reprinted, 1984.

Mangel, M. and Ludwig, D. 1977. Probability of extinction in a stochastic competition. *SIAM Journal of Applied Mathematics* 33, 256–67.

Minnis, P.E. 1985. *Social Adaptation to Food Stress*. Chicago: University of Chicago Press.

Morse, D.H. 1970. Ecological aspects of some mixed species foraging flocks of birds. *Ecological Monographs* 40, 119–68.

Nitecki, M.H. 1984. *Extinctions*. Chicago: University of Chicago Press.

Pauly, M.V. 1968. The economics of moral hazard: comment. *American Economic Review* 58, June, 531–6.

Pratt, J.W. 1964. Risk aversion in the small and in the large. *Econometrica* 32, 122–36.

Radner, R. 1981. Monitoring cooperative agreements in a repeated principal–agent relationship. *Econometrica* 49, 1127–48.

Riley, J.G. 1975. Competitive signalling. *Journal of Economic Theory* 10(2), April, 174–86.

Riley, J.G. 1979. Information equilibrium. *Econometrica* 47(2), March, 331–59.

Rothschild, M. and Stiglitz, J.E. 1976. Equilibrium in competitive insurance markets: an essay on the economics of imperfect information. *Quarterly Journal of Economics*, 90(4), November, 629–49.

Rubinstein, A. and Yaari, M.E. 1983. Repeated insurance contracts and moral hazard. *Journal of Economic Theory* 30, 74–97.

Seal, H.L. 1969. *Stochastic Theory of a Risk Business*. New York: Wiley.

Slobodkin, L.B. 1961. *Growth and Regulation of Animal Populations*. New York: Holt, Rinehart & Winston.

Spence, A.M. 1974. *Market Signalling: Informational Transfer in Hiring and Related Processes*. Cambridge, Mass.: Harvard University Press.

Zeckhauser, R. 1970. Medical insurance: a case study of the tradeoff between risk spreading and appropriate incentives. *Journal of Economic Theory* 2(1), March, 10–26.

integer programming. Integer programming is the youngest branch of mathematical programming: its development started in the late 1950s. A (linear or nonlinear) *integer program* is a linear or nonlinear program whose variables are constrained to be integer. We will consider here only the linear case, although there exist extensions of the techniques to be discussed to nonlinear integer programming.

The integer programming problem can be stated as

$$(P) \quad \min\{cx \,|\, Ax \geqslant b, x \geqslant 0, x_j \text{ integer}, j \in N_1 \subseteq N\},$$

where A is a given $m \times n$ matrix, c and b are given vectors of conformable dimensions, $N = \{1, \ldots, n\}$, and x is a variable n-vector. (P) is called a *pure integer program* if

$N_1 = N$, a *mixed integer program* if $\phi \neq N_i \neq N$. Integer programming is sometimes called *discrete optimization*.

MODELLING POTENTIAL. Integer programming is the most immediate and frequently needed extension of linear programming. Integrality constraints arise naturally whenever fractional values for the decision variables do not make sense. A case in point is the *fixed charge problem*, in which one wants to minimize a function of the form $\Sigma_i c(x_i)$, with

$$c(x_i) = \begin{cases} f_i + c_i x_i & \text{if } x_i > 0 \\ 0 & \text{if } x_i = 0 \end{cases}$$

subject to linear constraints. Such a problem can be restated as an integer program whenever x is bounded by setting

$$c(x_i) = c_i x_i + f_i y_i$$

$$x_i \leqslant U_i y_i, \qquad y_i = 0 \text{ or } 1$$

where U_i is an upper bound of x_i.

By far the most important special case of integer programming is the *0–1 programming problem*, in which the integer-constrained variables are restricted to 0 or 1. This is so because a host of frequently occurring nonlinearities, like logical alternatives, implications, precedence relations, etc., or combinations thereof, can be formulated via 0–1 variables. For example, a condition like

$$x > 0 \Rightarrow (f(x) \leqslant a \vee f(x) \geqslant b),$$

where a and b are positive scalars, x is a non-negative variable with a known upper bound M, $f(x)$ is a function whose value is bounded from above by $U > b$ and from below by $L < a$, while the symbol '$\vee$' means disjunction (logical 'or'), can be stated as

$$x \leqslant M(1 - \delta_1)$$

$$f(x) \leqslant a + (U - a)\delta_1 + (U - a)\delta_2$$

$$f(x) \geqslant b + (L - b)\delta_1 + (L - b)(1 - \delta_2)$$

$$\delta_1, \delta_2 = 0 \text{ or } 1$$

A linear program with *logical* conditions (conjunctions, disjunctions and implications) involving inequalities is called a *disjunctive program*, since it is the presence of disjunctions that makes these problems nonconvex. Bounded disjunctive programs can be stated as 0–1 programs and vice versa, but the disjunctive programming formulation has led to new methods.

Nonconvex optimization problems like bimatrix games, separable programs involving piecewise linear nonconvex functions, the general (nonconvex) quadratic programming problem, the linear complementarity problem and many others can be stated as disjunctive or 0–1 programming problems.

A host of interesting combinatorial problems can be formulated as 0–1 programming problems defined on a graph. The joint study of these problems by mathematical programmers and graph theorists has led to the recent development of a burgeoning area of research known as *combinatorial optimization*. Some typical problems studied in this area are: edge matching and covering, vertex packing and covering, clique covering, vertex colouring; set packing, partitioning and covering; Euler tours; Hamiltonian cycles (travelling salesman problem).

Applications of integer programming abound in all spheres of decision making. Some typical real-world problem areas where integer programming is particularly useful as a modelling tool include: facility (plant, warehouse, hospital, fire station) location; scheduling (of personnel, machines, projects); routing (of trucks, tankers, airplanes); design of communication (road, pipeline, telephone) networks; capital budgeting; project selection; analysis of capital development alternatives.

SOLUTION METHODS. Integer programs are notoriously hard to solve. Unlike linear programs, which are always solvable in a number of steps bounded by a polynomial function of the length of the data input, integer programs often require a number of steps that grows exponentially with problem size. However, sometimes an integer program (P) can be solved as a linear program; i.e., solving the linear program (L) obtained by removing the integrality constraints from (P), one obtains an integer solution. In particular, this is the case when all basic solutions of (L) are integer. For an arbitrary integer vector b, the constraint set $Ax \leqslant b$, $x \geqslant 0$ is known (Hoffman and Kruskal, 1958) to have only integer basic solutions if and only if the matrix A is totally unimodular (i.e. all nonsingular submatrices of A have a determinant of 1 or -1).

The best known instances of total unimodularity are the vertex-edge incidence matrices of directed graphs and undirected bipartite graphs. As a consequence, shortest path and network flow problems on arbitrary directed graphs, edge matching (or covering) and vertex packing (or covering) problems on bipartite graphs, as well as other integer programs whose constraint set is defined by the incidence matrix of a directed graph or an undirected bipartite graph, with arbitrary integer right-hand side, are in fact linear programs.

Apart from this important but very special class of problems, and a few other special classes, the difficulty in solving integer programs lies in the nonconvexity of the feasible set, which makes it impossible to establish global optimality from local conditions. The two main approaches to solving integer programs try to circumvent this difficulty in two different ways.

The first approach, which in the current state of the art is the standard way of solving general integer programs, is *enumerative* (branch and bound, implicit enumeration). It partitions the feasible set into successively smaller subsets, calculates bounds on the objective function value over each subset, and uses these bounds to discard certain subsets from further consideration. The procedure ends when each subset has either produced a feasible solution, or was shown to contain no better solution than the one already in hand. The best solution found during the procedure is a global optimum. Two early prototypes of this approach are due to Land and Doig (1960) and Balas (1965).

The second approach, known as the cutting plane method, is a *convexification* procedure: it approximates the convex hull of the set F of feasible integer points, by a sequence of inequalities that cut off (hence the term 'cutting planes') parts of the linear programming polyhedron, without removing any point of F. When sufficient inequalities have been generated to cut off every fractional point better than the integer optimum, then the latter is found as an optimal solution to the linear program (L) amended with the cutting planes. The first finitely convergent procedure of this type is due to Gomory (1958).

Depending on the type of techniques used to describe the convex hull of F and generate cutting planes, one can distinguish three main directions in this area. The first one uses algebraic methods, like modular arithmetic and group theory. Its key concept is that of subadditive functions. It is sometimes called the algebraic or group theoretic approach. The second one uses convexity, polarity and propositional calculus. Its main thrust comes from looking at the 0–1 programming problem as a disjunctive program. It is known as the convex analysis/disjunctive programming approach. Finally, the third direction applies to combinatorial programming problems, and it combines graph theory and matroid theory with mathematical programming. It is sometimes called polyhedral combinatorics.

Besides these two basic approaches to integer programming (enumerative and convexifying), two further procedures need to be mentioned, that do not belong to either category, but can rather be viewed as complementary to one or the other. Both procedures essentially *decompose* (P), one of them by partitioning the variables, the other one by partitioning the constraints. The first one, due to Benders (1962) eliminates the continuous variables of a mixed integer program (P) by projecting the feasible set F into the subspace of the integer-constrained variables. The second one, known as Lagrangean relaxation, removes some of the constraints of (P) by assigning multipliers to them and taking them into the objective function.

Each of the approaches outlined here aims at solving (P) exactly. However, since finding an optimal solution tends to be expensive beyond a certain problem size, approximation methods or *heuristics* play an increasingly important role in this area.

At present all commercially available integer programming codes are of the branch and bound type. While they can *sometimes* solve problems with hundreds of integer and thousands of continuous variables, they cannot be *guaranteed* to find optimal solutions in a reasonable amount of time to problems with more than 30–40 variables. On the other hand, they usually find feasible solutions of acceptable quality to much larger problems.

A considerable number of specialized branch and bound/implicit enumeration algorithms have been implemented by operations research groups in universities or industrial companies. They usually contain other features besides enumeration, like cutting planes and/or Lagrangean relaxation. Some of these codes can solve general (unstructured) 0–1 programs with up to 80–100 integer variables, and structured problems with up to several hundred (assembly line balancing, multiple choice, facility location), a few thousand (sparse set covering or set partitioning, generalized assignment), or several thousand (knapsack, travelling salesman) 0–1 variables.

At the current state of the art, while many real-world problems amenable to an integer programming formulation fit within the stated limits and are solvable in useful time, others substantially exceed those limits. Furthermore, some important and frequently occurring real-world problems, like job shop scheduling and others, lead to integer programming models that are almost always beyond the limits of what is currently solvable. Hence the great importance of approximation methods for such problems.

EGON BALAS

See also COMBINATORICS; INDIVISIBILITIES; LINEAR PROGRAMMING.

BIBLIOGRAPHY

Balas, E. 1965. An additive algorithm for solving linear programs with zero-one variables. *Operations Research* 13, 517–46.
Benders, J.F. 1962. Partitioning procedures for solving mixed-variables programming problems. *Numerische Mathematik* 4, 238–52.
Gomory, R.E. 1958. Outline of an algorithm for integer solutions to linear programs. *Bulletin of the American Mathematical Society* 64, 275–8.
Hoffman, A.J. and Kruskal, J.B. 1958. Integral boundary points of convex polyhedra. In *Linear Inequalities and Related Systems*, ed. H.W. Kuhn and A.W. Tucker, Princeton: Princeton University Press, 223–46.
Land, A.H. and Doig, A.G. 1960. An automatic method for solving discrete programming problems. *Econometrica* 28, 497–520.

integrability of demand. The lines of reasoning linking individual (ordinal) utility functions (or preference orderings) to individual demand functions run in both directions. Progressions from the former to the latter often begin with assumptions about the characteristics of a consumer's utility function and the requirement that he always chooses so as to maximize utility subject to a budget constraint, and then go on to derive the demand functions and the properties of these demand functions that logically ensue from such premises. Depending on context, certain of the properties of the demand functions so derived are expressed in differential terms (i.e., symmetry and negative definiteness of matrices of Slutsky substitution functions) or in revealed preference form. The reverse course takes the individual's demand functions and their properties as given and reconstructs a utility function from which, upon constrained maximization, the original demand functions could have been generated. In this second case, when the starting point includes the differential rather than revealed preference properties of demand, the argument usually involves (in part) the integration of a system of one or more differential equations. Hence the name 'integrability of demand' affixed to it.

There are at least two ways to structure an integrability of demand argument. Perhaps the more straightforward approach is simply to backtrack over the path that yields demand functions from utility functions via the theorem of Lagrange on maximization subject to constraint. This path may be summarized as follows: apply Lagrange's theorem to obtain

$$\frac{p_i}{p_I} = \frac{u_i(x)}{u_I(x)}, \qquad i = 1, \ldots, I-1, \tag{1}$$

$$\frac{m}{p_I} = x_I + \sum_{i=1}^{I-1} \frac{u_i(x)}{u_I(x)} x_i, \tag{2}$$

where $x_i > 0$ varies over quantities of commodity i, the index $i = 1, \ldots, I$, the vector $x = (x_1, \ldots, x_I)$, u_i is the partial derivative of the utility function u with respect to x_i, $p_i > 0$ is the price of good i, and $m > 0$ is the consumer's income. Equation (1) states that at a constrained maximum, the marginal rates of substitution or the negatives of the partial derivatives of indifference functions equal the price ratios, and equation (2) is a form of the budget constraint. Both equations together may be thought of as a system of inverse functions which are solved to secure demands as functions, h^i, of prices and income:

$$x_i = h^i\left(\frac{p_1}{p_I}, \ldots, \frac{p_{I-1}}{p_I}, \frac{m}{p_I}\right), \qquad i = 1, \ldots, I. \tag{3}$$

Of course, (3) may be written in the form

$$x_i = H^i(p, m), \qquad i = 1, \ldots, I,$$

where $p = (p_1, \ldots, p_I)$ and H^i is homogeneous of degree zero.

Consider now an integrability of demand argument that reverses the above steps. Start with the demand functions (3) having all the properties implied by utility maximization. The aim is to recover a utility function generator of these demand functions from the h^i. Backtracking from (3), solve for price–price and income–price ratios as functions, g^i, of x:

$$\frac{p_i}{p_I} = g^i(x), \qquad i = 1, \ldots, I-1, \tag{4}$$

$$\frac{m}{p_I} = g^I(x). \tag{5}$$

If the h^i are to be derivable from constrained utility maximization, then the g^i of (4) should indicate the negatives of the partial derivatives of appropriate indifference functions and g^I should be related to the budget constraint; that is, equation (4) should correspond to equation (1), and equation (5) to equation (2). But to say that the g^i are the negatives of the partial derivatives of indifference functions means that

$$\frac{\partial x_I}{\partial x_i} = -g^i(x), \qquad i = 1, \ldots, I-1. \tag{6}$$

Thus at every $x > 0$, the 'slopes' of the indifference surface through x in the direction of each of the coordinate axes are given by the g^i, for $i = 1, \ldots, I-1$. Integrating the differential equation system (6) 'fits' all of these slopes together to form an indifference map from which a utility function is deduced. (It is possible to integrate alternative, though related, systems of differential equations which yield the utility function directly.) Lastly, the appropriate characteristics of this utility function and the fact that its constrained maximization produces the original demand functions (3) is established.

Naturally, the properties of the demand functions h^i are crucial for such an integrability of demand argument to hold up. Among other things, these properties must permit the inversion of the h^i into the g^i and must ensure that the integration step can be carried out. Invertibility means that the h^i specify a 1–1 correspondence between values of the vectors $x = (x_1, \ldots, x_I)$ and $(p_1/p_I, \ldots, p_{I-1}/p_I, m/p_I)$. For the integration of (6) it is necessary that the g^i be continuous and, when $I > 2$, that a certain 'integrability' condition be satisfied. This guarantees that at least one indifference surface passes through every $x > 0$. To make certain that no more than one indifference surface passes through each x, a Lipschitz condition has to be in force. It turns out that the g^i are continuous as long as the h^i are continuous, that the integrability condition is equivalent to the symmetry of the matrix of Slutsky substitution functions, and that the Lipschitz condition is implied if certain partial derivatives of the g^i are bounded. All of these properties of demand functions except the last are derivable from utility functions that are twice continuously differentiable, increasing, strictly quasi-concave and whose indifference surfaces do not touch the boundaries of the commodity space. Even so, the properties of demand functions obtained from such utility functions are still 'roughly' sufficient to support the integrability of demand argument outlined above.

Problems arise when the properties of demand functions are derived from utility functions with modified characteristics. For example, the previously mentioned 1–1 correspondence may not appear in the h^i and hence invertibility from the h^i to the g^i may break down. In such a situation it is possible to restructure the integrability of demand argument to avoid the invertibility issue altogether. Since it turns out that the demand functions $H^i(p, m)$ may also be viewed as partial derivatives with respect to p_i of the expenditure or income compensation function

$$m = E(p, \mu),$$

where μ varies over utility levels, this is accomplished by integrating the system

$$\frac{\partial m}{\partial p_i} = H^i(p, m), \qquad i = 1, \ldots, I, \tag{7}$$

and converting the resulting expenditure function into a utility function. Once again the appropriate characteristics of the derived utility function have to be established, constrained maximization of it has to produce the given H^i, and enough properties of the H^i need to be present to sustain the argument.

Antonelli (1886) is usually credited with introducing economists to the integrability of demand argument. He began with the functions g^i and obtained a utility generator by integrating a system of differential equations related to (6). Many years later in a mathematical appendix, Samuelson (1950) inverted the h^i and then secured an indifference map by integrating another differential equation related to (6). In between, Antonelli's work seems to have been almost forgotten. Fisher (1892) independently 'rediscovered' the integrability problem in his doctoral dissertation, and various aspects of it were taken up subsequently by Volterra (1906), Pareto (1906), Allen (1932), Georgescu-Roegen (1936), Wold (1943, 1944), and others. More detailed history is given by Samuelson (1950) and Chipman (1971, introduction to part II). Hurwicz and Uzawa (1971) were the first to structure an integrability of demand argument based on the integration of (7).

Donald W. Katzner

See also DEMAND THEORY; REVEALED PREFERENCE; VOLTERRA, VITO.

BIBLIOGRAPHY

Allen, R.G.D. 1932. The foundations of a mathematical theory of exchange. *Economica* 12, May, 197–226.

Antonelli, G.B. 1886. On the mathematical theory of political economy. In *Preferences, Utility, and Demand*, ed. J.S. Chipman et al., New York: Harcourt Brace, Jovanovich, 1971, ch. 16.

Chipman, J.S. et al. (eds) 1971. *Preferences, Utility, and Demand.* New York: Harcourt Brace, Jovanovich, chs 7, 8, 9, and introduction to Part II.

Fisher, I. 1892. Mathematical investigations in the theory of value and prices. *Transactions of the Connecticut Academy of Arts and Sciences* 9, July, 1–124. Reprinted, New York: Augustus M. Kelley, 1965.

Georgescu-Roegen, N. 1936. The pure theory of consumer's behavior. *Quarterly Journal of Economics* 50, August, 545–93.

Hurwicz, L. and Uzawa, H. 1971. On the integrability of demand functions. In *Preferences, Utility, and Demand*, ed. J.S. Chipman, et al., New York: Harcourt Brace, Jovanovich, ch. 6.

Katzner, D.W. 1970. *Static Demand Theory.* New York: Macmillan, ch. 4.

Pareto, V. 1906. Ophelimity in nonclosed cycles. In *Preferences, Utility, and Demand*, ed. J.S. Chipman, et al., New York: Harcourt Brace, Jovanovich, 1971, ch. 18.

Samuelson, P.A. 1950. The problem of integrability in utility theory. *Economica* 17, November, 355–85.

Volterra, V. 1906. Mathematical economics and Professor Pareto's new manual. In *Preferences, Utility, and Demand*, ed. J.S. Chipman et al., New York: Harcourt Brace, Jovanovich, 1971, ch. 17.

Wold, H. 1943–4. A synthesis of pure demand analysis I, II, III. *Skandinavisk Aktaurietidskrift*, 1943, 85–118, 221–63; 1944, 69–120.

intelligence. From the latter half of the 19th century to the Great Depression and the rise of fascism in the 1930s, it was fashionable both in and out of scientific circles to stress the contribution of the genetic worth of individuals and groups to their economic success. This stress was to be as often found among progressives, who used the doctrine to affirm birth control, divorce, and equal educational and economic opportunity for women, as among conservatives, who relied upon eugenic arguments to justify the natural superiority of their favoured social classes, ethic groups, and races. Eugenics, for instance, was supported by such radicals as Havelock Ellis, Beatrice and Sydney Webb and George Bernard Shaw, as well as such conservatives as Francis Galton, Leonard Darwin and Charles Davenport.

Brought into disrepute by its association with Nazism, the notion of genetic destiny resurfaced in the United States in the 1960s as a conservative reaction to the civil rights movement of American blacks (Jensen, 1969; Herrnstein, 1971; Eysenck, 1971). The ensuing flurry of invective and empirical research has generated its proverbial quota of heat, and some light. My assessment of the evidence is that it provides no support for the notion that racial differences in economic success can be attributed to their genetic inferiority with respect to mental functioning, since no acceptable technique of correcting for environmental differences between distinct racial groups has been devised. This same evidence provides some positive evidence for the effect of genes on economic performance in general, but it is so difficult to separate genetic and environmental factors, even among such relatively restricted samples as monozygotic (identical) twins, that the extent of this effect is unknown. It certainly is not enough to justify the use of the notion of genetic differences in any serious way in the formulation of economic policy.

To illustrate this point, consider one of the most careful and powerful examinations of the role of genes in explaining earnings (Taubman, 1976a). Taubman uses a sample of 2468 pairs of monozygotic and dizygotic (fraternal) white male twins, attempting to explain differences in earnings at age 50 using such family background variables as parents' earnings and occupational status. This sample should provide the best possible evidence for or against the role of common genes, since monozygotic twins share all their genes, while dizygotic twins are no more genetically similar than two brothers. Assuming no assortive mating, no sex-linked genes, and no dominant and recessive genes, Taubman finds that the combined family environment explains 54% of the variance in earnings, while other influences explain the remaining 46%. However, depending on the extent to which twins share the same family environment more than two genetically unrelated individuals, the family contribution is apportioned so that the ratio of environmental to genetic factors ranges between 8% and 75%. In a related article (Taubman, 1976b) using this sample, Taubman shows that not correcting for family background leads to a severe upward bias in estimating the returns to years of education. But the extent to which this bias is due to genetic as opposed to social factors cannot be ascertained.

It has been suggested (Jensen, 1969) that individuals who perform badly on standardized cognitive tests be shunted out of the public educational system on grounds of the efficient application of economic resources. Certainly cognitive performance has been a central determinant of educational attainment in most modern societies. Yet one can show using a representative sample white American males (Bowles and Gintis, 1976, pp. 110–12) that the economic return to education does not fall appreciably when cognitive performance is controlled in a regression analysis of earnings and occupational status. Moreover, it can be shown that for the same sample, the observed relationship between social class background and earnings is only in small part due to the tendency of families to pass on IQ differences (Bowles and Gintis, 1976, pp. 120–22) either genetically or environmentally.

It is thus safe to say that if all differences in economic achievement were eliminated except for differences in IQ, and if the differences in the latter were maintained at their present level, there would be virtually perfect intergenerational economic mobility. In this sense, arguments which justify economic inequality, either among individuals or between races, on the basis of presumed intellectual differences, must be incorrect.

HERBERT GINTIS

See also EQUALITY; HUMAN CAPITAL; POVERTY.

BIBLIOGRAPHY
Bowles, S. and Gintis, H. 1976. *Schooling in Capitalist America: Educational Reform and the Contradictions of Economic Life.* New York: Basic Books.
Eysenck, H. 1971. *The IQ Argument.* New York: Modern Library.
Herrnstein, R. 1971. IQ. *Atlantic Monthly*, September.
Jensen, A. 1969. How much can we boost IQ and scholastic achievement? *Harvard Educational Review* 39(1), 1–123.
Taubman, P. 1976a. The determinants of earnings: genetics, family, and other environments; a study of white male twins. *American Economic Review* 66(5), December, 858–70.
Taubman, P. 1976b. Earnings, education, genetics, and environment. *Journal of Human Resources* 11(4), Fall, 447–61.

intensive rent. *See* EXTENSIVE AND INTENSIVE RENT.

interdependent preferences. Interdependent preferences arise in economic theory in the study of both individual decisions and group decisions. We imagine that a decision is required among alternatives in a set X and that the decision will depend on preferences between the elements in X. If the preferences represent different points of view about the relative desirability of the alternatives, of if they are based on multiple criteria that impinge on the decision, then we encounter the possibility of interdependent preferences.

There are two predominant approaches to interdependent preferences, the synthetic and the analytic. The synthetic approach begins with a set of preference relations on X and attempts to aggregate them into a holistic representative preference relation on X. This is done in social choice theory, where each original relation refers to the preferences of an individual in a social group. The aggregate relation is then referred to as a social preference relation. The synthetic approach also appears in studies of individual preferences, as when an individual rank-orders the alternatives for each of a number of criteria and then seeks a holistic ranking that combines the criteria rankings in a reasonable way.

In contrast, the analytic approach begins with a holistic preference relation on X and seeks to analyse its internal structure. This may involve a decomposition into components of preference, or it may concern trade-offs between factors that describe interactive contributions to overall preferences.

The synthetic approach often considers a list $(\gg_1, \gg_2, \ldots, \gg_n)$ of preference relations on X, where $x \gg_i y$ could mean that person i prefers x to y, or that an individual prefers x to y on the basis of criterion i. The problem may then be to specify a holistic relation $\gg = f(\gg_1, \gg_2, \ldots, \gg_n)$ for each possible n-tuple of individual relations.

The analytic approach often begins with X as a subset of the product $X_1 \times X_2 \times \cdots \times X_n$ of n other sets. It considers a holistic *is preferred to* relation $\gg$ on X and asks how $\gg$ depends on the X_i considered separately or in combination. Under suitably strong *independence* assumptions it may be possible to *define* $\gg_i$ for each i in a natural way from $\gg$ on X, and perhaps to establish a functional dependence of $\geqslant$ on the $\geqslant_i$. However, interdependencies among the factors will often preclude such a simple resolution.

HISTORICAL REMARKS. During the rise of marginal utility analysis in the latter part of the 19th century (Stigler, 1950), the utility of each commodity bundle in a set $X = X_1 \times X_2 \times \cdots \times X_n$ was thought of as an intuitively measurable quantity. Founders such as Jevons, Menger and Walras regarded x as preferred to y precisely when $u(x)$, the utility of x, is greater than $u(y)$. Their analytic approach ignored interdependencies since they used the independent additive utility form $u(x) = u_1(x_1) + \cdots + u_n(x_n)$.

Later writers such as Edgeworth, Fisher, Pareto and Slutsky discarded the additive decomposition for the general interdependent form $u(x_1, x_2, \ldots, x_n)$. Their ordinalist view of utilities as a mere reflection of a preference ordering remains dominant, and they considered interactive effects among goods, such as complementarities and substitutabilities. A fine example of interdependent analysis appears in Fisher (1892).

Fisher was also one of the first people to mention explicitly the interpersonal effect on individual utility (Stigler, 1950, p. 324). This occurs when one's utility and consequent demand depend on other people's consumption and could generally be expressed by $u_i(x_1, \ldots, x_n)$ as consumer i's utility when x_j denotes the commodity bundle of consumer j. Pigou (1903) considered the interpersonal effect in modest detail, and Duesenberry (1949) explored it in greater depth, but it has never been a prominent concern in economic theory.

Early examples of the synthetic approach in social choice theory come from Borda and Condorcet in the late 1700s. They asked: Given a list of voter preference rankings on a set X of $m \geqslant 3$ nominees, what is the best way of selecting a winner? Borda's answer was to assign $m, m-1, \ldots, 1$ points to each first, second, ..., last place nominee in the rankings and to elect the nominee with the largest point total. Condorcet advocated the election of a nominee who is preferred by a simple majority of voters to each other nominee in pairwise comparisons. Black (1958) contains an excellent review of their work and the proposals of later writers. The debate over good election methods continues today (Brams and Fishburn, 1983).

The turning point for social choice theory was Arrow's (1951) discovery that a few appealing conditions for aggregating individual preference orders on three or more candidates into social preference orders were jointly incompatible. The avalanche of research set off by Arrow's discovery is represented in part by Sen (1970, 1977), Fishburn (1973), Pattanaik (1971), and Kelly (1978).

In the area of risky decision theory, we envision a risky alternative as a probability distribution x on potential outcomes in a set C and observe that such decisions involve multiple factors since they entail both chances and outcomes. Bernoulli (1738) argued that a reasonable person will choose a risky alternative from a set X of distributions that maximizes his expected utility $\Sigma x(c)u(c)$. He proposed that u be assessed without reference to chance since he held an intuitive measurability view of utility. Consequently, his approach is wholly synthetic.

Little changed in the foundations of risky decisions during the next two centuries. Then, in a complete turnabout, von Neumann and Morgenstern (1944) introduced the analytic approach by beginning with a preference relation $\geqslant$ on X. Axioms for $\geqslant$ on X were shown to imply the existence of a real valued function u on C such that, for all x and y in X, $x \geqslant y$ precisely when x has greater expected utility than y, and u is to be assessed on the basis of comparisons between distributions. With a few exceptions, most notably Allais

(1953), subsequent research has adopted the von Neumann–Morgenstern approach.

In the rest of this essay we comment further on multiattribute preferences under 'certainty', interdependent preferences in risky decisions, and social choice theory.

MULTIATTRIBUTE PREFERENCES. We assume throughout this section that $\geqslant$ is a strict preference relation on $X = X_1 \times X_2 \times \cdots \times X_n$. A given X_i could represent amounts of commodity i, consumption bundles available to person i, levels of income and/or consumption in period i, or values that elements in X might have for criterion i. Also let u on X and u_i denote real valued functions.

A non-empty proper subset N of $\{1, 2, \ldots, n\}$ is defined to be $\geqslant$-*independent* if, for all x_N and y_N in the product of the X_i over N and for all $z_{(N)}$ in the product of the X_i over i not in N,

$$(x_N, z_{(N)}) \geqslant (y_N, z_{(N)}) \Leftrightarrow (x_N, w_{(N)}) \geqslant (y_N, w_{(N)}).$$

Most research for $\geqslant$ on X involves $\geqslant$-independence for some N, but this need not exclude elementary notions of preference interdependencies. Two models that presume all N to be $\geqslant$-independent are the additive model (see Krantz, Luce, Suppes and Tversky, 1971)

$$x \geqslant y \Leftrightarrow u_1(x_1) + \cdots + u_n(x_n) > u_1(y_1) + \cdots + u_n(y_n),$$

and the lexicographic model (Fishburn, 1974a) that places a value hierarchy on the factors.

Relationships between factors in the additive model and the more general model $x \geqslant y \Leftrightarrow u(x) > u(y)$ with u continuous, are often characterized by indifference maps or iso-utility contours. Interdependence arises in the lexicographic model from the fact that a small change in one factor overwhelms all changes in factors that are lower in the hierarchy.

Situations in which only some of the N are $\geqslant$-independent are reviewed by Keeney and Raiffa (1976, ch. 3) and Krantz, Luce, Suppes and Tversky (1971, ch. 7). Among other things, these models allow complete reversals in preferences over one factor at different fixed levels of the other factors. This, of course, is a very strong form of interdependence under which all N may fail to be $\geqslant$-independent.

Other general models for interdependent preferences are discussed by Fishburn (1972) for finite sets, and by Dyer and Sarin (1979) when u is viewed in the intuitive measurability way.

Models that explicitly incorporate the interpersonal effect in economic analysis have been investigated by Pollak (1976) and Wind (1976), among others. Pollak explores the influence of several versions of interdependence among individuals on short-run and long-run consumption within a group. Using models of demand that are locally linear in others' past consumption, he concludes that the distribution of income need not be a determinant of long-run per capita consumption patterns. Wind's work is representative of empirical approaches to the influence of others on an individual's choice behaviour.

RISKY DECISIONS. Interdependent preferences in risky decisions fall into two categories. The first concerns special forms for $u(c) = u(c_1, c_2, \ldots, c_n)$ in the context of von Neumann–Morgenstern expected utility theory when the outcome set C is a subset of a product set $C_1 \times C_2 \times \cdots \times C_n$. The second focuses on changes in the basic model that occur when the independence axiom that gives rise to the expected utility form $\Sigma x(c)u(c)$ is relaxed or dropped.

Decompositions of $u(c_1, c_2, \ldots, c_n)$ in the expected utility model have been axiomatized by various people. Reviews and extensions of much of this work appear in Keeney and Raiffa (1976) and Farquhar (1978). The simplest independent decompositions are the additive form and a multiplicative form. The first of these requires x and y to be indifferent whenever the marginal distributions of x and y on X_i are the same for every i. The multiplicative form arises when, for each non-empty proper subset N of $\{1, \ldots, n\}$, the preference order over marginal distributions on the product of the C_i for i in N, conditioned on fixed values of the other factors, does not depend on those fixed values.

An example of a more involved interdependent decomposition is the two-factor model (Fishburn and Farquhar, 1982) $u(c_1, c_2) = f_1(c_1)g_1(c_2) + \cdots + f_m(c_1)g_m(c_2) + h(c_1)$, which clearly allows a variety of interactive effects.

In the basic formulation for expected utility, assume that X is closed under convex combinations $\lambda x + (1 - \lambda)y$ with $0 < \lambda < 1$ and x and y in X. The *independence axiom* for expected utility asserts that, for all x, y and z in X and all $0 < \lambda < 1$.

$$x \gg y \Rightarrow \lambda x + (1 - \lambda)z \gg \lambda y + (1 - \lambda)z.$$

Systematic violations of this axiom uncovered in experiments by Allais (1953), Kahneman and Tversky (1979) and Mac-Crimmon and Larsson (1979) among others, have led to new theories of risky decisions (Kahneman and Tversky, 1979; Machina, 1982; Chew, 1983; Fishburn, 1982) that do not assume independence. Machina (1982) proposes a model that approximates expected utility locally but not globally. Fishburn (1982) weakens the usual transitivity and independence assumptions to obtain a non-separable model $x \gg y \Leftrightarrow \phi(x, y) > 0$ that allows preference cycles.

Related interdependent generalizations of Savage's subjective expected utility model for decisions under uncertainty are developed by Loomes and Sugden (1982) and Schmeidler (1984).

SOCIAL CHOICE. Many problems in social choice theory are related to Condorcet's phenomenon of cyclical majorities. This phenomenon occurs when voters have transitive preferences yet every nominee is defeated by another nominee under simple majority comparisons. The simplest example has three nominees and three voters with $x \gg_1 y \gg_1 z$, $z \gg_2 x \gg_2 y$ and $y \gg_3 z \gg_3 x$; x beats y, y beats z, and z beats x. Borda's point-summation procedure can fail to satisfy Condorcet's majority-choice principle, and it is notoriously sensitive to strategic voting. Moreover, all summation procedures based on decreasing weights for positions in voters' rankings are sensitive to nominees who have absolutely no chance of winning, but whose presence can affect the outcome.

Various problems and paradoxes for multicandidate elections that arise from combinatorial aspects of synthetic methods are discussed in Fishburn (1974b), Niemi and Riker (1976), Saari (1982) and Fishburn and Brams (1983). Analyses of strategic voting, which suggest that no sensible election method is immune from manipulation by falsification of preferences, are reviewed in Kelly (1978) and Pattanaik (1978).

Arrow's (1951) theorem offers a striking generalization of Condorcet's cyclical majorities phenomenon. Suppose X contains three or more nominees, each of n voters can have any preference ranking on X, and an aggregate ranking $\gg = f(\gg_1, \gg_2, \ldots, \gg_n)$ is desired for each list $(\gg_1, \gg_2, \ldots, \gg_n)$ of individual rankings. The question addressed by Arrow is whether there is any way of doing this that satisfies the following three conditions for all x and y in X:

(1) Pareto optimality: if $x \gg_i y$ for all i, then $x > y$;
(2) Binary independence: the aggregate preference between x

and y depends solely on the voters' preferences between x and y;

(3) Non-dictatorship: there is no i such that $x \gg y$ whenever $x \gg_i y$. Arrow's theorem says that it is impossible to satisfy all three conditions.

Several dozen related impossibility theorems have subsequently been developed by others. Many of these are noted in Kelly (1978) and Pattanaik (1978). As well as multi-profile theorems, like Arrow's, that use different lists of preference rankings to demonstrate impossibility, there are single-profile theorems (Roberts, 1980) that use only one list with sufficient variety in the rankings to establish impossibility.

Impossibility theorems, voting paradoxes, and results on strategic manipulation highlight the difficulty of designing good election procedures. Recent research to alleviate such problems (Dasgupta, Hammond and Maskin, 1979; Laffont and Moulin, 1982) focuses on the design of preference–revelation mechanisms (generalized ballots) and aggregation procedures that encourage people to vote in such a way that the outcome will agree with some theoretically best decision based on the true but unknown preferences of the voters. Other work, such as that on approval voting (Brams and Fishburn, 1983), continues to search for simple synthetic methods that minimize the problems that beset these methods.

PETER C. FISHBURN

See also ARROW'S THEOREM; EXTERNALITIES; REPRESENTATION OF PREFERENCES; UTILITY THEORY AND DECISION THEORY.

BIBLIOGRAPHY

Allais, M. 1953. Le comportement de l'homme rationnel devant de risque; critique des postulats et axiomes de l'école Américaine. *Econometrica* 21, 503–46.

Arrow, K.J. 1951. *Social Choice and Individual Values*. New York: Wiley. 2nd edn, 1963.

Bernoulli, D. 1738. Specimen theoriae novae de mensura sortis. *Commentarii Academiae Scientarium Imperialis Petropolitanae* 5, 175–92. Trans. by L. Sommer as 'Exposition of a new theory on the measurement of risk', *Econometrica* 22 (1954), 23–36.

Black, D. 1958. *The Theory of Committees and Elections*. Cambridge: Cambridge University Press.

Brams, S.J. and Fishburn, P.C. 1983. *Approval Voting*. Boston: Birkhäuser.

Chew, S.H. 1983. A generalization of the quasilinear mean with applications to the measurement of income inequality and decision theory resolving the Allais paradox. *Econometrica* 51, 1065–92.

Dasgupta, P., Hammond, P. and Maskin, E. 1979. The implementation of social choice rules: some general results on incentive compatibility. *Review of Economic Studies* 46, 185–216.

Duesenberry, J.S. 1949. *Income, Saving, and the Theory of Consumer Behavior*. Cambridge, Mass.: Harvard University Press.

Dyer, J.S. and Sarin, R.K. 1979. Measurable multiattribute value functions. *Operations Research* 27, 810–22.

Farquhar, P.H. 1978. Interdependent criteria in utility analysis. In *Multiple Criteria Problem Solving*, ed. S. Zionts, Berlin: Springer-Verlag, 131–80.

Fishburn, P.C. 1972. Interdependent preferences on finite sets. *Journal of Mathematical Psychology* 9, 225–36.

Fishburn, P.C. 1973. *The Theory of Social Choice*. Princeton: Princeton University Press.

Fishburn, P.C. 1974a. Lexicographic orders, utilities, and decision rules: a survey. *Management Science* 20, 1442–71.

Fishburn, P.C. 1974b. Paradoxes of voting. *American Political Science Review* 68, 537–46.

Fishburn, P.C. 1982. Nontransitive measurable utility. *Journal of Mathematical Psychology* 26, 31–67.

Fishburn, P.C. and Brams, S.J. 1983. Paradoxes of preferential voting. *Mathematics Magazine* 56, 207–14.

Fishburn, P.C. and Farquhar, P.H. 1982. Finite-degree utility independence. *Mathematics of Operations Research* 7, 348–53.

Fisher, I. 1892. Mathematical investigations in the theory of values and prices. *Transactions of the Connecticut Academy of Arts and Sciences* 9, 1–124. Reprinted, New York: Augustus M. Kelley, 1965.

Kahneman, D. and Tversky, A. 1979. Prospect theory: an analysis of decision under risk. *Econometrica* 47, 263–91.

Keeney, R.L. and Raiffa, H. 1976. *Decisions with Multiple Objectives: preferences and value tradeoffs.* New York: Wiley.

Kelly, J.S. 1978. *Arrow Impossibility Theorems.* New York: Academic Press.

Krantz, D.H., Luce, R.D., Suppes, P. and Tversky, A. 1971. *Foundations of Measurement.* Volume I: *Additive and Polynomial Representations.* New York: Academic Press.

Laffont, J.-J. and Moulin, H. (eds) 1982. Special issue on implementation. *Journal of Mathematical Economics* 10(1).

Loomes, G. and Sugden, R. 1982. Regret theory: an alternative theory of rational choice under uncertainty. *Economic Journal* 92, 805–24.

MacCrimmon, K.R. and Larsson, S. 1979. Utility theory: axioms versus 'paradoxes'. In *Expected Utility Hypotheses and the Allais Paradox,* ed. M. Allais and O. Hagen, Dordrecht: Reidel, 333–409.

Machina, M.J. 1982. 'Expected utility' analysis without the independence axiom. *Econometrica* 50, 277–323.

Niemi, R.G. and Riker, W.H. 1976. The choice of voting systems. *Scientific American* 234, 21–7.

Pattanaik, P.K. 1971. *Voting and Collective Choice.* Cambridge: Cambridge University Press.

Pattanaik, P.K. 1978. *Strategy and Group Choice.* Amsterdam: North-Holland.

Pigou, A.C. 1903. Some remarks on utility. *Economic Journal* 13, 58–68.

Pollak, R.A. 1976. Interdependent preferences. *American Economic Review* 66, 309–20.

Roberts, K.W.S. 1980. Social choice theory: the single-profile and multi-profile approaches. *Review of Economic Studies* 47, 441–50.

Saari, D.G. 1982. Inconsistencies of weighted summation voting systems. *Mathematics of Operations Research* 7, 479–90.

Schmeidler, D. 1984. Subjective probability and expected utility without additivity. Preprint #84, Institute for Mathematics and its Application, University of Minnesota.

Sen, A.K. 1970. *Collective Choice and Social Welfare.* San Francisco: Holden-Day.

Sen, A.K. 1977. Social choice theory: a re-examination. *Econometrica* 45, 53–89.

Stigler, G.J. 1950. The development of utility theory, Part I and II. *Journal of Political Economy* 58, 307–27, 373–96.

von Neumann, J. and Morgenstern, O. 1944. *Theory of Games and Economic Behavior.* Princeton: Princeton University Press. 2nd edn, 1947; 3rd edn, 1953.

Wind, Y. 1976. Preference of relevant others and individual choice models. *Journal of Consumer Research* 3, 50–7.

interest and profit. The analysis of the relationship between the rates of interest and profit deals with how to integrate the theory of money with the theory of value and distribution. Different views on this subject reflect alternative positions expressed in the debates over these theories. They describe how changes in the financial markets affect the production process and vice versa, and have different policy implications.

The dominant view on the relation between the rates of interest and profits since Adam Smith has been that while monetary factors affect the *daily* variations of the 'money' or 'market' interest rate, in equilibrium the interest rate can only be equal to its 'average' or 'natural' or 'real' value (as alternatively named in the literature), which is determined, independently of monetary factors, by the same forces that determine the general rate of profits on the capital invested in the production process. This view, though dominant, is not the

only one put forward in the literature. Indeed some outstanding economists (e.g., J.M. Keynes) have proposed alternative views according to which monetary factors are relevant, both temporarily and permanently, in determining the equilibrium level of economic variables.

The distinction between the notions of 'money' or 'market' interest rate and 'average' or 'natural' or 'real' interest rate can be traced in the writings of most economists dealing with this subject. Ricardo, for instance, drew a clear-cut distinction between these rates and provided a coherent analysis of how to relate them. The rate of profits was determined on the basis of a 'surplus' theory, in the tradition of the English classical political economists. He took as given the social product, the available technology and the real wage rate, disallowing any direct influence of monetary factors in the determination of the rate of profits. The 'average' or 'natural' interest rate was in his writings a *portion* of the rate of profits and was determined by the latter. As to the 'market' interest rate, it could undergo daily variations around its 'natural' level, on account of the changing conditions of competition between lenders and borrowers in the money markets.

Ricardo's writings clarify the role played by the analysis of the money markets in the treatment of this subject. He provided several examples to show how changes in the market interest rate occur and how this rate tends to move towards its natural level. The analysis of money markets, therefore, has to support the view that the rate of profits ultimately determines the interest rate by describing how competitive market forces make the natural level of the interest rate assert itself in the money markets.

Ricardo presented a coherent analysis of how the theory of the interest rate and of money has to be integrated with the classical theory of value and distribution, when the real wage rate is taken as given. Yet he did not provide a detailed analysis of the working of the money markets. Soon after his death some attempts to develop the latter analysis were presented by Tooke and J.S. Mill. Both used this analysis to criticize Ricardo's view and to claim that the interest rate can be determined both temporarily and permanently by causes which are independent of what happens to the rate of profits.

Tooke's and Mill's positions were stimulated by the sharp variations in the interest rate which occurred during and after the Napoleonic wars. These long-lasting variations in the interest rate were, according to them, the result of the policy followed to finance the Government debt, rather than the result of a change in the conditions of production implying a higher level of the rate of profits. The analysis of the interest rate they put forward had a strong influence on the economic literature. It failed, however, to give a correct account of the competitive market forces relating the rates of interest and profits. This led to the unconvincing view that the average interest rate and the rate of profits can move independently of each other.

A similar point of view was expressed some years later by Marx, who studied monetary issues at length both at a theoretical and at an empirical level, with particular reference to the experience of the British financial system. By looking at this system, Marx developed the view that the most powerful pressure-groups operating in the financial markets were able to affect the interest rate permanently (and consequently their share of surplus-value produced) through the introduction of financial innovations and through their influence on State interventions regulating the legal and institutional arrangements of the financial markets.

To support this idea Marx presented a detailed analysis of the working of these markets and of the determination of both the average and the market interest rates in terms of supply and demand for liquid means. He stressed the need to reject the notion of a 'natural' rate of interest determined on the basis of technological or material laws of production, and pointed out the analytical conditions allowing a determination of the interest rate independent of the rate of profits and based on historical, conventional elements.

Yet, like Tooke and Mill in the 1820s, Marx failed to correctly relate these two rates. He maintained Ricardo's determination of the rate of profits in terms of a given real wage rate within a surplus theory of value and distribution, and failed to work out the effects of the operation of competitive market forces coming into action when a divergence between the rates of interest and profits comes about.

However, from Marx's writings some insights can be derived for analysing these forces, even if he did not carry them out himself. He pointed out that the banking sector, like the other industrial sectors, has to earn at least the general rate of profits on the wages and capital anticipated to carry on its activity. Changes in the interest rates affect the revenues (interest received on bank loans and financial assets) and the costs (which include payments for wages, interest on deposits and the rate of profits on the capital advanced) of the banking firms. This produces adjustment processes tending to restore the conditions of equilibrium between revenues and costs, which set some constraints linking the movements of the rates of interest and profits.

The economic literature of the years during which Marx was developing Ricardo's surplus theory of value and distribution shows the progressive abandonment of this theory, which implied an inverse relationship between the rate of profits and the real wage rate. The general trend in this literature was to determine instead these two rates independently of each other, as the specific contributions of capital and labour to production. Leading historians of economic thought have argued that it is not possible to claim that a *new* theory of value and distribution was actually presented in those years. The analyses were not clearly spelt out, particularly those which dealt with the concept of capital.

This new trend was reflected in the analysis of the relationship between the rates of interest and profits. The tendency prevailed to identify interest and profit and use them as synonyms. Tooke, Fullarton and James Wilson claimed in those years that a *permanent* variation in the interest rate affects the costs of production and the prices in the same direction. For them a permanent change in the interest rate was *the same thing* as a change in the rate of profits. No one spoke any longer of independent movements of these two rates. Indeed, the whole analysis of the relation between the average interest rate and the rate of profits faded away. The only issue left for discussion was the temporary fluctuations of the market interest rate. Besides, the abandonment of the surplus theory of value and distribution, the confusions as to the definition of capital, and the tendency to identify demand for and supply of *money*- or *banking*-capital with demand for and supply of *real* capital – all these made impossible at that time the development of a monetary theory of the rate of profits, of a theory, that is, which recognizes the influence of monetary factors on this rate.

The concept of capital and its analytical role were more precisely spelt out by the economists who introduced the marginalist or neoclassical theory of value and distribution in the 1870s and later. In this new theory too, the natural interest rate and the general rate of profits were *the same thing*. The money rate of interest could vary independently only

temporarily. In equilibrium it had to be equal to its natural value determined as the rate of return to be made on the *real* capital employed in production.

Walras explicitly stated that money markets, so relevant in the real world, are a 'superfetation' in marginalist theory. Later on, Wicksell, having presented a rather developed analysis of the role played by monetary factors in disequilibrium, concluded that the money interest rate depends in the last analysis upon the supply of and the demand for real capital. In equilibrium no room can be allowed for the action of monetary forces.

An example of the marginalist view of the relationship between the rates of interest and profits can be found in *A Treatise of Money*, published by Keynes in 1930. This book is based upon the marginalist separation between the 'real' department and the 'monetary' department of economics. In the real department, in line with the marginalist theory of value and distribution, the equilibrium or natural level of the distributive variables, the relative prices and the level of output (which turns out to be full employment) are simultaneously determined. In the monetary department, as analysed in *A Treatise on Money*, equilibrium values are taken as given (or rather known from the real department). The fluctuations of the price level are then analysed by looking at the variations in the evaluations of the expected yields of investment goods and at the fluctuations of the money interest rate around its natural level.

In *A Treatise on Money* the instability of the demand for investment and the analysis of liquidity preference are both present. The latter analysis, presented in this book in the form of 'bear and bull positions', describes the working of the money markets and how changes in the interest rate come about. Their presence, however, does not imply the abandonment of the marginalist approach, which asserts itself in the determination of the natural interest rate.

An alternative view was presented by Keynes a few years later in the *General Theory*. In this book Keynes took a critical attitude towards the marginalist theory. From 1932 on, he denied the validity of the separation between real and monetary departments, proposing instead a 'monetary theory of production', where monetary factors were directly relevant to determine the equilibrium level of output, of the interest rate and of other distributive variables. According to this view, the traditional causal relation between the rates of interest and profits was reversed. The level of the latter rate depended upon the former.

The introduction of the concept of a 'monetary theory of production' coincided in Keynes's writings with the abandonment of the concept of 'natural interest rate'. A new 'monetary' theory of the interest rate was instead proposed to determine the 'average' or 'durable' (as Keynes named it: 1936, p. 203) level of this rate. Presenting this theory, Keynes stressed its historical, conventional character by claiming that *any* level of interest which is accepted with sufficient conviction as *likely* to be durable, *will* be durable (Keynes, 1936, p. 203). He pointed out that the policy of the monetary authority is a major determinant of the 'common opinion' as to the future value of the interest rate. But he also added that other elements of an economic or institutional character can affect this 'common opinion', for instance by persuading the public that the monetary authority will not be able to maintain its present policy.

However, the *General Theory* did not introduce any alternative analysis of how changes in the interest rate come about. The analysis of liquidity preference was represented in this book as dealing with the daily variations in the market interest rate and describing how the level of the interest rate, which is expected to be durable, tends to assert itself.

To support his new view as to the causal link between the rates of interest and profits, Keynes also presented in the *General Theory* an analysis of how competitive market forces tend to affect productive processes when temporary or persistent changes occur in the financial markets. This analysis was best framed in chapter 17 of this book. According to Keynes, investors in the financial and industrial sectors pay great attention to the rates of return of the different assets. They allow for certain differentials between these rates, which take into account the liquidity premium offered by the different assets. The equilibrium structure of the rates of return (that is the equilibrium differentials) is so determined. When the actual differentials do not correspond to the equilibrium ones, competitive market forces come into action, producing adjustment processes which affect the prices of the assets. Autonomous variations of the interest rates – as Keynes argued in the *General Theory* – if persistent can thus cause changes in the same direction in the rate of profits.

The analysis of the competitive forces tending to relate the rates of interest and profits, proposed by Keynes in the *General Theory*, can be considered as complementary to that hinted by Marx and described above. They point out two different market mechanisms which tend to relate the movements of these two rates. Combined together, these two analyses provide a base to argue for a monetary determination of the rate of profits. Those who accept a historical conventional determination of the interest rate, can claim the existence of a causal relationship moving from this rate to the rate of profits. Monetary factors can therefore be directly allowed in the determination of the rate of profits.

Sraffa's recent rehabilitation of the surplus theory of value and distribution moves along these lines. Taking advantage of the possibility offered by this theory to determine one distributive variable independently of the others, he suggests that it is preferable in order to analyse the present conditions of capitalist economies, to consider the rate of profits as an independent variable (determined by the level of the rates of interest on money), instead of following the classical political economists of the last century who took the real wage rate as independently determined (Sraffa, 1960, p. 33).

A new way to relate the rates of interest and profits, and consequently the theory of money, and that of value and distribution is therefore proposed within the recent rehabilitation of the surplus approach. This proposal can refer to the writings of outstanding economists to find theoretical support and to spell out its analytical implications.

The reconstruction of this analysis appears particularly relevant in the face of the present state of the neoclassical approach. As outstanding neoclassical economists have themselves recognized, no satisfactory integration between monetary and real variables has yet been presented within modern versions of neoclassical theory, that is, those developed after the work of Hicks (1939) and the subsequent works of Arrow, Debreu and Malinvaud. On account of this unsatisfactory integration, neoclassical economists still refer, as they themselves say, to the works of Wicksell and Fisher, that is, to those earlier versions of the neoclassical theory, whose internal consistency has been denied by the debate on capital theory of the 1960s and the 1970s.

The relationship between the rates of interest and profits can thus be considered one of the most open and controversial subjects of political economy.

CARLO PANICO

See also EQUAL RATES OF PROFIT; PROFIT AND PROFIT THEORY.

BIBLIOGRAPHY
Hicks, J.R. 1939. *Value and Capital*. Oxford: Clarendon Press.
Keynes, J.M. 1930. *A Treatise on Money*. 2 vols. London: Macmillan.
Keynes, J.M. 1936. *The General Theory of Employment, Interest and Money*. London: Macmillan.
Sraffa, P. 1960. *Production of Commodities by Means of Commodities*. Cambridge: Cambridge University Press.

interest rates. Interest is payment for use of funds over a period of time, and the amount of interest paid per unit of time as a fraction of the balance is called the interest rate. In some contexts, economists have found it conceptually useful to refer to a single number, the interest rate. In fact, at any point in time there are many prevailing interest rates. The rate actually charged will depend on such factors as the maturity of the loan, the credit-worthiness of the borrower, the amount of collateral, tax treatment of interest payments for both parties, and special features such as call provisions or sinking fund requirements.

A complete treatment of interest rates would account for all of these factors, but in fact it is hard enough to handle any one of them adequately. This entry considers one factor, the term to maturity for default-free bonds. This analysis of the *term structure of interest rates* will be approached from the partial equilibrium perspective of finance: the determinants of interest rates and the impact of changing short and long interest rates on the macroeconomy are not discussed.

Merton (1970 and other unpublished work) was the first to formulate the term structure problem using the continuous-time no-arbitrage framework exposited here. Cox, Ingersoll and Ross (1985a, 1985b), Dothan (1978), Richard (1978), and Vasicek (1977) solved term structure problems of this type for different stochastic processes.

INTEREST RATES IN A CERTAIN ENVIRONMENT. Historically, the theory of interest rates has been burdened with cumbersome notation designed to distinguish among spot rates, forward rates and rates of different terms. Notation and terminology will be kept to a minimum here. The instantaneous spot rate of interest at time t for a loan to be repaid an instant later will be denoted by $r(t)$. $R(t, T)$ will denote the continuously compounded interest rate for a zero coupon bond sold at t to be repaid at T, and $P(t, T)$ will be the price or present value per $1 face value of such a bond. The relation between these three quantities is

$$r(t) \equiv \lim_{T \downarrow t} R(t, T) \tag{1a}$$

$$P(t, T) \equiv e^{-R(t, T)(T - t)}. \tag{1b}$$

An investor who has funds to invest until time T could buy a T-period zero coupon bond with a guaranteed annualized return of $R(t, T)$. Alternatively, the investor could roll over a series of shorter bonds or buy a longer bond with the intention of selling it at time T. In the absence of uncertainty, all of these plans would have to realize the same final return to avoid the possibility of arbitrage. In particular

$$\frac{1}{P(t, T)} = \exp\left[\int_t^T r(s)\,ds\right] \tag{2}$$

or the continuously compounded long rate must be the average of the instantaneous rates,

$$R(t, T) = \frac{1}{T - t}\int_t^T r(s)\,ds. \tag{3}$$

(Note that with discrete compounding one plus the long rate is equal to the geometric mean of one plus the single-period rates. See Dybvig, Ingersoll and Ross (1986) for a catalogue of related results in both continuous and discrete time.)

ONE-FACTOR MODELS OF INTEREST RATES IN AN UNCERTAIN ENVIRONMENT. When future interest rates are not known in advance, these relations need not be realized, even on average, but the equilibrium that obtains will still depend on the trade-offs between different bond portfolio strategies. The resulting equilibrium relation among the interest rates will depend primarily on the information structure perceived by investors and, in particular, on the temporal resolution of uncertainty.

In this section we will assume that the information structure is Markov in the currently prevailing short rate, which is assumed to capture all currently available information relevant for pricing default-free bonds. $P(r, t, T)$ denotes the price at t of a zero coupon bond maturing at T with a face value of $1, given that the currently prevailing short rate is r. The evolution of the interest rate is assumed to follow a diffusion process.

$$\mathrm{d}r = f(r, t)\,\mathrm{d}t + g(r, t)\,\mathrm{d}\omega. \tag{4}$$

Here $f(\cdot)$ measures the expected change in the interest rate per unit time, $g(\cdot)$ measures the standard deviation of changes in the interest rate per unit time, and $\mathrm{d}\omega$ is the increment to a Wiener process.

The price of a zero coupon bond evolves according to

$$\mathrm{d}P(r, t, T)/P(r, t, T) = \alpha(r, t, T)\,\mathrm{d}t + \delta(r, t, T)\,\mathrm{d}\omega \tag{5}$$

where $\alpha(\cdot)$ is the bond's (endogenous) instantaneous expected rate of return and $\delta(\cdot)$ is its instantaneous standard deviation. By Itô's lemma

$$\alpha(r, t, T)P(r, t, T) = \tfrac{1}{2}g^2(r, t)P_{rr} + f(r, t)P_r + P_t \tag{6a}$$

$$\delta(r, t, T)P(r, t, T) = g(r, t)P_r \tag{6b}$$

where subscripts denote partial differentiation.

Equation (6a) is a partial differential equation relating the prices of a given bond at different points of time and with different prevailing short rates. Together with the known value of a bond at its maturity $[P(r, T, T) = 1]$ and mild regularity conditions, (6a) is equivalent to the integral

$$P(r, t, T) = E\left[\exp\left\{-\int_t^T \alpha[r(s), s, T]\,\mathrm{d}s\right\}\right] \tag{7}$$

(Friedman, 1975, Theorem 5.2). This integral demonstrates that when the source of uncertainty is a stochastically varying discount rate rather than a random cash flow, it is generally improper to discount by using the expected discount rate. Rather, we should use the geometric mean across states of the discount rate.

To price bonds using (6a) or (7), we must know $\alpha(\cdot)$. Intuitively, $\alpha(\cdot)$ will be equal to the risk-free rate plus a risk premium. As there is but a single source of uncertainty, the returns on all bonds will be perfectly correlated; therefore, the risk premium on any bond will be proportional to its exposure to the risk, and knowing $\alpha(\cdot)$ for one bond (for all interest rate levels) is sufficient. We can specify $\alpha(\cdot)$ by fiat, or it can be derived from an equilibrium model.

One equilibrium condition that is often imposed is the 'local' expectations hypothesis, $\alpha(r, t, T) = r$ (Cox, Ingersoll and Ross, 1981). Under the local expectations hypothesis the partial differential equation for bond pricing derived from (6a) and the integral in (7) are fully determined given the distribution of

interest rate changes. The local expectations hypothesis is a strong assumption, but for asset pricing purposes it is the only case that needs consideration. It can be shown that the absence of arbitrage implies that we can artificially reassign probabilities so that the local expectations hypothesis holds without changing any asset prices. The artificial probabilities are called the risk neutral probabilities or the equivalent martingale measure. Here is an informal proof in our context.

Consider any two zero coupon bonds. From (6a) their realized returns are perfectly correlated. Therefore, to ensure the absence of arbitrage possibilities, their expected excess returns must be proportional to their standard deviations

$$\alpha(r, t, T) - r = \pi(r, t)\delta(r, t, T)$$

$$= \pi(r, t)(P_r/P)g(r, t). \tag{8}$$

The risk premium term $\pi(\cdot)$ cannot depend on the bond in question, which is why it does not depend on T. We can now write the equivalent diffusion process under the martingale measure. Because we can infer the variance of any diffusion from its sample path and because the martingale measure has the same set of possible events as the original probability measure, the martingale standard deviations must be the same i.e., $g(\cdot)$. The drift under the martingale measure must equate expected returns across assets. Therefore, the drift term under the equivalent martingale measure must be $f^*(r, t) \equiv f(r, t) - \pi(r, t)g(r, t)$. Using the martingale measure (6a) therefore becomes

$$rP = \tfrac{1}{2}g^2(r, t)P_{rr} + f^*(r, t)P_r + P_t \tag{9}$$

and its solution analogous to (7) is

$$P(r, t, T) = E^*\left\{\exp\left[-\int_t^T r(s)\,\mathrm{d}s\right]\right\}. \tag{10}$$

where $E^*[\cdot]$ denotes expectation under the modified process with f^* as the drift term.

To illustrate how these tools are used, consider the simplest model in Cox, Ingersoll and Ross (1985b). The stochastic process for the interest rate is

$$\mathrm{d}r = \kappa(\mu - r)\,\mathrm{d}t + \sigma\sqrt{r}\,\mathrm{d}\omega. \tag{11}$$

For this process, the interest rate is attracted elastically toward its mean value μ and is influenced by a noise term whose variance is proportional to the prevailing level of the interest rate. As a consequence, the interest rate cannot become negative. Assuming logarithmic utility, the drift term for the equivalent process is $f^* = \kappa\mu - (\kappa + \lambda)r$.

With this specification of f^*, (9) or (10) is solved by

$$P(r, t, T) = A(T - t)\,\mathrm{e}^{-B(T - t)r} \tag{12}$$

where

$$A(\tau) \equiv \left[\frac{2\eta\,\mathrm{e}^{(\kappa + \lambda + \eta)\tau/2}}{b(\tau)}\right]^{2\kappa\mu/\sigma^2}, \qquad B(\tau) \equiv \left[\frac{2(\mathrm{e}^{\eta\tau} - 1)}{b(\tau)}\right]$$

$$b(\tau) \equiv 2\eta + (\kappa + \lambda + \eta)(\mathrm{e}^{\eta\tau} - 1), \qquad \eta \equiv \sqrt{(\kappa + \lambda)^2 + 2\sigma^2}.$$

The resulting zero coupon yield has a limit of

$$R_\infty = 2\kappa\mu/(\eta + \kappa + \lambda)$$

regardless of the current short rate. (The constancy of the long rate may seem like a severe restriction of this model. Actually, it is a property that is to be expected of any recurrent model. See Dybvig, Ingersoll and Ross, 1986.) If the current short rate is below R_∞, then the yield curve is upward sloping. If the current rate is greater than μ, then the yield curve slopes downward. For interest rates between these two levels the yield curve has a single hump.

Brown and Dybvig (1986) have estimated a reduced form of this model, yield curve by yield curve, to obtain a time series of parameter estimates. They found that the implied variance tracks actual interest rate volatility well. They also find some evidence of misspecification, including what appears to be a tax effect.

BOND PRICING WITH MULTIPLE SOURCES OF UNCERTAINTY. One criticism of previous models is that they permit only a single source of uncertainty. As a result of this assumption all bonds are perfectly correlated and all yield curves are characterized by a single parameter. Typically there would be factors that influence long and short rates differently. However, the basic techniques for bond pricing remain the same.

Suppose for simplicity that interest rates are determined by the short rate and one other state variable, x. The assumed dynamics for the two state variables are

$$dr = f(r, x, t) dt + g(r, x, t) d\omega_1$$
$$dx = \phi(r, x, t) dt + \gamma(r, x, t) d\omega_2. \qquad (13)$$

The resulting partial differential equation for bond pricing is

$$rP = \tfrac{1}{2}g^2 P_{rr} + \rho g\gamma P_{rx} + \tfrac{1}{2}\gamma^2 P_{xx} + f^* P_r + \phi^* P_x + P_t \qquad (14)$$

where as before $f^*(\cdot)$ and $\phi^*(\cdot)$ denote the modified (risk adjusted) drift terms under the equivalent martingale measure, and ρ is the correlation between the two Wiener processes. Solving this problem requires a specification of the joint stochastic process and the necessary modification of the equivalent martingale measure.

If the second state variable, x, is the price of some asset, then the risk premium, $-\pi_2(\cdot)\gamma(\cdot)$, associated with ω_2 is $rx - \phi(\cdot)$ giving $\phi^* = rx$ as in the Black–Scholes option pricing model. We can also infer the risk premium if x is functionally related to an asset's value and time. For example suppose that x is the yield-to-maturity on a zero coupon bond maturing at s. The price of this bond, $P(r, x, t, s)$, must itself satisfy the pricing equation. As $P(r, x, t, s) \equiv \exp[-x(s - t)]$, we know all of its required partial derivatives. Substituting them into (14) and solving for $\phi^*(\cdot)$ gives

$$\phi^*(r, x, t) = \frac{r - x - \tfrac{1}{2}\gamma^2(\cdot)(s - t)^2}{t - s} \qquad (15)$$

(Ingersoll, 1987). Of course, if x is not known to be related to a marketed asset, we can still specify the second risk premium arbitrarily or by reference to an equilibrium model.

Brennan and Schwartz (1979) used a finite difference numerical approximation to analyse a two factor model. In a test with US Government bonds, they concluded that the two factor model predicted bond prices much better than did a one factor model and on the whole was an adequate description of bond prices.

The traditional forecasting models using geometrically smoothed averages of past short rates as predictors of future rates can also be viewed as two (or more) factor models of this sort. (See Dobson, Sutch and Vanderford (1976) for a survey of the forecasting models.) For example, consider the model of Malkiel (1966) in which the short rate tends to return to a 'normal level', measured by a geometric average of past interest rates

$$x(t) \equiv \beta \int_0^\infty e^{-\beta s} r(t - s) ds. \qquad (16)$$

The dynamics of this model are

$$dr = \kappa(x - r)dt + \sigma(\cdot)d\omega \qquad (17a)$$
$$dx = \beta(r - x)dt. \qquad (17b)$$

As changes in the state variable x are locally deterministic, no risk adjustment is required. The modification to the stochastic process for r is handled in the usual fashion.

No closed form solution is known for this model when $\sigma(\cdot) = \sigma\sqrt{r}$ as in the Cox, Ingersoll and Ross (1985b) model. A solution for this and similar problems with other lag structures is given in Cox, Ingersoll and Ross (1981) when the variance is a constant.

Another form of two factor model uses real interest rates and expected inflation as its two state variables (Richard, 1978; Cox, Ingersoll and Ross, 1985b). Besides providing more flexibility for a better empirical fit, this formulation permits an identification of real and nominal effects for separate consideration.

APPLICATIONS OF INTEREST RATE MODELS. This continuous-time no-arbitrage method of pricing bonds is an outgrowth of the option pricing literature which gives it certain advantages over more traditional approaches. One advantage is the provision of a fully specified model of bond prices for empirical work. Another major advantage of term structure models of this type is that they provide a framework for valuation that is consistent with all of our other models based on the absence of arbitrage.

Thus, in addition to pricing zero coupon bonds and determining the term structure, this method can handle any other interest rate valuation problem. For example Cox, Ingersoll and Ross (1985b) value call options on interest rates. Applications to futures contracts, variable rate instruments, mortgages, loan commitments, etc. have also been published. Equation (9) or a multiple factor version such as (14) remains the fundamental relation among interest rate contingent claims. To value a particular claim the appropriate boundary condition is used in place of $P(r, t, T) = 1$.

Another advantage of such models is that they give an explicit measure of the risk characteristics of the priced assets. These risk measures can be used in immunizing bond portfolios or in relative performance measurement. Because they are derived in models based on the absence of arbitrage they are not subject to the same criticisms that have been made of traditional duration measures (see Ingersoll, Skelton and Weil, 1978).

Of theoretical interest is the relation between these models and the risk neutral or equivalent martingale valuation procedure. With a constant interest rate it can be shown that the value of a derivative asset that pays $H(S_T)$ at time T, contingent on the value S_T of its primitive asset and makes no other disbursements, is

$$V(S, t) = e^{-r(T-t)} E^*[H(S_T)]. \qquad (18)$$

The expectation $E^*[\cdot]$ is taken with respect to the risk neutral process for the primitive's price under which the actual expected rate of return is replaced by the risk-free interest rate. With a stochastic interest rate, the valuation is

$$V(S, t) = E^*\left\{\exp\left[-\int_t^T r(s) ds\right] H(S_T)\right\}. \qquad (19)$$

This equation generalizes both equations (18) and (10). The expectation in (19) is over the joint distribution under the martingale measure of interest rate paths and S_T. That is, the expectation assumes that

$$dS(t) = r(t)S dt + \sigma(\cdot)S d\omega_S \qquad (20a)$$
$$dr(t) = f^*(\cdot) dt + g(\cdot) d\omega_r \qquad (20b)$$

plus the modified processes of any other state variables that determine the term structure.

J.E. INGERSOLL

See also ARBITRAGE; CONTINUOUS-TIME STOCHASTIC PROCESSES; OPTION PRICING; TERM STRUCTURE OF INTEREST RATES; WIENER PROCESS.

BIBLIOGRAPHY
Brennan, M.J. and Schwartz, E.S. 1979. A continuous time approach to the pricing of bonds. *Journal of Banking and Finance* 3(2), September, 133–55.
Brown, S.J. and Dybvig, P.H. 1986. The empirical investigation of the Cox, Ingersoll, Ross theory of the term structure of interest rates. *Journal of Finance* 41(3), July, 616–30.
Cox, J.C., Ingersoll, J.E. and Ross, S.A. 1981. A re-examination of traditional hypotheses about the term structure of interest rates. *Journal of Finance* 36(4), September, 769–99.
Cox, J.C., Ingersoll, J.E. and Ross, S.A. 1985a. An intertemporal general equilibrium model of asset prices. *Econometrica* 53(2), March, 363–84.
Cox, J.C., Ingersoll, J.E. and Ross, S.A. 1985b. A theory of the term structure of interest rates. *Econometrica* 53(2), March, 385–407.
Dobson, S., Sutch, R. and Vanderford, D. 1976. An evaluation of alternative empirical models of the term structure of interest rates. *Journal of Finance* 31(4), September, 1035–65.
Dothan, L.U. 1978. On the term structure of interest rates. *Journal of Financial Economics* 6(1), March, 59–69.
Dybvig, P.H., Ingersoll, J.E. and Ross, S.A. 1986. Long forward rates can never fall. Unpublished working paper, Yale University.
Friedman, A. 1975. *Stochastic Differential Equations and Applications.* Vol. 1, New York: Academic Press.
Ingersoll, J.E. 1987. *Theory of Financial Decision Making.* Totowa, NJ: Rowman and Littlefield.
Ingersoll, J.E., Skelton, J. and Weil, R.L. 1978. Duration forty years later. *Journal of Financial and Quantitative Analysis* 13(4), November, 627–50.
Malkiel, B.G. 1966. *The Term Structure of Interest Rates: Expectations and Behavior Patterns.* Princeton: Princeton University Press.
Merton, R.C. 1970. A dynamic general equilibrium model of the asset market and its application to the pricing of the capital structure of the firm. Unpublished working paper, Sloan School of Management, MIT.
Richard, S.F. 1978. An arbitrage model of the term structure of interest rates. *Journal of Financial Economics* 6(1), March, 33–57.
Vasicek, O.A. 1977. An equilibrium characterization of the term structure. *Journal of Financial Economics* 5(2), November, 177–88.

interests. 'Interest' or 'interests' is one of the most central and controversial concepts in economics and, more generally, in social science and history. Since coming into widespread use in various European countries around the latter part of the 16th century as essentially the same Latin-derived word (*intérêt, interesse,* etc.), the concept has stood for the fundamental forces, based on the drive for self-preservation and self-aggrandizement, that motivate or should motivate the actions of the prince or the state, of the individual, and later of groups of people occupying a similar social or economic position (classes, interest groups). When related to the individual, the concept has at times had a very inclusive meaning, encompassing interest in honour, glory, self-respect, and even after-life, while at other times it became wholly confined to the drive for economic advantage. The esteem in which interest-motivated behaviour is held has also varied drastically. The term was originally pressed into service as a euphemism serving, already in the late Middle Ages, to make respectable an activity, the taking of interest on loans, that had long been considered contrary to divine law and known as the sin of usury. In its wider meanings, the term at times achieved enormous prestige as key to a workable and peaceful

social order. But it has also been attacked as degrading to the human spirit and corrosive of the foundations of society. An inquiry into these multiple meanings and appreciations is in effect an exploration of much of economic history and in particular of the history of economic and political doctrine in the West over the past four centuries.

The concept, moreover, still plays a central role in contemporary economics and political economy: the construct of the self-interested, isolated individual who chooses freely and rationally between alternative courses of action after computing their prospective costs and benefits to him or herself, that is, while ignoring costs and benefits to other people and to society at large, underlies much of welfare economics; and the same perspective has yielded important, if disturbing contributions to a broader science of social interactions, such as the Prisoner's Dilemma theorem and the obstacles to collective action because of free riding.

Two essential elements appear to characterize interest-propelled action: *self-centredness*, that is, predominant attention of the actor to the consequences of any contemplated action for himself; and *rational calculation*, that is, a systematic attempt at evaluating prospective costs, benefits, satisfactions, and the like. Calculation could be considered the dominant element: once action is supposed to be informed only by careful estimation of costs and benefits, with most weight necessarily being given to those that are better known and more quantifiable, it tends to become self-referential by virtue of the simple fact that each person is best informed about his *own* satisfactions and disappointments.

INTEREST AND STATECRAFT. Rational calculation also played the chief role in the emergence of the concept of interest-motivated action on the part of the prince in the 16th and 17th centuries. It accounts for the high marks interest-governed behaviour received during the late 16th- and early 17th-century phases of its career in politics. The term did duty on two fronts. First, it permitted the emergent science of statecraft to assimilate the important insights of Machiavelli. The author of *The Prince* had almost strained to advertise those aspects of politics that clashed with conventional morality. He dwelt on instances where the prince was well advised or even duty bound to practise cruelty, mendacity, treason, and so on. Just as, in connection with money lending, the term interest came into use as a euphemism for the earlier term usury, so did it impose itself on the political vocabulary as a means of anaesthetizing, assimilating and developing Machiavelli's shocking insights.

But in the early modern age, 'interest' was not only a label under which a ruler was given *new latitude* or was absolved from feeling guilty about following a practice that he had previously been taught to consider as immoral: the term also served to impose *new restraints* as it enjoined the Prince to pursue his interests with a rational, calculating spirit that would often imply prudence and moderation. At the beginning of the 17th century, the interests of the sovereign were contrasted with the wild and destructive passions, that is, with the immoderate and foolish seeking of glory and other excesses involved in pursuing the by then discredited heroic ideal of the Middle Ages and the Renaissance. This disciplinary aspect of the doctrine of interest was particularly driven home in the influential essay *On the Interest of Princes and States of Christendom* by the Huguenot statesman the Duke of Rohan (1579–1638).

The interest doctrine thus served to release the ruler from certain traditional restraints (or guilt feelings) only to subject him to new ones that were felt to be far more efficacious than the well-worn appeals to religion, morals, or abstract reason.

Genuine hope arose that, with princely or national interest as guide, statecraft would be able to produce a more stable political order and a more peaceful world.

INTEREST AND INDIVIDUAL BEHAVIOUR. The early career of the interest concept with regard to statecraft finds a remarkable parallel in the role it played in shaping behaviour codes for individual men and women in society. Here also a new license went hand in hand with a new restraint.

The new license consisted in the legitimation and even praise that was bestowed upon the single-minded pursuit of material wealth and upon activities conducive to its accumulation. Just as Machiavelli had opened up new horizons for the Prince, so did Mandeville two centuries later list a number of 'don'ts' for the commoner, in this case primarily in relation to money making. Once again, a new insight into human behaviour or into the social order was first proclaimed as a startling, shocking paradox. Like Machiavelli, Mandeville presented his thesis on the beneficial effects on the general welfare of the luxury trades (which had long been strictly regulated) in the most scandalous possible fashion, by referring to the activities, drives and emotions associated with these trades as 'private vices'. Here again, his essential message was eventually absorbed into the general stock of accepted practice by changing the language with which he had proclaimed his discovery. For the third time, euphemistic resort was had to 'interest', this time in substitution for such terms as 'avarice', 'love of lucre', and so on. The transition from one set of terms to the other is reflected by the first lines of David Hume's essay 'On the Independency of Parliament':

> Political writers have established it as a maxim, that, in contriving any system of government and fixing the several checks and balances of the constitution, every man ought to be supposed a *knave*, and to have no other end, in all his actions, than private interest. By this interest we must govern him, and, by means of it, make him, notwithstanding his insatiable avarice and ambition, cooperate to public good (Hume, 1742, vol. I, pp. 117–18, emphasis in the original).

Here interest is explicitly equated with knavishness and 'insatiable avarice'. But soon thereafter the memory of these unsavoury synonyms of interest was suppressed, as in Adam Smith's famous statement about the butcher, the brewer, and the baker who are driven to supply us with our daily necessities through their interest rather than their benevolence. Smith thus did for Mandeville what the Duke of Rohan had done for Machiavelli. His doctrine of the Invisible Hand legitimated total absorption of the citizen in the pursuit of private gain and thereby served to assuage any guilt feelings that might have been harboured by the many Englishmen who were drawn into commerce and industry during the commercial expansion of the 18th century but had been brought up under the civic humanist code enjoining them to serve the public interest *directly* (Pocock, 1982). They were now reassured that by pursuing gains they were doing so *indirectly*.

In fact, Adam Smith was not content to praise the pursuit of private gain. He also berated citizens' involvement in public affairs. Right after his Invisible Hand statement he wrote 'I have never known much good done by those who affected to trade for the public good' (1776, p. 423). Ten years before, Sir James Steuart had supplied an interesting explanation for a similar aversion toward citizens' involvement in public affairs.

> ... were everyone to act for the public, and neglect himself, the statesman would be bewildered ... were a people to

become quite disinterested, there would be no possibility of governing them. Everyone might consider the interest of his country in a different light, and many might join in the ruin of it, by endeavouring to promote its advantages (1767, vol. I, pp. 243–44).

In counterpart to the new area of authorized and recommended behaviour, these statements point to the important *restraints* that accompanied the doctrine of interest. For the individual citizen or subject as for the ruler, interest-propelled action meant originally action informed by rational calculation in any area of human activity – political, cultural, economic, personal and so on. In the 17th century and through part of the 18th, this sort of methodical, prudential, interest-guided action was seen as vastly preferable to actions dictated by the violent, unruly and disorderly passions. At the same time, the interests of the vast majority of people, that is of those outside of the highest reaches of power, came to be more narrowly defined as economic, material or 'moneyed' interests, probably because the non-elite was deemed to busy itself primarily with scrounging a living with no time left to worry about honour, glory, and the like. The infatuation with interest helped bestow legitimacy and prestige on commercial and related private activities, that had hitherto ranked rather low in public esteem; correspondingly, the Renaissance ideal of glory, with its implicit celebration of the public sphere, was downgraded and debunked as a mere exercise in the destructive passion of self-love (Hirschman, 1977, pp. 31–42).

THE POLITICAL BENEFITS OF AN INTEREST-BASED SOCIAL ORDER. The idea that the interests, understood as the methodical pursuit and accumulation of private wealth, would bring a number of benefits in the political realm took various distinct forms. There was, first of all, the expectation that they would achieve at the macrolevel what they were supposed to accomplish for the individual: hold back the violent passions of the 'rulers of mankind'. Here the best-known proposition, voiced early in the 18th century, says that the expansion of commerce is incompatible with the use of force in international relations and would gradually make for a peaceful world. Still more utopian hopes were held out for the effects of commerce on domestic politics: the web of interests delicately woven by thousands of transactions would make it impossible for the sovereign to interpose his power brutally and wantonly through what was called '*grands coups d'autorité*' by Montesquieu or 'the folly of despotism' by Sir James Steuart. This thought was carried further in the early 19th century when the intricacies of expanding industrial production compounded those of commerce: in the technocratic vision of Saint-Simon the time was at hand when economic exigencies would put an end, not just to *abuses* of the power of the state, but to any power whatsoever of man over man: politics would be replaced by administration of 'things'. As is well known this conjecture was taken up by Marxism with its prediction of the withering away of the state under communism. An argument that a century earlier had been advanced on behalf of emergent capitalism was thus refurbished for a new, *anti*-capitalist utopia.

Another line of thought about the political effects of an interest-driven society looks less at the constraints such as society will impose upon those who govern than at the difficulties of the task of governing. As already noted, a world where people methodically pursue their private interests was believed to be far more predictable, and hence *more governable*, than one where the citizens are vying with each other for honour and glory.

The stability and lack of turbulence that were expected to characterize a country where men pursue singlemindedly their material interests were very much on the minds of some of the 'inventors' of America, such as James Madison and Alexander Hamilton. The enormous prestige and influence of the interest concept at the time of the founding of America is well expressed in Hamilton's statement:

> The safest reliance of every government is on man's interests. This is a principle of human nature, on which all political speculation, to be just, must be founded (Hamilton [1784], cited in Terence Ball, 1983, p. 45).

Finally, a number of writers essentially extrapolated from the putative personality traits of the individual trader, as the prototype of interest-driven man, to the general characteristics of a society where traders would predominate. In the 18th century, perhaps as a result of some continuing disdain for economic pursuits, commerce and money-making were often described as essentially innocuous or 'innocent' pastimes, in contrast no doubt with the more violent or more strenuous ways of the upper or lower classes. Commerce was to bring 'gentle' and 'polished' manners. In French, the term innocent appended to commerce was often coupled with *doux* (sweet, gentle) and what has been called the thesis of the *doux commerce* held that commerce was a powerful civilizing agent diffusing prudence, probity and similar virtues within and among trading societies (Hirschman, 1977, 1982a). Only under the impact of the French Revolution did some doubt arise on the direction of the causal link between commerce and civilized society: taken aback by the outbreak of social violence on a large scale, Edmund Burke suggested that the expansion of commerce depended itself on the *prior* existence of 'manners' and 'civilization' and on what he called 'natural protecting principles' grounded in 'the spirit of a gentleman' and 'the spirit of religion' (Burke, 1790, p. 115; Pocock, 1982).

THE INVISIBLE HAND. The capstone of the doctrine of self-interest was of course Adam Smith's Invisible Hand. Even though this doctrine, being limited to the economic domain, was more modest than the earlier speculations on the beneficent *political* effects of trade and exchange, it soon came to dominate the discussion. An intriguing paradox was involved in stating that the *general* interest and welfare would be promoted by the self-interested activities of innumerable decentralized operators. To be sure, this was not the first nor the last time that such a claim of identity or coincidence or harmony of interests of a part with those of a whole has been put forward. Hobbes had advocated an absolute monarchy on the ground that this form of government brings about an identity of interest between ruler and ruled; as just noted, the writers of the Scottish Enlightenment saw an identity of interest between the general interests of British society and the interests of the middle ranks; such an identity between the interests of one class and those of society became later a cornerstone of Marxism, with the middling ranks having of course been supplanted by the proletariat; and finally, the American pluralist school in political science returned essentially to the Smithian scheme of harmony between many self-interests and the general interest, with Smith's individual economic operators having been replaced by contending 'interest groups' on the political stage.

All these *Harmonielehren* have two factors in common: the 'realistic' affirmation that we have to deal with men and women, or with groups thereof, 'as they really are', and an attempt to prove that it is possible to achieve a workable and progressive social order with these highly imperfect subjects,

and, as it were, behind their backs. The mixture of paradoxical insight and alchemy involved in these constructs makes them powerfully attractive, but also accounts for their ultimate vulnerability.

THE INTERESTS ATTACKED. The 17th century was perhaps the real heyday of the interest doctrine. Governance of the social world by interest was then viewed as an alternative to the rule of destructive passions; that was surely a lesser evil, and possibly an outright blessing. In the 18th century, the doctrine received a substantial boost in the economic domain through the doctrine of the Invisible Hand, but it was indirectly weakened by the emergence of a more optimistic view of the passions: such passionate sentiments and emotions as curiosity, generosity, and sympathy were then given detailed attention, the latter in fact by Adam Smith himself in his *Theory of Moral Sentiments*. In comparison to such fine, newly discovered or rehabilitated springs of human action, interest no longer looked nearly so attractive. Here was one reason for the reaction against the interest paradigm that unfolded toward the end of the 18th century and was to fuel several powerful 19th-century intellectual movements.

Actually the passions did not have to be wholly transformed into benign sentiments to be thought respectable and even admirable by a new generation. Once the interests appeared to be truly in command with the vigorous commercial and industrial expansion of the age, a general lament went up for 'the world we have lost'. The French Revolution brought another sense of loss and Edmund Burke joined the two when he exclaimed, in his *Reflections on the Revolution in France*, 'the age of chivalry is gone; that of sophisters, economists and calculators has succeeded; and the glory of Europe is extinguished for ever' (1790, p. 111). This famous statement came a bare 14 years after the *Wealth of Nations* had denounced the rule of the 'great lords' as a 'scene of violence, rapine and disorder' and had celebrated the benefits flowing from everyone catering to his interests through orderly economic pursuits. Now Burke was an intense admirer of Adam Smith and took much pride in the identity of views on economic matters between himself and Smith (Winch, 1985; Himmelfarb, 1984). His 'age of chivalry' statement, so contrary to the intellectual legacy of Smith, therefore signals one of those sudden changes in the general mood and understanding from one age to the next of which the exponents themselves are hardly aware. Burke's lament set the tone for much of the subsequent Romantic protest against an order based on the interests which, once it appeared to be dominant, was seen by many as lacking nobility, mystery, and beauty.

This nostalgic reaction merged with the observation that the interests, that is, the drive for material wealth, were not nearly as 'innocuous', 'innocent' or 'mild', as had been thought or advertised. To the contrary, it was now the drive for material advantage that suddenly loomed as a subversive force of enormous power. Thomas Carlyle thought that all traditional values were threatened by 'that brutish god-forgetting Profit-and-Loss Philosophy' and protested that 'cash payment is not the only nexus of man with man' (1843, p. 187). This phrase – cash-nexus – was taken over by Marx and Engels who used it to good effect in the first section of the *Communist Manifesto* where they painted a lurid picture of the moral and cultural havoc wrought by the conquering bourgeoisie.

Many other critics of capitalist society dwelt on the destructiveness of the new energies that were relased by a social order in which the interests were given free rein. In fact, the thought arose that these forces were so wild and out of control that they might undermine the very foundations on which the social order was resting, that they were thus bent on self-destruction. In a startling reversal, feudal society, which had earlier been treated as 'rude and barbarous' and was thought to be in permanent danger of dissolution because of the unchecked passions of violent rulers and grandees, was perceived in retrospect to have nurtured such values as honour, respect, friendship, trust and loyalty, that were essential for the functioning of an interest-dominated order, but were relentlessly, if inadvertently, undermined by it. This argument was already contained in part in Burke's assertion that it is civilized society that lays the groundwork for commerce rather than vice versa; it was elaborated by a large and diverse group of authors, from Richard Wagner via Schumpeter to Karl Polanyi and Fred Hirsch (Hirschman, 1982a, pp. 1466–70).

THE INTERESTS DILUTED. While the interest doctrine thus met with considerable opposition and criticism in the 19th century, its prestige remained nevertheless high, particularly because of the vigorous development of economics as a new body of scientific thought. Indeed, the success of this new science made for attempts to utilize its insights, such as the interest concept, for elucidating some non-economic aspects of the social world. In his *Essay on Government* (1820), James Mill formulated the first 'economic' theory of politics and based it – just as was later done by Schumpeter, Anthony Downs, Mancur Olson etc. – on the assumption of rational self-interest. But this widening of the use of the concept turned out to be something of a disservice. In politics, so Mill had to recognize, the gap between the 'real' interest of the citizen and 'a false supposition [i.e., perception] of interest' can be extremely wide and problematic (1820, p. 88). This difficulty provided an opening for Macaulay's withering attack in the *Edinburgh Review* (1829). Macaulay pointed out that Mill's theory was empty: interest 'means only that men, if they can, will do as they choose ... it is ... idle to attribute any importance to a proposition which, when interpreted, means only that a man had rather do what he had rather do' (p. 125).

The charge that the interest doctrine was essentially tautological acquired greater force as more parties climbed on the bandwagon of interest, attempting to bend the concept to their own ends. As so many key concepts used in everyday discourse, 'interest' had never been strictly defined. While individual self-interest in material gain predominated, wider meanings were never completely lost sight of. An extremely wide and inclusive interpretation of the concept was put forward at a very early stage in its history: Pascal's Wager was nothing but an attempt to demonstrate that belief in God (hence, conduct in accordance with His precepts) was strictly in our (long-term) self-interest. Thus the concept of *enlightened* self-interest has a long history. But it received a boost and special, concrete meaning in the course of the 19th century. With the contemporary revolutionary outbreaks and movements as an ominous backdrop, advocates of social reform were able to argue that a dominant social group is well advised to surrender some of its privileges or to improve the plight of the lower classes so as to insure social peace. 'Enlightened' self-interest of the upper classes and conservative opinion was appealed to, for example, by the French and English advocates of universal suffrage or electoral reform at mid-century; it was similarly invoked by the promoters of the early social welfare legislation in Germany and elsewhere toward the end of the century, and again by Keynes and the Keynesians who favoured limited intervention of the state in the economy through countercyclical policy and 'automatic stabilizers'

resulting from welfare state provisions. These appeals were often made by reformers who, while fully convinced of the intrinsic value and social justice of the measures they advocated, attempted to enlist the support of important groups by appealing to their 'longer-term' rather than short-term and therefore presumably *shortsighted* interests. But the advocacy was not only tactical. It was sincerely put forward and testified to the continued prestige of the notion that interest-motivated social behaviour was the best guarantee of a stable and harmonious social order.

Whereas enlightened self-interest was something the upper classes of society were in this manner pressed to ferret out, the lower classes were similarly exhorted, at about the same epoch but from different quarters, to raise their sights above day-to-day pursuits. Marx and the Marxists invited the working class to become aware of its *real interests* and to shed the 'false consciousness' from which it was said to be suffering as long as it did not throw itself wholeheartedly into the class struggle. Once again, the language of interests was borrowed for the purpose of characterizing and dignifying a type of behaviour a group was being pressed to adopt.

Here, then, was one way in which the concept of interest-motivated behaviour came to be diluted. Another was the progressive loss of the sharp distinction an earlier age had made between the passions and the interests. Already Adam Smith had used the two concepts jointly and interchangeably. Even though it became abundantly clear in the 19th century that the desire to accumulate wealth was anything but the 'calm passion' as which it had been commended by some 18th-century philosophers, there was no return to the earlier distinction between the interests and the passions or between the wild and the mild passions. Money-making had once and for all been identified with the concept of interest so that all forms of this activity, however passionate or irrational, were automatically thought of as interest-motivated. As striking new forms of accumulation and industrial or financial empire-building made their appearance, new concepts were introduced, such as entrepreneurial leadership and intuition (Schumpeter, 1911) or the 'animal spirits' of the capitalists (Keynes, 1936, pp. 161–63). But they were not contrasted with the interests, and were rather assumed to be one of their manifestations.

In this manner the interests came to cover virtually the entire range of human actions, from the narrowly self-centred to the sacrificially altruistic, and from the prudently calculated to the passionately compulsive. In the end, interest stood behind anything people do or wish to do and to explain human action by interest thus did turn into the vacuous tautology denounced by Macaulay. At about the same time, other key and time-honoured concepts of economic analysis, such as utility and value, became similarly drained of their earlier psychological or normative content. The positivistically oriented science of economics that flourished during much of this century felt that it could do without any of these terms and replaced them by the less value- or psychology-laden 'revealed preference' and 'maximizing under constraints'. And thus it came to pass that interest, which had rendered such long and faithful service as a euphemism, was now superseded in turn by various even more neutral and colourless neologisms.

The development of the self-interest concept and of economic analysis in general in the direction of positivism and formalism may have been related to the discovery, toward the end of the 19th century, of the instinctual-intuitive, the habitual, the unconscious, the ideologically and neurotically driven–in short, to the extraordinary vogue for the nonrational that characterized virtually all of the influential philosophical, psychological and sociological thinking of the time. It was out of the question for economics, all based on rationally pursued self-interest, to incorporate the new findings into its own apparatus. So that discipline reacted to the contemporary intellectual temper by withdrawing from psychology to the greatest possible extent, by emptying its basic concepts of their psychological origin–a survival strategy that turned out to be highly successful. It is of course difficult to prove that the rise of the non-rational in psychology and sociology and the triumph of positivism and formalism in economics were truly connected in this way. Some evidence is supplied by the remarkable case of Pareto: he made fundamental contributions both to a sociology that stressed the complex 'non-logical' (as he put it) aspects of social action and to an economics that is emancipated from dependence on psychological hedonism.

CURRENT TRENDS. Lately there have been signs of discontent with the progressive evisceration of the concept of interest. On the conservative side, there was a return to the orthodox meaning of interest and the doctrine of enlightened self-interest was impugned. Apart from the discovery, first made by Tocqueville, that reform is just as likely to unleash as to prevent revolution, it was pointed out that most well-meant reform moves and regulations have 'perverse' side effects which compound rather than alleviate the social ills one had set out to cure. It was best, so it appeared, not to stray from the narrow path of narrow self-interest, and it was confusing and pointless to dilute this concept.

Others agreed with the latter judgement, but for different reasons and with different conclusions. They also disliked the manoeuvre of having every kind of human action masquerade under the interest label. But they regarded as relevant for economics certain human actions and activities which cannot be accounted for by the traditional notion of self-interest: actions motivated by altruism, by commitment to ethical values, by concern for the group and the public interest, and, perhaps most important, the varieties of non-instrumental behaviour. A beginning has been made by various economists and other social scientists to take these kinds of activities seriously, that is, to abandon the attempt to categorize them as mere variants of interest-motivated activity (Boulding, 1973; Collard, 1978; Hirschman, 1985; Margolis, 1982; McPherson, 1984; Phelps, 1975; Schelling, 1984; Sen, 1977).

One important aspect of these various forms of behaviour which do not correspond to the classical concept of interest-motivated action is that they are subject to considerable variation. Take actions in the public interest as an example. There is a wide range of such actions, from total involvement in some protest movement down to voting on Election Day and further down to mere grumbling about, or commenting on, some public policy within a small circle of friends or family – what Guillermo O'Donnell has called 'horizontal voice' in contrast to the 'vertical' voice directly addressed to the authorities (1986). The actual degree of participation under more or less normal political conditions is subject to constant fluctuations along this continuum, in line with changes in economic conditions, government performance, personal development, and many other factors. As a result, with total time for private *and* public activity being limited, the intensity of citizens' dedication to their private interests is also subject to constant change. Near-total privatization occurs only under certain authoritarian governments, for the most repressive regimes do not only do away with the free vote and any open manifestation of dissent, but also manage to suppress, through their display of terrorist

power, all *private* expressions of inconformity with public policy, that is, all those manifestations of 'horizontal voice' that are actually important forms of public involvement.

An arresting conclusion follows. That vaunted ideal of predictability, that alleged idyll of a privatized citizenry paying busy and exclusive attention to its economic interests and thereby serving the public interest indirectly, but never directly, becomes a reality only under wholly nightmarish political conditions! More civilized political circumstances necessarily imply a less transparent and less predictable society.

Actually, this outcome of the current inquiries into activities not strictly motivated by traditional self-interest is all to the good: for the only certain and predictable feature of human affairs is their unpredictability and the futility of trying to reduce human action to a single motive–such as interest.

ALBERT O. HIRSCHMAN

See also ECONOMIC INTERPRETATION OF HISTORY; EXIT AND VOICE; INVISIBLE HAND; PROPERTY; SELF-INTEREST.

BIBLIOGRAPHY

Ball, T. 1983. The ontological presuppositions and political consequences of a social science. In *Changing Social Science*, ed. D.R. Sabia, Jr. and J.T. Wallulis, Albany: State University of New York Press.

Boulding, K.E. 1973. *The Economy of Love and Fear: A Preface to Grants Economics*. Belmont, California: Wadsworth.

Burke, E. 1790. *Reflections on the Revolution in France*. Chicago: Regnery, 1955.

Carlyle, T. 1843. *Past and Present*. New York: New York University Press, 1977.

Collard, D. 1978. *Altruism and Economy: A Study in Non-selfish Economics*. Oxford: Robertson.

Collini, S., Winch, D. and Burrow, J. 1983. *That Noble Science of Politics: A Study in Nineteenth-century Intellectual History*. Cambridge: Cambridge University Press.

Hamilton, A. 1784. Letters from Phocion, Number I. In *The Works of Alexander Hamilton*, ed. John C. Hamilton, New York: C.S. Francis, 1851, Vol. II, 322.

Himmelfarb, G. 1984. *The Idea of Poverty: England in the Early Industrial Age*. New York: Knopf.

Hirschman, A.O. 1977. *The Passions and the Interests: Political Arguments for Capitalism Before its Triumph*. Princeton: Princeton University Press.

Hirschman, A.O. 1982a. Rival interpretations of market society: civilizing, destructive, or feeble? *Journal of Economic Literature*, 20(4), December, 1463–84.

Hirschman, A.O. 1982b. *Shifting Involvements: Private Interest and Public Action*. Princeton: Princeton University Press.

Hirschman, A.O. 1985. Against parsimony: three easy ways of complicating some categories of economic discourse. *Economics and Philosophy*, 1, 7–21.

Hume, D. 1742. *Essays Moral, Political and Literary*. Ed. T.H. Green and T.H. Grose, London: Longmans, 1898.

Keynes, J.M. 1936. *The General Theory of Employment Interest and Money*. London: Macmillan.

Macaulay, T.B. 1829. Mill's Essay on Government. In *Utilitarian Logic and Politics*, ed. J. Lively and J. Rees, Oxford: Clarendon, 1978.

McPherson, M.S. 1984. Limits on self-seeking: the role of morality in economic life. In *Neoclassical Political Economy*, ed. D.C. Colander, Cambridge, Mass.: Ballinger, 71–85.

Margolis, H. 1982. *Selfishness, Altruism and Rationality*. Cambridge: Cambridge University Press.

Meinecke, F. 1924. *Die Idee der Staatsräson in der Neueren Geschichte*. Munich: Oldenburg.

Mill, J. 1820. *Essay on Government*. In *Utilitarian Logic and Politics*, ed. J. Lively and J. Rees, Oxford: Clarendon, 1978.

O'Donnell, G. 1986. On the convergences of Hirschman's *Exit, Voice and Loyalty* and *Shifting Involvements*. In *Development, Democracy and the Art of Trespassing: Essays in Honor of A.O. Hirschman*, ed. A. Foxley et al., Notre Dame, Ind.: University of Notre Dame Press.

Phelps, E.S. (ed.) 1975. *Altruism, Morality and Economic Theory*. New York: Russell Sage Foundation.

Pocock, J.G.A. 1982. The political economy of Burke's analysis of the French Revolution. *Historical Journal* 25, June, 331–49.

Rohan, H., Duc de. 1638. *De l'interêt des princes et états de la chrétienté*. Paris: Pierre Margat.

Schelling, T.C. 1984. *Choice and Consequence*. Cambridge, Mass.: Harvard University Press.

Schumpeter, J.A. 1911. *The Theory of Economic Development*. Cambridge, Mass.: Harvard University Press, 1951.

Sen, A. 1977. Rational fools: a critique of the behavioral foundations of economic theory. *Philosophy and Public Affairs* 6(4), Summer, 317–44.

Smith, A. 1776. *An Inquiry into the Nature and Causes of the Wealth of Nations*. Ed. E. Cannan, New York: Modern Library, 1937.

Steuart, J. 1767. *Inquiry into the Principles of Political Oeconomy*. Ed. A.S. Skinner, Chicago: University of Chicago Press, 1966.

Winch, D. 1985. The Burke–Smith problem and late eighteenth century political and economic thought. *Historical Journal* 28(1), 231–47.

intergenerational models. Recently it has been widely recognized by economists that the extension of the Arrow–Debreu model to a dynamic multi-period economy should not be restricted to models with a finite number of economic agents, each facing an infinite horizon. In analysing many economic problems, it seems natural to consider an open-ended horizon economy with individuals (or households) living for a finite number of periods; thus, at each date the economy consists of consumers of different ages (who interact with each other) and hence are inherently characterized by different economic parameters (such as their current income or planning horizon). In his seminal work, Samuelson (1958–9) attempts to explain, in an overlapping generations (OLG) equilibrium model, Irving Fisher's (1961) theory of interest. This simple model of a market economy characterized by an unbounded horizon, short-lived, overlapping, but essentially identical households, is different from the Arrow–Debreu model in various aspects. We shall concentrate upon similar models which have been successfully used to analyse microeconomic and macroeconomic problems. We shall focus upon the applications of these intergenerational models in: (1) efficient intergenerational and intertemporal allocation of resources, (2) intergenerational transfers (such as social security), and (3) optimal financing of government debt.

THE OVERLAPPING GENERATIONS MODEL: PARETO OPTIMALITY AND COMPETITIVE EQUILIBRIA. Let us describe the OLG model as it has been frequently used in the literature. Basically this is a formal generalization of Samuelson's (1958) model.

The economy starts in period 1 and continues over periods extending indefinitely into the future, $t = 1, 2, \ldots$. In each period there are l, $l \geq 1$, consumption goods, which are perishable, and an imperishable fiat 'money'. All households (or consumers) in this economy live for two periods, except those households which already exist at the inception of the economy (namely, were born in period 0) and who live out the balance of their lives in period 1. The tth generation, G_t, is the set of all households born at the beginning of period t, $t = 0, 1, 2, \ldots$. Therefore, in each period t, there are just two age groups

of households, the older generation, G_{t-1}, and the younger generation, G_t.

In each generation there are n households (or n 'types' of consumers). Each household will be denoted by a pair of indices (i, t), where $i = 1, \ldots, n$ and $t = 0, 1, \ldots$, and will be referred to as the ith household of the tth generation. Household (i, t), $t \geq 1$, has utility function u_i, $u_i : R_+^l \times R_+^l \to R$, defined over its lifetime consumption bundle $c_{it} = (c_{it}^y, c_{it}^o) \in R_+^l \times R_+^l$, where c_{it}^y is the consumption of (i, t), when 'young', and c_{it}^o is his consumption when 'old'. For $i \in G_0$ the utility function u_{i0} is defined over his consumption in period 1, c_{i0}, and $u_{i0} : R_+^l \to R$. Each household (i, t) is endowed with physical goods in each period of his life $w_i = (w_i^y, w_i^o)$ in $R_+^l \times R_+^l$ for $t \geq 1$. For $t = 0$, $w_{i0} \in R_+^l$ and $w_{i0} = w_i^o$. Households $(i, 0)$ of the oldest generation, G_0, may have initial endowments of money, $m_{i0} \geq 0$.

Each consumer can trade goods and money in markets in t at (present value) prices denoted as $p_t \in R_+^l$ for goods and $p_m \in R_+$ for money, respectively. We also assume that young households can borrow money, free of transaction cost, by means of issuing IOUs as long as there are households in the same generation which will accept the IOUs. Each household (i, t) faces the problem,

maximize $u_i(c_{it})$ s.t.

$$p_t c_{it}^y + p_{t+1} c_{it}^o \leq p_t w_i^y + p_{t+1} w_i^o, \qquad \text{for } t \geq 1 \text{ and } c_{it} \geq 0.$$

For i in G_0, the problem is

maximize $u_{i0}(c_{i0})$ s.t.

$$p_1 c_{i0} \leq p_1 w_{i0} + p_m m_{i0}, \qquad c_{i0} \geq 0.$$

Note that since the economy is stationary, the optimal consumption of (i, t), $c_i(p_t, p_{t+1})$, depends only on (p_t, p_{t+1}), and for each (i, o) the optimal consumption $c_{i0}(p_1, p_m)$ depends only on p_1 and p_m. The excess demand function for (i, t), $t \geq 1$, are

$$z_i(p_t, p_{t+1}) \equiv [z_i^y(p_t, p_{t+1}), z_i^o(p_t, p_{t+1})] = c_i(p_t, p_{t+1}) - w_i.$$

For (i, o) it is defined as $z_{i0}(p_1, p_m) = c_{i0}(p_1, p_m) - w_{i0}$.

A competitive equilibrium is a triplet of positive goods prices, $(p_t^*)_{t=1}^\infty$, a non-negative price of money, p_m^*, and optimal lifetime consumption profiles $[(c_{it}^*)_{i=1}^n]_{t=0}^\infty$ which are feasible, i.e.

(a) $\displaystyle\sum_{i=1}^n [c_{it}^{*y} + c_{i(t-1)}^{*o}] \leq \sum_{i=1}^n [w_i^y + w_i^o], \quad \text{for } t = 1, 2, \ldots$

and satisfies

(b) $c_{it}^* = z_i(p_t^*, p_{t+1}^*) + w_i,$ \qquad for all i, $t = 1, 2, \ldots$,

(c) $c_{i0}^* = z_{i0}(p_1^*, p_m^*) + w_{i0},$ \qquad for all i,

(d) $\displaystyle\sum_{i=1}^n [z_i^o(p_{t-1}^*, p_t^*) + z_i^y(p_t^*, p_{t+1}^*)] = 0$

\qquad for all $t = 2, 3, \ldots$,

(e) $\displaystyle\sum_{i=1}^n [z_{i0}(p_1^*, p_m^*) + z_i^y(p_1^*, p_2^*)] = 0.$

A competitive equilibrium is called a *monetary equilibrium* if $p_m^* > 0$, and a *non-monetary* equilibrium, if $p_m^* = 0$.

A feasible allocation $((c_{it})_{i=1}^n)_{t=0}^\infty$ is called *Pareto optimal* (P.O.) if there is no other feasible allocation $((\hat{c}_{it})_{i=1}^n)_{t=0}^\infty$ such that $u_i(\hat{c}_{it}) \geq u_i(c_{it})$ for all (i, t) and for some (h, τ), $1 \leq h \leq n$, $u_h(\hat{c}_{h\tau}) > u_h(c_{h\tau})$.

Samuelson analysed stationary allocations only and showed that without some extra-market institution, such as the fiat money already introduced in our presentation, competitive equilibria may not be Pareto optimal. Consider for example the case $l = 1$ $n = 1$, where the initial endowments are $w_i = (3, 1)$.

Let $u_i(c_i^y, c_i^o) = \ln c_i^y + \ln c_i^o$ and $u_{i0}(c_{i0}) = \ln c_{i0}$. Without fiat money the initial endowments allocation is the only competitive equilibrium when $p_t = 3^t$ for all t. This allocation is not P.O. since it is dominated by the P.O. allocation: $c_{i0}^* = 2$ and $c_i^* = (2, 2)$ for all i. If G_0 'contrives' fiat money, then a monetary equilibrium exists with this allocation and $p_t^* = 1$ for all t.

The main difference between an Arrow–Debreu model and the OLG model is the 'double infinity' of economic agents and commodities in the latter case. The reason for the failure of the second Theorem of Welfare Economics in the OLG models is the non–validity of Walras's Law in these infinite horizon models. This inherent property is due to the fact that, unlike the Arrow–Debreu economy, there is no possibility of trade between generations which do not overlap in their lifetime periods (see Gale, 1973).

Extending Samuelson's analysis to the non–stationary economies it was demonstrated by Cass et al. (1979) that when households are heterogeneous the following situations may occur: (a) there are both barter and monetary equilibria which are Pareto optimal; (b) there are both barter and monetary equilibria but none which is Pareto optimal. The non–optimality case, (b), highlights a fundamental difficulty: the mere creation of fiat money, i.e. the once-and-for-all augmentation of initial wealth, does not imply the second basic theorem of welfare economics; just the presence of money (trading for commodities at any conceivable positive price) may possibly not guarantee the Pareto optimality of competitive equilibrium.

In a non–stationary model, Okuno and Zilcha (1980) obtained a complete characterization of Pareto optimal competitive equilibria using the equilibrium prices. Also it is shown that certain monetary transfers (e.g. when the stock of money is increased at a uniform positive rate) can never achieve a Pareto efficient competitive equilibrium. Similar results have been attained by Balasko and Shell (1981). Cass and Shell (1983) introduced the concept of sunspot equilibria, that is equilibria in which uncertainty extrinsic to the economy operates through expectations to yield a fulfilled expectations competitive equilibrium in which the extrinsic randomness has real effects on prices and allocations. Cass and Shell examine assumptions on market structure and dynamics and find that sunspots can have no influence in a static world with complete markets, but can have effects when an OLG dynamics is introduced.

INTERGENERATIONAL TRANSFERS. The impact of annuity markets and social security upon savings, bequest and consumption has been studied recently in OLG models with uncertain lifetime. In these models there is a *continuum* of households in each generation. Each household maximizes its lifetime expected utility, where the utility from bequests appears explicitly. Uncertainty about lifetime concerns either the retirement period (as in Sheshinski and Weiss, 1981; Eckstein et al., 1985) or may extend to all periods (as in Karni and Zilcha, 1985). In the latter case it affects lifetime *earnings* and thus life insurance plays an important role in achieving efficient intergenerational allocations. In this model the first period income of each household depends upon the history of his family. In the presence of social security programmes and annuity markets agents can share death-related risks. When annuity markets operate, a non-discriminatory social security programme affects only the intergenerational allocation of resources. In the absence of private information regarding the survival probabilities, such a programme may lead to a non-optimal intragenerational allocation.

Kotlikoff and Summers (1981) use US data to estimate directly the contribution of intergenerational transfers to aggregate capital accumulation. They show that intergenerational transfers account for the vast majority of aggregate US capital formation; only a negligible fraction of actual capital accumulation can be traced to life-cycle savings.

Loury (1981) models the dynamics of earnings distribution among successive generations of workers as a stochastic process. The process arises from random assignments of abilities to individuals by nature together with the utility maximizing bequest decisions of their parents. In this OLG model parents cannot borrow to make human capital investments in their offspring. Consequently the allocation of training resources among the young generation depends upon the distribution of earnings among their parents. This implies in turn that the conflict between egalitarian and redistributive policies and economic efficiency is mitigated.

NATIONAL DEBT: ANALYSIS IN OLG MODELS. In recent years the OLG model has been applied to investigate the effects of national debt on real economic activity. Historically, attention has been focused on the question of whether or not individuals perceive government bonds as net wealth, the link between wealth and real activity being as given. Diamond (1965) analyses these questions in an OLG model with production (employing a durable capital good), in which individuals provide for their retirement years by lending to entrepreneurs. Diamond's framework has been used extensively in analysing optimal financing of government expenditures and debt; we shall describe it briefly here. Individuals live for two periods: working period and retirement period. The labour force in each period t, $t = 0, 1, 2, \ldots$, is $L_t = L_0 (1 + n)^t$. The output in period t is given by $Y_t = F(K_t, L_t)$ where K_t is the capital stock in period t. $K_t + Y_t$ should be divided between capital stock K_{t+1} available for production in the date $t + 1$ and aggregate consumption C_t. Unlike the optimal growth models, where the central planner determines the allocation between productive capital and consumption in each date (according to some social welfare function) Diamond considers allocations through the competitive mechanism. Individuals in G_t receive wage w_t which equals the marginal product of labour, $F_L(K_t, L_t)$. This wage is allocated between current consumption c_t^y and future consumption c_{t+1}^o, given the rate of interest on one period loans r_{t+1}, in a way that maximizes his lifetime utility $U(c_t^y, c_{t+1}^o)$. Therefore, $c_t^y = w_t - s_t, c_{t+1}^o = (1 + r_{t+1})s_t$ where $s_t = s(w_t, r_{t+1})$ is given by $U_1 = (1 + r_{t+1})U_2$. The equilibrium interest rate will equal the marginal product of capital, $r_{t+1} = F_K(K_{t+1}, L_{t+1})$, where the capital stock K_{t+1} is *the sum* of the individual's savings $S_t = L_t s(w_t, r_{t+1})$. The equilibrium condition in the capital market which relates the interest rate to the wage rate of the previous period is ($f(c)$ is the per-capita production function):

$$r_{t+1} = f'\left(\frac{S_t}{L_{t+1}}\right) = f'\left[\frac{s(w_t, r_{t+1})}{1 + n}\right]. \tag{1}$$

Given the factor–price frontier $w_t = \phi(r_t)$ and the initial (w_1, r_1) equations (1) define the equilibrium path. Diamond observes from these relations that the equilibrium need not occur at an interest rate exceeding the Golden Rule level. Thus the competitive solution may be dynamically inefficient. The open-ended nature of this economy is crucial to such analysis. Using this OLG model, Diamond demonstrates that external debt has two effects in the long run, both arising from the taxes needed to finance the interest payments. The taxes directly reduce available lifetime consumption of the individual taxpayer. Further, by reducing his disposable income taxes reduce his savings and thus the capital stock. Internal debt has both of these effects as well as a further reduction in the capital stock arising from the substitution of government debt for physical capital in individual portfolios. Barro (1974), using the same model, assumes that generation t's utility depends on its own consumption and leisure and upon the utility of its immediate offspring's utility. This actually connects generation t to all future generations since its utility depends on the entire future time path of consumption and leisure of its descendants. Barro argued that despite the limitation of finite lives of consumers in each generation, bonds will not be regarded as net wealth in a system characterized by intergenerational transfers. However, Barro admits that his neutrality result is valid for bonds that are redeemed at a known date. For a growing economy an increase in steady-state per-capita debt will generate net wealth if the rate of growth is greater than the interest rate. Further results concerning national debt were attained by Buiter (1979) and McCallum (1984).

Lucas (1972) uses Diamond's OLG framework, without population growth but including fiat money issued by the government and transferred to the old generation. Lucas shows that equilibrium prices and quantities exhibit what may be the central feature of the modern business cycle: a systematic relation between the rate of change in nominal prices and the level of real output. The relationship, essentially a variant of the well-known Phillips curve, is derived within a framework from which all forms of 'money illusion' are rigorously excluded: all prices are market clearing, all agents behave optimally in light of their objectives and expectations are formed optimally.

Tirole (1985) explores the interaction between productive and nonproductive savings in the model described above with capital accumulation. Some consequences of asset bubbles to asset pricing are derived. Once again the intergenerational interaction in an open-ended economy is important to such analysis.

Intergenerational models have been used in various other domains in economic theory due to their special dynamic characteristics. In international trade, for example, the examination of several questions related to exchange rate regimes has been carried out in an OLG model. Kareken and Wallace (1977) examine in various exchange rate regimes the differences that monetary-fiscal policy make. In the absence of capital controls the equilibrium exchange rate of the floating rate regime is indeterminate.

Bental (1985) applied an OLG model to two countries, two goods and two factors of production to show that in some cases the laissez-faire regime with free international trade and capital mobility is not necessarily a Pareto improvement over a regime in which capital is not internationally mobile. Furthermore, despite the fact that the laissez-faire regime is Pareto optimal and the restricted (portfolio autarky) regime is not, there do not exist simple temporal or intertemporal tax-transfer schemes which render the first allocation Pareto superior to the second.

ITZHAK ZILCHA

See also OVERLAPPING GENERATIONS MODEL; SOCIAL SECURITY; SUNSPOT EQUILIBRIA.

BIBLIOGRAPHY
Balasko, Y. and Shell, K. 1981. The overlapping generations model II: the case of pure exchange with money. *Journal of Economic Theory* 24(1), 112–42.
Barro, R. 1974. Are government bonds net wealth? *Journal of Political Economy* 82(6), 1095–117.

Bental, B. 1985. Is capital mobility always desirable? A welfare analysis of portfolio autarky in a growing economy. *International Economic Review* 26(1), 203–12.
Buiter, W. 1979. Government finance in an overlapping generations model with gifts and bequests. In *Social Security versus Private Savings*, ed. G.M. von Furstenberg, Cambridge: Ballinger.
Cass, D. and Shell, K. 1983. Do sunspots matter? *Journal of Political Economy* 91(2), 193–227.
Cass, D., Okuno, M. and Zilcha, I. 1979. The role of money in supporting the Pareto optimality of competitive equilibria in consumption loan models. *Journal of Economic Theory* 20(1), 41–80.
Diamond, P.A. 1965. National debt in a neoclassical growth model. *American Economic Review* 55(5), 1126–50.
Eckstein, Z., Eichenbaum, M.S. and Peled, D. 1985. Uncertain lifetimes and the welfare enhancing properties of annuity markets and social security. *Journal of Public Economics* 26(3), 303–26.
Gale, D. 1973. Pure exchange equilibrium of dynamic economic models. *Journal of Economic Theory* 6(1), 12–36.
Kareken, J.H. and Wallace, N. 1977. Portfolio autarky: a welfare analysis. *Journal of International Economics* 7(1), 19–43.
Karni, E. and Zilcha, I. 1986. Welfare and comparative statics implications of fair social security: a steady state analysis. *Journal of Public Economics* 30(1), 341–57.
Kotlikoff, L.J. and Summers, L.H. 1981. The role of intergenerational transfers in aggregate capital accumulation. *Journal of Political Economy* 89(4), 706–32.
Loury, G.C. 1981. Intergenerational transfers and the distribution of earnings. *Econometrica* 49(4), 843–67.
Lucas, R.E. 1972. Expectations and the neutrality of money. *Journal of Economic Theory* 4(2), 103–24.
McCallum, B. 1984. Are bond-financed deficits inflationary? A Ricardian analysis. *Journal of Political Economy* 92(1), 123–35.
Okuno, M. and Zilcha, I. 1980. On the efficiency of a competitive equilibrium in infinite horizon monetary economies. *Review of Economic Studies* 47(4), 797–807.
Samuelson, P.A. 1958. An exact consumption-loan model of interest with or without the social contrivance of money. *Journal of Political Economy* 66, December 1958, 467–82; 67, October 1959, 518–22.
Sheshinski, E. and Weiss, Y. 1981. Uncertainty and optimal social security systems. *Quarterly Journal of Economics* 96(2), 189–206.
Tirole, J. 1985. Asset bubbles and overlapping generations. *Econometrica* 53(5), 1071–100.

inter-industry analysis. *See* INPUT–OUTPUT ANALYSIS.

internal economies. The expression 'internal economies' is considered here solely in terms of Alfred Marshall's own formulation, which is quite different from current terminology referring to internal economies of scale. Modern terminology refers to a reduction in the average cost of production of a well specified commodity in relation to increases in the quantity produced, assuming, for every given quantity produced, the most appropriate utilization of the optimum productive plant. Marshall's concept of internal economies is analytically looser than this, but richer in empirical content and, possibly, in philosophical insight.

The twin terms, 'internal' and 'external' economies (and diseconomies) were first used by Marshall 'for indicating the fundamental distinction between the "internal" economies and wastes which come with an increase in the size of the individual representative firm; and those 'external' economies and wastes which come with an increase in the aggregate volume of a national or a local industry' (Marshall, [1890] 1961, vol. II, p. 347).

The main aim of the distinction was to help the applied economist in his attempts to disentangle the intricacies of contemporary socio-economic reality, rather than to provide an integral part of a formal theory of the relative values of the commodities; this is shown clearly enough by the ambiguous references to 'national or local industry' and by the use of such a fuzzy concept as the 'individual representative firm'. We must add that many times Marshall conveys the impression of confining external economies in the straightjacket of a single-product, homogeneous industry. Moreover, we must bear in mind, as Loasby aptly pointed out, that Marshall 'made no clear distinction between the theory of value and the theory of growth' (Loasby, 1978, p. 1, n. 1).

This vein of Marshallian thought derives from three sources: his vast and detailed knowledge of the literature on contemporary British and American industry; his own ruminations on the Smith–Babbage arguments on the division of labour and the internal organization of the firm, and, finally, his own early studies of mental science.

There is a passage in the *Principles*, that contains the kernel of the Marshallian ideas on the internal growth of the firm. 'Practice makes perfect', starts Marshall, taking up the well-known Smithian theme;

> physiology, [he continues], in some measure explains this fact. For it gives reasons for believing that the change is due to the gradual growth of new habits of more or less reflex or automatic action. Perfectly reflex actions ... are performed by the responsibility of the local nerve centre without any reference to the supreme central authority of the thinking power,... But all deliberate movements require the attention of the chief central authority: it receives information from the nerve centre or local authorities and perhaps in some cases direct from the sentient nerves, and sends back detailed and complex instructions to the local authorities, or in some cases direct to muscular nerves, and so co-ordinates their action as to bring about the required results ([1890] 1961, vol. I, pp. 250–51).

This quotation helps us put together the scattered pieces of the Marshallian theory of the growth of the firm under competitive conditions.

Under the spell of all the usual drives of the human mind (money-making propensity, 'instinct of the chase, desire for fame', etc), a business unit, working in a competitive context, is subject to a continual pressure to rationalize its most typical recurring operations and the tools used. So we have, simultaneously, both the development of 'skills' (a 'sort of capital of nerve force'), allowing the saving of time and of physical and, above all, nervous energies, and a rationalization of the process and the tools used. Alert to the danger of sliding to an abstract conception of the industrial process, Marshall makes room for historical and geographical peculiarities of the 'skilling' and 'rationalizing' processes.

But there comes a point 'when the action has thus been reduced to routine [that] it has nearly arrived at the stage at which it can be taken over by machinery' ([1890] 1961, vol. I, p. 254) At this point it is very probable that someone will invest the money and the inventive power required for the realization of the appropriate appliance.

When a machine is introduced into a manufacturing firm, its product becomes more uniformly specified and a cumulative process of mechanization and standardization could start. Marshall speaks of 'a great architectonic principle' according to which

> a well-driven machine tool could become the parent of new machine work more exact than itself ... and so on ... By successive steps larger and more delicate work is thrown upon the apparatus ... at last it becomes ... a thinking machine that changes its mode of action at the right time,

acting on hints given from within... . When all is in order, the machine is nearly self-sufficient. ([1890] 1961, vol. I, pp. 206–7).

The gradual introduction of specialized machinery results in more time and more nervous energies being made free at the hierarchical summit of the firm, in such a way that the entrepreneur can devote more of his time and energies to the 'broadest and most fundamental problems of his trade' (ibid., p. 284), that is to the collection and evaluation of information about general market trends and technological and organizational innovations.

The growing of a business unit above the other units of an industry, gives to it the opportunity of taking advantage of a better allocation of skills, of getting hold of 'big brains', of introducing innovations out of reach of the others, of obtaining better terms in buying, selling and borrowing. And consequently, in the words of Marshall: 'lowers the price at which he can afford to sell' (ibid., p. 315).

The basic constraint to the development of the individual firm lies in the conflict between the urge of the entrepreneur to decipher the environmental conditions of growth and the organizational requirements of the productive process. From this second viewpoint the best results can be attained by concentrating the entrepreneur's efforts on a narrow range of tasks. The simpler the work of direction, the larger the volume of output which can be efficiently controlled by a single mind, the greater the scope for the introduction of machines and uniform continuous processes. It would seem that the combined effect of these constraints would be a world composed of firms each one producing a great share of a very small range of commodities.

But this outcome would be apparently self-destroying for a world of competitive (albeit imperfectly) firms, like the Marshallian one. Marshall's answer to this challenge is both complex and stimulating. First of all, to make use of all the possible internal economies, a certain amount of individual volition is needed. The entrepreneur 'works hard and lives sparely ... subordinates trust him and he trusts them ... every improved process is quickly adopted ...'. If this behaviour 'could endure for a hundred years, he and one or two others like him would divide between them the whole of that branch of industry in which he is engaged'. But life is short and those who follow are not always fit to take over the task. The firms of many industries, at least before 'the great recent development of vast joint-stock companies, which often stagnate but do not readily die', like the trees of the forest 'gradually lose vitality and one after another ... give place to others'.

We must also remember that 'many of the lines of division between the trades which are nominally distinct are becoming narrower and less difficult to be passed ... A watch factory with those who worked in it could be converted without any overwhelming loss into a sewing-machine factory' (ibid., pp. 258–9). This continual trespassing of 'industrial' borderlines systematically frustrates the inner tendencies towards concentration and monopolization.

It must also be taken into account that the continual formation of economies, external to the single firm but internal, either to an industry or to some group of industries, in that they apply even to the smallest firms, systematically erodes part of the advantage of the bigger businesses. A particularly relevant example of it is given by the case of a localized population (the Marshallian 'industrial district') of medium-small sized firms, which, grouping together and specializing in various stages of the production process, achieve many of the large scale economies typical of the giant firms.

Marshall shares the classical view of a increasing average size of the business unit, but he is very careful to put it into a dynamical and historical context. Tendencies and counter-tendencies may result in different outcomes in terms of market structures. What is necessary for the process to be self-perpetuating is that the system should reproduce the complex of motivations which, given the structural character-istics of the industrial field, nourish the basic tendency of man towards liberation from purely mechanical tasks. In the words of R.A. Jenner: 'external and internal economies thus form counterbalanced forces of competition around which the disturbing thrusts of evolutionary change are held in control' (Jenner, 1964, p. 311).

GIACOMO BECATTINI

See also ECONOMIES AND DISECONOMIES OF SCALE; EXTERNAL ECONOMIES AND DISECONOMIES; INCREASING RETURNS TO SCALE.

BIBLIOGRAPHY

Jenner, R.A. 1964. The dynamic factor in Marshall's economic system. *Western Economic Journal* 3(1), 21–38.
Loasby, B.J. 1978. Whatever happened to Marshall's theory of value? *Scottish Journal of Political Economy* 1, 1–12.
Marshall, A. 1890. *Principles of Economics*. 9th (Variorum) edition, 2 vols, ed. C.W. Guillebaud, London: Macmillan, 1961.
Marshall, A. 1919. *Industry and Trade*. London: Macmillan.

internal migration. Among the determinants of population distribution within a country, migration is the most clearly responsive to economic opportunity. The history of the spatial structure of economic development can invariably be traced in patterns of migration. Migration also determines the size and flexibility of markets for labour and site-specific commodities, allowing for patterns of training, employment and consump-tion impossible in a restricted market. Even where migration does not result in net population redistribution, its impact on individual opportunities and the structure of exchange may be great.

In contrast to international movements, where direct state controls are frequently binding, internal migration is usually a function of choices by individuals or families. It is often taken as axiomatic that migration occurs in response to differences in location attractiveness, with wages and other aspects of employment opportunity playing a central role.

In the simplest framework, patterns of migration represent permanent moves to areas offering higher wages and away from those with less attractive opportunities; redistribution of population continues until opportunities across locations are equalized. However, since natural increase interacts with migration to influence population, migration patterns also reflect differences in natural increase. Wage differentials are expected to persist as a result of such differences, with continued net migration away from areas experiencing 'population pressure' and toward those with lower rates of natural increase. Such a characterization is at least roughly consistent with observed historical patterns. Net migration from rural to urban areas, and among regions, has tended to redistribute population toward higher wage areas (e.g., for the US, see Kuznets, 1957–64). Throughout much of the pre-industrial world, urban populations failed to reproduce themselves: cities survived only by offering opportunities that could attract migrants from rural areas.

Often, however, migration itself accelerates economic development. Where economies of scale and agglomeration are important, high levels of net migration or natural increase may improve economic opportunities, inducing increased migra-tion. The early growth of any urban area or the settlement of a new frontier depends on such a process.

Patterns of migration since World War II have caused researchers to expand the list of factors assumed to be important in attracting migrants. In a number of developed countries, net migration from rural to urban areas – which had continued over extended periods – has reversed, so that lower-density regions, with lower standards of living, have gained population by migration. Urban congestion and the availabil-ity of rural recreational and residential amenities are assumed to override differences in economic opportunity. Climate and various non-pecuniary benefits have been invoked to explain other movements that appear to go against monetary incentives.

The behavioural underpinnings of the observed patterns of net migration are less clear than has often been assumed. As a rule, net population redistribution is small relative to gross levels of migration. For example, in an area that grows due to migration, the numbers of migrants moving in and out are large compared to the net gain. Even where net migration is from low-opportunity to high-opportunity areas, a large number of migrants move in the 'wrong' direction. Migration is highly selective by a variety of demographic characteristics, and differentially selective according to the origin and destination.

To understand patterns of migration, it is useful to focus on the migration decision. A simple and fruitful way to examine such decisions is in terms of an investment model (Sjaastad, 1962). The move itself is usually costly, while benefits accrue over an extended period in a new location. In large part, this is why young adults are over-represented among migrants in general, and tend to predominate in those streams where employment opportunity is important. The returns of a permanent move are clearly maximized if the move takes place soon after labour market entry. Young workers also have fewer location-specific attachments, and will be able to make complementary human capital investments, for example, those associated with the adoption of new occupations.

Individuals who do not migrate early in their careers are less likely to move at a later point unless opportunities change, so that areas which have faced extended periods of economic depression often have relatively low rates of migrant departure. Migration from such areas is largely a function of the number of individuals reaching the age of labour force entry.

Migrant selectivity by other demographic characteristics varies according to the particular origin and destination, yet often the selection can be explained in terms of differential employment opportunities. For example, where wage differen-tials across locations differ by education, migrant selection will vary accordingly. Movement among locations often represents an exchange of migrants with different characteristics.

Migrants toward economically attractive areas tend to be selected within demographic groups as well. Those young adults who are most ambitious and productive, and who expect to be continuously employed, have the most to gain from a move. The transfer of economic activity that accompanies such migration may often exceed that indicated by simple numbers.

Although migrants may move in search of employment opportunity, non-pecuniary evaluations are critical in deter-mining the size and direction of migration streams. Psychic costs of leaving family, friends and familiar surroundings are appreciable. These latter costs differ not only across individuals, but also according to culture, as does the perceived desirability of potential destinations. Information channels also appear to delimit possible destinations. Although geographic distance is an important impediment to migration, movement between any pair of locations is much influenced by idiosyncratic ties. Moves are much more likely to occur between areas where there are personal contacts and like ethnic groups.

For any move to occur, there must always be some event that changes the migrant's evaluation of alternative locations. For the young adult who moves to find a job, entry into the labour market and the loosening of familial ties induce migration. Other life-cycle transitions, such as entry into college, marriage, the arrival of children, marital dissolution, or retirement, often induce migration as well. While the destination choice may be idiosyncratic to the individual, as where a retiree moves to be near friends or relatives, to some degree geographic areas are specialized according to the provision of amenities attractive to individuals in different stages. Migration then represents an exchange of populations as they move through the life cycle. The importance of such moves has increased in the 20th century; we see it reflected, for example, in the accelerated growth of college towns and retirement communities. In the US, by the 1960s more than a third of interstate migrants over age 60 moved into just three states, a much more concentrated pattern than that for migrants in general.

Despite the growth of such migration patterns, most migration continues to be related to labour market forces. In part, this is because the spatial distribution of employment opportunities has varied over time. The impact of such variation is clear where an economic decline in an area, or an improvement elsewhere, causes residents to move. But also, the net migration due to the movement of young migrants in search of employment may partly reflect a decline in the attractiveness of the area relative to that in previous generations. Although site-specific amenities are important in any one migrant's decision, preferences for these amenities are less likely to shift in such a way as to induce net migration than are employment opportunities. The recent urban–rural migration turnaround appears to confirm this pattern. While migrants to rural areas have always found rural residential amenities attractive – so that many of them make pecuniary sacrifices to move – the increased movement to rural areas appeared to reflect largely improved employment opportunities.

Clearly many of the apparent anomalies in aggregate migration would disappear with suitable disaggregation. Nonetheless, the volume and pattern of individual movements suggest that much migration represents decisions made with incomplete information. A few individuals who move repeatedly account for a disproportionate share of migration. For a migrant arriving in a new location, a repeat move, often a return to a previous residence, is common within the first two years. Such a pattern suggests that many migrants can only determine the attractiveness of a location after residing there for some minimum length of time, and so make multiple moves as they 'try out' alternative locations. Notably, those migrants who can migrate again at relatively low cost have an incentive to move initially to locations offering potentially high benefits even if they recognize that their chance of success is small (David, 1974). In fact, a few cities do attract large numbers of migrants, losing a disproportionate share of them after short periods.

The stochastic character of job search is particularly important for individuals whose skills are so specialized that they have relatively few opportunities in a restricted market. Educated workers are generally more likely to migrate longer distances, reflecting the value of obtaining a job that employs specific abilities. For such workers, as job tenure increases, intrafirm transfer becomes an important source of geographic mobility, but mobility due to changes of firm becomes less common, reflecting the cost of breaking a successful match (Schwartz, 1976).

Specialization is a force that induces movement among areas with similar industrial and occupational structure. Such patterns are superimposed on those which imply exchanges of migrants between areas with differing labour needs, and those that imply movements between areas with depressed and increasing overall opportunities. When the former forces dominate, as they frequently do even in less developed countries, it is common to find that migrants to a growth centre come not from the most depressed origins, but from regions with more advanced economic structures.

Migration must be placed in the context of an overall job and residence search strategy that incorporates the interests of families as well as individuals. In developing countries, where remittances are common, the migration of one member of a family is often a family decision; a rural family that sends a member to the city may raise its total income and diversify across sources of income. Where the family moves together, choice of destination must reflect the prospects of all members. While the chance of migration will decline for larger units, since gains that would induce one member to move may not be sufficient to cause the family to move, families that do move may move longer distances, reflecting the value of expanding the search space where acceptable destinations are scarce.

PETER MUESER

See also LABOUR MARKETS; REGIONAL DISTRIBUTION OF ECONOMIC ACTIVITY.

BIBLIOGRAPHY

David, P. 1974. Fortune, risk, and the microeconomics of migration. In *Nations and Households in Economic Growth*, ed. P. David and M. Reder, New York: Academic Press.

Kuznets, S. 1957–64. *Population Redistribution and Economic Growth: United States, 1870–1950.* Memoirs, Nos 45, 51, 61, Philadelphia: American Philosophical Society.

Schwartz, A. 1976. Migration, age and education. *Journal of Political Economy*, Pt I, 84(4), August, 701–19.

Sjaastad, L. 1962. The costs and returns of human migration. *Journal of Political Economy* 70(5), Pt 2, October, 80–93.

internal rate of return. The internal rate of return of an investment project is that discount rate or rate of interest Y which makes the stream of net returns x_t associated with the project equal to a present value of zero. It is the solution for i in the following equation in which θ indicates the physical lifetime of the investment project.

$$C(0, \theta) = \sum_{t=0}^{\theta} x_t (1 + i)^{-t} = 0.$$

The internal rate of return is compared with the market rate of interest in order to determine whether a proposed project should be undertaken or not.

Among the criteria to be used in determining the profitability of an investment project two others are frequently considered. Whereas the payout-period criterion is a crude rule of thumb

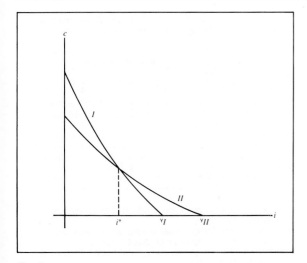

Figure 1

which ignores much of the time pattern of receipts, the net present value criterion is the most relevant 'rule' for optimal investment behaviour. If the present value (using the market rate of interest as the rate of discount) of a project's expected earnings is greater than its cost (including discounted future operating and maintenance costs), that is, if the net present value is positive, the investment project is potentially worth undertaking.

Whereas the net-present-value rule and the internal-rate-of-return rule lead to identical results in the two-period case and in the perpetuity case (which in essence is only a variant of the former), the two criteria may lead to different results in the multiperiod case. Figure 1 illustrates such a case in which the choice between two alternative investment options will lead to identical results for $i > i^*$ whereas the two criteria lead to different results for market rates of interest smaller than the cross-over rate i^* where the present value of I is higher while II has the higher internal rate of return. The failure of the internal rate of return criterion is the consequence of the implicit assumption that all intermediate receipts, positive or negative, are treated as if they could be compounded at the '*internal*' rate of return itself whereas the only appropriate *external* discounting rate is the market rate of interest (reinvestment problem).

When the investment projects are independent and with a perfect capital market (in which the lending and borrowing rates of interest are identical) the net present value is, in general, the only universally correct criterion of appraising investment projects (see Hirshleifer, 1958 and 1970, ch. 3). For the multiperiod case the internal-rate-of-return rule is not generally correct. Furthermore, there may be *multiple rates of return* that will equate the present value of a project to zero. A necessary condition for non-uniqueness of the internal rate of return is that there be more than one change of sign in the stream of receipts over the lifetime of a project.

The controversy about the multiplicity of the internal rate of return in the late Fifties led to the development of the *truncation theorem*. This theorem turns out to be important for the general problem of choosing the optimal investment period (for a historical survey of truncation theorems see Matsuda and Okishio, 1977). In 1969 Arrow and Levhari presented a new version of the truncation theorem which contrasted sharply with the other economists' method of choosing a truncation period so as to maximize the internal rate of return. They rightly pointed out that this criterion would not be adequate for the choice of the truncation period. Instead they advocated the maximization of the present value of the investment project as the proper criterion. It was demonstrated that the possibility of truncating investment projects at any age different from their physical lifetimes and at no extra costs leads to the following results:

(1) The maximized present value of the project is a monotonically decreasing function of the rate of interest. A corollary of this is that the internal rate of return is always unique.

(2) A rise (fall) in the rate of interest will always lower (raise) the present value of the remaining future net returns at all stages of the production process. Consequently the optimal economic lifetime, too, is a monotonically decreasing function of the rate of interest.

Flemming and Wright (1971) dropped the assumption of a constant rate of discount per unit of time and tried to generalize the theorem to the case of different interest rates over time, a case where the deficiency of the internal-rate-of-return rule is most obvious. However, the 'generalization' does not take us very far because the calculation would require perfect foresight of future rates. The authors emphasize that a 'slight relaxation' of this assumption is allowed because 'a change in expectation which causes' all rates 'to be revised in the same "direction" will alter the present values of all costlessly terminable projects ... in a common direction' (Flemming and Wright, 1971, p. 262). But even this proposition holds, in general, only when the change takes place uniformly, so that there is no change in the weights of the time pattern of the stream of net returns.

More interesting is the discussion of the impact of a consequence stream, that is, costs and benefits following from truncation. Whereas a positive *scrap value* can easily be incorporated the range of validity of the truncation theorem is severely limited in the case of *shut down costs*. Shut down costs can occur before and after truncation. Sen (1975) has shown that in the general case of a consequence stream following from truncation only minimal sufficiency conditions can be formulated: non-negative consequence sums (NCS) and non-negative consequence remainders (NCR), that is, the present value of the consequence stream for each t before and after the actual point θ of truncation has to be non-negative. Neither NCS nor NCR requires the present value of the consequence stream at θ to be non-negative, that is, a negative present value of the remaining process does not endanger the monotonicity result. But the conditions are very restrictive, because NCR is violated if the last item or the discounted value of the tail of the consequence stream is negative. This may be the case because of for example, redundancy payments, environmental protection or shut down costs of a nuclear power station.

The truncation theorem was originally developed in a *partial framework*. Nevertheless, Hicks (1973) and Nuti (1973) considered it applicable in a *general framework*. However, Eatwell's (1975) criticism of these authors has clarified that important propositions of the theorem do not carry over to the general framework (see also Hagemann and Pfister, 1978). At the partial level all prices in the economy are taken as given, that is, the individual's stream of net returns is considered not to be affected by changes in the discount rate. This assumption is impermissible when considering investment processes for society as a whole. At the general level the rate of profit is represented by the internal rate of return of the process as a whole for a given real wage. A variation of the discount factor, that is, the profit rate implies an opposite variation of the real wage rate. Because the present value of the whole process is both maximum and zero in competitive equilibrium the slope of the wage–profit curve is negative throughout. This is the only result one can draw under the conditions of the truncation theorem in the general setting. Neither the inverse relationship between the present value of the rest of the process and the rate of profit nor that between the optimal economic process length and the rate of profit invariably hold.

Furthermore, the analysis raises serious doubts as to the existence of an inverse monotonic relationship between interest and investment. The implication for Keynes's concept of the 'marginal efficiency of capital' is close at hand. As is well known, Keynes considered his concept 'identical' with Fisher's definition of the 'rate of return over cost' and stressed that there is no material difference 'between my schedule of the marginal efficiency of capital or investment demand-schedule and the demand curve for capital contemplated by some of the classical writers' (Keynes, 1936, pp. 140 and 178). To be sure, there are passages which indicate that Fisher was aware of the fact that prices and therefore not only the present values of the streams of net receipts but the net receipts themselves vary with variations in the rate of interest (see especially the 'more intricate than important' complication discussed in Fisher, 1930, pp. 170–71). However, the fixed-price assumption he commonly referred to implies a partial framework where the relationship between interest rates and prices is eliminated. It is therefore impossible to construct a demand curve for investment on the basis of a *ceteris paribus* clause for prices simply by variations of the rate of interest. An inverse macroeconomic relation between interest and investment cannot be derived from monotonicity results reached in a microeconomic framework. The difficulties encountered by Fisher and Keynes are discussed by Alchian and Garegnani from different points of view. Alchian (1955, p. 942) stresses that 'a schedule of investment demand at different market rates of interest requires that one compute the internal rates of return in terms of the prices that would prevail at each potential market rate of interest'. Garegnani (1978–9) brings into focus the problems involved in Keynes's concept of the schedule of the marginal efficiency of capital.

The return of the same truncation period and reswitching of techniques are closely linked phenomena occurring in a general framework. Some authors have tried to draw another analogy between the reswitching problem and the well-known possibility of the existence of multiple rates of return. Apparently the intention was to play down the importance of reswitching. This is reflected by the proposition that 'there is no new thing under the sun' (Bruno, Burmeister and Sheshinski, 1966, p. 553). However, multiple internal rates of return are a phenomenon related to the partial framework from which a generalization to the general level is not admissible. Truncation ensures the uniqueness of the internal rate of return but cannot rule out reswitching. Therefore, an analogy between the two phenomena does not exist.

HARALD HAGEMANN

See also INVESTMENT DECISION CRITERIA.

BIBLIOGRAPHY

Alchian, A.A. 1955. The rate of interest, Fisher's rate of return over costs and Keynes' internal rate of return. *American Economic Review* 45, December, 938–43.

Arrow, K.J. and Levhari, D. 1969. Uniqueness of the internal rate of return with variable life of investment. *Economic Journal* 79, September, 560–66.

Bruno, M., Burmeister, E. and Sheshinski, E. 1966. The nature and implications of the reswitching of techniques. *Quarterly Journal of Economics* 80, November, 526–53.

Eatwell, J. 1975. A note on the truncation theorem. *Kyklos* 28(4), 870–75.

Fisher, I. 1930. *The Theory of Interest*. New York: Macmillan.

Flemming, J.S. and Wright, J.F. 1971. Uniqueness of the internal rate of return: a generalisation. *Economic Journal* 81, June, 256–63.

Garegnani, P. 1978–9. Notes on consumption, investment and effective demand, I and II, *Cambridge Journal of Economics*, Pt. I, 2(4), December 1978, 335–53; Pt II, 3(1), March 1979, 63–82.

Hagemann, H. and Pfister, J. 1978. Zur Relevanz des Truncation-Theorems in partialanalytischer und totalanalytischer Sicht. *Jahrbücher für Nationalökonomie und Statistik* 193(4), August, 359–79.

Hicks, J. 1973. *Capital and Time. A Neo-Austrian Theory*. Oxford: Clarendon Press.

Hirshleifer, J. 1958. On the theory of optimal investment decision. *Journal of Political Economy* 66, August, 329–52.

Hirshleifer, J. 1970. *Investment, Interest, and Capital*. Englewood Cliffs, NJ: Prentice-Hall.

Keynes, J.M. 1936. *The General Theory of Employment, Interest and Money*. London: Macmillan.

Matsuda, K. and Okishio, N. 1977. Theorems of investment truncation. *Annals of the School of Business Administration*, Kobe University, 73–90.

Nuti, D.M. 1973. On the truncation of production flows. *Kyklos* 26(3), 485–96.

Sen, A. 1975. Minimal conditions for monotonicity of capital value, *Journal of Economic Theory* 11(3), December, 340–55.

international capital flows. A flow of international capital occurs when residents in one country (the capital exporter) extend loans to, or purchase the title to assets from, the residents of another country (the capital importer). By the principles of balance of payments accounting, which say that

Current account surplus

= Capital outflow + Increase in reserves

a net capital exporter must be providing a corresponding flow of goods and services to the capital importer unless it is simply drawing down its reserves. This is the flow of real capital corresponding to the flow of financial capital.

This flow of real capital enters the national income accounts as the trade surplus, $X - M$:

$$Y = C + I + G + (X - M).$$

Since income that is not consumed $(Y - C)$ is either paid in taxes (T) or saved, an alternative statement of the national income identity emphasizes that the sum of sectoral financial savings, including the foreign sector, must equal zero:

$$(I - S) + (G - T) + (X - M) = 0.$$

Thus capital flows finance imbalances between savings and investment by transferring real resources between one country and another.

Aspects of international capital flows that have received attention from economists include (a) facts regarding the magnitude, direction and form of capital flows; (b) the causes of capital flows; (c) the mechanisms by which a financial capital flow induces a corresponding real resource transfer; and (d) the consequences of capital flows. Writings on all these topics are scattered through the literature over space and time; the capital account of the balance of payments has never received a systematic synthetic treatment comparable to

Meade's (1951) analysis of the current account. From the classical writers to contemporary textbooks, the causes and consequences of capital flows have tended to receive less attention than the causes and consequences of trade flows. Indeed, the classical *definition* of international trade – of trade among areas between which the factors of production are immobile – was such as to preclude recognition of international capital flows.

FACTS. The first great age of international lending, in the half century prior to 1914, was effected primarily by the governments and railroads of borrowing countries issuing fixed-interest bonds on the capital markets of Britain and, to a lesser extent, France and Germany. From the late 19th century on, such 'portfolio investment' was increasingly supplemented by 'direct investment', principally from the United States, in which the creditor establishes a foreign subsidiary controlled through majority (or sole) shareholding. The mix of bond finance and direct investment was re-established in the 1920s though with the predominant flow from the United States to Europe. However, the international capital market virtually dried up after the widespread debt defaults provoked by the Great Depression. Capital movements in the 1930s largely took the form of hot-money flights of short-term funds resulting from expectations of devaluation or political persecution.

The disappearance of a private international capital market was assumed to be permanent by the post-war planners, who therefore created the World Bank to act as a public-sector substitute to channel capital to war-ravaged and developing countries. In the event the operations of the World Bank, though sizable and growing, were dwarfed by the revival of direct investment in the 1950s and subsequently the re-emergence of a market in financial capital. Unlike earlier times, however, capital flows primarily took the form of floating-interest bank loans rather than fixed-interest bond finance. A large part of this market was 'offshore', the 'Euromarket'. Following 20 years of explosive growth, the market showed signs of contracting in the 1980s, after a number of the heaviest net borrowers, mostly middle-income developing countries, had encountered difficulties in servicing their debts. The main capital-exporting areas in the postwar world were first the United States, subsequently OPEC, and now Japan, while the main capital-importing areas were first Europe, subsequently the developing countries, and currently the United States.

CAUSES. It is necessary to draw a sharp distinction between the causes of direct and portfolio investment. The former is driven by the desire of enterprises to exploit their 'intangible property' – patents, know-how, technology, trademarks, organizational or managerial expertise – in markets or sources of supply outside their home country. Direct investment flows tend to be relatively insensitive to short-run swings in macroeconomic conditions; it is flows of portfolio capital that vary sharply and therefore provide the focus of interest for economists modelling the macroeconomics of the open economy.

The basic assumption traditionally made by classical and most subsequent writers was that capital would flow abroad only if it were attracted by a higher rate of return (Iversen, 1935, ch. 2). From Adam Smith (1776) on, it was usual to argue that capital would not flow to the point of equalizing rates of return because disadvantages of foreign investment (the greater risks involved and lesser ability to exercise control) would need to be compensated by a higher prospective return. Early studies of the relative profitability of domestic versus foreign investment of Britain (by Lehfeldt, 1913–15) and France (by Harry Dexter White, 1933) cast doubt on the assumption that foreign investment had in fact been systematically and significantly more profitable to investors than domestic investment.

Explaining capital movements became in the postwar period a part of balance of payments theory. Harking back to the older literature, the writers who introduced the capital account into macroeconomic models of the open economy, notably J. Marcus Fleming (1962) and Robert Mundell (1968), assumed that a given differential in the interest rate between two countries would result in a given continuing capital flow. This became known as the 'flow theory of the capital account'.

Branson (1968) showed that this flow theory was inconsistent with portfolio theory, as developed in the 1950s by Markowitz (1952) and Tobin (1958). Portfolio theory argues that investors seek to distribute their stock of wealth among the available assets in such a way as to maximize utility, which depends on both expected income and its variance, explicitly introducing the notion that returns on most assets are subject to risk rather than perfectly foreseeable. It follows from portfolio theory that, if wealth were constant, a given interest rate differential would produce a given international distribution of the stock of capital, and thus a zero rather than a constant *flow*: hence this theory has been called the 'stock theory of the capital account'. (Under a floating exchange rate, of course, investors compare the domestic interest rate with the foreign interest rate plus the expected rate of appreciation of foreign exchange.) The stock theory implies that a once-for-all increase in the domestic interest rate would produce a once-for-all re-allocation of capital toward the domestic country, rather than a continuing inflow of capital.

In fact, capital flows are not as discontinuous as that might suggest. There are several explanations. (1) The stock of investors' wealth increases over time due to new saving, and thus maintenance of the same portfolio distribution implies a flow of capital from net creditors to net debtors. (2) Portfolio adjustment is not completed instantaneously but is spread out through time (at a rate determined by transactions and recontracting costs, which can imply very long lags where long-term assets are concerned). (3) Portfolio theory sets a *limit* to how much debtor countries with exchange controls may be able to borrow, not necessarily an *actual* level of borrowing. As long as that limit exceeds actual debt, capital inflow will be determined by demand. There is an interesting question as to whether that limit can be raised by offering lenders a higher interest rate: experience during the debt crisis has suggested that the scope for this is at best extremely limited, presumably because of the threat of adverse selection, which leads lenders to ration credit (Stiglitz and Weiss, 1981).

The flow and stock theories of the capital account are indistinguishable in one extreme case: that of perfect capital mobility, where the home country is able to borrow (or lend) unlimited sums from (to) the world capital market at a constant interest rate. This case has provided a popular assumption for model-builders, especially those in the monetarist tradition. Its attraction is, unquestionably, analytical tractability rather than empirical realism.

Empirical forecasts of capital flows need to disaggregate export credits and amortization payments from the flow of new portfolio capital. Export credits are explained principally by the flow of exports, especially of capital goods. Amortization payments depend on the level and maturity of past borrowing.

TRANSFER. Such interest as the classical economists showed in capital mobility revolved overwhelmingly around the question: what are the mechanisms by which a flow of financial capital induces a corresponding flow of real capital? Or, how does the transfer process operate? The initiating capital flow was treated as essentially autonomous. (In contrast, writers in the 1930s and 1940s emphasized that capital flows could be the *result* of a current account imbalance, or accommodating.)

The question provided a context in which to explain the operation of the gold-standard adjustment mechanism: a loan to a country would result in a transfer of gold, which would push the money supply up and thus induce enough inflation for the higher domestic prices to generate a trade deficit. In his doctoral thesis, Jacob Viner (1924) undertook a pioneering attempt to verify this theory empirically on the basis of Canadian data covering a period of massive foreign borrowing, 1900–13. He found that prices did indeed change as postulated by the theory, but also that gold movements were minimal and that trade flows seemed to respond too quickly to be explained by the theory – findings emphasized by Viner's teacher Taussig (1927), more than by Viner himself. Today we would explain these troubling findings as reflecting the fact that both the capital inflow and the current account deficit had a common cause, the improvement in investment opportunities as the prairies were opened to settlement by railroad construction.

Controversy over the 'transfer problem' flared up with the debate between Keynes and Ohlin in the *Economic Journal* in 1929. Keynes supported his pessimistic views concerning the impossibility of Germany paying reparations with an analysis predicated on the assumption that a real resource transfer could be induced only by relative price changes, as in the traditional gold-standard analysis. Ohlin's reply pointed to the complementary role of income effects – which, ironically, subsequently became the heart of what was thought of as Keynesian balance-of-payments theory. Indeed, subsequent writers asked whether the income effects induced by transfering a given amount of cash from country A to B might *alone* generate sufficient additional imports by B and a sufficient cut in the imports of A to effect the real transfer completely. The answer is that this would be true only under the implausible condition that one of the two countries be unstable in isolation (Metzler, 1942). Under normal circumstances, securing the remainder of the transfer will require either an exchange-rate change or allowing the reserve inflow (outflow) in B (from A) to increase (decrease) the money supply and thus further increase (decrease) nominal income.

EFFECTS. It is estimated that nowadays over 90 per cent of all transactions on the foreign exchange market are on capital rather than current account. The potential volume of capital flows that can be provoked in response to perceived profit opportunities is in turn vastly larger than the flows actually recorded on a typical day. This is why contemporary theory treats the exchange rate as an asset price which will be determined at a level where outstanding stocks of the different currencies are willingly held. The most glamorous research of recent years concerning the effects of capital flows has been focused on the implications for exchange rate behaviour.

The welfare effects of capital flows may be divided into two general categories, depending upon whether or not their attainment inherently requires a transfer of real resources. It may be surprising that off-setting flows of financial capital should be capable of enhancing welfare, but there are in fact three ways in which this is possible. (1) Off-setting flows of direct investment can allow multinationals to exploit their technological or other advantages in each other's markets. (2)

Because the returns on securities issued in different countries are imperfectly correlated, off-setting flows of portfolio investment enable investors in both countries to gain the benefits of risk diversification. (3) Where the tastes of investors differ between one country and another as regards the interest premium required to compensate for a sacrifice of liquidity, offsetting investment flows of assets with different maturities can be mutually beneficial (Despres, Kindleberger and Salant, 1966).

The second general category is where a capital flow does indeed finance a current account imbalance. There are two possible sources of welfare gain: from a productive loan that is used to finance an investment with returns above the cost of servicing the foreign borrowing, and from a consumption loan used to modify the time path of consumption. The classic case is of course that of the productive loan, whose impact on the time path of income and absorption is illustrated by Figure 1 (adapted from Lessard, 1981). The lower curve shows the time path that income, Y (equals absorption, A), would follow under capital-account autarky. However, when the return on investment exceeds the world interest rate, it pays to borrow, which enables A to exceed Y by the amount of the capital inflow. Because of the additional investment, GDP grows faster than it otherwise would, as shown by the curve 'Y with borrowing'. Absorption starts off above Y and remains above so long as capital inflows exceed debt service, but eventually debt service comes to exceed new borrowing and consequently absorption falls below income (though remaining above its level without foreign borrowing).

Conversely, a country where the marginal return on investment is lower than the world interest rate when all its savings are invested domestically has an incentive to invest abroad. There are sharp differences of view as to whether it is socially optimal to permit the uncontrolled export of capital. Some writers, such as Hobson (1914), in the first major treatise devoted exclusively to international capital flows, and Frenkel (1976), have eulogized the stages involved in progressing from foreign borrower to foreign lender to *rentier* as a natural and healthy result of development and the achievement of economic maturity. Others, like Jasay (1960) and MacDougall (1960), have catalogued potential sources of discrepancy between the private and social returns to (or costs of) foreign lending (borrowing), like tax revenues and induced changes in the level of wages and profits, without reaching strong conclusions as to the presumptive optimality of laissez-faire. Yet others, of whom Keynes (see especially his 1933 paper,

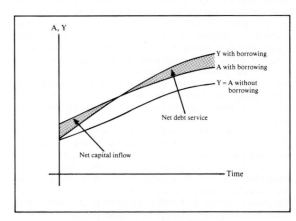

Figure 1 Impact of Productive Loans on Income and Absorption

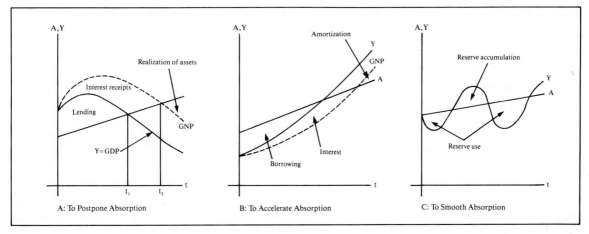

Figure 2 International Capital Flows to Modify the Time Path of Absorption

discussed by Crotty, 1983) was by far the most prominent, have argued vehemently that foreign investment resulted in a socially disastrous diversion of resources away from needed domestic investments.

Borrowing or lending to modify the time path of consumption is also capable of enhancing welfare when the time stream of income is more variable than the optimal time stream of consumption with a similar present value. Three possible cases, of lending to postpone consumption (the 1970s' OPEC low-absorber), of borrowing to accelerate consumption (the dream of the 1970s' NIC), and of borrowing and lending to smooth absorption (the primary producer with oscillating terms of trade), are illustrated in Figure 2 (also based on Lessard, 1981).

These potential sources of welfare gain have not always been realized in practice. Writers reflecting on the experience of the 1930s, notably Nurkse (1944), initiated analysis of 'disequilibrating capital flows': flows which aggravated a loss of reserves caused by a current account deficit rather than financing the deficit. Such flows were a characteristic of the adjustable-peg exchange rate mechanism adopted at Bretton Woods, and remained a source of concern at least down to the discussions of the Committee of Twenty on international monetary reform in 1972–4.

Neither have capital flows that did serve to finance current account imbalances always furthered economic welfare, as the debt crisis has demonstrated. In some countries, like Mexico, international borrowing had the effect of exaggerating rather than curtailing swings in income. Others, like Chile, accelerated consumption without the assurance of rising income that may make this policy rational. Still others, like Brazil, borrowed to make productive investments but found themselves unable to service the debts incurred (given the inflexible terms they had contracted) when world conditions changed for the worse, and were in consequence forced to cut back income – almost certainly by more than the gains previously realized from foreign borrowing.

No one doubts that the golden age of international lending prior to 1914, with capital flowing from the industrialized centre to the capital-hungry areas of recent settlement, brought gains to the world as a whole (though the optimality of exporting over half of British savings has often been contested from a nationalistic United Kingdom viewpoint). Since then, however, the record is dismal. Overlending to Germany in the 1920s thwarted adjustment; the capital inflow then dried up at

the time that world depression made adjustment excessively costly, so precipitating default. Hot money flows undermined economic management in the 1930s. Revival of an international capital market again led to excessive flows that thwarted adjustment in the 1970s, and then to a precipitate cutoff of new credit in the midst of a world recession that almost compelled a new round of defaults in the 1980s. Capital is now flowing on a vast scale to the most capital-rich country on earth and is fleeing areas where its real productivity is almost certainly higher. The net benefits of unrestricted capital mobility are indeed debatable.

JOHN WILLIAMSON

See also ABSORPTION APPROACH TO THE BALANCE OF PAYMENTS; CAPITAL FLIGHT; INTERNATIONAL FINANCE; INTERNATIONAL INDEBTEDNESS; MONETARY APPROACH TO THE BALANCE OF PAYMENTS.

BIBLIOGRAPHY

Branson, W.H. 1968. *Financial Capital Flows in the United States Balance of Payments.* Amsterdam: North-Holland.

Crotty, J.R. 1983. On Keynes and capital flight. *Journal of Economic Literature* 2(1), March, 59–65.

Despres, E., Kindleberger, C.P. and Salant, W.S. 1966. The dollar and world liquidity: a minority view. *The Economist,* 5 February 1966. Reprinted in C.P. Kindleberger, *International Money: A Collection of Essays,* London: Allen & Unwin, 1981.

Fleming, J.M. 1962. Domestic financial policies under fixed and floating exchange rates. *IMF Staff Papers* 9, November, 369–79.

Frenkel, J.A. 1976. A dynamic analysis of the balance of payments in a model of accumulation. In *The Monetary Approach to the Balance of Payments,* ed. J.A. Frenkel and H.G. Johnson, London: Allen & Unwin.

Hobson, C.K. 1914. *The Export of Capital.* London: Constable.

Iversen, C. 1935. *Aspects of the Theory of International Capital Movements.* Copenhagen: Levin and Munksgaard. Reprinted, New York: Augustus M. Kelley, 1967.

Jasay, A.E. 1960. The social choice between home and overseas investment. *Economic Journal* 70, March, 105–13.

Keynes, J.M. 1929. The German transfer problem. *Economic Journal* 39, March, 1–7. Reprinted in *The Collected Writings of John Maynard Keynes,* Vol. XI, London: Macmillan, 1983.

Keynes, J.M. 1933. National self-sufficiency. *The Yale Review* 22(4), June, 755–69. Reprinted in *The Collected Writings of John Maynard Keynes,* Vol. XXI, London: Macmillan, 1982.

Lehfeldt, R.A. 1913–15. The rate of interest on British and foreign investments. *Journal of the Royal Statistical Society.* Pt.I, 76, January 1913, 196–207; Pt II, 76, March 1913, 415–16; Pt III, 77, March 1914, 432–5; Pt IV, 78, May 1915, 452–3.

Lessard, D.R. 1981. Financial mechanisms for international risk sharing: issues and prospects. Paper presented to the Second International Conference on Latin American and Caribbean Financial Development, Caraballeda, Venezuela.

MacDougall, G.D.A. 1960. The benefits and costs of private investment from abroad: a theoretical approach. *Economic Record* 36, March, 13–35.

Markowitz, H. 1952. Portfolio selection. *Journal of Finance* 7, March, 77–91.

Meade, J.E. 1951. *The Theory of International Economic Policy.* Vol. I: *The Balance of Payments.* London: Oxford University Press.

Metzler, L.A. 1942. The transfer problem reconsidered. *Journal of Political Economy* 50, June, 397–414.

Mundell, R.A. 1968. *International Economics.* London: Macmillan.

Nurkse, R. 1944. *International Currency Experience: Lessons of the Inter-War Period.* Princeton: League of Nations.

Ohlin, B.G. 1929. The reparation problem: a discussion. *Economic Journal* 39, June, 172–83.

Smith, A. 1776. *An Inquiry into the Nature and Causes of the Wealth of Nations.* Ed. E. Cannan, London: Methuen, 1961.

Stiglitz, J.E. and Weiss, A. 1981. Credit rationing in markets with imperfect information. *American Economic Review* 71(3), June, 393–410.

Taussig, F.W. 1927. *International Trade.* New York: Macmillan.

Tobin, J. 1958. Liquidity preference as behavior towards risk. *Review of Economic Studies* 25, February, 65–86.

Viner, J. 1924. *Canada's Balance of International Indebtedness 1900–1913.* Cambridge, Mass.: Harvard University Press.

White, H.D. 1933. *The French International Accounts 1880–1913.* Cambridge, Mass: Harvard Economic Studies.

international finance. International finance is concerned with the determination of real income and the allocation of consumption over time in economies linked to world markets. Fundamental to international finance is the somewhat elusive idea of 'external balance', which in practice entails a path of external indebtedness that does not threaten a country's ability to meet its international obligations. Because the nature of the linkages among economies has varied across historical episodes, the requirements of external balance have varied as well. International finance studies the policies and market forces which may lead to external balance under various conditions. The history of the subject illustrates how the nature of world market linkages has itself been changed by national efforts to cope with external constraints.

The national income identity is the necessary groundwork for any discussion of external balance. The national income of an open economy equals domestic product plus net factor payments from abroad plus net international transfer payments; the current account equals net exports of goods and services (including all net factor payments) plus net transfers. If national expenditure is defined as the sum of consumption and investment (by both the public and private sectors), the national income identity asserts that national income less national expenditure equals the current account. When in surplus, the current account therefore measures the growth of the economy's external assets; when in deficit, it measures the growth of external debt.

THE CLASSICAL PARADIGM. The classical Ricardo–Mill barter trade theory shows how the terms of trade and international production pattern are determined in a stationary world economy with balanced trade. The classical analysis of the transition to balanced trade may be viewed as an account of the convergence process to the long-run barter equilibrium. As Ricardo noted in the *Principles* (1817):

> Gold and silver having been chosen for the general medium of circulation, they are, by the competition of commerce, distributed in such proportions amongst the different countries of the world as to accommodate themselves to the natural traffic which would take place if no such metals existed, and the trade between countries were purely a trade of barter.

Historically, however, the classical paradigm of external adjustment preceded Ricardo. Major elements of the theory had been expounded quite clearly by the early 18th century, but the most coherent and effective exposition was given by Hume in 1752.

Hume assumed a world economy that settles trade imbalances exclusively through imports or exports of precious metals that also serve as money. Building on the quantity theory of money, he constructed a full dynamic model of the balance of payments and the terms of trade. The famous price–specie–flow mechanism was put forth as an automatic market process that always works to restore balanced trade. Hume's goal was to refute mercantilist and protectionist arguments by showing that market forces would ensure in the long run a 'natural' distribution of specie among countries.

Hume invited his readers to imagine that four-fifths of Great Britain's money supply were 'annihilated in one night'. British prices would naturally fall, he argued, cheapening British exportables relative to foreign goods and creating a trade surplus. As a result of this surplus Britain would accumulate foreign wealth in the form of specie, seeing its money supply, and hence its prices, rise. Abroad, the drain of specie would lower prices. Britain's trade surplus would dwindle and eventually disappear once its terms of trade had improved sufficiently, and at this point, the natural distribution of specie would prevail. A hypothetical fivefold increase in Britain's money supply would set off the reverse process, involving an initial improvement in Britain's terms of trade and a trade balance deficit. Over time, specie would flow abroad as the terms of trade deteriorated and external equilibrium was restored.

There is little exaggeration in saying that issues raised by Hume's analysis dominated writing in international finance up until the inter-World War years. In a period that culminated in the classical gold standard, it was natural to take as the benchmark of external balance an absence of international specie movements. Hume had placed relative price movements at the centre of his account of how external balance would be attained, but subsequent writers asked whether direct income or wealth effects might also be operative, and whether external adjustment could take place in some cases without price changes. Such questions arose in the 1929 Keynes–Ohlin debate over the German transfer problem, but as Viner (1937) showed, the questions had been raised much earlier.

A simple model of a Humean world makes apparent some of the assumptions underlying the price–specie–flow mechanism. Such a model also serves as a springboard for understanding later developments in the analysis of external adjustment. (A more detailed exposition of a similar model is given by Dornbusch, 1973, whose analytical approach is, however, somewhat different from that taken here.)

Assume a world of two countries, each specialized in the production of a single commodity that is consumed in both countries. With given supplies of capital and labour within each country and perfect wage flexibility, home-country output is fixed at the full-employment level x and foreign-country output is fixed at y. Let q denote the price of y- goods in terms of x-goods (the terms of trade), z domestic expenditure measured in x-goods, and z^* foreign expenditure, also measured in x-goods. Then the domestic demands for the two goods are $c_x(q, z)$ and $c_y(q, z)$, while the foreign demands are $c_x^*(q, z^*)$ and $c_y^*(q, z^*)$.

Expenditure is determined by monetary conditions. The money supplies M and M^* are for simplicity taken to consist entirely of gold, and P and P^* denote the money prices of home and foreign goods, respectively. The exchange rate between domestic and foreign currency can be set at unity with no loss of generality, so the terms of trade, q, equal P^*/P. In each country there is a desired long-run (or 'natural') money supply this is proportional to nominal output, and saving behaviour is governed by discrepancies between natural and actual money supplies. Because a country's net saving here equals its current account, which by assumption is settled in specie, saving behaviour determines the evolution of national money supplies. These evolve according to the laws

$$dM/dt = \theta(\chi Px - M), \qquad dM^*/dt = \theta^*(\chi^* P^* y - M^*),$$

where χ (χ^*) is the reciprocal of the home (foreign) country's long-run monetary velocity and θ (θ^*) is the home (foreign) marginal propensity to dissave out of monetary wealth. Expenditure levels are therefore

$$z = (1 - \theta\chi)x + \theta M/P, \qquad z^* = (1 - \theta^*\chi^*)qy + \theta^* M^*/P,$$

where $\theta\chi$, $\theta^*\chi^* < 1$.

The model is closed by two equilibrium conditions. With a given world stock of monetary gold, M^w, home saving must equal foreign dissaving, that is, world expenditure must equal world output. In addition, the market for domestic goods must clear. By Walras's Law, these two equilibrium conditions imply equilibrium in the market for foreign goods.

The condition of zero desired world saving is $(dM/dt) + (dM^*/dt) = 0$, or

$$P = \frac{\theta M + \theta^*(M^w - M)}{\theta\chi x + \theta^*\chi^* qy}. \tag{1}$$

Equation (1) shows that, for given terms of trade and money supplies, the world price level adjusts to maintain consistency between the countries' saving plans. In equilibrium, this condition makes P a function of q and M, $P = P(q, M)$, with

$$\frac{q}{P}\frac{\partial P}{\partial q} = \frac{-\theta^*\chi^* qy}{\theta\chi x + \theta^*\chi^* qy} > -1,$$

$$\frac{M}{P}\frac{\partial P}{\partial M} = \frac{(\theta - \theta^*)M}{\theta\chi Px + \theta^*\chi^* P^* y} \lessgtr 0.$$

The market for x-goods clears when

$$c_x[q, (1 - \theta\chi)x + \theta M/P]$$
$$+ c_x^*[q, (1 - \theta^*\chi^*)qy + \theta^* M^*/P] = x. \tag{2}$$

Substitution of $P = P(q, M)$ and $M^* = M^w - M$ into (2) gives the curve describing combinations of M and q at which both goods markets clear and aggregate world saving is zero. The curve is labelled XX in figure 1 and is shown with a negative slope. The assumptions giving rise to this negative slope are crucial for analysing the Humean adjustment process. An increase in M (which necessarily implies an equal fall in M^*) causes an excess demand for x-goods equal to

$$\frac{\theta\theta^*(M + M^*)}{P(\theta M + \theta^* M^*)}[\partial c_x/\partial z - \partial c_x^*/\partial z^*]$$

near the system's long-run equilibrium (where $dM/dt = dM^*/dt = 0$). The term $\partial c_x/\partial z - \partial c_x^*/\partial z^*$ is the difference between the two countries' marginal propensities to

spend on home-country goods; if the home-country marginal propensity is larger – the 'orthodox' presumption in transfer analysis (Samuelson, 1971) – a redistribution of nominal balances in favour of the home country creates an incipient excess demand for its output. This excess demand is eliminated by a fall in q if the home-goods market is Walras stable, so XX slopes downward under standard assumptions concerning marginal spending propensities and Walrasian stability.

The curve in figure 1 labelled $dM/dt = 0$ describes points at which $M = \chi P(q, M)x$. This locus has a negative slope aigebraically smaller than that of XX. With goods markets continuously in equilibrium, the world economy travels along XX to its long-run equilibrium at point A, where international prices and the distribution of specie give rise to balanced trade.

In most respects the model confirms Hume's account of the external adjustment process. A (small) fall in M to M_0, for example, leads to terms of trade q_0, which are worse for the home country. The terms-of-trade change is a direct result of the transfer of purchasing power to foreigners, which produces an excess supply of home goods at the initial prices. The home balance-of-payments surplus that simultaneously emerges causes a gradual redistribution of money in favour of the home country, so the home terms of trade improve during the transition to external balance.

If $\theta = \theta^*$, equilibrium P is a function of q alone, with a negative elasticity greater than -1. The rise in q caused by a fall in M is thus accompanied by a less-than-proportional fall in P and a rise in P^* that are reversed as the economy returns to point A. These results are in accord with Hume's predictions, but they need not hold if the expenditure responses to real balances differ sufficiently in the two countries. If $\theta > \theta^*$, a transfer of money abroad raises world saving for given terms of trade, so P^* may *fall* along with P and then rise during the subsequent adjustment. Likewise, if $\theta < \theta^*$, a money transfer abroad may reduce world saving sufficiently that P must *rise*, along with P^*, to restore goods-market equilibrium in the short run. In this case, the initial response to the disturbance is followed by price deflation in both countries.

This stylized version of Hume's paradigm may be used to analyse the transfer problem. Suppose that ownership of a portion of the foreign country's endowment is given to the home country. Does the home trade deficit necessarily

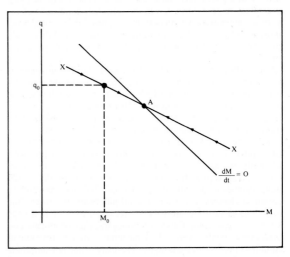

Figure 1

increase by the amount of the transfer, or is the transfer undereffected, requiring a flow of specie to the home country to balance international accounts? A second focus of debate in the literature is the possibility that the transfer imposes a 'secondary burden' on the paying country by adding an equilibrium terms-of-trade deterioration to the primary income burden. Keynes and Ohlin clashed on this point in 1929, with Keynes arguing that the secondary burden is inevitable.

To simplify, suppose that $\theta = \theta^*$ and $\chi = \chi^*$. Since long-run money demand rises with income, the $dM/dt = 0$ locus shifts to the right, implying that the transfer at first is undereffected and that the world's gold stock is redistributed toward the home country. Under the standard assumption regarding marginal spending propensities, the transfer also creates an excess demand for x-goods at the initial terms of trade, so XX shifts downward. A secondary burden is thus imposed on the paying country, and this burden worsens over time as balanced trade is reestablished.

THE INTERWAR PERIOD. The years between the World Wars saw a partial and ultimately unsuccessful return to the gold standard, followed by extensive experimentation with floating exchange rates and direct controls on international payments as means of attaining external balance. Nurkse's (1944) account of the period is probably the most influential one. Writers on international finance continued to conceptualize external balance in terms of reserve movements. The spread of the gold-exchange standard, under which central banks held as foreign reserves currencies tied to gold as well as gold itself, broadened the class of assets through which balance-of-payments deficits were financed.

International capital movements were discussed increasingly in the theoretical literature, but they were viewed for the most part as an adjunct to the classical balance-of-payments adjustment mechanism. The theoretical discussions merely formalized a mechanism that had long been exploited by the Bank of England to regulate gold flows. A country that suddenly developed a trade deficit would face declining international reserves, a declining money supply, and rising interest rates. Rising interest rates would, however, attract foreign capital inflows and thus dampen the resulting deficit in the balance of payments. On this view, interest-sensitive capital flows had a potentially stabilizing role to play in discouraging protracted reserve flows. Given the turbulent conditions of the period, contemporary writers fully recognized that capital flows motivated by fears of devaluation or political instability could just as well destabilize an already bad external payments problem.

Such 'short-term' or interest-sensitive capital movements were generally discussed separately from 'long-term' international capital movements which directly financed investment or government expenditures. Theoretical discussions of long-term capital movements focused mainly on the transfer mechanism, the balance-of-payments and terms-of-trade adjustments that would accompany an inter-country transfer of capital. Conspicuously absent from the literature were attempts to develop a normative intertemporal theory of international capital transfer. Such a theory naturally would have extended the prevailing external balance concept to comprise changes in nations' *overall* indebtedness rather than just changes in the central bank's foreign assets. It had been known, at least since Ricardo's *Principles*, that producers and consumers could gain if long-term foreign investment equalized profits internationally. The insight did not dominate thinking about the nature of external balance.

This gap in the literature is surprising in view of the developments in international capital markets over the previous century. Huge flows of long-term capital, primarily from Britain, had financed railroad construction and other investment in the Western Hemisphere. France and Germany also made significant foreign loans. In the early 1930s, widespread foreign debt default among the Latin American countries highlighted the need to analyse formally the sustainability of external debt paths. In the world assumed by Hume, specie flows had been the only means of settling current-account imbalances, and a concept of external balance based on balance-of-payments equilibrium had been defensible. Such a concept of external balance was outmoded, however, in a world where other types of asset trade could finance the current account.

The necessary change of perspective did not occur for several decades. Instead, the events and ideas of the interwar period led international financial theory to turn away sharply from the concern with the dynamics of international adjustment underlying the classical model. Emphasis shifted inward, to the interaction between the balance of payments and domestic economic conditions.

THE BRETTON WOODS PERIOD. The interwar experience had a profound influence on both the institutional framework of postwar international finance and the theoretical orientation of researchers. The international agreement reached at Bretton Woods in 1944 set up a world trading community linked by fixed dollar exchange rates, with a United States commitment to peg the dollar price of gold at $35 per ounce providing an anchor for the world price level. The agreement's provisions aimed to promote free trade in goods, but private capital movements were viewed as potentially disruptive and the widespread capital controls then in force were not discouraged. A prevailing view that flexible exchange rates had failed during the interwar period motivated the adoption of a fixed-rate system. Provision was made, however, for infrequent exchange-rate adjustment, after due consultation, in circumstances of 'fundamental disequilibrium' in the balance of payments.

Central to the design of the Bretton Woods system was a desire to avoid unemployment and ensure price-level stability. In the interwar years, many governments had resorted to competitive currency depreciations and trade restrictions aimed at reducing domestic unemployment. These 'beggar-thy-neighbour' moves made all countries worse off. Having recently experienced the hardships of the worldwide Great Depression, the Bretton Woods signatories recognized the goal of 'internal balance' – full employment with price stability – as a key aim of government policy. An International Monetary Fund was set up to reconcile the goals of internal and external balance. It was hoped that the availability of Fund credit would make it unnecessary for members to tolerate high unemployment in pursuing external balance, or to interfere with trade flows in pursuing internal balance.

In an environment of fixed exchange rates and extremely limited capital mobility, the overriding external consideration for governments was the available stock of foreign, particularly dollar, reserves. The operative external target was therefore the acquisition of as many dollars as possible through balance-of-payments surpluses. As the reserve centre, the United States enjoyed the privilege of being able to finance its own balance-of-payments deficits by borrowing dollars from foreign central banks. In reality, however, the United States was not totally free of a reserve constraint. Foreign central banks could, and did, use their dollars to buy gold from the US authorities at the official price. The problem of gold losses became important as the postwar period of 'dollar shortage' ended in the late 1950s. In 1960, Triffin put the American external dilemma in its most sombre light: Once foreign official dollar holdings exceed the official value of the US gold stock, it would become impossible to satisfy all foreign claims to US gold without a rise in the dollar price of metal. The resulting confidence problem, Triffin predicted, would undermine the stability of the Bretton Woods system.

As it developed immediately after World War II, international financial theory reflected the new institutional arrangements, along with the economic assumptions underlying Keynes's (1936) diagnosis of the unemployment of the 1930s. The new paradigm, set forth very effectively by Metzler (1948, pp. 212–213), assumed sticky price levels and wages along with fixed exchange rates, thus precluding the relative-price adjustments at the heart of the classical paradigm while opening the door to employment fluctuations:

> The important feature of the classical mechanism . . . is the central role which it attributes to the monetary system. The classical theory contains an explicit acceptance of the Quantity Theory of Money as well as an implied assumption that output and employment are unaffected by international monetary disturbances. In other words, the classical doctrine assumes that an increase or decrease in the quantity of money leads to an increase or decrease in the aggregate money demand for goods and services, and that a change in money demand affects prices and costs rather than output and employment ... . The essence of the new theory is that an external event which increases a country's exports will also increase imports *even without price changes*, since the change in exports affects the level of output and hence the demand for all goods. In other words, movements of output and employment play much the same role in the new doctrine that price movements played in the old.

An increase in external demand for a country's exports, for example, would raise the country's trade surplus in the first instance, but once the multiplier effect of the disturbance had raised income and hence import spending, the initial impact on the trade balance would be reduced. Metzler noted, however, that even if one assumed that investment spending responds positively to a rise in real income, it was unlikely that multiplier effects alone would ensure complete trade-balance adjustment in the short run.

The Keynesian account of external adjustment therefore contained an important gap. Private capital movements were largely ruled out in the Keynesian models, so incomplete trade-balance adjustment implied incomplete balance-of-payments adjustment and growing or shrinking central-bank foreign reserves. The models pushed monetary factors to the background, implicitly or explicitly assuming that central-bank sterilization operations were offsetting any monetary effects of the balance of payments. Only a few of the early postwar theorists, notably Meade (1951), assigned an important role to monetary factors.

Even if the sterilization assumption were granted, however, consideration of the system's inherent dynamics made clear the infeasibility of a *permanent* sterilization policy. Countries with persistent deficits would ultimately exhaust their available international reserves, including IMF credit; and even surplus countries might be unable to sterilize indefinitely if domestic financial markets were thin. How, then, could trade-balance equilibrium even be restored after a permanent external shock? Fiscal policy could be effective in situations where the needs of

internal and external balance were both served by the same measure. In dilemma situations where fiscal measures could move the economy toward external balance only at the cost of increasing its distance from internal balance, the 'fundamental disequilibrium' clause of the IMF Articles of Agreement could be invoked and the currency devalued. But no *automatic* market mechanism pushing the economy toward balance-of-payments equilibrium was featured in the early postwar writing.

In a series of remarkable papers published in the early 1960s, Mundell revived the explicit dynamic analysis of international adjustment. His models placed the monetary sector in the foreground, adopting a Keynesian liquidity-preference view of interest-rate determination. A prescient paper by Metzler (1960), written at about the same time, took a similar approach.

Mundell's paper on 'The International Disequilibrium System' (1961) criticized the Keynesian model's failure to account for the dynamic effects of payments imbalances. Even in a Keynesian world, Mundell argued, an income–specie–flow mechanism, analogous to Hume's price–specie–flow mechanism, ensures long-run balance-of-payments equilibrium. A 'fivefold increase' in a country's money supply, for example, depresses domestic interest rates, stimulates investment spending, and creates a deficit in the balance of payments. As the central bank loses reserves, however, the interest rate gradually rises and reduces investment, the process coming to an end (for a small country) only when the domestic money supply, the interest rate, investment, and output have returned to their original levels. The introduction of dynamic adjustment made it clear that sterilization could have only limited success as a policy response to permanent balance-of-payments disturbances. One source of dynamic effects, however, was not explicitly analysed in Mundell's work of the period. The omitted effect was the real-balance effect on expenditure, central to the classical account but possibly relevant (as Pigou had shown) under Keynesian conditions as well.

In line with the increasing international capital mobility that followed the European move toward currency convertibility in 1958, Mundell gave the capital account a prominent role in his models. The presence of capital mobility suggested a solution to the policy dilemmas that could arise under fixed exchange rates when the goals of internal and external balance appeared to conflict. Mundell showed that by gearing monetary policy to external balance and fiscal policy to internal balance, governments could simultaneously attain both goals. The key to the argument is the observation that monetary and fiscal expansion both raise output but have different effects on the capital account, monetary expansion causing capital outflows (by driving down the home interest rate) and fiscal expansion causing capital inflows (by raising the interest rate). With two independent instruments, both internal and external policy targets can be attained simultaneously.

While a major step forward, the Mundellian argument for a policy mix suffered from two drawbacks. First, the theoretical specification of the capital account as a function of international interest-rate levels was weak: it seemed unlikely that capital would flow at a uniform level forever even if the interest differential remained fixed. Missing was a discussion of stock equilibrium in international asset markets. The second problem with the policy mix was its definition of external balance. Would any policymaker view with satisfaction a permanently high interest rate that brought about balance-of-payments equilibrium by crowding out domestic investment and encouraging a build-up of external debt? Key considerations omitted from Mundell's model were the stock of net foreign claims and the associated flows of interest payments. Mundell himself (1968, p. 207) recognized that in many contexts, the definition of external balance as balance-of-payments equilibrium might be inadequate:

> Just as the composition of output is important (the division of output between investment and consumption affects additional growth targets), so an appropriate composition of the balance of payments is a legitimate target of policy.

Indeed, in spite of the continuing obligation to peg dollar exchange rates, the standard definition of external balance was becoming increasingly outmoded by the late 1960s. The balance of payments remained a legitimate concern, of course, in part because a large or persistent imbalance might look like 'fundamental disequilibrium' to the market and spark a speculative attack on the currency involved. But the increasing integration of national financial markets – a development epitomized by the growth of Eurocurrency trading – weakened the bite of the balance-of-payments constraint. In a hypothetical world of *perfect* capital mobility, a central bank short on reserves can essentially borrow them from abroad at no net cost simply by contracting domestic credit. Such an action, by causing an incipient rise in the home interest rate, leads to an instantaneous private capital inflow and an official reserve gain equal to the fall in domestic credit. The home interest rate, the money supply, output, and the national external debt are unchanged in the final equilibrium: the central bank holds more foreign assets and fewer domestic assets, while the home private sector, having made the mirror-image adjustment, holds fewer foreign assets and more domestic assets.

The case of perfect capital mobility is an extreme one that does not fit the facts of the late Bretton Woods period. None the less, the opportunities for central banks to borrow dollar reserves in the international capital market had grown since the early 1960s. The situation facing the United States was quite different. As the primary international reserve issuer, its responsibility was to peg the dollar price of gold, a responsibility that would have required the gearing of US monetary policy to that external commitment. In spite of such expedients as the two-tier gold market established by central banks in 1968, the US did not succeed in preserving the dollar's link to gold. Triffin had been right. After a series of violent speculative attacks, the US severed the dollar's gold link in August 1971 and in December 1971 devalued the dollar against major foreign currencies. The patchwork system of fixed exchange rates proved unstable, and in the first months of 1973 the postwar period of floating exchange rates began.

FLOATING EXCHANGE RATES. The industrialized countries adopted floating dollar exchange rates as an interim measure, but in fact a significant body of economists had come to advocate floating rates by 1973. Friedman's (1953) powerful case for flexible rates was the opening shot in a campaign to revise the then-prevailing view, expounded by Nurkse (1944), that the floating-rate experiments of the interwar years were disastrous. By the time Johnson wrote his well-known polemic of 1969, Friedman's views had gained many adherents.

The fundamental argument for floating rates was that they would free governments of the balance-of-payments constraint and allow them to use monetary policy to attain domestic economic goals. Equilibrium in the balance of payments would be automatic if central banks simply refrained from intervening in the foreign exchange market. At the same time,

floating rates would permit central banks to target their nominal money supplies without being frustrated by offsetting interest-sensitive foreign reserve flows. Widespread restrictions on trade and capital movements, motivated in part by a desire to impede reserve flows under the fixed-rate regime, could be dismantled.

Subsequent experience was to provide only partial vindication to the advocates of floating. In the decade after 1973, barriers to capital movement were reduced to insignificant levels in many of the industrial countries. This development helped spark unprecedented growth in international financial intermediation. Under the new exchange-rate regime, however, policymakers became more acutely aware that the traditional definition of internal balance as full employment cum price stability really involved two, quite distinct, goals. Under a floating exchange rate, monetary expansion aimed at domestic unemployment translates immediately into currency depreciation, higher import prices, and heightened inflationary expectations. Conversely, a rapidly adjusting exchange rate provides a powerful channel through which inflationary expectations can have a direct and immediate effect on inflation in an open economy. Any short-run tradeoff between inflation and unemployment would therefore be less favourable under a floating rate. Floating rates certainly allow countries to choose their own trend inflation rates. But it soon became evident that if disturbances to the economy originated predominantly outside the money market, the inflationary cost of using monetary policy to target employment could be quite high.

Sharp exchange-rate movements might also have adverse distributional effects in the economy, and these, together with a desire for price-level stability, led central banks to intervene, at times heavily, in the foreign exchange market. Correspondingly, the predicted drop in central banks' demand for international reserves did not materialize (although the composition of reserves did change over time as the Deutschmark and yen became important reserve currencies and the pound sterling retreated). Central banks' use of foreign reserves to manage exchange rates did not necessarily imply an operative balance-of-payments constraint, however, since in many countries the same exchange-rate effects could have been achieved at an unchanged reserve level through domestic credit measures.

Under conditions of limited capital mobility, such as those existing in the early 1950s when Friedman wrote, the automatic balancing of international reserve movements by a floating exchange rate amounted essentially to the automatic balancing of the current account. With means other than reserve flows available to settle current-account imbalances, however, there is no theoretical necessity for a floating rate to balance the current account in the short run. A current-account deficit, say, can be financed entirely through domestic borrowing abroad with no decline in the central bank's foreign assets. Experience was to show that floating exchange rates themselves could not prevent the emergence of large and persistent current-account imbalances. These imbalances were problematic not only because they usually entailed costs of shifting productive resources between the economy's tradable and nontradable sectors, but also because they implied changes in foreign debt and thus in sustainable future consumption levels.

Attention therefore shifted to the mechanism of current-account adjustment under floating exchange rates and capital mobility, with researchers asking, as Hume had, if market forces would automatically push economies toward current-account balance. The new generation of dynamic open-economy models produced in the mid-1970s built on a number of antecedents in the literature. One of these was the neoclassical monetary approach to the balance of payments, which stressed the real balance effect and the transition to long-run payments equilibrium (see, for example, Frenkel and Johnson, 1976). The second important antecedent was the closed-economy literature on money and growth, which had clarified the stock-flow distinction in multi-asset models with wealth accumulation. As suggested by the rational-expectations revolution in macroeconomics, many model builders endowed agents with forward-looking exchange-rate expectations that played a key role in clearing the asset markets.

The intrinsic dynamic mechanism in these models is fuelled by wealth, broadly defined to include not only real monetary balances, but also foreign assets and possibly capital, physical as well as human. (See Obstfeld and Stockman, 1985, for a survey.) In line with the long-run nature of the inquiry, the 'classical' conditions of price flexibility and full employment were generally assumed, giving a productions structure similar to the Humean model set out above. Where the models differed essentially from Hume was in the wider spectrum of marketable assets, and in the resulting portfolio problem of private agents. Each given configuration of world asset stocks determines a short-run equilibrium defined by the requirement of market clearing in asset as well as goods markets. The resulting equilibrium wealth levels and real interet rates determine consumption levels at home and abroad, but there is no necessary requirement of current-account balance in the short run: goods-market equilibrium implies only that one country's planned current-account surplus equals the other's planned current-account deficit. The international adjustment process can now be visualized. All else equal, the deficit country is running down its wealth by borrowing from abroad, so its consumption is falling and foreign consumption is rising. Under the orthodox transfer criterion, this redistribution of wealth between the countries causes the deficit country's terms of trade to deteriorate over time; if anticipated, the evolution of the terms of trade has further repercussions on world real interest rates and expenditure levels. The process comes to an end once the deficit country's consumption has fallen into line with its income, which is lower than initially because of the increased interest burden of the external debt. (A very similar adjustment process would take place with mobile capital and a fixed exchange rate, but reserve movements rather than exchange-rate movements would contribute to asset-market balance during the transition to long-run equilibrium.)

This simple picture of the adjustment process becomes more complicated once domestic capital accumulation is allowed. A current-account deficit may now finance an investment boom in which the deficit country's terms of trade improve over time. Eventually, however, the international wealth-flow mechanism restores a balanced current account. Further complications arise when the classical assumptions are dropped and Keynesian price stickiness in output markets is assumed. In such models, the approach to the long-run, full-employment equilibrium can be oscillatory.

For a single economy with Keynesian features, there is an analogue to the Mundellian idea of using monetary and fiscal policy simultaneously to attain internal and external targets. Figure 2, which is developed more fully in Obstfeld (1985, pp. 408–410), illustrates this approach. The downward sloping internal-balance schedule shows combinations of monetary and fiscal ease consistent with full employment. On the assumption that monetary ease improves the current account by depreciating the currency, the external-balance schedule, which shows policy settings consistent with some current-

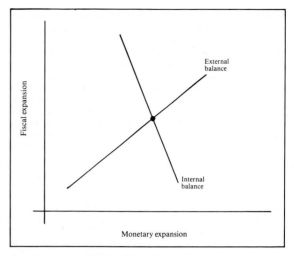

Figure 2

account target, slopes upward. The intersection of the two schedules shows how policies should be set to achieve both of the government's goals in the short run.

Even if one leaves aside the complex game-theoretic problems surrounding interactions between expectations and policy, the usefulness of the above framework as a normative guide is limited by its failure to incorporate some key dynamic elements. If the government can hit its targets only by running a budget deficit, its fiscal stance must eventually be reversed if the government debt is to be serviced. In addition, the policy equilibrium shown in Figure 2 may imply a domestic investment rate that is socially sub-optimal. Finally, the framework itself gives no guidance as to the appropriate external-balance criterion. The balanced current account reached in the hypothetical long-run equilibrium of a stationary world economy may be far off the mark in the short run in which policy decisions must be made. Recently, the theory of international finance has made partial progress in addressing these issues.

THE INTERTEMPORAL ANALYSIS OF EXTERNAL BALANCE. In the 1980s, it became increasingly common to analyse the dynamic behaviour of open economies in terms of the intertemporal maximization hypothesis applied by Fisher (1930) to the theory of saving and investment. As usual, this trend was the result of both new theoretical approaches in macroeconomics generally and of economic events that existing open-economy models seemed ill-equipped to analyse.

Lucas's (1976) influential critique of econometric policy evaluation was important in motivating the intertemporal approach. Lucas argued that the standard econometric models of the time would generally not be invariant to policy changes. Because the parameters estimated were not the 'deep' parameters describing preferences or technology, but instead reflected both deep structure and the policy environment prevailing over the estimation period, the models could not be used to analyse *changes* in the policy environment. Lucas's analysis suggested that more reliable policy conclusions might be drawn from open-economy models if demand and supply functions were derived from the optimal decision rules of maximizing households and firms.

Further impetus to develop an intertemporal approach came from events in the world capital market, particularly the international pattern of current accounts following the sharp oil-price increases of 1973–4 and 1979–80. The divergent patterns of a current-account adjustment by industrialized and developing countries raised the inherently intertemporal problem of characterizing the optimal response to external shocks. Neither classical nor Keynesian transfer analysis offered any reliable guidance on this question. Similarly, the explosion in bank lending to developing countries after the first oil shock sparked fears that some countries' external debt burdens would become unsustainable. The need to assess developing-country debt levels again led naturally to the notion of an intertemporally optimal current-account deficit.

Any intertemporal analysis of external balance must begin by specifying the economy's technological and market opportunities for shifting consumption over time. These opportunities are described by the economy's intertemporal budget constraint, which specifies the terms on which the economy can borrow or lend abroad, as well as the domestic investment technology. Separate analysis of the public and private sector's budget constraints illuminates the link between the public finances and external imbalance, as measured by the balance of payments or by the current account. The economy-wide budget constraint results from consolidation of the public- and private-sector constraints.

Assume for simplicity that a single good is consumed and produced on each date, and consider the position of a small open economy that can borrow or lend internationally at the real interest rate ρ. For each date t, the government of the economy chooses a level of real government consumption, $g(t)$, and a (possibly negative) level of real transfers to the private sector, $\tau(t)$. The government finances its outlays by issuing debt, by printing money, and by drawing on the interest paid by the central bank's foreign reserves. (For present purposes, the central bank's budget is best viewed as a component of the government's budget.) Let $b^G(t)$ denote real government bond holdings (other than central-bank foreign reserves), $D(t)$ the money value of central-bank domestic credit, $P(t)$ the money price level, and $r(t)$ real foreign reserves. If the government pays the interest rate ρ on the public debt $[-b^G(t)]$, then the path of government bond holdings satisfies the equation:

$$\mathrm{d}b^G(t)/\mathrm{d}t = \rho[b^G(t) + r(t)]$$
$$+ [1/P(t)]\,\mathrm{d}D(t)/\mathrm{d}t - g(t) - \tau(t). \quad (3)$$

Changes in the economy's money supply, $M^S(t)$, result from changes in the central bank's foreign or domestic assets. If the world price level P^* is constant (so that proportional changes in $P(t)$ equal proportional changes in the exchange rate), then the central-bank balance-sheet identity implies $\mathrm{d}M^S(t)/\mathrm{d}t = P(t)[\mathrm{d}r(t)/\mathrm{d}t] + \mathrm{d}D(t)/\mathrm{d}t$. Let $m(t)$ denote the private sector's desired real money balances and $\pi(t)$ the home inflation rate. On the assumption that the money market is continuously in equilibrium, $m(t) = M^S(t)/P(t)$ and equation (3) becomes

$$\mathrm{d}[b^G(t) + r(t)]/\mathrm{d}t = \rho[b^G(t) + r(t)]$$
$$+ \pi(t)m(t) + [\mathrm{d}m(t)/\mathrm{d}t] - g(t) - \tau(t). \quad (4)$$

Integrate (4) forward from $t = 0$ and impose the condition $\lim_{t \to \infty} \exp(-\rho t)[b^g(t) + r(t)] \geq 0$, which restricts the government to borrowing paths such that the public debt is asymptotically paid off. The result is the intertemporal budget constraint of the government,

$$\int_0^\infty [g(t) + \tau(t)]\exp(-\rho t)\,\mathrm{d}t$$
$$\leq \int_0^\infty [\pi(t)m(t) + \mathrm{d}m(t)/\mathrm{d}t]\exp(-\rho t)\,\mathrm{d}t + b^G(0) + r(0).$$

The inequality states that the present value of net government outlays must be less than the present value of the seigniorage from money creation plus the government's initial asset position. The latter quantity, in turn, equals central-bank foreign reserves less the public debt. For a world of perfect capital mobility, the constraint makes clear that it is the government's overall asset position that is relevant for assessing solvency. The level of foreign reserves $r(0)$ has little significance in itself. As noted earlier, the central bank can increase its reserves by selling other government assets (thus reducing $b^G(0)$ by an amount equal to the rise in reserves). The transaction requires no change in the path of planned government outlays, $g(t) + \tau(t)$.

Consider next the private sector. Let $b(t)$ denote net private real bond holdings and $k(t)$ real capital holdings. (By assumption capital's real price equals unity.) Foreigners do not hold domestic money or capital, although the analysis could easily be modified to account for these possibilities. Given an inelastic labour supply normalized at unity and a neoclassical production function $x[k(t), t]$, private-sector assets obey the equation

$$\mathrm{d}[b(t) + k(t) + m(t)]/\mathrm{d}t$$
$$= x[k(t), t] + \rho b(t) + \tau(t) - c(t) - \pi(t)m(t). \quad (5)$$

Define investment $i(t)$ as $\mathrm{d}k(t)/\mathrm{d}t$. The sum of (4) and (5) is

$$\mathrm{d}[b(t) + b^G(t) + r(t)]/\mathrm{d}t$$
$$= x[k(t), t] + \rho[b(t) + b^G(t) + r(t)] - c(t) - i(t) - g(t).$$

The sum $b(t) + b^G(t) + r(t)$ will be denoted by $f(t)$: $f(t)$ equals the economy's overall net claims on the rest of the world. Integrated forward and combined with the condition $\lim_{t \to \infty} \exp(-\rho t)f(t) \geq 0$, the above equation implies the economy's overall intertemporal budget constraint,

$$\int_0^\infty \{c(t) + i(t) + g(t) - x[k(t), t]\}\exp(-\rho t)\,\mathrm{d}t \leq f(0). \quad (6)$$

(The same constraint is relevant when the private sector is prohibited from transacting in the world capital market, but the paths of consumption, investment, and output would generally change if such a prohibition were imposed.)

Inequality (6) states that the present value of the economy's expenditures cannot exceed the present value of output plus initial net external assets. Alternatively, (6) constrains the present value of the economy's trade balance deficits to its initial foreign asset stock. The initial foreign asset stock thus limits the economy's ability to maintain absorption levels in excess of output.

An implication of the analysis is that the most appropriate indicator of flow disequilibrium in external transactions is the change in the economy's overall external assets – the current account. A surplus in the balance of payments may indicate low domestic credit expansion or growing domestic money demand; but when the government has unlimited access to the world capital market, a growing stock of foreign reserves is, in itself, neither a necessary nor a sufficient condition for a sound external position.

The important consequences of current-account flows do not imply that external balance and current-account balance are the same. In analogy with the idea of a high-employment government budget surplus, external balance could be defined roughly as a current account that maintains the highest possible steady consumption level consistent with the economy's expected intertemporal budget constraint. (A more exact definition would require a more explicit treatment of the preferences of households and the government.) Temporary unfavourable movements in output, world interest rates, or the terms of trade are appropriately offset by temporary current-account deficits, while temporary surpluses are an appropriate response to temporary favourable shocks. External balance in the face of a permanent shock, however, generally requires a rapid adjustment to current-account balance.

Similarly increases in the productivity of investment can justify a current-account deficit that is fully consistent with external balance in a long-run sense. In terms of equation (6), a technological innovation implying a gradual upward shift of the production function $x[k(t),t]$ generates higher levels of consumption and investment, and thus an initial current-account deficit. The ability to borrow abroad prevents the sharp rise in the interest rate that would occur initially in a closed economy; a higher investment level than under intertemporal autarky is supported by the foreign capital inflow. As productivity growth returns to normal, investment falls and current-account balance is restored with consumption and output at permanently higher levels.

These points can be made graphically in terms of a two-period Fisherian model (see Figure 3). The axes measure amounts of the two goods available, present and future consumption, and the indifference curves show preferences over those goods. Investment opportunities are described by the production-possibilities frontier, which indicates the amount of future consumption obtained from a given input of present consumption. With the opportunity to borrow abroad at an interest rate ρ, the economy chooses to invest at point A and consume at point B, both of these points lying on the economy's budget line, which has slope $-(1+\rho)$. Given preferences and technology, it is optimal for this economy to run a first-period current-account deficit equal to the horizontal distance between B and A; in period two, the country runs a surplus to repay its earlier borrowing. External balance thus entails an initial current-account deficit for the country shown, but surpluses for countries whose autarky interest rates are less than the equilibrium world rate ρ. The model is a parable of the development process.

When distortions in the economy cause the actual current account to diverge from its optimal level, governments may find it appropriate to adopt policies, such as taxes or subsidies on capital movement, that move the economy closer to the ideal external balance. Policies that operate directly on the distortions in question (if these can be identified) will, as usual, be best. Interesting problems arise when the countries being analysed are large enough that their governments can affect world real interest rates (and other world prices) through their actions. In this situation, the normative guidelines offered by the above approach are not directly applicable to policy analysis, and governments instead condition their actions on the conjectured responses of other governments. A Nash–Cournot equilibrium, in which each government maximizes over policy settings taking as *given* the policies of other governments, will in general be Pareto-inefficient from a global viewpoint. When governments recognize their policy interdependence, welfare in each country can be improved though policy cooperation. The practical difficulty lies in the negotiation process through which all parties agree to choose a particular point on the world contract curve.

SOVEREIGN BORROWING AND CREDIT CONSTRAINTS. The intertemporal analysis of external balance sketched above assumes a world in which individuals or at least governments can borrow unlimited amounts in the world capital market, subject only to their intertemporal budget constraints. Individual and sovereign borrowers alike, however, often appear to face binding credit constraints as a result of nonrepayment risk. After the early 1980s, the extreme difficulty for many industrializing countries of tapping world credit markets focused attention on how countries' borrowing possibilities are affected by the possibility of sovereign debt default. The problem is a central one because most developing-country debts are either contracted directly by government agencies or are government-guaranteed.

Eaton and Gersovitz (1981) presented an early explicit analysis of the sovereign repudiation problem in an international setting. Claims on sovereign debtors are usually not legally enforceable, so the analysis of sovereign default cannot be conducted in terms of bankruptcy laws that govern cases of individual default. Eaton and Gersovitz hypothesized that a sovereign debtor defaults whenever the present discounted benefit of doing so exceeds the present discounted cost. Potential lenders, understanding the debtor's decision rule, will never lend so much that a sure incentive to default is created. Accordingly, sovereign borrowers may find themselves credit-rationed, unable to borrow as much as would normally be optimal at the interest rate quoted by lenders.

There are several potential costs of sovereign default. A defaulting country's external assets, such as foreign reserves or goods in transit, can be seized. The country could, in addition, find itself unable to borrow in the future in response to unexpected changes in its income or technology. Continued participation in the world trade and payments system might become infeasible altogether.

This 'willingness to pay' hypothesis has radical implications for the analysis of external balance. The borrowing country shown in Figure 3 for example, would repudiate its foreign debt if that action were costless, thus avoiding the resource transfer it would otherwise have to make in the second period. As a result, period-one borrowing would take place at a country-specific interest rate reflecting the probability of default, with the extent of borrowing limited by the market's estimate of default costs. At interest rates so high that default was certain, no lending at all would occur.

The analysis of external balance becomes much more complex in such a setting. Not only is the allowable

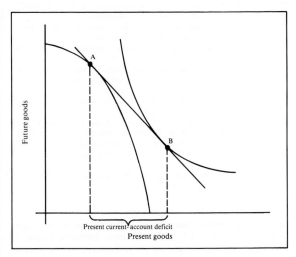

Figure 3

current-account deficit more severely circumscribed; in addition, the policymaker must consider how various policy actions will affect the costs of default and hence the availability of foreign credit. Trade liberalization measures that move the economy away from an autarkic production allocation increase the cost of default by making the economy more vulnerable to disruption of its foreign trade. Such measures will therefore ease international credit constraints at the same time as they improve the static allocation of national resources. Conversely, trade restrictions aimed at improving the current account may well reduce a country's creditworthiness.

The traditional balance-of-payments target has a rationale if the government believes that foreign credit lines may disappear unexpectedly. There is then a case for holding precautionary reserves to finance current-account deficits that may become necessary at times when credit happens to be tight or non-existent. The same purpose would be served, however, if foreign assets held by government agencies other than the central bank were run down at such times.

Internal and external balance may be irreconcilable for countries that seek to continue external debt service in the face of severe limitations on foreign borrowing. After the early 1980s, many developing countries were able to obtain private external finance only through 'forced' bank lending orchestrated by the IMF and central banks. Measures to reduce current-account deficits in line with the external funds available (and in line with IMF stabilization targets) pushed many economies into deep recession. As of this writing, it is unclear how long it will remain politically feasible for debtor governments to downplay internal-balance goals in order to continue avoiding default. There are increasingly-frequent calls for some form of debt relief. Such proposals amount to the *ex post* indexation of debt contracts to adverse contingencies that were not entirely under the debtors' control.

The debt crisis of the 1980s has raised deep and consequential questions about the types of assets traded between developed and developing countries. Before the debt crisis, the typical loan contract between banks and developing-country borrowers was indexed only to the London Inter-Bank Offered Rate, and not to other factors that might alter the borrower's ability to repay. Trade between developed and developing countries in a wider spectrum of state-contingent assets would improve the international allocation of risk, and thus help to avoid future debt crises. A greater share for equity in settling current-account imbalances is one possible step in this direction. Such reforms would not eliminate the sovereign-default problem entirely, nor would they eliminate the moral-hazard problem emphasized by critics of debt-relief proposals. The possibility of a widespread and synchronized default could be sharply reduced, however, under innovative external financing arrangments.

The structure of international financial intermediation also has implications for the mutual adjustment process of industrialized countries. Current-account imbalances are only one avenue through which countries can maintain long-run consumption levels in the face of real income fluctuations or changes in investment productivity. Similar consumption-smoothing can be obtained with smaller current-account imbalances if there is a greater degree of international portfolio diversification. Lucas (1982), for example, models a world of two exchange economies with perfect international risk sharing in which consumption levels can be perfectly correlated internationally even though current-account imbalances never take place. The problem of external balance therefore never arises in Lucas's idealized setting. In reality, the extent of international portfolio diversification seem to be much smaller than plausible financial models of an integrated world capital market would predict. Why this should be so is a major empirical puzzle, and a problem for policy as well.

MAURICE OBSTFELD

See also GOLD STANDARD; INTERNATIONAL TRADE; PURCHASING POWER PARITY; SPECIE-FLOW MECHANISM; SPOT AND FORWARD MARKETS.

BIBLIOGRAPHY

Dornbusch, R. 1973. Currency depreciation, hoarding, and relative prices. *Journal of Political Economy* 81(4), July/August, 893–915.

Eaton, J. and Gersovitz, M. 1981. Debt with potential repudiation: theoretical and empirical aspects. *Review of Economic Studies* 48, April, 289–309.

Fisher, I. 1930. *The Theory of Interest.* New York: Macmillan.

Frenkel, J.A. and Johnson, H.G. (eds) 1976. *The Monetary Approach to the Balance of Payments.* Toronto: University of Toronto Press.

Friedman, M. 1953. The case for flexible exchange rates. In M. Friedman, *Essays in Positive Economics,* Chicago: University of Chicago Press.

Hume, D. 1752. Of the balance of trade. In D. Hume, *Essays Moral, Political and Literary.* London: Longmans Green, 1898.

Johnson, H.G. 1969. The case for flexible exchange rates, 1969. *Federal Reserve Bank of St. Louis Monthly Review* 51(6), June, 12–24.

Keynes, J.M. 1936. *The General Theory of Employment Interest and Money.* London: Macmillan.

Lucas, R.E., Jr. 1976. Econometric policy evaluation: a critique. In *The Phillips Curve and Labor Markets,* ed. K. Brunner and A.H. Meltzer. Amsterdam: North-Holland.

Lucas, R.E., Jr. 1982. Interest rates and currency prices in a two-country world. *Journal of Monetary Economics* 10(3), November, 335–59.

Meade, J.E. 1951. *The Balance of Payments.* London: Oxford University Press.

Metzler, L.A. 1948. The theory of international trade. In *A Survey of Contemporary Economics,* ed. H.S. Ellis, Philadelphia: Blakiston.

Metzler, L.A. 1960. The process of international adjustment under conditions of full employment: a Keynesian view. In *Readings in International Economics,* ed. R.E. Caves and H.G. Johnson, Homewood: Irwin, 1968.

Mundell, R.A. 1961. The international disequilibrium system. In R.A. Mundell, *International Economics,* New York: Macmillan, 1968.

Mundell, R.A. 1968. The nature of policy choices. In R.A. Mundell, *International Economics,* New York: Macmillan.

Nurkse, R. 1944. *International Currency Experience: Lessons of the Inter-war Period.* Geneva: League of Nations.

Obstfeld, M. 1985. Floating exchange rates: experience and prospects. *Brookings Papers on Economic Activity* No. 2, 369–464.

Obstfeld, M., and Stockman, A.C. 1985. Exchange-rate dynamics. In *Handbook of International Economics,* ed. R.W. Jones and P.B. Kenen, Amsterdam: North-Holland.

Ricardo, D. 1817. *The Principles of Political Economy and Taxation.* London: J.M. Dent & Sons, 1911.

Samuelson, P.A. 1971. On the trail of conventional beliefs about the transfer problem. In *Trade, Balance of Payments, and Growth,* ed. J.N. Bhagwati et al., Amsterdam: North-Holland.

Triffin, R. 1960. *Gold and the Dollar Crisis.* New Haven: Yale University Press.

Viner, J. 1937. *Studies in the Theory of International Trade.* New York: Harper & Brothers.

international income comparisons. Despite the accumulating evidence of systematic error, the most common means of making international comparisons still remains the conversion of incomes expressed in own-currencies to a numeraire currency, most often the US dollar, via the exchange rate. The procedure has long been held suspect by travellers who could observe that some countries were dear and others inexpensive – i.e. that exchange rates did not reflect the purchasing power of currencies. As will be shown, the purchasing power of the currencies of poor countries tends to be understated by exchange rates. Exchange-rate conversions thus tend to exaggerate the dispersion of the real per capita incomes of the different nations.

Hence the unique character of making valid *international* income comparisons arises from the existence of different currency units and the need to compare their purchasing powers. For the most part, the other conceptual and empirical problems of international income comparisons are similar to those of within-nation comparisons across time or between persons at a given period.

Ignoring index number problems, the basic approach in recent international comparisons may be simply described as involving the derivation of a quantity (real income) comparison by dividing a price ratio into an expenditure ratio:

$$\frac{Q_j}{Q_b} = \frac{E_j}{E_b} \div \frac{P_j}{P_b}$$

where j and b are countries; Q's are physical quantities, E's are expenditures (for GDP or its components), and the P's are prices, the E's and P's being in own currencies. (P_j/P_b is the purchasing power parity.) Why not, it may be asked, make direct quantity comparisons of the commodities and services that make up the real incomes being compared? The answer is that it is more difficult for most kinds of goods to get a representative sample of country-to-country quantity ratios for the same or equivalent goods than it is to get a representative sample of price ratios. Also, the quantity ratios are more likely to be subject to greater sampling variations than price ratios.

THE EVOLVING SYSTEM OF COMPARISONS. The history of international income comparisons includes many ad hoc efforts based on exchange rate conversions and only a few careful attempts to compare the purchasing power of currencies (Kravis, 1984). The system of international income comparisons that is emerging towards the end of the 20th century has its origins in the study of the Organization for European Economic Cooperation (OEEC) by Gilbert and Kravis (1954). The OEEC study laid down a pattern that has been followed in the UN International Comparison Project (ICP) and related studies carried out by regional groups. In particular, the use of the price-times-quantity-equals-expenditure relationship and the breakdown of GDP in terms of its final product components rather than in terms of producing industries were carried over into the ICP work.

As of early 1986 the system included official benchmark studies for 60 countries with a 1980 reference date, and for diminishing numbers, for earlier reference dates as well, as far back as 1967. (UN and EC, 1985; Ward, 1985; Kravis, Heston and Summers [hereafter KHS], 1982). Unofficial estimates based on extrapolations, to be described presently, were available for most other countries (Beckerman, 1966; KHS, 1978b; Summers and Heston, 1984).

The evolution of the system has been greatly influenced by two major developments that had pervasive effects on the statistical description of the world economy. One was the emergence of national income accounting beginning in the years preceding World War II. Under the aegis of the United Nations a standardized system of national accounts was developed that was adopted by most nations of the world. The standardized system provided a common statistical framework for international comparisons with respect to such important matters as the definition of income (the gross domestic product concept) and the prices at which goods are to be valued (producers' market prices). Earlier international income comparisons tended to focus on the incomes of selected groups of wage earners or employees and to exclude incomes other than those arising from employment. The recent ones are therefore more comprehensive in scope, both with respect to the types of income and population coverage than the past wage-earner-oriented studies. The fact that the ICP studies offer separate estimates for personal consumption may offset the disadvantage that they are based on the inclusion of gross capital formation rather than on some net concept.

The other major development, the advent of the computer, has greatly expanded the availability of methods that meet the needs of comparisons involving many countries at once. In most studies preceding those of the ICP, small numbers of countries were involved and the method turned on a series of binary comparisons – i.e. comparisons between pairs of countries – sometimes, as in the OEEC studies, with all the other countries compared with a country selected as the centre country (star system). This approach has the advantages of simplicity and ease of understanding. An important disadvantage is that the quantitative relationships among the other countries, derived from the relation of each to the centre country, will vary with the choice of a country for the central role. Without the computer, it would have been infeasible to find and apply methods that were invariant to the selection of the base-country, and which at the same time had other desired properties. An important additional property for the real income comparisons that was sought and attained is matrix consistency. This property is akin to that afforded by a time-to-time national accounts table showing the income originating in different economic sectors in constant prices. That is, the figures in any one column (pertaining to a given country) may be added to yield aggregate GDP and subaggregates (e.g. consumption, food, etc.), and the figures on any row (each pertaining to a final expenditure category) show the correct quantity relationships between the different countries. (See, for example, KHS, 1982, p. 19.)

THE ACTUAL WORK OF PREPARING THE COMPARISONS. The actual work of the international income comparisons consists mainly of making price comparisons. The tasks involved are:

(1) Subdividing GDP into categories for which expenditure data and price comparisons can be obtained.

(2) Selecting and pricing a sample of specifications for each expenditure category.

(3) Aggregating the price relatives at the category level.

(4) Aggregation of the categories to form price and quantity indexes for GDP and its subaggregates.

In the literature on international income comparisons, the most intellectual effort has gone into devising index number formulas for the aggregation of the categories, a problem that has long fascinated economic statisticians. The formula chosen in the ICP is one which in principle values the quantities of goods in each country's GDP at world average prices. These values when summed yield the desired real income comparisons or comparisons for components of real income (e.g. consumption, food, beef). The formula, which was suggested by Robert Geary and amplified by S. Khamis, involves deriving the price comparisons and the average world prices simultaneously in two subsets of equations. Some statistical experiments have suggested that radically different results would not be produced by alternative formulas which have equally plausible claims to consideration (KHS, 1982, pp. 95ff).

It has been objected that the per capita quantity indexes for poor countries are inflated because the weights are dominated by the expenditure of the rich countries. However, some experimental work indicates that the results are not greatly changed when the weights of low income countries are greatly increased in calculating the world average prices (Kravis, 1984).

The quality of the income comparisons is, as a practical matter, more vulnerable to the care with which the price comparisons are carried out than to any other phase of the work. Not only must the sample of specifications of each detailed category be representative of price formation influences in each country but the items actually priced in the different countries must be equivalent in quality. Among the means used in the ICP to ensure such equivalents were international exchanges of samples, visits by price experts from one country to partner countries to consult with their counterparts and to examine goods in shops, and resort to informal and formal advice of merchants, manufacturers and engineers. Once specifications were identified, it was necessary to ensure that the price provided was the national average price; this was the responsibility of the country's statistical authorities to supply, sometimes from prices collected for other purposes and in other cases from special price surveys.

Certain services for which outputs are difficult to measure, including education, medical care and government, cannot always be treated in this standard specification approach. In some cases, as services of physicians, quality-adjusted inputs were used as proxies for measures for the international comparison of outputs. Some have claimed that the treatment of these services led to an overstatement of the real per capita GDP of low income countries in Phase III (Maddison, 1983) but sensitivity analysis indicates the possible impacts on real GDP per capitas of different treatments are small (Kravis, 1986).

SUBSTANTIVE FINDINGS. Comparisons based both on PPP and exchange rate conversions are shown in Table 1 for a selected set of countries. The countries are arranged by region and within region by ascending order of real GDP per capita; the exchange rate deviation index, the ratio of the PPP to the exchange rate conversion, is shown in column 3. It can be seen from column 3 that the estimates based on the PPP conversions tend to be higher for lower income countries. That is, poor countries tend to have lower price levels; their exchange rates understate the purchasing power of their currency. A consequence is that the spreads between the average incomes of the countries greatly diminishes when PPP conversions are used. For example, the ratio of the highest to the lowest per capita on the exchange rate basis is nearly 100 to 1 (Germany to Ethiopia) whereas on a PPP basis it is a little over 40 to 1 (Canada to Ethiopia).

Two lines of explanation have been offered for the tendency towards low prices in poor countries. In the productivity differential model, the productivity of poor countries is held to be lower in both traded and nontraded goods but by smaller differentials in nontraded goods (e.g. teaching). Prices of traded goods tend to be drawn to international levels; low productivity in a poor country thus means low wages. However, the same wage level will also prevail in the poor country's nontradables sector, but with productivity somewhat better, prices will be lower. An alternative explanation turns on the labour-abundant factor endowments of poor countries and assumes that nontraded goods (especially services) are

TABLE 1. Real Per Capital GDP, Selected Countries, 1980

	GDP per capita converted by		
	PPP (1)	Exchange rate (2)	Exchange rate deviation index (3) = (1) ÷ (2)
Africa			
Ethiopia	2.5	1.2	2.1
Kenya	5.6	3.7	1.5
Ivory Coast	12.0	11.2	1.1
Asia			
India	5.0	2.1	2.4
Korea	22.6	13.3	1.7
Japan	73.5	77.8	0.9
Europe			
Portugal	33.5	21.2	1.6
Spain	55.5	49.2	1.1
Italy	68.0	60.3	1.1
UK	72.1	81.6	0.9
France	85.4	106.0	0.8
Germany	89.1	116.2	0.8
Latin America			
Brazil	29.3	18.0	1.6
Argentia	33.6	47.4	0.7
North America			
Canada	101.5	94.5	1.1
US	100.0	100.0	1.0

Sources: U.N. and Commission of the European Communities, *World Comparisons of Purchasing Powers and Real Products for 1980; Phase IV of the International Comparison Project; Part I: Summary Results for 60 Countries*, New York, November 1985.

labour-intensive and therefore cheap in poor countries. The average price level is pulled down not only by low prices for nontradables but because tradables too, are cheaper since they almost always are sold with nontradable components (e.g. distribution costs).

Aside from the light shed on real per capita income, the ICP studies illuminate two other important aspects of the world economy. One is that it provides a comparison of the general (GDP) price level of the different countries. The price level is the ratio of the PPP to the exchange rate; it is the reciprocal of the exchange rate deviation index. Wide variations in price levels can be seen to exist even between different members of the European Common Market. These measures add to the existing information on *relative* movements of price levels, a measure of the *absolute* gap. They indicate for example that the German price level was 130 per cent of that of the US in 1980 and only 76 per cent of the US level in 1984 (Ward, 1985).

The other broad set of insights into the structure of the world economy arises from the price and quantity comparisons that are available for components of GDP. A host of questions, many arising in connection with basic economic analyses such as cross-country demand studies can be answered by these comparative price and quantity data. How do food prices differ in low and high income countries? The quantity of medical care? The extent of R&D in real terms? The amount of government services? and so on.

THE FUTURE OF INTERNATIONAL INCOME COMPARISONS. Since it is clear that benchmark estimates will not soon be available for many more than 60 to 70 countries, some less costly even if more approximate method of estimating real per capita GDP will have to be employed for the remaining 50 or so non-benchmark countries. Several approaches have been tried (Kravis, 1984, p. 18). One approach is based on an estimating equation that embodies the relationship between real GDP per capita of benchmark countries and certain physical indicators such as milk or steel consumption (Beckerman, 1966). Another uses certain widely available macroeconomic variables for the extrapolating equation (KHS, 1978a; Summers and Heston, 1984). (For example, real GDP per capita is taken as a function of exchange-rate-converted GDP per capita and the propensity to trade as measured by the ratio of exports plus imports to GDP.) The statistical margins of error surrounding these 'shortcut' estimates for non-benchmark countries make explicit the degree of uncertainty, in contrast to the seemingly unambiguous estimates produced by the exchange rate conversions. However, the exchange rate conversions are not error free; they are known to be biased. In fact, the shortcut estimates come closer than the exchange rage conversions to what full benchmark studies would yield, much closer for low and middle income countries (Kravis, 1986). Benchmark studies would be best, but given that they will not be available for all countries in the near future, a mixed set of benchmark and short cut PPP estimates should be used.

Irving B. Kravis

See also national income; purchasing power parity; real income.

BIBLIOGRAPHY
Beckerman, W. 1966. *International Comparisons of Real Income.* Paris: OECD Development Centre.
Gilbert, M. and Kravis, I.B. 1954. *An International Comparison of National Products and the Purchasing Power of Currencies: A Study of the United States, the United Kingdom, France, Germany, and Italy.* Paris: OEEC.
Kravis, I.B. 1984. Comparative studies of national incomes and prices. *Journal of Economic Literature* 22, March, 1–39.
Kravis, I.B. 1986. The three faces of the international comparison project. *World Bank Research Observer* 1, January, 3–26.
Kravis, I.B., Heston, A.W. and Summers, R. 1978a. *International Comparisons of Real Product and Purchasing Power.* Baltimore: Johns Hopkins University Press.
Kravis, I.B., Heston, A.W. and Summers, R. 1978b. Real per capita for more than one hundred countries. *Economic Journal* 88, June, 215–42.
Kravis, I.B., Heston, A.W. and Summers, R. 1982. *World Product and Income: International Comparisons of Real Gross Product.* Baltimore: Johns Hopkins University Press.
Maddison, A. 1970. *Economic Progress and Policy in Developing Countries.* New York: Norton.
Maddison, A. 1983. A comparison of the levels of GDP per capita in developed and developing countries, 1790–1980. *Journal of Economic History,* March, 27–41.
SOEC (Statistical Office of the European Community). 1977. *Comparisons in Real Values of the Aggregates of ESA, 1975.* Luxembourg: EEC.
SOEC (Statistical Office of the European Community). 1983. *Comparison in Real Values of the Aggregates of ESA, 1980.* Luxembourg: EEC.
SOEC (Statistical Office of the European Community). 1985. *Comparison of Price Levels and Economic Aggregates: The Results for African Countries.* Luxembourg: EEC.
Summers, R. and Heston, A.W. 1984. Improved international comparisons of real product and its composition: 1950–80. *Review of Income and Wealth* 30(2), June, 207–62.
United Nations 1985. *National Accounts Statistics, Main Aggregates and Detailed Tables, 1982.* New York: United Nations.
United Nations and Commission of the European Communities. 1986. *World Comparisons of Purchasing Powers and Real Product for 1980; Phase IV of the International Comparison Project:* Part I: *Summary Results for 60 Countries.* New York: United Nations.
Ward, M. 1985. *Purchasing Power Parities and Real Expenditures in the OECD.* Paris: OECD.

international indebtedness. After World War II many less-developed economies started industrialization programmes which contributed to the achievement of rapid and sustained growth of income. Although industrialization meant a continuous reduction of the import propensity of their economies, growth also implied that the level of imports tended to exceed that of exports. This gap was covered by external indebtedness and, to a lesser extent, by direct foreign investment. During the 1950s and the early 1960s, credit was granted mainly by the governments of advanced economies and multilateral financial organizations. Starting from the mid-1960s, however, international indebtedness was increasingly dominated by private banking, reducing the element of aid implicit in previous arrangements. More importantly, this shift in the nature of credit flows was less conducive to the coordination of policies between industrial and developing countries, the consequences of which became apparent later on.

The remarkable expansion of the world economy achieved during the 1960s was suddenly interrupted at the beginning of the 1970s, due mainly to the policy responses of advanced Western economies to the increase in commodity prices, especially oil. The dramatic increase in oil prices in 1973–4 and again in 1979–80 would have implied, *ceteris paribus,* a situation in which current-account surpluses of OPEC countries and other producers would have been mainly reflected in compensating deficits of the OECD economies – the major importers of oil – and to a lesser extent in deficits of non-oil producers in the Third World. In fact, however, the deficits of the latter swelled much more than what would have been expected on the basis of this assumption. The reason being that, in order to eliminate their own deficits, the advanced economies applied restrictive fiscal and monetary policies which reduced their growth rates (Llewellyn, Potter and Samuelson, 1985). This affected disproportionately the exports from developing economies, thereby increasing their need for foreign lending. Private banking was instrumental in 'recycling' large amounts of petro-dollars – indirectly, through financial intermediation – into the debit side of the balance-sheets of these economies.

Thus, between 1973 and 1980, the OECD economy was only required to shift real resources to oil producers in the form of extra exports – the only inevitable cost arising from oil price increases – of the order of half a percentage point of GDP. The rest of the oil earnings was used to accumulate financial assests. However, in spite of the small magnitude of this real transfer, the annual growth rate of income in these economies dropped two percentage points below its long-term trend. By 1980 the loss accumulated to 15 per cent of GDP (CEPG, 1980).

The maintenance of growth in many underdeveloped economies during the 1970s, plus the availability of international finance in great amounts created by the oil surpluses, increased enormously the level of their foreign debt. The dramatic rise in interest rates at the end of this period, which took place in creditor countries as a consequence of the application of monetarist policies, compounded the problems of debt servicing. Uncoordinated policies of the international banking community made this increase in debt possible, in spite of the heavy exposure of private institutions to sovereign borrowers in the Third World. Equally irrational was their reaction at the beginning of the 1980s, when general economic conditions deteriorated and a sudden awareness of the risks involved was regained. From lending in almost limitless quantities, abruptly the banks decided not to lend at all.

It is of some interest to analyse in more detail the costs of the adjustment process that this sudden change in the availability of finance implied for developing economies. Take the national accounts identity

$$Q \equiv D + X - M \qquad (1)$$

where Q is gross domestic product, D is domestic demand, and X and M are respectively exports and imports of goods and non-factorial services. The balance-of-payments identity can be represented as

$$B \equiv M - X + N \qquad (2)$$

where B is net foreign borrowing less capital movements abroad and N net interest payments. If m is the import propensity, then for simplicity one can assume that

$$M = m \cdot Q \qquad (3)$$

Taken together, these formulations imply that if B, X and N are given either by external circumstances or by history, then necessarily

$$D = [(B - N)(1 + m) + X]/m \qquad (4)$$

$$Q = (B - N + X)/m \qquad (5)$$

Under these assumptions, different mechanisms will be in operation in order to ensure that domestic demand and output attain the above values. Finally, transfers abroad of real resources are

$$T \equiv Q - D \equiv X - M \equiv -(B - N) \qquad (6)$$

Take now a typical situation before the debt crisis in which $B > N$; that is, in which transfers of resources were obtained by developing countries, say by $A = B - N > 0$. Then compare it with one where $B = 0$, prevailing after the crisis. This quantity can even be negative if capital flights increase, as is usually the case when doubts are cast about the creditworthiness of a country. Whereas before the economy was receiving A, now it is transferring resources abroad equivalent to the amount of N. The difference is $(A + N)$. More importantly, the difference in domestic demand is $(A + N)(1 + m)/m$ and $(A + N)/m$ for output. Since m is normally between 0.1 and 0.2, even if it is considerably reduced in the process of adjustment (through devaluation and other policies), this implies that the economy is not only transferring $(A + N)$, but that in order to be able to pay, domestic output and demand must drop by a large multiple of $(A + N)$. Whereas the former are resources obtained by creditor nations, the latter are simply wasted for the world economy as a whole. If this adjustment process takes place simultaneously in several economies, the wastage is even greater since X will also tend to fall.

Once this basic relationship between transfers abroad and the domestic levels of demand and output is grasped, it is straightforward to understand the nature of the policy recipes of international organizations such as the IMF. They are directed mainly towards the lowering of domestic demand through cuts in both the public sector deficit and consumption, the latter via a reduction of the real value of wages in terms of the exchange rate (i.e. devaluations). It is therefore not surprising that this adjustment process is normally accompanied by accelerating inflation, as a reflection of competing demands for shares in income, the level of which is drastically reduced. The external vulnerability of these economies is

further enhanced in the long-rung, and capacity to pay curtailed, since prolonged situations of depression are not conducive to capital formation and productivity growth. This makes it more difficult to reduce the import propensity and to stimulate exports. The import propensity is also likely to increase if the economy follows a trade liberalization policy, another element in the book of orthodox recipes.

The real resources which less-developed economies have been transferring abroad since 1982 is considerably greater in relative terms than the amounts involved for advanced countries following the oil crisis. For example, Argentina, Brazil and Mexico have had to expand in recent years their trade surpluses to around 6 per cent of GDP (ECLA, 1985). As a point of reference, Japan's trade surplus was not, at the time, greater than 2.5 per cent of output. Under these circumstances it is worth asking whether a policy, such as that proposed by the IMF and the international banks – which requires relatively poor countries to transfer abroad, year after year, a large proportion of resources, with a widening gap between potential and actual output – can be described as a permanent solution to the debt problem.

More likely than not, this situation will at some stage lead to an outright default by major debtors unless there is a change in the rules of the game. Defaults such as these have occurred in the past, and with little consequences to the debtors (Winkler, 1933; Wynne, 1951; Wood, 1980). They may become an attractive option given the magnitude of the resources which can be reclaimed for domestic uses. Due to their heavy exposure to sovereign nations, banks have a lot to lose in this event, and there is little they can do in terms of legal and other sanctions (Kaletsky, 1985). The costs of default can be minimized only to the extent that governments in advanced countries – once again, as during the 1950s and 1960s – play an active role in this field and the debt burden is shared equitably among the different parties involved. The incentive to do this lies in the fact that, given the private origin of international indebtedness, their own financial stability may be jeopardized.

VLADIMIR BRAILOVSKY

See also EXTERNAL DEBT; FISCAL AND MONETARY POLICIES IN DEVELOPING COUNTRIES; INTERNATIONAL LIQUIDITY.

BIBLIOGRAPHY

CEPG (Cambridge Economic Policy Group). 1980. World trade and finance: prospects for the 1980s. *Cambridge Economic Policy Review* 63, Department of Applied Economics, University of Cambridge and Gower Publishing Company.

ECLA (Economic Commission for Latin America). 1985. *Estudio Económico de América Latina y el Caribe, 1984*. United Nations, Economic and Social Council, LC/L.330.

Kaletsky, A. 1985. *The Costs of Default*. New York: Priority Press.

Llewellyn, J., Potter, S. and Samuelson, L. 1985. *Economic Forecasting and Policy: The International Dimension*. London: Routledge & Kegan Paul.

Winkler, M. 1933. *Foreign Bonds: An Autopsy*. Philadelphia: R. Swain.

Wood, P. 1980. *The Law and Practice of International Finance*. London: Sweet and Maxwell.

Wynne, W. 1951. *State Insolvency and Foreign Bondholders: Case Histories*. Vol. 2, New Haven: Yale University Press.

international inequality. *See* INEQUALITY BETWEEN NATIONS.

international liquidity. International liquidity may be defined as that stock of assets which is available to a country's monetary authorities to cover payments imbalances (when the exchange rate is fixed) or to influence the exchange value of the currency (when the exchange rate is flexible). A distinction may be drawn between unconditional liquidity, which is generally owned by the country concerned and may be used at its sole discretion, and conditional liquidity, which comprises access to borrowing facilities and is generally available only on conditions set by the lenders. Because of the obvious practical difficulties in measuring conditional liquidity, the operational measure of international liquidity that is generally used in discussion of the subject is that of gross international reserves.

The definition of international reserves used by the International Monetary Fund in compiling *International Financial Statistics* includes: gold; short-term foreign exchange holdings in convertible currencies; special drawing rights (SDRs); and reserve positions in the International Monetary Fund; As of December 1984, total holdings of reserves reported by member countries of the IMF amounted to SDR 438 billion (with gold valued at SDR 35 per ounce). Of this total, 8 per cent was represented by gold holdings (at SDR 35 per ounce); 10 per cent was reserve positions in the Fund; 4 per cent was SDRs; and the remainder was in holdings of foreign currencies (about 70 per cent of which was US dollars).

From an economic point of view, the most significant aspect of the subject of international liquidity is the relationship between the stock of reserves (whether for a country or the world as a whole) and other economic variables, such as the level of real output, the price level, and the pattern of balance of payments positions. It has been recognized that this relationship depends on the nature of the demand function for reserves, and of the arrangements governing reserve supply. Much of the literature on international liquidity has thus focused on these two aspects.

Concerning demand, the demand for reserves by countries, like that for national money by individual economic agents, rests on a desire to enhance welfare by cushioning fluctuations in absorption that might otherwise be made necessary by the non-synchronous nature of payments and receipts. A stock of reserves represents purchasing power that can be used to moderate the domestic economic impact of declines in foreign exchange receipts. Even where official reserve holdings are not in fact used for this purpose, their existence may facilitate the activation of international credits that serve the same purpose. Standard utility theory teaches that the optimum stock of reserves for a country will be that quantity at which the benefits of an additional unit of reserves (in terms of the flexibility it affords the monetary authorities) just balances the cost of acquiring and holding it. Key factors determining the demand for reserves are the amplitude of fluctuations to which an economy is subject in its external position; and the availability of alternative means of financing, or adjusting to, these payments disturbances. For an individual economy, therefore, reserve demand will depend on the structure of its balance of payments. Heavy dependence on exports of primary products subject to volatile supply and demand conditions will tend, *ceteris paribus*, to lead to a greater need for reserves. On the other hand, access to borrowing facilities to cover payments imbalances, or a willingness to allow the exchange rate or domestic policies to adapt so as to encourage accommodating capital flows, will reduce the demand for owned reserves.

For the world as a whole, it has generally been thought that payments disturbances would tend to grow with the underlying volume of world trade; however, there is less agreement about whether the relevant elasticity should, for practical purposes, be regarded as unity. (This debate parallels

that on the income elasticity of the transactions demand for cash within a national economy). A further important issue is the extent to which exchange rate flexibility alters (presumably reduces) the demand for reserves. While governments have always had alternatives to reserve use for financing payments disequilibria (e.g. official borrowing, or the manipulation of domestic policies to encourage capital inflows or outflows) the introduction of greater exchange rate variability has created much greater scope for economizing on reserves. The extent to which such scope has been used, has however, varied among countries and over time. As a result, economists have had much less success in estimating stable demand functions for international reserves than in finding stable money–demand functions in individual economies.

Concerning reserve–supply arrangements, there have been two alternative views of the mechanism at work, partly reflecting changing institutional conditions in the world economy. The traditional view, which was widely accepted until the latter 1960s, was that the stock of international liquidity was to a significant extent exogenously determined. There was little dispute that this was true of gold, where the price was fixed and the available physical quantity was being augmented at a relatively slow and predictable rate. It was also thought to be broadly characteristic of US dollars, with the payments deficit of the United States supplying foreign exchange which, under the then existing institutional arrangements, other countries felt obliged to accept and hold. The newer view of reserve supply arrangements sees the stock of international liquidity as being essentially demand-determined. With international capital markets having greatly expanded in size and efficiency, countries may collectively increase their stocks of owned liquidity through operations in domestic and international financial markets. In this view, reserve stocks can increase quite independently of the balance of payments position of reserve currency countries. If a country wishes to increase its reserve holdings, and is creditworthy, it may bid for the desired funds in international financial markets and hold them in the form of short-term securities. The liabilities which are the counterpart to the reserves thus created may be liabilities of the public or private sector, and may be issued by residents of any country with the creditworthiness (and exchange control permission) to do so.

These different views of reserve–supply arrangements, coupled with differences of opinion about the stability of the underlying demand for reserves, have major implications for the role of international liquidity in influencing developments in the world economy. At one extreme, if the demand for reserves was a stable function of variables that were closely linked to world output and trade, and if the available stock of reserves was externally fixed, there would be a tight linkage between international liquidity and economic activity. If reserves fell below the desired stock for a given level of economic activity, countries would, on average, seek to augment their liquidity through trade restrictions, exchange rate depreciation, or other measures aimed at strengthening their overall balance of payments. With a fixed stock of reserves, a deflationary bias would be imparted to the world economy until nominal incomes had been reduced sufficiently to correspond with the given reserve stock.

While it was never believed that such a tight linkage existed, the fear of reserve inadequacy became important during the 1960s. Gold was fixed in price and increasing only slowly in volume, while dollars were thought to be created by a process that would eventually prove unsustainable because of the effects of continued US deficits on countries' willingness to hold dollars. For this reason, it was felt that some other mechanism was required to meet the growth in demand for reserves. Initially, devices were employed to augment the supply of credit facilities (central bank 'swap' arrangements, General Arrangements to Borrow, increased drawing rights in the IMF) but it was generally felt that the system required a secular increase in owned reserves as well. The outcome of this debate was the decision (reached in 1967) to create a new international asset, Special Drawing Rights in the International Monetary Fund (SDRs). Since SDRs would not be the liabilities of any individual country, and would have a value linked to gold (later, to a basket of currencies), they would not be subject to the confidence factors that affected US dollars. At the same time, the volume of SDRs could be augmented (or reduced) by conscious decision in the light of the long-term needs of the world economy. At the time of writing, SDRs are valued in terms of a basket of the five major industrial country currencies and bear an interest rate (paid by countries whose holdings are less than their allocation and received by those countries whose holdings exceed their allocation) related to short-term market interest rates.

Since the shift to floating exchange rates, and partly as a result of it, a somewhat different view has developed of the factors influencing the demand for and supply of reserves. It has been recognized that the demand for reserves by major countries with flexible exchange rates cannot be easily identified, and may change over time. At the same time, the flexibility with which international capital markets have functioned has permitted changing reserve demand to be accommodated relatively easily. For these reasons, several major industrial countries have not thought it necessary for international liquidity to be deliberately augmented through allocations of SDRs.

On the other hand, most developing countries have continued to pursue some form of pegging arrangement, and many of them have experienced difficulty in preserving access to international capital markets. Their need for reserves has tended to increase in line with the volume of their international transactions, and their means of satisfying this need has relied heavily on action to improve the current account of their balance of payments. Many observers have therefore advocated continued creation of SDRs as a means of satisfying the reserve needs of these countries without requiring excessive adjustment on their part. At the same time, it has been pointed out that SDRs would preserve the 'seignorage' associated with reserve issuance for the international community at large. Proposals to 'link' liquidity creation to development assistance, by allocating SDRs in the first instance to developing countries or to development finance institutions, have enjoyed considerable popularity in the economic literature, but have not had the support of major countries.

A.D. CROCKETT

See also EXTERNAL DEBT; GOLD STANDARD; INTERNATIONAL MONETARY INSTITUTIONS; INTERNATIONAL MONETARY POLICY.

BIBLIOGRAPHY

Crockett, A.D. 1978. Control over international reserves. *IMF Staff Papers*, Washington, DC, March.
Heller, H.R. and Khan, M.S. 1978. The demand for international reserves under fixed and floating exchange rates. *IMF Staff Papers*, Washington, DC, December.
International Monetary Fund. 1970. *International Reserves: Needs and Availability*. Washington, DC.
Mundell, R.A. and Polak, J.J. (eds) 1977. *The New International Monetary System*. New York: Columbia University Press.
Von Furstenberg, G.M. (ed.) 1983. *International Money and Credit: The Policy Roles*. Washington, DC: International Monetary Fund.
Willett, T.D. 1980. *International Liquidity Issues*. Washington, DC: American Enterprise Institute.
Williamson, J. 1973. International liquidity: a survey. *Economic Journal* 83, September, 685-746.

international migration. The literature on international migration from developing countries has expanded considerably in recent years. Much of it tends to be empirical and policy-oriented with the majority of studies seeking to estimate the magnitude of inter-country migration flows and the associated benefits and costs. However, unlike the ever more abundant literature on the theoretical determinants and consequences of internal migration in developing nations, the literature on international migration is notable for its absence of theoretical models designed to analyse the causes and consequences of migration across borders. As a result, policy prescriptions for promoting or regulating international flows are often made without any underlying conceptual framework.

In part, the problem arises from the institutional context in which international migration occurs – placing legal, physical and cultural constraints on the unfettered locational choices of individual workers. The theoretical work to date, based mostly on neoclassical trade theory, skirts this difficulty by focusing on international migration in aggregate terms. By doing so, it has ignored the underlying basis for mass movements of labour – the individual migrant's decision-making process.

Furthermore, these economic models usually examine international migration within a perfect world of full employment, flexible wages and full factor mobility. Traditional trade theory views migration as a simple disequilibrium phenomenon in which labour seeks to equalize returns across countries. As long as real returns to labour are unequal, the models suggest that international migration will persist in the absence of physical constraint by governments. Only after factor returns are equalized will migration cease.

This view of international migration is implicit in many of the trade models ranging from the earlier simple neoclassical framework of Grubel and Scott (1966) to the more elaborate trade models of Berry and Soligo (1969), Rivera-Batiz (1982) and Ethier (1985), among others. Moreover, under the perfect world assumptions, these models often generate results that are not applicable to the not-so-perfect world of international migration. In the absence of an acceptable theoretical framework, therefore, most researchers have turned their attention to quantifying the direct benefits and costs of migration by focusing on the most obvious and easily obtainable information. Unfortunately, this approach has serious pitfalls, as the less obvious and indirect benefits and costs are typically overlooked.

A basic methodological requirement for theorization about international migration, therefore, is to include both macro- and micro-level factors and relationships within a single frame of reference. This essay will be a step in that direction. In presenting a model of the individual worker's decision to emigrate, it will lay a foundation for the neoclassical trade model which explains the mass migration of labour across borders as an attempt to equalize returns to factors. However, in contrast to traditional trade models, in our model, international migration will be shown to cease long before actual returns to labour are equalized across countries – i.e., it will cease when 'expected' returns are equalized. Perhaps, more importantly, by focusing simultaneously on three sectors – the domestic rural, the domestic urban and the foreign – our model will attempt to incorporate both internal and international migration choices within a single decision theoretic framework.

MIGRATION AND DOMESTIC UNEMPLOYMENT. One of the major purported benefits of organized, short-term labour emigration is that it contributes to the relief of unemployment in the labour-exporting country. This view will be disputed in the model to be presented here. While it may be true that labour emigration contributes to the relief of overall domestic unemployment, it will be shown that this favourable effect may be offset by a costly rise in urban unemployment provoked by increased rural-urban migration. Perhaps the most significant policy implication of this result is that policies designed to encourage labour exports in an effort to eliminate unemployment at home may be, in fact, exacerbating the domestic urban unemployment problem and increasing the rate of urbanization. The analysis also suggests that the promotion of labour exports may impose a significant cost on the rural sector as well. Such costs arise in the situation in which the rural sector does not possess a surplus of labour but large expected income differentials induce out-migration which, in turn, lowers rural output. In sum, our model of international migration suggests that the benefits and costs of labour exports must be examined and weighed more carefully than in previous studies, with particular emphasis on secondary and tertiary effects.

A preliminary review of empirical studies, both at the macro and micro level, suggests that the key variables underlying the decision to emigrate are the prospects for obtaining employment and the wage differentials between the country of immigration and that of emigration. Five well-known studies illustrate this point.

On a macro level, Hietala (1978) examined the labour flows between the Nordic countries in 1963–75 and found migration to be highly correlated with unemployment and wage differentials in almost all cases. This econometric study is particularly significant because it examined international migration in the absence of the usual physical and institutional constraints on migration – the Nordic countries have had a common labour market since 1954. Stahl (1984), in a major study of the current volume and characteristics of international migration in several ASEAN countries, also argued that wage differentials played a major role in the out-migration of workers from the leading labour-exporting countries.

On a micro level, Ulgalde (1979), in a survey of migrants from the Dominican Republic, found that 30% migrated because of unemployment and another 30% migrated in search of higher wages. Castano (1984), in a field study of Colombian workers migrating to Venezuela, found that 36% migrated in search of higher wages while another 26% cited the lack of jobs as their primary reason for moving. Finally, in a survey of Mexican migrants to the US, Cornelius (1978) found that 30% migrated due to the lack of jobs while another 31% went in search of higher wages.

Our model of international migration will therefore focus both on wage differentials and employment probabilities and thus will also be useful in analysing another type of major labour flow occurring in the world today – the flow of temporary workers, particularly from several Asian and a few Arab nations, to the oil-producing Persian Gulf countries. The prospect of substantial foreign exchange earnings, combined with the potential for reducing the high unemployment rates that often characterize the labour-sending countries, has led many developing countries to actually promote the export of their labour. The model will examine the various economic consequences of labour-export promotion policies and offer some insight into their possible costs.

A THREE-SECTOR MODEL OF LEGAL INTERNATIONAL MIGRATION. A major purported benefit of international migration is that it reduces the often high rates of unemployment in labour-exporting countries. While it may be true that labour emigration contributes to the relief of overall unemployment, one possibility ignored in the literature is that this favourable effect may be offset by a costly rise in urban unemployment provoked by increased rural-urban migration. For example, a number of studies have noted that labour export promotion has increased rural-urban migration and accelerated urbanization in several Asian and a few Arab labour-exporting countries. Stahl (1984) observed this effect in his study of the ASEAN countries, particularly with regard to the Philippines, Thailand and Indonesia, which export labour to the Middle East. The Population Information Center (1983) stated that the migration of skilled and semi-skilled Jordanian workers to Saudi Arabia and Kuwait opened up many jobs in the cities and provoked internal rural-urban migration to fill them. Finally, the United Nations Department of International Economic and Social Affairs (1982) declared that urbanization had been accelerated by emigration in Pakistan, also a major exporter of labour to the Middle East.

This is precisely the situation which arises when individuals seeking employment abroad must first come to their domestic urban areas in order to obtain foreign jobs. In many labour-exporting developing countries, such workers must register with urban modern-sector recruitment agencies (private or government run) which provide contracts with specific employers for employment abroad. Moreover, potential international migrants coming from outside the city must not only come to the urban centre to register for foreign jobs, they must also wait there in order to maximize their opportunity for foreign employment.

In a study of the major labour-exporting ASEAN countries, Stahl (1982) observed that recruiters, with registrants far in excess of the number of foreign jobs available, chose only those workers immediately present as delays in post and travel made it difficult to call in registrants residing outside the city. Under such circumstances, workers seeking foreign employment must migrate to the city and wait some period of time, usually in the informal sector, before obtaining a foreign job. For example, Martin (1984) noted that recruitment is concentrated in urban centres in the labour-exporting countries. Moreover, he argued that most potential international migrants enter the informal sector of the urban economy while waiting for the opportunity to migrate.

If one examines this situation within the context of a Todaro-type model of migration (Todaro, 1969; Harris and Todaro, 1970) then the offer of relatively higher-paying foreign jobs from recruiters in urban areas should increase the economic attractiveness of the urban economy and, hence, increase the rate of rural-urban migration and possibly also the rate of urban unemployment.

This section examines the effect of international labour emigration on the urban unemployment rate by extending the Todaro model of internal migration to rural-urban-international migration. In doing so, we arrive at a three-sector model of international legal migration based on assumptions that are in keeping with the observations made by researchers studying the major labour-exporting countries. Moreover, the model is unique within the current theoretical work on international migration because it focuses on the individual's decision to migrate.

The model. The model examines the behaviour of three types of migrants: rural workers seeking urban modern sector jobs; rural workers seeking foreign jobs; and urban employed workers seeking foreign jobs.

As within the framework of the original Todaro (1969) model, it is assumed that the decision to migrate depends on the expected relative income differential between the place of origin and the foreign destination. The expected income differential, in turn, depends on actual income differentials, the cost of migration and the probability of employment.

It is assumed that rural workers seeking urban jobs must wait in the informal sector some period of time before obtaining an urban modern-sector job. Those seeking foreign jobs must also come to the city in order to obtain those jobs. Furthermore, they must wait in the informal sector some period of time before obtaining foreign employment. Urban employed workers seeking foreign jobs are assumed to continue their employment in the urban sector until they have obtained jobs abroad.

For simplicity, it will be assumed that the natural rate of increase of the urban labour force is zero. Hence, any change is due to rural–urban migration and labour out-migration. This implies that at any point in time, the urban labour force consists of (1) rural immigrants looking for urban jobs, (2) rural immigrants looking for foreign jobs, with both groups residing in the informal sector of the urban economy, and (3) urban workers who are already employed in the modern sector of the economy. Under the assumption of a zero natural rate of increase, the employed 'urban' workers are actually former rural residents who have managed to obtain urban modern-sector jobs. Some of these urban employed will continue their jobs while others seek foreign employment. In terms of this analysis, it is the urban employed workers desiring foreign jobs that are important as their emigration affects the probability of employment of groups (1) and (2). Including a natural rate of increase of the urban labour force in the model would not change the results.

Formally, the model can be expressed as:

(I) IN-MIGRATION OF RURAL WORKERS. As in Todaro (1969), rural–urban migration of workers seeking domestic employment is a function (f) of the expected relative rural–urban income differential, $(\alpha_{r,u}^e)$:

$$f(\alpha_{r,u}^e(t)), \qquad f < 1, \quad f' > 0 \qquad (1)$$

which depends on the discounted value of average earnings in the rural and urban sectors (Y_r and Y_u, respectively), the probability of having an urban job (P_u), and the initial cost of migrating to the urban sector (C_u) such that:

$$\alpha_{r,u}^e(0) = \int_{t=0}^{t=n} P_u(t)[(Y_u(t) - Y_r(t))/Y_r(t)]\,e^{-rt}\,dt - C_u(0). \quad (1a)$$

By extending this framework to deal with international migration, rural–urban migration is also now a function (g) of the expected relative rural–foreign income differential $(\alpha_{r,f}^e)$:

$$g(\alpha_{r,f}^e(t)), \qquad g < 1, \quad g' > 0 \qquad (2)$$

which depends on the discounted value of average earnings in the rural and foreign sectors (Y_f), the probability of having a foreign job (P_f), the cost of migrating to the urban sector in order to find a foreign job (C_u), and the cost of migrating abroad (C_f) such that:

$$\alpha_{r,f}^e(0) = \int_{t=0}^{t=n} P_f(t)[(Y_f(t) - Y_r(t))/Y_r(t)]\,e^{-rt}\,dt$$
$$- C_u(0) - C_f(0). \quad (2a)$$

Since individuals obtain foreign jobs through recruiters and other foreign employment contacts in the urban sector, the probability of finding a foreign job is actually the probability of being selected for a foreign job by the recruiter with which the potential international migrant is registered. This probability corresponds to the random selection process described in Todaro (1969) and is less than one, since recruitment agencies have registrants in excess of the number of foreign jobs available.

It is assumed that once an individual succeeds in finding a foreign job through one of these domestic channels, he is guaranteed employment upon arrival at the foreign destination. In other words, upon having found a foreign job in the urban sector, the individual faces a probability equal to one of being employed in the country of immigration. This assumption seems reasonable in light of the fact that recruiters usually provide fixed time period employment contracts with specific employers in the labour importing country.

The cost of migrating abroad (C_f) may include recruiter's fees as well as transportation costs or any other costs associated with migrating abroad, including psychological costs. The important point is that in deciding whether to migrate, the individual must charge the cost to himself as if it were to apply the day that he leaves his rural home. In other words, the individual must have the amount C_f before leaving the rural sector as he will be unemployed some period of time before going abroad. The cost of migrating to the urban sector (C_u) is included in the potential migrant's calculations as he must go to the city in order to find a foreign job.

(II) OUT-MIGRATION OF URBAN EMPLOYED WORKERS. The rate of emigration of employed urban workers (h) is a function of the actual, rather than the expected, relative urban-foreign income differential ($\alpha_{u,f}$):

$$h(\alpha_{u,f}(t)), \qquad h < 1, \quad h' > 0 \qquad (3)$$

which depends on the discounted value of average income in the foreign and urban sectors less the costs of migration such that:

$$\alpha_{u,f}(0) = \int_{t=0}^{t=n} [(Y_f(t) - Y_u(t))/Y_u]\,e^{-rt}\,dt - C_f. \qquad (3a)$$

The intuition behind this formulation is that if, for example, the potential international migrant from the urban modern sector faces a one in three chance of finding a foreign job once he has decided to migrate, he will continue employment in the urban sector for three years prior to emigration. Thus, at the time of emigration, his expected income differential will be the same as his actual income differential.

(III) EXPLICIT EXPRESSIONS FOR THE PROBABILITIES OF EMPLOYMENT. In (Todaro, 1969), the probability of a rural–urban migrant being selected for an urban modern sector job in any given period is defined as the ratio of urban modern sector job openings relative to the number of urban unemployed. However, in this version of the model, the probability of being selected for a modern sector job is also affected by the emigration of employed modern sector workers which creates an additional flow of job openings. This probability can be expressed as:

$$\Pi_u(t) = (\gamma(t)N(t) + h(t)S(t))/(S(t) - N(t)) \qquad (4)$$

where $\gamma(t) = \dot{N}(t)/N(t)$ is the rate of urban job creation due to growth in modern sector output minus the growth of labor productivity. $N(t)$ is modern sector employment. $S(t)$ is the total urban labor force. Hence, $S(t) - N(t)$ is the number of unemployed in the urban sector (the informal sector).

The probability of a rural–urban migrant being selected for a foreign job is also affected by the emigration of employed modern sector workers who take some of the available foreign jobs:

$$\Pi_f(t) = (k(t)S(t) - h(t)S(t)/(S(t) - N(t)) \qquad (5)$$

where, $k(t)$ is the rate of foreign job creation and $k(t)S(t)$ yields total emigration opportunities for both employed modern sector workers and unemployed rural–urban migrants.

This formulation recognizes the fact that modern sector workers have an advantage over unemployed workers in obtaining foreign employment due to their work experience, wider contacts, and higher wage incomes to sustain themselves while waiting for the opportunity to emigrate. Thus, rural individuals seeking foreign jobs look at the net foreign jobs available after employed modern-sector workers have emigrated. To insure that net emigration opportunities are available for rural–urban migrants, we assume $k > h$.

In calculating the probability of obtaining urban (or foreign) jobs, rural individuals are assumed to compare the number of urban (or foreign) jobs available with the total number of urban unemployed $(S(t) = N(t))$ rather than with the number of of urban unemployed workers seeking urban jobs (or the number of urban unemployed workers seeking foreign jobs). This assumption seems more realistic from a behavioural point of view since rural individuals living in different locations would have great difficulty estimating the total number of urban (or foreign) jobs available with the total number of urban unemployed $(S(t) - N(t))$ rather than with the number of proxy for the number of other individuals seeking the same type of employment.

It is assumed further that the number of foreign jobs available grows with the size of the urban labour force $(S(t))$. This assumption stems from observations that the accumulation of a growing number of workers in the urban sector creates more opportunities for foreign employment as recruiters, employment agencies, and other traffickers in manpower expand to meet the growing demand for foreign jobs from domestic workers. The value of k, however, is a policy parameter, since governments may encourage or discourage labour export efforts.

(IV) URBAN LABOUR SUPPLY FUNCTION. The urban labour supply function in its most general form can be expressed as:

$$\frac{\dot{S}(t)}{S(t)} = f(\alpha^c_{r,u}(t)) + g(\alpha^c_{r,f}(t)) - k(t). \qquad (6)$$

For more explicit results and in keeping with Todaro (1969), a definite form of the labour supply function is assumed – that it is separable in income differentials (adjusted for the costs of migration) and the probability of employment. We also assume a one-period time horizon which allows us to treat the various income differentials as fixed. The basic results are unaffected by this assumption.

Letting $\alpha_{r,u} = (Y_u - Y_r)/Y_r$ and $\alpha_{r,f} = (Y_f - Y_r)/Y_r$, the labour supply function can now be expressed as:

$$\frac{\dot{S}}{S} = \Pi_u f(\alpha_{r,u}) + \Pi_f g(\alpha_{r,f}) - k. \qquad (6a)$$

Assuming that the f, g and h functions are continuous and monotonically increasing in their respective percentage real income differentials, we get unique values for these functions for any given set of differentials. Substituting for

$$\Pi_u \quad \text{and} \quad \Pi_f$$

from section (iii) into the above equation, the urban labour supply finally becomes:

$$\frac{\dot{S}}{S} = \left(\frac{\gamma N + hS}{S - N}\right)f + \left(\frac{kS - hS}{S - N}\right)g - k. \qquad (6b)$$

This equation simply says that the rate of increase of the urban labour force $(\dot{S}/S)$ is equal to rate of in-migration of rural workers seeking urban jobs in response to a given percentage rural–urban income differential (f) adjusted for the probability of finding an urban job $((\gamma N + hS)/(S - N))$ plus the rate of in-migration of rural workers seeking foreign jobs in response to a given percentage rural-foreign income differential (g) adjusted for the probability of finding a foreign job $((kS - hS)/(S - N))$ less the rate of emigration of both employed and unemployed workers from the urban sector.

(V) EQUILIBRIUM CONDITIONS. The equilibrium condition for our model is defined as the urban employment rate (E) such that:

$$\frac{\dot{E}}{E} = \frac{\dot{N}}{N} - \frac{\dot{S}}{S} = 0. \qquad (7)$$

Substituting for $\dot{N}/N$ from section (iii) and $\dot{S}/S$ from the final labour supply function into the above equation, we obtain the condition:

$$\gamma + k = \left(\frac{\gamma N + hS}{S - N}\right)f + \left(\frac{kS - hS}{S - N}\right)g. \qquad (8)$$

Thus, the equilibrium urban employment rate is one in which the total flow of migrants is just sufficient to fill the new modern sector jobs being created, replace the job vacancies left by emigrating employed workers, fill the emigration opportunities for unemployed workers and provide a net addition to the urban unemployed (in light of the higher modern sector employment). Dividing the numerator and denominator of the right hand side of the above equation by S, substituting $E = N/S$ and then solving for E we obtain:

$$E = \frac{\gamma + k + hg - kg - hf}{\gamma + k + \gamma f}. \qquad (9)$$

Taking the case where $\gamma > k > h$, we recall that: γ is the rate of urban job creation; k is the rate of foreign job creation, or total rate of emigration from the urban sector; f is the rate of in-migration of rural workers seeking urban jobs in response to a given rural–urban income differential; g is the rate of in-migration of rural workers seeking foreign jobs in response to a given rural–foreign income differential; h is the rate of out-migration of urban employed workers obtaining foreign jobs in response to a given urban–foreign income differential.

For meaningful results, it must be true that the equilibrium employment rate is positive and less than one. For $E > 0$, it must be true that:

$$\gamma + k + hg > kg + hf$$

and it is, since $k > kg$ and $\gamma > hf$. The condition for $E < 1$ is:

$$hg < \gamma f + hf + kg$$

which holds since $k > h$.

(VI) EFFECT OF AN INCREASE IN THE RATE OF FOREIGN JOB CREATION.

$$E_k = (\text{Dem}(1-g) - \text{Num}(1))/\text{Dem}^2 \qquad (10)$$

where 'Dem' is the denominator of equation (9) and 'Num' is its numerator.

The effect of an increase in emigration opportunities on the equilibrium urban employment rate depends on the relative values of f and g for a given γ and h. In the case where the value of g is such that:

$$g > f\,\frac{(\gamma + h)}{\gamma + h + \gamma f} \qquad (10a)$$

an increase in foreign job creation will lower the equilibrium rate of urban employment, or equivalently, it will raise the urban unemployment rate. That is, in the situation where g is sufficiently large (i.e. meets the above condition), the creation of additional emigration opportunities for unemployed workers will draw proportionally more rural individuals seeking foreign employment into the urban centre than is required to match emigration opportunities and hence, result in an increased urban unemployment rate. This implies that under such circumstances policies designed to eliminate unemployment through the promotion of labour exports will meet with increasing frustration, unless there is a simultaneous concentrated effort to raise rural incomes and hence, lower real income differentials.

(VII) EFFECTS OF AN INCREASED PROPENSITIES TO MIGRATE.

$$E_f = (\text{Dem}(-h) - \text{Num}(\gamma))/\text{Dem}^2 \qquad (11)$$

$$E_g = (\text{Dem}(h-k) - \text{Num}(0))/\text{Dem}^2 \qquad (12)$$

Equation (11) is clearly negative and since Dem > Num and $k > h$, equation (12) is also negative. Thus, an increased propensity to migrate on the part of rural workers seeking either urban or foreign jobs will raise the equilibrium rate of urban unemployment.

$$E_h = (\text{Dem}(g-f) - \text{Num}(0))/\text{Dem}^2 \qquad (13)$$

The net results depends on the relative values of (f) and (g). On one hand, the increased emigration of employed workers (h) creates more urban job opportunities as these workers leave and thereby raises the in-migration of rural workers seeking urban jobs via its impact on their probability function. On the other hand, it lessens emigration opportunities for rural workers seeking foreign jobs and hence lowers their migration into the urban sector. If f is greater than g, then the urban unemployment rate will rise. The urban unemployment rate rises because the increase in the flow of rural workers seeking urban jobs outweighs the decrease in the flow of rural workers seeking foreign jobs.

Recall that we are treating the percentage income differentials as given in this analysis. Hence, we are examining the effect of an increased desire to migrate reflected in an increase in f, g or h. This increased desire to migrate could be the result of any number of influences suggested by the literature including such non-economic factors as more information about working abroad from friends or relatives and hence, less insecurity on the part of the individual to emigrate or perhaps the result of 'modernized' tastes and hence, the desire to acquire foreign goods or perhaps the result of job frustration due to under-

employment or even an attraction to 'city lights'. However, keeping in mind the results of this section and recalling that f and g respond positively to increases in income differentials, it is immediately evident that if rural incomes fall, the values of both f and g will be larger in response to the larger rural-urban and rural-foreign income differentials, respectively, and the equilibrium rate of urban unemployment will be higher. Similarly, if foreign earnings rise, the values of both g and h will be larger in response to the larger rural-foreign and urban-foreign income differentials, respectively, and in the situation where $f > g$, the equilibrium rate of unemployment will be also higher.

(VIII) POLICY IMPLICATIONS. Most of the major labour-exporting countries are contending with the dual problems of rapid urbanization and high overall rates of urban unemployment. In light of our results, their efforts to lower the overall rates of unemployment through the promotion of labour exports (raising k) will increase rates of urbanization and urban unemployment in cases where the propensity to migrate of rural workers seeking foreign jobs (g) is sufficiently high. This implies that policy-makers will have to examine the benefits and costs of labour-export promotion more carefully. The social costs associated with rapid urbanization and high urban unemployment may offset to a considerable extent the benefits associated with increased foreign exchange earnings when one considers that in addition to the social costs of increased unemployment are the enormous costs of providing housing, education, health facilities, sanitation and public transportation for the new urban dwellers.

Moreover, as migration networks mature, one can expect an increase in the desire to migrate as individuals obtain more information about working abroad and become exposed to foreign consumption patterns from friends and relatives who have lived overseas. As was shown earlier, any increased propensities to migrate on the part of rural workers seeking foreign jobs (g), given existing income differentials, will result in higher urban unemployment rates. Furthermore, if one considers the trend of widening income differentials between the more advanced developing countries which are usually the labour-importers and the less developed countries which usually export labour, there exists the strong possibility that the urban unemployment problem will worsen as more individuals flock to the city to seek both the higher paying foreign jobs and the urban jobs that native emigrants leave behind.

Another implication of our model is for the rural sector. Although the impact of labour-export promotion on the rural sector was not examined directly, the model does show that the rate of out-migration from the rural sector would rise with increased opportunities for foreign employment. In the situation where the rural sector does not have a surplus of labour, this higher rate of out-migration may cause a decline in rural output. Another possible outcome is a costly rise in mechanization, which has already occurred in Jordan. Both possibilities would further exacerbate problems of urbanization and underemployment, the first by widening the rural-urban income gap and the second by displacing rural jobs and pushing more migrants toward the city.

As Stahl (1982 and 1984) has noted, one possible measure that policy-makers could take to mitigate this induced rural-urban migration effect is to disperse recruitment agencies throughout the country so as to give rural workers the opportunity to seek foreign employment without having to come to urban centres to obtain a foreign job. However, in cases where much of the recruitment is done through private agencies, it would be difficult to force private recruiters to relocate to areas that would be clearly inefficient from their point of view. The government might instead require that private recruiters obtain workers from a government-run central office that keeps files on registrants from various parts of the country (i.e., among those who have registered for foreign jobs at their local government foreign recruitment office) and draws on the available pool of registrants according to whatever criteria has been developed. However, as Stahl has pointed out, this measure may be also insufficient to deal with the problem due to the widespread existence of unlicensed recruitment offices which would be virtually impossible to regulate.

Perhaps the most important implication of the model arises out of the welfare evaluation of the foreign exchange remittances sent home by international migrants. Most researchers have concluded that the sheer size of these remittances indicates a large positive benefit for the labour-exporting nation, even in cases where remittances are used primarily for conspicuous consumption and/or real estate speculation. However, our model underlines an important negative element in the welfare equation that has been typically overlooked – namely, the social and private costs of increasing urban unemployment and possibly declining agricultural output generated by the induced migration of additional rural workers now seeking both modern sector and foreign higher wage jobs. Balancing these added costs against the presumed benefits of foreign remittances could conceivably tip the scales against the widespread belief in the positive association between increased international short-term migration and local economic gains.

M.P. Todaro and L. Maruszko

See also DEVELOPMENT ECONOMICS; HARRIS–TODARO MODEL.

BIBLIOGRAPHY
Berry, A.R. and Soligo, R. 1969. Some welfare aspects of international migration. *Journal of Political Economy* 77(5), September–October, 778–94.
Birks, J.S. and Sinclair, C.A. 1980. *International Migration and Development in the Arab Region*. Geneva: International Labor Office.
Castano, G.M. 1984. Migrant workers in the Americas: a comparative study of migration between Colombia and Venezuela and between Mexico and the United States. Center for U.S.–Mexican Studies, New York.
Cornelius, W.A. 1978. Mexican migration to the United States: causes, consequences, and U.S. responses. Migration Study Group, Massachusetts Institute of Technology.
Ethier, W.J. 1985. International trade and labor migration. *American Economic Review* 75, September, 691–707.
Grubel, H. and Scott, A. 1966. The international flow of human capital. *American Economic Review* 56, May, 268–74.
Harris, J. and Todaro, M.P. 1970. Migration, unemployment and development: a two sector analysis. *American Economic Review* 60(1), March, 126–42.
Hietala, K. 1978. Migration flows between Nordic countries in 1963–75. In International Union for the Scientific Study of Population, *Economic and Demographic Change: Issues for the 1980's*, Liège: IUSSP.
Martin, P.L. 1984. The economic effects of temporary workers. Mimeo, University of California, Davis, March.
Population Information Center, The Johns Hopkins University. 1983. *Population Report*, October.
Rivera-Batiz, F. 1982. International migration, non-traded goods and economic welfare in the source country. *Journal of Development Economics* 11(1), 81–90.
Stahl, C. 1982. *International Labor Migration and International Development*. Geneva: International Labour Organization.
Stahl. C. 1984. *International Labor Migration and the ASEAN Countries*. Geneva: International Labour Organization.
Todaro, M.P. 1969. A model of labor migration and urban unemployment in less developed countries. *American Economic Review* 59(1), March, 138–48.
Todaro, M.P. 1970. Labor migration and urban unemployment: reply. *American Economic Review* 60(3), September, 187–8.
Ulgalde, A.F. 1979. International migration from the Dominican Republic. *International Migration Review* 13(2), Summer, 235–54.
United Nations, Department of International and Social Affairs. 1982. International migration: policies and programs. *Population Studies* No. 80, New York: United Nations.

international monetary institutions. Domestic money is conceived of by society as a device to facilitate transactions in the marketplace, as a temporary store of value, and as a unit of account for contracts. Given the possibilities of fraud and counterfeiting, domestic monetary authorities have been established to regulate the quality of the domestic monetary unit in most countries. Such regulations attempt to guarantee the interchangeability of the different media, such as currency and the deposits of different banks, as well as stability in the value of the monetary unit, under conditions of prosperity.

International monetary arrangements are required under conditions of international trade, when residents of different countries must make payments to each other, and yet wish to hold most of their assets in terms of domestic currency. Such arrangements are designed to guarantee *convertibility* of assets denominated in different currencies, so that payments may be made independent of country of residence, thus facilitating a free and open trading system.

ALTERNATIVE EXCHANGE RATE MECHANISMS. Under a system of *pegged* exchange rates between different currencies, convertibility implies that domestic residents are free to obtain foreign currency at a *fixed* rate of exchange for the purchase of foreign goods and services, inclusive of normal trade credit. Likewise foreign residents are free to sell domestic currency obtained by sale of goods and services or to use it for purchase of domestic goods and services, at the same fixed rate of exchange. This definition may include but does not require free convertibility for capital account transactions (those arising from exchanges of financial assets only).

Under a *gold standard*, domestic residents and foreign residents may freely convert domestic currency into gold at a fixed rate of exchange. This type of convertibility was eliminated in the 1930s in favour of a gold *exchange* standard, which allowed only foreign monetary authorities to exchange domestic currency for gold. Gold convertibility of both types was ended as part of the Smithsonian Agreement of 1971 [see below].

Under a system of *floating* or *flexible* exchange rates, convertibility still implies that both domestic and foreign residents may freely convert domestic and foreign currency at the same rate of exchange for current account transactions, but the exchange rate at which this may be done is determined on a daily basis by market transactions, rather than being guaranteed by the domestic monetary authorities of the respective countries.

In 1985, only 45 out of the 148 member countries of the International Monetary Fund (IMF) maintained convertibility in any of the senses defined above. In all other countries, restrictions created differences in the exchange rates applying to exports and imports, leading to inefficient allocation of resources, as has been documented by Bhagwati (1978).

RESERVE ASSETS. In order to guarantee convertibility of the domestic currency into other convertible currencies, monetary authorities hold stocks of *reserve assets*, which are liquid assets held in readily accepted international media of exchange, such as gold, dollars, and a few other currencies. In addition, member countries have access to unconditional borrowing rights to obtain additional reserve assets in the form of IMF reserve positions and Special Drawing Rights. These, together with reserve asset holdings, make up *international liquidity*.

Since most international payments are handled by interbank transactions, banks have sought to minimize transactions costs by channelling their foreign exchange transactions through one or more *vehicle* currencies, the pound sterling in earlier days, but more recently the United States dollar. Because the dollar is so widely used in private exchange transactions, monetary authorities also find it convenient to operate in dollars to ensure the convertibility of their currencies.

ADJUSTMENT MECHANISMS. The existence of different national currencies and the need to achieve convertibility of the different currencies lead to the concept of balance of payments adjustment mechanism. At a given exchange rate, as long as the amount of foreign exchange earned through exports of goods and services and capital inflows just pays for imports and capital outflows, no external imbalance exists. If international capital markets were perfect and if investors were risk neutral so that assets denominated in different currencies were perfect substitutes for one another in private portfolios, there would in practice be a single world interest rate for short-term borrowing. Then imbalances between foreign exchange earnings and payments could simply be *financed* by borrowing in the international capital market. There would be no real distinction between the convertibility characteristics of the official liabilities of different borrowers.

But in fact, countries face very real limits on the amount of foreign currency they can borrow abroad in exchange for domestic currency because of *exchange rate* risk, which limits the willingness of risk-averse foreign lenders to acquire domestic currency assets. The ability to repay foreign currency debt is dependent on balance of payments adjustment as well. *Political* risk involves the possibility that exchange controls may be imposed in the future, preventing the repayment of foreign currency debt on the promised terms. Thus it is essential for countries to have access to a variety of adjustment mechanisms to eliminate external imbalances, as well as a variety of sources of official financing in the form of international liquidity. The primary mechanisms of balance of payments adjustment are through fluctuations of exchange rates and adjustments of income and price levels via monetary and fiscal policies. The need for adjustment can be postponed by imposition of tariffs and subsidies, quantitative restrictions on current account or capital account transactions, or controls over the allocation of foreign exchange. But, tariffs, quantitative restrictions, and exchange controls generally involve inefficiencies in the allocation of resources, including in the latter case loss of convertibility of the domestic currency. Changes in monetary and fiscal policies or exchange rates have their own costs in terms of domestic policy objectives foregone.

FINANCING. Thus a mixture of adjustment policies and financing are provided in a system of international monetary arrangements. *Official* financing is provided either by drawing on holdings of official reserve assets or by borrowing from foreign monetary authorities or international institutions. *Private* financing can be arranged by a monetary authority borrowing from foreign banks or the international bond market. Either provides the ability to postpone adjustment. The optimum mix of adjustment and financing for an individual country depends on the costs of the various alternatives. By setting the costs of these alternatives, international monetary arrangements influence the behaviour of the world economy.

A MODEL OF ADJUSTMENT VERSUS FINANCING. In the theory of adjustment versus financing, a country is faced with random balance of payments deficits and surpluses, which it may either finance by drawing on reserve assets or adjust by one of the adjustment mechanisms mentioned above. In one branch of the theory, due to Heller (1966) and others, the cost of adjustment is assumed to be a linear function of the size of the adjustment, so that any adjustments are postponed to the last minute, at which time full adjustment takes place. Alternatively, one may assume a nonlinear cost of adjustment, leading to a theory of partial adjustment. Kelly (1970) and Clark (1970) assume that the country's welfare function depends on the mean and variance of income, so that gradual adjustments are preferred. The analysis determines both the optimum level of reserve holdings, R^*, and the optimum rate of adjustment α to that level, according to the equation

$$\Delta R = \alpha[R^* - R_{-1}] + u, \qquad (1)$$

where u is normally distributed with mean zero and variance σ^2 and R_{-1} is the stock of reserves at the end of the previous period. This equation assumes that changes in the stock of reserves arise from both the random shocks in the balance of payments and the desired rate of adjustment to the optimal level of reserves. From equation (1) we find that the variance of reserve holdings decreases as the speed of adjustment α increases from zero to one.

Tchebychev's inequality then enables one to show that, for a given probability of not exhausting reserves, the optimum reserve holding R^* decreases with increasing α. As α increases, the need for more frequent adjustments raises the variance of income. Therefore the speed of adjustment should be chosen such that the welfare loss from increased variance in income due to a small increase in α is just counterbalanced by the welfare saving due to holding slightly smaller reserves.

According to this theory, international monetary institutions will strongly affect the behaviour of national policies concerning balance of payments adjustment and acquisition of reserves. Specifically, international money institutions will determine the opportunity cost of holding reserves, the penalty attached to running out of reserves, and the availability of different types of adjustment policies. By influencing countries' balance of payments adjustment policies, international institutions will also influence their domestic policies, since there is a tradeoff between internal and external objectives of policy.

THE ROLE OF MARKETS AND INSTITUTIONS. An optimal design for the international monetary system depends on balancing among a group of conflicting objectives: growth of real income and employment, stable prices, efficient allocation of resources, maintenance of convertibility of currencies, improving the distribution of income, and growth of world trade. The relevant tradeoffs can be understood in the context of an economic model. According to the model of adjustment and financing outlined above, reductions in the opportunity cost of holding reserves will lead to increased reserve holdings, a reduction in the speed of adjustment to imbalances, increased use of financing, and a decline in the variability of income. The slowdown in the speed of adjustment implies a change in the allocation of resources among countries. The increased use of financing may imply an increase in the rate of inflation. An optimal international system should balance these various considerations. For discussion of efforts to design such a system, see Solomon (1982) and the documents of the IMF's Committee of Twenty (1974).

In a purely laissez-faire system, market borrowing instead of official reserves would be the source of financing to postpone adjustment. Fluctuations in market interest rates would determine the terms of trade between adjustment and financing. As is usual in market solutions, the wealthy are in a better position to negotiate terms on loans. By contrast, a more institutionalized system provides access to financing at lower rates to those with a weaker market position, with more conditions on the use of the funds. Evaluating the difference between two such systems is a complex task. For an attempt, see Jones (1983).

THE EVOLUTION OF INTERNATIONAL MONETARY INSTITUTIONS. Between the close of the Napoleonic Wars and 1914, the international monetary system gradually moved onto the gold standard, which was fully achieved around 1880. Under the leadership of Great Britain, sterling operated as a vehicle currency during this period, allowing an efficient international payments mechanism to develop. The increasing substitution of bank deposits for currency allowed an ever-larger volume of payments to be supported by a gradually rising supply of gold. Despite the best efforts of the Bank of England and other central banks, periodic crises interfered with the continued convertibility of individual currencies. And the system was characterized by substantial fluctuations in employment and prices, albeit about a rising trend of employment with no trend in prices.

Following World War I, gold convertibility was resumed on a limited basis, until the Great Depression of 1929–33 brought it to an end. A period of fluctuating exchange rates, competitive devaluations, and increasing use of trade restrictions to promote domestic employment ensued. It is generally believed that the economic difficulties of the interwar period were major factors bringing on World War II.

The Bretton Woods system. The United States and Great Britain took the lead in constructing the postwar international monetary institutions, with John Maynard Keynes and Harry Dexter White drawing up rival designs for the new system. The Charter of the International Monetary Fund provided for a system based on pegged, but adjustable, exchange rates and an institution which would lend additional reserve assets to countries which were having temporary difficulties in maintaining convertibility. Resort to floating exchange rates, competitive devaluations, and trade restrictions to promote domestic employment were explicitly to be avoided, in the light of the problems of the 1930s.

The lending power of the International Monetary Fund was based on *quotas* of gold and domestic currency to be contributed by each member country. Only the gold was to be paid in initially, but if the Fund needed convertible currency to lend out, it would obtain it from any member whose currency was considered strong enough to be *usable*. Members could borrow automatically up to the amount of the gold portion or *tranche* of the quota, but only on demonstration of balance of payments need and subject to meeting conditions thereafter. For further discussion of IMF policies, see Williamson (1983).

The initial postwar problem involved the establishment of a payments system that would promote economic recovery and the growth of trade among the former combatants. The International Monetary Fund limited itself to establishing a set of agreed par values for the pegged exchange rates which could promote the growth of trade, leaving the provision of loans and grants for economic recovery to the United States, the strongest economy. Under this system, which was a form of *gold exchange* standard, countries declared their par values in terms of the United States dollar, which was convertible into gold at $35 an ounce. Thus the dollar became the *key currency* of the system, and most foreign exchange reserves came to be held in the form of dollars. Within Europe, convertibility remained limited until 1958, and the European Payments Union was established to facilitate intra-European payments. The re-establishment of convertibility led to fears that the International Monetary Fund might have inadequate resources to deal with the problems of large member countries. In 1960 the General Arrangements to Borrow were created, to enable the Fund to mobilize additional resources from its largest members, the Group of Ten.

With the recovery of the European economies in the 1950s and the achievement of convertibility in 1958, the United States dollar became gradually over-valued relative to gold and other currencies. As Robert Triffin (1960) pointed out, the key currency system required the United States to continue to run balance of payments deficits in order to supply other countries with increased foreign exchange reserves. As it did so, the gold reserve of the United States became increasingly inadequate to guarantee gold convertibility of dollars at $35 an ounce.

A variety of solutions to this problem were proposed, including the creation of an artificial reserve asset to substitute for dollars, an increase in the dollar price of gold, and the adoption of floating exchange rates. In 1968 the First Amendment to the Charter of the International Monetary Fund created Special Drawing Rights (SDRs), which have been distributed to member countries in proportion to their existing quotas in the Fund. SDRs, when utilized, permit the user to acquire convertible currencies from other members, upon the payment of interest and subject to certain rather mild conditions. They represent a centralized mechanism for increasing the stock of reserves. Nevertheless, by the early 1970s the gold convertibility of the dollar was under increasing pressure, for a variety of reasons. In August 1971 the dollar was unilaterally set loose from gold. The Smithsonian Agreement of December 1971 attempted to save the Bretton Woods system by multilateral realignment of exchange rates, including a devaluation of the dollar against gold and a widening of the narrow bands of fluctuation permitted around the newly fixed values. Some members of the European Communities (EC) agreed to maintain narrower margins of fluctuations versus each other's currency, in an arrangement that became known as the 'EC Snake'. Despite these efforts, the revised system lasted only a little more than a year.

Floating exchange rates. In April 1973, exchange rates of the major industrial countries began floating. At the same time, most developing countries continued to peg their currencies to the dollar or other developed country currency and the EC maintained the 'Snake'. About this time, a major effort to reconstruct international monetary institutions on the basis of pegged exchange rates began under the auspices of the International Monetary Fund's 'Committee of Twenty'. This effort collapsed in 1974, under the impact of the quadrupling of world oil prices by the Organization of Petroleum Exporting Countries.

In Jamaica in January 1976, the Interim Committee of the Board of Governors of the International Monetary Fund agreed on a Second Amendment to the Fund's Charter, ratifying the system of floating exchange rates. Stability of exchange rates was to be sought through stability of underlying monetary and fiscal policies rather than through pegging. Second, floating rates should be subject to a process of 'firm surveillance' by the International Monetary Fund. Third, it was hoped that the SDR would 'become the principal reserve asset', with the role of gold and the dollar being reduced. Fourth, the fixed official price of gold was abolished and one-third of the IMF's gold was disposed of. Acceptance of the *status quo* was all that could be accomplished. The result, according to Corden (1983), is an international laissez-faire system.

Since the Second Amendment, the most significant development has been the enlargement and strengthening of the EC 'Snake' in 1978, which was in the process renamed as the European Monetary System (EMS). The objectives of the enlarged EMS were to reduce intra-European exchange rate fluctuations, to promote convergence of macro-economic policies within Europe, and to reduce European dependence on US monetary policies. In addition to enlarging the number of members of the 'Snake' and the size of the short-term loan facilities, the EMS also created a nine-currency European Currency Unit (ECU) as a rival to the SDR. A proposed pooling of European reserves in a new European Monetary Fund was not implemented.

S.W. BLACK

See also GOLD STANDARD; INTERNATIONAL FINANCE; INTERNATIONAL MONETARY POLICY.

BIBLIOGRAPHY

Bhagwati, J. 1978. *Anatomy and Consequences of Exchange Control Measure.* Cambridge, Mass.: Ballinger.

Black, S.W. 1985. International money and international monetary arrangements. In *Handbook of International Economics*, Vol. 2, ed. R.W. Jones and P.B. Kenen, Amsterdam: North-Holland.

Clark, P.B. 1970. Optimum international reserves and the speed of adjustment. *Journal of Political Economy* 75, 356–76.

Corden, W.M. 1983. The logic of the international monetary non-system. In *Reflections on a Troubled World Economy*, ed. F. Machlup et al., London: Macmillan.

Heller, H.R. 1966. Optimal international reserves. *Economic Journal* 74, 333–52.

International Monetary Fund. 1974. *International Monetary Reform: Documents of the Committee of Twenty.* Washington, DC: International Monetary Fund.

Jones, M. 1983. International liquidity: a welfare analysis. *Quarterly Journal of Economics* 98, 1–23.

Kelly, M.G. 1970. The demand for international reserves. *American Economic Review* 60, 655–67.

Solomon, R. 1982. *The International Monetary System, 1945–1981.* New York: Harper & Row.

Triffin, R. 1960. *Gold and the Dollar Crisis.* New Haven: Yale University Press.

Williamson, J. (ed.) 1983. *IMF Conditionality.* Washington, DC: Institute for International Economics.

international monetary policy. One of the main characteristics of the international monetary system is the absence of an international monetary authority (central bank) with policy making powers comparable to those central banks have at the national level. Whereas national central banks typically regulate domestic money markets in one way or another, there is no comparable authority to regulate international money markets. As a result, international monetary conditions will be the outcome of a decentralized decision-making process, in which market forces play a role together with the policies of a few important countries. Ultimately, therefore, international monetary relations will be influenced by the nature of the

cooperation (or the lack of cooperation) among the central banks of the major countries.

There exist, of course, a number of international monetary institutions with important responsibilities. The most noteworthy are the Bank for International Settlements (BIS), and the International Monetary Fund (IMF). The latter has a major responsibility in providing credit to the countries with balance of payments and foreign debt problems. In addition, this role of the IMF has tended to increase during the last decade. Nevertheless it is fair to say that these institutions are far from having the powers and the responsibility a typical central bank has at the national level. It is also very unlikely that any of these institutions can be promoted to the position of a true world central bank in the foreseeable future.

The absence of a world central bank implies that the international monetary situation will be heavily influenced by the actions of the monetary authorities of the major countries. In fact, the nature of the domestic monetary policy regime in these countries is of crucial importance in the determination of the nature of the international monetary system. It is, therefore, useful to see how these domestic monetary policy regimes have changed over time, in particular in the United States which is the single most important country.

During the early postwar period the prevailing view in the US (and in other industrialized countries) was that the major responsibility of the central bank consisted in maintaining domestic price stability. This view, which originated in the writings of the classical economists, provided implicitly or explicitly the framework for monetary policy making in the major industrialized countries. The spillover of this view and of this policy attitude in the international sphere was a system of fixed exchange rates. The largest country, the US, was successful in stabilizing its price level. The other countries pegged their exchange rates to the US dollar, thereby also obtaining domestic price stability. As a result, during that period (which lasted roughly until the mid-Sixties, and which is usually called the Bretton Woods System) the world experienced stable exchange rates and the absence of inflation. This arrangement, however, could only work satisfactorily if the countries pegging to the dollar were willing to subordinate their domestic monetary policies to the maintenance of a fixed dollar rate of their currencies. As the US policy was predicated on maintaining price stability, the willingness of the other countries to impose on themselves the discipline of a fixed exchange rate was great.

This situation began to change when views about the responsibilities of the central bank altered. Instead of being the guardian of a stable purchasing power value of money, the central bank was increasingly seen as an institution responsible for the stabilization of economic activity. This led to problems with the stabilization of the price level, and undercut the basis of the fixed exchange rate system. Inevitably, as countries used monetary policies to stabilize output, inflation rates became more variable and also more different across countries. In the end, fixed exchange rates had to be abandoned, and from the early Seventies on the industrialized countries allowed their exchange rates to float.

The responsibilities of the US in the fundamental change of the international monetary environment have been widely discussed (see e.g. Triffin, 1968, Niehans 1974). By abandoning price stability as the major monetary policy objective, the US also transmitted inflationary shocks to those countries pegged to the dollar. As a result, these countries lost their willingness to maintain a fixed exchange rate with the dollar.

These changes in the domestic monetary policy regimes in the major industrialized countries have made exchange rates inevitably more variable than in the Bretton Woods period. This has also led to fundamental changes in the international monetary policy environment. Changes at three different levels can be identified.

A first change concerns the nature of the cooperation of the central banks. During the Bretton Woods period, the discipline of fixed exchange rates forced countries to coordinate their monetary policies closely. In fact, this coordination more or less automatically followed from these pegging arrangements. In addition, since pegging of the exchange rates occurred vis-à-vis the dollar, US monetary policies determined monetary conditions in the rest of the world. Thus, it can be said that the Bretton Woods system was a cooperative international monetary arrangement based on the leadership of the US.

The shift towards flexible exchange rates changed the nature of international monetary cooperation. First of all, as the movements of exchange rates tended to absorb monetary disturbances, the need to coordinate national monetary policies was generally felt to be less urgent. Second, when cooperation took place it tended to be of an ad hoc nature instead of automatic as in the Bretton Woods period.

It has been argued that this lack of explicit cooperative arrangements among the monetary authorities of the major industrialized countries is in itself a factor explaining the high volatility of the exchange rates observed since 1973 (see for example Williamson, 1984). Certainly, this volatility of the exchange rates came as a surprise to most academic economists, who had been influenced by the conventional wisdom of the Sixties stressing that a flexible exchange rate system would make a smooth adjustment of external equilibrium possible (see Friedman, 1953; Johnson, 1967; Sohmen, 1961). Those who read the old writers on the subject, however, knew better (see for example Bernholz, 1982, for a survey of older views about flexible exchange rates.)

A second major change since the inception of flexible exchange rates concerns the nature of monetary interdependence between nations. Academic opinion prior to the Seventies was that a system of flexible exchange rates would make individual countries more independent in setting domestic monetary policies than a system of fixed exchange rates. In particular, it was thought that flexible exchange rates would allow countries to determine their own inflation rates, so that even when other countries followed inflationary policies, those countries that wanted it could insulate themselves from these foreign inflationary shocks.

These predictions of the merits of flexible exchange rates have been fulfilled only partially. It turned out to be true that countries can select their long-run inflation rates more or less independently if they allow their exchange rate to vary. Thus it was possible for countries such as Switzerland to have an inflation rate of only four per cent per annum during 1973–84 whereas the average in the industrialized countries was more than eight per cent per annum. On the other hand, it also appears that the short-term movements of inflation rates have been more correlated across countries during the floating rate period than during the period prior to 1973. Thus, although countries now have a higher degree of independence in selecting their long-run inflation rates, the yearly movements in these inflation rates turn out to be more dependent on outside price shocks than during the Bretton Woods period.

The reasons for this unexpected phenomenon are twofold. First, the occurrence of flexible exchange rates coincided with major supply shocks during the Seventies which tended to raise the rate of inflation in all countries. Second, the exchange rate regime experienced since 1973 was not a pure floating

exchange rate system. Central banks continued to intervene heavily in the foreign exchange markets. They did this in (usually unsuccessful) attempts to mitigate the movements of the dollar. Thus, during the period 1973–78 the dollar declined substantially against the other major currencies, and central banks bought massive amounts of dollars in order to stem its slide. This had the effect of expanding the money stocks in all these countries and tended to accelerate inflation. Exactly the opposite occurred during the period 1979–84 when the dollar experienced an unprecedented surge, and when central banks sold dollars and deflated their own money stocks.

This system of managed floating produced the curious result that monetary expansions and contractions which originated in the US were transmitted to the other industrialized countries as they would have been under a fixed exchange rate arrangement. And yet the dollar continued to be highly volatile, as these interventions in the dollar exchange markets failed to have much effect on the movements of the dollar. In a sense it can be said that the international monetary arrangement of the Seventies and the early Eighties combined the disadvantages of fixed and flexible exchange rates.

The shift to more flexible exchange rates produced a third major change in the international monetary policy environment. During the Bretton Woods period a major concern of monetary policy makers was control of the creation of international reserves. It was then widely felt that the mechanism of international reserve creation which was implicit in the gold-exchange standard was deficient. The rate of growth of the stock of international reserves (gold and dollars) did not correspond to the needs of an expanding world trade. In addition, the system had the disadvantage of leading to a disproportionate growth of the dollar stock relative to the stock of gold. As a result, a confidence problem arose concerning the ability of the US to maintain the gold convertibility of the dollar (a problem analysed by Triffin, 1960, and Rueff, 1961).

The 'liquidity problem' of the Bretton Woods system led to numerous schemes to manage the creation of international reserves (for a survey, see Grubel, 1969, chapters 7 and 8). Without exaggeration it can be said that in those days the single most pressing issue of international monetary policy was thought to be this liquidity problem. Ultimately, this concern led to the creation of Special Drawing Rights, which were intended to substitute for gold, and which would enable the international community to regulate the creation of international reserves in a more rational way.

With the breakdown of the Bretton Woods system and of the gold exchange standard, these concerns about the creation of international reserves tended to fade away. Whereas in the Bretton Woods era the consensus was that the most important international monetary problem was how to create international reserves, there is now a growing consensus that the single most important issue faced by the international monetary system today is whether the degree of exchange rate variations has not become excessive. Concern that this may be the case has led some economists to propose schemes of coordination of monetary policies of the major industrialized countries, so as to stabilize exchange rates.

The need to come to such explicit cooperative agreements between central banks remains a controversial issue. There are essentially two schools of thought. The proponents of international monetary cooperation (e.g. McKinnon, 1984; Williamson, 1983) argue that the present flexible exchange rate regime leads to excessive movements of exchange rates, thereby making domestic macroeconomic management, and in particular the stabilization of the domestic price level, more difficult. In this view, cooperative agreements aimed at stabilizing the exchange rates must be given priority in order for countries to stabilize their economies in a more effective way.

A second school of thought turns the argument around and claims that domestic monetary stability comes first. In order to stabilize the exchange rate the monetary authorities of the major industrialized countries must follow more stable and predictable monetary policies. If this is done, the domestic price level can be stabilized so that the exchange rates can follow a more stable and predictable path. (Representative proponents of this view are Willett, 1983; Haberler, 1977.)

This debate has gone through many cycles in history. During the Twenties, after a period of strong fluctuations of the exchange rates, there was a widely held conviction that the paramount task in the field of international monetary cooperation was stabilizing the exchange rates of the major currencies. This was seen as a first step towards the successful stabilization of domestic economies (see Clarke, 1967, for a history of central bank cooperation during 1924–31). The whole cooperative effort underlying the Bretton Woods system was inspired by the same idea. During the 1960s, as major countries relaxed the monetary discipline needed to sustain a fixed exchange rate system, the view that domestic stability was a precondition for exchange rate stability gained respectability. In the early Seventies this view had become predominant among academic economists. Now, after many years of volatile exchange rate behaviour, the old view stressing the need to stabilize the exchange rates as a first step toward achieving domestic stability has regained some respectability.

The conflict between these two views, however, has not yet been settled. As a result, there is as yet no general agreement on how monetary policy should be conducted at the international level.

Paul De Grauwe

See also INTERNATIONAL FINANCE; MONETARY POLICY; SUPPLY SHOCKS; TRANSFER PROBLEM.

BIBLIOGRAPHY

Bernholz, P. 1982. *Flexible Exchange Rates in Historical Perspective.* Princeton Studies in International Finance No. 49, Princeton: Princeton University Press.

Clarke, S.V.O. 1967. *Central Bank Cooperation 1924–1931.* New York: Federal Reserve Bank of New York.

Friedman, M. 1953. The case for flexible exchange rates. In M. Friedman, *Essays in Positive Economics*, Chicago: University of Chicago Press.

Grubel, H. 1969. *The International Monetary System.* Harmondsworth: Penguin Books.

Haberler, G. 1977. The international monetary system after Jamaica and Manila. *Weltwirtschaftliches Archiv* 113.

Johnson, H. 1967. Theoretical problems of the international economy. *Pakistan Development Review.* Reprinted in R. Cooper, *International Finance*, Harmondsworth: Penguin Books, 1970.

McKinnon, R. 1984. *An International Standard for Monetary Stabilization.* Policy Analyses in International Economics 8, March, Washington, DC: Institute for International Economics.

Niehans, J. 1974. Reserve composition as a source of independence for national monetary policies. In *National Monetary Policies and the International Financial System*, ed. Robert Z. Aliber, Chicago: Chicago University Press.

Rueff, J. 1961. Gold exchange standard a danger to the west. *The Times*, London, 27–29 June.

Sohmen, E. 1961. *Flexible Exchange Rates, Theory and Controversy.* Chicago: University of Chicago Press.

Triffin, R. 1960. *Gold and the Dollar Crisis.* New Haven, Conn.: Yale University Press.

Triffin, R. 1968. *Our International Monetary System.* New York: Random House.

Willett, T. 1983. Functioning of the current international financial system: strengths, weaknesses and criteria for evaluation. In *International Money and Credit: The Policy Roles*, ed. G. von Furstenberg, Washington, DC: International Monetary Fund.

Williamson, J. 1983. *The Exchange Rate System.* Policy Analyses in International Economics, No. 5, Washington, DC: Institute of International Economics.

international trade. Edgeworth (1894) opened his survey of the theory of international values with the provocative statement: 'International trade meaning in plain English trade between nations, it is not surprising that the term should mean something else in Political Economy'. This could equally well be said today. What distinguishes international from domestic trade is the greater prevalence of barriers (both natural and artificial) to trade and factor movements in the former; different currencies; and (perhaps most important) autonomous governments, leading to a pattern of shocks which impact different countries in different ways. Because of these differences, a different type of theoretical model is called for. For example, international immobility of factors results in greater disparity in relative factor endowments among countries than among regions of the same country; these disparities may make it reasonable, as a first approximation, to ignore variations in supplies of factor services that come about in response to changes in factor rentals and commodity prices, if these variations are small in comparison with the differences in endowments. Likewise, great differences among resource endowments and productive techniques may make it reasonable to disregard differences in consumers' tastes within and across countries, even though this might be a very inappropriate type of simplification for purposes of analysing domestic trade.

The fact that national governments act independently leads to the need to analyse the effects of country-specific shocks, which take the form of intensification or liberalization of restrictions on trade or capital movements, unilateral transfers such as reparation payments, gifts, or loans, and disparities in monetary and fiscal policies. For this reason the emphasis in international-trade theory has from the beginning (Mill, 1848; Marshall, 1879) been on comparative statics: one wants to ascertain the qualitative, if not the quantitative, effect of a tariff or quota or transfer on the various quantities involved. To obtain unambiguous qualitative results one needs fairly drastic simplifications and strong assumptions. On the other hand, the emphasis in general-equilibrium theory (Walras, 1874; Pareto, 1896–97; Debreu, 1959) has been on proving the existence, stability, and Pareto-optimality of competitive equilibrium, for which much milder assumptions are required. A good definition of international-trade theory as it has evolved would therefore be: 'general-equilibrium theory with structure'.

The requirements of 'simplicity' in a theory are not absolute, but vary with the goals of the theory and the technical resources available to researchers at the time. There is not much virtue in simplicity if a result that holds in a model of two countries, two commodities, and two factors does not generalize in any meaningful way to higher dimensions. With the increasing possibilities of handling large-scale models and data sets and estimating their parameters numerically, it is natural to expect a movement of both general-equilibrium traditions towards each other.

Attention will be focused here on the neoclassical model developed by Haberler (1930, 1933), Lerner (1932, 1933, 1934), Ohlin (1928, 1933), Stolper and Samuelson (1941), Samuelson (1953), and Rybczynski (1955), which Baldwin (1982) has described as the 'Haberler–Lerner–Samuelson model' – an appellation which is more accurate than the usual 'Heckscher–Ohlin theory', since the model commonly employed makes the simplifying assumption – rejected by Ohlin (1933, ch. VII) except in his illustrative Appendix I – that factors of production are inelastic in supply and indifferent among alternative occupations, allowing one to define unambiguously a country's production-possibility frontier. This model has in recent years come to lose some of its hold on the profession – just as the Ricardian theory had in the 1930s – in favour of models that stress imperfect competition (see, e.g. Helpman and Krugman, 1985). However, these latter models have so far not been successfully formulated as general-equilibrium models, and are thus still in a formative stage. It goes without saying that, in the nature of the case, a partial-equilibrium model is incapable of explaining or predicting trade patterns or analysing the effect on prices and resource allocation of trade restrictions and transfers.

The material that follows is divided into two parts. Part 1 covers the mathematical foundations of the received theory, and deals with the duality between production functions and cost functions, the concept of a national-product function, the Stolper–Samuelson and Rybczynski relations between factor rentals and commodity prices and between commodity outputs and factor endowments, the concepts of trade-demand functions and trade-utility functions, world equilibrium and its dynamic stability. Part 2 covers the applications of these basic concepts to the most noteworthy problems that have been the object of attention in the theory of international trade since its beginnings: the explanation of trade flows, the effect of unilateral transfers on sectoral prices and resource allocation, and the effect of trade restrictions such as tariffs and quotas. The reader who is interested in substantive questions is advised to proceed directly to Part 2.

PART 1. THE MATHEMATICAL FOUNDATIONS

1. DUALITY OF COST FUNCTIONS AND PRODUCTION FUNCTIONS.

1.1. Let an industry produce a positive amount y of output of a particular product, with the aid of non-negative amounts v_j of m primary factors of production, determining the vector $v = (v_1, v_2, \ldots, v_m)$. A *production function* f is defined over the non-negative orthant E_m^+ of m-dimensional Euclidean space, with values $y = f(v)$ on the non-negative real line E_1^+. We assume that f has the following properties:

(a) *Upper semi-continuity:* for each y the set

$$A(y) = \{v : f(v) \geqq y\} \tag{1.1}$$

is closed;

(b) *Quasi-concavity:* for each y, the set $A(y)$ defined by (1.1) is convex;

(c) *Monotonicity:* if $v, v' \in E_m^+$ are such that $v' \geqq v$, then $f(v') \geqq f(v)$.

Further properties of f will be specified later on.

We shall denote by $w = (w_1, w_2, \ldots, w_m)$ a vector of *factor rentals*, i.e. prices of the services of the m factors of production. The following conventional notation will be adhered to:

$w \geqq 0$ means $w_i \geqq 0$ for all $i = 1, 2, \ldots, m$;

$w \geq 0$ means $w_i \geqq 0$ for all $i = 1, 2, \ldots, m$,

and $w_i > 0$ for some i;

$w > 0$ means $w_i > 0$ for all $i = 1, 2, \ldots, m$.

For each $y > 0$ and all $w \geqq 0$ we define the *minimum total cost function* G by

$$G(w, y) = \min_v \{w \cdot v : f(v) \geqq y\}, \tag{1.2}$$

where $w \cdot v$ denotes the inner product

$$\sum_{j=1}^m w_j v_j.$$

Mathematically, for each fixed y the function $G(\cdot, y)$ is the *support function* of the convex set $A(y)$ (cf. Fenchel, 1953). It has the following properties:

(a*) *Continuity* in w: for each y, $G(w, y)$ is continuous;

(b*) *Concavity* in w: if $0 < \theta < 1$ then

$$(1 - \theta)G(w^0, y) + \theta G(w^1, y) \leqq G[(1 - \theta)w^0 + \theta w^1, y];$$

(c*) *Monotonicity:* $y' \geqq y$ implies $G(w, y') \geqq G(w, y)$ and $w' \geqq w \geqq 0$ implies $G(w', y) \geqq G(w, y)$;

(d*) *positive homogeneity* in w: $G(\lambda w, y) = \lambda G(w, y)$ for all $\lambda > 0$.

Property (a*) follows from (a) and the definition of G; property (c*) follows the definition of G and the fact that $y' \geqq y$ implies $A(y') \subseteq A(y)$; property (d*) follows immediately from the definition of G. To prove (b*), let $w^0, w^1 \geqq 0$ and denote $w^\theta = (1 - \theta)w^0 + \theta w^1$; from the definitions of G and $A(y)$ in (1.2) and (1.1), we have

$$G(w^0, y) \leqq w^0 \cdot v \quad \text{for all} \quad v \in A(y);$$

$$G(w^1, y) \leqq w^1 \cdot y \quad \text{for all} \quad v \in A(y);$$

consequently,

$$(1 - \theta)G(w^0, y) + \theta G(w^1, y) \leqq w^\theta \cdot v \quad \text{for all} \quad v \in A(y).$$

Hence, in particular,

$$(1 - \theta)G(w^0, y) + \theta G(w^1 \cdot y)$$

$$\leqq \min_v \{w^\theta \cdot v : v \in A(y)\} \equiv G(w^0, y),$$

which is the result sought (cf. Uzawa, 1964b).

Of fundamental importance in international trade theory is the following *duality* theorem first proved by Shephard (1953). The formulation and proof contained in Theorem 1 to follow are due to Uzawa (1964b).

Theorem 1 (Duality Theorem). Define the set

$$B(y) = \{v : (\forall w \geqq 0) w \cdot v \geqq G(w, y)\}, \tag{1.3}$$

where G is defined by (1.2) and f satisfies properties (a), (b), (c). Then $B(y) = A(y)$, where $A(y)$ is defined by (1.1).

Proof: Let $v^0 \in A(y)$; then $f(v^0) \geqq y$, so for all $w \geqq 0$,

$$w \cdot v^0 \geqq \min_v \{w \cdot v : f(v) \geqq y\} \equiv G(w, y)$$

that is, $v^0 \in B(y)$.

Conversely, suppose $v^0 \notin A(y)$. Since $A(y)$ is closed and convex by properties (a) and (b) of f, it follows from the separating hyperplane theorem of closed convex sets (cf. Fenchel, 1953, p. 48) that there exists a vector $w^0 \neq 0$ such that

$$w^0 \cdot v^0 < \min_v \{w^0 \cdot v : v \in A(y)\} \tag{1.4}$$

(see Figure 1). Now if w^0 has a negative component, it follows from property (c) that the corresponding component of $v \in A(y)$ may be chosen to be arbitrarily large, hence no minimum of $w^0 \cdot v$ over $A(y)$ exists; consequently, $w \geqq 0$. But then the expression on the right of the inequality sign in (1.4) is just $G(w^0, y)$. From the definition of $B(y)$ in (3), it follows that $v^0 \notin B(y)$. q.e.d.

The duality theorem may be stated in words as follows: given the function G, the set $A(y)$ may be identified with the set of

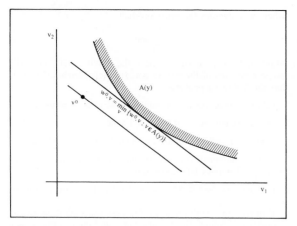

Figure 1

all factor combinations v which, at each constellation $w \geq 0$ of factor rentals, are at least as expensive as the minimal total cost of producing output y at factor rentals w.

1.2. Let us now explore the consequences of imposing a further condition on the production function f:

(d) *Positive homogeneity:* for all $\lambda > 0, f(\lambda v) = \lambda f(v)$.

From the definition of G in (1.2), we now have

$$G(w, y) = \min_v \left\{ w \cdot v : f\left(\frac{v}{y}\right) \geq 1 \right\}$$

$$= \min_b \left\{ yw \cdot b : f(b) \geq 1 \right\} \left(b = \frac{v}{y} \right)$$

$$= y \cdot \min_b \left\{ w \cdot b : f(b) \geq 1 \right\}.$$

Thus, $G(w, y)$ factors into two terms, of which the second depends only on $w \geq 0$ and may be denoted

$$g(w) = \min_v \left\{ w \cdot v : f(v) \geq 1 \right\}. \qquad (1.5)$$

We therefore have

Theorem 2. If f satisfies properties (a), (b), (c), (d), then the function G of (1.3) factors into

$$G(w, y) = yg(w) \qquad (1.6)$$

where g is defined by (1.5) and is continuous, concave, monotone, and positively homogeneous of first degree.

The properties of g specified in Theorem 2 follow directly from those of the function G.

We may now state a special form of the duality theorem for the case of homogeneous production functions.

Theorem 3. Let g be defined by (1.5) where f satisfies properties (a), (b), (c), (d), and let the function f^* be defined by

$$f^*(v) = \min_v \left\{ w \cdot v : g(w) \geq 1 \right\}. \qquad (1.7)$$

Then $f^* = f$.

Proof: Define the set

$$C(y) = \{ v : [\forall w \in A^*(1)] w \cdot v \geq y \} \qquad (1.8)$$

where for convenience we define

$$A^*(p) = \{ w : g(w) \geq p \}. \qquad (1.9)$$

(Since g is defined only for $w \geq 0$, $w \in A^*(p)$ implies $w \geq 0$).

First we shall show that $C(y) = B(y)$, where $B(y)$ is defined by (1.3). From (1.3) and (1.6), if $v^0 \in B(y)$ then for all $w \in A^*(1)$, $w \cdot v^0 \geq G(w, y) = yg(w) \geq y$, so $B(y) \subseteq C(y)$. Conversely suppose $v^0 \in C(y)$ and take any $w^0 \geq 0$. Then from the homogeneity of g we have $g[w^0/g(w^0)] = 1$, hence from the definition (1.8) of $C(y)$ it follows that

$$\frac{w^0}{g(w^0)} \cdot v^0 \geq y,$$

i.e., $w^0 \cdot v^0 \geq yg(w^0)$; thus $v^0 \in B(y)$. Therefore $B(y) = C(y)$ and by Theorem 1, $C(y) = A(y)$.

Now denote $r = w/g(w)$ and consider the set

$$C'(y) = \left[v : \min_r \left\{ r \cdot v : r \in A^*(1) \right\} \geq y \right]. \qquad (1.8')$$

If $r \cdot v \geq y$ for all $r \in A^*(1)$, then a fortiori $r \cdot v \geq y$ for the $r \in A^*(1)$ which minimizes $r \cdot v$; hence $C(y) \subseteq C'(y)$. Conversely, for all $r \in A^*(1)$ we have $r \cdot v \geq \min_r \{ r \cdot v : r \in A^*(1) \}$, so $C'(y) \subseteq C(y)$. Thus $C'(y) = C(y) = A(y)$. But from (1.7), (1.9) and (1.8') we have

$$C'(y) = \{ v : f^*(v) \geq y \}. \qquad (1.9)$$

Since $A(y) = C'(y)$ for all y, therefore f and f^* coincide.

q.e.d.

1.3. Let us consider the consequences of adding to the properties (a), (b), (c) of f given in §1.1 the following further properties:

(b_1) *Strict quasi-concavity:* for each y, the set $A(y)$ defined by (1.1) is strictly convex;

(e) *Differentiability:* f has continuous first-order partial derivatives.

For the time being, property (d) of §1.2 will not be used, but will be introduced again later on.

The problem of deriving the minimum total cost function $G(w, y)$ may be posed in terms of the following non-linear programming problem:

$$\text{minimize } \sum_{j=1}^{m} w_j v_j \text{ subject to } f(v) \geq y, v \geq 0. \qquad (1.10)$$

Form the Lagrangean function

$$L(p^*, v; y, w) = \sum_{j=1}^{m} w_j v_j - p^*[f(v) - y] \qquad (1.11)$$

where y, w are parameters and p^* is a Lagrangean multiplier. In accordance with the Kuhn–Tucker theorem (cf. Kuhn and Tucker, 1951, p. 486) in order for $v^0 = (v_1^0, v_2^0, \ldots, v_m^0)$ to be a solution of the minimum problem (1.10), it is necessary and sufficient that v^0 and some $p^* \geq 0$ satisfy

$$\left. \frac{\partial L}{\partial v_j} \right|_{v = v^0} = w_j - p^* \left. \frac{\partial f}{\partial v_j} \right|_{v = v^0} \geq 0; \qquad v_j \frac{\partial L}{\partial v_j} = 0 \qquad (1.12a)$$

and

$$\sum_{j=1}^{m} v_j^0 \left. \frac{\partial L}{\partial v_j} \right|_{v_j = v_j^0} = \sum_{j=1}^{m} v_j^0 \left(w_j - p^* \left. \frac{\partial f}{\partial v_j} \right|_{v = v^0} \right) = 0 \qquad (1.12b)$$

as well as

$$\frac{\partial L}{\partial p^*} = -[f(v^0) - y] \leq 0; \qquad p^* \frac{\partial L}{\partial p^*} = 0. \qquad (1.12c)$$

In the above we have used (e), but so far property (b_1) has not been used: Let us introduce the further properties:

(f) *indispensability:* $f(0) = 0$.

(f_1) *strict indispensability:* if v has a component $v_j = 0$ then $f(v) = 0$.

924

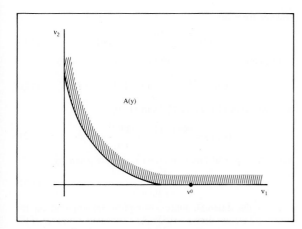

Figure 2

Now suppose the solution v^0 to (1.10) is such that $f(v^0) > y$ (see Figure 2). This violates (b$_1$), since strict quasi-concavity requires that if $v^0, v^1 \in A(y)$ and $0 < \theta < 1$, the point $v^\theta = (1 - \theta)v^0 + \theta v^1$ should be in the interior of $A(y)$. Suppose, however, that property (b$_1$) is not assumed, and that $f(v^0) > y > 0$; then $p^* = 0$ from (1.12c) hence $w \cdot v^0 = 0$ from (1.12b), and since $w \geqslant 0$ this implies that v^0 has a zero component. Thus, if (f$_1$) is assumed, we have $0 = f(v^0) > y > 0 - a$ contradiction. Thus, either (b$_1$) or (f$_1$) is sufficient – in conjunction with (a), (c), (e), to guarantee $f(v^0) = y$. If $w > 0$, a similar argument shows that (f) implies $f(v^0) = y$.

Now suppose that v^0 is such that strict inequality holds in (1.12a) for some j. Then $v_j^0 = 0$ from (1.12b) If (f$_1$) holds this would lead to a contradiction, since then $0 = f(v^0) \geqslant y > 0$. If (f$_1$) is not assumed, but if (b$_1$) holds, then strict inequality in (1.12a) implies that v^0 has a zero component, so v^0 is on the boundary of $A(y)$; but $2v^0$ is also on the boundary of $A(y)$, by property (c), and consequently the mid-point $1\frac{1}{2}v^0$ is as well, contradicting (b$_1$). Thus, if (a), (b), (c), (e) hold, then either (b$_1$) or (f$_1$) implies that equality holds in (1.12a) for all $j = 1, 2, \ldots, m$.

Consider a solution v^0 to (1.10) corresponding to a w^0 which has some zero components. Let $J = \{j : w_j^0 = 0\}$. Then if $w^0 \cdot v^0 = C^0$, certainly $A(y) \subseteq \{v : w^0 \cdot v^0 \geqslant C^0\}$. Let v^1 be such that $v_j^1 > v_j^0$ for $j \in J$ and $v_j^1 = v_j^0$ for $j \notin J$. Then $w^0 \cdot v^1 = w^0 \cdot v^0$,

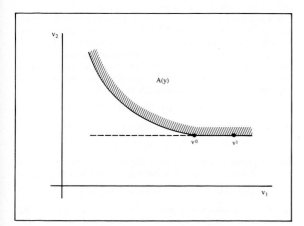

Figure 3

hence $v^1 \in \{v : w^0 \cdot v = C^0\}$. But by condition (c), $v^1 \in A(y)$; thus v^1 and v^0 are both on the boundary of $A(y)$, as is $(1 - \theta)v^0 + \theta v^1$ for $0 < \theta < 1$ (see Figure 3). This contradicts (b$_1$). Therefore under (b$_1$), a solution to (1.10) exists only if $w > 0$.

It should be noted that even if the function $G(w, y)$ of (1.2) is well defined in the sense

$$G(w, y) = \inf_v \{w \cdot v : f(v) \geqslant y\}, \qquad (1.2')$$

a solution of (1.10) need not exist. For example, if

$$f(v_1, v_2) = \frac{1}{\frac{1}{v_1} + \frac{1}{v_2}}$$

then

$$G(0, w_2; y) = yw_2$$

but the *infimum* is achieved as $(v_1, v_2) \to (\infty, y)$. On the other hand a solution to (1.10) always exists if $w > 0$; for, choosing any $v^0 \in \text{int } A(y)$ and $w^0 > 0$, the set

$$A(y) \cap \{v : w^0 \cdot v \leqslant w^0 v^0, v \geqslant 0\}$$

is compact by virtue of condition (a), and from (b) and (c) the minimum of $w^0 \cdot v$ over this set is the minimum over $A(y)$.

An immediate consequence of (b$_1$) is that if (1.10) has a solution, it is *unique*. Since (1.10) need not have a solution unless $w > 0$, it is of some advantage to replace (b$_1$) by a weaker condition which still ensures uniqueness provided $w > 0$. Such a condition is

(b$_2$) if $v^0 \neq v^1$ and neither $v^0 \geqslant v^1$ nor $v^1 \geqslant v^0$, and if $0 < \theta < 1$, then $f[(1 - \theta)v^0 + \theta v^1] > \min[f(v^0), f(v^1)]$.

The above discussion may now be summarized in the following theorem.

Theorem 4. Let conditions (a), (b), (c), (e), (f) hold. Then if either (b$_1$) or (f$_1$) holds, any solution v^0 to (1.10) has the property

$$w_j = p^* \left.\frac{\partial f}{\partial v_j}\right|_{v = v^0} \quad (j = 1, 2, \ldots, m); \qquad f(v^0) = y. \quad (1.12d)$$

If (b$_1$) holds, this solution is unique. If (b$_2$) holds and if $w > 0$, then a unique solution to (1.10) exists, and it satisfies (1.12d).

1.4. We now proceed with an analysis of the solution v of the programming problem (1.10) regarded as a function of the parameters $y > 0$, $w > 0$, when conditions (a), (b$_2$), (c), (e), (f) are assumed to hold.

In accordance with Theorem 4, the solution satisfies (1.12d) and is unique, given y and w. Thus we have the functions

$$v_j = \tilde{v}_j(w, y) \qquad (j = 1, 2, \ldots, m). \quad (1.13a)$$

It is shown in Fenchel (1953, pp. 102–4) that these functions are differentiable. Substituting (1.13a) into (1.12d) we obtain

$$p^* = w_j \left/ \frac{\partial}{\partial v_j} f[\tilde{v}_1(w, y), \quad \tilde{v}_2(w, y), \ldots, \tilde{v}_m(w, y)] \right.$$

$$\equiv \tilde{p}^*(w, y). \quad (1.13b)$$

The system of equations (1.12d) defines a mapping $\mathscr{F}$ from the non-negative orthant of $(m + 1)$-dimensional space into itself:

$$\mathscr{F}(v, p^*) = (w, y). \quad (1.14a)$$

Equations (1.13a) and (1.13b) define the inverse mapping:

$$\mathscr{F}^{-1}(w, y) = (v, p^*). \quad (1.14b)$$

In accordance with (1.2) we define

$$G(w, y) = \sum_{k=1}^{m} w_k \tilde{v}_k(w, y). \quad (1.15)$$

We shall also define the *indirect production function* $\tilde{f}$ by

$$\tilde{f}(w, y) = f[\tilde{v}_1(w, y), \tilde{v}_2(w, y), \ldots, \tilde{v}_m(w, y)] \quad (1.16a)$$

which satisfies the identity

$$\tilde{f}(w, y) = y \quad \text{for all} \quad w, y. \quad (1.16b)$$

Theorem 5. (*Fundamental Envelope Theorem of Production Theory*). The functions $G, \tilde{v}_j, \tilde{p}*$ of (1.15), (1.13a), (1.13b) are related by

$$\frac{\partial G(w, y)}{\partial w_j} = \tilde{v}_j(w, y) \quad j = 1, 2, \ldots, m \quad (1.17a)$$

and

$$\frac{\partial G(w, y)}{\partial y} = \tilde{p}*(w, y). \quad (1.17b)$$

Proof: Differentiating (15) with respect to w_j, we obtain

$$\frac{\partial G(w, y)}{\partial w_j} = \tilde{v}_j(w, y) + \sum_{k=1}^{m} w_k \frac{\partial \tilde{v}_k(w, y)}{\partial w_j}. \quad (1.18)$$

To prove (1.17a) we must show that the second term on the right of (1.18) vanishes. Differentiating (1.16a) with respect to w_j and making use of the identity (1.16b) and the chain rule, we obtain upon substitution of (1.13b),

$$0 = \frac{\partial \tilde{f}(w, y)}{\partial w_j} = \sum_{k=1}^{m} \frac{\partial f}{\partial v_k}\bigg|_{v_k = \tilde{v}_k(w, y)} \cdot \frac{\partial \tilde{v}_k(w, y)}{\partial w_j}$$

$$= \frac{1}{\tilde{p}*(w, y)} \sum_{k=1}^{m} w_k \frac{\partial \tilde{v}_k(w, y)}{\partial w_j}$$

and (1.17a) follows. Likewise, differentiating (1.16a) with respect to y and using the identity (1.16b) and the chain rule, we have upon making use once again of (1.13b),

$$1 = \frac{\partial \tilde{f}(w, y)}{\partial y} = \sum_{k=1}^{m} \left(\frac{\partial f}{\partial v_k}\right)_{v_k = \tilde{v}_k(w, y)} \cdot \frac{\partial \tilde{v}_k(w, y)}{\partial y}$$

$$= \frac{1}{\tilde{p}*(w, y)} \sum_{k=1}^{m} w_k \frac{\partial \tilde{v}_k(w, y)}{\partial y}.$$

Thus, from this result and (1.15),

$$\frac{\partial G(w, y)}{\partial y} = \sum_{k=1}^{m} w_k \frac{\partial \tilde{v}_k(w, y)}{\partial y} = \tilde{p}*(w, y),$$

establishing (1.17b).

q.e.d.

It may be noted immediately from (1.15) and (1.17a) that

$$G(w, y) = \sum_{k=1}^{m} w_k \frac{\partial G(w, y)}{\partial w_k},$$

providing the necessary and sufficient condition, by Euler's theorem, that G be homogeneous of degree 1 in w — a result already obtained in §1.1. Using (1.17a) again it follows that $\tilde{v}_j$ is homogeneous of degree zero in w.

Now let us introduce condition (d): the positive homogeneity (of degree 1) of the production function f. Using (1.15) and (1.13b) we have, by Euler's theorem,

$$G(w, y) = \sum_{k=1}^{m} w_k \tilde{v}_k(w, y)$$

$$= \tilde{p}*(w, y) \sum_{k=1}^{m} \left(\frac{\partial f}{\partial v_k}\right)_{v_k = \tilde{v}_k(w, y)} \cdot \tilde{v}_k(w, y)$$

$$= y\tilde{p}*(w, y)$$

whence from (6)

$$\tilde{p}*(w, y) = \frac{G(w, y)}{y} = g(w). \quad (1.19)$$

Defining

$$b_j(w) = \frac{\partial g(w)}{\partial w_j} \quad (j = 1, 2, \ldots, m) \quad (1.20)$$

we have from (1.17a), (1.19), and (1.20),

$$\tilde{v}_j(w, y) = \frac{\partial G(w, y)}{\partial w_j} = y \frac{\partial g(w)}{\partial w_j} = yb_j(w) \quad (1.21)$$

hence the optimal factor–product ratios are given by

$$\frac{v_j}{y} = b_j(w). \quad (1.22)$$

From the differentiability assumption (e) imposed on the function f we can derive a strict quasi-concavity property of the function g. For suppose $w^0 > 0, w^1 > 0$, and $w^0 \neq \lambda w^1$; then from (b_2) and (e), we have $b(w^0) \neq b(w^1)$, where

$$b(w) = [b_1(w), b_2(w), \ldots, b_a(w)]. \quad (1.23)$$

Now by definition of g [see (1.5)]

$$g(w^0) \leqq w^0 \cdot v \quad \text{for all} \quad v \in A(1)$$
$$g(w^1) \leqq w^1 \cdot v \quad \text{for all} \quad v \in A(1) \quad (1.24a)$$

and moreover

$$g(w^0) = w^0 \cdot v \quad \text{if and only if} \quad v = b(w^0)$$
$$g(w^1) = w^1 \cdot v \quad \text{if and only if} \quad v = b(w^1). \quad (1.24b)$$

Furthermore, $b(w^0) \neq b(w^1)$, so strict inequality must hold in one of the inequalities (1.24a); thus if $0 < \theta < 1$,

$$(1 - \theta)g(w^0) + \theta g(w^1)$$
$$< [(1 - \theta)w^0 + \theta w^1] \cdot v \quad \text{for all} \quad v \in A(1)$$

and therefore in particular

$$(1 - \theta)g(w^0) + \theta g(w^1)$$
$$< \min_v \{[(1 - \theta)w^0 + \theta w^1] \cdot v : v \in A(1)\}$$
$$= g[(1 - \theta)w^0 + \theta w^1].$$

So we have

$(b*)$ if $w^0 > 0, w^1 > 0$, and $w^0 \neq \lambda w^1$, and if $0 < \theta < 1$, then $g[(1 - \theta)w^0 + \theta w^1] > (1 - \theta)g(w^0) + \theta g(w^1)$.

If is not hard to see that a corresponding property (b_3) holds for f as well. Failure of (b_3^*) when f is not differentiable, allowing $b(w^0) = b(w^1)$ for $w^0 \neq \lambda w^1$, is illustrated in Figure 4.

In general, a flat segment on a production isoquant goes over into a kink on the dual cost isoquant, and vice versa. There is another still more subtle relationship, illustrated by the following function found in Katzner (1970, p. 54):

$$f(v_1, v_2) = (v_1^3 v_2 + v_1 v_2^3)^{1/4}.$$

Its dual minimum-unit-cost function is found to be

$$g(w_1, w_2) = 2^{-1/4}[(w_1 + w_2)^{4/3} - (w_1 - w_2)^{4/3}]^{3/4}.$$

The isoquants of f are extremely flat at $v_1 = v_2$, and as a result g is once but not twice differentiable at $w_1 = w_2$. A graph of

$$g(w_1, w_2) = w_1 b_1(w_1, w_2) + w_2 b_2(w_1, w_2)$$

for $w_2 = \bar{w}_2$ is shown in Figure 5. At $w_1 = \bar{w}_2$, $\bar{w}_2 b_2(w_1, \bar{w}_2)$ has a slope of $+\infty$ and $w_1 b_1(w_1, \bar{w}_2)$ has a slope of $-\infty$, yet their sum is differentiable. When the bordered Hessian of the

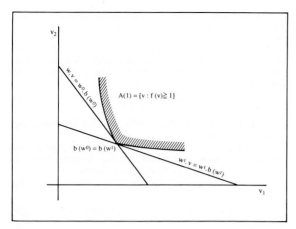

Figure 4

production function f is invertible, its inverse is the bordered Hessian of the cost function g; in the above example, it is not invertible at $v_1 = v_2$.

A useful illustration of the duality of cost and production functions is given by the case of CES (constant-elasticity-of-substitution) production functions (cf. Arrow et al., 1961; Uzawa, 1962):

$$f(v) = \left[\sum_{i=1}^{m} \alpha_i v_i^{1-1/\sigma} \right]^{\sigma/(\sigma-1)}$$

The corresponding cost functions have the form

$$g(w) = \left[\sum_{i=1}^{m} \alpha_i^{\sigma} w_i^{1-\sigma} \right]^{1/(1-\sigma)}$$

whose elasticity of substitution is $\sigma^* = 1/\sigma$.

2. THE PRODUCTION–POSSIBILITY SET.

Suppose a country to be capable of producing n commodities with the aid of m primary factors of production. Denoting the output of commodity j by y_j, and the input of factor i into the production of commodity j by v_{ij}, the production function may be written

$$y_j = f_i(v_{1j}, v_{2j}, \ldots, v_{mj}) = f_j(v_j) \qquad (j = 1, 2, \ldots, n), \quad (2.1)$$

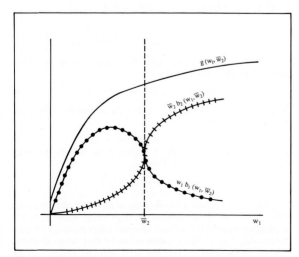

Figure 5

where

$$v_j = (v_{1j}, v_{2j}, \ldots, v_{mj}). \tag{2.2}$$

It will be assumed that f_j is:

(a) *Continuous;* i.e.,

$$\lim_{v_j \to v_j^0} f_j(v_j) = f_j(v_j^0);$$

(b) *Weakly monotone;* i.e., if $v_j^1 \geqq v_j^2$ (meaning that $v_{ij}^1 \geqq v_{ij}^2$ for $i = 1, 2, \ldots, m$) then $f_i(v_j^1) \geqq f_j(v_j^2)$, and if $v_j^1 > v_j^2$ (i.e., $v_{ij}^1 > v_{ij}^2$ for $i = 1, 2, \ldots, m$) then $f_j(v_j^1) > f_j(v_j^2)$;

(c) *Concave;* i.e., if v_j^0 and v_j^1 are any two vectors of primary inputs into the production of commodity j, then for any t in the interval $0 < t < 1$,

$$f_j[(1-t)v_j^0 + tv_j^1] \geqq (1-t)f_j(v_j^0) + tf_j(v_j^1); \tag{2.3}$$

(d) *Positively homogeneous of degree 1;* i.e., for any $\lambda > 0$,

$$f_j(\lambda v_j) = \lambda f_j(v_j). \tag{2.4}$$

It will be convenient to introduce the $m \times n$ allocation matrix

$$V = \begin{bmatrix} v_{11} & v_{12} & \cdots & v_{1n} \\ v_{21} & v_{22} & \cdots & v_{2n} \\ \cdot & \cdot & \cdot & \cdot \\ v_{m1} & v_{m2} & \cdots & v_{mn} \end{bmatrix} \tag{2.5}$$

The element v_{ij} is the input of factor i into the production of commodity j. The jth column of V will be denoted v_j; according to this notation, v_j is the transpose of v_j, denoted $v_j = v_j'$.

Let l_i denote the country's total endowment of factor i. Then for each i the following resource constraint holds:

$$\sum_{j=1}^{n} v_{ij} \leqq l_i \qquad (i = 1, 2, \ldots, m). \tag{2.6}$$

Using (2.5) this can be written in matrix notation as

$$\begin{bmatrix} v_{11} & v_{12} & \cdots & v_{1n} \\ v_{21} & v_{22} & \cdots & v_{2n} \\ \cdot & \cdot & \cdots & \cdot \\ v_{m1} & v_{m2} & \cdots & v_{mn} \end{bmatrix} \begin{bmatrix} 1 \\ 1 \\ \vdots \\ 1 \end{bmatrix} \leqq \begin{bmatrix} l_1 \\ l_2 \\ \vdots \\ l_m \end{bmatrix}, \tag{2.7}$$

or simply

$$V\iota \leqq l, \tag{2.8}$$

where ι is the column vector of n ones and $l = (l_1, l_2, \ldots, l_m)'$ is the column vector of factor endowments.

In the absence of any additional restrictions, condition (2.6) expresses the *perfectly mobility* of factors among industries.

The country's *production–possibility set* is the set of all possible output combinations $y = (y_1, y_2, \ldots, y_n)$ that can be produced with the production functions (2.1) under the resource constraints (2.6). Formally, it may be denoted

$$\mathscr{Y}(l) = \{y: \text{there exist allocations } v_{ij} \geqq 0 \text{ such that}$$

$$y_j = f_j(v_j) (j = 1, 2, \ldots, n) \quad \text{and}$$

$$\sum_{j=1}^{n} v_{ij} \leqq l_i \qquad (i = 1, 2, \ldots, m)\}. \tag{2.9}$$

For notational convenience we may define the function $f(V)$ as the vector-valued function

$$f(V) = (f_1(v_1'), \ f_2(v_2'), \ldots, f_n(v_n'))' \tag{2.10}$$

and write (2.9) in the more compact form

$$\mathscr{Y}(l) = \{y: (\exists V \geqq 0) y = f(V) \& V\iota \leqq l\}. \tag{2.11}$$

Note that with this notation, condition (2.3) can be written (for $t = t_j$) in the form

$$f(V^0(I - T) + V^1 T) \geqq (I - T)f(V^0) + Tf(V^1) \quad (2.12)$$

where $T = \operatorname{diag}(t_1, t_2, \ldots, t_n)$ is an $n \times n$ diagonal matrix with $0 < t_j < 1$. Likewise, (2.4) may be written (for $\lambda = \lambda_j$) in the form

$$f(V\Lambda) = \Lambda f(V), \quad (2.13)$$

where $\Lambda = \operatorname{diag}(\lambda_1, \lambda_2, \ldots, \lambda_n)$ is an $n \times n$ diagonal matrix with $\lambda_j > 0$.

Theorem 6. If assumptions (a), (b), and (c) hold, the production–possibility set $\mathscr{Y}(l)$ is convex.

Proof: Let y^0, y^1 both belong to $\mathscr{Y}(l)$; we are to show that for any t in the interval $0 < t < 1$, the output combination $y^t = (1 - t)y^0 + ty^1$ also belongs to $\mathscr{Y}(l)$ (see Figure 6).

Since $y^0, y^1 \in \mathscr{Y}(l)$, this means that there exist two allocation matrices V^0, V^1 each satisfying (2.8), such that $y^0 = f(V^0)$ and $y^1 = f(V^1)$. Denote $V^t = (1 - t)V^0 + tV^0$. Then from (2.8),

$$V^t l = (1 - t)V^0 l + tV^1 l \leqq (1 - t)l + tl = l, \quad (2.14)$$

so V^t is a feasible allocation, and by concavity,

$$f(V^t) \geqq (1 - t)f(V^0) + tf(V^1) = y^t, \quad (2.15)$$

i.e., for each $j = 1, 2, \ldots, n$, denoting $v_j^t = (1 - t)v_j^0 + tv_j^1$,

$$f_j(v_j^t) \geqq (1 - t)f_j(v_j^0) + tf_j(v_j^1) = y_j^t. \quad (2.15')$$

By continuity and monotonicity of f_j, there exist $\lambda_j^t \leqq 1$ such that

$$f_i(\lambda_j^t v_j^t) = y_j^t \quad (j = 1, 2, \ldots, n). \quad (2.16')$$

(In particular, (2.16') follows if the stronger homogeneity condition (d) holds, by taking $\lambda_j^t = y_j^t / f_j(v_j^t)$ if $y_t^t > 0$, and 0 otherwise.) Equivalently,

$$f(V^t \Lambda^t) = y^t. \quad (2.16)$$

It remains only to verify that the matrix $V^t \Lambda$ of allocations $\lambda_j^t v_{ij}^t$ satisfies the constraint (2.8). This is immediate from the fact that $0 \leqq \lambda_j^t \leqq 1$, whence from (2.14),

$$V^t \Lambda^t l = V^t \lambda^t \leqq V^t l \leqq l. \quad (2.17)$$

q.e.d.

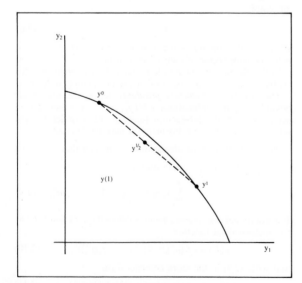

Figure 6

Note that homogeneity of production functions is not needed for the above result.

3. THE NATIONAL-PRODUCT FUNCTION. Let $p = (p_1, p_2, \ldots, p_n)'$ denote a vector of prices. The *national-product function* (cf. Samuelson, 1953; Chipman, 1972, 1974) is defined as the function

$$\Pi(p, l) = \max_{y \in \mathscr{Y}(l)} p \cdot y. \quad (3.1)$$

[See also Dixit and Norman (1980), who use the terminology 'revenue function'.]

For any fixed p, this has all the properties of a production function, but with some special peculiar features. These are illustrated in Figure 7 to be explained shortly.

For each commodity, $j = 1, 2, \ldots, n$, define the upper-contour set

$$A_j(y_j) = \{l^j = (l_1^j, l_2^j, \ldots, l_m^j) : f_j(l^j) \geqq y_j\}. \quad (3.2)$$

Then in particular,

$$A_j(Y/p_j) = \{l^j : p_j f_j(l^j) \geqq Y\} \quad (3.3)$$

is the set of factor–input combinations that will yield, at the given price p_j, an amount of commodity j worth at least Y. Throughout this section it will be assumed that each f_j satisfies properties (a)–(d) of the preceding section.

Let us now introduce a stronger monotonicity condition that refers to the entire vector-valued function (3.10). It may be stated as follows: f is

(e) *Strictly monotone*, i.e., for each $V = [v_{ij}]$ and each $i = 1, 2, \ldots, m$, there is a $j = 1, 2, \ldots, n$ such that $\delta > 0$ implies

$$f_j(v_{1j}, v_{2j}, \ldots, v_{ij} + \delta, \ldots, v_{mj})$$
$$> f_j(v_{1j}, v_{2j}, \ldots, v_{ij}, \ldots, v_{mj}). \quad (3.4)$$

In words, if there is an increase in the amount of any one of the m endowments, it is possible to find an industry where this additional input will lead to increased output.

For any family of sets $S_1, S_2, \ldots, S_n$, each a subset of m-dimensional Euclidean space E^m, the *arithmetic mean* of this family (which is, for convex S_j, also the *convex hull* of $\bigcup_{j=1}^n S_j$) is defined and denoted

$$\mathop{\mathbf{M}}_{j=1}^n S_j = \Big\{ s \in E^m : (\exists s^j \in S_j, \lambda_j \geqq 0, j = 1, 2, \ldots, n)$$

$$\sum_{j=1}^n \lambda_j = 1 \text{ and } s = \sum_{j=1}^n \lambda_j s^j \Big\}. \quad (3.5)$$

Analogously to (3.2) we define the upper-contour set of the national-product function by

$$A(p, Y) = \{l \in E_+^m : \Pi(p, l) \geqq Y\}. \quad (3.6)$$

The following theorem characterizes the isoquants of the function $\Pi(p, \cdot)$ (see Figure 7).

Theorem 7. Let all prices p_j be positive, $j = 1, 2, \ldots, n$, and let f satisfy conditions (a)–(d) of section 2, as well as the strict montonicity condition (e). Then

$$A(p, Y) = \mathop{\mathbf{M}}_{j=1}^n A_j(Y/p_j), \quad (3.7)$$

i.e., the upper-contour set consisting of all factor combinations l that give rise to a national product of at least Y, is the arithmetic mean of the n upper-contour sets consisting, for each commodity j, of all factor combinations l^j that, when allocated entirely to industry j, give rise to a national product of at least Y.

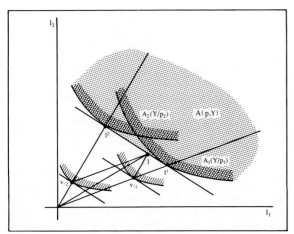

Figure 7

Proof: (a) let us first prove that

$$\mathop{M}_{j=1}^{n} A_j(Y/p_j) \subseteq A(p, Y). \tag{3.8}$$

Let

$$l \in \mathop{M}_{j=1}^{n} A_j(Y/p_j).$$

Then, by definition (3.5), there exist $l^j \in A(Y/p_j)$ and $\lambda_j \geqq 0$ such that

$$\sum_{j=1}^{n} \lambda_j = 1 \quad \text{and} \quad \sum_{j=1}^{n} \lambda_j l^j = l.$$

By definition (3.3), each l^j satisfies $p_j f_j(l^j) \geqq Y$, hence from the definition (3.1) of Π and the homogeneity of degree 1 of each f_j, we have

$$\Pi(p, l) \geqq \sum_{j=1}^{n} p_j f_j(\lambda_j l^j) = \sum_{j=1}^{n} \lambda_j p_j f_j(l^j) \geqq Y \sum_{j=1}^{n} \lambda_j = Y.$$

From definition (3.6) it follows that $l \in A(p, Y)$, and (3.8) follows.

(b) We now show that

$$A(p, Y) \subseteq \mathop{M}_{j=1}^{n} A_j(Y/p_j). \tag{3.9}$$

Let $l \in A(p, Y)$; then by definitions (3.6), (3.1) and (2.9), there exist allocations $v_j \in E_+^m$ such that

$$\sum_{j=1}^{n} v_j \leqq l \quad \text{and} \quad \sum_{j=1}^{n} p_j f_j(v_j) = \Pi(p, l) \geqq Y. \tag{3.10}$$

By the strict monotonicity of f, the first inequality of (3.10) must be an equality; for, if for some $i = i'$ we have

$$\sum_{j=1}^{n} v_{i'j} < l_{i'},$$

then for some

$$j = j' \quad \text{and} \quad 0 < \delta \leqq l_{i'} - \sum_{j=1}^{n'} v_{i'j}$$

the inequality (3.4) is satisfied, violating the definition (3.1) of $\Pi(p, l)$. Now define

$$\lambda_j = p_j f_j(v_j)/\Pi(p, l), \quad l^j = v_j/\lambda_j \quad (j = 1, 2, \ldots, n). \tag{3.11}$$

Then

$$\sum_{j=1}^{n} \lambda_j l^j = l \quad \text{where} \quad \sum_{j=1}^{n} \lambda_j = 1. \tag{3.12}$$

By homogeneity we have

$$p_j f_j(l^j) = p_j f_j(v_j)/\lambda_j = \Pi(p, l) \geqq Y,$$

hence $l^j \in A_j(Y/p_j)$ from (3.3). Together with (3.10) this implies that (3.9) holds.

q.e.d.

Since for each fixed p the national-product function $\Pi(p, \cdot)$ has the properties of a production function (i.e. it is continuous, concave, monotone, and positively homogeneous of degree 1), we may associate with it a corresponding minimum-unit cost function $\Gamma(p, \cdot)$ defined by

$$\Gamma(p, w) = \min_{l} \{w \cdot l : \Pi(p, l) \geqq 1\}. \tag{3.13}$$

This will be called the *national-cost function*. Letting $g_j(w) = \min_{v_j} \{w \cdot v_j : f_j(v_j) \geqq 1\}$ denote the minimum-unit cost function dual to the production function $f_j(v_j)\}$, we may define the upper-contour sets

$$A_j^*(p_j) = \{w : g_j(w) \geqq p_j\} \tag{3.14}$$

and

$$A^*(p) = \{w : \Gamma(p, w) \geqq 1\}. \tag{3.15}$$

The boundary of the intersection of all the sets (3.14) for $j = 1, 2, \ldots, n$ is known as the 'factor-rental frontier' (or 'factor-price frontier' – cf. Woodland, 1982, pp. 49–52). The following theorem shows that it is also the contour of the corresponding national-cost function. Its shape will be similar to that depicted in Figure 4.

Theorem 8. Let the prices p_j be positive, $j = 1, 2, \ldots, n$ and let f satisfy conditions (a) to (e) of section 1.2. Then

$$A^*(p) = \bigcap_{j=1}^{n} A_j^*(p_j). \tag{3.16}$$

Proof: Let $w \in A^*(p)$; then $\Gamma(p, w) \geqq 1$, i.e., $w \cdot l \geqq 1$ for all $l \in A(p, l)$. Choose such an l and let V be the optimal resource-allocation matrix; then

$$\Pi(p, l) = \sum_{j=1}^{n} p_j f_j(v_j) \geqq 1. \tag{3.17}$$

Defining λ_j and l^j as in (3.11), this gives (by homogeneity)

$$\sum_{j=1}^{n} p_j f_j(\lambda_j l^j) = \sum_{j=1}^{n} \lambda_j p_j f_j(l^j) \geqq 1, \tag{3.18}$$

and since

$$\lambda_j > 0 \quad \text{and} \quad \sum_{j=1}^{n} \lambda_j = 1$$

this implies $p_j f_j(l^j) \geqq 1$, i.e., $l^j \in A_j(1/p_j)$, for each j. Now by hypothesis, (3.17) implies $w \cdot l \geqq 1$ hence

$$\sum_{j=1}^{n} \lambda_j w \cdot l^j \geqq 1, \tag{3.19}$$

and by the same reasoning as above this implies $w \cdot l^j \geqq 1$ for all j, i.e.,

$$g_j(w)/p_j = \min_{l^j} \{w \cdot l^j : l^j \in A_j(1/p_j)\} \geqq 1 \tag{3.20}$$

or $g_j(w) \geqq p_j$. From the definition (3.14) this shows that $w \in A_j^*(p_j)$ for $j = 1, 2, \ldots, n$.

Conversely, let $w \in \bigcap_{j=1}^{n} A_j^*(p_j)$; then $g_j(w) \geqq p_j$ for $j = 1, 2, \ldots, n$. From the definition of g_j, this implies $w \cdot l^j \geqq 1$ for all $l^j \in A_j(1/p_j), j = 1, 2, \ldots, n$. Choosing $l^j \in A_j(1/p_j)$ such that

$$\sum_{j=1}^{n} \lambda_j l^j = l,$$

$$\Pi(p, l) \geqq \sum_{j=1}^{n} p_j f_j(\lambda_j l^j) = \sum_{j=1}^{n} \lambda_j p_j f_j(l^j) \geqq \sum_{j=1}^{n} \lambda_j = 1, \tag{3.21}$$

929

hence

$$w \cdot l = w \cdot \sum_{j=1}^{n} \lambda_j l^j = \sum_{j=1}^{n} \lambda_j w \cdot l^j \geqq 1. \qquad (3.22)$$

From the definition (3.13) this implies $\Gamma(p, w) \geqq 1$, and thus by (3.15) it follows that $w \in A^*(p)$.

q.e.d.

Let us introduce a further assumption, that each f_j is

(f) *Differentiable.*

Then from Theorem 7 it follows that $\Pi(p, \cdot)$ is differentiable. Its partial derivative with respect to l_i is defined as the *Stolper–Samuelson function*

$$\hat{w}_i(p, l) \equiv \frac{\partial}{\partial l_i} \Pi(p, l) (i = 1, 2, \ldots, m), \qquad (3.23)$$

and the corresponding vector-valued function $\hat{w}(p, l) = \partial \Pi(p, l)/\partial l$ is called the *Stolper–Samuelson mapping.* The values of this function are the shadow or implicit factor rentals of the respective factors.

Setting up the Lagrangean function

$$L(V, w; p, l) = \sum_{j=1}^{n} p_j f_j(v_j) - \sum_{i=1}^{m} w_i \left(\sum_{j=1}^{n} v_{ij} - l_i \right) \qquad (3.24)$$

corresponding to the definition of the national-product function, we obtain the Kuhn–Tucker conditions

$$\frac{\partial L}{\partial v_{ij}} = p_j \frac{\partial f_j}{\partial v_{ij}} - w_i \leqq 0, \quad \left(p_j \frac{\partial f_j}{\partial v_{ij}} - w_i \right) v_{ij} = 0; \qquad (3.25a)$$

$$\frac{\partial L}{\partial w_i} = l_i - \sum_{j=1}^{n} v_{ij} \geqq 0, \quad \left(l_i - \sum_{j=1}^{n} v_{ij} \right) w_i = 0. \qquad (3.25b)$$

It will be observed that conditions (3.25a) constitute, for each $j = 1, 2, \ldots, n$, precisely the Kuhn–Tucker conditions for cost-minimization in industry j, where w_i is the ith factor rental. The rentals defined by the Stolper–Samuelson mapping are therefore the market rentals that will obtain in competitive equilibrium.

Let us now explore the consequences of assuming that the function Π is differentiable with respect to p as well as l. Given p^0, l^0, let y^0 maximize $p^0 \cdot y^0$. Define the function

$$H(p, l^0) = \Pi(p, l^0) - p \cdot y^0.$$

Then $H(p^0, l^0) = 0$ and $H(p, l^0) \geqq 0$ for $p \neq p^0$ (by the definition of Π), hence H reaches a minimum with respect to p at $p = p^0$. Since differentiability of Π implies differentiability of H, we have

$$\frac{\partial H(p^0, l^0)}{\partial p_j} = \frac{\partial \Pi(p^0, l^0)}{\partial p_j} - y_j^0 = 0.$$

This shows that y^0 is the *unique* y which maximizes $p^0 \cdot y$ subject to $y \in \mathcal{Y}(l^0)$. This is equivalent to saying that the *production–possibility frontier* $\hat{\mathcal{Y}}(l)$ – i.e., the set of all $y \in \mathcal{Y}(l^0)$ which maximize $p \cdot y$ for some $p > 0$ – is *strictly concave to the origin.* The apparently innocuous assumption that Π is differentiable with respect to p has thus led to an important substantive conclusion.

When Π is differentiable with respect to p, the function

$$\hat{y}_j(p, l) = \frac{\partial}{\partial p_j} \Pi(p, l) \qquad (j = 1, 2, \ldots, n) \qquad (3.26)$$

is called the *Rybczynski function* for commodity j. The corresponding vector-valued function $\hat{y}(p, l)$ is called the *Rybczynski mapping.*

In general, we may define the *Rybczynski correspondence* by

$$\hat{y}(p, l) = \{ y \in \mathcal{Y}(l) : p \cdot y = \Pi(p, l) \}. \qquad (3.27)$$

The above result shows that if Π is differentiable with respect to p, this correspondence is a singleton-valued mapping. We shall now obtain a necessary and sufficient condition for this single-valuedness, i.e., for the strict concavity to the origin of $\mathcal{Y}(l)$.

Let the factor–output coefficients be denoted

$$b_{ij}(w) = \frac{\partial g_j(w)}{\partial w_i} (i = 1, 2, \ldots, m; \quad j = 1, 2, \ldots, n) \qquad (3.28)$$

where g_j is the minimum-unit-cost function dual to the production function f_j. The following result was obtained by Khang (1971) and Chipman (1972).

Theorem 9. Let p^0, l^0 be such that there exists a $y^0 > 0$ which maximizes $p^0 \cdot y$ subject to $y \in \mathcal{Y}(l^0)$, and let $w^0 = \hat{w}(p^0, l^0) = \partial \Pi(p^0, l^0)/\partial l$. Let f satisfy the strict monotonicity condition (e). Then in order that y^0 should be the unique maximizer of $p^0 \cdot y$ subject to $y \in \mathcal{Y}(l^0)$, it is necessary and sufficient that the n columns of the factor–output matrix

$$B(w^0) = \begin{bmatrix} b_{11}(w^0) & b_{12}(w^0) & \ldots & b_{1n}(w^0) \\ b_{21}(w^0) & b_{22}(w^0) & \ldots & b_{2n}(w^0) \\ & & & \\ b_{m1}(w^0) & b_{m2}(w^0) & \ldots & b_{mn}(w^0) \end{bmatrix}$$

be linearly indepenent.

Proof: For convenience, denote $B^0 = B(w^0)$. Then from strict monotonicity of f we have

$$B^0 y^0 = l^0. \qquad (3.29)$$

First we show that if rank $B^0 < n$ then y^0 is not unique. Since rank $B^0 < n$ there exists a vector $z^0 \neq 0$ such that

$$B^0 z^0 = 0. \qquad (3.30)$$

Choose $\epsilon^0 > 0$ such that

$$y^0 \pm \epsilon^0 z^0 > 0;$$

then $y^0 \pm \epsilon z^0 > 0$ for $0 < \epsilon < \epsilon^0$. From (3.29) and (3.30) we have $B^0(y^0 \pm \epsilon^0 z^0) = l^0$ whence $y^0 \pm \epsilon^0 z^0 \in \mathcal{Y}(l^0)$. Since y^0 maximizes $p^0 \cdot y$ over $\mathcal{Y}(l^0)$,

$$p^0 \cdot y^0 \geqq p^0 \cdot (y^0 \pm \epsilon^0 z^0),$$

i.e., $0 \geqq \epsilon^0 p^0 \cdot z^0 \geqq 0$. This implies $p^0 \cdot z^0 = 0$, hence $p^0 \cdot (y^0 \pm \epsilon z^0) = p^0 \cdot y^0$ for $0 < \epsilon < \epsilon^0$, i.e.,

$$y^0 \pm \epsilon z^0 \in \mathcal{Y}(l^0) \quad \text{for} \quad 0 < \epsilon < \epsilon^0.$$

This shows that y^0 is not unique.

Conversely we show that if y^0 is not unique then rank $B^0 < n$. Suppose $y^0, y^1 > 0$ both maximize $p^0 \cdot y$ subject to $y \in \mathcal{Y}(l^0)$, where $y^1 \neq y^0$. Then $B^0 y^0 = B^0 y^1 = l$, hence $B^0(y^0 - y^1) = 0$; since $y^0 - y^1 \neq 0$, this implies that rank $B^0 < n$.

q.e.d.

From this result it follows that a necessary condition for the production-possibility frontier to be strictly concave to the origin is that $m \geqq n$. If $m < n$, it is a ruled surface. However, the condition $m \geqq n$ is certainly not sufficient; one example is the case $m = n = 2$ when two isoquants for a dollar's worth of output are mutually tangent at a point along the endowment ray (cf. Lerner, 1933, p. 13). For further discussion of these points see Kemp, Khang and Uekawa (1978), and for an interesting characterization, see Inoue (1986) and Inoue and Wegge (1986).

To gain an intuitive understanding of the meaning of the differentiability of $\Pi(\cdot, l)$, let us assume that the f_j are differentiable and that the functions $\hat{v}_{ij}(p, l)$, obtained with the

$\hat{w}_i(p, l)$ by solving the above constrained-maximum problem, are also single-valued and differentiable. Then from

$$\Pi(p, l) = \sum_{j=1}^{n} p_j f_j(v_j) \qquad (3.31)$$

we have

$$\frac{\partial \Pi}{\partial p_k} = y_k + \sum_{j=1}^{n} \sum_{i=1}^{m} \left[p_j \frac{\partial f_j}{\partial v_{ij}} - w_i \right] \frac{\partial \hat{v}_{ij}}{\partial p_k} + \sum_{i=1}^{m} w_i \sum_{j=1}^{n} \frac{\partial \hat{v}_{ij}}{\partial p_k}. \quad (3.32)$$

If $w_i > 0$ then

$$\sum_{j=1}^{n} \hat{v}_{ij}(p, l) = l_i$$

and thus

$$\sum_{j=1}^{n} \partial \hat{v}_{ij} / \partial p_k = 0,$$

hence the last term of (3.32) must vanish. If $v_{ij} > 0$ then the bracketed term in (3.32) vanishes (by the Kuhn–Tucker conditions). If the bracketed term is negative then $\hat{v}_{ij} = 0$ by the Kuhn–Tucker conditions, and thus $\partial \hat{v}_{ij} / \partial p_k = 0$. In either case, the second term on the right in (3.32) vanishes. The trouble occurs in the intermediate case in which factor i is on the verge of being employed in industry j, hence $p_j \partial f_j / \partial v_{ij} - w_i = 0$ and $v_{ij} = 0$; it is precisely in this case that $\hat{v}_{ij}(\cdot, l)$ will not be differentiable at that point. Formula (3.26) therefore fails at switching points where factors are on the verge of being employed in particular industries; a small price change in one direction will lead to their continued unemployment, but in the other direction to their being employed. Thus, $\Pi(\cdot, l)$ is non-differentiable at such switching points. Likewise, it is non-differentiable when the conditions of Theorem 9 fail, in which case a small price change may lead to a country's switching from specialization in one commodity to specialization in another. All this would become clearer if the theory were to be recast in terms of subdifferentials (cf. Rockafellar, 1970).

Since $\Pi(p, \cdot)$ has the properties of a production function, from Theorem 7, it is concave; and since, as was seen above, $H(p, l^0) = \Pi(p, l^0) - p \cdot y^0$ is a minimum at $p = p^0$, where $\Pi(p^0, l^0) = p^0 \cdot y^0$, $H(\cdot, l)$ is convex, hence $\Pi(\cdot, l)$ is convex. That is, $\Pi(p, l)$ is convex in p and concave in l. If it is twice continuously differentiable then Samuelson's (1953) 'reciprocity theorem' holds:

$$\frac{\partial \hat{y}_j}{\partial l_i} = \frac{\partial^2 \Pi}{\partial p_j \partial l_i} = \frac{\partial^2 \Pi}{\partial l_i \partial p_j} = \frac{\partial \hat{w}_i}{\partial p_j}. \qquad (3.33)$$

4. THE STOLPER–SAMUELSON AND RYBCZYNSKI MAPPINGS. When a country diversifies its production, by which we shall mean that it produces all n consumable commodities, as long as it is not on the verge of specializing, its factor-endowment vector will lie in the interior of a diversification cone – the convex cone whose extreme rays pass through the factor-input vectors in the n industries which minimize costs at the given factor rentals (cf. McKenzie, 1955; Chipman, 1966). As is clear from Figure 7, the factor rentals will remain unchanged as the factor endowment vector varies within the interior of this cone; i.e. the function $\hat{w}(p, l)$ is independent of l for endowments l in this cone. Now if all n commodities are to be produced, costs cannot exceed prices; and competitive equilibrium requires that prices not exceed costs. Hence, from the homogeneity of degree 1 of the minimum-unit-cost functions, and by Theorem 5, we have

$$p_j = g_j(w) = \sum_{i=1}^{m} \frac{\partial g_j(w)}{\partial w_i} w_i = \sum_{i=1}^{m} b_{ij}(w) w_i, \qquad (4.1)$$

or in matrix notation, where w and p denote column vectors of m factor rentals and n commodity prices respectively,

$$p = g(w) = B(w)'w \qquad (4.2)$$

and thus

$$g[\hat{w}(p, l)] = p, \qquad (4.3)$$

i.e., $\hat{w}(\cdot, l)$ is a local inverse of the mapping g. Since the Jacobian matrix of g, $B(w)'$, must have rank m if the diversification cone has a non-empty interior (hence $n \geqq m$), the range of g is an m-dimensional manifold, hence (4.2) implies that the vector p of world prices cannot be varied arbitrarily (if the country is to continue to diversify) unless $n = m$.

Even when $n = m$, the mapping g is in general not globally univalent. Gale and Nikaido (1965) obtained strong sufficient conditions for global univalence, namely that the principal minors of $B(w)$ be positive (this condition can be slightly weakened). Inada (1971) obtained some alternative conditions. In a controversy with Pearce (1967), McKenzie (1967) showed that it did not suffice to assume that $B(w)$ had a non-vanishing determinant for all positive w. The condition that $|B(w)| \neq 0$ for some $w = w^0$ is of course sufficient for local invertibility of g, but this inverse mapping depends on l. If two countries with identical technologies have their endowment vectors l in the same diversification cone, their factor rentals will be equalized even if g is not globally univalent. Nikaido (1972) showed that a modification of conditions originally suggested by Samuelson (1953) is sufficient for global univalence of g.

Of particular interest is the nature of the Stolper–Samuelson mapping in regions where it is locally independent of l, i.e., the nature of the local inverses of g. For the reasons given above, discussion of this is effectively limited to the case $n = m$. Defining the diagonal matrices $W = \text{diag } w$ and $P = \text{diag } p$, and the matrix $\mathbf{B} = WBP^{-1}$, by dividing (4.1) through by p_j one sees that $\mathbf{B}$ is column-stochastic (i.e., has unit column sums in addition to having non-negative elements); denoting its elements by $\beta_{ij} = w_i b_{ij} / p_j = \partial \log g_j / \partial \log w_i$, these satisfy

$$\sum_{i=1}^{m} \beta_{ij} = 1.$$

Denoting the elements of B^{-1} by b^{ij} and those of $\mathbf{B}^{-1} = PB^{-1}W^{-1}$ by $\beta^{ij} = p_i b^{ij} / w_j$, these are equal to $\beta^{ij} = \partial \log \hat{w}_j / \partial \log p_i$. Denoting by ι_m the column vector of m 1s, from $\iota'_m \mathbf{B} = \iota'_m$ we have $\iota'_m \mathbf{B}^{-1} = \iota'_m \mathbf{B} \mathbf{B}^{-1} = \iota'_m$, hence $\mathbf{B}^{-1}$ also has unit column sums (cf. Chipman, 1969, p. 402).

In the case $m = n = 2$, if we follow the convention of numbering commodities and factors in such a way that, at the initial equilibrium, $|B(w)| > 0$, i.e.,

$$\begin{vmatrix} b_{11}(w) & b_{12}(w) \\ b_{21}(w) & b_{22}(w) \end{vmatrix} = b_{11}(w) b_{12}(w) \left[\frac{b_{22}(w)}{b_{12}(w)} - \frac{b_{21}(w)}{b_{11}(w)} \right] > 0 \quad (4.4)$$

(which means that industry 2 uses a higher ratio of factor 2 to factor 1 than industry 1), then B^{-1}, which has non-positive diagonal elements and unit column sums, must have diagonal elements greater than or equal to unity. If B has its elements all positive, then the off-diagonal elements of B^{-1} are negative and the diagonal elements greater than unity. This, in substance, is the Stolper–Samuelson (1941) theorem. In words, for some association of commodities and factors, a rise in one commodity price will lead to a more than proportionate rise in the corresponding factor rental.

A simple proof is illustrated in Figure 8, in the space of factor rentals. A rise in p_1 is shown by an upward shift in the isoquant $g_1(w_1, w_2) = p_1$ and a new intersection point with the isoquant $g_2(w_1, w_2) = p_2$ with lower w_2 and the rise in w_1 proportionately higher than that of p_1 (as long as the elasticities of substitution

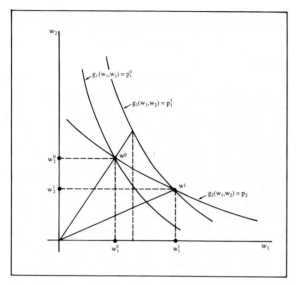

Figure 8

of the cost functions are positive, i.e., as long as the elasticities of substitution of the production functions are finite).

The Stolper–Samuelson theorem clearly does not generalize to higher dimensions. Either much stronger assumptions or much weaker conclusions are required. See Chipman (1969), Kuhn (1968), Inada (1971), Uekawa (1971), Ethier (1974), Jones and Scheinkinan (1977) and Neary (1985).

The Rybczynski functions $\hat{y}_j(p, l)$ exist as single-valued functions only for the case $m \geqq n$. If all n commodities are produced, and all m factors are fully employed, they satisfy the resource-allocation equation

$$B[\hat{w}(p, l)]\hat{y}(p, l) = l. \qquad (4.5)$$

When $m = n$, since then w is locally independent of l, $\hat{y}(p, l)$ is locally linear in l for any fixed p and may be written as

$$\hat{y}(p, l) = B[g^{-1}(p)]l \qquad (4.6)$$

(cf. Chipman, 1971, p. 214; 1972, p. 216). The curious shapes of the Rybczynski functions are illustrated in Figure 9.

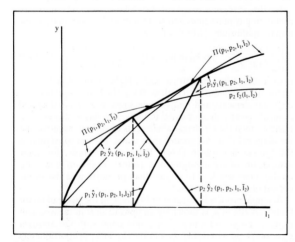

Figure 9

As in the case of the Stolper–Samuelson mapping, one can consider the elasticities of outputs with respect to factor endowments when $m = n$. Denoting $L = \text{diag } l$ and $Y = \text{diag } y$ (not to be confused with the national-income variable Y of section 3 above), we may define the matrix $\Lambda = L^{-1}BY$ with elements of $\lambda_{ij} = b_{ij}y_j/l_i = \partial \log l_i/\partial \log y_j$ (interpreting l_i in this relationship as requirements or demand for factor i, rather than supply). Its inverse $\Lambda^{-1} = Y^{-1}B^{-1}L$ has elements $\lambda^{ij} = b^{ij}l_j/y_i = \partial \log \hat{y}_i/\partial \log l_j$. From the resource-allocation constraint

$$\sum_{j=1}^{n} b_{ij}y_j = l_i \qquad (4.7)$$

it follows that

$$\sum_{j=1}^{n} \lambda_{ij} = 1,$$

i.e., Λ is row-stochastic (its elements are non-negative and its row sums are equal to unity). By the same reasoning as before, the row sums of Λ^{-1} are equal to unity. In the case $n = m = 2$, adhering to the convention (4.4) it follows that, when B has positive elements, the off-diagonal elements of Λ^{-1} are negative, and thus its diagonal elements are greater than unity. Thus, $\partial \log \hat{y}_i/\partial \log l_i > 1$ and $\partial \log \hat{y}_i/\partial \log l_j < 0$ for $j \neq i$; in words, a rise in the ith factor endowment will, at given world prices, lead to a more than proportionate rise in the output of the ith commodity, and a fall in the output of the jth commodity ($j \neq i$). As in the case of the Stolper–Samuelson theorem, this obviously does not generalize to higher dimensions unless stronger assumptions are made or weaker conclusions reached. A discussion of the nature of such generalized results will be found in Kemp and Wegge (1969), Wegge and Kemp (1969), Ethier (1974), Jones and Scheinkman (1977) and Neary (1985).

5. INTERINDUSTRIAL RELATIONSHIPS AND OTHER REFINEMENTS. The formal model treated so far assumes that production is completely integrated, contrary to fact. Indeed, a large part of international trade is in intermediate products. The main justification for not allowing for intermediate inputs at the very beginning is that it may obscure the logic of the analysis with inessential details. However, in view of the importance of the phenomenon it is desirable at this point to see how the formal framework needs to be modified (cf. McKinnon, 1966; Melvin, 1969a, 1969b; Khang and Uekawa, 1973).

In place of (2.1) one needs to substitute the production function

$$q_j = f_j(u_{1j}, u_{2j}, \ldots, u_{nj}, v_{1j}, v_{2j}, \ldots, v_{mj}) = f_j(u_j, v_j) \qquad (5.1)$$

(assumed homogeneous of degree 1) where q_j denotes gross output of commodity j, and u_{ij} denotes the amount of commodity i used as input to the production of commodity j. Its dual minimum-unit-cost function – equal to the price of commodity j when that commodity is produced – is denoted

$$p_j = g_j(p_1, p_2, \ldots, p_n, w_1, w_2, \ldots, w_m) = g_j(p, w), \qquad (5.2)$$

and the input-output and factor-output coefficients are, in accordance with Theorem 5,

$$a_{ij}(p, w) = \partial g_j(p, w)/\partial p_i; \quad b_{ij}(p, w) = \partial g_j(p, w)/\partial w_i. \quad (5.3)$$

The production–possibility set (2.9) is now replaced by the *net-output-possibility set* defined by

$$\mathcal{Y}(l) = \Big\{ y \in E^n : \text{there exist allocations } u_{ij} \geqq 0, v_{kj} \geqq 0$$

$$(i, j = 1, 2, \ldots, n; k = 1, 2, \ldots, m) \text{ such that}$$

$$y_j = f_j(u_j, v_j) - \sum_{k=1}^{n} u_{jk} \quad \text{and} \quad \sum_{j=1}^{n} v_j \leqq l \Big\} \qquad (5.4)$$

(cf. Khang and Uekawa, 1973). This set is convex. Khang and Uekawa (1973) developed a generalization of Theorem 9 (section 3 above); see also Kemp, Khang and Uekawa (1978) and Färe (1979). The concept of a national-product function can also be generalized to this case (cf. Chipman, 1985a, pp. 405–6), allowing one to define net supply (Rybczynski) functions.

When all commodities are produced, the net outputs, gross outputs, and factor endowments are related by

$$y = [I - A(p, w)]q, \quad B(p, w)q \leqslant l \quad (5.5)$$

where $A = [a_{ij}]$, $B = [b_{ij}]$ are the $n \times n$ and $m \times n$ input–output and factor–output matrices. Accordingly, the resource-allocation constraint may be expressed as

$$C(p, w)y \leqslant l \quad \text{where} \quad C = B(I - A)^{-1}. \quad (5.6)$$

When all commodities are produced and traded, minimum-unit costs are equal to the prices hence

$$p = g(p, w) = A(p, w)'p + B(p, w)'w. \quad (5.7)$$

If the Jacobian matrix of this transformation, $I - A(p, w)'$, satisfies the Hawkins–Simon (1949) conditions of having positive principle minors, then by the results of Gale and Nikaido (1965), (5.7) defines a set of consolidated cost functions $\psi(w)$ which satisfy

$$p = \psi(w) = [I - A(\psi(w), w)]'^{-1}B(\psi(w), w)'w. \quad (5.8)$$

The set of production functions dual to these consolidated cost functions defines a set of integrated production functions $\phi_j(v_j)$ corresponding to the unintegrated functions (5.1). These could be validly used in place of (5.1) provided net outputs were required to be nonnegative.

A concept which has proved very useful in analysing trade in intermediate products is that of a *value-added production function* introduced by Khang (1971, 1973) and Bruno (1973). If one assumes that the production functions (5.1) are twice continuously differentiable and that all intermediate inputs are used in positive amounts, by setting the partial derivatives $\partial f_j / \partial u_{ij} = p_i/p_j$ for $i = 1, 2, \ldots, n$, one may define implicitly the functions

$$u_{ij} = \hat{u}_{ij}(v_j, p) \quad (5.9)$$

which, when substituted back into (5.1) yield the *value-added production functions*

$$V_j(v_j, p) = p_j f_j[\hat{u}_j(v_j, p), v_j] - \sum_{i=1}^{n} p_i \hat{u}_{ij}(v_j, p). \quad (5.10)$$

These are shown to inherit (for given p) the homogeneity and concavity properties of the original production functions (5.1), and in particular to satisfy the envelope conditions $\partial V_j / \partial v_{ij} = w_i$ and $\partial V_j / \partial p_j = \delta_{ij} q_j - u_{ij}$, where δ_{ij} is the Kronecker delta.

From the mn equations

$$\frac{\partial V_j}{\partial v_{ij}} - \frac{\partial V_{j+1}}{\partial v_{i,j+1}} = 0 \quad (i = 1, 2, \ldots, m; j = 1, 2, \ldots, n - 1),$$

$$\sum_{j=1}^{n} v_{ij} = l_i \quad (i = 1, 2, \ldots, m), \quad (5.11)$$

provided $m \geqslant n$ one can in general solve for the functions $\hat{v}_{ij}(p, l)$. The system of equations determining the $\partial \hat{v}_{ij} / \partial p_k$ has the form

$$\begin{bmatrix} J_1 & -J_2 & 0 & \ldots & 0 & 0 \\ 0 & J_2 & -J_3 & \ldots & 0 & 0 \\ \cdot & & \cdot & & & \\ 0 & 0 & 0 & \ldots & J_{n-1} & J_n \\ I & I & I & \ldots & I & I \end{bmatrix} \begin{bmatrix} \partial \hat{v}_{\cdot 1} / \partial p_k \\ \partial \hat{v}_{\cdot 2} / \partial p_k \\ \ldots \\ \partial \hat{v}_{\cdot n-1} / \partial p_k \\ \partial \hat{v}_{\cdot n} / \partial p_k \end{bmatrix}$$

$$= - \begin{bmatrix} K_1 - K_2 \\ K_2 - K_3 \\ \ldots \\ K_{n-1} - K_n \\ 0 \end{bmatrix} \quad (5.12)$$

where

$$J_k = \left(\frac{\partial^2 V_k}{\partial v_{ik} \partial v_{jk}} \right)_{i, j = 1, 2, \ldots, m}, \qquad K_j = \left(\frac{\partial^2 V_j}{\partial v_{ij} \partial p_k} \right)_{i = 1, 2, \ldots, m}$$

are $m \times n$ and $m \times 1$ (cf. Bruno, 1973, p. 215).

As an illustration one may take the case $m = n = 2$, and pose the question: will an increase in the price of commodity 1 result in a reallocation of resources to industry 1? This question is of interest from the point of view of the theory of 'effective protection' (cf. Johnson, 1965; Corden, 1966, 1971). In this case (5.12) reduces to

$$\begin{pmatrix} J_1 & -J_2 \\ I & I \end{pmatrix} \begin{pmatrix} \partial \hat{v}_{\cdot 1} / \partial p_1 \\ \partial \hat{v}_{\cdot 2} / \partial p_1 \end{pmatrix} = - \begin{pmatrix} K_1 - K_2 \\ 0 \end{pmatrix} \quad (5.13)$$

and we find that, as long as the two value-added production functions use distinct factor ratios,

$$\begin{pmatrix} J_1 & -J_2 \\ I & I \end{pmatrix}^{-1} = -(J_1 + J_2)^{-1} \begin{bmatrix} I & J_2 \\ -I & J_1 \end{bmatrix} \quad (5.14)$$

where the product on the right is interpreted as scalar multiplication of the partitioned matrix by the scalar $(J_1 + J_2)^{-1}$. The matrix $J_1 + J_2$ is negative definite with positive off-diagonal elements, hence by the Hawkins–Simon (1949) theorem $(J_1 + J_2)^{-1} < 0$. The solution of (5.13) is then

$$\begin{pmatrix} \partial \hat{v}_{\cdot 1} / \partial p_1 \\ \partial \hat{v}_{\cdot 2} / \partial p_1 \end{pmatrix} = -(J_1 + J_2)^{-1} \begin{pmatrix} K_1 - K_2 \\ K_2 - K_1 \end{pmatrix}. \quad (5.15)$$

Thus, of course $\partial \hat{v}_{\cdot 2} / \partial p_1 = -\partial \hat{v}_{\cdot 1} / \partial p_1$.

Concentrating on industry 1 we obtain

$$\begin{pmatrix} \partial \hat{v}_{11} / \partial p_1 \\ \partial \hat{v}_{21} / \partial p_1 \end{pmatrix}$$
$$= -(J_1 + J_2)^{-1} \begin{pmatrix} \partial^2 V_1 / \partial v_{11} \partial p_1 - \partial^2 V_2 / \partial v_{12} \partial p_1 \\ \partial^2 V_1 / \partial v_{21} \partial p_1 - \partial^2 V_2 / \partial v_{22} \partial p_1 \end{pmatrix}. \quad (5.16)$$

Note that in the case of integrated production the vector on the right in (5.16) reduces to $(\partial f_1 / \partial v_{11}, \partial f_1 / \partial v_{21})'$, and since these two terms are positive it follows that both factors will move into industry 1. Is the same true in the case of non-integrated production? For example, if the two industries are steel and autos, and an import quota is imposed on steel, is it possible that, owing to the decline in the auto industry's use of the costlier steel, one of the factors will move out of steel into autos? [An example of this general kind was given by Ramaswami and Srinivasan (1971), but in relation to a model in which the sole intermediate input is an imported good not produced at home. See also Jones (1971).] It was shown by Bruno that a sufficient condition for both factors to move into industry 1 is that the production functions (5.1) be functionally separable, i.e. of the form $f_i(u_j, v_j) = f_j[\phi_j(u_j), \psi_j(v_j)]$. For technical reasons (to assure invertibility of the individual J_k matrices) Bruno assumed decreasing returns to scale for some of his results. For extensions not requiring this see Uekawa (1979).

Other generalizations and refinements may be briefly mentioned. An important restriction that needs to be relaxed is the assumption of constant returns to scale. Variable returns to scale have been analysed by Jones (1968), Herberg and Kemp (1969), Melvin (1971), Negishi (1972) and others. Inoue (1981) applied the concept of parametric external economies of scale introduced in Chipman (1970), allowing for economies of scale to be compatible with existence of competitive equilibrium, and obtained generalizations of the Samuelson reciprocity relations. In particular, if the production function facing a particular firm v, among the N_j firms in industry j, is

$$y_{jv} = k_j f_j(v_{jv}) = k_j f_j(v_{1jv}, v_{2jv}, \ldots, v_{mjv}) \qquad (5.17)$$

where f_j is concave and homogeneous of degree 1 and $k_j = \phi_j(y_j)$ where

$$y_j = \sum_{v=1}^{N_j} y_{jv}$$

(i.e., the coefficient k_j depends on industrial output y_j but is treated as a parameter by each firm), then since each cost-minimizing firm will hire its factor inputs in the proportion $v_{jv} = \lambda_{jv} v_j$ to the industry input vector v_j, where

$$\lambda_{jv} > 0 \quad \text{and} \quad \sum_{v=1}^{N_j} \lambda_{jv} = 1,$$

one obtains from (5.17)

$$y_j = \sum_{v=1}^{N_j} y_{jv} = \phi_j(y_j) f_j(v_j). \qquad (5.18)$$

Choosing ϕ_j to be of the form $\phi_j(y_j) = y_j^{1 - 1/\rho_j}$ this becomes

$$y_j = f_j(v_j)^{\rho_j}. \qquad (5.19)$$

Inoue showed that the reciprocity relation (3.33) is replaced by the condition $\partial \hat{y}_j / \partial l_i = \rho_j \partial \hat{w}_i / \partial p_j$ and that the symmetry condition $\partial \hat{w}_i / \partial p_j = \partial \hat{w}_j / \partial p_i$ is retained (as is to be expected from the parametric behaviour of producers), but that $\rho_i \partial \hat{y}_i / \partial p_i = \rho_j \partial \hat{y}_j / \partial p_j$. He also obtained sufficient conditions for the Stolper–Samuelson and Rybczynski theorems to generalize to the case of external economies or diseconomies of scale.

Alternative approaches have been followed by Ethier (1979, 1982), Helpman (1981), and others who do not distinguish between internal and external economies and therefore leave open the question of existence of general equilibrium.

Other types of refinements that have been introduced include the attempt to allow for joint production; one may refer especially to Chang, Ethier and Kemp (1980).

Finally, many authors have relaxed the assumption of international immobility of factors. The properties of the world production–possibility frontier in this case have been investigated by Chipman (1971), Uekawa (1972) and Otani (1973). The relation between capital mobility and technology transfer has been analysed by McCulloch and Yellen (1982) and Chipman (1982a). This is only a small part of a growing literature.

6. TRADE–DEMAND AND TRADE–UTILITY FUNCTIONS. A common simplifying assumption in international-trade theory is that consumer preferences can be aggregated. This assumption makes it possible to define a country's offer function in a simple way, as first indicated (in the case in which all goods are traded) by Meade (1953), and subsequently by Rader (1964), Chipman (1974a, p. 34n; 1979) and Woodland (1980). An extension to the case in which some goods are nontradable was derived by Chipman (1981).

Let us assume that a country produces n_1 tradable and n_3 nontradable commodities and imports an additional n_2 commodities which it does not produce. Let $\mathbf{x}_r$ denote the $n_r \times 1$ vector of quantities consumed in the rth category, and $\mathbf{p}_r$ the corresponding price vector, and let $\mathcal{X} \subseteq E^{n_1 + n_2 + n_3}$ denote the consumption set. If aggregate consumption $\mathbf{x} = (\mathbf{x}_1, \mathbf{x}_2, \mathbf{x}_3) \in \mathcal{X}$ is generated by maximization of an aggregate utility function

$U(\mathbf{x}_1, \mathbf{x}_2, \mathbf{x}_3)$ subject to a budget constraint

$$\sum_{r=1}^{3} \mathbf{p}_r \cdot \mathbf{x}_r \leqslant Y,$$

where Y is disposable national income, when the contours of U are strictly convex to the origin this yields a single-valued demand function $\mathbf{x} = \mathbf{h}(\mathbf{p}_1, \mathbf{p}_2, \mathbf{p}_3, Y)$. Let $\mathcal{Y}(l) \subseteq E^{n_1} \times \{0\}^{n_2} \times E_+^{n_3}$ denote the production–possibility set. The *trade set* may be defined as the set $\mathcal{Z}(l) = [\mathcal{X} - \mathcal{Y}(l)] \cap E^{n_1 + n_2} \times \{0\}^{n_3}$. For $z \in \mathcal{Z}(l)$ the *trade–utility function* may be defined as

$$\hat{U}(\mathbf{z}_1, \mathbf{z}_2; l) = \max_{\mathbf{x} - \mathbf{z} \in \mathcal{Y}(l)} U(\mathbf{x}), \qquad (6.1)$$

and the *trade-demand correspondence* by

$$\hat{\mathbf{h}}(\mathbf{p}_1, \mathbf{p}_2, D; l) = \{(\mathbf{z}_1, \mathbf{z}_2): \quad \mathbf{z} \in \mathcal{Z}(l)$$
$$\text{maximizes } \hat{U}(\mathbf{z}_1, \mathbf{z}_2; l)$$
$$\text{subject to } \mathbf{p}_1 \cdot \mathbf{z}_1 + \mathbf{p}_2 \cdot \mathbf{z}_2 \leqslant D\}, \qquad (6.2)$$

where D is the deficit in the balance of payments on goods and services.

When the production–possibility frontier $\hat{\mathcal{Y}}(l)$ is strictly concave to the origin, and the contours of $U(\mathbf{x})$ are strictly convex to the origin, the contours of $\hat{U}(\mathbf{z})$ are strictly convex to the origin, and $\hat{\mathbf{h}}$ becomes a single-valued function. When $n_3 = 0$,

$$\hat{\mathbf{h}}(\mathbf{p}, D; l) = \mathbf{h}[\mathbf{p}, \Pi(\mathbf{p}, l) + D] - \hat{\mathbf{y}}(\mathbf{p}, l). \qquad (6.3)$$

When this function is twice differentiable it has the properties that $\partial \hat{h}_i / \partial D = \partial h_i / \partial Y$ and

$$\hat{s}_{ij} \equiv \frac{\partial \hat{H}_i}{\partial p_j} + \frac{\partial \hat{h}_i}{\partial D} \hat{h}_j = \frac{\partial h_i}{\partial p_j} + \frac{\partial h_i}{\partial Y} h_j - \frac{\partial \hat{y}_i}{\partial p_j} \equiv s_{ij} - t_{ij}, \qquad (6.4)$$

where s_{ij} and $\hat{s}_{ij}$ denote the Slutsky substitution terms of the demand and trade–demand functions respectively, and t_{ij} is the transformation term.

If there are K countries, and a k superscript indicates the country, world equilibrium is defined by

$$\sum_{k=1}^{K} \mathbf{h}^k(\mathbf{p}_1, \mathbf{p}_2, D^k; l^k) = 0, \quad \text{where} \quad \sum_{k=1}^{K} D^k = 0. \qquad (6.5)$$

7. MARSHALLIAN OFFER FUNCTIONS AND DYNAMIC STABILITY. The Marshallian offer functions may be derived in a straightforward fashion from the trade–demand functions. Suppose two countries are trading two commodities, and that country k is exporting commodity k and importing commodity $j \neq k$. Assuming constant factor endowments (and thus ignoring the dependence of the functions on these endowments), each country's exports may be expressed as a single-valued function of its imports, provided the import good is non-inferior. In the case of country 1, denoting $r_2 = p_2 / p_1$ and $d^1 = D^1 / p_1$, provided commodity 2 is non-inferior, so that $\partial \hat{h}_2^1 / \partial p_2 < 0$, one may define its inverse trade-demand function $\hat{r}_2(z_2^1, d^1)$ implicitly by

$$\hat{h}_2^1[1, \hat{r}_2(z_2^1, d^1), d^1] = z_2^1, \qquad (7.1)$$

and its *trade function* (for $d^1 = 0$, its Marshallian reciprocal demand function) by

$$-z_1^1 = F^1(z_2^1, d^1) = \hat{r}_2(z_2^1, d^1) z_2^1 - d^1. \qquad (7.2)$$

Likewise for country 2, denoting $r_1 = p_1 / p_2$ and $d^2 = D^2 / p_2$, one defines in a similar way its inverse trade-demand function $\hat{r}_1$ by

$$\hat{h}_1^2[\hat{r}_1(z_1^2, d^2), 1, d^2] = z_1^2 \qquad (7.3)$$

and its trade function by

$$-z_2^2 = F^2(z_1^2, d^2) = \hat{r}_1(z_1^2, d^2) z_1^2 - d^2. \qquad (7.4)$$

Since $d^1 = -z_1^2 d^2/(d^2 + z_2^1)$ from the balance-of-payments constraint and the condition $D^1 = -D^2$, one may express the trade functions in the form

$$z_1^2 = Z_1(z_1^2, d^2) = F^1\left(z_2^1, \frac{-z_1^2 d^2}{d^2 + z_2^1}\right)$$

$$z_2^1 = Z_2(z_1^2, d^2) = F^2(z_1^2, d^2). \tag{7.5}$$

Following Alexander (1951), the *elasticities of trade* may be defined by

$$\alpha_1 = \frac{z_2^1}{Z_1}\frac{\partial Z_1}{\partial z_2^1}, \quad \alpha_2 = \frac{z_1^2}{Z_2}\frac{\partial Z_2}{\partial z_1^2}. \tag{7.6}$$

These may be related to the Marshallian elasticities of demand for imports

$$\eta^1 = -\frac{p_2}{\hat{h}_2^1}\frac{\partial \hat{h}_2^1}{\partial p_2}, \quad \eta^2 = -\frac{p_1}{\hat{h}_1^2}\frac{\partial \hat{h}_1^2}{\partial p_1} \tag{7.7}$$

in the following way. For country 2, (7.3) gives $\partial \hat{r}_1/\partial z_1^2 = 1/(\partial \hat{h}_1^2/\partial p_1)$, hence computing α_2 from (7.4) we find that

$$\alpha_2 = \left(1 + \frac{d^2}{z_2^1}\right)\left(1 - \frac{1}{\eta^2}\right) \tag{7.8}$$

hence

$$\eta^2 = \frac{1 + d^2/z_2^1}{1 + d^2/z_2^1 - \alpha_2} = \frac{z_2^1 + d^2}{z_2^1 + d^2 - z_1^2 \dfrac{\partial F^2}{\partial z_1^2}}. \tag{7.9}$$

The third expression in (7.9) provides a convenient way to read off the value of the elasticity from a diagram of the displaced Marshallian offer curve (see Figure 10).

Marshall (1879), following the outlines sketched by Mill (1848, vol. II, book III, ch. XXI, §1), introduced a dynamic process of adjustment which was formalized by Samuelson (1947, p. 266) as

$$\dot{z}_1^2 = \phi_1[Z_1(z_2^1, d^2) - z_1^2] \equiv P_1(z_1^2, z_2^1, d^2)$$

$$\dot{z}_2^1 = \phi_2[Z_2(z_1^2, d^2) - z_2^1] \equiv P_2(z_1^2, z_2^1, d^2) \tag{7.10}$$

(for the case $d^2 = 0$), where the ϕ_i are sign-preserving functions. Intuitively, if at a given level of import volume a country offers more units of the export good than it is currently exporting, it will increase its exports. According to Marshall (1923, p. 341), an excess of exports offered over current exports is an indication that profits are being made in the export industry, whence it will expand. Marshall's description of the rationale for this process was very terse, but a good discussion will be found in Amano (1968).

A complete description and classification of the stability properties of (7.10) appears to be lacking in the literature, and one will be briefly supplied here, for the case $d^2 = 0$. The methods can be extended readily to the case $d^2 \neq 0$, but the conditions become considerably more complex. Another simplification in the ensuing development (which does not affect stability conditions but only the shapes of the trajectories) will be to assume that $\phi_1 = \phi_2$, and for notational simplicity they will be taken to be identity functions $\phi_i(u_i) = u_i$.

An equilibrium is defined as a pair of values $\bar{z}_1^2, \bar{z}_2^1$ for which $P_i(\bar{z}_1^2, \bar{z}_2^1) = 0$, $i = 1,2$. Denoting deviations from these equilibrium values by $u_1 = z_1 - \bar{z}_1^2, u_2 = z_2^1 - \bar{z}_2^1$, and taking first-order Taylor approximations of the functions P_i around such an equilibrium, we obtain the system

$$\begin{pmatrix} \dot{u}_1 \\ \dot{u}_2 \end{pmatrix} = \begin{pmatrix} -1 & \gamma_1 \\ \gamma_2 & -1 \end{pmatrix}\begin{pmatrix} u_1 \\ u_2 \end{pmatrix} \tag{7.11}$$

where

$$\gamma_1 = \partial F^1/\partial z_2^1 \quad \text{and} \quad \gamma_2 = \partial F^2/\partial z_1^2 \tag{7.12}$$

and these derivatives are evaluated at the equilibrium point $(\bar{z}_1^2, \bar{z}_2^1)$.

Writing (7.11) in matrix notation $\dot{u} = Au$, it is well known and easily shown that if the characteristic values of A are distinct, its characteristic vectors are linearly independent. Letting V be the matrix whose columns are these characteristic vectors, we have $V^{-1}AV = \Lambda$, where Λ is the diagonal matrix of characteristic values (roots) λ_i. Defining $u^* = V^{-1}u$, we have $\dot{u}^* = V^{-1}\dot{u} = V^{-1}Au = V^{-1}AVu^* = \Lambda u^*$, whence $\dot{u}_i^* = \lambda_i u_i^*$. Each of these differential equations is solved to obtain $u_i^* = b_i e^{\lambda_i t}$, hence the desired solution is

$$u = Vu^* = \sum_{i=1}^{2} v^i b_i e^{\lambda_i t}$$

where v^i is the ith column of V. Assuming (to obtain a simple normalization) that the top row of V has no zero elements, they may be chosen equal to unity and we have

$$\begin{pmatrix} u_1 \\ u_2 \end{pmatrix} = \begin{pmatrix} 1 \\ v_1 \end{pmatrix} b_1 e^{\lambda_1 t} + \begin{pmatrix} 1 \\ v_2 \end{pmatrix} b_2 e^{\lambda_2 t}, \tag{7.13}$$

which substitutes a new coordinate system defined by the characteristic vectors (the columns of V as opposed to those of the identity matrix). These tilted coordinate axes spanned by the characteristic vectors are called 'separatrices' and the v_i's are called 'distribution coefficients' (cf. Andronov, Vitt and Khaikin, 1966, p. 258). Solving

$$\begin{pmatrix} \lambda - a_{11} & -a_{12} \\ -a_{21} & \lambda - a_{22} \end{pmatrix}\begin{pmatrix} 1 \\ v \end{pmatrix} = \begin{pmatrix} 0 \\ 0 \end{pmatrix} \tag{7.14}$$

we obtain for the characteristic values

$$\lambda = \frac{a_{11} + a_{22} \pm \sqrt{(a_{11} - a_{22})^2 + 4a_{21}a_{12}}}{2} = -1 \pm \sqrt{\gamma_1 \gamma_2} \tag{7.15}$$

and for the distribution coefficients

$$v = \frac{a_{22} - a_{11} \pm \sqrt{(a_{11} - a_{22})^2 + 4a_{21}a_{12}}}{2a_{12}} = \frac{\pm \sqrt{\gamma_1 \gamma_2}}{\gamma_1}. \tag{7.16}$$

First we may consider stability conditions. Asymptotic stability, in the sense $\lim_{t \to \infty} u_i(t) = 0$, requires that the real parts of both λ_i be negative. If the offer curves intersect with one positive and one negative slope, then $\gamma_1 \gamma_2 < 0$ and the real parts of both roots are -1. Instability can therefore only occur when both roots are real. (The intermediate case of repeated roots, for which special methods are required, will not be taken up here.)

If the offer curves intersect with both positive or both negative slopes, instability can occur only if $\gamma_1 \gamma_2 \geqslant 1$ hence the stability condition is, using (7.12), and (7.5) to (7.8),

$$1 > \gamma_1 \gamma_2 = \alpha_1 \alpha_2 = \left(1 - \frac{1}{\eta_1}\right)\left(1 - \frac{1}{\eta_2}\right) = \frac{(\eta^1 - 1)(\eta^2 - 1)}{\eta^1 \eta^2}. \tag{7.17}$$

In the form $\gamma_1 \gamma_2 < 1$, this stability condition was first obtained by Marshall (1923, p. 353). Since it has already been assumed that both goods are normal, the η^k are both positive and (7.17) reduces to the well-known condition

$$\eta^1 + \eta^2 - 1 > 0. \tag{7.18}$$

This condition was described by Hirschman (1949) as the 'Marshall–Lerner condition', but this must be characterized as one of the great misnomers of the theory of international trade. A condition formally equivalent to (7.18) was described by Lerner (1944, p. 378), but it referred to a model originated by

Bickerdike (1907, 1920) which may be interpreted as referring to economies which specialize in an exportable and a non-tradable good which are produced with a single factor of production (cf. Chipman, 1978, p. 67). This is described as the case of 'infinite elasticity of supply of exports'. As it happens, for that particular model the Bickerdike–Lerner elasticities coincide with the Marshallian ones, but in any case formula (7.18) was, for this case, already derived by Bickerdike (1920, p. 121) in his own idiosyncratic notation. As for Marshall, the passage for which (7.18) is attributed to him considers a case of near-neutral equilibrium (in which the offer curves approximately coincide with negative slope) and states concerning such conditions (1923, p. 354):

> they assume the total elasticity of demand of each country to be less than unity, and on the average to be less than one half, throughout a large part of its schedule.

A statement of conditions of instability no less explicit than this was already in Mill (1848, vol. II, book III, ch. XXI, §1, p. 158):

> until the increased cheapness of English goods induces foreign countries to take a greater pecuniary value, or until the increased dearness (positive or comparative) of foreign goods makes England take a less pecuniary value, the exports of England will be no nearer to paying for the imports than before...

The term 'Mill–Marshall condition' might be a more appropriate one to describe (7.18).

The various cases may now be classified. First if $\gamma_1 > 0$ and $\gamma_2 > 0$, then taking $\lambda_1 = -1 - \sqrt{\gamma_1 \gamma_2}$ and $\lambda_2 = -1 + \sqrt{\gamma_1 \gamma_2}$, where $\gamma_1 \gamma_2 < 1$ (i.e. $\gamma_2 < 1/\gamma_1$ – see Figure 11), we have $v_1 = -\sqrt{\gamma_2/\gamma_1}$ and $v_2 = \sqrt{\gamma_2/\gamma_1}$ and $\gamma_2 < v_2 < 1/\gamma_1$. Thus, country 1's offer curve is steeper than country 2's at the equilibrium point, and the separatrices pass between these two curves [this latter property no longer holds if $\phi_1' \neq \phi_2'$ in (7.10)]. Figure 11 displays the new coordinate axes given by the characteristic vectors of (7.13). The trajectory of u is a linear combination of trajectories of points moving along these axes towards the equilibrium point. From the above differential equations $\dot{u}_i^* = \lambda_i u_i^*$ we obtain

$$\frac{du_2^*}{du_1^*} = \frac{\lambda_2 u_2}{\lambda_1 u_1} \quad \text{with solution} \quad u_2^* = C|u_1^*|^{\lambda_2/\lambda_1} \quad (7.19)$$

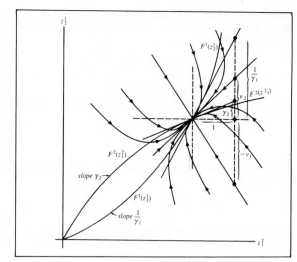

Figure 11

which gives the equation of family of a parabolas in the (u_1^*, u_2^*) plane. The transformation $u = Vu^*$ maps these into distorted parabolas as shown in Figure 11. This equilibrium is classified as a *stable node*.

The second case to be considered, corresponding to the northwest equilibrium point in Figure 12, has $\gamma_1 < 0$, $\gamma_2 < 0$, and $\gamma_1 \gamma_2 < 1$, so that $-1/\gamma_1 > -\gamma_2$ (i.e. country 1's offer curve is steeper than country 2's). With the λ_i as before we now have $v_1 = \sqrt{\gamma_2/\gamma_1}$ and $v_2 = -\sqrt{\gamma_2/\gamma_1}$ so that $-\gamma_2 < -v_2 < -1/\gamma_1$. This case is also one of a *stable node*. The difference is that in the previous case the base of the parabolas was the positively-sloped separatrix whereas in this case it is the negatively-sloped separatrix.

The third case is the intermediate unstable equilibrium shown in Figure 12, where $\gamma_1 < 0$, $\gamma_2 < 0$, and $-1/\gamma_1 < -\gamma_2$ so that $\gamma_1 \gamma_2 > 1$. The λ_i and v_i are as in the immediately preceding case, and the differential equation and solution (7.19) still hold, but this time it describes a family of hyperbolae. The separatrix with positive slope $v_1 = \sqrt{\gamma_2/\gamma_1}$ is stable, and the one with negative

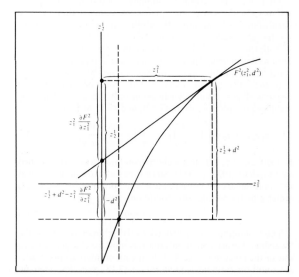

Figure 10

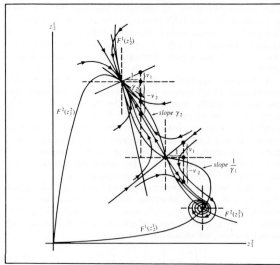

Figure 12

slope $v_2 = -\sqrt{\gamma_2/\gamma_1}$ is unstable. This unstable equilibrium is a *saddle-point*.

The fourth and final case is that in which $\gamma_1 > 0$ and $\gamma_2 < 0$, as in the southeast equilibrium point in Figure 12. The roots are complex, and the equilibrium is a *stable focus* or spiral, with movement towards equilibrium in the clockwise direction. In the case (not shown) $\gamma_1 < 0$ and $\gamma_2 > 0$, the direction of movement would be contraclockwise.

Marshall (1923, p. 353) considered another possible case of unstable equilibrium, in which $\gamma_1 > 0$, $\gamma_2 > 0$, and $\gamma_1\gamma_2 > 1$ hence $\gamma_2 > 1/\gamma_1$ (the country 2's offer curve is steeper than country 1's). Such a case, if it were possible, would be one of an *unstable node*.

The Marshallian dynamic adjustment mechanism is what is described as a 'non-tâtonnement process', in which trading takes place out of equilibrium. Alternative adjustment processes have been analysed by Jones (1961, 1974b), Kemp (1964, pp. 66–9), Chipman (1978) and others.

PART 2. THE APPLIED THEORY

8. THE EXPLANATION OF TRADE FLOWS. If there is any one thing that could legitimately be demanded of a theory of international trade, it is that it should be capable of explaining observed patterns and flows of trade among countries. The great strides in the subject have come with the development of new principles that provide such an explanation.

The principle of comparative advantage originated with Thornton (1802) whose problem was to explain why a country would lose gold reserves, i.e. why it would export gold. Gold outflows, he reasoned, would take place when (1802, p. 129; 1939, p. 150):

> goods, in comparison with gold coin, are made dear. The goods which are dear remain, therefore, in England; and the gold coin, which is cheap . . . , goes abroad.

The importance of *comparative* cheapness of gold was reiterated in Thornton's summary of the basic principles of his work (1802, pp. 76–7; 1939, p. 247):

> I would be understood to say, that in a country in which *coin alone* circulates, if, through any accident the quantity should become greater in proportion to the goods which it has to transfer than it is in other countries, the coin becomes cheap as compared with goods, or, in other words, that goods become dear as compared with coin, and that a profit on the exportation of coin arises.

This principle was absorbed into Ricardo's early work (1811; 1951, pp. 56–7), so much so that it caused Malthus (1811a, p. 341) to praise it for

> the doctrine, that excess and deficiency of currency are only *relative* terms; that the circulation of a country can never be superabundant, except in relation to other countries.

The principle was extended by Torrens (1815) and Ricardo (1817) to the explanation of trade in commodities other than money. This theory held sway for over a century, but since it was combined with a labour theory of value implying that countries' cost ratios were fixed, it led to the uncomfortable conclusion that either countries would specialize, or some countries' cost ratios would dictate world price ratios (cf. Graham, 1923). Moreover, the cost ratios of the various countries were taken as data, not in need of further explanation.

The first attempt to probe into the reasons for differences in comparative costs was evidently that of Heckscher, who stated (1919; 1949, p. 278):

A difference in the relative scarcity of the factors of production between one country and another is thus a necessary condition for a difference in comparative costs and consequently for international trade.

Heckscher did not say precisely what he meant by 'comparative costs', though he stressed the criterion of different relative factor rentals (presumably under autarky); nor did he make any specific statement concerning how differences in relative factor endowments (which he assumed fixed) would determine the precise pattern of trade. But he did completely anticipate Lerner's (1933) and Samuelson's (1949) theorems that with identical productive techniques trade would equalize factor rentals among countries.

Haberler (1930, 1933) liberated the classical theory from the labour theory of value by introducing the concept of a strictly-concave-to-the-origin production–possibility frontier, based on the allocation of factors (assumed in fixed total supply) among industries. Most of the subsequent formal development of the theory by Lerner (1932, 1933), Leontief (1933), and Meade (1952) was built on Haberler's concept.

Ohlin (1933, p. 29) summarized his theory as follows:

> The first condition of trade is that some goods can be produced more cheaply in one region than another. In each of them the cheap goods are those containing relatively great quantities of the factors cheaper than in other regions. These cheap goods make up exports, whereas goods which can be more cheaply produced in the other regions are imported. We may say, therefore, that exports are in each region composed of articles into the production of which enter large quantities of cheap factors.

The criterion of 'cheapness' was interpreted by Jones (1956) as referring to the pre-trade (autarkic) factor rentals in the respective countries. However, this appears to be directly contradicted by a statement of Ohlin's in the paragraph immediately following the above-cited passage:

> When reasoning like this we must, however, bear in mind one thing: whether a factor is cheaper or dearer in region A than in region B can be ascertained only when an exchange rate between the two countries has been established . . .

This appears to be a confusion, since it would lead to the conclusion that if the Heckscher–Lerner–Samuelson factor–rental–equalization theorem (which Ohlin rejected) holds, then there will be no trade!

Fruitful progress in any field requires one to filter truth out of error in theories that are carelessly stated yet contain important insights. Jones (1956) filtered out two logically distinct propositions both of which were given the name 'Heckscher–Ohlin theorem' and state that a country will export that commodity which is produced with relatively large amounts of its relatively abundant factor. The propositions differ in the definition of 'relative abundance': the *physical definition* (differences in relative factor endowments) and the *price definition* (differences in ratios of autarkic factor rentals) (see also Bhagwati, 1967).

The trouble with the 'price definition' of relative factor abundance is that it leads to a theory that is practically devoid of empirical content. In an internationally trading economy, autarkic prices are not observed. The theory would explain or predict observable trade patterns by data which cannot be observed. There is, moreover, a logical problem with the 'price definition', pointed out by Inada (1967): since competitive equilibrium is in general not unique, pre-trade factor rentals are not unique, hence the proposition is both ambiguous and false unless conditions are postulated that guarantee uniqueness (such as identical homothetic preferences).

The proposition under the 'physical definition' of factor abundance also holds under quite limited circumstances. A minimal set of sufficient conditions is the following: (1) there are two countries, two commodities – both tradable at zero transport costs – and two factors of production – both perfectly mobile between industries within countries but completely immobile between countries; (2) production functions are identical between countries, obey constant returns to scale and concavity, and use different ratios of factor inputs for all factor rentals; (3) preferences within and as between countries are identical and homothetic; and (4) trade is balanced. If any one of these conditions is omitted, the theorem can be shown to be false. It is not even clear how the theorem should be stated if condition (1) is generalized (but this will be discussed below). Only with (2) can one conclude that, at all price ratios, one country will produce a larger ratio of commodity 1 to commodity 2 than the other. Without (3), inhabitants of each country may have a strong relative preference for the commodity which that country produces in relative abundance, so that each country ends up importing rather than exporting that commodity. Without (4), a country that is borrowing or receiving a unilateral transfer from another may import both commodities from the other country.

A formal proof of the 'Heckscher–Ohlin theorem' (physical version) proceeds as follows (cf. Riezman, 1974). Let x_j^k, y_j^k, $z_j^k = x_j^k - y_j^k$ denote consumption, production, and net import of commodity j in country k and let l_i^k denote country k's endowment in factor i. Suppose $l_1^1/l_2^1 > l_1^2/l_2^2$, and assume that at all factor rentals w_1, w_2, the factor–output ratios satisfy $b_{11}(w)/b_{21}(w) > b_{12}(w)/b_{22}(w)$. Then from the Rybczynski theorem it follows that $\hat{y}_1(p_1, p_2, l_1^k, l_2^k)/\hat{y}_2(p_1, p_2, l_1^k, l_2^k)$ is an increasing function of l_1^k/l_2^k, hence at all (including equilibrium) prices, $y_1^1/y_2^1 > y_1^2/y_2^2$. From identical homothetic preferences, at all (including equilibrium) prices, $x_1^1/x_2^1 = x_1^2/x_2^2$. Now suppose by way of contradiction that country 1 does not export commodity 1, i.e., $z_1^1 \geqq 0$, or $x_1^1 \geqq y_1^1$. Then from balanced trade, $z_2^1 \leqq 0$, or $x_2^1 \leqq y_2^1$, hence $x_1^1/x_2^1 \geqq y_1^1/y_2^1$. Since $z_j^1 + z_j^2 = 0$ from free tradability of commodities at zero transport costs, a similar argument for country 2 shows that $y_1^2/y_2^2 \geqq x_1^2/x_2^2$. So $x_1^1/x_2^1 \geqq y_1^1/y_2^1 > y_1^2/y_2^2 \geqq x_1^2/x_2^2$, violating the equality $x_1^1/x_2^1 = x_1^2/x_2^2$. This contradiction proves the result.

Apparently the first attempt to subject the Heckscher–Ohlin theory to empirical test was that of Leontief (1953). It is apparent from the above that as long as a precise statement of the theory for more than two commodities, factors, and countries is lacking, the problem of testing it empirically is elusive at best.

The most prominent attempt to generalize the theory is that of Vanek (1968), which has been followed up by Leamer (1980). The ensuing summary will follow Leamer's treatment, somewhat generalized. Let A and B be $n \times n$ and $m \times n$ matrices (which in general depend on prices and factor rentals) of input–output and factor–output coefficients (assumed identical among countries), and let $C = B(I - A)^{-1}$ denote the integrated factor–output matrix. It is assumed that all n goods are traded at zero transport costs, and that a world equilibrium exists with equalization of factor rentals (which requires one effectively to assume $n \geqq m$), so that it makes sense to aggregate factor endowments over countries. The world consumption, production, and factor-endowment vectors are denoted

$$x = \sum_{k=1}^{K} x^k, \quad y = \sum_{k=1}^{K} y^k, \quad \text{and} \quad l = \sum_{k=1}^{K} l^k.$$

Finally, it is assumed that preferences are identical and homo-

thetic within and as between countries; this assures that whatever be the world prices, commodities will be consumed in the same proportion in all countries, and therefore $x^k = \alpha_k x$ where

$$\alpha_k > 0, \qquad \sum_{k=1}^{K} \alpha_k = 1.$$

Since world equilibrium requires $x = y$, we have $x^k = \alpha_k y$.

The vector of 'net imports of factor i by country k' is defined as

$$\hat{l}^k = Cz^k = C(x^k - y^k) = C(\alpha_k y - y^k) = \alpha_k l - l^k, \qquad (8.1)$$

where use is made of the full-employment condition $Cy^k = l^k$. Note that $\hat{l}^k$ is unique, even though y^k is not unique when $n > m$. Country k is said to be relatively well endowed in factor i relative to factor j if $l_i^k/l_j^k > l_i/l_j$. Leamer establishes a number of propositions:

Proposition 1: (Leamer's Corollary 2). If country k is a net exporter of factor i and a net importer of factor j, then it is relatively well endowed in factor i relative to factor j, i.e., $\hat{l}_i^k < 0$ and $\hat{l}_j^k > 0$ imply $l_i^k/l_j^k > l_i/l_j$.

This is proved by noting from (8.1) that $\hat{l}_i^k = \alpha_k l_i - l_i^k < 0$ implies $l_i^k > \alpha_k l_i$ and similarly $\hat{l}_j^k = \alpha_k l_j - l_j^k > 0$ implies $l_j^k < \alpha_k l_j$ hence $l_i^k/l_j^k > \alpha_k > l_j^k/l_j$.

Note that this generalizes not the Heckscher–Ohlin theorem but its converse. One would like to be able to say that $l_i^k/l_j^k > l_i/l_j$ implies $\hat{l}_i^k < 0$ and $\hat{l}_j^k > 0$, but this is in fact not true. An example used by Leamer to establish a different point can be used to establish this one as well. Let

$$C = \begin{pmatrix} 4 & 1 & 1 \\ 3 & 2 & 0.5 \\ 1 & 0 & 3 \end{pmatrix}, \quad y^k = \begin{pmatrix} 8 \\ 16 \\ 5 \end{pmatrix}, \quad y = \begin{pmatrix} 12 \\ 68 \\ 52 \end{pmatrix}, \quad p = \begin{pmatrix} 1 \\ 1 \\ 1 \end{pmatrix}.$$

Then country k and world factor endowments are

$$l^k = (53, 58.5, 23)' \quad \text{and} \quad l = (168, 198, 168)'$$

respectively, hence $l_1^k/l_1 = 0.32 > 0.30 = l_2^k/l_2$. However, national and world income are $p \cdot y^k = 29$ and $p \cdot y = 132$ respectively, giving $\alpha_k = 0.22$, hence $x^k = (2.64, 14.94, 11.42)'$ and thus $z^k = (-5.36, -1.06, 6.42)'$ (country k exports commodities 1 and 2 and imports commodity 3). It follows from (8.1) that $\hat{l}^k = (-16.1, -15.0, 13.9)'$ hence country k is a net exporter of both factors 1 and 2 and a net importer of factor 3.

The Vanek–Leamer 'generalization' of the Heckscher–Ohlin theorem is thus disappointing on two counts: first, it replaces the problem of explaining trade flows in actual commodities by that of explaining flows of abstract amounts of factors of production 'embodied' in the trade flows; secondly, it reverses the problem and uses the abstract embodied trade flows to explain or predict (or to 'reveal') the relative abundance of factors within countries. There is still a third cause for concern: the terminology 'amounts of factor services embodied in goods traded' to describe the empirically measurable entities (8.1) is justifiable only under the very special assumptions of the model, especially those of identical homothetic preferences and international equalization of factor rentals. For the description of the variables of a model to be valid only under the special assumptions of the model appears to be poor scientific practice. This is just a terminological objection, but words can be treacherous and terminology can lead one astray.

Proposition 2: (Leamer's Corollary 1). Country k is revealed to be relatively well endowed in factor i relative to factor j (i.e., $l_i^k/l_j^k > l_i/l_j$) if and only if $l_i^k/l_j^k > (l_i + \hat{l}_i^k)/(l_j^k + \hat{l}_j^k)$.

This follows very simply from the fact that $l_i^k + \hat{l}_i^k = \alpha_k l_i$, from (8.1). Since $l^k + \hat{l}^k = Cx^k$, the entities $l_i^k + \hat{l}_i^k$ are interpreted as 'the amount of factor i embodied in country k's consumption'. The significance of this proposition lies in the fact that one can infer a country's relative abundance (compared with the world) in one factor relative to another from data on this country's endowments, technology matrix and trade alone. It is rather remarkable that one should be able to do this without any data on factor endowments in the rest of the world. It shows the power of the assumptions of identical homothetic preferences and international equalization of factor rentals; but by the same token it leads one to be wary of assumptions that can provide so much information.

Leamer next considers Leontief's method of inferring a country's relative endowment ratios from trade, endowment, and technology data. Leontief considered imports and exports separately. Accordingly, one may define

$$z_j^k(+) = \max(z_j^k, 0), \qquad z_j^k(-) = -\min(z_j^k, 0) \qquad (8.2)$$

and

$$z^k(+) = [z_1^k(+), \ldots, z_n^k(+)]',$$
$$z^k(-) = [z_1^k(-), \ldots, z_n^k(-)]'. \qquad (8.3)$$

Then $z^k(+)$ is the vector of absolute values of imports, and $z^k(-)$ the vector of absolute values of exports. Then one may define

$$\hat{l}^k(+) = Cz^k(+), \qquad \hat{l}^k(-) = Cz^k(-), \qquad (8.4)$$

where $\hat{l}^k(+)$ and $\hat{l}^k(-)$ are the vectors of factor services embodied in gross imports and gross exports respectively. These vectors satisfy

$$z^k = z^k(+) - z^k(-), \qquad \hat{l}^k = \hat{l}^k(+) - \hat{l}^k(-). \qquad (8.5)$$

Leontief's criterion (1953, p. 343) for country k to be relatively well endowed in factor i relative to factor j was, in effect (for $i \neq j$),

$$\frac{\hat{l}_i^k(-)}{\hat{l}_j^k(-)} > \frac{\hat{l}_i^k(+)}{\hat{l}_j^k(+)}. \qquad (8.6)$$

This states that the ratio of amounts of factor i and factor j embodied in gross exports should be greater than the ratio of amounts of factor i and factor j embodied in gross imports.

Leamer's criticism is that the appropriate criterion is instead

$$\frac{l_i^k}{l_j^k} > \frac{l_i^k + \hat{l}_i^k}{l_j^k + \hat{l}_j^k}, \qquad (8.7)$$

i.e., that the endowment of factor i relative to that of factor j should exceed the ratio of the amounts of factors i and j embodied in consumption, since by Proposition 2 the fraction on the right in (8.7) is equal to the ratio of world endowments l_i/l_j. The numerical illustration referred to above was used by Leamer to show that these inequalities can conflict when, as in the case of Leontief's data, the amounts of $\hat{l}_i^k$, $\hat{l}_j^k$ of factors i and j embodied in net imports are both negative. Leamer showed that Leontief's procedure would be correct if $\hat{l}_i^k$ and $\hat{l}_j^k$ had opposite sign.

Proposition 3: (Leamer's Corollary 3). If $\hat{l}_i^k$, $\hat{l}_j^k$ have opposite sign then inequalities (8.6) and (8.7) are equivalent.

The method of proof is to show, first, that if $\hat{l}_i^k < 0$ and $\hat{l}_j^k > 0$ then both (8.6) and (8.7) hold, whereas if $\hat{l}_i^k > 0$ and $\hat{l}_j^k < 0$ then the inequalities opposite to both (8.6) and (8.7) hold. If $\hat{l}_i^k < 0$ and $\hat{l}_j^k > 0$ then $\hat{l}_i^k l_j^k < \hat{l}_j^k l_i^k$ hence $l_j^k \alpha_k l_i = l_i^k(l_j^k + \hat{l}_j^k) > l_j^k(l_i^k + \hat{l}_i^k) = l_i^k \alpha_k l_j$, establishing (8.7). Further, $\hat{l}_i^k = \hat{l}_i^k(+) - \hat{l}_i^k(-) < 0$ implies $\hat{l}_i^k(-)/\hat{l}_i^k(+) > 1$ and $\hat{l}_j^k = \hat{l}_j^k(+) - \hat{l}_j^k(-) > 0$ implies $\hat{l}_j^k(-)/\hat{l}_j^k(+) < 1$, whence (8.6) follows. The case $\hat{l}_i^k > 0$, $\hat{l}_j^k < 0$ is proved in exactly the same way, by reversing all the inequalities.

The following proposition provides a limited converse to Proposition 1, and thus a limited generalization of the Heckscher–Ohlin theorem:

Proposition 4: If there are two factors, both fully employed, and country k's trade is balanced, then $l_i^k/l_j^k > l_i/l_j$ implies $\hat{l}_i^k < 0$ and $\hat{l}_j^k > 0$.

This is proved by noting that the vector w of factor rentals must satisfy $w'C = p'$, hence from (8.1) $w'\hat{l}^k = w'Cz^k = p'z^k = 0$, the last equality being the condition of balanced trade. The full-employment condition $Cy^k = l^k$ implies that w has positive components, hence if $\hat{l}^k$ has only two components they must be of opposite sign. If $l_i^k/l_j^k > l_i/l_j$ then it follows from Proposition 1 that $\hat{l}_i^k < 0$ and $\hat{l}_j^k > 0$.

Leamer remarked (1980, p. 501) that in the unlikely world of two commodities Leontief's method was a correct method (when trade is balanced). But this was because Leamer assumed the number of products to be equal to the number of factors. However, it follows from Proposition 4 that Leontief's method is correct as long as there are two factors and trade is balanced.

Leamer showed that, on his method (which does not require any assumptions concerning balanced trade), the US was relatively well endowed in capital relative to labour on the basis of Leontief's 1947 data – the opposite of Leontief's conclusion. However, this deduction – correct and ingenious though it is – constitutes a Pyrrhic victory; for if the Heckscher–Ohlin theory were truly generalizable, it should follow from this conclusion that the US exported embodied capital services and imported embodied labour services, whereas it imported both. If one cannot draw such a conclusion, why is it useful to know that the US is relatively well endowed in capital relative to labour?

Leamer's conclusion appears to be that the culprit was natural resources – a factor left out of account by Leontief. This could well be true. But there is another possible explanation. In 1947 the US had an export surplus of $11.6 billion – roughly 5 percent of the gross national product. Even in the simple 2×2 case, such a violation of the conditions of the Heckscher–Ohlin theorem would be sufficient to invalidate the conclusion. A theory that purports to explain trade flows cannot afford to ignore trade imbalances.

Leontief's 1953 paper had an enormous impact on the theory of international trade. The foundations of the successor to the law of comparative advantage were questioned. Further empirical investigations were carried out and new hypotheses formulated. Some of the more notable of these will be discussed.

Grubel and Lloyd (1975) drew attention to the large proportion of international trade flows which constituted what they called 'intra-industry trade', that is, two-way trade in products belonging to the same industrial category. They also constructed an index to measure the intensity of intra-industry trade. To analyse this, one of course must examine the way statistics of international trade are aggregated. Suppose the n commodities entering international trade are partitioned into $\bar{n}$ groups, and let G by an $\bar{n} \times n$ 'grouping matrix', i.e. a matrix of ones and zeros, containing exactly one unit element in each column. Let $t^k = Pz^k$ be an $n \times 1$ vector of values of country k's net imports (= gross imports if positive, gross exports if negative – since cross-haulage may be ruled out at the finest levels of disaggregation), where $P = \operatorname{diag} p$. Then from (8.2) and (8.3) above we may define the vectors.

$$t^k(+) = Pz^k(+), \qquad t^k(-) = Pz^k(-). \qquad (8.8)$$

The components of the vector $t^k(+)$ are positive import values for commodities imported, and zeros otherwise; and those of $t^k(-)$ are positive export values for commodities exported and zeros otherwise. These vectors satisfy

$$t^k = t^k(+) - t^k(-); \qquad |t^k| = t^k(+) + t^k(-), \qquad (8.9)$$

where $|t^k|$ is the vector of absolute trade values.

Now from published trade statistics one will only observe the aggregate $\bar{n} \times 1$ vectors

$$\bar{t}^k(+) = Gt^k(+), \qquad \bar{t}^k(-) = Gt^k(-). \qquad (8.10)$$

The components of $\bar{t}^k(+)$ are aggregate imports, and those of $\bar{t}^k(-)$ aggregate exports, in the respective categories. Unlike the case with $t^k(+)$ and $t^k(-)$, whose inner product is zero, the inner product of $\bar{t}^k(+)$ and $\bar{t}^k(-)$ is in general not zero, that is, one will find imports and exports in both categories. From the data (8.10) one can obtain

$$\bar{t}^k(+) - \bar{t}^k(-) = G[t^k(+) - t^k(-)] = Gt^k \qquad (8.11)$$

which gives the net imports in each category, and

$$\bar{t}^k(+) + \bar{t}^k(-) = G[t^k(+) + t^k(-)] = G|t^k| \qquad (8.12)$$

which gives the total trade in each category. The Grubel–Lloyd index of intra-industry trade is essentially (except for multiplication by 100)

$$Q_{GL} = 1 - \frac{\iota'|Gt^k|}{\iota'G|t^k|} \qquad (8.13)$$

(cf. Grubel and Lloyd, 1975, p. 22). Related indices were earlier introduced by Hirschman (1945), Verdoorn (1960), Michaely (1962), Kojima (1964) and Balassa (1966).

The Grubel–Lloyd index satisfies $0 \leqslant Q_{GL} \leqslant 1$. The second inequality is immediate and the first follows from the fact that $|Gt^k| \leqslant G|t^k|$. The lower bound can be attained when aggregation is perfect in the sense that $\bar{t}^k(+)$ and $\bar{t}^k(-)$ are orthogonal, given that they are non-negative; for it follows then from (8.10)–(8.12) that $|Gt^k| = |Gt^k(+) - Gt^k(-)| = Gt^k(+) + Gt^k(-) = G|t^k|$. On the other hand for the upper bound to be attained requires $Gt^k = 0$ which implies $\iota'Gt^k = \iota't^k = 0$, i.e. balanced trade. An alternative index that surmounts this problem was suggested by Aquino (1978):

$$Q_{Aq} = 1 - \frac{1}{2}\iota'\left|\frac{Gt^k(+)}{\iota'Gt^k(+)} - \frac{Gt^k(-)}{\iota'Gt^k(-)}\right|. \qquad (8.14)$$

The lower bound of 0 is attained when aggregation is perfect, since then

$$\left|\frac{Gt^k(+)}{\iota'Gt^k(+)} - \frac{Gt^k(-)}{\iota'Gt^k(-)}\right| = \frac{Gt^k(+)}{\iota'Gt^k(+)} + \frac{Gt^k(-)}{\iota'Gt^k(-)} \qquad (8.15)$$

and the column sum of this vector is 2. The upper bound of 1 is attained when the two vectors on the right are equal to one another, and this does not require balanced trade. When trade is balanced, i.e., $\iota't^k = 0$, then $\iota'Gt^k = 0$ hence from (8.11) $\iota'Gt^k(+) = \iota'Gt^k(-)$ and from (8.12) each of these is equal to $\frac{1}{2}G|t^k|$, hence the Aquino index (8.14) reduces to the Grubel–Lloyd index (8.13).

Grubel and Lloyd (1975, pp. 86–7) distinguished two criteria for aggregation of commodities into groups: substitutability of products in consumption (e.g., wooden versus metal furniture, or natural versus artificial yarn), and similarly of input coefficients in production (e.g., petroleum tar and gasoline, or steel bars and steel sheets). They argued that intra-industry trade was quite compatible with the Heckscher–Ohlin theory in the case of goods in the former category, but not in the case of the latter. They presented the thesis that only increasing returns to scale and product differentiation could account for such trade, as well as trade in a third category of goods (such as automobiles and cigarettes) which were characterized by both

substitutability in consumption and similarity of input coefficients.

An outcome of this work was the development by Krugman (1979, 1980, 1982) of a model incorporating product differentiation and (internal) economies of scale, based on that of Dixit and Stiglitz (1977). A drawback of this model, however, is that it requires the imposition of strong 'symmetry' conditions in order to allow for existence of equilibrium, and as pointed out in Chipman (1982b) this in effect requires one to assume that preferences are functionally dependent on the technology. It is a good question, then, as to whether models which are logically overdetermined can play a useful role as descriptors of the real world.

An alternative and very interesting approach was introduced by Lancaster (1980) who recognizes the need for a model to be internally consistent and discussed the Nash equilibria of a trade model in which products, conceived as bundles of characteristics, are economic variables in the Chamberlinian sense. He concluded that intra-industry trade could be expected to occur even between countries which are identical in all respects.

One thing that has been overlooked in the literature on monopolistic competition in international trade is the possibility that intra-industry trade could be explained by the standard Haberler–Lerner–Samuelson model. The argument used by Grubel and Lloyd (1975, p. 88) was that if production functions were identical as between industries (as well as between countries), production-possibility surfaces would be flat (in accordance with Theorem 9 above – and as noted by Lerner (1933) in an observation he attributed to Joan Robinson). They also argued (p. 89) that these constant rates of transformation would be the same between countries (which would be true only if endowment ratios were the same). They concluded that there could be no trade under such circumstances; but in fact, since the supply (Rybczynski) correspondences are multi-valued in this case, the correct conclusion is that the amount of trade is indeterminate. If endowment ratios differed only slightly, it is quite apparent from the geometry of the situation that there would be a great amount of trade.

The situation can be depicted in terms of the well-known 'Lerner diagram' (Figure 13). Assume two countries to be mirror images of one another in their factor endowments, and to produce commodities whose production functions (identical as between countries) are mirror images of another. Assume further that preferences are identical and homothetic, and symmetric as between the two commodities. Then owing to the symmetry, prices of the products will be equal in equilibrium. The upper contour sets $A_j(Y^k/p_j)$ for the two countries (with $Y^1 = Y^2$) are shown in the figure, as are the factor endowment vectors l^1 and l^2. The equilibrium resource allocations v_j^k are as shown as in Figure 7. Now suppose that the two production techniques become more similar, and so as to preserve the equilibrium prices suppose the production functions remain mirror images of one another. The new upper contour sets are denoted $A'_j(Y^k/p_j)$, and the new resource allocations $v_j^{k\prime}$. It is clear that after the production techniques have become closer, each country will allocate a larger proportion of its factors to its export industry. If it is further assumed that the countries have Mill–Cobb–Douglas preferences, hence spend one-half of their incomes on each commodity, since the prices are equal each country will devote one-half of its income to its exportable both before and after the change, but a larger proportion of its national product will be composed of exportables after than before the change. It follows that each country will export a larger proportion of its export good when production functions are more similar than when they are dissimilar. If only one of the production functions becomes more similar to the other (no

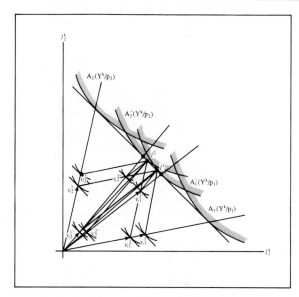

Figure 13

matter which one) it can be shown by a more complicated argument that the conclusion still holds.

What bearing does this have on intra-industry trade? If the criterion for aggregation is similarity of production functions, then when production functions are dissimilar, the commodities will likely be classified in different industries, whereas if they are similar they will be grouped into the same industry. By the above reasoning, one would then expect to observe more intra-industry than inter-industry trade. While extension to many commodities and factors requires careful analysis, this heuristic reasoning suggests that the HLS model would predict intra-industry trade, contrary to received opinion (cf. Lancaster, 1980; Helpman and Krugman, 1985).

Quite a different criticism of received opinion is suggested by Finger (1975), who builds upon the n-commodity 2-factor model developed by Jones (1956), Melvin (1968) and Bhagwati (1972). Bhagwati pointed out, by way of correcting Jones, that because of the production indeterminancy when $n > 2$ (the non-strict concavity to the origin of production–possibility frontiers), when factor rentals are equalized between countries it is not necessarily the case that the ranking of commodities by factor ratios will correspond to the actual trade pattern for some dividing line within the ranking. Finger did not rely on this argument, but assumed non-equalization of factor rentals and thus a unique ranking of commodities by 'exportability', to coin a term. Except for commodities near the dividing line, he noted that if commodities were grouped into industries according to factor intensities, there would be no intra-industry trade. His main point was that, in fact, commodities are not grouped into industries according to similarity of factor intensities.

Finger's argument does not carry over to the case of more than two factors. Here, then, there is a grey area since little is known concerning either (a) actual 'closeness' of vectors of technical coefficients within as opposed to between industries or (b) what in fact the theory would predict in the general case of more than two commodities and factors.

Generalizations of the 'Ricardian' (1-factor) and 'Heckscher–Ohlin' (2-factor) theories to the case of a continuum of commodities were developed by Dornbusch, Fischer and Samuelson

(1977, 1980). While these contributions provide interesting results and promising new techniques of analysis, they contain no surprises concerning the predicted pattern of trade.

The other noteworthy approach towards generalizing the Heckscher–Ohlin theorem is that introduced independently by Dixit and Norman (1980) and Deardorff (1980, 1982), and further developed by Dixit and Woodland (1982). If $\mathbf{p}^k$ is the price vector of country k under autarky and $\mathbf{z}^k$ its net-import vector under free trade, then by applying the revealed-preference criterion (assuming identical homothetic preferences) to country k's trade-preferences, we have $\mathbf{p}^k \cdot \mathbf{z}^k \geqq 0$. If there are two countries then from $\mathbf{z}^1 + \mathbf{z}^2 = 0$ we have $(\mathbf{p}^k - \mathbf{p}^i) \cdot \mathbf{z}^k \geqq 0$ for $i \neq k$, i.e., there is a positive correlation between net imports and differences in autarky prices. Deardorff (1982) obtained an analogous relation between autarky factor rentals and factor content of net imports, assuming identical tastes and technologies among countries. These are elegant results; but their practical usefulness is limited by the fact that autarky prices are not observed, and that one is usually interested in explaining actual trade patterns. Dixit and Woodland (1981) and Woodland (1982, p. 203) derived sufficient conditions for an increase in one (physical) endowment to cause an increase in a country's export of the corresponding commodity.

The contributions of Linder (1961), Posner (1961), Hufbauer (1966) and Vernon (1966) offer some important new ideas concerning the role of dynamics in the determination of trade patterns. Common to these is the idea that a country must develop an internal market first for a product, then realize economies of scale in it, before it can export it. This type of idea of course can be traced back to List (1841). Vernon (1966) in his interesting Schumpeterian analysis developed the concept of a 'product cycle': a country with a head-start can export a new product, but because of imitation by other countries must keep innovating to retain its technological lead and hence its comparative advantage in innovative products. For an interesting formalization see Jensen and Thursby (1986).

Finally one must note that there is a natural extension of laws of comparative advantage to the intertemporal sphere. Trade imbalances can be considered as intertemporal trades. In a model of one traded commodity and two periods, one may consider trade between present and future goods. A definite example has been presented in Chipman (1985b) in which each country has two factors (capital and labour) and produces two commodities (capital good and consumer good). The capital good is not traded, but used to augment the capital stock in the next period. If preferences as between the present and future good are identical and homothetic within and as between countries, and if production functions are identical between countries and have the property that the consumer–good industry uses a higher capital–labour ratio than the capital–good industry (a condition introduced by Uzawa, 1964a), then it follows that the country with the higher initial endowment of capital to labour will 'export' the present good (i.e. lend) to the other country, and 'import' the future good (i.e. be repaid by the other country in the next period). Obviously such models do not allow for rescheduling of debts. But in the simplest case they remind us that balanced trade is not to be expected, and is in fact far from optimal. It would be far more efficient for a theory of international trade to deal simultaneously with the problem of predicting trade flows and that of predicting trade balances as well. The field of 'international finance', or 'balance-of-payments theory', may be thought of as the field of intertemporal international-trade theory, but usually simplified to allow for only one commodity (and no production). A goal for the future is the development of a theory that simultaneously accommodates many commodities, factors, countries and periods.

9. THE TRANSFER PROBLEM. The term 'transfer problem' originated in the 1924 report of the Committee of Experts on Reparations chaired by General Charles G. Dawes (later that year elected Vice-President of the US), in charge of recommending the reparations payments of 1 billion gold marks to be paid by Germany to the Allies. The term initially referred to the problem of converting the German funds into foreign currency, as distinguished from the 'budgetary problem' of first raising the funds by taxation. Following Keynes's (1929) formulation, the term has come to encompass all the structural dislocations involved in carrying out the transfer, including changes in the exchange rate and the terms of trade.

It was observed by Smith (1776, book IV, ch. I; vol. II, p. 21) that a country could finance foreign expenditures by an export surplus, without the need to lose bullion reserves. On the other hand it was held by Thornton (1802, p. 139; 1939, p. 156) that:

the immediate cause . . . of the exportation of our coin has been an unfavourable exchange, produced partly by our heavy [foreign] expenditure, though chiefly by the super-added circumstance of two successively bad harvests.

And that in the case of the latter (1802, p. 131; 1939, p. 151):

In order . . . to induce the [foreign] country . . . to take all its payments in goods, and no part of it in gold, it would be requisite not only to prevent goods from being excessively dear, but even to render them excessively cheap.

This could be interpeted as saying that not only would the paying country's exchange rate have to depreciate but its terms of trade would have to deteriorate. It should be noted that this reasoning was applied by Thornton to the effect of a harvest failure (a decline in British production of importables) rather than to the effect of a transfer; a deterioration of Britain's terms of trade is to be expected in the former case but not necessarily in the latter.

King (1804) analysed the widely-held opinion that the depreciation of the Irish relative to the British pound could be attributed to the rent payments by Irish tenants to absentee landlords residing in England. He pointed out (p. 86; 1844, p. 108):

The residence of Irish proprietors in England has the necessary effect of diminishing the Irish imports, because the expenditure of revenue is transferred to another country; and it also increases the export of that produce which is no longer consumed at home.

The same reasoning was used by Foster (1804) – who referred to King (1804) – to argue that Britain's foreign expenditures produced the required export surplus (p. 18):

that part of the money lent, which was destined for foreign expenditure, was necessarily sent out either in specie or in bills of exchange, but, in each case, *necessarily forced* the exportation of British produce to that amount, to pay for these bills of exchange.

The same conclusion – but without the specific reasoning – was also reached by Ricardo (1811; 1951, II, p. 63), after a lengthy discussion of Thornton's case of the effect of a bad harvest:

If, which is a much stronger case, we agreed to pay a subsidy to a foreign power, money would not be exported whilst there were any goods which could more cheaply discharge the payment. The interest of individuals would render the exportation of the money unnecessary.

The qualifying phrase suggests that Ricardo thought Thornton was on firmer ground in the case of a bad harvest.

Malthus (1811a, pp. 344–5) took up Thornton's example of 'a bad harvest, or . . . a large subsidy to a foreign power' and argued against Ricardo that

if the debt for the corn or the subsidy . . . is paid by the transmission of bullion, . . . it is owing precisely to the cause mentioned by Mr. Thornton – the unwillingness of the creditor nation to receive a great additional quantity of goods not wanted for immediate consumption, without being bribed to it by excessive cheapness.

It should be noted that this would not be inconsistent with King's analysis – in the case of a bad harvest as opposed to that of a transfer. In Ricardo's later work (1817, ch. VI, pp. 162–72; 1951, ch. VII, pp. 137–42) – no doubt influenced by his discussions with Malthus – his views shifted somewhat in the direction of those of Thornton and Malthus, for he allowed for price divergences between countries – due necessarily to transport costs – and for movements of bullion in response to disturbances.

Although representing himself as a disciple of Ricardo, Mill (1844, p. 42) – see also Mill (1848, II, book III, ch. XXI, §4, pp. 166–7) – presented a doctrine that was much closer to those of Thornton and Malthus, or at least closer to late (1817) than to early (1811) Ricardo:

When a nation has regular payments to make in a foreign country, for which it is not to receive any return, its exports must annually exceed its imports by the amount of the payments which it is bound so to make. In order to force a demand for its exports greater than its imports will suffice to pay for, it must offer them at a rate of interchange more favourable to the foreign country, and less so to itself, than if it had no payments to make beyond the value of its imports.

Mill continued with his doctrine of the secondary burden (p. 43):

Thus the imposition of a tribute is a double burthen to the country paying it, and a double gain to that which receives it. The tributary country pays to the other, first, the tax, whatever be its amount, and next, something more, which the one country loses in the increased cost of its imports, the other gains in the diminished cost of its own.

Taussig (1917, 1927) in his writings attributed the above theory to Ricardo as well as Mill; this attribution was challenged by Viner (1924, p. 203), but Viner went too far in attributing the theory to Thornton. Mill held unequivocally that a transfer would worsen the paying country's terms of trade. Thornton held that a disturbance such as a transfer or a bad harvest would lower its exchange rate, and that a bad harvest would worsen its terms of trade; however, there is no unequivocal statement in his work to the effect that a transfer would worsen its terms of trade. Still, there is no question that Thornton, Malthus, and Mill were in the camp of those who believed that a transfer would entail changes in relative prices, whereas King (1804), Foster (1804) and Ricardo (1811) were in the camp of those who believed that it need not.

After a long hiatus, the topic of the transfer problem was taken up again by Bastable (1889, pp. 12–16) and Nicholson (1903, II, pp. 289–91), both of whom criticized Mill and essentially restated the arguments of King and Foster (but without reference to these authors). In view of these criticisms one could ask: was Mill's doctrine simply the result of a blunder? that of confusing the effect of a unilateral transfer – in which purchasing power is simply redistributed from one country to another – with that of a harvest failure – in which the world supply of our country's importable is diminished, resulting in its price rising relative to that of the exportable? In support of such an interpretation is the fact that Marshall (1923, p. 349) subsequently made precisely such a blunder – as was first noted by Robertson (1931, p. 179) – in depicting a transfer by a shift in the paying country's offer curve, forgetting to shift that of the receiving country. Against such an interpretation is that fact that the King–Foster–Bastable–Nicholson explanation tacitly assumes that all goods are tradable at zero transport costs, whereas Mill was not such a fool as to believe that, under that assumption, it would make sense to say of a transfer (made in money): 'This lowers prices in the remitting country, and raises them in the receiving' (Mill, 1848, II, book III, ch. XXI, §4, p. 167).

It was the great accomplishment of Taussig (1917) to introduce an idealization – the distinction between tradable goods ('international commodities') having zero transport costs, and nontradables ('domestic commodities') with effectively infinite transport costs – that could provide a definite interpretation of Mill's doctrine. In discussing the effect of a capital movement from Britain to the US, the following hint was thrown out (pp. 396–7):

> exporting industries in the United States . . . decline; . . . Less commodities are exported. More domestic commodities are made . . .

Taussig's theory was more fully developed in his book (1927, p. 35, ch. 26), and it led to an interesting response by Wicksell (1918) and to a number of empirical investigations and further conceptual developments including those of Williams (1920), Graham (1922, 1925), Viner (1924), Wilson (1931) and White (1933). Graham's analysis of the transfer problem (1925, pp. 213–14) in terms of adjustment of relative prices of tradable and nontradable commodities, and consequent resource reallocations between these sectors, foreshadowed much of Ohlin's subsequent treatment. Taussig's account was further developed by Viner (1937, pp. 323–65).

The most important contribution following Taussig's was that of Ohlin (1928), who followed the King–Foster–Bastable–Nicholson line of argument but with the noteworthy addition of nontradable commodities (which he called 'home-market goods', p. 6). He also assumed that each country produced importables as well as exportables and home-market goods, in contrast to Taussig whose analysis took account only of resource allocation between the export and domestic sectors. Ohlin showed that in this model a transfer would result in resource reallocation into home-market industries out of *both* export and import-competing industries in the receiving country, with the reverse movement in the paying country. He concluded, as had Graham (1925) before him, that there would be a tendency for the prices of home-market goods to rise relatively to those of international goods in the receiving country, and fall in the paying country (this would now be described as an appreciation of the receiving country's 'real exchange rate'). Thus, there need not be any change in the terms of trade. He went on to suggest that in a 'progressive country' the resource allocation would proceed smoothly so that in the long run no changes in relative prices of nontradables need take place (p. 10). A somewhat similar argument was presented by Cassel (1928, pp. 14–23) in the same year.

There followed the famous debate between Keynes (1929) and Ohlin (1929). Keynes noted:

> If £1 is taken from you and given to me and I choose to increase my consumption of precisely the same goods as those of which you are compelled to diminish ours, there is no transfer problem.

Arguing that this was not the correct representation of the facts, he stated that the transfer problem consisted (for the paying country) in 'the diversion of production out of other employments into the export trades (or to produce goods previously imported)', and he regarded the difficulties in accomplishing this as considerable. He expressed the opinion that this could be accomplished in the case of German reparations payments by a devaluation of the mark, but observed that this course of action was forbidden by the Dawes Committee. It followed that Germany would have to undergo a painful deflation. Keynes's view appeared to be closer to Thornton's than to Mill's, since the difficulty of resource reallocation between tradables and nontradables played a larger role than the terms of trade. In the *Treatise* ([1930, I, p. 334) he stressed the importance of the 'terms of trade', but this term was defined as the ratio of the 'money-rate of earnings' in the two countries, which is not the same as the ratio of the paying country's export to import prices.

The first formal model of the transfer problem was that of Pigou (1932), who considered the case of two tradable commodities and two countries each with separable Jevonian utility functions. He derived a condition for a transfer to worsen the paying country's terms of trade, expressed in terms of demand elasticities. Except for Metzler's (1942) treatment in terms of the Keynesian multiplier model of underemployment equilibrium – which lies completely outside the traditional discussion of the transfer problem – no further formalization took place until Samuelson's (1952, 1954) classic paper on the subject. Dispensing with Pigou's assumption of separable utilities, but retaining Pigou's formulation in terms of indirect (inverse) demand functions, Samuelson generalized Pigou's criterion and simplified it to the following form: for the transfer to worsen the paying country's terms of trade, it is necessary and sufficient (provided dynamic stability holds) that the paying country's marginal propensity to consume its export good be greater than the receiving country's marginal propensity to consume this same good (1954, p. 285).

Samuelson's result was obtained much more simply by Mundell (1960) using direct rather than indirect demand functions. The analysis may be most simply set out in terms of trade-demand functions (section 6 above). If two countries with a common currency are trading two commodities, and country 1 makes a transfer of T currency units to country 2, world equilibrium is defined by the equation

$$h_2^1(p_1, p_2, -T; l^1) + h_2^2(p_1, p_2, T; l^2) = 0, \tag{9.1}$$

which states that the world excess demand for commodity 2 is zero; a similar equation follows for commodity 1 by the balance-of-payments constraints $p_1^k h_1^k + p_2^k h_2^k = D^k (k = 1, 2)$. Supposing country 1 to be exporting commodity 1 to country 2, and choosing the price p_1 of commodity 1 as numeraire and setting it equal to $\bar{p}_1$, the above equation (for fixed endowment vectors l^1 and l^2) defines implicitly the function $p_2 = \bar{p}_2(T)$. We find that, assuming h_2^1 and h_2^2 to be differentiable,

$$\frac{d\bar{p}_2}{dT} = \frac{\partial h_2^1 / \partial D^1 - \partial h_2^2 / \partial D^2}{\partial h_2^1 / \partial p_2 + \partial h_2^2 / \partial p_2}. \tag{9.2}$$

Dynamic stability of equilibrium requires that if p_2 is above (or below) its equilibrium value $\bar{p}_2(T)$, it should fall (or rise); i.e., $\dot{p}_2 = dp_2/dt$ should have the opposite sign to $p_2 - \bar{p}_2(T)$. If we assume that $\dot{p}_2$ has the same sign as the world excess demand for commodity 2, then stability will hold provided world excess demand is negative for $p_2 > \bar{p}_2(T)$ and positive for $p_2 < \bar{p}_2(T)$. In the neighbourhood of $\bar{p}_2(T)$ this requires that the world excess demand for commodity 2 be a decreasing function of p_2, i.e., assuming differentiability,

$$\frac{\partial}{\partial p_2}[\hat{h}_2^1 + \hat{h}_2^2] = \frac{\partial \hat{h}_2^1}{\partial p_2} + \frac{\partial \hat{h}_2^2}{\partial p_2} < 0. \qquad (9.3)$$

In conjunction with (9.2) this implies that a transfer will worsen the paying country's terms of trade if and only if

$$\frac{\partial \hat{h}_2^2}{\partial D^2} > \frac{\partial \hat{h}_2^1}{\partial D^1}, \qquad (9.4)$$

i.e., if and only if country 2 will spend a larger amount of externally-received funds on its own export good than will country 2 on the same (its import) good. When there are no nontradables this reduces to Samuelson's criterion (1952, p. 286; 1954, p. 284).

The stability condition (9.3) is usually expressed in terms of elasticities. Defining

$$\eta^k = -\frac{p_j}{\hat{h}_j^k}\frac{\partial \hat{h}_j^k}{\partial p_j}(j \neq k), \qquad \hat{m}_j^k = p_j^k\frac{\partial \hat{h}_j^k}{\partial D^k},$$

$$\delta^k = -\frac{D^k}{\hat{h}_k^k}\frac{\partial \hat{h}_k^k}{\partial D^k}(k = 1, 2) \qquad (9.5)$$

and using (and differentiating) the balance-of-payments constraints and the homogeneity of degree 0 of the functions $\hat{h}_j^k$ in p_1, p_2, D^k, we may write (9.2) in the alternative form

$$\frac{d\bar{p}_2}{dT} = \frac{\hat{m}_2^2 - \hat{m}_2^1}{\hat{h}_2^1 \Delta} \qquad (9.6)$$

where

$$\Delta = -\frac{p_2}{\hat{h}_2^1}\frac{\partial(\hat{h}_2^1 + \hat{h}_2^2)}{\partial p_2} = \eta^1 + \frac{p_1 \hat{h}_1^2}{-p_2 \hat{h}_2^2}(\eta^2 - 1) + \delta^2. \qquad (9.7)$$

Condition (9.3) then translates into the stability condition $\Delta > 0$. Formula (9.7) generalizes the well-known formula $\Delta = \eta^1 + \eta^2 - 1 > 0$ obtained in (7.18) above, which Hirschman (1949) called the 'Marshall–Lerner condition'. For $T \neq 0$, (9.7) corrects Hirschman's expression, which omitted the term δ^2.

If there are no nontradables then as observed in section 6 $\partial \hat{h}_j^k/\partial D^k = \partial h_j^k/\partial Y^k$, and the condition that a transfer leave the terms of trade unchanged reduces to the condition that both countries have the same marginal propensities to consume each commodity. As Keynes noted, in this case there is no transfer problem.

Let us consider the general case in which we distinguish three categories of commodities in country k: n_1^k tradables and n_3^k nontradables produced and consumed in country k by means of m^k primary factors, and n_2^k tradables imported but not produced by country k. The country's trade-demand function may be obtained as follows. Let $\mathbf{p}_r^k = (p_{r1}^k, p_{r2}^k, \ldots, p_{rm_k}^k)'$ denote the vector of n_r^k prices of commodities in category $r(r = 1, 2, 3)$ and let $\mathbf{y}_r^k = (y_{r1}^k, y_{r2}^k, \ldots, y_{rm_k}^k)'$ denote the vector of n_r^k outputs of commodities in category $r(r = 1, 3)$, in country k. Let $\mathbf{h}_r^k(\mathbf{p}_1, \mathbf{p}_2, \mathbf{p}_3^k, Y^k)$ denote country k's aggregate demand function for the n_r^k commodities in category r, as a function of the three sets of prices and disposable income Y^k (including any transfers). Finally, let $\mathbf{w}^k$ denote the vector of country k's m^k factor rentals. Then the first set of n_3^k equations states that the demand for nontradables equals the supply:

$$\mathbf{h}_3^k[\mathbf{p}_1^k, \mathbf{p}_2^k, \mathbf{p}_3^k, \Pi^k(\mathbf{p}_1^k, \mathbf{p}_3^k, l^k) + D^k] = \mathbf{y}_3^k. \qquad (9.8a)$$

The second set of $n_1^k + n_3^k$ equations states that prices are equal to minimum unit costs for all produced commodities:

$$\mathbf{g}_k^k(\mathbf{w}^k) = \mathbf{p}_k^k$$
$$\mathbf{g}_3^k(\mathbf{w}^k) = \mathbf{p}_3^k. \qquad (9.8b)$$

The third set of m^k equations states that the demand for factors of production is equal to the supply:

$$\mathbf{B}_1^k(\mathbf{w}^k)\mathbf{y}_1^k + \mathbf{B}_3^k(\mathbf{w}^k)\mathbf{y}_3^k = l^k. \qquad (9.8c)$$

Here, $\mathbf{B}_r^k(\mathbf{w}^k)$ denotes the $m^k \times n_r^k$ matrix of factor–output ratios $b_{r, ij}^k = \partial g_{rj}^k/\partial w_i^k$, where $g_{rj}^k(\mathbf{w}^k)$ is the minimum-unit-cost function for the jth commodity in category $r(r = 1, 3)$.

Treating $\mathbf{p}_1, \mathbf{p}_2, D^k$, and l^k as parameters, the remaining variables $\mathbf{p}_3^k, \mathbf{w}^k, \mathbf{y}_1^k, \mathbf{y}_3^k$ constitute $n_3^k + m^k + n_1^k + n_3^k$ unknowns, which is equal to the number of equations in (9.8). This determines four 'reduced-form' functions $\tilde{\mathbf{p}}_3^k(\ )$, $\tilde{\mathbf{w}}^k(\ )$, $\tilde{\mathbf{y}}_1^k(\ )$, $\tilde{\mathbf{y}}_3^k(\ )$ whose arguments are $(\mathbf{p}_1, \mathbf{p}_2, D^k, l^k)$. The trade–demand functions are then defined by

$$\hat{\mathbf{h}}_r^k(\ ) = \mathbf{h}_r^k(\mathbf{p}_1, \mathbf{p}_2, \tilde{\mathbf{p}}_3^k(\ ), \Pi^k[\mathbf{p}_1, \tilde{\mathbf{p}}_3^k(\ ), l^k] + D^k) - \tilde{\mathbf{y}}_r^k(\ ),$$
$$(r = 1, 2), \qquad (9.9)$$

where $\tilde{\mathbf{y}}_2^k(\ ) = 0$. It was shown in Chipman (1981) that these are generated by maximizing a trade-utility function $\hat{U}^k(\mathbf{z}_1^k, \mathbf{z}_2^k)$ subject to the balance-of-payments constraint $\mathbf{p}_1 \cdot \mathbf{z}_1^k + \mathbf{p}_2 \mathbf{z}_2^k \leqslant D^k$, where $\mathbf{z}_j^k = \mathbf{x}_j^k - \mathbf{y}_j^k$. Defining $\hat{\mathbf{h}}^k(\ ) = [\hat{\mathbf{h}}_1^k(\ ), \mathbf{h}_2^k(\ )]'$, $\mathbf{p} = (\mathbf{p}_1, \mathbf{p}_2)'$, and $n = n_1^k + n_2^k$, a general solution of the transfer problem is obtained by solving $n - 1$ of the n equations $\hat{\mathbf{h}}^1(\ ) + \hat{\mathbf{h}}^2(\ ) = 0$, where $D^1 = -T, D^2 = T$, to obtain (subject to a normalization, e.g., setting the price of one tradable equal to unity) the function $\bar{\mathbf{p}}(T)$.

Some general remarks may be made about the solution of (9.8). In the case $m^k \geqslant n_1^k + n_3^k$ (there are at least as many primary factors as produced commodities), it follows from Theorem 9 that, save for exceptional cases, a single-valued Rybczynski function can be defined. Accordingly, substituting $\mathbf{y}_3^k = \hat{\mathbf{y}}_3^k(\mathbf{p}_1^k, \mathbf{p}_3^k, l^k)$ in (9.8a), one can solve immediately for the function $\tilde{\mathbf{p}}_3^k(\ )$ and then for the trade-demand function (9.9). At the other extreme, if $m^k = n_1^k$ then one can solve the first set of cost equations of (9.8b) for $\mathbf{w}^k = (\mathbf{g}_k^k)^{-1}(\mathbf{p}_k^k)$ and substitute into the second to obtain $\tilde{\mathbf{p}}_3^k(\ )$. These functions are then substituted in (9.8a) and (9.8c). The intermediate case $n_1^k < m^k < n_1^k + n_3^k$ presents much greater difficulties for explicit computation of solutions (cf. Chipman, 1980).

Some examples in the case of two tradables $(n_1^k + n_2^k = 2)$ and one nontradable $(n_3^k = 1)$ are of particular interest. To take the analytically simplest case first $(m^k \geqslant n_1^k + n_3^k)$ we may distinguish two subcases: $(m^k, n_1^k) = (2, 1)$ and $(m^k, n_1^k) = (3, 2)$. These are illustrated in Figures 14 and 15 respectively. In Case (2, 1) since each country specializes in its export good and its nontradable, and has two factors of production, each has a one-dimensional strictly-concave-to-the-origin production-possibility frontier as shown. In Case (3, 2), each country produces all three commodities with three factors, and thus each has a two-dimensional strictly-concave-to-the-origin production-possibility frontier. In Case (2, 1), when country 1 makes a transfer to country 2, resources are withdrawn from its nontradable sector and reallocated to its export sector, and the reverse movement takes place in country 2. World production of commodity 1 (country 1's export good) increases, and world production of commodity 2 (country 2's export good) decreases, so there is a general presumption that country 1's terms of trade will deteriorate, in accordance with the 'orthodox' Mill–Taussig presumption. In Case (3, 2), however, there is no reason to expect the terms of trade to change more in one direction than another; but there is a presumption that the price of the nontradable will fall relative to that of the tradables in

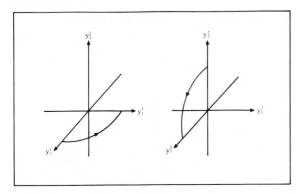

Figure 14

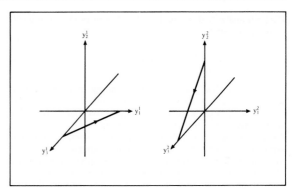

Figure 16

the paying country, and rise in the receiving country. This is in accordance with Graham's (1925) account, as well as Ohlin's (1928) – at least in the short run. It also appears fully consistent with Keynes's analysis (1929, 1930).

Before proceeding to the analytics, it is worth considering two other cases, in which $m^k = n_1^k$. Again, we may distinguish two subcases: $(m^k, n_1^k) = (1, 1)$ and $(m^k, n_1^k) = (2, 2)$. These are illustrated in Figures 16 and 17 respectively. In Case (1, 1), each country has a linear production-possibility frontier as between the exportable and the nontradable. As in Case (2, 1), there is a strong presumption in favour of the 'orthodox' doctrine; in fact, this case may be thought of as the one Mill had in the back of his mind. Case (2, 2), however, is quite different. Each country's production-possibility surface is ruled. As resources move between the tradables and nontradables sectors, it is possible for the movement to take place along the ruled rather than curved segments of the surface; in this case, there is no reason for any relative prices to change. This corresponds to Ohlin's view of the long-run outcome of a transfer, as well as with Cassel's (1928) view that capital movements will not disturb relative prices, and therefore purchasing-power parities. Contrasting Figures 15 and 17, we may say that the dispute between Keynes and Ohlin was really a dispute (if they had only realized it!) as to whether the production-possibility surface was or was not strictly concave to the origin.

Let us proceed to the analytics. In both the cases (2, 1) and (3, 2), the function $\tilde{p}_3^k(p_1, p_2, D^k, l^k)$ is defined implicitly by

$$h_3^k(p_1, p_2, \tilde{p}_3^k(\ ), \Pi^k[p_1, p_2, \tilde{p}_3^k(\ ), l^k] + D^k)$$
$$= \hat{y}_3^k(p_1, p_2, \tilde{p}_3^k(\ ), l^k). \quad (9.10)$$

Defining the Slutsky terms, transformation terms, and con-

sumption coefficients by

$$s_{ij}^k = \frac{\partial h_i^k}{\partial p_j^k} + \frac{\partial h_i^k}{\partial Y^k} h_j^k, \qquad t_{ij}^k = \frac{\partial \hat{y}_i^k}{\partial p_j^k}, \qquad c_i^k = \frac{\partial h_i^k}{\partial Y^k}, \qquad (9.11)$$

we find that $\partial \tilde{p}_2^k / \partial D^k = -(s_{33}^k - t_{33}^k)^{-1} c_3^k$ (hence an inward transfer raises the price of the nontradable provided it is a superior good), hence defining the trade-demand functions as in (9.9) we obtain

$$\hat{c}_i^k \equiv \partial \hat{h}_i^k / \partial D^k = c_i^k - (s_{i3}^k - t_{i3}^k)(s_{33}^k - t_{33}^k)^{-1} c_3^k,$$
$$(i = 1, 2) \quad (9.12)$$

where, in the case $(m^k, n_1^k) = (2, 1)$, $t_{i3}^k = 0$ for $i \neq k$. The condition for the 'orthodox' presumption (9.4) can therefore be stated as $\hat{c}_2^1 - \hat{c}_2^2 < 0$.

A set of sufficient conditions is readily obtained (cf. Chipman 1974). Suppose it is assumed that in the case of fixed production, hence pure exchange, a transfer will leave the terms of trade unchanged. In Chipman (1974a, p. 45) this was called the Hypothesis of Neutral Tastes. It states that $\hat{c}_2^1 = \hat{c}_2^2$ when $t_{ij}^k = 0$ for $i, j, k = 1, 2$, or

$$c_2^1 - s_{23}^1(s_{33}^1)^{-1} c_3^1 = c_2^2 - s_{23}^2(s_{33}^2)^{-1} c_3^2. \quad (9.13)$$

An example of preferences satisfying (9.13) is identical Mill–Cobb–Douglas preferences as between the two countries. Then sufficient conditions for $\hat{c}_2^1 < \hat{c}_2^2$ are

$$c_2^1 - (s_{23}^1 - t_{23}^1)(s_{33}^1 - t_{33}^1)^{-1} c_3^1 < c_2^1 - s_{23}^1(s_{33}^1)^{-1} c_3^1 \quad (9.14)$$

and

$$c_2^2 - s_{23}^2(s_{33}^2)^{-1} c_3^2 < c_2^2 - (s_{23}^2 - t_{23}^2)(s_{33}^2 - t_{33}^2)^{-1} c_3^2. \quad (9.15)$$

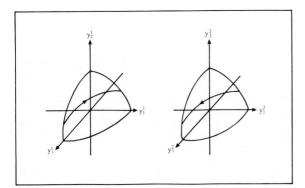

Figure 15

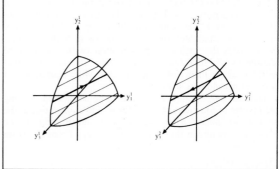

Figure 17

If $c_3^1 > 0$ and $c_3^2 > 0$, inequalities (9.14) and (9.15) are respectively equivalent to

$$\frac{s_{23}^1}{s_{33}^1} < \frac{t_{23}^1}{t_{33}^1} \quad \text{and} \quad \frac{s_{23}^2}{s_{33}^2} > \frac{t_{23}^2}{t_{33}^2}. \tag{9.16}$$

In the case $(m^k, n_1^k) = (2, 1)$ (corresponding to Figure 14), in which $t_{23}^1 = t_{13}^2 = 0$, these inequalities follow if $s_{23}^1 > 0$ and $s_{13}^2 > 0$, i.e. if in each country the importable and nontradable are Hicksian substitutes. In the case $(m^k, n_1^k) = (3, 2)$, conditions for (9.16) are more delicate, but under some special but symmetric assumptions it was found in Chipman (1974, p. 68) that the orthodox presumption holds. A special case of case (2, 1) was treated by Jones (1974b); see also Jones (1975).

Returning to the case $m^k = n_1^k$ let us first consider the case $(m^k, n_1^k) = (1, 1)$, corresponding to Figure 16. From the homogeneity of g_i^k, $g_i^k(w_1^k) = b_{ij}^k w_1^k$ where $b_{1i}^k = g_1^k(1)$, hence (9.8b) and (9.8c) yield

$$p_3^k = p_k b_{13}^k / b_{1k}^k \quad \text{and} \quad b_{1k}^k y_k^k + b_{13}^k y_3^k = l_1^k \tag{9.17}$$

hence from (9.8a)

$$\tilde{y}_k^k( ) = l_1^k / b_{1k}^k$$
$$- (b_{13}^k / b_{1k}^k) h_3^k(p_1, p_2, p_k b_{13}^k / b_{1k}^k, l_1^k p_k / b_{1k}^k + D^k). \tag{9.18}$$

The trade-demand function (9.9) then becomes, for $i = 1, 2$,

$$\hat{h}_i^k(p_1, p_2, D^k, l_1^k) = h_i^k(p_1, p_2, p_k b_{13}^k / b_{1k}^k, l_1^k p_k / b_{1k}^k + D^k)$$
$$- \delta_{ik} \tilde{y}_i^k(p_1, p_2, D^k, l_1^k) \tag{9.19}$$

where $\delta_{ik} = 1$ if $i = k$ and 0 if $i \neq k$. From this we derive

$$\frac{\partial \hat{h}_2^1}{\partial D^1} = \frac{\partial h_2^1}{\partial Y^1}, \qquad \frac{\partial \hat{h}_2^2}{\partial D^2} = \frac{\partial h_2^2}{\partial Y^2} + \frac{b_{13}^2}{b_{12}^2} \frac{\partial h_3^2}{\partial Y^2}. \tag{9.20}$$

In words (multiplying all these expressions by p_2): In country 1, which does not produce commodity 2, an additional dollar given or loaned to it has the same effect on the consumption of the importable as an equivalent increase in disposable national income. In country 2, however, which produces and exports commodity 2, an additional dollar received from country 1 has not only the direct effect on demand for exportables, but also an indirect effect brought about by the diversion of resources to the nontradable and the need to compensate for the fall in production of the exportable.

Of course, it is not enough to assume identical and homothetic preferences to assure $\partial h_2^1 / \partial Y^1 = \partial h_2^2 / \partial Y^2$, since these functions depend on prices of nontradables, which will in general be different in the two countries. However, if preferences are generated by utility functions of the separable form $U^k(x_1^k, x_2^k, x_3^k) = F[\bar{U}^k(x_1^k, x_2^k), x_3^k]$ and if the $\bar{U}^k$ are identical and homogeneous as between the two countries, then $\partial h_2^1 / \partial Y^1 = \partial h_2^2 / Y^2$ (cf. Chipman, 1974a, pp. 47–9), and provided the nontradables are superior goods in both countries, (9.20) yields the 'orthodox' presumption (9.4). This may be identified with Mill's doctrine, particularly since it is consistent with the labour theory of value and with Mill's assumption of constant expenditure shares.

The case $(m^k, n_1^k) = (2, 2)$ remains to be considered. Solving the first set of equations of (9.8b) for $\mathbf{w}^k$ and substituting in the second, we obtain $\tilde{p}_3^k(p_1, p_2)$; likewise, substituting in the $b_{ij}^k(w_1, w_2)$ we obtain the functions $\hat{b}_{ij}^k(p_1, p_2)$. From (9.8a) we then have

$$\tilde{y}_3^k(p_1, p_2, D^k, l^k)$$
$$= h_3^k(p_1, p_2, \tilde{p}_3^k(p_1, p_2), \Pi^k[p_1, p_2, \tilde{p}_3^k(p_1, p_2), l^k] + D^k) \tag{9.21}$$

whence $\partial \tilde{y}_3^k / \partial D^k = \partial h_3^k / \partial Y^k$. From (9.8c) we have

$$\begin{pmatrix} \tilde{y}_1^k( ) \\ \tilde{y}_2^k( ) \end{pmatrix} = \begin{pmatrix} \hat{b}_{11}^k( ) & \hat{b}_{12}^k( ) \\ \hat{b}_{21}^k( ) & \hat{b}_{22}^k( ) \end{pmatrix}^{-1} \left[\begin{pmatrix} l_1^k \\ l_2^k \end{pmatrix} - \begin{pmatrix} \hat{b}_{13}^k( ) \\ \hat{b}_{23}^k( ) \end{pmatrix} \tilde{y}_3^k( ) \right] \tag{9.22}$$

whence, using (9.9), we find that

$$\frac{\partial \hat{h}_2^k}{\partial D^k} = \frac{\partial h_2^k}{\partial Y^k} - \frac{\partial \tilde{y}_2^k}{\partial D^k}$$
$$= \frac{\partial h_2^k}{\partial Y^k} + [0 \ 1] \begin{pmatrix} \hat{b}_{11}^k( ) & \hat{b}_{12}^k( ) \\ \hat{b}_{21}^k( ) & \hat{b}_{22}^k( ) \end{pmatrix}^{-1} \begin{pmatrix} \hat{b}_{13}^k( ) \\ \hat{b}_{23}^k( ) \end{pmatrix} \frac{\partial h_3^k}{\partial Y^k}. \tag{9.23}$$

From this we may easily derive a sufficient condition for a transfer to leave all relative prices unaffected. Suppose production functions are identical in the two countries, and that factor rentals are initially equilized; then the $\hat{b}_{ij}^k( )$ are the same for $k = 1, 2$, and the prices of the nontradables are equal, i.e., $p_3^1 = p_3^2$. Accordingly,

$$\frac{\partial \hat{h}_2^1}{\partial D^1} - \frac{\partial \hat{h}_2^2}{\partial D^2} = \frac{\partial h_2^1}{\partial Y^1} - \frac{\partial h_2^2}{\partial Y^2}$$
$$+ [0 \ 1] \begin{pmatrix} \hat{b}_{11}( ) & \hat{b}_{12}( ) \\ \hat{b}_{21}( ) & \hat{b}_{22}( ) \end{pmatrix}^{-1} \begin{pmatrix} \hat{b}_{13}( ) \\ \hat{b}_{23}( ) \end{pmatrix} \left(\frac{\partial h_3^1}{\partial Y^1} - \frac{\partial h_3^2}{\partial Y^2} \right). \tag{9.24}$$

If it is assumed that preferences as between the two countries are identical and homothetic, given that the prices are the same the expression (9.24) vanishes. This may be identified with Ohlin's (1928) case – for the long run. In Figure 17, the direction of movement on the ruled production–possibility surfaces is along the ruled line segments.

Of course, if either production functions or preferences differ as between the two countries, the directions of movement in Figure 17 will be along the curved segments of the production–possibility surfaces, similar to those of Figure 15 except that the curvature and hence the price changes will not be so great. In this case there appears to be no presumption one way or the other with regard to the terms of trade (as opposed to the 'real exchange rate'). The outcome depends on the ranking of factor intensities by industry (cf. Chipman, 1974a, p. 61).

When preferences can be aggregated, the utility function which generates these preferences can usually be taken as a welfare indicator for some welfare criterion; for example, if preferences are identical and homothetic, the aggregate utility function is an indicator of potential welfare. In terms of the indirect trade–utility function

$$\hat{V}^k(p_1, p_2, D^k, l^k) = \hat{U}[\hat{\mathbf{h}}^k(p_1, p_2, D^k, l^k)] \tag{9.25}$$

we may define country 1's welfare as a function of the transfer by

$$W^1(T) = \hat{V}^1[\bar{p}_1, \bar{p}_2(T), -T, l^1]. \tag{9.26}$$

Then a simple computation shows that

$$\frac{\partial W^1}{\partial T} = -\frac{\partial \hat{V}^1}{\partial D^1} \left[1 + z_2^1 \frac{d\bar{p}_2}{dT} \right]. \tag{9.27}$$

The bracketed term indicates the primary and secondary burden of the transfer on the paying country. Note that even if there are great dislocations involving a change in the real exchange rate – of the kind Keynes envisaged – there is no secondary burden unless the terms of trade deteriorate.

The general case of n tradable commodities can be treated similarly, yielding the expression (where one of the $\bar{p}_i$ is a constant and the rest are functions of T):

$$\frac{\partial W^1}{\partial T} = -\frac{\partial \hat{V}^1}{\partial D^1} \left[1 + \sum_{i=1}^{n} z_i^1 \frac{d\bar{p}_i}{dT} \right]. \tag{9.28}$$

Here, the 'secondary burden' is measured by a change in the

difference between an import and an export price index, rather than a *ratio* of these – showing incidentally that the usual procedure of measuring a country's terms of trade as the ratio of its export to its import prices is inappropriate (a difference between two variables is never a monotone function of their ratio unless the variables are equal to one another, in which case the difference and ratio are both constant).

Analysis of the transfer problem in the multi-commodity case is fairly straightforward and need not be taken up here; cf. Chipman (1980, 1981). Space does not allow discussion of the multi-country transfer problem and the associated 'transfer paradoxes'; for a good summary of the state of the subject see Dixit (1983).

10. THE THEORY OF TARIFFS AND QUOTAS. The theory of the effect of tariffs and other trade barriers on the conditions of trading countries goes back to Torrens (1844, pp. 331–56) and Mill (1844, pp. 21–32) who showed that a country can improve its terms of trade by imposing a tariff or an export tax. Mill distinguished between a protecting and a non-protecting duty, the former being one sufficiently large to induce the country imposing it to start producing the import-competing good. He asserted (pp. 26–7) that there would be no advantage from a protecting duty; this followed from his assumption of constant costs, according to which a protective tariff would be equivalent to a prohibitive one.

The theory was briefly developed by Edgeworth (1894) in terms of Marshall's (1879) offer curves, but his treatment contained a flaw later uncovered by Lerner (1936) – the allegation of lack of symmetry between import and export taxes.

The next important step was the contribution by Bickerdike (1907), who established the proposition that a country could gain from a sufficiently small tariff, and could optimize its gains by a suitable choice of tariff rate. Bickerdike's theory – presented with extreme terseness – was greatly clarified by Edgeworth (1908). Bickerdike also noted what came later to be known as 'Lerner's symmetry theorem'.

Marshall (1923) analysed tariffs in terms of his offer curves, and noticed that if the foreign country's offer curve is inelastic a tariff will lead to an increase in amounts of both commodities available to consumers, and would thus constitute a clear gain to the country. In most essential respects this observation had already been made by Mill (1844, p. 22).

The modern theory of tariffs starts with Lerner's (1936) contribution, continues with Stolper and Samuelson's (1941) fundamental work, two important papers by Metzler (1949a, 1949b) and one by Bhagwati (1959), two major developments of Bickerdike's theory by Graaff (1949) and Johnson (1950), and an important contribution by Johnson (1960) further developed by Bhagwati and Johnson (1961) and Rao (1971). Also noteworthy are the expositions by Mundell (1960) and Jones (1969). The theory of quotas is less well developed – most of the theoretical analysis being of a partial-equilibrium nature; the primary reference is Bhagwati (1965). The traditional theory takes trade restrictions such as tariffs or quotas as exogenously-controlled instruments and examines their effects. In recent years there has been a great deal of interest in the opposite problem of explaining trade restrictions; it will not be possible to cover that literature here. However, mention should be made of the important model of retaliatory tariff behaviour and equilibrium originated by Johnson (1954) and developed by Gorman (1958), Panchamukhi (1961), Kemp (1964, ch. 15), Horwell (1966), Kuga (1973), Otani (1980), Mayer (1981), Riezman (1982) and Thursby and Jensen (1983), as well as the model of retaliatory quota behaviour studied by Rodriguez (1974) and Tower (1975).

The treatment to follow will consist of a synthesis of contemporary theory of trade restrictions, for the case of two tradable commodities, two factors, and two countries, divided into the theory of tariffs and the theory of quotas. And each will itself be divided into the classical aggregative treatment – in which each country is perceived as acting as a single rational agent – and the disaggregative treatment introduced by Johnson (1960) in which each factor of production, as well as the government, is treated as a rational agent.

10.1. Aggregative tariff theory. It will be assumed that country 1 is to impose a tariff on its imports of commodity 2 from country 2, and that commodity 1 in country 1 uses a larger proportion of factor 1 to factor 2 than commodity 2 in the initial equilibrium.

Country 1's excess demand for its import good is defined by

$$\hat{z}_2^1(p_1^2, p_2^2, T_2; l^1)$$
$$= \hat{h}_2^1[p_1^2, T_2 p_2^2, (T_2 - 1)p_2^2 \hat{z}_2^1(p_1^2, p_2^2, T_2; l^1); l^1] \quad (10.1)$$

where $\hat{h}_2^1(p_1^1, p_2^1, D^1; l^1)$ is country 1's trade-demand function [as in (6.3) above], and $T_2 = 1 + \tau_2$ is the tariff factor and τ_2 the tariff rate. Superscripts denote countries. It will be noted that the function $\hat{z}_2^1(\ )$ appears on both sides of the above equation, so it needs to be shown that a function $\hat{z}_2^1(\ )$ satisfying (10.1) exists and is unique. The existence of such a function defined locally (in a neighbourhood of initial equilibrium prices and tariff factor) follows from the implicit-function theorem provided $(T_2 - 1)p_2^2 \partial \hat{h}_2^1/\partial D^1 - 1 \neq 0$ at the initial equilibrium. More is needed for $\hat{z}_2^1(\ )$ to be defined globally, however. For each fixed p_1^2, p_2^2, T_2, l^1, (10.1) defines a mapping from the space of values z_2^1 of excess demand into itself, and from the principle of contraction mappings (cf. Kolmogorov and Fomin, 1957, p. 43), if $(1 - 1/T_2)\hat{m}_2^1$ lies in the interval $[0, 1)$, where $\hat{m}_2^1 = p_2^1 \partial \hat{h}_2^1/\partial D^1$ is country 1's marginal trade-propensity to consume commodity 2, then the mapping is a contraction and has a unique fixed point z_2^1. Since this argument has to be carried out for all prices and tariff factors, it is necessary to assume that $(1 - 1/T_2)\hat{m}_2^1$ is bounded below 1. Since $1 \leqslant T_2 < \infty$, $0 \leqslant 1 - 1/T_2 < 1$; it is thus sufficient to assume that $\hat{m}_2^1$ is bounded below 1. Since $\hat{m}_1^1 + \hat{m}_2^1 = 1$ this is equivalent to assuming that $\hat{m}_1^1$ is bounded above zero, i.e., the export good must be strongly superior.

An iterative procedure explained by Kolmogorov and Fomin (1957, p. 44), consisting of starting with a value of z_2^1 on the right to obtain a value on the left, and using this as the new value on the right, corresponds precisely to the procedure originally used by Metzler (1949b, p. 347) to define the tariff-inclusive offer function as Johnson (1960) has described it. See also Bhagwati and Johnson (1961, p. 230). Thus, Metzler used the principle of contraction mappings without apparently being aware of it. Of course, the above argument is still not entirely rigorous since the function (10.1) is not meaningful unless $z_2^1 > 0$; hence a more subtle argument is required. In the n-commodity case, the situation becomes still more complex. For some techniques of proof of existence of equilibrium under tariff distortions see Sontheimer (1971).

World equilibrium is defined by the condition

$$\hat{z}_2^1(p_1^2, p_2^2, T_2; l^1) + \hat{z}_2^2(p_1^2, p_2^2; l^2) = 0, \quad (10.2)$$

where $\hat{z}_2^2$ coincides with country 2's trade-demand function with $D^2 = 0$, and if commodity 1 is taken as numeraire and its price $\bar{p}_1^2$ held constant, for constant factor endowments (10.2) defines implicitly the function $\bar{p}_2^2(T_2)$. It yields the condition

$$\frac{d\bar{p}_2^2}{dT_2} = -\frac{\partial \hat{z}_2^1/\partial T_2}{\partial \hat{z}_2^1/\partial p_2^2 + \partial \hat{z}_2^2/\partial p_2^2}. \quad (10.3)$$

As in (9.3) above, dynamic stability requires that the denominator expression in (10.3) be negative, consequently the sign of $d\bar{p}_2^1/dT_2$ is the same as that of $\partial \tilde{z}_2^1/\partial T_2$. This illustrates the general principle of comparative statics: to find the effect of a tariff on the external price of the import good, one must ascertain the effect of the tariff on the import of that good when the external price is held constant. A tariff will improve country 1's terms of trade, then, if and only if it reduces country 1's demand for the import good when the terms of trade are held constant. It remains only to compute $\partial \tilde{z}_2^1/\partial T_2$.

Differentiating both sides of (10.1) with respect to T_2 one obtains after collecting terms

$$\frac{\partial \tilde{z}_2^1}{\partial T_2} = \frac{p_2^2 \hat{s}_{22}^2}{1 - (1 - 1/T_2)\hat{m}_2^1}. \tag{10.4}$$

This is negative, since the denominator has been assumed positive and the trade–Slutsky term $\hat{s}_{22}^2$ is necessarily negative – unless there is a kink in the trade–indifference curve at the initial equilibrium. This is the simple proof of the classical proposition, that can be traced back to Torrens and Mill, that a tariff will improve a country's terms of trade. The proof also suggests why the result need not be true in general – if preferences are not aggregable, for example if the government does not distribute the revenues and has preferences that differ from those of the public (cf. Lerner, 1936, and section 10.2 below).

In accordance with the above-mentioned principle of comparative statics, if one wishes to ascertain the effect of a tariff on the *internal* price of the import good, one must ask the question: What would be the effect of a tariff on demand for imports if the *internal* price were held constant?

To bring out an interesting aspect of the problem the formulation will be generalized (as in fact was done by Lerner, 1936) to leave open the question of what fraction of the tariff proceeds is collected by each of the two countries. Let ρ be the proportion collected by country 1 and $1 - \rho$ that collected by country 2, where $0 \leqslant \rho \leqslant 1$. Then the two countries' excess-demand functions $\tilde{z}_2^k(p_1^1, p_2^1, T_2, \rho; l^k)$ are defined implicitly by

$$\tilde{z}_2^1(\ ) = \hat{h}_2^1(p_1^1, p_2^1, \rho[1 - 1/T_2]p_2^1\tilde{z}_2^1(\ ); l^1)$$
$$\tilde{z}_2^2(\ ) = \hat{h}_2^2(p_1^1, p_2^1/T_2, -[1 - \rho][1 - 1/T_2]p_2^1\tilde{z}_2^2(\ ); l^2) \tag{10.5}$$

provided the $\hat{m}_k^k = p_k^k \partial \hat{h}_k^k/\partial D^k$ are both bounded above zero. (Note that the countries' respective balance-of-trade deficits, denominated in their own prices, satisfy

$$0 = p_1^1 z_1^1 + p_2^1 z_2^1 - D^1 = p_1^1 z_1^1 + p_2^1 z_2^1 - \rho(1 - 1/T_2)p_2^1 z_2^1$$
$$= -p_1^2 z_1^2 - p_2^2 z_2^2 - (1 - \rho)(1 - 1/T_2)p_2^1 z_2^1$$
$$= -p_1^2 z_1^2 - p_2^2 z_2^2 + D^2 \tag{10.6}$$

whence $D^1 + D^2 = \tau_2 p_2^2 z_2^1 \neq 0$ when $\tau_2 > 0$). The condition $\tilde{z}_2^1(\ ) + \tilde{z}_2^2(\ ) = 0$ of world equilibrium implicitly defines the function $\bar{p}_2(T_2, \rho)$ which satisfies

$$\frac{\partial \bar{p}_2^1}{\partial T_2} = -\frac{\partial \tilde{z}_2^1/\partial T_2 + \partial \tilde{z}_2^2/\partial T_2}{\partial \tilde{z}_2^1/\partial p_2^1 + \partial \tilde{x}_2^2/\partial p_2^1},$$
$$\frac{\partial \bar{p}_2^1}{\partial \rho} = -\frac{\partial \tilde{z}_2^1/\partial \rho + \partial \tilde{z}_2^2/\partial \rho}{\partial \tilde{z}_2^1/\partial p_2^1 + \partial \tilde{z}_2^2/\partial p_2^1}. \tag{10.7}$$

Consider first the effect of a tariff on the internal price of country 1's import good. From the previous stability argument, this has the sign of the effect of the tariff on demand for the import good when this internal price is held constant. When p_2^1 is held constant, it is apparent from (10.5) that there is only an income effect in country 1. Writing the derivatives as elasticities

one finds that

$$\frac{T_2}{\tilde{z}_2^1}\frac{\partial \tilde{z}_2^1}{\partial T_2} = \rho \hat{m}_2^{1'}; \qquad \frac{T_2}{\tilde{z}_2^2}\frac{\partial \tilde{z}_2^2}{\partial T_2} = \rho \hat{m}_2^{2'} - \hat{\sigma}_{22}^{2'} \tag{10.8}$$

where $\hat{\sigma}_{ij}^k = p_j^k \hat{s}_{ij}/z_i^k$ denotes the trade–Slutsky elasticity and where $\hat{m}_2^{1'}/\hat{m}_2^1 = \hat{\sigma}_{22}^{1'}/\hat{\sigma}_{22}^1 = T_2 - \rho(T_2 - 1)\hat{m}_2^1$ and $\hat{m}_2^{2'}/\hat{m}_2^2 = \hat{\sigma}_{22}^{2'}/\hat{\sigma}_{22}^2 = 1 + (1 - \rho)(T_2 - 1)\hat{m}_2^2$. Accordingly, the first equation of (10.7) can be expressed in elasticity form as

$$\frac{T_2}{\bar{p}_2^1}\frac{\partial \bar{p}_2^1}{\partial T_2} = \frac{\rho \hat{m}_2^{1'} + (1 - \rho)T_2\hat{m}_2^{2'} + \eta^2 - 1}{\eta^1 + \eta^2 - 1}$$
$$= \frac{\rho(\hat{m}_2^{1'} - \hat{m}_2^{2'}) + \hat{\sigma}_{22}^{2'}}{\eta^1 + \eta^2 - 1}. \tag{10.9}$$

The first equation of (10.9) is a generalization of Metzler's formula. There is a joint income effect from the countries given by the average of their adjusted marginal trade-propensities to consume commodity 2. For the tariff to raise the domestic price of the import good, the sum of this term and the elasticity of country 2's demand for imports must exceed unity. The reason for this is simple. If p_2^1 is held constant, in country 1 only the tariff revenues can be a source of its increased demand for imports. But the tariff lowers the price of commodity 2 in country 2, so there is both a general effect on country 2's demand from this source and the effect of the tariff revenues.

The second formula of (10.9) is perhaps even more instructive. It is obtained from the first from the decompositions

$$\eta^1 = \frac{[p + (1 - \rho)T_2]\hat{m}_2^1 - T_2\hat{\sigma}_{22}^1}{T_2 - \rho(T_2 - 1)\hat{m}_2^1},$$
$$\eta^2 = \frac{1 - \hat{m}_2^2 + \hat{\sigma}_{22}^2}{1 + (1 - \rho)(T_2 - 1)\hat{m}_2^2}. \tag{10.10}$$

Since the trade–Slutsky elasticity $\hat{\sigma}_{22}^2 = p_2^2 \hat{s}_{22}^2/z_2^2$ is positive (because $z_2^2 < 0$), a sufficient condition for a tariff to raise the internal price of country 1's import good is that $\hat{m}_2^{1'} \geqslant \hat{m}_2^{2'}$, and this of course will be recognized as Samuelson's (1952) condition for a transfer to a country to worsen or leave unchanged its terms of trade.

This can be explained by looking at the elasticities

$$\frac{1}{\tilde{z}_2^1}\frac{\partial \tilde{z}_2^1}{\partial \rho} = (T_2 - 1)\hat{m}_2^{1'}; \qquad \frac{1}{\tilde{z}_2^2}\frac{\partial \tilde{z}_2^2}{\partial \rho} = (T_2 - 1)\hat{m}_2^{2'}, \tag{10.11}$$

yielding for the second equation of (10.7) the expression

$$\frac{1}{\bar{p}_2^1}\frac{\partial \bar{p}_2^1}{\partial \rho} = \frac{(T_2 - 1)(\hat{m}_2^{1'} - \hat{m}_2^{2'})}{\eta^1 + \eta^2 - 1}. \tag{10.12}$$

The effects of a tariff imposed and collected by a country can be broken into two stages: In stage 1, the tariff revenues are allocated to the foreign country; consequently, this is equivalent to an export tax imposed by country 2. In accordance with Lerner's (1936) symmetry theorem, this is equivalent to an import tariff imposed by country 2, hence it improves country 2's terms of trade and thus (since p_1 is held constant) raises the domestic price of country 1's import good. In stage 2, the revenues from country 2's export tax are transferred to country 1, so that it becomes in effect an import tariff imposed by country 1. If the transfer has the 'orthodox' effect of improving country 1's terms of trade, it will lower the previously raised price of its import good. The net result will then be uncertain. If the transfer has the 'anti-orthodox' effect, the domestic price of country 1's import good will rise further.

The possibility that a tariff might lower the domestic price of the good on which it is imposed has come to be generally known as the 'Metzler case' (Johnson, 1960) or the 'Metzler paradox' (cf. Jones, 1974a), although the possibility was briefly noted by

Lerner (1936) in a footnote that also noted the other possibility that a tariff could worsen a country's terms of trade if the government (which purchases at external prices) has a sufficiently strong preference for importables. To avoid confusion between these two 'paradoxes', and because of the prominence of this effect in Metzler's work, the term 'Metzler paradox' will prove convenient.

In comparing an import tariff imposed by country 1 to an export tax imposed by country 2, one need only note that a small transfer of country 2's tax revenues to country 1 will raise country 1's potential welfare and lower country 2's. Integrating over a path from $\rho = 0$ to $\rho = 1$, since the integrand is positive so will be the integral. Hence an import tariff is preferable from country 1's point of view. Since for any quota equilibrium there is a corresponding tariff equilibrium (but not conversely – see Section 10.3 below), this reasoning establishes the superiority of an import quota to a 'voluntary export restraint', i.e., an export quota imposed by country 2.

10.2. Disaggregative tariff theory. Consideration of separate preferences as between the government and the public was introduced by Lerner (1936). Johnson (1960) went further and considered the separate preferences of the two factors of production. The following treatment encompases both, so that three separate classes of consumers are considered. The government will be called class 0, and factors 1 and 2 constitute classes 1 and 2. If preferences are homothetic and the proportional distribution of income within a class remains unchanged, then the class's aggregate demand is generated by an aggregate preference relation (cf. Chipman, 1974b); if preferences within a class are identical and homothetic, then the class's preferences can be aggregated regardless of how income is distributed within the class. In either case, the utility function of the class can be considered only as an indicator of its *potential* welfare (i.e. a rise in utility means that gainers could compensate losers).

In both Lerner's (1936) and Bhagwati and Johnson's (1961) treatment it is assumed that the government does not itself pay the tariff on imported goods. Of course, employees and many departments of government generally will purchase imported goods in domestic markets, but the Lerner assumption will be followed here, leading to the specification of the government's demand function for the jth commodity in country 1 by $x_{0j}^1 = h_{0j}^1(p_1^2, p_2^2, Y_0^1)$, where Y_0^1 is government revenue, assumed equal to the tariff revenues retained by the government. The demand on the part of factor i will instead be $x_{ij}^1 = h_{ij}^1(p_1^1, p_2^1, Y_i^1)$ for $i = 1, 2$, where factor i's income consists of its earned income plus its share in the tariff revenues. Dutiable imports are defined by $z_{d2}^1 = x_{12}^1 + x_{22}^1 - y_2^1$, where y_2^1 is the output of the importable in country 1. Letting δ_i^1 denote the share of class i in the tariff revenues, country 1's demand for imports $z_2^1 = x_{02}^1 + z_{d2}^1$ is given by

$$x_{02}^1 = h_{02}^1[p_1^1, p_2^1, \delta_0^1(T_2 - 1)p_2^2 z_{d2}^1]$$

$$z_{d2}^1 = \sum_{i=1}^{0} h_{i2}^1[p_1^1, T_2 p_2^1, l_i^1 \hat{w}_i^1(p_1^1, T_2 p_2^1, l_1^1, l_2^1)$$

$$+ \delta_i^1(T_2 - 1)p_2^2 z_{d2}^1] - \hat{y}_2^1(p_1^1, T_2 p_2^1, l_1^1, l_2^1) \tag{10.13}$$

where $\hat{w}_i^1$ is the Stolper–Samuelson function for the ith factor, which will generally (when both commodities are produced) depend only on the prices. The second equation of (10.13) implicitly defines the function $z_{d2}^1 = \hat{z}_{d2}^1(p_1^1, p_2^2, T_2, l^1)$; when this is substituted in the first equation of (10.13), the two summed equations define the function $\hat{z}_1^2(p_1^2, p_2^2, T_2; \delta^1, l^1)$ which determines country 1's demand for imports. It should be noted that the formulation is meaningful only if dutiable imports are positive.

There are two main questions of interest in this model: (1) under what conditions will a tariff improve country 1's terms of trade? (2) what will the effect of a tariff be on the welfare of the separate classes?

To answer the first question, by virtue of (10.3) one needs only to compute the partial derivative of country 1's demand for imports with respect to the tariff factor. This is found to be

$$\frac{\partial \hat{z}_2^1}{\partial T_2} = \delta_0^1 m_{02}^1 z_{d2}^1 + \frac{M_{02}^1}{T_2 M_{d2}^1}$$

$$\times \left\{ p_2^1 \left(\sum_{i=1}^{2} s_{i,22}^1 - t_{22}^1 \right) + \sum_{i=1}^{2} m_{i2}^1 a_i^1 \right\} \tag{10.14}$$

where

$$M_{02}^1 = 1 + (T_2 - 1)\delta_0^1 m_{02}^1,$$

$$M_{d2}^1 = 1 - (1 - 1/T_2) \sum_{i=1}^{2} \delta_i^1 m_{i2}^1,$$

and

$$a_i^1 = l_i^1 \frac{\partial \hat{w}_i^1}{\partial p_2^1} + \delta_i^1 z_{d2}^1 - x_{i2}^1. \tag{10.15}$$

The terms $s_{i,22}^1$ in (10.14) are the factors' Slutsky substitution terms, and the m_{i2}^1 are the classes' marginal propensities to consume the import good, defined as $m_{02}^1 = p_2^2 \partial h_{02}^1 / \partial Y_0^1$ and $m_{i2}^1 = p_2^1 \partial h_{i2}^1 / \partial Y_i^1$ for $i = 1, 2$.

Using Samuelson's reciprocity relation (3.33) and the homogeneity of degree 1 in factor endowments of the Rybczynski function one sees that

$$\sum_{i=1}^{2} a_i^1 = \sum_{i=1}^{2} l_i^1 \frac{\partial \hat{y}_2^1}{\partial l_i^1} + (\delta_1^1 + \delta_2^1) z_{d2}^1 - \sum_{i=1}^{2} x_{i2}^1$$

$$= \hat{y}_2^1 - \sum_{i=1}^{2} x_{i2}^1 + (1 - \delta_0^1) z_{d2}^1 = -\delta_0^1 z_{d2}^1 \leqslant 0. \tag{10.16}$$

By convention, country 1 is assumed to be relatively well endowed in factor 1 compared to factor 2, and commodity 1 uses a higher ratio of factor 1 to factor 2 than commodity 2 at the initial factor rentals. Accordingly, from the Stolper–Samuelson relation $\partial \hat{w}_2^1 / \partial p_2^1 > w_2^1 / p_2^1$ and the inequalities $0 \leqslant 1 - 1/T_2 < 1$ one finds from (10.15) that

$$p_2^1 a_2^1 > l_2^1 w_2^1 - \delta_2^1 p_2^2 z_{d2}^1 - p_2^1 x_{22}^1$$

$$\geqslant l_2^1 w_2^1 + \delta_2^1(T_2 - 1)p_2^2 z_{d2}^1 - p_2^1 x_{22}^1 = Y_2^1 - p_2^1 x_{22}^1 \geqslant 0 \tag{10.17}$$

hence from (10.16) $a_1^1 = -a_2^1 - \delta_0^1 z_{d2}^1 < 0$. It follows that

$$\sum_{i=1}^{2} m_{i2}^1 a_i^1 = (m_{22}^1 - m_{12}^1)a_2^1 - m_{12}^1 \delta_0^1 z_{d2}^1. \tag{10.18}$$

Substituting (10.18) in (10.14) one obtains the formula

$$\frac{\partial \hat{z}_2^1}{\partial T_2} = \delta_0^1 z_{d2}^1 \left(m_{02}^1 - \frac{M_{02}^1}{T_2 M_{d2}^1} m_{12}^1 \right)$$

$$+ \frac{M_{02}^1}{T_2 M_{d2}^1} \left\{ p_2^1 \left(\sum_{i=1}^{2} s_{i,22}^1 - t_{22}^1 \right) + (m_{22}^1 - m_{12}^1)a_2^1 \right\} \tag{10.19}$$

When $T_2 = 1$ initially, the expressions $M_{02}^1 / (T_2 M_{d2}^1)$ reduce to 1.

From (10.19) we can obtain the main results of Lerner (1936) and Johnson (1960). In Lerner's case, the two factors have identical preferences hence $m_{12}^1 = m_{22}^1$ and the second term on the right in (10.19) is negative. When the tariff rate is initially zero, a necessary condition for a tariff increase to worsen the terms of trade is that $m_{02}^1 > m_{12}^1$, i.e., that the government's marginal propensity to consume the import good be greater than that of the public. This is what Johnson (1960) called the 'Lerner case'; it may be called the 'Lerner paradox'. In Johnson's case, in which all tariff revenues are distributed to the

949

factors, hence $\delta_0^1 = 0$, a necessary condition for a tariff to worsen the terms of trade is that $m_{22}^1 > m_{12}^1$, i.e., that factor 2 have a greater marginal propensity to consume the importable than factor 1. This may be called the 'Johnson paradox'. When $\delta_0^1 = 0$ and $m_{12}^1 = m_{22}^1$, formula (10.19) reduces as it should to (10.4).

These paradoxes are easily explained. In Lerner's case, since the government makes its purchases at external prices, there is no substitution effect, only an income effect, hence if its marginal propensity to consume importables is higher than factor 1's this income effect may outweigh the substitution effect of the higher internal price. In Johnson's case, since at constant external prices factor 2 gains and factor 1 loses earned income by virtue of the Stolper–Samuelson theorem, if factor 2 has a relatively strong preference for the product in which it is used relatively intensively (commodity 2), the distributional income effect might outweigh the substitution effect.

Question (2) is also of great interest. From the relation $p_2^2 = \bar{p}_2^2(T_2)$ between the equilibrium price of commodity 2 in country 2 and the tariff factor (the price of commodity 1, equal in both countries, being held constant) one obtains the equilibrium value of the internal price $\bar{p}_2^1(T_2) = T_2 p_2^2(T_2)$, the amount imported $\bar{z}_2^1(T_2) = \hat{z}_2^1(\bar{p}_1^2, \bar{p}_2^2(T_2), T_2)$ and factor i's income $\bar{Y}_i^1(T_2) = l_i^1 \hat{w}_i^1(\bar{p}_1^2, \bar{p}_2^2(T_2)) + \delta_i^1(T_2 - 1)\bar{p}_2^2(T_2)\bar{z}_2^1(T_2)$. Defining factor i's potential welfare $W_i^1(T_2) = V_i^1[(\bar{p}_1^1, \bar{p}_2^1(T_2), \bar{Y}_i^1(T_2)]$ in terms of its indirect utility function, one has

$$\frac{\partial W_i^1}{\partial T_2} = \frac{\partial V_i^1}{\partial Y_i^1}\left[-h_{i2}^1 \frac{d\bar{p}_2^1}{dT_2} + \frac{d\bar{Y}_i^1}{dT_2}\right]. \quad (10.20)$$

After a series of steps one finds that (cf. Rao, 1971)

$$\frac{dW_i^1}{dT_2} = \frac{\partial V_i^1}{\partial Y_i^1}\left\{a_i^1 \frac{d\bar{p}_2^1}{dT_2} - \delta_i^1 z_2^1[1 - \tau_2(\eta^2 - 1)]\frac{d\bar{p}_2^2}{dT_2}\right\}. \quad (10.21)$$

The term in brackets is positive so long as $\tau_2 < 1/(\eta^2 - 1)$, i.e., the initial tariff rate is less than the 'optimal tariff' (Johnson, 1950). In the absence of either a 'Johnson paradox' or a 'Metzler paradox' under these circumstances, $d\bar{p}_2^2/dT_2 < 0$ and $d\bar{p}_2^1/dT > 0$, so factor 2 is a clear gainer. Since $a_i^1 < 0$, for $i = 1$ formula (10.21) indicates the conditions required, say when all

tariff proceeds are distributed to factor 1 ($\delta_1^1 = 1$), for factor 1 to be compensated for its loss of earnings.

Of great interest is the question of whether any factor in country 1 stands to gain by having the import tariff replaced by an export tax on commodity 2 imposed by country 2. An analysis similar to that of the previous subsection could be carried out, but it is enough to provide general indications. If country 1 makes a transfer to country 2, and if the 'orthodox presumption' holds, both p_2^1 and p_2^2 will rise, i.e. country 1's terms of trade will deteriorate and country 2's will improve. By the Stolper–Samuelson theorem, factor 2 in country 1 will gain and so will factor 2 in country 2; and in both countries factor 1 will suffer a decline in real earnings. Assuming factor 1 previously collected all the tariff revenues in country 1, it will now lose doubly: its real rental will fall and its share in the tariff proceeds will disappear. In country 2, factor 1 might come out even if it receives all the proceeds from the export tax. If this factor is neutral, it is then two to one in favour of the change. If and to the extent that quotas are equivalent to tariffs, this might provide an explanation for 'voluntary export restraints', even though from the point of view of the aggregative model such a policy on the part of country 1 would amount to 'shooting oneself in the foot'. (For other aspects of the problem see the interesting discussion in K. Jones, 1984.)

10.3. The aggregative theory of quotas.

If country 1 imposes a quota of q_2 on its imports of commodity 2, its demand for imports will be determined by

$$\hat{z}_2^1(p_1^2, p_2^2, q_2; l^1) = \min\{\hat{h}_2^1(p_1^2, p_2^2, 0; l^1), q_2\}. \quad (10.22)$$

This will be called the quota-constrained demand for imports. Owners of import licences will make a profit of $(p_2^1 - p_2^2)z_2^1$, and aggregate excess demand will be determined by

$$z_2^1 = \hat{h}_2^1[p_1^2, p_2^1, (p_2^1 - p_2^2)z_2^1; l^1]. \quad (10.23)$$

If the quota is ineffective [i.e., $q_2 > \hat{h}_2^1(p_1^2, p_2^2, 0; l^1)$] then $p_2^1 = p_2^2$; if it is effective, then setting $z_2^1 = q_2$ in (10.23) implicitly defines the function

$$p_2^1 = \hat{p}_2^1(p_2^2, q_2) \quad (10.24)$$

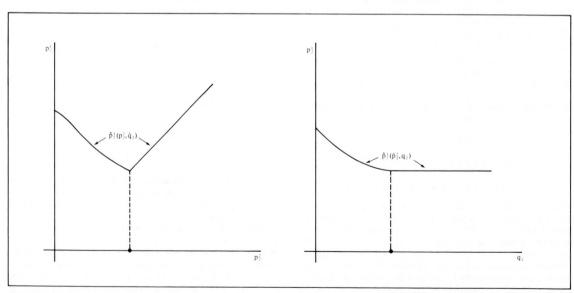

Figure 18

(where the arguments p_1^2, l^1 are suppressed, these being supposed constant). Defining $\hat{c}_j^k = \partial \hat{h}_j^k / \partial D^k$, $\hat{m}_j^k = p_j^k c_j^k$, and the implicit tariff factor $T_2 = p_2^1/p_2^2$, the derivatives of (10.24) when the quota is effective are found to be

$$\frac{\partial \hat{p}_2^1}{\partial p_2^2} = \frac{q_2 \hat{c}_2^1}{\hat{s}_{22}^1}, \qquad \frac{\partial \hat{p}_2^1}{\partial q_2} = \frac{1 - (1 - 1/T_2)\hat{m}_2^1}{\hat{s}_{22}^1}. \qquad (10.25)$$

When the quota is ineffective we have of course $\partial \hat{p}_2^1/\partial p_2^2 = 1$ and $\partial \hat{p}_2^1/\partial q_2 = 0.$, Figure 18 depicts the shape of $\hat{p}_2^1$ as a function of p_2^2 and q_2 separately. It is interesting in particular to note that when the quota is effective, a change in the external price of the import good will lead to a change in the internal price *in the opposite direction*. This is easily explained: if the external price falls, profits of holders of import licences will increase; as long as the import good is trade-superior (i.e., $\hat{c}_2^1 > 0$), this will lead to a rise in demand for imports which must be choked off by a price increase to maintain the level of the quota. This is one respect in which a quota is quite different from a tariff. It also makes it less than straightforward to extend the analysis of shared tariff revenues (section 10.1 above) to the case of profits from quota licences.

World equilibrium is defined by

$$\hat{z}_2^1(p_1^2, p_2^2, q_2; l^1) + \hat{z}_2^2(p_1^2, p_2^2; l^2) = 0 \qquad (10.26)$$

where $\hat{z}_2^2(p_1^2, p_2^2; l^2) = \hat{h}_2^2(p_1^2, p_2^2, 0; l^2)$, and when the quota is effective this leads to (holding p_1^2, l^1, l^2 constant)

$$\frac{d\bar{p}_2^2}{dq_2} = -\frac{1}{\partial \hat{z}_2^2/\partial p_2^2}, \quad \text{i.e.} \quad \frac{q_2}{p_2^2}\frac{d\bar{p}_2^2}{dq_2} = \frac{1}{\eta^2 - 1}. \qquad (10.27)$$

Since $\eta^1 = 0$ when the quota is binding, dynamic stability requires $\eta^2 > 1$. Figure 19 displays the 'catastrophic' effect of a quota when country 2's demand for imports is inelastic at the initial equilibrium (cf. Falvey, 1975). This is another respect in which a quota is different from a tariff.

From (10.27) and stability, a tightening of the quota (a decrease in q_2) will lead to a fall in the world price of commodity 2. One would expect it to lead to a rise in the domestic price of commodity 2 in country 1. From (10.24) we have

$$\frac{d\bar{p}_2^1}{dq_2} = \frac{\partial \hat{p}_2^1}{\partial p_2^2}\frac{d\bar{p}_2^2}{dq_2} + \frac{\partial \hat{p}_2^1}{\partial q_2} \qquad (10.28)$$

and from inspection of the expressions (10.25) and (10.27) it is

clear that as long as both goods are superior $d\bar{p}_2^1/dq_2 < 0$ as expected.

10.4. The disaggregative theory of quotas. Suppose class 0 is the group of holders of import licences and class i is factor i for $i = 1, 2$. A licence-holder facing given internal and external prices of the import good, the former greater than the latter, will optimize by taking the profit on the licence and consuming in domestic markets. Thus it is more appropriate to assume that all agents make their purchases at domestic prices. This is another difference between tariffs and quotas.

The disaggregative case differs from the aggregative one simply by replacing the aggregate trade–demand function by $\bar{z}_2^1(p_1^1, p_2^1, p_2^2, l^1, \delta^1)$ which is defined implicitly by

$$\bar{z}_2^1(\ ) = \sum_{i=1}^{2} h_{i2}^1[p_1^1, p_2^1, l_i^1 \hat{w}_i^1(p_1^1, p_2^1, l^1) + \delta_i^1(p_2^1 - p_2^2)\bar{z}_2^1(\ )] - \hat{y}_2^1(p_1^1, p_2^1, l^1), \qquad (10.29)$$

where by definition $l_0^1 = w_0^1 = 0$. (As in the disaggregative tariff case, this formulation is slightly less general than the aggregative one in that nontradables are excluded). Country 1's excess–demand function is defined by

$$\hat{z}_2^1(p_1^1, p_2^1, p_2^2, q_2; l^1, \delta^1)$$
$$= \min\{\bar{z}_2^1(p_1^2, p_2^1, p_2^2; l^1, \delta^1), q_2\}. \qquad (10.30)$$

and the function (10.24) is now defined implicitly by

$$\bar{z}_2^1(\bar{p}_1^2, \hat{p}_2^1, p_2^2; l^1, \delta^1) = \hat{z}_2^1(\bar{p}_1^2, \hat{p}_2^1, p_2^2, q_2; l^1, \delta^1). \qquad (10.31)$$

When the quota is effective the right side of (10.31) is just q_2. Then the derivatives of $\hat{p}_2^1$ satisfy

$$\frac{\partial \hat{p}_2^1}{\partial p_2^2} = -\frac{\partial \bar{z}_2^1/\partial p_2^2}{\partial \bar{z}_2^1/\partial p_2^1}, \qquad \frac{\partial \hat{p}_2^1}{\partial q_2} = \frac{1}{\partial \bar{z}_2^1/\partial p_2^1} \qquad (10.32)$$

and from (10.29) we have

$$\frac{\partial \bar{z}_2^1}{\partial p_2^1} = \frac{\sum_{i=0}^{2} s_{i,22}^1 - t_{22}^1 + \sum_{i=0}^{2} c_{i2}^1 a_i^1}{1 - \delta_i^1(1 - 1/T_2)\bar{m}_2^1},$$

$$\frac{\partial \bar{z}_2^1}{\partial p_2^2} = \frac{-q_1 \bar{m}_2^1}{1 - \delta_i^1(1 - 1/T_2)\bar{m}_2^1} \qquad (10.33)$$

where

$$\bar{m}_2^1 = \sum_{i=0}^{2} \delta_i^1 m_{i2}, \qquad T_2 = p_2^1/p_2^2,$$

and $a_i^1 = l_i^1 \partial \hat{w}_i^1/\partial p_2^1 + \delta_i^1 z_2^1 - x_{i2}^1$ is as in (10.15). Thus formulas (10.25) are replaced by formulas in which $\hat{m}_2^1$ is replaced by $\bar{m}_2^1$ and $\hat{s}_{22}^1$ is replaced by

$$\sum_{i=0}^{2} s_{i,22}^1 - t_{22}^1 + \sum_{i=0}^{2} c_{i2}^1 a_i^1. \qquad (10.34)$$

Because of possible differences in preferences this term need not be negative. However, a domestic stability argument could be used to show that it must be negative in the initial equilibrium, hence the disaggregative model does not introduce any essentially different features.

A great objective of the theory of commercial policy is to formulate the problem of international conflict in game-theoretic terms. Before one can do this one must have a payoff matrix. The types of computations that have been illustrated here constitute the raw material for such a strategic formulation. A beginning was developed by Scitovsky (1942), Johnson (1954) and Gorman (1958) in the case of tariffs and

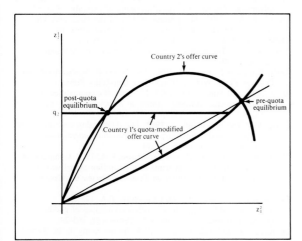

Figure 19

Rodriguez (1974) and Tower (1975) in the case of quotas. Many further promising developments have taken place (see the references cited at the beginning of this section as well as the excellent survey by McMillan, 1986). These models treat countries as aggregates. But we have seen that models of this type are unable to accommodate, let alone predict, the phenomenon of voluntary export restraints. A reasonable conjecture is that this tool provides the mechanism for a side-payment to otherwise injured parties in the foreign country, so as to avoid retaliation. Thus a strategic formulation must also consider the transfer problem. It is clear, also, that a proper formulation would require consideration of at least a four-person game. One may look forward to rich developments in this area in the future.

JOHN S. CHIPMAN

See also FOREIGN TRADE; HECKSCHER–OHLIN TRADE THEORY; LEONTIEF PARADOX; OFFER CURVES; TERMS OF TRADE; TRADEABLE AND NON-TRADEABLE COMMODITIES.

BIBLIOGRAPHY

Alexander, S.S. 1951. Devaluation vs import restriction as an instrument for improving trade balance. *International Monetary Fund Staff Papers* 1, April, 379–96.

Amano, A. 1968. Stability conditions in the pure theory of international trade: a rehabilitation of the Marshallian approach. *Quarterly Journal of Economics* 82, May, 326–39.

Andronov, A.A., Vitt, A.A. and Khaikin, S.E. 1966. *Theory of Oscillators*. Reading, Mass.: Addison-Wesley Publishing Co.

Aquino, A. 1978. Intra-industry trade and inter-industry specialization as concurrent sources of international trade in manufactures. *Weltwirtschaftliches Archiv* 114(2), 275–96.

Arrow, K.J., Chenery, H., Minhas, B.S. and Solow, R.M. 1961. Capital-labor substitution and economic efficiency. *Review of Economics and Statistics* 43, August, 225–50.

Balassa, B. 1966. Tariff reductions and trade in manufactures among the industrial countries. *American Economic Review* 56, June, 466–73.

Baldwin, R.E. 1982. Gottfried Haberler's contributions to international trade theory. *Quarterly Journal of Economics* 97, February, 141–59.

Bastable, C.F. 1889. On some applications of the theory of international trade. *Quarterly Journal of Economics* 4, October, 1–17.

Bhagwati, J.N. 1959. Protection, real wages and real incomes. *Economic Journal* 69, December, 733–48.

Bhagwati, J.N. 1965. On the equivalence of tariffs and quotas. In *Trade, Growth and the Balance of Payments*, ed. R.E. Caves, H.G. Johnson and P.B. Kenen, Amsterdam: North-Holland, 53–67.

Bhagwati, J.N. 1967. The proofs of the theorems on comparative advantage. *Economic Journal* 77, March, 75–83.

Bhagwati, J.N. 1972. The Heckscher–Ohlin theorem in the multi-commodity case. *Journal of Political Economy* 80, September–October, 1052–5.

Bhagwati, J.N. and Johnson, H.G. 1961. A generalized theory of the effects of tariffs on the terms of trade. *Oxford Economic Papers*, NS 13, October, 225–53.

Bickerdike, C.F. 1907. Review of *Protective and Preferential Import Duties* by A.C. Pigou. *Economic Journal* 17, March, 98–102.

Bickerdike, C.F. 1920. The instability of foreign exchange. *Economic Journal* 30, March, 118–22.

Bruno, M. 1973. Protection and tariff change under general equilibrium. *Journal of International Economics* 3, August, 205–25.

Cassel, G. 1928. The international movements of capital. In G. Cassel, T.E. Gregory, R.E. Kuczynski and H.K. Norton, *Foreign Investments*, Chicago: University of Chicago Press, 1–93.

Chang, W.W., Ethier, W.J. and Kemp, M.C. 1980. The theorems of international trade with joint production. *Journal of International Economics* 10, August, 377–94.

Chipman, J.S. 1965–66. A survey of the theory of international trade. *Econometrica* 33, July, 477–519; 33, October, 685–760; 34, January, 18–76.

Chipman, J.S. 1969. Factor-price equalization and the Stolper–Samuelson theorem. *International Economic Review* 10, October, 399–406.

Chipman, J.S. 1970. External economies of scale and competitive equilibrium. *Quarterly Journal of Economics* 84, August, 347–85.

Chipman, J.S. 1971. International trade with capital mobility: a substitution theorem. In *Trade, Balance of Payments, and Growth*, ed. J.N. Bhagwati, R.A. Mundell, R.W. Jones and J. Vanek, Amsterdam: North-Holland, 201–37.

Chipman, J.S. 1972. The theory of exploitative trade and investment policies: a reformulation and synthesis. In *International Economics and Development*, ed. L.E. Di Marco, New York: Academic Press, 881–916.

Chipman, J.S. 1974a. The transfer problem once again. In *Trade, Stability, and Macroeconomics*, ed. G. Horwich and P.A. Samuelson, New York: Academic Press, 19–78.

Chipman, J.S. 1974b. Homothetic preferences and aggregation. *Journal of Economic Theory* 8, May, 26–38.

Chipman, J.S. 1978. A reconsideration of the 'elasticity approach' to balance-of-payments adjustment problems. In *Breadth and Depth in Economics*, ed. J.S. Dreyer, Lexington, Mass.: Heath, 49–85.

Chipman, J.S. 1979. The theory and application of trade utility functions. In *General Equilibrium, Growth, and Trade*, ed. J.R. Green and J.A. Scheinkman, New York: Academic Press, 277–96.

Chipman, J.S. 1980. Exchange-rate flexibility and resource allocation. In *Flexible Exchange Rates and the Balance of Payments*, ed. J.S. Chipman and C.P. Kindleberger, Amsterdam: North-Holland, 159–209.

Chipman, J.S. 1981. A general-equilibrium framework for analyzing the responses of imports and exports to external price changes: an aggregation theorem. In *Methods of Operations Research*, Vol. 44, ed. G. Bamberg and O. Opitz, Königstein: Verlag Anton Hain, Meisenheim GmbH, 43–56.

Chipman, J.S. 1982a. Capital movement as a substitute for technology transfer: a comment. *Journal of International Economics* 12, February, 107–9.

Chipman, J.S. 1982b. Comment. In *Import Competition and Response*, ed. J.N. Bhagwati, Chicago: University of Chicago Press, 218–21.

Chipman, J.S. 1985a. Relative prices, capital movements, and sectoral technical change: theory and an empirical test. In *Structural Adjustment in Developed Open Economics*, ed. K. Jungenfelt and D. Hague, London: Macmillan, 395–454.

Chipman, J.S. 1985b. A two-period model of international trade and payments (MS).

Corden, W.M. 1966. The structure of a tariff system and the effective protective rate. *Journal of Political Economy* 74, June, 221–37.

Corden, W.M. 1971. The substitution problem in the theory of effective protection. *Journal of International Economics* 1, February, 37–57.

Deardorff, A.V. 1980. The general validity of the law of comparative advantage. *Journal of Political Economy* 88, October, 941–57.

Deardorff, A.V. 1982. The general validity of the Heckscher–Ohlin theorem. *American Economic Review* 72, September, 683–94.

Debreu, G. 1959. *Theory of Value*. New York: Wiley.

Dixit, A. 1983. The multi-country transfer problem, *Economics Letters* 13, 49–53.

Dixit, A.K. and Norman, V. 1980. *The Theory of International Trade*. Digswell Place, Welwyn: Cambridge University Press.

Dixit, A.K. and Stiglitz, J.E. 1977. Monopolistic competition and optimum product diversity. *American Economic Review* 67, June, 297–308.

Dixit, A.K. and Woodland, A. 1982. The relationship between factor endowments and commodity trade. *Journal of International Economics* 13, November, 201–14.

Dornbusch, R., Fischer, S. and Samuelson, P.A. 1977. Comparative advantage, trade and payments in a Ricardian model with a continuum of goods. *American Economic Review* 67, December, 823–9.

Dornbusch, R., Fischer, S. and Samuelson, P.A. 1980. Heckscher–Ohlin trade theory with a continuum of goods. *Quarterly Journal of Economics* 95, September, 203–24.

Drabicki, J.Z. and Takayama, A. 1979. An antinomy in the theory of comparative advantage. *Journal of International Economics* 9, May, 211–23.

Edgeworth, F.Y. 1894. The theory of international values. *Economic Journal* 4, March, 35–50; 4, September, 424–43; 4, December, 606–38.

Edgeworth, F.Y. 1908. Appreciations of mathematical theories. III. *Economic Journal* 18, September, 392–403; 18, December, 541–56.

Ethier, W. 1974. Some of the theorems of international trade with many goods and factors. *Journal of International Economics* 4, May, 199–206.

Ethier, W. 1979. Internationally decreasing costs and world trade. *Journal of International Economics* 9, February, 1–24.

Ethier, W.J. 1982. National and international returns to scale in the modern theory of international trade. *American Economic Review* 72, June, 389–405.

Falvey, R.E. 1975. A note on the distinction between tariffs and quotas. *Economica*, NS 42, 319–26.

Färe, R. 1979. On the flatness of the transformation surface, a counterexample. Discussion Paper No. 79–117, Department of Economics, Southern Illinois University, Carbondale, Ill.

Fenchel, W. 1953. *Convex Cones, Sets and Functions.* Princeton University, Department of Mathematics (mimeo).

Finger, J.M. 1975. Trade overlap and intra-industry trade. *Economic Inquiry* 13, December, 581–9.

Foster, J.L. 1804. *An Essay on the Principle of Commercial Exchanges, and More Particularly of Exchange between Great Britain and Ireland.* London: J. Hatchard.

Gale, D. and Nikaido, H. 1965. The Jacobian matrix and global univalence of mappings. *Mathematische Annalen* 49, 81–93.

Gorman, W.M. 1958. Tariffs, retaliation, and the elasticity of demand for imports. *Review of Economic Studies* 25, June, 133–62.

Graaff, J. de V. 1949. On optimum tariff structures. *Review of Economic Studies* 17(1), 47–59.

Graham, F.D. 1922. International trade under depreciated paper. The United States, 1862–79. *Quarterly Journal of Economics* 36, February, 220–73.

Graham, F.D. 1923. The theory of international values re-examined. *Quarterly Journal of Economics* 38, November, 54–86.

Graham, F.D. 1925. Germany's capacity to pay and the reparation plan. *American Economic Review* 15, June, 209–27.

Grubel, H.G. and Lloyd, P.J. 1975. *Intra-Industry Trade.* London: Macmillan.

Haberler, G. 1930. Die Theorie der komparativen Kosten und ihre Auswertung für die Begründung des Freihandels. *Weltwirtschaftliches Archiv* 32, July, 350–70. English translation: The theory of comparative costs and its use in the defense of free trade, in *Selected Essays of Gottfried Haberler*, ed. A.Y.C. Koo, Cambridge, Mass.: MIT Press, 3–19.

Haberler, G. 1933. *Die internationale Handel.* Berlin: Julius Springer. English translation (revised by the author): *The Theory of International Trade with its Applications to Commercial Policy.* London: William Hodge & Company, 1936.

Hawkins, D. and Simon, H.A. 1949. Note: some conditions of macroeconomic stability. *Econometrica* 17, July–October, 245–8.

Heckscher, E. 1919. The effect of foreign trade on the distribution of income (in Swedish). *Ekonomisk Tidskrift* 21, II, 1–32. English translation in *Readings in the Theory of International Trade*, ed. H.S. Ellis and L.A. Metzler, Homewood, Ill.: Irwin, 1950, 272–300.

Helpman, E. 1981. International trade in the presence of product differentiation, economies of scale and monopolistic competition: a Chamberlin–Heckscher–Ohlin approach. *Journal of International Economics* 11, August, 305–40.

Helpman, E. and Krugman, P.R. 1985. *Market Structure and Foreign Trade.* Cambridge, Mass.: MIT Press.

Herberg, H. and Kemp, M.C. 1969. Some implications of variable returns to scale. *Canadian Journal of Economics* 2, August, 403–15.

Hirschman, A.O. 1945. *National Power and the Structure of Foreign Trade.* Berkeley and Los Angeles: University of California Press. Expanded edn, 1979.

Hirschman, A.O. 1949. Devaluation and the trade balance: a note. *Review of Economics and Statistics* 31, February, 50–53.

Horwell, D.J. 1966. Optimum tariffs and tariff policy. *Review of Economic Studies* 33, April, 147–58.

Hufbauer, G.C. 1966. *Synthetic Materials and the Theory of International Trade.* London: Duckworth.

Inada, K. 1967. A note on the Heckscher–Ohlin theorem. *Economic Record* 43, March, 88–96.

Inada, K. 1971. The production coefficient matrix and the Stolper–Samuelson condition. *Econometrica* 39, March, 219–39.

Inoue, T. 1981. A generalization of the Samuelson reciprocity relation, the Stolper–Samuelson theorem, and the Rybczynski theorem under variable returns to scale. *Journal of International Economics* 11, February, 79–98.

Inoue, T. 1986. On the shape of the world production possibility frontier with three goods and two primary factors with and without capital mobility. *International Economic Review* 27, October, 707–26.

Inoue, T. and Wegge, L.L. 1986. On the geometry of the production possibility frontier. *International Economic Review* 27, October, 727–37.

Jensen, R. and Thursby, M. 1986. A strategic approach to the product life cycle. *Journal of International Economics* 21, November, 269–84.

Johnson, H.G. 1950. Optimum welfare and maximum revenue tariffs. *Review of Economic Studies* 19, No. 1, 28–35.

Johnson, H.G. 1954. Optimum tariffs and retaliation. *Review of Economic Studies* 21, No. 2, 142–53.

Johnson, H.G. 1960. Income distribution, the offer curve, and the effect of tariffs. *Manchester School of Economic and Social Studies* 28, September, 215–42.

Johnson, H.G. 1965. The theory of tariff structure, with special reference to world trade and development. In H.G. Johnson and P.B. Kenen, *Trade and Development*, Geneva: Librairie Droz, 9–29.

Jones, K. 1984. The political economy of voluntary export restraint agreements. *Kyklos* 37, Fasc. 1, 82–101.

Jones, R.W. 1956. Factor proportions and the Heckscher–Ohlin theorem. *Review of Economic Studies* 24, No. 1, 1–10.

Jones, R.W. 1961. Stability conditions in international trade: a general equilibrium analysis. *International Economic Review* 2, May, 199–209.

Jones, R.W. 1965a. Duality in international trade: a geometrical note. *Canadian Journal of Economics and Political Science* 31, August, 390–93.

Jones, R.W. 1965b. The structure of simple general equilibrium models. *Journal of Political Economy* 73, December, 557–72.

Jones, R.W. 1968. Variable returns to scale in general equilibrium theory. *International Economic Review* 9, October, 261–72.

Jones, R.W. 1969. Tariffs and trade in general equilibrium: comment. *American Economic Review* 59, June, 418–24.

Jones, R.W. 1970. The transfer problem revisited. *Economica*, NS 37, May, 178–84.

Jones, R.W. 1971. Effective protection and substitution. Journal of International Economics 1, February, 59–82.

Jones, R.W. 1974a. The Metzler tariff paradox: extensions to nontraded and intermediate commodities. In *Trade, Stability, and Macroeconomics*, ed. G. Horwich and P.A. Samuelson, New York: Academic Press, 3–18.

Jones, R.W. 1974b. Trade with non-traded goods: the anatomy of interconnected markets. *Economica*, NS 41, May, 121–38.

Jones, R.W. 1975. Presumption and the transfer problem. *Journal of International Economics* 5, August, 263–74.

Jones, R.W. and Scheinkman, J.A. 1977. The relevance of the two-sector production model in trade theory. *Journal of Political Economy* 85, October, 909–35.

Katzner, D.W. 1970. *Static Demand Theory.* New York: Macmillan.

Kemp, M.C. 1964. *The Pure Theory of International Trade.* Englewood Cliffs, NJ: Prentice-Hall.

Kemp, M.C. and Wegge, L.L.F. 1969. On the relation between commodity prices and factor rewards. *International Economic Review* 10, October, 407–13.

Kemp, M.C., Khang, C. and Uekawa, Y. 1978. On the flatness of the transformation surface. *Journal of International Economics* 8, November, 537–42.

Keynes, J.M. 1929. The German transfer problem. *Economic Journal* 39, March, 1–7. A rejoinder, 39, June, 179–82.

Keynes, J.M. 1930. *A Treatise on Money.* 2 vols, London: Macmillan.

Khang, C. 1971. An isovalue locus involving intermediate goods and its applications to the pure theory of international trade. *Journal of International Economics* 1, August, 315–25.

Khang, C. 1973. Factor substitution in the theory of effective protection: a general equilibrium analysis. *Journal of International Economics* 3, August, 227–43.

Khang, C. and Uekawa, Y. 1973. The production possibility set in a model allowing interindustry flows: the necessary and sufficient conditions for its strict convexity. *Journal of International Economics* 3, August, 283–90.

King, P., 7th Baron. 1804. *Thoughts on the Effects of the Bank Restrictions*. 2nd edn, London: T. Cadell and W. Davies.

Kojima, K. 1964. The pattern of international trade among advanced countries. *Hitotsubashi Journal of Economics* 5, June, 16–36.

Kolmogorov, A.N. and Fomin, S.V. 1957. *Functional Analysis*, Vol. 1. Translated from the 1st (1954) Russian edn, Rochester, NY: Graylock Press.

Krugman, P.R. 1979. Increasing returns, monopolistic competition, and international trade. *Journal of International Economics* 9, November, 469–79.

Krugman, P. 1980. Scale economies, product differentiation, and the pattern of trade. *American Economic Review* 70, December, 950–59.

Krugman, P. 1982. Trade in differentiated products and the political economy of trade liberalization. In *Import Competition and Response*, ed. J.N. Bhagwati, Chicago: University of Chicago Press, 197–208.

Kuga, K. 1973. Tariff retaliation and policy equilibrium. *Journal of International Economics* 3, November, 351–66.

Kuhn, H. 1968. Lectures on mathematical economics. In *Mathematics of the Decision Sciences*, Part 2, ed. G.B. Dantzig and A.F. Veinott, Providence, RI: American Mathematical Society, 49–84.

Lancaster, K. 1980. Intra-industry trade under perfect monopolistic competition. *Journal of International Economics* 10, May, 151–75.

Leamer, E.E. 1980. The Leontief paradox, reconsidered. *Journal of Political Economy* 88, June, 495–503.

Leamer, E.E. 1985. *Sources of International Comparative Advantage*. Cambridge, Mass.: MIT Press.

Leontief, W. 1933. The use of indifference curves in the analysis of foreign trade. *Quarterly Journal of Economics* 47, May, 493–503.

Leontief, W. 1953. Domestic production and foreign trade; the American capital position re-examined. *Proceedings of the American Philosophical Society* 97, September, 332–49.

Lerner, A.P. 1932. The diagrammatical representation of cost conditions in international trade. *Economica* 12, August, 346–56.

Lerner, A.P. 1933. Factor prices and international trade. *Economica* NS 19, February, 1952, 1–15.

Lerner, A.P. 1934. The diagrammatical representation of demand conditions in international trade. *Economica*, NS 1, August, 319–34.

Lerner, A.P. 1936. The symmetry between import and export taxes. *Economica* NS 3, August, 306–13.

Lerner, A.P. 1944. *The Economics of Control*. New York: Macmillan.

Linder, S.B. 1961. *An Essay on Trade and Transformation*. Stockholm: Almqvist & Wiksell, and New York: Wiley.

List, F. 1841. *Das nationale System der politischen Oekonomie*. Jena: Gustav Fischer, 1904.

Malthus, T.R. 1811a. Depreciation of paper currency. *Edinburgh Review* 17, February, 339–72.

Malthus, T.R. 1811b. Pamphlets on the bullion question. *Edinburgh Review* 18, August, 448–70.

Marshall, A. 1879. *The Pure Theory of Foreign Trade*. London: London School of Economics and Political Science, 1930.

Marshall, A. 1923. *Money Credit and Commerce*. London: Macmillan.

Mayer, W. 1981. Theoretical considerations on negotiated tariff adjustments. *Oxford Economic Papers*, N.S. 33, March, 135–53.

McCulloch, R. and Yellen, J. 1982. Can capital movements eliminate the need for technology transfer? *Journal of International Economics* 12, February, 95–106.

McDougall, I.A. 1965. Non-traded goods and the transfer problem. *Review of Economic Studies* 32, January, 67–84.

McKenzie, L.W. 1955. Equality of factor prices in world trade. *Econometrica* 23, July, 239–57.

McKenzie, L.W. 1967. The inversion of cost functions: a counter-example. *International Economic Review* 8, October, 271–8. Theorem and counter-example: 8, October, 279–85.

McKinnon, R.I. 1966. Intermediate products and differential tariffs: a generalization of Lerner's symmetry theorem. *Quarterly Journal of Economics* 80, November, 584–615.

McMillan, J. 1986. *Game Theory in International Economics*. London: Harwood Academic Publishers.

Meade, J.E. 1952. *A Geometry of International Trade*. London: George Allen & Unwin.

Melvin, J.R. 1968. Production and trade with two factors and three goods. *American Economic Review* 58, December, 1249–68.

Melvin, J.R. 1969a. Intermediate goods in production theory: the differentiable case. *Review of Economic Studies* 36, January, 124–31.

Melvin, J.R. 1969b. Intermediate goods, the production possibility curve, and gains from trade. *Quarterly Journal of Economics* 83, February, 141–51.

Melvin, J.R. 1971. International trade theory without homogeneity. *Quarterly Journal of Economics* 85, February, 66–76.

Metzler, L.A. 1942. The transfer problem reconsidered. *Journal of Political Economy* 50, June, 397–414.

Metzler, L.A. 1949a. Tariffs, the terms of trade, and the distribution of national income. *Journal of Political Economy* 57, February, 1–29.

Metzler, L.A. 1949b. Tariffs, international demand, and domestic prices. *Journal of Political Economy* 57, August, 345–51.

Michaely, M. 1962. *Concentration in International Trade*. Amsterdam: North-Holland.

Mill, J.S. 1844. *Essays on some Unsettled Questions of Political Economy*. London: John W. Parker.

Mill, J.S. 1848. *Principles of Political Economy with some of their Applications to Social Philosophy*. 2 vols, London: John W. Parker.

Mundell, R.A. 1960. The pure theory of international trade. *American Economic Review* 50, March, 67–110.

Neary, J.P. 1985. Two-by-two international trade theory with many goods and factors. *Econometrica* 53, September, 1233–47.

Negishi, T. 1972. *General Equilibrium Theory and International Trade*. Amsterdam: North-Holland.

Nicholson, J.S. 1903. *Principles of Political Economy*, 2nd edn. 3 vols, London: Adam and Charles Black.

Nikaido, H. 1972. Relative shares and factor price equalization. *Journal of International Economics* 2, August, 257–64.

Ohlin, B. 1928. The reparations problem. *Index* (Svenska Handelsbanken, Stockholm), No. 28, 2–33.

Ohlin, B. 1929. Transfer difficulties, real and imagined. *Economic Journal* 39, June, 172–8.

Ohlin, B. 1933. *Interregional and International Trade*. Cambridge, Mass.: Harvard University Press.

Oniki, H. and Uzawa, H. 1965. Patterns of trade and investment in a dynamic model of international trade. *Review of Economic Studies* 32, January, 15–38.

Otani, Y. 1973. Neo-classical technology sets and properties of production possibility sets. *Econometrica* 41, July, 667–82.

Otani, Y. 1980. Strategic equilibrium of tariffs and general equilibrium. *Econometrica* 48, April, 643–62.

Panchamukhi, V.R. 1961. A theory for optimum tariff policy. *Indian Economic Review* 9, October, 178–98.

Pareto, V. 1896–97. *Cours d'économie politique*. 2 vols, Lausanne: F. Rouge.

Pearce, I.F. 1967. More about factor price equalization. *International Economic Review* 8, October, 255–70. Rejoinder to Professor McKenzie. 8, October, 296–9.

Pigou, A.C. 1932. The effect of reparations on the real ratio of international interchange. *Economic Journal* 42, December, 532–43.

Posner, M.V. 1961. International trade and technical change. *Oxford Economic Papers* NS 13, October, 323–41.

Rader, J.T. 1964. Edgeworth exchange and general economic equilibrium. *Yale Economic Essays* 4(1), 133–80.

Ramaswami, V.K. and Srinivasan, T.N. 1971. Tariff structure and resource allocation in the presence of factor substitution. In *Trade, Balance of Payments, and Growth*, ed. J.N. Bhagwati,

R.W. Jones, R.A. Mundell, and J. Vanek, Amsterdam: North-Holland, 291–9.

Rao, V.S. 1971. Tariffs and welfare of factor owners: a normative extension of the Stolper–Samuelson theorem. *Journal of International Economics* 1, November, 401–15.

Ricardo, D. 1811. *The High Price of Bullion, a Proof of the Depreciation of Bank Notes*. 4th edn, London: John Murray. Reprinted in Sraffa (1951, Vol. III, 47–127).

Ricardo, D. 1817. *On the Principles of Political Economy, and Taxation*. London: John Murray. Reprinted in Sraffa (1951, Vol. I).

Riezman, R. 1974. A note on the Heckscher–Ohlin theorem. *Tijdschrift voor Economie* 19(3), 339–43.

Riezman, R. 1982. Tariff retaliation from a strategic viewpoint. *Southern Economic Journal* 48, January, 583–93.

Robertson, D.H. 1931. The transfer problem. In A.C. Pigou and D.H. Robertson, *Economic Essays and Addresses*, London: Macmillan, 170–81.

Rockafellar, F.T. 1970. *Convex Analysis*. Princeton: Princeton University Press.

Rodriguez, C.A. 1974. The non-equivalence of tariffs and quotas under retaliation. *Journal of International Economics* 4, August, 295–8.

Rybczynski, T.M. 1955. Factor endowment and relative commodity prices. *Economica* NS 22, November, 336–41.

Samuelson, P.A. 1947. *Foundations of Economic Analysis*. Cambridge, Mass.: Harvard University Press.

Samuelson, P.A. 1949. International factor-price equalisation once again. *Economic Journal* 59, June, 181–97.

Samuelson, P.A. 1952. The transfer problem and transport costs: the terms of trade when impediments are absent. *Economic Journal* 62, June, 278–304.

Samuelson, P.A. 1953. Prices of factors and goods in general equilibrium. *Review of Economic Studies* 21(1), 1–20.

Samuelson, P.A. 1954. The transfer problem and transport costs, II: analysis of effects of trade impediments. *Economic Journal* 64, June, 264–89.

Samuelson, P.A. 1971. On the trail of conventional beliefs about the transfer problem. In *Trade, Balance of Payments, and Growth*, ed. J.N. Bhagwati, R.W. Jones, R.A. Mundell and J. Vanek, Amsterdam: North-Holland, 327–51.

Scitovsky, T. 1942. A reconsideration of the theory of tariffs. *Review of Economic Studies* 9, No. 2, 89–110.

Shephard, R.W. 1953. *Cost and Production Functions*. Princeton: Princeton University Press. Reprinted, Berlin, Heidelberg, New York: Springer-Verlag, 1981

Smith, A. 1776. *An Inquiry into the Nature and Causes of the Wealth of Nations*. 2 vols, London: W. Strahan and T. Cadell.

Sontheimer, K.C. 1971. The existence of international trade equilibrium with trade tax-subsidy distortions. *Econometrica* 39, November, 1015–35.

Stolper, W. and Samuelson, P.A. 1941. Protection and real wages. *Review of Economic Studies* 9, November, 58–73.

Sraffa, P. 1951. *The Works and Correspondence of David Ricardo*. Vols I–IV, Cambridge: Cambridge University Press.

Taussig, F.W. 1917. International trade under depreciated paper. A contribution to theory. *Quarterly Journal of Economics* 31, May, 380–403.

Taussig, F.W. 1927. *International Trade*. New York: Macmillan.

Thornton, H. 1802. *An Enquiry into the Nature and Effects of the Paper Credit of Great Britain*. London: J. Hatchard. Reprinted, ed. F.A.v. Hayek, London: George Allen & Unwin, 1939; reprinted, New York: Augustus M. Kelley, 1962.

Thursby, M. and Jensen, R. 1983. A conjectural variation approach to strategic tariff equilibria. *Journal of International Economics* 14, February, 145–61.

Torrens, R. 1815. *An Essay on the External Corn Trade*. London: J. Hatchard.

Torrens, R. 1844. *The Budget. On Commercial and Colonial Policy*. London: Smith, Elder and Co.

Tower, E. 1975. The optimum quota and retaliation. *Review of Economic Studies* 42, October, 623–30.

Uekawa, Y. 1971. Generalization of the Stolper–Samuelson theorem. *Econometrica* 39, March, 197–217.

Uekawa, Y. 1972. On the existence of incomplete specialization in international trade with capital mobility. *Journal of International Economics* 2, February, 1–23.

Uekawa, Y. 1979. The theory of effective protection, resource allocation, and the Stolper–Samuelson theorem: the many-industry case. *Journal of International Economics* 9, May, 151–71

Uekawa, Y., Kemp, M.C. and Wegge, L.L. 1973. P- and PN-matrices, Minkowski- and Metzler-matrices, and generalizations of the Stolper–Samuelson and Samuelson–Rybczynski theorems. *Journal of International Economics* 3, February, 53–76.

Uzawa, H. 1962. Production functions with constant elasticity of substitution. *Review of Economic Studies* 29, October, 291–99.

Uzawa, H. 1964a. Optimal growth in a two-sector model of capital accumulation. *Review of Economic Studies* 31, January, 1–24.

Uzawa, H. 1964b. Duality principles in the theory of cost and production. *International Economic Review* 5, May, 216–20.

Vanek, J. 1968. The factor proportions theory: the N-factor case. *Kyklos* 21, Fasc. 4, 749–54.

Verdoorn, P.J. 1960. The intra-block trade of Benelux. In *Economic Consequences of the Size of Nations*, ed. E.A.G. Robinson, London: Macmillan, 291–329.

Vernon, R. 1966. International investment and international trade in the product cycle. *Quarterly Journal of Economics* 80, May, 190–207.

Viner, J. 1924. *Canada's Balance of International Indebtedness 1900–1913*. Cambridge, Mass.: Harvard University Press.

Viner, J. 1937. *Studies in the Theory of International Trade*. New York: Harper & Brothers.

Walras, L. 1874. *Éléments d'économie politique pure*. Lausanne: Corbaz.

Wan, H.Y. 1971. A simultaneous variational model for international capital movement. In *Trade, Balance of Payments, and Growth*, ed. J.N. Bhagwati, R.W. Jones, R.A. Mundell, and J. Vanek, Amsterdam: North-Holland, 261–87.

Wegge, L.L.F. and Kemp, M.C. 1969. Generalizations of the Stolper–Samuelson and Samuelson–Rybczynski theorems in terms of conditional input–output coefficients. *International Economic Review* 10, October, 414–25.

White, H.D. 1933. *The French International Accounts, 1880–1913*. Cambridge, Mass.: Harvard University Press.

Wicksell, K. 1918. International freights and prices. *Quarterly Journal of Economics* 32, February, 404–10.

Williams, J.H. 1920. *Argentine International Trade under Inconvertible Paper Money*. Cambridge, Mass.: Harvard University Press.

Wilson, R. 1931. *Capital Imports and the Terms of Trade*. Melbourne; Melbourne University Press in association with Macmillan.

Woodland, A.D. 1980. Direct and indirect trade utility functions. *Review of Economic Studies* 47, October, 907–26.

Woodland, A.D. 1982. *International Trade and Resource Allocation*. Amsterdam: North-Holland.

interpersonal utility comparisons. Suppose I am left with a ticket to a Mozart concert I am unable to attend and decide to give it to one of my closest friends. Which friend should I actually give it to? One thing I will surely consider in deciding this is which friend of mine would enjoy the concert *most*. More generally, when we decide as private individuals whom to help, or decide as voters or as public officials who are to receive government help, *one* natural criterion we use is who would derive the greatest benefit, that is, who would derive the *highest utility*, from this help. But to answer this last question we must make, or at least attempt to make, *interpersonal utility comparisons*.

At the common-sense level, all of us make such interpersonal comparisons. But philosophical reflection might make us uneasy about their meaning and validity. We have direct introspective access only to our *own* mental processes (such as our preferences and our feelings of satisfaction and dissatisfaction) defining our *own* utility function, but have only very indirect information about other people's mental processes. Many economists and philosophers take the view that our limited information about other people's minds renders it impossible for us to make meaningful interpersonal comparisons of utility.

COMPARISONS OF UTILITY LEVELS VS. COMPARISONS OF UTILITY DIFFERENCES. In any case, if such comparisons are possible at all, then we must distinguish between interpersonal comparisons of utility *levels* and interpersonal comparisons of utility *differences* (i.e. utility increments or decrements).

It is one thing to compare the utility level $U_i(A)$ that individual i enjoys (or would enjoy) in situation A, with utility level $U_j(B)$ that another individual j enjoys (or would enjoy) in situation B (where A and B may or may not refer to the same situation). It is a very different thing to make interpersonal comparisons between utility differences, such as comparing the utility increment

$$\Delta U_i(A, A') = U_i(A') - U_i(A) \qquad (1)$$

that individual i would enjoy in moving from situation A to situation A', with the utility increment

$$\Delta U_j(B, B') = U_j(B') - U_j(B) \qquad (2)$$

that individual j would enjoy in moving from B to B'. Either kind of interpersonal comparison might be possible without the other kind being possible (Sen, 1970).

Some ethical theories would require one kind of interpersonal comparisons; others would require the other. Thus, *utilitarianism* must assume the interpersonal comparability of utility *differences* because it asks us to maximize a social utility function (social welfare function) defined as the *sum* of all individual utilities. (There are arguments for defining social utility as the *arithmetic mean*, rather than the *sum*, of individual utilities (Harsanyi, 1955). But for most purposes – other than analysing population policies – the two definitions are equivalent because if the number of individuals can be taken for a *constant*, then maximizing the sum of utilities is mathematically equivalent to maximizing their arithmetic mean.) Yet, we cannot add different people's utilities unless all of them are expressed in the same utility units; and in order to decide whether this is the case, we must engage in interpersonal comparisons of utility *differences*. (On the other hand, utilitarianism does not require comparisons of different people's utility *levels* because it does not matter whether their utilities are measured from comparable zero points or not.)

Likewise, the interpersonal utility comparisons we make in everyday life are most of the time comparisons of utility *differences*. For instance, the comparisons made in our example between the utilities that different people would derive from a concert obviously involve comparing utility differences.

In contrast, the utility-based version of Rawls's *Theory of Justice* (1971) does require interpersonal comparisons of utility *levels*, but does not require comparisons of utility *differences*. This is so because his theory uses the *maximin principle* (he calls it the *difference principle*) in evaluating the economic performance of each society, in the sense of using the well-being of the *worst-off* individual (or the worst-off social group) as its principal criterion. But to decide which individuals (or social groups) are worse off than others he must compare different people's utility levels. (In earlier publications, Rawls seemed to define the worst-off individual as one with the lowest utility level. But in later publications, he

defined him as one with the smallest amount of 'primary goods'. For a critique of Rawls's theory, see Harsanyi, 1975.)

ORDINALISM, CARDINALISM AND INTERPERSONAL COMPARISONS. In studying comparisons between the utilities enjoyed by *one* particular individual i, we again have to distinguish between comparisons of utility *levels* and comparisons of utility *differences*. The former would involve comparing the utility levels $U_i(A)$ and $U_i(B)$ that i assigns to two different situations A and B. The latter would involve comparing the utility increment

$$\Delta U_i(A, A') = U_i(A') - U_i(A) \qquad (3)$$

that i would enjoy in moving from situation A to situation A', with the utility increment

$$\Delta U_i(B, B') = U_i(B') - U_i(B) \qquad (4)$$

that he would enjoy in moving from B to B'.

If i has a well-defined utility function U_i at all, then he certainly must be able to compare the utility *levels* he assigns to various situations; and such comparisons will have a clear behavioural meaning because they will correspond to the preference and indifference relations expressed by his choice behaviour. In contrast, it is immediately less obvious whether comparing utility *differences* as defined under (3) and (4) has any economic meaning (but see below).

A utility function U_i permitting meaningful comparisons *only* between i's utility levels, but *not* permitting such comparisons between his utility differences, is called *ordinal*; whereas a utility function permitting meaningful comparisons *both* between his utility levels and his utility differences is called *cardinal*.

As is well known, most branches of economic theory use only ordinal utilities. But, as von Neumann and Morgenstern (1947) have shown, cardinal utility functions can play a very useful role in the theory of risk taking. In fact, utility-difference comparisons based on von Neumann–Morgenstern utility functions turn out to have a direct behavioural meaning. For example, suppose that U_i is such a utility function, and let Δ_i^* and Δ_i^{**} be utility differences defined by (3) and by (4). Then, the inequality $\Delta_i^* > \Delta_i^{**}$ will be algebraically equivalent to the inequality

$$\tfrac{1}{2}U_i(A') + \tfrac{1}{2}U_i(B) > \tfrac{1}{2}U_i(B') + \tfrac{1}{2}U_i(A). \qquad (5)$$

This inequality in turn will have the behavioural interpretation that i *prefers* an equi-probability mixture of A' and of B to an equi-probability mixture of B' and of A. Of course, once von Neumann–Morgenstern utility functions are used in the theory of risk taking, they become available for possible use also in other branches of economic theory, including welfare economics as well as in ethical investigations. (It has been argued that von Neumann–Morgenstern utility functions have no place in ethics (or in welfare economics) because they merely express people's attitudes toward *gambling*, which has no moral significance (Arrow, 1951, p. 10; and Rawls, 1971, pp. 172 and 323). But see Harsanyi, 1984.)

Note that by taking an ordinalist or a cardinalist position, one restricts the positions one can consistently take as to interpersonal comparability of utilities:

(1) An *ordinalist* is logically free to *reject* both types of interpersonal comparisons. Or he may *admit* comparisons of different people's utility *levels*. But he *cannot* admit the interpersonal comparability of utility differences without becoming a cardinalist. (The reason is this. If the utility

differences experienced by one individual i are comparable with those experienced by *another* individual j, this will make the utility differences experienced by *one* individual (say) i likewise indirectly comparable with one another, which will enable us to construct a *cardinal* utility function for each individual.)

(2) A *cardinalist* is likewise logically free to *reject* both types of interpersonal comparisons. Or he may *admit* both. Or else he may admit interpersonal comparisons only for utility *differences*. (Though it is hard to see why anybody might want to reject interpersonal comparisons for utility levels if he admitted them for utility differences.) But he *cannot* consistently admit interpersonal comparisons for utility *levels* while rejecting them for utility *differences*. (This can be verified as follows. If utility levels are interpersonallly comparable, then we can find four situations A, A', B, and B' such that $U_i(A) = U_j(B)$ and $U_j(A') = U_j(B')$. But then we can conclude that

$$\Delta_i^* = U_i(A') - U_i(A) = \Delta_j^* = U_j(B') - U_j(B),$$

which means that at least the utility differences Δ_i^* and Δ_j^* are interpersonally comparable. But since U_i and U_j are *cardinal* utility functions, any utility difference Δ_i^{**} experienced by i is comparable with Δ_i^*, and any utility difference Δ_j^{**} experienced by j is comparable with Δ_j^*. Yet this means that *all* utility differences Δ_i^{**} experienced by i are comparable with *all* utility differences Δ_j^{**} experienced by j. Thus, cardinalism together with interpersonal comparability of utility levels *entails* that of utility differences.)

EXTENDED UTILITY FUNCTIONS. In what follows, I will use the symbols A_i, B_i, ... to denote the economic and non-economic resources available to individual i in situations $A, B, \ldots$ Moreover, I will use the symbol A_j to denote an arrangement under which j has the same resources available to him as were available to individual i under arrangement A_i. These entities $A_i, B_i, \ldots, A_j, B_j, \ldots$ I will call *positions*.

Interpersonal utility comparisons would pose no problem if all individuals had the *same* utility function. For in this case, any individual j could assume that the utility level $U_i(A_i)$ that another individual i would drive from a given position A_i should be the *same* as he himself would derive from a similar position. Thus, j could write simply

$$U_i(A_i) = U_j(A_j). \tag{6}$$

Of course, in actual fact, the utility of different people are rather *different* because people have different *tastes*, that is, they have different abilities to derive satisfactions from given resource endowments. I will use the symbols R_i, R_j, ... to denote the vectors listing the personal psychological characteristics of each individual $i, j, \ldots$ that *explain* the differences among their utility functions $U_i, U_j, \ldots$. Presumably, these vectors summarize the effects that the genetic make-up, the education, and the life experience of each incividual have on his utility function. This means that any individual j can attempt to assess the utility level $U_i(A_i)$ that another individual j would enjoy in position A_i as

$$U_i(A) = V(A_i, R_i), \tag{7}$$

where the function V represents the psychological laws determining the utility functions $U_i, U_j, \ldots$ of the various individuals $i, j, \ldots$ in accordance with their psychological parameters specified by the vectors $R_i, R_j, \ldots$. Since, by assumption, all differences among the various individuals' utility functions $U_i, U_j, \ldots$ are fully explained by the vectors $R_j, R_j, \ldots$, the function V itself will be the same for all individuals. We will call V an *extended utility function*. (See Arrow, 1978; and Harsanyi, 1977, pp. 51 – 60; though the basic ideas are contained already in Arrow, 1951, pp. 114 – 15.)

To be sure, we know very little about the psychological laws determining people's utility functions and, therefore, know very little about the true mathematical form of the extended utility function V. This means that, when we try to use equation (7), the best we can do is to use our – surely very imperfect – personal *estimate* of V, rather than V itself. As a result, in trying to make interpersonal utility comparisons, we must expect to make significant errors from time to time – in particular when we are trying to assess the utility functions of people with a very different cultural and social background from our own. But even if our judgements of interpersonal comparisons can easily be *mistaken*, this does not imply that they are *meaningless*.

Ordinalists will interpret both the functions U_i and the function V as *ordinal* utility functions and will interpret (7) merely as a warrant for interpersonal comparisons of utility *levels* (cf. Arrow, 1978). In contrast, cardinalists will interpret all these as *cardinal* utility functions and will interpret (7) as a warrant for *both* kinds of interpersonal comparison (cf. Harsanyi, 1977).

LIMITS TO INTERPERSONAL COMPARISONS. It seems to me that economists and philosophers influenced by *logical positivism* have greatly exaggerated the difficulties we face in making interpersonal utility comparisons with respect to the utilities and the disutilities that people derive from ordinary commodities and, more generally, from the ordinary pleasures and calamities of human life. (A very influential opponent of the possibility of meaningful interpersonal utility comparisons has been Robbins, 1932.) But when we face the problem of judging the utilities and the disutilities that other people derive from various *cultural* activities, we do seem to run into very real, and sometimes perhaps even unsurmountable, difficulties. For example, suppose I observe a group of people who claim to derive great aesthetic enjoyment from a very esoteric form of abstract art, which does not have the slightest appeal to me in spite of my best efforts to understand it. Then, there may be no way for me to decide whether the admirers of this art form *really* derive very great and genuine enjoyment from it, or merely *deceive themselves* by claiming that they do.

Maybe in such cases interpersonal comparisons of utility do reach unsurmountable obstacles. But, fortunately, very few of our personal moral decisions and of our public political decisions depend on such exceptionally difficult interpersonal comparisons of utility. (References additional to those listed below will be found in Hammond, 1977 and in Suppes and Winet, 1955.)

JOHN C. HARSANYI

See also HEDONISM; INTERDEPENDENT PREFERENCES; PIGOU, ARTHUR CECIL; VALUE JUDGEMENTS; WELFARE ECONOMICS.

BIBLIOGRAPHY
Arrow, K.J. 1951. *Social Choice and Individual Values.* 2nd edn, New York: Wiley, 1963.
Arrow, K.J. 1978. Extended sympathy and the possibility of social choice. *Philosophia* 7, 223–37.
Hammond, P.J. 1977. Dual interpersonal comparisons of utility and the welfare economics of income distribution. *Journal of Public Economics* 7, 51–71.

Harsanyi, J.C. 1955. Cardinal utility, individualistic ethics, and interpersonal comparisons of utility. *Journal of Political Economy* 63, 309–21. Reprinted as ch. 2 of Harsanyi (1977).

Harsanyi, J.C. 1975. Can the maximum principle serve as a basis for morality? A critique of John Rawls' theory. *American Political Science Review* 69, 594–606. Reprinted as ch. 4 of Harsanyi (1977).

Harsanyi, J.C. 1976. *Essays on Ethics, Social Behavior and Scientific Explanation*. Dordrecht: Reidel.

Harsanyi, J.C. 1977. *Rational Behaviour and Bargaining Equilibrium in Games and Social Situations*. Cambridge: Cambridge University Press.

Harsanyi, J.C. 1984. Von Neumann–Morgenstern utilities, risk taking, and welfare. In *Arrow and the Ascent of Modern Economic Theory*, ed. G.R. Feiwel, New York: New York University Press, 545–58.

Rawls, J. 1971. *A Theory of Justice*. Cambridge, Mass.: Harvard University Press.

Robbins, L. 1932. *An Essay on the Nature and Significance of Economic Science*. London: Macmillan.

Sen, A.K. 1970. *Collective Choice and Social Welfare*. San Francisco: Holden-Day.

Suppes, P. and Winet, M. 1955. An axiomatization of utility based on the notion of utility differences. *Management Science* 1, 259–70.

Von Neumann, J. and Morgenstern, O. 1947. *Theory of Games and Economic Behavior*. 2nd edn, Princeton: Princeton University Press.

intertemporal equilibrium and efficiency. People, corporations and governments take decisions for the future. What kind of consistency exists between these decisions? What role does the price system play in this respect? Is the resulting evolution efficient? How can economic organization be improved in order to permit a more satisfactory growth?

Confronted with such huge questions, economists have often answered quickly. Even when attention is limited to formal theory, which this article exclusively considers, many statements can be found which, taken as valid for a time, were later disproved. They had been obtained on special models and too easily given a broad validity. Indeed, the preliminary step should have been to find a general formal representation of economic activity through time, but this step was not given sufficient attention until the late 19th century (Böhm-Bawerk, 1888; Fisher, 1907). The central model with reference to which the whole theory can be built and developed clearly emerged only in the 1950s.

A survey on the subject must then start from first principles and note which major features of reality are still today neglected in main-stream theory. The significance of the most far-reaching results and the importance of some big question marks will then have to be assessed.

INTERTEMPORAL DECISIONS. Households save for future consumption, employees work overtime so as to have enough to enjoy their vacation, students strive to get a diploma so as to hold good jobs later, parents want to leave bequest to their children. Firms produce to inventories in the expectation to future sales, recruit and train staff that will later improve their competitiveness, install equipment to be used for many years, build new factories.

The main theories dealing with intertemporal economic problems see such decisions as parts of plans that the relevant agents make for all their future activities. Any household for instance is assumed not only to decide its present supply of labour and demand for goods, but also simultaneously to choose its plan for the labour to be later supplied and the goods to be later consumed, and this up to the end of its existence.

The notion of this plan can in principle be made richer by taking *uncertainties* into account; the future decisions are then conditional on events to be later observed, but they are already specified for all conceivable combinations of events. In principle again the structure of the plan must then depend on the structure of the *information* that the agent will receive. In the main intertemporal theories these complications coming from uncertainties and information are, however, neglected, so that the concept of a plan does not appear to be unduly abstract. When the relevance of these theories is assessed, one has to wonder about the consequences of the simplification, as will be seen in the sequel.

Analysis of intertemporal behaviour can adopt the familiar approach: the constraints to which the plan is subject and the objectives that it strives to achieve must be identified; then the optimization problem is solved. The purest of all theories simply transpose the classical analysis of consumer and producer behaviour (Debreu, 1959). They assume the existence of a full system of discounted prices, with one such price for each commodity at each present or future date, a price at which agents will be able to buy or sell as much of this commodity as they may wish. They then directly reinterpret as follows the constraints and objectives that static atemporal theories made familiar.

As between the many plans that he can think of, a consumer is assumed to have a system of preferences that is often conveniently represented by a utility function, whose argument is a consumption vector with as many components as there are commodities and dates. A budget constraint requires that the discounted value of the consumption vector does not exceed a given amount, the consumer initial wealth. The chosen plan maximizes the utility function subject to the budget constraint. It then follows that the consumption of the various commodities (and the supply of labour) depend on what are the discounted prices and the initial wealth. The present saving of the consumer may be said to be equal to the interest income earned on his initial wealth *minus* the value of his present consumption (labour income appears negatively in this value). It is immaterial in this theory to know how saving is invested. Hence, the consumption plan and the resulting saving plan are seen as involving the whole future *life cycle* of the consumer (Modigliani and Brumberg, 1954).

The plan of a producer is subject only to the constraints that technology imposes. The producer acts as a price taker. His objective is to maximize the discounted value of the plan. It follows that demand for inputs and supply of outputs are functions of the discounted prices. The balance between the value of present outputs and present inputs gives the financial surplus if positive or requirement if negative; this is subject to no direct constraint.

Such a theory of consumer and producer behaviour does not claim to apply to all problems concerning this behaviour. Clearly, analysis of the firm in particular must usually go far beyond the stylized description given above, even simply when investment behaviour is being studied (Nickell, 1978). But the theory is supposed to be appropriate for fitting into the discussion of the broad questions raised by intertemporal equilibrium and efficiency.

Even when it is so circumscribed, the intent cannot be considered as fully achieved. Significant limitations must be kept in mind, since they may forbid application of the theory to some of the problems raised by equilibrium and efficiency over time; indeed, some of these limitations have been the motivation for theoretical developments that will not be discussed at length here, but must be mentioned.

Full knowledge of the system of discounted prices for purchases of sales at all relevant future dates is of course an abstraction. Forward prices exist for only a few basic commodities and a limited horizon. Whereas the interest rates at which one can borrow or lend for more or less long durations are fairly well defined, with non-negligible transaction costs and fiscal interference, however, prices that will apply to future transactions have to be forecast by the agents. The uncertainties that their forecast necessarily contains are neglected. Among the many consequences of this major simplification, one particularly notes that it rules out fundamental problems concerning the characterization of decision criteria of business firms (Drèze, 1982).

Constraints on individual choices are also reduced to a minimum. No consideration is given to quantitative constraints, such as those following from mass unemployment on individuals looking for jobs or from business depression on firms looking for customers. When such constraints are binding, not only must the plans meet them, but also spill-over effects from one period to others occur, according to laws that follow from the theory of individual behaviour under rationing (Samuelson, 1947). In particular, consumers willing but unable to borrow are constrained by their current resources, a phenomenon that gives some justification to the Keynesian consumption function relating current consumption to current income.

Neglect of financial constraints may be considered as following from other theoretical simplifications, lack of uncertainty and full knowledge of discounted prices, which rule out insolvency; but it is often particularly restrictive. The role of financial constraints on investment behaviour indeed play a major part in the development of trade cycle theories (Haberler, 1937).

Another notable feature of the theory is the simplicity of the trading relations that it assumes. Consumers and producers buy from 'the market' or sell to 'the market'. A worker need not establish ties with a particular employer, nor a manufacturing firm to a particular supplier of raw material. Actually, intertemporal decisions are often subject to quite significant irreversibilities. Long-term commitments are frequent for easily understandable reasons, some of which having to do with the specificities that characterize many production processes (for instance, most equipment, once bought, cannot be resold). Long-term contracts are also predominant on the labour market, even though many of their clauses often remain implicit. This feature motivates significant research nowadays, under the heading of 'implicit contracts' (Rosen, 1985).

Limited as it is, the classical theory of individual intertemporal decisions is, however, indispensable as a starting point, from which the study of the many complexities of real life can proceed. It has moreover brought to light some quite relevant results, such as the fact that, contrary to common belief, the saving of a household need not be an increasing function of interest rates or that individual choice is bound to exhibit some degree of impatience (Koopmans, 1960).

AN INTERTEMPORAL ECONOMY. The theory of general intertemporal equilibrium can also transpose the more familiar static theory. But clearly when so doing it does not go very far; new complications, specific to intertemporal problems, must be faced.

The simple transposition of the general competitive equilibrium assumes the existence of a terminal date, 'the horizon', a given set of consumers and producers whose activities end at this date, if not before. They all decide their plans at the initial date, on the basis of a full system of discounted prices, and acting as price takers. Perfect competition is assumed to imply that discounted prices are such that all markets clear; more precisely for a given date and a given commoditiy, aggregate supply and demand are defined by addition of corresponding individual supplies and demands contained in individual plans, which may then be considered as fully announced; at equilibrium the aggregate supply is precisely equal to aggregate demand, and this applies for any date and commodity. Hence, all individual plans are, from the initial date, mutually consistent for all future dates.

The usefulness of such an abstract equilibrium concept cannot be judged independently of its application, in particularly for the discussion of properties linking discounted prices to the agents' individual characteristics. Before facing this discussion, it is enlightening to consider how the model can be revised; this was done in three ways.

First, the hypothesis of a full system of markets, one for each date and commodity, has been relaxed and the notion of a *temporary equilibrium* made explicit (Hicks, 1939; Arrow and Hahn, 1971; Grandmont, 1977). Markets then exist only for the exchange of commodities at the (initial) present date, as well as for the loans of one numeraire commodity from the present to the next future date. Thus, present prices and the interest rate of the first period are assumed to be determined by the law of supply and demand, individual plans being made mutually consistent for the initial data. But in deciding their plans, individual agents have to form anticipations about future prices. Nothing guarantees that these anticipations are correct, so that individual plans will be revised with the passage of time, as actual prices are found to differ from what was expected.

Formal properties of this more realistic model will not be discussed here. Cases can be defined in which anticipations are later realized. It is then possible, but not always necessary when the future is unbounded, that the sequence of temporary equilibria coincides with the equilibrium defined from the hypothesis of a full system of markets. Thus, two sources of difficulty can arise: false anticipations and on the other hand instability following from the myopic functioning of the market system (Hahn, 1968).

Second, coming back to the case of a full system of markets, one has relaxed the assumption of a finite horizon with a fixed set of agents. The problem of knowing which firms exist has not been considered as specific to the intertemporal models, and has not been discussed thus far in the framework of these models, given that infinitely lived firms have been assumed. But since the initial proposals of Allais (1947) and Samuelson (1958), consumers are more and more assumed to belong to overlapping generations, each generation living only for a finite time. Such a representation of the consumption sector is clearly more appropriate for long-term analysis than the assumption of a given set of consumers living for ever, but it raises new difficulties (Balasko–Shell, 1980–81).

Third, since long-term phenomena are often involved, it has been found natural and convenient to concentrate attention on specifications in which the exogenous conditions of economic activity, such as technology, tastes, size of the population, natural resources, remain the same through time or change in a simple way; for instance, population increasing at a constant rate while technology exhibits constant returns to scale and natural resources are unbounded. Within such specifications one has dealt with the particular case of a stationary equilibrium, or else with equilibria in which production and consumption all increase at the same constant rate, that is, the case of 'proportional growth'. The analytical usefulness of this assumption of stationarity was at the centre of an important debate on the building of the theory of capital during the 1930s (Knight, 1935; Hayek, 1936). It follows from the simple form that has the price system of a stationary equilibrium: all discounted prices can be computed from the prices of the present commodities using a single interest rate that applies to all future periods of unit duration. 'The interest rate' is then unambiguously defined (Malinvaud, 1953).

ANY GENERAL LAW? A clear formalization of intertemporal equilibrium not only serves to aid progress in the fundamental conceptualization of economic activity (hence indirectly in the rigour of the discussions concerning many particular questions) but should also lead to comparative statics properties, which, dealing with intertemporal equilibria, have also been called 'comparative dynamics properties'. Particular importance has been given to the question of knowing how the interest rate changes from one stationary equilibrium to another when some specific change is being brought to its exogenous determinants.

The study of this question concentrated on a number of conjectures, which turned out to be about as many disappointments for whose who had expected to find rigorous proofs of their general validity. It is now realized that the rate of interest is related in a very complex way to the many exogenous determinants of equilibrium and that changes of relative prices, which are associated with changes of interest, may be responsible for paradoxical effects. A brief survey of this theoretical search, that extended over many years, nevertheless reveals some basic issues.

Does a high preference of individuals for present consumption necessarily imply a high interest rate? The property was often asserted. When first publishing his *Theory of Interest* in 1907, Irving Fisher called it an impatience theory. Only later when he revised the book for the 1930 edition did he add the subtitle 'as determined by impatience to spend income and opportunity to invest it', which recognizes the role of the productivity of investment (Samuelson, 1967). Quite significant cases have indeed been found in the overlapping generation model for which changes of impatience leave the interest rate unchanged (Samuelson, 1958).

Does a decrease of the rate of interest mean a lengthening of the production process? The positive answer was taken for granted, at least as long as technology was given, by many economists and was at the head of the 'Austrian Theory' as developed mainly by Böhm-Bawerk (1888) and Hayek (1941). Actually, description of the production process was usually organized in such a way as to focus on the conjectured property, this being true also with such non-Austrian authors as Wicksell (1901). Final output, available for consumption at some date, was seen as resulting from a number of well identified primary inputs made at previous dates and having 'matured' since then. The notion of an average period of production looked natural; an inverse relationship between this period and the rate of interest was expected. However, it turned out that, even restricting attention to the case of one primary input and one final output, one could not prove the relationship unless a special definition was given to the production period and a special phrasing to the property (Hicks, 1939, 1973). Generalization to many primary inputs, many final outputs and many interdependent production processes raises the fundamental difficulty resulting from induced variations in relative prices; it is quite unlikely that a generalized property could be proved (Sargan, 1955).

A somewhat similar property was expected with another

formalization that seems to be much more appropriate for describing technology in modern industry. The property concerns the choice of techniques and the notion that different techniques should be selected at various stages of development, as relative scarcity of the two main factors, labour and capital, changes and the interest rate moves accordingly. Its formal specification actually requires a particular model. The production possibility set is seen as resulting from combination of a number of elementary processes, each one operating at constant returns to scale, with fixed input–output coefficients, and requiring a time just equal to one period. Specifying further this model and applying it to an economy with one primary factor (labour), n produced goods and no joint production (the 'Samuelson–Leontief technology'), one defines a technique as a selection of n processes, one for the production of each good.

In this model, given any value of the interest rate, one can determine one technique that is fully appropriate for production, no matter what is the consumption basket. It then seemed natural to conjecture that techniques thus appearing as efficient at different interest rates were ordered from the less capitalistic (high interest) to the most capitalistic ones (low interest). However, this conjecture is not generally valid, even in this special model: as the interest rate progressively declines, one may have to switch at some point away from some technique but have to switch back to it at a later point: this is the case of 'reswitching of techniques' (Morishima, 1966).

Is the interest rate systematically smaller when, with a given technology, one shifts from a stationary equilibrium to another one using the same labour input but more productive capital? Again, this looked like a natural property to be stated. Since in a perfect equilibrium with no uncertainty the net rate of profit must be equal to the interest rate, the property was associated with the notion that capital accumulation must depress profit rates.

The property holds in a purely aggregated model with just one produced commodity, used both for consumption and as productive capital (Solow, 1956). The significance of this model for a more general situation was at the heart of hot debates in the late 1950s and early 1960s, the main opponents being located in the two academic cities named Cambridge (Robinson, 1956; Lutz and Hague, 1961). A side issue was whether one could give unambiguous definitions to such aggregate notions as the volume of productive capital and the marginal productivity of capital. Eventually, both counterexamples and formal analysis of the problem showed that the property was not generally valid (Burmeister and Turnovsky, 1972).

The significance of these various negative theoretical results should of course not be overstated. While reflecting the basic complexity of the relationship between the full system of discounted prices and its determinants, the results do not prove that 'pathological cases' are often empirically relevant.

INTERTEMPORAL EFFICIENCY. In the same way as the classical theory of individual behaviour, the theory of the optimum allocation of resources can be transposed to the intertemporal framework. Pareto efficiency of a 'programme' made of a set of individual plans, also called 'Pareto optimality', is generalized in an obvious way that need not be spelled out. The two classical duality theorems directly apply as long as the horizon is bounded: the programme resulting from a competitive equilibrium of the type described above is Pareto efficient if no external effect occurs; conversely, under a convexity or atomicity assumption, to any Pareto efficient programme can be associated a set of discounted prices supporting this programme. Properties of this system of prices are similar to those of the competitive price system.

Interesting new applications of these properties may give insights on the evolution of prices through time. In particular it is easily found that, if extraction costs are negligible, the discounted efficiency price of an exploited exhaustible resource is the same for all future dates, which means that the undiscounted price increases at a rate equal to the interest rate of the numeraire (Hotelling, 1931). When forming decisions on the use of exhaustible resources, one should give as much weight to the distant future as to the present; discounting gives no comfort for such decisions.

Theoretical difficulties, however, occur when the more realistic case of an unbounded horizon is being considered. The most relevant of these difficulties concerns the Pareto efficiency of competitive equilibria; efficiency is still proved to hold if the discounted value of the productive capital that exists at date t decreases to zero when one lets t increase to infinity (Malinvaud, 1953); but examples of competitive equilibria that do not fulfil this condition and are not Pareto efficient can be found. Such examples may be characterized as cases of overcapitalization, an excessive capital stock being indefinitely maintained without this ever benefiting consumption.

When attention is limited to stationary equilibria, a negative interest rate reveals lack of efficiency, whereas a positive one implies efficiency (if no external effect exists). Similarly, the interest rate of the price system supporting an efficient proportional growth programme cannot be smaller than the rate of growth (Starrett, 1970). The borderline case of an interest rate equal to the growth rate corresponds to what was called 'the golden rule'. More precisely, a new notion of optimality has been defined as follows for proportional growth programmes: an optimal programme is feasible and no other feasible programme leads to larger consumptions (i.e. a larger consumption of some commodity at some date and no smaller consumption of any commodity at any date). This definition neglects the conditions at the initial date since an 'optimal' programme can require a large input of capital at this date, a larger than is required by other Pareto efficient proportional growth programmes. It was proved that a price system exists that supports such an optimal programme and contains an interest rate equal to the rate of growth (Desrousseaux, 1961; Phelps, 1961). This is another case in which discounting does not make the distant future negligible.

When it is considered in the preceding terms, the theory of intertemporal efficiency has a somewhat unrealistic aspect; or rather it seems to be quite partial in its treatment of the various questions that intertemporal efficiency raises both for planning and for the study of actual economic evolution. Indeed, the restrictions mentioned in the first section of this article are often serious.

For the theory of planning, even restricted to the medium and long terms, for which intertemporal choices are particularly important, problems concerning the gathering and exchange of information should not be neglected. If a system of discounted prices is to be used for supporting consistency of individual decisions with national objectives, its determination must be given very serious consideration. Moreover, planning often aims at correcting handicaps, distortions or market failures preventing economic development. Its long-term achievement then depends on how well it deals with problems that are not considered here but have motivated an important literature, dealing in particular with the determination of the best shadow discount rate to be used in project evaluation (Dasgupta et al., 1972).

Similarly, for assessing the performance of actual economic systems, one has still to face many questions that again often relate to problems of information. Three of them seem to deserve particular attention. First, the vision of agents exchanging in markets abstracts too much from the complexities of actual contractual arrangements, some of which deal precisely with intertemporal choices; one does not yet clearly see how these complexities react on the behaviour of the full economy, nor even how theory could approach the issue.

Second, the notion of an intertemporal competitive equilibrium should be replaced by that of a sequence of competitive temporary equilibria. It is then known that, even if anticipations are self-fulfilling along this sequence, intertemporal efficiency is not guaranteed; more precisely, the short-sightedness of equilibria seems to increase the likelihood of an overcapitalization of the type exhibited by the theory of the golden rule. This may occur because of too high saving propensities, because of risk aversion or because of oligopolistic market structures (Malinvaud, 1981). But the question of knowing whether and when this likelihood will materialize remains obscure.

Third, the dual assumption of permanent market clearing and permanently equilibrating prices rules out of consideration many issues, such as those arising from variations in the degree of unemployment or in the stimulus given by profitability. A rather common view among supporters of the market system sees these variations as negligible from a long-term perspective, economic evolution being supposed simply to oscillate around the long-term path determined by equilibrium analysis. But critics of the market system and some other economists have the opposite view: economic disequilibria would provide the main clue for an understanding of the comparative growth of nations (Schumpeter, 1934; Beckerman, 1966). Theory remains conspicuously weak with respect to solving this major debate.

E. MALINVAUD

See also ARROW–DEBREU MODEL; GENERAL EQUILIBRIUM; MULTISECTOR GROWTH MODELS; MYOPIC DECISION RULES; OPTIMAL CONTROL AND ECONOMIC DYNAMICS; OWN RATES OF INTEREST; RAMSEY MODEL; SEQUENCE ECONOMIES.

BIBLIOGRAPHY

Allais, M. 1947. *Economie et intérêt*. Paris: Imprimerie Nationale.

Arrow, K. and Hahn, F. 1971. *General Competitive Analysis*. San Francisco: Holden-Day.

Balasko, Y. and Shell, K. 1980–81. The overlapping generations model. *Journal of Economic Theory*. I. The case of pure exchange without money, December 1980, 23(3), 281–306; II. The case of pure exchange with money, February 1981, 24(1), 112–42.

Beckerman, W. 1966. The determinants of economic growth. In *Economic Growth in Britain*, ed. P.D. Henderson, London: Weidenfeld & Nicolson.

Böhm-Bawerk, E. von. 1889. *Positive Theories des Kapitales*. Trans. as Vol. II of *Capital and Interest*, South Holland, Ill.: Libertarian Press, 1959.

Burmeister, E. and Turnovsky, S.J. 1972. Capital deepening response in an economy with heterogeneous capital goods. *American Economic Review* 62(5), December, 842–53.

Dasgupta, P., Marglin, S. and Sen, A. 1972. *Guidelines for Project Evaluation*. New York: UNIDO, United Nations.

Debreu, G. 1959. *Theory of Value: An Axiomatic Analysis of Economic Equilibrium*. New York: Wiley.

Desrousseaux, J. 1961. Expansion stable et taux d'intérêt optimal. *Annales des Mines*, November, Paris.

Dreze, J. 1982. Decision criteria for business firms. In *Current Developments in the Interface: Economics, Econometrics, Mathematics*, ed. M. Hazewinkel and A. Rinney Khan, Dordrecht: D. Reidel.

Fisher, I. 1907. *The Rate of Interest*. New York: Macmillan, 2nd edn, 1930.

Grandmont, J.-M. 1977. Temporary general equilibrium theory. *Econometrica* 45(3), April, 535–72.

Haberler, G. 1937. *Prosperity and Depression*. Geneva: League of Nations. 3rd enlarged edn, 1941.

Hahn, F. 1968. On warranted growth paths. *Review of Economic Studies* 35, 175–84.

Hayek, F.A. von. 1936. The mythology of capital. *Quarterly Journal of Economics* 50, February, 199–228.

Hayek, F.A. von. 1941. *The Pure Theory of Capital*. London: Routledge & Kegan Paul.

Hicks, J. 1939. *Value and Capital*. Oxford: Clarendon.

Hicks, J. 1973. *Capital and Time*. Oxford: Clarendon.

Hotelling, H. 1931. The economics of exhaustible resources. *Journal of Political Economy* 39, 137–75.

Knight, F. 1935. The theory of investment once more: Mr. Boulding and the Austrians. *Quarterly Journal of Economics* 50, November, 36–67.

Koopmans, T. C. 1960. Stationary ordinal utility and impatience. *Econometrica* 28, April, 287–309.

Lutz, F. and Hague, D. (eds) 1961. *The Theory of Capital*. London: Macmillan.

Malinvaud, E. 1953. Capital accumulation and efficient allocation of resources. *Econometrica* 21, April, 233–68.

Malinvaud, E. 1962. Efficient capital accumulation: a corrigendum. *Econometrica* 30, July, 570–73.

Malinvaud, E. 1981. *Théorie macroéconomique*, Vol. 1. Paris: Dunod.

Modigliani, F. and Brumberg, R. 1954. Utility analysis and the consumption function: an interpretation of cross-section data. In *Post-Keynesian Economics*, ed. K. Kurihara, New Brunswick, NJ: Rutgers University Press; London: George Allen & Unwin, 1955.

Morishima, M. 1966. Refutation of the nonswitching theorem. *Quarterly Journal of Economics* 80, November, 520–25.

Nickell, S. 1978. *The Investment Decisions of Firms*. Cambridge: Cambridge University Press.

Phelps, E. 1961. The golden rule of capital accumulation. *American Economic Review* 51(4), September, 638–42.

Robinson, J. 1956. *The Accumulation of Capital*. London: Macmillan.

Rosen, S. 1985. Implicit contracts. *Journal of Economic Literature* 23(3), September, 1144–75.

Samuelson, P. 1947. *Foundations of Economic Analysis*. Cambridge, Mass.: Harvard University Press.

Samuelson, P. 1958. An exact consumption-loan model of interest with or without the social contrivance of money. *Journal of Political Economy* 66, December, 467–82.

Samuelson, P. 1967. Irving Fisher and the theory of capital. In *Ten Economic Studies in the Tradition of Irving Fisher*, ed. W.J. Fellner et al., New York: John Wiley & Sons.

Sargan, J.D. 1955. The period of production. *Econometrica* 23, April, 151–65.

Schumpeter, J. 1934. *The Theory of Economic Development*. Cambridge, Mass.: Harvard University Press.

Solow, R. 1956. A contribution to the theory of economic growth. *Quarterly Journal of Economics* 70, February, 65–94.

Starrett, D. 1970. The efficiency of competitive programmes. *Econometrica* 38(5), September, 704–11.

Wicksell, K. 1901. *Vorlesungen über Nationalökonomie*. Trans. as *Lectures on Political Economy*, London: Routledge and Kegan Paul, 1934, Vol. I.

intertemporal portfolio theory and asset pricing. The intent of this entry is to present intertemporal portfolio theory and asset pricing models, to explain their results and to illustrate the differences between multiperiod and single-period models. To appreciate intertemporal portfolio theory and asset pricing, it is necessary to understand the state of finance theory prior to the seminal intertemporal works of Merton (1969, 1971,

1973), Samuelson (1969), Fama (1970), Hakansson (1970) and Rubinstein (1974). Section I presents single-period theory and some general results on portfolio statistics. Section II presents intertemporal portfolio theory. Section III presents the intertemporal asset pricing model, and Section IV presents the consumption-oriented representation of it. Section V gives important extensions (without proof) and concludes the entry.

I. SINGLE-PERIOD PORTFOLIO THEORY AND ASSET PRICING

Portfolio choice in terms of means and variances of alternative portfolios' returns was rigorously modelled first in a single-period world by Markowitz (1952, 1959) and Tobin (1958). This theory was significantly extended by Sharpe (1964) and Lintner (1965). By requiring markets to clear in equilibrium, Sharpe and Lintner developed the well-known theory of equilibrium asset prices known as the capital asset pricing model (CAPM). This model was the premier general theoretical model of asset pricing, prior to Merton's (1973) development of the *intertemporal* capital asset pricing model (ICAPM). In fact, despite the development of the theoretically superior (more general) intertemporal asset pricing models, the single-period CAPM is widely used by investment practitioners today.

I.A. PORTFOLIO STATISTICS. In deriving both the single-period and the intertemporal CAPM, there are a few well-known facts about portfolio statistics that are used repeatedly to expedite the derivations. Those will be presented with the notational definitions that follow. First, let $\mathbf{w}^k$ be individual k's $A \times 1$ vector of portfolio weights for risky assets; the ith element represents the fraction of total wealth that is invested in the ith risky asset. From the investor's budget constraint, the amount placed in the riskless asset must be the residual fraction, i.e.,

$$w_0^k = 1 - \sum_i w_i^k.$$

The riskless asset's return is denoted r_f, and risky assets have normally distributed returns with an $A \times 1$ vector of means, $\boldsymbol{\mu}$, and a variance–covariance matrix $\mathbf{V}$. Two statistical results permit the mean and variance of any portfolio and the covariance between any two portfolios' returns to be found from the weights of the portfolios and from the joint distribution of individual assets' returns. (The reader may verify these results from elementary statistical theory on the mean and variance of a linear combination of random variables.)

Mean portfolio return

$$= \mu_p = w_0 r_f + \mathbf{w}' \boldsymbol{\mu} = r_f + \mathbf{w}'(\boldsymbol{\mu} - r_f \mathbf{1}). \quad (1)$$

Covariance of 2 portfolios' returns

$$= \sum_i \sum_j w_i^x w_j^y \sigma_{ij} = \mathbf{w}_x' \mathbf{V} \mathbf{w}_y = \sigma_{xy}, \quad (2)$$

where $\mathbf{w}_x'$ and $\mathbf{w}_y$ are the risky asset portfolios and $\mathbf{1}$ is an $A \times 1$ vector of ones. A useful special case of (2) is that the variance of any portfolio's return is $\sigma^2 = \mathbf{w}' \mathbf{V} \mathbf{w}$. Another useful special case of (2) is that, for any portfolio $\mathbf{w}$, the matrix product $\mathbf{V} \mathbf{w}$ gives the $A \times 1$ vector of covariance of all assets returns with the specified portfolio's return. To see this, view each row of the $A \times A$ identity matrix $\mathbf{I}$ as a 1-asset portfolio, and then apply fact (2) row by row to the matrix product $\mathbf{I} \mathbf{V} \mathbf{w} = \mathbf{V} \mathbf{w}$. For reference, these two special cases of (2) will be denoted (2') and (2''), respectively. Armed with these definitions and facts, we can now expeditiously derive the well-known single-period portfolio theory and CAPM of Sharpe (1964) and Lintner (1965).

I.B. OPTIMAL PORTFOLIO CHOICE. Each individual chooses at time 0 a portfolio that maximizes the expected value of a von Neumann–Morgenstern utility function for wealth at time 1, i.e., max $E[u^k(\tilde{W}_1^k)]$. Since the return on a portfolio is a linear combination of the returns on individual assets, and since the returns on individual assets are assumed to be normally distributed, wealth at time 1 is normally distributed. Thus, given initial wealth W^k, the entire probability distribution for wealth at time 1 is described by the mean and variance of the individual's portfolio return. Rewriting the individual's expected utility as a function of portfolio mean and variance and omitting superscripts for the individual's preferences and portfolio weights, the portfolio choice problem is:

$$\max_{\{\mathbf{w}\}} U(\mu_W, \sigma_W^2) \qquad (3)$$

where $\mu_W = r_f + \mathbf{w}'(\boldsymbol{\mu} - r_f\mathbf{1})$ and $\sigma_W^2 = \mathbf{w}'\mathbf{V}\mathbf{w}$.

Since individuals like higher mean and lower variance, each portfolio that is maximal for (3) will be 'mean-variance efficient'. Efficient portfolios are those with the highest mean for a given variance, or alternatively, are lowest variance for a given mean.

The choices of the portfolio weights for risky assets are unconstrained in the above problem, since the budget constraint is imposed by making the weight in the riskless asset the residual (negative amounts indicating borrowing). Implicitly differentiating (3) with respect to the vector of risky portfolio weights and setting the partials equal to zero gives a set of linear equations. Solving these by matrix inversion gives the following optimal risky asset portfolio:

$$\mathbf{w}^k W^k = T^k[\mathbf{V}^{-1}(\boldsymbol{\mu} - r_f\mathbf{1})], \quad \text{for all individuals } k, \qquad (4)$$

where $T^k = -(\partial U/\partial\mu)W^k/[2(\partial U/\partial\sigma^2)]$ is individual k's compensating variation in variance for a unit change in mean, holding utility constant. Thus, the higher T^k is, the higher k's risk tolerance. Dividing (4) by the sum of the risky asset weights eliminates the individual's wealth and risk tolerance from the new equation, giving the optimum mix of risky asset holdings relative to the total in risky assets.

Thus, we have a remarkable result (first attributed to Tobin, 1958): the optimal mix of risky assets in the individual's portfolio depends only upon the means, variances and covariances of risky returns (as perceived by that individual). The individual's current wealth and preferences only affect risky assets' demands through a scalar that is the same for all risky assets. This shows that an individual may separate the choice of the optimal risky portfolio mix from the choice of how much to place in that portfolio and how much in the riskless asset. Sharpe (1964) showed that if all individuals have the same probability beliefs $\{\boldsymbol{\mu}, \mathbf{V}\}$, then the optimal mix of risky assets is the same for all individuals. In fact, if there were a mutual fund that held all risky assets in the proportions given by $\mathbf{V}^{-1}(\boldsymbol{\mu} - r_f\mathbf{1})$, all individuals could achieve their optimal portfolios with that fund and a riskless asset holding. This property is known as 'two-fund portfolio separation'.

I.C. MARKET EQUILIBRIUM. CAPITAL ASSET PRICING MODEL. The aggregate values of individuals' asset holdings, divided by the aggregate market value of wealth of the economy (M), gives 'the market's' portfolio weights. Summing (4) over individuals k and dividing by aggregate wealth M gives the market portfolio. $\mathbf{w}^M$:

$$\mathbf{w}^M = T^M[\mathbf{V}^{-1}(\boldsymbol{\mu} - r_f\mathbf{1})], \qquad (5)$$

where

$$T^M = \left(\sum_k T^k\right)/M.$$

Since the market portfolio is a solution to (3) for an appropriate constant, the market portfolio is mean–variance efficient. Premultiplying (5) by $\mathbf{V}$ and using the statistical fact (2″), we have that the expected excess returns on assets in equilibrium are proportional to their covariances with the market's return, $\mathbf{V}_{aM}$:

$$\boldsymbol{\mu} - r_f\mathbf{1} = (1/T^M)\mathbf{V}_{aM}. \qquad (6)$$

Pre-multiplying (5) by $\mathbf{w}'\mathbf{V}$, using formulae for the mean and variance of a portfolio, and rearranging gives the value for the risk tolerance parameter: $(1/T^M) = (\mu_M - r_f)/\sigma_M^2$. The inverse of risk tolerance is termed risk aversion, so higher risk aversion among investors shows up as a higher expected excess return per unit of variance for the market portfolio. Substituting this into (6) gives the well known capital asset pricing model of Sharpe (1964) and Lintner (1965):

CAPM: $\qquad \boldsymbol{\mu} - r_f\mathbf{1} = \boldsymbol{\beta}_M(\mu_M - r_f), \qquad (7)$

where $\boldsymbol{\beta}_M = \mathbf{V}_{aM}/\sigma_M^2$ is the $A \times 1$ vector of assets' betas relative to the market portfolio. They are analogous to the slope coefficients in regressions of assets' returns on the market portfolio's return.

To this date, this single-period capital asset pricing model has been the most widely tested general model of asset prices under uncertainty. It makes the very strong prediction that the expected excess returns across assets are proportional in equilibrium to their betas relative to the market portfolio. Alternatively, it predicts that the market portfolio is mean–variance efficient, in that it gives the highest expected excess return per unit of standard deviation, considering all possible portfolio combinations. Empirical tests of the single-period CAPM usually reject it. Higher beta assets do have higher returns, but the CAPM of (7) is rejected as a representation of the data. Virtually every assumption used in the derivation of the CAPM has been weakened and empirically examined. What follows is the generalization to *multiperiod* or *intertemporal* consumption and investment decisions – probably the most important and productive generalization.

II. INTERTEMPORAL PORTFOLIO THEORY

Relaxation of the single-period assumption in portfolio theory has proceeded concurrently in two very similar types of models. First, discrete-time multiperiod models consider individuals who make consumption and investment decisions at fixed points in time, where the interval between decisions is a somewhat arbitrary choice. It is unlikely that an individual would choose only to revise at fixed dates in time, regardless of what happens in between, so these models initially cause concern. However, that concern is alleviated somewhat by the fact that the qualitative properties of optimal policies in many models are unaffected by the choice of updating interval. Key works in discrete-time multiperiod frameworks are those of Samuelson (1969), Hakansson (1970), Fama (1970), Rubinstein (1974, 1976), Long (1974), Dieffenbach (1975), Kraus and Litzenberger (1975), Lucas (1978), Breeden and Litzenberger (1978) and Brennan (1979).

The other model used for intertemporal portfolio theory and asset pricing is the continuous-time model pioneered by Merton (1969, 1971, 1973), and further developed by Cox, Ingersoll and Ross (1985a,b) and Breeden (1979, 1984, 1986). The continuous-time model assumes that individuals make consumption and portfolio decisions continuously. Although this is not realistic, since individuals do sleep and do things other than make economic decisions, it will not miss important consumption and portfolio adjustments due to the modelling of a fixed time between decisions.

In Merton's continuous-time model, the underlying random processes driving economic uncertainties are assumed to follow continuous-time stochastic processes with normally distributed increments and continuous sample paths. The underlying normality makes the continuous-time model a logical extension for the single-period CAPM and also gives it mathematical tractability that is often not found in discrete-time models. For example, with discrete-time models, a normally distributed stock return results in non-zero probability of a negative stock price. In the continuous-time model, the variance of the stock's return can approach zero as the stock's price approaches zero in such a way as to prevent negative stock prices, but have normally distributed increments at every instant in time. This entry will utilize the continuous-time model, but any important economic intuition found can also be derived in a discrete-time model.

In the intertemporal model, it is assumed that individuals choose consumption and investment policies that maximize their expected utilities across possible *lifetime* consumption paths. In both continuous-time and discrete-time models, preferences are typically assumed to be time-additive and state-independent, i.e., expected lifetime utility for individual k is: $E[\int u^k(c^k, t) \, dt]$. Although these preferences are not as general as theorists would like, much has been learned with them. It is assumed that the utility of consumption at any instant is monotonically increasing and strictly concave in consumption, in that partial derivatives are: $u_c^k > 0$ and $u_{cc}^k < 0$.

In using the techniques of stochastic dynamic programming to find the best consumption and portfolio policies, it is convenient to break the remaining utility of lifetime consumption into two parts and maximize the sum. At time t the first part is $u^k(c^k, t)$, the utility of the current consumption over the next period (or instant in time). The second part is the expected utility of consumption for all subsequent periods to that, $J^k(W^k, \mathbf{s}, t)$, which will be explained more fully below. Thus, the objective function is:

$$\max_{\{c, \mathbf{w}\}} \{u^k(c^k, t) + E_i[J^k(W^k, \mathbf{s}, t)]\} \qquad (8)$$

The current choice of consumption affects only the first part directly, but affects the budget constraint for investments made for future consumption. Differentiating (8) with respect to current consumption, taking into account that each additional unit of consumption today is a unit less of investible wealth, gives the standard condition that the marginal utility of consumption equals the marginal utility of wealth for an optimal policy:

$$u_c^k[c^k(W^k, \mathbf{s}, t), t] = J_W^k(W^k, \mathbf{s}, t). \qquad (9)$$

The key difference between single-period portfolio theory and its CAPM and the optimal results in an intertemporal equilibrium arises from the nature of the indirect utility function for wealth, $J(W^k, \mathbf{s}, t)$. The portfolio mix decision affects only the probability distribution of future wealth and therefore only affects J in (8) – the expected utility of future consumption that wealth will be used to buy. The $S \times 1$ vector $\mathbf{s}$ is a set of 'state variables' that describe consumption, investment and employment opportunities. When a person expects to live not just for an instant more, but for a period of time, the investment portfolio and consumption rate should be reviewed and adjusted continually. The utility that one expects to get during one's remaining lifetime depends positively on current wealth

(since higher wealth buys more goods), but also depends upon the state of investment opportunities. For example, a current wealth of \$100,000 provides a lower real consumption stream if the real riskless interest rate is 2 per cent, than if the real rate is 5 per cent. In this case, the real riskless rate is one of the state variables for investment opportunities. Examples of other economic state variables are the expected inflation rate(s) of goods, the expected productivity of capital or the expected return on the market portfolio, and the level of uncertainty about economic activity or of productivity. Of course, most of these would be considered as endogenous variables; more generally, the underlying exogenous variables could be substituted and the stochastic processes for the endogenous variables derived.

To see the effect of a stochastic investment opportunity set on the investment portfolio, consider a retiree who is relatively averse to risk and holds the single-period optimal portfolio – a little money in the market portfolio and a lot in riskless securities. With that portfolio, the investor has the same wealth if the market is up 10 per cent and the riskless rate is 2 per cent, as when the market is up 10 per cent and the riskless rate is 5 per cent. This may not be optimal, since this retiree has to reinvest his wealth and live off the income. The retiree is financially hurt in the state where the real riskless rate is 2 per cent, and is well off in the 5 per cent state. In addition to the market portfolio, this investor may optimally wish to buy some long-term bonds or interest rate futures contracts that go up in value as rates fall. Then the investor is hedged, by having more wealth to compensate for the poor reinvestment rate. If rates increase, the retiree has a capital loss on the bonds and, therefore, less wealth, but has a better reinvestment rate. Some investors may well prefer this to just holding the market portfolio. Thus, as we shall see, the single-period CAPM's two-fund theorem and the asset pricing model itself will not necessarily hold in a multiperiod economy. These are points all made clear in Merton's (1973) pathbreaking work.

Merton (1973) derived the optimal portfolio rules for an individual in an exchange economy, and Cox, Ingersoll and Ross (1985b) verified those same portfolio rules in a general equilibrium economy with production. Let subscripts of the indirect utility function be partial derivatives, and let $\mathbf{V}_{aa}$ now be the $A \times A$ variance–covariance matrix for assets' returns and $\mathbf{V}_{as}$ be the $A \times S$ covariance matrix of assets' returns with the various state variables. The optimal portfolio of risky assets in the intertemporal economy is:

$$\mathbf{w}^k W^k = T^k[\mathbf{V}_{aa}^{-1}(\boldsymbol{\mu} - r_f \mathbf{1})] + \mathbf{V}_{aa}^{-1}\mathbf{V}_{as}\mathbf{H}_s^k,$$

$$\text{for all individuals } k, \quad (10)$$

where $\mathbf{H}_s^k = -\mathbf{J}_{sW}^k / J_{WW}^k$. Notice that the first RHS term of (10) is the mean–variance efficient portfolio as in the single-period equations of (4). As for the other term, Breeden (1979) showed that each column j of the product matrix $\mathbf{V}_{aa}^{-1}\mathbf{V}_{as}$ represents the portfolio of assets that is most highly correlated in return with movements in state variable j. To see this, note that the portfolio that has the maximum correlation with state variable s_j is the one with the highest covariance with s_j, given a fixed portfolio variance. Mathematically:

Objective: $\quad \max_{\{\mathbf{w}_j\}} L = \mathbf{w}_j' \mathbf{V}_{a, sj} + \lambda[\sigma^2 - \mathbf{w}_j' \mathbf{V}_{aa} \mathbf{w}_j]$

Solution: $\quad \mathbf{w}_j = [\mathbf{V}_{aa}^{-1}\mathbf{V}_{a, sj}](1/2\lambda). \qquad (11)$

(The scalar does not matter, since all portfolios that are scalar multiples are perfectly correlated and have the same correlations with all other variables.) Thus, those S portfolios are the best hedge portfolios available for individuals to use in hedging opportunity set changes. The coefficient vector in (10),

$\mathbf{H}_s^k$, gives individual k's holdings of those hedge portfolios (which may be positive or negative).

Aggregating individuals' portfolios gives the market portfolio. Substituting this back into (10) gives:

$$\mathbf{w}^k W^k = (T^k/T^M)\mathbf{w}^M + [\mathbf{V}_{aa}^{-1}\mathbf{V}_{as}][\mathbf{H}_s^k - (T^k/T^M)\mathbf{H}_s^M],$$

for all individuals k,   (12)

where

$$T^M = \sum_k T^k \quad \text{and} \quad \mathbf{H}_s^M = \sum_k \mathbf{H}_s^k.$$

From this, it is clear that all individuals' portfolios can be obtained with $S + 2$ funds: (1) the market portfolio, (2) the riskless asset, (3) and the S best hedge portfolios for the state variables. No preferences are needed to set up the mutual funds. Breeden (1984) showed that if each of the S hedge portfolios is *perfectly* correlated with the state variable it hedges, then the allocation of contingent claims is an unconstrained Pareto-optimal allocation (ex ante, as in Arrow, 1951). If there is not a perfect hedge for some state variable, then preferences can be chosen so that the allocation is not unconstrained Pareto-optimal.

To complete the analysis, the $\mathbf{H}_s^k$ terms need to be examined, so we know what types of holdings different individuals should have in the hedge portfolios. Without stronger preference assumptions, analysis of the hedging terms is difficult. However, if one assumes that the vector of percentage compensating variations in k's wealth for state variables' changes $(\gamma_s^k = -\mathbf{J}_s^k/W^k J_W^k)$ are not a function of k's wealth, then Breeden (1984) has shown that:

$$\mathbf{H}_s^k = W^k(1 - T^{*k})\gamma_s^k,$$ (13)

where T^{*k} is k's Pratt–Arrow measure of relative risk tolerance. Since $\mathbf{H}_s^k$ give individual k's holdings of the hedge portfolios for opportunity set changes, an individual will attempt to hedge if and only if his or her relative risk tolerance is less than unity. Since unity represents the logarithmic utility case, those more risk averse than the log will tend to hedge, whereas those more tolerant than the log will tend to 'reverse hedge'. This type of result has been obtained by Merton (1969), Grauer and Litzenberger (1979), Dieffenbach (1975) and Breeden (1984).

The optimality of reverse hedging if relative risk tolerance is greater than unity is a very interesting result, since one certainly cannot rule out those preferences. To understand this result, consider a stochastic expected return on investments in the stock market. Apart from holding the market portfolio, one might wish to hedge or reverse hedge changes in the expected return on the market. For both hedger and reverse hedger, let us assume that an increase in expected return on the market is a good thing, in that expected lifetime utility is positively related to that opportunity. A hedger would say that when the expected return on the market is high, he needs less wealth; on the other hand, when the expected return is low, he needs more wealth to keep up his planned lifetime consumption level.

The person who would reverse hedge would view things differently, but not irrationally. That person would wish to have a lot of wealth to invest when the expected return on the market is high, in order to take advantage of the good returns. When returns are poor, our relatively risk tolerant person would wish to have little wealth to invest. Clearly, this strategy generates a higher multiperiod mean return and a higher multiperiod variance of return than does the hedging strategy. Neither strategy dominates the other for all risk averse individuals. Which is chosen depends upon the person's marginal rate of substitution function of mean for variance.

III. INTERTEMPORAL CAPITAL ASSET PRICING MODEL (ICAPM)

Given the general portfolio theory of the last section, this section derives the general intertemporal asset pricing model of Merton (1973). The first step shows that equilibrium expected returns on all assets are linear combinations of their covariances with the market portfolio and with the S portfolios that are most highly correlated with the opportunity set variables. To see this, aggregate individuals' asset demands (12) to get the market portfolio, pre-multiply that by $\mathbf{V}_{aa}(M/T^M)$, and rearrange to get:

$$\mu - r_f \mathbf{1} = [\mathbf{V}_{aM}\mathbf{V}_{as}]\begin{pmatrix} M/T_M \\ -\mathbf{H}_s^M/T_M \end{pmatrix}$$ (14)

It is easy to verify that the covariances of assets with the state variables are the same as their covariances with the returns on portfolios that are maximally correlated with the state variables (s^*), which have weights of $\mathbf{w}_{s^*} = \mathbf{V}_{aa}^{-1}\mathbf{V}_{as}$.

The next step is to derive the expected excess returns on the $S + 1$ mutual funds that individuals hold. Pre-multiplying (14) by the matrix of portfolio weights for the $S + 1$ funds, their expected excess returns are:

$$\begin{pmatrix} \mu_M - r_f \\ \mu_{s^*} - r_f\mathbf{1} \end{pmatrix} = \begin{pmatrix} \sigma_M^2 & \mathbf{V}_{Ms^*} \\ \mathbf{V}_{s^*M} & \mathbf{V}_{s^*s^*} \end{pmatrix}\begin{pmatrix} M/T_M \\ -\mathbf{H}_s^M/T^M \end{pmatrix}.$$ (15)

To see the implications of this, note that if the S hedging portfolios were uncorrelated with the market portfolio and with each other, their expected excess returns would be zero in the single-period CAPM. However, in the intertemporal model, the expected excess return on a hedge portfolio is negatively related to the aggregate hedging demand (opposite if reverse hedging), and proportional to the variance of the hedging portfolio's return. Thus, if individuals in aggregate wish to hedge investment opportunities with a portfolio, they bid up its price and bid down its expected return in equilibrium. As shown earlier, with normal hedging, those state variables with the largest compensating variations in wealth will have the largest hedging demands and will deviate the most from the single-period CAPM's return predictions, *ceteris paribus*.

The final step in Merton's intertemporal CAPM is to substitute expected excess returns on the $S + 1$ key portfolios from (15) for preference parameters in (14):

$$\mu - r_f\mathbf{1} = [\mathbf{V}_{aM}\mathbf{V}_{as}]\begin{pmatrix} \sigma_M^2 & \mathbf{V}_{Ms^*} \\ \mathbf{V}_{s^*M} & \mathbf{V}_{s^*s^*} \end{pmatrix}^{-1}\begin{pmatrix} \mu_M - r_f \\ \mu_{s^*} - r_f\mathbf{1} \end{pmatrix}$$

(ICAPM) $\quad = \boldsymbol{\beta}_{a, Ms^*}\begin{pmatrix} \mu_M - r_f \\ \mu_{s^*} - r_f\mathbf{1} \end{pmatrix}$ (16)

Thus, in the intertemporal economy, betas with respect to the market portfolio are not enough to describe the relevant risk of a security. Its covariances with the investment opportunity set also matter for both pricing and optimal portfolios.

IV. CONSUMPTION-ORIENTED ASSET PRICING MODEL (CCAPM)

Following seminal articles on asset pricing in discrete-time economies by Rubinstein (1976), Lucas (1978) and Breeden and Litzenberger (1978), Breeden (1979) showed that Merton's (1973) multi-beta intertemporal CAPM could be re-expressed with a single risk measure. The result found, which is derived below, is that Merton's multi-beta ICAPM reduces to a market price of risk multiplied by the asset's *consumption-beta*, which is its sensitivity of return to percentage movements in aggregate real consumption. This model is the consumption-based capital asset pricing model (CCAPM).

The optimal rate of current consumption in the continuous-time model is a function of the individual's current wealth and the state vector for investment opportunities, $c^k = c^k(W^k, \mathbf{s}, t)$. In the continuous-time model, the first-order Taylor series approximation is correct for the *stochastic* part of consumption movements. (In contrast, a second-order approximation is required to describe the *expected* change in consumption.) Thus, the stochastic movements in consumption, and the co-variances of assets' returns with k's consumption changes $\mathbf{V}_{a,ck}$, may be written as follows:

$$d\tilde{c}^k = c_W^k(d\tilde{W}^k) + \mathbf{c}_s^k(d\tilde{\mathbf{s}})$$
$$\mathbf{V}_{a,ck} = \mathbf{V}_{a,Wk}c_W^k + \mathbf{V}_{as}\mathbf{c}_s^k. \tag{17}$$

The risk aversion and hedging preference parameters that determine an individual's asset holdings can be rewritten in terms of an individual's direct utility function for consumption. To see this, implicitly differentiate the envelope condition [eqn (9), superscript k suppressed]:

$$T = -J_W/J_{WW} = -u_c/(u_{cc}c_W) = T_c/c_W \tag{18}$$
$$\mathbf{H}_s = -J_{sW}/J_{WW} = -\mathbf{c}_s/c_W, \quad \text{for each individual.} \tag{19}$$

Substituting these formulae into Merton's optimal asset demands, (10), pre-multiplying them by $(c_W^k\mathbf{V}_{aa})$ and using (17) to simplify gives:

$$\mathbf{V}_{a,ck} = T_c^k[\boldsymbol{\mu} - r_f\mathbf{1}], \quad \text{for each individual } k. \tag{20}$$

This shows that each individual holds assets in proportions that result in an optimal consumption rate that covaries with each asset in proportion to its expected excess return. The next step is to aggregate these individual optimality conditions, which shows that each asset's expected excess return is proportional to its covariance with *aggregate* consumption.

Define the 'consumption beta' for any asset or portfolio j, β_j, to be the covariance of j's return with percentage changes in aggregate consumption, divided by the variance of percentage changes in aggregate consumption. Thus, the consumption beta is the slope in the regression of the asset's return on percentage changes in (real, per capita) consumption. The consumption-oriented CAPM (CCAPM) follows easily from the aggregated version of (20), where the risk tolerance parameter is eliminated by using the expected excess return per unit of consumption beta for *any* portfolio M:

$$\boldsymbol{\mu} - r_f\mathbf{1} = [\boldsymbol{\beta}_C/\beta_{MC}](\mu_M - r_f). \tag{21}$$

Thus, Breeden (1979) showed that Merton's intertemporal CAPM, which required $S + 1$ betas to determine an asset's systematic risks and equilibrium return, can be collapsed into a consumption oriented CAPM, with only a single beta with respect to consumption. This helps the intuition in determining which types of assets should have equilibrium returns that are substantially different in multiperiod economies than in the single-period world of the original market-oriented CAPM. How much the CCAPM representation helps in the testing of the intertemporal model is the subject of much current debate.

In the intertemporal economy, the market portfolio is no longer mean-variance efficient. The portfolio that has the highest correlation of returns with aggregate real consumption is now mean-variance efficient. To see this, pre-multiply an aggregate version of (20) by $\mathbf{V}_{aa}^{-1}$: the LHS gives the maximum correlation portfolio for consumption, and the RHS shows that it satisfies the mean-variance efficiency property of eqn (4). The reason is simply that in the intertemporal economy one gets paid to take consumption-related risk, and no other. Any portfolio that is not highest correlation with consumption has wasted risk for no additional return.

Our understanding of these results is greatly enhanced by understanding the relation of asset prices to marginal utilities. Hirshleifer's (1970, ch. 9) presentation of the time-state preference model of Arrow (1964) and Debreu (1959) is used extensively by Fama (1970), Rubinstein (1974, 1976), Long (1974), Hakansson (1977), Lucas (1978), Breeden and Litzenberger (1978), Brennan (1979) and Cox, Ingersoll and Ross (1985b) in asset pricing models that were fundamental precursors to the developments here. They showed that the value and fair price of one more share of an asset is the expected marginal utility of its payoffs. The expected marginal utility of its payoffs depends on the expected sizes of the payoffs, the dates they are received and the covariances of their sizes with the marginal utilities at different dates of a unit of consumption or wealth [see (9)]. Assets that have their highest payoffs when consumption is high (positive consumption betas) are paying the most when least needed, i.e., when the marginal utility of consumption is low; they are less valuable and have higher equilibrium required returns than assets that pay most when consumption is down.

The consumption CAPM follows directly from the marginal utility insights, since with time-additive utility functions, consumption at any date t is a sufficient statistic for marginal utility at date t. Wealth is not a sufficient statistic for marginal utility in an intertemporal economy, since the quality of the investment opportunity set also affects the marginal utility of a unit payoff. The reason that covariance with the market (aggregate wealth) determines risk in the single-period CAPM is that with one period, consumption equals wealth. Since marginal utility is one-to-one with consumption, it is also one-to-one with wealth in the single-period model. In that case, consumption betas and market betas are the same and the CCAPM reduces to the CAPM.

V. EXTENSIONS AND CONCLUSIONS

Space prohibits the proof of other important results that have been proven or can be proven. For example, Breeden and Litzenberger (1978) showed that if the capital markets allocation is unconstrained Pareto-optimal in this intertemporal economy, then each individual's consumption is monotonically increasing in aggregate consumption and in every other person's consumption. All individuals' consumption rates should go up and down together, though not necessarily proportionally. Each individual's optimal portfolio is the one that results in the highest correlation of the individual's consumption with aggregate consumption. Breeden (1979) showed: (1) All assets have the same covariances with a portfolio that is maximally correlated with aggregate consumption as with consumption itself. As a result, the CCAPM may be stated and tested in terms of assets' betas with respect to that maximally correlated portfolio. (2) With commodity price uncertainty, the CCAPM holds in terms of expected real returns and the betas of real returns with real, per capita consumption. (3) The CCAPM holds without a riskless asset, by just replacing r_f with the expected return on a portfolio that is uncorrelated with real consumption movements. A similar result was shown earlier by Black (1972) for the single-period CAPM. Finally, Bergman (1985) showed that if preferences are time-multiplicative, rather than time-additive, Merton's intertemporal CAPM still holds, but Breeden's (1979) CCAPM extension does not. Thus, the ICAPM is more general than the CCAPM.

In conclusion, the past three decades have seen important developments in the modelling of consumption and portfolio choices under uncertainty. In my opinion, the intertemporal portfolio theory and asset pricing models presented here are the strongest and most useful theoretical models that we currently have. Finally, I must say that there were many more authors that were important to the development of this area than were described in this entry. The bibliography gives a more complete (but still abridged) listing of papers that serious students should read.

DOUGLAS T. BREEDEN

See also ARBITRAGE PRICING THEORY; ASSET PRICING MODEL; CAPITAL ASSET PRICING MODEL; FINANCE; PORTFOLIO ANALYSIS.

BIBLIOGRAPHY

Arrow, K.J. 1951. An extension of the basic theorems of classical welfare economics. In *Proceedings of the Second Berkeley Symposium on Mathematical Statistics and Probability*, ed. J. Neyman, Berkeley: University of California Press, 507–31.

Arrow, K.J. 1964. The role of securities in the optimal allocation of risk-bearing. *Review of Economic Studies* 31, 91–6.

Arrow, K.J. 1964. The theory of risk aversion. In K.J. Arrow, *Aspects of the Theory of Risk-Bearing*, Helsinki: Yrjö Jahnsson Foundation.

Banz, R.W. and Miller, M.H. 1978. Prices for state-contingent claims: some estimates and applications. *Journal of Business* 51, 653–72.

Beja, A. 1971. The structure of the cost of capital under uncertainty. *Review of Economic Studies* 38, 359–76.

Bergman, Y.Z. 1985. Time preference and capital asset pricing models. *Journal of Financial Economics* 14, 145–60.

Bhattacharya, S. 1981. Notes on multiperiod valuation and the pricing of options. *Journal of Finance* 36, 163–80.

Black, F. 1972. Capital market equilibrium with restricted borrowing. *Journal of Business* 45, 444–55.

Black, F. and Scholes, M.S. 1973. The pricing of options and corporate liabilities. *Journal of Political Economy* 81, 637–54.

Breeden, D.T. and Litzenberger, R.H. 1978. Prices of state-contingent claims implicit in option prices. *Journal of Business* 51, 621–51.

Breeden, D.T. 1979. An intertemporal asset pricing model with stochastic consumption and investment opportunities. *Journal of Financial Economics* 7, 265–96.

Breeden, D.T. 1984. Futures markets and commodity options: hedging and optimality in incomplete markets. *Journal of Economic Theory* 32, 275–300.

Breeden, D.T. 1986. Consumption, production, inflation, and interest rates: a synthesis. *Journal of Financial Economics* 16, 3–39.

Brennan, M.J. 1979. The pricing of contingent claims in discrete time models. *Journal of Finance* 34, 53–68.

Constantanides, G.M. 1982. Intertemporal asset pricing with heterogeneous consumers and without demand aggregation. *Journal of Business* 55, 253–67.

Cox, J.C., Ingersoll, J.E. and Ross, S.A. 1985a. A theory of the term structure of interest rates. *Econometrica* 53, 385–407.

Cox, J.C., Ingersoll, J.E. and Ross, S.A. 1985b. An intertemporal general equilibrium model of asset prices. *Econometrica* 53, 385–407.

Debreu, G. *Theory of Value*. New York: Wiley.

Dieffenbach, B.C. 1975. A quantitative theory of risk premiums on securities with an application to the term structure of interest rates. *Econometrica* 43, 431–54.

Fama, E.F. 1970. Multiperiod consumption-investment decisions. *American Economic Review* 60, 163–74.

Ferson, W.E. 1983. Expected real interest rates and aggregate consumption: empirical tests. *Journal of Financial and Quantitative Analysis* 18, 477–98.

Fischer, S. 1975. The demand for index bonds. *Journal of Political Economy* 83, 509–34.

Garman, M. 1977. A general theory of asset pricing under diffusion state processes. Working Paper No. 50, Research Program in Finance, University of California, Berkeley.

Grauer, F. and Litzenberger, R. 1979. The pricing of commodity futures contracts, nominal bonds and other risky assets under commodity price uncertainty. *Journal of Finance* 34, 69–83.

Grossman, S.J. and Shiller, R.J. 1982. Consumption correlatedness and risk measurement in economies with non-traded assets and heterogeneous information. *Journal of Financial Economics* 10, 195–210.

Hakansson, N.H. 1970. Optimal investment and consumption strategies under risk for a class of utility functions. *Econometrica* 38(5), September, 587–607.

Hakansson, N.H. 1977. Efficient paths toward efficient capital markets in large and small countries. In *Financial Decision Making under Uncertainty*, ed. H. Levy and M. Sarnat, New York: Academic Press.

Hall, R.E. 1978. Stochastic implications of the life cycle-permanent income hypothesis: theory and evidence. *Journal of Political Economy* 86, 971–87.

Hansen, L.P. and Singleton, K.J. 1983. Stochastic consumption, risk aversion, and the temporal behavior of asset returns. *Journal of Political Economy* 91, 249–65.

Hirshleifer, J. 1970. *Investment, Interest and Capital*. Englewood Cliffs: Prentice-Hall.

Huang, C.-F. 1985. Information structure and equilibrium asset prices. *Journal of Economic Theory* 34, 33–71.

Kraus, A. and Litzenberger, R.H. 1975. Market equilibrium in a state preference model with logarithmic utility. *Journal of Finance* 30(5), December, 1213–27.

Kydland, F.E. and Prescott, E.C. 1982. Time to build and aggregate fluctuations. *Econometrica* 50, 1345–70.

Lintner, J. 1965. Valuation of risk assets and the selection of risky investments in stock portfolios and capital budgets. *Review of Economics and Statistics* 47, February, 13–37.

Long, J.B. 1974. Stock prices, inflation, and the term structure of interest rates. *Journal of Financial Economics* 2, 131–70.

Lucas, R.E. 1978. Asset prices in an exchange economy. *Econometrica* 46, 14–45.

Marsh, T.A. and Rosenfeld, E.A. 1982. Stochastic processes for interest rates and equilibrium bond prices. *Journal of Finance* 38, 635–646.

Markowitz, H. 1952. Portfolio selection. *Journal of Finance* 12, 77–91.

Markowitz, H. 1959. *Portfolio Selection: Efficient diversification of investment*. New York: John Wiley.

Merton, R.C. 1969. Lifetime portfolio selection under uncertainty: the continuous-time case. *Review of Economics and Statistics* 51, 247–57.

Merton, R.C. 1971. Optimum consumption and portfolio rules in a continuous-time model. *Journal of Economic Theory* 3(4), 373–413.

Merton, R.C. 1973. An intertemporal capital asset pricing model. *Econometrica* 41, 867–87.

Mossin, J. 1966. Equilibrium in a capital asset market. *Econometrica* 34, October, 768–83.

Pratt, J.W. 1964. Risk aversion in the small and in the large. *Econometrica* 32(1–2), January–April, 122–36.

Pye, G. 1972. Lifetime portfolio selection with age dependent risk aversion. In *Mathematical Methods in Investment and Finance*, ed. G. Szego and K. Shell, Amsterdam: North-Holland, 49–64.

Richard, S.F. 1974. Optimal consumption, portfolio and life insurance rules for an uncertain lived individual in a continuous-time model. *Journal of Financial Economics* 2, 187–203.

Roll, R. 1977. A critique of the asset pricing theory's tests. Part I: On past and potential testability of the theory. *Journal of Financial Economics* 4, 129–76.

Ross, S.A. 1976. The arbitrage theory of capital asset pricing. *Journal of Economic Theory* 3, 343–62.

Rubinstein, M. 1974. A discrete-time synthesis of financial theory. In *Research in Finance*, Vol. 3, Greenwich, Conn.: JAI Press, 53–102.

Rubinstein, M. 1976. The valuation of uncertain income streams and the pricing of options. *Bell Journal of Economics and Management Science* 7, 407–25.

Samuelson, P.A. 1969. Lifetime portfolio selection by dynamic stochastic programming. *Review of Economics and Statistics* 57(3), August, 239–46.

Sharpe, W.F. 1964. Capital asset prices: A theory of market equilibrium under conditions of risk. *Journal of Finance* 19, 429–42.

Stulz, R.M. 1981. A model of international asset pricing. *Journal of Financial Economics* 9, 383–406.

Sundaresan, M. 1984. Consumption and equilibrium interest rates in stochastic production economies. *Journal of Finance* 39, 77–92.

Tobin, J. 1958. Liquidity preference as behavior towards risk. *Review of Economic Studies* 25, 65–86.

invariable standard of value. I. In Sections I–III, Chapter I of the *Principles*, Ricardo rejected Smith's labour-commanded theory of value in favour of an embodied-labour theory – for a justification of this, see Sraffa's Introduction to Ricardo (1951) or Garegnani (1984). However, in Sections IV and V, he was forced to modify his theory to take account of the effects of movements in income distribution. Thus, Ricardo had isolated two cases where the value of commodities would change – first, when there was an alteration in the amount of labour required, directly and indirectly, in production; and second, when there was a rise or fall in the value of labour, which operated through unequal capital–labour ratios in the different industries. On empirical grounds, Ricardo argued that the first would dominate the second:

> The greatest effect which could be produced on the relative prices of these goods from a rise of wages, could not exceed 6 or 7 per cent ... . Not so with that other great cause of the variation in the value of commodities, namely the increase or diminution in the quantity of labour necessary to produce them (Ricardo [1821], 1951, p. 36)

Thus, it was assumed in the remaining chapters of the *Principles* that changes in value were caused by changes in embodied labour.

This left the theory of value in an unsatisfactory state. Central to the Ricardian scheme since 'The Essay on Profits' (Ricardo, 1952) had been the rate of profits and its relation to the rate of growth; as a corollary, the determination of the laws determining income distribution was regarded as the principal problem in political economy (Ricardo [1821], 1951, p. 5). Yet, in the study of this problem, Ricardo found that the size of the national income, the quantity of capital and the amount of wages all varied with the distribution of income. Though fixed in physical composition, these variables changed because they were measured in terms of values, themselves functions of the distribution of income. What would be the effect then on the rate of profits, r, of an increase in the real wage rate, w? Would r necessarily decrease or could there be a sufficient rise in the value of net income to accommodate increases in both distributive parameters? It was in an attempt to answer such questions that Ricardo turned to the notion of an invariable standard of value and the associated distinction between relative value and real value.

Having identified the two causes of change in the values of commodities, Ricardo could define the characteristics of a standard measure of value: such a commodity would require a constant quantity of embodied labour in its production and have to be invariant with respect to changes in income distribution. 'Of such a measure, it is impossible to be possessed, because there is no commodity which is not itself exposed to the same variations as the things, the value of which is to be ascertained' (Ricardo [1821], 1951, pp. 43–4). Ricardo was defeated by this problem in *The Principles*, assuming gold to be invariable 'to facilitate the object of this enquiry, although I fully allow that money made of gold is subject to most of the variations of other things' (Ricardo [1821], 1951, p. 46). Nor was he able to achieve progress later, as evidenced by his paper on 'Absolute and Exchangeable Value' (Ricardo, 1952). II. For an attempt at a partial

solution to Ricardo's problem, we have to turn to the Standard Commmodity (§ 23 in Chapter IV of Sraffa (1960) is entitled 'An invariable measure of value'). In *Production of Commodities by Means of Commodities*, Sraffa was 'concerned exclusively with such properties of an economic system as do not depend on changes in the scale of production' (p. v). With a given technique of production, embodied labour cannot change; hence, when investigating the existence of an invariable standard of value, it suffices to consider invariance with respect to income distribution.

Consider a single-product industries, circulating capital model, as in Part I of Sraffa (1960), with price equations given by:

$$p' = wl' + (1 + r)p'A \tag{1a}$$

$$p'(I - A)x = 1 \tag{1b}$$

$p = (p_i)$ is the relative price vector, $l = (l_i)$ the vector of direct labour input coefficients, $A = (a_{ij})$ the matrix of input–output coefficients, w and r the uniform rates of wages and profits respectively, and x is the gross output vector. (1b) states that the actual net output of the economy is the *numéraire*. From (1a):

$$p' = wl'(I - (1 + r)A)^{-1} \tag{2}$$

so that, using (1b):

$$w = 1/l'(I - (1 + r)A)^{-1}(I - A)x \tag{3}$$

Then (2) and (3) imply that:

$$p' = l'(I - (1 + r)A)^{-1}/l'(I - (1 + r)A)^{-1}(I - A)x \tag{4}$$

i.e., equations (1) can be solved to yield w and p as functions of r in (3) and (4) respectively.

'The key to the movement of relative prices consequent upon a change in the wage rate lies in the inequality of the proportions in which labour and means of production are employed in the various industries' (Sraffa, 1960, §15, p. 12). It is a straightforward matter to show that relative prices are invariant with respect to income distribution if and only if there is a uniform value-capital/labour ratio in each industry, itself a manifestation of this underlying (mathematical) condition:

$$l'A = \lambda^*(A)l', \qquad \lambda^*(A) > 0 \tag{5}$$

i.e., l is a left characteristic vector of A corresponding to the Frobenius root, $\lambda^*(A)$. (For a proof of this statement, see Pasinetti (1977) or Woods (1985).)

Assuming that (5) does not hold, what would happen if, following Sraffa, we suppose that prices are constant as the wage, measured in terms of actual national income, is reduced from one and the rate of profits increased from zero. In those industries with a sufficiently low proportion of labour to means of production ('deficit'-industries), the reduction in wage payments is insufficient to met profit payments. On the other hand, there would be industries with a sufficiently high proportion of labour to means of production ('surplus'-industries) for the proceeds of the wage reduction to exceed profit payments.

> There would be a 'critical proportion' of labour to means of production which marked the watershed between 'deficit' and 'surplus' industries. An industry which employed that particular 'proportion' would show an even balance – the proceeds of the wage reduction would provide exactly what was required for the payment of profits at the general rate (Sraffa, 1960, §17, p. 13).

An industry characterized by that 'proportion' would also

exhibit it in its means of production, and in its means of production of means of production, etc. The output of such an industry would consist

> of the same commodities (combined in the same proportions) as does the aggregate of its own means of production – in other words, such that both product and means of production are quantities of the self-same composite commodity' (Sraffa, 1960, §24, p. 19).

Sraffa's Standard Commodity, which possesses that particular 'balancing proportion', is given by the semi-positive vector x^* which solves:

$$Ax^* = \lambda^*(A)x^* \qquad (6a)$$

$$l'x^* = 1 \qquad (6b)$$

As characteristic vectors are unique only up to scalar multiplication, (6b) is the normalization to fix the size of the Standard System. (The connection between the Standard Commodity and the right Frobenius characteristic vector was first perceived by Newman, 1962.) When measured in value terms, the ratio between gross output, x^*, and means of production, Ax^*, is invariant to changes in distribution. Calculating this ratio when the wage rate is zero, we obtain:

$$p'x^*/p'Ax^* = p'x^*/\lambda^*(A)p'x^* = 1/\lambda^*(A) = (1 + R) \qquad (7)$$

where R is Sraffa's 'balancing' ratio, identical to the maximum rate of profits. If A is productive, $\lambda^*(A) < 1$ so that $R > 0$. If, in addition, A is indecomposable (all commodities are basic), $x^* > 0$ and is, in fact, the only semi-positive characteristic vector of A. Thus, existence and uniqueness of the Standard proportions are derived straightforwardly from the Perron–Frobenius Theorem (see Pasinetti, 1977 or Woods, 1978).

Let the Standard Net Product be chosen as numeraire, i.e.

$$p'(I - A)x^* = 1 \qquad (8)$$

Then, from the price equations (1a) and (8), and the quantity equations (6a) and (6b), it can be shown that:

$$r = R(1 - w) \qquad (9)$$

(see Newman, 1962; Pasinetti, 1977; or Woods, 1978). That is, the real wage rate–rate of profits curve for the productive technique described by $\{A; l\}$ is a downward-sloping straight line.

It is a straightforward matter to show that (9) holds for a one-commodity model. In a multi-commodity model, prices in general vary with income distribution, as in (4), thereby obscuring the relation between distributive parameters. 'Particular proportions, such as the Standard ones, may give transparency to a system and render visible what was hidden' (Sraffa, 1960, §31, p. 23). The use of the Standard System is sufficient, not necessary, for the derivation of a downward-sloping w–r curve, for it can be shown from (3) that $dw/dr < 0$ in terms of any numéraire. 'It follows that if the wage is cut in terms of *any* commodity ... the rate of profits will rise; and vice versa for an increase of the wage' (Sraffa, 1960, §49, p. 40).

Thus, the Standard System always exists for the single-product industries, circulating capital model, implying a particularly simple form of the w–r curve.

There are three concluding points to be made.

(1) The Standard System does not necessarily exist for the pure joint production model of Part II of Sraffa (1960). For such a model, with input and output matrices A and B, the Standard Commodity would have to satisfy:

$$Bx^* = (1 + R)Ax^*. \qquad (10)$$

Sraffa considered the possibility that the Standard Commodity contains negative components. Manara (1980) has demonstrated, using simple numerical examples, that a productive multiple-product industries system does not necessarily have a maximum rate of profit; that is, the solution R, x^* of (10) may be negative or complex.

A reformulation of the system in terms of inequalities would give a simple von Neumann model; the von Neumann solution, which exhibits balanced growth, can be thought of as a generalization of the Standard proportions.

(2) A positive Standard System does exist for the single-product industries, fixed capital model in chapter X of Sraffa (1960). Furthermore, as demonstrated in Woods (1984), the Standard Commodity can be used to resolve the particular question of choice of technique which arises in this model – namely, the determination of the optimal economic lifetime of machinery.

(3) In any analysis of price and wage variation, some commodity must be chosen as *numéraire*. In chapter III, Sraffa (1960) operated with actual net output, as in (1b). When constructing the Standard Commodity, Sraffa supposes 'that there was an industry which employed labour and means of production in that precise proportion, so that with a wage-reduction, and *on the basis of the initial prices*, it would show an exact balance of wages and profits (Sraffa, 1960, §21, p. 16, emphasis added). Clearly, this is the intuitive argument underlying the construction of the Standard Commodity. However, it could not be expressed formally other than in terms of the Standard Commodity.

J.E. WOODS

See also STANDARD COMMODITY.

BIBLIOGRAPHY

Garegnani, P. 1984. Value and distribution in the classical economists and Marx. *Oxford Economic Papers* 36, 291–325.

Manara, C.F. 1980. Sraffa's model for the joint production of commodities by means of commodities. In *Essays on the Theory of Joint Production*, ed. L.L. Pasinetti, London: Macmillan, 1–15.

Newman, P. 1962. Production of commodities by means of commodities. *Schweizerische Zeitschrift für Volkswirtschaft und Statistik* 98, 58–75.

Pasinetti, L.L. 1977. *Lectures on the Theory of Production*. London: Macmillan.

Ricardo, D. 1821. *Principles of Political Economy and Taxation*. Vol. I of *Works and Correspondence of David Ricardo*, ed. P. Sraffa, Cambridge: Cambridge University Press.

Ricardo, D. 1952. *Papers and Pamphlets, 1815–1823*. Vol. IV of *Works and Correspondence of David Ricardo*, ed. P. Sraffa, Cambridge: Cambridge University Press.

Sraffa, P. 1960. *Production of Commodities by Means of Commodities*. Cambridge: Cambridge University Press.

Woods, J.E. 1978. *Mathematical Economics*. London: Longmans.

Woods, J.E. 1984. Notes on Sraffa's fixed capital model. *Journal of the Australian Mathematical Society*, Series B, *Applied Mathematics* 26, 200–232.

Woods, J.E. 1985. Notes on relative price invariance. *Giornale degli Economisti e Annali di Economia* 44, 135–52.

invariance principles. *See* MEANINGFULNESS AND INVARIANCE; TRANSFORMATIONS AND INVARIANCE.

inventories. Inventories consist of stocks of finished goods, goods-in-process, and raw materials and supplies held by business firms. Interest in inventories among economists stems primarily from the observation that inventory fluctuations are an important feature of business cycles.

This observation was first fully documented by Abramovitz (1950) who observed that, although the level of inventory investment is a small fraction of the level of GNP, changes in inventory investment are a large fraction of changes in GNP, especially in recessions. In particular, he pointed out that in the five business cycles in the US between the two world wars the average decline in inventory investment accounted for 47 per cent of the average decline in GNP during contractions. The post-World War II data for the US are even more dramatic. As reported by Blinder and Holtz-Eakin (1983), the average drop in inventory investment was 68 per cent of the drop in GNP during post-World War II recessions. In expansions, inventory movements account for a smaller fraction of movements in GNP, so that over the cycle as a whole inventory fluctuations are less influential than the data on recessions alone would indicate. Nevertheless, the point remains: inventory movements are a key feature of cyclical fluctuations in aggregate output.

To avoid confusion, it should be stressed that inventories appear to be a propagating mechanism, not a causal force, in business cycles. There is no evidence that exogenous shifts in inventory investment are an underlying cause of fluctuations in GNP.

MICROFOUNDATIONS. To provide a framework for analysing movements in inventories, economists have developed optimization models of the firm's behaviour. These borrow heavily from models in operations research, which are designed to guide actual firms in managing their inventory positions. The models of economists have been most fully developed for finished goods inventories held by manufacturers. The prime motive for holding inventories in these models is to serve as a buffer stock, that is, to absorb random fluctuations in demand. Among models of buffer stocks, two approaches may be distinguished: the quadratic criterion–linear constraint approach and the stochastic dynamic programming approach.

The former approach was developed in the pioneering book of Holt et al. (1960). There and in other uses of the theory in economics, for example, Belsley (1969), the firm was assumed to choose levels of output and inventories to minimize expected costs subject to an accumulation equation for inventories and an exogenous stochastic process for sales. To bring the price decision into the theory, Hay (1970), Blinder (1982), and others have assumed that the firm possesses monopoly power and maximizes discounted expected profits.

The basic model presumes that the firm sets its price and output before the random component of demand is revealed, and inventories are used to absorb any shocks to demand. Let P_t be price, Q_t output, N_t real sales or demand, H_t the stock of inventories at the end of the period, and ϵ_t a non-negative random variable. The firm is then assumed to maximize

$$E_0 \sum_{t=0}^{\infty} \rho^t [P_t N_t - c_0 - c_1 Q_t - c_2 Q_t^2 - v(H_t - H_t')^2] \quad (1)$$

subject to

$$H_t = H_{t-1} + Q_t - N_t$$

$$N_t = m_0 - m_1 P_t + \epsilon_t$$

where ρ is a discount factor, E_0 is an expectation operator, and c_0, c_1, c_2, m_0 are positive parameters.

The motive for the firm to hold inventories is essentially contained in the inventory-holding cost function, $v(H_t - H_t')^2$,

which is a U-shaped function of H_t that reaches a minimum at H_t'. It captures two forces: Higher inventories increase costs in the form of storage costs. But, lower inventories also increase costs in the form of lost sales, since lower inventories relative to sales increase the likelihood the firm will be caught out of stock. The holding cost function balances these forces.

The model has several advantages. It yields explicit solutions for the choice variables that are linear in the previous period's inventory stock and current and future sales. Further, certainty equivalence applies so that random variables may be replaced by their expected values. But there are disadvantages. The functional form assumptions are quite strong. The random component to demand must appear additively. And the discount rate must be constant, which limits the ability of the model to analyse the effects of changes in real interest rates through which monetary policy operates.

The stochastic dynamic programming approach to inventory-holding behaviour originates with the work of Arrow et al. (1951), who also developed a model of a firm that minimizes expected costs subject to an exogenous stochastic process for sales. Karlin and Carr (1962) later extended the model to allow for a price decision. The economic implications of the model have been drawn out by, among others, Mills (1962) and Zabel (1972).

Suppose the firm makes decisions over a planning horizon of T periods. Then, according to Bellman's Principle of Optimality, and using the earlier notation, the functional equation that describes the firm's optimal programme is

$$\Omega_T(H_{t-1}) = \max_{P_t, Q_t} \left\{ P_t m(P_t) - P_t D(b_t) - c(Q_t) \right.$$
$$\left. - A(b_t) + \rho_t \int_0^{\infty} \Omega_{T-1}(H_t) f(\epsilon_t) \, d\epsilon_t \right\} \quad (2)$$

where

$$b_t = H_{t-1} + Q_t - m(P_t)$$

$$H_t = b_t - \epsilon_t$$

$$D(b_t) = \int_{b_t}^{\infty} (\epsilon_t - b_t) f(\epsilon_t) \, d\epsilon_t$$

$$A(b_t) = \int_0^{b_t} a(b_t - \epsilon_t) f(\epsilon_t) \, d\epsilon_t$$

and where $m' < 0$, $m'' \leqslant 0$, $c' > 0$, $c'' \geqslant 0$, $a' > 0$, and $a'' \geqslant 0$. The expression, $\Omega_T(H_{t-1})$, denotes maximum discounted expected profits over T future periods.

Observe that $D(b_t)$ is expected shortages or stockouts, capturing the real sales that the firm can expect to lose if it is caught with too few inventories, while $A(b_t)$ is the expected storage cost of holding inventories. Unlike the quadratic cost-linear constraint approach, this model permits general functional forms to be used. Moreover, the decision rules will depend in general on moments higher than the mean of the probability distribution of demand. Finally, although the model has been formulated above with additive demand errors, it can be modified to allow for non-additive uncertainty. A disadvantage of the model is that explicit solutions are not possible.

In both approaches, the incentive for the firm to hold inventories is essentially the same. To see this, observe that expected shortages, $D(b_t)$, in (2) are inversely related to initial inventories and correspond to the cost savings to holding inventories – the downward-sloping section of the U-shaped cost curve – that appears in (1). In essence, both terms capture the benefits to the firm of holding a buffer stock in the form of finished goods inventories. Given this, it is not surprising that with additive demand uncertainty both approaches yield similar economic predictions, though of course the details will differ from model to model. Such predictions have served as the basis for much empirical work on inventories and have motivated the specification of inventory investment relationships in macroeconomic models.

One set of predictions concerns the response of decision variables to changes in initial inventories. As long as marginal production costs rise (i.e., $c_2 > 0$ or $c'' > 0$) so that there is some incentive to smooth production, both P_t and Q_t will be inversely related to H_{t-1}. Further, $-1 < (\partial H_t / \partial H_{t-1}) < 0$. This means that inventory adjustments will exhibit the characteristics of partial adjustment.

A second set of predictions refers to the effects of changes in exogenous variables. Generally speaking, P_t and Q_t will be positively related to exogenous shifts in anticipated demand (shifts in m_0 or the function $m(P_t)$) and inversely related to changes in inventory holding costs. Further, increases in current anticipated demand, declines in future anticipated demand, or increases in inventory holding costs will reduce inventory investment.

The basic models outlined above have been extended in numerous directions. A prominent one is that right from the beginning many authors have allowed production costs to depend on the change in the level of output as well as the level of output itself. This creates an additional incentive for the firm to smooth production beyond the incentive embodied in the assumption that $c'' > 0$. The difficulty with this notion is that the underlying rationale for including costs to changing the level of output is quite vague. A possible rationale that has been exploited in recent work – for example, Maccini (1984) – is that such costs reflect adjustment costs to changing quasi-fixed factors of production, for example, plant and equipment, or the stock of workers. But, then, a better theoretical procedure would be to incorporate decisions on quasi-fixed factors directly into the analysis. This more explicit specification of the economic forces at work permits an analysis of the important interaction between inventories and quasi-fixed factors of production.

Most inventory models that analyse price as well as output and inventory behaviour, have assumed that the firm possesses some monopoly power in output markets. This assumption facilitates a study of the relationship between price and inventory behaviour, but it is not essential. Models of a competitive industry can be formulated to yield similar predictions for movements in price, output and inventories – see, for example, Eichenbaum (1983).

So far we have concentrated, like the literature, on the holding of finished goods inventories. Theory that rationalizes the holding of inventories of goods-in-process and raw materials and supplies is much less well developed. It is common to adapt, rather casually, the earlier models designed for the study of finished goods inventories. But, this tactic fails to capture in a rigorous way the firm's motives in holding these other inventories, and may in fact generate specious predictions. This is an area that needs more research.

Finally, all the theories discussed so far are designed to explain the holding of manufacturers' inventories, and are not obviously applicable to the holding of inventories by wholesalers and retailers. The trouble is that for the latter agents the variable corresponding to Q_t – the variable that gives rise to additions to inventories – is orders of goods from manufacturers. But, the process of ordering and receiving goods from manufacturers may carry a substantial fixed cost which will induce the firm to 'bunch' rather than to 'smooth'

orders and deliveries. In recent work, Blinder (1981) has used ideas associated with S-s models of inventory behaviour – see, for example, Scarf (1960) – together with an aggregation procedure to undertake an analysis of retail inventories. This is an important development because retail inventories appear to exhibit at least as much volatility as manufacturers' inventories.

EMPIRICAL WORK. Empirical studies of inventories have been dominated by the use of the flexible accelerator model, first used in empirical work by Lovell (1961), and subsequently used by many authors, including Blinder and Holtz-Eakin (1983), Maccini and Rossana (1984), Irvine (1981), etc. The model takes the form

$$H_t - H_{t-1} = \lambda_1(H_t^* - H_{t-1}) + \lambda_2[N_t - E_t(N_t)] + u_t, \quad (3)$$
$$0 \leqslant \lambda_1 \leqslant 1, \qquad -1 \leqslant \lambda_2 \leqslant 0$$

where H_t^* is 'desired' inventories, $E_t(N_t)$ is expected sales so that $N_t - E_t(N_t)$ is a 'sales surprise', and u_t captures random forces other than sales that operate on inventories. Desired inventories should depend on current and future levels of expected demand, inventory holding costs, such as real interest rates, and factor input prices which shift marginal production costs.

This model embodies the essential ideas of the buffer stock models of inventory behaviour described above. As long as marginal production costs rise with output, $c_2 > 0$ or $c'' > 0$, so that there is an incentive for firms to smooth production, it follows that $\lambda_1 < 1$ so that firms will partially adjust inventory stocks to desired levels and that $\lambda_2 < 0$ so that sales surprises will be met partly by inventory adjustments. In the extreme, as $c_2 \to \infty$ or $c'' \to \infty$, it follows that $\lambda_1 \to 0$ and $\lambda_2 \to -1$, and vice versa. Moreover, the determinants of desired inventory stocks reflect the main exogenous variables that are embedded in the models of the firm's behaviour.

This model has been estimated over a wide variety of industries, sectors, countries, and sample periods. Different assumptions have been used for expectation formation schemes, and different econometric techniques have been used to handle statistical problems.

Despite the enormous amount of empirical work done with the model, a number of empirical puzzles remain. A major one surrounds the estimates of the parameters, λ_1 and λ_2, as several authors, most prominently Feldstein and Auerbach (1976) have pointed out. The estimates of λ_1 turn out to be very low, implying that the speed of adjustment of actual to desired inventories is very slow. This is implausible when wide savings in inventory investment amount to no more than a couple of weeks of production. Moreover, the estimates of λ_2 are also quite low, implying that sales surprises tend to be absorbed largely by production adjustments. But, this contradicts the low estimates of λ_1, which suggest that it is difficult, that is, costly, to adjust production levels in order to close gaps between H_t^* and H_{t-1}. Despite the use of relatively sophisticated econometric methods in recent work to deal with statistical problems, the puzzle remains unresolved.

Another puzzle is that real interest rates have generally performed poorly in inventory equations – see, for example, Blinder and Holtz-Eakin (1983) and Maccini and Rossana (1984) for recent studies. This is a surprise since it is widely believed that changes in credit conditions have a substantial effect on the inventory positions of business firms. This result may be owing to difficulties in measuring *ex ante* rates, or to a lack of variation in real rates, or to the presence of credit rationing which is not adequately captured in the price of credit.

Finally, a number of authors (e.g. Blinder and Holtz-Eakin, 1983), have observed that the variance of production actually exceeds the variance of sales in most industries. This fact appears to conflict with the above theoretical models which predict that inventories are used to smooth production relative to sales. To explain this phenomenon, the models need to be extended to allow for cost shocks in the form of, for example, real raw material prices, or more complex demand structures than serially uncorrelated random errors.

LOUIS J. MACCINI

See also ADAPTIVE EXPECTATIONS; ADJUSTMENT COSTS; BUFFER STOCKS; COBWEB THEOREM; INVENTORY CYCLES; INVENTORY POLICY; INVESTMENT; LAYOFFS; STOCHASTIC OPTIMAL CONTROL.

BIBLIOGRAPHY

Abramovitz, M. 1950. *Inventories and Business Cycles.* New York: National Bureau of Economic Research.

Arrow, K., Harris, T.B. and Marschak, J. 1951. Optimal inventory policy. *Econometrica* 19, July, 250–72.

Belsley, D. 1969. *Industry Production Behavior: The Order Stock Distinction.* Amsterdam: North-Holland.

Blinder, A. 1981. Retail inventory behavior and business fluctuations. *Brookings Papers on Economic Activity* No. 2,261G21 443–505.

Blinder, A. 1982. Inventories and sticky prices: more on the microfoundations of macroeconomics. *American Economic Review* 72(3), June, 334–48.

Blinder, A. and Holtz-Eakin, D. 1983. Inventory fluctuations in the United States since 1929. *Proceedings of the Conference on Business Cycles*, National Bureau of Economic Research.

Eichenbaum, M. 1983. A rational expectations equilibrium model of inventories of finished goods and employment. *Journal of Monetary Economics* 12(2), August, 259–77.

Feldstein, M. and Auerbach, A. 1976. Inventory behaviour in durable goods manufacturing: the target adjustment model. *Brookings Papers on Economic Activity* No. 2, 351–96.

Hay, G. 1970. Production, price and inventory theory. *American Economic Review* 60(4), September, 531–45.

Holt, C., Modigliani, F. Muth, J.B. and Simon, H. 1960. *Planning Production, Inventories and Work Force.* Englewood Cliffs: Prentice-Hall.

Irvine, O. 1981. Retail inventory investment and the cost of capital. *American Economic Review* 70(4), September, 633–48.

Karlin, S. and Carr, C. 1962. Prices and optimal inventory. In *Studies in Applied Probability and Management Science*, ed. K. Arrow, S. Karlin and H. Scarf, Palo Alto: Stanford University Press.

Lovell, M. 1961. Manufacturers' inventories, sales expectations and the acceleration principle. *Econometrica* 29, July, 293–314.

Maccini, L. 1984. The interrelationship between price and output decisions and investment decisions: microfoundations and aggregate implications. *Journal of Monetary Economics* 13(1), January, 41–65.

Maccini, L. and Rossana, R. 1984. Joint production, quasi-fixed factors of production, and investment in finished goods inventories. *Journal of Money, Credit, and Banking* 16(2), May, 218–36.

Mills, E. 1962. *Price, Output and Inventory Policy.* New York: Wiley.

Scarf, H. 1960. The optimality of (S-s) policies in a dynamic inventory problem. In *Mathematical Methods in the Social Sciences*, ed. K. Arrow, S. Karlin and P. Suppes, Palo Alto: Stanford University Press.

Zabel, E. 1972. Multiperiod monopoly under uncertainty. *Journal of Economic Theory* 5(3), December, 524–36.

inventory cycles.

I. THE CONTRIBUTION OF INVENTORY LIQUIDATION TO DECLINES IN GNP. Although inventory investment is a relatively small component of total GNP, even in boom years, the swing from inventory accumulation to massive liquidation is a fundamental factor in the propagation of cyclical reversals in the pace of economic activity. To illustrate, the 1981–2 decline in United States GNP of $31.8 billion dollars deserves to be called an inventory recession because the shift from positive inventory accumulation of $8.9 billion (1972 dollars) at the cycle peak to liquidation of stocks at an annual rate of $22.7, a decline in effective demand of $31.6 billion, greatly exceeded the collapse of any other component of real GNP.

As inspection of Table 1 makes clear, a drop off in inventory investment generally makes a major contribution to each recession's decline in effective demand. The short 1980 recession should probably not be counted as an exception to this rule in that inventory investment fell from $13.7 billion in the second quarter of 1979 to −$10.1 billion in the third quarter of 1980. However, the 1946 recession (which pre-dates the availability of quarterly GNP data) is atypical; this was not an inventory recession because the efforts of business enterprise to replenish stocks at the end of World War II served to soften the shock of postwar conversion. And in the Great Depression of the 1930s, the decline in inventories was overshadowed by a collapse of fixed investment that helped push the unemployment rate up to 25 per cent.

The critical importance of inventories has long been recognized, thanks in large measure to the fundamental empirical study of Moses Abramovitz (1950), who demonstrated that in the period between the two world wars a collapse of inventory investment contributed decisively to each recession's decline in effective demand. Inventories have continued to play a destabilizing role in each United States recession since the publication of Abramovitz's study. Also, substantial empirical evidence for a number of countries establishes that inventory recessions are a general characteristic of capitalist economies. Further, there is some evidence, reviewed by Attila Chikàn (1984), that socialist economies may also experience inventory cycles.

The phrase 'inventory recession' stresses the empirical fact that cyclical reversals in economic expansion are dominated by the liquidation of inventory stocks. But to characterize most cyclical reversals as 'inventory recessions' does not explain why the declines in the pace of economic activity come about or what policy measures, if any, should be applied to mitigate the sacrifice of jobs and output occasioned by recession. Empirical observation, such as the evidence of Table 1, cannot by itself establish that inventories are in any sense a fundamental cause of the business cycle rather than only a basic symptom or but one of a number of essential ingredients of the mechanism propagating business fluctuations. Both theory and empirical evidence are essential in the study of the inventory cycle mechanism.

II. MODELLING THE INVENTORY CYCLE. In the interwar period, members of both the psychological and the monetary schools of business cycle theory indicted inventory investment as a particularly critical factor in the generation of business cycles. R.G. Hawtrey (1928) argued that monetary factors had their primary effect on the economy through their influence on inventory investment. Pigou (1929) also stressed the impact of systematic errors of optimism and pessimism in explaining cyclical movements.

A major contribution to our understanding of inventory cycles was made by Erik Lundberg (1937), who showed how a set of quite simple assumptions suffices to generate cycles in economic activity, as illustrated by the data for a hypothetical inventory cycle reported on Table 2. Observe that a once and for all step increase in autonomous government spending from

TABLE 1. Contribution of Inventory Disinvestment to Cyclical Declines in GNP

GNP turning point date (Year & Quarter)	Gross National Product	Inventory Investment	Fixed Investment Nonresidential	Residential Investment	△ Inventory/ △ GNP (%)
Peak: 1948:4	497.9	5.3	51.9	24.1	
Trough: 1949:4	490.8	−7.7	43.5	26.9	
Change	*−7.1*	*−13*	*−8.4*	*2.8*	*183.10*
Peak: 1953:2	628.3	5.1	55.9	28.2	
Trough: 1954:2	608.1	−4.1	54.8	29	
Change	*−20.2*	*−9.2*	*−1.1*	*0.8*	*45.54*
Peak: 1957:3	688.5	3.7	67.3	28.9	
Trough: 1958:1	665.5	−6.8	61.5	28.2	
Change	*−23*	*−10.5*	*−5.8*	*−0.7*	*45.65*
Peak: 1960:1	740.7	12.7	67.4	37.3	
Trough: 1960:4	732.1	−5.3	66.3	32.7	
Change	*−8.6*	*−18*	*−1.1*	*−4.6*	*209.30*
Peak: 1969:3	1092	13.7	118.5	43.2	
Trough: 1970:1	1081.4	2.1	115.4	40.6	
Change	*−10.6*	*−11.6*	*−3.1*	*−2.6*	*109.43*
Peak: 1973:4	1266.1	23.7	140.7	57.4	
Trough: 1975:1	1204.3	−14.3	120.7	39.4	
Change	*−61.8*	*−38*	*−20*	*−18*	*61.49*
Peak: 1980:1	1496.4	−0.5	171.8	53	
Trough: 1980:2	1461.4	−2.1	162.2	42.4	
Change	*−35*	*−1.6*	*−9.6*	*−10.6*	*4.57*
Peak: 1981:2	1512.5	8.9	167.1	47.3	
Trough: 1982:4	1480.7	−22.7	181.3	40.6	
Change	*−31.8*	*−31.6*	*14.2*	*−6.7*	*99.37*
				Average	*94.81*

Note: All GNP magnitudes measured in 1972 dollars.

500 to 600 in period 5 (reported in column 1) disturbs the initial equilibrium, leading to cycles in output (column 5), sales (column 7), and inventory investment (column 8). Output rises from the initial equilibrium of 1200 to a peak of 1838 in period 8 and then slumps to a recession low of 1233 in period 13. Thus the attempts by business enterprises to use inventories as a buffer to insulate output from sales are frustrated in the aggregate; indeed, production fluctuates *more* than sales volume (compare columns 5 and 7) over the course of the inventory cycle.

The following details of this inventory cycle deserve notice: In each time period the entries are determined in accordance with the simple assumptions enumerated at the bottom of Table 2. Initially inventories serve as a buffer permitting business firms to meet the unanticipated increase in demand occasioned by the increase in government spending. The immediate impact of the increase in government spending is limited to a drawing down of inventories by 100 units in period 5, as reported in columns 8 and 9; there is no immediate change in output because business firms did not anticipate the increase in sales when scheduling production for period 5. For period 6 output of 1500 is scheduled in order to meet anticipated sales of 1350 (same as last period) plus 150 units to be added to inventory in order to both replace the items sold out of inventories in period 5 and to increase the inventory stock to the higher level of 675 (column 4) which is desired because of increased business volume. But in spite of

these adjustments, the economy does not achieve equilibrium in period 6; once again the immediate impact of an excess of sales over anticipations is met by drawing down inventories below the desired level. The boom is temporary, for eventually inventories catch up with the expanding economy. The economy gradually converges in a series of oscillations toward the equilibrium level presented in the bottom row of the table.

While the cycle displayed on Table 2 is stable, the economy converging in the limit to equilibrium, Lloyd Metzler (1941) showed analytically how the stability of the inventory cycle developed by Lundberg depended critically on the two parameters of the model, the Marginal Propensity to Consume (MPC) and the Marginal Desired Inventory Coefficient (MDIC). He proved that the model will converge toward equilibrium rather than explode only if the following stability condition is satisfied:

$$MPC \times (1 + MDIC) < 1.0$$

For the parameter values used in constructing Table 2 (MPC = 0.60 and MDIC = 0.5) we have $0.6 \times (1 + 0.5) = 0.9 < 1.0$, as required for stability; but if the reader recalculates the table with the MPC = 0.6 but MDIC = 0.7, a series of divergent cycles will be observed $(0.6 \times (1 + 0.7) = 1.02)$. Metzler also considered the implications of replacing the assumption that sales are expected to be the same as last period with extrapolative expectations.

TABLE 2. Lundberg–Metzler Inventory Cycle Model

	MPC = 0.60					MDIC = 0.50			
Time period	Government spending (1)	Anticipated sales (2)	Planned inventories (3)	Planned change in inventories (4)	Output (5)	Consumption (6)	Actual sales (7)	Actual inventory investment (8)	Inventory stock (9)
1	500	1250	625	0	1250	750	1250	0	625
2	500	1250	625	0	1250	750	1250	0	625
3	500	1250	625	0	1250	750	1250	0	625
4	500	1250	625	0	1250	750	1250	0	625
5	600	1250	625	0	1250	750	1350	−100	525
6	600	1350	675	150	1500	900	1500	0	525
7	600	1500	750	225	1725	1035	1635	90	615
8	600	1635	818	203	1838	1103	1703	135	750
9	600	1703	851	101	1804	1082	1682	122	872
10	600	1682	841	−30	1652	991	1591	61	932
11	600	1591	796	−137	1454	873	1473	−18	914
12	600	1473	736	−178	1295	777	1377	−82	832
13	600	1377	688	−144	1233	740	1340	−107	725
14	600	1340	670	−55	1285	771	1371	−86	639
15	600	1371	685	46	1417	850	1450	−33	606
16	600	1450	725	119	1569	942	1542	28	634
17	600	1542	771	137	1679	1007	1607	71	705
18	600	1607	804	98	1706	1023	1623	82	787
19	600	1623	812	24	1648	989	1589	59	846
20	600	1589	794	−52	1536	922	1522	15	861
New equilibrium	600	1500	750	0	1500	900	1500	0	750

Assumptions of the Lundberg–Metzler Inventory Cycle Model

1. Government spending (column 1) increases from 500 to 600 in period 5, generating cyclical movements in sales, output, consumption and inventory investment.
2. Expectations are "static", for anticipated sales in the current period (column 2) always equal actual sales in the preceding period (carried over from column 7).
3. Planned inventories (column 3) equal the marginal desired inventory coefficient (MDIC) times anticipated sales (column 2).
4. Planned inventory change (column 4) equals the excess of planned inventories over last period's inventory stock (column 9).
5. Output (column 5) is the sum of anticipated sales plus the planned change in +column 4.
6. Consumption (column 6) is the Marginal Propensity to Consume (MPC) times output.
7. Actual Sales (column 7) equal Government Spending plus Consumption (column 1 + column 6).
8. Actual Inventory Investment (column 8) equals output less sales (column 5–column 7).
9. The Inventory Stock (column 9) increases from the preceding period by Actual Inventory Investment (column 8).

While the Lundberg–Metzler model demonstrates how the inventory cycle can result from a few quite simple assumptions, simplicity has its costs: Lundberg and Metzler did not show that their assumptions about firm behaviour were compatible with the assumption of maximizing behaviour, they neglected problems of aggregation involved in moving from assumptions about firm behaviour to macro aggregates, they neglected the influence of monetary factors, and they assumed that the adjustment to surprise is met entirely through shifts in buffer stock inventories rather than through price reductions or adjustments in advertising expenditure. Their analysis did serve to inspire the investigation of these and several other issues by a number of authors. Holt and Modigliani (1961) showed how the Lundberg–Metzler assumptions about output determination could be derived from the assumption of firm optimizing behaviour, where the task faced by the entrepreneur involved the minimization of a dynamic quadratic loss function; of course, they also found that alternative specifications of the loss function would yield an embarrassing wealth of alternative behavioural equations. Lovell (1962) showed that the Lundberg–Metzler behavioural assumptions led to instability for any reasonable set of parameters when the problem of aggregating from the firm to the economy-wide level was addressed within the context of a multi-sector dynamic input–output model; while stability might still be obtained by replacing the Lundberg–Metzler assumption that firms attempt an immediate one-period correction of inventory imbalances with the flexible accelerator assumption that only a fraction of the gap between desired and actual inventories is eliminated each time period, he also showed that the system was necessarily unstable if firms had perfect expectations, correctly anticipating the volume of sales in the next period – that is to say, systematic expectational errors under certain circumstances contribute to stability. Lovell (1974) also showed how the real inventory cycle of Lundberg–Metzler could be influenced by monetary policy once desired inventories were assumed to depend on the interest rate as well as sales volume. In a more recent study, Blinder and Fisher (1981) find that the introduction of inventories modifies the standard rational expectations macroeconomic model in two important respects: first, cyclical rather than random fluctuations are generated when the economy is disturbed by unanticipated monetary shocks; second, if desired inventories are sensitive to real interest rates, even fully anticipated changes in money can affect real economic variables, which contradicts the policy impossibility result of rational expectations theory. Adding inventories to the simplified rational expectations model leads to an explanation of the cycle and simultaneously offers the hope for mitigating fluctuations through appropriate policy action.

III. EMPIRICAL RESEARCH. Recognition of the important role of inventories in recessions places a number of basic questions on the agenda for empirical research:

First, are the behavioural assumptions about firm behaviour made by Lundberg and Metzler consistent with the observations? Beginning in the mid 1950s, a number of studies established that the Lundberg–Metzler model is consistent with the evidence provided response lags are introduced; most investigators found that the flexible accelerator assumption that firms attempt only a partial adjustment of inventories toward their equilibrium level within a single period appears more appropriate than the assumption that firms engage directly in 'production smoothing'.

Second, are expectations of future sales volume subject to substantial error, are they rational, and do they have a substantial impact on inventory holdings? Investigators have in recent years been inclined to adopt the assumption of rational expectations as part of the maintained hypothesis rather than subjecting it to direct empirical test; where the rationality hypothesis has been tested, as in Hirsch and Lovell (1969), it has *not* dominated alternative models, such as the extrapolative model of Robert Ferber or the adaptive expectations model. Expectations may be subject to systematic error and forecast errors do affect inventory holdings; but the mechanism is not as simplistic as Lundberg and Metzler assumed in constructing their inventory cycle models.

Third, how big an impact do changes in nominal interest rates and anticipated price changes have on inventory holdings? Prior to the inflationary era of the 1970s, investigators were usually disappointed to find that their regressions yielded incorrect signs on interest rate variables approximately half the time, although these results often went unreported. Studies based on more recent data, of which that by Irvine (1981) may be the most notable, have found stronger indications that monetary conditions influence desired inventory stocks. If the great inflation of the 1970s sensitized business enterprises to the significance of interest rates as a component of inventory carrying costs, the inventory cycle may now be more of a monetary phenomenon than in the past.

MICHAEL C. LOVELL

See also ACCELERATION PRINCIPLE; BUSINESS CYCLES; COBWEB THEOREM; TRADE CYCLE.

BIBLIOGRAPHY

Abramovitz, M. 1950. *Inventories and Business Cycles.* New York: National Bureau of Economic Research.
Blinder, A.S. and Fisher, S. 1981. Inventories, rational expectations, and the business cycle. *Journal of Monetary Economics* 8(3), November, 277–304.
Chikàn, A. 1984. Inventory fluctuation in the Hungarian economy. In *New Results in Inventory Research,* ed. A. Chikàn, New York: Elsevier.
Hawtrey, R.G. 1928. *Trade and Credit.* London: Longmans, Green & Co.
Hirsch, A.G. and Lovell, M.C. 1969. *Sales Anticipations and Inventory Behavior.* New York: Wiley.
Holt, C.C. and Modigliani, F. 1961. Firm cost structure and the dynamic responses of inventories, production, work force and orders to sales fluctuations. In *Inventory Fluctuations and Economic Stabilization,* Joint Economic Committee, US Congress.
Irvine, F.O., Jr. 1981. Retail inventory investment and the cost of capital. *American Economic Review* 71(4), September, 633–48.
Lovell, M.C. 1962. Buffer stocks, sales expectations and stability: a multisector analysis of the inventory cycle. *Econometrica* 30, April, 267–96.
Lovell, M.C. 1974. Monetary policy and the inventory cycle. In *Trade Stability and Macroeconomics: Essays in Honor of Lloyd A. Metzler,* ed. G. Horwich and P.A. Samuelson, New York: Academic Press.
Lundberg, E. 1937. *Studies in the Theory of Economic Expansion.* London: P.S. King & Sons.
Metzler, L. 1941. The nature and stability of inventory cycles. *Review of Economic Statistics* 23, August, 113–29.
Pigou, A.C. 1929. *The Function of Economic Analysis.* London: Oxford University Press.

inventory policy under certainty. Inventories of raw materials, work-in-process, and finished goods are ubiquitous in firms engaged in production and/or distribution of one or more products. Indeed, in the United States alone, 1982 non-farm business inventories totalled over 500 billion dollars, or about 17 per cent of the gross national product that year. The annual cost of carrying these inventories, e.g., costs associated with capital, storage, taxes, insurance, etc., is significant – perhaps 25 per cent of the total investment in inventories, or about 125 billion dollars. Since the cost of carrying inventories is sizeable, a good deal of attention has been devoted to the problem of determining optimal or near-optimal inventory policies that properly balance the costs and benefits of carrying inventories. Moreover, since a firm's inventories are usually distributed among several facilities, e.g., plants, warehouses, retail outlets, effectively coordinating the inventory policies in multi-facility systems.

This essay has four goals. One is to discuss some of the main motives for carrying and distributing inventories in multi-facility inventory systems under certainty. Another is to explain how the form of optimal and/or near-optimal multi-facility inventory policies depend on the particular motive(s) present for carrying and distributing inventories. The third is to outline how the special structure of multi-facility inventory models can be exploited to carry out computations efficiently. The last is to give a brief historical perspective on some of the main developments in multi-facility inventory policy under certainty.

Since it is usually expensive to carry inventories, efficient firms would not do so without good reasons. Thus, it seems useful to ask what motivates a firm to carry inventories at a facility. For this purpose, it is convenient to differentiate *retailers*, i.e., facilities that face exogenous demands, from *wholesalers*, i.e., facilities that face only endogenous demands. Moreover, we shall assume that each facility produces a single product and that each unit output thereof consumes fixed amounts of the outputs of the other facilities that ship to it. Production at a facility is intended to have a broad interpretation, including procurement and shipments from another facility. There are known multi-period non-negative exogenous demands for the output of each facility in excess of the endogenous demands generated by the other facilities.

There are temporally varying production and storage costs at each facility that depend on the respective amounts produced and stored there. The goal is to find production and storage schedules at each facility that minimize the total system cost.

MOTIVES FOR RETAILERS TO CARRY INVENTORIES. Under these circumstances, two of the most significant motives for a retailer to carry inventories are the following.

(i) *Scale economies in supply.* Scale economies in supply occur for several reasons including procurement quantity discounts or set-up costs and production/transportation scale economies. When that is so, it is often economical to produce in a single period to satisfy exogenous and endogenous demands occurring over several periods with the aim of reducing the average unit cost of supply.

(ii) *Temporal increase in marginal cost of supplying demand.* The marginal cost of supplying demand in a period is the marginal cost of producing an amount equalling the demand in the period. A temporal increase in the marginal cost of supplying demand may arise in several ways. One is where the demands are stationary and there is a temporal increase in the marginal cost of production. The latter occurs, for example, when raw-materials prices, wage rates, or marginal transportation costs increase. A second is where there is a temporal increase in demand and diseconomies of scale in production. A temporal increase in demand may arise because of long-term growth or seasonality thereof. Production scale diseconomies occur when there are alternate sources of supply, each with limited capacity, or when production at a plant in excess of normal capacity must be deferred to a second shift or to over-time with an attendant increase in unit labour costs.

MOTIVES FOR WHOLESALERS TO CARRY INVENTORIES. Each of the above motives for retailers to carry inventories must be strengthened in order to motivate wholesalers to do likewise. Three significant ways to do this for a wholesaler whose output is directly or indirectly consumed by some retailer are the following.

(i) *Inter-facility storage-cost variations.* Inter-facility variations in storage costs are common, e.g., between retailers located on expensive city land and wholesalers located on inexpensive rural land. Such variations, when coupled with one of the motives for retailers to carry inventories, may also motivate wholesalers to carry inventories. To see why, consider the extreme case of a retailer with high enough storage costs to make storage there uneconomical. In that event, the retailer passes its demands on directly to any wholesaler whose output the retailer consumes – effectively making the wholesaler a retailer. Then, if either of the two motives for retailers to carry inventories is present at the wholesaler, the latter may be motivated to carry inventories. Of course, in the case of the second motive, the demand at the wholesaler is the demand passed on to it by the retailer.

(ii) *Inter-facility variations in supply scale economies.* Facility-dependent supply scale economies may motivate wholesalers to carry inventories. For example, if a wholesaler and a retailer are suppliers of a second retailer and there are no production costs at either retailer, then scale economies in production at the wholesaler may motivate it to carry inventories for the second retailer.

(iii) *Inter-facility variations and temporal increase in marginal cost of supplying demand.* Inter-facility variations in the marginal cost of supplying demand arise because of inter-facility variations in demands or production/transportation costs, capacity limitations, etc. Such variations, when coupled with a temporal increase in the marginal cost of

supplying demand, may motivate a wholesaler to carry inventories for a retailer that consumes its output. This can occur when the retailer's marginal production costs are temporally non-increasing and either there is a temporal increase in the marginal production cost at the wholesaler or there are diseconomies of scale in production at the wholesaler and rising demands at the retailer.

FORMULATION OF MULTI-FACILITY INVENTORY PROBLEM. In order to see how the different motives for carrying inventories influence the form of desirable inventory policies, it is necessary to formulate the problem more precisely. To that end, consider a collection of *facilities*, labelled $1, \ldots, f$, each producing a single product. Facilities can be interpreted in many ways, e.g., as plants, warehouses, retail outlets, machines in a plant, etc. the facilities are linked by the fact that the production of each unit at facility j *directly consumes* $e^{ij} \geqslant 0$ units of the output of facility i. The time-lags in shipments between facilities are negligible, i.e., production at facility j in a period consumes output at other facilities that may be produced at those facilities either before *or during* the period. There is a given exogenous non-negative demand d_{jt} in period $t = 1, \ldots, p$ for the ouput of facility j. The demands at each facility in each period are met as they occur. There is a real-valued cost $c_{jt}(w)$ (resp., $h_{jt}(w)$) of producing (resp., storing) $w \geqslant 0$ in (resp., at the end of) period t at facility j. The cost of producing w units at facility j in a period also includes the costs of transporting $e^{ij}w$ units of the output of each facility i to facility j in that period. We can and do assume without loss of generality that all $c_{jt}(0) = h_{jt}(0) = 0$. Let x_{jt} and y_{jt} be the respective amounts produced in and stored at the end of period t at facility j. Let $x_t = (x_{jt})$, $y_t = (y_{jt})$ and $d_t = (d_{jt})$ be respectively the f-element column vectors of *production. inventory and demand schedules in period t*, and let $x = (x_t)$, $y = (y_t)$ and $d = (d_t)$ be the $f \times p$ matrices of *p-period production, inventory and demand schedules*. Assume that $y_0 \equiv y_p \equiv 0$. The net production at each facility that is available to satisfy exogenous demands in, or to store at the end of, period t is $(I - E)x_t$ where $E \equiv (e^{ij})$ is the *consumption matrix*. The problem is to find a *production and storage schedule* $z = (x, y) \geqslant 0$ that minimizes the *cost*

$$\sum_{j,t} [c_{jt}(x_{jt}) + h_{jt}(y_{jt})] \tag{1}$$

subject to the *stock-conservation constraints*

$$(I - E)x_t + y_{t-1} - y_t = d_t, \qquad t = 1, \ldots, p. \tag{2}$$

FACILITY NETWORK. Associated with the consumption matrix E is a *facility network* **F** whose nodes are the facilities and whose arcs are the ordered pairs $i \rightarrow j$ of facilities i, j for which $e^{ij} > 0$. There is no loss of generality in assuming that **F** is *connected*, i.e., there is an undirected path from each facility to each other facility. For if not, the problem can be solved separately for the set of facilities in each connected component, i.e., maximal connected subnetwork, of **F**.

Facility j *directly* (resp., *indirectly*) *consumes* the output of facility i if there is a chain, i.e., directed path, from i to j with exactly one arc (resp., two or more arcs). Assume throughout that the facility network is *circuitless*, i.e., no facility directly or indirectly consumes its own output.

Several practical examples of circuitless facility networks are also *trees*, i.e., there is a unique simple path between each pair of facilities. Among the common special cases of trees are assembly, distribution, assembly-distribution, star and series networks. *Assembly* (resp., *distribution*) networks are rooted trees in which all arcs are directed towards (resp., away from) a distinguished facility called the root. The root is the facility at which final assembly of all products takes place in assembly networks and is the ultimate source of all product in distribution networks. *Star* networks are rooted trees in which all arcs are incident to the root, and so are assembly-distribution networks. *Series* networks are chains and so are at once assembly and distribution networks. Finally, *assembly-distribution* networks are rooted trees in which the root facility divides the tree into two subtrees, one an assembly network and the other a distribution network, both sharing the root facility. In trees, we can and do assume without loss of generality and without further mention that all products are measured in common units, i.e., $e^{ij} = 1$ for all arcs $i \rightarrow j$.

1. LINEAR COSTS

In this section, we consider the multi-facility inventory problem in its simplest setting, namely, where there are neither economies nor diseconomies of scale, so the cost function (1) is linear. This allows temporal increases in unit production costs and interfacility variations in unit storage or production costs. The former may motivate retailers to carry inventories and, when coupled with one of the latter, may motivate wholesalers to do likewise.

Extreme schedules of totally-Leontief-substitution systems. The problem is a totally-Leontief-substitution system. Moreover, at least one optimal schedule is an extreme point of the set of feasible schedules, and each extreme schedule satisfies

$$y_{j,t-1} x_{jt} = 0 \qquad (3)$$

for $1 \leqslant j \leqslant f$ and $1 \leqslant t \leqslant p$, i.e., facility j produces in period t only if that facility has no entering inventory in the period. This is so because production at a facility in a period and storage in the previous period are, in the terminology of Leontief-substitution systems, 'substitute' activities, and extreme schedules in such systems do not admit substitute activities. The condition (3) assures that whenever a facility produces, it satisfies all the endogenous and exogenous demands for its output in an interval of periods. Thus, there is no 'lot splitting', i.e., the endogenous demand at one facility created by production at another facility in a later period is entirely satisfied by production at the first facility in a *single* (no later) period. In this sense, each facility produces in 'larger' lots than do its followers.

Dynamic-programming computation of optimal dual prices. Moreover, the optimal dual price C_{jt} associated with the stock-conservation constraint for facility j in period t is the minimum cost of satisfying a unit of demand at facility j in period t. It can be shown that the C_{jt} satisfy, and can be calculated from, the dynamic-programming recursion

$$C_{jt} = \min \left(h_{j,t-1} + C_{j,t-1}, c_{jt} + \sum_{i \rightarrow j} e^{ij} C_{it} \right) \qquad (4)$$

for $1 \leqslant j \leqslant f$ and $1 \leqslant t \leqslant p$ where c_{jt} and h_{jt} are respectively the unit product and storage costs at facility j in period t and $C_{j0} \equiv h_{j0} \equiv 0$. Equation (4) expresses the fact that the optimal way to satisfy a unit of demand at facility j in period t is to choose the cheaper of two options. One is to provide a unit of product at facility j in period $t - 1$ as cheaply as possible and store it for one period. The other is to produce a unit at facility j in period t, thereby consuming e^{ij} units from each facility i in that period with those units being provided as cheaply as possible. Observe that it is optimal to produce a facility j in period t if and only $C_{jt} \leqslant h_{j,t-1} + C_{j,t-1}$, *independently* of the demand schedule d.

Once the optimal periods in which to produce each product are found, the desired optimal production schedule is obtained inductively as follows. Suppose the optimal production schedules $\mathbf{x}^j = (\mathbf{x}_{jt})$ have been found for all facilities j that directly consume the output of facility i. Then, if it is optimal for facility i to produce in period t and next in period $u + 1$, it follows from (2) and (3) that $\mathbf{x}_{it}$ can be calculated recursively from

$$\mathbf{x}_{it} = \sum_{k=t}^{u} d_{ik} + \sum_{i \rightarrow j} e^{ij} \sum_{k=t}^{u} \mathbf{x}_{jk}. \qquad (5)$$

Form of optimal production schedule. To sum up, some optimal schedule is extreme, and so a facility produces in a period only if the facility has no entering inventory in the period. If the consumption matrix and demand schedule are integer, then so is each extreme schedule. Optimal production at a facility in a period is a nondecreasing linear function of the present and future demand at the facility and all facilities that consume its output.

Running time. The running time of this dynamic-programming algorithm is $O(f^2 p)$ in general networks and falls to $O(fp)$ in planar networks, the last because the number of arcs in such networks does not exceed three times the number of nodes. Of course trees, and in particular. assembly-distribution networks, are planar, so the improved running time applies in such networks.

Nonlinear costs. The algorithm given above for solving the problem with linear costs is also useful when the costs are nonlinear. This is because many methods for solving problems with nonlinear costs, e.g., branch-and-bound, gradient methods, etc., entail solving a sequence of linear-cost problems, each of which can be solved in linear time by the recursion (4).

2. CONCAVE COSTS

In this section we generalize the linear-cost multi-facility inventory problem discussed in Section 1 to allow economies-of-scale by requiring the cost function (1) to be concave. Then the marginal costs of production and storage fall respectively the more one produces and stores. Since the class of additive concave cost functions contains the linear ones, all motives for carrying inventories at retailers and wholesalers with linear cost functions remain in force in this section. Beyond these, the introduction of scale economies in this section provides an added motive for carrying inventories at retailers. This fact, when coupled with either inter-facility variations in storage or production costs, both of which are allowed in this section, provide additional motives for carrying inventories at wholesalers.

Extreme schedules. As for the case of linear costs, if the minimum is attained, it is attained at an extreme schedule and so satisfies (3). However, because of the scale economies, optimal production is no longer linear in demand. For that reason, the dynamic-programming recursion (4) no longer solves the problem.

Dynamic-programming algorithm for series networks. However, there is a polynomial-time dynamic-programming algorithm for finding an optimal schedule in series networks in which facility f is the only retailer. To describe the algorithm, observe first that on iterating a representation like (5) of any extreme schedule $z = (x, y)$, that the sum $y_{i,t-1} + x_{it}$ of the initial inventory and production at facility i in period t equals the sum $d^{jl} \equiv \Sigma_{k=j}^{l-1} d_{fk}$ of the demands at facility f in periods $j, \ldots, l - 1$ for some $t \leqslant j < l$. Let $C_{it}(d^{jl})$ be the minimum cost of satisfying the demands at facility f in periods $j, \ldots, l - 1$ from the stock d^{jl} available at facility i at the beginning of period t. Then the $C_{it}(d^{jl})$ can be calculated from the dynamic-programming recursion.

$$C_{it}(d^{jl}) = \min_{\substack{j \leqslant k \leqslant l \\ t < k}} [c_{i+1, t}(d^{jk}) + C_{i+1, t}(d^{jk})$$
$$+ h_{it}(d^{kl}) + C_{i, t+1}(d^{kl})]$$

for $1 \leqslant i < f$, $1 \leqslant t \leqslant j < l \leqslant p + 1$ and $t < p$, together with fairly obvious boundary recursions where $i = f$ or $t = p$. This recursion expresses the fact that the minimum cost of satisfying the demands at facility f in periods $j, \ldots, l - 1$ from the stock d^{jl} on hand after production at facility i in period t is attained by dividing the stock d^{jl} into two parts, d^{jk} and d^{kl}, the former being sent to facility $i + 1$ for production in period t and the latter stored at facility i for one period.

Running time. The running time of the algorithm is $O(fp^4)$, which is $O(p^3)$ times that for the linear-cost case. If also the production and storage costs at each facility are respectively temporally non-increasing and non-decreasing in the facility index, the running time improves to $O(fp^3)$ because some optimal schedule is 'nested' in the sense to be defined shortly. For the case of a single facility, the running time drops further to $O(p^2)$.

General networks. The above algorithm can be generalized to arbitrary distribution networks, but the computational effort grows exponentially with the number of retailers. For that reason, we do not discuss this possibility. For the special case of one-warehouse multi-retailer networks (i.e., star distribution networks in which the root facility is a warehouse and the other facilities are retailers) in which the production and storage costs at the warehouse are respectively linear plus a set-up cost and linear, there is an algorithm for solving the problem whose running time is linear in the number of retailers, but exponential in the number of periods. However, no polynomial-time algorithm has been found for the general multi-facility problem, even for star networks.

EFFECTIVE HEURISTICS. This suggests the possibility that optimal schedules may be too difficult to find and that heuristics may instead be necessary. One heuristic for distribution systems is to optimize over the subclass of *nested* schedules, i.e., schedules x satisfying (3) for which $x_{it} > 0$ implies $x_{jt} > 0$ for each facility j that directly consumes the output of facility i. There is a dynamic-programming algorithm for finding an optimal nested schedule in a distribution system in $O(fp^3)$ time. Unfortunately, optimal schedules need not be nested – even in one-warehouse multi-retailer networks – because, for example, it may be optimal for a retailer with low demands to order less frequently than the warehouse. However, the nested-schedules heuristic can be adapted to give a reasonably effective heuristic when the demand schedules at each facility are *proportional* to one another.

STATIONARY CASE WITH SET-UP PRODUCTION AND LINEAR STORAGE COSTS. It turns out that there is an extraordinarily effective heuristic for the stationary (demands and costs) continuous-time infinite-horizon version of the problem in which there is a set-up production cost and a linear storage cost at each facility. The *effectiveness* of a heuristic for this problem is 100 per cent times the ratio of the infimum of the average cost per unit time over all policies to the average cost per unit time incurred by the heuristic.

The heuristic for this problem decomposes the multi-facility problem into a collection of single-facility problems, one for each facility. In each single-facility problem, there is a demand rate $r > 0$ per unit time, a set-up production cost $K > 0$ and a storage cost $h > 0$ per unit stored per unit time. Then one expects that a minimum-average-cost schedule will permit production only when stock runs out, which is the continuous-time analogue of (3). Thus, since the demand rates and costs are stationary, one anticipates that a minimum-average-cost schedule will entail producing every $T > 0$ periods an amount equalling the demand rT until the next instance of production. An easy calculation shows that the optimal value of T is given by the celebrated square-root formula $T° = (2K/hr)^{1/2}$. Now if $T > 0$ is any other production interval for which $2^{-1/2} \leqslant T/T° \leqslant 2^{1/2}$, then it is easy to show that the effectiveness of the new schedule is at least 94 per cent.

The heuristic for the multi-facility problem is constructed as follows. First, restrict attention to *power-of-two* schedules, i.e., those for which a facility produces only when it runs out of stock, the *production intervals* between successive times that a facility produces are identical, and the ration of the production intervals at distinct facilities is a (possibly negative) integer power of two. It is not difficult to find an expression $C(\mathbf{T})$ for the average cost of a power-of-two schedule $\mathbf{T} = (T^i)$ where T^i is the production interval used at facility i.

The *optimal power-of-two problem* is that of finding $\mathbf{T}$ that minimizes $C(\mathbf{T})$ subject to the power-of-two constraints and $\mathbf{T} \gg 0$. Instead of solving this problem, one next solves the *relaxation* thereof in which the power-of-two constraints are dropped. The relaxation is a minimum-convex-cost dual network-flow problem on a *cost network*. The nodes of the cost network correspond to the chains in the facility network that end at retailers. The arcs in the cost network join each node α to the node β that corresponds to the subchain formed by deleting the first node of the chain that corresponds to α. Denote by $\mathbf{T^*} = (T^{*i})$ the optimal solution of the relaxation of the optimal power-of-two problem. Remarkably and most important, the minimum average cost $C(\mathbf{T^*})$ of the relaxation is a lower bound on the average cost of an *arbitrary* (not necessarily power-of-two) schedule!

The cost-network problem entails reallocating the set-up costs and storage cost rates among the facilities in such a way that with the new cost parameters, each facility i's optimal production interval, when considered as a single-facility problem, is precisely T^{*i}. One then rounds off each T^{*i} to form a power-of-two schedule $\mathbf{T} = (T^i)$ satisfying $2^{-1/2} \leqslant T^i/T^{*i} \leqslant 2^{1/2}$ for each i. This assures that the effectiveness of the resulting power-of-two schedule is at least 94 per cent for each facility, and so is at least 94 per cent for the system. A somewhat more complex procedure guarantees that the system effectiveness is at least 98 per cent.

The cost-network problem can be reduced to solving a sequence of maximum-flow problems, each of which splits its predecessor into two smaller subproblems. In special cases, there are even more efficient algorithms. For example, the cost-network problem can be solved in $O(f \ln f)$ time in one-warehouse multi-retailer and assembly networks.

3. CONVEX COSTS

In this section we discuss the multi-facility inventory problem in the presence of diseconomies of scale by requiring the cost function (1) to be convex. Then the marginal costs of production and storage are non-decreasing in the amounts produced and stored respectively. This allows a temporal increase in the marginal cost of supplying demand, which may motivate retailers to carry inventories.

s-additive convex production cost function. Here we suppose that the production cost function is *s-additive convex,* i.e., $c_{ji}(w) = s_t c^j(w/s_t)$ for $w \in R$ with $c^j(\cdot)$ being a $+\infty$ or real-valued convex function on the real line and s_t being a positive *scale parameter* for each t. Also assume that there are no direct storage costs, so $h_{jt} \equiv 0$, but that it is possible to represent any such costs by absorbing them in the production costs with an appropriate choice of the scale parameters s_t, e.g., as we show below for capital costs. In particular, there is a motive to carry inventories at retailer j in period t if the marginal cost $\dot{c}(d_{jt}/st)$ of supplying the demand there in that period is less than that in some subsequent period k, say. This implies, and provided $\dot{c}^j(w)$ is strictly increasing in w, is implied by $d_{jt}/s_t < d_{jk}/s_k$. The last is so if $Od_{jt} \leqslant d_{jk}$ and $s_t \geqslant s_k$ with at least one inequality being strict. This formulation is rich enough to be useful because it provides for certain storage costs and allows temporal variations in the marginal cost of supplying demand, and yet is simple enough to admit a graphical solution.

Positively-homogeneous convex production cost function. As a particular example, suppose that the present value of the production cost at facility j is positively homogeneous of degree $q + 1 \geqslant 1$, i.e., $c_{jt}(w) = \beta^t |w|^{q+1} c_j$ for some discount factor $\beta > 0$ and $c_j > 0$. This accounts for the cost of capital invested in inventories. Then the production cost is s-additive convex with scale parameter

$$s_t = (1 + \rho)^{t/q} \tag{6}$$

where $\beta \equiv 1/(1 + \rho)$ and 100ρ per cent is the interest rate. Observe that if $\rho = 0$, then $s_t = 1$ for all t. If instead $\rho > 0$ (resp., $\rho < 0$), then s_t expands (resp., contracts) geometrically with the precise rate being greater than, equal to or less than $|\rho|$ according as $0 < q < 1$, $q = 1$ or $1 < q$.

Taut-string solution of the single-facility problem. Now return to the case of an arbitrary s-additive convex production cost function. Then the fundamental result for the single-facility problem is the Invariance Theorem which asserts that there is an optimal production schedule that is independent of the function $c \equiv c^1$, though it does depend on the demand schedule and scale parameters. Because of the Invariance Theorem, it suffices to solve the problem for any single strictly-convex function c. It turns out to be felicitous to put $c(w) = (w + 1)^{1/2}$ (which is strictly convex) because the problem can then be solved graphically. To see this, observe that the cost of a schedule (x, y) is then the length $\Sigma_t (x_t^2 + s_t^2)^{1/2}$ of the shortest polygonal path in the plane passing in order through the points (S_t, X_t) for $t = 0, \ldots, p$ where $S_t \equiv \Sigma_1^t s_k$ and $X_t \equiv \Sigma_1^t x_k$, or equivalently, the length of a taut string passing in order through those points. Now consider the taut string passing through the points $(S_o, D_o) = (0, 0)$ and (S_p, D_p), and lying above the points (S_t, D_t) for $t = 1, \ldots, p - 1$ where $D_t \equiv \Sigma_1^t d_k$. Let (S_t, X_t^*) be the coordinates of the taut string corresponding to S_t for each t. Then (X_t^*) is the least concave majorant of (D_t). Since the feasible cumulative production schedules are non-decreasing and majorize (D_t), it follows from the Invariance Theorem and the non-negativity of the demands that $x_t^* = X_t^* - X_{t-1}^* \geqslant 0$ is optimal for all t and convex c.

The above *taut-string solution* has the property that during any interval of periods in which inventories are held, optimal production is proportional to the scale parameter. Thus, if the scale parameter rises (resp., falls) over the interval, then so does optimal production. In particular, if the scale parameter is given by (6), then optimal production rises or falls geometrically in the interval according as $\rho > 0$ or $\rho < 0$.

The taut-string solution has another property that plays an

important role in solving the multi-facility problem, namely, it is positively homogeneous of degree one in the demand schedule. This is easily seen because, by the Invariance Theorem, there is no loss in taking the production cost function to be positively homogeneous.

Proportional demand schedules. In order to obtain a tractable solution to the multi-facility problem, we shall assume that the demand schedules at each facility are *proportional,* i.e., there is a p-period row vector d^* of non-negative demands and a column vector δ of non-negative *demand levels* associated with the f facilities such that $d = \delta d^*$. Thus the demand schedules at each facility exhibit a common pattern of temporal variation. This includes the cases in which the demands are facility-dependent and stationary, or there is a single retailer. Let π be the column vector of amounts that would have to be produced at the f facilities to satisfy the column vector δ of exogenous demands at those facilities. Evidently, $\delta = (I - E)\pi$. Let x^* be the (row) optimal production schedule (the taut-string solution) for the *standard* single-facility problem with demand schedule d^*, and let y^* be the corresponding row vector of inventories.

Optimal schedule for the multi-facility problem. Let $\mathbf{x}$ and $\mathbf{y}$ be the $f \times p$ matrices of optimal p-period production and inventory schedules at each facility. By combining the above results and the Invariance Theorem, one finds that optimal p-period production and inventory schedules for the multi-facility problem are given by the pair of rank-one matrices

$$\mathbf{z} = (\mathbf{x}, \mathbf{y}) = (\pi x^*, \delta y^*).$$

Thus, the optimal production and inventory schedules at different facilities are proportional to one another. Indeed, inventories are held at a retailer only to satisfy its exogenous demands, and are proportional to its demand level. No inventories are held at a facility to satisfy endogenous demands there because the production schedules at the facilities consuming the output of the given facility already smooth out the demand schedules. Hence, *just-in-time* scheduling, i.e. holding no inventories, is optimal at all wholesalers. This is consistent with the absence of either of the motives for wholesalers to hold inventories, namely, interfacility variations in storage costs or marginal costs of supplying demand. The last is fundamentally because there are no inter-facility variations in the scale parameters of the production cost functions and the demand schedules at each facility are proportional. Hence, up to a monotone transformation, all facilities exhibit the same temporal variation in their marginal costs of supplying demands.

Running time. The optimal schedule x^* for the standard single-facility problem can be found in $O(p)$ time. The vector π can be computed in $O(f^2)$ time in general networks and in $O(f)$ time in planar networks. Thus, $\mathbf{z}$ can be found in $O(f^2 + fp)$ time in general networks and in $O(fp)$ time in planar networks.

4. HISTORICAL PERSPECTIVE

The study of inventory problems has influenced and been influenced by the tools available for their analysis. The seventy odd years since the first economic-lot-size model was proposed by F.W. Harris (1915) can be reasonably divided into three phases. During the period 1915–1950, attention was focused mainly on the formulation and closed-form solution of relatively simple single-facility models under certainty using the calculus. The period 1950–1965 saw the introduction of dynamic, linear and nonlinear programming methods for the characterization and computation of optimal policies in the presence of certainty and/or uncertainty, again largely for single-facility problems.

During the period since 1960, attention has gradually shifted towards the problem of coordinating the inventory policies at several interrelated facilities. It was realized that many such multi-facility inventory problems could be formulated as dynamic nonlinear network-flow or Leontief-substitution models and that the special structure of these models permitted a unified theory of inventory control to begin to emerge. A qualitative theory of mathematical programming, namely, lattice programming and substitutes, complements and ripples, was developed to give a simple general method of studying the qualitative variation of optimal inventory policies with the problem parameters. The realization that optimal policies for multi-facility systems could be extraordinarily complex to compute and/or implement, coupled with the successful use of heuristics in several areas of combinatorial optimization, led to the development of provably effective and efficient heuristics for multi-facility inventory problems. The rapid recent development of computational geometry will no doubt stimulate further progress on multi-facility systems as we begin to be able to combine the symmetries of continuous space-time models with the computational efficiency of discrete space-time models.

We close by summarizing the main sources for the material in this entry. Motives for retailers to hold inventories are discussed by Arrow (1958) and Scarf (1963). The treatment of the multi-facility linear-cost problem is taken from Veinott (1969) who applied Dantzig's (1955) theory of Leontief-substitution systems. Earlier ad hoc treatments of the single-facility linear-cost case are reviewed by Arrow (1958). Clark and Scarf (1960) solved the series-network problem with stochastic demands at the last facility and linear costs at all facilities except the first one.

The results for the single-facility concave-cost problem begin with Wagner and Whitin (1958). The characterization of the extreme schedules for the multi-facility problem is due to Zangwill (1966), though the development using Leontief-substitution systems is taken from Veinott (1969). The algorithm for series networks is due to Zangwill (1969). The nested-schedules algorithm for distribution systems comes from Veinott (1969). Love (1972) gave conditions assuring that the last algorithm finds an optimal schedule for series networks. Erickson, Monma and Veinott (1981) have recently encompassed an improvement of the above algorithms for series networks in a send-and-split method for finding minimum-concave-cost network flows. This result reveals that the polynomial running time of these algorithms for series networks is explained by the planarity of the corresponding network-flow problem. The results on the 94 per cent and 98 per cent effective power-of-two heuristics for the multi-facility problem with stationary demands and costs are due to Roundy (1985a, 1986). He employed an algorithm of Maxwell and Muckstadt (1985) to solve the cost-network problem. The effectiveness of a 'nested-schedules' heuristic with proportional demand schedules is discussed by Roundy (1985b).

The Invariance Theorem for the single-facility convex-cost problem is due to Modigliani and Hahn (1955) for the case of stationary costs. The generalization to non-stationary costs and the taut-string solution are due to Veinott (1971) who encompassed this example in a theory of Invariant Network Flows. The results for the multi-facility convex-cost problem with proportional demand schedules are due to recent research of the author.

ARTHUR F. VEINOTT, JR.

See also NETWORKS; OPERATIONS RESEARCH.

BIBLIOGRAPHY

Arrow, K.J. 1958. Historical background. Chapter 1 in *Studies in the Mathematical Theory of Inventory and Production*, ed. K.J. Arrow, S. Karlin and H. Scarf, Stanford: Stanford University Press.

Clark, A.J. and Scarf, H. 1960. Optimal policies for a multiechelon inventory problem. *Management Science* 6(4), July, 475–90.

Dantzig, G.B. 1955. Optimal solution of a dynamic Leontief model with substitution. *Econometrica* 23(3), July, 295–302.

Erickson, R.E., Monma, C.L. and Veinott, A.F., Jr. 1981. Minimum-concave-cost network flows. Stanford, California, Department of Operations Research, Stanford University. Revised (1986) as Send-and-split method for minimum-concave-cost network flows. To appear in *Mathematics of Operations Research*.

Harris, F.W. 1915. Operation and Costs (The Factory Management Series). Chicago: A.W. Shaw Co., 47–52.

Love, S.F. 1972. A facilities in series inventory model with nested schedules. *Management Science* 18(5), January, 327–38.

Maxwell, W.E. and Muckstadt, J.A. 1985. Establishing consistent and realistic reorder intervals in production-distribution systems. *Operations Research* 33(6), November–December, 1316–41.

Modigliani, F. and Hohn, F.E. 1955. Production planning over time and the nature of the expectation and planning horizon. *Econometrica* 23(1), January, 46–66.

Roundy, R.O. 1985a. 98%-effective integer-ratio lot-sizing for one-warehouse multi-retailer systems. *Management Science* 31(11), November, 1416–30.

Roundy, R.O. 1985b. Efficient, effective lot-sizing for multiproduct multi-stage production/distribution systems with correlated demands. Technical Report 671, Ithaca, New York: School of Operations Research and Industrial Engineering, Cornell University.

Roundy, R.O. 1986. A 98%-effective rule lot-sizing for multi-product, multi-stage production/inventory systems. *Mathematics of Operations Research* 11(4), November, 699–727.

Scarf, H. 1963. A survey of analytical techniques in inventory theory. Chapter 7 in *Multi-Stage Inventory Models and Techniques*, ed. H. Scarf, D. Gilford, and M. Shelly. Stanford, California: Stanford University Press, 185–225.

Veinott, A.F., Jr. 1969. Minimum concave-cost solution of Leontief substitution models of multi-facility inventory systems. *Operations Research* 17(2), March–April, 262–91.

Veinott, A.F., Jr. 1971. Least d-majorized network flows with inventory and statistical applications. *Management Science* 17(9), May, 547–67.

Wagner, H.M. and Whitin, T.M. 1958. Dynamic version of the economic lot size model. *Management Science* 5(1), October, 89–96.

Zangwill, W.I. 1966. A deterministic multiproduct, multifacility production and inventory model. *Operation Research* 14(3), May–June, 486–507.

Zangwill, W.I. 1969. A backlogging model and a multi-echelon model of a dynamic economic lot size production system – a network approach. *Management Science* 15(9), May, 506–27.

investment. Investment is capital formation – the acquisition or creation of resources to be used in production. In capitalist economies much attention is focused on business investment in physical capital – buildings, equipment, and inventories. But investment is also undertaken by governments, nonprofit institutions, and households, and it includes the acquisition of human and intangible capital as well as physical capital. In principle, investment should also include improvement of land or the development of natural resources, and the relevant measure of production should include nonmarket output as well as goods and services produced for sale.

Thus, acquisition of an automobile by government or households is as much investment as acquisition of an automobile by a business firm. The car is used in all cases for the production of transportation services. Similarly, government construction of roads, bridges, and airports is as much investment as business acquisition of trucks and planes. Expenditures for research and development are investment whether undertaken by business, government, or nonprofit universities. And most important, education and training, wherever undertaken, are major forms of investment in human capital.

There is a widespread mythology that investment is good and the more investment the better. But investment may be good or bad and there may be too much as well as too little.

Classical and neoclassical economists have stressed the role of investment in providing for the future. Maintaining the current level of output requires keeping up the existing means of production. Economic growth, or the increase in the rate of output, is then seen as depending considerably on the acquisition of additional means of production, that is, investment in excess of the wearing away or depreciation of existing capital. Investment may also contribute to higher output where the new capital 'embodies' new and improved technology. That investment will contribute to economic growth presupposes, however, that the additional capital is useful. It must have a positive net product, which is to say that the additional capital must contribute more to future production than the value of the resources used to create it.

How far to go in allocating resources to investment depends upon our preferences for current consumption versus future consumption, or our preferences between our own consumption and that of our children and grandchildren. It also depends on the production function, that is, the terms under which additional capital can be converted into additional future output. It would hardly seem desirable to sacrifice $100 of current consumption to produce $100 of capital which would result in future production of only $90. The notion that this is not a relevant issue stems from the assumption that profit-seeking entrepreneurs would not freely undertake investment in which the costs are greater than the returns. It is not always perceived, however, that where governments offer subsidies or 'tax incentives' for investment to business to undertake investment which would not otherwise seem profitable, such unproductive capital formation is exactly what may be expected.

A second major role for investment has been seen in the achievement and maintenance of full employment. This requires that aggregate investment plus aggregate consumption equal the total output that would be produced if all individuals who wish to work could find employment. Investment may then be inadequate not only in failing to provide sufficient resources for future production. It may also be inadequate if it is insufficient to bring about the full utilization of existing resources. This latter problem has received major attention as a consequence of the work and influence of John Maynard Keynes (1936).

Another way of stating the condition necessary for full employment is that aggregate investment must equal aggregate saving out of the full-employment level of income. In national income accounts, measured investment and saving are always identically equal, owing to the identity of output and income which, aside from receipts from abroad, is earned only from production. That part of income not spent on consumption is saved. But that part of production not purchased by consumers must be acquired (or kept) by producers and hence is investment, though not necessarily intended investment. If we designate Y as income and output, C as consumption, S as saving, and I as investment, we then have $S = Y - C = I$.

While realized investment is thus identically equal to saving, investment and saving may be more or less than investment demand, i.e., intended investment. If investment demand is less than saving at the current level of income, producers will find that they cannot sell all that they produce. They will accumulate undesired inventories of finished goods (unintended investment in inventories), which should lead them to reduce production. Reduced production means less income and hence less consumption and saving. A shortfall of investment demand relative to saving therefore brings on a cumulative reduction of output and income until saving and investment are brought down to equality with the lesser investment demand. Insufficiency of investment demand has been identified with depression and recessions and tendencies toward chronic unemployment. Stimuli to investment, such as reductions in tax rates on income from capital, have thus seemed in order to bring investment demand up to the levels of saving that would be forthcoming with full employment. Conversely, excessive levels of investment demand can create inflationary pressures, calling for policies that would restrict investment.

These Keynesian perceptions as to the costs and benefits of investment are startlingly different from those of the classical models – old and new – which assume, implicitly or explicitly, that the economy is operating at full employment and full utilization of resources. In the classical models, more current production of capital must mean less current production of consumption goods and services. And more consumption now must mean less current investment and less output and consumption in the future.

In an economy with substantial unemployed resources, however, more investment need not and probably will not bring less consumption. Expenditures for additional investment will rather constitute additional income for their recipients, and this income will in turn be spent in large part on increased consumption. Thus the production of consumption goods and services will increase rather than decline. And more consumption may bring about more investment, as producers see a need for additional capital to increase the output of consumption goods and services.

Classical and Keynesian views also differ on the principal mechanism by which intended investment and saving are equated. In the classical view, changes in the rate of interest are presumed to perform this task. Investment demand is thought to be negatively related and very sensitive to the rate of interest, which is the cost of borrowing funds to finance capital spending. If investment demand is smaller than saving at the full-employment level of income, the classical analysis holds that the excess of funds in the credit market will depress interest rates, thereby inducing increases in investment demand (and possibly reductions in saving as the interest earned by savers falls) until intended investment and saving are equal. Thus, no change in the level of economic activity (output) need occur as in the Keynesian analysis. The Keynesian view of the equilibrating process has interest rates playing a smaller role than changes in output, because investment demand is thought to be relatively insensitive to interest rates, being dominated instead by producers' expectations of future demand for their products. Even if investment demand were sensitive to interest rates, expectations could be so pessimistic that even if the rate of interest were to fall to zero, there would be insufficient investment demand.

Empirical studies have attempted to measure the influence of interest rates, taxes, and expectations of future demand on investment decisions. Producers are presumed to acquire capital to increase their expected profits. The profitability of additional capital depends on its cost, on its expected productivity, and on expectations on the price at which additional output can be sold. Assuming that output is a fixed, 'well-behaved' function of capital and labour (strictly concave, with declining partial derivatives of output with respect to capital and labour and positive cross-partial derivatives), producers will acquire capital to the point where its declining marginal product equals its cost. This will then define both the desired, or equilibrium, capital–labour ratio and capital–output ratio. With the supply of labour and the rate of output fixed and no change in the relative price of capital and labour, investment in equilibrium will be equal to depreciation, or what is necessary to maintain the existing capital stock, and net investment will be zero. Positive net investment will then stem from increases in the demand for output or reductions in the relative price of capital. Increases in output will generate investment demand to maintain the equilibrium capital–output ratio. A reduction in the cost of capital would generate investment in order to increase the capital–labour and capital–output ratios. In either case, maintaining increased amounts of capital will generate further investment to cover increased depreciation.

In general, the desired capital stock may be written as:

$$K^* = f(p, c, Y^*), \qquad (1)$$

where p is the price of output, $c = q[i - (\dot{q}/q) + d]$ is the rental price or user cost of capital, q is the supply price of capital goods, i is the opportunity cost of capital, d is the rate of economic depreciation, and Y^* is desired output. If firms minimize expected costs of producing an exogenously given or expected output Y, then the wage rate, w, would be substituted for p. The rental price, or user cost of capital, c, is the cost per period of holding and maintaining one unit of capital. In the absence of taxes, it is the price of capital goods multiplied by the sum of the real interest rate and the rate of economic depreciation. The former measures the opportunity cost in terms of foregone net earnings from lending or otherwise investing money, plus the capital loss (or minus the capital gain) associated with changing prices of capital goods.

Building on this neoclassical theory of the firm developed by Haavelmo (1960), and assuming a Cobb–Douglas production function with elasticity of output with respect to capital, b, Jorgenson (1963, 1967) arrived at a demand function for capital with a particular form that has been employed in a large number of influential studies:

$$K^* = b(p/c)Y^* \qquad (2)$$

With an implicit unitary elasticity of K^* with respect to c, this formulation implies strong effects of monetary policy, via the rate of interest, and of tax policy so far as, by accelerated tax depreciation, investment subsidies, or exclusion of capital gains from taxation, it affects the value of c (see below).

The more general constant-elasticity-of-substitution (CES) production function may be used to generate a demand for capital having the form:

$$K^* = h(p/c)^s(Y^*)^r, \qquad (3)$$

where s, the elasticity of substitution between labour and capital, is the critical elasticity of demand for capital with respect to the relative price of capital, and r is the elasticity of demand for capital with respect to output. The elasticity, r, will be greater than, equal to, or less than unity as the returns to scale are decreasing, constant, or increasing.

If relative prices are constant, or if technology requires that capital and labour be used in fixed proportions (in which case

the elasticity of substitution is zero), then with constant returns to scale, desired capital is proportional to the demand for output. This form of the demand for capital leads to the 'acceleration principle', according to which net investment demand, arising from a desire to change the stock of capital, depends not on the level of demand for output, but on the *change* in demand for output (Clark, 1917). To induce firms to invest (acquire more capital), demand for output must be expected to rise. Both the original formulation by Jorgenson of the demand for capital (2) and the more general formulation (3) underly a 'flexible accelerator', where the desired capital–output ratio is not constant but depends on prices and on the scale of output and, as seen above, investment is subject to a distributed lag process (Koyck, 1954) affected by adjustment costs and the dynamic process governing the formation of expectations of future variables (Eisner and Strotz, 1963; Helliwell and Glorieux, 1970; Lucas, 1976; Eisner, 1978).

Many early econometric studies of investment behaviour tested the accelerator in various forms, but they generally did not allow for effects of prices on the desired capital–output ratio, which is the hallmark of Jorgenson's neoclassical approach. The major competing hypothesis was that investment depends on the level of profits, on the grounds that realized profits measure expected profits, or that capital market imperfections cause firms' capital expenditures to be constrained by the flow of internal funds (Meyer and Kuh, 1957). Reviews of these earlier investigations are found in Eisner and Strotz (1963) and Jorgenson (1971). The practice in recent studies has been to capture profit expectations by including expectations of the major determinants of profits, namely, sales, prices, and wages, or to approximate them by stock market valuations of firms. The flow of internal funds may play some role in investment decisions, not as a determinant of the desired capital stock but as a factor influencing the speed of adjustment of capital (Coen, 1971).

To study effects of tax policy on demand for capital, the rental price can be generalized to incorporate parameters of the tax system. For example, the after-tax cost of holding one unit of capital would be:

$$c = q[(1 - uv)i - (1 - uw)(\dot{q}/q) + d][1 - k - uz]/(1 - u) \qquad (4)$$

where u is the rate of taxation of business income; v is the proportion of the opportunity cost of capital (such as interest, dividends, and foregone earnings) that is tax deductible; w is the proportion of capital gains and losses effectively taxed; k is the effective rate of the investment tax credit or subsidy; and z is the present value of the tax depreciation expected from a dollar of investment (Hall and Jorgenson, 1967).

It can be seen, in this definition, that higher values of v, k and z (from accelerating tax depreciation) reduce the value of c, as does a higher value of w, provided that capital goods prices are expected to rise. The value of c would also be lowered by decreasing the rate of interest or other measure of the opportunity cost of capital. A higher rate of inflation of capital goods prices has two opposing effects on c. In so far as higher inflation reduces the real after-tax opportunity cost of capital, it reduces c. However, if tax depreciation is based on the historical cost of assets rather than on replacement costs, inflation reduces the present value of tax allowances, z, and thereby raises c (Feldstein, 1982). Finally, we may note that changes in the general rate of business taxation are ambiguous in their effects on c. If v and w are unity, and if the opportunity cost of capital is unaffected by a change in the business tax rate, then a decrease in u will reduce, leave unchanged, or increase c as the present value of tax allowances on a unit of investment (including the investment credit) is less than, equal to, or greater than the present value of economic depreciation (Hall and Jorgenson, 1971). But then, going back to equation (1), the effect of any of these parameters on K^* depends upon the elasticity of the latter with respect to c.

The desired capital stock does not in itself indicate the rate of investment, which is the rate of replacement of existing capital plus the rate of net additions. Both entail a combination of financial considerations and costs of adjustment, which will in turn relate to costs of acquiring information necessary to decisions, costs of planning, and the supply function for capital goods, all filtered through the expectations of agents. If adjustment costs are an increasing function of the rate of investment, it will generally prove optimal not to adjust capital to the desired level immediately, but instead to distribute changes in the capital stock over time (Eisner and Strotz, 1963).

The speed of adjustment of capital to changes in its desired or equilibrium level may depend on the causes and magnitudes of the changes. An increase in the demand for output may generate investment with all due speed as expectations become firm with regard to the permanence of the increased demand. If the increased demand for capital is due, however, to a fall in its relative price (due, let us say, to a reduction in the rate of interest), thus generating a demand for more durable and hence more substantial and expensive capital, the rate of investment may be slowed by the availability of existing capacity sufficient for current production. These considerations underlie the 'putty-clay' model in which the capital–labour ratio can be varied on newly installed capacity but cannot be altered on existing capacity. A demand for additional housing services will bring on investment in housing as rapidly as cost considerations permit. A lower rate of interest, causing substantial investment in more durable brick houses to replace less durable houses of wood or straw, would cause the rate of investment to increase only as existing houses of wood and straw wear out and are replaced.

Investment equations should thus in principle involve separate distributed lag responses to changes in relative prices and to changes in output. They should also admit the possibility that the lag distribution is not fixed and may vary with other economic parameters and the expectations function.

A logarithmic transformation of equation (3) yields

$$\ln K^* = \ln h + s \ln(p/c) + r \ln Y. \qquad (5)$$

Putting this in first difference form, we have:

$$\Delta \ln K^* = s \Delta \ln(p/c) + r \Delta \ln Y. \qquad (6)$$

Since the change in the logarithm of capital is the relative change in capital, we may treat the ratio of net investment to existing capital stock as approximately equal to $\Delta \ln K$, which may in turn be written as a distributed lag function of changes in the determinants of desired capital:

$$I_N/K_{-1} = s[q_1(L)\Delta \ln(p/c)] + r[q_2(L)\Delta \ln Y], \qquad (7)$$

where $q_1(L)$ and $q_2(L)$ are lag operators that indeed should be functions of such variables as the rate of interest, and the cost and availability of capital. Then finally, since investment equals net investment plus replacement, we may write

$$I = I_N + R = I_N + \mathrm{d}K_{-1}, \qquad (8)$$

where d, the replacement rate, may vary over time.

Estimates of investment functions of this type have often neglected influences of economic variables and expectations on adjustment processes and the replacement rate. Lag distribu-

tions are assumed to be of some fixed functional form, and d is assumed to be constant (for evidence that d may not be constant, see Feldstein and Foote, 1971; Eisner, 1972; Feldstein and Rothschild, 1974; and Coen, 1975).

Where production and lag parameters have not been unduly constrained by a priori specifications, estimates have generally yielded values of s, the elasticity of substitution, considerably less than unity, in some cases not substantially greater than zero (see Eisner and Nadiri, 1968; Coen, 1969; Lucas, 1969; Eisner 1978; and Chirinko and Eisner, 1982). Lag distributions estimated from time series and cross-section data have usually extended over a number of years (Eisner, 1978), and they often have inverted-U shapes. Where a putty-clay formulation has been employed with separate lags on relative prices and output, the mean lags on prices are typically much longer than those on output (Bischoff, 1971).

These findings of small price elasticities of demand for capital suggest some role, but a limited one, for monetary and tax policies to directly affect the general rate of investment through the rental cost of capital. The long lag distributions on relative prices suggest further difficulties in the use of monetary or fiscal policy to reduce cyclical fluctuations in investment. However, policy impacts may operate not only on the desired capital stock but also on the speed of adjustment of capital.

Repeated changes in tax parameters such as k, the rate of investment tax credit or subsidy, may be used to bring about intertemporal substitution of investment even if the effects on its long-run average are small. Thus, when investment is low, the marginal rate of subsidy, k, might be raised, while if investment were deemed too great, the value of k could be reduced to zero or indeed made negative (an investment tax instead of subsidy). Paradoxically, a fluctuating and uncertain investment subsidy/tax may have substantial effects on investment where permanent subsidies or taxes would not. There is thus an asymmetry between effects of changes in the cost of capital and changes in the demand for output, the effects of which on investment will be proportional to the permanence with which they are perceived (Eisner, 1978).

Most investment functions, with their *ad hoc*, fixed lag distributions and assumptions of static expectations, fail to capture accurately the effects of economic policies on the timing of investment or to distinguish properly between the effects of temporary and permanent policy changes (Lucas, 1976). To correct these shortcomings, adjustment costs must be explicitly introduced in the firm's optimization problem, so that instead of there being a desired stock of capital toward which the firm moves in a mechanistic way, there is a desired *path* of capital accumulation. Along such a path, the optimal rate of investment at each point, including the present, will in general depend on expected relative prices and output over the entire planning horizon.

Obtaining solutions to the firm's dynamic optimization problem under very general specifications of technology and expectations has proven difficult. To make such an approach empirically tractable, strong assumptions are usually made, for example, that the production function is quadratic, that adjustment costs are quadratic, symmetric, and separable from the rest of technology (the cost of adjusting capital, for example, does not depend on the quantities of capital and labour currently employed), and that expectations are characterized by relatively simple autoregressive processes.

The critical role in current investment of unobservable adjustment costs and of uncertain, shifting (and generally not directly observable) expectations of the future, stressed by Keynes, has sparked interest in a formulation of an investment function which directly relates demand prices and supply prices of capital. Going back to Keynes's *General Theory*, we have investment undertaken to the point where the expectation of marginal profit on investment (the 'marginal efficiency of investment') is equal to the rate of interest or, alternatively, the present value of expected returns from the marginal investment, using the rate of interest as the discount factor, is equal to the marginal supply price of newly produced capital goods. Building on this, Brainard and Tobin (1968) and Tobin (1969), presented a '*q*-theory' which sees investment as a positive function of the ratio, *q*, of the market value of capital to its replacement cost. The former may in principle be observed in the trading prices of stock shares along with bonded indebtedness of business firms. With proper adjustment for tax considerations, when the value of *q* is greater than unity, investment will take place because the cost of additional capital will be less than the market evaluation of the present value of returns from capital. Conversely, when *q* is less than unity, business demand for capital may be better satisfied by acquisitions taking over existing firms and their facilities than by new investment. In general, the rate of investment should be greater the greater the value of *q*.

Empirical estimation of '*q*' investment equations and predictions based on these estimates have not, however, proved very successful (Von Furstenberg, 1977; Abel, 1980; Summers, 1981; Hayashi, 1982; Abel and Blanchard, 1986). Suggested explanations of the difficulties include the fact that market values of firms may relate to much more than the tangible capital generally included in business investment, and the failure to distinguish marginal and average values of the cost of new capital versus the acquisition costs of existing firms (Chirinko, 1986).

Investment decisions are but one element of producers' plans for hiring or acquiring factors of production. Interrelationships between investment demand and demands for other inputs have been a subject of growing interest. Since factor demands are derived from a given production function, they share common technological parameters and may be estimated as a system of demand functions (Coen and Hickman, 1970). Such an approach calls attention to effects of investment stimuli that are often overlooked. For example, at a given level of output, the direct impact of an investment tax credit is to reduce the demand for labour, since it raises the relative cost of labour. Employment may eventually be raised, but only if the expansion in aggregate output induced by the increase in investment demand is large enough to offset the adjustment to a higher capital–labour ratio.

Additional interrelationships may arise when capital is not the only factor of production subject to adjustment costs. If labour input is also costly to change, then the rate of investment may depend not only on the desired adjustment in capital stock but also on the desired adjustment in employment (Nadiri and Rosen, 1969; Brechling, 1975; Epstein and Denny, 1983). Furthermore, since a firm must operate on its production function, factor adjustments cannot be entirely independent. If output is exogenously given and there are *n* inputs, *n* − 1 inputs can be independently adjusted, but the *n*th is determined by the production function, the level of output, and the quantities of the other inputs (Gould, 1969). It may be unreasonable to view the production function as a binding constraint, however, because it is difficult, if not impossible, to measure perfectly all inputs and their utilization rates.

With the development of dynamic optimization models of interrelated factor demands in which various types of capital and other inputs are subject to adjustment costs, and

expectations are not treated as static, it is possible to estimate the magnitude of adjustment costs for capital, to see how they affect and are affected by adjustments of other inputs, and to study the impacts of changes in producers' perceptions of the *processes* generating prices, output, and policy parameters. As we noted above, this approach necessitates strong restrictions on functional forms to obtain explicit decision rules for accumulation of capital and employment of other inputs (Meese, 1980). Where general forms of the production, adjustment cost, and expectations functions are assumed and the model cannot be solved completely, it is still possible to estimate the first-order conditions (Euler equations) which implicitly define the evolution of the optimal inputs (Pindyck and Rotemberg, 1983; Shapiro, 1986). Such estimates do not give a complete account of the dynamics of investment behaviour for any initial conditions and stochastic environment, but they do give insights about differing short- and long-run responses to, say, an unexpected increase in the price of energy starting today versus the same increase beginning five years from now but anticipated today.

We are brought back to the view that changes in the rate of aggregate demand and output are very likely to be of prime importance in the aggregate investment function. In equilibrium terms, the faster the rate of growth of aggregate demand, the greater the rate of investment demand. As long as the capital–output ratio remains unchanged, the absorption of a constant proportion of output in investment (a constant ratio of saving to income) requires a constant rate of growth of output and, in turn, of investment itself. For a greater relative rate of investment, there must be a proportionately faster rate of growth of output. Thus, the rate of growth, frequently seen by the classical economists as the essential result of investment, may be viewed, as well, as a critical determinant of investment.

If the capital–output ratio took on the not unreasonable value of three, corresponding almost precisely to estimates of fixed private capital and GNP in the United States for 1985, a rate of growth of output of 3 per cent per annum would account for net investment equal to 9 per cent of output. A reduction of the rate of growth of output from 3 per cent to 2 per cent, while maintaining the same capital–output ratio, would then entail (at least in the long run) a reduction in investment for purposes of expansion from 9 per cent of output to 6 per cent of output. With gross private domestic investment in the United States currently some 16 per cent of GNP, such a one percentage point drop in the rate of growth of output would thus imply a reduction of gross investment from the current 16 per cent of GNP to 13 per cent – a fall of some 19 per cent. Even mild economic downturns, as this relation predicts, have hence brought sharp declines in business investment. In the Great Depression of the 1930s, gross private domestic investment in the United States fell by some 90 per cent.

Asymmetry has been noted with regard to the determination of positive and negative net investment. In both cases, it has been argued that there is some form of constraint that sets bounds for the rate of investment attainable (Hicks, 1950). With positive investment, the limitation (and upper bound) is set by the amount of capacity available to create producers' goods. In regard to negative net investment, the limitation (or lower bound) relates to the speed at which existing plant and equipment can be worn out and inventories can be disposed of, with or without conversion to a form acceptable to non-producers. The upper bound is usually felt to be sufficiently high to permit a substantial investment boom if increases in aggregate demand or other factors are such as to bring about a great expansion of investment demand. A depression-induced fall in investment demand, however, may run quickly into a 'floor' to actual investment such that an economic slump, while cushioned, is prolonged by limitation of the rate at which excess stocks of capital can be worked off.

The rate of investment may be more influenced by government policy when the economy is slack than under conditions of full employment. For given the necessary identity between total saving and total investment, with output fixed on its full employment path, saving and investment can only be changed by altering the proportion of income and output going to saving. Private saving, however, may be determined essentially by private desires for wealth for purposes of retirement and bequests, and will relate essentially to rates of growth of the population and output rather than expected rates of return to wealth and saving.

It is widely argued that government fiscal policy and, particularly, the government budget surplus (government saving) or deficit (government dis-saving) affect total saving and hence total investment. In its crude form, this argument presupposes, however, that private saving is unaffected by government saving. In fact, for example, reduction of the public deficit, that is, reduction of public dis-saving, by reducing net income after taxes of the private sector, will reduce private saving. If income is reduced enough, private saving may be reduced as much or more than public saving is increased (dis-saving reduced).

If private saving and consumption propensities are essentially determined by life cycle and bequest motives, an increase in private wealth, brought on by increased private holdings of public debt created by a budget deficit, will generate more consumption demand now and in the future. Producers, operating in accord with rational expectations, will increase their investment demand in order to acquire capital to meet the anticipated increases in consumption. But this then, under conditions of full employment, unless blocked by unaccommodative monetary policy, will generate higher prices so that, despite the nominal budget deficit, there will be no increase in the real value of public debt, hence no real deficit and no change in either real consumption or real investment (Eisner, 1986).

Real deficits causing real increases in the value of public debt held by the private sector in excess of the rate of growth of real output would then occur only under conditions of less than full employment. In these conditions, resulting increased real consumption may well encourage rather than reduce real investment (Lange, 1938), that is, 'crowd-in' rather than 'crowd-out' investment. And under conditions of full employment, monetary policy or changes in tax parameters affecting the rental price of capital can only change investment to the limited extent they may be able to affect private saving and to the extent they affect foreign investment.

Government can, however, affect investment by its direct command over resources. To the extent that government buys goods and services, under conditions of full employment it must be reducing resources available for other output. Given private propensities to consume, some of the reduction in output, at least, will be in the production of capital. If the goods and services taken by government are for military purposes or public consumption, total investment will perforce be reduced. If the output purchased or produced by government is in the form of capital, tangible or intangible, total investment is likely to be increased.

Monetary and tax changes may have substantial, and in many instances unintended, impacts on the allocation of investment. Empirical studies generally reveal that investment

in residential buildings is more sensitive than most other forms of investment to the level and term structure of interest rates, so that monetary restraint, resulting in high short-term interest rates, tends to depress housing construction disproportionately. On the fiscal side, an investment tax credit available at a uniform rate on all types of capital will distort the allocation of capital in favour of shorter-lived assets. There are many common features of taxation – among them preferential treatment of capital gains, deductibility of nominal interest, and accelerated depreciation – that create disparities in effective tax rates on different types of investments and tend to lead to inefficient allocation of capital (King and Fullerton, 1984).

Where investments yield social returns that differ from returns to private investors, a case can be made for public subsidies or taxes. Capital expenditures on pollution abatement equipment, education, and research, for example, might qualify for subsidy. But the differential rates of investment subsidies found in national tax systems hardly seem defensible on grounds of economic externalities.

It is important that public policy toward investment relate to all of investment, human and intangible as well as tangible, and by households, government and government enterprises, and nonprofit institutions as well as by business. Business tangible investment (in plant, equipment and inventories) in fact comes to less that 20 per cent of total investment in the United States economy (Eisner, 1985). There is good reason to believe that the payoff in higher productivity and a higher path of consumption in the future is greater in all of this other investment – in research and development, in public infrastructure, and in the health and education of human beings.

ROBERT M. COEN AND ROBERT EISNER

See also ADJUSTMENT COSTS; NEOCLASSICAL SYNTHESIS; NEW CLASSICAL MACROECONOMICS; RATIONAL EXPECTATIONS; REPLACEMENT POLICY.

BIBLIOGRAPHY

Abel, A.B. 1980. Empirical investment equations: an integrative framework. In *On the State of Macro-Economics*, Vol. 12, ed K. Brunner and A.H. Meltzer, Carnegie-Rochester Conference on Public Policy, Amsterdam: North-Holland.

Abel, A.B. and Blanchard, O.J. 1986. The present value of profits and cyclical movements in investments. *Econometrica* 54(2), March, 249–73.

Bischoff, C.W. 1971. The effect of alternative lag distributions. In *Tax Incentives and Capital Spending*, ed. G. Fromm, Washington, DC: Brookings.

Brainard, W.C. and Tobin, J. 1968. Pitfalls in financial model-building. *American Economic Review, Papers and Proceedings* 58, May, 99–122.

Brechling, F. 1975. *Investment and Employment Decisions*. Manchester: Manchester University Press.

Chirinko, R.S. 1986. Business investment and tax policy: a perspective on existing models and empirical results. *National Tax Journal* 39(2), June, 137–55.

Chirinko, R.S. and Eisner, R. 1982. The effects of tax parameters in the investment equations in macroeconomic econometric models. In *Economic Activity and Finance*, ed. M.E. Blume, J. Crockett and P. Taubman, Cambridge, Mass.: Ballinger.

Clark, J.M. 1917. Business acceleration and the law of demand: a technical factor in economic cycles. *Journal of Political Economy* 25, March, 217–35. Reprinted in American Economic Association, *Readings in Business Cycle Theory*, Philadelphia: Blakiston, 1951.

Coen, R.M. 1969. Tax policy and investment behavior: comment. *American Economic Review* 59(3), June, 370–77.

Coen, R.M. 1971. The effect of cash flow on the speed of adjustment. In *Tax Incentives and Capital Spending*, ed. G. Fromm, Washington, DC: Brookings.

Coen, R.M. 1975. Investment behavior, the measurement of depreciation, and tax policy. *American Economic Review* 65(1) March, 59–74.

Coen, R.M. and Hickman, B.G. 1970. Constrained joint estimation of factor demand and production functions. *Review of Economics and Statistics* 52(3), August, 287–300.

Eisner, R. 1972. Components of capital expenditures: replacement and modernization versus expansion. *Review of Economics and Statistics* 54(3), August, 297–305.

Eisner, R. 1978. *Factors in Business Investment*. Cambridge, Mass.: Ballinger.

Eisner, R. 1985. The total incomes system of accounts. *Survey of Current Business* 65, January, 24–48.

Eisner, R. 1986. *How Real is the Federal Deficit?* New York: Free Press, Macmillan.

Eisner, R. and Nadiri, M.I. 1968. Investment behavior and the neo classical theory. *Review of Economics and Statistics* 50(3), August, 369–82.

Eisner, R. and Strotz, R.H. 1963. Determinants of business investment. In Commission on Money and Credit, *Impacts of Monetary Policy*, Englewood Cliffs, NJ: Prentice-Hall.

Epstein, L.G. and Denny, M.G.S. 1983. The multivariate flexible accelerator model: its empirical restrictions and an application to U.S. manufacturing. *Econometrica* 51(3), May, 647–74.

Feldstein, M.S. 1982. Inflation, tax rules and investment: some econometric evidence. *Econometrica* 50(4), July, 825–62.

Feldstein, M.S. and Foote, D.K. 1971. The other half of gross investment: replacement and modernization expenditures. *Review of Economics and Statistics* 53(1), February, 49–58.

Feldstein, M.S. and Rothschild, M. 1974. Towards an economic theory of replacement investment. *Econometrica* 42(3), May, 393–423.

Gould, J.P. 1969. The use of endogenous variables in dynamic models of investment. *Quarterly Journal of Economics* 83(4), November, 580–99.

Haavelmo, T. 1960. *A Study in the Theory of Investment*. Chicago: University of Chicago Press.

Hall, R.E. and Jorgenson, D.W. 1967. Tax policy and investment behavior. *American Economic Review* 58(3), June, 391–414.

Hall, R.E. and Jorgenson, D.W. 1971. Application of the theory of optimal capital accumulation. In *Tax Incentives and Capital Spending*, ed. G. Fromm, Washington, DC: Brookings.

Hayashi, F. 1982. Tobin's marginal q and average q: a neoclassical interpretation. *Econometrica* 50(1), January, 213–24.

Helliwell, J.F. and Glorieux, G. 1970. Forward-looking investment behavior. *Review of Economic Studies* 37(4), December, 499–516. Reprinted in *Aggregate Investment: Selected Readings*, ed. J.F. Helliwell, Harmondsworth: Penguin, 1976.

Hicks, J.R. 1950. *A Contribution to the Theory of the Trade Cycle*. Oxford: Clarendon Press.

Jorgenson, D.W. 1963. Capital theory and investment behavior. *American Economic Review, Papers and Proceedings* 53(2), May, 247–59.

Jorgenson, D.W. 1967. The theory of investment behavior. In *Determinants of Investment Behavior*, ed. R. Ferber, New York: Columbia University Press.

Jorgenson, D.W. 1971. Econometric studies of investment behavior: a survey. *Journal of Economic Literature* 9(4), December, 1111–47.

Keynes, J.M. 1936. *The General Theory of Employment, Interest and Money*. London: Macmillan.

King, M.A. and Fullerton, D.K. 1984. *The Taxation of Income from Capital*. Chicago: University of Chicago Press.

Koyck, L.M. 1954. *Distributed Lags and Investment Analysis*. Amsterdam: North-Holland.

Lange, O. 1938. The rate of interest and the optimum propensity to consume. *Economica* 5, February, 12–32. Reprinted in American Economic Association, *Readings in Business Cycle Theory*, Philadelphia: Blakiston, 1951.

Lucas, R.E. 1969. Labor-capital substitution in U.S. manufacturing. In *The Taxation of Income from Capital*, ed. A.C. Harberger and M.J. Bailey, Washington, DC: Brookings.

Lucas, R.E. 1976. Econometric policy evaluation: a critique. In *The Phillips Curve and Labor Markets*, ed. K. Brunner and A.H. Meltzer, Amsterdam: North-Holland.

Meese, R. 1980. Dynamic factor demand schedules for labor and capital under rational expectations. *Journal of Econometrics* 14, June, 141–58.

Meyer, J. and Kuh, E. 1957. *The Investment Decision*. Cambridge, Mass.: Harvard University Press.

Nadiri, M.I. and Rosen, S. 1969. Interrelated factor demand functions. *American Economic Review* 59(4), September, 457–71.

Pindyck, R.S. and Rotemberg, J.J. 1983. Dynamic factor demands and the effects of energy price shocks. *American Economic Review* 73(5), December, 1066–79.

Shapiro, M.D. 1986. The dynamic demand for capital and labor. *Quarterly Journal of Economics* 101(3), August, 513–42.

Summers, L.H. 1981. Taxation and corporate investment: a *q*-theory approach. *Brookings Papers on Economic Activity* No. 1, 67–140.

Tobin, J. 1969. A general equilibrium approach to monetary theory. *Journal of Money, Credit and Banking* 1(1), February, 15–29.

von Furstenberg, G.M. 1977. Corporate investment: does market valuation matter in the aggregate? *Brookings Papers on Economic Activity* No. 2, 347–97.

investment and accumulation. The standard view of accumulation goes something like this. In the short period, fraught with frictions and maladjustments, the demand for investment interacts with the supply of saving, more or less *à la* Keynes, to determine the growth of the capital stock. Keynesian policies may have fallen into disrepute, but for the short period the representative economist continues to use the tool box developed in the *General Theory* and its wake: accumulation falls out from the determination of national income.

In the longer period the same economist falls back on very different arguments: the mainstream of the profession takes accumulation to be determined by saving propensities, with nary a side glance at investment demand. That Japan has over the last quarter century devoted 30 per cent of gross output to fixed capital formation and Great Britain 20 per cent is conventionally explained in terms of higher Japanese saving propensities, not in terms of a greater propensity to invest.

In a still longer time frame, even saving propensities become irrelevant. In the asymptotic future beyond all future, accumulation is determined solely by population growth and technical change. Saving propensities may affect the steady-state capital: output ratio if the technology admits of substitution between labour and capital, but that is the limit of their influence.

Economists of a Marxian bent share the mainstream view, up to the asymptotic future, which they rightly dismiss as an irrelevant construct. The terminology may differ: difficulties of 'realization' is favoured for describing a shortfall of aggregate demand relative to aggregate supply, and hence (abstracting from foreign trade and government surpluses or deficits) for a shortfall of investment relative to saving. But for Marxians realization problems are generally confined to the short period; in the long run it is once again the saving propensities of capitalists, along with the rate of profit determined by class struggle, which determine the rate of accumulation. To be sure, neoclassical and Marxian theories of the determination of saving propensities differ, but for present purposes this is a secondary issue; the short run apart, the two theories agree that investment propensities are irrelevant to accumulation.

INVESTMENT IN THE GENERAL THEORY. Against these views stands the Keynesian view, which, applied to the long run, tells a very different story of the accumulation process. In the *General Theory* Keynes formalized his view of investment demand in terms of a 'marginal efficiency of capital' schedule which showed the amounts of investment that would be forthcoming at different rates of interest. The basic idea behind this schedule can be captured by supposing there to be a set of projects indexed by $i = 1, \ldots, n$, each requiring a unit of investment and returning, respectively, a cash flow of $r_1, \ldots, r_n$ in perpetuity. If the n projects are arrayed in descending order of r_i, then with the simplifying assumption that each investment costs one dollar, r_i is the *marginal* rate of return on the investment of i dollars. In Keynes's language, r_i is the marginal efficiency of capital.

Suppose now that the interest rate d, which represents the cost of capital to the investor, is also expected to be constant in perpetuity. Then the present value of the ith project's return is r_i/d. The profit-maximizing investor will go down the list until he reaches the project at which $r_i = d$, that is, the point at which the marginal rate of return just equals the cost of capital. In Keynesian terms, investment is determined by equating the marginal efficiency of capital to the cost of capital. More precisely, the array of projects being discrete, the profit-maximizing rule is to undertake all projects for which $r_i > d$ and to reject those for which $r_i < d$. If there is a project for which the relationship between r_i and d is one of exact equality, it is a matter of indifference whether the project is undertaken or not. The main point is that the discounted present values of returns, $r_i/d - d$ acts as the discount rate – exceed the assumed unit cost of the investment provided $r_i > d$.

So far there is nothing novel in this theory. Knut Wicksell would have had no problem making the argument his own, and indeed the schedule of the marginal efficiency of capital bears a close resemblance to the marginal productivity of capital schedule of mainstream theories of accumulation. Even the overall theoretical structure which Keynes builds by joining this schedule to schedules of consumption and liquidity preference would not have been uncongenial to Wicksell, particularly if its application is confined to the short period. This is presumably why Wicksell's Swedish followers chided Keynes for indulging 'the attractive Anglo-Saxon kind of unnecessary originality' (Myrdal, 1939, p. 8; the comment naturally refers to the *Treatise*, not to the *General Theory*).

Probably because of the affinity to a Wicksellian version of neoclassical economics, the mainstream of American economists has been able, as was suggested at the outset of this essay, to accept a version of the Keynesian analysis for the short period – much to the annoyance of Keynes's Cambridge disciples, like Joan Robinson and Nicholas Kaldor, who all their lives insisted that Keynes's main message was being lost in the translation. In the standard American view, the main point of Keynes was the limit of monetary policy, conceived of in terms of its affect on d, to affect investment demand, either because of a low elasticity of the marginal efficiency schedule, or, in the limit, because of the impossibility of reducing the interest rate (the famous 'liquidity trap', of which Keynes said in the *General Theory* that it was, as yet (1935), a theoretical possibility of which there had been no actual instances). The bottom line was the need not only for state intervention in the form of an activist monetary policy – this was fully present in Wicksell's analysis – but also in the form of fiscal policy. Indeed it does no disrespect to Keynes to accept that his influence owed as much to the intellectual justification he provided for an activist, interventionist state – for the end to *laissez faire* – as to the intellectual power of his ideas, certainly as these ideas were, reflected through the prism of the mainstream of the American economics profession.

AN ALTERNATIVE READING. But there is another reading of Keynes. The real departure of the Keynesian theory of investment from the orthodox one, in my judgment, starts from a recognition that the formalism of his theory of investment demand obscures its real content. The starting point is the recognition that the returns of any project, lying in the future, are inherently uncertain. The r_i's are not objectively given reality but a subjective construction of the investor. It has become fashionable to blur the old Knightian distinction between risk, objective and quantifiable, and uncertainty, subjective and qualitative, by means of the theory of subjective or personal probability. Even if the axioms underlying this theory are neither compelling (particularly the assumption of a complete ordering and the assumption of 'independence', what Leonard J. Savage called the 'sure thing' principle), nor borne out empirically in the behaviour of untutored individuals, subjective probability theory still has some heuristic value in modelling investment decisions, particularly in its emphasis on the psychology of the decision maker.

Subjective probability allows us to go behind the r_i's of present value calculations like the simple perpetuity formula r_i/d to more complex sums of the form

$$r_i^h = p_1^h u_1^h r_{i1}^h + p_2^h u_2^h r_{i2}^h + \cdots + p_m^h u_m^h r_{im}^h,$$

in which the generic term $p_j^h u_j^h r_{ij}^h$ is composed of these elements: p_j^h is Mr h's subjective probability of the occurrence of a particular complex of events (a 'state') in which his marginal utility of income (normalized) is u_j^h and his estimated return from project i is r_{ij}^h. The central point is that in the Keynesian view, each of the constituents – the probability p_j^h, the marginal utility of income u_j^h, and the state-specific return r_{ij}^h – owes as much to the imagination of the investor as it does to an objective reality. The more optimistic are investors, the higher the probabilities they will attach to states in which the returns and the marginal utilities of these returns are high, and the consequence will be higher values of r_i^h. Conversely, the more pessimistic are investors, the lower the r_i^h's that will be attached to the same projects. Thus the 'animal spirits' of investors play a crucial role in investment demand.

The recognition of a crucial role for animal spirits directs one's attention away from movements *along* the marginal efficiency schedule. The question becomes, what determines the position of this schedule? Evidently, according to what has just been said, it is, in the last instance, investors' evaluations of probabilities of various states, of the relative utility of income in different states, and of returns in different states.

It is equally evident that this makes a cumbersome theory. A more tractable model can be constructed by making the prospective returns the r_i^h's, depend on the general anticipations of capitalists. The higher is the general expectation of profits, the higher will be the anticipated rate of return on specific projects – a rising tide will be anticipated rate of return in specific projects – a rising tide will presumably lift all boats. In this view, movements of the entire marginal efficiency schedule, triggered by changes in expectations of profitability, not movements along a given schedule induced by changes in the interest rate, are the key to understanding the ups and downs of investment and output.

Such reasoning – this is of course a 'rational reconstruction' – permitted Keynes's heirs, Roy Harrod in Oxford and Robinson and Kaldor in Cambridge, who re-situated the *General Theory* in a long-period context, to recast the marginal efficiency schedule in terms of variables related to expected profits. Robinson's investment demand function, for example, and the argument of Keynes's own formulation, the rate of interest, disappears into the background of *ceteris paribus*.

The rationale for ignoring the cost of capital is the assumption of a highly insensitive responsiveness of investment and saving to plausible interest rate changes. Clearly, there are strong assumptions at work here about the relevant range of interest rate variation. As long as the anticipated returns are finite, there must be *some* level of the interest rate at which even the most attractive projects appear to all and sundry as uneconomic. Thus, unless saving is negatively related to the rate of interest *and* highly elastic, sufficient variation in interest rates could, with enough time, adjust the demand for investment to the supply of saving.

For most of the postwar period, however, the range of variation of interest rates has been too modest to test this possibility, at least if we identify the interest rate with the difference between the nominal rate on government or high grade corporate bonds and the rate of inflation. Indeed, from 1945 until 1980 this 'real' rate of interest never moved very far from zero in the United States. In the post-1980 disinflation and recession, real interest rates rose to levels which must give pause to even the most devoted neo-Keynesian. It is certainly too early at this writing to determine whether in the sweep of history this was a momentary aberration or the dawn of a new era; my own leaning is towards aberration, but that may reveal my neo-Keynesian predilections rather than a reasoned guess about the future.

The rate of profit generally expected on the capital stock as a whole (*re*) is itself no more observable than any other variable that lies in the future. In the neo-Keynesian literature *re*'s customarily are taken to be a function of a small number of variables which summarize the relevant information available to investors. The standard version of the theory simply extrapolates the current or immediate past rate of profit.

With simplifying assumptions like constant returns to scale, homogeneous capital, and a uniform, exogenously given rate of capacity utilization, it must be true *ex post* that the average rate of profit on new investment (which is the marginal rate of profit on the capital stock) turns out to equal the average rate of profit on all capital. Under these assumptions it might appear reasonable for the expected average rate of profit on the entire capital stock *re* to equal the expected rate of profit on new investment, which would impose, in the spirit of 'rational expectations', a consistency requirement for each investor

$$r^e \equiv \sum_i r_i^h \equiv \sum_i \sum_j p_j^h u_j^h r_{ij}^h.$$

But no such inference is warranted. Neo-Keynesian theory in my view is compatible with rational expectations – insofar as this notion makes any sense at all under conditions of subjective uncertainty. But an important measure of realism would be lost by constraining the theory in this way, for even under the stringent assumptions that lead to a uniform realized profit rate', there is no good reason to believe in rational expectations. The general expectation of a 4 per cent return is perfectly consistent with individual expectations that special gifts, opportunities, or kismet will result in a 10, 20 or even 50 per cent return on one's own projects. As P.T. Barnum would have it, a sucker is born every minute.

SELF-FULFILLING PROPHECY. Keynes made Barnum's sucker into a kind of hero: in Keynes's view it was doubtful that capitalism could function without such a 'spontaneous urge to action'. Neo-Keynesian theory makes an even stronger assertion, which has its roots in Keynes's *Treatise on Money*. As long as there are enough of them, the 'suckers' can turn the tables on the rest of us. Capitalists, as a class, have the power to shape conditions so that their expectations come true, at least in large enough part to maintain their confidence.

Unique among economic actors, this class has the power of self-fulfilling prophecy. Not only do actual profit rates affect capitalists' beliefs about future profits, capitalists' beliefs also have an impact on actual profits. This is not the same thing as the ability Joan Robinson imputed to capitalists to make the profit rate anything they liked, but it is a formidable power nonetheless.

There are two mechanisms by which capitalists' prophesies about profits become self-fulfilling, the distribution of income between capital and labour and the level of capacity utilization and employment. The first is the less familiar one, except to readers of Keynes's *Treatise on Money*. Suppose a closed economy with no government spending or taxation in which the starting point is a long run equilibrium characterized by equality both between desired saving and investment and between expected and actual (average) rates of profit. Imagine a change in 'animal spirits' which makes capitalists willing to undertake more investment at the going rate of profit than earlier was the case. In other words, imagine an outward shift in the investment demand function. In the simplest neo-Keynesian story, this addition to aggregate demand increases spending relative to income – remember *income* has remained unchanged – and drives the price level upward. Assuming money wages are fixed or at least sluggish, this reduces the real wage and shifts the income distribution in favour of profits. The process continues since higher realized profits lead to further expectations of higher profits and still more investment demand. But it does not continue indefinitely, because capitalists are assumed to save a higher proportion of their incomes than do workers. The upward spiral of prices and investment continues only until the extra saving induced by higher profits absorbs the extra investment, at which point the economy comes to a new equilibrium where desired saving and investment and actual and anticipated profits are agains equal, albeit at a higher level. In addition to the existence of an investment demand function that is distinct from the saving function, the sluggishness of money and wages and the difference in saving propensities between classes are crucial to this result.

Several observations are in order here. First, although the *fons et origo* of this theory is Keynes's *Treatise*, the theory has much in common with the model outlined in Josef Schumpeter's *Theory of Economic Development*. There are two important conceptual differences however. First, the neo-Keynesian equilibrium is formulated as a steady growth rather than stationary (zero growth) state that dominated in Schumpeter's time. Thus shifts from one equilibrium to another involve changes in the rate of growth rather than changes in the level of output.

A second difference is more fundamental. In both the Schumpeterian and the neo-Keynesian view, an outward shift in the investment demand function plausibly involves an expansion of the array of projects yielding returns in excess of the cost of capital. And in both views, the psychology of the capitalist class is crucial. Schumpeter no less than the neo-Keynesians recognized the subjective element in the estimation of returns. 'Invention', for Schumpeter, was necessary but not sufficient for 'innovation'. But here the resemblance ends. In the neo-Keynesian view invention is neither necessary nor sufficient for innovation. Investment can be a pure boot-strap operation; the theory requires nothing more than a change in business psychology to change investment demand, and a change in investment demand can lead the economy to a new equilibrium with a different rate of growth. Within the limits of saving propensities and the malleability of real wages, capitalists wishes are self-fulfilling.

Prices and profit rates change to validate the changes in capitalists' expectations!

THE CRUCIAL ASSUMPTIONS. There must be a trick. In fact, there are four crucial assumptions, two of which – the flexibility of real wages and the difference in propensities to save between capitalists and workers – have already been mentioned. The assumption that wages are set in money terms plays a central role in the analysis. Money wages need not be fixed once and for all, but Robinson's version of the neo-Keynesian story (and the Schumpeterian story for that matter) cannot be told at all without the assumption of sluggish money wages. Evidently it is the income distribution that adjusts saving and investment to each other. So if the income distribution is fixed, then it cannot do the job which neo-Keynesian theory assigns it.

A difference in saving propensities is equally important to the neo-Keynesian view of capitalism. There are various versions of the so-called Cambridge saving equation, and a fair amount of confusion about the content of alternative versions exists two decades after the most important contributions to this debate. But all versions of the theory take it for granted that capitalists' propensity to save exceeds workers'. By contrast, the principal neoclassical theory of saving, Franco Modigliani's life-cycle hypothesis, suggests that the propensity to save out of wages will exceed the propensity to save out of property income: wages will be disproportionately in the hands of people preparing for retirement and profits disproportionately in the hands of retirees. Theoretical dispute is of course nothing new in economics. What is more surprising is that we lack persuasive empirical evidence of one view or the other two decades after the theoretical battle was fairly joined.

The essential role that flexible real wages and differences in saving propensities play is relatively transparent, so perhaps 'trick' is not an appropriate description for either of these assumptions. But a third assumption is necessary to make capitalists' investment spending into a 'widow's cruse' (Keynes's metaphor), filling up with saving as fast as it is emptied in investment. This assumption is better hidden. We can see its role more clearly by asking how the process described for moving from one equilibrium in consequence of a shift in investment demand ever gets started. How is it that an increase in *desired* investment gets translated into *effective* demand? (The same question, it may be noted, might be asked about any displacement from macroeconomic equilibrium, for instance, the textbook displacement of a short-period equilibrium by a shift in the investment function.)

One possibility is that desired saving increases in line with desired investment, but this assumption in essence requires us to abandon the idea of separate saving and investment schedules. A second possibility that can be dismissed almost as easily is the just-so story fashionable in my youth: we were told to think in terms of cash 'hoards', with 'dishoarding' as the essential mechanism for initiating the disequilibrium transition from one steady state to another. That story won't wash because under contemporary conditions there simply aren't cash hoards of the requisite magnitude – if there ever were.

There *is* a better answer, and interestingly Wicksell, as well as Schumpeter and Keynes, give it in about the same way. Schumpeter is the clearest of the three, in contrast to whom Keynes is practically incoherent, perhaps because he thought the main point of the story so obvious that it did not require elaborate explanation. The main point, in two words, is credit money. The process of expanding investment can get started

with no accompanying increase in desired saving if capitalists are assumed to have access to an accommodating banking system which one way or another can create claims on scarce resources out of whole cloth. Here the psychology of financial capital joins the psychology of industrial capital, for unless the financiers share the optimism of industrialists, there is no way, absent those mythical cash hoards, by which investment can increase without a contemporaneous increase in desired saving, as the neo-Keynesian and the Schumpeterian story, with their reliance on price level and profit rate shifts to accommodate capitalists, would have it. Or for that matter, the Wicksellian story. Although ultimately it is the interest rate which adjusts desired investment and saving, Wicksellian disequilibrium is not a Walrasian virtual or hypothetical imbalance of *tâtonnement* with false trading, but an imbalance in real time which is sustained by credit money.

The importance of an accommodating banking system, or passive or endogenous money – these are all approximate equivalents – to the neo-Keynesian system explains why partisans of this view are necessarily hostile to the quantity of money theory of prices (the so-called 'quantity theory of *money*', but this is an obvious misnomer). The dispute is evidently not about the relationship between the quantity of money and the price level, a definitionally true relationship in its customary form, but about causality. In the neo-Keynesian view it is aggregate demand which drives *both* sides of the quantity equation, not MV (the product of money and velocity) which drives PX (the product of prices and quantities)

In an earlier essay on this subject (Marglin, 1984a), I suggested that in the neo-Keynesian story capitalists, as a class if not individually, were approximately in the position of the present Aga Khan, who in his student days is reputed to have asked his Harvard economics instructor how the theory of consumer choice worked *without* the budget constraint. V. Bhaskar has persuaded me that the analogy is misleading. The point is not that capitalists, even as a class, face no budget constraint: capitalists must be assumed to repay whatever debts are contracted to finance investment. The point is rather that, as a class, capitalists are able to change what are normally regarded as parameters of the budget constraint. By increasing relative demand, capitalists drive up the prices of the goods they sell relative to costs of production. The consequent increase in profits provides the wherewithal to retire debt as it becomes due.

There is a final assumption which must be introduced in order to make the neo-Keynesian argument that the longer run growth of the capitalist economy, as well as the share of investment in output and the profit rate, is sensitive to capitalists' animal spirits. This is the assumption of slack resources, specifically, slack labour-force growth must lead to continued capital deepening, which has finite limits both in the real world and in the textbook world of smooth substitution between capital the labour. The simplest, as well as the most realistic, justification of slack labour, is the argument that the capitalist sector of the economy is embedded in a larger entity from which it can, if needed be, draw labour. I refer here to a long run 'reserve army' constituted both by sectors of the national economy (the farm in an earlier day, the kitchen more recently) and by sectors of the international economy (the immigrants who provided labour for 19th- and 20th-century expansion in the United States and for the post-World War II boon in Europe).

Without this assumption, the power of capitalists to shape the history of capitalist economies would be much closer to the power any group has in a standard Arrow–Debreu model to influence relative prices through its preferences or, in a stochastic model, through its expectations. On the assumption of a limited quantity of land suitable for growing tobacco, a shift of smokers' preferences with respect to the pleasures of nicotine and the horrors of lung cancer will affect the equilibrium price of cigarettes. In a stochastic model, the belief that sunspots or any other exogenous variable matters may be sufficient to make the variable matter, even with the assumption of rational expectations (Azariadis, 1981; Cass and Shell, 1983). At issue in both these cases however is the distribution of a given pie, which stands in sharp contrast to neo-Keynesian theory, where distribution bears on the size of the pie itself.

Observe that the existence of slack resources is much more problematic in the long run than in the short period which was originally the focus of Keynes's analysis. For the short period, although controversy is not lacking even here, mainstream economists will generally accept it is the exception rather than the rule that capacity or manpower constrains output. It is in the long run that the true proportions of the neo-Keynesian departure from orthodoxy reveal themselves.

THE LEVEL OF REAL WAGES: EFFECT OR CAUSE OF ACCUMULATION? In a sense, this version of neo-Keynesian theory, like Keynes's own *General Theory*, proves too much. One cannot ask many of the more interesting questions about how changes in investment demand change the economy. For most of these questions turn out to be about movements along the demand curve rather than changes in its position, which mean that they are necessarily transitory if the initial position was one of equilibrium. For instance, one cannot ask what the effect of lower (or higher) real wages might be because the real wage, determined ultimately by the price level, is a consequence rather than a thermostat. Thus to find a thought experiment which illustrates the neo-Keynesian theory requires some care; it would not have done, for example, to take as the premise of a shift in the investment demand function that Mrs Thatcher and Mr Reagan had abolished collective bargaining. For while this might plausibly lead to the expectation of a higher profit rate, this expectation would translate into a movement *along* the existing investment demand schedule rather than a shift of the schedule. And supposing the economy was at equilibrium in the first place, a movement along the investment demand schedule is not sustainable unless the saving schedule shifts simultaneously.

In fact questions about the effect of changes in real wages on equilibrium come out of a very different approach to capitalism than that embodied in the *General Theory*. There Keynes calls this approach 'classical', even though the economists Keynes apparently had in mind as its exemplars were more neoclassical than classical. Classical may still be the right label since the Marxian strand of the classical theory, as typified by Michal Kalecki, even more than the neoclassical strand, makes the level of real wages a central determinant in the theory of accumulation, a thermostat rather than a thermometer. But if real wages are given exogenously, whether as a rate or a share, something else has to adjust if both investment and saving propensities are to continue to play a determining role in the accumulation process.

There are at least two ways out, besides the possibility of partitioning outcomes by 'regimes' in which one or another consideration – wages, investment, or saving – is a nonbinding constraint. One possibility is to allow each of the three determinants to operate with diminished force; this is the tack followed in Marglin (1984a, 1984b).

CAPACITY UTILITY AS AN ADJUSTMENT MECHANISM. Another possibility is to allow capacity utilization to enter the discussion more fully both as cause and as effect. This is not entirely unproblematic since it can be argued that the appropriate assumption for the long run (as distinct from Keynes's short period) is that capacity utilization settles down at some 'normal' rate, in other words, that it is not a variable. But it is hard to regard this issue as resolved; both the view that capacity utilization is endogenous and the view that it is exogenous in the long run can claim some empirical support.

The endogenous view, which Amit Bhaduri and the present author are currently developing, builds indirectly on the work of Kalecki and more directly on the work of Bob Rowthorn (1982). This view entails a significant reformulation of the investment demand function: investment demand is no longer a function of the expected profit rate but rather is a function of two of its constituents, the expected profit *margin qe* and the expected rate of capacity utilization *ze*. Definitionally $re = qezey$, where y represents the output:capital ratio at full capacity utilization, but with capacity utilization variable, there is no compelling reason why qe and ze should affect investment demand symmetrically, as they do in this formula. If, for example, the expected rate of profit is high because the profit margin is high and the expected rate of capacity utilization is low, the impact on investment demand may be very different from what a high expected rate of profit based on lower profit margins and higher capacity utilization would induce. In the second case there would be relatively little need for investment in order to have sufficient capacity to meet expected demand. Investment demand would be limited to new products, new processes, or the substitution of capital for labour. Thus the formulation of the expected rate of profit from a project as

$$r_i^h = p_1^h r_{i1}(q^e, z^e) + \cdots + p_m^h u_m^h r_{im}(q^e, z^e)$$

is at once more general and more plausible than the formulation in which r_i^h turns on re alone.

In this model, unlike the model in which aggregate investment turns on the expected rate of profit, one *can* ask questions about exogenous shifts in real wages. Profit margins are determined in large part by struggles over real wages, so an increase in the expected wage lowers qe and leads to a reduction in investment demand, that is, to a movement down the investment demand schedule. But the movement may be permanent since there is now another degree of freedom, capacity utilization, to accommodate the change.

A word of caution: the outcome of a change in real wages cannot be predicted solely on the basis of the qualitative structure of the model. Whether higher wages and lower profit margins will increase capacity utilization and accumulation depends on the relative strength of two opposing tendencies. On the one hand, from the capitalists' point of view, a decrease in profit margins makes investment less desirable at a given rate of capacity utilization. On the other hand, a shift in the distribution of income from capital to labour may be expected to increase consumption demand and thus to stimulate capacity utilization. Since investment is by assumption sensitive both to profit margins and to capacity utilization, the relative strength of these two influences becomes crucial. This framework is thus broad enough to accommodate both the 'stagnationist' interpretation of Keynes with its emphasis on high real wages as an engine of accumulation and an 'exhilarationist' interpretation, which emphasizes high profit margins.

Evidently a variety of theories, and an even wider variety of models, is possible within the broad Keynesian vision that aggregate demand matters. The essence of that vision is the role of investor psychology, the strength, in Keynes's words, of 'animal spirits' that 'urge to action rather than inaction'. Perhaps a paraphrase of Marx puts the main point best: capitalists make the history of capitalist economy, but not in circumstances of their own choosing. Whatever the mechanism of adjustment, whether income distribution or capacity utilization or some combination of the two, capitalists occupy a position of singular privilege in the neo-Keynesian conception, possessing the ability to impress their subjective construction of the future on the present functioning of the economy.

STEPHEN A. MARGLIN

See also ACCUMULATION OF CAPITAL; CLASSICAL GROWTH MODELS; NEOCLASSICAL GROWTH THEORY.

BIBLIOGRAPHY

Arrow, K. and Hahn, F. 1971. *General Competitive Analysis.* San Francisco: Holden-Day; Edinburgh: Oliver & Boyd.

Azariadis, C. 1981. Self-fulfilling prophecies. *Journal of Economic Theory* 25(3), December, 380–96.

Cass, D. and Shell, K. 1983. Do sunspots matter? *Journal of Political Economy* 91(2), April, 193–227.

Harrod, R. 1948. *Towards a Dynamic Economics.* London: Macmillan.

Kaldor, N. 1956. Alternative theories of distribution. *Review of Economic Studies* 23(2), 83–100.

Kaldor, N. 1957. A model of economic growth. *Economic Journal* 67, December, 591–624.

Kalecki, M. 1971. *Selected Essays on the Dynamics of the Capitalist Economy 1933–1970.* Cambridge: Cambridge University Press.

Keynes, J.M. 1930. *A Treatise on Money.* Vol. 1: *The Pure Theory of Money.* London: Macmillan.

Keynes, J.M. 1936. *The General Theory of Employment, Interest, and Money.* London: Macmillan; New York: Harcourt, Brace & Co.

Keynes, J.M. 1937a. Alternative theories of the rate of interest. *Economic Journal* 47, June, 241–52.

Keynes, J.M. 1937b. The ex-ante theory of the rate of interest. *Economic Journal* 47, December, 663–69.

Marglin, S.A. 1984a. *Growth, Distribution and Prices.* Cambridge, Mass.: Harvard University Press.

Marglin, S.A. 1984b. Growth, distribution, and inflation: a centennial synthesis. *Cambridge Journal of Economics* 8(2), June, 115–44.

Marglin, S.A. and Bhaduri, A. 1986. Distribution, capacity utilization, and growth. Harvard University.

Myrdal, G. 1939. *Monetary Equilibrium.* London: Hodge.

Robinson, J. 1956. *The Accumulation of Capital.* 2nd edn, London. Macmillan, 1965.

Robinson, J. 1962. *Essays in the Theory of Economic Growth.* London: Macmillan.

Rowthorn, B. 1982. Demand, real wages, and economic growth. *Studi Economici* 18.

Schumpeter, J.A. 1911. *The Theory of Economic Development: an Inquiry into Profits, Capital, Credit, Interest, and the Business Cycle.* Trans. R. Opie, Cambridge, Mass.: Harvard University Press, 1934.

Wicksell, K. 1901. *Lectures on Political Economy.* Vol. I: *General Theory.* Trans. E. Classen, London: Routledge & Kegan Paul, 1934.

investment decision criteria. Investment is present sacrifice for future benefit. Individuals, firms, and governments all are regularly in the position of deciding whether or not to invest, and how to choose among the options available. An individual might have to decide whether to buy a bond, plant a seed, or undertake a course of training; a firm whether to purchase a machine or construct a building; a government whether or not to erect a dam. Under the heading of investment decision criteria, economists have addressed the problem of how to choose rationally in situations that involve a tradeoff between present and future.

THE ECONOMIC THEORY OF INTERTEMPORAL CHOICE

The object of investment is taken to be to optimize one's pattern of consumption over time. The elements needed to determine an individual's investment decision are: (a) his *endowment*, in the form of a given existing income stream over time; (b) his *preference function*, which orders in desirability all possible time-patterns of consumption; and (c) his *transformation set*, which specifies the possibilities for transforming the original endowment into other time-combinations of consumption.

Figure 1 illustrates an artificially simple case of only two periods (say, this year and next) under conditions of certainty. Each point represents a combination of current consumption c_0 and future consumption c_1. The *endowment* combination Y has coordinates (y_0, y_1). *Time-preferences* are portrayed by the indifference curves $U_1, U_2, U_3, \ldots$, each such curve connecting combinations yielding equal satisfaction. The curve QQ' through the endowment position Y pictures the intertemporal *productive opportunities*. By sowing seed, for example, a person can sacrifice current consumption for future consumption – represented in the diagram by a movement from Y along QQ' to the northwest. (There may also be disinvestment opportunities, i.e., the individual might be able to draw upon the future so as to augment current consumption, which would be represented by a movement from Y along QQ' to the southeast.)

For a Robinson Crusoe, the optimum balance of present and future consumption – which in his isolated state must necessarily be identical to his provision for present and future production – occurs at point X^* along QQ'. In the situation pictured he achieves this optimum by *investing* the quantity $y_0 - x_0$ of current consumption claims. For example, having at hand a current corn endowment of y_0, he retains x_0 for current consumption and plants the remainder as seed. Next year he will reap as return from investment the amount $x_1 - y_1$ to augment his endowed availability of future corn.

If markets for trading between present and future income claims exist, however, in contrast with the Robinson Crusoe situation the individual will be able to disconnect the amount

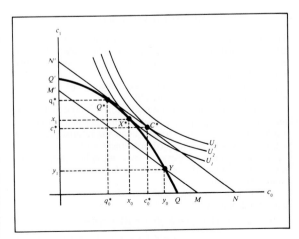

Figure 1 Investment and saving in a 2-period model

he *invests* from the amount he *saves*. These trading opportunities are shown in Figure 1 by the family of 'market lines' whose general equation is:

$$c_0 + c_1/(1 + r_1) = W_0 \qquad (1)$$

Here r_1 is the *interest rate* that discounts one-year future claims c_1 into their equivalent value in terms of c_0 claims. Along each market line the parameter W_0 represents the associated level of *wealth*. Put another way, wealth in equation (1) measures the *present worth* of any specified (c_0, c_1) vector – the future-dated element being 'discounted' at the given market interest rate r_1. In the diagram two market lines are shown: MM' through the endowment vector $Y = (y_0, y_1)$ indicates the individual's endowed wealth $W_0^Y = y_0 + y_1/(1 + r_1)$, while NN' represents the maximum attainable level of wealth $W_0^* = q_0^* + q_1^*/(1 + r_1)$.

If an individual has both productive and market opportunities, his optimizing decision in Figure 1 can be thought of as taking place in two stages. First he locates his 'productive solution' $Q^* = (q_0^*, q_1^*)$ by moving along QQ' so as to maximize attained wealth at the tangency with market line NN'. Second, he then transacts in the funds market, by lending or borrowing (exchanging current for future claims or vice versa) along NN' to find his 'consumptive solution' $C^* = (c_0^*, c_1^*)$ at the tangency of NN' with indifference curve U_2 in the diagram. Notice that his preferences do not at all affect the productive solution, but only how he chooses to 'finance' the investments made. Specifically, in the diagram here the amount he *invests* $(y_0 - q_0^*)$ exceeds the amount he *saves* $(y_0 - c_0^*)$. By borrowing on the market, in effect he has been able to get others to undertake part of the saving necessary to finance his projected investments.

This disconnection between the individual's productive and consumptive decisions in a regime of perfect markets is known as 'Fisher's Separation Theorem'. The essential implication is that individuals with diverging time-preferences can nevertheless come together and agree upon joint productive investments. Business firms and (to some extent) governments can be regarded as institutions designed for undertaking joint investments whose scale is too large for any single individual. The underlying principle is that those investment choices maximizing wealth value or present worth of the mutual undertaking will also maximize wealth for each and every participant therein.

THE PRESENT-VALUE RULE

The economic theory of intertemporal choice leads immediately to what is known as the *Present-Value Rule* for investment decision. This rule can be expressed in two essentially equivalent forms:

(i) Among the opportunities available, adopt the set of investments that maximizes wealth W_0.

(ii) Adopt any single investment project if and only if its present value V_0 is positive. (Taking into account, of course, any repercussions of that project upon the returns yielded by other members of the adopted investment set.)

As an obvious corollary, if two available projects are mutually exclusive, the one with the larger present value V_0 should be chosen.

Generalizing to the multi-period context, wealth as maximand becomes:

$$W_0 = q_0 + q_1/(1 + r_1) + q_2/[(1 + r_2)(1 + r_1)] + \cdots$$
$$\cdots + q_T/[(1 + r_T) \cdots (1 + r_2)(1 + r_1)] \quad (2)$$

Here the q_t are the coordinates of points along the $T + 1$-dimensional productive opportunity surface

$$\phi(q_0, q_1, \ldots, q_T) = 0,$$

a generalization of curve QQ' in Figure 1. T is the 'economic horizon', which may be infinite. And the r_t represent the successive short-term interest rates, each of which discounts prospective payments at any date into its wealth-equivalent at the next preceding date.

For a single project in the multi-date context, present value is defined as:

$$V_0 = z_0 + z_1/(1 + r_1) + z_2/[(1 + r_2)(1 + r_1)] + \cdots$$
$$\cdots + z_T/[(1 + r_T) \cdots (1 + r_2)(1 + r_1)] \quad (3)$$

Here the z_t are the dated payments or 'cash flows' associated incrementally with the project considered. Normally the z_t elements for earlier dates would include some with negative signs – or else the project could not be described as an investment – while those for later dates would have predominantly positive signs. In the special case where $r_1 = r_2 = \cdots = r_T = r$ – that is, where interest rates are expected to remain constant at the level r over time – the Present-Value formulas reduce to the more familiar forms:

$$W_0 = q_0 + q_1/(1 + r) + q_2/(1 + r)^2 + \cdots + q_T/(1 + r)^T \quad (2')$$
$$V_0 = z_0 + z_1/(1 + r) + z_2/(1 + r)^2 + \cdots + z_T/(1 + r)^T \quad (3')$$

The Present-Value solutions can also be formally generalized to allow for continuous rather than discrete time. As an illustrative simplified example, consider a project whose scale of current input or investment sacrifice i_0 is fixed while the output date is subject to choice (e.g., when to cut a growing tree). In Figure 2, horizontal distances represent time t and vertical distances value V_t at each date. Present Value V_0 is indicated by height along the vertical axis. The curve GG' represents productive growth of the asset – in the case of a tree, market value of the standing timber at any date. The 'discount curves' $D, D', D'', \ldots,$ are analogous to the 'market lines' of Figure 1. Each such curve represents the growth of a specific sum of present dollars by continuous compounding at a constant market rate of interest r, or alternatively the Present Value of any future payment continuously discounted at r. The optimal investment period $t = t^*$ is then the one that

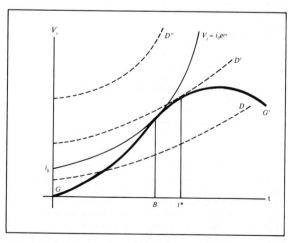

Figure 2 Optimal duration of investment

maximizes Present Value V_0, subject to the constraint on the available V_t described by the curve GG', in the equation:

$$V_0 = -i_0 + V_t e^{-rt} \qquad (4)$$

Geometrically, t^* is determined by the tangency of GG' with the highest discount curve (constant-wealth curve) attainable. The solution condition is then:

$$V_t'/V_t = r \qquad (5)$$

OTHER INVESTMENT CRITERIA

Certain investment criteria employed in business practice are definitely erroneous. One such is rapidity of 'payout' (the date when cash inflows first balance initial outlays), a formula that obviously fails to allow properly for time-discount. Controversy among theorists has centred upon a more interesting concept known variously as the 'internal rate' or the 'rate of return'. The internal rate for a project (or set of projects) is defined as ρ in the discrete discounting equation:

$$0 = z_0 + z_1/(1 + \rho) + z_2/(1 + \rho)^2 + \cdots + z_T/(1 + \rho)^T \qquad (6)$$

As before the z_t here are the successive terms, positive or negative, of the payments–receipts sequence associated incrementally with a particular project. In the special 'deepening' case illustrated in Figure 2, the corresponding concept under continuous compounding is defined implicitly in:

$$0 = -i_0 + V_t e^{-\rho t} \qquad (7)$$

where once again the V_t at any date is described by the productive opportunity curve GG'. Under these conditions ρ represents an average compounded rate of growth.

There has been some confusion between two quite different investment decision *rules* that both employ the internal-rate measure ρ: (i) choose projects so as to maximize ρ, versus (ii) adopt projects incrementally so long as $\rho > r$.

Maximum ρ Rule. If the internal rate ρ is interpreted as the average rate of growth, it may seem plausible that the investor should maximize ρ rather than wealth W_0. (Of course, maximizing a growth *rate* would scarcely make sense unless the initial outlay or scale of investment were held constant, which would not in general hold true.) The solution of (7) that maximizes ρ is shown in Figure 2 as $t = B$, notably earlier than the Present-Value solution $t = t^*$.

In favour of B over t^* it has been argued that, if the growth opportunity were to be replicated in perpetuity, returns from choosing the earlier 'rotation period' B must ultimately dominate those associated with cutting on each cycle at t^*. That is certainly true. However, if the decision problem concerns infinite rotation rather than a one-time cutting, for a valid comparison the relevant Present-Value measure would have to be a generalized one that allows for the associated infinite sequence of discounted returns. It can be shown that this generalized Present Value does coincide with B if the growth opportunity can be reproduced on an ever-broadening scale (e.g. on new land) – but only as funds are freed by cutting the tree or trees. This turns out to be an impossible or uninteresting case, because it implies that the productive opportunity must be of *infinite* market value if the maximized ρ exceeds the market interest rate r (and of zero value otherwise). In contrast, if the opportunity is a unique one which cannot be reproduced after cutting, as pictured in Figure 2, the simple $t = t^*$ solution remains correct. Another solution, $t = F$, found by the German forester Faustmann, is

appropriate when the opportunity can be reproduced over time by cutting and replanting but cannot be broadened in scale. F would be found by maximizing the Present Value V_0 of an infinite sequence of rotations, each being a constant-scale replication of the original opportunity. Like all the correct solutions, it is equivalent to maximizing the present worth of the opportunity under the stated assumptions. (F is not shown in Figure 2 but would lie between B and t^*.)

ρ vs. r Comparison Rule. The Comparison Rule says to adopt any project whose internal rate ρ exceeds the market rate of interest r. This rule remains popular in business practice, in part because it offers a convenient division of labour: calculation of the ρ's on individual projects might be delegated to subordinates, while top decision-makers choose the cutoff rate r that corresponds to the relevant market interest rate faced by the firm. Unfortunately, however convenient such a decision of labour may be, once again this is not in general a correct method of project selection.

The difficulty with the Comparison Rule first came to be appreciated when it was discovered that a sequence of positive and negative cash flows could have more than one ρ serving as solution of equation (6) above. A project represented by the annual payments sequence $-1, 5, -6$, for example, has two solutions: $\rho = 1$ and $\rho = 2$. (It can be shown that a project with $T + 1$ dated elements may have as many as T solutions.) This of course destroys the idea that the internal rate can generally be identified with a growth rate; an outlay of one dollar cannot be said to grow at both 100% and 200%. Various answers have been offered to the puzzle of which ρ to use in such cases. But the difficulty is immediately explained and resolved if we think instead in terms of Present Value. It turns out that the sequence $-1, 5, -6$ has positive V_0 (and is therefore worth adopting) for any constant market interest rate r *between* 100% and 200%, but at other values of r has negative Present Value (and should not be adopted). Perhaps even more illuminating is the project described by cash flows $-1, 3, -2\frac{1}{2}$. This sequence has no real solution for ρ in equation (6), the reason in Present-Value terms being that V_0 is negative for any *constant* r. Yet this is a perfectly respectable investment opportunity. After all, there is no justification for postulating (as is implicitly done by the Comparison Rule) that the anticipated sequence of market interest rates $r_1, r_2, \ldots, r_T$ *must* be constant over time (always equal to a common r). It turns out that the cash-flow pattern $-1, 3, -2\frac{1}{2}$ has positive Present Value (i.e., the project would be worth adopting) for many possible non-constant interest-rate sequences – for example, $r_1 = 100\%$ and $r_2 = 200\%$.

Summing up, therefore, the Present-Value Rule for investment decision – corresponding as it does to the principle of maximizing wealth within the opportunities available – is correct itself and also serves to define the range of validity of all the other rules considered.

GENERALIZATIONS AND EXTENSIONS

The preceding analysis needs to be extended in at least two important ways, so as to allow for: (1) uncertainty, and (2) imperfect and incomplete markets.

Uncertainty. Investment choices, involving as they do present sacrifice for future benefit, are peculiarly sensitive to uncertainty. However, so long as we can continue to assume a regime of complete and perfect markets, the Present-Value Rule is robust enough to retain validity even in a world of uncertainty. For, the proximate goal of any individual (or group of individuals organized in a firm or other joint enterprise) will still be to undertake productive activities so as to maximize wealth. Having achieved that goal, each and every individual investor will be in a position to distribute his attained wealth as desired over all possible dated contingencies in accordance with his time-preferences, degree of risk-aversion, and probability beliefs.

Economists use two main models for the analysis of uncertainty – state-preference and mean-versus-variability analysis. Since the latter, under certain assumptions, can be regarded as a special case of the former, for our purposes attention can be limited to the state-preference model. If markets for state-claims are complete and perfect, any pattern of varying returns over states of the world at a given date has a *certainty-equivalent* in value terms as of that date. In equations (3) and (3'), the z_t for any project can now be interpreted as certainty-equivalents (rather than as simple cash flows) defined by:

$$z_t = P_{t1}z_{t1} + P_{t2}z_{t2} + \cdots + P_{tS}z_{tS} \qquad (8)$$

Here z_{ts} represents the cash flow at date t contingent upon state of the world s obtaining – there being S distinguishable such states – while P_{ts} is the price at which a unit claim to income in state s at date t can be converted into (traded for) current certainty income.

Incomplete or imperfect markets. Markets are said to be *incomplete* if some objects of choice are non-tradeable. For example, futures markets for some commodities at far-distant dates do not exist, nor is it possible to trade in claims contingent upon each and every conceivable future uncertain event. Markets are said to be *imperfect* if there are costs of trading – for example, brokerage fees, transaction taxes, or expenses in locating exchange partners. Any real-world regime of markets will necessarily be both incomplete and imperfect, but for some purposes the assumption of complete and perfect markets may be a usable idealization. Unfortunately, once we depart from this idealization the problem of investment decision criteria becomes very difficult. The reason is that the Separation Theorem fails. Only under complete and perfect markets is the concept of wealth or Present Value unambiguously defined, so that the choice of productive investments can be entirely disconnected from individuals' personal time-preferences, risk-preferences, beliefs etc. Failure of the Separation Theorem particularly subverts the ability of investors to join together in undertaking large projects or groups of projects.

However, two different lines of analytical approach have yielded results of interest. (i) A number of techniques have been devised for locating 'utility-free' or 'efficient' investment choices. In general such techniques cannot determine an optimal project set, but they can serve to filter out options whose payoff patterns over dates and/or states are dominated by other available projects or project combinations. (ii) While investors' personal circumstances may diverge in innumerable ways, there should be some tendency for those similarly situated to group together. Thus, a firm whose investment opportunities yield far-future payoffs should tend to be owned by a 'clientele' consisting of individuals with moderate time-preferences, willing to forego current dividends in the hope of large long-term gain. It follows that unanimity as to the investment choices to be made may after all govern *within* the firm, for example as to the discount rate to employ in calculating Present Value, even in the absence of perfect and complete markets.

JACK HIRSHLEIFER

See also INTERNAL RATE OF RETURN; INVESTMENT PLANNING; PRESENT VALUE.

BIBLIOGRAPHY

The modern theory of investment and intertemporal choice was set down in classic form by Irving Fisher as part of his great works on interest (1907, chapters 8–9; 1930, chapters 10–13). The seminal works on uncertainty theory include Arrow (1953) for the state-preference approach and Markowitz (1959) and Sharpe (1964) for the mean-versus-variability model. Choice over time and choice under conditions of uncertainty are integrated in the treatise by Hirshleifer (1970) that builds upon these foundations. All these topics have been followed up in an enormous literature, of which only a few illustrative instances can be cited here: on investment decision formulas, Samuelson (1976); on utility-free or dominant choices, Pye (1966), Hanoch and Levy (1969), and DeAngelo (1981). A survey of investment decision criteria used in current business practice appears in Schall, Sundem, and Geijsbeek (1978).

Arrow, K.J. 1953. The role of securities in the optimal allocation of risk-bearing. Reprinted in K.J. Arrow, *Essays in the Theory of Risk-Bearing*, Chicago: Markham, 1971.
DeAngelo, H. 1981. Competition and unanimity. *American Economic Review* 7(1), March, 18–27.
Fisher, I. 1907. *The Rate of Interest*. New York: Macmillan.
Fisher, I. 1930. *The Theory of Interest*. New York: Macmillan.
Hanoch, G. and Levy, H. 1969. Efficiency analysis of choices involving risk. *Review of Economic Studies* 36(3), July, 335–46.
Hirshleifer, J. 1970. *Investment, Interest, and Capital*. Englewood Cliffs, NJ: Prentice-Hall.
Markowitz, H.M. 1959. *Portfolio Selection*. New York: Wiley.
Pye, G. 1966. Present values for imperfect capital markets. *Journal of Business* 39, January, 45–51.
Samuelson, P.A. 1976. Economics of forestry in an evolving society. *Economic Inquiry* 14(4), December, 466–92.
Schall, L.D., Sundem, G.L. and Geijsbeek, W.R. 1978. Survey and analysis of capital-budgeting methods. *Journal of Finance* 33(1), March, 281–7.
Sharpe, W.F. 1964. Capital asset prices: a theory of market equilibrium under conditions of risk. *Journal of Finance* 19(3), September, 425–42.

investment planning. The theories discussed here consist of two complementary formulations originating in India and in the United Kingdom in the 1950s and 1960s. Both deal with investment planning when development starts with virtually no capital goods industry. Thus they represent an expansion of the model of the Soviet economist Feld'man, since in the latter the economy did possess an investment sector albeit in a limited dimension (Feld'man, 1928a, 1928b).

The first approach, due to Dobb (1954, 1960) and to Sen (1960), deals with the choice of techniques and the sectoral distribution of investment and labour. The second approach, elaborated by a number of Indian scholars – Raj and Sen (1961), Naqvi (1963) – is more concerned with the sectoral allocation of investment goods under conditions of stagnant export earnings. The definition of sector is the same as in the Marx-based Feld'man model with the difference that the capital goods sector itself is divided into two branches. One branch consists of an intermediate sector producing equipment usable only in the consumption goods sector. The second branch is formed by *machine tools* which can reproduce themselves as well as be installed in the intermediate sector.

The emphasis on this kind of structural relations is aimed at providing analytical support to the view that sectoral investment planning by the State is a necessary, although not sufficient, condition for the emancipation from backwardness.

The starting point of both approaches is the historical consideration that colonialism has destroyed the traditional home industries, thereby making expansion dependent on the exports of primary products having low demand elasticities (Raj and Sen, 1961). It is this particular condition which justifies investment priority in the capital goods industry for a growth strategy oriented toward the home market (Dobb, 1967). Industrialization would then imply the creation of capital goods well in advance of any market demand for them, a process called by Dobb *the Accelerator in Reverse*.

Developing economies face the task of investing in a manner largely independent from the preexisting material structure. In this context, indivisibilities of capital equipment – which 'are likely to be significantly large (relatively to the scale of the economy) at early stages of development' (Dobb, 1960, pp. 11–12) – may make the expansion of a certain branch unprofitable although its growth can be of crucial importance for the formation of other industries. State planning of the sectoral allocation of investment performs the role of securing overtime the construction of complementary industries.

It must be noticed that some of the views put forward by Dobb and the Indian economists were part of the intellectual climate of the period. In the mid-1950s Prebisch started the debate over the terms of trade between industrialized and underdeveloped countries, arguing the long-term nature of the latter's unfavourable position. Politically, the first meeting of the non-aligned nations, held in the Indonesian city of Bandung in 1955, asserted the necessity to embark on a road privileging the domestic market. Institutionally, sectoral planning by the State seemed to have gained a firm hold also in a non-socialist country as important as India. Practically, the experience of the People's Republic of China suggested that a developing country could reduce the dependency on foreign exchange by building a machine tools industry (Raj, 1967).

Given this cultural and political framework, Dobb's pioneering work has a special place in the theories of planned development. It singled out the fact that the domestic economy of underdeveloped countries does not generate a surplus of wage goods large enough to allow a more or less smooth process of growth. Indeed, with most of the work force employed in subsistence activities, it would be impossible to set in motion the Accelerator in Reverse unless the bottleneck of a limited surplus is widened. The technical form of investment must therefore reflect this initial constraint. In setting forth the answer to the question of the choice of techniques, Dobb challenged the view that 'since a scarcity of capital relative to labour is a usual characteristic of underdeveloped economies, capital investment needs there to take the form of projects of "low capital intensity" ' (Dobb [1954], 1955, p. 139).

The gist of his and Sen's argument (Sen, 1960) can be presented as follows:

Consider an economy where fixed capital in the capital goods industry is so small that machines can be thought of as being produced by labour alone. Thus, employment in the capital goods sector multiplied by the productivity of labour – denoted by x – gives the total output of equipment. But employment in the capital goods sector is limited by the surplus produced in the wage goods sector. If 20 people work in the wage goods sector, where the productivity of labour (z) is 20 units per person and the real wage rate (w) is uniform throughout the economy and fixed at 10 units, then 20 people can be put to work in the capital goods sector. The crucial ratio is given by $(z-w)/w$, where $z-w$ is the surplus per unit of labour in the wage goods sector. If the bottleneck in the production of wage goods has to be widened without lowering the real wage, all newly produced machines should be installed in the wage goods sector. On the assumption that these do not depreciate and that each machine employs one worker, total output of capital goods will be equal to the increment in employment in the wage goods sector. The growth rate of the economy is therefore equal to the growth rate of employment in this sector. Given the above mentioned allocation policy, the growth rate is nothing but the productivity of labour in the capital goods sector multiplied by the ratio of the surplus to the wage rate. Hence:

$$g = x(s/w); \quad \text{where} \quad s = z - w. \tag{1}$$

Assuming no production lags, maximization of (1) yields:

$$-(dx/x) = (dz/z)(z/s). \tag{2}$$

According to equation (2), the growth rate would be maximized by using more costly methods of production in the capital goods industry, lowering the productivity of labour in this sector. At the same time, the delivery of improved and more expensive equipment would *ipso facto* raise labour productivity in the wage goods industry. With a positive wage rate – implying a z/s ratio greater than unity – this gain need not be as large as the loss of productivity in the capital goods industry. It is the asymmetrical change in the sectoral productivities of labour which leads to an overall increase in capital intensity.

The results do not change if unassisted labour builds machine tools for the intermediate investment sector. In this case the gains in the intermediate sector multiplied by z/s, should equal the losses in the machine tools industry.

With a construction based on a number of simplifying assumptions, Dobb and Sen provided the rationale for raising the capital intensity of production under conditions of abundant labour supply. Yet the assumptions turned out to be restrictive not so much in relation to traditional theory, but in relation to the scope and objective of the exercise.

Analytically the model does not succeed in giving a criterion for the choice of techniques when the economy embarks on a path of self expansion of the machine tools sector. The only possible observation is that this sector's productivity does not depend on any other branch of the economy, thus there is no constraint on the degree of capital intensity (Johansen and Ghosh, in Dobb, 1960). Dobb's and Sen's results depend very much on the assumptions of no production lags and of immortal machines. In macroeconomic terms, an increase in capital intensity generates a higher growth rate only if the share of investment in national income is raised more than proportionately, which may not be immediately feasible. In the interim period the economy will experience a lower growth rate and a lower share of consumption (Kalecki, 1972a). In turn, the notion of immortal machines becomes untenable whenever Dobb analyses the possibility of drafting the whole of the labour force in the two investment industries for the purpose of building the machine tools sector. If wear and tear is taken into account, as soon as no equipment flows to the wage goods sector its capital stock will shrink and so will the output of consumables. The wage rate will cease to be a parameter, becoming instead a variable conditioned by the proportions in which labour and machines are distributed. Hence, wear and tear and the socially minimum wage rate show the limit of the percentage in which machine tools can be reinvested in their own sector. This is a major structural and social aspect of any process of accelerated accumulation (Lowe, 1976; Halevi, 1981).

Dobb's contribution will remain a classic in the field because it introduced a novel perspective on the reasons for, and the modalities of, socialist-oriented development for the ex colonial countries. The fact that this approach is no longer followed can only in part be attributed to the limitations outlined above. Perhaps, in addition to the ever present ideological factor, one explanation lies in the change of the historical framework. There are, by now, significant instances in which a process of fast accumulation has taken place hand in hand with the persistence of phenomena such as landlessness and urban poverty. In countries like Brazil, Mexico and India, these are the problems that must be reflected in any planning strategy. The issue is not so much that of building a capital goods sector from scratch, but to conceptualize the economic and political nature of the phenomena (Kalecki, 1972b, 1972c; Taylor and Bacha, 1976).

The second approach, coming mainly from India, is a substantial improvement on the Mahalanobis variant of the Feld'man model (Mahalanobis, 1955). It uses the same hypothesis of two capital goods industries to discuss the sectoral allocation of machinery imported through a fixed sum of foreign earnings F. Raj and Sen (1961) assumed negligible amount of equipment in the intermediate investment sector I, and in the machine tools industry M. Furthermore, machine tools are used also for the extraction of raw materials R. The planners can freely choose the initial share of consumption over national income, production coefficients are given. In this context, if F is used to import I goods for the production of consumption goods C, the output of C goods will rise but its absolute increase will tend to nought because raw material requirements will also rise. A constant increment in C goods production can be obtained when F is used to import M goods for the production of I goods and for the extraction of R. In this case raw materials set a limit to the expansion of the I sector output. Finally, the output of consumption goods will grow at a constant absolute rate if M goods are imported in order to produce machine tools to be installed exclusively in the I and R sectors.

The original Raj–Sen paper did not discuss the proportions in which machine tools are reinvested in the M sector itself. In the literature that followed, the point was raised by Naqvi (1963) and later by Cooper (1983). Naqvi noted that reinvestment in the M goods sector would allow for a proportionate growth in C goods also in the presence of a limited amount of import earnings. Moreover he observed that central control of the M goods sector can be used to limit the creation of a luxury goods industry catering for the well to do. Cooper, on his part, argued that planners can more effectively influence the share of consumption by selecting the ratio in which M goods are to be replughed in their own sector. This is because the share of consumption over national income cannot be freely determined by planners, since it is fixed by the initial distribution of equipment. Planning models based on sectoral relations and on the principle of the Accelerator in Reverse, showed a greater longevity than choice of techniques models. The assumption of given production coefficients did not prevent the analysis of alternative growth paths and the introduction of limiting conditions such as minimum wage rate and stagnant export earnings (Das, 1974). The capital goods–consumption goods model has been used also as a framework for the application of optimal control theory in development planning (Stoleru, 1965), as well as for the analysis of unused capacity caused by a slow growing agricultural output (Patnaik, 1972; Raj, 1975).

Contributions to investment planning using analytically a Marxian sectoral approach have come mostly from Great Britain and from India. The Soviet mathematical economists seem to be more inclined toward generic multisectoral optimisation models. This may reflect a belief that a purely capital goods-consumption goods approach ceases to be relevant when a socialist economy possesses a developed industrial structure. Yet, as it emerges from reading the works of some Soviet economists of the mathematical school, generic multisector models cannot give a stylized picture of growth paths (Dadayan, 1981). Indeed, in the Western literature on growth, the crucial issue of the transition between two growth rates – a process called Traverse – is dealt with an analytical apparatus closer to Marx's sectoral characterization of the economy (Hicks, 1965; Lowe, 1976).

JOSEPH HALEVI

See also COST-BENEFIT ANALYSIS; DEVELOPMENT PLANNING; PLANNED ECONOMY; PROJECT EVALUATION.

BIBLIOGRAPHY

Cooper, C. 1983. Extensions of the Raj–Sen model of economic growth. *Oxford Economic Papers* 35(2), July, 170–85.

Dadayan, V. 1981. *Macroeconomic Models.* Moscow: Progress.

Das, R.K. 1974. *Optimal Investment Planning.* Rotterdam: Rotterdam University Press.

Dobb, M. 1954. A note on the so-called degree of capital-intensity of investment in under-developed countries. In M. Dobb, *On Economic Theory and Socialism, Collected Papers*, London, Routledge & Kegan Paul, 1955.

Dobb, M. 1960. *An Essay on Economic Growth and Planning.* London: Routledge & Kegan Paul.

Dobb, M. 1967. The question of 'investment-priority' for heavy industry. In M. Dobb, *Papers on Capitalism, Development and Planning*, New York: International Publishers.

Feld'man, G. 1928a. K teorii tempov narodnogo dokhoda. (On the theory of growth rates of the national income.) *Planovoe khoziaistvo*, November.

Feld'man, G. 1928b. K teorii tempov narodnogo dokhoda. (On the theory of growth rates of the national income.) *Planovoe khoziaistvo*, December.

Halevi, J. 1981. The composition of investment under conditions of non uniform changes. *Banca Nazionale del Lavoro Quarterly Review* 34(137), June, 213–32.

Hicks, J. 1965. *Capital and Growth.* Oxford: Clarendon Press.

Johansen, L. and Ghosh, A. 1960. Appendix: notes to chapters III and IV. In M. Dobb, *An Essay on Economic Growth and Planning*, London: Routledge & Kegan Paul.

Kalecki, M. 1972a. The problem of choice of the capital-output ratio under conditions of an unlimited supply of labour. In M. Kalecki, *Selected Essays on the Economic Growth of the Socialist and the Mixed Economy*, Cambridge: Cambridge University Press.

Kalecki, M. 1972b. Problems of financing economic development in a mixed economy. In M. Kalecki, *Selected Essays on the Economic Growth of the Socialist and the Mixed Economy*, Cambridge: Cambridge University Press.

Kalecki, M. 1972c. Social and economic aspects of 'intermediate regimes'. In M. Kalecki, *Selected Essays on the Economic Growth of the Socialist and the Mixed Economy*, Cambridge: Cambridge University Press.

Lowe, A. 1976. *The Path of Economic Growth.* Cambridge: Cambridge University Press.

Mahalanobis, P.C. 1955. The approach of operational research to planning in India. *Sankhya* 16, December, 3–131.

Naqvi, K.A. 1963. Machine-tools and machines: a physical interpretation of the marginal rate of saving. *Indian Economic Review* 6(3), February, 19–28.

Patnaik, P. 1972. Disproportionality crisis and cyclical growth. A theoretical note. *Economic and Political Weekly* 7, annual number, February, 329–36.

Raj, K.N. 1967. Role of the 'machine-tools sector' in economic growth. In *Socialism, Capitalism and Economic Growth, Essays Presented to Maurice Dobb*, ed. C. Feinstein, Cambridge: Cambridge University Press.

Raj, K.N. 1975. Linkages in industrialization and development strategy; some basic issues. *Journal of Development Planning* 8, 105–19.

Raj, K.N. and Sen, A.K. 1961. Alternative patterns of growth under conditions of stagnant export earnings. *Oxford Economic Papers* 13, February, 43–52.

Sen, A.K. 1960. *Choice of Techniques*. Oxford: Basil Blackwell.

Stoleru, L. 1965. An optimal policy for economic growth. *Econometrica* 33, April, 321–48.

Taylor, L. and Bacha E. 1976. The unequalizing spiral: a first growth model for Belindia. *Quarterly Journal of Economics* 90(2), May, 197–218.

invisible hand. 'The invisible hand' was a metaphor used by Adam Smith to describe the principle by which a beneficient social order emerged as the unintended consequences of individual human actions. Although Smith used the specific term 'invisible hand' in this sense only twice in his writings, once in the *Theory of Moral Sentiments* and once in *The Wealth of Nations*, the idea the metaphor connotes permeates all of his social and moral theories. Indeed, it was the notion of the invisible hand that enabled Smith to develop the first comprehensive theory of the economy as an interrelated social system. It is not much of an exaggeration to say that the invisible hand made theoretical social science itself possible.

In the *Theory of Moral Sentiments*, Smith, expounding on how the desire for wealth and luxury spurs men to great industry and production, points out that those who become rich from all this effort are not much better off in the things of this world that really count than the poor who work for them. The rich landlord, for example, desires many trivial luxuries, but can only consume a modest portion of the food his efforts produce; the rest must be paid to those who serve him. Rich landlords:

> ... in spite of their natural selfishness and rapacity, though they mean only their own conveniency, though the sole end which they propose from the labours of all the thousands whom they employ be the gratification of their own vain and insatiable desires, they divide with the poor the produce of all their improvements. They are led by an invisible hand to make nearly the same distribution of the necessaries of life which would have been made had the earth been divided into equal portions among all its inhabitants; and thus, without intending it, without making it, advance the interest of the society, and afford means to the multiplication of the species (*Moral Sentiments*, IV, 1, pp. 304–5).

In the *Wealth of Nations*, Smith uses the term 'invisible hand' in the context of explaining why restrictions of imports or on the use of one's capital are unnecessary:

> As every individual, therefore, endeavours as much as he can both to employ his capital in the support of domestick industry, and so to direct that industry that its produce may be of the greatest value; every individual necessarily labours to render the annual revenue of the society as great as he can. He generally, indeed, neither intends to promote the publick interest, nor knows how much he is promoting it. By preferring the support of domestick to that of foreign industry, he intends only his own security; and by directing that industry in such a manner as its produce may be of the greatest value, he intends only his own gain, and he is in this, as in many other cases, led by an invisible hand to promote an end which was no part of his intention (*Wealth of Nations*, IV, ii, p. 456).

The underlying notion of unintended orders that the invisible hand captures was not new in Smith. There were glimmerings of the idea in the 17th century in the writings of Petty and Locke. One of the earliest 18th-century forerunners to Smith's invisible hand was Bernard Mandeville. While there is some controversy as to whether Mandeville had any concept of a self-ordering system comparable to Smith's, it is clear that his infamous statement that private vices of greed, luxury and avarice lead to public benefits of abundant wealth (*Fable of the Bees*, 1714) stirred a great deal of debate. The philosophers of the Scottish Enlightenment, of whom Smith was one, rejected Mandeville's sensational equating of self-interest with greed, but they developed as a major theme of their writing the underlying idea that private actions can have beneficial public effects that were not intended by the actors. Adam Ferguson, for instance, described private property and political institutions in general as 'the results of human action, but not the execution of any human design' (*An Essay on the History of Civil Society*, 1767). David Hume appealed to the same notion when he explained how a system of justice emerges as the by product of a series of individual self-interested decisions about the disposition of particular disputes, and when he argued that human institutions like money and language arose from the actions of individuals directed toward another end (*Treatise of Human Nature*, 1740).

Adam Smith, like Mandeville, Ferguson and Hume, based his system on the observation that man is motivated by self-love. To Smith, however, self-love was potentially an admirable human characteristic that reflected a man's concern for his honour as well as his material welfare. Even more importantly, for Smith, self-love was the 'principle of motion' in social theory much as attraction is the principle of motion in Newton's physics. Those who believed that government was free to make any laws it chose to regulate society, Smith believed did not understand a most basic feature of human nature. The 'man of system', as Smith called him,

> seems to imagine that he can arrange the different members of a great society with as much ease as the hand arranges the different pieces upon a chess-board; he does not consider that the different pieces upon a chess-board have no other principle of motion besides that which the hand impresses upon them; but that, in the great chess-board of human society, every single piece has a principle of motion of its own, altogether different from that which the legislator might choose to impress upon it (*Theory of Moral Sentiments*, pp. 380–81).

The 18th century was, of course, the century immediately following Newton's great discovery. The scientific and moral consequences of Newton's universe were still being debated, and Newtonian ideas and patterns of thought were finding their way into all areas of intellectual activity. Not only was Adam Smith familiar with Newton's ideas but he had even written, early in his career, a history of astronomy, the last ten pages of which praised Sir Isaac Newton's system. Hence, it is plausible to see Adam Smith's economic system in a 'Newtonian' context as an attempt to explain a complex social order on the basis of a few simple principles of human action. The economic system that Smith in fact described is a product both of man's self-love and his peculiarly human propensity to 'truck, barter and exchange one thing for another' (*Wealth of Nations*, I, 2, p. 25). Exchange leads to the division of labour and the division of labour enables workers to take advantage of economies of scale with the unintended result that greater aggregate wealth is produced than if there were no exchange. Hence, the wealth of nations depends not on conscious governmental planning, but on the freedom of individuals to exchange, specialize and extend their markets. Furthermore, the overall beneficial nature of Smith's 'simple system of natural liberty' depends not on the benevolence of individuals, but upon the operation of self-love in a system of free exchange. Smith points out that in an exchange, we obtain our ends neither from coercing our partner nor from appealing to his sense of charity, but by engaging his own self-love in the exchange process. Two people trade because there are mutual gains from trade. Or as Smith puts it:

... man has almost constant occasion for the help of his brethren, and it is vain for him to expect it from their benevolence only. He will be more likely to prevail if he can interest their self-love in his favour, and show them that it is for their own advantage to do for him what he requires of them ... . It is not from the benevolence of the butcher, the brewer, or the baker, that we expect our dinner, but from their regard to their own interests (*Wealth of Nations*, I, 2, p. 26).

In general, the concept that the invisible hand so graphically captures – a concept Carl Menger restated as an 'organic understanding of social phenomena' ([1883] 1963, p. 127ff) and Hayek more recently referred to as a 'spontaneous order' (1973, vol. 1) – is composed of three logical steps. The first is the observation that human action often leads to consequences that were unintended and unforeseen by the actors. The second step is the argument that the sum of these unintended consequences over a large number of individuals or over a long period of time may, given the right circumstances, results in an order that is understandable to the human mind and appears as if it were the product of some intelligent planner. The third and final step is the judgement that the overall order is beneficial to the participants in the order in ways that they did not intend but nevertheless find desirable.

The first of these steps must have been obvious to human beings since they became capable of articulating their observations. The capriciousness of nature and the fallibility of human plans has been the theme of religious doctrine, philosophy and dramatic literature since the beginning of recorded time. It is only with the introduction of the notion that the independent actions of individual human beings can inadvertently give rise to an understandable and orderly social process that the unintended consequences of human plans becomes scientifically interesting. Clearly, without some notion of an invisible hand in human actions, social science would be impossible. The only alternative to describing social processes as the unintended by-products of purposeful human actions would be to view all social institutions and practices either as the predicted unfolding of conscious human plans or as the results of natural or supernatural phenomena beyond human experience. It is for this reason that Hayek has referred to spontaneous orders as a third category of phenomena between consciously planned organization and the physical world (1973, vol. 1, p. 20).

The third step in the 'invisible hand' formulation is not as uncontroversial as the other two nor is it even necessary to the description of a social order. To judge a spontaneous order to be benevolent, as Smith obviously did, is to judge it from a particular moral perspective and within a particular political and historical context. Smith's moral perspective was that of the participants in the system, and his judgement was that they would be better off under a predominantly free market system than under the system of mercantilist regulation that was still in force in 18th-century England. However, one could easily imagine a spontaneous order in which people were led as if by an invisible hand to promote a perverse and unpleasant end. The desirability of the order that emerges as the unintended consequences of human action depends ultimately on the kind of rules and institutions within which human beings act, and the real alternatives they face.

Spontaneous orders can be thought of in two, related ways. They can describe a set of regularities in a social system that is self-organizing in some way within the context of a set of social rules. In this interpretation, the constraints in the system could well be set by human design and can work for good or ill. Alternatively, spontaneous orders can be thought of as evolved orders where the rules themselves are the unintended products of human actions. For example, we think of a market economy as functioning according to a set of 'rules of the game' that permit allocative errors to be self-correcting within the system. The rules (laws, customs, the dictates of political organizations, and property rights) would be thought of as products of conscious human plans as in specific legislation and constitutional design, or alternatively, they could be thought of as themselves the unintended products of human action aimed at specific and narrow ends. In Adam Smith's writings, both interpretations of spontaneous orders can be found. In his moral philosophy, in the same manner as David Hume, Smith argues that our moral rules gradually evolve from the accumulation of individual experiences and judgements of individuals in concrete situations, while the rules themselves as evolved have the unintended results of promoting social stability. In his economic theory described in the *Wealth of Nations*, money prices and profit and loss provide the signals that lead to corrections in resource misallocations and to economic growth, while the economic institutions of markets, money, and division of labour all emerged in an evolutionary process. How one views the institutions of society makes a difference not only to one's political views, but also to how one evaluates an economic system.

The notion of spontaneous order in the sense of a self-ordering system continued to provide the foundation of economic science and especially general equilibrium theory throughout the 19th century and up to the present. In the general equilibrium formulation, self-love is translated into preference orders defined over all goods (but not benevolence or honour), and the political and social institutions of society are alterable by government corrective action. In this view, the invisible hand still makes the system run, but the optimality of the result is not necessarily guaranteed. Indeed, if one follows the logic of the argument to its conclusion, the invisible hand is palsied at best since it really only operates benevolently under conditions that are impossible to meet in the real world (Hahn, 1982).

An alternative formulation of the invisible hand that found its way from Adam Smith to Carl Menger views the economic institutions of a society as the unintended by-products of self-interested economic behaviour and sees these institutions as crucial to the self-ordering process. Hence, instead of asking what institutions would be necessary to make the invisible hand work perfectly. This view poses the question, what are the economic reasons why existing market institutions emerged and what unperceived purposes do they serve. Here, the attempt is to show how existing market institutions arise as the unintended consequences of human action. This, together with the argument that such institutions serve more ends than are known by planners, makes one more chary of 'fixing up' market arrangements.

KAREN I. VAUGHN

See also ECONOMIC HARMONY; INDIVIDUALISM; SMITH, ADAM.

BIBLIOGRAPHY

Ferguson, A. 1767. *An Essay on the History of Civil Society.* Edinburgh: Edinburgh University Press, 1966.

Hahn, F. 1982. Reflections on the invisible hand. *Lloyd's Bank Review* 144, April, 1–21.

Hayek, F. von. 1973. *Law, Legislation and Liberty.* Vol. I. Chicago: University of Chicago Press.

Hume, D. 1740. *Treatise of Human Nature.* Oxford: Oxford University Press, 1978.

Mandeville, B. 1714. *The Fable of the Bees: or, Private Vices, Public Benefits.* Oxford: Oxford University Press, 1924.

Menger, C. 1883. *Problems of Economics and Sociology.* Urbana: University of Illinois Press, 1963.

Smith, A. 1759. *The Theory of Moral Sentiments.* New York: Liberty Classics, 1969.

Smith, A. 1776. *An Inquiry into the Nature and Causes of the Wealth of Nations.* Ed. R.H. Campbell and A.S. Skinner, New York: Liberty Press, 1981.

involuntary unemployment. The most common and analytically useful definition of involuntary unemployment is based on the labour supply curve: if workers are off the labour supply curve – so that there is an excess supply of labour at the current real wage, then, by definition, there is involuntary unemployment. The amount of involuntary unemployment is equal to the amount of excess labour supply. If workers are on the labour supply curve, then, by definition, there is no involuntary unemployment. One could analogously define involuntary overemployment as a situation of insufficient supply of labour at the prevailing real wage (as may occur during wartime with wage and price controls), but the term is seldom used.

In a static, deterministic, utility maximization framework, the labour supply curve is simply the set of real wage and employment pairs for which the marginal rate of substitution of income for leisure is equal to the current real wage. Hence, involuntary unemployment can be equivalently defined using the utility function: if the real wage is greater than the marginal rate of substitution of income for leisure, then, by definition, there is involuntary unemployment. If the marginal rate of substitution of income for leisure is equal to the real wage, then there is no involuntary unemployment.

HISTORICAL EXAMPLES OF USAGE. This definition of involuntary unemployment is very close to that used by Keynes (1936). In Chapter II of the *General Theory*, Keynes writes ' ... the equality of the real wage to the marginal disutility of employment ... corresponds to the absence of "involuntary" unemployment' (p. 15). (Keynes makes the simplification that the marginal utility of income is constant, so that the marginal disutility of employment is the same as the marginal rate of substitution of income for leisure.) Keynes excluded frictional unemployment from involuntary unemployment. However, it is important to note that Keynes also excluded unemployment 'due to the refusal or inability of a unit of labour, as a result of legislation or social practices or of a combination for collective bargaining or of a slow response to change or of mere human obstinacy, to accept a reward corresponding to the value of the product attributable to its marginal productivity' (Keynes, 1936, p. 6). Thus, Keynes chose to exclude union wage differentials as well as minimum wage legislation as sources of involuntary unemployment. Clearly, Keynes wanted to focus on a particular type of involuntary unemployment.

Patinkin (1965, ch. 13) also used the static labour supply definition in his well-known analysis of involuntary unemployment:

> The norm of reference to be used in defining involuntary unemployment is the supply curve for labor ... as long as workers are 'on their labor supply curve' – that is, as long as they succeed in selling all the labor they want to at the prevailing real wage rate – a state of full employment will be said to exist in the economy (pp. 314–15).

Although Keynes developed and emphasized the idea of involuntary unemployment much more than economists had done before, the above definition based on the labour supply curve predates Keynes writings. In fact it was used by the 'classical' economists. For example, in 1914 Pigou proposed measuring involuntary unemployment of a group of persons by the number of hours' work by which employment ' ... falls short of the number of hours' work that these persons would have been willing to provide at the current rate of wages under current conditions of employment' (see Casson, 1983, p. 39). According to Keynes, however, classical theories (such as Pigou's) did not admit the possibility of involuntary unemployment. Unemployment of a particular group caused by union wage differentials or minimum wage legislation was admitted by the classical theory, but as mentioned above Keynes chose to classify this as voluntary.

CRITICISMS OF THE DEFINITION OF INVOLUNTARY UNEMPLOYMENT. Despite the analytical simplicity of the above definition based on labour supply, the term involuntary unemployment has resulted in many critiques and controversies. One of the criticisms stems from simple conflicts between the above technical definition and everyday non-technical usage of the term involuntary. For example, Fellner (1976) wrote, ' ... distinguishing elements of voluntariness from elements of involuntariness in the unemployment problem is a hopeless endeavour ... ' (p. 134) and that 'Keynes ' definition is unhelpful and so are all variants inspired by that definition' (p. 53). Fellner and others have been concerned that one can never determine the intentions of a given unemployed person so that the broad classification of unemployment into involuntary and voluntary is meaningless. Although the many connotations of the term involuntary may cause semantic difficulties (as may other concepts in economics such as 'rational' or 'marginal'), focusing on the technical definition given above would seem to avoid these difficulties.

A second criticism arises in the practical use of the concept of involuntary unemployment for public policy. From the above definition, one criterion of good macroeconomic performance would be zero, or very small, involuntary unemployment. (Strictly speaking, this is true only if the measured real wage is equal to the marginal productivity of labour, an equality that might not hold if optimal contracts of the type described below are important in the economy.) Since government unemployment statistics are commonly taken as an indicator of economic performance, one might hope that measured unemployment could be related to the concept of involuntary unemployment. However, this is very difficult and any attempt is bound to be criticized. Government unemployment statistics typically attempt to measure the number of unemployed who are looking for work, but who have not yet found work. However, aside from the problem of determining whether someone is looking for work, or how intensively, unemployment statistics obviously include frictional unemployment and other types of unemployment that would not be included as involuntary according to the above definition.

Even in a condition of relatively full employment, there exists some 'normal' unemployment, which government statistics need to be corrected for. Milton Friedman (1968) used the term 'natural' unemployment for the amount of unemployment that would exist, without excess supply, in equilibrium after wages and prices have adjusted. Another concept of normal unemployment is the non-accelerating inflation rate of unemployment (NAIRU), defined as the amount of unemployment that would exist when there is no tendency for wage or price inflation to rise or fall. Measuring the 'natural' rate or NAIRU in practice entails looking for an unemployment rate for which inflationary pressures are small and adjusting this rate for known changes in the demographic characteristics of the labour force. The natural rate of unemployment is not a constant, however, and these measurements have considerable error. Nevertheless a practical alternative to involuntary unemployment as a measure of economic performance is the difference between the actual unemployment rate and the natural unemployment rate. For policy purposes, this may serve as a reasonably close approximation to involuntary unemployment, but clearly it is a different concept. In particular, note that this measure can be negative, as when the unemployment rate falls below the natural rate in boom times. Fellner (1976) suggested focusing on this measure and hence on inflation stability, rather than on involuntary unemployment, and he argued that demand management (monetary and fiscal policy) should promote the maximum amount of employment that can be achieved without inflation instability. This measure is also the criterion used in stabilization studies that characterize a macroeconomic tradeoff in terms of the fluctuations of unemployment about the natural rate versus the fluctuations in inflation (see Taylor, 1980).

A third reason for criticism of the term involuntary unemployment is that the standard definition is essentially static and deterministic. In fact, the static, deterministic labour supply and demand model does not admit an explicit theory of frictional or natural unemployment. Without such a model it is difficult even to discuss whether a given level of unemployment is voluntary or optimal or not. Research on the microfoundations of unemployment (see for example Phelps et al., 1970), had as a major goal the development a model of equilibrium unemployment – using search and matching theory. Some search models generated unemployment that was Pareto optimal (see Lucas and Prescott, 1974, for example), but others included trading externalities and generated unemployment which could be non-optimal (see Diamond, 1983, for example). While not yet definitive, at the least this research shows that for many public policy questions it is necessary to go beyond the simplest model of labour supply, and thereby beyond the simple definition of involuntary unemployment.

In the *General Theory* Keynes presented a more convoluted definition of involuntary unemployment, and this has been a fourth source of controversy. According to Keynes (1936, p. 15),

> Men are involuntarily unemployed if, in the event of a small rise in the price of wage-goods relatively to the money-wage, both the aggregate supply of labour willing to work for the current money-wage and the aggregate demand for it at that wage would be greater than the existing volume of employment.

One can clearly envisage a point off the labour supply curve from this definition. However, there is much more. Embedded in the definition of involuntary unemployment are some of Keynes other ideas that were part of his *theory* of involuntary

unemployment, but logically distinct from the *definition* of involuntary unemployment. Within the definition it is noted that workers would be willing and able to have a reduction in their real wage (and still increase their work) if it occurred through an increase in the price level, but not if it occurred through a decline in the nominal wage. This 'stickiness' of nominal wages, which is generated as part of the market mechanism, is of course crucial to Keynes's theory. Also embedded in the definition is the assumption that firms are on their labour demand curve, so that a lower real wage would stimulate unemployment, an idea that is much less crucial for Keynes's ideas, as Leijonhufvud (1968) has emphasized. Why did Keynes emphasize this convoluted definition of involuntary unemployment? It seems clear that he wanted to highlight the crucial difference between his theory of unemployment and what he called classical theory. This difference centred on the inability, given the way labour markets and the whole economy interact, of individual workers to reduce unemployment simply by reducing nominal wages. As indicated above, Pigou based the definition of involuntary unemployment on the labour supply curve in much the same way that Keynes did, but the classical reason for its existence – simply that real wages were too high – was much different from the theory of deficient aggregate demand put forth by Keynes. In retrospect Keynes would have added clarity to his discussion by unbundling his theory and his definition of involuntary unemployment.

IMPLICATIONS OF RECENT TECHNICAL RESEARCH FOR THE CONCEPT OF INVOLUNTARY UNEMPLOYMENT. Five research developments during the last 20 years have had great relevance for the concept of involuntary unemployment: equilibrium macroeconomics, optimal contract theory, disequilibrium macroeconomics, efficiency or incentive wage theory, and staggered-wage setting theory. However, this relevance must be inferred from the research, because the term involuntary unemployment is seldom used explicitly, and perhaps avoided by many recent researchers.

Equilibrium macroeconomics. One strand of research macroeconomics has established a strategy of trying to explain the observed fluctuations in unemployment by equilibrium models in which workers are always on their labour supply curves. Wages and prices are perfectly flexible, and all markets clear in these models. Lucas and Rapping (1968) and Kydland and Prescott (1983) represent some of the seminal work in this strand of research. Clearly if these models turn out to be successful and to dominate other models, then the idea of involuntary unemployment would become useless for macroeconomics. Shifts of the labour supply curve – caused by intertemporal substitution of labour supply in response to temporary actual or perceived fluctuations in the real wage – are the main source of employment variability in these models. Research in this area is continuing and branching out into 'real business cycle' theory which ignores monetary factors in the cycle altogether. It appears, however, that a very high labour supply elasticity – by the standards of recent microeconomic empirical research (see MaCurdy, 1981) – is required for these models to be able to explain the observed fluctuations in employment.

Optimal contract theory. Studies by Azariadis (1975), Baily (1974) and others attempted to explain why involuntary unemployment would arise when there exist optimal contracts between firms and workers stipulating for fixed wage payments. However, when firms and workers have equal access to information, these studies have shown that, in the relevant sense, involuntary unemployment does not exist despite the fixed wage bill. In these optimal contract models the marginal rate of substitution of income for leisure is equal to the marginal productivity of labour – the condition for the optimality – in all possible states. Although workers are off their labour supply curve *ex post* (since the real wage is not necessarily equal to the marginal rate of substitution), this discrepancy has no welfare significance. Models in which firms have more information than workers about the nature of the shock can lead to a breakdown in the marginal conditions for optimality, but unless firms are more risk averse than workers the result is involuntary *over*-employment: the marginal productivity of labour is less than the marginal rate of substitution of income for leisure (see Green and Kahn, 1983, and Grossman and Hart, 1983). Viewed as an attempt to explain involuntary unemployment this research, therefore, has been unsuccessful. Taken literally, it shows that much of the unemployment that may have appeared as involuntary is, in fact, voluntary or at least efficient!

Disequilibrium theory. Malinvaud's (1977) careful examination of fix-price multimarket equilibria, following the tradition of Clower (1965) and Barro and Grossman (1971), has greatly helped to clarify the conceptual difference between Keynes's explanation of involuntary unemployment due to insufficient aggregate demand (where firms are constrained in product markets), and the classical unemployment associated with the real wage being too high (where firms are not constrained in product markets). This research also has had considerable policy relevance in the early 1980s because the high rates of unemployment in western Europe were diagnosed as classical rather than Keynesian by many economists.

Efficiency or incentive wages. Calvo (1979) and others have argued that involuntary unemployment can occur because high wages must be paid to give workers the incentive to work hard, to be productive, and not to shirk. As firms attempt to bid up their wages relative to other firms, an equilibrium is reached with all firms paying more than the wage in the absence of incentive effects and with involuntary unemployment: an excess supply of labour with unemployed workers willing to work at the going wage. This type of unemployment is not of the deficient demand type emphasized by Keynes, and given Keynes's willingness to lump other minimum wage unemployment in with frictional unemployment, it is likely that Keynes would have classified this type of unemployment as voluntary. Incentive wages would increase the normal unemployment (natural or NAIRU) rate, but there is little empirical evidence of how quantitatively important the effect is.

Staggered-wage setting theory. In these models (see Taylor, 1980, for example), wages are set with an aim to maintain relative wages unless there is a reason for relative wages to adjust. This relative wage setting leads average nominal wages to adjust with a lag described by a predictable dynamics to changes in demand. In these models prices are set as a mark-up over wages, and for this reason aggregate prices are almost as sticky as nominal wages. Combined with an elementary model of aggregate demand and an aggregate demand policy that does not fully accommodate inflation, these models are designed to be compared directly with the data and in fact lead to fluctuations in unemployment which have features similar to the real world. The unemployment in these models comes close to the usual definition of involuntary unemployment, but since explaining empirical regularities is a primary objective, unemployment enters the model directly as the deviation of unemployment from the natural rate – a more readily measurable quantity than involuntary unemployment. These models show that wage rigidities need not be very long to generate the type of fluctuations in unemployment that characterize the business cycle. Like the equilibrium models discussed above,

and unlike the other three research developments described above, these models are dynamic and can therefore be directly tested against time series data.

Although there has been a tendency for much recent research to avoid the term involuntary unemployment, and instead to define unemployment as appropriate to the theoretical or empirical objectives of the research itself, the term involuntary unemployment will probably continue to be used. Despite the criticism and controversy discussed above there is little harm in this usage, as long as the technical definition is emphasized. Its usage may encourage researchers to point out the connection of new results to past achievements.

JOHN B. TAYLOR

See also AGGREGATE SUPPLY FUNCTIONS; FULL EMPLOYMENT; IMPLICIT CONTRACTS; NATURAL RATE OF UNEMPLOYMENT; SEARCH THEORY; UNEMPLOYMENT; WAGE FLEXIBILITY.

BIBLIOGRAPHY
Azariadis, C. 1975. Implicit contracts and underemployment equilibria. *Journal of Political Economy* 83, 1183–202.
Baily, M.N. 1974. Wages and employment under uncertain demand. *Review of Economic Studies* 41, 37–50.
Barro, R.J. and Grossman, H. 1971. A general disequilibrium model of income and employment. *American Economic Review* 61(1), 82–93.
Calvo, G. 1979. Quasi-Walrasian theories of unemployment. *American Economic Review, Papers and Proceedings* 69, 102–7.
Casson, M. 1983. *Economics of Unemployment.* Oxford: Martin Robertson.
Clower, R. 1965. The Keynesian counter-revolution: a theoretical appraisal. In *The Theory of Interest Rates*, ed. F.H. Hahn and F.P.R. Brechling, London: Macmillan.
Diamond, P. 1982. Aggregate demand management in search equilibrium. *Journal of Political Economy* 90, 881–94.
Feliner, W. 1976. *Towards a Reconstruction of Macroeconomics: Problems of Theory and Policy.* Washington, DC: American Enterprise Institute.
Friedman, M. 1968. The role of monetary policy. *American Economic Review* 58, 1–17.
Green, J. and Kahn, C. 1983. Wage employment contracts. *Quarterly Journal of Economics* 98, Supplement, 173–87.
Grossman, S.J. and Hart, O. 1983. Implicit contracts, moral hazard, and unemployment. *American Economic Review* 71, 301–7.
Keynes, J.M. 1936. *The General Theory of Employment, Interest and Money,* London: Macmillan.
Kydland, F. and Prescott, E.C. 1982. Time to build and aggregate fluctuations. *Econometrica* 50, 1345–70.
Leijonhufvud, A. 1968. *On Keynesian Economics and the Economics of Keynes,* New York: Oxford University Press.
Lucas, R.E., Jr. and Prescot, E.C. 1974. Equilibrium search and unemployment. *Journal of Economic Theory* 7, 188–209.
Lucas, R.E., Jr. and Rapping, L. 1969. Real wages, employment, and inflation. *Journal of Political Economy* 77, 721–54.
MaCurdy, T.E. 1981. An empirical model of labour supply in a life cycle setting. *Journal of Political Economy* 89, 1059–85.
Malinvaud, E. 1977. *The Theory of Unemployment Reconsidered,* Oxford: Basil Blackwell.
Patinkin, D. 1965. *Money Interest and Prices.* 2nd edn, New York: Harper & Row.
Phelps, E.S. et al. (eds) 1970. *Microeconomic Foundations of Employment and Inflation Theory,* New York: W.W. Norton.
Taylor, J.B. 1980. Aggregate dynamics and staggered contracts. *Journal of Political Economy* 88, 1–23.

iron law of wages. The 'iron (or brazen) law of wages' is a term invented by Ferdinand Lassalle (1862) to describe the inexorable tendency of real wages under capitalism to adhere to a level just sufficient to afford the bare necessities of life. This law, he claimed, was not just a socialist indictment of capitalism but was authorized by leading 'bourgeois' economists such as Malthus and Ricardo. He failed to point out, however, that in Malthus and Ricardo the so-called 'subsistence' theory of wages was predicated on a theory of population growth according to which the supply of labour responds automatically to any gap between the going 'market price' and 'natural price' of labour, the latter being defined as a real wage sufficient to reproduce a working population of given size and composition. Lassalle, however, being a socialist, followed Marx in rejecting the Malthusian theory of population; what ensured the 'iron law of wages' for Lassalle, as for Marx, was the tendency for any rise in real wages to generate unemployment, thus setting in motion forces that reversed the rise. This threw the entire weight of argument for equilibrium adjustments in the labour market on the side of employers' demand; it provided no explanation of the supply of labour and thus failed to furnish a determinate theory of wages in long-run equilibrium. Ironically, therefore, there may be an 'iron law of wages' in Malthus and Ricardo, but there is certainly no such iron law in socialist economics. The question whether Malthus and particularly Ricardo can be said to have held the iron law or subsistence theory of wages was a favourite debating question in the latter half of the 19th century (see e.g. Marshall, 1890, pp. 508–9). There is no doubt that they held the view that real wages tend to fluctuate around a natural point of 'gravity', namely, the minimum level of food and other necessities required for existence. But, in the first place, these fluctuations, depending as they did upon decisions to marry and to have children, involved lag of at least 15–18 years, a point which Malthus (but not Ricardo) conceded explicitly. In the second place, the minimum-of-existence level of 'natural wages' was admitted to be a matter of custom and habit and therefore subject to a secular upward drift. It was therefore perfectly possible to argue for the existence of something like a normal long-run supply price of labour – a constant real wage, everything else being the same – while at the same time granting that the 'market price' of labour fluctuated around an ever-rising trend. In short, rising living standards under capitalism do not violate the iron law of wages, understood as a theory about the long-run equilibrium price of labour. But that is only to say that the iron law or subsistence theory of wages amounts for all practical purposes to accepting customary wages as an institutional datum (Schumpeter, 1954, p. 665).

In recent years there has been a revival of the old debate about whether Ricardo held the iron law of wages, but in an entirely new form: did Ricardo hold real wages to be constant at the subsistence level in stationary equilibrium or did he allow for an initial stage of increasing real wages followed by a final stage of declining wages alongside a secular fall in the rate of profit (Hollander, 1983)? It is doubtful whether this question yields one simple, neat answer, since it is clear that Ricardo operated with a number of different models regarding the determination of the 'natural price' of labour. In the very opening paragraph of the chapter on wages in Ricardo's *Principles of Political Economy of Taxation*, the 'natural price' of labour is defined as 'that price which is necessary to enable the labourers, one with another, to subsist and to perpetuate their race, without either increase or diminution'. This defines the natural price of labour to be the commodity wage that ensures a zero rate of population growth. But a page or two later, the natural price of labour is said to be that commodity wage which ensures a rate of growth of population equal to the rate of growth of the capital stock, so that market wages only rise above natural wages when capital accumulates faster than the growth of population. It is possible to make sense of

this in terms of modern growth theory, and many have done so (see Casarosa, 1978), but it is questionable whether Ricardo himself was aware of what he was doing, the more so as he frequently resorts to the constant-subsistence-wage assumption in the later tax chapters of the *Principles*.

MARK BLAUG

See also WAGES-FUND DOCTRINE.

BIBLIOGRAPHY

Casarosa, C. 1978. A new formulation of the Ricardian system. *Oxford Economic Papers* 30(1), March, 38–63.

Hollander, S. 1983. On the interpretation of Ricardian economics: the assumption regarding wages. *American Economic Review* 73(2), May, 314–18.

Lassalle, F.J.G. 1862. *Open Letter to the National Labor Association of Germany.* Trans. J. Ehmann and F. Bader, Cincinnati: Cincinnati Press, 1879.

Marshall, A. 1890. *Principles of Economics.* 9th (Variorum) edn, ed. C.W. Guillebaud, Vol. I, London: Macmillan, 1961.

Schumpeter, J.A. 1954. *History of Economic Analysis.* New York: Oxford University Press.

IS–LM analysis. The original IS–LM model was introduced by Sir John Hicks as a framework for clarifying the relationship between Keynes's theory and that of his predecessors. (In Hicks's famous paper, 'Mr Keynes and the "Classics" ' (1937), however, the now so familiar diagram bore the notation SI–LL.) Further attempts to define Keynes's theoretical contributions precisely within the basic IS–LM structure were made by Alvin H. Hansen (e.g. Hansen, 1953), Franco Modigliani (1944), Lawrence Klein (1947) and Don Patinkin (1948) among others. IS–LM became in this way not only the vehicle for popularizing Keynesian ideas and the mainstay of macroeconomic textbooks but, for several decades, the main organizing conception for macroeconomics in general. Even the very large macroeconometric models of several hundred equations were generally disaggregated IS–LM structures. When Modigliani (1963) surveyed the major developments in macroeconomics in the early 1960s, he did so by presenting an 'updated' IS–LM model. As late as 1971, Milton Friedman and his critics debated the issues between Monetarism and Keynesianism in accordance with IS–LM groundrules (cf. Gordon, 1973). From the late 1970s on, the grip of IS–LM on the macrotheoretical imagination began to loosen as the rational expectations movement came to rely on smallscale general equilibrium and game-theoretic models instead.

The equations of the basic model come in three blocks. The IS block, in the simplest version, consists of a consumption- (or savings-) function, an investment function, and a saving-equals-investment equilibrium condition. Government expenditures and taxation are optional features. The LM block, consists of a money demand (or liquidity preference) equation, a money supply equation, and an equilibrium condition for money. The employment block has an aggregate production function, from which is derived a labour demand function: the unemployment version of the model is usually completed by adding the restriction that the money wage cannot fall below a certain specified value ('rigid wages'); the full employment version has instead a labour supply function and an equilibrium condition for the labour market.

Each of the first two blocks can be reduced to a single relationship between income and the interest rate. The two

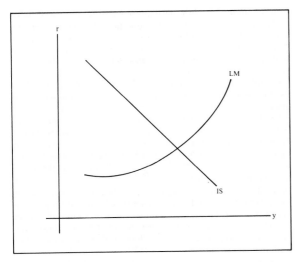

Figure 1 IS–LM

reduced forms in turn produce the familiar diagram shown in Figure 1.

The number of analytical uses of this Hicksian construction remains amazing. Generations of students have learned their macroeconomics by mastering the standard IS–LM exercises: a decline in the marginal efficiency of capital shifts IS leftwards and thus reduces both income and the interest rate; an increase in the money supply shifts LM rightwards and hence raises income while lowering the interest rate, etc. But while Hicks succeeded in compressing a lot of macroeconomics into these two dimensions, such simplification naturally came at a cost. In the later literature, it has often proven difficult to keep the inevitable limitations of the apparatus in clear perspective. Three sets of such problems deserve mention.

The stock–flow dimensional aspects of the model is one problem area and the one with which Hicks himself has been most preoccupied. The IS-schedule is a locus of alternative *flow*-equilibria. But a flow-equilibrium has to be defined over some interval of time and in the case of production this interval has to be fairly long – Hicks (1983) suggests that we think of it as a 'year'. The LM-schedule shows alternative *stock*-equilibria. These can be defined for a point in time. But to insist that realizations stay consistent with expectations so that stock-equilibrium is maintained over an entire 'year', leaves so little uncertainty in the model that the demand for liquidity part of the LM-curve becomes difficult to rationalize. Hicks, therefore, sees a basic tension in the IS–LM construction between the periods appropriate to the two reduced forms.

In still another reappraisal of IS–LM, Hicks (1986) points out that the IS–schedule summarizes the behaviour of the industrial sector and the LM that of the financial sector of the economy. This means that one of the characteristic simplifications of IS–LM analysis is, in effect, to disregard the balance-sheet of the industrial sector and the income-statement of the financial sector. This tends to direct attention away from cases where the two schedules are interdependent as, for instance, when an increase in current production is financed by bank credit so as to increase the money stock. It also makes IS–LM rather unsuitable for the analysis of many balance of payments problems.

A second set of problems has to do with a curious tendency for reliance on IS–LM to end up in a confusion of *nominal* and *real* (particularly real *intertemporal*) maladjustments. Consider, for instance, how the Phillips curve was used at one time to determine the expected price-level/output composition of a change in the level of money income. But a change in money income of some given magnitude can be brought about by either a real disturbance (an IS-shift) or a nominal disturbance (an LM-shift). Why the price/quantity 'trade-off' of a nominal shock should ever have been expected to be the same as for a real shock appears in retrospect as a riddle but the nominal/real distinction was seldom appropriately drawn in the Phillips curve controversy.

The most important instance of this nominal/real confusion, however, occurred much earlier and is embedded in the outcome of the 'Keynes and the Classics' debate. This debate made use of IS–LM by investigating what restrictions on the static form of the model would produce an unemployment solution. Initially, two hypotheses seemed of interest (e.g. Modigliani, 1944). One had the exogenous money supply set too low for nominal income to reach the level required for full employment at the given rigid money wage. the other had the liquidity preference function so specified that it kept the interest rate above the level required for saving and investment to be coordinated at full employment. In the course of the debate, however, this second possibility was eliminated. It was pointed out by Pigou, Patinkin and others that, if it were possible to reduce the money wage rate to an arbitrarily low value, then the so-called 'Pigou effect' would reduce the propensity to save to whatever extent is necessary in order to bring saving and investment into line at full employment no matter what the level of interest rate happens to be. This Pigou-effect argument was taken to dispose of Keynes's intertemporal disequilibrium case. Thus the debate came to the distinctly odd conclusion that Keynes had revolutionized economic theory by asserting the classic platitude that when money wages are too high for equilibrium in the labour market unemployment is the result.

Losing sight of Keynes's intertemporal coordination failure ('saving exceeds investment') has proved costly to the Keynesian tradition. It deprived Keynesians of their natural response to Friedman's hypothesis that the lag in the adjustment of money wages is the only obstacle to employment finding its natural rate. Instead of pointing out that the natural rate of unemployment hypothesis is true only when the interest rate equates saving and investment at full employment, Keynesians tended to reply that money wages are even less flexible than Monetarists would like to believe. But if unemployment is due to money wages being too high in relation to the money supply and if one cannot afford to wait for wages to adjust, inflation will come to seem the normal remedy for unemployment. This tendency is reinforced by the fact that the rationale for the traditional Keynesian fiscal remedies is also lost when removed from the original context of intertemporal coordination failure. If saving exceeds investment (so that there is an excess supply of present resources and an implicit excess demand for future ones) it makes sense for the government to spend now and tax later. If, however, it can be presumed that real interest rates efficiently coordinate intertemporal activities, it may also be presumed that Ricardian equivalence is likely to hold.

A third problem area concerns the precise nature of the 'short-run' for which the IS–LM 'equilibrium' is defined. Comparative statics exercises with IS–LM often produce solution states that obviously represent situations of *incomplete* adjustment. The assumptions that might motivate such incomplete adjustment are, however, often unstated and not always obvious. By varying the information assumptions of the model, it is easy, for instance, to make both schedules shift in response to the standard disturbances. Such interdependence rather undermines the basic modelling strategy which presumes that one reduced form will stay put while the other shifts, so that the IS–LM diagram can be used to generate predictions in the same straightforward way as the Marshallian supply and demand apparatus. Moreover, when the information assumptions are not spelled out, confusion easily arises over whether elasticities or interdependence of the schedules are at issue in a particular controversy (Leijonhufvud, 1983).

These points may be illustrated with reference to the theory of the monetary transmission mechanism. In vintage Keynesian theory, an increase in the money supply would shift LM rightwards while IS stayed put. If the interest elasticity of LM was high and that of IS low, monetary policy was seen to be 'ineffective' in the sense that, 'in the short run', the change in money income would be small. In a rational expectations model, a fully anticipated monetary impulse shifts both IS and LM in parallel fashion. In the short run, money income increases in full, constant-velocity proportion to the money injected into the system and this takes place independently of the interest elasticities of the two schedules. The older, Keynesian version can be rationalized in two distinct ways. Either one maintains that agents do not know of the nominal impulse or otherwise are unable fully to anticipate its eventual consequences; or else one interprets the increase in money, not as a pure nominal shock, but as an expansion of bank credit and inside money in, for instance, a regime with convertible money. IS–LM by itself, of course, will not help to settle the matter.

AXEL LEIJONHUFVUD

See also NEOCLASSICAL SYNTHESIS; NEW 'CLASSICAL' MACROECONOMICS.

BIBLIOGRAPHY

Hansen, A.H. 1953. *A Guide to Keynes.* New York: McGraw Hill.
Hicks, J. 1937. Mr Keynes and the Classics: a suggested interpretation. *Econometrica* 5, April, 147–59.
Hicks, J. 1983. IS–LM: an explanation. In *Modern Macroeconomics,* ed. J.P. Fitoussi, Oxford: Blackwell.
Hicks, J. 1986. Towards a more *General Theory.* Paper delivered at a Monetary Theory Symposium, Taipei, Taiwan, January.
Gordon, R.J. (ed.) 1973. *Milton Friedman's Monetary Framework: A Debate with His Critics.* Chicago: University of Chicago Press.
Klein, L.R. 1947. *The Keynesian Revolution.* New York: Macmillan.
Leijonhufvud, A. 1983. What was the matter with IS–LM? In *Modern Macroeconomics,* ed. J.-P. Fitoussi, Oxford: Blackwell.
Modigliani, F. 1944. Liquidity preference and the theory of interest and money. *Econometrica* 12, January, 45–88.
Modigliani, F. 1963. The monetary mechanism and its interaction with real phenomena. *Review of Economics and Statistics* 45(1), pt. 2, February, 79–107.
Patinkin, D. 1948. Price flexibility and full employment. *American Economic Review* 38, September, 543–64.

Isnard, Achylle Nicolas (1749–1803). French engineer and economist, Isnard was born at Paris on 25 February 1749; he died at Lyons on 25 February 1803. There are no details of his family history except that he had a devoted brother, J.L. Isnard, who was a lawyer and a judge, and who often interceded on his behalf. At the age of 17, Isnard entered the Ecole des Ponts et Chaussées which, even at this early date, inspired interest in political economy and exposed its students

to heavy doses of mathematics and statistics. On successfully completing his studies, Isnard began his career as an apprentice engineer in the district of Besançon. While engaged in various works of construction in these environs, he took the time to write his remarkable two-volume work, *Traité des richesses,* which was published in 1781.

Isnard's *Traité* is a highly original work, despite the fact that its theoretic core is embedded in otherwise unexceptional arguments against physiocratic doctrines. By this fact, we may infer that Isnard knew the physiocratic literature, but we can only speculate on his acquaintance with other writers. Given his background and training, the authors he would have most likely known are Boisguilbert and Vauban (Boisguilbert's ideas were represented in 19th-century course outlines at the Ecole des Ponts et Chaussées, and Vauban's views on the professionalization of engineers were largely responsible for the establishment of the Ecole). Boisguilbert certainly had a vision of an interconnected economy and of a kind of general equilibrium, although he failed to render his conception concrete by erecting any kind of formal, theoretic structure of a mathematical nature.

Isnard, on the other hand, was the first writer to attempt a mathematical definition and a mathematical proof of an economic equilibrium. Furthermore, he gave specific form to the general equilibrium concept by constructing a set of simultaneous equations which, in general form and content, anticipated the major elements of the Walrasian system, including the general interdependence of markets and quantities, the technical specifications of the exchange ratios, and the mathematical determination of the *numéraire.* It remained for Walras to add the engine of utility maximization and to adapt Isnard's model to his own purposes, something which, according to Jaffé (1969), he did with persistence, if not with ease. Isnard's pioneer efforts do not in any way denigrate Walras's monumental achievement, but they do lend force to the conviction that the development of economics was, and remains, a cumulative process.

Isnard's *Traité* is now extremely rare. However, the mathematics of his equilibrium analysis of exchange are partially accessible in Robertson (1949), Baumol and Goldfeld (1968), Jaffé (1969), and Theocharis (1983). The significance of Isnard's performance is that he discovered early on the truth that value is not an intrinsic thing but rather is a magnitude which necessarily varies in relation to other goods, whose worth is also interdependent. Specifically, Isnard anticipated the two-good world of Walras in which, for example, the demand for eggs is the supply of wheat and the demand for wheat is the supply of eggs. This elaboration of commodity interdependencies in real terms consumes approximately the first half of Walras's *Eléments.* Mathematically, Isnard treated value as an exchange ratio, moreover, and he worked out the equilibrium process of exchange both with and without money.

It is noteworthy that Isnard extended the subjects under his analytical purview to include, besides the theory of exchange, the theories of production, capital, interest, and foreign exchange. Jaffé (1969) has demonstrated that Walras's economic theory bears the imprint of Isnard in each of these areas. Underscoring merely the most striking example of the calibre of Isnard's analysis, Jaffé (1969, p. 40) emphasized his theory of capital and interest, which correctly laid down the rule for optimum resource allocation in the following terms:

> Capitals are distributed among different employments in agriculture, industry, and commerce in such a way that the ratios of their values to receipts from the sale of their

products less the costs of upkeep, repair, and replacement – that is, the ratios of [invested] funds to [net] returns – are everywhere the same in all enterprises. This uniformity is achieved and equilibrium established because funds flow to and abound in places where the yield [*intérêt*] is highest and because like things have one and the same value. When things have a higher price in one place than in another, they rush there and equilibrium is re-established. Let F be the value of the funds employed in agriculture and F′ that of the funds employed in industry; let B be the payments for the value of the products of agriculture less the cost of upkeep, repairs and replacement and B′ the payments for the products of industry less the same costs, then the ratio of F to F′ must be equal to the ratio of B to B′ for the ratio of F to B to be equal to the ratio of F′ to B′ or for the rate of interest [in the sense of rate of capitalization] to be everywhere the same. This uniformity [in the rate of capitalization] is realized not only between agriculture and industry in general, but also among individual enterprises.

It is, of course, necessary that perfect knowledge obtain for this conclusion to hold, but even without always making his assumptions explicit, Isnard anticipated much of modern microeconomic theory.

The scope and sweep of his analysis unquestionably entitle Isnard to a position of prominence in the history of economic thought. Yet appropriate recognition took a long time. Despite the filiation of ideas between Isnard and Walras, the 'father' of general equilibrium analysis mentioned Isnard's name in only one place, and that an obscure bibliographic article (a French reprint of Jevons's famous bibliography of mathematico-economic works) published in the *Journal des Economistes* in 1878. Add to this the ambiguity, idiosyncrasy and prolixity of Isnard's treatise. His definition and use of mathematical symbols is inconsistent and the essence of his arguments difficult to extract, nested as they are in a morass of other material that is neither very original nor very interesting. Such deficiencies were bound to handicap the recognition and acceptance of Isnard's contribution. In the final analysis, however, Isnard was simply a brilliant pioneer who wrote ahead of his time, and like so many other semi-tragic heroes of economic analysis (e.g. Cournot and Gossen), he failed to receive his due until long after departing the scene.

Isnard suffered in his personal life even as his ideas suffered (by neglect) in economics. Hot-tempered, yet not given to the intrigues apparently required to advance in the engineering ranks of a quasi-military public service, Isnard spent most of his career in a subordinate capacity. After he finally received a post worthy of his talents, his wife died, leaving him to raise three motherless children. At that point Isnard left government service and struggled in penury for some time. Recalled by Napoleon for the Egyptian Campaign in 1798, he was inexplicably left behind. Adding insult to injury, he was forced to take an oath of allegiance to the Republic even though he was an avowed royalist. He later became a member of the Tribunate under Napoleon and took an active part in the formation of public finance and conscription policies. But upon completing his term he resumed his engineer's career at Lyons, where he died soon after, 54 years to the day from his birth. Given his apparent influence on Walras, Theocharis (1983, p. 62) probably did not exaggerate much when he labelled Isnard's *Traité* 'one of the most important contributions in the history of the development of mathematical economics'.

R.F. HÉBERT

SELECTED WORKS

1781. *Traité des richesses*. 2 vols, London and Lausanne: F. Grasset.
1801. *Considérations théoriques sur les caisses d'amortissement de dette publique*. Paris: Duprat.

BIBLIOGRAPHY

Baumol, W.J., and Goldfeld, S.M. (eds) 1968. *Precursors in Mathematical Economics: An Anthology*. London: London School of Economics and Political Science.
Jaffé, W. 1969. A.N. Isnard, progenitor of the Walrasian general equilibrium model. *History of Political Economy* 1, 19–43.
Robertson, R.M. 1949. Mathematical economics before Cournot. *Journal of Political Economy* 57, December, 523–36.
Theocharis, R.D. 1983. *Early Developments in Mathematical Economics*. 2nd edn, London: Macmillan.

J

Jaffé, William (1898–1980).   Historian of economic thought whose important contributions were to the study of the work of Léon Walras, Jaffé was born in Brooklyn on 16 June 1898 and died in Toronto on i7 August 1980. He graduated from City College of New York with an AB degree in classics and English (1918), from Columbia University with an MA in history (1919), and from the University of Paris with a *Docteur en droit* in economics and political science (1924). He taught economics at Northwestern University (1928–66), and at York University in Ontario (1970–80). Jaffé translated Walras's *Eléments d'économie politique pure* into English (Walras, 1954), thereby providing a major stimulus to the study of his work; edited and exhaustively annotated Walras's scientific correspondence and related papers (Jaffé, 1965), thereby furnishing an encyclopedic storehouse of information about his writings; and wrote many essays on Walras's economic ideas (Walker, 1983). Jaffé believed that, even in its scientific aspects, a writer's work reveals the influence of his normative views and intellectual environment, and that to understand his work fully it is therefore necessary to study his biography and the era of which he was a part (Jaffé, 1965). He applied this thesis to the study of Walras's work, examining the aspects of his biography that had a bearing on his theories, explaining the antecedents of his scientific ideas and the philosophical sources of his normative conceptions, and interpreting and assessing his theories of demand, exchange, production, capital formation, money, *tâtonnement*, and general economic equilibrium.

In an extreme change of opinion, Jaffé came to believe, in the last seven years of his life, that Walras's theory of general equilibrium was intentionally a normative scheme, and that his theory of *tâtonnement* was intentionally a normative exercise in static analysis (Jaffé, 1980, 1981). It would be a disservice to Jaffé and a denial of his scholarship not to recognize that his soundest judgements on Walras were made during the first 43 years of his study of Walras' work, when he regarded Walras's economic theories as positive in intent and character and the theory of *tâtonnement* that Walras espoused during most of his career as an attempt to describe the general features of the dynamic adjustment of the market system toward equilibrium (Jaffé, 1967).

Donald A. Walker

SELECTED WORKS

1942. Léon Walras' theory of capital accumulation. In *Studies in Mathematical Economics and Econometrics*, ed. O. Lange et al., Chicago: University of Chicago Press, 37–49.
1965. Biography and economic analysis. *Western Economic Journal*, Summer.
1965. (ed.) *Correspondence of Léon Walras and Related Papers*. 3 vols, Amsterdam: North-Holland.
1967. Walras' theory of *tâtonnement*: a critique of recent interpretations. *Journal of Political Economy* 75(1), February, 1–19.
1972. Léon Walras's role in the 'marginal revolution' of the 1870s. *History of Political Economy*, Fall, 379–405.
1980. Walras's economics as others see it. *Journal of Economic Literature* 18(2), June, 528–49.
1981. Another look at Léon Walras's theory of *tâtonnement*. *History of Political Economy* 13(2), Summer, 313–36.

BIBLIOGRAPHY

Walker, D.A. 1981. William Jaffé, historian of economic thought, 1898–1980. *American Economic Review* 71(5), December, 1012–19. (This contains a complete bibliography of Jaffé's published writings).
Walker, D.A. 1983. William Jaffé, *Officier de Liaison intellectuel*. In *Research in the History of Economic Thought and Methodology*, ed. W.J. Samuels, Greenwich, Conn.: JAI Press.
Walker, D.A. (ed.) 1983. *William Jaffé's Essays on Walras*. New York: Cambridge University Press.
Walras, L. 1954. *Elements of Pure Economics*. Trans. and annotated by W. Jaffé, Homewood, Ill.: Richard D. Irwin.

Jaszi, George (born 1915).   Research administrator and expert in national accounts. Born in Budapest, Jaszi attended Oberlin College, the London School of Economics (BSc, 1936) and Harvard (PhD, 1946). He was employed by the Bureau of Economic analysis (or its predecessors) in its National Income Division 1942–59 (Division Chief 1949–59), as Assistant Director 1959–62, and as Director 1963–85.

Jaszi helped develop the US national income and product accounts that were introduced gradually during World War II and fully in 1947. The accounts for the government sector were his unique contribution. One aspect of this is that all government purchases, like other purchases not for resale, are counted as final products. For four decades Jaszi influenced the United Nations standardized system in addition to guiding the United States accounts. His firm grasp of the national income and product accounts as an integrated system and of the principle that they must rest upon quantifiable concepts made him particularly skilful in explaining and vindicating the 1947 system and its subsequent improvement. Jaszi's (1958) exposition of the accounts and responses to critics at a 1955 conference and his (1971) critique of comments by 46 economists were masterful. Elsewhere (1964), Jaszi showed that hedonic and conventional methods of allowing for quality change in output measurement are conceptually equivalent.

During 1963–85 Jaszi directed all the varied statistics and analyses of the Bureau of Economic Analysis – international, national, and regional. Many improvements were introduced during his tenure. He closely supervised BEA's *Survey of Current Business* and co-authored its 'Business Situation' section. His talks (e.g. Jaszi, 1972) helped balance exaggerated claims of (1) damage introduced into policy formulation by errors of estimate in the NIPA and (2) the possibility of greatly reducing such errors.

E. Denison

See also SOCIAL ACCOUNTING.

SELECTED WORKS

1943. (With M. Gilbert.) National income and national product in 1942. *Survey of Current Business*, March, 10–26.

1947. (With others in National Income Division.) National income and product statistics of the United States 1929–46. *Survey of Current Business*, July, Supplement.

1948. (With M. Gilbert, E. Denison. and C. Schwartz.) Objectives of national income measurement: a reply to Professor Kuznets. *Review of Economics and Statistics*, August, 179–95.

1956. Statistical foundations of gross national product. *Review of Economics and Statistics*, May, 205–14.

1958. The conceptual basis of the accounts: a reexamination. Reply and comments. *Studies in Income and Wealth*, Princeton: Princeton University Press, Vol. 22, 13–127, 140–45, 209–27, 300–22, 363–71, 402, 454–7, 521–35, 579–82.

1964. Comment. *Studies in Income and Wealth*, Vol. 28, 404–9.

1971. Review: an economic accountants' ledger. *Survey of Current Business*, July, Pt II, 183–227.

1972. Taking care of soft figures: reflections on improving the accuracy of the GNP. In University of Michigan Economics Department, *The Economic Outlook for 1972*, Ann Arbor: The Research Seminar in Quantitative Economics, 43–5.

1981. (With C. Carson.) National income and product accounts of the United States: an overview. *Survey of Current Business*, February, 22–34.

1986. An economic accountant's audit. *American Economic Review, Papers and Proceedings*, May, 411–17.

BIBLIOGRAPHY

Duncan, J. and Shelton, W. 1978. *Revolution in United States Government Statistics 1922–76*, Chapter 3. US Department of Commerce, Office of Federal Statistical Policy and Standards, Washington, DC: US Government Printing Office.

Jefferson, Thomas (1743–1826). The author of the Declaration of Independence and third President of the United States took a profound interest in the intellectual currents of his time that extended to economics. He was a friend of leading physiocrats, and energetically promoted the study of political economy in the United States. When in his seventies, the former president spent five hours a day over a period of three months revising the translation from the French of Destutt de Tracy's *Treatise on Political Economy*, which was eventually published in Georgetown, D.C., in 1817, after Jefferson had written more than 20 letters in search of a publisher.

Jefferson himself did not publish a systematic work on economics. His economic ideas can be gleaned from his *Notes on Virginia* (1785), his wide-ranging correspondence, and his activities in the service of the nation. His overriding aim was to perpetuate the rural economy characteristic of his native Virginia. He was deeply suspicious of banks, paper money and public borrowing, which he saw providing opportunities for financial manipulation and chicanery. He rejected government promotion of industrial development by subsidies or tariffs, and instead proposed to rely on imports of manufactures from abroad. His ideas ran counter to those of Alexander Hamilton, with whom he served in President Washington's cabinet. Although he outlived Hamilton by many years and attained the highest office in the land, the further development of the American economy was more in line with Hamiltonian than with Jeffersonian ideas. The force of circumstances and of incipient institutions caused Jefferson himself to assign to government an active role in economic development by imposing restrictions on foreign trade, nearly doubling the territory of the United States by the Louisiana Purchase, and promoting the development of the West with the Lewis and Clark Expedition. While the future course of the American economy was not shaped by Jefferson's ideals, his system of democratic values has served to this day as a principal factor integrating the American political community.

HENRY W. SPIEGEL

BIBLIOGRAPHY

Dorfman, J. 1946. *The Economic Mind in American Civilization 1606–1865*. New York: Viking, Vol. I.

Luttrell, C.B. 1975. Thomas Jefferson on money and banking; disciple of David Hume and forerunner of some modern monetary views. *History of Political Economy* 7(2), 156–73.

Spengler, J.J. 1940. The political economy of Jefferson, Madison and Adams. In *American Studies in Honor of William Kenneth Boyd*, ed. D.K. Jackson, Durham, North Carolina: Duke University Press.

Spiegel, H.W. 1960. *The Rise of American Economic Thought*. Philadelphia: Chilton.

Jenkin, Henry Charles Fleeming (1833–1885). Jenkin was a distinguished engineer whose wide interests and clarity of mind enabled him to make notable contributions to economic analysis.

Jenkin received his early education at Edinburgh Academy, but financial exigencies forced the family to move first to France and then to Italy. Consequently, Jenkin graduated from the University of Genoa in 1850. Returning to England in 1851, he spent ten years with various engineering firms working on the design and laying of submarine cables. In 1859 he became associated with William Thomson (Lord Kelvin) and frequently collaborated with him in later years, especially in contributing to the work of the British Association's Committee on Electrical Standards. In 1866 Jenkin was appointed Professor of Engineering at University College, London, and moved to a similar chair at the University of Edinburgh in 1868. Apart from his work in civil and electrical engineering, Jenkin distinguished himself as a critic of Darwin's theory of evolution, as an advocate of improved urban sanitation, and for the development of the system of monorail electric transport called telpherage.

Between 1868 and 1872 Jenkin published three economic papers whose theoretical quality and practical value have since earned him a deserved place in the history of economic thought. Recognizing that in current debates on trade unions 'the principles of political economy though often quoted are little understood', Jenkin set himself in his first paper to examine their application to the labour market. In the process he revealed the emptiness of the wages-fund concept, refuted the view that trade unions could not materially benefit their members and made the first clear statement in English economic writing of the concept of supply and demand as functions of price. These ideas he further developed and generalized in his 1870 paper, in which he analysed fully the determination of market price using diagrams to present the supply and demand functions in the form of intersecting curves. Jenkin specifically noted that in the long run cost of production chiefly determines the price of manufactured goods, but stressed 'how much the value of all things depends on simple mental phenomena, and not on laws having mere quantity of materials for their subject' (1887, II, p. 93).

In a third paper Jenkin applied his techniques of supply and demand analysis to the problem of tax incidence, stating the concept of consumers' surplus previously developed by Dupuit but apparently without knowledge of Dupuit's work.

Jenkin left two further essays on economic issues, which were published posthumously in his collected *Papers, Literary, Scientific, &c.* In 'Is one Man's Gain another Man's Loss?' (1884) he used a simple form of closed circuit diagram to illustrate the exchange process and its results. 'The Time-Labour System' contained an acute diagnosis of the differences between goods-markets and labour-markets with a proposal to improve the operation of the latter through what was in effect a system of guaranteed annual wages.

All Jenkin's economic writings were characterized by a striking combination of precise and lucid analysis with tolerant understanding of the facts of daily life in both the workshops and the counting-houses of the world he knew. In view of this their influence in his own time was surprisingly limited, although his 'Graphic Representation' (1870) does seem to have afforded the stimulus which led W.S. Jevons to publish his *Theory of Political Economy* in 1871.

R.D. COLLISON BLACK

SELECTED WORKS

1868. Trade-unions: how far legitimate? *North British Review*, March.
1870. The graphic representation of the laws of supply and demand, and their application to labour. In *Recess Studies*, ed. Alexander Grant, Edinburgh.
1872. On the principles which regulate the incidence of taxes. *Proceedings of the Royal Society of Edinburgh 1871–2*.
1879–81. The time-labour system. (Unpublished ms.)
1884. Is one man's gain another man's loss? (Unpublished ms.)

All published in *Papers, Literary, Scientific, &c*, ed. S.C. Colvin and J.A. Ewing, London: Longmans, Green & Co., 2 vols, 1887. Reprinted as No. 9 in *London School of Economics Series of Reprints of Scarce Tracts in Economics and Political Science*, 1931.

BIBLIOGRAPHY

Brownlie, A.D. and Lloyd Prichard, M.F. 1963. Professor Fleeming Jenkin, 1833–1885, pioneer in engineering and political economy. *Oxford Economic Papers*, NS 15, 204–16.

Jennings, Richard (1814–1891). Educated at Eton and Trinity College, Cambridge; called to the Bar 1838; afterwards Deputy Lieutenant and High Sheriff of Carmarthenshire.

Jennings was the author of two works (listed below) which are notable as early attempts to relate the study of psychology and physiology to political economy. In his *Natural Elements*, Jennings defined political economy as investigating 'the relations of human nature and exchangeable objects'. Consumption he defined as concerned with the contemplated effect of external objects upon man, and Production with the contemplated effect of man upon external objects. Jennings's treatment of the sensations attending consumption led Jevons to describe him as 'the writer who appears to me to have most clearly appreciated the nature and importance of the law of utility'. Hence it is as an early utility theorist that Jennings has been remembered, but his political economy had other interesting features. He forecast the use of mathematical methods, without himself employing them. In policy matters he was a sharp critic of laissez-faire and advocated the establishment of a Board of Public Economy, which might exercise control over the economy by adjustments of taxation and the rate of interest. Jennings's proposals for tax reform included a discriminatory tax to encourage women's employment and a provision for a 'considerable share' of the property of proprietors dying without close relatives to revert to the state. His work was characterized more by intriguing insights than consistently novel theorizing and was handicapped by a prolixity exceptional even by Victorian standards.

R.D. COLLISON BLACK

SELECTED WORKS

1855. *Natural Elements of Political Economy*. London: Longman, Brown, Green and Longmans.
1856. *Social Delusions concerning Wealth and Want*. London: Longman, Brown, Green and Longmans.

Jevons, William Stanley (1835–1882). Jevons was the ninth child of Thomas Jevons, a Liverpool iron merchant, and Mary Ann, daughter of William Roscoe, a noted banker, historian and art collector of the same city. The family were Unitarians and Stanley's background was thus that of a cultured and well-to-do Nonconformist family; but his childhood was shadowed by the death of his mother in 1845, the mental illness of his eldest brother, which began in 1847; and the failure of the family business in 1848.

Jevons's schooling, begun at the Mechanics Institute High School in Liverpool, was continued at University College School, London, and in 1851 he entered University College London itself to study chemistry and mathematics.

At this stage Jevons apparently intended to enter a business career without completing his degree but when a post as assayer to the newly established Mint in Sydney, Australia, was offered to him in 1853 he decided to take it, encouraged by his father, whose finances had never been restored after the family bankruptcy in 1848.

Jevons spent the years from 1854 to 1859 in Australia, applying his knowledge of chemistry at the Sydney Mint and studying, mainly botany and meteorology, in his spare time. From 1857 onwards his interest turned towards social and economic questions; he began to see his life-work as lying in 'the study of Man' and decided that this involved returning to England to improve his academic qualifications. Arriving home in September 1859 he re-enrolled at University College London, completing his BA in 1860 and then the MA course in 1862.

Jevons's attempts to make a career as a journalist in London met with little success and he followed up a suggestion made by his cousin, Henry Enfield Roscoe, who was already Professor of Chemistry at Owens College, Manchester, that he should apply for a vacancy there as a general tutor. When Jevons took up this very junior post in 1863 he was quite unknown even in academic circles, but he had already produced two works which were to prove of seminal importance in economics – his 'Brief Account of a General Mathematical Theory of Political Economy', first read before the British Association in 1862, and his outstanding applied research into changes in the value of gold. Within the same year, 1863, his first published contribution to the study of logic (*Pure Logic, or the Logic of Quality apart from Quantity*) also appeared, and during the next ten years he produced a series of works which established his standing as one of the leading thinkers of his time in both political economy and logic.

This was a period of rapid development in symbolic logic, primarily influenced by the work of George Boole and Augustus De Morgan. Following the lead which they had given, Jevons developed a system of what he called 'combinational' logic, which incorporated yet significantly simplified and improved upon the algebraic approach and notation used by Boole and De Morgan in attempting to set out the laws of reasoning. This system was essentially a logic of terms or classes, based upon a principle to which Jevons gave the name of 'the substitution of similars' – 'when we examine carefully enough the way in which we reason, it will be found in every case to consist in putting one thing or term in place of another to which we know it to have an exact resemblance in some respect' (*Primer of Logic*, 1876, p. 75). Thus Jevons conceived that substitution in logic could be carried out in exactly the same way as in algebra, and brought this out by expressing propositions in the form of equations.

Preoccupied with the development of these ideas, Jevons devoted less attention to other aspects of the subject, such as the logic of relations and quantification theory. Consequently after the contributions of Frege, Russell and Whitehead, Jevons's work had less influence on later logicians. On the other hand his was a type of logic which lent itself particularly to the mechanical performance of inference and Jevons devoted much time and energy between 1863 and 1874 to the working out of this system and to incorporating its principles into a 'logical machine' on which the processes of reasoning could be performed. This machine, which he demonstrated before the Royal Society in 1870, has been recognized as one of the ancestors of the modern computer.

In the years from 1867 to 1872 Jevons set out the whole of his ideas on formal logic and scientific method in a major treatise, the *Principles of Science* (1874). This was the first treatise of its kind to appear in England since John Stuart Mill's *System of Logic* (1843), which Jevons had come to regard as 'an extraordinary tissue of self-contradictions'. Jevons often made clear his resentment of Mill's authority, and this resulted chiefly from a low opinion of Mill's logic. Although Jevons unlike Mill gave no specific treatment of the methodology of the social sciences in his *Principles of Science*, the views which he set forth in it on the relation of mathematics and logic (which he regarded as the more fundamental of the two disciplines), on induction and deduction and on methods of measurement undoubtedly informed his economic work, both theoretical and applied.

That work was far from being neglected in the years when Jevons was developing his logical system. His book on *The Coal Question* (1865) had brought him a national reputation before his appointment as Cobden Professor of Logic, Mental and Moral Philosophy at Owens College in 1866, and he produced a significant series of papers on currency and coinage before returning to the central issues of value theory in his *Theory of Political Economy* (1871).

Jevons's achievements in logic and political economy were recognized by his election to Fellowship of the Royal Society in 1872. Yet combining this intense research programme with heavy teaching duties placed Jevons under great strain, which led to his leaving Manchester in 1876 to take up an appointment as Professor of Political Economy at University College, London. There his duties were considerably lighter, but in 1880 he gave up even this post in order to devote his energies entirely to research and writing. While on holiday with his family at Bulverhythe near Hastings in the summer of 1882, Jevons collapsed while swimming and was drowned.

Jevons's entire academic career thus spanned a period of just twenty years, but that twenty years witnessed the transition from classical political economy to neoclassical economics in Britain, and the fact that the discipline had in that time become a substantially different one was due in no small measure to the work of Jevons.

The fact that Jevons's best-known book, *The Theory of Political Economy* (1871) was first published in the same year as Menger's *Grundsätze der Volkswirtschaftslehre* and only three years before Walras's *Eléments de'économie politique pure* has led to his being bracketed with them as one of the founding fathers of what has come to be called 'The Marginal Revolution'.

Like all such generalizations, this contains an element of truth and an element of distortion. The element of truth is that in the *Theory of Political Economy* Jevons did begin from the point that 'value depends entirely upon utility', a viewpoint which was almost the polar opposite of that taken by many of the leading classical authors, and did emphasize the distinction between total utility and what he termed the 'final degree of utility' (du/dx) – 'that function upon which the theory of Economics will be found to turn' ([1871], 1970, p. 111). The element of distortion arises mainly from the fact that simply to regard Jevons as a pioneer of marginal analysis involves neglecting the relationship between the *Theory of Political Economy* and the rest of Jevons's economic work, but also partly from accepting Jevons's own assessment of the novelty of his approach too much at its face value – for his break with the past was not so sharp as he represented it to be.

Jevons's contributions to economics came at a time when there was much debate as to the appropriate method and future form of the science. Jevons's views on these points were clear and farsighted. He had no doubt that the deductive method which had served the classical authors so well did not need to be replaced, but rather reformed; and to a large extent that reform would consist in the explicit use of mathematical techniques. In the light of his work on logic and scientific method it was clear to Jevons that Economics belonged to the class of sciences 'which, besides being logical are also mathematical' – '... our science must be mathematical, simply because it deals with quantities' ([1871], 1970, p. 78). As to the form of the science, he was equally clear – 'it will no longer be possible to treat political economy as if it were a single undivided and indivisible science' (Jevons [1876], 1905, p. 197). Nevertheless, in whatever ways the subject might become divided and specialized, it would be pervaded by certain general principles. In the *Theory of Political Economy* Jevons set himself to the investigation of these principles – 'to the tracing out of the mechanics of self-interest and utility' ([1871], 1970, p. 50). But he also made substantial contributions to other divisions of the subject – especially in the study of movements of price level, the trade cycle, the economics of exhaustible resources and in questions of economic policy and the role of the state. The work of Jevons thus significantly affected both the foundations of economic theory and the superstructure of applied economics.

In the following sections, for convenience of reference, each of these will here be reviewed separately, but that should not be allowed to obscure the fact that Jevons himself recognized and understood the relationships between theory and measurement in economics more fully and more deeply than most of his contemporaries. He stressed that to say that economics must be a mathematical science did not imply that it must be an exact science, with accurately measurable data. Mathematical theories could be developed in advance of measurements; but the data were there to be measured – and Jevons set an outstanding example in both the painstaking collection and the inspired interpretation of masses of statistical information. He not only recognized the importance for theory of measuring economic quantities, but was among the first to bring out the significance for theory of the concept of the dimensions of such quantities, including notably the time dimension in many aspects of consumption, production and investment.

As to Jevons's relationship to his predecessors and contemporaries in the economic field, it is certainly true that, in the first edition of the *Theory of Political Economy* especially, he presented his work in such a way as to emphasize his break with the past, particularly with the political economy of Ricardo and J.S. Mill, and that he was disappointed by the comparatively slow recognition which his new ideas gained among his contemporaries. Nevertheless, he did not fail to realize and point out that there were elements of continuity as well as change in his work. He devoted considerable effort to research into the origins of mathematical economics and when he came to write the preface to the second edition of the *Theory of Political Economy* (1879) he was generous in conceding that many of his ideas had been anticipated by Cournot, Dupuit, von Thünen and Gossen. Jevons was equally receptive to, and appreciative of, the work of his contemporaries, notably Walras, and did much to introduce the work of continental European economists to English audiences.

I. Jevons himself asked readers of the *Theory of Political Economy* 'to bear in mind that this book was never put forward as containing a systematic view of economics. It treats only of the theory and is but an elementary sketch of elementary principles.' Jevons was not attempting the kind of task Mill or Marshall undertook in writing their *Principles*; nor was he even presenting the sort of outline of price determination in goods and factor markets which has since become the familiar form of neoclassical microeconomics. What he did attempt was a solution of what he saw as the problem of economics:

> Given, a certain population, with various needs and powers of production, in possession of certain lands and other sources of material: required, the mode of employing their labour which will maximise the utility of the produce (p. 254).

Jevons's solution to this problem was worked out in terms of Bentham's utilitarian philosophy with the aid of the differential calculus.

His opening chapter on 'The Theory of Pleasure and Pain' was avowedly based on Bentham's *Introduction to the Principles of Morals and Legislation* and developed the idea of treating pleasure and pain 'as positive and negative quantities are treated in algebra', the objective of each individual being 'to maximize the resulting sum in the direction of pleasure' (p. 97). Jevons moved on to set out the Theory of Utility, introducing first what he called 'the Law of the Variation of Utility' and distinguishing carefully between total utility and the degree of utility, 'the differential coefficient of u considered as a function of x' (p. 110).

Since it is by exchange that the commodities which provide utility are mainly obtained in suitable quantities and at appropriate times, Jevons's next step was to construct a Theory of Exchange, and this provides the keystone of the structure of the book. In it Jevons sets out the conditions under which two 'trading bodies' (a term devised by Jevons to cover individuals or groups), possessing and dealing in two commodities, will maximize their utilities, showing that in equilibrium final degrees of utility must be proportional to prices for each.

The trading bodies are here assumed to start dealing with fixed stocks of the two commodities; but stocks can be altered by production. Production in turn involves labour – 'painful exertion'. Hence the Theory of Labour which Jevons introduces after the Theory of Exchange is in effect the correlative of the Theory of Utility. Pleasure or utility results from the consumption of commodities, pain or disutility is involved in their production. In his discussion in this chapter of the 'Balance between Need and Labour' Jevons anticipated much that has figured in more recent discussions of the choice between income and leisure. But his Theory of Labour is also a theory of cost of production in disutility terms; and when Jevons came to integrate it with this theory of exchange he was able to show that, in equilibrium, prices would be proportional not only to final degrees of utility but also to costs of production (pp. 203–5).

Jevons added two other chapters before concluding his short book – one on the Theory of Rent, and one on the Theory of Capital – in which interest is explained in terms of the increment of produce derived from an increment of investment. The amount of investment of capital Jevons defined as a quantity of two dimensions – quantity of capital and the length of time for which it remains invested. In other words, interest is explained in terms of the marginal productivity of roundabout methods of production.

This has led many commentators to interpret these three concluding chapters as presenting a typical tripartite theory of factor prices, and hence to criticize Jevons for failing to see that the marginal productivity type of analysis which he developed to explain the return on capital could have been generalized to apply to rent and wages also. This is a misinterpretation with the benefit of hindsight. In fact the chapters on Rent and on Capital do not stand on the same footing as those which precede them; they are added to explain the role in the production of want-satisfying goods and services of non-human agents – which cannot experience the pleasures and pains dealt with in the earlier parts of the book.

There are other aspects of the *Theory of Political Economy* which are difficult to understand if it is approached from the standpoint of later neoclassical price theory – and attention has frequently been drawn to these. Among them are Jevons's failure to make use of demand and supply curves, the absence of any treatment of the entrepreneur or anything like a theory of the firm. The reasons for all these things become clear if the book is read as being exactly what Jevons said it was – an attempt 'to treat economy as a calculus of pleasure and pain' (p. 44). The result is as far removed from Marshall as it is from Ricardo, neither a classical nor a neoclassical text but a transitional work.

II. The most familiar dichotomy of the classical and neoclassical periods was that between value theory and monetary theory. In value theory, Jevons's position as a major innovator has long been recognized; in monetary theory it has been almost equally well recognized that his position was conventional, even conservative. Again there is an element of truth in this, but it must not be taken to imply that Jevons made no contribution to monetary analysis generally.

Like most of his contemporaries, and indeed successors before Keynes, Jevons did not question the broad validity of Say's Law and had no use for any of the current versions of under-consumptionism. Equally he accepted the Quantity Theory, supported the Bank Act of 1844 and was a convinced gold monometallist (Laidler, 1982, pp. 327 and 348). Yet monetary questions were a constant preoccupation of Jevons, and in some of the work which he did on them his remarkable qualities as an economic researcher were displayed to their best advantage. Jevons's theoretical and applied work here overlapped fruitfully; he displayed outstanding ability to work with large masses of factual data combined with equal capacity to frame novel hypotheses for interpreting them.

Statistical data in general, and time series in particular, had always a fascination for Jevons. With the instinct of a natural scientist he looked for rhythms and patterns in them, and one quality of time series which seems particularly to have intrigued him from his student days onwards was periodicity. In support of this assertion it may be pointed out that Jevons submitted two papers to Section F of the British Association in 1862. One was his 'Brief Account of a General Mathematical Theory of Political Economy', then ignored but now renowned as the first statement of his new ideas in the area of value theory; the other was entitled 'On the Study of Periodic Commercial Fluctuations'.

In this paper Jevons contended that 'every kind of periodic fluctuation ... must be detected and exhibited, not only as a subject of study in itself, but because we must ascertain and eliminate such periodic variations before we can correctly exhibit those which are irregular or non-periodic' (Jevons [1862], 1884, p. 4).

These ideas seem to have developed out of the work which Jevons began as early as 1860, when he 'undertook to form a Statistical atlas of say 30 plates exhibiting all the chief materials of historical stat[istics]' (Jevons [1861], 1972, Vol. I, p. 180) of which eventually only two diagrams were published, for reasons of cost. These diagrams seem to have brought home to Jevons the reality of a fall in the value of gold since the Australian and Californian discoveries of 1849, and he determined to investigate this long-run change in the price level and to disentangle it from other factors affecting that price level. The outcome was *A Serious Fall in the Value of Gold* (1863) – a piece of work which established new standards of sophistication for applied economic research and has earned the admiration of every commentator from then until now. As the most recent of them has written, 'Jevons had virtually to invent the concept of a price index for himself, as he also did a means of comparing changes in its value over time, a task which he accomplished with astonishing competence and originality' (Laidler, 1982, p. 334).

In fact, in this pamphlet Jevons covered virtually every aspect and implication of a long-term change in the price level with lucidity and accuracy. His final estimate of a fall in the value of gold between 9 and 15 per cent agreed closely with the results arrived at by J.E. Cairnes by more traditional *a priori* methods. It virtually settled a question about which there had been much uncertainty and controversy before he wrote and established his reputation as an economist firmly among his peers.

On the other hand, the outcome of Jevons's study of price level changes and their effects was certainly to support the prevailing monetary orthodoxy, and almost all of his later work in this field had the same tendency. The essence of a sound monetary system in his view was a currency convertible into gold and regulated by the state. Provision of the currency must, he thought, be a government monopoly, because 'if coining were left free, those who sold light coins at reduced prices would drive the best trade' (Jevons, 1875b, p. 54). Although he recognized that a well-managed inconvertible paper currency was theoretically possible, in practice he saw very real risks in leaving such a discretion to any government; nor did he see any advantages in a bimetallic standard or silver monometallism as against simple gold convertibility. In one respect only did his study of the variability of the value of gold lead him to propose a monetary novelty. When he came, in 1875, to set out his ideas for 'an ideally perfect System of Currency' he argued that 'gold must be employed only as the common denominator and temporary standard of value, in terms of which all prices will be expressed', being replaced as a permanent standard by a bundle of commodities. In other words this was to be a system of indexed contracts based on a 'tabular standard'. Earlier, in 1868, Jevons had drawn attention to the possibility of creating internationally acceptable money between countries on the gold standard by establishing uniform weights for their standard coins. Both these proposals involved what were to Jevons simple technical improvements in the operation of the existing monetary system which were easily attainable, but he tended to underestimate the political difficulties involved in bringing such changes into effect.

In *A Serious Fall*, as has already been noted, Jevons had urged that 'we must discriminate Permanent from Temporary Fluctuations of Prices' ([1863], 1884, p. 24). Among the causes of temporary fluctuations he had listed 'Variations of Permanent Investment', pointing out

that great commercial fluctuations, completing their course in some ten years, diversify the progress of trade, is familiar to all who attend to mercantile matters. The remote cause of these commercial tides has not been so well ascertained. It seems to lie in the varying proportion which the capital devoted to permanent and remote investment bears to that which is but temporarily invested soon to reproduce itself (pp. 27–8).

This passage has been noted, by Keynes amongst other commentators, as a remarkably prescient indication of the causes of the trade cycle. Yet when Jevons came to return to the question of the causes of 'commercial tides' some twelve years later he did not follow up his earlier idea. In his *Principles of Science* (1874) Jevons had written a section on 'Periodic Variations' in which he referred to a principle stated by the astronomer, Sir John Herschel:

the meaning of the proposition is that the effect of a periodic cause will be periodic, and will recur at intervals equal to those of the cause. Accordingly when we find two phenomena which do proceed, time after time, through changes of the same period, there is much probability that they are connected (Jevons [1874], 1958, p. 451).

It seems to have been about this time that Jevons began to frame the hypothesis that there might be a connection between the periodic fluctuations in sunspot activity which many 19th-century astronomers had observed and the periodic fluctuations in economic activity which he and other economists had noted.

Initially in a paper presented to the British Association in 1875 Jevons suggested that

the success of the harvest in any year certainly depends upon the weather, especially that of the summer and autumn months. Now, if this weather depends in any degree upon the solar period, it follows that the harvest and the price of grain will depend more or less upon the solar period, and will go through periodic fluctuations in periods of time equal to those of the sunspots (Jevons [1875a], 1884, pp. 194–5).

In support of this idea Jevons produced statistics of fluctuations of grain prices taken from Thorold Rogers's *History of Agriculture and Prices* and endeavoured to relate them to a solar period of 11.11 years, but 'did not venture to assert positively that the average fluctuations ... are solely due to variations of solar power' (p. 203). Indeed he withdrew this paper from publication on discovering that the same data would give other periods of variation equally well, but remained convinced that at the least 'the subject deserves further investigation' (ibid.).

Jevons continued that investigation in subsequent years, seeking confirmation of his hypothesis. In 1876 and 1877 he became aware of fresh evidence supporting the view of which he was becoming increasingly convinced – that the period in question was a decennial one. The astronomer J.A. Broun produced data which suggested that the sunspot period might be 10.45 rather than 11.11 years, while Sir William Hunter, the government statistician in India, published evidence of a decennial period in Indian famines. This led Jevons to suggest another possible linkage between the solar period and economic activity – 'it is the cheapness of food in India, which to a great extent governs the export trade from England to

India', hence high prices in India would be associated with low activity in Britain, and Jevons found evidence that 'the coincidence of commercial crises in Western Europe with high corn prices at Delhi is almost perfect'. Although 'fully alive to the weight of some of the difficulties and objections which have been brought forward against the theory', Jevons's last word on the subject was that 'the theory of the solar-commercial cycle and of the partially oriental origin of decennial crises has received such confirmation as time yet admits of' (Jevons [1882], 1981, Vol. VII, p. 111).

Nevertheless, Jevons's contemporaries remained for the most part unconvinced; and the general verdict of subsequent commentators has been that in this instance Jevons's normally sound instincts for handling and questioning statistical data failed him and that he allowed himself to be 'seduced by apparent regularities and associations, associations that would not survive the scrutiny of a less intoxicated eye' (Stigler, 1982, p. 364). On the other hand, it should be remembered that at the time when Jevons was writing, the whole study of business cycles was in its infancy. Even to recognize the existence of periodic cycles in economic indicators was itself an achievement; the hypothesis which Jevons put forward was no more far fetched than many others that were to be offered and, as Professor Laidler has recently stressed, the doctrine

> rested not just on some perhaps farfetched evidence of correlation, but on acute observation of the role of investment and credit market fluctuations in imparting an apparently decennial rhythm to the pace of business activity and to the accompanying time path of prices in Britain, and on a well articulated account of the link between British markets and an external source of disturbance in the shape of the Indian harvest which also seemed to fluctuate with a decennial rhythm (Laidler, 1982, p. 345).

III. The possibility of Britain's industrial supremacy being threatened by the exhaustion of her coal supplies had been raised in the House of Commons during debates on the Anglo-French trade treaty of 1860. Vague statements had been subsequently made about the size of British coal reserves but no firmly based forecasts of the demand for and supply of coal had been undertaken. In the summer of 1864 Jevons, casting about for a subject which might appeal to a wider audience than he had been able to reach with *A Serious Fall in the Value of Gold*, decided that 'the coming question' was that of coal. Working on the available statistical sources with his customary intensity he produced in a remarkably short space of time the book which first made his name nationally known, *The Coal Question* (1865).

In this Jevons's thesis was not that Britain's coal reserves would soon be exhausted but rather that the rapid growth of her population and industry in the 19th century had produced a *rate* of increase (of around 3.5 per cent per annum) in coal consumption which if maintained for long must compel the extension of mining to poorer or deeper seams, thereby greatly raising the cost of coal. To the extent that Britain's industrial position and progress depended on cheap coal it was consequently bound to be adversely and seriously affected within perhaps half a century.

As to the possibilities of escaping from this difficulty, Jevons was pessimistic. He was sceptical of the prospects of finding substitutes for coal, and with the benefit of hindsight it is evident that he seriously underestimated them, particularly in the cases of petroleum and natural gas. Likewise he discounted the possibility of diminishing the consumption of coal through more efficient use. Improved fuel economy by cutting costs

would, he argued, simply stimulate industrial expansion and so ultimately lead to increased demand for coal. Any attempt to restrict the export of coal, by taxes or prohibitions, would violate the principles of free trade to which Britain was committed and lay the country 'open to the imputation of perfidy' (Jevons [1865], 1906, p. 447). The only positive suggestion he offered 'towards compensating posterity for our present lavish use of cheap coal' was to reduce or pay off the National Debt, 'which would serve the three purposes of adding to the productive capital of the country, of slightly checking our present too rapid progress, and of lessening the future difficulties of the country' ([1865], 1906, p. 448).

This suggestion was in fact taken up by no less a person than Gladstone, then Chancellor of the Exchequer, who in his 1866 Budget speech referred to the prospects of coal exhaustion as among the grounds for his plans to reduce public debt. John Stuart Mill also supported Jevons's arguments and proposals. For a time Jevons had, as he wrote in his private journal, 'enough of newspaper fame' and what the newspapers referred to as 'The Coal Panic' led to the appointment of a Royal Commission on Coal. This did not report until 1871 and it did not really face up to the questions which Jevons had raised, but the estimates of total coal reserves which it produced were reassuring and the public forgot its fears.

Current concern with problems of energy reserves and costs inevitably gives a fresh interest to Jevons's concern with the coal question and there is an obvious temptation to treat him as a 'precursor' of modern ideas. In this connection it must be recognized that Jevons was not concerned with problems of the exhaustion of world energy reserves, or even British energy reserves, but merely with the effect which the rapid rate of consumption of cheap accessible coal must have on British economic growth and international competitiveness. Nevertheless, it remains true that he was one of the first economists to deal specifically with some of the problems presented by the exploitation of extractive resources and to identify the difficulties in which they might involve the early industrializing countries.

IV. One of the branches of political economy which Jevons foresaw arising as the discipline became more specialized and subdivided was that concerned with the principles of economic policy, but he never produced a general treatment of this subject. Perhaps his nearest approach to it was his last complete book, *The State in Relation to Labour* (1882), but this dealt with the role of the state only in a strictly delimited area. Nevertheless, when this is taken together with the variety of articles by Jevons which were collected and published after his death under the title of *Methods of Social Reform* (1883) it is possible to form a clear picture of his views on the principles which he considered should guide policy.

It has been shown above that Jevons's economic theory was largely based on Bentham's utilitarian philosophy, and the same is true of his approach to economic policy. Just as he employed the ideas of pleasure and pain in the former, so he employed the 'greatest happiness principle' in the latter. The effect of this was to lead Jevons to question whether

> according to the doctrine here upheld there is really any place at all for rules and general propositions. If a general law may be limited by a particular law, and that again by a further and more limited exception, we shall get down eventually to individual cases. When followed out, this is the outcome of the Benthamist doctrine. Every single act ought to be judged separately as regards the balance of good or evil which it produces (Jevons, 1882, pp. 17–18).

Consequently Jevons's attitude towards questions of public versus private action was always fundamentally undogmatic. Like John Stuart Mill he started from the position that 'the laissez-faire principle properly applied is the wholesome and true one' (Jevons [1876], 1905, p. 203) but found many exceptions to it among the various specific cases to which he devoted attention. Like most of his contemporaries his basic position was that whenever the parties to a bargain were of equal strength and competence and equally well informed the state should provide means to enforce the contract, but not interfere with its terms. Interference would, however, be legitimate when these conditions were not fulfilled, and Jevons therefore approved of state inspection and branding of commodities 'when the individual is not able to exercise proper judgement and supervision on his own behalf'. On similar grounds he endorsed the restriction of hours and conditions of labour, and argued that the Factory Acts should apply to agriculture and retail trade as much as to manufactures.

Following out these principles led Jevons into an attitude towards trade unions which was perhaps more logical than realistic; it was a critical but, he insisted, not a hostile one. He approved of unions acting as friendly societies and pressing for improvements in hours and conditions for their members, but was always opposed to any effort to settle wages by collective bargaining, which he saw as akin to any other combination in restraint of trade. Holding these two views led him to the somewhat naive conclusion that 'when workmen want to lessen their hours of work, they ought not to ask the same wages for the day's work as before' (Jevons, 1875b, p. 64).

In this respect Jevons lacked that lively appreciation of workers' values and attitudes which Fleeming Jenkin possessed. Yet there seems no adequate reason to doubt his own assertion that he was not 'involved in the prejudices of the capitalists' (Jevons [1868], 1883, p. 102). Jevons favoured a system of 'industrial partnership' and criticized capitalists who were unwilling to enter into such profit-sharing schemes with their employees; but he was prepared to go further than this by advocating the development of workers' cooperatives for production purposes (Jevons, 1882, p. 145).

In looking at all these cases Jevons was dealing with private contracts, particularly between employer and employee, and considering the desirability or otherwise of state interference with them. But even in his day economic policy involved more than this, and Jevons at one time or another made known his views as to the desirability or otherwise of the state itself undertaking various types of economic activity. They were typically eclectic: he accepted the case for state provision of many public goods such as external and internal security and law enforcement. Jevons also favoured generous public expenditure on education, museums and the amusements of the people, but deplored public provision of hospitals and dispensaries. The logic of this distinction was that education and cultural pursuits tended to improve 'the character of the people' whereas free medical services would tend to reduce the incentive to save and so would encourage improvidence.

On the question of public versus private enterprise Jevons's views were somewhat coloured by the fact that he tended to assume that public enterprise must be undertaken by something like the central government departments of his day. In the case of natural monopolies, like the postal service, this type of public enterprise could yield considerable economies, and Jevons therefore advocated a state parcel-post, which did not exist in his day. On the other hand 'work is always done more expensively by government' as a result of bureaucratic regulation and lack of competition, and this made a strong

case in favour of private enterprise in most cases. Eventually though, Jevons came back to his basic position here as elsewhere: 'nothing but experience and argument from experience can in most cases determine whether the community will be best served by its collective state action, or by trusting to private self-interest' (Jevons [1867], 1883, p. 278).

Despite his insistence on judging each case on its own merits according to Benthamite rules, a review of all his work on economic policy leaves some clear impressions. One is that while there are occasional traces of Victorian complacency, the sincerity and consistency of Jevons's sympathy with the less fortunate individuals and groups in society and of his desire to improve their position is not in doubt. Another is that Jevons favoured individualism more than collectivism; although his general position here was similar to that of John Stuart Mill, he was less disposed to accept the socialist case than was Mill. Nevertheless, over his lifetime the number of exceptions to the 'wholesome and true' principle of laissez-faire which he did allow, tended to increase as he recognized the changing character of economic life. After the words just quoted Jevons went on to add, 'as it seems to me, while population grows more numerous and dense, while industry becomes more complex and interdependent, as we travel faster and make use of more intense forces, we shall necessarily need more legislative supervision' (Jevons [1876], 1905, p. 204).

In matters of economic policy Jevons thus appears as a believer in incremental rather than radical change, but no defender of the status quo. His values were not those of a radical, whether of the left or of the right, but those of a liberal.

V. Jevons as portrayed here appears not so much a revolutionary as a transitional figure, but the change in economic thought which took place between 1860 and 1882 was profound and his contribution to that change was fundamental. In his comparatively short career Jevons was constantly breaking new ground in both theoretical and applied economics – marking out the trails which have often since become the well-used highways of the discipline. Coming into social science from natural science he brought to the subject an approach different from most of his predecessors, in Britain at least. He displayed a remarkable ability to combine painstaking collection and classification of data with a readiness to strike out daring interpretations of them and follow up new ideas – which sometimes, but very seldom, led him astray. History has identified Jevons with the origins of marginal utility economics, but Jevons was by nature not a one-subject man, and this was only one of his many achievements.

R.D. Collison Black

SELECTED WORKS

1863. *A Serious Fall in the Value of Gold ascertained, and its Social Effects set forth.* London: Edward Stanford. Reprinted in Jevons (1884), 13–118.

1865. *The Coal Question.* London: Macmillan. 3rd edn, revised and edited by A.W. Flux, 1906.

1871. *The Theory of Political Economy.* Pelican Classics edn, ed. R.D. Collison Black, Harmondsworth: Penguin Books, 1970.

1874. *The Principles of Science: A Treatise on Logic and Scientific Method.* London: Macmillan. Reprinted with introduction by Ernest Nagel, New York: Dover Publications, 1958.

1875a. The solar period and the price of corn. First published in Jevons (1884), 194–205.

1875b. *Money and the Mechanism of Exchange.* London: C. Kegan Paul & Co.

1875c. An ideally perfect system of currency. First published in Jevons (1884), 297–302.

1876. The future of political economy. *Fortnightly Review*, November. Reprinted in Jevons (1905), 187–208.

1882. *The State in Relation to Labour*. London: Macmillan.

1883. *Methods of Social Reform and other Papers*. London: Macmillan.

1884. *Investigations in Currency and Finance*. Ed., with an introduction by H.S. Foxwell, London: Macmillan.

1905. *The Principles of Economics: A Fragment of a Treatise on the Industrial Mechanism of Society, and Other Papers*. Preface by Henry Higgs, London: Macmillan.

1972–81. *Papers and Correspondence*, Vols. I–VII. Ed. R.D. Collison Black and R. Könekamp. London: Macmillan for the Royal Economic Society.

BIBLIOGRAPHY

Black, R.D.C. 1981. W.S. Jevons, 1835–82. Ch. 1 in *Pioneers of Modern Economics*, ed. D.P. O'Brien and J.R. Presley, London: Macmillan.

Hutchison, T.W. 1982. The politics and philosophy in Jevons's political economy. *The Manchester School* 50(4), 366–79.

Keynes, J.M. 1936. William Stanley Jevons: a centenary allocution on his life and work as economist and statistician. *Journal of the Royal Statistical Society* 99(3). Reprinted in J.M. Keynes, *Collected Writings*, Vol. X, London: Macmillan for the Royal Economic Society, 1972.

Laidler, D. 1982. Jevons on money. *The Manchester School* 50(4), 326–53.

Mays, W. and Henry, D.P. 1953. Jevons and logic. *Mind* 62, 484–505.

Robbins, L.C. 1936. The place of Jevons in the history of economic thought. *The Manchester School* 7(1), 1–7. Reprinted in *Manchester School* 50(4), (1982), 310–25 and in Robbins, *The Evolution of Modern Economic Theory*, London: Macmillan, 1970.

Stigler, S.M. 1982. Jevons as statistician. *The Manchester School* 50(4), 354–65.

Jevons as an economic theorist. The story goes that Jevons, along with his contemporaries, Menger and Walras, was a Founding Father of 'neoclassical' economics. Whether this claim has much substance could be, and has been, disputed. To mention just two issues: as Jevons candidly acknowledged, leading features of his theory had been anticipated, notably by Dupuit and Gossen (although his ideas were developed independently of theirs); and a marked, distinctively Jevonian imprint on mainstream 'neoclassical' thought is hard to distinguish, masked by the dominant influence of Alfred Marshall.

Credentials aside, the price exacted for Jevons's elevation to the patriarchy has been searching criticism from exponents of later 'neoclassical' economics (and also from adherents of other schools of thought). As well as receiving solemn praise he has been variously reproached, first, for theory he did produce, on account of its contamination by subsequently abandoned doctrines; secondly, for the absence of, or failure to develop, 'essential' 'neoclassical' doctrines; and thirdly, for lapses of internal consistency. Schumpeter's portmanteau judgement has been shared by many:

> his work on economic theory lacks finish. His performance was not up to his vision. Brilliant conceptions and profound insights ... were never properly worked out; they ... look[ed] almost superficial (1954, p. 826).

In what follows we will focus on Jevons's major theoretical work, *The Theory of Political Economy* (first edition 1871, second edition 1879; all references to the posthumous fifth edition of 1957). Our foremost aim is to describe the central content of Jevons's theory, remarking on its main, retrospectively perceived, weaknesses and strengths (this is merely to follow common practice, and does not reflect the author's

historiographical preferences). It will also be suggested that there are grounds for reconsidering certain commonly held views on the relationship between Jevons and his 'classical' predecessors.

1. MATHEMATICS, UTILITARIANISM AND METHODOLOGY. Lionel Robbins (1936) remarked in his classic piece on Jevons, 'If it were only for its apology for the mathematical method ... the *Theory of Political Economy* would still be memorable.' The *Theory* is indeed remarkable for Jevons's tireless advocacy of mathematical forms of reasoning, specifically of 'the fearless consideration of infinitely small quantities' (p. 3), i.e. differential calculus. He recognized, however, that he was not a 'skilful and professional mathematician', adding that when 'mathematicians recognise the subject as one with which they may usefully deal, I shall gladly resign it into their hands' (p. xiv)). But there was a caveat: it 'does not follow ... that to be explicitly mathematical is to ensure the attainment of truth' (p. xxiii). 'Truth' depended on the framing of economics as a 'Calculus of Pleasure and Pain' (p. vi), the problem of economics being 'to *maximise pleasure*' (p. 37, Jevons's italics; cf. p. 23).

This language signalled an acceptance of Jeremy Bentham's hedonistic psychology ('Utilitarianism') which has either been regarded as unfortunate by later 'neoclassicals', who believed themselves well rid of Utilitarianism, or as half-hearted to the point of non-existence: 'the Benthamite approach was thoroughly understood by Jevons and subtly rejected' (Robertson, 1951; cf. Young, 1912). This was probably wishful thinking: as Professor Collison Black has rightly said, 'Bentham's ideas permeated Jevons's *Theory* inescapably' (Black, 1972).

Just how far this was so is seen in the second chapter of the *Theory of Political Economy*, 'Theory of Pleasure and Pain', which considered 'how pleasure and pain can be estimated as magnitudes' (p. 28), the answer distilled from Bentham: a feeling of pleasure or its negative on the same scale, pain, is a function of its intensity, duration, the (un)certainty of its occurrence and its propinquity (proximity in time) or remoteness.

Taking pleasure and pain as 'undoubtedly the ultimate objects of the Calculus of Economics', Jevons tells us in the third chapter that 'it is convenient to transfer our attention ... to the physical objects or actions which are the source to us of pleasures and pains' (p. 37). Those which 'afford pleasure or ward off pain' are 'commodities' and those having the opposite effect are 'discommodities' (p. 58). 'Utility' is the 'abstract quality whereby an object serves our purposes, and becomes entitled to rank as a commodity' (p. 38) and there is an analogous relationship between 'disutility' and 'discommodity' (pp. 57–8).

Utility is not an *intrinsic* quality of commodities but is 'better described as a *circumstance of things* arising out of their relation to man's requirements' (p. 43, Jevons's italics). It 'must be considered as measured by, or ... actually identical with, the addition made to a person's happiness' (p. 45).

Jevons next distinguished between total utility and '*final degree of utility*', defined as 'the degree of utility of the last addition, or the next possible addition of a very small, or infinitely small, quantity to the existing stock' (p. 51). There followed a statement of the 'general law' that '*the degree of utility varies with the quantity of commodity, and ultimately decreases as that quantity increases*' (p. 53, Jevons's italics).

The 'general law' of diminishing final degree of utility was bracketed with, notably, the 'laws' that 'every person will choose the greater apparent good' and 'prolonged labour becomes more and more painful', collectively described as 'simple inductions on which we can proceed to reason deductively with great confidence' (p. 18). We have this confidence because the 'ultimate laws are known to us immediately by intuition, or, at any rate, they are furnished to us ready made by other mental or physical sciences' (ibid.).

Jevons's faith in his 'ultimate laws' was reflected in his endorsement of verificationism, a methodology which required him to find a way of measuring pleasure and pain/utility and disutility. He 'hesitate[d] to say' that they could be measured directly, arguing that '*it is from the quantitative effects of the feelings that we must estimate their comparative amounts*' (p. 11, Jevons's italics): 'quantitative effects' such as buying and selling, labouring and resting, producing and consuming. He also assured his readers that the focus would be on *incremental* variations: 'I never attempt to estimate the whole pleasure gained by purchasing a commodity' (p. 13); and he was just as emphatic that 'there is never, in any single instance, an attempt made to compare the amount of feeling in one mind with that in another', adding that 'I see no means by which such comparison can be accomplished' (p. 14).

Despite this stance, he did not restrict verification to the (incremental) behaviour of individuals. The 'laws of Economics', he claimed, 'will be theoretically true in the case of individuals, and practically true in the case of large aggregates' (pp. 89–90). The thorny problem here is that aggregate behaviour represents the 'quantitative effects' of feelings in different minds and we cannot meaningfully read back from effects to feelings, as verification of the 'ultimate laws' would require, if inter-personal comparisons are prohibited. As we shall see, however, Jevons made *explicit* inter-personal comparisons *and* drew attention to the pit-falls involved in his 'aggregative' approach.

2. THE THEORY OF EXCHANGE. Jevons's theory of exchange, developed in the fourth chapter of the *Theory*, is probably his best known piece of work. Jevons had no doubt of its significance, declaring that without a 'perfect comprehension' of the theory it was 'impossible to have a correct idea of the science of Economics' (p. 75).

The analysis rested on several assumptions. First, that of a 'perfect market' defined as 'two or more persons dealing in two or more commodities' with traders having 'perfect knowledge of the conditions of supply and demand, and the consequent ratio[s] of exchange' (pp. 85–7). Secondly, there are two 'trading bodies'. Jevons explained:

> The trading body may be a single individual in one case; it may be the whole inhabitants of a continent in another; it may be the individuals of a trade diffused through a country in a third. (p. 88).

But whatever its size, the trading body is treated as a single individual, aiming to maximise its utility and subject to an ultimately diminishing final degree of utility as it consumes more of any one commodity. Thirdly, each trading body is initially the sole possessor of a particular stock of commodity: one has beef and the other corn.

Within each class of commodity, individual units are homogeneous and therefore subject to the '*Law of Indifference*, meaning that, when two objects or commodities are subject to no important difference as regards the purpose in view, they will either of them be taken ... with perfect indifference by a purchaser' (p. 92, Jevons's italics). This 'general law of the utmost importance' implies that 'the price of the same commodity must be uniform at any one moment' (ibid.).

Commodities are also infinitely divisible, which is 'approximately true of all ordinary trade, especially international trade between great industrial nations' (p. 120): note the attempted justification of the analysis by an appeal to aggregate behaviour.

Finally, there is an implicit assumption that utility functions are additive: the (final degree of) utility from beef, for example, is uniquely determined by the quantity of beef alone. (The complications introduced by 'equivalent' commodities – Jevons's omnibus term for substitutes and complements – were considered, but only briefly: pp. 134–7.)

Jevons proceeded to a description of post-trade equilibrium. Using his notation:

a = corn initially held by the first trading body;
x = corn traded;
b = beef initially held by the second trading body;
y = beef traded;
$\phi_1(a-x)$ = the final degree of utility of corn remaining to the first trading body after trade;
$\psi_1 y$ = the final degree of utility of beef obtained by the first trading body;
$\psi_2(b-y)$ = the final degree of utility of beef remaining to the second trading body after trade;
$\phi_2 x$ = the final degree of utility of corn obtained by the second trading body.

The equilibrium position is:

$$\frac{\phi_1(a-x)}{\psi_1 y} = \frac{y}{x} = \frac{\phi_2 x}{\psi_2(b-y)} \tag{1}$$

the celebrated 'equation of exchange', described by Allyn Young (1912) as Jevons's 'most substantial contribution to distinctly mathematical analysis'.

Following Blaug (1985, p. 310), the equation may seen more familiar to modern readers if unit prices are introduced (as they were by Jevons at a later stage: the exchange analysis was obviously anterior to price formation). Letting p_c and p_b stand for the unit prices of corn and beef respectively, substitution into the equilibrium condition for the first trading body yields:

$$\frac{\phi_1(a-x)}{\psi_1 y} = \frac{p_c}{p_b} \tag{2}$$

And rearranging:

$$\frac{\phi_1(a-x)}{p_c} = \frac{\psi_1 y}{p_b} \tag{3}$$

Final degrees of utility are proportional to unit prices.

One criticism of this analysis has been that Jevons failed to appreciate the indeterminacy of bilateral exchange. As Edgeworth remarked using his own terminology, the equation of exchange (i.e. (1) above) holds for *all* points along the 'contract curve' and the equilibrium allocation cannot be determined 'in the absence of arbitration' (Edgeworth, 1881, p. 29). As a qualification, Jevons *did* provide a lucid discussion of indeterminacy and the need for arbitration, but only for the case of exchanging a perfectly divisible commodity with one that is indivisible (pp. 123–5).

More serious, some have contended, are the problems raised by 'trading bodies', which become acute when, in Jevon's demonstration of the 'real benefit derived from ... exchange' (p. 142), one body is transmogrified into 'Australia', complete with single, smooth and continuous utility functions for two commodities (p. 144). Given the assumed shapes of these functions, 'Australia' gains total total utility from trade (Jevons had earlier disavowed any attempt to measure total utility) although whether this benefit carries over to real Australians is unclear if inter-personal comparisons of utility are taboo.

A window on Jevons's reasoning is provided by his discussion of the numerical estimation of 'laws of utility'. He would proceed by *first* gathering statistics of the 'quantities of commodities purchased by the whole population at various prices' and *then* estimating the 'variation in the final degree of utility' (pp. 146–7). Taking the statistical 'demand function' as given, the transition to the utility function is made via the assumption of constancy in the final degree of utility for money (the utility from the increment of commodity purchased by, in Jevons's examples, one penny); granted this assumption as a 'first approximation' (p. 147), final degrees of utility will be directly proportional to prices.

Jevons may have thought that this procedure was legitimate because individual utility functions are not *directly* aggregated. But as we noted in section 1, there is still the problem of reading back from 'quantitative effects' to feelings. Ironically, by making an *explicit* inter-personal comparison, Jevons recognised some difficulties himself.

His generalized comparison was that the final degree of utility of money to poor families exceeds that to richer ones (pp. 140–1) and is prone to *change* when items of expenditure vary in price (p. 148). Hence his admission that there is a 'great difficulty' in the way of interpreting aggregate data stemming from 'vast differences in the condition of persons', to which he added that the difficulty is compounded by 'the complicated ways in which one commodity replaces or serves instead of another' (ibid.). On his own terms he had therefore exploded the implicit claim that multi-person trading bodies were meaningful entities.

3. THE THEORY OF LABOUR AND EXCHANGE-WITH-PRODUCTION.
The fifth chapter of the *Theory* is best known for the treatment of labour-supply, considered by Blaug to have been Jevons's 'most important contribution to the main stream of neoclassical economics' (1985, p. 313): a putative contribution which, however, has rather eclipsed an interesting discussion of production.

Taking the labour supply analysis first, Jevons's 'symbolic statement' (pp. 174-7) used the following notation:

t = duration of labour in clock-time;

x = commodity produced by an *independent* labourer (pp. 173, 176);

l = labour, conceived as '*painful exertion of mind or body undergone partly or wholly with a view to future good*' (p. 168, Jevons's italics); it was one of Jevons's 'ultimate laws' that labour eventually becomes increasingly painful as t is extended; and further note his conception of a *change* in l as that resulting from a change in an arbitrarily chosen interval of t, which allowed him to switch between what might be called an 'objective' rate of production (dx/dt) and a 'subjective' one (dx/dl);

u = utility derived by the labourer from the consumption of x or the commodities obtained in exchange for x; from the 'law' of diminishing final degree of utility this was taken to be a diminishing function of x.

With the 'reward to labour' given by $dx/dt \cdot du/dx$, work continues until $(-)dl/dt + dx/dt \cdot du/dx = 0$, described as the '*final equivalence of labour and utility*' (p. 177, Jevons's italics): work ceases when the utility obtained (directly or indirectly) from the commodity produced by an increment of labour equals the pain (disutility) incurred by supplying that labour. The analysis assumes that labour supply is continuously vari-

able and can stop at any moment, which Jevons acknowledged was unlikely when labourers are not self-employed (p. 181).

Having considered the supply of labour to a single production activity, Jevons extended the analysis to the case of a utility maximizing individual capable of producing two commodities, x and y. With $u_i(i = 1, 2)$ denoting the utility obtained from $x(i = 1)$ and $y(i = 2)$, the optimal distribution of labour is given by:

$$\frac{du_1}{dx} \cdot \frac{dx}{dl_1} = \frac{du_2}{dy} \cdot \frac{dy}{dl_2} \tag{4}$$

The 'increments of utility from the several employments' are equal (p. 184).

This brings us to the often neglected demonstration that 'the ratio of exchange of commodities will conform in the long run to the *ratio of productiveness*' (p. 186, Jevons's italics). It was assumed that production is still carried out by a single un-assisted labourer and (implicitly) that time periods of production are uniform. for dx/dl, described as the (subjective) 'rate of production', Jevons substituted the symbol ω; and for du/dx and du/dy he wrote, respectively, ϕx and ψy. Equation (4) then becomes:

$$\phi x \omega_1 = \psi y \omega_2. \tag{5}$$

The individual engages in exchange, gaining x_1 by giving up y_1. The 'equation of production' (5) is therefore modified, becoming $\phi(x + x_1) \cdot \omega_1 = \psi(y - y_1) \cdot \omega_2$, which, rearranged, gives:

$$\frac{\phi(x + x_1)}{\psi(y - y_1)} = \omega_2/\omega_1. \tag{6}$$

But from the equation of exchange [(10 in section 2], the left-hand side of (6) is equal to y_1/x_1; hence ω_2/ω_1; 'we have proved that commodities will exchange in any market in the ratio of the quantities produced by the same quantity of labour' (p. 187).

Jevons had demonstrated that under certain circumstances, a 'pure' labour theory of exchange ratios and his own 'subjective' theory of exchange could be harmonized (cf. Wicksteed). He also argued that under the 'general rule' of joint-production (p. 198) it becomes 'impossible to divide up the labour and say that so much is expended on producing [one joint-product] and so much on [the other(s)]' (p. 200); this showed 'all the more impressively' that it is 'demand and supply' which governs ratios of exchange and not 'ratios of productiveness' (p. 199). (Later work has shown that by solving a set of simultaneous equations it is possible to 'divide up the labour', although some of the imputed labour inputs may turn out to be negative.)

Finally, we must consider the contentious issue of whether Jevons's analysis supports the attribution to him of a 'marginal productivity' treatment of labour's reward. Schumpeter, for one, thought it did (1954, p. 940). But Jevons's analysis applies to a 'free labourer' (p. 176): one who is free, seemingly, from employer, 'capital' and land. Before concluding that he meaningfully applied marginal productivity analysis to labour we must await the introduction of other 'requisites of production'.

4. THE THEORY OF RENT.
The theory of (differential) rent, adopted and translated into calculus in the sixth chapter of the *Theory of Political Economy*, was credited to James Anderson (1777) and had been widely accepted since the time of Malthus and Ricardo. Jevons accepted the 'intensive' case (more intensive working of the same plot of land) and the 'extensive' one (the cultivation of separate plots differing in their yield). Commentators have taken greater notice not of Jevons's

treatment of rent *per se* but of the evidence it provides for a grasp of 'marginal productivity' analysis in relation to labour.

Let us take Jevons's treatment of the intensive case. Output from a plot of land is a function of 'labour' and is subject to eventually diminishing returns. When production stops the recompense is given by the (utility from) the produce obtained by the incremental input. But with all *previous* applications of 'labour' recompensed at the same rate – deduced from the 'law of indifference' – there arises a 'surplus', differential rent, because the previous applications were, by assumption, more productive.

According to the analysis, rent is not an element of cost-of-production, which agrees with the position of Ricardo and others. However, in his preface to the *Theory* (2nd edition) Jevons argued that this was not generally so: if land has alternative uses, the (differential) rent yielded in 'the most profitable employment' must be 'debited against the expenses of ... production' in other employments (p. xlix).

On the 'marginal productivity' issue, there is only a specious application of the analysis to labour: noting Jevons's assumption that 'increments of labour ... are equally assisted by capital' (p. 216), the marginal product of 'labour' is actually the joint marginal product of labour-and-capital; consequently, 'the separate elements of wages and interest become indeterminate' (Stigler, 1941, p. 21). We have yet to encounter a precise (or intentional) account from Jevons of wage determination in a context of multiple 'elements of production'.

5. THE THEORY OF CAPITAL AND INTEREST. Jevons's treatment of capital and interest – in Chapter 7 of the *Theory* – is often described as anticipating later 'Austrian' theory, particularly the variant developed by E. von Böhm-Bawerk. Commentators have therefore tended to praise or criticize according to their attitude towards the latter. However, even the favourably disposed have contended that there are weaknesses in Jevons's account.

'Capital' in its 'free' or 'uninvested' form is the '*aggregate of those commodities which are required for sustaining labourers of any kind or class engaged in work*' (p. 223, Jevons's italics). This is a 'real' conception but for purposes of aggregation Jevons used the 'transitory form of money' (p. 243), i.e. the monetary value of (aggregate) real wages.

Capital allows us to '*expend labour in advance*' (p. 226, Jevons's italics). It is therefore inextricably linked with time, its 'fixedness' depending on the relative time elapsing before the produce of 'supported' labour 'has returned profit, equivalent to the first cost, with interest' (p. 243). (The similarity between these views and Ricardo's was not lost on Jevons: pp. 222, 242.)

Jevons's general message was this:

> *whatever improvements in the supply of commodities* [which reduce the amount of 'painful labour'] *lengthen the average interval between the moment when labour is exerted and its ultimate result or purpose accomplished, such improvements depend upon the use of capital* (pp. 228–9, Jevons's italics).

But note that, for Jevons, 'improvements' always seem to involve a lengthening of the production process: a presumption which has been strongly criticised (see, for example, Stigler, 1941, p. 26).

In Jevons's examples, the cost of things in which capital is fixed is supposed to be repaid over their lifetimes. How is this cost calculated? Here, Jevons distinguished between the *amount of capital invested* (ACI) and the *amount of investment of capital* (AIC): the former is a quantity of free capital and the latter is the product of the ACI and the time for which it

remains invested. Letting t = the total time of investment and w = ACI, Jevons's formula is: AIC = $w \cdot \frac{1}{2} t$ (p. 236): it is this which must be repaid, with interest.

Providing only *simple* interest is involved, $\frac{1}{2}t$ for a given production technique – the average time of investment – only depends on the physical conditions of production and is therefore independent of income distribution; consequently, the simple interest rate can be expressed in terms of the average investment time. These results do not hold with *compound* interest (see, for instance, Steedman, 1972). In fact, Jevons seems to have recognised that his formula for the AIC would not hold with compound interest (p. 236) and that it is compound interst which is relevant (pp. 238–41). But the theoretical implications were not obviously comprehended.

We now consider Jevons's 'general expression for the rate of interest yielded by capital in any employment' (p. 245) which was clearly believed to have global relevance. Jevons assumed that 'the produce for the same amount of labour . . . [varies] as some continuous function of the time elapsing between the expenditure of the labour and the enjoyment of the result' (ibid). With t standing for time and Ft denoting produce at a point in time, the (instantaneous) interest rate is 'determined' by the ratio of an increment of produce $[F(t + \Delta t)]$ to the incremental AIC (which in this case is $\Delta t \cdot Ft$). In the limit, it is given by $F't/Ft$: '*the rate of increase of the produce divided by the whole produce*' (p. 246, Jevons's italics). Here, if nowhere else, there was an explicit and substantive application of marginal productivity analysis (but note that it pertained to the productivity of the ACI, or 'capitalization', and not capital *per se*).

Jevons thought his 'expression' could explain both the 'interest yielded by capital in any employment' and the (presumed) secular decline in the economy-wide interest rate: for each production process and in the aggregate, $F't/Ft$ 'must rapidly approach to zero, unless means can be found of continually maintaining the rate of increase' (p. 246, pp. 253–6). The analysis raises some knotty questions.

If, in the derivation of the 'expression', we credit Jevons with making the tacit assumption that ft and $F't$ are physically homogeneous, has reasoning *for that special case* was logically sound. But if he had in mind an example such as maturing wine, Ft and $F't$ are, *according to his own analysis* (pp. 238–9), value magnitudes, *dependent* on the (compounded) interest rate for their calculation. Whenever we step outside the ('as if') one commodity world and introduce compound interest, Jevons's 'expression'' will fail: by no stretch of the imagination is it 'general'. (In fact, the restrictions have been shown to be even more severe: see Steedman, 1972). When one considers the further difficulties involved in aggregating diverse production processes – not discussed by Jevons – the retrospective conclusion must be that his analysis was more productive of problems than of solutions. Needless to say, any expectation that Jevons *should* have spotted and resolved all these problems would be ludicrous.

6. DISTRIBUTION AND VALUE: FURTHER CONSIDERATIONS. Jevons's views on distribution and value merit further comment: it was above all in these (interrelated) areas that he believed himself to have radically departed from 'classical' teaching: a judgement widely shared.

On wages, he was scathing towards 'wage-fund theory' and 'natural wage doctrine'. The former, interpreted as the proposition that 'the average rate of wages is found by dividing the whole amount appropriated to the payment of wages by the number of those between whom it is divided', is dismissed in the Preface to the *Theory* (first edition) for being 'purely delusive' and 'a mere truism' (p. vi). As for 'natural wage doctrine', he inveighed against Ricardo's 'sweeping simplification' of 'a natural ordinary rate of wages for common labour ... [with] all higher rates ... merely exceptional circumstances, to be explained away on other grounds' (pp. 269–70). But whether he transcended these 'erroneous' views is questionable.

In the concluding chapter of the *Theory*, it emerges that wage-fund theory does have 'a certain limited and truthful application' (p. 268) when new business ventures are undertaken: the 'amount of capital which will be appropriated to the payment of wages ... will depend upon the amount of anticipated profits ... All workmen competent at the moment to be employed will be hired, and high wages paid if necessary' (p. 272).

What of the 'long-run' position? Supposing the new venture is successful, 'those who are first in the field make large profits' inducing other capitalists to enter the fray 'who, in trying to obtain good workmen, will [further] raise the rate of wages' (p. 271). Ultimately, 'only the market rate of interest is obtained for the capital invested ... [and] wages will have been so raised that the workmen reap the whole excess of produce, *unless ... the price of the produce has fallen*' (ibid., italics added). Whether *that* happens depends on the kind of labour involved, because 'the rate of wages ... of every species of labour will reduced to *the average proper to labour of that degree of skill*' (p. 272, italics added); so, if labour is skilled and educated, wages and therefore the 'price of produce' will remain high, whereas 'if only common labour' is required both will fall (p. 271).

Jevons had presented a 'wage-fund' explanation for the (entrepreneurial) 'short-run' and an explanation of 'long-run' wages that is hard to distinguish from 'natural wage' doctrine (especially in its Smithian form, where the 'natural' wage for 'common labour' is merely the current 'centre of gravity' and not necessarily a minimum 'subsistence' wage). The impression that he had done otherwise can probably be traced to ideas presented in his Theories of Labour and Rent which, if coupled with his statement that '*wages are clearly the effect not the cause of the value of the produce*' (p. 1, Jevons's italics; cf. p. 165), may seem to point towards an 'imputational' kind of 'marginal productivity' analysis. However, as shown in sections 3 and 4, the evidence to this effect from his Theories of Labour and Rent is suspect, and regarding the 'causal' claim, the argument documented in our previous paragraph implies an *interdependency* between wages and the 'value of produce'. The latter point is particularly relevant for an appreciation of Jevons's value analysis

Jevons complained about 'the thoroughly ambiguous and unscientific character of the term *value*' (p. 76, Jevons's italics) but after promising to 'discontinue the use of the word' (p. 81) he used it all the same, mainly in the sense of *exchange-value*, i.e. ratio(s) of exchange. He was keen to contrast his own opinion – '*value depends entirely upon utility*' (p. 1, Jevons's italics) with the (Ricardian) doctrine that 'value will be proportional to labour', which, he claimed, 'cannot stand for a moment, being directly opposed to facts' (p. 163). Four arguments were deployed.

First, as Ricardo had admitted, some commodities are not reproducible and so the 'labour theory' cannot apply to them. Secondly, again as Ricardo had allowed, market exchange ratios fluctuate around those given by comparative amounts of labour expended in production. Thirdly, '*labour once spent has no influence on the future value of any article*: it is gone and lost for ever' (p. 164, Jevons's italics). And fourthly, Jevons

objected to Ricardo's 'violent assumption' of homogeneous labour (p. 165).

The first of these arguments is true and need not detain us. The second would equally apply to Jevons's 'equation of exchange' (above, section 2) and is well answered by Jevons himself: 'We shall never have a Science of Economics unless we learn to discern the operation of law even among the most perplexing complications and apparent interruptions' (p. 111). Moving to the fourth argument and supposing that Ricardo *had* assumed homogeneous labour, this does not seem any less 'violent' than Jevons's preferred assumption of a *single labourer* (above, section 3).

That leaves the third argument, which Jevons clouded with his incisive demonstration that long-run values *might* conform to 'ratios of productiveness' (above, section 3). Ironically, the limitation that he chose not to highlight – stemming from joint-production – has recently been shown to create severe difficulties for deterministic labour theories (Steedman, 1977).

As for the claim about value depending 'entirely on utility', savaged by Marshall (1920, pp. 673–5), in the light of the whole *Theory of Political Economy* it did not accurately convey Jevons's position. For example, when a 'pure' labour theory holds, the 'ratio of exchange [value] governs the production as much as the production governs the ratio of exchange' (p. 188); in general it is the 'demand and supply of ... products' which 'rules value' (p. 199); and as we discovered earlier, wages will also influence long-run values.

We are now positioned to integrate Jevons's theory of value and distribution. In the 'short-run' wages are determined by his own version of 'wage-fund theory'; the price of produce is determined by supply and demand; and profit comes out of the residual or 'surplus'. In 'long-run equilibrium', supply of output is so adjusted relative to demand that revenue just covers what we might as well call (*pace* Jevons) natural wages and the natural rate of interest on capital (the uniform interest rate to which individual rates tend pp. 244–5); and if production is land-using, it must also cover (natural) rent as 'determined by the excess of produce in the most profitable employment' (p. xlix). In the 'long-run', then, there is no class of income which can properly be described as a 'surplus'.

8. CONCLUSION: JEVONS AND HIS CLASSICAL PREDECESSORS. Jevons accepted unreservedly both Malthusian-style population theory and the 'impossibility of general gluts' thesis (his strictures on gluts – pp. 202–3 – bear a close resemblance to Ricardo's). There was also a modified acceptance of differential rent theory with an allowance made for rent as an element of cost-of-production (a Smith-Ricardo 'meld'); and at least a 'family resemblance' between Jevons's treatment of capital and Ricardo's. A further point which has emerged is that Jevons was concerned with *aggregate* behaviour: a simplistic 'micro'/'macro' distinction between Jevonian and 'classical' analysis cannot be sustained.

A major contrast, however, is that Jevons's analysis was predominantly *static*, not because he believed 'dynamic' issues were uninteresting, rather that 'it would surely be absurd to attempt the more difficult [dynamic] question when the more easy [static] one is yet so imperfectly in our power' (p. 93). Despite its 'truth and vast importance' (p. 266) population theory was therefore excluded from the (static) 'problem of Economics' which Jevons stated as the allocation of *given* resources so as to '*maximise the utility of produce*' (p. 267, Jevons's italics).

His theory of value and distribution also contrasts with 'classical' analysis, though not for all the clichéd reasons. Jevons's treatment of both short and long-run wages could be described as 'classical' in substance. And in the long-run, *if* commodities sell at 'equilibrium' prices these can be decomposed into wages, interest and rent, all at their average (or 'natural') rates: to that extent, the analysis is reminiscent of Adam Smith's.

Where Jevons unambiguously and substantially departed from 'classical' teaching was with his 'marginal productivity' theory of 'capitalisation' and his application of (Bentham's) utility analysis to *all* aspects of economic behaviour, not just consumption–demand.

TERRY PEACH

BIBLIOGRAPHY

Black, R.D.C. 1972. Jevons, Bentham and De Morgan. *Economica* NS, 39, 119–34.

Blaug, M. 1985. *Economic Theory in Retrospect*. 4th edn, Cambridge: Cambridge University Press.

Edgeworth, F.Y. 1881. *Mathematical Psychics*. London: C. Kegan Paul & Co.

Jevons, W.S. 1957. *The Theory of Political Economy*. London: Macmillan. 5th edn, reprinted, New York: Augustus M. Kelley, 1965.

Marshall, A. 1920. *Principles of Economics*. 8th edn, London: Macmillan, 1949.

Robbins, L. 1936. The place of Jevons in the history of economic thought. *Manchester School of Economic and Social Studies* 7; reprinted in *Manchester School* 50, (1982), 310–25.

Robertson, R.M. 1951. Jevons and his precursors. *Econometrica* 19(3), 229–49.

Schumpeter, J.A. 1954. *History of Economic Analysis*. London: George Allen & Unwin.

Steedman, I. 1972. Jevons's theory of capital and interest. *Manchester School*. 40, 31–52.

Steedman, I. 1977. *Marx after Sraffa*. London: New Left Books.

Stigler, G.J. 1941. *Production and Distribution Theories*. New York: Macmillan.

Young, A.A. 1912. Jevons's *Theory of Political Economy*. *American Economic Review* 2, 566–89.

Jewkes, John (born 1902). Jewkes was educated at Barrow Grammar School and Manchester University. His first job was as Assistant Secretary of the Manchester Chamber of Commerce, 1925–26. He was then appointed Lecturer in Economics at Manchester University, and stayed there for three years. Following a period in the United States, he returned to Manchester as Professor of Social Economics in 1936. After holding this chair for ten years, he was appointed Stanley Jevons Professor of Political Economy at Manchester. In 1948 he became Professor of Economic Organization at Oxford, and a Fellow of Merton College, and held this chair until his retirement in 1969. His professional contacts, however, remained mainly outside Oxford. Jewkes had a distinguished wartime career. He became Director of the Economic Section of the War Cabinet Secretariat in 1941, and was appointed Director-General of Statistics and Programmes at the Ministry of Aircraft Production in 1943. This was followed by other posts, and after his return to university life he was a member of a number of Royal Commissions and other official committees.

Jewkes's Manchester roots, together with his wartime experience, made him a powerful advocate of free-market solutions. His first notable book on this subject was *Ordeal by Planning* (1948), followed by *Public and Private Enterprise* (1965), *New Ordeal by Planning* (1968), and *A Return to Free Market Economics?* (1978). In these works he advocated the virtues of the free market, as opposed to government ownership or government planning, as a fruitful background

for economic efficiency and individual initiative. He argued that government efforts to replace the market had produced one debacle after another, and also that economists claimed too much for their subject, thus reducing their potential usefulness. Before World War II Jewkes's work had concentrated on detailed studies of the economic and social problems of Lancashire – as, for example, in his *Wages and Labour in the Cotton Spinning Industry* (with E.M. Gray), 1935. Some of his work after the war also concentrated on detailed problems, but in a national or international context. For example, he published studies, jointly with his wife Sylvia, on medicine and the National Health Service, arguing that the state-operated National Health Service had displayed many weaknesses. A notable contribution to the literature on innovation was his *The Sources of Invention* (with David Sawers and Richard Stillerman), 1958. This was one of the earliest attempts at systematic investigation in this field. It successfully established the importance of the small-scale inventor, and showed that many notable 19th- and 20th-century inventions were essentially the work of one or two individuals, working with limited resources. This may well prove to be his most lasting contribution.

Z.A. SILBERSTON

SELECTED WORKS

1935. (With E.M. Gray.) *Wages and Labour in the Lancashire Cotton Spinning Industry*. Manchester: Manchester University Press.

1938. (With S. Jewkes.) *The Juvenile Labour Market*. London: Victor Gollancz.

1948. *Ordeal by Planning*. London: Macmillan.

1958. (With D. Sawers and R. Stillerman.) *The Sources of Invention*. London: Macmillan.

1961. (With S. Jewkes.) *The Genesis of the British National Health Service*. Oxford: Basil Blackwell.

1965. *Public and Private Enterprise*. London: Routledge & Kegan Paul.

1968. *The New Ordeal by Planning. The experience of the Forties and Sixties*. Revised edn of *Ordeal by Planning*, London: Macmillan; New York: St Martin's Press.

1978. *A Return to Free Market Economics? Critical essays on government intervention*. London: Macmillan.

Johannsen, Nicolas August Ludwig Jacob (1844–1928). Nicolas Johannsen was a brilliant outsider with insights into theoretical economics that were ahead of his time. He was born in Berlin but spent much of his life in New York, where he was active in the import–export business. He is best known for having anticipated Keynes's saving–investment relationship and the multiplier in *A Neglected Point in Connection with Crises* (1908). In other writings, some of which are quite elusive, having been published under pen names or in German, Johannsen developed a view of the economy in terms of circular flows of money and economic activities portrayed in the form of a wheel-of-wealth diagram (1903). This was not the first attempt of this kind, but was perhaps the first to provide a complete statistical underpinning. Like Silvio Gesell, but independently of him, Johannsen also proposed a tax on paper money, visualized as coming close to a single tax (1913).

As a forerunner of Keynes, Johannsen considered fluctuations in investment the strategic factor in business cycles. Depressions occur as a result of vanishing investment opportunities. If investment declined while saving stayed put, there would be an excess of saving over investment which Johannsen called 'impair saving'. This would go into 'impair investment', that is, purchases of existing assets or grants of loans to persons whose incomes were reduced in consequence of declining normal investment and who desired to maintain their consumption expenditure. Johannsen also drew attention to the fact that adverse effects suffered in one sector of the economy would spread and multiply through others. He estimated the propensity to save at one-seventh of income and the multiplier at five. The two estimates can be reconciled if allowance is made for negative saving, which Johannsen did not do. In the 1920s, Johannsen's concern with diminishing investment opportunities became more pronounced, and in 1926 he predicted that 'a depression seems due within an early year' (1926, p. 2).

Johannsen's views, highly unorthodox as they were at his time, were appreciated only among a handful of his contemporaries. He used pen names to hide his writings from his employers. Keynes referred to him with great condescension in the *Treatise on Money* and did not mention him in the *General Theory*. Johannsen seems to have been an iconoclast by habit. He also dabbled in astrophysics and there too, proffered views that offended orthodox opinion. His work in economics illustrates that important advances are often made by outsiders who do not suffer from the limitations of the expert.

About Johannsen, consult Dorfman (1949, vol. 3, pp. 408–13), the first comprehensive account of Johannsen's life and work, which may be supplemented by Dorfman's introduction to the 1971 reprint of Johannsen's *Neglected Point* as well as by his introductory essay to Clark (1970).

HENRY W. SPIEGEL

SELECTED WORKS

1903. (Under the pseudonym J.J.O. Lahn.) *Der Kreislauf des Geldes und Mechanismus des Sozial-Lebens*. Berlin: Puttkammer & Mühlbrecht.
1908. *A Neglected Point in Connection with Crises*. New York: The Bankers Publishing Co. Reprinted, New York: Kelley, 1971, with an introduction by Joseph Dorfman.
1913. *Die Steuer der Zukunft* Berlin: Puttkammer & Mühlbrecht.
1926. *Two Depression Factors*. Stapleton, NY: The Author.

BIBLIOGRAPHY

Dorfman, J. 1949. *The Economic Mind in American Civilization*. Vol. III, *1865–1918*. New York: Viking.
Dorfman, J. 1970. Some documentary notes on the relations among J.M. Clark, N.A.L.J. Johannsen and J.M. Keynes. In J.M. Clark, *The Costs of the World War to the American People*, reprinted New York: Kelley, 1970.
Patinkin, D. 1981. *Essays On and In the Chicago Tradition*. Durham, NC: Duke University Press.

Johansen, Leif (1930–1982). Johansen was born in Eidsvoll, Norway, on 11 May 1930 and died in Oslo on 29 December 1982. He entered the University of Oslo in 1948, and received the equivalent of a Master's degree in economics (*cand.oecon.*) in 1954. He was awarded a doctors degree (dr. philos) in 1961, for a dissertation with the title 'A Multi-Sectoral Study of Economic Growth'.

In 1951 Johansen became research assistant to Ragnar Frisch. After graduation the university awarded him a research fellowship. In 1958 he received a Rockefeller Fellowship, which he held until in 1959 he was appointed Associate Professor of Public Economics at the University of Oslo. On the retirement of Frisch in 1965, Johansen became Professor of Economics at the University of Oslo, with the special duty of lecturing on macro-economic planning.

Johansen's first important work is his doctoral dissertation, mentioned above. This book (1960) builds a bridge between the theory of economic growth, which had become fashionable in the 1950s, and Leontief's input–output model, which at the

time was widely used in economic planning and forecasting. The choice of dissertation topic was undoubtedly influenced by Johansen's two mentors, Frisch and Trygve Haavelmo, who were then both working in these fields.

In the dissertation Johansen presents a theoretical model, and applies it to Norwegian data. He analyses a 23-sector model of the economy, and it seems that at first this empirical part of the work was considered the more important.

After a few years, however, it became clear that the model, often referred to as the MSG-model, had considerable merits in itself. It became the basis for long-term planning by the Norwegian Ministry of Finance, and over the years it was developed and extended. Johansen took an active part in this work. It seems that the model also influenced planning methods in several countries, and a new and enlarged edition of the book was published in 1974.

The laws of production must play an important part in any growth model, and Johansen continued and extended the pioneering work of Frisch on production functions in a series of articles which had considerable influence. The main results in these papers are brought together and generalized in the book from 1972: *Production Functions: An Integration of Micro and Macro, Short Run and Long Run Aspects*. The subtitle indicates the high aspiration level of the book, and it did present production functions which were realistic and so general that they could be used in multi-sectoral planning models.

Johansen was a member of the Communist Party of Norway until he died, and he participated actively in some election campaigns. However, his political views have hardly left a trace in his professional writing. An uninformed reader will be at loss to divine which political opinions-if any-the author holds. Johansen seems to have written relatively few papers on planning in eastern Europe and on Marxist economic theory, and none after 1966. His objectives seem to be to inform and explain, rather than to convert, and often it seems that these papers are written on request – for instance his paper 'Labour Theory of Value and Marginal Utilities' (1963). This is an extension and clarification of some short comments he made in a discussion the year before. It shows that under certain circumstances the two theories can be reconciled. Johansen served on a number of expert committees appointed by different Norwegian governments, and was accepted as the objective scientist who would point out logical inconsistencies but never let his personal views influence the recommendations he made.

There is however little doubt that Johansen's political opinions had a marked effect on his career. Under the rules in force in the 1950s and 1960s it was impossible for him to obtain a visa to the USA. He therefore did not have the opportunity of spending some of his formative years at an American university. Such opportunities were regularly offered to bright young academics in Western Europe and usually had a profound influence on their later work. Johansen missed this experience, and in fact never visited the USA. He remained a European, and principally a Norwegian. Most of his work was published in Europe, and about half of it was written in Norwegian.

Johansen's political views did not affect his scientific work, but his views did inevitably influence his opinions as to which economic problems were important and which should be studied. His views naturally led him to study economic planning, and this subject remained Johansen's main interest during most of his professional life. His two-volume *Lectures on Macroeconomic Planning* (1977 and 1978) is a landmark. It is essentially a textbook which gives a balanced overview of the major issues in the economics of planning, integrating the results Johansen has reached over 25 years with those of the many others who have contributed to the development of the subject. As often Johansen appears as a master in reconciling different views and approaches. A third volume was in preparation at the time of Johansen's death, and this might have rounded off the work, and removed the many gaps and omissions which reviewers found in the presentation.

Economic planning is closely related to, if not a part of, the subject which has become known as 'public economics'. The subject may not be very well defined, and its contents have certainly changed over the years. The central topics however remain taxation, public expenditure and social welfare. When Johansen began to lecture on the subject at the University of Oslo in 1960, there was no single book which covered this heterogeneous subject. He published his own textbook in Norwegian in 1962. A revised and extended edition appeared in 1965, and was translated into English in the same year as *Public Economics*. The book did not give any clear definition, nor did it define the limits of the subject. Perhaps too tailor-made for his students at the University of Oslo, it deals very briefly with topics covered by other courses in the curriculum. The book did however have an impact, and helped to establish 'public economics' – suitably defined – as a recognized part of economics.

Johansen was one of the founders of the *Journal of Public Economics*, which first appeared in 1972. He served as co-editor from the beginning until his death, and he contributed the opening article of the new journal. Its title, 'On the optimal use of forecasts in economic policy decisions', indicates how broadly Johansen tended to view the subject of public economics.

In his later years Johansen developed a strong interest in game theory. He seems to have been led to this subject by Arrow's proof that non-dictatorial and efficient decisions were impossible, and he wrote a penetrating paper on the subject in 1969. A model of central planning will naturally be compared - for efficiency and fairness-with a model of free competition. The assumptions leading to neoclassical equilibrium are generally considered to be unrealistic, and game theory, with its different solution concepts based on compromises between coalitions, were developed as a generalization of the standard market model. The same idea can be applied to a central planning model. Plans are rarely drawn up and executed by a consistent single-minded dictator. Usually they appear as a compromise between different interest groups (coalitions) in society, or within a bureaucracy.

Johansen's first publication on game theory seems to be a short article in Norwegian with the title 'Plans and Games' from 1970, contributed to a 'Festschrift' with the general title 'Economics and Politics'. Here he shows that if there are several independent decision makers, with different preferences, the collective decision must necessarily be a compromise.

In the following years Johansen published a few papers in Norwegian along similar lines. His first paper on a game theory in English is 'A Calculus Approach to the Theory of the Core of an Exchange Economy', published in 1978. Debreu and Scarf (1963) proved that the core of a market game would, under certain conditions, shrink to the competitive equilibrium, as the number of players increased to infinity. Their proof, as well as the ones given by others, depends heavily on topological or measure theoretical arguments, which make the results inaccessible to most economists of the older generation. Johansen shows that the result can be reached by elementary methods, under the

assumptions conventionally made in neoclassical economic theory. The paper does not appear to be much cited, and its main effect may have been to give Johansen a deeper understanding of the subject.

Game theory is closely related to bargaining theory, and Johansen's next paper on the subject is 'The Bargaining Society and the Inefficiency of Bargaining' (1979). Here he wrote: 'I consider the game theory approach to economic problems to be the most appropriate paradigm as soon as we go beyond mere accounting and description of production technology and want to include various aspects of economic behaviour.' The conclusion of the essentially verbal discussion in the paper is that bargaining is not an efficient way of making social decisions. At the time of writing Johansen did not seem to be aware of the concept of 'bargaining sets' introduced by Aumann and Maschler (1964). The different bargaining sets include the core if it is not empty, and also some subsets corresponding to the cases in which the players fail to agree on a Pareto optimal outcome.

In one of his last papers, 'On the Status of the Nash Type of Noncooperative Equilibrium in Economic Theory' (1982) Johansen argues that the theorem of Nash (1950) has often been misinterpreted and misused in economic literature. The theorem just states that every *n*-person game has at least one equilibrium point in mixed strategies. In this purely mathematical context equilibrium point means what in mechanics is called a 'dead point', where the forces are in equilibrium. There are few reasons to assume that a point with this property should have any economic optimality property. Johansen observes, inter alia, that in the game known as the 'Prisoners' Dilemma', the only equilibrium point is the worst possible of all outcomes.

During his last years Johansen came to look at game theory as a general theory of economic behaviour, which contains as special cases the two extremes: completely centralized decision making and perfect competition. His work during these years showed that Johansen as always was a quick learner, and that at his death at the age of 52, he had gained mastery of the relevant parts of game theory. One can only make guesses about the general theories he might have developed if he had been given a few more years to live.

KARL H. BORCH

SELECTED WORKS
The obituary published by the Norwegian Academy of Science (*Yearbook* for 1983) lists 11 books and 138 articles written by Johannsen, about half of them in Norwegian. Some of the most important in English are:

1958. The role of the banking system in a macro-economic model. *International Economic Papers* 8, 91–110.
1959. Substitution versus fixed production coefficients in the theory of economic growth. *Econometrica* 27, 157–76.
1960. *A Multi-Sectoral Study of Economic Growth*. Amsterdam: North-Holland. 2nd enlarged edn, 1974.
1963. Labour theory of value and marginal utilities. *Economics of Planning* 2, 89–103.
1965. *Public Economics*. Trans. from Norwegian, Amsterdam: North Holland.
1969a. Ragnar Frisch's contributions to economics. *Swedish Journal of Economics* 71, 302–24.
1969b. An examination of the relevance of Kenneth Arrow's General Possibility Theorem for economic planning. *Economics of Planning* 9, 5–41.
1972a. *Production Functions*. Amsterdam: North-Holland.
1972b. On the optimal use of forecasts in economic policy decisions. *Journal of Public Economics* 1, 1–24.
1977. *Lectures on Macroeconomic Planning*. Vol. 1: *General Aspects*. Amsterdam: North-Holland.

1978a. *Lectures on Macroeconomic Planning.* Vol. 2: *Centralization, Decentralization, Planning under Uncertainty.* Amsterdam: North-Holland.

1978b. A calculus approach to the theory of the core of an exchange economy. *American Economic Review* 68, 813–20.

1979. The bargaining society and the inefficiency of bargaining. *Kyklos* 32, 497–522.

1982. On the status of the Nash type of noncooperative equilibrium in economic theory. *Scandinavian Journal of Economics* 84, 421–41.

BIBLIOGRAPHY

Aumann, R. and Maschler, M. 1964. The bargaining set for cooperative games. *Annals of Mathematical Studies* 52, 443–76.

Debreu, G. and Scarf, H. 1963. A limit theorem on the core of an economy. *International Economic Review* 4, 235–46.

Nash, J. 1950. Equilibrium points in n-person games. *Proceedings of the National Academy of Sciences* 36, 48–9.

Solow, R. 1983. Leif Johansen 1930–1982: a memorial. *Scandinavian Journal of Economics* 85, 445–59.

Johnson, Alvin Saunders

Johnson, Alvin Saunders (1874–1971). Alvin Johnson was born on 18 December 1874 near Homer, Nebraska, and died 7 June 1971 in Upper Nyack, New York. He received the BA (1897) and MA (1898) from the University of Nebraska and the PhD from Columbia University (1902). His varied teaching career included Bryn Mawr, Columbia, Nebraska, Texas, Chicago, Stanford (twice), Cornell and the New School for Social Research of which, in 1919, he was a founder and, beginning in 1923, director. He was president of the American Economic Association in 1936 and of the American Association of Adult Education in 1939. He was active in the struggle for academic freedom and other civil rights and in providing a haven, at the New School, for refugee scholars. His students included Walton Hale Hamilton, Frank H. Knight and James Harvey Rogers.

Johnson also had an active and varied editorial career. He was assistant editor of the *Political Science Quarterly*, founder and editor of *Social Research*, associate editor of the *Encyclopedia of the Social Sciences*, economics editor of the *New International Encyclopedia*, political science editor of the American edition of *Nelson's Encyclopedia*, and on the editorial council of the *Yale Review*. He also was a founder and member of the editorial staff of *The New Republic*.

Johnson, who also published novels and short stories, wrote as an economist on a wide range of theoretical and policy problems. He was also the author of a popular and respected principles text which went through several editions. As a student of (and secretary to) John Bates Clark, Johnson adhered to his marginalist approach to economic theory but combined his neoclassicism with social and institutionalist elements. His dissertation on rent theory stressed interproduct competition and tried to develop a non-Marxian conception of exploitation (thirty years prior to Joan Robinson's work). His early economic nationalism encompassed a limited pro-protectionist argument. In various writings he argued that labour saving machinery did not necessarily raise wages; that forward shifting of the corporate income tax requires price to be a function only of cost of production, which he deemed not prevalent; and that arguments against the minimum wage were based on static assumptions. He considered that prevailing theory offered only universalist, formal explanations to problems of price formation, whereas he found that price phenomena were also the product of a multiplicity of complex variables, and called for greater realism and empiricism. Following Clark, Johnson also anticipated Pigovian welfare economics arguing, in effect, that public ownership could be a solution to cases in which, because of nonappropriables,

marginal private benefits fell short of marginal social benefits. For many years he was active in the land reclamation movement.

In general, Johnson was a cautious reformer, advocating reform within the existing social order through the expansion of non-property rights as both a corollary to the security of property itself and a mark of a progressive economy.

WARREN J. SAMUELS

SELECTED WORKS

1903. *Rent in Modern Economic Theory: An Essay in Distribution.* New York: Columbia University Press.

1909. *Introduction to Economics.* Boston: Heath. 2nd edn, 1922.

1952. *Pioneer's Progress: An Autobiography.* New York: Viking.

1954. *Essays in Social Economics.* New York: New School for Social Research.

Johnson, Harry Gordon

Johnson, Harry Gordon (1923–1977). Harry G. Johnson was born in Toronto, Canada on 26 May 1923 and died in Geneva, Switzerland on 9 May 1977. Throughout his professional career he was a recognized leader of the economics profession in the United States, Britain and Canada, though his influence extended worldwide. He wrote prodigiously: 526 professional scientific articles, 41 books and pamphlets and over 150 book reviews. In addition, he edited 27 books and wrote numerous pieces of journalism. His writings are characterized by creative insights and by a unique capacity to synthesize; both clarify apparently untidy and unyielding masses of seemingly unrelated and abstruse contributions. His impact on the economics profession was enhanced by his ceaseless participations in conferences around the world, and by his willingness to lecture even at the smallest campus or institute, both of which he perceived as a professional obligation.

He graduated from the University of Toronto in 1943 and then spent a year at St. Francis Xavier University in Nova Scotia as Acting Professor of Economics (at the age of twenty). After military service in the Canadian Infantry, he proceeded to Cambridge, England, obtaining his BA in 1946. He taught in the following year at the University of Toronto, where he also received his MA, specializing in economic history. He then spent 1947–8 at Harvard, followed by a year at Jesus College, Cambridge and then election to a Berry-Ramsey Fellowship at King's College in 1949. He was to remain a Fellow of King's, teaching also at the London School of Economics, until he left Cambridge for the University of Manchester as Professor of Economic Theory in 1956. In 1959 he joined the University of Chicago as Professor of Economics, later becoming the Charles F. Grey Distinguished Service Professor of Economics, and remained there until his death. He was soon to combine the professorship at Chicago with a Chair at the London School of Economics (1966–74) and then with the Graduate Institute for International Studies in Geneva, Switzerland (1976–7).

These shifts in location and the associated changes in intellectual environment shaped his character as a cosmopolitan economist. The years in Cambridge and in Chicago were to be the most significant. For both campuses had, in addition to Johnson himself, remarkable figures in economic science such as Dennis Robertson, Richard Kahn, Nicholas Kaldor and Joan Robinson in Cambridge, and Milton Friedman, George Stigler and Theodore W. Schultz in Chicago. The strong professional and political views and interests of many of these economists must have deepened Johnson's interest in developing theory as a tool of policy making, and influenced the evolution of his own views and attitudes toward the various approaches to economics.

His writings span the entire range of the economics discipline: from the history of economic doctrines to the economics of the price of gold; from the theory of international commodity agreements to the theory of preferences and consumption; from an analysis of Keynesian economics to the theory of income distribution. They cover, too, the economics of reparations, the theory of productivity, growth and the balance of payments, the theory of tariffs, economic policies for Canada, Britain, the US and developing countries, the theory of excise taxes, the economics of public goods, the economics of common markets, the economics of monetary reform, the theory of inflation, the theory of index numbers, the theory of nationalism, the state of international liquidity, the theory of advertising, the relationship between planning and free enterprise, the theory of the demand for money, the choice between fixed and floating exchange rates, the economics of basic and applied research, the economics of the brain drain, the economics of poverty and opulence, the theory of distortions, the theory of money and economic growth, the theory of effective protection, the theory of human capital, the economics of bank mergers, an analysis of efficiency of monetary management, the economics of the North–South relationship, an analysis of minimum wages, the economics of student protest, an analysis of the infant-industry argument for protection, the economics of the multinational corporation, the economics of universities, the economics of libraries, the economics of international monetary union, the economics of dumping, an analysis of the role of uncertainty, the economics of smuggling, an analysis of income policy, the economics of speculation, an analysis of mercantilism, the economics of bluffing, an analysis of equal pay for men and women, an analysis of monetarism, an analysis of buffer stocks, the economics of patents, licenses and innovations, an analysis of legal and illegal migration, the economics of welfare and reversed international transfers, the monetary approach to the balance of payments, and the monetary approach to the exchange rate.

Four areas of interest and impact were clearly the most important and deserve to be highlighted: (i) the pure theory of international trade, (ii) macroeconomics, (iii) international monetary theory, and (iv) economic policies and issues of political economy.

Johnson's work on trade theory constitutes perhaps his most important scientific contribution. His early work in this area is collected in *International Trade and Economic Growth* (1958). This book contains his important and highly original papers in the theory of trade and growth (1953a, 1954). These articles, written at the time of the dollar shortage after the war, were to address the issues from the viewpoint of differential growth of productivity among trading countries, and were to put the whole theoretical discussion into a form that dominated the work of trade theorists for years.

His writings on the general equilibrium analysis of international trade include two influential companion papers on income distribution (1959b, 1960b). In addition, among his notable contributions are those that belong to what James Meade called the theory of trade and welfare. Four are particularly noteworthy. In chronological order, these are: his classic paper (1953b) on optimum tariffs and retaliation; the cost of protection and the scientific tariff (1960a), building on his earlier work measuring the gains from trade; optimal trade interventions in the presence of domestic distortions (1965a); and the possibility of income losses from economic growth of a small, tariff-distorted economy (1967d).

The paper on optimum tariffs and retaliation addresses the issue of whether a large country which exercises its monopoly power can be made worse off because of foreign retaliatory tariffs. Using a Cournot-type retaliation mechanism, Johnson showed that the country that initially imposes an optimal tariff can wind up better off than under free trade despite foreign tariff retaliation. From the viewpoint of Johnson's evolution as an economist, this paper is notable for two things. First, the early vintage Johnson was intrigued by analytical complexities of the kind that he found much less interesting later. Second, the policy implication of this early vintage analysis was to resurrect the classic case for the exercise of monopoly power by a large country; Johnson's later writings tended to go in the opposite direction, highlighting the great potential cost of departing from truly free trade.

The shift in Johnson's emphasis to the advantages of free trade is seen most directly in his work on the theory of optimal policy intervention in the presence of distortions and in his work on the theory of immiserizing growth. In both instances, Johnson opposed the use of tariffs, utilizing the insights of the theory of second-best as applied to problems of trade and welfare.

Finally, the impact of Johnson's paper on the scientific tariff (1960b) was in two areas: (i) the measurement of the cost of protection and (ii) the analytical propositions regarding optimal tariff structures. Johnson's theoretical contributions influenced empirical work on measuring the cost of protection, and on measuring the gains or losses to Britain from joining the EEC. Many of his contributions to the theory of tariffs and commercial policy are reprinted in his *Aspects of the Theory of Tariffs* (1971a).

Johnson's early contributions to macroeconomics were made during his tenure at Cambridge. In 'Some Cambridge Controversies in Monetary Theory' (1951b) he clarified the essence of the controversy between the Keynesian and the Robertsonian approaches to key issues like loanable funds versus liquidity preference, the savings–investment identity and the Gibson paradox, and he clearly demonstrated his talent for distilling and integrating complex issues into a coherent framework. His major contributions during that period, however, were his study of the implications of secular changes in the UK banks' assets and liabilities consequent on the replacement of private by public debt (1951a) and his active participation in the discussion surrounding the revival of monetary policy in the UK. Johnson was critical of the quality of British monetary statistics and in a series of articles attempted to make the case that improved monetary statistics were essential for well-managed monetary policy. In 'British Monetary Statistics' (1959a) he published his own labouriously constructed monetary aggregates for the period 1930–57, which stimulated further research.

Johnson's move to the University of Chicago (to which he was invited as the 'Keynesian') marked an increased research interest in monetary theory. His major contributions in the early 1960s are 'The *General Theory* After Twenty Five Years' (1961), the survey article 'Monetary Theory and Policy' (1962b) and 'Recent Developments in Monetary Theory' (1963a). These three contributions have since become classics in the field of monetary economics. They established Johnson's reputation as a scholar with a rare breadth of knowledge and with broad scientific and historical perspectives. The survey article is widely acclaimed as a masterpiece in scholarship and its contribution went far beyond surveying the 'state of the art'. Johnson's survey suggested a list of issues that would benefit from further research. In retrospect, this list seems to have served as the research agenda in the subsequent fifteen years. One of the notable issues on the list was his early skepticism on the stability of the Phillips curve in the face of changes in macroeconomic policies. His evaluations of the major developments in monetary economics (as of the early 1960s) have been influential and perceptive. These developments were the application of capital theory to monetary theory and the shift from static analysis to dynamic analysis. These contributions, along with others, are reprinted in *Money, Trade and Economic Growth* (1962c) and *Essays in Monetary Economics* (1967b) that also include his important contributions to the topic of money and economic growth.

As a result of his interest in the Keynesian revolution and his deep historical perspective, Johnson wrote his controversial article 'The Keynesian Revolution and the Monetarist Counter-Revolution' (1971b) which was first presented as the Richard T. Ely Lecture in 1970 and was reprinted in his *Further Essays in Monetary Economics* (1972a). This article is an exercise in the history of economic thought and scientific evolution. His interest in the various aspects of Keynes and his economic thought resulted in a series of provocative articles, some of which appeared posthumously in his joint book with his wife Elizabeth Johnson, *The Shadow of Keynes* (1978).

Johnson's major criticism of the Keynesian model was its failure to deal with the problem of inflation at the levels of both economic theory and economic policy. He was critical of the 'sociological' non-economic theories of inflation, as well as of price controls and incomes policy as remedies for inflation. His analysis of inflation was approached from the perspective of an international economist who views inflation (under a fixed exchange rate regime) as a global phenomenon, a proper analysis of which requires a shift of focus from the concept of monetary developments in individual countries to the concept of the aggregate world money supply. Johnson's view of world inflation is best exemplified in his *Inflation and the Monetarist Controversy* (1972b) which was delivered as the De Vries Lecture in 1971.

Throughout his professional life, Johnson continued his research on international monetary economics. Three articles in 1950 set the stage for what later on became the typical characteristics of his style of research: courage to take positions not always popular with others, the application of relatively simple economic techniques to a new range of problems with resultant important insights, and a passion for geometry as a tool of analysis. He took an early stand against raising the price of gold in terms of all other currencies (1950a), analysed the destabilizing effect of international commodity agreements on the prices of primary products (1950b) and produced an early diagrammatic analysis of income variations and the balance of payments (1950c) – an analysis which was conducted within the then typical Keynesian framework, a framework which he later criticized.

In his writings on the theory of the transfer problem, originally developed in the context of the postwar reparations, Johnson extended earlier work by P.A. Samuelson, L.A. Metzler, F. Machlup and J.E. Meade and demonstrated the potential provided by his philosophy that individual research effort is most productive when it utilizes the work of previous theorists as a foundation for new construction. Johnson's theme was that of 'continuity and multiplicity of effort'. In 'The Transfer Problem and Exchange Stability' (1956) he demonstrated that the problems of transfers and of exchange stability are formally the same and that all the possible methods of correcting balance of payments disequilibrium can be posed in terms of the analytical apparatus of the transfer problem. Almost two decades later (1974) he returned to the analysis of transfers with greater emphasis on the monetary aspects of the problem.

Johnson's most important contribution to the understanding of international monetary economics is 'Towards a General Theory of the Balance of Payments' printed in his *International Trade and Economic Growth* (1958). His insight was the emphasis on the monetary nature of a balance of payments surplus or deficit. '[A] balance-of-payments deficit implies *either* dishoarding by residents, *or* credit creation by the monetary authorities'; the former is inherently transitory and the latter is policy induced. As for policy, Johnson coined the distinction between 'expenditure reducing' policies and 'expenditure switching' policies. The insights contained in this important article are all the more remarkable considering the intellectual environment in the mid-1950s where to a large extent, the balance of payments was viewed as a 'real' (in contrast with 'monetary') phenomenon. This article may be viewed as the intellectual precursor of what would be termed fifteen years later 'the monetary approach to the balance of payments'.

Over the years, Johnson focused increasingly on policy issues with special reference to Canada (1962a; 1963b; 1965c). He supported the move to a flexible exchange rate regime (1969) but recognized, relying on the theory of optimum currency areas, that there are circumstances under which a small country (like Panama) might be better off maintaining a fixed parity.

His analysis of the international monetary system revealed his strength as a realistic political scientist. Monetary reform is not carried out in a vacuum. It is performed by representatives of independent nation states, to whom international commitments are likely to be secondary to national commitments. This view is reflected in his numerous commentaries on international monetary crises, in his doubts about the prospects of a stable European monetary union, in his appraisal of the Bretton Woods system and in his perceptive article 'Political Economy Aspects of International Monetary Reform' (1972d). He took a hard line on schemes designed to solve the international monetary problems by methods that channel resources to the less developed countries. He was aware that such a stance might be unpopular but his professional integrity determined his position; in his words, 'My reason for refusing to endorse such schemes is not that I am opposed to the less developed countries receiving more development assistance but I think that no useful purpose is served by misapplying economic analysis for political ends' (1967a, p. 8).

As world inflation accelerated in the 1960s Johnson recognized that in a world integrated through international trade in goods and assets, national rates of inflation cannot be fully analysed without a global perspective:

> I have become increasingly impressed in recent years with the conviction that the traditional division between closed-economy and open-economy monetary theory is a barrier to clear thought, and that domestic monetary phenomena for most of the countries with which economists are concerned can only be understood in an international monetary context (1972a, p. 11).

This perception of world inflation along with the analytical insights from his earlier work 'Towards a General Theory of the Balance of Payments' (printed in 1958) paved the way to his work on the monetary approach to the balance of payments which he viewed as the crowning achievement of his career. The intellectual roots of the monetary approach go back to the classic writers (David Hume and David Ricardo) and its early developments can be found in the work of economists associated with the International Monetary Fund (e.g. Jacques Polak). Johnson, however, along with Robert

A. Mundell and other members of the International Economics Workshop at the University of Chicago, introduced new and significant dimensions to the approach. Noting that the balance of payments is essentially a monetary phenomenon, he concluded that balance of payments policies will not produce an inflow of international reserves unless they increase the quantity of money demanded or unless domestic credit policy forces the resident population to acquire the extra money wanted through the balance of payments *via* an excess of receipts over payments. He saw himself as a missionary; and he was to take the lead in developing and disseminating the approach by encouraging and at times guiding the theoretical and empirical research in this field in various centres such as Chicago, London and Geneva. He coedited some of the results in *The Monetary Approach to the Balance of Payments* (Frenkel and Johnson, 1976). The evolution of the international monetary system into a regime of flexible exchange rates led to further extensions of the monetary approach and resulted in a new direction of theoretical and empirical research on the economics of exchange rates. Johnson stimulated much of the early research in the area and coedited *The Economics of Exchange Rates* (Frenkel and Johnson, 1978) which contains some of the resulting work.

In addition to his theoretical contributions, Johnson wrote profusely also on policy matters. His *Economic Policies Toward Less Developed Countries* (1967a) analyses proposals such as commodity schemes and preferential entry for manufactured exports of the less developed countries. Similarly, his work on the brain drain (1964, 1967c) propounded the view that the brain drain might be welfare-improving for the countries from which it occurred. This is one example of how, in his later years, his analyses increasingly questioned interventionist policies. Thus, the brain drain was beneficial rather than harmful; the multinational corporations were part of a non-zero-sum game and so on. The UNCTAD (United Nations Conference on Trade and Development), which addresses the less developed countries' problems and demands, and which he had looked on rather benignly in the early 1960s, came under his criticism in several writings as he came to feel that professional economists had allowed themselves to be influenced by their sympathies for the poor countries to the point of being led into empathetic and nonscientific research on trade and development.

It is impossible to conclude the brief survey of Johnson's prolific research without highlighting three other important aspects of his contribution. First, he was a humane social scientist who was interested in understanding social phenomena, in contributing to the improvement of welfare, and in understanding the development of knowledge and technological advances. These qualities are particularly evident in his *On Economics and Society* (1975a) and in *Technology and Economic Interdependence* (1975b). Second, he was a gifted teacher with a deep sense of mission and responsibility. He devoted great effort to the preparation of his lectures and always undertook an extremely heavy teaching load. Some of his lucid and insightful lectures are published in *Macroeconomics and Monetary Theory* (1972c) and *The Theory of Income Distribution* (1973). Third, he was widely respected as an editor, who demonstrated both considerable judgement and a talent for recognizing and encouraging the development of new and original lines of thought. He was devoted to his sustained role as an editor of the *Journal of Political Economy*. He also served on the editorial boards of the *Review of Economic Studies*, *Economica*, the *Journal of International Economics* and *The Manchester School of Economic and Social Research*.

Testifying to Johnson's impact on the economics profession is the number of articles devoted to the evaluation of his scientific contributions. Noteworthy in this respect are the special issues of the *Canadian Journal of Economics* (1978) and the *Journal of Political Economy* (1984) (which also contain a complete bibliography of Johnson's voluminous writings), as well as the entry in the *International Encyclopedia of the Social Sciences* (Bhagwati and Frenkel, 1979), on which this present entry draws.

Many honours came Johnson's way. He was invited to deliver many of the prestigious public lectures in economics: the Ely Lecture, the Wicksell Lectures, the De Vries Lecture, the Ramaswami Lecture, the Johansen Lectures and the Horowitz Lectures. He was elected to the Presidency of the Canadian Political Science Association (1965–6) and the Eastern Economic Association (1976–7), was Chairman of the (British) Association of University Teachers in Economics (1968–71), and was Vice-President of the American Economic Association (1976). He was a Fellow of the Econometric Society, the British Academy, the Royal Society of Canada, the American Academy of Arts and Sciences, a Distinguished Fellow of the American Economic Association and an honorary member of the Japan Economic Research Center. He was the holder of honorary degrees from St. Francis Xavier University, University of Windsor, Queen's University, Carleton University, University of Western Ontario, Sheffield University and the University of Manchester, and he was awarded the Innis-Gérin Medal of the Royal Society of Canada, the Prix Mondial Messim Habif by the University of Geneva, and the Bernhard Harris Prize by the University of Kiel, Germany just prior to his untimely death. The Canadian government named him an Officer of the Order of Canada in December 1976: a fitting tribute from his native country for a fully internationalist economist who had brought great distinction to his profession and his discipline.

JACOB A. FRENKEL

SELECTED WORKS

1950a. The case for increasing the price of gold in terms of all currencies – a contrary view. *Canadian Journal of Economics & Political Science* 16, May, 199–209.

1950b. The de-stabilising effect of international commodity agreements on the prices of primary products. *Economic Journal* 60, September, 626–9.

1950c. A note on diagrammatic analysis of income variations and the balance of payments. *Quarterly Journal of Economics* 64, November, 623–32.

1951a. Some implications of secular changes in bank assets and liabilities in Great Britain. *Economic Journal* 61, September, 544–61.

1951b. Some Cambridge controversies in monetary theory. *Review of Economic Studies* 19(2), 90–104.

1953a. Equilibrium growth in an international economy. *Canadian Journal of Economics & Political Science* 19, November, 478–500. Reprinted in (1958).

1953b. Optimum tariffs and retaliation. *Review of Economic Studies* 21(2), 142–53. Reprinted in (1958).

1954. Increasing productivity, income-price trends and the trade balance. *Economic Journal* 64, September, 462–85. Reprinted in (1958).

1956. The transfer problem and exchange stability. *Journal of Political Economy* 64(3), June, 212–25. Reprinted in (1958).

1958. *International Trade And Economic Growth*. London: George Allen & Unwin.

1959a. British monetary statistics. *Economica* 26, February, 1–17.

1959b. International trade, income distribution and the offer curve. *Manchester School* 27, September, 241–60. Reprinted in (1971a).

1960a. The cost of protection and the scientific tariff. *Journal of Political Economy* 68, August, 327–45.

1960b. Income distribution, the offer curve, and the effects of the tariffs. *Manchester School* 28, September, 215–42.

1961. The 'General Theory' after twenty-five years. *American Economic Review* 51, May, 1–17. Reprinted in (1962c).

1962a. *Canada In A Changing World Economy.* Alan B. Plaunt Memorial Lectures, Carleton University; Toronto: University of Toronto Press.

1962b. Monetary theory and policy. *American Economic Review* 52, June, 335–84. Reprinted in (1967b).

1962c. *Money, Trade And Economic Growth.* London: George Allen & Unwin.

1963a. Recent developments in monetary theory. *Indian Economic Review* 6(3), February, 29–69 and (4), August, 1–28.

1963b. *The Canadian Quandary: Economic Problems and Policies.* Toronto: McGraw-Hill.

1964. The economics of brain drain. In *Sixth Annual Seminar On Canadian-American Relations* (papers, Windsor: Windsor University Press, 37–50.

1965a. Optimal trade intervention in the presence of domestic distortion. In *Trade, Growth And The Balance Of Payments: Essays In Honor Of Gottfried Haberler*, ed. R.E. Baldwin et al., Amsterdam: North-Holland, 3–34.

1965b. The theory of tariff structure, with special reference to world trade and development. *Etudes Et Travaux De L'Institut Universitaire De Hautes*, Etudes Internationales De Geneva 4, 9–29. Reprinted in (1971a).

1965c. *The World Economy At The Crossroads.* Oxford: Clarendon Press.

1967a. *Economic Policies Toward Less Developed Countries.* Washington: Brookings Institution; London: George Allen & Unwin.

1967b. *Essays In Monetary Economics.* London: George Allen & Unwin; 2nd edn, 1969; Cambridge, Mass.: Harvard University Press.

1967c. Some economic aspects of brain drain. An appendix. *Notes On The Effects Of Emigration Of Professional People On The Welfare Of Those Remaining Behind. Pakistan Development Review* 7(3), Autumn, 379–411. Reprinted in (1975a).

1967d. The possibility of income losses from increased efficiency or factor accumulation in the presence of tariffs. *Economic Journal* 77, March, 151–4.

1969a. The case for flexible exchange rates, 1969. *Federal Reserve Bank of St. Louis Review* 51(6), June, 12–24. Reprinted in (1972a).

1971a. *Aspects Of The Theory Of Tariffs.* London: George Allen & Unwin.

1971b. The Keynesian revolution and the monetarist counter-revolution. *American Economic Review* 61, May, 1–14. Reprinted in (1972a).

1972a. *Further Essays In Monetary Economics.* London: George Allen & Unwin.

1972b. *Inflation And The Monetarist Controversy.* F. De Vries Lectures. Amsterdam and London: North-Holland.

1972c. *Macroeconomics And Monetary Theory.* Chicago: Aldine Publishing.

1972d. Political economy aspects of international monetary reform. *Journal of International Economics* 2, September, 401–24.

1973. *The Theory of Income Distribution.* London: Gray-Mills Publishing.

1974. The welfare economics of reversed international transfers. In *Trade, Stability and Macro-economics: Essays In Honor Of Lloyd A. Metzler*, ed. G. Horwich and P. Samuelson, New York and London: Academic Press, 79–110.

1975a. *On Economics And Society.* Chicago and London: University of Chicago Press.

1975b. *Technology And Economic Interdependence.* Trade Policy Research Centre Book, London: Macmillan, 1975.

1976. (With J.A. Frenkel, eds.) *The Monetary Approach To The Balance of Payments.* London: George Allen & Unwin, and Toronto: University of Toronto Press.

1978. (With J.A. Frenkel, eds.) *The Economics Of Exchange Rates.* Reading, Mass.: Addison-Wesley.

1978. (With E. Johnson) *The Shadow of Keynes.* Oxford: Basil Blackwell.

BIBLIOGRAPHY

Bhagwati, J.N. and Frenkel, J.A. 1979. Johnson, Harry G. *International Encyclopedia of the Social Sciences*, Biographical Supplement, Vol. 18, New York: The Free Press.

Canadian Journal of Economics, Supplement, 11(4), November 1978.

Journal of Political Economy 92(4), August 1984.

Johnson, William Ernest (1858–1931). English logician, philosopher, and economic theorist. The son of the headmaster of Llandaff House, a Cambridge academy, Johnson entered King's College in 1879 on a mathematical scholarship (11th wrangler, mathematics tripos 1882; first class honours, moral sciences tripos 1883). Initially a mathematics coach, then lecturer on psychology and education at the Cambridge Women's Training College, Johnson later held a succession of temporary positions at Cambridge (University Teacher in the Theory of Education, 1893 to 1898; University Lecturer in Moral Science, 1896 to 1901), until he was elected a Fellow of King's College in 1902 and appointed Sidgwick Lecturer in Moral Science in the University, where he remained until his death.

In the Cambridge of Johnson's day, economics was included among the moral sciences and, as C.D. Broad remarks, 'it was a subject in which Johnson's mathematical, logical, and psychological interests could combine with the happiest results' (Broad, 1931, pp. 500-501). Although he lectured on mathematical economics for many years, Johnson wrote only three papers on economics (Johnson 1891, 1894, 1913), of which only the last, 'The Pure Theory of Utility Curves', was published during his lifetime. This latter was, however, an important paper, representing 'a considerable advance in the development of utility theory' (Baumol and Goldfeld, 1968, p. 96), and 'contains several results that should secure for its author a place in any history of our science' (Schumpeter, 1954, p. 1063n). These include an analysis of utility based on marginal utility ratios, and a proof of the consistency of expenditure and convex indifference curves.

Johnson's aversion to publication has been variously ascribed to his 'ill health, diffidence, and a very high standard of achievement' (Broad, 1931, p. 505), and a 'rooted antipathy to publish anything until he was sure of everything' (Braithwaite, 1931). Indeed, between the publication of his treatise on *Trigonometry* in 1888, and his three volume work on *Logic* in the 1920s, he published only three papers on logic (Johnson, 1892, 1900, 1918) in addition to his paper on utility. Despite such a limited output Johnson retained his fellowship at King's (the continuance of which was periodically reviewed), due to the high regard in which he was held by his colleagues.

Johnson nevertheless exerted considerable influence on his colleagues and students at Cambridge through his lectures and personal interaction. One example, among many, is John Neville Keynes, the father of John Maynard Keynes and an eminent logician in his own right. When the senior Keynes was at work on the successive editions of his *Studies and Exercises in Formal Logic*, Johnson would come to lunch regularly to discuss the work; one result was that among the examples at the ends of chapters 'the hardest, neatest, and most ingenious problems are marked "J", which means that they were devised by Johnson' (Broad, 1931, p. 504). Among his students were John Maynard Keynes, Frank Ramsey, Ludwig Wittgenstein, C.D. Broad and Dorothy Wrinch (an early collaborator of Harold Jeffreys).

Nevertheless, it was only after the publication of his three volume *Logic* (Johnson 1921, 1922, 1924) – written only after the encouragement and assistance of his students, in particular Naomi Bentwich – that Johnson gained recognition outside Cambridge: honorary degrees from Manchester (1922) and Aberdeen (1926), and membership of the British Academy (1923). The third volume of the *Logic* concludes with a remarkable appendix on 'eduction', in which Johnson introduced his 'combination' and 'permutation' postulates. The latter of these was none other than the concept of exchangeability, soon to be independently rediscovered by Haag and de Finetti, and employed by the latter as a key element in his theory of subjective probability and statistical inference (Dale, 1985).

Johnson had, in fact, long been interested in the foundations of probability, but written nothing. Indeed Keynes remarked in his *Treatise on Probability* that 'when the following pages [Part II of the *Treatise*] were first in proof, there seemed little likelihood of the appearance of any work on Probability from his own pen ...' (Keynes, 1921, p. 116). Unfortunately, although Johnson now planned a fourth volume of his *Logic*, to be devoted to probability, he had 'left it too late': in 1927 he suffered a stroke and, although able to lecture again by the next year, he never fully recovered and died early in 1931. At his death only three chapters of the planned work had been completed (dating from about 1925); these were edited by R.B. Braithwaite and published posthumously (Johnson, 1932).

In these chapters Johnson introduced his 'sufficientness' postulate (the terminology is due to I.J. Good), a generalization and improvement on his earlier combination postulate. In a remarkable appendix (edited by Braithwaite, but the substance of whose argument was due to Johnson) the sufficientness postulate was shown to, in effect, characterize the use of symmetrical Dirichlet priors in the Bayesian analysis of multinomial sampling. This work anticipated by more than two decades Carnap's 'continuum of inductive methods', and indeed went further, technically, than Carnap (Zabell, 1982, p. 1098). Curiously, despite its seminal importance, the appendix was almost never published. (G.E. Moore, then editor of *Mind*, questioned whether such a technical result would be of interest to the journal's readership, and the appendix was left in only at Braithwaite's insistence (Braithwaite, 1982, personal communication).)

Johnson was one of a remarkable group of English intellectuals – most notably Jevons, Edgeworth, Keynes and Ramsey – who combined in varying proportions interests in economic theory and the philosophical foundations of logic, probability, statistics, and scientific inference. For further biographical details, see the obituary notices by C.D. Broad, (1931); R.B. Braithwaite, (1931); the unsigned A.D. (1932); and the entry on Johnson by Braithwaite in the *Dictionary of National Biography 1931–1940 (1949)*. R.F. Harrod (1951) contains scattered references to Johnson. For details of Wittgenstein's personally cordial but professionally strained relationship with Johnson, see Rush Rhees (1984).

Johnson's three papers on economics are reprinted, with brief commentary, in William J. Baumol and Stephen N. Goldfeld, (1968). The 1891 and 1894 papers were printed for private circulation, and are virtually unobtainable elsewhere. For a critical discussion of the 1913 paper, see F.Y. Edgeworth (1915). Due in part to what George Stigler has termed Johnson's 'concise and peculiar' style, and in part to the appearance of Slutsky's classic paper two years after the appearance of Johnson's *Economic Journal* paper, there has never been widespread recognition of Johnson's achievement in utility theory, and references to his work in the economic literature are few, brief, and scattered; see, for example Joseph A. Schumpeter (1954).

For a recent assessment of Johnson's philosophical work, see John Passmore (1968). Johnson's early work in logic is discussed in A.N. Prior (1949a). Discussions of the *Logic* include the critical notices by C.D. Broad (1922) as well as papers by H.W.B. Joseph (1927) and A.N. Prior (1949b).

The statistical implications of Johnson's 'sufficientness' postulate are discussed in I.J. Good (1965). Due to the circumstances of its publication, the argument showing that the postulate characterizes predictive probabilities was only outlined and contains several lacunae; for a complete version, historical background, and discussion of the connection with Carnap's work, see S.L. Zabell (1982). A.I. Dale (1986) summarizes and assesses Johnson's contribution to this area.

S.L. ZABELL

SELECTED WORKS

1888. *Trigonometry*. London: Macmillan.
1891. Exchange and distribution. *Cambridge Economic Club*, Lent Term, 1–6. Reprinted, with brief commentary in Baumol and Goldfeld (1968), 316–20.
1892. The logical calculus. *Mind* 17, 3–30, 235–50, 340–57.
1894. (With C.P. Sanger.) On certain questions connected with demand. *Cambridge Economic Club*, Easter Term, 1–8. Reprinted, with brief commentary, in Baumol and Goldfeld (1968), 40–48.
1900. Sur la théorie des équations logiques. *Bibliothèque du Congrès International de Philosophie* 3 (1901).
1913. The pure theory of utility curves. *Economic Journal* 23, 483–513. Reprinted, with brief commentary, in Baumol and Goldfeld (1968), 97–124.
1918. The analysis of thinking. *Mind* 27, 1–21, 133–51.
1921. *Logic*, Pt I. Cambridge: Cambridge University Press.
1922. *Logic*, Pt II: *Demonstrative Inference: Deductive and Inductive*. Cambridge: Cambridge University Press.
1924. *Logic*, Pt III: *The Logical Foundations of Science*. Cambridge: Cambridge University Press.
1932. Probability. *Mind* 41, 1–16 (The relations of proposal to supposal), 281–96 (Axioms), 409–23 (The deductive and inductive problems).
In addition to the above, Johnson wrote several critical reviews and a note for *Mind* during the years between 1886 and 1890, and contributed several entries to the original *Palgrave*.

BIBLIOGRAPHY

A.D. 1932. W.E. Johnson (1858–1931): an impression. *Mind* 41, 136–7.
Baumol, W.J. and Goldfeld, S.N. (eds) 1968. *Precursors in Mathematical Economics: An Anthology*. London: London School of Economics and Political Science; No. 19 in a series of Reprints of Scarce Works on Political Economy.
Braithwaite, R.B. 1931. W.E. Johnson. *The Cambridge Review* 52, 30 January, 220.
Braithwaite, R.B. 1949. W.E. Johnson. In *Dictionary of National Biography 1931–1940*. Oxford: Oxford University Press.
Broad, C.D. 1922. *Logic, Part I* by W.E. Johnson. *Mind* 31, 496–510.
Broad, C.D. 1924. *Logic, Parts II and III* by W.E. Johnson. *Mind* 33, 242–61, 369–84.
Broad, C.D. 1931. W.E. Johnson. *Proceedings of the British Academy* 17, 491–514.
Dale, A.I. 1986. A study of some early investigations into exchangeability. *Historia Mathematica* 12, 323–36.
Edgeworth, F.Y. 1915. Recent contributions to mathematical economics. *Economic Journal* 25, 36–63, 189–203 at 41–62.
Good, I.J. 1965. *The Estimation of Probabilities: An Essay on Modern Bayesian Methods*. Cambridge, Mass.: MIT Press.
Harrod, R.F. 1951. *The Life of John Maynard Keynes*. London: Macmillan.
Joseph, H.W.B. 1927. What does Mr Johnson mean by a proposition? *Mind* 36, 448–66.

Passmore, J. 1968. *A Hundred Years of Philosophy*. 2nd edn, New York: Penguin Books.

Prior, A.N. 1949a. Categoricals and hypotheticals in George Boole and his successors. *American Journal of Philosophy*.

Prior, A.N. 1949b. Determinables, determinates and determinants. *Mind* 53, 1–20, 178–94.

Rhees, R. (ed.) 1984. *Recollections of Wittgenstein*. Oxford: Oxford University Press, xvii, 51–2, 61–2, 103, 109.

Schumpeter, J.A. 1954. *History of Economic Analysis*. New York: Oxford University Press.

Zabell, S.L. 1982. W.E. Johnson's 'sufficientness' postulate. *Annals of Statistics* 10, 1090–99.

joint production. The network of cost relationships in a multi-product firm is much more complicated than in a single product one and the nature of these relationships has important implications for the structure and size of the firm, the organization and regulation of the industry, and the pattern and intensity of resource employment. The general neoclassical multi-product/multi-input decision framework presupposes that the firm is producing more than one output, but the central question is why do firms diversify into multiple outputs. The answer may lie in the characteristics of the cost functions and the nature of the demand facing the firm. We shall therefore briefly discuss: (1) the main causes of joint production; (2) the econometric techniques proposed for testing the presence of joint production and (3) the implication of joint production for the organization of the industry.

CAUSES OF JOINT PRODUCTION.

Joint production includes two cases: (1) when there are multiple products, each produced under separate production processes – i.e. the production function is non-joint; and (2) when several outputs are produced from a single production process. In the first case 'joint production' is a problem of aggregation while in the second case it is a technological phenomenon of 'intrinsic jointness'. Thus, writing a production or cost function with several outputs is by itself not an evidence of joint production; it is the absence of non-jointness which is a crucial test.

Recent literature has identified a variety of reasons for joint production but three causes stand out: economizing of some shareable inputs or economies of scope; jointness due to output interactions; and uncertainty on the demand side.

(1) *Economies of scope.* Suppose that a vector of outputs $y = (y_1, \ldots, y_n)$ and a vector of primary inputs $x = (x_1, \ldots, x_m)$ are technically related by the production structure characterized by its dual, the joint cost function $C = g(w, y)$ where $w = (w_1, \ldots, w_n)$ is the vector of input prices. Further assume that the cost function is *non-additive* with respect to all partitions of the commodity set. Economy of scope is defined for the partition of commodity set, h, as

$$\sum_{h=1}^{s} C(y_h, w) < \sum_{j=1}^{m} [C(y_j, w)],$$

$$y_j = (0, \ldots, y_j, \ldots, 0), \quad \text{if } s < m$$

where $\Sigma C(y_j, w)$ is the total cost when each commodity is produced separately (Lloyd, 1983). Economies (diseconomies) of scope will exist by this definition for *a given* partition of commodities.

Economies of scope may arise from fixed inputs such as physical and human capital that are shared or utilized without complete congestion. Some fixed inputs may be imperfectly divisible and could not easily be shifted from one production

to another so that the production of a subset of commodities may leave excess capacity in some stage of production. Another possibility is that some of the inputs may have a quasi-public characteristic which when purchased for use in one production process can be at least partially used in the production of other commodities.

(2) Economies of scope may also be due to interrelationships among products: two or more commodities may be produced jointly, at lower cost than if they were produced separately even in the absence of excess capacity and shareable inputs in the production process. An example will be a production process where $y_1 = f(x)$ but $y_2 = f_2(x, y_1)$ which characterizes many industrial and agricultural production processes. Another example is the case where it is not possible to produce zero quantities of the commodities produced jointly, i.e. the multiple output–input function $F(y, x) = 0$ is restricted to the combination (y, x) that precludes any element of the output vector y to be zero. Examples of such production can be found in agriculture (wool and mutton) and some chemical processes.

(3) Demand conditions are also important for the product structure of the firms; firms may avoid declines in revenue because of market saturation by producing new products, thereby substituting economies of scope for the economies of scale that the firm cannot achieve given the market conditions it faces. Another reason for joint output is attributed to uncertainty and risk aversion (Lloyd, 1983). Firms choose commodity diversification as a strategy to reduce risk in an environment of uncertainty though no jointness exists in their production process. Suppose a firm's profits from each commodity is random because the output and input prices are random variables; the firm maximizes the expected utility of aggregate profit given the joint probability distribution of the random variables. Under these sets of assumptions the firm will produce multiple outputs even though there is no technological reason for doing so. The presence of uncertainty plays the same role as shareability of input or intrinsic jointness of output in generating economies of scope.

ECONOMETRICS OF JOINT PRODUCTION. A major problem has been the difficulty of specifying a sensible and estimable functional form for the multi-product technology. The flexible functional form developed by Christensen et al. (1973), Diewert (1971), and Lau (1978) has made it possible to use the flexible production or cost functions, and particularly the translog cost or profit functions, to approximate multiple output technology. Other more suitable joint cost functions can be formulated but since the translog cost function is often used in empirical studies, we employ it for illustrative purposes. Consider the cost function

$$\ln C = \alpha_0 + \sum_i \alpha_i \ln w_i + \sum_k \beta_k \ln y_k + \frac{1}{2} \sum_i \sum_j \gamma_{ij} \ln w_i \ln w_j$$

$$+ \frac{1}{2} \sum_k \sum_l \theta_{lk} \ln y_l \cdot \ln y_2 + \sum_{ik} \sum_{ik} \delta_{ik} \ln w_i \ln y_k$$

$$i, j = 1, \ldots, m, \quad k, l = 1, \ldots, n \tag{1}$$

which is a quadratic approximation to an arbitrary multiple output cost function. The nature of the cost relationships can be tested by imposing the necessary parameter restrictions. For example, if it turns out that $\delta_{ik} \cdot \beta_l = \delta_{il} \cdot \beta_k$, then the cost function is separable, i.e. the ratio of any two marginal costs is independent of factor prices or factor intensities; then the cost function can be written as $C(y, w) = H(y)\omega(w)$. Another important feature of the production structure is non-jointness which is that total cost of producing all outputs be the same as the sum of the cost of producing each output separately, i.e.,

$C(y, w) = \Sigma_i g_i(w, y_i)$. This implies that the marginal cost of each output is independent of the level of any output. In terms of (1) the condition of non-jointness is $\theta_{kl} = -\beta_k \beta_l$ for $k \neq l$. Hall (1972) has shown that no multiple output technology with constant return to scale can be both separable and nonjoint; in fact all nontrivial separable technologies are inherently joint and cannot be used empirically to test hypotheses about jointness.

Ordinary translog cost function (1) (and cost function with logarithmic output variables) is inappropriate to measure economies of scope. By definition, if any of the outputs is zero the multiple product firm's cost will be zero, which suggests that if a firm specializes completely in one of the outputs it must incur no costs whatsoever. To overcome this problem it is necessary to modify (1) by performing a Box-Cox transformation on output variables, i.e. substitute $y^* = (y^\lambda - 1/\lambda)$, where λ is a parameter to be estimated. Another possibility is to formulate alternative cost functions such as the linear generalized Leontief joint cost function proposed by Hall (1972) or the CES multiple output cost function stated in the next section. Both are well defined for zero output levels.

Note two other issues: when allocatable fixed inputs are the sole cause of jointness the dual production models (multi-product costs or profit functions) are not very useful because they can recover the production function in the *sum* of the inputs and not in terms of individual allocations. This arises because all of the input allocations have the same market price, which enters the cost and profit functions. Appropriately specified *primal* model would permit identification of such allocations (Shumway et al., 1984). Also, a Giffen effect may arise in the case of multiproduct production. As the direct substitution effect of a change in price may not be negative for a factor in a single product line, although it will be over all lines. Moreover, the cross-substitution effect may not be equal in an individual product line (Hughes, 1981).

The measures of economies of scale and scope are, respectively.

$$S = \sum_{k=1}^{n} \left(\frac{\partial \log C}{\partial \log y_k^*} \right)^{-1}$$

and

$$S_c = \frac{C(y_1^*, 0) + \cdots + C(0, y_n^*) - C(y_1^*, \ldots, y_n^*)}{C(y_1^*, \ldots, y_n^*)}$$

and the relationship between them is shown to be

$$S = \frac{\sum_{i}^{n} \beta_i S_i}{1 - S_c} \quad \text{and} \quad \sum \beta_i = 1, \quad i = 1, \ldots, n$$

where S_i are measures of product specific economies of scale and β_i are roughly equal to the share of the variable cost of producing each output. If there is a sufficiently large economy of scope, it could result in economies of scale on the entire product set even if there is a constant return or some degree of diseconomies of scale in the separate products.

A number of econometric studies summarized by Bailey and Friedlaender (1982) and those by Denny and Pinto (1978), Brown, et al. (1979), Griffin (1977), Vincent et al. (1980), Just et al. (1983), and others have shown that at industry level (particularly in agriculture) multiple output production technologies prevail with differing degrees of jointness and economies of scope. However, further studies are required. Particularly, the role of technological progress in changing the intertemporal structure of the cost relations by unbundling some joint costs and giving rise to new ones requires considerable attention.

INDUSTRY STRUCTURE. Multi-product technology has important implications for the organization and regulation of industry. The characteristics of the underlying cost relationships could determine the optimal number of firms that may populate an industry; the industry may be dominated entirely by a single firm producing all of the output or may be characterized by duopolistic, oligopolistic or competitive forms. Baumol (1977) among others has formulated conditions for natural monopoly to prevail. When the cost function is *subadditive* the efficient supply condition is a single firm that can produce industry output at lower costs than two or more firms. The degree of contestability in many multiproduct industries depends to a great extent on the nature of the multiple output cost function. For example, consider the cost function

$$C(y_1, \ldots, y_n) = F + \left[\sum_{i=1}^{n} (y_i / a_i)^\beta \right]^{1/\alpha\beta} \quad (2)$$

where $F \geq 0$ is the fixed cost. Depending on the parameter values of (2) four market structure possibilities can be identified: (1) if ($F \geq 0$, $\alpha > 1$, $\beta > (1/\alpha)$), the industry is a natural monopoly; (2) if ($F = 0$, $\alpha < 1$, β arbitrary), the industry is competitive; (3) if ($F = 0$, $\alpha > 1$, $\beta < (1/\alpha)$), n specialized firms each producing the industry output of the specialized good will constitute the industry; (4) finally, if ($F > 0$, $\alpha < 1$, β arbitrary), at *small* levels of output, either a single firm ($\beta > 1$) or a number of specialized firms ($\beta < 1$) will populate the industry while at *large* levels of output, several smaller specialized firms will constitute the industry. Similar experiments can be carried out with the modified translog cost function (1).

The degree of contestability deduced from the characteristics of the cost relations (given sustainable prices) has policy implications for the entry of new firms, the degree of concentration in a market, and antitrust laws. In contestable markets, mergers may not be anticompetitive; the theory of joint production is also important for considering the boundary issue between regulated and unregulated portions of an industry and the related problem of cross-subsidization. Another policy concern is the potential pathological substitution effects in multiple-production processes that, at least in the short run, may lead to possible bottlenecks in factor utilization and may to some extent negate the effect of particular policies.

M. Ishaq Nadiri

See also COST AND SUPPLY CURVES; COST FUNCTIONS; DUALITY; VON NEUMANN TECHNOLOGY.

BIBLIOGRAPHY

Bailey, E.E. and Friedlaender, A.F. 1982. Market structure and multiproduct industries. *Journal of Economic Literature* 20(3), September, 1024–48.

Baumol, W.J. 1977. On the proper cost tests for natural monopoly in a multiproduct industry. *American Economic Review* 67(5), December, 809–22.

Brown, R.S., Caves, D.W. and Christensen, L.R. 1979. Modelling the structure of cost and production for multiproduct firms. *Southern Economic Journal* 46(1), July, 256–73.

Christensen, L., Jorgenson, D.W. and Lau, L.J. 1973. Transcendental logarithmic production frontiers. *Review of Economics and Statistics* 55(1), February, 28–45.

Denny, M. and Pinto, C. 1978. An aggregate model with multiproduct technologies. In *Production Economics: A Dual Approach to Theory and Applications*, Vol. 2, ed. M. Fuss and D. McFadden, Amsterdam: North-Holland.

Diewert, W.E. 1971. An application of the Shephard duality theorem: a generalized Leontief production function. *Journal of Political Economy* 79(3), May–June, 481–507.

Griffin, J.M. 1977. The econometrics of joint production: another approach. *Review of Economics and Statistics* 59(4), November, 389–97.

Hall, R.E. 1972. The specification of technology with several kinds of output. In *An Econometric Approach to Production Theory*, ed. D.L. McFadden, Amsterdam: North-Holland.

Hughes, J.P. 1981. Giffen inputs, the theory of multiple production. *Journal of Economic Theory* 25(2), October, 287–301.

Just, R.E., Zilberman, D. and Hochman, E. 1983. Estimation of multicrop production functions. *American Journal of Agricultural Economics* 65(4), 770–80.

Lau, L.J. 1978. Applications of profit functions. In *Production Economics: A Dual Approach to Theory and Applications*, Vol. 1, ed. M. Fuss and D.L. McFadden, Amsterdam: North-Holland.

Lloyd, P.J. 1983. Why do firms produce multiple outputs? *Journal of Economic Behavior and Organization* 4(1), March, 41–51.

Shumway, C.R., Pope, R.D. and Nash, E.K. 1984. Allocatable fixed inputs and jointness in agricultural production: implications for economic modeling. *American Journal of Agricultural Economics* 66(1), February, 72–78.

Vincent, D.P., Dixon, P.B. and Powell, A.A. 1980. The estimation of supply response in Australian agriculture: the CRESH/CRETH production system. *International Economic Review* 21(1), February, 221–42.

joint production in linear models. General *joint production* is defined as the simultaneous production of at least two commodities in one production process. This definition includes machines (fixed capital) that are not used up in one production period, and land which is by definition neither producible nor exhaustible.

Models are called *linear* in economic theory if a linear technology is used (implying constant returns to scale) and a finite number of production processes to produce a given number of goods. However, the classical theory, particularly in its modern form (Sraffa), does not presuppose constant returns; yet the formal tools of analysis are similar to those used in neoclassical theory in the linear case which justifies the inclusion of Sraffa below.

Linear models, as discussed here with regard to joint production, will be classified as (I) input–output models; (II) Von Neumann and Activity Analysis Models; and (III) Sraffa Models.

I JOINT PRODUCTION IN INPUT–OUTPUT ANALYSIS

The most widely used linear model in modern economics is undoubtedly Leontief's input–output model (see Leontief, 1936). As a theoretical model, it rules out joint production. In applied economics, joint production is found to be a ubiquitous phenomenon, and the methods used to cope with it are remarkable for their diversity. Hence, in applied input–output analysis one commonly aggregates or transforms industries and commodities in such a way that joint production does not appear. There are few publications which give a theoretical background to these constructions. Frequently joint production is discussed under the topic of 'secondary production' (see e.g. Armstrong, 1975; Chakraborty et al., 1984; Flaschel, 1983, pp. 333–58; Flaschel, 1982b; Gigantes, 1970; Stone, 1961). One usually assumes that there is a square matrix, the so-called 'use matrix', $Z = (z_{ij})$, where z_{ij} represents the quantity of a commodity i (rows) used in industry j (columns), and a square matrix $X = (x_{ij})$, often called 'make matrix', which represents the amount of the products j produced in each industry i, i,

$j = 1, \ldots, n$. X is separated into X_1 and X_2, where for X_1 all non-diagonal elements are zero. X_1 has the 'main products' on the diagonal. Several ways of dealing with 'by-products' shown in $X_2 = X - X_1$ have been suggested and used with the data to define an input–output matrix A.

(a) The *commodity technology model* simply assumes that an ascription of individual costs to commodities is possible and that one industry for each commodity is generated. Formally, the ascription is given by $A = Z(X')^{-1}$, where X' is the transposed matrix of X. (The matrix has to be transposed because the rows of the make matrix represent processes while the opposite convention is used for the use matrix). This model is used in applied input–output accounting (see e.g. UN, 1968, pp. 25–51). The assumption that the number of commodities equals the number of industries is essential here. Note that negative elements of the input–output matrix may occur.

(b) The *industry technology model* supposes that the technology is determined by the industry and that there exists a fixed commodity market share of each industry. Mathematically this model can be presented as follows:

$$A = [a_{ij}];$$
$$a_{ij} = \sum_{k=1}^{n} \left(x_{kj} \bigg/ \sum_{l=1}^{n} x_{lj} \right) \left(z_{ik} \bigg/ \sum_{l=1}^{n} x_{kl} \right)$$

or

$$A = Z(\overline{Xe})^{-1} X (\overline{X'e})^{-1},$$

where $\overline{Xe}$ ($\overline{X'e}$) is the diagonal matrix of vector Xe ($X'e$) and where $e' = (1, 1, \ldots, 1)$. This model is often thought to deal adequately with joint production. It is said that it can be extended to cover the case where the number of commodities exceeds the number of industries and has been applied by the Bureau of Economic Analysis (US Department of Commerce, 1980, pp. 37–51). It has, however, recently been attacked as being dependent on the base year prices chosen and on the distribution of the value added (see Chakraborty et al., 1984, pp. 89–90).

(c) The *by-product model* assumes that each industry produces outputs in fixed proportion and that one can say which of these outputs is the main one. The secondary products are then treated as negative inputs. Hence $Z - (X_2)' = AX_1$, $A = [Z - (X_2)'](X_1)^{-1}$. It is again essential that the number of commodities equals the number of industries. The matrix A may contain negative elements (see Stone, 1961, pp. 39–40; Chakraborty et al., 1984, p. 88).

(d) Gigantes (1970) and Chakraborty et al. (1984) distinguish between 'ordinary secondary products' and 'by-products'. In the case of 'ordinary secondary production' the application of the commodity technology model (case (a)) was proposed by both. While, however, Gigantes applied the industry technology model (case (b)) to by-products, Chakraborty et al. used the by-product model (case (c)) because of the deficiencies of the industry model pointed out by them.

(e) If the product is thought to be such that it is not the main product of any process it cannot be allocated to any industry of the economy as its single product. A *dummy industry* should then be introduced according to Stone (1961, pp. 41–2) which uses no input and shows the secondary product as output.

(f) Besides these constructs there are other models. For example, the Bureau of Economic Analysis supposed that, as in cases (a) and (c), to each by-product of an industry there is another industry of which that commodity is the main product. The amount of the commodity produced as a by-product is treated as if it were bought by the industry where it is the main product and added to the output of that industry (see US Department of Commerce, 1974).

All procedures have in common that they result in a system in which the number of processes is equal to that of the commodities, and usually it is assumed that the equality already holds at the level of the data, given in the form of square 'use' and 'make' matrices. This corresponds to what one should expect to hold in equilibrium from a classical point of view (Section 3).

II JOINT PRODUCTION IN VON NEUMANN MODELS AND ACTIVITY ANALYSIS

Joint production was first introduced in a linear model (with balanced growth and constant returns to scale) by John von Neumann (1937). In von Neumann models, joint production includes fixed capital but excludes land. The vast literature on the model (see Morishima, 1964) did not lead to a systematic analysis of different forms of joint production as special cases. Even the treatment of fixed capital as a joint product was first misunderstood; if a_{ij} is the input of a machine which depreciates by 10 per cent, it was thought that the output was $b_{ij} = 0.9\ a_{ij}$ (see Dorfman, Samuelson and Solow, 1958, pp. 382–3) while it is now generally accepted that the advantage of the joint production approach to fixed capital is based on the possibility of treating the old machine leaving a process as a different good from the one entering it so that depreciation has to be determined simultaneously with prices. Because von Neumann postulated that the rate of interest (or profit) is uniform and to be minimized and that the rate of balanced growth in the dual is maximized and equal to the rate of interest, the model chooses in general (but not always) $k \leqslant m$ production processes to be activated or $k \leqslant n$ positively priced commodities. The other m-k processes (or n-k commodities) are not used (or not produced). For the m-k processes are not profitable and the quantities of n-k commodities are overproduced and can be disposed of, that is, the prices of these will be zero. Therefore this model can also be a theoretical justification for the assumption mentioned above of a square 'make matrix' used in input–output analysis (see von Neumann, 1937; Schefold, 1978b, 1980a).

A similar argument was presented by Koopmans (1951) in his activity analysis. He studied the set of all possible baskets of commodities which are producible from efficient production processes. Like von Neumann he presupposed in his model constant returns to scale and that joint products which are socially not wanted and overproduced can be disposed at zero cost. Nevertheless, he considered briefly the problems that there might exist unwanted commodities which are necessarily produced along with the other socially desired ones. If the former endanger the consumption of the latter, they should be destroyed or transformed into desired commodities by additional processes. But the 'usefulness' will depend on the prices that result from the price determination of the whole economic system. Positively priced commodities are called 'useful'. If one assumes a linear objective function which chooses one technology out of the set of efficient technologies one will almost always obtain an economic system where the number of processes equals the number of commodities.

Activity analysis is much simpler in the case of single production where the so-called 'non-substitution theorem' holds: assuming (1) constant returns to scale, (2) a finite number of activities available to produce each commodity by means of other commodities and labour, and (3) a uniform rate of profit, one particular combination of activities will yield the highest rate of profit,

given the real wage (or, conversely, yield a higher real wage, given the rate of profit), *independently of the composition of output*. Prices will be positive if the economy produces a surplus of aggregate outputs over inputs and if the rate of profit is below the maximum rate of profit at which the entire surplus goes to profits and wages are zero. Formally $(1+r)pA + wl = p$, where A is the square input–output matrix which results from the choice of activities, l the labour vector, r the rate of profit, and w the wage rate.

This was called the '*non-substitution theorem*' by neoclassical economists, because prices here appear to be independent of demand changes, given distribution. However, there is really no room for this theorem in neoclassical analysis because a determination of factor-inputs and of relative outputs in terms of supply and demand necessarily links distribution with the demand for commodities (Garegnani, 1983).

The theorem more naturally corresponds to a classical approach, with prices reflecting costs and distribution being determined through other forces than supply and demand (Sraffa). The extension to joint production then poses a number of problems, in particular regarding the determination of prices. Early marginalists thought (erroneously) that prices of joint products could not be determined within a classical cost of production theory (see Kurz, 1984). The solution was provided by Sraffa who found a way to determine relative prices, given distribution, in a classical framework, by starting from square intput and output matrices.

III JOINT PRODUCTION IN SRAFFA MODELS

Unlike von Neumann, Koopmans, and Leontief, Sraffa (1960) did not assume constant returns to scale. His theory describes the technology, the composition of output and the state of distribution in a long period equilibrium of a closed economy which produces a surplus to be divided between profits, wages and – in the presence of land – rents. The variation of the activity levels which is essential to Koopmans's, von Neumann's, and Leontief's models was not the object of Sraffa's investigation. Assuming a given uniform rate of profit r, n processes in Sraffa's model determine prices of $n - 1$ commodities and the uniform wage rate w. He introduced joint production mainly but not exclusively because he wanted to consider fixed capital in his system. An industry can then no longer be characterized by the commodity it produces. But the number of industries can be expected to be equal to the number of commodities. For the socially wanted commodity basket is not producible if the number of industries is less than the number of produced commodities. In the converse case, prices would be overdetermined. Sraffa has thus shown that relative prices of joint products can be determined within a classical theory.

The model can be formulated as follows: $(1+r)Ap + wl = Bp$, where the elements $a_{i1}, \ldots, a_{in}$ of A, l_i, and $b_{i1}, \ldots b_{in}$ of B describe the quantities of the means of production, labour used, and the quantities of the produced commodities in the ith industry. Note that processes are now represented by rows and commodities by columns of the matrices A and B, following Sraffa, since the emphasis is on the determination of prices, given the structure of industries. The elements above can be normalized through $eB = e$ with $e(B - A) \geqslant 0$ (existence of a surplus), and $el = 1$, where $e = (1, 1, \ldots, 1)$.

(a) *Some contrasts between single product and joint production systems.* Most of the properties of single product systems do not hold for all possible joint production systems. For it is an essential property only of the former that the commodities are separately producible if one assumes constant returns to scale. With single production, output can adapt to all compositions demanded without changing the processes used (this is again the 'non-substitution theorem'). But in joint production systems there exists in general no vector of activity levels $q_i = e_i(B - A)^{-1}$, $e_i = (0, \ldots, 1, 0, \ldots, 0)$, necessary to produce one unit of one of the n commodities without running one of the industries at a negative level. It is, however, possible to single out joint production systems such that $(B - A)^{-1} \geq 0$ (see Schefold, 1978a). These systems are called all-productive systems. For if $(B - A)^{-1} \geq 0$, the activity levels are semi-positive ($q_i \geq 0$). If $(B - A)^{-1} > 0$, all processes are indispensable and the system is called all-engaging.

One can show that both, single product systems and joint production systems which are all-productive, have one and only one maximum rate of profit and standard commodity, and that prices are positive and rise monotonically in terms of the wage rate between zero and a maximum rate of profit. In general joint production systems, however, none of these properties necessarily holds. On the other hand, one can prove that in every basic (see Section IIIb) joint production system with a surplus and positive prices at $r = 0$, prices all turn negative and/or a maximum rate of profit will be reached where the wage rate is zero. Hence no basic joint production system is viable for all positive rates of profit.

Prices at $r = 0$ can be interpreted as 'labour values' or 'labour embodied', for if u is a vector of embodied labour, clearly $Au + l = Bu$ must hold. The same magnitudes can also be interpreted as employment multipliers for $u_i = q_i l$, where q_i is the vector of activity levels to produce one unit of commodity i in a 'subsystem': $q_i = e_i(B - A)^{-1}$. Some labour values may, however, be negative (see Sraffa, 1960, pp. 59–60; Schefold, 1971, pp. 24–6), if the joint production system is not all-productive, though prices at the ruling rate of profit are positive. It will then be possible to expand the system by a small amount without increasing total labour used. For if one assumes a joint production system with $p(r) > 0$, $r > 0$, and $u_i < 0$, producing a surplus $s = e(B - A) \geq 0$ at unit activity levels, ϵq_i is the vector of activity levels necessary for the additional production of a small amount of the commodity $i : \epsilon q_i = \epsilon e_i(B - A)^{-1}$. Hence the additional necessary 'quantity of labour' is $\epsilon q_i l = \epsilon e_i(B - A)^{-1} l$ or $\epsilon q_i l = \epsilon e_i u < 0$ for $u_i < 0$, while $(e + \epsilon q_i)(B - A) = s + \epsilon e_i$, $e + \epsilon q_i > 0$ for small ϵ. One can thus save labour by contracting an inefficient process and expanding an efficient one without reducing the surplus.

A reduction to dated quantities of labour analogous to the formula $p = \Sigma_{i=0}^{\infty}(1 + r)^i A^i l$ of single product models is not always possible in joint production systems. Steedman (1976) has shown that in general joint production models positive prices and a positive rate of profit are neither a sufficient nor a necessary condition for Marx's surplus value to be positive. Following this observation, some have begun to regard it as misleading to interpret u as a vector of 'labour values'.

(b) Basics and non-basics in joint production systems. In single production systems one can easily differentiate between basics, that is, commodities which are directly or indirectly enter the production of all other commodities, and non-basics. The system will be non-basic if and only if the input–output matrix is decomposable. If we assume joint production, however, a non-basic system is not necessarily decomposable as the following example shows: Consider a basic single product system (A, I, l) with n commodities and n industries. The nth industry of the system produces coke which is supposed to be basic, because it enters the production of steel. If the process produces gas, a by-product, and if we add a $(n + 1)$st industry

which also produces gas and if gas is only sold to consumers, the system remains indecomposable though gas then clearly is non-basic (see Schefold, 1971, pp. 7–8).

A system (A, B, l) is called non-basic if a linear combination of processes may be viewed as a decomposable system. Formally, if after a permutation of the columns of A and of B the matrix (A^2, B^2), formed of the last m columns of A and B, has at most rank m. Otherwise the system is called basic. If (A, B) is non-basic, $n - m$ industries have to be linearly dependent on at most m others (Sraffa, 1960, pp. 51–2). Without loss of generality, $A_{12} = HA_{22}$ and $B_{12} = HB_{22}$ for some $(m, n - m)$-matrix H (Manara, 1968). The system is transformed so that it falls into two parts:

$$(1 + r)(A_{11} - HA_{21})p_1 + w(l_1 - Hl_2) = (B_{11} - HB_{21})p_1,$$

$$(1 + r)(A_{21}p_1 - HA_{22}p_2) + wl_2 = B_{21}p_1 + B_{22}p_2.$$

According to this rather abstract definition the first part of the equations above can be solved without knowing the second. The basic commodities so obtained are uniquely defined (Schefold, 1971, pp. 12–23). The economic meaning is illustrated by the fact that a tax which affects the prices p_2 of the non-basic commodities will not affect the prices p_1 of the basic ones. Other possible distinctions between basics and non-basics were discussed by Schefold (1978a) and Flaschel (1982). For instance, all-engaging systems are always basic in the single product case, but not so with joint production.

(c) Fixed capital as a joint product. Sraffa introduced joint production mainly as a preliminary to the treatment of fixed capital (machines) (Sraffa, 1960, p. 43, fn. 1, p. 63). Sraffa assumed machines with constant efficiency, that is, the age of the machine does not affect the amount of the input and output of finished goods and labour. A machine enters the industry in the beginning of one production period as a mean of production and leaves it as a joint product at the end with the finished good which was intended to be produced. A finished good may be a consumption good (if not used as input), a new machine (if later transformed into an old one), a spare part (if used as input in conjunction with old machines only), a raw material, or some combination of the above. Sraffa shows that the value of the machines at different ages is dependent on the level of the rate of profit (Sraffa, 1960, p. 71). Sraffa's model can easily be extended to a model with machines which change their efficiency with their age (Schefold, 1971, pp. 48–80; Baldone, 1974; van Schaik, 1976). One can show that fixed capital systems, where other forms of joint production and the trade of used machines is excluded, behave very much like single product systems.

Mathematically an economic system with fixed capital of varying efficiency can be formulated with $\Sigma_{i=1}^{n} T_i$ equations: $(1 + r)(a_{it}p_1 + m_{i,t-1}p_2) + l_{it}w = b_{it}p_1 + m_{it}p_2$ for all ages $t = 1, \ldots, T_i$ and all industries $i = 1, \ldots, n$. The used machines m_{it}, $1 < t \leq T_i$, with price vector p_2, are produced jointly with a vector b_{it} of finished goods, i.e. new machines and other circulating capital, and consumption goods (price vector p_1). For this the quantities a_{ij}, $j = 1, \ldots, n$, of the commodities, an amount of labour l_{it}, and a one period younger machine $m_{i,t-1}$ is required. If $t = 0$ or $t = T_i$, $m_{it} = 0$, for new machines belong to finished goods and at the age T_i the machine is used up. Other joint production is excluded by the equations: $b_{ijt} = 0$ for all $i, j = 1, \ldots, n$, $t = 1, \ldots, T_i$, and $i \neq j$. Furthermore the output of final goods is normalized: $\Sigma_{it}b_{iit} = 1$. This system of $\Sigma_{i=1}^{n} T_i$ equations can be reduced to a system of n equations by eliminating the used machines: one combines the equations of each industry i by multiplying the ith equation by a factor $(1 + r)^{T_i - t}$ and summing over t for each industry i to get the reduced system:

$$(1 + r)A(r)p_1 + wl(r) = B(r)p_1$$

with

$$A(r) = [a_i(r)] = \left[\sum_{t=1}^{T_i} (1 + r)^{T_i - t} a_{it} \right],$$

$$B(r) = [b_i(r)] = \left[\sum_{t=1}^{T_i} (1 + r)^{T_i - t} b_{it} \right],$$

and

$$l(r) = [l_i(r)] = \left[\sum_{t=1}^{T_i} (1 + r)^{T_i - t} l_{it} \right].$$

This system is called an integrated system. It has an intuitive interpretation if the powers of $(1 + r)$ indicate the number of periods which pass between the use of the corresponding inputs (or the production of the outputs) and the end of the lifetime of the machine. The equations of the integrated system therefore show for each machine that total proceeds equal total costs over its lifetime, if one allows for interest at a rate equal to the rate of profit. The existence of the integrated system also shows that fixed capital, apparently a stock, can, in a sense, be reduced to a flow. The fact that it is possible to derive the prices of finished goods within the integrated system proves that prices of finished goods within the integrated system proves that prices of finished goods are determined independently of the existence of markets for old machines.

One can show (Schefold, 1971, 1974, 1980b) that in this fixed capital system, which is supposed to be basic, finished goods are separately producible and their prices are positive (exceptions may be due to basic finished goods which are used only as spare parts). All primary processes, except possibly those producing spare parts, are indispensable. A positive maximum rate of profit R and the standard commodity exist. One can also show by induction that: $p_{i0}(r) = \Sigma_{t=1}^{T_i} Y_{it}(1 + r)^{-t}$, that is that the price of the new machine $p_{i0}(r)$ equals the sum of the discounted 'expected' net returns: $Y_{it} = [b_{it} - (1 + r)a_{it}]p_1 - wl_{it}$. This equality is usually assumed as an equilibrium condition which links the 'past' with the 'future'. In a fixed capital model, however, this is not an assumption, but a theorem.

Prices of used machines p_2 can also be deduced indirectly through discounting from the integrated system which determines the prices p_1. Since the technical efficiency is allowed to vary with the age of the machines, one is able to describe the falling technical efficiency of a machine growing old, and rising efficiency which occurs, for example, because the machine itself is under construction. Moreover, one can prove: if net returns are positive at all ages of the machine in industry i at the ruling rate of profit r, the prices of the used machine are positive at all ages. If net returns of a machine are negative from age zero to age Θ, $t = 1, \ldots, \Theta$, prices are positive and rising up to age Θ. If a machine has negative returns from age Θ onwards up to age $T_i - 1$, prices are negative no later than at age Θ. If a machine is of falling efficiency, though net returns are positive, prices are falling from age Θ onwards (Schefold, 1974, 1980b, pp. 159–60). Note, however, that net returns are dependent upon changes of the rate of profit. It is possible that a machine of rising efficiency turns into a machine with falling efficiency if the rate of profit changes.

Old machines with negative prices can be eliminated by means of truncation (Nuti, 1973). If the first and all subsequent processes using an old machine with a negative price are truncated in the production of each finished good, a truncated fixed capital system can be reached such that the real wage is higher and all prices of finished goods in terms of the wage rate are lower at a given rate of profit and such that the processes of the truncated system, if used at prices of the untruncated system, will yield surplus profits during a transition (conversely, losses will be caused if the less efficient methods are used at prices of the efficient ones). This is again independent of the existence of markets for old machines, for the truncation found according to the criterion of the maximisation of the real wage in the integrated system is the same as that found by eliminating machines with negative prices.

(d) Counting of equations. The classical determination of prices through the structure of production and consumption (given distribution) leads, like input–output analysis dealing with joint production, to 'square' economic systems with as many commodities (with positive prices) as processes used. The intuitive argument why this should be so has been given above: fewer processes do not in general allow to produce the output in the proportions socially required, more processes lead to an overdetermination of prices.

A more rigorous formal treatment confirms this result for the case of constant returns to scale and balanced growth at a rate equal to the rate of profit (Golden Rule). On can establish an equilibrium with a von Neumann-type model (see section II above); there results a series of propositions which are analogous to those obtained for fixed capital systems:

Assuming a given basket of goods for final consumption d, square truncated systems can be defined which produce (or overproduce) the given basket at positive activity levels and prices; and optimal solutions (yielding the highest real wage at a given rate of profit) will be found with positive prices. The envelope of the wage curves of the truncations will be monotonically falling. In general, optimal solutions will be 'square systems' (number of commodities with positive prices equal to the number of processes used), and the last truncation appearing on the envelope will in general be all-productive ($[B - (1 + r)A]^{-1} \geqslant 0$) with a maximum rate of profit and a standard commodity (Schefold, 1978b). Much work has recently been done to determine the extent to which these properties also hold if the golden rule assumption is dropped, notably by N. Salvadori (1982) and B. Schefold.

There is a more direct economic argument which shows that 'counting of equations' works in the relevant cases within a classical framework. In fact, a 'square' system results from the forces of competition:

(1) It is true that even with single-product processes one commonly finds that several processes compete at any one moment in the production of the same commodity; these processes usually yield different rates of profit. But the 'socially necessary technique' determines the prices of production at the 'normal' rate of profit, relative to which obsolescent techniques make losses while more advanced ones yield extra profits. The same holds for multi-product industries. An excess of processes will, in the long run and in the absence of technical progress, in both cases be eliminated.

(2) Joint production now gives rise to another possibility: it may appear that there is only one multi-product process, or that there are 'too few', to determine relative prices in the classical fashion; however, it can be shown that incentives for the introduction of 'additional processes' will then arise. If, for instance, a new use is discovered for a byproduct of one process in a system which had been a single-product system, the byproduct can be sold initially (from the point of view of this theory) at an arbitrary price which, if the price is high,

may induce the introduction of a second process to manufacture the byproduct in a new process as an output or, if the price is low, to use the byproduct in a new process as an input. In both cases, the number of processes will again be equal to the number of commodities.

In the case of industrial production counting of equations thus leads to the postulate that there should be a second process which produces (or uses!) the byproduct of a process in a different proportion. Overdetermination of prices (too many processes, entailing quasi-rents) and underdetermination (excess of commodities with room left for new processes to be established) are therefore two forms of a disequilibrium which tends to be resolved in a 'square' equilibrium solution much in the same way as market prices tend towards prices of production (Schefold, 1985).

(3) The logic of the counting of equations allows one to predict a high degree of specialization in the presence of unproduced means of production. For example, land can be defined as a joint product which leaves a process unchanged (with improvements treated like fixed capital) so that the land price is equal to the rent, capitalized at the ruling rate of profit. There are two main forms of rent: extensive rent where the difference in rent is explained by the difference in production costs between two adjacent lands yielding the same crop, and intensive rent where a cost-intensive and a land-intensive method may coincide on one land to determine jointly the price (or rent) of the land, and the price of the crop.

Counting of equations shows that most lands will appear to be specialized if different crops could be grown on many lands, but by means of single product processes. For with, say, 10 crops and 100 types of land (differentiated according to location etc.), there will be room for 110 processes determining 100 rents and 10 crops prices, so that at least 90 lands will be specialized (Schefold, 1971, pp. 85–6), the choice being influenced by the rate of profit so that it does not necessarily reflect 'open efficiency' or 'fertility'. (Similar arguments can be made if one considers international trade.)

BERTRAM SCHEFOLD

See also LINEAR MODELS; VON NEUMANN TECHNOLOGY.

BIBLIOGRAPHY

Armstrong, A.G. 1975. Technology assumptions in the construction of U.K. Input–Output Tables. In *Estimating and Projecting Input-Output Coefficients*, ed. W.F. Gossling and R.I.G. Allen, London: Input-Output Publishing Co.

Baldone, S. 1974. Il capitale fisso nello schema teorico di Piero Sraffa. *Studi economici* 29, 45–106. Trans. as 'Fixed capital in Sraffa's theoretical scheme', in Pasinetti (1980).

Chakraborty, D., ten Raa, T. and Small, J.A. 1984. An alternative treatment of secondary production in input–output analysis. *Review of Economics and Statistics* 66(1), 88–97.

Dorfman, R., Samuelson, P.A. and Solow, R. 1958. *Linear Programming and Economic Analysis*. New York: McGraw-Hill.

Flaschel, K.P. 1982a. On two concepts of basic commodities for joint production systems. *Zeitschrift für Nationalökonomie* 42, 259–80.

Flaschel, P. 1982b. Input-output technology assumptions and the energy requirements of commodities. *Resources and Energy* 4, 359–89.

Flaschel, P. 1983. *Marx Sraffa und Leontief – Kritik und Ansätze zu ihrer Synthese*. Frankfurt: Peter Lang.

Garegnani, P. 1983. The classical theory of wages and the role of demand schedules in the determination of relative prices. *American Economic Review* 73, 309–13.

Gigantes, T. 1970. The representation of technology in input–output systems. In *Contributions to Input–Output–Analysis*, ed. A.P. Carter and A. Brody, Amsterdam and London: North-Holland.

Koopmans, T.C. 1951. Analysis of production as an efficient combination of activities. In *Activity Analysis of Production and Allocation*, ed. T.C. Koopmans, Proceedings of a Conference, New York: Yale University Press, 1976.

Kurz, H.D. 1984. Joint production in the history of economic thought. *Diskussionsbeiträge zur gesamtwirtschaftlichen Theorie und Politik*. Bremen: Universität Bremen.

Leontief, W.W. 1936. Quantitative input–output relations in the economic system of the United States. *Review of Economics and Statistics* 18, August, 105–25.

Manara, C.F. 1968. Il modello di Sraffa per la produzione congiunta di merci a mezzo di merci. *L'Industria* 1, 3–18. Trans. as 'Sraffa's model of the joint production of commodities by means of commodities,' in Pasinetti (1980), 1–15.

Morishima, M. 1964. *Equilibrium, Stability and Growth – A Multisectoral Analysis*. Oxford: Clarendon Press.

Neumann, J. von 1937. Über ein ökonomisches Gleichungssystem und eine Verallgemeinerung des Brouwerschen Fixpunktsatzes. In *Ergebnisse eines mathematischen Kolloquiums*, ed. K. Menger, Vol. 8, Leipzig: Verlag Franz Deuticke, 73–83. Trans. as 'A model of general equilibrium', *Review of Economic Studies* 13, 1945, 1–9.

Nuti, D. 1973. On the truncation of production flows. *Kyklos* 26(3), 485–96.

Pasinetti, L.L. (ed.) 1980. *Essays on the Theory of Joint Production*. London: Macmillan.

Salvadori, N. 1982. Existence of cost-minimizing systems within the Sraffa framework. *Zeitschrift für Nationalökonomie* 42, 281–98.

Schaik, A.B.T.M. van. 1976. *Reproduction and Fixed Capital*. Rotterdam: Tilburg University Press.

Schefold, B. 1971. *Theorie der Kuppelproduktion*. PhD. Basel, privately printed.

Schefold, B. 1974. Fixed capital as a joint product. Mimeo. Reprinted in Pasinetti (1980), 138–217.

Schefold, B. 1978a. Multiple product techniques with properties of single product systems. *Zeitschrift für Nationalökonomie* 38, 29–53.

Schefold, B. 1978b. On counting of equations. *Zeitschrift für Nationalökonomie* 38, 253–85.

Schefold, B. 1980a. Von Neumann and Sraffa: mathematical equivalence and conceptual difference. *Economic Journal* 90, 140–56.

Schefold, B. 1980b. Fixed capital as a joint product and the analysis of accumulation with different forms of technical progress. In Pasinetti (1980), 138–217.

Schefold, B. 1985. Sraffa and applied economics. *Political Economy: Studies in the Surplus Approach* 1(1), 17–40.

Sraffa, P. 1960. *Production of Commodities by Means of Commodities: Prelude to a Critique of Economic Theory*. Cambridge: Cambridge University Press.

Steedman, I. 1976. Positive profits with negative surplus value: a reply to Mr Wolfstetter. *Economic Journal* 86, 873–6.

Stone, R. 1961. *Input–Output and National Accounts*. Paris: OECD.

United Nations 1968. *A System of National Accounts*. New York: United Nations.

US Department of Commerce, Bureau of Economic Analysis. 1974. *Definitions and Conventions of the 1967 Input–Output Study*. Washington, DC.

US Department of Commerce, Bureau of Economic Analysis. 1980. *Definitions and Conventions of the 1972 Input–Output Study*. Washington, DC.

Jones, George Thomas (1902–1929). Born at Tunstall, Staffordshire, Jones entered Emmanuel College, Cambridge in 1921, took a First in Natural Sciences after two years and then, after an additional two years, a First in Economics. He continued at Christ's as a graduate student for a year, and won both the Adam Smith Prize and a two-year Rockefeller Fellowship, to Harvard in the first year and then in the second year to several other universities, including Stanford, where his thesis was written. He was awarded a PhD from Cambridge in the autumn of 1928 but soon afterwards was killed in a motorcar accident in Rouen.

Jones's career was bound up with that of Allyn Young. In the early 1920s, while Irving Fisher and Taussig were still prominent, Allyn Young at Harvard was becoming one of the best-known American economists. He was called upon to give economic advice to President Harding. There is an interesting legend to the effect that Harding did not like the advice he received, going on to say that he thought that the questions raised were not economic but statistical, and that he would send for the President of the American Statistical Association. 'Why do you want to send for me – I am here', growled Young.

In the 1920s the professorship at the London School of Economics, held by Edwin Cannan, fell vacant. Lionel Robbins was considered a leading candidate but too young (he was born in 1898), so a temporary seat holder was required. Allyn Young agreed to accept the position, but died from pneumonia in London in February 1929. Soon afterwards a new chair was created for Robbins and in fact Young's post was never filled (Robbins, 1971, p. 122). For the last few months of his life I was Young's part-time research assistant – not that we got much work done.

Young's principal interest at this time was in what we now call 'economies of scale', but were then called 'increasing returns', on which Young published a seminal paper (1928). In those days people thought that 'increasing returns', if they could be obtained at all, lay in vast self-contained organizations like Ford Motors, which won the admiration of the world in the 1920s. But Young pointed out that what mattered was increasing sub-division and specialization of the processes, and that average size of plant might actually fall. Young certainly startled British opinion when he states that if the population were doubled, British productivity would rise to the American level.

In contrast to Ford's attempt at self-sufficiency within one organization, its rival General Motors followed the opposite policy, contracting out the provision of components to specialized firms. This proved more successful than Ford's policy. (It is interesting to note that planners Soviet Russia have preferred the Ford model.)

At Harvard, Young had gathered around him some able research students, of whom G.T. Jones was outstanding. His early death left a large quantity of unedited text and tables, and Cambridge University Press gave me the job of reducing his tables and statistical writing to a publishable form – I cannot claim to have done it well. The theoretical part of his writing was referred to D.H. (subsequently Sir Dennis) Robertson, who replied that the question of 'increasing returns' was 'a foully difficult subject', on which however he thought there was some more important recent work. The book was eventually published by Cambridge University Press under the title *Increasing Return*.

Jones devised an ingenious method of comparing an index of the price of output of an industry with a weighted index of prices of its inputs. There was much difficulty, with the wholly inadequate statistical information of those days, in obtaining the necessary weights. The 'price' of capital he put at the current rate of interest; and obtained prices for labour and materials.

The industries covered by Jones were the cotton industry in Lancashire and in Massachusetts; iron production in Cleveland, one of the minor British iron producing districts; and the London building trade.

Jones found no evidence of increasing returns in Cleveland iron, and practically none in Lancashire cotton, but some in the growing Massachusetts industry. The London building trade showed almost constant returns to scale over the whole

period from 1840 to 1910, with one exception, namely the introduction of machinery in joinery workshops in the 1870s.

Cotton had been Britain's principal export industry up to the time of World War I. Its apparent inability to attain any economies of scale was a matter of cardinal importance, which contemporaries apparently failed to notice.

G.T. Jones left a posthumous son whose initials were also G.T., apparently with inherited skill – he is a highly capable economist in the Agricultural Economics Institute at Oxford.

COLIN G. CLARK

SELECTED WORKS

1933. *Increasing Return*. Ed. Colin Clark, Cambridge: Cambridge University Press.

BIBLIOGRAPHY

Robbins, L. 1971. *Autobiography of an Economist*. London: Macmillan.
Young, A.A. 1928. Increasing returns and economic progress. *Economic Journal* 38, December, 527–42.

Jones, Richard (1790–1855). Jones was born at Tunbridge Wells, Kent. After finishing his studies at Cambridge in 1816 he took holy orders and served as a curate at various places in England for the next decade and a half. During this time he developed an interest in political economy which culminated in his *Essay on the Distribution of Wealth: Vol. I. Rent* (1831a). Soon after publication he was appointed Professor of Political Economy at the newly established King's College, London. In 1835, following the death of Malthus, he was appointed Professor in the East India College at Haileybury and remained there until his death in 1855. He took an active part in the commutation of tithes and served as a commissioner of tithes from 1836 to 1851.

Jones never wrote the proposed second volume of his book and published very little else during his lifetime. The lectures he gave at King's College and East India College, together with other sundry essays and notes, were published soon after his death as *Literary Remains* (edited by W. Whewell, 1859). A persistent theme in Jones's work is a critique of the ahistorical, deductivist methods of the Ricardian school of political economy. He argued for a method he called 'inductivist' and was primarily concerned to overturn the Ricardian theory of rent with an historically based theory that distinguished between farmers' rents and various categories of peasant rents. He also developed a number of theoretical propositions concerning population and technology that contradicted the Malthusian orthodoxy.

Jones's iconoclastic theories were not well received by his contemporaries. McCulloch, in an extended review in *The Edinburgh Review* (1831), dismissed Jones's book as 'superficial', 'lacking in originality' and 'signally abortive' in its attempt to overthrow the Ricardian theory of rent. This opinion was generally held in the 19th century. However, he has acquired something of a reputation in the 20th century. Marx's favourable review of Jones in *Theories of Surplus–Value* (1905–10, ch. 24) has been a major contributing factor in this rehabilitation. Marx argued that Jones's theories were a substantial advance on Ricardo because, among other things, Jones had a sense of the historical differences in modes of production and was thus able to conceptualize rent as a form of surplus labour. Jones's theory of peasant rents has also attracted much attention. When his book was reprinted for the first time in 1914, for example, only the first half of his book on peasant rents was republished (see Jones, 1831b [1914], *Peasant Rents*). Historians of the role of British economic thought in India have shown that his theory of peasant rent had an important impact on policy debates in India in the latter part of the 19th century (Ambirajan, 1978, p. 175) and have assessed his theoretical contribution vis-à-vis the Ricardian school very favourably (Barber, 1975, ch. 12). Jones's approach to understanding the unfamiliar circumstances of rural India continues to have its advocates even today (Hill, 1982, pp. 14–15).

Miller, in two recent reviews of Jones's contribution to the history of economic thought, has attempted to assess the reputation to which Jones's orginality entitles him as distinct from the reputation that he has acquired. He finds that 'Jones did not really have a distinct inductive approach to offer' (1971, p. 206) and that his theory of rent 'largely deserved McCulloch's harsh judgement that it lacked originality' (1977, p. 360).

Originality is a difficult quality to assess because the theoretical perspective of the observer obviously affects any judgement made. Nevertheless it is clear that Jones's rehabilitation owes more to his advocacy of a method than to his theories. But this method is not 'inductivist'. Jones, as Miller (1971) correctly points out, employs both inductive and deductive reasoning. This is not evidence of a contradiction in Jones's thought, as Miller would argue. Jones's use of the term 'inductivist' is a simple misnomer. What is distinctive about Jones's method is the comparative and historical perspective he adopts. This method is now the basis of many non-neoclassical approaches to the economy. Not only does Jones deserve to be regarded as the founder of the English historical school (Edgeworth, 1899), he also deserves to be regarded as the founder of the English comparative economy school because of his contribution to the theory of peasant economy.

C.A. GREGORY

SELECTED WORKS

1831a. *An Essay on the Distribution of Wealth and on the Sources of Taxation*. Vol. I: Rent. New York: Augustus M. Kelley, 1964.
1831b. *Peasant Rents*. New York: Macmillan, 1914.
1859. *Literary Remains*. Ed. W. Whewell, New York: Kelley, 1964.

BIBLIOGRAPHY

Ambirajan, S. 1978. *Classical Political Economy and British Policy in India*. Cambridge: Cambridge University Press.
Barber, W.J. 1975. *British Economic Thought and India 1600–1858: A Study in the History of Development Economics*. Oxford: Clarendon Press.
Edgeworth, F.Y. 1899. Jones, Richard. *Dictionary of Political Economy*, ed. R.H.I. Palgrave, Vol. II, London: Macmillan.
Hill, P. 1982. *Dry Grain Farming Families*. Cambridge: Cambridge University Press.
Marx, K. 1905–10. *Theories of Surplus Value*. Pt III, Moscow: Progress, 1971.
McCulloch, J.R. 1831. Review of *An Essay on the Distribution of Wealth* by R. Jones. *Edinburgh Review* 54, 84–99.
Miller, W.L. 1971. Richard Jones: a case study in methodology. *History of Political Economy* 3, 198–207.
Miller, W.L. 1977. Richard Jones's contribution to the theory of rent. *History of Political Economy* 9, 346–68.

Joplin, Thomas (c1790–1847). Thomas Joplin's advocacy of joint stock banking at a time when the Bank of England enjoyed the exclusive right to that form of organization has earned him a place in banking history. Born about 1790 in Newcastle upon Tyne, he published a pamphlet there in 1822 in which he contrasted the many banking failures in England with the failure-free Scottish banks. It was their joint stock basis that according to Joplin accounted for their superior safety. He circulated the pamphlet widely, discussed it with Ricardo and, after Parliament repealed laws preventing the establishment of joint stock banks in Ireland, was instrumental in founding the Provincial Bank of Ireland in June 1824 with headquarters in London. Joplin was active in its management until 1828, when he left it to help launch the National Provincial Bank of England – his true memorial. The opposition of the central bank, country bankers, and London private banks, and the reluctance of investors delayed its formation for years (Ellis, 1953, 16).

National Provincial of England was established as a joint stock country bank in September 1833 under the Act of 1826 which repealed the six-partner rule for banks of issue situated at least 65 miles from London. It began operations in Gloucester in 1834. Although Joplin had discerned as early as 1823 that the Bank of England's note monopoly did not preclude deposit banking in London (explicitly authorized ten years later by the Banking Act of 1833), the head office in London, which administered the affairs of the outlying units, did no banking business there until 1866. As in the case of the Provincial Bank of Ireland, Joplin proposed that National Provincial consist not of branches but of a group of local banks with local shareholders 'deriving their profits, or incurring their losses (as the case may be) from or at and by means of such separate banks' (quoted from the bank's newspaper notice, Withers, 1933, p. 44). The scheme involved the parent bank in extending credit to the unit banks in return for a share of their profits. Given a trial in which it proved unworkable, the scheme was abandoned. Following this decision, a disgruntled Joplin quit the board of directors on which he served from 1833 to 1836, but persisted fruitlessly until 1842 in seeking compensation from the stockholders for past services. A biographer of the bank concludes, 'One can only suppose that he was a tactless and difficult person who got very much on the wrong side of his colleagues' sympathies' (Withers, 1933, p. 47). Joplin died in 1847 in Silesia, where he had gone for his health.

Joplin was a prolific writer. In the two decades following the publication of the 1822 pamphlet, he wrote a dozen works with the same core of ideas on currency and banking questions, the balance of payments, and the London money market. Many were issued in several editions. Some of his ideas about existing banking arrangements were perceptive. He was aware that the clearing process limited the expansion of an individual bank but did not so constrain the banking system as a whole (*Currency Reform*, 1844, pp. 43–5). He stated the doctrine of forced saving (*Views*, 1828, p. 146). Anticipating Bagehot, he understood that to stop a run, a source is needed 'of creating new money to meet the occasion' (*An Analysis*, 1832, 150, 248).

To reform the currency system, Joplin propounded over the years a plan to make the government the monopoly issuer of bank notes, replacing the Bank of England and the country banks. He alleged that Ricardo's posthumous *Plan for National Bank* was not original but taken from his work without acknowledgement and with details altered for the worse (*An Analysis*, 1832, pp. 178–81). Joplin made a contribution to the monetary debates of the period but his intellectual influence was limited.

ANNA J. SCHWARTZ

SELECTED WORKS

1822. *An Essay on the General Principles and Present Practice of Banking in England and Scotland*. Newcastle: E. Walker. 7th edn, London: Baldwin, Craddock & Joy.

1823. *Outlines of a System of Political Economy, written with a view to prove that the cause of the present Agricultural Distress is entirely artificial and to suggest a plan for the management of the currency.* London: Baldwin, Craddock & Joy.

1828. *Views on the Currency: in which the Connexion Between Corn and Currency is Shown: the Nature of our System of Currency Explained; and the Merits of the Corn Bill, the Branch Banks, the Extension of the Bank Charter, and the Small Note Act Examined.* London: James Ridgway and Baldwin & Craddock.

1832. *An Analysis and History of the Currency Question together with an Account of the Origin and Growth of Joint Stock Banking in England.* London: James Ridgway.

1839. *On Our Monetary System with an Explanation of the Causes by which the Pressures in the Money Market are Produced and a Plan for their Remedy.* London: James Ridgway.

1844. *Currency Reform: Improvement, Not Depreciation.* London: Richardson.

BIBLIOGRAPHY

Ellis, A. 1953. *Bold Adventure: The Pioneering Story of a Great Enterprise.* London: National Provincial Bank.

Withers, H. 1933. *National Provincial Bank, 1833 to 1933.* London: National Provincial Bank.

Juglar, Clément (1819–1905). Like his rather more illustrious compatriot François Quesnay, Juglar is an example of a physician turned economist. The circular flow of economic life – which it is often said Quesnay saw in terms of an analogy to the circulatory system – in Juglar's work seems have as its counterpart the view of the economic process as one of quasi-rhythmical variations between good and bad trade. This simple idea has been of profound importance in the study of alterations in the conditions of economic prosperity ever since. Both Wesley Clair Mitchell and Joseph Schumpeter in their classic studies of business cycles (in 1927 and 1939 respectively) credit Juglar's contribution as having been seminal in the field. For Mitchell, it was Juglar's recognition of the *cyclical* character of economic crises that established him as a pioneer (1927, p. 452); for Schumpeter it was Juglar's perception of how theory, statistics and history ought to contribute to the study of industrial fluctuations (1939, pp. 162–3). There is something to each of these claims, but it should not be forgotten that other authors had also done much in both of these areas – one may mention Samuel Jones Loyd, John Wade and Amasa Walker. As theorists of industrial fluctuations, of course, Sismondi, Rodbertus and Marx would also need to be mentioned.

Juglar practised as a physician until 1848. His first work in the social sciences was on the cyclical pattern of birth, death, and marriage rates in France, and it appeared in the *Journal des Economistes* in October–December 1851 and January–June 1852. He moved on to examine the discount policy of the Bank of France and published his findings in the *Annuaire de l'économie politique* for 1856 and in the *Journal des Economistes* for April–May 1857. In 1852 he was elected into the Société d'Economie Politique and he was one of the founders of the Société de Statistique de Paris in 1860. In 1868 he published an account of the policies and practices of the French monetary authorities and their effects on the exchanges.

There is, however, little doubt that Juglar's most important work on business cycles is his *Des crises commerciales et de leur retour périodique en France, en Angleterre, et aux Etats-Unis*, first published in 1860. Juglar's analysis of crises is essentially a monetary one – protracted periods of inflation and expansion are brought to an end when the banking system initiates a contraction in the face of unacceptable pressures on its specie reserves. This is very like the story Wicksell was later to tell, but without the sophistication of Wicksellian theory. Subsequent theories of the business cycle, which attributed the process to 'real' causes, were critical of this aspect of Juglar's argument. The observed periodicity of the cycle – of nine to ten years – is commonly known in the applied literature on business cycles as a Juglar cycle.

MURRAY MILGATE

SELECTED WORKS

1860. *Des crises commerciales et leur retour périodique en France, en Angleterre, et aux Etats-Unis.* 2nd edn, 1889.
1868. *Du change et de la liberté d'émission.*

BIBLIOGRAPHY

Mitchell, W.C. 1927. *Business Cycles: The Problem and its Setting.* New York: National Bureau of Economic Research.

Schumpeter, J.A. 1939. *Business Cycles: A Theoretical, Historical, and Statistical Analysis of the Capitalist Process.* 2 vols, New York: McGraw-Hill.

jurisprudence. Jurisprudence is the general theory of law: the study of what law is, what it is for, and how it comes into being. Until the 18th century there were essentially two approaches to this study. One regarded law as the expression of political power; the other considered it to be the expression of justice.

The first line of thought, today loosely known as positivism, holds that a rule is law because it has been laid down (*positum*) by whatever body has authority in the community. Already in Plato's *Republic*, Thrasymachus argued that 'just' and 'right' are merely names given by the power-holders in the state to the types of conduct they wish to impose on their subjects. With the rise of nation states in the 16th century, the theory was expressed in more sophisticated form. Sovereignty meant freedom from any kind of external limitation. Thus when the sovereign legislated, his will could not be subjected to any restriction. Even custom only became law when it had been confirmed by the sovereign's will. Thomas Hobbes developed the notion in *Leviathan* (1651), as part of his explanation of how civil society grew out of the state of nature. In his view, that condition was a war of all against all, in which each man sought only his own personal advantage. Men could enjoy the benefits of civilization only by submitting themselves to a ruler who would give them the security that they needed. Once established as sovereign, the ruler was subject to no control by his subjects, so long as he was able to offer them personal protection. Law was therefore nothing more than the commands of the sovereign.

The alternative line of thought holds that no rule can be considered to be law without regard to its moral content. Laws must be just and what is just in any given situation can be discovered from the nature of man as a rational and social animal. Whatever their cultural differences, all men are so constituted by nature (in the Stoic view set out, for example, by Cicero in *De legibus*) or by God (in the Christian view put forward, for example, by St Thomas Aquinas), that they share a common sense, or natural reason. It is this which tells them what is just and what is unjust. Until the 17th century those who adopted this approach, the natural lawyers, did not suggest that all laws were or could be natural. They accepted that many laws were merely positive, based on considerations of utility, but argued that natural law provided a criterion and that no rule could be law which actually contradicted it.

From the 17th century onwards there have been two versions of natural law theory. One is concerned to safeguard the position of the individual in the community. It exploited the ambiguity of the word for law in Latin and most European languages other than English (e.g., *ius, droit, Recht*), which means both the (objective) law in general and a (subjective) right enjoyed by an individual. Natural law came to be seen as concerned with the natural rights, particularly those of life, liberty and property, which belong to individual men in a state of nature. For John Locke, when men join together in civil society under a ruler, they cannot transfer to him more power than they have themselves. Since no one has absolute arbitrary power over himself, the ruler cannot receive from his subjects power to interfere with their natural rights, and any legislation which purports to do so is void. This natural rights theory provided the basis for the English settlement of 1688, which subordinated the Crown and its servants to the law, and a century later for the American Constitution under which any legislation which infringes the natural rights enshrined in it is void. Recently Ronald Dworkin (1978) has given new life to this line of thought.

The other version of natural law theory was largely motivated by the desire to present law as a science organized according to rational principles. Its principal exponents, Grotius and Pufendorf, held that the content of law must be justified in terms of principles which are as axiomatic as those of geometry and which have absolute validity in all times and places. Their 18th-century successors went further and argued that once these principles are established, complete systems of universal legal rules can be deduced from them by logic.

Although the obligatory character of natural law was derived from the divine will, discussion of its institutions was in terms of rational analysis of the needs of man in social life and they were generally justified by an often rather vague notion of maximizing the welfare of society. The removal of God from most accounts of natural law in the 18th century meant that its obligatory character had now to be provided by human will and its content justified by a secular utilitarianism. Several blueprints of 'natural' systems were offered to the enlightened rulers of the day to be enacted by them into law, foreshadowing the codification movement of the 19th century.

Jeremy Bentham and his disciple John Austin (*The province of jurisprudence determined*, 1832) insisted on a sharp distinction between the validity of law and its content. Building on Hobbes's notion, they held that laws are the commands of the sovereign, for whom utilitarianism provides a guide, and jurisprudence is merely their systematic exposition. In Austin's analysis, 'law properly so-called' requires an independent political society, the bulk of whose members must be in the habit of obedience to a person or body, not itself habitually obedient to any other person or body. Only the commands of such a 'sovereign' are law. The model was Parliamentary legislation and Benthamite–Austinian ideas had great influence on the spate of legislation enacted by Parliament after the Reform Act of 1832. However, the notion is not readily applicable either to the English common law, which was largely judge-made, although its rules could always be altered by Parliament, or to statutory rules enacted in a federal state where the powers of the different legislatures are limited by a constitution.

The positivist position has been forcibly re-stated in contemporary terms by H.L.A. Hart (1961), who concentrated on the obligatory character of the rules that make up the legal system. He developed the distinction between primary rules, which govern behaviour, and secondary rules, which specify the ways in which the primary rules can be identified ('the rules of recognition') or altered or applied through adjudication. But Hart is reluctant to accept the position that laws may

have any content at all and argues for a minimum form of protection for persons, property and promises, in any legal system.

The mid-18th century saw the beginnings of a third approach to jurisprudence. Montesquieu, in *De l'Esprit des lois* (1748) started from the apparently orthodox natural law position that law in general is human reason and that the laws of each nation should be the particular applications of that reason. He then demonstrated, however, that the laws most conformable to nature were those best adapted to the particular circumstances of each society. For the nature of things varies from one society to another and so laws must vary with the climate, soil, principal occupation of the people, their religion, manners and so on. Such factors together make up the spirit of a society to which the laws must conform. A few years later, in his *Lectures on Jurisprudence*, Adam Smith combined these ideas with that of the progress of societies from barbarism to civilization. Like the positivists, he was concerned to distinguish what a man can be compelled to do from what he ought morally to do, but argued that this changes as societies move from the hunting stage to the stages of shepherds, farmers and merchants. Like Grotius and Pufendorf, he thought of private law as primarily concerned with property and contract and showed that the relevant rules change according to what the 'impartial spectator' in a society considers to be property or according to what are his expectations on receiving a promise from another.

In the 19th century the German historical school, led by F.K. von Savigny, opposed the movement for codification of law with the romantic idea that a people's law, like its language, is inextricably linked to its traditional culture, so that it evolves not through the external will of a legislator but through 'internal silently operating forces'. In England, Sir Henry Maine (*Ancient Law*, 1861) adapted this notion to combat Benthamism, arguing that in 'progressive societies' law develops through recognized stages without the need for legislation. For example, he demonstrated the gradual substitution of the individual for the family as the unit with which the law is concerned, leading to the ordering of a person's social relations through free agreement rather than by his family position ('the movement from Status to Contract').

Although many of its propositions were incompatible with the findings of anthropology and legal history and were unacceptable to those concerned with reforming the law, historical jurisprudence did concentrate attention on the way law actually operates in society. Studies in the sociology of law have shown that it is misleading to see law as just a system of rules. People actually govern their behaviour by a variety of norms, such as custom or professional practice, and formal law is only one component. Jurisprudence should therefore be concerned with the interaction of the formal law with these other forms of social order.

This approach was developed, particularly in the USA, by the so-called legal realists. They focused attention on court decisions, and argued that law is nothing more than the prediction of what judges and officials concerned with the administration of the law will do in fact. They showed that although judges often rationalize their decisions in syllogistic form to emphasize legal certainty, they are in fact influenced by many factors other than existing rules. Policy preferences and unconscious social prejudices all affect decisions and must therefore form part of the subject matter of jurisprudence.

By contrast, the economic analysis of law is part of a larger movement to apply the economic model to an ever wider range of human behaviour and social institutions. Its main exponents, for example R.A. Posner (1977), argue that, when the legal regulation of non-market conduct in England and America is seen in economic terms, common law judges appear to have made their decisions in hard cases in such a way as to promote efficient resource allocation. It is not clear whether such findings are merely descriptive or also normative.

In recent years jurisprudence has been increasingly concerned with the process of legal reasoning, and with the question whether there is a particular legal logic. It is acknowledged that apart from legal rules, decisions are affected by certain fundamental notions, which must therefore be regarded as part of the system. Whereas traditionally jurisprudence was concerned mainly with private transactions between one citizen and another, it is today concerned more with the regulation of public power over the individual. There is a tension between those who see law as embodying certain values, such as individual liberty, procedural justice etc., which control the application of the rules, and those who see it instrumentally, as concerned only with the techniques of achieving certain goals settled by policies with the content of which the law itself is not concerned.

The latest current in American jurisprudence is the radical movement known as Critical Legal Studies, founded at a conference in Wisconsin in 1977. Its adherents reject the notion that legal practices derive from identifiable doctrines or principles and deny the existence of any specifically legal form of reasoning. Influenced by sociologists of knowledge, they argue that legal thought has claimed to be objective, when it is really dominated by ideologies and (oppressive) social orders. They reject the findings of the Economics of Law school and argue that plausible justifications can be found for any decisions which are desired on policy grounds.

P.G. STEIN

See also COMMON LAW; CONSTITUTIONAL ECONOMICS; LAW AND ECONOMICS; NATURAL LAW.

BIBLIOGRAPHY
Austin, J. 1832. *The Province of Jurisprudence Determined*. London: John Murray.
Dworkin, R.M. 1977. *Taking Rights Seriously*. 2nd edn, London: Duckworth, 1978.
Hart, H.L.A. 1961. *The Concept of Law*. Oxford: Clarendon Press.
Maine, H. 1861. *Ancient Law*. London: Murray.
Posner, R.A. 1973. *The Economic Analysis of Law*. 2nd edn, Boston, Mass.: Little, Brown & Co., 1977.
Smith, A. 1763. *Lectures on Jurisprudence*. Ed. R.L. Meek, D.D. Raphael and P.G. Stein, Oxford: Clarendon Press, 1978.

Justi, Johannn Heinrich Gottlob von (1720–1771). Born the son of a tax inspector in Thuringia in December 1720, Justi is best known as one of the architects of mid-18th-century Cameralism. He studied law at Wittenberg 1742–4, and after this embarked upon a career of literary activity and state service. During the early 1850s he taught 'commerce and public economics' in Vienna, and in 1755 published the substance of his teaching in his textbook *Staatswirthschaft*. By this time he was lecturing part-time in Göttingen, after 1757 resuming employment as an official with various rulers, in 1765 being appointed Prussian Inspector of Mines, Glass and Steel works. Accused of misusing state funds, he was imprisoned in 1768 and died in the fortress at Küstrin in 1771.

Justi's literary output and journalistic activity was extensive, ranging over aesthetics, philosophy, history, politics and economics. His major work is the *Staatswirthschaft* (1755, 1758), literally 'state economy', which details the manner in which a ruler should govern his lands to assure the 'happiness of the state' and a flourishing population. Cameralism had begun as a systematization of the principles followed by the administrators of the ruler's domains. In Justi these principles are identified with the management of the absolutist state, in which economic welfare is conceived as the path to political power. Welfare and wealth are produced by good government and the implementation of 'good police' – *Polizei* in the 18th-century sense of regulations covering all aspects of social action and public order. The 'science of police' is covered in a further textbook, *Grundsätze der Polizei-Wissenschaft*, 1756, 1759, 1782), which is claimed by Justi to provide the first systematic treatment of the subject, and which was in fact republished after his death in a revised edition. Justi's influence was strong during the later 18th century, and only diminished with the general decline of Cameralism at the turn of the century.

K. TRIBE

SELECTED WORKS
1755, 1758. *Staatswirthschaft*. 2 vols. Leipzig.
1756, 1759, 1782. *Grundsätze der Policey-Wissenschaft*. Göttingen.
1759. *Der Grundriss einer guten Regierung*. Frankfurt am Main.

justice.

1. JUSTICE AND UTILITARIANISM. The concept of justice is often invoked in economic discussions. Its relevance to economic evaluation is obvious enough. However, it is fair to say that in traditional welfare economics, when the notion of justice has been invoked, it has typically been seen only as a part of a bigger exercise, viz., that of social welfare maximization, rather than taking justice as an idea that commands attention on its own. For example, in utilitarian welfare economics (e.g. Pigou, 1952; Harsanyi, 1955) the problem of justice is not separated out from that of maximization of aggregate utility. This situation has been changing in recent years, partly as a result of developments in moral philosophy dealing explicitly with the notion of justice as a concept of independent importance (see especially Rawls, 1971, 1980).

In the utilitarian formulation the maximand in all choice exercises is taken to be the sum-total of individual utilities. The approach can be seen as an amalgam of three distinct principles: (1) welfarism, (2) sum-ranking, and (3) consequentialism. *Welfarism* asserts that the goodness of a state of affairs is to be judged entirely by the utility information related to that state, i.e., by information about individual utilities. All other information is either irrelevant, or only indirectly relevant as a causal influence on utilities (or as a surrogate for utility measures when such measurement cannot be directly done). The second principle is *sum-ranking*, which asserts that the goodness of a collection of utilities (or welfare indicators) of different individuals, taken together, is simply the sum of these utilities (or indicators). This eliminates the possibility of being concerned with inequalities in the distribution of utilities, and the overall goodness or 'social welfare' is seen simply as the aggregate of individual utilities. The third principle is *consequentialism*, which requires that all choice variables, such as actions, rules, institutions, etc., must be judged in terms of the goodness of their respective consequences. The overall effect of combining these three principles is to judge all choice variables by the sum-total of utilities generated by one alternative rather than another.

2. SUM-RANKING AND EQUALITY. A theory of justice can take issue with each of the principles underlying utilitarianism, and

in fact in the literature that has developed in recent decades, each of these principles has been seriously challenged (see the papers included in Sen and Williams, 1982). Some critiques have been particularly concerned with assessing and questioning the axiom of sum-ranking, and have considered the claims of equality in the distribution of well-being (see, for example, Phelps, 1973; Sen, 1973, 1977, 1982; Kern, 1978).

The summation formula can be defended either directly (e.g., in terms of attaching equal importance to everyone's 'interest': see Hare, 1981, 1982), or indirectly through invoking some model of 'impersonality' or 'fairness' (e.g., involving a hypothetical choice in a situation of primordial uncertainty, in which each person has to assume that he or she has an *as if* equal probability of becoming anybody else: see Vickrey, 1945; Harsanyi, 1955). Other routes to deriving sum-ranking involve independence or separability requirements of various kinds (see d'Aspremont and Gevers, 1977; Deschamps and Gevers, 1978; Maskin, 1978; Gevers, 1979; Roberts, 1980; Myerson, 1981; Blackorby, Donaldson and Weymark, 1984; d'Aspremont, 1985).

Whether the defences obtainable from these approaches are convincing enough has been a matter of some dispute. There have also been some interpretative discussions as to whether giving equal importance to everyone's 'interest' does, as alleged, in fact yield the formula of summing individual utilities *irrespective* of distribution, and also whether the additive formula that is obtained on the basis of hypothetical primordial choice is, in fact, a justification for adding individual utilities as they might be *substantively* interpreted in welfare economic exercises (see Pattanaik, 1971; Smart and Williams, 1973; Sen, 1982, 1985a; Blackorby, Donaldson and Weymark, 1984; Williams, 1985). It is not obvious that this debate has been in any way definitively concluded one way or the other.

3. THE DIFFERENCE PRINCIPLE AND LEXIMIN. Meanwhile, much attention has been paid to developing welfare-economic rules based on taking explicit note of inequalities in the distribution of utilities. A definitive departure on this came from Suppes (1966). Another major approach was developed in Rawls's (1971) *Theory of Justice*, even though Rawls himself was concerned not so much with the distribution of *utilities* but with that of the indices of primary goods (on which more later). The concern with the utility level of the worst-off individual has been formalized and reflected in various formulae suggested or derived in the rapidly growing welfare-economic literature on this theme. In particular, James Meade (1976) has provided an extensive treatment of this type of distributional issues, and it has also been penetratingly analysed by Kolm (1969), Phelps (1973, 1977), Atkinson (1975, 1983), Blackorby and Donaldson (1977), and others.

In fact, the Rawlsian 'Difference Principle', which judges states of affairs by the advantage of the least well-off person or group, has often been axiomatized in welfare economics and in the social-choice literature by equating advantage with utility. In this form, the 'lexicographic maximin' rule (proposed in Sen, 1970) has been axiomatically derived in different ways. The rule judges states of affairs by the well-being of the worst-off individual. In case of ties of the worst-off individuals' utilities, the states are ranked according to the utility levels of the *second* worst-off individuals respectively. In case of ties of the second worst-off positions as well, the third worst-off individuals' utilities are examined. And so on.

There is no necessity to interpret these axioms in terms of utilities only, and in fact the analytical results derived in this part of the social-choice literature can be easily applied

without the 'welfarist' structure of identifying individual advantage with the respective utilities. Various axiomatic derivations of lexicographic maximin – 'leximin' for short – can be found in Hammond (1976), Strasnick (1976), Arrow (1977), d'Aspremont and Gevers (1977), Sen (1977), Deschamps and Gevers (1978), Suzumura (1983), Blackorby, Donaldson and Weymark (1984), d'Aspremont (1985), among others. These can be seen as exercises that incorporate concern for reducing inequality, related to recognizing the claims of justice.

While the Rawlsian approach rejects the aggregation procedure of utilitarianism (i.e., ranking by sums), a major aspect of the Rawlsian theory involves the rejection of utility as the basis of social judgements (i.e., welfarism). Rawls (1971) argues for the priority of the 'principle of liberty', demanding that 'each person is to have an equal right to the most extensive basic liberty compatible with similar liberty for others'. Then, going beyond the principle of liberty, claims of efficiency as well as equity are both supported by Rawls's 'second principle' which *inter alia* incorporates his 'Difference Principle' in which priority is given to furthering the powers of the worst-off group. These powers are judged by indices of 'primary social goods' which each person wants (Rawls, 1971, pp. 60–65).

Primary goods are 'things that every rational man is presumed to want', including 'rights, liberties and opportunities, income and wealth, and the social bases of self-respect'. The Difference Principle takes the form, in fact, of maximin, or lexicographic maximin, based on interpersonal comparisons of indices of primary goods. This rule can be axiomatized in much the same way as the other 'lexicographic maximin' rule based on utilities, and all that is needed is a reinterpretation of the content of the axioms (with the objects of value being indices of primary goods rather than utilities).

The Rawlsian approach to justice, therefore, involves rejection both of welfarism and of sum-ranking. Furthermore, consequentialism is disputed too, since the priority of liberty might possibly go against judging all choice variables by consequences only. At least, in the more standard forms, consequentialism does involve such a conflict, even though it is arguable that the problem can be, to a great extent, resolved by a broader understanding of consequences, which takes into account the fulfilment and violation of liberties and rights, and also the agent's special role in the actions performed (Sen, 1985a).

4. UTILITIES, PRIMARY GOODS AND CAPABILITIES. The claim of primary goods to represent the demands of justice better than utilities is based on the idea that utilities do not reflect a person's advantage (in terms of well-being or powers) adequately. It is arguable that in making interpersonal comparisons of advantage, the metric of utilities (either in the form of happiness, or of desire fulfilment) may be biased against those who happen to be hopelessly deprived since the demands of unharrassed survival force people to take pleasure in small mercies and to cut their desires to shape in the light of feasibilities (see Sen, 1985a, 1985b). The status of 'preference' may be disputed in view of the need for critical assessment (see Broome, 1978; McPherson, 1982; Goodin, 1985, among others). Also, what types of pleasures should 'count' can itself be a matter for an important moral judgement. As Rawls (1971) points out, in the utilitarian formulation, we have the unplausible requirement that

if men take a certain pleasure in discriminating against one another, in subjecting others to a lesser liberty as a means of enhancing their self-respect, then the satisfaction of these desires must be weighed in our deliberations according to their intensity, or whatever, along with other desires (pp. 30–31).

These and other types of difficulties have been dealt with by some utilitarians through moving to less straight-forward versions of utilitarianism, for example Harsanyi's (1982) exclusion of 'all antisocial preferences, such as sadism, envy, resentment, and malice' (p. 56); see also the refinements proposed by Hare (1981, 1982), Hammond (1982) and Mirrlees (1982).

Recently, it has been argued that primary goods themselves may be rather deceptive in judging people's advantages, since the ability to convert primary goods into useful capabilities may vary from person to person. For example, while the same level of income (included among 'primary goods') may give each person the same command over calories and other nutrients, the nourishment of a person depends also on other parameters such as body size, metabolic rates, sex (and if female, whether pregnant or lactating), climatic conditions, etc. This indicates that a more plausible notion of justice may demand that attention be directly paid to the distribution of basic capabilities of people (see Sen, 1982, 1985b). The approach goes back to Smith's (1776) and Marx's (1875) focus on fulfilling needs.

The achievement of capabilities will, of course, be causally related to the command over primary goods, and the capabilities, in their turn, will also influence the extent to which utilities are achieved, so that the various alternative measures will not be independent of each other. However, the basic issue is the variable that should be chosen to serve as the proper metric for judging advantages of people – the equity and the distribution of which could form the foundations of a theory of justice. On this central issue several alternative views continue to flourish in the literature.

5. FAIRNESS AND ENVY. A view of justice that is not altogether dissimilar from Rawls's concerned with primary goods is captured by the literature on 'fairness', inspired by a pioneering contribution of Foley (1967). In this approach a person's relative advantage is judged by the criterion as to whether he or she would have preferred to have had the commodity bundle enjoyed by another person. This has been seen as a criterion of 'non-envy'. If no one 'envies' the bundle of anyone else, the state of affairs is described as being 'equitable'. If a state is both equitable and Pareto efficient, it is described as being 'fair' (even though the term fairness is also sometimes used interchangeably with only 'equitability').

There has been an extensive literature on existence problems, in particular whether equitability can be combined with efficiency in all circumstances. (The answer seems to be no, especially when production is involved: Pazner and Schmeidler, 1974.) There has also been considerable exploration of the effects of varying the criterion of equitability and fairness to reflect better the common intuitions regarding the requirements of justice. Various results on these problems and related ones have been presented, among many others, by Foley (1967), Schmeidler and Vind (1972), Feldman and Kirman (1974), Varian (1974, 1975), Svensson (1980), and Suzumura (1983).

It should be remarked that the fairness criterion does not provide a complete ranking of alternative states. It identifies some requirements of justice, which makes the states fair. Varian (1974) has argued, with some force, that 'social decision theory asks for too much out of the process in that it asks for an entire *ordering* of the various social states

(allocations in this case)', whereas 'the original question asked only for a "good" allocation; there was no requirement to rank all allocations' (pp. 64–5). While it is true that 'the fairness criterion in fact limits itself to answering the original question', the absence of further rankings may be particularly problematic if no feasible 'fair' allocation exists incorporating efficiency (as seems to be the case in many situations). Furthermore, while a 'pass-fail' criterion of justice may have attractive simplicity, it does not follow that two states, both passing this criterion, must be seen as being 'equally just'. Various 'finer' aspects of justice have indeed been discussed in the literature (see particularly Suppes, 1966; Kolm, 1969; Rawls, 1971; Meade, 1976; Atkinson, 1983).

It should also be noted that the 'fairness' literature deals with commodity allocations, or incomes, or some other part of the set of things that figures in Rawls's characterization of 'primary goods'. The list is, in fact, much less extensive than that of primary goods as defined by Rawls (1971), and as such it leaves out many considerations that are regarded as important in the Rawlsian framework (e.g., the social bases of self-respect). On the other hand the criticisms – discussed earlier – of the Rawlsian focus on primary goods (based on recognizing inter-individual variations in the ability to convert primary goods into capabilities) would apply *a fortiori* to the fairness approach as well.

6. LIBERTY AND ENTITLEMENTS. A different type of consideration altogether is raised by the place of liberty in a theory of justice. As was mentioned before, Rawls gives it priority. This priority has been questioned by pointing to the possibility that other things (e.g., having enough food) may sometimes be no less important than enjoying liberty without restriction by others. Rawls does, of course, attach importance to these other considerations, but in view of the priority of liberty, they may end up having too little impact on judgements regarding justice in many circumstances, and this might not be acceptable (on this see Hart, 1973).

On the other hand, in some other theories of justice, the priority of liberty has been given even greater importance than in the Rawlsian structure. For example, in Nozick's (1974) theory of 'entitlements', rights are given complete priority, and since these rights are characterized quite extensively, it is not clear whether or not much remains to be supported over and above the recognition of rights. Nozick argues against any 'patterning' of *outcomes*, indicating that any outcome that is arrived at on the basis of people's legitimate exercise of their rights must be acceptable because of the moral force of rights as such. These rights, in Nozick's analysis, include not only personal liberty, but also ownership rights over property, including the freedom to use its fruits, to use it freely for exchange, and to donate or bequeath it to others (thereby asserting the legitimacy of inherited property).

This type of approach has been criticized partly on grounds of what has been seen as its 'extremism', since the constraints imposed by rights can override other important considerations, for example reducing misery and promoting the well-being of the deprived members of the society. In fact, it has been argued that a system of entitlements of the kind specified by Nozick might well co-exist with the emergence and sustaining of widespread starvation and famines, which are often the result of legally sanctioned exercises of property rights rather than of natural calamities (on this see Sen, 1981). Although Nozick does refer to the possibility that in case of 'catastrophic moral horrors' rights may be compromised, it is not at all clear how his theory would accommodate such waiving of rights, in the absence of formulation of other, competing bases of moral judgements. On the other hand, there cannot be any doubt that Nozick's theory does capture some notions of justice that can be found in a less clear form in the literature. Nozick's analysis gives a well-formulated and illuminating account of an entitlement-based approach to justice.

7. SOURCES OF DIFFERENCE. To conclude, theories of justice explicitly or implicitly invoked in the literature show a variety of ways in which the demands of justice can be interpreted. There are at least three different bases of variation. One source of variation concerns the *metric* in terms of which a person's *advantage* is to be judged in the context of assessing equity and justice. Various metrics have been considered in this context, including utility (as under utilitarianism and other welfarist theories of justice), primary goods index (as in the Rawlsian theory of justice), capabilities index (as in theories emphasizing what people can actually do or be, e.g., Sen, 1985b), incomes or commodity bundles (as in the literature on 'fairness', and on statistical measures of poverty, e.g., Foster, 1984), various notions of command over commodity bundles and resources (as in some notions of 'equality' developed in the literature, e.g., Archibald and Donaldson, 1973; Dworkin, 1981), and so on.

A second source of difference relates to the *aggregating* of diverse information regarding the advantages of different individuals. One approach, best represented by utilitarianism, sees nothing being needed to be ascertained other than the *sum-total* of the overall utilities of different people. Insofar as distributional considerations come into this exercise, they enter in the conversion of goods to be distributed into the appropriate metric of individual utilities. For example, inequality in the distribution of incomes may be disvalued in the approach of utilitarian justice because it may lead to a reduction in the sum-total of individual utilities, through (interpersonally comparable) 'diminishing marginal utilities'. Other approaches are more concerned with distributional properties related to the different individuals' relative positions (vis-à-vis each other). The Rawlsian lexicographic maximin is one example of such a distributional concern, and there are others than can be considered, such as adding concave transformations of the individual utility indices (e.g., the additive formula used by Mirrlees, 1971, for his taxation assessment), and using various 'equity' axioms (e.g., Kolm, 1969, 1972; Sen, 1973; 1982; Atkinson, 1975, 1983; Hammond, 1976, 1979; d'Aspremont and Gevers, 1977; Roberts, 1980).

The third issue concerns the claimed *priority* of some particular aspect of a person's advantage (e.g., Rawls's insistence on the priority of liberty), *or* non-consequentialist priority of some processes over results (e.g., Nozick's, 1974, view of rights serving as unrelaxable constraints; or ideas of exploration based on counterfactual exercises of shared rights to social resources, e.g., Roemer, 1982).

Given the diversity of moral intuitions related to the complex notion of justice, which has been extensively used over centuries to arrive at normative assessment, it is not surprising that various theories of justice have been proposed in the economic and philosophical literature. The exercise of clearly understanding what the differences between distinct theories of justice consist of (and arise from) is, in some ways, the first task. This essay has been concerned with that task.

AMARTYA SEN

BIBLIOGRAPHY

Archibald, G.C. and Donaldson, D. 1979. Notes on economic equality. *Journal of Public Economics* 12, 205–14.

Arrow, K.J. 1951. *Social Choice and Individual Values.* New York: Wiley.

Arrow, K.J. 1963. *Social Choice and Individual Values.* 2nd edn, New York: Wiley.

Arrow, K.J. 1977. Extended sympathy and the possibility of social choice. *American Economic Review* 67, 219–25.

Atkinson, A.B. 1975. *The Economics of Inequality.* Oxford: Clarendon Press.

Atkinson, A.B. 1983. *Social Justice and Public Policy.* Brighton: Wheatsheaf; Cambridge, Mass.: MIT Press.

Bentham, J. 1789. *An Introduction to the Principles of Morals and Legislation.* London: Payne. Reprinted, Oxford: Clarendon Press, 1907.

Blackorby, C. and Donaldson, D. 1977. Utility versus equity: some plausible quasi-orderings. *Journal of Public Economics* 7(3), 365–82.

Blackorby, C., Donaldson, D. and Weymark, J.A. 1984. Social choice with interpersonal utility comparisons: a diagrammatic introduction. *International Economic Review* 25(2), 327–56.

Broome, J. 1978. Choice and value in economics. *Oxford Economic Papers* 30(3), 313–33.

d'Aspremont, C. and Gevers, L. 1977. Equity and the informational base of collective choice. *Review of Economic Studies* 44(2), 199–209.

d'Aspremont, C. 1985. Axioms for social welfare orderings. In *Social Goods and Social Organization: Essays in Memory of Elisha Pazner,* ed. L. Hurwicz, D. Schmeidler and H. Sonnenschein, Cambridge: Cambridge University Press.

Deschamps, R. and Gevers, L. 1978. Leximin and utilitarian rules: a joint characterisation. *Journal of Economic Theory* 17(2), 143–63.

Dworkin, R. 1981. What is equality? *Philosophy and Public Affairs* 10, 185–246.

Feldman, A. and Kirman, A. 1974. Fairness and envy. *American Economic Review* 64(6), 995–1005.

Foley, D. 1967. Resource allocation and the public sector. *Yale Economic Essays* 7(1), 45–98.

Foster, J. 1984. On economic poverty: a survey of aggregate measures. *Advances in Econometrics* 3.

Gevers, L. 1979. On interpersonal comparability and social welfare orderings. *Econometrica* 47(1), 75–89.

Goodin, R.E. 1985. *Protecting the Vulnerable.* Chicago: Chicago University Press.

Gottinger, H.W. and Leinfellner, W. (eds) 1973. *Decision Theory and Social Ethics: Issues in Social Choice.* Dordrecht: Reidel.

Hammond, P.J. 1976. Equity, Arrow's conditions and Rawls' Difference Principle. *Econometrica* 44(4), 793–804.

Hammond, P.J. 1979. Equity in two person situations: some consequences. *Econometrica* 47(5), 1127–36.

Hammond, P.J. 1982. Utilitarianism, uncertainty and information. In Sen and Williams (1982).

Hare, R.M. 1981. *Moral Thinking: Its Levels, Methods and Point.* Oxford: Clarendon Press.

Hare, R.M. 1982. Ethical theory and utilitarianism. In Sen and Williams (1982).

Harsanyi, J.C. 1955. Cardinal welfare, individualistic ethics, and interpersonal comparisons of utility. *Journal of Political Economy* 63, 309–21.

Harsanyi, J.C. 1982. Morality and the theory of rational behaviour. In Sen and Williams (1982).

Hart, H.L.A. 1973. Rawls on liberty and its priority. *University of Chicago Law Review* 40, 534–55.

Kern, L. 1978. Comparative distributive ethics: an extension of Sen's examination of the pure distribution problem. In Gottinger and Leinfellner (1978).

Kolm, S.C. 1969. The optimum production of social justice. In *Public Economics,* ed. J. Margolis and H. Guitton, London: Macmillan.

Kolm, S.C. 1972. *Justice et équité.* Paris: Edition du Centre National de la Recherche Scientifique.

Margolis, J. and Guitton, H. (eds) 1969. *Public Economics.* London: Macmillan.

Marx, K. 1875. *Critique of the Gotha Programme.* English translation, New York: International Publishers, 1938.

Maskin, E. 1978. A theorem of utilitarianism. *Review of Economic Studies* 45(1), 93–6.

McPherson, M.S. 1982. Mill's moral theory and the problem of preference change. *Ethics* 92(2), 252–73.

Meade, J.E. 1976. *The Just Economy*. London: Allen & Unwin.

Mirrlees, J.A. 1971. An exploration in the theory of optimum income taxation. *Review of Economic Studies* 38, 175–208.

Mirrlees, J.A. 1982. The economic uses of utilitarianism. In Sen and Williams (1982).

Myerson, R.B. 1981. Utilitarianism, egalitarianism, and the timing effect in social choice problems. *Econometrica* 49(4), 883–97.

Ng, Y.K. 1979. *Welfare Economics*. London: Macmillan.

Nozick, R. 1974. *Anarchy, State and Utopia*. Oxford: Blackwell.

Pattanaik, P.K. 1971. *Voting and Collective Choice*. Cambridge: Cambridge University Press.

Pattanaik, P.K. and Salles, M. (eds) 1983. *Social Choice and Welfare*. Amsterdam: North-Holland.

Pazner, E.A. and Schmeidler, D. 1974. A difficulty in the concept of fairness. *Review of Economic Studies* 41(3), 441–3.

Phelps, E.S. (ed.) 1973. *Economic Justice*. Harmondsworth: Penguin Books.

Phelps, E.S. 1977. Recent developments in welfare economics: Justice et Equité. In *Frontiers of Quantitative Economics*, vol. 3, ed. M.D. Intriligator, Amsterdam: North-Holland.

Pigou, A.C. 1952. *The Economics of Welfare*. 4th edn, London: Macmillan.

Rawls, J. 1971. *A Theory of Justice*. Cambridge, Mass.: Harvard University Press.

Rawls, J. 1980. Kantian construction in moral theory. The Dewey Lectures 1980. *Journal of Philosophy* 77.

Roberts, K.W.S. 1980. Interpersonal comparability and social choice theory. *Review of Economic Studies* 47, 421–39.

Roemer, J. 1982. *A General Theory of Exploitation and Class*. Cambridge, Mass.: Harvard University Press.

Schmeidler, D. and Vind, K. 1972. Fair net trades. *Econometrica* 40(4), 637–42.

Sen, A.K. 1970. *Collective Choice and Social Welfare*. San Francisco: Holden-Day, and Edinburgh: Oliver & Boyd; republished Amsterdam: North-Holland.

Sen, A.K. 1973. *On Economic Inequality*. Oxford: Clarendon Press; New York: Norton.

Sen, A.K. 1977. On weights and measures: informational constraints in social welfare analysis. *Econometrica* 45(7), 1539–72.

Sen, A.K. 1981. *Poverty and Famines: An Essay on Entitlement and Deprivation*. Oxford: Clarendon Press.

Sen, A.K. 1982. *Choice, Welfare and Measurement*. Oxford: Blackwell; Cambridge, Mass.: MIT Press.

Sen, A.K. 1985a. Well-being, agency and freedom: the Dewey Lectures 1984. *Journal of Philosophy* 82(4), 169–221.

Sen, A.K. 1985b. *Commodities and Capabilities*. Amsterdam: North-Holland.

Sen, A.K. and Williams, B. 1982. *Utilitarianism and Beyond*. Cambridge: Cambridge University Press.

Smart, J.J.C. and Williams, B.A.O. 1973. *Utilitarianism: For and Against*. Cambridge: Cambridge University Press.

Smith, A. 1776. *An Inquiry into the Nature and Causes of the Wealth of Nations*. London: Dent, 1954.

Strasnick, S. 1976. Social choice theory and the derivation of Rawls' Difference Principle. *Journal of Philosophy* 73(4), 85–99.

Suppes, P. 1966. Some formal models of grading principles. *Synthèse* 6, 284–306.

Suzumura, K. 1983. *Rational Choice, Collective Decisions and Social Welfare*. Cambridge: Cambridge University Press.

Svensson, L.G. 1980. Equity among generations. *Econometrica* 48(5), 1251–6.

Varian, H. 1974. Equity, envy and efficiency. *Journal of Economic Theory* 9(1), 63–91.

Varian, H. 1975. Distributive justice, welfare economics and the theory of fairness. *Philosophy and Public Affairs* 4, 223–47.

Vickrey, W. 1945. Measuring marginal utility by reactions to risk. *Econometrica* 13, 319–33.

Williams, B. 1973. A critique of utilitarianism. In Smart and Williams (1973).

Williams, B. 1985. *Ethics and the Limits of Philosophy*. Cambridge: Cambridge University Press.

just price. The idea of a just price goes back at least to Aristotle, but the term is principally associated with the writings of the scholastic philosophers of the Middle Ages. In considering their views, it is of interest to ask not only what they believed the just price was, but also how the doctrine was expected to affect people's actions and what function it served.

There is some disagreement among modern writers as to what the scholastics believed the just price of a good was, in part due to disagreement among the scholastic philosophers themselves. It appears that for many, including Aquinas, the just price of a good was normally its market price or, where price control existed, its legal price. For some, including Duns Scotus, the just price of a good meant its cost of production. For at least one, Henry of Langenstein, it appears to have been the price which allowed the producer to maintain his proper place in society. This last position was at one time widely believed to have been the dominant scholastic view. For evidence that it was not, see de Roover (1958).

According to Aquinas, the just price might be higher than the good's market value if the seller incurred some special cost in selling it, but not if 'a buyer derives great benefit from a transaction without the seller suffering any loss' (Aquinas, 1975, pp. 215–16). Hence the just price of a good appears to be its market value or its value to the seller, whichever is higher.

The doctrine of the just price was part of medieval moral philosophy, not medieval legal theory. The normal rule of both canon and Roman law was, with some exceptions, freedom of contract; a transaction was legitimate as long as both parties agreed to it. The doctrine of the just price was thus a statement about what was sinful, not about what was legal, and was normally enforced, if at all, by moral, not legal, sanctions.

To a modern economist, the idea of requiring people to buy and sell at the market price seems rather like the idea of compelling stones to fall when you drop them. The market price is, after all, the highest price at which a seller can sell his goods and the lowest price at which the buyer can buy them, hence the self-interest of the parties involved would seem to guarantee that goods will be sold at the market price, without the need for an additional moral sanction.

This is true under the circumstances where a market price is usually defined – on a market with many buyers and sellers. In a society where many markets are thin – where at many times and places there is at most one horse offered for sale and at most one person interested in buying a horse – the situation is more complicated. The typical transaction is a bilateral monopoly and may have a considerable bargaining range; the highest price the buyer is willing to pay for the horse may be significantly greater than the lowest price at which the seller is willing to sell it. If so, the requirement that the horse be sold at the market price – the price at which similar horses were sold when there was a competitive market, six months earlier in a town thirty miles away – would determined at what price within the bargaining range the horse sold.

This explanation seems consistent with much of the evidence. Medieval discussions of the just price, such as Aquinas's, describe its applications to two-party transactions.

Medieval writers on economic subjects seem to have been very much concerned with what we would describe as imperfect competition. Markets with many buyers and sellers certainly existed in the Middle Ages, but only at particular places and times. Transport costs were sufficiently high (wheat, transported by land, was estimated to double its price in twenty miles) to force many buyers and sellers to limit their transactions to those living near them. Hence we have the mixture of competitive markets and small-number transactions required by this interpretation of the doctrine. For a more detailed discussion, see Friedman (1980).

If the interpretation is correct, then the doctrine can be understood as (among other things) a device to promote economic efficiency. It provided, in effect, an arbitrated solution to what would otherwise be costly bilateral monopoly bargaining. Under most circumstances, the market price would be within the bargaining range. If the seller valued the good at above its market price, then the just price, according to Aquinas, would be raised accordingly, hence the transaction would still take place.

As a legal doctrine, the requirement that goods be sold at their market price unless they are worth more than that to the seller has serious difficulties. How can the court determine whether the potential seller's claim that the good is especially valuable to him is true? Making it a moral doctrine, to be enforced by the conscience of the seller (and possibly the threat of divine punishment), solves the problem; both the seller and God know whether the seller is telling the truth.

One explanation of the doctrine of the just price is as a solution to difficulties associated with imperfect competition.

Other explanations might view it as applying to other circumstances in which the ordinary market price does not automatically hold, such as fraud or duress. What seems clear is that for many, although not all, of the scholastics the just price was normally defined as the market price, and that the doctrine that individuals should buy and sell at the just price was a moral rule applied within a legal framework in which the usual rule was freedom of contract.

DAVID D. FRIEDMAN

See also AQUINAS, ST THOMAS; ARISTOTLE; SCHOLASTIC ECONOMIC THOUGHT.

BIBLIOGRAPHY

Aquinas, St. Thomas. *Summa theologiae*. Trans. and ed. M. Lefebure, New York: McGraw-Hill; London: Eyre Methuen, 1975.

de Roover, R. 1958. The concept of the just price: theory and economic policy. *Journal of Economic History* 18, December, 418–34.

Friedman, D. 1980. In defense of Thomas Aquinas and the just price. *History of Political Economy* 12(2), Summer, 234–42.